2

Lecture Notes in Computer Science 5558

Commenced Publication in 1973
Founding and Former Series Editors:
Gerhard Goos, Juris Hartmanis, and Jan van Leeuwen

AF881236

Editorial Board

David Hutchison
Lancaster University, UK
Takeo Kanade
Carnegie Mellon University, Pittsburgh, PA, USA
Josef Kittler
University of Surrey, Guildford, UK
Jon M. Kleinberg
Cornell University, Ithaca, NY, USA
Alfred Kobsa
University of California, Irvine, CA, USA
Friedemann Mattern
ETH Zurich, Switzerland
John C. Mitchell
Stanford University, CA, USA
Moni Naor
Weizmann Institute of Science, Rehovot, Israel
Oscar Nierstrasz
University of Bern, Switzerland
C. Pandu Rangan
Indian Institute of Technology, Madras, India
Bernhard Steffen
University of Dortmund, Germany
Madhu Sudan
Massachusetts Institute of Technology, MA, USA
Demetri Terzopoulos
University of California, Los Angeles, CA, USA
Doug Tygar
University of California, Berkeley, CA, USA
Gerhard Weikum
Max-Planck Institute of Computer Science, Saarbruecken, Germany

Massimo Tistarelli Mark S. Nixon (Eds.)

Advances in Biometrics

Third International Conference, ICB 2009
Alghero, Italy, June 2-5, 2009
Proceedings

Springer

Volume Editors

Massimo Tistarelli
Università di Sassari, Computer Vision Laboratory
Facoltà di Architettura di Alghero
Dipartimento di Architettura e Pianificazione (DAP)
Palazzo del Pou Salit, Piazza Duomo 6
07041 Alghero (SS), Italy
E-mail: tista@uniss.it

Mark S. Nixon
University of Southampton
School of Electronics and Computer Science
Southampton SO17 1BJ, UK
E-mail: msn@ecs.soton.ac.uk

Library of Congress Control Number: Applied for

CR Subject Classification (1998): I.5, I.4, K.4.1, K.4.4, K.6.5, J.1

LNCS Sublibrary: SL 6 – Image Processing, Computer Vision, Pattern Recognition, and Graphics

ISSN 0302-9743
ISBN-10 3-642-01792-4 Springer Berlin Heidelberg New York
ISBN-13 978-3-642-01792-6 Springer Berlin Heidelberg New York

This work is subject to copyright. All rights are reserved, whether the whole or part of the material is concerned, specifically the rights of translation, reprinting, re-use of illustrations, recitation, broadcasting, reproduction on microfilms or in any other way, and storage in data banks. Duplication of this publication or parts thereof is permitted only under the provisions of the German Copyright Law of September 9, 1965, in its current version, and permission for use must always be obtained from Springer. Violations are liable to prosecution under the German Copyright Law.

springer.com

© Springer-Verlag Berlin Heidelberg 2009
Printed in Germany

Typesetting: Camera-ready by author, data conversion by Scientific Publishing Services, Chennai, India
Printed on acid-free paper SPIN: 12670512 06/3180 5 4 3 2 1 0

Preface

It is a pleasure and an honour both to organize ICB 2009, the 3rd IAPR/IEEE International Conference on Biometrics. This will be held 2–5 June in Alghero, Italy, hosted by the Computer Vision Laboratory, University of Sassari. The conference series is the premier forum for presenting research in biometrics and its allied technologies: the generation of new ideas, new approaches, new techniques and new evaluations. The ICB series originated in 2006 from joining two highly reputed conferences: Audio and Video Based Personal Authentication (AVBPA) and the International Conference on Biometric Authentication (ICBA). Previous conferences were held in Hong Kong and in Korea. This is the first time the ICB conference has been held in Europe, and by Programme Committee, arrangements and by the quality of the papers, ICB 2009 will continue to maintain the high standards set by its predecessors.

In total we received around 250 papers for review. Of these, 36 were selected for oral presentation and 93 for poster presentation. These papers are accompanied by the invited speakers: Heinrich H. Bülthoff (*Max Planck Institute for Biological Cybernetics, Tübingen, Germany*) on "What Can Machine Vision Learn from Human Perception?", Sadaoki Furui (*Department of Computer Science, Tokyo Institute of Technology*) on "40 Years of Progress in Automatic Speaker Recognition Technology" and Jean-Christophe Fondeur (*SAGEM Security and Morpho, USA*) on "Large Scale Deployment of Biometrics and Border Control". In this conference we have also explored the evaluation of biometrics with a session of papers on challenges in evaluations: one on "Face Video Competition at ICB2009", on "Fingerprint and On-line Signature Verification Competitions at ICB 2009", on "Partial Face Matching between Near Infrared and Visual Images in MBGC Portal Challenge", and on "Overview of the Multiple Biometric Grand Challenge". The conference itself is followed by the annual Biometrics Summer School which has been held at Alghero for the past six years and its presenters, past and present, feature many of the programme and organising committee.

We are naturally very grateful for the efforts of many people. Firstly, we thank the entire conference committee for their valuable support in several stages of the organization. In particular we thank Anil Jain, Josef Kittler and Tieniu Tan who enthusiastically provided support and advice. The reviewing itself always appears a Herculean task and we were grateful that reviewers completed this in good time, and especially to those who came in at short notice. Equally, the other co-Program Chairs Arun Ross and Stan Li gave much of their time to finalise the final selection for the conference with major help in coordinating and actioning the reviewing process and developing the final program. The local organising committee have coordinated many of the local arrangements for the conference. Marinella Cardoni managed the review stage expertly. The Computer Vision Laboratory staff: Enrico Grosso, Linda Brodo, Massimo Gessa, Andrea Lagorio, Ajita Rattani and Elif Surer, are sorting the conference itself. Also, we are very grateful to the IAPR TC4 and the IEEE Biometrics Council, particularly to the Chairs Tieniu Tan and Rama Chellappa. The support from Accenture, GreenBit, Sagem, the European Union COST Action 2101 on "Biometrics in Identity

Documents", the foundation "Fondazione Banco di Sardegna", and the University of Sassari is also acknowledged. Finally, we are grateful to Springer for their production of the Proceedings, Alfred Hofmann and Wayne Wheeler in particular.

By papers, location and content this promises to be an excellent edition of the International Conference on Biometrics series. The conference continues to provide a snapshot of research in world-wide leading institutions on research in biometric recognition, and the many components if this fascinating research area, especially as it reaches world prominence given its increasing contribution to our societies' advancing security. By the program, papers and venue, we wish all delegates a most enjoyable conference

Massimo Tistarelli
Mark Nixon

Organization

General Chair

Massimo Tistarelli	University of Sassari, Italy

Conference Co-chairs

Davide Maltoni	University of Bologna, Italy
Javier Ortega-Garcia	Universidad Autonoma de Madrid, Spain

Advisory Board

Anil K. Jain	Michigan State University, USA
Joseph Kittler	University of Surrey, UK
B.V.K. Vijayakumar	Carnegie Mellon University, USA
Seong-Whan Lee	Korea University, South Korea
Tieniu Tan	Chinese Academy of Sciences, China

Competitions Chairs

Bernadette Dorizzi	Biosecure Foundation, France
Jonathon Phillips	NIST, USA

- Face competition: Norman Poh (University of Surrey, UK)
- Fingerprint competition: Raffaele Cappelli (University of Bologna, Italy)
- Signature verification competition: Sonia Garcia-Salicetti (GET-INT)
- Multimodal competition: Kryzstof Kryszczuk (EPFL)

Demo Chairs

Manuele Bicego	University of Sassari, Italy
Julian Fierrez	Universidad Autonoma de Madrid, Spain

Local Arrangement Chair

Enrico Grosso	University of Sassari, Italy

Program Chairs

Mark Nixon	University of Southampton, UK
Arun Ross	West Virginia University, USA
Stan Z. Li	Chinese Academy of Sciences, China

Biometric Competitions

Multibiometrics and Security

Gait

Iris

to face motion. The main contributions of this paper are (1) the hardware set-up used for data capture, (2) our inclusion of a face profile capture mechanism, and (3) an algorithm that we propose to optimise the overall 3D shape of the face.

The profile is obtained using a side-view camera that captures two images in close succession as the person passes by. The profile shape is then extracted from the images using an energy functional based on the difference between the two images in a 4D colour space. In order to combine the information from the two views, we apply an energy minimisation approach. The proposed method involves iteratively deforming the current estimate of the surface so that its profile matches that recovered from the side-view camera. The resulting field of surface normals is then used to synthesise new images and the difference between these and the raw images forms the energy functional. The minimisation is performed by adopting a field of normals that is intermediate between the initial estimate and the current estimate, as determined by the calculated energy.

There has been a vast amount of past research into shape reconstruction, and faces in particular, by the computer vision community. Amongst the best-known for 3D face recognition [2,3] are laser-triangulation [4], stereo vision [5], and morphable models [6]. Other, more general, methods that are relevant to our work include shape-from-shading [7] and photometric stereo [8,9]. Shape-from-shading is used by Zhao and Chellappa who exploit facial symmetry in order to reconstruct faces [10].

One of the most closely related contributions was proposed by Hernández et al. [11], who attempt to recover generic object shapes. In their work, they use varying views and illumination directions and assume a black background. The silhouettes of the objects are then extracted and used to form an initial estimate of the shape based on the visual hull. The estimate is then optimized based on an energy minimisation scheme similar to the one we present here. Another related work was by Nehab et al. [12], who use range data for the initial estimate of the shape and photometric stereo to reduce noise.

The principal advantages of our method are that (1) only inexpensive equipment is required, (2) the device is able to work in ambient illumination, (3) no assumptions are made about the symmetry of the face (which can be useful for face recognition applications [13]), (4) all images are captured almost instantaneously (unlike most other photometric stereo methods), and (5) the processing required for reconstruction is very efficient computationally. The primary disadvantages are the need for rapidly flashing lights and the space required for the two cameras and the four light sources.

2 Image Acquisition and Single-View Processing

Figure 1 shows a photograph of the device that we have constructed for data acquisition. The person walks towards camera on the back panel from the left. An ultrasound proximity switch detects the presence of the individual and triggers the acquisition procedure. Frontal images are captured using the high speed camera on the back panel with the face illuminated by the four flashguns in sequence.

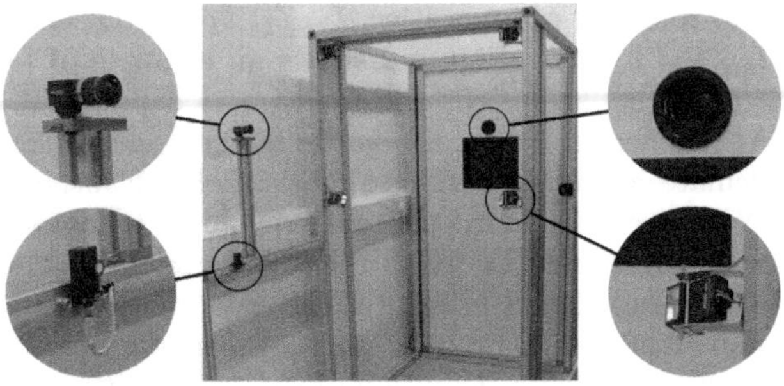

Fig. 1. Photograph of the image capture device. The following features are highlighted: Profile camera (top-left), frontal camera (top-right), ultrasound trigger (bottom-left) and one of the flashguns used in photometric stereo (bottom-right).

Profile images are taken using the camera to the left of Fig. 1 under ambient illumination. The rest of this section details the acquisition and processing of the images and the method used to align the data from each view.

2.1 Frontal Images

The first stage of our method is to capture frontal images of the face and make an initial estimate of the 3D geometry using photometric stereo. It is generally expected that the face will be moving at the time of acquisition. For this reason, it is necessary to use a high speed camera to rapidly acquire the images as the light sources change before significant motion is possible. We therefore use a Basler A504kc 1280×1024 pixel camera operating at 200 fps. It was found experimentally that this was the frame rate necessary to avoid face movement between images by more than one pixel. The light sources are Jessops 100M flashguns, which are synchronised to the camera frames. All interfacing and synchronisation is programmed in LabVIEW (although the image processing and shape estimation is performed in MATLAB). The sensor used to initiate the entire process is a highly directional Baumer ultrasound proximity switch.

Four greyscale images are captured by the camera with each corresponding to one of the four light sources shown in Fig. 1. The regions containing the actual face are extracted from the background using the method of Lienhart and Maydt

Fig. 2. Example of four raw images, $\mathcal{I}_1$ to $\mathcal{I}_4$

$$\mathcal{E} = \{E_i\} = \{\max(C_i E_{I_i}, 1)\} \quad (6)$$

where $\mathcal{E}_I$ and $\mathcal{C}$ are calculated using (4) and (5) respectively.

The energy now takes values in the interval $[0, 1]$. It is then used to modify the surface normals $\mathcal{V}'_n$ to arrive at the next estimate $\mathcal{V}_{n+1}$. For pixels where $E = 0$, then $\mathbf{v}'_n$ is used as the new surface normal $\mathbf{v}_{n+1}$. Where $E = 1$, then $\mathbf{v}_0$ is used. Otherwise a surface normal between the two is adopted for the new estimate.

Let the function $R(\mathbf{w}_1, \mathbf{w}_2, \theta)$ give a vector found by rotating $\mathbf{w}_1$ by an angle of θ towards $\mathbf{w}_2$. The next estimate of surface normal i is then given by the following:

$$\mathcal{V}_{n+1} = \{\mathbf{v}_{n+1,i}\} = \left\{R\left(\mathbf{v}'_{n,i}, \mathbf{v}_{0,i}, E \arccos\left(\mathbf{v}'_{n,i} \cdot \mathbf{v}_{0,i}\right)\right)\right\} \quad (7)$$

The set of refined surface normals, $\mathcal{V}_{n+1}$ is then re-integrated to form the refined height map $H_{n+1}(x, y)$.

The entire geometrical optimisation process described above is repeated until convergence. If M iterations are required for convergence, then $H'_M(x, y)$ is used for the final surface. Figure 6 shows an example of this. The iterative procedure takes about four or five seconds to complete on a typical modern personal computer, with cropped face image dimensions of 300×400 pixels.

4 Results

Collectively, Figs. 2 to 6 illustrate the various stages of our proposed algorithm for a typical face. As Fig. 3 demonstrates, standard photometric stereo is able to provide a qualitatively good reconstruction of the face without using any profile view information at all. However, as Fig. 5 shows, the reconstruction is only a very rough estimate of the actual shape. One major strength of our proposed technique is that the profile of the face is almost exactly correct when the side-view is considered. A slight undesirable consequence of this, is that the nose tends to be widened somewhat. The overall shape estimate is significantly better however for most faces that we considered (and at least a little better for all faces).

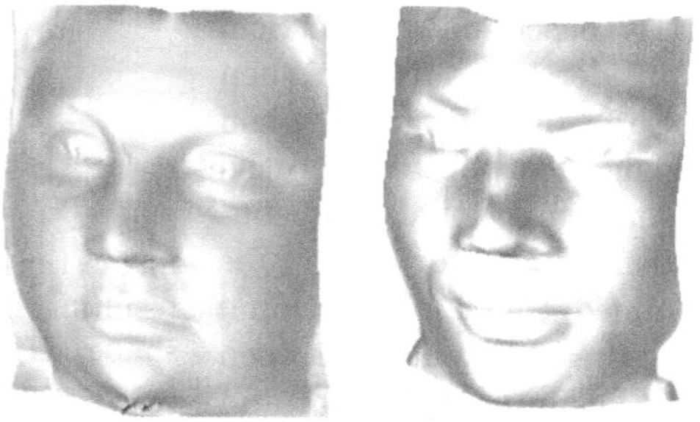

Fig. 7. Surface reconstructions for two of the subjects in Fig. 8. See Fig. 6 for the other reconstruction.

Three representative faces are shown in Figs. 6 to 8. For each face, two vertical cross-sections and two horizontal cross-sections are shown for the initial estimate, $H_0(x, y)$, the final estimate, $H'_M(x, y)$, and ground truth. The ground truth curves were obtained using a 3DMDface™ scanner [19]. Note that the right-hand vertical cross-sections do not pass through the centre of the face as such profiles match ground truth exactly.

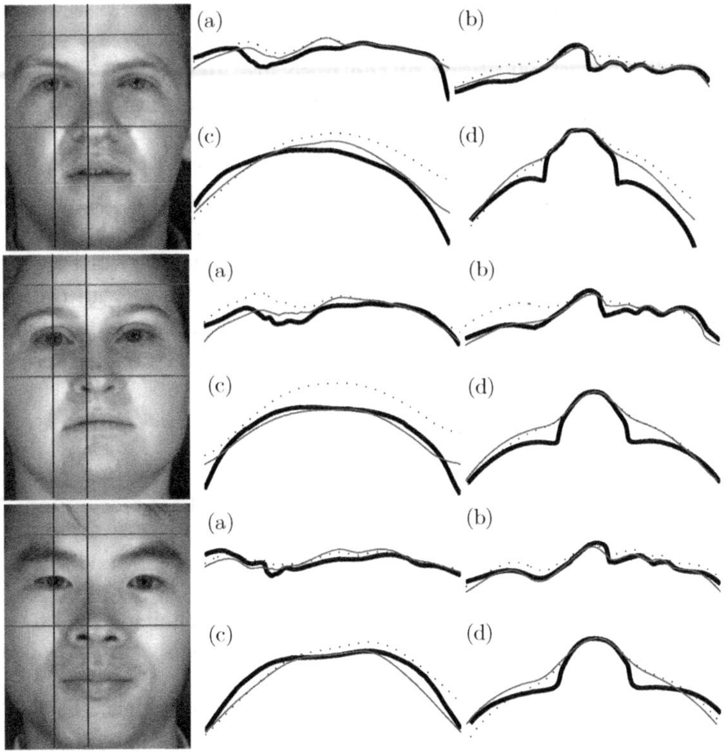

Fig. 8. Cross-sections of face reconstructions for three different subjects. Thick solid lines indicate ground truth, thin solid lines indicate the final estimate and broken lines indicate initial estimates. For each face, curve (a) corresponds to the left vertical line in the image, (b) the right vertical line, (c) the upper horizontal line, and (d) the lower horizontal line.

In all three cases, the final reconstructions match the ground truth reasonably well. In the first two cases, the final estimate is a major improvement over the initial estimate. The horizontal cross-sections across the forehead show particularly impressive results. In order to quantify the improvement to the estimate due the the incorporation of profile information, we have calculated the root-mean-square (RMS) error for the reconstructions when the nose tips are aligned. The reduction in RMS error when the profile information is used is from 26px to 17px for the first face, from 23px to 15px for the second face, and from 13px to 12px for the final face. Note that the small reduction for the final face reflects the fact that the initial estimate was already very good.

5 Conclusion

This paper has presented a new method for facial shape estimation using a combination of photometric stereo and profile view information. The paper clearly

2.2 Non-ICP-Based 3D Face Biometrics

The methods discussed in this section may utilize ICP in the overall process as we do; however, the key is that they do not use ICP as part of each match-score computation. So regardless of gallery size, ICP is utilized as part of probe preprocessing and is not utilized for each match-score computation.

Colbry et al. [9] recognized the issue of a need for a faster ICP technique that does not utilize ICP for each match score computation and presented an approach for faster matching. An input face is aligned to a stored model face and a depth map is rendered. After rendering the depth maps various sparse feature representations for analysis such as Gabor wavelets were utilized. These alternate match score computations were accomplished without using ICP during each match score calculation. Although eliminating the use of ICP as a part of each match score did dramatically decrease the running time, they found that it had a negative impact on biometric performance.

Kakadiaris et al. [10] produced depth maps utilizing a spin image. Next, the spin image was converted into a wavelet representation utilizing Haar and pyramid coefficients for comparison. Their approach performed well at 97.3% rank one, with performance comparable to that of Faltemier et al. [1]. Their approach did not require an alignment step, such as ICP, for each match score computation but did utilize ICP in the initial alignment wavelet-generating process. As such, their running times were much faster than those of Faltemier et al. [1] but utilized a distinctly different process. This process is distinct from much of the previous 3D face recognition approaches in that it is not based upon a 3D distance metric.

Russ et al. [4] proposed using a 3D PCA-based approach to calculate a unique 3D face vector that was used with PCA to perform recognition. The vectors were produced utilizing a combination of nearest neighbor and normal searches in order to ensure a complete vector. Next, the vectors were fed into principal component analysis to produce PCA coefficients utilized for recognition.

3 3D Signature Generation for Face Recognition

In this section, we will describe the process of generating 3D signatures. We employ a gallery $G = \{ X_1, X_2, \ldots X_{nG}\}$ of 3D face images captured for the purpose of enrollment, and a training set T of 3D images from fifty subjects not in T is used to bootstrap the generation of two generations of reference faces. Each 3D image is assumed to be roughly aligned to the same location and orientation.

3.1 Generation of Initial Reference Face

An initial reference face F_1 was generated by aligning, smoothing, and merging 3D images of ten distinct faces in T. We employed the Raindrop Geomagic [12] commercial software for this step. The reference face defines a canonical coordinate system. The reference face generated from this process can be seen in Figure 1.

A 3D signature is a vector representation $S = [\, p_1, \ldots p_n]$ (with p_n representing 3D points) of the 3D surface constructed from a structured traversal of face positions defined on a reference region.

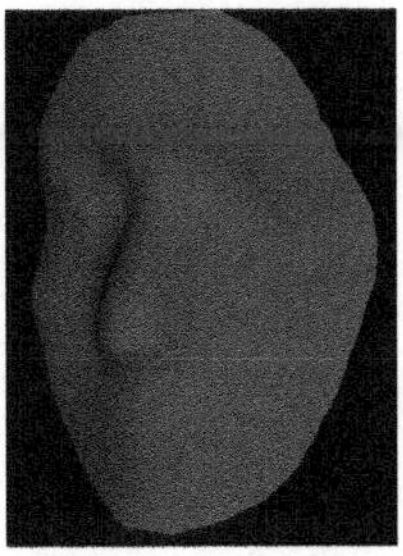

Fig. 1. Initial Reference Face Generated

Fig. 2. Complete Reference Face

After computation of the initial reference faces, a 3D signature is computed for each of the fifty face images in T using the following process.

- The reference face and the training face image T are aligned using ICP, with an initial coarse correspondence obtained from nose alignment.
- An indexed sequence of locations $S = \{p_1, \ldots p_{nS}\}$, $p_i = (x_i, y_i, z_i)$ consisting of all points on the initial reference face was defined using its canonical coordinate system. The initial reference face model consists of 15,000 points. We show this initial face in Figure 1. The initial distribution of points was made approximately uniform utilizing a uniform point sampling function in Raindrop Geomagic [12].
- Images of fifty images X_j in T were aligned with F_1 using ICP.
- *Calculation of 3D Signature*: Each of the aligned faces was sampled by searching along the normal vector on F_1 at each point p_i in S, retrieving the closest intersecting point on X_j. This yields an ordered set of 3D points on the aligned version of X_i, and a set of fifty correspondents $\{p_1, \ldots p_{50}\}$ for each p_i.
- The correspondents $\{p_1, \ldots p_{50}\}$ for each point p_i are averaged, yielding an average 3D signature which also serves as a second generation reference face. This face model is shown in Figure 2. We cropped the eye and nose region from this representation because the data from this area are typically noisy and contain holes due to sensor limitations.

To determine individual reference regions, the averaged reference face was manually cropped into different regions using Raindrop Geomagic [11]. This is done once during the initial reference region signature generation. Using this method, we produced eight random regions, as are shown in Figure 4. Regions 5 and 6 are subsets of region 2. However, the resulting 3D signatures are not simply subsets. The reason for this is that each reference region is aligned separately. The different alignments result in different 3D signatures. Each region utilized here yields an increase in the performance of the fusion result.

Selection of the reference region shape and point density plays a critical role in 3D signature generation. However, determining the ideal region shape and density to utilize is a problem that has not been addressed and most likely deserves more attention. Our method of reference-region generation has three main advantages over traditional hard-coded region-generation approaches.

- Ease of producing odd-shaped regions
- Ability to increase or decrease the number of points for comparison, regardless of the initial point density of the model
- Ability to control the point density for different parts of the same region

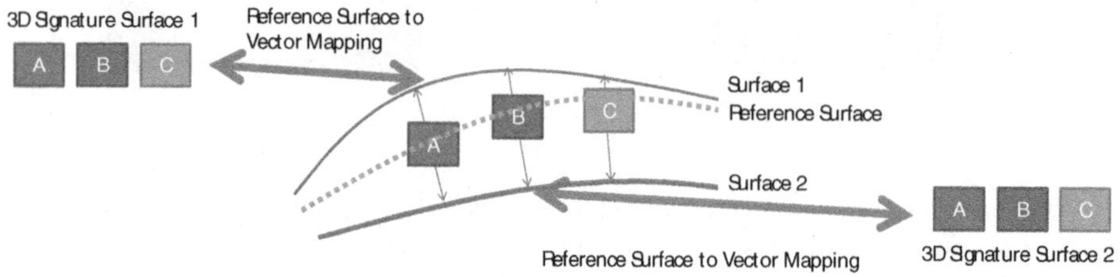

Fig. 3. 3D signature correspondence for two surfaces: Green dotted surface is the reference surface, and the red and blue surfaces are input surfaces; the double arrows represent correspondence

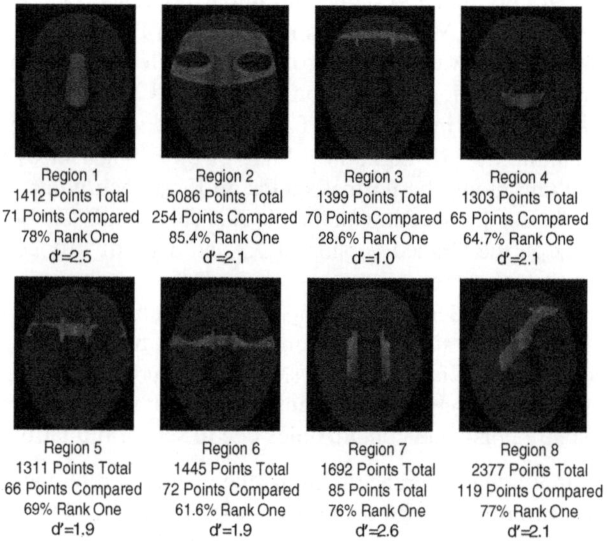

Fig. 4. 8 Reference Regions

The ability to customize region shape definition and point density has traditionally required more complex programming as the regions are defined via hard coding in methods such as Faltemier et al. [1]. Faltemier et al. [1] employed hard coded spherical regions (since they are relatively simple to define) and utilized the point densities of the existing gallery and probe face models as a basis for comparison. Our approach offers the ability to tune region shapes and point density to highly discriminatory face features by modifying the reference face regions.

3.2 Correspondence Search

Given a pair of 3D surfaces, one a reference region and the other an image of an unknown face, it is possible to determine 3D signature correspondence via a nearest neighbor (NN) search, a normal vector search, a hybrid of these two approaches, or a different correspondence method. The notion critical to a 3D signature is that a method of determining correspondence between two points is used.

The nearest-neighbor search assigns each point in the unknown face signature to the closest correspondent on the unknown face to that of the point in the reference face. Previous work [4] has documented that the NN approach does not pick the most representative points when areas of curvature differ between the reference and input surface.

A normal search searches along the normal vector to the reference region surface at each reference region point to find the intersection between the normal and the unknown face surface which is then the corresponded point used for the 3D signature. As such, it is possible that the reference region normal does not intersect the unknown face surface due to surface shape or holes in the 3D surface.

A hybrid technique proposed by Russ et al. [4] repairs missed correspondences by using a NN search to produce correspondences for probe points where the normal search fails. This is necessary for use with PCA as missing vector entries cannot be tolerated.

3.3 Individual Region Match Score Calculation

Calculation of match scores is simple when two 3D signatures are compared. No ICP step is needed during the match score calculation (as in [1][2][3]) since the signatures are pre-aligned. In our method, we calculate the average squared Euclidean distance between the vectors by accumulating the squared distance between each corresponding point pair. When utilizing a normal only search criteria we can ignore any points that lack a correspondent due to a failed normal search. No iterative techniques are used.

3.4 Score Fusion

After obtaining match scores from multiple regions, it is necessary to utilize all of the results to determine identity in a process commonly referred to as fusion. The most commonly used fusion methods include sum, product, minimum, and consensus. The concept of combining independent match scores into one result via fusion was originally proposed by Kittler et al [12].

Faltemier et al. [1] found optimal performance with a modified Borda count technique operating at the rank level. Each rank value was compared to a threshold. Ranks higher than the threshold were replaced by the threshold value and then all ranks were summed. This modified Borda approach performed 4.5% better than sum.

All rank-based fusion techniques, lose the proximity of match scores between in rank computation. We propose a new method named the Match Sum Thresholding (MST) approach. MST operates on the set of scores resulting from multiple 3D signature comparisons. As with the modified Borda count technique, MST replaces any region scores that exceed a predetermined matching score threshold with the value of that threshold.

The threshold used in the MST method is tuned for each 3D signature. Determining an individual MST score threshold for each region is necessary, because match score ranges vary from region to region. We precomputed the MST thresholds using the following approach. We limited the possible values of the thresholds to .01 increments searching for a biometric performance local maxima.

```
1. Set initial score thresholds to include all match
   scores
2. Continue until score thresholds do not change or
   change falls below a set threshold
   a. For each region
    i.  For a range of threshold values (such as
        i=0; i<2; i+=.01 )
       1. Find best threshold for current region
          based upon other region thresholds and MST
          calculation
       2. Set threshold for current region to best
```

4 Results

Our experiments used a subset of the FRGC 2.0 3D face data corpus [8]. Images were acquired at the University of Notre Dame using a Minolta Vivid 900 3D scanner. We used all subjects with two or more 3D face images, yielding a dataset comprising 410 subjects and 3939 total images, and conducted closed-set recognition trials. The data contains a variety of facial expressions. All experiments were performed with the earliest image as the gallery image and subsequent images as probes. This standard experiment has been commonly utilized in previous work and is referred to as Experiment 2 in the FRGC protocol. For more information on this dataset or experiment, please refer to Philips et al. [8].

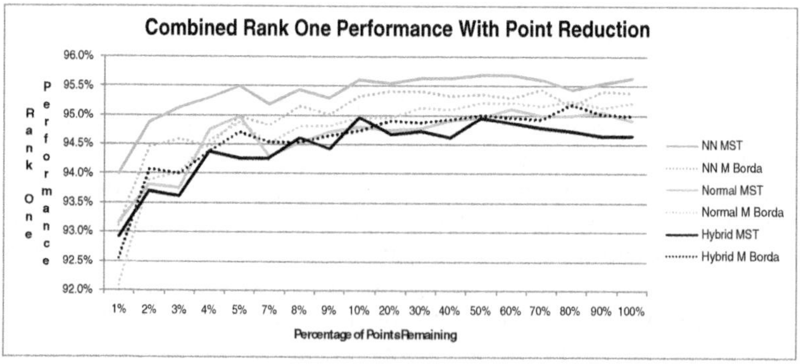

Fig. 6. Performance of Hybrid, Nearest Neighbor, and Normal 3D Signatures

4.1 3D Signature Search Method

Previous work [4] utilized a hybrid normal/NN search method to produce vectors for PCA input. We compared the performance of a hybrid approach, NN approach, and normal approach for generating vectors, for a range of 3D signature sizes between 1% and 100% of the original size. The rank one recognition rate recognition results are shown in Figure 6 with different correspondence methods and the MST fusion method.

Contrary to Russ et al. [4], we found that the NN (nearest neighbor) search criteria for 3D signature generation performed best. The difference in performance between the NN approach and the previously utilized hybrid search is even greater (approximately 1%), depending upon the number of points utilized for match score computation. It is possible that the use of individual regions as opposed to entire faces (as used in [4]) explains this difference.

4.2 Fusion Method

The previous highest performing fusion method for multiple regions was proposed by Faltemier et al. [1]. They utilized a modified Borda count to combine the results from each region for optimal performance. Their search criteria was a nearest neighbor approach. As seen in Figure 6, we found that the best-performing 3D-signature search criterion was the NN search method and examine only that correspondence method in this section.

Table 1. Threshold Values Used For Mst

	Region 1	Region 2	Region 3	Region 4	Region 5	Region 6	Region 7	Region 8
Region Thresh	0.55	1.2	0.4	0.6	0.85	0.8	0.85	0.8

Furthermore, our new MST fusion approach consistently outperformed the modified Borda approach by approximately 0.2% to 0.5%. The thresholds utilized for each region are shown in Table 1. We believe this demonstrates an advantage in this situation of using the matching score instead of the rank alone. The remainder of the experiments will be presented on the NN search criteria with MST fusion because those criteria obtained superior performance.

4.3 Running Time and Memory Requirements

In addition to a performance accuracy improvement, the 3D signature comparison technique yielded a decrease of nearly four orders of magnitude in matching execution time compared to the matcher proposed by Faltemier et al., due to the lack of an ICP step in each match pair comparison. The 3D-signature match-score computation running time and memory requirements are proportional to the total number of points in the 3D signature utilized for comparison.

Our test scenario utilized a single core of a dual-core laptop. The laptop utilized an Intel T7500 mobile processor operating at 2.2 GHz running a 32-bit version of Windows Vista. We found that on average we were able to compute 102,594 match scores per second for a single region. By comparison, Faltemier et al. [1] were able to compute 12 match scores per second on average for a single region. This makes the proposed approach approximately 8700 times faster.

Non-ICP–based 3D matchers, such as that described by Kakadiaris et al. [10], have been able to achieve speed improvements similar to those seen here. However, their approach is distinctly different from our 3D-signature approach in that it [10] does not utilize a surface distance comparison (its match scores are based upon wavelet representations). Traditionally the distance based match score approach as used here has been more utilized in the literature. The advantages or disadvantages of this type of an approach or the wavelet approach by Kakadiaris et al. [10] is an open research area.

The proposed 3D signature approach requires the storage of 802 3D points when 95% of the original signatures' points are removed by subsampling. This requires 2406 vertices (x, y, z) to be stored, or approximately 32 kilobytes per gallery image. By comparison, a densely sampled 3D face model of 20,000 vertices would require in excess of 240 kilobytes for vertex storage alone.

5 Conclusion

We have proposed to represent a 3D surface as a 3D signature that can be directly compared to another 3D signature in a process similar to ICP match-based distance biometrics. Both approaches represent match score as an error distance between two surfaces. However, our approach eliminates the need for ICP as a part of the match-score computation.

This process has advantages over other consistent representations, such as depth maps, because a 3D signature is not a 2D projection of the 3D surface; it is, instead, dependent upon a 3D reference region retaining the resulting 3D shape in the new representation. This allows for the data to be represented in a face based coordinate system that does not suffer any data loss as a projection to a depth map does.

The process of computing the signature and performing the comparison presented performance and accuracy advantages over an ICP based approach. The lack of an ICP invocation in each match yields a dramatic speed increase without encountering an accuracy penalty. We were able to calculate match scores 8,700 times faster than previous state of the art ICP based methods [1]. Each match score is computed by calculating the distance between two 3D signatures that are stored in vector form. Previous exploration of vector representations [4] utilized a hybrid normal search/NN approach. Here we show that a purely NN approach achieves improved performance.

In addition, we present a new region fusion approach named MST, which outperforms the modified Borda count approach [1], which was previously the state-of-the-art fusion method for region based 3D face biometrics. MST does not require computing match rank, as does the modified Borda count approach. The superior performance of MST is a result of it retaining the distance between match scores. Utilizing more complex comparison regions and improved fusion method we are able to achieve a 95.5% rank one identification utilizing 8 regions as opposed to 92.9% using 8 regions from the previous state of the art method [1].

In closing, the proposed 3D-signature method outperforms previous state-of-the-art ICP match score-based biometrics [1][2][3] in running time and biometric performance through the use of regions with more complex shapes and makes 3D face biometrics feasible for large applications with the increased speed and low memory requirements. In the future, we would like to further explore reference regions with a larger variety of shapes and densities. Although the regions here show the potential of complex regions, further exploration may yield further improvements in performance.

Acknowledgements

Biometrics research at the University of Notre Dame is supported by the Central Intelligence Agency, the National Science Foundation under grant CNS01-30839, by the US Department of Justice/National Institute for Justice under grant 2006-IJ-CX-K041, by iARPA and the Biometrics Task Force under US Army contract W91CRB-08-C-0093 and by UNISYS Corp.

References

1. Faltemier, T., Bowyer, K.W., Flynn, P.J.: A Region Ensemble for 3D Face Recognition. IEEE Transactions on Information Forensics and Security 3(1), 62–73 (2008)
2. Chang, K.I., Bowyer, K.W., Flynn, P.J.: Multiple Nose Region Matching for 3D Face Recognition under varying facial expression. IEEE Transactions on Pattern Analysis and Machine Intelligence 28(10), 1695–1700 (2006)
3. Lu, X., Colbry, D., Jain, A.K.: Three-Dimensional Model Based Face Recognition. In: 17th International Conference on Pattern Recognition (2004)
4. Russ, T., Boehnen, C., Peters, T.: 3D Face Recognition Using 3D Alignment for PCA. In: Computer Vision and Pattern Recognition, New York, pp. 1391–1398 (2006)
5. Besl, P., McKay, N.: A method for registration of 3-D shapes. IEEE Transactions on Pattern Analysis and Machine Intelligence, 239–256 (1992)
6. Bentley, J.L.: K-d Trees for Semidynamic Point Sets. In: SCG 1990: Proc. 6th Annual Symposium on Computational Geometry, pp. 187–197 (1990)
7. Phillips, P.J., Flynn, P.J., Scruggs, T., Bowyer, K.W., Chang, J., Hoffman, K., Marques, J., Min, J., Worek, W.: Overview of the face recognition grand challenge. In: IEEE Conf. on Computer Vision and Pattern Recognition, pp. 947–954 (2005)
8. Colbry, D., Folarin, O., Stockman, G.: 3D Face Identification—Experiments Towards a Large Gallery. Society of Photographic Instrumentation Engineers Defense & Security (2008)
9. Kakadiaris, I.A., Passalis, G., Toderici, G., Murtuza, N., Theoharis, T.: 3D Face Recognition. In: Proceedings of the British Machine Vision Conference, pp. 200–208 (2006)
10. Al-Osaimi, F.R., Bennamoun, M., Mian, A.: Integration of local and global geometrical cues for 3D face recognition. Pattern Recognition Letters, 1030–1040 (2008)
11. `http://www.raindropgeomagic.com/en/`
12. Kittler, J.V., Matas, J., Jonsson, K., Sanchez, M.U.R.: Combining Evidence in Personal Identity Verification Systems. Pattern Recognition Letters, 845–852 (1997)

at the borders. This cropping is performed using a binary mask. The cropped eigenvectors span a non-orthonormal subspace. Eqn. 1 describes the projection of the unseen 3D facial scans onto such a subspace.

$$\mathbf{c}_i = (\mathbf{E}^\top \mathbf{E})^{-1} \mathbf{E}^\top \mathbf{f}_i \tag{1}$$

where $\mathbf{f}_i$ is the projected face, $\mathbf{c}_i$ corresponds to the projection coefficients and the matrix $\mathbf{E}$ is formed from the cropped eigenvectors as column vectors.

The projection of a 3D facial scan described in Eqn. 1 loses a significant part when reconstructed using Eqn. 2. Whereas an expression residue is largely preserved during the same projection. This implies that the projection by itself significantly contributes to the decomposition process of the 3D facial scan. Further decomposition refinements are performed in the following stages (Section 2.2 and 2.3). Fig. 1.(b) shows example projections of 3D facial scans and shape residues on the expression residue subspace.

$$\mathbf{r}_{p_i} = \mathbf{E}\mathbf{c}_i \tag{2}$$

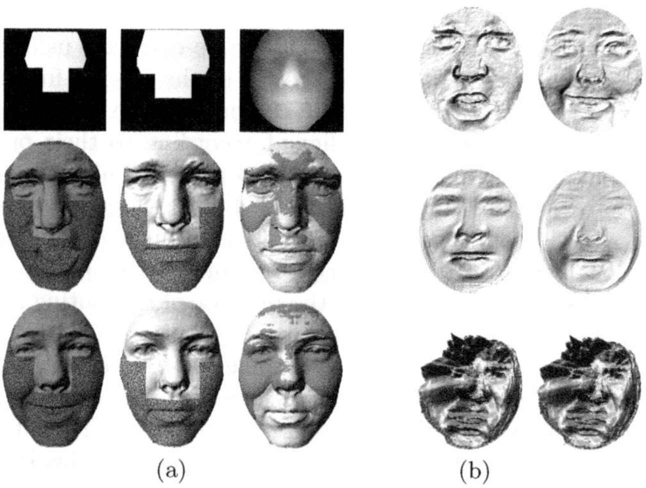

Fig. 1. (a) The binary masks (top row) which are used for scan pair (same subject) registration (second and third rows) according to the semi-rigid regions of the face using ICP. (b) The shapes of expressions residues (top row) are preserved during the projection on the PCA subspace (second row) whereas significant parts of 3D faces are lost during the same projection (third column).

2.2 Stage 2: Iterative Estimation of the Residue and Neutral Face

In this stage of the decomposition, the part of the 3D face which is preserved during the projection is further decomposed to expression and non-expression parts. The decomposition in this stage is data driven by training examples of 3D faces and expression residues. A large number of 3D faces and expression residues are projected onto the subspace using Eqn. 1. Let $\mathbf{f}$, $\mathbf{r}_e$ and $\mathbf{n}_e$ denote the projection coefficients of an unseen 3D face, expression residue estimate and neutral face estimate, respectively. Firstly, $\mathbf{r}_e$ is initialized to a vector of zero coefficients and $\mathbf{n}_e$ to $\mathbf{f}$. The N nearest neutral face projection to $\mathbf{n}_e$ and the M nearest neighbors of $\mathbf{r}_e$ are found, based on the city-block distance measure. $\mathbf{r}_e$ and $\mathbf{n}_e$ are updated using Eqns. 3 and 4. The process is repeated until convergence. The approach converges when the sets of nearest neighbors in an iteration are the same as in the previous iteration.

$$\mathbf{r}_{e_{i+1}} := \mathbf{r}_{e_i} + \frac{1}{2}(\bar{\mathbf{r}}_i - \mathbf{r}_{e_i} - \bar{\mathbf{n}}_i + \mathbf{n}_{e_i}) \tag{3}$$

$$\mathbf{n}_{e_{i+1}} := \mathbf{n}_{e_i} + \frac{1}{2}(\bar{\mathbf{n}}_i - \mathbf{n}_{e_i} - \bar{\mathbf{r}}_i + \mathbf{r}_{e_i}) \tag{4}$$

The vectors $\bar{\mathbf{r}}_i$ and $\bar{\mathbf{n}}_i$ are the mean projection coefficients of the nearest M residues and N neutral faces at stage number i, respectively. Note that at any iteration the sum of the residue estimate and the neutral estimate is equal to the unseen 3D face, $\mathbf{r}_{e_i} + \mathbf{n}_{e_i} = \mathbf{r}_{e_{i-1}} + \mathbf{n}_{e_{i-1}} = \mathbf{f}$. The updating equations also ensure that at convergence the nearest neighbors simultaneously represent reasonable residue and neutral face estimates.

2.3 Stage 3: Fitting to Nearest Neutral Faces

In the final stage, the subspace around the residue estimate from stage 2 is searched for an estimate that collectively fits the difference between the input 3D face $\mathbf{f}$ and the nearest neighboring neutral faces $\mathbf{n}_{1...N}$ found in stage 2 at convergence. The fitting applies an extra cost proportional to the deviation from stage 2 estimate. A matrix $\mathbf{E}_r$ is formed by concatenation of $\mathbf{E}$ N times followed by a diagonal square matrix $\mathbf{M}$, $\mathbf{E}_r = [\mathbf{E}^\top \ldots \mathbf{E}^\top \mathbf{M}^\top]^\top$. The diagonal elements of $\mathbf{M}$ are $\{\frac{S}{\sigma_1} \ldots \frac{S}{\sigma_k}\}$. The factor S determines the extent of the search portion of the subspace, a high value of S implies that only small deviations from stage 2 solution are allowed. The differences between $\mathbf{f}$ and $\mathbf{n}_{1...N}$ are concatenated in a vector $\mathbf{r}_r$ followed by a search restricting vector, $\mathbf{r}_r = [(\mathbf{f} - \mathbf{n}_1)^\top \ldots (\mathbf{f} - \mathbf{n}_N)^\top (\mathbf{M}\mathbf{r}_e)^\top]^\top$ where $\mathbf{r}_e$ is the residue estimate from stage 2. The final residue estimate and neutral face estimate are expressed as in Eqn. 5 and 6. Fig. 2 shows example 3D faces and their residue and neutral face decompositions.

$$\mathbf{r} = \mathbf{E}(\mathbf{E}_r^\top \mathbf{E}_r)^{-1} \mathbf{E}_r^\top \mathbf{r}_r \tag{5}$$

$$\mathbf{n} = \mathbf{f} - \mathbf{r} \tag{6}$$

3 Facial Expression Classification and Detection

Facial expression classification is performed based on the shape of the residue estimates and their normals. The normals are used because of their sensitivity to small shape variations which give the classifier a better ability to classify facial expressions that involve small surface deformations. The approach is to project

In the expression classification experiment, the non-neutral scans which were used to form the 1948 training residues were decomposed and their residue estimates were used for expression classification as described in Section 3. It is possible to use the training ground truth residues (difference between non-neutral and neutral scans) for classification rather than their estimates. However, experimental results have shown that the expressions are more accurately classified when the training residue estimates are used. Table 1 shows the results of the classifier. The detection rate of the fundamental expressions ranges from 96.22% to 83.02%. High detection rates were achieved especially for happy, surprise and neutral expressions. The rate tangibly decreases for the sad, angry and disgust expressions. This decrease is generally attributed to the scarcity of representing training data for these expressions, the small differences between them and their high variabilities.

Table 1. Expression detection and false detection rates for six fundamental expressions. The classification rate represents the rate of detecting all the fundamental expressions that exist in an image without any false detections.

	happy	surprise	sad	angry	disgust	neutral
Detection rate	94.3%	89.92%	94.14%	91.4%	83.02%	96.05%
False detection rate	2.56%	2.41%	0.64%	5.11%	0.54%	6.22%
Classification rate			86.3%			

The well-known PCA face recognition approach was applied to the decomposed neutral estimates. 406 estimates of neutral faces were extracted from non-neutral scans under severe facial expressions and used as probes. These probes were matched against a gallery of 100 subjects. The result was compared against the recognition performance when the decomposition was not used. The results shows about 6% increase in the identification performance when the proposed decomposition approach was used (39.9% for the non-decomposed scans and 46% for their neutral face estimates).

5 Conclusions

This paper presented a novel approach for decomposing an unseen 3D facial scans under facial expressions to neutral face estimates and expression residues. The residues were used for expression classification and gave a high expression classification results. This paper also demonstrated that the decomposition approach can boost the performance of existing face recognition algorithms that are designed for neutral facial scans.

References

1. Al-Osaimi, F., Bennamoun, M., Mian, A.: An Expression Deformation Approach to Non-rigid 3D Face Recognition. IJCV (2009)
2. Abboud, B., Davoine, F.: Bilinear factorisation for facial expression analysis and synthesis. Vision, Image and Signal Processing (2005)

3. Chandrasiri, N.P., Naemura, T., Harashima, H.: Interactive analysis and synthesis of facial expressions based on personal facial expression space. In: IEEE Conference on Automatic Face and Gesture Recognition (2004)
4. Chang, K., Bowyer, K., Flynn, P.: Adaptive Rigid Multi-region Selection for Handling Expression Variation in 3D Face Recognition. In: CVPR (2005)
5. Chen, Y., Medioni, G.: Object modeling by registration of multiple range images. In: ICRA (1991)
6. Chua, C., Han, F., Ho, Y.: 3D Human Face Recognition Using Point Signature. In: IEEE Conference on Automatic Face and Gesture Recognition (2000)
7. Faltemier, T., Bowyer, K., Flynn, P.: Using a Multi-Instance Enrollment Representation to Improve 3D Face Recognition. In: IEEE BTAS (2007)
8. Jollife, I.: Principal Component Analysis. Springer, Heidelberg (1986)
9. Lee, J., Lee, J.H., Kim, H., Moon, Y.: Automatic Synthesis of Realistic Facial Expressions. In: IEEE Symposium on Signal Processing and Information Technology (2007)
10. Matsugu, M., Mori, K., Mitari, Y., Kaneda, Y.: Subject independent facial expression recognition with robust face detection using a convolutional neural network. International Journal of Neural Networks (2003)
11. Mian, A., Bennamoun, M., Owens, R.: An Efficient Multimodal 2D-3D Hybrid Approach to Automatic Face Recognition. In: IEEE PAMI (2007)
12. Mpiperis, I., Malassiotis, S., Strintzis, M.: Bilinear Models for 3D Face and Facial Expression Recognition. IEEE Transactions on Information Forensics and Security (2008)
13. Phillips, P., Flynn, P., Scruggs, T., Bowyer, K., Chang, J., Hoffman, K., Marques, J., Min, J., Worek, W.: Overview of the Face Recognition Grand Challenge. In: CVPR (2005)
14. Ramachandran, M., Zhou, S., Jhalani, D., Chellappa, R.: A method for converting a smiling face to a neutral face with applications to face recognition. In: IEEE ICASSP (2005)
15. Ramanathan, S., Kassim, A., Venkatesh, Y., Wah, W.: Human Facial Expression Recognition using a 3D Morphable Model. In: IEEE Conference on Image Processing (2006)
16. Samper, S., Favreliere, H., Vacher, P., Coudert, T.: 3D measurement and synthesis of facial expressions. In: Symposium on 3D Analysis of Human Movement (2006)
17. Tang, H., Huang, T.: 3D facial expression recognition based on automatically selected features. In: IEEE Computer Vision and Pattern Recognition Workshops (2008)
18. Wang, H., Ahuja, N.: Facial Expression Decomposition. In: ICCV (2003)
19. Wang, S., Gu, X., Qin, H.: Automatic non-rigid registration of 3D dynamic data for facial expression synthesis and transfer. In: CVPR (2008)
20. Wang, J., Yin, L.: Static topographic modeling for facial expression recognition and analysis. Computer Vision ans Image Understanding (2007)
21. Zhan, Y., Ye, J., Niu, D., Cao, P.: Facial Expression Recognition Based on Gabor Wavelet Transformation and Elastic Templates Matching. In: International Conference on Image and Graphics (2004)

Pose Normalization for Local Appearance-Based Face Recognition

Hua Gao, Hazım Kemal Ekenel, and Rainer Stiefelhagen

Computer Science Department, Universität Karlsruhe (TH)
Am Fasanengarten 5, Karlsruhe 76131, Germany
{hua.gao,ekenel,stiefel}@ira.uka.de
http://isl.ira.uka.de/cvhci

Abstract. We focused this work on handling variation in facial appearance caused by 3D head pose. A pose normalization approach based on fitting active appearance models (AAM) on a given face image was investigated. Profile faces with different rotation angles in depth were warped into shape-free frontal view faces. Face recognition experiments were carried out on the pose normalized facial images with a local appearance-based approach. The experimental results showed a significant improvement in accuracy. The local appearance-based face recognition approach is found to be robust against errors introduced by face model fitting.

1 Introduction

Most of the current face recognition technologies are based on the appearance of face. One of the typical problems, where the appearance-based recognition approaches fail, is 3D pose variation. As the human face is a 3D object, its visual appearance under different view angles varies significantly. In [2], it has been shown that facial appearance variations caused by different head poses is larger than the ones caused by different identities. The large difference in appearance makes face recognition difficult when the face pose in the probe image differs from the one in the gallery image.

One way to compensate the pose variation is to add enough training samples that cover the pose-space of the gallery set. This sample augmentation scheme is adopted in many video-based face recognition systems [24] since the multi-pose samples can be obtained in video recordings. But in practice, it is not possible to cover the whole pose-space for each subject in the gallery. Moreover, enlarging the size of gallery set always slows down the system. For some recent application scenarios, such as identifying faces in the images from the Internet [16], this problem is even more significant because only limited number of gallery images are available. Pose normalization is one of the ways to solve the pose problem.

Chai et al. [5] presented a statistical affine transformation for pose normalization. They first estimated the face pose with a pose subspace algorithm, and divided the face region into three rectangles. An affine transformation was performed independently in the three rectangles and the transformation parameters

M. Tistarelli and M.S. Nixon (Eds.): ICB 2009, LNCS 5558, pp. 32–41, 2009.
© Springer-Verlag Berlin Heidelberg 2009

were statistically learned from the correspondence information between the specific pose and the frontal face. Experiments on the FERET database [20] showed that the recognition rate increases significantly compared to recognition without pose correction.

In [7], the active appearance model (AAM) was proposed as a model-based approach for face alignment and pose normalization. The shape of the model is optimized by minimizing the texture representation error. Once the model is fitted on an input image, the optimized model parameters can be used for face recognition [9]. Later in [8], the authors found that the shape parameters of AAMs can be used to estimate the pose angle. A frontal view of an input face image can be synthesized by configuring the shape parameters that control the pose. Instead of synthesizing a frontal face using the optimized model texture, the authors in [15] warp the texture inside the fitted model shape to the pose corrected shape. This texture warping approach has the advantage of preserving the textural information such as moles and freckles contained in the original image, which will be lost in the synthesis-based approach where the model parameters only represent the principal components of the appearance.

Blanz et al. [3] extended the 2D AAM to a 3D morphable model (3DMM). The vertices in a 3DMM shape are much more dense than an AAM shape. The dense shape model enables a high resolution face representation and the 3D shape enables linear transformation in 3D space in a more natural way. The 3DMM achieved promising results for illumination and pose-invariant face recognition. However, fitting a dense model requires much higher computational effort, which is not applicable for real-time face recognition systems.

In this paper, we investigated a pose normalization approach based on fitting 2D AAMs. The simultaneous inverse compositional (SIC) fitting algorithm [14] was employed for fitting generic models. To initialize the model more precisely and robustly against cluttered backgrounds, a progressive model fitting scheme was used. Different from [15], we generated shape-free face images as a pose normalization step, in which all the salient facial feature points were warped to a common coordinate framework. Different texture warping methods were studied and the local appearance-based face recognition approach was performed on the pose-normalized face images. We conducted our experiments on the FERET *b*-series database with a setup similar to the one in [5]. However, the profile face images with all rotation angles were evaluated, while in other studies not all of them were used [5,22,21]. Furthermore, we evaluated all the images including those where the AAM fitting failed, to show the robustness of the face recognition algorithms. Experiments show that local appearance-based approach is more robust against the errors introduced by model fitting and texture warping.

2 Active Appearance Models and Model Fitting

An active appearance model (AAM) is a generative parametric model which utilizes both shape and appearance information to represent a certain object such as the human face. A shape in AAM is defined as a set of normalized self-occluded face part is over-sampled during the warping, which enlarges the misalignment error. Another fact which we have noticed is that even if a semi-profile face is fitted perfectly, the warped frontal-view face still looks different from the real frontal face. Take the image labeled *bi* in Figure 3 for example: the left half-face is over-sampled and the right half-face is down-sampled after texture warping. This effect makes the left-eye wider and the right-eye narrower, which results in a different local appearance compared to the training image *ba* in Figure 3.

We carried out experiments with different face warping methods. First, the piecewise affine warp was considered. The original piecewise affine (PA) warp stretches zero-mean unit-length texture vector to intensity values in [0-255]. In addition to PA, piecewise affine warping with original pixel intensities was also considered and we named it PAN (Piecewise Affine warp No stretch). The third method (HIST) was also based on the piecewise affine warp, however, the histogram of the warped face was matched to the histogram of the mean face. In addition, the TPS-based warping technique was also evaluated. The face recognition results based on these four face warping approaches are plotted in Figure 4(a). Note that the recognition performances of these four methods are similar on series *be* and *bf*, where the face rotation angles are $\pm 15\,^{\circ}$. With the increasing degree of face rotation in depth, the recognition results differ from each other. Face warping with PAN outperforms the others in most cases. The performance of the other piecewise affine-based methods decreases dramatically when the fitting performance decreases due to large degree of face rotation. The reason for this behavior is that both methods change the histogram of the original input face image. In case of misalignment of the outline shape points, the background pixels may affect the histogram in the face area. The TPS warping, however, does not help in terms of face recognition. It achieves the worst results when testing on *bi*, *bh* and *bf*.

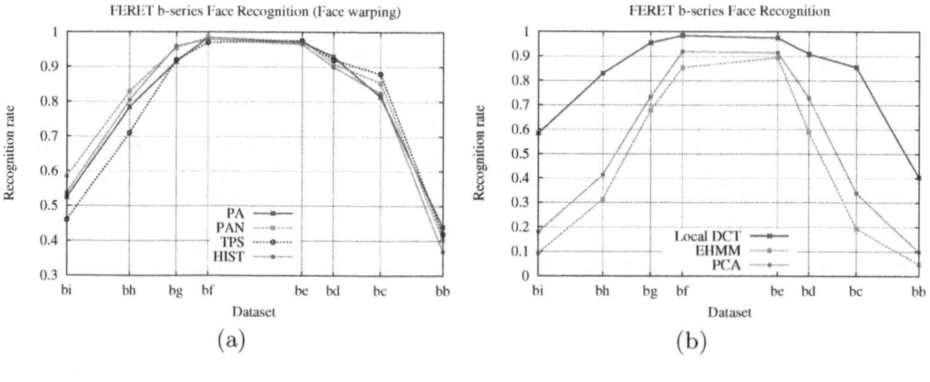

Fig. 4. Face recognition on the FERET *b*-series. (a) Face recognition rate with different face warping approaches. (b) Comparison of different face recognition techniques on pose-normalized face images with PAN face warping.

were statistically learned from the correspondence information between the specific pose and the frontal face. Experiments on the FERET database [20] showed that the recognition rate increases significantly compared to recognition without pose correction.

In [7], the active appearance model (AAM) was proposed as a model-based approach for face alignment and pose normalization. The shape of the model is optimized by minimizing the texture representation error. Once the model is fitted on an input image, the optimized model parameters can be used for face recognition [9]. Later in [8], the authors found that the shape parameters of AAMs can be used to estimate the pose angle. A frontal view of an input face image can be synthesized by configuring the shape parameters that control the pose. Instead of synthesizing a frontal face using the optimized model texture, the authors in [15] warp the texture inside the fitted model shape to the pose corrected shape. This texture warping approach has the advantage of preserving the textural information such as moles and freckles contained in the original image, which will be lost in the synthesis-based approach where the model parameters only represent the principal components of the appearance.

Blanz et al. [3] extended the 2D AAM to a 3D morphable model (3DMM). The vertices in a 3DMM shape are much more dense than an AAM shape. The dense shape model enables a high resolution face representation and the 3D shape enables linear transformation in 3D space in a more natural way. The 3DMM achieved promising results for illumination and pose-invariant face recognition. However, fitting a dense model requires much higher computational effort, which is not applicable for real-time face recognition systems.

In this paper, we investigated a pose normalization approach based on fitting 2D AAMs. The simultaneous inverse compositional (SIC) fitting algorithm [14] was employed for fitting generic models. To initialize the model more precisely and robustly against cluttered backgrounds, a progressive model fitting scheme was used. Different from [15], we generated shape-free face images as a pose normalization step, in which all the salient facial feature points were warped to a common coordinate framework. Different texture warping methods were studied and the local appearance-based face recognition approach was performed on the pose-normalized face images. We conducted our experiments on the FERET *b*-series database with a setup similar to the one in [5]. However, the profile face images with all rotation angles were evaluated, while in other studies not all of them were used [5,22,21]. Furthermore, we evaluated all the images including those where the AAM fitting failed, to show the robustness of the face recognition algorithms. Experiments show that local appearance-based approach is more robust against the errors introduced by model fitting and texture warping.

2 Active Appearance Models and Model Fitting

An active appearance model (AAM) is a generative parametric model which utilizes both shape and appearance information to represent a certain object such as the human face. A shape in AAM is defined as a set of normalized facial landmarks. To obtain enough variations in both shape and appearance, a set of representative facial images should be collected for training an AAM. Each facial image is manually labeled with a set of 2D landmarks, $[x_i, y_i], i = 1, 2, ...v$. A Procrustes analysis [13] is performed to make the shapes invariant to Euclidean transformation. Eigen-analysis is applied to reduce the redundancy in the normalized shape set and the resulting linear shape model represents a shape instance as $s = s_0 + \sum_{i=1}^{n} p_i s_i$, where s_0 is the mean shape, s_i is the ith shape basis, and $\mathbf{p} = [p_1, p_2, \ldots, p_n]$ are the shape parameters.

After building the shape model, each facial image is warped into the mean shape using a piecewise affine transformation. The transformation results in shape-free facial appearances which allows texture comparison in the coordinates of the mean shape. Eigen-analysis is performed on the shape-free appearances from all training images and the resulting model represents an instance appearance as $A = A_0 + \sum_{i=1}^{m} \lambda_i A_i$, where A_0 is the mean appearance, A_i is the ith appearance basis, and $\lambda = [\lambda_1, \lambda_2, \ldots, \lambda_m]$ are the appearance parameters.

The original AAM [7] learned the correlations between the shape model and appearance model by performing another eigen-analysis on the concatenated shape and texture parameters. This combined model is more compact. However, the choice of model fitting algorithms is restricted. Only numerical approaches are suitable for fitting the combined model. The authors in [18] proposed a more precise analytical approach for fitting AAMs using independent models in the framework of the inverse compositional (IC) image alignment algorithm. The IC algorithm is efficient and effective for fitting a person-specific AAM. However, the fitting accuracy degrades significantly on unseen faces. They extended their work in [14] for fitting a generic model to unseen persons with the simultaneous inverse compositional (SIC) algorithm, in which the shape parameter $\mathbf{p}$ and appearance parameter λ are optimized simultaneously during the search.

Given an input facial image $I(\mathbf{x})$, the goal of fitting an AAM is to find the optimal model parameters such that the synthesized model instance is as similar to the image observation as possible. This leads to a minimization of a cost function defined as,

$$E = \sum_{\mathbf{x} \in s_0} \left[I(\mathbf{W}(\mathbf{x}; \mathbf{p})) - A(\mathbf{x}, \lambda) \right]^2, \quad (1)$$

where $I(\mathbf{W}(\mathbf{x}; \mathbf{p}))$ is the warped input facial image, and $A(\mathbf{x}, \lambda)$ is the synthesized appearance instance. The minimization problem is usually solved by iteratively estimating the incremental update of the shape parameter $\Delta\mathbf{p}$ and the appearance parameter $\Delta\lambda$ and updating the current model parameters respectively. The IC and SIC algorithm, however, switch the role of the appearance template and the input image when computing $\Delta\mathbf{p}$. The shape is updated by composing an inverse incremental warping $\Delta W(\mathbf{x}; \mathbf{p})$ which is estimated where $\mathbf{p} = 0$. This IC framework enables the time-consuming steps of parameter estimation to be pre-computed and performed outside of the iterations.

It is known that the gradient-descent-based optimization problem usually requires a reasonable initialization. A poor initialization may cause the search to

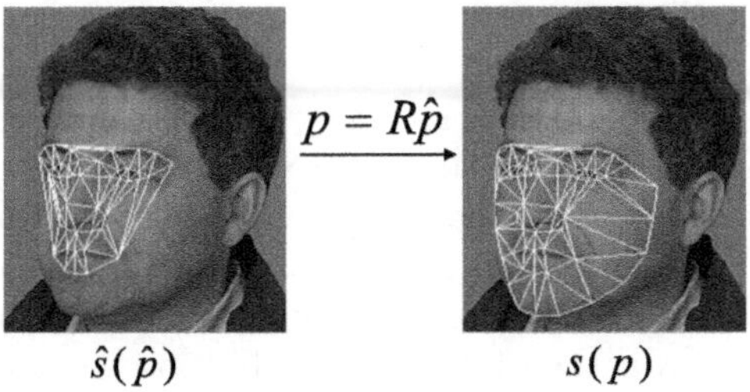

Fig. 1. Progressive AAM fitting

stuck in a local minimum. By utilizing some prior-knowledge of the face in question, the face shape can be initialized by transforming the mean shape close enough to the target. This initialization usually suffices for fitting frontal faces. However, when fitting semi-profile faces, part of the initialized shape cannot even cover the face. Thus the optimization can be affected by the included background pixels. Inspired by the work in [17], an inner-face model that only models the inner part of the face, i.e. eyes, nose, mouth, is built. The fitting is then performed progressively in two stages. In the first stage, only the inner-face model is initialized and fitted. In the second stage, the initial shape parameter $\mathbf{p}$ of the whole-face model is estimated from the obtained inner-face shape parameter $\hat{\mathbf{p}}$ with $\mathbf{p} = R\hat{\mathbf{p}}$, here we assume R is constant and it is obtained by performing linear regression on the training data of both models. A second fitting is carried out to fit the whole-face model using the initial parameters $\mathbf{p}$. Figure 1 depicts the bi-stage AAM fitting scheme, and it provides robust fitting accuracy in [17].

3 Pose Normalization

The most straightforward method to normalize the pose of a face image is the piecewise affine warping which is also used in the fitting iteration for sampling the texture inside the face mesh. The warp is realized by mapping the pixels in the fitted triangular mesh s to the base mesh s_0. For each pixel $\mathbf{x} = (x, y)^T$ in a triangle in the base mesh s_0, it can find a unique pixel $\mathbf{W}(\mathbf{x}; \mathbf{p}) = \mathbf{x}' = (x', y')^T$ in the corresponding triangle in the mesh s by an affine mapping in the barycentric coordinates. The implementation for the piecewise affine warp is detailed in [18].

The piecewise affine warping is simple but the deformation field is not smooth. Straight lines may usually be bended across triangle boundaries. To overcome this discontinuity, we investigated another warping technique based on thin-plate splines (TPS) [4] to improve the smoothness of the deformation. However, the behavior of the method is not always clear in-between feature points, especially in the case of large pose variations.

4 Face Recognition

The pose normalized shape-free facial images are masked with the AAM mean shape mask. All salient facial features are warped to the canonical locations. However, feature points around the chin area might be misaligned, which may create strong edges. As demonstrated in [11], the chin area does not contribute too much discriminative information compared to other facial features. For this reason, we decide to crop the chin area in the pose normalized facial image. Following the approach in [10], we scaled the cropped images to 64×64 pixels size and then divided them into 64 non-overlapped blocks of 8×8 pixels size. On each local block, the discrete cosine transform (DCT) is performed. The obtained DCT coefficients are ordered using a zig-zag scanning. The first component is skipped because it represents the average pixel intensity of the entire block. The following ten low frequency coefficients are retained which yields a ten dimensional local feature vector. Finally, the 64 local feature vectors are concatenated to construct the feature vector of a whole face image.

The DCT preserves the total image energy of the processed input block; therefore blocks with different brightness levels lead to DCT coefficients with different magnitudes. In order to balance each local block's contribution to the classification, the local feature vector is normalized to unit norm. To balance the contribution of each DCT frequency, each DCT coefficient is divided by its corresponding standard deviation [12] before the unit normalization. Nearest neighbor with L1 distance measure is used for classification.

5 Experiments

We built a generic AAM to fit it on all facial images of different subjects. The study in [14] showed that building generic AAMs is far more difficult than building person-specific AAMs. In order to reduce the effects of reconstruction errors, more training images are needed to cover all possible variations in pose, illumination, expression and identity.

In this work, we collected a set of 649 images from four publicly available databases: the IMM database [1], the ND1 database [6], the CMU PIE database [23], and the FERET database. The example images selected from these databases are displayed in Figure 2. Each training face image was manually labeled with 58 landmarks. The manual labeling is a time-consuming operation and is error-prone due to the accuracy limitations of a manual operation. Moreover, the interpretation of correct landmark locations differs between annotators. The inconsistent labeling introduces noise in the shape data which affects the face modeling and the performance of fitting, consequentially. To solve this problem, we employed an iterative data refitting method as suggested in [14] to eliminate the noise in the manual labels.

The face recognition experiments were carried out on the FERET *b*-series which contains 200 subjects, where each subject has nine images corresponding to different pose angles. Sample images of a single subject are shown in the first row of Figure 3. The label *ba* indicates the frontal series, which was used as the training set (gallery). The remaining eight series are non-frontal face images with different pose angles ($\pm 60\,^\circ$, $\pm 40\,^\circ$, $\pm 25\,^\circ$, and $\pm 15\,^\circ$). All except the *ba* series were used for testing (probe). None of the subjects used for training the AAM were used during the recognition experiments.

The example pose normalized images are depicted in the second row of Figure 3. The automatic AAM initialization is not always stable because of the large rotation angle in some series such as *bb* and *bi*. Automatic facial feature detectors fail to detect eyes and mouth, which are important for a precise AAM initialization. In this sense, we used a semi-automatic AAM initialization where the annotation of eye-centers and mouth-centers were used for model initialization.

In general, fitting near-frontal faces performs better than fitting semi-profile faces. When fitting semi-profile faces, even a small misalignment in the chin area may cause a large error in the warped face image. The reason is that the partially

Fig. 2. Sample images from IMM, ND1, FERET and CMU PIE face databases for training a generic model

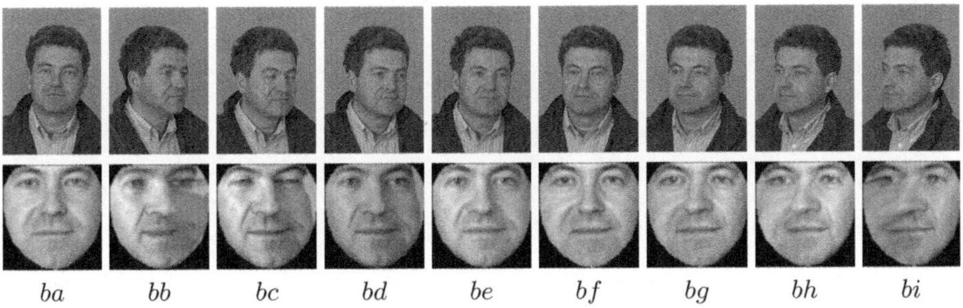

Fig. 3. The first row shows the example images from the FERET *b*-series. The face images in the series *ba* are all frontal. The series $b[b-i]$ contains faces in different yaw pose-angles: *bb* ($60\,^\circ$), *bd* ($40\,^\circ$), *bc* ($25\,^\circ$), *be* ($15\,^\circ$), *bf* ($-15\,^\circ$), *bg* ($-25\,^\circ$), *bh* ($-40\,^\circ$), and *bi* ($-60\,^\circ$). The second row shows the corresponding fitted faces with pose normalized.

self-occluded face part is over-sampled during the warping, which enlarges the misalignment error. Another fact which we have noticed is that even if a semi-profile face is fitted perfectly, the warped frontal-view face still looks different from the real frontal face. Take the image labeled *bi* in Figure 3 for example: the left half-face is over-sampled and the right half-face is down-sampled after texture warping. This effect makes the left-eye wider and the right-eye narrower, which results in a different local appearance compared to the training image *ba* in Figure 3.

We carried out experiments with different face warping methods. First, the piecewise affine warp was considered. The original piecewise affine (PA) warp stretches zero-mean unit-length texture vector to intensity values in [0-255]. In addition to PA, piecewise affine warping with original pixel intensities was also considered and we named it PAN (Piecewise Affine warp No stretch). The third method (HIST) was also based on the piecewise affine warp, however, the histogram of the warped face was matched to the histogram of the mean face. In addition, the TPS-based warping technique was also evaluated. The face recognition results based on these four face warping approaches are plotted in Figure 4(a). Note that the recognition performances of these four methods are similar on series *be* and *bf*, where the face rotation angles are $\pm 15\,^{\circ}$. With the increasing degree of face rotation in depth, the recognition results differ from each other. Face warping with PAN outperforms the others in most cases. The performance of the other piecewise affine-based methods decreases dramatically when the fitting performance decreases due to large degree of face rotation. The reason for this behavior is that both methods change the histogram of the original input face image. In case of misalignment of the outline shape points, the background pixels may affect the histogram in the face area. The TPS warping, however, does not help in terms of face recognition. It achieves the worst results when testing on *bi*, *bh* and *bf*.

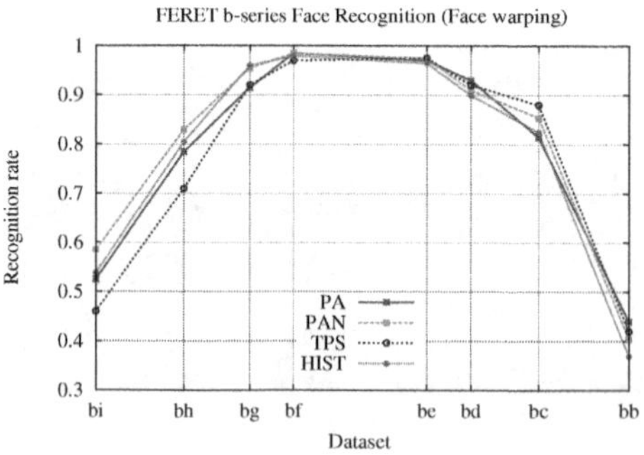

(a)

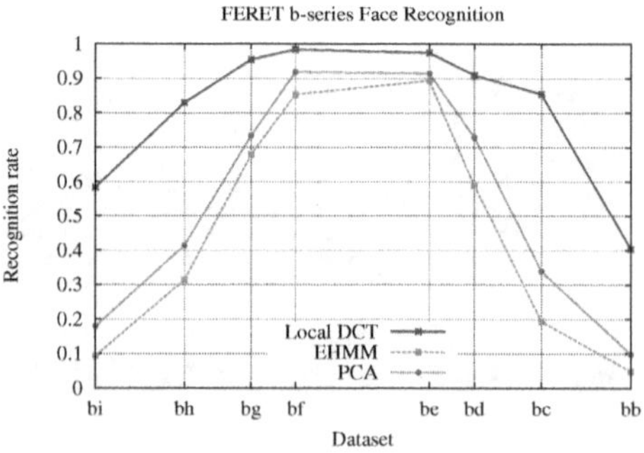

(b)

Fig. 4. Face recognition on the FERET *b*-series. (a) Face recognition rate with different face warping approaches. (b) Comparison of different face recognition techniques on pose-normalized face images with PAN face warping.

In addition to the different face warping methods, we also compared various face recognition techniques. Since the training set only contains one example per subject, the selection for other face recognition techniques is limited. Techniques which utilize intra-class information are not suitable for this evaluation. We selected two well-known face recognition approaches: the holistic Eigenfaces approach [25] and embedded hidden Markov models (EHMM) [19], which utilize DCT coefficients as local features as well. The selected algorithms were tested on face images warped with the PAN method. Figure 4(b) plots the corresponding recognition rates. Local DCT denotes the local appearance-based face recognition approach. For the Eigenfaces approach, 40 eigen-components were used, which yielded optimal performance. We noticed that Local DCT outperformed the other two approaches significantly over all the eight series. Moreover, the recognition rates of EHMM and PCA decreased drastically when the rotation angles increased. This demonstrates that the performance of the other approaches can be easily effected by local deformation of the face appearance.

We end this section by comparing the proposed face fitting and alignment method to a simple affine transform-based face alignment approach without pose correction. The results demonstrate that the face recognition performance increases significantly with the pose correction. As listed in Table 1, recognition with simple affine alignment only achieved 62.5% and 78.5% with $\pm 15\,^\circ$ degree of face rotation, and it did not work at all when the face rotation angles were larger than $40\,^\circ$.

Table 1. Face recognition results on the FERET series *bb*-*bi* with PAN and simple affine (SA) face warping. Recognition with local appearance-based approach.

Probe set	*bb*	*bc*	*bd*	*be*	*bf*	*bg*	*bh*	*bi*
Rec. rate (PAN)	44.0%	81.5%	93.0%	97.0%	98.5%	91.5%	78.5%	52.5%
Rec. rate (SA)	0.0%	5.5%	26.0%	62.5%	78.5%	26.5%	4.0%	1.0%

6 Conclusions

A face registration approach based on AAM fitting was studied in this work. The pose of an input face was normalized with a piecewise affine warp resulting in a frontal view of the input face. To initialize the model more precisely and to increase robustness against cluttered backgrounds, an inner-face AAM was built and the fitting was done progressively. Face recognition was performed on the pose-normalized face images using the local appearance-based approach. We extracted the local features from each block on a pose normalized image using the DCT, and then concatenated the normalized local features for classification.

We conducted experiments on the FERET *b*-series to evaluate the proposed approach. With the AAM-based approach we achieved 97.8% recognition rate on the data set with $\pm 15\,^\circ$ yaw pose-angle. The simple affine warp approach however, only recognized 70.5% of the testing images with this pose angle. When the rotation angle increased to $\pm 60\,^\circ$, recognition based on simple affine warp could hardly recognize any images in the probe set, while the AAM-based approach still achieved a recognition rate of 48.3%. The proposed pose normalization helped a lot to recognize faces with small angles of rotation as we observed in the experiments. We also compared the local appearance-based face recognition to other approaches such as the Eigenfaces and the EHMM. The local appearance-based approach significantly outperformed the other two approaches and it was more robust against the error introduced by AAM fitting and face warping.

Acknowledgments

This work was realized as part of the Quaero Programme, funded by OSEO, French State agency for innovation.

References

1. The IMM face database, http://www2.imm.dtu.dk/~aam/
2. Beymer, D., Poggio, T.: Face recognition from one example view. In: ICCV, Boston, MA, pp. 500–507 (1995)
3. Blanz, V., Vetter, T.: Face recognition based on fitting a 3D morphable model. IEEE Trans. on PAMI 25(9), 1063–1074 (2003)
4. Bookstein, F.L.: Principal warps: thin-plate splines and the decomposition of deformations. IEEE Trans. on PAMI 11(6), 567–585 (1989)
5. Chai, X., Shan, S., Gao, W.: Pose normalization for robust face recognition based on statistical affine transformation. In: Chai, X., Shan, S., Gao, W. (eds.) ICICS-PCM 2003, Singapore, vol. 3, pp. 1413–1417 (2003)
6. Chang, K., Bowyer, K., Flynn, P.: Face recognition using 2D and 3D facial data. In: Proc. ACM Workshop on Multimodal User Authentication, pp. 25–32 (2003)
7. Cootes, T.F., Edwards, G.J., Taylor, C.J.: Active appearance models. In: Burkhardt, H., Neumann, B. (eds.) ECCV 1998. LNCS, vol. 1407, pp. 484–498. Springer, Heidelberg (1998)
8. Cootes, T.F., Wheeler, G.V., Walker, K.N., Taylor, C.J.: View-based active appearance models. Image and Vision Computing 20(9-10), 657–664 (2002)
9. Edwards, G.J., Cootes, T.F., Taylor, C.J.: Face recognition using active appearance models. In: Burkhardt, H., Neumann, B. (eds.) ECCV 1998. LNCS, vol. 1407, pp. 581–595. Springer, Heidelberg (1998)
10. Ekenel, H.K., Stiefelhagen, R.: Local appearance based face recognition using discrete cosine transform. In: 13th European Signal Processing Conference (EUSIPCO 2005), Antalya, Turkey (2005)
11. Ekenel, H.K., Stiefelhagen, R.: Block selection in the local appearance-based face recognition scheme. In: CVPR Biometrics Workshop, New York (2006)
12. Ekenel, H.K., Stiefelhagen, R.: Analysis of local appearance-based face recognition: Effects of feature selection and feature normalization. In: CVPR Biometrics Workshop (2006)
13. Goodall, C.: Procrustes methods in the statistical analysis of shape. Journal of the Royal Statistical Society, Series B 53(2), 285–339 (1991)
14. Gross, R., Matthews, I., Baker, S.: Generic vs. person specific active appearance models. Image and Vision Computing 23(11), 1080–1093 (2005)

15. Guillemaut, J., Kittler, J., Sadeghi, M.T., Christmas, W.J.: General pose face recognition using frontal face model. In: Martínez-Trinidad, J.F., Carrasco Ochoa, J.A., Kittler, J. (eds.) CIARP 2006. LNCS, vol. 4225, pp. 79–88. Springer, Heidelberg (2006)
16. Huang, G.B., Ramesh, M., Berg, T., Learned-Miller, E.: Labeled faces in the wild: a database for studying face recognition in unconstrained environments. University of Massachusetts, Amherst, Technical Report 07-49 (October 2007)
17. Kim, D., Kim, J., Cho, S., Jang, Y., Chung, S.-T., Kim, B.-G.: Progressive AAM based robust face alignment. In: Proc. of World Academy of Science, Engineering and Technology, vol. 21, pp. 488–492 (2007)
18. Matthews, I., Baker, S.: Active appearance models revisited. IJCV 60(2), 135–164 (2004)
19. Nefian, A.: A hidden Markov model-based approach for face detection and recognition, Ph.D. dissertation, Georgia Institute of Technology (1999)
20. Phillips, P.J., Wechsler, H., Huang, J., Rauss, P.: The FERET database and evaluation procedure for face recognition algorithms. Image and Vision Computing 16(5), 295–306 (1998)
21. Sanderson, C., Shan, T., Lovell, B.C.: Towards pose-invariant 2D face classification for surveillance. In: Zhou, S.K., Zhao, W., Tang, X., Gong, S. (eds.) AMFG 2007. LNCS, vol. 4778, pp. 276–289. Springer, Heidelberg (2007)
22. Shan, T., Lovell, B.C., Chen, S.: Face recognition robust to head pose from one sample image. In: 18th Int'l. Conf. on Pattern Recognition, vol. 1, pp. 515–518 (2006)
23. Sim, T., Baker, S., Bsat, M.: The CMU pose, illumination and expression (PIE) database. In: Proc. IEEE Int'l. Conf. on Automatic Face and Gesture Recognition (May 2002)
24. Stallkamp, J., Ekenel, H.K., Stiefelhagen, R.: Video-based face recognition on real-world data. In: ICCV 2007, Rio de Jenario, Brasil (October 2007)
25. Turk, M., Pentland, A.: Eigenfaces for recognition. J. Cognitive Neuroscience 3(1), 71–86 (1991)

Bayesian Face Recognition Based on Markov Random Field Modeling

Rui Wang, Zhen Lei, Meng Ao, and Stan Z. Li*

Center for Biometrics and Security Research,
Institute of Automation, Chinese Academy of Sciences, Beijing 100190, China
{rwang,zlei,mao,szli}@cbsr.ia.ac.cn
http://www.cbsr.ia.ac.cn

Abstract. In this paper, a Bayesian method for face recognition is proposed based on Markov Random Fields (MRF) modeling. Constraints on image features as well as contextual relationships between them are explored and encoded into a cost function derived based on a statistical model of MRF. Gabor wavelet coefficients are used as the base features, and relationships between Gabor features at different pixel locations are used to provide higher order contextual constraints. The posterior probability of matching configuration is derived based on MRF modeling. Local search and discriminate analysis are used to evaluate local matches, and a contextual constraint is applied to evaluate mutual matches between local matches. The proposed MRF method provides a new perspective for modeling the face recognition problem. Experiments demonstrate promising results.

Keywords: Face Recognition, Markov random field, learning.

1 Introduction

Relations between image features provides important constraints for image analysis and face recognition. A face is seen as an array of pixel values. If we make an image darker or brighter or perform some other nonlinear transformation such as changing the lighting condition, the image is still a pattern of that particular face. However, if the pixels are shuffled, the image no longer contains a face pattern, even if the individual pixel values remain the same. This illustrates that contextual constraints between image features play a more critical role than the pixel values themselves.

Markov random field (MRF) theory provides a convenient and consistent way for modeling context-dependent constraints such as between image pixels and features derived from pixels. This is achieved through characterizing mutual influences among pixels and features using conditional MRF distributions. The practical use of MRF models is largely ascribed to a theorem that states the equivalence between MRFs and Gibbs distributions, as established by [1] and further developed by [2]. MRFs have been widely employed to solve image procesing and computer vision problems [3,4,5,6,7].

Recently, researchers have started investigations of face recognition problems from the MRF viewpoint. In [8], Qian and Huang presents a MRF-based method for face

* Corresponding author.

M. Tistarelli and M.S. Nixon (Eds.): ICB 2009, LNCS 5558, pp. 42–51, 2009.
© Springer-Verlag Berlin Heidelberg 2009

detection, where scale changes are dealt with by using image pyramid and orientation changes are handled by rotating the image pyramid. In [9], Dass and Jain use MRF to model the spatial distribution of pixel values for the face pattern. The MRF modeling is further enhanced by finding the optimal permutation of pixels that maximizes the discriminative power between face and nonface patterns [10]. Huang *et al.* [11] propose a hybrid method for face recognition using MRF and belief propagation. A face image is divided into non-overlapping patches, and the patches in the probe image are matched to those in the gallery images, giving a person ID for each patch. The final decision is made by voting IDs over patches. In [12], straight lines are extracted from a face image. By attaching properties and binary relations to the straight lines, a face is then represented as a graph. A partial matching is used to match two graphs and select the best match.

In this paper, we propose a novel approach for modeling face recognition problems using MRF and Bayesian statistics. Assuming that the face in an image has been properly normalized into a "canonical" frame, the following two issues are addressed: (1) effective representation of a face pattern using image properties and their contextual relations, and (2) matching with the image properties and their contextual relations.

The paper contains the following novel ideas. For issue (1), we explore mutual constraints between the images features in a neighborhood system, where the features used are the Gabor wavelet coefficients. Relationships between Gabor features at different pixel locations are used as contextual constraints. Discriminant analysis is performed on the Gabor features to select most effective ones.

To solve problem (2), MRF is used to constrain matching configurations. The posterior probability of MRF is formulated as the criterion function for the evaluation of MRF configurations for matching between features in the gallery and probe. The best matching is defined as the maximum a posteriori (MAP) solution. The MAP solution is the best match between the two faces in terms not only image features but also relationships between them.

The rest of the paper is organized as follows: Section 2 introduces basic MRF concepts and Gabor wavelets as base features. Section 3 presents the use of MRF modeling to constrain matching configurations and formulates the posterior probability function for face matching. Experimental results are presented in Section 4 and in Section 5, we conclude the paper.

2 Background

2.1 MRF Modeling

The problem of face recognition may be posed as (1) finding correspondence between features in gallery and probe, which can be posed as a labeling problem; (2) computing the score of matching (or distance) between the two, which may be posed as a probability estimation problem, and (3) making decision based on the distances. Here we introduce the notions for labeling problems and how MRF could constrain labeling configurations with contextual relations.

Assume that a face is divided into a number of regions for both probe and gallery faces. Here a region is centered at a pixel location, and some regions may overlap each other. Index the m regions in the probe by the set of sites $\mathscr{S} = \{1,\ldots,m\}$, and L regions in the probe by the set of labels $\mathscr{L} = \{1,\ldots,L\}$.

A labeling is then considered as realization $f = \{f_1,\ldots,f_m\}$ of a random field denoted $F = \{F_1,\ldots,F_m\}$. There, a random variable F_i takes a value f_i in $\mathscr{L}$, for which f is called a *configuration* of F, corresponding to a realization of the field.

The random variables F may be related to one another contextually via a *neighborhood system* $\mathscr{N} = \{\mathscr{N}_i \mid \forall i \in \mathscr{S}\}$ where $\mathscr{N}_i$ is the set of sites neighboring i. The neighboring relationship can be defined to include sites with the Euclidean distance from i.

A neighborhood system specifies a collection of *cliques*. A clique consists either of a single site $c = \{i\}$, or of a pair of neighboring sites $c = \{i,i'\}$. The collections of single-site and pair-site are denoted by $\mathscr{C}_1$ and$\mathscr{C}_2$, respectively.

F is said to be an MRF on $\mathscr{S}$ with respect to $\mathscr{N}$ if and only if $P(f) > 0$ (positivity), and $P(f_i \mid f_{\mathscr{S}-\{i\}}) = P(f_i \mid f_{\mathscr{N}_i})$ (Markovianity) are satisfied [1], where $\mathscr{S} - \{i\}$ is the set difference, $f_{\mathscr{S}-\{i\}}$ denotes the set of labels at the sites in $\mathscr{S} - \{i\}$ and $f_{\mathscr{N}_i} = \{f_{i'} \mid i' \in \mathscr{N}_i\}$ stands for the set of labels at the sites neighboring i. The Markovianity depicts the local characteristics of F.

The joint probability of a Markov random fields obeys a *Gibbs distribution*, which takes the following form

$$P(f) = Z^{-1} \times e^{-\frac{1}{T}U(f)} \quad (1)$$

where Z is a normalizing constant, T is the *temperature* constant, and $U(f)$ is the *energy function*. The energy

$$U(f) = \sum_{c \in \mathscr{C}} V_c(f) \quad (2)$$

is a sum of *clique potentials* $V_c(f)$ over all possible cliques $\mathscr{C}$. $P(f)$ measures the probability of the occurrence of a particular configuration, or "pattern", and is a decreasing function of the energy. A clique potential is defined on the labels on the clique c.

Contextual constraints are encoded in $V_c(f)$ for c containing more than one site. Consider up to two site cliques, $U(f)$ can be expressed as

$$U(f) = \sum_{\{i\} \in \mathscr{C}_1} V_1(f_i) + \sum_{\{i,i'\} \in \mathscr{C}_2} V_2(f_i, f_{i'}) \quad (3)$$

where it is V_2 that encode contextual constraints.

2.2 Gabor Wavelets

In this work, we use Gabor wavelets, instead of the raw face image pixels, as basic face features because Gabor wavelets features exhibit desirable characteristics of spatial locality and orientation selectively, and are optimally localized in the space and frequency domains.

A face is represented Gabor-based vector, derived as a set of convolutions with a family of complex Gabor filters:

$$\psi_{\mu,\nu} = \frac{k_{\mu,\nu}^2}{\sigma^2} \exp(-\frac{k_{\mu,\nu}^2 z^2}{2\sigma^2})[\exp(ik_{\mu,\nu}z) - \exp(-\frac{\sigma^2}{2})] \quad (4)$$

where μ and v define the orientation and scale of the Gabor kernels respectively, $z = (x, y)$, and the wave vector $k_{\mu,v}$ is defined as follows:

$$k_{\mu,v} = k_v e^{i\phi_\mu} \tag{5}$$

where $k_v = k_{max}/f^v$, $k_{max} = \pi/2$, $f = \sqrt{2}$, $\phi_\mu = \pi\mu/8$, with index $j = \mu + 8v$. This representation is chosen for its biological relevance and technical properties. The Gabor kernels resemble the receptive field profiles of simple cells in the visual pathway. They are localized in both space and frequency domains and achieve the lower bound of the space-bandwidth product as specified by the uncertainty principle. At each pixel location $X = (u, v)$, the Gabor coefficients are computed as

$$J_j(X) = \int I(X')\psi_j(X - X')\, d^2X' \qquad (j = 0, 1, 2 \ldots 40) \tag{6}$$

where $I(X)$ is the image grey level distribution.

Five scales, indexed by $\mu \in \{0, 1, 2, 3, 4\}$ and eight orientations, indexed by $v \in \{0, 1, 2, \ldots, 7\}$, are used. Convolving the filters with an image gives 40 complex coefficients, and we can then get 40 Gabor magnitudes as a more salient representation.

The Gabor filtering gives a set of 40 features for each region I of the gallery, denoted as a 40-dim vector $D(I)$. We can similarly compute $d(i)$ for each region i of the probe. These are the original data.

3 MAP-MRF Based Face Recognition

3.1 Formulation

The basic face matching problem is to compare a probe and a gallery by finding the best correspondences or region matches between the two. The regions in the probe that find no good matches in the gallery is labeled NULL ("no-match") indexed by 0. The problem of matching between probe and gallery using MRF modeling is formulated as labeling the sites in $\mathscr{S}$ in terms of the label set $\mathscr{L}^+ = \{0, 1, \ldots, L\}$.

The MAP-MRF matching maximizes the posterior probability, which is defined based on observation data and contextual constraints between sites

$$f^* = \arg\max_f P(f \mid d) \tag{7}$$

where d is the observation data and in this work corresponds to the set of Gabor features extracted from the probe face image. The posterior probability can be derived from the prior distribution $P(f)$ and the likelihood function $p(d \mid f)$. Assuming that f is a MRF, its distribution follows a Gibbs distribution of Equ.(1). Assuming that the likelihood distribution is also an exponential function $p(d \mid f) \propto \mathrm{e}^{-U(d \mid f)}$, then the posterior probability has the following form

$$P(f \mid d) \propto \mathrm{e}^{-U(f \mid d)} \propto \mathrm{e}^{-U(f) - U(d \mid f)} \tag{8}$$

The MAP estimate is equivalently found by minimizing the posterior energy function

$$f^* = \arg\min_f U(f \mid d) \tag{9}$$

The following derives $U(f)$ and $U(d \mid f)$.

The likelihood function $p(d \mid f)$ has the following characteristics:

1. It is conditioned on pure non-NULL matches $f_i \neq 0$,
2. It is independent of the neighborhood system $\mathcal{N}$, and
3. It depends on how the model object is observed in the scene which in turn depends on the underlying transformations and noise.

Assuming that the Gabor features in d are invariant under the considered class of transformations, and d of probe is related the corresponding features D of gallery via the observation model

$$d(i) = D(f_i) + e(i) \tag{10}$$

where e is additive independent zero mean Gaussian noise. The Gaussian assumptions may not be accurate but offers an approximation.

Then the likelihood function is a Gibbs distribution with the energy

$$U(d \mid f) = \Sigma_{i \in \mathcal{S}, f_i \neq 0} V_1(d(i) \mid f_i) \tag{11}$$

where the constraint $f_i \neq 0$ restricts the summations to take over the non-NULL matches. The likelihood potentials are

$$V_1(d(i) \mid f_i) = \begin{cases} \Delta(d(i), D(f_i)) & \text{if } f_i \neq 0 \\ 0 & \text{otherwise} \end{cases} \tag{12}$$

where $\Delta(d(i), D(f_i)) = |d(i) - D(f_i)|_e$ is some distance between the two random vectors respect to the noise e. In practice, we apply LDA (linear discriminant analysis) to find the discriminant subspace of the distribution of the training data, which is done over the whole image for all sites i, and use the Euclidean distance in the LDA subspace to replace the likelihood energy $U(d \mid f)$.

To model the prior distribution of Equ.(3), the single-site potentials are defined as

$$V_1(f_i) = \begin{cases} v_{10} & \text{if } f_i = 0 \\ 0 & \text{otherwise} \end{cases} \tag{13}$$

where v_{10} is a constant.

Between the probe and gallery, a pair of local matched regions (i, f_i) incurs a displacement vector $D(i, f_i) = [dx(i, f_i), dy(i, f_i)]$. The displacement image can be considered as a "flow map". The displacement in a neighborhood should be as consistent as possible. This is the smoothness constraint used as the contextual constraint.

We therefore define the pair-site potentials as

$$V_2(f_i, f_{i'}) = \begin{cases} v_{20} & \text{if } f_i = 0 \text{ or } f_{i'} = 0 \\ |D(i, f_i) - D(i', f_{i'})| & \text{otherwise} \end{cases} \tag{14}$$

where v_{20} is a constant and $|D(i, f_i) - D(i', f_{i'})|$ is the flow inconsistency. When the displacement vectors in the neighborhood are similar, the local flow inconsistency is small; otherwise they incur a large penalty. If either f_i or $f_{i'}$ is the NULL , it incurs a penalty of v_{20}; otherwise, the penalty is the flow inconsistency. The above clique potentials define the prior energy

$$U(f) = \sum_{i\in\mathscr{S}} V_1(f_i) + \sum_{i\in\mathscr{S}} \sum_{i'\in\mathscr{N}_i} V_2(f_i, f_{i'}) \tag{15}$$

The posterior energy is then obtained as $U(f \mid d) = U(f) + U(d \mid f)$.

3.2 Search Algorithm

Finding a MAP-MRF solution is to find the best matched correspondences between regions in the input probe and the gallery, according to both image features and their relationships. This amounts to finding the minimum of $U(f \mid d)$. While it is a combinatorial optimization problem and the global solution is computationally inhibitive, we present a heuristic local minimization algorithm to find a local solution as follows.

First, we do a local search over a neighborhood of i to minimize the simple Euclidean distance between $d(i)$ and $D(f_i)$. This gives the best matched region f_i^* in terms of the observation $d(i)$. For each i, if the distance is too large, we set $f_i^* = 0$. After this is done for all i, we calculate the Euclidean distance in LDA subspace between the data in d of the probe and D of the gallery. This gives the likelihood energy $U(d \mid f^*)$ of Equ.(11). After that, the displacements over the neighborhood can be computed. From these, the the flow inconsistency can be estimated according to Equ.(14). The final posterior energy $U(f^* \mid d)$ is evaluated as the cost for the best overall cost.

Supposing there are N gallery faces, then N costs can be obtained for the N galleries: $E(f^{(n)}) = U(f^{(n)} \mid d)$. Let the lowest cost among the N matches be

$$E_{\min} = \min_{n\in\{1,\ldots,N\}} E(f^{(n)}) \tag{16}$$

and the corresponding gallery image be

$$n_{\min} = \arg \min_{n\in\{1,\ldots,N\}} E(f^{(n)}) \tag{17}$$

A simplest scheme for face recognition is the following: If the overall minimum energy is small enough, *i.e.* $E_{\min} < E_{Thr}$, the probe is recognized as the person that gallery $n_{\min}$ belongs to, or otherwise it is considered as not in the gallery set. Some more sophisticated schemes are possible by considering all $E(f^{(n)})$s and the associated person identities.

4 Experiments

The experiments are presented to compare face recognition performance without (non-MRF methods) and with (the proposed MAP-MRF method) using contextual constraint. Six non-MRF algorithms are evaluated: PCA, Fisher face (FLDA) [13], regularized LDA (RLDA) [14], Gabor+PCA, Gabor+FLDA, Gabor+RLDA, which are popular methods and have achieved great success in face recognition. The Gabor+PCA, Gabor+FLDA, Gabor+RLDA and MAP-MRF methods use a selected set of 4000 Gabor features, for which 100 salient locations (regions) on a face are selected and 40 Gabor magnitude features are computed at each of the locations.

Two face databases, FERET [15] and FRGC ver 2.0 [16] are used for the experiments. All the images are rotated, scaled and cropped to 128×128 according to the provided eye positions succeeded by histogram equalization preprocessing. No further preprocessing is applied. For FERET database, the training set contains 731 images. In test phase, we use the gallery set containing 1196 images from 1196 subjects, and combine four provided probe sets together, totally including 2111 images to compose the probe set. So our test protocol should be more difficult than any of the four original protocols because we consider different factors (expression, illumination, aging etc.) together to evaluate the performance. For FRGC database, we select a subset from query set for experiment 4, which consists of still uncontrolled images including variations of illumination, expression, accessory and blurring. There are 316 subjects, each of which contains at least 10 images. We randomly select 10 images for each subject to get a total $316 \times 10 = 3,160$ images. These images are randomly divided into three sets. The training set consists of 116 persons, with 10 images per person. The left 200 subjects are divided into gallery and probe sets. For each person, 2 images randomly selected from the 10 images compose the gallery set and the left 8 images compose the probe set. The persons in training set are disjoint with those in gallery and probe sets. Fig. 1 illustrates some cropped face examples of FERET and FRGC databases.

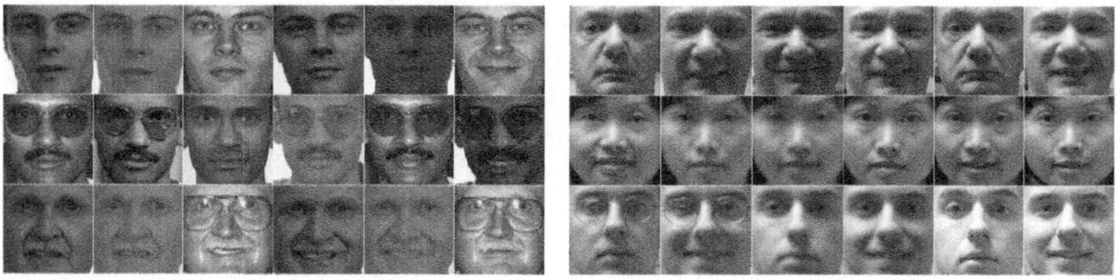

(a) (b)

Fig. 1. Face examples of FERET (a) and FRGC (b) databases

The results are reported in terms of three performance indices: rank-1 recognition rate, verification rate (VR) when the false accept rate (FAR) is 0.001, and equal error rate (EER).

Table 1 and 2 illustrate the results of different methods on FERET and FRGC databases respectively and Fig. 2 plots the corresponding ROC curves. For clarity, we only plot the results based on the Gabor features. In all experiments, the MAP-MRF method achieves the best accuracy. The results suggest that the use of contextual constraints is more robust to expression, aging, mis-alignment etc.

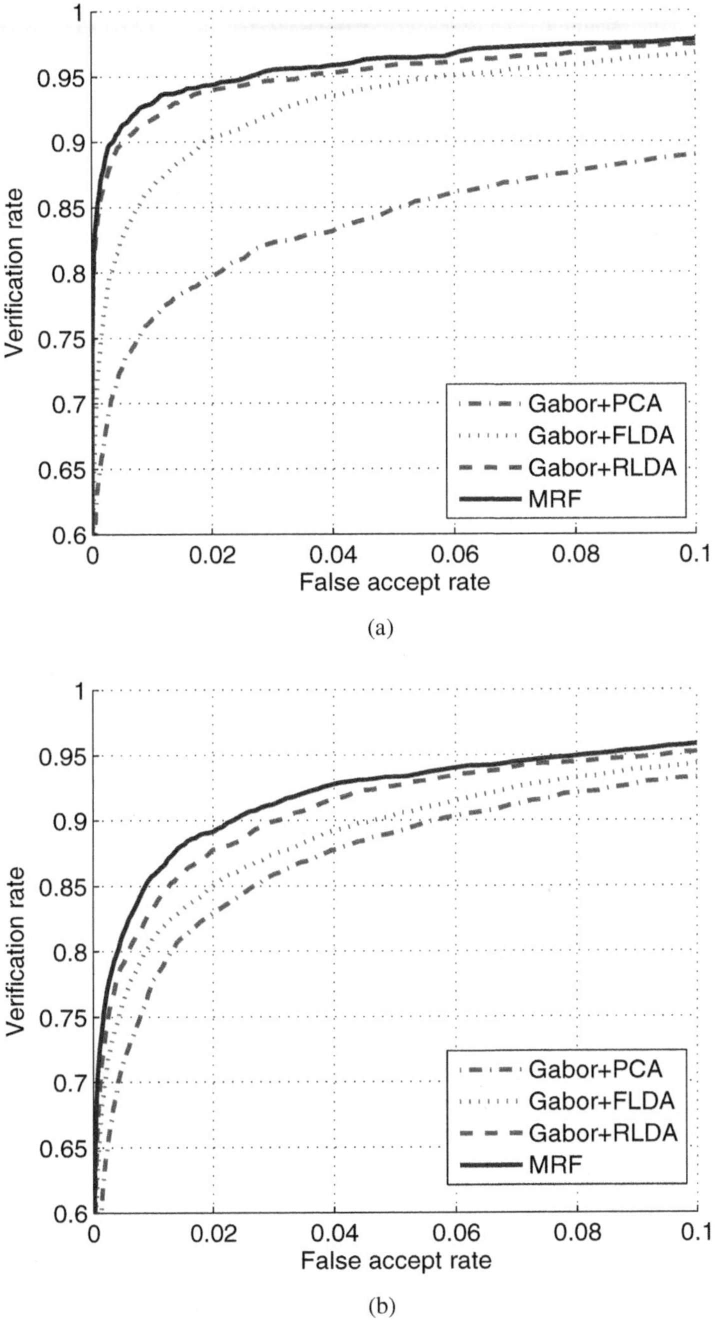

Fig. 2. Receiver operating characteristic (ROC) curves of different methods on FERET (a) and FRGC (b) databases

Table 1. Performance of different methods on FERET database

Method	Rank-1	VR@FAR=0.001	EER
PCA	0.5201	0.4713	0.1268
FLDA	0.6798	0.5585	0.1243
RLDA	0.7688	0.6750	0.0872
Gabor+PCA	0.7144	0.6400	0.1058
Gabor+FLDA	0.7575	0.6296	0.0732
Gabor+RLDA	0.8693	0.8167	0.0543
MAP-MRF	**0.8977**	**0.8541**	**0.0397**

Table 2. Performance of different methods on FRGC database

Method	Rank-1	VR@FAR=0.001	EER
PCA	0.6219	0.2759	0.1686
FLDA	0.7887	0.4172	0.1677
RLDA	0.8625	0.5394	0.1133
Gabor+PCA	0.8581	0.5759	0.0779
Gabor+FLDA	0.8962	0.6169	0.0785
Gabor+RLDA	0.9150	0.6678	0.0666
MAP-MRF	**0.9425**	**0.7250**	**0.0584**

5 Summary and Conclusions

In this paper, we present a MRF modeling method for face recognition. Contextual constraints, which should play an important role in recognition, are encoded in the formulation in the MAP-MRF framework, such that the objective function encodes not only constraints on image features but also relationships between them. An algorithm is provided to find an approximate optimal solution. The results show the advantage of using contextual constraints for face recognition in the MAP-MRF framework, and suggest MAP-MRF as a potential and competitive alternative for robust face recognition.

Acknowledgements. This work was supported by the following funding resources: National Natural Science Foundation Project #60518002, National Science and Technology Support Program Project #2006BAK08B06, National Hi-Tech (863) Program Projects #2006AA01Z192, #2006AA01Z193, and #2008AA01Z124, Chinese Academy of Sciences 100 people project, and AuthenMetric R&D Funds.

References

1. Hammersley, J.M., Clifford, P.: Markov field on finite graphs and lattices (unpublished, 1971)
2. Besag, J.: Spatial interaction and the statistical analysis of lattice systems (with discussions). Journal of the Royal Statistical Society, Series B 36, 192–236 (1974)
3. Chellappa, R., Kashyap, R.L.: Digital image restoration using spatial interaction models. IEEE Transactions on Acoustic, Speech and Signal Processing 30, 461–472 (1982)

4. Cross, G.C., Jain, A.K.: Markov random field texture models. IEEE Transactions on Pattern Analysis and Machine Intelligence 5(1), 25–39 (1983)
5. Derin, H., Elliott, H., Cristi, R., Geman, D.: Bayes smoothing algorithms for segmentation of binary images modeled by Markov random fields. IEEE Transactions on Pattern Analysis and Machine Intelligence 6(6), 707–720 (1984)
6. Geman, S., Geman, D.: Stochastic relaxation, Gibbs distribution and the Bayesian restoration of images. IEEE Transactions on Pattern Analysis and Machine Intelligence 6(6), 721–741 (1984)
7. Modestino, J.W., Zhang, J.: A Markov random field model-based approach to image interpretation. In: Proceedings of the IEEE Computer Society Conference on Computer Vision and Pattern Recognition, pp. 458–465 (1989)
8. Qian, R.J., Huang, T.S.: Objec detection using hierarchical mrf and map estimation. In: Proceedings of IEEE Computer Society Conference on Computer Vision and Pattern Recognition, December 1997, p. 186 (1997)
9. Dass, S.C., Jain, A.K.: Markov face models. In: Proceedings of IEEE International Conference on Computer Vision, Vancouver, July 9-12, pp. 112–116 (2001)
10. Dass, S.C., Jain, A.K., Lu, X.: Face detection and synthesis using Markov random field models. In: Proceedings of International Conference on Pattern Recognition, Quebec City, August 2002, pp. 112–116 (2002)
11. Huang, R., Pavlovic, V., Metaxas, D.: A hybrid face recognition method using markov random fields. In: Proceedings of International Conference on Pattern Recognition, Cambridge, UK (August 2004)
12. Park, B.G., Kyoung-Mu, Lee, S.U.: Face recognition using face-arg matching. IEEE Transactions on Pattern Analysis and Machine Intelligence 27(12), 1982–1988 (2005)
13. Belhumeur, P., Hespanha, J., Kriegman, D.: Eigenfaces vs. fisherfaces: recognition using class specific linear projection. IEEE Transactions on Pattern Analysis and Machine Intelligence 19(7), 711–720 (1997)
14. Ye, J., Xiong, T., Li, Q., Janardan, R., Bi, J., Cherkassky, V., Kambhamettu, C.: Efficient model selection for regularized linear discriminant analysis. In: CIKM, pp. 532–539 (2006)
15. Phillips, P.J., Moon, H., Rizvi, S.A., Rauss, P.J.: The FERET evaluation methodology for face-recognition algorithms. IEEE Transactions on Pattern Analysis and Machine Intelligence 22(10), 1090–1104 (2000)
16. Phillips, P.J., Flynn, P.J., Scruggs, W.T., Bowyer, K.W., Chang, J., Hoffman, K., Marques, J., Min, J., Worek, W.J.: Overview of the face recognition grand challenge. In: Proceedings of IEEE Computer Society Conference on Computer Vision and Pattern Recognition, pp. 947–954 (2005)

Pixelwise Local Binary Pattern Models of Faces Using Kernel Density Estimation

Timo Ahonen and Matti Pietikäinen

Machine Vision Group, University of Oulu, Finland
{tahonen,mkp}@ee.oulu.fi

Abstract. Local Binary Pattern (LBP) histograms have attained much attention in face image analysis. They have been successfully used in face detection, recognition, verification, facial expression recognition etc. The models for face description have been based on LBP histograms computed within small image blocks. In this work we propose a novel, spatially more precise model, based on kernel density estimation of local LBP distributions. In the experiments we show that this model produces significantly better performance in the face verification task than the earlier models. Furthermore, we show that the use of weighted information fusion from individual pixels based on a linear support vector machine provides with further improvements in performance.

1 Introduction

In the last years, different problems in face image analysis, such as face detection, face recognition and facial expression recognition have received very much attention in computer vision research. These problems are interesting from the viewpoints of basic research aiming to efficient descriptors for facial images and of applications such as surveillance and human-computer interaction [1].

Recently, a face descriptor based on Local Binary Patterns [2] was proposed [3]. In its original form, the method is based on dividing a facial image into small non-overlapping rectangular regions of equal size and describing each region using a local binary pattern histogram. A global description of the face is then formed by concatenating the histograms.

This representation has succesfully been applied to face recognition [3], face detection [4], facial expression recognition [5], demographic classification [6] and also other tasks such as general object recognition [7].

Some of the limitations of that method have also been discussed in the literature and different extensions and modifications have been proposed. In [8], Zhang *et al.* used AdaBoost learning algorithm for selecting a set of local blocks and their weights. Then, the LBP methodology was applied to the obtained blocks yielding in smaller feature vector length. Rodriguez and Marcel noticed that the method as such does not suit well for the face verification task and proposed an approach based on adapted, client-specific LBP histograms [9].

Even though the LBP labels are invariant to monotonic mappings of gray scale, illumination changes still pose a challenge to the local binary pattern

M. Tistarelli and M.S. Nixon (Eds.): ICB 2009, LNCS 5558, pp. 52–61, 2009.
© Springer-Verlag Berlin Heidelberg 2009

based representations. To alleviate this problem, Tan and Triggs proposed a preprocessing chain for illumination normalization. Furthermore, they introduced a three-level version of local patterns and a Hausdorff-like distance between query and gallery label images to replace histogram based approach. [10]

Other proposed enhancements include multi-scale block LBP which considers mean gray values from larger pixel blocks than original LBP [11], using patterns at multiple scales for representation [12] and combining LBP representation with other information such as Gabor filter based recognition [13], [14], [15].

In this work we propose a novel model for estimating local LBP distributions. This new method is based on kernel density estimation in the xy-space. We also propose the use of support vector machine for combining the information coming from different parts of the face as such learning method can be expected to result in higher performance than simple summation.

2 The Local Binary Pattern Operator

The local binary pattern operator [2] was originally designed for texture description. It has nevertheless shown very good performance also in many other application areas, and recently it has been used most widely in facial image description. LBP features are simple and fast to compute, yet LBP histograms have been shown to be discriminative features for face recognition even under some mis-alignment of the facial image [3], [12].

The basic LBP operator labels the pixels of an image by thresholding each 3×3 pixel neighborhood of the input image with the center value, multiplying the thresholded values by powers of two and summing them. Usually the histogram of the resulting labels is then used as an image descriptor.

The operator is extended to use neighborhoods of different sizes [2] by using a circular neighborhood around the center pixel and bilinearly interpolating the pixel values. This allows any radius and number of pixels in the neighborhood.

In the following, the general LBP operator is denoted by $LBP_{a,b}$, where a is the number of sampling points and b is the radius of the circular neighborhood. The operator outputs a labelled image, denoted by $f_l(x, y)$, whose value at integer coordinates (x, y) is one of the possible LBP labels $0, \ldots, N-1$. In this work we do not consider the uniform pattern and rotation invariant extensions of the LBP operator [2], thus the number of different labels produced by $LBP_{a,b}$ is $N = 2^a$.

2.1 Estimating LBP Histograms for Face Description

Most algorithms applying local binary patterns for face description make use of histograms of LBP labels computed within small image blocks. In [3], Ahonen *et al.* proposed dividing the facial image into R non-overlapping rectangular regions each of equal size and computing the histogram $H^r(i)$ independently within each block r

$$H^r(i) = n^r \sum_{x,y} I\{f_l(x,y) = i\} I\{(x,y) \in \text{Region}(r)\}, \tag{1}$$

where $r = 1, \ldots, R, i = 0, \ldots, N-1$ and

$$I\{A\} = \begin{cases} 1, & \text{A is true} \\ 0, & \text{A is false.} \end{cases}$$

The normalization constants n^r are set so that the sum of the histogram equals 1, and in case of rectangular regions of size of $w \times h$ pixels, the normalization constant is $n^r = 1/(wh)$.

As the amount of training data per client in face verification applications is limited in many cases, Rodriguez and Marcel [9] proposed using maximum a posteriori (MAP) adaptation [16] to achieve more reliable estimate of LBP distributions. In this context, MAP adaptation can be reduced to modelling the client-specific distribution of LBP labels by a weighted sum of histogram obtained from this client and a world model:

$$\hat{H}^r_C(i) = \alpha H^r_W(i) + (1-\alpha) H^r_C(i) \tag{2}$$

In this equation, $H^r_W(i)$ denotes the histogram for region r of the world model learned from training data independent of the client set, $H^r_C(i)$ is the model for client C and α is a weighting factor that is chosen through cross-validation in the development set.

The problem of model (1) is that finding the optimal division of face image area into R regions is a non-trivial task. Ahonen *et al.* proposed using a grid of rectangular block of equal size which was experimentally shown to be an effective way to encode facial information [3]. So far, also other works applying LBP histograms for face description have used this block based approach, either using a manually fixed grid (*e.g.* [3], [4], [5]) or by finding the division and weights denoting the relative importance of each region through some optimization procedure such as AdaBoost (*e.g.* [6], [8]) In some other works, the histogram based model has been omitted and LBP is used as preprocessing for linear discriminant analysis [17] or the query and gallery label images are compared directly using a Hausdorff like distance measure [10].

Another problem with model (1) is that when the values of $H^r(i)$ are used as estimates of probability distribution of LBP labels for pixels within the region r it is questionable whether the estimate is reliable near the borders of the region.

To address these two problems of the block based approach, we propose replacing it with Kernel Density Estimation (KDE) [18] in xy-space. KDE allows computing an estimate of probability distribution of LBP labels at any image coordinates (x, y) as a normalized sum of kernels positioned at occurrences of each LBP label. Here we use the symmetric Gaussian kernel, so the kernel density estimate of LBP distribution at location (x, y) becomes

$$H^{(x,y)}(i) = n^{(x,y)} \sum_{s,t} I\{f_l(s,t) = i\} \exp\left\{-\frac{(x-s)^2 + (y-t)^2}{2\sigma^2}\right\}. \tag{3}$$

The difference between block based and kernel density estimation of LBP local occurrence probability is illustrated in Fig. 1.

Again, the normalizing constants $n^{(x,y)}$ are set so that $\sum_i H^{(x,y)}(i) = 1$. If there are more than one label image available, as is the case in computing the

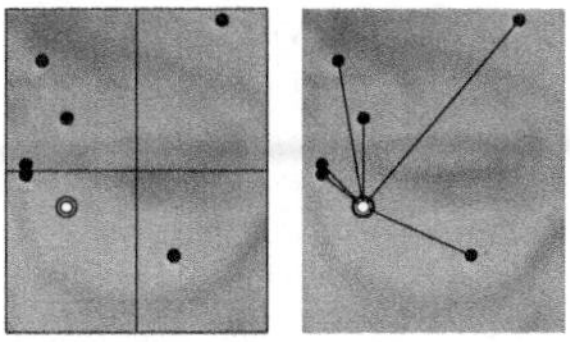

Fig. 1. Comparison of block based and kernel density estimation of LBP histograms in xy-space. Left: In block based estimation the probability of occurrence of a specific LBP label in arbitrary coordinates (x, y) is the number of this label in the region (x, y) belongs to divided by the total number of pixels in the region. Right: In kernel density estimation, the probability estimate is the sum of kernels positioned at occurrences of the label.

world model and sometimes also the client models, this means simply having more samples in the xy-space. In practice the resulting model can be computed by applying (3) to each of the images independently and then averaging over the obtained values.

2.2 Generative Model for LBP Based Face Verification

In this section we briefly review the generative model for LBP based face verification [9]. For full details and derivations, we refer to the original paper.

The biometric identity verification problem is the task of accepting or rejecting the claimed identity C of a subject using a biometric sample X from the subject. The decision is made based on a model θ_C of the client.

In the following, we assume that the sample X consists of the LBP labels computed from the facial image of the subject, i.e. $X = \{f_l(x, y)\}$, where the LBP label image $f_l(x, y)$ is obtained by applying a selected LBP operator to the input face image $f(x, y)$. Furthermore, if block based estimation of LBP probabilities is used, the LBP label image is further divided into local regions and we denote the k-th observation (pixel in the LBP label image) in r-th local region by l_k^r and in this case $X = \{l_k^r\}$. Now, assuming independence of the observations, the probability of observing features $X = \{l_k^r\}$ given model θ_C is

$$P(X|\theta_C) = \prod_{r=1}^{R} \prod_{k=1}^{K} \hat{H}_C^r(l_k^r), \tag{4}$$

The decision problem for accepting the claimed identity can then be expressed as a binary function of the log-likelihood ratio

$$\Lambda(X, C) = \log P(X|\theta_C) - logP(X|\theta_W) \tag{5}$$

$$= \sum_{r=1}^{R} \sum_{k=1}^{K} \left[\log(\hat{H}_C^r(l_k^r)) - \log(H_W^r(l_k^r)) \right] \tag{6}$$

The claim of identity C is accepted if $\Lambda(X, C) \geq \tau$ and rejected if $\Lambda(X, C) < \tau$. The decision threshold τ is usually learnt from a development set (see next section).

To increase the performance of the verification system, we propose replacing the block based client models $\hat{H}_C^r(i)$ and world model $H_W^r(i)$ of LBP distributions with the kernel density estimates $\hat{H}_C^{(x,y)}(i)$ and $H_W^{(x,y)}(i)$ computed using (3). The log likelihood ratio then becomes

$$\Lambda(X,C) = \sum_{x,y} \left[\log(\hat{H}_C^{(x,y)}(f_l(x,y))) - \log(H_W^{(x,y)}(f_l(x,y))) \right] \tag{7}$$

By denoting $\lambda_C(x,y) = \log(\hat{H}_C^{(x,y)}(f_l(x,y))) - \log(H_W^{(x,y)}(f_l(x,y)))$ we arrive at the decision function

$$D(X,C) = \text{sign}\left\{ \sum_{x,y} \lambda_C(x,y) - \tau \right\}, \tag{8}$$

and now the claim is accepted if $D(X,C) = 1$ and rejected if $D(X,C) = -1$.

2.3 Support Vector Machines for Combining Likelihoods

In essence, the decision function (8) is a sum of log likelihood ratios corresponding to individual pixels. As earlier studies with LBP based face recognition have shown, different facial areas contribute unequally in terms of recognition. In [3], a weighted sum of Chi square distances of LBP histograms from different facial regions outperformed unweighted sum. In that work the weights for individual regions were set based on the performance of each region when used for recognition alone. In [8], both the weights and the region parameters were obtained through Adaboost learning.

Inspired by those results, we propose replacing the sum in decision function (8) by a weighted sum, i.e. a general linear discriminant function, obtaining

$$D(X,C) = \text{sign}\left\{ \sum_{x,y} w^{(x,y)} \lambda_C(x,y) - \tau \right\}. \tag{9}$$

One can easily see that (8) is a special case of this decision function when $w^{(x,y)} = 1$ for all (x,y).

A convenient way to compute the weights $w(x,y)$ and the threshold τ is to resort to the linear Support Vector Machine [19]. Previously SVMs have been proposed for information fusion for person verification by, *e.g.* Ben-Yacoub *et al.* who used SVM to combine information from face and voice modalities in multi-modal person verification [20]. In face analysis, Heisele *et al.* used a support vector machine for fusing information from component-wise SVMs in face detection and recognition [21].

The linear SVM is obtained by finding such weights $w(x,y)$ that the minimum distance between the decision boundary and the training samples is maximized, i.e. it is a maximum margin classifier. The SVM can be seen as a structural risk minimization solution to the linear classification problem, and thus it is less prone to overtraining than some other learning methods. This is especially

Fig. 2. Example from the BANCA databases. From left: 1. controlled scenario 2. degraded scenario, 3. adverse scenario.

important in our case, since the input dimension is rather high with respect to the number of training samples. Despite that, in our experiments it was observed that the SVM did not show significant overfitting but it performed well also on unseen data.

3 Experiments

The performance of the proposed facial image representation in the face verification task was tested using the BANCA English database [22]. In the experimental setup we followed the procedures of [9] to make the results with kernel density estimation of LBP distributions comparable to block-based model used in that work.

3.1 BANCA Database

To test the performance of the proposed algorithm, we use the English section of BANCA database [22]. The BANCA English database consists of 52 subjects (26 male and 26 female) captured at 12 sessions. These 12 sessions were further divided into 3 scenarios: controlled, degraded and adverse. Fig. 2 shows example images from the different scenarios.

The images of the BANCA database can be used with different protocols defining which part of the data is used for training the subject models and which part is used for testing. We used the following five protocols in the experiments: Mc (Controlled data is used for client training and testing), Ud (Controlled data for training, degraded for testing), Ua (Controlled data for training, adverse for testing), P (Controlled data for training, all data for testing), G (All data for training, all data for testing). A more detailed description of the protocols can be found in [22].

The two possible cases that can occur in identity verification are true access, i.e. the claimed identity of subject is true and impostor access, which means that the claimed identity is false. Furthermore, the two possible error types are false acceptance, in which the system wrongly accepts an impostor and false rejection, in which the system wrongly rejects a true client.

Given a set of test true and impostor access cases, two error measures depending on the decision threshold τ can be defined. The false acceptance rate $P_{FA}(\tau)$ is $P_{FA}(\tau) = \frac{n_{FA}(\tau)}{n_I}$, where $n_{FA}(\tau)$ is the number of false acceptances given decision threshold τ and n_I is the total number of impostor access test cases. The false rejection rate $P_{FR}(\tau)$ is defined as $P_{FR}(\tau) = \frac{n_{FR}(\tau)}{n_T}$, where $n_{FR}(\tau)$ is the number of false rejections and n_T is the total number of true client test cases. In our experiments, we measure the system performance using half total error rate $HTER(\tau)$, which is $HTER(\tau) = \frac{P_{FA}(\tau)+P_{FR}(\tau)}{2}$.

The performance measure $HTER(\tau)$ depends on the decision threshold τ. To avoid bias in the reported results, the threshold should not be computed using the evaluation set but the performance should be assessed by first finding a threshold that produces lowest $HTER$ on a development set and then reporting the $HTER$ on an evaluation set obtained using this threshold. Also, other paramaters for the method to be evaluated should be selected in a similar fashion. For this reason, the BANCA database is divided into 2 groups of 26 subjects (13 women, 13 men) each. The decision threshold τ and other parameters can then be selected by optimizing the performance on group 1. Next, group 2 is used to evaluate the performance with these parameters and vice versa. The average $HTER$ of g1 (using g2 as development) and g2 (using g1 as development) is reported as performance measure for each protocol.

3.2 Experimental Setup

The images in the BANCA database were first geometrically normalized to size 84×68 pixels using the eye coordinates provided with the images.

In the 1st experiment, the $LBP_{4,2}$ operator was applied to the normalized images which resulted in labelled images of size 80×64 pixels. The world model $H_W^{(x,y)}(i)$ and client models $H_C^{(x,y)}(i)$ were then constructed using equation (3) and client model adaptation was done using (2). The parameters σ for KDE and α for model adaptation as well as the decision threshold τ were obtained by minimizing the $HTER$ in a development set. Using these parameters, the models for the evaluation set were classified using decision rule (8).

In the 2nd experiment the procedure was the same as in experiment 1 with the difference that the weighted fusion rule (9) was used instead of (8). In each protocol and group, the same parameters σ and α as in experiment 1 were used and the weights were obtained through training a SVM in the development set. Again, the $HTER$ obtained with these weights on the separate evaluation set was reported.

In experiments 3 and 4 we tested the effect of illumination normalization. It has been reported that the performance of LBP based face recognition can be enhanced with normalization procedure consisting of gamma correction, difference of gaussian filtering and contrast equalization [10]. In experiments 3 and 4 this preprocessing of [10] was applied to the images prior to applying LBP but otherwise the procedure was kept completely the same.

3.3 Results

The half total error rates of the proposed method under different protocols of the BANCA database are reported in Table 1.

Table 1. Half Total Error rates for different methods on BANCA protocols Mc, Ud, Ua, P and G

Method	Mc	Ud	Ua	P	G
0. LBP-block [9]	7.3	10.7	22.6	19.2	5.0
Best in [9]	4.9	10.7	17.6	14.8	5.0
1. LBP-KDE	4.3	6.4	18.1	17.6	4.0
2. Weighted LBP-KDE	3.7	6.6	15.1	11.6	3.0
3. PP+LBP-KDE	3.0	5.9	8.7	10.1	1.9
4. Weighted PP+LBP-KDE	3.2	7.1	10.4	9.1	1.5

For comparison, the error rates for LBP face verification with block based estimation of LBP distributions [9] is reported on the 1st row of the table. In that work, LBP was compared to two other methods (LDA/NC and DCT-mod2/GMM), and the 2nd row of the table shows the best result among these three methods.

The results from Experiment 1 show that switching from block based estimation (0) to kernel density estimation (1) improves the accuracy of LBP face verification in all the BANCA protocols. When it is further combined with weighted fusion rule trained with SVM (2), it produces better results than the comparison methods in all the protocols. With KDE the error rates drop in best cases to about half of those of LBP-block.

As expected, illumination normalization preprocessing (3 and 4) is useful also in case of BANCA images and especially in protocols Ua and P with test images containing severe illumination changes compared to training images. In the Mc protocol the effect of preprocessing is smaller as both the training and testing images have been taken under controlled conditions.

When assessing the result of support vector machine based weighted fusion of log-likelihoods from individual pixels, it can be noted that in case of non-preprocessed images, introducing weights gives better results in all the protocols except Ud. On the other hand, with preprocessed images, the weights obtained with SVM actually worsen the results in three protocols (Mc, Ud and Ua). In the P and G protocols the development set is larger than in MC, Ud and Ua, and in those two protocols the SVM performs best, which indicates that the SVM training clearly benefits from the higher amount of training data available.

4 Conclusions

In this work we proposed a novel method for estimating the local distributions of LBP labels for face description. The method is based on kernel density estimation in xy-space, and it provides much more spatial accuracy than the earlier, block-based methods.

The kernel density estimation based method alleviates two problems related to computing histograms within local blocks. First, finding an optimal way to divide the facial area into local blocks has proven to be a difficult problem. Second, the block histogram estimates of LBP label probabilities are likely to be unreliable near the block boundaries. Both these problems are avoided when using the proposed method.

We also proposed and evaluated the use of Support Vector Machine in information fusion of log-likelihoods corresponding to individual pixels for the binary classification task for identity verification. The SVM is a learning method for finding the weights corresponding to usefulness of information coming from that face area for the verification task. As it is less prone to overfitting than many other learning methods, it is well suited for this task as the input dimension is rather high.

In the experiments we showed that kernel density estimation clearly improves the verification performance in the BANCA dataset in comparison to block based method. Moreover, the SVM was shown to further enhance the results.

Here we applied kernel density estimation to LBP description of faces for the face verification task. However, we believe that this same methodology is applicable to other tasks such as face detection and alignment or in other application areas where local estimation of local binary pattern distributions are used. Also, the method is likely to be useful with other local descriptors than LBP. Further research includes exploring these areas, investigating other methods for fusing information from individual pixels, and experiments with larger datasets.

Acknowledgments. This work has been performed with partial support from the MOBIO project (contract number IST-214324), which belongs to the 7th Framework Research Programme of the European Union. The authors want to thank Dr. Abdenour Hadid for helpful comments in preparation of the manuscript.

References

1. Li, S.Z., Jain, A.K. (eds.): Handbook of Face Recognition. Springer, Heidelberg (2005)
2. Ojala, T., Pietikäinen, M., Mäenpää, T.: Multiresolution gray-scale and rotation invariant texture classification with local binary patterns. IEEE Trans. Pattern Analysis and Machine Intelligence 24(7), 971–987 (2002)
3. Ahonen, T., Hadid, A., Pietikäinen, M.: Face description with local binary patterns: Application to face recognition. IEEE Trans. Pattern Analysis and Machine Intelligence 28(12), 2037–2041 (2006)
4. Hadid, A., Pietikäinen, M., Ahonen, T.: A discriminative feature space for detecting and recognizing faces. In: Proc. Conf. Computer Vision and Pattern Recognition (CVPR 2004), vol. 2, pp. 797–804 (2004)
5. Shan, C., Gong, S., McOwan, P.W.: Robust facial expression recognition using local binary patterns. In: Proc. IEEE Int. Conf. on Image Processing (ICIP 2005), vol. II, pp. 914–917 (2005)

6. Yang, Z., Ai, H.: Demographic classification with local binary patterns. In: Lee, S.-W., Li, S.Z. (eds.) ICB 2007. LNCS, vol. 4642, pp. 464–473. Springer, Heidelberg (2007)
7. Zhang, H., Gao, W., Chen, X., Zhao, D.: Object detection using spatial histogram features. Image Vision Comput. 24(4), 327–341 (2006)
8. Zhang, G., Huang, X., Li, S.Z., Wang, Y., Wu, X.: Boosting local binary pattern (LBP)-based face recognition. In: Li, S.Z., Lai, J.-H., Tan, T., Feng, G.-C., Wang, Y. (eds.) SINOBIOMETRICS 2004. LNCS, vol. 3338, pp. 179–186. Springer, Heidelberg (2004)
9. Rodriguez, Y., Marcel, S.: Face authentication using adapted local binary pattern histograms. In: Leonardis, A., Bischof, H., Pinz, A. (eds.) ECCV 2006. LNCS, vol. 3954, pp. 321–332. Springer, Heidelberg (2006)
10. Tan, X., Triggs, B.: Enhanced local texture feature sets for face recognition under difficult lighting conditions. In: Zhou, S.K., Zhao, W., Tang, X., Gong, S. (eds.) AMFG 2007. LNCS, vol. 4778, pp. 168–182. Springer, Heidelberg (2007)
11. Liao, S., Zhu, X., Lei, Z., Zhang, L., Li, S.Z.: Learning multi-scale block local binary patterns for face recognition. In: Lee, S.-W., Li, S.Z. (eds.) ICB 2007. LNCS, vol. 4642, pp. 828–837. Springer, Heidelberg (2007)
12. Chan, C.H., Kittler, J., Messer, K.: Multi-scale local binary pattern histograms for face recognition. In: Lee, S.-W., Li, S.Z. (eds.) ICB 2007. LNCS, vol. 4642, pp. 809–818. Springer, Heidelberg (2007)
13. Yan, S., Wang, H., Tang, X., Huang, T.S.: Exploring feature descriptors for face recognition. In: Int. Conf. on Acoustics, Speech, and Signal Processing (ICASSP 2007) (2007)
14. Zhang, W., Shan, S., Gao, W., Chen, X., Zhang, H.: Local gabor binary pattern histogram sequence (LGBPHS): A novel non-statistical model for face representation and recognition. In: Proc. Tenth IEEE Int. Conf. on Computer Vision (ICCV 2005), vol. 1, pp. 786–791 (2005)
15. Tan, X., Triggs, B.: Fusing Gabor and LBP feature sets for kernel-based face recognition. In: Zhou, S.K., Zhao, W., Tang, X., Gong, S. (eds.) AMFG 2007. LNCS, vol. 4778, pp. 235–249. Springer, Heidelberg (2007)
16. Gauvain, J.L., Lee, C.H.: Maximum a posteriori estimation for multivariate gaussian mixture observations of markov chains. IEEE Trans. Speech and Audio Processing 2(2), 291–298 (1994)
17. Heusch, G., Rodriguez, Y., Marcel, S.: Local binary patterns as an image preprocessing for face authentication. In: Proc. IEEE Int. Conf. on Automatic Face and Gesture Recognition (FG 2006), pp. 9–14 (2006)
18. Scott, D.W.: Multivariate Density Estimation: Theory, Practice, and Visualization. Wiley-Interscience, Hoboken (1992)
19. Vapnik, V.N.: Statistical Learning Theory. Wiley-Interscience, Hoboken (1998)
20. Ben-Yacoub, S., Abdeljaoued, Y., Mayoraz, E.: Fusion of face and speech data for person identity verification. IEEE Trans. Neural Networks 10(5), 1065–1074 (1999)
21. Heisele, B., Serre, T., Poggio, T.: A component-based framework for face detection and identification. Int. Journal of Computer Vision 74(2), 167–181 (2007)
22. Bailly-Bailliére, E., Bengio, S., Bimbot, F., Hamouz, M., Kittler, J., Mariéthoz, J., Matas, J., Messer, K., Popovici, V., Porée, F., Ruíz, B., Thiran, J.P.: The BANCA database and evaluation protocol. In: Kittler, J., Nixon, M.S. (eds.) AVBPA 2003. LNCS, vol. 2688, pp. 625–638. Springer, Heidelberg (2003)

Improvements and Performance Evaluation Concerning Synthetic Age Progression and Face Recognition Affected by Adult Aging

Amrutha Sethuram, Eric Patterson, Karl Ricanek, and Allen Rawls

Face Aging Group, Computer Science Department, UNCW, USA
{sethurama,pattersone,ricanekk,rawlsa}@uncw.edu
http://www.faceaginggroup.com

Abstract. Aging of the face degrades the performance of face recognition algorithms. This paper presents recent work in synthetic age progression as well as performance comparisons for modern face recognition systems. Two top-performing, commercial systems along with a traditional PCA-based face recognizer are compared. It is shown that the commercial systems perform better than the baseline PCA algorithm, but their performance still deteriorates on an aged data-set. It is also shown that the use of our aging model improves the rank-one accuracy in these systems.

Keywords: Facial aging, synthetic age progression, performance evaluation.

1 Introduction

Of the many challenges to the performance of modern face-recognition (FR) algorithms, the robustness to age variation has perhaps been studied the least. Algorithms have made significant strides in the past decades toward dealing with real-world conditions such as pose and lighting.However, temporal variance particularly due to longer spans where aging changes are evident remains a difficulty. Overcoming this difficulty would bring useful improvements in robustness to biometrics and would also benefit related fields of forensics and medicine.

In this work, we demonstrate the severity of performance loss due to adult aging and discuss steps towards improving the related robustness of face-based biometrics. The organization of the paper is as follows. In section 2, a brief discussion of biometric-related work on aging of the human face is provided. In section 3, recent improvements concerning synthetic age progression are presented. Comparison and performance evaluation of two commercial face-recognition systems on longitudinal images are presented in section 4. The final section gives conclusions and a brief discussion on continued research.

2 Related Studies in Face Aging

Some attempts have been made at unifying anthropological and forensic studies that consider aspects of human aging and resulting changes in the face [3] [4], but

M. Tistarelli and M.S. Nixon (Eds.): ICB 2009, LNCS 5558, pp. 62–71, 2009.
© Springer-Verlag Berlin Heidelberg 2009

this knowledge has not been used in all cases to inform the relatively few studies conducted regarding the effects of growth and development and aging on synthetic progression and face recognition technologies [5] [6] [7] [8] [9]. Concerning simulation of aging in facial images or models, a few approaches have been considered. They can broadly be divided into geometric and/or physical-based [6] and analysis-synthesis model-based methods [5] [6] [7]. Some of the most promising analysis-synthesis methods have involved a variation of active appearance models (AAM) [10], [11] for growth and development [12], [13] or adult aging [8]. Several of these did not distinguish between childhood growth-and-development and adult effects of aging. It should be stressed, however, that these are distinct processes; pre-adult processes cause largely shape-based changes in the mid and lower face. Aging, however, occurring after roughly twenty years of age, consists of shape deformation from weight change, minor bone remodeling, and tissue degeneration combined with gravity effects along with significant textural change due to a variety of complex phenomenon [3]. It is this textural change that is critical to photo-realstic synthesis of age-progression and its use as a synthetic template to improve face recognition.

3 Improvements in Synthetic Age Progression

In this section, we discuss recent improvements made in synthetic age progression of images of adults. Our goal is to produce very accurate estimates, across human variation, of the changes in appearance of an individual due to the effects of aging over a number of years and to create synthetic images that realistically portray in a data-informed manner, the actual effects that occur in humans as documented by medical and anthropological literature. For the most part, all current approaches are still "general" models of aging because truly accurate models of individualized, idiosyncratic aging requires a large set of representative data for the same individual across adult lifespan. Such data is not currently available. Our ultimate aim, though, is to create models that may be altered specifically and automatically to improve accuracy of predictive images by accounting for personal individuality and lifestyle as possibly indicated by accompanying meta-data.

The improvements that we have achieved in our recent work are due in part to three areas of focus – improving quality and quantity of data, improving labeling of landmarks used for models, and improving methods for training and simulation in our synthetic age-progression technique. We include several images age-progressed from the most notable longitudinal databases FG-NET [16] and MORPH [1] for human comparison of recognizable images.

Our work uses an analysis-synthesis face-model approach, using Active Appearance Models (AAM) as the method of representation for both shape and texture information. Our recent methods have used support vector regression (SVR) to attempt to learn age-based properties of AAM parameters across the entire training set. Also, gradient-regression-based AAMs are used due to earlier work that suggests these may best represent texture information in faces.

Once the AAM is created, a Monte-Carlo simulation is performed (with 500,000 iterations) to garner the feature information learned by the support vectors, generating random face images within the range of acceptable face images, and classifying them based on their age estimated by the SVR solution. Bins are created and averaged for each age, ultimately creating a table of AAM parameters that may be indexed and differenced to synthetically age progress or regress a facial image. Particular care was taken to very accurately label (manually) the hundreds of images used to build these models. As our method for age-progression is data-driven, large numbers of clear images of sufficient resolution are necessary to build accurate image-based face models of what occurs over years of aging. We began by augmenting some of our past training sets and building a European-descent male data-set of approximately 200 images, demonstrating visually comparable performance improvements in the modes of the AAM, the simulated table of representative age parameters, and synthetic aging of images of similar individuals. Pictured first in Figure 1 are the representative faces for each decade of adult life as learned by the aging model. For comparison of synthesis improvements on an individual, Figure 2 shows an age-progression using the newer male model versus an earlier version shown in previously published work [15]. Better representative shape changes and finer texture changes are evident upon comparison of stills and movies of progression. Our most recent work has expanded the training set of images to that of 541 distinct individuals, currently of neutral expression and for the most part lacking facial hair, glasses, or other obstructions. Both the male-only model and the larger, varied model show an obvious representation of age changes in the first few modes of the AAM parameters. The larger, varied model also demonstrates observable gender and ethnicity representation in the modes. For example, one of the first few modes of the most recent, varied model is shown in Figure 3.

For additional comparison, images from FG-NET and MORPH are shown in Figures 4 and 5. These images are actually produced completely without previously incorporated information in either the AAM or the aging model, as they were not used in training at either level. It is possible that including them for the AAM training could improve the texture quality of the reconstructed and synthetically aged images, and future work may include comparisons of models aged that are not present in the AAM versus those that are (as this would certainly be possible in real-world scenarios to use all available information). In general, the images produced by the most recent methods appear to better illustrate texture and shape differences created by skin wrinkling and tissue sag over time, and these correspond well to those documented in the anthropological and medical literature [3] [4]. Minor shape changes, the appearance of rhytids, deepening of lines such as the nasio-labial, and ptosis of the orbital tissue are all well demonstrated. Some areas such as the forehead, though, do not demonstrate as noticeable and possibly not sufficiently accurate texture detail, but it is not clear at this point where the weakness may lie. In general, the progression of images demonstrate well the expected changes in shape and texture due to tissue and skin changes from adult aging and are better representative than past models.

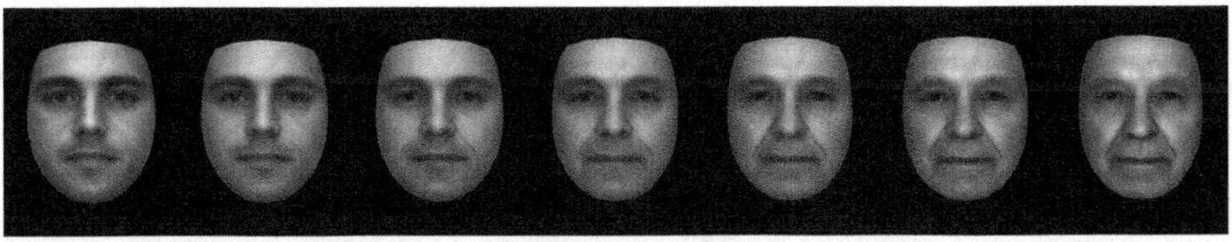

Fig. 1. The representative, average male of European-descent for each decade (2nd - 8th) as learned by the SVR training on approximately 200 images and generated by the Monte-Carlo simulation

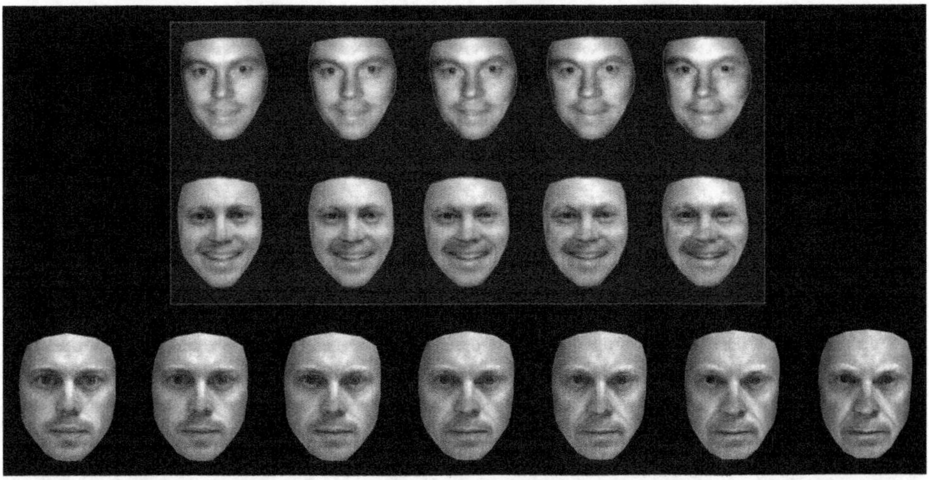

Fig. 2. Images from earlier work included for comparison, one of the authors age-progressed through the major decades. The top two rows are earlier versions (at original age 30 then 40,50,60,70), the bottom row aged using the aforementioned newer model (synthetic from 20 to 80).

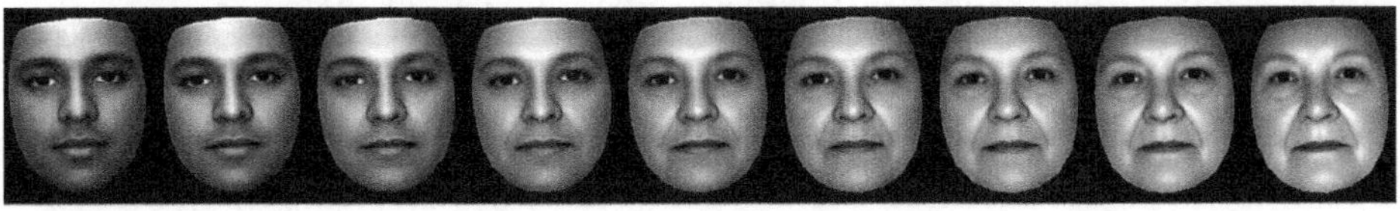

Fig. 3. One of the first principal modes of the AAM trained on 541 images representative of adult aging, demonstrating represention of age, gender, and ethnic variation

The current model also seems capable of reasonably reconstructing a wide variety of faces of different gender and ethnicity and representing characteristics of general aging within each of those. In the following sections, face-recognition performance and the application of these synthetic techniques will be discussed.

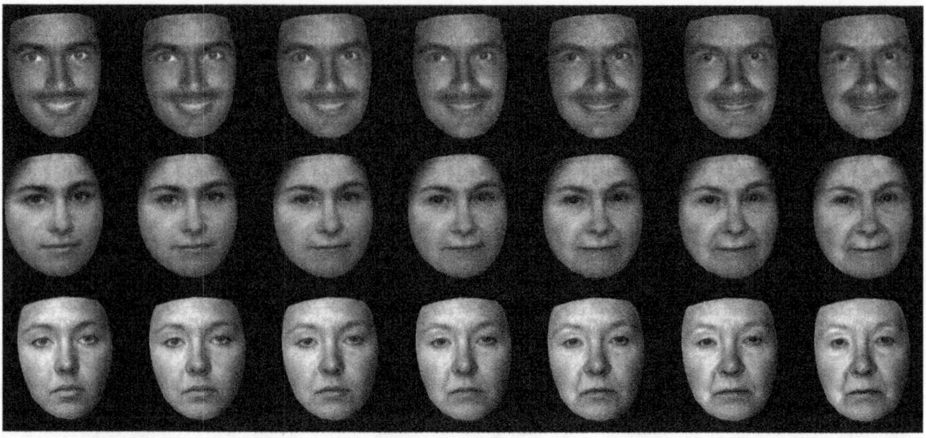

Fig. 4. Samples from the FG-NET database aged through the decades 20 to 80 years

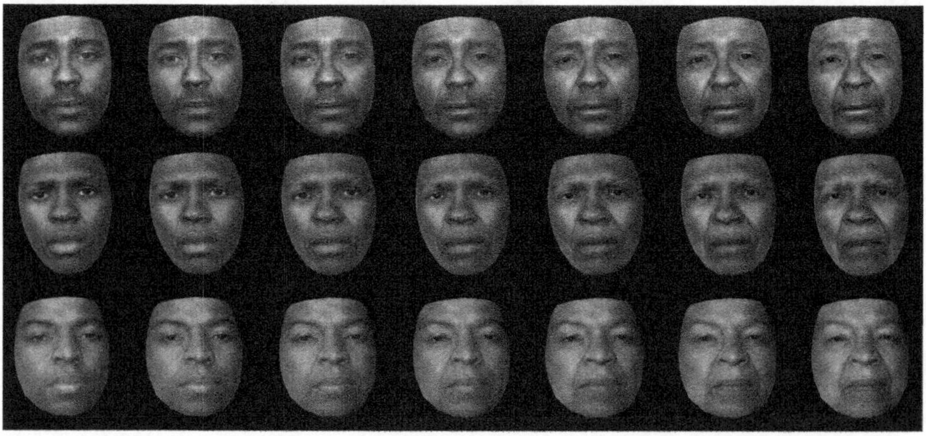

Fig. 5. Samples from the MORPH database aged through the decades 20 to 80 years

4 Performance Evaluation and Comparison

Two modern, high-performance, and commercially available face recognition systems were evaluated for performance on face images of individuals separated by several years of aging. For purposes of anonymity, the systems will be referred to as SDK *A* and SDK *B*. Methods and related results achieved using the SDKs provided by the two vendors are discussed in the following subsections.

4.1 Database

Currently, FG-NET [16] and MORPH [1] are the only two publicly available face-image databases that attempt to represent longitudinal changes. The FG-NET aging database consists of 1002 images of 82 subjects with a span of ages ranging from 0-69. Morph [17] is an ongoing data corpus in two albums that includes facial images of numerous individuals and with essential metadata, such as age, gender, ancestry, height, and weight. MORPH Album *1* contains digital scans (much like FG-NET) of photographs of 625 individuals with a total of 1690 images. MORPH Album *2* at present, contains 55,608 images of 13,673 individuals. In this work, a dataset comprising of both the FG-NET and MORPH Album *1* is used for training and testing the algorithms. 1299 images of individuals older than 18 and younger than 50 years of age were used to construct the training gallery and testing probes.

4.2 Experimental Setup

The goal of the experiments was to evaluate and compare the performance of advanced commercial systems relative to aging. A baseline PCA face recognition algorithm from the Colorado State University's Face Identification Evaluation System [18] was used as well. Two different experiments were conducted to evaluate the performance.

Experiment *1*. This experiment was set-up to evaluate the recognition rate of the SDKs and the baseline PCA algorithm on individuals as they aged. A combined dataset of Morph *1* and FG-NET databases was used for this experiment. The gallery and probes were constructed as outlined in Table. 1 w.r.t the age groups.

Table 1. Age range for the *gallery* and *probes* for Experiment 1

G/P	Age-Range(years)
Gallery (G)	18-30
Probe A (P_a)	18-30
Probe B (P_b)	31-40
Probe C (P_c)	41-50

The gallery data, G

$$G = [g_1, g_2, ..., g_n]; \tag{1}$$

was constructed such that youngest image of each person in the age range of 18-30 years available in the dataset was selected. Let P be the probe data defined by

$$P = [p_1, p_2, ..., p_m]; \tag{2}$$

which was constructed with aged images of the subjects from the Gallery. Three different probes P_a, P_b and P_c were generated to include aged images in different age-groups as shown in Table 1.

A quality rating of 80% or higher (based on the more rigorous SDK quality rating method) was used as a threshold in choosing images for both the gallery and probes. Input images were converted to 8-bit grayscale for all the experiments. Having set-up the data, SDK A and SDK B were trained on the images in the gallery G. For every experiment, scores were obtained for each image in the probe compared with every image in the gallery. The match list obtained was then ranked with decreasing order of the scores generated.

In keeping with a standard face evaluation system, Colorado State University's Face Identification Evaluation System (Version 5) [2] was used. This system provides a standard set of well known algorithms for face identification that assist in providing a baseline metric and are used to compare and contrast relative methods for identification. For this baseline comparison, Principle Component Analysis (PCA) was used. Images were prepared using CSU's preprocessing normalization techniques. The Euclidean distance measure was used to rank the probes against the gallery. To compare the performance directly with the scores generated with SDK A and SDK B, these distance measures were converted to similarity-like measures and were normalised using the following equation:

$$CSU\_Score = 1 - \left[\frac{Distance_i - min\,(Distance)}{max\,(Distance) - min\,(Distance)} \right] \tag{3}$$

4.3 Results and Discussion

Cumulative Match Characteristic (CMC) curves were obtained for the different probe sets. Performance of SDK A for different probes is shown in Figure 6, SDK B in Figure 7. The baseline PCA performance is shown in Figure. 8. For the same probe, (P_a), CMC curves were obtained for the two SDKs and baseline PCA algorithm. Results are shown in Figure 9. The experiment was repeated on probe, (P_b), the results of which are shown in Figure 10.

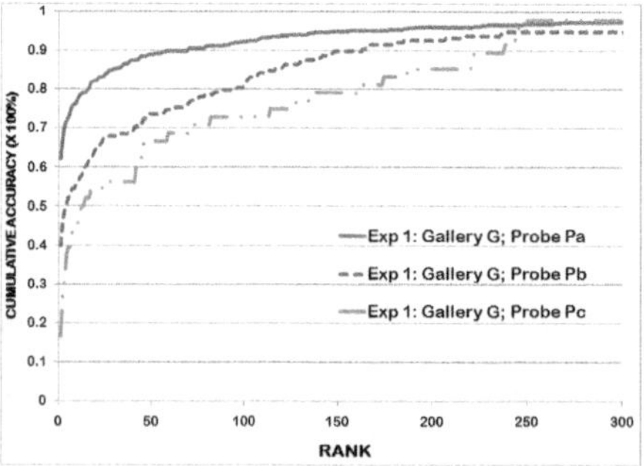

Fig. 6. CMC curves for SDK-A for different Probes

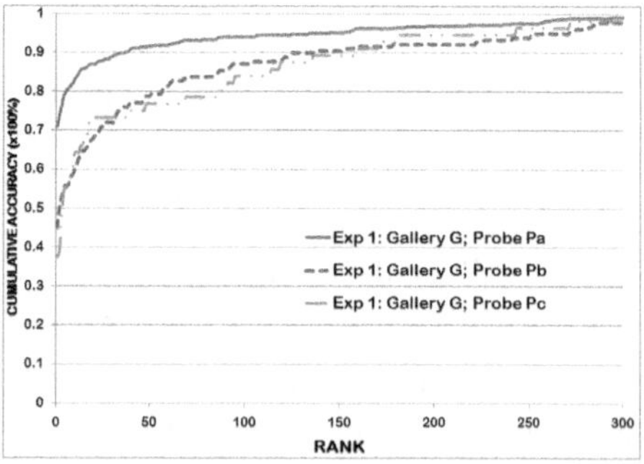

Fig. 7. CMC curves for SDK-B for different Probes

It is clear from figures 6- 8 that the performance of the algorithms decreases with increasing ages of the probes. In other words, the algorithms fail to identify subjects properly as they age versus their younger gallery images, indicating a

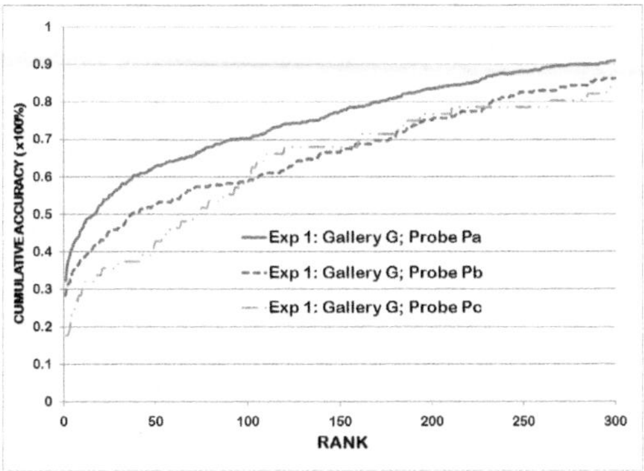

Fig. 8. CMC curves for PCA for different Probes

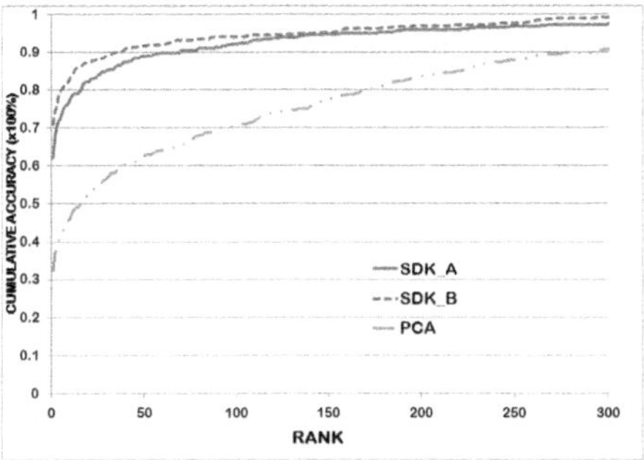

Fig. 9. Comparison of CMC curves of the three algorithms for Probe A

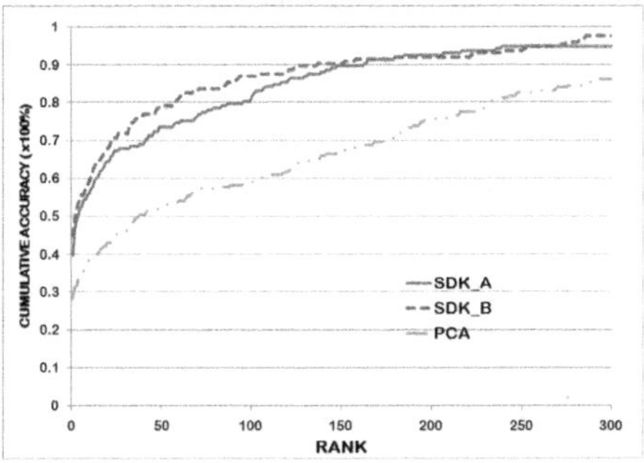

Fig. 10. Comparison of CMC curves of the three algorithms for Probe B

general lack of robustness to face aging even in these advanced systems. From Figure 9 and Figure 10, it can be concluded that performances of both SDK A and SDK B were expectedly better than the baseline PCA algorithm. In general, also, SDK B performed slightly better than SDK A.

Experiment *2*. The goal of this experiment was to quantify possible performance improvements with the introduction of synthetic aged images using our most recent techniques discussed earlier. This experiment was conducted on SDK A. One image of each of these subjects with an age-range between 18-24 years was used to build the gallery. The probe was created using an older image of each of these subjects whose actual ages were in the range of 35-50 years. Next, the images in the Gallery G_0 were age-progressed to match the approximate ages in the Probe P. These synthetic aged images were used to build another Gallery G_A. Rank-1 accuracy was computed for the probe matched with each of the galleries separately as shown in Table.2 This experiment, although still relatively small, demonstrates a notable gain of recognition performance by

Table 2. Rank-1 Accuracy for Probe P matched with the original younger gallery G_0 and synthesized older gallery G_A for Experiment 2

Exp	Rank-1 Accuracy
Original Images Only	18.75%
Including Age-Progressed Images	31.25%

injecting synthetically age-progressed images into a modern commercial system. With additional data and improved methods, future work should be able to further quantify this improvement and suggest specific techniques for the best performance gains.

5 Conclusions and Suggested Direction

In this paper, we have presented improved results in synthetic age-progression and also evaluated, relative to longitudinal data, the performance of two modern face recognizers. We have shown that even the performance of these advanced, commercially available systems degenerate with increasing age spans of test probes. We have also demonstrated the possibility of improvement, even in these systems, with the addition of synthetically age-progressed gallery images. We plan to extend and evaluate this experiment on a larger data-set when possible. Our future research will also focus on developing quantitative measures to evaluate performance in a more detailed manner. To conclude, this work has emphasized the need to develop robust algorithms invariant to facial aging as well as the need to continue improvements in synthetic age progression. Also, research efforts should be focused on collecting and processing necessary datasets to further improve and evaluate both age-progression and face-recognition technologies that seek to be age invariant.

References

1. Ricanek, K., Tesafaye, T.: Morph: A longitudinal image database of normal adult age-progression. In: 7th International Conference on Automatic Face and Gesture Recognition, April 2006, pp. 341–345 (2006)
2. Beveridge, et al: The CSU Face Identification Evaluation System User's Guide: Version 5.0. Colorado State University (May 1, 2003)
3. Albert, A.M., Ricanek, K., Patterson, E.: A review of the literature on the aging adult skull and face: Implications for forensic science research and applications. In: Forensic Science International (April 2007) (in Press)
4. Behrents, R.G.: Growth in the Aging Craniofacial Skeleton. University of Michigan, Ann Arbor (1985)
5. Lanitis, A., Taylor, C.J.: Towards automatic face identification robust to ageing variation. In: Proceedings of the Fourth IEEE International Conference on Automatic Face and Gesture Recognition, pp. 391–396 (2000)
6. Ramanathan, N., Chellapa, R.: Face verification across age progression. In: IEEE Conference Computer Vision and Pattern Recognition (2005)

7. Ricanek, K., Boone, E., Patterson, E.: Craniofacial aging impacts on the eigenface face biometric. In: Proceedings of the Sixth IASTED International Conference on Visualization, Imaging, and Image Processing, Palma de Mallorca, Spain, August 2006, pp. 249–253 (2006)
8. Patterson, E., Ricanek, K., Albert, A.M., Boone, E.: Automatic representation of adult aging in facial images. In: Proceedings of the Sixth IASTED International Conference on Visualization, Imaging, and Image Processing, Palma de Mallorca, Spain (August 2006)
9. Wang, J., Shang, Y., Su, G., Lin, X.: Age simulation for face recognition. In: 18th International Conference on Pattern Recognition (2006)
10. Lanitis, A., Taylor, C.J., Cootes, T.F.: Automatic identification and coding of human faces using flexible models. IEEE Transactions on Pattern Analysis and Machine Intelligence 19(7) (July 1997)
11. Edwards, G.J., Taylor, C.J., Cootes, T.F.: Interpreting face images using active appearance models. In: Proceedings of the Fifth European Conference on Computer Vision, pp. 581–595 (1998)
12. Lanitis, A., Taylor, C.J., Cootes, T.F.: Simulating aging on face images. In: Proceedings of the Second International Conference on Audio and Video-based Biometric Person Authentication (1999)
13. Lanitis, A., Taylor, C.J., Cootes, T.F.: Toward automatic simulation of aging effects on face images. IEEE Transactions on Pattern Analysis and Machine Intelligence 24(4) (April 2002)
14. Cootes, T.F., Kittipanya-ngam, P.: Comparing Variations on the Active Appearance Model Algorithm. In: Proc. BMVC 2002, vol. 2, pp. 837–846 (2002)
15. Patterson, E., Sethuram, A., Albert, M., Ricanek, K., King, M.: Aspects of Age Variation in Facial Morphology Affecting Biometrics. In: Proceedings of the IEEE Conference on Biometrics: Theory, Applications, and Systems, Washington, DC (September 2007)
16. FG-NET Aging Database, `http://www.fgnet.rsunit.com`
17. Face Aging Group, `http://www.faceaginggroup.com`
18. CSU FAce Identification Evaluation System, `http://www.cs.colostate.edu/evalfacerec/`

Binary Biometric Representation through Pairwise Polar Quantization

Chun Chen and Raymond Veldhuis

Signals and Systems Group, Electrical Engineering, University of Twente,
P.O. Box 217, 7500AE Enschede, The Netherlands
{c.chen,r.n.j.veldhuis}@utwente.nl

Abstract. Binary biometric representations have great significance for data compression and template protection. In this paper, we introduce pairwise polar quantization. Furthermore, aiming to optimize the discrimination between the genuine Hamming distance (GHD) and the imposter Hamming distance (IHD), we propose two feature pairing strategies: the long-short (LS) strategy for phase quantization, as well as the long-long (LL) strategy for magnitude quantization. Experimental results for the FRGC face database and the FVC2000 fingerprint database show that phase bits provide reasonably good performance, whereas magnitude bits obtain poor performance.

Keywords: Polar quantization, pairing, face, fingerprint recognition.

1 Introduction

Binary biometric representations have great significance for data compression and template protection [1]. A common way to extract binary strings is by quantizing and coding the real-valued biometric templates. The binary string has to suffice the following two requirements:

1. To obtain a verification performance with low false acceptance rate (FAR) and false rejection rate (FRR), the binary strings are desired to be not only discriminative, but also robust to intra-class variation;
2. Considering template protection, the bits generated by the imposters should be independent and identically distributed, in order to maximize the efforts of guessing the genuine template.

To achieve the requirements, many work focus on designing one-dimensional quantizers, relying on the feature's statistical properties, e.g. the genuine user probability density function (PDF) p_{g} and the background PDF p_{b} [2], [3], [4], [5], [6], [7]. Among them the fixed quantizer [5], [6] is global, constructed merely from the background PDF, whereas quantizers in [2], [3], [4], [7] are user-specific, constructed from both the genuine user PDF and the background PDF. Quantizers in [2], [3] and [4] have equal-width-intervals. Alternatively, quantizers in [5], [6], [7] have equal-probability-intervals. Furthermore, independent of

M. Tistarelli and M.S. Nixon (Eds.): ICB 2009, LNCS 5558, pp. 72–81, 2009.
© Springer-Verlag Berlin Heidelberg 2009

the one-dimensional qantizers, the DROBA principle [8] was proposed to assign various numbers of quantization bits to every feature. In this paper we concentrate on the quantizer design. Although one-dimensional quantizers yield reasonably good performances, quantizing every feature independently ends up with inflexible quantization intervals, for instance, orthogonal boundaries in the two-dimensional feature space. Therefore, two-dimensional quantization might bring more flexible quantizer structures.

In this paper, we propose quantization in polar coordinates, including phase and magnitude. Unlike in audio and image coding [9], where polar quantization is directly applied on predetermined complex variables, quantizing biometric features allows freedom to compose the pairwise features. Furthermore, we define a measure of the discrimination between the quantized bits, by computing the inter-class and intra-class scatters between the genuine and the imposter Hamming distances. To optimize such discrimination, we develop the long-short (LS) and the long-long (LL) pairing strategies for phase and magnitude, respectively.

In Section 2 the polar quantization is introduced. In Section 3 we propose the long-short and the long-long pairing strategies, to optimize the discrimination between the genuine and the imposter Hamming distances. In Section 4, some experimental results are given for the FRGC face database and the FVC2000 fingerprint database, and conclusions are drawn in Section 5.

2 Polar Quantization

Let $\mathbf{v} = \{v_1, v_2\}$ denote a two-dimensional feature vector. In polar coordinates, the phase θ and magnitude r are:

$$\theta = \text{angle}(v_1, v_2) \ , \tag{1}$$

$$r = \sqrt{v_1^2 + v_2^2} \ , \tag{2}$$

where θ is the counterclockwise angle from the v_1-axis, and r is the radial distance from the origin. We assume that biometric features have circularly symmetric background PDF, feasible for polar quantization. A n-bit phase quantizer is then constructed as:

$$\xi = \frac{2\pi}{2^n} \ , \tag{3}$$

$$Q_{\theta,i} = [(i-1)\xi \quad i\xi), \ i = 1, \ldots 2^n \ , \tag{4}$$

where $Q_{\theta,i}$ represents the ith quantization interval within boundaries $[(i-1)\xi \quad i\xi)$. When the background PDF is circularly symmetric, θ is uniformly distributed, leading to both equal-ξ-width and equal-2^{-n}-probability intervals.

A n-bit magnitude quantizer is constructed as:

$$B_0 = 0 \ , \tag{5}$$

$$B_i = \arg_B \left[\int_{B_{i-1}}^{B} \int_0^{2\pi} p_{\text{b}}(\theta, r) d\theta dr = 2^{-n} \right], \ i = 1, \ldots 2^n \ , \tag{6}$$

$$Q_{\text{r},i} = [B_{i-1} \quad B_i), \ i = 1, \ldots 2^n \ , \tag{7}$$

where $Q_{\mathrm{r},i}$ represents the ith quantization interval within boundaries $[B_{i-1} \ \ B_i)$. Determining these intervals depends on the background PDF p_{b}. The expression in (6) ensures equal-2^{-n}-probability intervals.

To summarize, both phase and magnitude quantization obtain equal background probability intervals. Thus, the imposters obtain independent and identically distributed bits.

3 Feature Pairing

3.1 Hamming Distance Discriminant Analysis

Often, binary biometric strings are matched via their Hamming distances. To design a verification system, a genuine Hamming distance (GHD) is computed when the query and the target share the same identity, otherwise an imposter Hamming distance (IHD) is computed. The decision is then made by applying a threshold T to both distances. In this paper, we aim to optimize the discrimination between the GHD and IHD densities.

We begin by defining the discrimination between GHD and IHD. Suppose we have P feature pairs. Each pair is quantized as a code $x_i, i = 1, \ldots, P$, and subsequently concatenated into the binary string $X = x_1, \ldots, x_P$. Considering a genuine user ω, with $\widehat{X}_\omega = \hat{x}_{\omega,1} \ldots \hat{x}_{\omega,P}$ as the enrollment template. Let $d_{\omega,i}$ and $d_{\bar{\omega},i}$ be the GHD and IHD for the ith feature pair, defined as:

$$d_{\omega,i} = h(x_i, \hat{x}_{\omega,i}), \ x_i \in \text{user } \omega \ ; \tag{8}$$

$$d_{\bar{\omega},i} = h(x_i, \hat{x}_{\omega,i}), \ x_i \notin \text{user } \omega \ , \tag{9}$$

where function h computes the Hamming distance between the two inputs. Then the GHD (D_ω) and the IHD ($D_{\bar{\omega}}$) for the entire binary string are:

$$D_\omega = \sum_{i=1}^{P} d_{\omega,i} \ ; \tag{10}$$

$$D_{\bar{\omega}} = \sum_{i=1}^{P} d_{\bar{\omega},i} \ . \tag{11}$$

Furthermore, for the ith feature pair, we define the expectation of the GHD ($\bar{d}_{\omega,i}$) and the IHD ($\bar{d}_{\bar{\omega},i}$) as:

$$\bar{d}_{\omega,i} = \mathcal{E}[d_{\omega,i}], \ x_i \in \text{user } \omega \ ; \tag{12}$$

$$\bar{d}_{\bar{\omega},i} = \mathcal{E}[d_{\bar{\omega},i}], \ x_i \notin \text{user } \omega \ . \tag{13}$$

Thus, for the entire binary string, the expectation of the GHD ($\overline{D}_\omega$) and the IHD ($\overline{D}_{\bar{\omega}}$) become:

$$\overline{D}_\omega = \sum_{i=1}^{P} \bar{d}_{\omega,i} \ ; \tag{14}$$

$$\overline{D}_{\bar{\omega}} = \sum_{i=1}^{P} \bar{d}_{\bar{\omega},i} \ . \tag{15}$$

A measure of separation between GHD and IHD densities is their intra-class scatter and the inter-class scatter. Thus, we would like to minimize the intra-class scatter $S_{\text{intra},\omega}$, defined as:

$$S_{\text{intra},\omega} = \mathcal{E}[(D_\omega - \overline{D}_\omega)^2] + \mathcal{E}[(D_{\bar{\omega}} - \overline{D}_{\bar{\omega}})^2] \ . \quad (16)$$

In the mean time, we want to maximize the inter-class scatter $S_{\text{inter},\omega}$:

$$S_{\text{inter},\omega} = (\overline{D}_{\bar{\omega}} - \overline{D}_\omega)^2 \ . \quad (17)$$

Substituting (10), (11), (14), (15) into (16) and (17), we have:

$$S_{\text{intra},\omega} = \mathcal{E}\Big[\big[\sum_{i=1}^{P}(d_{\omega,i} - \bar{d}_{\omega,i})\big]^2\Big] + \mathcal{E}\Big[\big[\sum_{i=1}^{P}(d_{\bar{\omega},i} - \bar{d}_{\bar{\omega},i})\big]^2\Big] \ ;$$

$$S_{\text{inter},\omega} = \big[\sum_{i=1}^{P}(\bar{d}_{\bar{\omega},i} - \bar{d}_{\omega,i})\big]^2 \ .$$

Assuming that the P feature pairs are independent, $S_{\text{intra},\omega}$ and $S_{\text{inter},\omega}$ can be approximated as:

$$S_{\text{intra},\omega} = \sum_{i=1}^{P}\Big[\mathcal{E}\big[(d_{\omega,i} - \bar{d}_{\omega,i})^2\big] + \mathcal{E}\big[(d_{\bar{\omega},i} - \bar{d}_{\bar{\omega},i})^2\big]\Big] = \sum_{i=1}^{P} S_{\text{intra},\omega,i} \ ; \quad (18)$$

$$S_{\text{inter},\omega} = \sum_{i=1}^{P}(\bar{d}_{\bar{\omega},i} - \bar{d}_{\omega,i})^2 = \sum_{i=1}^{P} S_{\text{inter},\omega,i} \ . \quad (19)$$

Hence, the intra-/inter-class scatter for the entire binary string is simplified as the sum of the intra-/inter-class scatter over all the feature pairs. Usually biometric features are not presented in pairs, allowing the freedom to compose the pairwise features to optimize $S_{\text{intra},\omega}$ and $S_{\text{inter},\omega}$. Therefore, we formulate the problem as: for a genuine user ω, to find a pairing configuration $\{C^*_{\omega,i}\}, i = 1, \ldots, P$, so that the discrimination between GHD and IHD is maximized:

$$\{C^*_{\omega,i}\} = \arg\max_{\{C_{\omega,i}\}} \frac{S_{\text{inter},\omega}}{S_{\text{intra},\omega}} \ , \quad (20)$$

$$= \arg\max_{\{C_{\omega,i}\}} \frac{\sum_{i=1}^{P} S_{\text{inter},\omega,i}}{\sum_{i=1}^{P} S_{\text{intra},\omega,i}} \ . \quad (21)$$

3.2 Pairing Solutions

In order to optimize (21), we first have to analyze how the genuine user PDF relates to $S_{\text{inter},\omega,i}$ and $S_{\text{intra},\omega,i}$ in case of the polar quantization. Afterwards, we could decide a strategy to pair features with specific p_{g} to optimize $S_{\text{inter},\omega,i}$ and $S_{\text{intra},\omega,i}$. However, it is difficult to analytically find an expression for the relation, due to lack of samples and complex integral calculation in polar coordinates. Therefore, we employ an empirical method to simplify the relation of $S_{\text{inter},\omega,i}$, $S_{\text{intra},\omega,i}$ and the genuine user PDF. We take two data sets: FRGC(version 1) [10] face database and FVC2000(DB2) fingerprint database [11].

- **FRGC:** It contains 275 users with various numbers of images, taken under both controlled and uncontrolled conditions. A set of standard landmarks, i.e. eyes, nose and mouth, are used to align the faces. The raw measurements are the gray pixel values, leading to a total of 8762 elements.
- **FVC2000:** It contains 8 images of 110 different users. Images are aligned according to a standard core point position. The raw measurements contain two categories: the squared directional field in both x and y directions, and the Gabor response in 4 orientations (0, $\pi/4$, $\pi/2$, $3\pi/4$). Determined by a regular grid of 16 by 16 points with spacing of 8 pixels, measurements are taken at 256 positions, leading to a total of 1536 elements [5].

We first apply PCA/LDA [12] to reduce both data sets into 50 features. Afterwards, for every genuine user, we randomly pair the features into 25 pairs. Following this pairing configuration, the entire data set, including the genuine user samples and the imposter samples are quantized via a 1-bit phase quantizer and a 1-bit magnitude quantizer, where the magnitude quantizer boundary is determined by a two-dimensional Gaussian density with zero mean and unit variance $p_{\mathrm{b}}(\mathbf{v}) = N(\mathbf{v}, 0, 1)$. The $S_{\mathrm{inter},\omega,i}$ and $S_{\mathrm{intra},\omega,i}$ for every feature pair is then computed based on the quantized bits. We repeat this process for all the genuine users in the data set. Eventually, we average $S_{\mathrm{inter},\omega,i}$ and $S_{\mathrm{intra},\omega,i}$ over all features as well as all genuine users, so that the averaged $\overline{S}_{\mathrm{inter},\omega,i}$ and $\overline{S}_{\mathrm{intra},\omega,i}$ are neither user nor feature biased. Intuitively, we speculate that the distance $r_{\omega,i}$, – distance between the feature pair mean and the origin – dominates the inter- and intra-class scatter. To analyze, in Fig. 1, we plot the value of $\overline{S}_{\mathrm{inter},\omega,i}$, $\overline{S}_{\mathrm{intra},\omega,i}$ as sorted by $r_{\omega,i}$, for both phase bits and magnitude bits. Both data sets reveal the same relations: Fig. 1(a) suggests that for phase quantization, when $r_{\omega,i}$ increases, the inter-class scatter increases and the intra-class scatter decreases; Fig. 1(b) suggests that for magnitude quantization, when the distance $\mid r_{\omega,i} - B \mid$ between the feature pair mean and the magnitude boundary

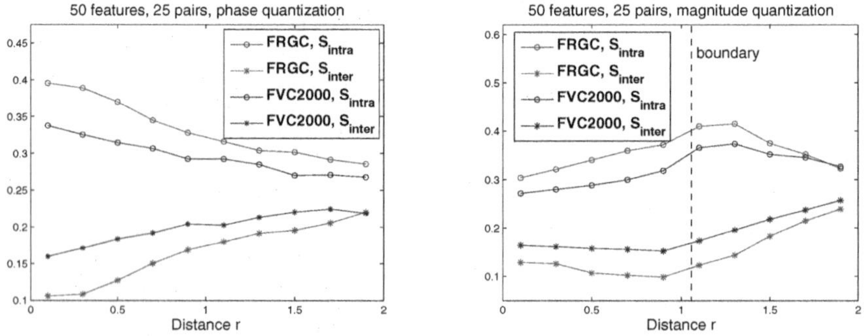

(a) (b)

Fig. 1. The averaged $\overline{S}_{\mathrm{inter},\omega,i}$ and $\overline{S}_{\mathrm{intra},\omega,i}$ for (a) phase and (b) magnitude, as sorted by $r_{\omega,i}$

increases, the inter-class scatter increases and the intra-class scatter decreases. Therefore, we simplify the problem (21) as:

$$\{C^*_{\omega,i}\} = \arg \max_{\{C_{\omega,i}\}} \sum_{i=1}^{P} r_{\omega,i}, \text{ for phase }, \tag{22}$$

$$\{C^*_{\omega,i}\} = \arg \max_{\{C_{\omega,i}\}} \sum_{i=1}^{P} \mid r_{\omega,i} - B \mid, \text{ for magnitude }. \tag{23}$$

Optimizing the inter- and intra-class scatter is now simplified as optimizing the overall distance of the feature pairs. To solve (22) and (23), the straightforward way is to conduct a brute force search of all possible pairing configurations and pick the one with the maximum overall distance. Unfortunately, the computational complexity is too high. Therefore, we propose the following two heuristic pairing strategies: Given $2P$ features, we first sort the mean of the $2P$ features $\{\text{abs}(\mu_k)\}$, $k = 1, \ldots, 2P$ from the smallest to the largest into a sequence of ordered feature indexes $\{I_1, I_2, \ldots, I_{2P}\}$.

1. **long-short (LS) strategy**: The index of the ith feature pair is:

$$C^*_{\omega,i} = (I_i, I_{2P+1-i}), \ i = 1, \ldots, P \ ; \tag{24}$$

2. **long-long (LL) strategy**: The index of the ith feature pair is:

$$C^*_{\omega,i} = (I_{2\times i-1}, I_{2\times i}), \ i = 1, \ldots, P \ . \tag{25}$$

Apparently, the long-short strategy selects features with a large mean and a small mean as a pair, keeping their distance $r_{\omega,i}$ large; thus, somehow maximizes the overall distances in (22) for phase quantization. Contrarily, long-long strategy selects features that either both have large means or both have small means, keeping their distance $r_{\omega,i}$ far away from the boundary; thus, maximizes the overall distances in (23) for magnitude quantization. The advantage of these two pairing strategies is that the computational complexity is low $O(P)$.

4 Experiments

We tested the polar quantization on the FRGC and the FVC2000 database, as described in Section 3.2. To first reduce the feature dimensionality, we applied PCA/LDA [12] on a training set, consisting of independent users from the enrollment and verification. The obtained transformation was then applied to both the enrollment and verification samples. In the enrollment, for every target user, assigned with Gray codes, the phase bits (phase+LS) were generated following the LS pairing strategy, while the magnitude bits (magnitude+LL) were generated following LL pairing strategy. By concatenating both the phase and the magnitude bits we obtained the total polar bits (polar+combined). The quantized codes, together with the pairing configuration $\{C^*_{\omega,i}\}$, was stored for each target user. During the verification, features of the query user were quantized according to the $\{C^*_{\omega,i}\}$ of the claimed identity, resulting in a query binary string. Eventually the verification performance was evaluated by a Hamming distance classifier. With, in total, n samples per user ($n = 8$ for FVC2000, and n ranges from 6 to 48 for FRGC), the division of the data is indicated in Table 1.

Table 1. Training, enrollment and verification data (number of users×number of samples per user) and the number of partitionings for FRGC and FVC2000

	Training	Enrollment	Verification	Partitionings
FRGC	$210 \times n$	$65 \times 2n/3$	$65 \times n/3$	5
FVC2000	$80 \times n$	$30 \times 3n/4$	$30 \times n/4$	20

Since both the phase and the magnitude have fixed equal-probability-intervals, we compared their performances with the one-dimensional fixed quantizer (1D fixed) [5], [6], which has the same property. We first investigated the 1-bit quantization ($n_\theta = n_\mathrm{r} = n_\mathrm{f} = 1$) performances of phase+LS, magnitude+LL and polar+combined. The EER results for the FRGC and the FVC2000 at various feature dimensionalities are shown in Fig. 2. In general, the magnitude bits give poor performances, whereas the phase bits consistently yield good performances and outperform 1D fixed quantization. Furthermore, since the magnitude bits are so poor, combining both phase and magnitude bits, as seen with polar+combined, does not show good performance.

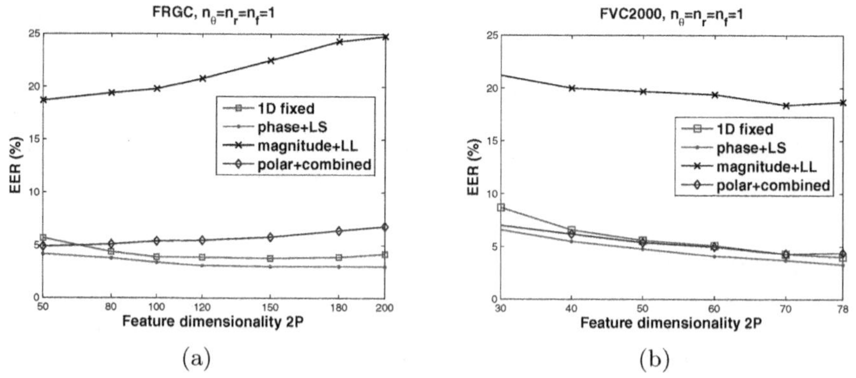

Fig. 2. The EER performances of phase+LS, magnitude+LL and polar+combined, as compared to the 1D fixed quantization, for (a) FRGC and (b) FVC2000

In Fig. 3 we further illustrate the GHD and IHD densities of the phase and the magnitude bits, at $P = 60$ for FRGC, as compared to the 1D fixed quantization. We observe that for the three types of bits, the mean of their IHD densities are all around 0.5, demonstrating the equal-probability-intervals. The IHD density

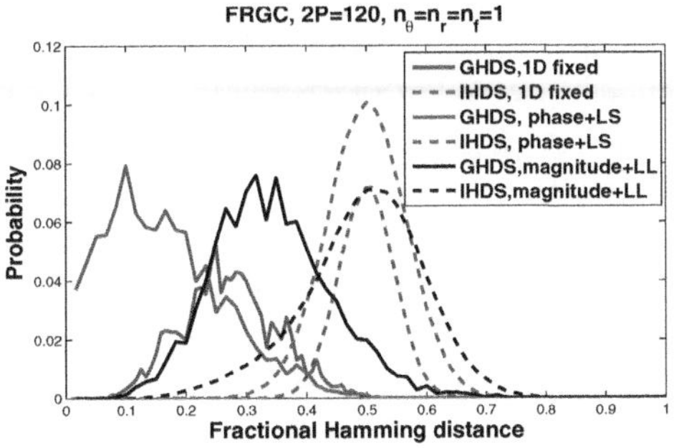

Fig. 3. The GHD and the IHD densities of phase+LS, magnitude+LL, and the 1D fixed quantization, at $2P = 120$ for FRGC

Table 2. The inter-class and intra-class scatter of phase+LS, magnitude+LL and the 1D fixed, for (a) FRGC and (b) FVC2000

(a)

FRGC	2P=50			2P=120		
	1D fixed	phase+LS	magnitude+LL	1D fixed	phase+LS	magnitude+LL
S_{inter} ($\times 10^{-2}$)	7.2	14.9	4.4	5.2	11.5	2.3
S_{intra} ($\times 10^{-2}$)	1.3	2.0	2.7	0.8	1.4	2.0

(b)

FVC2000	2P=50			2P=78		
	1D fixed	phase+LS	magnitude+LL	1D fixed	phase+LS	magnitude+LL
S_{inter} ($\times 10^{-2}$)	7.2	14.6	4.3	6.7	13.5	3.7
S_{intra} ($\times 10^{-2}$)	1.2	2.1	2.8	1.0	1.6	2.2

of the 1D fixed quantizer is relatively narrow, compared to those of the phase and magnitude. Additionally, the GHD density of the phase bits are located farther away from the IHD. Unfortunately, the GHD densities of the three types of bits are all quite wide.

The discrimination property explained above can be quantified as the inter- and intra-class scatter. Some examples computed on the fractional Hamming distances are given in Table 2. Consistent to what we observed in Fig. 3, for phase+LS, the S_{inter} is on average two times the 1D fixed. In the mean time, its S_{intra} does not increase much. For this reason, phase+LS yields better performance. On the other hand, magnitude+LL obtains smaller S_{inter} and larger S_{intra}, leading to poor performance. Based on these analysis, we could conclude that even with carefully paired features, the magnitude does not provide discriminative bits for classification. Contrarily, the phase provides reasonably discriminative bits for classification.

In fact, the 1-bit phase quantizer, with boundary at π, quantizes only the feature that has a larger mean in the pairs, leaving the other feature discarded. Thus it also acts as a feature selection procedure. Now we test the phase quantizer with more quantization bits at $n_\theta = 1, 2, 3, 4$, compared to the 1D fixed quantizer at $n_f = 1, 2$. Their EER results for the FRGC and the FVC2000 are shown in Fig. 4. Note that when $n_\theta = 2$, in the two-dimensional feature space, the phase quantizer has the same orthogonal boundaries as the 1D fixed quantizer at $n_f = 1$, leading to the same performances. Results show that at a given feature dimensionality, phase bits at $n_\theta = 1$ always give the best performances, while $n_\theta = 2, 3$ also yield reasonably good performances. Unfortunately, when $n_\theta = 4$, the performances turn poor. Generally, compared to the 1D fixed quantizer, the phase quantizer gives better performances at a lower bit length.

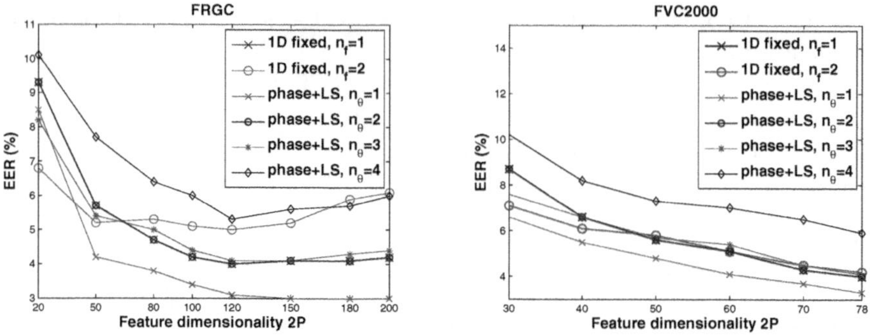

(a) (b)

Fig. 4. The EER performances of phase+LS and 1D fixed quantization at various feature dimensionalities with various quantization bits , for (a) FRGC and (b) FVC2000

To summarize, although the magnitude $r_{\omega,i}$ itself does not provide discriminative bits, it in fact facilitates generating better phase bits. Additionally, the phase quantization has the following properties: (1) The LS pairing strategy is universal and simple, without modeling the specific genuine feature PDF; (2) The phase quantizer boundaries are not necessarily orthogonal, allowing correlations between the two features. Furthermore, the phase is uniformly distributed as long as the background PDF is circularly symmetric, which easily fits many feature modalities.

5 Conclusions

In this paper, we propose extracting binary biometric bits through polar quantization. In addition to the polar quantizer structure, quantizing features in pairs allows feature pair configuration. Therefore, we propose the long-long and the long-short pairing strategies to optimize the discrimination between the genuine Hamming distance and the imposter Hamming distance. Experimental results on the FVC2000 and the FRGC database show that magnitude yields poor classification performances, whereas phase provides reasonably good performances.

Acknowledgments

This research is supported by the research program Sentinels (www.sentinels.nl). Sentinels is financed by Technology Foundation STW, Netherlands Organization for Scientific Research (NWO), and the Dutch Ministry of Economic Affairs.

References

1. Jain, A.K., Nandakumar, K., Nagar, A.: Biometric template security. EURASIP Journal on Advances in Signal Processing 8(2) (2008)
2. Vielhauer, C., Steinmetz, R., Mayerhofer, A.: Biometric hash based on statistical features of online signatures. In: Proc. 16th International Conference on Pattern Recognition (ICPR 2002), Quebec, Canada, vol. 1, pp. 123–126 (2002)
3. Hao, F., Wah, C.C.: Private key generation from on-line handwritten signatures. Information Management & Computer Security 10(4), 159–164 (2002)
4. Chang, Y., Zhang, W., Chen, T.: Biometrics-based cryptographic key generation. In: Proc. IEEE International Conference on Multimedia and Expo. (ICME 2004), Taipei, Taiwan, vol. 3, pp. 2203–2206 (2004)
5. Tuyls, P., Akkermans, A.H.M., Kevenaar, T.A.M., Schrijen, G.J., Bazen, A.M., Veldhuis, R.N.J.: Practical biometric authentication with template protection. In: Kanade, T., Jain, A., Ratha, N.K. (eds.) AVBPA 2005. LNCS, vol. 3546, pp. 436–446. Springer, Heidelberg (2005)
6. Kevenaar, T.A.M., Schrijen, G.J., van der Veen, M., Akkermans, A.H.M., Zuo, F.: Face recognition with renewable and privacy preserving binary templates. In: Proc. IEEE Workshop on Automatic Identification Advanced Technologies (AutoID 2005), NY, USA, pp. 21–26 (2005)
7. Chen, C., Veldhuis, R.N.J., Kevenaar, T.A.M., Akkermans, A.H.M.: Multi-bits biometric string generation based on the likelihood ratio. In: Proc. IEEE Conference on Biometrics: Theory, Applications and Systems (BTAS 2007) (2007)
8. Chen, C., Veldhuis, R.N.J., Kevenaar, T.A.M., Akkermans, A.H.M.: Biometric binary string generation with detection rate optimized bit allocation. In: IEEE Computer Society Conference on Computer Vision and Pattern Recognition Workshops (CVPR 2008) (2008)
9. Pobloth, H., Vafin, R., Kleijn, W.B.: Multivariate block polar quantization. IEEE Transactions on Communications 53(12), 2043–2053 (2005)
10. Phillips, P.J., Flynn, P.J., Scruggs, W.T., Bowyer, K.W., Chang, J., Hoffman, K., Marques, J., Min, J., Worek, W.J.: Overview of the face recognition grand challenge. In: Proc. IEEE Computer Society Conference on Computer Vision and Pattern Recognition (CVPR 2005), MD, USA, pp. 947–954 (2005)
11. Maio, D., Maltoni, D., Cappelli, R., Wayman, J.L., Jain, A.K.: FVC 2000: Fingerprint verification competition. IEEE Transactions on Pattern Analysis and Machine Intelligence 24(3), 402–412 (2002)
12. Veldhuis, R.N.J., Bazen, A., Kauffman, J., Hartel, P.: Biometric verification based on grip-pattern recognition. In: Proc. SPIE Security, Steganography, and Watermarking of Multimedia Contents VI (SSWMC 2004), CA, USA, vol. 5306, pp. 634–641 (2004)

Manifold Learning for Gender Classification from Face Sequences

Abdenour Hadid and Matti Pietikäinen

Machine Vision Group, P.O. Box 4500, FI-90014, University of Oulu, Finland

Abstract. We propose a novel approach to gender recognition for cases when face sequences are available. Such scenarios are commonly encountered in many applications such as human-computer interaction and visual surveillance in which input data generally consists of video sequences. Instead of treating each facial image as an isolated pattern and then combining the results (at feature, decision or score levels) as generally done in previous works, we propose to exploit the correlation between the face images and look at the problem of gender classification from manifold learning point of view. Our approach consists of first learning and discovering the hidden low-dimensional structure of male and female manifolds using an extension to the Locally Linear Embedding algorithm. Then, a target face sequence is projected into both manifolds for determining the gender of the person in the sequence. The matching is achieved using a new manifold distance measure. Extensive experiments on a large set of face sequences and different image resolutions showed very promising results, outperforming many traditional approaches.

1 Introduction

Determining the gender (i.e. man or woman) of a person in a given image or video is useful for many applications such as more affective Human-Computer Interaction (HCI), content-based image and video retrieval, restricting access to certain areas based on gender, collecting demographic information in public places, counting the number of women entering a retail store and so on.

First attempts of using computer vision based techniques to gender classification started in early 1990s. Since then, many approaches have been reported in literature. Among the most notable results to date are those obtained by Moghaddam and Yang [1], and also by Baluja and Rowley [2]. Moghaddam and Yang used raw pixels as inputs to Support Vector Machines (SVMs) and achieved a classification rate of 96.6% on FERET database of images scaled to 12×21 pixels [1]. Note that the considered FERET images were very clean and some persons may have appeared in both training and test sets. Comparable accuracy but at a higher speed was also reported by Baluja and Rowley who used AdaBoost to combine weak classifiers, constructed using simple pixel comparisons, into single strong classifier [2]. Note that both approaches are based on still images and assume well aligned faces. However, in many real-world applications (e.g. HCI, visual surveillance and content-based video retrieval) input

M. Tistarelli and M.S. Nixon (Eds.): ICB 2009, LNCS 5558, pp. 82–91, 2009.
© Springer-Verlag Berlin Heidelberg 2009

data generally consists of video sequences and it is not always obvious to hold the face alignment assumption. So, the question which arises then is how to efficiently perform gender classification from face sequences? We believe that this issue is understudied in literature as most works tend to process each frame (or some selected ones) and then fuse the results at feature, decision or score levels. Obviously, such approaches ignore the correlation between the face images as they are only treated as "single" or "isolated" patterns in the image space while, in fact, they lie on a low-dimensional manifold.

From these observations, we propose to look at the problem of gender classification from manifold learning point of view. The goal of face manifold learning is to discover the hidden low-dimensional structure of the face images. This is a very useful but also a challenging unsupervised learning problem. It is a useful task because mapping the high-dimensional faces into low-dimensional coordinates would facilitate and speed-up the further analysis (i.e. gender classification in our case) and also may avoid "the curse of dimensionality". It is also a challenging problem since the face manifold is highly nonconvex and nonlinear and so is the face manifold of any individual under changes due to pose, illumination and other factors.

Therefore, instead of treating each facial image as an isolated pattern and then fusing the results (at feature, decision or score levels) as generally done in previous works, we propose to exploit the correlation between the face images and consider the problem of gender classification from manifold learning point of view. Our approach consists of first learning and discovering, from a set of training face sequences, the hidden low-dimensional structure of male and female classes, thus yielding in male and female face manifolds. Then, a target face sequence can be projected into both manifolds for determining the gender of the person in the sequence. The "closest" manifold (in terms of a newly introduced manifold distance measure) will then determine the gender of the person. In the rest of this paper, we further develop the proposed approach and validate it through extensive experiments and comparisons against traditional approaches. Among the novel contributions in this work are:

→ A new method to gender classification from face sequences is presented and extensively evaluated;
→ An extension to the Locally Linear Embedding algorithm to handle face sequences is proposed;
→ A simple yet efficient manifold to manifold distance measure is introduced.

2 Locally Linear Embedding

The first key issue in our proposed approach is to learn the face manifolds by embedding the face images into low-dimensional coordinates. For that purpose, there exist several methods. The traditional ones are Principal Component Analysis (PCA) and Multidimensional Scaling (MDS). These methods are simple to implement and efficient in discovering the structure of data lying on or near linear subspaces of the high-dimensional input space. However, face images do not satisfy this constraint as they lie on a complex nonlinear and nonconvex manifold in the high-dimensional space. Therefore, such linear methods generally fail to discover the real structure of the face images in the low-dimensional space. As an alternative to PCA and MDS, one can consider some nonlinear dimensionality reduction methods such as Self-Organizing Maps (SOM) [3], talent variable models [4], Generative Topographic Mapping (GTM) [5], Sammon's Mappings (SM) [6] etc. Though these methods can also handle nonlinear manifolds, most of them tend to involve several free parameters such as learning rates and convergence criteria. In addition, most of these methods do not have an obvious guarantee of convergence to the global optimum. Fortunately, in the recent years, a set of new manifold learning algorithms have emerged. These methods are based on an Eigen decomposition and combine the major algorithmic features of PCA and MDS (computational efficiency, global optimality, and flexible asymptotic convergence guarantees) with flexibility to learn a broad class of nonlinear manifolds. Among these algorithms are Locally Linear Embedding (LLE) [7], ISOmetric feature MAPping (ISOMAP) [8] and Laplacian Eigenmaps [9].

We adopt in our present work the LLE approach for its demonstrated simplicity and efficiency to recover meaningful low-dimensional structures hidden in complex and high dimensional data such as face images. LLE is an unsupervised learning algorithm which maps high-dimensional data onto a low-dimensional, neighbor-preserving embedding space.

In brief, considering a set of N face images and organizing them into a matrix X (where each column vector represents a face), the LLE algorithm involves then the following three steps:

1. Find the k nearest neighbors of each point X_i.
2. Compute the weights W_{ij} that best reconstruct each data point from its neighbors, minimizing the cost in Equation (1):

$$\epsilon(W) = \sum_{i=1}^{N} \left\| X_i - \sum_{j \in neighbors(i)} W_{ij} X_j \right\|^2 \tag{1}$$

   while enforcing the constraints $W_{ij} = 0$ if X_j is not a neighbor of X_i, and $\sum_{j=1}^{N} W_{ij} = 1$ for every i (to ensure that W is translation-invariant).
3. Compute the embedding Y (of lower dimensionality $d << D$, where D is the dimension of the input data) best reconstructed by the weights W_{ij} minimizing the quadratic form in Equation (2):

$$\Phi(Y) = \sum_{i=1}^{N} \left\| Y_i - \sum_{j \in neighbors(i)} W_{ij} Y_j \right\|^2 \tag{2}$$

   under constraints $\sum_{i=1}^{N} Y_i = 0$ (to ensure a translation-invariant embedding) and $\frac{1}{N} \sum_{i=1}^{N} Y_i Y_i^T = 0$ (normalized unit covariance).

The aim of the first two steps of the algorithm is to preserve the local geometry of the data in the low-dimensional space, while the last step discovers the global structure by integrating information from overlapping local neighborhoods. LLE is an efficient approach to compute the low-dimensional embeddings of high-dimensional data assumed to lie on a non-linear manifold. Its ability to deal with large sizes of high-dimensional data and its non-iterative way to find the embeddings make it attractive. However, in its actual formulation, the LLE method can only be used with still images but not with video sequences. In the next section, we introduce extensions to LLE to also handle face sequences and then present our proposed approach to gender classification from face sequences.

3 The Proposed Approach to Gender Classification from Face Sequences

Given a set of training face sequences of males and females. First, let us denote the male face sequences as $\{Seq_1, Seq_2, ..., Seq_M\}$. In order to discover the male face manifold in the low-dimensional space, we modify and apply the LLE algorithm as follows:

- $1'$.a. Let $X_i^{(j)}$ be the column vector representing the ith face image of the jth sequence Seq_j.
- $1'$.b. In every sequence Seq_j, find the k nearest neighbors of each point $X_i^{(j)}$ among all points which do not belong to the sequence Seq_j.
- $2'$. Compute the weights on the newly constructed data and neighborhood in a similar way as in Step 2 of the original LLE algorithm.
- $3'$. Compute the embedding in a similar way as in Step 3 of the original LLE algorithm.

Note that, in this modified version of LLE, we enforce (in Step **$1'$.b.**) the neighbors of each point to be taken from other sequences than the one to which the data point belongs. The aim of this constraint is to avoid constructing disjoint manifolds, each corresponding to an individual. The other reason is that we are not only focusing on the relationships between the image instances of the same individual but more importantly we are also looking at learning the relationships between the images of different subjects from the entire male class. In other terms, we are interested in discovering the features and structure that are shared among the male faces. The introduced changes to LLE are somehow inspired by the recent works on semi-supervised locally linear embedding for classification (e.g. [10]). From now and on, we only consider LLE with the proposed changes. Therefore, for the clarity of the presentation, we continue using the term LLE to refer to the modified version of the algorithm in the rest of this paper.

After embedding the face sequences $\{Seq_1, Seq_2, ..., Seq_M\}$ using LLE, we obtain the coordinates of each face image in the low-dimensional space, thus defining the face manifold of the male class. Let us denote then the obtained embedding result as M_Y. Note that the calculation of M_Y involves only two free parameters which are the number of neighbors (k) and the dimension of the embedding space (d). We discuss the values of these two parameters in the experimental analysis in Section 4.

Similarly to the analysis of the male sequences, we also apply our described procedure to the female training face sequences to compute the face manifold F_Y corresponding to the female class. Thus, we finally obtain two embeddings (face manifolds M_Y and F_Y) that are extremely useful to gender classification of unseen individuals in target face sequences, as the next section explains.

Classifying a New Face Sequence?

To determine the gender of an unknown person in a given face sequence

$$\{Face_{frame(1)}, Face_{frame(2)}, ..., Face_{frame(L)}\}$$

we first project every face instance $Face_{frame(i)}$ into both male and female face manifolds in the low-dimensional space. The "closest" manifold will then determine the gender of the person in the sequence. The projection of the target face sequence into the male manifold, for instance, is done using the following steps:

a. Let now X_i be the column vector representing the face image ($Face_{frame(i)}$) from the new sequence.
b. Find the k nearest neighbors of each point X_i among the male training face samples.
c. Compute the weights W_{ij} that best reconstruct each data point X_i from its neighbors using Equation (1).
d. Use the obtained weights W_{ij} to compute the embedding Y_i of each point X_i (i.e. $Face_{frame(i)}$) as:

$$Y_i = \sum_{j \in neighbors(X_i)} W_{ij} \, M_Y^j \tag{3}$$

where M_Y^j refers to the embedding point in the male manifold of the jth neighbor of the point X_i.

As a result, we obtain the embedding Y of the new face sequence in the male manifold. Then, we compute how close is the embedding Y to the male manifold M_Y using:

$$D_{male} = \frac{1}{L} \sum_{i=1}^{L} \left\| Y_i - M_Y^{r(i)} \right\| \tag{4}$$

where L is the length of the target face sequence, Y_i is the embedding of the point X_i in the low-dimensional space and $M_Y^{r(i)}$ is the closest point (in term of Euclidean distance) from the manifold M_Y to Y_i as shown in Fig. 1.

In the same way as for the male manifold, we also compute the embedding of the new face sequence into the female manifold and then calculate the distance D_{female}. Finally, the smallest value between the two distances (D_{female} and D_{male}) defines the "closest" manifold and thus determines the gender of the person in the new sequence, as formulated in Equation (5).

$$Gender(Face_{frame(1)}, ..., Face_{frame(L)}) = \begin{cases} Man, & if \ D_{male} \prec D_{female} \\ Woman, & otherwise \end{cases} \tag{5}$$

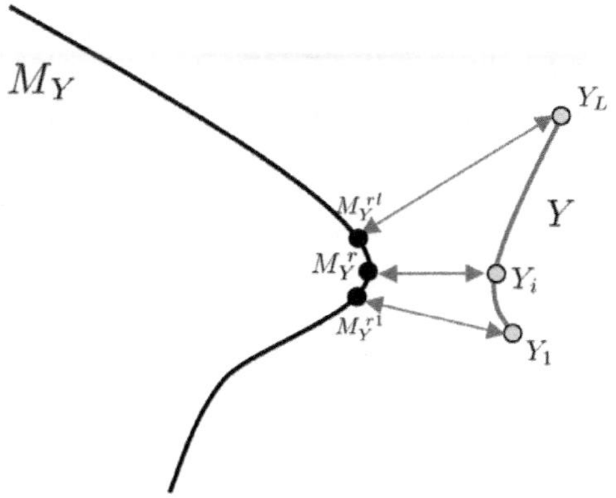

Fig. 1. Calculation of the distance between the manifold of a new face sequence (represented by its embedding Y) and the male manifold represented by its embedding M_Y

4 Experimental Analysis

4.1 Experimental Data

For experimental analysis of our proposed approach, we considered three different publicly available video face databases (namely CRIM [11], VidTIMIT [12] and Cohn-Kanade [13]) containing several face sequences subject to changes caused by different factors including face image resolution, illumination variations, head movements and facial expressions. CRIM is a large set of 591 face sequences showing 20 persons (10 female and 10 male) reading broadcast news for a total of about 5 hours. There are between 23 and 47 video sequences for each individual. The VidTIMIT database consists of audio recordings and video sequences of 43 subjects (19 female and 24 male), reciting ten short sentences

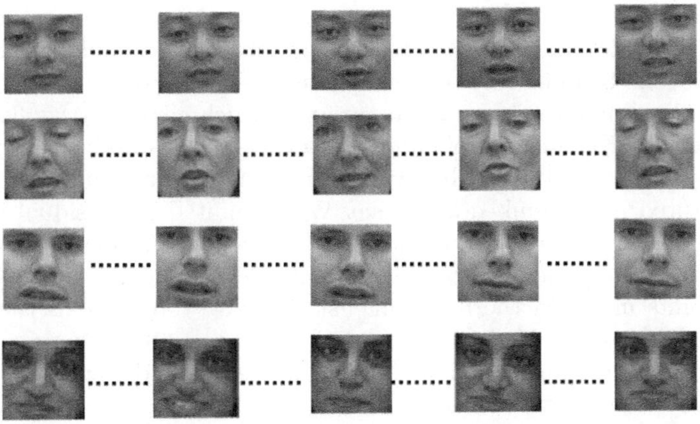

Fig. 2. Examples of face images from some sequences from the considered datasets

in three sessions with an average delay of a week between sessions, allowing for appearance and mood changes. Cohn-Kanade database consists of 100 subjects expressing different emotions like anger, disgust, fear, joy, sadness, and surprise. Sixty-five percent of the subjects were female, 15 percent African-American, and three percent Asian or Latino. We randomly segmented the datasets and extracted over 1000 video shots of 15 to 80 frames each. From each shot or sequence, we automatically detected the eye positions from the first frame. The determined eye positions are then used to crop the facial area in the whole sequence, yielding in nonaligned face images. Finally we scaled the resulted images into 3 different resolutions: 20×20, 40×40 and 60×60 pixels. Examples of face images from some sequences are shown in Fig. 2.

4.2 Results and Analysis

For evaluation, we adopted a 5-fold cross validation test scheme by dividing the 1000 sequences into five groups and using the data from four groups for training and the last group for testing. We repeated this process five times and we report the average classification rates.

As the proposed approach involves two free parameters which are the number of neighbors (k) and the dimension of the embedding space (d), we started then by analyzing the effect of these two factors on the system performance. Fig. 3 plots the gender classification results when varying k and d (These results are obtained using 40×40 pixels as face image resolution). For k, we can notice that the best rates are obtained using around 15 nearest neighbors and then the performance is decreasing for too small ($k \leq 6$) or too large ($k \geq 25$) values. This can be explained by the fact that too small values of k tend to make the reconstruction of each data point from its neighbors weak (see Equation (1)), while too large values tend to break the embedding as the locality assumption in LLE calculation may not hold. However, for a wide range of values between 10 and 20, the results remain good and stable. For the dimension of the embedding space (d), we can notice that the best results are obtained using values between $10 \leq d \leq 15$. Smaller values of d tend to decrease the performance while larger values ($d \geq 15$) do not alter the results. In fact, d corresponds to the intrinsic dimension of the training face sequences. An alternative to estimate the optimal value of d is to analyze the residual variance which characterizes how well the low-dimensional Euclidean embedding captures the geodesic distances of the input data. The lower the residual variance is, the better high-dimensional data are represented in the embedded space. We computed the residual variance as $1 - \rho^2(D_X, D_Y)$, where ρ is the standard linear correlation coefficient taken over all entries of D_X and D_Y; D_X is the matrix of geodesic distances between pairs of points in X (in the high-dimensional space) and D_Y is the matrix of Euclidean distances in Y (in the embedding space). Our experiments showed that the residual variance does not further decrease for dimensions ≥ 12 which consolidates our earlier findings.

Setting $k = 15$ and $d = 12$, we obtained gender classification rates of **96.8%**, **97.1%** and **97.2%** for resolutions of 20×20, 40×40 and 60×60 pixels,

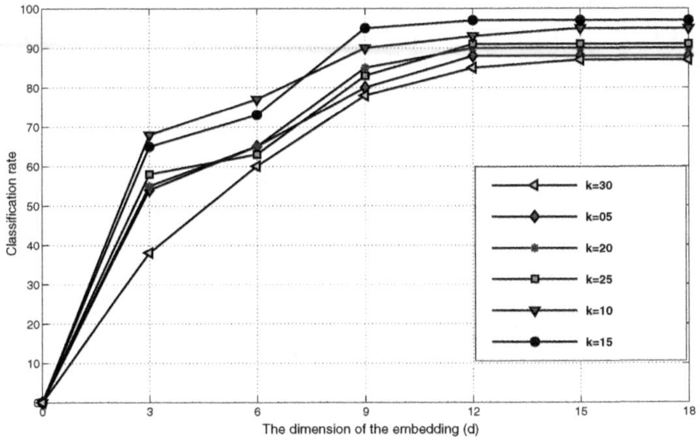

Fig. 3. Gender classification rates for different values of the number of neighbors (k) and the dimension of the embedding space (d)

Table 1. Comparison of gender classification results of several approaches on our test sequences with three different face image resolutions

Method	Average Gender Classification Rate (%)		
	20×20 Pixels	40×40 Pixels	60×60 Pixels
Pixels + SVM + Fusion	88.0	89.2	88.5
LBP + SVM + Fusion	90.5	91.0	92.1
VLBP + SVM	82.3	83.9	84.5
EVLBP + AdaBoost	83.1	84.4	84.6
Manifold Learning (Proposed Approach)	**96.8**	**97.1**	**97.2**

respectively. Although we did not perform any face alignment, the obtained results are very good and compare favourably against what is generally reported in literature even for well aligned faces [1][2]. For extensive and more objective comparison, we also implemented some state-of-the-art methods including **(i)** Use of raw pixels with SVM and fusion over the face sequence at score level; **(ii)** Use of Local Binary Pattern features (LBP) [14] with SVM and fusion over the face sequence at score level [15]; **(iii)** Use of Volume LBP with SVM [16] and **(iv)** Use of Extended Volume LBP with AdaBoost [15]. The obtained results using all these methods are summarized in Table. 1. From the table, we can clearly see that our proposed approach significantly outperforms all other considered methods.

Perhaps, the key reason behind the success of our approach lies in the effective use of the correlation between the face images through manifold learning. Our experimental analysis also showed that most state-of-the-art methods fail to handle faces under severe illumination changes and miss alignments, while our proposed approach seems to perform much better under such conditions.

Our results also indicated that image resolution does not affect very much gender classification performance and this confirms the conclusions of many other researchers.

5 Discussion and Conclusion

We proposed a novel approach to gender classification from face sequences. Our approach consisted of first learning the hidden low-dimensional structure of male and female manifolds using an extension to the original LLE algorithm. Then, a target face sequence is projected into both manifolds for determining the gender of the person in the sequence. The matching is achieved using a new manifold distance measure. Under natural conditions and without face alignment, a high classification rate of 97.2% is obtained for face sequences of 60×60 pixels, while most state-of-the-art methods fail to handle faces under miss alignments and illumination changes.

It is worth noting that, in addition to its efficiency, our approach involves only two free parameters which are quite easy to determine. Our future work consists of experimenting with much larger databases and further analyzing the few misclassification errors made by our system. It is also of interest to extend our approach to explicitly incorporate the temporal information during manifold learning.

Note that the proposed methodology can also be applied to other face analysis tasks such as face recognition, facial expression recognition and age classification from videos.

Acknowledgment

The financial support of the Academy of Finland is gratefully acknowledged.

References

1. Moghaddam, B., Yang, M.H.: Learning gender with support faces. IEEE Transactions on PAMI 24(5), 707–711 (2002)
2. Baluja, S., Rowley, H.: Boosting sex identification performance. International Journal of Computer Vision 71, 111–119 (2007)
3. Kohonen, T. (ed.): Self-Organizing Maps. Springer, Berlin (1997)
4. Bartholomew, D.J., Knott, M. (eds.): Latent variable models and factor analysis multidimensional scaling. Arnold Publication (1999)
5. Bishop, C.M., Svensen, M., Williams, C.K.I.: GTM: The generative topographic mapping. Neural Computation 10(1), 215–234 (1998)
6. Sammon, J.: A nonlinear mapping for data structure analysis. IEEE Transactions on Computers 18(5), 401–409 (1969)
7. Roweis, S.T., Saul, L.K.: Nonlinear dimensionality reduction by locally linear embedding. Science 290(5500), 2323–2326 (2000)
8. Tenenbaum, J.B., DeSilva, V., Langford, J.C.: A global geometric framework for nonlinear dimensionality reduction. Science 290(5500), 2319–2323 (2000)

9. Belkin, M., Niyogi, P.: Laplacian eigenmaps and spectral techniques for embedding and clustering. In: Dietterich, T.G., Becker, S., Ghahramani, Z. (eds.) Advances in Neural Information Processing Systems 14, pp. 585–591. MIT Press, Cambridge (2002)
10. Kouropteva, O., Okun, O., Pietikäinen, M.: Semisupervised visualization of high-dimensional data. Pattern Recognition and Image Analysis 15(4), 645–648 (2005)
11. CRIM, `http://www.crim.ca/`
12. Sanderson, C., Paliwal, K.K.: Noise compensation in a person verification system using face and multiple speech feature. Pattern Recognition 36(2), 293–302 (2003)
13. Kanade, T., Cohn, J.F., Tian, Y.: Comprehensive database for facial expression analysis. In: FG 2000, pp. 46–53 (2000)
14. Ojala, T., Pietikäinen, M., Mäenpää, T.: Multiresolution gray-scale and rotation invariant texture classification with local binary patterns. IEEE TPAMI 24, 971–987 (2002)
15. Hadid, A., Pietikäinen, M.: Combining motion and appearance for gender classification from video sequences. In: Proc. 19th International Conference on Pattern Recognition (ICPR 2008) (2008) (in press)
16. Zhao, G., Pietikäinen, M.: Dynamic texture recognition using local binary patterns with an application to facial expressions. IEEE TPAMI 29(6), 915–928 (2007)

A Random Network Ensemble for Face Recognition

Kwontaeg Choi[1], Kar-Ann Toh[2], and Hyeran Byun[1]

[1] Dept. of Computer Science, Yonsei University, Seoul, South Korea, 120-749
[2] School of Electrical & Electronic Engineering, Yonsei University
Seoul, South Korea, 120-749

Abstract. In this paper, we propose a random network ensemble for face recognition problem, particularly for images with a large appearance variation and with a limited number of training set. In order to reduce the correlation within the network ensemble using a single type of feature extractor and classifier, localized random facial features have been constructed together with internally randomized networks. The ensemble classifier is finally constructed by combining these multiple networks via a sum rule. The proposed method is shown to have a better accuracy(31.5% and 15.3% improvements on AR and EYALEB databases respectively) and a better efficiency than that of the widely used PCA-SVM.

Keywords: Face recognition, random projection, neural network.

1 Introduction

Face recognition is among the most active areas of research because it is a difficult and representative problem of high dimensionality with high commercial application values. Many methods have significantly advanced the field over recent years. However, the robustness remains an important issue under an unconstrained environment because the limited number of available training data can not well represent all possible facial variations such as lighting, pose and expression. A single classifier using such a training set may be biased and possesses a large variance, resulting in a poor performance[1][2].

An ensemble of classifiers or a multiple classifiers system(MCS) is becoming relevant in face recognition due to the remarkable improvements over the single classifier system. Since different representation of the input face images have different sensitivity to various face variations, an ensemble of different face classifiers which integrates the complementary information leads to an improved classification accuracy[3]. There are two types of ensemble. Intermodal ensemble combines different types of modalities such as audio, fingerprint, 3D face and 2D face images, and intramodal ensemble combines multiple classifiers within a single modality such as a 2D image. We shall focus on intramodal ensemble in this paper.

In [4], the resampling techniques was utilized to generate several subsets of samples from the original training dataset for LDA based classifiers. In [5], an

M. Tistarelli and M.S. Nixon (Eds.): ICB 2009, LNCS 5558, pp. 92–101, 2009.
© Springer-Verlag Berlin Heidelberg 2009

ensemble learning framework was developed based on a random sampling of three key components of a classification system namely the feature space, the training samples, and the subspace parameters. In [6], a random discriminant analysis was proposed to construct an optimal random subspace. In [7], a multi-classifier architecture was utilized using HMM, DCT, EigenFace and EigenObjects. In [3], two combination strategies namely, sum rule and RBF network, was adopted to integrate the outputes of three feature extraction methods, namely PCA, ICA, LDA. These works utilized various types of feature extractors and classifiers in order to increase the overall accuracy of the combined classifier.

These intramodal fusion networks appear somewhat complex due to a combination of heterogenous systems in terms of different kinds of feature extraction method and classifier. This is indeed a natural idea since a heterogeneous combination reduces the correlations among the classifiers. However, limitations of such a heterogenous system are the heavy computational cost and the tedious tuning of learning parameters. Moreover, they are not directly applicable in real-time problems such as video-based face tracking/recognition since most methods adopt the batch approach. An ensemble network with efficient online learning is thus an area worth exploring.

In this paper, we propose a homogeneous random network ensemble for face recognition particularly dealing with a large facial variation and with a limited number of training data. In order to reduce the correlation among classifiers in a homogeneous ensemble system, two ingredients namely, random projection and randomized network, have been utilized. The proposed method is differentiated from the above mentioned networks in the following ways: 1)the proposed network is a homogenous ensemble system whereas most other systems belong to the heterogeneous type, 2)an incremental(online) learning has been adopted whereas other ensemble networks adopt a batch learning.

The rest of this paper is organized as follows. Section 2 describes the concept of using localized random projection to extract facial feature by means of random basis. In Section 3, we present the proposed ensemble framework using randomized learning parameters selection. Section 4 describes an incremental learning formulation. In section 5, several experiments are presented to evaluate the proposed method in terms of its efficiency and generalization capabilities. Finally, our conclusion is given in Section 6.

2 Feature Extraction Using Random Basis

In this section, we present some backgrounds on random projection for immediate reference. This will be followed by an introduction of our proposed localized random projection method for face feature extraction.

2.1 Background of Random Projection(RP)

Unlike most training based feature extraction methods which require a lot of training samples and certain optimizing criteria, Random Projection(RP) has emerged as an efficient dimensional reduction method which does not require any training samples. In RP, the original high-dimensional data can be projected onto a low-dimensional subspace using a set of randomly generated basis where the orthonormal projection matrix consists of i.i.d. entries with zero mean and constant variance. According to [8], RP preserves approximately the pairwise distance of points in Euclidean space, volumes and affine distance and the structure of data without introducing significant distortion.

In order to significantly reduce the computational cost during the generation of projection matrix and projection onto a low-dimensional subspace, Achliopta[9] proposed a sparse random projection(SRP) with elements belonging to $\pm\sqrt{s}(s \in 1, 3)$ and 0. A projection matrix $R \in \mathbb{R}^{d \times D}$ with original face dimension D and reduced dimension d, consists of the following entities.

$$R_{ji} = \sqrt{s} \begin{cases} 1 & with\ prob\ \frac{1}{2s} \\ 0 & with\ prob\ 1 - \frac{1}{s} \\ -1 & with\ prob\ \frac{1}{2s} \end{cases}. \tag{1}$$

When sparsity s is 3, then $\frac{2}{3}$ of entities contain zero values. This leads to a threefold speedup. Moreover, the multiplications with $\sqrt{s}$ can be delayed and no floating point arithmetic is needed.

More recently, Li[10] proposed a very sparse random projection method. They showed that one can use $s >> 3$ (e.g. $s = \sqrt{D}$, or even $s = \frac{D}{logD}$) to significantly speedup the computation.

2.2 Proposed Localized Random Projection(LRP)

The relevance of SRP for feature extraction relies on how well the random distribution of non-zero entities (see Fig 1 (e) $\sim$ (f)) represents the distribution of facial features. While SRP focuses on efficient dimension reduction, several subspace methods such as PCA, LDA and ICA, consider both dimension reduction and effective feature representation. A uniform distribution of non-zero entities in SRP may not be representative for 2D face images. Hence, we cast a question, "How to change the distribution of the non-zero entities for face feature extraction while preserving the generation rule in Eq. (1)".

Inspired by localized feature extraction methods such as ICA, LFA and LNMF (see Fig. 1 (a) $\sim$ (c)), we propose a localized random projection in this work. Localized basis offers several advantages including stability to local deformations, lighting variations, and partial occlusion[11]. The proposed method modifies the uniform distribution of non-zero entities to localized distribution of non-zero entities as illustrated in Fig. 1 (g) $\sim$(i). The following provides more details.

Our main problem is about how to generate the sparse projection matrix with localized distribution of non-zero entities. In other words, we will need to determine the quantity, the position and the size(width, height) of localized blocks of non-zero entities. We will treat each basic vector as a 2D face image basis because we deal only with 2D images. Without loss of generality, we assume that each basis represents only one localized block. The position of a localized block can be set randomly within the size of the image template(we use 32×32 size).

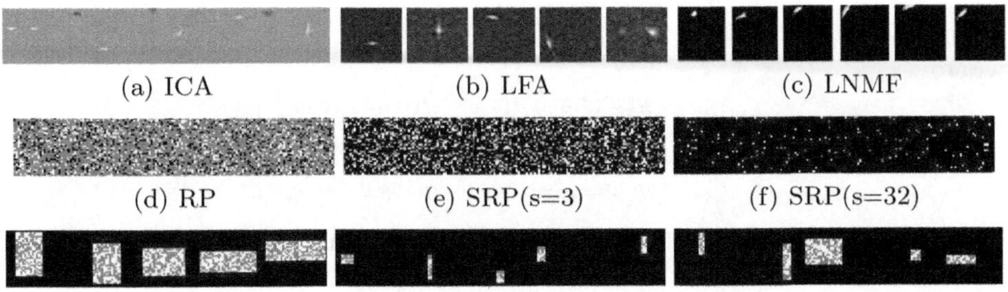

(a) ICA (b) LFA (c) LNMF
(d) RP (e) SRP(s=3) (f) SRP(s=32)
(g) LRP(s=3) (h) LRP(s=32) (i) LRP(s=random)

Fig. 1. Examples of basis obtained from various subspace methods. In (e) $\sim$ (i), black indicates 0, white indicates $sqrt(s)$ and gray indicates $-sqrt(s)$.

The size of each block is related to the amount of non-zero entities, given by sparsity s in Eq. (1). For example, when $s = 3$, $\frac{2}{3}$ of entities are non-zeros. In the case of a 32×32 template image, any size of rectangle block with $width \times height = 32 \times 32 \times \frac{2}{3}$ can be used. Fig. 1 (g) shows some possible localized basis and Fig. 1 (h) shows some possible localized basis when we use $s = 32 (= \sqrt{D} = \sqrt{32 \times 32})$.

One consideration here is regarding how to choose the best s for effective facial feature extraction. When s is small, many pixels are sampled for projection. When s is large, a small number of pixels are sampled for projection. Here we adopt the dynamic sparsity instead of fixed sparsity proposed by Achliopta and Li in order to capture small and large scale structures of face features as shown in Fig. 1 (i).

3 A Network Ensemble Using Randomized Parameters

In this section, we propose an ensemble framework of combining multiple networks with randomly selected learning parameters in order to reduce the correlation among classifiers.

3.1 Background of Single Layer Feedforward Network(SLFN)

Given n samples $x_i \in R^d$ and correspond target y_i, a standard SLFN classifier with h hidden nodes and activation function g can be mathematically modeled as

$$y_i = \sum_{j=1}^{h} \beta_j g(w_j x_i + b_j) \tag{2}$$

where $w_j = [u_{j1}, u_{j2}, ..., u_{jd}]^T$ is the weight vector connecting the jth hidden node to the input nodes, b_j is the threshold of the jth hidden node and $\beta_j \in R^h$ is the weight vector connecting the jth hidden node to the output nodes.

For k-class problem, denote $\Theta = [\boldsymbol{\beta}_1, ..., \boldsymbol{\beta}_k] \in \mathbb{R}^{h \times k}$ as the weight matrix which is a collection of weight vector $\boldsymbol{\beta}$ and $Y \in \mathbb{R}^{n \times k}$ as the indicator matrix. The n equations above can be written more compactly as

$$H\Theta = Y \quad (3)$$

where

$$\underset{n\times h}{H} = \begin{bmatrix} g(w_1x_1 + b_1) & \dots & g(w_hx_1 + b_h) \\ \vdots & \ddots & \vdots \\ g(w_1x_n + b_1) & \cdots & g(w_hx_n + b_h) \end{bmatrix} \quad (4)$$

The weight parameters Θ can be estimated by minimizing the Least-Squares Error giving the solution[12]:

$$\Theta = (H^TH)^{-1}H^TY \quad (5)$$

3.2 A Random Network Ensemble Framework

Face images with high dimensionality and a limited number of training examples for each class easily lead to overfitting, overtraining, small-sample effect and singularity particularly when using an RBF network for face recognition[13]. In order to solve these problems, we combine multiple SLFN classifiers instead of using a single Multi-Layered Perception(MLP) by the following strategies.

1. Reducing the number of empirically fixed learning parameters : A neural network has various learning parameters which affect the recognition accuracy and computation. In the ensemble approach, it is particularly a time consuming task to determine the empirically fixed learning parameters of all networks. The classifier with such parameters may be not proper when a small number of training samples are used. Here, we adopt the Extreme Learning Machine(ELM)[12] classifier which comes with a small number of learning parameters. In ELM, two weight parameters w and b are arbitrarily chosen and need not be adjusted at all unlike the conventional SLFN classifier. We train an individual ELM classifier using random number of features and hidden units. Thus, the proposed method does not have empirically fixed learning parameters.
2. Reducing the computational cost : The many number of classifiers is combined to increase the overall accuracy. However this leads to an increase in computational cost. Unlike MLP which uses an iterative search based on gradient descent approach, the ELM estimates deterministically only β parameter in closed-form for fast training.
3. Reducing the correlations between classifiers : In order to improve the overall accuracy, each classifier should be trained differently to reduce the redundant classification. In ELM, two weight parameters w and b are arbitrarily chosen and need not be adjusted at all. As we train multiple number of ELM classifiers, this generates different decision boundaries even though the same training set is used. Moreover, the facial features have been extracted using random basis. This, again, generates different decision boundaries among ELMs which used the similar weight parameters w and b within the activation function.

In short, we extract the face features using the proposed LRP and train multiple ELMs using randomly selected learning parameters, finally we combine these multiple number of ELMs using a sum rule. We will call this method LRP-ELM. When C number of LRP-ELM classifiers are combined via a sum rule, the class label of unseen sample x can be predicted using

$$class(x) = \arg\max_i \sum_{c=1}^{C} \Theta_{c,i}^T g(W_c R_c x + B_c), i = 1, 2, ..., k \tag{6}$$

where R_c is c^{th} projection matrix by LRP, W_c and B_c are c^{th} parameters of activation function, and $\Theta_{c,i}$ is i^{th} column vector of cth weight matrix Θ_c

4 Incremental Learning

Most classifiers adopt a batch based approach. This means that the classifiers need to be retrained when additional samples arrived. This could be a very time consuming task for ensemble approach.

The proposed LRP for facial feature extraction does not require the training set. Therefore an updating rule is not needed unlike the incremental PCA and incremental LDA. Moreover, the ELM can be retrained efficiently using only additional samples based on a recursive least square formulation. For this purpose, Huang[14] proposed an Online Sequential ELM(OSELM). When new feature block of data H_{t+1} and the corresponding indicator matrix Y_{t+1} are received, the parameter Θ_{t+1} can be estimated recursively as:

$$\begin{aligned} P_{t+1} &= P_t - P_t H_{t+1}^T (I + H_{t+1} P_t H_{t+1}^T)^{-1} H_{t+1} P_t \\ \Theta_{t+1} &= \Theta_t + P_{t+1} H_{t+1}^T (Y_{t+1} - H_{t+1} \Theta_t) \end{aligned} \tag{7}$$

Therefore the proposed random network ensemble can be applied efficiently once a new training set is available.

5 Experiment

Several experiments are conducted to evaluate the proposed method in terms of its generalization capability and efficiency. The experiments are repeated 10 times and the average test accuracy with the computational cost will be reported. The combination of snapshot based PCA and SVM[15] adopting polynomial and RBF kernels(PCA-SVM(Poly) and PCA-SVM(RBF), respectively) is used as the baseline classifier to compare the generalization capability and time complexity with the proposed method. We will use following three public face databases which contain many images per person and a large variation of imaging conditions.

1. AR database[16] (AR) : 126 identities/26 samples - different facial expressions, illumination conditions, and occlusions (sun glasses and scarf).
2. The extended Yale Face Database B[17] (EYALEB) : 28 identities / 1 frontal pose of 6 poses / 64 illumination conditions.
3. The Pose, Illumination and Expression Database [18](PIE) : 68 identities / 5 near frontal poses of 10 poses/ 5 illumination / 3 expression conditions.

We first evaluate the test accuracy of a single classifier using AR and EYALEB databases. Particularly, we evaluate the stability using a small number of training set, i.e. 30% of the dataset is used as training set with the remaining used as test set.

Fig. 2 (a) shows the test accuracy of compared methods at different number of feature dimensions using the AR database. In AR dataset with large face variation and with a limited number of training data in each class, the accuracy of PCA-SVM seems to be very low compared to the that of the proposed method because the training set is insufficient to predict the 70% test set which contains more face variations than that in the training set. The accuracy of PCA-SVM is seen to be 19.0% lower than the proposed method and seems to saturate at high dimensions. According to Fig. 2 (a), sparsity appears to affect the accuracy. For example, LRP-ELM(s=32) records a higher accuracy than that of LRP-ELM(s=3).

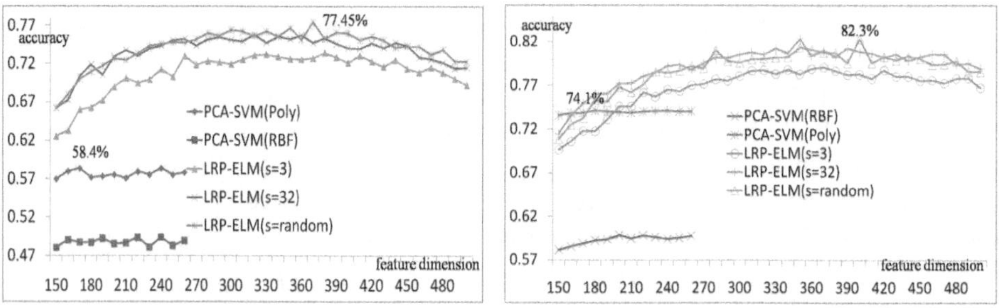

(a) AR database (b) EYALEB database

Fig. 2. Average test accuracy comparison on the proposed method and the PCA-SVM

Fig. 2 (b) shows the test accuracy of compared methods on the EYALEB database. The proposed method records a much higher accuracy(an improvement of 8.2%) than that of PCA-SVM. The accuracy of LRP-ELM is seen to be lower than that of PCA-SVM at low feature dimensions(< 150). However, the accuracy of the proposed method increases when more features have been used.

In the second experiment, the accuracy of an ensemble classifier is evaluated. Since there are three controllable parameters (feature dimension, the number of hidden units and the number of classifiers) that can affect the accuracy, we construct 5 cases of ensembles of classifiers according to the feature dimension($220 \sim 500$) and the number of hidden units($200 \sim 450$). These values are set randomly because we do not seek an empirically determined values by experiments. As seen from Fig. 3, as we increase the number of classifiers, the accuracy of the ensemble classifier via sum rule increases. And the ensemble classifier with large number of features and hidden units records good accuracy. This result shows a significant improvement of 12.5% in terms of test accuracy over that before fusion. In EYALB database, the best accuracy of the proposed method is 89.4%(7.1% accuracy improvement over that before fusion).

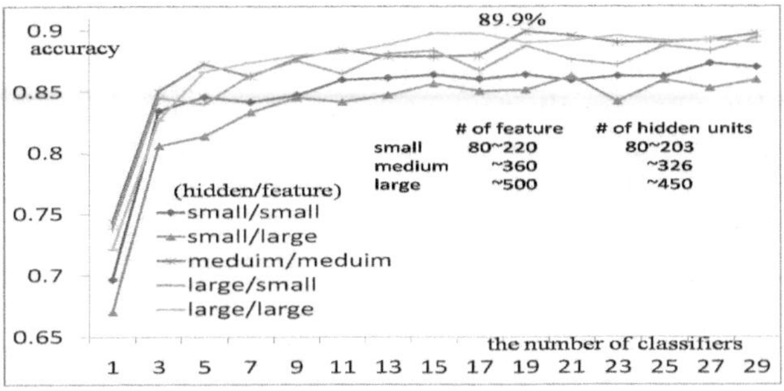

Fig. 3. Test accuracies of ensemble LRP-ELM classifiers(AR database)

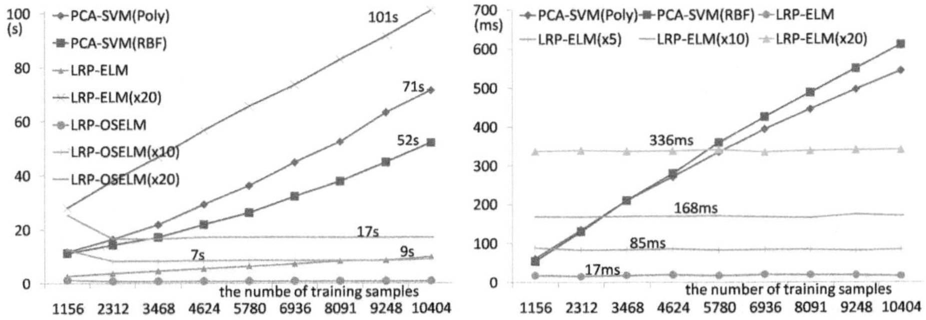

(a) Training (b) Test

Fig. 4. Execution time comparison on the proposed method and the PCA-SVM using PIE database. The number in parenthesis indicates the number of the combined classifiers.

To summarize, the accuracy of the combined classifier is 31.5% and 15.3% higher than that of a single PCA-SVM classifiers in AR and EYALEB databases respectively.

Finally, we compare the CPU execution times for both training and test using the PIE dataset which contains a large number of samples. Fig. 4 shows the CPU times for training and test. Here PCA-SVM used 100 eigen-faces and LRP-ELM used 400 localized random basis and 400 hidden units because PCA-SVM requires smaller dimension than that of LRP-ELM as shown in Fig. 2.

As seen from Fig. 4 (a), the proposed single LRP-ELM shows a much faster training speed than that of PCA-SVM(RBF). The ensemble classifier LRP-ELM(x20) using 20 classifiers is seen to be slower than single PCA-SVM(RBF). Since multiple LRP-ELM classifiers are combined via sum rule, LRP-ELM(x10) is 10 times slower than single LRP-ELM classifier. The computational cost of LRP-OSELM using Eq. (7) is evaluated. This approach records an almost constant execution time when the number of training samples increase, whereas the execution time of PCA-SVM and LRP-ELM increases due to retraining of the entire data set which includes the new and existing data.

Fig. 4 (b) shows the test execution time of the proposed method and PCA-SVM classifying one hundred samples. All methods show a low computational time within 610ms to classify one hundred samples. However, in mobile device or real-time applications, a fast test time is important. According to Fig. 4 (b), the proposed single LRP-ELM is seen to be 79 times faster than that of SVM when the number of training samples is 11,560. The ensemble classifier LRP-ELM(x20) using 20 classifiers is faster than single PCA-SVM when more than 5780 training samples have been used. Another benefit of the proposed LRP-ELM is that the execution time is almost independent of the total number of training samples because LRP-ELM is dependent on feature dimension and the number of hidden units, whereas PCA-SVM is dependent on the total number of training samples because the number of support vectors is proportional to the number of training samples.

6 Conclusion

In this paper, we proposed a random network ensemble for face recognition particularly for problems with a large appearance variation and with a limited number of training set. Unlike the conventional heterogeneous ensemble method, the proposed method is based on a homogenous system using random projection and randomized networks. In order to reduce the correlation between classifiers in the homogenous combination, we proposed localized random projections using sparse random basis. Next we train multiple number of single layer feedforward networks using randomly selected learning parameters. Finally we combined multiple number of randomized networks via a rum rule. The proposed ensemble network is subsequently extended to an incremental learning formulation. The proposed method is seen to improve the recognition accuracy of about 31.5% and 15.3% compared to that of SVM classifier using features extracted by PCA on AR and EYALEB databases, respectively. In terms of the computational cost, the proposed method has shown a better efficiency than that of PCA-SVM.

Acknowledgements

This work was supported by the Korea Science and Engineering Foundation (KOSEF) through the Biometrics Engineering Research Center(BERC) at Yonsei University. (R112002105070030(2008)(2008)).

References

1. Raudys, S.J., Jain, A.K.: Small Sample Size Effects in Statistical Pattern Recognition Recommendations for Practitioners. IEEE Trans on Pattern Analysis and Machine Intelligence 13(3), 252–264 (1991)
2. Skurichina, M., Duin, R.: Bagging, boosting and the random subspace method for linear classifiers. Pattern Anal. Appl., 121–135 (2002)

3. Lu, X., Wang, Y., Jain, A.K.: Combining classifiers for face recognition. In: ICME (2003)
4. Lu, X., Jain, A.K.: Resampling for Face Recognition. In: Kittler, J., Nixon, M.S. (eds.) AVBPA 2003. LNCS, vol. 2688. Springer, Heidelberg (2003)
5. Wang, X., Tang, X.: Random sampling face recognition. International Journal of Computer Vision (IJCV) (2005)
6. Zhang, X., Jia, Y.: A linear discriminant analysis framework based on random subspace for face recognition. In: Proceedings of Pattern Recognition, pp. 2585–2591 (2007)
7. Lemieux, A., Parizeau, M.: Flexible multi-classifier architecture for face recognition systems. In: The 16th International Conference on Vision Interface (2003)
8. Goel, N., Bebis, G., Nefian, A.: Face recognition experiments with random projection. In: SPIE (2005)
9. Achlioptas, D.: Database-friendly random projections. In: ACM Symposium on the Principles of Database Systems, pp. 274–281 (2001)
10. Li, P., Hastie, T., Church, K.W.: Very sparse random projections. In: KDD, pp. 287–296 (2006)
11. Feng, Li, S.Z., Shum, H.Y., Zhang, H.J.: Local non-negative matrix factorization as a visual representation. In: The Second International Conference on Development and Learning, pp. 178–183 (2002)
12. Er, M.J., Wu, S., Lu, J., Toh, H.L.: Face recognition with radial basis function (RBF) neural networks. IEEE Trans. Neural Netw. 13, 697–710
13. Huang, G.-B., Siew, C.-K.: Extreme learning machine with randomly assigned RBF kernels. International Journal of Information Technology 11(1) (2005)
14. Liang, N.Y., Huang, G.B., Saratchandran, P., Sundararajan, N.: A Fast and Accurate On-Line Sequential Learning Algorithm for Feedforward Networks. IEEE Trans. Neural Networks 17, 1411–1423 (2006)
15. `http://www.ece.osu.edu/osusvm`
16. Martinez, A., Benavente, R.: The AR Face Database (1998)
17. Georghiades, A.S., Belhumeur, P.N., Kriegman, D.J.: From Few to Many: Illumination Cone Models for Face Recognition under Variable Lighting and Pose. IEEE Trans. Pattern Anal. Mach. Intelligence 23(6), 643–660 (2001)
18. Sim, T., Baker, S., Bsat, M.: The CMU Pose, Illumination, and Expression Database. IEEE Trans. Pattern Anal. Mach. Intelligence 25(12), 1615–1618 (2003)

Extraction of Illumination-Invariant Features in Face Recognition by Empirical Mode Decomposition

Dan Zhang and Yuan Yan Tang

Department of Computer Science,
Hong Kong Baptist University,
Hong Kong SAR, China
{dzhang,yytang}@comp.hkbu.edu.hk

Abstract. Two Empirical Mode Decomposition (EMD) based face recognition schemes are proposed in this paper to address variant illumination problem. EMD is a data-driven analysis method for nonlinear and non-stationary signals. It decomposes signals into a set of Intrinsic Mode Functions (IMFs) that containing multiscale features. The features are representative and especially efficient in capturing high-frequency information. The advantages of EMD accord well with the requirements of face recognition under variant illuminations. Earlier studies show that only the low-frequency component is sensitive to illumination changes, it indicates that the corresponding high-frequency components are more robust to the illumination changes. Therefore, two face recognition schemes based on the IMFs are generated. One is using the high-frequency IMFs directly for classification. The other one is based on the synthesized face images fused by high-frequency IMFs. The experimental results on the PIE database verify the efficiency of the proposed methods.

Keywords: Empirical Mode Decomposition, Face recognition.

1 Introduction

Empirical mode decomposition (EMD) is an adaptive signal analysis method for nonlinear and non-stationary data. It has been originally introduced by Huang et al. in [1], which is the first step of Hilbert-Huang transform [2,3]. Essentially EMD decomposes a signal into a set of Intrinsic Mode Functions (IMFs), which recover the original input signal features in a multiscale sense from high frequency to low frequency. The IMFs heavily rely on the original signal rather than using predetermined filter or wavelet functions. Due to the non-parametric and data-driven advantages, EMD has been successfully applied in one dimensional signal analysis such as ocean waves, rogue water waves, sound analysis, and earthquake time records. Recently it has been more and more applied in image analysis. The first track can be retrospected to [4], in which S. Long et al. applied EMD to analyze digital slope images. The image data has been expressed in terms of an array of rows and columns, then the EMD is applied to these

M. Tistarelli and M.S. Nixon (Eds.): ICB 2009, LNCS 5558, pp. 102–111, 2009.
© Springer-Verlag Berlin Heidelberg 2009

arrays row by row. Linderhed et al. also adopted this row-by-row EMD method in the fusion of visible and infrared images [5,6]. The input images are vectorized in lexicographical order and EMD is performed on each channel vector separately. R. Bhagavatula et al. [7] addressed face recognition in the same way.

The EMD has been extended for two dimensional data in the literatures toward texture extraction [8] and image compression [10]. The two dimensional EMD (2DEMD) has a more wide application in various kinds of image analysis, such as image fusion [12], image compression [11], texture analysis [9], feature extraction [14], rainfall analysis [13], watermarking [15], and temperature variation [16]. Faces as two dimensional digital signal are generally processed by wavelets or other filters. However, these methods are not adaptive to the original faces and always affected by the predetermined functions. The facial features captured in this way sometimes are anamorphic, which always leads to low recognition rate. Though EMD based methods can capture the intrinsic feature of the iamges, few work done in face recognition.

Face recognition refers to technologies for human authentication or verification based on human facial characteristics. Though this biometric technique has been improved a lot, however, many present facial biometric systems are still inefficient when identifying the same person under varying illumination. Earlier studies [17,18,19] show that only the low-frequency component is sensitive to illumination changes. It indicates that the corresponding high-frequency components are more robust to the illumination changes. Therefore, the high-frequency components alone without the low-frequency components are sufficient for classification. In terms of EMD, due to its adaptiveness it is able to capture more representative features and especially more singular information in high-frequency IMFs. In this paper, two available face recognition schemes based on EMD are generated. One is using the high-frequency IMFs directly for classification. The other one is based on the synthesized face images fused by high-frequency IMFs. The proposed two methods efficiently address the illumination variance problem. The experimental results on the PIE database verify their efficiency.

The paper is organized as follows. Section 2 presents an overview of the EMD and its details of application in image. Section 3 presents the flow of the proposed two face recognition schemes. The experiments are demonstrated in Section 4. Finally, conclusions and future work outlook are presented.

2 Empirical Mode Decomposition and Its Applications in Images

EMD is a signal analysis technique for adaptive representation of non-stationary signals as sum of a set of IMFs. It captures information about local trends in the signal by measuring oscillations, which can be quantized by a local high frequency or a local low frequency corresponding to finest detail and coarsest content. Here we briefly introduce the sifting process of EMD, four main steps are contained. S_1, S_2, S_3 and S_4 are abbreviations of Step 1, Step 2, Step 3 and Step 4 respectively. Given a signal x(t),

S_1. Identify all the local minima and maxima of the input signals $x(t)$;

S_2. Interpolate between all minima and maxima to yield two corresponding envelopes $E_{max}(t)$ and $E_{min}(t)$. Calculate the mean envelope $m(t) = (E_{max}(t) + E_{min}(t))/2$;

S_3. Compute the residue $h(t) = x(t) - m(t)$. If it is less than the threshold predefined then it becomes the first IMF, go to S_4. Otherwise, repeat S_1 and S_2 using the residue $h(t)$, until the latest residue meets the threshold and turns to be an IMF;

S_4. Input the residue r(t) to the loop from S_1 to S_3 to get the next remained IMFs until it can not be decomposed further.

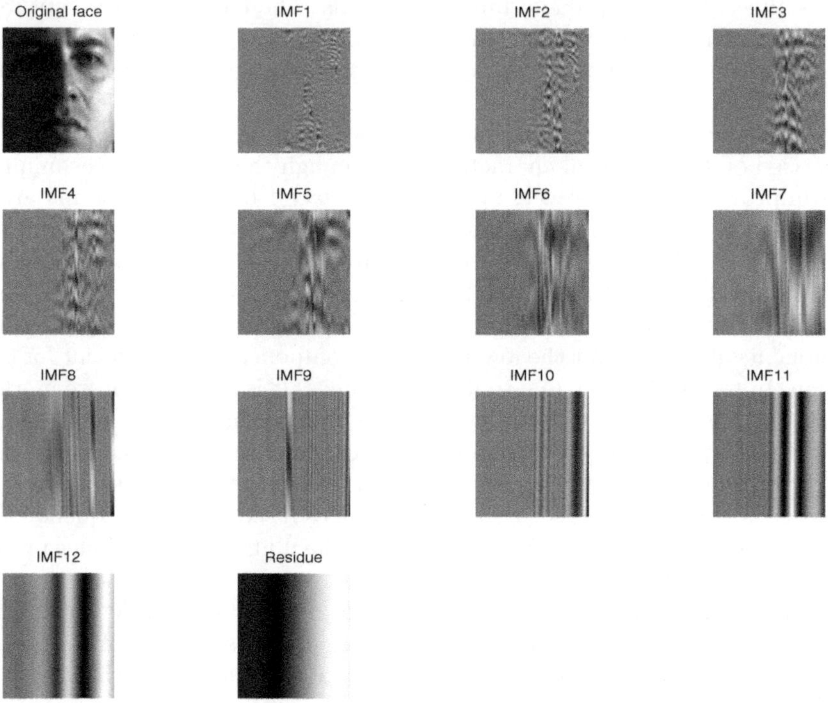

Fig. 1. Twelve IMFs and the residue decomposed by 1DEMDa on a face image with right point illumination from the PIE database. The IMFs captures the information of the original face from the finest to the coarsest. It is clear in this figure that the right illumination effects focus in the residue.

There are three key points in the algorithm implementation.

- Judge whether the residue signal $h(t)$ is an IMF or not. Theoretically, the judgement should accord two conditions [1]. First, the number of extrema and the number of zero-crossing must be at most differ by one. Second, the mean envelops obtained by the maximum envelop and the minimum envelop must equal to zero. However, it is difficult to achieve these two conditions

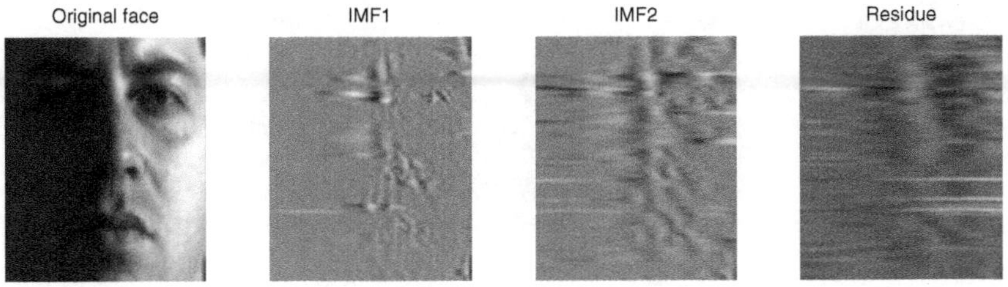

Fig. 2. Two IMFs and the residue decomposed by 1DEMDb on a face image with right point illumination from the PIE database

strictly in the sense of realistic implementation. Researchers always define a small threshold instead. If $h(t)$ is small enough then it is an IMF.

- Choose an appropriate interpolation functions, such as linear interpolation, cubic spline interpolation and so on.
- Stop criterion. How to judge the residue can be decomposed further or not? Generally, the decomposition stops when the residue becomes so small that it is less than a predefined threshold.

The EMD is originally proposed for one dimensional data. However, it has also been widely applied in image analysis. The idea is to express images as an array of one dimensional vectors. Generally two ways are adopted. One is to view the image as a long lengthened vector and then apply EMD on it. Another one is to apply on each row or column one by one as literatures [4,5,6,7] did. In order to discriminate these two methods, we name the former one as 1DEMDa and the latter one as 1DEMDb. Fig.1 and Fig.2 show all the IMFs and residues obtained by these two methods respectively. The IMFs captures the finest to the coarse information of the original face. The illumination effects are generally distributed in a few low level IMFs and the residues. Fig.1 shows an example. It is obvious that the right light effects mainly focus on the residue. By removing these low-frequency IMFs and residues, the remained high-frequency IMFs are less sensitive to the illumination effects, themselves alone are sufficient for classification. The extended 2DEMD has the similar sifting process as 1DEMD. Given the digital image $I = f(x, y)$, $x = 1, \cdots, M$, $y = 1, \cdots, N$, the process is summarized as follows.

S_1. Identify the extrema (both maxima and minima) of the image $I = f(x, y)$;

S_2. Generate the 2D envelopes by connecting maxima points (respectively, minima points) using surface interpolation. Determine the local mean m by averaging the two envelopes;

S_3. Subtract out the mean from the image to get an residue $h = I - m$, judge whether h is an IMF, if it is, go to S_4. Otherwise, repeat S_1 and S_2 using the residue h, until the latest residue turns to be an IMF;

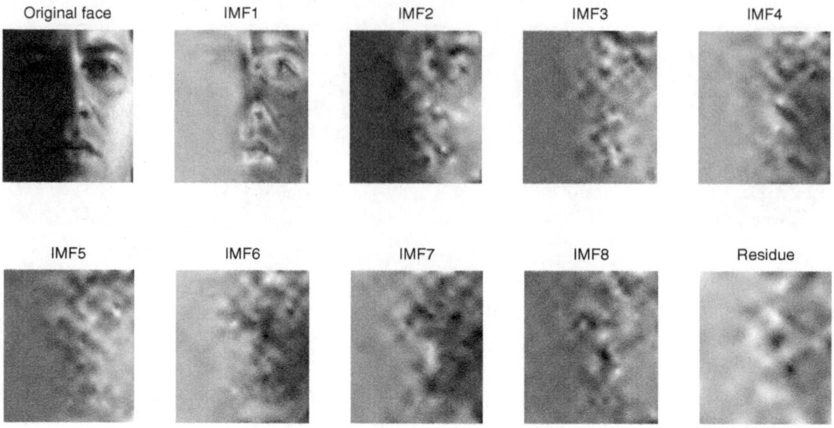

Fig. 3. Eight IMFs and the residue decomposed by 2DEMD on a face image with right point illumination from the PIE database

S_4. Input the residue h to the loop from S_1 to S_3 to get the next remained IMFs until it can not be decomposed further.

In the implementation of two dimensional case, the Stop Criterion (SD) should be defined to judge whether residue h is an IMF and when to stop the loops:

$$SD = \sum_{k=0}^{m} \sum_{i=0}^{n} [\frac{|h_{i(j-1)}(k,l) - h_{ij}(k,l)|^2}{h_{i(j-1)}(k,l)^2}].$$

Different with 1DEMD, the extrema detection and the surface interpolation seem more complicated. Choosing appropriate extrema detection method and surface interpolation functions are more important, otherwise, poor decomposition results are got. More details can be seen in [13]. Fig.3 demonstrates the decomposition results by the 2DEMD. The IMFs are from finest to coarsest as well, and the illumination effects focus on the residue.

3 Face Recognition Schemes

As described in the former sections, two face recognition schemes based on EMD are generated. One is using the high-frequency IMFs directly as facial features. The other one is based on the synthesized face images fused by high-frequency IMFs. Fig.4 and Fig.5 show the flowcharts of two methods respectively. In our implementation, the following four items should be illustrated.

- Down-sampling is adopted twice before the EMD process. It aims to decrease the dimension of the faces.
- We choose the 1st IMFs to represent the facial features since the 1st IMF is the highest-frequency component.

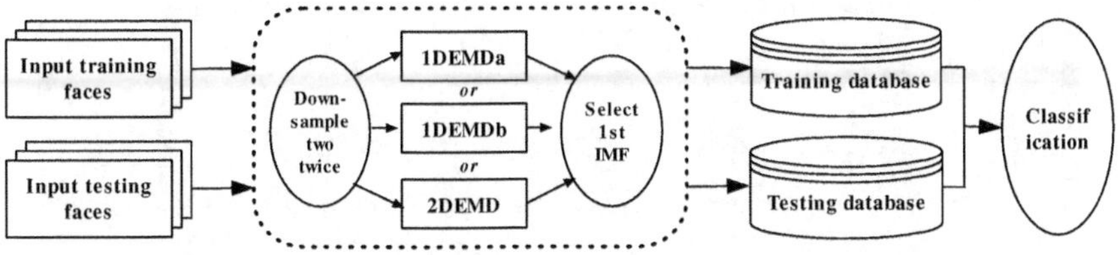

Fig. 4. Flowchart of frequency face recognition scheme

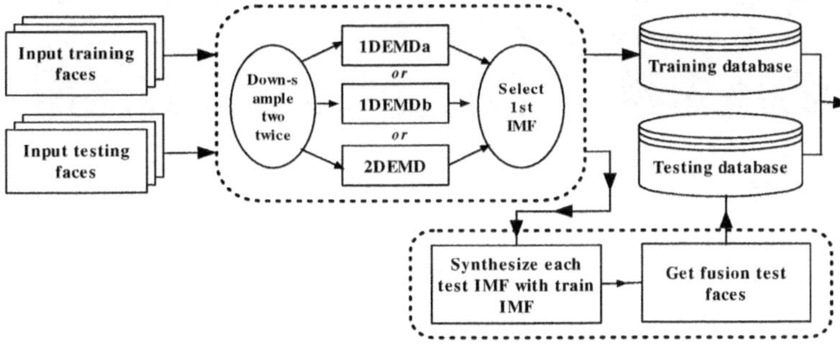

Fig. 5. Flowchart of fusion based face recognition scheme

- Fusion the 1st IMFs of the train faces and test faces. In our experiment, we adopt only one neutral illumination face for training, and aim to verify the other four faces with serious illumination effects. As described in former sections, illumination effects always exist in low frequency IMFs, i.e., the 1st IMF contains minimum illumination effects. The synthesized faces of the two 1st IMFs enhance their mutual features while the serious illumination effects are weakened a lot in the test faces.

4 Experimental Results

We evaluated the proposed two schemes using the PIE face database, which is accessible at http://www.ri.cmu.edu/projects/project_418.html. This database contains 41368 images of 68 people, each person under 13 different poses, 43 different illumination conditions, and 4 different expressions. Here, we only focused on the images with varying illuminations. Fig.6 showed 21 face samples of one person. All the face images were normalized into 112×92 size.

In our experiments, we have established two sub-database for evaluating the two schemes separately. The first one is designed for the frequency method. As the rectangles denoted in Fig.6, the first 6 images were chosen from each subject orderly. Thereinto 3 face images were used for training and the remained 3 face images for testing. There are total 20 cases, here we only evaluated 5 cases and the results listed in Table.1 are average ones. Another sub-database contains

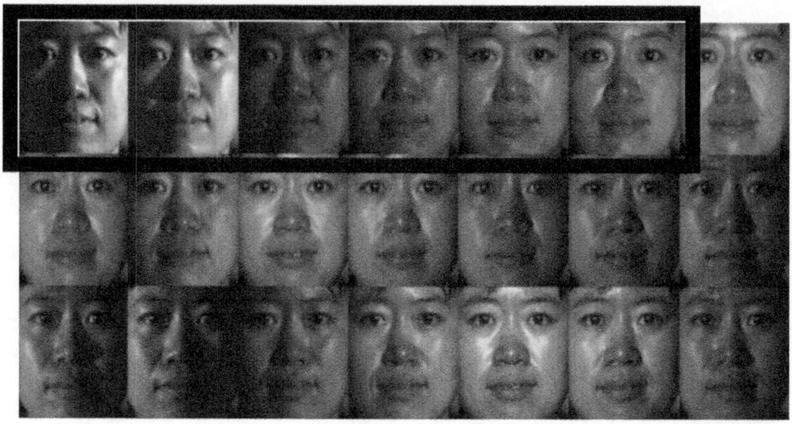

Fig. 6. Face image samples from the PIE database

5 faces per person. Each face is quite different from the other 4 in terms of illumination effects. We used the neutral one for training and the remained 4 for testing. Fig.7 showed an example.

Additionally, we have also compared our EMD based methods with the traditional wavelet based face recognition methods. In order to guarantee equitable comparisons, Daubechies wavelet 'db4' is adopted for wavelet decomposition, since 'db4' generally got the best performances as reported in [20,21]. Here we adopted Support Vector Machine (SVM) as classifier. Our experiments were implemented in a personal computer with Genuine Inete(R)T2300 CPU and 1.5G RAM and Matlab version 7.0 was used.

Table 1. Recognition rate versus different methods by scheme 1

Methods	1DEMDa	1DEMDb	2DEMD	'db4' wavelet
Correctness	99.51%	97.55%	93.63%	78.92%
	(203)	(199)	(191)	(161)

All the EMD based methods got much higher correctness than the wavelet based methods. Especially, 1DEMD applying on the lengthened image vector performed best, and the inferior one was 1DEMD applying row by row.

The next experiment we focused on testing the scheme 2. Fig. 7 demonstrated some face samples from the database and simultaneously showed the corresponding test fusion images. It is obvious that the illumination on test images have been removed a lot. Table. 2 showed the performances before fusion and after the fusion. The recognition rate are improved significantly comparing the poor correctness before fusion. All the EMD based methods performed nearly perfectly.

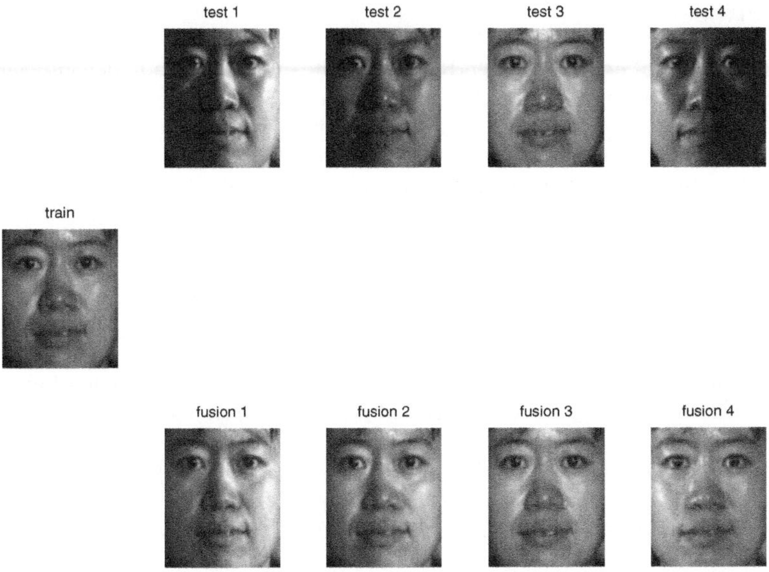

Fig. 7. Original testing face images and fusion face images by 1DEMDa

Table 2. Recognition rate versus different methods by scheme 2

Methods	1DEMDa	1DEMDb	2DEMD	'db4' wavelet
Correctness (before)	84.93% (231)	32.73% (89)	63.24% (172)	74.24% (202)
Correctness (after)	100% (272)	100% (272)	100% (272)	97.06% (264)

5 Conclusions

We have proposed two EMD based face recognition schemes in this paper. Both of the schemes have efficiently addressed the illumination effects problem. Compared with the traditional wavelet based methods, the proposed EMD based methods have much advantages in recognition rate. However, we also found that the EMD based methods have cost much time. Looking for the fast algorithm of EMD is one of our future work. Additionally, compared with 1DEMD, 2DEMD performed faster but poorer recognition rate. This is properly affected by the decomposition results which affected by the choices of surface interpolation methods. Finding the appropriate surface interpolation functions to improve the performances of 2DEMD is our remaining work in the coming days as well.

References

1. Huang, N.E., Shen, Z., Long, S.R., et al.: The empirical mode decomposition and the Hilbert spectrum for nonlinear and non-stationary time series analysis. Proceedings of the Royal Society A 454(1971), 903–995 (1998)
2. Huang, N.E., Shen, S.S.P.: Hilbert-Huang transform and its applications. Book in Interdisciplinary Mathematical Sciences, vol. 5 (2005)
3. Huang, N.E., Wu, M.L.C., Long, S.R., et al.: A confidence limit for the empirical mode decomposition and Hilbert spectral analysis. Proceedings of the Royal Society A 459(2037), 2317–2345 (2003)
4. Long, S.R.: Applications of HHT in image analysis. In: Huang, N.E., Shen, S.S.P. (eds.) Hilbert-Huang Transform and Its Applications. World Scientific, River Edge (2005)
5. Hariharan, H., Gribok, A., Abidi, B., Abidi, M.: Multi-modal Face Image Fusion using Empirical Mode Decomposition. In: The Biometrics Consortium Conference, Crystal City, VA (2005)
6. Hariharan, H., Koschan, A., Abidi, B., Gribok, A., Abidi, M.A.: Fusion of visible and infrared images using empirical mode decomposition to improve face recognition. In: IEEE International Conference on Image Processing, ICIP 2006, Atlanta, GA, pp. 2049–2052 (2006)
7. Bhagavatula, R., Savvides, M., Acoustics, M.: Analyzing Facial Images using Empirical Mode Decomposition for Illumination Artifact Removal and Improved Face Recognition. In: IEEE International Conference on Speech and Signal Processing, 2007 (ICASSP 2007), April 15-20, vol. 1, pp. 505–508 (2007)
8. Nunes, J.C., Bouaoune, Y., Delechelle, E., Niang, O., Bunel, P.: Image analysis by bidimensional empirical mode decomposition. Image and Vision Computing 21(12), 1019–1026 (2003)
9. Nunes, J.C., Guyot, S., Deléchelle, E.: Texture analysis based on local analysis of the Bidimensional Empirical Mode Decomposition. Machine Vision and Applications 16(3), 932–8092 (2005)
10. Linderhed, A.: 2-D empirical mode decompositions in the spirit of image compression. In: Wavelet and Independent Component Analysis Applications IX, Orlando, Fla, USA. Proceedings of SPIE, vol. 4738, pp. 1–8 (2002)
11. Linderhed, A.: Compression by image empirical mode decomposition. In: IEEE International Conference on Image Processing (ICIP 2005), vol. 1, pp. 553–556 (2005)
12. Hariharan, H., Gribok, A., Abidi, M., Koschan, A.: Image Fusion and Enhancement via Empirical Mode Decomposition. Journal of Pattern Recognition Research 1(1), 16–32 (2006)
13. Sinclair, S., Pegram, G.G.S.: Empirical Mode Decomposition in 2-D space and time: a tool for space-time rainfall analysis and nowcasting. Hydrol. Earth Syst. Sci. Discuss. 2, 289–318 (2005)
14. Wan, J., Ren, L., Zhao, C.: Image Feature Extraction Based on the Two-Dimensional Empirical Mode Decomposition. In: 2008 Congress on Image and Signal Processing, vol. 1, pp. 627–631 (2008)
15. Taghia, J., Doostari, M.A., Taghia, J.: An Image Watermarking Method Based on Bidimensional Empirical Mode Decomposition. In: 2008 Congress on Image and Signal Processing, vol. 5, pp. 674–678 (2008)
16. Fauchereau, N., Sinclair, S., Pegram, G.: 2-D Empirical Mode Decomposition on the sphere, application to the spatial scales of surface temperature variations. Hydrol. Earth Syst. Sci. Discuss. 5, 405–435 (2008)

17. Nastar, C.: The image shape spectrum for image retrieval. Technical report, INRIA, No. 3206 (1997)
18. Nastar, C., Moghaddam, B., Pentland, A.: Flexible images: matching and recognition using learned deformations. Computer Vision and Image Understanding 65(2), 179–191 (1997)
19. Zhang, Z.B., Ma, S.L., Wu, D.Y.: The application of neural network and wavelet in human face illumination compensation. In: Wang, J., Liao, X.-F., Yi, Z. (eds.) ISNN 2005. LNCS, vol. 3497, pp. 828–835. Springer, Heidelberg (2005)
20. Feng, G.C., Yuen, P.C., Dai, D.Q.: Human face recognition using PCA on wavelet subband. Journal of Electronic Imaging 9(2), 226–233 (2000)
21. Ekenel, H.K., Sanker, B.: Multiresolution face recognition. Image and Vision Computing 23, 469–477 (2005)

A Discriminant Analysis Method for Face Recognition in Heteroscedastic Distributions

Zhen Lei, Shengcai Liao, Dong Yi, Rui Qin, and Stan Z. Li*

Center for Biometrics and Security Research,
Institute of Automation, Chinese Academy of Sciences,
95 Zhongguancun Donglu, Beijing 100190, China
{zlei,scliao,dyi,rqin,szli}@cbsr.ia.ac.cn

Abstract. Linear discriminant analysis (LDA) is a popular method in pattern recognition and is equivalent to Bayesian method when the sample distributions of different classes are obey to the Gaussian with the same covariance matrix. However, in real world, the distribution of data is usually far more complex and the assumption of Gaussian density with the same covariance is seldom to be met which greatly affects the performance of LDA. In this paper, we propose an effective and efficient two step LDA, called LSR-LDA, to alleviate the affection of irregular distribution to improve the result of LDA. First, the samples are normalized so that the variances of variables in each class are consistent, and a pre-transformation matrix from the original data to the normalized one is learned using least squares regression (LSR); second, conventional LDA is conducted on the normalized data to find the most discriminant projective directions. The final projection matrix is obtained by multiply the pre-transformation matrix and the projective directions of LDA. Experimental results on FERET and FRGC ver 2.0 face databases show the proposed LSR-LDA method improves the recognition accuracy over the conventional LDA by using the LSR step.

Keywords: Least squares regression (LSR), discriminant analysis, face recognition.

1 Introduction

Subspace learning has attracted much attention and achieved great success in face recognition research area during the last decades. Among various methods, PCA and LDA [1]are the two most representative ones. PCA uses the Karhunen-Loeve transform to produce the most expressive subspace for face representation and recognition by minimizing the residua of the reconstruction. However, it does not utilize any class information and so it may drop some important clues for classification. LDA is then proposed and it seeks subspace of features best separating different face classes by maximizing the ratio of the between-class scatter matrix to the within-class scatter.

In theory, LDA is equivalent to Bayesian method if the distribution of samples in each class is obey to the Gaussian density with the same covariance matrix [2]. However, in practice, the covariance matrices from different classes are always heteroscedastic

* Corresponding author.

M. Tistarelli and M.S. Nixon (Eds.): ICB 2009, LNCS 5558, pp. 112–121, 2009.
© Springer-Verlag Berlin Heidelberg 2009

and it deteriorates the performance of LDA. There are usually two ways to handle this problem: One is to exploit specific heteroscedastic LDA [3,4,5,6] to solve the problem, not only taking into account the discriminatory information between class means, but also the differences of class covariance matrices. The other is to try to make the class covariance matrices consistent and thus to improve the performance of LDA on it.

In this paper, we follow the latter one and propose a least squares regression (LSR) based processing to normalize the distribution of samples before conventional LDA. In [7], researchers take into account the scale properties of each variable of feature on the whole data set and re-scale the variables to enhance the performance of LDA. In fact, each variable in different classes has different scale properties, and it is more reasonable to re-scale each variable in different classes individually. In our method, first, each variable of feature in the same class is normalized to be of unit variance so that the variables in every class have the same variance that makes the class distributions more consistent. Second, for its simplicity and effectiveness, we utilize LSR to learn the transform matrix from original data to the normalized one. If each variable of feature is independent, this normalization process would guarantee the covariance matrices from different classes identical. Even the variables are not independent, we argue this processing will still make the distributions of samples from different classes more consistent and therefore to improve the result of LDA. Besides, this normalization process could also make the sample distribution of each class more compact so that it is able to increase the separability of classes. Fig. 1 shows a toy example. The left one is the original data distribution and the right one is the data distribution after normalization. The line is the LDA projective direction. It explicitly illuminates the effect of the proposed normalization process for improving the separability of LDA result. Regarding the computational cost, compared to the heteroscedastic LDA [6], the proposed method only involves the LSR instead of the complex iterative optimization process, thus it is very efficient and can be applied in large scale data set.

The remainder of the paper is organized as follows. Section 2 reviews conventional LDA and details the LSR-LDA method. Section 3 compares the results of the proposed

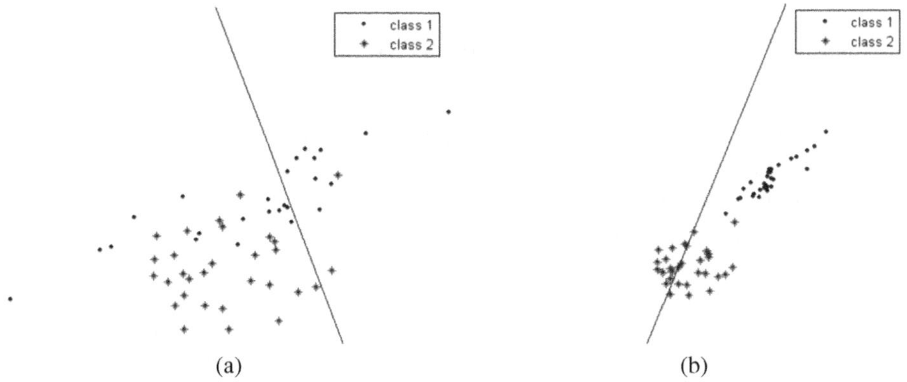

Fig. 1. A toy example that shows the advantage of the proposed normalization step, with data distribution and LDA projection before (left) and after (right) LSR normalization

method with other methods on FERET and FRGC databases and in Section 4, we conclude the paper.

2 LSR-LDA

2.1 Conventional LDA

Let $\mathbf{X} = [\mathbf{x}_1, \mathbf{x}_2, \ldots, \mathbf{x}_n]$ be a d-dimensional data set with n elements from $\{C_i|i = 1, 2, \ldots, L\}$ classes, where L is the number of the total classes. The within class scatter matrix S_w and the between class scatter matrix S_b in LDA are defined as follows:

$$S_w = \frac{1}{n} \sum_{i=1}^{L} \sum_{\mathbf{x}_j \in C_i} (\mathbf{x}_j - \mathbf{m}_i)(\mathbf{x}_j - \mathbf{m}_i)^T \tag{1}$$

$$S_b = \frac{1}{n} \sum_{i=1}^{L} n_i (\mathbf{m}_i - \mathbf{m})(\mathbf{m}_i - \mathbf{m})^T \tag{2}$$

where $\mathbf{m}_i = \frac{1}{n_i} \sum_{\mathbf{x}_j \in C_i} \mathbf{x}_j$ is the mean of data in class C_i, and $\mathbf{m} = \frac{1}{n} \sum_{i=1}^{L} \sum_{\mathbf{x}_j \in C_i} \mathbf{x}_j$ is the global mean vector. LDA searches such optimal projections that after projecting the original data onto these directions, the trace of the resulting between class scatter matrix is maximized while the trace of the within class scatter matrix is minimized. Let $\mathbf{W}$ denote a $d \times d'$ $(d' < d)$ projection matrix, LDA then chooses $\mathbf{W}$ so that the following object function is maximized:

$$J = \frac{tr(\tilde{S}_b)}{tr(\tilde{S}_w)} = \frac{tr(\sum_{i=1}^{L} n_i \mathbf{W}^T (\mathbf{m}_i - \mathbf{m})(\mathbf{m}_i - \mathbf{m})^T \mathbf{W})}{tr(\sum_{i=1}^{L} \sum_{\mathbf{x}_j \in C_i} \mathbf{W}^T (\mathbf{x}_j - \mathbf{m}_i)(\mathbf{x}_j - \mathbf{m}_i)^T \mathbf{W})} = \frac{tr(\mathbf{W}^T S_b \mathbf{W})}{tr(\mathbf{W}^T S_w \mathbf{W})} \tag{3}$$

The optimal projection matrix $\mathbf{W}_{opt}$ can be obtained by solving the following eigenvalue problem corresponding to the d' largest non-zero eigenvalues.

$$S_w^{-1} S_b \mathbf{W} = \mathbf{W} \mathbf{\Lambda} \tag{4}$$

where $\mathbf{\Lambda}$ is the diagonal matrix whose diagonal elements are the eigenvalues of $S_w^{-1} S_b$.

In real application, due to the usually high dimension of data and small size of samples, the within class scatter matrix S_w is often singular and the inverse of S_w does not exist, so the optimal solution of LDA in Eq. 4 cannot be found directly. To deal with this problem, many variants of LDA have been proposed such as PCA+LDA (Fisher-LDA or FLDA), Direct LDA (DLDA), Null space LDA (NLDA) [1,8,9] etc.. FLDA firstly utilizes PCA to select the most expressive subspace and reduces the dimension of feature to make S_w non-singular and then conducts LDA to derive the optimal projective directions. DLDA takes no account of the impact of S_w and finds the most discriminative projections in range space of S_b directly. NLDA is proposed to find the projections that maximize the between class scatter matrix S_b in the null space of S_w so that the singularity problem is settled.

2.2 LSR Normalization

LDA's performance highly depends on the sample distribution. Theoretically, under Gaussian assumption, it achieves its best performance with the same covariance matrix for different classes. Therefore, how to make data distributions from different classes as consistent as possible is a key point to achieve good result for LDA method.

Considering this, before LDA, we normalize the data distribution so as to make each dimension of data in every class to be of the same unit variance. In this way, we can make the distributions of different classes more consistent than original one and thus hope to improve the performance of LDA.

Let the sample set from the k-th class be $\mathbf{X}^k = [\mathbf{x}_1^k, \mathbf{x}_2^k, \cdots, \mathbf{x}_{n_k}^k]$ of d dimension, where n_k is the sample number, and $\mathbf{m}^k = 1/n_k \sum_{i=1}^{n_k} \mathbf{x}_i^k$ is the mean vector of the k-th class. Then we normalize each dimension of data to be of unit variance. That is

$$\sigma_j^k = \sqrt{\frac{1}{n_k}\sum_{i=1}^{n_k}(x_{ji}^k - m_j^k)^2}$$

$$x_{ji}'^k = \begin{cases} (x_{ji}^k - m_j^k)/\sigma_j^k + m_j^k & \sigma_j^k \neq 0 \\ x_{ji}^k & \sigma_j^k = 0 \end{cases} \qquad j = 1, 2, \ldots, d, \quad i = 1, 2, \ldots, n_k$$

The above normalization method is feasible in training set. When there comes a new sample, because of its unknown class label, we don't know how to normalize it. To handle this problem, we turn to learn the relationship between the original data and the normalized one so that the learned relationship can then be generalized to normalize the unseen data.

Suppose the normalized sample set be $\mathbf{X}'$. Our purpose is to learn the relationship f that describes the transform from the original data to the normalized one $\mathbf{x}_i' = f(\mathbf{x}_i)$. Under linear assumption, it is simplified to learn the transformation matrix $\mathbf{W}_1$ between the two spaces $\mathbf{x}_i' = \mathbf{W}_1^T\mathbf{x}_i$. Fortunately, this problem could be solved in means of least squares regression (LSR).

$$\mathbf{W}_1 = \arg\min_{\mathbf{W}_1} \sum_{i=1}^{n} \|\mathbf{W}_1^T\mathbf{x}_i - \mathbf{x}_i'\|^2 = \arg\min_{\mathbf{W}_1} tr((\mathbf{W}_1^T\mathbf{X} - \mathbf{X}')(\mathbf{W}_1^T\mathbf{X} - \mathbf{X}')^T) \tag{5}$$

By putting the derivative of above function with respect to $\mathbf{W}_1$ to zero, we obtain

$$2(\mathbf{W}_1^T\mathbf{X} - \mathbf{X}')\mathbf{X}^T = 0 \Rightarrow \mathbf{W}_1 = (\mathbf{X}\mathbf{X}^T)^{-1}\mathbf{X}\mathbf{X}'^T \tag{6}$$

In reality, the dimension of data is usually larger than the sample number and hence, the matrix $\mathbf{X}\mathbf{X}^T$ is deficient and its inverse does not exist. On the other hand, even the inverse of $\mathbf{X}\mathbf{X}^T$ exists, the derived result $\mathbf{W}_1$ may over-fit to the training set. In order to avoid the deficient problem and improve the generalization of the result, we impose regularized penalty, also known the prior knowledge [10] onto the objective function in Eq. 5 as

$$\mathbf{W}_1 = \arg\min_{\mathbf{W}_1} tr((\mathbf{W}_1^T\mathbf{X} - \mathbf{X}')(\mathbf{W}_1^T\mathbf{X} - \mathbf{X}')^T + \lambda\mathbf{W}_1\mathbf{W}_1^T) \tag{7}$$

Input: Let d-dimensional sample set be $\mathbf{X} = \{\mathbf{X}^1, \mathbf{X}^2, \cdots, \mathbf{X}^L\}$ from L classes $\{C_1, C_2, \ldots, C_L\}$, whose corresponding class means are $\{\mathbf{m}^1, \mathbf{m}^2, \ldots, \mathbf{m}^L\}$. $\mathbf{X}^k = [\mathbf{x}_1^k, \mathbf{x}_2^k, \cdots, \mathbf{x}_{n_k}^k]$ denotes the n_k samples from the k-th class.

(a) For every dimension of data, normalize the variance of variable in each class in training set to be unit.

for $k = 1 : L$

  for $j = 1 : d$

    $\sigma_j^k = \sqrt{\frac{1}{n_k} \sum_{i=1}^{n_k} (x_{ji}^k - m_j^k)^2}$

    for $i = 1 : n_k$

$$x_{ji}'^k = \begin{cases} (x_{ji}^k - m_j^k)/\sigma_j^k + m_j^k & \sigma_j^k \neq 0 \\ x_{ji}^k & \sigma_j^k = 0 \end{cases}$$

    end

  end

end

(b) Learn the pre-transformation matrix $\mathbf{W}_1$ using least squares regression according to Eq. 8.

(c) Compute the within and between scatter matrices based on the normalized samples according to Eq. 1 and 2.

(d) Compute the conventional LDA projection matrix $\mathbf{W}_2$ on the normalized samples according to Eq. 4.

Output:The two-step LDA projection matrix $\mathbf{W} = \mathbf{W}_1\mathbf{W}_2$.

Fig. 2. LSR based Two Step LDA algorithm

where λ controls the trade-off between the fitting accuracy in the training set and the generalization. We can then obtain the optimal result as

$$\mathbf{W}_1 = (\mathbf{X}\mathbf{X}^T + \lambda\mathbf{I})^{-1}\mathbf{X}\mathbf{X}'^T \tag{8}$$

In this way, when there comes a new data $\mathbf{x}_{new}$, we can normalize it by multiplying it with the learned transformation matrix $\mathbf{W}_1$ directly, $\mathbf{x}'_{new} = \mathbf{W}_1^T\mathbf{x}_{new}$.

2.3 LDA after LSR

After the normalization step, various LDA can be conducted on the normalized data to learn the most separable subspace and the final projective directions are obtained by multiplying the normalization transformation matrix with the LDA projections. The whole process of the LSR based two step LDA is illustrated in Fig. 2.

3 Experimental Results and Analysis

3.1 Data Preparation

Two face databases, FERET [11] and FRGC ver 2.0 [12] are tested. All the images are rotated, scaled and cropped to 44 × 40 according to the provided eye positions

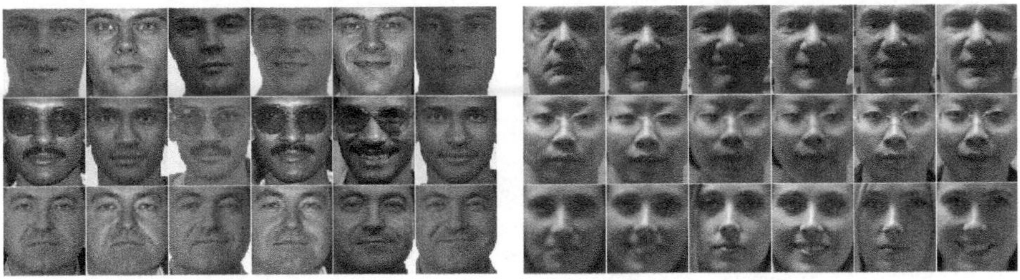

(a) (b)

Fig. 3. Face examples of FERET (a) and FRGC (b) databases

Table 1. The performance of different methods on FERET database

Methods	Rank-1	VR@FAR=0.001	EER
FLDA	0.6684	0.5557	0.1213
NLDA	0.7087	0.5751	0.1158
DLDA	0.7252	0.6390	0.1007
[7]+FLDA	0.6466	0.5012	0.1439
[7]+NLDA	0.6911	0.5396	0.1372
[7]+DLDA	0.7068	0.6087	0.1124
LSR-FLDA	0.6978	0.6594	0.0838
LSR-NLDA	0.7196	0.6793	0.0788
LSR-DLDA	0.7125	0.6618	0.0840

Table 2. The performance of different methods on FRGC v2.0 database

Methods	VR@FAR=0.001	EER
FLDA	0.5082	0.1153
DLDA	0.4366	0.1181
[7]+FLDA	0.5550	0.0875
[7]+DLDA	0.4386	0.1019
LSR-FLDA	0.5856	0.0552
LSR-DLDA	0.5585	0.0568

succeeded by histogram equalization preprocessing. For FERET database, the training set contains 731 images. In test phase, we use the gallery set containing 1196 images from 1196 subjects, and combine four provided probe sets (fb, fc, dupI, dupII) together, totally including 2111 images to compose the probe set. So our test protocol should be more difficult than any of the four original protocols because we consider different factors (expression, illumination, aging etc.) together to evaluate the performance. For FRGC database, the training set consists of 12776 face images from 222 individuals, including 6360 controlled and 6416 uncontrolled images. We choose experimental 4 setting which is considered as the most difficult case in FRGC to test the algorithm. In the test set, there are 16028 controlled images as the target set and 8014 query images which are uncontrolled ones, from 466 persons. The images are captured over several sessions. Fig. 3 illustrates some cropped face examples of FERET and FRGC databases.

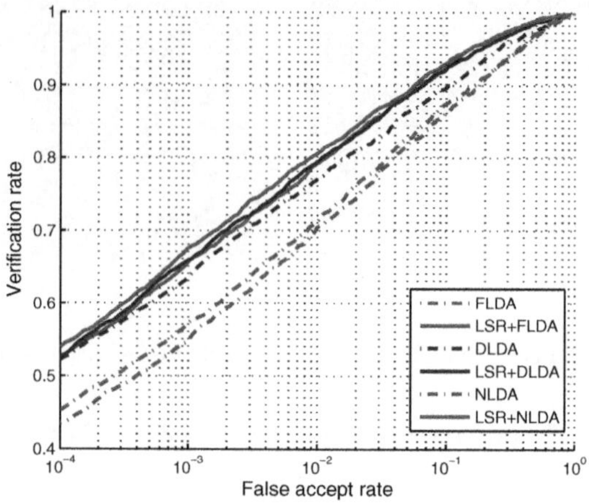

(a)

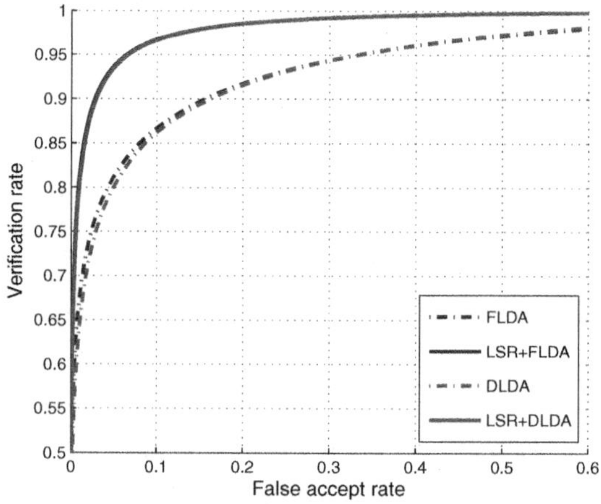

(b)

Fig. 4. Receiver operating characteristic (ROC) curves of different methods on FERET (a) and FRGC (b) databases

3.2 Performance Evaluation

In this experiment, the regularization parameter λ is set to be 1.0 empirically. The cosine distance (Eq. 9) is adopt to measure the dissimilarity of features and the nearest neighbor (NN) classifier is chosen to do the classification task.

$$d_{cos}(\mathbf{x}, \mathbf{y}) = -\frac{\mathbf{x}^T\mathbf{y}}{\sqrt{\mathbf{x}^T\mathbf{x}\mathbf{y}^T\mathbf{y}}} \tag{9}$$

The proposed LSR based normalization method is combined with various versions of LDA such as FLDA, DLDA, NLDA and compared with these LDAs without normalization stage. Moreover, we also compare the results of normalization method in [7]

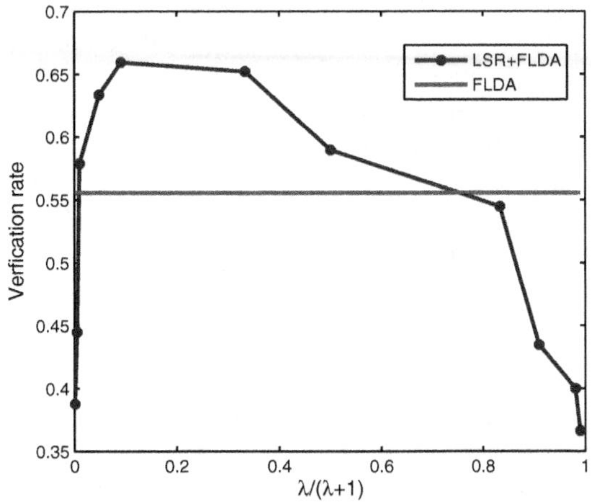

Fig. 5. Face verfication rates with respect to $\lambda/(\lambda + 1)$ on FERET database

Table 3. Computational cost of FLDA and LSR-FLDA on FRGC training set

	FLDA	LSR-FLDA
computational time (s)	23.69	31.71

on FERET and FRGC v2.0 respectively. For FERET, the results are reported as rank-1 recognition rate, verification rate (VR) when the false accept rate (FAR) is 0.001 and equal error rate (EER). For FRGC, we report the result as verification rate (VR) when the false accept rate (FAR) is 0.001 and equal error rate (EER) based on all matchings between query and target images.

Table 1 and 2 illustrate the recognition results of various methods on FERET and FRGC respectively and Fig. 4 shows the corresponding ROC curves. For clarity, we just plot the results of proposed method and conventional LDA in Fig. 4. For FRGC v2.0, because the sample number in training set is larger than the dimension of feature, there is no null space for within scatter matrix S_w, so that we don't have the results of NLDA. From the results, we can see that whether on FERET or FRGC database, whether with FLDA, DLDA or NLDA, the proposed LSR-LDA always obtains better results than conventional LDA, especially for FLDA, NLDA on FERET and FLDA, DLDA on FRGC database. It proves the proposed LSR based normalization step is an effective way to eliminate the affection of sample irregularity and could significantly improve the performance of LDA. It should be noted the normalization method in [7] doesn't always improve the performance of LDA. That may be because the databases in our experiments are larger and more challenge than that in [7].

The regularization coefficient λ is an adjustable parameter in our algorithm. We examine the impact of λ on recognition accuracy with LSR-FLDA on FERET database and plot the results with respect to different values of $\lambda/(\lambda + 1)$ in Fig. 5. The red line is the verification rate of original FLDA without the proposed normalization step. It can be seen the performance of LSR-FLDA keeps a higher accuracy during a large scale values of λ which indicates the robustness of the proposed algorithm.

Table 3 illustrates the experimental computational cost for FLDA and LSR-FLDA on FRGC training set. It is the average of 10 times running on a Core 2 Duo 2.4GHz and 2GB RAM PC with unoptimized matlab code. It shows that the proposed LSR-FLDA doesn't increase the computational burden too much, about 1.3 times than the cost of original FLDA. Considering the accuracy improvement of LSR-LDA, this additional spend is completely tolerable in practical applications.

4 Conclusions

In this paper, we propose a LSR based two step LDA for face recognition. Before LDA, each dimension of data from the same class are normalized to be of unit variance so that the distributions of different classes are imposed to be consistent. After that, we utilize least squares regression (LSR) to learn the pre-transformation from original data to normalized one. Thus, when there comes new data, it can be transformed to the normalized distribution directly. The final projective directions of LSR-LDA are obtained by multiplying the pre-transformation and the projections of conventional LDA on normalized data. Experimental results on FERET and FRGC databases show the proposed LSR based two step LDA significantly improves the performance compared to the conventional one.

Acknowledgements. This work was supported by the following funding resources: National Natural Science Foundation Project #60518002, National Science and Technology Support Program Project #2006BAK08B06, National Hi-Tech (863) Program Projects #2006AA01Z192, #2006AA01Z193, and #2008AA01Z124, Chinese Academy of Sciences 100 people project, and AuthenMetric R&D Funds.

References

1. Belhumeur, P., Hespanha, J., Kriegman, D.: Eigenfaces vs. fisherfaces: recognition using class specific linear projection. IEEE Transactions on Pattern Analysis and Machine Intelligence 19(7), 711–720 (1997)
2. Duda, R.O., Hart, P.E., Stork, D.G.: Pattern Classification, 2nd edn. John Wiley and Sons, Chichester (2001)
3. Loog, M., Duin, R.P.: Linear dimensionality reduction via a heteroscedastic extension of lda: The chernoff criterion. IEEE Transactions on Pattern Analysis and Machine Intelligence 26(6), 732–739 (2004)
4. Das, K., Nenadic, Z.: Approximate information discriminant analysis: A computationally simple heteroscedastic feature extraction technique. Pattern Recognition 41(5), 1548–1557 (2008)
5. Hsieh, P.F., Wang, D.S., Hsu, C.W.: A linear feature extraction for multiclass classification problems based on class mean and covariance discriminant information. IEEE Transactions on Pattern Analysis and Machine Intelligence 28(2), 223–235 (2006)
6. Kumar, N., Andreou, A.G.: Heteroscedastic discriminant analysis and reduced rank hmms for improved speech recognition. Speech Commun. 26(4), 283–297 (1998)

7. An, G., Ruan, Q.: Novel mathematical model for enhanced fisher's linear discriminant and its application to face recognition. In: Proceedings of International Conference on Pattern Recognition, pp. 524–527 (2006)
8. Yu, H., Yang, J.: A direct lda algorithm for high-dimensional data with application to face recognition. Pattern Recognition 34(10), 2067–2070 (2001)
9. Chen, L., Liao, H., Ko, M., Lin, J., Yu, G.: A new lda-based face recognition system which can solve the small sample size problem. Pattern Recognition 33(10), 1713–1726 (2000)
10. Hastie, T., Tibshirani, R., Friedman, J.: The Elements of Statistical Learning: Data Mining, Inference, and Prediction. Springer, Heidelberg (2001)
11. Phillips, P.J., Moon, H., Rizvi, S.A., Rauss, P.J.: The FERET evaluation methodology for face-recognition algorithms. IEEE Transactions on Pattern Analysis and Machine Intelligence 22(10), 1090–1104 (2000)
12. Phillips, P.J., Flynn, P.J., Scruggs, W.T., Bowyer, K.W., Chang, J., Hoffman, K., Marques, J., Min, J., Worek, W.J.: Overview of the face recognition grand challenge. In: Proceedings of IEEE Computer Society Conference on Computer Vision and Pattern Recognition, pp. 947–954 (2005)

Robust Face Recognition Using Color Information

Zhiming Liu and Chengjun Liu

Department of Computer Science, New Jersey Institute of Technology
Newark, New Jersey 07102, USA
zl9@njit.edu

Abstract. This paper presents a robust face recognition method using color information with the following three-fold contributions. First, a novel hybrid color space, the RC_rQ color space, is constructed out of three different color spaces: the RGB, YC_bC_r, and YIQ color spaces. The RC_rQ hybrid color space, whose component images possess complementary characteristics, enhances the discriminating power for face recognition. Second, three effective image encoding methods are proposed for the component images in the RC_rQ hybrid color space: (i) a patch-based Gabor image representation for the R component image, (ii) a multi-resolution LBP feature fusion scheme for the C_r component image, and (iii) a component-based DCT multiple face encoding for the Q component image. Finally, at the decision level, the similarity matrices generated using the three component images in the RC_rQ hybrid color space are fused using a weighted sum rule. The most challenging Face Recognition Grand Challenge (FRGC) version 2 Experiment 4 shows that the proposed method, which achieves the face verification rate of 92.43% at the false accept rate of 0.1%, performs better than the state-of-the-art face recognition methods.

1 Introduction

Robust face recognition methods require that algorithms be able to extract and exploit multiple facial features efficiently. Many current face recognition methods, however, start with one single feature extraction process, such as extracting the principal components by the Eigenfaces method [1] and extracting the linear disriminating components by the Fisherfaces method [2], and then utilize one classifer, such as the nearest neighbor classifier, for face recognition. In order to take advantage of the rich facial features resided in human face images, a better face recognition method should apply multiple feature extraction processes, and fuse the multiple face features for improving recognition performance. Fusion of multiple face features can take place by applying either various image features or various subspace transformed features.

The fusion of local and global features for improving face recognition performance has been studies recently [3], [4]. Generally speaking, local features such as those extracted by the Gabor [5] and the LBP [6] methods are different from

M. Tistarelli and M.S. Nixon (Eds.): ICB 2009, LNCS 5558, pp. 122–131, 2009.
© Springer-Verlag Berlin Heidelberg 2009

global features such as those derived by the Eigenfaces and the Fisherfaces methods, as different features deal with different scaled facial details. If the features are complementary to each other, the correlation between the outputs of the local and the global classifiers would be reduced. As a result, fusing the classification outputs at the decision level could lead to improved overall performance [7]. Currently, most methods extract the complementary features from the gray image only, which leads to fulfilling very limited complementary characteristics between the local and global features. A better method should seek a natural and much more powerful complementary face representation by capitalizing on the multiple images in a color space.

This paper presents a robust face recognition method integrating the local and global features derived from a novel hybrid color space. The motivation of our method is to explore the complementary facial information in order to boost face recognition performance by fusing their classification outputs. The novelty of our method is three-fold. First, a novel hybrid color space, the RC_rQ color space, is constructed out of three different color spaces: the RGB, YC_bC_r, and YIQ color spaces. The RC_rQ hybrid color space, whose component images possess complementary characteristics, enhances the discriminating power for face recognition. Second, three effective image encoding methods are proposed for the component images in the RC_rQ hybrid color space: (i) a patch-based Gabor image representation for the R component image, (ii) a multi-resolution LBP feature fusion scheme for the C_r component image, and (iii) a component-based DCT multiple face encoding for the Q component image. For each method, the Enhanced Fisher Model (EFM) [8] is applied to extract features for classification. Finally, at the decision level, the similarity matrices generated using the three component images in the RC_rQ hybrid color space are fused using a weighted sum rule. The most challenging Face Recognition Grand Challenge (FRGC) version 2 Experiment 4 shows that the proposed method, which achieves the face verification rate of 92.43% at the false accept rate of 0.1%, performs better than the state-of-the-art face recognition methods.

2 The Proposed Method

2.1 A Hybrid Color Space: RC_rQ

Color provides powerful information for object detection, indexing and retrieval, as "humans can discern thousands of color shades and intensities, compared to about only two dozen shades of gray" [9]. Color information is helpful for improving the performance of face recognition due to the complementary characteristics among the color component images. This paper addresses face recognition in a novel hybrid color space instead of the conventional color spaces. As the R component image in the RGB color space is more effective than other component images for face recognition [10], we define a new hybrid color space RC_rQ, where C_r and Q are from the YC_bC_r color space and the YIQ color space, respectively. Fig. 1 shows the component images in the RC_rQ color space. Note that the R component image in Fig. 1 has the fine face region, which is suitable for the

Fig. 1. The *RCrQ* hybrid color space. Note that the R component image has the fine face region while the C_r and Q component images contain partial face contour information.

extraction of Gabor features, while the C_r and Q component images contain partial face contour information.

2.2 The Patch-Based Gabor Image Representation for the *R* Image

The Gabor Image Representation (GIR) of an image captures salient visual properties such as spatial location, orientation selectivity, and spatial frequency characteristics [11]. Specifically, the GIR is the convolution of the image with a family of Gabor kernels that may be formulated as follows [11]:

$$\psi_{\mu,\nu}(z) = \frac{\|k_{\mu,\nu}\|^2}{\sigma^2} e^{-\frac{\|k_{\mu,\nu}\|^2 \|z\|^2}{2\sigma^2}} \left[e^{ik_{\mu,\nu}z} - e^{-\frac{\sigma^2}{2}} \right] \tag{1}$$

where μ and ν define the orientation and scale of the Gabor kernels, $z = (x, y)$, $\|\cdot\|$ denotes the norm operator, and the wave vector $k_{\mu,\nu}$ is defined as follows:

$$k_{\mu,\nu} = k_\nu e^{i\phi_\mu} \tag{2}$$

where $k_\nu = k_{max}/f^\nu$ and $\phi_\mu = \pi\mu/8$. k_{max} is the maximum frequency, and f is the spacing factor between kernels in the frequency domain. Let $R(x, y)$ represent the R component image, the convolution of $R(x, y)$ and a Gabor kernel $\psi_{\mu,\nu}$ may be formulated as follows:

$$O_{\mu,\nu}(z) = R(z) * \psi_{\mu,\nu}(z) \tag{3}$$

where $z = (x, y)$, $*$ denotes the convolution operator, and $O_{\mu,\nu}(z)$ is the convolution result corresponding to the Gabor kernel at orientation μ and scale ν. Commonly used Gabor kernels contain five different scales, $\nu \in \{0, ..., 4\}$, and eight orientations, $\mu \in \{0, ..., 7\}$. The set $\mathcal{S} = \{O_{\mu,\nu}(z) : \mu \in \{0, ..., 7\}, \nu \in \{0, ..., 4\}\}$, thus, forms the GIR of the image R.

The advantage of GIR stems from the integration of different spatial frequencies, spatial localities, and orientation selectivities. The GIR thus contains rich information for face recognition, which can be applied to extract features both locally and holisticly. Fig. 2 shows the outline of face recognition using the GIR. For local GIR feature extraction, we separate the GIR into an ensemble of patches. The GIR is disintegrated into 4 patches along the horizontal direction, with the adjacent scale images forming one group. As the GIR patch images reside in a very high dimensional space (16 times the original image size), DCT

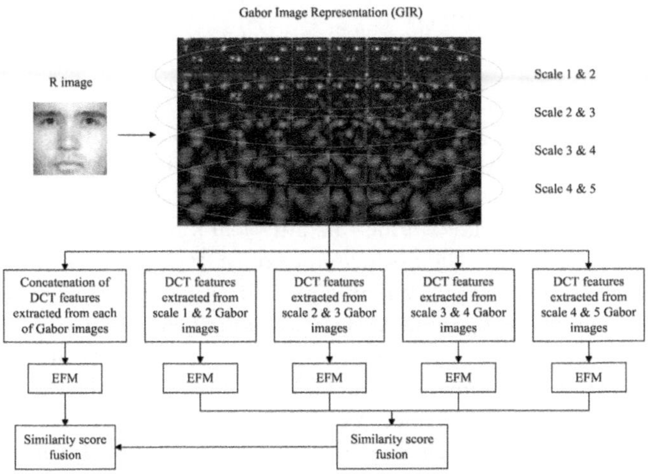

Fig. 2. The outline of face recognition using the GIR

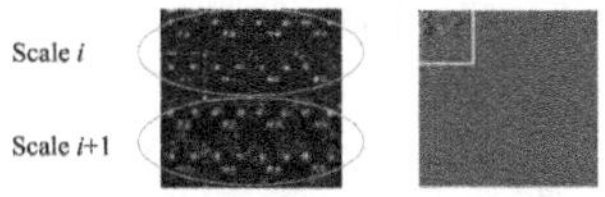

Fig. 3. A reshaped Gabor patch image and its DCT domain image, where $i \in \{1, ..., 4\}$. A frequency set selection scheme, which selects the low frequency set located in the upper-left corner, is used to reduce the dimensionality and decorrelate the redundancy of the GIR patch images.

is applied for dimensionality reduction for improving computational efficiency. To facilitate the DCT feature extraction, each GIR patch image is reshaped to a square, as shown in Fig. 3. After transforming the GIR patch image to the DCT domain, a frequency set selection scheme using a square mask is applied to select the low frequency feature set located in the upper-left corner. Then, the Enhanced Fisher Model (EFM) is used to classify these DCT feature sets. The four GIR patches generate four similarity matrices, which are fused by means of the sum rule. For holistic GIR feature extraction, the GIR is considered as a whole for classification. In particular, each of the Gabor convolved images is processed for dimensionality reduction using DCT. The DCT features derived from the 40 Gabor images are then concatenated to form an augmented vector, which is classified by the EFM. Previous research shows that Gabor kernels with different scales help improve performance differently, which implies that different number of DCT features should be used to capture the discriminant information corresponding to the Gabor kernels with different scales. The similarity matrix generated in this appraoch is subsequently fused with one generated from local approach by means of the sum rule, as shown in Fig. 2.

2.3 The Multi-resolution LBP Feature Fusion for the C_r Image

The success of Local Binary Patterns (LBP) [6] in face recognition is due to its robustness in terms of gray-level monotonic transformation. In practice, the face

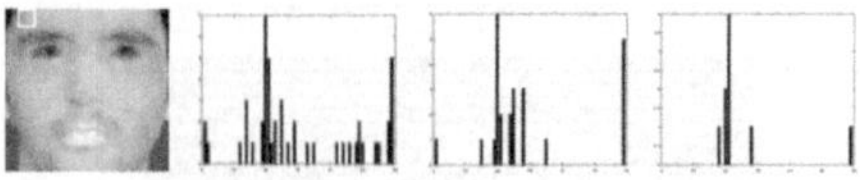

Fig. 4. A C_r image and three local histograms corresponding to the three scale operators: $LBP^{u2}_{8,1}$, $LBP^{u2}_{8,2}$, and $LBP^{u2}_{8,3}$, from a subwindow of 9×9 pixels

consists of uneven skin surface, which usually leads to nonmonotonic gray-level transformation. In this case, the performance of LBP degrades significantly, while Gabor kernel filters display excellent capabilities of resisting such variations. Compared with the R and Y images, the C_r image lacks the detailed information of skin surface, hence leads to less nonmonotonic gray-level transformation. Our research reveals that LBP outperforms Gabor filter in extracting discriminating features from the C_r image for face recognition.

In a 3×3 neighborhood of an image, the basic LBP operator assigns a binary label 0 or 1 to each surrounding pixel by thresholding at the gray value of the central pixel and replacing its value with a decimal number converted from the 8-bit binary number. Formally, the LBP operator is defined as follows:

$$LBP = \sum_{p=0}^{7} 2^p s(i_p - i_c) \tag{4}$$

where $s(i_p - i_c)$ equals 1, if $i_p - i_c \geq 0$; and 0, otherwise. Two extensions of the basic LBP were further developed [12]. The first extension allows LBP to deal with any size of neighborhoods by using circular neighborhoods and bilinearly interpolating the pixel values. The second extension defines the so called *uniform patterns*. When the binary string is considered circular, we can call LBP uniform if there are at most two bitwise transitions from 0 to 1 or vice versa. After extensions, LBP can be expressed as: $LBP^{u2}_{P,R}$, where P, R means P sampling points on a circle of radius R.

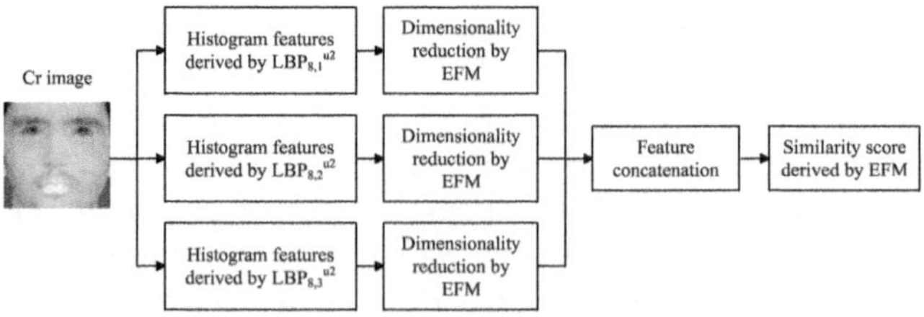

Fig. 5. Multiple resolution LBP feature fusion scheme

Note that the Gabor image representation encompasses the features corresponding to five scales for improving face recognition performance. Inspired by this idea, we combine the multiple-resolution information from the LBP operators. First, three LBP operators, $LBP^{u2}_{8,1}$, $LBP^{u2}_{8,2}$ and $LBP^{u2}_{8,3}$, are used to extract the multi-resolution histogram features from the C_r image. Fig. 4 shows three local histograms derived from three scale LBP operators in a subwindow of an image. It is easy to see that these histograms are complementary to one another. Smaller scale operators extract more detailed information (microstructure) and maintain the similar profile (macrostructure) as larger operators do. Second, three global histogram features are fused to form an augmented feature. One straightforward way is to concatenate the three global histograms, corresponding to $LBP^{u2}_{8,1}$, $LBP^{u2}_{8,2}$ and $LBP^{u2}_{8,3}$. However, this operation will result in the problem of high dimensionality. In this paper, we propose an LBP multiple-resolution feature fusion scheme, as shown in Fig. 5. For each global LBP histogram, the EFM is used to extract features and reduce dimensionality. Let $\mathbf{X}_{h_1}$, $\mathbf{X}_{h_2}$, and $\mathbf{X}_{h_3}$ be the reduced features after the EFM process. In particular, we first normalize and then concatenate the three reduced features and derive an augmented feature vector, $\mathbf{Y} = (\frac{\mathbf{X}_{h_1}-\mu_1}{\delta_1}; \frac{\mathbf{X}_{h_2}-\mu_2}{\delta_2}; \frac{\mathbf{X}_{h_3}-\mu_3}{\delta_3})$, where μ_i and δ_i are the mean and standard deviation of feature $\mathbf{X}_{h_i}$. By applying this fusion scheme, both the microstructures and the macrostructures of face image are utilized to extract the discriminating features, which contain much more face information than what a single LBP operator can provide.

2.4 The Component-Based DCT Multiple Face Encoding for the Q Image

In the YIQ color space, Y, I, and Q represent luminance, hue, and saturation, respectively. In terms of face image quality, the Q component image seems contain too much noise to be applied for face recognition by local feature extraction methods, such as GIR or LBP. The rough outlines of facial components in the Q image, such as nose, eyes and mouth, however, still contain valuable discriminating information. Based on this observation, we propose a holistic method to utilize the Q component image — the fusion of component-based DCT multiple face encoding.

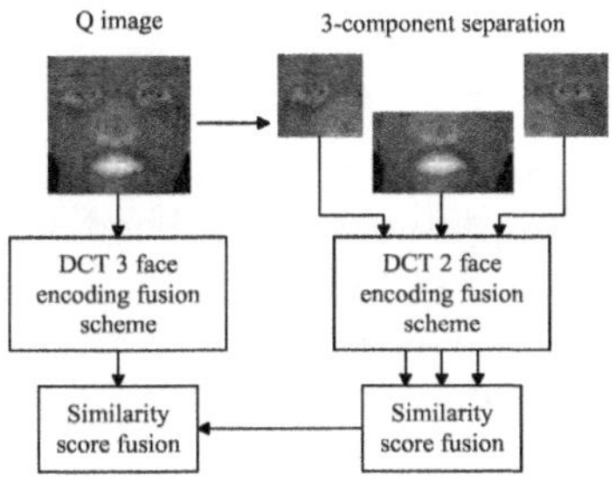

Fig. 6. Component-based DCT multiple face encoding fusion scheme

Component-based methods [13], [14] have been shown effective for improving face recognition performance, as the statistical variations caused by illumination and pose in each component image may be smaller than those in the whole face image [13]. We consider a simple seperation of three facial components shown in Fig. 6. As eyes and the vicinities have the most important discriminant information, they split up into the left eye component and the right eye component.

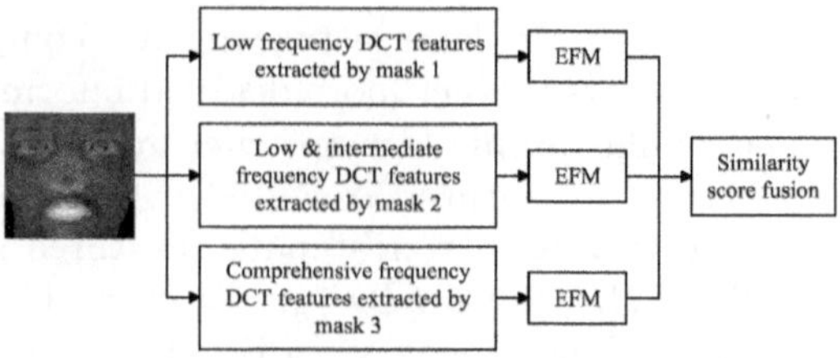

Fig. 7. DCT multiple face encoding fusion scheme

While the bottom half of face has weak discriminant capability, it is kept as an entity. Note that there are overlapping regions among the adjacent components, and each of three components is processed by the DCT 2 (using two DCT masks) face encoding fusion scheme to generate a similarity matrix. These three similarity matrices are further fused using the sum rule to generate a new similarity matrix.

The whole Q image is processed by the DCT 3 (using three DCT masks) face encoding fusion scheme to generate a similarity matrix. This similarity matrix is further fused with the similarity matrix derived using the facial components to derive the final similarity matrix.

The proposition of the DCT multiple face encoding fusion scheme is based on the observation that the reconstructed images of different DCT feature sets display different facial details. That is, these DCT feature sets are supposed to be complementary to each other. When fusing their classification outputs, the final classification result should be improved. Fig. 7 shows the outline of the three DCT face encoding fusion scheme for the whole Q image. Three masks, which are defined in the DCT domain shown in Fig. 3, are used to select three DCT feature sets. The decision of the mask size will be discussed in the experiment section.

3 Experiments

This section assesses the proposed method on a face recognition task using the FRGC version 2 Experiment 4 database [15]. The training set consists of 12,776 images that are either controlled or uncontrolled. The target set includes 16,028 controlled images and the query set contains 8,014 uncontrolled images. The sizes of images used in our experiments are 128×128 R images for Gabor feature extraction, and 64×64 C_r and Q images.

3.1 Effectiveness of the RC_rQ Hybrid Color Space

To evaluate the effectiveness of the new hybrid color space RC_rQ, we first conduct experiments on the Y, C_r, Q and R component images by applying the EFM with the 1,000 features and the cosine similarity measure. In particular, the R and Y images have the same face region as the C_r and Q images shown in Fig. 1. The size of all images is 64×64. At the decision level, prior to the similarity fusion, all similarity matrices are normalized using the z-score normalization.

Table 1. Face verification rates at 0.1% false accept rate of color component images

FVR (ROC III) at 0.1% FAR				
Y	C_r	Q	R	RC_rQ
56.41%	57.17%	53.97%	62.12%	79.98%

Table 2. Face verification rates at 0.1% false accept rate using the R images. Fusion I means the fusion of the four patches, and Fusion II means fusing the results of the whole GIR image and the Fusion I.

FVR (ROC III) at 0.1% FAR						
Scale 1&2	Scale 2&3	Scale 3&4	Scale 4&5	Fusion I	Whole GIR	Fusion II
68.27%	72.56%	73.99%	69.61%	82.81%	80.81%	84.12%

The results derived from the ROC III curves are listed in Table 1, which clearly indicates that the R component image carries much more discriminant information than the grayscale image (Y). Furthermore, by fusion at the decision level, the RC_rQ hybrid color space boosts the face recognition performance significantly. Although the C_r and Q images have lower than 60% verification rates, the overall verification performance is improved significantly when they are combined with the R component image. Therefore, the RC_rQ hybrid color space constitutes an excellent platform, from which one can focus on improving performance of each component image and expect to achieve good final performance by fusing their results.

3.2 Effectiveness of the Proposed Method

We first conduct experiments on R images using the proposed patch-based Gabor image representation. To alleviate the effect of illumination variations, an illumination normalization procedure [4] is applied to the R images. The illumination normalization is not applied to the C_r and Q images, because the irregular intensity values in C_r and Q images usually lead to unstable illumination normalization results that compromise face recognition performance.

When the GIR is considered as a whole, we reduce the dimensionality using the DCT domain mask defined in Fig. 3. As different Gabor kernel sacles contribute to the recognition performance differently, the size of masks is empirically chosen as follows to comply with such a characteristic: 8×8, 14×14, 17×17, 19×19, and 20×20, corresponding to scales 1 to 5, respectively. The resulting feature vector with size 10,480 is then processed by EFM (m = 1,100), producing the FVR (ROCIII) of 80.81% at FAR of 0.1%. For each of the GIR patch images, the 64×64 DCT features are chosen via masking. EFM, with m = 1,350, is used to classify these features. The face verification results using the R component image are listed in Table 2.

We then conduct experiments on C_r image using LBP features. To extract the LBP features, we divide a face image of 64×64 into 144 (12*12) overlapping windows of 9×9 pixels (3 pixel overlapping). The EFM (m = 1,300) is used

Table 3. Face verification rates at 0.1% false accept rate of C_r image

FVR (ROC III) at 0.1% FAR			
$LBP^{u2}_{8,1}$	$LBP^{u2}_{8,2}$	$LBP^{u2}_{8,3}$	Fusion
61.04%	62.63%	53.84%	69.38%

Table 4. Face verification rates at 0.1% false accept rate of Q image. LE and RE mean the left eye component and the right eye component.

<table>
<tr><td>Image</td><td colspan="3">Holistic</td><td colspan="2">LE</td><td colspan="2">RE</td><td colspan="2">Bottom half</td></tr>
<tr><td>Mask</td><td>$\mathbf{M}_{15\times15}$</td><td>$\mathbf{M}_{26\times26}$</td><td>$\mathbf{M}_{64\times64}$</td><td>$\mathbf{M}_{15\times15}$</td><td>$\mathbf{M}_{39\times39}$</td><td>$\mathbf{M}_{15\times15}$</td><td>$\mathbf{M}_{39\times39}$</td><td>$\mathbf{M}_{15\times15}$</td><td>$\mathbf{M}_{39\times64}$</td></tr>
<tr><td>Feature #</td><td>223</td><td>600</td><td>770</td><td>223</td><td>510</td><td>223</td><td>510</td><td>223</td><td>650</td></tr>
<tr><td rowspan="4">FVR
(ROC III)</td><td>41.48%</td><td>53.12%</td><td>54.46%</td><td>28.97%</td><td>32.79%</td><td>34.25%</td><td>36.41%</td><td>26.55%</td><td>32.97%</td></tr>
<tr><td colspan="3" rowspan="2">56.17%</td><td colspan="2">33.59%</td><td colspan="2">38.34%</td><td colspan="2">33.40%</td></tr>
<tr><td colspan="6">54.81%</td></tr>
<tr><td colspan="9">58.83%</td></tr>
</table>

Table 5. Comparison of the proposed method with the state-of-the-art

FVR (ROC III) at 0.1% FAR			
Method in [4]	Method in [3]	Method in [16]	Our method
83.6%	85.8%	87.5%	92.4%

to derive the discriminating features from each of three scale LBP histograms. After concatenation, another EFM ($m = 390$) is used to process the augmented feature vector. The face verification results using the C_r component image are listed in Table 3.

We finally conduct experiments on Q images using the proposed component-based DCT multiple face encoding. In particular, for the Q image of size 64×64, the sizes of the left eye, the right eye region and the bottom half of face components are 39×39, 39×39, and 39×64, respectively. For the DCT multiple face encoding, the selection of the DCT mask sizes affects the performance after fusion. As the number of subjects of training data is 222, the rank of the between-class scatter matrix is at most 221. In order to derive the 221 EFM features, the input feature vector should reside in a space whose dimensionality is larger than 221. We therefore choose $\mathbf{M}_{15\times15}$ as the smallest mask. The selection of the largest mask is determined by the size of images. For the holistic image, the size of the middle mask is empirically chosen to contain the low and intermediate frequencies. This size is 26×26 in our experiments. The experiment results are given in Table 4, where the numbers of features used by EFM are also included.

After generating three similarity matrices corresponding to the three component images in the RC_rQ color space, we fuse them by means of weighted sum. In our experiments, we empirically set the weights to 1.0, 0.6, and 0.8, respectively, based on the different roles of the R, C_r, and Q component images for face recognition. The final face verification results are shown in Table 5. Specifically, our proposed method, which achieves the face verification rate of 92.43% at the false accept rate of 0.1%, performs better than the state-of-the-art face recognition methods, such as [4] with the face verification rate of 83.6%, [3] with the face verification rate of 85.8%, and [16] with the face verification rate of 87.5% at the same false accept rate using the same face image database.

References

1. Turk, M., Pentland, A.: Eigenfaces for recognition. Journal of Cognitive Neuroscience 13, 71–86 (1991)
2. Belhumeur, P.N., Hespanha, J.P., Kriegman, D.J.: Eigenfaces vs. Fisherfaces: Recognition using class specific linear projection. IEEE Trans. Pattern Analysis and Machine Intelligence 19, 711–720 (1997)
3. Su, Y., Shan, S., Chen, X., Gao, W.: Hierarchical ensemble of global and local classifiers for face recognition. In: Proc. IEEE International Conference on Computer Vision (ICCV 2007) (2007)
4. Tan, X., Triggs, B.: Fusing gabor and lbp feature sets for kernel-based face recognition. In: Zhou, S.K., Zhao, W., Tang, X., Gong, S. (eds.) AMFG 2007. LNCS, vol. 4778, pp. 235–249. Springer, Heidelberg (2007)
5. Liu, C., Wechsler, H.: Gabor feature based classification using the enhanced Fisher linear discriminant model for face recognition. IEEE Trans. on Image Processing 11, 467–476 (2002)
6. Ahonen, T., Hadid, A., Pietikainen, M.: Face description with local binary patterns: Application to face recognition. IEEE Trans. Pattern Analysis and Machine Intelligence 28 (2006)
7. Kittler, J., Hatef, M., Robert, P., Matas, J.: On combining classifiers. IEEE Trans. Pattern Analysis and Machine Intelligence 20, 226–239 (1998)
8. Liu, C., Wechsler, H.: Robust coding schemes for indexing and retrieval from large face databases. IEEE Trans. on Image Processing 9, 132–137 (2000)
9. Gonzalez, R., Woods, R.: Digital Image Processing, 2nd edn. Prentice-Hall, Englewood Cliffs (2002)
10. Shih, P., Liu, C.: Comparative assessment of content-based face image retrieval in different color spaces. International Journal of Pattern Recognition and Artificial Intelligence 19, 873–893 (2005)
11. Daugman, J.: Uncertainty relation for resolution in space, spatial frequency, and orientation optimized by two-dimensional cortical filters. J. Optical Soc. Am. 2, 1160–1169 (1985)
12. Ojala, T., Pietikainen, M., Maenpaa, T.: Multiresolution gray-scale and rotation invariant texture classification with local binary patterns. IEEE Trans. Pattern Analysis and Machine Intelligence 24, 971–987 (2002)
13. Kim, T., Kim, H., Hwang, W., Kittler, J.: Component-based lda face description for image retrieval and mpeg-7 standardisation. Image and Vision Computing 23, 631–642 (2005)
14. Heisele, B., Serre, T., Poggio, T.: A component-based framework for face recognition and identification. International Journal of Computer Vision 74, 167–181 (2007)
15. Phillips, P., Flynn, P., Scruggs, T., Bowyer, K., Chang, J., Hoffman, K., Marques, J., Min, J., Worek, W.: Overview of the face recognition grand challenge. In: Proc. IEEE Conference on Computer Vision and Pattern Recognition (2005)
16. Kumar, V., Savvides, M., Xie, C.: Correlation pattern recognition for face recognition. Proceesings of the IEEE 94, 1963–1976 (2006)

Face Age Classification on Consumer Images with Gabor Feature and Fuzzy LDA Method

Feng Gao and Haizhou Ai

Computer Science and Technology Department,
Tsinghua University, Beijing 1000084, China
gaofeng03@mails.tsinghua.edu.cn, ahz@mail.tsinghua.edu.cn

Abstract. As we all know, face age estimation task is not only challenging for computer, but even hard for human in some cases, however, coarse age classification such as classifying human face as baby, child, adult or elder people is much easier for human. In this paper, we try to dig out the potential age classification power of computer on faces from consumer images which are taken under various conditions. Gabor feature is extracted and used in LDA classifiers. In order to solve the intrinsic age ambiguity problem, a fuzzy version LDA is introduced through defining age membership functions. Systematic comparative experiment results show that the proposed method with Gabor feature and fuzzy LDA can achieve better age classification precision in consumer images.

Keywords: Age classification, Gabor feature, membership functions, fuzzy LDA.

1 Introduction

As one of the main human facial attributes, aging plays a more complex role than other factors such as human identity, expression, gender or race. The progress of human aging is uncontrollable, with many internal and external influence factors such as one's health state, lifecycle and extreme weather conditions. Besides, because age is a temporal property of people, it's hard to collect the same person's face image across ages. It's also tedious and laborious to label the exact or approximate ages of collected faces. Due to these present difficulties, researches on human age are not as much as that on other face attributes.

However, researches on age progression and estimation have large potential in many applications, e.g. homeland security, parental control, age based Human-Computer interaction, passports renewal and finding missing individuals, and in particular face retrieval over internet or large scale face image database that is our targeted application area. In literature, different aspects of age progression are researched, including building complex models to predicate or simulate one's facial appearance in future [1], estimating or classifying the age of a given face image [2][4][5][6], age progression modeling to alleviate performance drop in face recognition [3][7].

Our work focus on the face age classification problem, the most related work with ours is Fu et al.'s work on age estimation [5] and Yang et al.'s work on age classification [6]. Fu et al. [5] applied linear dimensional reduction algorithm to map human

M. Tistarelli and M.S. Nixon (Eds.): ICB 2009, LNCS 5558, pp. 132–141, 2009.
© Springer-Verlag Berlin Heidelberg 2009

faces from pixel intensity space to a smaller space, which facilitated the following age regression using quadratic function or SVR [4], they achieved an average error of about 5 years in age estimation on a private large database. However, the age estimation variation across ages is large, without any result on coarse age classification task. Yang et al. used LBP feature and AdaBoost algorithm to construct a classifier to classify face as one of the three coarse categories: child, adult and old people. His training set is built from snapshot faces of Asian people taken under constraint illumination condition and without any expression and pose variations, thus the age classifier has limited performance in consumer images.

In our work, we partition the age into four categories, which are baby (0 to 1 approximately), child (2 to 16 approximately), adult (17 to 50 approximately), and old (after 50 approximately). We collected thousands of frontal or near frontal face images as training set from consumer images, there are variations in illumination and expression among those faces. One thing to mention, the age labeled in our collected data set is subjective one, not the objective one due to lack of the exact age information. Then we use Gabor features [9] as face representation and linear discriminant analysis (LDA) [10] to construct the final age classifier, achieving as much as 91% in precision on test set. Besides, to cope with the age ambiguity, we also employed the concept of fuzzy LDA classification by defining fuzzy age membership functions which not only utilize faces with vague ages to enlarge the training set, but also boost the estimation precision to a higher level significantly.

The rest of the paper is organized as follows: section 2 describes the Gabor features used for age classification; in section 3 fuzzy version LDA is presented; section 4 gives systematic comparative experiment results on a large consumer image dataset; and section 5 draws the conclusion.

2 Gabor Features

Gabor features [9] are popular in face representation; its effectiveness has been proved by many researches in fields like face recognition [9]. In face age classification, we also choose to use Gabor features, and in particular we extract Gabor features of 3 scales and 4 orientations that amount to 12 convolved face images of which only magnitude images are used as raw features as shown in Figure1. PCA is used to raw features for dimension reduction.

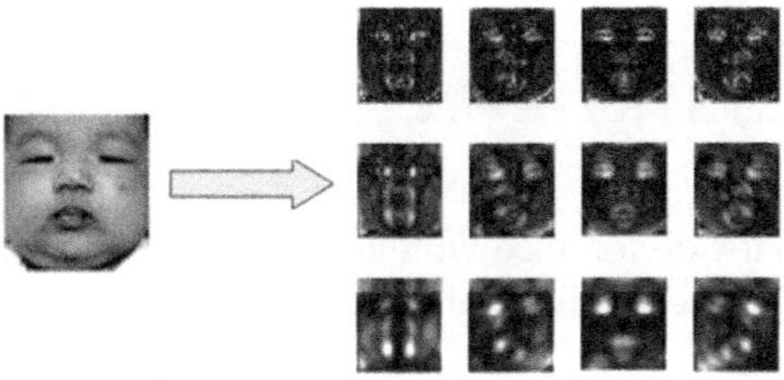

Fig. 1. The Gabor magnitude images of a baby face

Gabor filters are defined as below:

$$\psi_{\mu,\nu}(z) = \frac{\|k_{\mu,\nu}\|^2}{\sigma^2} e^{\left(-\|k_{\mu,\nu}\|^2 \|z\|^2 / 2\sigma^2\right)} \left[e^{ik_{\mu,\nu}z} - e^{-\sigma^2/2}\right]$$

where $k_{\mu,\nu} = k_\nu e^{i\phi\mu}$, $k_{\mu,\nu} = k_{\max} / \lambda^\nu$, $\phi_\mu = \pi\mu / 8$.

3 Fuzzy LDA

Linear discriminant analysis (LDA) as a classical dimension reduction method aims to find out optimal project directions to maximize the ratio of the between-class scatter and the within-class scatter. After finding the projected directions, data can be mapped to a low-dimensional subspace, and the nearest class center criteria can be used for classification.

In LDA method, each training sample is assigned to one class label exactly, this is easy in many other classification problems, because these classes have been clearly defined. But in age classification, mapping ages to age groups is somewhat intrinsically ambiguous. For example, it's usually quite easy to judge a face which is 6 years old as child, but how about the one with the label of 14, or what age group is appropriate for a 50 years old man. With labeled faces at hand, we need a way to map an age label to an appropriate age group to build a training set. There are three kinds of way to this goal: 1) Drop some vague ages; 2) Assign every age to only one age group. 3) Assign every age to age groups with the help of fuzzy age membership functions. In fact, the above three can all be expressed through age membership functions based on fuzzy mathematics. With the fuzzy age membership functions introduced, we modified the LDA method to a fuzzy version to suit the age classification problem.

3.1 Age Membership Functions

In order to cope with age ambiguity, we defined an age membership function $\mu_i(x)$ as:

$$0 \le \mu_i(x) \le 1, \quad i \in \{1,2,3,4\}, \quad 0 \le x, \quad \sum_{i=1}^{4} \mu_i(x) = 1$$

It describes to what extent a face with an age label x is a member of *i-th* age group. Note that the age ambiguity always happens at two adjacent age group's boundary, each age can belong to at most two age groups, therefore we elaborately designed the 3rd kind of age membership functions , as well as the other two non-fuzzy kind of membership functions as follows and their curves are shown in Figure 2.

(1) The 1st kind of age membership functions when some vague ages are dropped

$$\mu_1(x)=\begin{cases}1 & x=0\\ 0 & else\end{cases};\quad \mu_2(x)=\begin{cases}1 & 3\le x\le 12\\ 0 & else\end{cases};$$

$$\mu_3(x)=\begin{cases}1 & 20\le x\le 40\\ 0 & else\end{cases};\quad \mu_4(x)=\begin{cases}1 & x\ge 60\\ 0 & else\end{cases}$$

(2) The 2nd kind of age membership functions when every age is assigned to exactly one age group

$$\mu_1(x)=\begin{cases}1 & x=0\\ 0 & else\end{cases};\quad \mu_2(x)=\begin{cases}1 & 1\le x\le 17\\ 0 & else\end{cases};$$

$$\mu_3(x)=\begin{cases}1 & 18\le x\le 55\\ 0 & else\end{cases};\quad \mu_4(x)=\begin{cases}1 & x\ge 56\\ 0 & else\end{cases}$$

(3) The 3rd kind of fuzzy age membership functions

$$\mu_1(x)=\begin{cases}1 & x=0\\ 0.8 & x=1\\ 0.5 & x=2\\ 0 & else\end{cases};\quad \mu_2(x)=\begin{cases}0.2 & x=1\\ 0.5 & x=2\\ 1 & 3\le x<12\\ 1-1.5(\frac{x-12}{8})^2 & 12\le x<16\\ 1.5(\frac{x-20}{8})^2 & 16\le x<20\\ 0 & else\end{cases}$$

$$\mu_3(x)=\begin{cases}1.5(\frac{x-12}{8})^2 & 12\le x<16\\ 1-1.5(\frac{x-20}{8})^2 & 16\le x<20\\ 1 & 20\le x<36\\ 1-1.2(\frac{x-36}{24})^2 & 36\le x<48\\ 1.2(\frac{x-60}{24})^2 & 48\le x<60\\ 0 & else\end{cases};\quad \mu_4(x)=\begin{cases}1.2(\frac{x-36}{24})^2 & 36\le x<48\\ 1-1.2(\frac{x-60}{24})^2 & 48\le x<60\\ 1 & x\ge 60\\ 0 & else\end{cases}$$

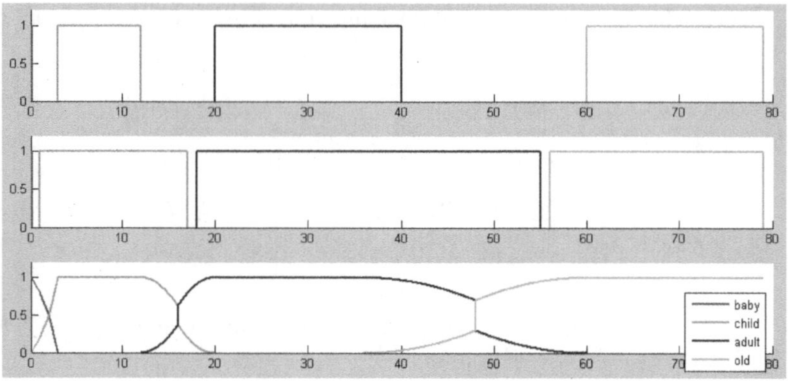

Fig. 2. Three kinds of fuzzy age membership functions: on the top, only non-vague ages are assigned to a certain age group, while in the middle, all ages are belonging to exactly one group, at the bottom is the more natural one being fuzzy because it takes into account of the age ambiguity

3.2 Fuzzy LDA Method

With age membership functions defined, they can be used in LDA by way of using age membership functions as class weighting values as follows:

$$S_B = \sum_{i=1}^{c} N_i^{'}(m_i - m)(m_i - m)^T,\ S_w = \sum_{j=1}^{C}\sum_{i=1}^{N} \mu_j(x_i)(x_i - m_j)(x_i - m_j)^T$$

where $m = \frac{1}{N}\sum_{i=1}^{N} x_i$, $m_i = \sum_{i=1}^{N}\mu_j(x_i)x_i \Big/ \sum_{i=1}^{N}\mu_j(x_i)$, $N_j^{'} = \sum_{i=1}^{N}\mu_j(x_i)$, N is the total number of training instances. As conventional LDA, the optimal projections are defined as below.

$$W_{opt} = \arg\max_{W} \frac{\left|W^T S_B W\right|}{\left|W^T S_W W\right|} = \left[w_1, w_2, \ldots w_m\right]$$

where $\left[w_1, w_2, \ldots w_m\right]$ is the set of generalized eigenvectors of $S_W^{-1} S_B$.

4 Experimental Results

A face image data set is collected from Internet, and divided into a training dataset and a test dataset. Using face detector and face alignment tool, these faces are automatically cropped and normalized in grey level and geometry as in [6], and each face is manually labeled with an age value estimated by human subjectively. The final training dataset consists of 5408 faces of 64 by 64 in resolution and the labeled ages range from 0 to 79 years old, while the test dataset consists of 57 babies, 350 children, 492 adults and 79 old people that amount to totally 978 photos. For the four-class classification, faces in the training dataset will be assigned to age groups according to their labeled age. Due to every face has an age value labeled, not only it can be assigned a certain age group according to the class membership functions, but also it can provide necessary data for building an age regression model. In our experiments, four aspects are approached: 1. what feature is most effective in age classification? 2. Does Fuzzy LDA help in improving age classification precision? 3. What's the performance of other classification methods? 4. Does age regression help in age classification?

4.1 Comparative Experiment on Different Features

Besides Gabor feature, we also used two other features: pixel intensity and LBP [8]. PCA is used for dimension reduction as preprocessing. We use the 1st kind of age membership functions defined in section 3.1 to label the whole collected face image set with age group attribute, which in fact is a subset of the original dataset containing 644 babies, 1427 children, 1691 adults and 1025 elder people by excluding some vague ages for better separation. The conventional LDA is used to extract a 3D discriminative feature space for each kind of feature. Figure 3 visualized the training data. It is clearly that Gabor feature is more discriminative than the other two. And the quantitative results in Table 1 (using the test set mentioned above) give more support to Gabor feature in its discriminant power and the generalization performance on the age classification problem.

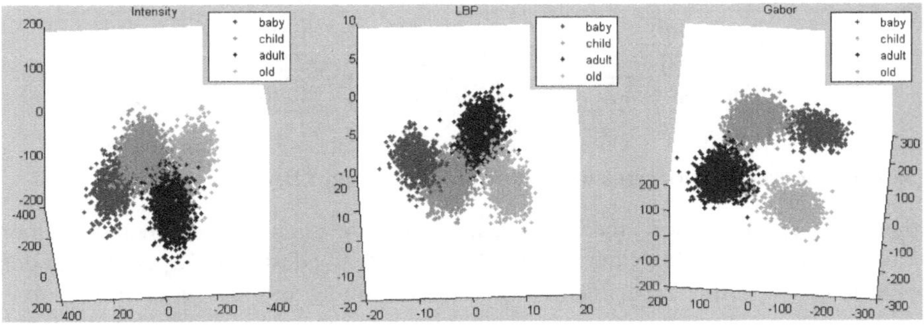

Fig. 3. The training data visualization in discriminative feature space (left for intensity feature, middle for LBP feature, right for Gabor feature)

Table 1. Age classification precision using various features

Dataset	Training set			Test set		
feature	intensity	LBP	Gabor	intensity	LBP	Gabor
Baby	95.34%	97.67%	98.29%	91.07%	96.43%	94.74%
Child	94.81%	97.06%	98.95%	89.14%	79.71%	92.57%
Adult	95.86%	96.87%	98.94%	88.38%	88.18%	91.46%
Old	95.12%	96.49%	99.71%	70.51%	75.64%	78.48%
Total precision	95.32%	96.95%	99.02%	87.39%	84.64%	91.00%

4.2 Comparative Experiments on Fuzzy LDA

In this experiment, we only use the Gabor feature which has been proved to be more effective than other features. We carry on our experiments using three kinds of membership functions defined in section 3.1, notice that the fuzzy LDA method using the 1st kind of age membership functions is equivalent to the LDA method with four explicit classes. Table 2 gives the result. By using the 2nd kind of age membership functions, although the training set is enlarged, the added faces are assigned to exactly one age group, the prediction performance on test set is dropped, especially for the baby class. While using the 3rd kind of age membership functions which are more smooth and natural, the prediction performance get improved to 92.54%, especially with old people class boosting from 78.48% to 88.61% and the children class from 92.57% to 94.57%, and the precision doesn't drop for adult and baby. From this result, it is obvious that the introduction of fuzzy LDA helps to utilize the face data with vague ages and improves the age classification generalization performance. In Figure 5 it shows some demo pictures containing the age classification result using our fuzzy LDA method with Gabor features.

Table 2. Comparative result using different age membership functions

Membership functions	(1)	(2)	(3)
Baby	94.74%	61.40%	94.74%
Child	92.57%	94.57%	94.57%
Adult	91.46%	90.04%	91.46%
Old	78.48%	77.22%	88.61%
Total precision	91.00%	88.96%	92.54%

4.3 Comparative Experiments with SVM, AdaBoost and LDA

Other than LDA method, there are many effective classification methods among which Support Vector Machines (SVM) [13] and Real AdaBoost [12] are two popular representatives. In this experiment, on the same training and test set as in section 4.1

Table 3. Comparative result using SVM, AdaBoost and LDA

Method	SVM		AdaBoost		LDA
Feature	One-vs-All (Gabor)	One-vs-Another (Gabor)	Haar	Gabor	Gabor
Baby	78.95%	85.96%	82.46%	75.44%	94.74%
Child	91.43%	91.71%	80.86%	71.14%	92.57%
Adult	88.41%	88.82%	80.08%	81.71%	91.46%
Old	86.08%	89.87%	73.42%	70.89%	78.48%
Total precision	88.75%	89.78%	79.96%	76.69%	91.00%

we report their results on age classification. For multi-class classification, we use a binary tree structure classifier as in [6] in which each node is trained by Real AdaBoost, and also pair-wise SVM classifiers using polynomial kernel trained with the "One-vs-Another" strategy that vote for the final decision [15] which got the best performance among different kernels including linear and Gaussian and also with the "One-vs-All" strategy [15]. In both methods Gabor features are used, and for AdaBoost we also used Haar [14] feature. Table 3 shows the results, in which we can see the LDA is comparative with the SVM, and it is much better than the AdaBoost in test set. For accuracy and efficiency, the LDA method is preferred in our age classification problem.

4.4 Experiment on Age Regression

Fu et al.'s work [4] on age estimation resulted in about 5 years old of mean absolute error. However, in their paper, the curve of the mean absolute errors in each age is heavily vibrating, with large errors in some ages even more than 10 years old. That means the precision of age estimation by regression has great variation over ages, together with a mean absolute error that is not so small for age classification requirement. Therefore we are not sure whether age regression method could help in age classification.

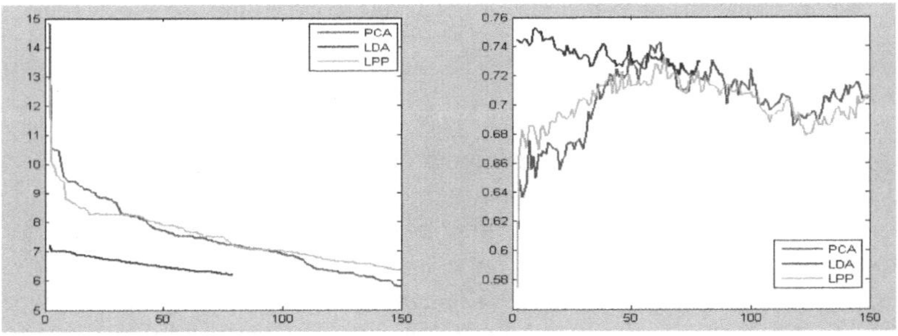

Fig. 4. Left side shows the age error in a 5-fold cross variation set, and right side shows the age classification precision in test set. The x-coordinate of both figures indicates the reduced dimension, and from left to right, the y-coordinate represents mean age error and classification precision respectively.

To be sure, we implemented the age regression method in which Gabor features are used as face representation, and then PCA, LDA and LPP (Locality Preserving Projections [11]) are used for linear dimensional reduction respectively and then Quadratic function [5] is used for age regression. Result in Figure 4 shows that the least age absolute error is 5.8061 and the best age classification precision on the test set is 75.26%, which is not comparable with that using the Fuzzy LDA. The reason is perhaps that variation related to age is too complicated to build a unified age estimation model. With the limited accuracy and vibrated performance, regression approaches for coarse classification task do not make sense.

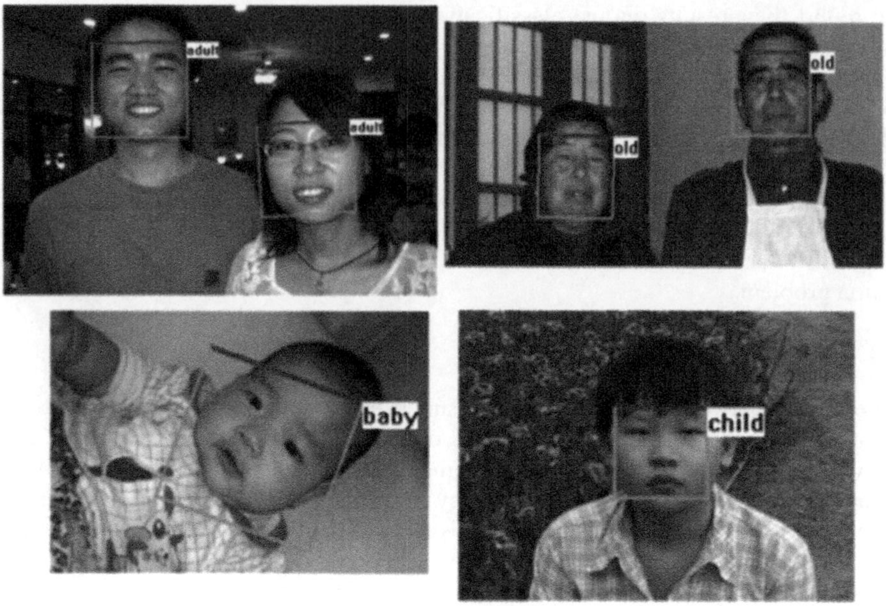

Fig. 5. Some age classification results using Gabor feature and Fuzzy LDA

5 Conclusion

In this paper, by introducing fuzzy membership functions, we proposed a Fuzzy LDA method using Gabor features for coarse age classification. Comparative experiments on different features and different age membership functions show that Gabor feature outperforms other features like pixel intensity and LBP, and the Fuzzy LDA can improve the classification precision even further. Besides, comparative experiments using SVM, AdaBoost and LDA have been done, to show that the LDA method works better for age classification. In addition, we have proved that for age classification it is more effective to use a discriminant classification method rather than to build a unified age regression model.

Acknowledgement

This work is supported by National Science Foundation of China under grant No.60673107, and it is also supported by a grant from HP Corporation.

References

1. Suo, J., Min, F., Zhu, S., Shan, S., Chen, X.: A Multi-Resolution Dynamic Model for Face Aging Simulation. In: CVPR 2007 (2007)
2. Geng, X., Zhou, Z.-H., Smith-Miles, K.: Automatic Age Estimation Based on Facial Aging Patterns. IEEE Transactions on Pattern Analysis and Machine Intelligence 29(12) (December 2007)

3. Wang, J., Shang, Y., Su, G., Lin, X.: Age simulation for face recognition. In: ICPR 2006 (2006)
4. Guo, G., Fu, Y., Dyer, C.R., Huang, T.S.: Image-Based Human Age Estimation by Manifold Learning and Locally Adjusted Robust Regression. IEEE Transactions on Image Processing 17(7) (July 2008)
5. Fu, Y., Huang, T.S.: Human Age Estimation With Regression on Discriminative Aging Manifold. IEEE Transactions on Multimedia 10(4) (June 2008)
6. Yang, Z., Ai, H.: Demographic classification with local binary patterns. In: Lee, S.-W., Li, S.Z. (eds.) ICB 2007. LNCS, vol. 4642, pp. 464–473. Springer, Heidelberg (2007)
7. Ling, H., Soatto, S., Ramanathan, N., Jacobs, D.W.: A Study of Face Recognition as People Age. In: ICCV 2007 (2007)
8. Ahonen, T., Hadid, A., Pietikäinen, M.: Face recognition with local binary patterns. In: Pajdla, T., Matas, J. (eds.) ECCV 2004. LNCS, vol. 3021, pp. 469–481. Springer, Heidelberg (2004)
9. Liu, C., Wechsler, H.: Gabor Feature Based Classification Using the Enhanced Fisher Linear Discriminant Model for Face Recognition. IEEE Transactions on Image Processing 11(4) (April 2002)
10. Belhumeur, P.N., Hespanha, J.P., Kriegman, D.J.: Eigenfaces vs. Fisherfaces: Recognition Using Class Specific Linear Projection. In: Buxton, B.F., Cipolla, R. (eds.) ECCV 1996. LNCS, vol. 1064. Springer, Heidelberg (1996)
11. He, X., Niyogi, P.: Locality Preserving Projections. In: NIPS 2004 (2004)
12. Schapire, S.: Improved Boosting Algorithms Using Confidence-rated Predictions. Machine Learning 37(3), 297–336 (1999)
13. Vapnik, V.N.: The Nature of Statistical Learning Theory. Springer, Heidelberg (1995)
14. Viola, P., Jones, M.: Rapid Object Detection using a Boosted Cascade of Simple Features. In: CVPR 2001 (2001)
15. Zhuang, L., Ai, H., Xu, G.: Training Support Vector Machines for Video Based Face Recognition. In: International Conf. on Image and Graphics. SPIE, vol. 4875, pp. 737–743 (2002)

The Big Brother Database: Evaluating Face Recognition in Smart Home Environments

Annalisa Franco, Dario Maio, and Davide Maltoni

C.d.L. Scienze dell'Informazione, Università di Bologna, via Sacchi 3, Cesena, Italy
DEIS – Viale Risorgimento, 2 – Bologna, Italy
{annalisa.franco,dario.maio,davide.maltoni}@unibo.it

Abstract. In this paper a preliminary study on template updating techniques for face recognition in home environments is presented. In particular a new database has been created specifically for this application, where the face images acquired are characterized by a great variability in terms of pose and illumination but the number of subjects is quite limited and a large amount of images can be exploited for intensive incremental learning. The steps of database creation and the characteristics of the data collected are described in detail. We believe such a database could be very useful to develop and optimize face recognition approaches for smart home environments. Moreover some preliminary results on incremental learning are provided and analyzed to evaluate the effects of incremental template updating on the recognition performance.

1 Introduction

Face is one of the most studied biometric characteristics and a huge literature exists on the topic [9] [14]. In recent years the attention of the scientific community has been focused on face recognition in real applications where the operating conditions are not optimal due to several factors such as the low quality of the acquired images and variable conditions of pose and illumination. The results of recent evaluations [12] prove that state-of-the-art algorithms perform well in controlled environments but still poorly work in the presence of critical conditions.

The focus of this paper is on a particular kind of applications, referred to as *smart home environment* in the literature, where some intelligence mechanisms personalize the environment (home, office, etc.) based on the presence of one or more subjects in the room. This scenario is characterized by specific constraints: the resolution of the images is typically low because they are usually acquired by low cost surveillance cameras that can be far from the subjects and the pose and illumination conditions are extremely variable since the subjects are observed during their daily activities. On the other hand the smart home environment offers some advantages: the number of users is usually limited and a huge amount of data can be continuously collected and exploited for incremental learning: in fact, the availability of so many images allows to progressively update and refine the user's template, initially created on the basis of a few training images. This continuous learning may help to better deal with the intrinsic variability of the face, the changes of look, and with the aging affect that, as well known, can significantly affect recognition accuracy.

M. Tistarelli and M.S. Nixon (Eds.): ICB 2009, LNCS 5558, pp. 142–150, 2009.
© Springer-Verlag Berlin Heidelberg 2009

In order to develop, test and optimize face recognition and template updating approaches for this particular context, specific datasets reflecting the above mentioned characteristics are needed: a huge set of images of a limited number of subjects, acquired daily for a long period while the subjects normally perform their daily activities. Actually several face databases are publicly available [9] [14], and some of them include images captured under variable pose and illumination conditions or in different sessions that allow to evaluate the aging effects on face recognition performance. For example, the BANCA [1] or the Notre Dame Human ID [4] databases have been acquired over a period of about 3 months, but the acquisition conditions are rather controlled and/or the number of different acquisition sessions is limited. To the best of our knowledge, none of the available datasets meets all the specified requirements. These motivations induced us to create a new ad-hoc database of images that we intend to make available to the scientific community. Acquiring from scratch such a large dataset is objectively difficult and very time consuming, so we decided to exploit an existing data source: the full video of one edition of the "Big Brother" reality show. A software tool has been developed to analyze the videos and to quickly detect and label the face images. The rest of this paper describes the database structure and its creation procedure: in particular the steps carried out and the software tools used are described in order to provide useful indications to create similar databases. Moreover some experimental results on template updating are given to evaluate the effects of incremental learning on the recognition performance.

The work is organized as follows. In section 2 the procedure for data extraction and labeling is presented; a detailed description of the database is reported in section 3 together with an analysis of the image quality, in section 4 some preliminary results on template updating are reported and, finally, in section 5 some concluding remarks are given.

2 Database Construction

The database has been created starting from the 2 DVDs made commercially available at the end of the 2006 "Big Brother" reality show produced for the Italian TV [7]. The videos contained in the DVDs summarize the 99 days of permanence of 20 participants in the house. In particular they represent the main events occurred (participants entering or leaving the house, loves, quarrels, weekly trials, etc), some of the participants' auditions and a brief presentation of all the participants, their work in the normal life, their family, etc. Most of the videos are labeled with the day of recording: this allowed us to organize the frames extracted from the videos in different subsets, each related to a specific day.

The steps carried out for the creation of the database are described in the following.

Conversion of the DVD video files to AVI files. A simple free conversion tool [3] was used to this purpose. Each video track was converted to an AVI file with the following settings: output frame rate 29.97, output size 720×576, video codec Indeo® Video 5.10. The number of AVI files extracted is 22, about 15 GB in total.

Frames extraction. About 3 frames per second have been extracted from each AVI file using the free software Image Grabber II [8]. The frames extracted (more than

42,000, ~3.3GB in jpg format) have been initially classified according to the day of recording.

Face detection and labeling. A software tool (see Fig. 1) has been developed to analyze the sequences of frames obtained in the previous steps. The tool provides the following functionalities:

- *Face detection.* This operation is fully automatic and consists of checking each image to verify the presence of one or more faces; this is performed by means of the face detector implemented in the OpenCv library [11], based on the AdaBoost approach proposed by Viola and Jones in [13]. A default classifier provided with the OpenCV library (haarcascade_frontalface_alt.xml) and trained for frontal images, has been used in this work.
- *Image labeling.* This procedure is semi-automatic and allows to assign an ID to the detected faces. The tool partially automates the operation, based on the idea that, within the same scene, the identity of the faces detected in similar positions does not change. A simple scene change detection method, based on the difference between subsequent images, has been implemented. At the beginning of a new scene the user is required to assign (by pointing and mouse clicking) the correct ID to the detected faces. In the subsequent frames of the same scene, if a face is detected approximately in the same position of a previously labeled face and the size of the two face windows is similar, the same ID is proposed. In case a new face is detected within an ongoing scene, the human intervention is required to assign the correct ID. When a scene change occurs, the labeling system is re-initialized. The false positives of the face location module can be simply ignored (i.e., no ID is assigned). The use of this tool allows to process/label a large number of frames in an acceptable amount of time.
- *Image storing.* The detected faces, independently of the original dimensions, are resized to 70×70 pixels and stored. The result is a set of very heterogeneous images in terms of quality since some faces, taken when the subject is close to the camera, have a higher resolution with respect to those normally acquired in the house. It is worth noting that only the false positives (images that do not correspond to a face) have been eliminated (actually a low percentage with respect to the faces correctly detected) and that no selection of the face images has been performed. For this reason some images are very hard to recognize, also for a human, due for example to partial occlusions, strong blurring effects or strange disguising.

The original Big Brother images, due to copyright protection, cannot be printed in this paper, but some example images, acquired in analogous conditions in a laboratory environment, are reported in Fig. 2. and Fig. 3. The two figures clearly demonstrate the high degree of variability observable in a home environment. In particular Fig. 2. shows that, even during a single acquisition session, the face aspect can drastically change due to pose and expression variations, occlusions or variable distance from the camera. The variations are even more evident when images of different acquisition sessions are compared (Fig. 3.).

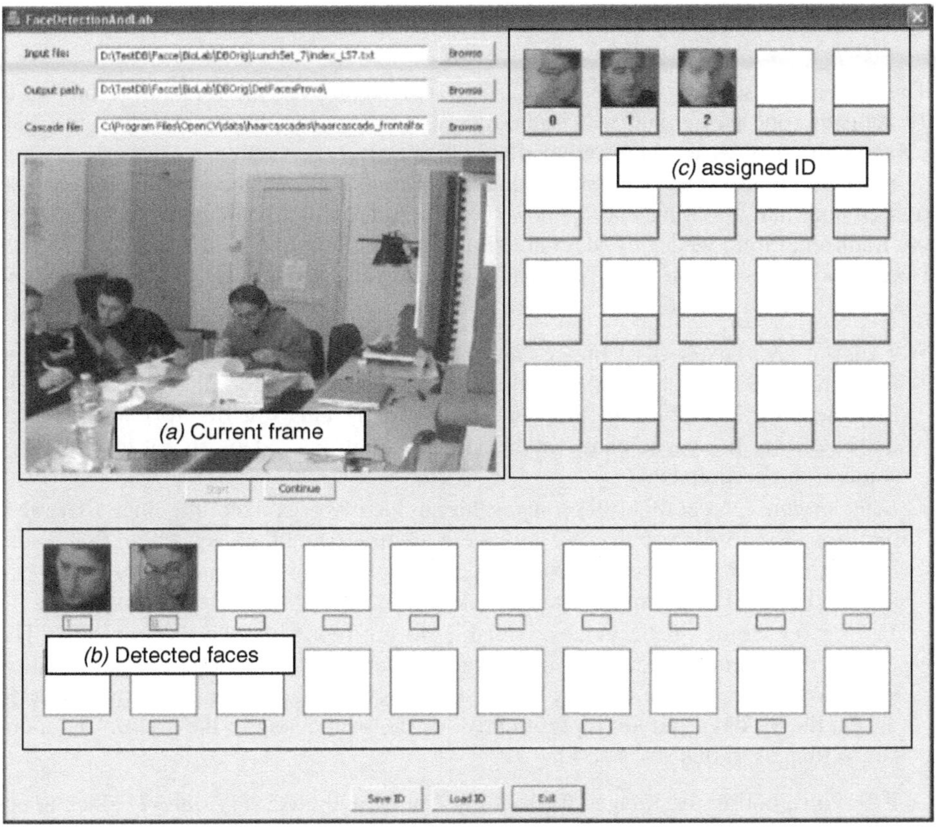

Fig. 1. A snapshot of the software tool developed for face detection and labeling. The three areas highlighted show the current frame analyzed *(a)*, the faces detected in the current frame *(b)* and the ID assigned to each individual *(c)*.

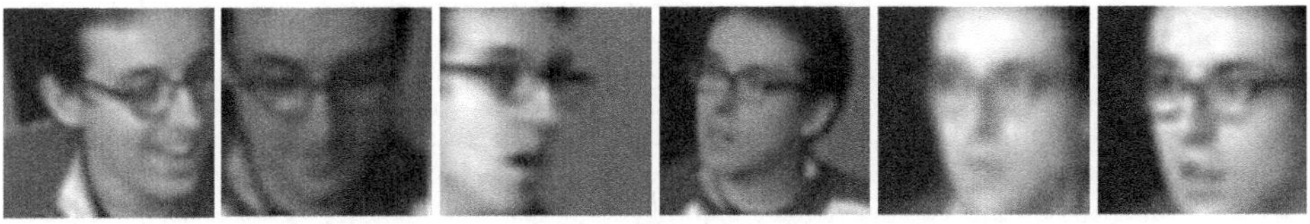

Fig. 2. Images of the same subject recorded during a single acquisition session

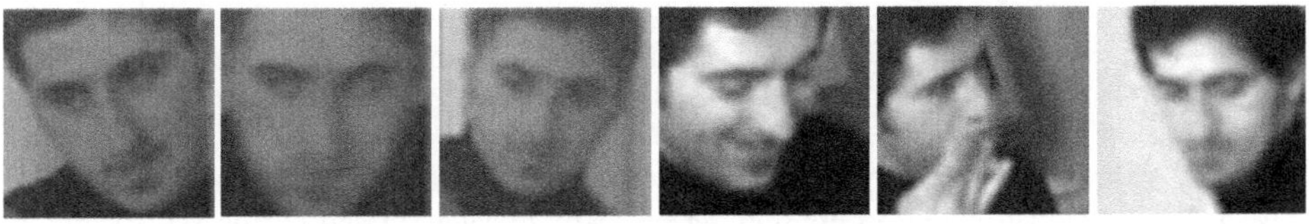

Fig. 3. Images of the same subject recorded during two different acquisition sessions

3 Database Description

The final database consists overall of 23842 images of size 70×70 pixels belonging to 18 subjects (one participant was immediately eliminated at the beginning of the reality show; no images are therefore available for him).

Since the aim of this work is to create a dataset that allows the training and evaluation of template updating approaches, the structure of the database is different from the traditional one, where the data are simply partitioned into *Training* and *Test sets*. Here an additional set of images, intended for template updating purposes, is provided so that the database is composed by the following subsets:

- *Training set*. Two sets of images are available for training: 651 images extracted from the video presenting the participants' biography (taken during the everyday life of each subject, at home or work, with friends, etc., low quality) and 5951 acquired during the participants' audition. The auditions video are not available for some of the participants.
- *Daily updates*. A set of 11898 images, hierarchically organized: the images are first partitioned according to the day they were recorded, and then according to the subject represented. Unfortunately only videos representing the main events have been included in the DVDs, so that images taken from 56 out of 99 days are available.
- *Testing set*. Some of the videos available do not report the day of recording. The images that cannot be included in the daily updates, since the temporal information is not available, can be used as testing set. These images refer to events occurred during the 99 days and are representative of the whole stay in the house. The number of images in this set is 5342.

It is worth noting the images for training can be gathered very quickly since about 200-300 images of a subject can be extracted from a video sequence of 1-2 minutes, if the acquisition conditions and the position of the individual are semi-controlled. On

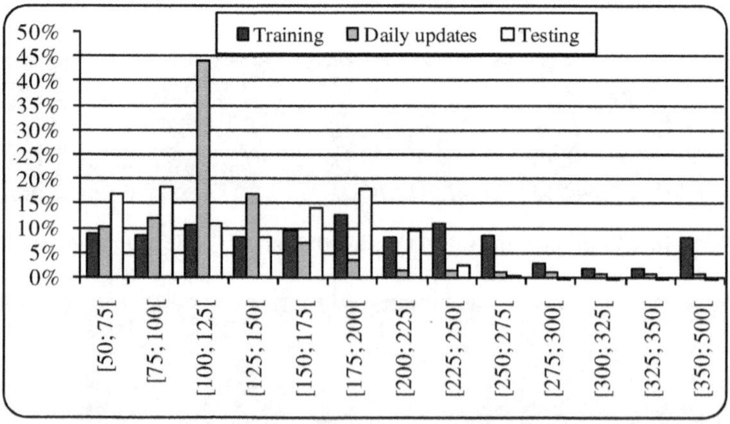

Fig. 4. Face size distribution in the training set, daily updates set and testing set. All the detected face windows are square and the size is simply the window side in pixel. Each bar represents the percentage of images of the related set having a size comprised in the specific interval.

the contrary, acquiring the images for updating is very hard since the acquisition conditions cannot be controlled and unfortunately the face detection module often fails to localize the individuals in the scene because they are too far from the camera or in particular poses.

To better understand the content of the database some statistics about the images are given. In particular the face size (before normalization) and the image quality distributions are reported in Fig. 4. and Fig. 5. respectively. In Fig. 4. the distribution of the face image size, as automatically detected, is reported for the training set, the daily updates and the testing set. It is worth noting that all the detected face windows are finally resized to a 70×70 pixels, that for most of the faces means a sub-sampling process. The graph shows that the original image size is quite variable: from 50 to more than 300 pixels in very few cases. However most of the detected face images have a size lower than 150 pixels.

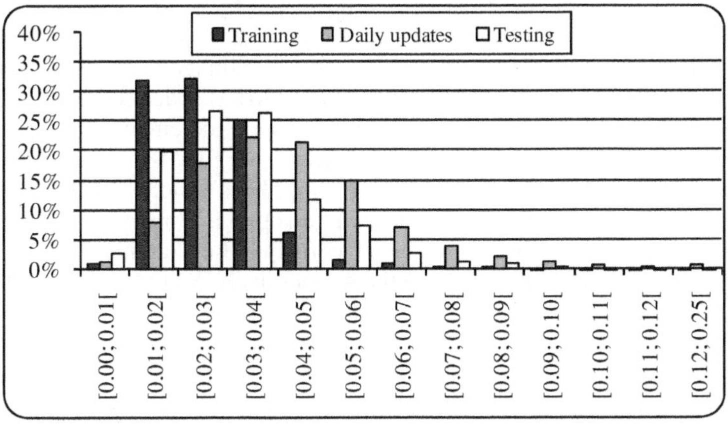

Fig. 5. Image quality distribution, evaluated according to the IQM measure, for the training set, daily updates set and testing set. Each bar represents the percentage of images of the related set having a quality score comprised in the specific interval.

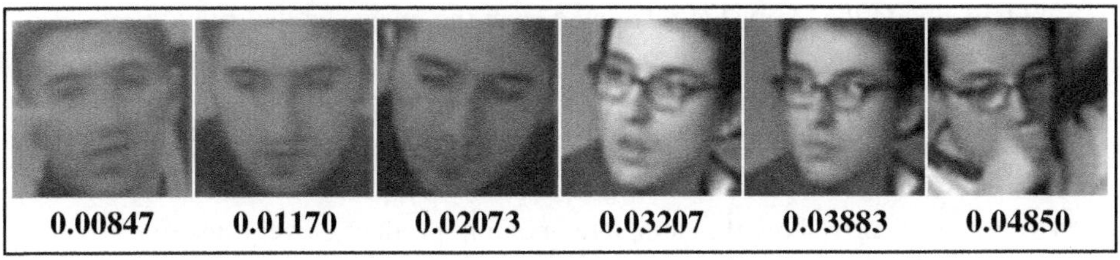

Fig. 6. IQM score of some example images taken an office environment

To characterize the image quality (see Fig. 5.), the Image Quality Measure (IQM) has been adopted. IQM is based on the analysis of the power spectrum and is well known in the literature [10]. It takes into account several factors that proved to be strictly related to the human visual quality assessment. The graph shows that the image quality is generally quite low, particularly for the images of the testing set. As a reference, some images taken in our laboratory with different IQM scores are shown in Fig. 6. Moreover it is worth noting that acceptable IQM scores do not necessarily guarantee that the image is easy to be recognized, since recognition can be seriously affected by other factors not measurable by IQM (e.g. particular poses or illumination, partial face occlusions, etc.).

4 Preliminary Learning Results

The final aim of this research is to design unsupervised learning techniques for face template updating; in this work some preliminary results obtained with a supervised updating technique are reported and analyzed to evaluate the effects of template updating on the recognition performance. Unsupervised leaning procedures will be studied in a future work.

In this experiment each individual is represented by a MKL space [2], i.e. a mixture of KL subspaces [6], created from the images in the *training set*. Recognition is base on a simple nearest neighbor criterion: the face image to be identified is associated to the individual represented by the closest MKL space according to the "distance from space" metric [2]. The initial templates are incrementally updated with the images in the *daily updates* set. An efficient procedure for MKL spaces merging [5] is used to avoid a complete recalculation of the templates which would be computationally very expensive. Updating is here done in a supervised manner: we incrementally update the templates by using images of corresponding identity. This constitute a simplification of a real scenario where the system itself has to decide whether to use a particular (unlabelled) image to update a specific template.

In this work we are focusing on a typical closed-set identification scenario (i.e., all the subjects are known to the system) and the performance are therefore measured in terms of recognition rate defined as the percentage of images in the testing set correctly recognized.

The results obtained are reported in Fig. 7 and Fig. 8. In particular in Fig. 7 the recognition rate is measured for each individual at different stages:

- *Initial RR*: represents the initial recognition rate, i.e. the baseline performance obtained with the templates calculated from the images in the *training set*.
- *RR last day*: refers to the recognition rate measured the last day of presence of each individual in the house.
- *RR max*: represents the maximum recognition rate measured for each subject during his/her whole presence in the house.

The results clearly show that template updating produces a very positive effect: the graph in Fig. 8 shows that the overall daily recognition rate rises from 33.60% (G0, initial template) to 56.27% (G99, templates updated with all the images in the *daily updates*). In most cases a considerable improvement is measured, in particular when a significant set of images is available. Moreover it is interesting to note the noticeable improvement measured for some individuals whose initial recognition rate was very low (e.g., subject 14), thus confirming the importance of template updating.

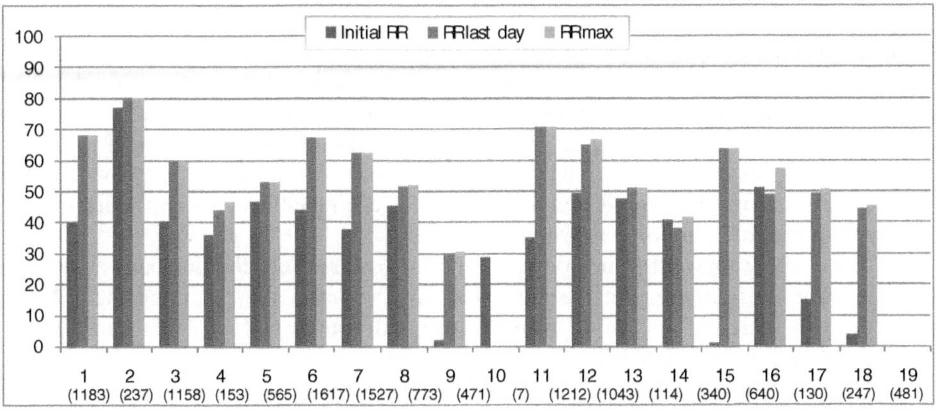

Fig. 7. Recognition rate of the 18 individuals, measured at different stages of the reality show. The x-axis reports the subject id and the total number of images available in the *daily updates* set for that individual.

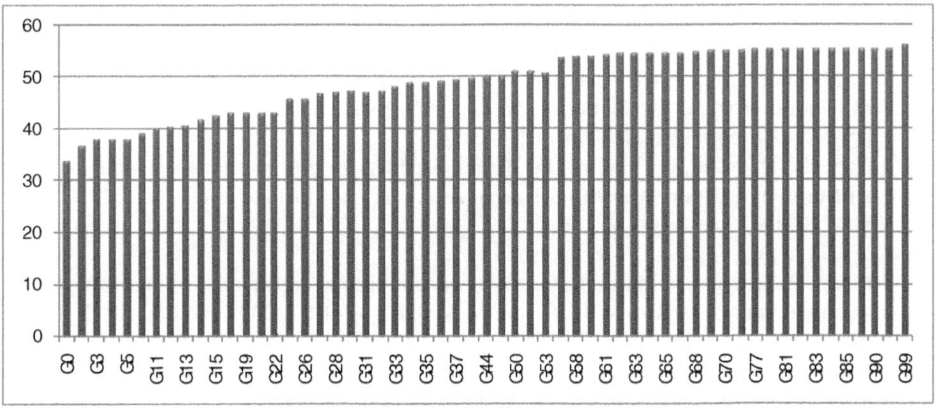

Fig. 8. Daily recognition rate of the images in the testing set

On the other hand the recognition accuracy (about 56% after updating) is quite low, even if for some subjects (for which more images are available) better results are achieved (about 70-80%). The poor performance is certainly due to the completely uncontrolled acquisition conditions that make face recognition very difficult in this specific scenario.

5 Conclusions

In this work the problem of template updating for face recognition in home environments is addressed. A new database, created with the aim of studying face recognition in this specific application scenario, is presented. The steps performed and the tools used or developed are described in detail to allow to the interested readers to create similar databases. In our opinion this can be very useful considering the difficulty of acquiring from scratch such a huge set of data, characterized by a great variability in terms of pose and illumination conditions, image quality and, above all, acquired from the same set of subjects observed daily for a long period. The analysis of the data obtained, on the basis of image quality and face image size, shows that the dataset created is complex and represents an interesting challenge for face recognition algorithms. Moreover some preliminary results on template updating are provided; the experiments carried out show that incremental learning allows to considerably improve the initial recognition performance.

At this stage copyright issues do not allow us to distribute the database created to the scientific community, but we are exploring the possibility of releasing a batch procedure to recreate it starting from the original source (DVDs) that can be purchased online for a few USD.

The work carried out to create the Big Brother database is of course just a first step towards the main objective of our research, which is to design effective unsupervised template updating techniques exploiting the peculiar information that can be obtained from a continuous observation of subjects in smart home environments.

References

[1] Bailly-Bailliere, E., Bengio, S., Bimbot, F., Hamouz, M., Kittler, J., Mariethoz, J., Matas, J., Messer, K., Popovici, V., Poree, F., Ruiz, B., Thiran, J.P.: The BANCA database and evaluation protocol. In: Kittler, J., Nixon, M.S. (eds.) AVBPA 2003. LNCS, vol. 2688, pp. 625–638. Springer, Heidelberg (2003)

[2] Cappelli, R., Maio, D., Maltoni, D.: Multi-space KL for Pattern Representation and Classification. IEEE Transactions on Pattern Analysis and Machine Intelligence 23(9), 977–996 (2001)

[3] DVDx, http://www.labdv.com/dvdx/

[4] Flynn, P., Bowyer, K., Phillips, P.J.: Assessment of time dependency in face recognition: an initial study. In: Kittler, J., Nixon, M.S. (eds.) AVBPA 2003. LNCS, vol. 2688, pp. 44–51. Springer, Heidelberg (2003)

[5] Franco, A., Lumini, A.: Mixture of KL Subspaces for relevance feedback. Multimedia Tools and Applications 37(2), 189–209 (2008)

[6] Fukunaga, K.: Statistical Pattern Recognition. Academic Press, San Diego (1990)

[7] Grande Fratello: Edizione 2006. Tutti i 92 giorni (2006), http://www.internetbookshop.it/dvd/8032442208456/grande-fratello-edizione.html

[8] Image Grabber II, http://www.videohelp.com/tools/ImageGrabber

[9] Li, S.Z., Jain, A.K.: Handbook of Face Recognition. Springer, Heidelberg (2005)

[10] Nill, N.B., Bouzas, B.H.: Objective Image Quality Measure Derived from Digital Image Power Spectra. Optical Engineering 31(4), 813–825 (1992)

[11] OpenCV Library, http://www.intel.com/technology/computing/opencv/

[12] Phillips, P.J., Scruggs, W.T., O'Toole, A.J., Flynn, P.J., Bowyer, K.W., Schott, C.L., Sharpe, M.: FRVT 2006 and ICE 2006 Large-Scale Results (2006), http://www.frvt.org/FRVT2006/docs/FRVT2006andICE2006LargeScaleReport.pdf2007

[13] Viola, P., Jones, M.J.: Rapid Object Detection using a Boosted Cascade of Simple Features. In: Proc. ICPR, vol. 1, pp. 511–518 (2001)

[14] Wechsler, H.: Reliable Face Recognition Methods. Springer, Heidelberg (2007)

A Confidence-Based Update Rule for Self-updating Human Face Recognition Systems

Sri-Kaushik Pavani[1,2], Federico M. Sukno[1,2], Constantine Butakoff[1,2], Xavier Planes[1,2], and Alejandro F. Frangi[1,2]

[1] Center for Computational Imaging & Simulation Technologies in Biomedicine (CISTIB), Information & Communications Technologies Department, Universitat Pompeu Fabra, Pg. Circumvallacio 8, Barcelona 08003, Spain
[2] Networking Biomedical Research Center on Bioengineering, Biomaterials and Nanomedicine (CIBER-BBN), Barcelona, Spain

Abstract. The aim of this paper is to present an automatic update rule to make a face recognition system adapt itself to the continuously changing appearance of users. The main idea is that every time the system interacts with a user, it adapts itself to include his or her current appearance, and thus, it always stays up-to-date. We propose a novel quality measure, which is used to decide whether the information just learnt from a user can be used to aggregate to what the system already knows. In the absence of databases that suit our needs, we present a publicly available database with 14,279 images of 35 users and 74 impostors acquired in a span of 5 months. Experiments on this database show that the proposed measure is adequate for a system to learn the current appearance of users in a non-supervised manner.

1 Introduction

In this paper, we describe an update rule to make a Face Recognition System (FRS) adapt itself to the continuously changing appearance of a human face. In previous research [10], it has been shown that there is an approximately linear decrease in the accuracy of a FRS with respect to the time elapsed between acquisition of training and testing images of a person. The decrease in the accuracy is the result of changes in facial appearance because of factors such as ageing, health, skin tanning, facial hair growth, scars, *etc.*

Adapting the FRS to the current appearance of users is vital because an un-updated FRS may have difficulties recognizing its rightful owner after a few years of use. A common way to maintain the FRS up-to-date is by learning the current appearance of a user whenever he/she interatcs with it. In this approach, when a person is identified by the FRS, the facial image goes through a selection process that decides whether the information learnt from that image is reliable or not. If the extracted information is deemed reliable, then the corresponding image is added to the training set of a particular user. Later, the FRS is re-trained with the recently updated training set, and thus, it remains up-to-date with the current appearance of the user.

M. Tistarelli and M.S. Nixon (Eds.): ICB 2009, LNCS 5558, pp. 151–160, 2009.
© Springer-Verlag Berlin Heidelberg 2009

A major risk in automatically incorporating new training images is the potential for a gradual corruption of the training set. Since images are added in a non-supervised manner, images on which one of the system components (*e.g.*, face detection or segmentation) failed could also be added. This leads to the corruption of the information that is used to recognize users. For example, the image shown in Fig. 1a, should not be added to the training set because the segmentation of the face into its constituent parts is inaccurate and therefore, the information learnt from this image will be incorrect.

In order to determine how accurately the facial texture that has been extracted from a test image represents the current appearance of a user, we propose a confidence measure called total confidence. If an image has high total confidence, it will be included in the training set, otherwise, it will be ignored. In our experiments, we used this measure to automatically update the FRS for a period of four months. We observed that the FRS maintained its accuracy after the updates and that the training set was not corrupted during the update process.

(a) (b)

Fig. 1. Examples of bad (a) and good (b) facial segmentation. Including inaccurately segmented images, such as (a), corrupts the training set.

2 Related Work

Over the years, two main approaches were developed to handle the ever-changing appearance of human faces. In [3] and [4], the FRS is made invariant to ageing by simulating the effect of ageing on the appearance of a face. Such approaches are generally failure-prone, as the future appearance of a face is not only dependent on ageing, but also on other unpredictable factors such as permanent scarring due to accidents and effects of cosmetic surgery. Such an approach is primarily intended for applications where recent images of a person are not available. For example, they are used in predicting the current appearance of a person who has been missing for a long time.

There are applications where simulation of ageing becomes unnecessary. For example, a FRS installed at home interacts with the users almost on a daily basis. In such scenarios, the appearance of the users can be learnt every time the system sees a user, and therefore, the system can remain up-to-date.

Aryananda [1] proposed and studied a humanoid robot (called Kismet) that has the ability to learn to recognize people during its interactions. The robot starts with an empty database and, as it interacts with people, it becomes a un-supervised FRS. In a reduced set of 9 individuals, the system was unable to learn 5 of them. It is not clear from their paper how the robot deals with cases when one of the elements of the recognition pipeline has been inaccurate.

Mou *et al.* [7] presented a technique to automatically create training databases for FRS in an unsupervised manner. The databases, once generated, can be used to automatically update a FRS. When a new face is input to the system, it is processed by the FaceVACS recognition technique which outputs the identity of a person and a similarity measure. If the similarity measure is greater than a preset threshold, the image is added to the training set. The authors produce results on 10 video sequences and they report that automatic construction of training sets occurred without corruption of data.

The main drawback of the above mentioned methods is that the results were reported on rather small databases, which questions the generalizability of their conclusions. One of the main reasons to use small databases is the unavailability of standard databases that have images of users captured on a daily basis over large periods of time. To this end, we present an image database containing 14,279 images of 35 users and 74 impostors acquired on a day-to-day basis over five months. The database, called Gradual Evolution of Facial Appearance database (GEFA), can be obtained free of cost from http://cilab2.upf.edu/gefa/.

3 Face Recognition System

A FRS was installed in a commonly visited area of our lab. The system, whose flowchart is shown in Fig. 2, has three main building blocks: face analysis, face identification and adaptive learning blocks.

3.1 Face Analysis Block

The face analysis block is responsible for detection and segmentation of human faces.

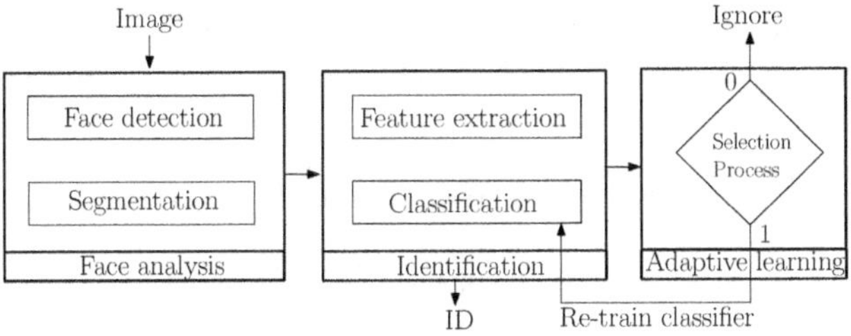

Fig. 2. Flow chart of the FRS. The input image is checked for the presence of faces. If any, the face is segmented and identified. In the selection process, the total confidence of the result is computed. This determines if the image is used to re-train the classifier.

Face Detection. To detect faces accurately in real-time, our system employs Viola and Jones face detection procedure [13]. The detector works at a speed of 15 frames per second at 12 resolution levels, while achieving low false acceptance rates.

Face Segmentation. For the segmentation of the prominent facial features our system employs Active Shape Models with Invariant Optimal Features (IOF-ASM) [12]. This algorithm can automatically locate the outlines of eyes, mouth, nose, brows and silhouette contour of the face with an average error of approximately 2% of the inter-eye distance.

3.2 Identification Block

The identification block extracts texture from a segmented face and uses it to identify a user.

Facial Feature Extraction. To make facial texture independent of the facial expression, each image was warped using a piece-wise affine mapping between the result of the segmentation step and the mean shape. The texture parameters for each image were extracted by constructing an image intensity model of an Active Appearance Model [2].

Classification. Once the texture parameters are obtained, they are whitened [8] and then a distance-weighted k-Nearest Neighbor classifier (k-NN) is used to identify the user in the image. The distance between two texture vectors is defined as the angle between them. The angle based distance metric was shown to be a good choice for PCA-based algorithms [9]. All the experiments in this paper were performed using the five nearest neighbors ($k = 5$).

3.3 Automatic Learning Block

Every time a facial image has been identified, it goes through an automatic selection process that determines whether it is suitable to be added to the training set of the appropriate user. The selection process deems two categories of images to be unsuitable for automatic learning: 1) images with outdated facial appearance, 2) images where the algorithms used in our system produce unreliable results.

4 Selection Process for Automatic Updates

The selection process is based on the following four confidence measures:

1. Temporal confidence (C_t). The accuracy of FRSs has been shown to decrease linearly with the time elapsed between enrollment and testing [10]. In our experiments, we assume that the maximum time limit of an image in representing a user's appearance is 5 years. The value of C_t is computed as in (1).

The quantity t_e stands for the time elapsed (in days) between the time of acquisition of the data and the current time.

$$\mathcal{C}_t = \max\left(0, 1 - \frac{t_e}{1825}\right) \tag{1}$$

2. Confidence of the face detector ($\mathcal{C}_d$). The quantity $\mathcal{C}_d$ is computed in two steps. Firstly, the likelihood p of a face is computed as the weighted ratio of the number of weak classifiers that label it as face to the total sum of weights of all the weak classifiers in the face detector. In the second step, the number of detection windows, n, fused during the clustering process is computed. We compute the confidence of the face detector as in (2).

$$\mathcal{C}_d = \sqrt[n]{\bar{p}} \tag{2}$$

where $\bar{p}$ is the mean of likelihood values, p, computed for all sub-windows that are fused to form a single detection region. It can be observed that as n increases, $\mathcal{C}_d$ also increases. Low number of detected sub-windows have been observed for faces that are non-frontal. High values of $\bar{p}$ and n were observed for frontal faces alone.
3. Confidence of the segmentation algorithm ($\mathcal{C}_{sg}$). The quantity $\mathcal{C}_{sg}$ is computed as the average of reliability estimate on each landmark of the segmented face. For details on how the reliability of each landmark is computed, readers are referred to [11].
4. Confidence of the classifier ($\mathcal{C}_c$). If $d_i, i \in \{1, ..., k\}$ are the distances of the test feature vector to the k nearest training feature vectors $\mathbf{b}_g^i$, then the confidence that the test image belongs to the class $\mathbb{C}_j$ is given by (3).

$$\mathcal{C}_c = \frac{\sum\limits_{\forall i \in \mathbb{D}_j} d_i^{-1}}{\sum\limits_{\forall i} d_i^{-1}}, \qquad \mathbb{D}_j = \{i | \mathbf{b}_g^i \in \mathbb{C}_j, i = 1, \ldots, k\} \tag{3}$$

All the above confidence values are key measures of the quality of the information obtained from a test image. A low value for any of the above mentioned factors indicates that the information obtained from an image does not accurately represent the current appearance of the user. Realizing this, we adopted the product rule to compute a global score, total confidence, as shown in (4).

$$\mathcal{C} = \mathcal{C}_t \cdot \mathcal{C}_d \cdot \mathcal{C}_{sg} \cdot \mathcal{C}_c \tag{4}$$

Further, we define a new confidence measure, system confidence ($\mathcal{C}_s$), that is an indicator of how confidently the algorithms we use in our FRS make their decisions. The system confidence is defined as in (5).

$$\mathcal{C}_s = \mathcal{C}_d \cdot \mathcal{C}_{sg} \cdot \mathcal{C}_c \tag{5}$$

The selection process proceeds as follows. Assume that a test image, with a total confidence $\mathcal{C}$ has been identified as belonging to class $\mathbb{C}_j$. It is added to the training set if and only if $\mathcal{C}$ is greater than $\mathcal{C}^{min}$, namely the minimum total confidence of all the images already enrolled as belonging to class $\mathbb{C}_j$.

Additionally, as images are continuously added to the training set during the update process, it is possible to overwhelm the FRS simply by accumulating a huge amount of training data. In order to limit the number of images per user, every time a new image is added to an already full training set, the image with the lowest total confidence, $\mathcal{C}^{min}$, is eliminated such that the total number of images per person is utmost $\mathcal{N}$.

5 Experimental Results

The following experiments aim to test the performance of the FRS when updated using the total confidence, $\mathcal{C}$. Before we present the results, the image database that was acquired to perform the tests is described.

5.1 Image Acquisition

The system described in Sec. 3, was used to obtain facial images of 35 individuals (belonging to 15 different nationalities) over a period of five months. A total of 11,918 images were collected. These images were separated into three categories:

1. `Training set`. The images in the `Training set` provide the ground-truth data to enroll users into the system. This set was obtained by selecting 472 images from the 3,498 images captured in the first month. The selection process was performed by a human expert who removed images on which the face detection and segmentation algorithms performed poorly. In other words, the 472 images that were selected for this set had high $\mathcal{C}_d$ and $\mathcal{C}_s$ values. Each individual in the training set had 11 ± 5 images.
2. `Update set`. This set contains the 6,420 images captured during the second, third and fourth months. The images from this set, in a chronological order, were used to automatically update the system. No image selection was performed, and therefore, this set may contain images where the detection and segmentation algorithms were inaccurate.
3. `Test set`. This set contains a total of 2000 images captured during the fifth month. As in the case of the `Update set`, no selection was performed.

To get an idea of the number of images in which the algorithms we use were inaccurate, we split the images in the `Update set` and `Test set` into five categories according to their $\mathcal{C}_s$ value. The first category contained images with

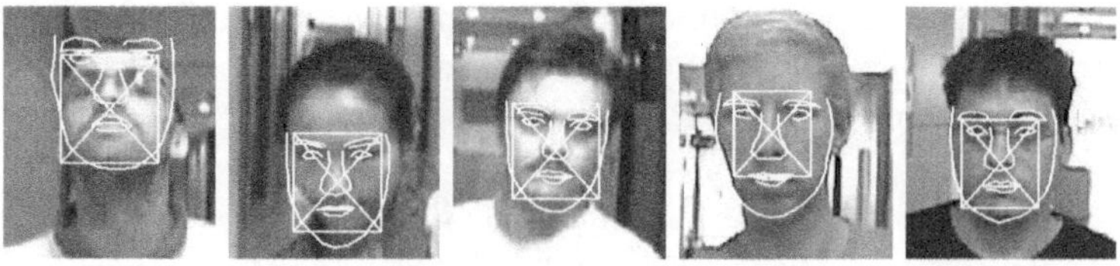

Fig. 3. Example images from the `Update set` and `Test set` arranged according to their $\mathcal{C}_s$ value. The first, second, third, fourth and fifth images (counted left to right) produced $\mathcal{C}_s$ values approximately equal to 0.1, 0.3, 0.5, 0.7 and 0.9 respectively.

$\mathcal{C}_s \in [0.0; 0.2)$, the second contained images with $\mathcal{C}_s \in [0.2; 0.4)$, and so on. The percentage of images that fell into each category was 5.5%, 16.5%, 16.5%, 18.5%, 43.0%, respectively. Some of the images belonging to each category are shown in Fig. 3. Images with $\mathcal{C}_s$ values lower than 0.8 have been observed to cause corruption of the texture data learnt for a user as the texture information obtained is not representative of the user.

5.2 Training and Update Procedures

The k-NN classifier was trained with the texture parameters obtained from the 472 images in the `Training set` and their corresponding user IDs. Using the images in the `Update set` (in chronological order), three types of updates were performed to the `Training set`:

1. $\mathcal{UC}$. In this case, updates were performed with the total confidence, C, as described in Sec. 4.
2. $\mathcal{UC}_c$. The update process is similar to that performed in [7]. In this scenario, if $\mathcal{C}_c$ computed for an image in the `Update set` is greater than $\mathcal{C}_c$ of any of the images in the `Training set`, then the image will be added to the `Training set`. Otherwise, the image is ignored.
3. $\mathcal{UC}_t$. This is a trivial update strategy where newer images from `Update set` replace older ones in the `Training set`. In other words, if an image has a $\mathcal{C}_t$ value greater than that of any of the images in the `Training set`, then it is added to the `Training set`.

During the update procedure, the number of training images belonging to a person is kept fixed to a preset value $\mathcal{N}$ as described in Sec. 4. We repeated our experiments for four different values of $\mathcal{N}$: 15, 30, 60 and 90. It was observed for $\mathcal{N} = 15$ case that $99.5 \pm 0.78\%$ of the original images were replaced after the updates. For the cases $\mathcal{N} = 30$, $\mathcal{N} = 60$ and $\mathcal{N} = 90$, the percentages of replaced images were $91.8 \pm 2.28\%$, $86.3 \pm 2.03\%$, and $69.3 \pm 1.1.5\%$, respectively.

5.3 Comparison of Accuracy

The accuracy of the FRS was computed in terms of the correct classification rate, which is a ratio of the number of correctly classified images to the total number of images that are tested. Fig. 4 shows the average scores of classification for the images in the `Test set` after the updates. Below are four main observations that can be made from Fig. 4.

1. Images with high $\mathcal{C}_s$ values are classified more accurately than those with low values. This is because, images with high $\mathcal{C}_s$ values are more accurately segmented, and therefore, the texture parameters obtained from these images are similar to the ground truth images in the `Training set`.
2. The accuracy of the FRS after $\mathcal{UC}$ (Fig. 4a) is higher than the accuracy obtained after $\mathcal{UC}_c$ (Fig. 4b) and $\mathcal{UC}_t$ (Fig. 4c). Since the `Test set` is the same for all three cases, the decrease in performance after $\mathcal{UC}_c$ and $\mathcal{UC}_t$ should be a result of a corrupted `Training set`. Analyzing the `Training set`

after $\mathcal{UC}_c$ and $\mathcal{UC}_t$, we observed that they contained misclassified images and images with inaccurate segmentation. Such corruption of the training sets never occurred in the training sets corresponding to $\mathcal{UC}$.

3. The standard error measurements (shown as error bars) corresponding to $\mathcal{UC}$ are the smallest (especially when $\mathcal{C}_s \geq 0.8$). Bigger error bars in Fig. 4b result from larger uncertainty in the measurement of correct classification rate. This means that there is higher dispersion in classification rates between the users: some of the users are classified better and some worse. This again is a result of the corruption of training images for some users.
4. In Fig. 4c, the accuracy of the FRS seems to decrease with smaller $\mathcal{N}$. This is because, FRS with low value of $\mathcal{N}$, is more susceptible to corruption. For example, one badly segmented image (belonging to a particular user) in a FRS with $\mathcal{N} = 15$ means that 6.6% of the data belonging to the user is corrupted. However, one badly segmented image in a FRS with $\mathcal{N} = 90$ means that only 1.1% of the data belonging to the user is corrupted. Thus FRS with bigger $\mathcal{N}$ are more robust to corruption. From Fig. 4(a,b), it can be noticed that the accuracy of the system does not change significantly with $\mathcal{N}$. This indicates that $\mathcal{UC}$ and $\mathcal{UC}_c$ are better update strategies than $\mathcal{UC}_t$.

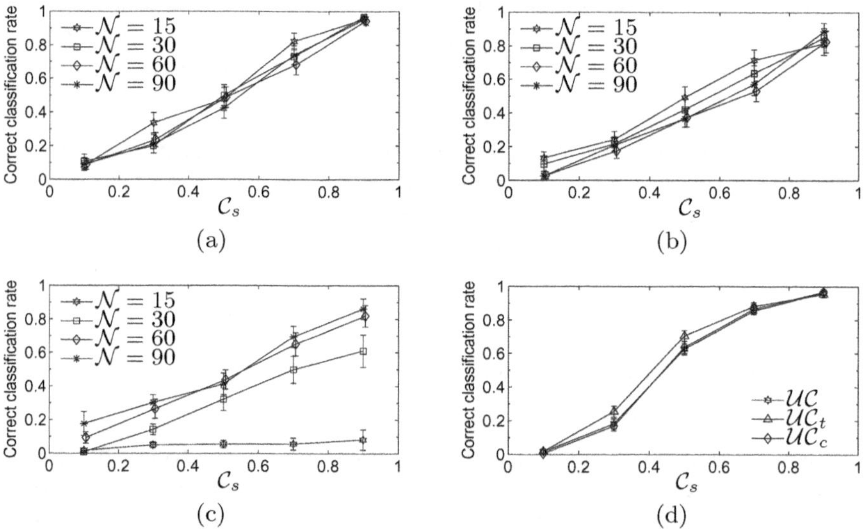

Fig. 4. Accuracy of the FRS after (a) $\mathcal{UC}$, (b) $\mathcal{UC}_c$, (c) $\mathcal{UC}_t$. (d) Accuracy of the FRS on the `Impostor set` with $\mathcal{N} = 15$. Error bars represent the standard error of the mean.

Limitations. GEFA does not contain images of users who exhibit abrupt appearance changes. Since images were captured on a day-to-day basis, appearance changes due to to skin tanning and facial hair occur gradually over a sequence of images belonging to a particular user. Abrupt appearance changes generally reduce the accuracy of the FRS. In the present set-up, if the FRS is having difficulties in recognizing a legitimate user, then the user is manually enrolled so that the FRS can learn his/her current appearance.

5.4 Robustness to Impostors

Robustness of the FRS to impostors was tested on an `Impostor set`, containing 2,379 images belonging to 74 individuals. These individuals were not part of the `Training set`, `Update set` or `Test set`. No manual selection of images was performed, and therefore, this database contained images where face detection or segmentation algorithms might have been inaccurate.

Robustness to impostors was tested as follows. Every image in the `Impostor set` was processed by the FRS, and its C_s value was calculated. If its C_s was lower than a preset threshold, then the image was labeled as impostor.

The correct classification rate on the `Impostor set` is calculated as the ratio of number of images classified as impostors to the total number of images in the `Impostor set`. Fig. 4d plots the correct classification rate of images in the `Impostor set` vs. the threshold value on C_s. It can be seen that if the minimum required C_s value for an image is set to 0.9, then approximately 95% of the impostors are classified correctly. Low threshold values on C_s cause misclassifications as the system confuses them with the enrolled users. High threshold values ($C_s > 0.9$) make FRS restrictive in the sense that even small inaccuracies during detection, segmentation or classification will not be tolerated.

In general, it seems that the system can be run in an optimal point by setting the minimum required C_s value to 0.9. At this operating point, the system classifies 95% of the legitimate users (from Fig. 4a) and 95% of the impostors correctly (from Fig. 4d). This is similar to the performance obtained by many other systems [5] when tested on the *XM2VTS* database [6].

6 Conclusions

We have presented an automatic update rule to adapt the FRS to the continuously varying appearance of faces. Every time the FRS identifies a person, and if it deems that its decision is reliable, the texture paremters obtained from the image are used to maintain the system updated.

The first contribution of this paper is the proposal of the total confidence metric to decide whether an image should be included in the training set or not. The proposed measure intends to quantify how accurately the facial parameters from a test image represent a user. We used the total confidence measure to discard two types of images which are known to cause corruption of training data: images with outdated facial appearance and images where the segmentation algorithm did not extract the facial texture reliably. Finally, this paper presents a new face database, GEFA, in which the appearance of a group of users was captured on a daily basis over five months. This database includes 11,918 facial images from 35 individuals. It also contains 2,349 images from 74 individuals who act as impostors. Detailed experiments performed on this database show that total confidence is an adequate measure for a system to update itself in an unsupervised manner while avoiding corruption of the training set.

References

[1] Aryananda, L.: Recognizing and remembering individuals: online and unsupervised face recognition for humanoid robot. In: IROS 2002, pp. 1202–1207 (2002)
[2] Cootes, T., Edwards, G., Taylor, C.: Active appearance models. IEEE Trans. Pattern Anal. Mach. Intell. 23(6), 681–685 (2001)
[3] Lanitis, A., Taylor, C., Cootes, T.: Modeling the process of ageing in face images. In: ICCV 1999, pp. 131–136 (1999)
[4] Lanitis, A., Taylor, C., Cootes, T.: Toward automatic simulation of aging effects on face images. IEEE Trans. Pattern Anal. Mach. Intell. 24(4), 442–455 (2002)
[5] Messer, K., et al.: Face verification competition on the XM2VTS database. In: Kittler, J., Nixon, M.S. (eds.) AVBPA 2003. LNCS, vol. 2688, pp. 964–974. Springer, Heidelberg (2003)
[6] Messer, K., Matas, J., Kittler, J., Luettin, J., Maitre, G.: XM2VTSDB: The extended M2VTS database. In: AVBPA 1999, pp. 72–77 (1999)
[7] Mou, D., Schweer, R., Rothermel, A.: Automatic databases for unsupervised face recognition. In: CVPRW 2004, pp. 90–97 (2004)
[8] Navarrete, P., Ruiz-del-Solar, J.: Comparative study between different eigenspace-based approaches for face recognition. In: Pal, N.R., Sugeno, M. (eds.) AFSS 2002. LNCS, vol. 2275, pp. 178–184. Springer, Heidelberg (2002)
[9] Perlibakas, V.: Distance measures for PCA-based face recognition. Pattern Recogn. Lett. 25(6), 711–724 (2004)
[10] Phillips, P., et al.: Face recognition vendor test 2002. Evaluation Report (2003), `http://www.frvt.org`
[11] Sukno, F., Frangi, A.: Exploring reliability for automatic identity verification with statistical shape models. In: AutoID 2007, pp. 80–86 (2007)
[12] Sukno, F., et al.: Active shape models with invariant optimal features. IEEE Trans. Pattern Anal. Mach. Intell. 29(7), 1105–1117 (2007)
[13] Viola, P., Jones, M.: Robust real-time face detection. Int. J. Comput. Vision 57(2), 137–154 (2004)

Facial Comparisons by Subject Matter Experts: Their Role in Biometrics and Their Training

Nicole A. Spaun

Federal Bureau of Investigation
Forensic Audio, Video and Image Analysis Unit
Building 27958-A, Pod E, Quantico VA 22135 USA, 703-985-1169
Nicole.Spaun@ic.FBI.gov

Abstract. The fingerprint biometrics community has accepted the involvement of human examiners as the ultimate verifiers of the output of automated systems, especially in latent fingerprint cases. Likewise, the facial biometrics community should recognize the importance of human experts working in association with the automated process, particularly when analyzing uncontrolled images. As facial biometric systems become more common, more facial image examiners will need to be trained. Currently there are no systematic training programs in place for these examiners. This paper outlines the knowledge needed to conduct facial examinations, and thus should be taught in a successful training program. A facial image comparison expert must be versed in many subjects including: comparative science, image science and processing, bones of the head, muscles of the face, properties of the skin, aging and alteration, legal issues and case law, and the history of facial identifications and photographic comparisons.

Keywords: Biometrics, forensics, facial recognition, facial identification.

1 Introduction

When automated fingerprint comparison systems, such as IAFIS, went into service, there were numerous human experts readily available to verify the results of those programs. In the United States, tens of state and local law enforcement agencies employ fingerprint examiners; this is in addition to the hundreds of experts employed within the US Departments of Justice, Defense, and Homeland Security alone. In contrast, there are comparatively few trained subject matter experts currently practicing facial comparisons for identification in justice systems around the world. Within the United States, there are less than 10 federal, state, and local agencies combined practicing forensic facial comparison today. Many agencies are welcoming the introduction of automated facial recognition systems as an answer to the comparison of people from surveillance images; however these agencies generally lack the experts to verify those results. As such, automated facial recognition will necessitate a substantial increase in the number of facial image examiners worldwide. Because there are currently no automated facial biometric systems with high enough accuracy to allow for "lights out" operation, forensic examiners will need to verify the resultant output

M. Tistarelli and M.S. Nixon (Eds.): ICB 2009, LNCS 5558, pp. 161–168, 2009.
© Springer-Verlag Berlin Heidelberg 2009

of these systems. Therefore, training and certification programs for facial image examiners need to be developed today for the coming wave of automated facial biometrics tomorrow.

This paper presents an outline of the history of facial identification in the legal system, the role for human subject matter experts in the automated facial recognition cycle, an outline of the levels of expertise needed to perform facial identifications for various purposes, and a summary of the spheres of knowledge that a facial image examiner would need to be versed in to perform scientific comparisons of faces.

1.1 History of Facial Identification, or Why There Are Few Trained Experts Today

The history of identification of people by their facial morphology in the legal system is a long and often punctuated one. In the late 1800's, Parisian anthropologist Alphonse Bertillon established a system for personal identification based on measurements of the face and body. Concurrently, fingerprinting was gaining popularity in India and England to identify people in the criminal justice system. At the turn of the century, the Bertillon system of identification fell out of favour when the uniqueness of the selected measurements was challenged by instances of misidentification [1]. Subsequently, fingerprinting quickly replaced the Bertillon system as the dominant means of identifying people to the exclusion of all others. The need for facial identification subject matter experts waned. The comparison of faces for official purposes was left to police, soldiers, guards, and other individuals, most of whom received more training in firearms than faces, if at all. As photography and surveillance became more prevalent through the 20th century, a small number of agencies began developing specialists to compare photographs of the faces of individuals. For example, during the Cold War the US Central Intelligence Agency compared photographs to identify Communist officials and their look-alikes [2].

Facial identification has suffered perception problems. While few people challenge the notion that experts are needed to compare fingerprints or footwear patterns to draw a proper conclusion, the majority of people believe that *anyone* can accurately identify faces in images without training and therefore experts are not needed. In the 1970's, Federal Bureau of Investigation (FBI) photographic examiners testified in several cases where the courts proposed that the conclusions of the facial comparison experts, while not harmful, may not be necessary except in the most difficult cases as the jury can view the images themselves and draw their own conclusions [3]. With such case law as precedent, it is apparent why few other agencies invested in establishing and training experts. Fortunately, subsequent cases have stated that subject matter expert testimony concerning facial comparisons is indeed beneficial.

Because people are accustomed to recognizing faces and not accustomed to comparing firearms, footwear, or fingerprints, it is assumed that experts are needed for analyzing the latter. However, the ability to conclusively identify people in images is different than recognizing individuals in person. Images are two-dimensional representations of a scene; while faces are highly three-dimensional involving many variables, including optics and perspective, that factor into facial image analysis. Thus it is the role of the facial image examiner to draw a scientific conclusion from the comparison of faces, acknowledging the complex variables, and their effects, in the process of imaging and comparing the human face.

1.2 The Current State of Forensic Facial Identifications

Unlike in fingerprint biometrics where the community developing automated systems realized the significant role that human experts will play in the course of action integrating with the automated systems, the facial biometric community is currently focused on developing new algorithms or attempting to fuse biometrics. The simple solution to enhance the accuracy of automated systems right now is to utilize human subject matter experts in the procedure. Also, it is essential that the output of automated facial recognition systems is verified by human examiners if it is to be used for legal purposes, just as is done with latent fingerprints. The availability of these subject matter experts, however, is an issue at the present time.

Currently there are numerous agencies using automated facial biometric systems to determine identities for purposes, such as access control and fraud detection, however often the people operating these systems have little or no facial and/or image analysis training. The agencies using these systems have generally created in-house training as external training is rarely available. Therefore, agencies that do have trained facial comparison subject matter experts are besieged with requests for their training programs and expert assistance. As such, today's subject matter experts on facial comparison must ensure the proficiency of the next generation of experts by sharing their training programs and science. A certification in facial image comparisons should be developed for the future and the first step is to define the topics of which an expert should have knowledge and practice.

2 Levels of Training

In an ideal world, every person comparing images of faces will be trained to the highest possible level of knowledge and practice in their field. In reality, this is not essential as the comparison of certain images involves fewer variables and, as a result, less training is needed to draw conclusions in such cases. Fingerprint comparison examinations are typically divided into two types: ten-print and latent print. Comparable to the differentiation of fingerprint comparisons by the type/quality of fingerprints under examination, facial comparison examinations can be divided by the type of images depicting faces that will be subject to facial comparison.

The terms *controlled* and *uncontrolled* have been used to describe the conditions under which the biometric sample has been obtained [4]. In facial biometrics, controlled images are front facial pictures with a subject bearing a neutral expression who is photographed under controlled, even lighting in front of a neutral background. Recommended standards exist for the taking of controlled images for the purpose of identification, which includes photographs taken for passports and arrest records (e.g. [5]). Uncontrolled images are simply that: not controlled. They are subject to great variations, including the pose and expression of the subject, the lighting and background of the environment, the distance between the camera and subject, etc. Examples of uncontrolled images include snapshots or surveillance images from both video and still cameras. In order for an examiner to understand and account for these complex, uncontrolled variables, they must have knowledge and expertise, and thus training, beyond that of examiners who are solely analyzing controlled, standardized images. Therefore, it can be suggested that examiners performing facial comparisons of controlled-to-controlled images are to be trained at a basic level and examiners performing facial comparisons of any uncontrolled images be trained to an advanced level. For the purposes of this paper, examiners trained to a basic level are referred to as Standardized Image Examiners (SIE) and those trained to an Advanced level are referred to as Uncontrolled Image Examiners (UIE) and cumulatively both types together shall be referred to as Facial Image Examiners (FIE).

3 Spheres of Training

The areas of training for facial image examiners can be divided into two categories: general and facial specific. An additional training component concerning legal issues and case law is essential for facial image comparison experts working within the justice system and/or are expected to present their conclusions in court. The general knowledge necessary for facial image examiners is similar to that taught for footwear, fingerprint, and other forensic disciplines involving the comparison of items. The science of comparing objects, including those depicted in images, has been documented in other forensic disciplines and therefore facial comparison training should receive similar training. Specifics of the face is a topic that should be covered in detail in any facial image training.

3.1 General Knowledge

History. Any training program in facial comparison, or any biometric for that matter, must first include a review of the history of personal identification. By exploring the past of any field, one learns about the advances and relapses that have occurred. A discussion of the early methods of individualizing persons would include an overview of the Bertillon system of personal measurements and a brief history of fingerprinting. Because facial image analysis is a subset of the discipline of Photographic Comparison, the history of the science of photographic comparisons should also be included, with a particular focus on photographic comparisons of faces.

Biometric Advances. An overview of automated facial biometrics should include focus areas such as the operation of common automated systems and the typical output of such automated systems. It is important for all facial image examiners to be familiar with the underlying methods of facial recognition systems, whether they are directly using these systems or not. The advanced level of training should include a brief review of the specific algorithms at use in most systems. Such a discussion would benefit the Uncontrolled Image Examiner by elucidating the sensitivities of these algorithms to the variable factors present in complex surveillance-type images. For example, some algorithms are most sensitive to facial geometry and therefore camera to subject distance variations would greatly affect those programs, while other systems are most affected by the choice of their training set of images.

Underlying Principles. The principles of photographic comparison are those of identification science: individuality and comparison. In the forensic community these principles are commonly referred to as ACE-V [6], which stands for Analyze, Compare, Evaluate, and Verify. Both the principles of individuality and of comparison are critical elements for the facial image examiner to learn.

The principle of individuality states that no two objects in the universe are the same [6]. Identical twins can be distinguished in images by observing features such as blemishes and scars; twins can also be individualized by ear structures and fingerprints. Because identical twins can be individualized, it can be inferred that any other randomly selected two people can be distinguished. A major challenge in identifying people by their facial morphology and geometry is the lack of statistical analyses available to document the percentage of uniqueness. The lack of specific statistics restricts the facial image examiner within the justice system to opinion based testimony. Automated facial recognition systems generate 'probabilities of match'. Without grounding in the actual statistical uniqueness of faces, however, those percentages can be considered subjective as they depend more on the chosen algorithm than observable traits. For instance, it is common for the true match in the output of a facial recognition system to be in a lower position in the gallery than the first position, implying that the system identified another individual as the best match to the probe. Future research must determine the statistics of facial uniqueness to explain the logic of match probabilities generated by the systems.

The principle of comparison is the scientific methodology for assessing the similarities and differences between items or persons. The ACE-V expression of this principle as applied to forensic science was articulated in 1973 for fingerprint, footwear and tire-tread analyses: to analyze the items, to compare their class and individual characteristics, to evaluate the significance of the similarities and dissimilarities, and to form an accurate conclusion that could be reached by a similarly trained examiner [6]. This method also applies to photographic comparisons of persons. Training in this area should be comprehensive and include hands on exercises to familiarize examiners with the methodology.

Image Science. Because the facial comparison is image based, it is important for facial comparison experts to know basic image science. Controlled facial imaging involves fewer variables than uncontrolled surveillance-type images and therefore the Uncontrolled Image Examiner will need to understand more factors than the Standardized Image Examiner. All examiners should be trained in the properties of digital images and cameras, including the basics of sensors and lenses. Additionally, one should understand compression and the effects it can have on an image. For example, the blocky artifacts created with JPEG compression could mask moles, freckles, and other smaller facial features. Also, understanding image perspective will allow facial image examiners to account for potential differences in facial size and shape; even in controlled images the subject is frequently asked to stand too close to the camera. If a person is less than 4 to 6 feet from a camera, regardless of lens or camera-make used, the perspective will distort the person's face, making their nose appear larger and their ears appear smaller than in reality. Thus it is always preferable to have the subject stand farther away from the camera and use an optical zoom with the camera, than to have a person stand very close to the camera.

Advanced training in image science will include more details about optics. It is not uncommon for surveillance-type images to exhibit barrel distortion, where the outer edge of the depicted scene can appear stretched. Should a face be present in this distortion area, the shape of the face can appear significantly different than in a comparison image. Likewise, uncontrolled images are subject to a wide variety of illumination differences which affect the appearance of the human face. For example, a controlled image should depict freckles on a face, where an uncontrolled image taken with a strong flash may depict a bright area on the person's cheeks, obscuring any facial freckles.

Image Processing. An important tool in image analysis is image processing. Basic image processing includes brightness and contrast adjustments, as well as rotations and cropping of images. Rotating images can be beneficial to align features, such as making the line of the eyes horizontal, and cropping can remove background information that is distracting to the facial depiction. Training should include these basic techniques; both knowledge of how they work and when to apply them.

Advanced level training should include several additional image processing techniques, such as sharpening and blurring [7]. Another tool is the separation of color channels in color images to detect subtle facial features. Additionally, an image examiner may be assisted by the creation of an overlay of questioned and known images. An overlay allows the examiner to switch between stacked images to ascertain differences in facial morphology and relative dimensions.

3.2 Facial Specifics

The human face is effectively an elastic sculpture built upon the foundation of bones and muscles, wrapped in skin, and subject to changes with expression, age, and damage. The specifics that need to be included in a training program are the properties of the face, the alteration of the face, and the methods of comparing faces.

Properties of the Face. Starting from the inside out is a logical way to become familiar with the human face. All facial image examiners should be trained in the structure of the bones and muscles of the head. Training should include a link between external expressions and the internal muscle groups involved in making such expressions, in order for the examiner to compare images depicting differing facial expressions. Furthermore, because wrinkles are typically formed perpendicular to the direction of muscle extension, knowledge of the muscles of the head is important to understanding the aging of the face. The outermost layer of the face, the skin, bears the texture that is most critical in individualizations. Commonly it is the co-location of facial blemishes, marks, and patterns thereof that allow a facial image examiner to identify a person to the exclusion of all others. A medical understanding of dermatology is not necessary at either the basic or advanced level of training, however a working knowledge of the visible features within the skin and their evolution is needed. For example, an examiner does not need to distinguish between a mole or a freckle but should know whether those dark marks will increase in number or fade in contrast through time.

Special attention should be given to the characteristics of the ear. Researchers have noted that they have not found any two ears that are alike in all parts and the ear is also stable throughout adulthood [8]. At a basic level, a facial comparison expert must be familiar with the names and nature of the features of the ear. At an advanced level, the history of ear individualization and ear comparisons should be reviewed. There are decades of research concerning the uniqueness of ears that could benefit a facial image expert analyzing uncontrolled images depicting views of ears.

To synthesize the physical properties of the face, training should include an overview of general facial shapes and the terminology used to describe them. Additionally, advanced training should include a review of known statistics regarding face shape.

Alteration of the Face. One of the drawbacks to facial biometrics is that the face is not stable over the life of an individual; nevertheless, it is one of the easiest biometrics to obtain. Alteration of the face can be temporary or permanent, to the texture or the structure. An additional type of alteration in facial image comparisons is alteration of the image itself. Facial image examiners must be aware of the possibilities of disguise and deception in imagery.

Aging of a face is the most common alteration depicted. Automated facial biometric systems are challenged by images of an individual taken years apart, therefore an expert verification can significantly improve the accuracy of the system. The aging process involves changes in both the texture and elasticity of the skin. Lifetime use of muscles will be reflected in the location and depth of facial wrinkles and furrows that develop. Many lifestyle factors will affect the aging of the face, including a history of drug usage, cigarette smoking, or weight changes.

Temporary, or transient, alteration of the face is easily achieved by the use of cosmetics and changes to hair. Makeup and cosmetics can mask the appearance of wrinkles and blemishes. Cosmetic medicine, such as the use of Botox, can also temporarily affect facial creases. Changes in length and color of hair are also transient alterations, although the hairline shape tends to remain stable until receding. Changes in facial hair are also common for both men and women. Women frequently alter their eyebrows and eyelashes, while men typically alter their beards and moustaches. Training should include comparisons of individuals with changes in hair and makeup.

Permanent facial alterations can be intentional or unintentional. Unintentional alterations include the formation of scars and the loss of hair. Intentional permanent alterations are can be achieved through plastic/cosmetic surgery. An examiner should be trained in recognizing these alterations to be able to account for such differences within comparisons.

The presence of possible image manipulation is a concern when individuals are allowed to submit their own images, such as visa applications, or within uncontrolled images obtained from unknown or unrestricted sources, such as second- or third-generation videos obtained from terrorist organizations. Training in the detection of image manipulation can be its own substantial training course; nonetheless some basics should be included in a training program for facial image examiners. Standardized image examiners should be able to recognize changes in the orientation of images, such as being flipped left to right or changes in color. When analyzing uncontrolled images, an examiner at the advanced level should be able to test for image changes due to a composite (i.e., "cut and paste"), such as differing perspective, focus, or lighting across an image.

3.3 Legal Issues

The legal issues relating to facial comparisons will vary by country and region. However, it is critical for any facial comparison expert who may testify in the judicial system to be aware of the legal issues and case law pertinent to facial comparisons within their region. All facial image examiners involved in the judicial system, i.e. forensic examiners, should be trained in the legal history of photographic comparisons, particularly facial comparisons, to include specific case law that helped, or hurt, the presentation of the science in court. Other legal issues to be covered involve the admissibility of facial comparisons in court. Within the United States, this involves familiarity with the Federal Rules of Evidence and specific rulings, such as Frye, Daubert, and Kumho Tire [9]. Lastly, a forensic facial image examiner should be trained in *how* to testify to their comparisons in court. Training should involve the creation of visual media to assist in presenting the comparison, as well as an overview of approaches to explaining the science to a layman.

4 Summary

Until a 'lights out' system can be developed for facial recognition, human examiners are a necessity. Similar to fingerprint examiners, human facial image examiners with sufficient training can be integrated into the facial recognition biometric chain to increase the accuracy and verify the results output from automated systems. Experts analyzing controlled images can be trained to a basic level because there are fewer variables in standardized images, while experts analyzing uncontrolled images must be aware of additional factors, such as variations in lighting, perspective and optics. The development of a successful training program is the first step in creating a certification program that will ensure a set level of knowledge is achieved by all facial image examiners.

References

1. Rogers, S.: The Personal Identification of Living Individuals, Charles C Thomas, Springfield Illinois (1986)
2. Brugioni, D.: Photo Fakery: The History and Techniques of Photographic Deception and Manipulation. Brassey's, Dulles Virginia (1999)
3. United States v. Cairns, 480 F.2d 927 (6th Cir. 1973); United States v. Brown, 501 F.2d 146 (9th Cir. 1974); United States v. Trejo, 501 F.2d 138 (9th Cir. 1974)
4. Woodward, J., Orlans, N., Higgins, P.: Biometrics. McGraw-Hill Professional, Blacklick Ohio (2003)
5. ANSI/NIST-ITL 1-2000
6. Tuthill, H., George, G.: Individualization: Principles and Procedures in Criminalistics, 2nd edn. Lightning Powder Company, Jacksonville Florida (2002)
7. Scientific Working Group on Imaging Technology (SWGIT), `http://www.swgit.org`
8. van der Lugt, C.: Earprint Identification. Elsevier, Gravenhage (2001)
9. Frye v. United States, 293 F. 1013 (D. C. Cir. 1923); Daubert v. Merrell Dow Pharmaceuticals, Inc., 509 U.S. 579 (1993); Kumho Tire Co. v. Carmichael, 526 U.S. 137, 154 (1999)

Face Gender Classification on Consumer Images in a Multiethnic Environment

Wei Gao and Haizhou Ai

Computer Science and Technology Department,
Tsinghua University, Beijing 100084, China
ahz@mail.tsinghua.edu.cn

Abstract. In this paper, we target at face gender classification on consumer images in a multiethnic environment. The consumer images are much more challenging, since the faces captured in the real situation vary in pose, illumination and expression in a much larger extent than that captured in the constrained environments such as the case of snapshot images. To overcome the non-uniformity, a robust Active Shape Model (ASM) is used for face texture normalization. The probabilistic boosting tree approach is presented which achieves a more accurate classification boundary on consumer images. Besides that, we also take into consideration the ethnic factor in gender classification and prove that ethnicity specific gender classifiers could remarkably improve the gender classification accuracy in a multiethnic environment. Experiments show that our methods achieve better accuracy and robustness on consumer images in a multiethnic environment.

Keywords: Boosting tree, gender classification, multiethnic environment.

1 Introduction

Face vision research has achieved significant advancement in the past decade especially on face detection and face alignment or facial feature location technologies that can readily provide effective tools to extract faces from raw images. With faces having being extracted, demography classification that includes gender, ethnicity and age become an interesting topic due to its potential applications in photo album management, shopping statistics for marketing, visual surveillance, etc. Unlike ethnicity and age estimation, gender classification has attracted more attention in face classification literature since it is the most basic information from a face which human can have a very clear division in perception.

In the early days, most of researches in gender classification are about human's perceiving for gender from a psychology point of view, where the computer is used just as an assistant tool and no automatic gender classification system is developed. More recently neural network methods were firstly used in gender classification. Golomb et al. [1] trained a gender classifier "SexNet" with a two-layer neural network on 90 facial images and achieved an accuracy of 91.9%. Gutta et al. [2] trained a neural network on 3000 faces from FERET dataset and decreased the error rate to 4%.

M. Tistarelli and M.S. Nixon (Eds.): ICB 2009, LNCS 5558, pp. 169–178, 2009.
© Springer-Verlag Berlin Heidelberg 2009

Balci [3] used eigenfaces and trained a Muti-Layer Perceptron on FERET Dataset to analyze which eigenface contributed to gender classification. Later, Moghaddam et al. [4] used SVM and achieved an accuracy of 96.6% on FERET's 1755 faces which was the best result on this set. However for human, only about 96% in accuracy can be achieved using face information only on this problem according to [5]. BenAbdlkader et al. [6] extracted both local and holistic features and used LDA and SVM on a dataset of about 13,000 faces that achieved 94.2% correct rate. Yang et al. [7] used LDA, SVM and Real AdaBoost respectively on 11,500 snapshots that achieved about 96.5% correct rate. Another interesting work, Lapedriza et al. [9] analyzed the external face's contribution to gender classification and developed a classifier [10] based on both internal and external faces that resulted in 91.7% correct rate on FRGC dataset.

In overview of most of previous works on gender classification, one thing in common is that all those face images used in the experiments are caught in constraint environments, and further they are using 5-fold CV verification method to evaluate performance that implies their test sets have the same distributions as their training sets. So the generalization ability on independent sets is still a problem. Shakhnarovich et al. [8] reported that gender classification on images from internet by AdaBoost algorithm can only achieve 78% accuracy and by SVM can only achieve 77% accuracy. Besides, the ethnic factor is less considered before. The gender classifier trained can not guarantee good generalization ability in a multiethnic environment.

In this paper we target at face gender classification on consumer images of which faces vary greatly in pose, illumination and expression. Active Shape Model (ASM) is used for normalization and a boosting tree is trained on about 10,100 face images from unconstrained environment. Comparative experiments with other methods including SVM and Real AdaBoost on independent test sets are reported to show its effectiveness. To handle a multiethnic environment, we treat face's ethnicity as a latent variable and the ethnicity specific gender classifiers are trained.

The rest of this paper is organized as follows: Section 2 gives an overview of our gender classification system, Section 3 describes the boosting tree classification method, Section 4 gives the gender classifier structure in multiethnic environment, and finally, Section 5 is the conclusion.

2 Gender Classification on Consumer Images

By consumer images we mean those digital photo images caught by popular users of digital cameras. Compared with faces caught from constraint environment, such as snapshots, faces in consumer images are more diverse in resolution, makeup, as well as in illumination, pose and expression (as shown in Fig.1), therefore they are more challenging to deal with in classification. In this situation, preprocessing and normalization become a critical issue.

As for gender classification methods, AdaBoost has been proved very effective in both accuracy and speed in the literature. Since AdaBoost is much faster than SVM, for potential practical applications we choose it to develop a boosting based method for gender classification. In fact, AdaBoost can mine discriminative features automatically from a large set by giving miss-classified samples more attention. Yang et al. [7] and Lapedriza et al. [9][10] showed that boosting algorithm achieved comparative accuracy with SVM in gender classification problem. But the main drawback of this algorithm is overfitting after over a certain number of iterations which means poor generalization ability in other dataset, especially on those with high intra-class variations. Since faces in consumer images are with great intra-class variations, it is found very difficult to learn a single boosting classifier as in [7] [9] [10] in our experiments, therefore divide and conquer strategy becomes necessary for better performance.

Fig. 1. Faces from Consumer Images

For a flowchart of our gender classification system, see Fig.2 First a face detection algorithm [12] is used to detect faces from consumer images and then a variation of ASM method [13] is used to locate 88 facial feature points for each detected face. For normalization, a shape free face texture is acquired by triangular warping from a shape aligned face to the 32×32 mean face shape. Compared with the conventional eye-center normalization in face recognition approaches, this method eliminates some pose and expression variations.

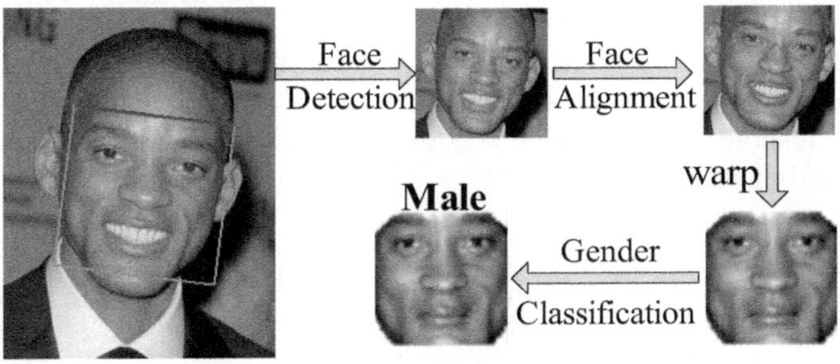

Fig. 2. Face Gender Classification Flowchart

3 Gender Classification by Boosting Tree

3.1 Probabilistic Boosting Tree

The probabilistic boosting tree (PBT) method is originally proposed by Tu [11] to deal with the problem of object categorization in natural scenes. It is a new divide-and-conquer strategy with soft probability boundary. The boosting tree method find the classification boundary step wisely by putting the ambiguous samples to both left and right sub-trees as shown in Fig.3 (left). Gradually, more similar samples will be sent to sub-tree nodes which results in a reduction of intra-variation. The boosting tree can approach the target posterior distribution by tree expansion.

In the boosting tree structure, each node is a strong classifier trained by AdaBoost algorithm. We adopt the LUT-based Real AdaBoost method in [14] and use simple Haar-like features [16] to construct weak classifiers. After T iterations of learning, the strong classifier has the form:

$$H(\boldsymbol{x}) = \sum_{t=1}^{T} \alpha_t h_t(\boldsymbol{x})$$

where $h_t(\boldsymbol{x})$ is the t-th weak classifier, α_t is the normalize coefficient and $H(\boldsymbol{x})$ is the output confidence.

To construct a boosting tree, the confidence output by the root node is further mapped to probability by sigmoid function as proposed in [15]:

$$q(+1 \mid x) = \frac{\exp\{2H(x)\}}{1+\exp\{2H(x)\}}, \quad q(-1 \mid x) = \frac{\exp\{-2H(x)\}}{1+\exp\{-2H(x)\}}$$

where $q(+1 \mid x), q(-1 \mid x)$ donates the sample's probability to be positive or to be negative respectively. Based on the probability above, we split the training set into sub-trees. This is done by choosing a threshold parameter ε to divided probability into three intervals as shown in Fig.3 (left), that is, the left tree with samples in $[0, \frac{1}{2} - \varepsilon)$, the right tree with samples in $(\frac{1}{2} + \varepsilon, 1]$ and the ambiguous samples in $[\frac{1}{2} - \varepsilon, \frac{1}{2} + \varepsilon]$ will be added into both the left and the right sub-tree (as show in Fig.3 (right)). In practice, instead of using a fixed threshold for every tree nodes as in [11], we choose a variable threshold for each node according to the distribution of samples to make the tree trained more balanced.

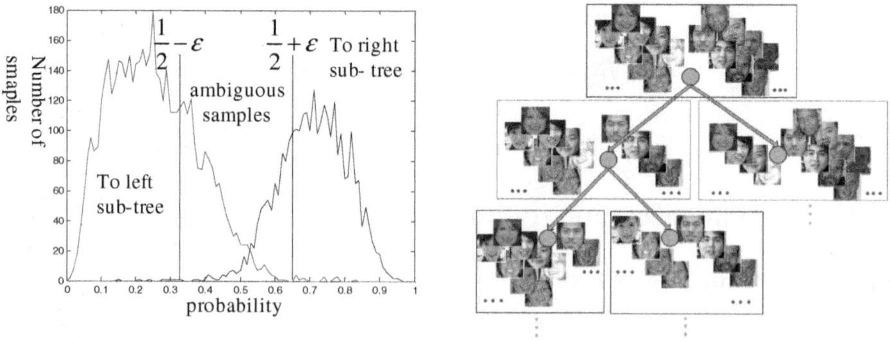

Fig. 3. (left) Histogram of probability distribution of positive and negative samples and three intervals divided. (right) Probabilistic boosting tree structure. Faces in the left and right of the nodes correspond to positive and negative samples.

The above procedure is repeated to construct a tree of which each node is a confidence-rated strong classifier learned by Real AdaBoost algorithm.

With the PBT trained, given a face sample, its normalized face is fed into the root node of the tree to start the decision procedure iteratively. At each node the probability to be a positive sample and that to be a negative sample are denoted as $q_P(+1 \mid x)$ and $q_N(-1 \mid x)$ respectively. And then it will be fed into both its left and right sub-tree to compute its corresponding probabilities $p_{right}(y)$ and $p_{left}(y)$. The final decision is computed as:

$$p(y \mid x) = q(+1 \mid x) p_{right}(y) + q(-1 \mid x) p_{left}(y) .$$

3.2 Experiment Result

Experiments are carried out on two kinds of face image datasets: snapshot datasets, and consumer image datasets. And for each kind of face images, two face sets are established: one for training and 5-fold CV test, while the other is totally independent from the training set which is used to judge the algorithms' generalization ability.

The snapshot training dataset (Snapshots) consists of 15,300 faces from controlled environment. The snapshot faces are all frontal with similar lighting condition and expressions. And the independent snapshot set consists of 1800 faces (Independent Snapshots). The consumer image training dataset consists of about 10,100 Mongoloid faces in real environment (Consumer Images) with significant changes in poses, illumination and expressions. Similarly another independent consumer image dataset is collected which consists of 1,300 faces (Consumer Images). All the face datasets collected above contain nearly equal number of samples for each gender. The boosting tree method is compared with SVM and Real AdaBoost on those datasets.

Table 1 gives both the results on the Snapshot dataset under the 5-fold CV verification protocol and the results tested on two other datasets of which 'All consumer images' means the sum of the two consumer image datasets. The SVM method uses Gaussian kernel. The Real AdaBoost method uses a strong classifier learned after 500 rounds. The boosting tree is composed of 15 strong classifier nodes and its depth is 4. The generalization ability is evaluated on the independent snapshots and consumer image dataset.

Table 1. Results on snapshot dataset under 5-fold CV and results tested on two independent datasets

	5-fold CV	Independent Snapshots	All consumer images
SVM	96.38	87.89	66.37
AdaBoost	96.50	90.41	80.50
PBT	97.13	93.48	82.07

Table 2. Results on Consumer Images under 5-fold CV and results tested on two independent datasets

	5-fold CV	Independent Consumer Images	All snapshots
SVM	90.24	88.13	90.72
AdaBoost	94.12	88.61	92.89
PBT	95.51	92.84	93.71

Table 2 gives both the results on the Consumer Image dataset and the results tested on two other datasets of which 'All snapshots' means the sum of the two snapshot image datasets. As before, the Real AdaBoost method uses a strong classifier learned after 500 rounds, and the PBT with 15 nodes and a depth of 4.

From the above results, we can see from Table 1 that all the three methods achieved comparative performance in snapshot datasets while their generalization ability on the consumer images is bad. However the PBT achieved better generalization ability than the other two methods on independent snapshot dataset. From the Table 2, we can see on the consumer images, PBT's generalization ability remarkably outperforms SVM and Real AdaBoost, and their generalization ability on the snapshot dataset is comparative with the classifier directly trained on snapshots. So, although there are variations between indoor controlled environments and unconstraint environments, the classifier trained on real consumer images from unconstraint environments can achieve better generalization ability. We can conclude that the PBT method can describe the classification boundary more accurately than the other two.

4 Gender Classification in a Multiethnic Environment

Compared with gender classification, ethnicity classification attracts less attention in demography classification. Intuitively ethnicity classification could be done almost in the same way as gender classification technically. But different from gender classification, ethnicity classification is much harder and sometimes even human can not have a very clear division for ethnicity in perception. In literature, G. Shakhnarovich et al. [8] divided ethnicity into two categories: Asian and Non-Asian, while in [7] [18] [19] three categories with Mongoloid, Caucasoid and African were adopted, and in [17] four ethnic labels with Caucasian, South Asian, East Asian, and African are used. In this paper, we use three ethnic labels with Mongoloid, Caucasian and African.

4.1 Generic Gender Classifier

We collect 2400 Mongoloid males and 2500 Mongoloid females, 2400 Caucasoid males and 2400 Caucasoid females, and 1800 African males and 1600 African females from consumer images for training. Another independent test set is collected which contains 400 faces for each ethnicity with half for each gender. We train two kinds of gender classifiers: first we train gender classifier for each ethnicity respectively and the results on test set are show in Table 3; and second, we train a gender classifier using all the males and females in the training set and the results on test set are shown in Table 4. All the gender classifiers in this section are trained in the same way as in Section 3.

Table 3. Gender classifier for each ethnicity respectively (MC, CC and AC mean gender classifier on Mongoloid, Caucasoid and African respectively)

	Mongoloid		Caucasoid		African	
	Male	Female	Male	Female	Male	Female
MC	90.1%	92.9%	95.7%	61.3%	93.5%	50%
CC	78.5%	89.1%	96.8%	88.7%	94%	69%
AC	52%	95.5%	55.5%	96%	94%	82%

Table 4. Generic gender classifier for all ethnicities

Mongoloid		Caucasoid		African	
Male	Female	Male	Female	Male	Female
84.5%	93.5%	86%	93%	95.5%	77%

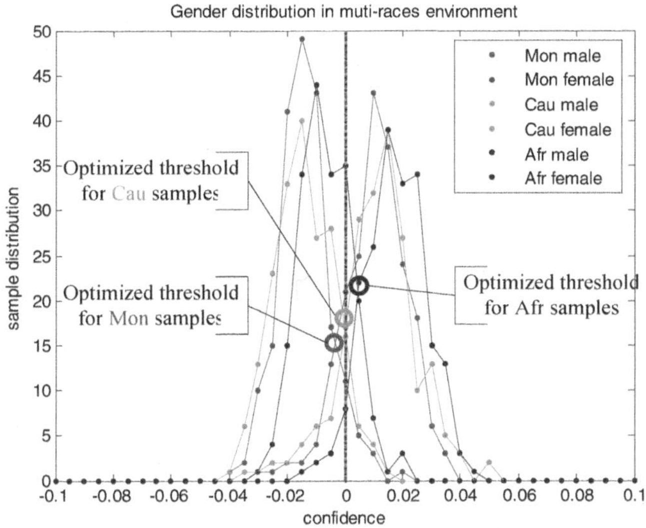

Fig. 4. Confidence distribution for different ethnicity in generic gender classifier (Positive for male and negative for female)

We can conclude from Table 3 that the gender classifier behaves well on the ethnicity it is trained on while can't achieve good results on other ethnicities. When we train a generic gender classifier for all ethnicities, the result as in Table 4 is not as good as training specific gender classifier as in Table 3. This can be explained by Fig.4, in generic gender classifier, we try to find the same threshold for all ethnicity faces, which in fact is not a best decision boundary for each ethnicity. As show in Fig.4, the decision boundary for Africans is apt to male side while the decision boundary for Mongoloid is apt to female side. That is why the generic gender classifier is inclined to classify Africans as males and Mongoloid as females as shown in Table 4.

4.2 Ethnicity Specific Gender Classifier

Enlightened by the analysis in Section 4.1, we propose an ethnicity specific gender classification framework as shown in Fig.5 for multiethnic environment. In the new framework, the ethnicity is treated as a latent variable for gender classification. We can formalize the procedure as:

$$P(G \mid F) = \sum_{E} P(G \mid E, F) P(E \mid F)$$

where G, E and F represent gender, ethnicity and face respectively.

We trained an ethnicity classifier with samples collected in Section 4.1 using AdaBoost.MH [20] and Haar-like features [16]. Gender classifiers on each ethnicity are from Section 4.1. The results of ethnicity specific gender classifier are compared with the generic gender classifier in Table 5. We can see that the ethnicity specific gender classifier performs better than the generic gender classifier, especially on Mongoloid males and African females, which is consistent with analysis of Fig.4. This experiment hints that faces from different ethnicity have different gender feature and in a multiethnic environment, gender classifier could be better by taking ethnicity as a latent variable. Some results are shown in Fig.6.

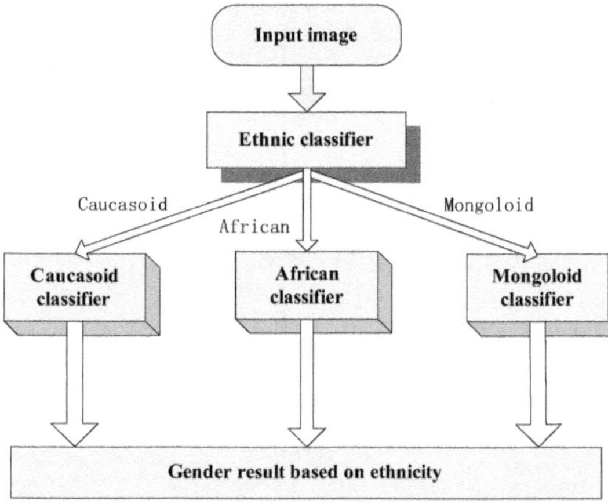

Fig. 5. Ethnicity specific gender classification framework

Table 5. Comparison of Generic Gender Classifier (GGC) and Ethnicity Specific Gender Classifier (ESGC)

	Mongoloid		Caucasoid		African	
	male	female	male	female	male	female
GGC	84.5%	93.5%	86%	93%	95.5%	77%
ESGC	89%	93%	86.3%	96.6%	94%	82%

Fig. 6. Gender classification results on consumer images in multiethnic environment

5 Conclusion

In this paper, a PBT approach for face gender classification on consumer images is presented. Faces on consumer images vary greatly in pose, illumination and expression that make it much more difficult than in the constrained environments. In this approach, Active Shape Model (ASM) is used for normalization and a PBT is trained for classification by which through divide and conquer strategy a more accurate classification boundary on consumer images is achieved. Experiments on both snapshots and consumer images show that the PBT method is better than the SVM and Real AdaBoost methods.

We also discussed the ethnicity factor in gender classification experimentally, to our best knowledge there is no such work before. We find that faces from different ethnicity have different gender feature, and gender classifier trained on a specific ethnicity could not get good generalization ability on other ethnicities. Finally, we improve the performance of gender classification in a multiethnic environment by treating ethnicity as a latent variable.

However, currently we can only deal with frontal or near frontal faces from consumer images. And the accuracy of gender classifier on Africans is not as high as on Mongoloid and Caucasoid. Another issue we have not considered is the impact of age on the face gender classification. Those are our future work.

Acknowledgement

This work is supported by National Science Foundation of China under grant No.60673107, and it is also supported by a grant from HP Corporation.

References

1. Golomb, B.A., Lawrence, D.T., Sejnowski, T.J.: SEXNET: A Neural Network Identifies Sex from Human Faces. In: NIPS 1990 (1990)
2. Gutta, S., Wechsler, H., Phillips, P.J.: Gender and Ethnic Classification of Face Images. In: FG 1998 (1998)
3. Balci, K., Atalay, V.: PCA for Gender Estimation: Which Eigenvectors Contribute? In: ICPR 2002 (2002)
4. Moghaddam, B., Yang, M.H.: Learning Gender with Support Faces. IEEE Trans. on Pattern Analysis and Machine Intelligence 24(5) (May 2002)
5. Hayashi, J., Yasumoto, M., Ito, H., Koshimizu, H.: Age and Gender Estimation based on Wrinkle Texture. In: ICPR 2002 (2002)
6. BenAbdelkader, C., Griffin, P.: A Local Region-based Approach to Gender Classification from Face Images. In: CVPR 2005 (2005)
7. Yang, Z., Li, M., Ai, H.: An Experimental Study on Automatic Face Gender Classification. In: ICPR 2006 (2006)
8. Shakhnarovich, G., Viola, P.A., Moghaddam, B.: A Unified Learning Framework for Real Time Face Detection and Classification. In: AFG 2002 (2002)
9. Lapedriza, A., Masip, D., Vitrià J.: Are External Face Features Useful for Automatic Face Classification. In: CVPR 2005 (2005)
10. Lapedriza, A., Manuel, M.J., Jiménez, J.M., Vitrià, J.: Gender Recognition in Non Controlled Environments. In: ICPR 2006 (2006)
11. Tu, Z.: Probabilistic Boosting-Tree: Learning Discriminative Models for Classification, Recognition, and Clustering. In: ICCV 2005 (2005)
12. Huang, C., Ai, H., Wu, B., Lao, S.: Boosting nested cascade detector for multi-view face detection. In: ICPR 2004 (2004)
13. Zhang, L., Ai, H., Xin, S., Huang, C., Tsukiji, S., Lao, S.: Robust Face Alignment Based on Local Texture Classifiers. In: ICIP 2005 (2005)
14. Wu, B., Ai, H., Huang, C.: LUT-Based AdaBoost for Gender Classification. In: Kittler, J., Nixon, M.S. (eds.) AVBPA 2003. LNCS, vol. 2688. Springer, Heidelberg (2003)
15. Schapire, R.E., Singer, Y.: Improved Boosting Algorithms Using Confidence-rated Predictions. Machine Learning 37, 297–336 (1999)
16. Viola, P., Jones, M.: Fast Multi-view Face Detection. In: Proc. of CVPR (2001)
17. Gutta, S., Huang, J.R., Jonathon, P., Wechsler, H.: Mixture of Experts for Classification of Gender, Ethnic Origin, and Pose of Human Faces. IEEE Transactions on Neural Networks
18. Yang, Z., Ai, H.: Demographic classification with local binary patterns. In: Lee, S.-W., Li, S.Z. (eds.) ICB 2007. LNCS, vol. 4642, pp. 464–473. Springer, Heidelberg (2007)
19. Hosoi, S., Takikawa, E., Kawade, M.: Ethnicity Estimation with Facial Images. In: FG 2004 (2004)
20. Schapire, R.E., Singer, Y.: Improved Boosting Algorithms Using Confidence-rated Predictions. Machine Learning (1999)

Multi-View Face Alignment Using 3D Shape Model for View Estimation

Yanchao Su[1], Haizhou Ai[1], and Shihong Lao[2]

[1] Computer Science and Technology Department, Tsinghua University
[2] Core Technology Center, Omron Corporation
ahz@mail.tsinghua.edu.cn

Abstract. For multi-view face alignment (MVFA), the non-linear variation of shape and texture, and the self-occlusion of facial feature points caused by view change are the two major difficulties. The state-of-the-art MVFA methods are essentially view-based approaches in which views are divided into several categories such as frontal, half profile, full profile etc. and each of them has its own model in MVFA. Therefore the view estimation problem becomes a critical step in MVFA. In this paper, a MVFA method using 3D face shape model for view estimation is presented in which the 3D shape model is used to estimate the pose of the face thereby selecting its model and indicating its self-occluded points. Experiments on different datasets are reported to show the improvement over previous works.

Keywords: Active Shape Model, face alignment, 3D face model.

1 Introduction

Automatically locating facial feature points on face images, i.e. face alignment (FA) is a critical task in many face related computer vision areas such as 3D face labeling, expression analysis and face recognition. For face alignment, there are two fundamental approaches, Active Shape Model (ASM) [1] and Active Appearance Model (AAM) [2]. Many variations of these two methods have been developed to improve their robustness and accuracy [3-5]. While for frontal FA there is already some robust algorithm can be used in practice [4], for MVFA where face is with large view change, it remains a challenging problem since both the shape and the texture of face in images change dramatically when the view changes.

In the literature of MVFA, non-linear modeling method such as Gaussian Mixture Model [6], kernel PCA [7], Bayesian Mixture Model with learning visibility of label points [8] and view-based methods such as view based DAM [9] and view-based ASM [10] are developed which are mainly 2D approaches with no appealing to 3D face information. Due to the intrinsic difficulties caused by face appearance changes in 2D face images of a 3D face, MVFA is still not a solved problem.

The state-of-the-art MVFA methods are essentially view-based approaches in which views are divided into several categories such as frontal, half profile, full profile etc. and each of them has its own shape and texture models. Since the texture

M. Tistarelli and M.S. Nixon (Eds.): ICB 2009, LNCS 5558, pp. 179–188, 2009.
© Springer-Verlag Berlin Heidelberg 2009

model used in local search of each label point of a particular shape model depends on its view category, these methods are very sensitive to the estimation of the view category. When the initial view is not predicted correctly, the results of local search become unreliable. And if the estimation of shape parameter does not deal with the potential outliers, ASM approach will fail.

In the original view-based ASM method [10] a non-linear optimization method for model selection is used in which each feature point is weighed dynamically so that only the feature points that are consistent with the shape model will have large weights, while the effect of outliers will be eliminated. Since this method does not completely depend on the local search of each label points, it is more robust against the initial view and cluttered background.

View based methods switch between different models of different views to cover the non-linear space of multi-view faces, thus, the selection of models, in other words, the estimation of view is a critical step in the algorithm. Although the overlapped definition of view ranges can mitigate the error caused by improper initialization of view, the automatic view estimation in the alignment procedure is still an important problem to be solved.

There are other MVFA approaches using 3D face model. In [11], view-based local texture models and a sparse 3D shape model which are trained using synthesized faces are integrated in an ASM-like framework to align faces with view changes. In [12], a parameterized 3D deformable face model is used to help with view based ASM but building its 3D face model is a very tough work.

In this paper, we combine a view-based ASM and a simple 3D face shape model built on 500 3D-scanned faces [13] to build a fully automatic MVFA system. Initialized by a multi-view face detector [14], we first use view based local texture model to local search the feature points around the initial shape [10], then a 3D face shape is reconstructed from those points using the 3D face shape model. According to the reconstructed 3D shape, we can get its view information from which self-occluded points can be indicated, and then the 2D shape model of this view is adopted to refine the observed non-occluded shape by non-linear parameter estimation.

2 View-Based Active Shape Model

In the case of MVFA, the shape and texture of faces in images change dramatically when the view changes. A single PCA model can only represent face shapes with limited view change due to the non-linear change of face shape. And further textures around the label points with large view changes are also hard to be characterized in a single model. So as in [10] we divide views into 7 categories and for each view we train a set of local texture models and a shape model. Therefore in MVFA, a face shape is represented by a PCA model of the view v it belongs to:

$$S = T_q\left(U_v \cdot p + \overline{S}_v\right) \tag{1}$$

So the objective of MVFA is to find the best PCA parameter p and pose q with some view v:

Fig. 1. Mean shapes of 7 view categories

$$(p,q) = \arg\max P\left(S\middle|v\right)\prod P_i(I \mid (x_i, y_i), v) \tag{2}$$

where $P(S|v) \sim N(\overline{S}_v, diag(\lambda_{1,v}, \ldots, \lambda_{m,v}))$ is the shape prior of specified view v ($\lambda_{i,v}$ is the i-th eigenvalue of the covariance matrix of shape samples with view v). And $P_i(I \mid (x_i, y_i), v)$ is the probability of the point (x_i, y_i) to be the i-th label point, which is determined by local texture model. The local texture model of each view is trained using Haar-like feature based boosted classifier [4] which given a texture patch can output the likelihood of this patch to be around the i-th label point.

The whole alignment procedure is as follow:

1. Given an image I, the bounding box and the estimated view v_0 is provided by the face detection module. And the initial shape S_0 is estimated by fitting the mean shape of v_0 into the bounding box of face. See Figure 1 for mean shape illustration.
2. For each label point, search locally around its current position for the best displacement (x_i^*, y_i^*) with the largest likelihood using the local texture models of current view.
3. Parameter estimation: for each view, estimate the parameter p_v' and q_v' using non-linear parameter estimation and then find the best view v' and its corresponding parameter p' and q' with the minimum reconstruction error of the shape.
4. Set the new shape $S = T_{q'}\left(U_{v'} \cdot p' + \overline{S}_{v'}\right)$, and current view $v = v'$
5. Iterates from step 2 until the shape S converged.

In the optimization of MVFA, the proper selection of the hidden view v, is a critical step which will severely affect its accuracy and robustness. So we have to develop a robust pose estimation method to select a proper view when an inaccurate initial view is given by the face detector.

3 3D Face Shape Model

While a 2D face model suffers when view changes, a 3D face model can easily overcome this obstacle. The 2D ASM deals both the intrinsic change (caused by the change of expression and different person) and the extrinsic change (caused by image projection) with a single linear model; while the 3D shape model reflects only the intrinsic change.

Similar as the 2D face shape model, a 3D face shape can be denoted by a list of 3D coordinates $S_{3d}=[x_1,y_1, z_1,\ldots,x_n,y_n,z_n]$ and here we use n=88 points in 3D-scanned faces which correspond to the 88 face alignment feature points in 2D face images. And our

3D face shape model is a PCA point distribution model of the 88 feature points built on a 3D-scanned BJUT-3D Face Database [13].

The first step to construct the 3D face shape model is to achieve the 3D locations of the 88 feature points in each 3D-scanned face as follows:

Each 3D-scanned face is rendered at various views through orthogonal projection, and then the 2D ASM is employed to obtain 2D feature point locations in the rendered images. And the corresponding 3D location of each feature point can be achieved by the following method:

Given a projection matrix (in fact, it is a coordinate transformation matrix) $P_i=(U_i^T\ V_i^T\ W_i^T)$, the projected coordinate of the i-th feature point $X=S_{3d,i}$ is $(x_i, y_i, z_i)^T=P_iS_{3d,i}$. The 2D ASM gives an observation of x_i as $\tilde{x}_i$. The depth channel of the rendered face gives an observation of Z_i as $\tilde{z}_i = \phi(\tilde{x}_i, \tilde{y}_i)$. We assume the error p.d.f. of the 2D ASM is Gaussian:

$$\begin{aligned} \tilde{x}_i &\sim N(x_i, \sigma_i^2) \\ \tilde{y}_i &\sim N(y_i, \sigma_i^2) \end{aligned} \tag{3}$$

For simplicity we assume the errors on different axes distribute independently. Thus the error of can be derived as

$$\begin{aligned} \tilde{z}_i &= \phi(\tilde{x}_i, \tilde{y}_i) \\ &= \phi(x_i + \delta x_i, y_i + \delta y_i) \\ &\approx \phi(x_i, y_i) + \delta x_i \phi_x(x_i, y_i) + \delta y_i \phi_y(x_i, y_i) \\ &= z_i + \delta x_i \phi_x(x_i, y_i) + \delta y_i \phi_y(x_i, y_i) \\ \tilde{z}_i &\sim N(z_i, \theta_i^2) \end{aligned} \tag{4}$$

where $\theta_i^2 = (\phi_x^2(x_i, y_i) + \phi_y^2(x_i, y_i))\sigma_i^2$

Rotating the 3D face and doing 2D ASM on the rendered faces give several sets of projection matrix and observed coordinates $(P_i, \tilde{x}_i, \tilde{y}_i, \tilde{z}_i)$. The log joint likelihood is:

$$\begin{aligned} L(X) &= \log\left(P((\tilde{x}_i, \tilde{y}_i, \tilde{z}_i), i = 1 \cdots n \mid X)\right) \\ &= \sum_i \left(\log P(\tilde{x}_i \mid X) + \log P(\tilde{y}_i \mid X) + \log P(\tilde{z}_i \mid X)\right) \\ &= \lambda - \sum_i \left(\frac{(\tilde{x}_i - U_i X)^2}{2\sigma_i^2} + \frac{(\tilde{y}_i - V_i X)^2}{2\sigma_i^2} + \frac{(\tilde{z}_i - W_i X)^2}{2\theta_i^2} \right) \end{aligned} \tag{5}$$

where λ is constant with respect to X.

The maximum likelihood estimation can be obtained analytically by letting the derivative to be zero.

$$X_{ML} = \arg\max_{X} L(X)$$
$$\nabla X = 0 \tag{6}$$

In practice the 2D ASM is not always accurate. So we use RANSAC algorithm for robust estimation of *X*.

After achieving the 3D locations of 88 feature points for each of the 500 3D-scanned faces in BJUT-3D Face Database [13], we select 389 3D face shapes among them which are good estimations to build the 3D point distribution model using PCA as the 3D face shape model.

$$S_{3d} = U \cdot p + \overline{S}_{3d} \tag{7}$$

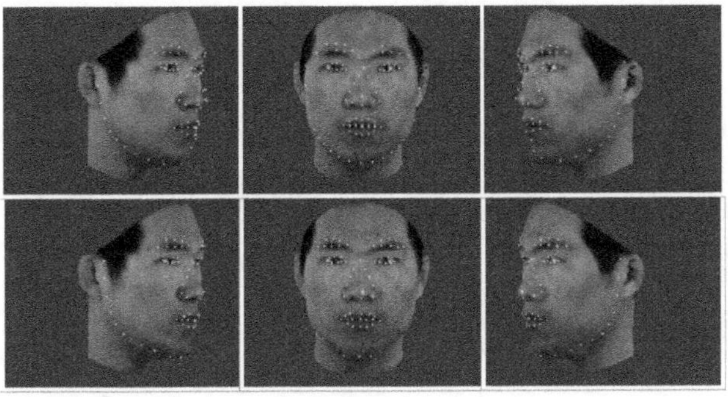

Fig. 2. Achieving 3D locations of feature points. First row shows 2D ASM results on different views. Second row shows reconstructed 3D position of feature points on those views.

4 View Estimation Using 3D Shape Model

Given an input image, suppose the orthogonal projection holds, the 2D shape of the face denoted as S2d in the image should be:

$$S_{2d,i} = M \cdot P \cdot S_{3d,i} + t \tag{8}$$

Each label point $S_{3d,i}$, which is determined by the PCA parameter p in equation (7), is first transformed (scaled and rotated) by the transform matrix P, then projected into the image plane orthogonally with translation t. Where P is an orthogonal projection matrix and $M = \begin{bmatrix} 1 & 0 & 0 \\ 0 & 1 & 0 \end{bmatrix}$ is the projection matrix.

According to equation (8), given a candidate 2D face shape S_{2d}, we could get the PCA parameter p and the pose information (*s*, *R*, *t*) using minimum square error estimation:

$$(P,t,p) = \arg\min \sum_{i} \left\| M \cdot P \cdot \left(U \cdot p + \overline{S}_{3d,i} \right) + t - S_{2d,i} \right\|^2 \tag{9}$$

We first reform the object function by denoting A=M·P, so that:

$$(A,t,p)=\arg\min\sum_i\left\|A\cdot\left(U\cdot p+\overline{S}_{3d,i}\right)+t-S_{2d,i}\right\|^2 \tag{10}$$

where A has the constraint:

$$a_{1,1}^2+a_{1,2}^2+a_{1,3}^2=a_{2,1}^2+a_{2,2}^2+a_{2,3}^2,\ a_{1,1}a_{2,1}+a_{1,2}a_{2,2}+a_{1,3}a_{2,3}=0 \tag{11}$$

which makes A an orthogonal projection.

We can solve the above optimization problem by optimizing pose parameter (A,t) and shape parameter p alternatively in an iteration procedure as follows:

1. Given p, solve (A,t)

$$(A,t)=\arg\min\sum_i\left\|A\left(U\cdot p+\overline{S}_{3d,i}\right)+t-S_{2d,i}\right\|^2 \tag{12}$$

An affine projection (A',t) can be estimated analytically and then we can get an orthogonal projection A by optimizing the object function using gradient descend initiated with the affine projection.

2. Given (A,t), solve p

$$p=\arg\min\sum_i\left\|A\left(U\cdot p+\overline{S}_{3d,i}\right)+t-S_{2d,i}\right\|^2 \tag{13}$$

It is a linear MSE and the solution goes straight forward.

Step 1 and Step 2 are iterated until the convergence of the reconstruction error.

5 Automatic MVFA

Given a face image, we initialize our algorithm by applying multi-view face detection [14] which provides a bounding box and a roll angle of the face. The roll angle corresponds to 5 view categories in {*-90°,-45°,0°,45°,90°*}. We select the initial view according to this angle and then compute the initial 2D shape by rotating and scaling the mean shape of initial view to fit the bounding box of face. Then the algorithm goes iteratively as follow:

1. Local Search: For the *i*-th label point, compute the likelihood $P(I|(x_i,y_i),v)$ using the local texture model of current view on every point around the current location of label point i, then select the best candidates $\{(x_i^*,y_i^*)\}$ with the largest likelihood as the new location. The observation shape is $S_{2d}^*=\{(x_i^*,y_i^*)\}$.
2. Pose Estimation using the 3D face shape model: use the observation 2D shape and the 3D shape model to estimate the pose parameter (A,t) and the shape parameter p of the 3D face model. Then compute the roll angle according to the projection matrix A and select current view. At the same time we can indicate the self-occluded label points by the reconstructed 3D shape.

3. 2D parameter estimation: estimate the shape and pose parameter using 2D shape model of current view. Given the observed shape S_{2d}^* and the visibilities of each label point, the new shape is reconstructed by minimizing the weighted reconstruction error of visible points. The dynamic weighting method used in [10] is still adopted in our algorithm to improve robustness.

Here is the flow chart of our automatic MVFA system.

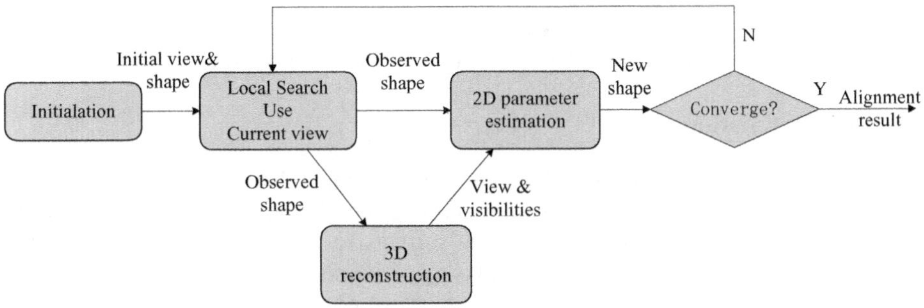

Fig. 3. Flowchart of MVFA

The whole alignment procedure is shown in Figure 4.

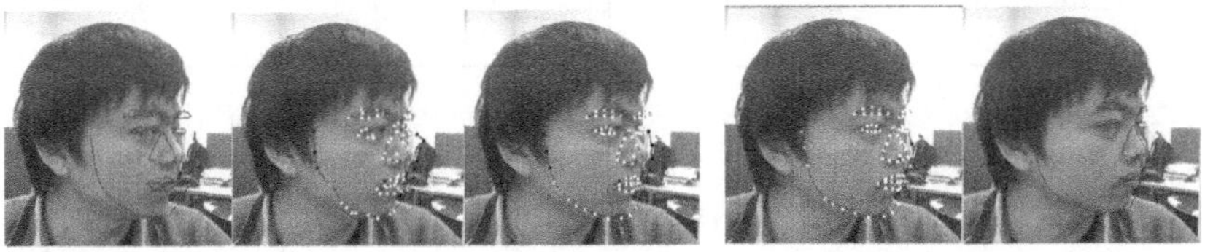

(a) (b) (c) (d) (e)

Fig. 4. Illustration of alignment procedure ((a) The algorithm is initialized by the mean shape of current view. (b) The observed shape is got by local search using local texture model. (c) With the observed shape, a 3D shape is reconstructed using 3D shape model and the pose is estimated. (d) The 2D shape is reconstructed from the observed shape. (e) The final shape when the iteration converged.)

6 Experiments

6.1 Training

A multi-view face database including totally 1800 images which are taken by a camera array with poses in the set {*-90°,-75°,-60°,-45°,-30°,0°*} is set up. The face images are about 250 by 250 in size each of which are manually labeled with 88 label points. 1500 images are used in training and the other 300 are used in testing. 4 ASMs of corresponding views shown in table 1 are trained using Haar-like feature based boosted classifiers which distinguish the textures around a feature point from the textures far away from the feature point [10] (here the view 5-7 are omitted since they are the mirrors of the view 2-4 which can use their mirrored models). Notice that the angle ranges of different views have overlaps in order to make each model more robust to view selection.

Table 1. Training models and roll angles

View	1	2	3	4
Angle	Frontal	0 to 45	30 to 60	60 to 90

6.2 View Estimation

View estimation results are tested on the 1500 training images. Table 2 gives the comparison results between the 3D approach and the 2D view-based approach. It can be seen that the 3D method can apparently improve the view estimation accuracy especially for those views with large off-image-plane (roll) angles which are very critical in MVFA since face alignment for faces of non-frontal views are much more sensitive to view selection.

Table 2. Comparison between the 3D approach and the 2D view-based approach

View	0	1	2	3
3D method	95%	93%	92%	95%
2D method	93%	90%	87%	85%

6.3 MVFA

On the 300 testing images, the performance of MVFA is measured by the average point-to-point errors between alignment result and the ground truth and is shown in Figure 5. It can be seen that the proposed approach outperforms the traditional view-based ASM algorithm [10]. In average, MVFA takes about 304ms to align a face.

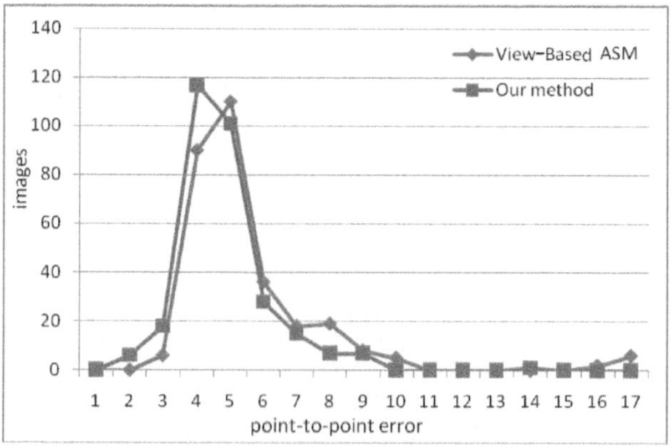

Fig. 5. Error distribution of alignment results

We also tested our method on the CMU-PIE database. Some results are shown in Figure 6. Since there are no ground truth data, we can only subjectively judge the correctness of alignment on a subset of CMU-PIE database. Among all the 1145 face images from the c02, c37, c27, c11, c14 view categories, our algorithm achieved 86.7% in correct rate, while the original method [10] can only achieve 74.5%.

Additional tests have also been taken on the Labeled faces in the wild database [16], which contains multi-view faces in unconstrained environments. Our method can deal with these faces rather well even though our training images are taken in constrained environments and do not cover such large variations in pose, illumination, background, focus and expression. See figure 7 for some results.

Fig. 6. Additional results on CMU-PIE database

Fig. 7. Additional results on Labeled faces in the wild database

7 Conclusion

In this paper, we presented an automatic MVFA framework by integrate 2D view based ASM with a 3D face shape model. Alignment is done in view-based ASM manner while during the iterations, the selection of models, in other words, the view estimation, is done using 3D face shape model. In addition, the 3D reconstructed shape is used to indicate invisible label points that can further improve the accuracy and robustness of the 2D view-based ASM method. Experiments show that view estimation using 3D model can help the view-based ASM method in both accuracy and robustness. Our future work will focus on extending the proposed method to more challenging datasets such as the Labeled faces in the wild database and consumer images over the internet.

Acknowledgement

This work is supported by National Science Foundation of China under grant No.60673107, and it is also supported by a grant from Omron Corporation.

References

1. Hill, A., Cootes, T.F., Taylor, C.J.: Active shape models and the shape approximation problem. In: BMVC 1995 (1995)
2. Cootes, T.F., Edwards, G.J., Taylor, C.J.: Active appearance models. IEEE Transactions on pattern analysis and machine intelligence 23(6) (June 2001)
3. Jiao, F., Li, S.Z., et al.: Face alignment using statistical models and wavelet features. In: CVPR 2003 (2003)
4. Zhang, L., Ai, H., et al.: Robust Face Alignment Based on Local Texture Classifiers. In: ICIP 2005 (2005)
5. Batur, A.U., Hayes, M.H.: A Novel Convergence for Active Appearance Models. In: CVPR 2003 (2003)
6. Cootes, T.F., Taylor, C.J.: A mixture model for representing shape variatio. In: BMVC 1997 (1997)
7. Romdhani, S., Gong, S., Psarrou, A.: A multi-view non-linear active shape model using kernel PCA. In: BMVC 1999 (1999)
8. Zhou, Y., Zhang, W., et al.: A Bayesian Mixture Model for Multi-view Face Alignment. In: CVPR 2005 (2005)
9. Li, S.Z., Yan, S.C., et al.: Multi-view face alignment using direct appearance models. In: AFG 2002 (2002)
10. Zhang, L., Ai, H.: Multi-View Active Shape Model with Robust Parameter Estimation. In: ICPR 2006 (2006)
11. Gu, L., Kanade, T.: 3D Alignment of Face in a Single Image. In: CVPR 2006 (2006)
12. Vogler, C., Li, Z.G., Kanaujia, A.: The Best of Both Worlds: Combining 3D Deformable Models with Active Shape Models. In: ICCV 2007 (2007)
13. The BJUT-3D Large-Scale Chinese Face Database. Technical Report No ISKL-TR-05-FMFR-001. Multimedia and Intelligent Software Technology Beijing Municipal Key Laboratory, Beijing University of Technology (2005)
14. Huang, C., Ai, H., et al.: High Performance Rotation Invariant Multiview Face Detection. IEEE Transactions on Pattern Analysis and Machine Intelligence 29(4), 671–686 (2007)
15. Sim, T., Baker, S., Bsat, M.: The CMU Pose, Illumination,and Expression (PIE) database of human faces. The roboticsinstitute, Carnegie Mellon University. Technical report (2001)
16. Huang, G.B., Ramesh, M., Berg, T., Miller, E.L.: Labeled faces in the wild: A database for studying face recognition in unconstrained environments. Technical Report 25(12), 7–49 (October 2007)

Analysis of Eigenvalue Correction Applied to Biometrics

Anne Hendrikse[1], Raymond Veldhuis[1], Luuk Spreeuwers[1], and Asker Bazen[2]

[1] Unversity of Twente, Fac. EEMCS, Signals ans Systems Group, Hogekamp Building, 7522 NB, Enschede, The Netherlands
a.j.hendrikse@utwente.nl
[2] Uniqkey Biometrics, The Netherlands
a.m.bazen@uniqkey.com

Abstract. Eigenvalue estimation plays an important role in biometrics. However, if the number of samples is limited, estimates are significantly biased. In this article we analyse the influence of this bias on the error rates of PCA/LDA based verification systems, using both synthetic data with realistic parameters and real biometric data. Results of bias correction in the verification systems differ considerable between synthetic data and real data: while the bias is responsible for a large part of classification errors in the synthetic facial data, compensation of the bias in real facial data leads only to marginal improvements.

1 Introduction

An important aspect of biometrics is data modeling. Modeling the statistics of data by covariance matrices is an example. Two techniques which rely on modeling by covariance matrices are Principle Component Analysis (PCA) and Linear Discriminant Analysis (LDA).

Because the covariance matrix of the data generating process, Σ, is usually unknown, it needs to be estimated from a training set. An often used estimate is the sample covariance matrix:

$$\hat{\Sigma} = \frac{1}{N-1} \mathrm{X} \cdot \mathrm{X}^T \tag{1}$$

where the columns of matrix X contain the training samples with the mean subtracted and N is the number of samples in the set.

In the modeling process we are often more interested in functions of the covariance matrix than in the covariance matrix itself. A commonly used function is the decomposition of the covariance matrix in eigenvectors and eigenvalues. The decomposition results we call population eigenvectors and population eigenvalues when derived from Σ and we call them by sample eigenvectors and sample eigenvalues when derived from $\hat{\Sigma}$. The ith population eigenvalue is denoted by λ_i and the ith sample eigenvalue is denoted by l_i. Though $\hat{\Sigma}$ is an unbiased estimate of Σ [1], the estimate of λ_i by l_i does have a bias.

M. Tistarelli and M.S. Nixon (Eds.): ICB 2009, LNCS 5558, pp. 189–198, 2009.
© Springer-Verlag Berlin Heidelberg 2009

In this article, we analyse the effect of this bias with two verification experiments. In the first experiment we use synthetic data so we can compare the verification performance of the system with and without the bias. In both the synthetic data and the real biometric data we compare performance improvement when applying several bias correction algorithms in several configurations.

An analysis of the bias is given in section 2.1. In section 2.2 we present a number of algorithms which reduce the bias. In section 3 we describe the verification system used in the experiments. We indicate where the bias will have its largest effect and where it should be compensated.

In section 4.1 we present an experiment with synthetic facial data, to determine the effect of the bias when the assumed model is correct. In section 4.2 we repeat the experiment with real facial data. In section 5 we present conclusions.

2 Eigenvalue Bias Analysis and Correction

2.1 Eigenvalue Bias Analysis

To find the statistics of estimators often Large Sample Analysis (LSA) is performed. The sample eigenvalues show no bias in this limit case where the number of samples is large enough that it solely determines the statistics of the estimator. However, in biometrics, the number of samples is often in the same order as the number of dimensions or even lower. Therefore, in the analysis of the statistics of the sample eigenvalues the following limit may be considered: $N, p \to \infty$ while $\frac{p}{N} \to \gamma$. Here N is the number of samples used, p is the number of dimensions and γ is some positive constant. Analysis in this limit are denoted General Statistical Analysis (GSA) [2]. In GSA the sample eigenvalues do have a bias.

To demonstrate GSA, we estimated sample eigenvalues of synthetic data with population eigenvalues chosen uniformly between 0 and 1. We kept $\gamma = \frac{1}{5}$ while we varied the dimensionality between 4, 20 and 100. In Figure 1 we show both the population eigenvalue probability function and the sample eigenvalue probability functions for 4 repetitions, given by

$$F_p(l) = p^{-1} \sum_{i=1}^{p} \mathrm{u}\left(l - l_i\right) \tag{2}$$

where $\mathrm{u}\left(l\right)$ is the step function. The empirical probability functions converge with increasing dimensionality, and they converge to a different probability function as the population probability function, due to bias. This example also shows that bias reduction is only possible for a minimum dimensionality, because only then the largest part of the error in l_i as estimate of λ_i is caused by the bias.

2.2 Eigenvalue Bias Correction Algorithms

The bias is a deterministic error and can therefore be compensated. In this section we present a number of correction algorithms we used in the verification experiments to reduce the bias. The correction algorithms provide new estimates of the population eigenvalues, which are denoted by $\hat{\hat{\lambda}}_i$.

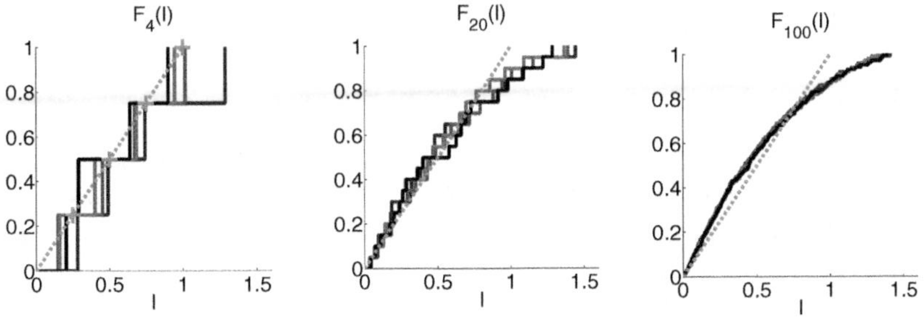

(a) 4 dimensions (b) 20 dimensions (c) 100 dimensions

Fig. 1. Examples of eigenvalue estimation bias toward the GSA limit. All lines indicate empirical probability functions based on sets of eigenvalues (see equation 2). The dashed line indicates the population distribution, the four solid lines are the empirical sample distribution.

1. The *Muirhead correction* [3] is given by a maximum likelihood estimate of the population eigenvalues:

$$\hat{\hat{\lambda}}_i = l_i - \frac{1}{n} l_i \sum_{j=1\ldots i-K, i+K\ldots p}^{p} \frac{l_j}{l_i - l_j} \tag{3}$$

   In the original formula K was set to one. However, to prevent strong fluctuations, we set $K = 50$, which is a simplified version of the Stein[4] algorithm.
2. The *Karoui correction* [5] is based on the Marčenko Pastur equation [6] which gives a relation between sample eigenvalues and the population eigenvalues in the limit considered in GSA. The algorithm finds an estimate of the empirical population eigenvalue probability function (Equation 2, with l replaced by λ) as a weighed sum of fixed probability functions, in our case a set of delta pulses and bar functions.
3. The *Iterative feedback* algorithm was developed by the authors and is new to our knowledge. To find the population eigenvalues the algorithm starts with an initial guess for the population eigenvalues, $\hat{\hat{\lambda}}_{i,1}$. In the mth iteration of the algorithm, synthetic data is generated with population eigenvalues equal to $\hat{\hat{\lambda}}_{i,m}$. The sample eigenvalues $\hat{l}_{i,m}$ of this synthetic data are determined. $\hat{\hat{\lambda}}_{i,m+1}$ is constructed via $\hat{\hat{\lambda}}_{i,m+1} = \hat{\hat{\lambda}}_{i,m} \cdot \frac{l_i}{\hat{l}_{i,m}}$. These steps are repeated until $\sum_{i=1}^{p} \left(\hat{l}_i - l_i\right)^2$ is below a preset threshold or $m > m_{\max}$.
4. The *Two Subset* correction is a classical technique in statistics to remove bias in estimates, where X is split in two subsets X_1 and X_2. From $(N/2 - 1)^{-1} X_1 {X_1}^T$ eigenvectors are estimated, denoted $\hat{\Phi}_1$. The variances in the second set along these estimated eigenvectors are used as $\hat{\hat{\lambda}}_i$'s, so $\hat{\hat{\lambda}}_i = \hat{\Phi}_{1,i}^T \cdot \frac{1}{N_2 - 1} X_2 X_2^T \cdot \hat{\Phi}_{1,i}$. The $\hat{\hat{\lambda}}_i$'s do not contain the bias of the original estimates. However, since the estimation is performed on half of the original set, the variance of the estimate increases. This might explain why this correction is not commonly used.

3 Verification System Description

3.1 System Setup

In our experiments we test the influence of the bias of eigenvalues in biometric systems, using a well known baseline PCA LDA system in our experiments. In this section we give a brief description of this system. For a more detailed discussion we refer to [7].

The input of the verification system are facial images. On these images some standard preprocessing is done, which results in a data sample $\boldsymbol{x}$ for each image. To transform these input vectors to a space where classification is possible, a transformation matrix T is determined in 3 steps based on a training set of example samples. In the first two steps we use PCA to reduce the dimensionality and whiten the data.

In the third step a projection to the most discriminating subspace is determined by modeling each data sample as $\boldsymbol{x} = \boldsymbol{x}_w + \boldsymbol{x}_b$. Variations between samples from the same class are modeled by $\boldsymbol{x}_w$, which is distributed as $N(0, \Sigma_w)$, a multi variate normal distribution with mean 0 and covariance matrix Σ_w. We model the variations between classes by $\boldsymbol{x}_b$, which is distributed as $N(\mu_t, \Sigma_b)$. Since the data is whitened, the most discriminating subspace is the subspace of the largest eigenvalues of Σ_b. Therefore the transformation matrix T is given by:

$$\mathrm{T} = \hat{\Phi}_{b,C_2}^T \cdot \hat{\Lambda}_{t,C_1}^{\frac{1}{2}} \cdot \hat{\Phi}_{t,C_1}^T \tag{4}$$

where $\hat{\Phi}_{t,C_1}$ are the first C_1 eigenvectors of $\hat{\Sigma}_t$, the covariance matrix of the training set, and $\hat{\Lambda}_{t,C_1}$ is a diagonal matrix with as diagonal the first C_1 eigenvalues of $\hat{\Sigma}_t$. $\hat{\Phi}_{b,C_2}$ are the first C_2 eigenvectors of $\hat{\Sigma}_b$.

After projecting samples in the classification space, we compare sample $\boldsymbol{x}$ with class c by calculating a matching score. We accept an identity claim if the score is above a certain threshold. The score is based on the log likelihood:

$$\mathrm{L}(\boldsymbol{x}, c) = -(\mathrm{T}\cdot\boldsymbol{x} - \mu_c)^T \cdot \hat{\Sigma}_w^{-1} \cdot (\mathrm{T}\cdot\boldsymbol{x} - \mu_c) + (\mathrm{T}\cdot\boldsymbol{x} - \mu_t)^T \cdot (\mathrm{T}\cdot\boldsymbol{x} - \mu_t) \tag{5}$$

3.2 Modifications for Eigenvalue Correction

In this verification system, there are two points where eigenvalue correction may improve results: in the whitening step, where the data is scaled based on eigenvalue estimates and in the matching score calculation, where the eigenvalues of the within covariance matrix in the classification space are needed. We perform eigenvalue correction after the dimensionality reduction, but before the whitening step.

At first sight, it seems that the eigenvalues of $\hat{\Sigma}_t$ need to be corrected. However, under the assumed model, the total covariance matrix Σ_t can be written as $\Sigma_b + \Sigma_w$. These matrices are estimated by $(C-1)^{-1}\sum_{c=1}^{C}\mu_c\mu_c^T$ and $(N-C)^{-1}\sum_{i=1}^{N}(\boldsymbol{x}_i - \mu_{\ell(\boldsymbol{x}_i)})(\boldsymbol{x}_i - \mu_{\ell(\boldsymbol{x}_i)})^T$ respectively, where C is the number of classes in the training set, μ_c is the mean of the training samples of class c, and

$\ell(\boldsymbol{x}_i)$ returns the class index of sample $\boldsymbol{x}_i$. Because both matrices are estimated with a different number of samples, their eigenvalues have a different bias. We therefore perform the correction in the following manner:

1. Estimate Σ_w and Σ_b.
2. Decompose both covariance matrices in eigenvectors and eigenvalues.
3. Construct new estimates of the covariance matrices using the original eigenvector estimates and the corrected eigenvalues.
4. Sum the two estimates to get a new estimate of Σ_t.

The corrected estimate of the covariance matrix is given by $\tilde{\Sigma}_r = \hat{\Phi}_r \cdot f_{N_r}(\hat{\Lambda}_r) \cdot \hat{\Phi}_r^T$, where r is either w or b and $f_{N_r}(\hat{\Lambda}_r)$ is an eigenvalue correction algorithm.

4 Experiments

In this section we describe two verification experiments with the system presented in the previous section. In the first experiment we used synthetic facial data while in the second experiment we used real facial data.

4.1 Synthetic Data Experiment

To generate synthetic data close to real facial data, we determined the data structure of a large set of face images in the FRGC database. The data contained 8941 facial images. All facial images were taken under controlled conditions with limited variations in pose and illumination. Also the faces in the facial images had a neutral expression and nobody wore glasses.

We model the facial data with the model in section 3. For generating synthetic data adhering to this model with parameters close to real facial data, we estimated the within class covariance matrix Σ_w and the between class covariance matrix Σ_b from the FRGC data. Since the eigenvalues of these estimates also contain a bias, we corrected their eigenvalues with the Two Subset correction, knowing from previous experiments that this correction led to better estimates of eigenvalues [8]. We kept μ_t zero.

We generated a small training set of 70 identities, with 4 samples per identity, so the bias should be comparable to small real face data sets. This training set was used to train a verification system. In the dimensionality reduction stage of the training the dimensionality was reduced to 150. In the LDA step, the 60 most discriminating features were retained.

We tested the following corrections: no correction, Muirhead correction, Karoui correction, Iterative Feedback correction, Two Subset correction and a lower bound correction. With the lower bound correction, we use the true covariance matrices of the synthetic data to calculate the actual variances along the estimated eigenvectors and use these values as $\hat{\hat{\lambda}}_i$'s. We assumed this correction would give an indication of the best possible error reduction.

We generated a test set with 1000 identities. For each identity 10 enrollment samples and 10 probe samples were generated. During the experiment 3 configurations were tested: correction of only the within class eigenvalues, correction of only the between class eigenvalues and correction of both the within and the between class eigenvalues. The DET curves of the three configurations are shown in Figure 2. In Figure 4a we show the relative EER improvement averaged over 5 repetitions.

The within class eigenvalues correction configuration shows a large difference between the no correction DET curve and the lower bound correction. Therefore the bias in the within class eigenvalues seems to have a large effect on the error rates. The Two Subset correction achieves on average slightly better results as the lower bound correction, but this is probably due to measurement noise. The performance of the Karoui correction fluctuates when the experiment is repeated. In some repetitions the Karoui correction reduces the error rates by half, but on average it increases the error rates as shown in Figure 4a.

The between class eigenvalues correction configuration shows hardly any difference between the different correction algorithms. It seems that the bias in the between class eigenvalues have little influence on the verification scores. The curve of both eigenvalue sets corrected shows no significant difference with the within only correction.

In Figure 3a and Figure 3b we show the corrected within class eigenvalues and between class eigenvalues respectively. The lower bound correction shows considerable fluctuations in the curve. This indicates that the ordering of the sample eigenvectors is wrong.

The lower bound curve is much flatter for the small eigenvalues in the within class correction than the no correction curve. The Two Subset correction also makes the curve much flatter for the smaller eigenvalues, although the eigenvalues are considerably larger than the lower bound correction. Considering the error rates are almost the same, the similarity in flatness seems more important than the actual value of the eigenvalues.

The Karoui correction shows a similar flatness until the 78th eigenvalue. After the 92th eigenvalue, all remaining eigenvalues are set to 0. This seems to have only a small effect on the error rates. This is remarkable since 0 within class variance would indicate very good features, while we know from the lower bound correction that the within class variance is non zero. However, if the between class variance is also zero, the direction will be neglected.

4.2 FRGC Facial Data Experiment

Eigenvalue correction with synthetic facial data caused a significant reduction of the error rates. In the next experiment we replaced the synthetic facial data with the face data set from the FRGC database. This data set is the set used in the previous experiment to determine the facial data structure.

The data set is split in a training set and a test set. The training set contained 70 randomly chosen identities, with a maximum of 5 samples per identity. The test set contained the remaining 445 identities. At most 5 samples per identity are used for enrolling, at least 1 sample is used as probe per identity.

In the training stage instead of reducing the dimensionality to 150, as described in section 3, only the null space is removed. After correction of the

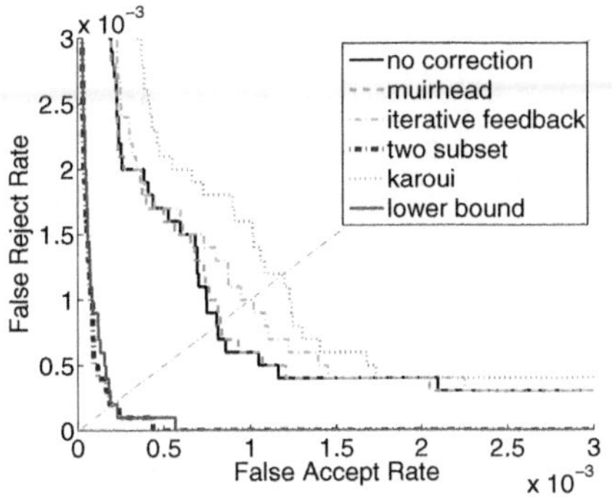

(a) Within eigenvalue correction only

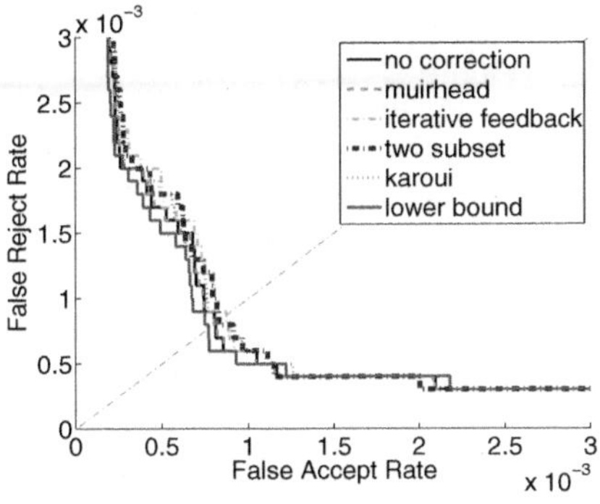

(b) Between eigenvalue correction only

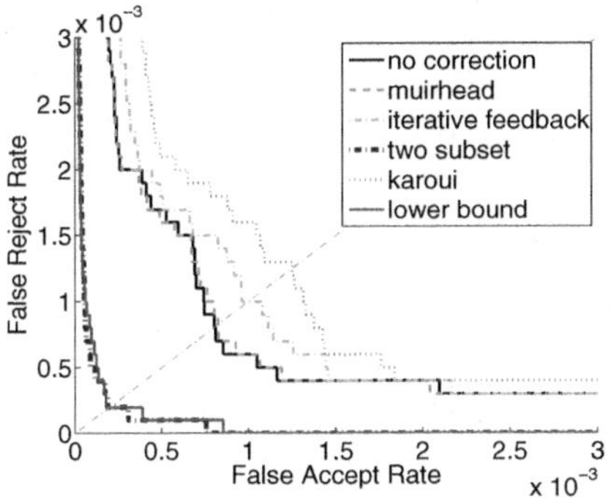

(c) Both within and between class eigenvalue correction

Fig. 2. DET curves for the synthetic data experiment

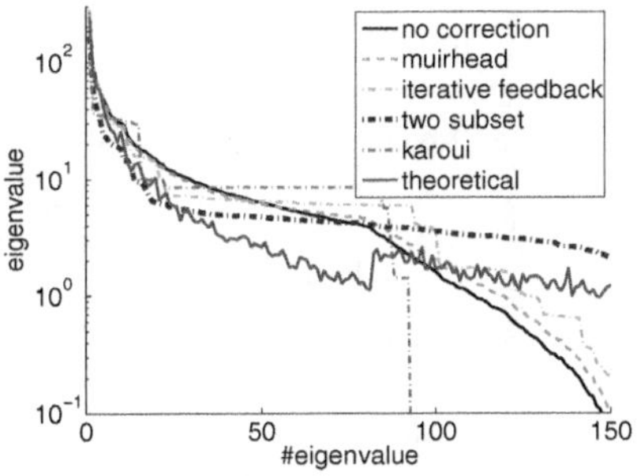

(a) Within class eigenvalues

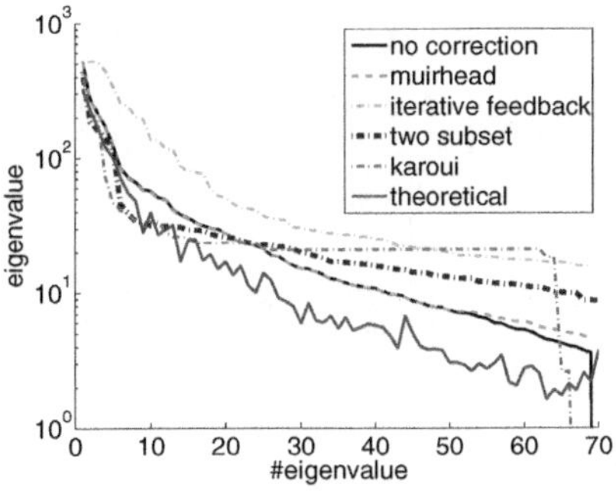

(b) Between class eigenvalues

Fig. 3. Scree plots of the corrected eigenvalues of synthetic data eigenvalues, the dimensionality is reduced to 150. The correction algorithms described in section 2.2 are compared.

The experiment is repeated 5 times for the same 3 configurations as in the synthetic data experiment. For each correction algorithm in each configuration we determined the Equal Error Rate (EER). This EER is compared with the no correction EER. The average over 5 repetitions of the relative improvement of EER is shown in figure 4b.

The results show that correcting only the between class eigenvalues increases the EER for all correction algorithms. The within correction decreases the EER for most algorithms. Correcting both eigenvalue sets decreases the EER for the iterative feedback algorithm and the Two Subset algorithm. But this decrease in EER is less than the decrease in EER if only the within class eigenvalues are corrected.

Comparing the different correction methods shows that in the within correction and both eigenvalue sets correction the Two Subset correction performs considerably better than the other corrections. The Karoui correction always increases the EER.

In Figure 5 we show the results of the first repetition. The Karoui corrections sets a large set of small eigenvalues to zero. This had remarkably little effect on the error rates. The Two Subset correction on the other hand assigns non zero values to eigenvalues which were originally zero.

Most correction algorithms show a trend: the largest eigenvalues are reduced while the smaller eigenvalues are increased. This effect is the strongest with the Two Subset correction. Since this correction method achieved the lowest error rates, it seems that in face recognition indeed the largest eigenvalues are over estimated while the smallest are under estimated, at least in the within class estimation.

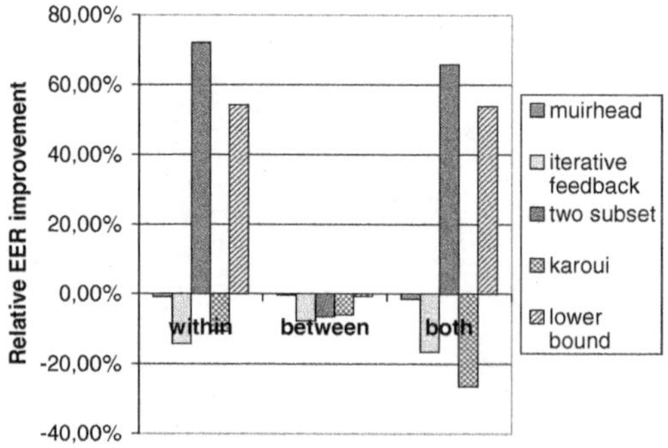

(a) synthetic facial data correction averaged over 5 repetitions

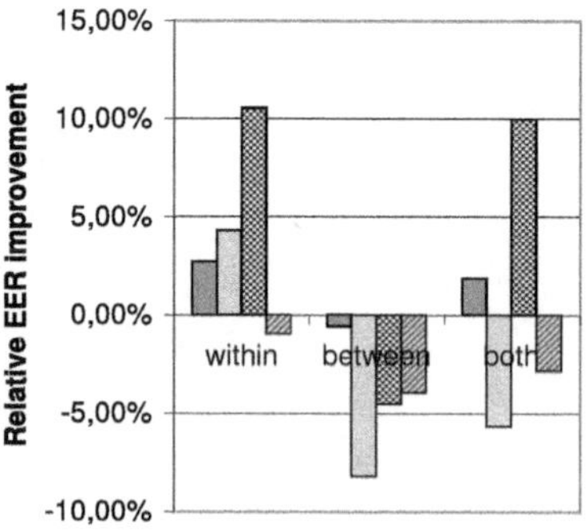

(b) real facial data correction averaged over 5 repetitions

Fig. 4. Relative Equal Error Rate improvement for each correction method. There are three configurations: only within class eigenvalues correction, only between class eigenvalues correction and both eigenvalue sets correction.

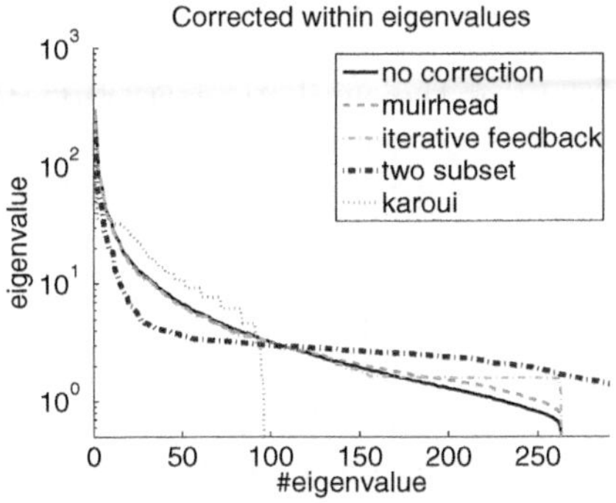

(a) Corrected within class eigenvalues

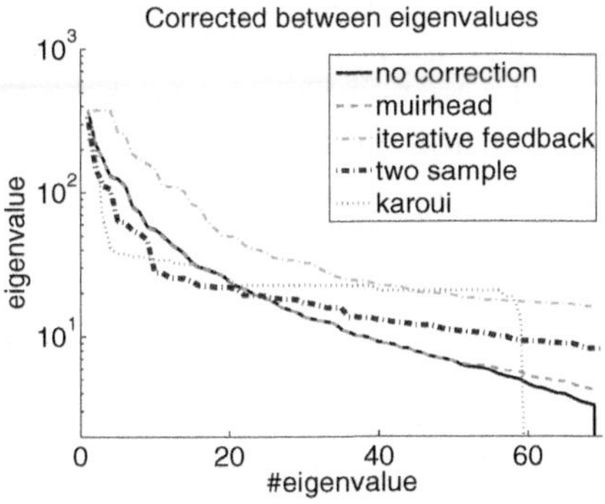

(b) Corrected between class eigenvalues

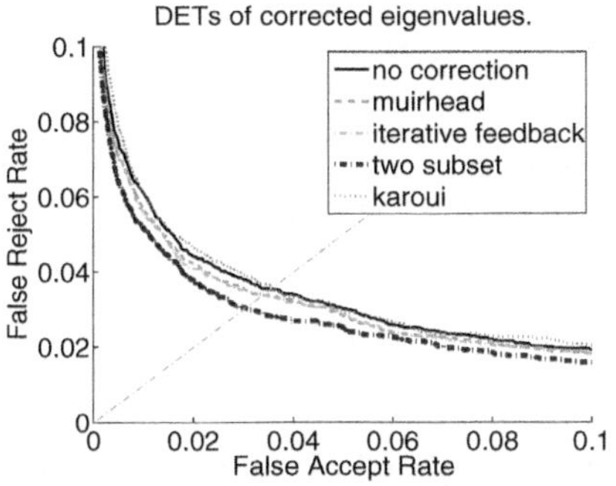

(c) DET curves of within correction only.

Fig. 5. Results of the first repetition of real facial data experiment

Comparing the results of the real facial data test with the results from the synthetic data shows that the EER's in real data are an order higher than the EER's in synthetic data. This suggests that the model we used is not sufficiently accurate for describing real facial data. However, in both experiments the Two Subset method showed the highest reduction in EER.

5 Conclusion

We showed that the GSA provides more accurate analysis of the sample eigenvalue estimator than LSA in biometrics: GSA on the estimator predicts that the estimates in biometrics will have a bias, which is observed in synthetic data, especially for the smaller eigenvalues.

Correcting only the within class eigenvalues has demonstrated the most effect. This is related to the previous conclusion: the best features are determined by the ratio of between class over within class variance. Therefore the best features probably lie in space spanned by the largest between class eigenvalues and the smallest within class eigenvalues. Since the smaller eigenvalues have more bias, within class correction has the most effect.

The Two Subset correction gave the best improvement of error rates in both the synthetic data experiment and the real facial data experiment. Although the performance of the correction was the same as the synthetic correction, the scree plots did differ. The corrections of the other algorithms did also significantly alter the eigenvalues, but this had little effect on the error rates for most of these corrections. Apparently the actual values of the eigenvalues do not have to be estimated very accurately.

The relative error reduction in the facial data is much lower as in the synthetic data by the Two Subset correction. Also the no correction error rates differ more than an order between the real facial data and the synthetic data. This suggest that the bias in the eigenvalue bias is only a moderate error factor in the real facial data.

References

1. Fukunaga, K.: Introduction to statistical pattern recognition, 2nd edn. Academic Press Professional, Inc., San Diego (1990)
2. Girko, V.: Theory of Random Determinants. Kluwer, Dordrecht (1990)
3. Muirhead, R.J.: Aspects of multivariate statistical theory. Wiley Series in Probability and Mathematical Statistics. John Wiley & Sons, Inc., Chichester (1982)
4. Stein, C.: Lectures on the theory of estimation of many parameters. Journal of Mathematical Sciences 34(1), 1371–1403 (1986)
5. El Karoui, N.: Spectrum estimation for large dimensional covariance matrices using random matrix theory. ArXiv Mathematics e-prints (September 2006)
6. Silverstein, J.W.: Strong convergence of the empirical distribution of eigenvalues of large dimensional random matrices. J. Multivar. Anal. 55(2), 331–339 (1995)
7. Belhumeur, P.N., Hespanha, J.P., Kriegman, D.J.: Eigenfaces vs. fisherfaces: Recognition using class specific linear projection. IEEE Transactions on Pattern Analysis and Machine Intelligence 19(7), 711–720 (1997)
8. Hendrikse, A.J., Spreeuwers, L.J., Veldhuis, R.N.J.: Eigenvalue correction results in face recognition. In: Twenty-ninth Symposium on Information Theory in the Benelux, pp. 27–35 (2008)

Multi-Region Probabilistic Histograms for Robust and Scalable Identity Inference

Conrad Sanderson[1,2] and Brian C. Lovell[1,2]

[1] NICTA*, PO Box 6020, St Lucia, QLD 4067, Australia
[2] The University of Queensland, School of ITEE, QLD 4072, Australia

Abstract. We propose a scalable face matching algorithm capable of dealing with faces subject to several concurrent and uncontrolled factors, such as variations in pose, expression, illumination, resolution, as well as scale and misalignment problems. Each face is described in terms of multi-region probabilistic histograms of visual words, followed by a normalised distance calculation between the histograms of two faces. We also propose a fast histogram approximation method which dramatically reduces the computational burden with minimal impact on discrimination performance. Experiments on the "Labeled Faces in the Wild" dataset (unconstrained environments) as well as FERET (controlled variations) show that the proposed algorithm obtains performance on par with a more complex method and displays a clear advantage over predecessor systems. Furthermore, the use of multiple regions (as opposed to a single overall region) improves accuracy in most cases, especially when dealing with illumination changes and very low resolution images. The experiments also show that normalised distances can noticeably improve robustness by partially counteracting the effects of image variations.

1 Introduction

When dealing with images obtained in surveillance contexts (e.g. via CCTV), automatic identity inference based on faces is considerably more difficult than in well controlled conditions (e.g. immigration checkpoints). The difficulties arise due to several concurrent and uncontrolled factors: pose (this includes both in-plane and out-of-plane rotations), expression, illumination and resolution (due to variable distances to cameras). Furthermore, an automatic face locator (detector) must be used which can induce further problems. As there are no guarantees that the localisation is perfect, faces can be at the wrong scale and/or misaligned [1].

A surveillance system may have further constraints: only one gallery image per person, as well as real-time operation requirements in order to handle large volumes of people (e.g. peak hour at a railway station). In this context the computational complexity of an identity inference system is necessarily limited,

* NICTA is funded by the Australian Government via the Department of Broadband, Communications and the Digital Economy, as well as the Australian Research Council through the ICT Centre of Excellence program.

M. Tistarelli and M.S. Nixon (Eds.): ICB 2009, LNCS 5558, pp. 199–208, 2009.
© Springer-Verlag Berlin Heidelberg 2009

suggesting that time-expensive approaches, such as the deduction of 3D shape from 2D images [2] (to compensate for pose variations), may not be applicable.

In this work we describe a Multi-Region Histogram (MRH) based approach[1], with the aim of concurrently addressing the above-mentioned problems. The MRH approach is an evolution of a method presented in [3], which in turn was inspired by 'visual words' used in image categorisation [4]. The method presented here builds on [3] primarily through **(i)** multiple regions to increase discrimination performance without adversely affecting robustness, **(ii)** a histogram approximation method in order to dramatically speed up calculation with minimal impact on discrimination performance, and **(iii)** a distance normalisation method to improve robustness in uncontrolled image conditions.

We continue the paper as follows. Section 2 describes the proposed MRH approach in detail, along with the associated histogram approximation and distance normalisation methods. In Section 3 the MRH approach is briefly contrasted to related methods, taking scalability into account. Results from evaluations and comparisons on the Labeled Faces in the Wild (LFW) [5] and FERET [6] datasets are given in Section 4. The main findings and an outlook are presented in Section 5.

2 Multi-Region Histograms of Visual Words

Each face is divided into several fixed and adjacent regions, with each region comprising a relatively large part of the face (see Fig. 1). For region r a set of feature vectors is obtained, $X_r = \{\boldsymbol{x}_{r,1}, \boldsymbol{x}_{r,2}, \cdots, \boldsymbol{x}_{r,N}\}$, which are in turn attained by dividing the region into small blocks (or patches) and extracting descriptive features from each block via 2D DCT [7] decomposition. Each block has a size of 8×8 pixels and overlaps neighbouring blocks by 75%. To account for varying contrast, each block is normalised to have zero mean and unit variance. Based on preliminary experiments we elected to retain 15 of the 64 DCT coefficients, by taking the top-left $4{\times}4$ submatrix of the $8{\times}8$ coefficient matrix and disregarding the first coefficient (as it carries no information due to the above normalisation).

For each vector $\boldsymbol{x}_{r,i}$ obtained from region r, a probabilistic histogram is computed:

$$\boldsymbol{h}_{r,i} = \left[\frac{w_1 p_1(\boldsymbol{x}_{r,i})}{\sum_{g=1}^{G} w_g p_g(\boldsymbol{x}_{r,i})}, \frac{w_2 p_2(\boldsymbol{x}_{r,i})}{\sum_{g=1}^{G} w_g p_g(\boldsymbol{x}_{r,i})}, \cdots, \frac{w_G p_G(\boldsymbol{x}_{r,i})}{\sum_{g=1}^{G} w_g p_g(\boldsymbol{x}_{r,i})} \right]^T \quad (1)$$

where the g-th element in $\boldsymbol{h}_{r,i}$ is the posterior probability of $\boldsymbol{x}_{r,i}$ according to the g-th component of a visual dictionary model. The visual dictionary model employed here is a convex mixture of gaussians [8], parameterised by $\lambda = \{w_g, \boldsymbol{\mu}_g, \mathbf{C}_g\}_{g=1}^{G}$, where G is the number of gaussians, while w_g, $\boldsymbol{\mu}_g$ and $\mathbf{C}_g$ are, respectively, the weight, mean vector and covariance matrix for gaussian g. The mean of each gaussian can be thought of as a particular 'visual word'.

[1] The source code for MRH can be obtained from http://arma.sourceforge.net/mrh/

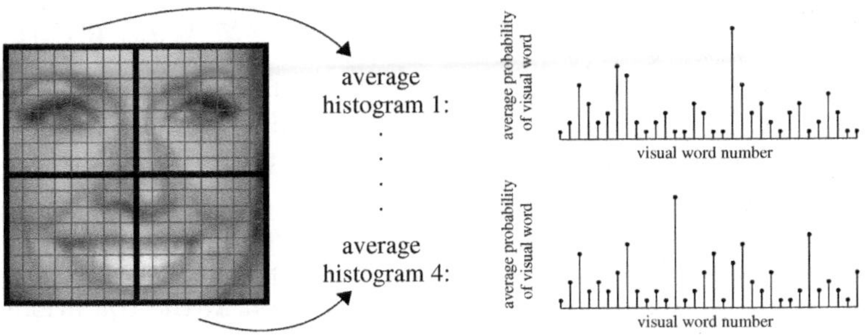

Fig. 1. Conceptual example of MRH face analysis using 2×2 regions, where each region is divided into small blocks. For each block descriptive features are placed into a vector. The posterior probability of the vector is then calculated using each gaussian in a visual dictionary, resulting in a histogram of probabilities. For each region the histograms of the underlying blocks are then averaged.

Once the histograms are computed for each feature vector from region r, an average histogram for the region is built:

$$\boldsymbol{h}_{r,\text{avg}} = \frac{1}{N} \sum\nolimits_{i=1}^{N} \boldsymbol{h}_{r,i} \tag{2}$$

The DCT decomposition acts like a low-pass filter, with the information retained from each block being robust to small alterations (e.g. due to in-plane rotations, expression changes or smoothing due to upsampling from low resolution images). The overlapping during feature extraction, as well as the loss of spatial relations within each region (due to averaging), results in robustness to translations of the face which are caused by imperfect face localisation. We note that in the 1×1 region configuration (used in [3]) the overall topology of the face is effectively lost, while in configurations such as 3×3 it is largely retained (while still allowing for deformations in each region).

The visual dictionary is obtained by pooling a large number of feature vectors from training faces, followed by employing the Expectation Maximisation algorithm [8] to optimise the dictionary's parameters (i.e. λ).

2.1 Normalised Distance

Comparison of two faces can be accomplished by comparing their corresponding average histograms. Based on [9] we define an L_1-norm based distance measure between faces A and B:

$$d_{\text{raw}}(A, B) = \frac{1}{R} \sum\nolimits_{r=1}^{R} \left\| \boldsymbol{h}_{r,\text{avg}}^{[A]} - \boldsymbol{h}_{r,\text{avg}}^{[B]} \right\|_1 \tag{3}$$

where R is the number of regions. To reach a decision as to whether faces A and B come from the same person or from two different people, $d_{\text{raw}}(A, B)$ can be compared to a threshold. However, the optimal threshold might be dependent on the image conditions of face A and/or B, which are not known a-priori. Inspired by cohort normalisation [10], we propose a normalised distance in order to reduce the sensitivity of threshold selection:

$$d_{\texttt{normalised}}(A,B) = \frac{d_{\texttt{raw}}(A,B)}{\frac{1}{2}\left(\frac{1}{M}\sum_{i=1}^{M} d_{\texttt{raw}}(A,C_i) + \frac{1}{M}\sum_{i=1}^{M} d_{\texttt{raw}}(B,C_i)\right)} \quad (4)$$

where C_i is the i-th cohort face and M is the number of cohorts. In the above equation cohort faces are assumed to be reference faces that are known not to be of persons depicted in A or B. As such, the terms $\frac{1}{M}\sum_{i=1}^{M} d_{\texttt{raw}}(A,C_i)$ and $\frac{1}{M}\sum_{i=1}^{M} d_{\texttt{raw}}(B,C_i)$ estimate how far away, on average, faces A and B are from the face of an impostor. This typically results in Eqn. (4) being approximately 1 when A and B represent faces from two different people, and less than 1 when A and B represent two instances of the same person. If the conditions of given images cause their raw distance to increase, the average raw distances to the cohorts will also increase. As such, the division in Eqn. (4) attempts to cancel out the effect of varying image conditions.

2.2 Fast Histogram Approximation

As will be shown in Section 4, the size of the visual dictionary needs to be relatively large in order to obtain good performance. Typically about 1000 components (gaussians) are required, which results in the calculation of histograms via Eqn. (1) to be time consuming. Based on empirical observations that for each vector only a subset of the gaussians is dominant, we propose a dedicated algorithm that adaptively calculates only a part of the histogram. The algorithm is comprised of two parts, with the first part done during training.

In the first part, the gaussians from the visual dictionary model are placed into K clusters via the k-means algorithm [8]. Euclidean distance between the means of the gaussians is used in determining cluster memberships. For each cluster, the closest member to the cluster mean is labelled as a *principal gaussian*, while the remaining members are labelled as *support gaussians*.

For a feature vector $\boldsymbol{x}$ an approximate histogram is then built as follows. Each of the K principal gaussians is evaluated. The clusters are then ranked according to the likelihood obtained by each cluster's principal gaussian (highest likelihood at the top). Additional likelihoods are produced cluster by cluster, with the production of likelihoods stopped as soon as the total number of gaussians used (principal and support) exceeds a threshold. The histogram is then constructed as per Eqn. (1), with the likelihoods of the omitted gaussians set to zero.

3 Related Methods and Scalability

The use of probabilistic histograms in MRH differs to the histograms used in [4] (for image retrieval/categorisation purposes), where a Vector Quantiser (VQ) based strategy is typically used. In the VQ strategy each vector is forcefully assigned to the closest matching visual word, instead of the probabilistic assignment done here.

For the purposes of face classification, MRH is related to, but distinct from, the following approaches: Partial Shape Collapse (PSC) [11], pseudo-2D Hidden Markov Models (HMMs) [12,13] and Probabilistic Local PCA (PLPCA) [14].

MRH is also somewhat related to the recently proposed and more complex Randomised Binary Trees (RBT) method [15], aimed for more general object classification. While both MRH and RBT use image patches for analysis, RBT also uses: (i) quantised differences, via 'extremely-randomised trees', between corresponding patches, (ii) a cross-correlation based search to determine patch correspondence, and (iii) an SVM classifier [8] for final classification.

The differences between MRH and PSC include: (i) the use of fixed regions for all persons instead of manually marked regions for each person, (ii) each region is modelled as a histogram rather than being directly described by a Gaussian Mixture Model (GMM), leading to (iii) MRH using only one GMM (the visual dictionary), common to all regions and all persons, instead of multiple GMMs per person in PSC. The use of only one GMM directly leads to much better scalability, as the number of gaussians requiring evaluation for a given probe face is fixed, rather than growing with the size of the face gallery. In the latter case the computational burden can quickly become prohibitive [3,10].

The MRH approach has similar advantages over PLPCA and HMM in terms of scalability and histogram based description. However, there are additional differences. In PLPCA each region is analysed via PCA instead of being split into small blocks. While the probabilistic treatment in PLPCA affords some robustness to translations, the use of relatively large face areas is likely to have negative impact on performance when dealing with other image transformations (e.g. rotations and scale changes). In HMM approaches the region boundaries are in effect found via an automatic alignment procedure (according to the model of each person the face is evaluated against) while in the MRH approach the regions are fixed, allowing straightforward parallel processing.

4 Experiments

The experiments were done on two datasets: LFW [5], and subsets of FERET [6]. We will show results of LFW first, where the number of face variations (as well as their severity) is uncontrolled, followed by a more detailed study on FERET, where each variation (e.g. pose) is studied separately.

The recent LFW dataset contains 13,233 face images which have several compound problems – e.g. in-plane rotations, non-frontal poses, low resolution, non-frontal illumination, varying expressions as well as imperfect localisation, resulting in scale and/or translation issues. The images were obtained by trawling the Internet followed by face centering, scaling and cropping based on bounding boxes provided by an automatic face locator. The original bounding boxes were expanded to include context. In our experiments we extracted closely cropped faces using a fixed bounding box placed in the same location in each LFW image[2]. The extracted faces were size normalised to 64×64 pixels, with an average distance between the eyes of 32 pixels. Examples are shown in Fig. 2.

[2] The upper-left and lower-right corners of the bounding box were: (83,92) and (166,175), respectively. Bounding box location was determined first via centering then shifting upwards to fit the eyes and mouth of 40 randomly selected LFW faces.

(i)

(ii)

Fig. 2. Examples from the LFW dataset: (i) master images (resulting from centering, scaling and cropping based on bounding boxes provided by an automatic face locator); (ii) processed versions used in experiments, extracted using a fixed bounding box placed on the master images. The faces typically have at least one of the following issues: in-plane rotations, non-frontal poses, low resolution, non-frontal illumination, varying expressions as well as imperfect localisation, resulting in scale and/or translation issues.

LFW experiments follow a prescribed protocol [5], where the task is to classify a pair of previously unseen faces as either belonging to the same person (matched pair) or two different persons (mismatched pair). The protocol specifies two views of the dataset: *view 1*, aimed at algorithm development & model selection, and *view 2*, aimed at final performance reporting (to be used sparingly). In view 1 the images are split into two sets: the training set (1100 matched and 1100 mismatched pairs) and the testing set (500 matched and 500 mismatched pairs). The training set is used for constructing the visual dictionary as well as selecting the decision threshold. The threshold was optimised to obtain the highest *average accuracy* (averaged over the classification accuracies for matched and mismatched pairs). In view 2 the images are split into 10 sets, each with 300 matched & 300 mismatched pairs. Performance is reported using the mean and standard error of the average accuracies from 10 folds of the sets, in a leave-one-out cross-validation scheme (i.e. in each fold 9 sets are used training and 1 set for testing). The standard error is useful for assessing the significance of performance differences across algorithms [5].

For consistency, experiments on FERET were designed to follow a similar pair classification strategy, albeit with manually found eye locations. The 'fa' and 'fb' subsets (frontal images) were used for training – constructing the visual dictionary as well as selecting the decision threshold. Using persons which had images in both subsets, there were 628 matched and 628 randomly assigned mismatched pairs. The 'b' subsets were used for testing, which contain controlled pose, expression and illumination variations for 200 unique persons. For each image condition there were 200 matched and 4000 mismatched pairs, with the latter obtained by randomly assigning 20 persons to each of the 200 available persons. Image transformations were applied separately to the frontal source images ('ba' series), obtaining the following versions: in-plane rotated (20°), scaled (bounding box expanded by 20% in x and y directions, resulting in shrunk faces), translated (shifted in x and y directions by 6 pixels, or 20% of the distance between the eyes), upsampled

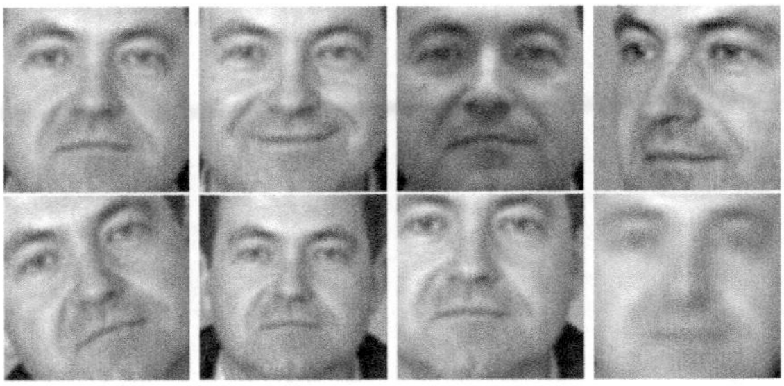

Fig. 3. *Top row:* examples of cropped images from FERET (neutral, followed by expression, illumination and pose change). *Bottom row:* transformed and cropped versions of the neutral source image (in-plane rotation, scale change, translation and upsampled low-res version).

from a low resolution version (with the low resolution version obtained by shrinking the original image to 30% of its size, resulting in an average eye distance of ~10 pixels). Example images are shown in Fig. 3.

In experiment 1 we studied the effect of increasing the size of the visual dictionary (from 2 to 4096 components) and number of regions (from 1×1 to 4×4) on the LFW dataset. As these variations constitute model selection, view 1 was used. The system used probabilistic histograms and normalised distances. Based on preliminary experiments, 32 randomly chosen cohort faces from the training set were used for the distance normalisation. The results, in Fig. 4(i), suggest that performance steadily increases up to about 1024 components, beyond which performance changes are mostly minor. Dramatic improvements are obtained by increasing the number of regions from 1×1 to 3×3. Using more regions (i.e. 4×4) shows no appreciable further performance gains.

In experiment 2 we fixed the number of regions at 3×3 and varied the size of the visual dictionary. The performance of systems using exact probabilistic histograms, approximate probabilistic and VQ based was compared. We also evaluated the performance of raw and normalised distances on both probabilistic and VQ based systems. Based on preliminary experiments, approximate histograms used $K=\frac{G}{10}$ clusters and a maximum of $\frac{G}{4}$ gaussians, where G is the size of the visual dictionary. The results, in Fig. 4(ii), point to the distance normalisation being helpful, with a consistent advantage of about 2 percentage points over raw distances (e.g. 72% vs 70%). The results further suggest that probabilistic histograms outperform VQ based histograms, also with an advantage of about 2 points. Finally, the performance of the computationally less expensive approximate probabilistic histograms is on par with exact probabilistic histograms.

In experiment 3 we used view 2 of LFW, allowing comparison with previously published as well as future results. Several configurations of MRH were evaluated as well as a baseline PCA system. Based on preliminary experiments, the baseline PCA based system used the euclidean distance as its raw distance and 61 eigenfaces (eigenfaces 4 to 64 of the training images, following the recommendation in [16] to skip the first three eigenfaces). The results, presented in Table 1, indicate that the performance of MRH based systems is consistent with

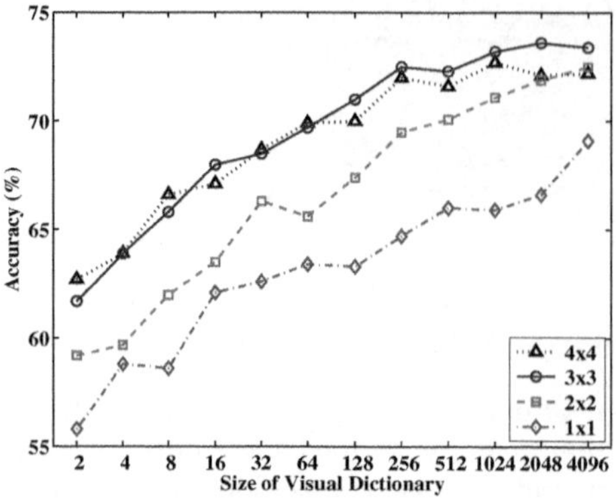

(i)

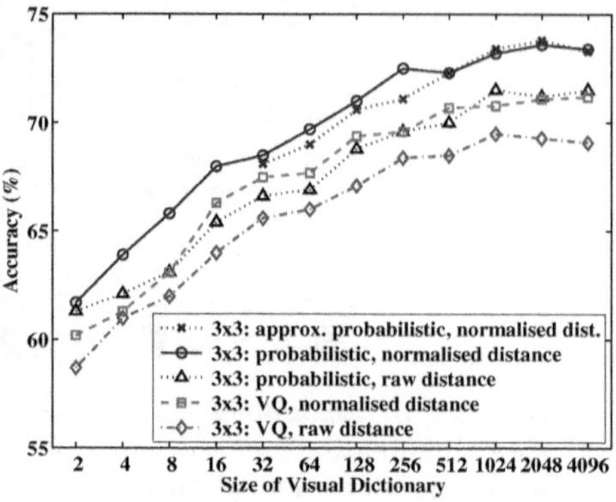

(ii)

Fig. 4. Accuracy rate for increasing size of the visual dictionary, on *view 1* of LFW (compound face variations). (i): MRH (probabilistic, normalised distance), with the number of regions varying from 1×1 to 4×4. (ii): 3×3 MRH, with either probabilistic or VQ based histogram generation, as well with and without distance normalisation.

Table 1. Results on *view 2* of LFW. Results for RBT obtained from http://vis-www.cs.umass.edu/lfw (accessed 2008-09-01), using the method published in [15]. MRH approaches used a 1024 component visual dictionary.

Method	Mean accuracy	Standard error
3×3 MRH (approx probabilistic, normalised distance)	72.35	0.54
3×3 MRH (probabilistic, normalised distance)	**72.95**	0.55
3×3 MRH (probabilistic, raw distance)	70.38	0.48
3×3 MRH (VQ, normalised distance)	69.35	0.72
3×3 MRH (VQ, raw distance)	68.38	0.61
1×1 MRH (probabilistic, normalised distance)	67.85	0.42
PCA (normalised distance)	59.82	0.68
PCA (raw distance)	57.23	0.68
Randomised Binary Trees (RBT)	72.45	0.40

experiments 1 and 2. Furthermore, the probabilistic 3×3 MRH method is on par with the more complex RBT method. The performance of PCA considerably lags behind all other approaches.

In experiment 4 images from FERET were used. The performances of probabilistic MRH with 3×3 and 1×1 configurations, as well as the baseline PCA based system, were compared. Both raw and normalised distances were evaluated. For testing, each image condition was evaluated separately. Moreover, for each image pair to be classified, the first image was always from the 'ba' series (normal frontal image). The results, presented in Fig. 5, indicate that increasing the number of regions from 1×1 to 3×3 improves accuracy in most cases, especially when dealing with illumination changes and low resolution images. The notable exceptions are faces with pose changes and in-plane rotations. We conjecture that the histogram for each region (3×3 case) is highly specific and that it has simply altered too much due to the pose change; in the 1×1 case, the

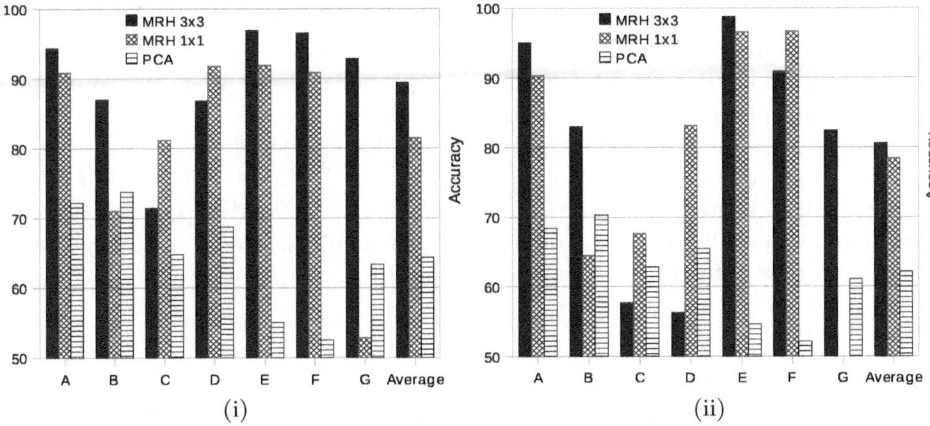

Fig. 5. Performance on FERET (separate face variations), using: (i) normalised distances, (ii) raw distances. A: expression change, B: illumination change, C: pose change, D: in-plane rotation, E: scale change, F: translation (shift), G: upsampled low resolution image.

single overall histogram is more general and parts of it are likely to be describing face sections which have changed relatively little. For the in-plane rotation, we conjecture that the performance drop is at least partially due to face components (e.g. eyes) moving between regions, causing a mismatch between the corresponding histograms of two faces; in the 1×1 case there is only one histogram, hence the movement has a reduced effect.

The use of normalised distances improved the average performance of all approaches. This is especially noticeable for MRH when dealing with pose changes and in-plane rotations. In all cases the 3×3 MRH system considerably outperformed the baseline PCA system, most notably for faces subject to scale changes and translations.

5 Main Findings and Outlook

In this paper we proposed a face matching algorithm that describes each face in terms of multi-region probabilistic histograms of visual words, followed by a normalised distance calculation between the corresponding histograms of two faces. We have also proposed a fast histogram approximation method which dramatically reduces the computational burden with minimal impact on discrimination performance. The matching algorithm was targeted to be scalable and deal with faces subject to several concurrent and uncontrolled factors, such as variations in pose, expression, illumination, as well as misalignment and resolution issues. These factors are consistent with face images obtained in surveillance contexts.

Experiments on the recent and difficult LFW dataset (unconstrained environments) show that the proposed algorithm obtains performance on par with the recently proposed and more complex Randomised Binary Trees method [15]. Further experiments on FERET (controlled variations) indicate that the use of multiple adjacent histograms (as opposed to a single overall histogram) on one hand reduces robustness specific to in-plane rotations and pose changes, while on the other hand results in better average performance. The experiments also show that use of normalised distances can considerably improve the robustness of both multiple- and single-histogram systems.

The robustness differences between multiple- and single-histogram systems suggest that combining the two systems (e.g. by a linear combination of distances) could be beneficial. Lastly, we note that the MRH approach is easily amenable to parallelisation: a multi-CPU machine can process regions concurrently, thereby providing a significant speed-up.

References

1. Rodriguez, Y., Cardinaux, F., Bengio, S., Mariethoz, J.: Measuring the performance of face localization systems. Image and Vision Comput. 24, 882–893 (2006)
2. Blanz, V., Vetter, T.: Face recognition based on fitting a 3D morphable model. IEEE Trans. Pattern Analysis and Machine Intelligence 25(9), 1063–1074 (2003)
3. Sanderson, C., Shan, T., Lovell, B.C.: Towards pose-invariant 2D face classification for surveillance. In: Zhou, S.K., Zhao, W., Tang, X., Gong, S. (eds.) AMFG 2007. LNCS, vol. 4778, pp. 276–289. Springer, Heidelberg (2007)
4. Nowak, E., Jurie, F., Triggs, B.: Sampling strategies for bag-of-features image classification. In: Leonardis, A., Bischof, H., Pinz, A. (eds.) ECCV 2006. LNCS, vol. 3954, pp. 490–503. Springer, Heidelberg (2006)
5. Huang, G.B., Ramesh, M., Berg, T., Learned-Miller, E.: Labeled Faces in the Wild: A database for studying face recognition in unconstrained environments. University of Massachusetts, Amherst, Technical Report 07-49 (October 2007)
6. Phillips, P., Moon, H., Rizvi, S., Rauss, P.: The FERET evaluation methodology for face-recognition algorithms. IEEE Trans. Pattern Analysis and Machine Intelligence 22(10), 1090–1104 (2000)
7. Gonzales, R., Woods, R.: Digital Image Processing, 3rd edn. Prentice-Hall, Englewood Cliffs (2007)
8. Bishop, C.: Pattern Recognition and Machine Learning. Springer, Heidelberg (2006)
9. Kadir, T., Brady, M.: Saliency, scale and image description. International Journal of Computer Vision 45(2), 83–105 (2001)
10. Sanderson, C.: Biometric Person Recognition — Face, Speech and Fusion. VDM Verlag (2008)
11. Lucey, S., Chen, T.: A GMM parts based face representation for improved verification through relevance adaptation. In: Proc. Computer Vision and Pattern Recognition (CVPR), vol. 2, pp. 855–861 (2004)
12. Nefian, A., Hayes, M.: Face recognition using an embedded HMM. In: Proc. Audio Video-based Biometric Person Authentication (AVBPA), pp. 19–24 (1999)
13. Cardinaux, F., Sanderson, C., Bengio, S.: User authentication via adapted statistical models of face images. IEEE Trans. Signal Processing 54(1), 361–373 (2006)
14. Martínez, A.M.: Recognizing imprecisely localized, partially occluded, and expression variant faces from a single sample per class. IEEE Trans. Pattern Analysis and Machine Intelligence 24(6), 748–763 (2002)
15. Nowak, E., Jurie, F.: Learning visual similarity measures for comparing never seen objects. In: Proc. Computer Vision and Pattern Recognition (CVPR), pp. 1–8 (2007)
16. Belhumeur, P., Hespanha, J., Kriegman, D.: Eigenfaces vs. fisherfaces: recognition using class specific linear projection. IEEE Trans. Pattern Analysis and Machine Intelligence 19(7), 711–720 (1997)

Heterogeneous Face Recognition from Local Structures of Normalized Appearance

Shengcai Liao, Dong Yi, Zhen Lei, Rui Qin, and Stan Z. Li*

Center for Biometrics and Security Research,
Institute of Automation, Chinese Academy of Sciences,
95 Zhongguancun Donglu, Beijing 100190, China
{scliao,dyi,zlei,rqin,szli}@nlpr.ia.ac.cn

Abstract. Heterogeneous face images come from different lighting conditions or different imaging devices, such as visible light (VIS) and near infrared (NIR) based. Because heterogeneous face images can have different skin spectra-optical properties, direct appearance based matching is no longer appropriate for solving the problem. Hence we need to find facial features common in heterogeneous images. For this, first we use Difference-of-Gaussian filtering to obtain a normalized appearance for all heterogeneous faces. We then apply MB-LBP, an extension of LBP operator, to encode the local image structures in the transformed domain, and further learn the most discriminant local features for recognition. Experiments show that the proposed method significantly outperforms existing ones in matching between VIS and NIR face images.

Keywords: Face Recognition, Heterogeneous, MB-LBP, DoG.

1 Introduction

In recent years, many face recognition systems are developed for practical use. Applications from the real world now faces an emergency problem that face images are always collected in quite different environments or even captured with different equipments. We call these images heterogeneous face images, and the corresponding new arisen problem as heterogeneous face recognition problem. Some examples of this problem appeared in real-world application are shown in [9,21]. A typical problem is that matching between near infrared (NIR) and visible light (VIS) face images in the situation that enrolment is finished with controlled indoor VIS face images, while authentication would be done using NIR face images to avoid the influence of the variational environment illumination.

Lin *et al.* proposed an algorithm called Common Discriminant Feature Extraction (CDFE) [9]. In the algorithm, two transforms were simultaneously learned to project the samples in both modalities respectively to a common feature space. However, though the authors claimed that they used a smooth factor to avoid overfitting, the algorithm still lacked generalization capability that they only got impressive result on an close-set experimental setting, while had poor performances on non-overlapping test set (*cf.* Section 4).

* Corresponding author.

M. Tistarelli and M.S. Nixon (Eds.): ICB 2009, LNCS 5558, pp. 209–218, 2009.
© Springer-Verlag Berlin Heidelberg 2009

Yi *et al.* [21] proposed another algorithm for heterogeneous face matching. There they also projected samples of two modalities onto a common subspace respectively, but instead they utilized the well-known CCA algorithm. Yang *et al.* [19] also proposed a CCA-based method for matching between 2D and 3D faces using sub-patches. Like CDFE, both methods failed to learn the essential property of heterogeneous faces that would tend to overfit on the training data (*cf.* Section 4).

In this paper, we present a novel algorithm and analyze the heterogeneous problems in NIR-VIS face recognition. It is obvious from the Lambertian model that the albedo term no longer serves as an ideal component for heterogeneous face recognition. Consequently, traditional approaches which directly match two kinds of face images would cause problems. Then what other representations could serve as an invariant property between heterogeneous faces? Finally we discover that the local image structures could be learned for a robust representation for the heterogeneous problem. The reason is that, no matter how the light source changes (homogeneous or heterogeneous), the local relationships of a face would not change too much. Therefore, it would be a robust representation if we learned the most intrinsic local image structures.

To achieve the goal, two steps are adopted in this work. First, we use Difference-of-Gaussian filtering to normalize the appearance of heterogeneous face images, so that they look similar with each other. This contributes further recognition. Second, after appearance normalization, we apply the Multi-scale Block LBP (MB-LBP) for feature representing, which is proved to be effective for encoding local image structures [8]. We call the resulting feature as Local Structure of Normalized Appearance (LSNA). Based on the representation we further learn the most discriminant features for recognition.

Through the above two stages, a robust local structure representation is learned for heterogeneous face recognition that generalizes well on unseen persons. Furthermore, the final model is universal that it is suitable for both homogeneous and heterogeneous faces.

The rest of this paper is organized as follows: In Section 2, we analyze problems in heterogeneous face recognition, present an appearance normalization method and give the explanation that why it is effective for heterogeneous face recognition problem. In Section 3 we apply MB-LBP to encode the intrinsic local structures between both kinds of source images, and learn a discriminant subspace based on it. The experiment results are given in Section 4. Finally, we summarize this paper in Section 5.

2 Appearance Normalization by Difference-of-Gaussian

2.1 Problem Analysis

According to the Lambertian model, an image $I(x, y)$ under a point light source is formed as follows.

$$I(x, y) = \rho_\omega(x, y)\mathbf{n}(\mathbf{x}, \mathbf{y})\mathbf{s} \quad (1)$$

where $\rho_\omega(x, y)$ is the albedo of the facial surface material at point (x, y) with the light source of wavelength ω, $\mathbf{n} = (\mathbf{n_x}, \mathbf{n_y}, \mathbf{n_z})$ is the surface normal in the 3D space, and $\mathbf{s} = (\mathbf{s_x}, \mathbf{s_y}, \mathbf{s_z})$ is the lighting direction. Here, albedo $\rho_\omega(x, y)$ reflects the spectral properties of facial skin, and $\mathbf{n(x, y)}$ is the geometric shape of the face.

Existing 2D face recognition researches mainly focus on visual face images, there the top-most factor that affects the face recognition performance is the direction of the incidence lighting relative to the face surface normal. With homogeneous lightings, $\rho_\omega(x, y)$ is the intrinsic property of the face , and is the ideal part for face recognition, while illumination variations are the extrinsic factor that should be removed from the product $\mathbf{n(x, y)s}$. In fact, over the past years much effort has been made to model illumination on visual faces and correct illumination directions (*cf.* [13], [4], [16], [3], [18]). Researches of this kind are all toward an aim to discover the intrinsic property: invariant $\rho_\omega(x, y)$ at point (x, y).

However, when dealing with heterogeneous images, such as matching between NIR and VIS images, the term $\rho_\omega(x, y)$ will not remain to be invariant any more. That is because the reflectance of the same face surface under heterogeneous light sources is different. Actually it is a function of the wavelength ω of the light source. Fig. 2 shows some examples of VIS vs. NIR face images (column (a) and (b)). The comparison demonstrates that the appearance of both kinds looks so different, particularly in the eyes part and the mouth part. It indicates that the reflectance of the same face surface under VIS and NIR light source is different. As a result, the above methods modelling illumination would fail to achieve their original goal under heterogeneous lighting condition, because the assumption of the invariant $\rho_\omega(x, y)$ does not hold any more.

Therefore, heterogeneous face recognition is confronted with more difficulties. From the Lambertian model we know that even with the same pose and lighting direction, one can not directly match two heterogeneous face images because of the distinct $\rho_\omega(x, y)$. As a result, traditional face recognition methods might encounter troubles when directly applied on heterogeneous face images.

From the above analysis, we know that $\rho_\omega(x, y)$ no longer serves as an ideal component for heterogeneous face recognition. Hence the problem could not be done directly upon image itself, even if we got the $\rho_\omega(x, y)$ part. The reason why previous researches [9,21,19] get poor performance on unseen data (*cf.* Section 4) is possibly because the learned model directly based on images lacks generalization.

Then what other representations are really essential for both homogeneous and heterogeneous face recognition? A good many of previous researches prove that local features can be used for robust representations, though they have not been examined on heterogeneous face recognition problem. It suggest us that local image structures might be the really intrinsic property of faces in despite of light sources. To testify this, our work adopts a two-stage procedure to deal with heterogeneous face recognition problem as described in the following materials.

2.2 Local Filtering

The first stage is to normalize the appearance of the input face images under both light sources, since they look quite different. In other words, we try to find a common image space where the appearance of both kinds of face images looks similar, so that it would be more easier for further recognition.

Unlike previous researches [9,21,19], in which they all learn a common subspace that lacks visual interpretation and the ability of generalization, here we try to first get a regularized appearance for heterogeneous face images using filtering without learning.

To achieve this purpose, we adopt the Difference-of-Gaussian (DoG) filtering to normalize the appearance, for its efficiency and effectiveness. It is successfully used in the well-known SIFT feature [11] for robustly identifying keypoint locations and scales. In [17], DoG filtering is also used in a preprocessing chain for illumination normalization.

Given an input image $I(x, y)$, DoG filtering is computed as

$$D(x, y|\sigma_0, \sigma_1) = (G(x, y, \sigma_0) - G(x, y, \sigma_1)) * I(x, y), \tag{2}$$

where $*$ is the convolution operator, and

$$G(x, y, \sigma) = \frac{1}{\sqrt{2\pi\sigma^2}} e^{-(x^2+y^2)/2\sigma^2}. \tag{3}$$

Here $\sigma_0 < \sigma_1$, so that it constructs a bandpass filter. This simple filtering is actually a subtraction of two Gaussian filtering, thus is efficient to compute. Fig. 1 shows an example of DoG filter.

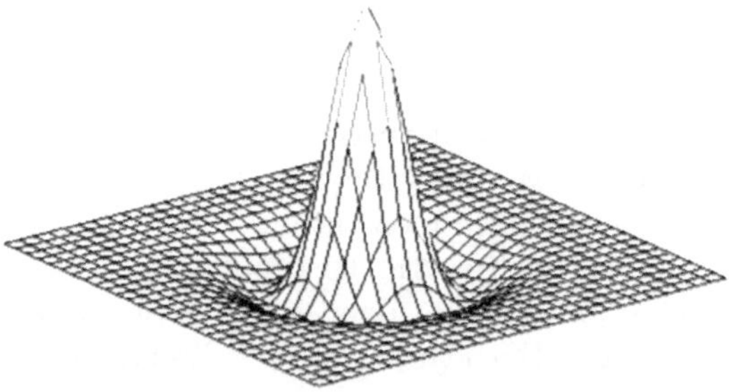

Fig. 1. An example of DoG filter

There are a number of reasons why we choose DoG filtering for normalizing appearance of heterogeneous face images, besides its computation efficiency.

Though the appearance of heterogeneous face images looks different (see Fig. 2 (a) and (b)), their local structures would be similar because they are all faces, particularly for the same person. DoG is a typical differential filter with excitatory and inhibitory lobes, like LoG (Laplacian-of-Gaussian) operator and Gabor filter. Differential filters are often used for representing image structures. They encode local relationships via comparing neighboring regions. We know that both DoG and LoG are scale-space operators with Gaussian kernel. Lowe [11] shows that the difference-of-Gaussian function provides a close approximation to the scale-normalized Laplacian of Gaussian, $\sigma^2\nabla^2 G$, which is studied by Lindeberg for representing visual structures [10]. Lindeberg also shows an interesting similarity with biological vision that these scale-space operators closely resemble receptive field profiles registered in neurophysiological studies of the mammalian retina and visual cortex [10].

Furthermore, DoG filtering also helps to reduce both illumination variation, image noise and aliasing, while preserving enough details for recognition. As we know, both image shading and highlight area caused by illumination lie in low frequency domain, while noise and aliasing are high frequency information. Hence suppressing both the lowest and highest spatial frequencies by bandpass filtering contributes to further recognition.

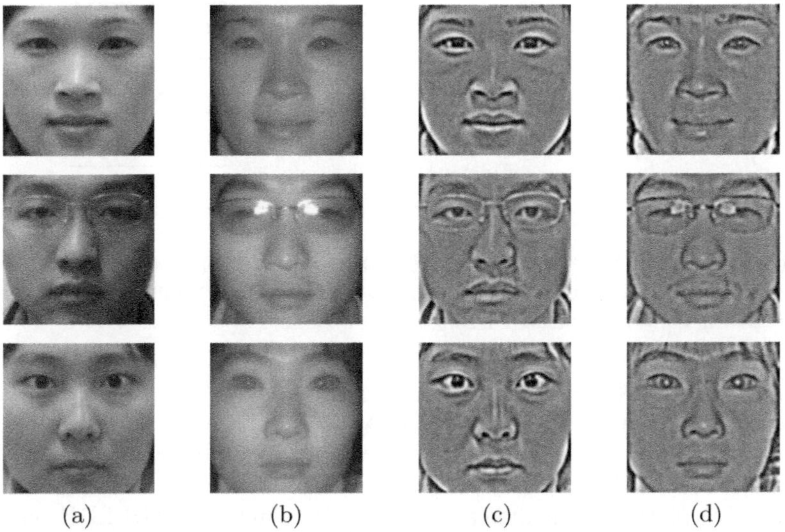

Fig. 2. (a) VIS face images of three people. (b) The corresponding NIR images of the same person as (a) in each row. (c) The corresponding DoG filtered images of (a). (d) The corresponding DoG filtered images of (b).

Fig. 2 shows some preprocessed VIS and NIR face images (column (c) and (d)) using DoG filtering. The parameters are set to $\sigma_0 = 1.0$ and $\sigma_1 = 2.0$ by default, as suggested in [17]. Compared with (a) and (b), we could clearly see that the distinct appearances of VIS and NIR images have been normalized to be similar after preprocessing. Besides, the image shadows and highlights are reduced after DoG filtering. The mouth part is normalized to be more similar, but the eyes part still looks different, so we still need to learn discriminant local structures for further recognition.

3 Learning Discriminant Local Structures

As mentioned before, the local structures of heterogeneous face images remain similar for the same person, which provides potential information for discriminating different persons. Therefore, we further apply MB-LBP, an extension of LBP operator, for multi-scale analysis of the local image structures after the appearance normalization.

Over the past years, Local Binary Patterns (LBP) is proved to be a powerful local descriptor for texture analysis and image structure representation [14,15,1]. The original LBP operator labels the pixels of an image by thresholding the 3×3-neighborhood of each pixel with the center value and considering the result as a binary string or a decimal number. Then the histogram of the labels can be used as a texture descriptor. An illustration of the basic LBP operator is shown in Fig. 3(a).

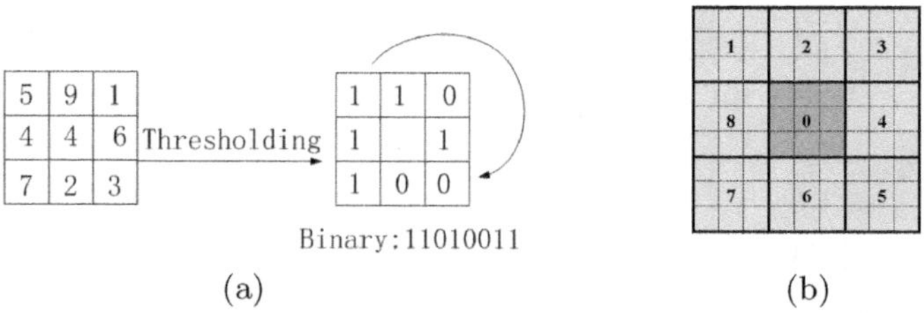

Fig. 3. (a) The basic LBP operator. (b) The 9×9 MB-LBP operator. In each sub-region, average sum of image intensity is computed. These average sums are then thresholded by that of the center block. MB-LBP is then obtained.

Recently, an extended LBP called MB-LBP [8,22] has been proposed for encoding multi-scale image structures. There, the computation is done based on average values of block subregions, instead of individual pixels (*cf.* Fig. 3(b)), and the size of the block indicates the scale. This way, MB-LBP only incurs a little more computation cost considering integral images, yet yields a more robust representation. Besides, the multi-scale representations encode not only microstructures but also macrostructures of image patterns, and hence provides a more complete image representation than the basic LBP operator.

In this work, we adopt the MB-LBP operator to encode local image structures on the normalized appearance, the resulting feature is denoted as Local Structure of Normalized Appearance (LSNA). we also use the histogram of statistically effective MB-LBP to represent the final feature set, as described in [8].

Considering the whole set of multi-scale features, MB-LBP would generate an over-complete representation. However, those excessive measures contain much redundant information, thus a further processing is needed to remove the redundancy and build effective classifiers. In this paper we use Gentle AdaBoost algorithm [2] to select the most effective MB-LBP features, as in [8].

Unlike in [9] and [21] where two kinds of heterogeneous face images are considered in two different spaces, we treat all the appearance normalized images as lying in the same space. Thus the intra-/extra-personal notation [12] is still suitable here, and the feature selection is then directly applied on both kinds of heterogeneous face images in the training set after appearance normalization. Finally, using the learned discriminant features, we apply R-LDA [20] on the whole training set to construct a universal subspace for distinguishing different persons, whether they are homogeneous or heterogeneous. Therefore, the learned model is suitable for both kinds of face recognition, in despite of the light sources.

In the test phase, for identifying two given images (homogeneous or heterogeneous), the test procedure of the proposed approach is as follows. First, both images are preprocessed using DoG filtering for appearance normalization. And then, the selected MB-LBP features are extracted on each output correspondingly. After that, the extracted feature vectors are both projected onto the learned subspace. Finally, the cosine distance is measured on the two projected vectors, and the output value is compared with a certain threshold to decide the identify result.

4 Experiments

4.1 Data Preparation

In order to evaluate the performance of the heterogeneous face recognition, we focus on the NIR vs. VIS problem and collect a corresponding NIR-VIS database for experiment. The whole database contains 4582 images from 202 subjects, including 2095 VIS and 3002 NIR images. All of them are frontal face images.

The database is divided into two distinct sets for training and testing.The training set contains 3464 images, including 1580 VIS and 1884 NIR images from the former 150 persons, and the left 52 persons constitute the test set. There is no intersection for both face image and person between training set and test set, for the purpose of constructing an open-set test protocol. In the test set, 515 VIS images compose the target set and 1118 NIR ones consist the probe set.

All the faces are cropped to 128×128 according to the automatically detected eye positions [5]. Fig. 2 has already shown some examples for VIS and NIR face images from this database.

4.2 Performance Evaluation

To verify whether an algorithm learns the essential property of the heterogeneous faces, the performance evaluation runs on an open-set test protocol as described in the previous subsection. Fig. 4 illustrates the performance evaluation result on the test set in ROC curve. Our proposed method (denoted as "LSNA") is compared with other existing methods. "CDFE" result is obtained using the approach of [9]. "PCA-CCA" and "LDA-CCA" are algorithms proposed by [21]. Besides, we also show the result of [8], denoted as "MB-LBP". Note that all parameters are carefully turned to give the best result for the compared methods, such as the smooth factor in CDFE.

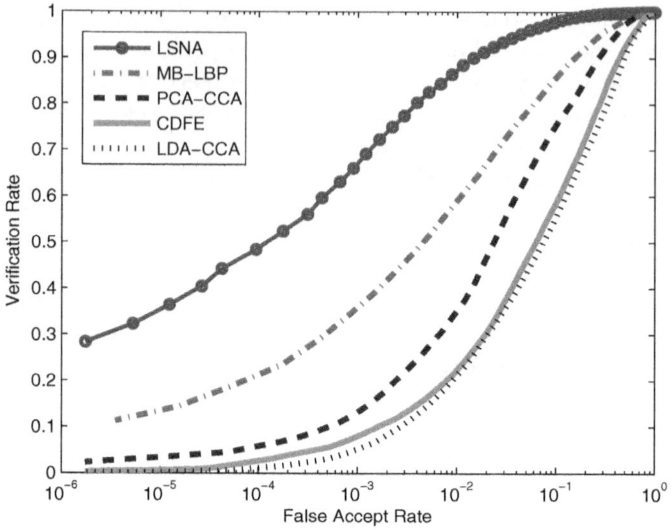

Fig. 4. Performance comparison with ROC curve

From the comparison it can be clearly seen that our approach significantly outperforms all other methods. Existing subspace projection based methods obtain poor results in the non-overlapping test set. As analyzed before, they are all applied directly on heterogeneous face images without appearance normalization and the intrinsic local-structure learning, as a result, they lack generalization ability and tend to overfit in the training data.

On the other hand, the proposed method learns the intrinsic local structure from the appearance-normalized heterogeneous face images, the resulting uniform model depends little on the training data, thus it generalizes well and gets a much better result on the test set. The verification rate of the proposed method at 0.1% false acceptance rate is 67.5%, and is 87.5% at false acceptance rate of 1%.

It can also be seen that the proposed approach outperforms MB-LBP method, which indicates that LSNA benefits a lot from the appearance normalization stage. The results also support our previous analysis that DoG filtering helps to normalize the heterogenous face appearance and contribute to recognition.

5 Summary and Conclusions

In this paper, we have presented an algorithm for heterogeneous face recognition and analyzed the particular problem of NIR to VIS face recognition. Because different types of face images have different skin spectra-optical properties, we first normalize the appearance of all heterogeneous face images using DoG filtering, and then we utilize MB-LBP to encode local image structures in the transformed space, and finally learn an universal model for both modalities of face recognition. Experiments have shown the effectiveness of the proposed method in matching between NIR and VIS faces.

Since heterogeneous face recognition is a difficult problem, which features are more suitable for this problem is still unknown till now. Future works would be investigating ordinal filters [7,6] for more flexible encodings to tackle the problem, and applying our method for other heterogeneous face recognition problems.

Acknowledgements. This work was supported by the following fundings: National Natural Science Foundation Project #60518002, National Science and Technology Support Program Project #2006BAK08B06, National Hi-Tech (863) Program Projects #2006AA01Z192, #2006AA01Z193, and #2008AA01Z124, Chinese Academy of Sciences 100 People Project, and AuthenMetric R&D Funds.

References

1. Ahonen, T., Hadid, A., Pietikainen, M.: Face recognition with local binary patterns. In: Pajdla, T., Matas, J. (eds.) ECCV 2004. LNCS, vol. 3021, pp. 469–481. Springer, Heidelberg (2004)
2. Friedman, J., Hastie, T., Tibshirani, R.: Additive logistic regression: a statistical view of boosting. Technical report, Department of Statistics, Sequoia Hall, Stanford Univerity (July 1998)
3. Gross, R., Brajovic, V.: An image preprocessing algorithm for illumination invariant face recognition. In: Proc. 4th International Conference on Audio- and Video-Based Biometric Person Authentication, Guildford, UK, June 9-11 (2003)
4. Jacobs, D., Belhumeur, P., Basri, R.: Comparing images under variable illumination. In: Proceedings of IEEE Computer Society Conference on Computer Vision and Pattern Recognition, pp. 610–617 (1998)
5. Li, S.Z., Chu, R., Liao, S., Zhang, L.: Illumination invariant face recognition using near-infrared images. IEEE Transactions on Pattern Analysis and Machine Intelligence 29(4), 627–639 (2007)
6. Liao, S., Lei, Z., Li, S.Z., Yuan, X., He, R.: Structured ordinal features for appearance-based object representation. In: Proceedings of the IEEE International Workshop on Analysis and Modeling of Faces and Gestures, pp. 183–192 (2007)
7. Liao, S., Lei, Z., Zhu, X., Sun, Z., Li, S.Z., Tan, T.: Face recognition using ordinal features. In: Zhang, D., Jain, A.K. (eds.) ICB 2006. LNCS, vol. 3832, pp. 40–46. Springer, Heidelberg (2006)
8. Liao, S., Zhu, X., Lei, Z., Zhang, L., Li, S.Z.: Learning multi-scale block local binary patterns for face recognition. In: Lee, S.-W., Li, S.Z. (eds.) ICB 2007. LNCS, vol. 4642, pp. 828–837. Springer, Heidelberg (2007)
9. Lin, D., Tang, X.: Inter-modality face recognition. In: Leonardis, A., Bischof, H., Pinz, A. (eds.) ECCV 2006. LNCS, vol. 3954, pp. 13–26. Springer, Heidelberg (2006)
10. Lindeberg, T.: Scale-space theory: A basic tool for analysing structures at different scales. Journal of Applied Statistics 21, 225–270 (1994)
11. Lowe, D.G.: Distinctive image features from scale-invariant keypoints. International Journal of Computer Vision 60, 91–110 (2004)

12. Moghaddam, B., Nastar, C., Pentland, A.: A Bayesain similarity measure for direct image matching. Media Lab Tech. Report No. 393, MIT (August 1996)
13. Nayar, S.K., Bolle, R.M.: Reflectance based object recognition. International Journal of Computer Vision 17(3), 219–240 (1996)
14. Ojala, T., Pietikainen, M., Harwood, D.: A comparative study of texture measures with classification based on feature distributions. Pattern Recognition 29(1), 51–59 (1996)
15. Ojala, T., Pietikainen, M., Maenpaa, M.: Multiresolution gray-scale and rotation invariant texture classification with local binary patterns. IEEE Transactions on Pattern Analysis and Machine Intelligence 24(7), 971–987 (2002)
16. Shashua, A., Raviv, T.R.: The quotient image: Class based re-rendering and recognition with varying illuminations. IEEE Transactions on Pattern Analysis and Machine Intelligence 23(2), 129–139 (2001)
17. Tan, X., Triggs, B.: Enhanced local texture feature sets for face recognition under difficult lighting conditions. In: Proceedings of the IEEE International Workshop on Analysis and Modeling of Faces and Gestures (2007)
18. Wang, H.T., Li, S.Z., Wang, Y.S.: Generalized quotient image. In: Proceedings of IEEE Computer Society Conference on Computer Vision and Pattern Recognition, pp. 498–505 (2004)
19. Yang, W., Yi, D., Lei, Z., Sang, J., Li, S.Z.: 2D-3D face matching using CCA. In: Proc. IEEE International Conference on Automatic Face and Gesture Recognition, Amsterdam, The Netherlands, September 17-19 (2008)
20. Ye, J., Xiong, T., Li, Q., Janardan, R., Bi, J., Cherkassky, V., Kambhamettu, C.: Efficient model selection for regularized linear discriminant analysis. In: Proceedings of the ACM Conference on Information and Knowledge Management, pp. 532–539 (2006)
21. Yi, D., Liu, R., Chu, R., Lei, Z., Li, S.Z.: Face matching from near infrared to visual images. In: Proceedings of the 2nd IAPR/IEEE International Conference on Biometrics, Seoul, Korea (August 2007)
22. Zhang, L., Chu, R., Xiang, S., Liao, S., Li, S.Z.: Face detection based on multi-block lbp representation. In: Lee, S.-W., Li, S.Z. (eds.) ICB 2007. LNCS, vol. 4642, pp. 11–18. Springer, Heidelberg (2007)

Sparse Representation for Video-Based Face Recognition

Imran Naseem[1], Roberto Togneri[1], and Mohammed Bennamoun[2]

[1] School of Electrical, Electronic and Computer Engineering
The University of Western Australia
imran.naseem@ee.uwa.edu.au, roberto@ee.uwa.edu.au

[2] School of Computer Science and Software Engineering
The University of Western Australia
bennamou@csse.uwa.edu.au

Abstract. In this paper we address for the first time, the problem of video-based face recognition in the context of sparse representation classification (SRC). The SRC classification using still face images, has recently emerged as a new paradigm in the research of view-based face recognition. In this research we extend the SRC algorithm for the problem of temporal face recognition. Extensive identification and verification experiments were conducted using the VidTIMIT database [1,2]. Comparative analysis with state-of-the-art Scale Invariant Feature Transform (SIFT) based recognition was also performed. The SRC algorithm achieved 94.45% recognition accuracy which was found comparable to 93.83% results for the SIFT based approach. Verification experiments yielded 1.30% Equal Error Rate (EER) for the SRC which outperformed the SIFT approach by a margin of 0.5%. Finally the two classifiers were fused using the weighted sum rule. The fusion results consistently outperformed the individual experts for identification, verification and rank-profile evaluation protocols.

1 Introduction

It is long known that appearance-based face recognition systems critically depend on manifold learning methods. A gray-scale face image of order $a \times b$ can be represented as an ab-dimensional vector in the original *image space*. However any attempt of recognition in such a high dimensional space is vulnerable to a variety of issues often referred to as the *curse of dimensionality*. Typically in pattern recognition problems it is believed that high-dimensional data vectors are redundant measurements of an underlying source. The objective of manifold learning is therefore to uncover this "underlying source" by a suitable transformation of high-dimensional measurements to low-dimensional data vectors. View-based face recognition methods are no exception to this rule. Therefore, at the feature extraction stage, images are transformed to low dimensional vectors in a *face space*. The main objective is to find a basis function for this transformation,

M. Tistarelli and M.S. Nixon (Eds.): ICB 2009, LNCS 5558, pp. 219–228, 2009.
© Springer-Verlag Berlin Heidelberg 2009

which could distinguishably represent faces in the face space. Linear transformation from the image space to the feature space is perhaps the most traditional way of dimensionality-reduction, also called "Linear Subspace Analysis".

A number of approaches have been reported in the literature including Principal Component Analysis (PCA) [3], [4], Linear Discriminant Analysis (LDA) [5] and Independent Component Analysis (ICA) [6], [7]. These approaches have been classified in two categories namely *reconstructive* and *discriminative* methods. Reconstructive approaches (such as PCA and ICA) are reported to be robust for the problem related to contaminated pixels, whereas discriminative approaches (such as LDA) are known to yield better results in clean conditions [8]. Nevertheless, the choice of the manifold learning method for a given problem of face recognition has been a hot topic of research in the face recognition literature. These debates have recently been challenged by a new concept of "Sparse Representation Classification (SRC)" [9]. It has been shown that unorthodox features such as downsampled images and random projections can serve equally well. As a result the choice of the feature space may no longer be so critical [9]. What really matters is the dimensionality of the feature space and the design of the classifier. The key factor to the success of sparse representation classification is the recent development of "Compressive Sensing" theory [10].

Due to the ever increasing security threats, video surveillance systems have been deployed on a large scale. With the additional temporal dimension, video sequences are much more informative than still images. As a result the person identification task is facilitated due to specific attributes of each subject such as head rotation and pose variation along the temporal dimension. Additionally more efficient face representations such as super resolution images can be derived from video sequences for further enhancement of the overall system. These motivations have urged researchers to look into the development of face recognition systems that can utilize the spatiotemporal information in video sequences. It is therefore becoming imperative to evaluate present state-of-the-art face recognition algorithms for video-based applications. With this understanding, this research is targeted to the extension of the recently proposed SRC classification for the problem of video-based face recognition. The primary objective is to critically analyze the new approach in comparison with state-of-the-art SIFT features based algorithm. The rest of the paper is organized as follows: Section 2 provides an overview of the SRC algorithm followed by a brief description of SIFT based recognition in Section 3. Experimental results and discussion are presented in Section 4, the paper concludes in Section 5.

2 Sparse Representation for Face Recognition

We now discuss the basic framework of the face recognition system in the context of sparse representation [9]. Let us assume that we have k distinct classes and n_i images available for training from the ith class. Each training sample is a gray scale image of order $a \times b$. The image is downsampled to an order $w \times h$

and is converted into a 1-D vector $\mathbf{v}_{i,j}$ by concatenating the columns of the downsampled image such that $\mathbf{v}_{i,j} \in \mathbb{R}^m$ $(m = wh)$. Here i is the index of the class, $i = 1, 2, \ldots, k$ and j is the index of the training sample, $j = 1, 2, \ldots, n_i$. All this training data from the ith class is placed in a matrix A_i such that $A_i = [\mathbf{v}_{i,1}, \mathbf{v}_{i,2}, \ldots\ldots, \mathbf{v}_{i,n_i}] \in \mathbb{R}^{m \times n_i}$. As stated in [9], when the training samples from the ith class are sufficient, the test sample $\mathbf{y}$ from the same class will approximately lie in the linear span of the columns of A_i:

$$\mathbf{y} = \alpha_{i,1}\mathbf{v}_{i,1} + \alpha_{i,2}\mathbf{v}_{i,2} + \cdots + \alpha_{i,n_i}\mathbf{v}_{i,n_i} \tag{1}$$

where $\alpha_{i,j}$ are real scalar quantities. Now we develop a dictionary matrix A for all k classes by concatenating A_i, $i = 1, 2, \ldots, k$ as follows:

$$\mathbf{A} = [A_1, A_2, \ldots, A_k] \ \in \mathbb{R}^{m \times n_i k} \tag{2}$$

Now a test pattern $\mathbf{y}$ can be represented as a linear combination of all n training samples $(n = n_i \times k)$:

$$\mathbf{y} = \mathbf{A}\mathbf{x} \tag{3}$$

Where $\mathbf{x}$ is an unknown vector of coefficients. Now from equation 3 it is relatively straight forward to note that only those entries of $\mathbf{x}$ that are non-zero correspond to the class of $\mathbf{y}$ [9]. This means that if we are able to solve equation 3 for $\mathbf{x}$ we can actually find the class of the test pattern $\mathbf{y}$. Recent research in compressive sensing and sparse representation [11,10,12,13,14] have shown that using the sparsity of the solution of equation 3, enables us to solve the problem using l_1-norm minimization:

$$(l^1): \quad \hat{\mathbf{x}}_1 = argmin \, \|\mathbf{x}\|_1 \quad ; \mathbf{A}\mathbf{x} = \mathbf{y} \tag{4}$$

Once we have estimated $\hat{\mathbf{x}}_1$, ideally it should have nonzero entries corresponding to the class of $\mathbf{y}$ and now deciding the class of $\mathbf{y}$ is a simple matter of locating indices of the non-zero entries in $\hat{\mathbf{x}}_1$. However due to noise and modeling limitations $\hat{\mathbf{x}}_1$ is commonly corrupted by some small nonzero entries belonging to different classes. To resolve this problem we define an operator δ_i for each class i so that $\delta_i(\hat{\mathbf{x}}_1)$ gives us a vector $\in \mathbb{R}^n$ where the only nonzero entries are from the ith class. This process is repeated k times for each class. Now for a given class i we can approximate $\hat{\mathbf{y}}_i = \mathbf{A}\delta_i(\hat{\mathbf{x}}_1)$ and assign the test pattern to the class with a minimum residual between $\mathbf{y}$ and $\hat{\mathbf{y}}_i$.

$$\underbrace{\min}_{i} r_i(\mathbf{y}) = \|\mathbf{y} - \mathbf{A}\delta_i(\hat{\mathbf{x}}_1)\|_2 \tag{5}$$

3 Scale Invariant Feature Transform (SIFT) for Face Recognition

The Scale Invariant Feature Transform (SIFT) was proposed in 1999 for the extraction of unique features from images [15]. The idea, initially proposed for a more generic object recognition task, was later successfully applied for the problem of face recognition [16]. Interesting characteristics of scale/rotation invariance and locality in both spatial and frequency domains have made the SIFT-based approach a pretty much standard technique in the paradigm of view-based face recognition. The first step in the derivation of the SIFT features is the identification of potential pixels of interest called "keypoints", in the face image. An efficient away of achieving this is to make use of the scale-space extrema of the Difference-of-Gaussian (DoG) function convolved with the face image [15]. These potential keypoints are further refined based on the high contrasts, good localization along edges and the ratio of principal curvatures criterion. Orientation(s) are then assigned to each keypoint based on local image gradient direction(s). A gradient orientation histogram is formed using the neighboring pixels of each keypoint. Contribution from neighbors are weighted by their magnitudes and by a circular Gaussian window. Peaks in the histogram represent the dominant directions and are used to align the histogram for rotation invariance. 4×4 pixel neighborhoods are used to extract eight bin histograms resulting in 128-dimensional SIFT features. For illumination robustness, the vectors are normalized to unity, thresholded to a ceiling of 0.2 and finally renormalized to unit length. Figure 1 shows a typical face from the VidTIMIT database [1,2] with extracted SIFT features.

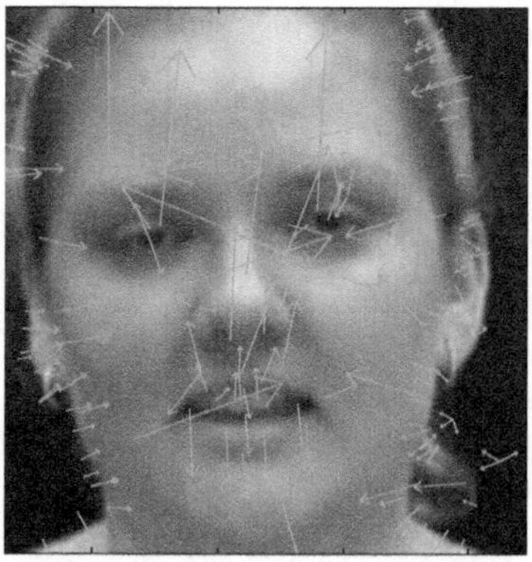

Fig. 1. A typical localized face from the VidTIMIT database with extracted SIFTs

During validation a SIFT feature vector from the query video f_q is matched with the feature vector from the gallery:

$$e = \arccos\left[f_q (f_g)^T\right] \quad (6)$$

where f_g corresponds to a SIFT vector from a training video sequence. All SIFT vectors from the query frame are matched with all SIFT features from a training frame using Equation 6. Pairs of features with the minimum error e are considered as matches. Note that if more than one SIFT vector from a given query frame happens to be the best match with the same SIFT vector from gallery (i.e. many-to-one match scenario), the one with the minimum error e is chosen. Other false matches were reduced by matching the SIFT vectors from only nearby regions of the two images.

In principle, for different image pairs we have different number of matches. This information is further harnessed to be used as an additional similarity measure between the two faces. The final similarity score between two frames is computed by normalizing the average error $\overline{e}$ between their matching pairs of SIFT features and the total number of matches z on a scale [0,1] and then using a weighted sum rule.

$$\overline{e}' = \frac{\overline{e} - \min(\overline{e})}{\max(\overline{e} - \min(\overline{e}))} \tag{7}$$

$$z' = \frac{z - \min(z)}{\max(z - \min(z))} \tag{8}$$

$$s = \frac{1}{2}(\beta_e \overline{e}' + \beta_z(1 - z')) \tag{9}$$

where β_e and β_z are the weights of normalized average error $\overline{e}'$ and normalized number of matches z' respectively. It has to be noted that $\overline{e}'$ is a distance (dissimilarity) measure while z' is a similarity score, therefore in Equation 9 z' is subtracted from 1 for a homogeneous fusion. Consequently s becomes a distance measure.

4 Results and Discussion

The problem of temporal face recognition using the SRC and SIFT feature based face recognition algorithms was evaluated on the VidTIMIT database [1], [2]. VidTIMIT is a multimodal database consisting of video sequences and corresponding audio files from 43 distinct subjects. The video section of the database characterizes 10 different video files from each subject. Each video file is a sequence of 512×384 JPEG images. Two video sequences were used for training while the remaining eight were used for validation. Due to the high correlation between consecutive frames, training and testing were carried out on alternate frames. Off-line batch learning mode [17] was used for these experiments and therefore probe frames did not add any information to the system.

Face localization is the first step in any face recognition system. Fully automatic face localization was carried out using a Harr-like feature based face detection algorithm [18] during off-line training and on-line recognition sessions. For the SIFT based face recognition, each detected face in a video frame was scale-normalized to 150×150 and histogram equalized before the extraction of the SIFT features. We achieved an identification rate of 93.83%. Verification experiments were also conducted for a more comprehensive comparison between

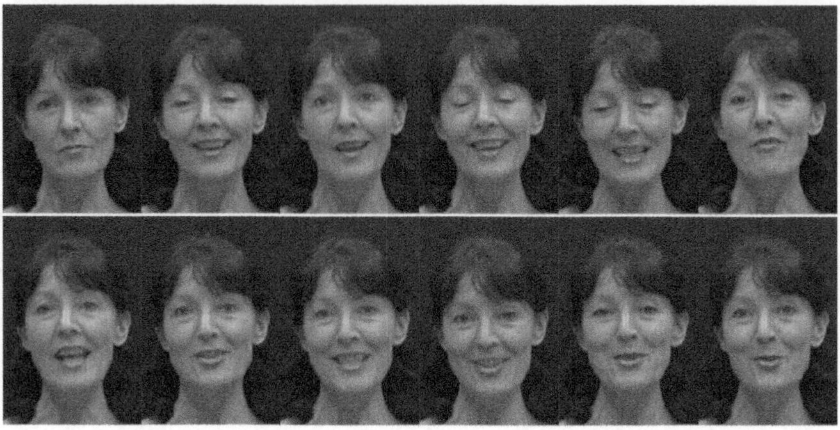

Fig. 2. A sample video sequence from the VidTIMIT database

the two approaches. An Equal Error Rate (EER) of 1.8% was achieved for the SIFT based verification. Verification rate at 0.01 False Accept Rate (FAR) was found to be 97.32%.

For the SRC classifier, each detected face in a frame is downsampled to order 10×10. Column concatenation is carried out to generate a 100-dimensional feature vector as discussed in Section 2. Off-line batch learning is carried out on alternate frames using two video sequences as discussed above. Unorthodox downsampled images in combination with the SRC classifier yielded quite comparable recognition accuracy of 94.45%. EER dropped to 1.3% with a verification accuracy of 98.23% at 0.01 FAR. The rank profile and ROC (Receiver Operating Characteristics) curves are shown in Figure 3 (a) and 3 (b) respectively.

We further investigated the complementary nature of the two classifiers by fusing them at the score level. The weighted sum rule is used which is perhaps

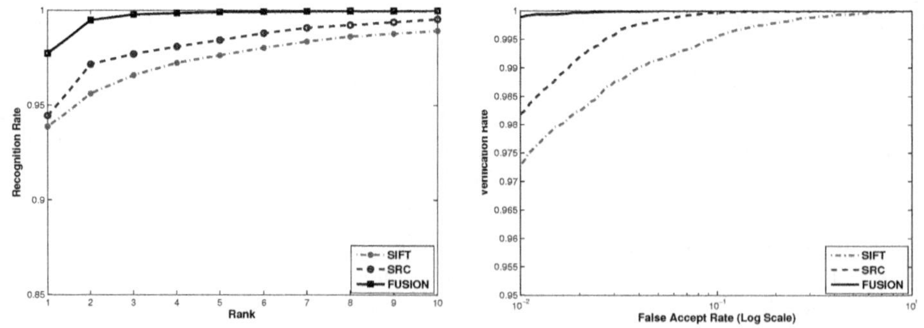

(a) (b)

Fig. 3. (a)Rank profiles and (b) ROC curves for the SIFT, SRC and the combination of the two classifiers the major work-horse in the field of combining classifiers [19]. Both classifiers were equally weighted and a high recognition accuracy of 97.73% was achieved which outperforms the SIFT based classifier and the SRC classifier by a margin of 3.90% and 3.28% respectively. Verification experiments also produced superior results with an EER of 0.3% which is better than the SIFT and the SRC based classification by 1.5% and 1.0% respectively. An excellent verification of 99.90% at an FAR of 0.01 is reported. Fusion of the two classifiers substantially improved the rank profile as well achieving 100% results at rank-5 only. A detailed comparison of the results is provided in Table 1.

Table 1. Summary of results

Evaluation Attributes	SIFT	SRC	Fusion
Recognition Accuracy	93.83%	94.45%	97.73%
Equal Error Rate	1.80%	1.30%	0.30%
Verification rate at 0.01 FAR	97.32%	98.23%	99.90%

Presented results certainly reflect a comparable performance index for the SRC classifier as compared to state-of-the-art SIFT based recognition. Extensive experiments based on identification, verification and rank-recognition evaluations consistently reflect better results for the SRC approach. Moreover the complementary information exhibited by the SRC method increased the verification success of the combined system to 99.9% for the standard 0.01 FAR criterion. Figure 4 shows variation in the recognition accuracy with the change in the normalized weight of the SRC classifier at the fusion stage. Approximately the highest recognition is achieved when both classifiers were equally weighted i.e. no prior information of the participating experts was incorporated in fusion.

Apart from these appreciable results it was found that the l_1-norm minimization using a large dictionary matrix made the iterative convergence lengthy and slow. To provide a comparative value we performed computational analysis for a randomly selected identification trail. The time required by the SRC algorithm for classifying a single frame on a typical 2.66 GHz machine with 2 GB memory was found to be 297.46 seconds (approximately 5 minutes). This duration is approximately 5 times greater than the processing time of the SIFT algorithm for the same frame which was found to be 58.18 seconds (approximately 1 minute). Typically a video sequence consists of hundreds of frames which would suggest a rather prolonged span for the evaluation of the whole video sequence. Noteworthy is the fact that experiments were conducted using an offline learning mode [17]. The probe frames did not contribute to the dictionary information. Critically speaking, the spatiotemporal information in video sequences is best harnessed using smart online [20] and hybrid [21] learning modes. These interactive learning algorithms add useful information along the temporal dimension and therefore enhance the overall performance. However, in the context of SRC classification, this would suggest an even larger dictionary matrix and consequently a lengthier evaluation.

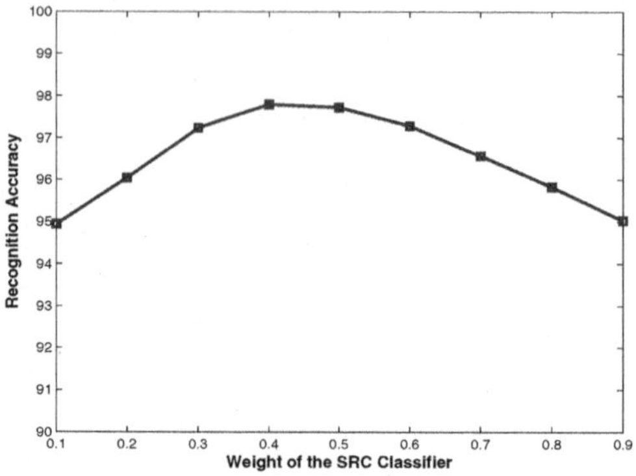

Fig. 4. Variation in performance with respect to bias in fusion

5 Conclusion

Sparse representation classification has recently emerged as the latest paradigm in the research of appearance-based face recognition. In this research we evaluated the approach for the problem of video-based face recognition. An identification rate of 94.45% is achieved on the VidTIMIT database which is quite comparable to 93.83% accuracy using state-of-the-art SIFT features based algorithm. Verification experiments were also conducted and the SRC approach exhibited an EER of 1.30% which is 0.5% better than the SIFT method. The SRC classifier was found to nicely complement the SIFT based method, the fusion of the two methods using the weighted sum rule consistently produced superior results for identification, verification and rank-recognition experiments. However since SRC requires an iterative convergence using an l_1-norm minimization, the approach was found computationally expensive as compared to the SIFT based recognition. Typically SRC required 5 minutes (approximately) for processing a single recognition trial which is 5 times greater than the time required by the SIFT based approach. To the best of our knowledge, this is the first evaluation of the SRC algorithm on a video database. From the experiments presented in the paper, it is quite safe to maintain that additional work is required before the SRC approach is declared as a standard approach for video-based applications. Computational expense is arguably an inherent issue with video processing giving rise to the emerging area of "Video Abstraction". Efficient algorithms have been proposed to cluster video sequences along the temporal dimension (for example [22] including others). These clusters are then portrayed by cluster-representative frame(s)/features resulting in a substantial decrease of complexity. Given the good performance of the SRC algorithm presented in this research, the evaluation of the method using state-of-the-art video abstraction methods will be the subject of our future research.

Acknowledgement

The authors would like to thank D. Lowe for providing the SIFT code. This research is partially funded by the Australian Research Council (ARC) grant No. DP0771294.

References

1. Sanderson, C., Paliwal, K.K.: Identity verification using speech and face information. Digital Signal Processing 14(5), 449–480 (2004)
2. Sanderson, C.: Biometric person recognition: Face, speech and fusion. VDM-Verlag (2008)
3. Jolliffe, I.T.: Pricipal Component Analysis. Springer, New York (1986)
4. Turk, M., Pentland, A.: Eigenfaces for recognition. Journal of Cognitive Neurosicence 3(1), 71–86 (1991)
5. Belhumeur, V., Hespanha, J., Kriegman, D.: Eigenfaces vs Fisherfaces: Recognition using class specific linear projection. IEEE Tran. PAMI 17(7), 711–720 (1997)
6. Comon, P.: Independent Component Analysis - A New Concept? Signal Processing 36, 287–314 (1994)
7. Bartlett, M., Lades, H., Sejnowski, T.: Independent component representations for face recognition. In: Proceedings of the SPIE: Conference on Human Vision and Electronic Imaging III, vol. 3299, pp. 528–539 (1998)
8. Duda, R.O., Hart, P.E., Stork, D.G.: Pattern Classification. John Wiley & Sons, Inc., Chichester (2000)
9. Wright, J., Yang, A., Ganesh, A., Sastri, S.S., Ma, Y.: Robust face recognition via sparse representation. IEEE Trans. PAMI (2008)
10. Donoho, D.: Compressed sensing. IEEE Trans. Inform. Theory 52(4), 1289–1306 (2006)
11. Candès, E., Romberg, J., Tao, T.: Robust uncertainty principles: Exact signal reconstruction from highly incomplete frequency information. IEEE Trans. Inform. Theory 52(2), 489–509 (2006)
12. Donoho, D.: For most large underdetermined systems of linear equations the minimal l_1-norm solution is also the sparsest solution. Comm. on Pure and Applied Math. 59(6), 797–829 (2006)
13. Candès, E., Romberg, J., Tao, T.: Stable signal recovery from incomplete and inaccurate measurements. Comm. on Pure and Applied Math. 59(8), 1207–1223 (2006)
14. Candès, E., Tao, T.: Near-optimal signal recovery from random projections: Universal encoding strategies? IEEE Tran. Infm. Theory 52(12), 5406–5425 (2006)
15. Lowe, D.: Object recognition from local scale-invariant features. In: Intl. Conf. on Computer Vision, pp. 1150–1157 (1999)
16. Bicego, M., Lagorio, A., Grosso, E., Tistarelli, M.: On the use of SIFT features for face authentication. In: CVPRW (2006)
17. Lee, K., Ho, J., Yang, M., Kriegman, D.: Visual tracking and recognition using probabilistic appearance manifolds. CVIU 99(3), 303–331 (2005)
18. Viola, P., Jones, M.: Robust real-time face detection. International Journal of Computer Vision 57(2), 137–154 (2004)

19. Kittler, J., Hatef, M., Duin, R.P.W., Matas, J.: On combining classifiers. IEEE Trans. on Pattern Analysis and Machine Intelligence 20(3), 226–238 (1998)
20. Liu, L., Wang, Y., Tan, T.: Online appearance model. In: CVPR, pp. 1–7 (2007)
21. Lee, K., Kriegman, D.: Online probabilistic appearance manifolds for video-based recognition and tracking. In: CVPR, vol. 1, pp. 852–859 (2005)
22. Chan, A.B., Vasconcelos, N.: Modeling, clustering, and segmenting video with mixtures of dynamic textures. IEEE Trans. PAMI 30, 909–926 (2008)

Face Image Quality Evaluation for ISO/IEC Standards 19794-5 and 29794-5

Jitao Sang, Zhen Lei, and Stan Z. Li*

Center for Biometrics and Security Research,
Institute of Automation, Chinese Academy of Sciences,
95 Zhongguancun Donglu, Beijing 100190, China
{jtsang,zlei,szli}@cbsr.ia.ac.cn

Abstract. Face recognition performance can be significantly influenced by face image quality. The approved ISO/IEC standard 19794-5 has specified recommendations for face photo taking for E-passport and related applications. Standardization of face image quality, ISO/IEC 29794-5, is in progress. Bad illumination, facial pose and out-of-focus are among main reasons that disqualify a face image sample. This paper presents several algorithms for face image quality assessment. Illumination conditions and facial pose are evaluated in terms of facial symmetry, and implemented based on Gabor wavelet features. Assessment of camera focus is done based on discrete cosine transform (DCT). These methods are validated by experiments.

Keywords: Face image quality, international standard, facial symmetry, out-of-focus.

1 Introduction

The approved ISO/IEC standard 19794-5[3] includes instructions for lighting, facial pose, focus, and so on, for taking face photos for applications such as E-passport. Illustrated by Fig 1, which is a standard face image, we can see besides normative requirements of size and proportion, the face is uniformly illuminated and captured from right ahead with no rotation or pitching, out-of-focus is not admissible either.

In ISO/IEC 19794-5, out-of-focus, non-frontal posture and side lighting are regarded as primary elements for poor face image quality. Clause 7.3.3 has regularization that captured image shall always be in focus from nose to ears and chin to crown, while clauses 7.2.2 and 7.2.7-7.2.10 require uniform illumination and fix the angel control for deviation from frontal head posture with respect to Pan/Tilt/Roll axis.

Face images of bad quality which do not accord with the requirements of the standards is a reason leading to face recognition performance degradation.

* Corresponding author.

M. Tistarelli and M.S. Nixon (Eds.): ICB 2009, LNCS 5558, pp. 229–238, 2009.
© Springer-Verlag Berlin Heidelberg 2009

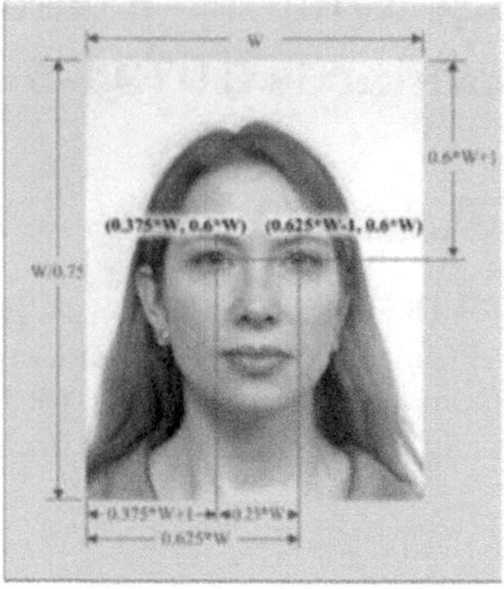

Fig. 1. Standard face image agreed with ISO/IEC 19794-5[3]

Table 1. Face image requirements in ISO/IEC-19794-5

Clause	Attribute	Constraint
Scene	Posture	control on deviation from frontal
	Illumination	Uniformly illuminated with no shadow
	Background	Plain light-colored
	Eyes	Open and clearly visible
	Glasses	No flash reflections, dark tint or heavy frames
	Mouth	Closed and clearly visible
Photographic	Head position	Placed in the center
	Distance to camera	Moderate head size
	Color	Color neutral and no red eye
	Exposure	Appropriate brightness
Digital	Focus	No out-of-focus and in Good sharpness
	Resolution	Width constraint of the head

Non-standard lighting or pose and out-of-focus are among the main reasons responsible for the performance degradation. One solution, where most researchers commit themselves, is to improve the algorithm itself by making it robust to possible degradation.

In virtue of face image quality evaluation method, face image quality are evaluated before subsequent enrollment and comparison and unqualified images will be abandoned so as to stabilize the systems performance. A framework for image quality evaluation was proposed in [1] and subsequently adopted in a draft of ISO/IEC 29794-5 [2].

Draft of ISO/IEC 29794-5[2] specifies methodologies for computation of objective and quantitative quality scores for facial images. Approaches for the determination of certain aspects, such as facial symmetry, resolution and size, illumination intensity, brightness, contrast, color, exposure, sharpness, etc, are introduced.

Moreover, draft of ISO/IEC 29794-5[2] suggests that facial quality be categorized into static subject and dynamic subject characteristics. In addition to the normative requirements presented in ISO/IEC 19794-5, open issue talking over objective metric for human perceived quality vs. metric for psFAR/psFRR are proposed in this draft, thus making it more practical and comprehensive. Measurements of face image quality are categorize into three aspects as scene requirements, photographic requirements and digital requirements, as in Table 1.

In this paper, we present methods for face image quality evaluation. Gabor wavelets are used as basis features to estimate the symmetry and then to evaluate lighting and pose conditions. Discrete Cosine Transform (DCT) and Inverse DCT (IDCT) are used to calculate a degree of out-of-focus. Experiment results are shown to illustrate the methods.

2 Gabor-Based Facial Symmetry

The illumination and pose variations are two main issues that cause severe performance degradation for most existing systems [9]. Uneven lighting and deviation from frontal posture give rise to severe facial asymmetry. So we can use facial symmetry to evaluate quality degradations caused by non-frontal lighting and improper facial pose. Difference between the left and right half regions of face give a visualized description of facial asymmetry, where the difference is zero when the face is strictly symmetric.

Suggested by the definition, difference between one face image and its mirror is a natural way to measure the symmetry. However, this method is too sensitive to face alignment which means it is available only when geometric centerline of the face image matches very well with the physical centerline.

Local filters which is more robust to alignment while sensitive to posture and illumination are needed. Gabor wavelets, Ordinal and Local Binary Patterns(LBP) filters provide effective local features met these requirements. Graft of ISO/IEC 29794-5[2] introduces a LBP-based method. LBP is sort of coding based on comparison with adjacent areas, thus will fail in extreme case when surrounding pixels brighten or darken in the same level. In this paper we propose to employ imaginary of Gabor filters as a solution to assess the facial symmetry.

Gabor feature is a popular descriptor in face recognition [4]. The Gabor kernels are defined as follows:

$$\psi_{\mu,\nu} = \frac{k_{\mu,\nu}^2}{\sigma^2} \exp(-\frac{k_{\mu,\nu}^2 z^2}{2\sigma^2})[\exp(ik_{\mu,\nu}z) - \exp(-\frac{\sigma^2}{2})] \quad (1)$$

where μ and ν define the orientation and scale of the Gabor kernels respectively, $z = (x, y)$, and the wave vector $k_{\mu,\nu}$ is defined as follows:

$$k_{\mu,\nu} = k_\nu e^{i\phi_\mu} \quad (2)$$

where $k_\nu = k_{max}/f^\nu$, $k_{max} = \pi/2$, $f = \sqrt{2}$, $\phi_\mu = \pi\mu/8$.

In face recognition, the magnitude of Gabor feature is usually adopted due to its robustness to illumination and expression. However, in this case, we require

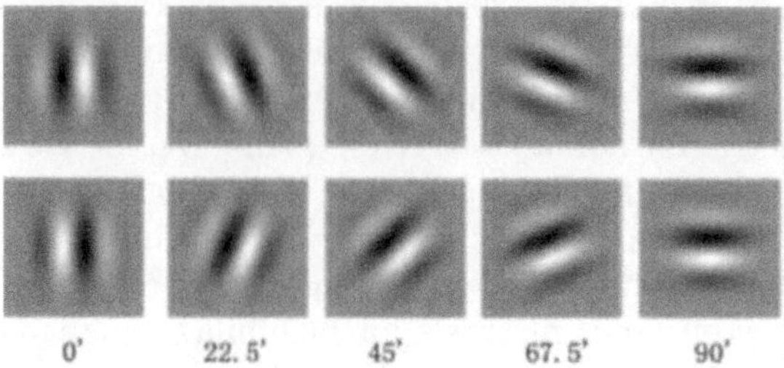

Fig. 2. Gabor filters of 5 orientations (the second row is the mirror filter corresponding to the first)

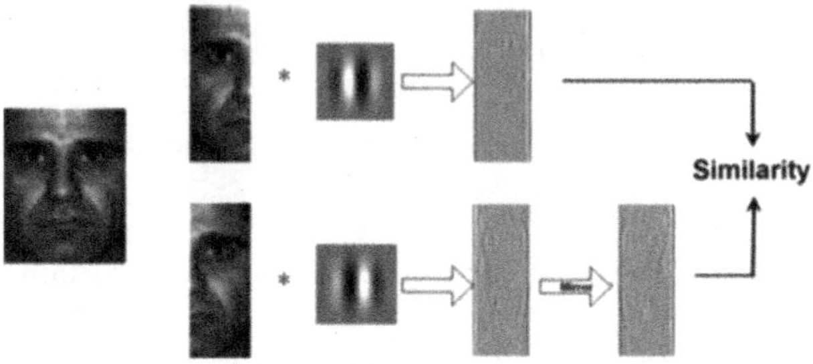

Fig. 3. Evaluation of facial symmetry using Gabor filters

the feature that is sensitive to illumination and pose. Therefore, the imaginary of Gabor filters, which not only has the characteristic of gradient operator, but also has orientation selectivity and stability, is utilized and believed to be competent for asymmetric. Fig. 2 shows the imaginary of Gabor filters in five orientations selected in this paper and Fig. 3 illustrates the whole process for face asymmetric measurement. Firstly, five pairs of mirror symmetric Gabor filters are acted on left and right half regions of a face image respectively. Secondly, the asymmetry is measured as the difference between the corresponding Gabor responses on left and right part calculated as:

$$Asymmetry = \sum_{i=1}^{M}\sum_{j=1}^{N}|L(i,j) - R(i,j)| \tag{3}$$

where M and N is the height and width of a image, L and R indicates the left and right half part of Gabor responses. Obviously, the bigger the value of *Asymmetry*, the worse the facial symmetry.

3 DCT-Based Sharpness

Most optical systems are equipped with components of automated out-of-focus detection and blur adjustment[6]. Blurriness, resulting from out-of-focus, is treated as important measurement for image quality evaluation[5]. In view of image processing, out-of-focus leads to lost in high-spatial frequency. Thus we proposed to use DCT to evaluate it on frequency domain.

Sharpness is another crucial factor affecting the performance of face recognition. The sharpness of a face image refers to the degree of clarity in both coarse and fine details in the face region.

Many measures for computing image sharpeness have been proposed in recent years[7,8]. Intuitively, N.K.Ratha et al. [5] proposes a variance based method to evaluate the sharpness of biometric images. In his method, an $M \times N$ image is divided into $m \times n$ regions. In every region, the variance is computed and if it is larger than a pre-defined threshold, this region is sharpness enough and considered as a good region. The final sharpness score is derived as the ratio of numbers of good regions to the total region number.

$$Sharpness1 = \frac{1}{Num} \sum_{i=1}^{Num} I(D_i > T) \tag{4}$$

where Num is the number of regions, D_i is the variance in region i and $I(\cdot) \in \{0, 1\}$ is an indication function of a boolean condition.

We propose a novel method in frequency domain to evaluate the sharpness of face image. The low-frequency information in an image corresponds to the global shape of components whereas the high-frequency counterpart corresponds to the details of skin. The out of focus face images usually lose high-frequency information. Fig. 4 illustrates two examples. The left is the blurring image captured in real world and the right is the simulated image by Gaussian convolution. It can be clearly seen the high frequency coefficients of blurring images (the lower images) are really small in their frequency domain.

Enlightened by the fact that little high-spatial frequency exists in most images, DCT are widely used in image compression. In this paper we utilize DCT and IDCT to measure the high-spatial frequency content, i.e., the sharpness of an image. In particular, with an input image I, it is first transformed into frequency domain by DCT operation. The coefficients which occupies the pre-defined $ratio$ ($ratio$ is chosen in sequence of run-length coding according to $rate$,

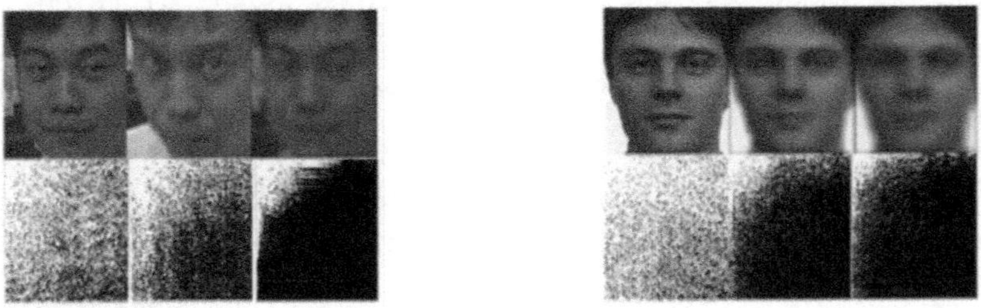

(a) (b)

Fig. 4. Images of different sharpness degrees and their DCT frequency results where $ratio = M \times N \times rate$) of total energy is preserved and then it is inverse-transformed to the image space by IDCT operation to obtain the recovered image R. The sharpness is finally evaluated by the difference of input image and the recovered one (Eq. 5). Obviously, the larger difference is, the sharper the input image is.

$$Sharpness2 = \frac{1}{M \times N}\sqrt{\sum_{i=1}^{M}\sum_{j=1}^{N}(R(i,j) - I(i,j))^2} \tag{5}$$

4 Experiments

The following experiments first examine on the facial asymmetry caused by non-frontal lighting and improper facial poses using the facial symmetry based methods presented in Section 2. Then DCT based method described in Section 3 are presented to measure the image sharpness.

CMU-PIE database is utilized to assess the facial symmetry problem. Images under different lighting conditions and different poses from 68 subjects are selected. In sharpness measurement experiment, 1850 images from FERET database are selected.

4.1 Lighting Symmetry

The proposed method is compared with the method that computes the difference on gray images directly, denoted as DDG and the LBP based method in [1]. All the images are cropped according to the automatically detected eye positions.

Fig. 5 shows the cropped face images under different lighting conditions and Fig. 6 illustrates the results of different facial symmetry assessment methods. It can be seen that the DDG method, due to its sensitive to misalignment and noise, is almost infeasible in lighting symmetry evaluation. The performance of the proposed Gabor based method and the LBP based method [1] are similar and they can be used to evaluate the lighting condition better.

4.2 Pose Symmetry

Four poses (27, 29, 11, 14) are selected to do the pose assessment. (Fig. 7). Similarly, the proposed Gabor based method is compared with DDG and LBP method in [1]. Fig. 8 shows the face quality asymmetry values for 4 pose categories. It can be seen the Gabor and LBP based methods are more applicable.

4.3 Sharpness Measurement

In this section, both real and simulated images are tested. In simulated case, the blurring images are generated by the convolution operation on original images with Gaussian filters (Eq. 6).

$$G(x,y) = \exp\{-\frac{x^2 + y^2}{2\sigma^2}\} \tag{6}$$

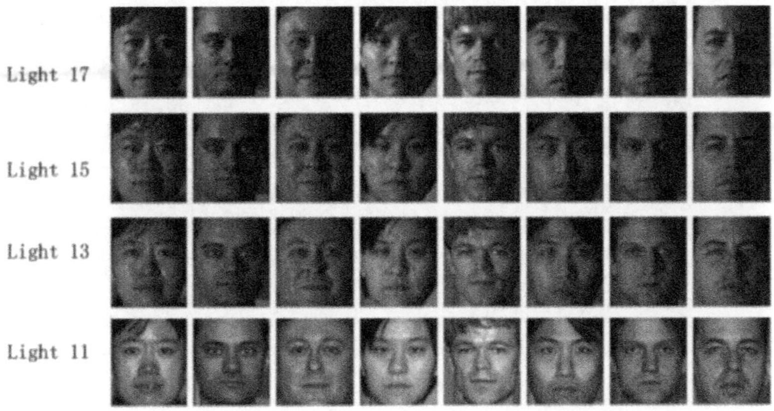

Fig. 5. Face examples under four different lighting conditions

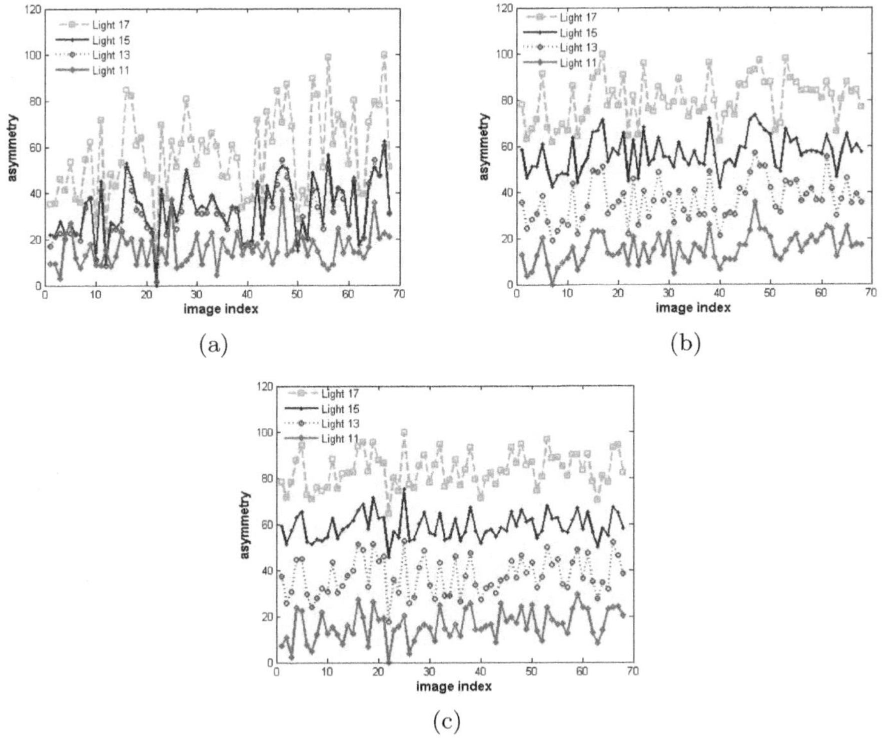

Fig. 6. Lighting asymmetry score using DDG (a), LBP (b) and Gabor (c) features

The proposed DCT based method is compared with the method in [5]. For method in [5], the image is divided into 14×12 and 7×6 respectively and the threshold is set as the mean of variances of all sub-regions. For the proposed

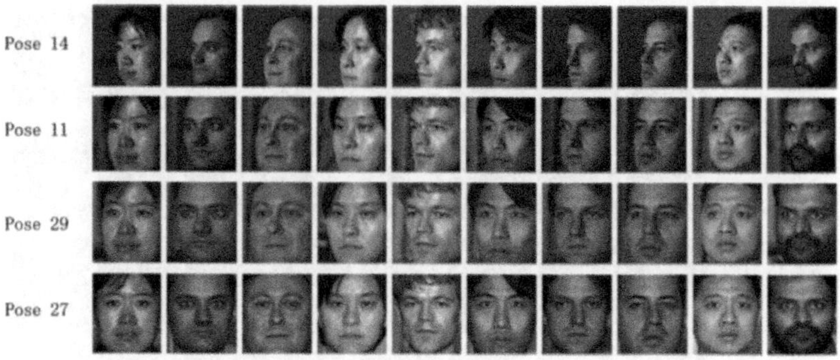

Fig. 7. Face examples under four poses

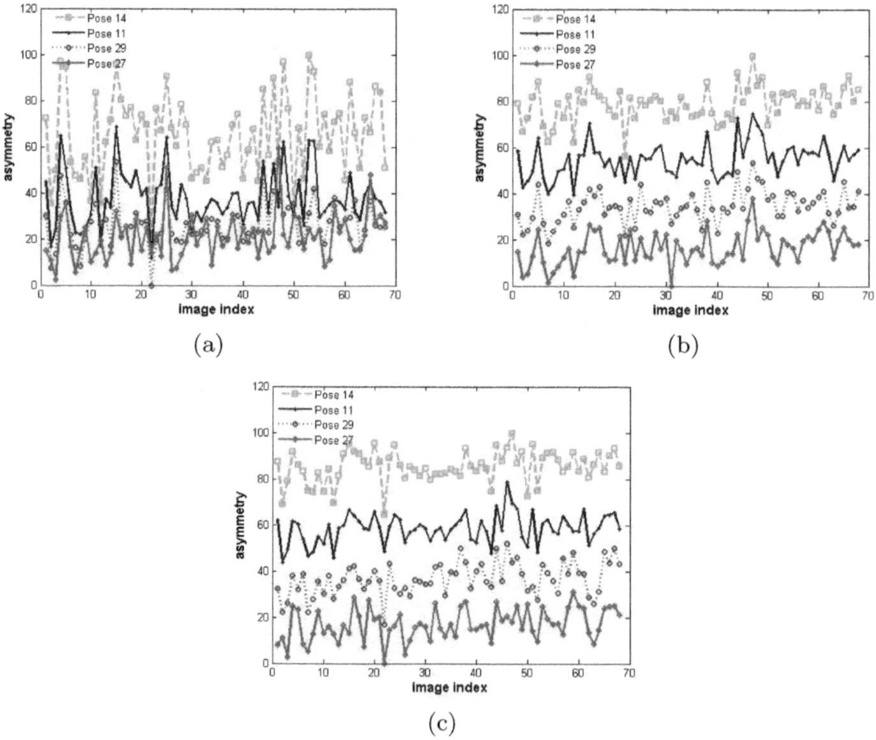

Fig. 8. Pose asymmetry score using DDG (a), LBP (b) and Gabor (c) features

method, the preserved DCT coefficient *rate* is set to 0.1 and 0.3 respectively. Fig. 9 illustrates the sharpness assessment results of these two methods on simulated images and real captured ones. It can be seen the proposed DCT based method achieves significantly better results to measure sharpness of images in both cases.

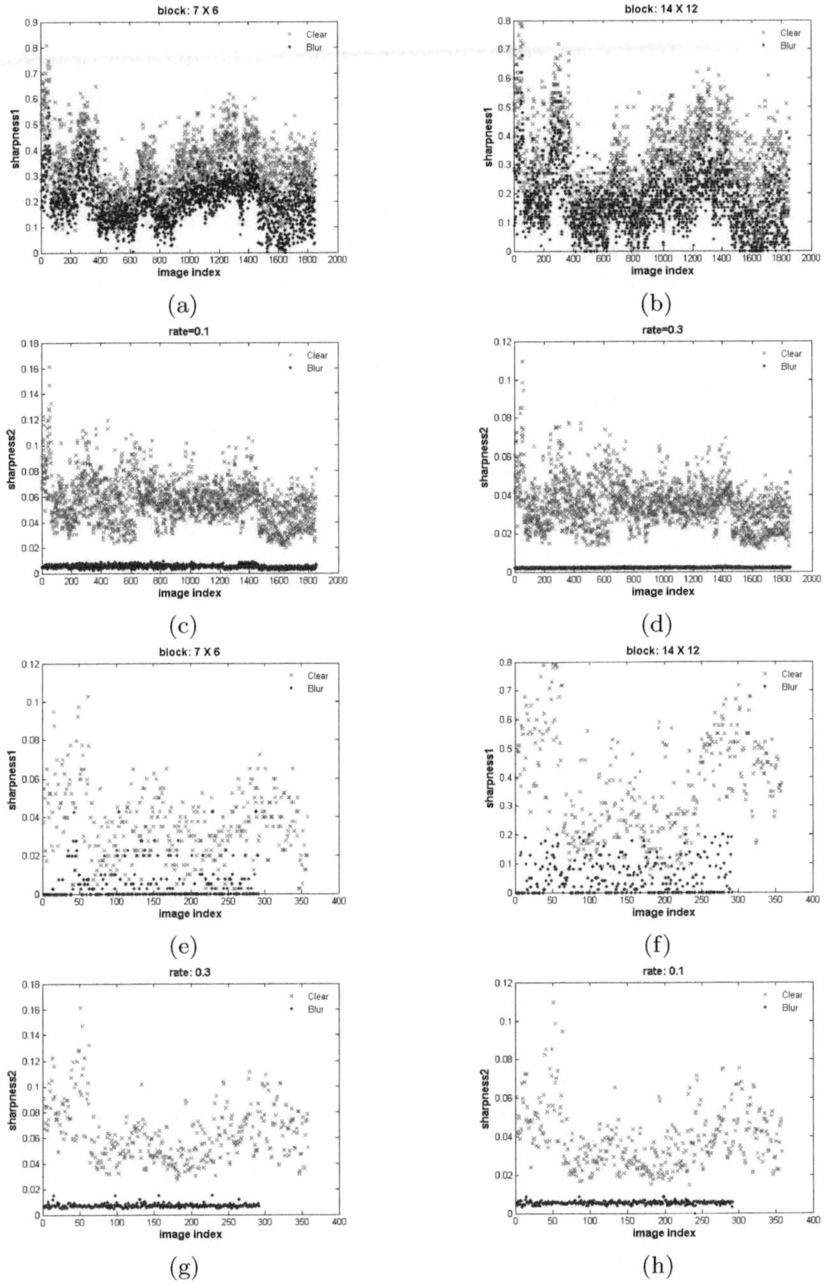

Fig. 9. Sharpness measurement results with different methods ((a)-(b) and (e)-(f) are results of the method in [5] with simulated blurring images and real captured blurring images. (c)-(d) and (g)-(h)are results of the proposed DCT method).

5 Conclusion

Enlightened by draft of ISO/IEC 29794-5 where standardization of face image quality is presented, we first introduce a novel Gabor-based facial symmetry measurement method to evaluate the changing illumination and improper posture. It provides comprehension from another aspect and offers alternative choice for evaluation. As face images are cropped according to the automatically detected eye positions, situations like misalignment and anatomically non-symmetric are not excluded intentionally, which indicates stability of the method. Also a DCT and IDCT combined algorithm is proposed to discern out-of-focus, which may replenish the standardization in ISO/IEC 29794-5.

Acknowledgement

This work was supported by the following fundings: National Natural Science Foundation Project #60518002, National Science and Technology Support Program Project #2006BAK08B06, National Hi-Tech (863) Program Projects #2006AA01Z192, #2006AA01Z193, and #2008AA01Z124, Chinese Academy of Sciences 100 People Project, and AuthenMetric R&D Funds.

References

1. Gao, X., Li, S.Z., Liu, R., Zhang, P.: Standardization of face image sample quality. In: Lee, S.-W., Li, S.Z. (eds.) ICB 2007. LNCS, vol. 4642, pp. 242–251. Springer, Heidelberg (2007)
2. ISO/IEC JTC 1/SC 37 N 1477. Biometric Sample Quality - Part 5: Face Image Data Sample Quality (Working Draft for comment) (February 12, 2007)
3. ISO/IEC JTC 1/SC 37 N 506. Biometric Data Interchange Formats Part 5: Face Image Data (March 22, 2004)
4. Liu, C., Wechsler, H.: Gabor feature based classification using the enhanced fisher linear discriminant model for face recognition 11(4), 467–476 (2002)
5. Ratha, N.K., Chen, S.Y., Jain, A.K.: Adaptive Flow Orientation-based Feature Extraction in Fingerprint Images. Pattern Recognition 28(11), 1657–1672 (1995)
6. Engeldrum, P.G.: Psychometric Scaling: A Toolkit for Imaging Systems Development. Imcotek Press (2000)
7. Shaked, D., Tastl, I.: Sharpness measure: towards automatic image enhancement 1(1), 937–940 (2005)
8. Wee, C., Paramesran, R.: Measure of image sharpness using eigenvalues. Inf. Sci. 177(12) (2007)
9. Zhao, W., Chellappa, R., Phillips, P., Rosenfeld, A.: Face Recognition: A Literature Survey. ACM Computing Surveys 35(4), 399–458 (2003)

Upper Facial Action Unit Recognition

Cemre Zor and Terry Windeatt

Centre for Vision, Speech and Signal Processing (CVSSP)
University of Surrey, Guildford, Surrey, United Kingdom GU2 7XH
{c.zor,t.windeatt}@surrey.ac.uk

Abstract. This paper concentrates on the comparisons of systems that are used for the recognition of expressions generated by six upper face action units (*AU*s) by using Facial Action Coding System (*FACS*). Haar wavelet, Haar-Like and Gabor wavelet coefficients are compared, using Adaboost for feature selection. The binary classification results by using Support Vector Machines (*SVM*) for the upper face *AU*s have been observed to be better than the current results in the literature, for example 96.5% for *AU2* and 97.6% for *AU5*. In multi-class classification case, the Error Correcting Output Coding (*ECOC*) has been applied. Although for a large number of classes, the results are not as accurate as the binary case, *ECOC* has the advantage of solving all problems simultaneously; and for large numbers of training samples and small number of classes, error rates are improved.

Keywords: FACS, ECOC, Adaboost.

1 Introduction

Face expression analysis and recognition have been one of the fastest developing areas due to its wide range of real life application areas such as emotion analysis, biometrics, computer graphics / simulations and image retrieval. The task is challenging and requires research into solving problems occurring under different illuminations, orientations and numerous other variations. In this paper, the aim is to compare binary and multi-class face expression analysis algorithms using the Facial Action Coding System (*FACS*) [5], which allows division of the main problem into sub-problems containing specific Action Units (*AU*s). Features based on Haar-Like, Gabor and Haar wavelet coefficients are compared using feature selection method based on Adaboost. Classification is performed by Support Vector Machines (*SVM*) and the multi-class problem is solved with Error Correcting Output Coding (*ECOC*).

Recognition rates for binary classification have been presented on three benchmark datasets and for binary and multi-class on the Cohn-Kanade frontal comprehensive database for facial expression analysis [20].

In Section 1.1, brief background of the problem has been presented. Section 2 describes the normalization steps, and in Sections 3 and 4, the components of the implemented systems; feature extraction, selection and classification have been explained in detail. Finally in Section 5, binary and multi-class classification results can be examined.

M. Tistarelli and M.S. Nixon (Eds.): ICB 2009, LNCS 5558, pp. 239–248, 2009.
© Springer-Verlag Berlin Heidelberg 2009

1.1 Background

In the literature, when facial expression analysis is considered, two main different approaches and two different parameterisations can be found. The first approach uses the whole frontal face image for classifying into six universal facial expression prototypes: disgust, fear, joy, surprise, sadness and anger. Ekman and Friesen have proposed in their related work that each of the six emotions has characteristic expression on the face and that's why recognition of them is necessary and sufficient [1][2]. Bartlett, Littlewort et al [3][4] have also used the method for fully automatic recognition systems.

As changes of features on specific sub-sections such as eyes and eyebrows are assumed to be more related to facial expression, dividing the face images into sub-sections for further processing is the main idea of the second approach. 'Facial Action Coding System', first developed by Ekman and Friesen [5]; for describing facial expressions by 44 different AUs widens the range of applications of face expression recognition and most current work on facial expression analysis depends on this decomposition into AUs.

There are also other methods that use neither the frontal face image as a whole nor all of the 44 AUs, but some other criteria such as the manually selected regions on face [6] or surface regions of facial features [7].

There are two main parameterisations that are used in both of the above approaches. Geometric based parameterization is an old way which consists of tracking and processing the motions of some spots on image sequences, firstly presented by Suwa [8] to recognize facial expressions. Cohn and Kanade later on tried geometrical modelling and tracking of facial features by claiming that each AU is presented with a specific set of facial muscles. In general, facial motion parameters [6][7] and the tracked spatial positioning & shapes of some special points [9] on face, are used as feature vectors for the geometric based method. These feature vectors are then used for classification. As for the disadvantages of this method; manual contour adjustments for the features on the initial image frames of each individual subject, lack of robustness occurring due to pose and illumination changes while the tracking is applied, and difficulties in the estimation of general movement / displacement parameters as a result of change in actions & expressions in morphological and dynamical senses [12], can be listed.

Rather than tracking spatial points and using positioning and movement parameters that vary within time, colour (pixel) information of related regions of the face are processed in appearance based parameterizations. Features such as Gabor, Haar wavelet coefficients, together with feature extraction and selection methods such as *PCA*, *LDA* and Adaboost are used within this framework. Example research can be found in [11][4].

The combination of the geometric and appearance based methods have also been used in some previous work. For example, Zhang [10] has tracked some fiducial points on the face images while also taking the Gabor wavelets of these points into account for the facial expression recognition.

In this paper, appearance based parameterization has been used together with *FACS*. Within the implementation of the binary classification scheme; Haar and

Gabor wavelet coefficients; and Haar-Like features have been used as extracted features. Feature selection has been applied by Adaboost, and *SVM* and Adaboost classifications have been performed in the final stage. When the multi-class classification problem is taken into consideration, the method used is the *ECOC* technique which is combined with the Adaboost feature selection and *SVM* classification techniques, together with an application of Bootstrapping on the training data.

2 Normalization

As the experiments were carried out on the upper face AUs, eye centres / pupils were localized on the input images as the first step of normalization. Localization was performed manually for the experiments so that the results were not dependent on the possible inaccuracies of an automated system. After the pupil coordinates were marked in each image, orientation and size normalizations were carried out through bilinear transforms. Images were made to have a distance of 32 pixels between the left and the right eye centres and a zero degree slope between the y-coordinates of these centres. Then, 32 by 32 square regions containing the eyes and the eyebrows were cropped around the pupils.

The previously detected, cropped, resized and rotated 32 by 32 gray scale images were then made to go through low pass filtering and a couple of histogram processing steps including contrast sketching and outlying pixels elimination. Elimination was accomplished by using the standard deviation information obtained from the histograms which look like scaled and shifted Gaussians. In addition to those, a log / power-law transformation was applied in order to have the average pixel value of each image equal to 128.

3 Feature Extraction and Selection

After the normalization step, Haar and Gabor wavelet coefficients, and Haar-Like features have been extracted in order to form up the feature vectors.

The superiority of Haar wavelet coefficients over the most commonly used Gabor wavelets and also the Haar-Like coefficients is their superiority in terms of extraction times. The mother wavelet function of Haar wavelets might be considered as a kind of step function:

$$\psi(x) = \left\{ \begin{array}{rl} 1, & 0 \leq t < 1/2 \\ -1, & 1/2 \leq t < 1 \\ 0, & otherwise \end{array} \right\} . \tag{1}$$

The wavelet decomposition of an image could therefore be defined as the combination of the resulting difference images calculated in different scales.

Haar-Like coefficients, introduced by Viola and Jones [13], are features which are indicatives of Haar basis functions. They consist of so-called two-rectangle, three-rectangle and four-rectangle features, which compose of differences between the sums of pixels of the same-size pairs of rectangles. Compared to Gabor and Haar wavelets, it takes a longer time to extract all the Haar-Like features of an input image. For example, a 24*24 image has 160000 Haar-Like features while a 32*32 one has more than 450000. This means that the set of features is many times over-complete; however, any rectangular sum can be computed in only four array differences by making use of the "integral image" method that is expressed by Viola and Jones [13]. Haar-Like features are sensitive to edges, boundaries and other important information hidden in pixel values such as the difference between the pixel values on the regions of motion on face.

Gabor wavelet transformation has been used in various kinds of signal and pattern processing / analysis areas both in spatial and in frequency domains and is found to give satisfactory results in application areas such as texture segmentation [16], fingerprint recognition [17] and face recognition [15]. The characteristics of Gabor wavelets such as their ability to get easily adjusted for detailed localization in spatial and frequency domains [18] and the similarity between their frequency & orientation representations and the human visual system components have made them popular for particular usage areas. Gabor wavelets are formed from the multiplication of a complex sinusoidal carrier with a Gaussian envelope.

$$g(x, y) = K \exp(-\pi(a^2 x'^2 + b^2 y'^2)) \exp(j(2\pi F_o(x \cos w_o + y \sin w_o) + P)) \quad (2)$$

where

$$x' = (x - x_0)\cos\theta + (y - y_0)\sin\theta; \;\; y' = (y - y_0)\cos\theta - (x - x_0)\sin\theta$$

and where K is the scale parameter for the magnitude of the envelope, a and b the parameters of the x and y axis of the envelope, θ the envelope rotation angle, F_o and w_o the magnitude and the direction of the sinusoidal carrier's frequency, P the phase of the carrier, and x_o and y_o the location of the peak value of the envelope.

The most commonly used values of the phase in the literature for facial expression recognition and also face recognition have been used here too: $P = \pi\frac{i}{8}; i = 0, ..., 7$. In case of frequencies, the upper limit is equal to 0.5 Hertz due the Nyquist sampling theory; and because having $\sqrt{2}$ as the spacing between the kernels in the frequency domain is found to give out experimentally good results on previous research, the values of F used are equal to: $\frac{fmax}{\sqrt{2}^u}, u = 0, ..., 4$ where $fmax = 0.5$.

For feature selection, the Adaboost algorithm has been used. Adaboost was originally a strong and fast classification algorithm, introduced by Freud and Schapire [14], making use of a weak binary classifier that strengthens its decisions in each iteration to end up with a final hypothesis with the lowest error rate. The weak binary classifier (weak learner) is any classifier for which the weighted classification error is expected to be better than chance. More recently, particularly in the Computer Vision community, Boosting has become popular as a feature selection routine, in which a single feature is selected in each Boosting iteration [13]. Specifically, the Boosting algorithm is modified so that, in each iteration, the individual feature which minimises the classification error on the weighted samples is chosen [13]. In our implementation, we use Adaboost with decision stump as weak learner.

4 Classification

Support Vector Machines (*SVM*), which were firstly developed from Statistical Learning Theory by Boser, Guyon and Vapnik [23], are used as binary classifiers in this paper. They aim to perform 2-class classification via optimally separating the data by making use of an N-dimensional hyper-plane. While dealing with the optimal hyper-planes that are to classify the data by minimizing the empirical classification error, *SVM* take also the maximization of the margin, in other words achievement of the maximum separation into account.

For multi-class classification, Error Correcting Output Coding (*ECOC*) has been used. In the *ECOC* matrix, every row represents the unique codeword (a binary string of n) of a class. For each of the n columns, a binary classifier, which is *SVM* here, has been trained. The features extracted are Gabor wavelets and the feature selection method is Adaboost. Also, in order to reduce correlation between individual columns, bootstrapping [21] is applied during training.

Each of the n trained *SVM* are applied to the new test pattern, and the resulting binary values are combined to create its binary codeword. This codeword is then compared to each of the r codewords and classified according to the codeword with closest distance, using Hamming distance [19].

One of the advantages of *ECOC* is the method's ability to correct bit errors up to a degree. If the minimum Hamming distance between any pair of codewords is d, then at least $\lfloor (d-1)/2 \rfloor$single bit errors can be corrected. Therefore, row separation in the design of the error correcting output code matrix is quite important for the classification results. Also, in order to create a good *ECOC* matrix, the Hamming distance between a pair of columns should be large enough so that the deterministic learning functions per each column are uncorrelated [19]. In our approach, the additional usage of bootstrapping helps further to de-correlate the columns.

5 Results

5.1 Binary Classification

Experiments were carried out on the Cohn-Kanade frontal face database [20] together with additional supplementary datasets from UCI Machine Learning Repository [22]. The images in the Cohn-Kanade dataset have been coded using *FACS*, which describes subject's expression in terms of *AU*s [20].

In the first part of the experiments, comparison of the two Adaboost algorithms, Adaboost as a classifier without feature selection (*Ada*); and with feature selection (*AdaFs*) have been carried out on small datasets, Breast Cancer Wisconsin (*Breast-w*), Sonar Mines vs. Rocks (*Sonar*), and Pima Indians Diabetes (*Diabetes*) which were taken from the UCI Machine Learning Repository. The

Breast-w dataset consists of 699 data with 10 features whereas the *Sonar* dataset consists of 208 data with 10 features, and the *Diabetes* of 768 data with 8 features. The number of features chosen by *AdaFs* was half of the number of total features within each dataset. In case of Diabetes and Sonar datasets, error rates were lower when *AdaFs* was used instead of *Ada*, whereas both of the rates for Breast-w dataset were similar. Due to these rates and also its ability to carry out feature selection at the same time with classification, *AdaFs* was considered to be superior to *Ada*.

The correct classification rates of *AdaFs* have therefore been compared to *AdaFs + SVM* combination in the second step. In *AdaFs + SVM*, half of the features were firstly selected from the data by *AdaFs* and the classification itself was carried out by *SVM*. *AdaFs + SVM* was better in terms of recognition rates, and Table 1 lists the error percentage results for the comparison of the three classification methods using 10-fold cross validation.

Table 1. Recognition error rates (%) of *Ada*, *AdaFs* and *AdaFs+SVM*, applied on UCI MLR

	Ada	*AdaFs*	*AdaFs + SVM*
Breast-w	4.0	4.6	3.5
Diabetes	28.1	26.7	22.5
Sonar	19.6	14.7	11.0

For the Cohn Kanade dataset *AdaFs + SVM* was found to give the best results, having an average of 5% increase over the performance of *Ada*, and 2% over *AdaFs*. Here, Haar wavelets were used as feature vectors and the number of features selected was set to 100 as there was a decrease in performance of about 1.5% when 200 and 300 were used. Also, there is the advantage of a decrease in training and run times for 100 features.

Though revealing good results, Haar wavelet coefficients depend on the differences between some determined pixel intensities; therefore even 1-2 pixel shifts that occur due to non-robust localizations of the eye centers may affect the overall results. Therefore Haar-Like features and Gabor wavelets were considered. Haar-Like features implement the idea of using the change in the contrast values between adjacent rectangular groups of pixels instead of the intensity values of individual pixels. Gabor wavelets are more robust to pixel intensity and lighting changes than Haar wavelets and are also meaningful in frequency domain. In Table 2, recognition rates for *AdaFs+SVM* (number of selected features being equal to 100) have been given for Haar wavelet, Haar-Like and Gabor wavelet coefficients on the upper face *AU*s of the 32 by 32 normalized Cohn-Kanade dataset images.

Although the resulting classification rates of Haar-Like features are not as successful compared to Haar wavelets, in real life applications where the localization of the centres are done automatically and where inaccurate localizations are more likely to occur, Haar wavelet features are expected to reveal poorer performance.

Table 2. Recognition error rates (%) obtained by Gabor, Haar wavelet coefficients, and Haar-Like features

	Haar Wavelet Coefficients	Haar-Like Features	Gabor Wavelet Coefficients
AU1	9.0	17.0	5.2
AU2	4.5	10.5	3.3
AU4	12.0	22.0	8.1
AU5	3.1	8.3	2.4
AU6	11.6	19.5	8.3
AU7	12.9	19.1	8.5

In case of Gabor wavelets, it was observed for all of the AUs that the first 10 out of 100 features used for classification have been extracted by using Gabor filters with mostly low complex sinusoidal frequencies. This is therefore an indication of the fact that useful frequency bands for facial AUs recognition are the low frequency bands in images. However, future work is needed to reveal why Gabor features do not make much use of edge information. One possible explanation might be that although convolution is taken into account, edge information might vary quite a lot in images even if they belong to the same class; and therefore this information may not be useful, and maybe even redundant. It was also found that both the number of features used and the parameters selected for Gabor wavelets play important roles in the final classification results. 100 features were found to give the optimal results together with the detected wavelet parameters: a=1/ 50, b=1/ 40 and the width and the height of the filter size being equal to half of the width and height of the image to be filtered. Experiments were also tested on 16 by 16 down-sampled images, and also on 1/10th down-sampled feature space; and the classification rates were found to be on average 5% less successful.

Although the classification rates for Gabor wavelets came out to be the most successful, Whitehill and Omlin [11] showed that extraction of Gabor wavelet coefficients is 300 times more costly than Haar. However, the down-sampled images or features might be used in real-time applications where speed is of importance.

5.2 Multi-class Classification

In order to carry out multi-class classification of the AUs, $ECOC$ is used. As Gabor features with *AdaFs* + SVM classification on 32 by 32 images were found to give the best binary classification, these components were inherited in the $ECOC$ system. The strategy consists of applying bootstrapping, as explained in Section 5, followed by *AdaFs* + SVM on Gabor features for each column. Tables 3, 4 and 5 show the results of experiments with ten-fold cross validation that were applied to three different sets of data, one having 12, the other 4 and the last one having 3 classes. In the set of experiments, in order to decrease the time needed for training, every 1 out of 10 Gabor features from a total of 40960 was taken to form the new feature set, to which *AdaFs* was applied to select 100 features.

Table 3. Recognition error and true positive rates (%) obtained by *ECOC* with 12 classes

Class No	*AU*s that are "on"	Number of Data	Overall Error / Tp Rate
1	1	7	5.3 / 0.0
2	4	26	12.1 / 17.8
3	6	65	14.6 / 62.9
4	7	5	3.0 / 0.0
5	1 & 2	23	8.4 / 20.0
6	1 & 4	20	6.6 / 42.0
7	4 & 7	47	16.0 / 61.1
8	6 & 7	13	6.5 /14.2
9	1 & 4 & 7	11	6.1 / 4.7
10	1 & 2 & 4	6	4.0 / 0.0
11	1 & 2 & 5	62	7.0 / 76.0
12	4 & 6 & 7	22	10.4 / 4.7

Table 4. Recognition error and true positive rates (%) obtained by *ECOC* with 4 classes

Class No	*AU*s	No. of Data	12 Class Err. / Tp	4 Class Err. / Tp
2	4	26	12.1 / 17.8	22.4 / 33.6
3	6	65	14.6 / 62.9	13.6 / 69.7
7	4 & 7	47	16.0 / 61.1	18.5 / 60.4
11	1 & 2 & 5	62	7.0 / 76.0	4.2 / 97.6

In Table 3, all the possible combinations of the upper face *AU*s that exist on the Cohn-Kanade dataset images were used to form 12 classes. The classes were created to be mutually exclusive and the ones having less than 5 data samples were discarded. Dealing with a 12 class problem is a complicated and challenging task and thus, in this stage we have ended up with an average 25.2% true positive (*tp*) rate, although the average error rate is 8.3%.

Apart from the number of classes, the effect of the number of training patterns on classification should also be considered. In the 12 class problem, the best *tp* rates were obtained by Class 11 and Class 3, and these classes are ones with the maximum number of patterns. Also, the worst ones were obtained by Class 1, Class 4 and Class 10, which have only 7, 5 and 6 patterns respectively. From Table 4, the four classes containing the maximum number of patterns, Class 11, Class 3, Class 7 and Class 2, can be seen to give better *tp* rates than the 12 class case. However, main reasons for some misclassifications are believed to be due to the low number of patterns in Class 2, and to the fact that two of the classes, Class 2 and Class 7 contain the same *AU*: *AU*7, which might cause confusion. Therefore the experiments shown in Table 5 were applied to the 3 class problem, which excludes Class 2. Since the classes now all have exclusive *AU*s, with only classes having the maximum number of patterns included, recognition and the *tp* rates are much improved on average.

Table 5. Recognition error and true positive rates (%) obtained by *ECOC* with 3 classes

Class No	*AU*s	No. of Data	12 Cl. Err. / Tp	4 Cl. Err. / Tp	3 Cl. Err./ Tp
3	6	65	14.6 / 62.9	13.6 / 69.7	12.9 / 76.5
7	4 & 7	47	16.0 / 61.1	18.5 / 60.4	6.4 / 95.4
11	1 & 2 & 5	62	7.0 / 76.0	4.2 / 97.5	8.0 / 97.5

6 Conclusion

In this paper both the binary and the multi-class classification results of the upper face action units have been presented through making use of several systems. In the binary case, the overall results were observed to be better than current results in the literature.

The system that was found to give out the best binary classification results was used in the multi-class case with *ECOC*. Although the recognition rates using twelve classes were low as expected due to classes have common *AU*s and insufficient training data, the results for smaller number of classes with large number of training data were much improved. Therefore the results are highly dependent on how the problem is decomposed and particularly if the classes are mutually exclusive. Future work is aimed at finding the best combination of feature selection and *ECOC* coding and decoding strategies for facial action unit recognition.

References

1. Ekman, P., Friesen, W.V.: Pictures of Facial Affect. Consulting Psychologist Press, Palo Alto (1976)
2. Izard, C., Dougherty, L., Hembree, E.A.: A System for Identifying Affect Expressions by Holistic Judgements. Univ. Of Delaware (unpublished manuscript) (1983)
3. Bartlett, M.S., Hager, J., Ekman, P., Sejnowski, T.: Measuring Facial Expressions by Computer Image Analysis. J. Psychophysiology 36, 253–263 (1999)
4. Bartlett, M.S., Littlewort, G., Lainscsek, C., Fasel, I., Movellan, J.: Machine Learning Methods for Fully Automatic Recognition of Facial Expressions and Facial Actions. In: IEEE International Conference on Systems, Men and Cybernetics, Netherlands, pp. 592–597 (2004)
5. Ekman, P., Friesen, W.V.: Facial Action Coding System: A Technique for the Measurement of Facial Movement. Consulting Psychologist Press, Palo Alto (1978)
6. Mase, K.: Recognition of Facial Expression from Optical Flow. IEICE Trans. E74(10), 3474–3483 (1991)
7. Yacoob, Y., Davis, L.S.: Recognizing Human Facial Expression from Long Image Sequences Using Optical Flow. IEEE Trans. Pattern Analysis and Machine Intelligence 18(6), 636–642 (1996)
8. Suwa, M., Sugie, N., Fujimora, K.A.: Preliminary Note on Pattern Recognition of Human Emotional Expression. In: Proc. International Joint Conf. Pattern Recognition, pp. 408–410 (1978)

9. Lanitis, A., Taylor, C., Cootes, T.: Automatic Interpretation and Coding of Face Images Using Flexible Models. IEEE Trans. Pattern Analysis and Machine Intelligence 19(7), 743–756 (1997)
10. Zhang, Z.: Feature-Based Facial Expression Recognition: Sensitivity Analysis and Experiments with a Multilayer Perceptron. Int'l. J. Pattern Recognition and Artificial Intelligence 13(6), 893–911 (1999)
11. Whitehill, J., Omlin, C.W.: Haar Features for FACS AU Recognition. In: Proceedings of the IEEE International Conference on Automatic Face and Gesture Recognition (2006)
12. Donato, G., Bartlett, M.S., Hager, J., Ekman, P., Sejnowski, T.J.: Classifying Facial Actions. IEEE Transactions on Pattern Analysis and Machine Intelligence 21(10), 974–988 (1999)
13. Viola, P., Jones, M.J.: Robust Real-Time Face Detection. International J. of Computer Vision 57(2), 137–154 (2004)
14. Freund, Y., Schapire, R.: A Decision-Theoretic Generalization of On-Line Learning and an Application to Boosting. Computer and System Sciences 55, 119–139 (1997)
15. Shen, L., Bai, L., Fairhurst, M.: Gabor Wavelets and General Discriminant Analysis for Face Identification and Verification. Image Vision Computing 25(5), 553–563 (2007)
16. Jain, A.K., Farrokhnia, F.: Unsupervised Texture Segmentation Using Gabor Filters. Pattern Recognition 24(12), 1167–1186 (1991)
17. Lee, C.J., Wang, S.D.: Fingerprint Feature Extraction Using Gabor Filters. Electronics Letters 35(4), 288–290 (1999)
18. Zhan, Y., Niu, D., Cao, P.: Facial Expression Recognition Based on Gabor Wavelet Transformation and Elastic Templates Matching. In: Third International Conference on Image and Graphics (ICIG 2004), pp. 254–257 (2004)
19. Dietterich, T.G., Bakiri, G.: Solving Multi-class Learning Problems via Error-Correcting Output Codes. J. Artificial Intelligence Research 2, 263–286 (1995)
20. Tian, Y., Kanade, T., Cohn, J.F.: Recognizing Action Units for Facial Expression Analysis. IEEE Transactions on Pattern Analysis and Machine Intelligence 23(2), 97–115 (2001)
21. Efron, B.: Bootstrap methods: Another Look at the Jackknife. The Annals of Statistics 7(1), 1–26 (1979)
22. Asuncion, A., Newman, D.J.: UCI Machine Learning Repository. School of Information and Computer Science, University of California, Irvine (2007), `http://www.ics.uci.edu/~mlearn/MLRepository.html`
23. Boser, B.E., Guyon, I.M., Vapnik, V.N.: A Training Algorithm for Optimal Margin Classifiers. In: COLT 1992: Proceedings of the Fifth Annual Workshop on Computational Learning Theory, pp. 144–152. ACM Press, New York (1992)

Automatic Partial Face Alignment in NIR Video Sequences*

Jimei Yang[1,2], Shengcai Liao[2], and Stan Z. Li[2,**]

[1] University of Science and Technology of China, Hefei 230027, China
[2] Center for Biometrics and Security Research, Institute of Automation, Chinese Academy of Sciences, Beijing 100190, China
{jmyang,szli}@cbsr.ia.ac.cn
http://www.cbsr.ia.ac.cn

Abstract. Face recognition with partial face images is an important problem in face biometrics. The necessity can arise in not so constrained environments such as in surveillance video, or portal video as provided in Multiple Biometrics Grand Challenge (MBGC). Face alignment with partial face images is a key step toward this challenging problem.

In this paper, we present a method for partial face alignment based on scale invariant feature transform (SIFT). We first train a reference model using holistic faces, in which the anchor points and their corresponding descriptor subspaces are learned from initial SIFT keypoints and the relationships between the anchor points are also derived. In the alignment stage, correspondences between the learned holistic face model and an input partial face image are established by matching keypoints of the partial face to the anchor points of the learned face model. Furthermore, shape constraint is used to eliminate outlier correspondences and temporal constraint is explored to find more inliers. Alignment is finally accomplished by solving a similarity transform. Experiments on the MBGC near infrared video sequences show the effectiveness of the proposed method, especially when PCA subspace, shape and temporal constraint are utilized.

Keywords: Face Alignment, Partial Faces, SIFT, MBGC.

1 Introduction

Face recognition is an important problem in both computer vision and biometrics. Most of researchers historically dealt with faces under constrained circumstances. However, with the development of state-of-the-art, researchers are shifting their interest to less constrained circumstances such as in surveillance video or portal video, where partial face recognition becomes a new challenge. Face alignment with partial face images is a prerequisite for solving this problem. Popular face alignment methods are mostly based on holistic face model,

* This work was performed at the CBSR, Chinese Academy of Sciences.
** Corresponding author.

M. Tistarelli and M.S. Nixon (Eds.): ICB 2009, LNCS 5558, pp. 249–258, 2009.
© Springer-Verlag Berlin Heidelberg 2009

such as Active Shape Model (ASM)[1], Active Appearance Model (AAM)[2]. When the integrality of faces cannot be guaranteed, these holistic models will lose their power. Facial part detectors are not subject to face incompletion. If we successfully detect two parts or even more, faces can be well aligned. However, in our setting, facial parts often uncertainly occur in the images and sometimes only one part occurs, so part detector based method will fail to obtain enough location information for alignment. Some test partial faces are shown in Fig.1. In this paper, we present a method for automatic partial face alignment based on scale invariant feature transform (SIFT)[3].

Fig. 1. Some typical partial faces in a video sequence

As a well-known local feature, SIFT has been used to perform face detection[4]. When performing in faces, SIFT keypoints have good repeatability in the same semantic region of different faces. By this property a set of uniform keypoints, called facial anchors, can be learned from a training database of holistic frontal faces. Descriptors attached to the same facial anchor together form its description. All of facial anchors and their respective descriptions compose a face model. When dealing with novel partial faces, we can establish point-wise correspondences between novel faces and the face model by matching keypoints to facial anchors. This sort of point-wise correspondences guarantee alignment robust to partial faces. Like the method in [3], we use the ratio of first best match and second best match to estimate the correctness of this correspondence. Some mismatches, however, still exist. Carneiro and Jepson[5] use shape context as semi-local feature integrated into SIFT to improve matching performance. Shape context[6] was originally proposed by Serge Belongie et al. to describe the object shape by shape point orientation histogram. In our setting, as far as sparsity is concerned, the orientation histogram of facial keypoints is not a stable feature so that we directly use shape constraint to prune outlier correspondences. A similarity transform can be solved from valid correspondences by using the method proposed in [7]. However, the number of valid correspondences established by one image is limited, which probably increase the risk of incorrect alignment. Temporal constraint in a video is further explored to enrich the pool of inlier correspondences. Two implications are derived from temporal constraint: pose continuity and identity consistency, both of which make it presumably easy to align faces to the same pose within a video. As a consequence, each of faces can contribute its inlier correspondences to one common similarity transform. It improves greatly alignment's robustness.

Our main contributions include: 1) SIFT based generative face model is learned and naturally overcomes the difficulties brought by face incompletion. 2) Shape constraint of face is used to prune correspondences, which guarantees that true matches are preserved and meanwhile most of false matches are eliminated. 3) Our method takes advantage of temporal constraint within a video to enrich correspondence inliers and thus obtain more robust results than frame-by-frame alignment.

The rest of this paper is organized as follows. In Section 2, a generative face model is built up. In section 3, partial face alignment with shape constraint and with temporal constraint are introduced in detail. In section 4, experiments are conducted on NIR video sequences released by Multiple Biometrics Grand Challenge (MBGC)[8] and show the performance of our method. This paper is finally concluded by discussion and future work in section 5.

2 Learning Face Model with SIFT

Scale Invariant Feature Transform (SIFT) developed by Lowe[3] combines a scale invariant keypoint detector and gradient histogram based descriptor. First, image scale-space is built and potential keypoints are identified by using difference of Gaussian (DoG) function. Final keypoints are selected based on measures of their stability. DoG as a close approximation of scale normalized Laplacian of Gaussian performed well in term of detection repeatability compared with other existing detectors[9]. The stability provides a prime basis for our face model. Second, the local coordinate system of keypoint is built in image scale-space. Based on the local coordinate system, The descriptor is represented by a 3D histogram of gradient locations and orientations. The contribution to the location and orientation bins is weighted by the gradient magnitude. The quantization of locations and orientations makes the descriptor robust to small geometric distortions and small errors in the keypoint detection. Mikolajczyk and Schimid[10] compared diverse existing descriptors and found SIFT best.

Using a set of holistic frontal faces as training database, we collect their SIFT keypoints and corresponding descriptors. All the keypoints are plotted in one typical face in Fig.2(a). Note that most of keypoints concentrate into small clusters around semantic facial regions such as eyes, nose and mouth. Considering location errors of DoG detector and feature displacements of different faces, we presume that each identifiable cluster of keypoints represent a facial semantic region and the mean of keypoints will be a good estimate of facial anchor. On the contrary, dispersive keypoints tend to be subject to some special features of certain faces and lack of generality and are thus removed before identifying facial anchors. We use following algorithm to remove dispersive keypoints: a keypoint, if the number of its neighbors within a small region R is less than certain number N_n, is considered dispersive and then removed. In our experiments, we set $R = 5$ and $N_n = 100$. After checking all keypoints, the remaining are shown in Fig.2(b). Finally, we identify facial anchors using Kmeans, which are represented by blue dots in Fig.2(c). Each facial anchor corresponds to a series of descriptors that are assigned by Kmeans. Comparing with the scheme of one keypoint with one descriptor, variance of descriptors coming from the same semantic region of different faces enrich feature representation and are thus less subject to some

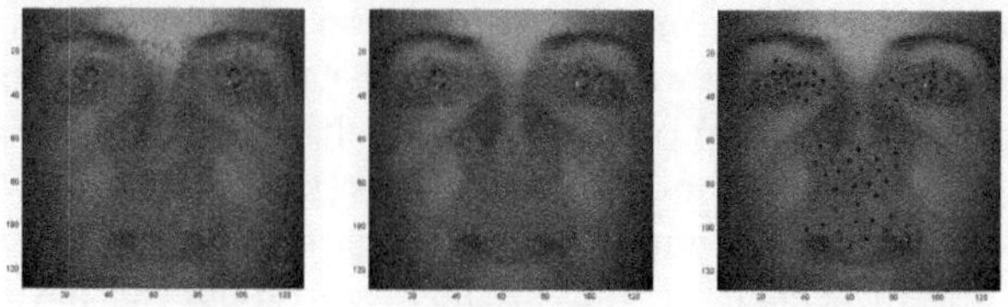

(a) (b) (c)

Fig. 2. The procedure of learning facial anchors by clustering. Red crosses represent keypoints and blue dots represent anchors.

special face. This property makes unseen face alignment possible. However, noises and even wrong descriptors brought by Kmeans could increase matching risk. Principle Component Analysis (PCA) is effective subspace learning technique and can be used here to represent the intrinsic structure of descriptors. The face model now consists of facial anchors $\{A_i, i = 1, 2, ..., T\}$ and their respective descriptor subspaces $\{S_i, i = 1, 2, ..., T\}$. We can also represent the facial anchor by a set of descriptor exemplars. The different methods of descriptions determine different feature matching criterion, which will be compared in section 4.

3 Partial Face Alignment

3.1 Matching with Shape Constraint

For each frame f of one video, its SIFT feature set consists of M keypoints $\{K_i, i = 1, 2, ..., M\}$ and their respective descriptors $\{D_i, i = 1, 2, ..., M\}$. For each K_i, we compute the distance from the descriptor D_i to each of descriptor subspaces S_j. The facial anchor related to the minimum distance d_1 is the best match and the one related to the second minimum distance d_2 is the second best match. We use the ratio: $r_i = d_1/d_2$ as criterion to estimate the correctness of this match. If $r_i < 0.85$, the match is accepted as a candidate correspondence, otherwise the match is discarded. When this procedure is finished, we sometimes will find that multiple keypoints are matched to a common anchor. In order to guarantee one-to-one correspondence we select the one which has minimum ratio r_i as the candidate correspondence. So we obtain a series of candidate correspondences: $\{C_i : K_i \longleftrightarrow A_j, r_i\}, i = 1, 2, ..., M', M' \leqslant M$.

Even if we adopt the ratio r to threshold matches, there still exist some mismatches due to some uncertain factors such as facial geometric distortion, non-facial features (hair, ear and clothing). We now utilize shape constraint to check candidate correspondences and further kill those outliers.

Given a set of keypoints or anchors, shape constraint is a collection of relative orientations between any two ones and represented as an index matrix. Centered in an anchor A_i, a polar coordinate system can be built so that each of remaining anchors $A_{j,j \neq i}$ gets an angle coordinate. In order to permit somewhat location errors of anchors, angles are quantified into 12 bins by 30°. Thus

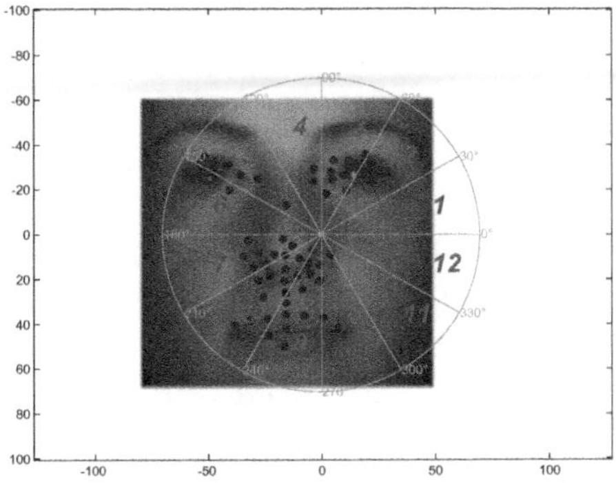

Fig. 3. A diagram for constructing shape constraint in face model. Centered in a given anchor (green dot), a polar coordinate system is plotted in green and orientation indices (red numbers) are assigned to other anchors (blue dots).

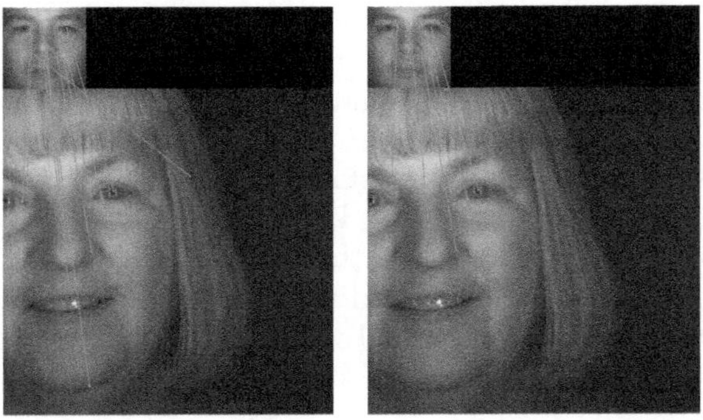

(a) (b)

Fig. 4. Before(a) and After(b) pruning with shape constraint. For each subfigure, the top left small 128 × 128 image is the template of face model and the bottom image is the novel 512 × 512 face to be aligned. Correspondences are represented by green line segments which connect the facial anchors and keypoints.

each of remaining anchors is labeled a bin index $b_{ij,j \neq i}$ that ranges from 1 to 12. Fig.3 illustrates the procedure. Arranging all these indices are arranged to form a matrix B. After computing the shape constraint matrix B of face model and shape constraint matrix D of novel face, the following algorithm is used to prune the outliers,

1. *Find the C_i which has minimum r_i as the first valid correspondence;*
2. *Set the keypoint K_i of the first valid correspondence as reference point in the novel face;*
3. *For the keypoint K_j of each remaining $C_{j,j \neq i}$;*
4. *By i as row coordinate and j as column coordinate, find index value b_{ij} in B and index value d_{ij} in D;*

5. *If* $d_{ij} == b_{ij}$*, then accept the current correspondence as the next valid correspondence, else eliminate it;*
6. *Go to the next* $C_{j,j \neq i}$.

An example is given in Fig.4 to illustrate the performance of this algorithm. In this way, the amount of correspondences from one partial face is relatively small. It probably reduces robustness of similarity transformation solution. In the following, we propose to explore the temporal constraint within videos to obtain more inlier correspondences.

3.2 Alignment with Temporal Constraint

The top row of Fig.5 lists some faces of one sequence. It is clear that their poses varies continuously and lightly. This kind of frame dependence or temporal constraint can help align the sequence. Consider that each of face has a certain amount of valid correspondences when matching to the face model. If we can make all these correspondences contribute to one uniform similarity transform, inliers will be greatly enriched. To achieve this, the first step is to ensure faces within a video have the same pose. Given a video sequence, $\{f_1, f_2, ..., f_N\}$, we extract SIFT features of each frame and then select the frame with most SIFT features as the reference frame f_r so that we can save different kinds of facial features as many as possible. For each of remaining frames f_i within sequence, we establish SIFT feature correspondences between f_i and the reference frame f_r. If the number of correspondences is less than two, then f_i is discarded; otherwise a similarity transform T_i can be solved with four parameters: scaling factor s_i, rotation angle θ_i, x translation t_{xi} and y translation t_{yi}. As a result, we obtain a series of new faces,

$$f_i' = T_i(f_i, \Theta_i), i = 1, 2, ..., N', N' \leqslant N, \tag{1}$$

where f_i' denotes new face and Θ_i denotes four similarity transform parameters: $\{s_i, \theta_i, t_{xi}, t_{yi}\}$. This matching procedure, called 'self-alignment', performs expectably well because of pose continuity and identity consistency. And shape constraint can also help improve the performance. The bottom row of Fig.5 are self-aligned faces with respect to the top row.

Each of self-aligned faces $f_i', i = 1, 2, ..., N'$ within a video can be matched to the face model and contributes a set of M_i valid correspondences $\{C_j^i\}, j = 1, 2, ..., M_i$ by using the method in section 3.1. From all of valid correspondences $\{C_j{}^i\}, j = 1, 2, ..., M_i, i = 1, 2, ..., N'$, an uniform similarity transform T_0 can be further solved by RANSAC algorithm. Final aligned faces are

$$f_i'' = T_0(f_i', \Theta_0) = T_0(T_i(f_i, \Theta_i), \Theta_0), i = 1, 2, ..., N', \tag{2}$$

where Θ_0 denotes four parameters of uniform similarity transform : scaling factor s_0, rotation angle θ_0, x translation t_{x0} and y translation t_{y0}.

(a) Some sample images from one video

(b) Aligned images

Fig. 5. Alignment within a video. The top row are some original sample images from one video with small pose variation and the bottom row are respective aligned images (in blue box) referred to the last image (in red box) which has most SIFT features in this video.

4 Experiments

In this section, we conduct experiments on MBGC NIR video sequences to evaluate our method's performance.

MBGC sponsored by multiple U.S. Government Agencies seeks to investigate, test and improve performance of face and iris recognition technology on both still and video imagery through a series of challenge problems. One of these problem is to recognize people from near infrared video sequences. There are together 139 sequences which consist of 2286 images with resolution 2048×2048. Sequences are acquired as people walk through a portal and consequently faces in sequences are partial, skewed and even missed. We select 249 holistic frontal faces as training set to learn face model and the remaining 2037 images are used as test set.

For the sake of computational efficiency, all the images are firstly compressed to the resolution 512 × 512 and 256 × 256, respectively. An important parameter that affects SIFT features is 'edge threshold'. Edges are poorly defined in

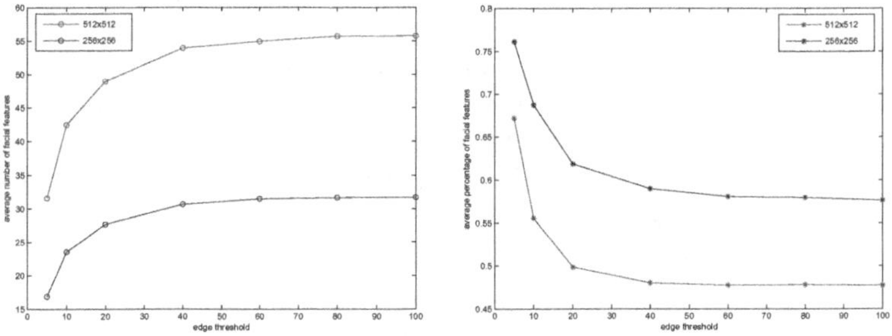

(a) The average numbers of facial features vs. edge threshold.

(b) The average percentages of facial features in total features vs. edge threshold.

Fig. 6. Edge threshold influence on facial SIFT features different-of-Gaussian function and are less stable to be candidate keypoints. The bigger the 'edge threshold' value is, the more edge-like SIFT features are accepted. As a result, more non-facial features are included, such as hair, ears, collar. In Fig.6(a) and Fig.6(b), we see that when edge threshold increases the average numbers of facial features accordingly increase but rather the average percentages of facial features decreases. When edge threshold passes 20 the average number and percentage tend to be stable. Thus, we set edge threshold to 20 in the following experiments.

In section 2, two kinds of description of facial anchor are introduced. These two methods result in different matching and further alignment performance. As four parameters should be solved, an alignment needs at less 2 valid correspondences. Therefore, if an image has more than 2 valid correspondences, the face is 'detected' and process to be aligned. In order to evaluate alignment performance, we define average square keypoint displacement d_i:

$$d_i = \sqrt{\frac{1}{M_i}\sum_{j=1}^{M_i}\|K'_j - A_j\|^2}, \tag{3}$$

where K'_j is aligned keypoint, A_j is its corresponding anchor and M_i is the number of correspondences. We use R, which has been defined as the smallest displacement tolerance in section 2, as threshold. If $d_i \leq R$, then the alignment is 'correct'; or else the alignment is 'incorrect'.

Table 1. Comparisons of different criterion in detection rate, correct rate and the number of incorrect alignments. Best results are red-marked.

	Resolution	Detection rate	Correct rate	Incorrect alignments
NN	512 × 512	49.3%	88.6%	129
	256 × 256	41.5%	95.7%	41
PCA	512 × 512	45.8%	98.3%	18
	256 × 256	38.1%	98.1%	17

Table 1 lists frame-by-frame alignment results under NN and PCA criteria. The best correct rate 98.3% is achieved under PCA criterion and the best detection rate 49.3% is achieved under NN criterion. Note that many images in test video database are lack of effective facial parts, like first and last several images in Fig.7(a). Thus, the detection rate is relatively low. NN seems to be more sensitive to the spatial resolution as the correct rate increase from 88.6% to 95.7% by increasing the spatial resolution from 512 × 512 to 256 × 256. PCA is less affected by this parameter as the correct rates in 512 × 512 and 256 × 256 are similar.

We now compare the performance of frame-by-frame alignment and alignment with temporal constraint in Table 2. The spatial resolution is set to 512 × 512 in this experiment. In the first step of alignment with temporal constraint, called

Table 2. Comparisons of alignment with and without temporal constraint in detection rate, correct rate and the number of incorrect alignments. Best results are red-marked.

	Temporal constraint	Detection rate	Correct rate	Incorrect alignments
NN	No	49.3%	88.6%	129
	Yes	49.4%	97.7%	26
PCA	No	45.8%	98.3%	18
	Yes	47.9%	99.9%	8

self-alignment, each frame is more easily matched to the reference frame than to an uniform face model and thus more facial images are correctly self-aligned and processed to the next step, called joint-alignment. In this step all the detected faces are matched to the face model and an uniform similarity transform is solved to align all self-aligned faces. Both of best detection rate and correct rate are achieved with temporal constraint. Note that the error in alignment with temporal constraint consists of two part. The one is introduced in the step of self-alignment and the other is generated in the step of joint alignment. For example, under PCA criterion and with temporal constraint, there are 8 incorrect alignments, 6 of which are self-alignment errors and 2 of which are coming from one video and are introduced in joint alignment. Finally, we show aligned images in the video plotted in Fig.7(a) by joint alignment in Fig.7(b).

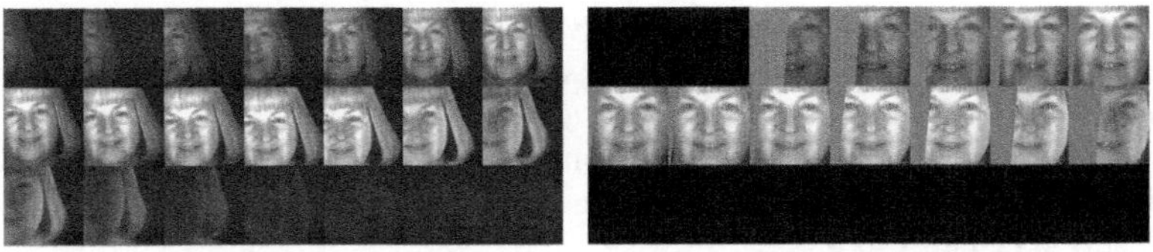

(a) Original video sequence (b) Aligned video sequence

Fig. 7. An example of video sequence alignment. Original video sequence has 21 frames. After alignment, 12 frontal partial faces are obtained and 9 images have been discarded which are illustrated by black ones.

5 Conclusions

In this paper, we introduce a novel problem of aligning partial faces in NIR video sequences and propose an effective solution. Our method has three novelties. First, an uniform face model is learned from a set of training faces by clustering analysis, which makes unseen face alignment possible. Second, shape constraint is used to eliminate outliers while matching the frames to model. Third, guaranteed by the temporal constraint we develop an scheme of joint alignment. Its results are shown to perform well. In the future, we will concentrate in decreasing the time complexity and develop more robust solution of similarity transform than RANSAC.

Acknowledgements

This work was supported by the following fundings: National Natural Science Foundation Project #60518002, National Science and Technology Support Program Project #2006BAK08B06, National Hi-Tech (863) Program Projects #2006AA01Z192, #2006AA01Z193, and #2008AA01Z124, Chinese Academy of Sciences 100 People Project, and AuthenMetric R&D Funds. We also would like to thank Andrea Vedaldi for his open source code of SIFT implementation.

References

1. Cootes, T., Cooper, D., Taylor, C., Gramham, J.: Active shape models - their training and application. Computer Vision and Image Understanding 61, 38–59 (1995)
2. Cootes, T., Adwards, G., Taylor, C.: Active appearance model. IEEE Transactions on Pattern Analysis and Machine Intelligence 23(6), 681–685 (2001)
3. Lowe, D.G.: Distinctive image features from scale-invariant keypoints. International Journal of Computer Vision 60(2) (2004)
4. Slot, K., Kim, H.: Keypoints derivation for object class detection with sift algorithm. In: Rutkowski, L., Tadeusiewicz, R., Zadeh, L.A., Żurada, J.M. (eds.) ICAISC 2006. LNCS, vol. 4029, pp. 850–859. Springer, Heidelberg (2006)
5. Carneiro, G., Jepson, A.D.: Pruning local feature correspondences using shape context. In: Proceedings of IEEE International Conference on Image Processing (2004)
6. Belongie, S., Malik, J., Puzicha, J.: Shape matching and object recognition using shape contexts. IEEE Transactions on Pattern Analysis and Machine Intelligence 24(4), 509–522 (2002)
7. Lowe, D.G.: Local feature view clustering for 3D object recognition. In: Proceedings of IEEE Computer Society Conference on Computer Vision and Pattern Recognition, pp. 682–688 (2001)
8. NIST: Multiple Biometric Grand Challenge, MBGC (2008), `http://face.nist.gov/mbgc`
9. Mikolajczyk, K., Tuytelaars, T., Schmid, C., Zisserman, A., Matas, J., Schaffalitzky, F., Kadir, T., Gool, L.V.: A comparison of affine region detectors. International Journal of Computer Vision 65(12), 43–72 (2005)
10. Mikolajczyk, K., Schmid, C.: A performance evaluation of local descriptors. IEEE Transactions on Pattern Analysis and Machine Intelligence 27(10), 1615–1630 (2005)

Parts-Based Face Verification Using Local Frequency Bands

Christopher McCool and Sébastien Marcel

Idiap Research Institute, Centre du Parc, CH-1920 Martigny, Switzerland
christopher.mccool@idiap.ch, sebastien.marcel@idiap.ch

Abstract. In this paper we extend the Parts-Based approach of face verification by performing a frequency-based decomposition. The Parts-Based approach divides the face into a set of blocks which are then considered to be separate observations, this is a spatial decomposition of the face. This paper extends the Parts-Based approach by also dividing the face in the frequency domain and treating each frequency response from an observation separately. This can be expressed as forming a set of sub-images where each sub-image represents the response to a different frequency of, for instance, the Discrete Cosine Transform. Each of these sub-images is treated separately by a Gaussian Mixture Model (GMM) based classifier. The classifiers from each sub-image are then combined using weighted summation with the weights being derived using linear logistic regression. It is shown on the BANCA database that this method improves the performance of the system from an Average Half Total Error Rate of 24.38% to 15.17% when compared to a GMM Parts-Based approach on Protocol P.

1 Introduction

The face is an object that we as humans know can be recognised. It is used to verify the identity of people on a daily basis through its inclusion in passports, drivers licences and other identity cards. However, performing automatic face verification has proved to be a very challenging task. This is shown by the fact that face recognition has been an active area of research for over 25 years [1], in fact the earliest research into face recognition was conducted by Bledsoe [2] in 1966.

Many techniques have been proposed to perform face verification ranging from Principal Component Analysis (PCA) [3] and Linear Discriminant Analysis (LDA) [4] through to feature distribution modelling techniques such as Hidden Markov Models (HMMs) [5] and Gaussian Mixture Models (GMMs) [6]. A recent advance in face verification has been the effective use of feature distribution modelling techniques. The first effective method of performing face verification using feature distribution modelling was in 2002 by Sanderson and Paliwal [6]; despite the earlier work of Samaria et al. [5,7] and Nefian and Hayes [8] who used HMMs.

The GMM Parts-Based approach, introduced by Sanderson and Paliwal, has been employed by several researchers [9,10]. This method consists of dividing

M. Tistarelli and M.S. Nixon (Eds.): ICB 2009, LNCS 5558, pp. 259–268, 2009.
© Springer-Verlag Berlin Heidelberg 2009

the face into blocks, or parts, and to then consider each block separately. The distribution of these parts is then modelled using Gaussian Mixture Modelling. This method is very different to other Parts-Based (or Component-Based) approaches which form an expert classifier each individual region or concatenate the information from the different Parts and then use a holistic-based classifier, for instance using SVMs [11].

In this paper an extension to the GMM Parts-Based approach (referred to as the Parts-Based approach from hereon) is proposed so that both Spatial and Frequency based decomposition can be performed. The frequency decomposition is achieved by collating the responses from each DCT coefficient from each block (observation) and forming a separate sub-image for each frequency. Each of these sub-images is then treated separately by a GMM based classifier. The classifiers from each sub-image are then combined using weighted summation with the weights being derived using linear logistic regression. Tests conducted on the BANCA database show that this extension is a significant improvement with the Average Half Total Error Rate being reduced from of 24.38% to 15.17% when compared to a baseline Parts-Based approach.

2 Related Work on GMM Parts-Based Face Verification

The Parts-Based approach divides the face into blocks, or parts, and treats each block as a separate observation of the same underlying signal (the face). In this method a feature vector is obtained from each block by applying the Discrete Cosine Transform and the distribution of these feature vectors is then modelled using GMMs. Several advances have been made upon this technique, for instance, Cardinaux et al. [9] proposed the use of background model adaptation while Lucey and Chen [10] examined a method to retain part of the structure of the face utilising the Parts-Based framework as well as proposing a relevance based adaptation.

2.1 Feature Extraction

The feature extraction algorithm is described by the following steps. The face is normalised, registered and cropped. This cropped and normalised face is divided into blocks (parts) and from each block (part) a feature vector is obtained. Each feature vector is treated as a separate observation of the same underlying signal (in this case the face) and the distribution of the feature vectors is modelled using GMMs. This process is illustrated in Figure 1.

The feature vectors from each block are obtained by applying the DCT. Even advanced feature extraction methods such as the DCTmod2 method [6] use the DCT as their basis feature vector; the DCTmod2 feature vectors incorporate spatial information within the feature vector by using the deltas from neighbouring blocks. The advantage of using only DCT feature vectors is that each DCT coefficient can be considered to be a frequency response from the image (or block). This property is exploited by the JPEG standard [12] where the coefficients are ranked in ascending order of their frequency.

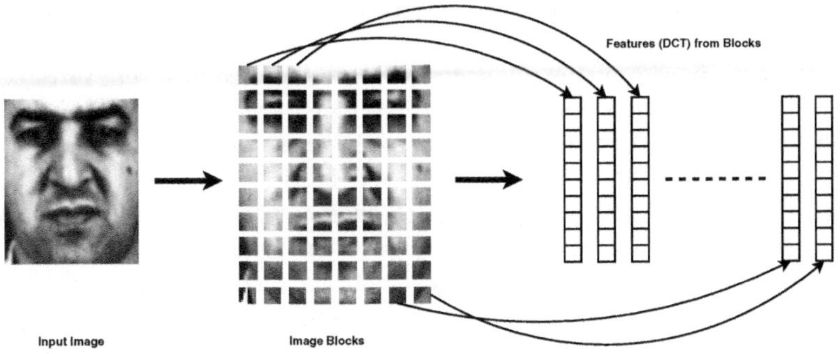

Fig. 1. A flow chart of describing the extraction of feature vectors from the face image for Parts-Based approaches

2.2 Feature Distribution Modelling

Feature distribution modelling is achieved by performing background model adaptation of GMMs [9,10]. The use of background model adaptation is not new to the field of biometric authentication in fact it is commonly used in the field of speaker verification [13]. Background model adaptation first trains a world (background) model Ω_{world} from a set of faces and then derives the client model for the ith client Ω^i_{client} by adapting the world model to match the observations of the client.

Two common methods of performing adaptation are mean only adaptation [14] and full adaptation [15]. Mean only adaptation is often used when there are few observations available because adapting the means of each mixture component requires fewer observations to derive a useful approximation. Full adaptation is used when there are sufficient observations to adapt all the parameters of each mode. Mean only adaptation is the method chosen for this work as it requires fewer observations to perform adaptation, this is the same adaptation method employed by Cardinaux et al. [9].

2.3 Verification

A description of the Parts-Based approach is not complete without defining how an observation is verified. To verify an observation, $\boldsymbol{x}$, it is scored against both the client (Ω^i_{client}) and world (Ω_{model}) model, this is true even for methods that do not perform background model adaptation [6]. The two models, Ω^i_{client} and Ω_{world}, produce a log-likelihood score which is then combined using the log-likelihood ratio (LLR),

$$h(\boldsymbol{x}) = \ln(p(\boldsymbol{x} \mid \Omega^i_{client})) - \ln(p(\boldsymbol{x} \mid \Omega_{world})), \quad (1)$$

to produce a single score. This score is used to assign the observation to the world class of faces (not the client) or the client class of faces (it is the client) and consequently a threshold τ has to be applied to the score $h(\boldsymbol{x})$ to declare (verify) that $\boldsymbol{x}$ matches to the ith client model Ω^i_{client} when $h(\boldsymbol{x}) \geq \tau$.

3 Local Frequency Band Approach

The method proposed in this paper is to divide the face into separate blocks and to then decompose these blocks in the frequency domain. This can be achieved by treating the frequency response from each block separately to form frequency sub-images. This method is applied to the DCT feature vectors obtained by applying the Parts-Based approach. Each coefficient can be considered independently because each coefficient of the DCT is orthogonal.

The technique is summarised as follows: (1) the face is cropped and normalised to a 68×68 image, (2) this image is divided into 8×8 blocks with an overlap of 4 pixels in the horizontal and vertical axes, (3) the DCT coefficients from each block are separated and used to form their own frequency sub-image, and (4) a feature vector is formed by taking a block from the frequency sub-image and vectorising the block. The way in which the frequency sub-images are formed is demonstrated in Figure 2.

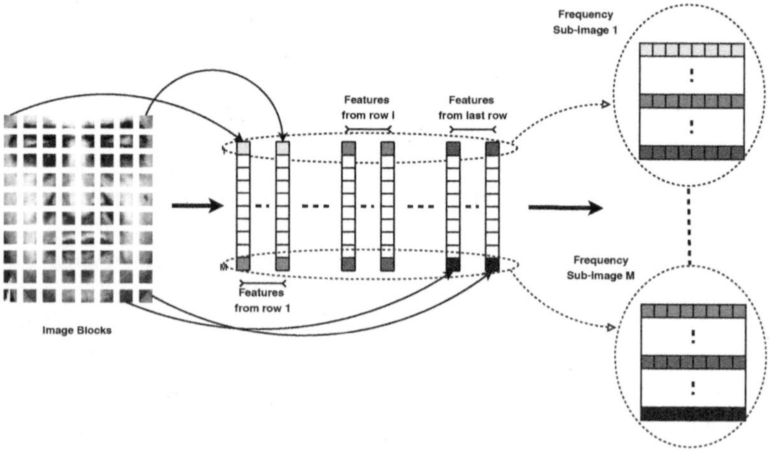

Fig. 2. The figure above describes how the face can be decomposed into separate frequency sub-bands (sub-images)

3.1 Motivation

To illustrate the differences between the frequency decomposition approach and the full Parts-Based approach the following statements are made. For the Parts-Based approach it is often stated that the face is broken into blocks and the distribution of each block is then modelled [6,10], however, another stricter statement would be that the frequency information from each block is simultaneously modelled since each dimension of the feature vector represents a different sampling frequency of the DCT. By contrast the frequency decomposition approach separates the frequency information from each local block and forms a feature vector from the resulting frequency sub-images. Many feature vectors are formed from a frequency sub-image and then modelled using background model adaptation, thus, the image is decomposed in both the spatial domain and the frequency domain.

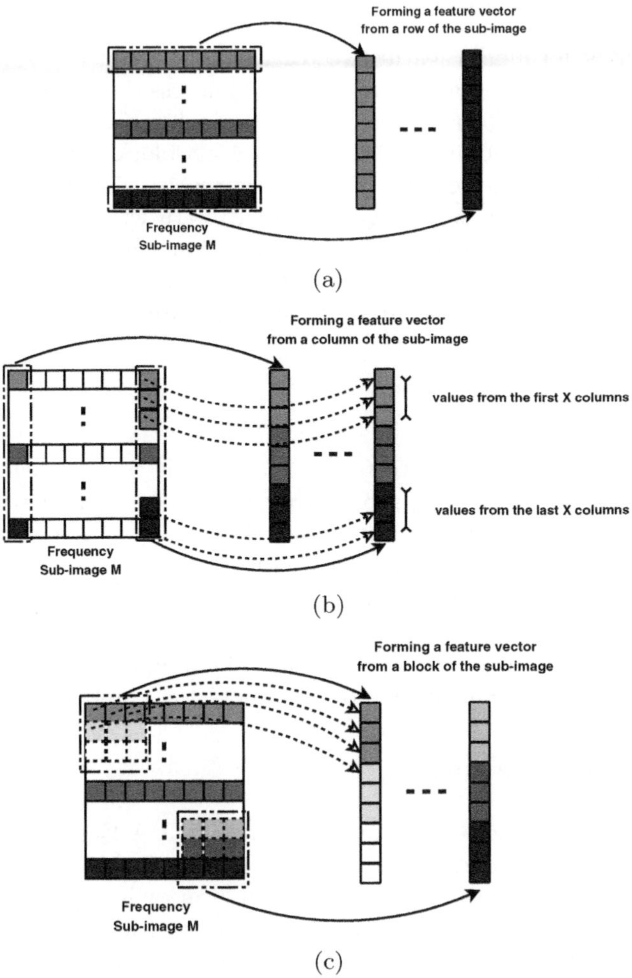

Fig. 3. Forming the feature vectors (a) along the row of the frequency sub-image, (b) along the columns of the frequency sub-image and (c) using blocks of the frequency sub-image

A side effect of working on the frequency sub-images is that the feature vectors formed from these sub-images will retain extra spatial information. This is because the Parts-Based approach obtains an observation from a block, however, the frequency decomposition approach gets the response from each block and then forms a feature vector using responses from several blocks. This means that the feature vectors formed from the frequency sub-images will actually span several blocks when compared to the Parts-Based approach, for instance the feature vector could be formed from a frequency sub-image by spanning an entire row or column of the image.

3.2 Feature Extraction

Three methods of forming a feature vector from the frequency sub-images are examined, these are to form a feature vector: (1) across the row of the frequency sub-image (row-based approach), (2) across the column of the frequency sub-image (column-based approach), or (3) from a 4×4 block of the frequency sub-image which is then vectorised (block-based approach). The choice of a 68×68 image results in frequency sub-images of size 16×16 which allows for the fair comparison of the three different feature extraction methods as each method will result in feature vectors of dimension $D = 16$ with $o = 16$ observations from each frequency sub-image. A visualisation of these three methods is provided in Figure 3.

3.3 Classifier

Having obtained these feature vectors a classifier is formed using the same background model adaptation approach that was used for the Parts-Based approach [9]. Each local frequency sub-band (k) produces a separate classifier (C_k) and these classifiers are then combined using weighted linear score fusion, $C_{w\_sum} = \sum_{k=1}^{K} \beta_k C_k$. This fusion technique is used as it was shown by Kittler et al. [16] that the sum rule (which is what weighted linear classifier score fusion abstracts to be) is robust to estimation errors. The weights, β_k, for the classifiers are derived using an implementation of linear logistic regression [17].

4 Results and Discussion

The database used for these experiments is the BANCA English database [18]. The images are cropped and scaled to a size of 68×68 pixels with a distance between the eyes of 33 pixels. Illumination normalisation is applied to each image as a two stage process, the image is histogram equalised and then encoded using a Local Binary Pattern (LBP) [19]. This is the same normalisation strategy employed by Heusch et al. [20] where the parameters used for the LBP are $R = 2$ and $P = 8$.

Results for the BANCA database are presented as the average Half Total Error Rate (HTER) using g_1 and g_2 in a cross-validated manner. Parameters such as the optimal number of mixture components M are derived using the development set of Protocol P and used as a constant throughout the remaining tests. The decision threshold and other parameters such as classifier weights, β_k, are derived on the independent development (tuning) set for each protocol.

Two baseline systems are considered for this work, one system uses DCT feature vectors and the other uses DCTmod2 feature vectors. DCTmod2 feature vectors are examined as it was previously found to be more robust than DCT feature vectors [6]. The size of the feature vectors, $D = 15$ for DCT feature vectors and $D = 18$ for DCTmod2 feature vectors, was chosen based on work conducted in [6]. Both baseline systems were developed using 68×68 face images with a varying number mixtures $M = [100, 200..., 500]$.

Results on the Development set of the P protocol found that, for both manual and automatic eye positions, the system using DCT feature vectors with $M = 100$ mixture components provides the best performance. This system was then used to produce full results for all of the BANCA protocols and are presented in Table 1. It can be seen that there is a consistent degradation in performance of, on average, 2.37% in absolute performance difference when automatic eye locations are used.

The effect of using manual and automatic eye positions is examined since any deployed face verification system will need to cope with errors introduced from an automatic face detection system. The manually annotated eye positions were provided with the BANCA database and the automatically annotated eye positions were obtained using a face detector based on a cascade of LBP-like features [21] [1]. There were 93 images (out of 6,540 images) where the automatic face detector could not find the face, these images were excluded from training, development and evaluation of the automatic systems.

The initial experiments indicated that all of the local frequency sub-band approaches, across all BANCA protocols, provided significantly improved performance when compared to the baseline system. It can be seen from the results in Table 2 that the optimal local frequency sub-band approach is the column-based approach followed by the block-based approach and finally the row-based approach. The column-based approach provides an absolute improvement of 9.21% for Protocol P with the HTER reducing from 24.38% to 15.17%; the frequency sub-band systems are optimised in a similar manner to the baseline systems, however, because there are fewer observations ($o = 16$ observations for each frequency sub-image whereas $o = 256$ for the Parts-Based approach) the number of mixtures were constrainted to $M = [5, 10, ..., 25]$.

Table 1. This table presents the average HTER for the baseline Parts-Based verification system across all of BANCA protocols

	P	G	Ud	Ua	Mc	Md	Ma
manual annotations	24.38%	12.38%	24.60%	29.46%	8.21%	16.35%	18.91%
automatic annotations	26.34%	15.33%	27.86%	30.51%	11.88%	18.63%	20.33%

The performance of the frequency sub-band approach when using automatic eye locations is superior to that of the baseline system. This is true across all test conditions except for the block-based approach on Protocol Mc, see Table 3 for full results. Examining the results in more detail it becomes obvious that the column-based approach degrades in a similar manner as the baseline system with an average absolute performance difference of 2.51%. This means that for both manual and automatic eye locations the column-based system significantly outperforms the baseline system. However, a different result appears for the row-based and block-based approaches whose performance degrades significantly,

[1] This detector has been implemented with torch3vision.idiap.ch

Table 2. This table presents the average HTER for the local frequency sub-band approaches on manually annotated eye locations for all of the BANCA protocols. Each system has $M_{row} = 15$, $M_{blk} = 20$ and $M_{col} = 5$ mixture components respectively. Highlighted are the best performing systems.

	P	G	Ud	Ua	Mc	Md	Ma
row-based	19.418%	7.24%	18.19%	23.06%	7.18%	9.98%	12.42%
block-based	16.74%	**5.72%**	14.46%	25.03%	**5.51%**	**7.20%**	**10.29%**
column-based	**15.17%**	5.80%	**13.40%**	**19.20%**	6.73%	7.88%	11.04%

Table 3. This table presents the average HTER for the local frequency sub-band approaches on automatically annotated eye locations for all of the BANCA protocols. Each system has $M_{row} = 15$, $M_{blk} = 20$ and $M_{col} = 5$ mixture components respectively. Highlighted are the best performing systems.

	P	G	Ud	Ua	Mc	Md	Ma
row-based	24.57%	9.97%	25.02%	27.81%	10.37%	16.66%	**15.00%**
block-based	21.86%	9.93%	23.04%	25.64%	12.34%	14.21%	16.79%
column-based	**16.64%**	**7.4%**	**17.50%**	**21.06%**	**8.07%**	**10.26%**	15.89%

Table 4. This table presents state-of-the-art results on the BANCA database using the average HTER (%) on the Mc protocol for Manual eye annotations

	SB_{Col}	BN	PSC-GMM	GMM
HTER on g1 (%)	10.13	9.01	11.31	N/A
HTER on g2 (%)	3.33	5.41	11.34	8.9

their HTER increases by an absolute average of 4.56% and 5.55% respectively when using automatic eye locations.

The result for the automatic eye locations demonstrates that the choice of feature vector formation has a significant impact on the local frequency sub-band approach. The column-based approach has empirically been shown to be more robust to localisation errors than either the row-based or block-based approaches. The column-based also performs better than either the row-based or block-based approaches for most of the test conditions. This fact could be explained by suggesting that the features of the face are more stable when scanned in a vertical manner, particularly when there is misalignment of the face image. However, this argument has not been explored fully and so it forms the basis of future work for this technique.

Finally, this system compares well to state-of-the-art techniques for the BANCA database. Previous state-of-the-art face verification systems, tested on the Mc protocol of BANCA, are taken from the work of Heusch and Marcel [22] who provide results from a Bayesian Network classifier (BN), a Partial Shape Collapse GMM classifier (PSC-GMM) and a state-of-the-art GMM Parts-Based classifier[2] are provided. The results are reproduced in Table 4 where it can be

[2] Results for the state-of-the-art GMM Parts-Based classifier are only available for g2.

seen that the column-based sub-band (SB_{Col}) approach performs very competitively when compared to the more complex BN approach.

5 Conclusions and Future Work

In this paper a novel extension to the Parts-Based approach has been introduced that results in an absolute improvement of the HTER of 9.21%. This novel extension decomposes the face into local frequency sub-bands. Feature vectors are extracted from these frequency sub-band images in one of three ways, in a: row-based manner, column-based manner or a block-based manner. It has been shown empirically that extracting the feature vectors in a column-based manner results in a robust and accurate method for performing face verification. However, it remains an open question why the column-based approach to feature formation is more robust and accurate than the row-based or block-based approaches.

Future work will examine potential reasons for the superior performance of the column-based approach. Work will also be conducted to determine if local frequency sub-band's can be merged, performing a feature level fusion, to simplify the task of classifier fusion. Finally, any future work will also need to address the possibility of applying a HMM approach to this local frequency sub-band technique.

Acknowledgements

This work has been performed by the MOBIO project 7th Framework Research Programme of the European Union (EU), grant agreement number: 214324. The authors would like to thank the EU for the financial support and the partners within the consortium for a fruitful collaboration. For more information about the MOBIO consortium please visit `http://www.mobioproject.org`

References

1. Zhao, W., Chellappa, R., Phillips, P., Rosenfeld, A.: Face recognition: A literature survey. ACM Computing Surveys 35(4), 399–459 (2003)
2. Bledsoe, W.W.: The model method in facial recognition. Technical report for Panoramic Research Inc. (1966)
3. Turk, M., Pentland, A.: Eigenfaces for recognition. Journal of Cognitive Neuroscience 3(1), 71–86 (1991)
4. Belhumeur, P.N., Hespanha, J.P., Kriegman, D.J.: Eigenfaces vs. fisherfaces: Recognition using class specific linear projection. IEEE Transactions on Pattern Analysis and Machine Intelligence 19(7), 711–720 (1997)
5. Samaria, F., Fallside, F.: Face identification and feature extraction using hidden markov models. Image Processing: Theory and Applications, 295–298 (1993)
6. Sanderson, C., Paliwal, K.K.: Fast feature extraction method for robust face verification. Electronic Letters 38(25), 1648–1650 (2002)
7. Samaria, F., Young, S.: Hmm-based architecture for face identification. Image and Vision Computing 12(8), 537–543 (1994)

8. Nefian, A., Hayes III, M.H.: Hidden markov models for face recognition. In: Proceedings of the 1998 IEEE International Conference on Acoustics, Speech, and Signal Processing, vol. 5, pp. 2721–2724 (1998)
9. Cardinaux, F., Sanderson, C., Marcel, S.: Comparison of mlp and gmm classifiers for face verification on xm2vts. In: International Conference on Audio- and Video-based Biometric Person Authentication, pp. 1058–1059 (2003)
10. Lucey, S., Chen, T.: A gmm parts based face representation for improved verification through relevance adaptation. In: Proceedings of IEEE Conference on Computer Vision and Pattern Recognition, vol. 2, pp. 855–861 (2004)
11. Heisele, B., Ho, P., Wu, J., Poggio, T.: Face recognition: component-based versus global approaches. In: Computer Vision and Image Understanding, vol. 91, pp. 6–21 (2003)
12. Pennebaker, W.B., Mitchell, J.L.: JPEG still image data compression standard. Van Nostrand Reinhold, New York (1993)
13. Doddington, G., Przybocki, M., Martin, A., Reynolds, D.: The NIST speaker recognition evaluation — overview, methodology, systems, results, perspective. Speech Communication 31(2-3), 225–254 (2000)
14. Reynolds, D.: Comparison of background normalization methods for text-independent speaker verification. In: Proc. European Conference on Speech Communication and Technology (Eurospeech), vol. 2, pp. 963–966 (1997)
15. Lee, C., Gauvain, J.: Bayesian adaptive learning and MAP estimation of HMM, pp. 83–107. Kluwer Academic Publishers, Boston (1996)
16. Kittler, J., Hatef, M., Duin, R.P.W., Matas, J.: On combining classifiers. IEEE Transactions on Pattern Analysis and Machine Intelligence 20, 226–239 (1998)
17. Brummer, N.: Tools for fusion and calibration of automatic speaker detection systems (2005), `http://www.dsp.sun.ac.za/~nbrummer/focal/index.htm`
18. Bailly-Bailliere, E., Bengio, S., Bimbo, F., Hamouz, M., Kittler, J., Mariethoz, J., Matas, J., Messer, K., Popovici, V., Poree, F., Ruiz, B., Thiran, J.P.: The banca database and evaluation protocol. LNCS, pp. 625–638. Springer, Heidelberg (2003)
19. Ojala, T., Pietikäinen, M., Mäenpää, T.: Multiresolution gray-scale and rotation invariant texture classification with local binary patterns. IEEE Trans. Pattern Anal. Mach. Intell. 24(7), 971–987 (2002)
20. Heusch, G., Rodriguez, Y., Marcel, S.: Local binary patterns as an image preprocessing for face authentication. In: International Conference on Automatic Face and Gesture Recognition, pp. 9–14 (2006)
21. Fröba, B., Ernst, A.: Face detection with the modified census transform. In: IEEE Conference on Automatic Face and Gesture Recognition, pp. 91–96 (2004)
22. Heusch, G., Marcel, S.: Face authentication with Salient Local Features and Static Bayesian network. In: IEEE / IAPR Intl. Conf. On Biometrics (ICB) (2007) IDIAP-RR 07-04

Local Gabor Binary Pattern Whitened PCA: A Novel Approach for Face Recognition from Single Image Per Person

Hieu V. Nguyen, Li Bai, and Linlin Shen

School of Computer Science, University of Nottingham,
Jubilee Campus, Wollaton Road, Nottingham, NG8 1BB, UK
{vhn,bai}@cs.nott.ac.uk, llshen@szu.edu.cn
http://www.nottingham.ac.uk/cs/

Abstract. One major challenge for face recognition techniques is the difficulty of collecting image samples. More samples usually mean better results but also more effort, time, and thus money. Unfortunately, many current face recognition techniques rely heavily on the large size and representativeness of the training sets, and most methods suffer degraded performance or fail to work if there is only one training sample per person available. This so-called "Single Sample per Person" (SSP) situation is common in face recognition. To resolve this problem, we propose a novel approach based on a combination of Gabor Filter, Local Binary Pattern and Whitened PCA (LGBPWP). The new LGBPWP method has been successfully implemented and evaluated through experiments on 3000+ FERET frontal face images of 1196 subjects. The results show the advantages of our method - it has achieved the best results on the FERET database. The established recognition rates are 98.1%, 98.9%, 83.8% and 81.6% on the fb, fc, dup I, and dup II probes, respectively, using only one training sample per person.

Keywords: Gabor Wavelet, Local Binary Pattern, Whitening, PCA, Feature Selection, Face Recognition.

1 Introduction

Face recognition is the identification of individuals from a database of labeled face images. The applications of face recognition can be found in surveillance and security, telecommunication, and smart environments [1]. The biggest challenge of face recognition comes from the numerous variations between images of the same face, such as changes in facial expression, illumination, aging or facial accessories. A robust face recognition system should recognize a face regardless of these intrapersonal facial variations [2].

Discriminating analysis based methods [3,4] have been widely used to suppress intrapersonal variations. This involves a projection into a subspace whose within-class scatter is small and the between-class scatter is large. A drawback of these methods is that they require large and representative training samples to

M. Tistarelli and M.S. Nixon (Eds.): ICB 2009, LNCS 5558, pp. 269–278, 2009.
© Springer-Verlag Berlin Heidelberg 2009

guarantee generalization. However, many face recognition tasks, such as airport security and law enforcement applications, can only offer small sized training samples. Consequently, discriminant analysis tends to over-fit the training data, and fail on future trials using novel (unseen) test data.

Recently, the local binary patterns (LBP) operator has been successfully used for face detection and recognition. LBP encodes both local and global information by a histogram [5]. Facial feature extracted by the LBP operator is robust to illumination variations because the LBP features are invariant to monotonic gray-scale changes. However, under varying lighting and aging effects, LBP performance is still not satisfactory [6].

Gabor based face representation is robust to illumination variations and efficient to describe local image features as multi-resolution histograms [7,8]. The combination of Gabor wavelets with LBP operator has been proposed to represent a face image to obtain robust feature against facial variations [9,10,11]. The combined operator is called Local Gabor Binary Patterns (LGBP) operator. Experimental results have demonstrated that the performance of face recognition with LGBP is superior to both of LBP-based approach and Gabor-based approach [9,10,11].

However, face representation with LGBP is high dimensional due to the multiple Gabor transformations of the LGBP operator [9]. Hence, further dimension reduction is necessary. There are two major categories of methods for dimension reduction: feature selection and feature transformation.

In this paper we introduce a robust face recognition scheme using a single image per person without using discriminant analysis. This scheme has three components: Gabor Filter, Local Binary Pattern (LBP) and Whitened PCA. Gabor and LBP have proven to have very powerful features that complement each other [11] while Standard PCA with whitening can effectively suppress the intrapersonal variation [12]. We combine these components in a novel way to solves the SSP problem.

2 Local Gabor Binary Pattern Whitened PCA Scheme

2.1 Face Representation

The representation approach based on a combination of Gabor Filters and Local Binary Patterns is illustrated in Figure 1 on page 271 [9]. A face image is modeled as a histogram sequence by the following steps:

(1) An input face image is normalized and transformed to obtain multiple Gabor Magnitude Images (GMIs) in frequency domain by applying multi-scale and multi-orientation Gabor filters;

(2) Each GMI is converted to Local Gabor Binary Pattern (LGBP) map;

(3) Each LGBP Map is further divided into non-overlapping rectangle regions with specific size, and histogram is computed for each region; and

(4) The LGBP histograms of all the LGBP Maps are concatenated to form the final histogram sequence as the face representation.

These 4 steps will be described in detail in the following sub-sections.

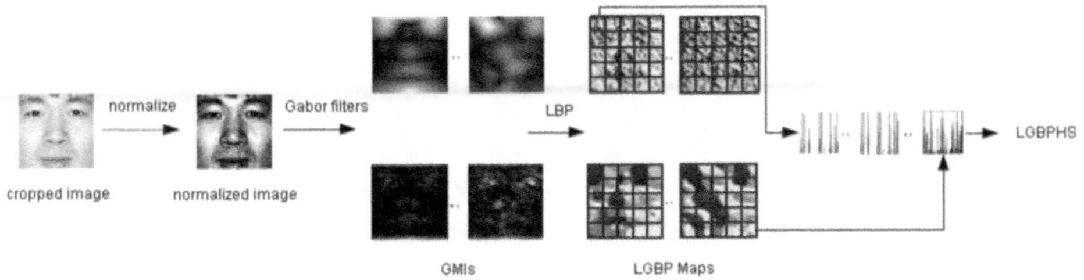

Fig. 1. The proposed LGBPHS face representation approach

Gabor Magnitude Image. Gabor filters most commonly used in face recognition have the form:

$$\psi_{\mu,\nu}(z) = \frac{\|k_{\mu,\nu}\|^2}{\sigma^2} e^{(-\|k_{\mu,\nu}\|^2\|z\|^2/2\sigma^2)} \left[e^{ik_{\mu,\nu}z} - e^{-\sigma^2/2}\right] \tag{1}$$

where μ and ν define the orientation and scale of the Gabor kernels, z = (x, y), $\|.\|$ denotes the norm operator, and the wave vector $k_{\mu,\nu} = k_\nu e^{i\phi_\mu}$, where $k_\nu = k_{max}/\lambda^\nu$ and $\phi_\mu = \pi\mu/8$. k_{max} is the maximum frequency, λ is the spacing factor between filters in the frequency domain.

The Gabor representation of a face image is derived by convolving the face image with the Gabor filters. Let $f(x, y)$ be the face image, its convolution with a Gabor filter $\psi_{\mu,\nu}(z)$ is defined as follows

$$G_{\psi f}(x, y, \mu, \nu) = f(x, y) * \psi_{\mu,\nu}(z) \tag{2}$$

where * denotes the convolution operator. Five scales $\nu \in \{0, \cdots, 4\}$ and eight orientations $\mu \in \{0, \cdots, 7\}$ Gabor filters are used. Convolving the image with each of the 40 Gabor filters can then generate the Gabor features. For each Gabor filter, one magnitude value will be computed at each pixel position, which will totally result in 40 Gabor Magnitude Images (GMIs).

Local Gabor Binary Pattern (LGBP). In order to further enhance the information in the Gabor Magnitude Images, we encode the magnitude values with LBP operator. The original LBP operator [13] labels the pixels of an image by thresholding the 3x3 neighborhood of each pixel $f_p(p = 0, 1, \ldots, 7)$ with the center value f_c, as shown in Figure 2 on page 272.

$$S(f_p - f_c) = \begin{cases} 1, & f_p \geqslant f_c \\ 0, & f_p < f_c \end{cases} \tag{3}$$

Then, by assigning a binomial factor 2^p to each $S(f_p - f_c)$, the LBP pattern at the pixel is

$$LBP = \sum_{p=0}^{7} S(f_p - f_c) 2^p \tag{4}$$

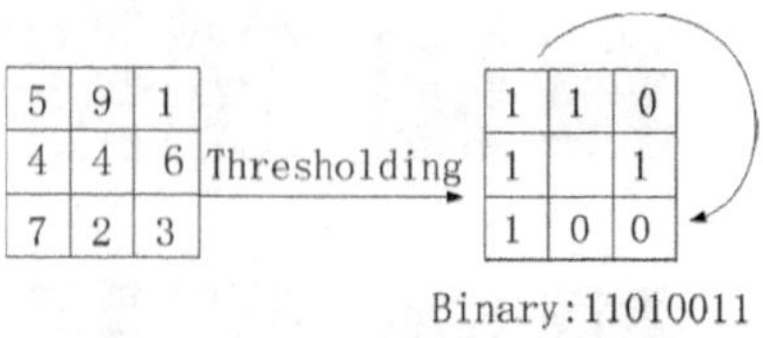

Fig. 2. The basic LBP operator

which characterizes the spatial structure of the local image texture. The operator LGBP denotes the LBP operates on GMI. We denote the transform result at position (x, y) of (μ, ν) - GMI as $G_{lgbp}(x, y, \mu, \nu)$, which composes the (μ, ν)-LGBP Map.

LGBP Histogram Sequence. The histogram **h** of an image $f(x, y)$ with gray levels in the range [0, L-1] could be defined as

$$\mathbf{h_i} = \sum_{\mathbf{x,y}} I\left\{f(x, y) = i\right\}, i = 0, 1, \ldots, L-1 \tag{5}$$

where i is the i-th gray level, h_i is the number of pixels in the image with gray level i and

$$I\left\{A\right\} = \begin{cases} 1, & A\, is\, true \\ 0, & A\, is\, false \end{cases} \tag{6}$$

Assume each LGBP Map is divided into m regions $R_o, R_1, \ldots, R_{m-1}$. The histogram of r-th region of the specific LGBP Map (from (μ, ν)-GMI) is computed by

$$H_{\mu,\nu,r} = (h_{\mu,\nu,r,0}, h_{\mu,\nu,r,1}, \ldots, h_{\mu,\nu,r,L-1}) \tag{7}$$

where

$$h_{\mu,\nu,r,i} = \sum_{(x,y)\in R_r} I\left\{G_{lgbp}(x, y, \mu, \nu) = i\right\} \tag{8}$$

Finally, all the histogram pieces computed from the regions of all the 40 LGBP Map are concatenated to a histogram sequence, R, as the final face representation $R = (H_{0,0,0}, \ldots, H_{0,0,m-1}, H_{0,1,0}, \ldots, H_{0,1,m-1}, \ldots, H_{4,7,m-1})$.

To solve the high dimensional problem of LGBP representation, we will discuss feature selection and feature transformation methods in the next section.

2.2 Whitened PCA

Standard PCA. A classical technique for dimensionality reduction, particularly in face recognition, is principle component analysis (PCA). In order to produce a compact representation, the feature vector is projected into a lower dimensional feature space found by principle components analysis

$$\mathbf{u} = \mathbf{W_{pca}x} \tag{9}$$

The input vectors are first transformed by subtracting the mean: $\Phi_i = x_i - m$. The principal components of the training data set are given by the eigenvectors of its covariance matrix $\sum = \frac{1}{n}\sum_{i=1}^{n}\Phi_i\Phi_i^T$. In practice, only $M(M < n-1)$ eigenvectors having the largest eigenvalues (and, hence, the largest variance in the data set) are kept empirically to form the projection matrix W_{PCA}.

PCA technique is optimal in maximizing the scatter of all the projected training samples, meaning PCA maximizes both intrinsic difference and intrapersonal variations in the training set. However, the latter is not useful for classification purposes. This problem is resolved using the LGBPWP scheme with the whitening process.

Whitening Process. PCA has two obvious shortcomings: (1) the leading eigenvectors encode mostly illumination and expression, rather than discriminating information [14]; and (2) Mean-Square-Error (MSE) principle underlying PCA favours low frequencies [15,16,17] and thus loses the discriminating information contained in the high frequency components. The whitening process normalizing the PCA based feature can directly counteract these disadvantages. Specifically, the PCA based feature, u is subjected to the whitening transformation and yields yet another feature set w:

$$\mathbf{w} = \mathbf{\Lambda_M^{-1/2}u} \tag{10}$$

where $\Lambda_M^{-1/2} = diag\{\lambda_1^{-1/2}, \lambda_2^{-1/2}, \ldots, \lambda_M^{-1/2}\}$. The integrated projection matrix $\Lambda_M^{-1/2}W_{PCA}$ treats variance along all principal component axes as equally significant by weighting components corresponding to smaller eigenvalues more heavily and is arguably appropriate for discrimination. Consequently, the negative influences of the leading eigenvectors are reduced while the discriminating details encoded in trailing eigenvectors are enhanced [12].

2.3 Feature Selection

Although LGBP features are very robust, they also cause the 'curse of dimensionality' problem [9,18] which can be solved by feature selection techniques. Feature selection algorithms such as Adaboost have been very effective [19,20,21]. Unfortunately, these techniques cannot be used in this situation because of the SSP assumption, or only one sample per person. Instead, we propose a simple but efficient selection method based on feature variance. Features are ranked based on their variance and highly variant features are selected. These features contain very useful information for classification but they also contain strong intrapersonal variation. This selection method is efficient because when used together with whitened PCA, the intrapersonal variation can be suppressed [12]. Selected features are visualized in Figure 3 on page 274. The gray level represents the number of selected features within each block. As can be expected, the selected features focus on discriminative areas such as eyes, eyebrows, mouth, and nose.

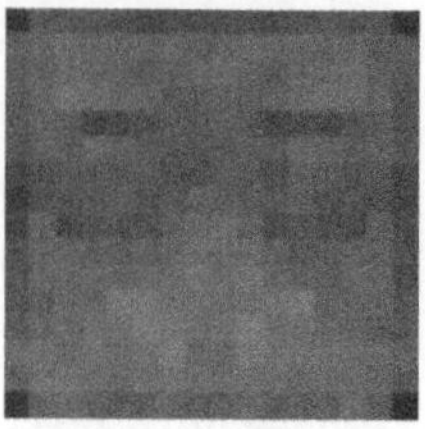

Fig. 3. Distribution of selected features based on their variances

2.4 Face Recognition

The original feature vector of the face image is first calculated as detailed in Section 2.1, and the low dimensional LGBPHS-based feature vector, w, is derived using (9). Let m_k^0, $k = 1, 2, \ldots, L$, be the prototype, the mean of training samples, for class ω_k in the feature space. The LGBPWP method applies the nearest neighbor (to the mean) rule for classification using similarity measure δ

$$\delta(w, m_k^0) = min_j\, \delta(w, m_j^0) \rightarrow w \in \omega_k \tag{11}$$

The image feature vector, w, is classified as belonging to the class of the closest mean, m_k^0, using the similarity measure δ.

Note that when the novel variation, unseen in the training set, is projected onto the feature space, most energy of it will distribute over all the eigenvectors. This is because such variations are somewhat independent of the variance retained by the feature space. In other words, novel variation, projected into the feature space, is incline to evenly affect the projected scale on each component, and thus take more effect on the $L1$ and $L2$ distance rather than the vector angle [12]. Therefore, the cosine similarity measure, δ_{CSM} , which is invariant to change in scale, is employed to perform nearest neighbor search in the feature space for face recognition.

$$\delta_{CSM}(w_1, w_2) = \frac{-w_1^T w_2}{\|w_1\|.\|w_2\|} \tag{12}$$

In the following section we will use experiments on the FERET database to back our claim that the LGBPWP method is robust for face recognition.

3 Experimental Evaluations

A large scale data set from The FacE REcognition Technology (FERET) database is used to assess the effectiveness and robustness of the new LGBPWP method on the face recognition task. This is a standard test for face recognition technologies [22]. The FERET facial database displays diversity across gender, ethnicity, and age. To obtain a strong and precise assessment of performance, the LGBPWP algorithm is evaluated against different categories of images due to varying lighting, expression, and acquisition time.

3.1 Implementation Parameters

In the preprocessing stage, firstly, face images are all cropped and resized to 128x128 pixels, and aligned using the locations of two eyes given in FERET database. Then histogram equalization is applied to reduce the illumination effect. Finally, all faces are normalized to zero mean and unit variance.

In the feature extraction stage, the window size of the local region will affect recognition performance. In order to keep more spatial information, in our experiments we empirically choose a small window size of 8x8 pixels. In order to reduce the dimension of LGBP features, we only use uniform local binary patterns [23]. As a result, the number of LGBP features is 40*(128/8)*(128/8)*59=604,160.

In the feature selection stage, top 25000 variance features, which is much fewer than the original number of features, are selected to feed to Whitened PCA for classification.

All images from the gallery in FERET database are used for PCA training.

3.2 Variance-Based versus Random Feature Selection

Intuitively, variance-based feature selection is efficient because highly variant features contain strong discriminant information. However, these features may also contain intrapersonal variation which is not good for recognition. Luckily, whitened PCA tends to suppress the intrapersonal variations. To demonstrate the effectiveness of the variance-based feature selection, we compare it with random feature selection method (LGBPRWP). As the rank curves in Figure 4 on page 275 show, the variance-based method gains much better performance.

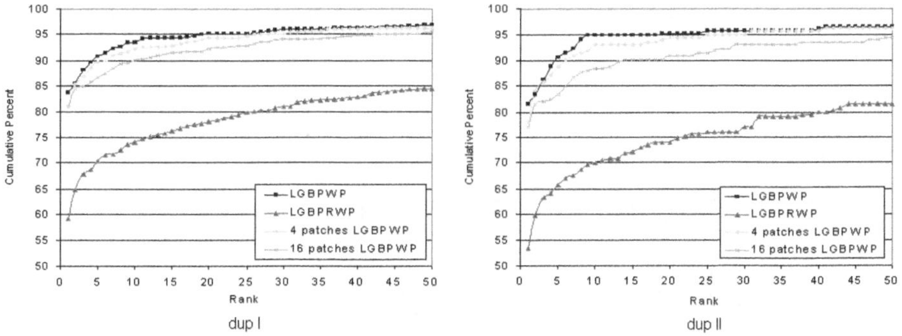

Fig. 4. Rank curves for the dup I (left) and dup II (right) probe sets

3.3 Holistic PCA versus Ensemble of Piecewise PCAs

Recently, patch-based and ensemble-based approaches [24] have gained a huge success in terms of recognition performance compared to traditional holistic approaches. However, in the case of LGBPWP, there is no clear difference between patch-based approach and holistic approach. Moreover, the holistic LGBPWP

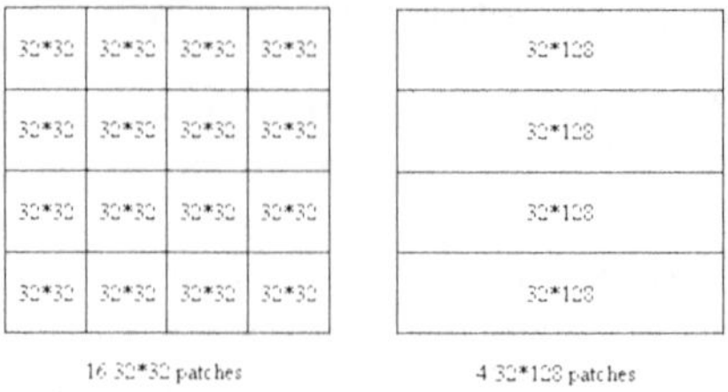

Fig. 5. 16 patches (left) and 4 patches (right) LGBPWP

achieves slightly better results in both dup I and dup II probe sets, as shown in Figure 4 on page 275. Because of the SSP problem, each patch does not contain enough discriminant information to rank matching faces correctly. PCA further compounds the problem. However, patch-based LGBPWP still achieved relatively good performance because of the robustness of Local Gabor Binary Pattern features. To demonstrate, we implemented two patch-based versions of LGBPWP. They are different in terms of patch sizes. As shown in Figure 5 on page 276, the first implementation contains 16 32x32 patches and the second one contains 4 32x128 patches. Rank curves in Figure 4 on page 275 clearly supports our judgement.

3.4 Results

In Table 1 on page 276, the rank-1 recognition rates of different algorithms on the FERET probe sets are listed for comparative purpose. As shown in the table, LGBPWP achieves similar results to top algorithms on fb and fc probe sets and a margin about 5% better than the second best algorithm on dup I and dup II probe sets.

Table 1. Rank-1 recognition rates of different algorithms on the FERET probe sets

Methods	fb	fc	dup I	dup II
Fisherface	94.0	73.0	55.0	31.0
Best Results of [25]	96.0	82.0	59.0	52.0
Results of [23]	97.0	79.0	66.0	64.0
Direct LGBPHS	94.0	97.0	68.0	53.0
HGPP	97.6	98.9	77.7	76.1
GEWC	96.3	99.5	78.8	77.8
LGBPWP	**98.1**	**98.9**	**83.8**	**81.6**

4 Conclusion and Discussion

In order to solve the problem of insufficient training data, a new face recognition scheme is proposed that uses single image per person without the need for discriminant analysis. The three major components of the proposed method are Gabor Filter, Local Binary Pattern (LBP) and Whitened PCA. The way these three techniques are combined forms a novel approach to solve the SSP problem. Gabor-based features and LBP-based features have proven to be very robust and complementary to each other while standard PCA with whitening can suppress intrapersonal variations very well. Variance-based feature selection is has also proven to be a very efficient when combined with whitened PCA. The results validated the arguments in this paper with the recognition rates of 98.1%, 98.9%, 83.8% and 81.6% on the fb, fc, dup I, and dup II probe using only one training image per person.

Acknowledgments. We acknowledge the support of this research by the University of Nottingham, the UK Engineering and Physical Science Research Council (EPSRC EP/F013477/1), and the China Science Foundation of Shenzhen City (CXQ2008019).

References

1. Chellappa, R., Wilson, C.L., Sirohey, S., et al.: Human and machine recognition of faces: a survey. Proceedings of the IEEE 83(5), 705–740 (1995)
2. Adini, Y., Moses, Y., Ullman, S.: Face Recognition: The Problem of Compensating for Changes in Illumination Direction. IEEE Transactions on Pattern Analysis and Machine Intelligence, 721–732 (1997)
3. Liu, C., Wechsler, H.: Gabor feature based classification using the enhanced fisher lineardiscriminant model for face recognition. IEEE Transactions on Image Processing 11(4), 467–476 (2002)
4. Swets, D.L., Weng, J.J.: Using Discriminant Eigenfeatures for Image Retrieval. IEEE Transactions on Pattern Analysis and Machine Intelligence, 831–836 (1996)
5. Ahonen, T., Hadid, A., Pietikäinen, M.: Face Description with Local Binary Patterns: Application to Face Recognition. IEEE Transactions on Pattern Analysis and Machine Intelligence, 2037–2041 (2006)
6. Ji, Q.(IEEE Senior Member), Zou, J.(IEEE Member), Nagy, G.(IEEE Fellow): A comparative study of local matching approach for face recognition. IEEE Transactions on Image Processing 16, 2617–2628 (2007)
7. Shan, S., Gao, W., Chang, Y., Cao, B., Yang, P.: Review the strength of Gabor features for face recognition from the angle of its robustness to mis-alignment. In: Proceedings of the 17th International Conference on Pattern Recognition, ICPR 2004, vol. 1 (2004)
8. Wiskott, L., Fellous, J.M., Krüger, N., von der Malsburg, C.: Face Recognition by Elastic Bunch Graph Matching. IEEE Transactions on Pattern Analysis and Machine Intelligence, 775–779 (1997)
9. Zhang, W., Shan, S., Gao, W., Chen, X., Zhang, H.: Local Gabor Binary Pattern Histogram Sequence (LGBPHS): A Novel Non-Statistical Model for Face Representation and Recognition. In: Proc. ICCV, pp. 786–791 (2005)
10. Zhang, W., Shan, S., Chen, X., Gao, W.: Local Gabor Binary Patterns Based On Mutual Information For Face Recognition. International Journal of Image and Graphics 7(4), 777–793 (2007)

11. Tan, X., Triggs, B., Vision, M.: Fusing Gabor and LBP Feature Sets for Kernel-Based Face Recognition. In: Zhou, S.K., Zhao, W., Tang, X., Gong, S. (eds.) AMFG 2007. LNCS, vol. 4778, p. 235. Springer, Heidelberg (2007)
12. Deng, W., Hu, J., Guo, J.: Gabor-Eigen-Whiten-Cosine: A Robust Scheme for Face Recognition. In: Zhao, W., Gong, S., Tang, X. (eds.) AMFG 2005. LNCS, vol. 3723, p. 336. Springer, Heidelberg (2005)
13. Ojala, T., Pietikäinen, M., Mäenpää, T.: Multiresolution Gray-Scale and Rotation Invariant Texture Classification with Local Binary Patterns. IEEE Transactions on Pattern Analysis and Machine Intelligence, 971–987 (2002)
14. Etcoff, N., Masoiu, N., Oliyide, O., Pentland, A., Starner, T., Turk, M.: Experiments with eigenfaces. In: Proc. Looking at People Workshop Int'l. Joint Conf. Artifical Intelligence (1993)
15. Sung, K.K., Poggio, T.: Example-based learning for view-based human face detection. IEEE Transactions on Pattern Analysis and Machine Intelligence 20(1), 39–51 (1998)
16. Moghaddam, B., Pentland, A.: Probabilistic visual learning for object representation. IEEE Transactions on Pattern Analysis and Machine Intelligence 19(7), 696–710 (1997)
17. Liu, C.: Gabor-Based Kernel PCA with Fractional Power Polynomial Models for Face Recognition. IEEE Transactions on Pattern Analysis and Machine Intelligence, 572–581 (2004)
18. Zhang, B., Shan, S., Chen, X., Gao, W.: Histogram of Gabor Phase Patterns (HGPP): A Novel Object Representation Approach for Face Recognition. IEEE Transactions on Image Processing 16(1), 57 (2007)
19. Bai, L., Shen, L.: InfoBoost for Selecting Discriminative Gabor Features. In: Gagalowicz, A., Philips, W. (eds.) CAIP 2005. LNCS, vol. 3691, pp. 423–432. Springer, Heidelberg (2005)
20. Shen, L., Bai, L.: MutualBoost learning for selecting Gabor features for face recognition. Pattern Recognition Letters 27(15), 1758–1767 (2006)
21. Shan, S., Yang, P., Chen, X., Gao, W.: AdaBoost Gabor Fisher Classifier for Face Recognition. In: Zhao, W., Gong, S., Tang, X. (eds.) AMFG 2005. LNCS, vol. 3723, pp. 279–292. Springer, Heidelberg (2005)
22. Phillips, P.J., Wechsler, H., Huang, J., Rauss, P.J.: The FERET database and evaluation procedure for face-recognition algorithms. Image and Vision Computing 16(5), 295–306 (1998)
23. Ahonen, T., Hadid, A., Pietikainen, M.: Face Recognition with Local Binary Patterns. LNCS, pp. 469–481. Springer, Heidelberg (2004)
24. Shan, S., Zhang, W., Su, Y., Chen, X., Gao, W., FRJDL, I.C.T.I., CAS, B.: Ensemble of Piecewise FDA Based on Spatial Histograms of Local (Gabor) Binary Patterns for Face Recognition. In: Proceedings of the 18th international conference on pattern recognition, pp. 606–609 (2006)
25. Phillips, P.J., Moon, H., Rizvi, S.A., Rauss, P.J.: The FERET Evaluation Methodology for Face-Recognition Algorithms. IEEE Transactions on Pattern Analysis and Machine Intelligence, 1090–1104 (2000)

3D Face Recognition Using Joint Differential Invariants

Marinella Cadoni[1], Manuele Bicego[1,2], and Enrico Grosso[1]

[1] Computer Vision Laboratory, DEIR, University of Sassari, Italy
[2] Computer Science Dept., University of Verona, Italy
{maricadoni,bicego,grosso}@uniss.it

Abstract. Stemming from a sound mathematical framework dating back to the beginning of the 20th century, this paper introduces a novel approach for 3D face recognition. The proposed technique is based on joint differential invariants, projecting a 3D shape in a 9-dimensional space where the effect of rotation and translation is removed. As a consequence, the matching between two different 3D samples can be directly performed in the invariant space. Thus the matching score can be effectively used to detect surfaces or parts of surfaces characterised by similar when not identical 3D structure. The paper details an efficient procedure for the generation of the invariant signature in the 9-dimensional space, carefully discussing a number of significant implications related to the application of the mathematical framework to the discrete, non-rigid case of interest. Experimental evaluation of the proposed approach is performed over the widely known 3D_RMA database, comparing results to the well established Iterative Closest Point (ICP)-based matching approach.

Keywords: Biometrics, 3D Face Recognition, Invariants.

1 Introduction

The typical approach in the automatic analysis of faces context consists in analysing the intensity image of a face – the so called 2D face. Nevertheless, in uncontrolled environments, illumination and pose variations may severely impair recognition systems based on sole 2D information. To overcome these problems, different alternatives have been proposed in recent years, based on 3D information. The approaches presented in the literature – see [1] for an comprehensive review – can be broadly divided into two classes: the first contains methods that perform recognition on the sole basis of shape, while the second comprises the so-called mutimodal algorithms which use both 3D and 2D data (namely the texture of the 3D shape) [2]. Here we concentrate on the former class: in this context, different are the features that may be used, ranging from the raw clouds of points to the curvatures or the face profiles, ending with the well known depth (range) images. In this paper we propose a novel approach for 3D face recognition, solely based on the analysis of the shape of a face represented as a raw cloud of points. The method presented here follows the typical scheme for point

M. Tistarelli and M.S. Nixon (Eds.): ICB 2009, LNCS 5558, pp. 279–288, 2009.
© Springer-Verlag Berlin Heidelberg 2009

cloud-based systems [3]: given two clouds of points representing two faces, the matching score is obtained by a registration process where the two clouds are aligned, and the matching score is given by the registration error. In this context, the widely employed registration technique is the general Iterative Closest Point (ICP) method [4], which aligns two shapes by iteratively minimizing the pair-wise distances between corresponding points. Even if some interesting and smart extensions have been proposed (e.g. to deal with non rigid variations [5,6,7]), the registration obtained with ICP is prone to be erroneous and time consuming. In fact the registration is obtained with an optimization process, which, starting from an initial alignment of the two shapes, iteratively minimises a closeness measure. Such process is a local optimiser, which converges to the nearest local minimum. It is therefore evident that a coarse pre-alignment is crucial to ensure a proper registration. Moreover, it should be noted that the registration time increases with the number of iterations, which could make the method unsuitable for real time recognition purposes. The face recognition approach presented in this paper aims to overcome these drawbacks, by proposing to perform the registration using joint differential invariants [13]. This sound mathematical framework provides an optimal way to project all points belonging to the two faces in a invariant space where the effects of rotation and translation are removed. The matching is then performed in such space. The resulting registration process is not based on an optimization procedure but directly derives from the invariant space. Moreover, no pre-alignment is required, since the effect of rigid variations is by definition removed. Finally, it should be emphasised that even in case of poor overlapping of the two faces (30% or less), the method has the potential to find the correct registration. It is important to note that such invariants are defined on smooth manifolds, assuming continuous surfaces. Clearly, the acquisition process samples the surface, and a too heavy sub-sampling may reduce the accuracy of the registration. Although this registration process reaches perfection in the ideal rigid-object noise-free case, in this paper we will show that it can be fruitfully exploited in the face case, where non rigid variations and noise may be present. In order to reduce the computational requirements, the matching is performed on specific points, detected around maxima of curvature. Preliminary experiments made on the well known 3D_RMA dataset [9] show promising results, also with respect to the standard ICP.

2 Joint Invariants for Surface Classification

In this section the theory that leads to Joint Invariants is presented, starting with an outline of the Moving Frames Theory of Olver [10], following with the description of the invariant signature adopted here and ending by detailing its application to the point cloud case.

2.1 Outline of the Moving Frame Theory

The classical differential invariant theory for surfaces embedded in the Euclidean space leads to invariants of second order derivative, the well known Gaussian and Mean curvatures (see [14]). These invariants, together with their 1st order derivatives with respect to the Frenet frame, parameterise a signature manifold that completely determines the surface up to Euclidean transformations (see [10]). This implies that two surfaces are the same (up to an Euclidean motion) if and only if their signature manifolds are the same and thus gives us a way to characterise surfaces up to Euclidean motions. In practice, a given a surface and a copy of it obtained by simply rototranslating the original one, share the same invariants signature. However, if we apply this methodology to experimental cases, where the surface is approximated by a sampling of points, any noise in the points will be amplified by the second order derivatives, making comparison of signatures difficult.

The theory developed by Olver in [10] is based on the classical theory of Moving Frames first introduced by Cartan [12] and provides us with an algorithm for building functionally independent sets of invariants that characterise a surface up to Lie Group transformations (which include Euclidean motions). By prolonging the group action to various cartesian copies of the surface, we can build invariants that depend on smaller order derivatives, the so called joint differential invariants. In particular, if taking one copy of the surface leads to the classical Gaussian and Mean curvature at each point of the surface, by considering three copies of the surface the invariants will depend on three points, and will consist of the three distance between the points and 6 first order invariants. If we prolong the action to enough copies (7 in the case of surfaces), we will find 21 invariants that depend only on the inter-point distances of 7 points of the surface at a time. As in the case of the classical differential invariants, joint invariants parameterise a signature that characterises the original surface up to the considered transformations.

The theory of joint differential invariants therefore gives a very elegant and powerful way of constructing a minimal set of invariants that are necessary to define a signature of the surface. The signature will live in a space of dimension equal to the number of invariants that parameterise it whereas its dimension will depend on the number of points the invariants are defined on (i.e. on the prolongation). This means that, through prolongation, we can have a representation of the original surface that is invariant to transformations and dependent on low order derivatives, at the expense of high dimensionality and computational complexity. In practical terms, a zero order signature is parameterised by the 21 inter-point distances between all ordered subsets of 7 points of the surface. If the surface is a point cloud consisting of n points, then we would have n^7 subsets each of which generates 21 invariants: if n is large, both the generation of the signature and any further processing become computationally challenging.

2.2 Invariants and Signature Generation

To compromise between computational time and robustness we choose a 3-fold prolongation, so that our invariants will depend on three points at one time. As we will see, this choice leads to three invariants of order zero plus six of order one. Let p_1, p_2 and p_3 be three points on the surface. If the surface is smooth we can define the normal vector n_i at each point p_i.

Furthermore, (see figure 2(a)) let r be the direction of the line between the first two points and n_t the normal to the plane through the 3 points:

$$r = \frac{p_2 - p_1}{\|p_2 - p_1\|} \quad \text{and} \quad n_t = \frac{(p_2 - p_1) \wedge (p_3 - p_1)}{\|(p_2 - p_1) \wedge (p_3 - p_1)\|}.$$

The zero order invariants we find are the 3 interpoint distances $I_k(p_1, p_2, p_3)$ for $k = 1, 2, 3$:

$$I_1 = \|p_2 - p_1\|, \; I_2 = \|p_3 - p_2\| \quad \text{and} \quad I_3 = \|p_3 - p_1\|$$

The first order invariants are the following:

$$J_k(p_1, p_2, p_3) = \frac{(n_t \wedge r) \cdot n_k}{n_t \cdot n_k} \quad \text{for } k = 1, 2, 3$$

and

$$\tilde{J}_k(p_1, p_2, p_3) = \frac{r \cdot n_k}{n_t \cdot n_k} \quad \text{for } k = 1, 2, 3.$$

To each triplet (p_1, p_2, p_3) on the surface we can then associate a point of the signature given by $(I_1, I_2, I_3, J_1, J_2, J_3, \tilde{J}_1, \tilde{J}_2, \tilde{J}_3)$. As the invariants depend on three points, each of which has two degrees of freedom on the surface, the signature will be a 6-dimensional manifold embedded in 9-dimensional space.

2.3 Point Cloud Implementation

Our aim is to adapt the general framework for constructing an optimal set of invariants to our case of interest: the registration of (possibly partially overlapping) clouds of points obtained by sampling surfaces. Let $\mathcal{F}_1 = \{p_1, \ldots, p_n\}$ and $\mathcal{F}_2 = \{q_1, \ldots, q_m\}$ be two clouds of points sampled from two faces. For $i = 1, 2$ we consider all unordered triplets of points in the sets $\mathcal{F}_i$ and calculate the invariants described in 2.2. This will produce two discrete sets of points, $\mathcal{S}_1$ and $\mathcal{S}_2$, that would lie on the signature manifold theoretically generated by the continuous surfaces of the faces, i.e. two sub-samplings of the signature manifolds. The next step is the comparison of the two signatures.

From the general theory it follows that if two signatures partially overlap (i.e. they intersect in a subset whose dimension equals their dimension), then also the surfaces that generated them will have the same property (they will overlap) after an Euclidean motion. In order to establish the intersection of $\mathcal{S}_1$ and $\mathcal{S}_2$, we need to define a metric in the embedding space R^9.

After normalising the values of the 9 invariants, we found the Euclidean metric was sufficient for reliably establishing matching points. Using a kd-tree algorithm [15], the search of the closest point $q_i \in \mathcal{S}_2$ for each point $p_i \in \mathcal{S}_1$ can be performed efficiently. Let $d_\star$ be a fixed threshold and M the set of pairs (p_i, q_i) that satisfy the inequality $\|p_i - q_i\| \leq d_\star$.

If we denote by $|M|$ the cardinality of M, the signatures intersection is defined to be $S_I = |M|/\min\{|S_1|, |S_2|\}$.

3 A Joint-Invariants Based Face Recognition Algorithm

In view of the theory outlined in 2, we could readily implement an algorithm for face recognition. Following the standard approach to point cloud-based 3D face recognition [3], given two faces to be compared, the idea is to register them and use the registration error as a matching score.

In details, suppose we have a training set G containing face scans of various subjects. Each face scan can be represented by its 6-dimensional signature embedded in E^9, which, as we have seen, characterises it up to Euclidean motion. When a test scan comes along and we want to compare it with the training scans we can build its signature and compare it to all signatures in the training set by evaluating the intersections S_I's. The unknown testing scan is then assigned to the subject with the most similar signature (highest value of the S_I's). The algorithm may be easily extended to the authentication scenario, where a testing face is authenticated if the template's signature is similar enough (given a threshold).

Using all points of the scans and invariants of maximum differential order equal to one, the matching would be robust and simple. Indeed, even considering facial expressions, it is reasonable to assume that there are enough stable points to distinguish an individual from another (since the method works also in case of partial overlapping) and no special metric is necessary to compare the signatures (see 2.3).

All this makes the framework very appealing. Unfortunately, computational complexity prevents us to readily apply it in this "full" form: the average cardinality of a face scan $\mathcal{F}$ in $3D$ databases can be beyond computationally capability. In the database we experimented on it is 4×10^3. Considering all these points would result in $4^3 \times 10^3$ signature points. This is computationally unfeasible, not to mention the processing time of such a bulk of data. To overcome this problem, we need a way to sub-sample the face scans. Since the invariants depend on distances and normals, we must ensure that, in common patches of different scans of the same subject, we sub-sample the same points.

3.1 Sub-sampling

Given a scan $\mathcal{F}$, we calculate an approximation of the mean curvature at each point p in this way. We fix a neighborhood radius r and consider all points within the sphere of radius r and centre p. We perform PCA analysis on the neighborhood points to find the principal directions and so the curvature value and the normal vector at the central point. Using the same neighborhood radius, we select the points of local maximum, and amongst them we take the 14 points of highest curvature (in norm). We found that this number of points was optimal to extract stable features like the nose tip and the eyes. In fact, to guarantee the presence of stable features, for each face we extract the maxima of curvature at different scales, by repeating the previous process 4 times using increasing neighborhood radii. As a result we have 4 sets of characteristic points on which to base the registration of the faces.

3.2 Signature Matching and Validation

For each set of characteristic points we consider all triplets, then for each triplet we determine the 9 invariants as in 2.2 and finally we collect them in a signature. At the end of the process we will have 4 signatures sets for each face. Given two faces $\mathcal{F}_1$ and $\mathcal{F}_2$, for each scale s_i, $i = 1, \ldots, 4$, we compare their signatures as in 2.3. For each s_i, we will have S_{Ii} matching pairs of signature points. As a final score we take the maximum of the S_{Ii}. In theory, this score should be enough for recognition purposes. In practice, however, the signatures are generated starting from a maximum of 14 points on the face scan, which leads to a heavy sub-sampling of the signature and so to the loss of certainty of properties that in the general theory follow from continuity and smoothness. Specifically, with no clues about proximity of matching pairs in the signature space, it might well happen that quite a few number of invariants are close enough to indicate a match, but this might just be the result of multiple transversal intersections of the two signatures. If our measure of signatures similarity solely consists on counting the number of matches over the total number of signature points we might be fooled by transversal matches. To prevent this, we need to validate the matches. Fortunately this almost comes for free: a match in the invariant space corresponds to a pair of matching triplets $\mathcal{T}_1 \subset \mathcal{F}_1$, $\mathcal{T}_2 \subset \mathcal{F}_2$. We find the roto-translation that takes $\mathcal{T}_2$ onto $\mathcal{T}_1$ and we apply it to all point of $\mathcal{F}_2$. To validate the match, we measure the "closeness" of the transformed scan to the other in this way. We start by looking for each point $q_i \in \mathcal{F}_2$ the closest point p_i in $\mathcal{F}_1$ and we save the Euclidean distance $d_i = \|q_i - p_i\|$ between the two. We get a set of distances $D = \{d_i\}_{i \in I}$ where $I = |F_2|$. However, since the acquisition viewpoint may change, even in the case of an accurate registration, some points might belong to only one of the scans and their distances from the closest points in the other scan could be relatively large. We therefore experimented two metrics. The first used a point to normal distance and is an approximation of volumetric distance: for each point $q_i \in \mathcal{F}_2$ we consider the normal line through it; if there is a point $p_i \in \mathcal{F}_1$ close enough to the line (where by close enough we mean comparable to the acquisition resolution), then we save their Euclidean distance d_i, otherwise q_i is ignored. In this way we remove outliers and we can evaluate the "closeness" of the transformed scan as the mean or median of distances (of remaining points). This measure proved to be reliable but time consuming for a practical experiment on a large number of scans. Also there are issues about the threshold on the point to normal distance since the face sampling might not be uniform (and it certainly is not if the cloud is the output of a single scan). A quicker but still reasonably robust alternative estimator of "closeness" of scans proved to be the median of distances computed over all the points of the clouds. This second metric has been used in extensive experiments.

4 Experimental Evaluation

We chose to test the proposed framework on the 3D_RMA database [9] which, despite being generally noisy, comprises scans of subjects taken from different viewpoints and with varying facial expressions. The database contains a total of 106 subjects whose faces where scanned using a structured light acquisition system during two different acquisition campaigns, each of 3 scans. In total there are 617 scans each of which contains on average 4×10^3 points. We divided the dataset into a training set $G = \{S_1, \ldots, S_{106}\}$, where each S_i consists of the first 3 scans of subject i, and a test set $P = \{V_1, \ldots, V_{106}\}$, where each V_i contains the remaining scans of subject i acquired in the second campaign. For all scans in the training set, we extracted the points of maximum curvature without any pre-processing on the points and calculated the signatures at 4 different scales (see 2.2). Then, for $j = 1, \ldots, 106$, we considered all test and train subjects subsets pairs $G_j = \{S_1, \ldots, S_j\}$, $P_j = \{T_1, \ldots, T_j\}$ and proceeded as follows: one test scan at a time from P_j was compared to all scans in G_j. The training scan that achieves maximum score after signatures matching, registration and validation through the median distance (see 3.2) has been taken as a match. If the matching test and train scans belong to the same subject the match is validated.

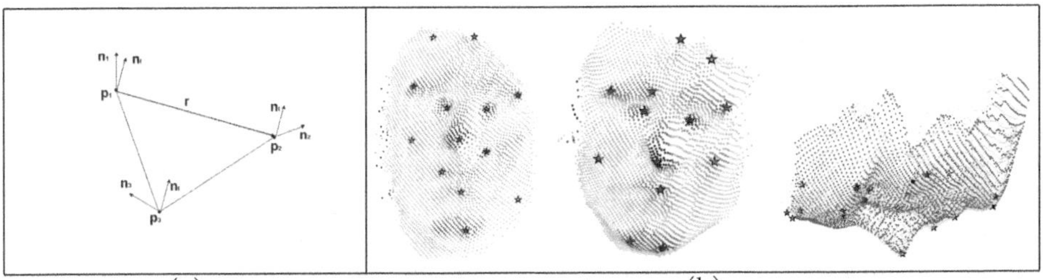

(a) (b)

Fig. 1. (a): Triplet with associated normals. (b): Characteristic points extraction and scans registration.

The first two images in figure 1(b) show the curvature variation in two scans of the same subject. The extracted points of local maximum absolute curvature values are represented by the stars. The third image is the result of the registration of the two scans; the profile view emphasises the accuracy of the registration.

To compare our results we set up the same experiment framework using Besl and McKay [4] original version of the ICP algorithm for registration and matching. This version of the ICP algorithm minimises point to point Euclidean distances (under a fixed threshold) through successive iterations.

To prevent ICP from converging to a local minima, a pre-registration has been provided by manual selection of nose tips. The final matching score is taken to be the mean or the median distance between corresponding points after the last iteration. The results of all tests are illustrated in figure 2(a). The figure should be read in this way: for each horizontal coordinate $j = 10, 11, \ldots, 106$, the vertical one expresses the accuracy (normalized to 1) of the recognition test

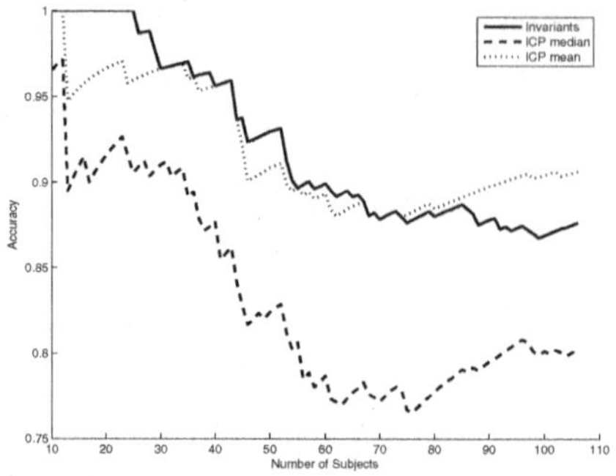

(a)

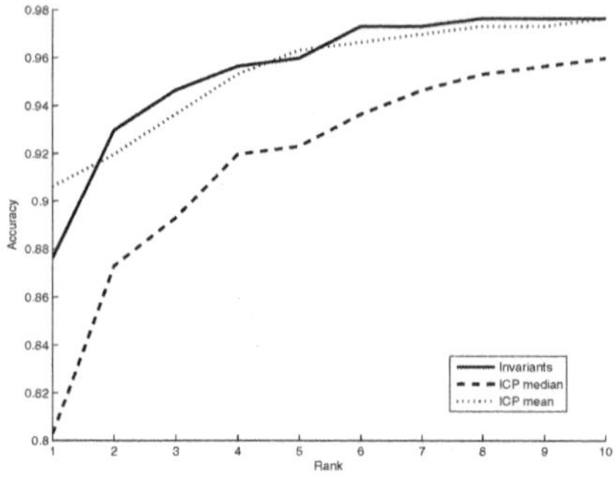

(b)

Fig. 2. (a) Matching scores of the different techniques as a function of database cardinality. (b) Cumulative match characteristic curves for Invariants and ICP methods.

carried out on the pair of subsets $G_j = \{S_1, \ldots, S_j\}$ and $P_j = \{T_1, \ldots, T_j\}$, namely the data subset obtained by considering the first j subjects.

The invariants method clearly outperforms the ICP (both with median and mean final error measurements) for $10 \leq j < 30$. At $j = 30$ the accuracy of both methods is 96.7%. For $j > 30$ the invariant method performance gracefully degrades to the advantage of ICP. At $j = 106$ the Invariant method score is 87.7% against the ICP score of 90.6%. The reason for this is due to the nature of the 3D_RMA database: in the first of the two acquisition campaigns and for the first 30 subjects only, the matching of corresponding points detected by the projected light pattern was manually aided, leading to low noise point coordinates. From subject 31 onwards, the correspondences were found automatically [9]. Since the extraction of curvatures suffers from noisy input data, the registration through invariants might lose in accuracy, leading to wrong matches. In fact, when we performed the recognition test on the test subset $P_{30} = \{T_1, \ldots, T_{30}\}$ paired with the whole train set $G = \{S_1, \ldots, S_{106}\}$, the matching score was exactly the same for both methods, indicating that the presence of impostors $\{S_{31}, \ldots S_{106}\}$ does not affect the result. This did not hold true for $j > 30$.

In figure 2(b), we can see the Cumulative Match Characteristic (CMC) curves for the Invariant and ICP methods applied to the whole database (on all 106 subjects). It represents the classification accuracy of the three methods at different ranks. Rank k means that a correct classification is assigned if the correct label is found within the first k best matches. From the figure we can observe that ICP performs better at rank 1, whereas from rank 2 onwards the Invariant method is almost always ahead.

The performance of ICP when the median is used to measure the final registration error instead of the mean (see figures 2 is also worthy of remark: we expected similar results with the use of the median, but surprisingly we had a substantial performance drop. Since the median and the mean are evaluated on the same set of matching closest points, the only way to explain the difference is that the distances were not normally distributed within the interval limited by zero and the ICP threshold value.

In the joint invariants test, to validate a registration, we did not set a threshold to discard closest points whose distances were quite big compared to acquisition resolution. This is because our purpose was not only to obtain a measure of closeness of the registered scans, but also to validate the registration, and so we would have had to set two thresholds, one on the distance and the other on the number of closest points under that distance. To avoid setting two thresholds we used the median on the set of distances $D = \{d_i\}_{i \in I}$ as explained in 3.2. In the light of the ICP results, however, we could improve the matching score by using a different metric to validate the registration. This is supported by some experiments carried out using the volumetric (point to normal) distance defined in 3.2, in which we have improvements on false subjects matching. Alas, due to time limits we could not get results on the whole database thus far.

The Joint invariant and ICP algorithms together with the tests were all implemented in MatLab and carried out on AMD Opteron of 2 GB of RAM and 2.5 GHz CPU speed. The joint invariant method took on average 2 seconds to extract the maximum curvature points, generate the signatures, and validate the matches of a pair of scans. In order to speed-up the procedure, the extraction of the maximum curvature points and the generation the signatures of the training set can been performed off line. This took about 45 minutes in total, and reduced the average pairwise matching time to 1.5 seconds. The ICP proved to be slower due to the iteration process, and took on average 4 seconds per pairwise matching.

5 Conclusions and Future Work

The main original contribution of this paper is the introduction of a novel approach for 3D face recognition, based on the sound mathematical framework of Moving Frames. In this context, a single signature for a cloud of 3D points is generated using joint differential invariants. Even if this concept of signature is defined for continuous surfaces, here it has been adapted to the considered discrete set of points, projecting a 3D shape in a 9-dimensional space where the effect of rotation and translation is removed.

In the paper we have presented an efficient procedure for the generation of an invariant signature in 9-dimensional space, suitable to be employed for registration-based matching. Experimental evaluation over the 3D_RMA database showed that the proposed method performance is in line with the well known ICP-based matching approach and outperforms it in the case of low noise input data. It should be noted that, contrary to the ICP, the proposed method does not require pre-registration and can work in case of limited overlap between surface scans.

Performance improvements, in terms of computational speed and robustness, are foreseeable with a more robust extraction of the points of maximum curvature. This is especially true in the case of noisy input data. Also, it is reasonable to think that pre-processing the data, e.g. cropping it to remove spikes due to hair or acquisition artefacts, would positively affect the results. Additional work will also be devoted to the implementation of a more sophisticated metric to validate the registration; in particular, an implementation of the volumetric distance described in 3.1 will be evaluated against the whole database. Tests will be run on other databases, in order to further evaluate the performance of the method in the presence of noise, facial expressions etc.

Finally, the ability to automatically capture and segment rigid parts of a face is expected to be a main outcome of this research effort in the short time.

References

1. Bowyer, K., Chang, K., Flynn, P.: A survey of approaches and challenges in 3d and multi-modal 2D+3D face recognition. Computer Vision and Image Understanding 101(1), 1–15 (2006)
2. Chang, K., Bowyer, K., Flynn, P.: An evaluation of multimodal 2d+3d face biometrics. IEEE Trans. Pattern Anal. Mach. Intell. 27(4), 619–624 (2005)
3. Gokberk, B., Akarun, L.: Comparative analysis of decision-level fusion algorithms for 3d face recognition. In: Proc. of Int. Conf. on Pattern Recognition, pp. 1018–1021 (2006)
4. Besl, P., McKay, N.: A method for registration of 3D shapes. IEEE Trans. on Pattern Analysis and Machine Intelligence 14, 239–256 (1992)
5. Chang, K., Bowyer, K., Flynn, P.: Adaptive rigid miltiregion selection for handling expression variations in 3D face recognition. In: IEEE Workshop on FRGC (2005)
6. Irfanoglu, M., Gokberk, B., Akarun, L.: 3D shape-based face recognition using automatically registered facial surfaces. In: Proc. of Int. Conf. on Pattern Recognition, pp. 183–186 (2004)
7. Lu, X., Jain, A.: Deformation analysis for 3d face matching. In: Proc. Int. Workshop on Applications of Computer Vision, pp. 99–104 (2005)
8. Lu, X., Jain, A., Colbry, D.: Matching 2.5D Face Scans to 3D Models. IEEE Trans. on Pattern Analysis and Machine Intelligence 28, 31–43 (2006)
9. Beumier, C., Acheroy, M.: Automatic 3d face authentication. Image and Vision Computing 18(4), 315–321 (2000)
10. Olver, P.J.: Joint Invariants Signatures. Found. Comput. Math. 1, 3–67 (2001)
11. Olver, P.J.: A survey of moving frames. In: Li, H., J. Olver, P., Sommer, G. (eds.) IWMM-GIAE 2004. LNCS, vol. 3519, pp. 105–138. Springer, Heidelberg (2005)
12. Cartan, É.: La Méthode du Repère Mobile, la Théorie des Groups Continue, et les Espaces Géneralisés. Hermann, Paris, France (1935)
13. Cadoni, M.I., Chimienti, A., Nerino, R.: Automatic Coarse Registration by Invariant Features. In: The 7th International Symposium on Virtual Reality, Archaeology and Cultural Heritage VAST (2006)
14. Do Carmo, M.: Differential Geometry of Curves and Surfaces. Prentice-Hall, Englewood Cliffs (1976)
15. de Berg, M., van Kreveld, M., Overmars, M., Schwarzkopf, O.: Computational Geometry, 2nd revised edn., pp. 99–105. Springer, Heidelberg (2000)

A Model Based Approach for Expressions Invariant Face Recognition

Zahid Riaz, Christoph Mayer, Matthias Wimmer,
Michael Beetz, and Bernd Radig

Department of Informatics, Technische Universität München
85748 Garching, Germany
{riaz,mayerc,matthias.wimmer,beetz,radig}@in.tum.de

Abstract. This paper describes an idea of recognizing the human face in the presence of strong facial expressions using model based approach. The features extracted for the face image sequences can be efficiently used for face recognition. The approach follows in 1) modeling an active appearance model (AAM) parameters for the face image, 2) using optical flow based temporal features for facial expression variations estimation, 3) and finally applying classifier for face recognition. The novelty lies not only in generation of appearance models which is obtained by fitting active shape model (ASM) to the face image using objective functions but also using a feature vector which is the combination of shape, texture and temporal parameters that is robust against facial expression variations. Experiments have been performed on Cohn-Kanade facial expression database using 62 subjects of the database with image sequences consisting of more than 4000 images. This achieved successful face recognition rate up to 91.17% using binary decision tree (BDT), 98.6% using Bayesian Networks (BN) with 10-fold cross validation in the presence of six different facial expressions.

Keywords: Active Appearance Models, Face Recognition, Facial Expressions Recognition, Binary Decision Trees, Bayesian Classifier.

1 Introduction

Since last three decades of face recognition technology there exists many commercially available systems to identify human faces, however face recognition is still an outstanding challenge against different kinds of variations like facial expressions, poses, non-uniform light illuminations, occlusions and aging effects. Meanwhile this technology has extended its role to Human-Computer-Interaction (HCI) and Human-Robot-Interaction (HRI). Person identity is one of the key tasks while interacting with the robots, exploiting the oblivious system security and authentication of the human interacting with the system. In such kinds of scenarios the acquired face images could contain various types of facial expressions along with above mentioned effects. This problem has been addressed in this paper resulting in a face recognition system which is robust against facial expressions. In [1, 2] authors give comprehensive survey of face recognition and some commercially available face recognition software.

M. Tistarelli and M.S. Nixon (Eds.): ICB 2009, LNCS 5558, pp. 289–298, 2009.
© Springer-Verlag Berlin Heidelberg 2009

As cameras are widely used and mounted on computer screens, embedded in mobiles and installed into everyday living and working environments, they have become valuable tools for human system interaction. A particular important aspect of this interaction is detection and recognition of faces and interpretation of facial expressions. These capabilities are deeply rooted in the human visual system and a crucial building block for social interaction. Consequently, these capabilities are an important step towards the acceptance of many technical systems. Although faces are the most important and natural way for human-human interaction but some outstanding challenges like uniqueness and accuracy made its market value a bit less than the other biometrics in year 2003. Later in 2006 Face recognition technology again raised up to 19% of the biometric market [23].

This publication focuses on one of the aspects of natural human-computer interfaces: our goal is to build a real-time system for face recognition that could robustly run in real-world environments. We develop it using model-based image interpretation techniques, which have proven its great potential to fulfill current and future requests on real-world image understanding. Our approach comprises methods that robustly localize facial features, seamlessly track them through image sequences, and finally infer the face identity.

2 Related Work

Traditional recognition systems have the abilities to recognize the human using various techniques like feature based recognition, face geometry based recognition, classifier design and model based methods. Linear subspace methods like Principal Components Analysis (PCA) was firstly used by Sirvovich and Kirby [3], which were latterly adopted by M. Turk and A. Pentland introducing the famous idea of eigenfaces [4,5]. This paper focuses on the modeling of human face using a two dimensional approach of shape, texture and temporal information and then utilizing this model for recognition purposes. This type of model using shape and texture parameters is called Active Appearance Models (AAMs), introduced by Edwards and Cootes [6]. In [7] authors have used weighted distance classifier called Mahalanobis distance measure. However, Edwards et al [8] isolated the sources of variation by maximizing the interclass variations using Linear Discriminant Analysis (LDA), the technique which was holistically used for Fisherfaces representation [9] which is similar to the eigenface approach resulting in out performance of previous approach. In [10] authors have utilized shape and temporal features collectively to form a feature vector for facial expressions recognition. Lanitis et al [11] used separate information for shape and gray level texture. These models utilize the shape information based on a point distribution of various landmarks points marked on the face image. In our approach a predefined shape model consisting of 134 points in two dimensional space, as was used by Wimmer et al [12], has been utilized.

3 Our Approach

This paper utilizes shape model fitting, extracting shape information, texture mapping for texture parameterization and estimating optical flow-based parameters for feature vector formation. The feature vector consists of the shape, texture and temporal variations, sufficient for considering local variations in shapes. All the subjects in the database are labeled for identification. A fitted face model, on the training images is then used for defining the reference shape in our experiments. This reference shape is calculated by finding the mean shape of the all shapes in the database.

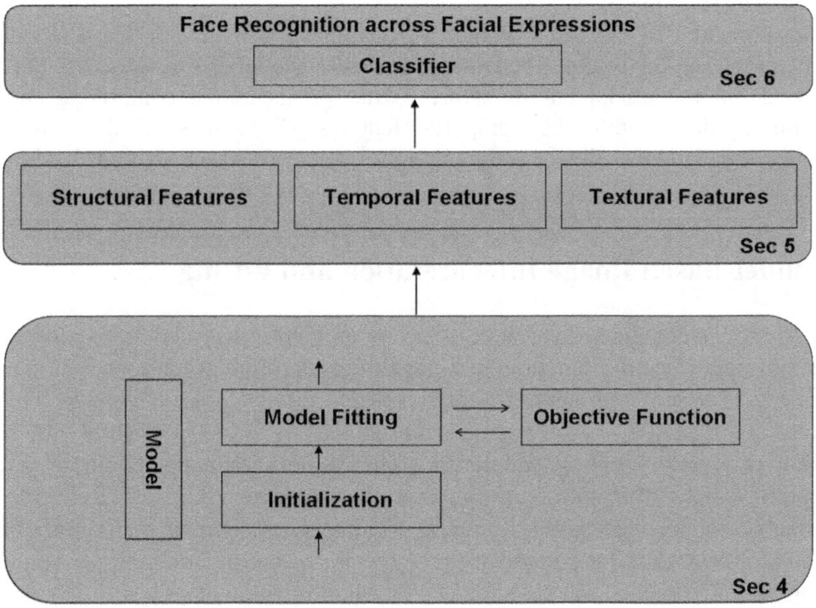

Fig. 1. Model-based techniques split the challenge of image interpretation into computationally independent modules. The lower right corners refer to the sections with detailed explanation.

In this paper an explicit 2D appearance model is used to develop a baseline for feature extraction. A point distribution model (PDM) is used as an active shape model. This model comprises of 134 points that prominently defines the location of local face features like eyes, nose and lips in 2D space. Face is localized in the image. An objective function is learned for fitting this model to the faces. After fitting the model to the example face image, texture information is extracted from the example image on a reference shape which is the mean shape of all the shapes of database. Image texture is extracted using planar subdivisions of the reference and the example shapes. Texture warping between the subdivisions is performed using affine transformation. This image texture is now normalized both in the sense of shape and varying illuminations effects, making the image robust for shape and illumination. Principal Component Analysis (PCA) is used to obtain the texture and shape parameters of the example image. This approach is similar to extracting Active Appearance Model (AAM) parameters. In addition to AAM parameters, temporal features of the facial changes are also calculated. Local motion of the feature points is observed using optical flow. We use reduced descriptors by trading off between accuracy and runtime performance. These features are then used for classifiers for face recognition. A detailed process flow of our approach is shown in Figure 1. Our approach achieves real-time performance and provides robustness against facial expressions for real-world applicability. This computer vision task comprises of various phases for which it exploits model-based techniques that accurately localize facial features, seamlessly track them through image sequences, and finally infer facial expressions visible. We specifically adapt state-of-the-art techniques to each of these challenging phases.

The remainder of this paper is divided in four sections. In section 4 model based image interpretation is described along with the model fitting process. Sections 5 discuss about the model based feature extraction technique comprising shape and appearance along with the temporal features. Section 6 deals with feature classification. Finally, the experimentation is performed in section 7 along with conclusions.

4 Model-Based Image Interpretation and Fitting

Model-based techniques consist of four components: the model, the initialization algorithm, the objective function, and the fitting algorithm. Our approach makes use of a statistics-based deformable model, as introduced by Cootes et al. [13]. The model contains a parameter vector *p* that represents its possible configurations, such as position, orientation, scaling, and deformation. Models are mapped onto the surface of an image via a set of feature points, a contour, a textured region, etc. Referring to [6], deformable models are highly suitable for analyzing human faces with all their individual variations. Its parameters $p = (tx, ty, s, \theta, b)^T$ comprise the translation, scaling factor, rotation, and a vector of deformation parameters $b = (b_{s,1}, \ldots, b_{s,m})^T$. The latter component describes the configuration of the face, such as the opening of the mouth, roundness of the eyes, raising of the eyebrows.

The initialization algorithm automatically starts the interpretation process by roughly localizing the object to interpret. It computes an initial estimate of the model parameters that needs to be further refined by the subsequent fitting algorithm. Our system integrates the approach of Viola and Jones, which is able to detect the affine transformation parameters (tx, ty, s, and θ) of frontal faces.

In order to obtain higher accuracy, we apply a second iteration of the Viola and Jones object detector to the previously determined image region of interest i.e. the face. This extension allows to roughly estimate the deformation parameters b_s as well, because we learn the algorithm to localize facial components, such as eyes and mouth. In the case of the eyes, our positive training examples contain the images of eyes, whereas the negative examples consist of image patches in the vicinity of the eyes, such as the cheek, the nose, or the brows. Note that the resulting eye detector is not able to robustly localize the eyes in a complex image, because it usually contains a lot of information that was not part of the training data. However, it is highly appropriate to determine the location of the eyes within a pure face image or within the face region of a complex image.

The objective function f(I, p) yields a comparable value that specifies how accurately a parameterized model *p* describes an image I. It is also known as the likelihood, similarity, energy, cost, goodness, or quality function. Without losing generality, we consider lower values to denote a better model fit. Traditionally, objective functions are manually specified by first selecting a small number of simple image features, such as edges or corners, and then formulating mathematical calculation rules. Afterwards, the appropriateness is subjectively determined by inspecting the result on example images and example model parameterizations. If the result is not satisfactory the function is tuned or redesigned from scratch. This heuristic approach relies on the designer's intuition about a good measure of fitness. Earlier works [12, 14] show that this methodology is erroneous and tedious. This traditional approach is depicted to the top in Figure 2.

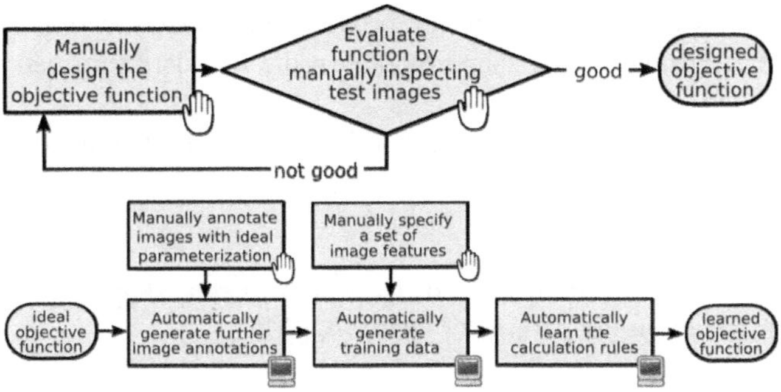

Fig. 2. The traditional procedure for designing objective functions (top), and the proposed method for learning objective functions (bottom)

To avoid these drawbacks, we recently proposed an approach that learns the objective function from annotated example images [14]. It splits up the generation of the objective function into several tasks partly automated. This provides several benefits: firstly, automated steps replace the labor-intensive design of the objective function. Secondly, this approach is less error prone, because giving examples of good fit is much easier than explicitly specifying rules that need to cover all examples. Thirdly, this approach does not rely on expert knowledge and therefore it is generally applicable and not domain-dependent. The bottom line is that this approach yields more robust and accurate objective functions, which greatly facilitate the task of the fitting algorithm.

The fitting algorithm searches for the model that best describes the face visible in the image. Therefore, it aims at finding the model parameters that minimize the objective function. Fitting algorithms have been the subject of intensive research and evaluation, e.g. Simulated Annealing, Genetic Algorithms, Particle Filtering, RANSAC, CONDENSATION, and CCD, see [15] for a recent overview and categorization. We propose to adapt the objective function rather than the fitting algorithm to the specifics of our application. Therefore, we are able to use any of these standard fitting algorithms, the characteristics of which are well-known, such as termination criteria, runtime, and accuracy.

5 Features Extraction

5.1 Active Shape Models

Different kind of shape models have been introduced by researchers depending upon the application. Some are landmark based models [16,17,18] defining some fixed points annotated on the images and then defining the boundaries around the objects. However some rely on the contour based approach. Different contours define the shape of the object for outlining it along with covering the feature inside an object [19]. Landmark based models however provide the exact location of the features inside the object. Fitting of this shape model on the face is performed by training an objective function (section 4).

The model is parameterized using PCA to form the shape feature vector.

$$x \approx x_m + P_s b_s$$

Where the shape x is parameterized by using mean shape x_m and matrix of eigenvectors P_s to obtain the parameter vector b_s [20].

5.2 Appearance Model

For the various images of the same person different types of variations are required to be modeled. For example, shape deformations including both facial expression changes and pose variations along with the texture variations caused by illuminations. For this reason, different normalizations are required to be performed at this stage. At first, shape variation is required to be controlled in order to record the texture. This can be achieved by defining a reference shape for a particular object. In our case, this reference image is mean shape, obtained by taking the mean of all the shapes of all persons in our database.

Once the texture is extracted it could be parameterized using PCA as,

$$g = g_m + P_g b_g$$

Where the texture g is parameterized by using mean texture g_m and matrix of eigenvectors P_g to obtain the parameter vector b_g [20].

Figure 3 (bottom-left) shows the mean shape of the subject in consideration. Since the points distribution defines a convex hull of points in space so a planar subdivision is defined for the reference shape to map image texture. Delaunay triangulation is used to divide the shape into a set of different facets. Figure 3 shows the delaunay triangulations of the reference shape.

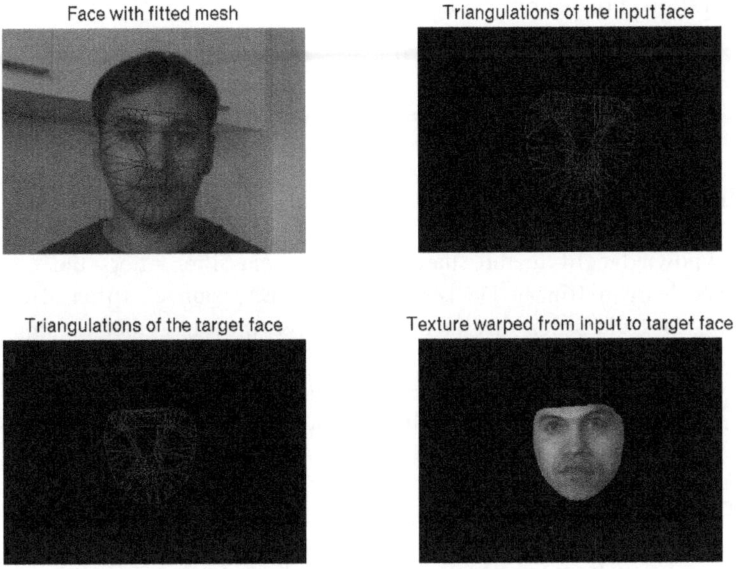

Fig. 3. Shape Subdivisions and Texture Warping

5.3 Temporal Features

Since facial expressions emerge from muscle activity, the motion of particular feature points within the face gives evidence about the facial expression. These features further help the classifier to learn the motion activity. Real-time capability is important, and therefore, a small number of feature points are considered only. The relative location of these points is connected to the structure of the face model. Note that we do not specify these locations manually, because this assumes a good experience of the designer in analyzing facial expressions. In contrast, we automatically generate *G* feature points that are uniformly distributed. We expect these points to move descriptively and predictably in the case of a particular facial expression. We sum up the motion $g_{x,i}$ and $g_{y,i}$ of each point $1 \leq i \leq G$ during a short time period. We set this period to 2 sec to cover slowly expressed emotions as well. The motion of the feature points is normalized by the affine transformation of the entire face (tx, ty, s, and θ) in order to separate the facial motion from the rigid head motion. In order to determine robust descriptors, PCA determines the H most relevant motion patterns (principal components) visible within the set of training sequences. A linear combination of these motion patterns describes each observation approximately correct. This reduces the number of descriptors ($H \leq 2G$) by enforcing robustness towards outliers as well. As a compromise between accuracy and runtime performance, we set the number of feature points to $G = 140$ and the number of motion patterns b_t to $H = 14$ containing.

The overall feature vector then becomes:

$$u = (b_{s,1},.....,b_{s,m},b_{t,1},.....,b_{t,H},b_{g,1},.....,b_{g,n})$$

Where b_s, b_t and b_g are shape, temporal and textural parameters respectively.

6 Feature Classification

With the knowledge of feature the vector u, a classifier infers the correct facial identity. We learn a Binary Decision Tree and Bayesian classifier. However, any other multi-class classifier that is able to derive the class membership from real valued features can be integrated as well, such as a k-Nearest-Neighbour classifier. We take 66% of the image sequences of the CKFE-DB as the training set and the remainder as test set for BDT, the evaluation on which is shown in the next section.

7 Experiments

Experiments have been performed on Cohn-Kanade-Facial-Expression database (CKFE-DB) for human faces. The CKFE-DB contains 488 short image sequences of 97 different persons performing the six universal facial expressions [21]. It provides researchers with a large dataset for experimenting and benchmarking purpose. Each sequence shows a neutral face at the beginning and then develops into the peak expression. Furthermore, a set of action units (AUs) has been manually specified by licensed Facial Expressions Coding System (FACS) [22] experts for each sequence. Note that this database does not contain natural facial expressions, but volunteers were asked to act. Furthermore, the image sequences are taken in a laboratory environment with predefined illumination conditions, solid background and frontal face views. Algorithms that perform well with these image sequences are not immediately appropriate for real-world scenes.

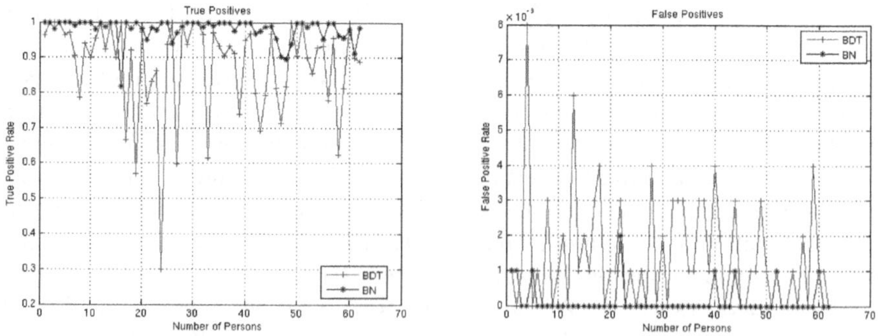

Fig. 4. True Positive and False Positive for 61 persons in experiments

For experimental purposes, image sequences of 62 persons have been used which consists of overall 4060 images. A binary decision tree is trained as classifier in 22.99 sec. We used 1381 images for testing the recognition results and successfully recognized 1259 images. The recognition rate achieved was 91.17% in the presence of facial expressions. Further, same set of the feature vectors are used to build a Bayesian Networks (BN) classifier. The classifier was built in 18.33 sec using 10-fold cross validation. This produces a better accuracy of 98.69% of accurate recognition. Figure 4 shows true positive and true negative for the database for BDT and BN respectively.

8 Conclusions

We introduced an idea to develop a feature vector which consists of three types of facial variations and is robust against the expressional changes in the human faces in real environments. Since the training set consists of the facial expressions information of a person, it can recognize the person even under various expressions. A Bayesian Network (BN) classifier is efficient to train and classify. However the benchmarked database consists of only frontal view of faces. This technique is capable of working in real time environment. This system is a constituent of HRI system. It can keep the person identity information even under the presence of facial expressions which could originate under human machine interaction scenarios. However, in real time environment the system can work by further improving it for light illuminations and using 3D information.

References

[1] Zhao, W., Chellapa, R., Rosenfeld, A., Philips, P.J.: Face Recognition: A Literature Survey, UMD CFAR Technical Report CAR-TR-948 (2000)

[2] Barrett, W.A.: A Survey of Face Recognition Algorithms and Testing Results. Proceeding of IEEE (1998)

[3] Sirovich, L., Kirby, M.: Low-dimensional procedure for the characterization of human faces. J. Opt. Soc. Am. A 4(3), 519–524 (1987)

[4] Turk, M.A., Pentland, A.P.: Face Recognition using Eigenfaces. In: IEEE Conference on Computer Vision and Pattern Recognition, pp. 586–591 (1991)

[5] Turk, M.A., Pentland, A.P.: Eigenfaces for Recognition. Journal of Cognitive Neuroscience 3(1), 71–86 (1991)

[6] Edwards, G.J., Taylor, C.J., Cootes, T.F.: Interpreting Face Images using Active Appearance Models. In: Proceedings of International Conference on Automatic Face and Gesture Recognition, pp. 300–305 (1998)

[7] Edwards, G.J., Cootes, T.F., Taylor, C.J.: Face Recognition using Active Appearance Models. In: Burkhardt, H., Neumann, B. (eds.) ECCV 1998. LNCS, vol. 1407, pp. 581–695. Springer, Heidelberg (1998)

[8] Edwards, G.J., Lanitis, A., Taylor, C.J., Cootes, T.: Statistical Models of Face Images: Improving Specificity. In: British Machine Vision Conference 1996, Edinburgh, UK (1996)

[9] Belheumeur, P.N., Hespanha, J.P., Kreigman, D.J.: Eigenfaces vs Fisherfaces: Recognition using Class Specific Linear Projection. IEEE Transaction on Pattern Analysis and Machine Intelligence 19(7) (July 1997)
[10] Wimmer, M., Riaz, Z., Mayer, C., Radig, B.: Recognizing Facial Expressions Using Model-Based Image Interpretation. In: Advances in Human-Computer Interaction, vol. 1, pp. 587–600 (I-Tech Book)
[11] Lantis, et al.: A unified Approach to Coding and Interpreting Face Images. In: 5th International Conference on Computer Vision, Cambridge USA (1995)
[12] Wimmer, M., Stulp, F., Tschechne, S., Radig, B.: Learning Robust Objective Functions for Model Fitting in Image Understanding Applications. In: Proceedings of the 17th British Machine Vision Conference, BMVA, Edinburgh, UK, pp. 1159–1168 (2006)
[13] Cootes, T.F., Taylor, C.J.: Active Shape models – smart snakes. In: Proceedings of the 3rd British Machine Vision Conference, pp. 266–275. Springer, Heidelberg (1992)
[14] Wimmer, M., Stulp, F., Pietzsch, S., Radig, B.: Learning local objective functions for robust face model fitting. IEEE Transactions on Pattern Analysis and Machine Intelligence (PAMI) 30(8) (2008)
[15] Hanek, R.: Fitting Parametric Curve Models to Images Using Local Self-adapting Seperation Criteria. PhD thesis, Department of Informatics, Technische Universität München (2004)
[16] Cootes, T.F., Edwards, G.J., Taylor, C.J.: Active Appearance Models. In: Burkhardt, H., Neumann, B. (eds.) ECCV 1998. LNCS, vol. 1407, pp. 484–498. Springer, Heidelberg (1998)
[17] Stegmann, M.B.: Active Appearance Models: Theory Extensions and Cases, Master Thesis, Technical University of Denmark (2000)
[18] Ahlberg, J.: An Experiment on 3D Face Model Adaptation using the Active Appearance Algorithm, Image Coding Group, Dept. of Electric Engineering, Linköping University
[19] Gupta, H.R., Chowdhury, A.K., Chellappa, R.: Contour-based 3D Face Modeling from a Monocular Video. In: British Machine Vision Conference 2004 (2004)
[20] Li, S.Z., Jain, A.K.: Handbook of Face recognition. Springer, Heidelberg (2005)
[21] Kanade, T., Cohn, J.F., Tian, Y.: Comprehensive database for facial expression analysis. In: Proceedings of Fourth IEEE International Conference on Automatic Face and Gesture Recognition (FG 2000), Grenoble, France, pp. 46–53 (2000)
[22] Ekman, P., Friesen, W.V.: The Facial Action Coding System: A Technique for the Measurement of Facial Movement. Consulting Psychologists Press, San Francisco (1978)
[23] Ross, A.: Introduction to Biometrics. Lecture in 5th International Summer School of Biometrics, Italy (June 2008)

Why Is Facial Occlusion a Challenging Problem?

Hazım Kemal Ekenel and Rainer Stiefelhagen

Computer Science Depatment, Universität Karlsruhe (TH)
Am Fasanengarten 5, Karlsruhe 76131, Germany
{ekenel,stiefel}@ira.uka.de
http://isl.ira.uka.de/cvhci

Abstract. This paper investigates the main reason for the obtained low performance when the face recognition algorithms are tested on partially occluded face images. It has been observed that in the case of upper face occlusion, missing discriminative information due to occlusion only accounts for a very small part of the performance drop. The main factor is found to be the registration errors due to erroneous facial feature localization. It has been shown that by solving the misalignment problem, very high correct recognition rates can be achieved with a generic local appearance-based face recognition algorithm. In the case of a lower face occlusion, only a slight decrease in the performance is observed, when a local appearance-based face representation approach is used. This indicates the importance of local processing when dealing with partial face occlusion. Moreover, improved alignment increases the correct recognition rate also in the experiments against the lower face occlusion, which shows that face registration plays a key role on face recognition performance.

1 Introduction

Partial face occlusion is one of the most challenging problems in face recognition. A face recognition system can confront occluded faces in real world applications very often due to use of accessories, such as scarf or sunglasses, hands on the face, the objects that persons carry, and external sources that partially occlude the camera view. Therefore, the face recognition system has to be robust to occlusion in order to guarantee reliable real-world operation. Sample occluded face images from a real world face recognition application are shown in Figure 1.

Several studies have been conducted in order to address this problem [1, 2, 3, 4, 5, 6]. In [1], face images are analyzed locally in order to handle partial face occlusion. The face image is first divided into k local regions and for each region an eigenspace is constructed. If a region is occluded, it is automatically detected. Moreover, weighting of the local regions are also proposed in order to provide robustness against expression variations. A similar approach is presented in [2], where a self-organizing map (SOM) is used to model the subspace instead of Gaussians or mixtures of Gaussians as in [1]. A face is represented by the face attributed relational graph (ARG) structure in [3]. This representation contains a set of nodes and binary relations between these nodes. In testing, first the correspondences between the ARG representations of the training and testing

M. Tistarelli and M.S. Nixon (Eds.): ICB 2009, LNCS 5558, pp. 299–308, 2009.
© Springer-Verlag Berlin Heidelberg 2009

Fig. 1. Sample occluded face images

samples are established. According to the distance between these representations, the classification is performed. In [4], robustness against occlusion is provided by combining the subspace methods that aim at best reconstruction, such as principal component analysis, with the subspace methods that aim at discrimination, such as linear discriminant analysis. A sparse signal representation is used to analyze partially occluded face images in [5]. Another representation based approach is proposed in [6]. Different from the studies [1,2,3,4,5] in which occluded images are only included in the testing set, in [6] they are included both in the training and testing sets. The occlusion problem is handled as a reconstruction problem and the classification is done according to the obtained reconstruction error on a test image.

All these studies have focused on finding a way to handle occlusion, especially upper face occlusion. However, until now, no study has been conducted on finding the main reason of low performance caused by the occlusion. Mainly, degraded performance is attributed to the fact that occlusion results in loss of discriminative information, especially in the case of upper face occlusion the eye region, which is known to be very discriminative, can not be utilized. In this study, we thoroughly analyzed the occlusion problem via various experiments and found that the main reason of low performance is not due to missing discriminative information, but the registration errors.

2 Local Appearance-Based Face Recognition

In this study, among the generic face recognition algorithms, local appearance-based face recognition algorithm [7,8] is chosen for face recognition. The reason to opt for this algorithm over other generic face recognition algorithms, such as eigenfaces [9] or Fisherfaces [10], is that it utilizes local facial regions for feature extraction, which is also a common characteristic of many face recognition algorithms that have been developed specifically to handle facial occlusion.

In the local appearance-based face recognition (LAFR) approach, a detected and registered face image is divided into blocks of 8×8 pixels size. Afterwards, the discrete cosine transform (DCT) is performed on each 8×8 pixels block. The obtained DCT coefficients are ordered using zig-zag scanning. From the ordered coefficients, according to a feature selection strategy, M of them are selected and normalized, resulting in an M-dimensional local feature vector.

Regarding feature normalization, there are two aspects. The first aspect is the value range of the DCT coefficients. The first coefficients have higher magnitudes than the later ones. Therefore, they contribute more to the calculated distance in a nearest neighbor classification scheme, hence having more importance in the classification. However, it is known that being able to represent more energy does not necessarily imply to have more discriminative power [11]. The second aspect is the total magnitude of each block's DCT coefficients. Since DCT is an orthonormal transformation and conserves all the energy of the processed input blocks, blocks with different brightness levels lead to DCT coefficients with different value levels. Because of this reason, the blocks with brighter content have more impact on the classification results. In order to balance the coefficients' contributions to the classification, the coefficients in each feature vector are first divided by their standard deviations that are learned from the training set, and then to balance the blocks' contributions to the classification, the local feature vector is normalized to unit norm [8].

Finally, the DCT coefficients extracted from each block are concatenated to construct the overall feature vector. The classification is done using a nearest neighbor classifier with L1 norm as the distance metric.

3 How Much Performance Loss Does Occlusion Cause?

Besides the face recognition algorithms that have been specifically developed to handle occlusion, face recognition algorithms attain very low performance when they are evaluated on occluded face images. For example, in [12], in the experiments on the AR face database [13], eigenfaces algorithm [9] obtained 48%, Fisherfaces [10] 45%, and FaceIt [14] 10% correct recognition rate when they were tested on the face images that contain upper face occlusion caused by sunglasses, whereas they attained 27%, 44%, and 81% respectively when they were tested on the face images that contain lower face occlusion caused by scarf.

In order to assess the drop in face recognition algorithm's performance due to partial face occlusion, we artificially occlude the eye region of the aligned faces and run the local appearance-based face recognition algorithm on the occluded faces, as well as on the face images without occlusion. The occlusion is imitated by painting the blocks in the second and third rows black. A sample aligned face image and the corresponding occluded face image are depicted in Figure 2.

The experiments have been conducted on the *Face Recognition Grand Challenge* (FRGC) database [15]. For each case, we run two different experiments.

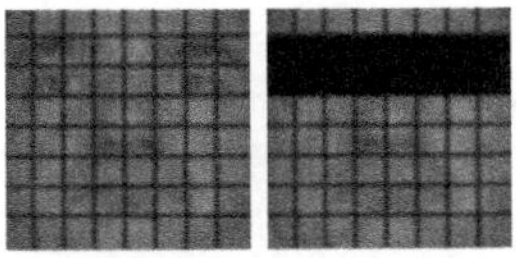

Fig. 2. Sample aligned face image and corresponding occluded face image

Table 1. Experiments on the FRGC database

Label of the experiment	Gallery	Probe
FRGC1	Face images collected under controlled conditions during fall 2003	Face images collected under controlled conditions during spring 2004
FRGC4	Face images collected under uncontrolled conditions during fall 2003	Face images collected under uncontrolled conditions during spring 2004

Fig. 3. Sample face images from the FRGC database. The first two images from the left are samples of face images collected under controlled conditions. The remaining two images are samples of face images collected under uncontrolled conditions.

Table 2. Correct recognition rates obtained on the FRGC database. The results in the first row is obtained using test face images that have no occlusion. The ones in the second row is attained using test face images that contain upper face occlusion as depicted in Figure 2.

	FRGC1	FRGC4
Without occlusion	97.9%	90.8%
With occlusion	95.9%	83.8%

One, *FRGC1*, with face images collected under controlled conditions and one, *FRGC4*, with face images collected under uncontrolled conditions. For controlled conditions, we selected 120 individuals from the FRGC experiment 1 data set, who have at least ten images both in spring 2003 and fall 2004 recordings. We used the images from spring 2003 for training and the ones from fall 2004 for testing. For uncontrolled conditions, we had the same experimental setup, but this time the images were selected from the FRGC experiment 4 data set. Table 1 shows the experimental setup. Sample images from the data sets can be seen in Figure 3.

The classification results with this experimental setup is given in Table 2. The results obtained on the test face images without occlusion are also presented in the table for comparison purposes. It is apparent that missing eye region information causes a decrease in the correct classification rate. Especially, if the experiment is more difficult, as in the FRGC4 experiment, where the training and testing data is collected under uncontrolled conditions, the decrease is more prominent. However, compared to the results obtained on the AR face database [12], the results are still high and the performance drop is not that dramatic.

This finding implies that, it is not only the missing eye region information that causes degradation in performance, but also there are some other factors which will be investigated in the next section.

4 What Is the Main Problem?

In this section, we conduct experiments on the occluded face images, compared the results with the ones from the previous section and try to find out the main problem. We used the AR face database [13] for the experiments. From the database, one image per subject is used from the first session for training. This image is annotated as "1: neutral expression". For testing we used five images per subject. Face images from the second session, which are annotated as "14: neutral expression", are used to assess the algorithm's performance when there exists no occlusion. For testing on occluded face images, two images per subject are used from each session, which are annotated as "8/21: wearing sunglasses" and "11/24: wearing scarf", where the first number corresponds to the label in the first recording session and the second one corresponds to the label in the second recording session. From these two images, the ones with annotations "8/21: wearing sunglasses" are used for testing against upper face occlusion and the ones with annotations "11/24: wearing scarf" are used for testing against lower face occlusion. In the data set, there are 110 subjects who have all these samples in both of the sessions. Overall, five separate experiments are conducted on this data set. Setups of the experiments are presented in Table 3. Sample input images and corresponding registered images are shown in Figure 4. Face images are aligned with respect to manually labeled eye center locations.

Table 3. Experiments on the AR face database

Label of the experiment	Gallery	Probe
AR1scarf	Face images without occlusion from session 1	Face images with scarf from session 1
AR1sun	Face images without occlusion from session 1	Face images with sunglasses from session 1
ARneutral	Face images without occlusion from session 1	Face images without occlusion from session 2
AR2scarf	Face images without occlusion from session 1	Face images with scarf from session 2
AR2sun	Face images without occlusion from session 1	Face images with sunglasses from session 2

The results of these experiments are listed in Table 4. As can be observed, upper face occlusion causes a significant drop in performance. The correct recognition rate decreases from 92.7% to 37.3%. However, the performance decrease

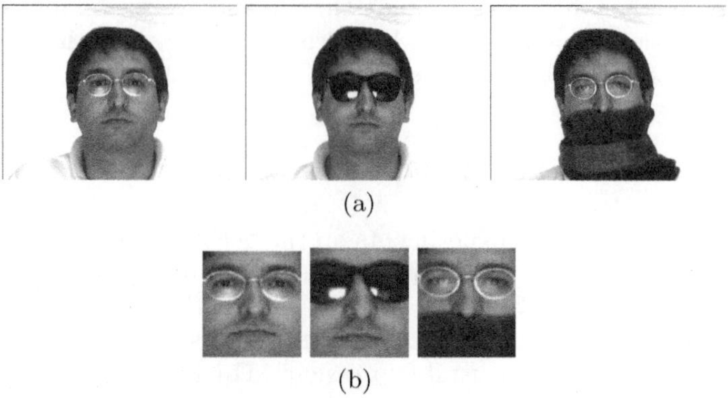

Fig. 4. Sample images from the AR face database. (a) Sample input images. (b) Corresponding registered images.

Table 4. Correct recognition rates obtained on the AR face database

	Correct recognition rate
AR1scarf	91.8%
AR1sun	38.2%
ARneutral	92.7%
AR2scarf	83.6%
AR2sun	37.3%

in the case of lower face occlusion is small. Compared to the results reported in [12], the performance obtained in the case of upper face occlusion is similar, on the other hand the results obtained on lower face occlusion is significantly higher. The performance increase in the case of lower face occlusion can be explained by the local representation approach, in which a change in a local region affects only the features that are extracted from the corresponding block while the features that are extracted from the other blocks remain unaffected, whereas in a holistic appearance-based face recognition approach, such as eigenfaces [9] or Fisherfaces [10], it can affect the entire feature representation. Nevertheless, this scheme does not help to improve the results obtained on the face images that contain upper face occlusion. Compared to the results in the previous section, the performance drop due to upper face occlusion on the AR face database is significantly higher. This implies that, there is another underlying problem, in addition to missing eye region information.

Recently, some studies have been conducted on the robustness of face recognition algorithms against registration errors [16, 17]. These studies have shown that face recognition algorithms' performance rely on face alignment accuracy. In [16], the robustness of eigenfaces algorithm [9] is analyzed against horizontal and vertical translations, as well as scale and rotation, which might occur due to alignment using erroneously detected facial feature locations. It is found that, the eigenfaces approach can tolerate variations only upto 5% of the face image size.

Table 5. Correct recognition rates obtained on the AR face database with the improved face alignment

	Correct recognition rate
AR1scarf	97.3%
AR1sun	98.2%
ARneutral	100%
AR2scarf	93.6%
AR2sun	95.5%

Similarly in [17], in addition to the eigenfaces approach, Fisherfaces [10], elastic bunch graph matching [18], Laplacianfaces [19], minimum average correlation energy (MACE) [20], and pseudo two-dimensional hidden Markov models [21] have been analyzed against registration errors. All these algorithms are observed to be sensitive to misalignment. To address this problem, in [22] a face alignment algorithm is proposed. In the proposed approach, face registration is done by minimizing the closest distance at the classification step. In this approach, once the face is located, positions of the facial features are roughly estimated. Search for the precise facial feature positions is conducted around the estimated positions. Various candidate facial feature positions are used to provide several aligned test face images, while comparing a test face image with an already aligned training face image. The facial feature positions, which lead to the aligned test face image that has the minimum distance to the training image, are selected as the facial feature locations. Thus for each training sample, separate eye center positions are determined for the test face image. In this way, inconsistencies across manual eye center labels of the training images are also handled.

Considering the significant performance drop in the case of upper face occlusion compared to the case of lower face occlusion and compared to the case of artificial occlusion, we hypothesize that the possible reason for this outcome is misalignment. Since the faces are occluded, even manual eye center labels can not be precise resulting misalignment. On the other hand, when we occlude already aligned face images we do not have this problem and the only factor that causes performance drop is the missing eye region information. In order to validate this hypothesis, we used the alignment algorithm in [22] which has been shown to be insensitive to misalignment due to imprecise facial feature localization. The results of this experiment are given in Table 5. It can be seen that the performance increases significantly. Even though a generic face recognition algorithm is used, the results are even better than the state-of-the-art algorithms that have been specifically developed to handle occlusion [1, 2, 3, 4, 5, 6]. It is also interesting to observe that, even the lower face benefit from better alignment, which implies that manual eye center labels may not be consistent, especially when the face images are not very high resolution. Another interesting point is that, slightly higher correct recognition rates are achieved on the upper face occlusion than the lower face occlusion. The reason can be the textured surface of the scarfs, which might affect the classification decision more than the black sunglasses.

5 Conclusions

In this paper, we investigated the main reason for the significant degradation of face recognition algorithms' performance in the case of partial face occlusion. We first artificially generate occluded face images by painting the eye region black and performed experiments both on the face images collected under controlled and uncontrolled conditions. We found that eye region contains valuable discriminative information, especially in the case of uncontrolled conditions. Although, in this experiment we have observed a performance drop due to occlusion, the results were not as low as the ones reported in the literature against upper face occlusion. Afterwards, we conducted experiments on the AR face database [13], both on the lower and upper face occlusion. We found that lower facial occlusion can be handled by local processing. However, against upper face occlusion we got very low performance. This was very intriguing, since the performance drop was not that severe on the artificially occluded face images. Comparing the experimental conditions between the artificially occluded face images and the occluded face images from the AR face database, there are two main differences. The first one is, in the artificial data the eye region is completely black, whereas in the natural one there are reflections. The second difference is, the manual eye center labels, which have been used for alignment is more precise in the artificially occluded face images, whereas due to sunglasses, manual eye center labels of the face images from the AR face database is not reliable. Recent studies on the impact of registration errors on face recognition performance [16,17] lead us to hypothesize that the second difference has more dominant effect. By utilizing an alignment approach [22], which is insensitive to imprecise facial feature localization, we run the experiments again and obtained very high increase in the performance indicating that the main reason causing the performance drop is not the missing eye region information, but the misalignment due to erroneously localized facial feature points. When it is handled, very high correct recognition rates can be obtained even with a generic local appearance-based approach.

Acknowledgement

This study is partially funded by the German Research Foundation (DFG) under Sonderforschungsbereich SFB 588 —Humanoid Robots— and by OSEO, French State agency for innovation, as part of the Quaero Programme. The authors would like to thank Mika Fischer for his contributions to the study.

References

1. Martinez, A.M.: Recognizing imprecisely localized, partially occluded and expression variant faces from a single sample per class. IEEE Transactions on Pattern Analysis and Machine Intelligence 24(6), 748–763 (2002)

2. Tan, X., Chen, S., Zhou, Z.H., Zhang, F.: Recognizing partially occluded, expression variant faces from single training image per person with SOM and soft k-NN ensemble. IEEE Transactions on Neural Networks 16(4), 875–886 (2005)
3. Park, B.G., Lee, K.M., Lee, S.U.: Face recognition using face-ARG matching. IEEE Transactions on Pattern Analysis and Machine Intelligence 27(12), 1982–1988 (2005)
4. Fidler, S., Skočaj, D., Leonardis, A.: Combining reconstructive and discriminative subspace methods for robust classification and regression by subsampling. IEEE Transactions on Pattern Analysis and Machine Intelligence 28(3), 337–350 (2006)
5. Wright, J., Ganesh, A., Yang, A., Ma, Y.: Robust face recognition via sparse representation. Technical report, University of Illinois, USA (2007)
6. Jia, H., Martinez, A.M.: Face recognition with occlusions in the training and testing sets. In: Proceedings of IEEE Int'l. Conf. on Automatic Face and Gesture Recognition (2008)
7. Ekenel, H.K., Stiefelhagen, R.: Local appearance-based face recognition using discrete cosine transform. In: Proc. of 13th European Signal Processing Conf. (EUSIPCO 2005) (2005)
8. Ekenel, H.K., Stiefelhagen, R.: Analysis of local appearance-based face recognition: Effects of feature selection and feature normalization. In: Conf. on Computer Vision and Pattern Recognition Workshop 2006, p. 34 (2006)
9. Turk, M., Pentland, A.: Eigenfaces for recognition. Journal of Cognitive Neuroscience 3(1), 71–86 (1991)
10. Belhumeur, P.N., Hespanha, J.P., Kriegman, D.J.: Eigenfaces vs. Fisherfaces: Recognition using class-specific linear projection. IEEE Transactions on Pattern Analysis and Machine Intelligence 19(7), 711–720 (1997)
11. Ekenel, H.K., Sankur, B.: Feature selection in the independent component subspace for face recognition. Pattern Recognition Letters 25(12), 1377–1388 (2004)
12. Gross, R., Shi, J., Cohn, J.F.: Que vadis face recognition? In: Proceedings of Workshop on Empirical Evaluation Methods in Computer Vision (2001)
13. Martinez, A.M., Benavente, R.: The AR face database. Technical Report 24, CVC (1998)
14. Penev, P.S., Atick, J.: Local feature analysis: a general statistical theory for object represntation. Network: Computation in Neural Systems 7, 477–500 (1996)
15. Phillips, P.J., Flynn, P.J., Scruggs, T., Bowyer, K.W., Chang, J., Hoffman, K., Marques, J., Min, J., Worek, W.: Overview of the Face Recognition Grand Challenge. In: Proc. 2005 IEEE Computer Society Conf. on Computer Vision and Pattern Recognition, Los Alamitos, CA, USA, vol. 1, pp. 947–954 (2005)
16. Lemieux, A., Parizeau, M.: Experiments on eigenfaces robustness. In: Proc. Int'l. Conf. on Pattern Recognition, vol. 1, pp. 421–424 (2002)
17. Rentzeperis, E., Stergiou, A., Pnevmatikakis, A., Polymenakos, L.: Impact of face registration errors on recognition. In: Maglogiannis, I., Karpouzis, K., Bramer, M. (eds.) Artificial Intelligence Applications and Innovations (AIAI 2006), pp. 187–194. Springer, Heidelberg (2006)
18. Wiskott, L., Fellous, J.M., Kruger, N., Malsburg, C.: Face recognition by elastic bunch graph matching. IEEE Transactions on Pattern Analysis and Machine Intelligence 19(7), 775–779 (1997)
19. He, X., Yan, S., Hu, Y., Niyogi, P., Zhang, H.J.: Face recognition using Laplacianfaces. IEEE Transactions on Pattern Analysis and Machine Intelligence 27(3), 328–340 (2005)

20. Xie, C., Kumar, B.V.K.V., Palanivel, S., Yegnanarayana, B.: A still-to-video face verification system using advanced correlation filters. In: Zhang, D., Jain, A.K. (eds.) ICBA 2004. LNCS, vol. 3072, pp. 102–108. Springer, Heidelberg (2004)
21. Samaria, F.S., Harter, A.C.: Parameterisation of a stochastic model for human face identification. In: Proceedings of the Second IEEE Workshop on Applications of Computer Vision, pp. 138–142 (1994)
22. Ekenel, H.K.: A Robust Face Recognition Algorithm for Real-World Applications. PhD thesis, Universität Karlsruhe (TH), Germany (2009)

Nasal Region-Based 3D Face Recognition under Pose and Expression Variations

Hamdi Dibeklioğlu[1], Berk Gökberk[2], and Lale Akarun[1]

[1] Boğaziçi University, Computer Engineering Department, Turkey
{hamdi.dibeklioglu,akarun}@boun.edu.tr
[2] University of Twente, Department of Electrical Engineering, The Netherlands
b.gokberk@utwente.nl

Abstract. In this work, we propose a fully automatic pose and expression invariant part-based 3D face recognition system. The proposed system is based on pose correction and curvature-based nose segmentation. Since the nose is the most stable part of the face, it is largely invariant under expressions. For this reason, we have concentrated on locating the nose tip and segmenting the nose. Furthermore, the nose direction is utilized to correct pose variations. We try both one-to-all and Average Nose Model-based methodologies for registration. Our results show that the utilization of anatomically-cropped nose region increases the recognition accuracy up to 94.10 per cent for frontal facial expressions and 79.41 per cent for pose variations in the Bosphorus 2D/3D face database.

1 Introduction

An effective face recognition system has to be fully automatic and robust enough for real life conditions where illumination, rotation, and expression variations are present. Although state-of-the-art 3D face recognition systems provide identification rates up to 99 per cent, they are mostly constrained by certain assumptions such as slight expression and pose variations [1]. Overcoming the problems due to *extreme* expression and pose changes still remains as a challenging problem.

A recently proposed approach for expression invariant 3D face recognition is a part-based system [2]. Part-based approach is useful to alleviate the pose, facial expression and partial occlusion effects on the recognition performance. To deal with these effects, the entire 3D model of the face is split into several regions and these regions are processed separately. Alyüz *et al.* [2] split the face into patches, and carry out an exhaustive search of all possible combinations of these surface to find the best subset of all the patches around the whole facial surface.

Moreno *et al.* [3] segment the 3D facial surface using signs of mean, Gaussian curvatures and several three dimensional descriptors. Cook *et al.* [4] use Log-Gabor Templates on range images to deal with occlusions, distortions and facial expressions. Recently, Kakadiaris et al. proposed matching of spin images prior to ICP for alignment. They used wavelet analysis for feature extraction and obtained good recognition results in the presence of facial expressions [5]. In [6], Chang *et al.* use multiple regions selected around the nose area which have the

M. Tistarelli and M.S. Nixon (Eds.): ICB 2009, LNCS 5558, pp. 309–318, 2009.
© Springer-Verlag Berlin Heidelberg 2009

least shape variation due to facial expressions. In [7] Faltemier *et al.* extend the use of multiple regions. Seven overlapping regions around the nose are extracted and the ICP algorithm [8] is used to align each segment separately. On the FRGC v.2 database, a rank-one accuracy of 94.9 per cent is reported using a voting mechanism. In their later work, Faltemier *et al.* [9] use 38 overlapping regions to find the best committee. By selecting multiple small regions on the face, any errors caused by a single region can be compensated for when combining the matching scores. These regions are extracted by spherical cropping of facial surface and aligned with the ICP. The best performance is provided by the fusion of 28 regions with a modified Borda count method, and the recognition accuracy is reported as 97.2 per cent on the FRGC v.2.

Most of these part-based systems first localizes some salient feature points over the facial surface such as nose tip and eye pits for component formation. There is a very broad range of facial landmark localization algorithms, ranging from statistical techniques to heuristics. For a comprehensive overview, the reader is referred to a recent survey [1]. In this paper we propose an automatic nose tip localization and nose segmentation method that can be applied to cases where significant pose variations are present. Extreme rotations create difficulties both for nose tip localization and nose segmentation because a significant part of the nose is missing due to self-occlusion. Our recognition scheme achieves significant success even under these conditions.

2 Nose Tip Estimation and Nose Segmentation

We propose a heuristic method to localize the nose tip and segment the nose region under pose and expression variations. The proposed algorithm estimates the regions with high protrusion and selects the most probable location based on curvature values. Principle normal curvatures are used since they are invariant to rotation and translation. We use an analytical method to estimate the principal curvatures by fitting a quadratic order surface. Details of the curvature estimation algorithm are given in [2].

2.1 Initial Estimation of the Nose Tip

An initial estimation of nose tip is required for segmentation of the nose region. First, Gaussian (K) and mean curvature (H) values of the 3D facial surface are estimated from the the principal curvatures as: $K = \kappa_1(p) \times \kappa_2(p)$, $H = \frac{\kappa_1(p)+\kappa_2(p)}{2}$ where, κ_1 and κ_2 denote the minimum and maximum principal curvatures, respectively. In theory, a HK map can be comprehended as a label map. Although this kind of a labeling would be useful to restrict the search area for statistical methods, it entails a certain loss of information. Our approach is based on the more informative continuous values of H and K maps [10]. We compose a difference map $\mathcal{D}_{i,j} = K_{i,j} - H_{i,j}$. In the proposed $\mathcal{D}$ map, the nose tip is revealed as a peaked cap. Then $\mathcal{D}$ map is smoothed by a Gaussian filter to cope with curvature sensitivity and surface noise. However, the peaks on the sides of the nose may be confused with the nose tip when there are large rotations. Therefore, we need a rotation invariant feature which yields higher values around the nose tip. To handle the rotation conditions, we estimate an Eigen map, $\mathcal{E}$. A sliding 3×3 pixels block is used for $\mathcal{E}$ map estimation. 3D coordinates of nine points on each block are taken as inputs and their covariance matrix is calculated. Then the eigenvalues and eigenvectors are estimated for the calculated covariance matrix. The biggest eigenvalue is selected and the ratio of the selected eigenvalue to sum of all eigenvalues is calculated. This is the value of the $\mathcal{E}$-map for the center point of the related block. Estimated $\mathcal{E}$-map gives relatively higher values for points with steep depth changes such as points on nose slopes. Additionally, we know that $\mathcal{D}$ has higher values for ridges and peaks. Eventually, all the nose region can be highlighted by the combination of these two indicators (see Fig. 1). A combination map $\mathcal{C}$ is formed such that:

$$\mathcal{C}_{i,j} = \frac{\mathcal{D}'_{i,j} + \mathcal{E}'_{i,j}}{2} \quad (1)$$

where $\mathcal{D}' = m(\mathcal{D})$, $\mathcal{E}' = m(\mathcal{E})$, and m denotes the min-max normalization function $m(x) = \frac{x - min(x)}{max(x) - min(x)}$.

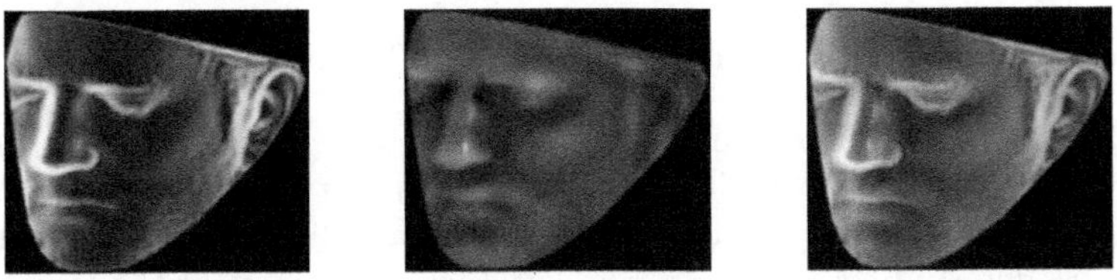

Fig. 1. The eigen, difference and the combination maps, respectively

We crop the difference and combination maps by 20 per cent on the top and bottom to focus on the center of the face. Left and right sides of the $\mathcal{D}$ map are eliminated by using $\mathcal{C}$ map. The vertical projection of the squared $\mathcal{D}$ map reveals a peak which corresponds to the nose region. We calculated the shape index (SI) in this interest region to locate the nose tip precisely. The shape index of the face image is the quantitative measure of the shape of a surface at a point [11], and it is defined as:

$$SI(p) = \frac{1}{2} - \frac{1}{\pi} tan^{-1} \frac{\kappa_1(p) + \kappa_2(p)}{\kappa_1(p) - \kappa_2(p)} \quad (2)$$

Then the interest region is thresholded by the mean value of the whole shape index map. Lastly, horizontal projection of the $\mathcal{D}$ map is calculated for remaining points and the point with the maximum value is selected as a close point to the nose tip.

2.2 Coarse Nose Cropping

After the estimation of the nose tip, points on the initial $\mathcal{D}$ map are cropped by a rectangular mask. Then, the minimum left and right side $\mathcal{D}$ values are found which are on the same y-coordinate with the nose tip. These minimum points are used as starting points for nose cropping by greedy search. Coarse cropping is based on finding the minimum energy path on the $\mathcal{D}$ map, by greedy search, as shown in Fig. 2.

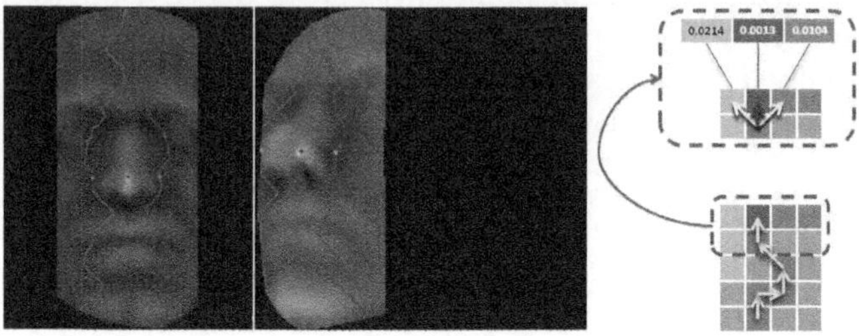

Fig. 2. Minimum energy path estimation: Green points show the initial minimum points. Red points denote the estimated nose tip.

2.3 Yaw Rotation Estimation and Pose Correction

Yaw rotations have to be estimated to rotate the face into a frontal pose. It can be assumed that the relative horizontal position of the nose tip gives a coarse estimation of the yaw rotation angle. Hence, normalized nose tip positions for subjects are determined relative to the leftmost valid point of the facial surface. In other words, minimum x coordinate of the facial surface is subtracted from the x coordinate of the nose tip, and the ratio of the calculated value to the width of the facial surface gives the normalized nose tip position. Since the nose tip position of the frontal poses is approximately at the middle, its value is 0.5. 90 degrees to the left and 90 degrees to the right rotated surfaces' nose tip positions are 0 and 1, respectively. Then, the estimated yaw angle is used to correct the rotation and to bring the facial points to a frontal pose. It should be noted that rotation compensation using nose tip is just a preliminary step. In the following registration steps, the ICP algorithm is utilized for more accurate pose normalization.

2.4 Nose Tip Correction and Fine Level Cropping

After coarse rotation compensation phase, $\mathcal{D}$ map is generated again, as explained in 2.1. Rotated nose region is cropped vertically in order to eliminate artifacts that may emerge around the previously invisible side of the nose. We prefer to eliminate these regions since interpolation is almost impossible considering the amount of missing surface data. In the next step, minimum energy

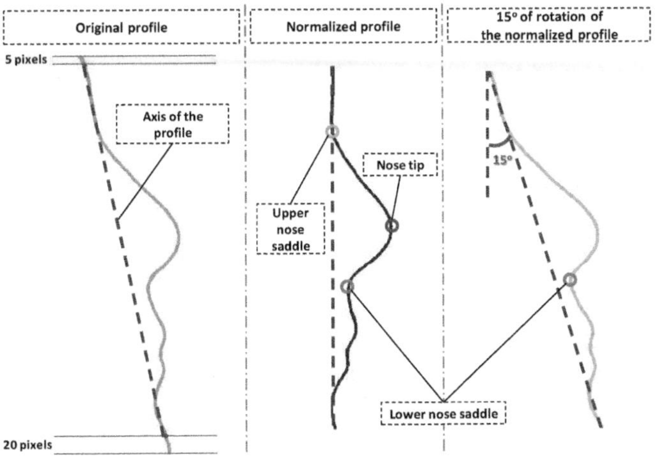

Fig. 3. Upper and lower limits estimation procedure of the nose region

paths are refined by updating the starting points and repeating the search. At the last step, upper and lower bounds of the nose are found as illustrated in Fig. 3. The profile is normalized, rotated, and local minima which corresponds to the upper and lower limits of the nose are found.

3 Registration and Recognition Methodology

Our registration method is based on the ICP algorithm. We compare two different approaches, namely, *one-to-all ICP* and Average Nose Model-based ICP (*ANM-based ICP*). In the one-to-all ICP approach, probe nose regions are registered to each gallery nose. In the ANM-based method, all of the gallery nose regions are registered to an ANM by the ICP. ANM is a generic 3D nose model computed from the gallery set. A modified version of the average face model construction algorithm presented in [12] is used to generate the ANM. Given all gallery nose regions, they are aligned to a consensus shape using manually located five fiducial points by the Procrustes analysis. After surface resampling, average surface is determined by taking the mean of depth coordinates. Details of the algorithm can be found in [12]. In the ANM-based approach, a probe nose is registered to the ANM at the recognition phase which provides dense point-to-point correspondence to every gallery nose. After registration, we resample the final nose surfaces by Delaunay trangulation-based interpolation and employ point set difference approach for dissimilarity calculation. If $\mathcal{N}$ and $\mathcal{N}'$ are the re-sampled depth maps of a registered training and test nose, respectively, a mean error function, E_{PSD} can be formulated as:

$$E_{PSD} = \frac{1}{G_h \times G_h} \sum_{i=1}^{G_h} \sum_{j=1}^{G_w} w_{i,j} \times \left| \mathcal{N}_{i,j} - \mathcal{N}'_{i,j} \right| \qquad (3)$$

where, G_h and G_w indicate height and width of the re-sampled depth maps, respectively. $w_{i,j}$ denotes the intersection of valid points in $\mathcal{N}$ and $\mathcal{N}'$. $w_{i,j}$ is zero for invalid points and one otherwise.

4 Experimental Results

We have used the Bosphorus 3D face database [13] in our experiments. The Bosphorus database is designed specifically to study extreme pose, expression, and occlusion variations (see Fig. 4). Additionally, the database incorporates realistic occlusions like eye glasses, hair tassel and eye rubbing. The Bosphorus database includes $+10°$, $+20°$, $+30°$, $\pm45°$, $\pm90°$ of yaw rotations, strong/slight upwards, strong/slight downwards, combinations of $\pm20°$ pitch and $+45°$ yaw rotations. Facial expressions include six basic emotions (happiness, surprise, fear, sadness, anger and disgust), as well as expressions based on facial action units. Database has two versions as v.1 and v.2. Bosphorus v.1 includes 34 subjects with only 10 expressions, 13 poses, four occlusions and four neutral faces, thus resulting in a total of 31 scans per subject. In version 2, there are 47 people with 53 different face scans per subject. Totally there are 34 expressions, 13 poses, four occlusions and one/two neutral faces. expression and pose variations in the Bosphorus Database are shown in Fig. 4.

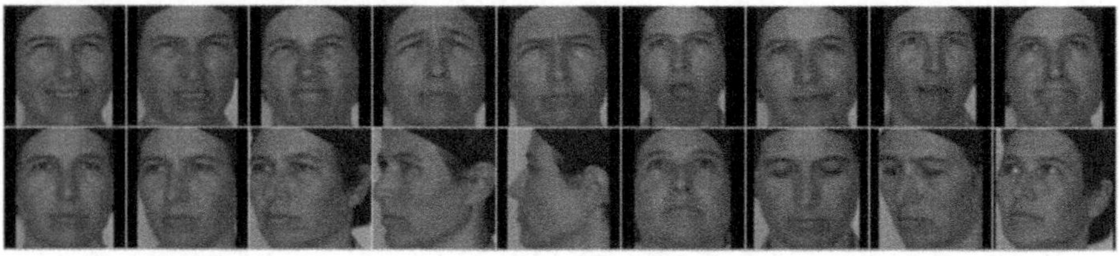

Fig. 4. Samples of expression and pose variations in the Bosphorus database

4.1 Nose Tip Localization and Segmentation

The proposed nose tip estimator has not been tested for facial images having 45° and 90° yaw rotations because these samples are not considered in our identification experiments. The reported success rates are obtained by accepting points within 10 per cent inter-eye distance to the ground truth (see Table 1 and Fig. 6). In the Bosphorus database, the average inter-eye distance is 64.9 millimeters. To control the generality of the used heuristics, we have also tested our algorithm on FRGC v.1 and achieved 100 per cent success with 10 per cent inter-eye distance acceptance. The main factors about the robustness and performance of our algorithm are the pose correction step and the refinement of the nose tip location during the segmentation process. Segmented nose regions are shown in Figure 6.

Table 1. The performance of the proposed node tip localization method

Type	Bosphorus v.1		Bosphorus v.2	
	Success Rate	# Samples	Success Rate	# Samples
Neutrals and Facial Expressions	100.00	475	99.68	1574
+Yaw Rotations (10°, 20°, 30°)	99.02	102	97.16	141
Strong and Slight Upwards	100.00	68	100.00	94
Strong and Slight Downwards	97.06	68	100.00	94
-20° Pitch and +45° Yaw	76.47	34	97.87	47
+20° Pitch and +45° Yaw	94.12	34	95.75	47

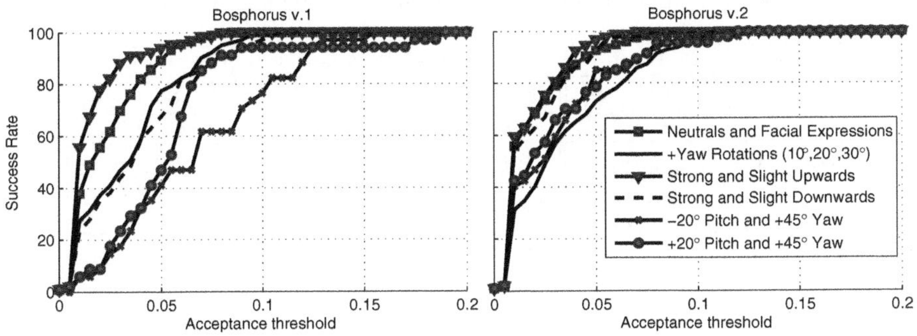

Fig. 5. The performance of the proposed node tip localization method on several acceptance thresholds of inter-eye distance

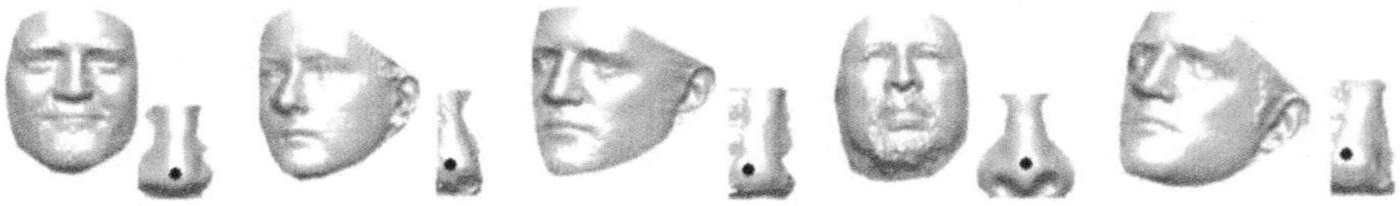

Fig. 6. Segmented nose regions under several expression and pose variations

4.2 Identification Experiments

In our identification experiments, the gallery set is composed of one neutral scan per subject. The remaining faces, probe images, are divided into two groups as i) frontal scans and ii) rotated faces. Poses with rotation of ±45° and ±90° are discarded in the rotation probe set because these images have insufficient amount of surface information for our identification algorithm. Number of samples in the gallery and probe sets for each version are given in Table 2.

Part-based identification rates are given in Table 3 and Table 4. Two different types of gallery sets are used in frontal tests of one-to-all methods. We refer to the one-to-all ICP and the ANM-based one-to-one ICP methods as $Natural_{one:all}$ and $Natural_{anm}$, respectively. The first one is composed of automatically cropped nose models, denoted as *Original*. The other one includes

Table 2. The gallery and probe sets for the Bosphorus v.1 and v.2

Version	Type	Gallery	Frontal Probe Set	Rotation Probe Set
Bosphorus v.1	Neutral	34	102	306
	Expression	-	339	-
	Total	34	441	306
Bosphorus v.2	Neutral	47	19	423
	Expression	-	1508	-
	Total	47	1527	423

Table 3. Rank-1 recognition rates of the one-to-all ICP approach for the frontal poses with expression in the Bosphorus database (* denotes the highest recognition rates)

Pose	Method	Gallery Type	Bosphorus v.1	Bosphorus v.2
Frontal	$Natural_{one:all}$	*Enlarged*	92.97	88.41
		Original	94.10*	89.19*
	$Spherical^{P1}_{one:all}$	*Enlarged*	73.71	65.62
		Original	89.57	80.48
	$Spherical^{P2}_{one:all}$	*Enlarged*	85.94	78.26
		Original	91.84	81.79

Table 4. Rank-1 recognition rates for the frontal poses with expression and poses with rotation in the Bosphorus database (* denotes the highest recognition rates)

Pose	Method	Gallery Type	Bosphorus v.1	Bosphorus v.2
Frontal	$Natural_{one:all}$	*Original*	94.10*	89.19*
	$Natural_{anm}$	*Original*	93.88	89.19*
Rotation	$Natural_{one:all}$	*Original*	79.41*	60.99
	$Natural_{anm}$	*Original*	73.20	62.55*

enlarged nose models, denoted as *Enlarged*, which are constructed by dilation with 10 pixels of the bounding boxes of *Original* set.

The region extraction method which is suggested in [9] is also implemented and tested for two different settings for comparative purposes. In this method, the nose is cropped by a sphere with a radius r. The center of the sphere is defined by offsets with respect to the tip of the nose, namely x-axis offset (x_o) and y-axis offset (y_o). Nose cropping parameters have been selected as in [9] which provide the best results in terms of rank-one recognition rate in the FRGC v.2. These tests are denoted as $Spherical^{P1}_{one:all}$ and $Spherical^{P2}_{one:all}$, respectively. The related parameters are set as $x_o = 0$, $y_o = +30$, $r = 40$ millimeters for $Spherical^{P1}_{one:all}$ and $x_o = 0$, $y_o = +20$ and $r = 35$ millimeters for $Spherical^{P2}_{one:all}$. Nose tip coordinates which are used in this method, are provided by our nose tip estimator. One-to-all ICP is used for the registration of the baseline method.

Test results indicate that proposed methods, $Natural_{one:all}$ and $Natural_{anm}$ with *Original* gallery type, provide the highest recognition rates for the frontal expressions in the Bosphorus v.1, 94.10 per cent and in the Bosphorus v.2, 89.19 per cent. The nose region which is segmented by using its natural boundaries is more discriminative than the point set of the manually cropped coarse nose region. As explained above, $Spherical^{P1}_{one:all}$ and $Spherical^{P2}_{one:all}$ use nose models which also covers an extra portion of the facial surface outside of the nose region. Because of that, the recognition success rates of the mentioned methods are degraded. If we compare the results for *Original* and *Enlarged* types of gallery set usages, it can be seen that *Enlarged* gallery set usage degrades the accuracy because of the same reason.

According to the results of the poses with rotations, the best rates achieved by $Natural_{one:all}$ on the Bosphorus v.1 and $Natural_{anm}$ on the Bosphorus v.2 are 79.41 per cent and 62.55 per cent, respectively. The rates are obviously decreased with respect to frontal poses by rotation variations. However the performance of the system is accurate enough for rotation conditions. It is needed to be noted that a very small part of a nose can be extracted under rotation variations. One final observation is related to one-to-one versus one-to-all methods: there is not a big performance difference between them despite very high computation complexity of the one-to-all method, i.e, registering a probe image with every gallery image without the use of the ANM.

5 Conclusion

A fully automatic part-based 3D face recognition system has been proposed in this work. Curvature-based nose localization/segmentation and part-based registration methods have been used to cope with rotations and facial expressions. The nose tip estimator is a curvature based heuristic method which includes pose correction. The results show that the nose tip is localized with a very good accuracy even under strong rotations and expressions. Our nose segmentation algorithm automatically finds the nose borders. According to the experimental results, recognition rates are significantly increased by using natural regions of the noses as opposed to enlarged nose regions. This is due to i) the shape of the segmented nose is discriminative and ii) the muscle deformations around the nose region increases the intra-class variations under expressions. Therefore fine segmentation of the nose region is crucial for identification. In our experiments, we show that high recognition performances can be attained under facial expression and rotation variations by a part-based method. Using only nose region, 94.10 and 79.41 per cent recognition accuracies can attained for expression and pose variations, respectively. As future work, the proposed system should be extended for pose variations larger than 45 degrees and other facial parts should also be consulted under such conditions.

References

1. Bowyer, K.W., Chang, K., Flynn, P.: A survey of approaches and challenges in 3D and multi-modal 3D + 2D face recognition. Computer Vision and Image Understanding 101, 1–15 (2006)

2. Alyüz, N., Gökberk, B., Dibeklioğlu, H., Akarun, L.: Component-based registration with curvature descriptors for expression insensitive 3D face recognition. In: International Conference on Automatic Face and Gesture Recognition (2008)
3. Moreno, A.B., Sanchez, A., Velez, J.F., Diaz, F.J.: Face recognition using 3D surface-extracted descriptors. In: Irish Machine Vision and Image Processing Conference (IMVIP) (2003)
4. Cook, J., Chandran, V., Fookes, C.: 3D face recognition using log-gabor templates. In: Biritish Machine Vision Conference, pp. 83–92 (2006)
5. Kakadiaris, I., Passalis, G., Toderici, G., Murtuza, N., Lu, Y., Karampatziakis, N., Theoharis, T.: 3D face recognition in the presence of facial expressions: an annotated deformable model approach. IEEE Trans. Pattern Analysis and Machine Intelligence 29, 640–649 (2007)
6. Chang, K.I., Bowyer, K.W., Flynn, P.J.: Adaptive rigid multi-region selection for handling expression variation in 3D face recognition. In: IEEE CVPR 2005, pp. 157–164 (2005)
7. Faltemier, T., Bowyer, K.W., Flynn, P.J.: 3D face recognition with region committee voting. In: Third International Symposium on 3D Data Processing, Visualization, and Transmission (3DPVT 2006), pp. 318–325 (2006)
8. Besl, P., McKay, N.: A method for registration of 3-D shapes. IEEE Trans. Pattern Analysis and Machine Intelligence 14, 239–256 (1992)
9. Faltemier, T., Bowyer, K., Flynn, P.: A region ensemble for 3-d face recognition. IEEE Transactions on Information Forensics and Security 3, 62–73 (2008)
10. Dibeklioğlu, H., Salah, A.A., Akarun, L.: 3D facial landmarking under expression, pose, and occlusion variations. In: IEEE BTAS 2008, Washington, DC, USA (2008)
11. Dorai, C., Jain, A.K.: Cosmos - a representation scheme for 3d free-form objects. IEEE Trans. PAMI 19, 1115–1130 (1997)
12. İrfanoğlu, M.O., Gökberk, B., Akarun, L.: 3D shape-based face recognition using automatically registered facial surfaces. In: International Conference on Pattern Recognition (ICPR 2004), vol. 4, pp. 183–186 (2004)
13. Savran, A., Alyüz, N., Dibeklioğlu, H., Çeliktutan, O., Gökberk, B., Sankur, B., Akarun, L.: Bosphorus database for 3D face analysis. In: The First COST 2101 Workshop on Biometrics and Identity Management (BIOID) (2008)

An Analysis-by-Synthesis Method for Heterogeneous Face Biometrics

Rui Wang, Jimei Yang, Dong Yi, and Stan Z. Li*

Center for Biometrics and Security Research,
Institute of Automation, Chinese Academy of Sciences, Beijing 100190, China
{rwang,jmyang,dyi,szli}@cbsr.ia.ac.cn

Abstract. Face images captured in different spectral bands, *e.g.* , in visual (VIS) and near infrared (NIR), are said to be heterogeneous. Although a person's face looks different in heterogeneous images, it should be classified as being from the same individual. In this paper, we present a new method, called *face analogy*, in the analysis-by-synthesis framework, for heterogeneous face mapping, that is, transforming face images from one type to another, and thereby performing heterogeneous face matching. Experiments show promising results.

Keywords: Heterogenous face biometrics, face analogy, face matching, analysis-by-synthesis.

1 Introduction

Different types face biometrics have been developed, including visual (VIS) [1], near infrared (NIR) [2], thermal infrared (TIR) [3], and 3D [4] image based. In any of these methods, it is assumed that both enrollment and query face images are of the same type. Face matching two different image types are said to be *heterogeneous* because the image formation characteristics are different.

Although heterogeneous face images of a given person have different face appearances by pixel values, the identify of the face images should be classified as the same whichever type of image the face is contained in. Face biometrics by matching between heterogeneous face images are collectively called *heterogeneous face biometrics* (HFBs).

Recent developments have led to several proposals of HFBs, that between VIS and face sketch [5], between VIS and NIR [6], and between 3D and NIR [7], and also reconstruction of the facial shape from an NIR image [8]. The MBGC (Multiple Biometric Grand Challenge) tests organized by NIST (National Institute of Standards and Technology) has devised the modality of matching between VIS and NIR face images as one of its experiments [9], with the purpose to examine how well NIR-VIS face biometric could be done and how its fusion with VIS-VIS face biometric could improve the overall performance.

In this paper, we present a new method, called *face analogy*, for HFBs, in the analysis-by-synthesis framework. The face analogy method transforms face images

* Corresponding author.

M. Tistarelli and M.S. Nixon (Eds.): ICB 2009, LNCS 5558, pp. 319–326, 2009.
© Springer-Verlag Berlin Heidelberg 2009

from one type to another. Heterogeneous face matching can thereby be performed. Experiments show promising results.

2 Image Modeling

2.1 Spectral Image Formation

Images used in face recognition is related to facial shape, skin and hair. A 3D face image is related to the shape only. It is captured by a range measuring system usually made from a laser range system or stereo vision system. Represent a range image taken from a viewpoint by $z(x,y)$. The pixel values measure the distances from the sensor to the facial surface points.

Spectral face images are different from 3D in nature. Fig. 1 illustrates ultra-violet (UV), VIS and infrared (IR) spectral bands in different wavelengths. The IR portion of electromagnetic wave can be divided into four spectral regions: near infrared (NIR), short-wave infrared, thermal infrared (TIR), and far infrared (FIR).

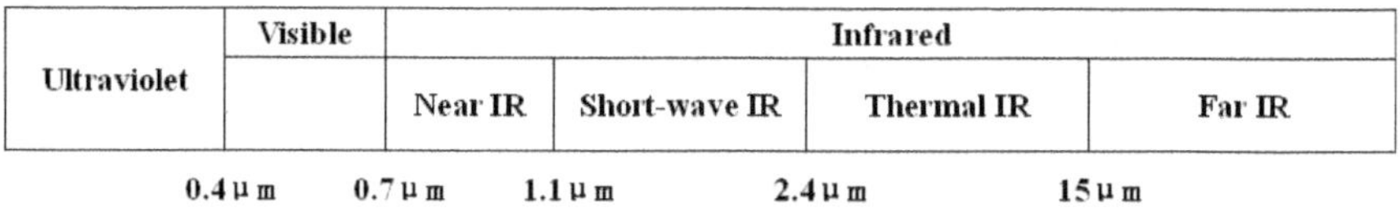

Fig. 1. Spectral bands of electromagnetic wave radiation

The Lambertian law provides an image formation model, relating a spectral image with the 3D shape of the sensed object, the object surface properties and the illumination source:

$$I(x,y) = \rho(x,y)\mathbf{n}(x,y)\mathbf{s} \tag{1}$$

where $I(x,y)$ is the spectral image, $\rho(x,y)$ is the albedo of the facial surface material at point (x,y) (also changing with wavelengths), $\mathbf{n} = (n_x, n_y, n_z)$ is the surface normal (a unit row vector) at the 3D surface point $z(x,y)$, and $\mathbf{s} = (s_x, s_y, s_z)$ is the point lighting direction (a column vector, with magnitude). The normal directions $\mathbf{n}(x,y)$ may be derived from the range image $z(x,y)$ but not vice versa.

2.2 Face Skin Spectroscopy

HFBs across spectral bands are related to facial skin and hair spectroscopy. Skin spectropy is a field of study on the interaction between electromagnetic wave and human skin. Essentially, skin spectroscopy studies the albedo ρ of skin as a function of wavelength. Understanding face spectroscopy helps formulating heterogeneous face image formation models and designing HFB algorithms.

The optics of skin is dependent on wavelength and dose of the incident light. These are the main bands that have been used practically for for non-invasive biometric applications. The spectrum of VIS is usually divided into three channels: red, green and blue.

In the VIS spectrum, melanin is the only pigment affecting the transmittance of epidermis. Shorter wavelengths, such as blue light, are highly absorbed by certain tissue components such as melanin and blood relative to longer wavelengths. In addition, the optical scattering increases as wavelength getting shorter in this spectral range.

The skin color is affected by the type and quantity of melanin in epidermis. This results in various skin colors from white to brown to black though the structure of human skin is similar across different race. The absorption of melanin decreases monotonously from the short wavelength of 0.25μm in the UV band (through the VIS band wavelength) to the long wavelength of 1.1μm in the NIR band.

Beyond 1.1μm, the absorption of melanin is negligible. Both transmittance and remittance of skin are uncorrelated with melanin and the most of IR is transmitted through epidermis into dermis. TIR of wavelength between 2.4 to 15 μm is often used passively to determine the temperature of skin. The amount of heat radiation emitted by skin increases with temperature, therefore thermography allows one to see variations in temperature.

The dermis has significantly different structure and composition with epidermis, In the dermis, the scattering effect dominates. Many observations show the transmittance and remittance are close to 100% across the spectrum from 0.3μm-2.4μm, indicating that very little radiation is absorbed there. Light at longer wavelengths can penetrate the dermis deeper than shorter wavelengths.

3 The Face Analogy Method

In the analysis-by-synthesis approach to HFBs, face analogy is a method for converting face images from one type to another by transforming facial textures. It shares the idea of "image analogies" [10] from the computer graphics community. For the NIR-VIS face analogy, an NIR face can be transformed analogously into a VIS face as if the lighting changes, based on a pair of well-registered NIR-VIS faces.

The NIR to VIS transform also has a practical motivation: We would like to take NIR images as the probe, take advantage that active NIR imaging can tackle the problem of variable illumination, and match against VIS faces of target.

Image analogy has two assumptions. One is that in any location of image, one type of texture pattern and the other type of texture pattern always co-occur. The other is that in any given training image pair, there are sufficient sorts of co-occurrent texture patterns so that only one pair of registered images are needed as training data.

However, these two assumptions tend to fail when applied to faces. Facial texture patterns are usually localized and thus co-occurrence just locally works. It means that texture pattern co-occurrence can only be trusted in the same facial semantic region, such as nose, eyes and corners of mouth. For example, if in an NIR image a texture pattern in the nose area is similar to one in the mouth area, their corresponding VIS texture patterns in the same areas cannot be guaranteed being similar.

We solve this problem in face analogy by introducing more pairs of well-registered NIR-VIS faces. NIR-VIS texture patterns coming from the same region of different face pairs together compose a training database for this region. Consequently, texture patterns in this region of a new NIR face can be found in the corresponding training database and then this NIR texture patterns can be transformed to VIS texture pattern by analogy. Further, by the analysis of NIR and VIS imaging models, we establish a common invariant using local normalization technique [11] and thereby parameterize the synthesis procedure.

As a result, the procedure of face analogy equivalently consists of a de-lighting (local normalization) step and a re-lighting (synthesis) step. In the following, we analyze VIS and NIR imaging models and then provide a procedure for face analogy.

3.1 Local Image Models

Assuming that the facial surface is Lambertian, then like Eq.1 a VIS face imaging model under spot lighting source is modeled by

$$I'(x,y) = \rho'(x,y)\mathbf{n}(x,y)\cdot\mathbf{s}' \tag{2}$$

When NIR lighting is near spot source, we have

$$I(x,y) = \rho(x,y)\frac{\mathbf{n}(x,y)\cdot\mathbf{s}}{r^2(x,y)} \tag{3}$$

where $r(x,y)$ is the distance from spot lighting to face. Note that from spectroscopy, the albedo functions $\rho'(x,y)$ for VIS and $\rho(x,y)$ for NIR are different. These two equations will be abbreviated as $I'(i) = \rho'(i)\mathbf{n}(i)\cdot\mathbf{s}'$ and $I'(i) = \rho'(i)\frac{\mathbf{n}(i)\cdot\mathbf{s}}{r^2(i)}$.

Without loss of generality, we assume that facial surface normal and lighting vary continuously, while more accurate formulation would take into account abrupt changes near boundaries. Then, in a small neighborhood $\mathcal{N}_i$ of pixel i, we have $\mathbf{n}(i)\cdot\mathbf{s}' = c'$ and $\mathbf{n}(i)\cdot\mathbf{s}/r^2(i) = c$, where c' and c are some constants, and the Lambertians become,

$$I'(i) = \rho'(i)c', i \in \mathcal{N}_i, \tag{4}$$

$$I(i) = \rho(i)c, i \in \mathcal{N}_i \tag{5}$$

3.2 Local Normalization

To construct common local invariants of NIR and VIS face images, we introduce an operator called *local normalization* as

$$f:\ h(i) \leftarrow \frac{I(i) - m(\mathcal{N}_i)}{\sigma(\mathcal{N}_i)}, \tag{6}$$

where $m(\mathcal{N}_i)$ and $\sigma(\mathcal{N}_i)$ denote the mean and the deviation within $\mathcal{N}_i$, respectively. Similarly define

$$g:\ h'(i) \leftarrow \frac{I'(i) - m'(\mathcal{N}_i)}{\sigma'(\mathcal{N}_i)}, \tag{7}$$

Taking the Lambertian equations into consideration, we get

$$h(i) = \frac{I(i) - m(\mathcal{N}_i)}{\sigma(\mathcal{N}_i)} = \frac{\rho(i) - \bar{\rho}(\mathcal{N}_i)}{\sqrt{\Sigma(\rho(i) - \bar{\rho}(\mathcal{N}_i))^2/n^2}}, \tag{8}$$

$$h'(i) = \frac{I'(i) - m'(\mathcal{N}_i)}{\sigma'(\mathcal{N}_i)} = \frac{\rho'(i) - \bar{\rho}'(\mathcal{N}_i)}{\sqrt{\Sigma(\rho'(i) - \bar{\rho}'(\mathcal{N}_i))^2/n^2}}, \tag{9}$$

Both $h(i)$ and $h'(i)$ are functions of the albedos.

We would like to model the relationship of $h(i)$ and $h'(i)$. Assume that there exists a local linear transformation $T(\mathcal{N}_i)$ from $\rho(i)$ to $\rho'(i)$, that is,

$$\rho'(i) = T(\mathcal{N}_i) \circ \rho(i), i \in \mathcal{N}_i. \tag{10}$$

Then, we can use the local linear transformations to equate

$$h'(i) = h(i) \tag{11}$$

3.3 Procedure

According to the analysis above, it is clear that we will accomplish the analogy from I to I' by applying a transformation $g^{-1} \circ f$, that is,

$$I' = g^{-1} \circ f \circ I, \tag{12}$$

where g^{-1} is,

$$g^{-1} : I'(i) \leftarrow h'(i) * \sigma'(\mathcal{N}_i) + m'(\mathcal{N}_i), \tag{13}$$

More specifically

$$I'(i) = \frac{I(i) - m(\mathcal{N}_i)}{\sigma(\mathcal{N}_i)} \cdot \sigma'(\mathcal{N}_i) + m'(\mathcal{N}_i). \tag{14}$$

Therefore, we can achieve face analogy when we can estimate these parameters.

Given a set of NIR faces $S = \{I_k\}$ and a set of VIS faces $S' = \{I'_k\}$, suppose that for each pair k, I_k and I'_k are mutually registered. Face Analogy takes S, S' and a new NIR face I as the input, and synthesizes the VIS counterpart I' of I as

$$I' \leftarrow \text{FaceAnalogy}(S, S', I). \tag{15}$$

The process of synthesizing a VIS face I' from an NIR face I is pixel-wise and in scan-line order. For each $i \in I'$, we first estimate $m(\mathcal{N}_i)$ and $\sigma(\mathcal{N}_i)$ by Eq.9, then search for the best match $I_k(\mathcal{N}_i)$ of $I(\mathcal{N}_i)$ among all NIR facial patterns defined in $\mathcal{N}_i$ and finally obtain the corresponding VIS facial pattern as

$$I'_k(\mathcal{N}_i) \leftarrow \text{BestMatch}(S, S', I, I', i). \tag{16}$$

So, we can estimate $\hat{m}'(\mathcal{N}_i)$ and $\hat{\sigma}'(\mathcal{N}_i)$ from $I'_k(\mathcal{N}_i)$ by Equ.13, and further obtain $I'(i)$ by Equ.(14). To reduce the error induced by scan-line, I' can be synthesized in four directions: from upper-left to bottom-right, bottom-right to upper-left, lower-left to upper-right, and upper-right to lower-left, and then their mean is computed as the final result.

4 Experiments

The following experiments are done to illustrate how to apply face analogy to synthesize a VIS face from a NIR probe face in terms of target VIS faces, and how to match the synthesized VIS face to the target face.

A database of NIR and VIS face image pairs of 100 persons was collected for the training and testing of face analogy and heterogeneous face matching. There are two NIR images and two VIS face images for each person, and so there are 200 NIR images and 200 VIS images in the database. The VIS and NIR images are captured in different time, containing variations in pose, expression, light condition, and so on. The face images are aligned with 68 landmarks manually into the size of 128×128 pixels. Fig.2 show some face images after alignment.

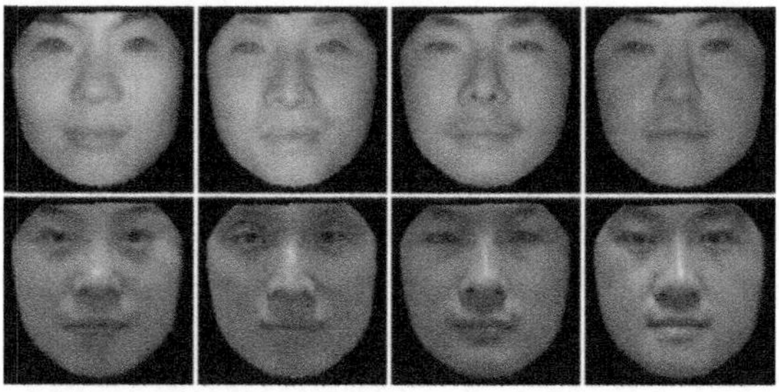

Fig. 2. Examples of aligned NIR (upper) and VIS (lower) face images

The 200 images in the NIR set are used as the source images from which the VIS (SYN) images are synthesized. The synthesis by face analogy is done in the leave-one-out manner, that is, by using all except one VIS images as the references. Fig. 3 shows some face analogy synthesized results. From the results, we can see the face analogy algorithm performs reasonably well for the NIR-VIS synthesis problem. The structure of NIR face is reserved meanwhile the appearance is transformed to the VIS type.

For the SYN-VIS matching, the 200 images in the NIR set are used as the probes, and the 200 images in the VIS set are used as the targets. No training is needed. The NIR probes are transformed to the counterpart VIS images (SYN) using face analogy. The SYN faces are then matched against the target of the VIS faces in the test set. Four matching methods are evaluated:

- NIR-VIS: face matching using the NIR-VIS image data directly.
- SYN-VIS: face matching using the SYN-VIS image data directly.
- NIR-VIS (LoG): face matching using Laplacian of Gaussian (LoG) filtered image data.
- SYN-VIS (LoG): face matching using LoG filtered image data.

The ROC curves of the four methods are shown in Fig. 4. In this experiment, the SYN-VIS methods outperform the rudimental NIR-VIS methods remarkably. Filtering NIR and VIS face images by LoG can further improve the performance.

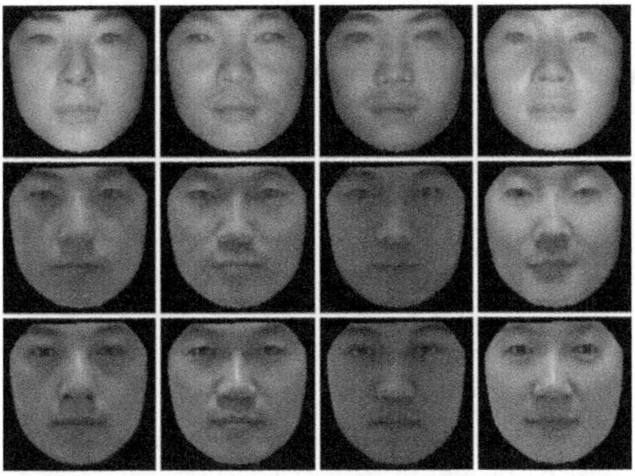

Fig. 3. Face analogy synthesized results: Source NIR (upper), synthesized VIS (middle), and true VIS (lower) face images

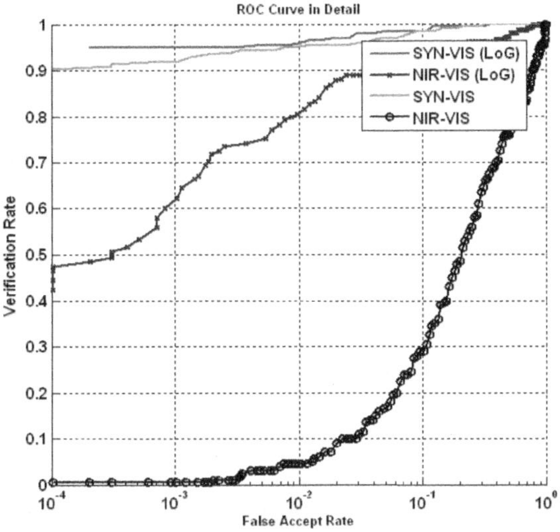

Fig. 4. The ROC curves of NIR-VIS heterogenous matchings and SYN-VIS matchings based on face analogy

5 Conclusions

Heterogeneous face biometrics (HFBs) perform biometric matching across heterogeneous face images. We have provided an analysis of problems in HFBs, and present a new method, face analogy, for transforming one type of face image to another, and use the synthesized image as a basis for NIR-VIS matching. Experiments are provided to illustrate the analysis-by-synthesis based algorithm.

Although the present analysis-by-synthesis method is for cross-spectrum face mapping between NIR and VIS types, cross-modality mapping such as between VIS and 3D could be tackled in a similar way. HFBs could be used standalone for biometric authentication or work as an added module to improve face recognition using homogeneous face images. Promising results are obtained.

Acknowledgements. This work was supported by the following funding sources: National Natural Science Foundation Project #60518002, National Science and Technology Support Program Project #2006BAK08B06, National Hi-Tech (863) Program Projects #2006AA01Z192, #2006AA01Z193, and #2008AA01Z124, Chinese Academy of Sciences 100 People Project, and AuthenMetric R&D Funds.

References

1. Zhao, W., Chellappa, R., Phillips, P., Rosenfeld, A.: Face recognition: A literature survey. ACM Computing Surveys, 399–458 (2003)
2. Li, S.Z., Chu, R., Liao, S., Zhang., L.: Illumination invariant face recognition using near-infrared images. IEEE Transactions on Pattern Analysis and Machine Intelligence 26 (2007) (Special issue on Biometrics: Progress and Directions)
3. Kong, S.G., Heo, J., Abidi, B., Paik, J., Abidi, M.: Recent advances in visual and infrared face recognition - A review. Computer Vision and Image Understanding 97(1), 103–135 (2005)
4. Bowyer, K.W., Chang, Flynn, P.J.: A survey of 3D and multi-modal 3D+2D face recognition. In: Proceedings of International Conference on Pattern Recognition, pp. 358–361 (2004)
5. Tang, X., Wang, X.: Face sketch recognition. IEEE Transactions on Circuits and Systems for Video Technology 14(1), 50–57 (2004)
6. Yi, D., Liu, R., Chu, R., Lei, Z., Li, S.Z.: Face matching between near infrared and visible light images. In: Lee, S.-W., Li, S.Z. (eds.) ICB 2007. LNCS, vol. 4642, pp. 523–530. Springer, Heidelberg (2007)
7. Yang, W., Yi, D., Lei, Z., Sang, J., Li, S.Z.: 2D-3D face matching using CCA. In: Proc. IEEE International Conference on Automatic Face and Gesture Recognition (2008)
8. Lei, Z., Bai, Q., He, R., Li, S.Z.: Face shape recovery from a single image using cca mapping between tensor spaces. In: Proceedings of IEEE Computer Society Conference on Computer Vision and Pattern Recognition (2008)
9. NIST: Multiple Biometric Grand Challenge (MBGC) (2008), `http://face.nist.gov/mbgc`
10. Hertzmann, A., Jacobs, C.E., Oliver, N., Curless, B., Salesin, D.H.: Image analogies. In: Fiume, E. (ed.) SIGGRAPH 2001, Computer Graphics Proceedings, pp. 327–340. ACM Press / ACM SIGGRAPH (2001)
11. Xie, X., Lam, K.M.: An efficient illumination normalization method for face recognition. Pattern Recognition Letters 27, 609–617 (2006)

Face Recognition with LWIR Imagery Using Local Binary Patterns

Heydi Méndez[1], Cesar San Martín[2], Josef Kittler[3], Yenisel Plasencia[1], and Edel García-Reyes[1]

[1] Advanced Technologies Application Center,
7a # 21812 e/ 218 y 222, Rpto. Siboney, Playa, C.P. 12200, La Habana, Cuba
{hmendez,ylpasencia,egarcia}@cenatav.co.cu
[2] Dep. of Electrical Eng. University of La Frontera, Casilla 54-D, Temuco, Chile
csmarti@ufro.cl
[3] University of Surrey, Guildford, Surrey, GU2 7XH, UK
J.Kittler@surrey.ac.uk

Abstract. In this paper, the merits of the Local Binary Patterns (LBP) representation are investigated in the context of face recognition using long-wave infrared images. Long-wave infrared images are invariant to illumination, but at the same time they are affected by a fixed-pattern noise inherent to this technology. The fixed-pattern is normally compensated by means of a non-uniformity correction method. Our study shows that the LBP approach is robust to the fixed-pattern noise, as well as to the presence of glasses. Not only no noise suppressing preprocessing is needed, but in fact if a non-uniformity correction method is applied, the image texture is amplified and the performance of the LBP degraded.

1 Introduction

The sensitivity to variations in illumination is one of the major limiting factors for face recognition system performance. Different methods have been proposed in the literature aiming to compensate for illumination variations, but the classification performance is still affected [1]. Among them, infrared (IR) imaging has been shown to be less affected by changes in ambient illumination. IR spectrum typically is divided in two groups: reflected IR ($0.7 - 2.4\mu$m) and thermal IR (2.4μm-14mm). Near-IR belongs to the reflected IR group. Being reflected by objects, it can serve as an active illumination source. The merits of using active Near-IR to provide constant illumination for face recognition have been demonstrated [2,3]. On the other hand, long-wave IR (LWIR), which belongs to the thermal IR, presents the following advantages: i) LWIR sensors collect the heat energy emitted by a body instead the light reflected; ii) it also exhibits invariant behavior under changes in illumination, being able to operate even in complete darkness; and iii) human skin has a high emissivity in $8 - 12\mu$m presenting a thermal signature which is unique to each individual.

In this paper, we focus on LWIR face recognition because of its benefits for outdoor applications where the environment illumination conditions are too variable. In the context of the previous work on face recognition using LWIR imagery

M. Tistarelli and M.S. Nixon (Eds.): ICB 2009, LNCS 5558, pp. 327–336, 2009.
© Springer-Verlag Berlin Heidelberg 2009

[4,5], some papers compare LWIR with visible images [6,7] and fuse these two kinds of data [8,9], showing promising results. In most of the cases, appearance based methods are used to represent and recognize LWIR images [7,10]. However, face recognition using LWIR imagery with Local Binary Patterns (LBP) has received little attention in the literature [9], in spite of its simplicity of implementation, efficiency and robustness already demonstrated in visible and Near-IR face recognition tasks.

An IR system consists of a group of detectors placed in an array located at the focal plane of an imaging optical system, known as focal-plane array (FPA) technology [11]. FPA is hindered by the non-uniformity (NU) effect, manifest in fixed-pattern noise (FPN), i.e. a pixel-to-pixel variation in the sensor array, caused by the differences between semiconductors. The FPN is superimposed on the true images degrading the quality of the captured data [12]. In order to solve this problem, several NU correction methods have been proposed to compensate for the NU noise in IR-FPA. For example, image-registration-based algorithms [13], algebraic scene-based algorithms [14], the constant statistics algorithms [15], and Kalman-filter-based algorithms [16,17], among others. But, to our knowledge, only in [18] these compensation methods are applied to LWIR face recognition, improving significantly the recognition performance.

In this paper we evaluate the performance of the LBP method applied to LWIR face recognition and its response to FPN. We show that LBP is very robust to the FPN and surprisingly its discriminatory capacity is degraded when the image texture is affected by random noise. In Section 2, we review the LBP method and discuss the origins, nature and effect of noise in IR images. Section 3 evaluates the performance of the proposed methodology. Section 4 concludes the paper.

2 Methodology

In this section the LBP method is described and basic notions about noise in LWIR images are presented.

2.1 Local Binary Patterns

The use of the LBP operator in face recognition was introduced in [19] and different extensions of the original operator have appeared afterwards [20]. As it can be appreciated in Figure 1, the original LBP operator represents each pixel of an image by thresholding its $3x3$- neighbourhood with reference to the center pixel value, g_c, and considering the result as a binary number, called the LBP code. The image is then divided into rectangular regions and histograms of the LBP codes are calculated over each of them. Finally, the histograms of each region are concatenated into a single one that represents the face image. A dissimilarity measure is used to compare the histograms of two different images.

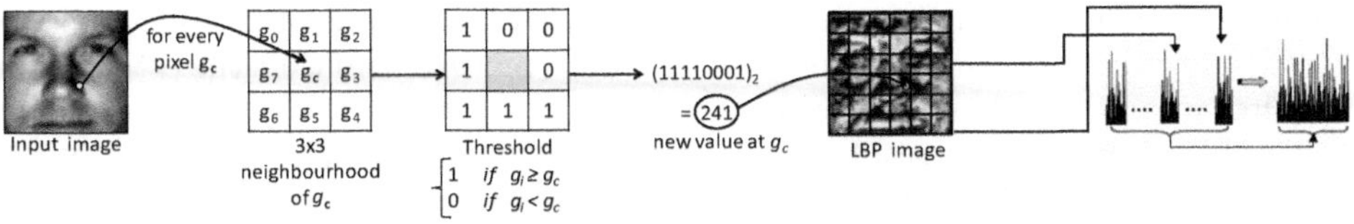

Fig. 1. Local Binary Patterns

2.2 Non-Uniformity Noise in LWIR Images and Its Correction

The IR detector has several known sources of noise, but the main contributing factors to signal corruption are the FPA temporal noise, the FPA NU noise, and the readout noise injected by the output electronics [12]. The FPN refers to any spatial pattern that does not change significantly from frame-to-frame, and it is due to differences in detector size, doping density and foreign matter getting trapped during fabrication. The multiplicative component of the FPN is due to the FPA photoresponse NU, which means that the responsivity changes from one detector to another. This component is produced by pixel-to-pixel differences in responsivity or gain, the detector size, spectral response and thickness in coatings of each detector. The additive component of the FPN, white noise, is mainly caused by the dark current, which is kept almost constant and does not vary from frame to frame because of its dependence on the temperature and the offset voltages of the amplifiers used. Taking all the noise process into consideration, each pixel on the FPA can be modelled at instant n, using a two-parameter linear equation given by:

$$Y_{ij}(n) = A_{ij}X_{ij}(n) + B_{ij} + v_{ij}(n), \tag{1}$$

where A_{ij} and B_{ij} are, respectively, the gain (multiplicative NU) and the offset (additive NU) associated with the $ijth$ detector. The random process $X_{ij}(n)$ is the IR photon flux collected at the $ijth$ detector. The term, $v_{ij}(n)$, corresponds to the additive noise, FPA temporal noise, which is commonly modelled as a zero mean white Gaussian random process called temporal random noise. $Y_{ij}(n)$ is referred to as readout data from the IR-FPA system.

In order to solve this problem, several NU compensation techniques have been developed. They can be divided into calibration techniques and scene-based correction methods. The first group requires two uniform references from blackbody radiator at different temperatures, and then, the gain and offset are obtained. The NU compensation is performed using the following equation:

$$\hat{X}_{ij}(n) = \frac{Y_{ij}(n) - \hat{B}_{ij}}{\hat{A}_{ij}}, \tag{2}$$

where $\hat{A}_{ij}$ and $\hat{B}_{ij}$ are the estimates of the gain and offset, and $\hat{X}_{ij}(n)$ is the estimated IR input irradiance.

The scene-based methods estimate gain and offset, but their performance is limited by the amount of spatio-temporal information and the diversity of temperature in the image sequence. The principal limitations of the scene-based method are that the camera must be in constant motion, and that the estimation requires a large number of frames.

A well-known scene-based NUC method is the constant statistics method proposed in [15]. The principal assumption of this method is that the input irradiance in mean and variances is the same for all sensors of the FPA. Applying the mean and variance to equation (1), and assuming that the mean and standard deviation of X_n are 0 and 1, respectively, the gain and offset can be obtained from:

$$\hat{A}_{ij} = \sigma_{Y_{ij}} - \sigma_{v_{ij}}, \tag{3}$$

$$\hat{B}_{ij} = \mu_{Y_{ij}}. \tag{4}$$

In order to obtain a solution with low estimation error, a good method of estimating of the mean and variance (large number of frames)is required. To avoid this conditions we assume that the readout data have a uniform distribution in a known range $[Y^{\min}, Y^{\max}]$. In such a case the mean and variance can be obtained respectively from:

$$\mu_{Y_{ij}} = \frac{Y_{ij}^{\max} + Y_{ij}^{\min}}{2} \quad \wedge \quad \sigma_{Y_{ij}} = \frac{Y_{ij}^{\max} - Y_{ij}^{\min}}{\sqrt{12}}, \tag{5}$$

and then, the correction is performed using equation (2) with the values obtained from (3) and (4).

3 Experimental Evaluation and Discussion

In our experiments we used the long-wave infrared images collected by Equinox Corporation [21]. This database is composed of three 40 frame sequences from 91 persons, acquired in two days with three different light sources: frontal, left lateral and right lateral. The frame sequences were recorded while people were uttering vowels standing in a frontal pose, and three more images from each person were taken to capture the expression of, respectively to smile, frown and surprise. In addition, the complete process was repeated for those persons who wore glasses. Since much of the data of one person is highly correlated, usually only a subset of the images is used for experimentation [6,7].

The LWIR images of Equinox are of 240x320 pixels size, they were radiometrically calibrated and represented as gray-scale images with 12 bits per pixels. The calibrated images from the database present a dead-pixel distortion besides the typical FPN. This source of noise means that the detector always gives the same readout value independent of the input irradiance. In order to remove it, it is possible to assume that the IR irradiance collected by the sensor ij is to be close to the neighbors around the sensor ij, then we assumed this value as the readout data.

The different face images of a person are not aligned as can be appreciated in Figure 2, however the displacement is not too great. Taking into account this

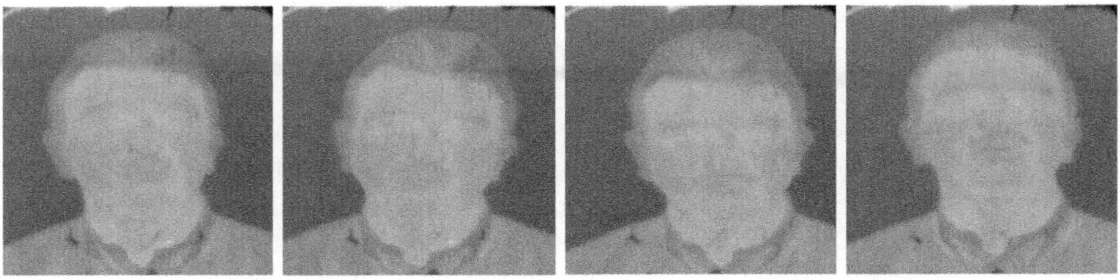

Fig. 2. Examples of LWIR images of one person

fact and that it is not easy to precisely detect the facial features in LWIR, the images were not geometrically normalized. On the other hand, the size of the LBP regions was selected not too small neither too big, $18x21$ pixels as in [19].

Because the use of glasses changes the thermal emission perceived in LWIR face images, we split our experiments into two cases: with and without glasses.

3.1 The No Glasses Case

In this case, we use the same procedure as other authors [7,8] who constructed the following multiple subsets using only three images of the vocal pronunciation frame sequence (vowel frames) and the three expression frames of each subject in each illumination condition:

VA: Vowel frames, all illuminations.
EA: Expression frames, all illuminations.
VF: Vowel frames, frontal illumination.
EF: Expression frames, frontal illumination.
VL: Vowel frames, lateral illuminations.
EL: Expression frames, lateral illuminations.

The performance of the LBP based matching is evaluated by using each time one set as gallery and another as a test set. Some of the subset combinations are not considered in the experiments since one subset is included in the other. Table 1 below, where the columns represent the gallery sets and the rows the test sets, shows the top-match recognition performance of the LBP method.

Table 1. Correct classification percent with LBP method

	VA	**EA**	**VF**	**EF**	**VL**	**EL**
VA		98.13	97.75	98.50	99.25	95.13
EA	97.67		93.77	98.44	97.67	100.0
VF		98.13		97.00	97.38	95.51
EF	99.24		99.62		99.24	95.08
VL		98.13	96.63	96.63		96.63
EL	97.74		92.83	95.09	98.49	

Although the alignment of the faces in Equinox images is very similar, even a small localization error usually affects the appearance based methods. The obtained results, giving an average of 97.3 percent of correct classification are very promising, taking into account that images were not geometrically normalised or cropped. This performance is comparable with the best performance of an appearance based method obtained earlier with the same LWIR images previously geometrically transformed, by means of the Linear Discriminant Analysis (LDA) method [22], reported in [7] and summarized in Table 2.

Table 2. Classification results obtained with the LDA method [7]

	VA	EA	VF	EF	VL	EL
VA		99.6	98.3	96.2	99.6	99.3
EA	97.4		94.0	98.1	96.8	99.2
VF		100.0		97.0	98.8	98.6
EF	97.2		94.6		95.6	97.9
VL		99.5	97.4	95.8		99.6
EL	97.4		93.7	97.1	97.4	

Note that the results are very similar, with the LDA achieving an average of 97.5 percent of correct classification. However, besides the fact that no geometric normalisation was needed, LBP has the advantage over the LDA method that it does not need more than one image per person in the gallery set, neither it requires a training set, which are very important properties of a face recognition system.

In order to improve the results and following the idea in [18], prior to computing the LBP representation we applied a NU correction method, estimating the FPN as was described in section 2.2 since the blackbody images are not available from [21]. Table 3 shows the identification rates obtained when the correction was applied.

Table 3. Classification results with the LBP method after NU correction

	VA	EA	VF	EF	VL	EL
VA		88.89	95.63	87.89	95.38	90.14
EA	97.67		94.94	93.77	98.05	97.67
VF		85.77		92.51	90.26	86.02
EF	98.48		98.86		98.48	94.70
VL		91.26	93.63	83.65		90.76
EL	96.60		93.96	93.21	98.11	

Surprisingly, with an average value of 93.3 percent, the performance with the NU correction was lower than without it. Inspecting the original and NU corrected images in Figure 3, it is apparent that although the NUC method

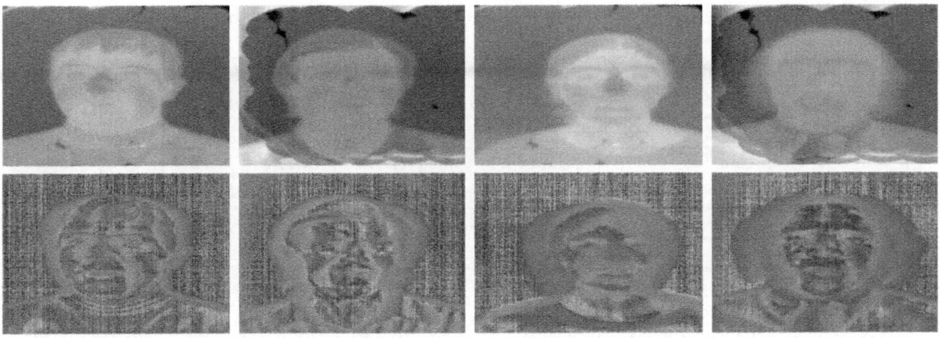

Fig. 3. Examples of LWIR original (first row) and NU corrected (second row) images

suppresses fixed-pattern noise in the IR images, the random noise is magnified and the image texture is affected. Since LBP is a texture descriptor, it is sensitive to this kind of noise.

In order to support the hypothesis that the LBP method is sensitive to random noise in LWIR images, we conducted the same experiments adding some random noise artificially to the original images, as can be seen in Figure 4.

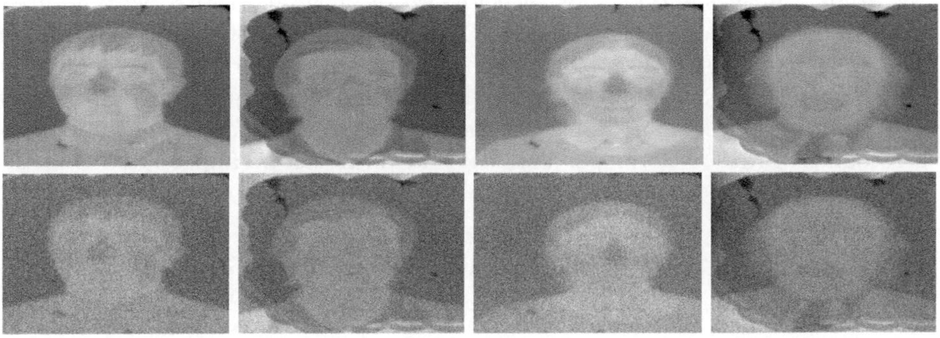

Fig. 4. Examples of LWIR original (first row) and random noise affected (second row) images

Table 4 displays the results of the experiments adding the random noise to the original LWIR images, with an average of 86.3 percent they confirm that LBP method decreases its performance in the presence of this kind of noise.

3.2 The Glasses Present Case

It is well known that in the presence of glasses the performance of the recognition algorithms for thermal images is very low [5]. There is some work addressing this problem, e.g. in [6] where they detect and remove the eyeglass area in the

Table 4. Classification results with the LBP method adding random noise

	VA	EA	VF	EF	VL	EL
VA		90.26	86.64	86.77	92.26	90.01
EA	87.03		81.06	80.16	87.55	91.05
VF		82.27		89.89	86.14	85.27
EF	89.14		88.76		85.61	83.33
VL		89.01	83.02	82.77		88.64
EL	86.79		80.50	82.64	87.92	

LWIR images in order to improve the recognition performance. We conducted an experiment to test the performance of the LBP method in LWIR face images when people are wearing glasses.

In this case, we used the VA subset as a gallery set and constructed three new test sets using only the LWIR images of those people who are wearing glasses in the following manner:

GF: all glasses images with frontal illumination.
GL: all glasses images with left lateral illumination.
GR: all glasses images with right lateral illumination.

Table 5 shows the results of the LBP method in the presence of glasses, for the original images and also for the same set of images corrupted with additive random noise. As can be seen, in general, the performance of the method is lower than for the no glasses case. However the results are comparable and even better on average than those obtained with the earlier methods which need to detect and preprocess eyeglass regions. The method used in [6], for example, achieves a recognition rate of 85.77 percent. Once again it can be appreciated that the results are affected by random noise.

Table 5. Correct classification with LBP in the presence of glasses

	GF	GL	GR	Average
VA-original images	83.67	85.41	89.58	86.20
VA-with added noise	61.22	64.58	66.67	64.14

4 Conclusions

This paper applies the Local Binary Pattern method to LWIR face imagery, in order to perform person identification tasks. We observed through experimentation that LBP performs well on LWIR images, as compared to the state of the art methods for these kind of images, having some advantages compared with other appearance based methods in its simplicity of implementation and not needing any training images. It is important to note that the good results were obtained although the images were miss-aligned, which is an important property for thermal face recognition where it is easy to detect the face region but not the face features to make a geometric normalisation. Even in the presence of glasses the LBP method achieves good results without applying any preprocessing step. On the other hand we note that a magnified random noise affects the identification rate of the LBP method and the application of NU compensation magnifies this kind of noise, although it reduces the fixed-pattern noise. In conclusion it can be said that the LBP method is robust to the fixed-pattern noise and it may be applied directly to the LWIR face imagery with an identification rate average of 97.3 percent, achievable on good quality images.

References

1. Zou, X., Kittler, J., Messer, K.: Illumination Invariant Face Recognition: A Survey. In: First IEEE International Conference on Biometrics: Theory, Applications, and Systems, pp. 1–8 (2007)
2. Zou, X., Kittler, J., Messer, K.: Face recognition using active near-ir illumination. In: British Machine Vision Conference Proceedings (2005)
3. Li, S.Z., Chu, R., Liao, S., Zhang, L.: Illumination invariant face recognition using near-infrared images. IEEE Trans. Pattern Anal. Mach. Intell. 29(4), 627–639 (2007)
4. Socolinsky, D.A., Wolff, L.B., Neuheisel, J.D., Eveland, C.K.: Illimination invariant face recognition using thermal infrared imagery. In: Proc. IEEE CS Conf. Comp. Vision and Pattern recognition, vol. 1, pp. 527–534 (2001)
5. Kong, S.G., Heo, J., Abidi, B.R., Ki Paik, J., Abidi, M.A.: Recent advances in visual and infrared face recognition - a review. Computer Vision and Image Undertanding 97(1), 103–135 (2005)
6. Heo, J., Savvides, M., Vijaya Kumar, V.K.: Performance Evaluation of Face Recognition using visual and thermal imagery with advanced correlation filters. In: Conference on Computer Vision and Pattern Recognition, pp. 9–14. IEEE Computer Society, Los Alamitos (2005)
7. Socolinsky, D.A., Selinger, A.: A Comparative Analysis of Face Recognition Performance with Visible and Thermal Infrared Imagery. In: ICPR 2002: Proceedings of the 16Th International Conference on Pattern Recognition, vol. 4(4) (2002)
8. Bebis, G., Gyaourova, A., Singh, A., Pavlidis, I.: Face recognition by fusing thermal infrared and visible imagery. Image and Vision Computing 24(7), 727–742 (2006)
9. Singh, R., Vatsa, M., Noore, A.: Integrated multilevel image fusion and match score fusion of visible and infrared face images for robust face recognition. Pattern Recogn. 41(3), 880–893 (2008)
10. Chen, X., Flynn, P.J., Bowyer, K.w.: PCA-based face recognition in infrared imagery: baseline and comparative studies. In: IEEE International Workshop on Analysis and Modeling of Faces and Gestures, AMFG, pp. 127–134 (2003)
11. Pron, H., Menanteau, W., Bissieux, C., Beaudoin, J.L.: Characterization of a focal plane array (FPA) infrared camera. Quantitativa Infrared Thermography QIRT Open Archives, QIRT, 2000–2061 (2000), http://qirt.gel.ulaval.ca
12. Milton, A., Barone, F., Kruer, M.: Influence of nonuniformity on infrared focal plane array performance. Opctical Engineering 24, 855–862 (1985)
13. Hardie, R., Hayat, M., Armstrong, E., Yasuda, B.: Scene-based nonuniformity correction using video sequences and registration. Applied Optics 39, 1241–1250 (2000)
14. Ratliff, B., Hayat, M., Tyo, J.: Generalized algebraic scene-based nonuniformity correction algorihtm. The JOSA-A Opt. Soc. of America 22, 239–249 (2005)

15. Harris, J., Chiang, Y.: Nonuniformity correction of infrared image sequences using constant statistics constraint. IEEE Trans. on Image Processing 8, 1148–1151 (1999)
16. Pezoa, J., Hayat, M., Torres, S., Rahman, M.: Multimodel kalman filtering for adaptive nonuniformity correction in infrared sensors. The JOSA-A Opt. Soc. of America 23, 1282–1291 (2006)
17. Averbuch, A., Liron, G., Bobrovsky, B.Z.: Scene based non-uniformity correction in thermal images using Kalman filter. Image and Vision Computing 25, 833–851 (2007)
18. San Martin, C., Meza, P., Torres, S., Carrillo, R.: Improved Infrared Face Identification Performance Using Nonuniformity Correction Techniques. In: Blanc-Talon, J., Bourennane, S., Philips, W., Popescu, D., Scheunders, P. (eds.) ACIVS 2008. LNCS, vol. 5259, pp. 1115–1123. Springer, Heidelberg (2008)
19. Ahonen, T., Hadid, A., Pietikäinen, M.: Face recognition with local binary patterns. In: Pajdla, T., Matas, J. (eds.) ECCV 2004. LNCS, vol. 3021, pp. 469–481. Springer, Heidelberg (2004)
20. Marcel, S., Rodriguez, Y., Heusch, G.: On the Recent Use of Local Binary Patterns for Face Authentication. International Journal on Image and Video Processing Special Issue on Facial Image Processing (2007)
21. http://www.equinoxsensors.com/products/HID.html
22. Belhumeur, P.N., Hespanha, J.P., Kriegman, D.J.: Eigenfaces vs. Fisherfaces: Recognition using class specific linear projection. IEEE Trans. Pattern Anal. Machine Intell. 7(19), 711–720 (1997)

A Classification Framework for Large-Scale Face Recognition Systems

Ziheng Zhou, Samuel Chindaro, and Farzin Deravi

Department of Electronics, University of Kent
Canterbury, UK CT2 7NT
{z.zhou,s.chindaro,f.deravi}@kent.ac.uk

Abstract. This paper presents a generic classification framework for large-scale face recognition systems. Within the framework, a data sampling strategy is proposed to tackle the data imbalance when image pairs are sampled from thousands of face images for preparing a training dataset. A modified kernel Fisher discriminant classifier is proposed to make it computationally feasible to train the kernel-based classification method using tens of thousands of training samples. The framework is tested in an open-set face recognition scenario and the performance of the proposed classifier is compared with alternative techniques. The experimental results show that the classification framework can effectively manage large amounts of training data, without regard to feature types, to efficiently train classifiers with high recognition accuracy compared to alternative techniques.

Keywords: Classification framework, face recognition and kernel Fisher discriminant.

1 Introduction

In the past two decades, there has been a great deal of research and development in the field of face recognition (FR) [14]. To develop a large-scale practical FR system, it is essential to use a large facial image database for training and testing so that real-world scenarios that may be faced in target applications can be effectively represented. Fortunately, some large databases have been built up for testing various FR technologies [8-10].

It is well known that the task of face recognition can be turned into a simple and effective two-class classification problem [4,6,7]. To do that, facial features are extracted from an image pair (instead of a single image) and then classified into the intra- and extra-personal categories. Here, the intra-personal features represent those calculated from two images of the same persons while the extra-personal features are calculated from two images from different persons.

Although adopting the classification problem is not new in the literature, using it to build up an FR system based on a large facial database (e.g., the Face Recognition Grand Challenge (FRGC) database [9]) still remains challenging. The first major challenge is how to handle the large amount of training data. For example, if the FRGC database is used and only the controlled frontal images are exploited,

M. Tistarelli and M.S. Nixon (Eds.): ICB 2009, LNCS 5558, pp. 337–346, 2009.
© Springer-Verlag Berlin Heidelberg 2009

according to [9], there are more than 20 thousand such images from which more than 400 million image pairs can be sampled for training and testing. Moreover, the number of intra-personal image pairs and extra-personal image pairs are extremely imbalanced. Fig. 1 shows an example that explains such imbalance. In the figure, the black squares represent images of Subject i and rest of the squares stand for other images in the database. The stars mark all possible intra-personal image pairs that can be sampled for Subject i, while the circles locate the extra-personal image pairs. It is obvious that in a large database, there will be much more extra-personal image pairs than the intra-personal pairs for every Subject i, resulting in the huge imbalance. Here, the question turns to be how to sample image pairs to form a balanced training dataset with a reasonable size.

The way that the training and test data are defined depends on whether a FR application is a close-set or an open-set problem. For an open-set problem, the system will be expecting in the test data some subjects which have not been encountered at the training phase. For a close-set scenario, all subjects in the test are also included in the training set. In recent evaluation campaigns [10], it has been observed that the open-set scenario could show a much lower FR performance compared to a closed-set scenario. Very often, a real-world large-scale FR system is required to deal with an open-set scenario. Therefore, the second challenge will be how to properly select and train a classifier using the available training data to classify the test data from some unknown subjects.

In this paper, we propose a generic classification framework to tackle the above-mentioned challenges. The framework consists of an image-pair sampling strategy for preparing a balanced set of samples for training and a kernel-based classifier that can perform reasonably in an open-set scenario and can be trained using a large training dataset. The framework is essentially "feature agnostic" – meaning that it is independent of the choice feature-extraction algorithms used to produce training and test vectors for classification. The multi-resolution local binary pattern (LBP) [1] features are used to test the system on the FRGC database. Experiments are designed to simulate the open-set scenario and results indicate that the framework significantly improves the performance.

The rest of the paper is organised as follows: Section 2 and 3 describes the sampling strategy and the classifier, respectively. The experiments and results are presented in Section 4. Section 5 provides a summary and conclusions.

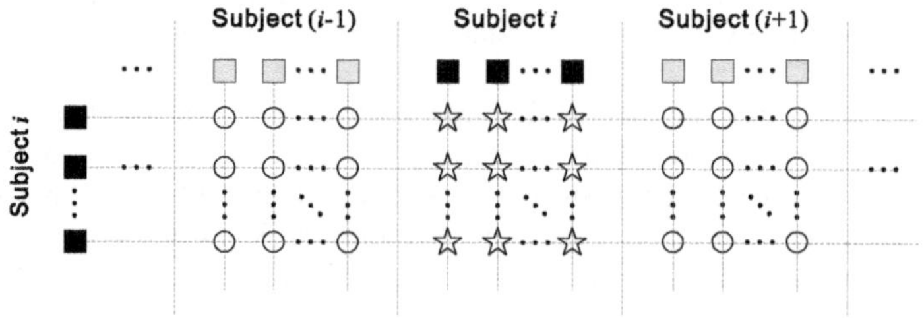

Fig. 1. An example showing the imbalance between the number of intra-personal image pairs and the number of extra-personal image pairs that can be sampled for Subject i. The black squares represent images of Subject i and rest of the squares stand for other images in the database. The stars mark all possible intra-personal image pairs that can be sampled for Subject i, while the circles locate the extra-personal image pairs.

2 Data Sampling Strategy

As mentioned above, a large facial database (e.g., the FRGC database) could contain thousands of face images from which millions of image pairs can be sampled for training and test purposes. The huge amount of data available and the imbalance between the numbers of intra- and extra-personal image pairs make the preparation of a suitable training dataset non-trivial for any FR system.

To cope with the large number of training samples and to make each subject evenly weighted in the training dataset, one common way [13] is to select equal number of images for each subject and sample all the intra- and extra-personal image pairs from the selected images. By evenly weighted, we mean that the numbers of intra- and extra-personal image pairs including images of a particular subject should be the same for each subject. Although the method is intuitively straightforward, the obtained training dataset is not suitable for the classification problem as explained in the example below.

Suppose that there is a database containing $N_{\mathrm{sub}} = 300$ subjects (a large facial database could include many more subjects) and for each subject, $n = 4$ images are randomly selected. Assume that there are ten images from Subject i and seven images from Subject j. Fig. 2(a) and Fig. 3(a) show the intra-personal pairs sampled for Subject i and extra-personal pairs sampled for Subject i and j. (Note that without loss of generality, images belonging to a subject can always be organised in the order that starts with the selected images represented by the black boxes.) Here the image pair (I_1, I_2) is considered to be the same as (I_2, I_1) and the pairs formed by the same images are only included once. The total number of intra-personal pairs N_{int} can be calculated as $\frac{1}{2}n(n-1)N_{\mathrm{sub}} = 1800$ and the total number of extra-personal pairs N_{ext} can be calculated as $\frac{1}{2}n^2 N_{\mathrm{sub}}(N_{\mathrm{sub}} - 1)$ which is more than 7×10^5. The disadvantage of the sampling method is that N_{int} and N_{ext} are decided by n given N_{sub} and $N_{\mathrm{int}} \ll N_{\mathrm{ext}}$ always holds no matter how n changes. Besides the large imbalance between N_{int} and N_{ext}, from the figures, it can been that the image pairs are not randomly sampled from all the possible positions.

In this paper, we propose a strategy for sampling a balanced dataset in which each subject is evenly weighted. Here the subjects involved in the sampling are denoted as $\{s_1, s_2, \dots, s_{N_{\mathrm{sub}}}\}$ and the images from s_i as $\{I_1^i, I_2^i, \dots, I_{N_{s_i}}^i\}$ where N_{s_i} is the number of images. Instead of selecting images for each subject, we first list all the intra- and extra-personal pairs. For s_i, the intra-personal pairs can be expressed as:

$$\mathcal{P}_{\mathrm{int}}^i = \{(I_u^i, I_v^i) | u < v, 1 \le u \le (N_{s_i} - 1), 2 \le v \le N_{s_i}\}. \quad (1)$$

In Eq. (1), the image index u is always smaller than v to avoid including an image pair twice and to exclude those formed by the same images. For a subject pair (s_i, s_j) $(i < j)$, the extra-personal pairs can be expressed as:

$$\mathcal{P}_{\mathrm{ext}}^{i,j} = \{(I_u^i, I_v^j) | 1 \le u \le N_{s_i}, 1 \le v \le N_{s_j}\}. \quad (2)$$

In Eq. (2), the subject index i is always smaller than j to avoid considering a subject pair twice. We then randomly choose n_{int} pairs from each $\mathcal{P}_{\text{int}}^{i}$ and n_{ext} pairs from each $\mathcal{P}_{\text{ext}}^{i,j}$ to form the training dataset. In this way, the total numbers of the intra- and extra-personal pairs can be computed as:

$$N_{\text{int}} = n_{\text{int}} N_{\text{sub}}$$
$$N_{\text{ext}} = \frac{1}{2} n_{\text{ext}} N_{\text{sub}} (N_{\text{sub}} - 1) \tag{3}$$

It can be seen that given fixed N_{sub}, N_{int} and N_{ext} are controlled by n_{int} and n_{ext}, respectively. The number $n_{\text{int}}/n_{\text{ext}}$ can be customised to enlarge/limit the number of intra-/extra-personal image pairs in the training dataset. Following the sampling example mentioned above, Fig. 2(b) and 3(b) show the image pairs obtained using the proposed sampling strategy where $n_{\text{int}} = 15$ and $n_{\text{ext}} = 5$. It can be seen that not only twice more intra-personal pairs are selected, but more images of Subject i are involved in the sampled image pairs, which in some sense makes the image pairs

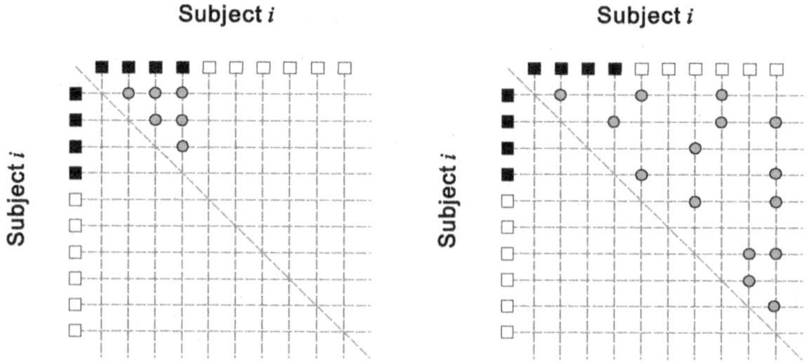

(a) (b)

Fig. 2. The intra-personal image pairs sampled for Subject i using (a) the normal image-based sampling method and (b) the proposed image-pair-based sampling strategy

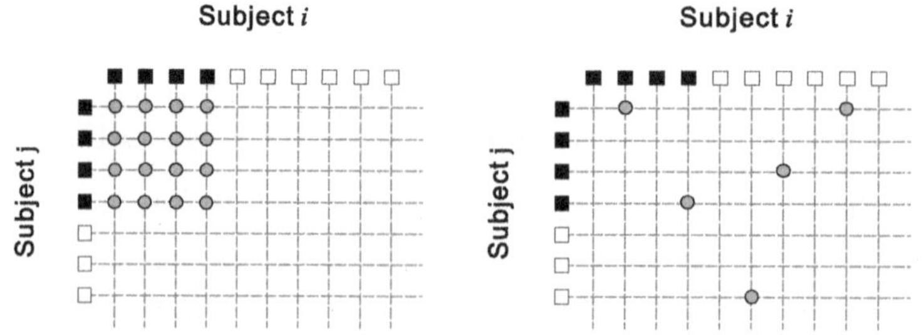

(a) (b)

Fig. 3. The extra-personal image pairs sampled for Subject i and Subject j using (a) the normal image-based sampling method and (b) the proposed image-pair-based sampling strategy more representative for Subject i. On the other hand, because of $n_{\text{ext}} = 5$, the number of extra-personal pairs is significantly reduced to one third of the previous size. Note that subjects in a large database often have more images and a larger n_{ext} can be chosen to sample more intra-personal pairs.

3 Modified Kernel Fisher Discriminant

For an effective FR system, the choice of the classifier is equally important as the selection of informative and discriminatory features, especially for an open-set problem. In some cases, a simple classifier (e.g., a K-nearest neighbour classifier or a Bayesian classifier [6]) can do the job. However, if sufficient training data is available, a more sophisticated classifier could significant improve the system performance particularly when the extracted features have a complex distribution in the feature space. Recently, the kernel-based classification methods (e.g., the support vector machine (SVM) [11] and the kernel Fisher discriminant (KFD) [5]) have been widely used to solve some non-linear classification problems [4,5,7,13]. If two classes of data points cannot be separated sufficiently in the original space, the kernel methods provide a way to map the points into a higher-dimensional feature space in which they could distribute more sparsely and therefore, more easily be separated. The merit of these methods is that the computation of the mapping can be avoided by the 'kernel trick' [11] which makes it computationally feasible and affordable.

Let Φ be the mapping to the new feature space $\mathcal{F}$ and $\Phi(\boldsymbol{x}) \in \mathcal{F}$ be the point mapped from a data sample $\boldsymbol{x}$. The kernel $\mathscr{k}$ is a function defined in the original space to calculate the dot product of two mapped samples $\Phi(\boldsymbol{x})$ and $\Phi(\boldsymbol{x}')$, that is, $\mathscr{k}(\boldsymbol{x},\boldsymbol{x}') = \langle\Phi(\boldsymbol{x}),\Phi(\boldsymbol{x}')\rangle$. In both SVM and KFD, the gram matrix $\boldsymbol{K}$ [11] is calculated from all the training samples to solve some optimisation problem. Each entry $\boldsymbol{K}(i,j)$ is defined as the value of the kernel $\mathscr{k}(\boldsymbol{x}_i,\boldsymbol{x}_j)$ where $\boldsymbol{x}_i$ and $\boldsymbol{x}_j$ are the ith and j^{th} samples in the training dataset. According to the definition, the dimension of matrix $\boldsymbol{K}$ will be $N \times N$ where N is the total number of samples for training. Although a sampling strategy has been proposed to significantly reduce the size of the training data, as there are possibly hundreds of subjects included in the database, there could still be tens of thousands of image pairs selected for training according to Eq. (**3**), which would make $\boldsymbol{K}$ very large. Keeping such a large matrix not only requires a large amount of memory, but makes the optimisation problem computationally very expensive, sometimes even infeasible. To solve the problem, Joachims [3] developed the SVM^{light} system, an implementation of an SVM learner which addresses the problem of having large training dataset. In this paper, a modified KFD (MKFD) is proposed to tackle the problem in a more efficient way.

Let $\{\boldsymbol{x}_1,\boldsymbol{x}_2 \dots,\boldsymbol{x}_N\}$ be the training samples. The classic KFD algorithm tries to search a vector $\boldsymbol{w}^* \in \mathcal{F}$ that maximises the ratio of the intra-class variations of the mapped feature points over the extra-class variations. Using the theory of reproducing kernels [11], $\boldsymbol{w}$ can be written as a linear combination of all the samples, that is, $\boldsymbol{w} = \sum_{n=1}^{N} \alpha_n \, \Phi(\boldsymbol{x}_n)$. Using the kernel trick, the problem can be converted to find the optimal parameters $\boldsymbol{\alpha}^*$ that maximises $\mathcal{J}(\boldsymbol{\alpha}) = \frac{\boldsymbol{\alpha}^{\text{T}}\mathbf{M}\,\boldsymbol{\alpha}}{\boldsymbol{\alpha}^{\text{T}}\boldsymbol{N}\,\boldsymbol{\alpha}}$ where $\boldsymbol{\alpha} = [\alpha_1,\alpha_2,\dots,\alpha_N]^{\text{T}}$. The definitions of matrices $\boldsymbol{M}$ and $\boldsymbol{N}$ can be found in [5]. In this case, $\boldsymbol{M}$ and $\boldsymbol{N}$ can be easily computed from $\boldsymbol{K}$ and to keep both matrices makes the computation more difficult.

A similar situation is encountered in the SVM where a set of parameters $\boldsymbol{\alpha}$ is to be optimised. It is known that $\boldsymbol{w}$ only depends on a subset of training samples (the support vectors). The support vectors are found by solving an optimisation problem which itself involves the calculation of $\boldsymbol{K}$ and only those parameters α_{n} corresponding to the support vectors have non-zero values. Inspired by the finding from the SVM, we modify the KFD algorithm by selecting a subset $\mathcal{X}$ of the training samples (obtained by the sampling strategy) to construct the linear combination for $\boldsymbol{w}$. The samples in $\mathcal{X}$ are half intra-personal and half extra-personal and are randomly selected from the training dataset. Let $Q = \{q_m\}_{m=1}^{M}$ be the indices of the selected samples. Zero values are assigned to all the α_{n} where n is not in Q. In this way, only $\{\alpha_{q_1}, \alpha_{q_2}, \dots, \alpha_{q_M}\}$ need to be optimised, which makes the KFD computationally feasible for a large training dataset. Although not using the full set of the training samples might cause some information loss, the experimental results show that significant improvement can still be achieved by the MKFD method when compared with alternative techniques.

4 Experiments and Results

4.1 Experimental Setting

Experiments have been designed to test the system in a face verification scenario. A subset of the FRGC database was used. In total, the subset contains 16 thousands face images all taken under a controlled environment (e.g., with a static clear background and controlled lighting) during the 2003-2004 academic year. The large size of the set of images makes it suitable for demonstrating the proposed framework. The normalisation procedure described in [2] was used to pre-process images in the experiments.

Results can be influenced by the choice of particular training and test data. To reduce this effect, we prepared seven image groups each of which contained images from 50 subjects. In total, there were 12992 face images in the groups. Note that the subjects in each group were unique and did not appear in the other groups thereby simulating an open-set scenario very often encountered in a practical large-scale FR system. Cross validation was adopted to test the system, using one image group for validating and the rest of the images for training. The validating set was changed from one group to another until all the groups had been used.

Since the proposed classification framework does not specify any feature extraction method, the local binary pattern (LBP) technique described in [1] was used to extract facial features for testing the framework. To have different kinds of facial features, images were partitioned into 3×4, 5×5, 7×7, 10×10 and 14×14 local regions and three different LBP operators $\mathrm{LBP}_{8,2}^{u2}$, $\mathrm{LBP}_{8,4}^{u2}$ and $\mathrm{LBP}_{16,4}^{u2}$ (see [1] for details) were exploited, resulting in totally 15 kinds of facial features. For an image pair, the chi-square distances were computed from the local LBP histograms as the facial features. Note that all the features were zero-score normalised [12] in the experiments.

To prepare the training datasets, we need to decide the values of n_{int} and n_{ext}. Some experiments were carried out using training datasets with different sizes and based on the results we set $n_{int} = 100$ and $n_{ext} = 5$. For the test datasets, all the intra- and extra-personal image pairs were used to calculate the test samples. To employ MKFD, two most commonly used kernels were tested: the RBF kernels, $\mathscr{k}(\boldsymbol{x}, \boldsymbol{x}') = \exp(-\|\boldsymbol{x} - \boldsymbol{x}'\|/p)$ and the polynomial kernels, $\mathscr{k}(\boldsymbol{x}, \boldsymbol{x}') = (\boldsymbol{x} \cdot \boldsymbol{x}')^d$ where p and d are positive constants. The RBF kernels largely outperformed the polynomial kernels and were used in the experiments. Finally, the size of the subset $\mathcal{X}$ in the MKFD was chosen to be 5000 based on some exploratory experiment.

4.2 Experimental Results

In the first experiment, following [1], face images were partitioned into 7×7 local regions and the LBP features were extracted using $\mathrm{LBP}_{8,2}^{u2}$. The classification framework was performed on the seven image groups defined for the cross-validation. To test the robustness of the framework, in each turn of the cross-validation, we sampled three different training datasets using the sampling strategy. The MKFD classifier was then trained on them and tested on the same test dataset. To compared with the framework, we implemented the LBP system developed in [1], the linear Fisher discriminant (LFD) classifier, and the SVM^{light} [3]. Here the LFD and the SVM^{light} are trained on the same training datasets as used by the MKFD classifier.

Table 1 and Fig. 4 show the results in terms of the equal error rates (EERs) and the true acceptance rates (TARs) at the false acceptance rates (FARs) of 0.1% and 1% which are two important rates for evaluating system performance [7,9]. Based on the same facial features, the proposed classification framework and the SVM^{light}system significantly outperform the other two methods. Considering the EERs, the proposed framework performed slightly better than the SVM^{light} system. Table 2 summarises the CPU time used to train the MKFD classifier and the SVM^{light} system.

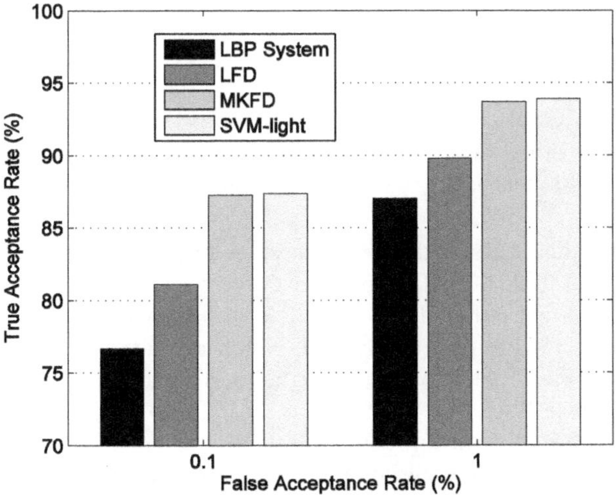

Fig. 4. Average TARs reported at the FARs of 0.1% and 1%

(Experiments were carried out on a PC with an Intel Core 2 Duo 2.4Hz CPU and 4G memory.) It can be seen that using the same training datasets, the SVM^{light} required eight times more training time than the MKFD. Moreover, there were averagely 9.57×10^3 support vectors learned by the SVM^{light} system each time, while the size of the subset $\mathcal{X}$ used by the MKFD was set to be 5000 all the time.

In the second experiment, we investigated how the size of the subset $\mathcal{X}$ could affect the performance of the MKFD. Using the same training and test datasets in the first experiment, different sizes of $\mathcal{X}$ were chosen to train and test the MKFD.

Table 1. Means and STDs of the EERs for different algorithms tested in the first experiment

	LBP System	LFD	MKFD	SVM^{light}
Mean (%)	6.10	4.98	3.52	3.87
STD (%)	1.43	1.33	1.11	1.23

Table 2. Means and STDs of the CPU time used to train the MKFD and SVM^{light}

	MKFD	SVM^{light}
Mean (sec)	1.0×10^3	8.0×10^3
STD (sec)	1.2	1.1×10^2

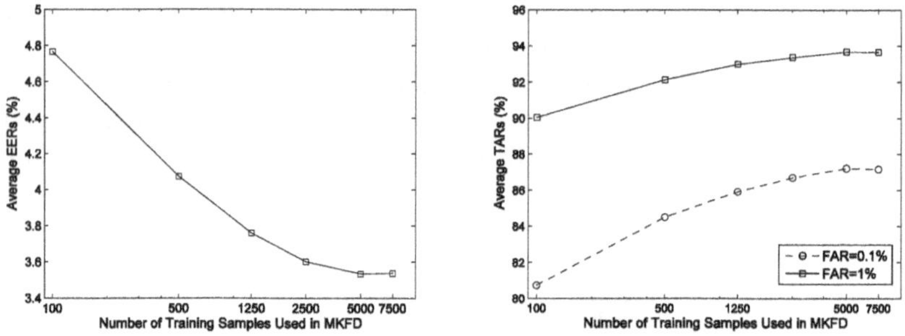

(a) (b)

Fig. 5. Different sizes of the subset $\mathcal{X}$ are chosen to train and test the MKFD classifier. The average EERs and TARs at the FARs of 0.1% and 1% are reported in (a) and (b), respectively.

Fig. 5(a) shows the EERs at the different sizes of $\mathcal{X}$. The rates dropped quickly from the size of 100 to 2500. After that, the curve reached the bottom at the size of 5000. It can be seen that although the size of $\mathcal{X}$ was increased by half from 5000 to 7500, the EER remained almost unchanged, which indicates that increasing the size of $\mathcal{X}$ does not always help to increase the performance. Fig. 5(b) shows the corresponding average TARs at the FARs of 0.1% and 1%. Once again, the two curves confirmed the finding in Fig. 5(a).

In the final experiment, we explored the capability of the classification framework for fusing different sources of facial information. To do that, various image partitionings and LBP operators were used to extract a range of feature sets. The

Table 3. Average EERs (%) calculated using different image partitioning and LBP operators

Image Partitioning	$LBP^{u2}_{8,2}$	$LBP^{u2}_{8,4}$	$LBP^{u2}_{16,4}$
3 × 4	11.36	5.06	4.43
5 × 5	4.59	3.68	3.20
7 × 7	3.52	3.49	3.20
10 × 10	3.27	3.73	3.91
14 × 14	3.01	6.15	8.27

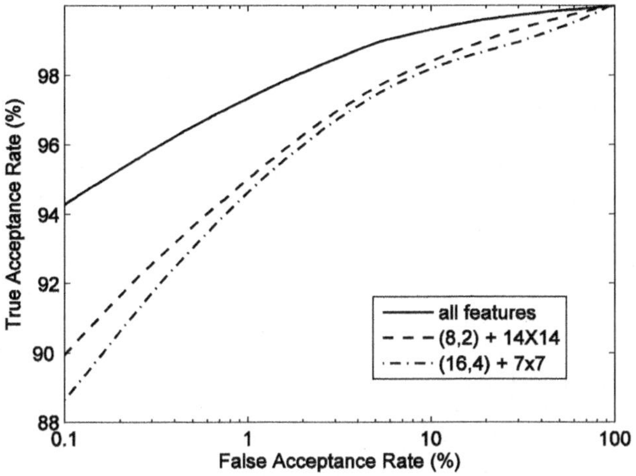

Fig. 6. ROC curves for the fused features and two individual features with the smallest EERs

framework was then tested for all the three LBP feature types on each of the seven image partitionings. Table 3 shows the average EERs calculated for each combination. The resulting features were then concatenated to produce fused features used for classification. Fig. 6 presents the results in terms of ROC curves. Besides the ROC for all features, the figure also shows the ROC curves for the two kinds of features with the smallest EERs. It can be seen that the classification framework not only can handle each of the different kinds of features, but is capable of fusing the facial information carried by the features to give a much better performance.

5 Conclusions

In this paper, a generic classification framework has been proposed for large-scale face recognition systems. The framework consists of two components: a data sampling strategy and a modified kernel Fisher discriminant classifier. The sampling strategy is aimed at dealing with the imbalance within the large amount of training data when image pairs are sampled for feature extraction. The modified KFD provides a simple solution for using the KFD algorithm given a large training dataset. Various experiments have been carried out in an open-set scenario and results suggest that the classification framework can provide an effective solution in terms of accuracy and computational efficiency.

References

1. Ahonen, T., Hadid, A., Pietikäinen, M.: Face description with local binary pattern: Application to face recognition. TPAMI 28(12), 2037–2041 (2006)
2. Beveridge, J.R., Bolme, D.S., Draper, B.A., Teixeira, M.: The CSU face identification evaluation system: its purpose, features, and structure. Machine Vision and Applications 16(2), 128–138 (2005)
3. Joachims, T.: Making large-scale SVM learning practical. In: Advances in Kernel Methods - Support Vector Learning. MIT Press, Cambridge (1998)
4. Jonsson, K., Kittler, J., Li, Y., Matas, J.: Support Vector Machines for Face Recognition. In: BMVC 1999, Nottingham, UK, pp. 543–553 (1999)
5. Mika, S., Rätsch, G., Weston, J., Schölkopf, B., Muller, K.R.: Fisher discriminant analysis with kernels. In: Proceedings of the 1999 IEEE Signal Processing Society Workshop, pp. 41–48. IEEE Press, Piscataway (1999)
6. Moghaddam, B., Wahid, W., Pentland, A.: Beyond Eigenfaces: probabilistic matching for face recognition. In: FG 1998, pp. 14–16 (1998)
7. Phillips, P.J.: Support vector machines applied to face recognition. In: Proceedings of Advances in Neural Information Processing Systems II, pp. 803–809 (1999)
8. Phillips, P.J., Grother, P., Micheals, R.J., Blackburn, D.M., Tabassi, E., Bone, M.: Face recognition vendor test 2002: overview and summary (2003)
9. Phillips, P.J., Flynn, P.J., Scruggs, W.T., Bowyer, K.W., Chang, J., Hoffman, K., Marques, J., Min, J., Worek, W.: Overview of the face recognition grand challenge. In: CVPR, San Diego, CA, pp. 947–954 (2005)
10. Phillips, P.J., Scruggs, W.T., O'Toole, A.J., Flynn, P.J., Bowyer, K.W., Schott, C.L., Sharpe, M.: FRVT 2006 and ICE 2006 Large-Scale Results. TR-NISTIR 7408 (2007)
11. Schölkopf, B., Smola, A.J.: Learning with Kernels: Support Vector Machines, Regularization, Optimization, and Beyond. MIT Press, Cambridge (2001)
12. Snelick, R., Uludag, U., Mink, A., Indovina, M., Jain, A.: Large-scale evaluation of multimodal biometric authentication using state-of-the-art systems. TPAMI 27(3), 450–455 (2005)
13. Yang, J., Frangi, A.F., Yang, J., Zhang, D., Jin, Z.: KPCA Pluse LDA: a complete kernel Fisher discriminant framework for feature extraction and recognition. TPAMI 27(2), 230–244 (2005)
14. Zhao, W., Chellappa, R., Rosenfeld, A., Phillips, P.J.: Face Recognition: A Literature Survey. ACM Computing Surveys, 399–458 (2003)

Synthesizing Frontal Faces on Calibrated Stereo Cameras for Face Recognition

Kin-Wang Cheung, Jiansheng Chen, and Yiu-Sang Moon

The Department of Computer Science and Engineering,
The Chinese University of Hong Kong, Shatin, N.T., Hong Kong
{kwcheung,jschen,ysmoon}@cse.cuhk.edu.hk

Abstract. Current automatic face recognition systems often require users to face towards the capturing camera. To extent these systems for user non-intrusive application scenarios such as video surveillance, we propose a stereo camera configuration to synthesize a frontal face image from two non-frontal face images. After the head pose has been estimated, a frontal face image is synthesized using face warping and view morphing techniques. Face identification experiments reveal that using the synthetic frontal images can achieve comparable performance with real frontal face images.

1 Introduction

Advances in the automatic face recognition technology have led to its extensive deployments in security applications around the world. In Britain, face scanners have been installed in airports so that passengers can perform self-help immigration clearance. Passengers who use this automated machine are expected to face towards the scanner and wait inside the unmanned gates, until their faces match with the data stored in their biometric passports. In the United States, visitors entering the border will have their face images taken at the entry ports by the immigration officers. The photographs are then matched with criminal face databases to check if the visitors are suspects. During the 2008 Beijing Olympic Games, security checkpoints with face verification capability were allocated in venue entrances to ensure the audiences were the genuine ticket holders, as shown in Figure 1.

A common characteristic of these face recognition systems is that users have to follow the instructions strictly so as to co-operate with the system. For example, in the British airport face scanner system, users are expected to stand upon a fixed location, and turn their faces directly towards the capturing camera so that clear frontal face images can be captured. Such a user intrusive approach will probably fail when the user does not offer full co-operation, say by not facing the camera during authentication. Therefore, solving the non-frontal facial image problem is critical for user non-intrusive application scenarios such as video surveillance.

One approach to recognize non-frontal faces is to use gallery images with the same pose as the captured live image for face matching. Head pose can be

M. Tistarelli and M.S. Nixon (Eds.): ICB 2009, LNCS 5558, pp. 347–356, 2009.
© Springer-Verlag Berlin Heidelberg 2009

Fig. 1. The face recognition systems adopted in British (Top Left), the United States (Bottom Left), and China (Right)

approximately estimated from a single face image through statistical learning [1]. After that, algorithms such as the view-based eigenface proposed by Pentland *et al.* [2] can be applied to do the non-frontal face recognition. The main drawback of this approach, however, is that face images of each user in various viewpoints need to be collected beforehand.

Another approach is to synthesize frontal faces based on 2D image statistics or 3D head models. In [3], Beymer and Poggio generated faces in virtual views based on prior knowledge of faces in 2D. In [4], Ashraf *et al.* divided a non-frontal face image into several non-overlapping 2D patches, and aligned the patches based on the correspondences learned from examples. In [5], Gross *et al.* re-rendered face images across viewpoint using eigen light-fields. In [6], Blanz *et al.* generated a frontal face by fitting one non-frontal image into the 3D morphable model proposed in [7]. Once the frontal faces are generated, traditional face recognition algorithms like PCA [8] and LDA [9] can be applied directly. One common problem of these methods is that facial texture extracted from a single non-frontal image does not necessarily cover the whole face region, since some parts of the face are occluded in non-frontal view. To solve this problem, Li *et al.* [10] proposed a method in which two sides of a face are captured simultaneously using two cameras. A mean 3D facial model is fitted to the two images respectively so that they can both be rotated to the frontal view. The synthesized frontal face image is computed by averaging the two rotated images. By adopting the mean 3D face model, this method actually ignores the valuable stereo information which can be utilized to estimate the individual facial structure of the user, leading to the possible unrealistic frontal face synthesis. One straightforward way of utilizing the stereo information is to incorporate 3D surface reconstruction. However, this process is usually computational intensive and slow. Facing these problems, we propose a novel scheme for frontal face synthesis using two cameras. The stereo information provided by the cameras is fully utilized to make this approach simple, fast and realistic.

The key technique of the proposed method is *view morphing*, which was proposed by Seitz and Dyer for synthesizing a novel view of an object from two images of the same object in two different viewpoints [11]. The main idea is the view of a virtual camera in between the two real cameras can be interpolated by the two input images. This technique works by first aligning two input images using projective transform. After that, a virtual view is created by interpolating the intensity values of the two aligned images. Finally, the interpolated image is re-projected to produce the novel view. View morphing works well when most surfaces are visible in both images, because this technique does not recover the 3D shape of the object throughout the procedures. To deal with occlusions from different viewpoints, the interpolation may have to be guided by explicitly-defined control points. In the proposed method, two calibrated cameras are placed side-by-side in converging directions as shown in Figure 2. When the user is inside the overlapping field-of-view region, two face images are captured. The head pose is estimated by calculating 3D locations of facial points using stereo geometry. If the head is facing between the two cameras, a frontal face is synthesized following the procedures in Figure 3. The face recognition system can then perform verification or identification using this synthesized frontal face image.

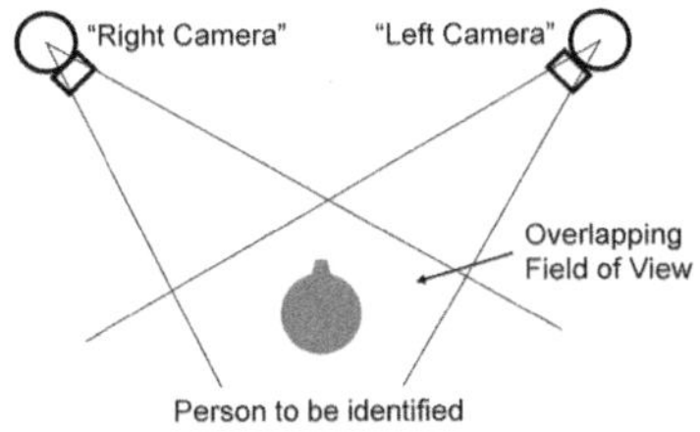

Fig. 2. The camera configuration of our approach

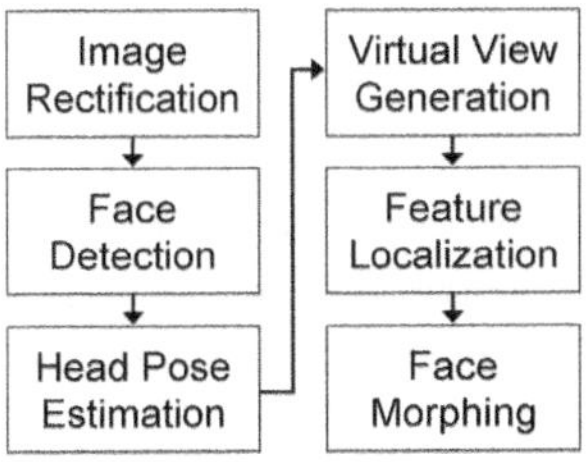

Fig. 3. The overflow of our approach

The remainder of this paper is organized as follows: In Section 2, we describe how to align two images and estimate the head pose. Section 3 explains how "*view morphing*" is applied to generate a virtual view. Section 4 illustrates how a frontal face is synthesized by morphing. Section 5 shows some case studies on face identification experiments. The proposed method is summarized in section 6.

2 Image Alignment and Head Pose Estimation

When two face images are captured, we have no information about the head location and orientation in the 3D world. Therefore, we have to calculate the head pose first, using the 3D locations of facial points. Since our cameras are placed in converging directions, we need to transform the two captured images and switch the stereo camera configuration back to parallel views.

To begin with, the stereo cameras are first calibrated [12] such that the intrinsic and extrinsic parameters are known. The captured images are rectified as shown in Figure 4, i.e. the epipolar lines become collinear and parallel to the baseline. The optical centers of the two cameras are kept unchanged after image rectification. It is because any two views which share the same optical center can be related by a planer projective transformation. In fact, this procedure is equivalent to the pre-warping step in view morphing. Further details can be found in Seitz and Dyer's paper [11].

Fig. 4. (Left & Right) Two captured images I_L from the "*left camera*" and I_R from the "*right camera*". (Middle) The rectified stereo images.

A simple method to estimate the head pose is to assume the eye centers and the mouth center are co-planar, and the orientation of this 3D plane is close to the real head pose. The orientation of the "*face plane*" can be determined from three non-collinear points in the 3D space. The nose tip is not chosen as mentioned in [1] because the appearance of the nose is heavily distorted when the head is rotated (particularly in yaw angle). It makes the nose tip much harder to be located when the actual head pose is still unknown.

Fig. 5. Face extraction from the top right image in Figure 4. The result of (Left) background subtraction [13] and (Middle) skin color model. (Right) The final face detected.

A course-to-fine approach is adopted to locate the facial features. First, the face area is extracted using background subtraction [13] and skin color model in YCrCb channel, as illustrated in Figure 5. Next, the approximate eye and mouth regions are predicted using the eye map and mouth map proposed in [14]. Finally, the eye and mouth centers can be located using machine learning approaches such as [15], assuming that the face is almost upright in the images after rectification.

After the correspondences of the eye and mouth centers are found, we can calculate the 3D locations of these feature points using calibrated stereo rig. Thus, the normal of the "*face plane*" can be determined. The head pose is resolved using the face normal on the mid-point of the eye and mouth centers P. The estimated head pose is close to the real pose as far as the eye and mouth centers are visible in both images, i.e. the features are not self-occluded by the face.

3 View Morphing in Parallel Views

Once we have estimated the head location and orientation in the 3D world, we can continue to generate a virtual view using *view morphing* technique in parallel views. After rectification, the problem of computing a shape-preserving morph from two non-parallel perspective views is reduced to the case of parallel views. To synthesize a frontal face image, we create a virtual camera in between the two real cameras. The virtual camera center C_V is defined as the intersection of the baseline and the light-ray from P parallel to the projection vector of the face normal on the C_L-C_R-P plane, as illustrated in Figure 6.

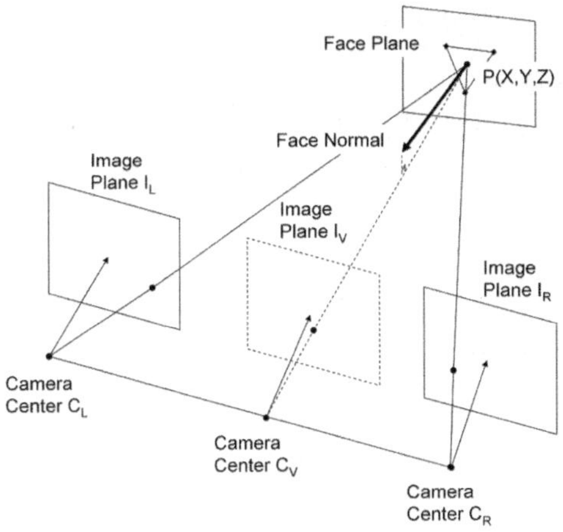

Fig. 6. The parallel stereo rig in which a virtual view is generated

For each pixel in the virtual image plane I_V, the light-ray which passes through the camera center C_V and the pixel can be extended to the "*face plane*". Suppose the intersection of the light-ray and the face plane is the 3D location of this particular pixel, we can backtrack the image projections and find out the subsequent pixels in left and right image planes I_L and I_R. This assumption is essential to align the face images in I_L and I_R to the virtual image plane I_V, especially to ensure the eye and mouth centers are aligned to the same locations in I_V.

4 Facial Feature Detection and Face Morphing

Now we have aligned the two non-frontal face images so that the eye and mouth centers are in the same location on the image plane of the virtual camera, as shown in Figure 7. A straight forward approach to blend the two images is to interpolate the intensity values linearly. However, the result is far from satisfactory. The blended image appears to be "*ghosted*" especially in the nose region. The underlying reason is facial features are distorted in different magnitudes over the change of head pose. The human nose tip, which is prominent on the face, shifted more rapidly than other feature points such as the eye corners when the pose is varied in yaw angle. Therefore, we have to guide the way feature points are shifted between the two non-frontal faces.

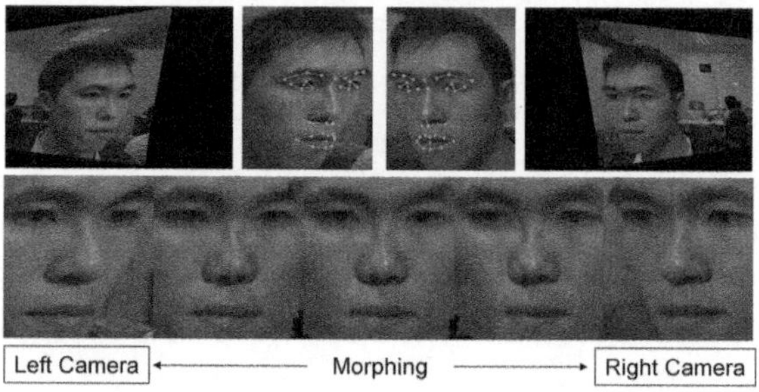

Fig. 7. (Top Left & Right) The non-frontal face images from the virtual camera view after face alignment. (Top middle) The facial features detected by the set of pose-specific ASM. (Bottom) The face morphing results using the non-frontal face images from the virtual camera view.

Certainly using only three feature points we have detected, the eye and mouth centers, is not sufficient to guide the warping of the whole face. As we have estimated the head pose from stereo vision, we can apply a set of view-specific Active Shape Model (ASM) [16]. Each ASM is trained from face images of the same pose, for example, using face images in the Facial Recognition Technology (FERET) database [17] with 0° (frontal), ±15°, ±25°, or ±40° varied in yaw angle. The ASM with the training face pose nearest to the estimated head pose is chosen to detect facial features. The eye and mouth centers are used to initialize the ASM, such that the search of scale and orientation of ASM is bounded. After applying the ASM, the locations of 31 additional facial features are found as shown in Figure 7.

Facial features extracted from the ASM and the eye centers are selected as the control points to regulate the mesh warping procedure. To avoid the "*ghosting*" effect from linear blending, the bottom part of the face, which includes the nose and the mouth, is combined using the more apparent half of each warped non-frontal face image. The upper part of face, which included the eyes and eyebrows, are blended linearly instead. This face morphing method produces a

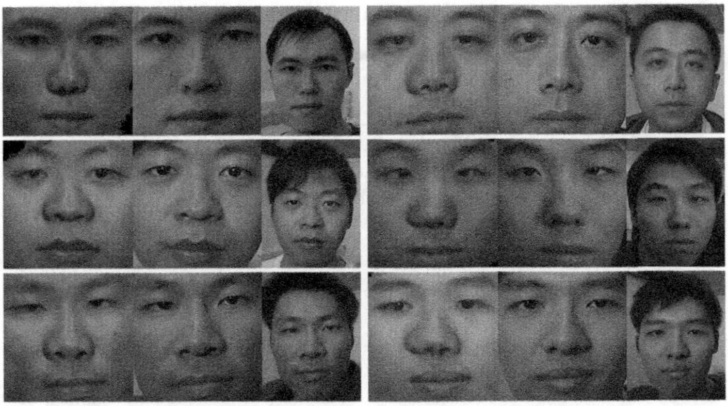

Fig. 8. Some synthesized results using two non-frontal face images. (Column A & D) The synthesized frontal face regions. (Column B & E) The real frontal face regions cropped from the person's head images (Column C & F).

more visually-plausible synthesized frontal face than using linear blending alone, as demonstrated in Figure 7. In addition, the final synthesized frontal faces are shown in Figure 8.

5 Face Recognition with Synthesized Frontal Faces

Now we have synthesized visually-plausible frontal face images using the proposed method. Still, we have to investigate whether these synthesized frontal faces can be applied in automatic face recognition algorithms. We adopted the PCA method used in [18]. Cosine angle was used as the distance measure. The face images were normalized to 130x150 grayscale with the same intraocular (eye-to-eye) distance. The gallery consisted of frontal face images with regular expression from 1196 individuals in the FERET database [17]. The probe set consisted of frontal face images with alternative expression from 1195 individuals in the FERET database. Around 500 frontal face images in the FERET database were randomly chosen for eigenface training. The identification result on gallery vs. probe set (Experiment 1) was 71.5% at rank 1, as shown in Table 1.

We conducted a similar experiment (Experiment 2) by including the 6 synthesized frontal face images to the probe set and the 6 real frontal face images to the gallery. The identification result was shown in Figure 9. The result was encouraging indeed since all synthesized faces could be identified correctly within the first 7 matches. How about we use the respective non-frontal face images for face recognition? We replaced the synthesized frontal face images in the probe set by the 12 non-frontal face images captured by the left and right cameras (Experiment 3). The identification result was worse than Experiment 2, in which only 4 non-frontal face images were identified correctly within the first 20 matches, as shown in Table 1.

Table 1. Identification results on the FERET database

	Experiment 1 (Original)	Experiment 2 (Synthesized)	Experiment 3 (Non-frontal)
Rank 1	71.5%	50.0% (3/6)	8.3% (1/12)
Rank 5	84.9%	83.3% (5/6)	16.7% (2/12)
Rank 10	89.1%	100.0% (6/6)	25.0% (3/12)
Rank 20	92.6%	–	33.3% (4/12)
Rank 50	96.2%	–	66.7% (8/12)

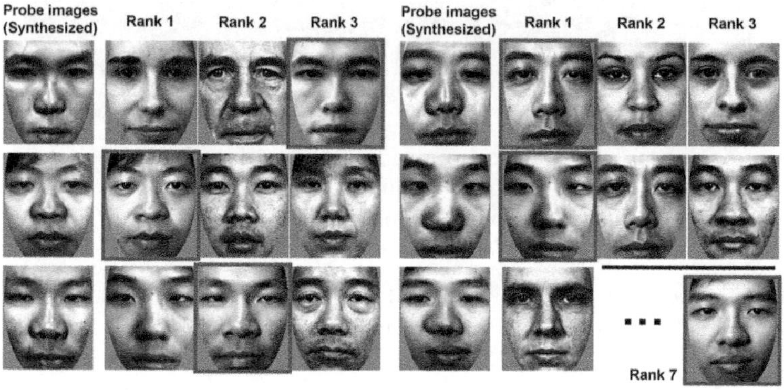

Fig. 9. The identification result of the synthesized frontal faces on the FERET database. The probe images (the synthesized frontal face images) are shown in the leftmost column. The first 3 matches with the highest scores in the PCA recognition algorithm are shown on the right. The correct matches are squared in red, indicating the correct individuals have been identified.

Since racial characteristics might take a role in the identification tests on the FERET database, further experiments were conducted using the CAS-PEAL database [19]. The CAS-PEAL database was collected from the north-east China, in which nearly all subjects were native Chinese. The setup of our identification test was as follows: The gallery consisted of frontal face images with regular expression (normal) from 377 individuals in the database. The probe set consisted of frontal face images with alternative expression (frown) from the same 377 individuals in the database. Frontal face images from the other 663 individuals in the database with regular expression were chosen for eigenface training. The identification result on gallery vs. probe set (Experiment 4) was 74.3% at rank 1, as shown in Table 2.

We conducted a similar experiment (Experiment 5) by including the 6 synthesized frontal face images to the probe set and the 6 real frontal face images to the gallery. The identification result was shown in Table 2. All synthesized faces could be identified correctly within the first 3 matches. In addition, the identification result using respective non-frontal face images (Experiment 6) was worse than using synthesized face images in Experiment 5, in which only 4 non-frontal

Table 2. Identification results on the CAS-PEAL database

	Experiment 1 (Original)	Experiment 2 (Synthesized)	Experiment 3 (Non-frontal)
Rank 1	74.3%	50.0% (3/6)	8.3% (1/12)
Rank 5	89.4%	100.0% (6/6)	33.3% (4/12)
Rank 10	94.4%	–	33.3% (4/12)
Rank 20	96.8%	–	33.3% (4/12)
Rank 50	99.2%	–	50.0% (6/12)

face images were identified correctly within the first 20 matches, as shown in Table 2.

The identification results on the CAS-PEAL database agree with the results on the FERET database, which is we can identify users correctly using their synthesized frontal faces. Furthermore, the experimental results demonstrate our proposed method is capable to synthesize frontal faces which are visually-plausible and also machine recognizable.

6 Conclusion

In this paper, we proposed a method to extend automatic face recognition systems to non-intrusive applications, since users in such scenarios will not face towards the capturing camera *exactly* for the system to perform face recognition. This goal is achieved by synthesizing a frontal face image from two non-frontal face images captured by a pair of stereo cameras. The head pose is first estimated using the 3D locations of the eye and mouth centers. If the head is facing between the two cameras, a virtual view can be created using view morphing. Since facial features are distorted in different magnitudes over the change of head pose, control points for face warping are extracted using a set of pose-specific ASM. Afterwards, a frontal face can be synthesized by combining the two warped non-frontal face images.

The synthesized frontal faces are visually plausible. Moreover, we conducted some case studies to evaluate whether these synthesized face images can be identified using automatic face recognition algorithms. The results are encouraging as all synthesized frontal faces can be identified correctly within the first 7 ranks in both the FERET and the CAS-PEAL databases, which is comparable to the performance using real frontal face images. The results also reveal that synthesized frontal faces can be used in automatic face recognition. To conclude, this method can provide a flexible manner to recognize faces without explicit user co-operation.

Acknowledgements. The work described in this paper was substantially supported by a grant from the Research Grants Council of the Hong Kong Special Administrative Region, China (Project No. 415207). Portions of the research in this paper use the FERET database of facial images collected under the FERET program, sponsored by the DOD Counterdrug Technology Development Program Office. The research in this paper use the CAS-PEAL-R1 face database collected under the sponsor of the Chinese National Hi-Tech Program and ISVISION Tech. Co. Ltd.

References

1. Murphy-Chutorian, E., Trivedi, M.: Head pose estimation in computer vision: A survey. IEEE Trans. on PAMI (accepted)
2. Pentland, A., Moghaddam, B., et al.: View-based and modular eigenspaces for face recognition. In: CVPR, pp. 84–91. IEEE Press, Los Alamitos (1994)
3. Beymer, D., Poggio, T.: Face recognition from one example view. In: ICCV, pp. 500–507. IEEE Press, Los Alamitos (1995)
4. Ashraf, A., Lucey, S., et al.: Learning patch correspondences for improved viewpoint invariant face recognition. In: CVPR, pp. 1–8. IEEE Press, Los Alamitos (2008)
5. Gross, R., Matthews, I., et al.: Eigen light-fields and face recognition across pose. In: Fifth IEEE Int. Conf. on FG, pp. 1–7. IEEE Press, Los Alamitos (2002)
6. Blanz, V., Grother, P., et al.: Face recognition based on frontal views generated from non-frontal images. In: CVPR, vol. 2, pp. 454–461. IEEE Press, Los Alamitos (2005)
7. Blanz, V., Vetter, T.: A morphable model for the synthesis of 3D faces. In: SIGGRAPH, pp. 187–194. ACM Press, New York (1999)
8. Moghaddam, B., Pentland, A.: Probabilistic visual learning for object representation. IEEE Trans. on PAMI 19(7), 696–710 (1997)
9. Zhao, W., Chellappa, R., et al.: Discriminant analysis of principal components for face recognition. In: Third IEEE Int. Conf. on FG, pp. 336–341. IEEE Press, Los Alamitos (1998)
10. Li, C., Su, G., et al.: Frontal face synthesis based on multiple pose-variant images for face recognition. In: Lee, S.-W., Li, S.Z. (eds.) ICB 2007. LNCS, vol. 4642, pp. 948–957. Springer, Heidelberg (2007)
11. Seitz, S.M., Dyer, C.R.: View morphing. In: SIGGRAPH, pp. 21–30. ACM Press, New York (1996)
12. Zhang, Z.: Flexible camera calibration by viewing a plane from unknown orientations. In: ICCV, pp. 666–673. IEEE Press, Los Alamitos (1999)
13. Agrawal, A., Raskar, R., et al.: Edge suppression by gradient field transformation using cross-projection tensors. In: CVPR, vol. 2, pp. 2301–2308. IEEE Press, Los Alamitos (2006)
14. Hsu, R., Abdel-Mottaleb, M., et al.: Face detection in color images. IEEE Trans. on PAMI 24(5), 696–706 (2002)
15. Smeraldi, F., Bigun, J.: Retinal vision applied to facial features detection and face authentication. Pattern Recognition Letters 23, 463–475 (2002)
16. Cootes, T., Cooper, D., et al.: Active shape models - their training and application. Computer Vision and Image Understanding 61(1), 38–59 (1995)
17. Phillips, P.J., Wechsler, H., et al.: The feret database and evaluation procedure for face recognition algorithms. Image and Vision Computing J. 16(5), 295–306 (1998)
18. Delac, K., Grgic, M., et al.: Independent comparative study of pca, ica, and lda on the feret data set. Int. J. of IST 15(5), 252–260 (2006)
19. Gao, W., Cao, B., et al.: The cas-peal large-scale chinese face database and baseline evaluations. IEEE Trans. on SMC (Part A) 38(1), 149–161 (2008)

Nasal Region Contribution in 3D Face Biometrics Using Shape Analysis Framework

Hassen Drira[1], Boulbaba Ben Amor[1,2], Mohamed Daoudi[1,2], and Anuj Srivastava[3]

[1] LIFL (UMR USTL/CNRS 8022), Université de Lille 1, France
[2] Institut TELECOM/TELECOM Lille 1, France
[3] Departement of Statistics, Florida State University, Tallahassee, FL 32306, USA
{hassen.drira,boulbaba.benamor,mohamed.daoudi}@telecom-lille1.eu,
anuj@stat.fsu.edu

Abstract. The main goal of this paper is to illustrate a geometric analysis of 3D facial shapes in presence of varying facial expressions using the nose region. This approach consists of the following two main steps: *(i)* Each nasal surface is automatically denoised and preprocessed to result in an indexed collection of nasal curves. During this step one detects the tip of the nose and defines a surface distance function with that tip as the reference point. The level curves of this distance function are the desired nasal curves. *(ii)* Comparisons between noses are based on optimal deformations from one to another. This, in turn, is based on optimal deformations of the corresponding nasal curves across surfaces under an elastic metric. The experimental results, generated using a subset of FRGC v2 dataset, demonstrate the success of the proposed framework in recognizing people under different facial expressions. The recognition rates obtained here exceed those for a baseline ICP algorithm on the same dataset.

Keywords: 3D face/nose biometrics, shape analysis, automatic preprocessing.

1 Introduction and Motivations

The various tools that are called *Biometric Technologies* are simply means physiological characteristics, human body parts and their appearances, used to point-point individual human beings in the course of daily activities. The appearances of body parts, especially in imaged data, have a large variability and are influenced by their shapes, colors, illumination environment, presence of other parts, and so on. Therefore, the biometrics researchers have focused on body parts and images that try to minimize this variability within class (subjects) and maximize it across classes. *3D face* has recently emerged as a major trend in facial biometric which illustrates this idea. Since 2D (visible light) images of faces are greatly susceptible to variations in the imaging environments (camera pose, illumination patterns, etc), the researchers have argued for the need to use 3D face data, typically collected by laser scanners, for studying shapes of peoples' faces

M. Tistarelli and M.S. Nixon (Eds.): ICB 2009, LNCS 5558, pp. 357–366, 2009.
© Springer-Verlag Berlin Heidelberg 2009

and using this shape analysis for biometrics. The output from laser scanners are minimally dependent on the external environmental factors and provide faithful measurements of shapes facial surfaces. It's the case the only remaining variability that is manifested within the same class, i.e. within the measurements of the same person, is the one introduced by changes in facial expressions. Facial expressions, such as smile, serious, fear, and anger, are prime indicators of the emotional state of a person and, thus, are important in estimating mood of a person, for example in developing intelligent ambient systems, but may have a lesser role in biometric applications. In fact, variations in facial expressions change the shapes of facial surfaces to some extent and introduce a nuisance variability that has to be accounted for in shape-based 3D face recognition. We argue that the variability introduced by facial expressions has become one of the most important issues in 3D face recognition. The other important issue is related to data collection and imperfections introduced in that process. It is difficult to obtain a pristine, continuous facial surface, or a mesh representing such a surface, with the current laser technology. One typically gets holes in the scanned data in locations of eyes, lips, and outside regions. For instance, scans of people with open mouths result in holes in the mouth region.

To handle these issues – shape variability due to facial expressions and presence of holes in mouth, we advocate the use of *nose region* for biometric analysis. At the outset the shape of the nose seems like a bad choice of feature for biometrics. Indeed, the shapes of noses seem very similar to a human observer but we will support this choice using real data and automated techniques for shape analysis. We do not assert that this framework will be sufficient for identifying human subjects across a vast population, but we argue for its role in shortlisting possible hypotheses so that a reduced hypothesis set can be evaluated using a more elaborate, multi-modal biometric system. The stability of nose data collection, the efficiency of nasal shape analysis, and the invariance of nasal shape to changes in facial expressions make it an important biometric.

The rest of this paper is organized as follows, section 2 gives a brief description of FRGC database and the preprocessing step. In section 3, we explain the differentiel-geometric framework of curves and its extension to 3D surfaces that we used to analyse 3D shapes of nasal region. Finally, in section 4, we show the experimental protocol and some preliminary results on a subset of FRGC v2 database containing expressive faces.

2 Automatic Data Preprocessing

In order to assess the recognition performance of the proposed framework, we use a subset of FRGC v2 dataset. This benchmark database [9] includes 4007 3D frontal scans for 466 subjects and is considered as a challenging database as it contains sessions with both neutral and non-neutral expressions. Moreover, the laser-based 3D scanner used in the acquisition process introduces noise in the data. In fact, some of 3D face scans suffer from missing data (holes), spikes, artefacts specially in teeth region, occlusions caused by the hair, etc. We focus in this work on designing a complete solution for 3D face analysis and recognition using only the nose region. For that purpose, it is crucial to begin by denoising the data by removing spikes, filling holes and extract only the useful part of face and then nose from original depth image. Figure 1 shows different steps of our preprocessing solution to overcome these problems. Starting from an original range image of a face, we firstly apply a 2D median filter in order to remove spikes while preserving edges. Secondly, using a smooth 2D interpolation, we fill holes by adding points in parts where the laser has been completely absorbed (e.g. eyes, eyebrows and opend mouth). Next, we use the Delaunay triangulation to generate a triangulated mesh from the point cloud. On this mesh, we localize the nose tip, necessary for cropping the useful region of the face and nose region segmentation. For this purpose, a sphere function having center the nose tip and radius $R = 100mm$ is constructed and the part inside the sphere is keeped. Finally, a collection of geodesic level curves are extracted by locating iso-geodesic points from the reference vertex (nose tip) using the Dijkstra algorithm [2]. In order to segmentate nasal region, we consider the N first level curves. For more details about data proprocessing please refer to [1]. This algorithm has successfully processed 3971 faces in FRGC v2, which means a success rate of 99.1%. Actually, it is the nose detection step that fails more than other steps. For these faces, we have fixed manually the nose tip and so we have cleaned all the FRGC v2 faces, 99.1% automatically and 0.9% manually.

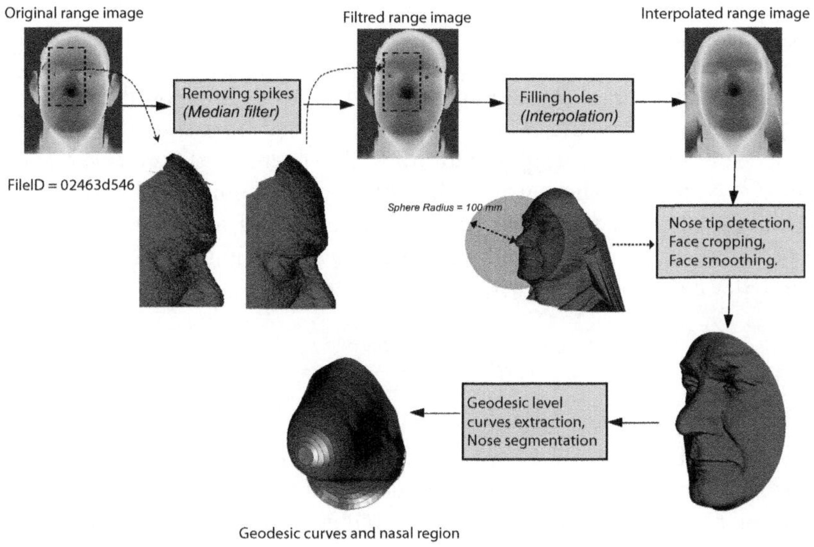

Fig. 1. Automatic FRGC data preprocessing and nose curves extraction

3 A Geometric Framework for Nose Analysis

As indicated earlier, our goal is to analyse shapes of facial surfaces using shapes of facial curves. In other words, we divide each surface into an indexed collection of simple, closed curves in $\mathbb{R}^3$ and the geometry of a surface is then studied using the geometries of the associated curves. Since these curves, previously called *facial curves*, have been defined as level curves of an intrinsic distance function on the surface, their geometries in turn are invariant to the rigid transformation (rotation and translation) of the original surface. At least theoretically, these curves jointly contain all the information about the surface and one can go back-and-forth between the surface and the curves without any ambiguity. In practice, however, some information is lost when one works with a finite subset of these curves rather than the full set. Later, through experiments on real data, we will demonstrate that the choice of facial curves for studying shapes of facial surfaces is both natural and convenient. In the following section, we will describe a differential-geometric approach for analysing shapes of simple, closed curves in $\mathbb{R}^3$. In recent years, there have been several papers for studying shapes of continuous curves. The earlier papers, including [12,6,7,8], were mainly concerned with curves in $\mathbb{R}^2$, while the curves in higher dimensions were studied later. In this paper, we will follow the theory laid out by Joshi et al. [3,4] for *elastic shape analysis* of continuous, closed curves in $\mathbb{R}^n$ and particularize it for facial curves in $\mathbb{R}^3$. The mathematical framework for using elastic shape analysis of facial curves was first presented in [11].

3.1 Nose Curves

We start by considering a closed curve β in $\mathbb{R}^3$. Since it is a closed curve, it is natural to parametrize it using $\beta : \mathbb{S}^1 \to \mathbb{R}^3$. We will assume that the parametrization is non-singular, i.e. $\|\dot{\beta}(t)\| \neq 0$ for all t. The norm used here is the Euclidean norm in $\mathbb{R}^3$. Note that the parametrization is not assumed to be arc-length; we allow a larger class of parametrizations for improved analysis. To analyse the shape of β, we shall represent it mathematically using a *square-root velocity function* (SRVF), denoted by $q(t)$, according to:

$$q(t) \doteq \frac{\dot{\beta}(t)}{\sqrt{\|\dot{\beta}(t)\|}} . \tag{1}$$

$q(t)$ is a special function that captures the shape of β and is particularly convenient for shape analysis, as we describe next. Firstly, the squared $\mathbb{L}^2$-norm of q, given by: $\|q\|^2 = \int_{\mathbb{S}^1} \langle q(t), q(t) \rangle \, dt = \int_{\mathbb{S}^1} \|\dot{\beta}(t)\| dt$, which is the length of β. Therefore, the $\mathbb{L}^2$-norm is convenient to analyse curves of specific lengths. Secondly, as shown in [3], the classical elastic metric for comparing shapes of curves becomes the $\mathbb{L}^2$-metric under the SRVF representation. This point is very important as it simplifies the calculus of elastic metric to the well-known calculus of functional analysis under the $\mathbb{L}^2$-metric. In order to restrict our shape analysis to closed curves, we define the set: $\mathcal{C} = \{q : \mathbb{S}^1 \to \mathbb{R}^3 | \int_{\mathbb{S}^1} q(t)\|q(t)\|dt = 0\} \subset \mathbb{L}^2(\mathbb{S}^1, \mathbb{R}^3)$.

For more details about the proposed curve shape analysis framework, reader can refer to [11]. Figure 2 shows some illustrations of this idea. The top two rows show nasal surfaces of same subject under different expressions and two level curves extracted from each of these surfaces. Between them we display geodesic paths between the corresponding level curves of the two noses, obtained using the path-straightening approach. The remaining two rows display nasal surfaces and curves of two different subjects. In each case, the first and the last curves are the ones extracted from the two surfaces, and the intermediate curves denote equally-spaced points on the corresponding geodesic α. These curves have been scaled to the same length to improve display of geodesics. We will use the notation $d(\beta_1, \beta_2)$ to denote the geodesic distance, or the length of the geodesic in $\mathcal{S}$, between the two curves β_1 and β_2.

Fig. 2. Examples of geodesic between curves

Why do we expect that shapes of facial curves are central to analysing the shapes of facial surfaces? There is plenty of psychological evidence that certain facial curves, especially those around nose, lips and other prominent parts, can capture the essential features of a face. Our experiments support this idea in a mathematical way. We have computed geodesic distances between corresponding nasal curves of different faces – same people different facial expressions and different people altogether. We have found that the distances are typically smaller for faces of the same people, despite different expressions, when compared to the distances between nasal curves of different people.

3.2 Nose Surfaces

Now we extend ideas developed in the previous section for analysing shapes of nasal curves to the shapes of nose region surfaces. As mentioned earlier, we are going to represent a nose region surface S with an indexed collection of the level curves of the D function (geodesic distance from the nose tip).

That is,

$$S \leftrightarrow \{c_\lambda, \lambda \in [0, L]\} ,$$

Fig. 3. Geodesic path between source and target noses (a) First row: intra-class path, source and target with different expressions (b) Three last rows: inter-class path

where c_λ is the level set associated with $D = \lambda$. Through this relation, each facial surface has been represented as an element of the set $\mathcal{C}^{[0,L]}$. In our framework, the shapes of any two noses are compared by comparing their corresponding nasal curves. Given any two surfaces S^1 and S^2, and their nasal curves $\{c^1_\lambda, \lambda \in [0, L]\}$ and $\{c^2_\lambda, \lambda \in [0, L]\}$, respectively, our idea is to compare the facial curves c^1_λ and c^2_λ, and to accumulate these differences over all λ. More formally, we define a distance: $d_s : \mathcal{C}^{[0,L]} \times \mathcal{C}^{[0,L]} \rightarrow \mathbb{R}_{\geq 0}$, given by

$$d_s(S^1, S^2) = \int_0^L d(c^1_\lambda, c^2_\lambda) d\lambda \ . \tag{2}$$

Here, the distance inside the integral is the geodesic distance function between the shapes of any curves, described in the last section. It has been shown in [10] that this distance is actually a proper distance between the elements of the $\mathcal{C}^{[0,L]}$. In addition to the distance $d_s(S^1, S^2)$, which is useful in biometry and other classification experiments, we also have a geodesic path in $\mathcal{C}^{[0,L]}$ between the two points representing by S^1 and S^2. This geodesic corresponds to the optimal elastic deformations of facial curves and, thus, facial surfaces from one to other. Shown in Figure 3 are examples of such geodesic paths between given facial surfaces – One example (top row) involves faces of same people but with different facial expressions while the other examples show geodesics between faces that belong to different people.

4 Experiments and Evaluations

we focus in our experiments on the common biometric scenarions which are authentication (or verification) and identification (or recognition). As far as verification scenarion is concerned, performance is reported on a *Receiver Operator*

Characteristic (ROC) that shows the trade-off between verification and false accept rates. In the identification scenarion, however, the results for facial identification are often displayed using a *Cumulative Match Characteristic* (CMC) curve. This curve displays the cumulative identification rates as a function of the rank distribution. This provides an indication of how close one may be to getting the correct match if the rank-one match was incorrect. In order to produce results for both scenarions and to explore effect of the presence of facial expressions on performance, a similarity matrix between a gallery and a probe datasets is computed. The gallery contains 125 sessions for 125 different subjects acquired with neutral expressions selected from FRGC v2 dataset. The probe dataset includes completely different sessions of these subjects under non-neutral facial expressions. Due to sensitivity of our algorithm (on full face) to opened mouth, expressions in probe dataset include only scans with closed mouths. In this matrix, the diagonal terms represent match scores (or *Genuine Access*) contrary to non-diagonal terms which represent Non-match scores (or *Imposter Access*). These scores allow us to produce the ROC and the CMC curves for this protocol. We compare results of our algorithm with a standard implementation of ICP which is considered as a baseline in 3D face recognition. Baseline performance serves to demonstrate that a challenge problem can be executed, to provide a minimum level of performance, and to provide a set of controls for detailed studies. The same protocol was followed to compute similarity matrices for both algorithms on the same preprocessed data.

Figure 4 shows ROC curves for our approach on face, on nose region and ICP algorithm on nose and on face. As shown in this figure, the ROC curves (for face

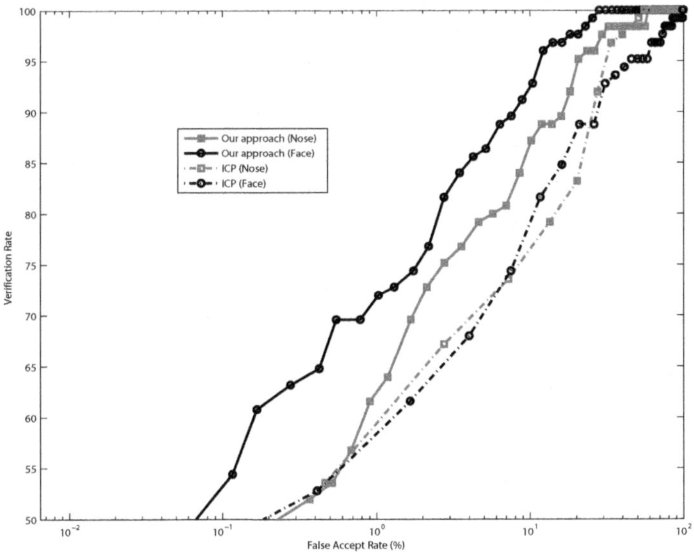

Fig. 4. ROC curves for our approach and ICP (baseline) on face and nose surfaces

and nose region) of our approach are almost always above the ICP ones. Which means that our verification rates at each false accept rate are greater than ICP ones. This result is expected as our approach is more robust in presence of facial expressions. As far as ICP is concerned, ROC curves for face and nose region are overlapped. ROC curve of our approach applied only on nose region is often above ICP ROC curves, applied on nose region and on the full face. Therefore, using only nose region in our approach gave a better verification results than ICP applied on the full face. Although, using the full face with our approach give always the better verification results.

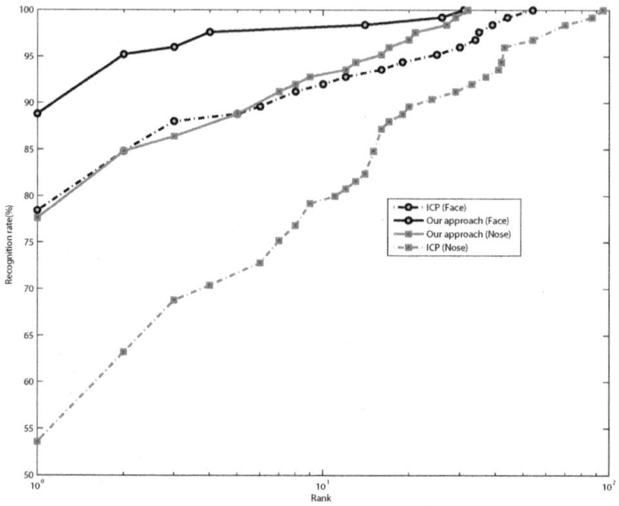

Fig. 5. CMC curves for our approach and ICP (baseline) on face and nose surfaces

These observations are also confirmed with identification scenarion which is illustrated with CMC curves of the two algorithms applied on nose region and on the full face. As shown in Figure 5, rank-one recognition rate given by our algorithm applied on the full face is the best one, it's about 88.8%. Then results of rank-one recognition for our approach applied on nose region and ICP applied on the full face are similar, about 78%. The worst result is given by ICP applied on nose region. At the rank-four, our approach on the full face is able to recognize 97.8% of the subjects in contrast with ICP on the full face which gives a similar result with our approach on the nose region, it's about 89%. After this rank, the recognition rates given by our approach on the nose region are always better than the ones given by ICP applied on the full face. So, using the full face for our approach gives always the best recognition and authentification results. However, appling our approach only on the nose region gives better results than ICP applied on the full face. Figure 6 shows some examlpes of some noses (with their faces) which are recognized by our approach but not by the baseline (ICP). This result can be explicated by the nature of the approaches themselves. Actually, as known, ICP algorithm search for the best rigid transformations to apply on a 3D points cloud (which represents the first face or nose) to be the closest to the second 3D points cloud (the second face or nose). As shown in this figure, 3D points positions have widely changed and it is different to ICP to find a smallest rigid transformation between these noses in probe and their correspondant in gallery. Although our approach succes to recognize them as it is based on shape analysis which is more robust to facial deformations in comparison with ICP.

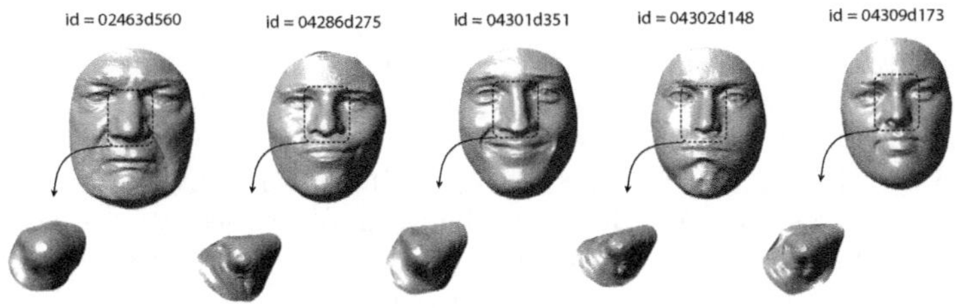

Fig. 6. Examples of noses that are recognized by our approach and not by ICP

5 Conclusions

In this paper, we have illustrated a geometric analysis of 3D nasal shapes in presence of both neutral and non-neutral facial expressions. In this analysis, the preprocessing is completely automated - the algorithm processes the face scan data, detects the tip of the nose, extracts a set of facial curves and extract nasal surface. The main tool presented in this paper is the construction of geodesic paths between arbitrary two nasal surfaces. The length of a geodesic between any two noses is computed as the geodesic length between a set of their nasal curves. This length quantifies differences in their shapes ; it also provides an optimal deformation from one to the other. In order to validate our approach in presence of facial expressions, a similarity matrix between 125 probe images with facial expressions and 125 gallery images with the neutral expression is computed. Authentication and recognition scores are produced and compared with a standard implementation of ICP as a baseline. The results of our algorithm outperform the baseline ICP algorithm which prove robustness of the proposed framework to deformations caused by facial expressions.

Acknowledgements

This research is supported in part by the ANR under the project ANR-07-SESU-004 and the Contrat de Projet Etat-Rgion (CPER) Rgion Nord-Pas-De calais Ambient Intelligence and partially supported by the following grants : ARO W911NF-04-01-0268 and AFOSR FA9550-06-1-0324 to Anuj Srivastava. Additionally, Anuj Srivastava was supported by visiting professorships from University of Lille I in summers of 2007 and 2008.

References

1. Ben Amor, B., Drira, H., Ballihi, L., Srivastava, A., Daoudi, M.: An experimental illustration of 3D facial shape analysis under facial expressions. Annals of telecommunications (accepted for publication)
2. Dijkstra, E.W.: A note on two problems in connection with graphs. Numerische Math. 1, 269–271 (1959)
3. Joshi, S.H., Klassen, E., Srivastava, A., Jermyn, I.H.: A novel representation for efficient computation of geodesics between n-dimensional curves. In: IEEE CVPR (2007)
4. Joshi, S.H., Klassen, E., Srivastava, A., Jermyn, I.H.: Removing shape-preserving transformations in square-root elastic (SRE) framework for shape analysis of curves. In: Yuille, A.L., Zhu, S.-C., Cremers, D., Wang, Y. (eds.) EMMCVPR 2007. LNCS, vol. 4679, pp. 387–398. Springer, Heidelberg (2007)
5. Klassen, E., Srivastava, A.: Geodesics between 3D closed curves using path-straightening. In: Leonardis, A., Bischof, H., Pinz, A. (eds.) ECCV 2006. LNCS, vol. 3951, pp. 95–106. Springer, Heidelberg (2006)
6. Klassen, E., Srivastava, A., Mio, W., Joshi, S.: Analysis of planar shapes using geodesic paths on shape spaces. IEEE Pattern Analysis and Machine Intelligence 26(3), 372–383 (2004)
7. Michor, P.W., Mumford, D.: Riemannian geometries on spaces of plane curves. Journal of the European Mathematical Society 88, 1–48 (2006)
8. Mio, W., Srivastava, A., Joshi, S.: On shape of plane elastic curves. International Journal of Computer Vision 73(3), 307–324 (2007)
9. Phillips, P.J., Flynn, P.J., Scruggs, T., Bowyer, K.W., Chang, J., Hoffman, K., Marques, J., Min, J., Worek, W.: Overview of the face recognition grand challenge. In: CVPR 2005: Proceedings of the 2005 IEEE Computer Society Conference on Computer Vision and Pattern Recognition (CVPR 2005), vol. 1, pp. 947–954 (2005)
10. Samir, C., Srivastava, A., Daoudi, M., Klassen, E.: An intrinsic framework for analysis of facial surfaces. International Journal of Computer Vision (accepted for publication)
11. Srivastava, A., Samir, C., Joshi, S., Daoudi, M.: Elastic shape models for face analysis using curvilinear coordinates. Journal of Mathematical Imaging and Vision (2008) (accepted for publication)
12. Younes, L.: Computable elastic distance between shapes. SIAM Journal of Applied Mathematics 58, 565–586 (1998)

Generic versus Salient Region-Based Partitioning for Local Appearance Face Recognition

Hazım Kemal Ekenel and Rainer Stiefelhagen

Computer Science Depatment, Universität Karlsruhe (TH)
Am Fasanengarten 5, Karlsruhe 76131, Germany
{ekenel,stiefel}@ira.uka.de
http://isl.ira.uka.de/cvhci

Abstract. In this paper, we investigate different partitioning schemes for local appearance-based face recognition. Five different salient region-based partitioning approaches are analyzed and they are compared to a generic partitioning scheme. Extensive experiments have been conducted on the AR, CMU PIE, FRGC, Yale B, and Extend Yale B face databases. The experimental results show that generic partitioning provides better performance than salient region-based partitioning schemes.

1 Introduction

Since the beginning of 1990s, appearance-based holistic approaches have been dominating the face recognition research [1,2,3]. Although local appearance information, in other words, using salient regions, has been shown to be superior to the holistic information in [4,5], interestingly face recognition research has focused on holistic approaches and local appearance based face recognition has been ignored in a great extent. It has not had as much impact as the holistic approaches, and compared to the plethora of the holistic methods, only a few techniques have been proposed to perform local appearance-based face recognition. The main reason for this is that the initial local appearance based approaches [4,5] require the detection of salient features i.e. eyes- which may not be an easy task. Moreover, erroneous detection of these local regions leads to severe performance drops.

Recently, local appearance-based face recognition approaches have been attracting growing interest [6,7,8,9,10,11,12,13,14]. In [6,14], the local facial regions are located by a support vector machine (SVM) and the combined local features are classified again with SVM. In [7], the face image is divided into rectangular sub-images without considering any salient regions, and the eigenfaces approach is then performed on each of these sub-images. In [8], the features extracted from local parts are modelled using GMM which are estimated by benefiting relevance adaptation (RA). In [11], the face image is partitioned into several local regions and each local region is represented by linear discriminant analysis (LDA). To combine the features extracted from each local region, another LDA

M. Tistarelli and M.S. Nixon (Eds.): ICB 2009, LNCS 5558, pp. 367–375, 2009.
© Springer-Verlag Berlin Heidelberg 2009

is used. A combination of DCT and LDA is applied on facial components in [9]. In [13], local binary pattern based face recognition is proposed. In this approach the image is divided into rectangular blocks and each block is represented by histogram of the local binary patterns (LBP). In [10,12], a generic face representation approach is introduced for local appearance-based face recognition. Local features, that are derived from DCT coefficients, are used for representing the local regions. The obtained local features are combined by conserving the spatial information. The proposed approach has been also utilized for 3-D and video-based face recognition [15,16].

In this study, we investigate the effect of different salient region-based partitioning schemes to the performance of the local appearance based face recognition (LAFR) approach. Normally, the proposed LAFR technique partitions an aligned face image into 8×8 pixels resolution non-overlapping blocks without considering any salient regions [10,12]. In addition to this generic partitioning, we test five different salient region-based partitioning schemes. Experimental results on AR [17], CMU PIE [18], FRGC [19], Yale B [20] and Extended Yale B [21] face databases show that generic partitioning provides better performance than salient region-based partitioning schemes.

The organization of the paper is as follows. In Section 2, local appearance-based face recognition is explained. Utilized partitioning approaches are introduced in Section 3. Experimental results are presented and discussed in Section 4. Finally, in Section 5, conclusions are given.

2 Local Appearance-Based Face Recognition

Local appearance-based face recognition is based on statistical representations of the non-overlapping local facial regions and their combination at the feature level. The underlying idea is to utilize local information while preserving the spatial relationships. In [10], the discrete cosine transform (DCT) is proposed to be used to represent the local regions. Its compact representation ability is superior to that of the other widely used input independent transforms like Walsh-Hadamard transforms. Although Karhunen-Loève transform (KLT) is known to be the optimal transform in terms of information packing, its data dependent nature makes it infeasible to use it for some practical tasks. Furthermore, DCT closely approximates the compact representation ability of the KLT, which makes it very useful for representation both in terms of information packing and in terms of computational complexity. In addition, DCT has been shown to be a better representation method for modeling the local facial appearance compared to KLT in terms of face recognition performance [10].

Local appearance-based face recognition approach is proposed as a generic, practical and robust face recognition algorithm. Feature extraction using local appearance-based face representation can be summarized as follows: A detected and normalized face image is divided into blocks of 8×8 pixels resolution. The reason for having 8×8 pixels block size is to provide sufficient compactness on one hand, and to keep stationarity within the block on the other hand. The DCT is applied on each block. Then, the obtained DCT coefficients are ordered using zig-zag scan pattern. From the ordered coefficients, M of them are selected and normalized according to a feature selection and feature normalization strategy resulting in an M-dimensional local feature vector [12]. Finally, the DCT coefficients extracted from each block are concatenated to construct the global feature vector. For details of the algorithm please see [10,12].

3 Generic vs. Salient Region-Based Partitioning

In this study, five different salient region-based partitioning schemes, that are derived from the previous modular/component/patch based studies [4,5,9,11,14], are compared to the generic partitioning of the local appearance-based face recognition approach. In the implementation, the salient regions are divided into 8×8 pixels resolution non-overlapping blocks and the DCT is applied on each block. From the DCT coefficients that are ordered according to the zig-zag scan pattern, ten of them are selected by omitting the first DCT coefficient and selecting the following first ten of them. The selected coefficients are divided to their standard deviations. Afterwards, the local feature vector is normalized to the unit norm. The overall feature vector for a salient region is constructed by concatenating the local feature vectors that are extracted from the blocks of the corresponding salient region. The feature vector of the combined regions is generated by concatenating the local feature vectors of each region.

The generic partitioning is performed by dividing the face image into 8×8 pixel resolution non-overlapping blocks as depicted in Figure 1. The first salient region-based partitioning scheme (P1) is similar to the one in [4]. It consists of three regions: Eyes, nose and mouth. A sample image, illustrating this partitioning scheme, is given in Figure 2. The second partitioning scheme (P2) is from [5]. Four salient regions are used for face recognition: Left eye, right eye, nose and mouth. The partitioning on a sample image is shown in Figure 4. The third partitioning scheme (P3) is derived from [9]. Larger left eye and right eye regions that contain partially the nose and some parts below the eyes, and the nose region are the salient regions used in this partitioning scheme. The partitioning on a sample image is illustrated in Figure 6. The fourth partitioning scheme is an approximation of the one in [11]. It has five regions: forehead, left eye, right eye, lower left and right parts of the face. The partitioning on a sample image can be seen in Figure 8. The fifth partitioning scheme is derived from [14]. There are 14 learned components as shown in Figure 10.

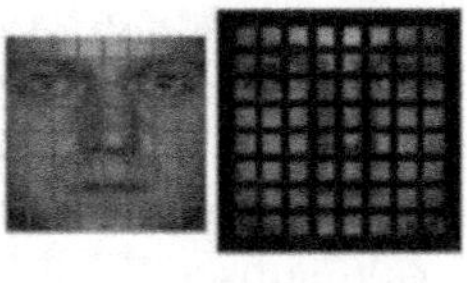

Fig. 1. Generic partitioning scheme

4 Experiments

The performance of the algorithm is tested on five different benchmarks, that are widely used in the face recognition research community. The face recognition grand challenge (FRGC) version 2 data set [19] is used to test the algorithm under controlled and uncontrolled conditions. For controlled conditions, 120 individuals from the FRGC experiment 1 data set are selected, who have at least ten images both in fall 2003 and spring 2004 recordings. The images from fall 2003 recordings are used for training and the ones from spring 2004 are used for testing. For uncontrolled conditions, the same experimentation setup is used, but this time the images are selected from the FRGC experiment 4 data set. The AR face database [17] is utilized for testing against partial face occlusion. 550 face images of 110 subjects are selected from this database. Separate experiments were conducted for upper and lower face occlusion. For both cases, a single face image that does not contain any occlusion is used for training. To assess the performance against upper face occlusion, a single face image that contains sun glasses is used. For lower face occlusion experiments, a single face image that contains a scarf is used. In order to test the algorithm against illumination variations, experiments are conducted on the CMU PIE [18] and the Yale/Extended Yale B face databases [20,21]. There are 68 subjects in the CMU PIE database. 21 images per subject are chosen from the illumination set for the experiments. From these images, only the frontally illuminated face image is used for training. The remaining twenty images are used for testing. The Yale face database B contains ten subjects under 64 different illumination conditions. The first subset that has close to frontal illumination is used for training. For testing, subsets 2, 3, 4 and 5 are used. These subsets contain 12, 12, 14, 19 images per person, respectively. With increasing subset number, the illumination variations become stronger. The extended Yale face database B contains Yale face database B and has the same organization. It has 28 additional subjects making 38 subjects in total. Overall, there are 15 training-testing combinations. Two from the FRGC ver. 2 database, one for controlled and one for uncontrolled case; one from the CMU PIE; four from the AR, two for lower face occlusion and two for upper face occlusion -one of them is within session and the other one is between session experiment. Finally, eight from the Yale B and Extended Yale B face databases. Four from each one.

The obtained results using P1 partitioning scheme with the individual components and the combined representation on each experimental setup are shown in Figure 3. The correspondences between the abbreviations on the x-axis and the experiments are as follows: F1: FRGC under controlled conditions, F4: FRGC under uncontrolled conditions, CP: CMU PIE, A1sc: AR, occlusion with scarf, within session, A1sun: AR, occlusion with sunglasses, within session, Aisc: AR, occlusion with scarf, between session, Aisun: AR, occlusion with sunglasses, between session, Y2,Y3,Y4,Y5: Yale subset 2,3,4,5 and EY2,EY3,EY4,EY5: Extended Yale subset 2,3,4,5, respectively. The best results are obtained with the combined representation except on the Aisun experiment, where mouth region provides the highest correct classification rate. Eye region is found to be the

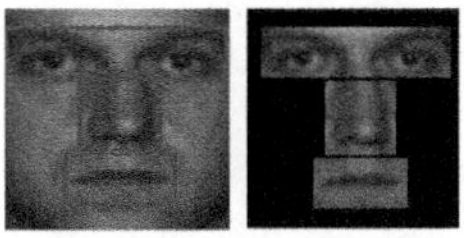

Fig. 2. Salient regions obtained with the P1 partitioning scheme

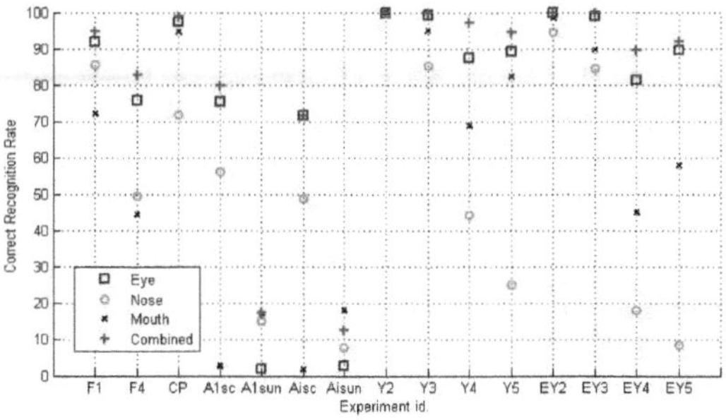

Fig. 3. Correct identification rates obtained with the P1 partitioning scheme

second best performing region, except in the experiments where upper face occlusion exists. On Y2 and EY2, 100% correct recognition rate is achieved just using the eye region. Depending on the experimental setup, either nose region or mouth region comes the third. In the experiments with high illumination variations, such as CP, Y4, Y5, EY4 and EY5, mouth region is found to be more useful for identification than the nose region. This is expected, since in the case of illumination variation due to cast shadows the appearance of nose region is affected severely. The other reason for this outcome is the lack of expression variations in the used data sets for these experiments. Mouth region works also better in the experiments that contain upper face occlusion, namely, A1sun and Aisun experiments. Having sunglasses decreases the amount of the discriminative information that the nose region contains. In F1, F4, A1sc and Aisc experiments nose region reaches higher recogniton rates than the mouth region. The expression variations in F1 and F4 experiments deteriorate the performance of the mouth region. Obviously, in the case of lower face occlusion, the mouth region has no use.

The results obtained by using P2 partitioning scheme can be seen from Figure 5. The outcomes are similar to the ones obtained with the P1 partitioning scheme. Combined representation achieves the best results. Eye regions have the second place. There is no big difference in left and right eye regions' correct classification rates. The same observations are valid for the nose and mouth regions. The only difference is observed on the CP experiment where the mouth region performs slightly better than the eye regions. It can be also observed that the performance difference between the mouth region and the eye regions is less than the one attained on the experiments that contain illumination variations with P1 partitioning scheme. The reason is, the region that contains both of the eyes have more discriminative power than the individual eye regions.

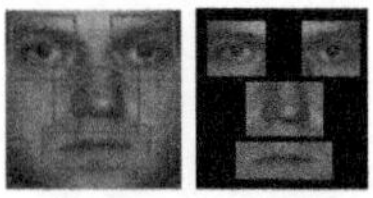

Fig. 4. Salient regions obtained with the P2 partitioning scheme

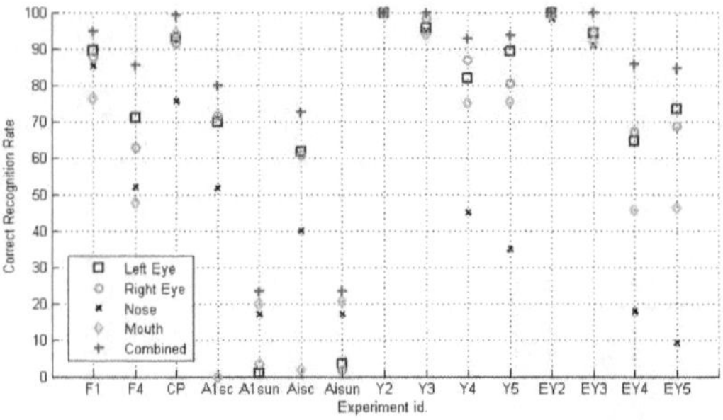

Fig. 5. Correct identification rates obtained with the P2 partitioning scheme

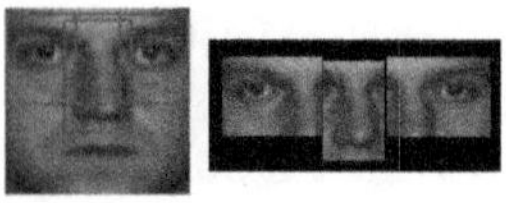

Fig. 6. Salient regions obtained with the P3 partitioning scheme

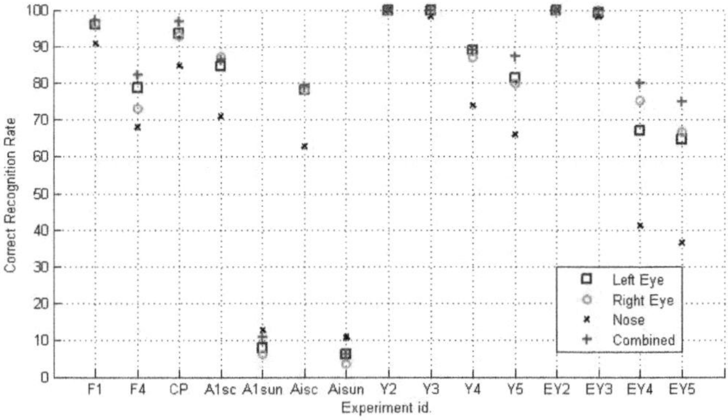

Fig. 7. Correct identification rates obtained with the P3 partitioning scheme

The correct identification rates that are achieved with the P3 partitioning scheme are presented in Figure 7. Combined representation attains the highest correct recognition rates in most of the experiments. On A1sun and Aisun nose region achieves the best results, whereas on A1sc right eye region outperforms the others. Once more time, it has been observed that, except upper face occlusion, eye regions contain more disciminative power than the nose region.

Figure 9 shows the correct identification rates obtained by using the P4 partitioning scheme. The best performance is always achieved with the combined representation. On the experiments with lower face occlusion, as expected, lower face regions perform poorly and on the experiments with upper face occlusion, eye regions perform poorly. In most of the cases forehead region achieves higher correct recognition rates compared to the other salient regions on the experiments that contain large illumination variations, since this region is less affected

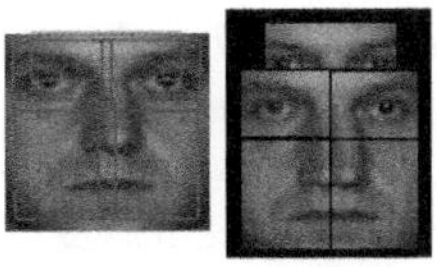

Fig. 8. Salient regions obtained with the P4 partitioning scheme

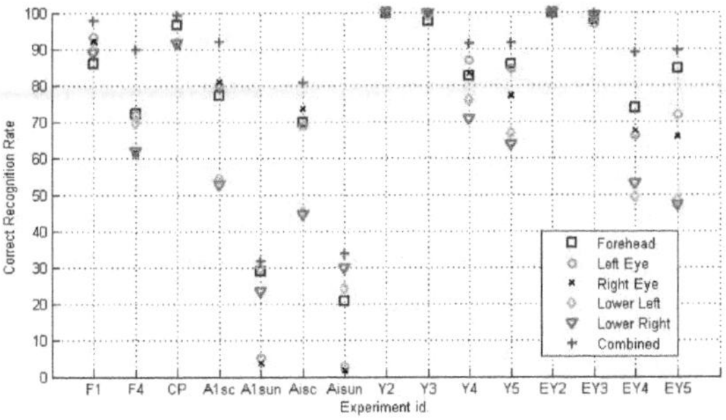

Fig. 9. Correct identification rates obtained with the P4 partitioning scheme

from the changes in lighting. Both the eye regions and lower facial parts contain partially the nose region which makes them sensitive to the changes in appearance due to cast shadows.

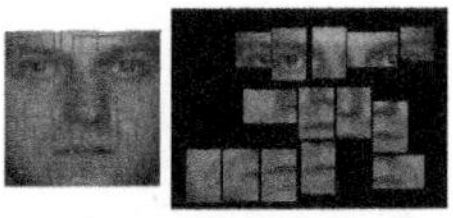

Fig. 10. Salient regions obtained with the P5 partitioning scheme

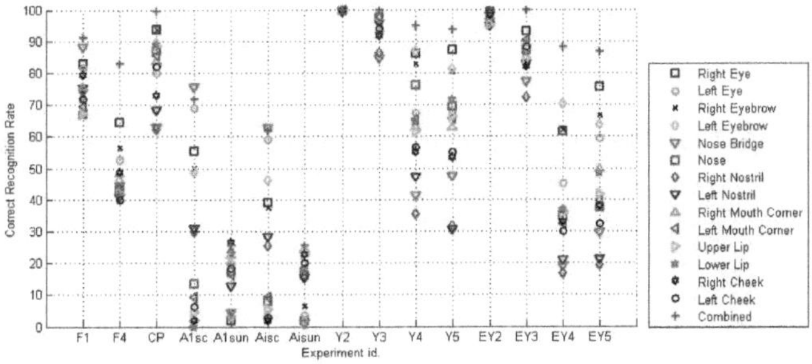

Fig. 11. Correct identification rates obtained with the P5 partitioning scheme

Correct identification rates attained by using the P5 partitioning scheme are given in Figure 11. Most of the time the best performance is achieved with the combined representation. Only on A1sc, Aisc experiments nose bridge region, and on A1sun right cheek region performs better. Depending on the experimental condition, performance order of the facial parts changes. Besides the experiments with upper face occlusion, right eye, left eye, right eyebrow and left eyebrow regions consistently achieve high classification rates.

The comparison of combined representation of different partitioning schemes is shown in Figure 12. Generic partitioning is found to be superior to the salient region-based partitioning in most of the cases. For example, on F4, the performance is 83.0% with P1, 85.6% with P2, 82.3% with P3, 89.8% with P4, 82.8% with P5 and 90.8% with generic partitioning. Only on the Y4 experiment P1 partitioning scheme outperforms generic partitioning. However, on EY4, which contains Y4 as a subset, generic partitioning provides better results. P4 partitioning scheme also provides consistently high results. On the experiments that contain large illumination variations, P3 partitioning scheme is found to be the poorest performing one. The reason is on each part nose is included to some extend, which makes it sensitive to the cast shadows. P1 performs better than P2 on these experiments again due to its less sensitivity to the cast shadows.

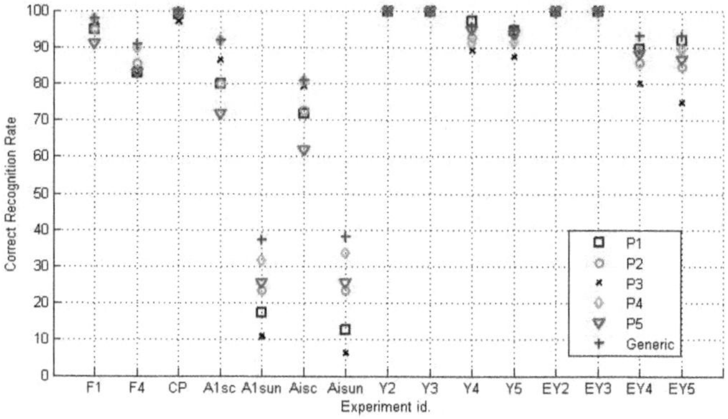

Fig. 12. Correct identification rates obtained with the combined representations

5 Conclusions

In this study, we evaluate different salient region-based partitioning schemes for local appearance-based face recognition algorithm, which is originally proposed as a generic approach and partitions an aligned face image into 8×8 pixels resolution non-overlapping blocks without considering any salient regions. We compare five different salient region-based partitioning schemes. It has been found that generic partitioning provides higher correct recognition rates than salient region-based partitioning approaches. Thus, justifying that there is no need to focus on salient regions and perform salient region-based partitioning.

Acknowledgement

This study is partially funded by the German Research Foundation (DFG) under Sonderforschungsbereich SFB 588 —Humanoid Robots— and by OSEO, French State agency for innovation, as part of the Quaero Programme. The authors would like to thank Dr. Bernd Heisele and Dr. Tae-Kyun Kim for providing information about the partitioning schemes that they have used.

References

1. Turk, M., Pentland, A.: Eigenfaces for recognition. Journal of Cognitive Neuroscience 3(1), 71–86 (1991)
2. Belhumeur, P.N., Hespanha, J.P., Kriegman, D.J.: Eigenfaces vs. fisherfaces: Recognition using class specific linear projection. IEEE Trans. on PAMI 19(7), 711–720 (1997)
3. Moghaddam, B., Jebara, T., Pentland, A.: Bayesian face recognition. Pattern Recognition 33(11), 1771–1782 (2000)
4. Brunelli, R., Poggio, T.: Face recognition: Features versus templates. IEEE Trans. on PAMI 15(10), 1042–1052 (1993)
5. Pentland, A., Moghaddam, B., Starner, T.: View-based and modular eigenspaces for face recognition. In: Proc. of IEEE CVPR, pp. 84–91 (1994)
6. Heisele, B., Ho, P., Wu, J., Poggio, T.: Face recognition: component-based versus global approaches. Computer Vision and Image Understanding 91(1-2), 6–21 (2003)
7. Gottumukkal, R., Asari, V.K.: An improved face recognition technique based on modular PCA approach. Pattern Recognition Letters 25(4), 429–436 (2004)
8. Lucey, S., Chen, T.: A GMM parts based face representation for improved verification through relevance adaptation. In: Proc. of IEEE CVPR, pp. 855–861 (2004)
9. Lee, H.J., Kim, H.J., Kim, W.Y.: Face recognition using component-based DCT/LDA. In: Proc. of IWAIT (2005)
10. Ekenel, H.K., Stiefelhagen, R.: Local appearance based face recognition using discrete cosine transform. In: Proc. of EUSIPCO (2005)
11. Kim, T.K., Kim, H., Hwang, W., Kittler, J.: Component-based LDA face description for image retrieval and MPEG-7 standardisation. Image and Vision Computing 23(7), 631–642 (2005)
12. Ekenel, H.K., Stiefelhagen, R.: Analysis of local appearance-based face recognition: Effects of feature selection and feature normalization. In: Proc. of IEEE CVPR Biometrics Workshop (2006)
13. Ahonen, T., Hadid, A., Pietikainen, M.: Face description with local binary patterns: application to face recognition. IEEE Trans. on PAMI 28(12), 2037–2041 (2006)
14. Heisele, B., Serre, T., Poggio, T.: A component-based framework for face detection and identification. International Journal of Computer Vision 74(2), 167–181 (2007)
15. Ekenel, H.K., Gao, H., Stiefelhagen, R.: 3-D face recognition using local appearance-based models. IEEE Trans. on Information Forensics and Security 2(3), 630–635 (2007)
16. Stallkamp, J., Ekenel, H.K., Stiefelhagen, R.: Video-based face recognition on real-world data. In: Proc. of IEEE ICCV, pp. 1–8 (2007)
17. Martinez, A.M., Benavente, R.: The AR face database, CVC Tech. Report #24 (1998)
18. Sim, T., Baker, S., Bsat, M.: The CMU pose, illumination, and expression (PIE) database. IEEE Trans. on PAMI 25(12), 1615–1618 (2003)
19. Phillips, P.J., Flynn, P.J., Scruggs, T., Bowyer, K.W., Chang, J., Hoffman, K., Marques, J., Min, J., Worek, W.: Overview of the face recognition grand challenge. In: Proc. of IEEE CVPR, pp. 947–954 (2005)
20. Georghiades, A.S., Belhumeur, P.N., Kriegman, D.J.: From few to many: Illumination cone models for face recognition under variable lighting and pose. IEEE Trans. on PAMI 23(6), 643–660 (2001)
21. Lee, K.C., Ho, J., Kriegman, D.J.: Acquiring linear subspaces for face recognition under variable lighting. IEEE Trans. on PAMI 27(5), 684–698 (2005)

Near Infrared Face Based Biometric Key Binding

Meng Ao and Stan Z. Li*

Center for Biometrics and Security Research,
Institute of Automation, Chinese Academy of Sciences,
95 Zhongguancun Donglu, Beijing 100190, China
{mao,szli}@cbsr.ia.ac.cn

Abstract. Biometric encryption is the basis for biometric template protection and information security. While existing methods are based on iris or fingerprint modality, face has so far been considered not reliable enough to meet the requirement for error correcting ability. In this paper, we present a novel biometric key binding method based on near infrared (NIR) face biometric. An enhanced BioHash algorithm is developed by imposing an NXOR mask onto the input to the subsequent error correcting code (ECC). This way, when combined with ECC and NIR face features, it enables reliable binding of face biometric features and the biometric key. Its ability for template protection and information cryptography is guarantied by the theory of encryption. The security level of NIR face recognition system is thereby improved. Experimental results show that the security benefit is gained with a sacrifice of 1-2% drop in the recognition performance.

Keywords: BioHash, ECC, Key Binding, NIR Face, Biometric Encryption.

1 Introduction

Cryptology of security keys plays an important role in information security [14]. When the key and its owner are separated, the computer know whether the key is correct but not the user. If an attacker gets the key, the information will be decrypted easily. This presents a problem in security.

A biometric based system gets personal identity, but relies on the operating system or access key to control access to the information stored on the system. So, biometrics alone can solve the security problem only partially.

In addition, biometric has its inherent shortcomings. Biometric characteristics (especially physiological biometrics) are largely immutable, resulting in permanent biometric compromise when stolen or leaked. Therefore, how to protect biometric templates becomes an important issue. In traditional systems using passwords, Message-Digest Algorithm 5 (MD5) is widely used to store passwords

* Corresponding author.

M. Tistarelli and M.S. Nixon (Eds.): ICB 2009, LNCS 5558, pp. 376–385, 2009.
© Springer-Verlag Berlin Heidelberg 2009

[13]. The system only stores the hashed password, which ensures the system will not leak the password.

A good solution to these problems is to bind the access key with its owner's identity for each piece of restricted information. There are two approaches to biometric based information security. One is to convert a biometric template into a stable vector and use MD5 or other one-way algorithm to protect the biometric template. A problem with this approach is that the biometric template is still unchangeable. The other is to bind a biometric feature to a key. If the system is cracked by brutal force attack, the system administrator just needs to change a binding key.

However, there is a problem in applying biometrics to security key generation: Biometric features are subject to certain degree of variation and so may not be 100% the same for all feature extraction session. To tackle such inherent variability of biometrics, error correcting code (ECC) is used to correct errors or noise in biometric features, to finally enhance the reliability of biometric key binding [12,29,28,27]. Since the ECC algorithms only work in binary system with hamming distance. In order to use ECC algorithms, BioHash[25,23,26,32,31] method can be introduced to convert the biometric feature vectors into binary strings. Therefore, BioHash and ECC theory can be combined to enhance key binding to a biometric system.

Most of existing works in this direction are mainly based on fingerprint and iris modalities [5,10,11,18,9,20,4,16,3,2,19,7]. This is because fingerprint and iris features are more stable and can be easily incorporated with ECC [8,24,30,1].

There has been some literature reporting key binding with face biometric features[17,15]. These works are based on a visible light face recognition. The experiments in these papers are close-set face recognition. That means these works can not be used in real application. The main reason is that face has thus far been considered not stable enough to be error-corrected by ECC. Face biometric is affected by lighting condition, expression, and other factors [34]. The variability in face features often exceeds the correcting ability of ECC.

Near infrared (NIR) face recognition introduced recently has achieved great success. Not only it overcomes the illumination problem but also achieved significant higher accuracy performance than conventional visual face biometrics [21]. It can be used reliably for 1-to-many face identification. This new technology makes face biometric key binding possible.

In this paper, we present a novel biometric key binding method based on near infrared (NIR) face biometric. The contributions are the following: First, we develop an enhanced BioHash algorithm by imposing an NXOR mask onto the input to the subsequent error correcting code (ECC). Second, we present and NIR face based key binding for improving the security level of NIR face recognition system. the BioHash is combined with ECC and NIR face features to enables reliable binding of face biometric features and the biometric key.

While the ability of the BioHash algorithm for template protection and information cryptography is guarantied by the theory of encryption, we present experimental results show that the security benefit is gained with a sacrifice of 2% drop in the recognition performance.

The rest of the paper is organized as follows. Section 2 introduces BioHash method and the enhanced BioHash. Section 3 describes the NIR face based encryption algorithm. Section 4 shows the experiment result.

2 Enhanced BioHash

BioHash represents a series of operations which combine a high dimensional feature vector with a user-specific tokenised random numbers (TRN) to produce a binary bit string, so that the similarity score between two feature vectors can be measured by the hamming distance [26,32,6]. The basic idea of BioHash is to use threshold calculation to obtain a bit 0 or 1. The following describe three BioHash procedures for converting a feature vector to a binary string.

BioHash 1. By repeatedly converting a feature vector to a binary string by comparing it with a random vector with same length.

1. Use a feature extraction technique to extract the biometric feature. The biometric feature is represented in a vector form, $x \in \mathscr{R}^n$, with n denoting the feature length of x. Repeating x m times to obtain a new feature vector $X = \{x, ..., x\}$.
2. Use token to generate a pseudo-random vectors $r \in \mathscr{R}^{n \times m}$ based on a seed. m is a integer. The entire distribution of r is same as x.
3. Use the following threshold calculation to obtain a binary string $b_1 b_2 ... b_{n \times m}$ with length $n \times m$

$$b_i = \begin{cases} 1 \; if \; X_i > r_i \\ 0 \; if \; X_i \leqslant r_i \end{cases} i = \{1, ..., n \times m\} \tag{1}$$

The n-dimensional biometric feature x is thus converted to a binary string b of length $n \times \; m$.

BioHash 2. By calculating the dot product of the feature vector and several random vectors, and then comparing the dot product results with a threshold.

1. Employ the input token to generate a set of pseudo-random vectors, $r_i \in \mathscr{R}^M$ for $i = 1, \ldots, m$ based on a seed.
2. Apply the Gram-Schmidt process to $r_i \in \mathscr{R}^M$ for $i = 1, \ldots, m$ to obtain a set of orthonormal vectors $p_i \in \mathscr{R}^M | i = 1, ..., m$, also called Tokenised Random Number, TRN.
3. Calculate the dot product of v, the feature vector obtained from first step and each orthnonormal vector in TRN, p_i, such that $\langle v, p \rangle$.
4. Use a threshold τ to obtain a binary string, b whose elements are defined as

$$b_i = \begin{cases} 1 \; if \; \langle v, p \rangle > \tau \\ 0 \; if \; \langle v, p \rangle \leqslant \tau \end{cases} i = \{1, ..., m\} \tag{2}$$

BioHash 3. Converting a feature vector to a binary string by calculating the range of arguments of the complex, which is generated by adding a random imaginary to the feature vector.

1. Use token to generate a set of pseudo-random vectors, $r_i \in \mathscr{R}^M | i = 1, ..., m$ with the entires distributed according to $N(0,1)$ and apply the Gram-Schmidt process to transform the basis $r_i \in \mathscr{R}^n | i = 1, ..., m$ into an orthonormal set of $r, r_{\perp i} \in \mathscr{R}^n | i = 1, ..., m$.
2. Mix the x with $r_{\perp i}$ iteratively to form the complex number, $\{z_i = x_i + r_{\perp i} j | i = 2, ..., n\}$, where $j = \sqrt{-1}$ and calculate their complex arguments, $\{arg(z_i) \in \mathscr{R}^n | i = 1, ..., n\}$.
3. Average the complex arguments, $\left\{\alpha_i = \frac{1}{n}\sum_{j=1}^{n} arg(z_i) \in \mathscr{R}^m | i = 1, ..., m\right\}$ where $-\pi \leqslant \alpha_i < \pi$ and $m < n$. Then we get the binary string b

$$b_i = \begin{cases} 1 \; if \; -\pi < \alpha < 0 \\ 0 \; if \; 0 \leqslant \alpha < \pi \end{cases} i = \{1, ..., m\} \tag{3}$$

Enhanced BioHash

Among the three BioHash algorithms, BioHash 1 is advantageous in that binary strings it generates is much longer than the others and easy to calculate. With a longer binary string, we can bind into longer key, which increases the security of system.

However, this incurs a problem that we cannot using an ECC algorithm to correct the binary string: The binary string obtained can be any one in the space $\{0,1\}^n$; however, ECC cannot decode some of these. ECC coding and decoding algorithms are generally used in pair. There are some strings in the space $\{0,1\}^n$ which are out of ECC decoding space.

Here we present an enhanced BioHash method to solve this problem. Suppose that b is a binary string with length n which is converted from a biometric feature vector. For any binary string b' of length n, we generate a mask string $M = b \; NXOR \; b')$ by

$$M_i = b_i \; NXOR \; b'_i, i = 1, ..., n \tag{4}$$

Here NXOR (Not eXclusive OR) is a logical operation. The NXOR is defined as follows

$$1 \; NXOR \; 1 = 1, 0 \; NXOR \; 0 = 1, 1 \; NXOR \; 0 = 0, 0 \; NXOR \; 1 = 0. \tag{5}$$

An NXOR between two strings is defined as NXOR on each bit respectively. The binary string b is then converted to b' using the operation $b' = b \; NXOR \; M$ by

$$b_i \; NXOR \; M_i = b'_i, i = 1, ..., n \tag{6}$$

With the same mask M, the hamming distance between b_1 and b_2 does not change because the property of NXOR operator, that is,

$$D_{Hamming}(b_1, b_2) = D_{Hamming}((b_1 \; NXOR \; M), (b_2 \; NXOR \; M)) \tag{7}$$

3 NIR Face Based Key Binding

The motivation of NIR face based key binding is the following: We need a face encryption system. Error correcting code (ECC) is usually needed to solve the problem incurred by variation in biometric features. The ECC based method requires that the biometric module be able to achieve a minimum performance that is associated with the ability ranger of ECC. The stability of NIR face features can satisfy this requirement of ECC, as will be explained.

When the key binding method is introduced into a face recognition system, the enrollment and recognition processes are both changed. So, a description of face enrollment and recognition for biometric key binding will be provided. How to use multi-image for enrollment to improve the system performance will be described. The security level of the key binding NIR face recognition system and the external function of the system will be discussed, too.

3.1 Error Correcting Code

Error Correcting Code is a code in which each data signal conforms to specific rules of construction so that departures from this construction in the received signal can generally be automatically detected and corrected. The basic idea is for the transmitter to apply one or more of the above error detecting codes; then the receiver uses those codes to narrow down exactly where in the message the error (if any) is. Shannon's theorem points out that the maximum attainable efficiency of an error-correcting scheme versus the levels of noise interference expected.

Two main categories are convolutional codes and block codes. Examples of the latter are Hamming code, BCH (Bose-Chaudhuri-Hocquenghem) code, Reed-Solomon code, Reed-Muller code, Binary Golay code, and low-density parity-check codes [33]. Here we use BCH code to convert the unstable BioHash binary string to a stable binary string.

In the error correcting theory the ratio of error bits to total bits has an upper limit. Suppose b_1 and b_2 are the two BioHash strings from the same person. If we use the error correcting code to transform the BioHash binary string to a stable binary string, the difference of two binary strings from the same person should be less than the upper limit. As we know, face images usually vary large due to various lighting changes which may make the difference of features from the same person larger than the upper limit. To deal with this problem, here we choose NIR face recognition as the basic system which has an advantage on face encryption. See Fig.1.

The use of NIR imaging brings a new dimension for applications of invisible lights for face detection and recognition [21]. It not only provides appropriate active frontal lighting but also minimizes lightings from other sources. The following figure compares NIR face images with visible light (VL) face images. The VL images contain large performance-deteriorating lighting changes whereas the NIR are good for face recognition. The fixed lighting direction much simplifies the problem of face recognition. Face recognition algorithms are more stable in under controlled lighting. That meets the requirement of error correcting code theory.

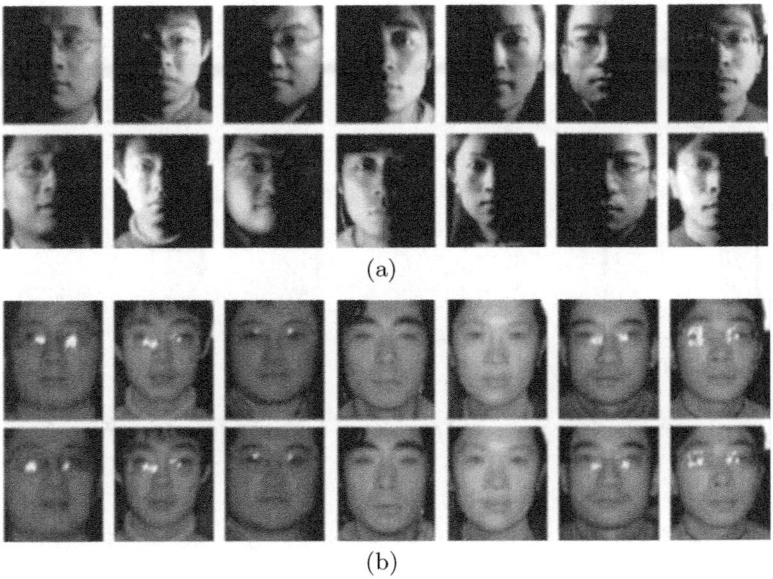

Fig. 1. Comparison of VL images (a) vs. NIR images (b) taken in the same environmental lighting conditions

3.2 Face Enrollment

The enrollment process is showed in Fig.2(a). Suppose the length of face feature is n. The length of BioHash binary string is $n \times m$. The threshold of the hamming distance is τ. The the enroll process is:

1. Face feature extraction. Extract a feature vector x from a NIR face image.
2. Use the basic BioHash method to convert the face feature vector x to a binary string b. The seed of the pseudo-random vector in BioHash is s.
3. Generate a random integer vector. Convert the random vector to binary system. Here we get a random binary string k with the length of t. This is the binded key.
4. Use BCH coding algorithm to transform k to k_{BCH}. The length l of k_{BCH} is a little smaller than $n \times m$ or equal. Use the first l bits of b to get the mask $M = k_{BCH}\ NXOR\ b$.
5. Record the M , s and k values in the database. k should be protected by MD5 or other one-way algorithm.

Item 5 refers the storage in the database. It is obvious that we can no nothing about the face feature just from M and s. However, if k can be obtained by the attacker, the attacker may concoct a new feature vector which can pass the system. So the binded key k should be protected by MD5.

Note that in item 1 in BioHash 1, the face feature vector x is repeated m times. This brings multiimages for the enrolled person to improve the performance of

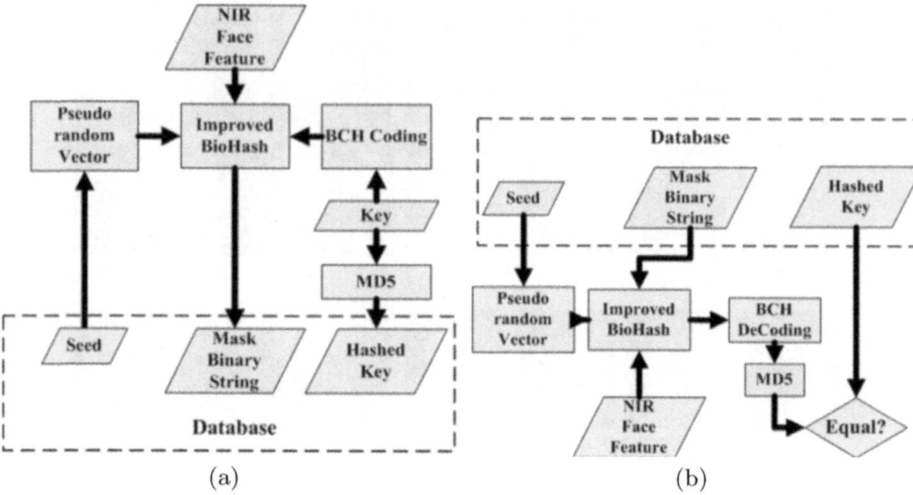

Fig. 2. (a) The enrollment process. (b) The recognition process.

the system. Suppose there are several images from one person with corresponding feature vectors $x_1, x_2, ..., x_s$. We use $\{x_1, x_2, ..., x_s, ..., x_1, x_2, ..., x_s\}$ instead of $\{x, x, ..., x\}$ in item 1.

3.3 Face Recognition

The recognition, shown in Fig.2(b), is described as follows:

1. Face feature extraction. Extract a feature vector x' from a NIR face image.
2. Get s from the database. Use s to generate a seudo-random vector for Bio-Hash. Convert x to the binary string b'.
3. Get M from the database. Calculate the masked string $k'_{BCH} = b' \, NXOR \, M$.
4. Use BCH decoding algorithm to transform k'_{BCH} to k'.
5. Check if k' is k by MD5. if $k' = k$, then x' and x are from the same person.

Because the probability of collision of MD5 is very small, the performance of the recognition mainly depends on the result of BioHash.

We can see in enrollment and recognition processes that the face template stored in the system are protected by MD5 or other one-way algorithm. Here the template protection depends on the security of encryption algorithms. The security of the face recognition system is guaranteed by the encryption theory [13].

The NIR face base key binding can be used in information cryptography. The binding key can be the same key for encryption and decryption of a message. This application has a wide range of prospects.

4 Experiments

Experiments are done to compare the performance of the original NIR face features algorithm and the enhanced BioHash binary strings with face key binding incorporated. The purpose is to evaluate how much sacrifice in recognition accuracy the system has to pay for the gained security.

The original algorithm converts a 142×120 face image to a 256-dimensional feature vector by the algorithm in paper [22]. There are no intersection between the training database and testing database. The testing database contains 1176 images, which are from 294 people with 4 images per person. Every image is compared with other images. So the number of similarity scores is $C_{1176}^{2} = 690900$.

The original system uses $L2$ distance as the similarity score. The key banding system checks whether the result of MD5 is equal. BCH decode algorithm ensures the two binary strings getting different codes when the hamming distance is bigger than BCH threshold. So we only need to calculate ROC of the BioHash binary string with hamming distance. The result is shown in Fig.3(a).

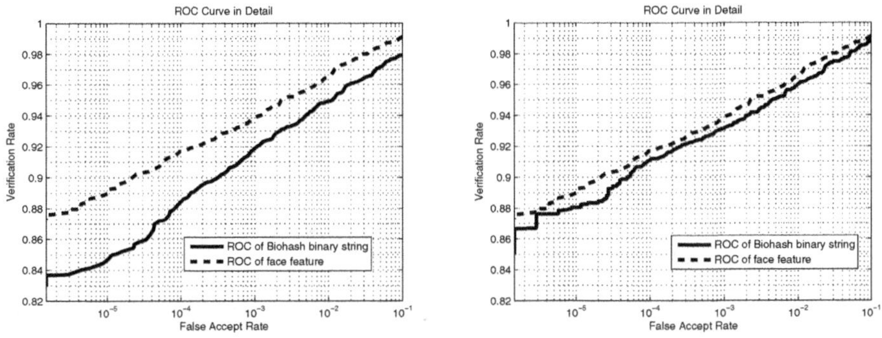

(a) (b)

Fig. 3. The comparation of ROC curve between BioHash binary string and raw face feature: (a) When Biohash binary string length is 4095, and (b) When Biohash binary string length is 66175

We set the length of BioHash binary string to be 4095, threshold 285, and corresponding key length as 707. The enhanced BioHash binary string achieves verification rate (VR) of 92.01% at FAR = 0.11%. This decreased VR by 2% from the original system. However, this method offers security to the face recognition system, and information protection ability for the face recognition system.

In BCH algorithm, when the total number is 4095, the corresponding error bits can be 1 to 343. That means in our method, the range of FAR is from 0 to 8.32%, and the range of VR is from 0 to 97.79%.

The BCH code length can be longer in order to make a long key. When the code length increases, BCH algorithm takes more time to code and decode. That also makes the brute force search very difficult. See Fig.3(b). When we set the length of BioHash binary string to be 66175, threshold 3407, and corresponding key length as 11131. And with the long biohash binary string, the VR also increases to 93.15%.

5 Conclusion

This paper has presented a novel biometric key binding method, enhanced BioHash. The method, when incoporated with NIR face biometric, enables face biometric based template protection, file encrypting and many other applications. The gain in security is proved by existing theory, with a little drop in recognition accuracy.

Acknowledgements. This work was supported by the following funding resources: National Natural Science Foundation Project #60518002, National Science and Technology Support Program Project #2006BAK08B06, National Hi-Tech (863) Program Projects #2006AA01Z192, #2006AA01Z193, and #2008AA01Z124, Chinese Academy of Sciences 100 people project, and AuthenMetric R&D Funds.

References

1. Alabbadi, M., Wicker, S.B.: A digital signature scheme based on linear error-correcting block codes (1995)
2. Bakhtiari, S., Pieprzyk, J.: On the weakness of gongs collisionful hash function. Journal of Universal Computer Science 3, 185–196 (1997)
3. Bakhtiari, S., Safavi-naini, R., Pieprzyk, J.: On password-based authenticated key exchange using collisionful hash functions (1996)
4. Berson, T., Gong, L., Lomas, T.: Secure, keyed, and collisionful hash functions. December 1993. Included in Technical Report SRI-CSL-94-08, Computer Science Laboratory, SRI International, Menlo Park, California (May 1994)
5. Bodo, A.: Method for producing a digital signature with aid of a biometric feature, German Patent, DE 4243908A1 (1994)
6. Cheung, K.H., Kong, A.W.-K., Zhang, D., Kamel, M., You, J.: Revealing the secret of facehashing. In: Zhang, D., Jain, A.K. (eds.) ICB 2006. LNCS, vol. 3832, pp. 106–112. Springer, Heidelberg (2006)
7. Clancy, T.C., Kiyavash, N., Lin, D.J.: Secure smartcardbased fingerprint authentication (2003)
8. Crepeau, C., Kilian, J.: Achieving oblivious transfer using weakened security assumptions (1988) (extended abstract)
9. Daugman, J.: How iris recognition works. IEEE Transactions on Circuits and Systems for Video Technology 14, 21–30 (2004)
10. Davida, G., Frankel, Y., Matt, B.: On enabling secure applications through off-line biometric identification. In: 1998 IEEE Symposium on Security and Privacy, Proceedings, May 1998, pp. 148–157 (1998)
11. Davida, G.I., Frankel, Y., Matt, B.J., Peralta, R.: On the relation of error correction and cryptography to an offline biometric based identification scheme (1999)
12. Dodis, Y., Ostrovsky, R., Reyzin, L., Smith, A.: Fuzzy extractors: How to generate strong keys from biometrics and other noisy data. SIAM J. Comput. 38(1), 97–139 (2008)
13. Feldmeier, D.C., Karn, P.R.: Unix password security - ten years later (1990)
14. Garrett, P.B.: Making, Breaking Codes: Introduction to Cryptology. Prentice Hall PTR, Upper Saddle River (2000)

15. Goh, A., Ling, D.N.C., Goh, A.: Computation of cryptographic keys from face biometrics. In: Lioy, A., Mazzocchi, D. (eds.) CMS 2003. LNCS, vol. 2828, pp. 1–13. Springer, Heidelberg (2003)
16. Gong, L.: Collisionful keyed hash functions with selectable collisions. Inf. Process. Lett. 55(3), 167–170 (1995)
17. Jin, A.T.B., Ling, D.N.C., Goh, A.: Personalised cryptographic key generation based on facehashing. Computers & Security 23(7), 606–614 (2004)
18. Juels, A.: A fuzzy vault scheme (2002)
19. Juels, A.: Fuzzy vaults: Toward secure client-side matching (2002)
20. Juels, A., Wattenberg, M.: A fuzzy commitment scheme (1999)
21. Li, S., Chu, R., Liao, S., Zhang, L.: Illumination invariant face recognition using near-infrared images. IEEE Transactions on Pattern Analysis and Machine Intelligence 29(4), 627–639 (2007)
22. Liao, S., Zhu, X., Lei, Z., Zhang, L., Li, S.Z.: Learning multi-scale block local binary patterns for face recognition. In: Lee, S.-W., Li, S.Z. (eds.) ICB 2007. LNCS, vol. 4642, pp. 828–837. Springer, Heidelberg (2007)
23. Lumini, A., Nanni, L.: An improved biohashing for human authentication. Pattern Recogn. 40(3), 1057–1065 (2007)
24. McEliece, R.J.: A public-key cryptosystem based on algebraic coding theory. Deep Space Network Progress Report 44, 114–116 (1978)
25. Nanni, L., Lumini, A.: Empirical tests on biohashing. Neurocomputing 69(16-18), 2390–2395 (2006)
26. Nanni, L., Lumini, A.: Random subspace for an improved biohashing for face authentication. Pattern Recogn. Lett. 29(3), 295–300 (2008)
27. Nichols, R.K.: ICSA Guide to Cryptography. McGraw-Hill Professional, New York (1998)
28. Soutar, C., Roberge, D., Stoianov, A., Gilroy, R., Kumar, B.V.: Biometric encryption: enrollment and verification procedures, vol. 3386, pp. 24–35. SPIE (1998)
29. Soutar, C., Roberge, D., Stoianov, A., Gilroy, R., Vijaya Kumar, B.V.K.: Biometric Encryption using image processing (April 1998)
30. Stern, J.: A new identification scheme based on syndrome decoding (1994)
31. Teoh, A., Jin, B., Connie, T., Ngo, D., Ling, C.: Remarks on biohash and its mathematical foundation. Inf. Process. Lett. 100(4), 145–150 (2006)
32. Teoh, A.B.J., Kuan, Y.W., Lee, S.: Cancellable biometrics and annotations on biohash. Pattern Recogn. 41(6), 2034–2044 (2008)
33. Wikipedia. Error detection and correction — Wikipedia, the free encyclopedia (2008) (Online; accessed November 6, 2008)
34. Zhao, W., Chellappa, R., Phillips, P., Rosenfeld, A.: Face recognition: A literature survey. ACM Computing Surveys, 399–458 (2003)

Fuzzy 3D Face Ethnicity Categorization

Cheng Zhong, Zhenan Sun, and Tieniu Tan

National Laboratory of Pattern Recognition, Institute of Automation, Chinese Academy of Sciences, Beijing, 100080, P.R. China
{czhong,znsun,tnt}@nlpr.ia.ac.cn

Abstract. In this paper, we propose a novel fuzzy 3D face ethnicity categorization algorithm, which contains two stages, learning and mapping. In learning stage, the visual codes are first learned for both the eastern and western individuals using the learned visual codebook (LVC) method, then from these codes we can learn two distance measures, merging distance and mapping distance. Using the merging distance, we can learn the eastern, western and human codes based on the visual codes. In mapping stage, we compute the probabilities for each 3D face mapped to eastern and western individuals using the mapping distance. And the membership degree is determined by our defined membership function. The main contribution of this paper is that we view ethnicity categorization as a fuzzy problem and give an effective solution to assign the 3D face a reasonable membership degree. All experiments are based on the challenging FRGC2.0 3D Face Database. Experimental results illustrate the efficiency and accuracy of our fuzzy 3D face ethnicity categorization method.

Keywords: 3D Face, Fuzzy Ethnicity Categorization, Learned Visual Codebook.

1 Introduction

Automatic identification of human faces is a very challenging research topic, which has gained much attention during the last few years [1]. However, human faces can not only provide the identities, but also the demographic information, such as gender and ethnicity, which are also necessary for human face perception.

Some effort has been made on the ethnicity categorization area over the last two decades. Gutta et al. [2] treated ethnicity categorization as a multi-class problem including Caucasian, Asian, Oriental and African origins categories. Based on the intensity faces, they adopted a hybrid classifier, ensemble of radial basis functions and a decision tree, for the categorization. Qiu et al. [3] considered ethnicity categorization as a two-class problem, including asian and non-asian. Based on the iris image, they combined the Gabor features and the Adaboost classifier to achieve categorization. Lu et al. [4] also viewed ethnicity categorization as a two-class problem. In their solution, they adopt SVM for both the intensity and depth modalities, and the matching scores were combined

M. Tistarelli and M.S. Nixon (Eds.): ICB 2009, LNCS 5558, pp. 386–393, 2009.
© Springer-Verlag Berlin Heidelberg 2009

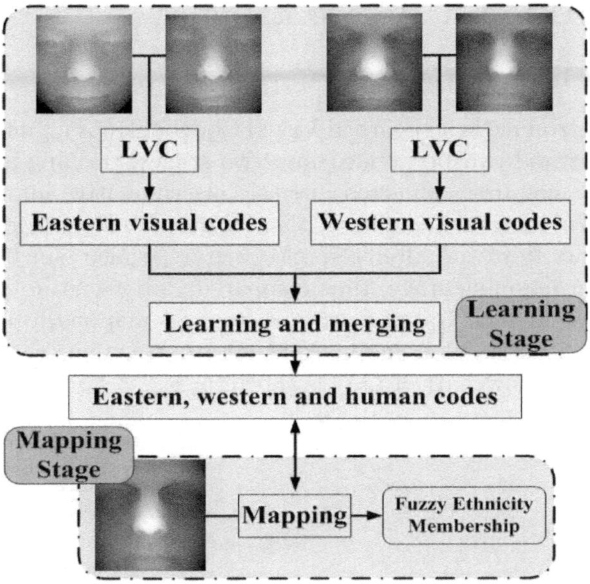

Fig. 1. The flowchart of our fuzzy 3D face categorization

using the sum rule. The drawback of these works [3][4] is that the ethnicity categorization is not a simple two-class problem (asian and non-asian), and there are also some other ethnic categories, such as Latin and Indian. Although [2] proposed four ethnic categories, it can not cover all the ethnic groups.

In this paper, we propose a fuzzy 3D face ethnicity categorization algorithm to solve the ethnic problems. The flowchart is shown in Fig.1. In learning stage, we first learn the eastern and western visual codes using the training eastern and western 3D faces based on the LVC method [5], then the merging and mapping distances are learned based on all the visual codes. When the distance between two different visual codes (eastern and western) is below merging distance, they are assigned to the human codes, which are viewed as the common properties of humans. The remaining eastern and western visual codes become the eastern and western codes (the specific properties of eastern and western individuals). In mapping stage, for each 3D face, we compute its probabilities for both eastern and western using the mapping distance and determine its membership degree by our membership function.

The main contribution of this paper is the introduce of fuzzy membership degree into the ethnicity categorization. Therefore, we need not to pay much attention on whether we can cover all of the ethnic groups. And how to accurately extract the ethnic information embedded in the 3D faces becomes our main task.

The remainer of this paper is organized as follows. In Section 2, we give the detail information of our fuzzy 3D face categorization. We describe our experimental results and discussions in Section 3. Finally, this paper is concluded in Section 4.

2 Fuzzy 3D Face Categorization

2.1 A Brief Introduction of LVC

In [5], Zhong introduced the Learned Visual Codebook (LVC) into 3D face recognition. This method can be divided into two stages, training and mapping. In training stage, they first extracted intrinsic discriminative information embedded in 3D faces using Gabor filters, then K-means clustering was adopted to learn the centers from the filter response vectors, which can be viewed as the visual codes. In mapping stage, they compared Gabor feature vectors with the learned codes. And each Gabor feature vector was mapped to the learned code with the nearest distance. A mapping histogram was constructed to represent each 3D face. The flowchart of LVC is shown in Fig.2. For detailed information of this algorithm, you can refer to [5].

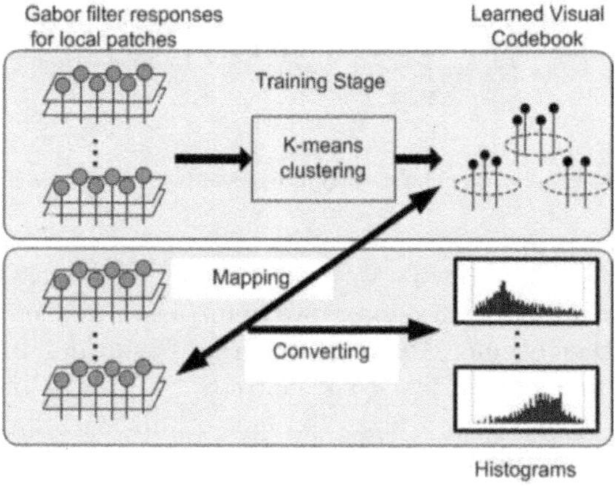

Fig. 2. The flowchart of LVC algorithm

2.2 Eastern Code, Western Code and Human Code

Learned Visual Codebook (LVC) has achieved good performance in 3D face recognition [5]. However, in this paper our task is 3D face ethnicity categorization. Intuitively, we first learn the visual codes for both eastern and western individuals, which are shown in Fig.3(a). Here, we adopt the Gabor features with only 1 scale and 2 orientations and QuadTree clustering to learn the visual codes [6]. From the figure, we find that there is a large overlap area between the eastern and western visual codes, which we can viewed as the common properties of humans (human codes). And there are also some codes corresponding to the properties of eastern or western individuals (eastern or western codes). Therefore, we can learn the eastern codes, western codes and human codes based on the obtained eastern and western visual codes.

Fig.3(b) shows three kinds of distance measures extracted from visual codes distribution. Because in LVC framework, each 3D face is divided into some local

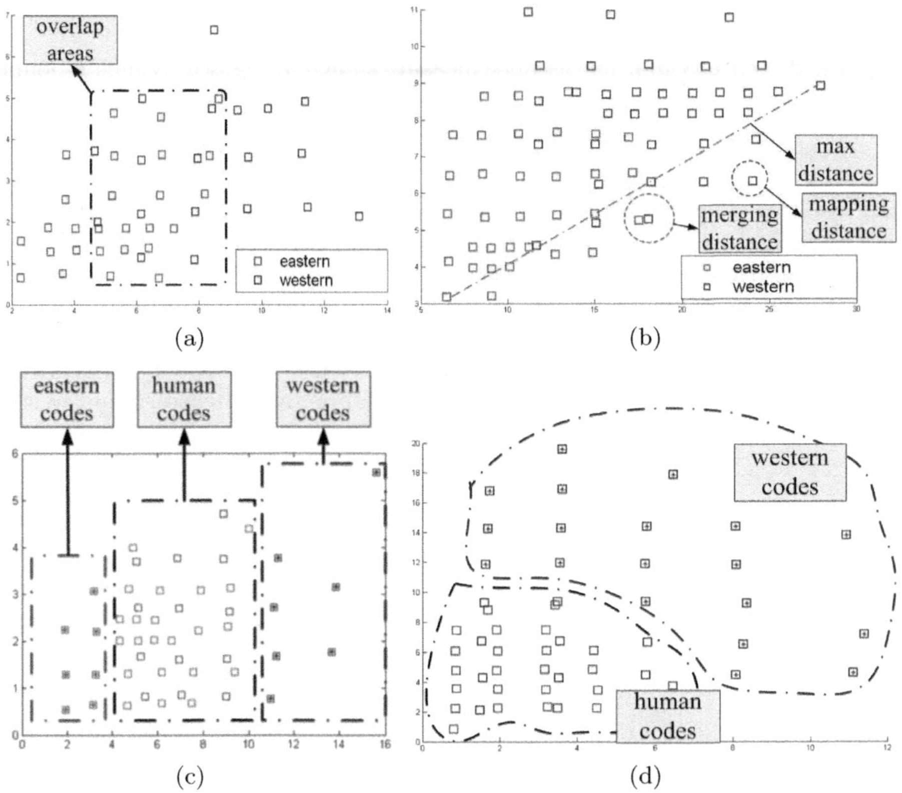

Fig. 3. How to learn the eastern, western and human codes. (a) is the centers of visual codes from both eastern and western individuals, (b) is the distance measures for the visual codes learning, (c) is the local patch with all three kinds of codes, (d) is the local patch without the eastern codes.

textures [5]. Thus for each local texture, we compute their maximum distance between two visual codes (max distance). Then we can obtain the merging distance as follows:

$$D1 = ratio1 * dmax \tag{1}$$

where $D1$ is the merging distance, $ratio1$ is a predefined value and $dmax$ is the max distance. Merging distance is defined to generate eastern, western and human codes. When the distance between a western and an eastern visual code is below $D1$, these two codes are categorized as the human codes. For a western visual code, when there is no eastern visual codes with the distance below $D1$, this code can be categorized as western codes. For an eastern visual code, when there is no western visual codes with the distance below $D1$, this code can be categorized as eastern codes. Some examples are shown in Fig.3(c) and Fig.3(d) In some local textures, there will be no learned eastern codes. But there will be learned western codes for all local textures.

2.3 Membership Function

Mapping distance is another distance measure shown in Fig.3(b), which is defined as

$$D2 = ratio2 * dmax \tag{2}$$

where $D2$ is the mapping distance, $ratio2$ is a predefined value and $dmax$ is the max distance. The definition of mapping distance is to compute the number of the Gabor feature vectors belonging to eastern and western categories, which is represented as $eastnum$ and $westnum$. The initial values for $eastnum$ and $westnum$ are both zero. For each 3D face image, first we convert it to many Gabor feature vectors. Then if the distance between a vector and one of the eastern (western) codes is below mapping distance, the $eastnum$ ($westnum$) adds one. The probability of the 3D face corresponding to eastern and western individuals are defined as follows:

$$probability_{eastern}(I) = eastnum/num \tag{3}$$

$$probability_{western}(I) = westnum/num \tag{4}$$

where I is the input 3D face image, num is the total number of the Gabor feature vectors. Then our membership function is simply defined as

$$membership(I) = probability_{eastern}/probability_{western} \tag{5}$$

The obtained value is to represent the membership degree. When the value is small, the face image is similar to western individuals. And when the value is large, the face image is similar to eastern individuals.

3 Experiments and Discussion

Our proposed method is evaluated based on the FRGC2.0 3D Face Database and the expression division protocol proposed by Geometrix, which divides the FRGC2.0 data into three subsets: neutral, small expression and large expression [7] [8]. To achieve our ethnicity categorization task, we manually select the

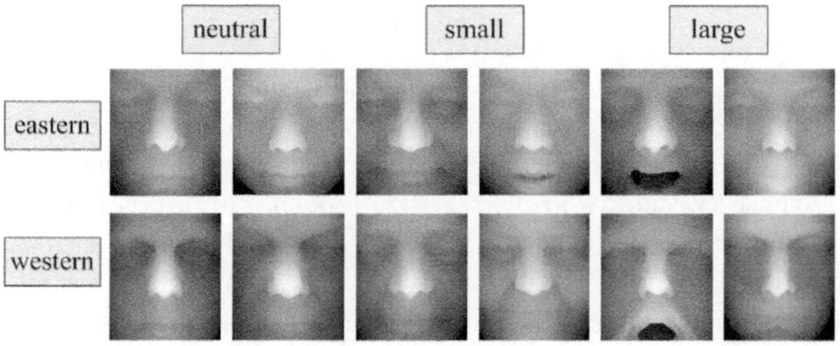

Fig. 4. Some example images in FRGC2.0 database (neutral subset). The first row shows the images of eastern people, the second row shows the images of western people.

eastern individuals and western individuals for all these three subsets, in which we assign the ethnic categories such as Latin, Indian or African origins, into the western group. Then for the neutral expression subset, we have 807 eastern individuals and 1662 western individuals. For the small expression subset, we have 247 eastern individuals and 549 western individuals. For the large expression subset, we have 163 eastern individuals and 579 western individuals. Some example images are shown in Fig.4.

In our experiments, we only adopt 50 eastern individuals and 50 western individual with neutral expression as training set to learn the eastern codes, western codes and human codes. The testing stage contains three situations: neutral expression, small expression and large expression. The categorization performance is measured by the percentage of the eastern or western data located into the western, middle and eastern areas, which are shown in Table.1 and Table.2. From Table.1, we find that the ratios of eastern individuals categorized into eastern area are 75.59%, 80.16% and 74.85% respectively in the three situations. From Table.2, we find that the ratios of western individuals categorized into western area are 89.65%, 79.05% and 77.20% respectively in the three situations.

Table 1. Eastern performance

Category	Western	Middle	Eastern
Neutral	5.95 %	18.46 %	75.59 %
Small	5.26 %	14.57 %	80.16 %
Large	7.36 %	17.79 %	74.85 %

Table 2. Western performance

Category	Western	Middle	Eastern
Neutral	89.65 %	9.45 %	0.90 %
Small	79.05 %	16.94 %	4.01 %
Large	77.20 %	18.48 %	4.32 %

4 Discussion

Experimental results illustrate our fuzzy 3D face categorization algorithm gives a comparable performance in all three data sets. Even in the large expression subset, more than 74% images are correctly categorized for both eastern and western individuals. Compared to previous methods, our method only adopt the simple Gabor features (1 scale and 2 orientations), no strong classifiers, such as SVM, Adaboost or Neural Network, are involved in the categorization procedures [9]. Therefore, our method is very promising and has great potential for the ethnicity categorization task, which is not limited in the 3D face modality.

From our experiments, we find that: First, only a small number of the individuals from Latin, Indian and African origins are mis-categorized. Because in data preparation section, we manually select these ethnic groups into the western individuals. Therefore, they have the western-similar faces. Second, although the western codes covers more space than the eastern codes. The distribution of the western individuals are more compact than the eastern individuals, as shown in Fig.5. In addition, the expression variations have little influence on the categorization performance for eastern individuals, while for western individuals, there is a distinct drop for the categorization performance when the situations change from neutral to small or large expressions. The main reason is that the data distribution of *westnum* is compact, but the data distribution of *eastnum*

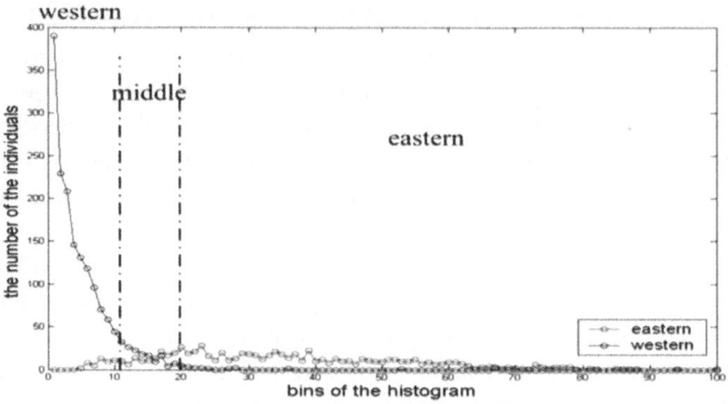

Fig. 5. The data distributions in the neutral subset

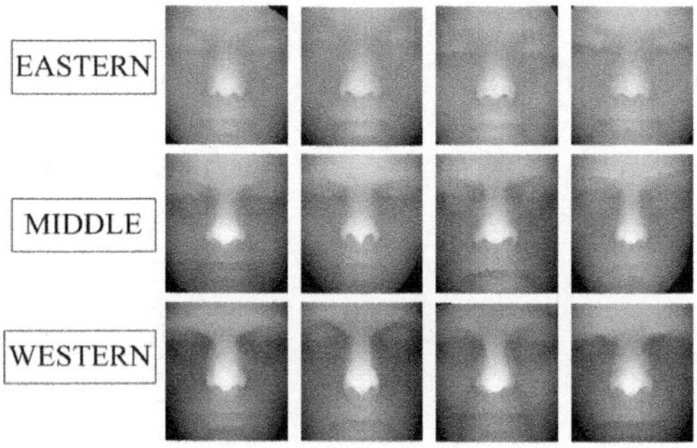

Fig. 6. The representative images in the categorized eastern, middle and western areas

covers a large area. This illustrates the appearances of western individuals are more similar and distribute tightly around some prominent facial structures, and the appearances of eastern individuals are more diversified and cover more smooth facial structures. Finally, from Fig.6 we can find that the representative eastern individuals have a flat 3D face with no prominent structures, while the representative western individuals have a 3D face with sharp contrast and some prominent structures, such as the high-bridged nose and the deep-sunken eyes. The representative individuals in middle area are similar to the mixtures of the eastern and western individuals, which can be viewed as the transition between these two areas. Therefore, it is reasonable to use the fuzzy value for ethnicity categorization.

5 Conclusion

In this paper, we have proposed an efficient and accurate fuzzy 3D face categorization algorithm. Because we can learn the characteristics of the eastern and western individuals, for a 3D face, the membership degree can be obtained based on its similarity to the eastern or the western characteristics. Experimental results illustrate the effectiveness and robustness of our proposed fuzzy method for categorization.

Acknowledgement

This work is supported by the National Basic Research Program of China (Grant No. 2004CB318100), the National Natural Science Foundation of China (Grant No. 60736018, 60723005), the National Natural Science Foundation of China (Grant No. 60702024), the National Hi-Tech Research and Development Program of China (Grant No.2006AA01Z193, 2007AA01Z162) and the Chinese Academy of Sciences.

References

1. Zhao, W., Chellappa, R., Rosenfeld, A.: Face recognition: a literature survey. ACM Computing Surveys 35, 399–458 (2003)
2. Gutta, S., Wechsler, H., Phillips, P.J.: Gender and ethnic classification of face images. In: IEEE International Conference on Automatic Face and Gesture Recognition (1998)
3. Qiu, X., Sun, Z., Tan, T.: Global texture analysis of iris images for ethnic classification. In: Zhang, D., Jain, A.K. (eds.) ICB 2005. LNCS, vol. 3832, pp. 411–418. Springer, Heidelberg (2005)
4. Lu, X., Chen, H., Jain, A.K.: Multimodal facial gender and ethnicity identification. In: Zhang, D., Jain, A.K. (eds.) ICB 2006. LNCS, vol. 3832, pp. 554–561. Springer, Heidelberg (2006)
5. Zhong, C., Sun, Z., Tan, T.: Robust 3D face recognition using learned visual codebook. In: IEEE Conference on Computer Vision and Pattern Recognition (2007)
6. Zhong, C., Sun, Z., Tan, T.: Learning efficient codes for 3D face recognition. In: International Conference on Image Processing (2008)
7. Phillips, P.J., Flynn, P.J., Scruggs, T., Bowyer, K.W., Chang, J., Hoffman, K., Marques, J., Min, J., Worek, W.: Overview of the face recognition grand challenge. In: Proceedings of IEEE Conference on Computer Vision and Pattern Recognition (2005)
8. Maurer, T., Guigonis, D., Maslov, I., Pesenti, B., Tsaregorodtsev, A., West, D., Medioni, G.: Performance of geometrix $activeid^{TM}$ 3d face recognition engine on the frgc data. In: Proceedings of IEEE Conference on Computer Vision and Pattern Recognition (2005)
9. Duda, R.O., Hart, P.E., Stork, D.G.: Pattern Classification, 2nd edn. John Wiley & Sons, Inc., Chichester (2001)

Faceprint: Fusion of Local Features for 3D Face Recognition

Guangpeng Zhang and Yunhong Wang

School of Computer Science and Engineering, Beihang University
gpzhang@cse.buaa.edu.cn, yhwang@buaa.edu.cn

Abstract. 3D face recognition is a very active biometric research field. Due to the 3D data's insensitivity to illumination and pose variations, 3D face recognition has the potential to perform better than 2D face recognition. In this paper, we focus on local feature based 3D face recognition, and propose a novel Faceprint method. SIFT features are extracted from texture and range images and matched, the matching number of key points together with geodesic distance ratios between models are used as three kinds of matching scores, likelihood ratio based score level fusion is conducted to calculate the final matching score. Thanks to the robustness of SIFT, shape index, and geodesic distance against various changes of geometric transformation, illumination, pose and expression, the Faceprint method is inherently insensitive to these variations. Experimental results indicate that Faceprint method achieves consistently high performance comparing with commonly used SIFT on texture images.

Keywords: 3D face recognition, local feature, fusion.

1 Introduction

As an important biometrics face recognition has a long history, and enormous amount of face recognition algorithms have been proposed. Most of these algorithms are 2D-image-based, that is to identify or verify subjects through their face images captured by cameras or video recorders. A literature survey of these methods can be found in [1]. Despite its popularity, 2D face recognition is sensitive to illumination, pose and expression variations. In order to deal with these limitations, researchers are paying more and more attentions to 3D face recognition [2], as 3D model captures the exact shape of facial surface, thus is invariant to illumination and pose variations. But 3D-based methods are more sensitive to expression variations than 2D-based, and influenced by occlusion. Actually 2D images and 3D models are two different modalities, they are complementary, and fusion of these two modalities may benefit face recognition [2].

Zhao et al. [1] divided still 2D face recognition techniques into holistic, feature-based, and hybrid categories. This categorization can also be extended to 3D face recognition. Holistic method uses the whole face region as the input to face recognition algorithm. Typical holistic methods include PCA [3] and LDA [4] based methods. Most of these popular 2D face recognition methods were also

M. Tistarelli and M.S. Nixon (Eds.): ICB 2009, LNCS 5558, pp. 394–403, 2009.
© Springer-Verlag Berlin Heidelberg 2009

extended to 3D face recognition. Many 3D face recognition methods such as ICP-based matching [22], annotated deformable model [23], isometry-invariant representation [17] also belong to this category. In feature-based methods, local (geometric or appearance) features are extracted from facial feature points or regions such as eyes, nose, and mouth, and fed into classifiers. Many earlier works are feature-based face recognition [5]. They measured distances and angles between fiducial points to do matching. The most successful feature-based method is graph matching [6], which is based on Dynamic Link Architecture (DLA). It's also extended to 3D recognition [7]. Point signature method proposed by Chua et al. [8] belongs to this category. Hybrid methods use both local features and the whole face region to do recognition. Typical methods include modular eigenfaces [10], component-based [11] et al.

Among these three categories, we pay special attention to feature-based methods. This kind of methods extracts features from local points or regions, and has the potential to deal with expression variations and occlusion. As no matter when expression variation or occlusion occurs, there always exists some local points or regions remain invariant. If we can match these invariant points or regions, recognition is done. SIFT is used in this paper to extract robust local features from feature points. SIFT proposed by Lowe [12] is a robust feature extraction and matching method against image scaling, translation, and rotation, and partially invariant to illumination changes and affine or 3D projection. Because of its great property, SIFT spreads rapidly into various applications since proposed, including face recognition. M. Bicego et al [9] first utilized SIFT on 2D face recognition, matching scores of feature descriptors was used for verification. The method was further improved by incorporating graph composed of the detected SIFT points [16]. Mian et al [13] utilized SIFT on 2D texture images together with SFR (Spherical Face Representation) on 3D data to form a rejection classifier. As far as we know, all the proposed method of using SIFT for face recognition are performed on 2D domain (texture images), and none of the results reported in the above publications is very high. We think the main reason for the unsatisfying performance is that, unlike artificial objects such as buildings, face is a smooth surface, and there are not so many distinctive feature points. As reported in [13], the average number of detected feature points is 80. Another reason is that images change greatly with illumination variations, while SIFT is only partially insensitive to illumination variations.

In order to deal with the limitations of utilizing SIFT for face recognition, we resort to 3D data, and shape index is extracted from range images, as shape index is derived from 3D curvatures, so it's invariant to illumination and pose variations. Compared with 2D texture images, shape index images have much more details as shown in Fig. 2(c), thus more feature points can be extracted from them. In order to deal with sensitivity of 3D data to expression variations, geodesic distances in between matched feature points are also computed and matched. The final matching score is obtained by fusion of the number of matched 2D and 3D SIFT feature points, and geodesic distance ratio between matched models using likelihood ratio test strategy. We call this proposed method "Faceprint". Just like minutia in fingerprint, Faceprint describes local facial features, and is robust against many kinds of variations.

The paper is organized as follows. SIFT, shape index and geodesic features are briefly introduced in section 2, and then Faceprint method is described in detail in section 3. Experiments are carried out on FRGC v2.0 database [24], and results are shown in section 4. Section 5 concludes the paper.

2 SIFT, Shape Index and Geodesic distance

2.1 Scale Invariant Feature Transform

SIFT (Scale Invariant Feature Transform) was proposed by Lowe [12], [14] for object recognition. The original image is repeatedly convolved with Gaussians of different scales separated by a constant factor k to produce an octave in scale space. Once an octave is processed, the coarsest image of this octave is down sampled by 4 to be the start image of next octave. After convolution with Gaussian, neighboring images in each octave are subtracted to get DOGs (Difference of Gaussian). DOG function provides a close approximation to scale-normalized Laplacian of Gaussian [15]. Extremas are detected by comparing a pixel with its 26 neighbors in $3 * 3$ regions at the current and adjacent scales.

$$G(x, y, k\sigma) - G(x, y, \sigma) \approx (k - 1)\sigma^2\nabla^2 G \tag{1}$$

After detection of extremas, a threshold is used to eliminate key points with low contrast. Later key points that have a ratio between the principle curvatures greater than a threshold are removed, because these points are located along edges and poorly defined.

For each key point, an orientation histogram is calculated from the gradient orientations of its neighboring sample points. Each sample is weighted by its gradient magnitude and a Gaussian-weighted circular window. Highest local peak and peak that's within 80% of the highest peak are detected, and the corresponding orientations are assigned to the key point.

For each orientation of a key point, a feature vector is extracted as a descriptor from the gradients of its neighboring sample points. In order to achieve orientation invariance, the coordinates and gradient orientations or neighboring points are rotated relative to the key point orientation. Then a Gaussian function is used to assign a weight to the gradient magnitude of each point. Points that are close to the key point are given more emphasis than those far from it. Orientation histograms of $4 * 4$ sample regions are calculated, each with eight orientation bins. Thus a feature vector with dimension of $4 * 4 * 8 = 128$ is formed. To achieve illumination insensitive, the feature vector is normalized to unit length and thresholded such that no element is larger than 0.2, and normalized to unit length again. This normalized feature vector is the descriptor of the central key point.

Given the descriptors, matching between key points on two images can be done. A matching of one key point to another is only accepted if the matching distance is less than a threshold t times the distance to the second closest match, t is set to 0.6 in this work. The SIFT based key point detection and matching is done with the help of Lowe's code.

2.2 Shape Index

Shape Index feature was proposed by Dorai et al. [25] to represent surface concave and convex attributes. The Shape Index at point p is defined as:

$$S(p) = \frac{1}{2} - \frac{1}{\pi} \arctan \frac{\kappa_1(p) + \kappa_2(p)}{\kappa_1(p) - \kappa_2(p)} \tag{2}$$

where k_1 and k_2 represent maximum and minimum principal curvatures respectively. As principal curvatures are invariant to pose variations, so is Shape index.

The value of Shape Index lies in the range of $[0, 1]$. Local shape at point p is a spherical cup when $S(p) = 0$, and a spherical cap when $S(p) = 1$. When Shape Index changes from 0 to 1, local shape changes from spherical cup to spherical cap.

Fig. 2(c) shows an example of shape index image, in which dark pixels represent lower shape index, while bright pixels represent higher shape index.

2.3 Geodesic Distance

In the specific case of 3D mesh, geodesic distance between two vertices is defined as the shortest path connecting them. As shown in [17], geodesic distance is invariant to deformation of 3D model caused by expression variations. A numerically consistent algorithm for distance computation on triangulated mesh was proposed by Kimmel and Sethian [18] as a generalization of the fast marching method [19]. Using this method, the geodesic distances between a surface vertex and the rest of the n surface vertices can be computed in $O(n)$ operations. We use this method for computing geodesic distances between SIFT key points on 3D mesh.

3 Faceprint

The flow chart of Faceprint method is shown in Fig. 1. A 3D model of one subject contains one texture image and one range image, which are densely registered. Unlike the careful normalization including face pose correction, resampling, and segmentation performed in reference [13], faces are simply cropped out of the background using a bounding box on both texture and range images, by which computational complexity is reserved. As SIFT is very robust to image scaling, translation and rotation, we can fully benefit from advantages of SIFT, and do not need to do pose correction and resampling. After normalization, the size of the texture and range images is different from one model to another. Histogram equalization is carried out on normalized texture images.

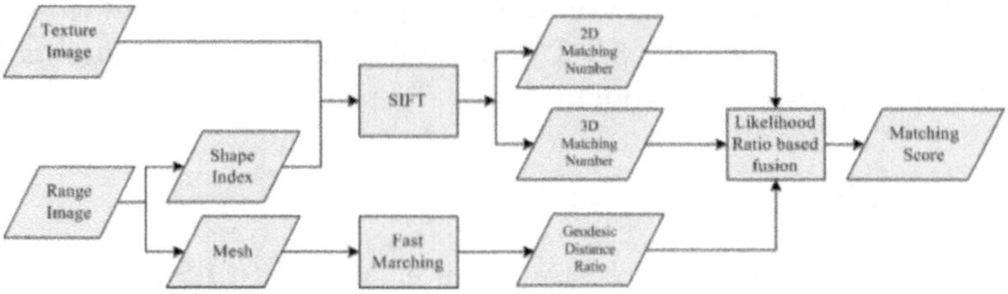

Fig. 1. Flow chart of Faceprint method

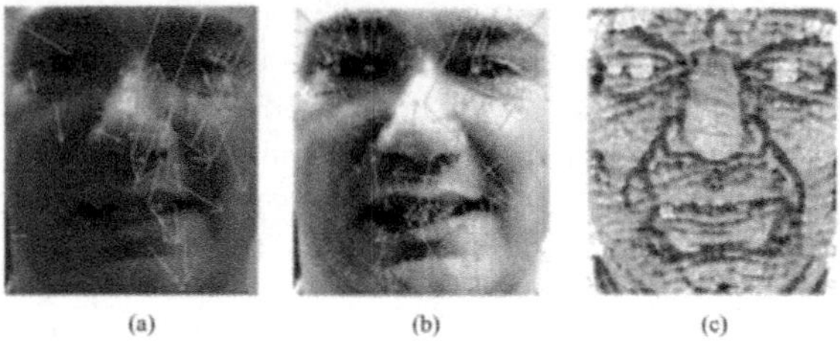

Fig. 2. SIFT features extracted from (a) texture images, (b) histogram equalized texture images, and (c) shape index images

From range images, shape index images as shown in Fig. 2(c) are calculated, SIFT is performed not only on texture images but also on shape index images. Shape index images provide robust features against pose and illuminations, thus repeatability is guaranteed, and these features are more sufficient and with more diversity than those of texture images. In our experiments, the average number of descriptors extracted from shape index images is 862, while that from texture images is 37, and that from histogram equalized texture images is 128. Fig. 2 shows the SIFT features extracted from texture images, histogram equalized texture images, and shape index images, all of which were captured at the same time for the same subject. From Fig. 2, we can see that compared with texture and histogram equalized texture images, shape index image provides the most evenly spaced and largest number of descriptors.

Then the range image is triangulated and mesh is reconstructed. Triangulation is done by dividing each square in the image grid with the shortest diagonal edge. As texture and range images are densely registered, so each detected SIFT key point on texture image corresponds to a vertex on the generated mesh, unless those located in regions such as eyes where no range data is captured. Fast marching is performed for each key point on the mesh, and the geodesic distances between key points are obtained after that.

On matching, detected SIFT key points from texture images and shape index images are matched separately, and the matching number of key points is treated as matching score. The other matching score is the ratio of geodesic distance between corresponding pairs of matched points.

Given a set of key points $\{p_i|i = 1, \cdots, N\}$ on mesh X, there is a set of corresponding matching points $\{p'_i|i = 1, \cdots, N\}$ on mesh X', then the geodesic distance ratio $GDR(X, X')$ between X and X' is defined as below.

$$GDR(X, X') = \frac{1}{N(N-1)/2} \sum_{i=1}^{N-1} \sum_{j=i+1}^{N} \frac{\min(gd(p_i, p_j), gd(p'_i, p'_j))}{\max(gd(p_i, p_j), gd(p'_i, p'_j))} \tag{3}$$

where $gd(p_i, p_j)$ means the geodesic distance between points p_i and p_j. According to its definition, the geodesic distance ratio lies in the range of $[0, 1]$.

Therefore, after the above process, we get three kinds of matching scores, namely, the matching number of key points from texture images MNT and shape index images MNS, and the geodesic distance ratio GDR between models. The final matching score is obtained by fusion of these three kinds of matching scores.

Score level fusion can be divided into three categories [20]: transformation-based, classifier-based, and density-based fusion. A comparison of eight biometric fusion methods conducted by NIST [21] favors the density-based fusion, specifically the likelihood ratio based fusion. In order to reduce the computational complexity, Gaussian Mixture Model (GMM) is used in [20] instead of kernel density estimator (KDE) for density estimation during likelihood ratio based fusion. However, in our case, the matching scores MNT and MNS are both of discrete values, which are inconvenient to be modeled with GMM. For the models with no matching key points, the GDR is set to be zero, and as a result there are many zero valued GDR, mostly for those matching between different subjects' models, which is also hard to be handled by GMM. Considering these inconvenience, we choose KDE for density estimation instead of GMM.

The likelihood ratio based fusion is formulated as below. Given a vector of K matching scores $\mathbf{s} = [s_1, s_2, \cdots, s_K]$, and estimated genuine density $\widehat{f}_{gen}(\mathbf{s})$ and impostor density $\widehat{f}_{imp}(\mathbf{s})$, compute the likelihood ratio $LR(\mathbf{s}) = \widehat{f}_{gen}(\mathbf{s})/\widehat{f}_{imp}(\mathbf{s})$, assign $\mathbf{s}$ to the genuine class if $LR(\mathbf{s}) \geq \eta$. We assume the independence of the three matching scores, therefore the density function $\widehat{f} = \prod_{k=1}^{3} \widehat{f}_k(s_k)$.

4 Experiments and Results

4.1 Dataset

Experiments are conducted on FRGC v2.0 database [24], which contains 4007 3D models of 466 subjects, and each model contains a pair of registered texture image and range image. Among these subjects, 57% are male and 43% are female, with the age distribution: 65% 18-22 years old, 18% 23-27 and 17% 28 years or over. The database was collected during 2003-2004 academic year, thus contains time variations. It also contains neutral and non-neutral expression variations.

All the texture and range images are simply cropped using a bounding box determined by two eye-outer corners and chin point, so that faces are separated from the background. These simply cropped images have pose, illumination, resolution, and expression variations, and also contain occlusions and holes.

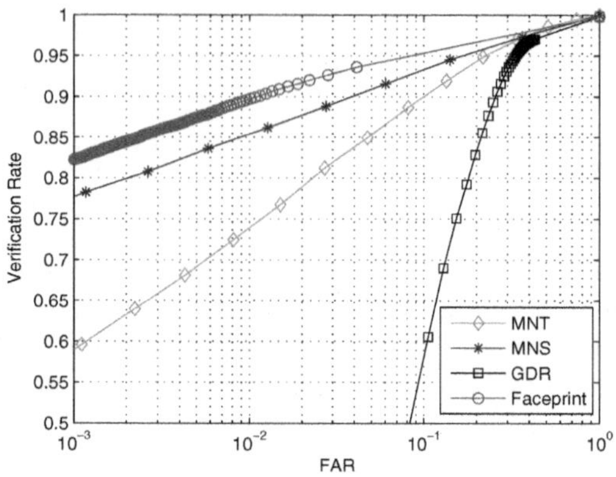

Fig. 3. ROC curves of All vs. All experiment

The database is divided into a training set and a testing set. The training set contains 977 models of randomly selected 100 subjects, and 3030 models of the rest 366 subjects form the testing set. The training set is used to estimate densities for likelihood ratio based fusion.

Four experiments are carried out, namely, All vs. All, and Neutral vs. Neutral, Neutral vs. Non-Neutral, and Neutral vs. All. All vs. All experiments are conducted to be consistent with FRGC experiment 3. All possible pairs of 3D models int the testing set are compared, so that a 3030*3030 similarity matrix is calculated. In Neutral vs. Neutral, Neutral vs. Non-Neutral, and Neutral vs. All experiments, 366 neutral expression faces are selected as the gallery set. The rest 2664 images are treated as probe set in Neutral vs. All experiment, while only neutral and non-neutral face models are used in Neutral vs. Neutral and Neutral vs. Non-Neutral experiments respectively. The reason to do Neutral vs. Neutral, Neutral vs. Non-Neutral, and Neutral vs. All experiments is to analyze robustness against expression variations.

4.2 All vs. All

As described above, in the All vs. All experiment, Faceprint method is used to match all possible pairs of 3030 3D models, and a 3030*3030 similarity matrix is obtained. From the similarity matrix ROC curves can be derived. We compare the ROC curves of the proposed Faceprint (fusion) method with those of MNT, MNS, and GDR in Fig. 3.

From Fig. 3, we can see that the proposed Faceprint method performs the best, verification rate of 82.3% is achieved at false accept rate of 0.1%. MNS performs much better than MNT which demonstrates the robustness of Shape index images is better than that of texture images. GDR performs not so well, but still shows some distinguishability.

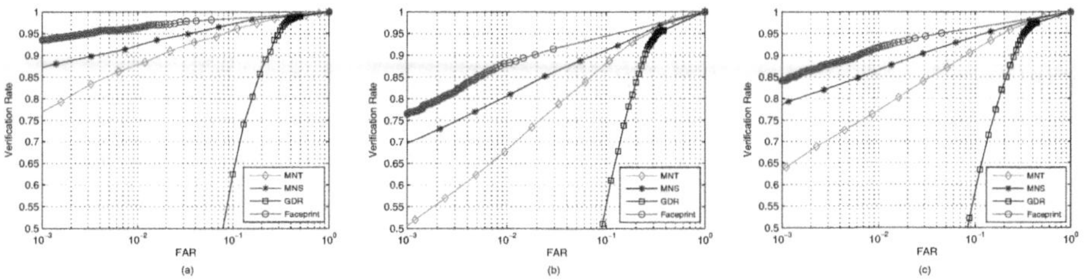

Fig. 4. ROC curves of experiments (a) Neutral vs. Neutral, (b) Neutral vs. Non-Neutral, and (c) Neutral vs. All

4.3 Neutral vs. Neutral, Non-Neutral, and All

In order to evaluate the robustness against expression variations, three experiments are conducted, namely, Neutral vs. Neutral, Neutral vs. Non-Neutral, and Neutral vs. All. ROC curves of these three experiments are shown in Fig. 4.

As can be seen from Fig. 4, no matter with or without expression variations, the order of the performance of Faceprint, MNT, MNS, and GDR remains the same. The performance degradation when non-neutral expressions are included in the experiment is calculated, and shown in Table 1. From Table 1, we can see that performance of MNT degrades most seriously (34.3%), followed by MNS (20.2%), and Faceprint (18.3%), while performance of GDR degrades by only 11.5%. To be noted, as it's hard and unreasonable to get the verification rate at FAR of 0.1%, the performance degradation of GDR is calculated at FAR of 1%, which is different from the other three matching scores. The results show that SIFT features extracted from shape index images are more robust than those from texture images, and geodesic distance ratio is most insensitive to expression variations. Fusion of these three kinds of matching scores increases the performance while retaining most of the robustness against expression variations.

Table 1. Performance degradation of Faceprint, MNT, MNS, and GDR when expression variations are included in the experiments

	Faceprint FAR=0.1%	MNT FAR=0.1%	MNS FAR=0.1%	GDR FAR=1%
Neutral vs. Neutral	93.6%	77.0%	87.2%	62.6%
Neutral vs. Non-Neutral	76.5%	50.6%	69.6%	55.4%
Performance degradation	18.3%	34.3%	20.2%	11.5%

From all the above four experiments, we can see that SIFT features extracted from shape index images (MNS) perform much better than the commonly used texture SIFT features (MNT), and the proposed Faceprint method performs the best by fusion the three kinds of matching scores.

5 Conclusions

In this paper, we pay special attention to local feature based 3D face recognition, and propose a novel Faceprint method. SIFT features are extracted from texture and range images and matched, the matching number of key points together with geodesic distance ratios between models are used as three kinds of matching scores, likelihood ratio based score level fusion is conducted to calculate the final matching score. Because of the invariance of SIFT to image scaling, translation and rotation, the robustness of shape index against illumination and pose changes, and the insensitivity of geodesic distance to expression variations, the Faceprint method is inherently robust against all these variations. Systematical experiments are carried out on FRGC v2.0 database, and the experimental results indicate that Faceprint method achieves consistently high performance comparing with commonly used SIFT on texture images.

Acknowledgements

This work was supported by Program of New Century Excellent Talents in University, National Natural Science Foundation of China (No. 60575003, 60332010, 60873158), Joint Project supported by National Science Foundation of China and Royal Society of UK (60710059), and Hi-Tech Research and Development Program of China (2006AA01Z133).

References

1. Zhao, W., Chellappa, R., Rosenfeld, A., Phillips, P.J.: Face Recognition: A Literature Survey. ACM Computing Surveys, 399–458 (2003)
2. Bowyer, K.W., Chang, K., Flynn, P.: A survey of approaches and challenges in 3D and multi-modal 3D + 2D face recognition. Computer Vision and Image Understanding 101(1), 1–15 (2006)
3. Turk, M., Pentland, A.: Eigenfaces for recognition. Journal of Cognitive Neuroscience 3(1), 71–86 (1991)
4. Belhumeur, P., Hespanha, J., Kriegman, D.: Eigenfaces vs. Fisherfaces: Recognition Using Class Specific Linear Projection. IEEE Transactions on pattern analysis and machine intelligence 19(7), 711 (1997)
5. Kanade, T.: Computer Recognition of Human Faces. Interdisciplinary Systems Research 47 (1977)
6. Wiskott, L., Fellous, J., Kruger, N., von der Malsburg, C.: Face Recognition by Elastic Bunch Graph Matching. IEEE Transactions on Pattern Analysis and Machine Intelligence 19(7), 775–779 (1997)
7. Husken, M., Brauckmann, M., Gehlen, S., von der Malsburg, C.: Strategies and benefits of fusion of 2D and 3D face recognition. In: IEEE Workshop on Face Recognition Grand Challenge Experiments (2005)
8. Chua, C., Han, F., Ho, Y.K.: 3D human face recognition using point signature. In: Proc. IEEE International Conference on Automatic Face and Gesture Recognition, pp. 233–238 (2000)

9. Bicego, M., Lagorio, A., Grosso, E., Tistarelli, M.: On the use of SIFT features for face authentication. In: Proc. IEEE International Conference on Computer Vision and Pattern Recognition Workshop, pp. 35–41 (2006)
10. Pentland, A., Moghaddam, B., Starner, T.: View-Based and modular eigenspaces for face recognition. In: Proc. of the IEEE Conf. on Computer Vision and Pattern Recognition, pp. 84–91 (1994)
11. Huang, J., Heisele, B.: Blanz. V.: Component-based Face Recognition with 3D Morphable Models. In: Proc. of the 4th International Conference on Audio- and Video-Based Biometric Person Authentication, pp. 27–34 (2003)
12. Lowe, D.G.: Object recognition from local scale-invariant features. In: Proc. of the International Conference on Computer Vision 1999, pp. 1150–1157 (1999)
13. Mian, A.S., Bennamoun, M., Owens, R.A.: An Efficient Multimodal 2D-3D Hybrid Approach to Automatic Face Recognition. IEEE Transactions on pattern analysis and machine intelligence 29(11), 1927–1943 (2007)
14. Lowe, D.G.: Distinctive image features from scale-invariant keypoints. International Journal of Computer Vision 60(4), 91–110 (2004)
15. Lindeberg, T.: Scale-space theory: A basic tool for analysing structures at different scales. Journal of Applied Statistics 21(2), 224–270 (1994)
16. Kisku, D.R., Rattani, A., Grosso, E., Tistarelli, M.: Face Identification by SIFT-based Complete Graph Topology. In: 5th IEEE Workshop on Automatic Identification Advanced Technologies, Alghero, Italy (2007)
17. Bronstein, A.M., Bronstein, M.M., Kimmel, R.: Three-dimensional face recognition. International Journal of Computer Vision 64(1), 5–30 (2005)
18. Kimmel, R., Sethian, J.A.: Computing geodesic on manifolds. Proc. US National Academy of Science 95, 8431–8435 (1998)
19. Sethian, J.A.: A review of the theory, algorithms, and applications of level set method for propagating surfaces. Acta numerica (1996)
20. Nandakumar, K., Chen, Y., Dass, S.C., Jain, A.K.: Likelihood Ratio Based Biometric Score Fusion. IEEE Transactions on Pattern Analysis and Machine Intelligence 30(2), 342–347 (2008)
21. Ulery, B., Hicklin, A.R., Watson, C., Fellner, W., Hallinan, P.: Studies of Biometric Fusion. NIST, Tech. Rep. IR 7346. (2006)
22. Lu, X., Jain, A.K., Colbry, D.: Matching 2.5D Face Scans to 3D Models. IEEE Transactions on pattern analysis and machine intelligence 28(1), 31–43 (2006)
23. Kakadiaris, I.A., Passalis, G., Toderici, G., Murtuza, N., Lu, Y., Karampatziakis, N., Theoharis, T.: Three-Dimensional Face Recognition in the Presence of Facial Expressions: An Annotated Deformable Model Approach. IEEE Transactions on pattern analysis and machine intelligence 29(4), 640–649 (2007)
24. Phillips, P.J., et al.: Overview of the Face Recognition Grand Challenge. In: Proc. Of IEEE Conf. on Computer Vision and Pattern Recognition, pp. I:947–954 (2005)
25. Dorai, C., Jain, A.K.: COSMOS - A Representation Scheme for 3D Free- Form Objects. IEEE Transactions on Pattern Analysis and Machine Intelligence 19(10), 1115–1130 (1997)

Combining Illumination Normalization Methods for Better Face Recognition

Bas Boom, Qian Tao, Luuk Spreeuwers, and Raymond Veldhuis

University of Twente, EEMSC, Signals & Systems,
P.O. box 217, 7500 AE, Enschede, The Netherlands
{b.j.boom,q.tao,l.j.spreeuwers,r.n.j.veldhuis}@utwente.nl

Abstract. Face Recognition under uncontrolled illumination conditions is partly an unsolved problem. There are two categories of illumination normalization methods. The first category performs a local preprocessing, where they correct a pixel value based on a local neighborhood in the images. The second category performs a global preprocessing step, where the illumination conditions and the face shape of the entire image are estimated. We use two illumination normalization methods from both categories, namely Local Binary Patterns and Model-based Face Illumination Correction. The preprocessed face images of both methods are individually classified with a face recognition algorithm which gives us two similarity scores for a face image. We combine the similarity scores using score-level fusion, decision-level fusion and hybrid fusion. In our previous work, we show that combining the similarity score of different methods using fusion can improve the performance of biometric systems. We achieved a significant performance improvement in comparison with the individual methods.

1 Introduction

One of the major problems with face recognition under uncontrolled conditions is the illumination variation, which is often larger than the variations between individuals. Using illumination normalization methods, we want to correct the illumination variations in a single face image. In literature, several methods have been proposed to make face images invariant for illumination. These methods can be divided into two categories. The first category normalizes the face image by applying a preprocessing step on the pixel values using information from the local region around that pixel. Examples of these approaches are Histogram Equalization [1] or (Simplified) Local Binary Patterns [2], [3]. These approaches are direct and simple, but fail to model the global illumination conditions. The second category estimates a global physical model of the illumination mechanism and its interaction with the facial surface. In this category falls for instance the Quotient Image [4], Spherical harmonics [5], 3D morphable models [6]. These methods are able to estimate the global illumination condition, but are also more complicated and require training to model the illumination conditions.

In practise, we have observed that both categories of illumination normalization algorithms have their advantages and disadvantages. The methods in the

M. Tistarelli and M.S. Nixon (Eds.): ICB 2009, LNCS 5558, pp. 404–413, 2009.
© Springer-Verlag Berlin Heidelberg 2009

first category have problems with regions which are not illuminated because of hard shadows. These shadow regions have a large signal to noise ratio which makes the correction for local methods almost impossible. However, these local methods work well on the illuminated parts of the image and on lightly shadowed (soft shadows) areas. The second category is able to reconstruct the parts with hard shadows, using statistical models. But our current implementation of a global method does not model face variations like glasses and expressions. To summarize, the local methods work well on illuminated part of the image, also the parts which are not modelled by the global methods. The global methods are however able to reconstruct parts which contain hard shadows, which is not possible using a local method. By combining methods from both categories, we aim to improve the performance in face recognition under different and uncontrolled illumination conditions.

Combining the two different illumination correction methods can be done on three different levels, namely at the feature level, the score level and the decision level. We will concentrate on the last two levels of fusion because of its simplicity. To achieve this, the preprocessed images are individually classified and the scores are fused using score-level [7], decision-level fusion [8] and a combination of both these methods named hybrid fusion [9].

Our paper is organized as follows, in Section 2 we describe the two illumination correction algorithms. Section 3 explains how we combine these algorithms for face recognition. In Section 4, we show the experiments and results and Section 5 gives the conclusions.

2 Illumination Normalization

For illumination normalization, we use two methods which come from different categories. The method from the first category is the Local Binary Patterns [10] preprocessing, where different papers [2], [3] claim its invariance to illumination conditions. In the second category, we use the illumination correction approach in [11], which is able to correct illumination variations using a single 2D facial image as its input. If we compare this method with [4] or [12], it is more advanced using a 3D shape model and a shadow and reflection model, but in comparison with 3D morphable models [6] it is still computational efficient. In the following subsections, we describe both methods in more detail.

2.1 Local Binary Patterns

Local Binary Patterns method (LBP) is proposed in [10] and are often used as features in face recognition. The standard LBP give the 3×3-neighbors the value 0 if they are smaller than the center pixel value and 1 otherwise. This result in a 8 bit string, which represents the pattern at the center point. We can also obtain from it a decimal representation between 0 and 255. LBP is a relative measure which makes it invariant against monotonic gray-scale transformations. A big range of illumination changes consist of monotonic gray-scale transformations in large regions in the image.

There are also extensions to the original LBP, which use a bigger radius and more spacing for the neighbors or use a different weighting scheme for the different bits. In this paper, we use the simplest LBP as preprocessing to obtain the illumination invariant images, i.e. based on a 3×3-neighborhood.

2.2 Model-Based Face Illumination Correction

The method from the second category [11] uses the Lambertian model to correct illumination in face images. It assumes that a face image is illuminated from a single light source at an infinit distance, which gives the following Lambertian equation:

$$b(x) = c(x)\mathbf{n}(x)^T \mathbf{s} i + e(x; \mathbf{s}) \tag{1}$$

In this case, x is the pixel position and $b \in \mathcal{R}$ is the pixel intensity in the image. The pixel intensity is determined by the shape and the light condition. The shape $\mathbf{h}(x) = c(x)\mathbf{n}(x)^T$ is defined by the surface normals $\mathbf{n} \in \mathcal{R}^3$ and the albedo $c \in \mathcal{R}$. The light condition $\mathbf{v} = \mathbf{s}i$ is defined by the normalized vector $\mathbf{s} \in \mathcal{R}^3$ which denotes the light direction and the intensity of the light given by $i \in \mathcal{R}$. We added the error term $e \in \mathcal{R}$, which allows us to handle shadows and reflections, which are not modelled in the Lambertian model. We can also vectorize the face image which gives the following notation: $\mathbf{b} = H\mathbf{v} + \mathbf{e}(\mathbf{s})$. Using an image of M pixels, we have a vectorized face images, $\mathbf{b} \in \mathcal{R}^{M \times 1}$ and a matrix $H \in \mathcal{R}^{M \times 3}$ which contains the face shape and the error term $\mathbf{e}(\mathbf{s}) \in \mathcal{R}^{M \times 1}$. The goal of this method is to correct the illumination in a face image $b(x)$, by estimating the shape $\mathbf{h}(x)$ and light $\mathbf{v}$. This approach uses a face shape model, so that we are able to estimate a face shape given that we know the light direction. We can also evaluate how well the face shape fit under a certain light direction. Because the light direction is unknown, we calculate multiple face shapes for a set of light direction and evaluate these shapes. Using kernel regression, a sort of weighted interpolation using the evaluations of the face shapes, we can calculate a final shape using the best evaluated shapes. The pseudo-code of this approach is given below:

- Learn a model of the face shape and the error term (offline)
- For a set of different light direction $\mathbf{s}_j$
  - Calculate a shadow and reflection term $e_j(x; \mathbf{s}_j)$
  - Estimate the light intensity i_j which gives us light condition $\mathbf{v}_j$
  - Fit the face shape model to the face image which gives us a shape $\mathbf{h}_j(x)$
  - Evaluate the shape $\mathbf{h}_j(x)$ which gives us a distance measure d_j
- Calculate the final shape $\mathbf{h}(x)$ using kernel regression
- Refine the albedo of the final shape $c(x)$ to obtain more details

In this paper, j is the index for a set of different light directions, where we calculate for J light directions error terms $e_j(x; \mathbf{s}_j)$, light conditions v_j, face shapes $\mathbf{h}_j(x)$ and evaluations of the face shapes. The final shape is used to calculate frontal illumination on the face, giving us an illumination normalized face image. In the next sections, we describe the different steps of the pseudo-code.

Learn the Face Shape and the Error Model. A set of face images and 3D range maps allows us to calculate the face shapes H using a image and a range map (see [11]). Using multiple face shapes, we can determine with Principal Component Analysis (PCA) a mean shape $\overline{H}$ and k variations from this mean shape denoted by $T_k \in \mathcal{R}^{M \times 3}$. This allows us to describe the face shape H as follows $\overline{H} + \sum_{k=1}^{K} T_k y_k$.

The Lambertian model is unable to deal with shadows and reflections. For this reason, we added the error term $e(x; \mathbf{s})$, which depends on the light direction. We use a 2D face database with labelled light direction to learn the error term for each light directions. We determine the face shape under the frontal illumination conditions. Next, we determine for the same face under different illumination conditions the error with respect to the Lambertian model. We calculate for every pixel position a mean and variance of the error using this dataset. This gives us a face independent error term to model shadows and reflections.

Estimate and Evaluate the Face Shape given a Light Direction. Given the light direction $\mathbf{s}_j$, we estimate and evaluate a face shape using the face image and our face shape model. We obtain the mean error $e(x; \mathbf{s})$ as described in the previous section using kernel regression for a given light direction (see [12]). We still have to estimate the light intensity i_j and the face shape H_j. By replacing the unknown face shape H_j by the mean shape $\overline{H}$, we estimate the light intensity i_j using the following equation:

$$i_j = \arg\min_{i_j} \|(\overline{H}\mathbf{s}_j)i_j - (\mathbf{b} - \mathbf{e}_j(\mathbf{s}_j)\|^2 \tag{2}$$

Because this is a overcomplete system, we can use the mean face shape $\overline{H}$ to estimate the light intensity i_j, which still gives a very accurate estimation. This might however normalize the skincolor, because the mean shape contains the mean albedo instead of the user-specific skincolor. The light conditions are now given by $\mathbf{v} = \mathbf{s}i$, so we only have to obtain the face shape H_j. To obtain the face shape H, we substitute $\overline{H} + \sum_{k=1}^{K} T_k y_k$ into the Lambertian equation as follows:

$$\overline{H}\mathbf{v}_j + \sum_{k=1}^{K} T_k \mathbf{v}_j y_{j,k} = \mathbf{b} - \mathbf{e}_j(\mathbf{s}_j) \Rightarrow \sum_{k=1}^{K} T_k \mathbf{v}_j y_{j,k} = \mathbf{b} - \mathbf{e}_j(\mathbf{s}_j) - \overline{H}\mathbf{v}_j \tag{3}$$

We write $T_k \mathbf{v}_j = A_k$ and $\mathbf{b} - \mathbf{e}_j(\mathbf{s}_j) - \overline{H}\mathbf{v}_j = \mathbf{c}$, which allows us to calculate the variations $\mathbf{y}_j$:

$$\mathbf{y}_j = \arg\min_{\mathbf{y}_j} \|A\mathbf{y}_j - \mathbf{c}\|^2 \tag{4}$$

To calculate the shape H_j from the parameters $\mathbf{y}_j$, we use model obtain in section 2.2. This shape allows us to calculate a reconstructed image $\mathbf{b}_j = H_j \mathbf{v}_j + \mathbf{e}_j(\mathbf{s}_j)$. To evaluate the face shape, we use two conditions: first the found variations $\mathbf{y}_j$ to the mean shape should be small and second the reconstructed image $\mathbf{b}_j$ should be similar to the real image $\mathbf{b}$. This gives the following distance measure:

$$d_j = \sum_{k=1}^{K} \frac{y_k^2}{\lambda_k} + \frac{(\mathbf{b} - \mathbf{b}_j)^T(\mathbf{b} - \mathbf{b}_j)}{\rho} \tag{5}$$

The calculation for variations which are not model by the shape model (ρ) is described in [11]. Using this distance measure d_j, we can easily evaluate the quality of the found shape for a certain light direction.

Calculate Final Shape Using Kernel Regression. In the previous section, we explained how we obtain a face shape using only the face image $\mathbf{b}$ and the light direction $\mathbf{s}_j$. We perform this procedure for a set of light directions, where we use a similar grid of light directions as is used in the Yale B database. We calculated the face shape parameters $\{\mathbf{y}_j\}_{j=1}^{J}$ for each light direction $\mathbf{s}_j$. We evaluate the shapes which gives us the distance measures $\{d_j\}_{j=1}^{J}$. Using kernel regression on the obtain parameters $\mathbf{y}_j$, we calculate the final face shape parameters:

$$\mathbf{y} = \sum_{j=1}^{J} w_j \mathbf{y}_j / (\sum_{j=1}^{J} w_j) \qquad\qquad w_j = \exp[-\frac{1}{2}(d_j)/\sigma)^2] \tag{6}$$

In the above equation, σ is determined such that 5 percent of the distances lie within $1 \times \sigma$. Using these final face shape parameters, we can calculate the final face shape and estimate the light conditions in the image.

Refinement of Albedo of the Final Shape. The final face shape does not contain all the details present in the original image, which can be explained by the fact that we perform dimension reduction and kernel regression. To recover these details, we recalculate the albedo $c(x)$ of the shape, because the albedo usually contains the details, while the surface normals contain the larger structures of the shape. The albedo can be calculated using a MAP estimate given by the following equation:

$$c(x)_{MAP} = \arg\max_{c} P(b(x)|c(x))P(c(x)) \tag{7}$$

$$c(x)_{MAP} = \arg\max_{c(x)} \mathcal{N}(b(x) - (c(x)\mathbf{n}(x)^T \mathbf{s}i + \mu_e(x, \mathbf{s})), \sigma_e^2(x, \mathbf{s})) \times \mathcal{N}(\mu_c(x), \sigma_c^2(x)) \tag{8}$$

We assume that the probabilities are normal distributions ($\mathcal{N}$). The mean (μ_e) and variance (σ_e) of the error term are calculated 2D face database with labelled light direction and the mean (μ_c) and variance (σ_c) of the albedo are determined from the set of the face images and 3D range maps used to learn the face shape. We find the minimum by taking the derivative and set it equal to zero. The new albedo c_{MAP} contains more details than the albedo in the final shape.

3 Fusion to Improve Recognition

In the previous section, we proposed two methods to obtain illumination invariant face images. The resulting images of these methods can be seen in Figure 1.

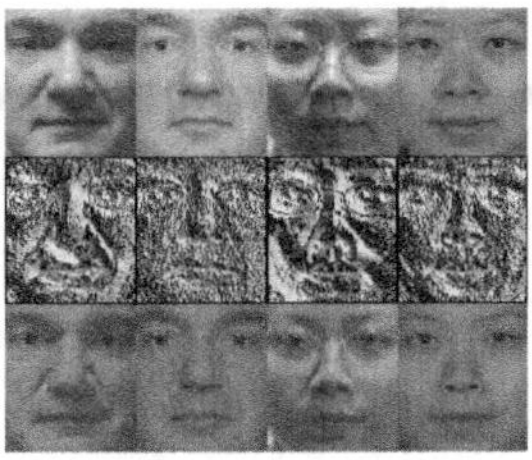

(a) FRGCv1 database

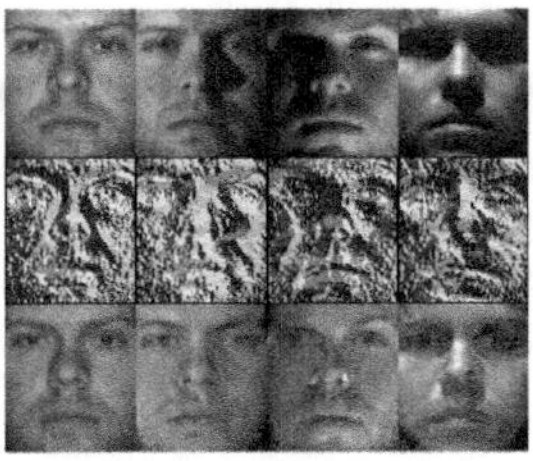

(b) Yale B databases

Fig. 1. Face Images from the Yale B database and FRGCv1 database, upper without correction, middle preprocessed with LBP, lower preprocessed with Model-based Illumination Correction

In this section, we combine these methods to improve the face recognition under different illumination conditions. We train two face classifiers, one with the LBP and one with Model-based Face Illumination correction. The face classifier (log-likelihood ratio after a feature reduction using a PCA and LDA transformation) gives us a similarity score, which we are able to fuse. For fusion, we use the following methods: SUM rule score-level fusion [7], OR rule decision-level fusion [8] and hybrid fusion [9].

In the case of score-level fusion, we can take the joint likelihood ratio, which in our case means that we can sum the scores obtained from the log-likelihoods ratios. This gives us the advantage that we do not have to estimate the different density functions or perform a normalization step to the similarity scores. This method of fusion, we denote as SUM rule fusion.

Although theoretically, score-level fusion should achieve the optimal performance, it is not very robust to outliers. For this reason, we also use decision-level fusion with the OR rule to combine the reciever operation characteristics (ROC). The ROC is determined from the similarity scores of the face classifiers and can be obtained by varying the threshold, thus producing a different false reject rate β and false accept rate $\alpha = 1 - p_r$. This specific pair (α, β) is called an operation point, which corresponds to a threshold t in the similarity scores. In the case of fusion, there can be N classifiers and each is characterized by its ROC, $p_{r,i}(\beta_i), i = 1, ..., N$. By assuming that our classifiers are independent, the final performance of the OR rule can be estimated as $\beta = \prod_{i=1}^{N} \beta_i$ and $p_r(\beta) = \prod_{i=1}^{N} p_{r,i}(\beta_i)$. By searching for the optimal operation points, the fusion with the OR rule can be formulated as:

$$\hat{p}_r(\beta) = \max_{\beta_i | \prod_{i=1}^{N} \beta_i = \beta} \{ \prod_{i=1}^{N} p_{r,i}(\beta_i) \} \tag{9}$$

We can prove that the estimated $\hat{p}_r(\beta)$ is never smaller than any of the components $p_{r,i}(\beta), i = 1, ..., N$ at the same β. Because we do not have the ROC $\hat{p}_r(\beta)$

in analytical form, we estimate a ROC from evaluation data. The ROC $\hat{p}_r(\beta)$ is therefor characterized by discrete values and can be solved numerically [8].

The hybrid fusion is a combination between score-level fusion and decision-level fusion. In hybrid fusion, we first perform the SUM rule score-level fusion to combine the ROCs of both classifiers. The ROCs of both classifiers together with ROC given by the SUM rule are then fused using OR rule decision-level fusion. Using hybrid fusion, we hope to combine the advantages of score-level fusion and the decision-level fusion.

4 Experiments and Results

The purpose of the illumination corrections is to improve the verification rates in face recognition. We performed a recognition experiment on the Yale B databases (the Yale B [13] and extended Yale B [14] database), which contain face images under different labelled illumination conditions created in a laboratory. This experiment tests the ability to make illumination invariant images under all kinds of illumination conditions (also with hard shadows). We also perform an experiment on the Face Recognition Grand Challenge version 1 (FRGCv1) database [15] which contain face images taken under controlled and uncontrolled conditions. This experiment allows us test the ability to correct the uncontrolled illumination conditions and compare them to the controlled images in the gallery.

The Yale B Databases are created to model and test the effects of illumination on face images. In our correction algorithm, we use the Yale B databases to obtain our error model for shadows and reflections. Because we trained our error model on the Yale B databases, we performed a leave one person out experiment for the Model-based Illumination Correction.

In our face recognition experiment, we correct all the face images with both correction methods. We use only the face images with a azimuth and elevation angle below ± 90 degrees in this experiment. For face recognition, we trained on the face images of thirty persons and performed a one-to-one verification experiment on the remaining eight persons, leaving out the face images taken under similar illumination conditions. We repeated this experiment until we compared the face images of all person in the Yale B databases with each other. To train the fusion methods, we used the scores from the Yale B database (10 persons). For testing, we used the scores of Extended Yale B database (29 persons). In Figure 2, we show all the results on the Extended Yale B database.

The experiment which we perform on the Yale B databases is a difficult experiment, because sometimes two images illuminated from opposite positions are compared. We observe from Figure 2(b) that the Model-based Illumination Correction works better at a FAR $< 25\%$ than the Local Binary Patterns. By fusing the two methods, we can improve the recognition results significantly. In Figure 2(a), we observe that a diagonal line is probably the best separation between imposter scores and genuine scores. This explains why the SUM rule performs slightly better than the OR rule and hybrid fusion (see Figure 2(b)).

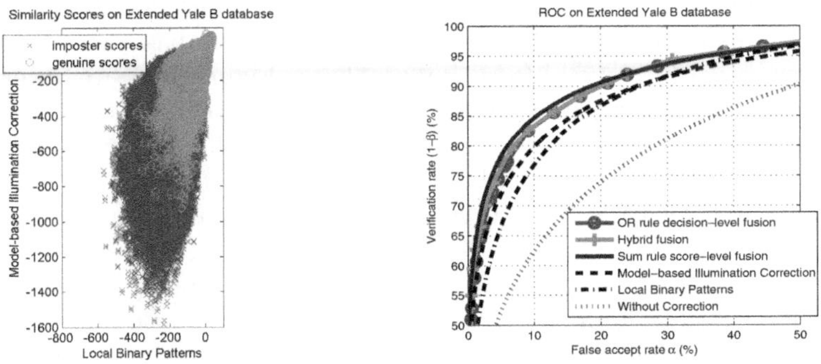

Fig. 2. Score Plot and ROC of the two illumination correction methods on the Extended Yale B database, ROCs also contains the fusion results and the results without illumination correction

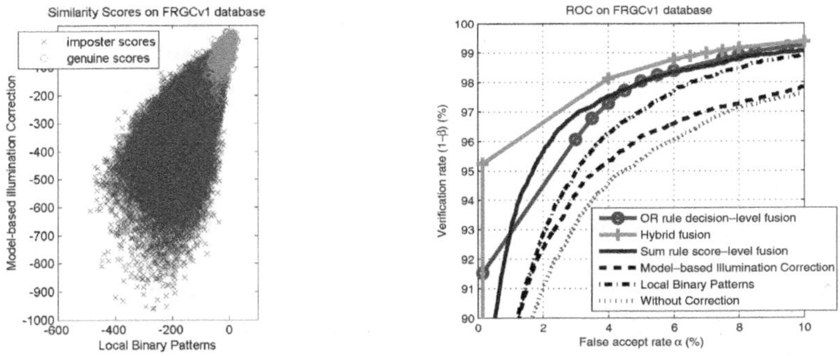

Fig. 3. Score Plot and ROC of the two illumination correction methods on the FRGCv1 database, ROCs also contains the fusion results and the results without illumination correction

The FRGCv1 Database contains frontal face images taken under both controlled and uncontrolled conditions. In our experiment, we randomly divided the uncontrolled and controlled face images into two parts, each containing approximately half of the face images. We used the first halves of both sets to train our face classifiers and the fusion methods, the second half of the controlled images are used for the enrollment of the one user template for every person and the second half of the uncontrolled images are used as probe images. We repeat this experiment 20 times using different random splits of the database to become invariant for statistical fluctuations. This experiment simulates a video surveillance scenario, where we usually have a gallery of high quality images, but the probe images are obtained under uncontrolled conditions. Both our illumination correction algorithms preprocess all the images, also the controlled images. The recognition results are shown in Figure 3.

Although, we have in this experiment less extreme illumination conditions, there are also other challenges in the FRGCv1 database beside illumination, like expressions and out of focus images. From Figure 3(b), we observe that the Local Binary Patterns work better on this database than the Model-based Illumination Correction. The main reason for this difference is that Model-based Illumination Correction has a larger amount of outliers, due to glasses and expressions (see also Figure 3(a) where relatively many genuine scores (circles) have larger negative values for the Model-based Illumination Correction). Using the simple SUM rule to combine the face classifiers already improves the overall recognition results. In figure 3(b), we observe that the recognition results of the OR rule are similar to the sum rule, and the hybrid fusion clearly outperforms the other fusion methods on this database.

5 Conclusions

We combine two different methods to correct for illumination in face images and obtain better results in face recognition. We show that both methods are able to correct for illumination in face images. The Local Binary Patterns method corrects pixel values based on the local neighborhood in the image. This method shows good results on uncontrolled images, but cannot recover large regions with hard shadows. The Model-based Illumination Correction shows that it can deal with these shadow regions, but it has problems in uncontrolled conditions which contain unmodelled effects, like glasses and expressions. Because both methods have different strengths and weaknesses, we combine the illumination normalization methods using fusion. We use three different fusion methods: SUM rule score-level fusion, OR rule decision-level fusion and hybrid fusion. The performance of the simple SUM rule fusion already improves the results significantly and works best for large variations in illumination. The performance of the OR-rule is in both experiments slightly worse than the SUM rule. The hybrid fusion, which tries to combine the advantages of both fusion methods, gives the largest improvement in performance when we correct for uncontrolled illumination conditions occurring in a video surveillance environment.

References

1. Shan, S., Gao, W., Cao, B., Zhao, D.: Illumination normalization for robust face recognition against varying lighting conditions. In: IEEE International Workshop on Analysis and Modeling of Faces and Gestures, AMFG 2003, October 17, 2003, pp. 157–164 (2003)
2. Heusch, G., Rodriguez, Y., Marcel, S.: Local binary patterns as an image preprocessing for face authentication. In: 7th International Conference on Automatic Face and Gesture Recognition, FGR 2006, April 10-12, pp. 6–14 (2006)
3. Tao, Q., Veldhuis, R.N.J.: Illumination normalization based on simplified local binary patterns for a face verification system. In: Biometrics Symposium 2007 at The Biometrics Consortium Conference, Baltimore, Maryland, USA, pp. 1–6. IEEE Computational Intelligence Society, Los Alamitos (2007)

4. Shashua, A., Riklin-Raviv, T.: The quotient image: class-based re-rendering and recognition with varying illuminations. IEEE Transactions on Pattern Analysis and Machine Intelligence 23(2), 129–139 (2001)
5. Basri, R., Jacobs, D.: Lambertian reflectance and linear subspaces. IEEE Transactions on Pattern Analysis and Machine Intelligence 25(2), 218–233 (2003)
6. Blanz, V., Vetter, T.: A morphable model for the synthesis of 3D faces. In: SIGGRAPH 1999: Proceedings of the 26th annual conference on Computer graphics and interactive techniques, pp. 187–194 (1999)
7. Dass, S.C., Nandakumar, K., Jain, A.K.: A principled approach to score level fusion in multimodal. In: Kanade, T., Jain, A., Ratha, N.K. (eds.) AVBPA 2005. LNCS, vol. 3546, pp. 1049–1058. Springer, Heidelberg (2005)
8. Tao, Q., Veldhuis, R.N.J.: Optimal decision fusion for a face verification system. In: Lee, S.-W., Li, S.Z. (eds.) ICB 2007. LNCS, vol. 4642, pp. 958–967. Springer, Heidelberg (2007)
9. Tao, Q., Veldhuis, R.: Hybrid fusion for biometrics: Combining score-level and decision-level fusion. In: IEEE Computer Society Conference on Computer Vision and Pattern Recognition Workshops, CVPR Workshops 2008, June 2008, pp. 1–6 (2008)
10. Ojala, T., Pietikainen, M., Harwood, D.: Comparative study of texture measures with classification based on feature distributions. Pattern Recognition 29, 51–59 (1996)
11. Boom, B., Spreeuwers, L., Veldhuis, R.: Model-based reconstruction for illumination variation in face images. In: 8th International Conference on Automatic Face and Gesture Recognition (FGR 2008) (2008)
12. Sim, T., Kanade, T.: Combining models and exemplars for face recognition: An illuminating example. In: Proc. CVPR Workshop on Models versus Exemplars in Computer Vision (December 2001)
13. Georghiades, A., Belhumeur, P., Kriegman, D.: From few to many: Illumination cone models for face recognition under variable lighting and pose. IEEE Transactions on Pattern Analysis and Machine Intelligence 23(6), 643–660 (2001)
14. Lee, K., Ho, J., Kriegman, D.: Acquiring linear subspaces for face recognition under variable lighting. IEEE Transactions on Pattern Analysis and Machine Intelligence 27(5), 684–698 (2005)
15. Phillips, P., Flynn, P., Scruggs, T., Bowyer, K., Chang, J., Hoffman, K., Marques, J., Min, J., Worek, W.: Overview of the face recognition grand challenge. In: IEEE Computer Society Conference on Computer Vision and Pattern Recognition, CVPR 2005, June 2005, vol. 1, pp. 947–954 (2005)

Bayesian Networks to Combine Intensity and Color Information in Face Recognition

Guillaume Heusch[1,2] and Sébastien Marcel[1]

[1] Idiap Research Institute, Centre du Parc, rue Marconi 19, P.O. Box 592, 1920 Martigny, Switzerland
[2] Ecole Polytechnique Fédérale de Lausanne (EPFL), 1015 Lausanne, Switzerland
{guillaume.heusch,sebastien.marcel}@idiap.ch

Abstract. We present generative models dedicated to face recognition. Our models consider data extracted from color face images and use Bayesian Networks to model relationships between different observations derived from a single face. Specifically, the use of color as a complementary observation to local, grayscale-based features is investigated. This is done by means of new generative models, combining color and grayscale information in a principled way. Color is either incorporated at the global face level, at the local facial feature level, or at both levels. Experiments on the face authentication task are conducted on two benchmark databases, XM2VTS and BANCA. Obtained results show that integrating color in an intelligent manner improves the performance over a similar baseline system acting on grayscale only, but also over an Eigenfaces-based system were information from different color channels are treated independently.

1 Introduction

Face recognition is an active research area, probably because of its numerous applications, ranging from video surveillance to human-computer interaction for instance. Hence, there exists numerous systems allowing to recognize people based on their face image. The vast majority of such existing approaches typically act on grayscale images only, since color is usually considered to introduce high variability. Nevertheless, it was shown that color plays an important role in human face recognition [1] [2]. It is thus likely that it may also carry useful information for computer-based systems. Surprisingly, only a few studies are using color in automatic face recognition systems. Torres et al. [3] developed a color Eigenfaces system, where Principal Component Analysis (PCA) is independently applied on each color channels and results are then combined for final classification. They showed that an improvement is obtained over traditional Eigenfaces acting on grayscale images [4]. This result was later confirmed in a study by Gutta et al. on the larger FERET database [5]. Another interesting study is due to Sadeghi et al. [6]: different channels from numerous colorspaces are first classified independently thanks to Linear Discriminant Analysis (LDA). An optimal subset of such classifiers is then found, and selected classification scores

M. Tistarelli and M.S. Nixon (Eds.): ICB 2009, LNCS 5558, pp. 414–423, 2009.
© Springer-Verlag Berlin Heidelberg 2009

are combined using Support Vector Machines. Another approach proposed in [7] consists in extracting color features to use them as input to an Elastic Graph Matching algorithm. Again, color features were shown to perform better than grayscale-based ones.

Bayesian Networks provide an elegant framework to describe relationships (and hence correlations) between different pieces of information. In this work, our aim is to derive models describing the process that generates observations of different nature extracted from face images. Such generative models will then be used for recognition purposes. Specific generative models for face recognition were recently proposed by Heusch and Marcel [8]. The authors proposed a tree-structured Bayesian Network to describe data extracted from grayscale face images. More precisely, they assumed that observations derived from salient facial features are related to each other, and hence tried to model correlations between such observations. Going one step further, we believe that such models are suitable to model correlations between local, grayscale-based features and other information, such as color. Hence, in this contribution, new models integrating color at the global face level, the local facial feature level but also at both levels are derived. Experimental evaluation on face authentication is carried out on the XM2VTS [9] and the BANCA [10] databases. Results show that integrating color intelligently into dedicated generative models may help at reducing the authentication error rate, at least when the training and testing acquisition conditions are quite similar. The best proposed model, correlating grayscale and color at both the local and the global level show a significant improvement in performance as compared to a similar baseline model acting on the luminance channel only [8], but also performs better than the Eigenfaces-based system acting on color channels independently [3]. Obtained results thus suggest that color is of valuable information when combined to grayscale in a coherent manner.

In the next section, Bayesian Networks are briefly introduced, before describing the proposed models and the features in Section 3. Section 4 details the experimental framework and section 5 describes the databases and discusses the obtained results. Finally, a conclusion is drawn in Section 6.

2 Bayesian Networks

A Bayesian Network is a probabilistic graphical model representing the joint probability distribution over a set of random variables, and having the ability to encode dependencies among these variables [11]. It is specified as a directed acyclic graph, where nodes represent random variables and directed links represent causal relationships between these variables. Defining a set of random variables $\mathbf{U} = (x_1, ..., x_n)$, the joint probability defined by a Bayesian Network is given by the following chain rule:

$$P(\mathbf{U}) = \prod_{i=1}^{n} P(x_i|Parents(x_i)) \quad (1)$$

where $Parents(x_i)$ denotes the set of parents of the node representing the variable x_i. Hence, a Bayesian Network is fully defined by the structure of the graph, and by its parameters: the conditional distribution of each variable given its parents. Computing probabilities in the network is referred to as *inference*. It is typically done to update the state knowledge of a set of hidden variables when other variables, referred to as *evidence*, have been observed. In our case, inference is carried out thanks to the Junction Tree Algorithm [12]. Another important issue is how to learn such models from data. Learning in Bayesian Networks may either refer to structure learning, parameters learning or both. Since in this work, the structure is derived according to prior domain knowledge, the focus is made on learning the parameters. As the proposed models contain hidden variables, a natural choice is the Expectation-Maximization (EM) algorithm [13].

3 Proposed Models

3.1 Reference Model

The model presented in [8] relies on the assumption that facial features are related to each other. To model such relationships, a tree-structured Bayesian Network is proposed. This generative model assumes that there exists pairwise relationships between observations derived from grayscale images. Namely, relationships between eyebrows and eyes, eyes and nose and nose and mouth are considered. This model performs better than a simpler generative model (based on Gaussian Mixtures Model) were independence between facial features is assumed. However, it only acts on local features derived from intensity images and does not take advantage of the Bayesian Networks framework to integrate other source of information in a smart manner.

3.2 Color at the Global Level

The first proposed model is depicted in Figure 1 and should be understood as follow: the root node is used to relate various information describing the face. A face thus consists in a relationship between different 'types' of facial features (nodes **B**: eyebrows, **E**: eyes, **N**: nose and **M**: mouth). In addition, a 'type' of color is also modelled through node **C**. This hidden node *causes* observations derived in each color channel: it is hence assumed that information coming from different color channels are explicitly correlated. Finally, the different types of facial feature, as well as the type of color, generates the corresponding observations extracted from the face image. Note also that, unlike the reference model [8], a single hidden node is used to model the relationship between the different observations. Actually, global color information has to be related to the whole face rather than to pairwise relationships between facial features.

3.3 Color at the Local Level

To model the face more accurately, we also propose to incorporate color information at the local facial feature level. For this purpose, every type of facial

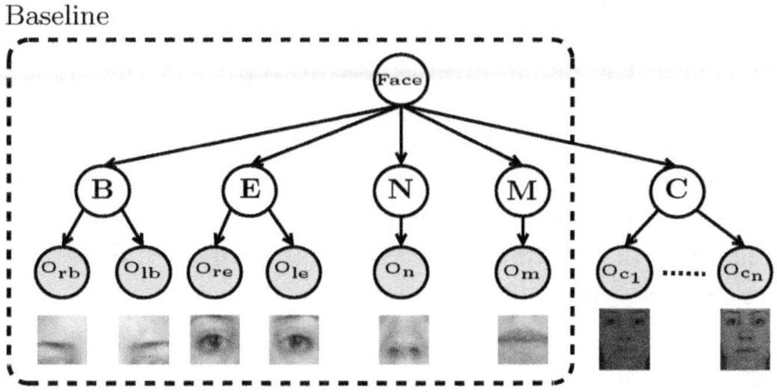

Fig. 1. Bayesian Network model for the face incorporating color information at the global face level. Gray nodes represent the observations extracted from the face image. White nodes are the hidden variables describing 'types' of observations.

feature is not only *explained* using grayscale appearance, but also with color information, as depicted in Figure 2. Note that two additional layers of hidden nodes were introduced in this model. The first one aims at separating the left and right part of 'symmetric' observations (such as eyebrows and eyes, the separation is made through nodes **Br**, **Bl**, **Er** and **El**). This is done to incorporate color information directly at the facial feature level: we choose to correlate the grayscale appearance of a particular facial feature with its color information instead of correlating color of symmetric features together. The purpose of the second additional layer is to separate the appearance of the facial feature from its color.

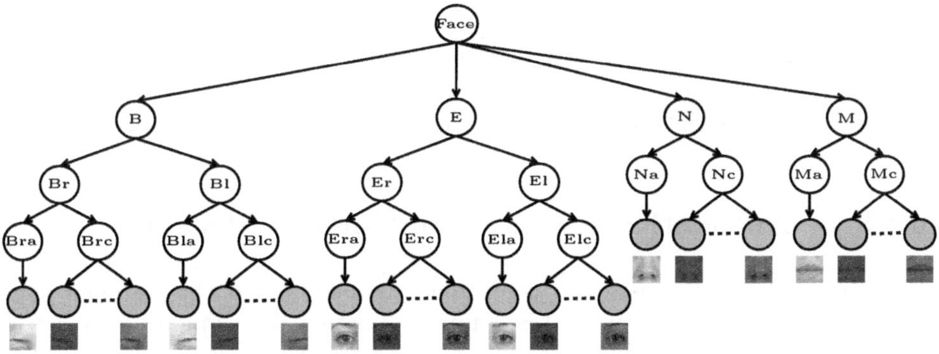

Fig. 2. Bayesian Network model for the face incorporating color information at the local facial feature level

3.4 Color at Both Levels

Color information at the global level is used to represent the global skin color of the face. On the other hand, color information at the local level aims at modelling the color of the associated facial feature. Hence, these two different observations may provide complementary information. As a consequence, we also propose a model combining both global and local information. This model consists in the local model depicted in Figure 2, where the branch of the global model (Figure 1) corresponding to color information is added.

3.5 Feature Extraction

As observations are derived around facial features, they are first located in the face image using an Active Shape Model (ASM) [14]. For local grayscale observations, the same feature extraction scheme as in [8] is applied here. The original color image is converted to grayscale. Multiple squared windows are then cropped around each facial feature, by adding shifts of a variable amount of pixels. Each extracted window is preprocessed using histogram equalization in order to enhance its contrast. Finally, a feature vector is obtained by applying a two-dimensional Discrete Cosine Transform (2D-DCT) on each preprocessed window.

Global Color Observations. In this framework, our aim is to extract skin color information. Hence, a bounding box containing only the inner part of the face is cropped from the color image, based on eyes position. After being preprocessed by histogram equalization, it is subsampled to yield a low-resolution representation of the face, which discards details and thus mainly contains skin-colored pixels. Finally, feature vectors representing color are obtained by decomposing each color channel in terms of 2D-DCT.

Local Color Observations. Regarding the color observations at the local feature level, the same windows as for the grayscale observations are cropped (using shifts as well), but from the original color image. Then, each extracted window is preprocessed by histogram equalization. Finally, feature vectors are obtained by decomposing each color channel in each window in terms of 2D-DCT.

4 Face Authentication and Performance Measures

Face authentication consists in confirming (or denying) a client's claim supported by its face image. In such a framework, either the claimant provides its real identity, either it is trying to fool the system (it is then referred to as an *impostor*). The system has thus to make a decision on whether the claimant is a true client or an impostor. Since modelling all possible impostors is not feasible, a so-called *world model* is trained thanks to the EM algorithm with the Maximum Likelihood (ML) criterion [13] using data coming from different identities. In face authentication, there are usually few training examples available for each client, and hence Maximum Likelihood estimates of the parameters for the client specific models may be inaccurate. To tackle this problem, a form of Maximum A Posteriori (MAP) adaptation [15] is used to adapt client models from a nearby distribution, given by the world model. This approach was already successfully applied to this task [16] [8].

When using generative models, authentication decision is typically performed by taking the likelihood ratio between the model corresponding to the claimed identity and the world model, which is used to represent arbitrary impostors. Given a client's claim supported by its face representation $\mathbf{X}$ (i.e. the set of observations derived from the face image, as depicted on Figures 1 and 2), the decision is made according to:

$$\Lambda(\mathbf{X}) = \log p(\mathbf{X}|\theta_C) - \log p(\mathbf{X}|\theta_{world}) \tag{2}$$

where $p(\mathbf{X}|\theta_C)$ is the probability that the client's model θ_C has generated the data $\mathbf{X}$ and $p(\mathbf{X}|\theta_{world})$ is the probability that the data were generated by an impostor. Based on a threshold τ, the claim is accepted if $\Lambda(\mathbf{X}) \geq \tau$.

In a face authentication framework, two kinds of error can occur: either the true claimant is rejected (false rejection), or an impostor is accepted (false acceptance). Hence, authentication results are typically presented using the Half Total Error Rate, which combines the False Acceptance Rate (FAR) and the False Rejection Rate (FRR) and is defined as:

$$HTER(\tau) = (FAR(\tau) + FRR(\tau))/2 \quad [\%] \tag{3}$$

5 Experiments and Results

In this work, we are interested in seeing if color is a valuable source of information for the face authentication task and, more importantly, we also would like to know if combining color and grayscale intelligently is better than treating such information independently. Color is here encoded in the HSV colorspace, since it was previously shown to be suitable for various computer vision tasks. Results are reported using the three proposed models (referred to as *Global-HSV*, *Local-HSV* and *Both-HSV*), but also with three baseline algorithms. The first one is similar to [8] and consists in the first proposed model, but where the color part has been discarded (see Figure 1) and is referred to as *Local-gray*. The second and third one are our own implementations of the traditional Eigenfaces algorithm [4] (*PCA-gray*) and of the color-based Eigenfaces (*PCA-HSV*) proposed in [3].

Results presented throughout this work are obtained using the following settings for feature extraction: windows of size 24x24 pixels were extracted from the original images based on the results given by the ASM, with shifts of 2, 4 and 6 pixels in each directions. Regarding global color information, faces windows of size 64x80 pixels were first cropped and then subsampled to 24x24 pixels. We kept the first 64 DCT coefficients to build the final feature vectors. For the Eigenfaces-based system, 95% of the variance was kept, and the standard correlation was used as the metric. All these hyper-parameters, as well as the threshold τ, were selected by minimizing the Equal Error Rate (EER: when the FAR equals the FRR) on a separate validation set.

5.1 Experiments on the XM2VTS Database

The XM2VTS database [9] contains 295 identities, among which 200 are used as true clients and the remainder are used to simulate impostors. Recordings were acquired during four sessions under controlled conditions and covering a time period of five months. Along with the database, two experimental protocols, specifying which images have to be used for training, validation and testing have been defined. In Table 1, HTER performance with its 95% confidence interval is reported for the various systems using both XM2VTS protocols.

Table 1. HTER Performance on the test set of XM2VTS

System		HTER [%]	
		LP1	LP2
Baselines	Local-gray	2.74 (0.96)	2.43 (0.84)
	PCA-gray	5.32 (1.07)	4.28 (1.07)
	PCA-HSV	3.09 (0.87)	1.94 (0.49)
Proposed	Global-HSV	2.33 (0.90)	1.25 (0.69)
	Local-HSV	1.98 (0.87)	0.98 (0.60)
	Both-HSV	**1.22** (0.59)	**0.59** (0.42)

One can first remark that color is indeed a valuable information. Actually, results obtained with the color-based Eigenfaces algorithm significantly outperforms the classical Eigenfaces, as expected. However, it reaches almost the same performance as the reference generative model acting on grayscale only. This suggest that treating color channels as independent sources of information is not really successful. Note that, on the other hand, the proposed generative models integrating color in a principled way yields better performance than the similar model based on grayscale features only (*Local-gray*), again emphasizing the usefulness of color information in face processing.

Correlating color and grayscale information (*Global-HSV*) seems to be better than treating such information independently (*PCA-HSV*). This is again evidenced by the results obtained with the proposed generative models: they all perform better than the color-based Eigenfaces system. Regarding the different proposed generative models, we can see that modelling color at the facial feature level consistently yields better results than using information derived from the whole face image. This result is not surprising since the local model is inherently more accurate than the global one. Note finally that the best performance is achieved with the model taking both global and local color information into account (*Both-HSV*), suggesting that both clues are valuable and complementary to describe an identity.

5.2 Experiments on the BANCA Database

To assess the validity of our approach, experiments were also carried out with the baseline systems and the proposed generative models on the more challenging

BANCA database [10]. This database contains 52 clients (English corpus), equally divided into two groups $g1$ and $g2$ used for validation and test respectively. An additional set with 10 images of 30 other subjects is also provided as the world model. In this database, image acquisition was performed according to three different scenarios: Controlled (high-quality camera, uniform background, controlled lighting), Degraded (webcam, non-uniform background) and Adverse (high-quality camera, arbitrary conditions). Examples of acquisition conditions can be seen on Figure 3. There exists several experimental protocols defining which scenarios and which images have to be used for enrollment and testing. In this study, the protocols Mc, Ua, Ud, P and G have been considered.

(a) controlled (b) degraded (c) adverse

Fig. 3. Example of the different scenarios in the BANCA database

In Table 2, HTER performance with its 95% confidence interval for the different protocols is reported on the test set $g2$. Obtained results show that when the acquisition conditions are not well-controlled, global skin color is not a reliable clue anymore. This is evidenced by the performance obtained with the Eigenfaces-based system, but also with generative models taking this information into account (i.e. *Global* and *Both*). Indeed, the baseline generative model based on grayscale features (*Local-gray*) performs better when there is a strong mismatch between training and testing conditions. However, modelling the color at the local facial feature level achieves good results: when the training and testing conditions are the same (protocols Mc and G), local color information

Table 2. HTER on the test set $g2$ of the BANCA database

System		HTER [%]				
		Mc	Ua	Ud	P	G
Baselines	Local-gray	2.24 (0.93)	20.51 (2.64)	**19.90** (2.59)	**16.52** (1.38)	6.85 (0.95)
	PCA-gray	20.38 (2.62)	41.60 (3.17)	37.02 (3.17)	34.51 (1.80)	21.84 (1.56)
	PCA-HSV	14.71 (2.30)	34.07 (3.11)	32.34 (3.08)	29.78 (1.72)	18.48 (1.47)
Proposed	Global-HSV	5.19 (1.41)	28.11 (2.94)	32.08 (3.04)	24.22 (1.63)	11.50 (1.22)
	Local-HSV	**1.89** (0.87)	**17.24** (2.43)	20.77 (2.65)	18.80 (1.41)	**5.58** (0.86)
	Both-HSV	3.21 (1.09)	21.79 (2.72)	22.02 (2.66)	19.94 (1.52)	6.31 (0.93)

improves the performance, and clearly outperforms the Eigenfaces-based systems. Another interesting result is obtained with protocol Ua: in this case, even if the training/testing conditions are different, the model integrating color performs better than the baseline system. This can be explained by the fact that the same acquisition device was used in both scenarios, and thus color seems to remain consistent across controlled and adverse conditions (see Figure 3).

6 Conclusion

In this contribution, new generative models based on Bayesian Networks were proposed to tackle the face authentication task. The purpose of these models was to integrate color in a principled way into a local-feature based model acting on grayscale observations only. To do so, new models were derived: they combine color and grayscale information either at the global face level, at the local facial feature level or at both levels. Face authentication experiments were conducted on two different benchmark databases. Obtained results showed that improvement can be gained when color is combined to grayscale as additional information. Namely, we showed that the proposed models are suitable for the face authentication task, at least when the acquisition conditions between enrollment and testing are quite similar. In particular, the model taking color into account at both the global and the local level significantly outperforms a similar baseline system acting on grayscale-based features, as well as a color-based Eigenfaces algorithm [3]. Correlating different sources of information thus seems to be more effective than treating them independently. However, when there is a strong mismatch between training and testing conditions, color information may become confusing. An obvious possible future direction is hence the investigation of other colorspaces, and particularly the one taking the illuminant into account (i.e. CIE-XYZ and its derivatives). Besides, it would also be interesting to combine more than one colorspace representation, since this approach was shown to yield good results [6].

Acknowledgements

This work has been funded by the GMFace project of the Swiss National Science Foundation (SNSF) and by the MOBIO project 7th Framework Research Programme of the European Union (EU). Softwares were implemented using the TorchVision library[1] and experiments were carried out using the PyVerif framework[2].

References

1. Russell, R., Sinha, P., Biedermann, I., Nederhouser, M.: Is Pigmentation Important For Face Recognition? Evidence From Contrast Negation. Perception 35, 749–759 (2006)

[1] http://torch3vision.idiap.ch

[2] http://pyverif.idiap.ch

2. Sinha, P., Balas, B., Ostrovsky, Y., Russel, R.: Face Recognition by Humans: Nineteen Results All Computer Vision Researchers Should Know About. Proceedings of the IEEE, Special Issue on Biometrics: Algorithms and Applications 94(11), 1948–1962 (2006)
3. Torres, L., Reutter, J.Y., Lorente, L.: The Importance of the Color Information in Face Recognition. In: IEEE Intl. Conf. on Image Processing (ICIP), vol. 3, pp. 627–631 (1999)
4. Turk, M., Pentland, A.: Face Recognition Using Eigenfaces. In: IEEE Intl. Conf. on Computer Vision and Pattern Recognition (CVPR), pp. 586–591 (1991)
5. Gutta, S., Huang, J., Chengjun, L., Wechsler, H.: Comparative Performance Evaluation of Gray-Scale and Color Information for Face Recognition Tasks. In: Bigun, J., Smeraldi, F. (eds.) AVBPA 2001. LNCS, vol. 2091, pp. 38–43. Springer, Heidelberg (2001)
6. Sadeghi, M., Khoshrou, S., Kittler, J.: SVM-Based Selection of Colour Space Experts for Face Authentication. In: Lee, S.-W., Li, S.Z. (eds.) ICB 2007. LNCS, vol. 4642, pp. 907–916. Springer, Heidelberg (2007)
7. Jones, C.I., Abott, A.L.: Color Face Recognition by Hypercomplex Gabor Analysis. In: IEEE Intl. Conf. on Automatic Face and Gesture Recognition (AFGR), pp. 126–131 (2006)
8. Heusch, G., Marcel, S.: Face Authentication with Salient Facial Features and Static Bayesian Network. In: Lee, S.-W., Li, S.Z. (eds.) ICB 2007. LNCS, vol. 4642, pp. 878–887. Springer, Heidelberg (2007)
9. Messer, K., Matas, J., Kittler, J., Lüttin, J., Maitre, G.: XM2VTSDB: The Extended M2VTS Database. In: Intl. Conf. Audio- and Video-based Biometric Person Authentication (AVBPA), pp. 72–77 (1999)
10. Bailly-Baillière, E., et al.: The Banca Database and Evaluation Protocol. In: Kittler, J., Nixon, M.S. (eds.) AVBPA 2003. LNCS, vol. 2688, pp. 625–638. Springer, Heidelberg (2003)
11. Pearl, J.: Probabilistic Reasoning in Intelligent Systems: Networks of Plausible Inference. Morgan Kaufmann, San Francisco (1988)
12. Cowell, G., Dawid, P., Lauritzen, L., Spiegelhalter, J.: Probabilistic Networks and Expert Systems. Springer, Heidelberg (1999)
13. Dempster, A., Laird, N., Rubin, D.: Maximum Likelihood From Incomplete Data via the EM Algorithm. The Journal of Royal Statistical Society 39, 1–37 (1977)
14. Cootes, T., Taylor, C., Cooper, D., Graham, J.: Active Shape Models: Their Training and Applications. Computer Vision and Image Understanding 61(1), 38–59 (1995)
15. Gauvain, J.L., Lee, C.H.: Maximum A Posteriori Estimation for Multivariate Gaussian Mixture Observations of Markov Chains. IEEE Trans. on Speech and Audio Processing 2(2), 291–298 (1994)
16. Cardinaux, F., Sanderson, C., Bengio, S.: User Authentication via Adapted Statistical Models of Face Images. IEEE Trans. on Signal Processing 54(1), 361–373 (2005)

Combining Facial Skin Mark and Eigenfaces for Face Recognition

Zhi Zhang, Sergey Tulyakov, and Venu Govindaraju

Center for Unified Biometrics and Sensors (CUBS)
University at Buffalo, USA

Abstract. In this paper we investigate the use of facial skin mark information for biometric person verification. We performed statistical analysis of the facial skin mark information. The position, size and color intensity of the skin marks are considered as features for skin mark based face matching. Developed facial skin mark matcher has good performance, but can not be applied to faces with no detected skin marks. Due to non-universality of skin mark information, a novel combination algorithm of traditional Eigenfaces matcher with skin mark matcher has been proposed. The resulting combined face matcher has universality property and delivers better performance than either single matcher. The AR Face Database was used in experiments.

1 Introduction

In order to process ultraviolet radiation human skin produces a pigment, melanin. A localized concentration of melanin in the skin can form into a mole or a freckle, termed as a melanocytic naevus. Nevi, as well as, other small scale structures on the skin surface such as wrinkles or scars, have different appearance from normal skin regions. With their relatively stable appearance, these irregularities, especially prominent moles, freckles or scars, in the face region present potential valuable information, and therefore can facilitate the task of face recognition. In particular, the skin irregularities have been used by law enforcement agencies for purpose of personal identification. In this paper we study the benefits of using skin mark information for the automatic face recognition.

Most of the current face recognition systems are designed to process faces holistically. One of the most prominent examples of holistic matching algorithms, the Eigenfaces approach [1], which introduces PCA on the raw image data to reduce feature dimension, therefore implicitly treats local variations as noise. Similar approaches include Fisherfaces [2]. Other model based systems, such as Active Appearance Model in $2D$ [3], or the Morphable Model in $3D$ [4] use PCA to model intra-class variations. Small unexpected details cannot be captured by these methods. Many existing algorithms utilize textons, DCT coefficients or Gabor wavelet features to represent local facial skin information, but there are no explicit representations of aforementioned facial skin marks.

Only limited research has been done so far on exploiting skin marks for face recognition. The work by Lin et al. [5] utilizes a multilayer representation of a face with global appearance, facial features, skin texture and irregularities. A SIFT framework

M. Tistarelli and M.S. Nixon (Eds.): ICB 2009, LNCS 5558, pp. 424–433, 2009.
© Springer-Verlag Berlin Heidelberg 2009

is utilized to detect and describe details of irregular skin regions, which are combined into an elastic graph for recognition. Based on the *3D* Morphable Model developed in [4], the work of Pierrard et al. [6] presented a framework to localize prominent skin irregularities, mainly moles, in facial skin regions. Their system detects potential moles with a sensitive multi-scale template matching algorithm. Using skin segmentation and local saliency measurement, the candidates are filtered based on their discriminative potential.

The goal of this paper is to present study of facial skin marks as a possible biometric trait for the task of personal identification. Statistical analysis has been done on the characteristics of facial skin marks, such as universality and stability. We developed a facial skin mark matching algorithm, which has good performance. However, due to the nature of facial skin mark, it does not have perfect universality. To address this issue, one novel approach combining conventional Eigenfaces algorithm with our facial skin mark matcher was presented, which has universality property and delivers better performance than either one of the algorithm alone.

The organization of this paper is as follows: In Section 2, the statistical analysis on the characteristics of facial skin marks and their detection, localization, and feature presentation will be presented. Proposed approach will be presented in Section 3. Experiment setups, results and evaluations will be given in Section 4, including face recognition tests using facial skin marks alone and tests using facial skin marks combined with Eigenfaces algorithm. Finally the concluding remarks will be given in Section 5.

2 Study of Facial Skin Marks

2.1 Dataset

No public dataset of skin marks, especially facial skin marks, was found. Instead, generic face databases have to be considered. In this work, *The AR Face Database* [7] is used to build a facial skin mark dataset. In this publicly available dataset, most of the subjects had two sessions of images taken, with two weeks in between these sessions, which provides possible changes of appearance in the facial skin regions of the same subject, such as changes of make-up, facial hair, facial skin irritations (for example acne, razor burns) and spectacles. Based on these changes, the characteristics of facial skin marks, especially intra-class stability, can be analyzed to a certain extent.

In this work, a facial skin mark dataset was built on the frontal images with neutral expression of each subject in the AR face database. As shown in Table 1 240 images of 120 subjects in total were included in this dataset with exactly 2 images per subject.

The automatic labeling of facial skin marks appears to be a difficult problem due to their diversity and the frequent similarity to the neighboring background pixels. In

Table 1. The facial skin marks dataset built on *The AR Face Database*

	Number of subjects	Number of images
Female	55	110
Male	65	130

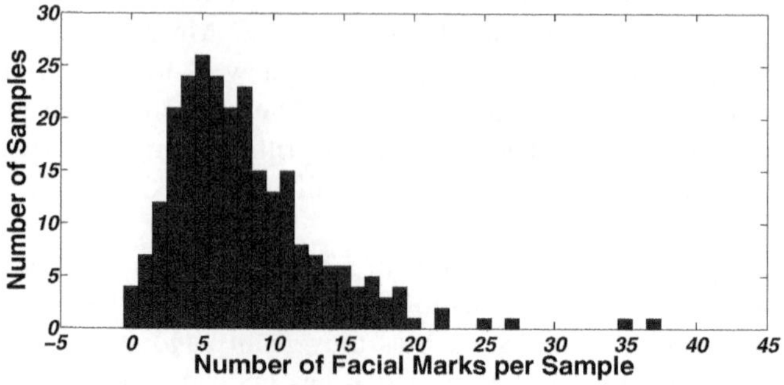

Fig. 1. Histogram of numbers of skin marks labeled for all the facial images

this work we utilize a semi-automatic approach for building facial skin mark database. First, all the images were studied by human observer closely in the full size and then one seed pixel for each facial skin mark was labeled manually. The inner eye corners of each face in the images were also marked for reference purpose. All of the 240 facial images aforementioned images were processed accordingly. Seed pixels were subsequently used for automatic extraction of features.

2.2 Statistical Analysis

With the ground-truthed dataset, statistical analysis was done on the characteristics of facial skin marks, especially their universality and intra-class stability.

Figure 1 shows the histogram of the number of facial skin marks in each of the available facial images. As shown in Table 2, the minimum number of facial skin marks in one facial image is 0 while the maximum number is 37 with a standard deviation of 5.5. From Figure 1, we can conclude that these extreme cases happen far less frequently compared with the cases where number of facial skin marks lies between these two numbers. The total number of facial images that contain 3 to 12 facial skin marks comprises 74.5% of all the available images. And 98.8% of all the availabe facial images have at least one facial skin mark.

Table 2. Statistics of the number of facial skin marks in all the facial images and the intra-class differences in number of facial skin marks for all the subjects

	Number of facial skinmarks for all the facial images	Intra-class differences in number of facial skin marks
Min	0	0
Max	37	10
Average	8.2	1.7
Stdev	5.5	1.7

Intra-class differences are measured by calculating the differences between the number of facial skin marks of the two images from the same subject. The right column in Table 2 summaries the analysis on this matter. Within the 120 subjects, the minimum intra-class difference is 0, while the maximum difference is 10, with the average of 1.7 and the standard deviation of 1.7. Causes of this discrepancies could be one or more of the following conditions:

- Occlusion caused by change of
  - Hair
  - Facial hair
  - Cosmetics
  - Spectacles
- Short term changes of skin appearance caused by
  - Acne
  - Freckles
  - Razor burns
- Environment changes such as changes of lightings.

Proper compensation can be done to eliminate environment changes to certain level. However, it is rather difficult a task to counteract the other changes since the appearance of facial skin region changes over time. Among types of facial skin marks, moles and scars are more stable than small dark regions caused by acne, freckles, razor burns or other skin irritations, however, it is difficult to recognize them completely separately since they all tend to have the similar appearance in contrast to their surrounding normal skin regions.

Although, this analysis of facial skin marks was only conducted on a dataset with limited size, it can, to some extent, be used to justify the characteristics of facial skin marks as a potential biometric trait. First, the universality of facial skin marks is satisfying with over 90% availability. In order to test the intra-class stability, experiments were conducted and results were shown in Section 4.

2.3 Feature Presentation of Facial Skin Marks

Based on the seed pixel labeled for each facial image, a region growing algorithm was utilized on each of region of interest that contains one of the facial skin marks, such that each of the facial skin marks would grow from one pixel into a patch containing a group of pixels that have similar intensity. After the region growing algorithm, the centers of the facial skin marks were re-calculated accordingly.

Two types of features were defined for facial skin marks, namely distance feature and local features. For distance feature, one polar coordinate system was built on each facial image, with the middle point of the two inner eye corners, O, as the pole and the ray from O to the left corner of the right eye as polar axis. Then, the distance feature of each facial skin mark, M_i, can be represented by the polar coordinates of its newly calculated center point, (r_i, θ_i). Two local features were introduced in this work, the area of the facial skin mark, A_i, and the average intensity of all the pixels included in this skin mark, I_i, for facial skin mark M_i.

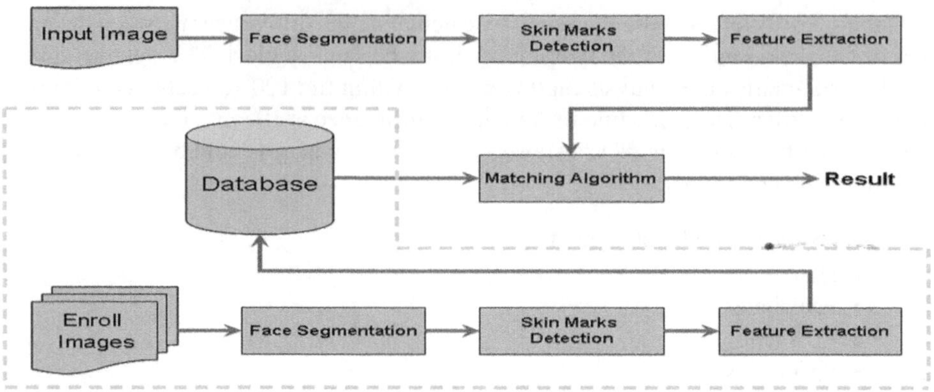

Fig. 2. The framework of personal identification system using facial skin marks as sole biometrics

3 Facial Skin Marks as Biometric Trait

To verify the feasibility of using facial skin marks alone to perform personal identification, one framework was proposed as shown in Figure 2. Facial images are enrolled by face segmentation, facial skin marks detection, and feature extraction for all the detected marks, then the features are stored into the database. For a probe image, the similar process would be conducted to extract its features, then the probe features would be matched against all the record in the database. Finally, the match with the highest score would be returned as identification result.

3.1 Matching Algorithm

Since it is possible for the images of the same subject to have different number of facial skin marks extracted, the matching of any two facial images based on facial skin marks could result in matching N marks in one image against M marks in the other, where $N \neq M$. To match two images, G_1 and G_2, based on their facial skin marks, one matching algorithm was proposed as following.

For the ith facial skin mark in G_1, and for the jth facial skin mark in G_2, the weighted distance between these two facial skin marks is calculated as:

$$D_{i,j} = \sqrt{\sum_{f=1}^{fn} \left(\omega_f \times Dist(F_{i,f}, F_{j,f})\right)^2}, \quad \begin{array}{l} 0 \leq i < M \\ 0 \leq j < N \end{array} \qquad (1)$$

In Equation 1, fn is the number of features for each facial skin mark, and as aforementioned, there are 3 features defined for each skin mark in this work. ω_f is the weight for the fth feature, with $\sum_{f=1}^{fn} \omega_f = 1$. $Dist(F_i, F_j)$ calculates the distance between two features F_i and F_j. Depend on the type of these features, $Dist(F_i, F_j)$ could calculate Euclidean distance or simple subtraction of these two features. If this $D_{i,j}$ is smaller than a preset threshold, then these two facial skin marks are considered matched. After processing all the possible matching pairs of facial skin marks in G_1 and G_2, the

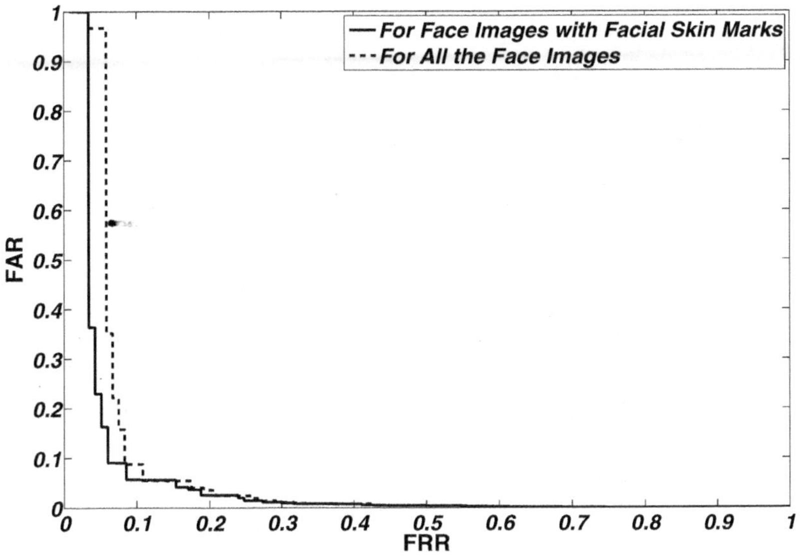

Fig. 3. ROC curves for facial skin mark matcher

number of matches, $N_{matches}$, are counted. Then the final matching score of matching these two images is calculated as:

$$S = \frac{N_{matches}}{N_{matching\_pairs}} \tag{2}$$

If this score S is higher than a threshold, then these two images are considered matched, otherwise not matched. Another scenario need to be considered is when at least one of the matching faces has no or too few facial skin marks detected. A flag was introduced for these cases, where the flag would be set as "*FAIL*". For normal cases, this flag is "*OK*".

4 Experiment Results and Discussion

Two set of experiments were conducted in this work. One set of experiment is using facial skin marks as single biometric trait to perform personal identification. The other is using the combination of facial skin marks and a conventional face recognition algorithm, in this work Eigenfaces algorithm to perform personal identification. For these two set of experiments, the same dataset, as mentioned in Section 2, was used, and of the two images of each subject, one is used as gallery image and the other as probe image.

4.1 Facial Skin Marks as Single Biometric Trait

In this set of experiments, facial skin marks alone was used as a biometric trait to perform personal identification. A system was build as described in Figure 2. Due to the

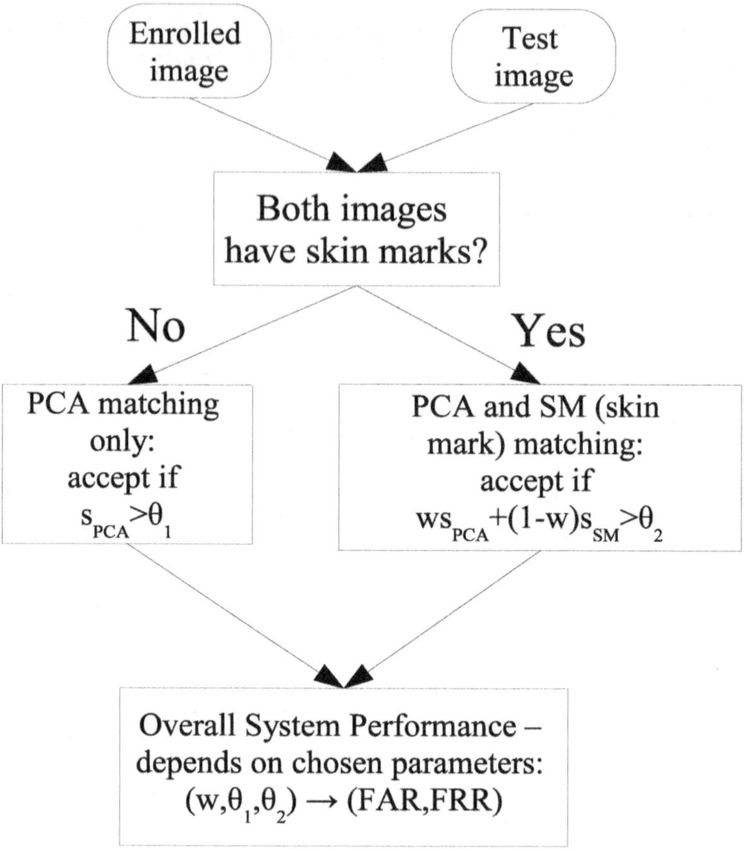

Fig. 4. Combination algorithm for branched system

nature of this trait, it is possible that the images of some of the subjects may not show any facial skin marks at all, which is the reason we introduced a flag as mention above. In order to evaluate the system under various conditions, two modes of experiments were conducted. One is to only process the images with certain number, F_n, of facial skin marks. In this work, F_n was set to 1. The other mode is to process all the available images, which contains both images with some facial skin marks extracted and images with no facial skin marks. With these two modes of experiments conducted on the available dataset, the results is shown in Figure 3.

4.2 Combine Facial Skin Marks with Eigenfaces

We used the weighted sum combination method to combine the scores produced by the Eigenfaces matching and skin mark matching algorithms. Usually this combination method has reasonably good performance [8], but the weights should be trained with the help of training data. Since not all faces possess skin marks and skin mark matching

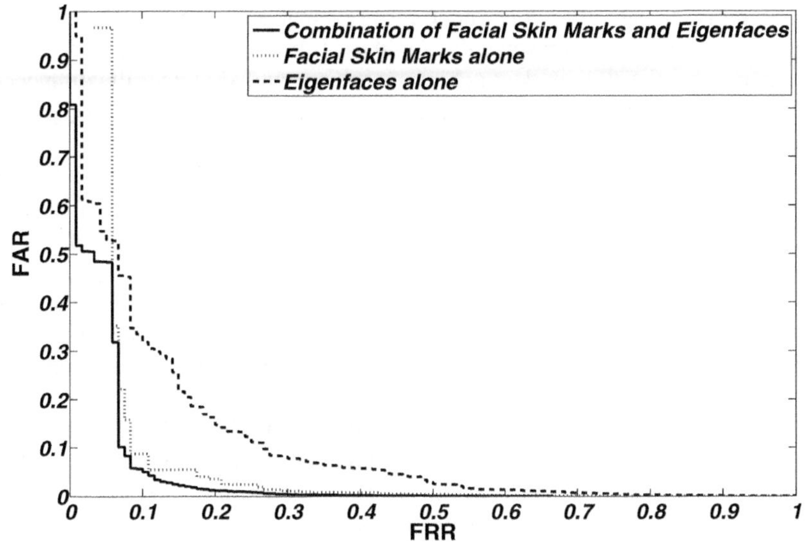

Fig. 5. ROC curves for Eigenfaces, skin mark matcher and for combined algorithm

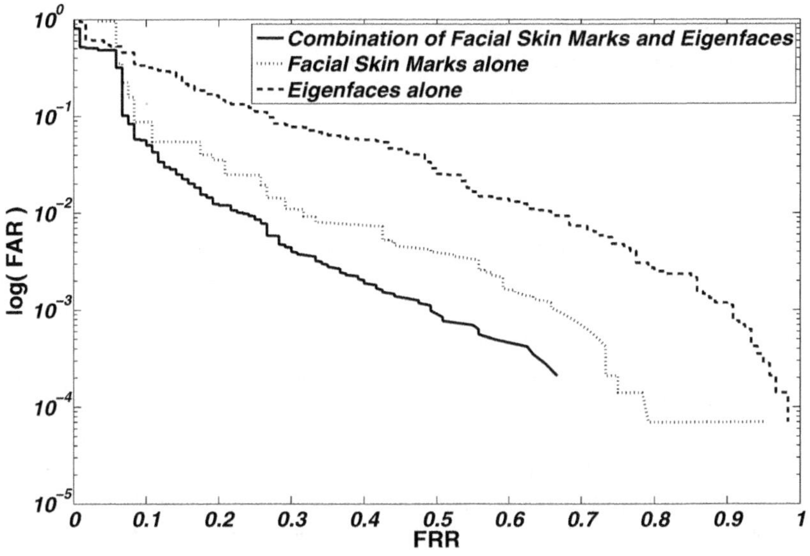

Fig. 6. ROC curves for Eigenfaces, skin mark matcher and for combined algorithm. The FAR axis was log-scaled in order to better show the performance of these algorithms for small FAR values.

score can not be calculated, we need to provide a separate processing for such cases. The general structure of the combination algorithm is shown in Figure 4.

The system has two execution branches depending on whether both matched face images have skin marks detected. If there are no skin marks in any of the images we can only use Eigenfaces matching score and accept verification if the Eigenfaces matching score $S_{Eigenfaces}$ is better than some threshold θ_1. If both images have skin marks, we additionally possess the skin mark matching score S_{SM} and accept verification decision if the weighted sum of two scores is better than than some threshold θ_2: $wS_{Eigenfaces} + (1-w)S_{SM} > \theta_2$.

The particular choice of operating parameters w, θ_1 and θ_2 will result in specific values of *FAR* and *FRR*. We call the choice of parameters $\{w, \theta_1, \theta_2\}_0$ locally optimal, if corresponding system's FAR_0 and FRR_0 are optimal: no other set of parameters $\{w, \theta_1, \theta_2\}_1$ delivers better FAR_1 and FRR_1, i.e. $FAR_1 \leq FAR_0$ and $FRR_1 \leq FRR_0$. The set of locally optimal values *FAR* and *FRR* gives us the ROC curve for optimal combinations of the type shown in Figure 4. Note, that different points on produced ROC curve will generally have different values of parameters w, θ_1 and θ_2.

The presented combination algorithm is different from both serial and parallel combination architectures investigated previously in classifier combination field. It is also different from combination methods based on classifier selection methods. The difference stems from the fixed branch selection which is determined by the presence of skin marks in compared images. We can call this type of combinations as "*branch combinations*".

From the implementation side, the training of the presented combination method might rely on exhaustive search of locally optimal pairs (FAR, FRR) with respect to system parameters w, θ_1 and θ_2. In our case we applied some additional optimizations for this search: efficient calculation of *FAR* and *FRR* based on sorted arrays of genuine and impostor matching scores and utilizing discrete nature of calculated *FAR* and *FRR* determined by the limited number of total training samples.

4.3 Results of Combination

Due to branching nature of combination algorithm we can analyze the performance improvements due to each branch. If compared face images have no skin marks, we rely only on Eigenfaces matcher. Consequently, we can view the combination algorithm as an improvement over skin mark matcher algorithm, especially in the area where skin mark matcher can make only reject decisions due to absence of skin marks. The areas of ROC curves with small FRR of figure 5 indicate that the performance of combined matcher is dominated by the performance of a Eigenfaces matching algorithm.

On the other hand, in the area of small FAR we can see in Figure 6, uniform improvement of combined algorithm over both Eigenfaces and skin mark matcher. This improvements can be explained by the good ability of skin mark matcher to reject false matches based on skin mark information. Such information seems to be not utilized in Eigenfaces matching algorithm.

5 Conclusion

This paper investigated the use of facial skin mark information for biometric person verification and presented a novel combination algorithm of conventional algorithm Eigenfaces and skin mark matcher. The resulting combined face matcher has universality property and delivers better performance than either single matcher. Future work comprises expanding current study of facial skin marks by incorporating more face image data, especially images with high resolution, and introducing more detailed classification and analysis of different types of facial skin marks. Study of the pattern of long term changes of facial skin marks can be useful in further evaluation of the stability of facial skin mark as a biometric trait. Facial skin marks combined with other conventional models could also be an interesting problem to investigate.

References

1. Turk, M., Pentland, A.: Face recognition using eigenfaces. In: IEEE Conf. on Computer Vision and Pattern Recognition, June 1991, pp. 586–591 (1991)
2. Belhumeur, P.N., Hespanha, J.P., Kriegman, D.J.: Eigenfaces vs. fisherfaces: recognition using class specific linearprojection. IEEE Trans. Pattern Anal. Mach. Intell. 19(7), 711–720 (1997)
3. Cootes, T.F., Edwards, G.J., Taylor, C.J.: Active appearance models. IEEE Trans. Pattern Anal. Mach. Intell. 23(6), 681–685 (2001)
4. Blanz, V., Vetter, T.: Face recognition based on fitting a 3D morphable model. IEEE Trans. Pattern Anal. Mach. Intell. 25(9), 1063–1074 (2003)
5. Lin, D., Tang, X.: Recognize high resolution faces: From macrocosm to microcosm. In: IEEE Conf. on Computer Vision and Pattern Recognition, vol. 2, pp. 1355–1362 (2007)
6. Pierrard, J.S., Vetter, T.: Skin detail analysis for face recognition. In: IEEE Conf. on Computer Vision and Pattern Recognition, June 2007, pp. 1–8 (2007)
7. Martinez, A.M., Benavente, R.: The AR Face Database. CVC Technical Report #24 (June 1998)
8. Ivanov, Y., Serre, T., Bouvrie, J.: Error weighted classifier combination for multi-modal human identification. Technical Report AI Memo 2005-035, Massachusetts Institute of Technology Computer Science and Artificial Intelligence Laboratory (2005/12/14 (2005)

Analysis of the Utility of Classical and Novel Speech Quality Measures for Speaker Verification

Alberto Harriero, Daniel Ramos, Joaquin Gonzalez-Rodriguez, and Julian Fierrez

ATVS – Biometric Recognition Group,
Escuela Politecnica Superior, Universidad Autonoma de Madrid,
C. Francisco Tomás y Valiente 11, 28049 Madrid, Spain
{alberto.harriero,daniel.ramos,joaquin.gonzalez,
julian.fierrez}@uam.es

Abstract. In this work, we analyze several quality measures for speaker verification from the point of view of their utility, i.e., their ability to predict performance in an authentication task. We select several quality measures derived from classic indicators of speech degradation, namely ITU P.563 estimator of subjective quality, signal to noise ratio and kurtosis of linear predictive coefficients. Moreover, we propose a novel quality measure derived from what we have called Universal Background Model Likelihood (UBML), which indicates the degradation of a speech utterance in terms of its divergence with respect to a given universal model. Utility of quality measures is evaluated following the protocols and databases of NIST Speaker Recognition Evaluation (SRE) 2006 and 2008 (telephone-only subset), and ultimately by means of error-vs.-rejection plots as recommended by NIST. Results presented in this study show significant utility for all the quality measures analyzed, and also a moderate decorrelation among them.

Keywords: Speaker verification, quality, utility, SNR, degradation indicator.

1 Introduction

Speaker recognition is nowadays a mature field with multiple applications in security, access control, intelligence and forensics. The State of the Art is based on the use of spectral information of the speech signal, combining such information in multiple ways, and compensating the inter-session variability of speech recordings [1,2].

Despite the significant advance on the performance of the technology in the field, partly due to the efforts of NIST and their successful periodic Speaker Recognition Evaluations [3], the field of speaker recognition faces important challenges. Among them, performance of comparisons when there is a high mismatch between enrollment and testing speech conditions is far from being solved, although the improvements in this sense in the last years have been remarkable [1]. Moreover, the mismatch in the conditions of the speech databases for system tuning and for operational work (the so-called database mismatch problem [4]) has a strong impact in the performance of the systems, and attenuates the beneficial effects of compensation techniques.

M. Tistarelli and M.S. Nixon (Eds.): ICB 2009, LNCS 5558, pp. 434–442, 2009.
© Springer-Verlag Berlin Heidelberg 2009

In order to solve the problems associated to session variability in speech, the speaker recognition scientific community continues their efforts on improving the existing compensation algorithms [1]. These methods are mainly based on data-driven approaches modeled with statistical techniques such as factor analysis [1]. Although their demonstrated success, such techniques are sensitive to the existence of a rich development corpus, desirably in similar conditions to those of the operational framework, which may not be available in general. Moreover, there is other knowledge about the speech signal which can be efficiently extracted from excerpts and used as information about the variability of the speech signal and its impact on the performance of speaker recognition systems. Among such knowledge are the quality measures, as recently proposed by NIST [5].

In this work, we present an analysis of several quality measures from the point of view of their utility, i.e., their usefulness as a predictor of system performance. Some of the analyzed quality measures are derived from classical indicators of speech degradation, namely Signal to Noise Ratio (SNR), statistics from Linear Predictive Coefficients (LPC) and estimators of subjective quality (such as ITU P.563 recommendation [6]). Moreover, we propose a quality measure with an attractive interpretation, derived from what we have called Universal Background Model Likelihood (UBML). The work also presents a framework for the obtaining of the proposed quality measures from speech. The paper is completed with experimental results using telephone speech and protocols from recent NIST Speaker Recognition Evaluation Evaluations (SRE), where the utility of quality measures is shown by the performance measures recommended by NIST [5].

The paper is organized as follows. In section 2, we define the quality measurement framework according to previous work in the literature [6,7], We also present three classical quality measures derived from classical indicators of speech degradation. In section 3, we present a novel quality measure based on what we have called the Universal Background Model Likelihood (UBML). Results showing the analysis of the four analyzed quality measures, including the proposed one derived from UBML, are described in section 4, where the utility of the proposed measures is analyzed using two different databases from NIST Speaker Recognition Evaluations (2006 and 2008). Experiments allow the identification of the most useful quality measures for predicting performance, based on protocols recommended by NIST [5]. Finally, conclusions are drawn in section 5.

2 Quality Measures for Speaker Verification

The idea that the quality of the speech signal affects the ability of an automatic system to distinguish among people from their voices is somewhat intuitive, as it happens in other biometric traits [8]. In fact, the measurement of speech quality has been a major topic of research during the last decades [9]. The need to monitor the quality of speech signals on telephone networks has lead to the development of several algorithms to estimate the subjective quality of a speech signal [9], understood as the quality perceived by a given user. The recommendation P.563 of the International

[1] The last research workshop on the topic at John Hoskins University deserves special attention (http://www.clsp.jhu.edu/workshops/ws08/groups/rsrovc/).

Telecommunications Union (ITU) [6] is an estimation method of the subjective speech quality which includes the effects of the majority of existing impairments in modern telephony networks. Its output is computed from 51 parameters, which are indicators of different possible degradations. The quality measures from this study are mainly based on degradation indicators found in ITU P.563 as well as other work in the literature [10].

According to previous work in the literature [6,8], we define a quality measure as a scalar magnitude which predicts the performance of a given biometric system. Under such a definition, utterances with poor quality are more likely to be misclassified than those of good quality. A quality measure is defined to be bounded in the range between 0 and 1, where 0 corresponds to the worst possible quality value and 1 to the best one. As this scalar is based on parameters which, in general, do not belong to this range, a mapping function has to be applied, in such a way that for every possible value of a degradation indicator x, the mapping assigns a quality value $Q(x) \in [0,1]$.

The evaluation of quality measures is carried out following the recommendations given by NIST [5], according to which a quality measure is considered useful if as we reject scores with the lowest quality values, the system performance improves.

2.1 Classical Quality Measures

Quality measures defined in this section have been used before with the purpose of evaluating speech degradation [6,10].

Signal to Noise Ratio (SNR). The SNR degradation indicator has been calculated as follows: making use of a energy-based voice activity detector, each utterance is separated in non overlapping voiced and un-voiced frames of 20 ms. Then, average energy is calculated for both types of frames. Finally, SNR is computed as:

$$SNR = 10 \times \log\left(\frac{E_{voiced}}{E_{unvoiced}}\right). \tag{1}$$

where E_{voiced} and $E_{unvoiced}$ are the mean energies of the voiced and unvoiced sections. This method for measuring SNR has one main drawback: as it depends on the VAD accuracy, it may have problems to differentiate voiced from un-voiced sections for noisy or very high activity utterances.

We defined the SNR quality mapping function as:

$$Q_{SNR}(x) = \frac{x}{60}. \tag{2}$$

where x is the SNR value, which is supposed to belong to the range 0-60 dB. Values outside this range will be limited prior to mapping to quality.

Kurtosis LPC (KLPC). Kurtosis is a 4th order statistic which measures the degree of fat tails of a distribution. In this case, kurtosis is applied to the LPC coefficients distribution, as is done in ITU P.563 recommendation [6]. For every 20 ms frame, 21 LPC coefficients are obtained. Then, kurtosis is calculated as:

$$k = \frac{1}{P}\sum_{p=1}^{P}\left(\frac{a_p - \frac{1}{P}\sum_{p=1}^{P} a_p}{\sigma}\right)^4. \tag{3}$$

where σ represents the standard deviation of LPC coefficients, a_p. Finally, all kurtosis values from all the voiced frames are averaged.

As it will be shown later, the system performance decreases as KLPC increases. According to this, we defined its mapping function as:

$$Q_{KLPC}(x) = 1 - \left(\frac{x-3}{8}\right). \tag{4}$$

where x is the KLPC value, which based on our experiments, is supposed to belong to the range 3-11.

ITU P.563 Recommendation (P.563). ITU provides an implementation of the algorithm defined on this recommendation. The algorithm generates a Mean Opinion Score (MOS) [11] for each utterance, which is representative of the utterance subjective quality. The MOS belongs to the range 1-5, where 1 corresponds to the worst possible quality value, and 5 to the best one. The input utterance must have a length between 3 and 20 seconds. All utterances duration were between 2 and 5 minutes long, so they had to be divided in smaller fragments and their MOSs were averaged.

The mapping function has been defined according to the MOS scale:

$$Q_{P563}(x) = \frac{(x-1)}{4}. \tag{5}$$

3 UBML: A Novel Quality Measure for Speaker Verification

In this work we propose a degradation indicator in the context of speaker verification based on Gaussian Mixture Models (GMM) [13], although the approach can be used in any possible system, no matter the modeling scheme. The proposed measure is motivated by a simple idea. Given that a Universal Background Model (UBM) from a GMM represents the common distribution of speaker features for a given expected operational database, degraded signals are more likely to differ from a UBM than non-degraded signals. Thus, the likelihood between any utterance and the UBM can be used as a measure of speech degradation. Moreover, it is well-known that speech utterances not matching a given UBM in a GMM system will tend to perform poorly, and therefore a simple measure of the match between a given speech utterance and the UBM like UBML will predict performance for any utterance. Although it may be argued that the likelihood with respect to a UBM may represent many other speaker-dependent information non related to speech degradation, experiments with UBML showed a strong relationship between system performance and this indicator, supporting the assumed hypothesis. In section 5, the validity of this measure is further discussed.

Obtaining UBM likelihood is a mandatory step when using a GMM system, and therefore if such a system is used, the obtention of UBML indicators is costless. However, for other systems UBML can be previously computed and its quality measure used as well. Given a speaker GMM model λ_t and any utterance O for which feature vectors have been extracted, a similarity score is typically computed as:

$$S(O, \lambda_t) = \log p(O, \lambda_t) - \log p(O, \lambda_{UBM}). \tag{6}$$

where $p(.,\lambda)$ is the probability density function for any model λ. The last term gives the likelihood between any utterance and the UBM:

$$\text{UBML} = \log p(O, \lambda_{UBM}) \,. \tag{7}$$

We define the mapping function based on the typical distribution of UBML according to the experiments performed in this work, whose values lay within the range (-13,-5). It is expected that for a given GMM system configuration this value will not significantly change its range among databases. Thus, we map the quality measure as follows:

$$Q_{UBML}(x) = \frac{(x+13)}{8} \,. \tag{8}$$

4 Experiments

4.1 Databases, Systems and Protocols

In order to evaluate the utility of quality measures, we have used telephone databases and protocols from NIST Speaker Recognition Evaluations 2006 and 2008, which represents a real challenge in terms on session variability [3]. We have selected both corpuses for experiments in order to show the general behavior of the proposed quality measures among different telephone databases. This fact allows a general strategy of training quality mappings from degradation indicators using a given database (namely NIST SRE 2006) and using such mapping on a different one (namely NIST SRE 2008). For NIST SRE 2008, we have selected the telephone-only subtask of the core condition, namely *short2- short3 tlf-tlf*. For NIST SRE 2006, the whole core condition is used, namely *1conv4w-1conv4w*. For both conditions in the different evaluations, each conversation (coined *short2* for training and *short3* for testing) has an average duration of 5 minutes, with 2.5 minutes of speech on average after silence removal. Variability due to different transmission channels, languages and environmental conditions is present, but even more accused in SRE 2008. Although there are speakers of both genders in the corpus, no cross-gender trials are defined.

For score computation, the ATVS GMM system has been used, where speech data known to come from a given speaker is represented using Gaussian Mixture Models adapted from a Universal Background Model. The front-end consists of the extraction of 19 MFCC plus deltas, and processed with rasta filtering and feature warping. Channel factors at feature level have been used for channel compensation [1]. GMM of 1024 mixtures have been used for modeling. Finally, T-Norm has been used for score normalization. The background set for T-Norm cohorts, channel compensation and background modeling is a subset of databases from previous NIST SRE.

4.2 Degradation Indicators Evaluation

Experiments presented in this section were carried out for 12 different degradation indicators, from previous work in the literature [6,7,10] . They were intended to show the variations of the system performance depending on the magnitude of each indicator, which is useful in order to determine the mapping function from indicator to quality measure. From the whole set of 12, we selected those which showed a clearer relationship with the system performance, namely SNR, ITU P.563 and KLPC.

The experiment was carried out as follows:

1. For every utterance in the databases, each degradation indicator was computed.
2. Scores from the experimental set-up were computed for the described protocols and using the ATVS GMM system.
3. For every score i, a mean degradation indicator μ_i is generated computing the arithmetic mean of the indicators for the training utterance and the test segment.
4. Scores are arranged according to their mean degradation indicator μ_i.
5. The first 20% of ordered scores are selected. This is known as set k. For each score set k, the EER_k is computed, as well as the mean degradation indicator.
6. The last step is repeated 100 times for each set of scores k=1,...100. Each time selecting a set of scores with higher degradation indicator. The last set will correspond to the 20% scores with highest degradation indicator.
7. As a result, we obtain 100 EER values and 100 mean degradation values, which correspond to 100 overlapped sets of scores. EER is then represented with respect to its corresponding mean set degradation value.

The following plots show the result of the best performed degradation indicators from the 12 analyzed. We also show the results for the proposed UBML.

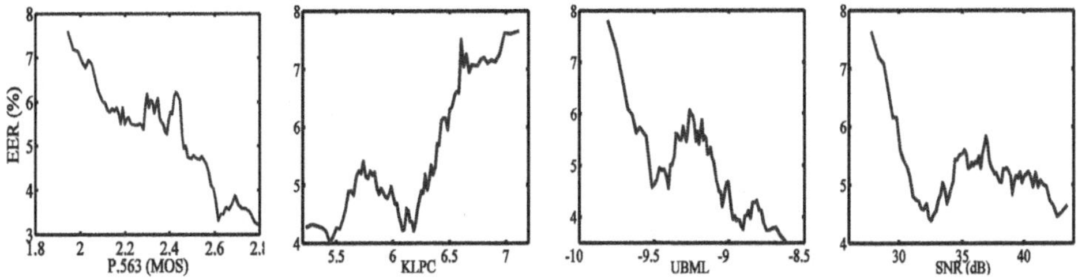

Fig. 1. EER (%) for every set of scores with a given mean indicator value: P563, KLPC, UBML and SNR, for the NIST SRE 2006 database

As we can observe, all of them show a clear relationship with the system performance, particularly UBML and P563, for which the EER decreases roughly from 8% to 4% for the set of scores with highest qualities.

4.3 Correlation Experiments

Given any two quality measures, the linear correlation coefficient among them gives an estimate of how similar is the information they provide about speech degradation in each utterance. This may be interesting in order to combine different quality measures and to optimize the available information to discriminate degraded quality samples. On the following tables we show the correlation coefficients for the five quality measures for both SRE 2006 and 2008 databases.

As we can observe, in general all correlation values are moderate. It can be observed a remarkable correlation between UBML and the measures P.563 and SNR. Since P.563 and SNR are well-known degradation indicators, this fact confirms the hypothesis stated in Section 4: UBML is an indicator of signal degradation.

Table 1. Correlation coefficients for the four quality measures: snr, klpc, p563 and ubml

	SRE 2006			SRE 2008		
	snr	klpc	ubml	snr	klpc	ubml
p563	0.136	0.192	0.223	-0.005	0.145	0.097
snr		0.182	0.386		-0.132	0.536
klpc			-0.034			-0.281

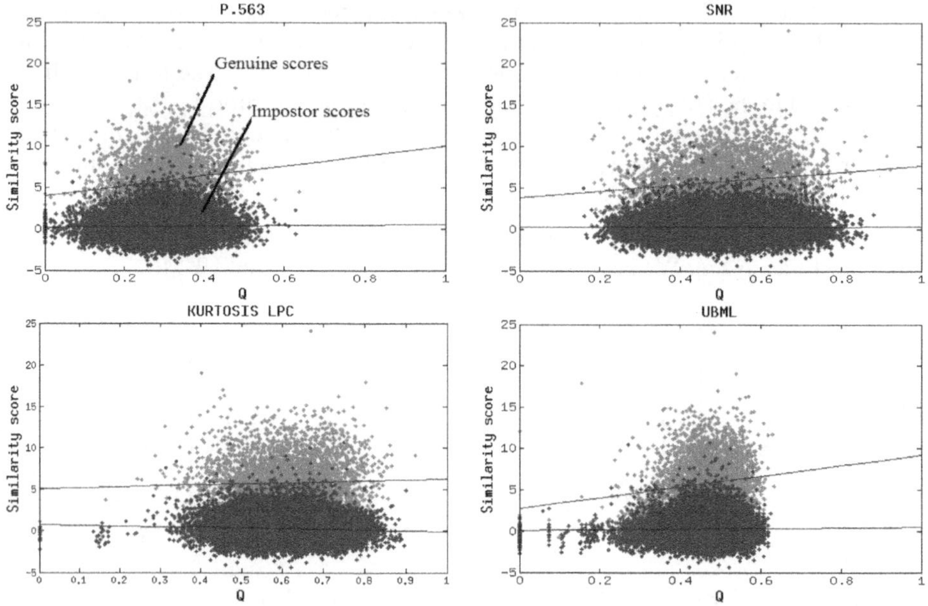

Fig. 2. Similarity scores against Q, for every quality measure for the SRE 2008 database

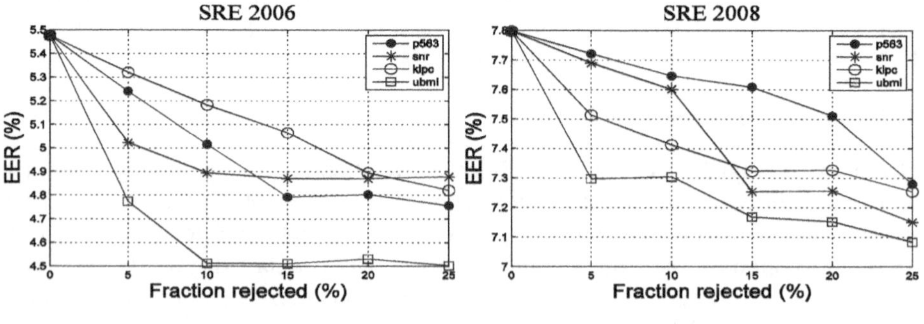

Fig. 3. EER (%) against rejected scores (%), for both NIST SRE 2006 and 2008 databases

SNR measure presents a low correlation with P.563. This may be due to the low noise level of both databases, since P.563, which selects the strongest of several degradation indicators, is not considering SNR a dominant one. However, SNR has a clear correlation with UBML, which means that the likelihood between any utterance and the UBM is quite sensitive to the noise contained in the utterance.

4.4 Utility Experiments

In this section we try to show the effectiveness of the quality measures as predictors of the system performance. We make use of two kinds of graphic representations: scores-vs-quality scatter plots and error-vs-rejection plots. On the first one, we represent the similarity score against their corresponding quality values (Q), which are obtained combining the qualities of the two involved utterances as:

$$Q = \sqrt{Q_t \cdot Q_{tr}}. \tag{9}$$

where Q_t and Q_{tr} are the quality measures of the test and train utterances.

Since better quality values are supposed to predict better results, target and non-target scores should get more separated as Q gets more close to 1. Regression lines fitted on the plots are intended to show this tendency.

As we can observe, for the quality measures P563, SNR and UBML, scores show a clear tendency to get separated for higher values of Q.

Finally, EER vs reject plots are used as recommended by NIST to show the utility of quality measures [5]. In these plots, the EER is represented against a given percentage of scores rejected with lowest quality values. The curve is supposed to decrease if the quality measure is useful as the rejection percentage increases. We have represented the results for the rejection fractions: 5, 10, 15, 20 and 25%.

We can observe that EER decreases for all the quality algorithms as we reject scores. In general, all measures perform better for the 2006 database. It is worth noting that UBML is the best performed measure for both databases, especially for 2006, where the EER decreases a 20% after rejecting the 10% of the scores.

5 Conclusions

In this paper we have analyzed the utility of several quality measures obtained from different indicators of speech degradation typically used in speech processing, namely ITU P.563 estimator of subjective quality, signal to noise ratio (SNR) and LPC Kurtosis (KLPC). We have also proposed a novel quality measure based on the likelihood of a speech segment with respect to a universal model (UBML), which measures degradation in a speech segment by its divergence with respect to such a model. Performance of the quality measures has been presented following the recommendations by NIST, and also using different databases and protocols from NIST Speaker Recognition Evaluations. In all cases, a remarkable utility has been obtained, and a moderate correlation has been observed among different quality measures. Thus, we can argue that the analyzed measures are predictors of speaker verification performance, and therefore they can be used as information in order to compensate for performance drops due to speech degradation.

Future work is mainly related with the use of the obtained quality measures for improving speaker verification performance, and also as complementary information to other data-driven approaches for session variability compensation or fusion in speaker recognition. The potential uses of the promising UBML-based quality measure will be also explored in depth. Finally, a more complete classification of quality measures for speaker verification will be also addressed, including the utility analysis of other different quality measures.

Acknowledgements

This work has been supported by the Spanish Ministry of Education under project TEC2006-13170-C02-01. The authors thank Fernando Alonso-Fernandez and Ignacio Lopez-Moreno for fruitful discussions and suggestions.

References

1. Kenny, P., Ouellet, P., Dehak, N., Gupta, V., Dumouchel, P.: A Study of Inter-Speaker Variability in Speaker Verification. IEEE Transactions on Audio, Speech and Language Processing 16(5), 980–988 (2008)
2. Brümmer, N., Burget, L., Černocký, J., Glembek, O., Grézl, F., Karafiát, M., van Leeuwen, D., Matějka, P., Schwarz, P., Strasheim, A.: Fusion of heterogeneous speaker recognition systems in the STBU submission for the NIST speaker recognition evaluation 2006. IEEE Transactions on Audio, Speech and Language Processing 15(7), 2072–2084 (2007)
3. Przybocki, M.A., Martin, A.F., Le, A.N.: NIST Speaker Recognition Evaluations Utilizing the Mixer Corpora—2004, 2005, 2006. IEEE Transactions on Audio, Speech and Language Processing 15(7), 1951–1959 (2007)
4. Ramos, D., Gonzalez-Rodriguez, J., Gonzalez-Dominguez, J., Lucena-Molina, J.J.: Addressing database mismatch in forensic speaker recognition with Ahumada III: a public real-casework database in Spanish. In: Proc. Interspeech 2008, vol. 1, pp. 1493–1496 (2008)
5. Grother, P., Tabassi, E.: Performance of Biometric Quality Measures. IEEE Transactions on Pattern Analysis and Machine Intelligence 29(4), 531–543 (2007)
6. Malfait, L., Berger, J., Kastner, M.: P.563-The ITU-T Standard for Single-Ended Speech Quality Assessment. IEEE Trans. On audio, speech and language processing 14(6)
7. Garcia-Romero, D., Fierrez-Aguilar, J., Gonzalez-Rodriguez, J., Ortega-Garcia, J.: Using Quality Measures for Multilevel Speaker Recognition. Computer Speech and Language 20(2,3), 192–209 (2006)
8. Alonso-Fernandez, F., Fierrez, J., Ortega-Garcia, J., Gonzalez-Rodriguez, J., Fronthaler, H., Kollreider, K., Bigun, J.: A comparative study of fingerprint image-quality estimation methods. IEEE Trans. on Information Forensics and Security 2(4), 734–743 (2007)
9. Grancharov, V., Kleijn, W.B.: Speech Quality Assessment. Springer Handbook of Speech Processing. Springer, Heidelberg (2008)
10. Richiardi, J., Drygajlo, A.: Evaluation of speech quality measures for the purpose of speaker verification. In: Proc. of Odyssey 2008, the ISCA Speaker and Language Recognition Workshop, Stellenbosch, South Africa (2008)
11. Mean opinion score (MOS) terminology, ITU-T Rec. P.800.1 (2003)
12. Reynolds, D.A.: Speaker Verification Using Adapted Gaussian Mixture Models. Digital Signal Processing 10, 19–41 (2000)

Impact of Prior Channel Information for Speaker Identification

C. Vaquero[1,2,*], N. Scheffer[2], and S. Karajekar[2]

[1] University of Zaragoza
Maria de Luna 1, 50018 Zaragoza, Spain
cvaquero@unizar.es
[2] SRI International,
333 Ravenswood Avenue, Menlo Park, CA 94025-3493, USA
nicolas.scheffer@sri.com,
sachin@speech.sri.com

Abstract. Joint factor analysis (JFA) has been very successful in speaker recognition but its success depends on the choice of development data. In this work, we apply JFA to a very diverse set of recording conditions and conversation modes in NIST 2008 SRE, showing that having channel matched development data will give improvements of about 50% in terms of Equal Error Rate against a Maximum a Posteriori (MAP) system, while not having it will not give significant improvement. To provide robustness to the system, we estimate eigenchannels in two ways. First, we estimate the eigenchannels separately for each condition and stack them. Second, we pool all the relevant development data and obtain a single estimate. Both techniques show good performance, but the former leads to lower performance when working with low-dimension channel subspaces, due to the correlation between those subspaces.

1 Introduction

Cepstral features with the Gaussian mixture model (GMM) is a very commonly used configuration for a speaker recognition system in NIST speaker recognition evaluations. Channel mismatch is a major problem in performance degradation and the most successful approach to addressing this problem has been joint factor analysis (JFA) [1]. JFA models the variability in the features into speaker and channel variability. As with any other statistical modeling technique, the choice of development data used for JFA is crucial for the best performance.

In this work we explore different approaches to estimating channel variability, testing them with NIST 2008 SRE data. The data contains telephone conversations recorded over telephone and over different microphones as well as interviews recorded over different types of microphones. There is a lot of development data for telephone conversations, but there is very limited data for interviews. In

* This work has been supported in part by the program FPU from MEC of the Spanish Government.

M. Tistarelli and M.S. Nixon (Eds.): ICB 2009, LNCS 5558, pp. 443–453, 2009.
© Springer-Verlag Berlin Heidelberg 2009

addition, the microphones used in telephone conversations and interviews are similar, so the development data for the former can be reused for the latter.

In our first approach we divide the development data into different sets, each belonging to a particular microphone and communication mode. JFA is performed separately for each set and the resulting estimates are stacked as a single estimate of channel variability.

In our second approach we use all the development data to obtain a single estimate of channel variability. Both approaches work well when using a dimension high enough to model all channel subspace variability, but with a lower dimension the single estimate obtains better results, since the first approach shows correlation between different channel subspaces. Further analysis on correlation is done to show the importance of selecting the dimension for every channel subspace when stacking channel estimates.

2 System and Experimental Protocol

We describe the JFA system and the experimental protocol.

2.1 Joint Factor Analysis System

Let us define the notations that will be used throughout this discussion. The JFA framework uses the distribution of an underlying GMM, the universal background model (UBM) of mean m_0 and diagonal covariance Σ_0. Let the number of Gaussians of this model be N and the feature dimension be F. A supervector is a vector of the concatenation of the means of a GMM: its dimension is $N \times F$. The speaker component of the JFA model is a factor analysis model on the speaker GMM supervector. It is composed of a set of eigenvoices and a diagonal model. Precisely, the supervector m_s of a speaker s is governed by

$$m_s = m_0 + Vy + Dz \tag{1}$$

where V is a tall matrix of dimension $NF \times R_S$, and is related to the eigenvoices (or speaker loadings), that spans a subspace of low-rank R_S. D is the diagonal matrix of the factor analysis model of dimension NF. Two latent variables y and z entirely describe the speaker and are subjected to the prior $N(0, 1)$. The nuisance (or channel) supervector distribution also lies in a low-dimensional subspace of rank R_C. The supervector for an utterance h with speaker s is

$$m_h = m_s + Ux \tag{2}$$

The matrix U, known as the eigenchannels (or channel loadings), has a dimension of $NF \times R_C$. The loadings U, V, D are estimated from a sufficiently large dataset, while the latent variables x, y, z are estimated for each utterance.

To train the matrices, several iterations of the expectation maximization (EM) algorithm of the factor analysis framework are used. An alternative minimum divergence estimation (MDE) is used at the second iteration to scale the latent variables to a $N(0,1)$ distribution. To train a speaker model, the posteriors of x, y, z are computed using a single iteration (via the Gauss-Seidel method as in [2]).

The verification score for each trial was a scalar product between the speaker model mean offset and the channel-compensated first-order Baum-Welch statistics centered around the UBM. This scalar product was found to be simple yet very effective [3] and was subsequently adopted by the JHU fast scoring group [4].

2.2 System Description

We have used a gender-dependent JFA system for all experiments. The UBM consists of a GMM of 512 gaussians trained on telephone conversations of about 3 minutes duration each from the NIST SRE04 database [5]. The male UBM was trained on 1970 conversations, while the female UBM was trained on 2741 conversations. A 300-rank eigenvoice matrix was trained to model the speaker subspace, using telephone conversations from NIST SRE04 and Switchboard II databases. Male speaker subspace was trained on recordings from 1368 speakers, while female speaker subspace was trained on recordings from 992 speakers.

For training channel subspace, three different datasets were considered: for tackling conditions involving telephone data, 1675 conversations from 118 male speakers and 2409 conversations from 183 female speakers from the NIST SRE04 were used. For dealing with conditions involving microphone data, 1207 and 1414 telephone conversations from 44 male and 52 female speakers, respectively, recorded over different microphones from the alternate microphone data of the NIST SRE05 were used. To have some prior information of interview data type, the interview development data from NIST SRE08 was used. This dataset contains long interview recordings over several microphones from three male (138 recordings) and three female speakers (141 recordings).

For scoring normalization zt-norm was used. We used 207 male and 292 female speaker models from the NIST SRE04 database for t-norm and 1374 male and 1770 female speaker segments for z-norm from the same database.

2.3 Experimental Protocol

Results are reported on the NIST SRE08 required condition, short2-short3. This condition takes just one session of the target speaker for enrollment and one session for testing. Short2-short3 is divided into several conditions, and we are interested in four of them:

- Two conditions using interview sessions for enrollment:
  - Condition 1, involving 34181 microphone channel matched trials, testing on interview sessions
  - Condition 4, involving 11741 microphone-telephone trials, testing on telephone calls

- Two conditions using telephone calls for enrollment:
  - Condition 5, involving 8454 telephone-microphone trials, testing on telephone calls recorded over different microphones.
  - Condition 6, involving 35896 telephone channel matched trials, testing on telephone calls.

Results are reported in terms of both equal error rate (EER) and detection cost function (DCF) as described in NIST SRE.

3 Channel Compensation Using Different Types of Data

We investigate the importance of having matching data to perform channel compensation in a JFA system.

3.1 Baseline System without Channel Compensation

To analyze the improvement obtained in performance when using channel compensation, we consider as baseline the JFA system without channel compensation. For this purpose, a classic Maximum a Posteriori (MAP) system was used.

The results for interview and telephone conditions are shown in Table 1 where, for two reasons, the former are worse than the latter. First, the interview condition uses microphones that are different from telephone and second, the background data is made up entirely of telephone conversation.

Table 1. Results of a classic MAP on NIST-SRE-2008, where results with interview data (condition 1) are worse than telephone data (condition 6)

Train-Test (Condition)	DCF	EER
Interview-Interview (1)	0.639	16.898
Interview-Telephone (4)	0.793	19.276
Telephone-Interview (5)	0.404	11.549
Telephone-Telephone (6)	0.523	11.725

3.2 JFA with Telephone Data

Table 2 show the results for four conditions with JFA performed with telephone development data. The results show that condition 1, which has the biggest mismatch with the development data, improves by only about 5% and is almost insensitive to the rank of the eigenchannel matrix. On the other hand, condition 6, which is the most matched condition, improves by almost 40% and increased the rank of about 500 gives the best performance on this condition.

The mixed data conditions show an interesting trend. When telephone data is used in testing (condition 4), the improvement is about 15%. In addition, the effect of using more eigenchannels is negligible. When telephone data is used in training (condition 5), the improvement is more. It is around 30% with improvement in performance with as many as 200 eigenchannels. There is no improvement from adding more eigenchannels. We discuss the mixed results further in Section 5.

Table 2. Results for several ranks on all conditions, using telephone development data for training channel compensation. Best results according to a trade-off between performance and matrix rank are emphasized. DCF (x10)/EER(%).

Rank	Condition 1	Condition 4	Condition 5	Condition 6
50	0.609/15.771	0.824/16.742	0.348/8.560	0.356/7.506
100	**0.582/15.355**	0.827/17.104	0.309/7.609	**0.350/6.908**
200	0.579/15.572	**0.821/16.471**	**0.298/6.997**	0.347/6.834
300	0.578/15.607	0.821/17.195	0.305/7.201	0.337/6.796
500	0.576/15.797	0.821/17.376	0.311/7.337	0.336/6.684

3.3 JFA with Microphone Data

Table 3 shows the results for four conditions with JFA performed with microphone data. Note that this data is more matched to condition 1 than condition 6. The difference between condition 1 and the development data is that the former is an interview and the latter is a conversation. As expected, the performance of condition 1 improves by almost 75% and it gets better with higher rank. The performance of condition 6 improves only slightly and with smaller-rank eigenchannels.

Table 3. Results for several ranks on all conditions, using microphone development data for training channel compensation. Best results according to a trade-off between performance and matrix rank are emphasized. DCF (x10)/EER(%).

Rank	Condition 1	Condition 4	Condition 5	Condition 6
50	0.341/6.352	0.652/13.394	0.346/8.288	0.537/11.202
100	0.319/5.789	0.612/13.032	0.321/8.084	0.536/11.016
200	**0.278/5.199**	**0.582/11.493**	0.302/7.473	**0.521/10.680**
300	0.269/5.017	0.575/11.222	0.290/7.745	0.52/10.792
500	0.270/4.991	0.568/10.679	**0.279/7.405**	0.518/10.754

The same trend is seen with mixed conditions. This trend is opposite of the trend seen with JFA using only telephone data (Table 2). The results show that there is about a 50% improvement in the performance when interview data is used in training (condition 4) with the results getting better with more eigenchannels. The latter can be also seen with condition 5 (interview data used in testing only), but the gains are about 30%. As mentioned earlier we will elaborate on the mixed conditions in Section 5.

3.4 JFA with Interview Data

Table 4 shows the results for four conditions with JFA performed with interview development data. Although it is matched to condition 1, it was very sparse. It had only six speakers (three male and three female) with different microphones. This is reflected in the experiments by the reduced rank of eigenchannels that

Table 4. Results for several ranks on all conditions, using interview development data for training channel compensation. Best results according to a trade-off between performance and matrix rank are emphasized. DCF (x10)/EER(%).

Rank	Condition 1	Condition 4	Condition 5	Condition 6
50	**0.446/9.194**	**0.728/15.656**	**0.428/10.734**	**0.558/12.360**
75	0.441/9.09	0.739/15.837	0.416/10.598	0.559/12.397
100	0.435/9.004	0.731/15.747	0.417/10.734	0.555/12.360
125	0.434/8.917	0.732/15.656	0.415/10.598	0.552/12.360

could be estimated from this data. Results are similar to Table 3; performance of condition 1 improves with increased rank. Performance of condition 6 is worse with these eigenchannels. Among mixed conditions, there is about a 20% improvement when interview data is used for training. When interview data is used for testing there is a very small improvement.

4 Importance of Matched Development Data

Some important conclusions can be made by comparing different results from the previous section. First is the importance of matched development data for getting the best performance. The results show that the best results on the telephone data are obtained with JFA trained with telephone data. It is the same with interview data where JFA trained with microphone and interview data gives better performance than JFA trained with telephone data. The results are not obvious in the sense that any statistical technique relies on the match between development and evaluation data. The results are interesting in the case of mismatch. These results show that JFA is not very effective with the mismatched development data.

Second, the mixed conditions (4 and 5) prefer matched development data for training condition. For example, when telephone development data is used, the best performance is obtained on condition 5. When microphone development data is used, the best performance is obtained on condition 4. The results on mixed conditions also show an interesting trend. These results improve with the addition of eigenchannels but only up to about 200 of them.

Third, note that interview data differs from telephone data in two aspects: recording microphones and communication mode. As mentioned earlier, the interview development data is very small compared to microphone development data. However, the results show that the microphone data is more representative of the interview condition than the interview data. One hypothesis about these results is that interview does not differ from conversation as a communication mode but the difference between microphones is what makes two conditions different. This effect was seen in FRTIV data collection [6]. It will be interesting to investigate further with this hypothesis.

5 Building a Robust System to Compensate for Unseen Channels

The JFA approach can take advantage of having different datasets in order to compensate for unseen channels as linear combinations of known channels. This can be achieved in two ways: it is possible to train different eigenchannel matrices on every dataset separately and then stack them as a single matrix, or channel compensation can be trained on the whole merged dataset.

5.1 Stacking Channel Matrices

Stacking channel matrices is an interesting approach to modeling different channels with one eigenchannel matrix, as it is a modular way to deal with many datasets. In this approach, one eigenchannel matrix is trained for every dataset, and all these matrices are stacked together. This approach is nondestructive and enables the model to select the dimension that best fits the data.

However, selecting for every database the dimension that best fits the data will lead us to obtain large eigenchannel matrices, so in order to avoid impacting the system speed when stacking many matrices, we choose for every data set that dimension obtaining best performance according to a trade-off between performance and matrix rank. Finally, we choose a 100-rank telephone (see Table 2), a 200-rank microphone (see Table 3) and a 50-rank interview eigenchannel matrix (see Table 4) to be stacked, and they will be stacked in two steps: first we will stack telephone and microphone matrices, obtaining a matrix of rank 300, and then we will add the interview matrix, obtaining a final rank of 350.

Table 5. Results obtained stacking eigenchannel matrices trained on three different datasets: telephone (phn), microphone (mic), interview (int). Results in terms of DCF(x10)/EER(%).

System trained on	Condition 1	Condition 4	Condition 5	Condition 6
Best single dataset	0.278/5.199	0.582/11.493	0.309/7.609	0.336/6.684
Phn, mic	0.264/5.277	0.483/9.321	0.203/4.959	0.339/6.311
Phn, mic, int	**0.211/4.263**	**0.460/9.050**	**0.198/4.775**	**0.332/6.263**

Table 5 shows results obtained for channel-matched conditions as channel matrices are stacked. For every condition, results for the best-performing system among the 100-rank telephone, 200-rank microphone and 50-rank interview are shown for comparison. We can see that stacking matrices does not decrease performance for matched conditions, while it brings a great improvement in mismatched conditions.

We can see as well that the interview data provides additional information, which helps to improve performance on conditions involving interview. Indeed, a 1% absolute gain is observed at the EER for condition 1. This highlights again the importance of having prior knowledge on the data for NIST evaluation and for realistic scenarios.

5.2 Retraining Channel Compensation

Retraining channel compensation is the easiest way to assure good performance across every channel, but it has an important disadvantage: it takes a lot of time to retrain an eigenchannel matrix. Indeed, matrices with very high rank should be trained to model all variability contained in several databases.

Table 6. Results obtained retraining channel compensation merging different datasets: telephone (phn), microphone (mic), interview (int). Results in terms of DCF(x10)/EER(%).

System trained on	Condition 1	Condition 4	Condition 5	Condition 6
Best single dataset	0.278/5.199	0.582/11.493	0.309/7.609	0.336/6.684
Phn, mic	0.248/5.442	0.476/9.683	**0.202**/5.095	0.334/6.161
Phn, mic, int	**0.212/4.255**	**0.444/9.321**	0.206/**4.823**	**0.330/6.049**

Table 6 shows results when retraining a 300-eigenchannel matrix using telephone and microphone datasets, and a 350-eigenchannel matrix on all data available. Comparing these results with those obtained in Table 5 we can see that it is possible to obtain similar performance by stacking eigenchannel matrices as by retraining a new eigenchannel matrix. Moreover, stacking matrices is a practical answer for enriching channel modeling without having to retrain on all data.

However, when low-rank matrices are used, stacking is not as good as retraining, especially on cross-channel conditions. This can be seen on Table 7, where results of stacking 50 phone, 50 microphone, and 50 interview eigenchannels are compared to those obtained by retraining 150 eigenchannels.These results show the importance of rank selection for every channel subspace before stacking, as has been done in this work.

This behavior is explained by the correlation between channel subspaces.

Table 7. Comparison between stacking and retraining for low dimension channel subspaces. Results in terms of DCF(x10)/EER(%).

Channel compensation	Condition 1	Condition 4	Condition 5	Condition 6
Stacking (all datasets)	0.229/4.835	0.501/10.226	0.226/5.571	**0.325**/6.647
Retraining (all datasets)	**0.228/4.645**	**0.463/9.140**	**0.212/4.891**	0.339/**6.497**

5.3 Correlation between Channel Subspaces

When stacking eigenchannel matrices, some of the vectors stacked may be correlated, thus providing redundant information as well as very high ranks. To analyze the correlation between channel subspaces, an orthonormal base of the subspace is estimated for every eigenchannel matrix. We call these subspaces *phone*, *microphone*, and *interview* subspace. We use Singular Value Decomposition (SVD), as every low-rank $m \times n$ matrix M can be represented as:

$$M = U \Sigma V^* \tag{3}$$

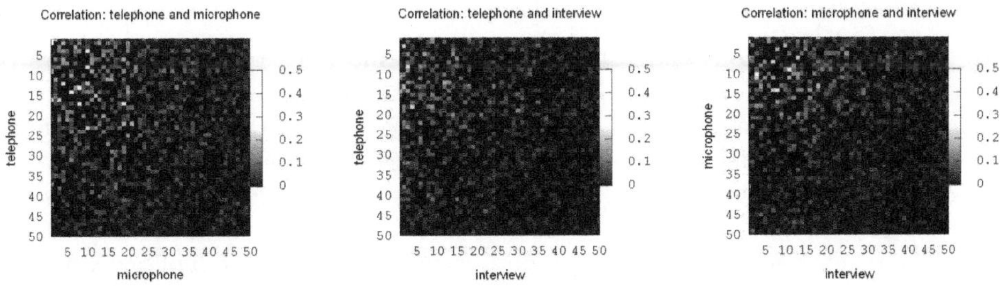

Fig. 1. Correlation between channel subspaces, trained on male data. The importance of the eigenvectors is ranked with respect to their singular values.

where U is an $m \times m$ unitary matrix containing a set of orthonormal basis vectors for the output space, Σ is a diagonal $m \times n$ matrix containing the singular values of the transformation defined by M, and V is an $n \times n$ matrix containing a set of orthonormal basis vectors for the input space.

Modeling the channel in this way allows us to get the correlation between channel matrices by just projecting one orthonormal channel matrix onto another. Figure 1 shows the correlation between phone, microphone, and interview subspaces, trained on male data, on a dimension reduced to 50. This correlation is higher between microphone and interview, and between telephone and microphone, while lower between telephone and interview. We can see that correlation appears mainly in those directions corresponding to the highest singular values.

When pooling all data together, the training method can take advantage of this correlation and model the channel directions more precisely. However, as more variability directions are modeled (i.e., rank is increased), the stacked approach tends to perform as well as the retraining approach.

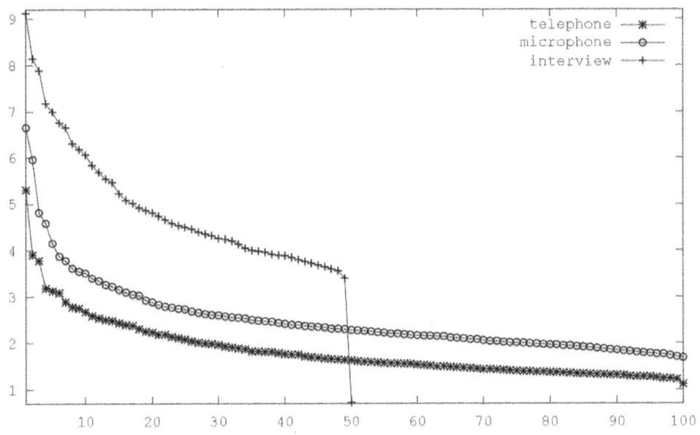

Fig. 2. First 100 singular values for every channel subspace, trained on male data. Interview and microphone subspaces shadow the telephone subspace.

5.4 Variability of Channel Subspaces

To remove correlation when stacking different matrices to obtain a low-rank stacked matrix, we can stack higher-rank matrices that are known to obtain good performance and then reduce the rank by means of SVD, keeping those directions showing higher variability (higher singular values).

However, analyzing Figure 2, which shows the first 100 singular values for every channel subspace, trained on male data, it can be noticed that microphone and interview channels have more variability, so that telephone variability directions will be shadowed and removed when reducing the rank, decreasing performance on those conditions involving telephone channel.

6 Conclusions

The problem of missing prior knowledge on the channels involved in the speaker verification task was studied. We showed the major impact of this information as without any information. A state-of-the art system would have the performance of a system not using channel compensation.

Then, we investigated two approaches for merging information from several datasets to build a more robust system. The first approach consists of stacking matrices trained with different datasets, which is a modular solution, allowing us to deal with multiple datasets with different types of recording for training channel compensation. The second approach consists of retraining the whole eigenchannel matrix.

The stacking approach was as good as the retraining approach when using high-rank eigenchannel matrices, but worse, especially on channel mismatched conditions, when using low-rank eigenchannel matrices. A further analysis has shown that this effect is probably due to the correlation present between different channel subspaces, correlation that the retraining approach may detect and take into account to obtain the final channel variability directions but that the stacking approach cannot get rid of.

Finally, a study on the singular values of telephone, microphone, and interview channel subspaces has shown that it is not possible to take advantage of the correlation for reducing high-rank stacked matrices to a desired rank. Directions from channels having high variability will shadow directions from channels showing less variability, losing performance in conditions having low variability.

As a final conclusion, we can assert that in order to assure performance as good for stacking matrices as for retraining channel compensation, a previous analysis on the optimal rank for every matrix to be stacked is needed. Stacking will lead to higher rank matrices than retraining, but it will assure good performance for every condition.

References

1. Kenny, P., Ouellet, P., Dehak, N., Gupta, V., Dumouchel, P.: A Study of Inter-Speaker Variability in Speaker Verification. IEEE Trans. Audio, Speech and Language Processing 16(5), 980–988 (2008)

2. Vogt, R., Baker, B., Sridharan, S.: Modelling session variability in text-independent speaker verification. In: Ninth European Conference on Speech Communication and Technology, ISCA (2005)
3. Brümmer, N.: SUN SDV system description for the NIST SRE 2008 evaluation, Montreal, Canada (2008)
4. JHU: Johns Hopkins University, Summer workshop, Robust Speaker ID, Fast scoring team, Baltimore, MD (2008)
5. NIST: The NIST year 2005 speaker recognition evaluation plan (April 2004), http://www.nist.gov/speech/tests/spk/2004/SRE-04_evalplan-v1a.pdf
6. Shriberg, E., Graciarena, M., Bratt, H., Kathol, A., Kajarekar, S., Jameel, H., Richey, C., Goodman, F.: Effects of Vocal Effort and Speaking Style on Text-Independent Speaker Verification. In: Proceedings of Interspeech, Brisbane, Australia (2008)

Minimising Speaker Verification Utterance Length through Confidence Based Early Verification Decisions

Robbie Vogt and Sridha Sridharan*

Speech and Audio Research Laboratory,
Queensland University of Technology, Brisbane, Australia
{r.vogt,s.sridharan}@qut.edu.au

Abstract. This paper presents a novel approach of estimating the confidence interval of speaker verification scores. This approach is utilised to minimise the utterance lengths required in order to produce a confident verification decision. The confidence estimation method is also extended to address both the problem of high correlation in consecutive frame scores, and robustness with very limited training samples. The proposed technique achieves a drastic reduction in the typical data requirements for producing confident decisions in an automatic speaker verification system. When evaluated on the NIST 2005 SRE, the early verification decision method demonstrates that an average of 5–10 seconds of speech is sufficient to produce verification rates approaching those achieved previously using an average in excess of 100 seconds of speech.

1 Introduction

A number of practical issues inevitably arise in the process of deploying a speaker verification system. Typically these difficulties involve determining system parameters such as the required quantities of speech for adequately trained models and for accurate verification trials, as well as deciding an appropriate decision threshold to achieve the required verification error rates. Despite the importance of such decisions, very limited speaker verification research has been published that specifically address these issues. This work focuses on the issue of test utterance length.

Ideally, a verification system would produce a verification confidence from a trial, as this is the most useful and usable result from a system designer perspective: Knowing that there is a 96% probability that an utterance was produced by speaker s makes it easy for a designer to employ Bayesian logic to produce the best possible system. There are two distinct impediments to this: Firstly, accurately estimating the prior probability of a true trial is problematic due to the difficulties in identifying and quantifying the non-target class, and secondly, scores produced by verification systems would need to be representational of

* The authors would like to acknowledge the collaborative contribution of Torqx Pty. Ltd. on this research. This research was supported by the Australian Research Council Discovery Grant No DP0877835.

M. Tistarelli and M.S. Nixon (Eds.): ICB 2009, LNCS 5558, pp. 454–463, 2009.
© Springer-Verlag Berlin Heidelberg 2009

true likelihoods ratios, which is rarely the case for automatic speaker recognition systems.

Prompted by the importance of presenting meaningful results in forensic applications, recent work has begun to address the production of accurate likelihood ratios [1] and the interpretation of scores that are *not* likelihood ratios [2]. Also, the analysis and evaluation of speaker verification systems based on the accuracy of output likelihood ratios is also a topic of recent interest [3]. Regardless, speaker verification systems do not in general produce scores that should be interpreted as true likelihood ratios.

Given these difficulties with determining an accurate verification confidence, an alternative approach pursued in this work is to determine a method by which one can state that the "true" verification score for a trial lies within the interval $\Lambda_S = a \pm b$ at, for example, the 99% confidence level. Here the "true" verification score is defined as the score that the verification system would produce given an infinite quantity of testing speech. The Early Verification Decision (EVD) method, first proposed in [4], exploits this verification score confidence interval to make confident verification decisions with minimal speech based on a specified threshold. This paper expands substantially on [4] and additionally investigates operating at the minimum DCF threshold and the interaction of the EVD method with Z-norm score normalisation.

The following section describes the baseline speaker verification system used in this paper and explores the effect on performance of reducing the available test data. Section 3 then presents the EVD method for minimising test utterance length by estimating confidence intervals on the speaker verification score. Several methods of estimating the verification score confidence interval are then developed including an extension to incorporate Z-norm score normalisation. Experimental evaluation of these estimates are presented in Section 4.

2 Baseline System and Experimental Setup

The verification system used in this study is a GMM-UBM system with inter-session variability modelling as described in [5]. The verification score used for this system is the expected log-likelihood ratio of the target speaker to the UBM. The expectation is taken over the individual frame-based log-likelihood ratios for the test utterance,

$$\Lambda_S = \frac{1}{T}\sum_{t=1}^{T} \ell_S(t) = \frac{1}{T}\sum_{t=1}^{T} \log\left(\frac{p(\boldsymbol{x}_t|\lambda_S)}{p(\boldsymbol{x}_t|\lambda_{ubm})}\right) \quad (1)$$

where, $p(\boldsymbol{x}|\lambda)$ is the standard GMM density.

This system uses explicit inter-session variability modelling [5] in the training procedure to mitigate the effects of mismatch, however session variability was not considered during testing. This configuration was chosen to have performance representative of the current state-of-the-art but avoiding the complication of estimating the session conditions of the testing utterance. Additionally, Z-Norm score normalisation [6] was applied to this system.

Table 1. The effect of shortened test utterances on verification performance

System	No Normalisation		Z-Norm Normalisation	
	Min. DCF	Act. DCF	Min. DCF	Act. DCF
Reference	.0293	.0293	.0249	.0249
20 sec	.0391	.0422	.0368	.0406
10 sec	.0489	.0601	.0482	.0636
5 sec	.0616	.0976	.0626	.1031
2 sec	.0794	.1770	.0810	.1851

Experiments were conducted on the 2005 NIST SRE protocol using conversational telephony speech drawn from the Mixer corpus [7]. The focus of these results is on the 1-side training, common evaluation condition of this corpus.

2.1 The Effect of Short Verification Utterances

While researchers typically prefer as much data as possible to make the most reliable verification decision possible, system designs desire utterances to be as short as possible to minimise the inconvenience for the user. Compromise is usually necessary. Thus, an understanding of the impact of limiting verification utterance lengths is important. Table 1 assesses this impact for the baseline system. These results demonstrate that utterance length, predictably, has a significant effect on overall system performance in the range that is typically of interest for a system designer, as previously observed [8].

Table 1 presents both the minimum DCF value as well as the actual DCF value if the optimal threshold of the reference system is chosen. The substantial difference between the minimum and actual detection costs can be seen to be increasing as the utterance length is reduced, to the extent that it is more costly to use the 2-second system with the best threshold for the reference system than reject every verification claim *a priori* (this gives a DCF of 0.1). These numbers also highlight the difficulty of choosing a suitable threshold as this choice is evidently affected by the choice of utterance length.

Results including Z-Norm score normalisation are included in the rightmost columns of Table 1. This application shows a clear advantage for the reference system with 15% reduction in DCF. This advantage, however, is less apparent when shortened utterances are used. This is particularly apparent in the case of the actual DCF results using 2 and 5 second utterances where the application of Z-Norm has a detrimental effect.

3 The Early Verification Decision Method

The aim of the Early Verification Decision (EVD) method is to minimise the amount of speech required to make a verification decision. This is achieved by making a verification decision as soon as we are confident the "true" verification score is above or below the specified threshold based on the *confidence interval* of the current estimated score.

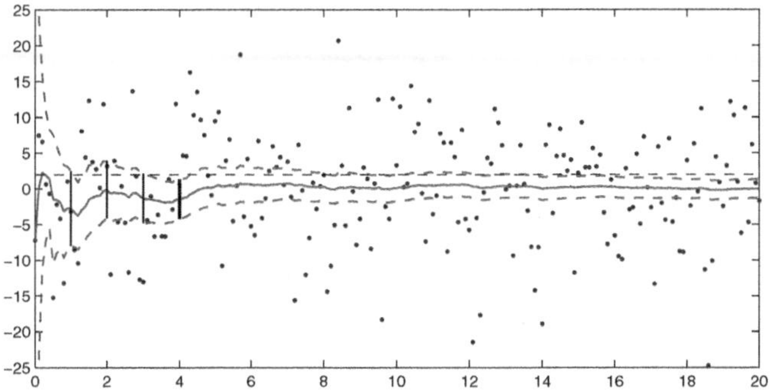

Fig. 1. Example verification trial using the early decision method. After observing only 4 seconds of speech a *reject* decision can be made.

The current verification score estimate is assumed to be a random variable drawn from a distribution with the "true" score as mean. To determine the confidence interval it is thus necessary to determine the variance of this distribution. This variance is usually dependent on many factors such as whether a trial is a genuine or impostor trial (which we can not know *a priori*), the length of a particular verification utterance and the noise and other environmental conditions of the recording. Consequently, the variance must be estimated individually for each verification trial using the observed sequence of frame scores as the fundamental statistics for this estimation. This estimation forms the basis of the EVD method and is addressed in the next section.

An example of the early verification decision process is presented in Fig. 1. In this figure, the samples (frame scores) used to estimate the distribution are represented as dots, the evolving mean verification score estimate is shown as a thick red line and the 99% confidence interval of this estimate depicted with dashed lines above and below the estimate. The verification threshold is shown as a horizontal line through the centre of the figure. After a couple of seconds of the trial the estimate of the verification score is quite eratic, which is reflected in the wide confidence interval, but looks to be converging to a point below the threshold. By four seconds the estimate seems to be more stable as more samples become available and the width of the confidence interval has narrowed to be entirely below the threshold. At this point, *after only four seconds*, we can be confident that the verification score will continue to lie below the threshold and thus make a reject decision for this trial. The subsequent part of the trial confirms that the verification score does in fact continue to lie below the threshold and the confidence interval continues to narrow, even though the entire confidence interval does not necessarily lie below the threshold at all times.

3.1 Variance Estimation Approaches

As detailed in [4], the crux of confidence-based EVD method is the ability to estimate confidence intervals on the ELLR score. This ability in turn relies on estimating the variance of the ELLR estimate distribution from the sequence of observed frame scores. To do this, it is assumed that the observed verification score is a random *process* that evolves over time. It is assumed that this random process is Gaussian at time t, has a fixed mean (the "true" score) and a time-dependent variance, that is

$$\Lambda_S(t) \sim \mathcal{N}\left(\mu_S, \sigma_S^2(t)\right) . \tag{2}$$

Presented in [4] were several methods for estimating $\sigma_S^2(t)$ of this process. These methods are summarised here.

Firstly, the *Naïve* approach exploits the central limit theorem and the fact that the verification score is a sum of the frame scores, which in this case are assumed to be i.i.d. random variables. Thus, if $\ell_S(t)$ has sample mean m_ℓ and variance s_ℓ^2, the ELLR verification score will have a mean and variance approximated by

$$\mu_S = m_\ell \qquad\qquad \sigma_S^2 = \frac{s_\ell^2}{T-1} \tag{3}$$

The *Decorrelated* variance estimate attempts to compensate for the high level of correlation between consecutive acoustic feature vectors and, consequently, frame scores. This compensation is achieved through a transformation approach to reduce the correlation by producing a series of ELLR estimates $\boldsymbol{y}_S$ from short, fixed-length, non-overlapping frame sequences,

$$y_S(i) = \frac{1}{N} \sum_{t=Ni}^{N(i+1)-1} \ell_S(t) \tag{4}$$

where N is the length of the short frame sequences. If N is sufficiently large, the correlation between successive $y_S(i)$ drops to a negligible level.

From $\boldsymbol{y}_S$, it is then possible to estimate the overall ELLR mean and variance

$$\mu_S = m_y \qquad\qquad \sigma_S^2 = \frac{s_y^2}{T/N-1} \tag{5}$$

where m_y and s_y^2 are the sample mean and sample variance of y_S respectively.

Finally, for the EVD approach to be effective, it is particularly important to robustly estimate the variance of the frame scores with a very limited number of samples. This issue is also exacerbated by the correlated nature of these scores. In this work a more robust variance estimate is produced through Bayesian estimation and introducing *a priori* information. This *With Prior* estimate is given by

$$\hat{s}^2 = \frac{\tau\kappa^2 + (M-1)s^2}{\tau + (M-1)}, \tag{6}$$

where s^2 is unbiased sample variance from M samples and κ^2 and τ are hyper-parameters of the prior distribution, which takes the form of a Dirichlet distribution [9]. This estimate can then be used to produce more robust estimates of the ELLR variance using either (3) or (5) above.

Table 2. Results at the Actual DCF operating point for the EVD method

System	Act. DCF	Trial Length		Shortcut Errors	
		Median	Mean	Impostor	Target
Reference	.0293	103.4	103.4	–	–
Naïve					
90% Conf.	.1032	2	2.9	7.2%	22.1%
99% Conf.	.0600	3	5.4	2.9%	13.2%
99.9% Conf.	.0427	4	8.4	1.3%	7.7%
Decorrelated	$N = 10$				
90% Conf.	.0701	2	4.4	3.9%	15.6%
99% Conf.	.0369	5	11.3	0.7%	4.8%
99.9% Conf.	.0314	9	17.7	0.2%	1.4%
With Prior	$\tau = 100, \kappa^2 = 0.25$				
90% Conf.	.0583	3	5.4	2.7%	12.9%
99% Conf.	.0325	7	13.1	0.3%	3.0%
99.9% Conf.	.0302	11	19.8	0.1%	0.9%

3.2 Verification Score Normalisation

Typically, raw scores output by speaker verification systems are further processed to normalise for factors such as the quality of the trained speaker model, mismatch between the training and testing conditions and the linguistic content in the test utterance. Z-Norm [6] is an example of a score normalisation technique that normalises the verification score by the mean and variance of the speaker model's response to a set of impostor trials.

It is straight forward to apply Z-Norm to the applications described above as it can be characterised as a simple linear transform of the frame-based scores. If the Z-Norm statistics are given by μ_Z and σ_Z then the normalised ELLR score is given by,

$$\Lambda_Z(s) = \frac{\Lambda(s) - \mu_Z(s)}{\sigma_Z(s)} = a\Lambda(s) + b \tag{7}$$

where $a = 1/\sigma_Z(s)$ and $b = -\mu_Z(s)/\sigma_Z(s)$. As the ELLR score is a scaled sum of the frame scores, this transform can alternatively be applied directly to the individual frame scores,

$$\ell'_S(t) = a\ell_S(t) + b; \qquad \Lambda_Z(s) = \frac{1}{T}\sum_{t=1}^{T} \ell'_S(t). \tag{8}$$

It is then straightforward to apply any of the estimates in (3) through (6) above using the transformed frame scores.

4 Experimental Results

The results of using the EVD scoring approach are presented in Table 2 for the *Naïve*, *Decorrelated* and *With Prior* variance estimates with the threshold set to minimise the NIST Detection Cost Function (DCF) for the reference system. Performance is shown at three confidence levels, 90%, 99% and 99.9%, which are the minimum confidence with which the "true" verification score must be above or below the DCF threshold for the system to make an early verification decision. Also included are the results for the reference system, replicated from Table 1. These results do not include Z-norm score normalisation at this stage. The performance of the systems in Table 2 are measured by the *actual* DCF value achieved at the specified threshold. Also included are measures of the average utterance length for each system, measured by both the mean and median[1] statistics.

It can be seen from the actual DCF results that the performance of the EVD approach drops behind that of the reference system in all cases—sometimes dramatically so—but this drop is both expected and actually quite small when the utterance lengths are also taken into consideration. This can be readily seen by comparing the results of Tables 1 and 2. For example, the *Decorrelated 99.9%* system shows a 7% relative drop in actual DCF but achieves this performance with less than a tenth of the utterance length for most trials (median length of 9 seconds). In contrast, using a similar but fixed utterance length of 10 secnods results in an actual DCF of .0601 (Table 1); this is more than a 100% increase in DCF.

Further analysing the utterance length statistics for the EVD systems, there is a consistent discrepancy between the mean and median statistics as the mean length are considerably longer in each case. This indicates a signifcantly skewed distribution of utterance lengths and that the *mean* test utterance lengths are dominated by a relatively small number of long trials. For the *Naïve* EVD systems, the majority of trials provide a result within 2, 3 or 4 seconds, with increasing confidence level, as indicated by the *median* trial lengths in Table 2.

This last point has an amazing implication: For the majority of trials a text-independent speaker verification system will produce the same decision with only *a few seconds* of speech that it will with almost 2 *minutes* of speech.

Fig. 2 is a DET plot of the *Naïve* EVD systems at differing confidence levels. Also shown is the DET curve for the baseline reference system using all available speech and a system using a fixed 2-second utterance length (dotted curve) as a "worst case" system.[2] For all systems the operating point at the specified threshold is highlighted with a circle.

Interestingly, the DET curves for these systems veer away from the reference system the farther they are from the DCF operating point. The performance curves of the early decision systems drop back toward the 2-second worst-case system in these areas. This effect is even more dramatic at the DCF operating point than for the EER, as explored in [4]. This characteristic is a direct consequence of the EVD method as the system is only interested in the performance at the specified threshold and essentially trades performance in other areas for shorter test utterances.

By comparing the Tables 1 and 2 it can be seen that the EVD method is effective in trading performance at a specific operating point for shorter trials. It is also evident increasing the required confidence level provides an improved DCF for each of the EVD methods, demonstrating that setting the confidence level

[1] The median utterance length for the EVD systems always falls on a whole-second increment as the EVD implementation used in these experiments only tests the stopping criteria at 1-second intervals.

[2] The EVD systems were restricted to a 2 sec minimum utterance length.

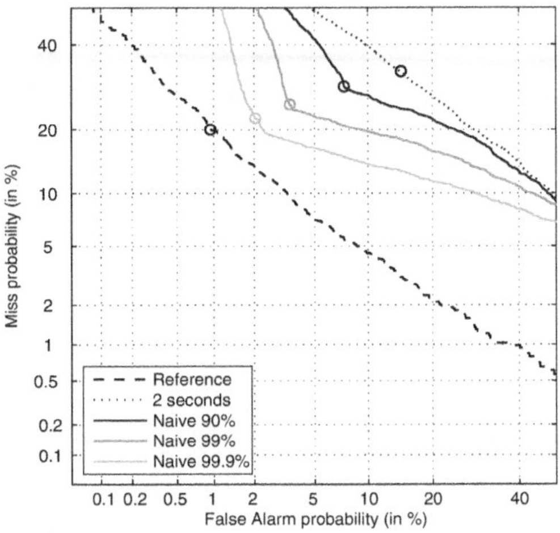

Fig. 2. DET plot using the naïve method at the minimum DCF threshold

is a viable method of controlling the trade-off between verification performance and utterance length.

The two rightmost columns of Table 2 quantify the errors introduced by the early decision criteria for impostor and target trials, respectively. These represent the trials for which the reference system and the EVD system have produced differing decisions. This is the approximate loss introduced by the early decision methods and, if the distribution assumptions and estimates are accurate, should closely match the confidence levels specified.

It can be seen from these results that the error rates for the *Naïve* system do not match the specified confidence levels well, particularly as the confidence is increased. The fact that the error rates don't reflect the desired confidence levels suggests that the *Naïve* variance estimates are not sufficiently accurate, particularly when based on a small number of frames.

It is also evident that, unlike in [4], the errors introduced by the EVD method are not evenly distributed between the target and impostor trials at the DCF operating point, with the target trial errors far outweighing the low rate of impostor trial errors. It is hypothesised that this discrepancy is due to the threshold lying much closer to the centre of the target trial score distribution (at approximately 20% miss rate) compared to near the tail of the impostor scores distribution (approvimately 1% false alarms) at this threshold. Hence it is simpler to dismiss a larger proportion of the impostor trials due to the increased distance of the score to the threshold.

This situation is improved significantly with the introduction of the *Decorrelated* variance estimation. With $N = 10$ and a typical frame rate of 100 frames per second, this method averages the frame scores over approximately 0.1 seconds of active speech. It can be seen from these results that decorrelating the samples used to estimate the ELLR score distribution does in fact reduce the proportion of errors introduced by the early decision scoring method, resulting

Table 3. Results with Z-Norm score normalisation at the Actual DCF operating point for the EVD method

System	Act. DCF	Trial Length		Shortcut Errors	
		Median	Mean	Impostor	Target
Reference	.0249	103.4	103.4	–	–
Naïve					
90% Conf.	.1078	2	2.9	7.7%	20.8%
99% Conf.	.0585	3	5.5	3.0%	12.5%
99.9% Conf.	.0407	4	8.7	1.4%	7.3%
Decorrelated	$N = 10$				
90% Conf.	.0701	2	4.5	4.1%	14.7%
99% Conf.	.0331	5	11.7	0.7%	4.4%
99.9% Conf.	.0271	10	18.4	0.2%	1.6%
With Prior	$\tau = 100, \kappa^2 = 0.25$				
90% Conf.	.0565	3	5.5	2.8%	12.1%
99% Conf.	.0283	7	13.6	0.3%	2.5%
99.9% Conf.	.0257	12	20.5	0.1%	0.8%

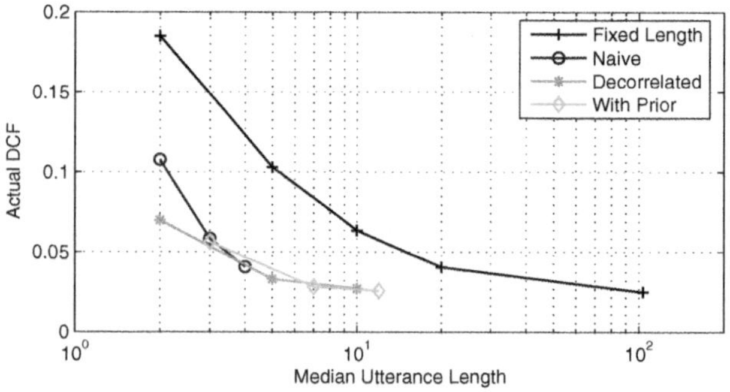

Fig. 3. Median utterance length versus Actual DCF for the fixed short utterance and EVD systems

in performance closer to that of the reference system. The *Decorrelated* approach also produces errors at a rate much closer to the specified confidence level. While the rate at 99.9% confidence is still an order of magnitude too high for target trials, this result at least demonstrates that the variance estimated is more accurate with the data correlations diminished. There is an increase in both the mean and median utterance length associated with the decorrelated estimation method, however, despite this increase the median utterance lengths required are still very short at less than 10 seconds even at the 99.9% confidence level.

By incorporating a prior in the variance estimate it is possible to reduce the performance discrepancy between the reference system and the early decision version to be insignificant. This improved performance unfortunately comes at the cost of longer verification utterances both in terms of the mean and median length statistics (last three rows of Table 2). The hyperparameter τ was fixed at the equivalent of 1 sec while a value of $\kappa^2 = 0.25$ was determined empirically for this system.

Table 3 reproduces the results of Table 2 with the combination of the EVD method and Z-Norm score normalisation by applying the transform described in Section 3.2. These results demonstrate that the EVD method is just as effective with the application of Z-norm, showing much the same trends as described above. As with the fixed length systems, it can be seen that the efficacy of Z-Norm with the EVD method is reduced with shorter utterance lengths. Notably, though, this effect is not as severe with the EVD approach as only the *Naïve 99.9%* system is degraded through the application of Z-Norm.

Fig. 3 graphically summarises the performance of the early verification decision approach by comparing the actual DCF to the median utterance length. Also presented are the fixed utterance-length systems as a reference. All systems have Z-Norm score normalisation applied. It is evident that the EVD method demonstrates consistently and significantly superior performance compared to specifying a fixed utterance length.

5 Summary

This paper introduced a novel method for estimating the confidence interval for speaker verification scores based on estimating the variance of individual frame scores. Several enhancements to this estimate were proposed to increase its robustness and accuracy for the peculiarities of GMM-based speaker verification. The Early Verification Decision (EVD) method, based on this confidence interval estimates, demonstrated that as little as 5–10 seconds of active speech on average was able to produce verification results approaching that of using an average of over 100 seconds of speech.

References

1. Gonzalez-Rodriguez, J., Drygajlo, A., Ramos-Castro, D., Garcia-Gomar, M., Ortega-Garcia, J.: Robust estimation, interpretation and assessment of likelihood ratios in forensic speaker recognition. Computer Speech & Language 20(2-3), 331–355 (2006)
2. Campbell, W.M., Brady, K.J., Campbell, J.P., Granville, R., Reynolds, D.A.: Understanding scores in forensic speaker recognition. In: Odyssey: The Speaker and Language Recognition Workshop (2006)
3. Brümmer, N., du Preez, J.: Application-independent evaluation of speaker detection. Computer Speech & Language 20(2-3), 230–275 (2006)
4. Vogt, R., Sridharan, S., Mason, M.: Making confident speaker verification decisions with minimal speech. In: Interspeech, pp. 1405–1408 (2008)
5. Vogt, R., Sridharan, S.: Explicit modelling of session variability for speaker verification. Computer Speech & Language 22(1), 17–38 (2008)
6. Auckenthaler, R., Carey, M., Lloyd-Thomas, H.: Score normalization for text-independent speaker verification systems. Digital Signal Processing 10(1/2/3), 42–54 (2000)
7. Martin, A., Miller, D., Przybocki, M., Campbell, J., Nakasone, H.: Conversational telephone speech corpus collection for the NIST speaker recognition evaluation 2004. In: International Conference on Language Resources and Evaluation, pp. 587–590 (2004)
8. Martin, A., Przybocki, M.: The NIST 1999 speaker recognition evaluation—an overview. Digital Signal Processing 10(1-3), 1–18 (2000)
9. Gauvain, J.L., Lee, C.H.: Bayesian adaptive learning and MAP estimation of HMM. In: Lee, C.H., Soong, F., Paliwal, K. (eds.) Automatic Speech and Speaker Recognition: Advanced Topics, pp. 83–107. Kluwer Academic, Boston (1996)

Scatter Difference NAP for SVM Speaker Recognition

Brendan Baker, Robbie Vogt, Mitchell McLaren, and Sridha Sridharan

Speech and Audio Research Laboratory
Queensland University of Technology
GPO Box 2434, Brisbane, AUSTRALIA, 4001
{r.vogt,bj.baker,m.mclaren,s.sridharan}@qut.edu.au

Abstract. This paper presents Scatter Difference Nuisance Attribute Projection (SD-NAP) as an enhancement to NAP for SVM-based speaker verification. While standard NAP may inadvertently remove desirable speaker variability, SD-NAP explicitly de-emphasises this variability by incorporating a weighted version of the between-class scatter into the NAP optimisation criterion. Experimental evaluation of SD-NAP with a variety of SVM systems on the 2006 and 2008 NIST SRE corpora demonstrate that SD-NAP provides improved verification performance over standard NAP in most cases, particularly at the EER operating point.

1 Introduction

Automatic speaker verification technology has advanced considerably in recent years, and remains to be a highly active research area. This interest is evidenced by strong and continuing international participation by leading research groups in recent NIST Speaker Recognition Evaluations (SRE). These evaluations have fostered important developments in the technology, and have provided a public forum for participating institutions to publicise these advancements.

In recent NIST SRE's, there has been an increase in the popularity of support vector machine (SVM) approaches, with techniques based around SVM's achieving widespread success. A large variety of SVM speaker verification implementations have been proposed using differing feature representations. including cepstral polynomials [1], MLLR transform coefficients [2], recognised phonetic sequences [3] and adapted GMM mean supervectors [4].

An instrumental development that has attributed to the success of SVM techniques is that of nuisance attribute projection (NAP). Originally proposed by Solomonoff, *et al.* [5,6], NAP has been shown to be an effective method of reducing the performance degradation introduced by mismatch between the training and testing utterances of a speaker. Through a modification of the kernel function, NAP allows for the removal of dimensions of the feature space dominated by "nuisance variation." These removed dimensions are generally determined through a data-driven approach over a large background population database. The most common form of NAP seeks to remove within-class variation which

M. Tistarelli and M.S. Nixon (Eds.): ICB 2009, LNCS 5558, pp. 464–473, 2009.
© Springer-Verlag Berlin Heidelberg 2009

can be observed through the differences between examples of the same speaker in the background population. A possible sub-optimality of this approach is that there is no mechanism to prevent desirable speaker information from also being removed along with the session variability. That is to say, NAP does not explicitly avoid removing between-class variability, while it's assumed that it is this variability that is useful for discriminating between speakers.

Building upon previous work in this area [7], this paper addresses this potential sub-optimality in the NAP training by adopting an approach that explicitly avoids the incorporation of speaker information in the discarded dimensions. The alternate training method, termed *scatter difference nuisance attribute projection* (SD-NAP), is examined in detail. As described in [7], SD-NAP seeks to de-emphasise the speaker information in the NAP projection by incorporating between-class scatter information in the projection training. This paper provides further exploration of this alternate formulation, assessing its performance across different corpora and using varied feature sets and implementations.

The following sections describe both the standard NAP kernel function, and scatter difference analysis that underpins the modified training method. Following this, a series of experiments are performed to compare both the modified and standard NAP approaches. Initial experimentation is performed using a GMM mean supervector system on both 2006 and 2008 NIST SRE corpora. Following this, further results are presented on alternate SVM-based speaker verification systems, utilising different feature sets and kernel functions.

2 Nuisance Attribute Projection

NAP is used to combat errors introduced as a result of inter-session variation, or more simply, mismatch between training and testing utterances of a speaker. Through a modification of the kernel function, NAP allows for the removal of dimensions of the feature space dominated by this inter-session variation. These dimensions are generally determined through a data-driven approach using a large background database consisting of typically 100's or 1000's of speakers, each with numerous recordings or sessions. By examining the differences between examples of the same speaker in the background population, the within-class variation can be used as a model of the nuisance or inter-session variation.

More specifically, NAP attempts to remove the unwanted within-class variation of the observed feature vectors [5,6]. This is achieved by applying the transform

$$\boldsymbol{y}' = \boldsymbol{P}_n \boldsymbol{y} = \left(\boldsymbol{I} - \boldsymbol{V}_n \boldsymbol{V}_n^T \right) \boldsymbol{y} \tag{1}$$

where $\boldsymbol{I}$ is the identity matrix and $\boldsymbol{V}_n$ is an $R_z \times R_y$ orthogonal projection matrix. $\boldsymbol{P}_n$ therefore introduces a null space of dimension R_z into the transformed features that corresponds to the range of $\boldsymbol{V}_n$.

As the purpose of NAP is to remove unwanted variability, $\boldsymbol{V}_n$ is trained to capture the principal directions of within-class variability of a training dataset, that is, it finds the vectors $\boldsymbol{v}$ that maximise the criterion

$$J(\boldsymbol{v}) = \boldsymbol{v}^T \boldsymbol{S}_w \boldsymbol{v} \tag{2}$$

where $\boldsymbol{S}_w$ is the within-class scatter of the training data. This is equivalent to finding the eigenvectors corresponding to the largest eigenvalues satisfying

$$\boldsymbol{S}_w \boldsymbol{v} = \lambda \boldsymbol{v}. \quad (3)$$

As the dimension of the input space is very large (the dimensions of the GMM mean supervectors is $12,288 \times 1$) and the number of background data samples is relatively small (approximately 3,400 utterances from 430 speakers extracted from 2004 and 2005 NIST SRE data), the correlation matrix method [8] is used to determine the principal components. Determining the eigenvalues and eigenvectors of the $3,400 \times 3,400$ correlation matrix is evidently more practical and efficient than the direct eigen decomposition of the covariance matrix $\boldsymbol{S}_w$. This is essentially equivalent to kernel PCA.

2.1 Speaker Information Removed with NAP

As highlighted in [7], the original NAP formulation proposed by Solomonoff, *et al.* [5] does not explicitly avoid removing between-class variability, while it's assumed that it is this variability that is useful for discriminating between speakers. In Figure 1 a plot of the variability captured in the leading NAP dimensions is provided, calculated by measuring the variance of the supervector observations projected onto these dimensions for the 2,800 observations in the training database. It is this information that is discarded by the NAP kernel.

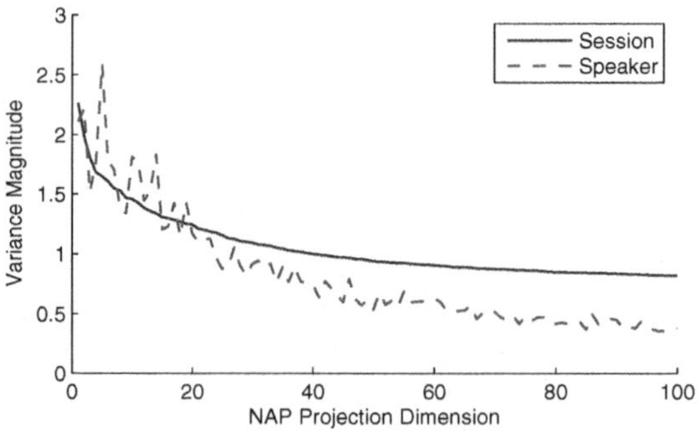

Fig. 1. Session and speaker variability magnitude of the SRE 2004 training data captured by the first 100 dimensions of the NAP projection

It can be seen from Figure 1 that there is a considerable amount of speaker variability removed along with the session variability using the NAP method and, in fact, for many of the first 20 dimensions the speaker variability is *greater* than the amount of session variability removed. This observation certainly motivates a NAP training algorithm that is more selective in the information it removes.

2.2 Scatter Difference Analysis

In [7], an alternate NAP formulation was proposed that seeked to reduce the speaker variability removed. This was achieved by incorporating the between class scatter information in the projection matrix optimisation criterion, and using the difference between both within- and between-class matrices such as in [9]. The criterion with this approach can be expressed as

$$J(\boldsymbol{v}) = \boldsymbol{v}^T \left(\boldsymbol{S}_w - m\boldsymbol{S}_b\right) \boldsymbol{v} \tag{4}$$

where $\boldsymbol{S}_b$ is the between-class scatter of the training data, and m controls the influence of the between-class scatter statistics. It should be noted that this approach introduces a database-dependent tuning parameter to weight the relative importance of $\boldsymbol{S}_b$ and $\boldsymbol{S}_w$.

The scatter difference criterion is optimised in the same manner as the standard NAP method, that is, by solving the eigenvalue problem. As with standard NAP, correlation matrices are used to avoid the issues caused by the very high dimensionality of the supervector features.[1]

To suppress the speaker information in the resulting transform, m is typically set to be in the range 0 to 1. The bounds of this range correspond to the special cases of standard NAP ($m = 0$), and equal weighting between session and speaker information ($m = 1$).

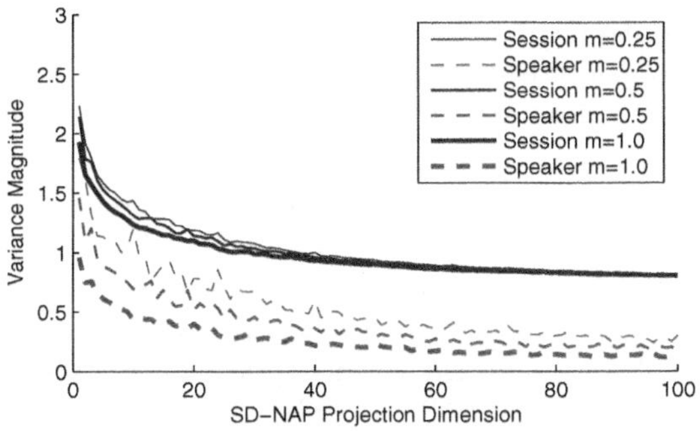

Fig. 2. Session and speaker variability magnitude of the SRE 2004 training data captured by the first 100 dimensions of the scatter difference NAP projection with different values of m

Figure 2 shows the session and speaker variance in the leading dimensions of the projection trained with $m = 1$, that is weighting the within- and between-class scatter statistics equally, as well as with $m = 0.5$ and $m = 0.25$, corresponding to a reduced influence of the between-class scatter statistic.

[1] The scatter difference: $\boldsymbol{S}_w - m\boldsymbol{S}_b$, is not necessarily positive definite, so generalised eigenvalue decomposition must be used rather than singular value decomposition.

Comparing these results to Figure 1, it can be seen that the scatter difference criterion has significantly reduced the speaker variability captured by the NAP transform, as desired, with only a small drop in the session variance magnitude. Furthermore, as m increases the reduction in captured speaker variability becomes more pronounced, as expected.

3 Database and Evaluation

The SD-NAP technique was evaluated using a number of different SVM speaker verification implementations, with varied feature sets, and across two separate evaluation corpora. A description of each of the system configuration and evaluation corpora follows.

3.1 Evaluation Corpora

The NIST 2006 and 2008 SRE corpora were used to evaluate and compare the performance of the proposed method. Results were derived from the "all trials" condition of the official evaluation protocol which includes trials spoken in all-languages.

3.2 GMM Mean Supervector SVM System

The mean of a MAP adapted GMM [10] in the form of a supervector provides a suitable representation of an utterance for modelling with an SVM classifier [4]. For the system used in this study, a GMM mean supervector is formed by concatenating the component mean vectors of a MAP-adapted GMM that is $\boldsymbol{\mu}(s) = \left[\boldsymbol{\mu}_1(s)^T \dots \boldsymbol{\mu}_C(s)^T\right]^T$ where $\boldsymbol{\mu}_c(s)$ are the component means. The GMM-UBM system used in this work is based around the system described in [11] with the resulting supervectors having dimension of $24 \times 512 = 12,288$. Each dimension of the SVM feature space is normalised by the mean and standard deviation of the corresponding observations in the background dataset. T-Norm score normalisation [12] was also used on the resulting verification scores using a T-Norm dataset constructed from the NIST 2004 SRE data.

3.3 English Phonetic Lattice N-Gram SVM System

This system used phonetic transcripts produced by an English phone recogniser. The English phone recogniser was trained using data from the Fisher corpus [13]. Gender-dependent phone recognisers were used. 100 hours of speech per gender were extracted from the Fisher database to train three state, 16 mixture component, and gender-dependent HMMs using HTK. The system is capable of recognising a total of 43 phonetic labels. In a similar manner to the approach described in [14], rather than using 1-best transcriptions phone lattices were utilised. The expected frequencies of unigrams, bigrams and trigrams in each utterance were concatenated and used to form a feature vector. Only the 10,000 most frequently occurring trigrams (determined on the background set) were included. The n-gram frequencies were weighted according to their posterior probability of occurrence in the recognition lattice. T-Norm score normalisation was also applied to this system.

3.4 PPRLM Lattice N-Gram SVM System

The PPRLM (parallel phone recogniser with language modelling) system uses phone transcriptions obtained from multiple open-loop phone recognisers (OLPR) each trained on one of 6 languages; English, German, Hindi, Japanese, Mandarin and Spanish. This parallel stream architecture was first described for the speaker verification task in [15]. The multi-stream decoding was performed using QUTs HMM based OLPR - trained on the OGI multi-lingual database. The same support vector feature extraction process was used for all six language streams. The expected frequencies of unigrams, bigrams and trigrams in each utterance were concatenated and used to form a feature vector. Only the 10,000 most frequently occurring trigrams (determined on the background set) were included. Once again, the n-gram frequencies were weighted according to their posterior probability of occurrence in the recognition lattice. Scores from each stream were fused through a linear combination, with weightings calculated via logistic regression. T-Norm score normalisation was applied to each stream prior to combination.

3.5 MLLR SVM System

The MLLR system makes use of the same English phone recogniser as described in Section 3.3. The male and female English phone recogniser HMM models served as reference models for computing speaker-dependent MLLR transforms. Using the alignments produced by the phonetic decoder, a five class regression tree was used (4 data driven classes + silence) to obtain a set of MLLR transforms for each training segment. The acoustic features contain 39 components, resulting in a transform vector for each class of dimension 39×40 (39×39 transform matrix + 39×1 vector bias). The transform components for each class (excluding the silence class) were concatenated to form a single feature vector for each conversation side resulting in a total feature vector length of 6,240. Based on the suggestion in [2], transforms from both male and female models were generated for each conversation side and concatenated to form a final 12,480 length supervector. Once again, the SVM space was constructed with a linear kernel and was rank normalised. During development, T-score normalisation was found to provide little benefit for this system, and as such this process was excluded. This conclusion was also drawn in [2].

4 Results and Discussion

Initial experimentation was performed using the GMM mean supervector system (see Section 3.2). Table 1 presents results of the SD-NAP method in comparison to conventional NAP and a baseline system without session variability compensation on both the 2006 and 2008 NIST SRE protocol. For the scatter difference approach to NAP matrix training, results are presented for a range of values for the matrix weighting term m.

It is clear from the results in Table 1 that significant performance improvements over the baseline can be achieved by adopting a form of nuisance attribute projection. As expected, the original NAP formulation, proposed by Solomonoff, *et al.* achieves a significant improvement over the baseline system.

The SD-NAP results indicate that further improvements can be achieved by minimising the amount of speaker variability removed. Almost all SD-NAP configurations trialled showed improvement over the standard NAP approach. The best results across 2006 and 2008 evaluations for both equal error rate and minimum detection cost criteria resulted when the SD-NAP technique was employed with $m = 0.2$.[2]

Table 1. Performance of the GMM Mean Supervector System evaluated on all trials (combined male and female) of the 2006 and 2008 NIST SRE's. 50 dimensions of session variability were removed using NAP/SD-NAP.

	NIST SRE 2006		**NIST SRE 2008**	
System	**EER**	**Min. DCF**	**EER**	**Min. DCF**
Baseline	7.30%	.0363	10.26%	.0542
Standard NAP ($m = 0$)	4.82%	.0232	7.14%	.0375
SD-NAP $m = 0.05$	4.73%	.0230	7.10%	.0373
SD-NAP $m = 0.1$	4.67%	.0225	6.96%	.0372
SD-NAP $m = 0.2$	**4.54%**	**.0221**	**6.80%**	**.0366**
SD-NAP $m = 0.5$	4.73%	.0235	7.05%	.0372
SD-NAP $m = 1$	5.34%	.0261	7.46%	.0388

The SD-NAP technique was evaluated for the remaining verification systems, with comparisons made again to both standard NAP and baseline systems. Table 2 provides a performance summary of the trialled configurations. For the SD-NAP configurations, the weighting factor m was optimised on the 2006 data. Once again, it is immediately clear from these results that incorporating nuisance attribute projection of some form allows for significant improvements in performance to be achieved. Both NAP and SD-NAP configurations provide marked improvements over the baseline systems for all systems and evaluations.

Comparing the NAP and SD-NAP configurations, for most cases, the SD-NAP approach achieves superior error rates to that of standard NAP. This is with the exception of the multi-stream PPRLM n-gram system. In this case, adopting the SD-NAP approach in substitution of standard NAP resulted in a slight degradation across both evaluations.

[2] While negative values for the weighting m were investigated in [7] and gave improved performance, these same negative values gave only degraded performance in this study. It is hypothesized that the results in the previous study were anomalis and possibly related to differences or details in the system implementatations.

Table 2. Performance of the SVM speaker verification systems evaluated on all trials of the 2006 and 2008 NIST SRE's. 50 dimensions of session variability were removed using NAP/SD-NAP.

	NIST SRE 2006		NIST SRE 2008	
System	**EER**	**Min. DCF**	**EER**	**Min. DCF**
GMM SVM				
Baseline	7.30%	.0363	10.26%	.0542
Standard NAP	4.82%	.0232	7.14%	.0375
SD-NAP $m = 0.2$	**4.54%**	**.0221**	**6.80%**	**.0366**
English N-gram				
Baseline	13.82%	.0630	18.90%	.0910
Standard NAP	11.42%	.0496	14.49%	**.0794**
SD-NAP $m = 0.2$	**10.04%**	**.0480**	**13.42%**	**.0794**
PPRLM N-gram				
Baseline	15.38%	.0631	17.24%	.0764
Standard NAP	**8.24%**	**.0426**	**11.85%**	**.0719**
SD-NAP $m = 0.2$	8.30%	.0442	11.96%	**.0719**
MLLR				
Baseline	10.34%	.0442	14.34%	.0626
Standard NAP	9.43%	.0418	12.30%	.0542
SD-NAP $m = 0.2$	**9.26%**	**.0409**	**11.85%**	**.0535**

Table 3. Comparison of NAP and SD-NAP for individual streams of PPRLM system on all trials (combined male and female) of the 2006 NIST SRE

	Standard NAP		SD-NAP	
System	**EER**	**Min. DCF**	**EER**	**Min. DCF**
English	11.75%	.0521	11.01%	.0522
German	12.36%	.0549	11.48%	.0542
Hindi	11.64%	.0521	10.87%	.0522
Japanese	12.64%	.0547	12.00%	.0543
Mandarin	11.89%	.0521	11.23%	.0520
Spanish	12.61%	.0539	11.86%	.0536
Combined	8.24%	.0426	8.30%	.0442

It is also interesting to note that the optimal value for the SD-NAP weighting m remained fairly stable across the different systems and evaluations. Using a constant value $m = 0.2$ for all systems gave the best performance.

4.1 Analysis of Individual PPRLM Streams

Further analysis was performed to gain a better indication as to why the SD-NAP approach did not outperform the standard NAP method when used for the PPRLM system. Table 3 provides a performance summary breakdown for the six individual language streams when tested on the 2006 SRE, along with the combined result.

Interestingly, a number of the individual language streams display equal or better performance when the SD-NAP approach is used rather than the standard NAP formulation. Particularly, improvement is apparent for the EER performance statistic, where using SD-NAP lowers the error rate across all languages. Unfortunately, as already presented, once the streams are combined using logistic regression, this improvement is negated, with the standard NAP approach surpassing SD-NAP. These results warrant further investigation in future studies. A number of factors, including the order in which T-Norm score normalisation and fusion is performed, and the callibration of input scores prior to logistic regression combination may need to be re-considered when using SD-NAP.

5 Conclusions and Future Work

This paper examined an alternate NAP training approach that explicitly avoids the incorporation of speaker information in the discarded dimensions, The method proposed uses an alternate training criterion based on scatter difference analysis. The SD-NAP method seeks to de-emphasise important speaker information in the NAP projection by incorporating a weighted version of the between-class scatter.

The SD-NAP technique was evaluated and compared to both baseline and standard NAP approaches using a number of different SVM speaker verification implementations and evaluation corpora. Results demonstrated that consistent improvements over the standard NAP approach could be achieved. Although the SD-NAP method introduces an additional database-dependent tuning parameter m, experiments revealed that the optimal value for this weighting term remained fairly stable across the different systems and evaluations.

The only exception to the observed improvement achieved through use of SD-NAP was for the multi-stream PPRLM-based speaker verification system. Further analysis of this system revealed that for the individual component streams, equal or better performance was generally achieved over NAP. This trend, however, was reversed after the application of T-Norm and stream combination. The effect that SD-NAP has on output score distributions along with its interaction with score normalisation schemes such as T-Norm warrants further investigation.

Acknowledgements. This research was supported by the Australian Research Council (ARC) Discovery Grant Project ID: DP0877835.

References

1. Campbell, W.: Generalized linear discriminant sequence kernels for speaker recognition. In: IEEE International Conference on Acoustics, Speech, and Signal Processing, vol. 1, pp. 161–164 (2002)
2. Stolcke, A., Ferrer, L., Kajarekar, S.: Improvements in MLLR-transform-based speaker recognition. In: Odyssey: The Speaker and Language Recognition Workshop (2006)

3. Campbell, W., Campbell, J., Reynolds, D., Jones, D., Leek, T.: Phonetic speaker recognition with support vector machines. In: Advances in Neural Information Processing Systems, vol. 16 (2004)
4. Campbell, W., Sturim, D., Reynolds, D., Solomonoff, A.: SVM based speaker verification using a GMM supervector kernel and NAP variability compensation. In: IEEE International Conference on Acoustics, Speech and Signal Processing, vol. 1, pp. I–97–I–100 (2006)
5. Solomonoff, A., Quillen, C., Campbell, W.: Channel compensation for SVM speaker recognition. In: Odyssey: The Speaker and Language Recognition Workshop, pp. 57–62 (2004)
6. Solomonoff, A., Campbell, W., Boardman, I.: Advances in channel compensation for SVM speaker recognition. In: IEEE International Conference on Acoustics, Speech and Signal Processing, vol. I, pp. 629–632 (2005)
7. Vogt, R., Kajarekar, S., Sridharan, S.: Discriminant NAP for SVM speaker recognition. In: Odyssey: The Speaker and Language Recognition Workshop (2008)
8. Fukunaga, K.: Introduction to Statistical Pattern Recognition. Academic Press, San Diego (1990)
9. Liu, Q., Tang, X., Lu, H., Ma, S.: Face recognition using kernel scatter-difference-based discriminant analysis. IEEE Transactions on Neural Networks 17(4), 1081–1085 (2006)
10. Reynolds, D., Quatieri, T., Dunn, R.: Speaker verification using adapted Gaussian mixture models. Digital Signal Processing 10(1/2/3), 19–41 (2000)
11. McLaren, M., Vogt, R., Baker, B., Sridharan, S.: A comparison of session variability compensation techniques for SVM-based speaker recognition. In: Interspeech 2007, pp. 790–793 (2007)
12. Auckenthaler, R., Carey, M., Lloyd-Thomas, H.: Score normalization for text-independent speaker verification systems. Digital Signal Processing 10(1/2/3), 42–54 (2000)
13. Cieri, C., Miller, D., Walker, K.: The Fisher Corpus: a Resource for the Next Generations of Speech-to-Text. In: International Conference on Language Resources and Evaluation, pp. 69–71 (2004)
14. Hatch, A., Peskin, B., Stolcke, A.: Improved phonetic speaker recognition using lattice decoding. In: IEEE International Conference on Acoustics, Speech and Signal Processing (2005)
15. Andrews, W., Kohler, M., Campbell, J., Godfrey, J., Hernández-Cordero, J.: Gender-dependent phonetic refraction for speaker recognition. In: IEEE International Conference on Acoustics, Speech and Signal Processing, vol. 1, pp. 149–152 (2002)

Data-Driven Impostor Selection for T-Norm Score Normalisation and the Background Dataset in SVM-Based Speaker Verification

Mitchell McLaren, Robbie Vogt, Brendan Baker, and Sridha Sridharan

Speech and Audio Research Laboratory, QUT, Brisbane, Australia
{m.mclaren,r.vogt,bj.baker,s.sridharan}@qut.edu.au

Abstract. A data-driven background dataset refinement technique was recently proposed for SVM based speaker verification. This method selects a refined SVM background dataset from a set of candidate impostor examples after individually ranking examples by their relevance. This paper extends this technique to the refinement of the T-norm dataset for SVM-based speaker verification. The independent refinement of the background and T-norm datasets provides a means of investigating the sensitivity of SVM-based speaker verification performance to the selection of each of these datasets. Using refined datasets provided improvements of 13% in min. DCF and 9% in EER over the full set of impostor examples on the 2006 SRE corpus with the majority of these gains due to refinement of the T-norm dataset. Similar trends were observed for the unseen data of the NIST 2008 SRE.

1 Introduction

An issue commonly faced in the development of speaker verification systems is the selection of suitable datasets. Several recent studies have highlighted the importance of selecting appropriate training, development or impostor data for SVM-based speaker verification to match the evaluation conditions [1,2]. Two datasets fundamental to SVM-based classification that must be appropriately selected to maximise classification performance are the background dataset and the dataset used for test score normalisation (T-norm) [3].

The background dataset is a collection of negative or impostor observations used in the training of an SVM in which discrimination between the background dataset and the speaker examples is maximised [4]. Often, the number of impostor observations significantly outweighs that of speaker examples such that the SVM system relies heavily on the background observations to provide most of the observable discriminatory information. The background dataset must, therefore, consist of suitable impostor examples to ensure good classification performance.

Similar to the background dataset, the T-norm dataset is a collection of impostor examples from which a set of T-norm models are trained. T-normalisation uses these models during testing to estimate an impostor score distribution with

M. Tistarelli and M.S. Nixon (Eds.): ICB 2009, LNCS 5558, pp. 474–483, 2009.
© Springer-Verlag Berlin Heidelberg 2009

which classification scores are normalised [3]. Consequently, the reliable estimation of impostor score distributions, and ultimately, the potential gains offered through T-norm are dependent on the appropriate selection of T-norm dataset.

Data-driven background dataset refinement [5] was a recently proposed technique to individually assess the suitability of each candidate impostor example, from a large and diverse dataset, for use in the background dataset. The support vector frequency of an example was used as a measure of its relative importance in the background dataset to rank the set of impostor examples. The top N examples from this ranked set were then used as a *refined* background dataset to provide improved classification performance. The system configuration in this study refined a single dataset to be used as both the background and the T-norm dataset. Of interest, however, is the way in which classification performance is affected when the size of the refined background and T-norm datasets are allowed to vary independently.

The following study investigates how sensitive SVM-based classification performance is to the selection of a suitable T-norm dataset compared to the selection of the background dataset. This is analysed by observing performance when independently varying the number of highest-ranking candidate impostor examples in the refined background and T-norm datasets. The ranking of candidate impostor examples is performed using the same refinement procedure outlined in [5].

The recently proposed data-driven background dataset refinement technique is presented in Section 2 followed by a discussion in Section 3 on T-norm score normalisation in SVM-based classification. Section 4 details the experimental protocol with results presented in Section 5.

2 Data-Driven Background Dataset Selection

Data-driven background dataset refinement [5] was recently shown to be an effective method for the selection of a highly informative background dataset from a set of candidate impostor examples such that it exhibited performance gains over the best heuristically-selected background from the same initial resources. The technique makes use of a development dataset to systematically drive the selection of the impostor dataset based on the relevance of each example in the background dataset. In this approach, the ranking of impostor observations is performed using a criterion that involves exploiting the information possessed by the support vectors of a trained SVM.

2.1 Support Vector Frequency

The support vector machine is a discriminative classifier trained to separate classes in a high-dimensional space. A kernel is used to project input vectors into this high-dimensional space where a separating hyperplane is positioned to maximise the margin between the classes [4]. The training of a speaker SVM results in the selection of a subset of both positive and negative examples from the training dataset termed *support vectors* that are used to construct the separating hyperplane. Examples that were selected as support vectors in the SVM hold a common property of being the most difficult to classify, lying on, or within, the margin between classes. In contrast, those training examples that were not selected as support vectors provided no information in the training of the SVM.

The process of determining a subset of support vectors during SVM training can be considered a data selection process in which the most informative examples are chosen from the training dataset. In light of this, it can be stated that the impostor support vectors are the most informative set of background examples with respect to the client data.

Based on this observation, the *support vector frequency* of an example provides a measure of its relative importance in the background dataset. The support vector frequency of an example is defined as the number of times that it is selected as a support vector while training a set of SVMs on a development dataset.

2.2 Background Dataset Refinement

Given a diverse set of vectors B, compiled from a number of available resources, this dataset can be refined into a suitable background dataset using a set of *development* client vectors S. The speakers and vectors in the set S should be disjoint from those in B.

1. Using the full set of impostors B as the background dataset, train SVMs for each vector in the set of development client models S.
2. Calculate the support vector frequency of each impostor example in B as the total number of instances in which it was selected as a support vector for the development client models.
3. The refined impostor dataset R_N is chosen as the top N subset of B ranked by the support vector frequency ($R_N \subset B$).
4. For several values of N, use R_N in the evaluation of a development corpus to determine the optimal number of examples to be included in the refined background dataset.

It is important to note that the support vector frequencies are likely to be heavily dependent on the characteristics found in the development set S. For this reason, S should be selected based on the knowledge of the broad characteristics (such as gender, language and audio conditions) expected to exist in the corpus for which the impostor dataset is intended to be used.

3 Test Score Normalisation

Test score normalisation (T-norm) [3] is a technique used to counteract the statistical variations that occur in classification scores and was found to be an integral part of most speaker verification systems submitted to the recent NIST speaker recognition evaluations (SRE) [6]. T-norm aims to normalise the score distributions of all test segments to a single scale so as to improve performance when applying a global threshold to test classification scores. This involves scoring each test utterance on a set of impostor models, trained from the T-norm dataset, producing an impostor score distribution with mean μ_I and standard deviation σ_I. The score s, obtained when comparing the test segment to a client model, is then normalised using,

$$\overline{s} = \frac{s - \mu_I}{\sigma_I} \quad (1)$$

The reliable estimation of the normalisation parameters μ_I and σ_I is dependent on the observable characteristics of the T-norm dataset. Consequently, the objective of normalising scores to a global scale will be more attainable if the selection of the T-norm dataset is tailored toward the expected evaluation conditions.

3.1 T-Norm Dataset Selection for SVM-Based Classification

The desired characteristics of the T-norm dataset closely match those wanted of a background dataset, in being a set of examples that appropriately represents the impostor population. Campbell et al. demonstrated this commonality of requirements by comparing the use of a single dataset for both the background and T-norm datasets to the use of disjoint datasets [7]. It was found that performing T-norm using the background dataset provided an improvement in performance over unnormalised scores, while the disjoint T-norm dataset resulted in somewhat degraded performance. It is unclear why this degradation occured, however, possible explanations include mismatch between the T-norm and evaluation conditions and also the limited size of the T-norm dataset.

The use of a single impostor dataset for the background and T-norm datasets was further explored in the recent study on data-driven background dataset refinement [5]. The evaluation of a development corpus demonstrated consistent performance gains as the large and diverse background dataset was more extensively refined. Corresponding gains were also observed in the evaluation of an unseen corpus.

The following study extends on the research in [5] by investigating the degree that classification performance depends on the suitable selection of the T-norm dataset compared to that of the background dataset. In contrast to the selection of a single dataset, *intersecting* datasets will be formed through the *independent* selection of the top N examples for the T-norm and background datasets from a single, ranked impostor dataset. In this way, the smaller dataset will be a subset of the other. Analysis of performance over a range of dataset sizes is expected to provide insight as to how sensitive SVM classification performance is to the selection of each of these datasets and whether the characteristics of the best refined background dataset are in fact similar those of the best T-norm dataset.

While this study will focus on the data-driven selection of *intersecting* background and T-norm datasets, the use of refined, *disjoint* datasets will also be evaluated for completeness.

4 Experimental Protocol

4.1 GMM-SVM System

SVM classification in the following experiments was based on GMM mean supervectors using the associated GMM mean supervector kernel [8]. The GMM system used in this study was based on 512-component models and was previously described in [9].

The SVM implementation uses the open source LIA_MISTRAL package [10] based on the libSVM library [11]. Nuisance attribute projection (NAP) [8] was employed to reduce session variation with the 50 dimensions of greatest session variation being removed from all supervectors.

4.2 Evaluation Datasets

Gender-dependent background datasets were collected from NIST 2004 and NIST 2005 databases and a random selection of 2000 utterances[1] from each of Fisher and Switchboard 2 corpora giving a total of 6444 male and 7766 female observations. The number of impostor examples from each of these data sources can be found in Table 1. The limited amount of data from the NIST 2005 corpus is due to the intentional exclusion of utterances from any speakers that also appear in the NIST 2006 corpus. For this study, these datasets consisted only of telephony data. Conversations were spoken in a range of languages with the majority in English. Large gender-dependent background datasets B were compiled from all available resources as listed in Table 1.

The gender-dependent development client dataset S used to calculate support vector frequencies was compiled from the training and testing utterances in the all-language, 1conv4w condition of the NIST 2006 SRE. Consisting of 1950 male and 2556 female client vectors, this provided a moderate degree of resolution in the support vector frequency statistic.

The NIST 2008 SRE corpus was used to observe how well the refined background and T-norm datasets generalised to unseen data. All NIST 2008 results were derived from condition 6 as specified in the official evaluation protocol [12] which includes trials spoken in all-languages while being restricted to telephony data, matching the conditions found in the development dataset S.

5 Results

5.1 Development Evaluations

Figure 1 depicts 3-D plots of the min. DCF and EER obtained on the NIST 2006 development corpus when using a range of refined T-norm and background dataset sizes selected as the highest-ranking impostor observations from the full dataset B. The darker peaks in the plots designate improved performance.

Figure 1(a) indicates that the min. DCF was more sensitive to the selection of a suitable T-norm dataset than the background dataset. This is evident in the higher

[1] Selected randomly due to memory limitations restricting the full dataset size.

Table 1. Number of impostor examples from each data source

Gender	Fisher	SWB2	NIST04	NIST05
Male	2000	2000	1901	543
Female	2000	2000	2651	1115

performance variation along the dimension corresponding to the size of the T-norm dataset compared to that of the background dataset. In contrast to the min. DCF, the EER plot in Figure 1(b) appears to exhibit less consistent dataset dependencies. As the background dataset is increasingly refined, however, the sensitivity of performance to the selection of the T-norm dataset appears to become clearer. Interestingly, the EER begins to degrade quicker than the min. DCF as the T-norm and background datasets are refined too extensively.

Maximum classification performance was found when using the top 1000 ranking impostor examples as the T-norm dataset and the subset of the top 250 observations as the background dataset. These datasets will be designated by the notation Bkg=B_{250} and T-norm=B_{1000}. Results from development evaluations using these refined datasets and the full impostor set B are detailed in Table 2. The use of both refined datasets provided a relative gain of 13% in min. DCF and 9% in EER over the full dataset B which is a statistically significant improvement at the 99% and 95% confidence level[2], respectively. With T-norm=B, the refined background dataset offered performance improvements over the full background, however, superior gains were observed from the refined T-norm dataset over the full T-norm set when Bck=B. These results demonstrate, firstly, that background dataset refinement can sucessfully be applied to the task of T-norm dataset selection, and secondly, that SVM-based classification is more dependent on the selection of a suitable T-norm dataset than the background dataset.

5.2 NIST 2008 Evaluations

The objective of this section was to determine whether the dataset-dependent performance trends, observed in the previous section (Section 5.1), were also reflected in the evaluation of the unseen data of the NIST 2008 SRE. Figure 2 depicts the 3-D plot of the min. DCF from these evaluations as the full set of impostor examples B was refined after being ranked using the NIST 2006 SRE corpus as development data.

Figure 2 indicates that the NIST 2008 SRE min. DCF performance was more sensitive to the selection of a suitable T-norm dataset than that of background dataset selection, thus supporting the observations from development evaluations in Section 5.1. In contrast to the background dataset, the T-norm dataset appears to provide relatively high gains through increased refinement as observed by the darker regions of the plot.

Results from NIST 2008 SRE when using the datasets selected based on NIST 2006 development evaluations (Bkg=B_{250} and T-norm=B_{1000}) are detailed in Table 3 along with results obtained using the full impostor set. The use of the

[2] Based on the proposed method in [13] (independent case).

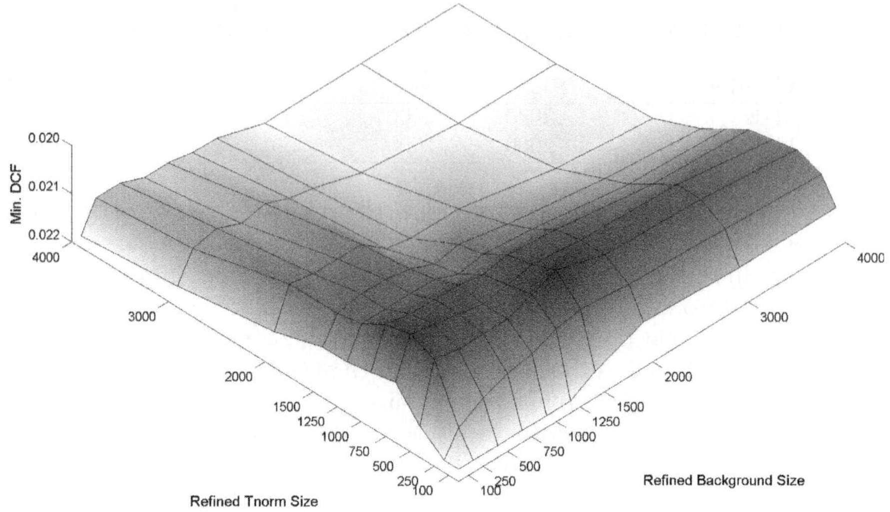

(a) Min. DCF

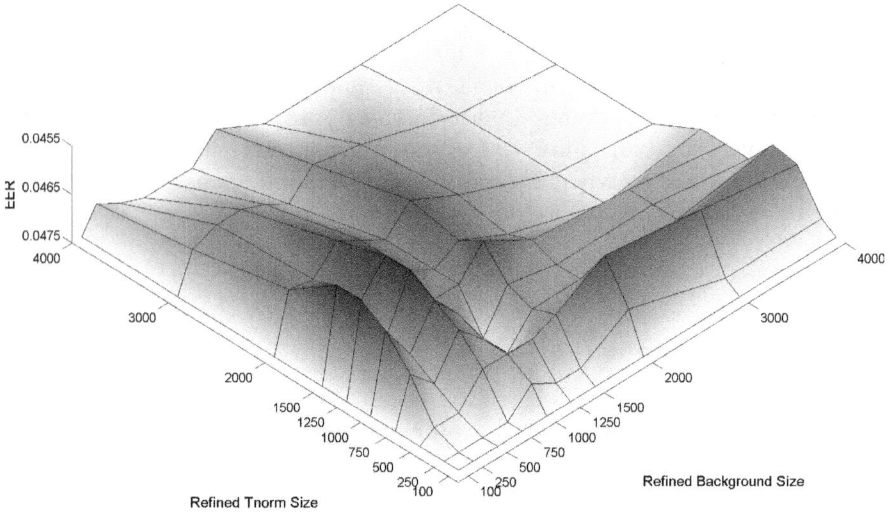

(b) EER

Fig. 1. Min DCF. and EER on NIST 2006 SRE when performing data-driven impostor selection of intersecting background and T-norm datasets

refined T-norm dataset offers substantial improvements over the full dataset while, surprisingly, the refined background dataset achieves comparable performance to the full dataset. These results demonstrate that the selection of a suitable T-norm dataset can have more impact on potential classification performance than the background dataset in the evaluation of an unseen corpus. When using both the refined datasets, a statistically significant improvement of 8% was observed in the min. DCF at the 99% confidence level[2] over the full datasets, however, no gain was found in EER.

Table 2. Performance on NIST 2006 SRE when using full dataset B and best refined intersecting T-norm and background datasets

Config. (Bck / T-norm)	Bck	T-norm	Min. DCF	EER
Full / Full	B	B	.0234	5.06%
Full / Refined	B	B_{1000}	.0215	4.70%
Refined / Full	B_{250}	B	.0223	4.84%
Refined / Refined	B_{250}	B_{1000}	.0203	4.59%

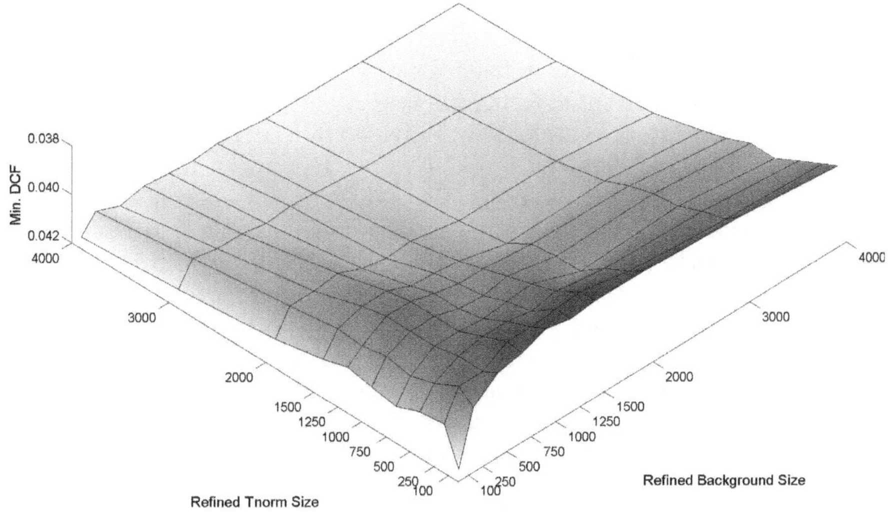

(a) Min. DCF

Fig. 2. Min DCF. on NIST 2008 SRE when using refined intersecting background and T-norm datasets ranked using NIST 2006 data

Table 3. Performance from NIST 2008 SRE using full and best refined T-norm and background datasets selected based on NIST 2006 evaluations

Config. (Bck / T-norm)	Bck	T-norm	Min. DCF	EER
Full / Full	B	B	.0435	8.34%
Full / Refined	B	B_{1000}	.0408	8.18%
Refined / Full	B_{250}	B	.0421	8.57%
Refined / Refined	B_{250}	B_{1000}	.0399	8.35%

5.3 Refinement of Disjoint Datasets

The refinement of a single dataset to form intersecting background and T-norm datasets provided a suitable means of investigating the dependence of SVM-based classification performance to their selection. The most common dataset configuration, however, involves the use of disjoint datasets as observed in recent NIST SRE submissions [6]. This section endeavours to determine firstly,

Table 4. Performance on NIST 2006 and NIST 2008 SRE using full and best refined *disjoint* T-norm and background datasets

Config. (Bck / T-norm)	Bck	T-norm	2006 SRE DCF	2006 SRE EER	2008 SRE DCF	2008 SRE EER
Full / Full	I	T	.0230	4.93%	.0432	8.34%
Full / Refined	I	T_{750}	.0214	4.73%	.0408	8.44%
Refined / Full	I_{250}	T	.0222	4.86%	.0428	8.37%
Refined / Refined	I_{250}	T_{750}	.0202	4.48%	.0394	8.30%

whether similar T-norm-dependence trends are observed when using disjoint datasets, and secondly, whether refined disjoint datasets provide performance improvements over refined intersecting datasets.

The full set of impostor examples B was divided to form the unrefined disjoint T-norm and background datasets T and I respectively, such that the speakers and vectors in these sets were separate. These disjoint subsets contained similar proportions of examples from each of the data sources listed in Table 1. Ranking of these sets was performed independently using the NIST 2006 corpus.

The refined disjoint datasets providing maximum performance in the NIST 2006 development evaluations were Bkg=I_{250} and T-norm=T_{750}. Results from trials on both NIST 2006 and NIST 2008 SRE using these datasets are detailed in Table 4. In the evaluation of both corpora, the refined T-norm dataset demonstrated substantial improvements over the full T-norm dataset. In contrast, the refined background provided somewhat improved results in the development evaluations, however, these benefits were only reflected in the NIST 2008 SRE when used in conjunction with the refined T-norm dataset. These results demonstrate that, even in the case of disjoint datasets, SVM-based speaker verification performance is more dependent on the suitable selection of the T-norm dataset than that of the background dataset.

Comparing the results in Table 4 to those detailed in Tables 2 and 3, the refined disjoint datasets (Bkg=I_{250} and T-norm=T_{750}) were found to provide marginal performance improvements over the refined intersecting datasets (Bkg=B_{250} and T-norm=B_{1000}). This performance gain may also bring to light an underlying characteristic of background dataset refinement in that ranking of impostor examples may become more robust as the ratio of development SVMs in set S to the size of the full impostor set B is increase. Future work will investigate the impact that the size of B has on dataset refinement.

6 Conclusion

This study investigated the dependence of SVM-based classification performance to the selection of suitable background and T-norm datasets. The recently proposed background dataset refinement technique [5] was used to rank a large set of candidate impostor examples from which the top N highest-ranking observations were independently selected to form refined intersecting background and T-norm datasets. Evaluations were performed on both the NIST 2006 SRE development corpus and the unseen NIST 2008 SRE using a range of refined dataset sizes.

It was determined that SVM-based speaker verification classification performance is more sensitive to the selection of a suitable T-norm dataset than of background dataset selection. The best refined T-norm dataset, as determined by NIST 2006 development evaluations, provided substantial gains in both NIST 2006 and 2008 SRE irrespective of background choice. In contrast, the best refined background dataset offered only marginal performance improvements unless used in conjunction with the refined T-norm dataset, in which case maximum performance was obtained. Likewise, the refinement of disjoint background and T-norm datasets further demonstrated the high dependence of SVM-based speaker verification performance on the choice of T-norm dataset.

Acknowledgments. This research was supported by the Australian Research Council (ARC) Discovery Grant Project ID: DP0877835.

References

1. Kajarekar, S.S., Stolcke, A.: NAP and WCCN: Comparison of approaches using MLLR-SVM speaker verification system. In: Proc. IEEE ICASSP, pp. 249–252 (2007)
2. Stolcke, A., Kajarekar, S.S., Ferrer, L., Shriberg, E.: Speaker recognition with session variability normalization based on MLLR adaptation transforms. IEEE Trans. on Audio, Speech, and Language Processing 15, 1987–1998 (2007)
3. Auckenthaler, R., Carey, M., Lloyd-Thomas, H.: Score normalization for text-independent speaker verification systems. Digital Signal Processing 10(1), 42–54 (2000)
4. Burges, C.: A tutorial on support vector machines for pattern recognition. Data Mining and Knowledge Discovery 2(2), 121–167 (1998)
5. McLaren, M., Baker, B., Vogt, R., Sridharan, S.: Improved SVM speaker verification through data-driven background dataset selection. To be presented in Proc. IEEE ICASSP (2009)
6. Nation Institute of Standards and Technology: NIST speech group website (2006), `http://www.nist.gov/speech`
7. Campbell, W., Reynolds, D., Campbell, J.: Fusing discriminative and generative methods for speaker recognition: Experiments on switchboard and NFI/TNO field data. In: Proc. Odyssey, pp. 41–44 (2004)
8. Campbell, W., Sturim, D., Reynolds, D., Solomonoff, A.: SVM based speaker verification using a GMM supervector kernel and NAP variability compensation. In: Proc. IEEE ICASSP, pp. 97–100 (2006)
9. McLaren, M., Vogt, R., Baker, B., Sridharan, S.: A comparison of session variability compensation techniques for SVM-based speaker recognition. In: Proc. Interspeech, pp. 790–793 (2007)
10. Bonastre, J., Wils, F., Meignier, S.: ALIZE, a free toolkit for speaker recognition. In: Proc. IEEE ICASSP, pp. 737–740 (2005)
11. Chang, C., Lin, C.: LIBSVM: A library for support vector machines (2001), `http://www.csie.ntu.edu.tw/~cjlin/libsvm`
12. NIST: The NIST Year 2008 Speaker Recognition Evaluation Plan (2008), `http://www.nist.gov/speech/tests/sre/2008/sre08_evalplan_release4.pdf`
13. Bengio, S., Mariéthoz, J.: A statistical significance test for person authentication. In: Proc. Odyssey, pp. 237–244 (2004)

Support Vector Machine Regression for Robust Speaker Verification in Mismatching and Forensic Conditions

Ismael Mateos-Garcia, Daniel Ramos, Ignacio Lopez-Moreno, and Joaquin Gonzalez-Rodriguez

ATVS – Biometric Recognition Group, Escuela Politecnica Superior, Universidad Autonoma de Madrid, C. Francisco Tomás y Valiente 11, 28049 Madrid, Spain
{ismael.mateos,daniel.ramos,ignacio.lopez, joaquin.gonzalez}@uam.es

Abstract. In this paper we propose the use of Support Vector Machine Regression (SVR) for robust speaker verification in two scenarios: *i)* strong mismatch in speech conditions and *ii)* forensic environment. The proposed approach seeks robustness to situations where a proper background database is reduced or not present, a situation typical in forensic cases which has been called *database mismatch*. For the mismatching condition scenario, we use the NIST SRE 2008 core task as a highly variable environment, but with a mostly representative background set coming from past NIST evaluations. For the forensic scenario, we use the Ahumada III database, a public corpus in Spanish coming from real authored forensic cases collected by Spanish Guardia Civil. We show experiments illustrating the robustness of a SVR scheme using a GLDS kernel under strong session variability, even when no session variability is applied, and especially in the forensic scenario, under database mismatch.

Keywords: Speaker verification, forensic, GLDS, SVM classification, SVM regression, session variability compensation, robustness.

1 Introduction

Speaker verification is currently a mature technology which aims at determine whether a given speech segment of unknown source belongs to the identity of a claimed individual or not. Among the most important challenges of a speaker verification system is the robustness to the mismatch in conditions between training and testing utterances, being its compensation a main factor for the improvement of system performance. Recently, this task has been carried out by the use of data-driven session variability compensation techniques based on factor analysis, which have become the state of the art in these technologies as can be seen in the periodic NIST Speaker Recognition Evaluations (SRE) [1]. Such techniques can be applied to the best-performing systems working at the spectral level, mainly based on Gaussian Mixture Models (GMM) [2] and Support Vector Machines (SVM) [3], increasing their robustness and accuracy. Among all the different compensation variants, the Nuisance

M. Tistarelli and M.S. Nixon (Eds.): ICB 2009, LNCS 5558, pp. 484–493, 2009.
© Springer-Verlag Berlin Heidelberg 2009

Attribute Projection (NAP) [4] has been used for SVM modelling techniques, presenting the advantages of simplicity and efficiency with respect to other more sophisticated approaches [5]. In particular, NAP has demonstrated its usefulness in systems based on SVM Classification (SVC) using Generalized Linear Discriminant Sequence (GLDS) kernel [3]. Although SVC-GLDS performance is slightly worse than other modelling approaches such as GMM or GMM-SVM [6], it constitutes an additional source of information about speaker identity, and can be combined with other systems by means of fusion [7].

Despite of their unquestionable success, factor analysis techniques still present important challenges to face. The use of such compensation techniques is strongly conditioned to the availability of databases for training the algorithms involved. In real applications the availability of development data in desirable conditions is unfortunately unfrequent. In many situations the technology developers tune their systems with databases coming from a different environment from the conditions of the operational data. This is very typical in forensics, where in each case the conditions of the recordings to analyze are extremely variable in terms of acoustic environment, channel, speaking style, emotional state, language, etc. It is almost impossible to think in the availability of a background database for all the combination of conditions in a possible case. This mismatch in the conditions between background data for system tuning and operational data has been coined *database mismatch* in a recent work [8], and constitutes an important challenge to face in the current state of the art.

In this paper we propose the use of Support Vector Machine Regression (SVR) using a GLDS kernel for robust speaker verification under strong mismatch and forensic conditions. In order to show the adequacy of our approach, we use two different speech databases: *i)* NIST SRE 2008, presenting strong mismatching conditions; and *ii)* Ahumada III, a public database in Spanish coming from authored real forensic cases and collected by Spanish Guardia Civil, which also presents different conditions than NIST databases typically used for background modelling and session variability compensation. This paper is organized as follows. First, the new approach SVM regression is introduced in Section 2. Section 3 presents the proposed SVR-GLDS system for speaker verification. In Section 4, experiments are presented in the two proposed scenarios. Results show the adequacy of SVR-GLDS for robust speaker verification, even when no session variability compensation is performed. Finally, conclusions are drawn in Section 5.

2 Support Vector Machine (SVM) Regression

SVR approach for GLDS speaker verification has been recently proposed by the authors in [9]. In the SVR case the goal is more general than in the widely extended SVC approach. Regression aims at learning a n-dimensional function from the data and classification aims at obtaining a classification boundary. In regression, the vector labels, y_i, are seen as a function of x_i, $g_n(x_i) = y_i$. In a binary classification problem, such as speaker verification, $g_n(\cdot)$ is a discrete function with just two values: $g_n(x_{\text{target}}) = +1$ and $g_n(x_{\text{nontarget}}) = -1$. SVR will try to find the discrete function $f(\cdot) \simeq g_n(\cdot)$.

The main difference between SVC and SVR is the loss function. SVC penalizes the situation where $f(\cdot) < g_n(\cdot)$, but as SVR aims at estimating a function, it also penalizes $f(\cdot) > g_n(\cdot)$. The loss function should consider such effect, and there are different options in the literature. A popular choice is the ε-insensitive loss function [10], where vectors are penalized when $|f(\cdot) - g_n(\cdot)| > \varepsilon$. The objective hyperplane in the SVR case will then be:

$$w = \min\left(\frac{1}{2}w^T \cdot w + C\frac{1}{m}\sum \xi_{c,i} + \xi'_{c,i}\right) .$$
$$\text{subject to: } \begin{cases} 0 \le f(x_i) - y_i \le \xi_{c,i} + \varepsilon \\ 0 \le y_i - f(x_i) \le \xi'_{c,i} + \varepsilon \end{cases} \tag{1}$$

If we compare these criteria with SVC in Equation (2), we observe some differences. We have the SVC penalty variable, $\xi_{c,i}$, for those vectors for which $f(x_i) > g_n(x_i) + \varepsilon$, and a new variable $\xi'_{c,i}$ for those ones for which $f(x_i) < g_n(x_i) - \varepsilon$.

$$w = \min\left(\frac{1}{2}w^T \cdot w + C\frac{1}{m}\sum \xi_{c,i}\right) .$$
$$\text{subject to: } 0 \le \xi_{c,i} \le 1 - y_i f(x_i) \tag{2}$$

The loss functions, $f'_{loss}(x_i)$ (SVR) centered at $f(x_i) = g_n(x_i)$ and $f_{loss}(x_i)$ (SVC) at $f(x_i) = y_i$, are defined in (3) and shown in Fig. 1.

$$f'_{loss}(x_i) = \max\{0, |y_i \cdot f(x_i)| - \varepsilon\} .$$
$$f_{loss}(x_i) = \max\{0, 1 - y_i \cdot f(x_i)\} . \tag{3}$$

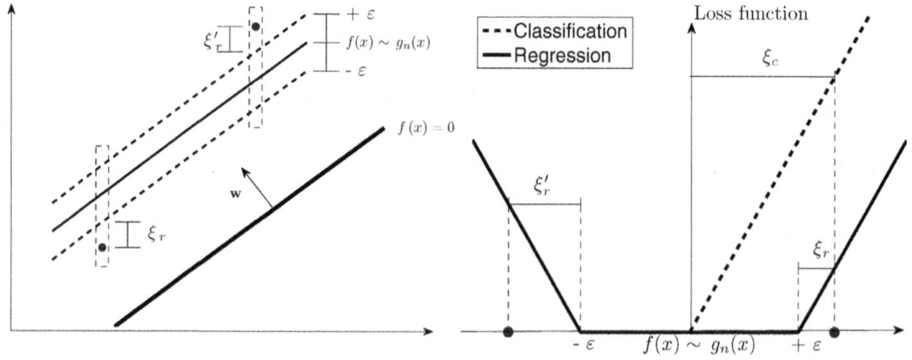

Fig. 1. SVR vs. SVC: boundaries and loss functions

3 SVR-GLDS for Speaker Verification

We propose to use SVR with a ε-insensitive loss function for the speaker verification task. Recently, the authors showed the performance of this novel approach over the core task of NIST SRE 2006 [9], a telephone scenario, obtaining good results in comparison with SVC.

One of the main advantages of using the SVR approach in the GLDS space relates to the use of support vectors for SVM training. On the one hand, SVC uses support vectors which are near the boundary between classes, where the vectors use to be scarce. Moreover, variability in the conditions of speech may significantly change the final hyperplane, introducing undesired variability and therefore performance degradation. On the other hand, SVR selects support vectors from areas where there is a higher concentration of vectors. Thus, the SVC hyperplane may be more sensitive than SVR to outliers, noisy vectors, etc. In this sense, SVR can present a more robust performance than SVC against outlier support vectors due to extreme conditions in some speech utterances.

Another advantage of the SVR approach relies on the use of the ε parameter. There are some works in the literature [10] that relate the ε parameter to the noise or variability of the function estimate. Following such assumptions, we proved in a previous work [9] that tuning ε allows us to adapt the SVR training process to the variability in the expanded feature space.

4 Experiments

4.1 SVM-GLDS Systems

Both ATVS SVC-GLDS and SVR-GLDS systems are based on a GLDS kernel as described in [3]. Feature extraction is performed based on audio files processed with Wiener filtering[1]. The front-end consists on the extraction of 19 MFCC plus deltas. As a first stage to avoid session variability compensation, CMN *(Cepstral Mean Normalization)*, RASTA filtering and feature warping are performed. A third degree polynomial expansion GLDS kernel is performed on the whole observation sequence, and a separating hyperplane is computed between the training speaker features and the background model. NAP is applied for session variability compensation according to [4]. Finally, the T-Norm score normalization technique is applied. We have used the LibSVM library[2] for training both SVM algorithms.

The background set for system tuning is a subset of databases from previous NIST SRE evaluations, including telephone and microphone channels. The T-Norm cohorts were extracted from the NIST SRE 2005 target models, 100 telephone models and 240 microphone models. NAP channel compensation was trained using recordings belonging to NIST SRE 2005 speakers which are present in both telephone and microphone data.

---

[1] A Wiener filtering implementation is available at Berkeley Webpage: http://www.icsi.berkeley.edu/ftp/global/pub/speech/papers/qio

[2] Software available at LibSVM webpage: http://www.csie.ntu.edu.tw/~cjlin/libsvm

4.2 Databases and Experimental Protocol

Experiments have been performed using two different databases. First, the NIST SRE 2008 [1] constitutes a highly mismatching environment. Second, Ahumada III represents real forensic casework speech in conditions different to those of the background data [8].

NIST SRE 2008 database and protocol represents a real challenge in terms on session variability. The training and test conditions for the core task include not only conversational telephone speech data but also conversational speech data recorded over microphone channels involving an interview scenario, and additionally, for the test condition, conversational telephone speech recorded over a microphone channel. The evaluation protocol defines the following training conditions: 10 seconds, 1 *(short2)*, 3 and 8 conversation sides and long conversation; and the following test condition: 10 seconds, 1 *(short3)* conversation side and long conversation. Each "short" conversation, either recorded over a telephone or a microphone, has an average duration of 5 minutes, with 2.5 minutes of speech on average after silence removal. Interview segments contain about 3 minutes of conversational speech recorded by a microphone, most of the speech generally spoken by the target speaker. In our case the experiments followed the core task, namely *short2* training conditions, and *short3* test condition *(short2-short3)*.

Taking into account the test and train channel types, the evaluation protocol can be divided in 4 conditions: *tlf-tlf* (37050 trials), *tlf-mic* (15771 trials), *mic-mic* (34046 trials) and *mic-tlf* (11741 trials).

Ahumada III consists of authorized conversational speech acquired by the Acoustic and Image Processing Department of Spanish Guardia Civil from real forensic cases. The acquisition procedure uses two of the systems and procedures followed by Guardia Civil. As its present release, the recording procedure considered consists of digitalized analog magnetic recordings from GSM mobile calls, from those recordings of this type received in the last ten years, those authorized (case by case) by the corresponding judge after a trial and added to a database registered in the Spanish Ministerio del Interior, known as Base de Datos de Registros Acústicos (BDRA)[3]. In future releases of the database, speech will be included from digital wiretaps recorded directly from Spanish mobile telephone operators, the system known as SITEL (nationwide digital interception system).

Ahumada III Release 1 (Ah3R1)[4] consists of 61 speakers from a number of real cases with GSM BDRA calls across Spain, with a variety of country of origin of speakers, emotional and acoustic conditions, and dialects in the case of Spanish speech. There is no variability dimension is gender, as all of them are male speakers. All 61 speakers in Ah3R1 have two minutes of speech available from a single phone call to be used as unquestioned (control) recording, with the purpose of model enrollment or voice characterization. Additionally, ten speech segments for 31 speakers and five segments for speakers are included for testing issues, each one from a different call. Such fragments present between 7 and 25 seconds of speech, with an average

[3] With reference public scientific file number 1981420003 from Spanish Guardia Civil, Orden Ministerial INT/3764/2004 de 11 de noviembre.

[4] Ahumada III is publicly available for research purposes under license agreement and conditions (contact: http://atvs.ii.uam.es).

duration of 13 seconds. An evaluation protocol has been generated consisting in computing all possible scores from models trained with the enrollment utterances and test segments in the database (27084 trials).

4.3 Results

Strong Mismatching Conditions in NIST SRE 2008. The performance of SVC-GLDS over NIST SRE 2008 is first evaluated with two different configurations: *i)* without including any compensation technique, and *ii)* including a NAP compensation scheme. This shows the effect of compensating variability using NAP with a suitable background database. Table 1 shows the performance of the system detailed per condition. Results are presented both as EER *(Equal Error Rate)* and DCF_{min} as defined in NIST SRE [1]. It is observed that the performance of the system significantly improves when NAP is added, both for EER and DCF_{min} values. The improvement is bigger when strong channel mismatch occurs (*tlf-mic* or *mic-tlf* conditions).

Table 1. EER and DCF_{min} in NIST SRE 2008 short2-short3, for SVC-GLDS and SVR-GLDS with and without NAP session variability compensation

		tlf-tlf	tlf-mic	mic-mic	mic-tlf
SVC	EER	13.8	24.1	17.4	23.5
	DCF_{min}	0.054	0.075	0.075	0.078
SVC + NAP	EER	10.2	**13.9**	**13.0**	15.3
	DCF_{min}	0.047	**0.053**	**0.057**	**0.059**
SVR	EER	10.0	15.1	15.4	16.4
	DCF_{min}	0.045	0.055	0.065	0.064
SVR + NAP	EER	**9.6**	14.3	13.8	**15.0**
	DCF_{min}	**0.045**	0.053	0.060	0.062

In order to use the proposed SVR-GLDS system, tuning the ε parameter is firstly required, and the variation of its performance with respect to such parameter is presented in Table 2. As we saw in [9] the system performance significantly changes as a function of this parameter.

Table 2. EER and DCF_{min} in NIST SRE 2008 short2-short3, for different values of ε in SVR-GLDS without NAP session variability compensation

		$\varepsilon = 0.05$	0.1	0.2	0.4	0.8
tlf-tlf	EER	**9.9**	10.0	10.9	13.5	13.9
	DCF_{min}	0.046	**0.045**	0.047	0.052	0.054
tlf-mic	EER	16.9	**15.1**	16.6	23.8	24.0
	DCF_{min}	0.059	**0.055**	0.063	0.074	0.075
mic-mic	EER	15.7	**15.4**	15.9	16.8	17.4
	DCF_{min}	**0.064**	0.065	0.067	0.074	0.075
mic-tlf	EER	17.0	**16.4**	18.8	22.8	23.6
	DCF_{min}	**0.063**	0.064	0.066	0.078	0.078

In most cases $\varepsilon = 0.1$ significantly improves the system performance, which is very similar to the optimum value in cases where it is seen at $\varepsilon = 0.05$. The optimal value of the parameter is coherent with the experiments presented in [9] using telephone speech in NIST SRE 2006 database and protocol. Thus, without NAP compensation, system tuning of the ε parameter seems robust over different databases, and should be performed one time and not for each one of the four conditions.

Next approach shows the performance of applying the same NAP compensation scheme to SVC-GLDS and SVR-GLDS systems. As the NAP transformation changes the properties of the expanded space a ε tuning is required before using the proposed system, the compensated parameters vectors will be significantly different to the previous ones. Table 3 shows the performance for different values of ε.

In this case the optimal value for the ε parameter varies depending on the condition. The optimal value observed for the non-compensated feature space was $\varepsilon = 0.1$, we will use this value in the rest of experiments. Fig. 2 *a)* presents a comparison between the performance of SVR-GLDS + NAP with $\varepsilon = 0.1$ and the optimal selection of ε for each one of four the conditions. The performance is similar.

Table 3. EER and DCF_{min} in NIST SRE 2008 short2-short3, for different values of ε in SVR-GLDS with NAP session variability compensation

		$\varepsilon = 0.05$	0.1	0.2	0.4	0.8
tlf-tlf	EER	9.7	**9.6**	10.1	10.2	10.2
	DCF_{min}	0.046	**0.045**	0.046	0.047	0.047
tlf-mic	EER	17.0	14.3	**13.3**	13.9	13.9
	DCF_{min}	0.059	0.053	**0.052**	0.053	0.053
mic-mic	EER	15.5	13.8	13.4	**13.0**	**13.0**
	DCF_{min}	0.062	0.060	0.057	**0.057**	0.057
mic-tlf	EER	17.1	**15.0**	15.7	15.3	15.3
	DCF_{min}	0.062	0.062	0.061	**0.059**	0.059

Finally, we compare the performance of the two approaches, SVC-GLDS and SVR-GLDS, with and without NAP compensation scheme. Table 1 shows the comparison in EER and DCF_{min} values for each condition and Fig. 2 *b)* shows the global DET curves of the systems. The system with the best performance in most part of the cases is SVC-GLDS + NAP, obtaining a relative improvement in EER of 31% and 19% in DCF_{min} value. However, the proposed system, SVR-GLDS, presents a similar performance before and after channel compensation. This has the advantage that there is no need of using NAP to obtain similar performance as SVC-GLDS + NAP. It is worth noting that if no channel compensation could be applied because the non-availability of a background database, the SVC-GLDS performance worsens significantly, especially when strong session mismatch occurs (*tlf-mic* and *mic-tlf*). If a suitable database is available, NAP may significantly improve the performance, but if such database is not available or the representative data is scarce, SVR-GLDS seems a convenient option for obtaining robustness. The latter may be the case in many real applications, such as the forensic environment. Moreover, if a suitable database is available SVR-GLDS + NAP provides just a reduced improvement, in both EER and DCF_{min} values (5% and 3% respectively), with respect to SVR-GLDS.

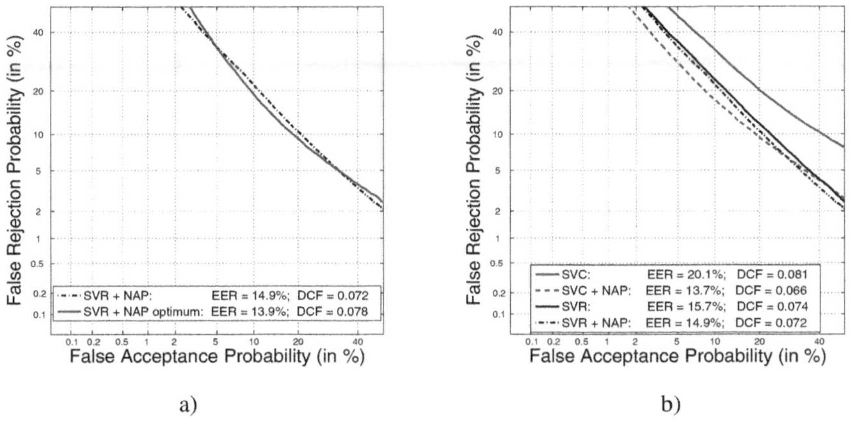

Fig. 2. DET curves in NIST SRE 2008 short2-short3 task: *a)* SVR + NAP ($\varepsilon = 0.1$) and SVR + NAP (ε optimum); *b)* SVC, SVC + NAP, SVR and SVR + NAP

Real Forensic Conditions in Ahumada III. In order to show the performance of the proposed system in similar conditions to those found in real forensic cases, Fig. 3 *b)* shows the SVC-GLDS performance with and without including NAP compensation over Ahumada III. As we observed in NIST SRE 2008, the performance of the system improves when NAP is added, but in this case the relative improvement is significantly lower (13% versus 31% in EER). Moreover, it is observed a degradation in DCF_{min} performance after NAP compensation. The loss in NAP compensation effectiveness can be attributed to the lack of background data in operational conditions. Thus, when a high *database mismatch* is observed among the background and the operational databases, session variability compensation techniques are not only less efficient, but can also even degrade performance [8].

In order to be robust to such lack of background data, the proposed SVR-GLDS approach is used. First, we perform an experiment to show the variability of performance with respect to the ε value. Table 4 presents such results.

Table 4. EER and DCF_{min} in Ahumada III, for different values of ε in SVR-GLDS with and without NAP session variability compensation

		$\varepsilon = 0.05$	0.1	0.2	0.4	0.8
SVR	EER (%)	**14.6**	14.8	15.5	17.4	17.6
	DCF_{min}	0.055	**0.055**	0.058	0.058	0.059
SVR + NAP	EER (%)	15.1	**14.8**	15.6	15.3	15.3
	DCF_{min}	**0.054**	0.056	0.059	0.062	0.062

The system performance with and without NAP is similar, as we saw in NIST SRE 2008 (Table 2 and Table 3). The optimal ε value lays between 0.05 and 0.1, Fig. 3 *a)* shows a comparison between the performance of SVR-GLDS + NAP with these ε values, the technique is not very sensitive. The system performance is similar. Finally, the DET curves of the two approaches with and without session variability compensation are showed in Fig. 3 *b)*.

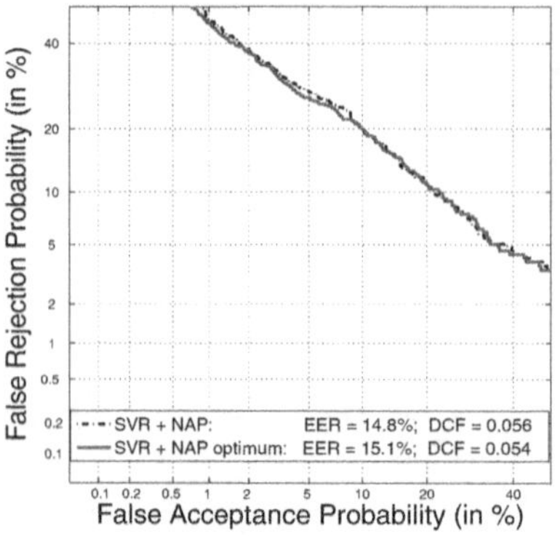

a)

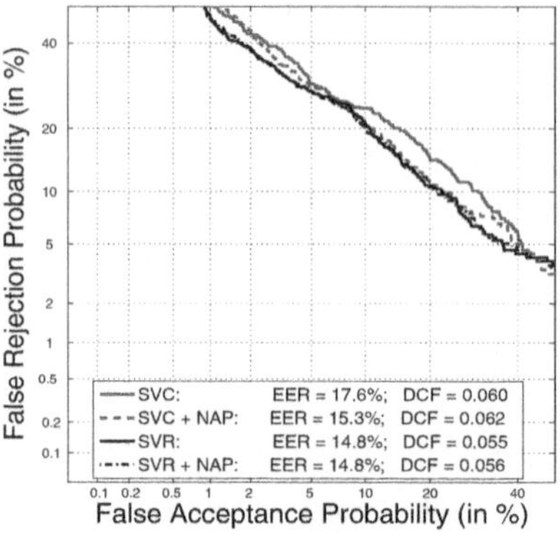

b)

Fig. 3. DET curves in Ahumada III: *a)* SVR + NAP ($\varepsilon = 0.1$) and SVR + NAP (ε optimum); *b)* SVC, SVC + NAP, SVR and SVR + NAP

We observe that using a forensic corpus under database mismatch conditions without any compensation scheme the SVR performance is better than SVC (relative improvement of 16% in EER and 8% in DCF_{min} value), a similar situation can be seen in Fig. 2 b) over NIST SRE 2008. Once we have included NAP the performance of SVR and SVC is similar, but slightly better for SVR. These results are different than those presented for NIST SRE 2008, where in general SVC-GLDS + NAP outperformed SVR-GLDS + NAP. In forensic case, where suitable databases are difficult to obtain SVR seems a more convenient option for obtaining robustness.

5 Conclusions

In this paper we propose a robust approach for speaker verification by means of Support Vector Machine Regression (SVR). The presented work shows that SVR using a GLDS kernel is robust to the lack of a proper background set for NAP session variability compensation, clearly outperforming Support Vector Machine Classification (SVC) in such a situation. This is in accordance with previous work of the authors, where telephone-only speech was used [9]. In this work, two much harder scenarios are proposed. First, NIST SRE 2008 core task is used as a highly mismatching database with multichannel data. Results in this scenario show similar performance among SVC and SVR when NAP is trained with a proper background dataset. However, we simulate the lack of such a database by eliminating the compensation step, and SVR clearly outperforms SVC, showing a much higher robustness. Second, Ahumada III database is used, which consists of speech from real forensic cases. In this scenario, where a background database is not available (i.e., under database mismatch), results show a much lesser effectivity of the NAP compensation technique. Moreover, SVR performs better than SVC, confirming the robustness simulated in NIST SRE 2008.

This work shows that, if a suitable background database for NAP is not available, SVR outperforms SVC, being also a better option in order to obtain robustness to unseen conditions. Moreover, NAP may significantly improve the performance of the system, but under database mismatch its effectiveness is significantly reduced. This is especially important in forensic scenarios, where the availability of a proper database adapted to the case at hand may be almost impossible in many situations.

Future work includes the exploration of different SVR approaches for the GLDS space, such non-linear loss functions and different kernels. We will also explore the complementarity and correlation of SVR with respect to other approaches in the state of the art in speaker verification such as GMM and GMM-SVM.

Acknowledgements. This work has been supported by the Spanish Ministry of Education under project TEC2006-13170-C02-01. We also thank Lt. Cln. Jose Juan Lucena and people from the Acoustics and Image Processing Department from Guardia Civil for their important effort in collecting data for forensic purposes.

References

1. National Institute of Standards and Technology (NIST), 2008 speaker recognition evaluation plan (2008), `http://www.nist.gov/speech/tests/sre/2008/index.html`
2. Reynolds, D.A.: Speaker Verification Using Adapted Gaussian Mixture Models. Digital Signal Processing 10, 19–41 (2000)
3. Campbell, W.M., Quatieri, T.F., Dunn, R.B.: Support Vector Machines for Speaker and language Recognition. Computer Speech and Language 20, 210–229 (2006)
4. Solomonoff, A., Campbell, W.M., Boardman, I.: Advances in Channel Compensation for SVM Speaker Recognition. In: Proc. Of ICASSP, pp. 629–632 (2005)
5. Kenny, P., Oullet, P., Dehak, N., Gupta, V., Dumouchel, P.: A Study of Inter-Speaker Variability in Speaker Verification. IEEE Transactions on Audio, Speech and Language Processing 16(5), 980–988 (2008)
6. Campbell, W.M., Campbell, J.P., Reynolds, D.A., Singer, E., Torres-Carrasquillo, P.A.: Support Vector Machines using GMM Supervectors for Speaker Verification. Signal Processing Letters 13(5), 308–311 (2006)
7. Brümmer, N., et al.: Fusion of Heterogeneous Speaker Recognition Systems in the STBU Submission for the NIST Speaker Recognition Evaluation 2006. IEEE Transactions on Audio, Speech and Language Processing 15(7), 2072–2084 (2007)
8. Ramos, D., Gonzalez-Rodriguez, J., Gonzalez-Dominguez, J., Lucena-Molina, J.J.: Addressing Database Mismatch in Forensic Speaker Recognition with Ahumada III: a Public Real-Casework Database in Spanish. In: Proc. Of Interspeech, pp. 1493–1496 (2008)
9. Lopez-Moreno, I., Mateos-Garcia, I., Ramos, D., Gonzalez-Rodriguez, J.: Support Vector Regression for Speaker Verification. In: Proc. Of Interspeech, pp. 306–309 (2007)
10. Smola, A.J., Schoelkopf, B.: A Tutorial on Support Vector Regression. Tech. Rep. NeuroCOLT2 Technical Report NC2-TR-1998-030, Royal Holloway College (1998)

Scores Selection for Emotional Speaker Recognition*

Zhenyu Shan and Yingchun Yang**

College of Computer Science and Technology, Zhejiang University,
Hangzhou, China, 310027
{shanzhenyu,yyc}@zju.edu.cn

Abstract. Emotion variability of the training and testing utterances is one of the largest challenges in speaker recognition. It is a common situation where training data is the neutral speech and testing data is the mixture of neutral and emotional speech. In this paper, we experimentally analyzed the performance of the GMM-based verification system with the utterances in this situation. It reveals that the verification performance improves as the emotion ratio decreases and the scores of neutral features against his/her model are distributed in the upper area than other three scores(neutral against the model of other speakers, and non-neutral speech against the model of himself/herself and other speakers). Based on these, we propose a scores selection method to reduce the emotion ratio of the testing utterance by eliminating the non-neutral features. It is applicable to the GMM-based recognition system without labeling the emotion state in the testing process. The experiments are carried on the MASC Corpus and the performance of the system with scores selection is improved with an EER reduction from 13.52% to 10.17%.

Keywords: Speaker Recognition, Emotional Speech, Feature Selection.

1 Introduction

Emotion variability, which means the emotion states mismatch between the training and testing utterances, is one of the dominant factors that affect the performance of speaker recognition. In most papers, such kind of recognition is called emotional speaker recognition or affective speaker recognition. Different emotion states will affect speech production mechanism in different ways, which leads to acoustical changes in his/her speech[3, 7]. These changes will induce the distribution variability of training and testing features, thus aggravate the performance.

In recent years, many researches have been devoted to the emotional speaker recognition. In general, two situations are considered based on the emotion states of the testing/training utterances. In the first situation, all kinds of emotion states in the testing utterances can be found in the training utterances. Scherer[6] presented an

* This work is supported by the foundings: NCET-04-0545, NSFC_60525202/ 60533040, 863 Program 2006AA01Z136, PCSIRT0652, ZPNSF Y106705, National Key Technology R&D Program (No. 2006BAH02A01).

** Corresponding author.

M. Tistarelli and M.S. Nixon (Eds.): ICB 2009, LNCS 5558, pp. 494–502, 2009.
© Springer-Verlag Berlin Heidelberg 2009

effective structured training approach which aims at making the system get familiar with the emotion variation of the user's voice. As to the second, a restriction is added that the training utterances contain only one emotion state(usually neutral), while the testing utterances involve other emotion states. The adding of this restriction makes it unnecessary to ask the registered users to provide emotional speech, which is user friendly in applications. However, it will affect the system performance. To alleviate the negative effects of the restriction, Wu *et al.*[2] developed feature modification rule to convert neutral features to emotion features which were used to train the emotion model. Shan *et al.* [10, 11] presented the natural-emotion GMM transformation algorithm to obtain the emotion speech distribution from its neutral speech directly. Wu *et al.*[5] proposed an emotion-dependent score normalization method derived from H-norm to alleviate the effect of speaker-emotion variability. In these methods, the emotion states involved in testing should be available in advance and the testing utterance is of single emotion state.

In most circumstances, the emotional element, if there is any, always lasts ephemeral when users provide testing utterance, since they tend to provide neutral speech for testing in real application(as discussed in [12]). Normally, the emotional testing utterance is a mixture of neutral and emotion speech. A parameter named the emotion ratio is introduced to define the lasting time of emotional element in one sentence. It is calculated as the emotional speech length against the whole length of the sentence. According to this definition, the emotion ratio of testing utterance in the second situation is 100%. And in this new situation, the focus of our research, the emotion ratio is between 0% and 100%. The utterances of different emotion ratios will affect the recognition performance in different ways. The relationship between the emotion ratio and the system performance is analyzed. Two results are concluded from the experiments: the performance is improved as the emotion ratio decreases and scores of the neutral speech against his/her own model mainly are distributed in the upper area. Based on these, a scores selection method is proposed to reduce the emotion ratio of testing utterances for improving the performance. According to the distribution of the scores, it sets a threshold to eliminate emotional features. Experimental results show that the performance of different emotion ratios is improved by the scores selection method.

The remainder of this paper is organized as follows. In Section 2, the emotional speech corpus and the speaker recognition based on GMM are introduced. In Section 3, the system performance on the testing speech with different emotion ratios is evaluated and four scores distributions are analyzed. In Section 4, the scores selection method is proposed and the experimental results are given. Conclusions are drawn in Section 5.

2 Corpus and System Description

2.1 Emotional Speech Corpus

An emotional speech database MASC (Mandarin Affective Speech Corpus) is used in our experiments. It is a corpus provided by linguistic data consortium(LDC) and contains recordings of 68 native speakers (23 female and 45 male) and five kinds of emotional states: neutral, anger, elation, panic and sadness. Each speaker reads 5 phrases, 20 sentences three times for each emotional state and 2 paragraphs only for neutral. These materials cover all the phonemes in Chinese. It is constructed for prosodic and linguistic investigation of emotion expression in Mandarin. Only the sentences in the corpus are applied in the experiments. These sentences include all the phonemes and most common consonant clusters in Mandarin. The types of sentences are: simple statements, a declarative sentence with an enumeration, general questions (yes/no question), alternative questions, imperative sentences, exclamatory sentences, special questions (wh-questions). More details can be found in [4].

2.2 System Description

The speaker verification system used for experiments in this paper is based on the traditional GMM-based method [1]. The diagonal 32-order GMM is applied for modeling. The speech is segmented into frames by a 20-ms window progressing at a 10-ms frame rate and the 13-dimension MFCCs[8] are extracted from the speech frames. T-norm[9] technique is used to normalize the dissimilarity scores.

3 Experiments with Different Emotion Ratios

In this section, two sets of experiments are designed to analyze the influence of different emotion ratios on a GMM-based speaker verification system.

Table 1. The composition of the testing utterances with different emotion ratios

Emotion Ratio (%)	Non-Neutral Speech Length (Sentence)	Neutral Speech Length (Sentence)	Testing Speech Length (Sentence)	Testing Speech Number for each speaker
0	0	3	3	15
20	0.6	2.4	3	45
33.3	1	2	3	45
50	1.5	1.5	3	45
66.6	2	1	3	45
80	2.4	0.6	3	45
100	3	0	3	15

In the first set of experiments, the system performance is evaluated with the testing utterances of different emotion ratios. The training data is the first 5 sentences read for three times and the testing data is the left (15*3) sentences, which is applied to compose the testing utterances with different emotion ratios. The emotion ratio of all the non-neutral utterances in the Corpus is viewed as 100% and the neutral 0%. The testing utterance with certain emotion ratio is obtained by mixing the neutral utterance with anger, elation, panic or sadness one respectively. The details of the testing utterances are shown in Table 1. In the experiment, each sentence in the Corpus is treated as a length unit. The testing utterances contains 3 sentences (units) lasting about 6-12s. Each utterance is verified against the models of all 68 speakers. The result of the experiments is shown in Fig. 1. The EER rises in direct proportion to the emotion ratio in each emotion state. It indicates that the system achieves a better performance when the emotion ratio is lower.

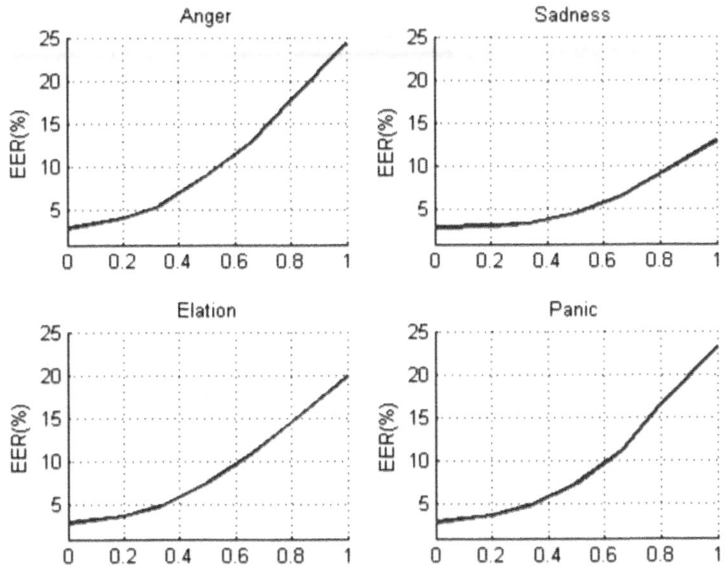

Fig. 1. The performance of emotional speaker recognition with different emotion ratios. X-axis is the emotion ratio and Y-axis is the EER(%).

The aim of the second set of experiments is to analyze the relationship between the four types of scores. The first and the second type is the user's neutral and non-neutral speech scored against the model of him/her own respectively(S_N, S_E), while the third and the fourth is the neutral and non-neutral speech scored against others' model(S_N', S_E'). The training utterances are the same as the above experiment and the left utterances of each emotion state are used for scoring. For each speaker, the number of features of each emotion state exceeds 10500 frames. All features are scored against his/her own model and other 67 speakers' models.

The score distribution of one speaker is shown in Fig. 2. It is a typical distribution of the four scores and almost the same for other speakers. To the same speaker, S_N is mainly distributed in the upper score area and S_E in the relatively lower score area, while the distributions of S_N' and S_E' are similar to each other in the lower area. The mean of all speakers' S_N is larger than other three scores and the mean of [$avg(S_N)$-$avg(S_E)$] is 2.5484. Most speakers (66 of 68) satisfies $avg(S_E) > avg(S_N')$ and the mean of [$avg(S_E)-avg(S_N')$] is 1.4729. The 47 of all speakers satisfies $avg(S_N') > avg(S_E')$ and the left 21 speakers show the opposite. So these results can be concluded as the following formula: $avg(S_N) > avg(S_E) > avg(S_N') \approx avg(S_E')$. It is a statistical result of the four scores. It can be concluded that the scores of the neutral speech against his/her model are distributed in the highest area. That is to say, most neutral features can be distinguished from his/her emotion features when scoring against his/her model. It also indicates that the system performs better when the utterance is long enough.

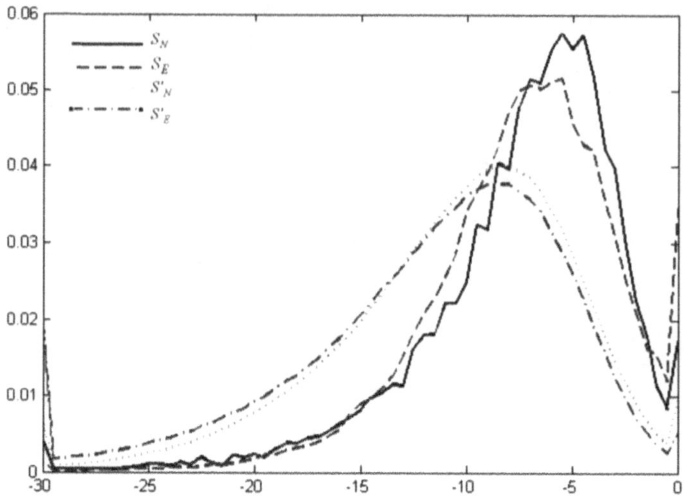

Fig. 2. The 0.5-bin histogram shows the distribution of four types of scores. X-axis is the score value and Y-axis is the possibility.

The results of the experiment elicit two methods to improve the performance of emotional speaker recognition. One is applying the test utterances with low emotion ratio, which suggests the user to provide pure neutral speech for testing or eliminating the non-neutral part from the testing speech. The other is increasing the testing utterances length, yet it can't be satisfied in most applications.

4 Scores Selection

4.1 Method

In the above section, it can be inferred that decreasing the emotion ratio of the testing utterances is preferable to improve the emotional speaker recognition. Thus, we propose a simple method to distinguish the non-neutral features from the neutral ones in the scoring processing of testing with the purpose to reduce the emotion ratio of testing utterances.

In the GMM-based method, the score is calculated as the average of all testing frames' log-posteriori probability. Formally,

$$S = \sum_{t=1}^{T} \log G(x_t) / T \ . \tag{1}$$

where $G(x_t)$ is the posteriori probability for one feature scored against the GMM and T is the number of feature frames. It only adopts the average of all frames' posteriori probability (scores) for reasoning while their distribution information is ignored. In the above section, we discover that the neutral features can be distinguished from emotion features by their score distribution.

A scores selection method with two steps is proposed to eliminate the non-neutral features. First, all features' scores and the average of these scores are calculated. Second, the scores higher than the average are selected for the final score computation as in Equation (1). In other words, not all features are effective in the score computation thus some of them are treated as non-neutral features and eliminated to decrease the emotion ratio.

In this method, the eliminated features from the same utterance are different when scored against the different models. When the testing utterance and the model are obtained from the same speaker, more non-neutral features will be eliminated because $avg(S_N)>avg(S_E)$. This means that the emotion ratio of testing utterance is reduced. While from different speakers, the eliminated neutral and non-neutral features are almost the same because $avg(S_N') \approx avg(S_E')$. Therefore, the emotion ratio changes very little. Though both scores rise up, the score of utterances against his/her own model increases more, so it is advantageous for improving the performance.

4.2 Performance Evaluation

One set of experiments was carried out to evaluate the performance of this method. The setting of the experiments was the same as the ones in Section 3. The result of the system with or without scores selection is shown in Fig.3. The system using scores selection outperforms the traditional method in all emotion ratios, which shows this method is effective for speaker recognition with different emotion ratios. Out of our expectation, the performance is also improved when the whole testing utterance only involves neutral or non-neutral state. The performance is markedly improved when the testing utterance is mixed with anger, elation and panic speech, while not obviously better for sadness. The possible reason is that the discrepancy between neutral and sadness features is smaller than other emotion states. It can be found in Fig.1 that the testing utterances blended with sadness speech are the best of the four verification performances. It indicates the score selection method performs better when the non-neutral features are distinguished from neutral ones. In further, all the testing utterances with four emotion states and different emotion ratios are applied for another verification experiment. The result is shown in Fig.4. The performance is improved with the EER reduction from 13.52% to 10.17%. It indicates its effectiveness for speaker recognition with different emotion ratios.

This experiment is designed to evaluate the effectiveness of the scores selection method by calculating the eliminated non-neutral features. The emotion ratio is reduced after using the selection method as shown in Table 2. When the original emotion ratio is 50%, the emotion ratio deceases to about 43% after the feature selection. And its outcome is almost the same as the 33% emotion ratio when not applying it. For sadness, the reduction of emotion ratio is less, so the performance is not so good. This result shows the scores selection method can eliminate the non-neutral features from the testing utterance to decrease the emotion ratio despite it can't remove all non-neutral features.

In the experiment, the average of the testing utterance's scores is used as the threshold. The scores above the average are applied in the final score computation. Based on the rule of scores distribution, the threshold should be set possibly high. However, the reasoning method is on Bayesian theory and its precondition is that there is enough observed data. If the threshold is set too high, there are not enough observed features when the testing utterances length is fixed. Therefore, in the future, further research is still required to get a better threshold.

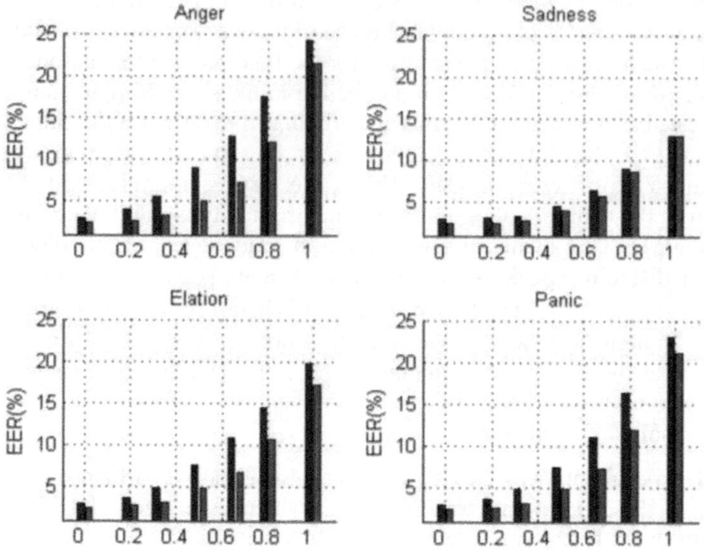

Fig. 3. Comparison of GMM-based system with(right) or without(left) scores selection method in MASC Corpus for four emotion states. X-axis is emotion ratio and Y-axis is EER(%).

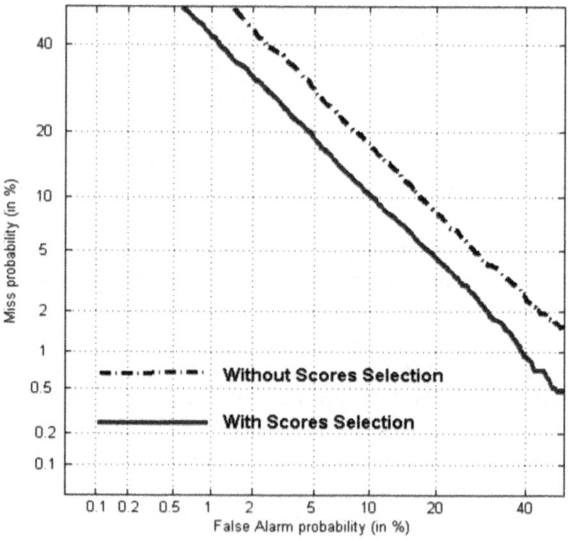

Fig. 4. Comparison of GMM-based system with or without scores selection method in MASC Corpus. The EER of the system with scores selection is 10.17% and without is 13.52%.

Table 2. The emotion ratio(ER) after feature selection. The testing utterances and the model are from the same or different speaker. The first/second row is the proportion of the number of changed non-neutral/neutral features against the original features.

	Anger same/different	**Sadness** same/different	**Elation** same/different	**Panic** same/different
Non-Neutral(%)	38.41/42.69	40.81/44.55	40.11/43.81	40.71/45.43
Neutral(%)	53.27/43.66	47.16/44.66	51.45/47.13	51.32/46.48
ER (original)	50.00/50.00	50.00/50.00	50.00/50.00	50.00/50.00
ER (changed)	41.90/49.44	46.40/49.94	43.81/48.17	44.24/49.43

5 Conclusions

In the emotional speaker recognition, it is a common situation where users provide neutral utterance mixed with emotion element. In this paper, we analyze the influence of the different emotion ratios on the verification performance and two results are concluded. The first is that the performance of emotional speaker recognition is improved as the emotion ratio declines. The second is that the scores of neutral speech against his/her model are distributed in the upper area, compared with other three scores. The four types of scores are neutral and non-neutral speech against the model of himself/herself and other speakers. On the basis of the results, a scores selection method is proposed to eliminate the non-neutral element from testing speech. The experiment on MASC shows that this method can eliminate the non-neutral features from the testing utterance to decrease the emotion ratio. The system performance is improved with the EER reduction from 13.52% to 10.17%.

Acknowledgements

We would like to thank two anonymous referees for their valuable comments and suggestions.

References

1. Reynolds, D.A., Rose, R.C.: Robust Text-Independent Speaker Identification Using Gasussian Mixture Speaker Models. IEEE Ttransactions on Speech and Audio Processing 3(1), 72–83 (1995)
2. Wu, Z., Li, D., Yang, Y.: Rules Based Feature Modification for Affective Speaker Recognition. In: ICASSP 2006, vol. 1, pp. 661–664 (2006)
3. Scherer, K.R., Johnstone, T., Klasmeyer, G.: Can automatic speaker verification be improved by training the algorithms on emotional speech? In: Proceedings of ICSLP 2000, Beijing, China, pp. 807–810 (2000)
4. Wu, T., Yang, Y., Wu, Z., Li, D.: MASC: A Speech Corpus in Mandarin for Emotion Analysis and Affective Speaker Recognition. In: Odyssey 2006, June 2006, pp. 1–5 (2006)
5. Wu, W., Zheng, T.F., Xu, M.-X., Bao, H.-J.: Study on Speaker Verification on Emotional Speech. In: ICSLP 2006, September 2006, pp. 2102–2105 (2006)

6. Scherer, K.R.: A cross-cultural investigation of emotion inferences from voice and speech: implication for speech technology. In: Proc. ICSLP 2000 (2000)
7. Scherer, K.R., Johnstone, T., Bänziger, T.: Automatic verification of emotionally stressed speakers: The problem of individual differences. In: Proc. of SPECOM 1998 (1998)
8. Vergin, R., O'Shaughnessy, D., Farhat, A.: Generalized Mel Frequency Cepstral Coefficients for Large-Vocabulary Speaker-Independent Continuous-Speech Recognition. IEEE Ttransactions on Speech and Audio Processing 7(5), 525–532 (1999)
9. Auckenthaler, R., Carey, M., Lloyd-Thomas, H.: Score normalization for text-independent speaker verification systems. Digital Signal Processing 10, 42–54 (2000)
10. Shan, Z., Yang, Y., Wu, Z.: Natural-Emotion GMM Transformation Algorithm for Emotional Speaker Recognition. In: InterSpeech 2007, pp. 782–785 (2007)
11. Shan, Z., Yang, Y.: Polynomial Function Based Neutral-Emotion GMM Transformation for Speaker Recognition. In: ICPR 2008 (2008) (accepted)
12. Shan, Z., Yang, Y., Wu, Z.: SCS: A Speech Check-in System. In: The 8th International Conference on Signal Processing, vol. 4, pp. 752–756 (2006)

Automatic Cross-Biometric Footstep Database Labelling Using Speaker Recognition

Rubén Vera-Rodríguez[1], John S.D. Mason[1], and Nicholas W.D. Evans[1,2]

[1] Speech and Image Research Group, Swansea University, Singleton Park, Swansea, SA2 8PP, UK
[2] Institut Eurécom, 2229 route des Crêtes, 06904 Sophia-Antipolis, France
{r.vera-rodriguez.405831,j.s.d.mason}@swansea.ac.uk,
nicholas.evans@eurecom.fr

Abstract. The often daunting task of collecting and manually labelling biometric databases can be a barrier to research. This is especially true for a new or non-established biometric such as footsteps. The availability of very large data sets often plays a role in the research of complex modelling and normalisation algorithms and so an automatic, semi-unsupervised approach to reduce the cost of manual labelling is potentially of immense value.

This paper proposes a novel, iterative and adaptive approach to the automatic labelling of what is thought to be the first large scale footstep database (more than 10,000 examples across 127 persons). The procedure involves the simultaneous collection of a spoken, speaker-dependent password which is used to label the footstep data automatically via a pre-trained speaker recognition system. Subsets of labels are manually checked by listening to the particular password utterance, or viewing the associated talking face; both are recorded with the same time stamp as the footstep sequence.

Experiments to assess the resulting label accuracy, based on manually labelled subsets, suggest that the accuracy of the automatic labelling is better than 0.1%, and thus sufficient to assess a biometric such as footsteps, which is anticipated to have a much higher error rate.

Keywords: Automatic database labelling, speaker verification, score normalisation, footstep biometric, multimodal biometrics.

1 Introduction

When developing a new biometric one of the first considerations entails the collection of a representative dataset of meaningful size. Data collection is notoriously expensive and problematic but instrumental to the success of a new project and confidence in the results. Many fundamental questions need to be addressed. Among them are: the number of samples; the number of clients; the enrolment and labelling procedures. To get these wrong would devalue the database and any results derived from it.

Doddington's 'rule of 30' [1] gives some guidance regarding the number of samples, the expected error rate and the confidence in the result. He states that

M. Tistarelli and M.S. Nixon (Eds.): ICB 2009, LNCS 5558, pp. 503–512, 2009.
© Springer-Verlag Berlin Heidelberg 2009

'to be 90 percent confident that the true error rate is within +/- 30 percent of the observed error rate, there must be at least 30 errors'. Thus if we expect a relatively higher error rate we may be satisfied with a smaller database than if we expect a relatively lower error rate; there is a trade-off between database size and expected error rate. Of course if we are researching a new biometric then we cannot know the expected error rate; it is likely that one of the fundamental goals of the research is to establish precisely this. We might opt to conduct an initial trial on a small dataset to help us decide upon the required database size, however to extract the best value the database should be sufficient for both today's and tomorrow's research. Advances in biometrics research often come from very large databases designed to facilitate the learning of complex modelling and normalisation strategies which may not have been possible on smaller datasets. Thus in order to prepare for the research of tomorrow it is in any case always advantageous to collect as large a database as possible within economic and practical constraints.

The financing of database collection can however be difficult to obtain especially when the research involves a new biometric for which a baseline error rate does not exist and we cannot reliably predict the potential of the biometric under investigation. It is sometimes possible to reduce the cost of collection through automated collection systems and this can go someway to help the labelling of the collected data, namely the assignment of ownership to each collected sample.

In the collection system described here there are two modes, supervised and unsupervised. The initial enrolment of each person participating in the database collection (donor) is supervised and hence the allocation of an identity label to this enrolment data is also supervised. Subsequently, and for the large majority of the data collection, the process is unsupervised. Thus a strategy is necessary to assign the correct donor identity to each of the recorded signals. This paper describes such a strategy using a combination of automation plus human cross-checking. The automation itself uses a biometric approach based on person specific spoken utterances captured at the same time as each of the footstep signals is captured and recorded.

2 Concept of Automatic Labelling

The cost of human resources can represent a potential barrier to research: we might have difficulty in justifying and financing the collection and manual labelling of a large database unless the commercial potential of a new biometric is proven. However, we might not be able to demonstrate the true potential of the biometric without a large database and we have something of a chicken and egg situation. Over recent years we have been investigating a relatively new and little researched biometric and in this paper we describe our approach to collect a large database without the full cost of manual labelling. The idea is to collect a multimodal biometric database, where the primary mode of interest is footsteps and the secondary modes are speech, talking face and gait. Each set (comprising footstep, speech, talking face and gait sequence) is assumed to be consistent, coming from just one person, by a time stamp assigned at the time of capture from a single clock. Of course, in the absence of supervision, more than one person could mischievously combine to give anomalous data sets but this is thought to be a small risk under the given conditions. The secondary modes, particulary speech and talking face, are included specifically to assist in labelling the database in order to subsequently carry out biometric research based on footsteps. This labelling process is a combination of enrolment, automation and human cross-checking. This paper describes the automation process based on (acoustic) speaker recognition with the goal of accurately labelling the database. Speech is chosen for convenience and is also one of very low proven error rates [2]. Speech also perhaps reduces the need for sequestering when compared to other biometrics with lower EER, like fingerprints or iris. The assignment of a speaker specific PIN gives a text-dependent characteristic and importantly provides a means of human based cross-checking of the labels. The PIN is allocated at the (supervised) enrolment stage.

Naturally, even when the speaker recognition has proven low error rates, there will nonetheless be concerns over the labelling accuracy and its unpredictable repercussions. This though, has to be seen in the context of (i) international evaluation campaigns which have been shown to contain labelling anomalies, and (ii) the trade-off between a large database (with a small number of labelling errors) which provides the richness required for the development of complex modelling and normalisation algorithms and a smaller database with (possibly) no labelling errors. In any case for a larger database perhaps inevitably with some labelling anomalies, we suppose that (i) the potential of the biometric can be assessed and (ii) we have greater confidence in the results than we would otherwise have for a smaller database. Even the sceptic has to accept that it would be unwise to suppose that a database does not have labelling anomalies and their potential for occurrence is a function of the size of the database. In the approach described in this paper we accept that labelling errors are possible, and this paper describes the efforts to minimize the number of errors using a range of strategies.

Speech is used for the automatic labelling as shown in Figure 1. There is a large data set captured in an unsupervised mode and hence unlabelled. This is introduced into an speaker recognition system, which is trained on manually labelled data as ground truth. Then a decision based on the speech signal is taken to obtain new labels. These new labels apply not only to the speech but also to the other contemporaneous signals in the data set, namely the footstep signals together with the talking face and gait image sequences. We refer to one example of these four signals as a *set* with the primary interest here being the footsteps and the speech. The talking face has benefit in the manual labelling and cross-checking for anomalies of a set.

At the time of carrying out the following experiments the database was comprised of a total of 11,537 sets. The collection of the database was unsupervised, apart from the initial enrolment session of each person in the system, where normally around 10 sets were manually labelled. Apart from the enrolment data

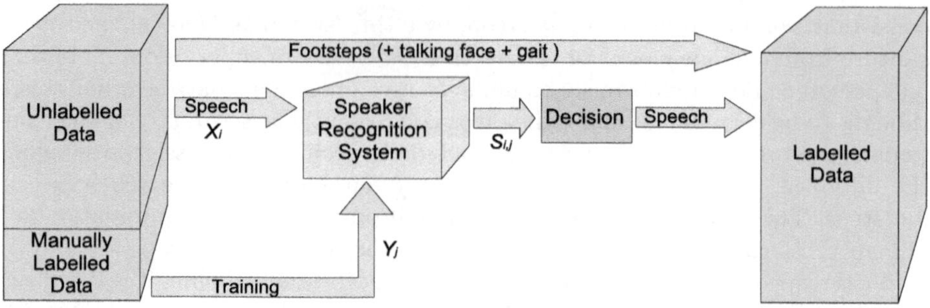

Fig. 1. Speech-based automatic labelling system. The speaker recognition system trained on labelled data receives unlabelled data. The decision process considers all combination of scores from the data on the left and systematically labels the most likely sets, passing them across to the right repeating the process until all data is labelled or discarded.

(1,123 sets in total from 127 clients), more data was manually labelled (1,385 sets in total). This labelling exercise took part during the collection stage itself and proved the enormity of the task, amply demonstrating the need for automation.

In the speaker recognition system, each X_i test data, left of Figure 1, being i = 1....9,029, is tested against the Y_j models, being j = 1....127, created from the manually labelled data. This could be seen as a form of 1-in-N identification if there was confidence that X_i definitely came from one of the Y_j persons in the enrolled set. Alternatively, the task may be viewed as a specific case of verification applying acceptance to the most likely pairing across all X, Y combinations. The benefit of this interpretation is (at least) two-fold, both being of critical importance. The first covers the case when X_i is outside of set Y. The second is in terms of score normalization. Before any test-to-model score is assessed it is normalized using standard techniques well established in the speaker recognition world [8,9,10].

The key point is that, the assignment of any data set is prioritized in an order of confidence. This means that the most likely assignments take place first. Also once X_i is assigned, to say Y_j, then it is possible to re-train the model for person Y_j in an adaptive manner, potentially improving the model as a representation of person Y_j. Of course such adaptation can be dangerous in the case of false acceptance. We address this issue in our experimental procedures (Sect. 5).

3 Speaker Recognition System

The speaker recognition system shown in Figure 1 is based on a linear frequency cepstral coefficient (LFCC) front-end and a Gaussian mixture model (GMM) system [4], using SPro[1] and ALIZE[2] open source toolkits. The GMM system

[1] http://gforge.inria.fr/projects/spro/
[2] http://www.lia.univ-avignon.fr/heberges/ALIZE/

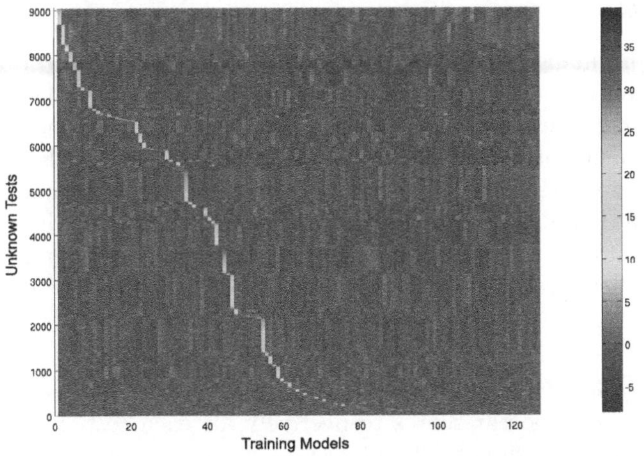

Fig. 2. Representation of the score matrix after using the speaker verification system. Test signal scores against trained models with manually labelled data (enrolment plus human labelling).

is close to the description in [5]. The signal is characterised by 33 coefficients including 16 LFCC, their first derivative coefficients and the energy derivative (16LFCC + 16Δ + ΔE).

The first experiment relates to 9,029 unlabelled sets tested against each of the 127 models trained on the manually labelled sets (2,508). The recogniser scores $S_{i,j}$ can form a score matrix as represented in Figure 2 with scores for the 9,029 unlabelled sets plotted against the 127 models. The colour scale in the figure shows the range of the scores. The 127 models are sorted according to the amount of training data, with the largest on the left to models with the smallest amount of training data on the right. The correlation with the amount of training data and high scores is distinct, with almost half of the 127 models having very few high scores. The 9,029 test signals have been sorted to have high scores together per model, this way we could have an idea of the number of tests that belong to each model. As Figure 2 shows, there are not many high scores for models on the right; this suggests that there are not many test signals that belong to those models. It is clear that score normalisation is essential in order to remove the influence of the amount of data in the training models.

4 Score Normalisation

Score normalisation is widely used in biometrics, for example it is a key factor when fusing different biometric modalities, and here in the context of speaker recognition it is required to balance inherent test data and model variation. The most popular methods for score normalisation are T-norm [8,9] and Z-norm [10].

4.1 Test Normalisation

First, a test normalisation is applied to the scores from each of the Y models for a given test utterance (the rows of the score matrix in Figure 2). This normalisation attempts to align test scores by using scores from impostor models. In the case where a particular unknown test belongs to one of the models, only one of the scores can be a true score while the rest correspond to impostor scores. Of course, in the present context the one particular model is unknown, hence here all 127 scores are used with the one potentially true score assumed to be swamped by the remaining 126 scores. A general equation for the test normalisation is given by:

$$STn_{i,j} = (S_{i,j} - \mu_i)/\sigma_i \tag{1}$$

where $STn_{i,j}$ are the normalised scores and μ_i and σ_i the mean and standard deviation of the impostor scores respectively for each test. Figure 3 (a) shows an example of a test signal. The plot shows scores against the 127 models corresponding to a horizontal trajectory across Fig 2), and indicating a very high likelihood of ownership of the given test signal by model number 105. The profile after the test normalisation can be observed in Figure 3 (a) with red dashed line. As observed, the effect of this normalisation makes a wider range for the scores and also gives a common zero threshold. Figure 3 (b) shows the same but for one of a bad test example where there is no score clearly higher than the rest.

4.2 Model Normalisation

Second, model normalisation is applied which attempts to align between-speaker differences by producing statistical parameters for each model to align the scores to zero. Figure 3 (c) shows an example of the score distribution for a good model. As the figure shows there are some high scores on the right (zoom of the distribution on the top of Figure 3 (c)) that are likely to correspond to true scores for this model, and a large number of low scores on the left, which would be likely to belong to other models or not within the 127 set. To carry out a normalisation similar to Z-norm, it is necessary to obtain the distribution for the out of class data to calculate its mean and standard deviation. Therefore, the score distribution was approximated by two gaussian distributions, using the one with lower mean and higher weight as the distribution of the out of class data. A general equation for the model normalisation is given by:

$$SZTn_{i,j} = (STn_{i,j} - \mu_j)/\sigma_j \tag{2}$$

where $SZTn_{i,j}$ are the normalised scores and μ_j and σ_j are the mean and standard deviation of the out of class scores respectively for each model. Figure 3 (e) shows the effect that this normalisation makes to the distribution, i.e. aligning the out of class distribution to zero. Figure 3 (d) shows an example of a bad model, one with either small amount of training data, or few high scores. The effect of the model normalisation can be observed in Figure 3 (f). This normalisation is called ZT-norm and in [11] a reduction of 20% EER is reported when compared with standard Z-Norm.

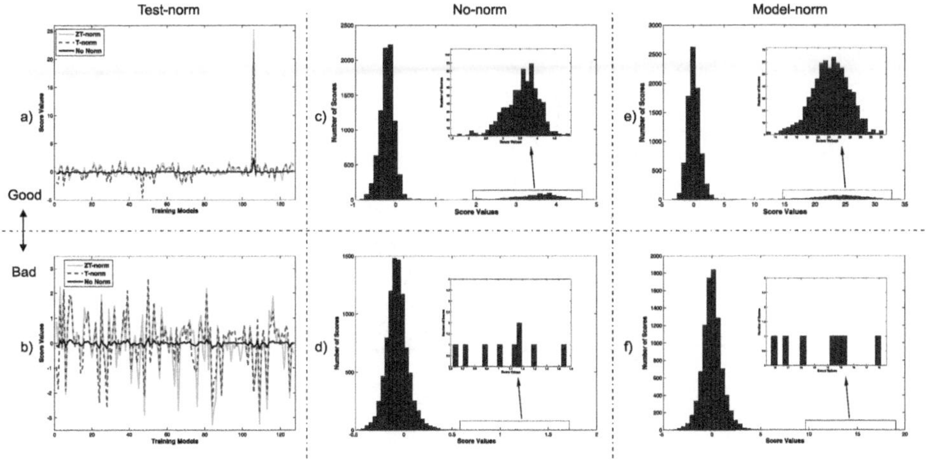

Fig. 3. (a)Example of scores for a good test utterance against 127 models. Black solid profile before normalisation. Red dashed profile after T-norm. Solid light blue profile after ZT-norm. (b) Same as (a) but for a bad test utterance. (c)Example of score distribution for a good model and 9,029 test sets before normalisation. (d) Same as (b) but for a bad model. (e) Same model as (c) but after ZT-norm. (f) Same model as (d) but after ZT-norm.

5 Automatic Labelling, Implementation and Results

This section describes the iterative process followed to carry out the labelling of the database. Figure 4 shows the structure of the database, which is a ragged array with different amounts of data per person and per labelling class: enrolment, manually labelled, automatically labelled (by speaker recognition) and unlabelled. The diagram represents a state during the iterative recognition and label allocation process. The enrolment session is the only one that is square as there are consistently 10 data sets per model. Then, there is another extra set of manually labelled data, followed by the automatically labelled set and finally the unlabelled set. Unlabelled data on the top is iteratively moved down as it becomes labelled by an iterative process. The automatic labelling process may be summarised by the following steps:

- Test all unlabelled data against all models (obtaining scores as per Figure 2).
- ZT-norm the scores.
- For each unassigned set find the most likely model.
- Sort all unassigned sets.
- Label the best set(s).
- Update model(s) using newly labelled data.

This process is repeated until the label confidence threshold is reached shown as the knee point in Figure 5 (a). As indicated above, the assignment of the unknown sets to the models is an iterative process. In each iteration unlabelled

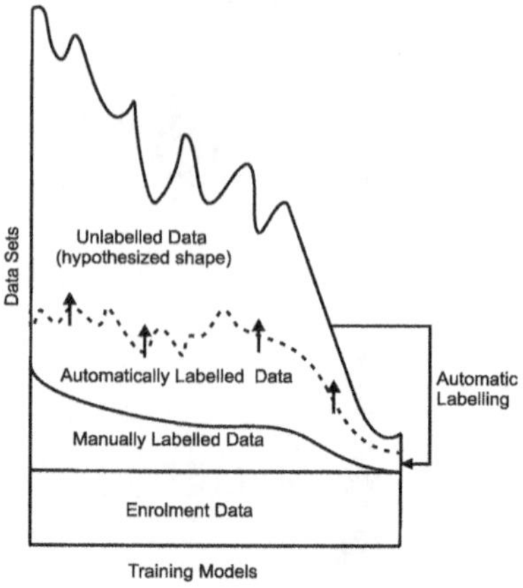

Fig. 4. Database structure. Enrolment session data on the bottom, then manually labelled data, assigned data and unlabelled data on the top. In each iteration the best sets of the unlabelled data are assigned to the respective models according to the criteria defined.

data is tested against 127 models producing a score matrix as in Figure 2. Then a ZT-norm is carried out as described in Section 4. At each iteration, data with the highest overall score statistics (best sets) are assigned to their respective models, and then are used to re-train the models in an adaptive manner. To prioritize the tests to be assigned to the models in an order of confidence, the tests are sorted considering statistics of the peak score and variance of the remaining 126 scores for each test (rows of Figure 2). In this way, tests with less confidence remain until the last iterations. This strategy is independent to the number of models, which in a dynamic database can be variable.

At each iteration it is possible to estimate a figure of the relative merit of the assignment. Figure 5 (a) shows the EER for the tests assigned at each iteration. The left part of the figure is a region with high confidence due to the sorting of the tests assigned (tests have similar profiles as example in Figure 3 (a)). The middle part of Figure 5 (a) is a region with less confidence, and the right part is a region of very low confidence suitable for manual labelling or rejection, where examples such as those in Figure 3 (b) can be found. Figure 5 (a) shows how the EER stays in a range of 0.06% - 0.1% until the last 400 tests are assigned. Then the error increases to 0.58%. This suggests that either this data are different in some way to data previously assigned either due to within class variation or alternatively the data comes from outside of the set of the 127 speakers. Therefore, this data in this region and to the right should be labelled manually or rejected. Using the above strategy, 8,867 out of the 9,029 unlabelled sets were each labelled to one of the 127 models.

In order to assess the labelling accuracy, a further speaker recognition experiment was conducted using manually labelled sets. This represents a meaningful evaluation of the speaker recognition system and hence a reasonable assessment of the automatic labelling of the database. The results are shown in Figure 5 (b). The profiles relate to two systems trained on a manually labelled set of data common to both, namely a smaller set of further manually labelled data and a much larger set (8,867) of automatically labelled data. most reassuringly, the error rate for the later are smaller than for the manually labelled set. This might be attributed to differences in the gender ratio: in the manually labelled set a balance across gender was sought; in the automatically labelled set there are proved to be more males than females given that is well known that speaker recognition error rates tend to be higher for females. This could account for the difference in the two profiles.

Finally, having lower confidence levels, data in the vicinity of the knee point in Figure 5 (a) could be manually labelled. Even so, the numbers requiring manual labelling have been massively reduced by the automatic procedure described with a predicted error rate in the region of 0.1% well within limits for study of footsteps as a biometric, the ultimate goal of this work.

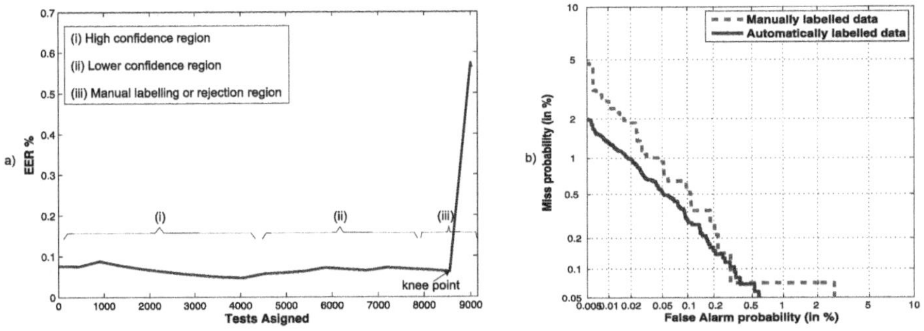

Fig. 5. (a) EER against test assigned in the iterations. (b) DET curve estimating the expected error of the speaker recognition system. Manually labelled data in dashed profile, and automatically labelled data in solid profile.

6 Conclusions

This paper describes an automatic system to label a database designed to assess footsteps as a biometric. The novel contribution is the way in which the data has been collected and labelled. A total of four modes were collected simultaneously namely footsteps, speech, talking face and gait. All four modes within a set were linked by the same time stamp. Of principal importance are the footsteps, followed by the speech, the later included for labelling both manually and automatically. The large majority of the 9,000 plus signals have been labelled automatically using speech as a biometric. This has significantly reduced the manual effort and therefore cost of creating the database. Most large scale collections such as this one, are likely to have some form of data anomalies. Here we estimate the labelling errors to be less than 1% and thus, sufficient for the objectives of studying footsteps as a biometric. Finally it is clear that this system could be applied to the collection of other large scale biometric databases, where supervision and labelling is likely to prove expensive.

References

1. Doddington, G.R., Przybocki, M.A., Martin, A.F., Reynolds, D.A.: The NIST speaker recognition evaluation: Overview methodology, systems, results, perspective. Speech Communication 31, 225–254 (2000)
2. Toledano, D.T., Esteve-Elizande, C., Gonzalez-Rodriguez, J., Fernandez-Pozo, R., Hernandez-Gomez, L.: Phoneme and Sub-Phoneme T-Normalization for Text-Dependent Speaker Recognition. In: Proc. IEEE Speaker and Language Recognition Workshop (Odyssey) (2008)
3. Vera-Rodriguez, R., Lewis, R.P., Mason, J.S.D., Evans, N.W.D.: A Large Scale Footsteps Database for Biometric Sutdies Created using Cross-Biometrics for Labelling. In: Proc. 10th IEEE International Conference on Control, Automation, Robotics and Vision, ICARCV, Vietnam (2008)
4. Reynolds, D.A., Quatieri, T.F., Dunn, R.: Speaker Verification Using Adapted Gaussian Mixture Models. Digital Signal Processing 10(1-3), 19–41 (2000)
5. Bonastre, J.-F., Scheffer, N., Fredouille, C., Matrouf, D.: NIST 2004 Speaker Recognition Evaluation Campaign: New LIA Speaker Detection Platform Based on ALIZE Toolkit. In: Proc. NIST SRE 2004 Workshop, Spain (2004)
6. Fauve, B., Evans, N.W.D., Mason, J.S.D.: Improving the Performance of Text-Independent Short Duration GMM and SVM Based Speaker Verification. In: Proc. Odyssey: the Speaker and Language Recognition Workshop (2008)
7. Fauve, B., Bredin, H., Karam, W., Verdet, F., Mayoue, A., Chollet, G., Hennerbert, J., Lewis, R., Mason, J., Mokbel, C., Petrovska, D.: Some Results From The Biosecure Talking Face Evaluation Campaign. In: Proc. ICASSP (2008)
8. Auckenthaler, R., Carey, M.J., Lloyd-Thomas, H.: Score normalisation for text-independent speaker verification system. Digital Signal Processing (DSP), a review journal - Special issue on NIST 1999 Speaker Recognition Workshop 10(1-3), 42–54 (2000)
9. Navrati, J., Ramaswamy, G.N.: The awe and mystery of T-Norm. In: Proc. Eurospeech, Geneva, pp. 2009–2012 (2003)
10. Li, K.P., Porter, J.E.: Normalizations and selection of speech segments for speaker Recognition Scoring. In: Proc. ICASSP, pp. 595–598 (1988)
11. Zhang, S., Zheng, R., Xu, A.: A Comparative Study of Feature and Score Normalization for Speaker Verification. In: Proc. IEEE Odyssey Speaker and Language Recognition Workshop, pp. 531–538 (2005)

Towards Structured Approaches to Arbitrary Data Selection and Performance Prediction for Speaker Recognition

Howard Lei

The International Computer Science Institute, Berkeley, CA
hlei@icsi.berkeley.edu
http://www.icsi.berkeley.edu

Abstract. We developed measures relating feature vector distributions to speaker recognition (SR) performances for performance prediction and potential arbitrary data selection for SR. We examined the measures of mutual information, kurtosis, correlation, and measures pertaining to intra- and inter-speaker variability. We applied the measures on feature vectors of phones to determine which measures gave good SR performance prediction of phones standalone and in combination. We found that mutual information had an -83.5% correlation with the Equal Error Rates (EERs) of each phone. Also, Pearson's correlation between the feature vectors of two phones had a -48.6% correlation with the relative EER improvement of the score-level combination of the phones. When implemented in our new data-selection scheme (which does not require a SR system to be run), the measures allowed us to select data with 2.13% overall EER improvement (on SRE08) over data selected via a brute-force approach, at a fifth of the computational costs.

Keywords: Text-dependent speaker recognition, mutual information, relevance, redundancy, data selection.

1 Introduction

Conversational speaker recognition (SR) involves the task of determining whether a certain speaker spoke in a certain conversation. SR systems have historically relied on GMM speaker models [1], and involves scoring test utterances against target speaker models to determine if the target speaker spoke in the test utterance. The Equal Error Rate (EER), which represents the rate at which false accepts equal false rejects, is a common SR system evaluation standard.

Unit-based text-dependent speaker recognition (UTSR) is the speaker recognition approach where only certain units (i.e. words, phones, syllables) found in speech data are used to construct entire speaker recognition systems [2]. These approaches have been successfully applied in conversational speaker recognition tasks, where the data consists of lengthy conversations between speakers, and the speech is not lexically constrained [2][3]. While discarding much of the speech,

M. Tistarelli and M.S. Nixon (Eds.): ICB 2009, LNCS 5558, pp. 513–522, 2009.
© Springer-Verlag Berlin Heidelberg 2009

the advantages of UTSR for conversational speaker recognition (SR) are threefold: to focus speaker modeling power on more informative regions of speech, to reduce intra-speaker lexical variability, and to reduce the total amounts of data required for faster processing.

The units examined in the past include word N-grams, syllables, phones, and Automatic Language Independent Speech Processing (ALISP) units [4] (which are designed to mimic the phones) and MLP-based phonetic units [5]. Many of the units, such as the words and phones, are used only because their transcripts are readily available via Automatic Speech Recognition, and are incorporated without regard to their actual speaker discriminative abilities. Moreover, there has been no evidence suggesting that words, phones, and/or syllables are ideal sets of units for UTSR. The eventual aim of this work is to allow one to step beyond the use of these units, and to examine the speaker discriminative capabilities of all possible speech segments that can act as units.

This work involves the development of measures as computationally inexpensive ways of determining which units are speaker discriminative based solely on feature vectors of the units. The measures would allow for a quick determination of SR performances of each unit without having to run the SR system, which could take days depending on the units used. For an arbitrary set of units, one task is to compute the measures on the feature vectors of each unit separately. Measures computed in this matter (referred to as relevance measures) would give an indication of the relevance of the unit with respect to the SR task. Measures that have high correlation (in magnitude) with the SR EERs of the units would have good predictive value for SR, and would eventually be good measures for arbitrary data selection.

In UTSR, the units are usually combined at either the feature-level or score-level. To get a good prediction of the effectiveness of unit combination, another task is to compute the measures on pooled features for sets of units, so that a correlation between the measures and the EER achieved via the combination of the set of units is obtained. Measures computed in this manner (referred to as redundancy measures) give an indication of the redundancy of the units amongst one another, whereby units that combine well are less redundant, and vice versa.

Finding effective relevance and redundancy measures will allow for the eventual selection of arbitrary sets of units that produce the best SR performances. Note that the task of data selection is more difficult than the related task of feature selection, in that there are typically many more feature vectors than the number of feature dimensions in a speech utterance, and there are no pre-defined orderings of the feature vectors as opposed to the feature dimensions.

This paper is organized as follows: Section 2 describes the database and our SR system for computing the EERs, section 3 describes the measures, section 4 describes our data-selection scheme, section 5 describes the units used, section 6 describes the experiments and results and provides a brief discussion, and section 7 provides a summary of the current work and describes the applicability of this work to future research in UTSR.

2 Data, Preprocessing, and Speaker Recognition

We used the Switchboard II and Fisher corpora for universal background speaker model training, SRE06 for development, and SRE08 for testing. All corpora consists of telephone conversations between two unfamiliar speakers. A conversation side (roughly 2.5 minutes for non-Fisher and 5 minutes for Fisher) contains speech from one speaker only. 1,060 conversation sides with 128 speakers are used for SRE06, and 1,108 conversation sides with 160 speakers for SRE08. 1,553 background conversation sides are used from Switchboard II and Fisher. Only female English telephone conversation sides are used for this work. There are ~55,000 total trials for SRE06 with ~7,000 true speaker trials, and ~47,000 trials for SRE08 with ~6,500 true speaker trials. We are provided with force-aligned phone ASR decodings for all conversation sides by SRI, obtained via the DECIPHER recognizer [6].

A 512-mixture GMM-UBM system [1] with MAP adaptation and MFCC features C0-C19 (with 25 ms windows and 10 ms intervals) with deltas is used for computing the EERs of units. The ALIZE implementation is used [7], and the MFCC features are obtained via HTK [8].

3 The Measures

We have implemented various measures for determining the relevance and redundancy of units. For the relevance task, we want to determine how well the measure(s), when computed using the feature vectors of a unit, correlate with the SR performance of the unit. For the redundancy task, we want to determine how well the measure(s), when computed using the feature vectors of a pair of units, correlate with the EER improvement of the MLP-based combination of SR scores of the unit pairs. The measures include mutual information, kurtosis, intra- and inter-speaker variances, Fisher's ratio, and Pearson's correlation.

3.1 Mutual Information as Relevance Measure

Mutual information, which measures the mutual dependence of two variables, has historically been used successfully in the related area of feature selection, such as in [9], [10], and [11]. Typical feature selection algorithms involving mutual information select features with high mutual information with respect to a classification label or class, such that the features are relevant to the classification task. For the case of SR, the classification classes (which are discrete) are the distinct speakers. The mutual information between a continuous vector $\boldsymbol{X}$ and a discrete classification label Y (with distributions $p(\boldsymbol{x})$, $p(y)$, and $p(\boldsymbol{x}, y)$), is given as follows:

$$I(\boldsymbol{X}; Y) = \sum_{y} \int_{\boldsymbol{x}} p(\boldsymbol{x}, y) \log \frac{p(\boldsymbol{x}, y)}{p(\boldsymbol{x})p(y)} d\boldsymbol{x}. \tag{1}$$

For this particular work, the mutual information is computed between the feature vectors ($\boldsymbol{X}$) constrained by a particular unit, and the speakers (Y). The technique described in [9] for computing the mutual information via approximation of equation 1 is used here. The Parzen windowing technique is used to model the distribution of the continuous-valued feature vectors:

$$\hat{p}(\boldsymbol{X}) = \frac{1}{S}\sum_{i=1}^{S}\delta(\boldsymbol{x}-\boldsymbol{x}_i, h). \tag{2}$$

where $\delta(\cdot)$ is the Parzen window function [12]. A Gaussian window is used, where h represents the standard deviation.

3.2 Kurtosis as Relevance Measure

Kurtosis is a measure of peakiness and/or non-Gaussianity of a random variable. Kurtosis mismatches between training and test conversation sides have been shown to adversely affect speaker recognition performance, and kurtosis feature normalization is an effective way to improve speaker recognition performance [13]. Kurtosis is defined for random variable X as:

$$Kurtosis(X) = \frac{E(x^4)}{E(x^2)^2} - 3 \tag{3}$$

For this work, the kurtosis measure is evaluated on the entire set of feature vectors for each unit.

3.3 Fisher's Ratio, Intra- and Inter-speaker Variances as Relevance Measures

Fisher's ratio and intra- and inter-speaker variances all give measures of class-separability, whereby features/data with high Fisher's ratio, high inter-speaker variances, and low intra-speaker variances have high relevance with respect to the classification task. For this work, Fisher's ratio is the ratio of the inter- to intra-speaker variances of the feature vectors of a unit, where we estimated the inter-speaker variance as follows:

$$\sum_{speaker:s}(\boldsymbol{\mu}_s-\boldsymbol{\mu})^T(\boldsymbol{\mu}_s-\boldsymbol{\mu}). \tag{4}$$

and the intra-speaker variance as follows:

$$\sum_{speaker:s}\frac{1}{N_s}\sum_{i\in s}(\boldsymbol{x}_i-\boldsymbol{\mu}_s)^T(\boldsymbol{x}_i-\boldsymbol{\mu}_s). \tag{5}$$

where N_s and $\boldsymbol{\mu}_s$ are the number and average of feature vectors respectively for speaker s, $\boldsymbol{\mu}$ is the overall average of the feature vectors, and $\boldsymbol{x}_i$ is feature vector i.

3.4 Pearson's Correlation as Redundancy Measure

For a pair of units, Pearson's correlation is computed using the average MFCC feature values of each unit for each conversation side. Specifically, for each conversation side, the average values of the MFCC feature vectors for each unit are computed. Pearson's correlation between the averaged values of each unit is computed across all conversation sides. Note that the correlation is computed separately for each dimension of the feature vectors, and an overall correlation is obtained by averaging the correlations of each dimension. Fig. 1 illustrates this computation.

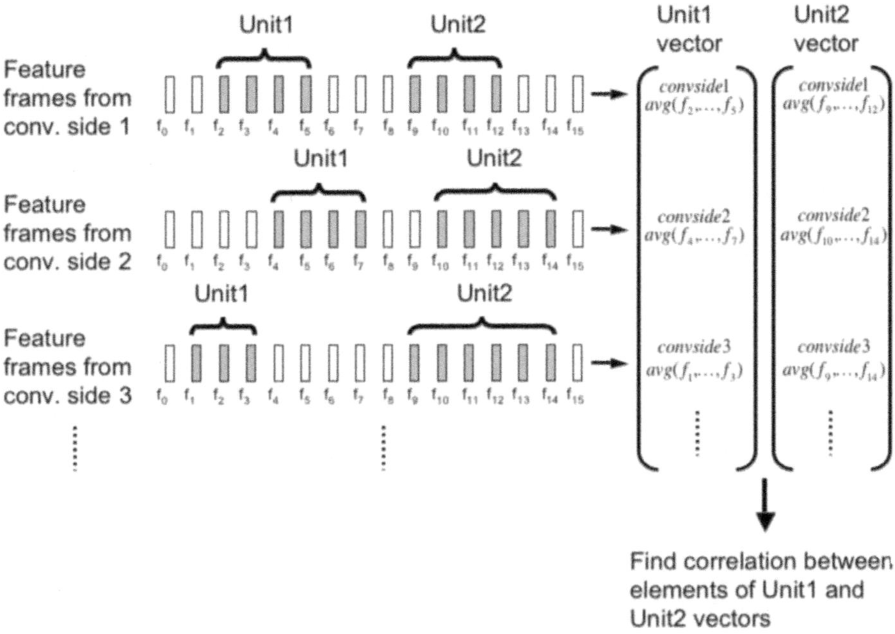

Fig. 1. Illustration of the procedure for computing Pearson's correlation as a redundancy measure

Hence, a Pearson's correlation value is associated with each pair of units. The correlation between this correlation and the relative MLP-based score level combination improvement of the unit pair is obtained to determine how well the measure predicts the redundancy of the unit pair. The relative MLP-based score level combination is determined by the relative score-level combination EER improvement over the average EER of the units standalone.

Note that we've also implemented mutual information as a redundancy measure, but found that Pearson's correlation is more effective.

4 Data Selection Scheme Involving the Measures

Our data selection scheme involving the measures is based off of the feature selection approach in [10]. Specifically, given a set of units, the task is to select

N units that produce the best SR result in combination. Given the relevance measures for each unit and redundancy measures for unit pairs, our data selection approach is the following: for a given set of pre-selected units P, determine if an additional unit Q should be selected by maximizing the following objective OBJ:

$$OBJ(Q) = Rel(Q) - \alpha \sum_{p \in P} Red(Q, p). \tag{6}$$

where $Rel(Q)$ is the value of the relevance measure for unit Q, $Red(Q, p)$ is the value of the redundancy measure between Q and p, and α is a weight between the relevance and redundancy factors. This objective allows one to select units that have good standalone speaker discriminative power (according to $Rel(Q)$) and are not redundant in their speaker discriminative characteristics with pre-selected units.

5 The Units

The following set of 30 phones represent the units used for this work: /ɑ/, /æ/, /ʌ/, /ɔ/, /ə/, /ɑʸ/, /b/, /d/, /ð/, /ɛ/, /ɝ/, /e/, /f/, /h/, /I/, /i/, /k/, /l/, /m/, /n/, /o/, /p/, PUH, /r/, /s/, /t/, /u/, /v/, /w/, /j/, /z/, where PUH is the vowel in a filled pause, and the remaining phones are denoted by their IPA symbols. These phones are selected from the set of all phones because they occur most frequently in the SRE06 conversation sides.

Phones intuitively represent a good starting point for the evaluation of measures because they span the vast majority of the acoustic space of speech. Each phone represents a small section of the acoustic space, largely separate from the acoustic spaces of other phones. Hence, the use of phones allows the measures to be computed on many different parts of the acoustic space, and the value obtained for a measure using a particular phone would be largely specific to the section of acoustic space represented by the phone. Note that for each conversation side, some phone instances are removed to ensure roughly equal numbers of frames for all phones in the conversation side.

6 Experiments and Results

6.1 Mutual Information as Relevance Measure

Mutual information as a relevance measure is implemented for each of the 30 phones on SRE06 with 128 female speakers, and is our most effective measure. A -0.8352 correlation is obtained between the mutual informations and EERs of the phones. This correlation implies that in general, phones with good SR performance (low EER) also have high mutual information, and that mutual information is an effective measure for SR performance prediction. The phones with the lowest EER and highest mutual information involve the nasals, some consonants, and no vowels: /t/, /k/, /s/, /p/, /f/, /v/, /d/, /ð/, /z/, /b/, /m/, /n/. It is hypothesized that the use of delta features, which capture transitions into and out of the consonants, may have improved the speaker discriminative abilities of the consonants.

The following 6 phones (according to their IPA symbols): /t/, /d/, /b/, /m/, /I/, /ε/ resulted in a -0.9969 correlation between mutual information and EER on the SRE06 corpus. The same phones show a -0.9819 correlation on the SRE08 corpus, suggesting that if all speech data are comprised of the 6 phones, a good indication of which phones are speaker discriminative based on their individual mutual information values can be achieved.

6.2 Kurtosis, Fisher's Ratio, Intra- and Inter-speaker Variances as Relevance Measures

Kurtosis, Fisher's ratio, and intra- and inter-speaker variances are also computed on each of the 30 phones, and their values are compared to the EERs of the phones. SRE06 results for the correlations of kurtosis, Fisher's ratio, and intra- and inter-speaker variances for each phone with respect to the EERs are shown in Table 1. The result for mutual information is shown alongside these results.

Table 1. Correlations of kurtosis, Fisher's ratio, intra- and inter-speaker variances, and mutual information for each phone with their EERs. Results obtained on SRE06.

Measure	Value
Kurtosis	0.715
Fisher's ratio	0.363
Intra-speaker variance	0.580
Inter-speaker variance	0.539
Mutual information	-0.835

According to Table 1, mutual information and kurtosis have the most significant correlations (-0.835 and 0.715 respectively) with the EERs of the 30 phones. Note that the correlation between inter-speaker variance and EER is positive, which is counterintuitive, since the inter-speaker variance should be high for phones with good speaker discriminative ability (and hence low EER). While this is rather strange, past results on Nuisance Attribute Projection (NAP) have suggested that minimizing inter-speaker variance helps SR performances [14][15]. One possible explanation for this is that features with high inter-speaker variance also have high intra-speaker variance in general (this has been shown by examining plots of the feature vectors along the top 2 PCA dimensions for speaker pairs). Nevertheless, these results demonstrate a significance in the correlations between all measures and EER (with the possible exception of Fisher's ratio, which only has a correlation of 0.363). Thus, the measures are useful for SR performance prediction.

6.3 Pearson's Correlation as Redundancy Measure

The approach described in section 3.4 is implemented on the 128 female speakers of SRE06. Correlations are obtained between the feature vectors of all distinct pairs of the 30 phones, along with the relative improvement in the MLP-based score-level combinations of the pairs (two SRE06 splits are created; MLP weights are trained using one split and tested on the other). The latter is obtained by computing the EER improvements of phones in combination over the average of the standalone phone EERs.

The optimal correlation between the correlation of feature vectors and the EER relative improvements of phone pairs is -0.486, which is obtained by considering only C0 and C1 of the MFCC feature vectors without their deltas (a -0.409 correlation is obtained when considering all MFCC coefficients). This result suggests that if the correlation between feature vectors of two phones is high, then the relative improvement of their score-level combination is low, and vice versa. Hence, Pearson's correlation is a suitable measure of unit redundancy.

6.4 Data Selection Investigation and Discussion

We've applied the mutual information relevance measure (our best measure) and Pearson's correlation redundancy measure to the data selection scheme described in section 4. Obtaining the mutual information and Pearson's correlation measures requires roughly a fifth of the computational costs of running the SR system for all phones. We've also used the standalone EERs of the individual phones as the baseline relevance measure. Only C0 and C1 are used for Pearson's correlation measure, which produces the optimal correlation according to section 6.3. All measures (including the standalone EERs) are obtained on SRE06. Two splits of SRE06 are used to train the α parameter from equation 6.

The data selection scheme in section 4 is used to select the top 5 phones for MLP-based score-level combination on SRE08 (with MLP weights trained on SRE06). We've also selected the top 5 phones with the lowest standalone EERs for SRE08, and compared the phone combination EERs obtained via the two approaches. Table 2 shows the EER results on SRE08 for α equal to its optimal value (where both relevance and redundancy are used) and zero (where only relevance is used), along with the phones selected.

According to Table 2, selecting the top 5 phones in combination using our data selection approach with mutual information relevance measure and optimal α gives a 13.8% EER on SRE08, which is a 2.13% improvement over selecting the top 5 phones with the best EERs (14.1% EER). Note that even though the improvement is not significant, we've shown that we can select an effective set

Table 2. MLP score-level combination of top 5 phones selected according to relevance and redundancy measures with optimal α, and standalone EERs. Results obtained on SRE08.

Data selection approach	Relevance measure	Phones selected	EER (%)
Relevance and redundancy	Mutual information	/d/, /h/, /k/, /t/, /v/	13.8
Relevance only	Mutual information	/f/, /k/, /p/, /s/, /t/	14.5
Relevance and redundancy	Standalone EERs	/b/, /k/, /n/, /t/, /z/	13.5
Relevance only	Standalone EERs	/b/, /d/, /k/, /t/, /z/	14.1
Top standalone EERs	–	/b/, /d/, /k/, /t/, /z/	14.1

of units without having to run the actual SR system. Our result also achieves a 4.83% improvement over using mutual information and no redundancy measure (14.5% EER), and is within 2.22% of the result using standalone phone EERs as the baseline relevance measure (13.5% EER).

Table 2 also suggests that using only the mutual information relevance measure (with no redundancy measure) for data selection does not improve results over using phones with top standalone EERs (14.5% EER vs. 14.1% EER). The latter can be expected, since the mutual information relevance measure has only an imperfect correlation with EER (-0.835). Note from Table 2 that the top 5 phones are all consonants.

According to the results, we have demonstrated that it is possible to select effective units for SR without running the actual SR system, by obtaining relevance and redundancy measures (mutual information and Pearson's correlation in our case) from acoustic feature vectors with good SR performance predictions. Note that using the phone EERs as a baseline relevance measure requires running the SR system, but improves only insignificantly over using mutual information. Our results indicate that taking both relevance and redundancy (as opposed to just relevance) into consideration for SR data selection leads to better unit selection. Interestingly, only the MFCC C0 and C1 coefficients are sufficient for computing our redundancy measure.

7 Conclusion and Future Work

In this work, we've investigated the feasibility of obtaining measures for data selection and performance prediction for unit-based text-dependent speaker recognition. As a starting point, we've used a set of 30 phones as units, and obtained various measures having significant correlations with EERs of the phones. We've shown that it is possible to select a set of units based off of relevance and redundancy measures, which gives equal or better speaker recognition results in combination than the combination of units with best standalone EERs, and does not require a speaker recognition system to be run.

In the future, we will attempt to develop more effective measures, investigate data selection using other types of units, and investigate other types of features. Once we are satisfied with the effectiveness of our measures, we will use the measures to select an arbitrary set of units which would have the globally optimal speaker recognition result in combination for particular types of systems. The arbitrary selection of units would be computationally feasible via the use of our measures.

Acknowledgements

The author wishes to thank Andreas Stolcke of SRI for providing speech recognition decodings. This research is funded by NSF grant number 0329258.

References

1. Reynolds, D.A., Quatieri, T.F., Dunn, R.: Speaker Verification using Adapted Gaussian Mixture Models. Digital Signal Processing 10, 19–41 (2000)
2. Sturim, D., Reynolds, D., Dunn, R., Quatieri, T.: Speaker Verification using Text-Constrained Gaussian Mixture Models. In: ICASSP, vol. 1, pp. 677–680 (2002)
3. Lei, H., Mirghafori, N.: Word-Conditioned Phone N-grams for Speaker Recognition. In: ICASSP, vol. 4, pp. 253–256 (2007)
4. Hannani, A., Toledano, D., Petrovska-Delacrétaz, D., Montero-Asenjo, A., Hennebert, J.: Using Data-driven and Phonetic Units for Speaker Verification. In: IEEE Odyssey (2006)
5. Gerber, M., Beutler, R., Pfisher, B.: Quasi Text-Independent Speaker-Verification based on Pattern Matching. In: Interspeech, pp. 1993–1996 (2007)
6. Stolcke, A., Bratth, H., Butzberger, J., Franco, H., Rao Gadde, V., Plauche, M., Richey, C., Shriberg, E., Sonmez, K., Weng, F., Zheng, J.: The SRI March 2000 Hub-5 Conversational Speech Transcription System. In: NIST Speech Transcription Workshop (March 2000)
7. Bonastre, J.F., Wils, F., Meignier, S.: ALIZE, a free Toolkit for Speaker Recognition. In: ICASSP, vol. 1, pp. 737–740 (2005)
8. HMM Toolkit (HTK), `http://htk.eng.cam.ac.uk`
9. Kwak, N., Choi, C.: Input Feature Selection by Mutual Information Based on Parzen Window. IEEE Transactions on Pattern Analysis and Machine Intelligence 24 (2002)
10. Peng, H., Long, F., Ding, C.: Feature Selection Based on Mutual Information: Criteria of Max-Dependency, Max-Relevance, and Min-Redundancy. IEEE Transactions on Pattern Analysis and Machine Intelligence 27 (2005)
11. Ellis, D., Bilmes, J.: Using Mutual Information to Design Feature Combinations. In: ICSLP (2000)
12. Parzen, E.: On Estimation of a Probability Density Function and Mode. Annals of Math. Statistics 33 (1962)
13. Xie, Y., Dai, B., Yao, Z., Liu, M.: Kurtosis Normalization in Feature Space for Robust Speaker Verification. In: ICASSP, vol. 1 (2006)
14. Vogt, R., Kajarekar, S., Sridharan, S.: Discriminant NAP for SVM Speaker Recognition. In: IEEE Odyssey (2008)
15. Lei, H.: NAP, WCCN, a New Linear Kernel, and Keyword Weighting for the HMM Supervector Speaker Recognition System. Technical report, International Computer Sciences Institute (2008)

Beyond Minutiae: A Fingerprint Individuality Model with Pattern, Ridge and Pore Features

Yi Chen[1,2] and Anil K. Jain[1]

[1] Michigan State University
[2] DigitalPersona Inc.

Abstract. Fingerprints are considered to be unique because they contain various distinctive features, including minutiae, ridges, pores, etc. Some attempts have been made to model the minutiae in order to get a quantitative measure for uniqueness or individuality of fingerprints. However, these models do not fully exploit information contained in non-minutiae features that is utilized for matching fingerprints in practice. We propose an individuality model that incorporates all three levels of fingerprint features: pattern or class type (Level 1), minutiae and ridges (Level 2), and pores (Level 3). Correlations among these features and their distributions are also taken into account in our model. Experimental results show that the theoretical estimates of fingerprint individuality using our model consistently follow the empirical values based on the public domain NIST-4 database.

Keywords: Fingerprint individuality, statistical models, probability of random correspondence (PRC), minutiae, ridges, pores.

1 Introduction

A number of court challenges have brought into question the validity and reliability of using fingerprints for personal identification [1,2,3]. These challenges are based on, among other factors, the lack of (i) conclusive evidence to support the claim of fingerprint uniqueness, and (ii) scientific evaluation of criteria used to determine a match between two fingerprints. These legal challenges have generated substantial interest in studies on fingerprint individuality.

Fingerprint individuality can be formulated as the probability that any two prints from different fingers will be "sufficiently" similar. Because similarity between fingerprints is often quantitatively defined based on the similarity of fingerprint features, it is equivalent to finding the *probability of random correspondence (PRC)*, or the probability of matching k features, given that two impostor fingerprints contain m and n features, respectively. The key to calculating PRC is to model the distribution of fingerprint features, as it is not only the number of features, but also their spatial distributions that account for individuality. The variety of features that are incorporated in the individuality model is also important. Latent experts use three levels of features, namely, Level 1 (e.g., pattern), Level 2 (e.g., minutiae and ridges) and Level 3 (e.g., pores) in

M. Tistarelli and M.S. Nixon (Eds.): ICB 2009, LNCS 5558, pp. 523–533, 2009.
© Springer-Verlag Berlin Heidelberg 2009

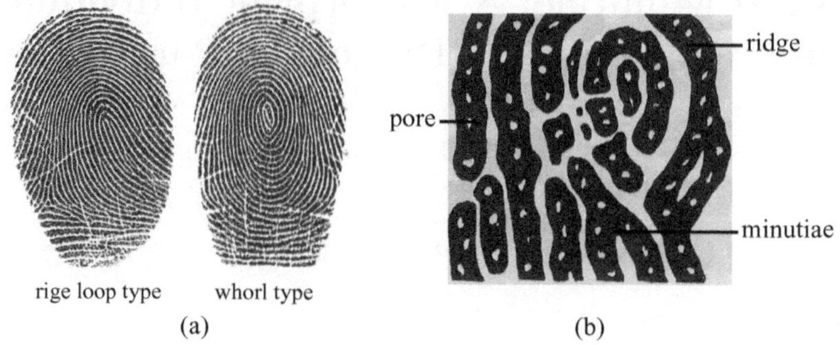

Fig. 1. Fingerprint features (a) two different pattern types (b) minutiae, ridges and pores

fingerprint matching [4]. All of these features are believed to be distinctive and unique and should be considered in the individuality model (see Fig. 1).

To learn the spatial distribution of fingerprint features, a couple of generative models have been proposed [5,6]. Pankanti et al. [5] modeled the minutiae as uniformly and independently distributed, however, it has been empirically determined that minutiae tend to cluster where ridge directions change abruptly, with higher density in the core and delta regions [7,8]. To account for this clustering tendency of minutiae, Zhu et al. [6] proposed a finger-specific mixture model, in which minutiae are first clustered and then independently modeled in each cluster. In modeling features other than minutiae, Fang et al. [9] modeled ridges by classifying the ridge segment associated with each minutia into one of sixteen ridge shapes. However, most of the ridge shapes used in this model were quite rare as the ridge segments associated with minutiae either terminate or bifurcate. Roddy and Stosz [10] proposed to approximate the spatial distribution of pores using a discrete grid, and consequently, their formulation is sensitive to distortion as well as the size and the position of the grid cells.

2 Proposed Model

We propose to evaluate fingerprint individuality by modeling the distribution of minutiae, ridge and pore features. In particular, the minutiae and ridge features are separately modeled for each fingerprint pattern type (whorl, left loop, right loop, arch and tented arch). This is because minutiae and ridge feature distributions are highly correlated with the fingerprint pattern [8], which was also demonstrated by Zhu et al [6]. Compared to finger-specific models, pattern-specific models can be easily generalized to different target populations and are less prone to overfitting, especially when the number of minutiae in a fingerprint is small. To model ridges and pores, transform-invariant features such as ridge period, ridge curvature, and pore spacing associated with each minutia are derived.

2.1 Pattern-Specific Minutiae Modeling

Given a large database of fingerprints, we assign each image to one of the five major pattern types [11]: whorl, left loop, right loop, arch and tented arch. For each fingerprint class, minutiae from fingerprints in that class are consolidated and their spatial distribution is estimated. These minutiae are grouped into clusters based on their positions (X) and directions (O) using the EM algorithm [12]. In each cluster, the minutiae position X is modeled by a bivariate Gaussian distribution and the minutiae direction O is modeled using a Von-Mises distribution [6]. Each minutia $m(X, O)$ in a fingerprint of class G has the following mixture density:

$$f(x, o|\Theta_G) = \sum_{g=1}^{N_G} \tau_g \cdot f_X(x|\mu_g, \Sigma_g) \cdot f_O(o|\nu_g, \kappa_g), \tag{1}$$

where N_G is the number of clusters in the mixture for class G, τ_g is the weight for the g-th cluster, Θ_G is the set of parameters describing the distributions in each cluster, $f_X(x|\mu_g, \Sigma_g)$ is the p.d.f. over minutiae position in the g-th cluster, and $f_O(o|\nu_g, \kappa_g)$ is the p.d.f. over minutiae direction in the g-th cluster. Minutiae in the same cluster have the same distribution of directions, establishing a dependence between minutiae position and direction.

Let T be a fingerprint belonging to class G with minutiae density $f(x, o|\Theta_G)$. Similarly, let Q be a fingerprint belonging to class H with minutiae density $f(x, o|\Theta_H)$. Let $m(X^T, O^T)$ and $m(X^Q, O^Q)$ be two minutiae from T and Q, respectively. The probability that these two minutiae would match is defined as

$$P(|X^T - X^Q| \leq x_0, |O^T - O^Q| \leq o_0|\Theta_G, \Theta_H) \tag{2}$$

$$= \sum_{g=1}^{N_G} \sum_{h=1}^{N_H} \tau_g \cdot \tau_h \cdot P(|X^T - X^Q| \leq x_0|\mu_g, \mu_h, \Sigma_g, \Sigma_h) \cdot P(|O^T - O^Q| \leq o_0|\nu_g, \nu_h, \kappa_g, \kappa_h),$$

where parameters $x_0 = 15$ pixels and $o_0 = 22.5$ degrees are used as tolerances. Note that this probability can be directly computed since $(X^T - X^Q)$ follows a 2D Gaussian distribution with mean $(\mu_g - \mu_h)$ and covariance $(\Sigma_g + \Sigma_h)$; and $(O^T - O^Q)$ can be approximated by a Von-Mises distribution with mean $(\nu_g - \nu_h)$ and variance $\kappa_{g,h}$ defined as [13]:

$$A(\kappa_{g,h}) = A(\kappa_g)A(\kappa_h), \tag{3}$$

$$A(x) = 1 - \frac{1}{2} - \frac{1}{8x^2} - \frac{1}{8x^3} + o(x^{-3}). \tag{4}$$

Finally, the PRC, or the probability of matching k pairs of minutiae between Q and T, is calculated as [6]

$$p(m, n, k) = \frac{e^{-\lambda} \cdot \lambda^k}{k!}, \lambda = mnl, \tag{5}$$

where m and n are the number of minutiae in Q and T, respectively, and $l = P(|X^T - X^Q| \leq x_0, |O^T - O^Q| \leq o_0|\Theta_G, \Theta_H)$ is the probability calculated in Eq. (2). This PRC calculation corresponds to the Poisson probability mass function with mean $\lambda = mnl$, which can be interpreted as the expected number of matches from the total number of mn possible pairings between Q and T with the probability of each match being l.

2.2 Pattern-Specific Ridge Modeling

To incorporate ridge features in the model, we extend the density function in Eq. (1) to the following mixture density for class G:

$$f(x, o, r, c|\Theta_G) = \sum_{g=1}^{N_G} \tau_g \cdot f_X(x|\mu_g, \Sigma_g) \cdot f_O(o|\nu_g, \kappa_g) \cdot f_R(r|\omega_g, \sigma_g^2) \cdot f_C(c|\lambda_g), \tag{6}$$

where $f_R(r|\omega_g, \sigma_g^2)$ and $f_C(c|\lambda_g)$ are the probability density functions of ridge period and ridge curvature associated with each minutia in the g-th cluster, respectively. We use a Gaussian distribution to model ridge period because ridge period only fluctuates towards the tip (smaller) and flexion crease (larger) regions. Ridge curvature, on the other hand, is usually low except near the core and delta regions, and is modeled by a Poisson distribution. Note that minutiae in the same cluster have the same distribution of direction, ridge period and curvature, hence, establishing a dependency among these features.

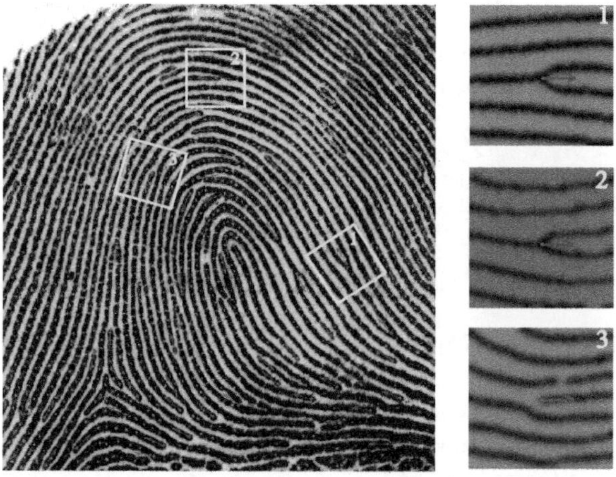

Fig. 2. Converting local ridge structure to a canonical form such that matching two minutiae and their associated ridges is equivalent to matching the position and direction of minutiae and the period and curvature of the ridges

Let T be a fingerprint belonging to class G with minutiae density $f(x, o, r, c|\Theta_G)$. Similarly, let Q be a fingerprint belonging to class H with minutiae density $f(x, o, r, c|\Theta_H)$. Let $m(X^T, O^T, R^T, C^T)$ and $m(X^Q, O^Q, R^Q, C^Q)$ be two minutiae from T and Q, respectively. The probability that these two minutiae would match is then defined as

$$\begin{aligned} &P(|X^T - X^Q| \leq x_0, |O^T - O^Q| \leq o_0, |R^T - R^Q| \leq r_0, |C^T - C^Q| \leq c_0|\Theta_G, \Theta_H) = \\ &\sum_{g=1}^{N_G} \sum_{h=1}^{N_H} \tau_g \cdot \tau_h \cdot P(|X^T - X^Q| \leq x_0|\mu_g, \mu_h, \Sigma_g, \Sigma_h) \cdot P(|O^T - O^Q| \leq o_0|\nu_g, \nu_h, \kappa_g, \kappa_h) \\ &\qquad \cdot P(|R^T - R^Q| \leq r_0|\omega_g, \omega_h, \sigma_g^2, \sigma_h^2) \cdot P(|C^T - C^Q| \leq c_0|\lambda_g, \lambda_h), \end{aligned} \tag{7}$$

where $r_0 = 2$ and $c_0 = 2$. Here, matching the ridges associated with two minutiae is equivalent to matching the ridge period and curvature after converting the ridge structure associated with each minutia to a canonical form centered at the minutia and rotated by the minutia direction (see Fig. 2). Similar to Eq. (2), this probability can be directly computed since $(R^T - R^Q)$ follows a Gaussian distribution with mean $(\omega_g - \omega_h)$ and variance $(\sigma_g^2 + \sigma_h^2)$ and $(C^T - C^Q)$ follows a Skellam distribution [14] with mean $(\lambda_g - \lambda_h)$ and variance $(\lambda_g + \lambda_h)$. The final PRC is again computed as in Eq. (5), except that l is replaced by Eq. (7).

2.3 Pore Modeling

Since pores are almost evenly spaced along the ridges [10], modeling pores can be approximated by modeling the intra-ridge spacing of pores on modeled ridges. As a result, each minutia and its local features (position X, direction O, ridge period R, ridge curvature C and pore spacing S) have the following mixture density for fingerprints from class G:

$$f(x, o, r, c, s|\Theta_G) = \tag{8}$$
$$\sum_{g=1}^{N_G} \tau_g \cdot f_X(x|\mu_g, \Sigma_g) \cdot f_O(o|\nu_g, \kappa_g) \cdot f_R(r|\omega_g, \sigma_g^2) \cdot f_C(c|\lambda_g) \cdot f_S(s|\mu_p, \sigma_p^2),$$

where $f_S(s|\mu_p, \sigma_p^2)$ is the probability density function for pore spacing. Because there is no evidence that intra-ridge pore spacing is dependent on minutiae location or ridge flow patterns, pore spacing is not clustered.

Let T be a fingerprint belonging to class G with minutiae density $f(x, o, r, c, s|\Theta_G)$. Similarly, let Q be a fingerprint belonging to class H with minutiae density $f(x, o, r, c, s|\Theta_H)$. Let $m(X^T, O^T, R^T, C^T, S^T)$ and $m(X^Q, O^Q, R^Q, C^Q, S^Q)$ be two minutiae from T and Q, respectively. The probability of matching the two minutiae features can be calculated as follows:

$$P(|X^T - X^Q| \le x_0, |O^T - O^Q| \le o_0, |R^T - R^Q| \le r_0, |C^T - C^Q| \le c_0, |S^T - S^Q| \le s_0$$
$$|\Theta_G, \Theta_H) = \sum_{g=1}^{N_G} \sum_{h=1}^{N_H} \tau_g \cdot \tau_h \cdot P(|X^T - X^Q| \le x_0 | \mu_g, \mu_h, \Sigma_g, \Sigma_h)$$
$$\cdot P(|O^T - O^Q| \le o_0 | \nu_g, \nu_h, \kappa_g, \kappa_h) \cdot P(|R^T - R^Q| \le r_0 | \omega_g, \omega_h, \sigma_g^2, \sigma_h^2)$$
$$\cdot P(|C^T - C^Q| \le c_0 | \lambda_g, \lambda_h) \cdot P(|S^T - S^Q| \le s_0 | \sigma_p^2), \tag{9}$$

where $s_0 = 2$. Again, this probability can be directly calculated since $(S^T - S^Q)$ follows a Gaussian distribution with mean 0 and variance $2\sigma_p^2$. Again, PRC is still computed as in Eq. (5), except that l is now replaced by Eq. (9). Note that after incorporating ridge and pore features in our individuality model, PRC represents the probability of matching k minutiae not only in their positions and directions, but also with respect to the local ridge period and curvature as well as pore spacing in the neighborhood of the matching minutiae.

3 Model Evaluation and Validation

In order to demonstrate the utility of our individuality model, we perform the following evaluation protocol: (i) learn the mixture density of minutiae and their local ridge and pore features, (ii) compute the theoretical probability of random correspondence (PRC), and (iii) compare the theoretical PRC with the empirical values obtained on a public database. The database used for evaluation is the NIST Special Database 4 (NIST-4) [16], which contains 2,000 pairs (4,000 images) of inked rolled prints at 500 ppi. Rolls are appropriate for fingerprint individuality study because they provide a complete representation of fingerprints. Each fingerprint in NIST-4 comes with a class label (28.2% right loop, 26.6% left loop, 21.5% whorl, 19% arch and 4.7% tented arch) assigned by forensic experts. These prints are manually aligned by the author based on the locations of core(s). For example, fingerprints of the left loop, right loop and tented arch classes are aligned at the core point; fingerprints of the whorl class are aligned at the centroid of the two cores; and fingerprints of the arch class are aligned at the highest curvature point on the most upthrusting ridge. Figure 3 shows the empirical distribution of minutiae positions from fingerprints in each of the five classes. It can be observed that the minutiae distribution is highly correlated with the fingerprint class. Note the higher minutiae density in the core and delta regions, which is consistent with Champod's finding [8].

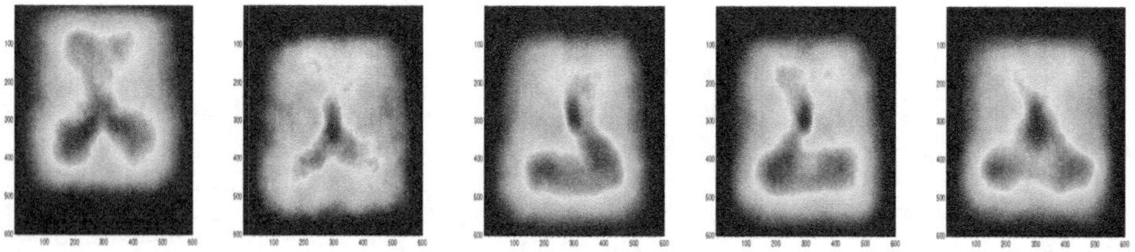

(a) (b) (c) (d) (e)

Fig. 3. Distribution of minutiae positions (empirically extracted [20]) in each of the five fingerprint classes (a) arch, (b) tented arch, (c) left loop, (d) right loop, and (e) whorl, obtained from 4,000 rolled fingerprints in NIST-4. The darker the area, the higher the minutiae density. These plots (with size 600 × 600) were smoothed using a disk filter with a radius of 25 pixels. The center of each plot corresponds to the alignment origin.

The theoretic probability of matching a minutiae pair between impostor fingerprints of each class is calculated using Equation 2. As shown in Table 1, impostor fingerprints from the same class have a higher matching probability (highlighted in gray) than those from different classes. This is consistent with the results of Jain et al.'s study [18], which revealed that fingerprints from the same class are more likely to be matched (have higher False Accept Rate) than fingerprints from difference classes.

For comparison, we also calculate the empirical probabilities of matching a minutiae pair between impostor fingerprints based on NIST-4. An in-house fingerprint matcher [15] was used to automatically establish minutiae correspondences between the impostor pairs in NIST-4. To be compatible with our model, the matcher is restricted to register fingerprints within a 50 × 50 neighborhood

Table 1. Theoretical probabilities of matching a minutiae pair between impostor fingerprints belonging to class A=arch, TA=tented arch, L=left loop, R=right loop and W=whorl based on the proposed model

Type	A	TA	L	R	W
A	13.20×10^{-4}	6.05×10^{-4}	6.33×10^{-4}	3.95×10^{-4}	4.65×10^{-4}
TA	6.05×10^{-4}	12.76×10^{-4}	6.85×10^{-4}	5.01×10^{-4}	6.40×10^{-4}
L	6.33×10^{-4}	6.85×10^{-4}	10.95×10^{-4}	7.44×10^{-4}	4.92×10^{-4}
R	3.95×10^{-4}	5.01×10^{-4}	7.44×10^{-4}	11.59×10^{-4}	4.74×10^{-4}
W	4.65×10^{-4}	6.40×10^{-4}	4.92×10^{-4}	4.74×10^{-4}	10.01×10^{-4}

Table 2. Empirical probabilities of matching a minutiae pair between impostor fingerprints belonging to class A=arch, TA=tented arch, L=left loop, R=right loop and W=whorl based on NIST-4

Type	A	TA	L	R	W
A	14.33×10^{-4}	12.25×10^{-4}	7.73×10^{-4}	8.31×10^{-4}	2.35×10^{-4}
TA	12.25×10^{-4}	12.54×10^{-4}	8.33×10^{-4}	9.17×10^{-4}	2.61×10^{-4}
L	7.73×10^{-4}	8.33×10^{-4}	11.46×10^{-4}	5.92×10^{-4}	4.36×10^{-4}
R	8.31×10^{-4}	9.17×10^{-4}	5.92×10^{-4}	11.54×10^{-4}	4.64×10^{-4}
W	2.35×10^{-4}	2.61×10^{-4}	4.36×10^{-4}	4.64×10^{-4}	6.36×10^{-4}

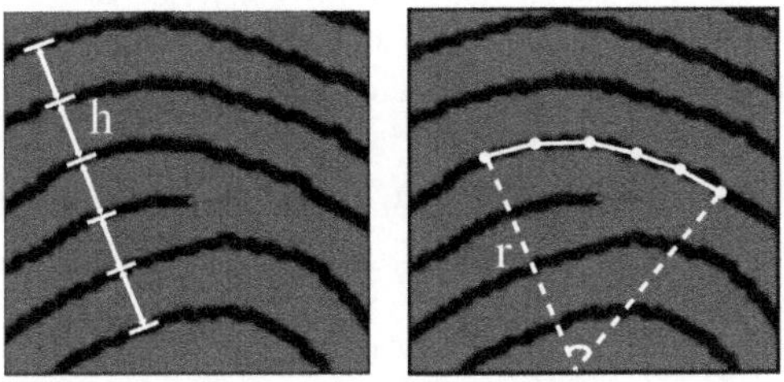

Fig. 4. Ridge features in a neighborhood (30 × 30 pixels) of a minutia. (a) ridge period (average of inter-ridge distance h), and (b) ridge curvature (inverse of radius r).

of the manually aligned origin. A total of 7,998,000 impostor matches were conducted and the empirical probability of matching a minutiae pair between two impostor prints is computed by the average of $k/(m \times n)$, where k is the number of matched minutiae and m and n are the number of minutiae in the two fingerprints. This probability is tabulated by the fingerprint class information, resulting in all possible intra-class and inter-class probability values (see Table 2). Note that the theoretical probabilities based on our model are closer to empirical probabilities (with correlation coefficient 0.59) compared to those calculated by Zhu et al.'s model (empirical probability $= 1.8 \times 10^{-3}$, theoretical probability $= 6.50 \times 10^{-4}$) [6].

Table 3. Theoretical probabilities of matching a minutiae pair and their local ridge features between impostor fingerprints belonging to class A=arch, TA=tented arch, L=left loop, R=right loop and W=whorl based on the proposed model

Type	A	TA	L	R	W
A	6.21×10^{-4}	2.59×10^{-4}	2.62×10^{-4}	1.74×10^{-4}	1.88×10^{-4}
TA	2.59×10^{-4}	6.06×10^{-4}	3.36×10^{-4}	2.40×10^{-4}	2.92×10^{-4}
L	2.62×10^{-4}	3.36×10^{-4}	5.52×10^{-4}	3.71×10^{-4}	2.24×10^{-4}
R	1.74×10^{-4}	2.40×10^{-4}	3.71×10^{-4}	5.99×10^{-4}	2.21×10^{-4}
W	1.88×10^{-4}	2.92×10^{-4}	2.24×10^{-4}	2.21×10^{-4}	4.56×10^{-4}

Table 4. Empirical probabilities of matching a minutiae pair and their local ridge features between impostor fingerprints belonging to class A=arch, TA=tented arch, L=left loop, R=right loop and W=whorl based on NIST-4

Type	A	TA	L	R	W
A	5.86×10^{-4}	4.70×10^{-4}	2.91×10^{-4}	3.10×10^{-4}	0.84×10^{-4}
TA	4.70×10^{-4}	5.68×10^{-4}	3.68×10^{-4}	3.89×10^{-4}	1.07×10^{-4}
L	2.91×10^{-4}	3.68×10^{-4}	5.05×10^{-4}	2.39×10^{-4}	1.79×10^{-4}
R	3.10×10^{-4}	3.89×10^{-4}	2.39×10^{-4}	5.12×10^{-4}	1.93×10^{-4}
W	0.84×10^{-4}	1.07×10^{-4}	1.79×10^{-4}	1.93×10^{-4}	2.78×10^{-4}

To incorporate ridge features, we extract ridge period and curvature in a 30 $\times$ 30 neighborhood of each minutia (see Fig. 4). Ridge period is calculated as the average inter-ridge distances, and ridge curvature, defined as the inverse of its radius, is calculated by the second derivative of sampled ridge points. In the model, these features are used to retrain the clustering algorithm and reevaluate the theoretical probability in Equation 7. In the empirical case, minutiae correspondences that disagree in ridge period or curvature (with difference larger than r_0 and c_0, respectively) are removed. The resulting number of matched minutiae pairs k' is used to recalculate the empirical probability as the average of $k'/(m \times n)$. The theoretical and empirical matching probability matrices tabulated by fingerprint class information after incorporating the ridge features are shown in Tables 3 and 4, respectively. As we can see, higher probabilities are still observed among impostor fingerprints from the same class than those from different classes. The use of ridge features reduces the probability of random minutiae correspondence both empirically and theoretically; the correlation coefficient between the two probability matrices is increased to 0.67.

When pores are also incorporated in the model, we reevaluate the theoretical probability of matching a minutiae pair using Equation 9. Because pores are more reliably captured at 1000 ppi than 500 ppi, we use the NIST-30 database [17] (720 rolled images at 1000 ppi) for extracting [19] and modeling the distribution of pore spacing. The empirical distribution and a Gaussian fit to the distribution with mean 12.67 and standard deviation 0.82 are shown in Figure 5. This is consistent with the estimates of Roddy and Stosz [10], who suggested that the most frequently observed pore spacing is 13 pixels at 1100 ppi. The theoretically

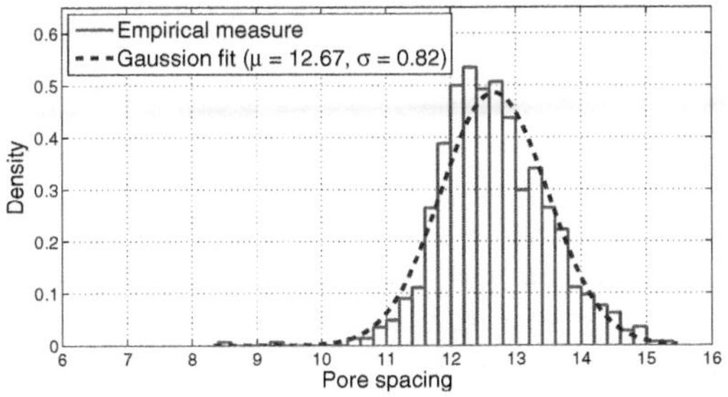

Fig. 5. Distribution of intra-ridge pore spacings (distance between two neighboring pores on the same ridge) obtained from 720 fingerprints in NIST-30 (1000 ppi) using an automatic pore extraction algorithm [19]

Table 5. Theoretical probabilities of matching a minutiae pair, their local ridge and pore features between impostor fingerprints belonging to class A=arch, TA=tented arch, L=left loop, R=right loop and W=whorl based on the proposed model

Type	A	TA	L	R	W
A	5.68×10^{-4}	2.37×10^{-4}	2.40×10^{-4}	1.59×10^{-4}	1.72×10^{-4}
TA	2.37×10^{-4}	5.55×10^{-4}	3.08×10^{-4}	2.20×10^{-4}	2.67×10^{-4}
L	2.40×10^{-4}	3.08×10^{-4}	5.05×10^{-4}	3.40×10^{-4}	2.05×10^{-4}
R	1.59×10^{-4}	2.20×10^{-4}	3.40×10^{-4}	5.48×10^{-4}	2.02×10^{-4}
W	1.72×10^{-4}	2.67×10^{-4}	2.05×10^{-4}	2.02×10^{-4}	4.17×10^{-4}

Table 6. An example comparison among PRCs obtained from Pankanti's uniform model, Zhu's mixture model and the proposed model with empirical values on NIST-4. Empirical values were calculated using Eq. (5) by substituting λ with the mean number of matches (weighted by the class distribution) found empirically.

(m,n,k)	Uniform [5]	Mixture [6]	Proposed	Empirical
(52,52,12)	4.3×10^{-8}	4.4×10^{-4}	5.77×10^{-7}	7.72×10^{-7}
(62,62,12)	2.9×10^{-7}	4.1×10^{-4}	1.79×10^{-5}	2.34×10^{-5}

probability matrix of matching a minutiae and their local ridge and pore features tabulated by fingerprint class information is shown in Table 5.

Having obtained the probability of matching one minutiae pair, we can calculate the PRC, or the probability of matching k minutiae pairs, given m and n minutiae in the two fingerprints using Equation 5. For example, when only minutiae position and orientation are incorporated, the theoretic PRC of having, say, $k = 12$ minutiae matches given $m = 52$ and $n = 52$ is 5.77×10^{-7}. This is more consistent with the empirical probability (7.72×10^{-7}) than the previously published models (see Table 6). When ridge period and curvature are incorporated in addition to minutiae, both theoretical and empirical probabilities (for $k = 12, m = 52, n = 52$) are reduced to 1.86×10^{-10} and to 5.36×10^{-11}, respectively. When pore spacing is incorporated, the theoretic probability drops further to 6.93×10^{-11}.

4 Summary

We have proposed a mixture model to evaluate fingerprint individuality based on five major fingerprint classes using minutiae and non-minutiae features. Experimental results show that the estimated matching probabilities between impostor fingerprints are highly correlated with the fingerprint class information and, as expected, the probabilities reduce when additional features (ridges and pores) are incorporated in the model. Our theoretical estimates of PRCs are consistent with those based on empirical matching. In practice, however, both theoretical and empirical estimates can be affected by factors such as image quality and the feature extraction and matching algorithms. In future work, we would like to account for these factors in our individuality model.

Acknowledgments. This work was supported by research grants NIJ 2007-RG-CX-K183, NIJ-2007-DN-BX-0005, and ARO W911NF-06-1-0418.

References

1. Plaza, U.S. v., et al.: 179 F. Supp. 2d 492, E.D. Pa (2002)
2. Mitchell, U.S. v.: 365 F. 3d 215, 3d Cir. (2004)
3. Hearing Document of State of Maryland vs. Bryan Rose, Case No.: K06-0545, Circuit Court for Baltimore County (2007)
4. Ashbaugh, D.R.: Quantitative-Qualitative Friction Ridge Analysis: An Introduction to Basic and Advanced Ridgeology. CRC Press, Boca Raton (1999)
5. Pankanti, S., Prabhakar, S., Jain, A.K.: On the Individuality of Fingerprints. IEEE Trans. on Pattern Analysis and Machine Intelligence 24(8), 1010–1025 (2002)
6. Zhu, Y., Dass, S., Jain, A.K.: Statistical Models for Assessing the Individuality of Fingerprints. IEEE Trans. on Information Forensics and Security 2, 391–401 (2007)
7. Kingston, C.: Probabilistic Analysis of Partial Fingerprint Patterns. University of California, Berkeley (1964)
8. Champod, C., Lennard, C., Margot, P., Stoilovic, M.: Fingerprints and Other Ridge Skin Impressions. CRC Press, Boca Raton (2004)
9. Fang, G., Srihari, S., Srinivasan, H.: Generative Models for Fingerprint Individuality Using Ridge Types. In: Proc. International Workshop on Computational Forensics, Manchester, UK (2007)
10. Roddy, A.R., Stosz, J.D.: Fingerprint Features - Statistical Analysis and System Performance Estimates. Proc. IEEE 85(9), 1390–1421 (1997)
11. Henry, E.: Classification and Uses of Fingerprints. Routledge and Sons, London (1900)
12. Figueiredo, M., Jain, A.K.: Unsupervised Learning of Finite Mixture Models. IEEE Trans. on Pattern Analysis and Machine Intelligence 24(3), 381–396 (2002)
13. Mardia, K.V.: Statistics of Directional Data. Academic Press, London (1972)
14. Skellam, J.G.: The Frequency Distribution of the Difference Between Two Poisson Variates Belonging to Different Populations. Jounal of the Royal Statistical Society 109(3), 294 (1946)

15. Jain, A.K., Hong, L., Bolle, R.: On-line Fingerprint Verification. IEEE Trans. on Pattern Analysis and Machine Intelligence 19(4), 302–314 (1997)
16. NIST Special Database 4 (2009), http://www.nist.gov/srd/nistsd4.htm
17. NIST Special Database 30 (2009), http://www.nist.gov/srd/nistsd30.htm
18. Jain, A.K., Prabhakar, S., Pankanti, S.: On the Similarity of Identical Twin Fingerprints. Pattern Recognition 35(8), 2653–2663 (2002)
19. Jain, A.K., Chen, Y., Demirkus, M.: Pores and Ridges: High-Resolution Fingerprint Matching Using Level 3 Features. IEEE Trans. on Pattern Analysis and Machine Intelligence 29(1), 15–27 (2007)
20. Neurotechnologija Verifinger 4.2 SDK, http://www.neurotechnology.com (2009)

Active Fingerprint Ridge Orientation Models

Surinder Ram[1], Horst Bischof[1], and Josef Birchbauer[2]

[1] Institute for Computer Graphics and Vision
University of Technology Graz, Austria
[2] Siemens Austria, Siemens IT Solutions and Services
Biometrics Center

Abstract. This paper proposes a statistical model for fingerprint ridge orientations. The active fingerprint ridge orientation model (AFROM) iteratively deforms to fit the orientation field (OF) of a fingerprint. The OFs are constrained by the AFROM to vary only in ways according to a training set. The main application of the method is the OF estimation in noisy fingerprints as well as the interpolation and extrapolation of larger OF parts. Fingerprint OFs are represented by Legendre Polynomials. The method does not depend on any pre-alignment or registration of the input image itself. The training can be done fully automatic without any user interaction. We show that the model is able to extract the significant appearance elements of fingerprint flow patterns even from noisy training images. Furthermore, our method does not depend on any other computed data, except a segmentation. We evaluated both, the generalisation as well as the prediction capability of the proposed method. These evaluations assess our method very good results.

1 Introduction

Automatic Fingerprint Identification Systems (AFIS) have evolved to a mature technique and are becoming part of the daily lives of millions of people all over the world. For example, in many countries fingerprints are taken as part of the visa application process. Therefore, in the recent years we observe a step into a new dimension with respect to the size and complexity of automatic fingerprint identification systems. Large attention has been paid to the emerging problems, but still there is an ever-increasing need for better recognition rates. The latter is especially true for fingerprint images of poor quality.

The attractiveness of fingerprints results from their uniqueness which does not change through the life of individuals [10]. Three types of characteristic features [7] can be extracted from a fingerprint image: a) patterns, which are the macro details of a fingerprint such as ridge flow and pattern type. b) minutiae, which are points where ridges bifurcate or end. c) pores, edge contours, incipient ridges, breaks, creases and other permanent details.

The extraction of the key features (fingerprint patterns, minutiae) is highly depending on the correct estimation of local ridge orientation. Ridge orientation is inevitably used for detecting, describing and matching fingerprint features such as minutiae and singular points (SPs). For minutiae detection, special filtering

M. Tistarelli and M.S. Nixon (Eds.): ICB 2009, LNCS 5558, pp. 534–543, 2009.
© Springer-Verlag Berlin Heidelberg 2009

schema (see references in [7]) are available to enhance fingerprint images in order to extract the last bit of information available in the image. Note that the use of the mentioned filtering methods can only be successful if the correct ridge orientation is available.

Therefore, large efforts are made in order to extract reliable orientation data from fingerprints. Many methods for ridge orientation estimation exist in the literature (see [7]). The described methods proceed locally, and extract the orientation in a given area. Typically, this is done by estimating the gradients in the considered area. Unfortunately, determination of ridge orientation becomes more difficult when image quality is low (typically caused by noise, smudges, scars, weat or dry fingers, etc.). Thus even the 'best' orientation estimation algorithm will fail in regions of low image quality. The classic solution to solve this problem, is to smooth the ridge orientation. Such filtering methods are mostly based around an approach described by Witkin and Kass in [8]. This method splits orientation into vectorial parts (x-part, y-part) and then smooths this parts using low pass filtering schema. Note that this method is identical to the method of local orientation by tensor smoothing [2]. A representative example of a low pass filtering scheme is described by Bazen and Gerez in [1]. It is noteworthy to mention, that using such filtering methods, only small regions can be re-estimated successfully.

1.1 Related Work

The limitations of the above mentioned filtering schema gave rise to more sophisticated, model-based methods. Model-based approaches attempt to re-estimate OFs of larger areas in the image.

Early attempts of fingerprint ridge orientation modelling are described by Sherlock et al. [12] who model the orientation using a so-called zero-pole model. This orientation model is far too simple and fails describing the ridge orientation accurately. Vizcaya and Gerhardt improve on this model in [13] by using a piecewise linear model around the SP. This model is applied for synthesis of fingerprints as described in [7].

A combination method is described by Zhou and Gu in [16,17]. These methods first describe the global orientation field using power series and than model SPs locally. Unfortunately, the algorithm is difficult to apply in practice, since combining the two parts of the model involves heuristics. Furthermore, the algorithm requires reliable detection of SPs.

In [9], Li et al. model the orientation of fingerprints using higher order phase portraits. Therefore, the method divides the fingerprint into several predefined regions and approximates them using piecewise linear phase portraits. In a further step this method computes a global model using the piecewise linear phase portraits. Similar problems as described above apply also to this algorithm, namely the required separation of fingerprints into predefined regions and the robust detection of SPs.

Wang et al. [14] present a OF model based on trigonometric polynomials. Their approach (coined FOMFE) does not require the detection of SPs. The application includes orientation interpolation, SP detection and database indexing based on the model parameters. For OF smoothing, we [11] found that this method does not perform significantly better in comparison to classical low pass filtering schema.

Another method is described in [11], where we propose the use of Legendre Polynomials for modelling fingerprint OFs. We argue that SPs yield discontinuities in the OF which are difficult to model using polynomials. Instead we propose to use a fractional term, where the numerator and denominator parts are computed from the orientation (vectorial x-part, y-part). A non-linear optimization scheme enables this fractional function to approximate high curvature areas (especially SPs) without the necessity to model discontinuities. This is achieved by exploiting the zero-poles of the polynomials for modelling SPs and enables the method to perform better than other methods.

Very recently, Huckemann et al. [6] proposed a global OF model-based on quadratic differentials. This model can approximate fingerprint OFs using only five coefficients. These coefficients are geometrically interpretable and have a clear meaning. One drawback of this method is that it can not model every fingerprint type. Furthermore, the model is clearly not 'flexible' enough for a precise approximation of fingerprint OFs.

1.2 Prior Knowledge within Fingerprint Orientation Models

All the above mentioned fingerprint OF models do not contain the possibility of adding prior knowledge to aid the process of orientation estimation. The nomenclature of the term prior knowledge in this context implies that the model 'knows' patterns of valid fingerprints. Prior knowledge of valid fingerprint patterns can be used to provide tolerance to noisy or missing data. There are two main characteristics which the model should possess. First, the model should be general, meaning it should be capable of generating any plausible fingerprint pattern. Second, and crucially, it should be specific, which means the model should only be capable of generating 'naturally occurring' fingerprint patterns.

Typical examples of active models in literature are active shape models (ASM) used for modelling shape variations and active appearance models (AAM) for matching a statistical model of object shape and appearance to a new image. These models are built during a training phase. The latter method is widely used for matching and tracking faces and for medical image interpretation. For a more thorough overview see [3].

2 Training the Model

We use commercial fingerprint software from Siemens (Siemens IT Solutions and Services, Biometrics Center) for local OF estimation and for the segmentation of the image into foreground/background pixels. Note that no other processing, i.e. registration or alignment has been employed. For the training phase, the raw OF is smoothed using the method described in [11], this step is essential as it is necessary to estimate the OF also in the corners of the image (background).

2.1 Representation of Fingerprint Flow Patterns

For the representation of fingerprints OFs, we use 12th order Legendre Polynomials as described in [11]. Alternatively, one could also use the parametric OF representation as described in [14] by Wang et al. In the following, we give a short overview of the used OF approximation method. Let $2O(x,y)$ be the doubled orientation and $\Phi(\mathbf{x}) = [\phi_0(\mathbf{x}) \ldots \phi_n(\mathbf{x})]$ the row vector containing the set of basis functions $\phi(\mathbf{x})$ evaluated for a given coordinate $\mathbf{x} = (x, y)$. The system matrix is given as $\mathbf{V}$ and consists of the row vectors $\Phi(\mathbf{x})$. $\mathbf{f_x}$ and $\mathbf{f_y}$ contain the vectorial orientation data (computed using sine/cosine function from $2O(x,y)$). Then one can compute the parameter vector $\mathbf{c} = [\mathbf{a}, \mathbf{b}]$ for the vectorial approximation as described in the following:

$$\mathbf{a} = \mathbf{V}_\mathbf{w}^+ \mathbf{W} \mathbf{f_y} \qquad \mathbf{b} = \mathbf{V}_\mathbf{w}^+ \mathbf{W} \mathbf{f_x} \tag{1}$$

Where $\mathbf{V}_\mathbf{w}^+ = (\mathbf{V}^T \mathbf{W} \mathbf{V})^{-1} \mathbf{V}^T$ is the pseudo-inverse of the system matrix $\mathbf{V}$. The diagonal weighting matrix $\mathbf{W}$ is computed using fingerprint segmentation, where the diagonal elements are $\omega = 0$ for background and $\omega = 1$ for foreground pixels. For further details regarding the construction of the system matrix $\mathbf{V}$ we refer the reader to [11].

2.2 Computing a Subspace

Suppose now we have s sets of parameters $\mathbf{c}_i = [\mathbf{a}_i, \mathbf{b}_i]$ which were generated from s fingerprints as described above. These vectors form a distribution in the n dimensional space. If one can model this distribution, one can generate new examples similar to those in the original training set. Furthermore, one can decide whether a given OF is a plausible fingerprint flow patterns. We apply Principal Component Analysis (PCA) to the set of parameters in order to find a linear subspace where realistic fingerprints 'reside'. Therefore, we compute the mean $\overline{\mathbf{c}}$ and the covariance $\mathbf{S}$ of the data, followed by the eigenvectors $\mathbf{e} = [e_1, e_2, \ldots, e_t]$ and the corresponding eigenvalues $\boldsymbol{\lambda} = [\lambda_1, \lambda_2, \ldots, \lambda_t]$ of $\mathbf{S}$ (sorted largest first). Let Ω be the space of all possible parameters and Ψ the linear subspace spanned by the PCA. Then we can project parameters from Ω to Ψ using the linear projection φ:

$$\mathbf{d}_i = \varphi(\mathbf{c}_i) = \mathbf{e}^T(\mathbf{c}_i - \overline{\mathbf{c}}) \qquad \text{projection } \varphi \tag{2}$$

$$\mathbf{c}_i = \varphi^{-1}(\mathbf{d}_i) = \overline{\mathbf{c}}_i + \mathbf{e}\mathbf{d}_i \qquad \text{inverse projection } \varphi^{-1} \tag{3}$$

here $\mathbf{c}_i$ represents a point in the high dimensional space Ω and $\mathbf{d}_i$ the same point projected in to the linear subspace Ψ. The number of eigenvectors t to retain should be chosen so that the model represents a sufficiently large proportion of the total variance. Thus, the original high dimensional data can be approximated using a model with much fewer parameters. In Figure 1 the eigenvalue spectrum of 2000 fingerprint vectors (NIST4, f-prints database) is shown. Note that these fingerprints were not registered nor aligned in any other form. Only image cropping according to a segmentation has been performed.

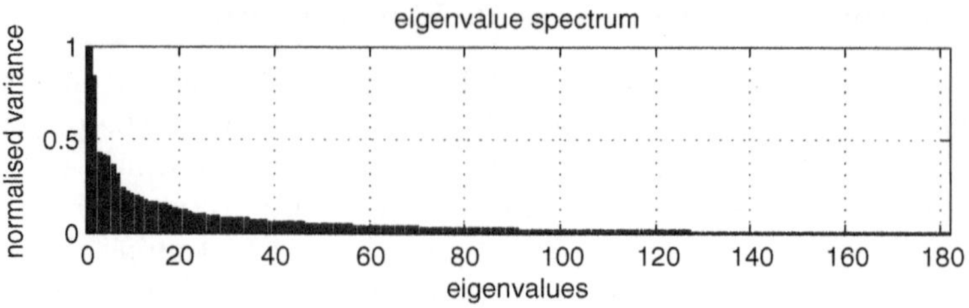

Fig. 1. Eigenvalue spectrum of 2000 fingerprint vectors (NIST4 f-prints database). Note how the first 40 eigenvalues correspond to more than 95% of the data's variance.

3 Fitting the Model to the Nearest Plausible Fingerprint Flow Pattern

In order to only generate examples similar to the training set, we have to choose a parameter $\mathbf{d} \in \Psi$. Therefore, we have to minimize the following cost function:

$$\underset{\mathbf{d}}{min} \quad \sum_{j=1}^{i} \omega_i \left[\sin\Big(M(\mathbf{x}_j) - O(\mathbf{x}_j) \Big) \right]^2 + \mu \left[\frac{1}{P(\mathbf{x}_j)} - P(\mathbf{x}_j) \right]^2 \tag{4}$$

The cost function compares the model's orientation estimation $M(\mathbf{x}_j)$ with the observed function value $O(\mathbf{x}_j)$ (obtained from local image gradients). We use the sin-function in order to resolve the discontinuity problem at zero and π. Then, one can compute $M(\mathbf{x}_j)$ as described in Equation 5:

$$M(\mathbf{x}_j) = \frac{1}{2} \arctan \frac{\Phi(\mathbf{x}_j)\mathbf{a}^T}{\Phi(\mathbf{x}_j)\mathbf{b}^T} \tag{5}$$

$$P(\mathbf{x}_j) = (\Phi(\mathbf{x}_j)\mathbf{a}^T)^2 + (\Phi(\mathbf{x}_j)\mathbf{b}^T)^2 \tag{6}$$

Note that $\mathbf{a}$ and $\mathbf{b}$ can be computed by the inverse mapping $\mathbf{c} = [\mathbf{a}, \mathbf{b}] = \varphi^{-1}(\mathbf{d})$ The second term of Equation 4 is a penalty function which regularizes the orientation vector to unit length ($\sin^2 + \cos^2 = 1$). This regularization (given in Equation 6 is necessary since the the training was done exactly with this condition fulfilled. On the other hand, allowing a minor deviation from unit length provides tolerance to rotation and translation.

3.1 Optimization

The minimization of the cost function in Equation 4 is done by using the Levenberg-Marquard (LM) algorithm. Note that, as described above, each iteration of the LM uses the inverse mapping $\mathbf{c}_i = \varphi^{-1}(\mathbf{d}_i)$ in order to evaluate the cost function as given in Equation 4. The factor μ is set to $3 * 10^{-4}$ in all our experiments. The initial value $\mathbf{d_0}$ for the LM is set to the null vector. This corresponds to the mean OF ($\overline{\mathbf{c}}$). The LM algorithm stops when a minima is reached or when the number of iterations exceeds 40.

4 Evaluation

This section presents the experimental results. For training the model, we used the NIST4 special database [15]. This database contains 2000 fingerprints evenly distributed among the five Henry classes. The number of eigenmodes is limited to 80. For evaluation of the proposed method, we used the NIST4 s-prints (all 2000 images) and the FVC2006 2a [4] (all 1680 images) database.

4.1 Generalisation Test

In this subsection we test how well the proposed model generalises to a given test database. Therefore the model is fitted to the raw (unsmoothed) OF of the given fingerprint. To measure the quality of the fit, we compute the absolute mean deviation between the ground truth OF and the fitted orientation field in degrees, where a error is only computed for foreground pixels. The ground truth OF is computed using the mentioned fingerprint software. The figures depicted in Figure 2(a) show cumulative distribution functions of the absolute mean deviation in degrees summarized over all images of the database. Most of the images show a mean deviation of smaller than five degrees. A large fraction of this error can be adhered to the block-wise processing of the commercial fingerprint matcher. Furthermore, we want to point out that the ground truth OF contains errors and thus a possible improvement using the proposed method is impossible to measure.

The reader should note that this evaluation procedure is exactly the same as described in [6]. The experiments and datasets are identical, but the authors of [6] deleted 20% of the images (due to missing SPs). In direct comparison our model generalises significantly better to fingerprints than the one proposed by Huckemann et al. [6]. We can report almost all of the images to have smaller than 10 degrees absolute mean deviation, in comparison [6] can only report 50%.

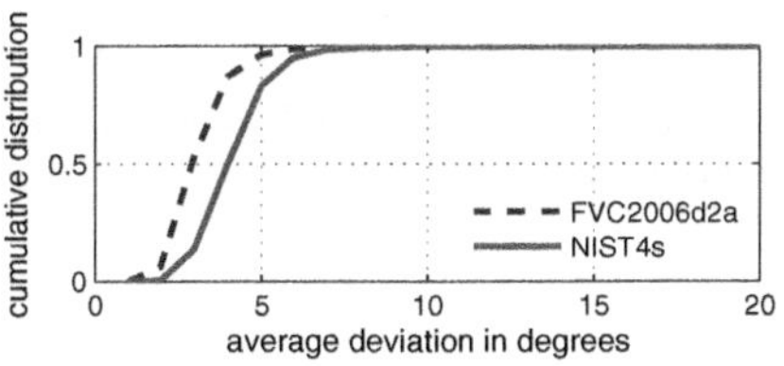

(a) *generalisation*

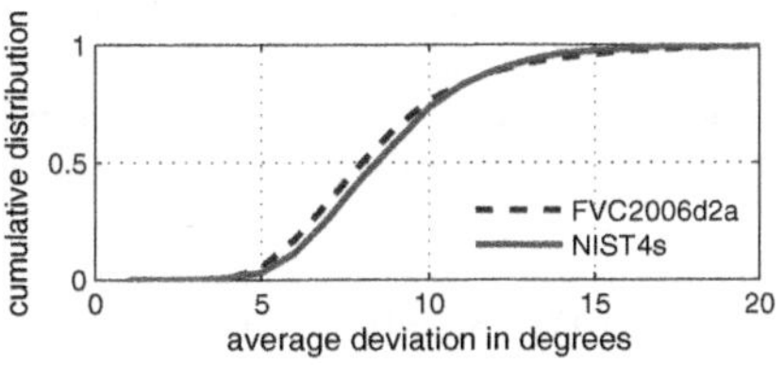

(b) *prediction*

Fig. 2. Generalisation and Prediction Evaluation

4.2 Prediction Test

The orientation interpolation capability of the proposed algorithm is tested in a simulated scenario where we remove 70% of the OF (see Figure 3). The OF remains in a rectangle with 40% area size, except a smaller rectangle with 10%

of the total image size in pixels. Both rectangles are centred in the middle of the image and exhibit the same aspect ratio as the image. Using this scenario, we tested the extrapolation as well as interpolation ability of the proposed algorithm. The figures are computed for the predicted OF only. Additionally, the background is removed from the input OF. This prediction evaluation is done using the same database configuration as mentioned above. It should be noted that not all predictions with a large absolute mean deviation are wrong in terms of plausibility.

The only comparable work with a significant large evaluated database is available from Hotz [5] (co-author of [6]). In his evaluation scenario the prediction was performed for only 5% occlusion (compared to 70% of our testing scenario). Unfortunately, this makes a possible comparison meaningless.

In Figure 6 one can see the translation and rotation invariance of the proposed method. Illustration 6(c) shows a 180 degree rotated image of a loop type fingerprint (the straight example is given in Figure 5). As can be seen, the model corrects the OF to a whorl type fingerprint - the most plausible valid pattern. Figure 3 shows comparisons with other methods available in literature.

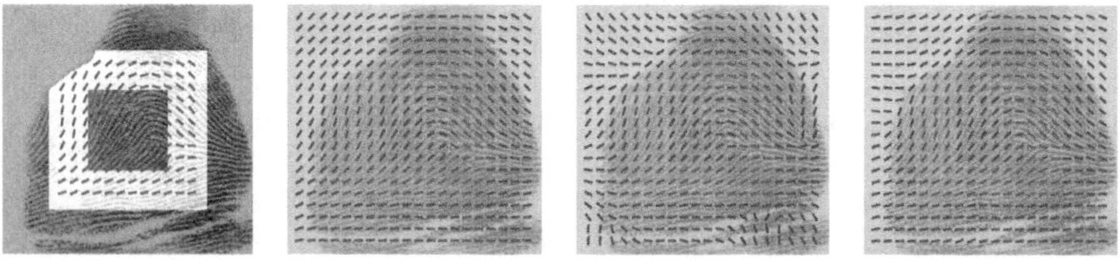

(a) *input OF highlighted* (b) *proposed* (c) *FOMFE, 162 parameters* (d) *Legendre Polynomials*

Fig. 3. Prediction ability of various methods proposed in literature. Input OF as shown in 3(a). Green color is used to display the interpolated/extrapolated OF. Subfigure 3(b) shows the results of the proposed method. The prediction cabability of the FOMFE approach (exactly as in [14]) is shown in image 3(c). Subfigure 3(d) shows results of the approach described in [11] (exactly as described in the paper). Note that the proposed method (Figure 3(b)) generates the most plausible OF.

4.3 Estimating the Number of Modes

In order to estimate the best number of Eigenmodes we performed the above mentioned prediction and generalisation experiments for a varying number of Eigenmodes. The evaluation criteria for the prediction and generalisation figures was the relative number of fingerprints with less than eight degrees average error. Due to the computational burden, only the first 100 images of the NIST4 database (s-prints) were used. We performed this evaluation for two scenarios. In the first scenario (shown in Subfigure 4(a)) one can see how the number of coefficients affects the model without regularisation. Furthermore, it shows the trade off between generalisation and prediction capability of the method. In

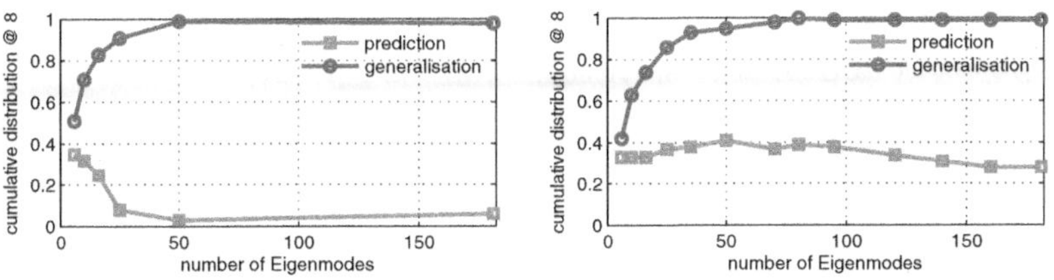

(a) *no regularization* (b) *with regularization*

Fig. 4. Regularisation. Applying a regularisation on the cost function, where we force the orientation vectors to unit length, we can significantly improve the results.

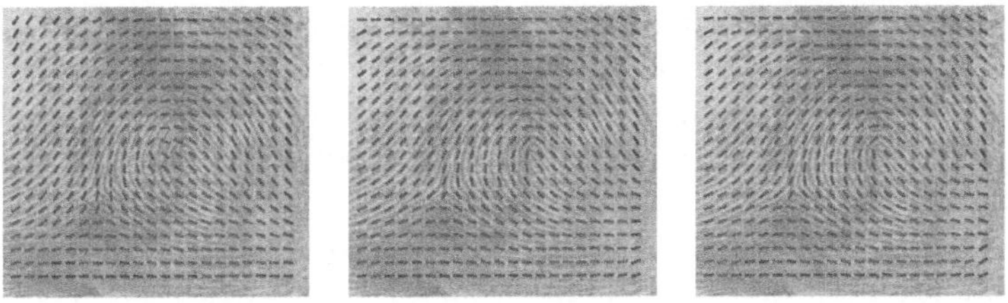

(a) *5 Eigenmodes* (b) *80 Eigenmodes* (c) *182 Eigenmodes*

Fig. 5. Number of Eigenmodes. In case of to few Eigenmodes (Subfigure 5(a)) the model fails to generalize, especially visible at SPs. The application of a higher number of Eigenmodes allows the model to fit the shown OF more precisely.

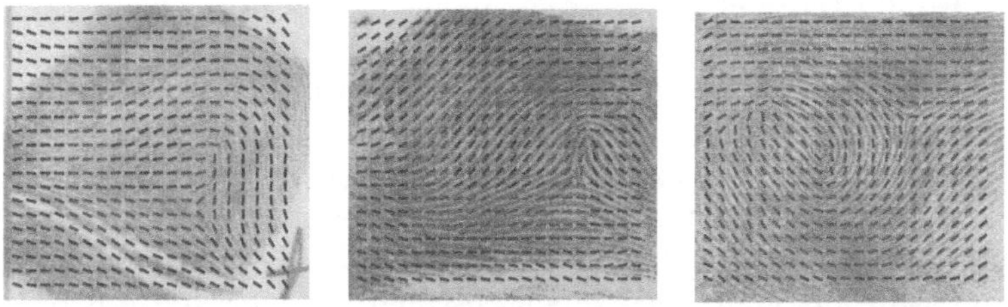

(a) *rotation* (b) *translation* (c) *upside down of 5*

Fig. 6. Subfigure 6(a) shows a 45 degree rotated fingerprint image. In Subfigure 6(b) an uncentered loop type fingerprint is shown. Figure 6(c) displays an upside down loop which has been corrected to a whorl. Predicted OF is shown in green.

general, a lower number of Eigenmodes results in good prediction figures but bad generalisation capability of the model - and vice versa. The second scenario (see Subfigure 4(b)) shows the model as proposed with the regularisation. It is clearly visible that the regularisation leads to a significant improvement.

4.4 Conclusion

In this paper we presented a statistical model for fingerprint ridge orientation. The fingerprint orientation field (OF) can be constrained by the Active Fingerprint Ridge Orientation Model (AFROM) to vary only in ways seen in a training set. The OF of fingerprints is represented by a vectorially linear regression using Legendre Polynomials. Fitting parameters to a given fingerprint is done using the Levenberg-Marquard (LM) algorithm. During the optimization procedure the parameters are limited to a previously learned linear subspace, where only 'legal' fingerprints reside. Using the proposed method, the AFROM iteratively deforms to fit an OF of a fingerprint. Our method does not depend on any prealignment or registration of the considered images. The training can be done fully automatic without any user interaction. Furthermore, our method does not depend on any other computed data, except a segmentation.

In the evaluation section of this paper, we perform generalisation and prediction tests of the proposed method. A generalisation test is done in order to evaluate how well the model 'fits' to a large number of OFs. Using the presented prediction test, we asses how specific the model is. This is the ability to constrain unknown or noisy regions of the OF to valid fingerprint flow patterns. All experiments are performed on public databases (from which one is fairly different to the learning dataset). These experiments, comparable with a very recent paper [6], assess our method a very good performance. Furthermore, it should be noted that our method is the first fingerprint OF model making use of prior knowledge for OF estimation. The major conception behind existing methods (e.g. [6]) is a hand crafted model which fits only to valid fingerprint OFs, without the possibility for machine based training. Our approach can also be seen as a method to find those elements (Eigenmodes comply to 'Eigen-Orientations') which, when (linear) combined, give biological valid patterns of fingerprints. We want to point out, that we used the full NIST4 f-prints database for training, including many noisy fingerprint images.

Future work includes the experimentation with other subspace methods than PCA (e.g. ICA, K-PCA, etc.). Moreover, the regularization term of the cost functional accommodates a large potential for future improvements. Another topic is the inclusion of an image quality estimation algorithm, where the model adjusts the amount of prior knowledge depending on the local image quality.

Acknowledgments. This work has been funded by the Biometrics Center of Siemens IT Solutions and Services, Siemens Austria.

References

1. Bazen, A.M., Gerez, S.H.: Systematic methods for the computation of the directional fields and singular points of fingerprints. IEEE Trans. Pattern Anal. Mach. Intell. 24(7), 905–919 (2002)
2. Bigün, J., Granlund, G.H.: Optimal orientation detection of linear symmetry. In: Proceedings of the IEEE First International Conference on Computer Vision, London, Great Britain, June 1987, pp. 433–438 (1987)

3. Cootes, T., Taylor, C.: Statistical models of appearance for computer vision (1999)
4. Fierrez, J., Ortega-Garcia, J., Torre, Gonzalez-Rodriguez, J.: Biosec baseline corpus: A multimodal biometric database. Pattern Recognition 40(4), 1389–1392 (2007)
5. Hotz, T.: Modelling and analysing orientation field of fingerprints. Phd thesis (2007)
6. Huckemann, S., Hotz, T., Munk, A.: Global models for the orientation field of fingerprints: An approach based on quadratic differentials. PAMI 30(9), 1507–1519 (2008)
7. Jain, A.K., Maltoni, D.: Handbook of Fingerprint Recognition. Springer, New York (2003)
8. Kass, M., Witkin, A.: Analyzing oriented patterns. Comput. Vision Graph. Image Process. 37(3), 362–385 (1987)
9. Li, J., Yau, W.Y., Wang, H.: Constrained nonlinear models of fingerprint orientations with prediction. PR 39(1), 102–114 (2006)
10. Pankanti, S., Prabhakar, S., Jain, A.K.: On the individuality of fingerprints. IEEE PAMI 24(8), 1010–1025 (2002)
11. Ram, S., Bischof, H., Birchbauer, J.: Curvature preserving fingerprint ridge orientation smoothing using legendre polynomials. In: IEEE CVPR Biometrics Workshop (2008)
12. Sherlock, B.G., Monro, D.M.: A model for interpreting fingerprint topology. PR 26(7), 1047–1055 (1993)
13. Vizcaya, P.R., Gerhardt, L.A.: A nonlinear orientation model for global description of fingerprints. PR 29(7), 1221–1231 (1996)
14. Wang, Y., Hu, J., Phillips, D.: A fingerprint orientation model based on 2d fourier expansion (fomfe) and its application to singular-point detection and fingerprint indexing. PAMI 29(4), 573–585 (2007)
15. Watson, C.I., Wilson, C.L.: Nist special database 4: Fingerprint database. National Institute of Standards and Technology (1992)
16. Zhou, J., Gu, J.: A model-based method for the computation of fingerprints' orientation field. IP 13(6), 821–835 (2004)
17. Zhou, J., Gu, J.: Modeling orientation fields of fingerprints with rational complex functions. PR 37(2), 389–391 (2004)

FM Model Based Fingerprint Reconstruction from Minutiae Template

Jianjiang Feng and Anil K. Jain

Department of Computer Science and Engineering
Michigan State University
{jfeng,jain}@cse.msu.edu

Abstract. Minutiae-based representation is the most widely adopted fingerprint representation scheme. The compactness of minutiae template has created an impression that the minutiae template does not contain sufficient information to allow the reconstruction of the original fingerprint image. This belief has now been shown to be false; several algorithms have been proposed that can reconstruct fingerprint images from minutiae templates. However, these reconstruction techniques have a common weak point: many spurious minutiae, not included in the original minutiae template, are generated in the reconstructed image. Moreover, some of these techniques can only reconstruct a partial fingerprint. In this paper, a novel fingerprint reconstruction algorithm is proposed, which not only reconstructs the whole fingerprint, but the reconstructed fingerprint contains very few spurious minutiae. A fingerprint image is modeled as a 2D Frequency Modulation (FM) signal whose phase consists of the continuous part and the spiral part (which corresponds to minutiae). An algorithm is proposed to reconstruct the continuous phase from minutiae. The proposed reconstruction algorithm has been evaluated with respect to the success rates of type-I attack (match the reconstructed fingerprint against the original fingerprint) and type-II attack (match the reconstructed fingerprint against the different impressions of the original fingerprint) using a commercial fingerprint recognition system. Both types of attacks were shown to be successful in deceiving the fingerprint system.

Keywords: Fingerprint synthesis, fingerprint reconstruction, minutiae, AM-FM, orientation field.

1 Introduction

Fingerprint recognition systems play a crucial role in many situations where a person needs to be verified or identified with high confidence. A fingerprint is an oriented texture consisting of interleaving ridges and valleys. The pattern of minutiae, which are the endpoints and merging points of ridges, is believed to be unique to each finger. Due to its distinctiveness and compactness, minutiae-based representation has become the most widely adopted fingerprint representation scheme. Since minutiae template is a compact representation of a fingerprint, it has been assumed that it is not possible to reconstruct the original fingerprint from a minutiae template. In other words, minutiae extraction algorithms have been traditionally viewed as a "one-way" transform. Recently, however, this belief has been challenged by some researchers [1, 2, 3] who were successful in reconstructing a fingerprint image from the given minutiae template.

M. Tistarelli and M.S. Nixon (Eds.): ICB 2009, LNCS 5558, pp. 544–553, 2009.
© Springer-Verlag Berlin Heidelberg 2009

Hill [1] proposed a technique to determine the fingerprint structure from minutiae template (including singular points). In this work, orientation field was generated based on singular points according to the model in [4]. A line drawing algorithm was used to generate a sequence of splines passing through the minutiae. Ross et al. [2] first estimated the orientation field using selected minutiae triplets in the template. Streamlines were then traced starting from minutiae and border points. Linear Integral Convolution was used to impart texture-like appearance to the ridges. Finally, the image was smoothed to obtain wider ridges. This reconstruction algorithm can only generate a partial fingerprint. In addition, streamlines that terminated due to distance constraint between adjacent streamlines generate spurious minutiae. Cappelli et al. [3] estimated the orientation field by fitting a modified model proposed in [5] to the minutiae directions. Gabor filtering is iteratively performed starting from minutiae on an image initialized by local minutiae pattern. A rendering step is performed to make the reconstructed fingerprint image appear more realistic. However, this algorithm also generates many spurious minutiae in the reconstructed fingerprints.

Fingerprint reconstruction from minutiae (hereinafter simply referred to as fingerprint reconstruction) is very similar to fingerprint synthesis [6] except that the goals and the inputs of the two techniques are different. The goal of fingerprint reconstruction is to obtain an artificial fingerprint that resembles the original fingerprint as much as possible, while the goal of fingerprint synthesis is to generate artificial fingerprints that are as realistic as possible. For fingerprint reconstruction, the minutiae from a given fingerprint must be provided, while for fingerprint synthesis, no input is needed (except for statistical models learned from many real fingerprints).

The well-known SFINGE fingerprint synthesis method of Cappelli et al. [6] performs Gabor filtering on a seed image according to the orientation and frequency images; minutiae automatically emerge during the filtering procedure. Some intraclass variations, such as spatial transformation, touching area, nonlinear distortion, ridge dilation/shrinking and noise, are simulated to generate realistic impressions of the master fingerprint. One main limitation of SFINGE is that minutiae can not be controlled. As a result, SFINGE may generate problematic fingerprints that contain too few minutiae or very long ridges. It is well known that the distribution of minutiae in fingerprints is not random and fingerprints of different pattern types have different minutiae distributions [2]. The minutiae distribution of fingerprints generated by SFINGE may not conform to such distributions since these minutiae are automatically generated during the image filtering process. Similar fingerprint synthesis methods have also been proposed in [7, 8]. Reaction-diffusion technique described in [9] can also be used for synthesizing fingerprints. Bicz [10] described a fingerprint synthesis technique based on 2D FM model. The phase of FM model consists of the continuous component and the spiral component, which corresponds to minutiae. A fingerprint is synthesized by generating each component separately and combining them. Separation of the continuous phase and the spiral phase makes minutiae controllable. However, the most important step, generating continuous phase component, was not described in [10]. According to the demo software provided by the author, only a partial fingerprint (around the core) can be generated and the orientation field of each of the four fingerprint patterns is fixed. Larkin and Fletcher [11] proposed a fingerprint analysis framework based on AM-FM model where a fingerprint image is decomposed into the four components of the model. Separate compression of each component leads to a compressed fingerprint with an impressive compression factor of 239.

In this paper, a novel approach to fingerprint reconstruction from minutiae template is proposed, which uses FM model as the fingerprint representation scheme. Specifically, a new algorithm is proposed to reconstruct the continuous phase of FM model from minutiae. The advantages of our approach over existing approaches to fingerprint reconstruction [1, 2, 3] are: (i) the complete fingerprint can be reconstructed, and (ii) the reconstructed fingerprint contains very few spurious minutiae. The proposed reconstruction algorithm has been quantitatively assessed by matching reconstructed fingerprints against the original fingerprints (termed as type-I attack) and against different impressions of the original fingerprints (termed as type-II attack) using a commercial fingerprint SDK, Neurotechnology VeriFinger 4.2 [12]. Type-I attack was found to have a high chance in deceiving the fingerprint recognition system in both the verification and identification experiments. Type-II attack also has a significantly higher accept rate than that of impostor match.

2 Fingerprint Representation

Larkin and Fletcher [11] proposed to represent a fingerprint image as a 2D amplitude and frequency modulated (AM-FM) signal:

$$I(x, y) = a(x, y) + b(x, y)\cos(\Psi(x, y)) + n(x, y), \quad (1)$$

which is composed of four components: the intensity offset $a(x, y)$, the amplitude $b(x, y)$, the phase $\Psi(x, y)$, and the noise $n(x, y)$. Here we are only interested in the phase $\Psi(x, y)$, since ridges and minutiae are totally determined by the phase; the other three components just make the fingerprint appear realistic. Therefore, an ideal fingerprint is represented as a 2D FM signal:

$$I(x, y) = \cos(\Psi(x, y)). \quad (2)$$

According to the Helmholtz Decomposition Theorem [13], the phase can be uniquely decomposed into two parts: the continuous phase and the spiral phase:

$$\Psi(x, y) = \Psi_C(x, y) + \Psi_S(x, y). \quad (3)$$

The gradient of the continuous phase $\Psi_C(x, y)$ is termed as instantaneous frequency $G(x, y)$. The direction of instantaneous frequency is normal to ridge orientation. The amplitude of instantaneous frequency is equal to the ridge frequency.

The spiral phase $\Psi_S(x, y)$ corresponds to minutiae:

$$\Psi_S(x, y) = \sum_{n=1}^{N} p_n \arctan\left(\frac{y - y_n}{x - x_n}\right), \quad (4)$$

where x_n and y_n denote the coordinates of the nth minutia, and N denotes the total number of minutiae. The direction of a minutia is determined by its polarity $p_n \in$

(a) (b)

Fig. 1. (a) Synthetic fingerprint without minutiae and (b) synthetic fingerprint with 7 minutiae

$\{1, -1\}$ and the local ridge orientation $O(x_n, y_n)$, which is defined in the continuous phase. Assume the ridge orientation is in the range $[0, \pi]$. The direction of a minutia is equal to $O(x_n, y_n)$ when it has positive polarity, or $O(x_n, y_n) + \pi$ when it has negative polarity. Adding spiral to a continuous phase generates minutiae. Figure 1(a) shows a synthesized fingerprint that only contains continuous phase. Adding 7 spirals to the continuous phase leads to the synthesized fingerprint in Fig. 1(b).

3 Fingerprint Reconstruction

3.1 Problem Statement

A set of N minutiae $\{x_n, y_n, \alpha_n\}, 1 \leq n \leq N$ of a fingerprint is given, where (x_n, y_n) and α_n denote the location and direction of the nth minutia, respectively. In terms of FM model, the spiral phase is given and the direction of instantaneous frequency of the continuous phase is known at the locations of the N minutiae. The problem is to reconstruct the original fingerprint image modeled by Eq. (1). This is obviously an ill-posed problem, since the important information required to reconstruct the continuous phase of fingerprints, ridge frequency, is unknown. Information needed to reconstruct realistic fingerprints, such as brightness, contrast, the background noise of fingerprint sensor, and detailed ridge features (pores, contours) etc., is also not available. A more practical goal is to estimate the FM representation of the original fingerprint, $cos(\Psi(x, y))$. To obtain the phase $\Psi(x, y)$, the following three steps are performed: orientation field reconstruction, continuous phase reconstruction, and combination of the spiral phase and the continuous phase. The flow chart of the proposed fingerprint reconstruction algorithm is depicted in Fig. 2.

3.2 Orientation Field Reconstruction

Ross et al. [2] used selected minutiae triplets to estimate the orientation field in triangles. Cappelli et al. [3] estimated orientation field by fitting an orientation field model to the orientations at minutiae. Both these methods have a minimum requirement on the number of minutiae. We propose a novel orientation field reconstruction algorithm that

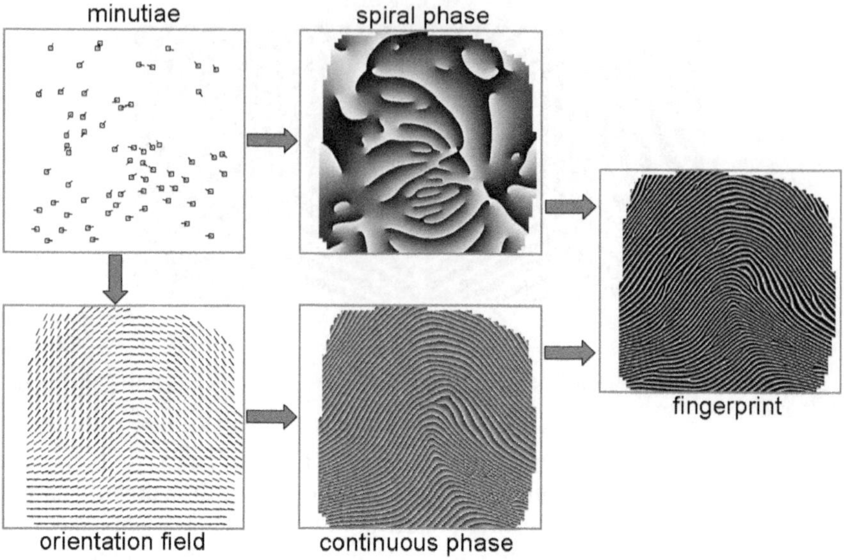

Fig. 2. Flow chart of the proposed fingerprint reconstruction algorithm

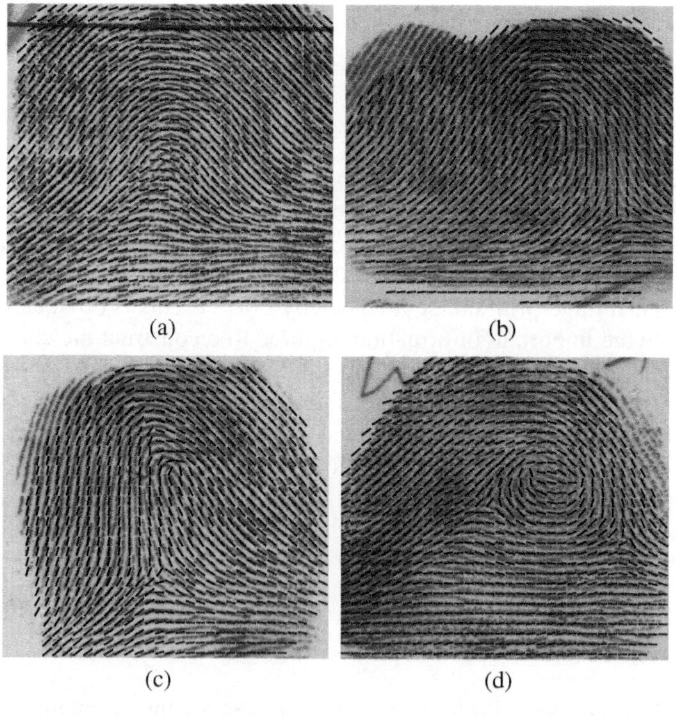

Fig. 3. Reconstructed orientation field for the four main types of fingerprints: (a) arch, (b) left loop, (c) right loop, and (d) whorl

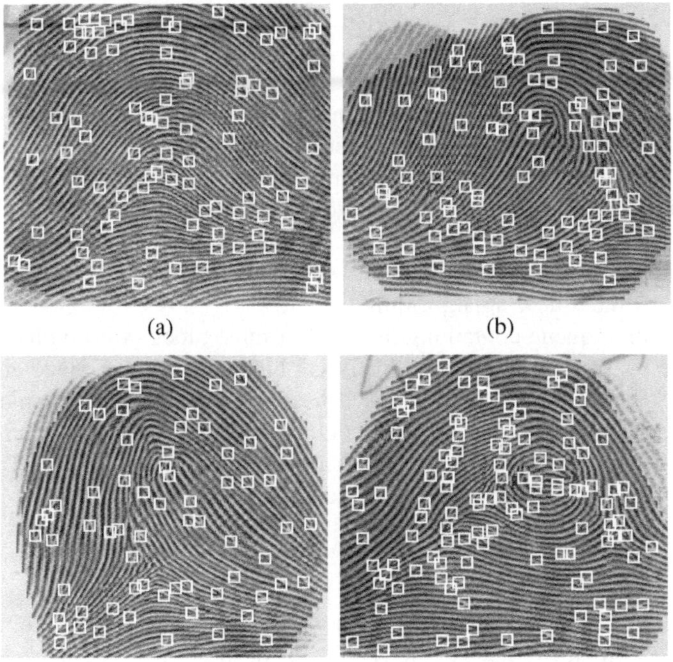

(a) (b) (c) (d)

Fig. 4. Reconstructed fingerprints (overlaid on the original fingerprints) for the four main types of fingerprints: (a) arch, (b) left loop, (c) right loop, and (d) whorl. The original minutiae are denoted by white squares.

can work even when only one minutia is available. Figure 3 shows the reconstructed orientation field for the four main types of fingerprints.

The image is divided into non-overlapping blocks of 8×8 pixels. Foreground mask for the fingerprint image is obtained by dilating the convex hull of minutiae using a disk-shaped structuring element of 8×8 pixels. The local ridge orientation at block (m, n) is predicted by using the nearest minutia in each of the 8 sectors. The minutia direction α_k is doubled to make α_k equivalent to $\alpha_k + \pi$. The cosine and sine components of $2\alpha_k$ of all the K selected minutiae are summed:

$$u = \sum_{k=1}^{K} \frac{\cos(2\alpha_k)}{d_k}, v = \sum_{k=1}^{K} \frac{\sin(2\alpha_k)}{d_k}, \tag{5}$$

where d_k denotes the Euclidean distance between the block center and the kth minutia. Then the orientation at block (m, n) is computed as: $O(m, n) = \frac{1}{2} \arctan(\frac{v}{u})$.

3.3 Continuous Phase Reconstruction

The continuous phase of a fingerprint is modeled by piecewise planes at each foreground block (m, n) of 8×8 pixels:

$$\Psi_C(x,y) = 2\pi \cdot F(m,n) \cdot (\cos(O(m,n))x + \sin(O(m,n))y) + P(m,n),$$
$$8(m-1) \le x < 8m, 8(n-1) \le y < 8n, \quad (6)$$

where $F(m,n)$, $O(m,n)$ and $P(m,n)$ denote the ridge frequency, the ridge orientation and the phase offset at block (m,n), respectively. Since it is not possible to estimate the ridge frequency from minutiae (if the ridge count information between minutiae is provided, then it is possible to estimate the ridge frequency), we have used a constant frequency value 0.1 for the whole image, which corresponds to a ridge period of 10 pixels in 500 ppi images. The only unknown value in Eq. (6), the phase offset $P(m,n)$, is estimated by the following algorithm.

Starting with a queue containing the top left-most block (whose phase offset is assumed to be 0), in each iteration, a block is obtained from the queue and each of its four-connected neighbors is checked if it has been reconstructed (namely, the phase offset has been estimated). If one of the neighboring blocks has not been reconstructed, the phase offset of this block is estimated and it is put into the queue. This procedure is performed until the queue is empty (which means that the continuous phase has been reconstructed at all the foreground blocks). An ancillary image is used to record the reconstructed blocks.

Here we describe how to estimate the phase offset of a block using all of the already reconstructed four-connected neighbors. Consider one of the neighbors, say block $(m-1,n)$, of block (m,n). The phase images of these two blocks are required to be continuous at the border pixels $\{(x,y) : x = 8(m-1), 8(n-1) \le y < 8n\}$. For each border pixel (x,y), a phase offset of block (m,n) is estimated:

$$\psi = 0.2\pi \cdot (\cos(O(m-1,n)) \cdot x + \sin(O(m-1,n)) \cdot y) + P(m-1,n)$$
$$-0.2\pi \cdot (\cos(O(m,n)) \cdot x + \sin(O(m,n)) \cdot y). \quad (7)$$

The mean value is used as the phase offset of block (m,n).

After the continuous phase is reconstructed, the reconstructed fingerprint can be obtained by combining the continuous phase and the spiral phase. In Fig. 4, the reconstructed fingerprint for each of the four types of fingerprints in Fig. 3 is overlaid on the original fingerprint. The reconstructed fingerprints match the original fingerprints well. But, the reconstructed fingerprints do contain a few spurious minutiae, especially in the region of singularity. The main cause for the spurious minutiae is the assumption of a fixed ridge frequency for the whole image, which is not true for real fingerprints.

4 Experiments

A reconstructed fingerprint may be used to attack the system that contains the original fingerprint template (termed as type-I attack) or the systems where the same finger has also been enrolled (termed as type-II attack). Such fingerprint recognition systems may work in the verification mode or in the identification mode. To evaluate the performance of the proposed reconstruction algorithm in these situations, VeriFinger 4.2 by Neurotechnology [12] was assumed to be the attacked fingerprint recognition system.

The verification experiment was performed on FVC2002 DB1_A [14], which consists of 100 fingers and 8 impressions per finger. Fingerprints were reconstructed from

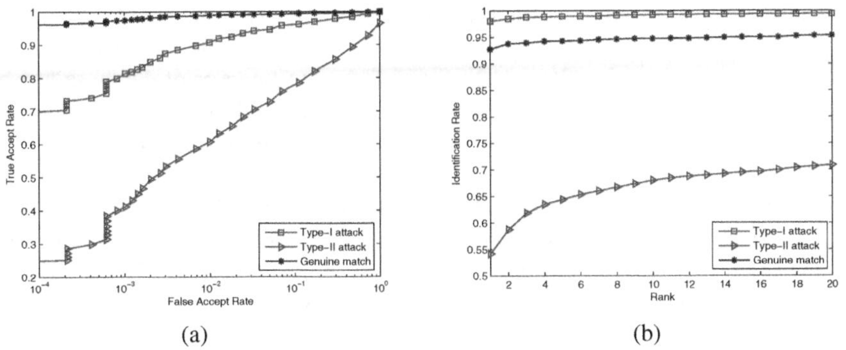

Fig. 5. Performance of the proposed reconstruction algorithm. (a) ROC curves of type-I attack, type-II attack and genuine match against impostor match on FVC2002 DB1_A. (b) CMC curves of type-I attack, type-II attack and genuine match on NIST SD4.

the minutiae templates of all the 800 images. Each reconstructed fingerprint was matched against all the 8 impressions of the same fingerprint in the database, which produced 800 type-I attacks and 5,600 type-II attacks. For the purpose of comparison, 2,800 genuine matches were obtained by cross matching 8 impressions of the same fingerprint and 4,950 impostor matches were obtained by cross matching the first impression of different fingerprints. By changing decision thresholds, the accept rates of type-I attack, type-II attack, genuine match and impostor match were obtained. The three Receiver Operating Characteristic (ROC) curves in Fig. 5(a) are plots of the true accept rates (termed as TAR) of type-I attack, type-II attack and genuine match against the false accept rates (termed as FAR) of impostor match. It can be observed that even under a very secure setting (0.1% FAR), both type-I attack and type-II attack have a much higher accept rate (70% and 25%, respectively) than the accept rate of impostor match, namely 0.1% FAR. As expected, type-I attack has more chance than type-II attack in deceiving a fingerprint verification system. But the impact of type-II atack should not be underestimated, since it can be launched against all fingerprint systems where the same finger has been enrolled. It should be noted that this performance can not be directly compared to the performance reported in [3], since (i) only type-I attack was considered in [3]; (ii) only the first impression (with large image size and good image quality) of each finger was used for reconstruction in [3]; and (iii) different fingerprint matchers have very different ability in dealing with artificial fingerprints, as shown in [3].

The identification experiment was performed on NIST SD4 [15], which consists of 2,000 pairs of ink-on-paper fingerprints (called file and search fingerprints, respectively). Fingerprints were reconstructed from the templates of 2,000 file fingerprints. The reconstructed fingerprints were matched against 2,000 file fingerprints and 2,000 search fingerprints to obtain 2,000 type-I attacks and 2,000 type-II attacks, respectively. For the purpose of comparison, 2,000 search fingerprints are also matched against 2,000 file fingerprints to obtain 2,000 genuine matches. The Cumulative Match Characteristic (CMC) curves of type-I attack, type-II attack and genuine match are given in Fig. 5(b). The rank-1 identification rate of 98.1% indicates that type-I attack has a very high identification rate in deceiving a fingerprint identification system, even higher than the identification rate of a genuine match. It should be noted that the rank-1 identification rate of type-I attack is only 23% in [2]. Since the same fingerprint database has been used and the same minutiae extraction and matching algorithms have been employed, we can conclude that the proposed reconstruction algorithm performs better than the algorithm of Ross et al. [2].

The reconstruction of fingerprints in FVC2002 DB1 and NIST SD4 takes around 3 and 8 seconds per image, respectively, when tested on a PC with 3GB of RAM and a 3GHz Pentium 4 CPU. The algorithm is currently implemented in MATLAB and we expect the computational costs to be significantly reduced after optimization.

5 Conclusions and Future Work

A FM model based fingerprint reconstruction scheme has been proposed. The phase of FM model is composed of the continuous phase and the spiral phase. A reconstructed fingerprint is obtained by reconstructing the orientation field, reconstructing the continuous phase and combining the continuous phase with the spiral phase. The experimental results show that the reconstructed images are very consistent with the original fingerprints and that there is a high chance to deceive a state-of-the-art commercial fingerprint recognition system.

The reconstructed fingerprints still contain a few spurious minutiae, especially in the high curvature regions. The spurious minutiae can be avoided by (i) detecting and subtracting spirals from the continuous phase, or (ii) permitting the ridge frequency to vary in a reasonable range. To obtain reconstructed images more consistent with the original fingerprints, ridge frequency and minutiae type should be utilized. To make the reconstructed fingerprints more realistic, brightness, ridge thickness, pores and noise should be modeled. The accept rate of the reconstructed fingerprints can be further improved by reducing the image quality around the spurious minutiae. To reduce the risk of attacks using reconstructed fingerprints, robust fingerprint template security [16] and spoof detection techniques [17] should be developed.

Acknowledgments

We would like to thank Dr. Kieran G. Larkin, Canon Information Systems Research Australia (CiSRA), for his help in generating the synthetic fingerprint in Fig. 1. This work was supported by ARO grant W911NF-06-1-0418 and NIJ grant 2007-RG-CX-K183.

References

1. Hill, C.: Risk of masquerade arising from the storage of biometrics. Master's thesis, Australian National University (2001)
2. Ross, A., Shah, J., Jain, A.K.: From template to image: Reconstructing fingerprints from minutiae points. IEEE Trans. Pattern Analysis and Machine Intelligence 29(4), 544–560 (2007)

3. Cappelli, R., Lumini, A., Maio, D., Maltoni, D.: Fingerprint image reconstruction from standard templates. IEEE Trans. Pattern Analysis and Machine Intelligence 29(9), 1489–1503 (2007)
4. Sherlock, B.G., Monro, D.M.: A model for interpreting fingerprint topology. Pattern Recognition 26(7), 1047–1055 (1993)
5. Vizcaya, P.R., Gerhardt, L.A.: A nonlinear orientation model for global description of fingerprints. Pattern Recognition 29(7), 1221–1231 (1996)
6. Maltoni, D., Maio, D., Jain, A.K., Prabhakar, S.: Handbook of Fingerprint Recognition. Springer, Heidelberg (2003)
7. Novikov, S.O., Glushchenko, G.N.: Fingerprint ridges structure generation models. In: Proc. SPIE (International Workshop on Digital Image Processing and Computer Graphics), vol. 3346, pp. 270–274 (1997)
8. Araque, J.L., Baena, M., Chalela, B.E., Navarro, D., Vizcaya, P.R.: Synthesis of fingerprint images. In: Proc. 16th International Conference on Pattern Recognition, pp. 422–425 (2002)
9. Witkin, A., Kass, M.: Reaction-diffusion textures. SIGGRAPH Computer Graphics 25(4), 299–308 (1991)
10. Bicz, W.: The idea of description (reconstruction) of fingerprints with mathematical algorithms and history of the development of this idea at optel (2003), `http://www.optel.pl/article/english/idea.htm`
11. Larkin, K.G., Fletcher, P.A.: A coherent framework for fingerprint analysis: are fingerprints holograms? Optics Express 15(14), 8667–8677 (2007)
12. Neurotechnology Inc., VeriFinger, `http://www.neurotechnology.com`
13. Ghiglia, D.C., Pritt, M.D.: Two-Dimensional Phase Unwrapping: Theory, Algorithms, and Software. John Wiley and Sons, New York (1998)
14. FVC 2002: The Second International Competition for Fingerprint Verification Algorithms, `http://bias.csr.unibo.it/fvc2002/`
15. NIST SD4: NIST 8-Bit gray scale images of fingerprint image groups (FIGS), `http://www.nist.gov/srd/nistsd4.htm`
16. Nandakumar, K., Jain, A.K., Pankanti, S.: Fingerprint-based fuzzy vault: implementation and performance. IEEE Trans. Information Forensics and Security 2(4), 744–757 (2007)
17. Nixon, K.A., Rowe, R.K.: Multispectral fingerprint imaging for spoof detection. In: Proc. SPIE (Biometric Technology for Human Identification II), vol. 5779, pp. 214–225 (2005)

Robust Biometric System Using Palmprint for Personal Verification

G.S. Badrinath and Phalguni Gupta

Dept. of Computer Science and Engineering, Indain Institute of Technology Kanpur, India
{badri,pg}@iitk.ac.in

Abstract. This paper describes a prototype of robust biometric system for verification. The system uses features extracted using Speeded Up Robust Features (SURF) operator of human hand. The hand image for features is acquired using a low cost scanner. The extracted palmprint region is robust to hand translation and rotation on the scanner. The system is tested on IITK database and PolyU database. It has FAR 0.02%, FRR 0.01% and an accuracy of 99.98% at original size. The system addresses the robustness in the context of scale, rotation and occlusion of palmprint. The system performs at accuracy more than 99% for scale, more than 98% for rotation, and more than 99% for occlusion. The robustness and accuracy suggest that it can be a suitable system for civilian and high-security environments.

Keywords: Robust, Scale, Occlusion, Rotation, Translation, Scanner.

1 Introduction

Biometrics establishes identity of a person by physiological and/or behavioral characteristics. Identification / verification systems based on biometrics have found its applications widely in commercial and law enforcement applications. Systems based on fingerprint features are most widely used and biometric features generated from iris are considered to be the most reliable biometric features [22]. Using human hand as biometric feature is relatively a new approach. Features like principle lines, wrinkles, ridges, minutiae points, singular points, and texture pattern in the palmprint region between wrist and fingers can be used for its representation [20]. Furthermore, systems based on hand features are the most acceptable to users [8].

Limited work has been reported on palmprint identification and verification, despite the importance of palmprint features. Some of the recent research efforts in [1-3], [11] address the problem of palmprint recognition in large databases, and achieve very low error rates. There exist verification/identication system based on palmprint extracted features using various transforms such as Fourier Transform [10], Discrete Cosine Transform [3], Karhunen-Loeve transform [1], Wavelet transform [2],[4],[11] Fisher Discriminant Analysis [16], Gabor filtering [2], Neural networks [3,7] and Independent component Analysis [15]. Also there are systems based on datam points [6], local binary pattern histograms [5], statistical signatures [4], hand geometry features [3]. Furthermore, there exist multimodal systems fusing feature from

M. Tistarelli and M.S. Nixon (Eds.): ICB 2009, LNCS 5558, pp. 554–565, 2009.
© Springer-Verlag Berlin Heidelberg 2009

hand–geometry and face [12], eigenpalm and eigenfinger [1] and fusing palmprint features extracted using haar wavelet and principle component anlysis [11].

In [6], [7], systems using ink marking to capture the palmprint patterns have been presented. The systems are not widely accepted due to considerable attention and high cooperation is required in providing a biometric sample. In recent papers on palmprint based recognition system, palmprint images are captured from a digital camera [2] and users hand is placed in constrained environment using pegs for palmprint image acquisition. The problem with peg based system is that the user may not able to enroll or verify themselves if not able to place his/her hand in between the pegs due to injuries or physically challenged. Hence, there is a need to build a system which is

- Robust to translation and rotation of hand image on the scanner.
- Robust to spatial resolution (scale) of the image.
- Robust to occlusion (partially exposed) of the hand image.
- Robust to rotation of the palmprint image given for verification.

Further, it should be highly accurate, available at reasonable cost so that it is suitable for civilian and high end security applications.

This paper has proposed a novel method to extract features in palmprint which can be used in personal verification system. The proposed method tries to incorporate some of the above said characteristics in the palmprint based biometric system. It makes use of Speeded Up Robust Features (SURF) [14] to detect key-points in a given palmprint and determines local features around the key-points as its descriptors. Number of matching key-points for verification is carried out using nearest neighbour ratio method [21]. The hand images are acquired using low cost flat bed scanner.

2 Speeded Up Robust Features (SURF)

The SURF [14] is recently emerged cutting edge methodology for pattern recognition, and has been used in general object recognition and for other machine vision applications [17], [18]. SURF has been designed for extracting highly distinctive invariant features from images. The following major stages are used to compute the SURF feature vectors of a given image.

2.1 Key-Point Detectors

The SURF key-point detector is based on hessian matrix. Given a point $P = (x, y)$ in an image I, the hessian matrix $H = (x, \sigma)$ in P at scale σ is defined as follows

$$H(P,\sigma) = \begin{bmatrix} L_{xx}(P,\sigma) & L_{xy}(P,\sigma) \\ L_{xy}(P,\sigma) & L_{yy}(P,\sigma) \end{bmatrix} \quad (1)$$

where $L_{xx}(P,\sigma)$, $L_{xy}(P,\sigma)$ and $L_{yy}(P,\sigma)$ are the convolution of the Gaussian second order derivatives $\frac{\partial^2}{\partial x^2} g(\sigma)$, $\frac{\partial^2}{\partial x \partial y} g(\sigma)$ and $\frac{\partial^2}{\partial y^2} g(\sigma)$ with the image I at the point P respectively. The second order Gaussian derivatives are approximated using box filters. Image convolutions with box filters are computed rapidly using integral images [19]. Key-points are localized in scale and image space by applying a non maximum suppression in a $3 \times 3 \times 3$ neighborhood.

2.2 Key-Point Descriptor

This stage consists of two steps. In first step, a circular region is constructed around the extracted key-points. Dominant orientation of the circular region is computed using the Haar wavelets response in both x and y directions. The resulting maximum is considered as dominant orientation and used to generate the key-point feature vector. Since, the feature vector of the key-point are measured relative to the dominant orientation, hence the generated feature vectors are invariance to image rotation.

In second step a square region is constructed around each extracted key-point and aligned along the dominant orientation. The square region is partitioned into sixteen smaller sub-regions of size 4×4. Haar wavelet responses are computed for each sub-region. The sum of the wavelet responses d_x and d_y for each sub-region are used as feature values. Furthermore, the absolute values $|d_x|$ and $|dy|$ are summed to obtain the polarity of the image intensity changes. Thus, the feature vector *Vec* of the sub-image is given by

$$Vec = \left\{ \sum d_x, \sum d_x, \sum |d_x|, \sum |d_y| \right\} \tag{2}$$

The SURF feature vector of the key-point is formed concatenating the feature vectors *Vec* from all sixteen 4x4 sub-regions around key-point. So the feature vector of key-point contains 16 x 4 = 64 elements as the SURF feature vector.

3 Proposed System

This section presents a robust biometric system for verification using SURF features of palmprint. Like other biometric systems, it consists of four major stages, described in the following sections.

3.1 Image Acquisition

Hand images are obtained at a spatial resolution of 200 dots per inch in gray scale using a flatbed scanner. The device is constraints (pegs) free. Typical gray level image obtained from the scanner is shown in Fig. 1a.

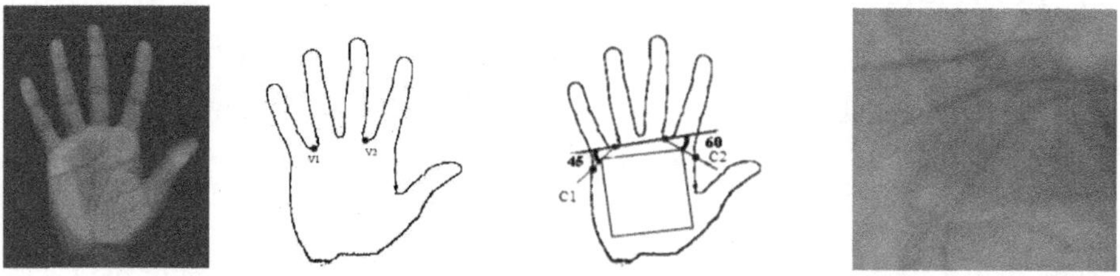

(a) (b) (c) (d)

Fig. 1. (a) Scanned Image, (b) Hand Contour and reference point, (c) Relevant points and Region of interest (palmprint), (d) Extracted Region of Interest (palmprint)

3.2 Pre-processing and Region of Interest Extraction

In this phase hand image is pre-processed and palmprint region is extracted. Global thresholding has been applied to extract the hand from the background. The contour of the hand image as shown in Fig. 1b is extracted applying contour-tracing algorithm [13] on the binarised image. Two reference points (*V1, V2*) between the fingertips and the valleys are determined on the contour of hand image as shown in Fig 1b.

Square area as shown in Fig. 1c with two of its corners placed on middle of line segments *C1-V1* and *V2-C2* is considered as region of interest or palmprint. The line segments *C1-V1* and *V2-C2* are inclined at an angle of 45^0 and 60^0 respectively to the line joining *V1* and *V2*. The extracted region of interest of the original gray scale image is shown in Fig. 1d.

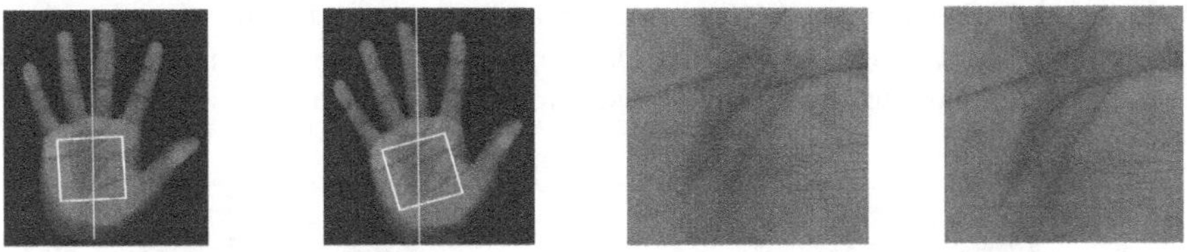

(a) (b)

Fig. 2. (a) Images of same subject with different orientation of placement relative to the symmetry (Yellow Line) of work surface, (b) Extracted region of interest for images shown in Fig. 5a

Since placement of palm on the scanner is pegs free, orientation of placement would vary for every incident. Two images of same subject with different orientations of placement are shown in Fig. 2a. The extracted region of interest or palmprint is relative to the reference points *V1* and *V2* which are stable for the subject. So the extracted palmprint region remains the same as shown in Fig. 2b for the subject independent to the orientation of placement. Hence the proposed extraction procedure of the system for extracting palmprint makes the system robust to rotation. From the experiments it is observed that the system is robust to rotation for about $\pm 35^0$.

3.3 Palmprint Feature Extraction

After pre-processing of hand image, and region of interest or palmprint extraction, feature are extracted for verification. SURF is used to extract the features which provides good discrimination ability. The SURF extracted features of the image are

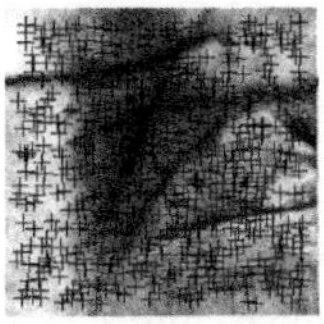

Fig. 3. Detected SURF key-points of the Palmprint

found to be invariant to scaling, rotation, and translation of the image [14]. Thus extracted feature from an image can be matched correctly with high probability against features from a large image database. The detected SURF key-points for the palmprint image is shown in Fig. 3.

3.4 Matching

To verify the live palmprint, the SURF features computed for the enrolled image should be matched with SURF features of the live palmprint image. Matching between SURF key-points of live and enrolled palmprint image is done using nearest neighbor ratio matching [21]. Based on the number of matching points between the live image and enrolled image, a decision is made. Thus more the number of matching points between two images, greater is the similarity between them. To illustrate the matching of extracted palmprint using SURF operator, an example is shown in Fig 4a. Matching between palmprint images of different users is called as imposter matching. An example of imposter matching between two users in IITK database is shown in Fig 4b.

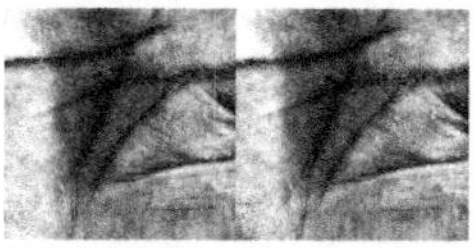

(i) Extracted palmprint (Id: 87_00 and 87_01) as input pair

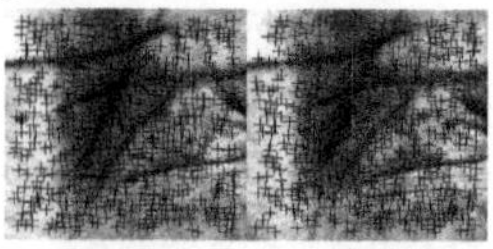

(ii) 627 and 607 Key-points detected for the input pair

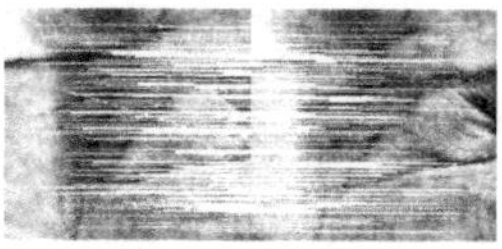

(iii) 210 SURF Matching Key-points obtained for the input pair

(a)

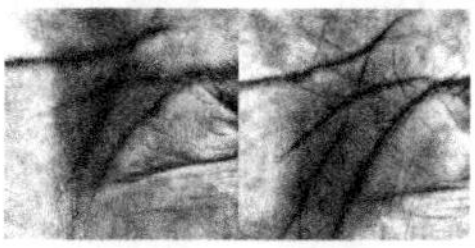

(i) Extracted palmprint (Id: 87_00, and 21_01) as input pair

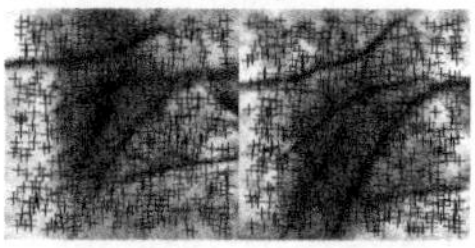

(ii) 627 and 656 Key-points detected for the input pair

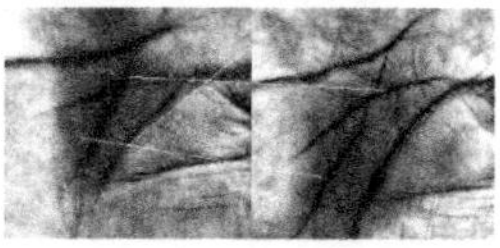

(iii) 3 SURF Matching key-points obtained for the input pair

(b)

Fig. 4. (a) Illustration of genuine palmprint matching using SURF operator, (b) Illustration of imposter matching using SURF operator

4 Experimental Results

The proposed system is tested on two sets of image databases. (1) The Indian Institute of Technology Kanpur (IITK) database, and (2) The Hong Kong Polytechnic University (PolyU) database.

4.1 Datasets

IITK has collected a database comprised of 200 images from 100 subjects. All hand images are collected at a spatial resolution of 200 dots per inch using a low cost flat bed scanner, and 256 gray levels. The database is divided into training set and testing set. One image per subject is considered for training, while other image of the subject is used for testing.

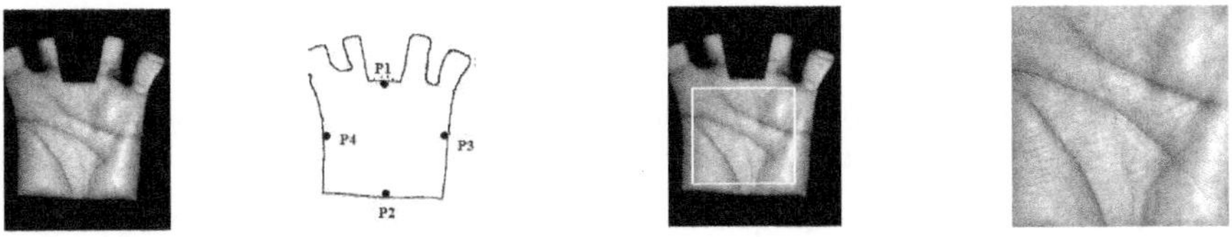

(a) (b) (c) (d)

Fig. 5. (a) PolyU sample image, (b) Reference points, (c) Region of interest in gray scale image, (d) Extracted palmprint

The proposed system has also been tested on the database from PolyU [9]. The images are collected at spatial resolution of 75 dots per inch, and 256 gray levels using CCD [2]. Images are captured placing pegs. Around 17 images per palm are collected in two sessions. 3200 grayscale images corresponding to 200 different palms are used for the experiment. Fig. 5a shows the sample of database. The database is divided into testing set and training set. Four images per palm are considered for training, while remaining twelve images of the palm is considered for testing.

Table 1. Accuracy, FAR and FRR of the proposed system and [11]

	IITK database			PolyU Database		
	FAR	FRR	Accuracy	FAR	FRR	Accuracy
Eigenpalm [11]	9.65	9.43	90.45	6.44	8.38	92.58
Haarpalm [11]	5.00	13.8	90.57	4.56	12.8	91.28
Fusion [11]	4.73	5.66	94.80	4.17	8.10	93.85
Proposed	**0.02**	**0.0**	**99.98**	**0.0**	**0.0**	**100.0**

In order to extract region of interest for PolyU database following method is proposed. Four reference points $P1$, $P2$, $P3$ and $P4$ are located on the contour of palm as shown in Fig. 5b. In gray-scale image extract 200 x 200 pixels palm area with its center coinciding with intersecting point of line segments $P1$–$P2$ and $P3 - P4$. Fig. 5c shows the region of interest in gray scale image, and the extracted palmprint image is shown in Fig. 5d.

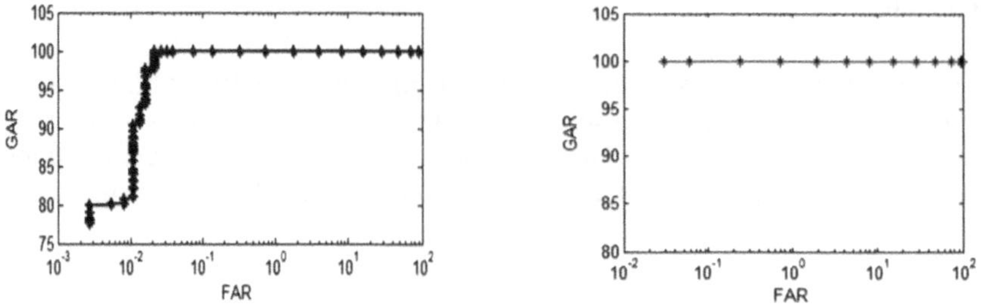

(a) (b)

Fig. 6. (a) ROC curve for IITK database, (b) ROC curve for PolyU database

The experiment has been performed for both the datasets using SURF. The Receiver Operating Characteristic (ROC) curves of the proposed system for is shown in Fig. 6a and Fig. 6b respectively, The proposed system has been compared with known available system [11]. Table 1 shows the accuracy, FAR and FRR of the proposed and previous [11] systems for both the datasets.

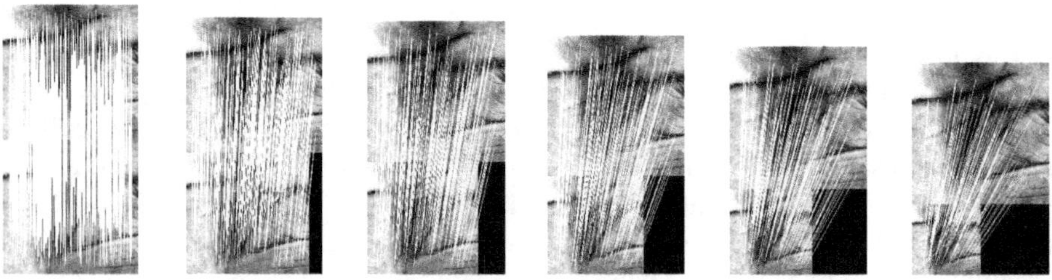

(a) 1.0*W* x 1.0*H* (b) 0. 9*W* x 0.9*H* (c) 0.8*W* x 0.8*H* (d) 0.7*W* x 0.7*H* (e) 0.6*W* x 0.6*H* (f) 0.5*W* x 0.5*H*

Fig. 7. Surf matching key-points between enrolled image with downscaled palmprint image

4.2 Scale Invariant

Since key-point features which are extracted using SURF approach is invariant to scale of the image, the system is robust to be spatial resolution. In order to investigate the robustness to scale, palmprint images in the testing set are down scaled using bicubic interpolation. The testing set of palmprint images are scaled to *0.9 W* x *0.9 H, 0.8 W* x *0.8 H, 0.7 W* x *0.7 H, 0.6 W* x *0.6 H,* and *0.5 W* x *0.5 H* to test the system, where *W* and *H* are the width and height of the enrolled images respectively. Matching points between the enrolled image and scaled testing palmprint images of the same subject are shown in Fig. 7. The ROC curve for matching with different scales of test images on different datasets are shown in Fig. 8a and Fig. 8b respectively.

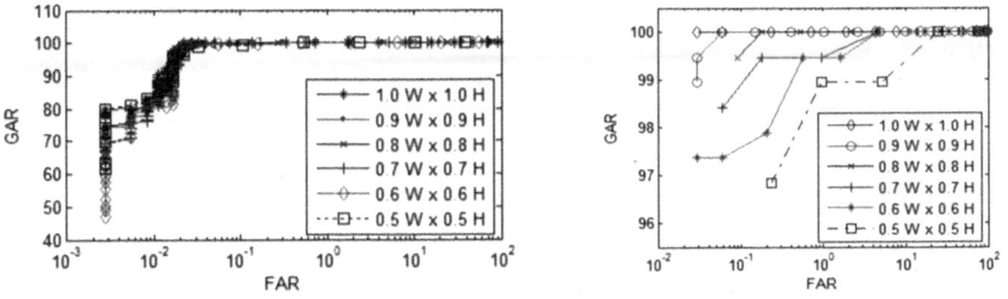

(a) (b)

Fig. 8. (a) ROC curve for IITK database on different scales of the test images, (b) ROC curve for PolyU database on different scales of the test images

Table 2. Accuracy obtained at different scales of testing image

	1.0 *W* x 1.0 *H*	0.9 *W* x 0.9 *H*	0.8 *W* x 0.8 *H*	0.7 *W* x 0.7 *H*	0.6 *W* x 0.6 *H*	0.5 *W* x 0.5 *H*
IITK	99.98%	99.98%	99.95%	99.92%	99.83%	99.74%
PolyU	100.0%	99.97%	99.91%	99.65%	99.46%	98.99%

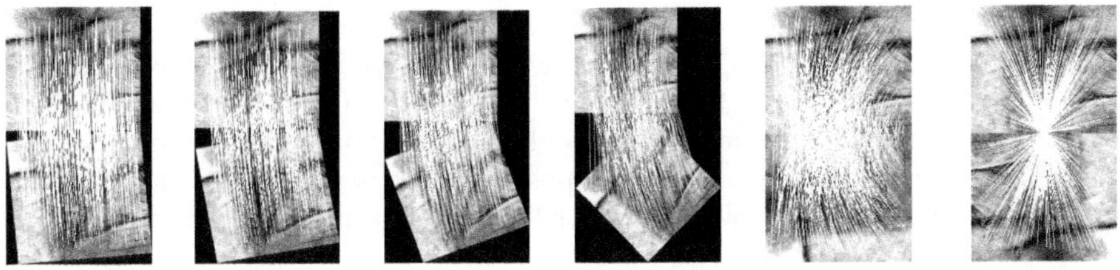

(a) 5^0 (b) 10^0 (c) 20^0 (d) 45^0 (e) 90^0 (f) 180^0

Fig. 9. Surf matching key-points between enrolled image and rotated palmprint image

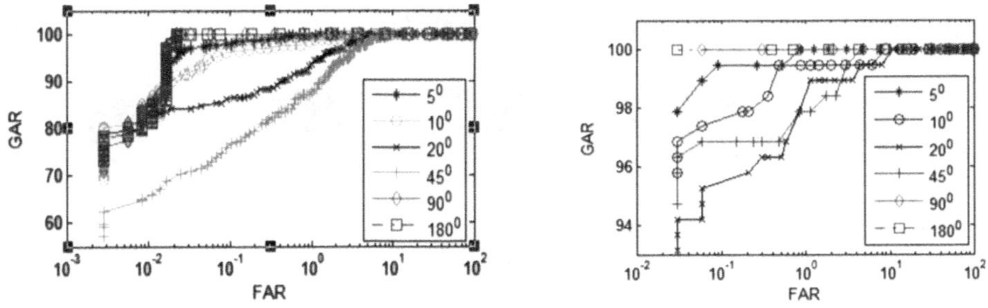

(a) (b)

Fig. 10. (a) ROC curve for IITK database on different rotation of the test images, (b) ROC curve for PolyU database on different rotation of the test images

The accuracy obtained for the different scales of testing images on IITK and PolyU database is shown in Table 2. It is observed that the accuracy of the system came down below 99% for scale of the testing images less than *0.5 W* x *0.5 H* of enrolled image size. From Table 2, it can be infered that the proposed system is highly robust to scale.

4.3 Rotation Invariant

The SURF feature vectors of a key-point are relative to the dominant orientation of the key-point. Hence the key-point featuress remain the same irrespective to the orientation of the palmprint image. So the proposed system is robust to the rotation of palmprint image.

Table 3. Accuracy obtained at different rotation of testing image

	5^0	10^0	20^0	45^0	90^0	180^0
IITK	99.59%	98.82%	98.52%	98.08	99.98	99.98
PolyU	99.69%	99.50%	98.92%	98.54%	100.0%	100.0%

In order to investigate the robustness to rotation, palmprint images are rotated synthetically using bi-cubic interpolation. The palmprint images in the testing set rotated by 5^0, 10^0, 20^0, 45^0, 90^0, and 180^0 are used to test the system. Matching points between the enrolled palmprint image of a subject with different orientation of testing palmprint image are shown in Fig. 9. ROC curve for IITK and PolyU database at different orientations of test images is shown in Fig. 10a and Fig. 10b. The accuracy for different orientation of test image on IITK and PolyU is shown in Table 3. From the table it can be seen that the accuracy of the proposed system is above 98% independent to rotation of the testing image, which implies the system is highly robust to rotation.

4.4 Occlusion Invariant

The proposed system extracts features using SURF which describes the image with local region around key-points. Hence they are independent to area of occlusion. So the proposed system can work in the places where user is able to expose partial palm region to the scanner and remaining region occluded.

(a) 0.1 *W* x 0.1*H* (b) 0.2 *W* x 0.2 *H* (c) 0.3 *W* x 0.3 *H* (d) 0.4 *W* x 0.4 *H* (e) 0.5 *W* x 0.5 *H*

Fig. 11. Surf matching key-points between enrolled image and occluded palmprint image

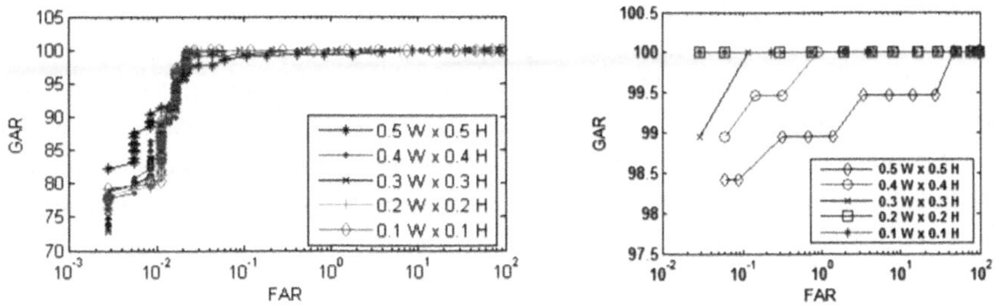

(a) (b)

Fig. 12. (a) ROC curve on IITK database for occlusion of the test images, (b) ROC curve for occlusion of the test images on PolyU database

In order to investigate the robustness to occlusion, palmprint images in the testing set are synthetically occluded with black square region. The typical palmprint images used to test the system are occluded with *0.1 W* x *0.1 H, 0.2 W* x *0.2 H, 0.3 W* x *0.3 H, 0.4 W* x *0.4 H,* and *0.5 W* x *0.5 H* of the image. Fig. 11 shows an example for matching key-points between the enrolled palmprint image and occluded testing image of the same subject. ROC curves for both the datasets at different size of occlusions are shown in Fig. 12a and Fig. 12b respectively. The accuracy obtained at different sizes of occlusion on testing images is shown in Table 4. From the results in Table IV it can be inferred that the proposed system at occlusion of *0.5 W* x *0.5 H* performs with accuracy more than 99%. Hence it implies the proposed system is highly robust to occlusion.

Table 4. Accuracy obtained for different size of occlusion in the testing image

	0.1 *W* x 0.1 *H*	0.2 *W* x 0.2 *H*	0.3 *W* x 0.3 *H*	0.4 *W* x 0.4 *H*	0.5 *W* x 0.5 *H*
IITK	99.98%	99.97%	99.96%	99.80%	99.48%
PolyU	100.0%	100.0%	99.94%	99.66%	99.31%

From results in Table 1, it can be inferred that the proposed system performs better than the recent best known system [11] which fuses the matching scores of the haar wavelet and eigenpalm classifiers. The proposed system also addresses the robustness in context of scale, rotation and occlusion of palmprint, and its performance is shown in Table 2, 3, and 4.

5 Conclusions

In this paper the use of SURF features in the context of palmprint verification has been investigated. A technique to extract palmprint from hand-image image which is robust to rotation, translation on the scanner surface has been presented. The extracted palmprint is found to be invariant to orientation and translation of palm on scanner, which makes the system robust to rotation and translation. The proposed system is tested on IITK and PolyU databases. The proposed key-point descriptors extracted using SURF outperform the earlier known system which fuses the matching scores of the haar wavelet and eigenpalm classifiers [11]. The recognition accuracy of the system is 99.98%, along with FAR 0.021% and FRR 0.01% at original sizes. The system is found to be robust to scale, rotation and occlusion. Thus the design of the system with robustness, performance and use of low cost scanner for acquisition of palm image has demonstrated the possibility of using this system for high-end security, and civilian applications.

References

1. Ribaric, S., Fratric, I.: A biometric identification system based on Eigenpalm and Eigenfinger features. IEEE Trans. on Pattern Analysis Machine Intelligence 27(11), 1698–1709 (2005)
2. Zhang, D., Kong, W.K., You, J., Wong, M.: Online palmprint identification. IEEE Transaction on Pattern Analysis and Machine Intelligence 25(9), 1041–1050 (2003)
3. Kumar, A., Zhang, A.: Personal recognition using hand shape and texture. IEEE Transaction on Image Processing 15(8), 2454–2461 (2006)
4. Zhang, L., Zhang, D.: Characterization of Palmprints by Wavelet Signatures via Directional Context Modeling. IEEE Transaction on Systems, Man, and Cybernetics 34(3), 1335–1347 (2004)
5. Wang, X., Gong, H., Zhang, H.: Palmprint Identification using Boosting Local Binary Pattern. In: 18th International Conference on Pattern Recognition, vol. 3, pp. 503–506 (2006)
6. Zhang, D., Shu, W.: Two novel characteristics in palmprint verification: Datum point invariance and line feature matching. Pattern Recognition 32(4), 691–702 (1999)
7. Han, C.-C., Cheng, H.-L., Lin, C.-L., Fan, K.-C.: Personal authentication using palmprint features. Pattern Recognition 36, 371–381 (2003)
8. International Committee for Information Technology Standards. Technical Committee M1-Biometrics (2005), http://www.incits.org/tc_home/m1.htm
9. The PolyU palmprint database, http://www.comp.polyu.edu.hk/~biometrics
10. Wenxin, L., Zhang, D., Xu, Z.: Palmprint Identification by Fourier Transform. Intl. Journal of Pattern Recognition and Artificial Intelligence 16(4), 417–432 (2002)
11. Badrinath, G.S., Gupta, P.: An Efficient Multi-algorithmic Fusion System based on Palmprint for Personnel Identification. In: Intl. Conf. on Advanced Computing, pp. 759–764 (2007)
12. Ross, A., Jain, A.K.: Information fusion in biometrics. Pattern Recognition Letters, 2115–2125 (2003)
13. Pavlidis, T.: Algorithms for graphics and image processing. Springer, Heidelberg (1982)
14. Bay, H., Tuytelaars, T., Van Gool, L.: SURF: Speeded up robust features. In: Leonardis, A., Bischof, H., Pinz, A. (eds.) ECCV 2006. LNCS, vol. 3951, pp. 404–417. Springer, Heidelberg (2006)
15. Lu, G., Wang, K., Zhang, D.: Wavelet based independent component analysis for palmprint identification. In: Intl. conf. on Machine Learning and Cybernetics, pp. 3547–3550 (2004)
16. Wang, Y., Ruan, Q.: Kernel Fisher Discriminant Analysis for Palmprint Recognition. In: 18th Intl. Conf. on Pattern Recognition, pp. 457–460 (2006)
17. Bay, H., Fasel, B., Van, L.: Interactive museum guide: Fast and robust recognition of museum objects. In: First Intl. workshop on mobile vision (2006)

18. Murillo, A.C., Guerrero, J.J., Sagues, C.: SURF features for efficient robot localization with omnidirectional images. In: IEEE Intl. Conf. on Robotics and Automation, pp. 3901–3907 (2007)
19. Viola, P., Jones, M.: Rapid object detection using a boosted cascade of simple features. In: IEEE Conf. on Computer Vision and Pattern Recognition, pp. 511–518 (2001)
20. Shu, W., Zhang, D.: Automated Personal Identification by Palmprint. Optical Engineering 37(8), 2359–2362 (1998)
21. Mikolajczyk, K., Schmid, C.: A performance evaluation of local descriptors. Pattern Analysis and Machine Intelligence 27, 1615–1630 (2005)
22. Independent Testing of Iris Recognition Technology Final Report. International Biometric Group (2005)

Accurate Palmprint Recognition Using Spatial Bags of Local Layered Descriptors

Yufei Han, Tieniu Tan, and Zhenan Sun

Center for Biometrics and Security Research, National Laboratory of Pattern Recognition
Institute of Automation, Chinese Academy of Sciences
{yfhan,tnt,znsun}@nlpr.ia.ac.cn

Abstract. State-of-the-art palmprint recognition algorithms achieve high accuracy based on component based texture analysis. However, they are still sensitive to local variations of appearances introduced by deformation of skin surfaces or local contrast variations. To tackle this problem, this paper presents a novel palmprint representation named Spatial Bags of Local Layered Descriptors (SBLLD). This technique works by partitioning the whole palmprint image into sub-regions and describing distributions of layered palmprint descriptors inside each sub-region. Through the procedure of partitioning and disordering, local statistical palmprint descriptions and spatial information of palmprint patterns are integrated to achieve accurate image description. Furthermore, to remove irrelevant and attributes from the proposed feature representation, we apply a simple but efficient ranking based feature selection procedure to construct compact and descriptive statistical palmprint representation, which improves classification ability of the proposed method in a further step. Our idea is verified through verification test on large-scale PolyU Palmprint Database Version 2.0. Extensive experimental results testify efficiency of our proposed palmprint representation.

1 Introduction

The past decade has witnessed a great progress of biometrics technology, which identifies different people by their physiological and behavioral difference, such as face, iris, retinal, gait and etc [1]. Palmprint recognition is a novel but promising member of biometrics. Large region of palm supplies amounts of discriminative line and texture patterns, which could be captured using a low resolution imaging device, such as a low-cost CCD camera [2]. Thanks to its high recognition accuracy and convenience for practical use, palmprint recognition provides a promising way for identity check. Palmprints normally refer to appearance properties of structure patterns of epidermis on palm regions, which are formed due to movement of fingers, structures of tissue and muscles gradually after birth and spread over different areas of palmprint images. Therefore, a key issue in palmprint recognition is to design an efficient image representation to depict the skin textures accurately. In previous work, component based texture analysis provides a reasonable framework for palmprint recognition [2-8]. Most state-of-the-art approaches [5-8] can be ranged into this framework. For one

M. Tistarelli and M.S. Nixon (Eds.): ICB 2009, LNCS 5558, pp. 566–576, 2009.
© Springer-Verlag Berlin Heidelberg 2009

palmprint image, characteristics of palmprints inside each local patch are described using image filters, like multi-directional Gabor filter set [5][6] or orthogonal line ordinal filters [7]. Either phase [5] or magnitude [6][7] of filtering responses are used as local texture descriptors, which are rearranged into a feature matrix or template according to their spatial order. Feature matching is conducted by comparing texture features of corresponding patches pair-wisely in two palmprint images. Differences between local regions are accumulated to form a similarity measure between the two images. The component based framework makes good use of intrinsic properties of palmprints. As a result, this framework obtains good performance. However, it also suffers from variations of palmprint appearances between intra-class samples. The reason for such variation can be concluded into two aspects, as shown in Fig.1. For one thing, due to different postures of hands in practical conditions, captured palmprint images usually contain contraction or stretching of skin surfaces caused by movement of muscle tissues, which leads to deformation of skin textures. For another thing, variations of illumination settings over local region also change contrast information of palmprint appearances. For two palmprint images belonging to the same class, the variations of appearances easily lead to discrepancies between component based texture descriptions in corresponding regions, which consequently limits further increasing of descriptive power.

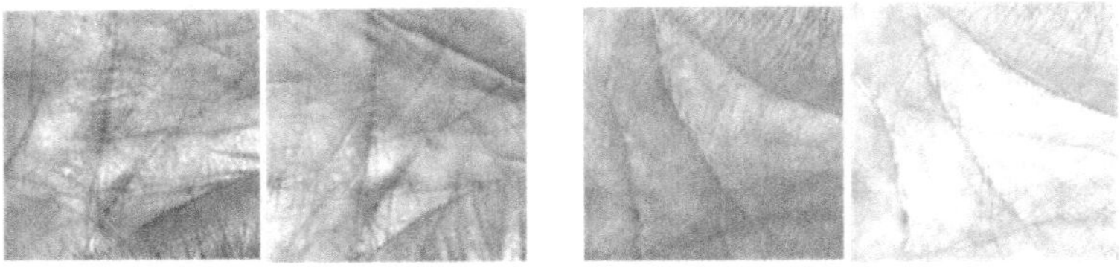

(a) (b)

Fig. 1. Variations of palmprint appearances caused by (a) contraction or stretching of skin surfaces and (b) variations of illumination settings

Unlike component based methods, Bag-of-Feature model is a holistic statistical image representation [11][13]. General scheme of this approach is to represent an image by an orderless collection of local features, disregarding their spatial information. Bag-of-Feature model of one specific image is constructed by computing a histogram of votes, which counts occurrence frequency of local features in the image [11][13]. It makes an improvement in robustness of image description against local appearance deformations in nature. This model is popular in object recognition or scene categorization. Motivated by Bag-of-feature model, in this paper, we propose a statistical palmprint representation, called Spatial Bags of Local Layered Descriptors (SBLLD), as illustrated in Fig.2. We firstly construct local layered descriptors for palm lines and textures using multi-directional quadrature Gabor filters, which are both discriminative palmprint features. A tree-structured visual vocabulary of local layered descriptors is then generated and pruned without prior clustering on training data. Finally, the whole image is subdivided into non-overlapped sub-regions. Distributions of visual vocabulary within each sub-region are described following Bag-of-Feature framework and aggregated into a histogram sequence according to spatial order of sub-regions.

Furthermore, we select the most discriminative attributes from histogram based palmprint features. With this helpful procedure, the proposed method achieves even higher accuracy with more compact feature subset. Our goal of this paper can be concluded into two aspects. For one thing, by compensating loss of spatial information with the subdividing scheme, we improve descriptive ability of statistical image model to be suitable for palmprint classification. For the other thing, by explicitly introducing the statistical image description into palmprint representation, more robust recognition against local variations of appearances are achieved compared with component based approaches.

Remainder of this paper is organized as follows: In Section.2, we give detailed description of the proposed method. Section.3 illustrates procedure of ranking based feature selection. Comprehensive experimental results are listed and analyzed in Section.4. Section.5 concludes the whole paper.

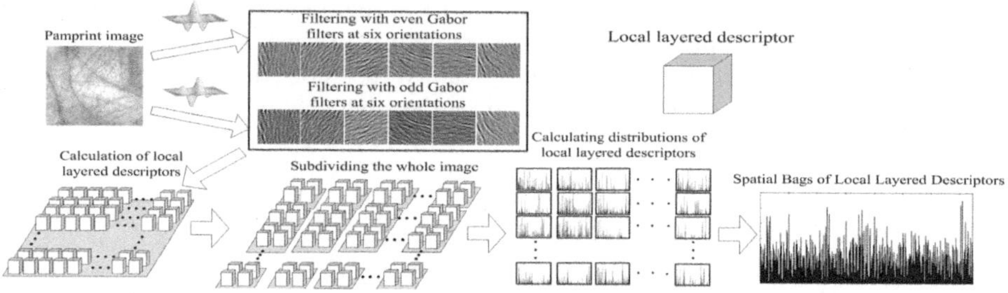

Fig. 2. Construction of SBLLD based palmprint representation

2 Description of SBLLD Scheme

2.1 Local Layered Palmprint Descriptor

In low resolution palmprint images, line segments and textures are observable and discriminative features [2]. Theoretically, quadrature Gabor filter provides accurate time-frequency location and good orientation selectivity to describe both of palmprint patterns within local image patches. A complex quadrature Gabor function is shown in Eq.1

$$G(x,y,\theta,u,\sigma)=\frac{1}{2\pi\sigma^2}\exp(-\frac{x^2+y^2}{2\sigma^2})(\cos(2\pi u\tilde{\theta})+i\sin(2\pi u\tilde{\theta})) \quad (1)$$

where $\tilde{\theta}=x\cos\theta+y\sin\theta$ and $i=\sqrt{-1}$. u and σ are the frequency of sinusoid wave and the standard deviation of Gaussian envelope. The real part of the function, namely even Gabor filter, optimally models directional line-like patterns such as wrinkles and principle lines in palmprints. Meanwhile, the imaginary part, namely odd Gabor filter, approximates the 1st-order directional derivatives of intensities. By constructing odd Gabor filter templates, we can obtain good estimation of image gradient, which is an efficient indicator of local image textures [8][9]. In previous work, Zhang and

Kong et al [5][6] use Gabor filter sets to extract quantified phase features of palmprint images. Kong et al [6] combines multi-directional even Gabor filters and a "winner-take-all" scheme to describe direction of line segments in palmprints. Their results denote efficiency of quadrature Gabor filter in palmprint analysis. Motivated by properties of the filter, we construct a two-layered descriptor *T* to depict both of discriminative palmprint patterns inside a local image patch, using quadrature Gabor filters at different orientation bands. Fig.3 illustrates structure of *T*.

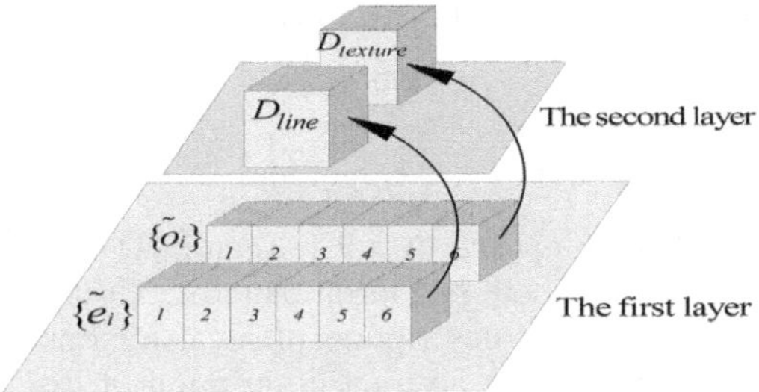

Fig. 3. The structure of two-layered descriptor

Before construction of the local descriptor, we adopt quadrature Gabor kernels with zero DC values at six orientations $\theta \in \{0, \pi/6, \pi/3, \pi/2, 2\pi/3, 5\pi/6\}$. The parameter u and σ of them equal to 0.0916 and 5.6179. The size of each filter is 35×35. Empirical analysis testifies that this setting is suitable for palmprint recognition [5][6]. By performing convolution on a local patch with those Gabor kernels, totally 12 filtering responses $\{e_i, o_i\ (i = 1,2\ldots6)\}$ are obtained. e_i and o_i are filtering responses of the even and odd Gabor filter respectively, as defined in Eq.2.

$$\begin{aligned} e_i &= P(x, y) * \mathrm{Re}\{G(x, y, \frac{(i-1)\pi}{6}, u, \sigma)\} \\ o_i &= P(x, y) * \mathrm{Im}\{G(x, y, \frac{(i-1)\pi}{6}, u, \sigma)\} \end{aligned} \tag{2}$$

where $\mathrm{Re}\{G\}$ and $\mathrm{Im}\{G\}$ represent even and odd Gabor filter. Elements of the first layer in the descriptor is then derived by applying ordinal encoding procedure [7] on original filtering responses, as shown in Eq.3.

$$\begin{cases} \tilde{e}_i \text{ or } \tilde{o}_i = 1 & \text{if } e_i > 0 \text{ or } o_i > 0 \\ \tilde{e}_i \text{ or } \tilde{o}_i = 0 & \text{if } e_i < 0 \text{ or } o_i < 0 \end{cases} \tag{3}$$

where $\tilde{e}_i$ and $\tilde{o}_i$ are ordinal codes of corresponding filtering results. Ordinal encoding scheme [10] improves robustness of filter based appearance descriptions against changes of illumination settings and indicates intrinsic structures of lines and texture patterns. Thus, as we see in Fig.3, the first layer is defined as a 2nd-order tensor (matrix), with $\{\tilde{e}_i\}$ and $\{\tilde{o}_i\}$ $(i = 1,2\ldots6)$ as its row vectors. The former one is used as robust line description in the descriptor, while the latter one is utilized for depicting textures in palmprints. Compared with the first layer, the second one contains more abstract characteristics of appearances within local regions. Based on $\{e_i\}$ and $\{o_i\}$, we further calculate quantified orientation of prominent line patterns D_{line} (ranging from 1 to 6), which equals to the index of response e_i that achieves the minimum magnitude among $\{e_i\}$ (i=1,2..6) [6]. This orientation information plays an important role in accurate description of palm lines [6]. For describing texture patterns of palmprints, the approximated direction of image gradient $D_{texture}$(ranging from 1 to 6) is estimated from $\{o_i\}$ (i=1,2..6) [8]. According to Eq.4, $D_{texture}$ is defined as the index of the angle band which has the strongest contrast response.

$$D_{line} = \arg\min_i e_i$$
$$D_{texture} = \arg\max_i |o_i| \tag{4}$$

The second layer is then composed by a 2-D vector {D*line*, D*texture*}, which corresponds to directional feature of both palmprint patterns. Through the layered scheme, we can achieve a comprehensive description of local regions in palmprint images. Firstly, in the top layer, we extract abstract description of discriminative palmprint patterns based on the effective directional information. Secondly, in the bottom layer, we make use of robust ordinal features to depict local appearances of line segments and textures in a more comprehensive manner. In our work, we sample a palmprint image densely. As a result, for a h×w image, totally h×w local layered descriptors are derived.

2.2 Formulation of Tree-Structured Codebook

In our application, we adopt an efficient coding scheme to map layered descriptors to leaf nodes of a tree-structured codebook, which is consistent with formation of the descriptors. Based on this scheme, we can obtain visual codes fast using a look-up table, without prior clustering [10] performed in classic Bag-of-Feature model. The whole codebook contains three stages, as shown in Fig.4. On the root stage, the codebook C is divided into two non-overlapped sub-codebooks S_{line} and $S_{texture}$, which correspond to representations of line segments or texture patterns in palmprints. They are used in parallel during encoding visual patterns. In the following stage, each sub-codebook is partitioned into six branches, indexed by integers from 1 to 6, which correspond to values of D_{line} or $D_{texture}$ in the local descriptors. Within a given branch, totally 63 leaf nodes transverse all possible combinations of six ordinal codes { $\tilde{e}_i$ } or { $\tilde{o}_i$ }. Given a fixed D_{line} or $D_{texture}$, a descriptor is mapped to a leaf node by computing a label L_{leaf} following Eq.5.

$$L_{leaf} = \sum_{i=1}^{6} 2^{i-1}\tilde{e}_i \text{ or } L_{leaf} = \sum_{i=1}^{6} 2^{i-1}\tilde{o}_i \tag{5}$$

According to structure of the codebook, each local layered descriptor corresponds to two leaf nodes in the codebook, which belong to S_{line} and $S_{texture}$ respectively. Moreover, directional information of lines and texture patterns has a dominant position in the palmprint description, which is consistent with verified conclusions in palmprint recognition [5][6]. With the same orientation label, quantified outputs of quadrature Gabor kernels represent finer differences of appearances within local patches, in order to provide helpful supplements for accurate palmprint representation.

As we can see in Fig.4, there are totally 768 leaf nodes in the codebook, which cover all possible visual patterns that one local layered descriptor can represent, without considering practical significance of each pattern in palmprint analysis. Actually, some patterns in the sub-codebooks S_{line} are surplus. For example, assuming that D_{line} equals to 1, filtering respones $\tilde{e}_1$ is negative for sure. Thus, any pattern with the first element in { $\tilde{e}_i$ } equaling to 1 is invalid for palmprint description. We remove such surplus patterns from the codebook. As a result, there are 582 nodes left to form a more reasonable and compact codebook for palmprint description.

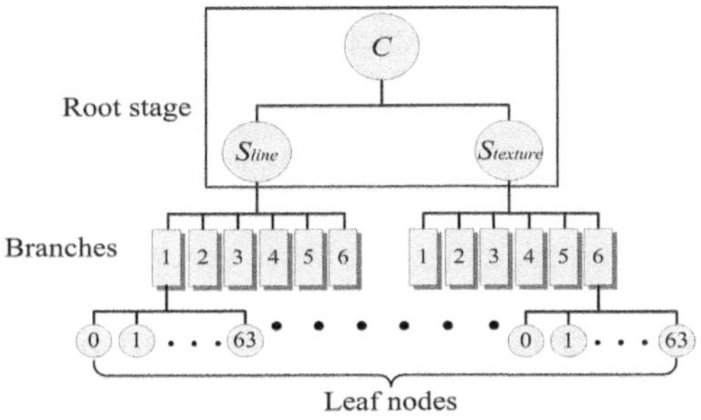

Fig. 4. The hierarchical structure of visual codebook

2.3 Subdividing Scheme

Although holistic image model is robust enough to local appearance deformation, it loses all spatial information of local features, making the image representation not suitable for accurate palmprint classification. In our method, we partition the whole image into sub-regions or spatial bags, and compute histograms of layered descriptor inside each sub-region. The proposed SBLLD is then constructed by aggregating histograms into a sequence according to their spatial order. Chi-square distance between SBLLD features is taken as the similarity measure. This subdividing scheme [12] has been practiced efficiently in computer vision, both for global and local image model. By dividing images into sub-blocks, we obtain distribution information of local descriptors at finer scale, which improves the ability of the statistical image representation to depict local details. Scale of partitioning is the key factor in the scheme. With increasing finer scale, the proposed representation becomes more sensitive to local appearance variations, which leads reduction of robustness to local appearance variations and higher dimensionality in the obtained histogram sequences. Therefore, to achieve an optimal trade-off, we should choose a proper size of individual sub-block. Experiments in Section 4 will give a detailed analysis on this issue.

3 Ranking Based Feature Selection

A typical SBLLD usually has high dimensional histogram based representation. In theory, higher dimension can provide more discriminating power. However, it can also introduce non-informative attributes into the histogram vector of SBLLD, which are irrelevant with category information. In SBLLD, each dimension of the histogram corresponds to one visual pattern in a specific region of palmprints. Nevertheless, not all patterns provide efficient information for classification. For example, patterns appearing in different palmprint images with almost equal frequencies contain little discriminative power. Moreover, random image noise can also generate meaningless patterns. This is not a completely brand-new problem. Similarly in text categorization, large amount of vocabulary in corpus always leads to high dimensional histograms of term (word or phrase) occurrence [11]. To remove non-informative elements, all terms in texts are sorted and listed according to a term-goodness based criterion, such as information gain and document frequency [11]. Then a desired degree of terms at back of the list are eliminated from the vocabulary. The ranking based approach is suitable for high dimensional data due to its linear time complexity in terms of dimensionality. Simple as it is, it is efficient in selecting a well suited feature subset for classification [11]. Motivated by this idea, we perform feature selection based on a classification oriented criterion, rather than those general ones. Each dimension of the histogram vector is treated as one feature in SBLLD model. For each feature i, we define a stump classifier $C(i)$ based on normalized difference of corresponding dimension in two different SBLLD feature vectors A and B, as denoted by Eq.6. In the expression, T is a per-determined threshold. By performing an optimal threshold on outputs of $C(i)$, we can obtain the minimum classification error (MCE) of $C(i)$, which is used to evaluate predicting power of each single feature. With a set of labeled training data, the optimal threshold $t_{optimal}$ is obtained following Eq.7.

$$C(i) = \mathrm{sgn}\left(\frac{|A(i) - B(i)|}{A(i) + B(i)} - t\right)$$
$$C(i) = \begin{cases} 1 & A \text{ and } B \text{ belong to the same class} \\ -1 & A \text{ and } B \text{ belong to different classes} \end{cases} \tag{6}$$

$$E = \frac{\text{False rejected items} + \text{False accepted items}}{\text{Total amounts of matching}}$$
$$t_{optimal} = \arg\min_t E \tag{7}$$

where E is classification error corresponding to a specific threshold. All features are sorted in an ascending order according to MCE of each $C(i)$. The lower the minimum error is, the more relevant the specific feature is to palmprint classification. The optimal subset consists of top K items in the ranking list. Lower K leads to more compact representation. However, it may also exclude useful terms. In the following section, we choose a proper value of K to achieve the best performance.

4 Experiments on PolyU Palmprint Database Version 2.0

We compare performances of the proposed method on PolyU Palmprint Database Version 2.0 [13]. This database contains 7,752 384*286 palmprint images, which are captured from 386 palms. Each palm has two sessions of images, either of which contains more than 9 images. Average time interval between two sessions is two months, which introduce changes of light conditions, focus of the imaging device and deformation of skin textures due to varied hand postures between two occasions of image capturing [2], as shown in Fig.5. In our experiments, our proposed method is compared with four state-of-the-art component based algorithms [5][6][7] through one-to-one verification test.

After preprocessing, regions of interests, with the size of 128×128 are obtained. In this section, we implement SBLLD by dividing the whole ROI into 8*8,16*16 and 32*32 non-overlapped blocks, named as $SBLLD_8$, $SBLLD_{16}$,$SBLLD_{32}$ respectively, which cover descriptions from coarse scale to fine spatial scale. As we see in Table.1, although $SBLLD_8$ achieves the highest accuracy, improvement of performance is limited compared with $SBLLD_{16}$. Too fine spatial scale in division leads to increasing sensitivity to local appearance changes, which conversely deteriorates descriptive power of features. It also results in too high dimensional (148,992 bins) histogram vectors of $SBLLD_8$ requiring much computational cost. Thus, we choose $SBLLD_{16}$ to obtain a good trade-off. With the optimal division scheme, the SBLLD achieves satisfied performance, which is similar as the highest recognition accuracy of the state-of-the-art methods. The histogram vector of $SBLLD_{16}$ has overall 37,248 bins, with each bin treated as an attribute in the representation. As described in Section 3, we further

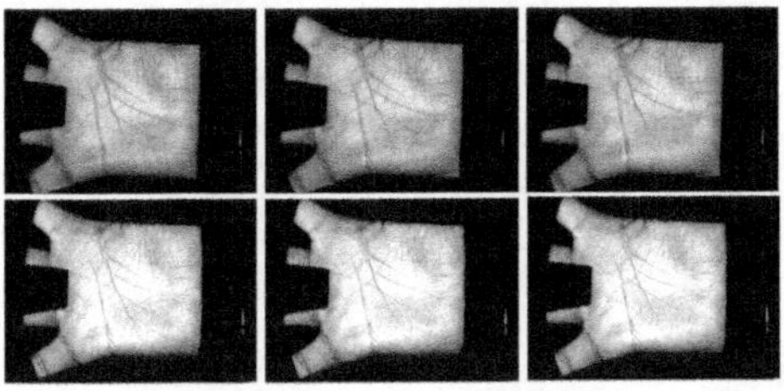

Fig. 5. Six examples of captured images in PolyU Palmprint Database Version 2.0

Table 1. Performances on the PolyU Database with different sub-dividing schemes

	d'[14]	EER[14]
$SBLLD_8$	4.16	0.038%
$SBLLD_{16}$	4.22	0.049%
$SBLLD_{32}$	3.99	0.581%
Ordinal code [7]	6.90	0.052%
Competitive code [6]	5.84	0.045%
Fusion code [5]	5.40	0.217%

Table 2. Verification performances with different choices of K on the testing set

	d'[14]	EER[14]
$K = 5000$	4.78	0.048%
$K = 7000$	4.69	0.039%
$K = 10000$	4.57	0.026%
$K = 13000$	4.49	0.016%
$K = 15000$	4.43	0.017%
$K = 17000$	**4.40**	**0.011%**
$K = 20000$	4.34	0.015%
$K = 23000$	4.30	0.012%
$K = 25000$	4.27	0.017%
$K = 27000$	4.25	0.020%
$K = 30000$	4.22	0.025%
$K = 33000$	4.21	0.028%
$K = 37248$	4.22	0.051%

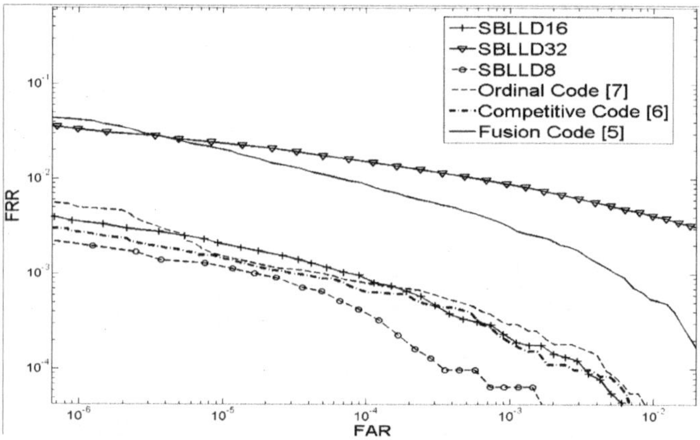

Fig. 6. ROC curves with different sub-dividing schemes

sort all attributes in descending order according to their predicting power measured on a training data set which is composed by four images randomly selected from each class in PolyU Database. Left 6,208 images form a testing set. The top K attributes in the sorted list compose a subset S. Chi-square distance between tailored histogram vectors, which only reserve attributes in S, is defined as the similarity measure. We test the proposed tailored palmprint representation on the testing set. As we can see in Table.2, less K leads to more strict selection. Aggressive choice of K, e.g. less than 5000, even excludes informative features from the representation, which results in much performance deterioration of the proposed method. By increasing K gradually from 1000 to 10000, performance of our method is improved greatly. During further increasing from 10000 to 17000, improvement of recognition accuracy becomes gentler, which confirms that leading portions of the ranking list contribute much more in classification than the posteriors. Conversely, K larger than 20000 doesn't enhance classification ability stably or even worse. It verifies that attributes at the back of the list are usually irrelevant with classification task. They are generated by non-informative patterns in palmprint images. As a result, they provide no more stable clues for recognition or even decrease differences between inter-class samples. As we can see in Table.2, K equaling to 17000 leads to the best performance on test data set, much higher than using all original attributes. By proper removal of irrelevant terms, the proposed SBLLD model can achieve efficient palmprint recognition, even more accurate than state-of-the-art methods.

5 Conclusion

In this paper, we have proposed a novel statistical palmprint representation, called Spatial Bags of Local Layered Descriptor (SBLLD). In this method, by integrating local distributions of local layered descriptors and their spatial order, we obtain accurate description of discriminative palmprint patterns, such as lines and textures on skin surfaces of palm regions, which consistently achieves accurate palmprint recognition. By performing the ranking based feature selection on the proposed statistical image representation, our proposed method obtains great improvement of performance with more compact histogram feature vectors, even better than most of state-of-the art methods using component based texture analysis. However, the proposed model still needs additional storage and computation cost to construct and store high dimensional features. This problem has not been addressed well. In the future work, we will focus on this issue following the proposed framework.

Acknowledgement

This work is supported by research grants from the National Basic Research Program (Grant No.2004CB318110), the Natural Science Foundation of China (Grant No.60723005, 60736018, 60702024), NLPR 2008NLPRZY-2, the National Hi-Tech Research and Development Program of China (2006AA01Z193,2007AA01Z162).

References

[1] Jain, A.K., Bolle, R.M., Pankanti, S. (eds.): Biometrics: Personal Identification in Networked Society. Kluwer, Norwell (1999)

[2] Zhang, D., Kong, W.K., You, J., Wong, M.: Online Palmprint Identification. IEEE Transactions on Pattern Analysis and Machine Intelligence 25(9), 1041–1050 (2003)

[3] Wu, X.Q., Wang, K.Q., Zhang, D.: Palmprint Recognition using Directional Line Energy Feature. In: Proceedings of the 17th ICPR, vol. 4, pp. 475–478 (2004)

[4] You, J., Kong, W.K., Zhang, D., Cheung, K.: On Hierarchical Palmprint Coding with Multi-features for Personal Identification in Large Databases. IEEE Transactions on Circuit Systems for Video Technology 14(2), 234–243 (2004)

[5] Kong, W.K., Zhang, D.: Feature-Level Fusion for Effective Palmprint Authentication. In: Zhang, D., Jain, A.K. (eds.) ICBA 2004. LNCS, vol. 3072, pp. 761–767. Springer, Heidelberg (2004)

[6] Kong, W.K., Zhang, D.: Competitive Coding Scheme for Palmprint Verification. In: Proceedings of the 17th ICPR, vol. 1, pp. 520–523 (2004)
[7] Sun, Z.N., Tan, T.N., Wang, Y.H., Li, S.Z.: Ordinal Palmprint Representation for Personal Identification. In: Proceedings of CVPR 2005, vol. 1, pp. 279–284 (2005)
[8] Han, Y.F., Sun, Z.N., Tan, T.N.: Palmprint Recognition Based on Directional Features and Graph Matching. In: Lee, S.-W., Li, S.Z. (eds.) ICB 2007. LNCS, vol. 4642, pp. 1164–1173. Springer, Heidelberg (2007)
[9] Chen, H.F., Belhumeur, P.N., Jacobs, D.W.: In Search of Illumination Invariants. In: Proceedings of CVPR 2000, pp. I:254–261 (2000)
[10] Leung, T., Malik, J.: Representing and recognizing the visual appearance of materials using three-dimensional textons. International Journal of Computer Vision 43, 29–44 (2001)
[11] Yang, Y., Pedersen, J.O.: A Comparative Study on Features Selection in Text Categorization. In: Proceedings of the 14th ICML, pp. 412–420 (1997)
[12] Lazebnik, S., Schmid, C., Ponce, J.: Beyond Bags of Features: Spatial Pyramid Matching for Recognizing Natural Scene Categories. In: Proceedings of CVPR 2006, vol. 2, pp. 2169–2178 (2006)
[13] PolyU Palmprint Database, `http://www.comp.polyu.edu.hk/~biometrics/`
[14] Daugman, J., Williams, G.: A Proposed Standard for Biometric Decidability. In: Proceedings of CardTech/ SecureTech Conference, Atlanta, GA, pp. 223–234 (1996)

Pose Invariant Palmprint Recognition

Chhaya Methani and Anoop M. Namboodiri

International Institute of Information Technology, Hyderabad, India
chhaya@research.iiit.ac.in, anoop@iiit.ac.in

Abstract. A palmprint based authentication system that can work with a multi-purpose camera in uncontrolled circumstances, such as those mounted on a laptop, mobile device or those for surveillance, can dramatically increase the applicability of such a system. However, the performance of existing techniques for palmprint authentication fall considerably, when the camera is not aligned with the surface of the palm. The problems arise primarily due to variations in appearance introduced due to varying pose, but is compounded by specularity of the skin and blur due to motion and focus. In this paper, we propose a method to deal with variations in pose in unconstrained palmprint imaging. The method can robustly estimate and correct variations in pose, and compute a similarity measure between the corrected test image and a reference image. Experimental results on a set of 100 user's palms captured at varying poses show a reduction in Equal Error Eate from 22.4% to 8.7%.

1 Introduction

The use of digital imaging, specifically camera based imaging, for biometric authentication has changed the applicability of biometric authentication for personal security. Camera based imaging is fast, inexpensive, convenient to use, and is easily available to the common man. In addition to biometric modalities such as face, gait, etc., one can employ camera based imaging for fingerprints, palmprints, or even handwriting. It also facilitates the use of multi-modal biometrics, which alleviates many concerns on accuracy and applicability, as it is possible to capture different biometric features using the same imaging setup and at the required resolution.

Despite all the advantages, there are many challenges to overcome to realize biometric authentication systems that use unconstrained camera-based imaging. The major problems include background clutter or noise that make segmentation difficult, illumination changes that can affect the feature extraction process, scale and resolution changes, and the increased flexibility offered to the subject in terms of pose changes. The onus then lies on the algorithm to accommodate all these variations in the best possible way. This adversely affects the recognition rate that is much higher when limited degree of freedom is allowed. As a result, many camera based recognition systems employ a capturing setup that constrains the pose and scale, and provides uniform illumination and constant background. Such a setup may not add much in terms of cost, but makes the apparatus cumbersome to use, liable to frequent damage, less intuitive and user friendly, and very specific to a particular biometric. Hence, there is a potential to develop new algorithms that are more robust to environmental variations and also

M. Tistarelli and M.S. Nixon (Eds.): ICB 2009, LNCS 5558, pp. 577–586, 2009.
© Springer-Verlag Berlin Heidelberg 2009

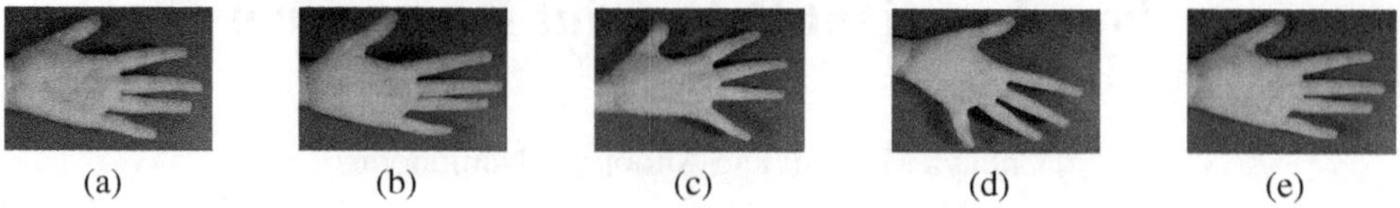

Fig. 1. Variations in hand pose due to unconstrained imaging

allow more flexibility to the subject. Efforts in this direction include achieving pose and illumination invariance for face recognition [1], view invariance for gait recognition [2] and handling different poses for hand geometry based authentication [3].

In this paper, we look at the problem of achieving pose invariance in palmprint based authentication. Initial palmprint matching systems used scanned images from inked palms. A detailed image having the major lines, wrinkles and even creases of the palm was obtained. Approaches developed for these images involved matching datum points, and the line orientation at them [4]. This constituted offline palmprint matching. Real-Time Online Palmprint matching was made possible with the use of electronic imaging by capturing low resolution images using devises such as web cameras. The assumptions of controlled environment and restricted palm pose were convenient, and only prominent palm lines and their structure were used for matching [5].

As both hand-geometry and palmprint use the same imaging setup, one could also combine them both in a multi-biometric setup [6]. The common preprocessing step computes the two valley points p_1 and p_2 as shown in Figure 2. As the palm is cut out with respect to the coordinate axes formed by joining these two stable points on the palm, the extraction process becomes invariant to in-plane rotations and translation. Focus of current research is, hence, to find better ways for representation of palm [7] and discovering new ways of feature representation for improved and robust matching. Popular representations include Gabor filter responses [5], wavelet basis representation, Fourier representation, orientation histogram based etc. [7],.

The partial rotation and translation invariance to in-plane rotation and translation is achieved by constrained pose hand imaging. This suffers from problems relating to user inconvenience, being less intuitive, unsatisfactory user cooperation, a cumbersome setup and limiting the possibilities of extending the system. These issues must be addressed to make the system more robust and better adept to practical scenarios. Approaches to deal with varying poses either try to compute pose-invariant features, or apply a transformation to the image to normalize pose variations.

G. Zheng et al in [3] proposed projective invariant hand geometry features. In [8], Doublet et al. propose novel way of contact less hand recognition. The approach primarily concentrates on extracting the palm from an image with cluttered background. Their hand detection method is based on a combination of skin color modeling and shape models. For the purpose of recognition, the approach assumes that the palm is being held parallel to the camera's imaging plane.

In this work, we extend their approach to include matching images with hand variations in 3D space(see Figure 1). We deal with projective variations that will make palmprint recognition possible using a truly unconstrained imaging setup. Pose variations distort the absolute position and relative distance between the lines, and change

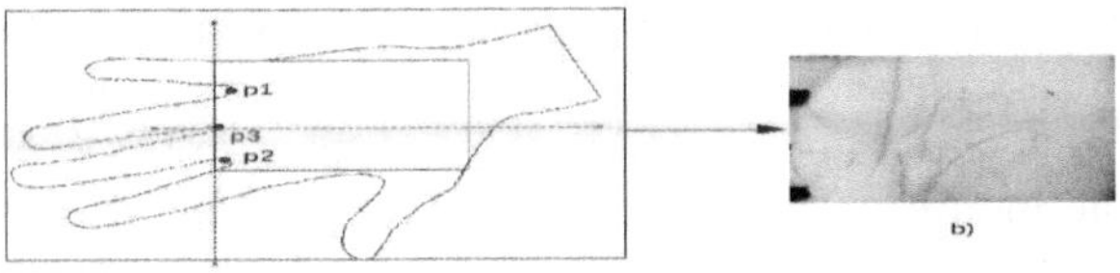

Fig. 2. Extraction of palm region based on finger valleys

their characteristic properties like length, width, curvature, aspect ratio etc. Hence, traditional matching can not be applied directly to images having different poses.

In this paper, we approach the problem of unconstrained palmprint imaging and the related pose variations for the first time. We propose a novel algorithm for pose invariant palmprint matching. As a part of the algorithm, we propose a robust image alignment technique and a method to detect the orientation of the hand. We show the results of our approach on a synthetic and a real dataset. The synthetic dataset gives us an opportunity to isolate, quantify and incrementally study the effects of pose changes, while the real dataset accurately models the practical problems such as skin deformations, specular reflections, etc. Some images from our dataset are shown in Figure 5.

The rest of the paper is organized as follows. Section 2 explains the approach chosen to attain pose invariance and reasons for it. Section 3 describes the practical challenges in image alignment and the solution. Section 4 gives a step wise formulation of the solution proposed by us, followed by experimental results in Section 5.

2 Unconstrained Palmprint Imaging

In our work, we consider images taken with unrestricted and mostly unsupervised hand motion. This may lead to a change in hand pose. Typical set of images to be matched is shown in Figure 1(a) and 1(b). Matching can be done in two ways.

Finding Projective Invariant Features in Palm. As pointed out in [3], when subjected to geometrical projective transformations, cross ratios created by 5 coplanar points with a specific configuration are invariant to pose change. The approach assumes the planarity of hand, and the landmark points are chosen from the the finger creases.

The bottleneck of this is that it requires us to find 5 stable landmarks with specific geometric relationship on the palm images. However, considering the structure of palm, which consists of three major lines: the heart, head and life lines, the number of distinctive points that can be reliably identified are very limited. Points formed by intersection of two lines are not reliably detected due to blur, illumination and resolution changes, and the problem increases further due to introduction of wrinkles during pose changes. In short, the palm images are weakly textured and do not have enough acceptable landmark points to consider the computation of projective invariant features.

Pose Correction. A second approach is that of explicitly determining the pose and then correcting for it. A fundamental matrix generally models the transformation between objects in 3D world ($[x', y', z']\mathcal{F}[x, y, z]^T = 0$). However, we can restrict the set of possible transformations, by assuming the thickness of the palm to be negligible, approximating it to a planar surface. It should be noted here that the palm, when in different poses, becomes prone to deformations of the skin because of the movement of the fingers and the thumb. Hence, additional line like impressions are introduced. However, these deformations can be ignored if the palm is relatively flat, and for our current purpose, we choose to ignore the skin deformations. If we assume the palm to be a rigid plane, correction of pose variation becomes well posed. From multi-view geometry of imaging, two images of a planar object are always related by a linear transform, characterized by 8 parameters [9]. Given any point in one image, in the homogeneous coordinate system, $[x, y, 1]^T$, the corresponding point in the second image is given by:

$$[x', y', c] = \mathcal{H}[x, y, 1]^T, \tag{1}$$

where the actual image points are x'/c and y'/c. The Homography matrix, $\mathcal{H}$, can be estimated from correspondences between two views, up to scale. Hence $\mathcal{H}$ has 8 unknowns and we need 4 point correspondences to solve it. But, as mentioned earlier, it is difficult to find robust correspondence points in the weakly textured palm image. Although the flexibility offered by RANSAC in choosing the initial point set, makes it possible. We call this correction the Image Alignment stage, which is described below.

3 Image Alignment

Our goal is to align two palm images with each other so as to match them properly. For this, we need to find 4 point correspondences for applying Homography for pose correction. In the absence of robust correspondences, we use a set of interest points as the initial correspondence and try to improve it using RANSAC. The two most stable points on the palm are the valley points p_1 and p_2 shown in Figure 2, and we directly know the correspondence between them. For the other two correspondences, we search the palm lines for any points with a strong surrounding context in both images.

The process of interest point detection proceeds as follows: We find an edge map of the lines using a curvature based method proposed in [10]. The approach is more robust to noise, and extracts prominent lines in the image. However, in many cases the edge map is still noisy. So, we adopt a two-step line detection algorithm. First, two edge images are computed with a high and a low threshold(Figure 3(b) and (c)). The image with higher threshold has parts of all the major lines and prominent wrinkles. We use these points as seeds to search in the low-threshold image using morphological operations to get the final edge map. The final image obtained is shown in Figure 3(d).

In the second step, we richly sample points from edge map as indicated by the cross marks in Figure 3(e). These points form the initial candidates that have the highest probability of being the interest points. The initial point set is culled by removing those without significant correspondence in the second image. The correspondence level is estimated by using correlation(See Algorithm 1). The list of corresponding points selected by this process is referred to as putative matches.

There are possible scenarios where the correspondences found are incorrect, since we have used the local context only. Another possibility is the introduction of additional interest points due to skin deformation or illumination difference. To compute a reliable

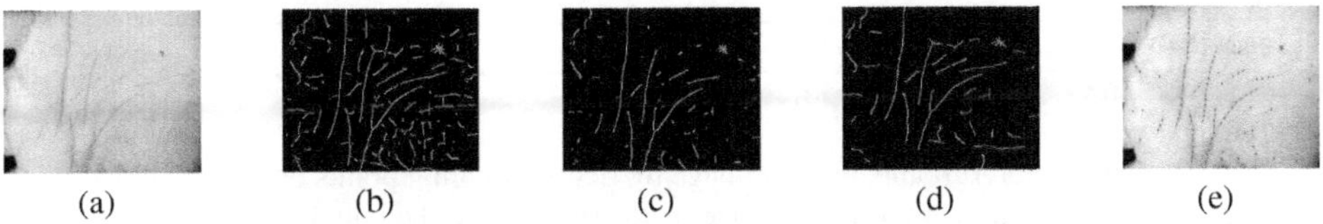

Fig. 3. Steps being followed to get interest points from the image. a) Palm Image, b) Edge map with a lower threshold, c) Edge map with a higher threshold, d) Edge map after enhancing image in c, e) Sampled interest points marked on the palm image.

estimate of homography, we use a RANSAC based method [11]. The input to RANSAC is n pairs of putative matches, including the 2 points computed from the finger valleys. We modify the traditional RANSAC algorithm slightly to always include the two valley points in the set of 4 points, as we have high confidence in them.

Finally, we transform the test image with the computed Homography matrix. This results in the sample image being modified as per the dimensions of the template. Homography computation is sensitive to even small errors in point correspondences. The efficacy of the entire algorithm finally depends on how good correspondences turn out to be. The overall process of image alignment is presented in Algorithm 1. A typical set of input images and the aligned output image are shown Figure 4.

4 Proposed Solution

The overall matching algorithm includes preprocessing, extraction of the region of interest (ROI), image alignment, feature extraction, and matching.

Preprocessing. Consists of three steps, a)Extracting hand from the background and b)Determining the orientation of the hand, and c)locating the two finger valleys.

To segment out the hand from a cluttered background, an ada boosting based approach was proposed by Doublet *et al.* [8]. We use a uniform background in our experiments as our focus is to solve the problem of unknown hand pose for matching. Next, we need to estimate the approximate orientation of the palm. We do this by computing the direction of highest variation among the foreground pixels(largest Eigen vector). If I is the palm image with dimensions $m \times n$, then we take a matrix $M = [X\ Y]$, where

$$X = x_i, Y = y_i \mid 1 \leq x_i \leq m, 1 \leq y_i \leq n \;\&\; I[x_i, y_i] \neq 0 \qquad (2)$$

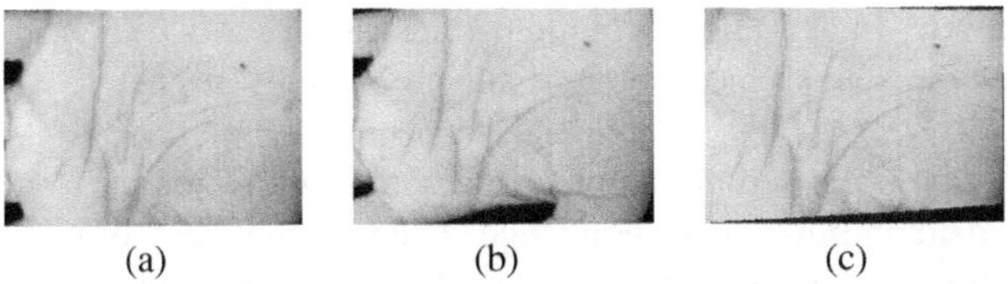

Fig. 4. Result of aligning Image a to Image b, using Homography, has been shown in Figure c

Algorithm 1. Algorithm for Image Alignment

Input: $Image_1, Image_2$ with different poses.
Output: *pose corrected* $Image_2$.

1. Compute interest points in each image by richly sampling points on the edge map.
2. Use correlation, with a window size of 11, to compute the putative correspondences. If the number of correspondences obtained is less than some threshold, label it as imposter.
3. *RANSAC based robust Homography estimation:*
   (a) **for** $k = 1$ to N **do**
   Selection of Point Set: Select the two point pairs randomly from the putative matches and compute the Homography H_k by including the two valley points.
   Distance Computation: Calculate the distance d_i for each pair of putative correspondences after Homography transformation.
   Compute Inliers: Given a threshold t, compute the total number of inliers, n_k among putative matches, such that $d_i < t$.
   (b) **end for**
   (c) Choose the final Homography, H as $\arg\max_{H_k}(n_k)$.
4. *Output the image obtained by transforming* $Image_2$ *with* H.

The largest eigen value of the scatter matrix, S, of M above gives us the orientation of the palm. This method corrects large degrees of in-plane rotations as opposed to previous approaches. Finally, we extract the valley points as in [5].

Extraction of ROI. The region of interest in the hand image is defined with respect to the two valley points detected above. We assume the middle valley point, p_3 to be the origin, and compute the line passing through p_3 and parallel to the line joining p_1 and p_2(Figure 2). We then extract the maximal rectangular region aligned to the above line, containing foreground pixels. This allows us to consistently extract the entire visible area of the palm, irrespective of the change in view. Note that this is an approximate region of interest, and the matching is done after image alignment.

The palm extraction process indirectly solves the problem of variations in palm size. In approaches assuming a fixed palm to camera distance, palm area of fixed size is cut out from all the subjects. In such a case, the subjects with palm size smaller than the fixed size are either rejected or are able to infiltrate the system. This problem was raised by Poon *et al.* [12], who proposed the use of variable sized elliptical ROI to address it. By extracting variable sized palm areas, we are able to handle this problem as well.

Image Alignment. Image alignment algorithm was described in Section 3. In most cases, we get two images of the same size that are suitably aligned to each other. However, in certain cases, it fails to compute an accurate match. RANSAC is a robust algorithm for Homography computation, and any failure to compute a valid transformation indicates errors in point correspondence. The reason is lack of correlation at local windows of points. So, we assume that palm textures are different and assign a high dissimilarity value, effectively labeling them as belonging to different individuals.

Feature Extraction. Once the images are aligned, one can use any one of the existing approaches for matching palmprints. In our experiments, we match responses of Gabor Filters from corresponding windows and their neighborhood, as proposed in [5].

Matching. A typical pose corrected image is shown in Figure 4(c). To match the images, we only concentrate on the overlapping parts of the two palms. This is referred to as matching with masking, and the distance is computed as the sum of hamming distances between the binarized real and imaginary filter responses [5]. For certain pairs of genuine matches, error in image alignment can reduce the resulting matching score. To overcome this, we define the final dissimilarity score as $d_{final} = min(d_{fixed}, d_{corrected})$, where the individual dissimilarity scores are computed assuming fixed pose, and after pose correction. If Homography computation is accurate, the dissimilarity score after pose alignment should reduce for genuine pairs.

In short, we compute the best available matching score between any two image pairs. This reduces the variability in genuine matching scores. However, it also improves the matching scores of imposter pairs. The result of this bias towards better match is clearly visible in the results presented in the following section.

5 Experimental Results and Discussion

We present results of the proposed approach on PolyU dataset with synthetically generated pose variations along with a real dataset with varying palm poses.

Synthetic Dataset. We selected 50 individuals from the PolyU database, and generated various poses. For each subject, we randomly pick 1 image, apply out of plane rotations, and get images from the resulting view. We generated 9 different poses for each subject by rotating the image about X and Y axes (Z being the camera axis), with angles varying from 0° to 45°. Figure 5 shows some of the examples from two classes.

To compare our approach with the traditional approach, which assumes a fixed pose (except in-plane rotation and translation), we obtained two values of matching score between every pair, with and without image alignment. Note that the fixed pose approach always extracts a fixed-sized square ROI. Each image was matched with 4 registered

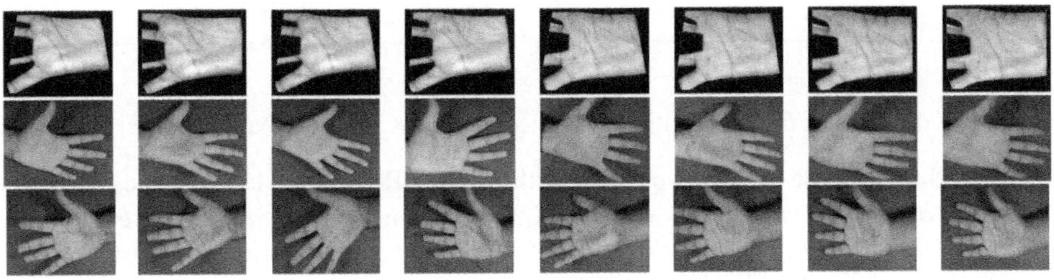

Fig. 5. A set of images from the Dataset. First row shows 4 images each from 2 users in Synthetic Dataset. 4 left hand images each from 2 users in Real Dataset are shown in the second row and right hand is shown in third row.

images, and the minimum score is taken as the matching score. Comparison of ROC curves between our algorithm and fixed pose approach, for Synthetic Dataset, is given in Figure 6(a). A semilog plot of the same is given in figure 6(c), to highlight the low FAR regions.

Real Dataset. The synthetic dataset does not model the complexities of unconstrained real world imaging. In reality, variations arising from illumination changes, skin deformation, concave nature of the palm, and specular reflection from the skin, should be considered to check the effectiveness of a matching algorithm(Figure 7). We thus created a real world dataset. The setup consists of a fixed camera looking at a uniform background. We collect 5 different palm images with varying poses from each subject. We consider the right and left palms of a user as two different classes, and the dataset consists of 100 classes coming from 50 individuals. However, for the purpose of computation of imposter distances, we do not compare an image from a right palm to a left palm. Figure 5 shows four examples each from four classes from the dataset.

For every user, 2 images were used as the registration images. And the rest were used for testing. Two matching scores for each pair were computed using the fixed pose and the proposed approaches. The ROC curves for the two have been presented in Figure 6(b) and (d).

Table 1 compares the Equal Error Rate while using the two approaches. For synthetic data, we categorized pose variations in 5 groups to analyze performance incrementally. We also present the results while blindly applying the pose correction using Homography along with the complete matching algorithm. From the results on synthetic data, we note that the pose correction step is inaccurate and introduces additional errors in case of data with small pose variations. However, as the extent of pose variation increases, the correction phase in the proposed approach becomes beneficial, increasing the overall accuracy of the matcher.

Table 1. Equal Error Rate(EER) on two datasets with the fixed pose and proposed approach

Method	Synthetic Data					Real Data
	$0°$-$20°$	$20°$-$30°$	$30°$-$35°$	$35°$-$40°$	$40°$-$45°$	
Fixed Pose Approach	0.01%	3.24%	3.71%	16.93%	30.92%	22.4%
Blind Pose Correction	16.48%	12.40%	11.14%	14.98%	11.92%	16.51 %
Proposed Approach	0.47%	4.19 %	11.14%	14.98%	11.92%	8.71%

A few interesting characteristics to be noted in ROC plot of real data in Figure 6(d):

- *(p):*Note that the GAR drops even with high FAR scores. This indicates the presence of some genuine pairs with low similarity. This can arise in uncontrolled imaging scenarios from variations due to blur, wrinkles, illumination, etc.(Figure 7)
- *(q):* The second region of interest is to the left with low GAR, which indicates the presence of image pairs from different palms with very high similarity. This arises mainly due to presence of images with most of the pixels being saturated due to specular reflections from the skin(Figure 7(c)).

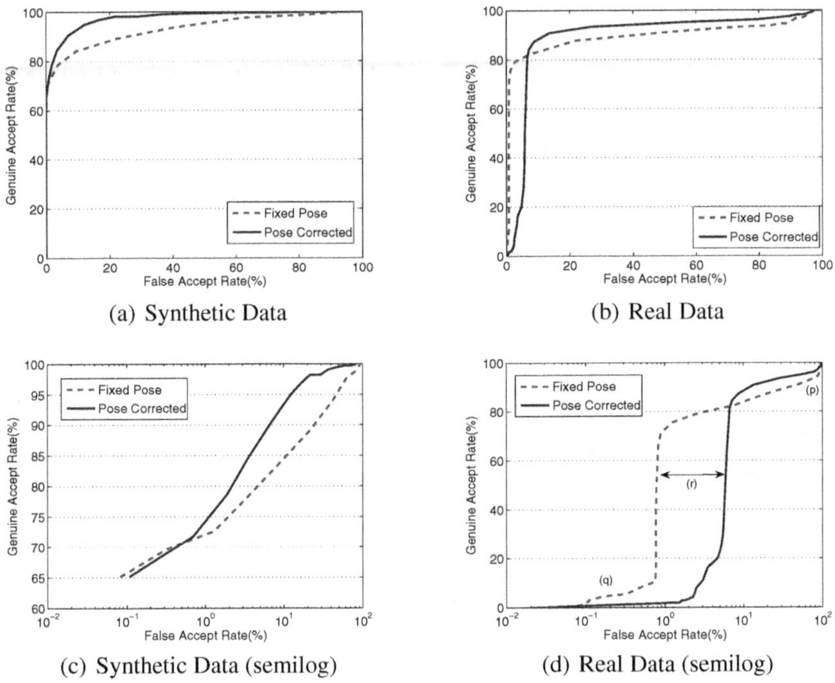

Fig. 6. ROC curves for the Synthetic and Real Datasets in *linear* and *logarithmic* scales

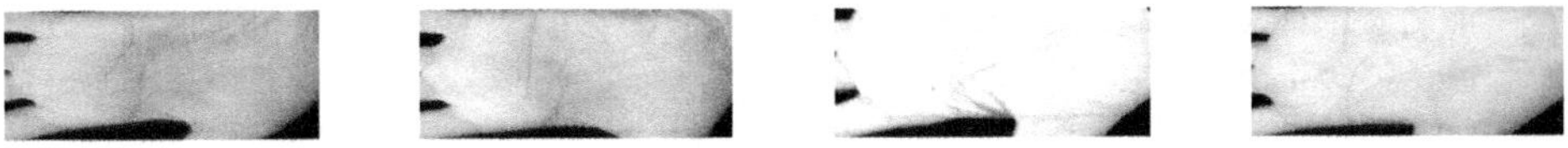

(a) (b) (c) (d)

Fig. 7. Images from our dataset with blur, specular reflection, and skin deformations

- *(r):* Note the drop in GAR in pose corrected version is earlier than the uncorrected one.The reason being that pose correction reduces distance between pairs from different palms too, making some of them closer to genuine pair distance.

One can rectify some of the above problems by improving the capturing device. For example, to reduce the effects of specular highlights(p), use a polarizing filter and improve the gain control of the camera. Improvements in camera sensor and lens can also lead to a reduction in blur(q). The last problem is inherent to the approach of pose correction, and requires better matching algorithm to overcome.

6 Conclusion and Future Work

In this paper, we presented a novel approach for enabling unconstrained palmprint recognition over a variety of poses. The dataset we collected contains significant amount of variations due to pose, illumination and specularity. We describe the challenges faced by such a system and proposed a mechanism to address the pose changes. We found that recognition accuracy is better for well-focused images as it is easier to find stable corresponding points across poses. The accuracy of the system is mostly affected by the imposter matches turning good because of a flexible alignment process. Our method achieves considerable reduction in the EER from 22.4% to 8.7% for a set of 100 users.

One could imagine the use of additional information such as skin color, hand shape, and positions of finger creases, etc. to improve the overall accuracy of matching. One could also improve the accuracy of the approach using better imaging, and using more accurate methods for correspondence computation. One could also employ skin deformation models or using articulated or concave models for the palm surface.

References

1. Kahraman, F., Kurt, B., Gokmen, M.: Robust face alignment for illumination and pose invariant face recognition. CVPR (November 2007)
2. Kale, A., Chowdhury, A.R.: Towards a view invariant gait recognition algorithm. In: IEEE Conference on AVSS, pp. 143–150 (2003)
3. Zheng, G., Wang, C.J., Boult, T.E.: Application of projective invariants in hand geometry biometrics. IEEE Transactions on Information Forensics and Security 2(4), 758–768 (2007)
4. Duta, N., Jain, A.K., Mardia, K.V.: Matching of palmprints. Pattern Recognition Letters 23, 477–485 (2002)
5. Zhang, D., Kong, W.K., You, J., Wong, M.: Online palmprint identification. IEEE Trans. Pattern Anal. Mach. Intell. 25(9), 1041–1050 (2003)
6. Kumar, A., Wong, D.C.M., Shen, H.C., Jain, A.K.: Personal verification using palmprint and hand geometry biometric. In: Kittler, J., Nixon, M.S. (eds.) AVBPA 2003. LNCS, vol. 2688, pp. 668–678. Springer, Heidelberg (2003)
7. Sun, Z., Tan, T., Wang, Y., Li, S.Z.: Ordinal palmprint representation for personal identification. In: Proc. IEEE Computer Vision and Pattern Recognition (CVPR), pp. 279–284 (2005)
8. Doublet, J., Lepetit, O., Revenu, M.: Contact less hand recognition using shape and texture features. ICSP Proceedings 3 (2006)
9. Hartley, R., Zisserman, A.: Multiple view geometry in computer vision (2000)
10. Garg, S., Sivaswamy, J., Chandra, S.: Unsupervised curvature-based retinal vessel segmentation. In: Proc. of IEEE International Symposium on Bio-Medical Imaging(ISBI), pp. 344–347 (2007)
11. Xiaowei, L., Yue, L., Yongtian, W., Dayuan, Y.: Computing homography with ransac algorithm: a novel method of registration. In: Proceedings of the SPIE, vol. 5637, pp. 109–112 (2005)
12. Poon, C., Wong, D., Shen, H.: A new method in locating and segmenting palmprint into region-of-interest. In: Proc. of 17th International Conference on Pattern Recognition (ICPR 2004), pp. 1051–1054 (2004)

Palmprint Recognition Based on Regional Rank Correlation of Directional Features

Yufei Han, Zhenan Sun, Tieniu Tan, and Ying Hao

Center for Biometrics and Security Research, National Laboratory of Pattern Recognition
Institute of Automation, Chinese Academy of Sciences
{yfhan,znsun,tnt,yhao}@nlpr.ia.ac.cn

Abstract. Automatic personal identification based on palmprints has been considered as a promising technology in biometrics family during recent years. In pursuit of accurate palmprint recognition approaches, it is a key issue to design proper image representation to describe skin textures in palm regions. According to previous achievements, directional texture measurement provides a powerful tool for depicting palmprint appearances. Most of successful approaches can be ranged into this framework. Following this idea, we propose a novel palmprint representation in this paper, which describes palmprint images by constructing rank correlation statistics of appearance patterns within local image areas. Promising experimental results on two large scale palmprint databases demonstrate that the proposed method achieves even better performances than the state-of-the-art approaches.

1 Introduction

Palmprint recognition technology distinguishes one individual from others based on differences of skin appearances in central palm regions of hands [1]. According to previous work, most of discriminating line-like image patterns in palmprints can be captured using only a low-resolution imaging device, like web cameras or low-cost CCD cameras. Due to its applicability, palmprint recognition has attracted increasing attention during recent years, especially in civil use, such as airport and custom.

In this paper, we focus on palmprint analysis based on low-resolution (<= 100 dpi) images. In low resolution palmprint images, there are amount of line segments and irregularly distributed patches on skin surfaces, which are formed due to movement of fingers, structures of tissue and muscles after birth. They are discriminative patterns for personal identification task [1]. Early research in palmprint recognition borrows idea from fingerprint recognition [2][3]. There are many algorithms proposed to makes use of geometry information of principal lines, datum points, delta point features or even minutiae. However, they all require relatively high computation cost and are sensitive to appearance variations. In most state-of-the-art approaches [4-10], palmprints are treated as directional image textures spread in images, which are well characterized by a variety of local texture descriptors, such as wavelet coefficients [5], Gabor feature [1][7][8] and so on. Local texture analysis provides comprehensive and accurate descriptions for both spatial information and texture properties of palmprints. Thus it achieves the most efficient palmprint recognition by far.

M. Tistarelli and M.S. Nixon (Eds.): ICB 2009, LNCS 5558, pp. 587–596, 2009.
© Springer-Verlag Berlin Heidelberg 2009

Motivated by the framework of local texture analysis, we propose to utilize correlation statistics of local distribution of texture features to represent palmprints. In our approach, directional DoG (Difference of Gaussian) filters are performed on palmprint images straightly to extract basic structures of palmprint appearances. Inside each local region, characteristics of skin textures are further encoded by spatial enhanced correlation statistics of distribution of the low-level texture descriptors, named regional appearance correlation (RAC) in this paper. This statistical representation describes distribution properties of skin textures accurately. Since the proposed correlation features does not lie on Euclidean space. Similarity between the proposed correlation based features is evaluated based on correlation matrix distance which is previously used in [14].

The remainder of this paper is organized as follows. Section 2 introduces the correlation based local texture descriptor. Section 3 describes details of the proposed RAC scheme. In Section 4, we demonstrate performances of the proposed approaches on two large-scale palmprint databases. Section 5 concludes the whole paper.

2 Regional Appearance Correlation Based Palmprint Descriptor

2.1 Rank Correlation Statistics as Local Texture Descriptor

In the proposed palmprint representation, we utilize the first and second order DoG filters (Differential of Gaussian) to extract information about structures of skin textures in palmprint images. In Eq.1 and Eq. 2, we illustrate the general expressions of the first and second order DoG filters f^{θ} and g^{θ} respectively, where $x_1 = x\cos\theta + y\sin\theta$, $y_1 = -x\sin\theta + y\cos\theta$

$$f^{\theta} = -\frac{2x_1}{\delta_x^2} exp(-(\frac{x_1}{\delta_x})^2 - (\frac{y_1}{\delta_y})^2) \tag{1}$$

$$g^{\theta} = \{4\frac{x_1^2}{\delta_x^4} - \frac{2}{\delta_x^2}\} exp(-(\frac{x_1}{\delta_x})^2 - (\frac{y_1}{\delta_y})^2) \tag{2}$$

In Eq.1 and Eq.2, θ is the rotation angle of the filters in 2D Cartesian coordinates. In scale space theory [11][12], DoG filters have been used as efficient indicators of differential image structures. The first order Gaussian derivate, also known as canny operator, is widely used for estimating image gradients and describing edges or contrast variations. The directional second order Gaussian derivate is close to optimal for detecting orientation of line-like patterns, like Gabor filters [13]. Benefited from Gaussian kernel, both of them can provide noise suppression in differentiating operation. In the proposed method, we perform convolution with DoG filters on palmprint images pixel by pixel. The size of the proposed filters is fixed to 35*35. We set $\delta_y/\delta_x =$ 3 for the scale to achieve good orientation selection. Empirical results testify that this setting is suitable for describing palmprints patterns. Furthermore, based on X-Y separable property of Gaussian derivates [12], the 2D convolution procedure is computationally efficient. At each pixel, we can obtain two 6D filtering response vectors

v_f and v_g containing filter results on six orientations $\{\theta = i\pi/6\ (i = 0,1\ldots5)\}$. The ith components of v_f and v_g expose and enhance basic first and second order differential structures of palmprint appearances along the orientation $i\pi/6$ respectively, which corresponds to directional edges and line-like segments in palm regions. Thus, the filtering response vectors $\{v_f\}$ and $\{v_g\}$ inside a local region form a comprehensive description of image structures along multiple orientations. In this paper, the proposed palmprint representation is designed based on distribution properties of the vectors over local regions, which can capture intrinsic texture characteristics of palmprint images in a statistical way.

In each local region, we firstly calculate rank correlation statistics of the 6D vectors $\{v_f\}$ and $\{v_g\}$ aggregated inside the region respectively to construct a local texture descriptor. This idea originates from use of covariance as a successful region descriptor in object recognition and detection [14]. From the aspect of image description, covariance statistics evaluate correlation between multiple image features, which achieves information fusion of different image descriptors. In our work, we firstly calculate covariance matrix COV of k 6D response vectors $\{v^i\ (i = 1,2\ldots k)\}$ ($v = v_f$ or v_g) inside a local block following Eq.3, where $\bar{v}$ is the average of : $\{v^i\ (i = 1,2\ldots k)\}$:

$$COV_{i,j} = \frac{1}{k-1}\sum_{n=1}^{k}(v_i^n - \overline{v_i})(v_j^n - \overline{v_j}) \tag{3}$$

Nevertheless, as we can see in Eq.3, elements of covariance have close relation with the scalar value of the filtering responses that can be easily affected by appearance variations due to local contrast variations and deformation of skin surface in palm regions. As a result, it reduces robustness of covariance based feature against the appearance changes between intra-class samples and limits improvement of accuracy. To attack this problem, we propose to use spearman correlation (rank correlation) instead. Firstly, for each filtering vector v_f or v_g in a local region, we sort components of the vector in a descending order. Rank indexes of the six components form a rank statistics vector S_f or S_g, corresponding to v_f or v_g respectively. This procedure is illustrated in Fig.1. After that, we compute a covariance matrix D on all rank statistics vectors $\{S_f\}$ or $\{S_g\}$ inside the region following Eq.3. Finally, a 6*6 rank correlation matrix R is then derived based on the covariance matrix D, which is used as a local texture descriptor in the proposed method. Eq.4 illustrates the procedure:

$$R = \begin{pmatrix} 1 & r_{12} & \cdots & r_{16} \\ r_{21} & 1 & \cdots & r_{26} \\ \vdots & \vdots & & \vdots \\ r_{61} & r_{62} & \cdots & 1 \end{pmatrix} \quad r_{ij} = \frac{d_{ij}}{\sqrt{d_{ii}}\sqrt{d_{jj}}} \tag{4}$$

In Eq.4, d_{ij} is the corresponding elements of the covariance matrix D. In our scheme, we replace scalar filtering responses with their rank statistics. This method has been also known as ordinal encoding procedure [9]. For image representation, qualitative ordinal relation or rank statistics between image features presents robustness to local monotonous appearance variations or outlier observations. The ordinal feature reflects intrinsic structures of image patterns or descriptions. Notably, the rank statistics are widely used as discriminative indicators of directional skin textures in palm regions [7][8][9]. In the rank correlation matrix, non-diagonal elements are normalized correlation coefficients. Compared with covariance matrix, the normalized correlation coefficients remove affection of variation of original filtering results in a further step. They evaluate correlation measure between multiple components in the rank statistics vector, which is the intrinsic source of descriptive power in correlation based features. Furthermore, rank correlation coefficient is a non-parametric robust statistic. It doesn't require any prior assumption about function forms of multivariate distributions. As a matter of fact, in palmprint recognition, we seldom know about the exact form of distribution of local filtering responses. Thus, rank correlation provides more flexible tool to describe palmprint appearances than parametric statistics, such as covariance measure in Eq.3, which is not distribution-free.

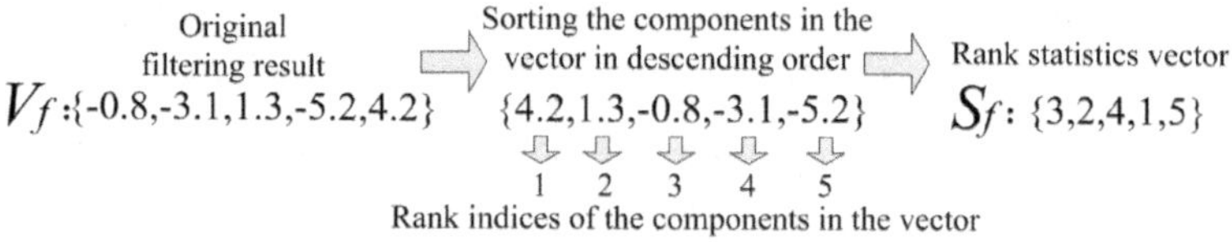

Fig. 1. Construction of rank statistics vector

3 Construction of Regional Appearance Correlation

Before further processing, we regulate geometrical deformation in palmprint images and crop square regions of interests (ROI) for feature extraction. To construct descriptions of local texture characteristics, we divide the whole ROI image into non-overlapped local blocks with a specific spatial resolution, as illustrated in Fig.3. The texture description of the whole palmprint images is formed by rearranging the proposed statistical palmprint descriptors of each block, named Regional Appearance Correlation (RAC), according to spatial arrangement of the blocks.

Construction of the proposed RAC descriptor is shown in Fig.2. Each local block is further divided into four sub-regions. In the ith sub-region, as described in Section.2, after convolution with the first and second order DoG filters at six orientations, we obtain 6D filtering response vectors v_f and v_g at each pixel. Then two 6*6 rank correlation matrices M_i and M_{i+5} are calculated based on all filtering response vectors $\{v_f\}$ and $\{v_g\}$ in this sub-block respectively. This procedure is shown in Fig.2. They indicate characteristics of distribution of local image structures in the ith sub-region. Furthermore, we obtain correlation matrix M_5 and M_{10}, which are calculated based on all vectors $\{v_f\}$ and $\{v_g\}$ inside the whole local block respectively. Both of them evaluate global statistical texture properties in the block. In our method, they are used as complementation to correlation features that focus on finer texture structures within sub-regions. Thus, for each block in the image, we derive totally ten correlation matrices, which form a 6*6*10 tri-tensor $\{M_i\}$ (i = 1,2…,9,10), named as Regional Appearance Correlation (RAC) in this paper. The first five matrices in this tri-tensor are derived based on all $\{v_f\}$ in this block, while left five are calculated from all $\{v_g\}$ in the block, as we can see in Fig.3. Based on the proposed scheme, we can find that the overall palmprint representation of the whole image involves spatial information of palmprint textures on two successive levels. For two RAC descriptor A and B derived from two separate local blocks, their orders of arrangement in the overall image representation denote geometric relation between two blocks, which represents topological structures of palmprint textures at block-level. Similarly, within each block, the precedence relation between different components M*i* in a RAC descriptor indicates spatial layout of palmprint image structures at finer sub-region level. Through this way, we can combine statistical texture descriptions on both coarse and fine spatial scales to achieve a comprehensive description of palmprint appearances within the local block. In our method, because the proposed correlation matrix doesn't lie on Euclidean space, we utilize a distance metric for symmetric correlation matrix as in Eq.5, which is proposed in [14]. Similarity between two RAC A and B is evaluated by calculating the mean of distance measures between corresponding A*i* and B*i* in the RAC descriptors.

$$Dis(A_i, B_i) = \sqrt{\sum_{k=1}^{n} ln^2 \lambda_k (A_i, B_i)} \quad (i = 1,2\ldots10) \tag{5}$$

$\{\lambda_k\}$ are the generalized eigen values of A*i* and B*i*. To testify efficiency of correlation based feature, we also construct RAC descriptor based on covariance matrices of original filtering results in our work, named as "covariance based RAC". Comparison of performances between two types of RAC descriptors is involved in the following section.

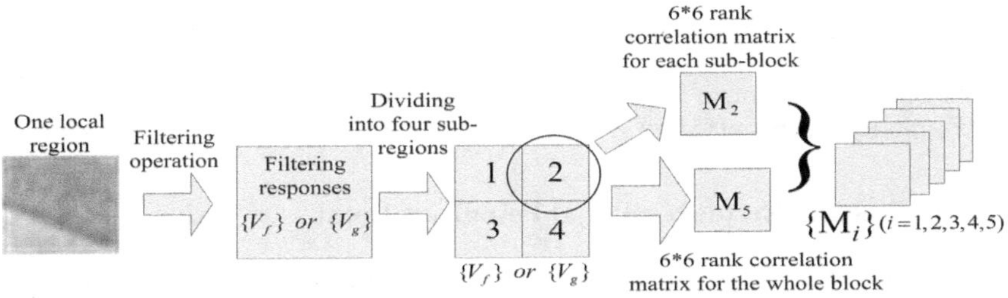

Fig. 2. Generation of rank correlation matrices inside a local image block

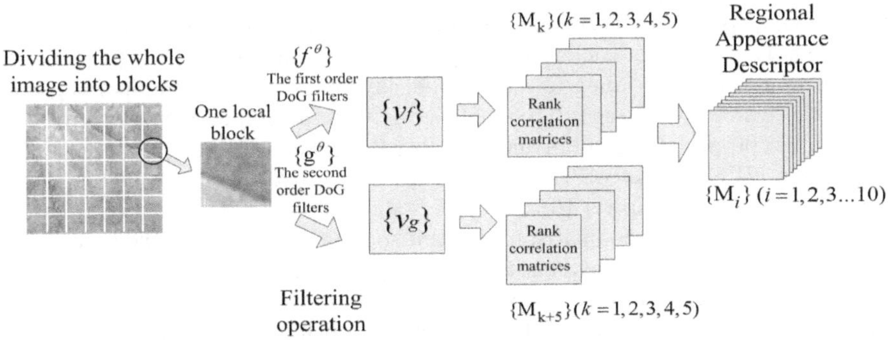

Fig. 3. Construction of Regional Appearance Descriptor

4 Experimental Results

We test the correlation based representation on two large-scale palmprint databases, CASIA Palmprint Database [16] and PolyU Palmprint Database Version 2.0 [15]. Both of them contain deformation of skin textures due to variations of hand postures during image capturing, which is shown in Fig.4. In PolyU Palmprint Database Version 2.0 [15], each palm contain two different sessions. The average time interval between them is about two months. Light conditions and focus of the imaging device are changed between two occasions of image capture [1], which is challengeable to recognition algorithms. Palmprint images in CASIA Palmprint Database [16] are captured without pegs or plates to fix users' hands during image capturing [16]. Orientations and postures of hands are varied in the captured images, as we can see in Fig.7. It leads to contraction and stretching of skin surface in palm regions. In our experiments, we implement the other three state-of-the-art component based recognition algorithms [7][8][9] and evaluate performances on the two databases. The proposed correlation based features is compared with them to testify validity of the proposed method.

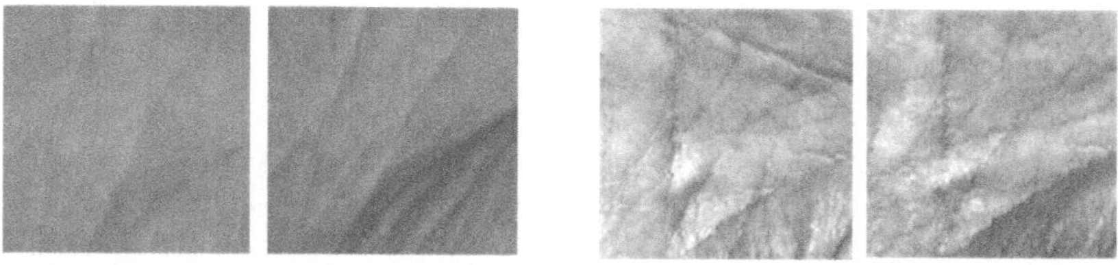

(a) (b)

Fig. 4. Examples of deformation of skin textures in CASIA Palmprint Database [16] (a) and PolyU Palmprint Database [15] (b)

4.1 Experimental Results on PolyU Palmprint Database Version 2.0

There are totally 7,752 palmprint images from 385 palms in the PolyU database [15]. Six examples of captured images in this database are shown in Fig.5. After preprocessing, regions of interests, with the size of 128×128 are obtained from captured image. To construct the proposed RAC description, we divide the whole ROI into 16*16 local blocks firstly. For each local region, we further construct the proposed RAC descriptor. Therefore, for each ROI image, we can obtain totally 8*8 RAC descriptors which are arranged according to their spatial layouts to form appearance representation of the whole ROI image. Fig.6 and Tab.1 illustrate comparison of performances between our proposed algorithm and the other three state-of-the-art component based approaches [7][8][9]. As we can see in experimental results, we can find that our method achieves much lower EER than the state-of-the-art component based methods [7][8][9]. Furthermore, comparison between roc curves of RAC scheme and "RAC based on covariance" testifies that the local appearance variations result in deterioration of performances of covariance based feature, as we described in Section.2. The proposed rank correlation based texture description improves robustness to appearance change to a great extent.

Table 1. Performances comparisons on PolyU Palmprint Database Version 2.0

Algortihms	*d'*[17]	EER[17]
RAC	4.10	0.01%
RAC based on covariance	3.81	1.01%
Ordinal Code [9]	6.19	0.05%
Competitive Code [8]	5.84	0.04%
Fusion code [7]	5.40	0.21%

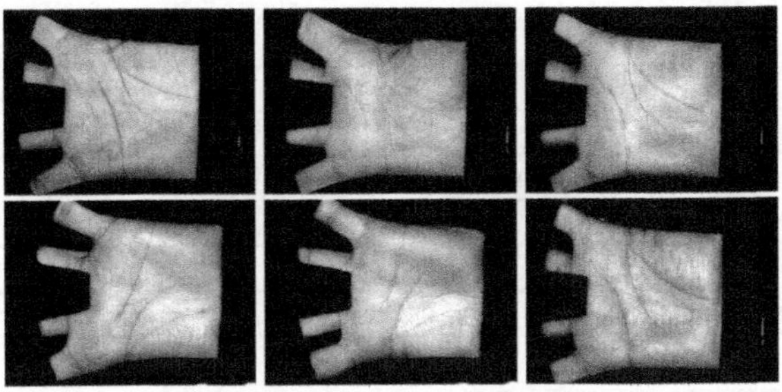

Fig. 5. Six examples of captured images of PolyU Palmprint Database Version 2.0

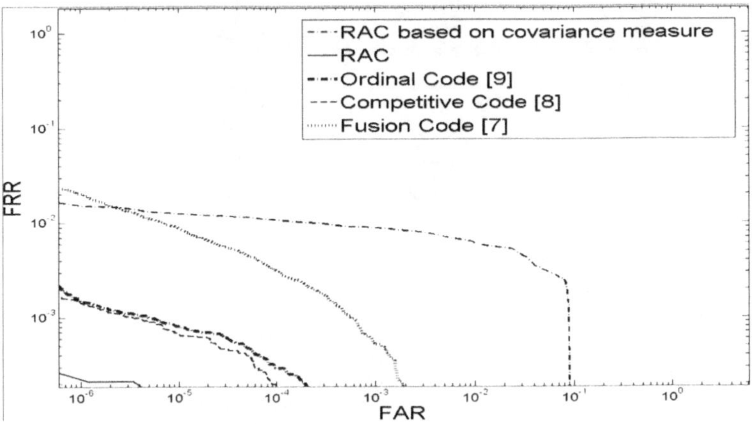

Fig. 6. ROC curves of all algorithms on PolyU Palmprint Database Version 2.0

4.2 Experimental Results on CASIA Palmprint Database

CASIA Palmprint Database contains totally 4,512 palmprint images from 282 palms which are captured with common cameras and peg-free palmprint imaging devices. The derived images are all 640*480 RGB color images [16], as shown in Fig.7. After transforming them into gray images and performing normalization procedure on them, 176*176 ROI are obtained. For each ROI, we divide it into 22*22 blocks. For each local block, we further sub-divide it into 11*11 sub-regions and construct local RAC palmprint descriptors based on local blocks. As we can see in Fig.8 and Tab.2, we can draw similar conclusion that the rank correlation really extract stable and discriminative directional characteristics of palm textures, while covariance based scheme suffers from local appearance variations. It is worth to note that intra-class samples in this dataset are almost captured in a same time session. Thus, the CASIA database contains less appearance variations, such as changes of contrast and deformation of skin textures than the PolyU database, which are the main reasons for deterioration of performances of covariance based RAC scheme. Therefore, as we can see in the experimental results, the covariance based feature performs relatively better than the state-of-the-art approaches in the CASIA database.

Table 2. Performances comparisons on CASIA Palmprint Database

Algortihms	*d'* [17]	EER[17]
RAC	4.77	0.03%
RAC based on covariance	5.69	0.05%
Ordinal Code [9]	5.65	0.08%
Competitive Code [8]	3.82	0.19%
Fusion code [7]	3.73	0.57%

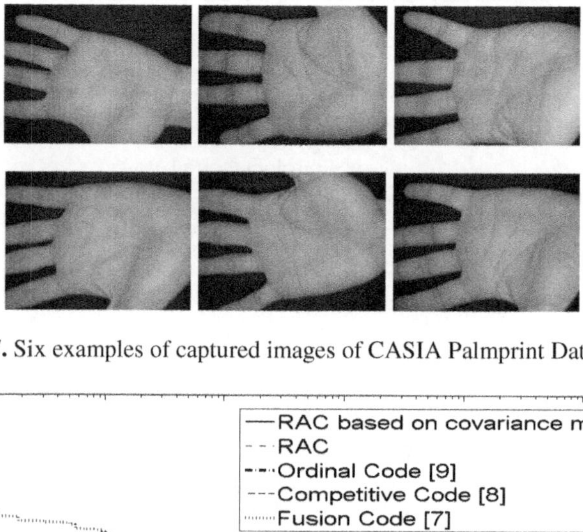

Fig. 7. Six examples of captured images of CASIA Palmprint Database

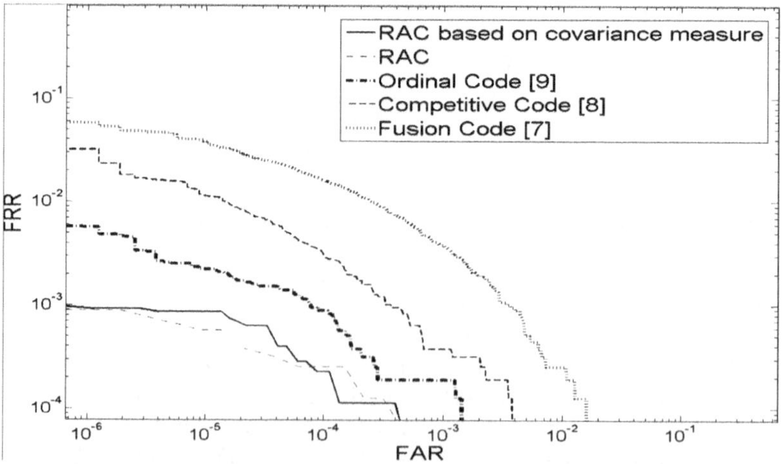

Fig. 8. ROC curves of all algorithms on CASIA Palmprint Database

5 Conclusion

In this paper, we propose a rank correlation based palmprint image representation. It is composed by two main elements. We firstly divide the whole image to local blocks and enhance image structures within each block along multiple orientations using the first and second order directional DoG filter banks over the whole palmprint image. Characteristics of palmprints inside each local region are then evaluated by rank correlation statistics of filtering responses aggregated in the region, which actually form a description of non-isotropic image textures. By utilizing the rank correlation statistics, we obtain robustness to local appearance deformation due to qualitative coding of filtering responses. More importantly, the rank statistics extracts intrinsic characteristics of directional skin textures within local regions. In a further step, correlation measure between qualitative descriptions along different orientations provides accurate descriptions of local image structures. Experimental results on CASIA Palmprint Database [16] and PolyU Palmprint Database Version 2.0 [15] illustrate effectiveness of our method. Compared with the state-of-the-art methods [7][8][9], the proposed method achieves even higher accuracy. The proposed correlation measure based image descriptor can be applied not only in palmprint recognition, but also in face and iris image analysis in biometrics.

Acknowledgement

This work is supported by research grants from the National Basic Research Program (Grant No.2004CB318110), the Natural Science Foundation of China (Grant No.60723005, 60736018, 60702024), NLPR 2008NLPRZY-2, the National Hi-Tech Research and Development Program of China (2006AA01Z193,2007AA01Z162).

References

[1] Zhang, D., Kong, W.K., You, J., Wong, M.: Online Palmprint Identification. IEEE Transactions on Pattern Analysis and Machine Intelligence 25(9), 1041–1050 (2003)
[2] Zhang, D., Shu, W.: Two Novel Characteristics in Palmprint Verification: Datum Point Invariance and Line Feature Matching. Pattern Recognition 32(4), 691–702 (1999)
[3] Duta, N., Jain, A.K., Mardia, K.V.: Matching of Palmprint. Pattern Recognition Letters 23(4), 477–485 (2001)
[4] Wu, X.Q., Wang, K.Q., Zhang, D.: Palmprint Recognition using Directional Line Energy Feature. In: Proceedings of the 17th ICPR, vol. 4, pp. 475–478 (2004)
[5] Zhang, L., Zhang, D.: Characterization of Palmprints by Wavelet Signatures via Directional Context Modeling. IEEE Trans on SMC-B 34(3), 1335–1347 (2004)
[6] You, J., Kong, W.K., Zhang, D., Cheung, K.: On Hierarchical Palmprint Coding with Multi-features for Personal Identification in Large Databases. IEEE Transactions on Circuit Systems for Video Technology 14(2), 234–243 (2004)
[7] Kong, W.K., Zhang, D.: Feature-Level Fusion for Effective Palmprint Authentication. In: Zhang, D., Jain, A.K. (eds.) ICBA 2004. LNCS, vol. 3072, pp. 761–767. Springer, Heidelberg (2004)

[8] Kong, W.K., Zhang, D.: Competitive Coding Scheme for Palmprint Verification. In: Proceedings of the 17th ICPR, vol. 1, pp. 520–523 (2004)

[9] Sun, Z.N., Tan, T.N., Wang, Y.H., Li, S.Z.: Ordinal Palmprint Representation for Personal Identification. In: Proceedings of CVPR 2005, vol. 1, pp. 279–284 (2005)

[10] Han, Y.F., Sun, Z.N., Tan, T.N.: Palmprint Recognition Based on Directional Features and Graph Matching. In: Lee, S.-W., Li, S.Z. (eds.) ICB 2007. LNCS, vol. 4642, pp. 1164–1173. Springer, Heidelberg (2007)

[11] Koenderink, J., van Doorn: Generic neighbourhood operators. IEEE Transactions on Pattern Analysis and Machine Intelligence 14, 597–605 (1992)

[12] Daugman, J.G.: Two-dimensional spectral analysis of cortical receptive field profiles. Vision Research 20, 847–856 (1980)

[13] Florack, L.: Image Structure. Kluwer Academic Publishers, Dordrecht (1997)

[14] Tuzel, O., Porikli, F., Meer, P.: Region covariance: A fast descriptor for detection and classification. In: Leonardis, A., Bischof, H., Pinz, A. (eds.) ECCV 2006. LNCS, vol. 3952, pp. 586–600. Springer, Heidelberg (2006)

[15] PolyU Palmprint Database, http://www.comp.polyu.edu.hk/~biometrics/

[16] CASIA Palmprint Database, http://www.cbsr.ia.ac.cn/english/Palmprint%20Databases.asp

[17] Daugman, J., Williams, G.: A Proposed Standard for Biometric Decidability. In: Proceedings of CardTech/ SecureTech Conference, Atlanta, GA, pp. 223–234 (1996)

Direct Pore Matching for Fingerprint Recognition

Qijun Zhao, Lei Zhang, David Zhang, and Nan Luo

Biometrics Research Centre, Department of Computing,
The Hong Kong Polytechnic University, Kowloon, Hong Kong
{csqjzhao,cslzhang,csdzhang,csnluo}@comp.polyu.edu.hk

Abstract. Sweat pores on fingerprints have proven to be useful features for personal identification. Several methods have been proposed for pore matching. The state-of-the-art method first matches minutiae on the fingerprints and then matches the pores based on the minutia matching results. A problem of such minutia-based pore matching method is that the pore matching is dependent on the minutia matching. Such dependency limits the pore matching performance and impairs the effectiveness of the fusion of minutia and pore match scores. In this paper, we propose a novel direct approach for matching fingerprint pores. It first determines the correspondences between pores based on their local features. It then uses the RANSAC (RANdom SAmple Consensus) algorithm to refine the pore correspondences obtained in the first step. A similarity score is finally calculated based on the pore matching results. The proposed pore matching method successfully avoids the dependency of pore matching on minutia matching results. Experiments have shown that the fingerprint recognition accuracy can be greatly improved by using the method proposed in this paper.

Keywords: Fingerprint recognition, pore matching, level-3 features, fusion.

1 Introduction

Automatic fingerprint recognition systems (AFRS), as well-known biometric techniques, are nowadays widely used in various applications such as forensics and access control [1-2]. Most existing AFRS utilize the minutiae (the endings and bifurcations of ridges, belonging to level-2 features) to recognize fingerprints and to identify persons [11-14]. However, as people's desire for higher security levels keeps increasing, it is highly necessary to base the recognition of fingerprints on more features, but not merely minutiae.

An example non-minutia feature is the sweat pores on fingerprints [3] as shown in Fig. 1. They have been used for a long time by forensic officers [4], and proven to be very distinctive features [10]. However, as level-3 fingerprint features, pores can not be reliably extracted from normal low-resolution fingerprint images [1]. Instead, a minimum of 1000dpi is required [3]. Recently, thanks to the development of high resolution imaging techniques, some pore-based AFRS have been proposed in literature [5-9].

In the earliest pore-based AFRS developed by Stosz and Alyea [5], fingerprints are first aligned by searching for the best alignment in a discretized transformation parameter space. The correlation between manually marked regions is used to choose the best alignment. Such correlation based method was later used by Kryszczuk et al. [6-7]

M. Tistarelli and M.S. Nixon (Eds.): ICB 2009, LNCS 5558, pp. 597–606, 2009.
© Springer-Verlag Berlin Heidelberg 2009

in their study of using pores to recognize fragmentary fingerprints. After aligning the fingerprints, the pores on them can be matched by simply comparing their coordinates in the aligned fingerprint coordinate system. These pore matching methods are limited in the following factors. First, their accuracy highly depends on the way of discretizing the transformation parameters. Second, they have to search through all possible rotations and translations, which is computationally very expensive.

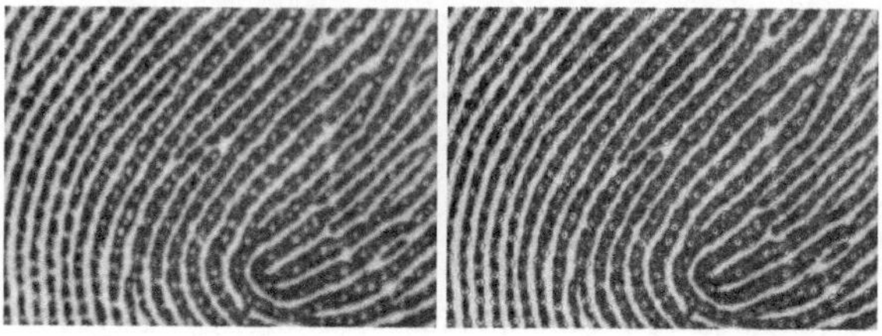

Fig. 1. Left: an example fingerprint image. Right: the extracted pores on it (marked by circles).

The state-of-the-art pore matching method was recently proposed by Jain et al. [8-9] In the method, the fingerprint images were first aligned based on the minutia features on them by using a string-matching algorithm. Minutiae on the fingerprints were then matched and paired. Pores lying in a rectangular neighborhood to each pair of matched minutiae were cropped and rotated according to the directions of the two minutiae. Afterwards, they were matched by using the iterative closest point (ICP) algorithm which is capable to handle sets of points with different numbers of points and can compensate for non-linear deformation between them. The average distance between matched pores was taken as the pore match score. This score was then fused with the minutia match score by using the weighted summation scheme. Compared with previous pore matching methods, this method can cope with fingerprint transformation more efficiently. However, it matches pores based on the minutia matching results. Consequently, the pore matching accuracy is limited by the minutia matching accuracy and the match scores of minutiae and pores will be not independent. Such dependency will impair the effectiveness of the subsequent fusion of the match scores.

In order to decouple the matching processes of minutiae and pores and to make the fusion of minutia and pore match scores more effective, this paper proposes a novel direct approach to matching pores on fingerprints. The proposed method incorporates the alignment into the matching process and can handle complicated transformations. It successfully avoids the dependency of pore matching on minutia matching. As a result, the fusion of the match scores of minutiae and pores is expected to be more effective because they are independent. Section 2 will introduce the proposed method in detail. Section 3 then gives our experimental results and section 4 concludes the paper.

2 Direct Pore Matching

In this paper, the pores are extracted from fingerprint images by using a state-of-the-art method proposed by Zhao et al. [15] Fig. 1 gives an example fingerprint image and the pores extracted from it. The extracted pores are recorded by their locations on the fingerprint images and described by feature vectors defined later in this section. In order to match the pores on two fingerprint images, they are first pair-wise compared and initial correspondences between them are established based on their local features. The initial pore correspondences are then refined by using the RANSAC (RANdom SAmple Consensus) algorithm to give the final pore matching results. A pore match score is finally calculated for the two fingerprint images based on both the initial and final pore correspondences. Below, we introduce the proposed method step by step.

2.1 Initial Pore Correspondence Establishment

The first step of our pore matching approach is to pair-wise compare the pores and establish the initial correspondences between them. In order to fulfill this purpose, we have to associate each pore with a descriptor and present a method to compare the pores based on their descriptors. In the literature of computer vision, lots of local descriptors have been proposed to describe point features [16]. Most of them are based on the gradients in the local neighborhood to the points. Unfortunately, from our experiments, we observed that the gradients in the local neighborhood to pores on fingerprints are not very distinctive. In the context of fingerprint recognition, Jain et al. [17] proposed to use a bank of Gabor filters to generate a feature vector from the region surrounding the reference point to describe the fingerprint. However, this method failed to cope well with different transformations on fingerprints.

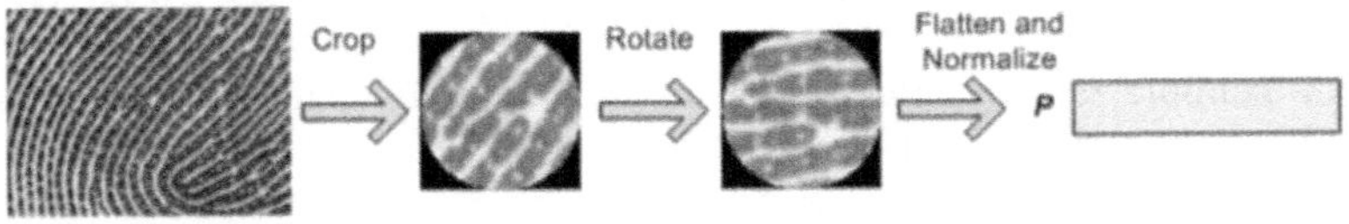

Fig. 2. The flowchart of constructing the descriptor for a pore (marked by circle). *P* is a row vector whose entries are pixel values on the rotated circular neighborhood.

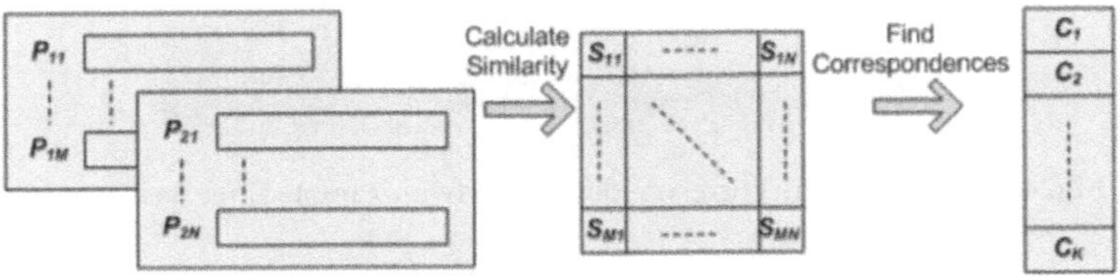

Fig. 3. The flowchart of establishing initial pore correspondences

Our descriptor in this paper is directly built from the pixel values in the local neighborhood to the pore. Specifically, we first smooth the fingerprint image with a Gaussian filter. This is to remove noise in the fingerprint. Then we set up a circular neighborhood to each pore. The circular neighborhood is rotated such that the ridge orientation at the location of the pore becomes horizontal. By using this rotated circular neighborhood, the descriptor is rotation invariant. A feature vector is then obtained by flattening the neighborhood and it is normalized to have zero mean and unit length.

The normalization makes the feature vector invariant to monotone contrast changes and simplifies the subsequent computation of correlation between the feature vectors. This feature vector is defined as the descriptor for the pore. Fig. 2 shows the flowchart of the construction of the descriptor.

Given the descriptors for pores on two fingerprints, we can now compare them pair-wise and establish initial correspondences between them. Suppose $\{P_{1i} \mid i=1,2,\cdots,M\}$ and $\{P_{2j} \mid j=1,2,\cdots,N\}$ are the sets of pores on two fingerprints, $P_{1i}, P_{2j} \in R^n$ are the descriptors of the pores. We define the similarity between two pores P_i and P_j as

$$S_{ij} = P_i P_j^T , \tag{1}$$

where 'T' denotes the transpose. Note that this similarity is equivalent to the correlation between the two feature vectors because they have zero mean and unit length after the normalization. After comparing all pores pair-wise, we get a similarity matrix $S \in R^{M \times N}$. We find for each pore on the first fingerprint a pore which is the most similar one to it among all the pores on the second fingerprint. Similarly, a pore is also found for each pore on the second fingerprint from the pores on the first fingerprint. If P_{1i_k} has its most similar pore as P_{2j_k} and P_{1i_k} is also the most similar pore of P_{2j_k}, then a correspondence is established as $C_k = (P_{1i_k}, P_{2j_k})$. Finally, we will get a number of pore correspondences. Let us assume there are K pairs of corresponding pores and in the next subsection, we will further refine these correspondences to remove the false ones. Fig. 3 shows the flowchart of the initial pore correspondence establishment. An example result of initial pore correspondences determined on two fingerprint images is shown in Fig. 4.

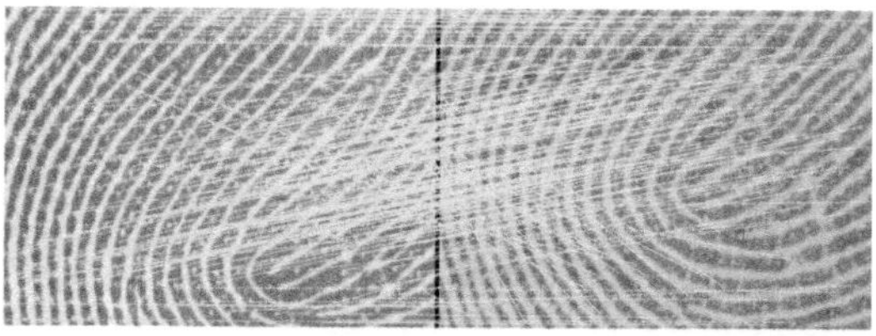

Fig. 4. The initial pore correspondences on two example fingerprint images

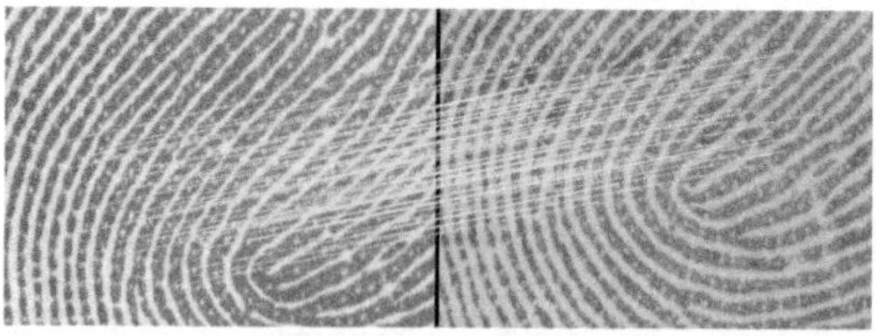

Fig. 5. The refined pore correspondences on the two example fingerprint images

2.2 Pore Correspondence Refinement

In the above step of initial pore correspondence establishment, we consider only the local features and the translation and rotation on fingerprints. As a consequence, there are still lots of false pore correspondences among the initial pore correspondences as can be seen in Fig. 4. In this refinement step, we will cope with the pores in a global context and consider the non-rigid deformation between fingerprints.

Among all the pore correspondences, we call the correct correspondences as inliers and the false ones as outliers. These outliers can severely disturb the estimated transformation between fingerprints, and consequently make the matching results unreliable. In order to exclude the large number of outliers, we have to use a method which is robust to outliers. The ICP algorithm previously used for pore matching requires a good coarse alignment and is sensitive to outliers [20]. Therefore, we take the RANSAC algorithm proposed by Fishler and Bolles [18]. It is a widely used robust estimator for model fitting and has now become a standard in the field of computer vision. Based on the RANSAC algorithm, the pore correspondence refinement proceeds as follows.

First, we choose at random a set C_R of three pairs of corresponding pores from the initial results. We choose three pairs because they are sufficient to exactly determine an affine transformation that we assume occurring to fingerprints. Let (x, y) be a pore on the first fingerprint image and $(\tilde{x}, \tilde{y})$ its corresponding one on the second print after affine transformation. They are associated by

$$\begin{pmatrix} \tilde{x} \\ \tilde{y} \end{pmatrix} = \begin{bmatrix} a_{11} & a_{12} \\ a_{21} & a_{22} \end{bmatrix} \begin{pmatrix} x \\ y \end{pmatrix} + \begin{bmatrix} t_x \\ t_y \end{bmatrix}. \tag{2}$$

There are six parameters in this model and therefore at least three pairs of corresponding pores are needed to uniquely determine the transformation. Based on the chosen three pairs of pores, we estimate the affine transformation simply by solving a set of linear equations.

Second, we evaluate the obtained transformation. The pores on the first fingerprint are transformed according to the obtained transformation. They are then compared with the pores on the second fingerprint. If the distance between two pores is below a given threshold, they are then thought to be matched. If two corresponding pores in the initial results are still matched after the transformation, the correspondence is then a correct one under the transformation. The set C_I of these correct correspondences with respect to the transformation is the consensus set of the chosen correspondences and defines the inliers of C_R.

Third, we check the termination conditions. We set two conditions. One is that the maximum number of iterations N_m has been reached and the other is that the sufficient number of iterations N_s has been reached. N_m is pre-specified, and N_s is calculated as follows. Suppose we want to ensure with a probability p that at least one chosen correspondence set C_R in the iterations is free from outliers, in other words, all the correspondences and thus the transformation estimated based on them are correct. Let ε be the percentage of outliers over the initial correspondence set with respect to the transformation obtained in the current iteration. According to [19], the sufficient number of iterations is then given by

$$N_s = \log(1-p)/\log(1-(1-\varepsilon)^3) . \tag{3}$$

If either of the two conditions is met, the algorithm is stopped and the largest set of inliers, C_I^*, is output as the final result of pore correspondences. Otherwise, go back to the first step and iterate until either of the termination conditions is met. Fig. 5 gives the result of refining the initial pore correspondences shown in Fig. 4. Obviously, the false matches are removed successfully.

2.3 Match Score Computation

In sub-sections 2.1 and 2.2, we show our approach to matching pores. In the process, we get the initial and refined correspondences between the pores on two fingerprints. Based on these results, we can define a similarity between the two fingerprints, i.e. the pore match score of the two fingerprints. In this paper, we define it as

$$MS_{pore} = \frac{|C|+|C^*|}{2\times\min\{M,N\}} , \tag{4}$$

where C and C^* are the set of initial pore correspondences and the set of refined pore correspondences respectively, '$|C|$' denotes the number of elements in the set C, and M and N are the numbers of pores on the two fingerprints.

3 Experiments

In order to objectively assess the performance of the proposed method, we need a dataset of high resolution (at least 1000dpi) fingerprint images. Because there is no such a free high resolution live-scan fingerprint image database in the public domain, we built by ourselves a high resolution fingerprint scanner (approx. 1200dpi). Using this scanner, we collected a set of 1,480 fingerprint images of 148 fingers (see Fig. 1 for an example image). Five images were scanned for each finger in each of two sessions (about two weeks apart). These images have a spatial size of 320 by 240 pixels. They capture a fingerprint area of about 6.5mm by 4.9mm. Therefore, they are very small partial fingerprints and it is very challenge to recognize such fingerprints. According to [6-7], the fingerprint recognition benefits more from the pores when the used fingerprint images cover small fingerprint areas. In order to emphasize the contribution of pores, we thus conducted our experiments in this paper with the above mentioned fingerprint images.

We compared our direct pore matching method (denoted as DP) with the state-of-the-art minutia-based pore matching method (denoted as MICPP) proposed in [8]. In the implementation, the minutiae on two fingerprints are first extracted and matched by using the algorithm in [21]. After matching the minutiae, the pores lying in the neighborhoods of each pair of matched minutiae are matched by using the ICP algorithm, resulting in *N* match scores (*N* is the number of pairs of matched minutiae), which are defined as the summation of two terms: the mean distance between all matched pores and the percentage of unmatched pores. The pore match score between the two fingerprints is finally defined as the average of the first three smallest match scores.

We consider the recognition accuracy in five cases: using only minutia match scores (denoted as MINU), using only pore match scores of DP, using only pore match scores of MICPP, using fused scores of MINU and DP (MINU-DP), and using fused scores of MINU and MICPP (MINU-MICPP). The fused score MS is calculated by using the weighted summation scheme defined as

$$MS = \omega \bullet MS_{pore} + (1-\omega) \bullet MS_{minu} , \tag{5}$$

where MS_{pore} is the pore match score, MS_{minu} is the minutia match score, and $\omega \in [0,1]$ is the weight for MS_{pore}. Before fusion, both the match scores of pores and minutiae are normalized to [0, 1] by using the maximum-minimum normalization.

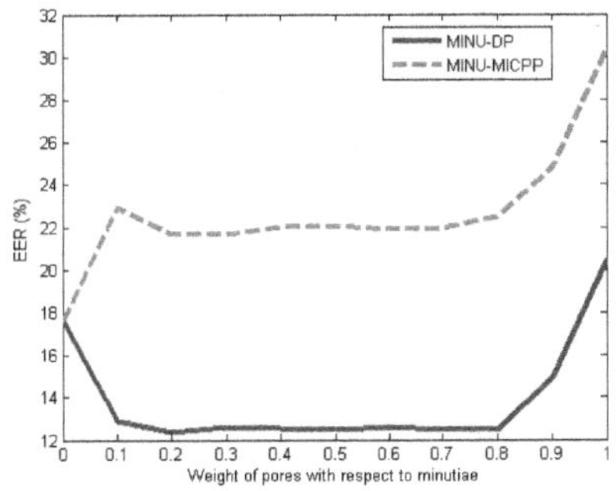

Fig. 6. The EER of MINU-DP and MINU-MICPP when different weights are used

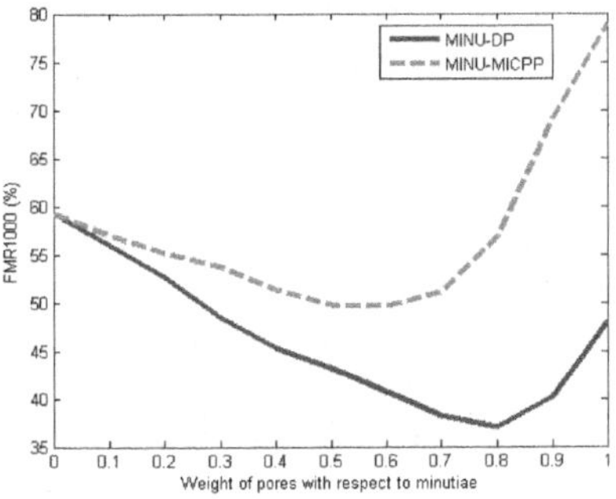

Fig. 7. The FMR1000 of MINU-DP and MINU-MICPP when different weights are used

In the fingerprint recognition experiments, the following matches were carried out: 1) Genuine matches: Each of the fingerprint images in the second session was matched with all the fingerprint images in the first session, leading to 3,700 genuine matches, and 2) Imposter matches: the first fingerprint image of each finger in the second session was matched with the first fingerprint image of all the other fingers in the first session, resulting in 21,756 imposter matches. Based on these match scores, we calculated the equal error rates (EER) and FMR1000 [22] of the methods MINU-MICPP and MINU-DP when different weights were used. The results are presented in Figs. 6 and 7. As shown in Fig. 6, the EER of using only minutiae is 17.67%, and the EER of the proposed DP and its counterpart MICPP are respectively 20.49% and 30.45%. By fusing the minutia and pore match scores, our proposed method MINU-DP gets the lowest EER of 12.4% when $\omega = 0.2$, which improves the EER of using only minutiae by 29.82%. In contrast, the MINU-MICPP method fails to make any improvement at all. As for the FMR1000, according to the curves shown in Fig. 7, using only minutiae, the obtained FMR1000 is 59.24%, whereas using only pores are 48.11% and 78.95% by DP and MICPP respectively. After fusing the minutia and pore match scores, the FMR1000 can be improved to 37.05% and 49.68% respectively by MINU-DP and MINU-MICPP. The best accuracy of each case is listed in Table 1.

Table 1. The best EER and FMR1000 of the five cases

Method	MINU	DP	MICPP	MINU-DP	MINU-MICPP
EER	17.67%	20.49%	30.45%	12.4%	21.66%
FMR1000	59.24%	48.11%	78.95%	37.05%	49.68%

From these experimental results, we can see that when only using the pore match scores, the EER is worse than that of minutia match scores, but the FMR1000 could be better. This demonstrates that pores, if advanced pore matching methods are used, are good choices for applications where a very low false acceptance rate is required (or the cost of false acceptance is much higher than that of false rejection). When using only pores, our proposed method obtains much better recognition accuracy (32.71% improvement on EER and 39.06% improvement on FMR1000) compared with the MICPP method. When fusing the match scores of minutiae and pores, the accuracy can be further improved by our method. When the MICPP method is used for pore matching, compared with the results of using only minutiae, no improvement is observed on EER and 16.14% improvement on FMR1000. But when using our proposed method, the fusion can achieve 29.82% and 37.46% improvement respectively upon the EER and FMR1000 of using only minutia match scores. We believe that this improvement owes to the following factors. First, our proposed direct pore matching method incorporates the alignment into the matching process and can better cope with complicated transformations between fingerprints. Second and more important, it enables the independency between minutia and pore matching processes and thus makes the fusion between the minutia and pore match scores more effective.

4 Conclusions

This paper presents a direct pore matching approach for fingerprint recognition. It incorporates the alignment into the pore matching process and can cope with complicated transformations between fingerprints. By making the pore matching independent from the minutia matching, it enables the fusion of the minutia and pore match scores more effective in improving the fingerprint recognition accuracy. Experiments testified that better recognition accuracy can be obtained by using the proposed method. Compared with the state-of-the-art minutia-based pore matching method, our method can achieve more than 30% improvement on the recognition accuracy when using only pores. By fusing minutia and pore match scores, our method improves the recognition accuracy with respect to that of using only minutiae by 29.82% and 37.46% respectively in terms of EER and FMR1000, whereas the counterpart method makes no improvement at all on EER and only 16.14% improvement on FMR1000. A shortcoming of the proposed method is its complexity in describing the pores. Considering the large number of pores on fingerprints, it is worthy to reduce the dimension of the pore descriptors while not impairing the discrimination power and to speed up the comparison of pores.

Acknowledgements

The work is partially supported by the CERG fund from the HKSAR Government, the central fund from The Hong Kong Polytechnic University, the Natural Science

Foundation of China (NSFC) under Contract No. 60620160097, and the National High-Tech Research and Development Plan of China (863) under Contract No. 2006AA01Z193 and 2007AA01Z195.

References

1. Maltoni, D., Maio, D., Jain, A.K., Prabhakar, S.: Handbook of Fingerprint Recognition. Springer, New York (2003)
2. Ratha, N., Bolle, R.: Automatic Fingerprint Recognition Systems. Springer, New York (2004)
3. CDEFFS: Data Format for the Interchange of Extended Fingerprint and Palmprint Features, Working Draft Version 0.2 (2008), `http://fingerprint.nist.gov/standard/cdeffs/index.html`
4. Ashbaugh, D.R.: Quantitative-Qualitative Friction Ridge Analysis: An Introduction to Basic and Advanced Ridgeology. CRC Press, Boca Raton (1999)
5. Stosz, J.D., Alyea, L.A.: Automated System for Fingerprint Authentication Using Pores and Ridge Structure. In: SPIE Conference on Automatic Systems for the Identification and Inspection of Humans, vol. 2277, pp. 210–223 (1994)
6. Kryszczuk, K., Drygajlo, A., Morier, P.: Extraction of Level 2 and Level 3 Features for Fragmentary Fingerprints. In: Second COST Action 275 Workshop, pp. 83–88 (2004)
7. Kryszczuk, K., Morier, P., Drygajlo, A.: Study of the Distinctiveness of Level 2 and Level 3 Features in Fragmentary Fingerprint Comparison. In: Maltoni, D., Jain, A.K. (eds.) BioAW 2004. LNCS, vol. 3087, pp. 124–133. Springer, Heidelberg (2004)
8. Jain, A.K., Chen, Y., Demirkus, M.: Pores and Ridges: Fingerprint Matching Using Level 3 Features. In: 18th International Conference on Pattern Recognition, vol. 4, pp. 477–480 (2006)
9. Jain, A.K., Chen, Y., Demirkus, M.: Pores and Ridges: Fingerprint Matching Using Level 3 Features. IEEE Trans. Pattern Analysis and Machine Intelligence 29(1), 15–27 (2007)
10. Roddy, A., Stosz, J.: Fingerprint Features - Statistical Analysis and System Performance Estimates. Proceedings of the IEEE 85(9), 1390–1421 (1997)
11. Maio, D., Maltoni, D.: Direct Gray-Scale Minutiae Detection in Fingerprints. IEEE Trans. Pattern Analysis and Machine Intelligence 19(1), 27–40 (1997)
12. Jiang, X., Yau, W.Y., Ser, W.: Minutiae Extraction by Adaptive Tracing the Gray Level Ridge of the Fingerprint Image. In: ICIP 1999, vol. 2, pp. 852–856 (1999)
13. Chen, X., Tian, J., Cheng, J., Yang, X.: Segmentation of Fingerprint Images Using Linear Classifier. EURASIP Journal on Applied Signal Processing 2004(4), 480–494 (2004)
14. Wu, C., Tulyakov, S., Govindaraju, V.: Robust Point-Based Feature Fingerprint Segmentation Algorithm. In: Lee, S.-W., Li, S.Z. (eds.) ICB 2007. LNCS, vol. 4642, pp. 1095–1103. Springer, Heidelberg (2007)
15. Zhao, Q., Zhang, L., Zhang, D., Luo, N., Bao, J.: Adaptive Pore Model for Fingerprint Pore Extraction. In: ICPR 2008 (2008)
16. Mikolajczyk, K., Schmid, C.: A Performance Evaluation of Local Descriptors. IEEE Trans. Pattern Analysis and Machine Intelligence 27(10), 1615–1630 (2005)
17. Jain, A.K., Prabhakar, S., Hong, L., Pankanti, S.: Filterbank-based Fingerprint Matching. IEEE Trans. Image Processing 9(5), 846–859 (2000)
18. Fishler, M., Bolles, R.: Random Sample Consensus: A Paradigm for Model Fitting with Applications to Image Analysis and Automated Cartography. Comm. ACM 24(6), 381–395 (1981)

19. Hartley, R., Zisserman, A.: Multiple View Geometry in Computer Vision, 2nd edn. Cambridge Univ., Cambridge (2003)
20. Phillips, J.M., Liu, R., Tomasi, C.: Outlier Robust ICP for Minimizing Fractional RMSD. In: 6th International Conference on 3-D Digital Imaging and Modeling, pp. 427–434 (2007)
21. Feng, J.: Combining Minutiae Descriptors for Fingerprint Matching. Pattern Recognition 41, 342–352 (2008)
22. Maio, D., Maltoni, D., Cappelli, R., Wayman, J.L., Jain, A.K.: FVC 2002: Second Fingerprint Verification Competition. In: ICPR 2002, pp. 811–814 (2002)

A Novel Fingerprint Matching Algorithm Using Ridge Curvature Feature*

Peng Li, Xin Yang, Qi Su, Yangyang Zhang, and Jie Tian

Institute of Automation, Chinese Academy of Sciences, Beijing 100190 China
tian@ieee.org, jie.tian@ia.ac.cn
http://www.fingerpass.net

Abstract. Fingerprint matching based on solely minutiae feature ignore the abundant ridge information in fingerprint images. We propose a novel fingerprint matching algorithm which integrates minutiae feature with ridge curvature map(RCM). The RCM is approximated by a polynomial model which is computed by Least Square(LS) method. In the matching stage, phase-only correlation matching method is employed to match two RCMs. Then sum fusion rule is selected to combine the minutiae matching score and the RCM matching score. Experiments conducted on FVC2002 and FVC2004 databases show that proposed algorithm can obtain more promising performance than solely minutiae-based algorithm and several other multi-feature fusion algorithms.

Keywords: Fingerprint matching, Polynomial model, RCM, Phase-only correlation, Sum rule.

1 Introduction

Most classical Automatic Fingerprint Identification Systems(AFIS) employ minutiae information, including the position, orientation and type, as the distinctive feature. However, using solely minutiae, which belongs to micro feature in fingerprint representation, misses the abundant macro features, for example orientation field, ridge frequency, ridge curvature and so on. Pankanti et al.[1] analyzed and computed the performance's theoretical upper bound of solely minutiae-based fingerprint recognition systems. So more and more researchers begin to focus on exploring the supplementary features for minutiae feature in AFIS[2][3][4].

Fingerprint pattern is constituted by a group of ridge and valley curves. The ridge and valley have the same trend in nature, so only the ridge curve is taken into consideration in this paper. Ridge curvature is a kind of important macro

* This paper is supported by the Project of National Natural Science Foundation of China under Grant No. 60875018 and 60621001, National High Technology Research and Development Program of China under Grant No. 2008AA01Z411, Chinese Academy of Sciences Hundred Talents Program, Beijing Natural Science Foundation under Grant No. 4091004.

M. Tistarelli and M.S. Nixon (Eds.): ICB 2009, LNCS 5558, pp. 607–616, 2009.
© Springer-Verlag Berlin Heidelberg 2009

feature in fingerprint representation and has been widely used in singular point detection[5][6], classification[7], alignment[8] and matching[9] of fingerprints. Liu et al.[5] and Koo et al.[6] compute fingerprint's ridge curvature using the orientation field information and multi-resolution method. They define the ridge point with the maximum curvature value as the reference point of fingerprint. Wang et al.[7] apply the fingerprint ridge curvature into classification with fixed class number. In their algorithm, 25 sampling points are selected and the corresponding curvature values are computed. Then the curvature values are used for composing a 25-dimension feature vector, which are input into the Radial Basis Function Neural Networks(RBFNN) to obtain the classification result. Yager and Amin[8] proposed a two-stage fingerprint alignment method, in the first stage of which the ridge curvature is employed as one of the features. The other two are fingerprint orientation field and ridge frequency and all of them belong to macro feature. Saleh and Adhami transformed all the entire fingerprint feature, including minutiae, ridge curvature, relative position etc., into the so-called Angel-Curvature domain. In their method, the curvature of some ridge points are employed.

In this paper, we explore the fingerprint ridge curvature feature for supplementing the minutiae-based matching algorithm, which is first proposed to our knowledge. The fingerprint ridge curvature is approximated using a polynomial model, whose coefficients are stored, with little extra memory cost(typically 50 bytes), for generating the whole curvature map in the matching phase. We compute the matching score of curvature maps by using correlation matching method based on Fourier-Mellin Transformation[10]. The curvature maps' matching score is fused with the minutiae matching score using sum rule in the decision level.

2 Fingerprint Ridge Curvature Map

Fingerprint ridge contains abundant information representing the flow direction and the orientation's change amount to a length-fixed ridge segment. Introducing ridge curvature into minutiae-based matching algorithm can help improve the accuracy of AFIS. We compute the curvature values of every pixel in the fingerprint image and display all of them using density in another gray scale image, which we call Ridge Curvature Map(RCM). The RCM we propose is obtained by two steps: coarse curvature value extraction and polynomial model fitting.

2.1 Coarse Curvature Value Computing

Ridge curvature is defined as the orientation's change amount along an length-fixed arc and commonly computed by tracing the thinned ridge. Yager et al.[8] represented each thinned ridge with a cubic B-Spline to approximately calculate the change rate on some pixel's tangent direction. And then a Gaussian low-pass filter is used to obtain the whole RCM. The disadvantage of Yager's method lies on the difficulty of computing the minutiae's curvature and coarse curvature value's separative distribution due to 1-pixel width ridge. In our method, the ridge contour line, rather than thinned ridge, is used to compute the coarse curvature value. And we define the curvature as the orientation's difference along a arc with fixed-number pixels. Thus the effective coarse curvature data increase by about one time and the minutiae's curvature can also be computed correctly.

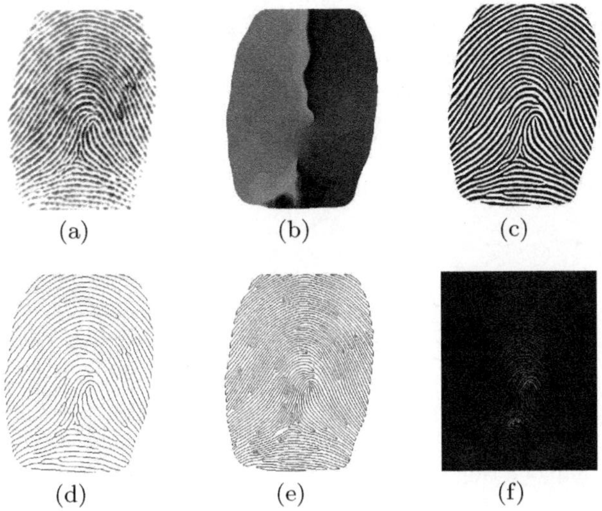

Fig. 1. Intermediate images in coarse ridge curvature value computing procedure((a) is from FVC2002 DB1). (a) input fingerprint image; (b) orientation image; (c) binary image; (d) thinned image; (e) contour image; (f) contour curvature image.

The coarse curvature is calculated by the following steps:

1) Gradient-based orientation filed computation(See Fig.1(b));
2) Image enhancement and binarization. Here we adopt a structural anisotropic filter[11] to enhance the input fingerprint image I_{IN}. The filter kernel is defined in space domain as:

$$h(x, y, \psi) = c_1 + c_2 \cdot exp(-\frac{x_\psi^2}{2\sigma_1^2} - \frac{y_\psi^2}{2\sigma_2^2}) \cdot \frac{\sin f \cdot x_\psi}{f \cdot x_\psi} \tag{1}$$

where,

$$\begin{aligned} x_\psi &= x \cos \psi + y \sin \psi \\ y_\psi &= -x \sin \psi + y \cos \psi. \end{aligned} \tag{2}$$

The parameters c_1, c_2, σ_1^2 and σ_2^2 are all selected empirically as $c_1 = -1$, $c_2 = 2$, $\sigma_1^2 = 4$ and $\sigma_2^2 = 2$. And the parameter f is defined by the fingerprint ridge frequency. After obtaining the enhancement image I_E, global threshold based binarization method is performed and the threshold equals 128. The binarization image is shown in Fig.1(c).

3) Extracting fingerprint ridge contour image using Sobel operator. The resultant image I_C is shown in Fig.1(e). It doubles the curvature information amount compare to Fig.1(d), which shows the thinned ridge image I_T. The spacing between lines of I_T is much larger than I_C.
4) Coarse curvature value computation. We approximately compute the ridge contour line pixel's curvature using the orientation difference of a arc with fixed-number pixels. As is shown in Fig.2, a contour pixel p_0's curvature is decided by the orientation difference of pixel p_1 and p_2, i.e. $c(i_0, j_0) = |o(i_1, j_1) - o(i_2, j_2)|$. For the minutiae p_3, the same computation method is applied to obtain the curvature, $c(i_3, j_3) = |o(i_4, j_4) - o(i_5, j_5)|$. In Fig.2, the fixed pixel length $\triangle L$ is 10 pixels. The resultant coarse curvature map is shown in Fig.1(f).

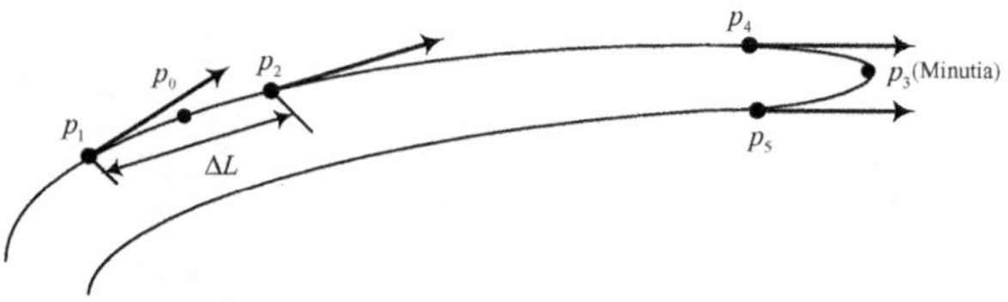

Fig. 2. Sketch map of a ridge contour line and the ridge curvature computation method

2.2 RCM Generation Based on Polynomial Model

The coarse curvature data of all the contour pixels are huge and very memory-costing to store them, which is not necessary at all because there exists a mass of data redundancy. In fingerprint feature analysis and representation, polynomial model is an effective tool to fit the macro feature[12]. We employ a bivariate polynomial function to model the fingerprint ridge curvature. The Least Square(LS) algorithm is used to compute the model's coefficients. The polynomial model is written as Equ.3.

$$\tilde{p}(x, y) = \sum_{i=0}^{k} \sum_{j=0}^{k} a_{ij} x^i y^j \tag{3}$$

where k is the polynomial degree. Let $A = \{a_{ij}\}$ denotes the coefficient matrix, $i = 1, 2, \cdots, k, j = 1, 2, \cdots, k$. A can be obtained by minimizing the square error between the polynomial value and the coarse curvature computed from the given fingerprint image, i.e., minimizing the following function:

$$\varepsilon = \sum_{i=1}^{n} [\tilde{p}(x_i, y_i) - p(x_i, y_i)]^2 \tag{4}$$

where n is the number of sampling points and $p(x_i, y_i)$ denotes the coarse curvature value of the i-th sampling point. The element $a_{i,j}$ of A can be computed by solving the following equation:

$$\frac{\partial \varepsilon}{\partial a_{i,j}} = 0, i \in [1, k], j \in [1, k] \tag{5}$$

In order to avoid ill solution, Singular Value Decomposition(SVD) method is used to solve Equ.5. In addition, k in Equ.3 is selected as 4 according to the tradeoff between the computing cost and the accuracy.

In the end, the 25 coefficients($k = 4$) of A are stored into the fingerprint template to generate RCM in the matching phase by using Equ.3. The total memory cost is $2bytes \times 25 = 50bytes$. Fig.3 compares the RCMs generated by polynomial model and by Yager's method[8], respectively. It shows that the polynomial model can generate smoother RCMs.

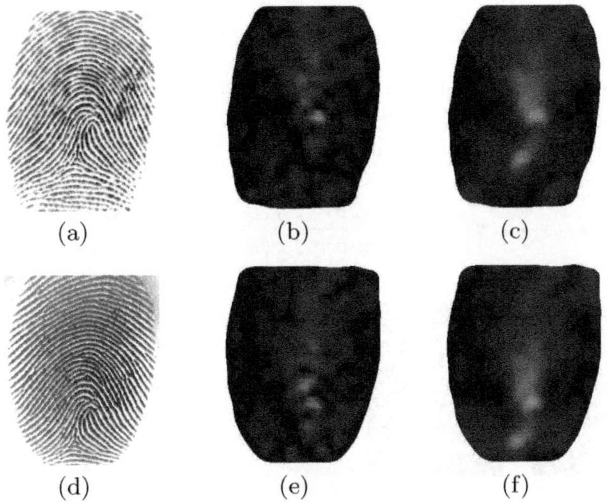

Fig. 3. RCMs. (a)&(d) input fingerprint images(from FVC2002 DB1); (b)&(e) RCMs extracted by method[8]; (c)&(f) RCMs extracted by proposed method.

3 Fingerprint Matching Fusing Ridge Curvature and Minutiae

In this section, we will clarify the matching scheme combining RCM and minutiae information. Fourier-Mellin Transform(FMT) based correlation matching algorithm is utilized for the RCM matching. And its matching score will be combined with the minutiae matching algorithm's with sum rule in the decision level.

3.1 RCM Matching Algorithm

An important method in fingerprint matching is phase-only correlation matching [13], which employs gray-scale images to compute the affine transform's translation parameter and the correlation matching score of two images. The translation parameter and matching score can be computed in either space domain or frequency domain. However, due to large time consuming in space domain, correlation matching is commonly performed using Fourier Transform in frequency domain.

In this paper, we compute the correlation matching score by performing Fourier-Mellin Transform on RCMs. Let f_1 and f_2 denote two RCMs in space domain and their corresponding FMT maps are F_1 and F_2 respectively. Let the relation between f_1 and f_2 be written as(assuming two images have the same size here):

$$f_2(x, y) = f_1[(x \cos\theta_0 + y \sin\theta_0) - x_0, (-x \sin\theta_0 + y \cos\theta_0) - y_0] \quad (6)$$

where (x_0, y_0) and θ_0 denote the translation and rotation parameter, respectively. Correspondingly, the relation between F_1 and F_2 can be written as:

$$F_2(u, v) = F_1(u \cos\theta_0 + v \sin\theta_0, -u \sin\theta_0 + v \cos\theta_0) \cdot exp(-j2\pi(ux_0 + vy_0)) \quad (7)$$

Equ.7 shows that the translation parameter (x_0, y_0) and the rotation parameter θ_0 can be computed separately. According to the property of Fourier Transform, (x_0, y_0) can be indicated with cross-power spectrum of f_1 and f_2, as shown in Equ.8.

$$\hat{R}_{F_1 F_2}(u, v) = \frac{F_1(u, v) \cdot F_2^*(u, v)}{|F_1(u, v) \cdot F_2^*(u, v)|} = e^{-j2\pi(ux_0 + vy_0)} \quad (8)$$

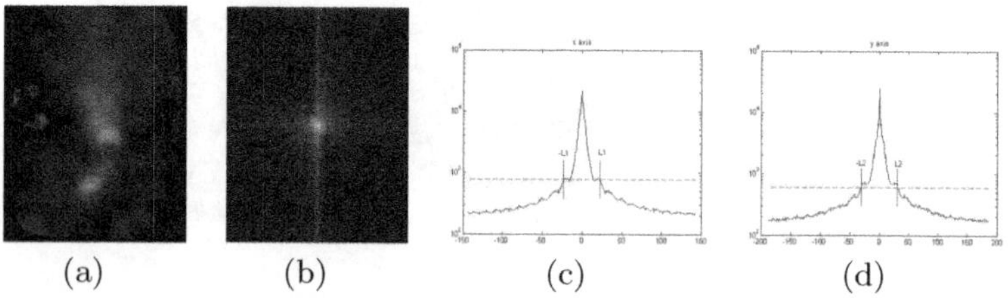

Fig. 4. RCMs in band-limited frequency domain. (a) A fingerprint RCM; (b) Amplitude spectrum of (a); (c) Band limitation in x-axis projection; (d) Band limitation in y-axis projection.

Inverse Fourier Mellin Transform(IFMT) is performed on Equ.8 to obtain a correlation image $I_{cor} = \hat{r}_{f_1 f_2}(x, y)$ in space domain. If f_1 and f_2 are from the same finger, in I_{cor} there will be an aculeated single peak value, similar to monoimpulse, which denotes the position of (x_0, y_0). Otherwise, the peak value's distribution is relatively gentle. This is so-called Phase-Only Correlation(POC) matching. In order to sharpen the peak value, we utilize Band-limited POC(LDPOC)[13] to compute the phase of the cross power spectrum density of f_1 and f_2. LDPOC selects the mean of the power spectrum as the threshold, based on which domain energy region is decided. Fig.4 illuminates the decision of main energy region. After LDPOC, the peak value $v_p (0 < v_p < 1)$ could be viewed as the matching score S_c. If there are several peaks, the sum of them will be S_c.

3.2 Minutiae Matching Algorithm and Fusion Strategy

We use similarity histogram based minutiae matching algorithm, proposed by Zhang et al.[14], to obtain the minutiae matching score. Let a minutiae $m_i =$

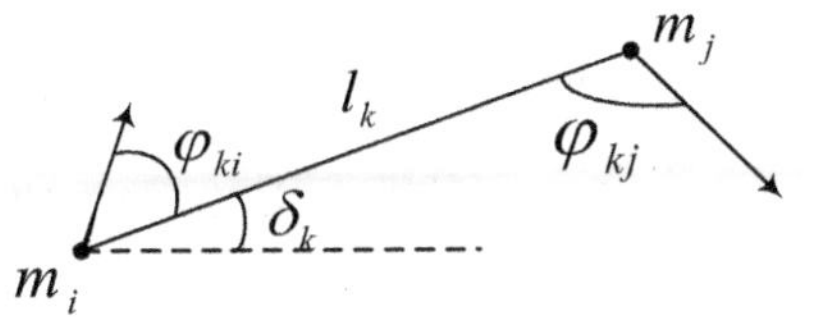

Fig. 5. Structure of minutiae pair

$(x_i, y_i, \theta_i, t_i)$, where (x_i, y_i) denotes the position, θ_i the orientation and t_i the type. All the minutiae in a fingerprint constitute the template $M = \{m_i | i \in [1, n]\}$($n$ is the minutiae's total number). The basic local structure of minutiae pair, illuminated as Fig.5, is utilized in minutiae matching algorithm.

We search the minutiae template to find all the basic structures, which is subject to Equ.9.

$$\begin{aligned} l(m_i, m_j) &= \sqrt{(x_i - x_j)^2 + (y_i - y_j)^2} \\ d_{min} &< l(m_i, m_j) < d_{max}, \end{aligned} \tag{9}$$

where $l(m_i, m_j)$ is the distance between m_i and m_j; d_{min} and d_{max} are two distance thresholds.

Given the template fingerprint T and the query fingerprint Q, we extract two local structure sets $P^T = \{p_k^t(l_t^t, \delta_k^t, \phi_{ki}^t, \phi_{kj}^t) | k \in [1, N_1]\}$ and $P^Q = \{p_k^q(l_t^q, \delta_k^q, \phi_{ki}^q, \phi_{kj}^q) | k \in [1, N_2]\}$, where N_1 and N_2 both are the number of local structure. For $p_m^t \in P^T$ and $p_n^q \in P^Q$, the similarity between them is defined as:

$$\begin{aligned} s(m,n) &= \begin{cases} 1 - \frac{\Delta p_{mn}}{c_0} & \Delta p_{mn} < c_0 \\ 0 & otherwise \end{cases} \\ \Delta p_{mn} &= \eta_1 \cdot |l_m^t - l_n^q| + \eta_2 \cdot |\delta_m^t - \delta_n^q| + \eta_3 \cdot |\varphi_{mi}^t - \varphi_{mi}^q| + \eta_4 \cdot |\varphi_{ni}^t - \varphi_{ni}^q|, \end{aligned} \tag{10}$$

where, empirically the constants are selected as $c_0 = 100, \eta_1 = 0.2, \eta_2 = 0.2, \eta_3 = 0.3, \eta_4 = 0.3$.

Before matching is performed, the alignment procedure and minutiae structure pairing between the template and the query need to be conducted first[14]. For arbitrary $p_m^T \in P^T$ and $p_n^Q \in P^Q$, all the local similarity $s(m, n)$ are computed and the similarity histogram[14] is plotted to decide the reference minutiae structure and the corresponding relation of all local minutiae structures. Afterwards, all the corresponding minutiae structure's similarity degrees are computed and accumulated to get the overall similarity score S_{TQ}(where k is the total number of basic structures).

$$S_{TQ} = \sum_{m,n} s(m,n)/k \tag{11}$$

Fusing different features and their corresponding classifiers can obtain better matching performance than using single feature. We will fuse the ridge curvature matching score S_c and the minutiae matching score S_{TQ} in the decision level, where several fusion strategies can be considered, including max rule, min rule, product rule, sum rule. Experiments in section 4 show the sum rule performs best among all of them.

4 Experimental Results and Analysis

In this section, we select fingerprint data sets from FVC2002 and FVC2004 to test the performance of proposed fusion matching algorithm. Both FVC2002 and FVC2004 include 4 data sets(3 real fingerprint sets from different sensors and 1 synthetic set), which we denote as DB1-DB4. Each of these 8 sets consists 800 fingerprints, i.e., 8 samples for each of 100 fingers, respectively. All the real samples are collected under natural condition and the synthetic ones simulate the fingerprints' natural quality and distribution. For each set, we conduct $\frac{8\times7}{2} \times 100 = 2800$ genuine matches and $\frac{100\times99}{2} = 4950$ imposter matches. The indicator EER(Equal Error Rate) is used to evaluate the performance and the test platform is a P4 2.4G PC with 512M RAM. The algorithm is implemented with C and Matlab computing engine hybrid programming.

Table 1. Accuracy comparison of fusion strategies on FVC2002

	EER_c	EER_{TQ}	EER_{max}	EER_{min}	$EER_{product}$	EER_{sum}
DB1	9.7%	2.1%	11.2%	1.4%	1.3%	1.1%
DB2	8.7%	1.3%	7.8%	0.69%	0.64%	0.59%
DB3	15%	4.2%	16%	3.9%	3.6%	3.3%
DB4	12%	1.7%	13.5%	1.5%	1.3%	1%

4.1 Fusion Rule Selection

Using multi-classifier correlation computation method proposed by Kuncheva et al.[15], we calculate the correlation coefficient between RCM matching and minutiae matching on DB1 and DB2 of FVC2002. Respectively, they are 0.18 and 0.19, which are lower than the ones(0.28 and 0.24) reported by[12]. This shows the independence between RCM matching and minutiae matching is higher than the two classifiers used in[12].

We test the performance of the four fusion rules(maximum rule, minimum rule, product rule and sum rule) on 4 sets from FVC2002. Table 1 shows the results, which can tell that sum rule performs best among all the fusion rules. So we select sum rule to test the final performance of proposed fusing matching method.

4.2 Results on Data Sets from FVC2002

Table 2 shows the performance comparison of proposed fusing algorithm and solely minutiae-based matching method on data sets from FVC2002. Combining modeled RCM with minutiae leads average EER decrease by 0.83% on 4 data sets. Due to using of C and Matlab computing engine hybrid programming, average enroll time and match time consumed by proposed method are somewhat longer than solely minutiae-based method. However, if the entire algorithm is transplanted to an AFIS and implemented with C, the time consuming can be much shorter.

Table 2. Performance comparison of fusing method and minutiae-base method on FVC2002

	Image description	Algorithm	EER	Enroll time	Match time
DB1	388 × 374,500dpi	Minutiae	2.1%	0.09s	0.019s
		RCM+Minutiae	1.1%	0.7s	0.26s
DB2	296 × 560,569dpi	Minutiae	1.3%	0.1s	0.019s
		RCM+Minutiae	0.59%	0.9s	0.27s
DB3	300 × 300,500dpi	Minutiae	4.2%	0.07s	0.019s
		RCM+Minutiae	3.3%	0.5s	0.26s
DB4	288 × 384,500dpi	Minutiae	1.7%	0.08s	0.019s
		RCM+Minutiae	1.0%	0.5s	0.26s

Table 3. Accuracy comparison of different matching algorithms on FVC2002

	DB1	DB2	DB3	DB4
Zhang et al.[14]	2.6%	1.32%	4.55%	3.13%
Yager et al.[8]	2.6%	3.0%	–	–
Wan and Zhou[12]	4%	3.5%	–	–
Proposed method	1.1%	0.59%	3.3%	1.0%

We also compare proposed fusing matching method with other minutiae-based or fusing method on data sets from FVC2002. The results are shown in Table 3 and our proposed method performs best.

4.3 Results on Data Sets from FVC2004

Data sets from FVC2004 are more difficult to deal with than ones from FVC2002. We select FVC2004 DB1 and DB2 to test proposed method's performance on large distortion or low quality data set. Fingerprints of DB1 are captured by CrossMatch V300 optical sensor, and the image size is 640 × 480 with 500dpi. There exists relatively large distortion in images of DB1. Fingerprints of DB2 are captured by URU4000 optical sensor, and the image size is 328 × 364 with 500dpi. Much images in DB2 fall into low quality images and difficult to deal with. Respectively, proposed fusing matching method obtains EER 7.3% and 5.9% on DB1 and DB2, compared with 7.6% and 6.4% obtained by solely minutiae-based method. The results show that performance improvement on FVC2004 data sets is weaker than FVC2002 data sets. We can conclude that proposed fusing method is not able to deal with large distortion and low quality fingerprint images effectively, i.e., RCM is easy influenced by the distortion and image quality.

5 Conclusion

Fingerprint matching algorithm combining minutiae feature and ridge curvature feature is proposed in this paper. More refined ridge curvature is extracted from fingerprint ridge contour map. Then a polynomial model to approximate global ridge curvature is used to obtain resultant RCMs, which are matched by phase-only correlation method in the matching phase. Experiments conducted on FVC2002 and FVC2004 databases show the promising performance of proposed fusing matching algorithm.

References

1. Pankanti, S., Prabhakar, S., Jain, A.K.: On the Individuality of Fingerprints. IEEE Trans. Pattern Anal. Mach. Intell. 24(8), 1010–1025 (2002)
2. He, Y., Tian, J., Luo, X., Zhang, T.: Image Enhancement and Minutiae Matching in Fingerprint Verification. Pattern Recognition Letters 24(9), 1349–1360 (2003)
3. Tico, M., Kuosmanen, P.: Fingerprint Matching Using an Orientation-Based Minutia Descriptor. IEEE Trans. Pattern Anal. Mach. Intell. 25(8), 1009–1014 (2003)
4. Jain, A.K., Prabhakar, S., Lin, H., Pankanti, S.: Filterbank-based Fingerprint Matching. IEEE Trans. Image Processing 9(5), 846–859 (2000)
5. Liu, M., Jiang, X., Kot, A.C.: Fingerprint Reference-point Detection. EURASIP Journal on Applied Signal Processing 2005(4), 498–509 (2005)
6. Koo, W.M., Kot, A.: Curvature-Based Singular Points Detection. In: Proceedings of International Conference on Audio-and Video-Based Biometric Person Authentication, pp. 229–234. Springer, Heidelberg (2001)
7. Wang, X., Li, J., Niu, N.: Fingerprint Classification Based on Curvature Sampling and RBF Neural Networks. In: Wang, J., Liao, X.-F., Yi, Z. (eds.) ISNN 2005. LNCS, vol. 3497, pp. 171–176. Springer, Heidelberg (2005)
8. Yager, N., Amin, A.: Fingerprint Alignment Using a Two Stage Optimization. Pattern Recognition Letters 27(5), 317–324 (2006)
9. Saleh, A.A., Adhami, R.R.: Curvature-based Matching Approach for Automatic Fingerprint Identification. In: Proceedings of the 33rd Southeastern Symposium on System Theory, pp. 171–175 (2001)
10. Chen, Q., Defrise, M., Deconinck, F.: Symmetric Phase-only Matched Filtering of Fourier-Mellin Transforms for Image Registration and Recognition. IEEE Trans. Pattern Anal. Mach. Intell. 16(12), 1156–1168 (1994)
11. Chen, X., Tian, J., Zhang, Y., Yang, X.: Enhancement of Low Quality Fingerprints Based on Anisotropic filter. In: Zhang, D., Jain, A.K. (eds.) ICB 2006. LNCS, vol. 3832, pp. 302–308. Springer, Heidelberg (2006)
12. Wan, W., Zhou, J.: Fingerprint Recognition Using Model-based Density Map. IEEE Transactions on Image Processing 15(6), 1690–1696 (2006)
13. Ito, K., Nakajima, K., Kobayashi, K., Aoki, T., Higuchi, T.: A Fingerprint Matching Algorithm Using Phase-Only Correlation. IEICE Trans. Fundamentals E87-A(3) (2004)
14. Zhang, T., Tian, J., He, Y., Cheng, J., Yang, X.: Fingerprint Alignment Using Similarity Histogram. In: Proceedings of International Conference on Audio- and Video-Based Biometric Person Authentication, pp. 854–861. Springer, Heidelberg (2003)
15. Kuncheva, L.I., Whitaker, C.J.: Measures of Diversity in Classifiers Ensembles and Their Relationship with the Ensemble Accuracy. Mach. Learn. 51, 182–207 (2003)

Fingerprint Matching Based on Neighboring Information and Penalized Logistic Regression*

Kai Cao, Xin Yang, Jie Tian, Yangyang Zhang, Peng Li, and Xunqiang Tao

Institute of Automation, Chinese Academy of Sciences, Beijing 100190 China
tian@ieee.org, jie.tian@ia.ac.cn
http://www.fingerpass.net

Abstract. This paper proposes a novel minutiae-based fingerprint matching algorithm. A fingerprint is represented by minutiae set and sampling points on all ridges. Therefore, the foreground of a fingerprint image can be accurately estimated by the sampling points. The similarity between two minutiae is measured by two parts: neighboring minutiae which are different in minutiae pattern and neighboring sampling points which are different in orientation and frequency. After alignment and minutiae pairing, Nine features are extracted to represent the matching status and penalized logistic regression (PLR) is adopted to calculate the matching score. The proposed algorithm is evaluated on fingerprint databases of FVC2002 and compared with the participants in FVC 2002. Experimental results show that the proposed algorithm achieves good performance and ranks 5th according to average equal error rate.

Keywords: Minutia, fingerprint matching, penalized logistic regression, ridge.

1 Introduction

A fingerprint is the pattern of ridges and valleys on the surface of a fingertip. Due to their strongpoint, such as uniqueness and permanence, fingerprint recognition becomes one of the most important biometric technologies. Previous researchers have proposed many kinds of fingerprint matching algorithms [1,2,3]. According to the fingerprint representation, the existing fingerprint algorithms can be classified into the following categories: image or correlation based [12], ridge feature based [13], orientation feature based [14] and minutiae based [15]. The performance of image based algorithm is largely affected by non-linear distortion. Ridge image is an effective and compact representation of the fingerprint image. However, the correspondences between ridges are difficult to establish due to break and wrong connection in low quality fingerprint images. The difference between two orientation fields is not evident if two fingerprints have similar singular locations. A minutia represents local discontinuities and marks

* This paper is supported by the Project of National Natural Science Foundation of China under Grant No. 60875018 and 60621001, National High Technology Research and Development Program of China under Grant No. 2008AA01Z411, Chinese Academy of Sciences Hundred Talents Program, Beijing Natural Science Foundation under Grant No. 4091004.

M. Tistarelli and M.S. Nixon (Eds.): ICB 2009, LNCS 5558, pp. 617–626, 2009.
© Springer-Verlag Berlin Heidelberg 2009

position where the ridge comes to an end or bifurcates into two, which is one of the discriminative and reliable features. However, the information provided by only minutiae set is limited since it discards the global ridge information. Therefore, researchers proposed to make use of other discriminatory features to establish the correspondence among minutiae and enhance the performance [1,2,3]. Different from above methods, we combine minutiae and all ridges to represent the fingerprint and propose a novel similarity measure methodology based on neighboring minutiae and neighboring sampling points on the ridges. The neighboring minutiae capture the difference of local minutiae pattern, while the neighboring sampling points capture the difference of ridge orientation and ridge frequency around two minutiae.

After minutiae pairing the matching score computation is another challenging problem. A traditional way to calculate the matching score is $n^2/(N_T \times N_I)$, where N_T and N_I represent the numbers of minutiae in template and query minutiae sets, respectively, and n is the number of matched minutiae in both sets. Bazen and Gerez [4] claimed that using $n^2/(N_T + N_I)$ to compute the matching score will give better results. However, both of these two formulas can not deal with low quality fingerprint images and small overlapping region. Recently, Feng[5] tried to solve the problem of minutiae-based matching as a two class pattern recognition problem and used support vector machine (SVM) to give the classification result. From Bayesian decision theory we can get the least error if we know the probability distribution of the features. In this paper we adopt penalized logistic regression (PLR) [9] to give the matching result, which has the advantage of providing the estimation of the underlying probability. To our knowledge it is the first work that tries to solve the minutiae-based matching as a probability estimation problem.

The rest of this paper is organized as follows. In section 2, fingerprint representation is described. In section 3 similarity measure method between two minutiae is discussed. In section 4 the penalized logistic regression model is introduced. In section 5 experimental results are reported and the conclusion is drawn in section 6.

2 Fingerprint Representation

For an input gray-scale fingerprint image, the method proposed by Hong et al. [6] is used to enhance the image. Then the thinned image is obtained and postprocessed by the knowledge-based enhancement method [7]. The original image and the thinned image are displayed in Fig.1(a)-(b).

In this paper we combine minutiae with all sampling points on the ridges to represent a fingerprint. The minutiae points are detected from the thinned image by Hong's method [6]. For poor fingerprint images, some spurious minutiae may exist. Thus, it is necessary to deal with the spurious minutiae. A minutia that satisfies one of the following two conditions will be regarded as a spurious minutia and discarded: (1) the ridge (s) associated with the minutia is (are) shorter than a threshold; (2) the number of minutiae circled the minutia within a small radius is larger than a threshold. The valid minutiae set is represented as $\{(x_i, y_i, \theta_i, t_i)\}_{i=1}^{n}$, where i is the index of the minutia, (x_i, y_i) is its coordinate, θ_i is the direction in the range $(-\pi, \pi]$, and t_i is the minutia type (termination or bifurcation).

Each ridge in the thinned image is traced. The ridges associated with a bifurcation are treated as three different ridges. To make the tracing process efficient, we use the following tracing order: (1) ridges associated with a bifurcation; (2) ridges associated with a termination; (3) ridges without associated minutiae. If the length of a ridge is shorter than a threshold (shorter than 12 pixels in our experiment) then the ridge is regarded as an invalid one and discarded. Each valid ridge is sampled at a constant interval. Then all of the ridges in a fingerprint are represented as $SP = \{\{x_{ij}, y_{ij}, \theta_{ij}\}_{j=1}^{nr_i}\}_{i=1}^{N_r}$, where x_{ij}, y_{ij} and θ_{ij} denote the x, y coordinates and tangential direction in the range $(-\pi/2, \pi/2]$ of the *j*th point on the *i*th ridge respectively, N_r denotes the number of valid ridges in the thinned image, nr_i denotes the number of sampling points of the *i*th ridge. An example of ridge tracing and sampling is illustrated in Fig.1(c)-(d).

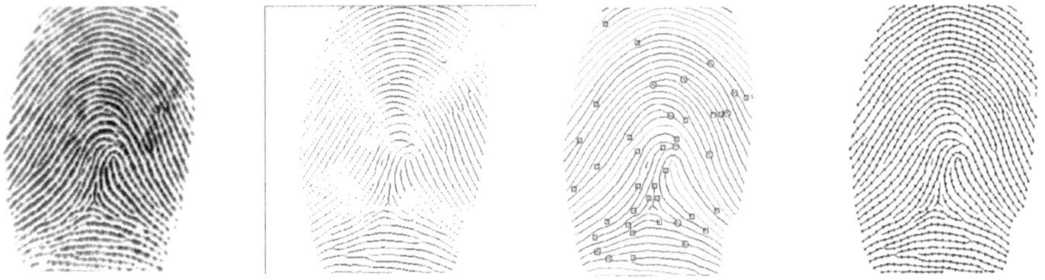

(a) (b) (c) (d)

Fig. 1. (a) Intensity image (500 dpi); (b) thinned image; (c) minutiae and an example of ridge tracing (red curves denote ridges associated with a bifurcation, blue curves denote ridges associated with a termination, green curves denote ridges without associated minutiae and black curves denote invalid ridges); (d) thinned image and sampling points

3 Similarity Measure between Two Minutiae

In this section, we propose an approach to obtain the foreground of the fingerprint from the sampling points and a novel similarity measure between two minutiae using neighboring minutiae and neighboring sampling points.

3.1 Foreground of Fingerprint

Tico and Kuosmanen [1] proposed an effective matching score calculation approach by only considering the minutiae that fall inside the common region of both fingerprints. The similarity calculation between two minutiae based descriptors defined by Feng[5] was also confined to the common region of two fingerprints. The common region plays an important role in fingerprint matching and it can be regarded as the intersection of two fingerprint foregrounds after registration. However, the foreground is hard to estimate by the minutiae set. In this paper, we use Graham's algorithm [8] to determining the convex hull of the sampling set which can be viewed as the boundary of the fingerprint. The points inside the convex hull are regarded as the foreground of the fingerprint. Fig. 2 illustrates the process.

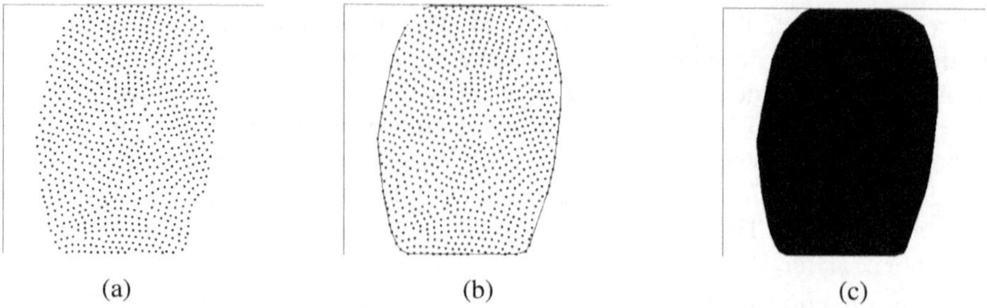

Fig. 2. (a) Sampling points; (b) convex hull; (c) foreground of fingerprint

3.2 Similarity Measure Based on the Neighboring Minutiae

Suppose p is a minutia in the template fingerprint and q is a minutia in the input fingerprint, the similarity calculation process can be divided into two stages: (1) minutia p and its neighbors are mapped to the coordinate system of q, (2) minutia q and its neighbors are mapped to the coordinate system of p.

In the stage 1, let $N(p,r) = \{p_i\}_{i=1}^{n_p}$ denote the set of the neighboring minutiae circled p within r radius, $N(q, r+\Delta r) = \{q_i\}_{i=1}^{n_q}$ denote the set of the neighboring minutiae circled q within $r + \Delta r$ radius and T represent the corresponding rigid transformation from p to q. Each minutia p_i in $N(p,r)$ is mapped to p_i' using T. The distance difference between p_i' and q_j is calculated as follows,

$$dLen_1(p_i', q_j) = \sqrt{(x'_{p_i} - x_{q_j})^2 + (y'_{p_i} - y_{q_j})^2}. \quad (1)$$

The orientation difference between p_i' and q_j is calculated as

$$d\phi_1(p_i', q_j) = \begin{cases} 2\pi - \left|\theta_{p_i'} - \theta_{q_j}\right| & \text{if } \left|\theta_{p_i'} - \theta_{q_j}\right| > \pi, \\ \left|\theta_{p_i'} - \theta_{q_j}\right| & \text{otherwise.} \end{cases} \quad (2)$$

Then the contribution of p_i with respect to the minutia p is computed as

$$C_{p_i} = \max_{q_j} f(dLen_1(p_i', q_j), disThr1, disThr2) \cdot f(d\phi_1(p_i', q_j), oriThr1, oriThr2), \quad (3)$$

where $disThr1$ and $disThr2$ are two distance threshold, $oriThr1$ and $oriThr2$ are two orientation distance threshold, and function f is defined as

$$f(x, th1, th2) = \begin{cases} 1 & \text{if } x \le th1, \\ 0 & \text{if } x > th2, \\ \frac{x - th1}{th2 - th1} & \text{otherwise.} \end{cases} \quad (4)$$

In the stage 2, we define another two neighboring structures: $N(q,r)$ and $N(p, r+\Delta r)$, the meaning of which is similar as stage 1. We use the same symbol T to represent the relative rigid transformation from q to p. Each minutia q_j is mapped to q_j' using T. Similar to the stage 1, the contribution of q_j to the minutia q is computed as follows

$$C_{q_j} = \max_{p_i} f(dLen_(p_i, q_j'), disThr1, disThr2) \cdot f(d\phi_1(p_i, q_j'), oriThr1, oriThr2). \quad (5)$$

The similarity based on neighboring minutiae between p and q is measured using the following formula

$$MS_{pq} = \frac{1+\sum_{p_i \in N(p,r)} C_{p_i}}{M_p+bias} \cdot \frac{1+\sum_{q_j \in N(q,r)} C_{q_j}}{M_q+bias}, \quad (6)$$

where M_p is the number of the minutiae in $N(p, r)$ whose transformed version are in the foreground of input fingerprint, M_q is the number of the minutiae in $N(q, r)$ whose transformed version are in the foreground of the template fingerprint and $bias$ is a parameter.

3.3 Similarity Measure Based on the Neighboring Sampling Points

In this section, we propose a similarity measure method which can capture both ridge orientation and ridge frequency difference. Similar to the section 3.2, there are two stages in this similarity measure process for minutiae pair p and q from template and input fingerprint, respectively.

In the stage 1, let $SN(p, sr) = \{s_i\}_{i=1}^{sn_p}$ denote the sn_p sampling points circled p within rs radius, $SN(q, sr + \Delta sr) = \{r_j\}_{j=1}^{sn_q}$ denote the sn_q sampling points circled q within $sr + \Delta sr$ radius and T denote the relative transformation. Each sampling point s_i in $SN(p, sr)$ is mapped to s_i' using T. The Euclidean distance is not appropriate to this condition because position of the sampling points will be slight different due to the different initial tracing point. In this work, we adopt digital distance map [16] to measure the distance between the sampling point pair, formally

$$dLen_2(s_i', r_j) = \left|(s_s' - r_j)^T \eta_{s_i'}\right|, \quad (7)$$

where $\eta_{s_i'}$ is the normal to the direction of sampling point s_i'. Here we use s_i' and r_j to represent their coordinates for convenience. This formula gives a "point to line" distance, which is appropriate to reflect the ridge frequency. Fig.3 illustrates this case. The orientation distance between s_i' and r_j is computed using the following formula:

$$d\phi_2(\theta_1, \theta_2) = \begin{cases} \pi - |\theta_1 - \theta_2| & \text{if } |\theta_1 - \theta_2| > \frac{\pi}{2}, \\ |\theta_1 - \theta_2| & \text{otherwise.} \end{cases} \quad (8)$$

Then the contribution of s_i with respect to the minutia p is computed as

$$C_{s_i} = f(dLen_2(s_i', r_{j^*}), sdThr1, sdThr2) \cdot f(d\phi_2(s_i', r_{j^*}), soThr1, soThr2), \quad (9)$$

where

$$j^* = \arg\min_{j, dLen_1(s_i', r_j)) < Thr} dLen_2(s_i', r_j) \quad (10)$$

In stage 2 the calculation process is similar to the stage 1. Then the similarity based on neighboring sampling points between p and q can calculated as the following

$$SS_{pq} = \frac{\sum_{s_i \in SN(p,sr)} C_{s_i}}{s_p+1} \cdot \frac{\sum_{r_j \in SN(q,sr)} C_{r_j}}{s_q+1}, \tag{11}$$

where S_p is the number of the sampling points in $SN(p, sr)$ whose transformed version are in the foreground of input fingerprint, S_q is the number of the sampling points in $SN(q, sr)$ whose transformed version are in the foreground of the template fingerprint.

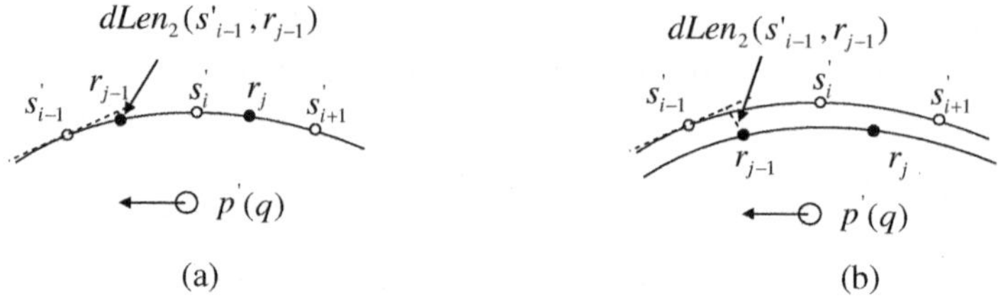

Fig. 3. (a) Frequency around p and q is same, then $dLen_2(s'_{i-1}, r_{j-1})$ is close to 0; (b) frequency around p and q is different then $dLen_2(s'_{i-1}, r_{j-1})$ is larger than that in (a)

3.4 Combined Similarity Measure

The similarity measure based on the neighboring minutiae capture local minutiae pattern while the similarity based on the neighboring sampling point capture the local information of the orientation and the location distribution. Therefore, the two similarity functions can be combined to measure the similarity between the minutiae pair. Weighted mean rule and product rule are two widely used fusion rules, shown as follows:

$$S_{pq} = \eta \cdot SS_{pq} + (1 - \eta) \cdot MS_{pq}, \tag{12}$$

$$S_{pq} = SS_{pq} \cdot MS_{pq}, \tag{13}$$

where η is a weighting coefficient. Experiment in section 5.1 will give the performances of the two formulas.

4 Matching Score Computation

Let $\{p_i\}_{i=1}^{N_p}$ and $\{q_j\}_{j=1}^{N_q}$ denote two minutiae sets from template and input fingerprint, respectively, $\{s_{ij}\}_{i=1,j=1}^{N_p,N_q}$ denote the similarities between minutiae pair. The minutiae matching algorithm proposed in Ref.[5] is adopted to establish the correspondences. It contains the following steps: (1) normalize the similarities and sort them in decreasing order; (2) one of the top N_α minutiae pairs is chosen as reference pairs, using which two minutiae sets are aligned; (3) Nine features in section 4.1 are extracted and PLR is used to compute the score;(4) this process repeat N_α times and the largest score is used as the matching score. In the following, we describe the feature extraction and PLR.

4.1 Feature Representation

Nine features are selected to represent the matching status:

(1) The number of matched minutiae pairs (n);
(2) The ratio of matched minutiae ($mRatio$) in the common region, which is defined as:

$$mRatio = (2n)/(IN_T + IN_I),$$

where IN_T and IN_I denote the number of minutiae from the template and input fingerprint located inside the common region, respectively.
(3) The mean similarity ($simi$) of the matched minutiae pair.
(4) The mean difference of the length ($lenDiff$) between two minutiae pairs.
(5) The mean difference of the minutiae direction between two minutiae pairs
(6) The average number of sampling points ($aveSNum$) in the common region, which is defined as

$$aveSNum = (S_T + S_I)/2.$$

where the S_T and S_I are the number of sampling points from the template and input fingerprint, which are located in the common region, respectively. This feature describes the size of the common region.
(7) The ratio of the difference $SNumDiff$ of the sampling points, which describes the frequency difference in the common region and is defined as

$$SNumDiff = |S_T - S_I| / (S_T + S_I).$$

(8) The corresponding sampling points are considered to be matched if their associated minutiae are matched. If the number of the sampling points associated with one minutia is difference from the other of the pair. We choose the smaller number as the number of the matched sampling points. Suppose there are *si* matched sampling points for minutiae pair *i*. Then the total number of the matched sampling points ($spNum$) is calculated as follows:

$$spNum = \sum_{i=1}^{n} s_i.$$

(9) The mean difference between the minutiae and the sampling points ($MSLenDiff$) is adopted to represent the matching status of the texture. Suppose that q_i and p_i are the ith minutiae pair, sq_j and sp_j are the last matched sampling points for the jth minutiae pair q_j and p_j, then the $MSLenDiff$ can be calculated as follows:

$$MSLenDiff = \frac{1}{n} \sum_{i=1}^{n} |dLen_1(q_i, sq_j) - dLen_2(p_i, sp_j)|.$$

4.2 Penalized Logistic Regression

The PLR can be divided into two processes: training and validation. We assume there are N samples in the training set. Each sample can be in one of two classes: class 0 and class 1. Class 0 means the two fingerprints are from two different fingers while class 1 means the two fingerprints are from the same finger. Let

$x_i^T = (x_{i_0}, x_{i_1}, x_{i_2}, \cdots, x_{i_d})$ be the ith feature vector with $d+1$ features, where x_{i_0} is the constant term with the value 1, y be the class label of the feature vector. The task of the training is to estimate parameters β in the probability model $p(x;\beta)$, where $p(x;\beta) = P(y=1|X=x;\beta)$ is the conditional probability of a feature vector belonging to class 1 given $X = x$. Since only two classes are considered, the conditional probability of x representing class 0 is consequently $1 - p(x;\beta)$. The normal logistic regression model for a two class problem is

$$f(x;\beta) = \log \frac{p(x;\beta)}{1-p(x;\beta^T)} = \beta^T x, \tag{14}$$

where $\beta = [\beta_0, \beta_1, \cdots, \beta_d]^T$ are the $d+1$ parameters that need to be estimated and $f(x;\beta)$ is formally, Then the penalized binomial log-likelihood for N observation is

$$L(\beta) = \sum_{i=1}^{N} y_i \log p(x_i;\beta) + \sum_{i=1}^{N} (1-y_i) \log(1-p(x_i;\beta))) - \frac{\lambda}{2} \sum_{j=1}^{d} \beta_j^2. \tag{15}$$

Since the sizes of the samples from two classes is greatly different, we modify the objective function as follows to reduce the impact of the imbalance

$$L'(\beta) = C_1 \sum_{i=1}^{N} y_i \log p(x_i;\beta) + C_2 \sum_{i=1}^{N} (1-y_i) \log(1-p(x_i;\beta)) - \frac{\lambda}{2} \sum_{j=1}^{d} \beta_j^2 \tag{16}$$

where $C_1 = 1/\sum_{i=1}^{n} y_i$ and $C_2 = 1/\sum_{i=1}^{n}(1-y_i)$ are two weighting parameters. The Newton-Raphson algorithm [10] is used to find the optimal value of β which minimizes the objective function (16). The parameter λ is selected by cross-validation.

5 Experimental Results

Experiments are performed over the fingerprint databases provided by FVC2002. Our experiments check the discriminating ability of the similarity measure between minutiae pairs and evaluate the final matching performance.

5.1 Validation of Similarity Measure between Minutiae Pairs

In this experiment two kinds of minutiae pairs are selected to evaluate the performance of the similarity measure: corresponding minutiae pairs and non corresponding minutiae pairs. The minutiae pairs matched in the genuine matches are called corresponding minutiae pairs otherwise are called non-corresponding minutiae pairs. This experiment is conducted over FVC2002 DB1_B. Totally 6992 corresponding minutiae pairs are obtained by minutiae matching algorithm proposed by He et al [2]. The same number of non-corresponding minutiae pairs are randomly selected from the imposter matches. Four kinds of similarity degrees are calculated for all minutiae pairs. For each similarity degree false match rate (FMR) and false non-match rate (FNMR) at a fixed threshold t are recorded as FMR(t) and FNMR(t). By varying the threshold t from 0 to 1, a Receiver Operating Characteristics (ROC) curve is obtained. Fig.4 shows that the performance of two combined similarity measures is similar and better than other two. In the following experiment we use the product rule as the similarity measure for convenience.

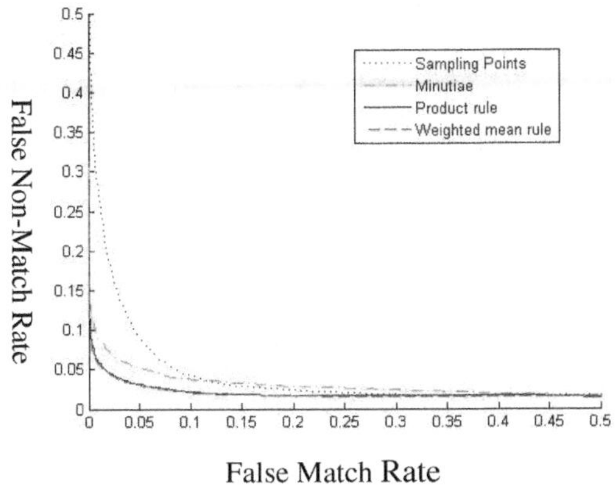

Fig. 4. The ROC curves of various similarity measures including neighboring minutiae based, neighboring sampling points based, combined by product rule and combined by weighted mean rule

5.2 Performance on FVC2002

The experiments reported in this section have been conducted on the four databases of FVC 2002. Database B is used to select penalized λ and train the parameter β while database A is used to validate the proposed algorithm. For a genuine match or imposter match from database B, totally N_α attempts are performed, and the feature vector corresponding to the attempt that maximizes $mRatio \cdot simi$ is selected as a training feature vector. We compare our result with that of algorithm PB15 and PB05 in Table 1. Both of PB15 and PB05 have participated in the competition of FVC2002, and got the 4th and 5th place ranked by the average equal error rate (AveEER), respectively. According to the ranking rule our algorithm is in the 5th place.

Table 1. Equal error rate on four databased of PB15, PB05 and ours

	DB1 (%)	DB2 (%)	DB3 (%)	DB4 (%)	Average (%)
PB15	0.63	1.03	0.81	0.61	0.77
PB05	0.52	0.69	1.48	0.98	0.92
ours	0.34	0.51	1.70	0.88	0.86

6 Conclusion

In this paper we represent the fingerprint by minutiae and ridge sampling points. This representation holds the uniqueness of the fingerprint and is compact and effective. In the light of this representation we propose a novel similarity measure method to capture neighboring minutiae pattern, ridge orientation and ridge frequency. After the alignment and minutiae pairing nine features are extracted to represent the matching status of the input and query fingerprint, and penalized logistic regression is adopted to estimate the probability of the two fingerprints coming from a same finger. Experimental results on FVC2002 show that the proposed algorithm achieves good performance and ranks 5th place according to the average equal error rate.

References

[1] Tico, M., Kuosmanen, P.: Fingerprint Matching Using an Orientation-Based Minutia Descriptor. IEEE Trans. Pattern Anal. Mach. Intell. 25(8), 1009–1014 (2003)
[2] He, Y., Tian, J., Luo, X., Zhang, T.: Image Enhancement and Minutiae Matching in Fingerprint Verification. Pattern Recognition Letters 24(9), 1349–1360 (2003)
[3] He, Y., Tian, J., Li, L., Chen, H., Yang, X.: Fingerprint Matching Based on Global Comprehensive Similarity. IEEE Trans. Pattern Anal. Mach. Intell. 28(6), 850–862 (2006)
[4] Bazen, A.M., Gerez, S.H.: Fingerprint Matching by Thin-Plate Spline Modelling of Elastic Deformations. Pattern Recognition 36(8), 1859–1867 (2003)
[5] Feng, J.: Combining Minutiae Descriptors for Fingerprint Matching. Pattern Recognition 41(1), 342–352 (2008)
[6] Lin, H., Wan, Y., Jain, A.K.: Fingerprint Image Enhancement: Algorithm and Performance Evaluation. IEEE Trans. Pattern Anal. Mach. Intell. 20(8), 777–789 (1998)
[7] Luo, X., Tian, J.: Knowledge Based Fingerprint Image Enhancement. In: Proc. 15th ICPR, pp. 4783–4786 (2000)
[8] Graham, R.L.: An Efficient Algorithm for Determining The Convex Hull of a Finite Planar Set. Information Processing Letters 26, 132–133 (1972)
[9] Shen, L., Tan, E.T.: Dimension Reduction-Based Penalized Logistic Regression for Cancer Classification Using Microarray Data. IEEE/ACM Trans. On Computational Biology and Bioinformatics 2(2), 166–175 (2005)
[10] Hastie, T., Tibshirani, R., Friedman, J.: The Elements of Statistical Learning. Springer, Heidelberg (2001)
[11] The 2nd fingerprint verification competition, `http://bias.csr.unibo.it/fvc2002/`
[12] Wilson, C.L., Watson, C.I., Paek, E.G.: Effect of Resolution and Image Quality on Combined Optical and Neural Network Fingerprint Matching. Pattern Recognition 33(2), 317–331 (2000)
[13] Feng, J., Ouyang, Z., Cai, A.: Fingerprint Matching Using Ridges. Pattern Recognition 39(11), 2131–2140 (2006)
[14] Gu, J., Zhou, J., Yang, C.: Fingerprint Recognition by Combining Global Structure and Local Cues. IEEE Trans. On Image Processing 15(7), 1952–1964 (2006)
[15] Jain, A.K., Hong, L., Bolle, R.: On-Line Fingerprint Verification. IEEE Trans.Pattern Anal. Mach. Intell. 19(4), 302–314 (1997)
[16] Borgefors, G.: Distance Transformations in Digital Images. Computer Vision, Graphics, and Image Processing 34(3), 344–371 (1986)

A Novel Region Based Liveness Detection Approach for Fingerprint Scanners

Brian DeCann, Bozhao Tan, and Stephanie Schuckers

Clarkson University, Potsdam, NY 13699 USA
{decannbm,tanb,sschucke}@clarkson.edu

Abstract. Biometric scanners have become widely popular in providing security to information technology and entry to otherwise sensitive locations. However, these systems have been proven to be vulnerable to spoofing, or granting entry to an imposter using fake fingers. While matching algorithms are highly successful in identifying the unique fingerprint biometric of an individual, they lack the ability to determine if the source of the image is coming from a living individual, or a fake finger, comprised of PlayDoh, silicon, gelatin or other material. Detection of liveness patterns is one method in which physiological traits are identified in order to ensure that the image received by the scanner is coming from a living source. In this paper, a new algorithm for detection of perspiration is proposed. The method quantifies perspiration via region labeling methods, a simple computer vision technique. This method is capable of extracting observable trends in live and spoof images, generally relating to the differences found in the number and size of identifiable regions per contour along a ridge or valley segment. This approach was tested on a optical fingerprint scanner, Identix DFR2100. The dataset includes a total of 1526 live and 1588 spoof fingerprints, arising from over 150 unique individuals with multiple visits. Performance was evaluated through a neural network classifier, and the results are compared to previous studies using intensity based ridge and valley liveness detection. The results yield excellent classification, achieving overall classification rates greater than 95.5%. Implementation of this liveness detection method can greatly improve the security of fingerprint scanners.

Keywords: Biometrics, fingerprint, liveness detection, neural network.

1 Introduction

Biometric systems are becoming popularized for their ability to grant security to restricted individuals, without them needing to carry small objects such as keys, or memorize long and abstract passwords[1]. However, fingerprint scanners have been shown to be vulnerable to attack at the sensor level, granting access to well made spoof fingers, or even dismembered live fingers [2]. Liveness detection is a measure that determines whether or not the source of the image presented to a biometric scanner is from a living individual. Its purpose is to provide extra security to the system by working cooperatively with matching algorithms that recognize an enrolled user.

M. Tistarelli and M.S. Nixon (Eds.): ICB 2009, LNCS 5558, pp. 627–636, 2009.
© Springer-Verlag Berlin Heidelberg 2009

Fig. 1. Region Labeling of a Fingerprint Image. Left) Binary image with foreground (white) and background (black) pixels. Right) Labeled image. Note each region is assigned a different color.

Software-based liveness detection has been an active research area in recent five years, which uses the information already captured by traditional fingerprint systems to detect life signs, for example, skin deformation or elasticity, perspiration pattern or characteristics of spoof and live images. Skin deformation [3,4,5,6] technique uses the information regarding how the fingertip's skin deforms when pressed against the scanner surface. Perspiration pattern detection method [13,14,7] quantifies the perspiration phenomenon inherent only to live fingers due to the existence of pores and perspiration using two time series images or a single image. Also researchers investigate the characteristics of spoof and live fingers, like power spectrum [12] or band-selected Fourier Spectrum [9], skin coarseness [11], valley noise [15], gray-level profile and frequency patterns [8] or integration of multiple static features [10].

The aim of this study is to build upon the previous research and propose a new liveness detection method, region labeling technique, which provides different characteristics of live and spoof features which may improve upon classification. In general, region labeling is a simple computer vision technique, which identifies areas of foreground from the background. Refer to the following figure for an example. Here, region labeling can be applied to quantify observable differences between live and spoof images. Regarding the capture of a single image, there exists more identifiable regions along both ridge and valley segments than contained in spoof images. This would be expected because ridge segments in live images tend to have consistent intervals of high and low intensity, corresponding to pore locations, a trait not common in spoof images. In the valleys, spoof images are often prone to pressure deformation, causing ridge segments to bleed into the valley. Live images remain relatively clean, though due to the effects of noise, several small areas of intensity should be present. Difference images, a type of image where two captures are subtracted from one another can also expose traits which can be exploited by region labeling. In this format, the element of control an imposter has over a fingerprint scanner is exposed. Here, the slight variance of pressure and rotation distorts ridge segments which would otherwise remain intact, resulting in more identifiable regions in spoof images than live images. Additionally, tracking region size can also be taken into account to verify this claim. This method is advantageous in that it looks at the perspiration pattern in a different sense, as well as isolating traits of spoof images. Another advantage is simplicity where the required computations are presently available and well known.

2 Region Labeling Algorithm

The most naturally evident form of vitality in a fingerprint is its perspiration. Since perspiration can only occur from sweat glands of living tissue, the ability to quantify it makes for an excellent liveness classification method. Fig. 2 illustrates the differences in time between a living fingerprint, and one that has been spoofed.

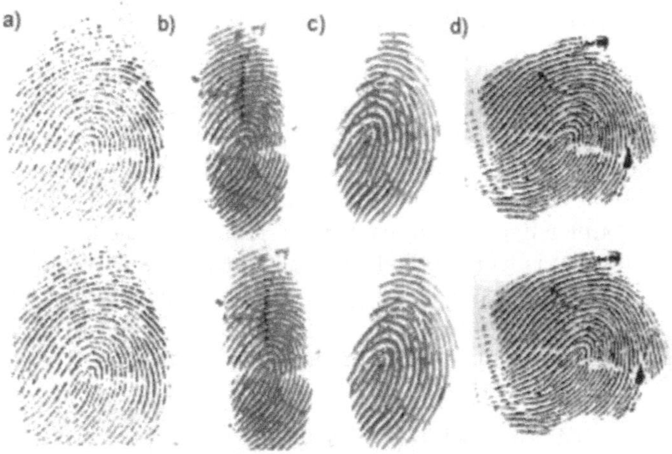

Fig. 2. Fingerprint images for a particular subject. The top row represents time zero, and the bottom 2 seconds following. a) Live, b) Silicone, c) Gelatin, d) PlayDoh.

From Fig. 2, it is seen that the live image darkens, as well as thickens along the ridges with time. The spoof images however, remain relatively consistent, and are noticeably darker in intensity. The valleys of the live image as well, are relatively clean, and free of pixels, while this does not appear to be the case for the spoof images. Furthermore, the lengths of the ridge segments along the live signal are smaller, and spotty in appearance. This differs from spoof images in the sense that the segments for those images are rather long and consistent. From a statistical standpoint, the motives of the previous work are clear, as through observation one can visibly see evident intensity based differences, and expect differences in the frequency domain.

While these images alone make it difficult to visually witness the perspiration effect, the subtraction of two images, can illustrate this much more efficiently, as expressed in Fig.3.

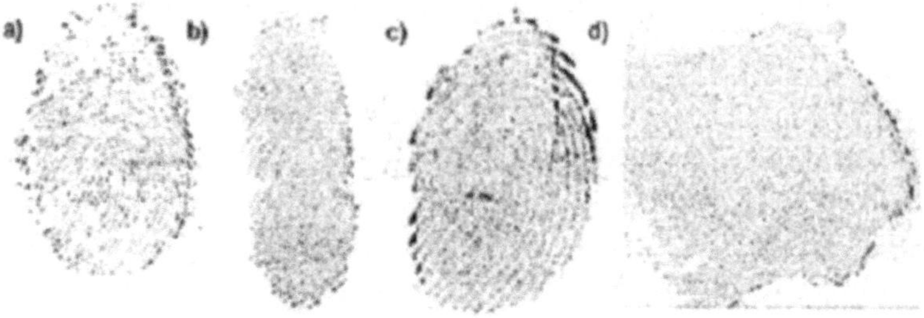

Fig. 3. Absolute difference between images at t=2 and t=0. a) Live. b) Silicone. c) Gelatin. d) PlayDoh (Darkness indicates significant change).

2.1 Image Capture and Preprocessing

Images are captured sequentially at an interval of one second. Difference images are then created from the subtraction of the final image from the initial. Preprocessing of captured images begins with a cropping of a rectangular region whose center is the center of the scanner. This reduces the computational effort to a degree by assuming that the user will not place a piece of their finger along the corners of the scanner itself in an attempt to gain access. The first step of preprocessing the images is to adaptively account for irregularities in regard to the scanner itself. This is done through use of a program that compares the recorded images to a blank capture, and uses a 2% threshold in which to decide if a pixel should reflect the intensity presented in the image, otherwise it is discarded.

2.2 Mask Generation and 1-D Signal Extraction

Veridicom SDK is used to convert the gray image to binary image. Once the image is binarized, a simple thinning algorithm can thin the binarized image down to a width of 1 pixel. The remaining binary image then has all of its Y intersections removed, so that a signal has a well defined start and endpoint. Following a shift procedure, the resulting contours are aligned with either the center of the ridges or valleys, depending on operation. A length check removes contours that are too small for computation, on the basis that their pixel distance is shorter than the length between pores found on the skin.

2.3 Region Identification

At this point, the filtered initial and subsequent images, along with a contour mask, representing pixel locations of the ridges or valleys are prepared for use. In order to account for the entirety of intensity values that may lie within a ridge or valley, the contour mask is dilated to a width of three pixels. Fig. 4 illustrates superposition of the mask along both ridge and valley signals, before and after dilation.

The next step is to identify the size and number of regions present along the dilated contours. Here, the methodology diverges into application for both single and double image capture. For multiple image capture, a difference image is created based on pixel intensity change between the initial and secondary capture. Three types of difference images can be observed. Pixel locations whose change is negative reflects an area that has decreased in intensity, or became darker. This would arise from true perspiration, increasing pressure, or movement. Additionally, an image can be created to reflect positive change in pixel values. This sort of intensity change naturally arises from motion. Finally, an image reflecting the absolute difference illustrates the combination of any sort of pixel change.

Each individual contour along the difference image is initially checked for a length threshold. Assuming the thresholding criterion is met, the labeling operation can then be performed on an isolated contour. As a requirement for

Fig. 4. Live Region Identification. Note the variance of region size. Left) Negative Difference Image with Contour of Interest Colored (Ridge Signal). Right) Labeled Contour of Interest.

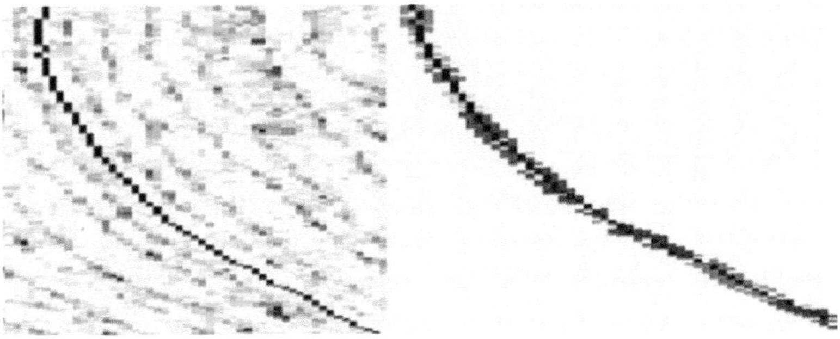

Fig. 5. Spoof Region Identification. Note the consistency of region size. Left) Negative Difference Image with Contour of Interest Colored (Ridge Signal). Right) Labeled Contour of Interest.

region labeling algorithms, the image itself must first be converted into its respective foreground and background. In this case, the conversion is trivial since the only pixels of interest are those that have changed along the contour being examined. The labeling algorithm itself is very simple. It operates by isolating an individual pixel, and any 8 connected neighborhood pixels are assigned a number corresponding to its region identifier. The result of this labeling operation, along with a newly labeled image, is the total number of regions found within the contour itself. A visual of the labeling process is process is provided in Fig. 4 and 5.

Fig. 4 and 5 help to illustrate a specific property in classification of live and spoof images. The live image for example, shows about 6 regions of differing size. Also evident are the gaps between each region. These gaps are indicative of pore locations. The spoof image in Fig. 5 however, has many more regions, with the majority smaller in size than represented in Fig. 4. The lack of gaps, as evident in Fig. 4, illustrates how pressure deformation removes pore locations, and instead replaces them with regions with a size of a few pixels.

For single image capture, the only difference is the steps involving the process to convert the image into a binary format. In this case, the pixels of interest are those that are deemed to be within a specific tolerance range. In the algorithm, foreground pixels are determined by being within 90% of the maximum intensity value along the contour of interest. This rationality for the threshold is to prevent pixels that have weak intensity values to contribute, and possibly join adjacent regions.

2.4 Feature Extraction

Following the labeling procedure for each specific image type, statistical properties can be extracted. Features include the total number of regions found (per image), regions found per contour, average region size (in pixels) and the standard deviation of region size. Use of the time series difference images (absolute, negative, positive) and the first capture images yield a total of 16 total measures. Recalling that the measures can be repeated for both ridge and valley analysis results in potentially 32 region based measures available for classification.

Fig. 6 illustrates a few examples of features extracted between live and spoof fingerprints. As we can see from Fig. 6-a, spoof difference images on average contain more regions per contour along ridge segments, and fewer along the valleys, when compared to a live difference image, supporting the original claim. As previously mentioned, this occurs as a result of the lack of control an imposter has over a fingerprint scanner when using a spoof. Minor inconsistencies in the ridge segment are reflected as pixel change and will subsequently result in a higher region count. As there is little apparent change in the valley of a live image (Fig. 2-a), the effects of noise result in several small regions, which are typically absorbed in a spoof image due to unintended pressure deformation (Fig 3-b,c,d). Fig. 6-b aims to measure the consistency of region size in ridge segments. As can be inferred, region size in spoof images is much less random than live images. This result continues to support the image presented in Fig. 5,

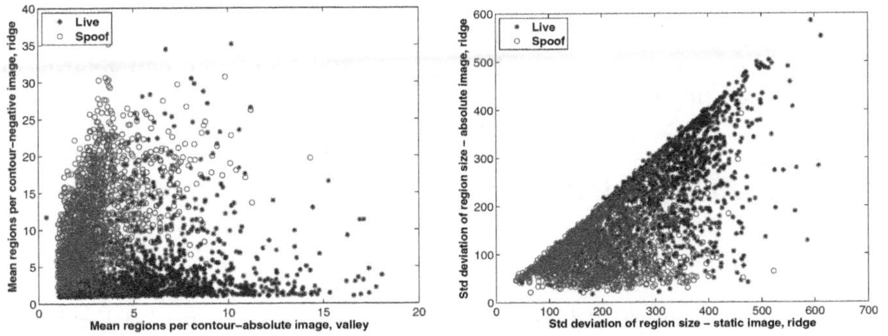

(a) Mean regions per contour, valley absolute image vs ridge negative image

(b) Std deviation of region size, ridge static image vs ridge absolute image

Fig. 6. Feature comparison between live and spoof fingerprints

where the illustrated contour contains many small regions with little variation. In a single image, the result verifies the randomness property of region size in a live image because pixel intensity is much more variant (Fig. 2-a).

2.5 Classification

After region labeling features are extracted, standard pattern classification tools (neural network, nearest neighbour, classification tree, etc.) can be used to separate non-live subjects from the live subjects. The non-live category includes the spoof data made from Play-Doh, gelatin and silicone fingers. The software we used for classification is from WEKA (Waikato Environment for Knowledge Analysis) software tool [16] that provides different classification techniques for large data sets. 10-fold cross-validation test was used to test the new method.

3 Data Collection

Identix DFR 2100 was used to collect all fingerprint data for this study. Time series data of at 0s and 1s are primarily used. The International Review Board (IRB) at A University approved all protocols for data collection. Fingerprints that are collected consist of the Right thumb and Right index fingers. There are 160 live subjects involved in the data collection, each subject with average of 4.6 visits with an average of one month between visits.

In order to create the Spoof fingers, the material which is used to create a spoof finger is placed and conformed to a fingerprint cast created from dental impression materials, which contains the fingerprint of an individual. Spoof images are created from common household items such as PlayDoh, Gelatin (Knox brand), and commercial grade Silicone (Dow Corning Sylgard 184 Silicone Elastomer Kit). Spoof variability is introduced into the spoof library through time.

In the process of collecting data, several images are created of the same spoof material over approximately ten minutes as the spoof materials dry.

In total, we have collected 1526 live images, 800 Play-Doh, 649 gelatin and 139 silicone spoof fingerprints.

4 Experiments and Results

4.1 Performance Evaluation

To evaluate our system, we utilize terms for evaluating biometric identification or verification systems. Liveness classification is the process of determining whether the input finger is live or spoof.

Because of the threshold selection or decision logic, liveness classification makes two type of errors, i.e., *spoof false acceptance* (SFA), where a spoof finger is accepted as live and *live false rejection* (LFR), where a live finger is rejected. Two other terms are also often used, *true acceptance*, where a live finger is recognized as live and *true rejection*, where a spoof finger is recognized as spoof. Normalized SFA and LFR are used, which are called spoof false acceptance ratio (SFAR) and live false rejection ratio (LFRR), respectively. They can be calculated as

$$SFAR = \frac{SFA}{NS} \tag{1}$$

$$LFRR = \frac{LFR}{NL} \tag{2}$$

where SFA and LFR is the number of SFA and LFR samples, respectively; and NS and NL are the total number of spoof finger samples and live finger samples, respectively. Equal error rate (EER) is when SFAR equals LFRR. To evaluate the performance of our system, we use Receiver Operator Characteristic (ROC) curve, which plots the change of true positive rate (1-LFRR) with SFAR for varying thresholds.

4.2 Results and Discussion

10-fold cross-validation was used to evaluate the performance of the proposed algorithm. To make a comparison, we also implemented the liveness detection algorithms described in [14,15]. Ridge signal algorithm [13] quantifies the perspiration pattern along the signal along ridges, while valley noise algorithm [15] characterize the difference between live and spoof fingerprints along valleys. Fig. 7 shows the ROC curve. As seen from this figure, the new region label technique can reach an EER of 4.5%, while ridge signal algorithm can reach 5.1% and valley noise algorithm can reach 3.2%, respectively. Also, the fusion of three approaches is also implemented at feature level, the outcome can reach an EER of as low as 1.2%. The proposed method explores the perspiration pattern in more detail, investigating the positive, negative and absolute difference images

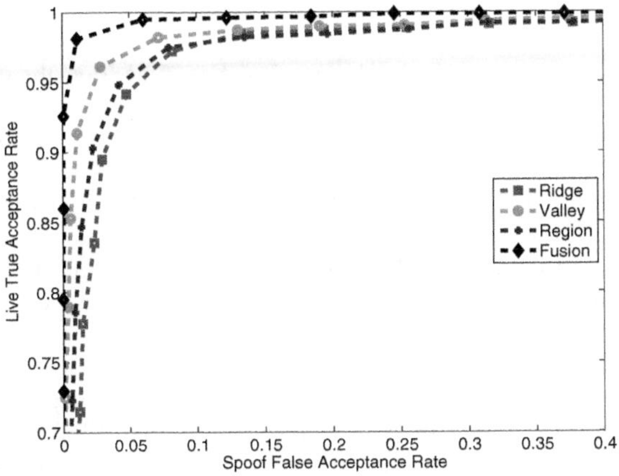

Fig. 7. ROC comparison between ridge signal algorithm, valley noise algorithm and new region label algorithm

between 0 and 1 second images, in addition to the first single image. The addition of the new region label technique can be used to fuse with ridge signal or valley noise algorithms to improve the overall liveness classification.

Furthermore, another standout feature of these measures is that they do not require great lengths of computation. The majority of the calculations are simple logical relations (thresholding), basic mathematical operations (creating difference images), and use of basic image processing algorithms.

Future work includes testing on other types of scanners, collecting more live data with environmental changes, like temperature or humidity. Also it would be necessary to investigate more spoofing techniques, like trying different spoof materials or techniques, which can cause a spoof finger to break the algorithm. These information will be used to tune the proposed algorithm to a more robust liveness classification agent.

5 Conclusion

This study demonstrated a new method to quantify liveness detection through region labeling of the first capture and three difference images of first and second capture (absolute, positive and negative). This approach differs from previous methods in the sense that it aims to interpret liveness based on its visual characteristics that characterize live and spoof images. The test database for this study consisted of over 1,500 live and spoof images for Identix fingerprint scanner. The size of the database helps to ensure there exists sufficient testing variability, to encompass situations that would arise in everyday use.

References

1. Jain, A., Bolle, R., Pankanti, S.: Biometrics: Personal Identification in Networked Society. Springer, Heidelberg (1999)
2. Schuckers, S.A.C.: Spoofing and anti-spoofing measures. Information Security Technical Report 7(4), 56–62 (2002)
3. Maltoni, D., Maio, D., Jain, A., Prabhakar, S.: Handbook of Fingerprint Recognition. Springer, Nu (2003)
4. Antonelli, A., Cappelli, R., Maio, D., Maltoni, D.: Fake Finger Detection by Skin Distortion Analysis. IEEE Transactions on Information Forensics and Security 1(3), 360–373 (2006)
5. Chen, Y., Jain, A., Dass, S.: Fingerprint Deformation for Spoof Detection. In: Proc. of Biometrics Symposium (BSYM 2005), Arlington, VA, September 19-21 (2005)
6. Zhang, Y., Tian, J., Chen, X., Yang, X., Shi, P.: Fake Finger Detection Baed on Thin-late Spline Distortion Model. In: Lee, S.-W., Li, S.Z. (eds.) ICB 2007. LNCS, vol. 4642, pp. 742–749. Springer, Heidelberg (2007)
7. Jia, J., Cai, L.: Fake Finger Detection Based on Time-series Fingerprint Image Analysis. In: Huang, D.-S., Heutte, L., Loog, M. (eds.) ICIC 2007. LNCS, vol. 4681, pp. 1140–1150. Springer, Heidelberg (2007)
8. Uchida, K.: Image-Based Appraoch to Fingerprint Acceptability Assessment. In: Zhang, D., Jain, A.K. (eds.) ICBA 2004. LNCS, vol. 3072, pp. 294–300. Springer, Heidelberg (2004)
9. Jin, C., Kim, H., Elliott, S.: Liveness Detection of Fingerprint Based on Band-Selective Fourier Spectrum. In: Nam, K.-H., Rhee, G. (eds.) ICISC 2007. LNCS, vol. 4817, pp. 168–179. Springer, Heidelberg (2007)
10. Choi, H., Kang, R., CHoi, K., Kim, J.: Aliveness Detection of Fingerprints using Multiple Static Features. In: Proc. of World Academy of Science, Engineering and Technology, July 2007, vol. 22 (2007)
11. Moon, Y.S., Chen, J.S., Chan, K.C., So, K., Woo, K.C.: Wavelet based fingerprint liveness detection. Electronic Letters 41(20), 1112–1113 (2005)
12. Coli, P., Marcialis, G.L., Roli, F.: Power spectrum-based fingerprint vitality detection. In: IEEE Workshop on AutoID (June 2007)
13. Derakhshani, R., Schuckers, S.A.C., Hornak, L., O'Gorman, L.: Determination of vitality from a noninvasive biomedical measurement for use in fingerprint scanners. Pattern Recognition 36(2) (2003)
14. Tan, B., Schuckers, S.: Liveness Detection for Fingerprint Scanners Based on the Statistics of Wavelet Signal Processing. In: IEEE CVPRW (June 2006)
15. Tan, B., Schuckers, S.: A New Approach for Liveness Detection in Fingerprint Scanners Based on Valley Noise Analysis. SPIE Journal of Electronic Imaging 17(1) (2008)
16. Witten, I.H., Frank, E.: Data Mining: Practical machine learning tools and techniques, 2nd edn. Morgan Kaufmann, San Francisco (2005)

Focal Point Detection Based on Half Concentric Lens Model for Singular Point Extraction in Fingerprint

Natthawat Boonchaiseree and Vutipong Areekul

Kasetsart Signal & Image Processing Laboratory (KSIP Lab),
Department of Electrical Engineering, Kasetsart University, Bangkok, 10900, Thailand
g4965124@ku.ac.th, vutipong.a@ku.ac.th

Abstract. A focal point is a kind of singular points, closely related to a core point, which can be derived from curvature of fingerprint ridges and valleys. It is expected that the focal point is more reliable than the core point in case of low quality fingerprint. This paper proposes a new efficient focal point localization method based on a half concentric lens model. The half concentric lens window, with directional adaptive, accelerates convergence of a focal point localization process rapidly. Moreover, concentric lens similarity factor is introduced in order to measure orientation and stability of an extracted focal point. From experimental results, the proposed scheme is out-performed most of singular point detection schemes in literature in term of location accuracy and consistency. For computational complexity, algorithm requires average 75 millisecond execution-time from original fingerprint to a unique focal point.

Keywords: Singular Point, Focal Point, Half Concentric Lens, Concentric Lens's Similarity, Fingerprint Registration.

1 Introduction

Large scale fingerprint identification requires extremely fast and efficient algorithm in order to satisfy practical demands such as national identification. To achieve this goal, algorithm should capture strong and distinguished global features. Consider pivotal features existing in general fingerprint, singular points are some of these global features which are easily noticed by human eyes. In general, singular point can be classified into two types; core point and delta point. A focal point [1], closely related to a core point, can be included as one of a kind of singular points.

In practical fingerprint identification, singular points are not easily detected. There exist many problems caused by finger conditions and live-scan fingerprint sensors. For example, finger conditions can be too dry or too wet resulting in fingerprints with low quality. It becomes worse if dust, sweat, and old prints are included. Moreover, different fingerprint sensor types can cause different types of noises and quality. Somehow, size of sensor is matter due to where people press their fingers, resulting in partial fingerprint with missing singular point. On the other hand, arch type of fingerprint has no core and most algorithms have run into the difficulties of localizing its core. Finally, non-linear distortion and elastic distortion is occurred between finger

M. Tistarelli and M.S. Nixon (Eds.): ICB 2009, LNCS 5558, pp. 637–646, 2009.
© Springer-Verlag Berlin Heidelberg 2009

and sensors with various pressures. These problems drastically reduce accuracy of most singular point detection schemes reported in literature.

Existing works on locating singular points can be classified into three broad approaches; pattern-based, partitioning-based, and projection-based methods. The pattern-based method employs fixed patterns such as core and delta models. Then these patterns are correlated with orientation fields or ridges of fingerprint in order to find the locations of similar patterns. For example, Kawagoe and Tojo proposed the classic and popular method, called Poincare′index in 1984 [2]. This method searches for fixed patterns of core, delta, and whorl in fingerprint. A successful improvement of this classic method from Bazen and Gerez can be found in [3]. Park et.al. [4] analyzed directional field pattern and proposed filtering rules to detect singular point including arch type. Another sub-classes of pattern-based approach is called complex symmetrical filters, introduced by Nilsson and Bigun [5]. These filter shapes are similar to core and delta. Singular point detection is obtained by convolving fingerprint orientation fields with these filter models. Hence computational complexity depends on sizes of filters and fingerprint image.

The second approach, called partitioning based method, utilize boundary of quantized directional fields which leads to singular point. For example, Hung and Huang [6], introduced fault-line, a borderline which separate two adjacent regions of the same quantized directional fields. These fault-lines converge to singularities. Another example, by Ramo et al. [7], used transition lines of orientation fields and detected reference points by inspection of line intersections.

The third approach is referred to as projection-based method. The idea is to project fingerprint information, such as orientation fields, into some mathematic models or patterns, and analyze these models and patterns according to a priori knowledge, resulting in singular point position. For example, Jiang et al. introduced algorithm based on hierarchical analysis of the orientation coherence [8]. Liu et al. proposed a new way to project orientation fields into another field called local axial symmetry fields, then these fields is used to find a reference point location [9]. By projection-based approach, the focal point was first introduced by Rerkrai and Areekul [10]. The focal point is a centroid of crossings, the intersection of two normal lines or gradients of curved ridges. The original focal point algorithm was demonstrated experimentally to be quite stable and a proof of concept, but it was of a very high computational complexity. Later Areekul et al. [11] introduced the second algorithm. Instead of calculating all enormous crossings to find a reliable centroid, this algorithm employs mean-shift approach by allowing iterative search along highly curved ridges. This approach reduces amount of computational time into practical implementation. However, this algorithm still needs fingerprint enhancement in order to effectively track local ridges. Recently Areekul and Boonchaiseree [12] introduced the third algorithm by analyzing a focal point from orientation fields instead of true ridges in order to reduce computational complexity without losing accuracy. In addition, this previous algorithm introduced a fixed half circle window which is suitably designed for orientation field selection, resulted in algorithm having a good performance.

In this paper, the concept of a half concentric lens or a half circle window from [12] is extended with directional adaptation. Moreover, concentric lens similarity factor is introduced in order to control algorithm's convergence.

2 Concentric Lens Similarity Factor

This proposed algorithm allows a half concentric lens to be rotated in order to find suitable directional field's curves in each iteration step. The half concentric lens or a geometric half circle window, with radius of R (pixels) and a center point at (x_c, y_c) (pixels), contains some number of partitioning blocks as shown in Fig. 1(a). Note that all blocks with partial parts inside this half circle window are included.

In order to measure pattern of directional fields in a half circle window or a half concentric lens, we define perfect directional fields in form of a concentric pattern as shown in Fig 1(d). Then we define concentric lens similarity factor (CLS), α, as following equation.

$$\alpha = \frac{N}{M}\left(1 - \sum_{(i,j)\in CL}^{N} \frac{\left|\theta_{fh}(i,j) - \theta_{cl}(i,j)\right|}{N\pi/2}\right) \tag{1}$$

where $\theta_{fh}(i, j)$ and $\theta_{cl}(i, j)$ are actual DFs in half circle window and ideal DFs of perfect concentric lens pattern, respectively. Each DF of block i^{th} column and j^{th} row should be member of concentric lens, CL, or in area of half circle window. N is a number of reliable DF (discussed in 3.1) and M is total number of DF in half circle window. Note that CLS factor ranges from zero to one where one the most similarity and zero the least similarity. Fig 1 (b,c,d) shows CLS factor with various DF patterns.

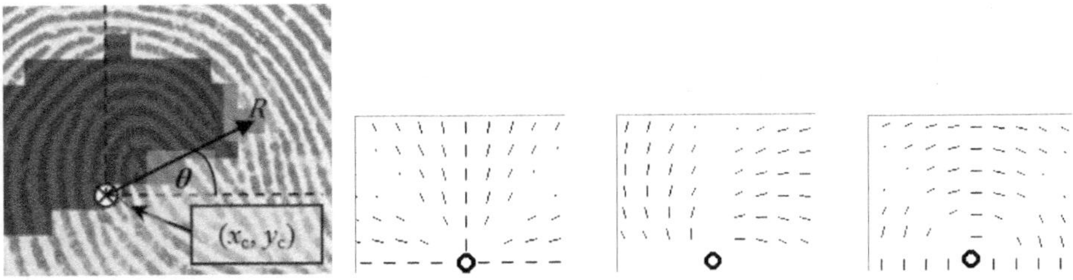

(a) Half Concentric Lens (b) α = 0 (c) α = 0.5 (d) α = 1

Fig. 1. (a) Half Concentric Lens or Geometric Half Circle Window. (b), (c) and (d) Concentric Lens Similarity Factor (CLS) with various DF patterns.

3 Focal Point Extraction Algorithm

The proposed focal point detection algorithm consists of the following steps: preprocessing, initial point localization, intermediate centroid iterative, and focal point selection. These steps are explained as follows.

3.1 Preprocessing

The input fingerprint image is partitioned into $N \times N$ blocks (N=16 in this experiment), and its directional field of each block is carefully calculated by using Hong's method [13]. For higher precision of directional field estimation at pixel resolution, Bazen and

Gerez's approach [3] is more suitable. The $\varphi_f(i, j)$ is a gradient, perpendicular to its directional field $\theta_f(i, j)$ in the same block with position (i, j). Then $Coh(i, j)$, coherence of directional field, is calculated using method in [3]. The $Coh(i, j)$ is employed as a strength of directional field in each block, ranging from zero to one (0, 1). This $Coh(i,j)$ value is used to indicate reliability of each directional field. In this experiment, DF with $Coh(i, j) > 0.3$ consider reliable and use in localization process. The rests are ignored. Note that DF smoothing method is not applied in our scheme in order to preserve and protect original DFs as possible.

3.2 Initial Point Localization

Let $x_B(i, j)$ and $y_B(i, j)$ be a center of block coordinate at column i^{th} and rows j^{th}, respectively. $\varphi_f(i, j)$ represents a gradient orientation or an orientation with perpendicular to its directional field. A linear equation through the center of block coordinate is

$$y - y_B = m_B(x - x_B); \quad \text{where } m_B = \tan(\varphi_f(i, j)) \tag{2}$$

where m_B is a slope of gradient orientation. Our goal is to find all crossings of these gradients. However, some parallel DFs or low curvature may spread gradient crossings far apart resulting in focal point's divergence. Hence a gradient orientation constraint between two DFs is defined as

$$\Delta\varphi = \left|\varphi_{f1} - \varphi_{f2}\right| > \varphi_{threshold}, \tag{3}$$

where $\Delta\varphi$ is absolute gradient difference between gradient φ_{f1} and gradient φ_{f2}. In order to obtain a crossing point from two gradients, the above constraint must be satisfied. In experiment, if $\Delta\varphi$ is less than $\varphi_{threshold} = 30$ degree ($\pi/6$), then a gradient crossing point could be calculated. Assume that (x_P, y_P) is a coordinate of the gradient crossing and it can be obtained by

$$\left(x_p = \frac{m_{B2}x_{B2} - m_{B1}x_{B1} + y_{B1} - y_{B2}}{m_{B2} - m_{B1}},\ y_p = \frac{m_{B2}(y_{B1} - m_{B1}x_{B1}) + m_{B1}(y_{B2} - m_{B2}x_{B2})}{m_{B2} - m_{B1}} \right) \tag{4}$$

To control quality of a gradient crossing point contributed to a focal point, another constraint, related to distance between crossing (x_P, y_P) and center of block (x_{B1}, y_{B1}) and (x_{B2}, y_{B2}), is defined as

$$\Delta d = \left|d_{B1C} - d_{B2C}\right| < \Delta d_{threshold} \tag{5}$$

where d_{B1C} and d_{B2C} are distance between crossing (x_P, y_P) and a center of block (x_{B1}, y_{B1}) and (x_{B2}, y_{B2}), respectively. This constraint helps filtering some overrated crossings, and $\Delta d_{threshold}$ is equal to 128 pixels (8 blocks) in this experiment.

Once all gradients crossings are calculated, the number of crossings in $N \times N$ blocks (16×16 in our experiment) is accumulated to create 2-D histogram of crossings as shown in Fig 2 (c). Then a center of the maximum block is a candidate for an initial point. Note that a number of initial points can be more than one in order to capture all possible local maximum points.

(a) Original (b) Gradient Crossings (c) Crossing Histogram

Fig. 2. Demonstrate Detection of Initial Point Localization in Crossing Histogram

3.3 Intermediate Centroid Iterative

This iterative process is a key of focal point localization and it composes of three major steps. Details of each step are discussed as follows.

Step 1: Find Orientation with Maximum CLS
Given initial point or an initial block, a half concentric lens window is located with radius R blocks (6 blocks in this experiment). This window is rotated by 22.5 degree step size, with maximum range ± 45 degree, resulting in only 5 orientations as shown in Fig 3. Then concentric lens similarity factor is calculated for each step size. If the initial point is outside image with non-overlapped area between concentric lens window and image, iteration will stop.

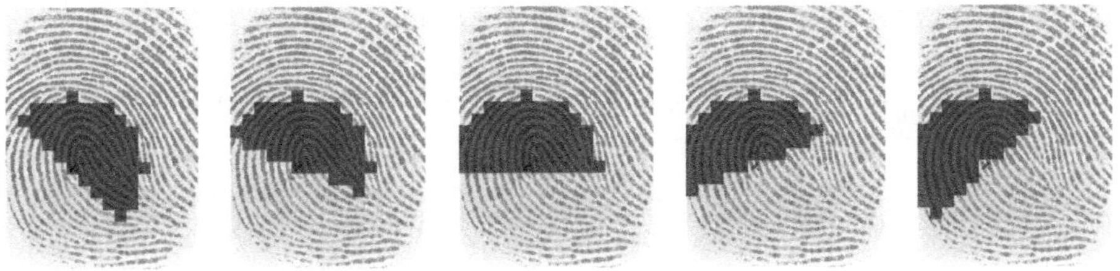

(a) $\alpha = 0.6597$ (b) $\alpha = 0.7626$ (c) $\alpha = 0.8185$ (d) $\alpha = 0.8278$ (e) $\alpha = 0.8069$

Fig. 3. Half Concentric Lens in various Orientations $-\pi/4$, $-\pi/8$, 0, $\pi/8$, and $-\pi/4$, respectively

Step 2: Find Intermediate Centroid
In this step, all gradient crossings by equations (2) to (5), generated from concentric lens window, are calculated. Then an current intermediate centroid, $(x_{C(k)}, y_{C(k)})$, at k^{th} iterative, can be derived by

$$(x_{C(k)}, y_{C(k)}) = \left(\frac{1}{N(k)} \sum_{i=1}^{N(k)} x_{p(i,k)}, \frac{1}{N(k)} \sum_{i=1}^{N(k)} y_{p(i,k)} \right) \tag{6}$$

where $N(k)$ is the total number of gradient crossings (x_P, y_P), created in this k^{th} iteration.

Step 3: Check Convergence
Focal point converges when consecutive intermediate centroids do not move in iterative process. The ε distance error and focal point convergence condition can be written by

$$\varepsilon = \sqrt{\left(x_{C(k)} - x_{C(k-1)}\right)^2 + \left(y_{C(k)} - y_{C(k-1)}\right)^2} < \varepsilon_{Threshold} \tag{7}$$

To be precise, if distance error between a current intermediate centroid $(x_{C(k)}, y_{C(k)})$ and a previous intermediate centroid $(x_{C(k-1)}, y_{C(k-1)})$ is less than $\varepsilon_{Threshold}$, this iterative process will be terminated. Otherwise, the current intermediate centroid is a new initial point and iterative process continues (go back to Step 1 again). Note that iterative process has limited number of iterations to 10 rounds in order to protect limit cycle situation. The $\varepsilon_{Threshold}$ is set to zero (pixel) in this experiment.

3.4 Focal Point Selection

In general, if convergence condition is satisfied, the focal point is selected from the last intermediate centroid of the last iteration. Moreover, step 1 is employed again to find maximum CLS factor α which represented the best orientation for the detected focal point. Overall flowchart of the new focal point localization algorithm is concluded in Fig 4.

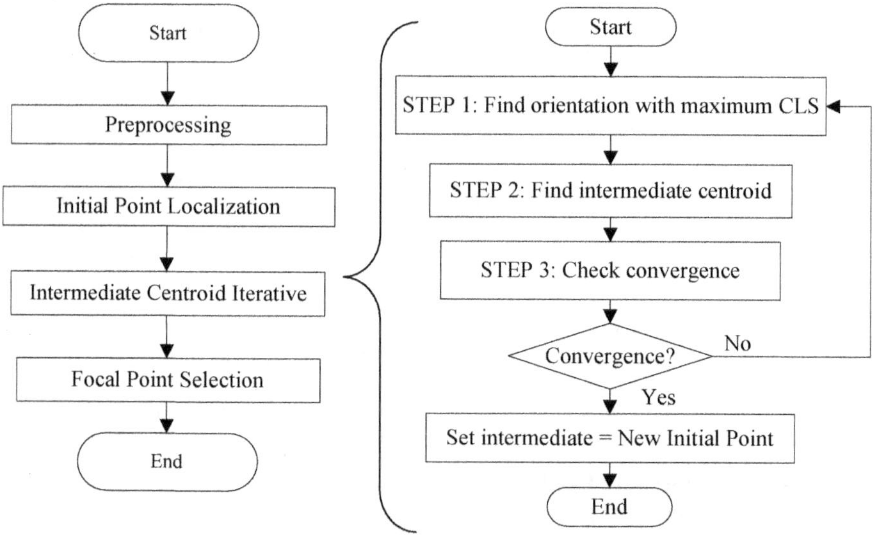

Fig. 4. Flowchart of the Proposed Focal Point Localization Algorithm

4 Focal Point's Error Measurement and Evaluation

Evaluation and error measurement of a focal point must be carefully designed and tested because the focal point cannot be visually detected by human eyes then it cannot be manually marked. Two approaches are proposed for distance error measurement. Then orientation error measurement is also concluded.

4.1 Distance Error Measurement

In general, most singular point detection algorithms compare location of detected singular points with manually marked cores and deltas. Unfortunately, a pinpoint location of a focal point cannot be manually located because of invisible location. However, an expected area can be estimated. Therefore, in [12], the method "A" is introduced by linearly projecting an obtained focal point into another fingerprint image with 3 manually reference points. The second method "B" manually projects all obtained focal points into one fingerprint image. Both methods are discussed in detail as follows.

a) Evaluation Method A: Approximately Linear Projection Method [12]
In this evaluation method, three reference points are manually marked around singularity area of each finger, 8 fingerprint images, of all 800 images in FVC2000 Db2a database. Then linear projection can be applied to relocate an obtained focal point from one image to another image. A location of the focal point is projected onto two sides of a triangle, formed by 3 manually marked points [12]. By keeping the same projection ratio of triangle sides, the projected focal point can be relocated from one image to another image, and distance error can be measured. In Facts, there are three possible projection ways by selecting two out of three sides of triangle combinations. All possible ways are used and error distance is averaged. Detail is discussed in [12].

This method suffered from two major causes; linear projection error and manually 3-points marked error. Unfortunately, linear projection is not good estimation because of nonlinear distortion of fingerprint. Moreover, distance error depends on three marked points. If a focal point is inside a marked triangle, distance error is reasonable. However, if the focal point is outside the marked triangle, error is not reasonable. This method is very useful in case of computerized evaluation for optimizing parameters. However, experimental results may not be reliable to compare with other's results.

b) Evaluation Method B: Manually Marked Projection Method
To solve elastic problems in Method A, for each finger with 8 fingerprint images, other seven focal points are manually marked into one selected fingerprint image. Then distance error can be calculated by averaging distance error between the reference focal point to other 7 focal points. All focal points are rotated to be the reference focal point, resulted in distance error for each reference focal point. Finally, distance error from 8 reference points is averaged, resulted in average distance error for this finger. Clearly this method should result less error than Method A because nonlinear projection error is excluded. Only manually marked error may exist.

4.2 Orientation Error Measurement

Orientation error, *OE*, can be measured by using Liu's method [8] is defined as

$$OE = \min\{|\theta_P - \theta_R|, \pi - |\theta_P - \theta_R|\}, \qquad -\frac{\pi}{2} < \theta_P, \theta_R < \frac{\pi}{2} \tag{8}$$

where θ_P is manually measured orientation from 3 marked points, and θ_R is extracted orientation, referred to the focal point. (The total result of *OE* will be quantized by 5 orientations with range ± 45 degree for evaluation in this experiment.).

5 Experimental Results

Experimental results tested on FVC2000Db2A (100 fingers, 8 fingerprints /finger, capacitive sensor size 364×256 pixels, 500 dpi) which is also tested in [4,5,6,7,10,11]. Accuracy performance comparisons of various algorithms, referred to distance error and orientation error, are shown in Table 1 and Table 2, respectively. Computational complexity in term of execution time can be compared among limited algorithms using Intel Centrino Duo, CPU 1.6GHz, and 2MBs of RAM, as shown in Table 3.

Table 1. Position Accuracy Comparison of Singular Point Detection Algorithms in each Literature on FVC2000Db2A Database

Algorithms	Distance Error, *DE* (pixels)					Accuracy (%)	
	[0, 10]	(10, 20]	(20, 40]	> 40	Fail	≤ 10 pixel	≤ 20 pixel
Proposed Alg./Eval.B	668	101	26	5	-	83.50	96.12
M. Liu et.al. [8]	659	90	25	13	13	82.37	93.62
T. Liu et.al. [9]	654	91	← 46 →		9	81.75	93.12
Proposed Alg./Eval.A	624	143	20	13	-	78.00	95.87
Areekul [12]/Eval.B	610	124	44	21	1	76.25	91.75

Algorithm	Distance Error, *DE* (pixels)		Accuracy (%)
	[0, 15)	> 15	< 15 pixels
Proposed Alg./Eval.B	751	49	93.875
K. Nilsson et. al. [5]	749	51	93.625
Proposed Alg./Eval.A	730	70	91.250

Algorithm	Magnitude Standard Deviation (pixels)	
	All FVC2000 Db2a Database	Arch-type Only (144 Archs in *)
C.H. Park et.al. [4]	6.385	10.690
*Proposed Alg./Eval.B	7.301 (Means = 6.481)	10.667 (Means = 11. 885)
*Proposed Alg./Eval.A	9.327 (Means = 10.532)	14.614 (Means = 14.412)

Table 2. Orientation Accuracy Comparison of Singular Point Detection Algorithms on FVC2000Db2A Database

Algorithm	Orientation Error, *OE* (degree)				Accuracy (%)
	[0,11.25]	(11.25,22.5]	> 22.5	Fail	< 22.5 pixels
T. Liu. et.al. [9]	690	47	50	13	93.647
Proposed Alg.	521	136	143	-	82.125

Table 3. Compare Computational Time among Focal Point Localization Algorithms, FVC2000Db2A Database, with Intel Centrino Duo 1.6GHz and 2MBs of RAM

Process	Average Execution Time (millisecond)		
	Algorithm in [11]	Algorithm in [12]	Proposed Algorithm
Segmentation	26.98	-	-
DF Extraction	13.10	13.10	13.10
Enhancement	535.72	-	-
Focal Point Localization	30.10	29.52	61.94
Total	**605.90**	**42.62**	**75.04**

In Table 1, our proposed scheme with measurement method B yields the best experimental results in singular point localization accuracy. However, Table 2 shows problem with orientation accuracy of detected reference point. The main reason is that orientation is quantized into 5 steps with $\pi/4$ apart. Hence accuracy may not be good but it can be improved in the near future. Finally, in Table 3, the proposed algorithm is compared with two previous algorithms in term of execution time.

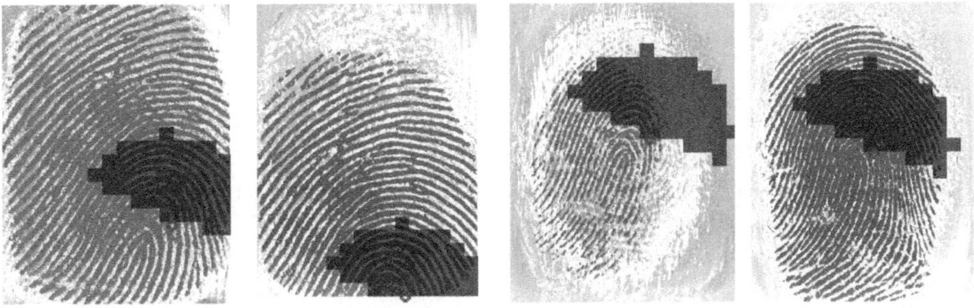

(a) #38_1 (*DE* = 9.53 pixels) (b) #38_4 (*DE* = 10.72 pixels) (c) #81_7 (*DE* = 2.07 pixels) (d) #81_8 (*DE* = 2.06 pixels)

Fig. 5. Good Results from Proposed Algorithm on FVC2000Db2A (DE by Method B Eval)

(a) #22_1 (*DE* = 60.64 pixels) (b) #69_4 (*DE* = 43.55 pixels) (c) #100_1 (*DE* = 90.93 pixels)

Fig. 6. Failure from the Proposed Algorithm on FVC2000Db2A (DE by Method B Eval)

Fig 5 shows good results from the proposed algorithm. Fig 6 reported failure from the proposed algorithm. For example, without most-north curve area in Fig 6(a), without right-side curve area in Fig 6(b), and low quality core in Fig 6 (c).

6 Conclusion and Future Research

In this paper, projection-base method (Focal Point Projection Concept) and template-base method (Concentric Lens Model) are combined for a new proposed algorithm. The proposed algorithm provides fast convergence with high accuracy and good consistency, comparing with other singular point detection methods in literature. In the near future, applications of focal point could be proposed for fingerprint classification. The proposed algorithm can be further improved by multi-resolution approach. In addition, orientation accuracy must be improved and tested with other FVC databases such as thermal slide sensor or FVC2004 with high elastic distortion.

Acknowledgments. This work was partially supported by Department of Electrical Engineering, Kasetsart University, the Thailand Research Fund and the Commission on Higher Education through Grant No. RMU4980027, and JON KEE's Scholarship.

References

1. Maltoni, D., Maio, D., Jain, A.K., Prabhakar, S.: Handbook of Fingerprint Recognition. Springer, Heidelberg (2003)
2. Kawagoe, M., Tojo, A.: Fingerprint Pattern Classification. Pattern Recognition 17, 295–303 (1984)
3. Bazen, A.M., Gerez, S.H.: Systematic Methods for the Computation of the Directional Fields and Singular Points of Fingerprints. IEEE Trans. Pattern Anal. Machine Intell. 24, 905–919 (2002)
4. Park, C.H., Lee, J.J., Smith, M.T.J., Park, K.H.: Singular Point Detection by Shape Analysis of Directional Fields in Fingerprint. Pattern Recognition 39, 839–855 (2005)
5. Nilsson, K., Bigun, J.: Registration of Fingerprints by Complex Filtering and by 1D Projections of Orientation Images. In: Kanade, T., Jain, A., Ratha, N.K. (eds.) AVBPA 2005. LNCS, vol. 3546, pp. 171–183. Springer, Heidelberg (2005)
6. Hung, D.C.D., Huang, C.: A Model Optimized Singular Point Detection Algorithm for Fingerprint Images. In: Proc. Florida Artificial Intelligence Research Symposium (9th), pp. 444–448 (1996)
7. Rämö, P., Tico, M., Onnia, V., Ṣaarinen, J.: Optimized Singular Point Detection Algorithm for Fingerprint Images. In: Proc. Int'l. Conf. Image Processing, vol. 3, pp. 242–245 (2001)
8. Liu, M., Jiang, X.: Kot. A.C.: Fingerprint Reference-Point Detection. EURASIP Journal on Applied Signal Processing, 498–509 (2005)
9. Liu, T., Zhang, C., Hao, P.: Fingerprint Reference Point Detection Based on Local Axial Symmetry. In: Proc. Int'l. Conf. Pattern Recognition, vol. I, pp. 1050–1053 (2006)
10. Rerkrai, K., Areekul, V.: A New Reference Point for Fingerprint Recognition. In: Proc. Int'l. Conf. Image Processing, vol. 2, pp. 499–502 (2000)
11. Areekul, V., Suppasriwasuseth, K., Jirachaweng, S.: The New Focal Point Localization Algorithm for Fingerprint Registration. In: Proc. Int'l. Conf. Pattern Recognition, vol. IV, pp. 497–500 (2006)
12. Areekul, V., Boonchaiseree, N.: Fast Focal Point Localization Algorithm for Fingerprint Registration. In: Proc. 3rd IEEE ICIEA, pp. 2089–2094 (2008)
13. Hong, L., Wang, Y., Jain, A.K.: Fingerprint Image Enhancement: Algorithm and Performance Evaluation. IEEE Trans. on Pattern Analysis and Machine Intelligence 20(8), 777–789 (1998)

Robust Fingerprint Matching Using Spiral Partitioning Scheme

Zhixin Shi and Venu Govindaraju

Center for Unified Biometrics and Sensors(CUBS)
State University of New York at Buffalo
Buffalo, NY 14228, USA
zshi@cedar.buffalo.edu
http://cubs.buffalo.edu

Abstract. Fingerprint matching for low quality or partial fingerprint images is very challenging. It is mainly because the features such as minutia points can not be extracted reliably. In the case of partial fingerprint images captured using solid state sensors, enough number of minutia points may not be included. In this paper, we introduce a novel fingerprint representation that combines information from each extracted minutia with detected ridges in its neighborhood. The proposed algorithm first enhances a fingerprint image and generates a binary image. Then instead of using thinning-based algorithms, the ridges are extracted using a chaincode scheme, which retains the original thickness of the ridges and precise local orientations. The minutia points are detected by tracing the ridge lines. Finally the enriched local structural features are built for each minutia by a spiral coding using the ridge line orientations around the minutia. The new features are translation and rotational invariant. Each feature vector represents a minutia and its neighboring ridge structures. Matching of two fingerprints is performed by calculating the Euclidean distances between pairs of corresponding feature vectors. Preliminary experiments show that the proposed algorithm is effective.

1 Introduction

A fingerprint is a graphical pattern of ridges and valleys on the surface of a human finger. Due to the uniqueness and permanence of fingerprints, they are among the most reliable human characteristics that can be used for personal identification [1,2]. The uniqueness of a fingerprint is determined by the local ridge characteristics and their relationships [1,3]. Ridge endings and ridge bifurcations, called minutiae, are generally considered to be the two most prominent characteristics. Most Automatic Fingerprint Identification Systems (AFIS) are based on minutiae matching. Each minutia is represented by attributes such as the type of the minutia point (ridge ending or bifurcation), its location and orientation. With detected minutia points, fingerprint matching is then reduced to a point matching problem. There are several minutia-based matching algorithms proposed in the literature. Besides approaches that apply direct point matching,

M. Tistarelli and M.S. Nixon (Eds.): ICB 2009, LNCS 5558, pp. 647–655, 2009.
© Springer-Verlag Berlin Heidelberg 2009

many minutia-based matching algorithms take into account the translational and rotational variations. A coupled approach is often used to match minutiae in clusters by constructing a translation and rotation invariant structure. A local structure composed of two or more minutiae close to each other serves to obtain minutiae correspondences[4,5,6,7,8]. These minutia-based matching approaches use only the minutia points and rely on the interdependencies between minutia details. The matching performance is evaluated by counting the matched minutiae.

Thus, the performance of a fingerprint verification/identification system using minutia-based approaches depends heavily on the performance of the corresponding minutia extraction algorithms. In practice, due to rotation and displacement of the finger placed on the sensor, there is often only a partial overlap between the template and query fingerprint images. This problem becomes more severe in the case of small solid-state sensors that are increasingly popular. Furthermore the fingerprint image quality is also affected by several factors such as the presence of scars, variations of the pressure between the finger and acquisition sensor, worn artifacts, and the environmental conditions during the acquisition process. Poor quality image affects the extraction of minutiae.

Another class of matching approaches are correlation-based using other features besides minutia points such as features from ridge orientations, phase-based features from Fourier transforms or statistic features from local intensity variations[9,10]. These additional features provide more discriminative information. Matching these new features requires fingerprint registration[11]. But such alignments can only be obtained using other key features such as core points[9] or minutiae[11], which may be elusive. Typically, minutiae-based methods perform better than correlation-based ones.

In this paper, we propose a novel fingerprint representation that combines information from each extracted minutia with detected ridges in its neighborhood. The proposed algorithm first enhances a fingerprint image to generate a binary fingerprint image. Then instead of using thinning-based algorithms, the ridges are extracted using a chaincode scheme, which retains the ridges along with their original thickness and precise local orientations. The minutia points are detected by tracing the ridge lines. Finally, the enriched local structural features are built for each minutia by a circular coding using the ridge lines around the minutia. The new features are translation and rotational invariant. Each feature vector represents a minutia and its neighboring ridge structures. Matching of two fingerprints is by calculating the Euclidean distances between pairs of corresponding feature vectors.

The key aspects of our proposed representation are that the enriched minutiae features have stronger discriminative power and they can be used independently in matching fingerprints. The minutia and non-minutiae information extracted from the surrounding ridges are used together in building correspondences between fingerprints. In contrast to the minutia-based approaches[4,5,6,7,8], matching using the proposed representation requires fewer minutiae and therefore more practical with small images or partial images with small overlap areas. The proposed representation for one minutia is independent of that for any other minutia detected in the fingerprint hence it is more tolerant of erroneous outcomes of the minutia detection algorithm such as missing and spurious minutiae. The idea is about using rich local characterization for shape representation in order to reduce the complexity of feature correspondence and matching problems. The translation and rotation invariant properties also make the matching algorithm efficient.

Closely related to our approach is the work of Qi et al.[12] and Tico et al.[13] Both use a sampling scheme to choose sample points in a circular neighborhood around a minutia and they record ridge orientations at the sample points. The feature vector is then constructed by the differences between the orientations at the samples points and the orientation of the center minutia. The sampling scheme employed by Qi et al. chooses sample points along three 120° ray lines starting from a center minutia point. The sampling points in the work of Tico et al. are taken from a group of concentric circles. The distance between two points on the same circle is fixed. Minutiae extraction and ridge orientation field computation in both papers are thinning based and rely on an enhanced binary fingerprint image. Our minutia extraction is based on chaincode contours. The non-minutia features are extracted from the detected ridge lines which provide more reliable information than what can be obtained from the rough orientation field. For extracting ridge features, we apply a spiral tessellation scheme, which divides a neighborhood region around a minutia into sectors. Within each sector, ridge information is collected, which provides information that is more reliable than that collected from a number of random sampling points.

The circular tessellation is used in Jain's work[9] on reference points such as singular points to collect global features. Our spiral tessellation is constructed using an Archimedes' spiral centered at a minutia point. The spiral starts from the minutia's orientation direction. A number of ray lines starting from the center minutia point together with the spiral partition the neighborhood area of the minutia into sectors. The spiral tessellation is the key of our method, which makes the construction of translation and rotation invariant feature vectors possible.

The rest of the paper is organized as follows. Section 2 describes the minutia extraction algorithm used in our experiments. The definition of the proposed hybrid local structure and the matching algorithm are presented in section 3. The experiment is presented in section 4. Finally, section 5 summarizes our work and provides pointers for future work along this direction.

2 Chaincode-Based Image Enhancement and Minutia Extraction

2.1 Fingerprint Image Enhancement by Directional Filters

An ideally sensed or scanned fingerprint image has clear and distinct ridges and valleys. For example, an ideal fingerprint image could be a rolled ink impression on a fingerprint card. The finger skin profile made of the ridges is evenly pressed on the flat paper card, which leaves the ink impression of ridges as continuous flow of foreground passes, and in between the ridges are the valleys as background with white color. But in practice, the fingerprint scanning devices are far from ideal. Even the NIST fingerprint images scanned from the inked fingerprint cards are not perfect. An image quality improvement algorithm is almost always needed. Our earlier work[14] employed a chaincode based approach in computing the ridge orientation field efficiently. The algorithm combines both the binary and the gray-scale image enhancement methods. A locally adaptive algorithm is used to obtain a binary fingerprint image of sufficient quality. The local direction field is estimated using a fast chaincode-base algorithm and a 15×15 mask, followed by applying an elliptically shaped filter[15] with its major axis aligned parallel to the local ridge direction. This increases the connectivity along the ridge direction.

2.2 Minutiae Extraction Using Chaincode

Most fingerprint minutia extraction methods are thinning-based where the skeletonisation process converts each ridge to one pixel wide. Minutia points are detected by locating the end points and bifurcation points on the thinned ridge skeleton based on the number of neighboring pixels.

Our chaincode based method is obtained by scanning the image from top to bottom and right to left. The transitions from white (background) to black (foreground) are detected. The contour is then traced counterclockwise and expressed as an array of contour elements. Each contour element represents a pixel on the contour. It contains fields for the x,y coordinates of the pixel, the slope or direction of the contour into the pixel, and auxiliary information such as curvature.

In a binary fingerprint image, ridge lines are more than one pixel wide. Tracing a ridge line along its boundary in counterclockwise direction, a termination minutia (ridge ending) is detected when the trace makes a significant left turn. Similarly, a bifurcation minutia (a fork) is detected when the trace makes a significant right turn[14].

The chaincode based algorithm extract minutiae and ridge lines much more reliably. The integrity of the ridge lines is preserved with respect to their locations and thickness. Figure 1 shows an example of minutia and ridge line result.

3 Definition of the Proposed Hybrid Local Structure

Regarding a shape as a discrete set of sample contour points, which propose a representation scheme that comprises of a coarse distribution of the shape with respect to each point. Since the contour sample points can be chosen in any number, one can always pick enough points in order to accurately represent the underlying shape. In contrast, the minutia points of a fingerprint capture only a very limited amount of information from the rich information content present in the fingerprint pattern. Consequently, a representation of the fingerprint pattern with respect to each minutia detail must employ non-minutia information like

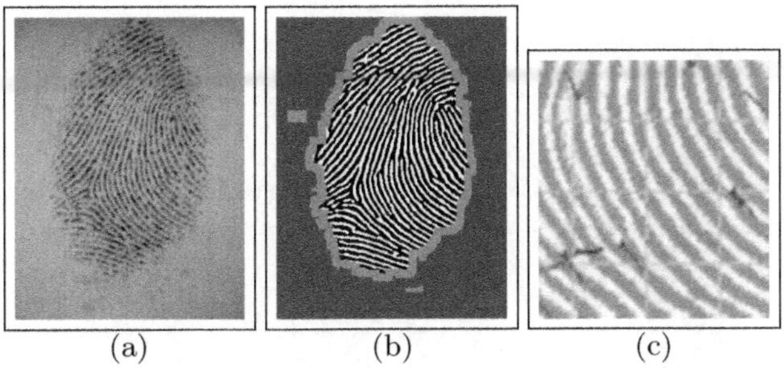

Fig. 1. Sample result showing fingerprint minutia and ridge line extractions

brightness, ridge orientation, transform features, etc. In this work,we investigate the use of orientation information for minutia representation in the context of the fingerprint pattern.

3.1 Compute the Correlation Position for a Minutia

We use an Archimedes' spiral centered at a minutia point, which is started from the minutia's orientation direction. Assume the angle between the orientation direction and the horizontal direction is θ_0. Then the polar equation of the Archimedes' spiral is

$$\rho = a(\theta - \theta_0).$$

A number of ray lines starting from the center minutia point together with the spiral curve partition the neighborhood of the point. We draw N ray lines that evenly divide the range of 0 to 2π, which give us N sectors in each turn of the spiral (Figure 2). Since the first full turn covers the smallest area near the center, we start collecting features from the sectors starting from the second turn. The indices of the feature vector are arranged corresponding to the sectors along the spiral turning counter-clock-wise outward (Figure 3).

In the neighborhood of a center minutia covered by the spiral tessellation, if any other minutia point is included in a sector, the minutia contributes to the feature vector a strong correlation feature. The minutia information such as its location, orientation and type are added to the structure of the corresponding feature entry. Instead of searching every pixel in a sector for possible minutia points, we go with a more efficient approach to incorporating the surrounding minutiae in the feature vector.

For each such minutia point we compute its feature index that indicates which sector the minutia point belongs to. Assume that the current central minutia point is at $P_c = (x, y)$ and its orientation is represented by a vector $v_c = (X, Y)$ (Figure 3). For a minutia point at $P_m = (x_m, y_m)$, we compute a vector from P_c to P_m as $v_m = (X_m, Y_m)$, where $X_m = x_m - x$ and $Y_m = y_m - y$. To determine the angle θ_m between v_c and v_m, we first compute

$$\beta = \arccos\left(\frac{v_c \cdot v_m}{|v_c||v_m|}\right) = \arccos\left(\frac{XX_m + YY_m}{\sqrt{X^2+Y^2}\sqrt{X_m^2+Y_m^2}}\right)$$

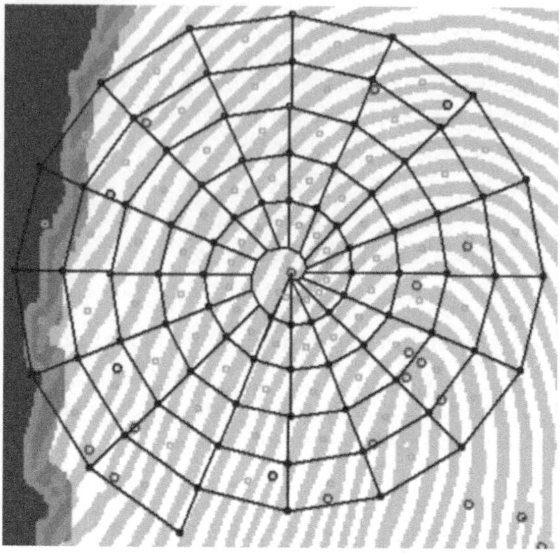

Fig. 2. Spiral tessellation around a minita point

and

$$D = XY_m - YX_m.$$

Then the angle $\theta_m \in [0, 2\pi)$ is

$$\theta_m = \begin{cases} \beta & \text{if } D \geq 0, \\ 2\pi - \beta & \text{if } D < 0. \end{cases}$$

If the distance between P_m and P_c is $d = \sqrt{(x_m - x)^2 + (y_m - y)^2}$, and the successive turnings of the spiral have a constant separation distance of $2a\pi$, then the index of the feature vector corresponding to the sector that P_m belongs to is

$$index = \left\lfloor \frac{d - a\theta_m}{2a\pi} \right\rfloor N + \left\lceil \frac{\theta_m}{\Delta\theta} \right\rceil,$$

where $\Delta\theta = \frac{2\pi}{N}$. If we call the spiral the circular boundary of the sectors, and the sector including P_m is S_m, then the first term is the index of the turn that makes the insider circular boundary of S_m. The second term is the angular index indicating which sector in the turn P_m is in. We use the ceiling instead of floor in the second term because our regular indexing starts from 1. The first feature entry indexed at 0 is from the entire inner area inside the first turn. Therefore around each minutia a feature vector is constructed which includes a list of indexed feature entries extracted from the above sectors. Since the features are calculated with respect to the center minutia, the features are rotation invariant.

3.2 Add Local Orientation Features

An orientation field is calculated for the fingerprint image before the constructions of each feature vector. When a feature vector is constructed around a center

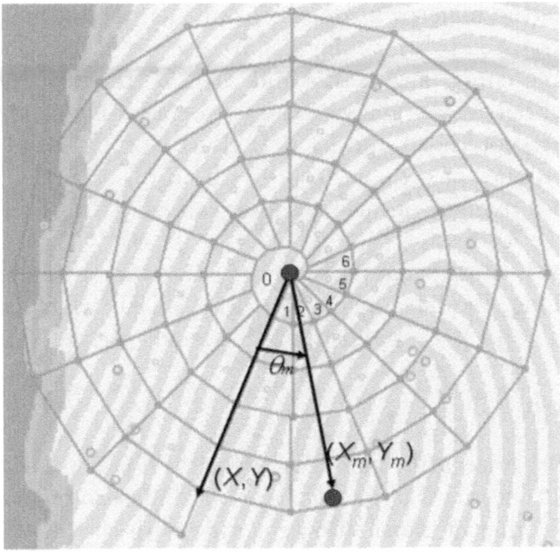

Fig. 3. Feature vector indexing arrangement and relative angle from a center minutia to another minutia point with respect to the direction of the center minutia

minutia, an average orientation direction in each sector of the spiral tessellation is calculated using the orientation field. The difference of the average orientation direction and the direction of the center minutia is the relative orientation of the sector, which is saved as one feature for the sector in the feature vector.

3.3 Matching Two Structure Feature Vectors

A Match of two fingerprints is by finding the best match of feature vectors of one fingerprint with those of the other fingerprint. Because the proposed feature vector construction is based on a spiral tessellation, a match between a pair feature vectors can be done by simply matching the features in the corresponding entries, which eliminates the need of an alignment between the two fingerprints. For each feature entry, the matching score is calculated from the matchings in the corresponding sector. The score is a weighted sum of the following values: (1) the number matched minutiae of same type over the total number of minutiae in the sector, (2) the difference between the orientations, and (3) overlapping sectors for common areas between two fingerprints.

4 Experiment and Results

Our algorithm has been implemented and tested primarily on the DB1 fingerprint databases of FVC2002 [16]. The DB1 database contains 110 different fingers and 8 impressions of each finger yielding a total of 880 fingerprints (388 pixels 374 pixels) at 500 dots per inch. The database has two different sets: A and

B. Set A contains the fingerprint images from the first 100 fingers, while Set B has the images from the other 10 fingers. We use Set B of each database as our training set for matching parameters and then we perform the experiment on the fingerprints of Set A. We followed the protocols of FVC2002 [16] to evaluate the FAR (False Accept Rate) and FRR (False Reject Rate) of our system. For FRR, the total number of genuine tests (with no rejection) is (87)/2100=2800. For FAR, the total number of false acceptance tests (with no rejection) is (100 x 99)/2 = 4950.

The experiment is meant for showing the effectiveness of the new feature for fingerprint matching. We have achieved a 1.94% EER which is better than the result presented in [8] (2.13%) and the NIST result presented in [17] (3.35%).

5 Conclusion

We present a novel fingerprint matching algorithm based on hybrid local structural similarity. The local structural features are built around a detected minutia by combining the minutia information together with the local ridge orientation information. A feature structure is then constructed by sampling the minutiae and local ridge orientations using a spiral partition scheme. The new features are translation and rotational invariant. Each feature vector represents a minutia and its neighboring ridge structures. Matching of two fingerprints is done by calculating the distances between each pair of corresponding feature vectors from the two fingerprints. The final matching score is converted from the scores of the best matching pairs. We presented an experimental evaluation of the proposed approach and showed that the proposed feature structure is effective and it exceeds the performance of the earlier matching algorithms.

References

1. Lee, H.C., Gaensslen, R.E.: Advances in Fingerprint Technology. CRC Press, New York (1991)
2. O'Gorman, L.: Fingerprint verification. In: Jain, A.K., Bolle, R., Pankanti, S. (eds.) Biometrics-Personal Identification in Networked Society, pp. 43–64. Kluwer Academic, The Netherlands (1999)
3. Moenssens, A.: Fingerprint Technology. Chilton Book Company, London (1971)
4. Wahab, A., Chin, S., Tan, E.: Novel approach to automated fingerprint recognition. In: IEE Proceedings Vision, Image and Signal Processing, vol. 145(3), pp. 160–166 (1998)
5. Chen, Z., Kuo, C.H.: A topology-based matching algorithm for fingerprint authentication. In: 125th Annual IEEE Int. Carnahan Conference on Security Technology, pp. 84–87 (1991)
6. Ratha, N.K., Bolle, R.M., Pandit, V.D., Vaish, V.: Robust fingerprint authentication using local structural similarity. In: Fifth IEEE Workshop on Applications of Computer Vision, December 2000, pp. 29–34 (2000)
7. Chikkerur, S., Cartwright, A.N., Govindaraju, V.: K-plet and coupled BFS: A graph based fingerprint representation and matching algorithm. In: Zhang, D., Jain, A.K. (eds.) ICB 2006. LNCS, vol. 3832, pp. 309–315. Springer, Heidelberg (2006)

8. Jea, T., Govindaraju, V.: A minutia-based partial fingerprint recognition system. Pattern Recognition 38(10), 1672–1684 (2005)
9. Jain, A.K., Prabhakar, S., Hong, L., Pankanti, S.: Filterbank-based fingerprint matching. IEEE Trans. Pattern Analysis and Image Processing 9(5), 846–859 (2000)
10. Ito, K., Nakajima, H., Kobayashi, K., Aoki, T., Higuchi, T.: A fingerprint matching algorithm using phase-only correlation. IEICE Trans. Fundamentals E87-A, 886–894 (2004)
11. Ross, A., Jain, A., Reisman, J.: A hybrid fingerprint matcher. Pattern Recognition 36(7), 1661–1673 (2003)
12. Qi, J., Yang, S., Wang, Y.: Fingerprint matching combining the global orientation field with minutia. Pattern Recognition Letters 26(15), 2424–2430 (2005)
13. Tico, M., Kuosmanen, P.: Fingerprint matching using an orientation-based minutia descriptor. IEEE Trans. Pattern Analysis and Machine Intelligence 25(8), 1009–1014 (2003)
14. Shi, Z., Govindaraju, V.: A chaincode based scheme for fingerprint feature extraction. Pattern Recognition Letters 27(5), 462–468 (2006)
15. Greenberg, S., Aladjem, M., Kogan, D.: Fingerprint image enhancement using filtering techniques. Real-Time Imaging 8, 227–236 (2002)
16. Maltoni, D., Maio, D., Jain, A., Salil, P.: Handbook of Fingerprint Recognition. Springer, New York (2003)
17. Watson, C.I., Garris, M.D., Tabassi, E., Wilson, C.L., McCabe, R.M., Stanley, J.: User's guide to nist fingerprint image software 2 (nfis2). National Institute of Standards and Technology (2004)

Performance and Computational Complexity Comparison of Block-Based Fingerprint Enhancement

Suksan Jirachaweng, Teesid Leelasawassuk, and Vutipong Areekul

Kasetsart Signal & Image Processing Laboratory (KSIP Lab),
Department of Electrical Engineering, Kasetsart University, Bangkok, 10900, Thailand
g4885038@ku.ac.th, g4765176@ku.ac.th, vutipong.a@ku.ac.th

Abstract. Performance and computational complexity comparisons of various block-based fingerprint enhancement schemes are tested and reported in this literature. Enhancement performance is evaluated by comparing equal error rates, which obtained by a proposed fingerprint matching algorithm using local and global features. Various enhancement methods are tested; i.e. three types of spatial Gabor filtering, short-time Fourier transform filtering, and discrete cosine transform filtering. These enhancement schemes also tested with various databases such as FVC2000, FVC2002, and FVC2004. Finally, computational complexity of enhancement implementation is analyzed and concluded.

Keywords: Fingerprint Enhancement, Fingerprint Enhancement Performance Comparison, Spatial-Domain/Frequency-Domain Fingerprint Enhancement, Fingerprint Enhancement Computational Complexity.

1 Introduction

Inevitably, many fingerprint identification applications are playing an important role in our everyday life from personal access control to global border control. To pursue this goal, automatic fingerprint identification system (AFIS) must be proved to be highly reliable and flexible. Since most AFISs are based on minutiae matching, these systems rely on good quality of input fingerprint images for minutiae and ridge extraction. Unfortunately, poor quality of fingerprint is now a major problem for most AFISs. In order to reduce this problem, fingerprint enhancement is required to improve quality of fingerprint in order to precisely extract minutiae [1].

The need for superb enhancement is inevitable for poor quality fingerprints, where recoverable regions contain necessary features for matching. However, processing limitation of devices, such as embedded system or smart card with limited memory size, confines computational complexity of enhancement algorithm. Therefore, trade-off between enhancement performance and computational complexity is also one of interesting problems for practical implementation of fingerprint recognition.

There are many fingerprint enhancement algorithms in literature. Scope of this paper is focused on widely used block-based contextual filtering, which can be roughly classified into two major types; i.e. spatial-domain and frequency-domain. In spatial-domain approaches, Gabor filtering plays an important role [2]. This method employs a set of pre-computed Gabor filters that selected by local ridge orientation and local

M. Tistarelli and M.S. Nixon (Eds.): ICB 2009, LNCS 5558, pp. 656–665, 2009.
© Springer-Verlag Berlin Heidelberg 2009

ridge frequency in order to perform 2-D convolution with an entire image. In addition, computational complexity of this Gabor filtering can be reduced by applying separable Gabor filters as reported in [3,4]. In frequency-domain approaches, Gabor filter can be implemented in Short Time Fourier Transform (STFT) domain, resulting in better enhancement around high curvature zone [5]. Enhancement can be applied in Discrete Cosine Transform (DCT) domain [6]. The STFT and DCT approaches take advantage of frequency domain, which could design arbitrary 2-D filter shape, corresponding to improve quality of fingerprint in high curvature zones.

Goal of this paper is to compare block-based fingerprint enhancement of various schemes, of various domains, and among various fingerprint databases, in term of recognition performance and computational complexity comparison.

2 Fingerprint Enhancement Problems

To compare enhanced quality of fingerprint subjectively, 4 important properties of block-based fingerprint enhancement should be investigated; i.e. high curvature ridge preservation, intrinsic estimation accuracy, poor quality ridge recoverability, and spurious minutia avoidance. These four cases are discussed as follows.

Firstly, high curvature area always suffers from partitioning. In spatial domain, Gabor filter enhancement assumes that each block contains only unidirectional ridges, resulting in blocking artifacts at boundary around high curvature area. On the other hand, frequency domain approaches (STFT & DCT) take advantage of 2-dimensional filter design with arbitrary shaping, as shown in Fig 1/(1)/(b, c, d).

Secondly, ridge intrinsic plays an important role in fingerprint recognition such as filtering, global orientation modeling, fingerprint classifying, and fingerprint matching. Ridge intrinsic estimation can be done in both spatial and frequency domains. For spatial-domain approach, ridge orientation is estimated based on the relative between horizontal gradient and vertical gradient, and then ridge frequency is estimated based on gray ridge projection along ridge's direction. Hence, accuracy of the ridge frequency will depend on accuracy of the ridge orientation estimation that computed previously. For frequency-domain approach, ridge orientation and ridge frequency parameters are simultaneously estimated from the highest peak coefficients of spectrum. However, with low quality fingerprint, the highest peak could not pinpoint. For example, Fig 1/(2 & 3)/(b, c, d) show error results from ridge orientation estimation and ridge frequency estimation, respectively.

Thirdly, poor quality ridge recoverability is one of significant indicator of fingerprint enhancement performance. Consider original oily fingerprint, shown in Fig 1/(4)/(a), enhanced with 3 different methods, resulting in Fig 1/(4)/(b, c, d). The Fourier approach performed best enhancement in oily ridge region because of two reasons; i.e. 1) Fourier domain approach preserves energy normalization, resulting in adaptive ridge's contrast without interfering DC coefficient. 2) 2D Fourier bases are based on parallel lines pattern of different frequencies in vertical and horizontal orientation. This domain is very suitable for parallel ridges shaping with Gabor filter. Unlike Fourier, 2-D DCT bases are similar to chess board patterns, and it is not suitable for ridge enhancement in general.

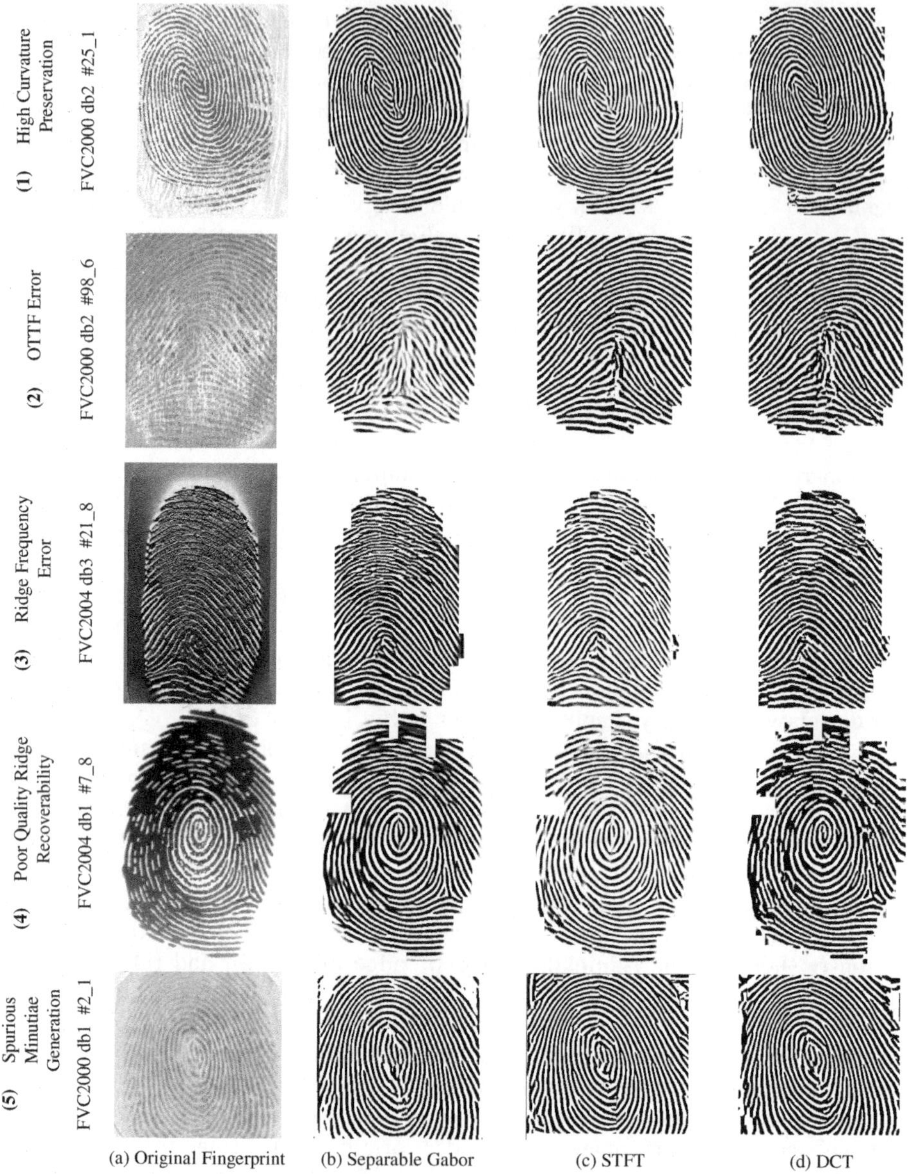

Fig. 1. Various problems of fingerprint enhancement with different domain approach (a) Original fingerprint from FVC2000 and FVC2004 (b) Enhanced results from separable Gabor in spatial domain[4], (c) Enhanced results from STFT approach modified from [5] in frequency domain, (d) Enhanced results from DCT approach method [6] in frequency domain

The last problem relates to capability of spurious minutia avoidance. In case of segmentation error or noisy and dirty fingerprint, frequency domain approaches, which has advantage of poor ridge recoverability, generates fake ridges and spurious minutia in this area, as shown in Fig 1/(5)/(b, c, d). This is clearly a drawback of frequency domain approaches. In conclusion, subjective performance comparison of three fingerprint enhancement approaches due to four major problems is summarized in the Table 1.

Table 1. Summary of Subjective Performance Comparison among three Fingerprint Enhancement Algorithms due to four Important Properties of Fingerprint Enhancement

Capabilities	**Gabor Filtering (Spatial)**	**STFT (Frequency)**	**DCT (Frequency)**
High Curvature Ridge Preservation	*Poor*	*Good*	*Good*
Accuracy of Intrinsic Parameter Estimation	*Fair*	*Good*	*Good*
Poor quality Ridge Recoverability	*Fair*	*Good*	*Poor*
Spurious Minutia Avoidance	*Good*	*Poor*	*Poor*

3 Effect of Enhancement on Fingerprint Recognition

The simply way to test enhancement performance is by testing their results with the same fingerprint matching algorithm. A fingerprint matching algorithm was developed based on three major features; focal point as singular point [7], orientation fields, and minutiae. This matching algorithm has been locally reported in [8,9] and its concept is reviewed for clear explanation as the following.

A focal point, similar to a core point, can be used as a reference point for fingerprint registration and recognition. The concept of this matching algorithm is to reorganize minutiae features around the focal point as 1-Dimensional feature sequence. Minutiae features can be categorized into *single minutiae* feature and *couple minutiae* feature. The single minutia feature consists of (x_i, y_i) coordinate of the ith minutia in fingerprint image, θ_i minutia-tail direction, and t_i minutia type (end or bifurcation), as shown in Fig 2 (a). The couple minutiae feature consists of d_{ij} distance between minutiae i^{th} and j^{th}, α_I and α_j relative angles between a minutiae pair, and r ridge count. The double minutiae features are demonstrated in Fig 2 (b).

To arrange minutiae into a 1-D feature sequence, in case of focal point existence, minutiae features are arranged around the focal point. The order of minutiae depends on their position and orientation, respected to the focal point. By using the vertical axis as a zero degree reference as shown in Fig. 2 (c), the 1st single minutia is a head of sequence, following by the 2nd minutia with counterclockwise search, resulting in a looped minutiae sequence in Fig. 2(d). In case of missing focal point, single minutia can be ordered by its tail direction. By using the vertical axis as a zero degree reference, each minutia tail direction can be arranged in order from zero to 360 degree. This single minutia sequence should be unique for each fingerprint if tail direction is perfectly detected. Fig. 2(e) shows actual links of a looped minutiae vector sequence without the focal point.

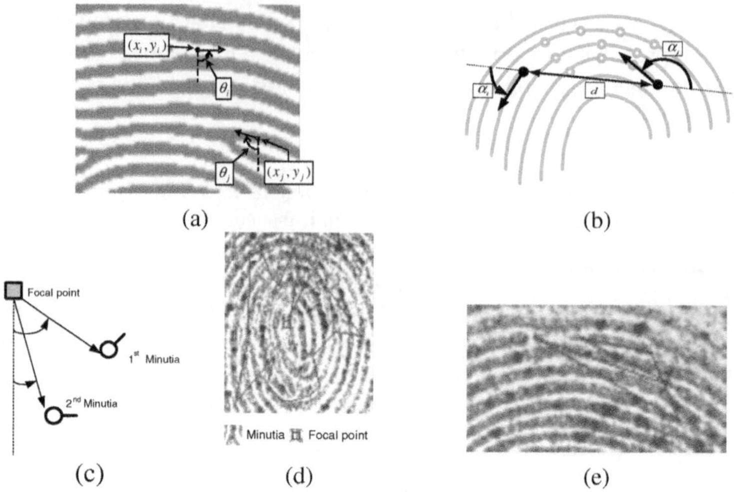

Fig. 2. Minutiae features and ordering: (a) Single minutia features. (b) Double minutiae features (c) Two minutiae ordering. (d) Actual looped minutiae sequence with a focal point. (e) Actual looped minutiae sequence without focal point.

After finishing looped minutiae arrangement, the next step is to generate the double minutiae features from the looped sequence of single minutia. By coupling two nearest neighbor minutiae in the looped sequence, a pair of minutiae can be formed and the double minutiae feature can be generated as shown in Fig 2(b). For example, the first minutia is paired with the second minutia in the looped sequence. The second minutia is coupled with the third, and so on.

A matching process composes of two stages; initial sequence alignment and minutiae sequence matching. Detail of this matching algorithm is fully explained in [9]. Matching concept is briefly explained as follows.

Step 1) Initial Sequence Alignment

- Select each minutia from both sequences, use a minutia point as a reference point, and correlate all orientation fields around minutia point and select a few minutiae couple from both sequences with highly correlative scores.
- From these few selected minutiae, find a few nearest neighbor minutiae and test with similarity score of single minutiae features and double minutiae features. In case of focal point existence, few double minutiae feature vectors can be traced and checked with a focal point. In case of no focal point, all minutiae vectors are tested with similarity score.
- Select the couple minutiae with the highest similarity score to be an initial head vector.

Step 2) Minutiae Vector Sequence Matching

- In case of no focal point input comparison, template sequence must pre-generate minutiae loop sequence without a focal point in order to prepare comparison to non-focal point input sequence.

- Use the initial head vector from step 1 for synchronizing both sequences. Then apply dynamic programming through both sequences and count matching minutiae vectors. Convert it to matching score.

The experimental results have been evaluated on public fingerprint database FVC2000, FVC2002, and FVC2004 [1,10] (100 fingers, 8 fingerprints each) in term of equal error rate. The fingerprint image is partitioned into blocks of 16×16 pixels, and a basic segmentation scheme using mean and variance is employed. Five fingerprint enhancement filtering types are evaluated as follows; Traditional Gabor filtering with non-quantized orientation (TG) [2], Separable Gabor filtering with 8-quantized orientation (SG8) [3], Separable Gabor filtering with non-quantized orientation (SG) [4], Short Time Fourier Transform approach (STFT) [5], and Discrete Cosine Transform approach (DCT) [6]. In the spatial domain approaches, the discrete Gabor filters are the same 25×25 fixed-window size. For the STFT and the DCT approaches in frequency domain, fingerprint image is also partitioned into 16×16 blocks but each block is transformed with 32×32 overlapped window in order to reduce blocking artifacts. Note that the probability estimation in [5] is not included. Finally, enhancement results are tested with our minutiae matching algorithm, and equal error rate (EER) is reported in Table 2.

Table 2. Summary of verification performance comparison among five fingerprint enhancement algorithms with FVC 2000, FVC 2002, and FVC 2004 Fingerprint Databases

Enhanced Algorithms	Average Equal Error Rate in Percent (%)											
	FVC2000				FVC2002				FVC2004			
	DB1a	DB2a	DB3a	DB4a	DB1a	DB2a	DB3a	DB4a	DB1a	DB2a	DB3a	DB4a
TG [2]	N/A	1.63	6.28	**4.38**	5.09	3.89	9.71	3.85	**6.30**	8.14	6.84	**6.48**
SG [3]	N/A	2.04	6.40	4.56	5.89	3.92	8.57	**3.83**	6.53	7.98	6.95	6.50
SG8 [4]	N/A	2.98	6.93	6.17	5.43	5.09	12.1	5.38	8.36	9.28	8.63	6.84
STFT [5]	N/A	1.23	5.75	7.06	5.28	3.46	7.45	4.92	6.75	**7.50**	**5.39**	7.73
DCT [6]	N/A	**1.11**	**5.06**	5.98	**4.38**	**3.10**	**6.84**	4.08	6.75	10.3	6.74	6.82

Note that N/A is not available, due to our technical execution problem with segmentation.

4 Computational Complexity Analysis and Comparison

In order to compare computational complexity among various enhancement algorithms (Spatial Gabor Filtering, Short Time Fourier Transform (STFT), and Discrete Cosine Transform (DCT)), computational cost is measured in term of a number of additions (N^{+}) and multiplications ($N^{\times}$) per one partitioned block, with size $W \times W$ pixels. Assume that one partitioned window has only one dominant ridge orientation. Enhancement consists of intrinsic estimation process and filtering process. However, focal point localization process is not included.

4.1 Spatial Gabor Filtering Approach

1-D band-pass Gabor filtering for 8 directions (SG8) [3] is analyzed. This algorithm is composed of three processes.

4.1.1 Ridge Orientation Estimation

Ridge orientation estimation is obtained by summation of V_x and V_y gradients in horizontal and vertical directions respectively. These two gradients are obtained by convolution with horizontal and vertical mask filtering (3×3 pixels). Then ridge orientation can be estimated by $\theta = \tan^{-1}(V_x/V_y)$. Computational complexity is summarized in Table 3.

Table 3. Number of the Additions and Multiplications in Gradients Operating

	G_x	G_y	$V_x=2\sum\sum(G_x G_y)$	$V_y=\sum\sum(G_x^2 - G_y^2)$	Total
N^+	$9W^2$	$9W^2$	W^2+1	W^2	$20W^2+1$
$N^\times$	$9W^2$	$9W^2$	W^2	$2W^2$	$21W^2$

4.1.2 Ridge Frequency Estimation

Ridge frequency can be directly estimated in frequency domain. Computational complexity is reported in term of $[N^+, N^\times]$, a number of addition and multiplication respectively. Algorithm consists of three main steps: *Step1)*: all gray ridge pixels in the block are projected into 1D sequence along ridge orientation, which requires $[W^2, 0]$. *Step 2)*: the projected sequence is transformed to frequency domain, which requires $[\mathrm{FFT}_{1D}^{+}(W), \mathrm{FFT}_{1D}^{\times}(W)]$, where $\mathrm{FFT}_{1D}^{+}(n)$ and $\mathrm{FFT}_{1D}^{\times}(n)$ denote the number of addition and multiplication of FFT_{1D} (Radix 2) as described in Table 4. *Step 3)*: ridge frequency is estimated from a distance between origin and the highest peak spectrum, which requires $[W, 2W]$.

4.1.3 Filtering Process

In order to estimate Gabor filtering computational operation, filtering process is split into two consecutive phases. For the first phase, an 1-D band-pass Gabor filter, size L, convolved with all pixels in extended block zone, size $W(W+L-1)$, along the direction perpendicular to ridge orientation. The second phase performs convolution with an 1-D Gabor low-pass filter for all pixels in the block, total W^2 pixels, along ridge direction. Implementation detail is clearly explained in [3]. Since all Gabor filters are symmetry, filtering requires only $\lfloor L/2 \rfloor$ additions and $\lfloor L/2 \rfloor+1$ multiplications. Total numbers of additions and multiplications are equal to $(W(W+L-1)+W^2)\times\lfloor L/2 \rfloor$ and $(W(W+L-1)+W^2)\times(\lfloor L/2 \rfloor+1)$, respectively.

4.2 Frequency Approach

In general, fingerprint enhancement in frequency domain utilizes overlapping window in order to preserve ridge continuity, and to eliminate blocking artifacts. Computation of each stage is explained as follows,

4.2.1 Overlapping Decomposition/Reconstruction

In general, spectrum of fingerprint's ridges locates in a specific frequency band. Other frequency bands are considered as noise, which could be eliminated in filtering process. In practice, only specific ridge frequency band is transformed and other frequency bands are ignored in order to reduce unnecessary computation, and to preserve memory.

To further reduce computational complexity, redundancy in overlapping window of FFT^{H}_{2D} and DCT^{H}_{2D} can be removed. Basically, two dimensional FFT_{2D} and DCT_{2D} can be implemented by using separable Fast Fourier transform (FFT_{1D}) and separable Fast Discrete Cosine transform (DCT_{1D}). The separable transform applies

1-D transform along the vertical as shown in Fig 3(a) (left), and then another same 1-D transform along the horizontal as shown in Fig 3(a) (right). Assume that overlapping window size is $2W \times 2W$ pixels, only a quarter at the center of total FFT coefficient's area, $(W \times W)$, is related to ridge intrinsic estimation as shown in Fig 3(b), and only $W \times W$ pixels on the top-left DCT coefficients are significant as shown in Fig 3(c).

Therefore, redundancy can be further reduced by performing convolution along gray zone of all blocks in a row as shown in Fig. 3(d). For example, consider the block B_2, a half number of transformed columns can be shared with the previous block B_1 and the next block B_3 as shown in Fig. 3(e) for $STFT^H_{2D}$ and Fig. 3(f) for DCT^H_{2D}. This technique reduces computation by 4 times, $\{FFT_{2D}(2n \times 2n) = (4)FFT^H_{2D}(2n \times 2n)\}$, as shown in Table 4.

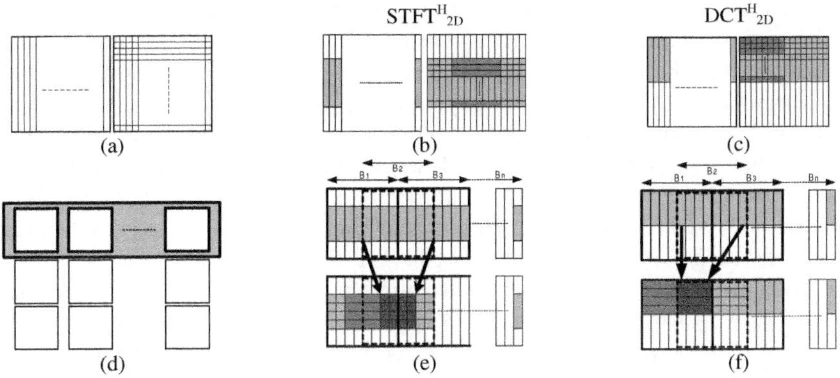

Fig. 3. Separable transforms (a) Traditional separable transform by vertical and horizontal 1-D transform, (b) Separable FFT, (c) Separable DCT transform, (d) Block transform in row, (e) Implemented $STFT^H_{2D}$ for fingerprint enhancement, and (f) Implemented DCT^H_{2D} for fingerprint enhancement

Table 4. Number of addition and multiplication requirement for different transform types

FFT, DCT (Radix 2)	+ (Addition)	× (Multiplication)
$FFT_{1D}(n)$	$(6n)(\log_2 n)$	$(4n)(\log_2 n)$
$FFT_{2D}(n \times n)$	$(2n)\ FFT^{+}_{1D}(n)$	$(2n)\ FFT^{\times}_{1D}(n)$
$DCT_{1D}(n)$	$(n)(\log_2 n)$	$(n)(\log_2 n)$
$DCT_{2D}(n \times n)$	$(2n)\ DCT^{+}_{1D}(n)$	$(2n)\ DCT^{\times}_{1D}(n)$
Overlap	+	×
$FFT_{2D}(2n \times 2n)$	$(4n)\ FFT^{+}_{1D}(2n)$	$(4n)\ FFT^{\times}_{1D}(2n)$
$FFT^H_{2D}(2n \times 2n)$	$(n)\ FFT^{+}_{1D}(2n)$	$(n)\ FFT^{\times}_{1D}(2n)$
$DCT_{2D}(2n \times 2n)$	$(4n)\ DCT^{+}_{1D}(2n)$	$(4n)\ DCT^{\times}_{1D}(2n)$
$DCT^H_{2D}(2n \times 2n)$	$(n)\ DCT^{+}_{1D}(2n)$	$(n)\ DCT^{\times}_{1D}(2n)$

4.2.2 Intrinsic Parameter Estimation

Since ridge frequency and ridge orientation parameters can be easily estimated from a distance and an angle between zero frequency coefficient and highest peak frequency position. For FFT approach, each spectrum magnitude calculation requires one addition and two multiplications because of complex number coefficient. Since Fourier coefficients are symmetry. Therefore, a total required number of addition and multiplication for spectrum magnitude calculation are only $(W^2)/2$ and $(2W^2)/2$, respectively. For DCT approach, coefficient is real and no computation needs because spectrum can be referred directly to its absolute value. However, DCT approach needs an additional phase analysis to classify ridge orientation [6]. This step requires two 2-D filters with size of 5×3 pixels for calculating average directional strength. Then, it requires totally 2(5×3) additions and 2 multiplications for normalization.

4.2.3 Filtering

For STFT approach, it requires two steps; frequency filtering and angular filtering. A band-pass filter in circular shape, with r-pixel radius, is applied directly by point to point multiplication in frequency domain. Outside filter region is forced to be zero without any operation. Therefore, the total number of multiplications is $2\times2\times\lfloor \pi r^2 \rfloor/2$. The first term represents a number of filtering time (frequency filtering and directional filtering). The second term represents a number of complex dimension (Real and Imaginary). The $\lfloor \pi r^2 \rfloor$ is an approximate size of band region. The last term represents computational reduction from a symmetry property.

For DCT approach, filtering process consists of two arithmetic operations; multiplication and convolution. The multiplication needs for frequency-domain filtering and its computational complexity is similar to the FFT approach. The convolution composes of $(r+l-1) \times r \times \lfloor l/2 \rfloor$ additions and $(r+l-1) \times r \times (\lfloor l/2 \rfloor+1)$ multiplications, where l represents a diagonal filter length. The first term $(r+l-1)$ and second term r represent a number of filtering convolution in a major direction and a minor direction of phase flow (45 or 135 degree) in DCT domain, respectively. The last term with $\lfloor l/2 \rfloor$ and $(\lfloor l/2 \rfloor+1)$ represents a number of additions and multiplications in each step of translation in convolution [6].

Finally, the total number of additions and multiplications of each stage are summarized in the Table 5. Note that (.)* shows computational complexity results under our experimental condition with $W = 16$, $L = 25$, $r = 10$ and $l = 5$.

Table 5. Number of addition and multiplication requirement for various fingerprint enhancement processes

	G1D8		STFT		DCT	
	+	×	+	×	+	×
Decomposition & Reconstruct	-	-	$(24W^2)(\log_2 2W)$ *(30720)	$(16W^2)(\log_2 2W)$ *(20480)	$(4W^2)(\log_2 2W)$ *(5120)	$(4W^2)(\log_2 2W)$ *(5120)
Gradient Operation	$20W^2+1$ *(5121)	$21W^2$ *(5376)	-	-	-	-
Frequency estimation	$W^2+(6W)(\log_2 W)+W$ *(656)	$(4W)(\log_2 W)+2W$ *(288)	-	-	-	-
Peak Searching	-	-	$W^2/2$ *(128)	W^2 *(256)	-	-
Quadratic Detection	-	-	-	-	30	2
Multiplication Filtering	-	-	-	$\lfloor 2\pi r^2 \rfloor$ *(628)	-	$\lfloor 2\pi r^2 \rfloor$ *(628)
Convolution Filtering	$(W(W+L-1)+W^2)\lfloor L/2 \rfloor$ *(10752)	$(W(W+L-1)+W^2) \times(\lfloor L/2 \rfloor+1)$ *(11648)	-	-	$r(r+l-1)\lfloor l/2 \rfloor$ *(280)	$r(r+l-1)(\lfloor l/2 \rfloor+1)$ *(420)
Total	16529	17312	61560	21364	5430	6170

5 Conclusion

Performance comparison of block-based fingerprint enhancement was reported in this paper. From Table 1, Short Time Fourier enhancement approach seems to tolerate to most problems except spurious minutiae and fake ridge generation. From Table 2, DCT enhancement approach successfully gives minimum equal error rate (EER) in 5 out of 11 tested databases. Note that DCT approach works well for fair fingerprint quality, but it fails to all FVC2004 databases, which contain many low quality fingerprints with elastic distortion. On the other hand, STFT achieves good results in FVC2004 databases, with minimum EER in 2 out of 4. In addition Spatial Gabor filtering works very well with all synthetic fingerprints (all Db4a). Surprisingly, DCT approach consumes the least computational complexity comparing with STFT and spatial Gabor filtering.

Acknowledgments. This work was partially supported by Department of Electrical Engineering, Kasetsart University, Thailand Research Fund (TRF) through the Royal Golden Jubilee Ph.D. Program (Grant No.PHD/0017/2549), and the Commission on Higher Education through the TRF Research Scholar (Grant No. RMU4980027).

References

1. Maltoni, D., Maio, D., Jain, A.K., Prabhakar, S.: Handbook of Fingerprint Recognition. Springer, Heidelberg (2003)
2. Hong, L., Wang, Y., Jain, A.K.: Fingerprint Image Enhancement: Algorithm and Performance Evaluation. IEEE Trans. on Pattern Analysis and Machine Intelligence 20(8), 777–789 (1998)
3. Areekul, V., Watchareeruetai, U., Tantaratana, S.: Fast Separable Gabor Filter for Fingerprint Enhancement. In: Zhang, D., Jain, A.K. (eds.) ICBA 2004. LNCS, vol. 3072, pp. 403–409. Springer, Heidelberg (2004)
4. Areekul, V., Watchareeruetai, U., Suppasriwasuseth, K., Tantaratana, S.: Separable Gabor filter realization for fast fingerprint enhancement. In: Proc. Int. Conf. on Image Processing (ICIP 2005), pp. III-253–III-256 (2005)
5. Chikkerur, S., Cartwright, A.N., Govindaraju, V.: Fingerprint Enhancement Using STFT Analysis. Pattern Recognition 40, 198–211 (2007)
6. Jirachaweng, S., Areekul, V.: Fingerprint Enhancement Based on Discrete Cosine Transform. In: Lee, S.-W., Li, S.Z. (eds.) ICB 2007. LNCS, vol. 4642, pp. 96–105. Springer, Heidelberg (2007)
7. Areekul, V., Suppasriwasuseth, K., Jirachawang, S.: The New Focal Point Localization for Fingerprint Registration. In: Proc. Int. Conf. on Image Pattern Recognition (ICPR 2006), vol. 4, pp. 497–500 (2006)
8. Leelasawassuk, T., Areekul, V.: Looped Minutiae Matching in Fingerprint Verification. In: Proceedings of International Workshop on Advanced Image Technology (IWAIT 2007), pp. 924–928 (2007)
9. Leelasawassuk, T.: Fingerprint Matching Using a Focal Point as a Reference Point. Master of Engineering Thesis, Department of Electrical Engineering, Kasetsart University, Thailand (2007)
10. `http://bias.csr.unibo.it`

Reference Point Detection for Arch Type Fingerprints

H.K. Lam[1], Z. Hou[1], W.Y. Yau[1], T.P. Chen[1], J. Li[2], and K.Y. Sim[2]

[1] Computer Vision and Image Understanding Department
Institute for Infocomm Research, A*STAR
(Agency for Science, Technology and Research),
Fusionopolis, Singapore
{hklam,zhou}@i2r.a-star.edu.sg
[2] School of Electrical and Electronic Engineering
Nanyang Technological University
Singapore

Abstract. Reference point detection is an important task in the design of an automated fingerprint identification system. Many algorithms have emerged with acceptable results but are mostly suitable for non-arch type fingerprint. It still remains as a challenging problem to reliably identify reference points for fingerprints of the arch type. A topological method is presented in this paper to detect reference points in arch type fingerprint images. To evaluate the performance, 400 arch type fingerprint image pairs in the NIST DB4 database are utilized. The alignment accuracy on average is about 35 pixels in distance and 9 degrees in orientation, which is very well comparable with respect to state-of-the-arts as designed for non-arch type fingerprints.

1 Introduction

In the design of an automated fingerprint identification system (AFIS), a frequently encountered problem is to reliably locate a reference point. An accurately located reference point could substantially improve the efficiency of an AFIS [1]. Usually the reference point is selected from a list of singular points (core/delta point) [2] located in the acquired fingerprint image. Singular points usually refer to discontinuities in the fingerprint orientation field and are important features for fingerprint images as they can effectively be used for fingerprint classification, registration, alignment and etc.

However, it is not trivial to locate singular points accurately and precisely as the result can be affected by a number of factors such as noise, wetness and dryness of the finger, and the type of fingerprint class (arch or non-arch). Nevertheless, the singular point can be said to be more easily located in a non-arch type fingerprint compared to an arch type fingerprint. This is because in the case of arch type fingerprint, the ridges are almost parallel across the whole fingerprint image and as a result, the conventional concept of singular points is not well defined.

Thus it is not surprising to witness that most efforts for reference point detection focus on fingerprints of non-arch types and ignore the arch types [3-6]. Although compared with other fingerprint types, the percentage of people with arch type

M. Tistarelli and M.S. Nixon (Eds.): ICB 2009, LNCS 5558, pp. 666–674, 2009.
© Springer-Verlag Berlin Heidelberg 2009

fingerprint is smaller, in absolute numbers, they are still quite large. For techniques on reference point detection to be practically useful in an AFIS, there is no reason to ignore the arch type fingerprint, in particular when considering the high demand on security in the era that terrorism remains a severe threat on people's daily life and economic prosperity.

The rest of the paper is organized as follows: Section 2 will detail our method of locating reference points in arch type fingerprint images, Section 3 will show our results and compare with some of the well known singular point detection algorithm as designed for non-arch type fingerprint images and Section 4 concludes the paper with a brief discussion on future direction.

2 Methodology

The acquired fingerprint image is firstly segmented to extract the foreground where the fingerprint structure is present. Then fingerprint orientation is estimated using the method in [7] to obtain the orientation map. Fig. 1 shows the original image and the orientation image where the dot (in orientation image) indicates background. The region of interest in the following processing will be limited to the foreground where the fingerprint structure is present.

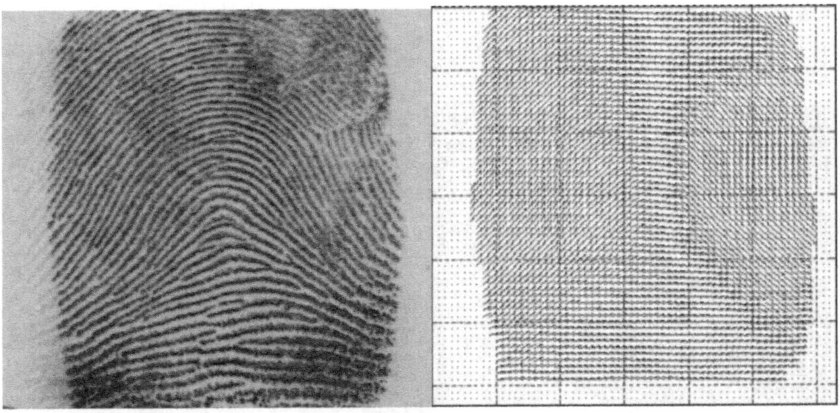

(a) (b)

Fig. 1. (a) An arch type fingerprint image, (b) orientation image

2.1 Finding Symmetry Line of Fingerprint Structure

Compared with non-arch type fingerprints, the most distinctive feature of the arch type ones is the line symmetry for the overall fingerprint structure. To utilize this topological feature for characterizing the arch type fingerprint, the orientation of the fingerprint image is further processed to separate it into two levels and to detect the symmetry line from this segmentation map.

Toward that end, a threshold to segment the orientation map of the fingerprint image must be calculated. This is done automatically by first obtaining an orientation histogram from the orientation map and then finding a point such that the area of each region of the thresholded orientation histogram is equal. Fig. 2 shows the orientation histogram of the fingerprint image in Fig. 1 and the automatically calculated threshold drawn in red. The threshold is applied to the orientation map to obtain a two level segmentation map. Then morphological processing is applied to derive a smooth segmentation map, as shown in Fig. 3(a) for example.

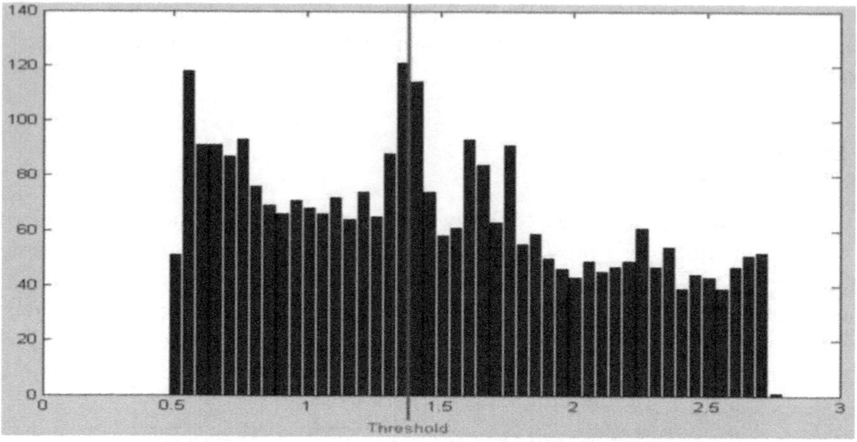

Fig. 2. Two level thresholding on the orientation histogram

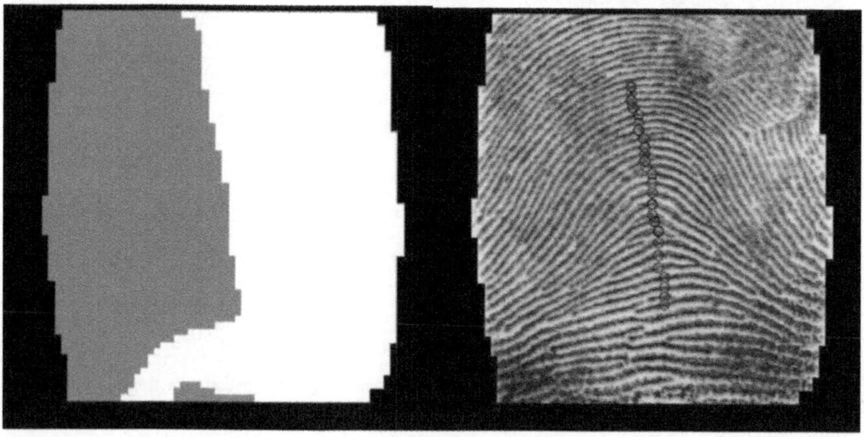

(a) (b)

Fig. 3. (a) A two level segmentation of the orientation field for an arch type fingerprint image. (b) The major symmetric line as identified from boundaries between the segmented regions in (a) and represented as blue circles.

Very often, the segmented orientation map does not produce two perfectly symmetric regions, as seen from Fig. 3(a) for instance. Through intensive investigation, we observe that frequently there will be one major boundary line that runs through the segmentation map. Thus, we can confidently select the longest boundary line as the symmetry line of the fingerprint image.

Firstly, the coordinates of the boundary points are traced according to 8-neighbourhood connectivity and labeled as a group. For instance, there are two boundary lines in Fig. 3(a). After all boundary points are labeled, the line that is longest and cuts closest along the centre of the fingerprint structure will be the symmetry line of the fingerprint image. Fig. 3(b) shows the result of the identified symmetric line from fig. 3(a). For convenience, the point set in the line is denoted as $\ell = \{P_i \mid i = N_\ell\}$, where N_ℓ is the number of the points.

2.2 Locate Reference Point and Orientation

After identifying the symmetric line, we are ready to define the reference point for arch type fingerprints. As aforementioned, line symmetry is the most significant feature of arch type fingerprints, which distinguishes arch type fingerprints from non-arch type ones in terms of global pattern characterization. For non-arch type fingerprints, singular points represent the most evident global features, contain rich information for fingerprint classification and can reliably be used as candidate of reference point.

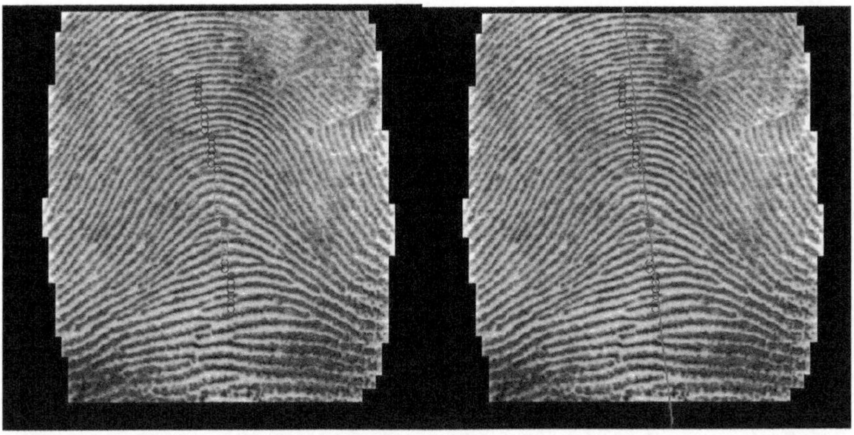

(a) (b)

Fig. 4. (a) The located reference point (in red) has the greatest change of orientation angle along the symmetric line. (b) The green circles are used to calculate the orientation of the reference point. The magenta line represents the orientation of the reference point.

Although singular points are not well defined for arch type fingerprints, we argue that candidate of reference point detection should be determined from the points that most fundamentally characterize the arch type fingerprints. With this in mind, we define the reference point of an arch type fingerprint as the point in the above identified symmetric line with the greatest change in direction represented using:

$$P^{*} = \max_{i \in N_{\ell}} \|\nabla \varphi_i\| \tag{1}$$

where φ_i is the orientation at the i-th pixel. Fig. 4(a) shows the detected reference point for an arch type fingerprint.

Besides finding the location of the reference point, the orientation of the reference point is also estimated from the identified symmetric line. Towards that end, the points within a distance from the located reference point P^{*} are selected, as denoted by $P_k, k = 1, \cdots, n$. Then a straight line $y = \beta_0 + \beta_1 x$ is fitted among the points and the parameters are estimated using the least square error method. The orientation of the reference point is estimated as

$$\varphi^{*} = \arctan \beta_1 \tag{2}$$

Fig. 4(b) shows the result of obtaining the orientation of the reference point from the symmetric line.

3 Results

3.1 Evaluation Criteria

To evaluate the performance of a method for fingerprint reference point detection, one usually compares the detected results with respect to the ground truth which is derived from experts' labeling. However, this kind of strategy is problematic for the case of arch-type fingerprints. As aforementioned, the concept of singular point is not well defined for arch-type fingerprints, thus there is a large degree of uncertainty for fingerprint experts to manually identify a point as the reference point. To tackle this issue, we propose here to validate the performance through matching as detailed in the following:

1. Two fingerprint images from the same subject are matched and the correspondence of matched minutia pairs is established;
2. The same two fingerprint images are aligned using the reference points;
3. The average matching minutiae distance difference and angle difference of the normalized set is calculated and used as performance index.

In order to have a fair comparison of our algorithm and due to the lack of good reference point locating algorithm for arch type fingerprint images in the literature, we compare our result for arch type fingerprint images to that of well know algorithm for ***non-arch*** type fingerprint images. For non-arch type fingerprint images, the reference point was located using the complex filter [8]. For arch type fingerprint images, our method was used to locate the reference point.

The database used for testing is NIST DB4 where there are 2000 pairs of images. 20% of the images are arch-type images. Out of the rest of the 1600 pairs of images, only 1280 pairs of the non-arch type fingerprint images are usable because the complex filter is unable to find a reference point in one of either image. The 1680 pairs of fingerprint images are matched using software [9] to obtain the matching minutiae for use in computing the performance of the reference point locating algorithms.

Table 1. Results of our reference point detection algorithm for arch type fingerprint images and comparison with respect to the complex filter results on non-arch type fingerprint images

	Complex filter method on non-arch type image	Our proposed method on arch type image
Average matching minutiae normalized distance difference	28.58 pixels	35.18 pixels
Average matching minutiae normalized angle difference	9.41 degrees	8.79 degrees

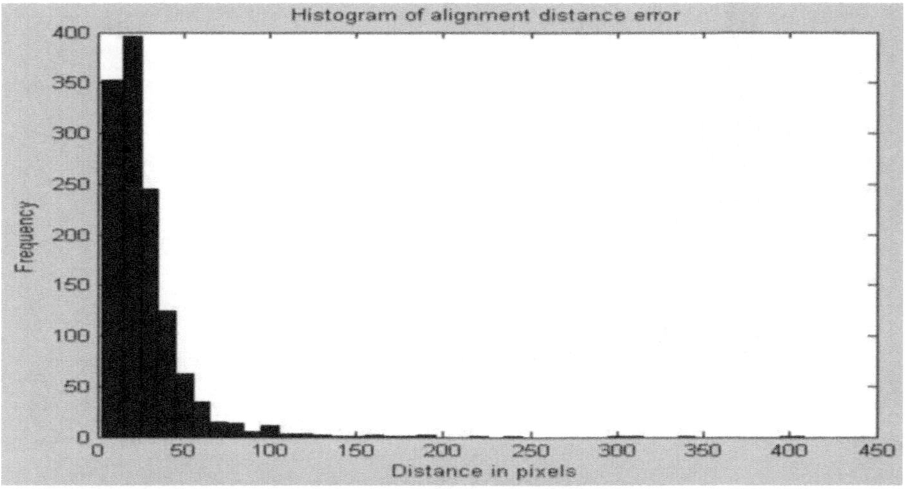

(a)

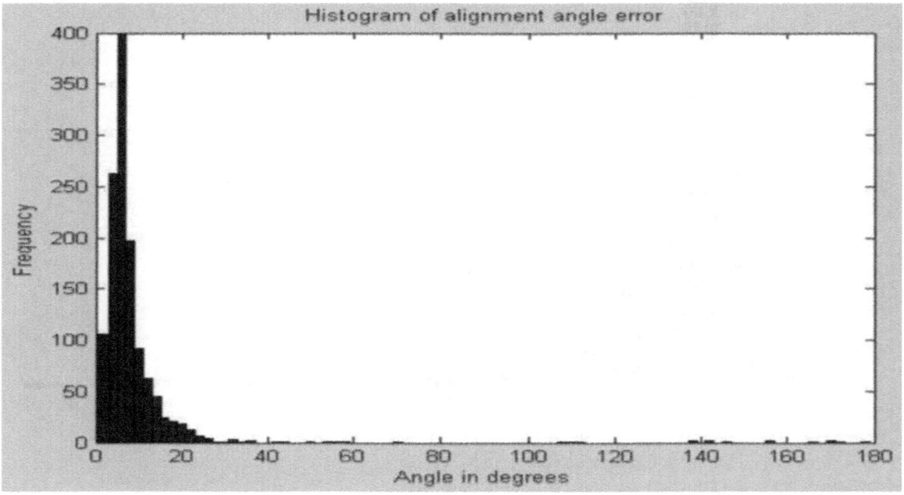

(b)

Fig. 5. (a) Histogram showing the distribution of distance difference. (b) Histogram showing the distribution of angle difference. Algorithm used is complex filter for non-arch type images only.

3.2 Experimental Results

The result is summarized in Table 1 (3rd column). The alignment accuracy of the proposed method is about 35 pixels in distance and 9 degrees in orientation. Table 1 also shows the result of the complex filter method (2nd column). From the table, it is clear that our algorithm is able to accurately determine the position and orientation of the

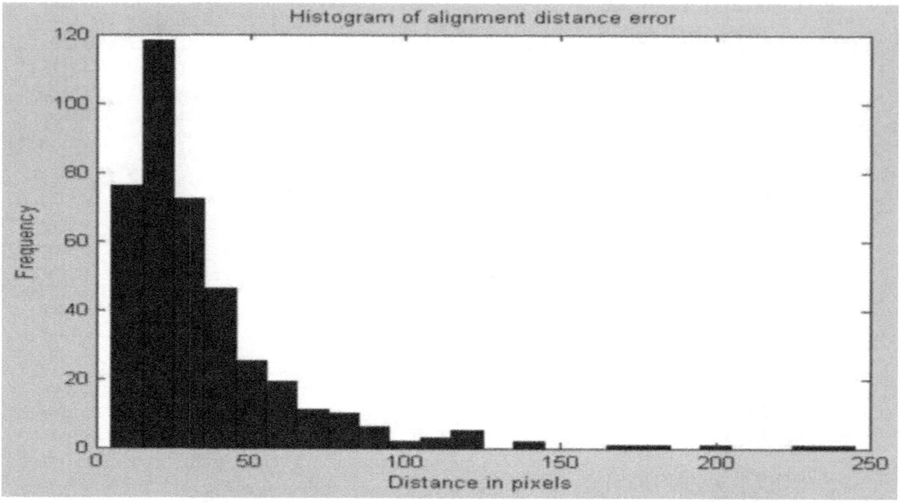

(a)

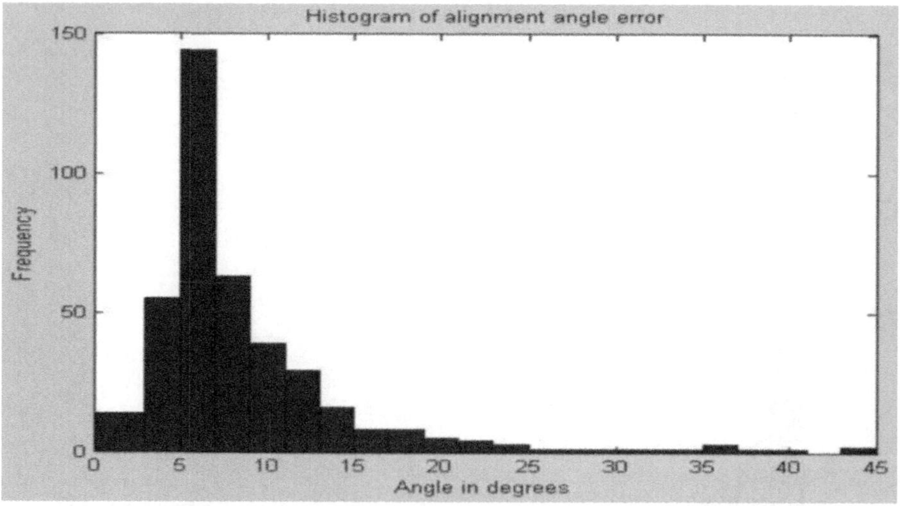

(b)

Fig. 6. (a) Histogram showing the distribution of distance difference. (b) Histogram showing the distribution of angle difference. Algorithm used is our proposed method for arch type images only.

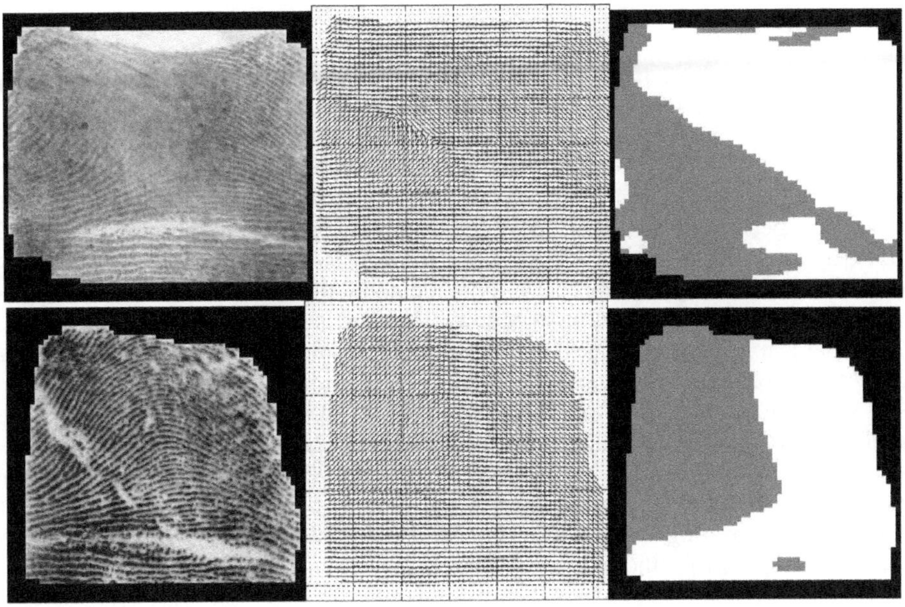

Fig. 7. Two instance of the same finger. The series on the top shows noise in the image causing the orientation map to be computed incorrectly and subsequently segmented wrongly. The series on the bottom shows a proper segmented orientation map and thus correctly detected reference point. Alignment error is large in this case.

reference points in arch type fingerprint images. The result is at least comparable to well known algorithm for locating reference points in non-arch type fingerprint images. The result of the proposed method is also comparable with respect to methods in [10,11] (as evaluated using the FVC2000 DB2 database) when considering that the image quality of FVC2000 DB2 is much better than the NIST DB4 used here and either [10] or [11] rejects about 10 images in their result.

For a more detailed comparison, the distributions of the error in both distance and angle are plotted in Fig. 5 (for the complex filter method) and Fig. 6 (for the proposed method) respectively. From Fig. 5, one can see that the alignment error is fairly small for most images and the occurrence of large error is mainly due to the poor image quality, thus it is reasonable to align non-arch type fingerprints using the complex filter method. Similar observation can be found in Fig. 6, where proposed method is satisfactory for most arch type fingerprints with reasonable quality. From Figs. 5 and 6, it can be expected that the performance of fingerprint alignment can substantially be improved by fingerprint enhancement, which is out of the scope of this presentation.

For the proposed method, a reasonable image quality means that the segmentation of the orientation pattern can be reliably derived. Otherwise, large alignment error will happen. Fig. 7 shows a typical example on how the image quality affects the estimation of the orientation field as well as the result of segmentation.

Both methods are implemented in Matlab. The complex filter takes around 2.0s to complete the reference point detection, while our algorithm takes about 0.05s.

4 Conclusion

To summarize, we have presented a method to locate reference points in arch type fingerprint images. The method starts from the detection of the symmetric line, which is topologically the most remarkable feature in arch type fingerprints, and defines the reference point as the point in the symmetric line with largest change in orientation. The orientation of the reference point is also estimated from the identified symmetric line. The method has been evaluated using the 400 arch type fingerprint image pairs in the NIST DB4 database. Our alignment accuracy is about 35 pixels in distance and 9 degrees in orientation, which is well comparable to the complex filter method as targeted for non-arch type fingerprint images. The experiment demonstrates that the proposed method is simple but is it fast and accurate. With this method, it can be expected that the technique of reference point detection can be integrated into practical AFIS to improve efficiency.

It should be pointed out that the results for our algorithm are based on a coarse level analysis (8x8 blocks). Thus, further processing can be done to locate the reference points to pixel accuracy. In addition, some images are too wet or very noisy, and significantly degrade the performance of the proposed method. Therefore more reliable methods for fingerprint enhancement and orientation calculation may be researched on to improve the location accuracy.

References

1. Chikkerur, S., Ratha, N.: Impact of singular point detection on fingerprint matching performance. In: 4th IEEE workshop on Automatic Identification Advanced Technologies, pp. 207–212 (2005)
2. Jain, A.K., Prabhakar, S., Hong, L., Pankanti, S.: Filterbank-based fingerprint matching. IEEE Trans. Image Processing 9, 846–859 (2000)
3. Kitiyanan, N., Havlicek, J.P.: Modulation domain reference point detection for fingerprint recognition. In: 6th IEEE Southwest Symp. on Image Analysis and Interpretation, pp. 147–151 (2004)
4. Zhang, H., Yin, Y., Ren, G.: An improved method for singularity detection of fingerprint images. In: Li, S.Z., Lai, J.-H., Tan, T., Feng, G.-C., Wang, Y. (eds.) SINOBIOMETRICS 2004. LNCS, vol. 3338, pp. 516–524. Springer, Heidelberg (2004)
5. Ohtsuka, T., Kondo, A.: A new core and delta detection for fingerprints using the extended relation graph. In: IEICE Trans. on Fundamentals of Electronics, Communications and Computer Sciences, pp. 2587–2592 (2005)
6. Huang, C., Liu, L., Douglas Hung, D.C.: Fingerprint analysis and singular point detection. Pattern Recognition Letters 15, 1937–1945 (2007)
7. Wang, Y., Hu, J., Phillips, D.: A Fingerprint Orientation Model Based on 2D Fourier Expansion (FOME) and Its Application to Singular-Point Detection and Fingerprint Indexing. Trans. on Pattern Analysis and Machine Intelligence 29(4), 573–585 (2007)
8. Nilsson, K., Bigun, J.: Complex filters applied to fingerprint images detecting prominent symmetry points used for alignment. In: Tistarelli, M., Bigun, J., Jain, A.K. (eds.) ECCV 2002. LNCS, vol. 2359, pp. 39–47. Springer, Heidelberg (2002)
9. `http://www.fingerprint.nist.gov/NBIS/index.html`
10. Jiang, X., Liu, M., Kot, A.C.: Reference point detection for fingerprint recognition. In: Intl. Conf. in Pattern Recognition, pp. 540–543 (2004)
11. Liu, T., Zhang, C., Hao, P.: Fingerprint reference point detection based on local axial symmetry. In: Intl. Conf. in Pattern Recognition, pp. 1050–1053 (2006)

Palmprint Verification Using Circular Gabor Filter

Azadeh Ghandehari[1] and Reza Safabakhsh[2]

[1] Faculty of Engineering & Technology, Islamic Azad University (Saveh Branch)
[2] Computer Engineering Department, Amirkabir University of Technology
(Tehran Polytechnic)
ghandehari@iauscd.ir, safa@aut.ac.ir

Abstract. Recently, researchers have been paying an excessive amount of attention to biometric palmprint which has gained popularity and received significant prominence due to its high stability and uniqueness. In this study, two certain filters have been taken into consideration, namely Gabor filter and Circular Gabor filter which are used to obtain feature information from two distinguishing regions of interest, square and inscribed circle areas as the central part sub-images and the two palmprint images are compared with one another in terms of their hamming distance. The outcomes of the experiment gave an indication to the fact that circular Gabor Filter had a comparatively better performance than traditional one in extracting distinct feature information.

Keywords: Biometrics, Feature extraction, Gabor filter, Texture analysis.

1 Introduction

A plentitude of distinguishing traits that have a wide variety of usage for personal authentication exist, which include fingerprint, face, voice, iris, hand geometry and so forth. Palmprint can be undoubtedly referred to as one of the highly reliable means of making a distinction between two individuals, since the print patterns are not duplicated even in monozygotic twins [1]. More significantly, the details of these patterns are perpetual and the rich structures of the palmprint offer plenty of helpful information for the purpose of recognizing one individual from others [2-4].

Recently, most researches in biometrics have been focused on palmprint authentication, and many approaches have been developed. According to the operating space in which palmprint features extracted, those approaches can be categorized into spatial domain and frequency domain. Due to the type of palmprint features, the above mentioned domains can be categorized into structural feature-based approaches and statistical feature-based approaches. The first approach is a key strategy to extracting structural information, like principal lines and creases, from the palm for recognition [7-11]. The second approach indicates the works that appear in the literature, including eigenpalm [5], fisherpalms [11], global texture energy [13], Gabor filters [13], Fourier Transform [6], and local texture energy [15]. Each approach demonstrates its strengths and weaknesses, and the choice depends on the temperament of application: operational mode, processing speed, memory storage and quality of the image acquired. Generally, a palmprint has some principal lines, wrinkles, singular points and

M. Tistarelli and M.S. Nixon (Eds.): ICB 2009, LNCS 5558, pp. 675–684, 2009.
© Springer-Verlag Berlin Heidelberg 2009

minutiae. Some algorithms such as the stack filter [17] can extract the principal lines. However, these principal lines do not contribute adequately to high accuracy because of their similarity amongst different people. Thus, wrinkles play an important role in palmprint authentication. However, in low-resolution palmprint images, some images do not have clear wrinkles. So, in this paper texture analysis has been applied to palmprint authentication.

2 Overview of System Architecture

The proposed system is divided into two phases, namely the enrollment and verification, as shown in Fig 1.

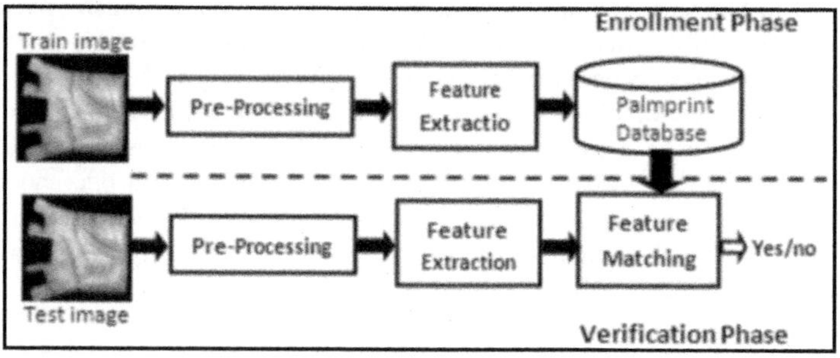

Fig. 1. Block diagram of the proposed palmprint verification phases

The important tasks in the system include the pre-processing, feature extraction and feature matching. In the pre-processing stage, the alignment and orientation of the hand images are corrected for use in the successive tasks. In the feature extraction stage, the most discriminating features of the palms are extracted for representation, and finally in the feature matching stage, comparisons are performed and decisions are made to find out whether two palmprint features are from the same person. The details of each of these components are discussed in the subsequent sections.

2.1 Pre-processing

In pre-processing stage, it is necessary to obtain a sub-image from palmprint image and to eliminate the variations caused by rotation and translation. There are two common approaches for the processing of palm images, which differ from each other because of the shape of the sub-image. In the first one, called square-based approach, the sub-images are obtained by cutting a square from the main image, while in the second one the sub-images are extracted by a circular cutting of the main palm image; his is called inscribed circle-based approach. There are many articles reporting the use of square-based approach whereas the inscribed circle-based approach has hardly attracted the attention of researchers. In this paper, we employed both approaches to create two different data bases for feature extractions.

2.1.1 Square-Based Approach

The five main steps of palmprint image preprocessing to extract a square with the fixed size from predefined position under the coordinate system are as follows (see Fig. 2):

Step 1: Apply a low-pass filter, such as Gaussian, to the original image. Then use a threshold to convert this original image into a binary image as shown in Fig. 2(b).

Step 2: Extract the boundaries of the holes between fingers using a boundary-tracking algorithm. The start points, (Sx_i, Sy_i) and end points, (Ex_i, Ey_i) of the holes are then marked in the process (see Fig. 2(c)).

Step 3: Calculate the center of gravity (Cx_i, Cy_i) of the holes. Then construct a line that passes through (Cx_i, Cy_i) and the midpoint of (Sx_i, Sy_i) and (Ex_i, Ey_i). Based on these lines, two key points (K_1, K_2) can easily be detected (see Fig. 2(d)).

Step 4: Line up K_1 and K_2 to get the Y-axis of the palmprint coordinate system and make a line through their midpoint which is perpendicular to the Y-axis, to determine the origin of the coordinate system (see Fig. 2(e)). This coordinate system can align different palmprint images.

Step 5: Extract a sub-image with the fixed size on the basis of coordinate system, which is located at the certain part of the palmprint for feature extraction (see Fig. 2(f)).

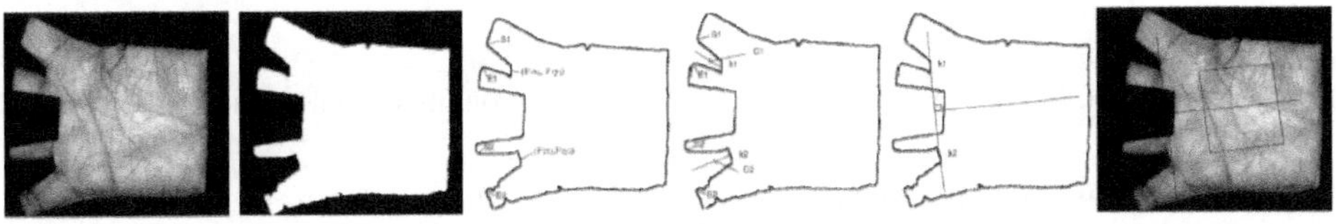

(a) (b) (c) (d) (e) (f)

Fig. 2. Main steps of preprocessing: (a)original image, (b)binary image, (c)boundary tracking, (d)key points (k1 and k3) detecting, (e)the coordinate system and (f)the central part of a palm

2.1.2 Inscribed Circle-Based Approach

In contrast to the square-based approach, this approach is based on an inscribed circle to extract the central part of a palmprint; therefore a round image is obtained rather than a square one. The followings describe the details of this approach (Fig. 3):

Step 1: convert the gray scale image into a binary image. Then, smooth the boundary between the white and black pixel and obtain the contour of a palmprint (see Fig. 3(c)).

Step 2: calculate the biggest inscribed circle for the contour of the palmprint and obtain its center and radius. Because different palms are different sized, the radii may be different (see Fig. 3(c)).

Step 3: once the circle is determined, the extraction only invokes obtaining all pixels inside the circle (see Fig. 3(d)).

(a) (b) (c) (d)

Fig. 3. Main steps of preprocessing: (a)original image, (b)binary image, (c)biggest inscribed circle and (d) circle area central part sub-image

2.1.3 Comparison between Two Approaches

There are cases in which a palmprint cannot be segmented properly using the square-based approach while it can be done so using the circle-based approach. Fig. 4 shows a sample that the ring finger cannot be placed correctly due to its extreme darkness which cannot be separated from the background. As a result, this palmprint cannot be correctly segmented using the square-based approach. On the other hand, by using the circle-based segmentation approach, the central part palmprint sub-image can be extracted successfully (Fig. 4(c).)

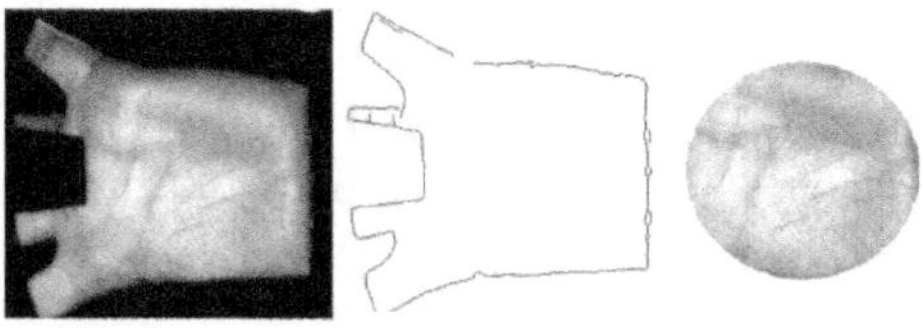

(a) (b) (c)

Fig. 4. Palmprint central part sub-image extraction for special palmprint

The circle-based approach can extract a larger area for the feature extraction, making it possible to include more features. It is worth noting that though the circle-based approach provides a larger surface area for the palmprint feature extraction, it takes longer for feature extraction. Hence, the conclusion is that square-based approach has the advantage of faster processing on the feature extraction while circle-based approach has more features obtained.

Samples from the same palm may have different circle sizes, but the difference is usually within 2-3 pixels, which does not cause a problem for palmprint matching. Different palms may have different circle sizes. It provides a useful means on palmprint classification, i.e. use the circle size as the first step of the classification criteria.

2.2 Palmprint Feature Extractions

In this section, the extraction of texture features has been fulfilled through using two types of Gabor filters. The first type of Gabor filter, the conventional one, is used in the texture analysis and the second type, the improved version of the conventional one, is rotation invariant which has been suggested for feature extraction.

2.2.1 Traditional 2D Gabor Filter

The Gabor filter is an effective tool for texture analysis, and has the following general form:

$$G(x,y,\theta,u,\sigma)=\frac{1}{2\pi\sigma^2}\exp\left\{-\frac{x^2+y^2}{2\sigma^2}\right\}\times\exp\{2\pi iu(xCos\theta+ySin\theta)\} \tag{1}$$

where $i=\sqrt{-1}$; u is the frequency of the sinusoidal wave; θ controls the orientation of the function and σ is the standard deviation of the Gaussian envelope.

Gabor function, $G(x,y,\theta,u,\sigma)$ with a special set of parameters (σ,θ,u), is transformed into a discrete Gabor filter $G[x,y,\theta,u,\sigma]$. The parameters are chosen from 12 sets of parameters listed in Table 1. In this paper, the parameter selection is similar to [12]. In order to provide more robustness to brightness, the Gabor filter is turned to zero DC (direct current) with the application of the following formula:

$$\tilde{G}[x,y,\theta,u,\sigma]=G[x,y,\theta,u,\sigma]-\frac{\sum_{i=-n}^{n}\sum_{j=-n}^{n}G[i,j,\theta,u,\sigma]}{(2n+1)^2} \tag{2}$$

where $(2n+1)^2$ is the size of the filter. In fact, the imaginary part of the Gabor filter automatically has zero DC because of odd symmetry. This adjusted Gabor filter will convolute with a sub-image defined in Section 2.1. The sample point in the filtered image is coded to two bits, (b_r,b_i) by the following inequalities,

$$b_r=1 \;\; if \;\; \mathrm{Re}[\tilde{G}[x,y,\theta,u,\sigma]*I]\geq 0 \tag{3}$$

$$b_r=0 \;\; if \;\; \mathrm{Re}[\tilde{G}[x,y,\theta,u,\sigma]*I]<0 \tag{4}$$

$$b_i=1 \;\; if \;\; \mathrm{Im}[\tilde{G}[x,y,\theta,u,\sigma]*I]\geq 0 \tag{5}$$

$$b_i=0 \;\; if \;\; \mathrm{Im}[\tilde{G}[x,y,\theta,u,\sigma]*I]<0 \tag{6}$$

where I is the sub-image of a palmprint. Using this coding method, only the phase information in palmprint images is stored in the feature vector. Fig. 5 shows the features generated by the 12 filters listed in Table 1.

Table 1. The traditional Gabor filter's parameters and their definitions

Levels	No	Sizes	θ	F	σ
1	1	9×9	0	0.3666	1.4045
	2	9×9	45	0.3666	1.4045
	3	9×9	90	0.3666	1.4045
	4	9×9	135	0.3666	1.4045
2	5	17×17	0	0.1833	2.8090
	6	17×17	45	0.1833	2.8090
	7	17×17	90	0.1833	2.8090
	8	17×17	135	0.1833	2.8090
3	9	35×35	0	0.1833	2.8090
	10	35×35	45	0.1833	2.8090
	11	35×35	90	0.1833	2.8090
	12	35×35	135	0.1833	2.8090

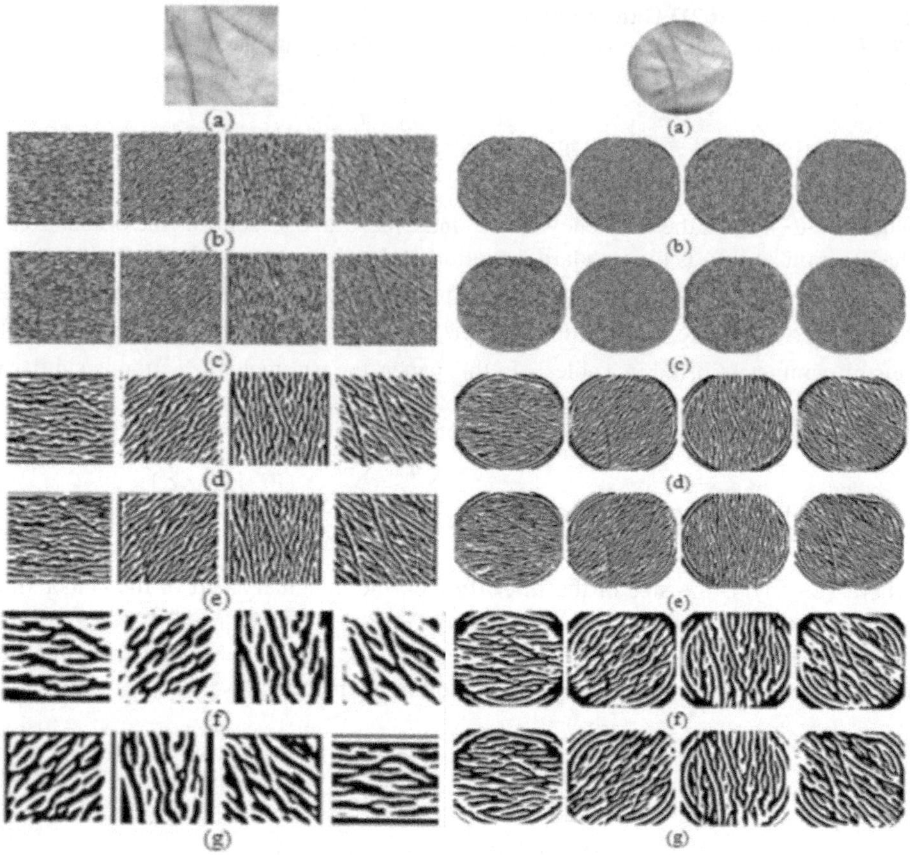

Fig. 5. Original image from DBI and DBII and their features generated by 12 filters listed in the Table1: (a)original image, (b), (d) and (f)real parts of features from Levels 1–3 filters, respectively, (c), (e) and (g)imaginary parts of features from Levels 1–3 filters, respectively

2.2.2 Circular Gabor Filter

Circular Gabor-based filter is a modified version of traditional Gabor filter. It is isotropic which means its characteristics are equal in every direction. It has been successfully used for invariant texture segmentation [15]. The circular Gabor filter is:

$$cG(x, y, F, \sigma) = \frac{1}{2\pi\sigma^2}\exp\left\{-\frac{(x^2+y^2)}{2\sigma^2}\right\}.\exp\left\{2\pi i F(\sqrt{x^2+y^2})\right\} \tag{7}$$

where F is the central frequency of the circular Gabor filter. The frequency domain representation of the circular Gabor filter is given by

$$F(u,v) = \frac{\sqrt{2\pi}}{2}\alpha\exp(-\frac{(\sqrt{u^2+v^2}-F)^2}{2\alpha^2} \tag{8}$$

where $\alpha = \frac{1}{2\pi\sigma}$. The parameters are chosen from 3 sets of parameters listed in Table 2. Palmprint presentation is exactly identical to section 2.2.1. Fig. 6 shows the features generated by the filter sets listed in Table 2.

Table 2. The circular Gabor filter's parameters and their definitions

Levels	No	Sizes	F	σ
1	1	9×9	0.3666	1.4045
2	5	17×17	0.1833	2.8090
3	9	35×35	0.0916	5.6179

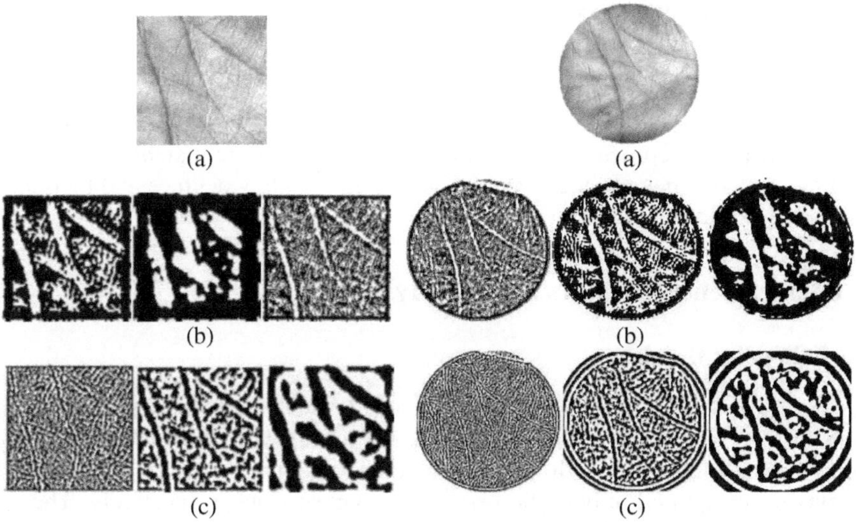

Fig. 6. Original image from DBI and DBII and their features generated by 3 filters listed in the Table2: (a) original image, (b) real parts of features from Levels 1–3 filters, (c), imaginary parts of from Levels 1–3 filters

2.3 Feature Matching/Classification Procedures

In order to describe clearly the matching process, each feature vector is considered as two 2-D feature matrices, real and imaginary. Palmprint matching is based on a normalized hamming distance. Let P and Q be two palmprint feature matrices. The normalized hamming distance can be defined as

$$D_0 = \frac{\sum_{i=1}^{n}\sum_{j=1}^{n}(P_R(i,j)\otimes Q_R(i,j) + P_I(i,j)\otimes Q_I(i,j))}{2N^2} \tag{9}$$

where $P_R(Q_R)$ and $P_I(Q_I)$ are the real part and the imaginary part of $P(Q)$, respectively; the Boolean operator, "$\otimes$", is equal to zero if and only if the two bits, $P_{R(I)}(i, j)$ and $Q_{R(I)}(i, j)$ are equal and the size of the feature matrices is N × N. It is noted that D_0 is between 1 and 0. The hamming distance for perfect matching is zero. In order to provide translation invariance matching, Eq. (9) can be improved as

$$D_{\min} = \min_{|s|<S,|t|<T} \frac{\sum_{i=\max(1,1+s)}^{\min(N,N+s)} \sum_{j=\max(1,1+t)}^{\min(N,N+t)} (P_R(i+s,j+t) \otimes Q_R(i,j) + P_I(i+s,j+t) \otimes Q_I(i,j))}{2H(s)H(t)} \quad (10)$$

where S = 2 and T = 2 control the range of horizontal and vertical translation of a feature in the matching process, respectively and

$$H(s) = \min(N, N+s) - \max(1+1+s) \quad (11)$$

The hamming distance, $D_{\min}$ can support translation matching; nevertheless, because of unstable preprocessing, it is not a rotational invariant matching. Therefore, in enrollment mode, the coordinate system is rotated by a few degrees and then the sub-images are extracted for feature extraction.

3 Experimental Results and Analyses

3.1 Palmprint Database

The proposed system has been tested on the PolyU-Palmprint-Database [16] that contains 7752 grayscale images corresponding to 386 different individuals. Around twenty samples from each individual were collected in two sessions, where around 10 samples were captured in both the first and second sessions, respectively. The size of all images is 384×284 with 75 dpi resolution. The square central parts of each image extracted with size 128×128 are named DBI and the circular central parts of each image extracted with average size of 100 pixels named DBII. Fig. 7 shows four typical images from this database.

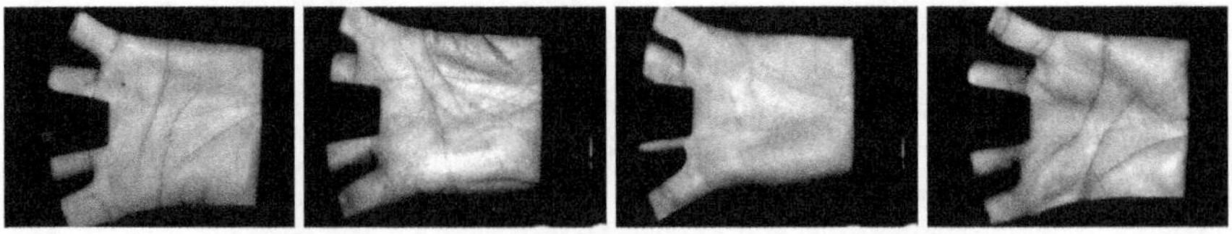

Fig. 7. Six typical images from the database

3.2 Verification Test

To obtain the verification accuracy of our palmprint system and better parameters for our system, 12 different sets of parameters listed in Table 1 and 3 different sets listed in Table 2 are used to test the method. Each of the images in DBI (DBII) is matched with all other palmprint images in the same database. A matching is counted as a correct matching if two palmprint images are collected from the same palm; otherwise it is an incorrect matching. The classification accuracy of the system based on Gabor features is presented in Tables 3.

Table 3. Summary of accuracy rate for the different filters on two databases

<table>
<tr><th colspan="2" rowspan="3">Filter No.</th><th colspan="4">DBI: Square sub-image</th><th colspan="4">DBII: Circle sub-image</th></tr>
<tr><th colspan="2">Traditional Gabor Filter</th><th colspan="2">Circular Gabor Filter</th><th colspan="2">Traditional Gabor Filter</th><th colspan="2">Circular Gabor Filter</th></tr>
<tr><th>FAR (%)</th><th>FRR (%)</th><th>FAR (%)</th><th>FRR (%)</th><th>FAR (%)</th><th>FRR (%)</th><th>FAR (%)</th><th>FRR (%)</th></tr>
<tr><td rowspan="4">Level 1</td><td>1</td><td>0</td><td>54.4</td><td rowspan="4">0</td><td rowspan="4">15.0</td><td>0</td><td>57.2</td><td rowspan="4">0</td><td rowspan="4">10.2</td></tr>
<tr><td>2</td><td>0</td><td>38.0</td><td>0</td><td>48.0</td></tr>
<tr><td>3</td><td>0</td><td>40.8</td><td>0</td><td>37.6</td></tr>
<tr><td>4</td><td>0</td><td>39.2</td><td>0</td><td>43.6</td></tr>
<tr><td rowspan="4">Level 2</td><td>5</td><td>0</td><td>12.4</td><td rowspan="4">0</td><td rowspan="4">1.2</td><td>0</td><td>12.4</td><td rowspan="4">0</td><td rowspan="4">0.9</td></tr>
<tr><td>6</td><td>0</td><td>5</td><td>0</td><td>5.8</td></tr>
<tr><td>7</td><td>0</td><td>14.0</td><td>0</td><td>8.8</td></tr>
<tr><td>8</td><td>0</td><td>7.6</td><td>0</td><td>8.4</td></tr>
<tr><td rowspan="4">Level 3</td><td>9</td><td>0</td><td>6.2</td><td rowspan="4">0</td><td rowspan="4">3.8</td><td>0</td><td>2.2</td><td rowspan="4">0</td><td rowspan="4">1.2</td></tr>
<tr><td>10</td><td>0</td><td>1.4</td><td>0</td><td>0.8</td></tr>
<tr><td>11</td><td>0</td><td>6.8</td><td>0</td><td>1.4</td></tr>
<tr><td>12</td><td>0</td><td>2</td><td>0</td><td>0.6</td></tr>
</table>

4 Conclusion

Three key issues are involved in palmprint verification: preprocessing and segmentation, feature extraction and matching. The Gaussian filter was used for preprocessing. Whereas, noise reduction and the smoothing of palmprint images are done in preprocessing, palmprint segmentation involves correcting distortions and putting all palmprints under the same coordinate system so that the expected area of each palmprint (ROI) can be extracted for feature extraction and matching. Two kinds of palmprint segmentation approaches were presented. One was designed to extract 100×100 fixed size square area as the central part sub-image for the feature extraction, while the other one suggested using an inscribed circle (with the mean average radius of 100 pixel for database images).

Two different methods were employed to extract features from two sub-images of the central area of the palm. The first method used was the Gabor Filter feature extraction. Two kinds of Gabor Filter, traditional and circular, were used for this purpose, and two palmprint images are compared in terms of a criterion called hamming distance. The accuracy of 99.90% was obtained for identity verification through applying the circular Gabor filter on the square sub-images from the central area of the palm for the purpose of extracting feature.

References

1. Kong, A., Zhang, D., Lu, G.: A Study of Identical Twins' Palmprints for Personal Authentication. In: Zhang, D., Jain, A.K. (eds.) ICB 2006. LNCS, vol. 3832, pp. 668–674. Springer, Heidelberg (2006)
2. Jain, A., Bolle, R., Pankanti, S.: Biometrics: Personal Identification in Networked Society. Kluwer Academic Publishers, Dordrecht (1999)
3. Miller, B.: Vital Signs of identity. IEEE Spectrum 2, 22–30 (1994)
4. Ross, A., Jain, A.K.: Multimodal Biometrics: An Overview. In: Proceedings of the 12th European Signal Processing Conference, pp. 1221–1224 (2004)
5. Lu, G., Zhang, D., Wang, K.: Palmprint Recognition using Eigenpalms Features. Pattern Recognition Letters 24, 1473–1477 (2003)
6. Li, W., Zhang, D., Xu, Z.: Palmprint Identification by Fourier Transform. International Journal of Pattern Recognition and Artificial Intelligence 16, 417–432 (2003)
7. Funada, J., Ohta, N., Mizoguchi, M., Temma, T., Nakanishi, K., Murai, A., Sugiuchi, T., Wakabayashi, T., Yamada, Y.: Feature Extraction Method for Palmprint Considering Elimination of Creases. In: Proceedings of the 14th International Conference on Pattern Recognition, vol. 2, pp. 1849–1854 (1998)
8. Zhang, D., Shu, W.: Two Novel Characteristics in Palmprint Verification: Datum Point Invariance and Line Feature Matching. Pattern Recognition 32, 691–702 (1999)
9. Wu, X., Wang, K., Zhang, D.: Fuzzy Directional Element Energy Feature (FDEEF) based Palmprint Identification. In: Proceedings of the 16th International Conference on Pattern Recognition, vol. 1, pp. 95–98 (2002)
10. Han, C.C., Cheng, H.L., Lin, C.L., Fan, K.C.: Personal Authentication using Palmprint Features. Pattern Recognition 2, 281–371 (2003)
11. Wu, X., Zhang, D., Wang, K.: Fisherpalms based Palmprint Recognition. Pattern Recognition Letters 24, 2829–2838 (2003)
12. Li, W., You, J., Zhang, D.: Texture-based Palmprint Retrieval using a Layered Search Scheme for Personal Identification. IEEE Transactions on Multimedia 7, 891–898 (2005)
13. Kong, W.K., Zhang, D., Li, W.: Palmprint Feature Extraction using 2-D Gabor Filters. Pattern Recognition 10, 2339–2347 (2003)
14. You, J., Li, W., Zhang, D.: Hierarchical Palmprint Identification via Multiple Feature Extraction. Pattern Recognition 35, 847–859 (2002)
15. Zhang, J., Tan, T.: Rotation Invariant Texture Segmentation via Circular Gabor Filters. In: IEEE International Conference on Pattern Recognition (2002)
16. Biometrics Research Centre (BRC), `http://www4.comp.polyu.edu.hk/~biometrics/`
17. Wu, P.S., Li, M.: Pyramid Edge Detection based on Stack Filter. Pattern Recognition Letters 18, 329–348 (1997)

Kernel Principal Component Analysis of Gabor Features for Palmprint Recognition*

Murat Aykut and Murat Ekinci

Computer Vision & Pattern Recognition Lab.
Department of Computer Engineering, Karadeniz Technical University
61080 Trabzon, Turkey
{murat_aykut,ekinci}@ktu.edu.tr

Abstract. This paper presents Gabor-based kernel Principal Component Analysis (KPCA) method by integrating the Gabor wavelet and the KPCA methods for palmprint recognition. The intensity values of the palmprint images extracted by using an image preprocessing method are first normalized. Then Gabor wavelets are applied to derive desirable palmprint features. The transformed palm images exhibit strong characteristics of spatial locality, scale, and orientation selectivity. The KPCA method nonlinearly maps the Gabor wavelet image into a high-dimensional feature space and the matching is realized by weighted Euclidean distance. The proposed algorithm has been successfully tested on the PolyU palmprint database which the samples were collected in two different sessions. Experimental results show that this method achieves 97.22% accuracy for PolyU dataset using 3850 images from 385 different palms captured in the first session as train set and the second session im0061ges as test set.

Keywords: Palmprint recognition, Biometrics, Gabor-wavelet, Kernel PCA.

1 Introduction

Over the last decade, palmprint recognition has attracted growing attention in biometrics [1]. The palmprint is a relatively new biometric feature, and can be used to recognize a person based on unique features in his palm, such as the principal lines, wrinkles, ridges, and texture etc. Furthermore, the several advantages of the palmprint identification can be (1) low-resolution imaging; (2) low-cost capture devices; (3) low intrusiveness; (4) difficulties to a fake palmprint; (5) stable line features, and so forth. In the literature, various techniques have been proposed for palmprint recognition. These techniques have focused on the extraction of structural [2], statistical [3], textural [4], algebraic [1] and multiple biometrics features [5].

In this paper, a novel Gabor-based kernel PCA method, by integrating the Gabor wavelet and the kernel PCA is presented for palmprint recognition. In the proposed method, Gabor wavelets first derive desirable palm features characterized by spatial frequency, spatial locality, and orientation selectivity to cope with the variations of

* This work has fully been supported by the TUBITAK Research Project 107E212.

M. Tistarelli and M.S. Nixon (Eds.): ICB 2009, LNCS 5558, pp. 685–694, 2009.
© Springer-Verlag Berlin Heidelberg 2009

illumination and location. The kernel PCA method [9] is then applied to project palmprints from the high dimensional palmprint space to a significantly lower-dimensional feature space, in which the palmprints can be discriminated much more efficiently. Finally, weighted Euclidean distance (WED) and the nearest neighbor classifier are used for feature matching and classification.

This paper also expands on [10] by including new experimental studies. The feasibility of the proposed method has been successfully tested on PolyU palmprint dataset that contains 7,752 images corresponding to 386 different palms. In the dataset, the images were acquired under variable illumination conditions and collected on two separate occasions. A large number of papers in the literature reported their recognition results on either the earlier or new version of PolyU databases, but least of them made informed comparisons among different algorithms. Here, we provide some comparative experiments to examine the performance of the proposed algorithm.

This paper is organized as follows. Section 2 introduces briefly Gabor wavelets and Gaborpalm extraction. Description of recently popular feature extraction method kernel PCA is given in Section 3. Experimental studies and results on palmprint database are summarized in Section 4, followed by conclusions in Section 5.

2 Gabor Feature Representation

Gabor (wavelet, kernel, or filters) functions are Gaussian modulated by complex sinusoids. The Gabor wavelets-based features were introduced to image analysis due to their biological relevance, performance of capturing salient visual properties [12] and computational properties [11]. The Gabor wavelets have been widely applied to texture segmentation [13], fingerprint recognition [15] and face recognition [8].

In the Gabor wavelets, basis function is the Gabor function defined as:

$$g(x,y)=\left(\frac{1}{2\pi\sigma_x\sigma_y}\right)\exp\left[-\frac{1}{2}\left(\frac{x^2}{\sigma_x^2}+\frac{y^2}{\sigma_y^2}\right)+2\pi jWx\right] \tag{1}$$

where σ_x and σ_y are the horizontal and vertical standard deviation. Then Gabor wavelets can be obtained with dilations and rotations of this function like this:

$$\begin{aligned} &g_{mn}(x,y)=a^{-m}g(x',y'),\ a>1,\\ &x'=a^{-m}(x\cos\theta+y\sin\theta),\ \ and\ \ y'=a^{-m}(-x\sin\theta+y\cos\theta), \end{aligned} \tag{2}$$

where $\theta = n\pi/K$, a^{-m} is the scale factor, and K is the total number of orientations. Because Gabor functions form a non-orthogonal basis set, filtered images has redundant information. For this purpose, Manjunath and Ma [14] developed a strategy to reduce this redundancy. In this strategy, horizontal and vertical standard deviation values are computed according to lower and upper center frequencies of interest.

The Gabor feature representation of a palm image is obtained by convolving the family of Gabor kernels which can be produced by selecting different orientations and scales, with the palm image. Therefore, the Gabor wavelet representation formed by the set $S=\{P_{mn}(x,y): m\in\{0,\cdots,7\},\ n\in\{0,\cdots,4\}\}$.

In our approach, to reduce the computational costs of calculation for palmprint representation, five different scales and four orientations of Gabor wavelets are used. Parameters of the Gabor function are selected as U_{low}=0.2, and U_{high}=0.5, empirically.

3 Kernel PCA

The KPCA is a technique for non-linear feature extraction [9]. It is used to nonlinearly transform an input space into a high-dimensional feature space to achieve linearly separable with high probability for nonlinearly separable patterns in the input space.

Let $x_1, x_2, \cdots, x_M \in \mathbb{R}^N$ be the data in the input space (Gabor coefficients in this work), and let Φ be a nonlinear mapping between the input space and the feature space: $\Phi : \mathbb{R}^N \rightarrow F$, and then performing a linear PCA in *F*.

Assuming the mapped data is centered on the feature space, kernel PCA diagonalizes the estimate of the covariance matrix of the mapped data, defined as:

$$C = \frac{1}{M} \sum_{i=1}^{M} \Phi(x_i) \cdot \Phi(x_i)^T \tag{3}$$

For KPCA, the nonlinear mapping, Φ, usually defines a kernel function. The most often used kernel functions are polynomial, Gaussian and sigmoid kernels [9]. Gaussian kernel was used for the experiments in this work, and it is defined as:

$$k(x_i, x_j) = \exp\left(-\frac{\left\| x_i - x_j \right\|^2}{2\sigma^2} \right), \tag{4}$$

The eigenvalues, $\lambda_1, \lambda_2, \cdots, \lambda_M$, and the eigenvectors, $V_1, V_2, \cdots, V_M$ of kernel PCA can then be derived by solving the following eigenvalue equation as given in [9]:

$$MA\Lambda = KA, \tag{5}$$

where $A = (\alpha_1, \cdots, \alpha_M)$ is an M x M orthogonal eigenvector matrix, Λ is a diagonal eigenvalue matrix and *M* is the number of training samples. $K \in \mathbb{R}^{M \times M}$ define a kernel matrix by means of dot product in the feature space. Since the eigenvalue equation is solved for *A*, we will have to normalize *A* to ensure that eigenvalues of kernel PCA have unit norm in the feature space, therefore $\alpha_j = \alpha_j \big/ \sqrt{\lambda_j}$. After the normalization, the eigenvector matrix, V, of kernel PCA is computed as follows:

$$V = DA, \quad D = \left[\Phi(x_1) \Phi(x_2) \cdots \Phi(x_M) \right] \tag{6}$$

where *D* is the data matrix in the feature space. Now let x be a test example whose map in the higher dimensional feature space is $\Phi(x)$. The kernel PCA features for this example are derived as follows:

$$F = V^T \Phi(x) = A^T B, \quad B = \left[\Phi(x_1) \cdot \Phi(x) \quad \Phi(x_2) \cdot \Phi(x) \quad \cdots \quad \Phi(x_M) \cdot \Phi(x) \right] \tag{7}$$

4 Experiments

4.1 PolyU Palmprint Database

The performance of the proposed method is evaluated on PolyU [3] palmprint database. In the literature, PolyU database is most-well known palmprint database which includes palm samples captured from different sessions. PolyU palmprint database was obtained by collecting palmprint images from 193 individuals using a palmprint capture device. This database contains 7752 gray scale images corresponding to 386 different palms. The samples were collected in two sessions, where ten samples were captured in each session. In addition, they changed the light source and adjusted the focus of the CCD camera so that the images collected on the first and second occasions could be regarded as being captured by two different palmprint devices [3].

At the experiments, we use the preprocessing technique described in [3] to align the palmprints. In this technique, the tangent of the two holes (they are between the fore and the middle finger, and between the ring and the little finger) are computed and used to align the palmprint. The central part of the image, which is 128 x 128, is then cropped to represent the whole palmprint. Such preprocessing greatly reduces the translation and rotation of the palmprints captured from the same palms. We then use z-score normalization on the gray-level distribution of the image. After the discrete transforms and feature extraction stages the WED is used to cluster the features [16].

4.2 Palmprint Identification

Palmprint identification searchs to answer the question "who is the person?". In identification the image is matched against each template and, finally, the most similar template is obtained as the identification result.

In our experiments, firstly region of interest is cropped from the palm images. In order to reduce the computational cost of the Gabor wavelet, each subimage is reduced to 32 x 32 for Gabor-based experiments. We then obtain the Gabor feature representation of the palm images. Then kernel PCA works and nonlinearly derives low-dimensional features. WED-based matching is finally used to cluster those features. In our recognition experiments, first session samples are selected as training, and the second session samples are selected as test set. In the experiments, different feature lengths (50, 100, 200, 300, and 380) are used for comparative results.

The most well-known discrete transforms, such as FFT, DCT and DWT are alternatively employed in the proposed algorithm in order to examine the effects of these transform. In the FFT, the coefficients correspond to the lower frequencies than 3 x 3, and higher frequencies than 16 x 16, are discarded by filtering. In DCT, lower frequencies correspond to the 12.5% coefficients are also selected as useful features. In DWT, the wavelet transform of the palm image is extended to the fourth level. We then choose lowest frequency subimage with a matrix of 16 x 16 as the feature vectors obtained by Daubechies-8 filter banks [17]. In discrete transforms, parameters were empirically determined to achieve highest accuracies for palmprint recognition.

The experimental results are also summarized in Table 1. The number given in Table 1 represents the correct recognition rates. As can be seen from Table 1, the presented approach has achieved highest recognition rate. To improve the experimental results by doing more experiments, the image size 64 x 64 has not been able to choosen because of the memory limitation constrained by the operating system. Hence, 32 x 32 image size will be used for Gabor based approach. The other important point summarized in Table 1 is that, kernel PCA gives higher performance than PCA for all feature lengths used.

Table 1. Comparative performance evaluation for different methods on the PolyU database with different feature lengths

Method	Feature Length (Eigenvectors)				
	50	75	100	200	300
PCA	88.597%	90.311%	90.857%	91.246%	91.246%
KPCA	88.597%	90.415%	90.857%	91.116%	91.168%
DCT+KPCA	89.74%	91.636%	92.311%	93.376%	93.454%
FFT+KPCA	71.324%	76.181%	78.805%	82.411%	84.493%
DWT+KPCA	89.376%	91.324%	92.493%	93.428%	93.584%
Gabor+PCA	82.597%	88.025%	90.701%	93.61%	94.311%
Gabor+KPCA	83.038%	88.285%	90.987%	93.714%	94.415%
Gabor+KPCA (64 x 64)	86.103%	90.259%	92.051%	94.259%	94.805%

Table 2. Comparison of correct recognition rates (%) of different methods with the different number of the samples per palm in the kernel matrix calculation

Method	Number of samples in each class									
	1	2	3	4	5	6	7	8	9	10
DWT+KPCA	93.58	93.69	93.77	94.0	94.05	94.23	94.36	94.47	94.52	94.60
Gabor+PCA	94.41	94.65	94.78	94.78	95.84	96.0	96.36	96.57	96.86	96.91
Gabor+KPCA	94.57	94.65	95.58	96.54	96.57	96.75	96.88	97.09	97.12	97.22

The results presented so far are for the kernel matrix, calculated by using only one sample for each palm. To make further improving on the recognition accuracies, the different number of the samples for each palm are used in the following experiment. This is also one of the contribution in this paper. In this experiment we will discuss results for only 32 x 32 image sizes. As mentioned above, while the image size is reduced to 64 x 64 instead of 32 x 32, the application of the presented approach was not successful for the all samples used for per palm due to typical constraints on the memory available in term of the operating system. Operating systems based Microsoft Windows NT provide applications with a 4GB virtual address space. The virtual address space is divided such that 2GB is available to the application and the other 2GB is available only to the system [20]. Table 2 presents the comparison of correct recognition rates using KPCA on DWT features, PCA and KPCA on gabor-wavelet features. A high recognition rate (97.22%) was achieved for the proposed method, with the all samples usage in calculation of the kernel matrix. It is evident that the number of the samples for each class play an important role in the kernel trick. In the experiments given in Table 2, the feature dimensions of the samples were dynamically determined with PCA or KPCA method by ratio of the sum of chosen largest eigenvalues to all (about 98%).

When the palmprints are collected in different sessions, direction and amount of stretching of a palm may vary so that even palmprints from the same palm may have a little rotation and translation. Furthermore, the lighting, translation, and orientation conditions in both sessions are very different. Hence they will effect the accuracy. As conclusion, the experiments achieved on the training and test images collected in different sessions are becoming more realistic. Figure 1 shows typical samples and problems in the database. The palms shown in the first column were used as the train set, and the corresponding to the last two samples were also employed as test set.

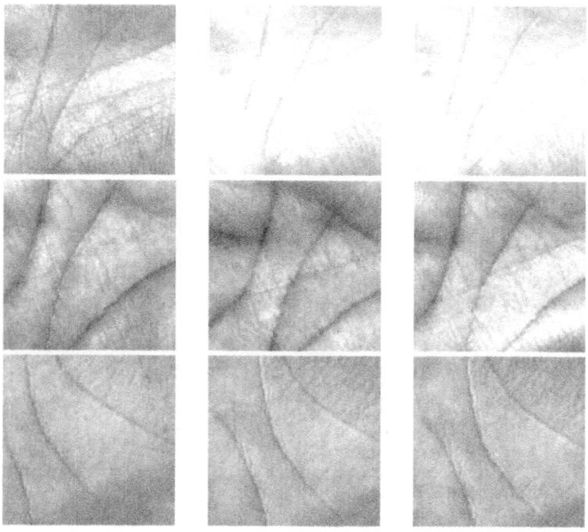

Fig. 1. Experimental results by different lighting (top), orientation (middle) and rotation (bottom) conditions

In biometric systems, the recognition accuracy will decrease dramatically when the number of image classes increase [1]. In our previous work [10], we have achieved 100% accuracy for the earlier version of PolyU palmprint database which contains 600 images from 100 different palms. 300 images collected in the first session were selected for training, the rest samples were also used as test set.

4.2.1 Comparisons

Most published papers in the literature has presented their experimental results on the samples collected in the same session. For instance, the presented results in the published papers [6][7][18] were achieved by using training and test sets contain the palmprint images collected from one session only. In the other paper [4], they selected some of the images collected in the first session as training set, and then used the rest images as test set. As different from them, in our experiments, given in Table 1 and 2, first session palm images were chosen as training set and the second session palm images were used as the test set.

To compare the performance of the proposed algorithm, the experimental results given in Table 3 were obtained from the first session samples only. We took the four samples of each person as training set, and the remaining six samples as test set. In this test before the Gabor transform images are sampled to 64 x 64 resolution. The numbers given in Table 3 represent the corresponding recognition rate. A high recognition rate (99.697%) was achieved for kernel PCA with Gabor-wavelet and WED-based classifier approach. The experimental results of different discrete transforms are also summarized in Table 3 to show their performance.

Table 3. Testing results of different approaches with different feature lengths

Method	Feature Length				
	50	100	200	300	380
PCA	98.747%	99.179%	99.093%	99.05%	98.963%
KPCA	98.877%	99.222%	99.05%	99.006%	98.92%
DWT+PCA	98.834%	99.309%	99.352%	99.352%	99.395%
Gabor+PCA	98.056%	99.222%	99.438%	99.438%	82.728%
DWT+KPCA	98.747%	99.309%	99.568%	99.654%	99.654%
Gabor+KPCA	98.1%	99.265%	99.525%	99.697%	99.697%

Table 4 comparatively presents the experimental results achieved by different palmprint recognition approaches. The works in [6][7][18][4] used the early versions or some samples of the PolyU palmprint database. In other works [5][19], they also collected their palmprint images on one occasion only.

Table 4. Comparisons of different palmprint recognition methods

	The approaches						
	This paper	In [6]	In [7]	In [18]	In [4]	In [5]	In [19]
Different palms	386	300	382	50	190	100	160
Total samples	3860	3000	3056	200	3040	1000	1600
Recog. Rate (%)	99.697	99.2	99.149	98.0	98.13	95.8	97.25

In the work [6], a linear projection-based on Fisher's linear discriminant was used. Then a nearest-neighbor classifier was employed for classification. In [7], they proposed an eigenpalm method and applied to recognition with an Euclidean distance. In another work [18], they use the wavelet transform and identify the predominant structures using context modeling. The work in [4] proposed an approach based on DCT and linear discriminant technique. In their experiments, they used five images of each person as the training samples, and the remainder (11 images of each person) as the test samples. Kumar and Zhang [5] proposed a method of hand-shape and palmprint image segmentation, and the combination of features from these images.

Their approach uses DCT and some classification algorithms to evaluate the performance. In the study [19], kernel FDA is proposed for palmprint recognition. Although our proposed method is tested on the public database includes more different palms, the proposed approach achieves highly competitive performance with respect to the other published papers.

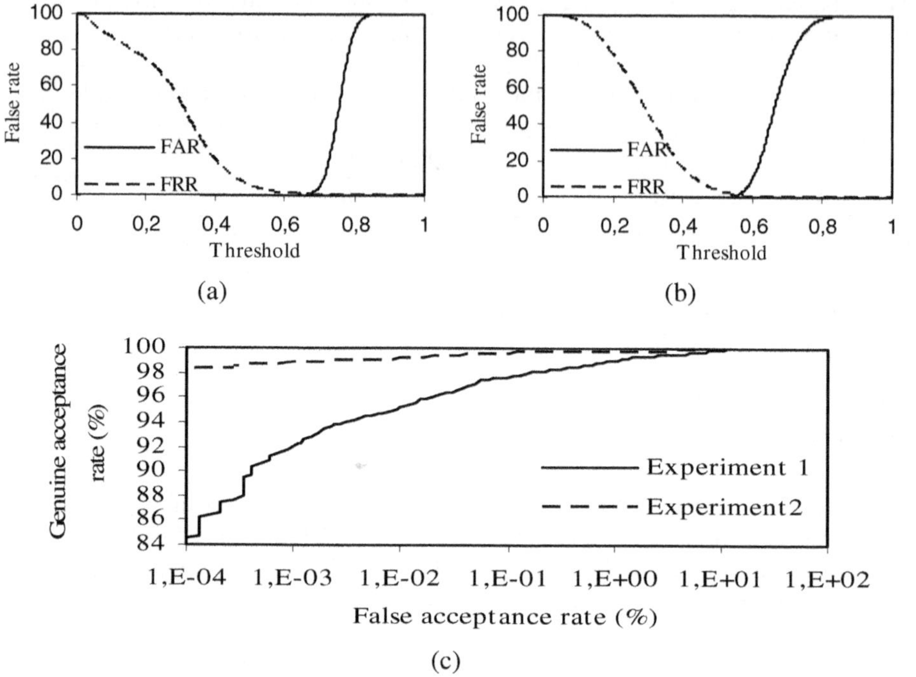

Fig. 2. Verification results of the proposed method (a) FAR-FRR curves of first experiment (b) FAR-FRR curves of second experiment (c) ROC curves of the experiments

Table 5. Experimental results near cross-over points of the FAR and FRR curves

First verification experiment			Second verification experiment		
Threshold	FAR (%)	FRR (%)	Threshold	FAR (%)	FRR (%)
0.737	27.866	0	0.609	13.915	0
0.721	14.282	0.051	0.591	7.166	0.285
0.705	6.255	0.129	0.573	3.188	0.467
0.673	0.804	0.337	0.551	0.960	0.987
0.661	0.337	0.337	0.533	0.321	1.636
0.645	0.102	0.441	0.515	0.099	2.337
0.613	0.009	0.909	0.478	0.009	4.909
0.579	4.7×10^{-4}	1.428	0.434	4.05×10^{-4}	9.610
0.561	6.7×10^{-5}	1.844	0.398	6.76×10^{-5}	15.662
0.553	0	2.155	0.396	0	16.103

4.3 Palmprint Verification

Biometric verification answers the question of "whether the person is whom he claims to be" by examining his biometric features. To obtain the verification accuracy, we perform two experiments which research impact of the time lag. In both of them, Gabor wavelet transform is applied for the palmprint images at 32 x 32 resolutions. Feature extraction and matching procedures are achieved by using KPCA and WED. In the KPCA, kernel space is formed using 10 first session samples for each class.

In the first experiment on the verification, the palm images in the first session only were used as database. Each palmprint image is then matched with all the other palmprint images in the database. The total number of comparisons is 14,818,650, where the number of intra-class comparisons is 34,650 and the number of inter-class comparisons is 14,784,000. In the verification system, the performance of the system is often measured by the false accept rate (FAR) and false reject rate (FRR). Fig. 2.a and Fig. 2.c show the FAR and FRR curves and corresponding Receiver Operating Characteristic (ROC) curve, which is a plot of genuine acceptance rate against false acceptance rate for all possible operating points, respectively. The experimental results near the cross-over point of the FAR and FRR curves are tabulated in Table 5. From Fig. 2 and Table 5, it can be seen that EER of this experiment is 0.338% and genuine acceptance rate is 98.16% while FAR is about 6×10^{-5} %.

In the second experiment, the palm images collected in the first session were used as training set; the images accumulated in the second session were used as test set. Each palmprint image in the training set is matched with all the palmprint images in the test set. The total number of comparisons is 14,822,500, where the number of intra-class comparisons is 38,500 and the rest are inter-class comparisons. Fig. 2.b and Fig. 2.c show the FAR and FRR curves of this experiment and corresponding ROC curve, respectively. The experimental results near the cross-over point of the FAR and FRR curves are finally tabulated in Table 5. From Fig. 2.b and Table 5, it can be seen that EER of this experiment is 0.982% and genuine acceptance rate is 90.39% while FAR is about 5×10^{-4} %.

5 Conclusions

In this paper, Gabor-based KPCA method by integrating the Gabor wavelet representation of palm images and the kernel PCA method are proposed for palmprint recognition. The Gabor wavelets are particularly aggressive at capturing the features of the local structures corresponding to spatial frequency, spatial localization, and orientation selectively and Gaborpalms can be robust against variations due to illumination and pose changes. The kernel PCA overcomes many limitations of its linear counterpart by nonlinearly mapping the input space to a high-dimensional feature space. The proposed algorithm has been successfully tested on public data set from the PolyU palmprint database for which the samples were collected in two different sessions. In the experiments, the samples captured in the first session were used as training set, the others collected in the second session were chosen as test set. Experimental results show that 1) Gabor-wavelet is an effective and efficient palmprint representation, and 2) the proposed recognition approach achieves highly competitive performance with respect to the published major palmprint recognition approaches.

References

1. Zhang, D., Jing, X., Yang, J.: Biometric Image Discrimination Technologies. Computational Intelligence and Its Application Series. Idea Group Publishing (2006)
2. Zhang, D., Shu, W.: Two novel characteristics in palmprint verification: Datum point invariance and line feature matching. Pattern Recognition 32(4), 691–702 (1999)
3. Zhang, D., Kongi, W., You, J., Wong, M.: Online Palmprint Identification. IEEE Trans. on Pattern Analysis and Machine Intelligence 25(9), 1041–1049 (2003)
4. Jing, X.Y., Zhang, D.: A Face and Palmprint Recognition Approach Based on Discriminant DCT Feature Extraction. IEEE Trans. on Systems, Man, and Cybernetics 34(6) (2004)
5. Kumar, A., Zhang, D.: Personal Recognition Using Hand Shape and Texture. IEEE Transactions on Image Processing 5(8), 2454–2460 (2006)
6. Wu, X., Zhang, D., Wang, K.: Fisherpalms Based Palmprint Recognition. Pattern Recognition Letters 24(15), 2829–2838 (2003)
7. Lu, G., Zhang, D., Wang, K.: Palmprint Recognition Using Eigenpalms Features. Pattern Recognition Letters 24(9-10), 1463–1467 (2003)
8. Liu, C.: Gabor-Based Kernel PCA with Fractional Power Polynomial Models for Face Recognition. IEEE Transactions on PAMI (5), 572–581 (2004)
9. Scholkopf, B., Smola, A.: Learning with Kernels: Support Vector Machine, Regularization, Optimization and Beyond. MIT Press, Cambridge (2002)
10. Ekinci, M., Aykut, M.: Gabor-Based Kernel PCA for Palmprint Recognition. IET Electronics Letters 43(20), 1077–1079 (2007)
11. Daugman, J.G.: Two-Dimensional Spectral Analysis of Cortical Receptive Field Profile. Vision Research 20, 847–856 (1980)
12. Liu, D., Wechsler, H.: Independent Component Analysis of Gabor Features for Face Recognition. IEEE Transactions on Neural Networks 14(4), 919–928 (2003)
13. Weldon, T.P., Higgins, W.E., Dunn, D.F.: Efficient Gabor Filter Design for Texture Segmentation. Pattern Recognition 29(12), 2005–2015 (1996)
14. Manjunath, B.S., Ma, W.Y.: Texture Feature for Browing and Retrieval of Image Data. IEEE Transaction on Pattern Analysis and Machine Intelligence 18(8), 837–842 (1996)
15. Lee, C.J., Wang, S.D.: Fingerprint Feature Extraction Using Gabor Filters. Electronic Letters 35(4), 288–290 (1999)
16. Zhu, Y., Tan, T., Wang, Y.: Biometric Personal Identification Based on Handwriting. In: IEEE Int. Conference on Pattern Recognition, vol. 2, pp. 797–800 (2000)
17. Daubechies, I.: Ten Lecture on Wavelets. Capital City Press, Philadelphia (1992)
18. Zhang, L., Zhang, D.: Characterization of Palmprints by Wavelet Signatures via Directional Context Modeling. IEEE Trans. on Systems, Man, and Cybernetics 34, 1335–1347 (2004)
19. Wang, Y., Ruan, Q.: Kernel Fisher Discriminant Analysis for Palmprint Recognition. In: The 18th International Conference on Pattern Recognition (ICPR 2006) (2006)
20. Microsoft Windows Platform SDK Windows XP SP2, Memory Management, © (2004)

Latent Fingerprint Matching: Fusion of Rolled and Plain Fingerprints

Jianjiang Feng, Soweon Yoon, and Anil K. Jain

Department of Computer Science and Engineering
Michigan State University
{jfeng,yoonsowo,jain}@cse.msu.edu

Abstract. Law enforcement agencies routinely collect both rolled and plain fingerprints of all the ten fingers of suspects. These two types of fingerprints complement each other, since rolled fingerprints are of larger size and contain more minutiae, and plain fingerprints are less affected by distortion and have clearer ridge structure. It is widely known in the law enforcement community that searching both rolled and plain fingerprints can improve the accuracy of latent matching, but, this does not appear to be a common practice in law enforcement. To our knowledge, only rank level fusion option is provided by the vendors. There has been no systematic study and comparison of different fusion techniques. In this paper, multiple fusion approaches at three different levels (rank, score and feature) are proposed to fuse rolled and plain fingerprints. Experimental results in searching 230 latents in the NIST SD27 against a database of 4,180 pairs of rolled and plain fingerprints show that most of the fusion approaches can improve the identification performance. The greatest improvement was obtained by boosted max fusion at the score level, which reaches a rank-1 identification rate of 83.0%, compared to the rank-1 rate of 57.8% for plain and 70.4% for rolled prints.

Keywords: Latent fingerprint, rolled fingerprint, plain fingerprint, fusion, minutiae matching.

1 Introduction

Automated Fingerprint Identification Systems (AFIS) have played an important role in forensic, law enforcement and many civilian applications. Fingerprint images in AFIS can be broadly classified into three categories, namely, (i) rolled, (ii) plain/flat and (iii) latent. Figure 1 shows these three types of fingerprint images from the same finger. Rolled fingerprint images are obtained by rolling a finger from one side to the other ("nail-to-nail") in order to capture all the ridge details of a finger. Plain fingerprints are those in which the finger is pressed down on a flat surface but not rolled. Rolled and plain impressions are obtained either by scanning the inked impression on paper or by directly using livescan devices. In AFIS, rolled and plain fingerprints are generally acquired in an attended mode and are subject to a recapture if the image quality is poor. In

M. Tistarelli and M.S. Nixon (Eds.): ICB 2009, LNCS 5558, pp. 695–704, 2009.
© Springer-Verlag Berlin Heidelberg 2009

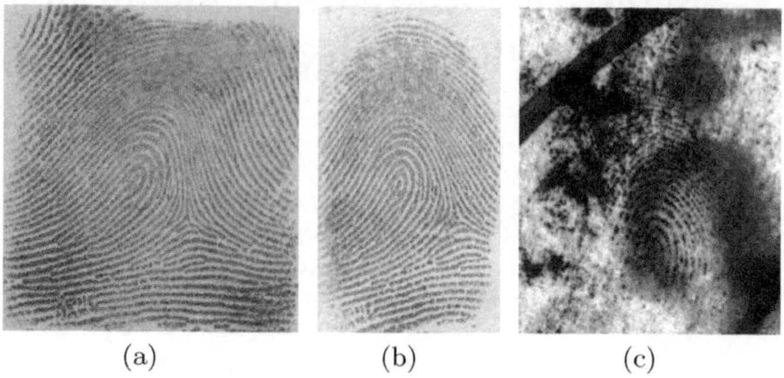

Fig. 1. Three types of fingerprint images. (a) Rolled fingerprint, (b) plain fingerprint and (c) latent fingerprint.

contrast, latent fingerprints are lifted from surfaces of objects that are inadvertently touched or handled by a person through a variety of means ranging from simply photographing the print to more complex dusting or chemical processing [1]. With small area, unclear ridge structure, complex background and strong distortion, latent fingerprints generally have the worst image quality among the three types of fingerprints. However, it is the matching of a latent fingerprint against a database of rolled/plain fingerprints that is of utmost importance in forensics and law enforcement to apprehend suspects.

AFIS may work in automatic or semi-automatic mode depending on specific applications. In civil background check application, 10 finger impressions (rolled or plain) are submitted and the AFIS automatically returns the mated subject or reports no matches found. In suspect identification application, latents are submitted and the system generally returns a list of top candidates, which are then reviewed by latent experts.

Fingerprint image quality has a significant impact on system accuracy. Since the quality of latents is not controllable, it is important to ensure that enrolled fingerprints in the database have as clear ridge structure as possible. During the enrollment, the rolled and plain fingerprints of the ten fingers of a person are obtained through fourteen impressions. These impressions can be captured by using traditional tenprint cards or livescan devices. Traditional tenprint cards (see Fig. 2) contain the rolled impressions of the ten fingers as well as four slap impressions: the left slap (four fingers of the left hand), the right slap (four fingers of the right hand) and the thumb slaps (left and right thumbs). A segmentation algorithm is used to automatically segment the slaps into individual plain fingerprints. There are two reasons for including plain fingerprints in tenprint cards: (i) while rolled fingerprints contain larger size and larger number of minutiae, plain fingerprints are less distorted and often have clearer ridges; and (ii) out-of-sequence rolled fingerprints can be easily detected by matching rolled images to slap images [2]. With an aim to avoid the time-consuming and error prone rolling process, a touchless 3D fingerprint sensor was proposed in [3], which reconstructs 3D fingerprint using stereo vision techniques and then

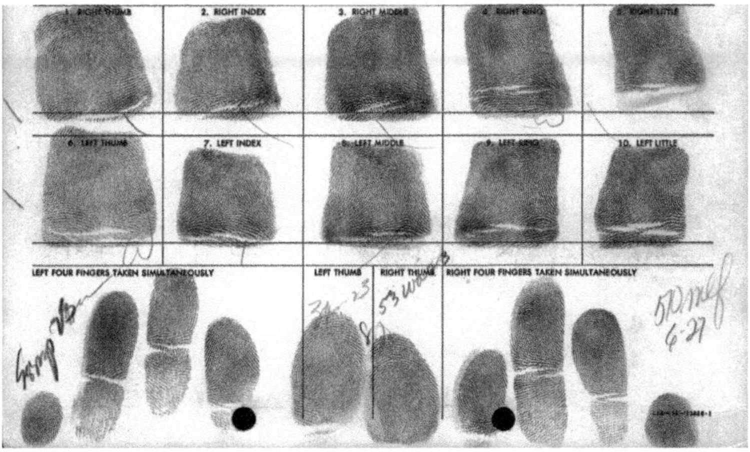

Fig. 2. A tenprint card from NIST SD29 consisting of fourteen fingerprint impressions

creates 2D rolled-equivalent fingerprints. While such a scanner is very desirable, the image quality provided by the current prototype of the scanner still needs significant improvement [4].

There have been some studies that compare matching accuracy using plain and rolled fingerprints. In the plain-to-rolled fingerprint matching experiments conducted by NIST [5], the identification rate of 10-finger plain-to-rolled and rolled-to-rolled is 97.5% and 97.9%, respectively. The Federal Bureau of Investigation (FBI) [6] has conducted an experiment to evaluate the accuracy of matching latents against plain and rolled fingerprints. The hit rate of searching 250 latents in NIST SD27 against plain and rolled fingerprints in the FBI's Integrated Automated Fingerprint Identification System (IAFIS) is 38.8% and 54.4%, respectively. Fusion at rank level leads to a hit rate of 61.2%. Given this relatively low performance of latent fingerprint matching [7,8], there is a strong need for the improvement in the fusion of plain and rolled fingerprints for latent fingerprint matching.

Fusion of plain and rolled fingerprints is a type of multibiometric technique [9]. A multibiometric system can combine evidences from different sources, such as multiple fingers [10], multiple sensors [11], multiple samples [12] and multiple representations [13]. Fusing plain and rolled fingerprints can be viewed as fusing multiple samples of the finger. It can also be viewed as combining the impressions of the same finger captured by different sensors or sensing procedures. The evidences provided by different sources can be combined at different levels, such as score level [10,13], feature level [12] and image level [12,14]. Although it is generally believed that early stage fusion (image level) should provide better performance, fusion at the score level has received much more attention than fusion at the other levels. There are relatively few studies on fusion at the feature and image level in the literature [12,14]. In addition, no systematic work has been done to compare different fusion approaches at different levels for fingerprint identification, especially in the case of fusing plain and rolled fingerprints for latent matching.

In this study, multiple fusion approaches at each of the three different levels (rank, score and feature) are proposed to fuse rolled and plain fingerprints. These fusion approaches are evaluated according to the matching accuracy in searching latent fingerprints against rolled and plain fingerprints. Experimental results indicate that most of the fusion methods can improve the matching accuracy; the greatest improvement was obtained by boosted max fusion at the score level.

(a) (b) (c)

Fig. 3. (a) A latent fingerprint, (b) mated plain fingerprint, and (c) mated rolled fingerprint. Reliable minutiae are marked by green squares and unreliable ones are marked by red circles.

2 Fusion Experiments

2.1 Database and Matcher

Experiments were conducted on the new NIST SD27 database, which consists of 230 latents and 202 tenprint cards. The link between latents and the mated fingerprints was provided. The old NIST SD27 was not used, since it does not contain plain fingerprints. To increase the background database size, all 216 tenprint cards in set A of NIST SD29 were also used. Totally, the background database consists of 4,180 pairs of plain and rolled prints.

The features in the latents were manually marked, which is a common practice in the law enforcement community. Three types of features are used: minutiae, ridge orientation map and quality map. A minutia has four attributes: x, y, direction and quality. The minutia quality has two levels: reliable and unreliable. Ridge orientation map and quality map are defined on non-overlapping blocks of 16×16 pixels. The quality of a block has three levels: background, bad quality and good quality. The features of the plain and rolled fingerprints were automatically detected using the feature extraction algorithm in [8]. Figure 3 shows a latent and the mated plain and rolled fingerprints (with minutiae marked).

The minutiae matching algorithm in [8] was used to match the 230 latents against the database containing 4,180 pairs of plain and rolled fingerprints. The

Cumulative Match Characteristic (CMC) curves of of latent-to-plain and latent-to-rolled matching are shown in Fig. 4. These two CMC curves are also included in all the subsequent figures as a baseline to evaluate fusion performance. A CMC curve plots the rank-k identification rate against k, for $k = 1, 2, \cdots, 20$. The rank-k identification rate indicates the proportion of times the mated fingerprint occurs in the top k matches. The rank-1 identification rates of latent-to-plain and latent-to-rolled are 57.8% and 70.4%, respectively. As expected, the accuracy of seaching latents against rolled images is much higher than against plain images. This phenomenon was also observed in the experiments conducted by FBI [6], where the rank-1 rates of latent-to-plain and latent-to-rolled matching are 38.8% and 54.4%, respectively.

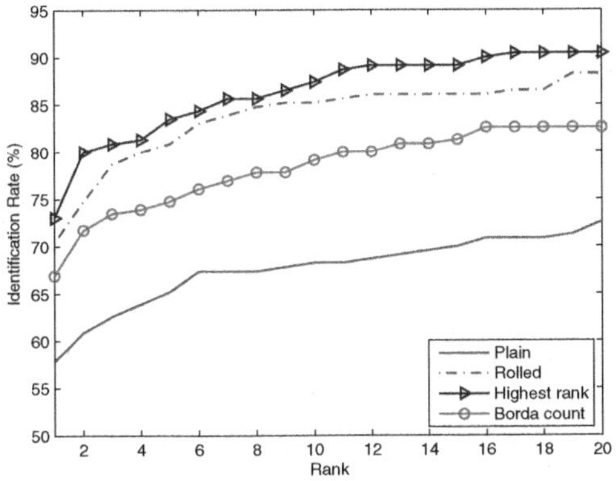

Fig. 4. The CMC curves of two rank level fusion methods: highest rank and Borda count

2.2 Rank Level

Two rank level fusion methods are adopted: highest rank and Borda count [15]. In the highest rank method, fingers are sorted with respect to the higher rank of plain and rolled fingerprints. If two fingers have the same higher rank, they are sorted by the lower rank of plain and rolled fingerprints. If both the higher and the lower ranks are the same, the two fingers are randomly sorted. The Borda count method uses the sum of the ranks of plain and rolled fingerprints to sort fingers. If the sums are the same, the two fingerprints are sorted by the higher rank of plain and rolled fingerprints. If both the higher and the lower ranks are the same, the two fingers are randomly sorted. The CMC curves of these two rank level fusion methods are given in Fig. 4. It can be observed that highest rank method is much better than Borda count. The accuracy of Borda count is even lower than that of latent-to-rolled matching. This is not surprising, since if either one of plain and rolled fingerprints has poor quality, the fused rank according to Borda count will be very low. In our experiments, we found that weighted Borda count also does not provide better performance.

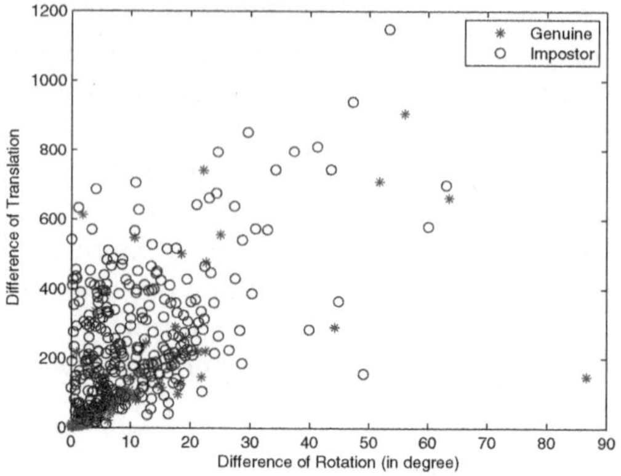

Fig. 5. The scatter plot of the rotation and translation differences of genuine matches and impostor matches

2.3 Score Level

Five score level fusion rules are tested: min, max, sum, product and boosted max. The purpose of the boosted max is to boost the score of genuine matches. Generally, for genuine matches, the spatial transformations among latent, plain and rolled fingerprints are consistent. But for impostor matches, the spatial transformations are generally not consistent. Let T_{LP}, T_{LR} and T_{PR} denote the rigid transformation matrices from latent to plain, latent to rolled and plain to rolled, respectively. These transformations are computed by our minutiae matcher. If the rigid transformation represented by $T_{LPR} = T_{LP} \times T_{PR}$ is similar to the rigid transformation represented by T_{LR}, the transformations among latent, plain and rolled fingerprints are viewed as consistent; otherwise they are viewed as inconsistent. Two rigid transformations, T_{LPR} and T_{LR}, are said to be similar if the difference between the two rotations is less than $\pi/8$ and the Euclidean distance between the two translations is less than 50 pixels. Figure 5 shows the scatter plot of the rotation and translation differences between T_{LPR} and T_{LR} for genuine matches and impostor matches. In the scatter plot, the genuine matches form a cluster near the origin, while the impostor matches scatter in the entire plane. The fused score according to boosted max rule is computed as:

$$s_F = \begin{cases} \max(s_P, s_R) + w \cdot \min(s_P, s_R) & \text{if consistent} \\ \max(s_P, s_R) & \text{if inconsistent} \end{cases}, \tag{1}$$

where s_P and s_R denote the scores of latent-to-plain and latent-to-rolled, respectively, and w is a weight assigned to the minimum score, which has been set as 0.3 empirically. The CMC curves of these five score level fusion methods are given in Fig. 6. It can be observed that (i) all score level fusion rules provide improved performance except for the min rule, and (ii) boosted max gives the largest improvement, improving the rate-1 identification rate from 70.4% (rolled) and 57.8% (plain) to 83.0%. The good performance of boosted max can be explained by the scatter plot in Fig. 5.

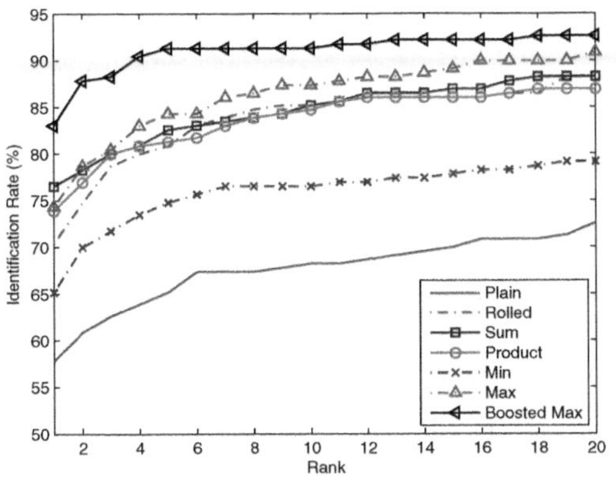

Fig. 6. The CMC curves of five score level fusion methods: sum, product, min, max and boosted max

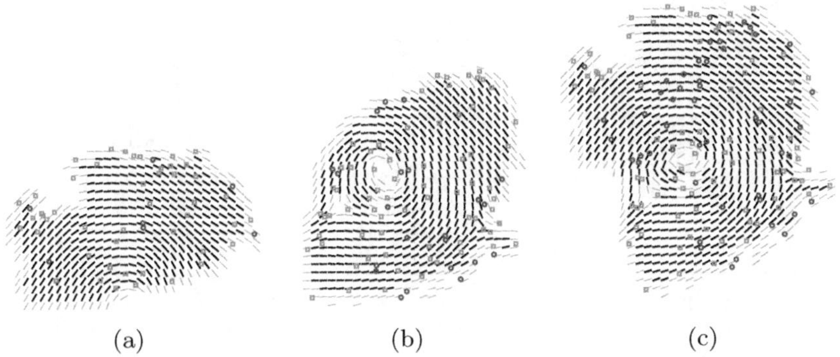

Fig. 7. Features of (a) the plain fingerprint, (b) the rolled fingerprint and (c) the composite fingerprint. Reliable minutiae are marked by green squares and unreliable ones are marked by red circles. Orientations in good quality regions are marked by dark lines and orientations in bad quality regions are marked by bright lines.

2.4 Feature Level

The plain and rolled fingerprints of the same finger are matched using our matcher. The output of the matcher includes a matching score, a set of matched minutiae, and the spatial transformation between the two images. The features of the plain fingerprint are mapped into the coordinate system of the rolled fingerprint. Three types of regions are identified: common region, plain-only region and rolled-only region. In the common region and the rolled-only region, the ridge orientation and quality of the rolled fingerprint are adopted. In the plain-only region, the ridge orientation and quality of the plain fingerprint are adopted.

Now we consider merging two sets of minutiae, minutiae from the plain fingerprint and minutiae from the rolled fingerprint, into one composite minutiae

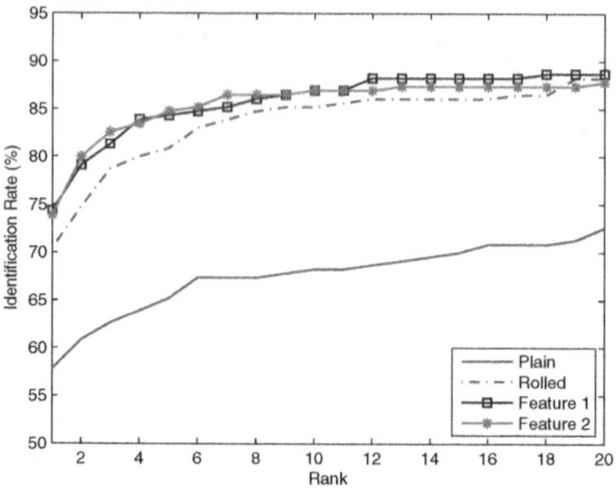

Fig. 8. The CMC curves of two different feature level fusion methods

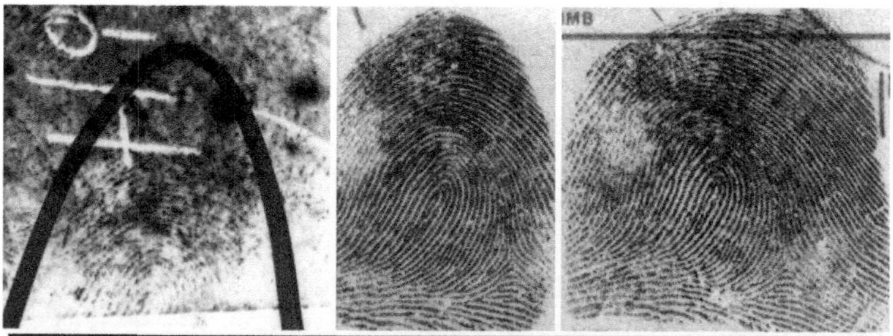

	Plain	Rolled	Highest Rank	Boosted Max	Feature Level
Score	0.39	0.30	-	0.46	0.42
Rank	16	119	31	1	3

Fig. 9. Fusion performance for a latent fingerprint and the mated plain and rolled fingerprints. Boosted max fusion gives the best result.

set. Minutiae that do not lie in the common region of the two fingerprints are directly used in the composite minutiae set. The minutiae in the common region may be either matched or unmatched. For matched minutiae, the minutiae in the rolled fingerprints are used and set as reliable if either of the two matched minutiae is reliable. The unmatched minutiae in the common region are treated in two different ways: (i) use all the unmatched minutiae, or (ii) if an unmatched minutia is reliable, set its quality as unreliable and include it in the composite minutiae set; if an unmatched minutia is unreliable, do not include it in the composite minutiae set. Figure 7 shows the features of a plain fingerprint, the mated rolled fingerprint and the composite fingerprint. The CMC curves of these two feature level fusion methods are given in Fig. 8. It can be observed that (i) both the methods provide improved performance and (ii) the first minutiae fusion method is slightly better than the second one.

An example is given in Fig. 9 to compare the different fusion methods. When searching for the latent in Fig. 9 in the databases, the mated plain fingerprint gets a matching score of 0.39 and a rank of 16, and the mated rolled fingerprint gets a matching score of 0.30 and a rank of 119. The higher rank of the plain image is because in the region that corresponds to the latent, the plain image has clearer ridge structures than the rolled image. Boosted max rule improves the mated finger to rank one and the feature level fusion improves it to rank three. The highest rank method gives a poor result due to the effect of ties.

3 Conclusions and Future Work

The problem of fusing rolled and plain fingerprints to improve the latent matching accuracy has been considered. Different fusion methods at three different levels (namely rank, score and feature) are studied. Experimental results in searching 230 latents against a database consisting of the plain and rolled fingerprints of 4,180 fingers indicate that boosted max provides the greatest improvement in identification accuracy.

The eventual goal of fusing rolled and plain fingerprints is to obtain full fingerprints of high quality. On one hand, new sensors or combination of sensors should be considered and on the other hand, robust fusion schemes should be explored based on existing data and sensors. Fingerprint mosaicking is a promising direction; however, simple image mosaicking using minutiae-based alignment is not likely to provide good performance. To develop a robust fingerprint mosaicking algorithm, the distortion between rolled and plain fingerprints must be accurately estimated, which is a challenging problem. We have only focused on fusing exemplar fingerprints in this work. Fusion of latent fingerprints is another opportunity for improving matching accuracy. For instance, a fingerprint mosaicking algorithm can be used to merge multiple latents left by the same finger into a single latent with higher image quality and larger image size.

Acknowledgments

We would like to acknowledge the support of Sagem Morpho and would like to thank Austin Hicklin, Noblis for providing us the new NIST SD27 database. This work was supported by ARO grant W911NF-06-1-0418 and NIJ grant 2007-RG-CX-K183.

References

1. Lee, H.C., Gaensslen, R.E. (eds.): Advances in Fingerprint Technology. CRC Press, New York (2001)
2. Komarinski, P.: Automated Fingerprint Identification Systems (AFIS). Academic Press, London (2004)

3. Parziale, G., Diaz-Santana, E.: The Surround Imager: a multi-camera touchless device to acquire 3D rolled-equivalent fingerprints. In: Zhang, D., Jain, A.K. (eds.) ICB 2006. LNCS, vol. 3832, pp. 244–250. Springer, Heidelberg (2006)
4. NIST: Evaluation of Fast Tenprint Capture Devices, http://fingerprint.nist.gov/FTcapture/index.html
5. Wood, S.S., Wilson, C.L.: Studies of plain-to-rolled fingerprint matching using the NIST algorithmic test bed (ATB). NISTIR 7112 (2004)
6. Swann, S.: Needs and applications of latents at FBI/CJIS, http://www.itl.nist.gov/iad/894.03/latent/workshop/proc/P6_Swann_LatentOverview_2.pdf
7. NIST: Summary of results from ELFT07 phase I testing (2007), http://fingerprint.nist.gov/latent/elft07/phase1_aggregate.pdf
8. Jain, A.K., Feng, J., Nagar, A., Nandakumar, K.: On matching latent fingerprints. In: Proc. CVPR Workshop on Biometrics, pp. 1–8 (2008)
9. Ross, A., Nandakumar, K., Jain, A.K.: Handbook of Multibiometrics. Springer, Heidelberg (2006)
10. Wilson, C., et al.: Fingerprint vendor technology evaluation 2003: Summary of results and analysis report. NISTIR 7123 (2004)
11. Marcialis, G.L., Roli, F.: Fingerprint verification by decision-level fusion of optical and capacitive sensors. In: Proc. ECCV Workshop on Biometric Authentication, pp. 307–317 (2004)
12. Jain, A.K., Ross, A.: Fingerprint mosaicking. In: Proc. International Conference on Acoustic Speech and Signal Processing (ICASSP), pp. 4064–4067 (2002)
13. Ross, A., Jain, A.K., Reisman, J.: A hybrid fingerprint matcher. Pattern Recognition 36(7), 1661–1673 (2003)
14. Hara, M., Toyama, H.: Print image synthesizing device and method, and print image synthesizing program. US Patent Application Publication No. 2006/002595A1 (2006)
15. Ho, T.K., Hull, J.J., Srihari, S.N.: Decision combination in multiple classifier systems. IEEE Trans. Pattern Analysis and Machine Intelligence 16(1), 66–75 (1994)

Massimo Tistarelli Mark S. Nixon (Eds.)

Advances in Biometrics

Third International Conference, ICB 2009
Alghero, Italy, June 2-5, 2009
Proceedings

Springer

Volume Editors

Massimo Tistarelli
Università di Sassari, Computer Vision Laboratory
Facoltà di Architettura di Alghero
Dipartimento di Architettura e Pianificazione (DAP)
Palazzo del Pou Salit, Piazza Duomo 6
07041 Alghero (SS), Italy
E-mail: tista@uniss.it

Mark S. Nixon
University of Southampton
School of Electronics and Computer Science
Southampton SO17 1BJ, UK
E-mail: msn@ecs.soton.ac.uk

Library of Congress Control Number: Applied for

CR Subject Classification (1998): I.5, I.4, K.4.1, K.4.4, K.6.5, J.1

LNCS Sublibrary: SL 6 – Image Processing, Computer Vision, Pattern Recognition, and Graphics

ISSN 0302-9743
ISBN-10 3-642-01792-4 Springer Berlin Heidelberg New York
ISBN-13 978-3-642-01792-6 Springer Berlin Heidelberg New York

This work is subject to copyright. All rights are reserved, whether the whole or part of the material is concerned, specifically the rights of translation, reprinting, re-use of illustrations, recitation, broadcasting, reproduction on microfilms or in any other way, and storage in data banks. Duplication of this publication or parts thereof is permitted only under the provisions of the German Copyright Law of September 9, 1965, in its current version, and permission for use must always be obtained from Springer. Violations are liable to prosecution under the German Copyright Law.

springer.com

© Springer-Verlag Berlin Heidelberg 2009
Printed in Germany

Typesetting: Camera-ready by author, data conversion by Scientific Publishing Services, Chennai, India
Printed on acid-free paper SPIN: 12670512 06/3180 5 4 3 2 1 0

Preface

It is a pleasure and an honour both to organize ICB 2009, the 3rd IAPR/IEEE International Conference on Biometrics. This will be held 2–5 June in Alghero, Italy, hosted by the Computer Vision Laboratory, University of Sassari. The conference series is the premier forum for presenting research in biometrics and its allied technologies: the generation of new ideas, new approaches, new techniques and new evaluations. The ICB series originated in 2006 from joining two highly reputed conferences: Audio and Video Based Personal Authentication (AVBPA) and the International Conference on Biometric Authentication (ICBA). Previous conferences were held in Hong Kong and in Korea. This is the first time the ICB conference has been held in Europe, and by Programme Committee, arrangements and by the quality of the papers, ICB 2009 will continue to maintain the high standards set by its predecessors.

In total we received around 250 papers for review. Of these, 36 were selected for oral presentation and 93 for poster presentation. These papers are accompanied by the invited speakers: Heinrich H. Bülthoff (*Max Planck Institute for Biological Cybernetics, Tübingen, Germany*) on "What Can Machine Vision Learn from Human Perception?", Sadaoki Furui (*Department of Computer Science, Tokyo Institute of Technology*) on "40 Years of Progress in Automatic Speaker Recognition Technology" and Jean-Christophe Fondeur (*SAGEM Security and Morpho, USA*) on "Large Scale Deployment of Biometrics and Border Control". In this conference we have also explored the evaluation of biometrics with a session of papers on challenges in evaluations: one on "Face Video Competition at ICB2009", on "Fingerprint and On-line Signature Verification Competitions at ICB 2009", on "Partial Face Matching between Near Infrared and Visual Images in MBGC Portal Challenge", and on "Overview of the Multiple Biometric Grand Challenge". The conference itself is followed by the annual Biometrics Summer School which has been held at Alghero for the past six years and its presenters, past and present, feature many of the programme and organising committee.

We are naturally very grateful for the efforts of many people. Firstly, we thank the entire conference committee for their valuable support in several stages of the organization. In particular we thank Anil Jain, Josef Kittler and Tieniu Tan who enthusiastically provided support and advice. The reviewing itself always appears a Herculean task and we were grateful that reviewers completed this in good time, and especially to those who came in at short notice. Equally, the other co-Program Chairs Arun Ross and Stan Li gave much of their time to finalise the final selection for the conference with major help in coordinating and actioning the reviewing process and developing the final program. The local organising committee have coordinated many of the local arrangements for the conference. Marinella Cardoni managed the review stage expertly. The Computer Vision Laboratory staff: Enrico Grosso, Linda Brodo, Massimo Gessa, Andrea Lagorio, Ajita Rattani and Elif Surer, are sorting the conference itself. Also, we are very grateful to the IAPR TC4 and the IEEE Biometrics Council, particularly to the Chairs Tieniu Tan and Rama Chellappa. The support from Accenture, GreenBit, Sagem, the European Union COST Action 2101 on "Biometrics in Identity

Documents", the foundation "Fondazione Banco di Sardegna", and the University of Sassari is also acknowledged. Finally, we are grateful to Springer for their production of the Proceedings, Alfred Hofmann and Wayne Wheeler in particular.

By papers, location and content this promises to be an excellent edition of the International Conference on Biometrics series. The conference continues to provide a snapshot of research in world-wide leading institutions on research in biometric recognition, and the many components if this fascinating research area, especially as it reaches world prominence given its increasing contribution to our societies' advancing security. By the program, papers and venue, we wish all delegates a most enjoyable conference

Massimo Tistarelli
Mark Nixon

Organization

General Chair

Massimo Tistarelli	University of Sassari, Italy

Conference Co-chairs

Davide Maltoni	University of Bologna, Italy
Javier Ortega-Garcia	Universidad Autonoma de Madrid, Spain

Advisory Board

Anil K. Jain	Michigan State University, USA
Joseph Kittler	University of Surrey, UK
B.V.K. Vijayakumar	Carnegie Mellon University, USA
Seong-Whan Lee	Korea University, South Korea
Tieniu Tan	Chinese Academy of Sciences, China

Competitions Chairs

Bernadette Dorizzi	Biosecure Foundation, France
Jonathon Phillips	NIST, USA

- Face competition: Norman Poh (University of Surrey, UK)
- Fingerprint competition: Raffaele Cappelli (University of Bologna, Italy)
- Signature verification competition: Sonia Garcia-Salicetti (GET-INT)
- Multimodal competition: Kryzstof Kryszczuk (EPFL)

Demo Chairs

Manuele Bicego	University of Sassari, Italy
Julian Fierrez	Universidad Autonoma de Madrid, Spain

Local Arrangement Chair

Enrico Grosso	University of Sassari, Italy

Program Chairs

Mark Nixon	University of Southampton, UK
Arun Ross	West Virginia University, USA
Stan Z. Li	Chinese Academy of Sciences, China

Publications Chair

Annalisa Franco	University of Bologna, Italy

Publicity Chairs

Josef Bigun	Halmstad University, Sweden
Robert Rowe	Lumidigm, USA
Zhenan Sun	Chinese Academy of Sciences, China

Registration Chair

Marinella Cadoni	University of Sassari, Italy

Sponsorship Chair

Giuseppe Parziale	iNVASIVE CODE, USA

Tutorials Chairs

Farzin Deravi	University of Kent, UK
Harry Wechsler	George Mason University, USA

Program Committee

Andy Adler	Canada
J.-L. Alba-Castro	Spain
George Bebis	USA
Samy Bengio	Switzerland
Bir Bhanu	USA
Frederic Bimbot	France
Kevin Bowyer	USA
Rama Chellappa	USA
Timothy Cootes	UK
Larry Davis	USA
Andrzej Drygajlo	Switzerland
Mike Fairhurst	UK
Miguel A. Ferrer-Ballester	Spain
Patrick Flynn	USA
Sadaoki Furui	Japan
Carmen Garcia-Mateo	Spain
Dominique Genoud	Switzerland
Shaogang Gong	UK
Venu Govindaraju	USA
Patrick Groether	USA

Steve Gunn	UK
Bernd Heisele	USA
Xudong Jiang	Singapore
Kenneth Jonsson	Sweden
Behrooz Kamgar-Parsi	USA
Takeo Kanade	USA
Jaihie Kim	Korea
Naohisa Komatsu	Japan
Ajay Kumar	India
Xiaoguang Lu	USA
Jean-Francois Mainguet	France
Dario Maio	Italy
John Mason	UK
James Matey	USA
Karthik Nandakumar	Singapore
Larry O'Gorman	USA
Alice O'Toole	USA
Sharath Pankanti	USA
Matti Pietikainen	Finland
Ioannis Pitas	Greece
Tomaso Poggio	USA
Salil Prabhakar	USA
Ganesh N. Ramaswamy	USA
Nalini Ratha	USA
Fabio Roli	Italy
Albert Ali Salah	The Netherlands
Ben Schouten	The Netherlands
Tamas Sziranyi	Hungary
Jie Tian	China
Doroteo T. Toledano	Spain
Patrick Verlinde	Belgium
Alessandro Verri	Italy
Yunhong Wang	China
Wei Yun Yau	Singapore
Pong Chi Yuen	Hong Kong
David Zhang	Hong Kong

Additional Reviewers

Gholamreza Amayeh
Manuele Bicego
Josef Bigun
John Daugman
Farzin Deravi
Bernadette Dorizzi
Ali Erol
Shengcai Liao
Chengjun Liu
Davide Maltoni
Jiri Navratil
Javier Ortega-Garcia
Roberto Paredes
Giuseppe Parziale

Nicholas Evans
Mike Fairhurst
Jianjian Feng
Julian Fierrez
Annalisa Franco
Kazuhiro Fukui
Enrico Grosso
Anil K. Jain
Joseph Kittler
Zhen Lei
Yongping Li
Jonathon Phillips
Norman Poh
Zhenan Sun
Tieniu Tan
Massimo Tistarelli
Kar-Ann Toh
Bhagavatula Vijayakumar
Junxian Wang
Harry Wechsler
Yong Xu
Junping Zhang

Local Organizing Committee

Linda Brodo	University of Sassari, Italy
Marinella Cadoni	University of Sassari, Italy
Pietro Coli	University of Cagliari, Italy
Andrea Lagorio	University of Sassari, Italy
Gian Luca Marcialis	University of Cagliari, Italy
Ajita Rattani	University of Cagliari, Italy
Elif Surer	University of Sassari, Italy
Roberto Tronci	University of Cagliari, Italy

Table of Contents

Face

Speech

Fingerprint and Palmprint

Biometric Competitions

Multibiometrics and Security

Gait

Iris

Other Biometrics

Overview of the Multiple Biometrics Grand Challenge

P. Jonathon Phillips[1], Patrick J. Flynn[2], J. Ross Beveridge[3], W. Todd Scruggs[4], Alice J. O'Toole[5], David Bolme[3], Kevin W. Bowyer[2], Bruce A. Draper[3], Geof H. Givens[6], Yui Man Lui[3], Hassan Sahibzada[1], Joseph A. Scallan III[7], and Samuel Weimer[5]

[1] National Institute of Standards and Technology, Gaithersburg, MD 20899, USA
[2] Computer Science & Engineering Department, U. of Notre Dame, Notre Dame, IN 46556, USA
[3] Department of Computer Science, Colorado State U., Fort Collins, CO 80523, USA
[4] SAIC, Arlington, VA 22203, USA
[5] School of Behavioral and Brain Sciences, The U. of Texas at Dallas, Richardson, TX 75083, USA
[6] Department of Statistics, Colorado State U., Fort Collins, CO 80523, USA
[7] Schafer Corp., Arlington, VA 22203, USA

Abstract. The goal of the Multiple Biometrics Grand Challenge (MBGC) is to improve the performance of face and iris recognition technology from biometric samples acquired under unconstrained conditions. The MBGC is organized into three challenge problems. Each challenge problem relaxes the acquisition constraints in different directions. In the Portal Challenge Problem, the goal is to recognize people from near-infrared (NIR) and high definition (HD) video as they walk through a portal. Iris recognition can be performed from the NIR video and face recognition from the HD video. The availability of NIR and HD modalities allows for the development of fusion algorithms. The Still Face Challenge Problem has two primary goals. The first is to improve recognition performance from frontal and off angle still face images taken under uncontrolled indoor and outdoor lighting. The second is to improve recognition performance on still frontal face images that have been resized and compressed, as is required for electronic passports. In the Video Challenge Problem, the goal is to recognize people from video in unconstrained environments. The video is unconstrained in pose, illumination, and camera angle. All three challenge problems include a large data set, experiment descriptions, ground truth, and scoring code.

1 Introduction

Since 1993, progress in automatic face recognition has been assisted by a series of "challenge problems" and "independent evaluations" organized by the U.S. Government [1] [2] [3] [4] [5] [6]. The challenge problems were designed to advance the evolving state-of-the-art in the field of automatic face recognition.

M. Tistarelli and M.S. Nixon (Eds.): ICB 2009, LNCS 5558, pp. 705–714, 2009.
© Springer-Verlag Berlin Heidelberg 2009

The independent evaluations were designed to measure progress made in automatic face recognition. A common thread connecting these challenge problems and evaluations was benchmarking performance of frontal face recognition from still images taken under controlled illumination. Additionally, more difficult aspects of face recognition have been incorporated into most of these efforts. A fundamental result of the Face Recognition Vendor Test (FRVT) 2006 was its demonstration of significant progress, relative to earlier evaluations, on frontal face recognition from still images taken under controlled illumination [6].

Challenge problems and third party evaluation in iris recognition began in 2004 [5] [6] [7] [8] [9]. To better understand the state-of-the-art in iris recognition, Newton and Phillips [10] performed a meta-analysis on the ICE 2006, the ITIRT, and the Iris 06 evaluations. The finding of the meta-analysis was that despite the differences in the testing protocols, sensors, image quality, subject variability and failures to enroll and acquire, the performance results from all three evaluations were comparable[1].

The Multiple Biometrics Grand Challenge (MBGC), currently underway, is aimed at examining a set of biometrics in a single evaluation program. An important part of this challenge is to relax the controlled imaging constraints used in previous tests. Again, the emphasis is to pose challenge problems that motivate improved recognition from biometric samples acquired under unconstrained conditions. In this paper, we provide an overview of the MBGC. The MBGC is organized into three challenge problems: the Portal Challenge Problem, the Still Face Challenge Problem, and the Video Challenge Problem. We begin with a brief statement of the goals of each of the three challenge problems. Next, we present a description of the data used in the challenge problems. Finally, we describe the protocol and experiments being applied in each challenge problem and sketch out the projected schedule for the MBGC.

2 Challenge Problems

In the Portal Challenge Problem, the goal is to recognize people from near-infrared (NIR) and high definition (HD) video as they walk through a portal. Iris recognition can be performed from the NIR video and face recognition from the HD video. Person recognition can be augmented beyond face or iris recognition alone by fusing the NIR and HD modalities. The Portal Challenge Problem is designed to examine identification from biometric data captured passively as a person passes through a portal. The Portal Challenge Problem for the first time makes NIR video clips available to researchers to support iris recognition research. Thus, solving the Portal Challenge Problem will require the development of a number of new technologies. First is the development of algorithms that can process the NIR video clips to locate ocular and iris regions and perform recognition from these regions. The second is the development of fusion algorithms to perform recognition from both the NIR and HD video clips. Because the biometric samples in the NIR and HD video sequences are acquired in

[1] A survey of iris recognition can be found in Bowyer et al [11].

a portal environment, it is not guaranteed that high quality iris or face samples will be captured. The fusion algorithms need to be able to handle cases where an iris is not present or one of the samples is poorly imaged. The requirement for recognition from NIR video clips has the potential to extend recognition from the iris to the entire ocular region of the face.

The Still Face Challenge Problem builds on the Face Recognition Grand Challenge (FRGC) adding non-frontal face images to the task and increasing the size and scope of experiments with images acquired under uncontrolled illumination. A survey of work addressing face recognition from unconstrained environments can be found elsewhere [12] and provides some baseline estimates of the state-of-the-art for this problem. The increase in the scope of the Still Face Challenge Problem includes recognition from highly compressed images.

The Video Challenge Problem breaks new ground by being the first challenge problem to address face recognition from unconstrained video. It makes available information that is commonly included in surveillance video and so provides a look at the level of recognition that can be achieved in security applications where video is available.

The MBGC is being conducted in two parts with two versions of the challenge problems. One of the goals of the version 1 is to introduce participants to the challenge problems and the MBGC protocol. The goal of version 2 is to encourage the development of algorithms that can handle large datasets. The version 1 challenge problems were released starting in May 2008 and the version 2 challenge problems will be released starting in February 2009. This paper describes the MBGC as of January 2009 and primarily focuses on the version 1 challenge problems. The version 2 challenge problems builds on version 1. Version 2 will consist of the same three challenge problems with the same types of biometric samples. The primary difference between the two versions is that larger datasets will be included in version 2.

3 Data

The MBGC challenge problems consist of data collected at the U. of Notre Dame specifically to support the MBGC and a previously collected dataset at the U. of Texas at Dallas [13]. The data collected at the U. of Notre Dame are included in all three challenge problems. The data collected at the U. of Texas at Dallas are included in only the Video Challenge Problem.

The data in the Portal Challenge Problem consists of five recording types. Two of the recording types were collected from an LG 2200 sensor: still iris images and video iris sequences. The still iris images intentionally represent a broader range of quality than the LG 2200 sensor would normally acquire, see Phillips et al [5] [6] [9] for details. The LG 2200 iris video sequences were digitized from the National Television System Committee (NTSC) video signal from the sensor. The NTSC video signal produced by the LG 2200 was digitized by a DayStar XLR8 USB video digitizer attached to a Macintosh host system, and stored in uncompressed format. Initial processing edited each subject's original

video, which contained both eyes, into two clips, one containing the left eye only and one containing the right eye only. These clips were transcoded to MPEG-4 format with a high bit rate allowance, thus yielding near-lossless encoding.

The third recording is very high resolution NIR video acquired from a Sarnoff Iris on the Move (IoM) system [14], see Figure 1. The IoM system was designed to capture iris imagery as a person walks through a portal.

The fourth recording type was acquired with a High Definition (HD) video camera, The HD video camera was bore-sighted beside the NIR video cameras in the IoM chassis. This camera captured HD video as a subject walked through the portal, see Figure 1. The HD video sequences are distributed in portrait mode, and the dimension of the video are 1440 pixels high by 1080 pixels wide. Because the camera was in portrait mode, the video was compressed vertically and the interlace artifacts show up in the vertical direction. The fifth recording type was a medium resolution still digital frontal image.

The data for the Still Face Challenge Problem were collected with high resolution digital cameras, 4 and 6 mega-pixels. Images were collected with both controlled and uncontrolled illumination. The images with controlled illumination were collected in a studio environment with controlled lighting. The images with uncontrolled illumination were collected in hallways and outdoors. Both frontal and non-frontal face images were collected.

The data for the Video Challenge Problem were collected with HD and Standard Definition (SD) video cameras. The data were collected in hallways, atria, and outdoors with unconstrained pose and illumination, see Figure 2. In the Video Challenge Problem version 1, the data from the U. of Notre Dame consisted of two types of video sequences: walking and activity. In the walking sequences, the subjects walked towards the camera. In the activity sequences, the subjects performed an action that required subjects to look away from the camera. An example action is picking up or looking at a book. The data from U. of Texas at Dallas consist of two types of video sequences: walking and conversation. In the walking sequences the subjects walked towards the camera. In the conversation sequences, a camera is looking down on a conversation between two subjects. One subject's back is to the camera and this subject is to be ignored in the experiments. The other subject's face is facing the camera at an off angle. This is the subject to be recognized.

4 Portal Challenge Problem

The Portal Challenge Problem version 1 consists of five experiments. Each experiment was designed to address different aspects of the Portal Challenge Problem data. These aspects address single mode face and iris recognition, and multiple biometric recognition. The five experiments in the Portal Challenge Problem are briefly described below and summarized in Table 1.

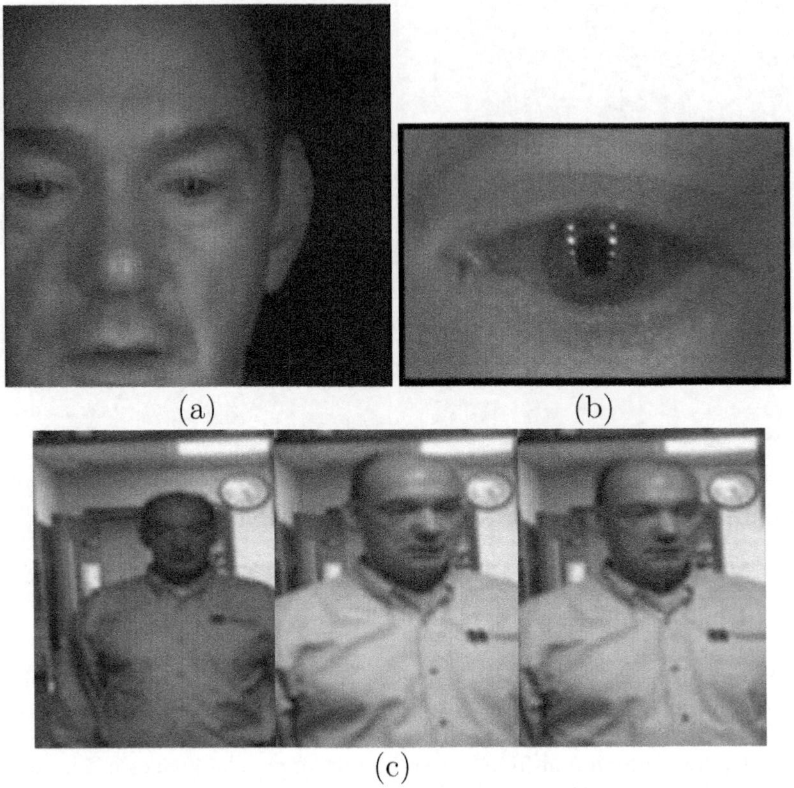

Fig. 1. Example of imagery collected as a subject walks through the portal. The image in (a) is a full 2,000 by 2,000 pixel NIR frame acquired by the IoM and (b) is the left ocular region from the frame in (a). There are approximately 120 pixels across the iris in (a) and (b). The images in (c) are three frames from an HD video sequence of a subject walking through the portal.

Still Iris versus NIR: The target image is an iris still from an LG 2200 camera. The query image is a NIR video sequence. This experiment tests iris recognition.

Video Iris versus NIR: In this experiment both the target and query are video sequences. The target is an LG 2200 iris video and the query is a NIR video sequence. This experiment tests iris matching using video in both the target and query sets.

Still Face versus HD Video: The target is a still face images and the query images is an HD video sequence. This experiment tests facial recognition.

Still Iris and Still Face versus NIR and HD Video: This experiment combines the elements from the pervious two experiments (Video Iris vs. NIR, and Still Face vs. HD Video Face). The target images consist of a face still and an LG 2200 still iris. The query images consists of two video sequences. The

(a) (b) (c) (d)

Fig. 2. Example frames from video sequences in the Video Challenge Problem. The images in (a) and (b) are from the U. of Notre Dame collection. The image in (a) is from a video sequence of a subject walking towards the cameras in an atrium and (b) is a subject performing a non-frontal activity outdoors. The images in (c) and (d) are from the U. of Texas at Dallas collection. The image in (c) is from a video sequence of a subject walking towards the cameras in an atrium. The image in (d) is acquired from a video camera looking down on a conversation. In (d), the subject whose face can be seen is the subject to be recognized.

first is an HD video sequences, and the second is a NIR video sequence. This experiment tests recognition from face and iris imagery.

Video Iris and Still Face versus NIR and HD Video: This experiment is the same as previous experiment except that LG 2200 video iris replaces the LG 2200 still iris images.

5 Still Face Challenge Problem

The goal of the Still Face Challenge Problem is to improve face recognition performance on images taken under conditions that are representative of a broad range of real-world scenarios. One set of experiments uses high resolution still face images taken in both controlled and uncontrolled illumination environments. This set of experiments challenges participants to match 'controlled to uncontrolled' face images and to match 'uncontrolled to uncontrolled' face images.

Table 1. A summary of the experiments in the Portal Challenge Problem version 1. The first column is the name of the experiment. The second and third columns respectively list the type of imagery in the target and query sets. The last column lists the primary biometric modality in the experiment.

Experiment	Target image	Query image	Problem
still iris vs NIR	still iris	NIR video	iris
video iris vs NIR	video Iris	NIR video	iris
still face vs HD	still face	HD video	face
still iris & still face vs NIR & HD	still iris & still face	NIR & HD video	multiple biometrics
video iris & still face vs NIR & HD	video iris & video face	NIR & HD video	multiple biometrics

The target sets consist of frontal face images and the query sets consist of both frontal and non-frontal face images.

A second set of experiments in the Still Face Challenge Problem is designed to model the size of face images to meet the International Civil Aviation Organization's (ICAO) standard for electronic passports [15]. Two of the primary requirements of this ICAO standard are the size of the face and compression of the face image. The face is required to be frontal with a size range from 90 to 120 pixels between the centers of the eyes. The face image is required to be compressed to between 8 KBytes and 20 KBytes. The images can be compressed by either JPEG or JPEG2000. To measure the range of possible performance, two sets of images were generated from the frontal images in the Still Face Challenge Problem, see Figure 3. In the first set, the images were rescaled to 90 pixels between the eyes and compressed to 8 KBytes. In the second set, the images were rescaled to 120 pixels between the eyes and compressed to 20 KBytes. Both sets of images were compressed using JPEG2000.

6 Video Challenge Problem

The Video Challenge Problem is designed to encourage the development of face recognition from video taken in hallways, atria, and outdoors under unconstrained illumination, pose, and camera angle. Version 1 of the Video Challenge Problem examined the effects of HD and SD video on performance. Version 1 also examined the effects of matching walking to walking sequences, activity to activity sequences, and walking to activity sequences, see Figure 1. All comparisons in the Video Challenge Problem are between video sequences taken on different days. This is to avoid matching based on clothes. Human performance on recognizing people in video sequences will be measured from data in the Video Challenge Problem. Human performance experiments will be designed to allow for direct comparison of human and machine recognition performance on the same pairs of video sequences.

Fig. 3. Example of face images in the Still Face Challenge Problem. The image in (a) is an original still face image taken outdoors. The size of the image is 3008 pixels wide by 2000 pixels high, and the size of the face is 232 pixels between the centers of the eyes. The image in (b) is the original that has been rescaled to 120 pixels between the centers of the eyes, cropped, and compressed to 20Kbytes. The image in (c) is the original that has been rescaled to 90 pixels between the centers of the eyes, cropped, and compressed to 8Kbytes. The dimensions of the image in (b) are 408 pixels wide by 528 pixels high, and the image in (c) is 306 pixels wide by 396 pixels high.

7 Protocol

The MBGC version 1 data and challenge problems were made available to participants starting in May 2008[2]. Each challenge problem consisted of a data set, description of the experiments, ground truth, and code to score the experiments.

[2] Instructions for getting the MBGC challenge problems and participating in the MBGC can be found at `http://face.nist.gov/mbgc/`

Results from the MBGC version 1 were presented at the Second Multiple Biometrics Grand Challenge Organizational Workshop held on 5 December 2008. The MBGC version 2 data and challenge problems will be made available to participants starting in February 2009.

Participants could submit results for any experiment in one or more of the three challenge problems. Participants could submit results for multiple algorithms for an experiment. All groups that submitted results agreed in advance that their performance results could be publicly attributed to them.

The MBGC is open to academic institutions, research laboratories and to companies worldwide. Groups granted access to the MBGC data are not required to submit results for analysis. Participation in MBGC version 1 is not required for participation in MBGC version 2.

8 Conclusion

The MBGC challenge problems are designed to advance face and iris recognition performance in unconstrained scenarios of clear practical importance. For the Portal Challenge Problem, the goal is to recognize people as they walk through a portal. The challenge is to develop algorithms that can process both the iris and the ocular regions of the face, and to fuse both the NIR and HD video sequences. The goal of the Still Face Challenge Problem is to recognize people in uncontrolled still images, and to test the effects of compression on face recognition. The Video Challenge Problem is the first challenge problem to address recognition of people in unconstrained video. The MBGC and its participants are developing technology to support the next generation of biometric applications.

Acknowledgements

We acknowledge the support of the the Biometric Task Force, the Department of Homeland Security's Directorate for Science and Technology , the Intelligence Advanced Research Projects Activity (IARPA), the Federal Bureau of Investigation (FBI), and the Technical Support Working Group (TSWG). The identification of any commercial product or trade name does not imply endorsement or recommendation by the Colorado State U., the National Institute of Standards and Technology, SAIC, Schafer Corp., the U. of Notre Dame or the U. of Texas at Dallas.

References

1. Phillips, P.J., Wechsler, H., Huang, J., Rauss, P.: The FERET database and evaluation procedure for face-recognition algorithms. Image and Vision Computing Journal 16(5), 295–306 (1998)
2. Phillips, P.J., Moon, H., Rizvi, S., Rauss, P.: The FERET evaluation methodology for face-recognition algorithms. IEEE Trans. PAMI 22, 1090–1104 (2000)

3. Phillips, P.J., Grother, P.J., Micheals, R.J., Blackburn, D.M., Tabassi, E., Bone, J.M.: Face recognition vendor test 2002: Evaluation report. National Institute of Standards and Technology, Tech. Rep. NISTIR 6965 (2003), http://www.frvt.org
4. Phillips, P.J., Flynn, P.J., Scruggs, T., Bowyer, K.W., Chang, J., Hoffman, K., Marques, J., Min, J., Worek, W.: Overview of the face recognition grand challenge. In: IEEE Computer Society Conference on Computer Vision and Pattern Recognition, pp. 947–954 (2005)
5. Phillips, P.J., Scruggs, W.T., O'Toole, A.J., Flynn, P.J., Bowyer, K.W., Schott, C.L., Sharpe, M.: FRVT 2006 and ICE 2006 large-scale results, National Institute of Standards and Technology, Tech. Rep. NISTIR 7408 (2007), http://iris.nist.gov/
6. Phillips, P.J., Scruggs, W.T., O'Toole, A.J., Flynn, P.J., Bowyer, K.W., Schott, C.L., Sharpe, M.: FRVT 2006 and ICE 2006 large-scale results. IEEE Trans. PAMI (in Press)
7. Authenti-Corp, Iris recognition study 2006 (IRIS06), Authenti-Corp, Tech. Rep. version 0.40 (March 2007), http://www.authenti-corp.com/iris06/report/
8. International Biometric Group, Independent testing of iris recognition technology, International Biometric Group, Tech. Rep. (May 2005), http://www.ibgweb.com/reports/public/ITIRT.html
9. Phillips, P., Bowyer, K.W., Flynn, P.J., Liu, X., Scruggs, W.T.: The Iris Challenge Evaluation 2005. In: Second IEEE International Conference on Biometrics: Theory, Applications, and Systems (2008)
10. Newton, E.M., Phillips, P.J.: Meta-analysis of third-party evaluations of iris recognition. IEEE Trans. on SMC-A 39(1), 4–11 (2009)
11. Bowyer, K.W., Hollingsworth, K., Flynn, P.J.: Image understanding for iris biometrics: a survey. Computer Vision and Image Understanding 110(2), 281–307 (2008)
12. Zhao, W., Chellappa, R., Phillips, P.J., Rosenfeld, A.: Face recognition: A literature survey. ACM Computing Surveys 35, 399–458 (2003)
13. O'Toole, A.J., Harms, J., Snow, S.L., Hurst, D.R., Pappas, M.R., Ayyad, J.H., Abdi, H.: A video database of moving faces and people. IEEE Trans. PAMI 27, 812–816 (2005)
14. Matey, J.R., Naroditsky, O., Hanna, K., Kolcynksi, R., LoIancono, D.J., Mangru, S., Tinker, M., Zappa, T.M., Zhao, W.Y.: Iris on the move: Acquisition of images for iris recognition in less constrained environments. Proceedings of the IEEE 94, 1936–1947 (2006)
15. ICAO New Technologies Working Group, ICAO Document 9303, Tech. Rep., http://www2.icao.int/en/MRTD/Pages/Doc9393.aspx

Face Video Competition

Norman Poh[1], Chi Ho Chan[1], Josef Kittler[1], Sébastien Marcel[2], Christopher Mc Cool[2], Enrique Argones Rúa[3], José Luis Alba Castro[3], Mauricio Villegas[4], Roberto Paredes[4], Vitomir Štruc[5], Nikola Pavešić[5], Albert Ali Salah[6], Hui Fang[7], and Nicholas Costen[7]

[1] CVSSP, University of Surrey, Guildford, GU2 7XH, Surrey, UK
normanpoh@ieee.org, c.chan@surrey.ac.uk, j.kittler@surrey.ac.uk
[2] Idiap research institute, Marconi 19, Martigny, CH
[3] Signal Technologies Group, Signal Theory and Communications Dept., University of Vigo, 36310, Spain
[4] Universidad Politécnica de Valencia, Instituto Tecnológico de Informática, Camino de Vera s/n, 46022 Valencia, Spain
[5] Faculty of Electrical Engineering, University of Ljubljana, Tržaška 25, SI-1000 Ljubljana, Slovenia
[6] CWI, Science Park 123, 1098 XG Amsterdam
[7] Department of Computing and Mathematics, Manchester Metropolitan University, UK M1 5GD

Abstract. Person recognition using facial features, e.g., mug-shot images, has long been used in identity documents. However, due to the widespread use of web-cams and mobile devices embedded with a camera, it is now possible to realise facial video recognition, rather than resorting to just still images. In fact, facial video recognition offers many advantages over still image recognition; these include the potential of boosting the system accuracy and deterring spoof attacks. This paper presents the first known benchmarking effort of person identity verification using facial video data. The evaluation involves 18 systems submitted by seven academic institutes.

1 Introduction

With an increasing number of mobile devices with built-in web-cams, e.g., PDA, mobile phones and laptops, face is arguably the most widely accepted means of person verification. However, the biometric authentication task based on face images acquired by a mobile device in an uncontrolled environment is very challenging. One way to boost the face verification performance is to use multiple samples.

Previous attempts at assessing the performance of face verification algorithms have been restricted to matching still images, e.g., the three FERET evaluations[1] (1994, 1995 and 1996), the face recognition vendor tests (FRVTs 2000, 2002 and 2006)[2], and assessment on XM2VTS and BANCA databases [1,2]. The well known Face Recognition Grand Challenge [3] includes queries with multiple still images but this is far from the vast amount of data available in video matching.

[1] http://www.itl.nist.gov/iad/humanid/feret/feret_master.html

[2] http://www.frvt.org

M. Tistarelli and M.S. Nixon (Eds.): ICB 2009, LNCS 5558, pp. 715–724, 2009.
© Springer-Verlag Berlin Heidelberg 2009

The evaluation exercise presented here is the first known effort in assessing *video-to-video* matching, i.e., in both enrolment and verification phases, the data captured is in the form of video sequence. This is different from still-image-to-video matching, one of the evaluation scenarios currently examined by the NIST Multiple Biometric Grand Challenge[3] (MBGC). Note that NIST MBGC aims at "portal application" where the task is to verify the identity of person as he/she walks through an access control check point. The video-to-video matching adopted here has a slightly different application, with a focus on mobile devices, where a sequence of unconstrained (talking) face images can be expected.

The video-to-video face verification assessment has several objectives, among which are:

- to promote the development of algorithms for analysing video sequences (e.g., exploring the talking face dynamics)
- to assess the merit of multi-template face representation
- to compare whether early integration (e.g., feature-level fusion) is better than late integration (e.g., decision-level fusion) in dealing with sequences of query images.

2 Database, Protocols, Facial Video Annotations

Towards the above goal, we have opted to use the publicly available BANCA database [4][4]. It has a collection of face and voice biometric traits of up to 260 persons in 5 different languages, but only the English subset is used here. It contains a total of 52 persons; 26 females and 26 males. The 52 persons are further divided into two sets of users, which are called g1 and g2, respectively. Each set (g1 or g2) is designed to be balanced in gender, i.e., having 13 males and 13 females. According to the experimental protocols reported in [4], when g1 is used as a development set (to build the user's template/model), g2 is used as an evaluation set. Their roles are then switched. This corresponds to a two-fold cross-validation procedure.

The BANCA database was designed to examine matching under the same recording conditions (as the enrolment session) and two different challenging conditions: recording under a noisy (adverse) environment and with a degraded device. In each of the three conditions, four recordings were performed. The clean conditions apply to sessions 1–4; adversed conditions to sessions 5–8; and degraded conditions to sessions 9–12. There are altogether seven experimental protocols specifying the sessions to be used for enrolment and for testing in an exhaustive manner. In this face video recognition evaluation, we focused on two protocols, namely the match controlled (Mc) and unmatched adversed (Ua) protocols. The first protocol was intended as a vehicle to design and tune their face verification systems. The second protocol aims at testing the systems under more realistic and challenging conditions.

In the Mc protocol, session 1 data is used for enrolment whereas the data from sessions 2–4 are reserved for testing. In the Ua protocol, the session 1 data again is used

[3] http://face.nist.gov/mbgc

[4] http://www.ee.surrey.ac.uk/CVSSP/banca

for enrolment but the test data is taken from session 5–8 (recorded under adversed conditions). The ICB2009 face video competition was thus naturally carried out in two rounds, with the first round focusing on the Mc protocol and the second round on the Ua protocol.

In order to be consistent with the previous BANCA evaluations [1,2], we also divided a query video sequence into 5 chunks, each containing 50 frames for convenience; the remaining frames were simply not used.

In order to standardise the evaluation, we provided a pair of eye coordinates, based on the face detector provided by the OmniPerception's SDK[5]. However, the participants could use their own face detectors. For each image in a video sequence, the SDK also annotated the following quality measurements. Note that the entire processes from detection to annotation were done automatically. No effort was made to fine tune the system parameters, and in consequence, some imperfectly cropped images were observed. The image quality measures assessed.

1. Overall reliability
2. Brightness
3. Contrast
4. Focus
5. Bit per pixel
6. Spatial resolution (between eyes)
7. Illumination
8. Background uniformity
9. Background brightness
10. Reflection
11. Presence of glasses
12. In-plane rotation
13. In-depth rotation
14. Frontalness

In the above list, "frontalness" quantifies the degree of similarity of a query image to a typical frontal (mug-shot) face image. The overall reliability is a compounded quality measure obtained by combining the remaining quality measures. Two categories of quality measures can be distinguished: face-specific or generic. The face-specific ones strongly depend on the result of face detection, i.e., frontalness, rotation, reflection, between-eyes spatial resolution in pixels, and the degree of background uniformity (calculated from the remaining area of a cropped face image). The generic ones are defined by the MPEG standards. All the annotation data (including eye coordinates and quality measures) has been published on the website "http://face.ee.surrey.ac".

A preliminary analysis shows that when the frontalness measure is 100%, the detected face is always frontal. On the other hand, any value less than 100% does indeed suggest an imperfect face detection, or else a non-ideal (non-frontal) pose.

3 Summary of Submitted Systems

The submitted face verification systems can be categorised according to whether they are image-set-based or frame-based (comparison) approach. In the image-set based approach, a video sequence is analysed and treated as a set of images. When comparing two video sequences, this approach, in essence, compares two *sets* of images. On the other hand, the frame-based approach directly establishes similarity between two images, each obtained from their respective video sequence. If there are P and Q images in both sequences, there will be at most PQ similarity scores. The frame-based approach

[5] http://www.omniperception.com

Table 1. Overview of the submitted face verification systems

	Systems	Pre-processing	Face rep.	Feature Extraction	Classifier	Quality measure used	Process all images
Holistic	idiap-pca-pearson	HEQ		PCA	Pearson	No	Yes
	idiap-pca-nc	HEQ		PCA	NC	No	Yes
	idiap-pca-cor	HEQ		PCA	StdCor	No	Yes
	idiap-lda-pearson	HEQ		PCAxLDA	Pearson	No	Yes
	idiap-lda-nc	HEQ		PCAxLDA	NC	No	Yes
	idiap-lda-cor	HEQ		PCAxLDA	StdCor	No	Yes
	mmu		AM	LDA	Avg(NC)	No	Yes
Local	idiap-dcthmmt-v1	HEQ		DCT	HMM	No	Yes
	idiap-dcthmmt-v2	HEQ		DCT	HMM	No	Yes
	idiap-dctgmm	HEQ		DCTmod2+xy	GMM	No	Yes
	idiap-LBP-dctgmm		LBP	DCTmod2+xy	GMM	No	Yes
	cwi-Cq			DCT	Max(NC)		Yes
	cwi-Eq			DCT	Max(NC)		Yes
	cwi-Cr			DCT	Max(NC)		No
	cwi-Er			DCT	Max(NC)		No
	upv	Local-HEQ	LF	PCA	Avg(KNN)	Yes	No
	uni-lj	ZMUV + HEQ	Gb2	KDA+PCA	WNC	Yes	No
	uvigo	Ani	Gb1		GMM	Yes	No

The following keys are used: AM = Appearance model, ZMUV = zero mean and unit-variance, Ani = Anisotropic+local mean subtraction, LF = Local feature Gb1 = Gabor(magnitude) Gb2 = Gabor(phase+magnitude, NC = Normalised correlation, WNC = Sum of whitened NC Note: OmniPerception's face detector was used by all systems.

would select, or otherwise combine these similarity scores to obtain a final similarity score. Among the systems, only the MMU system belongs to the image-set based approach, while the remaining systems are the frame-based approach.

Face verification systems can also be further distinguished by the way a face image is treated, i.e, either holistic or local (parts-based) appearance approach. In the former, the entire (often cropped) image is considered as input to the face classifier. In the latter, the face images are divided into (sometimes overlapping) parts which are then treated separately by a classifier.

Table 1 summarises the systems by this categorisation. Principal component analysis (PCA), or Eigenface, and local discriminant analysis (LDA), or Fisherface, are perhaps the most representative (and popular) examples of the holistic approach due to the pioneer work of Turk and Pentland [5]. Many of these systems were submitted by IDIAP as baseline systems, tested on the Mc protocol (and not the Ua protocol). Recent face verification research has been dominated by the local appearance approach, as exemplified by *most* of the submissions in this competition. The details of each system can be found in "http://face.ee.surrey.ac.uk/data/face_icb2009.pdf".

4 Evaluation Metrics

We use two types of curves in order to compare the performance: the Detection Error Trade-off (DET) curve [6] and the Expected Performance Curve (EPC) [7]. A DET

curve is actually a Receiver Operator Curve (ROC) curve plotted on a scale defined by the inverse of a cumulative Gaussian density function, but otherwise similar in all aspects. We have opted to use EPC because it has been pointed out in [7] that two DET curves resulting from two systems are not comparable. This is because such comparison does not take into account how the decision thresholds are selected. EPC turns out to be able to make such comparison possible. Furthermore, the performance across different data sets, resulting in several EPCs, can be merged into a single EPC [8]. Although reporting performance in EPC is more meaningful than DET as far as performance comparison is concerned, it is relatively new and has not gained a widespread acceptance in the biometric community. As such, we shall also report performance in DET curves, but using only a subset of operating points.

The EPC curve, however, is less convenient to use because it requires two sets of match scores, one used for tuning the threshold (for a given operating cost), and the other used for assessing the performance. In our context, with the two-fold cross-validation defined on the database (as determined by g1 and g2), these two match scores can be conveniently used.

According to [7], one possible, and often used criterion is the weighted error rate (WER), defined by:

$$\mathrm{WER}(\beta, \Delta) = \beta\,\mathrm{FAR}(\Delta) + (1 - \beta)\,\mathrm{FRR}(\Delta), \tag{1}$$

where FAR is the false acceptance rate, FRR is the false rejection rate at a given threshold Δ and $\beta \in [0, 1]$ is a user-specified coefficient which balances FAR and FRR. The WER criterion generalises the criterion used in the annual NIST's speaker evaluation [9] as well as the three operating points used in the past face verification competitions on the BANCA database [1,2]. In particular the following three coefficients of β are used:

$$\beta = \frac{1}{1+R} \text{ for } R = \{0.1, 1, 10\}$$

which yields approximately $\beta = \{0.9, 0.5, 0.1\}$, respectively.

The procedure to calculate an EPC is as follows: Use g1 to generate the development match scores; and g2, the evaluation counterpart. For each chosen β, the development score set is used to minimise (1) in order to obtain an operational threshold. This threshold is then applied to the evaluation set in order to obtain the final pair of false acceptance rate (FAR) and false rejection rate (FRR). The EPC curve simply plots half total error rate (HTER) versus β, where HTER is the average of FAR and FRR. Alternatively, the generalisation performance can also be reported in WER (as done in the previous BANCA face competitions). To plot the corresponding DET curve, we use the pair of FAR and FRR of all the operating points, as determined by β. Note that this DET curve is a *subset* (in fact discrete version) of a conventional continuous DET curve because the latter is plotted from continuous empirical functions of FAR and FRR. By plotting the discrete version of the DET curve, we establish a *direct correspondence* between EPC and DET, satisfying both camps of biometric practitioners, while retaining the advantage of EPC which makes performance comparison between systems less biased.

5 Results

The DET curves of all submitted systems for the g1 and g2 data sets, as well as for the Mc and Ua protocols, are shown in Figure 1. By merging the results from g1 and g2, we plotted the EPCs for Mc and Ua in Figure 2 (plotting β versus HTER). To be consistent with the previous published BANCA evaluations [1,2], we also listed the individual g1 and g2 performance, in terms of WER, in Table 2 for the Mc protocol and in Table 3 for the Ua protocol.

The following observations can be made:

- **degradation of performance under adversed conditions:** It is obvious from Figure 2 that all systems systematically degrade in performance under adversed conditions.

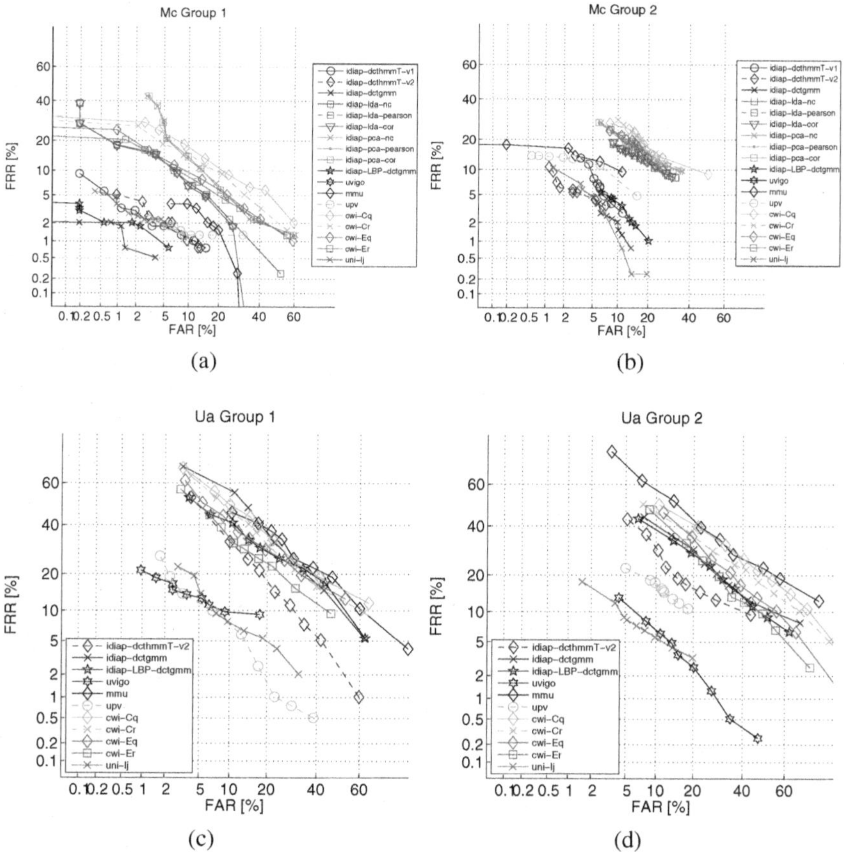

Fig. 1. DET curves of the submitted systems evaluated on the g2 (evaluation set) of the BANCA video based on the Mc protocol. Note that the uvigo system achieved zero EER on the Mc g2 datasets. As a result, its DET curve reduces to a single point at the origin ((∞, ∞) in the above normal inverse scales.

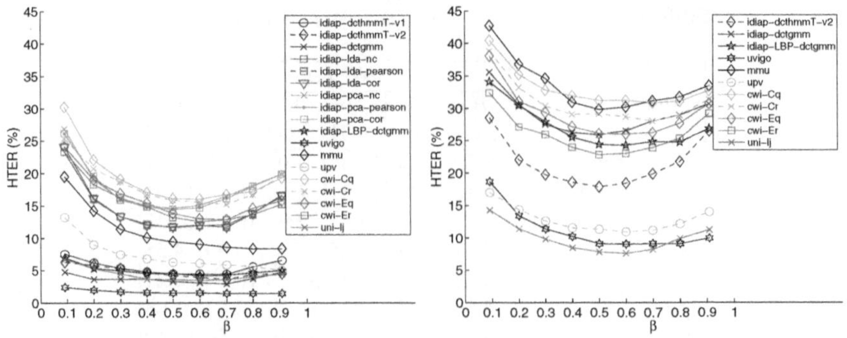

(a) Mc (b) Ua

Fig. 2. EPC curves of the submitted systems evaluated on the g2 (evaluation set) of the BANCA video based on the Mc protocol

Table 2. Performance of g1 and g2 based on the Mc protocol using video sequences

systems	WER (%)					
	$R = 0.1$		$R = 1$		$R = 10$	
	G1	G2	G1	G2	G1	G2
idiap-dcthmm†	7.52	4.90	5.45	0.64	2.56	0.12
idiap-dcthmm‡	7.78	3.76	5.13	2.08	1.17	2.74
idiap-dcthmmT-v2	1.34	2.03	4.20	4.29	1.92	3.93
idiap-dctgmm	0.82	5.14	1.12	5.48	0.82	1.96
idiap-LBP-dctgmm	0.75	6.26	1.63	7.37	1.22	2.77
uvigo	1.05	0.42	0.77	2.31	0.45	4.20
mmu	5.94	2.14	9.84	9.07	5.21	9.64
upv	3.01	1.81	5.06	7.50	4.00	5.86
cwi-Cq	3.80	9.84	14.20	18.14	7.28	12.76
cwi-Cr	3.66	11.72	13.14	18.69	6.49	12.40
cwi-Eq	2.84	9.51	10.90	16.83	6.32	11.49
cwi-Er	2.59	9.73	9.87	16.63	6.25	11.68
uni-lj	0.86	2.18	2.34	4.81	2.32	2.02

†: Experimental results on *still* images, taken from [1] with automatic localisation. ‡: Similar to †, except with manual localisation.

- **holistic vs. local appearance methods:** From Figure 1(a) and (b) as well as Figure 2(a), we observe that the performance of the holistic appearance methods (PCA and LDA) is worse than that of the local appearance methods, except for the CWI classifier (where photometric normalisation was not performed). Thus, we can expect that the performance of CWI to be similar to the performance of other local appearance methods in the raw image space, such as idiap-dctgmm, idiap-dcthmmt-v2 and upv if photometric normalisation were to be performed.
- **still vs. video comparison:** Among the submitted systems, only IDIAP's DCT-HMM system was involved in the previously reported results for the Mc protocol

Table 3. Performance of g1 and g2 based on the Ua protocol

systems	WER (%)					
	$R = 0.1$		$R = 1$		$R = 10$	
	G1	G2	G1	G2	G1	G2
idiap-dcthmmT-v2	8.52	8.66	18.65	17.08	6.37	12.61
idiap-dctgmm	9.10	11.03	27.31	24.49	10.54	13.31
idiap-LBP-dctgmm	8.34	10.08	23.85	24.94	10.58	11.47
uvigo	2.81	5.06	8.75	9.49	10.00	4.55
mmu	13.61	9.88	27.72	31.96	10.97	18.21
upv	4.00	6.60	9.29	13.46	3.98	11.45
cwi-Cq	9.06	14.18	28.08	34.46	16.54	11.19
cwi-Cr	9.43	11.41	26.60	31.79	14.50	11.79
cwi-Eq	8.72	14.73	24.23	27.98	16.50	8.48
cwi-Er	8.00	12.23	21.38	24.29	12.86	8.80
uni-lj	4.67	3.03	8.78	6.99	4.78	4.83

[1] which was based on 5 still images taken from a video sequence (as opposed to five video chunks as done here). The results for this classifier are shown in Table 2 (comparing rows 1-2 with row 3). In theory, one would expect the classifier tested on video sequence to be better than still images. Unfortunately, such conclusion cannot be made except for $R = 0.1$.

- **Pre-processing:** In dctgmm methods, the performance of applying HEQ is better than that of applying LBP as a pre-processing method for Mc protocol. However, the case is reversed for Ua protocol because HEQ enhances shadows while LBP features are invariant to such monotonic transformation (in relation to the neighbourhood pixels cast under shadows). In other words, the selection of the pre-processing methods should be dependent on the environmental conditions.
- **Sample size:** Cwi's submission has four variations: depending on the dichotomies: system complexity, i.e., Cheap (C) versus Exepnsive (E); and strategy for choosing the query samples, i.e, random (r) versus quality-based (q). Two observations can be noted: First, the performance of cwi-Eq and cwi-Er are better than that of cwi-Cq and cwi-Cr. Second, using *more* template and query features can improve the cwi system. A rigorous and systematic design of experiments is still needed to find out the usefulness of the provided quality measures, and more importantly, the most effective ways of using such auxilliary information. This is a challenging problem for two reasons. First, not all 14 quality measures provided are relevant to a face matching algorithm, e.g., an algorithm that is robust to illumination changes would, in principle, be invariant to some photometric measures used here (brightness, contrast, etc). This implies that a quality measure selection strategy is needed. Second, quality measures are themselves not discriminatory for distinguishing subjects but discriminatory in distinguishing environmental conditions.
- **Multi resolution Contrast Information:** The best algorithm of this competition for MC protocol is UVigo where the WER at R=1 is 0.77% for G1 and 2.31% for G2. For UA protocol, the best algorithm is uni-lj where WER at R=1 is 8.78% for G1 and 6.99% for G2. In fact, the performance of these two systems is very close

but uni-lj is slightly better overall as the average of WER at different R is 3.96% for G1 and 3.98% for G2, while the result of UVigo is 3.97% for G1 and 4.34% for G2. The success of these two algorithms derives from the use of multi resolution contrast information.

6 Discussion and Future Evaluation

Because the target application scenario of this assessment is on mobile devices, computational resources are crucial. For this reason, when benchmarking a face verification algorithm, the cost of computation has to be considered. For instance, a fast and light algorithm, capable of processing all images in a sequence, may be preferred over an extremely accurate algorithm only capable of processing a few selected images in a sequence. However, the former algorithm may be able to achieve better performance since it can process a much larger number of images within the same time limit and memory requirement. The above scenario highlights that the performance of two algorithms cannot be compared on equal grounds, unless both use comparable computation costs, taking the time, memory and computational resources into consideration.

The current evaluation has not taken this cost factor into consideration, but this will be carried out in future. The idea is to request each participant to run a benchmarking program, executable in any operating system. The time registered by the program will be used as a *standard unit time* for the participant's system. Thus the time to process a video file for a participant, for instance, will be reported in terms of multiples (or fractions) of the participant's standard unit time.

7 Conclusions

This paper presents a comparison of video face verification algorithms on BANCA database. Eighteen different video-based verification algorithms from a variety of academic institutions participated in this competition. The results show that the performance of the local appearance methods is better than that of the holistic appearance methods. Secondly, using more query and selected template features to measure similarity improve the system performance. Finally, the best algorithm in this competition clearly shows that multi resolution contrast information is important for face recognition.

Acknowledgement

The work of NPoh is supported by the advanced researcher fellowship PA0022_121477 of the Swiss NSF; NPoh, CHC and JK by the EU-funded Mobio project grant IST-214324; NPC and HF by the EPSRC grants EP/D056942 and EP/D054818; VS and NP by the Slovenian national research program P2-0250(C) Metrology and Biometric System, the COST Action 2101 and FP7-217762 HIDE; and, AAS by the Dutch BRICKS/BSIK project.

References

1. Messer, K., Kittler, J., Sadeghi, M., Hamouz, M., Kostyn, A., Marcel, S., Bengio, S., Cardinaux, F., Sanderson, C., Poh, N., Rodriguez, Y., Kryszczuk, K., Czyz, J., Vandendorpe, L., Ng, J., Cheung, H., Tang, B.: Face authentication competition on the BANCA database. In: Zhang, D., Jain, A.K. (eds.) ICBA 2004. LNCS, vol. 3072, pp. 8–15. Springer, Heidelberg (2004)
2. Messer, K., Kittler, J., Sadeghi, M., Hamouz, M., Kostin, A., Cardinaux, F., Marcel, S., Bengio, S., Sanderson, C., Poh, N., Rodriguez, Y., Czyz, J., Vandendorpe, L., McCool, C., Lowther, S., Sridharan, S., Chandran, V., Palacios, R.P., Vidal, E., Bai, L., Shen, L.-L., Wang, Y., Yueh-Hsuan, C., Liu, H.-C., Hung, Y.-P., Heinrichs, A., Muller, M., Tewes, A., vd Malsburg, C., Wurtz, R., Wang, Z., Xue, F., Ma, Y., Yang, Q., Fang, C., Ding, X., Lucey, S., Goss, R., Schneiderman, H.: Face authentication test on the BANCA database. In: Int'l. Conf. Pattern Recognition (ICPR), vol. 4, pp. 523–532 (2004)
3. Phillips, P.J., Flynn, P.J., Scruggs, T., Bowyer, K.W., Chang, J., Hoffman, K., Marques, J., Min, J., Worek, W.: Overview of the Face Recognition Grand Challenge. In: IEEE Computer Society Conference on Computer Vision and Pattern Recognition, pp. 947–954 (2005)
4. Bailly-Baillière, E., Bengio, S., Bimbot, F., Hamouz, M., Kittler, J., Marithoz, J., Matas, J., Messer, K., Popovici, V., Porée, F., Ruiz, B., Thiran, J.-P.: The BANCA Database and Evaluation Protocol. In: Kittler, J., Nixon, M.S. (eds.) AVBPA 2003. LNCS, vol. 2688. Springer, Heidelberg (2003)
5. Turk, M., Pentland, A.: Eigenfaces for Recognition. Journal of Cognitive Neuroscience 3(1), 71–86 (1991)
6. Martin, A., Doddington, G., Kamm, T., Ordowsk, M., Przybocki, M.: The DET Curve in Assessment of Detection Task Performance. In: Proc. Eurospeech 1997, Rhodes, pp. 1895–1898 (1997)
7. Bengio, S., Marithoz, J.: The Expected Performance Curve: a New Assessment Measure for Person Authentication. In: The Speaker and Language Recognition Workshop (Odyssey), Toledo, pp. 279–284 (2004)
8. Poh, N., Bengio, S.: Database, Protocol and Tools for Evaluating Score-Level Fusion Algorithms in Biometric Authentication. Pattern Recognition 39(2), 223–233 (2005)
9. Martin, A., Przybocki, M., Campbell, J.P.: The NIST Speaker Recognition Evaluation Program, ch. 8. Springer, Heidelberg (2005)

Fingerprint and On-Line Signature Verification Competitions at ICB 2009

Bernadette Dorizzi[1], Raffaele Cappelli[2], Matteo Ferrara[2], Dario Maio[2], Davide Maltoni[2], Nesma Houmani[1], Sonia Garcia-Salicetti[1], and Aurélien Mayoue[1]

[1] Institut TELECOM; TELECOM & Management SudParis, Evry, France
[2] Biometric System Laboratory, DEIS, University of Bologna, Italy
{cappelli,ferrara,maio,maltoni}@csr.unibo.it,
{Bernadette.Dorizzi,Nesma.Houmani,
Sonia.Salicetti,Aurelien.Mayoue}@it-sudparis.eu

Abstract. This paper describes the objectives, the tasks proposed to the participants and the associated protocols in terms of database and assessment tools of two competitions on fingerprints and on-line signatures. The particularity of the fingerprint competition is to be an on-line competition, for evaluation of fingerprint verification tools such as minutiae extractors and matchers as well as complete systems. This competition will be officialy launched during the ICB conference. The on-line signature competition will test the influence of multi-sessions, environmental conditions (still and mobility) and signature complexity on the performance of complete systems using two datasets extracted from the BioSecure database. Its result will be presented during the ICB conference.

Keywords: Fingerprint Verification, On-line signature Verification, Technical Evaluations, Signature Categorization, Entropy.

1 Introduction

Conducting biometric technical evaluations is important in order to stimulate research in the challenging and open issues left unsolved in the biometrics at hand. Comparative evaluations also provide to the stakeholders, information about the state of the art of the algorithms and the progress of the technology as well as tools such as databases and protocols for future benchmarking.

In this paper, we concentrate on two evaluations, which are now open in fingerprint and on-line signature. Their results will be ready for a presentation at the ICB 2009 conference. The first competition, FVC-onGoing is in the continuity of the FVCs competitions (Fingerprint Verification Competition) organized by University of Bologna in the past years. The second one, BSEC 2009 (BioSecure Signature Evaluation Campaign) follows the road opened last year by BMEC 2009 (BioSecure Multimodal Evaluation Campaign) except that this year, it is focused on one modality only, namely on-line signature and it will use the newly available BioSecure Database.

The rest of the paper describes in two separate sections (Section 2 and 3) the two competitions, their objectives, the tasks proposed to the participants and the associated protocols in terms of database and assessment tools.

M. Tistarelli and M.S. Nixon (Eds.): ICB 2009, LNCS 5558, pp. 725–732, 2009.
© Springer-Verlag Berlin Heidelberg 2009

2 FVC-onGoing

Performance evaluation is important for all biometric modalities and particularly so for fingerprint recognition, which is receiving widespread attention for citizen identity verification and identification in large-scale applications. Unambiguously and reliably assessing the current state of the fingerprint recognition technology is mandatory for understanding its limitations and addressing future research.

After the success of the first four fingerprint verification competitions (FVC2000 [1], FVC2002 [2], FVC2004 [3] and FVC2006 [4]), the Biometric System Laboratory (University of Bologna) has decided to organize a new online evaluation campaign for fingerprint recognition technologies: *FVC-onGoing*.

FVC-onGoing will offer web-based automatic evaluation of fingerprint recognition algorithms on a set of sequestered datasets, reporting results using well known performance indicators and metrics.

The aim is to track the advances in fingerprint recognition technologies, through continuously updated independent testing and reporting of performances on given benchmarks. The benchmark datasets will not evolve over time; in case new datasets will be added in the future, they will form a different benchmark or a new version of an existing one: in this way, only results obtained on the same data will be compared.

The algorithms will be evaluated using strongly supervised approaches (see [3]), to maximize trustworthiness of the results.

While previous FVC initiatives were organized as "competitions", with specific calls and fixed time frames, FVC-onGoing will be:

- an "on going competition" always open to new participants;
- an evolving online repository of evaluation metrics and results.

Furthermore, the evaluation will be not only limited to fingerprint verification algorithms: ad hoc metrics and datasets for testing specific modules of fingerprint verification systems will be available. In fact, since results are always reported as FMR/FNMR values of the entire fingerprint recognition system developed, it is practically impossible to understand if an advancement in performance is due to a specific matching technique or is in large part due to a minor change in an existing feature extraction method. For example, the only way to objectively compare fingerprint matchers is to start from the same set of features (i.e., the set of minutiae for minutiae based matchers). This will allow to better understand the limits and the challenges not only of the whole recognition problem, but also of its building blocks, with clear benefits for researchers and algorithms' developers.

Benchmarks (with specific datasets and testing protocols) for the following (sub)problems are currently being developed, and others may be added in the future:

- Fingerprint Verification (assessment of the accuracy of one-to-one fingerprint matching algorithms);
- Orientation Image Extraction (assessment of the accuracy of orientation image extraction algorithms);
- Minutiae Extraction (assessment of the accuracy of minutiae extraction algorithms);
- Minutiae Matching (assessment of the accuracy of minutiae matching algorithms on datasets of minutiae templates).

One of the main goals of FVC-onGoing is to fully automate the main steps of the evaluation: participant registration, algorithm submission, performance evaluation, and reporting of the results. To this purpose, the Biometric System Laboratory is currently developing a web-based evaluation framework, whose architecture and typical workflow are shown in Fig. 1. The development of this software system is expected to be completed by January 2009; during the first months of 2009, a beta testing phase will be carried out with some invited participants; the official start of FVC-onGoing is planned to be held in conjunction with ICB2009 (the reader may refer to [5] for the most updated information).

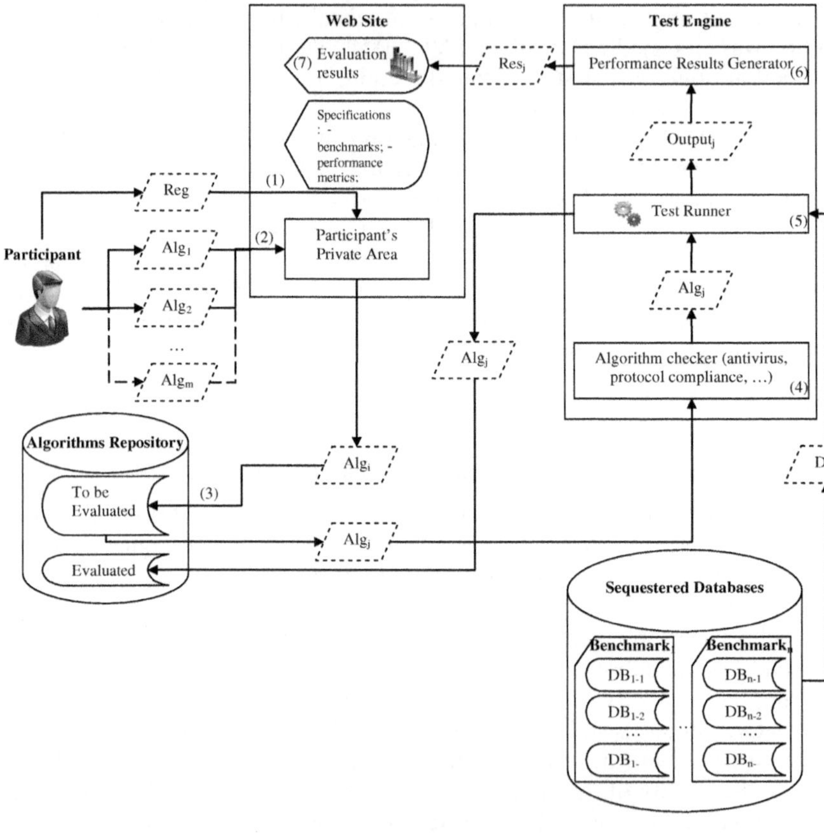

Fig. 1. The diagram shows the architecture of the FVC-onGoing evaluation framework and an example of a typical workflow: a given participant, after registering to the Web Site (1), submits some algorithms (2) to one or more of the available benchmarks; the algorithms (binary executable programs compliant to a given protocol) are stored in a specific repository (3). Each algorithm is evaluated by the Test Engine that, after some preliminary checks (4), executes it on the dataset of the corresponding benchmark (5) and processes its outputs (e.g. matching scores) to generate (6) all the results (e.g. EER, score graphs, ...), which are finally published (7) on the Web Site.

3 BSEC2009: The BioSecure Signature Evaluation Campaign

3.1 Motivation

Pen-based interfaces on mobile platforms give nowadays another possible applicative context for signature verification algorithms. In the previous competitions, the First International Signature Verification Competition (SVC'2004) [6] and the BioSecure Multimodal Evaluation Campaign (BMEC'2007) [7,8], data came from a single sensor in each case: a digitizing tablet at SVC'2004 [6] (although 2 tasks with different parameters were tested) and a Personal Digital Assistant at BMEC'2007 [7, 8]. Also, in both cases performance was evaluated globally on all the persons of the database independently of any quality measured on the data. Finally, the impact of temporal variability was not studied at all in SVC'2004 and BMEC'2007; we thus aim in BSEC'2009 at doing a first step in this direction, by evaluating algorithms performance on two sessions spaced of 5 weeks.

BSEC2009 (BioSecure Signature Evaluation Campaign) [9] will be performed on the two existing largest databases containing the same persons, acquired on two different devices : indeed, BioSecure Signature Corpus DS3 and DS2 are the first on-line signature multisession databases acquired in a mobile scenario (on a PDA) for DS3, and on a digitizing tablet for DS2 [7,8]. BSEC2009 aims at measuring the real impact of a mobile platform on algorithms' performance using databases of large size. Another main goal of BSEC2009 is to evaluate algorithms' performance according to information content in signatures [9].

3.2 Objectives

The BioSecure Signature Evaluation Campaign has several objectives:

- To measure the real impact of mobility acquisition conditions on algorithms performance, by using the two largest existing online signature databases containing the same persons, BioSecure Signature Databases DS2 and DS3.
- To measure the impact of the information content of signatures on algorithms' performance, thanks to a protocol categorizing the data of both DS2 and DS3 in subsets. Indeed, performance will be measured by the evaluator not only globally (on the whole test database), but also on categories of signatures in terms of information content. To this end, we will exploit the notion of Client-Entropy, introduced in [10,11] to categorize users depending on the quality of the signature (Complexity, Variability). This quality measure will not be available to participants; it will be only used by the evaluator in order to carry out performance assessment. In Section 3.3, we present in details the method provided by the organizer for categorizing the signatures.
- To do a first evaluation of the impact of time variability in the field.

3.3 Entropy Measure and Client Categories

3.3.1 Measuring Client-Entropy with a Hidden Markov Model

The entropy of a random variable only depends on its probability density values; thus, a good estimation of this probability density is important. As there are local dependencies in the dynamics of the hand-draw, a local paradigm for density estimation is natural.

We consider each signature as a succession of portions, generated by its segmentation via the client-Hidden Markov Model (HMM) [12]. Therefore, we obtain as many portions in each signature as there are states in the client-HMM. Then we consider each point *(x,y)* in a given portion as the outcome of one random variable Z that follows a given probability mass function $p(z) = Pr(Z = z)$ where z belongs to the Alphabet A of ordered pairs *(x,y)*.

Indeed, such random variable is discrete since its alphabet A has a finite number of values, nevertheless, the hand-draw as a time function is naturally a continuous process from which we retrieve a sequence of discrete values by the digitizer. For this reason, although $Z=(x,y)$ is discrete, its entropy in bits is defined as

$$P(Z) = -\sum_{z \in A} p(z).\log_2(p(z)) \tag{1}$$

To compute the Client-Entropy measure, we first train the Client-HMM on 10 genuine signatures of such writer, after computing a personalized number of states, as follows:

$$N = \frac{T_{Total}}{M * 30} \tag{2}$$

where T_{Total} is the total number of sampled points available in the genuine signatures, and M=4 is the number of Gaussian components per state. We consider that on average 30 sampled points are enough to estimate the mean vector and diagonal covariance matrix of each Gaussian component.

Then we exploit the Client-HMM to generate by the Viterbi algorithm the portions on which entropy is computed for each genuine signature. On each portion, we consider the probability density estimated by the Client-HMM to compute locally the entropy. We then average the entropy over all the portions of a signature and normalize the result by the signing time of the signature. Averaging this measure across the 10 genuine signatures on which the probability density was estimated by the HMM, allows generating a measure of the Client- Entropy. This measure is expressed in bits per second. Time-normalization allows comparing users between them in terms of entropy.

3.3.2 Client Categories with Entropy Measure

We performed on the two BioSecure Subsets DS2 and DS3 containing the same 382 persons, a *K*-Means on the Client- Entropy measure values for different values of *K* and reached a good separation of signatures with *K=3* as shown in Fig 1 and Fig 2, respectively on DS2 and DS3.

We notice visually that on the two Subsets of DS2 and DS3, the first category of signatures, those having the highest Client-Entropy measure, contains short, simply drawn and not legible signatures, often with the shape of a simple flourish. At the opposite, signatures in the third category, those of lowest Client-Entropy measure, are the longest and their appearance is rather that of handwriting, some are even legible. In between, we notice that signatures with medium Client-Entropy measure (second category) are longer and sometimes become legible, often showing the aspect of a complex flourish.

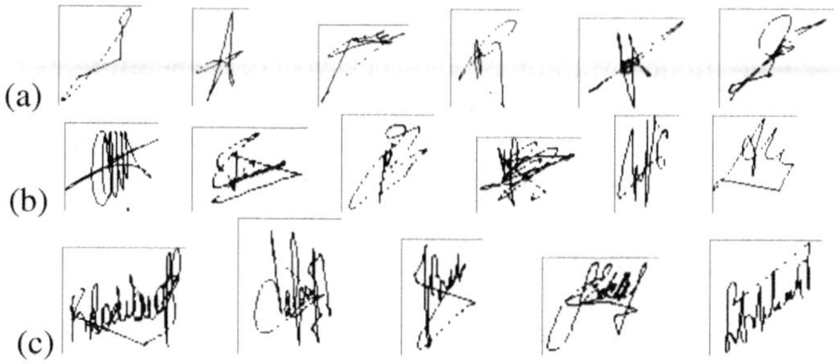

Fig. 1. Examples of signatures from DS2 of (a) highest, (b) medium and (c) lowest Client-Entropy measure

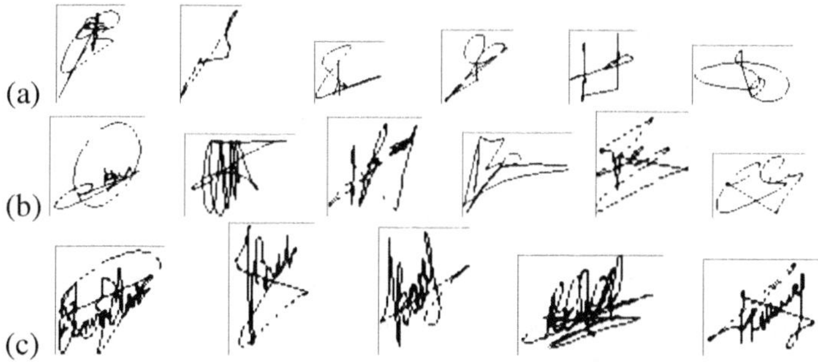

Fig. 2. Examples of signatures from DS3 of (a) highest, (b) medium and (c) lowest Client-Entropy measure

3.4 Protocol and Databases

3.4.1 Description of Databases

We use two databases in this evaluation: the two subsets of the Online Signature Databases acquired in the framework of the BioSecure Network of Excellence [7], DS3, acquired on a mobile platform (Personal Digital Assistant (PDA)) and DS2, acquired on a digitizer [8].

The whole BioSecure Signature Subcorpus DS3 and DS2 are not yet publicly available but, acquired on several sites in Europe, they are the first on-line signature multisession databases acquired in a mobile scenario (on a PDA) for DS3 and on a digitizer for DS2.

DS3 contains the signatures of 713 persons, acquired on the PDA HP iPAQ hx2790, at the frequency of 100Hz and a touch screen resolution of 1280*960 pixels. Three time functions are captured from the PDA: x and y coordinates and the time elapsed between the acquisition of two successive points. The user signs while standing and has to keep the PDA in his or her hand. Two sessions were acquired spaced of around 5 weeks, each containing 15 genuine signatures. The donor was asked to perform, alternatively, three times five genuine signatures and twice five forgeries. For skilled forgeries, at each session, a donor is asked to imitate five times the signature of two other persons. In order to imitate the dynamics of the signature, the forger

visualized on the PDA screen the writing sequence of the signature he/she had to forge and could sign on the image of such signature in order to obtain a better quality forgery both from the point of view of the dynamics and of the shape of the signature.

As original data acquired on the PDA have the property of variable time elapsed between two consecutive points captured, the organizer will perform time interpolation.

On the other hand, DS2 contains data from 667 persons acquired in a PC-based offline supervised scenario and the digitizing tablet WACOM INTUOS 3 A6. The pen tablet resolution is 5080 lines per inch and the precision is 0.25 mm. The maximum detection height is 13 mm and the capture area is 270mm (width) x 216mm (height). Signatures are captured on paper using an inking pen. At each sampled point of the signature, the digitizer captures at 100 Hz sampling rate the pen coordinates, pen pressure (1024 pressure levels) and pen inclination angles (azimuth and altitude angles of the pen with respect to the tablet). This database contains two sessions, acquired two weeks apart. Fifteen (15) genuine signatures were acquired at each session as follows: the donor was asked to perform, alternatively, three times five genuine signatures and two times five skilled forgeries. For skilled forgeries, at each session, a donor is asked to imitate five times the signature of two other persons.

Two development sets of 50 users from respectively BioSecure Subsets DS2 and DS3 have been distributed to participants. Note that the 50 users are not the same in the two development sets.

Two test sets of 382 users from respectively BioSecure Subsets DS2 and DS3 are used and they will be kept sequestered. Note that the two test sets contain the same 382 users, in order to measure the real impact of mobility acquisition conditions on algorithms performance.

3.4.2 Protocol

Two tasks will be defined on DS2: Task1 considering the five available time functions captured by the digitizer (pen coordinates, pen pressure and pen inclination) and Task 2 with only pen coordinates.

For each individual enrolled in the signature database, we describe the enrolment and test sets valid for development sets as well as tests on sequestered data: five genuine signatures, acquired during the first session, are used as reference signatures for distance-based approaches and as training set for statistical approaches. The tests are done on the remaining 10 genuine signatures acquired during the first session, 15 genuine signatures from the second session, 20 skilled forgeries acquired during both sessions, and 30 random forgeries.

Submitted systems should compare each signature of the test set to the five signatures of the training set or to a model constructed from these five signatures. For each test, only one score should be provided by the system. This score must be normalized between 0 and 1.

3.4.3 Participants and Organizer Tasks

Participants are free to propose classifiers coupled to feature extraction of their choice. Their objective should be to optimize their approaches on both sets of development data.

Participants are required to send an executable code following precise guidelines that will be provided by the organizer. The organizer will provide comparative global

results (on all 382 persons) as well as on categories of writers as defined in Section 3.3. The organizer will also test the executables proposed by the participants in conditions of time variability, thanks to the protocol given in Section 3.4.2.

Up to now, 11 academic participants are registered to BSEC‘2009.

4 Conclusion

We have presented in this paper two competitions that are on-going, on fingerprints and on-line signatures. Both present original protocols which are representative of the actual state of the art in their domain and which will provide interesting comparative results. FVC on-going will be officially launched at ICB 2009 in June while the results of BSEC 2009 will be presented during the conference.

References

1. Maio, D., Maltoni, D., Cappelli, R., Wayman, J.L., Jain, A.K.: FVC 2000: Fingerprint Verification Competition. IEEE Transactions on Pattern Analysis Machine Intelligence 24(3), 402–412 (2002)
2. Maio, D., Maltoni, D., Cappelli, R., Wayman, J.L., Jain, A.K.: FVC 2002: Second Fingerprint Verification Competition. In: Proceedings 16th International Conference on Pattern Recognition (ICPR 2002), Québec City, August 2002, vol. 3, pp. 811–814 (2002)
3. Cappelli, R., Maio, D., Maltoni, D., Wayman, J.L., Jain, A.K.: Performance Evaluation of Fingerprint Verification Systems. IEEE Transactions on Pattern Analysis Machine Intelligence 28(1), 3–18 (2006)
4. Cappelli, R., Ferrara, M., Franco, A., Maltoni, D.: Fingerprint verification competition 2006. Biometric Technology Today 15(7-8), 7–9 (2007)
5. FVC-onGoing web site, `http://bias.csr.unibo.it/fvcongoing/` (accessed, November 2008)
6. Yeung, D.Y., Chang, H., Xiong, Y., George, S., Kashi, R., Matsumoto, T., Rigoll, G.: SVC 2004: First international signature verification competition. In: Zhang, D., Jain, A.K. (eds.) ICBA 2004. LNCS, vol. 3072, pp. 16–22. Springer, Heidelberg (2004)
7. `http://www.biosecure.info`
8. `http://www.int-evry.fr/biometrics/BMEC2007`
9. `http://biometrics.it-sudparis.eu/BSEC2009/`
10. Garcia-Salicetti, S., Houmani, N., Dorizzi, B.: A Client-entropy Measure for On-line Signatures. In: Proceedings of IEEE Biometrics Symposium (BSYM), Tampa, USA, September 2008, pp. 83–88 (2008)
11. Houmani, N., Garcia-Salicetti, S., Dorizzi, B.: A Novel Personal Entropy Measure confronted with Online Signature Verification Systems Performance. In: Proceedings of IEEE Second International Conference on Biometrics: Theory, Applications and Systems (BTAS), Washington (September 2008)
12. Rabiner, L., Juang, B.H.: Fundamentals of Speech Recognition. Prentice Hall Signal Processing Series (1993)

Partial Face Matching between Near Infrared and Visual Images in MBGC Portal Challenge

Dong Yi, Shengcai Liao, Zhen Lei, Jitao Sang, and Stan Z. Li*

Center for Biometrics and Security Research,
Institute of Automation, Chinese Academy of Sciences,
95 Zhongguancun Donglu, Beijing 100190, China
{dyi,scliao,zlei,jtsang,szli}@cbsr.ia.ac.cn

Abstract. The latest multi-biometric grand challenge (MBGC 2008) sets up a new experiment in which near infrared (NIR) face videos containing partial faces are used as a probe set and the visual (VIS) images of full faces are used as the target set. This is challenging for two reasons: (1) it has to deal with partially occluded faces in the NIR videos, and (2) the matching is between heterogeneous NIR and VIS faces. Partial face matching is also a problem often confronted in many video based face biometric applications.

In this paper, we propose a novel approach for solving this challenging problem. For partial face matching, we propose a local patch based method to deal with partial face data. For heterogeneous face matching, we propose the philosophy of enhancing common features in heterogeneous images while reducing differences. This is realized by using edge-enhancing filters, which at the same time is also beneficial for partial face matching. The approach requires neither learning procedures nor training data. Experiments are performed using the MBGC portal challenge data, comparing with several known state-of-the-arts methods. Extensive results show that the proposed approach, without knowing statistical characteristics of the subjects or data, outperforms the methods of contrast significantly, with ten-fold higher verification rates at FAR of 0.1%.

Keywords: Multiple biometric grand challenge (MBGC), MBGC portal challenge, video based face recognition, near infrared (NIR), heterogeneous face biometrics.

1 Introduction

The latest multiple biometric grand challenge (MBGC 2008) [1] sets up a series of experiments to evaluate recent advances of biometric technologies. In the MBGC portal challenge, near infrared (NIR) face videos are used as one of the probe sets whereas the target set consists of full faces in visual (VIS) images, as illustrated in fig.1. The NIR videos contain partial faces due to out-of-view during video capturing. Such partial occlusions of faces are often confronted in video surveillance and other face based applications.

Many methods have been proposed to deal with the occlusion problem in face recognition, which mainly based on patch or local feature strategies. For example, in [2], local nonnegative matrix factorization is used to learn local feature subspaces for face

* Corresponding author.

M. Tistarelli and M.S. Nixon (Eds.): ICB 2009, LNCS 5558, pp. 733–742, 2009.
© Springer-Verlag Berlin Heidelberg 2009

Fig. 1. Examples of full VIS face targets (upper) and partial NIR face probes (lower) from MBGC portal challenge

recognition with occlusions. In [3], a face image is divided into local parts, and each part is modeled by a Gaussian distribution. The probability of a given image is calculated with the sum of all Gaussian models. In [4], instead of dividing face image into patches, the authors create six subparts of images through masking some regions (eg. left eye, right eye, nose or mouth) and learn a PCA subspace for each subset. A given face image is projected to these six PCA subspaces separately, then the final score is fused by weighted sum rule.

As face recognition technology advances, active near infrared (NIR) imaging has been used to effectively solve the illumination problem. On the other hand, many applications require visual (VIS) image based enrollment. To take advantage of latest technology while meeting requirements from applications, heterogeneous face biometrics by matching face images from different sources has become a new direction.

Recent years have seen developments in non-visual image based face biometrics, including 3D [5], thermal infrared (TIR) [6], and NIR [7] based methods. In many face based biometric applications, such as access control and E-Passport, the enrollment of face templates is done using VIS face images. Such images are normally acquired in controlled environment where the lighting is approximately homogeneous and frontal.

However, face biometrics has to deal with changes in lighting conditions. In dark environment, the quality of VIS face images can be too poor to be used for face recognition. Recent NIR technology well overcomes problems arising from lighting changes and has been successfully applied to face recognition in normal or dark environments [7]. So many face recognition systems mount NIR imaging device to adapt dark environment. While the NIR based method requires that enrollment face images be acquired using the NIR as well, how to take advantage of the new technology of NIR face to match against visual (VIS) face as required in many applications becomes an interesting problem.

Heterogeneous face biometrics by matching face images from different image sources has become a new direction for face recognition research. Several methods have

been proposed to solve this problem. In [8], Tang and his colleagues developed a PCA based method for face sketch recognition against face photos. In [9], the authors propose a unified framework for inter-modality face matching called common discriminant feature extraction (CDFE). CDFE can be considered as a modified linear discriminant analysis (LDA) method for heterogeneous face matching, in which scatter matrices are defined on two different types of images, and local consistency of manifold is imposed to regularize the dimension reduction. CDFE is applied to sketch-photo image matching and NIR-VIS face image matching, and obtains good results. In [10], a method is proposed for NIR-VIS face image matching problem based on canonical correlation analysis (CCA) [11]. Given that NIR-VIS image pairs of subjects are available for training, a CCA based correlation mechanism can be learned from corresponding NIR-VIS face images. The CCA learning is performed between features in PCA or LDA subspaces, rather than between images, by which better classification performance is gained.

In this paper, we propose an approach for solving the new challenging problem of matching between partial NIR faces and VIS faces. This is challenging for two reasons: (1) it has to deal with partially occluded faces in the NIR videos, and (2) the matching is between heterogeneous NIR and VIS faces. We adopt the philosophy of enhancing common features while reducing differences between heterogeneous image sources. A Laplacian of Gaussian (LoG) filter is applied to NIR and VIS face images. Edge points or low level feature points are extracted and converted into binary images. This significantly reduces heterogeneities between the NIR and VIS face images. The filtered images are segmented into small patches, and then template matching is done between these patches. Because the proposed method is local patch and edge-enhancing feature based, it is able to solve the problem of partial matching in heterogeneous images. Another advantage of the method is that it is non-learning based and basically requires no training data in the design of the classifiers. Hence, it largely overcomes the overfitting problem in learning based methods.

Experiments are performed using the MBGC data sets. The proposed method is compared with several state-of-the-art methods, such as CDFE [9], PCA-CCA and LDA-CCA [10], on the same basis of patch segmentation. Extensive results show that the proposed approach, without knowing statistical characteristics of the subjects or data, outperforms the methods of contrast significantly, with ten-fold higher verification rates at FAR of 0.1%.

The rest of the paper is organized as follows. Partial face processing and analysis are discussed in Section 2. Local, edge-enhancing features are described in Section 3 for partial and heterogeneous face matching. The NIR-VIS face matching algorithm is described in Section 4. Section 5 describe experimental set up and extensive results.

2 Partial Face Processing

MBGC portal challenge includes two kinds of face data: the full VIS face images as the target set, and the NIR videos containing full or partial faces as the probe set. Some partial NIR faces and full VIS faces are shown in Fig. 1. We treat the problem as still face matching since temporal information is difficult to explore with such data sets, that is, departing the NIR videos into frames and then converting the problem into

matching between sets of still frames. Before matching, all face images need to be properly aligned.

In conventional methods with full faces, face images are usually aligned by four steps: (1) face Detection; (2) face landmark point localization, usually of two eyes, and sometime with nose and mouth; (3) face alignment according to the eye centers and sometime with references to nose and mouth; (4) cropping of aligned face.

However, for partial face images, as in MBGC NIR portal videos, not all landmarks are present. Face detection becomes difficult and landmark localization become inaccurate. To solve this problem, we developed an automated algorithm for landmark localization and thereby face alignment with full faces. According to located landmark points, a full or partial NIR face is cropped; this also gives the corresponding mask of the valid face area. An example is shown in Fig.2.

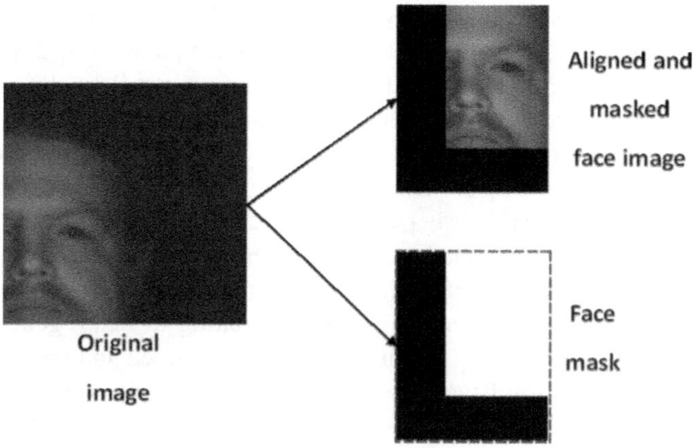

Fig. 2. A partial NIR face image, its corresponding aligned and masked version, and the face mask

While this paper is focused on the the problem of partial face matching in heterogeneous images, we leave the presentation of automated localization of landmarks in partial faces in a separate paper. However, we will evaluate partial face recognition performance with manual versus automated alignment.

3 Face Descriptors for NIR-VIS Matching

For heterogeneous face matching, we adopt the philosophy of enhancing common features while reducing differences between heterogeneous image sources. By visual inspection (Fig. 3), we can see although all pixel values of NIR and VIS face images are different, the structure of points, edges or contours are very similar. These low level features (such as nose location, eye shape, mouth contour, etc.) could supply common and essential information for NIR-VIS matching. By experiments, we found that features common to NIR and VIS images are mostly contained near edges. Therefore, we

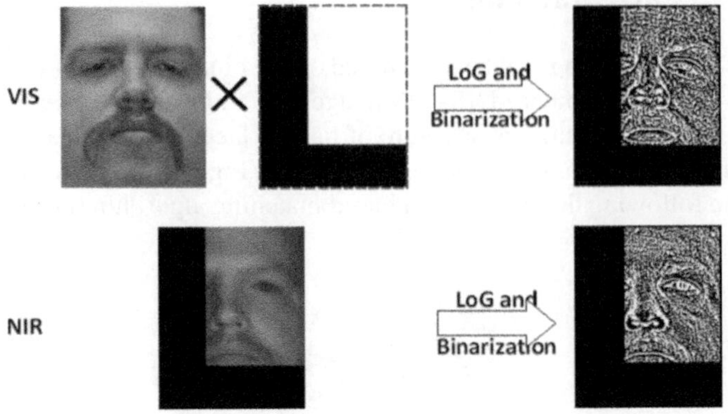

Fig. 3. NIR and VIS face images and their corresponding LoG filtered binary images

investigated into various edge enhancing features and found the Laplacian of Gaussian (LoG) is a good choice.

Laplacian filter is usually used to highlight the regions of rapid intensity change and is therefore often used for edge detection. Because Laplacian filter is approximating a second derivative measurement on a image, it is very sensitive to noise. To suppress noise, the image is smoothed by a Gaussian kernel before applying the Laplacian filter. This is equivalent to applying a LoG filter, which reduces high frequency noise components prior to the edge detection. The 2-D LoG function centered on zero and with Gaussian standard deviation σ has the following form:

$$LoG(x,y) = -\frac{1}{\pi\sigma^4}\left[1 - \frac{x^2+y^2}{2\sigma^2}\right] e^{-\frac{x^2+y^2}{2\sigma^2}} \tag{1}$$

Assuming we use square 2D-LoG kernel to filter face images, there will be only two parameters of a 2D-LoG kernel: kernel size and Gaussian standard deviation. In practice, these two parameters are selected according the size of source image.

In this paper, LoG filter is used to detect the edge information of NIR and VIS face images. Firstly, using mask image to remove the redundant pixels of VIS image comparing to NIR image. Then NIR and VIS face images are filtered by a LoG filter and thresholded into binary images. Thresholded binary images are more robust to noise than gray images.

Fig. 3 illustrates a NIR and VIS face image and their corresponding LoG filtered binary images. We can see the difference between NIR-VIS image pair is remarkably reduced by binarized LoG filtering. This processing step provides a solid basis for the subsequent face matching. Surely, other filters also can be used to replace LoG filter, such as Canny operator, Sobel operator, and etc. Some kinds of local feature invariant to NIR-VIS spectrum are good choices too. Although this paper does not provide a thorough evaluation of all these filters (Gabor filters is proved not good), it directs to the choice of edge enhancing filters.

4 NIR-VIS Face Matching

The NIR-VIS face matching will be done based on the binary face features and the mask of the face area. Given a binary NIR-VIS image pair and mask, we need to compute a similarity score. Considering the structure of human face and possible occlusions, we divide the whole face into local patches of 16×16 (Fig. 4) and perform patch based matching. The following describes patch based matching algorithm for the face matching problem.

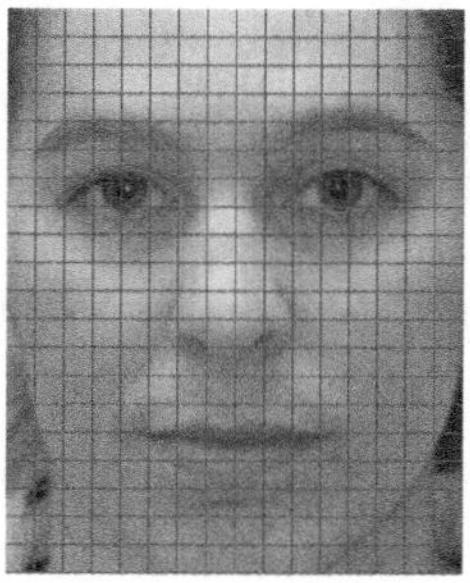

Fig. 4. The 256×320 pixel area of a face is divided into 16×16 local patches

First, segment the face from the background and obtain the mask of valid area of the face. Then compute the descriptor for each of the patches in the masked area. The binary features of corresponding patches are compared and distance computed using the Hamming distance. While the Hamming distance is suited for comparing two binary images, it's very sensitive to occlusion, expression and mis-alignment. Therefore, a local search is performed and the minimum Hamming distance is finally found to measure the difference between two patches. These are illustrated in Fig. 5. All the minimum distances are fused by the sum rule to obtain the overall distance.

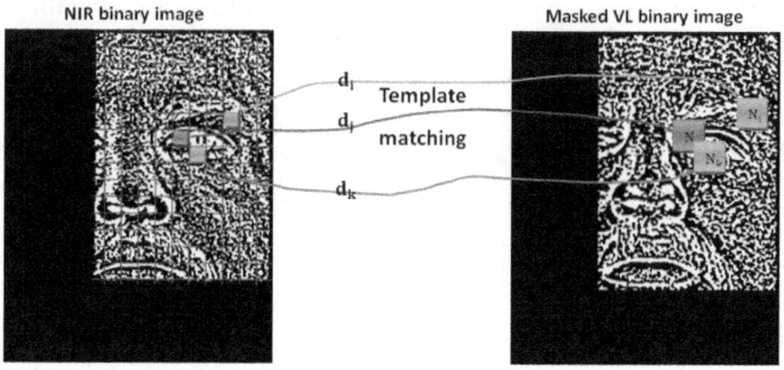

Fig. 5. The NIR-VIS face image matching strategy. The 256×320 pixel area of a face is divided into 16×16 local patches.

5 Experiments

The experiments are aimed to evaluate the performance of the proposed method in MBGC portal challenge, as compared to existing methods,including PCA-CCA, LDA-CCA, CDFE. Because PCA-CCA, LDA-CCA, CDFE cannot process partial data directly, for fair comparison, they are all implemented in the patch based way. That means these methods are applied to image patches and then patch matching scores are fused into a final score. Besides, in order to evaluate the influence of face alignment, all the experiments are conducted based on results of manual and automated face alignment. Therefore, the following 8 schemes are evaluated:

1. PCA-CCA-M: Patch based PCA-CCA on manually aligned face data;
2. PCA-CCA-A: Patch based PCA-CCA on automatically aligned face data;
3. LDA-CCA-M: Patch based LDA-CCA on manually aligned face data;
4. LDA-CCA-A: Patch based LDA-CCA on automatically aligned face data;
5. CDFE-M: Patch based CDFE on manually aligned face data;
6. CDFE-A: Patch based CDFE on automatically aligned face data.
7. Proposed-M: Proposed method on manually aligned face data;
8. Proposed-A: Proposed method on automatically aligned face data;

5.1 Database Description

For the portal challenge, MBGC only supply a database for testing. It contains 148 NIR videos and 148 VIS images of 114 persons. The still VIS face images are aligned and normalized into the size of 256×320 pixels. Frames are extracted from the NIR videos and masked and aligned in manual and automated way separately. Some aligned face images in MBGC database are shown in Fig.6. All the compared methods have the same basis of alignment.

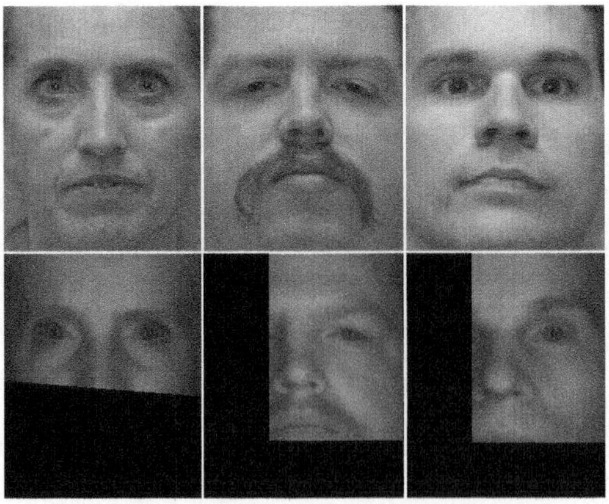

Fig. 6. Examples of aligned VIS face images (upper) and NIR face images (lower) from the MBGC testing set

Because the CDFE and CCA based methods, unlike the proposed method, need a training set, a database of NIR and VIS face images of 200 persons was collected by ourselves. All the faces and eyes were detected in the images, aligned with eye coordinates and cropped into 256×320 pixels too, using an automated procedure, with results of acceptable quality. Some example training face images were shown in Fig.7.

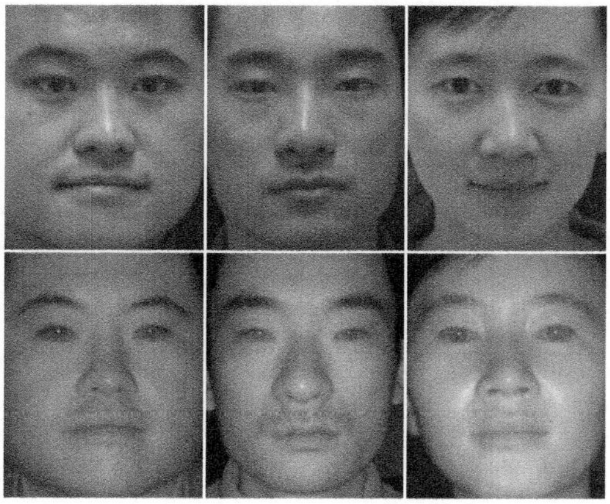

Fig. 7. Examples of aligned VIS face images (upper) and NIR face images (lower) from the self-prepared training set

From these, we have the training set and testing set as follows:

1. NIR training set: Collected by ourselves, consisting of 200 persons, 6 NIR images per person;
2. VIS training set: Collected by ourselves, consisting of the same 200 persons with NIR training set, 6 VIS images per person;
3. NIR testing set: Coming from MBGC database, consisting of 114 persons not overlapping with training set, 1-2 NIR vidoes per person;
4. VIS testing set: Coming from MBGC database, consisting of the same 114 persons with NIR testing set, 1-2 VIS images per person.

Data sets (1) and (2) are used for training PCA-CCA, LDA-CCA and CDFE, whereas (3) and (4) are for testing. Apparently, the testing sets did not include any of persons in the training sets.

5.2 Performance Evaluation

The training of PCA-CCA, LDA-CCA and CDFE is performed as follows: PCA and LDA projection matrices are computed from the NIR and VIS patches of the training set, respectively. The projection reduces the dimensionality from 16×16 to 199. The CCA and CDFE projections further reduce the 199 to 100 dimensions to reduce overfitting.

The ROC curves of the compared methods are shown in Fig. 8. At FAR=0.001, the proposed Proposed-M and The Proposed-M achieves impressive 77% verification rate at FAR=0.1%, more than ten-fold higher than the compared PCA-CCA, LDA-CCA and CDFE methods of PCA-CCA-M, PCA-CCA-A, LDA-CCA-M, LDA-CCA-A, CDFE-M, and CFFE-A. Understandably, the manual alignment leads better performance than the automated alignment, with about 5% higher verification rate at FAR=0.001 for the proposed method. Evidently, we can see CCA based methods and CDFE performed poorly in portal challenge, which may be due to their poor generalization ability while they did well on the training data. The proposed method is very stable even without training.

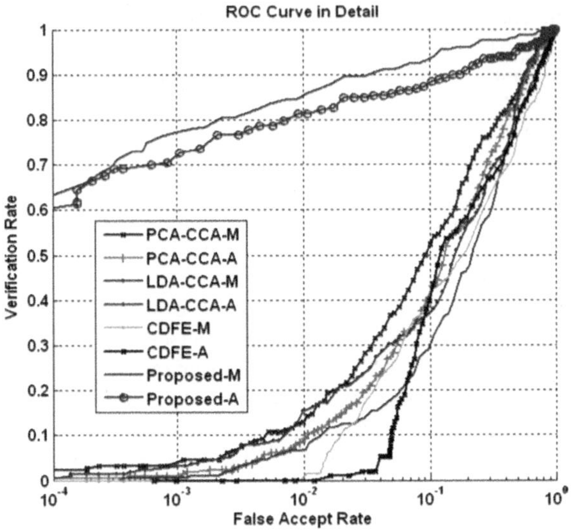

Fig. 8. ROC curves for 8 schemes in the proposed method

6 Conclusions

In this paper, we proposed a novel approach for partial face matching in NIR and VIS images for MBGC portal challenge. To deal with partial occlusion, local patch based matching is devised. To deal with NIR-VIS face matching, effective features are extracted using binarized LoG filters, which also help matching of partial faces. The experimental results on the MBGC portal challenge data illustrate the proposed method obtains significantly higher verification rate than the compared methods. The Proposed-M achieves impressive 77% verification rate at FAR=0.1%, more than ten-fold higher than the compared PCA-CCA, LDA-CCA and CDFE methods. These results support that the proposed approach is an effective solution for the new challenging problem.

Acknowledgements. This work was supported by the following funding resources: National Natural Science Foundation Project #60518002, National Science and Technology Support Program Project #2006BAK08B06, National Hi-Tech (863) Program Projects #2006AA01Z192, #2006AA01Z193, and #2008AA01Z124, Chinese Academy of Sciences 100 people project, and AuthenMetric R&D Funds.

References

1. NIST: Multiple Biometric Grand Challenge (MBGC) (2008), `http://face.nist.gov/mbgc`
2. Li, S.Z., Hou, X.W., Zhang, H.J.: Learning spatially localized, parts-based representation. In: Proceedings of IEEE Computer Society Conference on Computer Vision and Pattern Recognition, Hawaii, December 11-13, vol. 1, pp. 207–212 (2001)
3. Martinez, A.: Recognizing imprecisely localized, partially occluded, and expression variant faces from a single sample per class. IEEE Transactions on Pattern Analysis and Machine Intelligence 24(6), 748–763 (2002)
4. Tarres, F., Rama, A.: A novel method for face recognition under partial occlusion or facial expression variations. In: 47th International Symposium on ELMAR, June 2005, pp. 163–166 (2005)
5. Bowyer, K.W., Chang, Flynn, P.J.: A survey of 3D and multi-modal 3D+2D face recognition. In: Proceedings of International Conference on Pattern Recognition, August 2004, pp. 358–361 (2004)
6. Kong, S.G., Heo, J., Abidi, B., Paik, J., Abidi, M.: Recent advances in visual and infrared face recognition - A review. Computer Vision and Image Understanding 97(1), 103–135 (2005)
7. Li, S.Z., Chu, R., Liao, S., Zhang, L.: Illumination invariant face recognition using near-infrared images. IEEE Transactions on Pattern Analysis and Machine Intelligence 26 (April 2007) (Special issue on Biometrics: Progress and Directions)
8. Tang, X., Wang, X.: Face sketch recognition. IEEE Transactions on Circuits and Systems for Video Technology 14(1), 50–57 (2004)
9. Lin, D., Tang, X.: Inter-modality face recognition. In: Leonardis, A., Bischof, H., Pinz, A. (eds.) ECCV 2006. LNCS, vol. 3954, pp. 13–26. Springer, Heidelberg (2006)
10. Yi, D., Liu, R., Chu, R., Lei, Z., Li, S.Z.: Face matching between near infrared and visible light images. In: Lee, S.-W., Li, S.Z. (eds.) ICB 2007. LNCS, vol. 4642, pp. 523–530. Springer, Heidelberg (2007)
11. Hotelling, H.: Relations between two sets of variates. Biometrika 28, 321–377 (1936)

Fusion in Multibiometric Identification Systems: What about the Missing Data?*

Karthik Nandakumar[1], Anil K. Jain[2], and Arun Ross[3]

[1] Institute for Infocomm Research, A*STAR, Fusionopolis, Singapore
knandakumar@i2r.a-star.edu.sg
[2] Michigan State University, East Lansing, MI, USA
jain@cse.msu.edu
[3] West Virginia University, Morgantown, WV, USA
arun.ross@mail.wvu.edu

Abstract. Many large-scale biometric systems operate in the identification mode and include multimodal information. While biometric fusion is a well-studied problem, most of the fusion schemes have been implicitly designed for the verification scenario and cannot account for missing data (missing modalities or incomplete score lists) that is commonly encountered in multibiometric identification systems. In this paper, we show that likelihood ratio-based score fusion, which was originally designed for verification systems, can be extended for fusion in the identification scenario under certain assumptions. We further propose a Bayesian approach for consolidating ranks and a hybrid scheme that utilizes both ranks and scores to perform fusion in identification systems. We also demonstrate that the proposed fusion rules can handle missing information without any ad-hoc modifications. We observe that the recognition performance of the simplest rank level fusion scheme, namely, the highest rank method, is comparable to the performance of complex fusion strategies, especially when the goal is not to obtain the best rank-1 accuracy but to just retrieve the top few matches.

1 Introduction

Biometric systems can operate in two modes, namely, *verification* and *identification*. In verification, the query is compared only to the template of the claimed identity and the identity claim is accepted as "genuine" if the degree of similarity is sufficiently high. On the other hand, the goal in identification systems is to determine the identity of an individual from a large set of possible identities. Here, the user's biometric input is compared with the templates of all the persons enrolled in the database and the system outputs either the identity of the person whose template has the highest degree of similarity with the user's input (closed set ID) or a decision indicating that the user presenting the input is not an enrolled user (open set ID). Many large-scale biometric applications such as the FBI-IAFIS and US-VISIT IDENT program work in the open-set identification mode. Since the number of enrolled users in the database can be quite

* This research was supported by the Center for Identification Technology Research at West Virginia University.

M. Tistarelli and M.S. Nixon (Eds.): ICB 2009, LNCS 5558, pp. 743–752, 2009.
© Springer-Verlag Berlin Heidelberg 2009

large (e.g., more than 80 million subjects in the FBI-IAFIS system [1]), the identification task is significantly more challenging than verification.

Since multibiometric systems provide several advantages over unimodal systems such as increased population coverage and lower error rates, the databases in many large scale ID systems are increasingly becoming multimodal (e.g., fingerprint, face and palmprint in Next Generation IAFIS, multiple fingers and face in US-VISIT). Biometric fusion has been extensively studied in the literature and a number of fusion schemes have been proposed [2]. However, most of the existing fusion schemes have been implicitly designed for the verification scenario. For instance, the density-based and classifier-based approaches typically model biometric fusion as a two-class (genuine and impostor) classification problem. Hence, such schemes cannot be directly applied to the identification scenario. In [3], Tulyakov and Govindaraju identified four possible strategies in the context of combining multi-class classifiers and described a method for improving the verification performance of a biometric system by utilizing scores obtained by matching the query to all the templates in the database. However, they fail to provide a technique for optimizing the identification performance.

Moreover, the problem of missing data is frequently encountered in large scale multibiometric ID systems. The missing information may be in the form missing modalities in the template or the query or incomplete score/rank information from the individual matchers. In many pattern classification systems, the problem of missing feature values is handled either through imputation (predicting the missing values based on the available features) or by designing dynamic classification rules. One such example is the approach proposed by Dinerstein et al. [4], which involves learning a separate Support Vector Machine (SVM) classifier for each possible combination of input modalities and selecting the appropriate SVM based on the query presented during authentication. However, such a solution is clearly inefficient and not scalable.

Our goal is to develop a biometric fusion scheme specifically for the identification mode that can seamlessly handle partial information without any need for designing rules on a case-by-case basis. Towards this end, we have made the following three contributions. Firstly, we present a mathematical framework for score fusion in identification systems. Secondly, we propose a Bayesian scheme for rank-level fusion in multibiometric systems, which can be combined with the score fusion framework to obtain a hybrid score-rank fusion scheme. Finally, we also demonstrate the effectiveness of the proposed schemes in handling missing information.

2 Score Fusion in Multibiometric Identification

A likelihood ratio (LR)-based framework for score level fusion was presented in [5] for the verification scenario. This LR framework is based on the Neyman-Pearson theorem and it maximizes the genuine accept rate (GAR) at any desired false accept rate (FAR), provided the genuine and impostor match densities are known or can be estimated accurately. By following a Bayesian approach, we now show that the likelihood ratios computed in the LR framework can also be used for fusion in the identification scenario provided the following three assumptions are satisfied: (i) prior probabilities are equal for all users, (ii) the match scores for different persons are independent of one another, and (iii) genuine (impostor) match scores of all users are identically distributed.

Table 1. Transpose of a sample match score matrix, $\boldsymbol{S}$, from the NIST-Multimodal database. Since the number of enrolled users in this database is 517 and the number of modalities is 4, $\boldsymbol{S}$ is a 517×4 matrix.

User #	1	2	3	4	⋯	514	515	516	517
Left Index Finger	29	17	8	8	⋯	9	6	4	9
Right Index Finger	84	7	7	5	⋯	8	5	4	7
Face Matcher 1	0.58	0.57	0.46	0.51	⋯	0.52	0.55	0.49	0.52
Face Matcher 2	73.6	73.9	65.7	65.4	⋯	66.3	66.7	66.8	72.2

Let K denote the number of matchers in the multibiometric system and N be the number of persons enrolled in the system. Suppose for a given query, we observe the $N \times K$ score matrix $\boldsymbol{S} = \left[s_n^k\right]$ (see Table 1). Note that s_n^k represents the match score output by the kth matcher for the nth template in the database, $k = 1, \cdots, K$; $n = 1, \cdots, N$. Our goal is to determine the true identity I of the given query based on $\boldsymbol{S}$. According to the Bayes decision theory [6], the query should be assigned to the identity I_{n_0} that maximizes the posteriori probability, i.e.,

Assign query to identity I_{n_0} if

$$P(I_{n_0}|\boldsymbol{S}) \geq P(I_n|\boldsymbol{S}), \forall\, n = 1, \cdots, N. \tag{1}$$

The above decision rule applies only to closed set identification. For open set identification, the query is assigned to identity I_{n_0} only when equation (1) holds and $P(I_{n_0}|\boldsymbol{S}) \geq \tau$, where τ is a threshold. We can estimate the posteriori probabilities $P(I_n|\boldsymbol{S})$ in the following manner. According to the Bayes theorem,

$$P(I_n|\boldsymbol{S}) = \frac{p(\boldsymbol{S}|I_n)P(I_n)}{p(\boldsymbol{S})}, \tag{2}$$

where $p(\boldsymbol{S}|I_n)$ is the likelihood of observing the score matrix $\boldsymbol{S}$ given that the true identity is I_n and $P(I_n)$ is the prior probability of observing the identity I_n. Under the assumption of equal priors for all users, the posteriori probability $P(I_n|\boldsymbol{S})$ is proportional to the likelihood $p(\boldsymbol{S}|I_n)$. Ideally, we would like to estimate the conditional density of $\boldsymbol{S}$ individually for each user because it captures the complete information about dependencies between the scores assigned to the different users and the user-specific characteristics of the match scores. However, directly estimating the conditional density of $\boldsymbol{S}$ is not practical due to the following two reasons: (i) since $\boldsymbol{S}$ is a $N \times K$ dimensional matrix and N is usually quite large in identification scenarios, estimating the density of $\boldsymbol{S}$ requires a significant number of training samples for each user, which is not generally available in multibiometric databases; and (ii) the density of $\boldsymbol{S}$ needs to be re-estimated frequently due to changes in the list of enrollees.

If match scores for different persons are independent of one another, the likelihood $p(\boldsymbol{S}|I_n)$ can be simplified as

$$p(\boldsymbol{S}|I_n) = \prod_{j=1}^{N} p(\boldsymbol{s}_j|I_n) = p(\boldsymbol{s}_n|I_n) \prod_{j=1, j\neq n}^{N} p(\boldsymbol{s}_j|I_n). \tag{3}$$

Here, $p(\boldsymbol{s}_n|I_n)$ represents the density of genuine match scores corresponding to user I_n (denoted by $f_{gen,n}(\cdot)$) and $p(\boldsymbol{s}_j|I_n), j \neq n$ represents the densities of the impostor scores (denoted by $f_{imp,n}(\cdot)$). However, when the genuine (impostor) match scores of all users are assumed to be identically distributed ($f_{gen,n}(\cdot) = f_{gen}(\cdot)$ and $f_{imp,n}(\cdot) = f_{imp}(\cdot), \forall\ n$), equation (3) can be further simplified as

$$p(\boldsymbol{S}|I_n) = f_{gen}(\boldsymbol{s}_n) \prod_{j=1,j\neq n}^{N} f_{imp}(\boldsymbol{s}_j) = \frac{f_{gen}(\boldsymbol{s}_n)}{f_{imp}(\boldsymbol{s}_n)} \prod_{j=1}^{N} f_{imp}(\boldsymbol{s}_j). \tag{4}$$

Thus, the likelihood of observing the score matrix $\boldsymbol{S}$ given that the true identity is I_n is proportional to the likelihood ratio that was used by the authors in [5] for the verification scenario. Furthermore, if we assume that the scores of different matchers are conditionally independent, we can estimate the joint density of the genuine (impostor) match scores by the product of the marginal densities. Hence, the decision rule in equation (1) can be restated as

Assign query to identity I_{n_0} if

$$\prod_{k=1}^{K} \frac{f_{gen}^k(s_{n_0}^k)}{f_{imp}^k(s_{n_0}^k)} \geq \prod_{k=1}^{K} \frac{f_{gen}^k(s_n^k)}{f_{imp}^k(s_n^k)}, \forall\ n = 1, \cdots, N, \tag{5}$$

where $f_{gen}^k(\cdot)$ and $f_{imp}^k(\cdot)$ are the marginal densities of the genuine and impostor match scores, respectively, output by the kth matcher.

3 Bayesian Approach for Rank Level Fusion

When a biometric system operates in the identification mode, the output of the system may also be a set of ranks (possible matching identities sorted in a decreasing order of match scores). Although the ranks are derived from the match scores, the rank information captures the relative ordering of the scores corresponding to different users. Suppose for a given query, we observe the $N \times K$ rank matrix $\boldsymbol{R} = \left[r_n^k\right]$, where r_n^k represents the rank output by the kth matcher for the nth template in the database, $k = 1, \cdots, K$; $n = 1, \cdots, N$. Let $r_n^{'}$ be a statistic computed for user n such that the user with the lowest value of $r^{'}$ is assigned the highest consensus (or reordered) rank. For example, in the highest rank method [7], each user is assigned the highest rank (minimum r value) as computed by different matchers, i.e., the statistic for user n is

$$r_n^{'} = \min_{k=1}^{K} r_n^k. \tag{6}$$

Ties are broken randomly to arrive at a strict ranking order based on the new statistic $r^{'}$. Ho et al. [7] proposed other methods such as Borda Count and logistic regression, which compute the statistic $r^{'}$ as a linear combination of the ranks. Melnik et al. [8] proposed the use of non-linear functions to combine the ranks of the individual matchers.

We now propose a new rank combination statistic based on Bayes decision theory. Let $P_k(r)$ be the probability that the identity which is assigned rank r by the kth

matcher is the true identity, $r = 1, \cdots, N$; $k = 1, \cdots, N$. Note that the cumulative distribution function of the discrete rank distribution $P_k(r)$ is nothing but the Cumulative Match Characteristic (CMC) curve [9], which is commonly used to depict the performance of an identification system. Grother and Phillips [10] and Bolle et al. [11] show that the rank distribution $P_k(r)$ can be estimated provided the marginal genuine and impostor match score densities $f^k_{gen}(\cdot)$ and $f^k_{imp}(\cdot)$ are known. This estimation again requires two of the assumptions used in section 2, namely, (i) scores of the individual users are independent and (ii) score distributions of different users are identical.

For a given query, suppose that the identity I_n is assigned the rank r^k_n by the kth matcher. From the definition of the rank distribution $P_k(r)$, $P_k(r^k_n)$ is the posteriori probability that I_n is the true identity given r^k_n. Further, if we assume that the matchers are independent, we can compute the new rank combination statistic as the product of the posterior probabilities of the individual matchers.

$$r^{'}_n = \prod_{k=1}^{K} P_k(r^k_n), \text{ for } n = 1, \cdots, N. \tag{7}$$

For the rank statistic computed using equation (7), the user with the largest value of $r^{'}$ should be assigned the highest consensus rank. So, the rank posterior based fusion rule can then be defined as follows.

Assign query to identity I_{n_0} if

$$r^{'}_{n_0} \geq r^{'}_n, \forall\, n = 1, 2, \cdots, N. \tag{8}$$

Note that likelihood ratio based score fusion rule shown in equation (5) utilizes only the match scores corresponding to a particular user, when computing the likelihood ratio for that user. In other words, the relative information between the scores of different users is ignored when computing the score likelihood ratio. On the other hand, the rank posterior based fusion rule in equation (8) considers only the relative order information between the scores of different users and the actual score values are ignored. Therefore, we can treat the score and rank information as two different pieces of evidence and define a hybrid fusion scheme that utilizes both the match scores and the ranks. Let R be the combined score and rank statistic, defined as

$$R_n(\boldsymbol{S}, \boldsymbol{R}) = P(I_n|\boldsymbol{S})r^{'}_n, \tag{9}$$

where the posterior probability based on the match score matrix $\boldsymbol{S}$, $P(I_n|\boldsymbol{S})$, is computed by substituting equation (4) in equation (2) and the posterior probability based on the rank matrix $\boldsymbol{R}$ is obtained using equation (7). The hybrid score-rank fusion rule can then be defined as

Assign query to identity I_{n_0} if

$$R_{n_0} \geq R_n, \forall\, n = 1, 2, \cdots, N. \tag{10}$$

4 Fusion with Missing Data

In a multibiometric ID system, three types of missing data can occur. Firstly, not all the biometric information corresponding to an individual may be available in the database. This leads to missing modalities (templates) for some of the users. Secondly, all the modalities may not be available from the user during authentication, resulting in incomplete queries. Finally, the individual matchers may output a partial list of scores or ranks corresponding to only the top m matching identities. Missing data needs to be accounted for, both when training the fusion algorithm and during identification.

The likelihood ratio-based score fusion technique presented in equation (5) has the ability to handle the three kinds of missing data. Since we assume that the score distributions are identical for all users and the matchers are independent, we need to estimate only the marginal genuine and impostor match score densities for each matcher. Therefore, even if some of the scores in the score matrix $\boldsymbol{S}$ are not available for a fraction of the users, we can still estimate the required densities by pooling match scores from all the users. During identification, when a match score s_n^k is not available, it implies that the kth matcher does not provide any evidence in support of the nth identity. This fact can be easily incorporated in equation (5) by setting the corresponding likelihood ratio $f_{gen}^k(s_n^k)/f_{imp}^k(s_n^k)$ to the value 1 with no change in the decision rule required.

Missing data affects the rank posterior-based fusion scheme more severely. When there are missing templates in the database, the rank information provided by a matcher is not quite meaningful. An incorrect identity may be assigned rank 1 (even when the match score is very low) if the template corresponding to the true identity is not available. Hence, it is not possible to estimate the rank distribution accurately when there are missing templates. Consequently, the rank posterior-based fusion scheme and the hybrid score-rank fusion scheme cannot be employed in this scenario. However, in the case of a missing modality in the query sample presented during authentication, the rank posterior-based technique can still be used by assigning equal posterior probabilities to all the users, i.e., if the kth modality is not available in the query sample, $P(I_n|r_n^k) = 1/N, \forall\ n = 1, \cdots, N$. Partial rank lists can be handled by randomly assigning lower ranks to the identities that were not originally assigned a rank by the matcher. For example, if a matcher outputs only the ranks corresponding to the top m matching identities, the other identities can be randomly assigned the ranks from $(m + 1)$ to N. In most biometric matchers, the probability $P_k(r)$ is high only for the top few matches and is very small for large r. Therefore, the random assignment of lower ranks does not affect the performance of the system.

5 Performance Evaluation

The identification performances of various score and rank level fusion strategies were evaluated on the three partitions of the NIST-BSSR1 database. The first partition is the NIST-Multimodal database, consisting of 517 users with two fingerprints (left and right index fingers) and two face scores (obtained from two different face matchers). The second partition of NIST-BSSR1 is the NIST-Fingerprint database, which is an example of multi-instance(finger) biometric system. This partition consists of scores

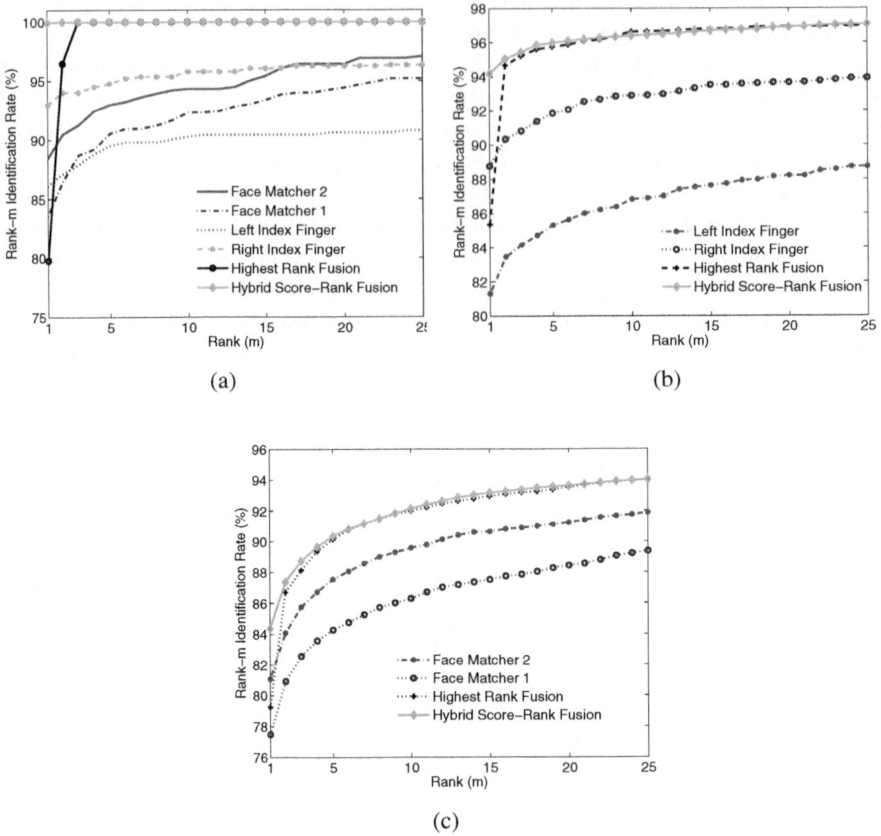

Fig. 1. Cumulative Match Characteristic (CMC) curve of highest rank fusion and the hybrid score-rank fusion rules on (a) NIST-Multimodal ($K = 4, N = 517$), (b) NIST-Fingerprint ($K = 2, N = 6,000$) and (c) NIST-Face ($K = 2, N = 3,000$) databases. The CMC curves shown here are the average recognition rates obtained over 20 cross-validation trials.

from left and right index fingerprint matches of $6,000$ individuals. The third partition is the NIST-Face database, which consists of scores from two face matchers applied on three frontal face images from $3,000$ individuals.

The cumulative match characteristic (CMC) curves of the individual matchers and the highest rank and hybrid score-rank fusion rules on the NIST-BSSR1 database are shown in Figures 1(a), 1(b) and 1(c). In each experiment, half the users were randomly selected to be in the training set for estimating the marginal densities and the rank distribution. The remaining half of the database was used for evaluating the fusion performance. The above training-test partitioning was repeated 20 times and the reported CMC curves correspond to the mean identification rates over the 20 trials. The Gaussian mixture model based scheme [5] was used for estimating the score densities.

Among the various rank level fusion schemes such as highest rank, Borda count and logistic regression, we observed that the highest rank method achieves the best rank-m

recognition rate when $m \geq K$, where K is the number of matchers. Hence, only the recognition rates of the highest rank method are reported here. It is well-known that the highest rank method works well when the number of users is large compared to the number of matchers [7], which is usually the case in biometric identification systems. This is because the highest rank method utilizes the strength of each matcher effectively. Even if only one matcher assigns a high rank to the correct user, it is likely that the correct user will receive a high rank after reordering. However, there can be up to K ties at rank 1 due to conflicting decisions output by the K matchers. Since the ties are broken randomly without considering the relative accuracies of the matchers, the identification rates at ranks 1 to $K-1$ are not very high. In fact, the rank-1 accuracy of the highest rank method is usually less than that of the best individual matcher.

The recognition rates of likelihood ratio-based score fusion, rank posterior fusion and hybrid score-rank fusion rules were observed to be quite similar on all the three partitions of NIST-BSSR1. While the hybrid score-rank fusion rule achieves a marginal improvement in the recognition rates over the other two fusion rules, the differences in the recognition rates of the three fusion rules is less than 1% at all ranks. Therefore, only the performance of the hybrid score-rank fusion rule is reported in Figures 1(a), 1(b) and 1(c). In the case of the NIST-Multimodal database, the hybrid score-rank fusion rule provides 100% rank-1 accuracy, while the rank-1 accuracy of the best single matcher (right index finger) was only 93.7%. The hybrid score-rank fusion rule improves the rank-1 accuracy from 88.9% for the best single matcher (right index finger) to 94% on the NIST-Fingerprint database. Finally, on the NIST-Face database the improvement is comparatively lower (81.2% for the best face matcher and 84.1% for the score-rank fusion rule) due to the strong correlation between the two face matchers.

The results also indicate that the performance of the simplest rank level fusion scheme, namely, the highest rank method, is comparable to performance of the more complex score and rank fusion strategies for ranks greater than or equal to K, where K is the number of matchers. Therefore, in practical multibiometric ID systems with a large number of users, it may be sufficient to use the highest rank method if the goal is to retrieve the top few matches. However, if the best rank-1 accuracy is desired and if the score information is available, then the hybrid score-rank fusion rule can be employed.

We have also evaluated the robustness of the fusion schemes under the missing data scenario. We simulate missing data by randomly removing scores from the score matrices obtained from NIST-BSSR1 with probability m_p (whose value is set to 0.05, 0.1 or 0.25). When the template corresponding to the kth modality of user n is deemed to be missing, only the score s_n^k is removed. When the kth modality is deemed to be missing in the query, the kth column in the score matrix $\boldsymbol{S}$ is removed (see Table 2). While simulating missing data, care is taken to ensure that each user has a template corresponding to at least one modality and each query provides at least one modality to be used for identification. To avoid cases where there are no common modalities between the query and the user templates in the database (e.g., only the left index fingerprint of a user is available in the database, whereas the query contains only the right index fingerprint) we study the missing template and missing query scenarios separately.

Figure 2 shows the CMC curves for different fusion strategies under the missing data scenario on the NIST-Face database (similar results were observed for the other two

Table 2. Transpose of a sample match score matrix with missing data (denoted by x) simulated from the NIST-Multimodal database. In this example, the query multimodal sample does not contain the fingerprint from the left index finger, the templates corresponding to the right index finger are not available for users 3 and 517 and face templates are not available for user 515.

User #	1	2	3	4	···	514	515	516	517
Left Index Finger	x	x	x	x	···	x	x	x	x
Right Index Finger	84	7	x	5	···	8	5	4	x
Face Matcher 1	0.58	0.57	0.46	0.51	···	0.52	x	0.49	0.52
Face Matcher 2	73.6	73.9	65.7	65.4	···	66.3	x	66.8	72.2

databases). As explained in section 4, the rank-posterior and hybrid score-rank fusion schemes cannot be applied in the case of missing templates. Only the likelihood ratio based score fusion rule is used in this scenario (see figure 2(a)). On the other hand, the hybrid score-rank fusion rule can be used in the missing query modalities scenario (see figure 2(b)). Similar to the case, where is there is no missing data, the performance of the highest rank fusion rule is comparable to that of the score fusion rules for larger ranks. However, the highest rank fusion rule has a significantly worse rank-1 recognition rate in the case of missing templates, because the ranks assigned by the matchers are not very meaningful when there are missing templates.

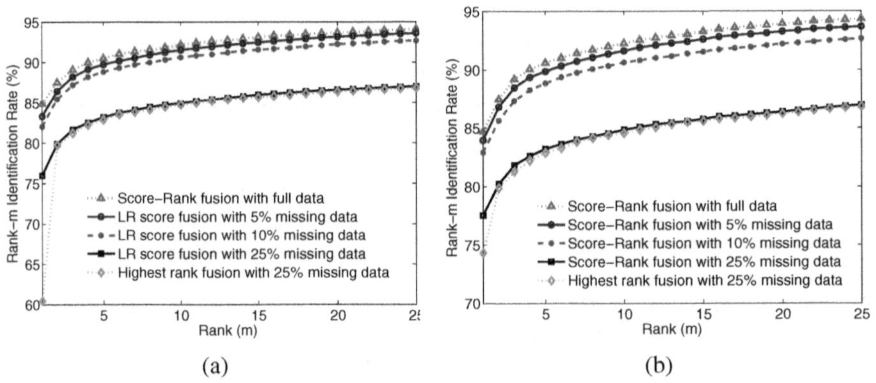

Fig. 2. Cumulative Match Characteristic (CMC) curve of different fusion rules on the NIST-Face databases under the missing data scenario. (a) Missing templates and (b) missing queries.

From Figure 2, we observe that there is a decrease in the recognition rates with increasing amount of missing data. For this particular database, the reduction in the accuracy is marginal when amount of missing data is less than 10%. However, when the amount of missing data is set to 25%, there is a significant reduction in the accuracy. This decrease in accuracy can be attributed to two reasons. During training, due to missing modalities in the template or query, there are lesser number of match scores available for estimating the marginal score densities. During identification, the identity

is decided based on fewer evidences compared to the case where there is no missing data. Since the density estimation is sensitive to the number of match scores available during training, the decrease in accuracy depends on the size of the training database. Since the NIST-Face and NIST-Fingerprint databases have a larger number of genuine match scores ($6,000$), the densities can be reliably estimated even when 10%-15% of the genuine scores are missing. However, the NIST-Multimodal database has only 517 genuine scores. As a result, we observed that even when 10% of the scores are unavailable, the decrease in the resulting recognition rate is significant.

6 Conclusions

Fusion in the multibiometric identification scenario is a critical problem in many large-scale practical biometric systems. In this work, we have shown that the likelihood ratio-based score fusion algorithm that was originally designed for the verification mode, can be extended for fusion in the identification scenario. Moreover, we have shown that the likelihood ratios computed based on the match scores can be combined with the rank-based posterior probabilities and the hybrid rank and score level fusion scheme achieves high recognition performance in multibiometric identification systems. The proposed fusion rules can also easily deal with missing data that is commonly encountered multibiometric identification systems without any need for ad-hoc modifications.

References

1. Federal Bureau of Investigation: Integrated Automated Fingerprint Identification System, http://www.fbi.gov/hq/cjisd/iafis.htm
2. Ross, A., Nandakumar, K., Jain, A.K.: Handbook of Multibiometrics. Springer, Heidelberg (2006)
3. Tulyakov, S., Govindaraju, V.: Combining Matching Scores in Identification Model. In: Proceedings of ICDAR, Seoul, South Korea, August-September 2005, pp. 1151–1155 (2005)
4. Dinerstein, S., Dinerstein, J., Ventura, D.: Robust Multi-Modal Biometric Fusion via Multiple SVMs. In: IEEE Intl. Conf. on Systems, Man and Cybernetics, October 2007, pp. 1530–1535 (2007)
5. Nandakumar, K., Chen, Y., Dass, S.C., Jain, A.K.: Likelihood Ratio Based Biometric Score Fusion. IEEE Trans. on PAMI 30(2), 342–347 (2008)
6. Duda, R.O., Hart, P.E., Stork, D.G.: Pattern Classification. John Wiley & Sons, Chichester (2001)
7. Ho, T.K., Hull, J.J., Srihari, S.N.: Decision Combination in Multiple Classifier Systems. IEEE Trans. on PAMI 16(1), 66–75 (1994)
8. Melnik, O., Vardi, Y., Zhang, C.H.: Mixed Group Ranks: Preference and Confidence in Classifier Combination. IEEE Trans. on PAMI 26(8), 973–981 (2004)
9. Moon, H., Phillips, P.J.: Computational and Performance Aspects of PCA-based Face Recognition Algorithms. Perception 30(5), 303–321 (2001)
10. Grother, P., Phillips, P.J.: Models of Large Population Recognition Performance. In: Proceedings of CVPR, Washington DC, USA, June/July 2004, vol. 2, pp. 68–75 (2004)
11. Bolle, R., Connell, J., Pankanti, S., Ratha, N., Senior, A.: The Relationship Between the ROC Curve and the CMC. In: Proceedings of Fourth IEEE Workshop on Automatic Identification Advanced Technologies (AutoID), Buffalo, USA, October 2005, pp. 15–20 (2005)

Challenges and Research Directions for Adaptive Biometric Recognition Systems

Norman Poh[1], Rita Wong[1], Josef Kittler[1], and Fabio Roli[2]

[1] University of Surrey, Guildford, GU2 7XH, Surrey, UK
{n.poh,s.wong,j.kittler}@surrey.ac.uk
[2] Department of Electrical and Electronic Engineering,
University of Cagliari Piazza d'Armi 09123 Cagliari, Italy
roli@diee.unica.it

Abstract. Biometric authentication using mobile devices is becoming a convenient and important means to secure access to remote services such as telebanking and electronic transactions. Such an application poses a very challenging pattern recognition problem: the training samples are often sparse and they cannot represent the biometrics of a person. The query features are easily affected by the acquisition environment, the user's accessories, occlusions and aging. Semi-supervised learning – learning from the query/test data – can be a means to tap the vast unlabeled training data. While there is evidence that semi-supervised learning can work in text categorization and biometrics, its application on mobile devices remains a great challenge. As a preliminary, yet, indispensable study towards the goal of semi-supervised learning, we analyze the following sub-problems: model adaptation, update criteria, inference with several models and user-specific time-dependent performance assessment, and explore possible solutions and research directions.

1 Introduction

Portable electronic devices such as mobile phones and PDAs are becoming important means to provide wireless access to the Internet and other telecommunication networks anytime, anywhere. Very often, such access requires the verification of the user's identity in order to ensure that the person is really whom he/she claims to be. While knowledge-based authentication such as PINs or passwords can be used, they can be forgotten, or easily compromised when shared, copied or stolen. In comparison, biometrics is a more effective alternative because it is by far a more natural, reliable and friendly means of authentication. Thanks to the availability of cameras and microphones in today's mobile devices, audio- and visual-based biometrics such as face and speech can be readily used for this purpose.

In this context, we aim to develop and evaluate new mobile services that are secured by bi-modal speech and face biometrics. We shall call this problem "mobile biometry" (Mobio). Due to the device mobility, the problem of biometric authentication is much more challenging for at least two reasons: First, it has to deal with changing and often uncontrolled environments, e.g., external noise and varying illumination conditions. Under such conditions, a biometric query data can appear very differently from the one

M. Tistarelli and M.S. Nixon (Eds.): ICB 2009, LNCS 5558, pp. 753–764, 2009.
© Springer-Verlag Berlin Heidelberg 2009

acquired during enrollment (template). As a consequence, the device performance can degrade drastically. Second, mobile devices have limited memory and CPU resources. This provides a natural constraint on the size of biometric template/model[1] and the type of processing algorithms (which favor those of low computation).

One possible way to improve the device authentication performance is by automatically labeling the abundant query (test) data and incorporating them as part of the training data [3,9]. Such a strategy belongs to a category of processes known as *semi-supervised learning* [14]. While initially developed for text classification, its application is now extended to speech verification [6] (termed incremental enrollment by the authors), face recognition [8] (called "Eigenspace updating") and multimodal biometrics [13] (known as template co-updating).

In [13], two distinct methods have been examined: self-training and co-training. In self-training, an initially trained classifier based on labeled data attempts to give labels to an unlabeled data set. The newly labeled data are then incorporated into the training set. The algorithm iterates until a convergence criterion is satisfied. In co-training, two initially trained classifiers based on labeled data attempt to provide labels to an unlabeled data set. Since two classifiers are involved, two labeled data sets are produced. The union of these two newly labeled data sets are then used as part of the training set for each of the two classifiers.

The study in [13] is particularly relevant to our problem here, i.e., mobile biometry. Indeed, it has been shown that both self-training and co-training are promising solutions to overcome the lack of biometric training samples at enrollment. However, mobile biometry presents a significantly different challenge, which can be summarized by the following four issues.

First, mobile biometry requires that data be labeled online (or in small batch of data) instead of offline (i.e., processing in batch), as is always practiced in experiments involving co-training/self-training. The online or small-batch processing is the consequence of the limited storage of mobile devices.

Second, there is no distinctive co-training and test phases as mobile devices are continuously updated as needed. This implies that using the usual three-partition experimental designs – involving training or enrollment (with only labeled data), co-training (with unlabeled data) and test data sets – is inappropriate. An ideal assessment would reflect the actual application scenario, which should contain only two partitions of data set: one for enrollment and a separate one for both co-training and testing (where error is estimated). As a result, there is a need to design an error estimator capable of tracking the error dynamics.

Third, there is a need to use quality measures in model adaptation[2]. Until now, procedures on co-training depend uniquely on the confidence of classifier decision, which is

[1] We use the term "template" when referring to the stored features representing a person's biometric trait whereas the term "model" as a more general concept in order to refer to the parameters of a discriminative classifier or a statistical model, or that of an intermediate feature extraction process such as eigenspace analysis.

[2] Model adaptation is a more general concept than semi-supervised learning such as co-training. The latter refers uniquely to training or adaptation on self-labeled test data. In general, adaptation refers to the update of *model parameters* in various contexts, e.g., quality of samples.

often derived directly from the classifier output. In the biometric community, an important and rapidly developing research area is concerned with the use of quality measures. For example, if a quality measure is designed to annotate the head orientation of a face image (giving pitch, tilt and yaw angles), it would make sense to cluster the images according to different head pose and create a new template for each cluster. Conversely, without quality measures, a direct update to the old model, without considering head orientations can be catastrophic. The same argument applies to other quality measures annotating the presence of glasses, the type of emotional states and lighting conditions. This motivates us to use quality-dependent model adaptation, in addition to the confidence of the classifier output, as a criterion to create a new model or update an existing one.

Last but not least, when there are several models/templates (recalling that the number is bounded by the memory of mobile devices), inference with several models requires a special treatment. Drawing on the work in [10], a possible framework is also proposed in this paper for mobile biometry.

Section 2 begins with a more formal presentation of the concept of co-training according to the original paper of Blum and Mitchell [3]. The remaining sub-sections analyze the above mentioned four issues related to applying semi-supervised learning to biometric authentication in general, but using mobile biometry as a test field. For each issue, a possible solution is also elaborated. Section 3 then concludes the paper.

2 Open Issues and Research Directions for Adaptive Biometric Systems

We wish to learn the concept $f : X \to Y$ given some labeled L and unlabeled U data set drawn from $P(X)$, where X is a feature vector and Y is a class label. The features describing X can be partitioned into $X^{(1)}$ and $X^{(2)}$, i.e., $X = X^{(1)} \times X^{(2)}$, such that f can be computed from either $X^{(1)}$ or $X^{(2)}$. Let g_1 and g_2 be the trained classifier from $X^{(1)}$ and $X^{(2)}$, respectively. Then, the objective of co-training can be expressed as finding g_1 and g_2 such that:

$$\exists_{g_1,g_2}(\forall_x \in X)g_1(x^{(1)}) = f(x) = g_2(x^{(2)}).$$

from U and L. An example of co-training procedure for a bimodal face and speech biometric authentication problem, which is a binary classification task, is shown in Algorithm 1.

An important result from [3] is that if $X^{(1)}$ and $X^{(2)}$ are *conditionally independent* given Y, and the concept f is "PAC learnable" from noisy labeled data, then, f is PAC learnable from weak initially labeled data plus unlabeled data. Probably approximately correct learning or PAC learning refers to the fact that one can train a classifier in order to learn the concept f, giving low generalization error in finite time and space.

In biometric authentication, $X^{(1)}$ can be a feature vector extracted, for instance, from a face image, whereas $X^{(2)}$ can be a feature vector extracted from a speech recording. Y is a person's identity. Not only that $X^{(1)}$ and $X^{(2)}$ are *conditionally independent* given Y, a condition that must be fulfilled in order for co-training to work, they are simply (unconditionally) independent in our application. Because of this independence,

Algorithm 1. The co-training algorithm

- Given: labeled data L and unlabeled data U
- Loop:
  - Train $g1$ (face classifier) using L
  - Train $g2$ (speech classifier) using L
  - Allow $g1$ to label p positive, n negative examples from U
  - Allow $g2$ to label p positive, n negative examples from U
  - Add these self-labeled examples to L

for instance, it is not possible to predict a person's face features given his/her speech features without any prior information. We therefore have a strong case here supporting the conjecture that co-training will work for bimodal person authentication. However, in practice, updating the wrong samples, i.e., from other users, may result in degraded performance [6]. Therefore, despite the sound theory, over-updating a model with the wrong person's biometric data will eventually be counter-productive.

A second obstacle to the successful deployment of semi-supervised learning in biometrics is that the intra-person variability is larger than inter-person variability. For instance, the lighting conditions can cause two face images of the same person appear much more differently than two images of different persons taken in the same lighting condition. This implies that trying to incorporate all types of variability into a single model will only reduce its discriminative power, hence, decreasing the recognition performance. A simple, yet, effective solution is to maintain several models, each capturing only local variation as gauged by quality measures (lighting conditions being one example here).

The following four sections will address open issues related to adaptive biometric systems. The first section provides a Bayesian interpretation of model adaptation. An example based on fingerprint minutia-based classifier is also shown. The second section addresses training and inference with several models. Each model differs from the others as it captures an aspect of a biometric template based on some observed quality measurements. The third section proposes a simple model creation/update criterion, also based on quality measurements. Finally, the last section addresses the issue of performance estimation as models adapt through time.

2.1 A Bayesian Interpretation of Biometric Model Adaptation

There are basically three operations that are essential to manipulating biometric models/templates; they are defined as follow:

- Model creation

$$Create : data \rightarrow new\ model$$

- Model adaptation:

$$Update : model, data \rightarrow updated\ model$$

- Model deletion

$$Delete : model \rightarrow \emptyset$$

While adding and deleting models is straightforward, it is not so for model adaptation because the latter requires combination of several biometric samples. Classifiers based on statistical models can often be implemented in the form of online learning, i.e., "old" model parameters can be updated with new ones after observing a training sample. This implies that learning can be done incrementally. A state-of-the-art classifier in speaker verification known as Gaussian mixture model (GMM) with maximum *a posteriori* (MAP) adaptation is a good example:

$$p(\theta|x) \propto p(x|\theta)p(\theta)$$

where $p(x|\theta)$ is the likelihood of the data (given a model with parameter θ) and $p(\theta)$ is the prior probability over the parameter θ. One maximizes the right hand side in order to find the most probable value of θ, i.e., θ_*. This value becomes an initial estimate that can be updated when a new sample becomes available.

If there are $x_1, \ldots, x_T$ observations, also denoted as $x_1 : x_T$, one finds the value of θ that maximizes $p(\theta|x_1 : x_T)$, as follows (ignoring the normalizing factor in each step since we are only interested in maximizing the function with respect to θ):

$$\begin{aligned} p(\theta|x_1 : x_T) &\propto \prod_{i=1}^{T} p(x_i|\theta)p(\theta) \\ &\propto \prod_{i=2}^{T} p(x_i|\theta)p(\theta|x_1) \\ &\propto \prod_{i=3}^{T} p(x_i|\theta)p(\theta|x_1, x_2) \\ &\propto \vdots \\ &\propto p(x_T|\theta)p(\theta|x_1 : x_{T-1}) \end{aligned} \tag{1}$$

where $p(\theta|x_1) \propto p(x_1|\theta)p(\theta)$. The recursive formulation of (1) implies that in order to calculate the optimal value of θ given all previously observed T samples, one only needs to use the parameter calculated up to $T-1$ to do so. This dispenses with the need to keep all previous training samples, which is a memory demanding requirement. The above recursive formulation of MAP implies that density-based classifiers can benefit from continuous training as this leads to finding the optimal value of θ. In the terms used in [4, Chap. 3], the recursive formulation of (1) is known as *true recursive Bayesian learning*, its right hand term, $p(\theta|x_1 : x_T)$, as a *reproducing density*; and the term $p(\theta)$ as a *conjugate prior*.

For non-statistical model, adaptation may seem not obvious at first. For instance, Jiang and Ser [7] showed that it is possible to combine several minutiae-based fingerprint templates to form a "super-template". It turns out that the super-template approach can be seen as a *simplified* statistical model taking the form of a multivariate Gaussian distribution with isotropic covariance. Let x_m^t be a minutia at the m-th location of the t-th fingerprint. We require here that all minutia locations of different templates are

aligned first. A minutia can contain location and/or directional information. Assuming that the minutiae of a fingerprint is independent, a fingerprint template can then be represented by $p(X|\theta) = \prod_m p(x_m|\theta)$ where $p(x_m|\theta)$ is a Gaussian and $\theta = \epsilon$ is common to all observations t and locations m, i.e.,

$$p(X|\theta, t) = \prod_{m=1}^{M} p(x_m|\theta, t) = \prod_{m=1}^{M} \mathcal{N}(x_m|\mu_m^t, \epsilon)$$
$$= \frac{1}{(2\pi\epsilon)^{M/2}} \exp\left\{-\frac{1}{2\epsilon}\sum_{m=1}^{M} ||x_m - \mu_m^t||^2\right\}$$

where μ_m^t is the location (and orientation) of a minutiae taken from the t-th template. Note that ϵ corresponds to the variance of x_m. The posterior is then:

$$p(\theta|X_1 : X_T) \propto p(X_1 : X_T|\theta)p(\theta) \quad (2)$$
$$= \prod_{t=1}^{T} p(X_t|\theta)p(\theta)$$
$$\propto \exp\left\{-\frac{1}{2\epsilon}\sum_{t=1}^{T}\sum_{m=1}^{M} \| \underbrace{x_m - \mu_m^t} \|^2\right\} \quad (3)$$

with equality only if we consider the normalizing factor $\frac{1}{(2\pi\epsilon)^{MT/2}}$. Here, $p(\theta)$ is a uniform distribution over the θ space.

When $\epsilon \to 0$, the minutia x_m^t that is close to μ_m^t will give a peak value and near zero otherwise. In this limit, we see that the counting approach proposed by Jiang and Ser [7] and (3) converge. According to (3), with a large number of samples T, spurious minutiae will receive low weights, whereas frequently occurred minutiae will receive high weights, hence, playing a more important role during recognition. This behavior is very similar to the k-nearest neighbor algorithm [4].

A related study on face recognition, called "Eigenspace-updating", found in [8] can also be considered a special case of our recursive MAP framework. In this case $p(x|\theta)$ is a multivariate Gaussian whose parameters θ are a mean vector and a covariance matrix. It is well known that mean and covariance can be incrementally updated by observing one example at a time. This is, again, a realization of (1). However, slightly different from this formulation, the authors introduced the concept of weight decay whose aim is to give more weight to the latest test samples rather than giving all samples equal weights.

A second concept called "twin-subspace updating", also found in [8], is to maintain two models (classifiers) instead of one. The motivation is that one model may capture a frontal view whereas another may capture another slightly different view (a profile view would be another extreme). When there are more than one models, it is necessary to decide which model will contribute more to the final match score. This issue is treated in the next section.

2.2 Quality-Based Training and Inference with Multiple Models

In this section, we will first treat the subject of training with several models and then consider how inference conditioned on quality can be approached using a Bayesian framework.

Before doing so, it is instructive to categorize templates/models into the following three primary types:

- single-template, where only a single template is available
- multi-template, where several templates are used
- super-template, where several templates are combined to form a single one.

Higher level types of templates are possible, for example, using multiple super-templates where each super-template captures an aspect of biometric features, acquired in a particular condition (e.g., same head orientation, lighting conditions, etc) or using a particular device.

Let q denote a sample quality measurement and θ_* be the optimal value that maximizes (2), i.e., $\theta_* = \arg\max_\theta p(\theta|X_1 : X_T)$. In this case, it is reasonable to expect that samples of similar quality to be comparable; otherwise, they are not. This suggests that one should choose a model/template that matches a particular acquisition condition measurable by $q \in \mathbb{R}^{\mathtt{q}}$, i.e., a vector containing q measurements. For instance, for face verification, these measures can be illumination, focus, reliability of face detection, etc. It is possible to cluster q into several states. Let Q be the cluster indices of q. We expect that images of good quality will cluster together and similarly for those of moderate and bad quality. A well known clustering algorithm that can suitably be used for this purpose is a Gaussian mixture model (GMM) [1]. For a pre-specified number of clusters $\mathtt{Q}$, a GMM models the density $p(q) = \sum_{Q=1}^{\mathtt{Q}} p(q|Q)p(Q)$. The parameters of this model are found using the Expectation-Maximization (EM) algorithm. The posterior probability of a cluster Q given an observation q, also known as *responsibility* in EM, is then given by $P(Q|q) = \frac{p(q|Q)P(Q)}{p(q)}$. Now, we can estimate the optimal parameter θ using the recursive MAP framework, but in a quality-dependent manner:

$$p(\theta|x_1 : x_T, Q) \propto p(x_T|\theta, Q)p(\theta|x_1 : x_{T-1}, Q) \tag{4}$$

which is very similar to (1), except that all terms are dependent on the cluster index Q. Thanks to (4), we can estimate:

$$p(\theta|x_1 : x_T, q_1 : q_T) = \sum_Q p(\theta|x_1 : x_T, Q)P(Q|q_1 : q_T) \tag{5}$$

which is the actual goal. One maximizes (5) with respect to θ and this can be done in a recursive MAP framework, i.e., when samples arrive one at a time. Using a derivation very similar to (1), except that all terms are dependent on Q, one can show that (5) is proportional to:

$$p(\theta|x_1 : x_T, q_1 : q_T) \propto \underbrace{\sum_Q p(x_T|\theta, Q)P(Q|q_T)}\cdot p(\theta|x_1 : x_{T-1}, q_1 : q_{T-1}). \tag{6}$$

(6) shows that new parameters (the left hand side of equation) can be updated from old parameters (represented by $p(\theta|x_1 : x_{T-1}, q_1 : q_{T-1})$) using the underbraced term. Therefore, in order to maximize $p(\theta|x_1 : x_T, q_1 : q_T)$ with respect to θ, one does not need to keep the previous samples $(x_1 : x_{T-1}, q_1 : q_{T-1})$ but only needs the last estimate of parameters from these samples. This shows that one can design an online algorithm not only for a conventional model (without quality information), but also for the conditional model based on quality.

Note that the responsibility term $P(Q|q_T)$ only appears in the underbraced term, implying that $P(Q|q_T)$ is only effective on the latest T-th observation. Conditioning the parameter estimation based on Q, as in (4), therefore makes the estimation of (5) (being the real objective) a tractable proposition.

When there are several models/templates, it is reasonable to expect that some models may contribute more in making the accept/reject decision than the others. Let θ_*^Q be the optimal value that maximizes (5), and $Q = 1, \ldots, \mathtt{Q}$. In this case, using the Neyman Pearson theorem, the optimal classifier should give the following output:

$$y = \log \frac{\sum_Q p(x|\theta_*^Q, \mathtt{C}) P(Q|q)}{\sum_Q p(x|\theta_*^Q, \bar{\mathtt{C}}) P(Q|q)} \tag{7}$$

where $p(x|\theta_*^Q, \mathtt{C})$ is the density of a client (the reference user) model whereas $p(x|\theta_*^Q, \bar{\mathtt{C}})$ is the density of an anti-client model, i.e., one that represents all other users. This is a concept borrowed from speaker verification, also known as a universal background model [12]. It can be observed that the contributions of these two terms are appropriately weighted by the responsibility term $P(Q|q)$. The effectiveness of the above formulation was shown in [10], in the context of intramodal fusion involving several face verification systems, with x being a vector of system outputs and q being a vector of some face related quality measures such as reliability of face detection, background uniformity, presence of glasses, head orientation, etc. Note that these factors are used because they are known to affect the performance of a face verification system.

Instead of using (7), in a simplified case, we can also choose to use only a single model that is the *most* appropriate (according to the responsibility term):

$$y = \log \frac{\sum_Q p(x|\theta_*^Q, \mathtt{C}) f(Q, q)}{\sum_Q p(x|\theta_*^Q, \bar{\mathtt{C}}) f(Q, q)},$$

where

$$f(Q, q) = \begin{cases} 1 & Q = \arg\max_Q P(Q|q) \\ 0 & \text{otherwise,} \end{cases}$$

2.3 Quality-Based Model Update and Creation

In the previous section, we showed that it is possible to make inference with several models. Each model is different because it gauges the feature density for a given Q which can be the acquisition condition, or a different presentation. In both cases, an observed biometric feature set will appear very differently from the previously stored biometric template.

A criterion is indeed needed to identify when a model should be updated or created. If x' is a newly labeled sample having q' as its quality measurements, one chooses the quality state Q_* that maximizes the responsibility $P(Q|q')$, i.e.,

$$Q_* = \arg\max_{Q=1}^{\mathtt{Q}} P(Q|q'). \quad (8)$$

However, in reality, the model $p(x'|\theta, Q_*)$ may not exist. In fact, right after enrollment, in a typical scenario where enrollment is done with a single template, there is only one quality state, i.e., $\mathtt{Q} = 1$. In the absence of the model $p(x|\theta, Q_*)$, one should naturally create the model $p(x|\theta, Q_*)$ using the observation x'.

However, if the model $p(x|\theta, Q_*)$ exists, one can update the model using (6). The new parameter can be obtained as follows:

$$\theta^Q_{new} = \arg\max_\theta \sum_Q p(x'|\theta, Q) p(\theta^Q_{old}) P(Q|q') \quad (9)$$

A simpler approach, taking the hidden variable Q as an observed one, is to use the following update rule:

$$\theta^{Q_*}_{new} = \arg\max_\theta p(x'|\theta, Q_*) p(\theta^{Q_*}_{old}) \quad (10)$$

Both (9) and (10) constitute two variants of the maximization step in a typical EM algorithm. In comparison, in both cases, (8) corresponds to the expectation step in an EM algorithm [2].

$P(Q|q)$ represents a prior knowledge of all possible combinations of factors affecting the system performance. This function was obtained by modeling $p(q|Q)$ using a clustering algorithm. This suggests that one should train $p(q) = \sum_Q P(Q)p(q|Q)$ from as much data as possible from a separate development database containing many more persons, environmental conditions, presentation styles, and even different devices. An accurate estimation of $P(Q|q)$ would guarantee the success of model adaptation. Designing a good set of quality measures, therefore, cannot be overemphasized.

2.4 Person-Specific and Time Dependent Performance Evaluation

There are, in general, two approaches to performance assessment with evolving biometric models. We shall refer to one of these approaches as a *separate* adapt-and-test strategy and another as a *joint* adapt-and-test strategy. Both are shown in Figure 1. In the first strategy, a partition of data is used to adapt a biometric model and another disjoint partition is used for evaluating its performance. Such an approach is inefficient in terms of data usage since the data used for adaptation cannot be used for testing, and vice-versa. Worst still, it does not reflect our actual application scenario where a mobile device is allowed to update biometric models on the test data. However, by using a unique test set, one can use the widely accepted error counting approach. In this approach, one first determines a threshold and then calculates the number of false match and false non-match events.

In contrast, the joint adapt-and-test strategy, which allows one to adapt (with unlabeled data) and test (with the known labels), has two advantages. First, it corresponds

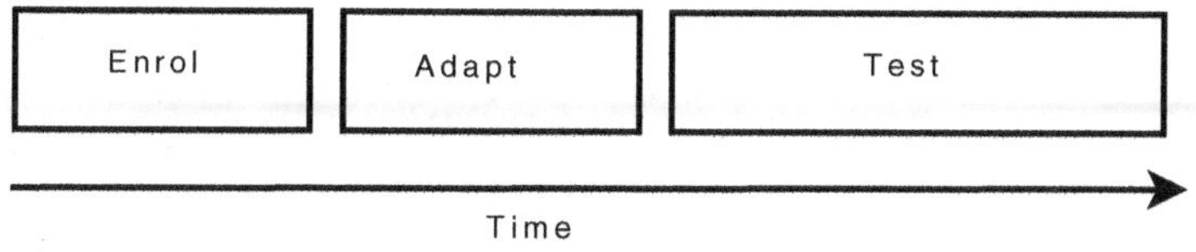

(a) Separate adapt-and-test strategy

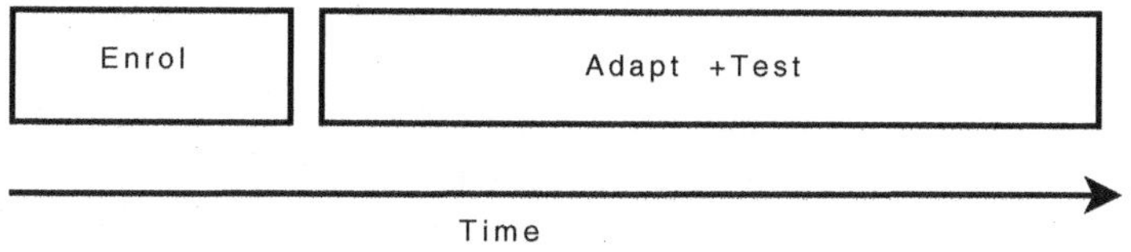

(b) Joint adapt-and-test strategy

Fig. 1. Two assessment strategies for co-training: (a) is a usual strategy and (b) is our proposal

better to our mobile application scenario. Second, it is more efficient in terms of data usage, since the data used in model adaptation can, at the same time, be used to asses the model performance. Unfortunately, its major disadvantage is the absence of an error estimator.

We shall elaborate on an error estimator that can be used for joint adapt-and-test strategy mentioned above, which is adapted from [11]. This procedure is able to track performance change in terms of false match rate (FMR) and false non-match rate (FNMR) over time and on a per person basis. Estimating this error is difficult because of the paucity of data, especially the genuine user scores. However, it is possible if one imposes the constraint that the user-specific class-conditional (genuine user or impostor) scores follow a particular parametric family of distributions (Gaussian in [11]) and that it is continuous in time. In so doing, one can estimate the performance to an arbitrary time precision. This method compares favorably with the conventional error-counting approach which utilizes a sliding window, e.g., [5], and as a result suffers from the dilemma between precision in performance and the time resolution, i.e., higher performance precision entails lower time resolution and vice-versa.

In the context of [11], it was found that even without any model adaptation, some biometric models can degrade over time, while others improve with use. This phenomenon justifies the adaptation of biometric model by using the test data. However, they also suggest that one should consider the adaptation on a per user basis. In our case, when one regularly adapts a model through time, it is reasonable to expect that the performance of a biometric model will evolve with time.

We shall summarize the procedure to estimate the error trend on a per person basis below [11]:

1. Fit a regression line to each of the genuine and impostor match scores. An example of regression fits for both genuine and impostor scores can be found in Figure 2[3].

[3] In principle, one does not expect any trend for the impostor scores, i.e., the regression line for the impostor should be parallel to the x-axis. Therefore, any deviation from this can be attributed to estimation error.

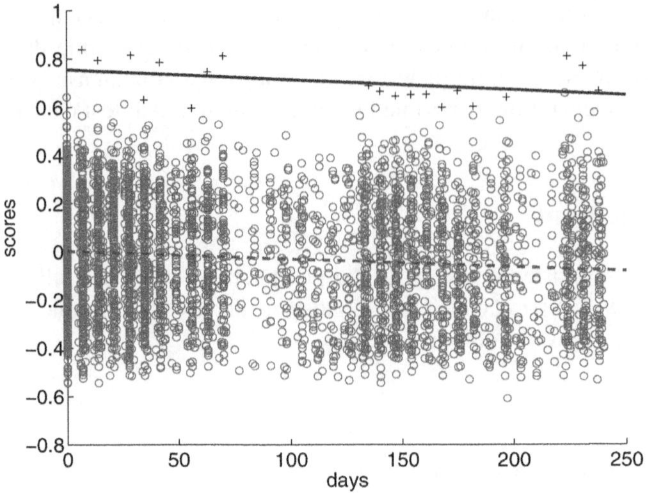

Fig. 2. Scatter plot of genuine user ("+") and impostor ("o") match scores for a single user's template over 250 days (the X-axis). Higher match scores imply genuine user class. The interruption in genuine match scores around the 100-th day is due to no observations being made during the term break. The straight lines are the regression fits on the data (continuous line for the genuine user match scores and dashed line for the impostor ones).

2. Obtain the first and second order statistics of the two regression lines for a fixed time interval. Let these parameters be $(\mu^I_{j,t}, \sigma^I_{j,t})$ for the impostor regression line and $(\mu^C_{j,t}, \sigma^C_{j,t})$ for the client counterpart, where j denotes the identity of a user and $t \in [1, T]$ is a time index.
3. Estimate the false match rate and false non-match rate as follows:

$$\mathrm{FMR}_{j,t}(\Delta) = \Phi\big(\Delta|\mu^I_{j,t}, (\sigma^I_{j,t})^2\big) \tag{11}$$

and

$$\mathrm{FNMR}_{j,t}(\Delta) = 1 - \Phi\big(\Delta|\mu^G_{j,t}, (\sigma^G_{j,t})^2\big) \tag{12}$$

for a given threshold Δ in the score space, where $\Phi\big(\Delta|\mu, (\sigma)^2\big)$ is a cumulative density function with mean μ and standard deviation σ for a chosen distribution.

Once $\mathrm{FNMR}_{j,t}(\Delta)$ and $\mathrm{FMR}_{j,t}(\Delta)$ are calculated, they can be plotted for each user j and through time $t = 1 : T$. In this way, one can plot a user-specific detection error trade-off (DET) or receiver operating characteristic (ROC) curves that evolves with time.

3 Conclusions

In this paper, we highlighted four challenges as well as provided solutions to semi-supervised learning in the context of biometric person authentication. The issues are:

online model updating, training and inference with several models, quality-based criteria to control the creation of a new model, and person-specific time-dependent performance evaluation. At the time of writing, an audio-visual database is being collected with mobile devices. Future experiments will be conducted to test each of the proposed solutions.

Acknowledgement

This work was supported partially by the prospective researcher fellowship PA0022_121477 of the Swiss National Science Foundation, and by the EU-funded Mobio project grant IST-214324 (www.mobioproject.org).

References

1. Bishop, C.: Neural Networks for Pattern Recognition. Oxford University Press, Oxford (1999)
2. Bishop, C.M.: Pattern Recognition and Machine Learning. Springer, Heidelberg (2007)
3. Blum, A., Mitchell, T.: Combining labeled and unlabeled data with co-training. In: COLT: Proceedings of the Workshop on Computational Learning Theory, pp. 92–100. Morgan Kaufmann Publishers, San Francisco (1998)
4. Duda, R.O., Hart, P.E., Stork, D.G.: Pattern Classification and Scene Analysis. John Wiley and Sons, New York (2001)
5. Flynn, P.J., Bowyer, K.W., Phillips, P.J.: Assessment of Time Dependency in Face Recognition: An Initial Study. In: Kittler, J., Nixon, M.S. (eds.) AVBPA 2003. LNCS, vol. 2688, pp. 44–51. Springer, Heidelberg (2003)
6. Fredouille, C., Mariéthoz, J., Jaboulet, C., Hennebert, J., Mokbel, C., Bimbot, F.: Behavior of a bayesian adaptation method for incremental enrollment in speaker verification. In: ICASSP 2000 - IEEE International Conference on Acoustics, Speech, and Signal Processing, Istanbul, Turkey, June 5–9 (2000)
7. Jiang, X., Ser, W.: Online fingerprint template improvement. IEEE Tran. Pattern Analysis and Machine Intelligence 24(8), 1121–1126 (2002)
8. Liu, X., Chen, T., Thornton, S.M.: Eigenspace updating for non-stationary process and its application to face recognition. Pattern Recognition 36(9), 1945–1959 (2003)
9. Nigam, K., McCallum, A.K., Thrun, S., Mitchell, T.M.: Text classification from labeled and unlabeled documents using EM. Machine Learning 39(2/3), 103–134 (2000)
10. Poh, N., Heusch, G., Kittler, J.: On Combination of Face Authentication Experts by a Mixture of Quality Dependent Fusion Classifiers. In: Haindl, M., Kittler, J., Roli, F. (eds.) MCS 2007. LNCS, vol. 4472, pp. 344–356. Springer, Heidelberg (2007)
11. Poh, N., Kittler, J.: A Method for Estimating authentication Performance Over Time, with Applications to Face Biometrics. In: Rueda, L., Mery, D., Kittler, J. (eds.) CIARP 2007. LNCS, vol. 4756, pp. 360–369. Springer, Heidelberg (2007)
12. Reynolds, D.A., Quatieri, T., Dunn, R.: Speaker Verification Using Adapted Gaussian Mixture Models. Digital Signal Processing 10(1–3), 19–41 (2000)
13. Roli, F., Didaci, L., Marcialis, G.L.: Template co-update in multimodal biometric systems. In: Lee, S.-W., Li, S.Z. (eds.) ICB 2007. LNCS, vol. 4642, pp. 1194–1202. Springer, Heidelberg (2007)
14. Zhu, X.: Semi-supervised learning literature survey, University of Wisconsin-Madison (2008) (online publication)

Modelling FRR of Biometric Verification Systems Using the Template Co-update Algorithm

Luca Didaci, Gian Luca Marcialis, and Fabio Roli

Dept. of Electrical and Electronic Eng., University of Cagliari, Italy
{luca.didaci,marcialis,roli}@diee.unica.it

Abstract. The decrease of representativeness of available templates during time is due to the large intra-class variations characterizing biometrics (e.g. faces). This requires the design of algorithms able to make biometric verification systems adaptive to such variations. Among others, the template co-update algorithm, which uses the mutual help of two complementary biometric matchers, has shown promising experimental results. The present paper is aimed to describe a theoretical model able to explain the co-update behaviour. In particular, the focus is on the relationships between error rate and gallery size increase. Preliminary experimental results are shown to validate the proposed model.

1 Introduction

Personal verification systems based on biometrics reached very notable improvements in the last years [1-2]. In particular, multi-modal verification systems allow to obtain an increase of performance by exploiting advantages of two or more biometrics.

However, many open issues still remain. Among others, the dependency of performance from the biometric variations related to short-time or medium-time causes. For example, a scratch on the finger skin or the face, or an unknown expression, or the so-called "aging" effect, which is crucial in performance of face recognition systems [1]. With regard to this problem, the topic called "template update" has received an increasing interest. The template(s) is a sample, or a set of samples, which is considered "representative" of the subject appearance [1]. Templates are usually captured during the enrolment session, under the supervision of a human expert. However, it is widely acknowledged that capturing all possible subject variations during an enrolment session is impossible. On the other hand, the system performance strongly depends on such representativeness. Accordingly, updating a template, by additional enrolment sessions, is considered as a good solution in order to follow the intra-class variations of each biometric [3].

Recently, "semi-supervised" [4-9] template update has been proposed. In this approach, unlabelled samples (that is, collected during system operations) showing very strong similarity to existing templates are added into the gallery of the related subject, thus increasing the representativeness of related set of templates. Template update solutions involve both mono-modal and multi-modal verification systems [3-10]. In particular, the so-called "co-update" algorithm showed that the mutual "help" of two weakly correlated biometrics (e.g., fingerprints and faces) can increase the gallery

M. Tistarelli and M.S. Nixon (Eds.): ICB 2009, LNCS 5558, pp. 765–774, 2009.
© Springer-Verlag Berlin Heidelberg 2009

representativeness more quickly than systems adopting only one biometric [7-8], thus improving the overall performance.

In [10], a statistical framework modelling the co-update behaviour has been proposed. This model starts from hypotheses which fit the practical scenarios involving fingerprint and face biometrics, providing a first analytical framework able to explain the relationship between the amount of unlabelled data collected and the gallery size for a certain application. However, that model is lack of a convincing explanation of the verification performance of the system, in particular, the False Rejection Rate, when the estimated FAR is zero. Moreover, that model underestimates the FRR, especially in case of clients not "easily" recognizable, as we show in this paper. The aim of the present work is to fill such gap by adopting a more realistic model for genuine users distribution. This allows to point out the relationships between the amount of unlabelled data collected and the error rate reduction achieved. The impact is practical, since this model can be adopted by designers for the preliminary assessment of template co-update algorithms in real environments. Experimental results on a chimerical data set show the reliability of the framework by comparing the gallery sizes and errors predicted by the model with the gallery sizes and errors obtained on test data.

The paper is organized as follows. Section 2 describes the current version of the model. Section 3 describe our proposal for the modelling of verification performance as function of the gallery size increase. Section 4 reports the experimental results which validate the model, and also show the difference with the previous one. Section 5 concludes the paper.

2 Recent Advances in Modelling the Template Co-update Algorithm

The "complementary" performances of biometric matchers using distinct biometric traits, for example, face and fingerprint, are one of the fundamental motivations for multi-modal biometrics [7,11]. Intuitively, each recognizer is expected to assign correct labels to certain input data which are difficult for the other. In the case of template co-update, two matchers are used (bi-modal system).

Description of co-update algorithms is given in Figure 1. Let D_l be a set of labelled biometric data. These data are the templates enrolled by human supervision, thus they are labelled with users identities (clients names). It is supposed that a batch D_u (usually, much larger than D_l) of data is acquired during system operations. Each element of D_u is a pair of biometric samples, i.e. face and fingerprint, from the same individual. During the off-line co-update phase, each matcher is applied to the batch D_u to verify the identity of each couple of biometric samples. A biometric sample is added to the set D_l if and only if the identity of the other biometric sample of the couple has been verified.

In this paper, we adopted the same terminology used in [10] for sake of uniformity with the state-of-the-art. Let us call "master" the biometric that assumes the supervisor's role, and "slave" the biometric whose gallery is augmented thanks to the master biometric. During the co-training process both biometrics assume, alternatively, master and slave roles. The process is repeated a specified number of times.

As for the majority of semi-supervised template update algorithms, the core of co-update is the insertion of new samples without external supervision. This might introduce misclassified samples in gallery. For this reason, only "confident examples" are added by a very high co-update threshold value at which no false matches occur, namely, FAR = 0%.

Co-update pros and cons have been preliminary evaluated in [8]. But only in [10] a first trial to theoretically explain why it works has been made. In the following, we summarize the main achievements of framework proposed in Ref. [10], and in Section 3 we specify what we added in order to improve that framework.

In the enrolment (supervised) session, collect a set D_l of face and fingerprint images. A couple of face and fingerprint images is acquired for each user.
Create the fingerprint and face templates using the set D_l
Loop for H iterations:
  Collect a set D_u without supervision. Each element of D_u is a pair of face and fingerprint samples $\{\mathbf{x}_1, \mathbf{x}_2\}$ from the same individual
  For each biometric i, $i \in \{0,1\}$
    Let i be the 'master' biometric and $|1-i|$ be the 'slave' biometric
    For each couple of elements in D_u
      If the matcher 'master' verifies the claimed identity, the sample of the 'slave' biometric is added to the set D_l
  Update face and fingerprint templates using the augmented labelled set D_l

Fig. 1. Co-updating algorithm. The main steps of the algorithm are shown.

Let n be the number of different templates contained in a gallery, for a given identity. Let N_{TOT} be the maximum number of templates that a gallery can contain. The role, master or slave, of the two biometrics, is highlighted by superscripts M and S.

Some assumptions are made in order to propose a mathematically tractable model [10]. The first assumption regards the finite discrete space used to represent different biometric impressions. Each identity can produce a maximum of $N_{TOT}^{(face)}$ and $N_{TOT}^{(fingerprint)}$ different observations of the face and fingerprint biometrics, respectively. This assumption can be motivated easily in strongly controlled environments, where a small amount of substantial intra-class variations is admitted, but not collectable by the supervised enrolling. The main consequence of this assumption is that a new example, very similar to an existing one, can be considered already present in the gallery.

The second assumption states that all biometric impressions are equally probable (as usually stated in biometric applications), and the two traits are conditionally independent given the identity. The motivation is that no correlation is supposed to be between the face appearance and the related client's fingerprint image [1-2].

In the co-update stage, a set D_u of $|D_u|=k$ pairs of biometric samples (i.e. face and fingerprint) are presented to the system. We indicate with k' the number of samples of the master biometric whose identity has been verified (i.e. is over the updating threshold in practice).

The value *k'* can be easily computed by considering that each of *k* biometric samples of the master biometric can be drawn with replacement from a homogeneous population of $N_{TOT}^{(M)}$ elements. The probability that a sample will be verified is $p = 1 - FRR^{(M)}$, where $FRR^{(M)}$ is the False Rejection Rate for the master biometric. The problem can be modeled using a binomial distribution [10]. Therefore, on average:

$$k' = k\left(1 - FRR^{(M)}\right). \tag{1}$$

It is worth noting that the effectiveness of the master biometric as supervisor towards the 'slave' biometric is related to its FRR value. If FRR=0 (*k'*=*k*) the master biometric is equivalent to a "true" supervisor because it will verify the claimed identity of all 'genuine' samples in D_u. If FRR>0, then only *k'*<*k* samples can be verified. Thus, only *k'*<*k* samples will be added to the slave gallery.

Now, let *d* be the random variable representing the number of elements in *k'* that are different among themselves, and *dn* the random variable representing the number of elements in *k'* that are different and not present in gallery.

The pdf $p_d(d)$ can be modeled using a multinomial distribution:

$$p_d(d) = \binom{N_{TOT}}{d} \sum_{i=0}^{d} (-1)^i \binom{d}{i} \left(\frac{d-i}{N_{TOT}}\right)^{k'}. \tag{2}$$

The conditional pdf $p_{dn/d}(dn/d)$ – the probability to have *dn* different samples not present in gallery if *d* samples are different among themselves – can be modeled using an hyper-geometric distribution:

$$p_{dn|d}(dn \mid d) = h\left(dn; N_{TOT}, (N_{TOT} - n), d\right). \tag{3}$$

where N_{TOT} is the number of items in the population, $(N_{TOT} - n)$ is the number of items in the population that are classified as 'highly confident genuine users', *d* is the number of items in the sample, *dn* is the number of items in the sample that are classified as 'highly confident genuine users'.

By using Eqs. (2) and (3), the pdf $p_{dn}(dn)$ is derived [10]. Here, we are interested to the expected value $E[dn]$. This represents the expected enlargement of the gallery due to the collection of *k'* elements. By assuming that for each co-update step the number $\Delta n^{(S)}$ of new and distinct samples added to the slave gallery is about equal to its expected value $E[dn]$:

$$\Delta n^{(S)} \approx E[dn] = \sum_{dn=0}^{k'} \left\{ dn \cdot p_{dn}(dn) \right\} = \left(1 - \left(1 - \frac{1}{N_{TOT}^{(S)}}\right)^{k\left(1-FRR^{(M)}\right)}\right)\left(N_{TOT}^{(S)} - n^{(S)}\right). \tag{4}$$

Let $n_0^{(S)}$ be the initial size of the gallery, and $n_{(i)}^{(S)}$ the size of the gallery at the *i*-th co-update step. The length of the gallery at the *i*-th step can be written as:

$$n_{(0)}^{(S)} = n_0^{(S)} \quad n_{(i)}^{(S)} = n_{(i-1)}^{(S)}\, w + N_{TOT}^{(S)}\,(1-w) \;;\quad w = \left(1 - 1/N_{TOT}^{(S)}\right)^{k\left(1-FRR^{(M)}\right)} . \tag{5}$$

The weight w takes into account the effectiveness of the master gallery as a supervisor.

The $FRR^{(M)}$ value determine the system's dynamics, so it is worth to relate its value to the length of the master gallery.

3 The Problem of Modeling FRR

In order to model FRR, in [10] has been introduced, for each client i, a parameter called "connection degree" m_i, correspondent to the average number of examples that produce a 'score' over the updating threshold. Therefore, "difficult" clients can be modeled by a low value of m_i, whilst high value of m_i is adopted for "easy" clients. In [10], the probability that no examples among the m_i are in the gallery, which correspond to FRR value, has been computed:

$$FRR^{(M)} = \begin{cases} 0 & if\ m_i > \left(N_{TOT}^{(M)} - n^{(M)}\right) \\ h(r=0; N_{TOT}^{(M)}, m_i, n^{(M)}) & if\ m_i \le \left(N_{TOT}^{(M)} - n^{(M)}\right) \end{cases} \tag{6}$$

Where h is a hypergeometric distribution on a random variable r equal to the number of examples, in the gallery, among the m_i .

This approach is too rough, because it does not take into account that, actually, each sample has his own "m_i". In other words, each j-th biometric sample is characterized by a $m_{ij} \ge 0$ connection degree with other samples, and the i-th client is characterized by the sequence $\{m_{i1}, m_{i2}, \ldots, m_{iN_{TOT}}\}$ of connection degrees. In particular, eq. (6) neglects that there is a subset of samples for which this connection degree is zero. We refer to such samples with the term "isolated". As a consequence, isolated samples contribute to the FRR with a constant value equal to 1, because they cannot "match" to other samples.

Therefore, we propose to correct eq. (6) by considering the fraction of samples that are isolated ($m_{ij} = 0$), namely, f_I, and the fraction of connected samples ($m_{ij} > 0$), f_C. Eq. (6) becomes as follows:

$$FRR^{(M)} = \begin{cases} f_I \cdot 1 + f_C \cdot 0 & if\ n^{(M)} > N_{TOT}^{(M)} - m_i \\ f_I \cdot 1 + f_C \cdot h(0; N_{TOT}^{(M)}, m_i, n^{(M)}) & if\ n^{(M)} \le N_{TOT}^{(M)} - m_i \end{cases} . \tag{7}$$

When the gallery reaches its maximum size, $n=N_{TOT}$, the FRR value depends only on the isolated samples, for the reason we explained above. In other words, there is a "saturation" value for FRR, that is $FRR(n=N_{TOT})=f_I$. It is worth noting that saturation value is 0 for the model according to eq. (6).

Eq. (5) allows us to model the gallery size increase during the co-update iterations, whilst eq. (7) models the FRR of the master biometrics in a more realistic manner than eq. (6). Notice that, in order to simplify the model, we adopted the average of m_{ij} for connected samples. Now, this average value does not take into account isolated samples, but only connected samples. The above relationships can be used to predict the behaviour of co-update. In the next Section the experimental validation of the proposed model by eqs. (5, 7) is done by co-updating a face and a fingerprint matcher on real data.

4 Experimental Results

4.1 The Data Set

The data set adopted consists of 42 individuals composed of 20 face and fingerprint images for each individual, by keeping in mind the independence of face and fingerprint traits. The time span of both the collected data sets spans over one year. Forty-two clients with 20 instances representing significant illuminations changes and variations in facial expressions per person were used from the Equinox corporation database [12]. The fingerprint data set has been collected by the authors using Biometrika Fx2000 optical sensor. The images are acquired with variations in pressure, moisture and time interval to represent large intra-class variations. The results are computed on five random coupling of face and fingerprint datasets and are averaged. Whilst minutiae are simply extracted from the fingerprint images after commonly used processing algorithms [2,13], PCA is computed on the whole data set and applied to face images in order to reduce the size of the overall feature space. 95% of energy is retained according to the current literature [1,14].

It is worth noting that the face data set is notably "simpler" than the fingerprint one. We adopted so different data sets in order to show the effect of intra-class variations on the model prediction ability.

4.2 Experimental Protocol

First of all, we implemented a simple bi-modal identification system made up of a PCA-based face matcher and a fingerprint matcher using the "String" matching algorithm ("String" is based on minutiae points). We used the standard versions of these two recognition algorithms [13-14]. Then, we implemented the template co-update algorithm in Figure 1. Both the eigenspace of the PCA-based face matcher and the co-update threshold value at FAR = 0% are computed using the whole dataset, and has not been updated during the co-update process. This approach is common to some template update practices [4-5]. The update of face and fingerprint templates is performed simply by adding new examples to the user's gallery.

In the experiments the initial gallery has been set as follows. We selected, as the initial template in D_l for the i-th client, a sample whose connection degree is exactly m_i. In other word, the initial template is a connected sample 'near' to other m_i samples. In the proposed model the i-th client is characterized by (f_C, m_i) so this initial template is representative for the client. The rationale for this choice is to exclude outliers from the initial gallery, likewise to what happens in real situations, where the initial template is chosen in completely supervised fashion.

We simulated the acquisition of a batch set D_u by generating several sets of k=10 couples – face and fingerprint – of 'genuine' examples, drawn with replacement from a homogeneous population of $N_{TOT}^{(face)} = N_{TOT}^{(fingerprint)} = 20$ samples. We are aware that adopted database size may not be very well appropriate for the task, but it respects, on average, the size adopted in other template update works reported in literature [4-9]. Since duplicated samples may provide high match scores, which is not realistic, we disregard the first highiest match score when such duplicates are present in the gallery.

In order to set the correct parameters in the proposed model, for each client and for each biometric we computed a) the value f_c, that is, the fraction of samples that produce a 'score' over the threshold S^*, and b) the value m_i, that is, the integer nearest to the average connection degree of the connected samples. Table 1 shows the computed parameters. For each client, results are averaged on ten trials (experimental values) and predicted using parameters in Table 1.

Table 1. List of parameters used to model the chimerical dataset. Fraction of connected samples (f_c) and average connection degree characterizing the connected samples (m_i) are reported for each client and for each biometric.

ID	face		fingerprint	
	m_i	f_c	m_i	f_c
1	1	0.4	0	0
2	3	0.8	2	0.4
3	3	0.85	2	0.2
4	4	0.85	2	0.2
5	3	0.75	1	0.25
6	3	0.85	1	0.1
7	2	0.65	1	0.2
8	3	0.85	1	0.2
9	2	0.65	0	0
10	3	0.85	1	0.15
11	**4**	**1**	**1**	**0.35**
12	4	0.5	1	0.25
13	3	0.9	1	0.35
14	5	0.9	1	0.15

ID	Face		fingerprint	
	m_i	f_c	m_i	f_c
15	3	0.65	0	0
16	3	0.8	2	0.35
17	2	0.55	1	0.3
18	4	0.85	1	0.1
19	2	0.75	3	0.3
20	3	0.85	2	0.5
21	3	0.75	1	0.1
22	4	0.85	0	0
23	4	0.8	1	0.15
24	3	0.8	0	0
25	4	0.75	2	0.45
26	4	0.85	2	0.4
27	3	0.9	1	0.35
28	1	0.75	2	0.25

ID	face		fingerprint	
	m_i	f_c	m_i	f_c
29	3	0.75	0	0
30	**2**	**0.65**	**1**	**0.2**
31	4	0.8	0	0
32	2	0.35	3	0.2
33	5	0.9	1	0.15
34	2	0.6	1	0.1
35	2	0.95	2	0.2
36	3	0.85	0	0
37	4	0.75	1	0.1
38	4	0.85	0	0
39	5	0.9	1	0.15
40	3	0.8	0	0
41	3	0.8	2	0.4
42	2	0.65	1	0.1

4.3 Experimental Results

FRR values and gallery sizes obtained on test data are compared to correspondent values by eqs. (5-7). Fig. 2 compares the values obtained on test data with the values predicted by the model proposed in [10] and by ours. Reported trend is obtained by averaging values over all the clients. For each iteration of the algorithm depicted in Fig. 1, the FRR value evaluated at zeroFAR and the gallery sizes are reported, whilst theoretical lines follow eqs. (5, 6) and eqs (5, 7). In order to highlight the peculiarity of the model, values regarding two typical clients, 'easy' and 'difficult', are also show in Figs. 3 (client 11) and 4 (client 30).

As we can deduce from Table 1, client 11 is an 'easy' client. The face biometric presents m_{11}=4, f_c=1. This means that every sample is connected, thus producing a score over the threshold S^* with at least another sample. Being m_{11}=4, each sample is averagely connected to other four ones. Worth noting, the fingerprint biometric presents m_{11}=1, f_c=0.35: only 35% of the samples are connected, with connection degree m=1. Remaining samples are isolated. They can be introduced in the gallery thanks to the master, face, biometric only. However, they don't contribute to the expressive power of the fingerprints gallery, since they cannot match with other samples.

Referring to the same Table 1, client 30 is a difficult client. The face biometric presents m=2, f_c=0.65: only 65% of the samples are connected, with connection degree m=2. Moreover, the fingerprint biometric exhibits m=1, f_c=0.2. Other face and fingerprint samples are isolated, and, even if they are in the gallery, cannot contribute to its expressive power.

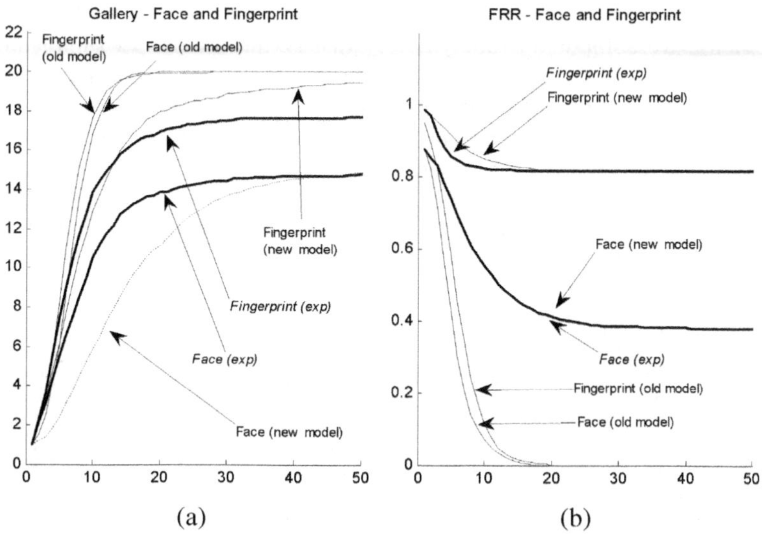

Fig. 2. Experimental trend and predicted values of gallery sizes (a) and FRR (b) for fingerprint and face biometrics. Reported trend is obtained averaging values over all the clients. Values predicted by the previous model and the new proposed model are shown.

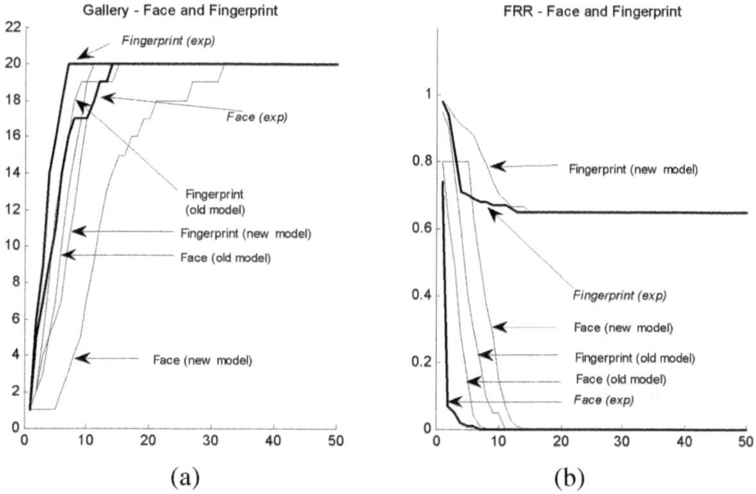

Fig. 3. Client n.11. Experimental trend and predicted values of gallery sizes (a) and FRR (b) for fingerprint and face biometrics. Values predicted by the previous model and the new proposed model are shown.

The following observations can be drawn from Figures 2-4.

The model is able to predict the experimental performance, with a negligible difference. Predictions are more accurate than those obtained in [10], due to improvements in the model. That model estimates FRR by neglecting the differences in

expressive power between isolated and connected samples. Consequently, if a gallery reaches its maximum size thanks to the other biometric (co-update scenario), the previous model estimates FRR=0. However, in the real scenario, due to the presence of isolated samples, the true value of FRR will be greater than 0 (FRR=1 if all the samples in gallery are isolated). As we can notice from Figs. 2-4, this drawback is completely eliminated in the present model.

The saturation of theoretical and experimental curves is obviously due to the form of eqs.(5, 7). In particular, the weight w tends to decrease as the gallery size of both matchers increases, depending on the value of their FRR. Thus, the size of galleries must converge to N_{TOT} if m_i is not zero. According to the model, at the end of co-update process, FRR($n=N_{TOT}$)=f_I. It is worth noting that the correct prediction of FRR values is matter of primary importance both for the validation of the model and for the designers, that need relevant information about the performance of the system.

Some issues have not yet been investigated: (1) in real environments N_{TOT} is unknown. This problem can be overcame by considering N_{TOT} as the number of samples in the batch exhibiting, for each client, a score above the zeroFAR threshold adopted; (2) the same for the estimation of m_i. These assumptions do not consider the problem of impostors insertion, a still open issue as remarked in [9, 10]. Therefore, in a future work, we will investigate the impact of N_{TOT} and m_i predicted on a batch of samples including even impostors, on a larger, non chimerical data set.

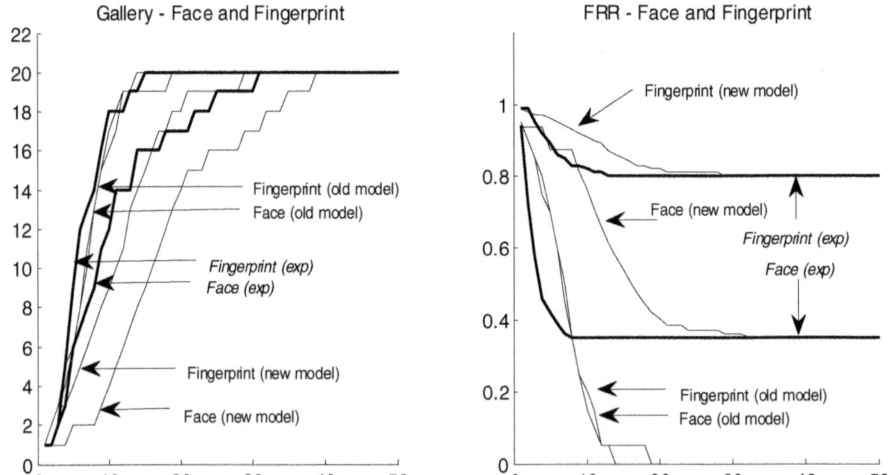

(a) (b)

Fig. 4. Client n. 30. Experimental trend and predicted values of gallery sizes (a) and FRR (b) for fingerprint and face biometrics. Values predicted by the previous model and the new proposed model are shown.

5 Conclusions

In this paper, we improved the theoretical framework explaining the template co-update behaviour proposed in a state-of-the-art work, by a more effective and realistic modelling of the verification performance in terms of FRR. The model still assumes that no impostors are collected during system operation, by setting the updating threshold at zeroFAR. Therefore, the assessment of verification performance is only related to that operational point.

Although we pointed out some issues still remain, we believe that the template co-update algorithm is well explained by proposed model. Therefore, it is suitable for designing bi-modal verification systems which can improve with use.

References

1. Li, S.Z., Jain, A.K. (eds.): Handbook of face recognition. Springer, Heidelberg (2005)
2. Maltoni, D., Maio, D., Jain, A.K., Prabhakar, S.: Handbook of fingerprint recognition. Springer, Heidelberg (2003)
3. Uludag, U., Ross, A., Jain, A.K.: Biometric template selection and update: a case study in fingerprints. Pattern Recognition 37(7), 1533–1542 (2004)
4. Jiang, X., Ser, W.: Online Fingerprint Template Improvement. IEEE Trans. PAMI 24(8), 1121–1126 (2002)
5. Ryu, C., Hakil, K., Jain, A.: Template adaptation based fingerprint verification. In: Proc. of ICPR, Hong Kong, vol. 4, pp. 582–585 (2006)
6. Roli, F., Marcialis, G.L.: Semi-supervised PCA-based face recognition using self-training. In: Yeung, D.-Y., Kwok, J.T., Fred, A., Roli, F., de Ridder, D. (eds.) SSPR 2006 and SPR 2006. LNCS, vol. 4109, pp. 560–568. Springer, Heidelberg (2006)
7. Roli, F., Didaci, L., Marcialis, G.L.: Template co-update in multimodal biometric systems. In: Lee, S.-W., Li, S.Z. (eds.) ICB 2007. LNCS, vol. 4642, pp. 1194–1202. Springer, Heidelberg (2007)
8. Roli, F., Didaci, L., Marcialis, G.L.: Adaptive biometric systems that can improve with use. In: Ratha, N., Govindaraju, V. (eds.) Advances in Biometrics: Sensors, Systems and Algorithms, pp. 447–471. Springer, Heidelberg (2008)
9. Rattani, A., Marcialis, G.L., Roli, F.: Capturing large intra-class variations of biometric data by template co-updating. In: IEEE Workshop on Biometrics, Int. Conf. on Vision and Pattern Recognition CVPR 2008, Anchorage, Alaska, USA (2008) (in press)
10. Didaci, L., Marcialis, G.L., Roli, F.: A Theoretical and Experimental Analysis of Template Co-update in Biometric Verification Systems. In: da Vitora-Lobo, N., Kasparis, T., Roli, F., Kwok, J.T., Georgiopoulos, M., Anagnostopoulos, G.C., Loog, M. (eds.) S+SSPR 2008. LNCS, vol. 5342, pp. 765–774. Springer, Heidelberg (2008)
11. Ross, A., Nandakumar, K., Jain, A.K.: Handbook of Multibiometrics. Springer, Heidelberg (2006)
12. http://www.equinoxsensors.com/products/HID.html
13. Jain, A.K., Hong, L., Bolle, R.: On-line Fingerprint Verification. IEEE Transactions on Pattern Analysis and Machine Intelligence 19(4), 302–314 (1997)
14. Turk, M., Pentland, A.: Eigenfaces for Face Recognition. Journal of Cognitive Neuroscience 3(1), 71–86 (1991)

Bipartite Biotokens: Definition, Implementation, and Analysis*

W.J. Scheirer[1,2] and T.E. Boult[1,2]

[1] Univ. of Colorado at Colorado Springs, Colorado Springs, CO - 80918
[2] Securics Inc, Colorado Springs, CO - 80918

Abstract. Cryptographic transactions form the basis of many common security systems found throughout computer networks. Supporting these transactions with biometrics is very desirable, as stronger non-repudiation is introduced, along with enhanced ease-of-use. In order to support such transactions, some sort of secure template construct is required that, when re-encoded, can release session specific data. The construct we propose for this task is the *bipartite biotoken*. In this paper, we define the bipartite biotoken, describe its implementation for fingerprints, and present an analysis of its security. No other technology exists with the critical reissue and secure embedding properties of the bipartite biotoken. Experimental results for matching accuracy are presented for the FVC 2002 data set and imposter testing on 750 Million matches.

1 Introduction

Template protection schemes solve an important problem inherent in biometrics: the threat of permanent feature compromise. Biometrics, unlike passwords or PINs, cannot be changed during the course of an individual's life. Many different schemes have been proposed in the literature [1] for template protection. Certain classes of these schemes support key release upon successful matching. *Key-binding* biometric cryptosystems bind key data with the biometric data. *Key-generating* biometric cryptosystems derive the key data from the biometric data. Both classes support a key release that may be used for cryptographic applications, including standard symmetric key cryptography, where key storage is problematic. Biometrics coupled with traditional cryptography presents several advantages, including ease-of-use and stronger non-repudiation properties. Unfortunately, the work to date has not been able to support cryptographic transactions as described in [13]. Further, the actual security and matching accuracy of even the most popular schemes is questionable.

The *fuzzy vault* scheme [2] is a key-binding biometric cryptosystem that hides a secret κ within a large amount of chaff data. Briefly explained, Alice places κ in a fuzzy vault and locks it using a set A of elements from some public universe U. To unlock the vault, and retrieve κ, Bob must present a set B that substantially overlaps with A. To protect κ, it is encoded as coefficients of a

* Work supported in part by NSF STTR 0750485 and NSF PFI 0650251.

M. Tistarelli and M.S. Nixon (Eds.): ICB 2009, LNCS 5558, pp. 775–785, 2009.
© Springer-Verlag Berlin Heidelberg 2009

polynomial p. A set of points R is constructed from A and $p(A)$. In addition to these points, chaff points C are randomly generated and inserted into R. The subset matching problem is solved with an error correction code. To decode κ, if Bob's B approximately matches A, he can isolate enough points in R that lie on p so that applying the error correcting code he can reconstruct p, and hence κ. Several implementations of biometric fuzzy vaults have been produced, including a fingerprint implementation [3], a password hardened implementation [4], and a multi-modal fingerprint & iris implementation [5].

Multiple serious attacks have questioned the security of fuzzy vaults. The work of [11] introduces three attacks against a variety of secure template technologies. For fuzzy vaults, the attack via record multiplicity (ARM), surreptitious key inversion (SKI) attack, and substitution attacks all apply. The authors of [5] concede that the fuzzy vault "is not a perfect template protection scheme" because of the attacks of [11], yet the security analysis presented in [5] does not consider their impact. Password hardened fuzzy vaults [4] were introduced in response to the ARM attack, but still fall prey to the SKI attack, facilitating recovery of the original biometric data, and the substitution attack, allowing the placement of a backdoor into the template. Other, brute-force oriented, attacks against fuzzy vaults have included CRC checks [6], and chaff point identification [7]. On the issue of performance, the published results have been promising, albeit achieved with very limited testing. How the matching accuracy of fuzzy vaults scales to realistic amounts of data has yet to be shown.

The *fuzzy extractor* scheme [8] is a key-generating cryptosystem that binds some random data with the biometric data to produce a unique key. A fuzzy extractor incorporates a *secure sketch* construct to allow the precise reconstruction of a noisy input w given an instance of the sketch s and a sample w'. A secure sketch SS bound with a random number i forms the basis of the fuzzy extractor instance P, which returns a key R, when approximate input matching is successful. Given (questionable) assumptions, [8] shows that in an information theoretical sense, that fuzzy extractors could achieve entropic security, with P and R leaking no information that helps to predict w. The security analysis of [8] is largely constrained to modeling the probability of an attacker guessing R, and the effects of key generation on this probability.

While theoretical security analyses may be important, in biometrics, the operational security is tied to the GAR and FAR. For effective security a system needs the FAR to be less than 1 in *millions* or *billions*. Despite the formal models of security in [8], an impostor may be able to achieve a false match releasing the key. This security is a constraint of the matching algorithm, not just the template protection scheme. To date, there is no published work on the GAR/FAR performance of fuzzy extractors. Moreover, fuzzy extractors may suffer from practical constraints during error-prone data collection [10], making it difficult to generate a key that is both stable and highly random.

Revocable biotokens [12] have emerged as a different solution to the template protection problem, and have been described as being able to support key release [13]. For any biometric data that can be split into stable and unstable

components, the stable portion can be encrypted in a reliable fashion, while the unstable portion is left in the clear. This provides for the definition of a biotoken transform that scales/translates the data, and then separates it into a quotient q and modulus or remainder, r. Since q is stable, it can be encrypted or hashed for both probe and gallery data, and require an exact match. This transform induces a distance measure in encoded space: first test if the encoded q values are identical; if they are, then the residuals r are then used to compute distance. In this paper, we analyze secure key release from revocable biotokens.

This paper introduces the implementation details of the bipartite biotoken construct. In Sec. 2, we review the definition of bipartite biotokens, as introduced as a general theoretical construct in [13]. With this definition, we go on to summarize an implementation of fingerprint bipartite biotokens in Sec. 3, and present a security analysis of this implementation in Sec. 4. Finally, in Sec. 5, we experimentally show that bipartite biotokens outperform existing secure template data release mechanisms, and have *useful genuine accept rates when set for zero false accepts in over 750 Million imposter trials.*

2 The Definition of Bipartite Biotokens

The notion of data splitting to support revocable fingerprint biotokens was introduced in [12]. Using this knowledge, and the concept of public key cryptography, we can develop the re-encoding methodology for revocable biotokens. The re-encoding property, introduced in [13], is essential for supporting a viable transactional framework - tokens with unique data must be generated quickly and automatically to support the transaction. Bipartite biotoken generation from a stored biotoken allows the required data release when matching against tokens generated from original biometric features during the course of the transaction.

Assuming the biometric produces a value v that is *transformed* via scaling and translation to $v' = (v-t)*s$, the resulting v' is split into the overall stable component q, and the the residual component r. In the base scheme, for a user j, their residual $r_j(v')$ is left in the clear. The amount of stable & unstable data is a function of the modality being considered. For the initial transformation $w_{j,1}(v', P)$ of q, a public key P is required. For nested re-encodings, w_j is re-encoded using some transformation function T (which may be a hash function, or another application of public key cryptography) creating a unique new transformation for each key that is applied: $w_{j,1}(v', P),\ w_{j,2}(w_{j,1}, T_2), \ldots, w_{j,n}(w_{j,n-1}, T_n)$.

Using public key cryptography, the nesting process can be securely invertible if the private key associated with the first stage of encoding is available. With this nesting in mind, we can define three properties for the bipartite biotoken:

1. Let B be a secure biotoken, as described in [12]. A bipartite biotoken B_p is a transformation $bb_{j,k}$ of user j's kth instance of B. This transformation supports matching in encoded space of any bipartite biotoken instance $B_{p,k}$ with any secure biotoken instance B_k for the biometric features of a user j and a common series of transforms P, T_2, ..., T_k.

2. The transformation $bb_{j,k}$ must allow the embedding of some data D into B_p, represented as: $bb_{j,k}(w_{j,k}, T_k, D)$.
3. The matching of B_k and Bp_k must release D if successful, or a random string r if not successful.

3 The Implementation of Bipartite Biotokens

The implementation of the bipartite token, Fig. 1, is an extension of the concepts of revocable biotokens [12] and fuzzy vaults [2], which are prerequisite for a solid understanding as in the limited space we focus on the key advances. There are four major advances in the bipartite biotoken implementation:

1. The bipartite representation implements Reed-Solomon for error correction
2. The bipartite representation uses biotoken encoded "pair rows", which are rotation and translation independent
3. The bipartite representation does *not* store the points at which the embedded polynomial is evaluated
4. The bipartite representation allows for multiple simultaneously embedded polynomials, supporting larger keys with lower numbers of matching pairs.

While the original fuzzy vault work suggested the use of Reed-Solomon (RS) codes, we are unaware of any fingerprint fuzzy vaults that have actually implemented them, probably because of the inherent difficulty of alignment, ordering issues, and the high potential error rate. Our implementation uses an RS code with varying levels of error correction selectable at encoding time. For efficiency,

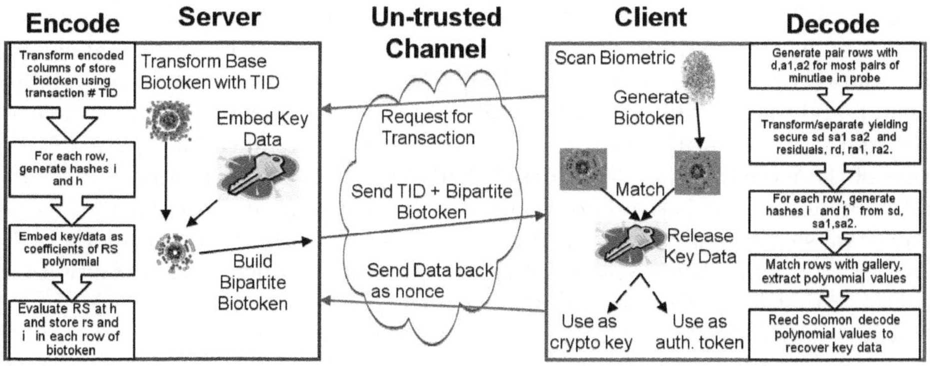

Fig. 1. Sequence diagram for the bipartite biotoken. Since the embedded data can be unique on a transactional basis, a variety of cryptographic protocols can be supported [13]. The embedded data can be a nonce that is sent back to the server for validation. It can also be a one-time token that is used for authentication. Or, in a more traditional application of key-binding schemes, it can be a symmetric or public cryptographic key. All are advantageous when the communications channel is un-trusted; only a legitimate party can unlock the embedded secret.

we choose to work over $GF(2^8)$, where the coefficients and evaluation points are all 8 bit quantities. We represent the data D to be stored as a K-byte block, with E bytes of error correction, yielding a total payload block $B = K + E$. The polynomial encodes the B bytes of data. The RS polynomial representing the B byte payload body is then evaluated at a set of points, with the value of the resulting polynomial being stored. This allows for a very fast implementation, with the average matching and key extraction attempt requiring less than 1 millisecond on a 3Ghz processor, where we use pre-computed gallery files and start from minutiae for the probe. With this, we can easily vary both the key size, up to 1024 bits, and the level of error correction, with little impact on speed.

Using the "pair row" representation of the Bozorth-like matcher of [12], we have a representation that is inherently rotation and translation invariant. With the biotoken encoding of a row pair we have the raw distance and angles separated and the stable parts of those numbers are protected. Let d, a_1 and a_2 be the distance and angle fields of the row, and let sd, sa_1 and sa_2 be the stable components of these with rd, ra_1 and ra_2 the reflected modulus [12] residuals.

For polynomial evaluation, we hash the 24 bits of sd, sa_1 and sa_2 into i, an 8 bit quantity that is stored in the gallery. The value i is then hashed, per transaction, a second time to define the point at which the polynomial is evaluated. To support multiple key columns, we evaluate this hash h for different polynomials yielding values $rs_1 \ldots rs_4$. Note the evaluation point/hash value h is not stored.

The result is an "encoded bipartite row" that contains the unprotected fields and 6 protected fields (the encoded stable field w used for matching, index i and 4 columns of evaluated polynomials). We require at least 14 rows, padding the key if it does not require 4 columns to represent it. The location of the w is randomized per row. The evaluated RS polynomials for the 4 key columns, $rs_1 \ldots rs_4$, follow w using a circular mapping of the 6 slots. For example, if the random index was 3, then the sequence would be: $[rs_3, rs_4, w, rs_1, rs_2, i]$.

When matching a probe, the system creates all the fields for each of its rows, including the "un-stored" hash value (h) for polynomial evaluation. A probe row potentially matches a gallery row if it finds a matching w among the encoded fields and the residuals (rd, ra_1, ra_2) are within threshold. This test is necessary, but not sufficient, for a correct match. With w identified, the algorithm can then extract the evaluated polynomial values, $rs_1 \ldots rs_4$. If w is incorrectly identified, if the row is an accidental match, or if the underlying hash value (h) is incorrect (because of a random collision in generating/matching w), some values labeled $rs_1 \ldots rs_4$ will be extracted, but will be incorrect. Prints will produce many potentially matching rows, usually (determined empirically) 200-800 if a true match and 50-600 if a non-matching print. The second stage of our Bozorth-like matching is generation of a consistent subgraph from the potentially matching rows. This results in selection of a set (20-70) of mostly correct matched rows. Without an effective way to select probable rows from the set of potentially matching rows, the level of error correction or search needed would be impractical (e.g., $\binom{200}{20}$ is 10^{27}). We extract the k values for each of the j key columns and

obtain a set of hash evaluation points h_j and their Reed-Solomon polynomial evaluations $rs_{j,k}$ at the associated points.

Now comes one of the important implementation details, addressing both security and efficiency. One could effectively improve robustness by increasing the level of ECC, but doing so increase the ease with which an attacker can crack the key. Instead we use a two level hashing to improve robustness. Our two level mapping will, in general, map multiple $sd, sa1, sa2$ sets to the same index. We implemented a procedure to collect the multi-values during the mapping, check for consistency and use that redundancy to help resolve any conflicts that arise when noisy data is mapped. The result of the mapping and consistency check is a vector of length B polynomial values (some of which may be missing) that holds the values of the evaluated RS polynomial for each location. The vector B, with gaps marked, is fed into the RS decode function, which allows us to recover D with up to g gaps and e errors, as long as $2g + e < E$, where E is the number of ECC bytes used. Each key column is recovered separately, with larger keys being the concatenation of multiple columns. For added security, a checksum is compute over the 6 unprotected columns the gallery biotope. The data D are XORed with a checksum before embedding, and again after decoding, and this prevents any tampering with the biotoken.

4 Security Analysis of Bipartite Biotokens

A security analysis of the underlying revocable biotokens was presented in [12]. The security analysis of bipartite extension is twofold - the analysis of the impact of the attacks of [11] and the brute force attack necessary to recover D.

The bipartite approach prevents the ARM, SKI, and the blended substitution attacks of [11]. For the ARM attack, we refer back to the implementation details of Sec. 3, where we introduced the hashing methodology to protect sd, sa_1 and sa_2 through data reduction ambiguity and transaction specific hasing, and the further RS polynomial encoding of the resulting hash. Thus, an attacker with access to the RS polynomial encodings cannot correlate between different biotokens - the evaluation points are never found within the encoding. For the SKI attack, we again note that the encoding binding D is tied to a hashed form of the secure data, which does not change the security analysis of [12]. The blended substitution attack is solved through the use of the specially crafted embedded data wherein D is XORed with a checksum of the stored biotoken. Thus, the system automatically checks for tampering by XORing the computed checksum of the biotoken with the released data before returning the key. If the checksum is correct, the valid D will result. Bipartite biotokens are still susceptible to a straight substitution attack, whereby an attacker replaces the columns of the biotoken with their own. However, this is effectively a denial of service for the legitimate user and hence detectable.

To recover D, an attacker would have to estimate B with a sufficiently small number of errors. At a minimum this requires K correct rows, where for each row the attacker would have to "label" the fields in each row, they would have to

guess the hashes/indices which serve as the evaluation point for the polynomial. For each row there are 5 choices for ordering, and 2^8 possible hash values per key for a total ambiguity of 1280 per-row per key. The attacker would have to simultaneously recover and order K of these. It is likely there may be a reduction from the full hash space to the evaluation index of the RS polynomial, which may reduce the per-row ambiguity down to K^{5*B}, so if $K = 16$ and $E = 6$, it may reduce the attack effort to 2^{480} attempts. Thus, as we shall see, this is not the limiting aspect of the bipartite biotokens security.

To date, security analyses of protected templates presumed the attacker cannot find added constraints to relate the data or exploit non-uniformity in the distributions. An analogy would be discussing string entropy for password strength, knowing that the distribution of actual passwords is far from uniform; dictionary attacks and rainbow tables are often very successful. Prior works [8] [9] have used "entropy" models to address this and suggest security bounds. The problem with this is that measuring entropy depends on how the well the data is "coded" and the models of interdependence. It is difficult to accurately estimate or bound entropy from below, which is what is needed for a security estimate. In practice a likely more intelligent attack exists against any biometric protection scheme than simple brute force, which is to use a large amount of biometric data to see if any of the imposters can release the encoded data. This *doppleganger attack* is the biometric equivalent of a dictionary attack. The hundreds of thousands of prints publicly available provide at least a basic doppleganger dictionary.

If the False Accept Rate (FAR) of a system is 1 in X attempts, then a doppleganger attack consists of trying sufficiently more than X different attempts. When papers such as [8] [9] prove they are "secure" or others [3] [4] [5] claim they have N bits of security but have measurable FARs greater than 1 in 2^N, then one can only conclude the assumptions underlying the proofs of the security models are fundamentally flawed. A FAR of zero, tested on $< 2^N$ items (for example, [3] [4] tested on $< 2^{14}$) only documents N-bit security. Since this depends on experimental analysis we return to our analysis of the doppleganger attack against bidirectional biotokens after presenting the experimental results.

5 Experimental Results

To test the matching accuracy and security of the implemented bipartite biotokens, we ran a series of large-scale tests for the FVC 2002 [14] data set, varying a series of parameters. These parameters include bytes of error correction, released data size, size of the probe biotoken, and size of the gallery biotoken. Most importantly, many different embedded data sizes were tried, from 112 bits to 1024 bits. Support for "large key" sizes has not been available for any published template protection scheme to this point. Large key sizes here include 512 bits and 1024 bits, both of which are suitable for public key cryptography. All results we report are at single points of the ROC curve, where the FAR is 0, in order to mitigate doppleganger attacks described in Sec. 4. Our focus in this paper is high security; other points on the ROC curve represent opportunities for attack,

as the FAR grows larger. In order to facilitate this requirement, all experiments reported use 6 bytes of ECC or less, except for the 1024 bit experiments, which allow larger ECC and still maintain a 0 FAR on large sets. Many different probe and gallery size combinations produce the same GAR, for the same ECC level.

First, we compare revocable biotokens to three fuzzy vault approaches presented in [3] and [4] for three different released data sizes. As in those papers, part of FVC2002 DB2 was used, providing 100 distinct probe/gallery pairs, yielding 100 genuine matches and 9,900 impostors trials. Table 1 contains the results, with bipartite biotokens showing significant improvement. Moreover, the best performing fuzzy vault scheme, the mosaic with 2 queries, incorporated information from 4 different print impressions (2 to build the mosaic, and 2 for the queries), while bipartite biotokens outperforms it with 1 probe and 1 gallery. GAR numbers reported for the fuzzy vaults have been adjusted to count reported "failure to capture rates" as mis-detections (as they operationally are).

Support for standard cryptographic key sizes is of primary interest. Thus, in Table 2 we show results for two common symmetric key sizes (192 bits and 256 bits) and two common public key sizes (512 bits and 1024 bits), for the same protocol as in Table 1 using FVC2002 for DB1 and DB2. A drop in performance is noted for the 1024 bit data size, but we note the GAR for DB2 is better than the results reported by [4] for a much smaller embedded data size. While the small test presented in Tables 1 & 2 indicates promising performance, it is inadequate to gauge real operational performance. For instance, any "0% FAR" reported is really just FAR $< 2^{-14}$.

Table 1. A comparison of different template protection schemes on the FVC2002 DB2 data set for three small released data/key sizes. The password vault experiments of [4] presume the attacker does not know the password; the FAR rates when the attacker does know this information (as we presume in our tests) are not given but are known not to be zero. Bipartite Biotokens provide significant improvements over the three fuzzy vault schemes of [3] and [4]. Bipartite biotoken error correction parameters were 5 bytes for 112 bits, 2 bytes for 128 bits, and 6 bytes for 160 bits.

	112 Bits		128 Bits		160 Bits	
	GAR %	FAR %	GAR %	FAR %	GAR %	FAR %
F.P. Fuzzy Vault	89	0.13	89	0.01	84	0
F.P. FV, Mosaic with 2 Queries	96	0.24	95	0.04	89	0
Password Vault	88	?	86	?	79	?
Bipartite Biotokens	97	0	97	0	97	0

Table 2. Results for larger key sizes appropriate for public key cryptography. Best GAR and ECC sizes yielding FAR=0 for FVC02 DB1 & DB2.

FVC02 DB#	192 Bits		256 Bits		512 Bits		1024 Bits	
	GAR %	ECC	GAR %	ECC	GAR %	ECC	GAR %	ECC
1	97	5	94	2	95	5	77	10
2	97	2	97	2	92	6	82	9

Table 3. Multiple recovery attempts for a test gallery of 200 unique fingers from FVC 2002 DB1 and DB2. 6 bytes of error correction was used with both the probe and gallery biotoken size of 8000 bytes, except for right column, which used 9 bytes ECC, 8000 byte gallery and 20,000 byte probe sizes. These are the same parameters as our large test below, with 0 FAR over more than 750 Million imposter trials.

Recovery Attempts	128 Bits GAR %	256 Bits GAR %	512 Bits GAR %	1024 Bits GAR %	1024 Bits 9B ECC GAR %
1	93.0	93.0	91.0	63.0	78.0
2	97.5	97.5	97.0	74.0	87.0
3	98.0	98.0	97.5	77.0	89.0
4	98.5	98.5	98.5	77.5	89.0
5	99.0	99.0	99.0	79.0	89.5
6	99.0	99.0	99.0	81.0	89.5

For a meaningful security analysis we need orders of magnitude larger imposter testing, and multiple attempts at entry. Thus, we created a much larger test out of DB1 and DB2, including all available images per finger. For true matches, we have 200 unique fingers. For the impostor test data DB1 and DB2 provide a total of 158,400 impostor attempts from the same collection. To show the operational security of bipartite biotokens and protection from doppleganger attacks, we extended our tests with mixed data from FVC 2002 and 2004, the rolled prints in NIST DB29, and NIST DB14. This provides a doppleganger dictionary of over 63,000 images used to attack each gallery entry, yielding potentially billions of non-match attempts of which we have completed 750 Million so far. Scanned/rolled prints are justified for use in the attack, even if not operationally relevant, as they provide more minutiae. Imposter/FAR testing also included rolled against rolled. Data was processed with *mindct* to extract minutiae, keeping the highest 150 quality minutiae found.

For a more realistic model of usage, we are interested in assessing the effect of multiple attempts to match on the GAR - a realistic scenario, as the user is given multiple attempts to match in an operational system. For multiple matching attempts using the 7 other prints for each gallery, the prints were attempted in order of "quality," as defined in [14], from best to worst. [3] uses a similar methodology for allowing 2 attempts to match, but also fuses 2 images for probe and gallery. Results in table 3 show good initial performance but major operational improvement with multiple attempts (the 7th did not produce different results from the 6th). Since 1024 bit keys require more matching, we can improve the GAR by increasing ECC and token size, while maintaining its zero FAR rate. Thus we report 1024 bits using 9 bytes of error correction and larger probes, yielding an acceptable GAR rates on 2 or more attempts.

When using appropriately chosen keys this experiment shows an acceptable GAR with *Zero False Accepts from processing 750 Million impostor trials to date*, with 6 Bytes of ECC, 128 bit, 256 bit, or 512 bit keys and 8000 bytes for the probe and gallery. This unprecedented "single finger" performance can be attributed to the biotoken transform dimensional lifting [12] combined with

the novel key embedding/recovery process with error correction. Multi-finger implementations would further enhance the security. As any biometric security analysis should, our testing presumes the attacker has all passwords and all non-embedded keys. Operationally these are generally unknown. Including the security gains of the multiple non-embedded keys and passwords would increase the effective operational security well beyond the tested FAR.

This experiment also highlighted the need to analyze weak pseudo-random number generators and *weak keys* - well known issues in crytposystems.[1] Any biometric template protection with keys or random data, especially if using error correction, must address these issues as weak keys/seeds can drive up the FAR and reduce security. For example, the FAR drops to about 1 in 10 million with random keys/seeds. Our results reported are for appropriately chosen keys. As keys and random number usage impacts the biotope transforms, it is very complex to analyze; weak keys will be the subject of a future paper.

6 Conclusion

The key-binding biometric cryptosystem problem is a challenging, yet essential aspect of the template protection domain. In this paper, we introduced the implementation details that are necessary to build a bipartite biotoken, which supports many different cryptographic protocols, as well as a thorough security analysis covering a range of common attacks. Our experiments show a significant improvement in accuracy and embedding capacity over the most recent published results for several implementations of the fingerprint fuzzy vault. With an 8KByte tokens securely storing a 256bit embedded key, 93% GAR and 97.5% "two try" GAR was shown. At those setting testing with a doppleganger dictionary of over 60,000 attacks per print, totalling over 750 Million impostor tests to date (and still counting), there have been no false accepts when using appropriate keys. Bipartite biotokens provide an extremely attractive secure template technology, ready for large scale use.

References

[1] Jain, A., Nandakumar, K., Nagar, A.: Biometric Template Security. EURASIP Journal on Advances in Signal Processing 2008, Article ID 579416
[2] Juels, A., Sudan, M.: A Fuzzy Vault Scheme. In: Proc. of the IEEE Intl. Symposium on Information Theory (2002)
[3] Nandakumar, K., Jain, A.K., Pankanti, S.: Fingerprint-based Fuzzy Vault: Implementation and Performance. IEEE Trans. on Info. Forensics and Security 2(4), 744–757 (2007)
[4] Nandakumar, K., Nagar, A., Jain, A.K.: Hardening Fingerprint Fuzzy Vault Using Password. In: Lee, S.-W., Li, S.Z. (eds.) ICB 2007. LNCS, vol. 4642, pp. 927–937. Springer, Heidelberg (2007)

[1] See Sec. 14.10 of B. Schneier, *Applied Cryptography* J.Weily&Sons, 1996.

[5] Nandakumar, K., Jain, A.K.: Multibiometric Template Security Using Fuzzy Vault. In: IEEE Conf. Biometric Theory, Application and Systems (2008)
[6] Mihailescu, P.: The Fuzzy Vault for Fingerprints is Vulnerable to Brute Force Attack (2007), http://arxiv.org/abs/0708.2974v1
[7] Chang, W., Shen, R., Teo, F.W.: Finding the Original Point Set Hidden Among Chaff. In: Proc. of the ACM Sym. on Info. Computer And Comm. Security (2006)
[8] Dodis, Y., Reyzin, L., Smith, A.: Fuzzy Extractors. In: Tuyls, P., Skoric, B., Kevenaar, T. (eds.) Security with Noisy Data: Private Biometrics, Secure Key Storage and Anti-Counterfeiting, ch. 5, pp. 79–99. Springer, Heidelberg (2007)
[9] Boyen, X., Dodis, Y., Katz, J., Ostrovsky, R., Smith, A.: Secure Remote Authentication Using Biometrics. In: Cramer, R. (ed.) EUROCRYPT 2005. LNCS, vol. 3494, pp. 147–163. Springer, Heidelberg (2005)
[10] Ballard, L., Kamara, S., Reiter, M.: The Practical Subtleties of Biometric Key Generation. In: USENIX Security Symposium, August 2008, pp. 61–74 (2008)
[11] Scheirer, W., Boult, T.: Cracking Fuzzy Vaults and Biometric Encryption. In: Proc. of the 2007 IEEE Biometrics Symposium, held in conjunction with the Biometrics Consortium Conference (BCC 2007), Baltimore, MD (2007)
[12] Boult, T., Scheirer, W., Woodworth, R.: Secure Revocable Finger Biotokens. In: Proc. of the IEEE Conf. on Computer Vision and Pattern Recognition (2007)
[13] Scheirer, W., Boult, T.: Bio-cryptographic Protocols With Bipartite Biotokens. In: Proc. of the IEEE 2008 Biometrics Symposium, held in conjunction with the Biometrics Consortium Conference (BCC 2008), Tampa, FL (2008)
[14] Maio, D., Maltoni, D., Wayman, J., Jain, A.K.: FVC 2002: Second Fingerprint Verification Competition. In: Proc. of the 2002 Int. Conf. Pattern Recognition (2002)

Fusion of LSB and DWT Biometric Watermarking Using Offline Handwritten Signature for Copyright Protection

Cheng-Yaw Low[1], Andrew Beng-Jin Teoh[2], and Connie Tee[1]

[1] Faculty of Information Science and Technology
Multimedia University, Jalan Ayer Keroh Lama, 75450 Melaka, Malaysia
{cylow,tee.connie}@mmu.edu.my
[2] Electrical and Electronic Engineering Department
Yonsei University, Seoul, South Korea
bjteoh@yonsei.ac.kr

Abstract. Biometric watermarking was introduced as the synergistic integration of biometrics and digital watermarking technology. This paper proposes a novel biometric watermarking technique, which embeds offline handwritten signature in host image for copyright protection. We propose to combine the conventional LSB-based and DWT-based watermarking techniques into a unison framework, which is known as LSB-DWT in this paper. The proposed LSB-DWT technique is evaluated against various simulated security attacks, i.e. JPEG compression, Gaussian low-pass filtering, median filtering, Gaussian noise, scaling, rotation and cropping. The experimental results demonstrate that the proposed LSB-DWT technique exhibits remarkable watermark imperceptibility and watermark robustness.

Keywords: Biometric watermarking, offline handwritten signature, copyright protection.

1 Introduction

Digital watermarking is the practice of concealing additional information in digital document, e.g. image, audio, video, etc, which is generally termed as host. The host is traditionally embedded with originator's name, graphical logo, serial number, binary bitstring, etc, which are coined as conventional watermark henceforward, to prosecute copyright infringements. The major drawback of the conventional watermark is that it lacks the non-repudiation property as anyone including malicious users can embed a particular name or logo in the host.

Biometrics, which measures human physiological and behavioural characteristics, e.g. fingerprint, iris, hand geometry, face, handwritten signature, etc, provides unique means to recognise a person. In recent years, biometrics is synergistically merged into the digital watermarking technology to replace the conventional digital watermarking techniques. This advancement can be categorized as follows:

M. Tistarelli and M.S. Nixon (Eds.): ICB 2009, LNCS 5558, pp. 786–795, 2009.
© Springer-Verlag Berlin Heidelberg 2009

i. Watermarked Biometrics - The host is a biometrics, whereas the watermark can either be a biometrics, or other proprietary notice. For example, biometrics, e.g. fingerprint, iris, or demographic information, e.g. name, etc, can be embedded in the biometric host to protect against biometric template theft. In addition, there are some cases where the biometric host only acts as a carrier of genuine biometrics. In such cases, the genuine biometrics can be highly protected as the intruders might not be aware that the biometric host is invisibly attached with the genuine biometrics.
ii. Biometric Watermarking - This category employs biometrics as the watermark, whereas the host can be any copyrighted documents. Since biometrics provides uniqueness which can hardly be counterfeited, misplaced, or shared, biometric watermarking promises stringent security against fraudulent watermark.

Biometric watermarking was first proposed by Jain in year 2002. Jain and Uludag [5] suggested an amplitude modulation-based biometric watermarking technique for two application scenarios. In the first scenario, fingerprint minutiae were embedded in fingerprint, face and arbitrary images, which were merely acting as a carrier to secure the genuine fingerprint minutiae. In the next application scenario, fingerprint minutiae were embedded in face image, and the watermarked face image was then encoded in smart card. To authenticate the smart card holder, his/her fingerprint was captured and matched against the fingerprint minutiae stored in the smart card. On the other hand, Jain et al. [4] proposed a biometric watermarking technique to embed a person's face code in his/her fingerprint image. The extracted face code was used as an additional source, besides the fingerprint image, to verify the system users. Namboodiri and Jain [6] recommended a biometric watermarking technique to embed online handwritten signature in host image. The Equal Error Rate (EER) of Namboodiri's technique was reported to be approximately 7%. In spite of that, Namboodiri's technique is highly fragile towards further image manipulations, e.g. JPEG compression, image filtering, etc. Hassanien [7] introduced a Discrete Wavelet Transform (DWT)-based biometric watermarking technique to embed iris code in host image. However, the credibility of Hassanien's technique was only evaluated against JPEG compression.

The conventional Least Significant Bit (LSB)-based and DWT-based techniques are recently fused into a unison framework, which is abbreviated as LSB-DWT. Chen and Wang [3] applied integer-to-integer DWT to transform host image into equivalent DWT domain of integer coefficients, and watermark was embedded in 5-bit of DWT coefficients based on exclusive-OR operation. The shortcoming of integer-to-integer DWT is that it involves round-off function which might eliminate the watermark. The proposed LSB-DWT technique exploits conventional DWT to decompose host image into DWT domain, and offline handwritten signature, i.e. still signature image, is then repeatedly embedded in 5-bit of DWT coefficients for copyright protection. The most significant advantage of handwritten signature over other biometric attributes is that it has traditionally been used for authenticating official documents and thus it is socially accepted.

2 Biometric Watermarking

Fig. 1 shows the generic diagram of the proposed LSB-DWT technique. In general, it encompasses of four main modules:

i. Pre-processing, Feature Extraction, and Discretisation - To use signature image as watermark, it is first transformed into binary bitstring, which is referred to as signature code in this paper. Since the signature image might contain scratches, speckles, smears, or other unwanted artefacts that can thwart feature extraction, it is smoothed using median filtering. The signature image is then converted to binary image of 300 x 200 pixels to isolate the signature from the background. Subsequently, the signature image is projected into feature space via Discrete Radon Transform (DRT). As DRT produces feature space of high dimension, Principle Component Analysis (PCA) is applied to compress the DRT feature space while retaining the major characteristics. The PCA feature space is lastly discretised into the signature code of N bits, where N is set to 10, 50, and 100, based on Kevenaar et al. [8].
ii. Watermark Embedding - The signature code is embedded in the host image to establish the host authenticity (refer Section 3.3). Typically, one or more secret keys are required to safeguard the signature code. This is due to the reason that the watermarked host image can be exposed to a wide range of security attacks, e.g. JPEG compression, image filtering, rotation, scaling, etc, through insecure distribution channels.
iii. Watermark Extraction - The signature code is extracted from the watermarked host image based on the secret keys used in the watermark embedding module (refer Section 3.4).
iv. Matching - The extracted signature code is compared to the original template stored in the database to validate the host authenticity. The extracted signature code is accepted as genuine, or rejected as fraudulent based on an empirical threshold.

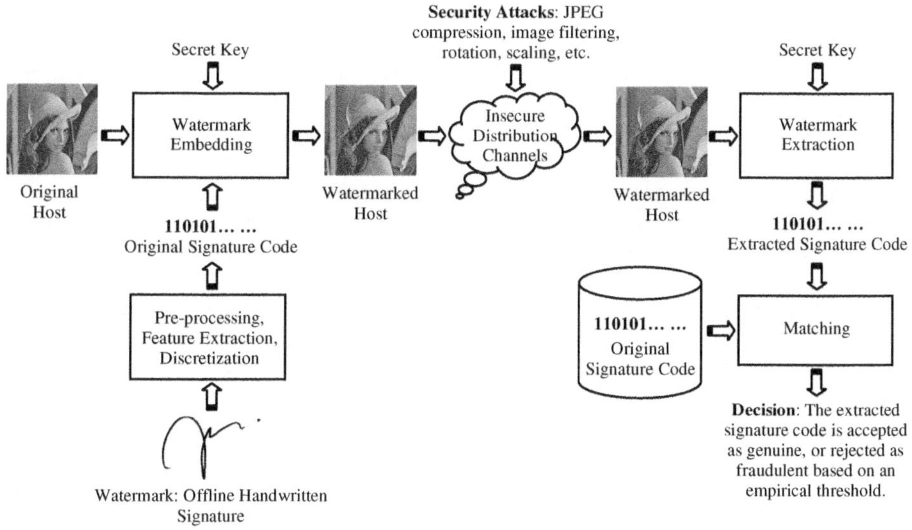

Fig. 1. Generic diagram of the proposed LSB-DWT technique

In accordance with Hartung and Martin [2], a fundamental biometric watermarking technique should at least possess the following requirements:

i. Watermark Imperceptibility - The watermark should be transparent from visual inspection, where both of the original and the watermarked host images should be indistinguishable in terms of host fidelity.
ii. Watermark Robustness - The watermark should be resistant to security attacks, where the extracted watermark should contain zero, or negligible distortion.
iii. Watermark Capacity - The watermark should convey as much information as possible.
iv. Watermark Security - The watermark should only be accessible by authorized parties. This can be achieved using cryptographic keys.

3 Proposed LSB-DWT Technique

3.1 Least Significant Bit

The least significant bit (LSB) denotes the right-most bit of a given binary number, which conveys the least important information; on the other hand, the most significant bit (MSB) is the left-most bit of the binary number, which conveys the most important information. Fig. 2 depicts the 8-bit binary representation of 149, where the MSB and the LSB correspond to the values of 128 and 1, respectively.

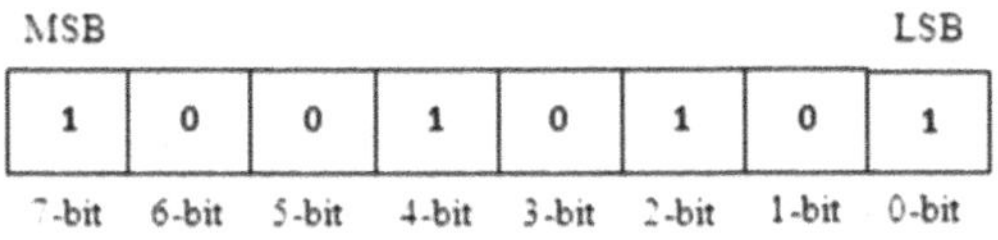

Fig. 2. 8-bit binary representation of decimal 149

The conventional LSB-based watermarking technique embeds the watermark in the LSB of the host pixels. As the LSB is the most insignificant as compared to other bits, the watermark can be embedded in the host image without corrupting the host fidelity. In spite of that, LSB-based technique is highly fragile against security attacks. This is because the watermark can be distorted or damaged if the watermarked host image is exposed to the common image manipulations, e.g. JPEG compression, image filtering, etc. Instead of the LSB, the watermark can be embedded in the l-bit, except the MSBs (the bits closest to, and including the MSB), to improve watermark robustness. This is due to the fact that the MSBs are the most significant and thus there are very sensitive to modification.

3.2 Discrete Wavelet Transform

Discrete Wavelet Transform (DWT) transforms the host image into four frequency sub-bands of equal sizes at each decomposition level n, namely an approximation sub-band (LL_n) and three detailed sub-bands (HL_n, LH_n and HH_n). The approximation sub-band refers to the low frequency sub-band, which can be down-sampled into multiple levels to obtain the next coarser domains. In contrast, the detailed sub-bands represent

the middle and the high frequency sub-bands, which are the finest and thus additional down-sampling is prohibited.

As the approximation sub-band consists of the salient attributes of the host image, e.g. smooth variation region, the watermark can be embedded in this sub-band to gain extra watermark robustness. However, a trade-off exists where the host fidelity can be degraded, or corrupted at the worst. The detailed sub-bands contain the image details, e.g. edges and textured regions, which are less significant and therefore less sensitive to the host fidelity. Rather than HH_n, the watermark is typically embedded in HL_n and LH_n as HH_n is susceptible to security attacks. Depending on the intended applications, the watermark can still be embedded in HH_n.

3.3 Proposed LSB-DWT Watermark Embedding Technique

The proposed LSB-DWT technique is an adaptive fusion of the conventional LSB-based and DWT-based technique. In general, the watermark, i.e. signature code in this paper, is embedded in the selected DWT coefficients of the host image using the LSB substitution technique. Let the signature code S is of N bits as follows:

$$S = [s_0, s_1, s_2, \dots, s_{N-1}],\ s_i \in \{0,1\}.$$

where N is set to 10, 50 and 100 bits; the host image is decomposed into equivalent 3-level DWT domain and a pseudorandom number generator is initialised using a secret key k to select N random coefficients C from HL_3 and LH_3 sub-bands with respect to an empirical threshold t as follows:

$$C = [c_0, c_1, c_2, \dots, c_{N-1}],\ C > t,\ C \in \{HL_3, LH_3\}.$$

The proposed LSB-DWT technique embeds s_i in the integer part of c_i. Assume c_i is the selected coefficient of value 149.1258; the integer part, i.e. 149, is converted to 8-bit binary representation, and the fractional part, i.e. 0.1258, is temporary stored (see Fig. 3(a)). Assume also s_i is a bit '1' (see Fig. 3(b)); the bit '1' is inserted in the 5-bit of the binary number to construct the watermarked binary number. The watermarked binary number is then reconverted back to the decimal representation, and the original fractional part, i.e. 0.1258, is reinstated to obtain the watermarked coefficient of value 181.1258. This process is repeated until the signature code is completely embedded in the host image. In the experiments, s_i is repeatedly embedded in the host image for 10 times to improve watermark robustness. At last, 3-level inverse DWT is performed to generate the watermarked host image.

3.4 Proposed LSB-DWT Watermark Extraction Technique

To extract the signature code, the watermarked host is decomposed into its 3-level DWT domain. After that, the watermarked coefficients C' are identified from HL_3' and LH_3' sub-bands by reusing the secret key k as follows:

$$C' = [c_0', c_1', c_2', \dots, c_{N-1}'],\ C' \in \{HL_3', LH_3'\}.$$

Let c_i' be the watermarked coefficient of value 181.1258 (see Fig. 4); the integer part, i.e. 181, is converted to 8-bit binary representation, and the 5-bit is assumed to be the extracted signature code s_i'. As the signature code is embedded in the host image for multiple times, s_i' is determined by finding the most frequent occurrence bit (mode).

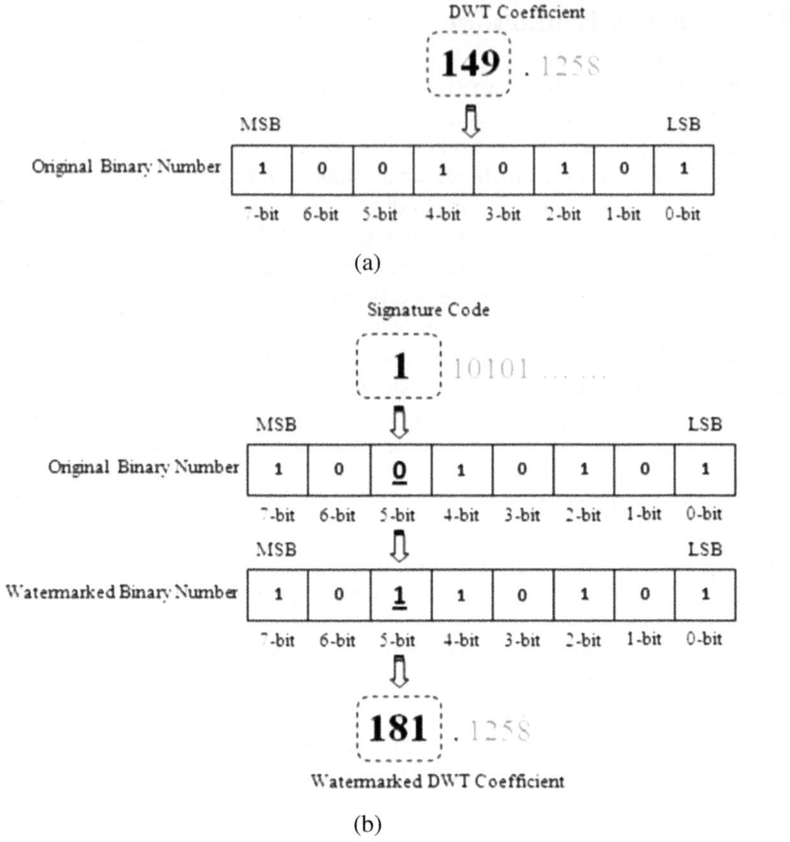

Fig. 3. Generic diagram of the proposed LSB-DWT watermark embedding technique

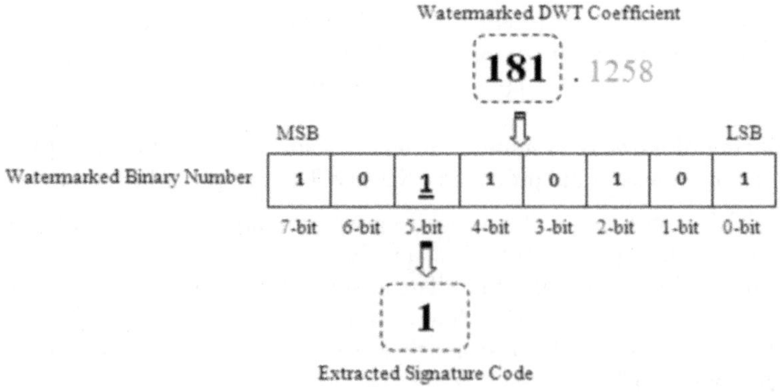

Fig. 4. Generic diagram of the proposed LSB-DWT watermark extraction technique

4 Performance Evaluation

A signature database, which consists of 500 offline handwritten signatures obtained from 50 signatories, has been constructed for performance evaluation. Unlike other biometric attributes such as fingerprint, iris, hand geometry, face, etc, the handwritten signature is a behavioural attribute, which has certain degree of intra-class variation due to age, illness, emotional state and geographical location of the signatory. This implies that there is no handwritten signature with the exact signing style although it can be provided by the same signatory. Hence, signature acquisition was completed within two contact sessions. During the first contact, each signatory was requested to provide 5 handwritten signatures. One week after the first contact, each signatory was again requested to provide another 5 handwritten signatures. All acquired handwritten signatures were then scanned and transformed into signature code of 10, 50, and 100 bits. Throughout the experiments, a gray-level Lena image was adopted as the host image. It was formatted in bitmap with the size of 512 x 512 pixels.

4.1 Performance Criteria

The performance of the proposed LSB-DWT technique was investigated in terms of watermark imperceptibility and watermark robustness.

Watermark Imperceptibility. Due to the fact that human eyes can tolerate different degree of distortion, a quantitative index, i.e. Peak Signal to Noise Ratio (PSNR), is adopted to estimate the dissimilarity between the original and the watermarked host images. Let I_{ori} and I_w denote the original and the watermarked host images of m x n pixels, respectively; PSNR can be defined via Mean Square Error (MSE) in the unit of logarithmic decibel (dB) as follows:

$$PSNR = 10\log_{10}\left(\frac{255^2}{MSE}\right), \text{ where } MSE = \frac{1}{mn}\sum_{i=0}^{m-1}\sum_{j=0}^{n-1}\left[I_{ori}(i,j) - I_w(i,j)\right]^2 .$$

PNSR only provides a rough estimation of the watermarked host image, irrespective of its host fidelity. A higher PSNR indicates that the watermarked host image is closer to the original one from the perspective of host contents. In accordance with Chen et al. [1], the host fidelity is acceptable for any PSNR greater than 30 dB.

Watermark Robustness. A number of security attacks were simulated to validate the watermark robustness of the proposed LSB-DWT technique, which include:

i. JPEG Compression - The host image was compressed using JPEG compression of JPEG quality ratio 80%, 50% and 20%.
ii. Gaussian Low-pass Filtering - The host image was filtered using Gaussian low-pass filter of 3 x 3 neighbourhood and standard deviation of 0.5.
iii. Median Filtering - The host image was distorted using median filtering of 3 x 3 neighbourhood.
iv. Gaussian Noise - The host image was added with Gaussian noise of zero-mean and variance of 0.01.
v. Scaling - The host was shrunk to 128 x 128 pixels using the nearest neighbour interpolation method, and is then rescaled back to its original size.

vi. Rotation - The host image was rotated by 10^0 in counter-clockwise direction using the nearest neighbour interpolation method, and is then rotated by 10^0 in clockwise direction to restore the host image back to its initial orientation.
vii. Cropping - A pre-determined area was cropped from the host image to simulate data reduction attack.

The distortion rate of the extracted signature code was measured after the host image was subjected to the simulated security attacks using normalized Hamming distance:

$$\varepsilon(S,S')=\frac{\sum S\oplus S'}{N}.$$

where *S* and *S'* denote the original and the extracted signature code, respectively, and *N* refers to the signature code length of 10, 50, or 100 bits. In the experiments, the extracted signature code is assumed to be damaged if the distortion rate exceeds 0.50.

5 Experimental Results

Table 1 summarizes the experimental results of the proposed LSB-DWT technique. It reveals that PSNR is inverse-proportional to the signature code length *N*. However, the proposed LSB-DWT technique is demonstrated to be capable of safeguarding the host fidelity from severe degradation as PSNR for N = 10; 50; 100 are at least 45dB. Furthermore, Fig. 5 illustrates that the watermarked host image of the proposed LSB-DWT technique for N = 100 is still close the original one. Table 1 also reveals that the proposed LSB-DWT technique is extremely robust against JPEG 80%, JPEG 50%, Gaussian low-pass filtering, median filtering, rotation and cropping, as the signature code can be fully extracted. Meanwhile, JPEG 20%, Gaussian noise, and scaling are proven to be able to deform the signature code insignificantly.

The proposed LSB-DWT technique is compared with the conventional LSB-based technique, Hassanien's technique in [7], and Chen's techniques in [3]. In general, Hassanien technique is a DWT-based method, which embeds the iris code in the host image based on the proposed quantization rules. On the other hand, Chen's technique is also a fusion of the conventional LSB-based and DWT-based techniques, which

Table 1. Experimental results of the proposed LSB-DWT technique

The Proposed LSB-DWT Technique			
Signature Code Length (*N* bits)	**10**	**50**	**100**
PSNR (dB)	55.8463	49.0162	45.9280
Simulated Security Attacks		**Distortion Rate**	
No Attack	0.0000	0.0000	0.0000
JPEG 80%	0.0000	0.0000	0.0000
JPEG 50%	0.0000	0.0000	0.0000
JPEG 20%	0.0000	**0.0200**	**0.0100**
Gaussian Low-Pass Filtering	0.0000	0.0000	0.0000
Median Filtering	0.0000	0.0000	0.0000
Gaussian Noise	0.0000	0.0000	**0.0200**
Scaling	0.0000	0.0000	**0.0300**
Rotation	0.0000	0.0000	0.0000
Cropping	0.0000	0.0000	0.0000

Fig. 5. Watermarked host image of the proposed LSB-DWT technique for N = 100

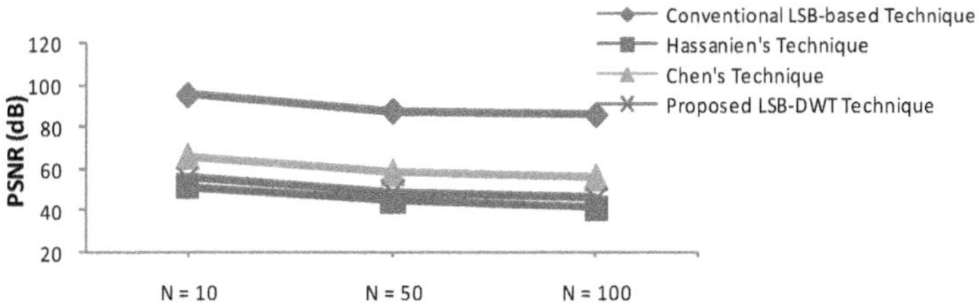

Fig. 6. PSNR comparison between the conventional LSB-based, Hassanien's, Chen's, and the proposed LSB-DWT techniques

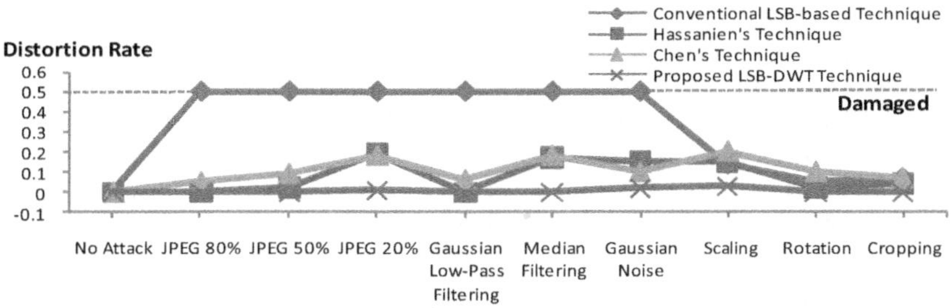

Fig. 7. Distortion comparison between the conventional LSB-based, Hassanien's, Chen's, and the proposed LSB-DWT techniques for $N = 100$

exploits integer-to-integer DWT to transform the host image into DWT domain. The watermark is then embedded in the 5-bit of the selected DWT coefficients based on exclusive-OR operation.

Fig. 6 depicts the PSNR distributions of the conventional LSB-based, Hassanien's, Chen's, and the proposed LSB-DWT-based techniques. It is clearly demonstrated that the conventional LSB-based technique is outstanding in safeguarding the host fidelity.

According to Fig. 7, the conventional LSB technique is revealed to be fragile against JPEG compression, Gaussian low-pass filtering, median filtering, and Gaussian noise. This is because the LSB consists of the least significant components which

are highly vulnerable to frequency-based operations. Despite Hassanien's technique outperforms the conventional LSB technique, it still lacks the desired robustness to protect the signature code from distortion. In addition, Fig. 7 also reveals that the proposed LSB-DWT technique is superior to Chen's technique. This is due to the reason that Chen's technique uses integer-to-integer DWT, which involves round-off function. Thus, the signature code can be unintentionally eliminated. Apart from that, the proposed LSB-DWT technique uses redundant watermarking scheme, which is capable of improving the watermark robustness.

6 Conclusion

This paper proposes a novel biometric watermarking technique, which is referred to as LSB-DWT technique, to embed offline handwritten signature in copyrighted host image as the claim of rightful ownership. A variety of security attacks are simulated to investigate the watermark robustness of the proposed LSB-DWT technique, which include JPEG compression, Gaussian low-pass filtering, median filtering, Gaussian noise, scaling, rotation, and cropping. Experimental results show that the proposed LSB-DWT technique is proficient in protecting the host fidelity, and is sufficiently robust against the simulated security attacks.

Acknowledgements

This work was supported by the Korea Science and Engineering Foundation (KOSEF) through the Biometrics Engineering Research Center (BERC) at Yonsei University (Grant Number: R112002105080020(2008)).

References

1. Chen, T.S., Chang, C.C., Hwang, M.S.: A Virtual Image Cryptosystem based upon Vector Quantization. IEEE Transactions on Image Processing 7, 1485–1488 (1998)
2. Hartung, F., Kutter, M.: Multimedia Watermarking Technique. IEEE Transactions - Invited Paper 87(7), 1079–1107 (1999)
3. Chen, T., Wang, J.C.: Image Watermarking Method using Integer-to-Integer Wavelet Transform. Tsinghua Science and Technology 7(5), 508–512 (2002)
4. Jain, A.K., Uludag, U., Hsu, R.L.: Hiding a Face into a Fingerprint Image. In: IEEE International Conference on Pattern Recognition (2002)
5. Jain, A.K., Uludag, U.: Hiding Biometric Data. IEEE Transactions on Pattern Analysis and Machine Intelligence 25(11), 1494–1498 (2003)
6. Namboodiri, A.M., Jain, A.K.: Multimedia Document Authentication using On-line Signatures as Watermarks. In: Proceedings of SPIE on Security, Steganography and Watermarking of Multimedia Content VI, vol. 5306, pp. 653–662 (2004)
7. Hassanien, A.E.: Hiding Iris Data for Authentication of Digital Images using Wavelet Theory. Pattern Recognition and Image Analysis 16(4), 637–643 (2005)
8. Kevenaar, T.A.M., Schrijen, G.J., Van der Veen, M., Akkermans, A.H.M.: Face Recognition with Renewable and Privacy Preserving Binary Templates. In: IEEE Workshop on Automatic Identification Advanced Technologies, pp. 21–26 (2005)

Audio-Visual Identity Verification and Robustness to Imposture

Walid Karam[1,2], Chafic Mokbel[1], Hanna Greige[1], and Gérard Chollet[2]

[1] University of Balamand, Deir El-Balamand, Al-Kurah, Lebanon
{Walid.Karam,Chafic.Mokbel,Hanna.Greige}@balamand.edu.lb
[2] CNRS-LTCI, TELECOM ParisTech, 46 rue Barrault, 75634 Paris, France
{karam,chollet}@telecom-paristech.fr

Abstract. The robustness of talking-face identity verification (IV) systems is best evaluated by monitoring their behavior under impostor attacks. We propose a scenario where the impostor uses a still face picture and a sample of speech of the genuine client to transform his/her speech and visual appearance into that of the target client. We propose *MixTrans*, an original text-independent technique for voice transformation in the cepstral domain, which allows a transformed audio signal to be estimated and reconstructed in the temporal domain. We also propose a face transformation technique that allows a frontal face image of a client to be animated, using principal warps to deform defined MPEG-4 facial feature points based on determined facial animation parameters. The robustness of the talking-face IV system is evaluated under these attacks. Results on the BANCA talking-face database clearly show that such attacks represent a serious challenge and a security threat to IV systems.

Keywords: Identity verification, audio-visual forgery, talking-face imposture, voice conversion, face animation, biometric verification robustness.

1 Introduction

Biometric identity verification (IV) systems are starting to appear on the market in various commercial applications. However, these systems are still operating with a certain measurable error rate that prevents them from being used in a full automatic mode, and still require human intervention and further authentication. This is primarily due to the variability of the biometric traits of humans over time because of growth, aging, injury, appearance, physical state, etc. Impostors attempting to be authenticated by an IV system to gain access to privileged resources could take advantage of the non-zero false acceptance rate of the system by imitating, as closely as possible, the biometric features of a genuine client.

The purpose of this paper is threefold. 1– It evaluates the performance of IV systems by monitoring their behavior under impostor attacks. These attacks include the transformation of the face and the voice biometric modalities. 2– It introduces *MixTrans*, a novel mixture-structure bias voice transformation technique in the cepstral domain, which allows a transformed audio signal to be estimated and reconstructed in

M. Tistarelli and M.S. Nixon (Eds.): ICB 2009, LNCS 5558, pp. 796–805, 2009.
© Springer-Verlag Berlin Heidelberg 2009

the temporal domain. 3– It proposes a face transformation technique that allows a 2D face image of the client to be animated. This technique employs principal warps to deform defined MPEG-4 facial feature points based on determined facial animation parameters (FAP). The talking-face BANCA database is used to test the effects of voice and face transformation on the IV system.

The rest of the paper is organized as follows. Section 2 describes the imposture techniques used, including *MixTrans* and the MPEG-4 face animation with thin-plate spline warping technique. Section 3 discusses the experimental results on the BANCA talking-face database. Section 4 wraps up with a conclusion.

2 Audiovisual Imposture

Audiovisual imposture is the deliberate modification of both speech and face of a person so as to make that person sound and look like someone else. The goal of such an effort is to analyze the robustness of biometric identity verification systems to deliberate forgery attacks. An attempt is made to increase the acceptance rate of an impostor. Talking-face impostor has been reported in [12] with a single modality conversion, i.e. face animation, using a commercial animation software, CrazyTalk [1]. Other types of deliberate impostor evaluations are reported in [6] and [4], all based on face conversion. In this work, techniques for conversion of both the speech and the visual appearances of clients are developed and are treated below.

2.1 Speaker Transformation

Speaker transformation, also referred to as voice transformation, is the process of altering an utterance from a speaker (impostor) to make it sound as if it were articulated by a target speaker (client.) Such transformation can be effectively used to replace the client's voice in a video to impersonate that client and break an IV system.

MixTrans. A linear time-invariant transformation in the temporal domain is equivalent to a bias in the cepstral domain. However, speaker transformation may not be seen as a simple linear time-invariant transformation. It is more accurate to consider it as several linear time-invariant filters, each operating in a part of the acoustic space. This leads to the following form of the transformation:

$$\mathcal{T}_\theta(\mathbf{X}) = \sum_k \Pi_k(\mathbf{X} + \mathbf{b}_k) = \sum_k \Pi_k \mathbf{X} + \sum_k \Pi_k \mathbf{b}_k = \mathbf{X} + \sum_k \Pi_k \mathbf{b}_k \,. \tag{1}$$

where $\mathbf{b}_k$ represents the kth bias and Π_k is the probability of being in the kth part of the acoustic space given the observation vector $\mathbf{X}$. Π_k is calculated using a universal GMM modeling the acoustic space.

Once the transformation is defined, its parameters have to be estimated. Assume that speech samples are available for both the source and the target speakers, but do not correspond to the same text. Let λ be the stochastic model for a target client. λ is a GMM of the client. Let $\mathbf{X}$ represent the sequence of observation vectors for an impostor (a source client.) Our aim is to define a transformation $\mathcal{T}_\theta(\mathbf{X})$ that makes the source client vector resemble the target client. In other words, we would like to have

the source vectors be best represented by the target client model λ through the application of the transformation $\mathcal{T}_\theta(\mathbf{X})$. In this context the Maximum likelihood criterion is used to estimate the transformation parameters.

$$\hat{\theta} = \arg\max_{\theta} \mathcal{L}(\mathcal{T}_\theta(\mathbf{X})|\lambda)\,. \tag{2}$$

Since λ is a GMM, and $\mathcal{T}_\theta(\mathbf{X})$ is a transform of the source vectors $\mathbf{X}$, and $\mathcal{T}_\theta(\mathbf{X})$ depends on another model λ_w, then $\mathcal{L}(\mathcal{T}_\theta(\mathbf{X})|\lambda)$ in (2) can be written

$$\begin{aligned}\mathcal{L}(\mathcal{T}_\theta(\mathbf{X}|\lambda) &= \prod_{t=1}^{T} \mathcal{L}(\mathcal{T}_\theta(\mathbf{X}_t)|\lambda)\\ &= \prod_{t=1}^{T}\sum_{m=1}^{M} \frac{1}{(2\pi)^{\frac{D}{2}}\left|\sum_m\right|^{\frac{1}{2}}} e^{-\frac{1}{2}(\mathcal{T}_\theta(\mathbf{X}_t)-\mu_m)^T \sum_m^{-1}(\mathcal{T}_\theta(\mathbf{X}_t)-\mu_m)}\\ &= \prod_{t=1}^{T}\sum_{m=1}^{M} \frac{1}{(2\pi)^{\frac{D}{2}}\left|\sum_m\right|^{\frac{1}{2}}} e^{-\frac{1}{2}\left(\mathbf{x}_t+\sum_{k=1}^{K}\Pi_{kt}b_k-\mu_m\right)^T \sum_m^{-1}\left(\mathbf{x}_t+\sum_{k=1}^{K}\Pi_{kt}b_k-\mu_m\right)}\,.\end{aligned} \tag{3}$$

Finding $\{b_k\}$ such that (3) is maximized is found through the use of the EM algorithm. In the expectation "E" step, the probability α_{mt} of component m is calculated. Then, at the maximization "M" step, the log-likelihood is optimized dimension by dimension for a GMM with a diagonal covariance matrix:

$$ll = \sum_{t=1}^{T}\sum_{m=1}^{M}\alpha_{mt}\left[\log\frac{1}{\sigma_m\sqrt{2\pi}} - \frac{1}{2}\frac{\left(\mathbf{X}_t+\sum_{k=1}^{K}\Pi_{kt}\mathbf{b}_k-\mu_m\right)^2}{\sigma_m^2}\right].$$

Maximizing,

$$\frac{\partial ll}{\partial b_l} = 0 \Rightarrow -\sum_{t=1}^{T}\sum_{m=1}^{M}\alpha_{mt}\frac{\left(\mathbf{X}_t+\sum_{k=1}^{K}\Pi_{kt}\mathbf{b}_k-\mu_m\right)\Pi_{lt}}{\sigma_m^2} = 0, \qquad \text{for } l = 1\ldots K$$

then, in matrix notation,

$$-\left(\sum_m\sum_t\frac{\alpha_{mt}\Pi_{lt}\Pi_{kt}}{\sigma_m^2}\right)\left(\mathbf{b}_k\right) = \left(\sum_m\sum_t\frac{\alpha_{mt}\Pi_{lt}(\mathbf{X}_t-\mu_m)}{\sigma_m^2}\right). \tag{4}$$

This matrix equation is solved at every iteration of the EM algorithm.

Speech Signal Reconstruction. It is known that the cepstral domain is appropriate for classification due to the physical significance of the Euclidean distance in this space. However, the extraction of cepstral coefficients from the temporal signal is a nonlinear process and the inversion of this process is not uniquely defined. Therefore, a solution has to be found in order to take advantage of the good characteristics of the cepstral space while applying the transformation in the temporal domain. Several techniques have been proposed to solve this problem. In [10], harmonic plus noise analysis has been used for this purpose. Instead of trying to find a transformation allowing the passage from the cepstral domain to the temporal domain, we adopt a different strategy. Suppose that an intermediate space existed where transformation could be directly transposed into the temporal domain. Fig. 1 shows the process where

the temporal signal goes through a two-step feature extraction process leading to the cepstral coefficients that may be easily transformed into target speaker-like cepstral coefficients by applying the transformation function $\mathcal{T}_\theta(\mathbf{X})$ as discussed previously.

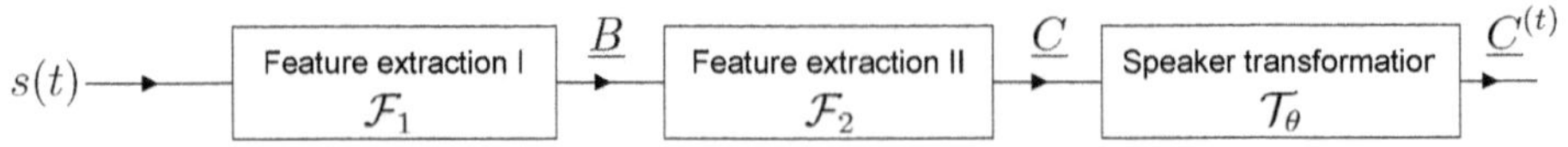

Fig. 1. Signal to transformed cepstral coefficients

The transformation that is trained in the cepstral domain cannot be directly projected on the temporal domain since the feature extraction module $(\mathcal{F}_1 \circ \mathcal{F}_2)$ is highly nonlinear. However, a speaker transformation determined in the $\underline{B}$ space may be projected on the signal space, e.g. $\underline{B}$ space may be the spectral domain. But, for physical significance it is better to train the transformation in the cepstral domain. Therefore, we suppose that another transformation $\mathcal{T}'_\theta(\mathbf{X})$ existed in the $\underline{B}$ space leading to the same transformation in the cepstral domain satisfying thereby the two objectives; transformation of the signal and distance measurement in the cepstral domain. This is shown in Fig. 2.

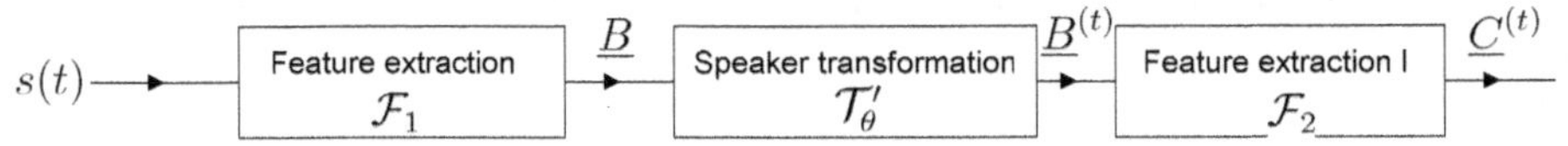

Fig. 2. Transformation in a signal-equivalent space

The remaining issue is estimating parameters θ of the transformation $\mathcal{T}'_\theta(\mathbf{X})$ in order to get the same transformation result as in the cepstral domain. This is detailed next.

Estimating Signal Transformation is Equivalent to Calculating Cepstral Transformation. The transformation in the cepstral domain is presumably determined; the idea is to establish a transformation in the $\underline{B}$ space leading to cepstral coefficients similar to the one resulting from the cepstral transformation.

Let $\underline{\hat{C}}^{(t)}$ represents the cepstral vector obtained after applying the transformation in the $\underline{B}$ domain; let $\underline{C}^{(t)}$ represents the cepstral vector obtained when the transformation is applied in the cepstral domain. The difference defines an error vector: $\underline{e} = \underline{C}^{(t)} - \underline{\hat{C}}^{(t)}$ and the quadratic error $E = |e|^2 = \underline{e}^T \underline{e}$. Starting from a set of parameters for $\mathcal{T}'_\theta$, the gradient algorithm may be applied in order to minimize the quadratic error E. At each iteration of the algorithm the parameter θ is updated using the equation $\theta^{(i+1)} = \theta^{(i)} - \mu \frac{\partial E}{\partial \theta}$, where μ is the gradient step.

The gradient of the error with respect to parameter θ is given by:

$$\frac{\partial E}{\partial \theta} = -2\underline{e}^T \frac{\partial \underline{\hat{C}}^{(t)}}{\partial \theta}. \tag{5}$$

Finally, the derivative of the estimated transformed cepstral coefficient with respect to θ, can be obtained using a gradient descent:

$$\frac{\partial \hat{\underline{C}}^{(t)}}{\partial \theta} = \frac{\partial \hat{\underline{C}}^{(t)T}}{\partial \underline{B}^{(t)}} \frac{\partial \underline{B}^{(t)}}{\partial \theta} \,. \tag{6}$$

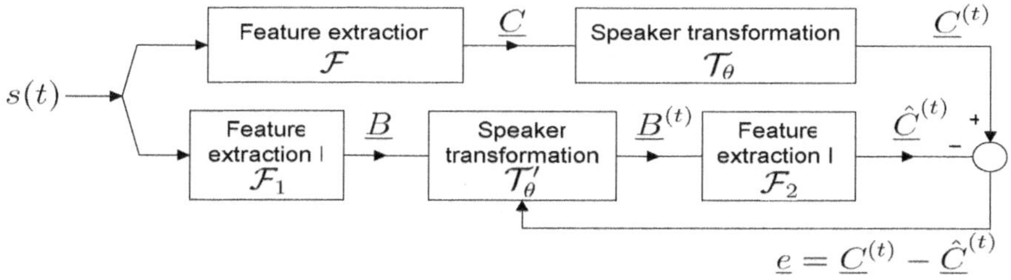

Fig. 3. Signal-level transformation parameters tuned with a gradient descent algorithm

In order to illustrate this principle, let us consider the case of MFCC analysis leading to the cepstral coefficients. In this case, $\mathcal{F}_1$ is just the Fast Fourier Transform (FFT) followed by the power spectral calculation (the phase being kept constant). $\mathcal{F}_2$ is the filterbank integration in the logarithm scale followed by the inverse DCT transform. We can write:

$$\hat{C}_l^{(t)} = \sum_{k=1}^{K} log\left(\sum_{i=1}^{N} a_i^{(k)} B_i^{(t)}\right) cos\left(2\pi l \frac{f_k}{F}\right) . \tag{7}$$

$$B_i^{(t)} = B_i.\theta_i \,. \tag{8}$$

where $\{a_i\}$ are the filter-bank coefficients, f_k the central frequencies of the filter-bank, and θ_i is the transfer function at frequency bin *i* of the transformation $\mathcal{T}'_\theta(\mathbf{X})$.

Using (7) and (8), it is straightforward to compute the derivatives in (6):

$$\frac{\partial \hat{C}_i^{(t)}}{\partial B_j^{(t)}} = \sum_{k=1}^{K} \frac{a_j^{(k)}}{\sum_{i=1}^{N} a_i^{(k)} B_i^{(t)}} cos\left(2\pi l \frac{f_k}{F}\right) . \tag{9}$$

$$\frac{\partial B_i^{(t)}}{\partial \theta_j} = B_j \delta_{ij} \,. \tag{10}$$

(5), (6), (9), and (10) allow the implementation of this algorithm in the case of MFCC.

Once $\mathcal{T}'_\theta(\mathbf{X})$ completely defined, the transformed signal may be determined by applying an inverse FFT to $B(t)$ and using the original phase to recompose the signal window. In order to consider the overlapping between adjacent windows, the Overlap and Add (OLA) algorithm is used.

Initializing the Gradient Algorithm. The previous approach is computationally expensive. Actually, for each signal window, i.e. 10ms to 16ms, a gradient algorithm is to be applied. In order to alleviate this high computational algorithm, a solution consists in finding a good initialization of the gradient algorithm. This may be obtained by using as initial value for the transformation $\mathcal{T}'_\theta(\mathbf{X})$, the transformation obtained for the previous signal window.

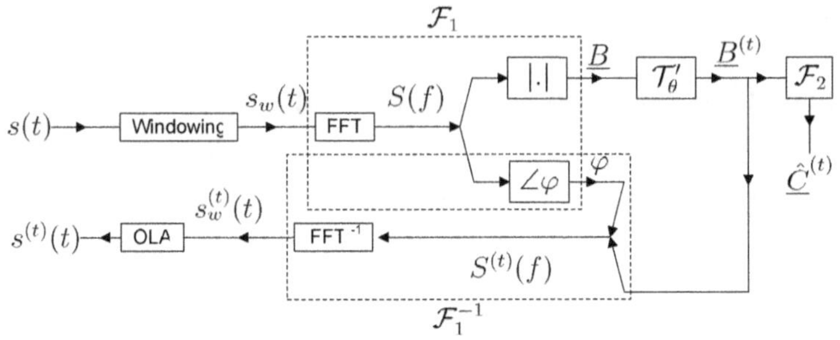

Fig. 4. Speech signal feature extraction, transformation, and reconstruction

2.2 Face Animation

To complete the scenario of audiovisual imposture, speaker transformation is coupled with face transformation. It is meant to produce synthetically an "animated" face of a target person, given a still photo of his face and some animation parameters.

The face animation technique used in this paper is MPEG-4 compliant, which uses a very simple thin-plane spline warping function defined by a set of reference points on the target image, driven by a set of corresponding points on the source image face.

MPEG-4 2-D Face Animation. MPEG-4 is an object-based multimedia compression standard, which defines a standard for face animation [**11**]. It specifies 84 feature points that are used as references for Facial Animation Parameters (FAPs). 68 FAPs allow the representation of facial expressions and actions such as head motion and mouth and eye movements. Two FAP groups are defined, visemes (FAP group 1) and expressions (FAP group 2). Visemes (FAP1) are visually associated with phonemes of speech; expressions (FAP2) are joy, sadness, anger, fear, disgust, and surprise.

An MPEG-4 compliant system decodes a FAP stream and animates a face model that has all feature points properly determined. In this paper, the animation of the feature points is accomplished using a simple thin-plate spline warping technique.

Thin-Plate Spline Warping. The thin-plate spline (TPS), initially introduced by Duchon [5], is a geometric mathematical formulation that can be applied to the problem of 2D coordinate transformation. The name *thin-plate spline* indicates a physical analogy to bending a thin sheet of metal in the vertical z direction, thus displacing x and y coordinates on the horizontal plane.

Given a set of data points $\{w_i, i = 1, 2, ..., K\}$ in a 2D plane – for our case, MPEG-4 facial feature points – a radial basis function is defined as a spatial mapping that

maps a location x in space to a new location $f(x) = \sum_{i=1}^{K} c_i \phi(\|x - w_i\|)$, where $\{c_i\}$ is a set of mapping coefficients and the kernel function $\phi(r) = r^2 \ln r$ is the thin-plate spline [3]. The mapping function $f(x)$ is fit between corresponding sets of points $\{x_i\}$ and $\{y_i\}$ by minimizing the "bending energy" *I*, defined as the sum of squares of the second derivatives

$$I[f(x,y)] = \iint\limits_{\mathbb{R}^2} \left[\left(\frac{\partial^2 f}{\partial x^2}\right)^2 + 2\left(\frac{\partial^2 f}{\partial xy}\right)^2 + \left(\frac{\partial^2 f}{\partial y^2}\right)^2 \right] dx\, dy \cdot \tag{11}$$

3 Effects of Imposture on Verification – Experimental Results

To test the robustness of IV systems, a state-of-the-art baseline audio-visual IV system is built. This system follows the BANCA [8] "P" protocol and is based on a classical GMM approach for both speech and face modalities. It is completely independent from the voice and face transformations described above.

BANCA defines "random" impostures, where a speaker proclaims in his/her own voice and face to be someone else. This "zero-effort" imposture is unrealistic and IV systems detect easily the forgery by contrasting the impostor model against that of the claimed identity. To make the verification more realistic, deliberate attacks are modeled with the transformation of both the voice and the face of the impostor.

To verify a claimed identity, audio and face feature vectors are matched against the claimed speaker model and against the world model. GMM client training and testing is performed on the open-source speaker verification toolkit BECARS [2].

3.1 Verification Experiments

Speaker Verification. To process the speech signal, a feature extraction module calculates relevant feature vectors from the speech waveform. On a signal "FFT" window shifted at a regular rate, cepstral coefficients are derived from a filter bank analysis with triangular filters. A Hamming weighting window is used to compensate for the truncation of the signal. Then GMM speaker classification is performed with 256 Gaussians. The world model of BANCA is adapted using MAP adaptation, and its parameters estimated using the EM algorithm.

A total of 234 true client tests and 312 "random impostor" tests per group were performed. Fig. 7 (a) shows the DET curves for speaker verification on g1 and g2, with an EER of 4.38% and 4.22% respectively.

Face Verification. Face verification is based on processing a video sequence in four stages: 1– Face detection, localization and segmentation, 2– Normalization, 3– Facial Feature extraction and tracking, and 4– Classification.

The face detection algorithm used in this work is a machine learning approach based on a boosted cascade of simple and rotated haar-like features for visual object detection [7]. Once a face is detected, it is normalized (resized to 48×64, cropped to 36×40, gray-scaled, and histogram equalized) to reduce the variability of different aspects in the face image such as contrast and illumination, scale, translation, and rotation. The face tracking module extracts faces in all frames and retains only 5 per video for training and/or testing.

The next step is face feature extraction. We use DCT-*mod2* proposed in [9].

In a similar way to speaker verification, GMM's are used to model the distribution of face feature vectors for each person.

For the same BANCA "P" protocol, and a total of 234 true clients and 312 "random impostor" tests (per group per frame – 5 frames per video) the DET curves for face verification are shown in Fig. 7 (b) with an EER of 23.5% (g1) and 22.2% (g2).

Score Fusion. A final decision on the claimed identity of a person relies on both the speech-based and the face-based verification systems. To combine both modalities, a fusion scheme is needed. The simple weighted sum rule fusion technique is used in this study. The sum rule computes the audiovisual score s by weight averaging: $s = w_s s_s + w_f s_f$, where w_s and w_f are speech and face score weights computed so as to optimize the equal error rate on the training set. The speech and face scores must be in the same range (e.g. $\mu = 0, \sigma = 1$) for the fusion to be meaningful. This is achieved by normalizing the scores as follows: $s_{norm(s)} = (s_s - \mu_s)/\sigma_s$ and $s_{norm(f)} = (s_f - \mu_f)\sigma_f$.

Fig. 7 (c) shows an improvement of the verification by score fusion of both modalities, with an EER of 4.22% for g1, and 3.47% for g2.

3.2 Transformation Experiments

Voice Conversion Experiments. BANCA has total of 312 impostor attacks per group in which the speaker claims in his own words to be someone else. These attempts are replaced by the transformed voices as described in section 2.1 above. For each attempt, MFCC analysis is performed and transformation coefficients are calculated in the cepstral domain using the EM algorithm. Then the signal transformation parameters are estimated using a gradient descent algorithm. The transformed voice signal is then reconstructed with an inverse FFT and OLA as described in section 02.1. The pitch of the transformed voice had to be adjusted to better match the target speaker's pitch. Verification experiments are repeated with the transformed voices. The result is an increase of the EER from 4.38% to 10.6% on g1, and from 4.22% to 12.1% on g2 (Fig. 7 (a)).

Face Conversion Experiments. Given a picture of the face of a target person, the facial feature points are first annotated as shown in Fig. 5. A total of 61 feature points out of the 83 of MPEG-4 are used, the majority of which belong to the eyes and the mouth regions. Others have less impact on FAP's or do not affect them at all.

The FAP's used in the experiments correspond to a subset of 33 out of the 68 FAP's defines by MPEG-4. Facial actions related to head movement, tongue, nose, ears, and jaws are not used. The FAP's used correspond to mouth, eye, and eyebrow movements, e.g. horizontal displacement of right outer lip corner (`stretch_r_cornerlip_o`). Fig. 6 shows frames animating the noted expressions.

A synthesized video sequence is generated by deforming a face from its neutral state according to determined FAP values. For the experiments presented in this work, these FAP's are selected so as to produce a realistic talking head. The detection and the measure of the level of audiovisual speech synchrony is not treated in this work, but has been reported in [4] to improve the verification performance.

(a) Original sample video frame (b) Annotated face

Fig. 5. Feature point annotation on the BANCA database (client number 9055)

BANCA has total of 312 impostor attacks per group in which the speaker claims in his own words and facial expressions to be someone else. These are replaced by the synthetically animated videos with the transformed speech. The experiments have shown a deterioration of the performance from an EER of [23.5%, 22.2%] on [g1, g2] to [37.6%, 33.0%] (Fig. 7 (b)) for face, and from [4.22%, 3.47%] to [11.0%, 16.1%] for the audio-visual system (Fig. 7 (c)).

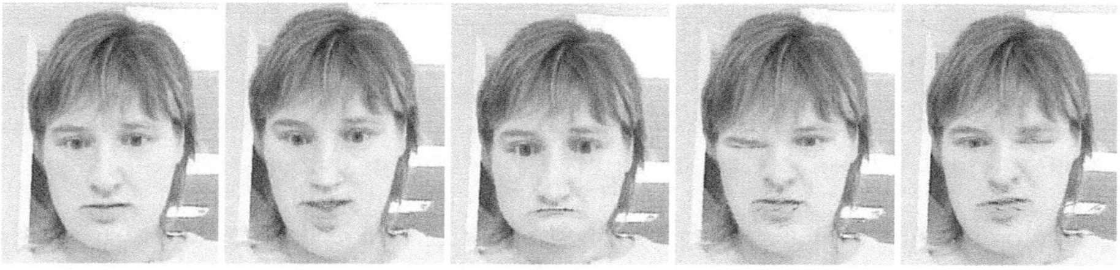

(a) neutral face (b) joy expression (c) sad expression (d) right eye wink (e) left eye wink

Fig. 6. Selected frames from an animated face with various expressions

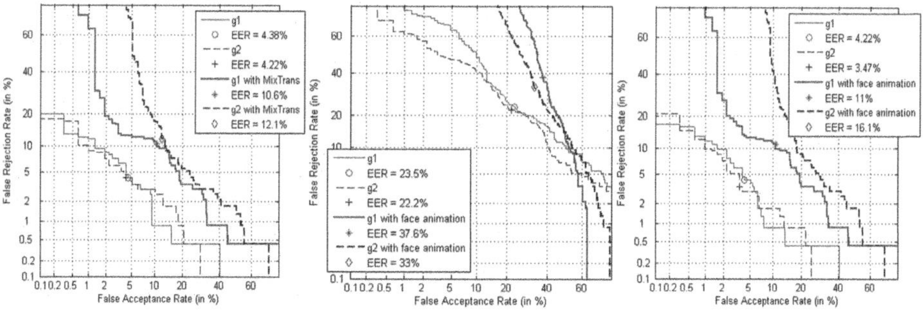

(a) Speech (b) Face (c) Audio-visual (fused results)

Fig. 7. Verification and imposture results on BANCA

4 Conclusion

This paper evaluates the robustness of audio-visual imposture on biometric identity verification systems. It proposes *MixTrans*, a mixture-structure bias voice transformation technique in the cepstral domain, which allows a transformed audio signal to be estimated and reconstructed in the temporal domain. It also couples the audio conversion with an MPEG-4 compliant face animation system that warps facial feature points using a simple thin-plate spline. The proposed audiovisual forgery is completely independent from the baseline audiovisual IV system, and can be used to attack any other audiovisual IV system. The Results drawn from the experiments show that state-of-the-art IV systems are vulnerable to forgery attacks, which indicate more impostor acceptance, and, for the same threshold, more genuine client denial.

References

1. Reallusion crazytalk animation studio software, `http://www.reallusion.com/crazytalk/`
2. Blouet, R., Mokbel, C., Mokbel, H., Soto, E.S., Chollet, G., Greige, H.: Becars: A free software for speaker verification. In: Proc. ODYSSEY 2004, pp. 145–148 (2004)
3. Bookstein, F.: Principal warps: Thin-plate splines and the decomposition of deformations. IEEE Transactions on Pattern Analysis and Machine Intelligence 11(6), 567–585 (1989)
4. Bredin, H., Chollet, G.: Making talking-face authentication robust to deliberate imposture. In: IEEE International Conference on Acoustics, Speech and Signal Processing (ICASSP 2008), pp. 1693–1696 (2008)
5. Duchon, J.: Interpolation des fonctions de deux variables suivant le principe de la flexion des plaques minces. R.A.I.R.O. Analyse numérique 10, 5–12 (1976)
6. Fauve, B., Bredin, H., Karam, W., Verdet, F., Mayoue, A., Chollet, G., Hennebert, J., Lewis, R., Mason, J., Mokbel, C., Petrovska, D.: Some results from the biosecure talking face evaluation campaign. In: IEEE International Conference on Acoustics, Speech and Signal Processing (ICASSP 2008), vol. 1, pp. 4137–4140 (2008)
7. Lienhart, R., Maydt, J.: An extended set of haar-like features for rapid object detection. In: Proceedings of the International Conference on Image Processing, vol. 1, pp. I–900–I–903(2002)
8. Popovici, V., Thiran, J., Bailly-Bailliere, E., Bengio, S., Bimbot, F., Hamouz, M., Kittler, J., Mariethoz, J., Matas, J., Messer, K., Ruiz, B., Poiree, F.: The BANCA database and evaluation protocol. In: Kittler, J., Nixon, M.S. (eds.) AVBPA 2003. LNCS, vol. 2688, pp. 625–638. Springer, Heidelberg (2003)
9. Sanderson, C., Paliwal, K.K.: Fast feature extraction method for robust face verification. IEE Electronics Letters 38(25), 1648–1650 (2002)
10. Stylianou, Y., Cappe, O., Moulines, E.: Continuous probabilistic transform for voice conversion. IEEE Transactions on Speech and Audio Processing 15(6), 131–142 (1998)
11. Tekalp, A., Ostermann, J.: Face and 2-d mesh animation in mpeg-4. Image Communication Journal 15(4-5), 387–421 (2000)
12. Verdet, F., Hennebert, J.: Impostures of talking face systems using automatic face animation. In: Proceedings of the IEEE Conference on Biometrics: Theory, Applications and Systems (BTAS 2008) (2008)

Theoretical Framework for Constructing Matching Algorithms in Biometric Authentication Systems

Manabu Inuma[1,2], Akira Otsuka[1,2], and Hideki Imai[1,2]

[1] Research Center for Information Security (RCIS)
National Institute of Advanced Industrial Science and Technology (AIST)
Akihabara-Daibiru Room 1003, 1-18-13, Sotokanda, Chiyoda-ku
Tokyo 101-0021, Japan
[2] Department of Electrical, Electronic, and Communication Engineering
Faculty of Science and Engineering, Chuo University
1-13-27 Kasuga, Bunkyo-ku Tokyo 112-8551, Japan

Abstract. In this paper, we propose a theoretical framework to construct matching algorithms for any biometric authentication systems. Conventional matching algorithms are not necessarily secure against strong intentional impersonation attacks such as wolf attacks. The wolf attack is an attempt to impersonate a genuine user by presenting a "wolf" to a biometric authentication system without the knowledge of a genuine user's biometric sample. A "wolf" is a sample which can be accepted as a match with multiple templates. The wolf attack probability (WAP) is the maximum success probability of the wolf attack, which was proposed by Une, Otsuka, Imai as a measure for evaluating security of biometric authentication systems [UOI1], [UOI2]. We present a principle for construction of secure matching algorithms against the wolf attack for any biometric authentication systems. The ideal matching algorithm determines a threshold for each input value depending on the entropy of the probability distribution of the (Hamming) distances. Then we show that if the information about the probability distribution for each input value is perfectly given, then our matching algorithm is secure against the wolf attack. Our generalized matching algorithm gives a theoretical framework to construct secure matching algorithms. How lower WAP is achievable depends on how accurately the entropy is estimated. Then there is a trade-off between the efficiency and the achievable WAP. Almost every conventional matching algorithm employs a fixed threshold and hence it can be regarded as an efficient but insecure instance of our theoretical framework. Daugman's algorithm proposed in [Da2] can also be regarded as a non-optimal instance of our framework.

1 Introduction

Biometric authentication systems automatically identify or verify individuals by physiological or behavioral characteristics. They are used in various services such as the immigration control at an airport, the banking transactions at an ATM,

M. Tistarelli and M.S. Nixon (Eds.): ICB 2009, LNCS 5558, pp. 806–815, 2009.
© Springer-Verlag Berlin Heidelberg 2009

the access control to restricted areas in a building, and so on. The increase in the need of biometric authentication systems makes it important to explicitly evaluate the security of them.

The false acceptance rate (*FAR*) (see the definition (3) in Section 2.2) is traditionally used as a security measure against the zero-effort impersonation attack. The zero-effort approach assumes that an attacker will present his/her own biometric data. But, it is clearly not a rational assumption, since an attacker attempting to impersonate a genuine user will try to present a biometric data of the genuine user or its imitation.

Ratha et al. approximately calculate the success probability of a brute-force attack in a typical fingerprint-minutiae matching algorithm [RCB1]. The brute-force approach assumes that an attacker blindly selects an input value. However, if an attacker has some information about the algorithm employed in the system, the attacker might be able to find a sample which shows high similarity to most of the templates. Such a biometric sample is called a **wolf** (cf. [ISO1]). An attacker could impersonate a genuine user with much higher probability than *FAR* by presenting a wolf to a biometric authentication system.

With regard to the artefact attack, Matsumoto et al. showed that some biometric authentication systems often falsely accepts some artefacts [MMYH1]. Therefore we should assume that an attacker may find a special input value not only from biometric samples but also from non-biometric samples. Une, Otsuka, Imai extended the definition of a wolf to include a non-biometric input value and defined the **wolf attack probability (*WAP*)** (see Definition 3) [UOI1], [UOI2]. *WAP* can be regarded as the upper bound of the success probability of attacks without the knowledge of a genuine user's biometric sample. Une, Otsuka, Imai proposed that *WAP* can be used as a security measure to evaluate the lower bound of a security level in a biometric authentication system.

Our goal is to propose a theoretical framework to construct matching algorithms for biometric authentication systems. Almost every conventional matching algorithm employs a fixed threshold determined based on *FAR* and the false rejection rate (*FRR*) (see the definition (1) in Section 2.1). It is not necessarily secure against the wolf attack. Une, Otsuka, Imai showed that in some of such matching algorithms, there actually exist strong wolves and *WAP* can be extremely higher [UOI1], [UOI2]. Surprisingly, as far as we know, no research have been conducted on security of matching algorithms until now. This is the first paper which studies the security characteristics of matching algorithms and gives a theoretical framework how to construct them securely.

Suppose a matching algorithm employs a threshold determined by the entropy of the probability distribution for each input value. We prove that if the entropy for each input value is perfectly given, then the matching algorithm is secure against the wolf attack (Theorem 1, 2).

In the real world, it might be difficult to perfectly calculate the entropy for each input value, however, a more accurate computation of the entropy can achieve a lower *WAP*. Then there is a trade-off between the efficiency of the matching algorithm and the achievable *WAP* in the matching algorithm.

Previous results can be regarded as instances of our theoretical framework. Almost every previous matching algorithm employs a fixed threshold. In our theoretical framework, it can be regarded as an efficient instance which assumes all input values have the same entropy. However, as mentioned above, it is not exactly secure against the wolf attack.

Daugman proposed a matching algorithm in which a threshold is determined for each match by taking account the number of bits available for comparison [Da2]. His method can also be regarded as an instance of our framework, which assumes every bit of a sample independently and identically contributes to the entropy of the probability distribution. *WAP* in his algorithm can be relatively lower than that in an ordinary algorithm employing a fixed threshold. However, his matching algorithm is not necessarily secure against the wolf attack (see details in Section 4), since we have to assume that an attacker knows more accurate information about the probability distributions.

This paper continues as follows. In Section 2, we will briefly introduce a typical model of biometric authentication systems and give explicit definitions of *FAR*, *WAP*, and security against the wolf attack. Our proposal can be easily adapted to all matching algorithms of all modalities that employ symmetric prametric functions such as the ordinary (Hamming) distance as the dissimilarity measure. We will construct matching algorithms in the general case (Theorem 1) and in the normal distribution case (Theorem 2) and show that these matching algorithms are secure against the wolf attack. They give a theoretical framework to construct secure matching algorithms for any biometric authentication systems. In Section 4, we will reconsider previous results in our theoretical framework.

2 Model (Preliminaries)

A biometric authentication system can be used for verification or identification of individuals. In verification, a user of the system claims to have a certain identity and the biometric system performs a one to one comparison between the offered biometric data and the template which is linked to the claimed identity. In identification, a one to all comparison is performed between the offered data and all available template stored in the database to reveal the identity of an individual. In this paper, we will discuss verification systems.

Let $\mathcal{U}$ be a set of all possible users of the biometric authentication system. Namely $\mathcal{U}$ is a set of all human individuals. For each user $u \in \mathcal{U}$, the identity of u can be denoted by u, namely the identities of users can be identified with $\mathcal{U}$. Let $\mathcal{M}$ be a finite set with a symmetric prametric function $d : \mathcal{M} \times \mathcal{M} \to \mathbb{R}$, namely $d(x,y) = d(y,x)$, $d(x,y) \geq 0$, $d(x,x) = 0$ for all $x, y \in \mathcal{M}$.

In an enrollment phase, for any user $u \in \mathcal{U}$, an acquisition device measures a biometric data of u. After processing the measurement data and extracting relevant features, the features are represented as an element t_u of $\mathcal{M}$. Then the template t_u of $u \in \mathcal{U}$ is stored in the database of the system. In a verification phase (matching phase) *match*, a user $v \in \mathcal{U}$ claims an identity $w \in \mathcal{U}$ and a biometric measurement is acquired from v. This measurement is also transformed

into an element s of $\mathcal{M}$. A matching process compares s with t_w and *match* generates a message, *accept* or *reject*, by a predetermined threshold $\tau \in \mathbb{R}_{\geq 0}$ as follows:

$$match(v,w) = \begin{cases} accept \text{ if } d(s,t_w) < \tau \\ reject \text{ if } d(s,t_w) \geq \tau \end{cases} .$$

Each user $u \in \mathcal{U}$ enrolls and offers a certain biometric sample of u in an enrollment phase and a verification phase, respectively. Therefore $\mathcal{U}$ can be regarded as a set of the biometric samples of users. For each biometric sample $u \in \mathcal{U}$, let X_u be a random variable on $\mathcal{M}$ representing noisy versions of u, namely $P(X_u = s)$ denotes the probability that biometric data of u will be transformed into $s \in \mathcal{M}$. Assume that the X_u, $u \in \mathcal{U}$, are independent.

2.1 The False Rejection Rate

The false rejection rate (FRR) is the probability that a genuine user is rejected, namely it is defined by

$$\begin{aligned} FRR &= \operatorname*{Ave}_{u\in\mathcal{U}} P(\,match(u,u) = reject\,) \\ &= \frac{1}{n}\sum_{u\in\mathcal{U}} \sum_{\substack{(s,t)\in\mathcal{M}\times\mathcal{M} \\ d(s,t)\geq\tau}} P(X_u = s)P(X_u = t) \\ &= 1 - \frac{1}{n}\sum_{u\in\mathcal{U}} \sum_{\substack{(s,t)\in\mathcal{M}\times\mathcal{M} \\ d(s,t)<\tau}} P(X_u = s)P(X_u = t) \end{aligned} \tag{1}$$

where $n = \#\mathcal{U}$. For each user $u \in \mathcal{U}$, let FRR_u denote the probability that the user u with the correct identity claim u will be rejected. Namely, FRR_u is defined by

$$\begin{aligned} FRR_u &= \sum_{\substack{(s,t)\in\mathcal{M}\times\mathcal{M} \\ d(s,t)\geq\tau}} P(X_u = s)P(X_v = t) \\ &= 1 - \sum_{\substack{(s,t)\in\mathcal{M}\times\mathcal{M} \\ d(s,t)<\tau}} P(X_u = s)P(X_v = t) \quad . \end{aligned} \tag{2}$$

It is easy to check that $FRR = \frac{1}{n}\sum_{u\in\mathcal{U}} FRR_u$.

2.2 The False Acceptance Rate

The false acceptance rate (FAR) is the probability that an offer of a user with a wrong identity claim will be incorrectly accepted, namely FAR is defined by

$$\begin{aligned} FAR &= \operatorname*{Ave}_{\substack{(u,v)\in\mathcal{U}\times\mathcal{U} \\ u\neq v}} P(\,match(u,v) = accept\,) \\ &= \frac{1}{n(n-1)} \sum_{\substack{(u,v)\in\mathcal{U}\times\mathcal{U} \\ u\neq v}} \sum_{\substack{(s,t)\in\mathcal{M}\times\mathcal{M} \\ d(s,t)<\tau}} P(X_u = s)P(X_v = t) \quad . \end{aligned} \tag{3}$$

The measure *FAR* is traditionally used to express a recognition accuracy of biometric systems. It is also used as a measure to evaluate the security of systems against the zero-effort impersonation attack.

The zero-effort approach assumes that an attacker attempting to impersonate a genuine user will present his/her own biometric data. This assumption is clearly so far from reality, since an attacker will try to present a biometric data of a genuine user or its imitation.

2.3 The Wolf Attack Probability

Une, Otsuka, Imai proposed a new security measure for biometric authentication systems [UOI1], [UOI2]. If an attacker can find an input value which matches many templates, then he succeed in impersonating a genuine user with a higher probability than *FAR* by presenting the input value to the biometric authentication system. Such an input value obtained from a biometric sample is called a **wolf** by many authors (cf. [ISO1]). However, such an input value might be obtained not only from biometric samples but also from non-biometric samples. Matsumoto et al. show by experimentation that some artefacts can be falsely accepted in some biometric authentication systems [MMYH1].

Considering these facts, we will extend the definition of a wolf as follows.

Let $\mathcal{A}$ be a set of all possible samples including non-biometric samples such as artefacts or synthetic samples. For each $w \in \mathcal{A}$, let FAR_w denote the probability that the sample w with a wrong identity claim $v \neq w$ will be incorrectly accepted and let AR_w denote the probability that the sample w with random claim will be accepted. Namely, FAR_w and AR_w are respectively defined by

$$\begin{aligned} FAR_w &= \underset{v \in \mathcal{U}\setminus\{w\}}{\mathrm{Ave}} P(match(w,v) = accept) \\ &= \frac{1}{\#\,(\mathcal{U}\setminus\{w\})} \sum_{v \in \mathcal{U}\setminus\{w\}} \sum_{\substack{(s,t) \in \mathcal{M}\times\mathcal{M} \\ d(s,t) < \tau}} P(X_w = s)P(X_v = t)\,, \end{aligned} \tag{4}$$

$$\begin{aligned} AR_w &= \underset{v \in \mathcal{U}}{\mathrm{Ave}}\, P(match(w,v) = accept) \\ &= \frac{1}{n} \sum_{v \in \mathcal{U}} \sum_{\substack{(s,t) \in \mathcal{M}\times\mathcal{M} \\ d(s,t) < \tau}} P(X_w = s)P(X_v = t) \quad . \end{aligned} \tag{5}$$

It is easy to check that $FAR = \frac{1}{n} \sum_{u \in \mathcal{U}} FAR_u$. The following theorem describes the relation between FRR_w, FAR_w and AR_w.

Lemma 1

$$AR_w = \begin{cases} FAR_w & \text{if } w \in \mathcal{A}\setminus\mathcal{U} \\ \frac{1}{n}(1 - FRR_w) + \left(1 - \frac{1}{n}\right) FAR_w & \text{if } w \in \mathcal{U}\,. \end{cases} \tag{6}$$

Therefore it immediately follows that

$$\frac{1}{n}\sum_{u\in\mathcal{U}} AR_u = \frac{1}{n}(1 - FRR) + \left(1 - \frac{1}{n}\right) FAR \ . \tag{7}$$

Proof. For any $w \in \mathcal{A}\backslash\mathcal{U}$, it is clear that $AR_w = FAR_w$, since $\mathcal{U}\backslash\{w\} = \mathcal{U}$. For any $w \in \mathcal{U}$, from the definitions (2), (4) and (5) of FRR_w, FAR_w and AR_w, respectively, we have

$$\begin{aligned} AR_w &= \frac{1}{n} \sum_{\substack{(s,t)\in\mathcal{M}\times\mathcal{M} \\ d(s,t)<\tau}} P(X_w = s)P(X_w = t) \\ &\quad + \frac{n-1}{n}\cdot\frac{1}{n-1} \sum_{v\in\mathcal{U}\backslash\{w\}} \sum_{\substack{(s,t)\in\mathcal{M}\times\mathcal{M} \\ d(s,t)<\tau}} P(X_w = s)P(X_v = t) \\ &= \frac{1}{n}(1 - FRR_w) + \left(1 - \frac{1}{n}\right) FAR_w \ . \end{aligned}$$

Therefore the results follow. □

Put

$$AR = \frac{1}{n}\sum_{u\in\mathcal{U}} AR_u = \frac{1}{n}(1 - FRR) + \left(1 - \frac{1}{n}\right) FAR \ .$$

Note that usual biometric authentication systems obviously satisfy $FAR \leq 1 - FRR$, namely $FAR \leq AR$.

Definition 1. (cf. [UOI1, Definition 3]) *A* **wolf** *is defined as a sample* $w \in \mathcal{A}$ *such that* $AR_w > AR$.

For any $AR < p \leq 1$, a wolf w such that $AR_w = p$ is called a p-**wolf**. In particular, 1-wolf is called a **universal wolf**.

Definition 2. [UOI1, Definition 4] *Assume the following two conditions.*

(i) *The attacker has no information of a biometric sample of a genuine user to be impersonated.*
(ii) *The attacker has complete information of the algorithms employed in the enrollment phase and the verification phase.*

The **wolf attack** *is defined as an attempt to impersonate a genuine user by presenting p-wolves with large p's to minimize the complexity of the impersonation attack.*

Definition 3 (Wolf attack probability (WAP)). (cf. [UOI1, Definition 5]) *The* **Wolf attack probability** *is defined by*

$$WAP = \max_{w\in\mathcal{A}} \operatorname*{Ave}_{v\in\mathcal{U}} P(\mathit{match}(w,v) = \mathit{accept}) = \max_{w\in\mathcal{A}} AR_w \ . \tag{8}$$

It is clear that $AR \leq WAP$. Therefore, if $FAR \leq 1 - FRR$, then we have $FAR \leq AR = \operatorname{Ave}_{u \in \mathcal{U}} AR_u \leq WAP$.

Definition 4 (Security against the wolf attack). *For any $\delta > 0$, a biometric authentication system is δ-secure against the wolf attack if $WAP < \delta$, namely there exist no wolf $w \in \mathcal{A}$ such that $AR_w \geq \delta$.*

If we use only FAR as a security measure against impersonation attacks, then we cannot explicitly evaluate the security against wolf attacks. Une, Otsuka, Imai proposed to evaluate the security level against the wolf attack by computing WAP [UOI1], [UOI2].

3 Matching Algorithms Secure against the Wolf Attack

For each $s \in \mathcal{M}$ and $x \in \mathbb{R}_{\geq 0}$, the probability $P_s(x)$ that a template $t \in \mathcal{M}$ obtained from a biometric sample of a user will be at a distance less than x from s is defined by

$$P_s(x) = \frac{1}{n} \sum_{v \in \mathcal{U}} \sum_{\substack{t \in \mathcal{M} \\ d(s,t) < x}} P(X_v = t) \quad . \tag{9}$$

Then we have

$$AR_w = \sum_{s \in \mathcal{M}} P(X_w = s) P_s(\tau) \quad .$$

3.1 General Case

Fix $\delta > 0$. Then we will construct a matching algorithm δ-secure against the wolf attack as follows.

Almost every conventional matching algorithm employs a fixed threshold τ predetermined based on FRR and FAR. However, we will employ a threshold τ_s determined for each element $s \in \mathcal{M}$ obtained from the sample $w \in \mathcal{A}$ offered in the verification phase. For each $s \in \mathcal{M}$, put

$$\tau_s = \max\{x \in \mathbb{R}_{\geq 0} \mid P_s(x) < \delta\} \ .$$

Note that a set $S = \{x \in \mathbb{R}_{\geq 0} \mid P_s(x) < \delta\}$ is a non-empty closed subset of $\mathbb{R}_{\geq 0}$ and therefore there exists the maximum of S.

For the implementation, we need to gather enough templates from each $v \in \mathcal{U}$ and estimate the probabilities $P(X_v = t)$ for all $t \in \mathcal{M}$. Then we can determine the threshold τ_s for each $s \in \mathcal{M}$ by doing the exhaustive search of all possible $x \geq 0$ such that $P_s(x) < \delta$.

It is clear that

$$WAP = \max_{w \in \mathcal{A}} \sum_{s \in \mathcal{M}} P(X_w = s) P_s(\tau_s) < \delta \quad .$$

The above discussion gives the following theorem.

Theorem 1. *If the information about the probability distribution $P_s(x)$ for each $s \in \mathcal{M}$ is completely given, then, for any $\delta > 0$, we can construct a matching algorithm δ-secure against the wolf attack.*

3.2 Normal Distribution Case

We assume that the distribution $P_s(x)$ is normal with mean m_s and standard deviation σ_s for each $s \in \mathcal{M}$, namely

$$P_s(x) = \int_{-\infty}^{x} \frac{1}{\sqrt{2\pi}\sigma_s} \exp\left(-\frac{1}{2}\left(\frac{x - m_s}{\sigma_s}\right)^2\right) dx \tag{10}$$

for any $x > 0$. More strictly, we assume that $P_s(x)$ can be approximately estimated by the above equation. The distributions of Hamming distances for Daugman's iriscode satisfy this assumption (cf. [Da1], [Da2]). Some authors use the Gaussian assumption as the basis of their analysis (cf. [AYL1], [Ka1], [Wa1]). In general, the real-valued features will tend to approximate a Gaussian distribution when they are obtained by a linear combinations of many components, e.g. feature extraction techniques based on the principle component analysis (PCA) or the linear discriminant analysis (LDA) (cf. [AYL1]). Under this assumption, we can construct a secure and simple matching algorithm and show that the matching algorithm is optimal, namely WAP is minimized to the (almost) same value as AR.

Define the entropy $H(P)$ of the probability distribution P by

$$H(P) = \int_{-\infty}^{\infty} -P(x) \log_2 P(x)\, dx\ .$$

By the assumption (10), it can be easily checked that

$$H(P_s) = \log_2 \left(\sqrt{2\pi e} \cdot \sigma_s\right)\ . \tag{11}$$

We work with entropies $H(P)$ of continuous probability distributions P. Then the entropy $H(P)$ is not always non-negative. It is clear from (11) that if $\sigma_s < \frac{1}{\sqrt{2\pi e}}$, then $H(P_s) < 0$. Note that if a fixed threshold is employed, then an input value $s \in \mathcal{M}$ which has higher entropy $H(P_s)$ and therefore larger deviation σ_s can be accepted with higher probability.

Fix a real number α. For each $s \in \mathcal{M}$, put

$$\tau_s = \alpha\sigma_s + m_s = \frac{\alpha 2^{H_s}}{\sqrt{2\pi e}} + m_s \tag{12}$$

where $H_s = H(P_s)$. By the assumption (10), we have

$$P_s(\tau_s) = \int_{-\infty}^{\tau_s} \frac{1}{\sqrt{2\pi}\sigma_s} \exp\left(-\frac{1}{2}\left(\frac{x - m_s}{\sigma_s}\right)^2\right) dx = \int_{-\infty}^{\alpha} \frac{1}{\sqrt{2\pi}} \exp\left(-\frac{z^2}{2}\right) dz \tag{13}$$

for all $s \in \mathcal{M}$. Put

$$\delta(\alpha) = \int_{-\infty}^{\alpha} \frac{1}{\sqrt{2\pi}} \exp\left(-\frac{z^2}{2}\right) dz \quad .$$

The following theorem can be immediately proved.

Theorem 2. *Assume that the standard deviation σ_s (or the entropy H_s) and the mean m_s are perfectly given for each $s \in \mathcal{M}$. Then the matching algorithm employing the thresholds τ_s, $s \in \mathcal{M}$, defined by (12) is $\delta(\alpha)$-secure against wolf attacks. Moreover, we have $AR_w = AR = WAP = \delta(\alpha)$ for all $w \in \mathcal{A}$.*

Proof. By the calculation (13), for all $w \in \mathcal{A}$, we have

$$AR_w = \sum_{s \in \mathcal{M}} P(X_w = s) P_s(\tau_s) = \delta(\alpha) \ .$$

Therefore the results follow. □

Our generalized matching algorithm gives a theoretical framework for constructing matching algorithms secure against the wolf attack for any biometric authentication system. Under the ideal condition that for each $s \in \mathcal{M}$, the distribution $P_s(x)$ is completely calculated, our matching algorithm is optimal against the wolf attack.

In the real world, it might be difficult to explicitly calculate the distribution $P_s(x)$ for all $s \in \mathcal{M}$, however, a more accurate computation of σ_s, H_s, or m_s for each $s \in \mathcal{M}$ can achieve a lower *WAP*. Consequently, there is a trade-off between the efficiency of the matching algorithm and the security evaluated by the achievable *WAP*. In the next section, we will reconsider previous results as instances of our theoretical framework.

4 Previous Results in Our Framework

In this section, we will review previous results in the context of our theoretical framework.

A conventional matching algorithm employing a fixed threshold can be viewed as an efficient instance of our framework, which assumes every input value has a constant entropy instead of computing the entropy for each input value. Such a matching algorithm is not secure against the wolf attack.

Daugman proposes a matching algorithm which employs a variable threshold in place of a fixed threshold as follows [Da2]. He employs a fractional Hamming distance $d = fHD$ defined by $fHD(s,t) = \dfrac{HD(s,t)}{k}$ for any $s, t \in \mathcal{M} = \{0,1\}^{2048}$, where k is the number of bits available for comparison. He determines a threshold depending on k as follows:

$$\tau(s,t) = \frac{\alpha'}{\sqrt{k}} + \frac{1}{2} \tag{14}$$

where $\frac{1}{2}$ is the average of $fHD(s,t)$ estimated from his database. His algorithm can also be regarded as an instance of our framework, which assumes every bit of each sample independently and identically contributes to the probability distribution.

However, his algorithm is not necessarily secure against the wolf attack, since every bit is not exactly independent and identical and the distributions $P_s(x)$, $s \in \mathcal{M}$, can be considerably different from each other. We assume that an attacker has more accurate information about the distributions $P_s(x)$, $s \in \mathcal{M}$. If the attacker can successfully find a smart input value $s \in \mathcal{M}$ such that the entropy $H(P_s(x))$ is extremely high, then he can be incorrectly accepted with much higher probability than AR.

Daugman's matching algorithm is not always secure against the wolf attack, however, it motivated us to research a theoretical framework to construct secure matching algorithms.

References

[AYL1] Adler, A., Youmaran, R., Loyka, S.: Towards a measure of biometric feature information. In: Pattern Analysis and Applications. Springer, London (Online first), doi:10.1007/s10044-008-0120-3

[Da1] Daugman, J.: High Confidence Visual Recognition of Persons by a Test of Statstical Independence. IEEE Trans. on Pattern Anal. Mach. Intell. 15(11) (November 1993)

[Da2] Daugman, J.: Probing the uniqueness and randomness of IrisCodes: Results from 200 billion iris pair comparisons. Proceedings of the IEEE 94(11), 1927–1935 (2006)

[ISO1] International Organization for Standardization (ISO), International Electronical Commission (ICE): ISO/IEC CD 19792: Imformation technology - Security techniques - Security evaluations of biometrics (2006)

[Ka1] Kevenaar, T.: Protection of Biometric Information. In: Tuyls, P., Skoric, B., Kevenaar, T. (eds.) Security with Noisy Data: On Private Biometrics, Secure Key Storage and Anti-Counterfeiting, ch. 7, pp. 113–125. Springer, London (2007)

[MMYH1] Matsumoto, T., Matsumoto, H., Yamada, K., Hoshino, S.: Impact of Artificial 'Gummy' Fingers on Fingerprinting Systems. In: Opt. Sec. and Count. Det. Tech. IV, Proc. of SPIE, vol. 4677, pp. 275–289 (2002)

[RCB1] Ratha, N.K., Connell, J.H., Bolle, R.M.: Enhancing security and privacy in biometrics-based authentication systems. IBM Syst. J. 40, 614–634 (2001)

[UOI1] Une, M., Otsuka, A., Imai, H.: Wolf Attack Probability: A New Security Measure in Biometric Authentication Systems. In: Lee, S.-W., Li, S.Z. (eds.) ICB 2007. LNCS, vol. 4642, pp. 396–406. Springer, Heidelberg (2007)

[UOI2] Une, M., Otsuka, A., Imai, H.: Wolf Attack Probability: A Theoretical Security Measure in Biometrics-Based Authentication Systems. IEICE, Transactions on Information and Systems 2008 E91-D (5), 1380–1389 (2008)

[Wa1] Wayman, J.S.: The cotton ball problem. In: Biometrics Conference, Washington, DC, USA, September 20-22 (2004)

A Biometric Menagerie Index for Characterising Template/Model-Specific Variation

Norman Poh and Josef Kittler

CVSSP, University of Surrey, Guildford, GU2 7XH, Surrey, UK
normanpoh@ieee.org, j.kittler@surrey.ac.uk

Abstract. An important phenomenon influencing the performance of a biometric experiment, attributed to Doddington et al (1998), is that the match scores (whether under genuine or impostor matching) are strongly dependent on the model or template from which the match scores have been derived. Although there exist studies to classify the characteristic of the template/model, as well as the query data, into animal names such as sheep, goats, wolves and lambs – so-called Doddington's menagerie, or higher semantic categories considering simultaneously both genuine and impostor match scores, due to Yager and Dunstone (2008), there is currently absence of means to characterise the extent of Doddington's menagerie. This paper aims to design such an index, called the biometric menagerie index (BMI). It is defined as the ratio of the between-client variance and the expectation of the total variance. BMI has three desirable properties. First, it is invariant to shifting and scaling of the match scores. Second, its value lies between zero and one, with zero implying the absence of Doddington's menagerie effect, and one signifying its strong presence. Third, it is experimentally verified that BMI generalizes to *different choices* of impostor population. Our findings based on the XM2VTS benchmark score database suggest the followings: First, the BMI of genuine match scores is generally higher than that of the impostor match scores. Second, two different matching algorithms observing the same biometric data may have significantly different BMI values, hence suggesting that the biometric menagerie is algorithm-dependent.

1 Motivation

An automatic biometric system works by first building a model or template specific to a user. During the operational phase, the system compares a scanned biometric sample with the registered model to decide whether an identity claim is genuine or not (from a different person). As a result, each user model/template exhibits different behaviours in terms of output scores when being presented genuine and impostor biometric samples. The consequence is that some user models are better than others in representing the user's identity. Doddington et al's initial study [1] attempted to characterise the user models by how easy or difficult they can be recognised, and how easy query samples can imitate others, hence, introducing user categories by names such as sheep, goats, lambs and wolves. A sheep is a person who can be easily recognised; a goat is a person who is particularly difficult to be recognised; a lamb is a person who is easy to imitate; and a wolf is a person who is particularly successful at imitating others. A more recent

M. Tistarelli and M.S. Nixon (Eds.): ICB 2009, LNCS 5558, pp. 816–827, 2009.
© Springer-Verlag Berlin Heidelberg 2009

study by Yager and Dunstone [2] further distinguishes four other semantic categories of users by considering both the genuine and impostor match scores, for *each* claimed identity, simultaneously. However, their approach considers only the *client-specific* first order moments (i.e., for each claimed identity) of the match scores. Poh and Kittler [3] further considered the second order moments from which several client-specific class-sepabarability criteria can be derived. Thanks to these criteria, the authors deomnstrated empirically that client models/templates (in the gallery) can be ranked according to their performance, hence separating well-behaved models from badly behaved models (in terms of performance).

Referring to Doddington's menagerie, sheep are characterised by high genuine similarity match scores whereas goats are characterised by low genuine similarity match scores. Lambs are defined as a symmetry of goats, i.e., having high impostor similarity match scores. Finally, wolves are persons who can consistently give high impostor similarity scores when matched against the client models (enrolled templates in the gallery). While sheep dominate the population of client models, goats constitute only a small fraction of the population. However the latter category constitutes disproportionately a large portion of false rejection error. Although the original Doddington's study was applied to speaker verification, the same phenomenon was independently observed in [3] using the face, fingerprint and iris biometric modalities; [4] using the fingerprint modality; [5] using the face modality; and many others, e.g., [2]. Using finger-vein and fingerprint as case studies, Une et al [6] proposed a measure known as the wolf attack probability, which quantifies the maximum probability of success of impersonating a victim by feeding wolves in a biometric system. These studies provide a mounting evidence that the biometric menagerie is a general phenomenon inherent in all biometric experiments.

As a result of these menagerie studies, it is beginning to be recognised that fine-tuning the system parameters (including feature extraction and classifier or distance matching parameters) and the decision threshold for each individual client model (classifier) can greatly boost the recognition (identification and verification) performance further. For instance, *lowering* the similarity decision threshold (than a globally pre-set value) for the goats is likely to compensate for their disproportionately high false rejection errors, and thus *increasing* the similarity decision threshold for the lambs in order to compensate for their disproportionately high false acceptance error. This strategy is called client (model/template) specific decision. Examples of such strategies abound: [7,8,9,10,11]. Rather than adjusting the thresholds, one can instead transform the match score distribution. This alternative strategy is called *client-specific score normalisation*. Examples are Z-norm [12], F-norm [13], EER-norm [14] and model-specific log-likelihood ratio (LLR)-based normalisation [15]. Both categories of approaches have been discussed and summarised in [16].

Clearly, a fundamental understanding of Doddington's menagerie has an important impact on designing and optimising a biometric system as a whole. There is, however, a certain lack of understanding of this phenomenon. For instance, we ignore the reason for the existence of wolves, as well as of lambs and goats. However, we know that it is certainly dependent on the choice of biometric device, the result of the user's interaction with the device and the acquisition environment, hence related to the quality of

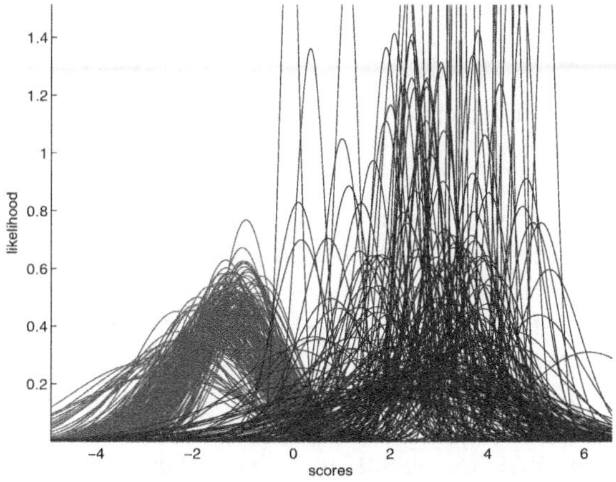

Fig. 1. An example of (client) model-specific likelihood plots (conditioned on client/genuine (blue) and impostor (red) classes, for all the 200 users using the first data set. The zoo index for the genuine match scores is 0.96 whereas for the impostor match scores, it is 0.23.

biometric samples. Although Grother and Tabassi [17] have attempted to define *quality* as a scalar summary of a sample's appropriateness to be used for matching, quality itself is clearly multi-facetted. There is currently a missing link between template-related quality measures and the Doddington's menagerie.

The goal of this paper is to advance the understanding of Doddington's menagerie by proposing a statistical measure to quantify the extent of this phenomenon. To date, there simply exists no such measure. The only means thus far has been applying a client-specific score normalisation/decision threshold procedure and observe if the procedure indeed yields an improved performance, or not, e.g., [12,13,14,15]. Attempting to diagnose such procedure is difficult since this can be due to inaccurate estimation of the parameters, especially when dealing with the genuine match score distribution (which has very few samples compared to its impostor counterpart).

The proposed measure is called the *biometric menagerie index* (BMI) or simply the *zoo index*, as a means to directly measure the dependency of a match score on a user model/template. In order to motivate our approach, we illustrate in Figure 1 the model/template-specific class-conditional score distribution, i.e., one distribution for each template and for each class, subject to being involved in genuine (same-person) and impostor (different-person) matching. An important observation here is that the genuine and impostor match scores have very different characteristics. In particular, the genuine match scores exhibit higher model-dependency than the impostor match scores for this example. This figure also shows that the compounded genuine (impostor) similarity score distribution is a mixture of *client-specific* genuine (impostor) similarity score distributions, hence, can have multiple modes.

This paper is organised as follows: Section 2 explains how BMI was derived. Section 3 shows the BMI values for face and speech biometrics and this is followed by conclusions.

2 Deriving a Zoo Index

In this section, we shall analyse the match scores conditioned on each class and each claimed identity, which is necessarily drawn from a distribution specified by $p(y|k, j)$. This distribution can be estimated from the score set $\mathcal{Y}_j^k$ introduced in the previous section. The goal of this section is to derive a zoo index. Section 2.1 will first introduce the notation. Section 2.2 defines BMI mathematically. We then present the properties of BMI in Section 2.3. Finally, the estimation of the relevant statistics is described in Section 2.4.

2.1 Notation

We shall denote $j' \in \mathcal{J}'$ to be the *true* identity whereas $j \in \mathcal{J}$ to be the *claimed* identity which must come from a client set (gallery) $\mathcal{J} = [1, \ldots, J\}$ and there are J clients (for whom templates/models are built). The set $\mathcal{J}'$ is a set of identities/users not in $\mathcal{J}$, i.e., the two populations do not overlap: $\mathcal{J} \cap \mathcal{J}' = \emptyset$. Thus, throughout this study, we deal with *open-set* recognition, and an extension to closed-set recognition is straightforward.

In order to simplify the notation, we also assume that there is only a single model associated with each client. In a usual identification setting, one compares all samples belonging to j' against the model of j in order to obtain a score set $\mathcal{Y}(j, j')$. When $j = j'$, the match score set is genuine (client), whereas when $j \neq j'$, it is impostor. We further introduce two *client-specific* score sets, both dependent on the claimed identity: $\mathcal{Y}_j^{\mathtt{C}} \equiv \mathcal{Y}'(j, j)$ for the client class, and $\mathcal{Y}_j^{\mathtt{I}}$ for the impostor class. In this study, for the impostor class, the scores are a *union* or aggregation of all other users (from $\mathcal{J}'$) claiming to be j, i.e., $\mathcal{Y}_j^{\mathtt{I}} \equiv \bigcup_{j' \in \mathcal{J}, j' \neq j} \mathcal{Y}'(j, j')$.

2.2 Deriving the Zoo Index Using the Within- and Between-Class-Variances

Using the above notations, we shall use the score variable y to reprensent an element in the set $\mathcal{Y}_j^k$ for a given class $k = [\mathtt{C}, \mathtt{I}\}$ (client or impostor) and a given claimed identity j. The unknown disribution from which $\mathcal{Y}_j^k$ was generated is denoted by $p(y|k, j)$. Thus, the unconditional distribution of y is

$$p(y) = \sum_{k,j} p(y|k, j) P(j|k) P(k)$$

where $P(j|k)$ is the prior class-conditional probability claiming identity j and $P(k)$ is the prior class probability.

The class-conditional expected value (global mean) of y is defined by:

$$\begin{aligned} \mu^k &= \sum_{j=1}^{J} \left(\int_y p(y|k, j) y \, dy \right) P(j|k) \\ &\equiv \mathbb{E}_j \left[\mathbb{E}_y \left[y|k, j \right] |k \right] \end{aligned} \quad (1)$$

where $P(j|k) = P(j, k)/P(k)$ and we have used the following notation

$$\mathbb{E}_y \left[y|k, j = j_* \right] \equiv \int_y y p(y|k, j = j_*) \, dy$$

to denote the expectation of y conditioned on a particular claimed identity $j = j_*$ as well as the class k, and

$$\mathbb{E}_j[\bullet|k] \equiv \sum_{j=1}^{J} \bullet P(j|k)$$

to denote the expectation over all clients, conditioned on k.

In contrast to (1), the (class-conditional) client-specific mean is defined as:

$$\mu_j^k \equiv \mathbb{E}_y[y|k,j]$$

which is related to the (class-conditional) global mean by:

$$\mu^k = \mathbb{E}_j\left[\mu_j^k|k\right]$$

as is evident from (1).

The (class-conditional) global score variance, for a given claimed identity j_* can be calculated as follows:

$$\begin{aligned}
\mathbb{E}_y\left[\left(y-\mu^k\right)^2|k,j_*\right] &= \mathbb{E}_y\left[\left(y-\mu_{j_*}^k+\mu_{j_*}^k-\mu^k\right)^2|k,j_*\right] \\
&= \mathbb{E}_y\left[\left(y-\mu_{j_*}^k\right)^2|k,j_*\right]+\mathbb{E}_y\left[\left(\mu_{j_*}^k-\mu^k\right)^2|k,j_*\right] \\
&\quad+\underbrace{2\left(\mu_{j_*}^k-\mu^k\right)\mathbb{E}_y\left[\left(y-\mu_{j_*}^k\right)|k,j_*\right]}
\end{aligned}$$

where we introduced the term $\mu_{j_*}^k$. Note that since under the expectation $\mathbb{E}_y[\bullet|k,j_*]$, the second term is invariant, and the third (underbraced) term vanishes, we have:

$$\underbrace{\mathbb{E}_y\left[\left(y-\mu^k\right)^2|k,j_*\right]}_{V_{k,j*}^{tot}} = \underbrace{\mathbb{E}_y\left[\left(y-\mu_j^k\right)^2|k,j_*\right]}_{V_{k,j*}^{W}}+\underbrace{\left(\mu_j^k-\mu^k\right)^2}_{V_{k,j*}^{B}} \qquad (2)$$

where V_{k,j_*}^{tot} is the *total variance*, V_{k,j_*}^{W} is the *within-client* variance and $V_B^k(j_*)$ is the *squared bias* (or squared distance) between the global mean and the client specific mean, all conditioned on the class label k (which can be genuine or impostor matching) and a particular claimed identity j_*. If one takes the expecation $\mathbb{E}_j[\bullet|k]$ on both sides of (2), then the two terms on the right-hand sides are known as variance and squared bias, respectively. The resultant two components constitutes the bias-variance decomposition of Krogh and Vedelsby [18].

Using the fact that j is a variable, and j_* being its realisation, we shall similarly define $V_{k,j}^{tot}$ to be the variable of V_{k,j_*}^{tot} (the latter being a realization). We shall also define $V_{k,j}^{W}$ and $V_{k_j}^{B}$ in a similar manner. We then propose to define the *client-specific* biometric menagerie index (BMI) (for a given client with index j) to be:

$$\text{BMI}_{k,j} \equiv \frac{V_{k,j}^{B}}{V_{k,j}^{tot}}.$$

Likewise, the BMI index for the whole population of clients is:

$$\bar{\text{BMI}}_k \equiv \mathbb{E}_j\left[\text{BMI}_{k,j}|k\right]$$

with variance $\mathbb{E}_j\left[\left(\text{BMI}_{k,j}-\bar{\text{BMI}}_k|k\right)^2\right]$.

2.3 Properties of BMI

Because of (2), $V^B_{k,j} \leq V^{tot}_{k,j}$ since $V^W_{k,j}$ is always *positive*. As a result, the client-specific BMI index is always bounded:

$$0 \leq \text{BMI}_{k,j} \leq 1,$$

and so is the BMI index for the whole population

$$0 \leq \bar{\text{BMI}}_k \leq 1.$$

When $\text{BMI}_{k,j} = 0$, this implies that $V^B_{k,j} = 0$. This can only happen when the client-specific mean μ^k_j coincides with the global mean μ^k. If this is true over the whole population, then $\bar{\text{BMI}}_k = 0$. This strongly implies that the Doddington's menagerie contains only one species.

On the other hand, when $\bar{\text{BMI}}_k = 1$, this implies that $\mathbb{E}_j[V^W_{k,j}|k] = 0$. In other words, the total variance is fully determined by the expected squared bias $\mathbb{E}_j[V^B_{k,j}|k]$ (or the between class variance), and the expected within-client variance has little importance. This signifies the presence of widely different species in the Doddington's menagerie. In practice, we have never observed that the BMI takes on the extreme values of zero or one. Therefore, we emphasize that the BMI value should *always* be expressed along with its confidence interval (characterised by its variance).

Apart from the bounded property, BMI is also invariant to shifting and scaling. It is trivial to show the invariance of BMI to shifting. Suppose y is shifted by a constant. Then, its expected value will be shifted by the same amount. This constant simply gets canceled out when calculating the variance of the shifted score, because this operation involves only finding the difference between the shifted score and shifted expected value. Hence, all the terms in (2) are not affected by the constant shift. It follows that BMI is invariant to shifting.

BMI is also invariant to scaling. Suppose that y is scaled by a constant a, i.e., ay. Then, all the terms in (2) will be scaled by a^2. The resultant BMI will be:

$$\text{BMI}_{k,j} = \frac{a^2 V^B_{k,j}}{a^2 V^{tot}_{k,j}} = \frac{V^B_{k,j}}{V^{tot}_{k,j}}$$

hence invariant to the constant scaling.

We simulated four different possible scenarios giving different BMI values, all as a function of the following *client-specific* means and variances (arranged in a vector):

$$\boldsymbol{\mu} = [-7.00, -3.50, 0, 3.50, 7.00]' \text{ and } \boldsymbol{\sigma} = [0.20, 1.00, 3.00, 0.70, 4.00]'.$$

The results are shown in Figures 2 (a)–(d). The actual mean and variance used for each sub-figure are shown in the caption. In (a), all client-specific means are aligned to the global mean (zero in our case), so that the between-client variance is set to zero, hence, giving BMI=0. The other extreme, giving BMI=1 is shown in (d). This was done by decreasing the variance to very small values so that the between-client variance dominates the total variance. Figures (b) and (c) are two intermediate cases sampled from the range of possibilities.

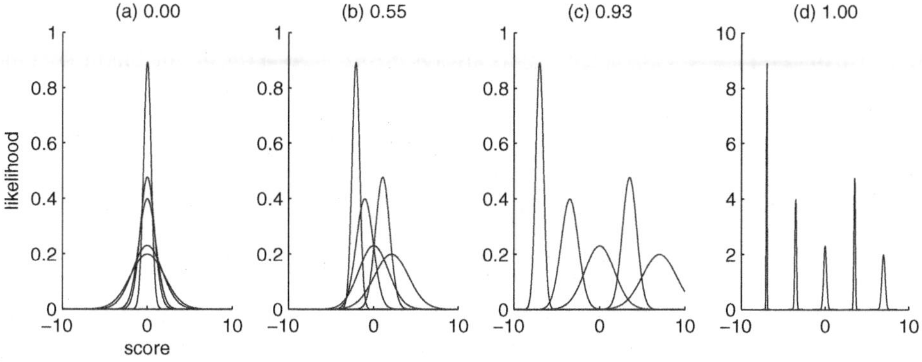

Fig. 2. Examples of client-specific distributions for various BMI values (in the subtitle). The following means and variances are used: (a) $[\boldsymbol{\mu} \times 0, \boldsymbol{\sigma}]$, (b) $[\boldsymbol{\mu} \times 0.3, \boldsymbol{\sigma}]$, (c) $[\boldsymbol{\mu}, \boldsymbol{\sigma}]$, and (d) $[\boldsymbol{\mu}, \boldsymbol{\sigma} \times 0.01]$.

2.4 Empirical Estimates

In the followings, we will work with the sample $y_{ij}^k \in \mathcal{Y}_j^k$ for any claimed identity j, conditioned on the class k, and $i = 1, \ldots, N^k$, and there are N^k samples. The estimated global and client-specific means are:

$$\hat{\mu}^k = \frac{1}{JN^k} \sum_{j=1}^{J} \sum_{i=1}^{N^k} y_{ij}^k \quad \text{and} \quad \hat{\mu}_j^k = \frac{1}{N^k} \sum_{i=1}^{N^k} y_{ij}^k$$

The estimated (squared) bias term is:

$$\hat{V}_{k,j}^B = (\hat{\mu}_j^k - \hat{\mu}^k)^2$$

The estimated total variance is:

$$\hat{V}_{k,}^{tot} = \frac{1}{N^k} \sum_{i=1}^{N^k} (y_{ij}^k - \hat{\mu}^k)^2.$$

Assuming all hypotheses to be equiprobable, the expected mean and variance of BMI index can be estimated as follow:

$$\widehat{\text{BMI}}^k = \frac{1}{J} \sum_{j=1}^{J} \left(\frac{\hat{V}_B^{k,j}}{\hat{V}_{k,j}^{tot}} \right)$$

$$\widehat{var}(\text{BMI}^k) = \frac{1}{J} \sum_{j=1}^{J} \left(\frac{\hat{V}_B^k(j)}{\hat{V}_{tot}^k(j)} - \widehat{\text{BMI}}^k \right)^2$$

The variance of BMI is a useful indicator of the confidence of the estimated BMI. The smaller the variance, the higher the confidence of the estimate.

3 Experiments

3.1 Data-Set

We use the publicly available[1] XM2VTS multimodal score-level fusion benchmark database [19] for this purpose. 13 sets of experiments were available, among which 6 (resp. 7) are on the speech (resp. face) biometrics. The distribution of match scores, in general, exhibits high central tendency except for two comparison subsystems which are based on Multi-layer Perceptrons (MLPs) having tangent hyperbolic tanh) output. For these systems, their outputs are mapped into a linear scale using the inverse of $\tanh$, hence, denoted by "MLPi". This one-to-one transformation is non-linear and is essential to guarantee the central tendency of the match scores. Since the BMI index is invariant to shifting and scaling, no further prepossessing is needed, and the experiments can be compared directly using BMI.

There are 200 client models that are assessed on two data sets, termed dev (for development) and eva (for evaluation) sets. There are 3 genuine samples per client for the dev set under Lausanne Protocol I (LP1) and 4 for Lausanne protocol II (LP2). These protocols define what data the expert systems should use for enrolling the clients. Although these protocols entail changes in the number of genuine match scores for the dev set, they remain the same for the eva set, which is two genuine match scores per client model, hence a total of 200×2 genuine match scores for eva but 200×3 for dev LP1 and 200×4 for dev LP2.

It is worth noting that the impostor population of dev and eva sets are different, thus, $\mathcal{J}'_{dev}$ (the former) consists of 25 impostors and $\mathcal{J}'_{eva}$ (the latter) consists of 70 impostors. Each impostor contributes 8 samples. This results in 25×8 impostor scores for the impostor dev and 70×8 for the impostor eva set. Using different impostor population is beneficial for testing the stability of BMI under changes in impostor population.

3.2 Results

Figure 3 shows the BMI values and its variance for the 13 score data sets conditioned on the class label (genuine or impostor), and on the data set used, i.e., dev or eva. We can observe the following:

(i) The variances of impostor BMI match scores are generally much higher than their genuine counter parts. This is expected since there are many more impostor match scores than the genuine match scores. For instance, $200 \times 75 \times 8$ impostor match scores versus 200×2 genuine match scores for the eva set. As a result, the impostor BMI can be estimated with *higher* confidence.

(ii) The *expected* genuine match scores have, in general, higher BMI values than their corresponding impostor match scores.

(iii) The BMI values between the development and the evaluation sets are somewhat consistent in pattern. In order to verify this futher, we plotted BMIs of the dev set versus their corresponding values on the eva set in Figure 4.

[1] http://personal.ee.surrey.ac.uk/Personal/Norman.Poh/web/fusion

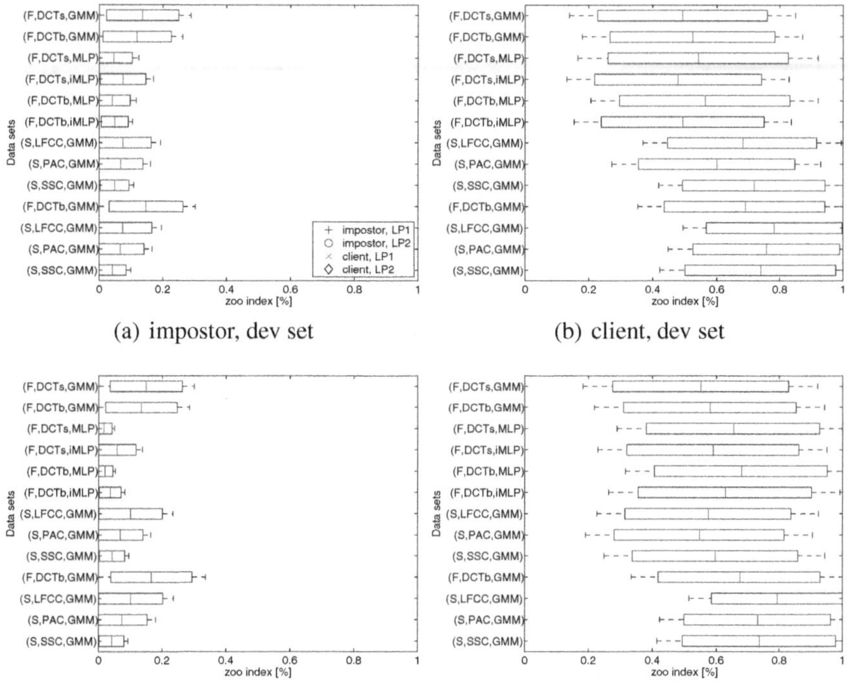

(a) impostor, dev set

(b) client, dev set

(c) impostor, eva set

(d) client, eva set

Fig. 3. The BMI of 13 systems calculated on the four data sets formed by the following dichotomies: client vs impostor, *dev* vs. *eva*. Each figure shows 13 BMI values – with the first 9 (from top to bottom) from LP1 systems whereas the remaining 4 from LP2 – conditioned on *dev* and *eva* sets. Each BMI was estimated from 200 client-specific BMI values (since there are 200 client models) in each system.

Concerning observation (iii), it is worth recalling that the impostor populations in the *dev* and the *eva* sets consist of *different* subjects (persons), and that the client models (in the gallery) remain the same. By comparing BMI with different impostor data sets, this experiment confirms that BMI is indeed *insensitive* to the choice of impostor population.

The client BMI for LP1 is *inconsistent* between *dev* and the *eva* sets, with correlation as low as -0.11. In comparison, the client BMI for LP2 is more consistent (giving correlation as high as 0.95; see Figure 4 caption). This is because the genuine match scores between the *dev* and *eva* sets for LP1 are *biased*. In LP1, the *dev* score sets were derived from the same session (in the same visit) of data set as the one used to build the base-classifier model, but this is not the case for LP2 (where a different session of data was used). As a result, for LP1, the genuine match scores in *dev* is *positively biased* (resulting in better recognition than expected). For LP2, there is no such bias, since the *dev* and *eva* sets contains similar *cross session* variability, with *dev* scores derived from the third session and *eva* from the fourth session. Recall also that in both LP1 and LP2, the data used to generate the *eva* score set is the same (the forth session/visit). Two

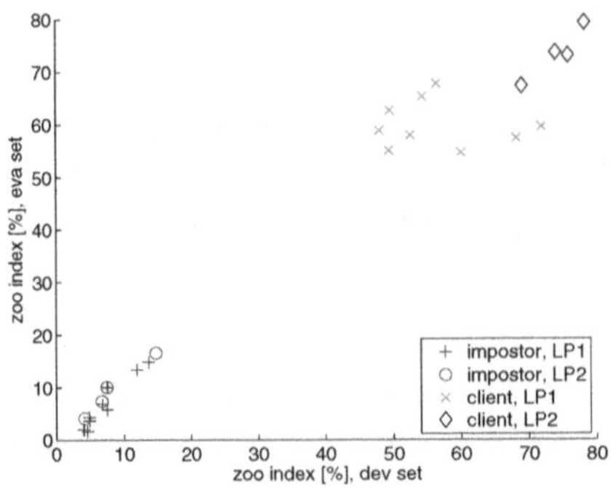

Fig. 4. Comparison of BMI calculated on the development set (X-axis) and the same statistic calculated on the evaluation set (Y-axis) for (a) impostor BMI (on LP1 data set), (b) impostor BMI (LP1), (c) client BMI (LP1), and (d) client BMI (LP1). (a) and (b) consists of 9 systems and LP2 consists of 4 systems. The correlation of BMI between the *dev* and *eva* sets for these four data sets are 0.96, 0.98, −0.11 and 0.95, respectively.

consecutive sessions were separated by about one month's interval, but this is certainly not consistent across all the users. This observation highlights the importance of using *unbiased* (cross-session) data set to estimate the BMI values.

4 Conclusions

In this paper, we proposed a biometric menagerie index (BMI), based on the expected within-client variance and expected squared bias (or between-client variance). The current study is somewhat preliminary; the following extensions are currently being investigated:

- **Wider range of biometrics:** While the current study is limited to face and speech, more biometric modalities will be considered
- **Multivariate extension of BMI:** The current study is limited to calculating BMI on a single modality. A future study will include an analysis of user-specific class-conditional BMI *across different modalities/systems*. This will allow to resolve whether two BMI values, each obtained from their respective user model, are correlated or not. If they are correlated, then the presence of lamb may be attributed to data of insufficient quality. On the other hand, if the BMI values are not correlated, then the presence of lamb may not be caused by the data, but by the choice of classifier or feature representation, which may systematically fail to discriminate a certain group of users.
- **Pre-processing techniques:** The current study relies on the central tendency of class-conditional match scores, essentially relying on the good estimates of the first and second order moments. It is therefore vital to ensure that the data adhere to this assumption. A promising technique to achieve this is by means of Gaussian Copula [20,21], which has been used successfully to estimate the effective biometric sample size. An extension of BMI to multimodal biometrics by means of multivariate Gaussian Copula is straightforward.

Acknowledgement

This work was supported partially by the advanced researcher fellowship PA0022_121477 of the Swiss National Science Foundation and by the EU-funded Mobio project (www.mobioproject.org) grant IST-214324.

References

1. Doddington, G., Liggett, W., Martin, A., Przybocki, M., Reynolds, D.: Sheep, Goats, Lambs and Wolves: A Statistical Analysis of Speaker Performance in the NIST 1998 Speaker Recognition Evaluation. In: Int'l. Conf. Spoken Language Processing (ICSLP), Sydney (1998)
2. Yager, N., Dunstone, T.: Worms, chameleons, phantoms and doves: New additions to the biometric menagerie. In: IEEE Workshop on Automatic Identification Advanced Technologies, June 2007, pp. 1–6 (2007)
3. Poh, N., Kittler, J.: A Methodology for Separating Sheep from Goats for Controlled Enrollment and Multimodal Fusion. In: Proc. of the 6th Biometrics Symposium, Tampa, pp. 17–22 (2008)
4. Hicklin, A., Ulery, B.: The myth of goats: How many people have fingerprints that are hard to match? Tech. Rep. NISTIR 7271, National Institute of Standards and Technology (2005)
5. Wittman, M., Davis, P., Flynn, P.J.: Empirical studies of the existence of the biometric menagerie in the frgc 2.0 color image corpus. In: Conf. on Computer Vision and Pattern Recognition Workshop, June 2006, p. 33(2006)
6. Une, M., Otsuka, A., Imai, H.: Wolf attack probability: A theoretical security measure in biometric authentication systems. IEICE-Transactions on Info and Systems E91-D(5), 1380–1389 (2008)
7. Furui, S.: Cepstral Analysis for Automatic Speaker Verification. IEEE Trans. Acoustic, Speech and Audio Processing / IEEE Trans. on Signal Processing 29(2), 254–272 (1981)
8. Pierrot, J.-B.: Elaboration et Validation d'Approaches en Vérification du Locuteur, Ph.D. thesis, ENST, Paris (September 1998)
9. Chen, K.: Towards Better Making a Decision in Speaker Verification. Pattern Recognition 36(2), 329–346 (2003)
10. Saeta, J.R., Hernando, J.: On the Use of Score Pruning in Speaker Verification for Speaker Dependent Threshold Estimation. In: The Speaker and Language Recognition Workshop (Odyssey), Toledo, pp. 215–218 (2004)
11. Genoud, D.: Reconnaissance et Transformation de Locuteur, Ph.D. thesis, Ecole Polythechnique Fédérale de Lausanne (EPFL), Switzerland (1998)
12. Auckenthaler, R., Carey, M., Lloyd-Thomas, H.: Score Normalization for Text-Independant Speaker Verification Systems. Digital Signal Processing (DSP) Journal 10, 42–54 (2000)
13. Poh, N., Bengio, S.: F-ratio Client-Dependent Normalisation on Biometric Authentication Tasks. In: IEEE Int'l. Conf. Acoustics, Speech, and Signal Processing (ICASSP), Philadelphia, pp. 721–724 (2005)
14. Fierrez-Aguilar, J., Ortega-Garcia, J., Gonzalez-Rodriguez, J.: Target Dependent Score Normalisation Techniques and Their Application to Signature Verification. In: Zhang, D., Jain, A.K. (eds.) ICBA 2004. LNCS, vol. 3072, pp. 498–504. Springer, Heidelberg (2004)
15. Poh, N., Kittler, J.: On the Use of Log-likelihood Ratio Based Model-specific Score Normalisation in Biometric Authentication. In: Lee, S.-W., Li, S.Z. (eds.) ICB 2007. LNCS, vol. 4642, pp. 614–624. Springer, Heidelberg (2007)
16. Poh, N., Kittler, J.: Incorporating Variation of Model-specific Score Distribution in Speaker Verification Systems. IEEE Transactions on Audio, Speech and Language Processing 16(3), 594–606 (2008)

17. Grother, P., Tabassi, E.: Performance of Biometric Quality Measures. IEEE Trans. Pattern Analysis and Machine Intelligence 29(4), 531–543 (2007)
18. Krogh, A., Vedelsby, J.: Neural Network Ensembles, Cross-Validation and Active-Learning. In: Advances in Neural Information Processing Systems, vol. 7 (1995)
19. Poh, N., Bengio, S.: Database, Protocol and Tools for Evaluating Score-Level Fusion Algorithms in Biometric Authentication. Pattern Recognition 39(2), 223–233 (2005)
20. Song, P.X.-K.: Multivariate dispersion models generated from gaussian copula. Scandinavian Journal of Statistics 27(2), 305–320 (2000)
21. Dass, S.C., Zhu, Y., Jain, A.K.: Validating a Biometric Authentication System: Sample Size requirements. IEEE Trans. on Pattern Analysis and Machine 28(12), 1319–1902 (2006)

An Usability Study of Continuous Biometrics Authentication

Geraldine Kwang[1], Roland H.C. Yap[1], Terence Sim[1], and Rajiv Ramnath[2]

[1] School of Computing, National University of Singapore, Singapore
u0407321@alumni.nus.edu.sg, {ryap,tsim}@comp.nus.edu.sg
[2] Temasek Laboratories, National University of Singapore, Singapore
tslrr@nus.edu.sg

Abstract. We present an usability study for a bi-modality Continuous Biometrics Authentication System (CBAS) that runs on the Windows platform. Our CBAS combines fingerprint and facial biometrics to authenticate users. As authentication is continuous, CBAS constantly contributes a computational overhead of up to 42% to the computer system. This usability study seeks to investigate (a) whether this overhead will have an impact on the performance of users to complete tasks; and (b) whether the users deem the responsiveness of the system to be acceptable. The results of our study are encouraging, indicating that the runtime cost of a CBAS system has no measurable statistical impact on the task completion by users. We found that user acceptance of CBAS to be good and they did not perceive the CBAS to degrade system response. This suggests that continuous biometrics for authentication is viable – the CBAS benefits outweighs system impact drawbacks.

1 Introduction

Most computer systems require authentication in order to use the system. Traditionally, the authentication is done at sign-on or login time either with a password and/or biometric authentication. Combining a password with biometrics can give additional assurance that the user logging in is the authorized user rather than an intruder. However, once the user has successfully signed-on or logged-in, most systems assume that the user continues to be legitimate.

In critical or high assurance environments, we would want to ensure that the user continues to be the legitimate user. Thus, authentication of the user needs to performed in a "*continuous*" fashion over the entire time interval when the user is actively using the system. It is also desirable that the authentication process be transparent so that users do not need to do anything special, e.g. periodic password challenges are not transparent. Continuous authentication using biometrics fits these needs since many biometrics can be monitored in a non-intrusive way at a sufficient frequency to approximate continuous authentication.

Ideally, a Continuous Biometric Authentication System (CBAS) should be transparent so that the user is hardly aware of its presence. This paper investigates the usability issues for a Windows-based CBAS. We believe this is the

M. Tistarelli and M.S. Nixon (Eds.): ICB 2009, LNCS 5558, pp. 828–837, 2009.
© Springer-Verlag Berlin Heidelberg 2009

first such study to evaluate the viability of CBAS on a general user population. Although the usage of a CBAS can be transparent, the overheads of continuous biometric verification and fusion can be significant which impacts on usability. We study the user perception of CBAS on first time users and whether the run-time overhead of CBAS affects their performance on common Windows tasks.

Our study on a moderate sized user population shows that a CBAS is quite acceptable to users. For interactive programs, such as Microsoft Office productivity applications, the system overhead of the CBAS, while significant, did not cause significant statistical difference on user performance. Although not the focus of this paper, we also show that the false reject rates for continuous biometrics are quite low. This paper gives the first successful use of continuous biometrics with a realistic user study which shows that continuous biometrics is usable for a general user population and also the viability of continuous authentication as security mechanism. Our results shows the potential of CBAS to achieve transparent continuous authentication without causing significant impact on overall usability.

2 Related Work

Recently there has been interest in continuous authentication. In response to the 11 September 2001 attacks, two design proposals using continuous biometrics authentication to safeguard the aircraft cockpit against unauthorized control were proposed by Carrilo [1].

Altinok and Turk [2] experimented with using voice, face, and fingerprint biometrics for continuous authentication. They articulated two key issues in continuous authentication, that is, the need for integration across both modality and time, and the requirement that the system must be able to determine "authentication certainty" at any point in time, even in the absence of observations. However, their work focused only on multimodal fusion and they did not study a real system with realistic workloads – their experiments used only simulated individuals because of the difficulty of getting real users.

In our previous work [3], we developed a bi-modal continuous biometrics authentication system integrated with the Linux kernel. This paper is based on a newer version of the CBAS which has been integrated with Windows.

Existing biometrics usability studies are based on one-off authentication systems which focus on the system's ability to enroll and verify individuals effectively. One-off systems are not relevant in a continuous authentication setting. Furthermore, CBAS has the additional factor of continuous runtime overhead which can impact on usability. We believe that this is the first paper to measure CBAS usability in realistic environment and task setting using task completion time [4] as an objective performance metric.

3 The Windows CBAS

This paper does not go into the details of the Windows XP-based CBAS system. Instead, we refer the interested reader to our previous paper [3] in which the

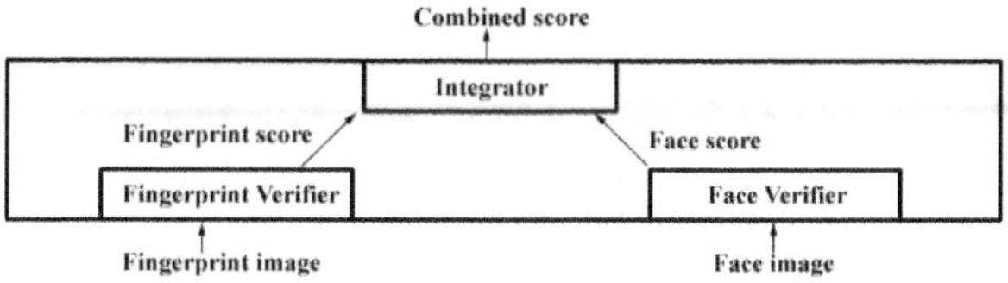

Fig. 1. CBAS architecture

decision-making method (using a Hidden Markov Model (HMM)) is described for the Linux operating system. We largely implemented the same method in Windows XP, with appropriate changes to take into account the differences in the operating system internals [5]. Here, we briefly describe the overall CBAS architecture, which suffices to understand the rest of the paper.

Our CBAS uses a normal webcam and a special mouse with a fingerprint sensor located at where the thumb would normally be placed. Face images captured by the webcam (at one image per second) are sent to a *Face Verifier* to compute a face score, while thumbprint images captured asynchronously by the fingerprint mouse (at one image every two seconds) are sent to the *Fingerprint Verifier* to compute a fingerprint score. Both scores are fused by the *Integrator* to produce a *Combined Score*. Fig. 1 shows these three components. The *Combined Score*, which reflects the probability of the continued presence of the user, determines whether the user is legitimate or an imposter. When CBAS determines the user is an intruder, it can be configured to disable the keyboard and mouse (thus locking out the user) to prevent further usage of the computer.

As our CBAS system fuses video and fingerprint biometrics, it allows for transparent monitoring. However, the challenge is that the continuous authentication requires frequent verification of both biometrics (between 0.5-2s) which itself leads to substantial overheads. We ran micro- and macro-benchmarks and found that overheads ranged from 26-42% which is consistent with [3]. This overhead is significant and highlights the need for usability studies on the performance impact of CBAS.

4 CBAS Experimental Design

The focus of this paper is to determine whether the overheads of CBAS will have an impact on overall usability on a user population. Participants were invited to our usability lab to carry out four work tasks for 1.5 to 2 hours on a workstation that had CBAS running in the background. They were told that the purpose of their participation was to help us try out a novel continuous biometrics authentication system.

Participants are first enrolled into the CBAS to train the face and fingerprint verifiers. After enrollment, participants were instructed to open the CBAS application called *BioMonitor* before beginning their tasks. They were informed that continuous authentication starts the minute they opened the BioMonitor

application. In reality, the BioMonitor could be either the actual CBAS BioMonitor or a placebo BioMonitor which resembled and behaved exactly like the real one. The overhead of the placebo BioMonitor was negligible. This allows us to subject the same participant to two different test conditions: (i) with the real BioMonitor running; and (ii) with the placebo BioMonitor running.

Out of the four tasks, the last task actually repeated a task similar to the first. Participants were asked to leave for a five-minute break after the third task was finished. While they were outside the lab, the version of the BioMonitor used was switched with the other BioMonitor (real with placebo, and vice versa). The experiment is designed to measure the effect of the real BioMonitor on task completion time: (i) for the real BioMonitor group, the effect of removing the BioMonitor overhead; and (ii) for the placebo BioMonitor group, the effect of adding the BioMonitor overhead. In the experimental setup, the placebo and real BioMonitor look alike, thus any difference in task completion times between the real and placebo can be attributed to the BioMonitor overhead [6].

Only one of the three tasks was repeated because it was difficult to make participants repeat more than one task; they may get bored, tired, annoyed. The first task was chosen to be the repeat task since it was done furthest away in time. According to Chambliss [7], repeating a task raises the problem of learning effects transfer. This means that the participant will become more familiar with the task after each repetition and therefore would be able to finish it more quickly. However, the repeated task should be one which should take about the same time, hence, small changes were made to the contents of the task (transposed text, different values and pictures in the sample task documents) to reduce learning effects. Despite this, we found that we still had to adjust for some learning effects (see Sec. 5.1).

4.1 Task Set

Tasks were designed to resemble possible real-life office work using Microsoft Word, Excel and PowerPoint. They were predefined so that users were only required to replicate the sample documents which they were given. These tasks used applications which were familiar to the participants. To accommodate participants with varying levels of expertise, the tasks required only the use of basic application features.

Each task was given in the form of a realistic scenario where the participant played the role of an employee having to do some work for his supervisor. Not only does this help the participant take on the role of actual end-users, it also means that the experiment results will be a better predictor of CBAS's performance in the workplace [8]. Participants may also find it easier to "stay in role" and overcome any latent hesitation and self-consciousness if the scenario reflected familiar situations, with realistic reasons (motivation) for performing the tasks [8]. During the pre-experiment briefing, participants were also reminded to perform the tasks as they normally would.

According to Dumas and Redish [4], usability testing is an emotional experience for participants. Many of them will be tense and they will need frequent

breaks to retain their concentration if they have to perform for more than a few hours. Hence, participants were asked to play computer games or watch short videos for about 10 minutes in between tasks.

We permuted the three tasks (Word, PowerPoint, Excel) to obtain six possible task order permutations. Each participant was then randomly assigned one permutation. The different permutations allow us to determine whether or not the results will vary when different types of tasks are repeated. It also reduces the chance that task order will affect completion time.

To keep track of the time-taken to complete individual tasks accurately, we used *Process Logger* [9] to track when programs on Windows are started and finished. It was chosen because of its low overhead (less than 1%). The timings used were real time since we want to measure task performance and overhead rather than CPU usage.

Dumas and Redish [4] highlight two problems related to timing how long it takes to complete tasks: (i) the participant has completed the task but is not ready to say that the task is done; and (ii) the participant says that the task is done, but in reality it is not. As such, the Experiment Administrator has to play the role of a moderator (to consistently apply the policy on when a task is considered done) and was present at all times to:

1. Observe the participants and intervene if they appeared to be having trouble with an application feature (e.g. inserting a table into a slide).
2. Ensure that the tasks were completed correctly (accuracy) before allowing them to continue. Since we want to time the tasks separately, we need a way to get participants to stop between tasks [4]. Hence, participants were given one task at a time. After checking that the task had been completed satisfactorily, the Experiment Administrator would then instruct the participant to quit the current application before beginning the next task.
3. Stop them if they have finished the task but still want to check their work. This is because prolonged checking should not be considered as task time.

4.2 Participants

The experiment was open to all students from our university since Microsoft Office 2003 is the most commonly-used productivity suite for their regular academic work. Over a two-week period, we had a total of 58 volunteers participating in the study coming from various faculties (29 males and 29 females).

5 Experimental Results

All participants, regardless of whether they used the real or placebo BioMonitor, underwent enrollment to train CBAS to recognize their face and fingerprint biometrics. This was needed for the placebo participants since their interaction with the system had to be identical to that of the real BioMonitor participants. Besides, they were also tested with the real BioMonitor for the final task.[1] The

[1] Recall that the placebo and real BioMonitor participants switch roles for the repeat (fourth) task.

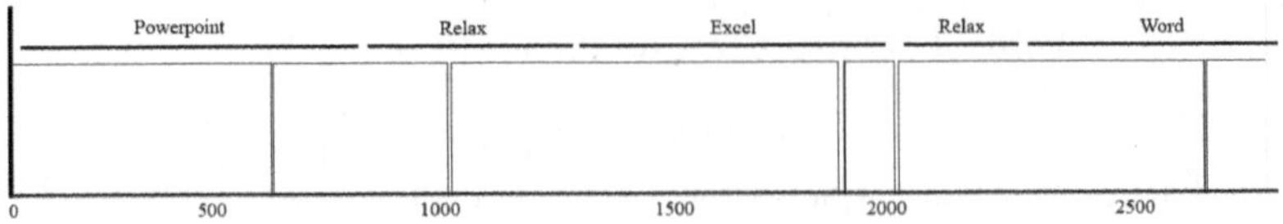

Fig. 2. A plot of *Combined Score* (vertical axis) vs. time (horizontal axis, in seconds) for a user. The dips which drop to 0 in the plot are False Rejects, i.e. where the *Combined Score* drops below the legitimate user/imposter threshold. The horizontal lines at the top indicate which application the user is running at the given times.

mean time for face enrollment was 101s, and that for fingerprint enrollment was 43s. Only 24% needed to repeat face enrollment, and only one person had to repeat fingerprint enrollment. Since all participants were first time users of the system, the enrollment process was rather smooth with the mean total enrollment time of 144s.

Although this study is primarily concerned with task performance under a CBAS, we also report the False Reject Rates (FRR). The FRR is the percentage of time the system falsely determines that the true user is an imposter. The BioMonitor continuously identified 48% of the participants correctly all the time (FRR of 0%) and the remainder (52%) had a non-zero but small FRR. For the real BioMonitor group, the mean FRR was 0.86%. In the case of the placebo BioMonitor group, the mean FRR was a little larger at 3.1%. This is not unexpected since the placebo group use the real BioMonitor during the fourth task which was only for a short period. Overall, the FRR is low, which suggests that most of the time the CBAS is *correctly identifying the user.*

In many cases, there were periods between the switching of tasks where no biometric observations were obtained and thus the BioMonitor could not verify the presence of the user. In other cases, users needed some time to settle down. Fig. 2 is an example of a graph of a user session where FRR > 0. There are a few small blips on the graph which is reasonable over the total period of 2800s. We expect that that the user might not always be using the mouse or have his face adequately captured by the webcam.

Two remarks about FRR and FAR (False Accept Rates) are in order: (1) Because we are primarily concerned with the impact of overheads on task completion time, we do not lock the user out when the *Combined Score* falls below the threshold. Doing so necessitates re-authentication, which will introduce delays not directly caused by overheads. Moreover, if the FRR is high, frequent lock-outs will frustrate the user, and may even cause him to abort work altogether. While these effects are worth studying, it is not the primary goal of this paper. (2) It is well-known that the FRR can be traded off with the FAR. Again, while it may be interesting to study such trade-offs, this is not the goal of our current work. Thus no FAR is reported here.

5.1 Accounting for Learning Effects

We first studied whether there are learning effects (see Sec. 4) in the completion time for repeating the last task. We invited 28 more participants for this learning

Table 1. Paired t-test results for learning effects

Task	N	Paired Differences					t	df	Sig- 2 tailed
		Mean Time	Std.Dev	Std. Error Mean	95% Confidence Interval of the Difference				
					Lower	Higher			
Word	10	59.100	48.884	15.459	24.130	94.070	3.823	9	.004
Powerpoint	8	233.500	140.495	49.672	116.043	350.957	4.701	7	.002
Excel	10	212.100	132.752	41.980	117.135	307.065	5.052	9	.001

effects experiment. These participants did not have to go through the enrollment process and they used a normal Windows setup with neither real nor placebo BioMonitor running.

Paired t-tests were carried out on the difference between the first versus the second completion time for all task types. Table 1 shows a summary of the paired t-test results. The mean times are measured in seconds. Our paired-samples t-test analysis indicates that at the $p < .01$ level:

1. For the 10 participants who repeated the *Word Task*, the mean time taken to complete it the first time ($518.20s$) was *significantly higher* ($p = .004, t(9) = 3.823$) than the mean time to do it the second time ($459.10s$).
2. For the 8 participants who repeated the *PowerPoint Task*, the mean first task completion time ($775.38s$) was *significantly higher* ($p = .002, t(7) = 4.701$) than the mean second task completion time ($541.88s$).
3. For the 10 participants who repeated the *Excel Task*, the mean first task completion time ($917.70s$) was *significantly higher* ($p = .001, t(9) = 5.052$) than the mean second task completion time ($705.60s$).

This means that despite our best efforts to reduce learning effects (the repeat task is similar to the first so that the task completion time should be similar), participants were still able to repeat the tasks more quickly for all task types.

To account for learning effects, the mean time differences (Word = 59.1s, PowerPoint = 233.5s, Excel = 212.1s) were used to adjust the completion time of the repeat tasks for the placebo and real BioMonitor groups. Although this is admittedly crude, it is unclear how else to account for the learning effect so that the participants do not suffer from experimental fatigue and other complications (more repetitions would have to account for more learning effects).

5.2 Completion Time with and without CBAS

To determine whether the BioMonitor had any effect on completion time, we carried out paired t-tests on the difference between the completion time of the BioMonitor group versus the placebo group ($n = 58$). The combined paired t-test results in Table 2 show that at the $p < .01$ level:

- There was *no significant difference* between time taken to complete the *Word* task with and without the real BioMonitor ($p = .135$)[$t(21) = -1.554$, $p > .05$]).

Table 2. Paired t-test results for task completion time

Task	N	Paired Differences: BioMonitor - Placebo Completion Time					t	df	Sig- 2 tailed
		Mean Time	Std.Dev	Std. Error Mean	95% Confidence Interval of the Difference				
					Lower	Higher			
Word	22	-21.955	66.278	14.131	-51.341	7.432	-1.554	21	.135
Powerpoint	18	121.500	320.883	75.633	-38.071	281.071	1.606	17	.127
Excel	18	17.056	171.686	40.467	-68.322	102.433	.421	17	.679

- There was *no significant difference* between time taken to complete the *PowerPoint* task with and without the real BioMonitor ($p = .127$)[$t(17) = 1.606, p > .05$]).
- There was *no significant difference* between time taken to complete the *Excel* Task with and without the real BioMonitor ($p = .679$)[$t(18) = .421, p > .05$]).

We see that there is no significant difference in completion times for all tasks. This gives evidence that even though the CBAS has significant impact on system overhead, for the interactive tasks tested in this experiment, we cannot conclude that there is any significant impact on task performance.

6 Post Usability Experiment Survey

As part of our usability study, we were also interested in finding out the subjective satisfaction of the participants with using CBAS. This is because there may be psychological factors at play, even with small system overheads the users might feel that the system is less usable or less responsive. On the other hand, there might be significant system overheads, yet the users could be quite happy with using the system.

Since human perception is to a large extent subjective, post test surveys are an essential part of usability studies because they collect subjective evaluation data about the usability of the system [8]. Our survey was self formulated; using five-point Likert scales for ratings and open-ended questions for participants to share their views in their own words. Due to space constraints, we list only four sample survey questions here, and discuss the findings in the next section:

- I felt comfortable using this system. (rating)
- Overall, I am satisfied with the responsiveness of the system whenever I type using the keyboard. (rating)
- Overall, I was comfortable using the fingerprint mouse throughout the session. (rating)
- Please share with us how you feel this biometrics monitoring security system can be improved. (open-ended)

7 Discussion

Our experimental study shows that there are no significant differences in the time taken to complete a task by the same participant, whether or not the CBAS is

continuously running in the background. We were pleasantly surprised by these results as we had expected that the high CBAS overhead (up to 42%) would have more impact in increasing the overall task time.

Analysis of the answers from the post experiment survey also show that participants were generally satisfied with the system's responsiveness to mouse and keyboard input. We had similar positive responses to questions on comfort level, satisfaction with the CBAS and overall ease of use. Thus, both the experimental results and the responses from the post experiment survey are consistent. We conclude that the CPU overhead of running the CBAS did not have a significant effect on the users' task performance or in their perception of system usability for all the tasks tested in the experiment.

We wish to point out that we do not claim that there is no difference between having the CBAS running versus not having it. System measurements show there is system impact but the question is whether this is significant given a realistic system use scenario. As CBAS is for humans, we focus on appropriate human-centric measures. Rather than CPU-bound tasks, we focus on interactive tasks. Furthermore, we measure our results based on the user performance metrics rather than system overhead.

When asked to share with us what they did not like about CBAS, many participants said that because the webcam was "too obvious", it made them feel uncomfortable. They also suggested that it would be better to have the camera hidden away. This shows that the discomfort may be not really be due to the fact that they were subjected to surveillance, but more so due to them being able to see webcam. This suggests that CBAS systems need to have more discreet camera design/placement. As for the fingerprint mouse, all participants said they were comfortable using it.

Finally, although our objective is to determine whether or not CBAS has an impact from a user's perspective, we also have some interesting results on enrollment and FRR. Even though our participants were all first time users to the CBAS, we had no problems with enrollment and the mean time needed was small, just 144s. We did expect the FRR to be non zero. However, in many cases, it turned out to be either zero for the entire session or for most of the session. Both of these findings also suggest that actual implementation and deployment of a CBAS may be easy.

8 Conclusion and Future Work

Although there have been proposals for CBAS design and prototype systems, the important human factor issues have not been previously addressed. We show using a moderately large scale usability experiment on real users that the effect of CBAS on users is small. They can perform tasks just as well even with a large CBAS system overhead. From a user perspective, the presence of the CBAS is not readily perceived. It is also relatively easy to enroll new users to the CBAS. The authentication error from continuous authentication is also small as the false reject rate is under 1%. Overall our results show that the key objective of

a transparent CBAS is realizable. Furthermore, we did not experience any user acceptance problems.

In the near future, we plan to: (a) gather participants from a wider age group / education background, so as to better generalize our findings to the larger population; (b) allow for multitasking to better approximate the working environment in the real world (our current study restricts users to one task at a time); (c) account for the novelty factor of using biometrics, which could have caused users to be more tolerant of any degradation in system performance; and (d) explore whether multi-core CPUs can reduce system overheads.

Acknowledgments

We acknowledge the support of Microsoft Research Asia (R-252-000-260-720) and the Defence Science Technology Agency of Singapore (R-252-000-261-422).

References

1. Carrillo, C.: Continuous biometric authentication for authorized aircraft personnel: A proposed design. Master's thesis, Naval Postgraduate School (2003)
2. Snelick, R., Indovina, M., Yen, J., Mink, A.: Multimodal biometrics: issues in design and testing. In: International conference on Multimodal interfaces, pp. 68–72 (2003)
3. Kumar, S., Sim, T., Janakiraman, R., Zhang, S.: Using continuous biometric verification to protect interactive login sessions. In: Annual Computer Security Applications Conference, pp. 441–450 (2005)
4. Dumas, J., Redish, J.: A Practical Guide to Usability Testing. Greenwood Publishing Group Inc., USA (1993)
5. Yap, R., Sim, T., Kwang, G., Ramnath, R.: Physical access protection using continuous authentication. In: IEEE Conference on Technologies for Homeland Security, pp. 510–512 (2008)
6. Alreck, P., Settle, R.: The Survey Research Handbook, 3rd edn. McGrawHill Publishers, New York (2004)
7. Chambliss, D.: Making Sense of the Social World. Sage Publications Inc., USA (2003)
8. Rubin, J.: Handbook of Usability Testing: How to Plan, Design and Conduct Effective Tests. John Wiley and Sons Inc., USA (1994)
9. Process logger, `http://keleos.h11.ru/proclog`

A First Approach to Contact-Based Biometrics for User Authentication

Athanasios Vogiannou[1,2], Konstantinos Moustakas[2], Dimitrios Tzovaras[2], and Michael G. Strintzis[1,2]

[1] Electrical & Computer Engineering Department, Aristotle University of Thessaloniki
[2] Informatics and Telematics Institute, CERTH

Abstract. This paper presents the concept of contact-based biometric features, which are behavioral biometric features related to the dynamic manipulation of objects that exist in the surrounding environment. The motivation behind the proposed features derives from activity-related biometrics and the extension of them to activities involving objects. The proposed approach exploits methods from different scientific fields, such as virtual reality, collision detection and pattern classification and is applicable to user authentication systems. Experimental results in a dataset of 20 subjects show that the introduced features comprise a very efficient and interesting approach in the research of biometric features.

1 Introduction

Biometrics have recently gained significant attention from researchers while they have been rapidly developed for various commercial applications ranging from access control against potential impostors to the management of voters to ensure no one votes twice [1,2]. These systems require reliable personal recognition schemes to either confirm or determine the identity of an individual requesting their services. A number of approaches have been described in the past to satisfy the different requirements of each application such as unobtrusiveness, reliability, permanence, etc.

Biometric methods are categorized to physiological and behavioral [3], depending on the type of used features. Physiological biometrics are based on biological measurements and inherent characteristics of each human. The fingerprint is a typical example of physiological biometrics which is widely used in law enforcement for identifying criminals [4]. Other recent applications are based on identification with the iris [5] or the face [6].

On the other hand, behavioral biometrics are related to specific actions and the way that each person executes them. The most known example of behavioral biometrics is gait recognition [7]. Behavioral biometrics are the newest technology in biometrics and they have yet to be researched in detail. They are supposed to be less reliable than physiological biometrics, however they are less obtrusive and simpler to implement [3,8].

M. Tistarelli and M.S. Nixon (Eds.): ICB 2009, LNCS 5558, pp. 838–846, 2009.
© Springer-Verlag Berlin Heidelberg 2009

Both types of biometric features are based on human characteristics of everyday life. One such interesting characteristic is the hand and the way people used it to manipulate objects. Although significant amount of research has been performed on various aspects of dynamic hand gestures [9,10], such features have not been employed for biometrics yet.

The present paper introduces contact-based biometric features and investigates the feasibility of such features in user authentication applications. The presented concept derives from the simple observation that every person handles the objects of the surrounding environment quite differently. For example, the action of picking up a glass or holding a knife depends on the physiological characteristics of each person and the way that this person is used to manipulate objects. Contact biometrics belong to the general category of behavioral biometrics and can also been thought as a specialization of activity-related biometrics [11,7].

The general overview of the framework is given in the section 2, while sections 3, 4 and 5 describe in detail the setup, the feature extraction and the authentication modules of the proposed approach respectively. Experimental results are provided in section 6 and the conclusions are drawn in section 7.

2 Overview

Although humans interact with objects in various ways, the present work concentrates on the hand as it is the major human part for object manipulation. It is important to point out that contact-based features of the hand are not the same as hand biometrics which have already been employed for human recognition [12]. Although certain hand characteristics, such as the size of the palm or the length of the fingers, have an effect in the way humans manipulate objects, contact-based biometrics are primarily concerned with the *behavioral features* and the *dynamics* of the specific action. In particular, except from the gesture posture of the hand, contact based features include the information about the collisions between the hand and the object.

In summary the proposed scheme consists of four major parts:

- Setup of a 3D virtual environment
- Registration of the user's hand and the objects in the virtual environment
- Extraction of features during an action performed by the user using an object
- Final classification and authentication

The 3D environment and the registration of the objects in virtual environment is necessary because collision detection algorithms can only be used with knowledge of the 3D geometry of each object. Based on this setup, the next step is to extract different types of features and produce the final authentication result.

3 3D Environment Setup and Model Registration

The most important, and at the same time difficult, task of the registration procedure is the 3D reconstruction of the user's hand. In the proposed method, the

hand is modeled as a set of 5 fingers connected to the palm, which is represented by a simple rectangle (Figure 1(a)). Each finger has 4 Degrees of Freedom (DOF) and consists of 3 phalanxes which are modeled as simple ellipsoids. Other 3 DOF describe the orientation of the whole hand.

For the registration of the hand, the CyberGlove® (www.immersion.com) device was employed. The CyberGlove® (Figure 1(b)) provides the angles between the phalanxes of the hand, and therefore it is possible to reconstruct a 3D representation of the hand. Note, that the virtual representation of the hand is not perfectly accurate because the size of the fingers and the phalanxes are not known. This inaccuracy is considered as noise and does not significantly affect the results.

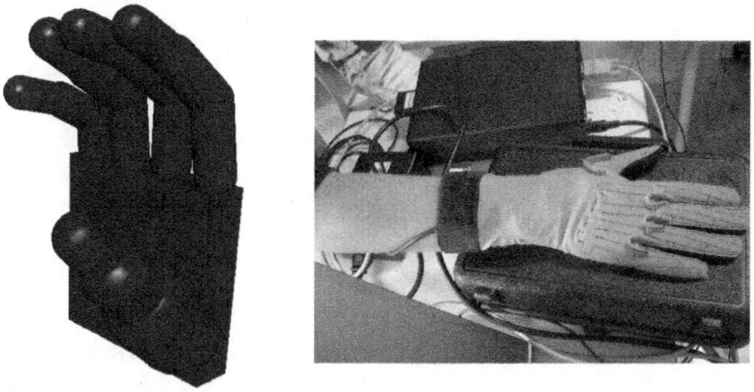

(a) (b)

Fig. 1. The 3D representation of the hand (a) and the CyberGlove® (b)

The objects of the environment can be registered using computer vision techniques for object tracking [13]. Since object tracking is out of the scope of this paper, a simple color based tracker [14] was implemented. However, it is not absolutely necessary to have an accurate representation of the object in the virtual environment. Regarding rigid objects, which are the most common case, the geometry of each object can be simplified using a priori information. This simplification is possible as the real shape of each object is usually not directly related to the way it is handled. For example, a glass can be represented by a simple cylinder since the user grabs only the outer surface of the glass.

4 Feature Extraction

The classification features of contact-based biometrics are twofold. From one hand there is the feature set of the hand posture (static, user oriented) and on the other the collision information (dynamic, user-object interaction). The first set is related to the hand posture and consists of the data extracted from the CyberGlove® while the second one is based on collision detection algorithms

and the information that we can retrieve from them. Although collision features are indeed related to the hand posture, the transformation between these two feature spaces is unknown and impossible to define for general shapes due to the algorithmic nature of collision detection methods and the dynamic position of each object.

4.1 Hand Posture Features

Any representation of the hand posture can be linearly defined by the angles produced by the CyberGlove® device. Therefore, these angles constitute a feature space of the hand posture. Figure 2 displays the position and the orientation of the joints and the respective measured angles in the hand. Every DOF of each joint corresponds to a feature. The final feature $\boldsymbol{F}_a$ vector is a 23-dimensional vector consisting of

$$\boldsymbol{F}_a = \{\boldsymbol{W}_0, \boldsymbol{W}_1, ..., \boldsymbol{W}_{22}\}$$

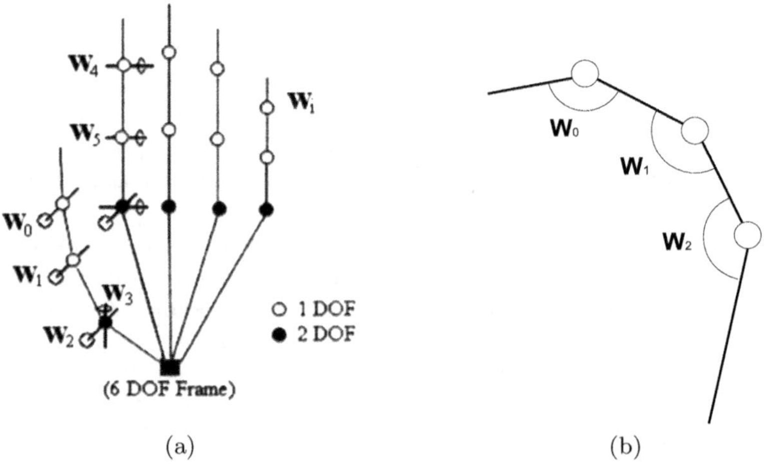

Fig. 2. Posture Features. Each DOF of the joints between the phalanxes and the palm corresponds to a feature. Figure (a) illustrates the measured angles for all the hand while figure (b) focuses on a single finger.

4.2 Collision Features

The collision features consist of any information that can be acquired by employing state-of-the-art algorithms for proximity queries. These include penetration depth[15], closest distance [16,17], contact points etc. The literature in the field is vast and there are numerous algorithms to accurately perform queries in real-time. The interested reader is directed to [15,16,17,18,19,20,21,22] for further details. For the purposes of the presented work, the algorithms for rigid convex objects [23,19] of the software package SOLID (http://www.dtecta.com/) were used.

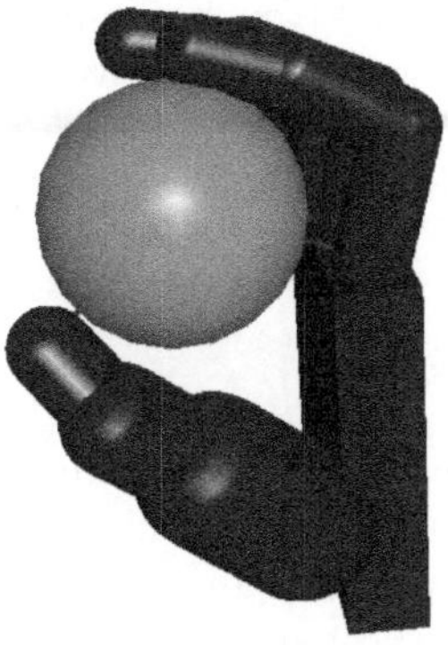

(a)

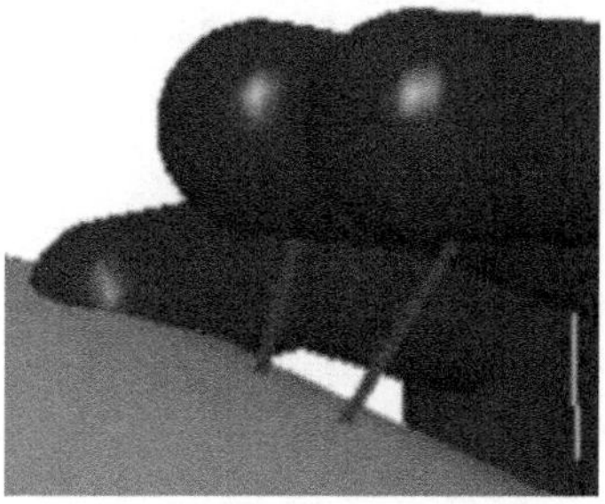

(b)

Fig. 3. Collision Features. Either the closest distance or the penetration depth is used, depending on the collision state between each finger and the object.

Proximity queries are performed between the object and every finger of the user's hand. Each query refers to either of two states, collision or no collision between the two virtual shapes. For example penetration depth can only be calculated when two objects intersect since it is always 0 otherwise. However, in a user-object interaction scheme it is necessary to continuously produce discriminant feature samples. Thus, any proximity query as a single feature would not provide adequate information to a classifier.

In order to overcome this difficulty, the combination of the penetration depth and the closest distance is used, depending on the collision state, to define the feature space. The penetration depth and the closest distance are usually described as 3D vectors in virtual simulations. However, in our case it is preferred to describe them as the pair of points $(\mathbf{p}_{finger}, \mathbf{p}_{object})$, one on the finger and the other one on the object, that define the respective vector $\mathbf{v} = \mathbf{p}_{finger} - \mathbf{p}_{object}$. This way the 3D position of each finger affects the values of the feature vector, while $\mathbf{v}$ would only describe the relative direction which is usually similar even for different fingers. The 3D vectors are represented in the local coordinate system of the hand. Let $\mathbf{pd}_k$ and $\mathbf{cd}_k$ denote the points of the penetration depth and the closest distance respectively where k is either a finger or the object. The feature sample $f_e(i, O)$ for finger e and object O on the i-th frame is

$$f_e(i, O) = \begin{cases} (\mathbf{pd}_e, \mathbf{pd}_O), & \text{if e and O collide} \\ (\mathbf{cd}_e, \mathbf{cd}_O), & \text{if e and O do not collide} \end{cases}$$

Figure 3 displays the closest distance between two fingers and the surface of a 3D object. The final collision feature vector $F = \bigcup_e \{f_e\}$ is formed using the collision information from all the 5 fingers and is a 30-dimensional vector.

The feature vectors of both collision and angle data are calculated during an object-related action at every frame update of the virtual environment. Note that in a realistic application we would also need an activity recognition module to detect object-related actions in a free environment [24]. However, the presented

work focuses on the evaluation of contact-biometrics and not in the general setup of a specific application. Therefore, it is considered that the extracted feature data refer only to a specific object-related action.

5 Classification and User Authentication

In general, authentication systems involve an enrollment stage where the system calculates and stores a template of the subject's feature measurement. For most biometric features, this is a single feature vector. However, in the case of the presented method, the enrolment sessions generate a significant number of samples due to the high sampling frequency (90 smples/sec) of the CyberGlove® and the dynamics of the action. The set of enrollment samples is used for training a classifier so that it recognizes impostors from genuine users.

More specifically, let F_i denote the training set of feature samples of subject s_i. The enrollment sample ${f_i}^e$ is selected as the average feature of the enrollment session. Then, the feature sets of genuine user G_i and impostor users I_i are generated by

$$f \in G_i : f = a - {f_i}^e \quad \text{where} \quad a \in F_i$$

and

$$f \in I_i : f = a - {f_i}^e \quad \text{where} \quad a \in F_{k \neq i}$$

The above 2-class classification problem is solved with Support Vector Machines (SVMs) in the proposed method. In particular, a SVM M_i is trained for each subject s_i so that it separates G_i from I_i.

When a user claims to be the subject s_i, the system samples the respective feature vector f and calculates $f_d = f - {f_i}^e$. If f_d is classified in G_i, the user s_i is authenticated. Otherwise the user is considered to be an impostor.

6 Experimental Results

The experimental setup includes one testing action that is performed by 20 subjects. In particular, the right hand of each user and a glass were registered in the virtual environment for the action notated as "grabbing the glass". Each person performed the action 10 times which produced around 1000 feature samples in total for each subject.

In the following, the results are presented in two parts. The first part evaluates the distinctiveness while the second part the performance of the proposed authentication approach. The results are also separated according to the type of features, i.e. only collision features, only angle features and the combination of the two, and according to the type of the kernel used with the SVM.

6.1 Authentication Distinctiveness

Table 1 displays the False Rejection and False Acceptance rates for the 8 test sessions of the 20 subjects. All the sessions were performed with the same calibration of the glove for each subject, i.e. the subjects did not put off the glove

Table 1. FRR and FAR for 20 subjects

Kernel Type	Linear			RBF		
Features	Posture	Collision	Combined	Posture	Collision	Combined
FRR	0.01 %	0.5 %	0 %	0 %	0.1 %	0 %
FAR	0 %	0.08 %	0 %	0 %	0 %	0 %

between sessions. From the acquired samples, the first session was used for the enrollment and the rest were used for testing.

The results demonstrate the high distinctiveness of the proposed features since FRR and FAR are practically 0. Although these results confirm our assumption, there is also the concept of consistency and the reliability of the measurements which is evaluated in the next section.

6.2 Authentication Performance

The term performance used here is generally defined in [3]. More specifically, what is of concern in our case is the consistency and the accuracy of the measurements. In order to evaluate the performance of the proposed approach, data acquisition were performed with two different calibrations. More specifically, each subject performed another set of 10 sessions but in the meantime it put off the glove and put it on again. No restrictions or guidelines were given on how to wear the glove at either of the session sets.

Table 2 displays the results. The efficiency of the method was reduced, when compared to the first data-set (Figure 1). This is caused by the difference in the way that each subject put on the glove at the second calibration, introducing this way a significant amount of undefined noise. Nevertheless, the combination of the features is very robust, especially for the FRR. This fact illustrates the high authentication efficiency of the proposed approach in the presence of noise.

Table 2. FRR and FAR for 20 subjects with different calibrations of the glove

Kernel Type	Linear			RBF		
Features	Posture	Collision	Combined	Posture	Collision	Combined
FRR	2.5 %	4.2 %	0.6 %	2.3 %	3.13 %	0.6 %
FAR	4.3 %	10.76 %	4.5 %	4.3 %	5.8 %	4.5 %

7 Conclusions and Future Work

In this paper we presented the novel contact-based biometric features that are related to activity-related biometrics and include the dynamic characteristics of interactions with objects. The results show that the proposed method can achieve very high rates of authentication performance and therefore comprises a very interesting approach for further research in activity-related biometrics. The proposed method can be integrated along with other types of features in a user authentication system so as to improve the overall efficiency.

Acknowledgements

This work was supported by the EU funded ACTIBIO IST STREP (FP7-215372).

References

1. Qazi, F.A.: A survey of biometric authentication systems. In: Security and Management, pp. 61–67 (2004)
2. Xiao, Q.: Security issues in biometric authentication. In: Information Assurance Workshop, IAW 2005, June 2005, pp. 8–13 (2005)
3. Jain, A., Ross, A., Prabhakar, S.: An introduction to biometric recognition. IEEE Transactions on Circuits and Systems for Video Technology 14(1), 4–20 (2004)
4. Maltoni, D., Maio, D., Jain, A.K., Prabhakar, S.: Handbook of Fingerprint Recognition. Springer, Heidelberg (2003)
5. Chowhan, S., Shinde, G.: Iris biometrics recognition application in security management, vol. 1, pp. 661–665 (2008)
6. Chang, K.I., Bowyer, K.W., Flynn, P.J.: An evaluation of multimodal 2d+3d face biometrics. IEEE Trans. Pattern Analysis and Machine Intelligence 27(4), 619–624 (2005)
7. Boulgouris, N., Chi, Z.: Gait recognition using radon transform and linear discriminant analysis. IEEE Transactions on Image Processing 16(3), 731–740 (2007)
8. Delac, K., Grgic, M.: A survey of biometric recognition methods. In: Electronics in Marine, 2004. Proceedings Elmar 2004. 46th International Symposium, pp. 184–193 (2004)
9. Moustakas, K., Strintzis, M., Tzovaras, D., Carbini, S., Bernier, O., Viallet, J., Raidt, S., Mancas, M., Dimiccoli, M., Yagci, E., Balci, S., Leon, E.: Masterpiece: Physical Interaction and 3D Content-based Search in VR Applications. IEEE Multimedia 13(3), 92–100 (2006)
10. Tzovaras, D., Nikolakis, G., Fergadis, G., Malasiotis, S., Stavrakis, M.: Design and Implementation of Haptic Virtual Environments for the Training of the Visually Impaired. IEEE Transactions on Neural Systems and Rehabilitation Engineering 12(2), 266–278 (2004)
11. Kale, A., Cuntoor, N., Chellappa, R.: A framework for activity specific human recognition. In: International Conference on Acoustics, Speech and Signal Processing, Orlando, FL (2002)
12. Kukula, E., Elliott, S.: Implementation of hand geometry: an analysis of user perspectives and system performance. IEEE Aerospace and Electronic Systems Magazine 21, 3–9 (2006)
13. Yilmaz, A., Javed, O., Shah, M.: Object tracking: A survey. ACM Comput. Surv. 38(4) (2006)
14. Ren, Y., Chua, C.S., Ho, Y.K.: Color based tracking by adaptive modeling. In: 7th International Conference on Control, Automation, Robotics and Vision, ICARCV 2002., vol. 3, pp. 1597–1602 (2002)
15. Kim, Y.J., Lin, M.C., Manocha, D.: Incremental penetration depth estimation between convex polytopes using dual-space expansion. IEEE Transactions on Visualization and Computer Graphics 10(2), 152–163 (2004)
16. Larsen, E., Gottschalk, S., Lin, M., Manocha, D.: Fast distance queries with rectangular swept sphere volumes, vol. 4, pp. 3719–3726 (2000)

17. Lin, M.C., Canny, J.F.: A fast algorithm for incremental distance calculation, pp. 1008–1014 (1991)
18. Lin, M.C., Gottschalk, S.: Collision detection between geometric models: A survey. In: IMA Conference on Mathematics of Surfaces, pp. 37–56 (1998)
19. van den Bergen, G.: Collision Detection in Interactive 3D Environments. Morgan Kaufmann, San Francisco (2003)
20. Ericson, C.: Real-Time Collision Detection. Morgan Kaufmann, San Francisco (2004)
21. Thomas, F., Torras, C.: 3D collision detection: A survey. Computers and Graphics 25, 269–285 (2001)
22. Teschner, M., Kimmerle, S., Zachmann, G., Heidelberger, B., Raghupathi, L., Fuhrmann, A., Cani, M.P., Faure, F., Magnetat-Thalmann, N., Strasser, W.: Collision detection for deformable objects. In: Eurographics State-of-the-Art Report (EG-STAR), pp. 119–139. Eurographics Association (2004)
23. van den Bergen, G.: Efficient collision detection of complex deformable models using AABB trees. Journal on Graphics Tools 2 (1997)
24. Gavrila, D.M.: The visual analysis of human movement: a survey. Computer Vision and Image Understanding 73(1), 82–98 (1999)

Template Update Methods in Adaptive Biometric Systems: A Critical Review

Ajita Rattani, Biagio Freni, Gian Luca Marcialis, and Fabio Roli

Department of Electrical and Electronic Engineering,
University of Cagliari, Italy
{ajita.rattani,biagio.freni,marcialis,roli}@diee.unica.it

Abstract. Template representativeness is a fundamental problem in a biometric recognition system. The performance of the system degrades if the enrolled templates are un-representative of the substantial intra-class variations encountered in the input biometric samples. Recently, several template updates methods based on supervised and semi-supervised learning have been proposed in the literature with an aim to update the enrolled templates to the intra-class variations of the input data. However, the state of art related to template update is still in its infancy. This paper presents a critical review of the current approaches to template updating in order to analyze the state of the art in terms of advancement reached and open issues remain.

1 Introduction

Biometrics refers to the use of physiological or biological characteristics to recognize the identity of an individual. These characteristics are unique to each individual and remain unaltered during the individual's lifetime [1].

A typical personal biometric authentication system operates in two distinct modes: enrolment and verification. In enrolment mode, individual's biometric sample(s) are captured, processed, features extracted and labels are assigned to them to establish identity, representing the template(s) of an individual. Verification mode, compares query biometric samples of the respective individual with the enrolled template(s) to verify an identity. It is quite difficult to have a perfect match ('100%') between template and a genuine query image, due to the presence of several factors: human-sensor interaction, environmental conditions, temporal variations (short term like scars in fingerprint surface and long term like aging in face) and other temporary variations like change in facial expression or affine transformation in fingerprints. These factors introduce variations in the query image, resulting in performance degradation and making enrolled templates, usually acquired in controlled environment, un-representative [2].

The solution lies in making the biometric system "adaptive" to the intra-class variations of the input data. The standard approaches to deal with this problem are acquiring multiple impressions of each individual, making the threshold parameters adpative, or adopting multi- biometrics [1].

Recently, novel solutions have been introduced in the form of template updating where the classifiers adapts itself to the intra-class variations based on learning

M. Tistarelli and M.S. Nixon (Eds.): ICB 2009, LNCS 5558, pp. 847–856, 2009.
© Springer-Verlag Berlin Heidelberg 2009

methodologies like supervised or semi-supervised learning. Although these methods are promising, the state-of-art related to them is still in its infancy. A unified taxonomy is missing along with the lack of clear statement of the problem [3]. Therefore, to further stimulate research on this important topic of template un-representativeness and their solution as update procedures, this paper critically reviews the template update methods to biometrics. The aim of this survey is to highlight current state-of-the-art with respect to current approaches, in particular, the learning methodology adopted and experimental evaluation followed. Our goal is also to give an insight into the open issues that still remain to be addressed.

Thus, we propose a taxonomy in which the current template update methods can be classified (Section 2). We discuss and compare some of the key features of these methods (Section 2.1 and 2.2), and the usual template update and management procedure adopted (Section 3). We have also analysed different protocols followed for the performance evaluation with respect to the threshold parameter set, the number of unlabelled data used for updating the templates, properties of database used, and percentage recognition accuracy enhancement reported in the literature, traits covered so far etc. As all these factors are very vital information to be taken into consideration for a fair evaluation of each update method (Section 4). Section 5 concludes the paper and discusses some open issues that still remain to be considered.

2 Biometric Template Update Methods: A Taxonomy

Template representativeness is a crucial problem in the biometrics, as the input biometric data is subject to on-going changes due to presence of intra-class variations thus making the initial enrolled templates non representative of them. As a result, performance degradations arise. To deal with this issue, adaptive biometric systems, based on appropriate "template update" methods, have been proposed [3]. The basic step in these methods are the assignment of identity labels, considering which a possible taxonomy is proposed in Figure 1. Identity label assignment can be done either in a completely supervised manner (i.e. by human expert intervention), or by using automatic "learning" methodology ("semi-supervised"). The key difference between "supervised" [2, 4-5] and semi-supervised learning [6-13, 15] is the technique followed for the data labelling. In supervised template update methods, the label assignment is manual, whilst in semi-supervised methods, it is automatic.

Supervised methods (Fig. 1, Vertex 1) proposed so far are usually offline [2, 4-5] as they operate on a batch of collected samples. On the other hand, semi-supervised methods (Fig. 1, Vertex 2) are automated methods that assigns identity labels to the unlabelled data on the basis of their own knowledge, derived through current enrolled templates without the intervention of human supervisor [6-13, 15]. These methods avoid the cost related to manual assignment of labels to unlabelled data used for updating. Therefore, they can also operate online.

The second basic step is, which is also followed in the taxonomy, is the template selection procedure. Among all templates, manually or automatic labelled, it is necessary to select the most "representative" ones, in terms of intra-class variations. Accordingly, supervised methods can be divided on the basis of template selection techniques (Fig. 1, Vertex 1.1) employed to reduce the gallery size, that is, clustering

(Fig. 1 Vertex 1.1.1) [2], and editing [4] (Fig. 1 Vertex 1.1.2). On the basis of the same selection principle, semi-supervised methods (Fig. 1 Vertex 2) can be further subdivided into: single modality [6-9, 12] (Fig. 1 Vertex 2.1) and multiple modality (Fig. 1 Vertex 2.2) [10, 11, 15] on the basis of whether the selection and update process operates in a uni-modal or multimodal environment. The single modality methods are further categorized as based on self-training [6-9] and graph-based [12]. Methods based on self-training (Fig. 1 Vertex 2.1.1) updates themselves iteratively using the highly confident input data [6-9] and graph based methods [12] (Fig. 1 Vertex 2.1.2) work by analysing the structure of the input data in the form of a graph [7]. The self-training based methods are further categorized as online (Fig. 1 Vertex 2.1.1.1) [6-8] or offline methods (Fig. 1 Vertex 2.1.1.2) [9] on the basis if the update is performed as soon as the input data arrives or after the batch of data has been collected during the system operation. Feature editing has also been employed for online methods [6-7] (Fig. 1 Vertex 2.1.1.1.1). The proposed graph based [12] and multimodal update methods [10, 11, 15] are offline methods (Fig. 1 Vertex 2.1.1.2). According to the presented taxonomy, the proposed template update methods in the scientific literature have been critically reviewed.

Sections 2.1 and 2.2 discuss methods in the proposed taxonomy in more detail.

2.1 Supervised Methods

Supervised methods (Fig. 1 Vertex 1) to template update have been introduced in [2, 4-5]. These methods update the system to the batch of unlabelled data collected during the system operation. They require the intervention of supervisor for assigning identity labels to the collected data which are used for updating. The labels assigned during the normal system operation are stripped off and the positively labeled samples by the human expert are added to the template set for updating. These are offline methods as the updating task is performed when the batch of input data has been collected after a fixed time interval. The supervisor should have a knowledge of the correct intra-class variations as noisy biometric data. For example, a noisy fingerprint image due to dirt accumulation on fingerprint sensor is also a difficult sample, but does not represent a genuine intra-class variation. Therefore, supervised methods which exploit properties of fingerprints in terms of features used for their comparison (e.g. minutiae), perform automatic template selection (Fig.1 Vertex 1.1.).

These methods can be classified on the basis of the type of techniques used for selecting the representative templates: clustering-based [2] (Fig.1 Vertex 1.1.1) and editing-based [4] (Fig.1 Vertex 1.1.2). The former are based on applying standard clustering algorithms to each updated template gallery, and selecting an appropriate template from each cluster. To this aim, two algorithms, namely, MDIST, based on finding the templates which minimize the intra-distances among samples in the gallery, or DEND, based on the so-called "dendogram" [17], in order to select the templates exhibiting large intra-class variations, have been proposed. In latter [4], each gallery is viewed as a training set of a NN classifier. Accordingly, if a large batch of samples is available, the aim is to "edit" it in order to find the subset exhibiting the minimum size and also keeping the same classification accuracy of the training set. In [4], several existing editing algorithm have been applied with promising results. In particular, it has been noticed that editing based approaches are

not dependent on the "number of clusters" (i.e. number of templates), an input parameter for clustering-based approaches, set manually. Moreover, it has been argued that editing approach could help in retaining variable number of templates for each individual, by detecting easy and difficult clients, for which a small number of templates are enough or which require large number of templates, for a good classification accuracy [4].

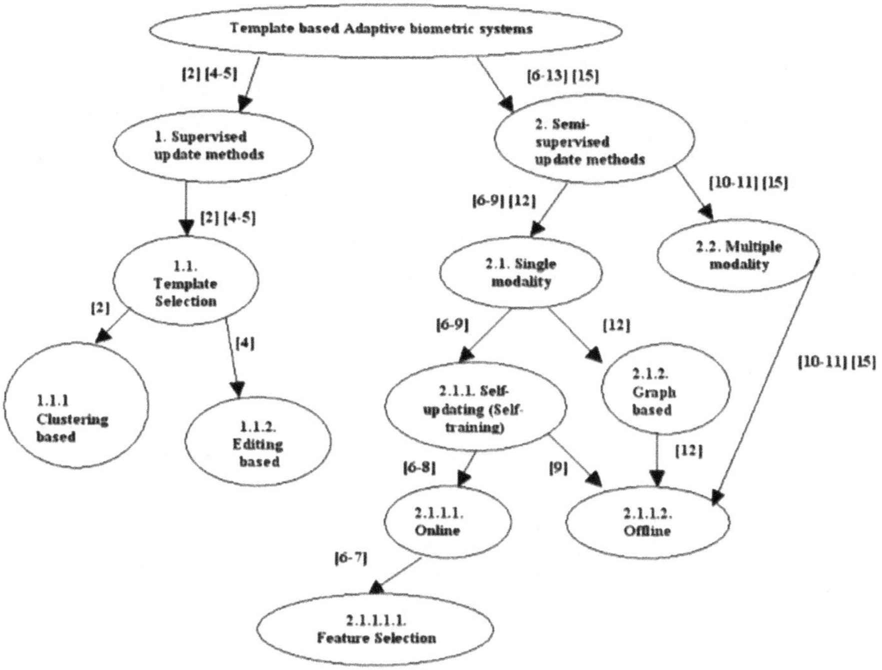

Fig. 1. The proposed taxonomy for the template update methods. The edges are labeled by the reference number of the proposed methods.

2.2 Semi-supervised Methods

As an alternative to supervised approaches, semi-supervised methods (Fig.1 Vertex 2) have been introduced [3, 6-13, 15]. "Semi-supervised" methods derive their name from the machine learning approaches related to the joint use of labeled and unlabelled data for improving the classification accuracy, named as "semi-supervised learning" [18]. In this topic, term "labeled data" is used to denote initial training samples, and "unlabelled data" to the samples available during system operation and "pseudo-labeled" by the classifier. In biometric applications, the labeled data are the initial enrolled templates and the unlabeled ones are the input biometric data collected during the system operation. Since classification approach is threshold-based in biometrics, labeled and unlabelled data are jointly used to modify such threshold and the composition of galleries by adding "pseudo-labeled" data to the related client's gallery.

Semi-supervised methods can be subdivided into "single-modality" (Fig. 1 Vertex 2.1) and "multi-modality" (Fig. 1 Vertex 2.2) based approaches.

2.2.1 Single Modality

Single modality based methods update the biometric system without using any external help or knowledge [2-5, 7]. They can be based on "self-training" (Fig.1 Vertex 2.1.1) or graph-mincut (Fig.1 Vertex 2.1.2). These methods can be either online [6-8] or offline [9-13, 15-16], depending on whether the templates are updated as soon as the input data arrives or after a batch of data has been collected over a specific time-period.

a) Self-Updating "Self-training" based methods

These methods to template update, update themselves iteratively (Fig.1 Vertex 2.1.1) [6-9]. Specifically, these methods iteratively add "highly genuine" unlabelled data to the template set of the respective client. Highly genuine data is recognized if matching score between input data and the nearest template in the gallery, is over a fixed "updating threshold". In online updating (Fig.1 Vertex 2.1.1.1), templates are updated as soon as the input data is available [6-8]. Some online approaches [6-7] also proposed an appropriate method for managing templates into the gallery (see Section 3). Offline methods (Fig.1 Vertex 2.1.1.2) [9-13, 15] to template update operate on the batch of unlabelled data (labels assigned during the system operations are stripped off). It is worth mentioning that both of these types of method are dependent on the threshold value selected for updating, as only the input data whose matching score is above the selected threshold is added for updating. They operate at stringent acceptance threshold and exploit (i.e. add to the clients' galleries) only the confidently classified samples in order to avoid the introduction of impostors into the gallery set of the client. But it has been shown that these methods can exploit only the patterns very similar ("near") to the enrolled templates which leads to non-exploitation of many difficult and informative intra-class variations but relaxing the threshold results in more probability of impostor's introduction [11]. They are also dependent on the representativeness of initial enrolled templates. Online methods are also dependent on the order of the sequence of input data as it affects the matching score [7]. Ref. [21] have compared the performance of offline methods with online methods and no significant differences were reported in the performance.

b) Graph-based methods

Graph based methods [12] (Fig.1 Vertex 2.1.2) try to overcome the limitation of self update, which is only able to exploit data very near to initial templates, by organizing labeled and unlabelled data in a graph-like structure, where nodes are all the samples and edges are labelled with similarity values among them. The aim is assign labels to input data by studying the underlying structure between the labeled and unlabelled data.

Among the various graph based methods in semi-supervised learning [18], graph-mincut optimization has been proposed in [12]. The aim of graph mincut is to find, minimum weight set of edges, if cut, partitions the graph in two parts related to two opposite classes, the unlabelled data belonging to genuine class partition are labelled positive [20]. This method has the advantage of bypassing the selection of the optimum threshold, the method is not iterative hence not time consuming. Preliminary

experiments on comparison of graph-mincut with self update states that the former results in better assignment of labels and also reduces the creep in of impostors. On the other hand, these methods can only be offline (Fig. 1 Vertex 2.1.1.2).

2.2.2 Multi-modality

These methods utilize multimodal biometrics for updating (Fig.1 Vertex 2.2) [10,11,15]. They are based on the "co-training" concept of semi-supervised learning technique [19]. Specifically, two classifiers help each other in the updating process on the assumption that they work on i.i.d independently and identically distributed data [10-11]. The main intuition is that, if matchers are complementary, one matcher operating at high confidence helps other to identify "difficult patterns" [10-11, 15].

The offline version of this method (Fig.1 section 2.1.1.2) has been proposed in [10] with preliminary experiments. These showed that co-update is better than self-update in capturing large intra-class variations of the input data without relaxing acceptance threshold. Another work [11] investigated the conditions under which the co-update can update the system better than self-update specifically. Reported results pointed out that co-update can outperform self-update when the initial enrolled templates are non-representative of the input data as can happen in an uncontrolled environment. However, in a controlled environment when the input data is representative, performances of both self-update and co-update are comparable [11]. This work also presented for the first time in literature the effects of impostor's introduction and studied the performance degradation of self-update due to this problem [11].

3 Template Management Strategies

In this section, it is discussed the strategy of managing selected input samples in the template gallery. An appropriate strategy is necessary since, in real applications, all possible collected intra-class variations are not virtually retainable, due to the constraints of the available memory size or verification time required, which obviously increases with the number of templates in the gallery. With regard to this problem, the procedure of updating of template galleries so far are: (1) super template based, in which the input data is always fused to a common single template called "super template" embedding all the information together [6-8], or (2) instance based, in which the input data is always added as a separate instance to the gallery set of the respective client [9-13, 15]. Most of the self-update online methods (Fig.1 Vertex 2.1.1.1) have followed super template based approach. However, the offline based self-update (Fig.1 Vertex 2.1.1.2), template co-update (Fig.1 Vertex 2.2) and graph-based methods (Fig.1 Vertex 2.1.2) have followed instance based approach. Recently, an intermediate approach has been followed too by always fusing the data which is similar to enrolled templates and adding intra-class variations as a separate instance to the template gallery set [11].

The above mentioned approaches, does not take into account that, in real applications, the available memory space is quite limited. Thus, due to space and time constraints may be some fixed number, for example, only M templates (with M small) could be stored. In order to take advantage of benefits of template update in constraint environment, template replacement criteria [5] have been proposed. They

treat templates as as a set of samples where obsolete templates can be "replaced" by novel and more representative ones. Accordingly, approaches like FIFO, LFU, and clustering algorithms, used for different contexts [2], have been modified [5] in order to be employed for template replacement. This approach has been so far tested online, and has shown the property of strongly decreasing misclassification errors during updating, by limiting the problem of impostors insertion in client's galleries [5].

4 Experimental Protocols and Performance Evaluation

The effectiveness of a template update method is dependent on the operational factors such as the number of verification attempts, the prior probability of an impostor attempt and on the availability of intra-class variation over the operational timescale, thus the experimental protocol adopted for the evaluation of these methods are worthy to be examined.

An experimental protocol defines the number of initial images to be used for training the system, how is the threshold set, is it re-estimated or kept fixed, how many samples are used for updating the templates, what is the stopping criteria of these update methods, how is the performance enhancement evaluated for each method. The evaluation of these methods are also affected by the properties of the dataset in respect of the number of samples per client, more the better, and the period of the collection span of the database to account for temporal intra-class variations too.

Different protocols have been adopted in the literature of template update methods. Also a separate paper proposing flow chart of template-update methods and a protocol for performance evaluation has also been introduced in the literature [16]. The aim of this protocol is to evaluate the performance of the update procedure ir-respective of the order of sequence of input unlabelled data arrival. Accordingly, three ordered sequence of unlabelled data are used for updating the templates i.e.,(i) impostor matching first, in which the impostor unlabelled samples are followed by genuine ones, (ii) random matching, random presentation of genuine or impostor samples, (iii) genuine matching first, in which genuine samples are followed by impostor ones. The performance is evaluated accordingly for each sequence individually and is averaged, giving overall enhancement. Some self-update online methods (Fig.1 Vertex 2.1.1.1.) [7] have followed this protocol. Other self-update methods have only operated only at random matching scenario [6, 8]. All the self-update methods have operated at fixed threshold which is not re-estimated after each update procedure by an unlabelled input data.

However, most of the template self-update offline methods [9] (Fig.1 Vertex 2.1.1.2) and multi-modality based co-update methods [10-11, 15] (Fig.1 Vertex 2.2) have followed the protocol of training using an initial template and have operated at threshold fixed at 0 or 1% FAR and is re-estimated after each iteration of updating [7-10, 12] [10-11, 15]. The performance of online update procedures are reported to progress as the input data are used for updating without evaluation on a separate test set [6-8]. However, offline methods have used separate test set [9, 10-11, 15] for the performance evaluation due to updated templates. The graph based methods [12] (Fig.1 section 2.1.2) have also followed the protocol of offline methods but the difference is in the technique, as these methods assign labels to all the unlabelled data together, the performance on the test set has been evaluated after all the unlabelled samples being labeled and the template gallery updated, instead of progressive

Table 1. Description of the dataset used, their sizes, number of unlabelled data and if separate test set used, the size of it is mentioned for fair evaluation and understanding. Multiple entries denote multiple datasets used and accordingly multiple information are mentioned.

Ref.	Dataset & Traits used(dataset, trait)	Size of the database	#Unlabeled samples	% Performance enhancement	Separate Test set
[2] *	Home-made, finger	50 x 100	25	9.99% 31.39%	75
[4]	Equinox, face	50 x 100	supervised	97.6 %	50
[5]	FVC2002 Db1-4 , finger	100 x 8	6	53%	50
[6]	FVC2000 DB1-4, finger	100 x 8	6	31.39 %	No
	Home-made, finger	12 x 200	199	50	
[7]	Home-made, finger	41	10	32.81%	No
[8]	Home-made, face	20 x 220 30 x 75 7 x 24	9 5 3	75 % 80% 60%	210 70 21
[9]	AR, face	100 x 8	6	74 %	7
[10]	AR/FVC-2002 DB2, face & fingerprint	100 x 8	7	46%	No
[11]	Equinox and DIEE, face & fingerprint	42 x 20	9	59 %	11
[12]	Equinox, face	57x129	50	40 % - 27%	71

performance evaluation. A separate study in template update has evaluated the impact of template self-update at different threshold conditions and security level [13] and reported results indicated more performance enhancement at user-specific threshold with re-estimation at security level of 1% FAR [13].

As mentioned above, efficiency of an update method is also a function of the properties of dataset employed for its experimental evaluation. Table 1 gives the database description, the size of the database (number of people x samples per person), number of unlabelled data (samples used for template updating), the % performance enhancement reported and test set size, consisting of genuine and impostors samples (to evaluate the performance of the updated system), if separate test set is used for performance evaluation (Table 1 reports information only for the references proposing different update methods). Note that mostly the systems are trained with single initial enrolled template before the application of update procedures. Worth mentioning, stop criteria too, plays an important role in offline methods as it effects the number of unlabelled data used for the template updating thus effecting the gained performance. At present, some methods have operated on fixed number of unlabelled data used for updating [7][9] and some self update offline [13] (Fig.1 Vertex 2.1.1.2) and co-update methods [10-11] (Fig.1 Vertex 2.2) are based on updating the templates till no more samples from unlabelled batch can be used for updating. As the number of samples used for updating effects the performance reached, making the stop criteria very important issue.

All these information together gives complete detail about the proposed template update methods in literature and results in fair comparison of them. It can be seen that on an average the performance enhancement is significant but the size of unlabelled data and fixed number of iterations for self-update offline methods [9] (Fig.1 Vertex 2.1.1.2) does not provide a fair evaluation of these methods. Thus a fair stop criteria is still needed to be formulated for these offline methods. For the graph-based methods (Fig.1 Vertex 2.1.2) [12] no stop criteria is required as they are based on partitioning the graph and assigning the labels.

Moreover so far only face and fingerprint traits have been studied for template updating using PCA or EBGM classifier for face and Minutiae/String based method for fingerprint [2-13, 15] respectively. May be template updating can be studied for other trait too. Since the protocol is different, the methods cannot be compared directly. In future, these methods can be evaluated on the common test bed for the better comparison of the proposed template update methods. * Table 1, Ref [2] denotes two performances improvement due to batch and augment updating.

5 Discussions and Conclusions

Template update is a very challenging topic. In this paper, we critically reviewed the main approaches proposed in the literature. To the best of our knowledge, this is the first survey about template update methods. We proposed a possible taxonomy for summarizing the current state-of-the-art, highlighting the advancements and limitations of the proposed methods.

Although template update methods have shown to be promising, some open issues still need to be addressed for their effective implementation. In particular, the existence, if any, of the tradeoff between performance enhancement and gallery size maintained due to updating. Worth mentioning, all the template update methods are prone to impostors introduction and the attraction of more samples of it which may gradually lead to "creep in" of identity, when the genuine person loses its identity. Very recently, the problem of impostor's introduction has been stated in [11]. Ref. [13] has reported that even with operation of update procedures at stringent threshold condition, the introduction of impostor cannot be avoidable. As apart from factors like incorrect estimation of threshold or basic FAR of the system, these methods are very much prone to impostors introduction due to the presence of "very difficult" clients, "wolves and lambs" according to the Doddington's zoo [13]. Wolves are the clients having the ability to imitate others irrespective of stringent threshold conditions and lambs are the clients vulnerable to impostors attack and the presence of these characteristic clients result in impostor introduction. To model the early stoppage of impostor introduction due to these client is still an open issue.

References

1. Ross, A., Nandakumar, K., Jain, A.K.: Handbook of Multi-biometrics. Springer, Heidelberg (2006)
2. Uludag, U., Ross, A., Jain, A.: Biometric template selection and update: a case study in fingerprints. Pattern Recognition 37(7), 1533–1542 (2004)
3. Roli, F., Didaci, L., Marcialis, G.L.: Adaptive biometric systems that can improve with use. In: Ratha, N., Govindaraju, V. (eds.) Advances in Biometrics: Sensors, Systems and Algorithms, pp. 447–471. Springer, Heidelberg (2008)

4. Freni, B., Marcialis, G.L., Roli, F.: Template Selection by Editing Algorithms: a case of Study in Face Recognition. In: da Vitoria Lobo, N., Kasparis, T., Roli, F., Kwok, J.T., Georgiopoulos, M., Anagnostopoulos, G.C., Loog, M. (eds.) S+SSPR 2008. LNCS, vol. 5342, pp. 745–754. Springer, Heidelberg (2008)
5. Freni, B., Marcialis, G.L., Roli, F.: Replacement algorithms for fingerprint template update. In: Campilho, A., Kamel, M.S. (eds.) ICIAR 2008. LNCS, vol. 5112, pp. 884–893. Springer, Heidelberg (2008)
6. Jiang, X., Ser, W.: Online Fingerprint Template Improvement. IEEE Trans. PAMI 24(8), 1121–1126 (2002)
7. Ryu, C., Hakil, K., Jain, A.: Template adaptation based fingerprint verification. In: Proc. of ICPR, Hong Kong, vol. 4, pp. 582–585 (2006)
8. Liu, X., Chen, T., Thornton, S.M.: Eigenspace updating for non-stationary process and its application to face recognition. Pattern Recognition, 1945–1959 (2003)
9. Roli, F., Marcialis, G.L.: Semi-supervised PCA-based face recognition using self training. In: Yeung, D.-Y., Kwok, J.T., Fred, A., Roli, F., de Ridder, D. (eds.) SSPR 2006 and SPR 2006. LNCS, vol. 4109, pp. 560–568. Springer, Heidelberg (2006)
10. Roli, F., Didaci, L., Marcialis, G.L.: Template co-update in multimodal biometric systems. In: Lee, S.-W., Li, S.Z. (eds.) ICB 2007. LNCS, vol. 4642, pp. 1194–1202. Springer, Heidelberg (2007)
11. Rattani, A., Marcialis, G.L., Roli, F.: Capturing large intra-class variations of biometric data by template co-updating. In: IEEE Workshop on Biometrics, Int. Conf. on Vision and Pattern Recognition CVPR 2008, Anchorage, Alaska, USA (2008)
12. Rattani, A., Marcialis, G.L., Roli, F.: Biometric template update using the graph mincut: a case study in face verification. In: 6th IEEE Biometric Symposium, Tampa, USA (2008)
13. Marcialis, G.L., Rattani, A., Roli, F.: Biometric template update: an experimental investigation on the relationship between update errors and performance degradation in face verification. In: da Vitoria Lobo, N., Kasparis, T., Roli, F., Kwok, J.T., Georgiopoulos, M., Anagnostopoulos, G.C., Loog, M. (eds.) S+SSPR 2008. LNCS, vol. 5342, pp. 684–693. Springer, Heidelberg (2008)
14. Doddington, G., Liggett, W., Martin, A., Przybocki, M., Reynolds, D.: Sheeps, Goats, Lambs and Wolves: A Statistical Analysis of Speaker Performance in the NIST 1998 Speaker Recognition Evaluation. In: Proc. ICSLD, Sydney, Australia (1998)
15. Didaci, L., Marcialis, G.L., Roli, F.: A theoretical and experimental analysis of template co-update in biometric verification systems. In: da Vitoria Lobo, N., Kasparis, T., Roli, F., Kwok, J.T., Georgiopoulos, M., Anagnostopoulos, G.C., Loog, M. (eds.) S+SSPR 2008. LNCS, vol. 5342, pp. 745–754. Springer, Heidelberg (2008)
16. Ryu, C., Kim, H.: Fingerprint Verification Testing Scenarios for Multi- impression Enrollment and Template Adaptation. In: Proc. of Biometric Symposium, pp. 19–21, 39–40 (2005)
17. Jain, A.K., Dubes, R.C.: Algorithms for clustering data. Prentice-Hall, Englewood Cliffs (1988)
18. Zhu, X.: Semi-supervised learning literature survey, Computer Sciences TR 1530 (2006)
19. Blum, A., Mitchell, T.: Combining labelled and unlabeled data with co-training. In: Proc. of the Workshop on Computational Learning Theory, pp. 92–100 (1998)
20. Blum, A., Chawla, S.: Learning from labeled and unlabeled data using graph mincuts. In: Int. Conf. on Machine Learning, pp. 19–26 (2001)
21. Freni, B., Marcialis, G.L., Roli, F.: Online and Offline Fingerprint Template Update Using Minutiae: An Experimental Comparison. In: Perales, F.J., Fisher, R.B. (eds.) AMDO 2008. LNCS, vol. 5098, pp. 441–448. Springer, Heidelberg (2008)

Simulating the Influences of Aging and Ocular Disease on Biometric Recognition Performance

Halvor Borgen[1], Patrick Bours[1], and Stephen D. Wolthusen[1,2]

[1] Norwegian Information Security Laboratory, Gjøvik University College, P.O. Box 191, N-2802 Gjøvik, Norway
[2] Information Security Group, Department of Mathematics, Royal Holloway, University of London, Egham Hill, Egham TW20 0EX, United Kingdom

Abstract. Many applications of ocular biometrics require long-term stability, yet only limited data on the effects of disease and aging on the error rates of ocular biometrics is currently available. Based on pathologies simulated using image manipulation validated by opthalmology and optometry specialists, the present paper reports on the effects that selected common ocular diseases and age-related pathologies have on the recognition performance of two widely used iris and retina recognition algorithms, finding the algorithms to be robust against many even highly visible pathologies, permitting acceptable re-enrolment intervals for most disease progressions.

1 Introduction

High levels of accuracy combined with being relatively difficult to forge are making ocular (iris and retina) biometrics attractive for identification and authentication in a number of areas. These include border control or other machine readable identification documents where, unlike in the case of access control to restricted areas, it is highly desirable to have long intervals between re-enrolment. While long-term stability of features such as fingerprints is well understood and limited research has also been conducted on the impact e.g. of dermal pathologies [1], this has not been the case for ocular features, particularly for the robustness of ocular biometrics to pathologies in general and the stability of the features used for recognition over time given aging and pathologies. Unlike in the case of fingerprints and other external features such as facial images, where medical and police records provide longitudinal records of feature stability and in many cases also disease progression, we are not aware of such data sets existing for iris and retina images, tracking disease and age progression from a healthy baseline and particularly using the multiple baseline images required for accurate initial enrolment of the ocular features. We have therefore chosen to acquire high-quality images directly in case of retinal images and using an existing database for iris images [2] as a baseline images for enrolment in biometric systems, simulating selected pathologies. To ensure that the simulations provided a faithful representation of these pathologies, all simulated stages were validated by opthalmology and optometry specialists.

M. Tistarelli and M.S. Nixon (Eds.): ICB 2009, LNCS 5558, pp. 857–867, 2009.
© Springer-Verlag Berlin Heidelberg 2009

2 Biometric Recognition Techniques Based on Ocular Features

For the purposes of this paper, a single algorithm and parameterization was chosen for both retina and iris recognition techniques. While different algorithms, signal acquisition mechanisms, and tuning are likely to yield different quantitative results, the main purpose of this paper is to explore whether these biometric techniques exhibit sufficient long-term stability, which is a sufficiently coarse requirement so as to allow qualitative extrapolation to other ocular biometric systems.

Retina Recognition. The suitability of the retinal vascular pattern for identification purposes was posited by Simon and Goldstein, later studies by Tower established experimentally that even identical twins exhibited randotypical retinal vascular patterns [3,4,5]. While technical solutions for capturing retinal images and their use in biometrics was already developed in the 1970s [6], it has not seen widespread adoption. A retina scanner must illuminate an annular region of the retina through the pupil either in the infrared or visible spectrum. This region centered on the fovea is approx. 10° off the visual axis of the eye. For recognition, the reflected vasculature contrast is recorded, capturing the pattern of blood vessels on the retina, the choroidal vasculature, and surrounding contrast. Successful acquisition requires alignment and fixation as well as a unobstructed optical pathway between sensor and the retina. The latter implies that the optical system must be able to accommodate different focal lengths so subjects can reliably focus on the target since eyeglasses would introduce distortions. Moreover, as the remainder of this paper will argue, it is also important to have a well-characterized optical pathway in the eye itself as certain diseases and age-related changes may also affect signal acquisition. However, this pathway and dynamically adjustable illumination also provides opportunities for liveness tests (e.g. dilation, depth effects, detection of moisture). In addition, a counterfeit retina or image must not only replicate the same vascular pattern in multiple focal layers but also simulate reflectivity of the retina and optical pathway as well as the flexible lens, the focusing of incoming and reflected beams, and its dynamic changes in response to external stimuli. Early recognition mechanisms include [7,8,9]. Here we present a system based on Hill's original design adapted to readily available acquisition as described in [10].

Iris Recognition. The randotypical patterns emerging during the growth of the eye yield a large feature space and are not only distinct among indiviuals but also different for each eye of the same individual [11,6]. This was identified as a suitable feature for biometric identification and verification by ophthalmologists Flom and Safir [12] while Daugman developed the dominant algorithm for encoding and matching [13,14]. The iris is a part of the central, or uveal, coat of the eye and consists of a trabecular meshwork [5,15] of elastic connective tissue. This trabecular meshwork pattern is completed during the first eight months of gestation, and remains stable apart from possible depigmentation [16] and the

effects of disease and trauma. Clinical evidence suggests that iris color change in adolescence has no effect on trabecular patterns while post-adolescence, depigmentation, and shrinking of the pupillary opening can occur with advanced age in healthy irises [11]. An iris scan requires a high-resolution grey-scale image (approx. 200 pixels across the diameter of the iris), preferably illuminated by infrared light. Since signal acquisition typically requires the capture of multiple frames, several dynamic liveness tests can be integrated in the acquisition, e.g. tracking pupil to iris diameter ratios. The diameter of the pupil is constantly oscillating due to the complex interplay of the muscles of the iris [11], which can also be induced by changes in illumination. Similarly, the fact that the lens floats outside the body of the eye and that the iris lies beneath the cornea creates a detectable difference in geometry if a custom contact lens is used to replicate the pattern of the iris itself. Depending on the acquisition mechanism used, a segmentation algorithm is required to identify the iris region as well as to characterize images of sufficient quality that are e.g. not excessively occluded by eyelids. In addition, distortions introduced by eyelashes, the presence of specularity (e.g. owing to poor illumination characteristics) or the iris image being out of focus can also degrade the acquired signal. Moreover, changes in lighting condition (as well as other stimuli) can result in pupillary dilation, which in turn will result in a nonaffine transformation of the iris. For the purposes of this paper, the algorithm by Daugman [13] in the implementation by Masek was chosen [17] (modified only to handle a slightly different image size). This algorithm must first find both the pupillary boundary and the outer boundary of the iris, the limbus. Once the searches for both boundaries have reached the single pixel precision, a similar approach to detect curvilinear edges is used to localize the eyelid boundaries. The path of contour integration in the operator is changed from circular to arcuate to fit the statistical estimation methods with the parameters to describe optimally the available evidence for each eyelid boundary. Each of the isolated iris patterns is then demodulated to extract its phase information using quadratic 2D Gabor wavelets.

3 Ocular Diseases and Their Simulation

A number of diseases and age-related developments of the eye can have a negative impact on the features and characteristics relevant to iris and retina recognition; some features may affect only one technique in particular while others can affect both approaches at the same time. The following section provides a brief overview of selected common pathologies and their potential impact.

Glaucoma. [18] causes the pressure of the fluids inside the eye to rise slowly, resulting in loss of visual acuity or blindness. In some severe cases, the increase in pressure causes severe damage to the optic nerve. Two types of adult glaucoma can be distinguished, open and closed angle glaucoma. Here, canals providing drainage to the eye become clogged over time, causing the pressure in the inner eye (intraocular pressure, IOP) to slowly rise owing to insufficient drainage of fluids. Closed angle (acute or narrow angle) glaucoma, is more rare. The main

difference is that the pressure rises very quickly due to severe clogging or blocking of the drainage canals. Here, the peripheral iris outer edges block the anterior chamber angle and the drainage canals because the pupil dilates too quickly or too much as may happen when entering a dark room. High pressure can result in a corneal edema, which in turn can create corneal scarring. Also of interest in this context is that treatment of severe narrow angle closure is accomplished by removing a segment of the outer iris edge (trabeculectomy) while less severe cases respond to medication (latanoprost), which in turn may result in pigmentation changes [19]. *Macular degeneration* [20,21] is a (generally age-related, hence AMD) disease of the retina. The macula is the central portion of the retina responsible for fine details in the vision. Loss of vision occurs when photoreceptors in the macula degenerate. In atrophic (dry) AMD, also referred to as geographic atrophy, irregular pigmentation of the macular region occurs, but no hemorrhage or exudation is evident in the macular region. Here, yellow-white deposits accumulate in the retinal pigment epithelium (RPE) tissue beneath the macula (see [22]). These deposits, called drusen, are waste products from the photoreceptor cells while exudative AMD is characterized by subretinal choroidal neovascularization. *Cataracts* are the result of a breakdown of cellular waste products in the lens, resulting in a blurred or cloudy lens. Three types can be distinguished: Nuclear, cortical, and subcapsular cataracts. Nuclear cataracts are the most common form in the center of the lens while cortical cataracts (common in diabetic patients) form in the lens cortex, gradually extending to the center of the lens. Subcapsular cataracts start at the back of the lens working its way forward and is associated with retinitis pigmentosa, diabetes, and high dosage of steroids [23]. *Retinopathy* refers to several types of retinal diseases affecting the fine retinal vasculature. In case of hypertensive retinopathy, high blood pressure damages and causes hemorrhaging of the vasculature; this may also be accompanied by exudates and cotton-wool spots. In case of ocular arteriosclerosis, retinal arteries harden, causing local hypertension in arteries and also in the capillaries. This results in a change in the capillary structure and may also cause hemorrhages, drastically changing the blood vessel pattern. Other possible signs of the disease are exudates and macular edema [24]. Diabetic retinopathy is the most common diabetic eye disease and one of the leading causes of blindness [24]. Here, the vasculature either swells, resulting in a leak of fluids, or pathological angiogenesis can ensue. Leaking vasculature results in blurry vision or, if affecting the macula, result in a macular edema. Two types of diabetic retinopathy are distinguished, proliferative and non-proliferative, the former is the more severe form. Non-proliferative retinopathy does not include vascular growth, but hemorrhages and exudates will also occur. *Pathological angiogenesis* is the abnormal rapid proliferation of blood vessels which is caused by cells receiving insufficient supplies of oxygen, as can e.g. be the case in tumor growth. This results in the release of angiogenic molecules that attract inflammatory and endothelial cells. The inflammatory cells secrete molecules that intensify the angiogenic process [25]. While angiogenesis can occur in any location, is retinal pathological angiogenesis that is of particular interest. *Keratitis* refers to an inflammation of the

cornea, which can have a number of different etiologies. Common symptoms for these are dullness of the cornea that sometimes turn into a grey-white connective tissue called macula cornea. Bacterial keratitis is caused by a bacterial infection and introduces a small grey-white collection of leukocytes (white blood cells) causing the surface of the cornea to turn dull because of the edema, with fungal keratitis yielding similar symptoms. Further variations can be caused by viral keratitis and photokeratitis [26].

4 Disease Simulation

Lacking a well-characterized longitudinal data series for ocular pathologies described in section 3, disease progression had to be simulated followed by subject expert validation in each case. The source images for healthy retinae and iridae were the UBIRIS database [2] in case of iridae and a data set collected by the authors in collaboration with Sykehuset Innlandet Lillehammer, Norway. For all diseases, three progression stages simulations were produced, with the first simulating disease onset and the final one advanced stages of a given disease, with the simulated images being validated by ophthalmologists.

4.1 Diseases Affecting the Iris and Cornea

Many pathologies, particularly those affecting the cornea, impact both iris and retina images. The following is a selection of common pathologies and symptoms likely to be encountered in aging populations and long-term biometric identification and authentication environments. The UBIRIS images are from 241 subjects and two sessions with a size of 200×150 pixels and the difference between the two series being the addition of noise in the second series. For the purposes of the present paper only the noise-less image series was used (see figure 1a for an example of an original image from this series).

Keratitis and Infiltrates. To simulate several types of keratitis, infiltrates of different sizes were spread over the corneal area. For severe cases, this will result in a whitening of the entire corneal area. Figures 1b through 1d illustrate different stages of central keratitis and infiltrates.

Blurring and Dulling of the Cornea. Corneal bleaching or clouding can be caused by diseases such as Glaucoma, Maroteaux-Lamy syndrome, Hurler's syndrome, KID syndrome, and a number of diseases similar to keratitis and scarring of the cornea (see fig. 1e for advanced-stage corneal bleaching).

Scarring and Surgery of the Cornea. Corneal scarring can occur in patients suffering from glaucoma or other diseases where the intraocular pressure builds up and creates corneal hemorrhaging. Corneal scarring is also common after injuries where a physical object has been in contact with the cornea. Glaucoma surgery is performed by removing a small part of the iris, at the outer edge, to release the pressure. This leaves a black mark on the iris, with unclear edges. Change

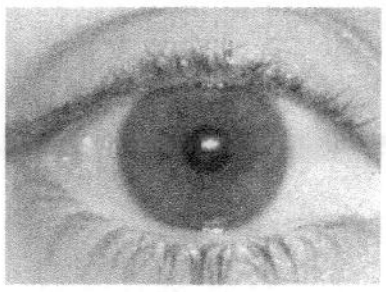

(a) Original image prior to manipulation (from [2])

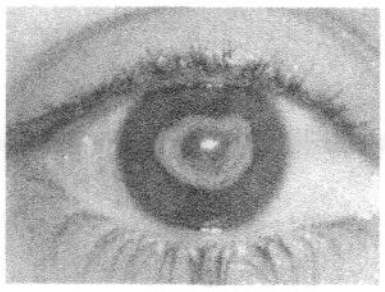

(b) Central keratitis, advanced stage simulation

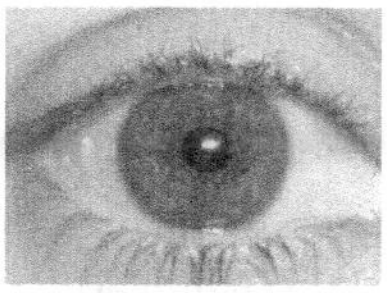

(c) Simulated infiltrates across entire cornea, low density

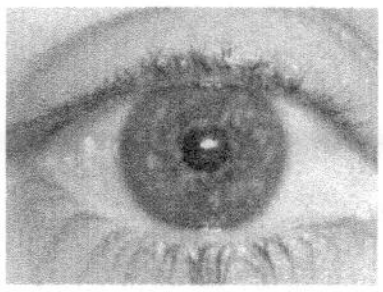

(d) Simulated infiltrates across entire cornea, high density

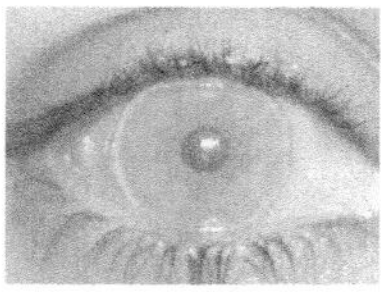

(e) Simulated corneal bleaching

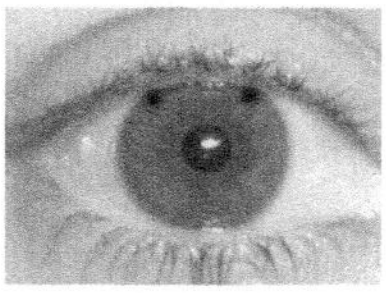

(f) Simulated corneal scars from glaucoma surgery

Fig. 1. Keratitis and infiltrates, corneal bleaching and scarring

in iris color is not normally observed, but can occur when using glaucoma medication [19]. Figure 1f shows an example of scars typical of glaucoma surgery at intermediate size.

Angiogenesis. Abnormal vascularization can occur whenever a subject suffers from vein occlusion, or lack of oxygen to the cornea. The latter is often related to wearing contact lenses but can also be caused by tumor growth.

Tumors and Melanoma. In addition to the vascular growths noted in the preceding paragraphs, deformations and larger-scale occlusions owing to hemorrhaging or melanoma are of particular interest. To simulate such pathologies, entire sections of the iris images were colored uniformly either black or white (given that hemorrhages and tumors in the iris are dark while conjunctival tumors spreading to the iris are light and pale). While tumors and hemorrhages can also affect the pupil, the simulation did not modify the pupil as this tends to affect image segmentation algorithms, not the recognition algorithms that are of interest in the present paper (see [22] for examples of a hemorrhages).

4.2 Diseases Affecting the Retina

Retinal images used in the experiments were acquired at Sykehuset Innlandet, Lillehammer, Norway using a standard Topcon TRC-50IX mydriatic retinal camera. The digital image sensor affixed to the TRC-50IX yielded a color (RGB) image at 768 by 576 (PAL resolution, square pixels), with the images stored in an uncompressed TIFF format under illumination by visible light provided by two integral light sources, a 100 *W* (max.) halogen lamp for observation, and a max. 300 *Ws* Xenon flash for photography. The camera had an angular coverage of 50°, 35°, and 20° and total observation magnification of $10x$ at 50°, $13.3x$ at 35°, and $23.3x$ at 20°, respectively as well as photographic magnification of

1.84x at 50°, 2.45x at 35°, and 4.28x at 20°, respectively (at zero diopter) [10]. For the symptoms described in the remainder of this section, it should be noted that these are not specific or exclusive to particular diseases, the latter are provided mainly as examples. *Cataracts and Lens Blurring* Cataracts can result in a blurring of the lense, which affects the ability of the lens to focus light onto the retina. Figure 2a illustrates a severe cataract that would also significantly affect a patient's vision. *Hemorrhaging* Retinal hemorrhaging is common in diseases involving hypertension, and in diabetes-related ocular diseases. Only hemorrhages within the recognition algorithm's scan circle were considered in the simulation. Hemorrhages can appear as both small red dots, and large collections of blood, both of which were simulated, although figure 2b only shows large-area hemorrhaging.

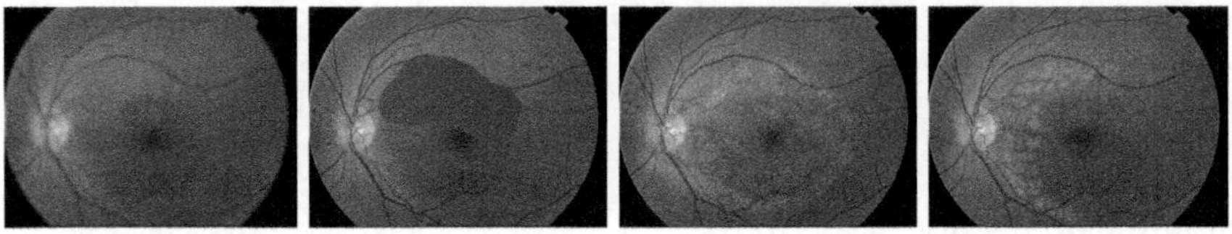

(a) Cataract (b) Hemmorrhaging (c) Drusen (d) Cotton-wool spots

Fig. 2. Retinal hemorrhaging, cataracts, Drusen and cotton-wool spots

Exudates and Drusen. Cellular waste materials called drusen are e.g. found in macular degeneration while exudates and cotton-wool spots are found in several different retinopathies, including hypertensive and diabetic retinopathy, macular degeneration, and arteriosclerosis. Figure 2c shows a simulation of drusen while figure 2d illustrates simulated cotton-wool spots.

5 Experimental Results

For the iris recognition stage, 17 subjects were chosen based on a combination of the intra-class and inter-class Hamming distances. These Hamming distances indicated that for our data set, a threshold between 0.30 and 0.35 was optimal, and the threshold was set at 0.35 for matching (i.e. a Hamming distance over 0.35 between the healthy template and the simulated diseased image led to a rejection; for these experiments only the FRR was of interest). The simulations were carried out on a set of manipulated images from one healthy image, and the remaining four healthy iris images as templates in four separate simulations. Table 1a summarizes the findings for all simulations. We note that the main reason for the false acceptances observed was faulty segmentation of the iris. Here, interactions with eye color (and hence contrast with the pupil) were observed empirically as dark eyes yielded lower match rates. While the data set used in the present simulation was too small to draw statistically significant conclusions, similar observations for the relevance of eye color can be found in related research [27]. This interplay was particularly observed for selected pathologies:

Corneal bleaching was only accepted for subjects with very bright iris color. Further bleaching had only limited effects at the initial and intermediate stages. However, advanced corneal bleaching led to rejections in most cases as iris segmentation (in particular, the detection of the outer iris boundary) becomes problematic. *Central keratitis* was frequently rejected in subjects with a bright iris color, again owing to segmentation problems. Although this occurred for all subjects, the FRR was higher with bright-eyed subjects. *High density infiltrates* was frequently rejected in subjects with a dark iris color. This effect was is similar to corneal bleaching, but results in an uneven spread of high density infiltrates.

Table 1.

(a) Iris FRR for different pathologies

Pathology	Total FRR
Corneal bleaching	65.2%
Central keratitis	61.7%
Change in iris color	0.5%
Infiltrates, high density	61.8%
Infiltrates, low density	32.8%
Glaucoma surgery scar	0.0%
Corneal scarring	72.5%
Corneal scarring w/bleaching	86.8%
Vessel growth	6.6%

(b) Retinal FRR for different pathologies

Pathology	Total FRR
Blurred lens	0.0%
Cotton-wool spots	11.7%
Drusen	43.3%
Exudates	5.0%
Hemorrhage, large	38.3%
Hemorrhage, small	38.3%
Vessel growth	68.3%

For the retina portion, 20 subjects were chosen, and the threshold for matching was kept at a correlation coefficient of 0.7. Tests were carried out during the adjustment phase of our algorithm, and they proved that a threshold of 0.7 would work most of the time, with a data set of images taken with visible light. Only one healthy image per retina test subject was available, the same that was used for the manipulations, which shifted results towards lower correlation coefficients (and hence FRR). Table 1b summarizes the findings for all simulations of pathologies affecting the retina. As shown in table 1b, neovascularization had the biggest impact on FRR among the simulated pathologies with an average FRR of 68.33%. The other blood-related signs of disease, i.e. hemorrhaging, resulted in an average FRR of 38.3%. The noise from these signs of disease were darker, and covered a larger area than exudates, drusen, and cotton-wool spots with higher density. The results indicate that darker eye color or illumination results in a lower FRR in general but our results do not allow the clear distinction between individual factors. This is the case for the signals acquired for the experiments discussed here (using visible-spectrum light) but must be expected to be even more pronounced in case near-infrared light is used for signal acquisition as near-IR light is absorbed effectively by blood vessels and hence also by hemorrhage artifacts. Finally, as was the case for iris recognition, we also observed a similar possible connection between retina color and illumination and the number of matches in general. The full results of the experimental study reported here can be found in [22].

6 Related Work

As noted in section 1, the extent of literature on the subject of the present paper appears to be somewhat limited. Notably, however, Roizenblatt *et al.* [28] performed a quantitative study on the impact of cataract surgery on the texture of affected irises and their effects on a specific iris recognition system and algorithm while Smith discusses the effects of iris pigmentation on recognition performance as well as the effects and implications of ocular illumination for signal acquisition [27]. As noted in section 2, the lack of an available retina recognition system has resulted in the authors of the present paper implenting a variant of Hill's system; details on the image acquisition system used as well as on the retina recognition algorithm developed for the experiments described in this paper can be found in [22,10].

7 Conclusion

While the results of the present study can only be seen as a preliminary study given the limitations imposed by the small data set and the need to resort to the simulation of pathologies which should be followed up by a longitudinal study with a sufficiently large population to allow statistical analysis over factors such as average disease progression times, the qualitative results from the simulation indicate that for application areas where biometric identification and authentication may occur only infrequently and on an irregular basis (e.g. in border control applications), disease progression can occur sufficiently rapidly to be problematic for recognition performance in aggressive forms. As an example, corneal bleaching at disease progression stages between 3 and 6 months after onset will result in a FRR of 67%, 78%, and 87%, respectively, for iris recognition (for these results, the full UBIRIS database was used with 231 subjects, resulting in a 95% confidence interval for the values given). Moreover, future work should also investigate the effect of iris color on recognition performance for both iris and retina recognition as well as other morphological features that can affect recognition performance, particularly the ability to accurately segment eye images. This in turn may yield further insights for the optimization of segmentation heuristics and hence result in improved recognition performance.

Acknowledgments. The authors would like to thank Robert "Buzz" Hill for advice and discussions on the development of retina recognition algorithms, Dr. Farshad Heybaran of the Opthalmology Department at Sykehuset Innlandet Lillehammer and Dr. Vibeke Sundling at the Department of Optometry and Visual Science at Buskerud University College for discussions on physiological properties of the ocular system and retina. Nurse Nina Løvli Wangen at Sykehuset Innlandet Lillehammer kindly assisted with the acquisition of retinal images.

References

1. Bundesamt für Sicherheit in der Informationstechnik: Public Final Report: Evaluation of Fingerprint Recognition Technologies — BioFinger. Technical report, BSI and BKA, Bonn, Germany (August 2004)
2. Proença, H., Alexandre, L.A.: UBIRIS: A Noisy Iris Image Database. In: Roli, F., Vitulano, S. (eds.) ICIAP 2005. LNCS, vol. 3617, pp. 970–977. Springer, Heidelberg (2005)
3. Simon, C., Goldstein, I.: A New Scientific Method of Identification. New York State Journal of Medicine 35(18), 901–906 (1935)
4. Simon, C.: The Retina Method of Authentication. Series 4: Unpublished Writings by Carleton Simon, 1900–1925, 1935–1938, 1940–1945, Files A-S. M. E. Grenander Department of Special Collections & Archives, State University of New York at Albany (1936)
5. Bolle, R., Pankanti, S., Jain, A.K.: Biometrics: Personal Identification in a Networked World. Kluwer Academic Publishers, Dordrecht (1998)
6. Woodward, J.D., Orlans, N.M., Higgins, P.T.: Biometrics: Identity Assurance in the Information Age. Osborne McGraw Hill, New York (2003)
7. Hill, R.B.: Apparatus and method for identifying individuals through their retinal vasculature pattern. U.S. Patent 4109237 (1978)
8. Hill, R.B.: Fovea-centered eye fundus scanner. U.S. Patent 4620318 (1986)
9. Samples, J.R., Hill, R.V.: Use of infrared fundus reflection for an identification device. American Journal of Ophthalmology 98(5), 636–640 (1984)
10. Borgen, H., Bours, P., Wolthusen, S.D.: Visible-Spectrum Biometric Retina Recognition. In: Proc. 4th Int'l. Conf. IIH-MSP, Harbin, China, pp. 1056–1062. IEEE Press, Los Alamitos (2008)
11. Wildes, R.P.: Iris Recognition: An Emerging Biometric Technology. Proc. IEEE 85(9), 1348–1363 (1997)
12. Flom, L., Safir, A.: Iris Recognition System. U.S. Patent 4641349 (1987)
13. Daugman, J.: How Iris Recognition Works. IEEE Transactions on Circuits and Systems for Video Technology 14(1), 21–30 (2004)
14. Daugman, J.: Probing the Uniqueness and Randomness of IrisCodes: Results From 200 Billion Iris Pair Comparisons. Proc. IEEE 94(11), 1927–1935 (2006)
15. Nanavati, S., Thieme, M., Nanavati, R.: Biometrics: Identity Verification in a Networked World. John Wiley & Sons, New York (2002)
16. Wayman, J.L., Jain, A.K., Maltoni, D., Maio, D.: Biometric Systems: Technology, Design and Performance Evaluation. Springer-Verlag, Berlin (2005)
17. Masek, L., Kovesi, P.: MatLab Source Code for a Biometric Identification System Based on Iris Patterns. B.Sc. thesis, U. of Western Australia (May 2003)
18. Cole III, R.M.: Glaucoma. Review of Optometry 143(5) (May 2006)
19. Teus, M.A., Arranz-Márquez, E., Lucea-Suescun, P.: Incidence of iris colour change in latanoprost treated eyes. British Journal of Ophthalmology 86(10), 1085–1088 (2002)
20. Nowak, J.Z.: Age-Related Macular Degeneration (AMD): Pathogenesis and Therapy. Pharmacological Reports 58(3), 353–363 (2006)
21. Gehrs, K.M., Anderson, D.H., Johnson, L.V., Hageman, G.S.: Age-related macular degeneration – Emerging pathogenetic and therapeutic concepts. Annals of Medicine 38(7), 450–471 (2006)
22. Borgen, H.: The Effects of Eye Disease and Aging of the Eye on Biometric Authentication. Master's thesis, Gjøvik University College, Gjøvik, Norway (July 2007)

23. Allen, D., Vasavada, A.: Cataract and surgery for cataract. British Medical Journal 333(7559), 128–132 (2006)
24. Torpy, J.M., Glass, T.J., Glass, R.M.: Retinopathy. Journal of the American Medical Association 293(1), 128 (2005)
25. Gariano, R.F., Gardner, T.W.: Retinal angiogenesis in development and disease. Nature 438(7070), 960–966 (2005)
26. Pavan-Langston, D.: Diagnosis and therapy of common eye infections: Bacterial, viral, fungal. Comprehensive Therapy 9(5), 33–42 (1983)
27. Smith, K.N.: Analysis of Pigmentation and Wavefront Coding Acquisition in Iris Recognition. Master's thesis, West Virginia U., Morgantown, WV, USA (May 2007)
28. Roizenblatt, R., Schor, P., Dante, F., Roizenblatt, J., Belfort Jr., R.: Iris recognition as a biometric method after cataract surgery. BioMedical Engineering OnLine 3(2), 1–7 (2004)

Cancelable Biometrics with Perfect Secrecy for Correlation-Based Matching

Shinji Hirata and Kenta Takahashi

Hitachi Ltd., Systems Development Laboratory,
292, Yoshida-cho, Totsuka-ku, Yokohama-shi, Kanagawa-ken, 244-0817, Japan
{shinji.hirata.sb,kenta.takahashi.bw}@hitachi.com

Abstract. In this paper, we propose a novel method of Cancelable Biometrics for correlation-based matching. The biometric image is transformed by Number Theoretic Transform (Fourier-like transform over a finite field), and then the transformed data is masked with a random filter. By applying a particular kind of masking technique, the correlation between the registered image and the input matching image can be computed in masked domain (i.e., encrypted domain) without knowing the original images. And we proved theoretically that in our proposed method the masked version does not leak any information of the original image, in other words, our proposed method has perfect secrecy. Additionally, we applied our proposed method to finger-vein pattern verification and experimentally obtained very high verification performance.

1 Introduction

Biometric authentication has the advantage of security and usability compared to traditional authentication methods like password or token. Biometrics cannot be stolen, forgotten, or shared. But recently, protecting biometric templates has become an issue. In a client/server-type biometric authentication system, biometric templates are stored in a database on the authentication server. In this case, it is difficult to prevent internal fraud by server's administrator, such as taking out biometric templates from the server. Furthermore, user's psychological resistance against centralized control of biometric is high. Besides, it is impossible to revoke biometric unlike password or token, and therefore if biometric is leaked out once and threat of forgery has occurred, the user cannot securely use his biometric anymore. The only remedy is to replace the template with another biometric feature. However, a person has only a limited number of biometric features.

Cancelable Biometrics [1] is a biometric verification scheme which was introduced to address this problem. This scheme enables the system to store and match templates while keeping them secret. The biometric is transformed using a parametrized distortion function. This preserves user's privacy and enhances security since it is impossible to recover the original biometric from the transformed version. A compromised template can be revoked using another transformation. Many authors have proposed methods for realizing Cancelable Biometrics. The

M. Tistarelli and M.S. Nixon (Eds.): ICB 2009, LNCS 5558, pp. 868–878, 2009.
© Springer-Verlag Berlin Heidelberg 2009

followings are some of the works in this direction. Savvides et al. [2] proposed a method that encrypts the training images used to synthesize the correlation filter for face recognition. Connie et al. [3] proposed a method that hashes palmprint templates with a set of pseudo-random keys to obtain a unique code called palmhash. Ratha et al. [4] proposed several transformations for cancelable fingerprint, such as cartesian, radial and functional transformation. In order to realize Cancelable Biometrics, to design the transform function properly is important. First, it is important to preserve the accuracy. Secondly, it is required to prevent the attacker from recovering the original biometric feature from the transformed feature. Ideally, the transformed feature itself does not leak any information about the original one. But none of the existing methods meets both requirements at the same time.

In this paper, we propose a novel method of Cancelable Biometrics which meets both requirements at the same time. It is applicable to correlation-based matching and utilizes "Number Theoretic Transform" [5]. We show theoretically that the accuracy is preserved and the transformed version (i.e, our cancelable template) does not leak any information of the original image Additionally, we apply our proposed method to finger-vein pattern verification and show experimentally that the verification performance is high enough.

2 Preliminary

2.1 Motivation

Here, we explain our motivation to propose a new method of Cancelable Biometrics. Savvides et al. [2] developed a method of generating cancelable template for face recognition using the minimum average correlation energy (MACE) filter. They encrypt the training face images by multiplying the Fourier transform of the face images with the Fourier transform of a random convolution kernel (that is equivalent to convolving the images with random convolution kernel) and synthesize the encrypted MACE filter, which is their cancelable template. Their cancelable template $\boldsymbol{h}'$ is defined as follows: Let $\boldsymbol{h} = \{h_i\}$ be the unencrypted original MACE filter (i.e., original template). Let $\boldsymbol{l} = \{l_i\}$ be the random filter. Then the encrypted MACE filter, that is, their cancelable template $\boldsymbol{h}' = \{h'_i\}$ is defined as $h'_i = l_i h_i$ for i-th component.

They show that the verification performance is preserved. However, their cancelable template leaks partial information of the original template, thus there is a possibility to recover the original template from the cancelable template. The reason is as follows: The range of l_i is given as a system parameter and naturally we can assume that the attacker knows this range. So, suppose $l_i \in (0, l_{max}]$. And we can also assume that the attacker knows the range of h_i. This is because she can estimate theoretically the range of h_i from the specification of their algorithm and the image used. Here, suppose $h_i \in (0, h_{i,max}]$. Note that the assumption of $h_i > 0$ does not cause loss of generality. Now assume that the attacker obtains the cancelable template $\boldsymbol{h}'$. Since $h_i = \frac{h'_i}{l_i}$, $l_i \in (0, l_{max}]$ and $h_i \in (0, h_{i,max}]$, then she gets to have the inequality of $\frac{h'_i}{l_{max}} \leq h_i \leq h_{i,max}$ For

the attacker, from the knowledge of h'_i, the range of h_i is confined to $[\frac{h'_i}{l_{max}}, h_{i,max}]$ compared to the original range of $(0, h_{i,max}]$. This means that h'_i leaks partial information about h_i.

We can address this problem if l_i, h_i and h'_i are elements of a finite field. The reason is as follows: Let $GF(p)$ be a finite field with order p, where p is prime. Assume $l_i, h_i, h'_i \in GF(p)(l_i \neq 0, h_i \neq 0, h'_i \neq 0)$ and l_i is uniformly random. By definition, we have $h_i = h'_i l_i^{-1}$. The mapping $\texttt{Inverse}; l_i \mapsto l_i^{-1}$ is bijective since $l_i \in GF(p)(l_i \neq 0)$. And for given h'_i, the mapping $\texttt{Multiply}_{h'_i}; l_i^{-1} \mapsto h'_i l_i^{-1}$ is bijective since $l_i^{-1}, h'_i \in GF(p)(l_i \neq 0, h'_i \neq 0)$. Thus, the composite mapping $\texttt{Multiply}_{h'_i} \circ \texttt{Inverse}$ for given h'_i is bijective. From this and the assumption that l_i is uniformly random, we conclude h_i is independent from h'_i, that is to say, h'_i cannot leak any information about h_i.

To make h_i be an element of $GF(p)$, the process of creating h_i must be performed over $GF(p)$. In their method, Fourier Transform (FT) is utilized to create h_i. In fact, there is a Fourier-like transform over $GF(p)$, which is called "Number Theoretic Transform (NTT)" [5]. If it is possible to create $\boldsymbol{h}$ by using NTT instead of FT, we can solve the problem. But unfortunately, we cannot create $\boldsymbol{h}$ by using NTT. Their method utilizes the MACE filter. To make the MACE filter, computing the spatial-frequency power spectrum of the images is required. The number theoretic transform of the image means nothing physical, while the fourier transform of the image means frequency component. Thus the power spectrum cannot be obtained through NTT. This is why we cannot create $\boldsymbol{h}$ by using NTT.

But it is possible to compute the simple correlation between two images by using NTT. Hence, not for the advanced correlation filter (like the MACE filter), but for a matching algorithm based on the simple correlation, we can construct a method of Cancelable Biometrics which multiplies the number theoretic transform of the image by a random filter. Thus, in the following , we propose a new method of Cancelable Biometrics for a simple correlation-based matching algorithm using NTT.

2.2 Correlation-Based Matching

Before presenting our proposed method, we describe the correlation-based matching algorithm. The following is a well-known algorithm of correlation-based matching [6], which is applicable to image-based biometric verification (e.g., fingerprint, iris, finger-vein and so on).

$f(x, y)$ and $g(x, y)$ are the values at position (x, y) of the registered image and the input matching image, respectively. The size of the registered image $f(x, y)$ is $W_f \times H_f$ and the size of the input matching image is $W_g \times H_g$, where $W_f < W_g$ and $H_f < H_g$. And $f(x, y)$ and $g(x, y)$ are taken from integers. $w_{f,g}(p, q)$ is the correlation between $f(x, y)$ and $g(x, y)$ at the relative displacement (p, q). $w_{f,g}(p, q)$ is defined as follows:

$$w_{f,g}(p, q) = \sum_{(x,y) \in S(p,q)} f(x - p, y - q) g(x, y), \tag{1}$$

where $S(p,q)$ is the region that the registered image $f(x,y)$ overlaps the input matching image $g(x,y)$ at the displacement (p,q).

$w_{f,g}(p,q)$ is not directly used as a measure of match, instead we use *Peak-to-Mean*. First, let *peak* be the maximum value of $w_{f,g}(p,q)$. Second, on the correlation plane, let us consider the sidelobe region (excluding a central rectangular mask) centered at the peak. Let *mean* be the mean of $w_{f,g}(p,q)$ in this sidelobe region. Then, *Peak-to-Mean* is defined as follows: *Peak-to-Mean* $=$ *peak* $-$ *mean*. *Peak-to-Mean* means the peak's height to the sidelobe and this indicates the similarity of two images. Now note that *Peak-to-Mean* is computed based on only the correlation $w_{f,g}(p,q)$.

3 Proposed Method

In this section, we propose a new method of Cancelable Biometrics for the typical correlation-based matching which is described in Section 2.2. If the correlation can be computed while keeping the two images secret, we can construct Cancelable Biometrics for correlation-based matching. In order to realize this, we utilize "Number Theoretic Transform (NTT)" [5].

NTT Ψ is a Fourier-like transform over $GF(p)$ and is denoted as follows: $\Psi; (a_1, \cdots, a_n) \mapsto (A_1, \cdots, A_n)$, where $a_i, A_i \in GF(p)$. For details of NTT, please see [5]. Although here for convenience described for 1-D data arrays, it is easily applicable to 2-D image [7]. NTT Ψ has following property: $\Psi(\boldsymbol{a} * \boldsymbol{b}) = \Psi(\boldsymbol{a})\Psi(\boldsymbol{b})$, where $\boldsymbol{a} = \{a_i\}, \boldsymbol{b} = \{b_i\}$ and $*$ is convolution. This property means that the convolution of data arrays corresponds to the component-wise multiplication in the NTT domain, and this property is called "Cyclic Convolution Property (CCP)". In fact, Fourier Transform has the same property. And CCP implies that a convolution can be computed by:

$$\boldsymbol{a} * \boldsymbol{b} = \Psi^{-1}\{\Psi(\boldsymbol{a})\Psi(\boldsymbol{b})\}. \tag{2}$$

Note that in order to compute the correlation between $\boldsymbol{a}$ and $\boldsymbol{b}$, the order of the one data array has to be inverted.

By using CCP of NTT, we can compute the correlation while keeping the two images secret, thus we can construct Cancelable Biometrics for correlation-based matching. In order to apply NTT, the size of the registered image must be equal to the size of the input matching image. Thus, we pad the value to the pixels outside the registered image $f(x,y)$. The value of padding in the extended region is 0. By doing this, the registered image $f(x,y)$ is extended to the same size as the input matching image $g(x,y)$. Let $\tilde{f}(x,y)$ be the extended version of $f(x,y)$. Note that the correlation is invariant under this extension since the padding value is 0, that is, $w_{\tilde{f},g}(p,q) = w_{f,g}(p,q)$.

Figure 1 depicts the block diagram of our proposed method. $F(u,v)$ and $G(u,v)$ are the number theoretic transform of $\tilde{f}(x,y)$ and $g(x,y)$, respectively, where (u,v) is the position in the NTT domain. $F(u,v) \in GF(p)$ and $G(u,v) \in GF(p)$. $R(u,v)$ is a random filter. Let $R(u,v)$ be uniformly random over $GF(p)$, but $R(u,v) \neq 0$. $R^{-1}(u,v)$ is a random filter which is the inverse element of

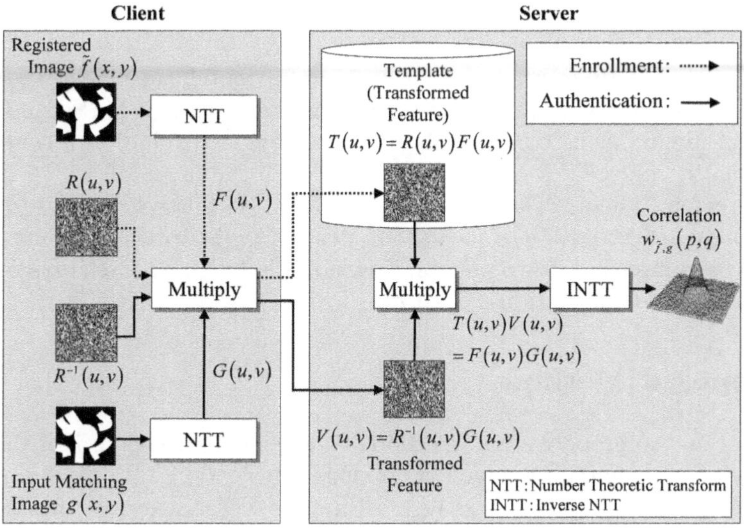

Fig. 1. Block diagram of the proposed method

$R(u,v)$. Thus $R^{-1}(u,v) \in GF(p)$, $R^{-1}(u,v) \neq 0$ and $R(u,v)R^{-1}(u,v) = 1$. In what follows, the transformed biometric feature is referred to as the transformed feature and the data to be used for transforming the feature is referred to as the parameter.

In enrollment stage, we first perform NTT to $\widetilde{f}(x,y)$ and get $F(u,v)$, and then multiply $F(u,v)$ by $R(u,v)$ (i.e., mask $F(u,v)$ with a random filter $R(u,v)$). Our cancelable template $T(u,v)$ is defined as $T(u,v) = R(u,v)F(u,v)$ and $R^{-1}(u,v)$ is the parameter. These processes are performed in the client. Then $T(u,v)$ is sent to the server and stored into the database and $R^{-1}(u,v)$ is stored in the client.

In authentication stage, we first perform NTT to $g(x,y)$ and get $G(u,v)$, and then multiply $G(u,v)$ by $R^{-1}(u,v)$ (i.e., mask $G(u,v)$ with a random filter $R^{-1}(u,v)$). Our transformed feature $V(u,v)$ is defined as $V(u,v) = R^{-1}(u,v)G(u,v)$. These processes are performed in the client and then $V(u,v)$ is sent to the server. The matching process is performed in the masked domain (i.e., encrypted domain) on the server. This process is as follows: we first multiply $T(u,v)$ by $V(u,v)$. Then we compute the correlation $w_{\widetilde{f},g}(p,q)$ from $T(u,v)V(u,v)$ by using Inverse-NTT. This is because $T(u,v)V(u,v) = F(u,v)G(u,v)$ must hold since $R(u,v)R^{-1}(u,v) = 1$ and we can obtain the correlation according to Eq.(2). This means that we can compute the correlation between $f(x,y)$ and $g(x,y)$ in the masked domain (i.e., encrypted domain), while keeping $F(u,v)$ and $G(u,v)$ secret, that is, keeping $f(x,y)$ and $g(x,y)$ secret. Finally we compute *Peak-to-Mean* from the correlation $w_{\widetilde{f},g}(p,q)$ and authenticate the user.

In addition, $T(u,v)$ can be changed by varying $R(u,v)$, thus revocation of the template is enabled.

4 Analysis

4.1 Accuracy Preservation

In general, an error in computing the score would occur due to applying the transformation and the matching accuracy may degrade compared to the version without transformation [4]. It is important to reduce the accuracy degradation and to preserve the accuracy.

Here, we discuss the accuracy preservation of our proposed method when applied to the algorithm described in Section 2.2. NTT is a transform over $GF(p)$, thus pixels of image must be elements of $GF(p)$. For this, if a pixel f is negative, then we convert f to $p - |f|$. Besides, any pixel of the correlation plane obtained through Inverse-NTT is not negative, that is, negative pixels are not computed correctly. To compensate this, the following is performed: let p be larger than about twice as large as the maximum of the correlation. If a pixel w of the correlation plane is larger than the maximum of the correlation, then we convert w to $w - p$. By this, we can compute the correlation plane correctly. Hence, the matching score is invariant when applying our proposed method, thus the accuracy is preserved.

4.2 Recovery Resistance

From the point of view of template protection, it is required to prevent the attacker from recovering the original biometric feature from the transformed feature without knowledge of the parameter. Ideally, the transformed feature itself does not leak any information about the original biometric. Furthermore, it is also required to prevent the attacker from recovering the original biometric feature from the parameter without knowledge of the transformed feature. Also ideally, the parameter itself does not leak any information about the original biometric.

(1) Resistance against Recovery from Template

We here discuss the resistance against recovery of the original images $f(x, y)$ or $g(x, y)$ from the cancelable template $T(u, v)$ or the transformed feature $V(u, v)$. In what follows, we will prove that it is impossible to recover $F(u, v)$ from $T(u, v)$ without knowledge of $R(u, v)$ (we will treat the case of $G(u, v)$ later).

Since the transform of masking $F(u, v)$ with $R(u, v)$ is component-wise in the NTT domain, it is sufficient to prove it for only one single component. Let us consider just one component at the position of (u, v). Let $r \in GF(p)$ $(r \neq 0)$ be the (u, v)-th component of $R(u, v)$, $s \in GF(p)$ be the (u, v)-th component of $F(u, v)$ and $t \in GF(p)$ be (u, v)-th component of $T(u, v)$. Then we have $t = rs$. We will prove that it is impossible to recover s from t without knowledge of r.

We now define the transform function $\phi_r(s) = rs$. And let $\mathbf{\Phi}$ be the family of the transform functions ϕ_r. We can here consider $\mathbf{\Phi}$ as an encryption algorithm, where r is a encryption key, s is a plaintext and t is a ciphertext. And we define the transform function $\phi_r^{-1}(t) = r^{-1}t$. Decryption algorithm $\mathbf{\Phi}^{-1}$ is defined as the family of the transform functions ϕ_r^{-1}.

There is a formal definition about secrecy in cryptography, that is, Perfect Secrecy [8]. Before presenting the definition of Perfect Secrecy, let us introduce the definition of Cryptosystem according to [8].

Definition 1. *Cryptosystem is a tuple* $(\mathbf{M}, \mathbf{C}, \mathbf{K}, \mathbf{E}, \mathbf{D})$ *with the following properties:*

1. $\mathbf{M}$ *is a set and called the plaintext space.*
2. $\mathbf{C}$ *is a set and called the ciphertext space.*
3. $\mathbf{K}$ *is a set and called the key space.*
4. $\mathbf{E} = \{E_k; k \in \mathbf{K}\}$ *is a family of functions* $E_k; \mathbf{M} \rightarrow \mathbf{C}$.
5. $\mathbf{D} = \{D_k; k \in \mathbf{K}\}$ *is a family of functions* $D_k; \mathbf{C} \rightarrow \mathbf{M}$.
6. *For each* $e \in \mathbf{K}$, *there is* $d \in \mathbf{K}$ *such that* $D_d(E_e(m)) = m$ *for all* $m \in \mathbf{M}$

In our case, the cryptosystem is $(S, T, R, \mathbf{\Phi}, \mathbf{\Phi}^{-1})$, where $S = \{0, 1, \cdots, p-1\}$ is the set of plaintexts s, $R = \{1, 2, \cdots, p-1\}$ is the set of encryption keys r (Note that R does not include 0) and $T = \{0, 1, \cdots, p-1\}$ is the set of ciphertexts t.

We now present the Shanon's definition of Perfect Secrecy:

Definition 2. *Cryptosystem* $(\mathbf{M}, \mathbf{C}, \mathbf{K}, \mathbf{E}, \mathbf{D})$ *has Perfect Secrecy if the events that a particular ciphertext occurs and that a particular plaintext has been encrypted are independent (i.e.,* $\Pr(m|c) = \Pr(m)$ *for all plaintexts* m *and all ciphertexts* c*).*

Definition 2 implies that it is impossible even to estimate m from c without knowledge of k. Thus, in order to prove that it is impossible to recover s from t without knowledge of r, it is sufficient to prove that $(S, T, R, \mathbf{\Phi}, \mathbf{\Phi}^{-1})$ has Perfect Secrecy.

But it is easy to see that one can recover s without knowledge of r if $t = 0$ in $(S, T, R, \mathbf{\Phi}, \mathbf{\Phi}^{-1})$. This is because according to definition of ϕ_r if $t = 0$, then $s = 0$ must hold since $r \neq 0$. In order to avoid this, let us eliminate the case of $s = t = 0$ and define another set of plaintexts and ciphertexts; $\widetilde{S} = \{1, \cdots, p-1\}$ is the set of plaintexts $\widetilde{s}$, and $\widetilde{T} = \{1, \cdots, p-1\}$ is the set of ciphertexts $\widetilde{t}$. As we describe later, we can prove that the cryptosystem $(\widetilde{S}, \widetilde{T}, R, \mathbf{\Phi}, \mathbf{\Phi}^{-1})$ has Perfect Secrecy.

Now only the case of $s = 0$, that is, the case of $F(u, v) = 0$ is a problem. But if the case of $F(u, v) = 0$ rarely happens, this would not be a problem in practice. In order to examine how often the case of $F(u, v) = 0$ happens, as an example, we applied NTT to finger-vein pattern images and experimentally obtained the rate of the case of $F(u, v) = 0$ in all $F(u, v)$s among the whole dataset. The rate is 0.016%. The rate is low enough, thus the amount of leaked information is so small that in practice one cannot recover the entire $F(u, v)$ from it. We will see the details of this experiment in Section 5.

In order to prove that the cryptosystem $(\widetilde{S}, \widetilde{T}, R, \mathbf{\Phi}, \mathbf{\Phi}^{-1})$ has Perfect Secrecy, we use the following Shanon's theorem [8]:

Theorem 1. *Let* $|\mathbf{M}| = |\mathbf{K}| = |\mathbf{C}| < \infty$ *and* $\Pr(m) > 0$ *for any plaintext* m. *Cryptosystem* $(\mathbf{M}, \mathbf{C}, \mathbf{K}, \mathbf{E}, \mathbf{D})$ *has perfect secrecy if and only if the probability distribution on the key space is the uniform distribution and if for any platintext* m *and any ciphertext* c *there is exactly one key* k *with* $E_k(m) = c$.

Now we can prove the following theorem using Theorem 1:

Theorem 2. *Cryptosystem* $(\widetilde{S}, \widetilde{T}, R, \mathbf{\Phi}, \mathbf{\Phi}^{-1})$ *has Perfect Secrecy.*

Proof. By definition, $|\widetilde{S}| = |\widetilde{T}| = |R| < \infty$, $\Pr(\widetilde{s}) > 0$ for any $\widetilde{s}$ and the probability distribution of r is uniform.

Let us confirm that for any $\widetilde{s}$ and any $\widetilde{t}$ there is exactly one key r with $\phi_r(\widetilde{s}) = \widetilde{t}$. If we assume that for arbitrary $\widetilde{s}$ and $\widetilde{t}$, $\widetilde{t} = \phi_{r_1}(\widetilde{s}) = \phi_{r_2}(\widetilde{s})$ where $r_1 \neq r_2$, then $r_1\widetilde{s} = r_2\widetilde{s}$ must hold. There exists $\widetilde{s}^{-1}$ and we can multiply both parts of this equation by $\widetilde{s}^{-1}$. Then we have $r_1 = r_2$. But this contradicts the assumption that $r_1 \neq r_2$. Hence, for any $\widetilde{s}$ and $\widetilde{t}$ there is exactly one key r with $\phi_r(\widetilde{s}) = \widetilde{t}$. This proves the theorem. □

Therefore, $T(u, v)$ does not leak any information of $F(u, v)$, hence it is impossible to recover $F(u, v)$ from $T(u, v)$ without knowledge of $R(u, v)$ (note that there is the exceptional case of $F(u, v) = 0$, but in practice this case is rare and not a problem).

In the same way, we can prove that $V(u, v)$ does not leak any information of $G(u, v)$, hence it is impossible to recover $G(u, v)$ from $V(u, v)$ without knowledge of $R^{-1}(u, v)$ (also note that there is the exceptional case of $G(u, v) = 0$, but this case is not a problem in practice).

(2) Resistance against Recovery from Parameter

We here discuss the resistance against recovery of the original image $f(x, y)$ from the parameter $R^{-1}(u, v)$ which is stored in the client. $R^{-1}(u, v)$ is uniformly random and independent from $f(x, y)$. Hence, it is impossible to recover $f(x, y)$ from $R^{-1}(u, v)$ without knowledge of $T(u, v)$. In addition, even if the attacker collects two or more different $R^{-1}(u, v)$s which are used to match against the same original image $f(x, y)$, he cannot recover $f(x, y)$ because $R^{-1}(u, v)$ does not include any information of $f(x, y)$.

(3) Resistance against Recovery from Correlation

We here discuss another possible approach to recover the original images $f(x, y)$ and $g(x, y)$. The server knows the correlation between $f(x, y)$ and $g(x, y)$ in matching phase. If the server administrator is malicious, she would try to take advantage of this correlation to recover $f(x, y)$ and $g(x, y)$.

In order to recover $f(x, y)$ and $g(x, y)$, the attacker regards Eq.(1) as a simultaneous equation and try to solve it. This simultaneous equation has $H_f W_f + H_g W_g$ variables and $H_g W_g$ equations. The solution of the equation is underspecified because the number of variables is larger than the number of equations. If the authentication is repeated n times, the number of variables is $H_f W_f + nH_g W_g$ and the number of equations is $nH_g W_g$. The solution of the equation is also underspecified because the number of variables is larger than the number of equations. Thus, we conclude that to recover the original images from the correlation is impossible.

5 Experiments

In our experiment, we applied the proposed method to finger-vein pattern matching and examine the verification performance. We used the infrared finger images obtained from 17 volunteers in our company. For each person, three fingers of each hand are used. The dataset contains infrared images of 102 different fingers, with a pair of images per finger. The images are captured by the infrared sensing device which was developed as a prototype in our company (for details of the device, see [9]). Figure 2 shows an example of infrared finger image. By using the algorithm described in [9], the finger-vein pattern images are extracted from the infrared finger images.

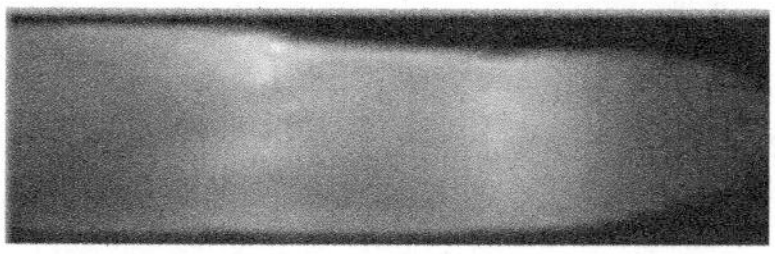

Fig. 2. Example of infrared finger image

We applied our proposed method to finger-vein pattern images and measured matching score (*Peak-to-Mean*) for assessing the verification performance. We generated 102 genuine scores and 10302 (= 102 × 101) impostor scores by performing all-against-all matches. Figure 3 shows the normalized frequency distribution of score for genuine and impostor matching. There is a clear margin of separation between the scores of the genuine class and the impostor class, that is, the verification performance is high enough.

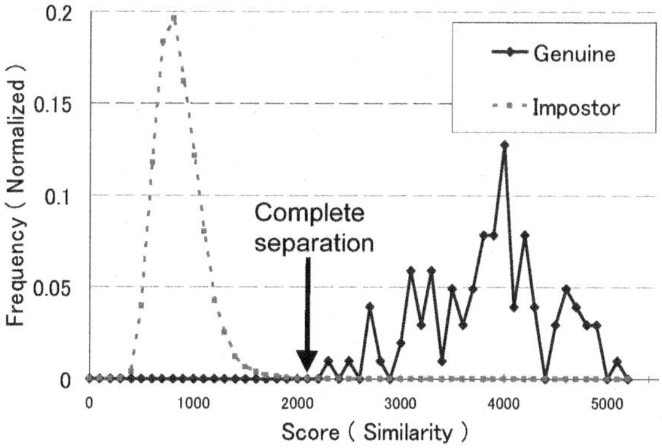

Fig. 3. Normalized frequency distribution of score for genuine and impostor matching

In Section 4.2, we analyzed the resistance against recovery of the original image from our cancelable template and noticed that only the case of NTT component being 0 is a problem. But if this case rarely happens, this would not

be a problem in practice. In order to see how often this case happens, we applied NTT to finger-vein pattern images and examined the rate of the case of NTT component being 0. We investigated all the NTT components among the whole dataset of finger-vein pattern images ($102 \times 2 = 204$ images) and then obtained a rate of 0.016% of NTT components being 0. Table 1 shows the number of images that include NTT component being 0. 70%(= 144/204) of the images include no NTT component being 0 and the number of NTT component being 0 is at most 3 per image. Thus, the rate is low enough and there is no particular image that include many NTT components being 0. Hence, the amount of leaked information is so small that in practice one cannot recover the entire $F(u, v)$ or $G(u, v)$ from it.

Table 1. Number of images that include NTT components being 0

Num of NTT component being 0 per image	0	1	2	3
Num of image (Total 204)	144	52	6	2

6 Conclusion

In this paper, we proposed a novel method of Cancelable Biometrics for correlation-based matching. The main idea is to transform the image by Number Theoretic Transform and mask the transformed data with a random filter. By applying a particular kind of masking technique, the correlation between the registered image and the input matching image can be computed in masked domain (i.e., encrypted domain) without knowing the original images. Thus, the matching accuracy is invariant when applying our proposed method. And we proved theoretically that in our proposed method the masked version does not leak any information of the original image, in other words, our proposed method has perfect secrecy. Additionally, we applied our proposed method to finger-vein pattern verification and experimentally obtained very high verification performance.

Acknowledgement

This paper partially contains research achievements of a national project funded by Ministry of Internal Affairs and Communications in Japan, "R&D for advancement of functionality and usability in information history management".

References

1. Ratha, N.K., Connell, J.H., Bolle, R.M.: Enhancing security and privacy in biometric-based authentication systems. IBM System Journal 40(3) (2001)
2. Savvides, M., Vijayakumar, B.V.K., Khosla, P.K.: Cancelable Biometric Filters for Face Recognition. In: 17th International Conference on Pattern Recognition (ICPR 2004), vol. 3, pp. 922–925 (2004)

3. Connie, T., Teoh, A., Goh, M., Ngo, D.: PalmHashing: a novel approach for cancelable biometrics. Information Processing Letters 93(1), 1–5 (2005)
4. Ratha, N.K., Connell, J.H., Bolle, R.M., Chikkerur, S.: Cancelable Biometrics: A Case Study in Fingerprints. In: 18th International Conference on Pattern Recognition (ICPR 2006), vol. 4, pp. 370–373 (2006)
5. Agarwal, R.C., Burrus, C.S.: Number theoretic transforms to implement fast digital convolution. Proc. IEEE 63(4), 550–560 (1975)
6. Rosenfeld, A., Kak, A.C.: Digital Picture Processing, 2nd edn., vol. 2. Academic Press, London (1982)
7. Reed, I.S., Truong, T.K., Kwoh, Y.S., Hall, E.L.: Image Processing by Transforms Over a Finte Field. IEEE Transactions on Computers C-26(9), 874–881 (1977)
8. Buchmann, J.A.: Introduction to Cryptography, 2nd edn. Springer, Heidelberg (2004)
9. Miura, N., Nagasaka, A., Miyatake, T.: Feature extraction of finger-vein patterns based on repeated line tracking and its application to personal identification. Machine Vision and Applications 15(4), 194–203 (2004)

An Information Theoretic Framework for Biometric Security Systems*

Lifeng Lai[1], Siu-Wai Ho[2], and H. Vincent Poor[1]

[1] Department of Electrical Engineering,
Princeton University, Princeton, NJ, 08512. USA
{llai,poor}@princeton.edu
[2] Department of Information Engineering,
The Chinese University of Hong Kong, Shatin, N.T., Hong Kong
swho@ie.cuhk.edu.hk

Abstract. An information theoretic framework is established to analyze the performance of biometric security systems. Two performance metrics, namely privacy, measured by the normalized equivocation rate of the biometric measurements, and security, measured by the rate of the key generated from the biometric measurements, are first defined. A fundamental tradeoff between these two metrics is then identified. The scenario in which a potential attacker does not have side information is considered first. The privacy-security region, which characterizes the above-noted tradeoff, is derived for this case. An important role of common information among random variables is revealed in perfect privacy biometric security systems. The scenario in which the attacker has side information is then considered. Inner and outer bounds on the privacy-security tradeoff are derived in this case.

Keywords: Biometric, Entropy, Mutual Information, Privacy, Security.

1 Introduction

Addressing reliability and privacy issues properly is important for the widespread application of biometric security systems. Due to measurement noise or other factors such as aging or injury, two measurements of the same biometric characteristics will not produce the same result. Hence, biometric measurements cannot be directly used for encryption in the same way that a secret key would typically be used. The issue of *privacy* arises from the fact that biometric characteristics are stored in a certain form in database, which creates a security threat. Unlike passwords, biometric characteristics cannot be changed. Hence, if the database is compromised, irreversible identity theft is possible.

In recent years, there has been increasing research interest in addressing these issues. A number of interesting approaches have been proposed (see, e.g., [1,2]

* This research was supported in part by the National Science Foundation under Grants CNS-06-25637 and CCF-07-28208.

M. Tistarelli and M.S. Nixon (Eds.): ICB 2009, LNCS 5558, pp. 879–888, 2009.
© Springer-Verlag Berlin Heidelberg 2009

and [3] for overviews). The basic idea of these approaches is to generate a secret key and helper data during the initial enrollment stage. The key is used for encryption. The helper data is stored in the database. In the release stage, by combining the noisy measurements with the helper data, one can recover the key which is then used to decrypt the message. The helper data can be viewed as the parity-check bits of an error correcting code, and the effects of noise can be mitigated by such error correction. The existing approaches focus on maximizing the rate of the key that can be recovered successfully from the noisy measurements. This approach is motivated by the fact that in an encryption system, the equivocation of the encrypted message is limited by the entropy of the key [4]. From an information theoretic perspective, these existing approaches can be modelled as a problem of generating a secret key from common randomness [5], and hence the largest rate of the key can be characterized [6]. On the other hand, although the biometric measurements are not stored in the database in plain form, the helper data still contains information about the biometric measurements.

While the existing approaches maximize the key rate, they do not address the privacy issue adequately. In practice, the protection of the biometric measurements themselves is at least as important as maximizing the key rate. To increase the security level of the encrypted messages, we would like to make the key rate as large as possible. On the other hand, to preserve the privacy, we need to ensure that information leakage about the biometric measurements themselves is as small as possible. One question naturally arises: can we maximize the rate of the generated key while simultaneously minimizing the information leakage? In this paper, by establishing an information theoretic foundation[1] for biometric security systems, we show that there exists a fundamental tradeoff between security, measured by the rate of the generated key, and privacy, measured by the normalized equivocation of the biometric measurements, in any biometric security system. Thus, we cannot achieve both goals simultaneously. More specifically, we first rigorously formulate the privacy-security tradeoff in biometric security systems. We then identify and characterize this fundamental tradeoff for several different scenarios. In the first scenario, we require perfect security (rigorous definition will be given in the sequel) of the generated key. In this scenario, we consider two systems differentiated by whether the user is allowed to select the key or not. In each system, we characterize the security-privacy tradeoff. Furthermore, we propose schemes that fully achieve any particular point on the tradeoff curve. We show that the performance of the existing approach is one particular point on the derived tradeoff curve. We further show that the freedom of selecting a key does not affect the privacy-security tradeoff. In the second scenario, we require perfect privacy of the biometric measurements. We identify a close relationship between the common randomness between the biometric characteristics obtained during the enrollment and release stages and the rate of a secret key that can be generated. Finally, we study the scenario in which an attacker has side information about the biometric measurements. Again both types of systems are considered. Inner and outer bounds on the privacy-security

[1] Please refer to [7] for basic background on information theory.

region are derived for these situations. These bounds are shown to match under certain conditions of interest.

Due to space limitations, we include only the outline of the proofs of our results. Details of these proofs can be found in [8].

2 Model

We denote the biometric measurements taken during the enrollment stage by X^n and the biometric measurements taken during the verification stage by Y^n. Here, we assume that X^n and Y^n are sequences with length n taking values from n-fold product sets $\mathcal{X}^n$ and $\mathcal{Y}^n$, respectively. We assume that these measurements are generated according to a joint distribution $P_{X^nY^n}(x^n, y^n) = \prod_{i=1}^{n} P_{XY}(x_i, y_i)$. Specific models for the distribution of biometric measurements can be found, for example, in [9].

During the enrollment stage both the key K, ranging over $\mathcal{K}$, and the helper data V, ranging over $\mathcal{V}$, are generated. The key K is used to perform various tasks, such as message encryption. The helper data V is stored in the database to assist the recovery of the key from the noisy measurements Y^n during the release stage. Regarding the generation of key K, we consider two types of systems: namely non-randomized systems and randomized systems. In non-randomized systems, as shown in Figure 1 (a), both V and K are generated from X^n by functions h_n and $\tilde{h}_n$, respectively, so that $V = h_n(X^n)$ and $K = \tilde{h}_n(X^n)$. In randomized systems, a key K, which is independent with X^n, is randomly generated during the enrollment stage. Then V is generated from the randomly chosen key K and the biometric measurements X^n by a function h_n^* so that $V = h_n^*(X^n, K)$. The randomized system is illustrated in Figure 1 (b).

During the release stage, by providing the noisy measurement Y^n and data stored in the database V, we generate an estimate $\hat{K}$ of the key. Let g_n be the recovery function, and thus $\hat{K} = g_n(Y^n, V)$. In order to perform decryption, we require an arbitrarily small error probability during the key recovery stage.

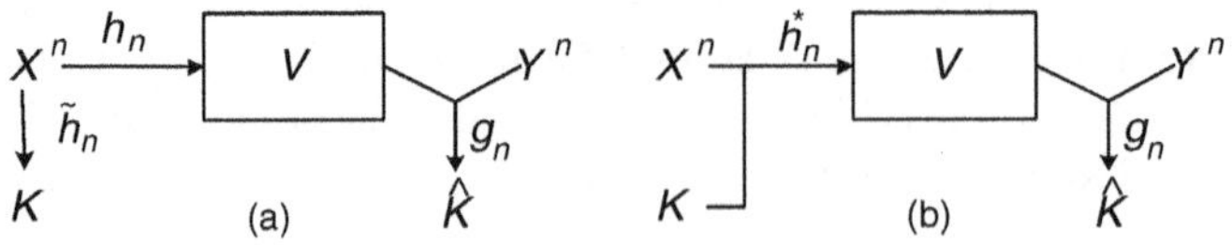

Fig. 1. Two different approaches for generating keys in biometric encryption systems: (a) a non-randomized approach; (b) a randomized approach

We first consider perfect security systems, in which we require that V does not contain any information about the generated key. More specifically we require that for every $\epsilon > 0$, $n^{-1}I(K;V) \leq \epsilon$ for sufficiently large n. Here $I(\cdot;\cdot)$ denotes

the mutual information between its two arguments. As mentioned before, the security level of the encrypted message is related to the rate of the generated key, and hence we measure the security level of the system by $R = n^{-1}H(K)$. Here $H(\cdot)$ denotes the entropy of its argument. The privacy of the biometric measurements is defined as the normalized equivocation rate $\Delta_P = H(X^n|V)/H(X^n)$. The larger this quantity, the greater the degree of privacy of the biometric measurements. If this quantity can be made arbitrarily close to 1, then we can achieve perfect privacy, which means that V does not leak any information about X^n, since $\Delta_P = 1$ implies $I(X^n; V) = 0$.

Definition 1 (perfect security system). *In a perfect security biometric encryption system, a privacy-security pair (Δ_P, R) is said to be achievable if, for each $\epsilon > 0$, there exist an integer n, coding functions, namely h_n and $\tilde{h}_n$ in non-randomized systems (i.e. $K = \tilde{h}_n(X^n)$, $V = h_n(X^n)$) and h_n^* in randomized systems (i.e., $V = h_n^*(X^n, K)$), and a decoding function, namely g_n (i.e., $\hat{K} = g_n(V, Y^n)$), satisfying the following conditions:*

$$n^{-1}H(K) \geq R, \tag{1}$$

$$H(X^n|V)/H(X^n) \geq \Delta_P, \tag{2}$$

$$n^{-1}I(V; K) \leq \epsilon, \tag{3}$$

$$\mathbb{P}[K \neq \hat{K}] \leq \epsilon \quad \textit{and} \tag{4}$$

$$n^{-1}\log|\mathcal{K}| \leq R + \epsilon. \tag{5}$$

Here, we offer some explanations of this definition. In this definition, (1) implies that the rate of the key should be larger than R; (2) implies that the privacy level of the system should be at least Δ_P; (3) says that the data stored in the database does not leak any information about the generated key; (4) implies that we can recover the key with high probability; and the combination of (1) and (5) says that the key is nearly uniformly generated.

For the second scenario, we consider a perfect privacy system, in which we require that the data stored in the database does not leak any information about the biometric measurements, that is for each $\epsilon > 0$, we require $I(X^n; V) \leq \epsilon$ for sufficiently large n. At the same time, we relax the requirement on the generated key, namely to allow $I(V; K)$ to range from 0 to $H(K)$. Of course, the smaller $I(V; K)$ the better. We measure the performance of a perfect privacy system by 1) the rate of the generated key $n^{-1}H(K)$, and 2) the normalized equivocation of the generated key $\Delta_s = H(K|V)/H(K)$. If $\Delta_s = 1$, we have $I(V; K) = 0$.

Definition 2 (perfect privacy system). *In a perfect privacy biometric security system, a rate-equivocation pair (R, Δ_s) is achievable if, for any $\epsilon > 0$, there exist an integer n, coding functions, namely h_n and $\tilde{h}_n$ in non-randomized systems (i.e. $K = \tilde{h}_n(X^n)$, $V = h_n(X^n)$) and h_n^* in randomized systems (i.e., $V = h_n^*(X^n, K)$), and a decoding function, namely g_n (i.e., $\hat{K} = g_n(V, Y^n)$), satisfying the following conditions:*

$$n^{-1}H(K) \geq R, \tag{6}$$

$$I(X^n; V) \leq \epsilon, \tag{7}$$
$$H(K|V)/H(K) \geq \Delta_s, \tag{8}$$
$$\mathbb{P}[K \neq \hat{K}\} \leq \epsilon \quad and \tag{9}$$
$$n^{-1} \log |\mathcal{K}| \leq R + \epsilon. \tag{10}$$

Another situation of interest is that in which, besides the data V stored in the database, an attacker of the system has side-information about the biometric characteristics. This models the situation in which the attacker obtains side-information from other sources, such as biometric characteristics stored in other databases or biometric characteristics from the relatives of the user. We denote the side observation at the attacker by Z^n, ranging in the set $\mathcal{Z}^n$, and assume that it is correlated with (X^n, Y^n). Furthermore, we assume $P_{X^nY^nZ^n}(x^n, y^n, z^n) = \prod_{i=1}^{n} P_{XYZ}(x, y, z)$. Since the attacker has both V and Z^n, the privacy level is now measured as $H(X^n|VZ^n)/H(X^n)$, and the generated key is required to be independent of V and Z^n.

Definition 3 (side-information at attacker). *In a biometric system with side-information Z^n available to an attacker, a privacy-security pair (Δ_P, R) is said to be achievable if, for any $\epsilon > 0$, there exist an integer n, coding functions, namely h_n and $\tilde{h}_n$ in non-randomized systems (i.e. $K = \tilde{h}_n(X^n)$, $V = h_n(X^n)$) and h_n^* in randomized systems (i.e., $V = h_n^*(X^n, K)$), and a decoding function, namely g_n (i.e., $\hat{K} = g_n(V, Y^n)$), satisfying the following conditions:*

$$n^{-1}H(K) \geq R, \tag{11}$$
$$H(X^n|VZ^n)/H(X^n) \geq \Delta_P, \tag{12}$$
$$n^{-1}I(VZ^n; K) \leq \epsilon, \tag{13}$$
$$\mathbb{P}[K \neq \hat{K}\} \leq \epsilon \quad and \tag{14}$$
$$n^{-1} \log |\mathcal{K}| \leq R + \epsilon. \tag{15}$$

3 Perfect Key Case

In this section, we study perfect security systems, in which the data stored in the database contains limited information about the generated key. Our goal is to characterize the relationship between the key size and the information leakage about the biometric measurements.

We first consider non-randomized systems. As discussed in Section 2, in a non-randomized system, both the key K and data V are generated from the biometric measurements X^n. Some existing schemes, for example, the secure sketch approach of [2] and the coding approach of [9], belong to this category. The theorem below establishes the performance limits of this type of biometric security system. The basic idea of the achievability scheme behind this theorem is to construct a compressed version U^n of X^n, and then generate the key K and helper data V as functions of U^n. Roughly speaking, we generate approximately $2^{nI(U;X)}$ U^n sequences. For each $x^n \in \mathcal{X}^n$, we find a u^n that is jointly

typical (in this paper, we use strong typicality defined in [7]) with x^n and assign this u^n as the compressed version of x^n. Since the number of X^n sequences is approximately $2^{nH(X)}$, which is larger than the number of U^n sequences in the codebook, each U^n will correspond to more than one X^n. We further reduce the information required to be stored in the database by using source coding with side-information [7], in which U^n is the source sequence at the encoder and Y^n is the side information present at the decoder. Roughly speaking, we divide all sequences U^n into approximately $2^{n(I(U;X)-I(U;Y))}$ bins, each containing approximately $2^{nI(U;Y)}$ sequences. Thus, each U^n sequence has two indices: a bin index and an index within each bin. We store the bin index in the database as helper data, and set the key value as the index of U^n within each bin. Hence, the rate of the key is approximately $I(U;Y)$. With the bin index and noisy measurements Y^n, we can recover U^n during the release stage with high probability. We can then further recover the key. Furthermore, it can be shown that the mutual information between the data stored in the database (i.e. the bin index) and the key (i.e. the index of the sequence within the bin) can be made arbitrarily small. Thus this scheme guarantees the perfect security of the generated key. By different choices of U, we control the leakage of information about the biometric measurements and the rate of the generated key. Furthermore, we are able to prove a converse result, and thus show that the above mentioned scheme is optimal.

Theorem 1. *Let $\mathcal{C}_N$ be the set of the privacy-security pairs (Δ_P, R) satisfying the following conditions:*

$$\Delta_P \leq 1 - \frac{I(U;X) - I(U;Y)}{H(X)} \quad \text{and} \tag{16}$$

$$R \leq I(U;Y), \tag{17}$$

for some auxiliary random variable U such that (U, X, Y) satisfies the Markov chain condition $U \to X \to Y$. Then a privacy-security pair (Δ_P, R) is achievable if and only if $(\Delta_P, R) \in \mathcal{C}_N$.

We note that a similar scheme was used in [10] with a different purpose. From this theorem, we know that in order to maximize the rate of the key, we should set $U = X$. The rate of the key is then $I(X;Y)$. Correspondingly, the privacy level is $1 - H(X|Y)/H(X)$. This recovers the existing results of [11,12]. On the other hand, in order to achieve perfect privacy, the auxiliary random variable U in (16) should be chosen such that $I(U;X) = I(U;Y)$. The maximal rate achievable is then

$$\max_U I(U;Y) \quad s.t. \quad U \to X \to Y \quad \text{and} \quad I(U;X) = I(U;Y). \tag{18}$$

In randomized systems, during the enrollment stage, users have the freedom to choose the values of the keys but they are not required to remember them. For example, the fuzzy vault scheme studied in [1] belongs to this category. Here, the key K can be viewed as a source of additional randomness. It is reasonable

to conjecture that this additional randomness could help in achieving better performance, at least for the privacy of the biometric measurements. The theorem below disproves this conjecture. The basic idea of the achievablity scheme is as follows. We first use the scheme in the proof of Theorem 1 to generate a key J, choosing from a set $\mathcal{J}$ with size $|\mathcal{J}|$. Then for a uniformly generated key K from a set $\mathcal{K}$, we store $J \oplus K$ in the database, along with other information required to be stored for the achievability scheme described above Theorem 1. Here $\oplus$ denotes mod-$|\mathcal{J}|$ addition. If we set $\mathcal{K} = \mathcal{J}$, $J \oplus K$ will be approximately uniformly distributed over $\mathcal{J}$, and is independent of other random variables of interest. Hence, this additional information stored in the database will not provide any information about the generated key and biometric measurements. In the release stage, we first obtain an estimate $\hat{J}$ of J using the same achievability scheme as that of Theorem 1. We then recover K via $J \oplus K \oplus \hat{J}$. Since $\hat{J} = J$ with high probability, $\hat{K}$ is equal to K with high probability. We show in the converse that the performance of the above mentioned scheme is optimal.

Theorem 2. *Let $\mathcal{C}_R$ be the set of the privacy-security pairs (Δ_P, R) satisfying the following conditions:*

$$\Delta_P \leq 1 - \frac{I(U;X) - I(U;Y)}{H(X)} \quad \text{and} \tag{19}$$

$$R \leq I(U;Y), \tag{20}$$

for some auxiliary random variable U such that (U, X, Y) satisfies the Markov chain condition $U \to X \to Y$. Then a privacy-security pair (Δ_P, R) is achievable if and only if $(\Delta_P, R) \in \mathcal{C}_R$.

From here, we can see that $\mathcal{C}_N = \mathcal{C}_R$, and hence, the additional randomization does not increase the region. But one advantage of this randomized approach is that the system is revocable.

4 Perfect Privacy for Biometric Measurements

In this section, we consider the perfect privacy case. As discussed in Section 3, if we consider both perfect privacy and perfect secrecy, i.e. both $I(X^n; V)$ and $n^{-1}I(V; K)$ can be made arbitrarily small, the problem can be solved by looking for suitable auxiliary random variables U. Thus in this section we generalize the requirement on the generated key by allowing $I(V; K)$ to be nonzero, as specified in Definition 2.

First consider non-randomized systems, in which $H(K|X^n) = 0$ since K is a function of X^n. Thus in this case, $I(X^n; V) \leq \epsilon$ implies that $I(K; V) \leq \epsilon$. Hence in non-randomized systems, perfect privacy means perfect security. This case has been considered in Section 3. Therefore, it is sufficient to discuss only randomized systems in the remainder of this section.

In the following, we show a close relationship between perfect privacy and common random information, which is defined as follows.

Definition 4. *For two random processes X^n and Y^n, there exists a common random process between them with entropy rate not less than α if for each $\eta > 0$, there exist n and functions ψ_n of X^n and ϕ_n of Y^n such that*

$$\mathbb{P}[\psi_n(X^n) \neq \phi_n(Y^n)] \leq \eta \quad \text{and} \tag{21}$$

$$n^{-1}H(\psi_n(X^n)) \geq \alpha - \eta. \tag{22}$$

This definition says that if X^n and Y^n have a common random process with entropy rate α, then one can generate two random variables: $\psi_n(X^n)$ solely based on X^n and $\phi_n(Y^n)$ solely based on Y^n, with the property that each of these two random variables has entropy $n\alpha$ and equals to the other one with high probability. Now, if there exists a common random process between the biometric measurements X^n and Y^n with entropy rate R, we can construct a system with perfect privacy. We first generate a random variable $J = \psi_n(X^n)$ during the enrollment stage, and store a function $f(K, J)$ in the database. Now, as long as $H(K) \geq nR$, there exists a function f such that $I(X^n; f(K, J)) = 0$, which means that there is no privacy leakage. During the release stage, based on the biometric measurements, we can first generate $\hat{J} = \phi_n(Y^n)$, and then recover the key K. Based on Definition 4, $\hat{J} = J$ with high probability, and hence $K = \hat{K}$ with high probability. The following theorem makes these ideas precise.

Theorem 3. *A privacy-rate pair (R, Δ_s) is achievable if and only if there exists a common random process between X^n and Y^n with entropy rate not less than $R\Delta_s$.*

5 Side-Information at an Attacker

In this section, we consider a situation in which, besides the data V stored in the database, an attacker has side-information about the biometric characteristics. This models the situation in which the attacker obtains side-information from other sources, such as biometric characteristics stored in other databases or biometric characteristics from the relatives of the user.

We first consider the non-randomized approach, in which both V and K are functions of the biometric measurements X^n, i.e., $V = h_n(X^n)$ and $K = \tilde{h}_n(X^n)$. We begin with a scheme that provides an inner bound on the set of all achievable privacy-security pairs. The basic idea is based on that of Theorem 1. We first generate a compressed version U^n of X^n, and then perform source coding with side information (U^n as the source sequence at the source coding encoder, and Y^n as the side information present at the decoder). That is we divide the U^ns into bins, and store the bin index in the database. In Theorem 1, we set the key value as the index of U^n in each bin. Now the attacker has additional information, the key rate should be reduced accordingly in order to guarantee that the attacker does not obtain any information about the generated key. We fulfill this goal by further partitioning each bin into subsets. We set the key value as the subset index. Using ideas from the analysis of the wiretap channel [13], it can

be shown that there exists a partition such that even with the side information at the attacker and bin index, the attacker will not be able to infer too much information about the generated key. We then characterize the privacy leakage of this scheme. With the bin index and noisy information Y^n, we can recover U^n, and then recover the key by looking at the subset index of the recovered sequence U^n. Using information inequalities, we also provide an upper bound on the performance achievable by any scheme.

Theorem 4. *Let $C_{s,in}$ be the set of (Δ_P, R) satisfying the following conditions:*

$$\Delta_P \leq 1 - \frac{I(X;UZ) - I(U;Y|W) + I(U;Z|W)}{H(X)} \quad \text{and} \tag{23}$$

$$R \leq I(U;Y|W) - I(U;Z|W), \tag{24}$$

and $C_{s,out}$ be the set of (Δ_P, R) satisfying the following conditions:

$$\Delta_P \leq 1 - \frac{I(X;UZ) - I(U;Y) + I(U;Z|W)}{H(X)} \quad \text{and} \tag{25}$$

$$R \leq I(U;Y|W) - I(U;Z|W), \tag{26}$$

in which W and U are auxiliary random variables such that (W, U, X, Y, Z) satisfies the following Markov chain condition $W \to U \to X \to (Y, Z)$. Any pair in $C_{s,in}$ is achievable, while any pair outside of $C_{s,out}$ is not achievable.

Now consider randomized systems, in which K is randomly generated and independent of X^n. The helper data V is a function of K and X^n; that is $V = h_n^*(K, X^n)$. An achievable region is described by the following scheme. The basic idea is to first generate a key J, choosing from a set $\mathcal{J}$ with size $|\mathcal{J}|$, using the scheme in the proof of Theorem 4. Then for a randomly generated key K, we store $J \oplus K$ in the database, along with other information required to be stored in Theorem 4. Here $\oplus$ denotes mod-$|\mathcal{J}|$ addition. If we set $\mathcal{K} = \mathcal{J}$, $J \oplus K$ will be approximately uniformly (these terms can be made rigorous, but these details are omitted here due to space limitations) distributed over $\mathcal{J}$, and is independent of other random variables of interest. In the release stage, we first obtain an estimate $\hat{J}$ of J using the same scheme as that of Theorem 4. We then recover K via $J \oplus K \oplus \hat{J}$. Since $\hat{J} = J$ with high probability, $\hat{K}$ is equal to K with high probability. Using information theoretic inequalities, we provide an upper-bound on the achievable privacy-security pairs.

Theorem 5. *Let $C_{sr,in}$ be the set of (Δ_P, R) pairs satisfying the following conditions:*

$$\Delta_P \leq 1 - \frac{I(X;UZ) - I(U;Y|W) + I(U;Z|W)}{H(X)} \quad \text{and} \tag{27}$$

$$R \leq I(U;Y|W) - I(U;Z|W), \tag{28}$$

and let $C_{sr,out}$ be the set of (Δ_P, R) pair satisfying the following conditions:

$$\Delta_P \leq 1 - \frac{I(X;Z|U) - I(U;Y|W) + I(U;Z|W)}{H(X)} \quad \text{and} \tag{29}$$

$$R \leq I(U;Y|W) - I(U;Z|W), \tag{30}$$

in which W and U are auxiliary random variables such that (W, U, X, Y, Z) satisfies the following Markov chain condition $W \rightarrow U \rightarrow X \rightarrow (Y, Z)$. Then any pair in $C_{sr,in}$ is achievable, while any pair outside of $C_{sr,out}$ is not achievable.

6 Conclusions

An information theoretic framework has been established to study the performance of biometric security systems. More specifically, biometric security systems have been studied under a privacy-security tradeoff framework. Two different scenarios, in which the attacker either has side-information about the biometric measurements or not, have been considered. In the scenario for which the attacker does not have side-information, we have considered two cases of perfect security and perfect privacy. In both cases, the complete privacy-security region has been identified. For the scenario in which the attacker has side-information about the biometric measurements, the perfect security case has been considered, for which inner and upper bounds on the privacy-security region have been derived.

References

1. Juels, A., Sudan, M.: A fuzzy vault scheme. In: Proc. IEEE Intl. Symposium on Information Theory, pp. 293–297 (2002)
2. Sutcu, Y., Li, Q., Memon, N.: Protecting biometric templates with sketch: Theory and practice. IEEE Trans. Inf. Forensics and Security 2, 503–512 (2007)
3. Jain, A.K., Nandakumar, K., Nagar, A.: Biometric template security. EURASIP Journal on Advances in Signal Processing, 1–17 (2008)
4. Shannon, C.E.: Communication theory of secrecy systems. Bell System Technical Journal 28, 656–715 (1949)
5. Maurer, U.M.: Secret key agreement by public discussion from common information. IEEE Trans. Inf. Theory 39, 733–742 (1993)
6. Tuyls, P., Goseling, J.: Biometric Authentication. Springer, Berlin (2004)
7. Cover, T.M., Thomas, J.A.: Elements of Information Theory. Wiley, New York (1991)
8. Lai, L., Ho, S.-W., Poor, H.V.: Privacy-security tradeoffs in biometric security systems. IEEE Trans. on Inf. Theory (submitted, 2008)
9. Draper, S., Khisti, A., Martinian, E., Vetro, A., Yedidia, J.: Using distributed source coding to secure fingerprint biometrics. In: Proc. IEEE Intl. Conf. on Acoustics, Speech, and Signal Processing, pp. 129–132 (2007)
10. Csiszár, I., Narayan, P.: Common randomness and secret key generation with a helper. IEEE Trans. Inf. Theory 46, 344–366 (2000)
11. Cohen, G., Zemor, G.: The wire-tap channel applied to biometrics. In: Proc. IEEE Intl. Symposium on Information Theory and its Applications (2004)
12. Ignatenko, T., Willems, F.: On privacy in secure biometrics authentication systems. In: Proc. IEEE Intl. Conf. on Acoustics, Speech, and Signal Processing, pp. 121–124 (2007)
13. Wyner, A.D.: The wire-tap channel. Bell System Technical Journal 54, 1355–1387 (1975)

Constructing Passwords from Biometrical Data

Vladimir B. Balakirsky, Anahit R. Ghazaryan, and A.J. Han Vinck*

Institute for Experimental Mathematics, Ellernstr. 29, 45326 Essen, Germany
v_b_balakirsky@rambler.ru, a_ghazaryan@rambler.ru, vinck@iem.uni-due.de

Abstract. We propose a probabilistic model for constructing passwords on the basis of outcomes of biometrical measurements. An algorithm for the transformation of biometrical data to passwords is given. The performance of the authentication scheme is evaluated by the compression factor, the false acceptance/rejection rates, the probability distribution over the set of passwords, and the probability of a correct guess of the input biometrical data mapped to the known password. An application of the results to DNA measurements is presented.

1 Probabilistic Model for Biometrical Data and Their Use for Authentication

Suppose that there is a non–stationary memoryless source generating a vector $\mathbf{b} = (b_1, \ldots, b_n)$, where $b_t \in \mathcal{B}_t$ and $\mathcal{B}_t = \{0, \ldots, Q_t - 1\}$, $t = 1, \ldots, n$. We also denote $\mathcal{B}^{(n)} = \mathcal{B}_1 \times \ldots \times \mathcal{B}_n$ and write $\mathbf{b} \in \mathcal{B}^{(n)}$. Let the probability distribution (PD) over the set $\mathcal{B}_t$ be given as

$$P_t = \Big(\Pr_{\text{bio}}\{B_t = b\},\, b \in \mathcal{B}_t \Big). \tag{1}$$

Thus, the probability of receiving the vector $\mathbf{b}$ is computed as

$$\Pr_{\text{bio}}\Big\{B^{(n)} = \mathbf{b}\Big\} = \prod_{t=1}^{n} \Pr_{\text{bio}}\Big\{B_t = b_t\Big\}, \tag{2}$$

where $B^{(n)} = (B_1, \ldots, B_n)$ denotes the vectors of random variables.

The vector $\mathbf{b}$ will be referred to as the biometric vector. For example, this vector can represent outcomes of $n = 28$ DNA measurements of a person [1]. Some basic characteristics of the PD's $P_1, \ldots, P_n$ can be found in [2] where different additive block coding schemes using the DNA data were developed. We will also use these data to illustrate the main points of the present contribution. In particular, to store the vector $\mathbf{b}$ in the database (DB), one needs $\sum_{t=1}^{n} \lceil \log Q_t \rceil$ bits, and this sum is equal to 140 for the DNA data.

The authentication of people is one of the reasons for processing biometrical data [3]. In this case, names and passwords of a certain group of people, called the users, are stored in the DB. Having received a pair (name, password), the

* This work was partially supported by the DFG.

M. Tistarelli and M.S. Nixon (Eds.): ICB 2009, LNCS 5558, pp. 889–898, 2009.
© Springer-Verlag Berlin Heidelberg 2009

authentication scheme checks whether the pair is valid or not. In biometrical authentication procedures, passwords are formed on the basis of the outcomes of biometrical measurements of the users. We understand such a transformation as a mapping

$$F : \mathbf{b} \in \mathcal{B}^{(n)} \to \mathbf{z} \in \mathcal{Z}^{(n)}, \tag{3}$$

where $\mathcal{Z}^{(n)} = \mathcal{Z}_1 \times \ldots \times \mathcal{Z}_n$ and $\mathcal{Z}_t = \{0, \ldots, q_t - 1\}$, $t = 1, \ldots, n$, are fixed sets. We will also assume that $q_1 \leq Q_1, \ldots, q_n \leq Q_n$ and write $\mathbf{z} = F(\mathbf{b})$.

Let us discuss the basic requirements to the strings that can be effectively used as passwords, which are taken into account in our considerations.

1. The length of passwords is usually rather small (about 8 bytes), and passwords of different users "have to look as completely random sequences" to create difficulties to an attacker, who wants to guess the password of a certain user. This requirement to an algorithm of assigning passwords is not easy to formalize, unless a pseudo–random number generator is used. However, biometrical authentication assumes that designers of the system try to extract randomness from user's personal data and do not include any external source, i.e., the function F in (3) is a deterministic function. The problem, which appears in this context, is caused by the point that components of the biometric vector are generated by a non–uniform source. In particular, the probability of the most likely vector of outcomes of the DNA measurements is equal to $2^{-78.8}$, while the vectors have length 140 bits.
2. The biometrical characteristics of a fixed user can be non–typical. Nevertheless, the system has to provide the authentication for such a user as well. Therefore, we also have to study the performance of the authentication scheme for fixed biometrical vectors.
3. The outcomes of the biometrical measurements of the same person can be hardly exactly repeated. The absence of external randomness also implies that basic probabilistic characteristics of the authentication scheme, such as the false rejection rate (the probability that the identity claim of a user is rejected) and the false acceptance rate (the probability that the identity claim of a different user is accepted), have to be computed over the ensemble of biometric vectors and their noisy observations. We will present a probabilistic model for the noise of observations, which is needed for computing the false rejection rate. In our opinion, introduction of a proper model is the key point of processing biometrical data for authentication purposes.
4. The passwords of users are usually stored in a highly protected part of the DB where an attacker is not supposed to have any access. However, the biometrical direction in authentication also generates another types of attackers. Namely, suppose that the attacker knows the password of a certain user and he is interested in the biometrical characteristics of the person instead of passing through a fixed authentication test with the acceptance decision. The reasons for this interest come from the point that some parameters, like the DNA data, are very difficult to receive and, being compromised, these data are compromised forever.

The structure of a general biometrical authentication scheme under our considerations is given in Figure 1. The vector $\mathbf{b}^*$ and the corresponding password $\mathbf{z}^*$ are fixed. Another input to the verifier is a biometric vector $\mathbf{b}$, and we consider two alternatives denoted by "ind" and "dep":

$$\text{“ind”} \Rightarrow \Pr\Big\{B^{(n)} = \mathbf{b} \,|\, B^{(n)*} = \mathbf{b}^*\Big\} = \Pr_{\text{bio}}\Big\{B^{(n)} = \mathbf{b}\Big\}, \tag{4}$$

$$\text{“dep”} \Rightarrow \Pr\Big\{B^{(n)} = \mathbf{b} \,|\, B^{(n)*} = \mathbf{b}^*\Big\} = \Pr_{\text{err}}\Big\{B^{(n)} = \mathbf{b} \,|\, B^{(n)*} = \mathbf{b}^*\Big\}. \tag{5}$$

The conditional probabilities at the left–hand sides specify the source in Figure 1, which either knows the vector $\mathbf{b}^*$ or not. In the "ind" case, when the verifier has to reject the identity claim, the vector $\mathbf{b}$ is formed by a memoryless source introduced in (1), (2). In the "dep", when the verifier has to accept the identity claim, the vector $\mathbf{b}$ is formed by a memoryless source parameterized by the vector $\mathbf{b}^*$ in such a way that

$$\Pr_{\text{err}}\Big\{B^{(n)} = \mathbf{b} \,|\, B^{(n)*} = \mathbf{b}^*\Big\} = \prod_{t=1}^{n} \Pr_{\text{err}}\Big\{B_t = b_t \,|\, B_t^* = b_t^*\Big\},$$

where, for all $t = 1, \ldots, n$,

$$\left(\left(\Pr_{\text{err}}\Big\{B_t = b \,|\, B_t^* = b^*\Big\}, \, b \in \mathcal{B}_t \right), b^* \in \mathcal{B}_t \right) \tag{6}$$

is a collection of the conditional PD's satisfying the inequalities

$$\Pr_{\text{err}}\Big\{B_t = b^* \,|\, B_t^* = b^*\Big\} \geq 1 - \varepsilon \tag{7}$$

for all $b^* \in \mathcal{B}_t$ and a fixed $\varepsilon \in (0, 1/2)$.

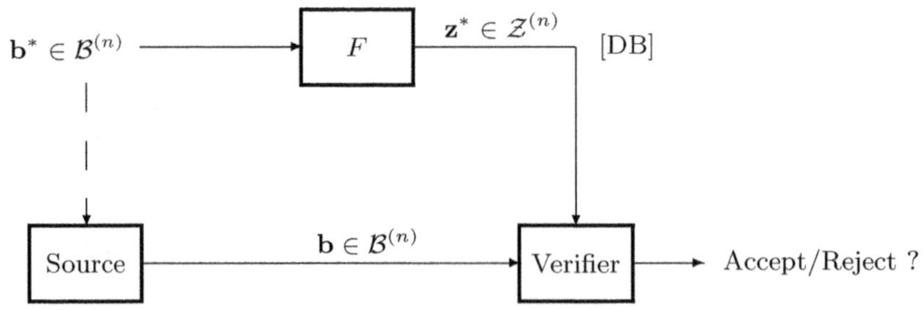

Fig. 1. The structure of a general biometrical authentication scheme

The model for the noise of observations above can be directly used for practical DNA measurements [1]. Some details of this application will be described in Section 5. We also think that this model is general enough in a sense that it can be used for other biometrical data in the following way. Since finding a probabilistic description of the noise of observations is usually very difficult (for example, in

the case when locations of minutiae points of the fingerprint are measured, the errors are caused by shifts and rotations of the finger, the light, the pressure, etc.), in practical biometrical systems, designers include an algorithm that tries to match two outcomes of the measurements. The acceptance/rejection decision is made on the basis of the comparison of the total number of observations and the number of observations that are matched by the program. Our model for the noise, where we only specify the probability that the input symbol is unchanged, assumes that the matching program is included into the channel between the sample and its observation.

In the following considerations, we concentrate on constructing passwords under the model described above. This problem is non–trivial because of the requirement that the strings stored in the DB have to look as randomly chosen strings, while the non–stationary source (1), (2) can be arbitrary. These considerations generalize some ideas developed for the non–stationary continuous sources [4] to the discrete case. In the continuous case, we assumed that there is a memoryless source given by the probability distribution functions $\Phi_1(y), \ldots, \Phi_n(y)$, $y \in \mathcal{R}$, where $\Phi_t(y) \in [0,1]$ is a non–decreasing function with $\Phi_t(-\infty) = 0$ and $\Phi_t(+\infty) = 1$ for all $t = 1, \ldots, n$. Then the transformation of an observed vector $\mathbf{y} = (y_1, \ldots, y_n) \in \mathcal{R}^n$ to $(\Phi_1(y_1), \ldots, \Phi_n(y_n)) \in [0,1]^n$ creates the vector whose components are independent random variables uniformly distributed over the $[0,1]$ interval. The quantization of each component into 2^L levels by using the thresholds $i2^{-L}$, $i = 0, \ldots, 2^L$, where L is a fixed integer, generates a binary vector of length Ln uniformly distributed over the set $\{0,1\}^{Ln}$, and this statement is true for any L. Furthermore, the distance between two arguments y and y' of the function Φ_t defined as $|y - y'|$ is translated into the distance defined as $|\Phi_t(y) - \Phi_t(y')|$, by the properties of the function Φ_t. In the discrete case we cannot use this approach one–by–one, but its ideas are relevant.

2 Restricted Biometrical Authentication Schemes and Their Parameters

Let us restrict the considerations to the following case.

1. Let
$$\max_{b \in \mathcal{B}_t} \Pr_{\text{bio}}\Big\{ B_t = b \Big\} < 1 - \varepsilon. \tag{8}$$
2. Let F be a component–wise mapping determined by functions $f_t : \mathcal{B}_t \to \mathcal{Z}_t$, $t = 1, \ldots, n$, in such a way that
$$F(\mathbf{b}) = (f_1(b_1), \ldots, f_n(b_n))$$
for all $\mathbf{b} \in \mathcal{B}^{(n)}$.
3. Let the verifier make the acceptance/rejection decision on the basis of the pair $(F(\mathbf{b}^*), F(\mathbf{b}))$ instead of the pair $(F(\mathbf{b}^*), \mathbf{b})$.

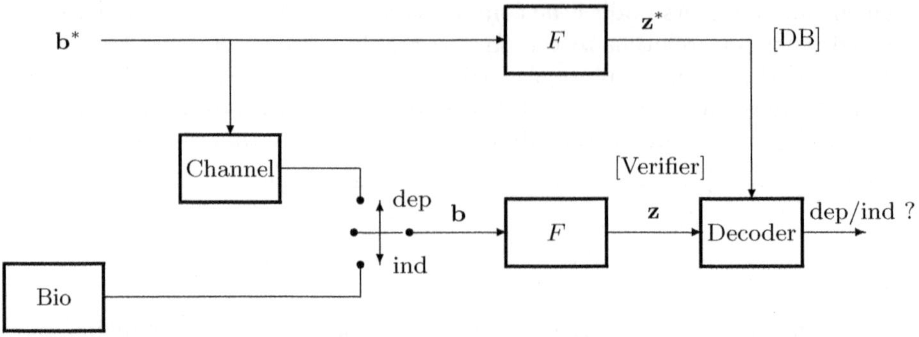

Fig. 2. Representation of the restricted biometrical authentication scheme

The structure of the restricted biometrical authentication scheme is illustrated in Figure 2. We represent the relationships given in (4), (5) as two possibilities: the vector $\mathbf{b}$ is either received as a result of transmission of the vector $\mathbf{b}^*$ over a channel or it is independently generated by the biometric source. The verification is split in two blocks. At first, the password $F(\mathbf{b})$ is computed and then the decoder processes the pair of passwords to make a decision. In general case, the decoding algorithm is specified by the sets $\mathcal{A}(\mathbf{z}^*) \subseteq \mathcal{Z}^{(n)}$, $\mathbf{z}^* \in \mathcal{Z}^{(n)}$, in such a way that the decision is "dep"(accept) if $F(\mathbf{b}) \in \mathcal{A}(\mathbf{z}^*)$ and "ind"(reject) if $F(\mathbf{b}) \notin \mathcal{A}(\mathbf{z}^*)$. Our probabilistic model for the noise assumes that the verifier has to make a decision on the basis of the value of the Hamming distance between received vectors, $\mathcal{A}(\mathbf{z}^*) = \mathcal{A}_\rho(\mathbf{z}^*)$, where

$$\mathcal{A}_\rho(\mathbf{z}^*) = \Big\{ \mathbf{z} \in \mathcal{Z}^{(n)} : d(\mathbf{z}, \mathbf{z}^*) \leq n\rho \Big\}, \tag{9}$$

where $\rho \in [0, 1]$ is a specified parameter and

$$d(\mathbf{z}, \mathbf{z}^*) \triangleq \Big| \Big\{ t \in \{1, \ldots, n\} : z_t \neq z_t^* \Big\} \Big|.$$

Thus,

$$\text{Decision} = \begin{cases} \text{"ind"}, \text{ if } d(\mathbf{z}, \mathbf{z}^*) > n\rho, \\ \text{"dep"}, \text{ if } d(\mathbf{z}, \mathbf{z}^*) \leq n\rho, \end{cases}$$

The probabilities of incorrect decisions (the false acceptance and the false rejection rates) for the vector $\mathbf{b}^*$ that is mapped to the vector $\mathbf{z}^*$ are expressed as

$$\text{FAR}(\mathbf{b}^*) = \sum_{\mathbf{b}:\, F(\mathbf{b}) \in \mathcal{A}_\rho(\mathbf{z}^*)} \Pr_{\text{bio}} \Big\{ B^{(n)} = \mathbf{b} \Big\},$$

$$\text{FRR}(\mathbf{b}^*) = \sum_{\mathbf{b}:\, F(\mathbf{b}) \notin \mathcal{A}_\rho(\mathbf{z}^*)} \Pr_{\text{err}} \Big\{ B^{(n)} = \mathbf{b} \,|\, B^{(n)*} = \mathbf{b}^* \Big\}.$$

One of main parameters characterizing the performance of the algorithm is

$$C \triangleq \frac{\sum_{t=1}^n \lceil \log Q_t \rceil}{\sum_{t=1}^n \lceil \log q_t \rceil},$$

which can be understood as the compression factor. The other parameters introduced below are relevant to the analysis of the false acceptance/false rejection rates and the cryptographic properties. One can easily see that if an attacker wants to guess the biometric vector of a person, then the best estimate is the vector having the maximum probability. The probability that this guess is correct is equal to $\hat{\omega} \triangleq \prod_{t=1}^{n} \hat{\omega}_t$, where $\hat{\omega}_t \triangleq \max_{b \in \mathcal{B}_t} \omega_t(b)$ and

$$\omega_t(b) \triangleq \Pr_{\text{bio}}\Big\{B_t = b\Big\}, \quad b \in \mathcal{B}_t.$$

The transformation of biometric vectors to passwords and changing the attacker's task as guessing the password makes the probability of success equal to $\hat{\pi} \triangleq \prod_{t=1}^{n} \hat{\pi}_t$, where $\hat{\pi}_t \triangleq \max_{z \in \mathcal{Z}_t} \pi_t(z)$ and

$$\pi_t(z) = \sum_{b:\ f_t(b)=z} \Pr_{\text{bio}}\Big\{B_t = b\Big\}, \quad z \in \mathcal{Z}_t, \tag{10}$$

is the PD over the t-th component of the passwords. Furthermore, the best prediction of the input biometric vector, given the password $\mathbf{z}^*$, is the biometric vector having the maximum probability among the vectors mapped to $\mathbf{z}^*$. As it is easy to see, the probability that this prediction is correct can be computed as $\hat{\gamma}(\mathbf{z}^*) \triangleq \prod_{t=1}^{n} \hat{\gamma}_t(z_t^*)$, where $\hat{\gamma}_t(z_t^*) \triangleq \max_{b \in \mathcal{B}_t} \gamma_t(b|z_t^*)$ and

$$\gamma_t(b|z_t^*) \triangleq \frac{1}{\pi_t(z_t^*)} \begin{cases} \omega_t(b), & \text{if } f_t(b) = z_t^*, \\ 0, & \text{if } f_t(b) \neq z_t^*, \end{cases} \quad b \in \mathcal{B}_t,$$

is the z_t^*-conditional PD over the t-th component of the passwords.

The notation above is illustrated in Table 1, where

$$\overline{\gamma}_t \triangleq \sum_{z^* \in \mathcal{Z}_t} \pi_t(z)\hat{\gamma}_t(z^*), \quad \hat{\gamma} \triangleq \max_{z^* \in \mathcal{Z}_t} \hat{\gamma}_t(z^*)$$

denote the average and the maximum probability of the successful guess of the input biometric vector. In the stationary case, when the biometric source is described by the presented PD for all $t = 1, \ldots, n$, the probabilities of the correct guess of the biometric vector and the password are equal to $(0.2401)^n$ and $(0.2596)^n$, i.e., they are close enough. The guessing algorithm for the attacker, who has access to the password and wants to find the biometric vector, depends on the password. The average and the maximum probabilities of success are equal to $(0.7401)^n$ and $(0.9674)^n$, respectively.

The false acceptance/false rejection rates can be expressed using the notation above. These expressions and proofs can be found in [5]. In the present correspondence we restrict ourselves to the description of the algorithm for constructing the functions $f_1, \ldots, f_n$ and to the application of the verification procedure to the biometric systems with the DNA measurements.

Table 1. Example of the mapping $f_t : \{0,\ldots,9\} \to \{0,\ldots,3\}$

$b =$	9	7	3	5	6	0	8	1	2	4
$\omega_t(b) =$	.2401	.0081	.2254	.0342	.1862	.0529	.0882	.0874	.0414	.0361
$z = f_t(b) =$	0		1		2		3			
$\pi_t(z) =$	.2482		.2596		.2381		.2531			
$\gamma_t(b\|z) =$	.9674	.0326	.8682	.1318	.7820	.2180	.3485	.3453	.1636	.1426
$\hat{\gamma}_t, \overline{\gamma}_t =$	.9674,.7401									
$\hat{\pi}_t =$	.2596									
$\hat{\omega}_t =$	.2401									

3 Constructing the Functions $f_1, \ldots, f_n$

In the following considerations, we omit the index $t \in \{1,\ldots,n\}$ for a formal brevity and extend the ideas presented in [4] for the continuous case to the discrete case.

Let us determine the function f by a partitioning of the set $\mathcal{B}$ by q pairwise disjoint subsets $\mathcal{F}(0),\ldots,\mathcal{F}(q-1)$ in such a way that $f(b) = z$ is equivalent to $b \in \mathcal{F}(z)$. For example, the function f with $f(0) = f(2) = 0$, $f(1) = 1$, $f(3) = 2$ is specified by the sets $\mathcal{F}(0) = \{0,2\}$, $\mathcal{F}(1) = \{1\}$, $\mathcal{F}(2) = \{3\}$. "A greedy algorithm" for constructing the sets $\mathcal{F}(0),\ldots,\mathcal{F}(q-1)$ is presented below.

F1: Set $z = 0$.
F2: Set $\mathcal{F}(z) = \emptyset$, $S = 0$, $\Delta = 2^{-q}$.
F3: Denote

$$\mathcal{B}_0 = \mathcal{F}(z) \bigcup \Big[\bigcup_{z'=0}^{z-1} \mathcal{F}(z') \Big],$$

$$\Delta_0 = \min_{b \in \mathcal{B} \setminus \mathcal{B}_0} \Big| 2^{-q} - (S + \omega(b)) \Big|,$$

$$b_0 = \arg \min_{b \in \mathcal{B} \setminus \mathcal{B}_0} \Big| 2^{-q} - (S + \omega(b)) \Big|.$$

If $\Delta_0 > \Delta$, then go to 5.
F4: Include b_0 into the set $\mathcal{F}(z)$, increase S by $\omega(b_0)$, and substitute Δ_0 for Δ.
F5: Increase z by 1. If $z \leq q-2$, then go to 2.
F6: Set

$$\mathcal{F}(q-1) = \mathcal{B} \setminus \Big[\bigcup_{z=0}^{q-2} \mathcal{F}(z) \Big].$$

F7: Output the sets $\mathcal{F}(0),\ldots,\mathcal{F}(q-1)$. End.

4 Application of the Authentication Algorithm to the DNA Measurements

We will use the following mathematical model for the DNA measurements [2]. Suppose that there are n sources. Let the t-th source generate a pair of integers according to the PD

$$\Pr_{\mathrm{DNA}}\Big\{ (R_{t,1}, R_{t,2}) = (r_{t,1}, r_{t,2}) \Big\} = p_t(r_{t,1}) p_t(r_{t,2}),$$

where $r_{t,1}, r_{t,2} \in \mathcal{R}_t = \{c_t, \ldots, c_t + k_t - 1\}$ and integers $c_t, k_t > 0$ are given. The outcome of the t-th measurement is defined as

$$R_t \triangleq \Big(\min\{R_{t,1}, R_{t,2}\}, \max\{R_{t,1}, R_{t,2}\} \Big). \tag{11}$$

Hence, for all $i \in \mathcal{R}_t$,

$$\Pr_{\mathrm{DNA}}\Big\{ R_t = (i,j) \Big\} = \begin{cases} 0, & \text{if } j \in \{0, \ldots, i-1\}, \\ p_t^2(i), & \text{if } j = i, \\ 2p_t(i)p_t(j), & \text{if } j \in \{i+1, \ldots, c_t + k_t - 1\}. \end{cases}$$

We assume that $R_1, \ldots, R_n$ are mutually independent pairs of random variables, i.e.,

$$\Pr_{\mathrm{DNA}}\Big\{ R^{(n)} = \mathbf{r} \Big\} = \prod_{t=1}^{n} \Pr_{\mathrm{DNA}}\Big\{ R_t = r_t \Big\},$$

where $R^{(n)} = (R_1, \ldots, R_n)$ and $\mathbf{r} = (r_1, \ldots, r_n)$, $r_t \in \mathcal{R}_t \times \mathcal{R}_t$. To make the notation consistent with the notation of Section 2, let us map $Q_t = k_t(k_t+1)/2$ pairs $r_t = (i_t, j_t)$, where $j_t \geq i_t$, that can occur with positive probability to integers $b \in \mathcal{B}_t = \{0, \ldots, Q_t - 1\}$ in a lexicographic order.

The formalization above appears because the DNA measurements are usually understood as measurements of the numbers of repeats of certain motifs in the paternal and the maternal allele where the measuring device cannot distinguish between data coming from different allele. Therefore, the outcomes $r_{t,1}, r_{t,2}$ can be represented as observations of the sets $\{r_{t,1}, r_{t,2}\}$. This information can be equivalently presented as the value of the random variable R_t defined in (11).

We present parameters of the PDs obtained from the DNA measurements, which are relevant to the evaluation of the performance of the authentication algorithm, for the **TH01** allele and the total sums obtained for $n = 28$ allele in Table 2. A more complete version of this table can be found in [5]. One can see that the storage of biometric vectors requires 140 bits and the PD over these vectors is non–uniform, as the probability of the most likely vector is equal to $2^{-78.8}$. The encoding with parameters $q_t = \lceil \log 1/\hat{\omega}_t \rceil$ creates passwords of length 68 with the PD close to the uniform PD (the probability of the most likely

Table 2. Parameters of three variants of the encoding for the biometrical authentication with the DNA measurements when repeats of certain motifs in $n = 28$ allele, numbered by $t = 1, \ldots, n$, are measured

				$C = 140/140$			$C = 140/68$			$C = 140/28$		
t	Name	$\log Q_t$	$\log \hat{\omega}_t$	$\log q_t$	$\log \hat{\pi}_t$	$\log \hat{\gamma}_t$	$\log q_t$	$\log \hat{\pi}_t$	$\log \hat{\gamma}_t$	$\log q_t$	$\log \hat{\pi}_t$	$\log \hat{\gamma}_t$
12	**TH01**	3.32	−2.07	4	−2.07	0	2	−1.93	−0.04	1	−1.00	−1.07
	$\sum_n$	128.6	−78.8	140	−78.8	0	68	−66.7	−9.0	28	−27.8	−49.0

vector is equal to $2^{-66.7}$). However, the maximum probability of the correct guess of the biometric vector is equal to $2^{-9.0}$. This probability can be decreased to $2^{-49.0}$ by assigning $q_t = 2, t = 1, \ldots, n$, which creates passwords of length 28 bits. However, the false acceptance rate will be also increased, as it is illustrated in the end of the section. A possible implementation of our transformations of input biometrical data is presented by the example below.

Example. (the quantities below describe the **TH01** allele in Table 2, $t = 12$). Let $c_t = 6$, $k_t = 4$, and $(p_t(6), \ldots, p_t(9)) = (0.23, 0.19, 0.09, 0.49)$. Then

$$\Big[p_t(i)p_t(j)\Big]_{i,j=6,\ldots,9} =$$

	$j = 6$	$j = 7$	$j = 8$	$j = 9$
$i = 6$	.0529	.0437	.0207	.1127
$i = 7$	.0437	.0361	.0171	.0931
$i = 8$	.0207	.0171	.0081	.0441
$i = 9$	.1127	.0931	.0441	.2401

To construct the PD ω_t, we transform this matrix to the right triangular matrix below. The entries above the diagonal are doubled, and the entries below the diagonal are replaced with zeroes,

	$j = 6$	$j = 7$	$j = 8$	$j = 9$
$i = 6$	.0529	.0874	.0414	.2254
$i = 7$		.0361	.0342	.1862
$i = 8$			.0081	.0882
$i = 9$				.2401

Let $q_t = 4$ and let the sets $\mathcal{F}_t(0), \ldots, \mathcal{F}_t(3)$ be constructed by the F1–F7 algorithm. Then $\mathcal{F}_0(z) = \{9, 7\}$, $\mathcal{F}_1(z) = \{3, 5\}$, $\mathcal{F}_2(z) = \{6, 0\}$, $\mathcal{F}_3(z) = \{8, 1, 2, 4\}$. The PDs obtained using this partitioning of the set $\{0, \ldots, 9\}$ were presented in Table 1.

To implement the encoding, the authentication scheme needs the table below.

$(i, j) =$	(6, 6)	(6, 7)	(6, 8)	(6, 9)	(7, 7)	(7, 8)	(7, 9)	(8, 8)	(8, 9)	(9, 9)
$b_t =$	0	1	2	3	4	5	6	7	8	9
$z_t =$	3	3	2	1	2	1	3	0	3	0

The outcome of the measurements $(i_{t,1}, j_{t,2})$ has to be found in the first row, and the symbol z in the corresponding column is sent to the output. Notice that the index of the pair $(i_{t,1}, j_{t,2})$ can be easily computed from $c_t = 6$ and $k_t = 4$. Therefore, the storage of the first and the second rows of the table is not necessary.

Some numerical results are given in Table 3 where we show the values of the false acceptance rate as a function of ρ and the length of passwords. These data illustrate the point that our approach is robust in a sense that it can reach a desired trade–off between different parameters characterizing the performance.

Table 3. The DNA measurements: the values of the false acceptance rate and the parameter ε(FAR) such that FRR $\leq$ FAR for all $\varepsilon < \varepsilon$(FAR)

	$C = 140/140$		$C = 140/68$		$C = 140/28$	
np	FAR	ε(FAR)	FAR	ε(FAR)	FAR	ε(FAR)
3	1.3e−17		7.0e−15		3.8e−02	0.0376
4	5.9e−16	0.0001	2.4e−13	0.0003	6.7e−02	0.0672
5	2.0e−14	0.0006	6.2e−12	0.0016	1.1e−01	0.1524
6	5.0e−13	0.0024	1.2e−10	0.0053		

References

[1] Korte, U., Krawczak, M., Merkle, J., Plaga, R., Niesing, M., Tiemann, C., Han Vinck, A.J., Martini, U.: A cryptographic biometric authentication system based on genetic fingerprints. Sicherheit, 263–276 (2008)

[2] Balakirsky, V.B., Ghazaryan, A.R., Han Vinck, A.J.: Additive block coding schemes for biometric authentication with the DNA data. In: Schouten, B., Jul, N.C., Drygajlo, A., Tistarelli, M. (eds.) BIOID 2008. LNCS, vol. 5372, pp. 160–169. Springer, Heidelberg (2008)

[3] Bolle, R.M., Connell, J.H., Pankanti, S., Ratha, N.K., Senior, A.W.: Guide to Biometrics. Springer, NY (2004)

[4] Balakirsky, V.B., Ghazaryan, A.R., Han Vinck, A.J.: An Algorithm for Biometric Authentication Based on the Model of Non-Stationary Random Processes. In: Lee, S.-W., Li, S.Z. (eds.) ICB 2007. LNCS, vol. 4642, pp. 319–327. Springer, Heidelberg (2007)

[5] Balakirsky, V.B., Ghazaryan, A.R., Han Vinck, A.J.: Mathematical model for constructing passwords from biometrical data. In: Security and Communication Networks, vol. 2(1), pp. 1–9. Wiley, Chichester (2009)

Efficient Biometric Verification in Encrypted Domain

Maneesh Upmanyu, Anoop M. Namboodiri, K. Srinathan, and C.V. Jawahar

Center for Visual Information Technology
International Institute of Information Technology, Hyderabad
{upmanyu@research.,anoop@,srinathan@,jawahar@}iiit.ac.in

Abstract. Biometric authentication over public networks leads to a variety of privacy issues that needs to be addressed before it can become popular. The primary concerns are that the biometrics might reveal more information than the identity itself, as well as provide the ability to track users over an extended period of time. In this paper, we propose an authentication protocol that alleviates these concerns. The protocol takes care of user privacy, template protection and trust issues in biometric authentication systems. The protocol uses asymmetric encryption, and captures the advantages of biometric authentication. The protocol provides non-repudiable identity verification, while not revealing any additional information about the user to the server or vice versa. We show that the protocol is secure under various attacks. Experimental results indicate that the overall method is efficient to be used in practical scenarios.

1 Introduction

The primary advantages of biometrics over other authentication mechanisms are its convenience, security, and non-repudiable nature [1]. However, the assertions on security and non-repudiation are valid only if the integrity of the overall system is maintained [2] A hacker who gains physical or remote access to an authentication server can steal the stored templates, which are non-replaceable in case of plain templates. Concerns are also on the privacy as many biometrics reveal personal information beyond just identity. Widespread use of biometric authentication also provides the ability to track a person through every activity in his life, which introduces another significant privacy concern.

The primary concerns in widespread use of biometrics for remote and onsite authentication are in i) template protection, ii) privacy of the user, iii) trust between user and server, and iv) network security. For civilian applications, these concerns are often more serious than the accuracy of the biometric itself.

The ideal solution to overcoming all the privacy and security concerns would be to apply a strong encryption (say RSA) on the biometric samples as well as the classifier parameters, and carry out all the computations in the encrypted domain. However, the primary goal of a strong encryption algorithm is to destroy any pattern that would be present in the data. We now need to carry out a pattern classification task (identity verification) in the encrypted domain. These two goals are contradictory. In other words, security/privacy and accuracy seems to be opposing objectives. Different secure authentication solutions achieve their goal through a compromise between privacy and accuracy or by making restrictive assumptions on the biometric data.

M. Tistarelli and M.S. Nixon (Eds.): ICB 2009, LNCS 5558, pp. 899–908, 2009.
© Springer-Verlag Berlin Heidelberg 2009

The primary difference in our approach is that we are able to design the classifier in the plain feature space, which allows us to maintain the performance of the biometric itself, while carrying out the authentication on data with strong encryption, which provides high security/privacy. However, such an approach would require an *algebraic homomorphic encryption* scheme [3], which is not known to exist till date. We show that a specific distribution of work between the client (sensor) and the server (authenticator), coupled with a novel randomization scheme can achieve the goal.

Over the years a number of attempts have been made to address the problem of template protection and privacy concerns and despite all efforts, *a template protection scheme with provable security and acceptable recognition performance has thus far remained elusive* [4]. Jain et al. [4] classifies the existing approaches into two groups: *feature transformation-based* and *biometric cryptosystems*. We will look at the two groups, in light of this security-accuracy dilemma. Detailed reviews of the work on template protection can be found in Jain *et al.* [4], Uludag *et al.* [5], and Ratha *et al.* [6].

The first class of approaches that use feature transformation, such as *Salting and Non-invertible Transform* [4] offers security using a transformation seeded by a user specific key. A classifier is designed in the encrypted feature space. Hence, one cannot employ strong encryption here, which necessarily leads a compromise made between security and the performance. Moreover salting based solutions are usually specific to a biometric trait [7,6]. Kong *et al.* do a detailed analysis of the current biohashing based approaches [8], and concludes that the zero EER reported is obtained in carefully set experimental conditions and unrealistic under assumptions from a practical view point.

Biometric cryptosystems use the biometric as a protection for a secret key (Key Binding approach [9]) or to directly generate a secret key (Key Generation approach [10]). The authentication is done using the key, which is unlocked/generated by the biometric. We note that to provide template protection, key is to be unlocked/generated at the client's end. However, this would become a key based authentication scheme, thus losing the non-repudiable nature of biometric authentication. According to Jain *et al.* [4], biometric cryptosystems such as Fuzzy Vault and Fuzzy extractor, in their true form, lack diversity and revocability, and results in performance degradation as the matching is done using error correction schemes. Biometric cryptosystems, along with salting based approaches introduce diversity and revocability in them. However, one can recover the plain biometric from multiple secrets secured using the same key [11].

The approach that is closest to our proposed one is termed *ZeroBio* authentication, proposed by Nagai *et al* [12]. It makes use of client side computation and communication between the client and the server to classify a biometric feature vector using a 3-layer neural network. The client computes the outputs of the hidden layer and transfers it to the server, which completes the authentication by computing the output values of the neural network. The mechanism of zero-knowledge proof using communication is used to ensure honesty. The method is both efficient and generic, however, the server can estimate the weights at the hidden layer from multiple observations over authentications. Once the weights are known, the server can also compute the feature vector of the biometric, thus compromising both security and privacy. The system could also be compromised if an attacker gains access to the client computer, where the weight information is available in plain.

This paper proposes an approach that is generic in the sense that we can implement a generic and powerful classifier such as support vector machines(SVM). Moreover, we achieve complete privacy, as the biometric that is passed to the server is encrypted using strong asymmetric encryption. We also achieve efficiency in computation using interaction with the client along with a novel randomization scheme, while maintaining the security of the server templates. In short, we addresses all the concerns mentioned in the introduction. Specifically, 1) the use of strong encryption addresses privacy concerns, 2) a third party based enrollment scheme takes care of the trust issues. 3) provable protection is provided against replay and client-side attacks, 4) user tracking is avoided by using different templates for different servers.

The framework provides the ability to classify any feature vector, and hence is applicable to multiple biometrics. Moreover, as the authentication is directly based on the biometric, non-repudiable nature of biometrics is fully utilized. Note that the proposed approach does not fall into any of the categories discussed before, and opens a new direction of research to look at privacy preserving biometric authentication.

2 Authentication in Encrypted Domain

To explain the protocol in a simple setting, we consider the problem of verification as that of classifying genuine and imposter samples using a perceptron. The protocol can be extended to more generic and powerful classifiers, such as *Support Vector Machines* (SVMs). The only restriction to the approach is that the feature vector is assumed to be of fixed length n. Note that even for biometrics such as fingerprints, one can define fixed length feature representations [13].

2.1 Authentication

Let ω be the parameters of the linear classifier, such that the user is accepted if $\omega \cdot x < \tau$, where τ is a threshold. As we do not want to reveal the parameter vector (ω) or the test sample (x) to the server, we need to carry out the computations in the encrypted domain. To achieve this, we use a class of encryptions that are multiplicative homomorphic [14]. An encryption scheme, $E(x)$ is said to be multiplicative homomorphic, if $E(x)E(y) = E(xy)$ for any two numbers x and y. We use the popular RSA encryption scheme, which satisfies this property. Note that if we have an encryption scheme that is homomorphic to both addition and multiplication (algebraic homomorphic), we can carry out the computation, directly at the server side. However, such an encryption has not been discovered till date.

During enrollment, the server receives the client's public key, E, as well as the classifier parameters vector ω in the encrypted form, i.e., $E(\omega)$. The authentication happens over two rounds of communication between the client and the server. To perform authentication, the client locks the biometric test sample using her public key and sends the *locked ID* ($E(x_i)$) and the username to the server. We note that the computation of: $\omega \cdot x$ requires a set of scalar multiplications, followed by a set of additions. As the encryption used (RSA) is homomorphic to multiplication, we can compute, $E(\omega_i x_i) = E(\omega_i)E(x_i)$, at the server side. However, we cannot add the results to compute the authentication function. Unfortunately, sending the products to the client

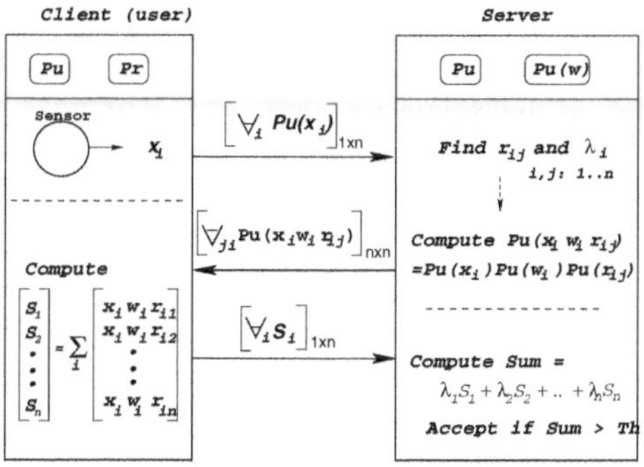

Fig. 1. The proposed authentication process using linear classifier

for addition will reveal the classifier parameters to the user, which is not desirable. We use a clever randomization mechanism that achieves this computation without revealing any information to the user. The randomization makes sure that the client can do the summation, while not being able to decipher any information from the products. The randomization is done in such a way that the server can compute the final sum to be compared with the threshold. The overall algorithm of the authentication process is given in Algorithm 1. Note that all the arithmetic operations that we mention in the encrypted domain will be $modulo-$ operations.

In the algorithm, the server carries out all its computation in the encrypted domain, and hence does not get any information about the biometric data (x) or the classifier parameters (ω). A malicious client also cannot guess the classifier parameters from the products returned as they are randomized by multiplication with r_{ji}. The reason why the server is able to compute the final sum S in *Step 8* of Algorithm 1 is because we impose the following condition on r_{ji}s and λ_js during its generation:

$$\forall_i, \sum_{j=1}^{k} \lambda_j \, r_{ji} = 1 \tag{1}$$

Substituting equation 1 in the expansion of the final sum (S) in Algorithm 1, we get:

$$S = \sum_{j=1}^{k} \lambda_j \, S_j = \sum_{j=1}^{k} \lambda_j \sum_{i=1}^{n} \omega_i \, x_i \, r_{ji} = \sum_{i=1}^{n} \sum_{j=1}^{k} \lambda_j \, \omega_i \, x_i \, r_{ji} \tag{2}$$

$$= \sum_{i=1}^{n} \omega_i \, x_i \sum_{j=1}^{k} \lambda_j \, r_{ji} = \sum_{i=1}^{n} \omega_i \, x_i$$

We note that the server is unable to decipher any information about the original products in the whole process, and directly obtains the final sum-of-products expression. This quantity measures the confidence that the test biometric belongs to the claimed identity, and does not reveal any information about the actual biometric itself. The authentication process thus maintains a clear separation of information between the client and the server and hence provides complete privacy to the user, and security to the biometric. Moreover, the clear biometric or parameters are never stored at any place, thus

Algorithm 1. Authentication

1: Client computes feature vector, $x_{1..n}$, from test data
2: Each feature x_i is encrypted ($E(x_i)$) and sent to server
3: Server computes $kn + k$ random numbers, r_{ji} and λ_j, such that,$\forall_i$, $\sum_{j=1}^{k} \lambda_j \, r_{ji} = 1$
4: Server computes $E(\omega_i \, x_i \, r_{ji}) = E(\omega_i) \, E(x_i) \, E(r_{ji})$
5: The kn products thus generated are sent to the client
6: The client decrypts the products, and returns their sum $S_j = \sum_{i=1}^{n} \omega_i \, x_i \, r_{ji}$ to the server
7: Server computes $S = \sum_{j=1}^{k} \lambda_j \, S_j$
8: **if** $S > \tau$ **then**
9: return *Accepted* to the client
10: **else**
11: return *Rejected* to the client
12: **end if**

avoiding serious losses if the server or the client computer is compromised. We will take a detailed look at the related security aspects in section 4.

As noted before, the linear classification was used for illustration of the algorithm, as it is instructive and easy to understand. The extension of this approach to compute more complex functions such as the kernelized inner products are given in section 3. One can also deal with variable length features and warping based matching techniques using a similar approach. However, a complete treatment of such solutions are beyond the scope of this paper.

Applicability: We have not made any assumptions on the specific biometric being used in the framework. One could use any biometric as long as the feature vector embeds the samples in a Euclidean space. The classifier itself was assumed to be a linear classifier. However, one can extend it to work with kernel based methods (as we show in the next section) and hence any verification problem that can be carried out using a generic SVM-based classifier can be modeled by this protocol. One could also extend the protocol to work with the Neural Networks.

3 Extension to Kernels and Other Variations

Even though the linear classifier model can support some of the simple template matching approaches, it does not generalize to other model based classifiers. We will now extend the proposed approach to deal with the kernel form of the linear classifier, the support vector machine (SVM).

Kernel-based classification: In the linear case, we described a procedure, *secure–Product*, to compute the inner product of two encrypted vectors without revealing its contents. However, in order to use a kernel based classifier at the server for verification, one needs to compute a discriminating function of the form: $S = \sum_{i=1}^{N} \alpha_i d_i \kappa({v_i}^T x) =$

$\alpha \cdot \kappa(v, x)$ where the rows of v are the support vectors and $\kappa()$ is referred to as the kernel function.

We can extend the $secureProduct$ procedure to deal with kernel based classification as well. We note that the parameter of the kernel function is a set of inner products of vectors. This could be calculated in a similar fashion as the $secureProduct$. Once we obtain the individual inner products, we can compute the kernel functions, κ, at the server side. The discriminant function to be computed is once again the dot product of the vector of κ values and the α vector. This could again be computed, securely using the $secureProduct$ procedure. We note that this procedure allows us to compute any kernel function at the server side.

The above approach is more generic and secure than any of the secure authentication protocols in the literature. Moreover, it does not reveal any information about the classifier to the client. However, as the results of the intermediate inner products are known to the server, this simple extension is not completely blind in the information theoretic sense. One can solve this problem using another round of communication with the client and define a completely blind kernel-based verification protocol.

Security Extensions: The client end contains the biometric acquisition device as well as keys for encryption and decryption. Security at the client end is critical, especially when using a public terminal to access any service. One could move the private key to a card or carry out the decryption operation, completely in a smart card in case of insecure environments. Another option would be to secure the private keys at the client end using a fuzzy vault [9], which is unlocked only with the biometric that is provided for authentication. This provides a double layer of security through the biometric provided by the user.

RSA is just one of the public key encryption algorithms that is multiplicative homomorphic. This can be replaced by other similar encryptors. One could analyze the computation cost and security issues for each encryption method. A further speed up is possible by reducing the number of support vectors [15].

4 Security, Privacy, and Trust Issues

Security of the system refers to the ability of the system to withstand attacks from outside to gain illegal access or deny access to legitimate users. Security is hence a function of the specific biometric used as well as the overall design of the system.

Privacy on the other hand is related to the amount of user information that is revealed to the server. Ideally, one would like to reveal only the identity and no additional information. Most of the current systems provides very little privacy, and hence demands trust between the user and the server. We now take a closer look at the security and privacy aspects of the proposed system.

4.1 System Security

Biometric systems are known to be more secure as compared to passwords or tokens, as they are difficult to reproduce. As the authentication process in the proposed system is directly based on biometrics we gain all the advantages of a generic biometric system.

The security is further enhanced by the fact that an attacker needs to get access to both the user's biometric as well as her private key to be able to pose as an enrolled user.

Server Security: Let us assume that the hacker gains access to the template database. In this case, all the templates (or classifier parameters) in the server are encrypted using the public key of the respective clients. Hence gaining access to each template is as hard as cracking the public key encryption algorithm. Moreover, if by any chance a template is suspected to be broken, one could create another one from a new public-private key.

In case the hacker is in the server *during* the authentication, he can try to extract information from his entire "view" of the protocol, i.e. the encrypted classifier parameters $E(\omega_i)$, encrypted test vector $E(x_i)$ and other intermediate data such as random numbers r_{ij}'s, λ's etc. We ask what hacker can learn about the critical data, viz., ω_i's and x_i's? The hacker only obtains k linear equations over the n variables $y_1, y_2, \ldots, y_n$, namely, $S_j = \sum_{i=1}^{n} r_{ji} y_i$ for all $j \in [1, k]$, where $y_i = \omega_i x_i$. Thus, if at all the hacker obtains some information, it is only about the y_i's and not about the ω_i's or x_i's. Notwithstanding, nothing additional is revealed even about the y_i's, by choosing k to be such that $|\mathbb{D}| \geq |\mathbb{Y}|^{\frac{n}{n-k}}$, where $\mathbb{Y}$ is the domain of y_i's and $\mathbb{D}$ is the domain of r_{ji}'s.

Client-End Security: A hacker having access to the user's computer, will not be able to carry out the authentication, as the biometric is not stored on the client's computer. The private key itself is often hard wired into the decryption hardware and often cannot be read out. However, in the worst case, the attacker may be able to decrypt messages in a black-box fashion. Even in such a case, the ω_i's cannot be obtained by the attacker from his entire view of the protocol, as long as $|\mathbb{D}| \geq |\mathbb{Y}|^{\frac{n}{d(k-n)+n}}$, where d is the number of fake authentication requests made to the server. Reconciling this case with the previous one, and taking the limit $d \rightarrow \infty$, we get the optimal value of $k = n$.

Network Security: An attacker having control over the insecure network can watch the traffic on the network, as well as modify it. However, all the traffic on the network are encrypted either using the clients public key or using the random numbers generated by the server. Hence the attacker will not be able to decipher any information. A replay attack is also not possible as the data communicated during the second round of communication is dependent on the random numbers generated by the server.

4.2 Privacy

Privacy, as noted before deals with the amount of user information that is revealed to the server, during the process of enrollment and authentication.

We noted that there are two aspects of privacy to be dealt with. i) Concern of revealing personal information: As the template or test biometric sample is never revealed to the server, the user need not worry that the use of biometrics might divulge any personal information other than her identity. ii) Concern of being tracked: One can use different keys for different applications (servers) and hence avoid being tracked across uses. In fact, even the choice biometric or real identity of the user itself is known only to the enrolling server. The authenticating server knows only the user ID communicated by the enrollment server and the biometric is obtained in the form of an encrypted feature vector.

As the user and server need not trust each other, the framework is applicable to a variety of remote and on-site identity verification tasks.

5 Implementation and Analysis

An authentication protocol based on a client-server model was implemented that can perform verification over an insecure channel such as the Internet. The following experiments and analysis evaluates the accuracy and performance of the proposed approach for verification.

5.1 Implementation

For the evaluation purpose a generic SVM based verifier based on a client-server architecture was implemented in GNU/C and the *XySSL* RSA library. All experiments are conducted on AMD X2 Dual Core 4000+ processor, 750MB DDR2 RAM and over the network.

Both exponentiation operations in encryption and decryption assumes that the data consists of positive integers. The feature vectors and the SVM parameters are mapped to integers in the twos complement form after scaling to retain the precision. Thus all our computations are now done in two's complement arithmetic. If x_i is a parameter to be encrypted, the forward mapping is defined as: $x_i' = compl(\lfloor s.x_i + 0.5 \rfloor)$, where s is a scale factor, depending on the range of values for x_is, and $compl()$ maps the integral numbers to its twos complement. The corresponding reverse mapping is done by the server, once the results are obtained. Figure 2(a) shows a hand-geometry based authentication tool that we have implemented based on the proposed method.

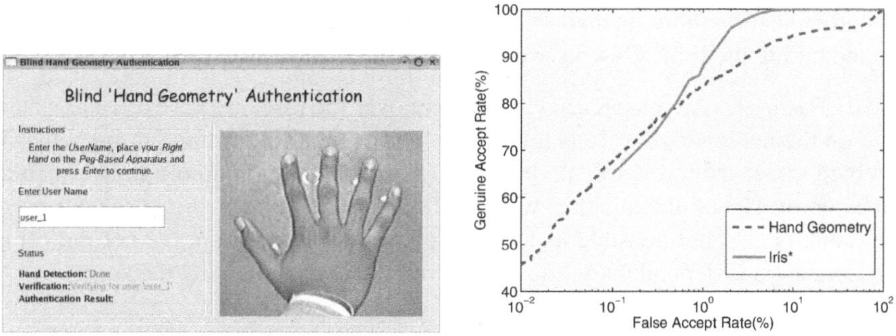

(a) Hand geometry based authentication tool (b) ROC Curves for verification

Fig. 2.

As the protocol implements a generic classifier, without making any simplification assumptions, the accuracy of the classifier should be identical to that of the original classifier. One could expect small variations in accuracy due to the round off errors used in the mapping function described above. To verify the effectiveness of using SVMs as a classification model for biometric verification problems, we tested it on two different modalities. The verification accuracies after 3-fold cross validation on each of the datasets is presented in Table 1.

The first set of experiments was done on the CASIA IRIS database. The Version 1 of the dataset consists of 108 users with 7 images per user (the seven images are

Table 1. Verification accuracy on biometric datasets

Dataset	# of Features	Avg num of Support Vectors	Accuracy
Hand Geometry	20	310	98.38%
CASIA Iris	9600	127	98.24%

collected over two separate imaging sessions). The iris code consists of 9600 binary features. 3 samples per user were used for training and 4 sample per user were used for testing purpose in each experiment. For the second set of experiments, we used a hand-geometry dataset that was collected in-house. The dataset consisted of 149 users with 10 hand images each. The features consists of the 14 finger length and width features described by Jain *et al.* [16]. For each experiment 4 images per user were used for training purpose and the remaining 6 were used for testing. Figure 2(b) shows the *receiver operating characteristic (ROC)* plots for the biometrics using fixed length feature vector representation[1]. The accuracies are similar to those obtained by running SVM in the plain domain, and hence shows the effectiveness of the proposed scheme for biometric based verification problems.

5.2 Computation and Communication Overheads

The additional computation that needs to be carried out can be divided into two parts: i) Modulo multiplications to be done for encryption/decryption and inner product, and ii) the additional time spent in the computation of random numbers, products and sums. As the modulo multiplications and encryption decryption operations can be done efficiently using dedicated hardware available [18], we analyze the time required for both, separately. Consider a biometric with feature vector of length n. In the protocol, the client needs to do n encryptions for the test vector x.

For the linear classifier, the server needs to do kn encryptions of the random numbers and $2kn$ multiplications, so as to compute $E(\omega_i x_i r_{ji})$, where k≤n, each of which is an integer. The client needs to do kn decryptions. Additional computations at the server includes $n + kn$ modulo multiplications of encrypted numbers at the server end, and kn non-encrypted additions at the client end. In addition, the server needs to generate kn random numbers. For most practical biometrics, the total run time required for all these (non-encrypted) computations together on current desktop machines is less than 10 milliseconds. The communication overhead, in addition to regular authentication, includes sending kn numbers from the server to the client and sending k numbers from the client back to the server for evaluation of the final result.

Extending the analysis to a kernel based classifier with n_v support vectors, one would need to repeat the *secure product* n_v times, once for each support vector. In addition, there is one round of *secure product* to compute the final result. Hence the time required will be $n_v + 1$ times that required for the linear classifier. In practice the total time taken (other than those implemented in hardware) is less than one second.

[1] Wang *et al.* [17].

6 Conclusions

The proposed method for biometric authentication is extremely secure under a variety of attacks and can be used with a wide variety of biometric traits. As the verification can be done in real-time with the help of available hardware, the approach is also practical in many applications. The use of smart cards to hold encryption keys enables applications such as biometric ATMs and access of services from public terminals. Possible extensions to this work includes secure enrollment protocols and encryption methods to reduce computations. Efficient methods to do dynamic warping based matching of variable length feature vectors can also enhance the utility of the approach.

References

1. Jain, A.K., Ross, A., Prabhakar, S.: An introduction to biometric recognition. IEEE Transactions on Circuits and Systems for Video Technology 14(1), 4–20 (2004)
2. Ratha, N.K., Connell, J.H., Bolle, R.M.: Enhancing security and privacy in biometrics-based authentication systems. IBM Systems Journal 40(3), 614–634 (2001)
3. Fontaine, C., Galand, F.: A survey of homomorphic encryption for nonspecialists. EURASIP J. Inf. Secur. 2007(1), 1–15 (2007)
4. Jain, A.K., Nandakumar, K., Nagar, A.: Biometric template security. EURASIP J. Adv. Signal Process. 8(2), 1–17 (2008)
5. Uludag, U., Pankanti, S., Prabhakar, S., Jain, A.K.: Biometric cryptosystems: Issues and challenges. Proceedings of the IEEE 92(6), 948–960 (2004)
6. Ratha, N., Chikkerur, S., Connell, J., Bolle, R.: Generating cancelable fingerprint templates. IEEE Trans. on PAMI 29(4), 561–572 (2007)
7. Teoh, A., Jin, B., Connie, T., Ngo, D., Ling, C.: Remarks on BioHash and its mathematical foundation. Information Processing Letters 100(4), 145–150 (2006)
8. Kong, A., Cheung, K., Zhang, D., Kamel, M., You, J.: An analysis of biohashing and its variants. Pattern Recognition 39(7), 1359–1368 (2006)
9. Juels, A., Sudan, M.: A fuzzy vault scheme. DCC 38(2), 237–257 (2006)
10. Dodis, Y., Reyzin, L., Smith, A.: Fuzzy extractors: How to generate strong keys from biometrics and other noisy data. In: Cachin, C., Camenisch, J.L. (eds.) EUROCRYPT 2004. LNCS, vol. 3027, pp. 523–540. Springer, Heidelberg (2004)
11. Walter, J.S., Boult, T.E.: Cracking fuzzy vaults and biometric encryption. In: Biometrics Symposium (2007)
12. Nagai, K., Kikuchi, H., Ogata, W., Nishigaki, M.: ZeroBio: Evaluation and development of asymmetric fingerprint authentication system using oblivious neural network evaluation protocol. In: Proceedings of ARES 2007, pp. 1155–1159 (2007)
13. Farooq, F., Bolle, R.M., Jea, T.Y., Ratha, N.: Anonymous and revocable fingerprint recognition. In: Proceedings of the Biometrics Worshop (CVPR 2007), pp. 1–7 (2007)
14. Menezes, A., van, O., Paul, C., Vanstone, S.A.: Handbook of Applied Cryptography (1996)
15. Abe, S.: Support Vector Machines For Pattern Classification. Springer, Heidelberg (2005)
16. Jain, A.K., Ross, A., Pankanti, S.: A prototype hand geometry-based verification system. In: Proceedings of the AVBPA 1999, pp. 166–171 (1999)
17. Wang, Y., Han, J.: Iris recognition using SVM. In: Yin, F.-L., Wang, J., Guo, C. (eds.) ISNN 2004. LNCS, vol. 3173, pp. 622–628. Springer, Heidelberg (2004)
18. Blum, T., Paar, C.: High-radix montgomery modular exponentiation on reconfigurable hardware. IEEE Transactions on Computers 50(7), 759–764 (2001)

A New Approach for Biometric Template Storage and Remote Authentication

Neyire Deniz Sarier

Bonn-Aachen International Center for Information Technology
Computer Security Group
Dahlmannstr. 2, D-53113 Bonn Germany
denizsarier@yahoo.com

Abstract. In this paper, we propose a new remote biometric based authentication scheme, which is designed for distributed systems with a central database for the storage of the biometric data. For our scheme, we consider the recently introduced security notions of Identity and Transaction privacy and present a different storage mechanism for biometrics resulting in a reduced database storage cost. Besides, the components of the system do not need to store any biometric template in cleartext or in encrypted form, which affects the social acceptance of the system positively. Finally, we compare our results with existing schemes satisfying the current security notions and achieve improved computational complexity.

Keywords: Remote authentication, Biometric template security, Identity privacy, Distributed systems.

1 Introduction

Biometric-based identification provides unforgeable authentication without requiring the user to store any secret identification data or remember long passwords. Biometric information is unique, unforgettable, non-transferable and it could be easily integrated with password-based and/or token-based authentication techniques. As each authentication technique has its weaknesses, a multi-factor authentication scheme with the correct design and security model results in a reliable system.

Currently, the secrecy of biometric data is viewed with skepticism since it is very easy to obtain biological information such as fingerprint, iris or face data through fingerprint marking or using a camcorder. However, biometrics is a sensitive information, thus it should not be easy to obtain the biometric data by compromising the server, where the biometrics of each user is often associated with his personal information. This also affects the social accpetence of the biometric systems especially when biometric data are stored in a central database which can be vulnerable to internal or external attackers.

Biometric authentication could be categorized broadly as remote server or client end authentication, where in the first case, the remote server stores the

M. Tistarelli and M.S. Nixon (Eds.): ICB 2009, LNCS 5558, pp. 909–918, 2009.
© Springer-Verlag Berlin Heidelberg 2009

reference biometric data and performs the matching. In a typical biometric based remote authentication scheme, the user registers his identity information and biometrics to the server end that stores this information either in cleartext or in encrypted form. When the user wants to authenticate himself, the user provides a fresh biometric, which is compared to the previously stored biometric information after decryption or in the encryption domain by exploiting the homomorphic properties of the underlying encryption scheme.

The security and privacy protection of remote biometric-based authentication systems is enhanced by implementing distributed biometric systems, where the goal is to detach the biometric data storage from the service provider and to guarantee the notions of user privacy and database privacy, which have been recently introduced as a new security model for biometric authentication. Current systems implementing this approach provide provable security in this new model, but they are not suitable for weak computational devices such as smart cards and RFID's, since the user has to encrypt at each authentication request his fresh biometric information using public key encryption. Moreover, the employment of homomorphic encryption schemes and Private Information Retrieval (PIR) systems results in high communication and computational costs. Consequently, one has to design a secure and efficient remote biometric authentication scheme for a distributed system, where the service provider, the database and the client end with a smart card collaborate during the authentication process.

1.1 Related Work

Remote biometric-based authentication systems could be classified based on the employment of cryptographic encryption schemes or lightweight computational primitives. The systems described in [1,2,3,4,5] combine homomorphic encryption techniques with biometrics in a distributed environment. Specifically, the user U registers its biometric template in cleartext or in encrypted form at the database DB. Besides, U registers his personal information and the index of the database storage location of his biometrics at the service provider SP. To authenticate himself, U encrypts his biometrics using a homomorphic encryption scheme and sends this to SP, which retrieves the index of U to be used in a PIR protocol between SP and DB. In [1,5], an independent verification unit is additionally required for the matching operation and the final decision. In [1], the biometric template is stored as a plaintext and a user sends the encryption of each singe bit using Goldwasser-Micali scheme resulting in a high transmission and computation cost. Also, the relationship between the user's identity and his biometrics is kept private by employing a PIR with the communication cost linear in the size N of the database. Besides, protocols using Paillier Public Key System are described in [5,6], where the authors of [5] present an attack against the scheme of [1] that reveals the user's biometric data to SP. Furthermore, the scheme of [1] is improved in terms of communication cost by combining a PIR, a secure sketch and a homomorphic encryption scheme [3,4,2].

For the second case that consider biometric data as a secret information, a biometric authentication system for weak computational devices is presented [7],

which requires the use of a fixed permutation stored by the server and the client. Additionally, a number of secret values are stored in the client's smart card, whose loss results in the compromise of the biometric template by evasedropping on the communication channel [1]. Moreover, in [8] efficient biometric authentication schemes for mobile devices are presented, where as opposed to [7], a reference biometric template should be stored on the mobile device. Similarly, other systems are designed using error-correction procedures [9,10,11,12,13,14,15,16]. In [17], the author describes a brute force attack that extracts both the secret and the biometric template from the fuzzy vault [15] with $O(10^{17})$ binary operations and thus he suggests the use of strong cryptographic techniques for a secure biometric authentication.

Finally, a combination of the two approaches is presented in [18], where a multi-factor biometric authentication system is described using a public key pair that is generated by combining a secret key and a biometric based key. The system requires the selection of a representative template that is fixed to one pattern and a Public Key Infrasturacture (PKI) for the public key certificates. However, the template should be stored in the user's smart card and a correct matching on card will activate the keys to be used in the authentication.

1.2 Motivation and Contributions

The privacy protection and the secure storage of the biometric templates are the main concerns for the biometric-based authentication schemes. As it is noted in [4], privacy protection not only means the attackers inability to compromise the biometric template but also the protection of the sensitive relationship between the identity and the biometric information of the user. To achieve this property, we separate the storage of personal identity information from the storage of biometrics using the distributed structure of [3,4,2], which is composed of the user U_i, the sensor client SC, the service provider SP and the database DB. Here, SC, SP and DB are independent of each other and the latter two are assumed to be malicious whereas the sensor client is always honest. This way, SP cannot obtain the biometrics of the user and can have business agreements with different parties that make the sensor client available to users at different locations. Also, DB could function as a trusted storage for different SP's.

Since SC captures the biometric data and performs the feature extraction, this component could be installed as a Trusted Biometric Reader as in [19]. Alternatively, a special smartcard biometric reader [7] could be used to capture biometrics and perform the necessary computations for session key generation and AES, which would provide the highest security for the user since no transfer of biometric data will take place between the user's smart card and SC. However, as it is noted in [20], the computational cost for feature extraction is very high compared to AES and Elliptic Curve Digital Signature (ECDS) computation on a smart card. Thus, we only require a smart card that implements AES and an efficient Identity based encryption (IBE) system such as the Boneh/Franklin scheme [21] to perform a short session key generation using the stored private key. This way, the session keys could be constructed in an anonymous, authenticated

and efficient way. Specifically, an IBE scheme is called as anonymous if the ciphertext does not reveal the identity of the recipient [22]. Besides, the users should not store any biometric data in their smart cards as relying only on tamper proofness is not a wise assumption [7]. Even if the smart card is lost or stolen, the compromise of the secret values and keys should not reveal the biometrics since the revocation of these values is possible as opposed to biometric data. Finally, we aim to design an efficient system that minimizes the costs of storage, encryption and communication. For this purpose, we propose a different approach for the storage of the biometrics in the database and describe a new remote biometric-based authentication system.

2 Preliminaries and Definitions

Our system consists of four independent entities: A human user U_i with identity ID_i, the sensor client SC with ID_{SC}, the service provider SP with ID_{SP} and the database SB with ID_{DB}. Similar to existing authentication schemes, our system is composed of two phases: the registration and the verification phase, which have the following workflow.

1. In the registration phase, the human user U_i presents its biometrics to SC, which captures the raw biometric data and extracts the feature vector $B_i = (\mu_1, ..., \mu_n)$. Next, U_i registers each feature at a randomly selected storage location in DB and registers his personalized username ID_i at SP. Here, the size of the database is denoted as N and the dimension of the feature vector is denoted as n.
2. In the verification phase, the user U_i presents its biometrics to SC, which computes the feature set B'_i. Using cryptographic techniques, SP communicates with U_i and DB and decides based on a distance metric and a predefined threshold to accept or to reject U_i. In our scheme, we will consider set overlap as the distance metric [23,24], where the value d represents the error tolerance in terms of minimal set overlap.

2.1 Assumptions on the System

- Liveliness Assumption: This is an indispensable assumption for any biometric system as it guarantees with high probability that the biometrics is coming from a live human user.
- Security link Assumption: To provide the confidentiality and integrity of sensitive information, the communication channel between U_i, SC, SP and DB should be encrypted using standard protocols. Specifically, the session key generation should be performed in an anonymous and authenticated way.
- Collusion Assumption: Due to the distributed system structure, we assume that DB and SP are malicious but they do not collude. Additionally, the sensor client is always honest.

2.2 Security Requirements

Identity Privacy: Informally, this notion guarantees the privacy of the sensitive relationship between the user identity and its biometrics against a malicious service provider or a malicious database even in case of multiple registrations of the same user with different personalized usernames. Briefly, it means that the service provider or the database (or an attacker that has compromised one of them) cannot recover the biometric template of the user [4].

Transaction Privacy: Informally, transaction anonymity means that a malicious database cannot learn anything about the personal identity of the user for any authentication request made to the service provider.

The formal definitions of the security notions could be found in [2,4,1,3].

3 A New Biometric Authentication Scheme

In this section, we present a new remote biometric-based authentication scheme using a different approach for storing the biometric features resulting in a secure and more efficient protocol compared to the existing protocols. For this purpose, we use Boneh/Franklin IBE [21] scheme to encrypt a random session key for AES and an efficient PIR protocol [25] which allows SP to retrieve an item from the DB without revealing which item SP is retrieving. Due to page limitations, the reader is referred to [21,27,25] for a detailed discussion of IBE and PIR.

3.1 Registration Phase

The registration phase consists of the following initialization of the components.

1. The four components of the system, namely, U_i, SC, SP and DB are initialized by the Private Key Generator (PKG) of the IBE system with the private keys $d_i, d_{SC}, d_{SP}, d_{DB}$, respectively.
2. The user U_i presents its biometrics to the sensor client which extracts the feature set $B_i = (\mu_1, ..., \mu_n)$ of the user. Here, each of the features of arbitrary length can be hashed using some collision-resistant hash function [24] or mapped to an element of a finite field [23].
3. The user picks some random indexes $i_k \in Z$ where $1 \leq k \leq n$ and registers each feature at the locations i_k of the database.

   Remark 1. If the location of the database is already occupied by another feature, then the user selects another random index for the feature. Also, if some of the features of the user are already stored in the database, then the database returns the indexes of the common features. Thus, common features are not stored more than once, which decreases the total storage cost of the database.

4. The user U_i registers its personalized username at the service provider and stores the index list $Index_i = (i_1, ..., i_n)$ in his smart card.

3.2 Verification Phase

The verification phase has the following workflow.

1. U_i inserts his smart card into the terminal of SC and presents its biometrics. SC performs feature extraction to compute the feature set B_i'. U_i and SC agree on a session key using Boneh-Franklin IBE [21] scheme and SC encrypts the feature set using AES. To provide non-malleability, SC also signs the hash of the feature set B_i' using an efficient identity based signature scheme based on bilinear pairings [26]. Finally, SC communicates with the smart card of U_i by sending the $E(B_i') = Enc_{AES}(B_i')||Sign(H(B_i'))$, where H is a cryptographic hash function.
2. U_i decrypts the first part of $E(B_i')$ using the session key to obtain the feature set $B_i' = (\mu_1', ..., \mu_n')$ and verifies the signature.
3. U_i makes an authentication request to SP and both entities agree on a short session key using IBE. U_i sends the encryption of the index list as $E_k = Enc_{AES}(i_k)||H(Enc_{AES}(i_k), H(i_k))$ for $1 \leq k \leq n$.
4. SP decrypts each E_k and obtains the $Index_i$ of U_i.
5. For $1 \leq t \leq N$, DB randomly selects r_t and computes $E_t^1 = r_t \oplus \mu_t$ and $E_t^2 = H(r_t \oplus \mu_t, H(r_t))$.
6. SP runs a PIR protocol [25] to obtain each masked feature and the corresponding hash value of $Index_i$ from the database. Next, SP stores the hash values in the set S_1 to be used later in the matching stage.
7. For $1 \leq l \leq n$, SP computes $E_l = Enc_{AES}(M_l, Sign(H(M_l)) = Enc_{AES}(r_l \oplus \mu_l), Sign(H(r_l \oplus \mu_l))$ and sends each E_l to U_i.
8. For $1 \leq l \leq n$, U_i decrypts each E_l using the session key and computes $r_l' = M_l \oplus \mu_l'$. Lastly, U_i sends each $R_l = Enc_{AES}(H(r_l'))$ to SP.
9. SP decrypts each R_l and computes $H(M_l, H(r_l'))) = H(r_l \oplus \mu_l, H(r_l \oplus \mu_l \oplus \mu_l'))$, which are then stored in the set S_2. Finally, using the threshold d, SP checks $|S_1 \cap S_2| \geq d$. If this condition is satisfied, then the user is authenticated, otherwise rejected.

4 Analysis of the Protocol

In the first part, we evaluate the major security criteria that should be satisfied in a biometric authentication system

- Identity-biometric template relation: At the registration phase, a user selects a random number for each feature of his biometrics and each feature is stored as a separate entry using the randomly selected index. Hence, even if the database is compromised, the attacker would not be able to find an index that points to a biometric template stored as cleartext or encrypted. This also provides security against the database since it only stores a randomly ordered pool of features from different users, where each feature is hashed using a specific cryptographic hash function before it is stored in the database. Besides, when the same user registers at the service provider using different personalized (pseudorandom) usernames, than the service provider is not even aware of this situation since it does not store any index number corresponding to the database storage location as opposed to [2,4,5,1].

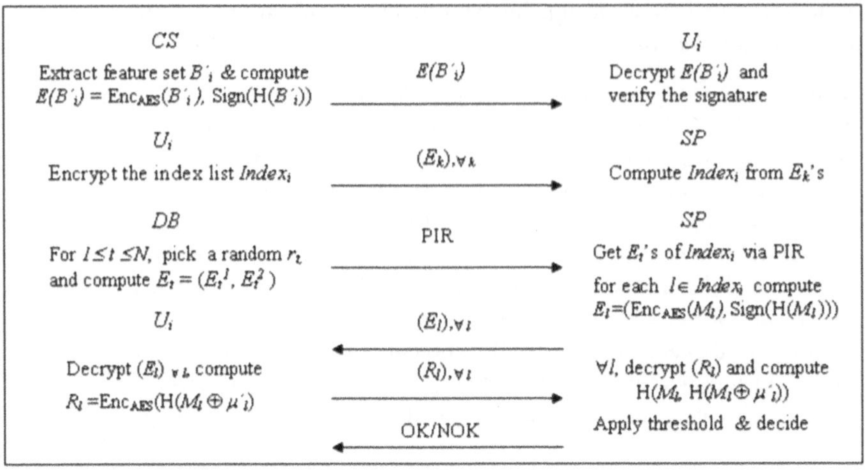

Fig. 1. Verification phase of the new Protocol

- No single point of failure: In order to impersonate a user, the attacker needs to obtain both the biometrics and the smart card that stores the private key and the index list of the user.
- Loss of the secret does not compromise the template: The user has to store only a private key and some index numbers in the smart card instead of storing his biometric template and the compromise of these values do not reveal the biometric information in any step of the authentication. When the user's smart card is lost or stolen, then the user can obtain a new secret key from PKG and the index list by reregistering to the database. However, most of the user's features are already stored in DB, so DB only returns the corresponding index list after the user presents his feature vector.
- No replay attack possible: At each authentication request, the database xores each feature in the database with a new random number and sends these values to the service provider. Also, the service provider and the user agrees on a different session key for each application and the authentication requires the collaboration of each party. So, it is not possible to send replayed data.
- No need for PKI: Our scheme uses an efficient Anonymous Identity Based Encryption (IBE) scheme such as Boneh/Franklin IBE [21] for the generation of a short session key, hence, an eavesdropper (or a malicious database) on the communication channel cannot discover the identity of the user since the ciphertext does not reveal anything about the identity of the recipient (and the sender for authenticated Boneh/Franklin scheme [27]) of the ciphertext [22]. This is a vital point in the privacy of the user identity in a biometric authentication system, however, in many schemes, this property is not considered. Also, our design does not require a PKI and public key certificates as in [18].
- Encryption of the channel: The communication channel between the components of the system is encrypted, hence eavesdropping on the channel is prevented.

In addition, our new design has the following advantages in terms of computation and storage costs.

- Efficient memory storage: Since each feature is stored as a separate entry in the database, there could be common features belonging to different users. Thus, during registration phase, the database could check for this situation and could return the index of the previously stored feature. This way, the size of the registered feature set and the total storage in the database could be smaller. Besides, since no biometric template is stored as an entry, there is no need to apply a public key encryption scheme such as ElGamal to store the biometric data as encrypted, where the ciphertext size is twice the plaintext size as in [4,2]. Finally, the choice of the system parameters of [3,1] result in a constraint on the size of the database, whereas our design is also suitable for a large scale central database that stores biometric data.
- Lower computational cost: In [3,1], the database performs $O(N)$ exponentiations modulo q^2 [3] and modulo q [1], where q is an RSA modulus with $|q|$=2048 bits. Similarly, the schemes of [4,2] require $O(N)$ exponentiations in group G, on which the ElGamal public key scheme is defined. The computational cost of our scheme is dominated by the $O(N)$ random number selections and $O(N)$ hash computations in order to encrypt each feature stored in the database using one time pad. Except for the session key generations, we use symmetric key encryption and lightweight cryptographic primitives, hence, our scheme is suitable for energy constrained devices. In the following table, we summarize various remote biometric-based authentication schemes that satisfy the security model described in section 2.

Table 1. Comparison of various biometric authentication systems

Scheme	Communication Cost	Computation Cost	Storage Cost*
System 1 [1]	$O(N)$	M exponentiations + $(MN)/2$ multiplications	M bits
System 2 [3]	$O(\log^2(N))$	$O(N)$ exponentiations	128 Kbytes
System 3 [4]	$O(k+2M)$	$O(N)$ exponentiations	$2M$ bits
System 4 [2]	$O(k+2M)$	$O(N)$ exponentiations	$2M$ bits
Our System	$O(n(k+\|\mu\|))$	$O(N)$ random number selections + $O(N)$ Hash computations	$\|\mu\|$

*At each entry of the database
Abbreviations: N=total number of entries in the database; n=dimension of the feature vector of a user; M= size of the biometric template; $|\mu|$=size of a feature; $k \geq \log(N)$

5 Conclusion

In this paper, we presented a new design for a remote biometric based authentication protocol, where the entities of the system are independent of each other. The system follows the state-of-the-art security model for biometric authentication systems with an improved computational complexity. Besides, a different storage

mechanism for the biometric data is introduced, which could be of independent interest for the biometrics and information security community. In addition to the increased efficiency in the database storage, this approach also affects the social acceptance of biometric systems operating with central databases positively since the compromise of the database (namely, a random pool of features) would not help any attacker in the recovery of a user's template, which could otherwise only be guaranteed by storing the biometrics as encrypted. Finally, an open problem is to decrease the communication cost of the distributed biometric authentication systems, which is caused by the use of PIR systems.

Acknowledgement

The author is grateful to her supervisor Prof. Dr. Joachim von zur Gathen for his valuable support, encouragement and guidance.

References

1. Bringer, J., Chabanne, H., Izabachène, M., Pointcheval, D., Tang, Q., Zimmer, S.: An application of the goldwasser-micali cryptosystem to biometric authentication. In: Pieprzyk, J., Ghodosi, H., Dawson, E. (eds.) ACISP 2007. LNCS, vol. 4586, pp. 96–106. Springer, Heidelberg (2007)
2. Bringer, J., Chabanne, H., Pointcheval, D., Tang, Q.: Extended private information retrieval and its application in biometrics authentications. In: Bao, F., Ling, S., Okamoto, T., Wang, H., Xing, C. (eds.) CANS 2007. LNCS, vol. 4856, pp. 175–193. Springer, Heidelberg (2007)
3. Bringer, J., Chabanne, H.: An authentication protocol with encrypted biometric data. In: Vaudenay, S. (ed.) AFRICACRYPT 2008. LNCS, vol. 5023, pp. 109–124. Springer, Heidelberg (2008)
4. Tang, Q., Bringer, J., Chabanne, H., Pointcheval, D.: A formal study of the privacy concerns in biometric-based remote authentication schemes. In: Chen, L., Mu, Y., Susilo, W. (eds.) ISPEC 2008. LNCS, vol. 4991, pp. 56–70. Springer, Heidelberg (2008)
5. Barbosa, M., Brouard, T., Cauchie, S., de Sousa, S.M.: Secure biometric authentication with improved accuracy. In: Mu, Y., Susilo, W., Seberry, J. (eds.) ACISP 2008. LNCS, vol. 5107, pp. 21–36. Springer, Heidelberg (2008)
6. Schoenmakers, B., Tuyls, P.: Efficient binary conversion for paillier encrypted values. In: Vaudenay, S. (ed.) EUROCRYPT 2006. LNCS, vol. 4004, pp. 522–537. Springer, Heidelberg (2006)
7. Atallah, M.J., Frikken, K.B., Goodrich, M.T., Tamassia, R.: Secure biometric authentication for weak computational devices. In: Patrick, A.S., Yung, M. (eds.) FC 2005. LNCS, vol. 3570, pp. 357–371. Springer, Heidelberg (2005)
8. Yoon, E.J., Yoo, K.Y.: A secure chaotic hash-based biometric remote user authentication scheme using mobile devices. In: Chang, K.C.-C., Wang, W., Chen, L., Ellis, C.A., Hsu, C.-H., Tsoi, A.C., Wang, H. (eds.) APWeb/WAIM 2007. LNCS, vol. 4537, pp. 612–623. Springer, Heidelberg (2007)
9. Boyen, X., Dodis, Y., Katz, J., Ostrovsky, R., Smith, A.: Secure remote authentication using biometric data. In: Cramer, R. (ed.) EUROCRYPT 2005. LNCS, vol. 3494, pp. 147–163. Springer, Heidelberg (2005)

10. Crescenzo, G.D., Graveman, R.F., Ge, R., Arce, G.R.: Approximate message authentication and biometric entity authentication. In: Patrick, A.S., Yung, M. (eds.) FC 2005. LNCS, vol. 3570, pp. 240–254. Springer, Heidelberg (2005)
11. Dodis, Y., Reyzin, L., Smith, A.: Fuzzy extractors: How to generate strong keys from biometrics and other noisy data. In: Cachin, C., Camenisch, J.L. (eds.) EUROCRYPT 2004. LNCS, vol. 3027, pp. 523–540. Springer, Heidelberg (2004)
12. Juels, A., Sudan, M.: A fuzzy vault scheme. Des. Codes Cryptography 38(2), 237–257 (2006)
13. Juels, A., Wattenberg, M.: A fuzzy commitment scheme. In: ACM Conference on Computer and Communications Security, pp. 28–36. ACM, New York (1999)
14. Tuyls, P., Goseling, J.: Capacity and examples of template-protecting biometric authentication systems. In: Maltoni, D., Jain, A.K. (eds.) BioAW 2004. LNCS, vol. 3087, pp. 158–170. Springer, Heidelberg (2004)
15. Uludag, U., Pankanti, S., Jain, A.K.: Fuzzy vault for fingerprints. In: Kanade, T., Jain, A., Ratha, N.K. (eds.) AVBPA 2005. LNCS, vol. 3546, pp. 310–319. Springer, Heidelberg (2005)
16. Uludag, U., Jain, A.: Securing fingerprint template: Fuzzy vault with helper data. In: Computer Vision and Pattern Recognition Workshop. IEEE Computer Society, Los Alamitos (2006)
17. Mihailescu, P.: The fuzzy vault for fingerprints is vulnerable to brute force attack. CoRR abs/0708.2974 (2007)
18. Itakura, Y., Tsujii, S.: Proposal on a multifactor biometric authentication method based on cryptosystem keys containing biometric signatures. Int. J. Inf. Sec. 4(4), 288–296 (2005)
19. Salaiwarakul, A., Ryan, M.D.: Verification of integrity and secrecy properties of a biometric authentication protocol. In: Chen, L., Mu, Y., Susilo, W. (eds.) ISPEC 2008. LNCS, vol. 4991, pp. 1–13. Springer, Heidelberg (2008)
20. Park, B., Moon, D., Chung, Y., Park, J.W.: Impact of embedding scenarios on the smart card-based fingerprint verification. In: Lee, J.K., Yi, O., Yung, M. (eds.) WISA 2006. LNCS, vol. 4298, pp. 110–120. Springer, Heidelberg (2007)
21. Boneh, D., Franklin, M.K.: Identity-based encryption from the weil pairing. SIAM J. Comput. 32(3), 586–615 (2003)
22. Boneh, D., Gentry, C., Hamburg, M.: Space-efficient identity based encryptionwithout pairings. In: FOCS 2007: Proceedings of the 48th Annual IEEE Symposium on Foundations of Computer Science, pp. 647–657. IEEE Computer Society, Los Alamitos (2007)
23. Sahai, A., Waters, B.: Fuzzy identity-based encryption. In: Cramer, R. (ed.) EUROCRYPT 2005. LNCS, vol. 3494, pp. 457–473. Springer, Heidelberg (2005)
24. Baek, J., Susilo, W., Zhou, J.: New constructions of fuzzy identity-based encryption. In: ASIACCS 2007, pp. 368–370. ACM, New York (2007)
25. Gentry, C., Ramzan, Z.: Single-database private information retrieval with constant communication rate. In: Caires, L., Italiano, G.F., Monteiro, L., Palamidessi, C., Yung, M. (eds.) ICALP 2005. LNCS, vol. 3580, pp. 803–815. Springer, Heidelberg (2005)
26. Bellare, M., Namprempre, C., Neven, G.: Security proofs for identity-based identification and signature schemes. In: Cachin, C., Camenisch, J. (eds.) EUROCRYPT 2004. LNCS, vol. 3027, pp. 268–286. Springer, Heidelberg (2004)
27. Pan, J., Cai, L., Shen, X.: Promoting Identity-Based Key Management in Wireless Ad Hoc Networks. In: Xiao, Y., Shen, X., Du, D. (eds.) Wireless/Mobile Network Security - Signals and Communication Technology, pp. 83–102. Springer, Heidelberg (2007)

A Biometric Key-Binding and Template Protection Framework Using Correlation Filters

Vishnu Naresh Boddeti[1], Fei Su[2], and B.V.K. Vijaya Kumar[1]

[1] Carnegie Mellon University, Pittsburgh PA 15213, USA
naresh@cmu.edu, kumar@ece.cmu.edu
[2] Beijing University of Posts and Telecommunications, Beijing, China
sufei@bupt.edu.cn

Abstract. We describe a new framework to bind cryptographic keys with biometric signatures using correlation filters. This scheme combines correlation filter based biometric recognition with biometric key-binding while offering template protection, revocability, diversity and security. We demonstrate the effectiveness of our scheme via numerical results on the CMU-PIE face database.

Keywords: Encryption, Key-Binding, Correlation Filters, Face Recognition, Template Protection, Biometrics.

1 Introduction

The fundamental idea of cryptography is to encipher a "secret" or message into an intermediate form, also called "cipher text", in which the original message exists in a hidden state. The same message can be transformed into many intermediate forms by using different ciphers chosen by a key, called "cipher key". The original message can be recovered accurately by reversing this process only by using the correct decryption key. Most often these keys are very long and nonsensical making it very difficult for humans to remember. Hence they are stored either on a smart card or in a computer database and are released using some other form of authentication (usually a password).

While current cryptographic algorithms provide high security, they suffer from some limitations not due to weaknesses in the algorithms themselves but due to the setup of the whole security system. The decryption keys are only as secure as the password or smart card used for authentication to release the key. These passwords and smart cards can be lost or compromised, effectively nullifying the security provided by the cryptographic algorithms. Another concern is that there is no way of knowing if the password is being provided by a genuine person or an impostor.

Linking the decryption keys to the physiological and behavioral traits of a person like face, iris, fingerprint, gait etc. would address some of the above mentioned shortcomings of cryptographic authentication systems. These biometric characteristics cannot be lost or forgotten like passwords, thereby reducing the chances of being compromised. It is also not easy to copy, share or distribute the biometric signatures and more importantly they enable us to tie the key directly to the user. Hence biometric-based authentication systems are good candidates to either replace normal password-based systems or can be used in conjunction with them to provide two-factor authentication.

M. Tistarelli and M.S. Nixon (Eds.): ICB 2009, LNCS 5558, pp. 919–929, 2009.
© Springer-Verlag Berlin Heidelberg 2009

Having said that, biometric authentication systems come with their own challenges. The main concern is that cryptographic systems require exactitude while biometric measurements are inherently noisy due to both natural variation and noise in acquiring biometric data. Hence combining cryptography with biometric recognition is challenging. Some requirements of biometric cryptographic systems are as follows:

1. ***Revocability:*** Lost or stolen passwords are easy to revoke and new passwords can be issued. Biometric authentication systems should have the same capability.
2. ***Security:*** Biometric authentication systems need to be as secure as password based systems. Also, it should be computationally hard to reverse engineer either the secret key or the raw biometric from the biometric template.
3. ***Performance:*** In order that only an authentic user be able to release the key, the error rates i.e., in terms of false accepts and false rejects should be low.
4. ***Diversity:*** Should have the ability to issue diverse templates, obtained from the same biometric, to be used for multiple applications.

The correlation filter based framework that we propose in this work has been designed keeping the above in mind. Correlation filters can be designed to tolerate a variety of sources of appearance variability in the image. Due to this property, correlation filters are suitable candidates for combining biometrics with cryptography. The filters used in this work, while performing their primary task of biometric matching, have the functionality of key-binding built into them. This is achieved by extending the traditional correlation filter design as explained in Section 2.

1.1 Related Work

Biometric security and privacy are important factors in the adoption and deployment of biometric authentication systems. There have been many approaches to address the issues in bringing together biometric authentication and traditional cryptography.

Soutar et al. [1][2][3] proposed a correlation-based key binding algorithm for fingerprint based authentication. They design a correlation filter from representative training images and then set the filter magnitude to 1 for all frequencies, while multiplying the filter by a random phase mask. Key-binding is done by linking the key to the binarized correlation plane. Though we also use correlation filters in our framework, our work differs from their's in both the correlation filter design and the key-binding algorithm.

Another popular technique is the Fuzzy Vault scheme proposed by Juels and Sudan [4]. This scheme has been used for fingerprint based recognition [5] and even in iris recognition [6]. While fuzzy vaults offer tolerance to biometric variability, they have many limitations [7][8], some of which are being addressed. The primary limitation with these schemes is the amount of security they provide, which is not on par with the present day encryption techniques like AES-128, triple DES etc.

Ratha et al. [9] and Savvides et. al. [10] proposed techniques for obtaining cancelable templates for fingerprints and faces respectively, which helps address the issues of revocability and diversity. In this work, we incorporate the later cancelability scheme.

Another approach to combining cryptography with biometrics has been the use of Error Correction Coding (ECC) to overcome the natural variability in biometrics. Hao et al. [11] use Hadamard and Reed Solomon codes for iris recognition while Sutcu et

al. [12][13] use low density parity check (LDPC) codes on fingerprints. These methods require binary representation of the features to use ECC, which may not always be possible for biometrics like faces, gait, voice etc.

The rest of the paper is organized as follows. In Section 2, we describe the correlation filters used for matching. Our framework is presented in Section 3 followed by security analysis in Section 4. Experimental results are given in Section 5.

2 Correlation Filters

A correlation filter can be represented by a template that is specifically designed from a set of patterns (e.g., iris images) that are representative of a particular class. This template is matched against a query by performing a cross-correlation between the two. For computational efficiency this is done in the frequency domain,

$$C(x,y) = FT^{-1}\{FT\{I(x,y)\} \cdot F^{*}(u,v)\} \tag{1}$$

where *I(x,y)* is a query pattern and *F(u,v)* is the frequency domain representation of the correlation filter with *u* and *v* denoting the spatial frequencies.

The filter is constrained to give a sharp peak at the center of the correlation plane $C(x,y)$ for an authentic match and no such peak for an impostor as shown in Fig. 1. The degree of match is measured by the peak-to-correlation energy (PCE) ratio defined as:

$$PCE = \frac{peak - \mu}{\sigma} \tag{2}$$

μ and σ are the mean and the standard deviation, respectively of the correlation plane.

Correlation filters have been found to perform well in biometric recognition problems like face, iris, fingerprint and palm print recognition [14][15][16]. A variety of advanced correlation filters are available [17] and in this work we use a modified Minimum Average Correlation Energy (MACE) [18] filter. The MACE filter is given by,

$$\mathbf{h} = \mathbf{D}^{-1}\mathbf{X}(\mathbf{X}^{+}\mathbf{D}^{-1}\mathbf{X})^{-1}\mathbf{u} \tag{3}$$

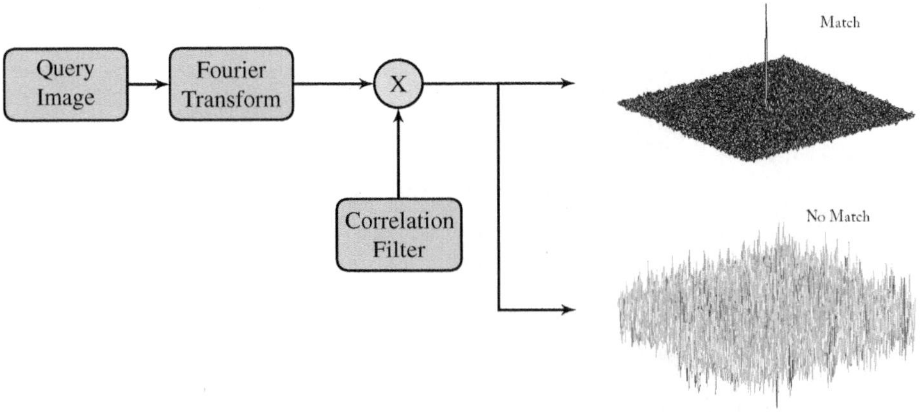

Fig. 1. Application of correlation filter on a query image

where $\mathbf{X} = [\mathbf{z_1} \ .... \ \mathbf{z_n}]$ contains the vectorized versions of the two-dimensional Fourier transforms (FT) of the training images, $\mathbf{h}$ is the frequency domain correlation filter, $\mathbf{u}$ contains the correlation plane origin constraints and $\mathbf{D}$ is a diagonal matrix with the average power spectrum of the training images along its diagonal.

2.1 Multi-peak Correlation Filter

Correlation filters have been traditionally designed to produce just one peak at the center of the correlation plane for an authentic match and no such peak for an impostor. We extend this by adding more constraints to the filter design to produce an arbitrary number of peaks at arbitrary locations for an authentic query and no discernible peaks for an impostor. This can be thought of as summing together multiple correlation planes, each having one peak at a pre-specified location which in turn can be obtained by shifting the peak in the traditional design (at the origin) to the specified location. See Fig. 2 for an illustration. Even though the absolute locations of the peaks may vary depending on the relative displacements between the gallery and probe images, the relative locations of these multiple peaks are, in general stable when using good quality images, thus approaching the exactness required by cryptography. This property together with the flexibility in designing the filter is exploited to bind the message with the biometric to give one template. The multi-peak MACE filter designed with n training images and m unique peak locations is of the same form as the MACE filter and is given by,

$$\mathbf{h} = \mathbf{D}^{-1}\mathbf{X}(\mathbf{X}^{+}\mathbf{D}^{-1}\mathbf{X})^{-1}\mathbf{u} \tag{4}$$

where $\mathbf{X} = [\mathbf{z_1 e}^{-\mathbf{j}(\mathbf{ux_1}+\mathbf{vy_1})} \ .... \ \mathbf{z_n e}^{-\mathbf{j}(\mathbf{ux_1}+\mathbf{vy_1})} \ .... \ \mathbf{z_1 e}^{-\mathbf{j}(\mathbf{ux_m}+\mathbf{vy_m})} \ .... \ \mathbf{z_n e}^{-\mathbf{j}(\mathbf{ux_m}+\mathbf{vy_m})}]$ now contains in the columns, the vectorized versions of the two-dimensional FT of the modulated training images. The modulation for each peak is $e^{-j(ux+vy)}$ where (x,y) is the desired location of the correlation peak. $\mathbf{D}$ is a diagonal matrix with the average power spectrum of the modulated training images along its diagonal.

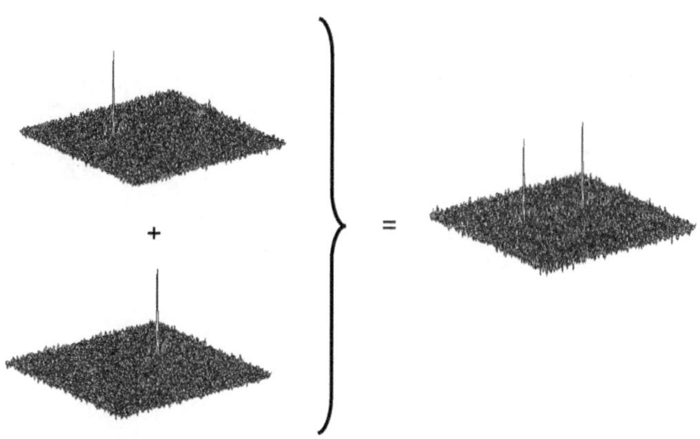

Fig. 2. A multi-peak correlation plane can be thought of as the sum of many correlation planes each producing one peak, but at distinct locations

3 Proposed Framework

The main idea behind our approach is to use correlation filters for the dual purpose of biometric matching and information hiding thereby binding the hidden information to the training images used to design the filter. During authentication this hidden information is released automatically if the match is determined to be an authentic match.

In an encryption system, two inputs, "secret" and "cipher key", are algorithmically combined to give out one output, "cipher text". The framework that we present here is flexible enough that we can hide either the "secret" or the "cipher key" in the filter. Each of the alternatives has its own advantages and disadvantages. Hiding the message directly would impose constraints on the maximum length of the message since there is a trade-off between how many peaks can be packed and how reliably we can recover the information. However it must be noted that recovering the information by brute force is as hard as decrypting a conventionally encrypted message. When hiding the encryption key, the message can now be of any length although the length of the encryption key is limited by the constraints in the performance of the filters. To allow for flexibility in the key length, we can use one more level of encryption to encrypt the "cipher key". Now the key required for the second encryption is hidden in the correlation filter and the first "cipher key" now becomes the message in our framework. Such a scenario would arise when the key length required for the first encryption is beyond the capability of our framework (e.g., RSA-1024). In this initial work, we only consider the case where the key is hidden in the correlation filter.

Template protection is achieved by convolving the images with a random kernel as proposed by Savvides et.al [10]. Fig. 3 shows an example.

Fig. 3. Image encryption with a random convolution kernel

The resulting image looks very different from the original image, effectively securing the original images. It also helps us achieve revocability and diversity by assigning different kernels for different people and different applications. One way to do this would be by asking the user for a password and using it as a seed for the random number generator from which the random kernel is obtained. It was shown theoretically and empirically in [10] that matching in this encrypted domain does not lead to any loss of performance. We design the modified MACE filter introduced in the previous section with these encrypted images.

3.1 Enrollment

Fig. 4 shows the block diagram of the enrollment process. During enrollment the user provides multiple samples of his/her biometric (shown as "Training Data" in Fig. 4), the "secret" that is sought to be protected and a password to create the random kernel to encrypt the images. The number of peaks in the output correlation plane for a user

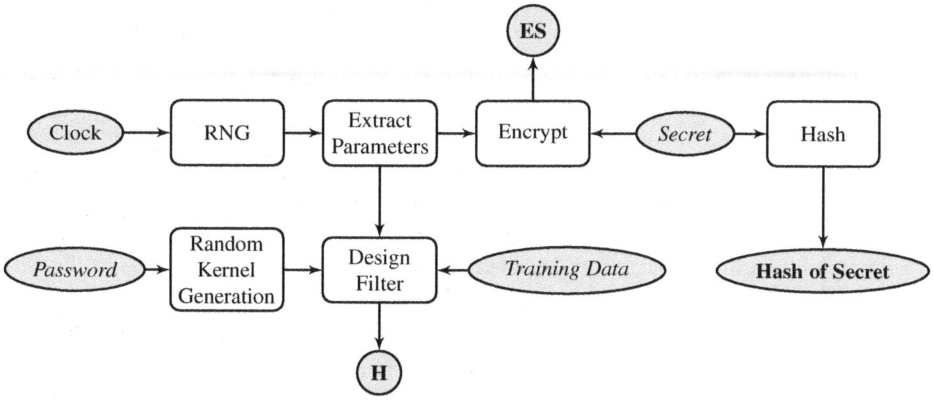

Fig. 4. Block diagram of the enrollment process. Uncolored blocks denote internal processing. Colored blocks with *italicized* text denote input and **bold** text denote output.

is selected at random between some minimum and maximum number depending on the desired level of security. Then the locations of the peaks are selected at random depending on the resolution of the images. We further choose the polarity of each peak (± 1) at random. This set of numbers representing the number of peaks, the polarity of the peaks and the locations of the peaks are concatenated to form the encryption key for that user and is used to encrypt the message. This whole process is captured in the blocks labeled "clock", "RNG" and "Extract Parameters".

The peak locations are used as constraints while designing the multi-peak MACE filter described in the previous section. The filter is the cancelable template for the user and is stored in the database along with the encrypted message. We also store the hash value of the message using a one-way hash function. Thus each user has their own cancelable template designed with a random number of peaks, polarity of peaks and peak locations.

To get stable correlation planes with sharp peaks, the peak locations for designing the filter are chosen such that there is at least a 3 pixel separation between adjacent peaks. We also do not allow the peaks to be within 5 to 10 pixels from the boundary of the image. Sharper peaks corresponds to better recognition performance and greater number of peaks provides more security. Both of these cannot be achieved simultaneously since there exists a trade-off between how many peaks one can pack in the correlation plane and the sharpness of the peaks.

3.2 Authentication and Identification

Fig. 5 shows a block diagram of the authentication process. In the authentication stage, the user provides the biometric (denoted as "probe" in Fig. 5) along with the password for his/her random kernel. The biometric is convolved with the random kernel derived from the password. The resulting image is then correlated with the filter stored in the database. If the match is authentic, the resulting correlation plane would have sharp peaks at the locations that the filter was constrained to produce peaks at. From the peak locations, the key is retrieved and is used to decrypt the stored encrypted message. For

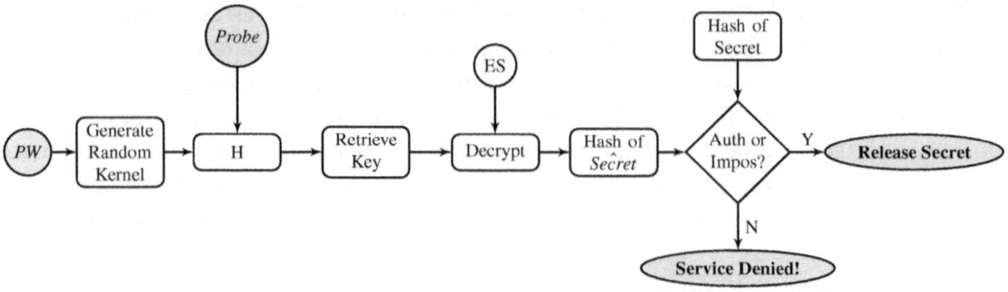

Fig. 5. Block diagram of the authentication process. Uncolored blocks denote internal processing. Colored blocks with *italicized* text denote input and **bold** text denote output.

an authentic match, the retrieved key would be correct while for an impostor, it would be wrong leading to incorrect message decryption. This is confirmed by comparing the stored hash value of the original message with the hash value of the recovered message. Only when the two match, the "secret" is released. For an impostor the decryption would fail because there would not be any sharp peaks in the correlation result when either the password is wrong or the biometric is wrong or both are wrong.

4 Security Analysis

In the proposed framework, protection is provided at two levels, password and biometric. Both the biometric and the password have to be correct for correct decoding of the protected message. We consider three kinds of possible attack strategies. In all the scenarios, we assume that the attacker has access to the database i.e., the encrypted message, the biometric template and the hash value of the message.

4.1 Brute Force

This corresponds to the scenario when the attacker does not have access to both the password and the biometric. The attacker now has to guess the decryption key which means he must guess the number of peaks, polarity of the peaks and the peak locations. Once the number of peaks and their peak locations are correctly guessed, the attacker gets the decryption key and can decrypt the encrypted message. For an image size of MxN pixels and a filter design with Q peaks, the entropy $L(Q)$ in bits is given by:

$$L(Q) = \sum_{i=1}^{Q} [\log_2\{(M-q)(N-q)-(i-1)p^2\}-1] \tag{5}$$

where q is the size of the boundary region that is ignored and p is the minimum separation between the peaks in the correlation plane. The $(M-q)(N-q)$ term quantifies the number of available peak locations for the first peak. Now since there is a minimum required separation of p in either direction between the peaks, the number of peak locations for the subsequent peaks reduces by p^2. We also allow for the retrieved peak location to be off by 1 in either x or y direction which is accounted for by subtracting 2 bits per peak and adding one bit for the sign of the peak.

To provide equal amount of security to all the users of the system in the same application, each user is assigned the same number of peaks depending on the security and performance trade-off. Thus the amount of security is $L(Q)$ with Q peaks per user. In situations where more flexibility is needed and equal security for all users is not required, each user can be assigned different number of peaks. It must be noted that there is no storage or computational cost associated with using different number of peaks. However, the amount of security and the recognition performance would be different.

4.2 Compromised Biometric

This is the scenario where the attacker does not know the password but obtains the biometric, for example a face image from the subject's web site or fingerprints from the objects the subject touches etc. In this case, the security depends on how guessable the password is and also on the quality of the biometric stolen by the hacker. Without the password, the random mask would be wrong and the decoding will fail. So the amount of security is equal to the strength of the user's password or the complexity of the encryption used, whichever is lower.

4.3 Compromised Password

This is the scenario where the attacker knows the password but does not have the biometric. In this case since the biometric would be wrong, the result of the correlation with the filter would either not produce any peaks or produce them at the wrong location which causes the decryption to fail. The amount of security in this case would depend on the complexity associated with guessing the biometric (e.g., by hill climbing technique [19]) or the complexity of the encryption used, whichever is lower.

5 Numerical Results

We used the CMU-PIE database to evaluate the effectiveness of the proposed scheme and we allowed for illumination changes only in our experiments. We believe that since biometric key-binding would be used for applications such as access control, e-commerce etc., it is reasonable to expect user cooperation while presenting their biometric. Illumination is something that can't really be controlled very well (even for cooperative users) and hence becomes one of the sources of variability in appearance. Further we use both the PIE-lights and PIE-nolights for our evaluation and present the verification results. The PIE-lights database consists of 68 classes with 24 images per class while PIE-nolights consists of 66 classes with 21 images per class. In our experiments we used 3 images for training the MACE filter and the rest of the images for testing. The 3 training images were chosen such that one has illumination from the left, another has illumination from the right and the third has illumination from the front.

We conducted numerical experiments corresponding to two different levels of security. In one case we use both the password and the biometric for authentication (referred to as whole), while in the other case we use only the biometric i.e., the images are not convolved with a random mask which is the scenario when the password is compromised (referred to as part). Table 1 summarizes our results for both subject verification

Table 1. Performance with 5 Peaks and Image Size 256

Exp. Type	Recognition		Decryption			
	Lights	No Lights	Lights		No Lights	
	EER (%)	EER (%)	DFR (%)	FDR (%)	DFR (%)	FDR(%)
Whole	0.0	0.083	1.4	0.0	3.2	0.0
Part	0.072	0.65	2.3	3.2	3.5	1.6

and successful message decryption. We show equal error rates (EER) for verification performance and to quantify key-retrieval performance we look at the decryption failure rate (DFR) for the authentic users and the false decryption rate (FDR) for the impostors i.e., impostors who could retrieve the key of another user.

When the whole system is used, i.e., when both password and biometric are available, the FDR is very close to zero, while in the case where only the biometric is used, the FDR is higher at about 3.2% in the case of PIE-lights. This is because of the presence of users who look similar to each other. When the password is used, since the images are convolved with a random mask the effect of the similarity in the face images is reduced by the randomness of the mask thereby reducing the FDR. Hence the use of the random mask not only protects the template but also helps reduce the FDR.

The results in Table 1 used images of resolution 256x256 with 5 peaks for each user. A region of 10 pixels near the boundary was ignored and the minimum distance between the pixels was set to 5. The entropy for 5 peaks is 74 bits which is the amount of security of the system since all the users have the same number of peaks.

Table 2 summarizes results for both subject verification and successful message decryption when the number of peaks is increased to 7 using images of size 256x256 giving 103 bits of security. Increasing the number of peaks from 5 to 7 increases the amount of security. However there is a performance degradation in both DFR and FDR. The FDR when using the whole system is still 0% which is very desirable.

Table 2. Performance with 7 Peaks and Image Size 256

Exp. Type	Recognition		Decryption			
	Lights	No Lights	Lights		No Lights	
	EER (%)	EER (%)	DFR (%)	FDR (%)	DFR (%)	FDR(%)
Whole	0.0	0.12	2.0	0.0	3.4	0.0
Part	0.1	1.0	2.6	3.8	3.9	3.0

Table 3 shows results for both subject verification and successful message decryption with 6 peaks with the resolution of the images being 128x128 which amounts to 75 bits of security. Using images of smaller size reduces the security and also since the peaks are more closely packed there is a performance degradation.

Table 3. Performance with 6 Peaks and Image Size 128

Exp. Type	Recognition		Decryption			
	Lights	No Lights	Lights		No Lights	
	EER (%)	EER (%)	DFR (%)	FDR (%)	DFR (%)	FDR(%)
Whole	0.0	0.08	0.21	0.0	4.3	0.0
Part	0.07	1.2	0.14	0.25	4.4	0.05

6 Limitations

To understand the true effectiveness of our scheme, one must think of all possible attack strategies. However due to space constraints we will briefly mention only some limitations of the proposed framework.

- Key retrieval is tied strongly to subject verification which means this scheme is effective only when the verification performance is acceptable.
- Stealing biometrics is not too difficult (e.g., one can easily obtain face images from a personal web site), in which case it may turn out that our system is only as strong the password used to encrypt the images. However all biometric-based authentication systems would also face the same problem.
- Once the password in our scheme is compromised, it might become vulnerable to a hill climbing attack. However, it must be noted that the hidden message can be recovered only by a very good estimate of the biometric via hill climbing.
- If the database is not very secure the stored template can be tampered with leading to a denial of service attack or more seriously, a blended substitution attack is also possible. A simple scheme to prevent both of these attacks is to check if the template has been tampered before using it or by making the database read-only.

7 Conclusions

We presented a novel framework for biometric key-binding and template protection using correlation filters and demonstrated its effectiveness in terms of recognition performance and security. Our framework provides great flexibility and scope for expansion. The presented scheme also addresses other important issues like cancelability and diversity and offers two-factor authentication.

Acknowledgements

This work was supported in part by US Army Research Office (ARO) support to CyLab at Carnegie Mellon University.

References

1. Soutar, C., et al.: Proc. SPIE, Optical Security and Counterfeit Deterrence Techniques II, vol. 3314, pp. 178–188 (1998)
2. Soutar, C., et al.: Proc. SPIE, Optical Pattern Recognition IX, vol. 3386, pp. 24–35 (1998)

3. Soutar, C., et al.: Biometric encryption. In: Nichols, R.K. (ed.) ICSA Guide to Cryptography. McGraw-Hill, New York (1999)
4. Juels, A., Sudan, M.: Proc. of IEEE Intl. Symp. on Info. Theory, Lausanne, Switzerland, p. 408 (2002)
5. Uludag, U., et al.: Fuzzy vault for fingerprints. In: Kanade, T., Jain, A., Ratha, N.K., et al. (eds.) AVBPA 2005. LNCS, vol. 3546, pp. 310–319. Springer, Heidelberg (2005)
6. Lee, Y.J., et al.: IEEE Transactions on Systems, Man, and Cybernetics-Part B 38, 1302–1313 (October 2008)
7. Scheirer, W.J., Boult, T.E.: Proc. of Biometrics Symposium, September 2007, pp. 1–6 (2007)
8. Kholmatov, A., Yanikoglu, B.: Proc. of SPIE Symposium on Security, Forensics, Steganography, and Watermarking of Multimedia Contents X, San Jose, USA, vol. 6819, pp. 68190O–68190O-7 (2008)
9. Ratha, N.K., et al.: IEEE Trans. Pattern Anal. Mach. Intell., 561–572 (2007)
10. Savvides, M., et al.: International Conference on Pattern Recognition, pp. 922–925 (2004)
11. Hao, F., et al.: IEEE Trans. Computers, 1081–1088 (2006)
12. Sutcu, Y., et al.: IEEE International Symposium on Information Theory (ISIT), pp. 2297-2301 (July 2008)
13. Sutcu, Y., et al.: IEEE Computer Society Conference on Computer Vision and Pattern Recognition (CVPR), June 2008, vol. 23, pp. 1–6 (2008)
14. Vijaya Kumar, B.V.K., et al.: Proceedings of International Conference of Image Processing, September 2002, pp. 53–56 (2002)
15. Savvides, M., Vijaya Kumar, B.V.K.: Proceedings of IEEE Conference on Advanced Video and Signal Based Surveillance, July 2003, pp. 45–52 (2003)
16. Thronton, J., et al.: Proceedings of International Conference on Image Analysis and Recognition, September 2005, pp. 1098–1105 (2005)
17. Vijaya Kumar, B.V.K.: Applied Optics 31, 4773–4801 (1992)
18. Mahalanobis, A., et al.: Applied Optics 26, 3633–3640 (1987)
19. Adler, A.: Audio and Video Based Biometric Person Authentication, pp. 1100–1109 (2005)

Security-Enhanced Fuzzy Fingerprint Vault Based on Minutiae's Local Ridge Information*

Peng Li, Xin Yang, Kai Cao, Peng Shi, and Jie Tian

Institute of Automation, Chinese Academy of Sciences, Beijing 100190 China
tian@ieee.org, jie.tian@ia.ac.cn
http://www.fingerpass.net

Abstract. Fuzzy vault is a practical and promising fingerprint template protection technology. However, this scheme has some security issues, in which cross-matching between different vaults may be the most serious one. In this paper, we develop an improvement version of fuzzy vault integrating minutiae's local ridge orientation information. The improved fuzzy fingerprint vault, a two factor authentication scheme to some extent, can effectively prevent cross-matching between different fingerprint vaults. Experimental results show that, if and only if the fingerprint and the password of users are simultaneity obtained by the attacker, the fuzzy vault can be cracked. Results under three scenarios indicate that, although the authentication performance of Scena.1 decreases a little in term of GAR, the security of Scena.2 and Scena.3, hence the security of the whole scheme, is enhanced greatly.

Keywords: Fuzzy vault, Cross-matching, Minutiae descriptor, Hermit's interpolation polynomial, SHA-1.

1 Introduction

Traditional biometric systems expose some disadvantages, for example, template security issue due to raw data storage and irrevocability issue because of the inherence property of biometric (e.g., fingerprint, iris, face). We can hardly find solutions for these disadvantages solely by means of biometric theory itself. Therefore, the technology combining biometrics and cryptography, called biometric encryption, biometric template protection or biometric cryptosystem, has attracted remarkable attention, because it may provide potential solutions for the above problems. A good survey for this field can be found in [1].

Over past years, many technologies which integrate biometrics with cryptography were proposed. Souter et al. [2] performs Fourier Transform upon fingerprint

* This paper is supported by the Project of National Natural Science Foundation of China under Grant No. 60875018 and 60621001, National High Technology Research and Development Program of China under Grant No. 2008AA01Z411, Chinese Academy of Sciences Hundred Talents Program, Beijing Natural Science Foundation under Grant No. 4091004.

M. Tistarelli and M.S. Nixon (Eds.): ICB 2009, LNCS 5558, pp. 930–939, 2009.
© Springer-Verlag Berlin Heidelberg 2009

images and then combines the phase information with a random-selected key to make a Lookup Table. If the decryption step succeeds, the correct key is released from the Lookup Table, otherwise a reject signal is given. This is the first practical algorithm in this field but no performance results have been reported. Biohashing [3] is a two-factor authentication method, which uses WFMT(Wavelet Fourier Mellin Transform) feature of fingerprints and performs iterative inner product upon WFMT and a group of random number vectors. The 0-EER can be achieved providing the key (random number vector) is not stolen. However, if the key is stolen, the EER may be much higher than the plain biometric system [4]. Juels and Sudan [5] proposed the fuzzy vault scheme to try to bridge the gap between the fuzziness of biometrics and the exactitude of cryptography. Lee et al. [6] proposed a cancelable fingerprint template technology, using the local minutiae information. This method provided some promising results using transformed fingerprint templates.

Among all the technologies above, fuzzy vault(FV) is the most practical one in the term of security and authentication performance. Juels and Sudan [5] first proposed the fuzzy vault scheme. Clancy et al.[7] proposed the implementation of fuzzy vault for fingerprint, using the cartesian coordinates of minutiae. Holding the assumption that the query and template fingerprints were prealigned, they reported FRR$\approx$20-30%(FAR$\approx$0). Uludag et al.[8][9] and Nandakumar et al.[10] proposed more robust and effective implementation of fuzzy fingerprint vault. They also developed an automatic alignment method in the encryption domain, using ridge curvature maximum values(i.e., so-called helper data). The performance they reported is FRR$\approx$10%(FAR$\approx$0%), making significant improvement compared to Clancy's implementation. Although fuzzy fingerprint vault becomes more and more effective, there remains some security and privacy issues involved with it. As pointed out in [11], fuzzy fingerprint vault leaks some information about the original fingerprint minutiae template. Chang et al.[12] exploited the chaff points' statistical property to distinguish real minutiae from the chaff. Kholmatov and Yanikoglu [13] realized the correlation against fuzzy fingerprint vault, obtaining 59% success rate for two correlated fingerprint databases. Nandakumar et al.[11] tried to solve this problem using password as an additional authentication layer. However, if the password is obtained by the attacker, he/she can perform cross-matching just as [13] shows. In addition, Li et al.[14] pointed that Reed-Solomon correction code is not appropriate for fuzzy vault. After analyzing CRC, we think that it is not suitable for fuzzy vault either, so SHA-1 is used to replace CRC in our security-enhanced scheme to check the key's correction. Minutiae descriptor has been used for enhancing the security of fuzzy fingerprint vault [16], but it can not prevent cross matching described in [13].

In this paper, we propose a security-enhanced fuzzy fingerprint vault integrating the local minutiae information into the original scheme. From each minutia, we derive the corresponding invariant value based on the descriptor proposed in [15]. Afterwards the original minutiae are transformed into deformation domain using the invariant values. The procedures of encoding and decoding for fuzzy fingerprint vault are performed in the deformation domain.

2 Methods Enhancing FV's Security

The security of fuzzy fingerprint vault is mainly enhanced by transforming all the original minutiae into deformation domain. Like [6], the transforming method extracts from each minutia a translation and rotation invariant value, which is obtained by using a user-specific random vector and the orientation information of the neighboring regions around each minutia. And then we compute the transformation amount for each minutia by using the designed changing function. However, we compute the local ridge orientation vector using minutiae decriptor[15], rather than equal angel sample in all the concentric circles[6]. And the changing function is generated using Hermit's Interpolation Polynomial, in place of piece-wise linear interpolation[6]. The transformation accuracy is improved through resorting to minutiae descriptor and Hermit's Interpolation Polynomial. The encoding and decoding phases of fuzzy vault are carried out in the transformed domain. Respectively, the encoding and decoding procedures of security-enhanced fuzzy fingerprint vault are shown in Fig. 1 and Fig. 2.

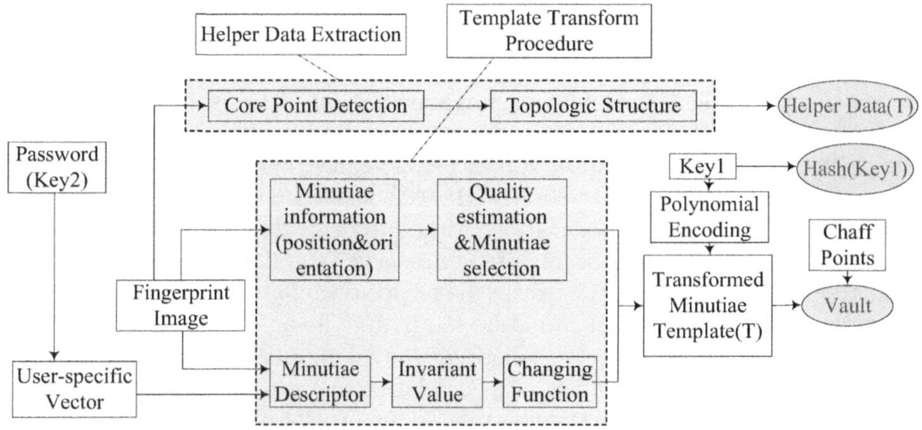

Fig. 1. Encoding procedure of security-enhanced fuzzy fingerprint vault

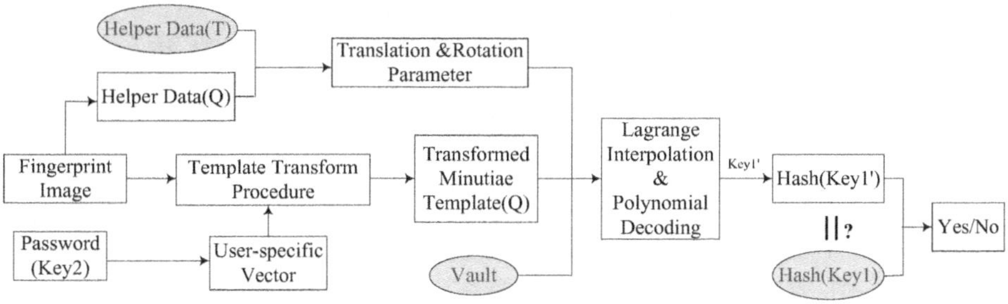

Fig. 2. Decoding procedure of security-enhanced fuzzy fingerprint vault

2.1 Invariant Value Extraction

We compute each minutia's invariant value using a similar method described in [6]. However, in [6], the authors conduct equal angle sample in all the concentric circles, which may miss some ridge information in outer layer circles because the arcs connecting two adjacent sample points may cross two or more adjacent ridges. So we adopt ridge orientation-based minutiae descriptor[15] to compute the invariant value. Fig. 3 shows the difference between sample structure in [6] and minutiae descriptor sample structure. We adopt the same parameter setting as [15]. The minutiae descriptor consists four concentric circles and the radius are 27, 45, 63 and 81 pixels. Respectively, they contain 10, 16, 22 and 28 points, uniformly sampled in the corresponding circle. In all 76 sample points are obtained. Fig. 3 shows that minutiae descriptor can acquire more information than equal angle sample method. One difference is that we extract the minutiae descriptor in thinned ridge image. For the sample point `sp_i`(denoting its position with `co_i` and its 8-neighborhood with `nh_i`), we compute its orientation `O_i` as follows:

```
If sp_i locates in any thinned ridge
  Then co_i is recorded;
Else If sp_i locates in no thinned ridge AND a ridge point is
within nh_i
  Then the nearest ridge point rp_i is recorded;
Else If sp_i locates in no thinned ridge AND no ridge point is
within nh_i
  Then sp_i is labeled as a background point;
If a ridge point is recorded
  Then O_i is computed by the ridge orientation;
Else If the background point
  Then O_i is the same as the corresponding minutia's orientation;
```

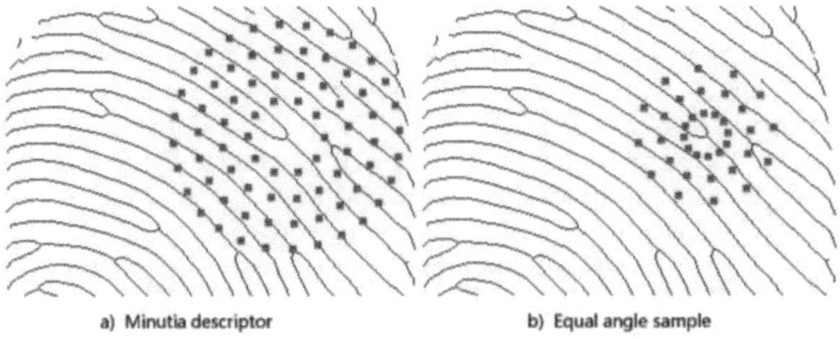

Fig. 3. Minutia descriptor vs. Equal angle sample

Denote the i-th minutia's orientation θ_i and the j-th sample point's orientation θ_{ij}. Thus we can obtain a 76-dimension translation and rotation invariant vector, depicted in (1):

$$T_i = [d(\theta_{i,1} - \theta_i), d(\theta_{i,2} - \theta_i), ..., d(\theta_{i,76} - \theta_i)]$$
$$d(\theta_1, \theta_2) = \begin{cases} \theta_1 - \theta_2, & \text{if } -\frac{\pi}{2} < (\theta_1 - \theta_2) < \frac{\pi}{2} \\ \theta_1 - \theta_2 + \pi, & \text{if } -\pi < (\theta_1 - \theta_2) < -\frac{\pi}{2} \\ \theta_1 - \theta_2 - \pi, & \text{if } \frac{\pi}{2} < (\theta_1 - \theta_2) < \pi \end{cases} \quad (1)$$

The subsequent procedure of generating an invariant value is the same as [6]. First, generate an user-specific random vector R_{pwd} with the same length as T_i, i.e., 76. Then two 76-dimen vectors are normalized to obtain t_i and r_{pwd}. Finally, the i-th invariant value corresponding to i-th minutia is computed using inner product, as shown in (2).

$$m_i = t_i \circ r_{pwd} \quad (2)$$

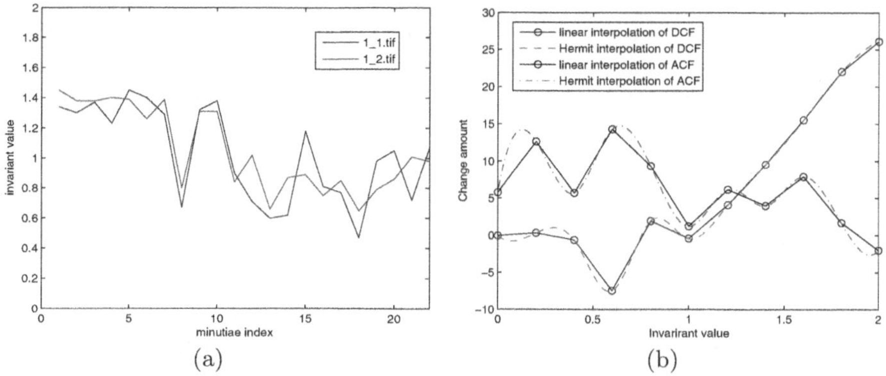

Fig. 4. (a) "Invariant" values of corresponding minutiae in "1_1.tif" and "1_2.tif" in FVC2002 DB2 set A. The X axis is the minutiae index (22 in all) and the Y axis is the invariant value; (b) Hermit's Interpolation Polynomial is smoother than piece-wise linear interpolation.

In theory, m_i is translation and rotation invariant, but it changes a little sample to sample due to the presence of noise and feature extraction error. We compute all the corresponding minutiae's invariant values of image "1_1.tif" and "1_2.tif" in FVC2002 DB2 set A. Our minutiae extraction algorithm finds 22 couples of corresponding minutiae. Fig. 4(a) shows the difference of the invariant values computed in "1_1.tif" and "1_2.tif", using the same password(i.e., random generator). From Fig. 4(a), we can see that 18 couples of corresponding minutiae have "invariant" values with absolute error less than 0.2, and 12 couples have the ones with absolute difference less than 0.1. The tiny difference can be compensated in the following minutiae matching procedure, using bounding box matching method as used in [10].

2.2 Changing Function Design

Compared to [6], the biggest improvement that our design method makes lies in the interpolation method. We employ Hermit's Interpolation Polynomial, a smoother interpolation curve, to replace the piece-wise linear interpolation used in [6].

The transformation (translation and rotation) amount of each minutia is determined by the output of the changing function, whose input is the invariant value corresponding to the minutia. In theory, if the same "invariant" values are obtained from differently impressed fingerprint images, the transformation amount will be accordant. So the geometric relation between original fingerprint templates is preserved after transformation. That is why the encoding and decoding procedures of fuzzy fingerprint vault can be conducted in the transformed domain. Designing a group of changing functions which can derive coherent output from the invariant value is a challenging problem. Our change function is designed as follows:

1) Create two random number sequences X and Y, whose ranges are respectively limited in $[-\beta, -\alpha] \cup [\alpha, \beta]$ and $[-\eta, -\gamma] \cup [\gamma, \eta]$, using the user's password as seed;
2) Sum the outputs of X and Y as the control points of two changing function, respectively. The control points of distance change function(DCF) and angle change function(ACF) are generated as following equation (3):

$$\begin{aligned} L_{pwd}(nT) &= x_0 + x_T + ... + x_{(n-1)T} + x_{nT} = \sum_{i=0}^{n} x_{iT} \\ \Theta_{pwd}(nT) &= y_0 + y_T + ... + y_{(n-1)T} + y_{nT} = \sum_{i=0}^{n} y_{iT} \end{aligned} \tag{3}$$

3) Perform Hermit's Interpolation Polynomial to obtain the values of $L_{pwd}(p)$ and $\Theta_{pwd}(p)$ between $(k - 1)T$ and kT. Hermit's Interpolation Polynomial is smoother than linear interpolation used in [6]. Fig. 4(b) can show this point (typically, $\alpha = 5, \beta = 10, \eta = 5, \gamma = 10$). The contrastive minutiae maps before and after transformation of a fingerprint from FVC2002 DB2 are shown in Fig.5(b) and Fig.5(c). Fig.5(d) lays transformed minutiae map upon original minutiae map and shows that there are few couples of minutiae identified as corresponding.

2.3 Encoding

In the security-enhanced fuzzy fingerprint vault, besides the key concealed in the Vault, we also need a password, which is used to generating invariant value and the changing function, as elaborated in the two subsections above. So our scheme is somewhat two-factor authentication scheme: password+fingerprint. If and only if correct fingerprint and password(Key2) are presented, Key1 will be released, otherwise a reject signal is given. We will confirm this point in the subsequent experiments. The encoding procedure is as shown in Fig.1.

During vault encoding phase, we first extract all the minutiae and use minutiae quality estimation method, as employed in[10], to select 20-40 top-ranking and well-separated minutiae. At the same time, the invariant value corresponding to each minutia is computed using the Key1(password) and the ridge orientation based minutiae descriptor, hence the change function. Afterwards the selected minutiae are changed into the transformation domain by using the user-specific

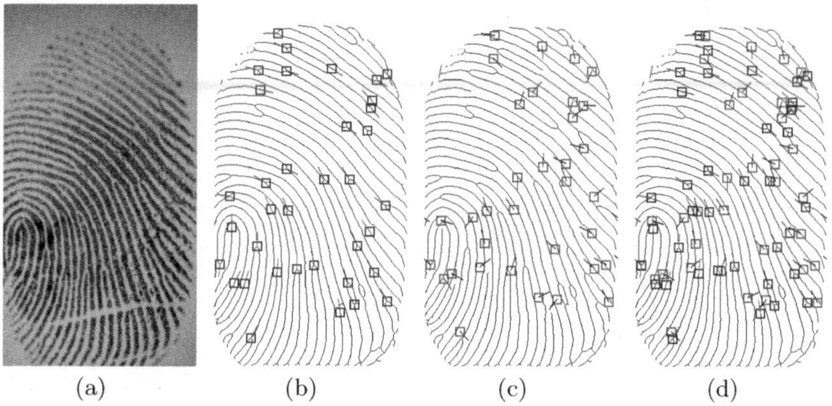

Fig. 5. (a) A fingerprint from FVC2002 DB2; (b) Minutiae map before transformation; (c) Minutiae map after transformation; (d) Transformed minutiae map overlays original minutiae map

changing function. Then we generate in the Galois Field $GF(2^{16})$ a polynomial f, whose coefficient is determined by the Key2 with length 16*n(n-1 is the polynomial degree). The transformed minutiae's location and orientation $x_i(i = 1, 2, ..., r)$ are encoded and projected in the polynomial f to obtain $f(x_i)$, and then $\{(x_i, f(x_i), i = 1, 2, ..., r\}$ are stored into the vault. A group of chaff points $\{(y_j, z_j)(j = 1, 2, ..., s)\}$, which don't lie in f, are added into the vault. Now the vault consisting $r+s$ elements has been created. In addition, topological structure based helper data extraction method [17] is used in our scheme and the helper data are also stored in encoding procedure to be used for automatic alignment in encrypted domain. Based on Li et al.[14]'s analysis, we replace the CRC-16 with SHA-1 check, that is to say, we compute *SHA-1(Key2)* and store it for key check in the decoding phase.

2.4 Decoding

In decoding phase(see Fig.2), a query fingerprint image is presented with the user's password. The same template transformation method is performed by using the query's invariant values to obtain Template(Q). Meanwhile, the query's helper data(Q) are also extracted and compared with helper data(T) to acquire the translation and rotation parameter, according to which Template(Q) is aligned to obtain Template(Q'). Afterwards a bounding-box minutiae matcher, as used in [10], is adopted to search the corresponding minutiae in the vault. Thus an unlocking set, in which each element consists $n(n-1$ denotes the polynomial degree) candidate points, is decoded to get the candidate $Key2'$. Then SHA-1($Key2'$) is computed and compared with SHA-1($Key2$). If these two hash values equal, it proves that $Key2' = Key2$ and the user's key is restored correctly, otherwise a failure will be reported.

3 Experiments

In order to validate the authentication performance and security of our improved fuzzy vault scheme, experiments are conducted under three scenarios: normal scenario(Scena.1), password-stolen scenario(Scena.2) and fingerprint-stolen scenario(Scena.3). The 1st and 2nd samples of each finger in FVC2002 DB2 Set A are selected for experiments because they have less transformation than other samples. The control point range parameters are set empirically and typically as: $\alpha = 5, \beta = 10, \gamma = 5, \eta = 10$, which makes a trade-off between authentication performance and security.

For Scena.1, assuming the attacker has no knowledge of the user's password, fingerprint minutiae templates from the same finger are transformed using the same password. For genuine test, the first sample is used for template and the second for query, which yields 100 trials in all. For imposter test, the first samples of the first 10 fingers are used for template respectively, the first sample of fingers, whose index is larger than the template finger, is used for query. This branch experiment yields 945 trials in all. The terms GAR(Genuine Accept Rate) and FAR(False Accept Rate) are used to indicate the performance. Table 1 gives the GAR and FAR corresponding to original fuzzy vault and our proposed scheme. Results show that the GARs of security-enhanced FV scheme have tiny decrease than original scheme, while FARs of two schemes approximately equals. In spite of this point, please note that the most valuable point of our proposed scheme lies in that it can eliminate the possibility of cross-matching between two vaults from the same finger. However it could satisfy the need of typical cryptography protocol, for example typical 128-bit AES key(n=8).

Table 1. Performance comparison of proposed security-enhanced FV scheme and the original FV scheme in Scena.1

	Degree	**7**	**8**	**9**	**10**	**11**
Proposed scheme	**GAR(%)**	89	88	86	84	82
	FAR(%)	0.07	0.03	0.04	0.03	0.02
Original scheme	**GAR(%)**	92	92	90	89	87
	FAR(%)	0.3	0.06	0.02	0.01	0

For Scena.2, given the user's password is stolen by the attacker, i.e., using the same password and fingerprints from different fingers to crack our scheme. We conduct 198 trials in all in this branch experiments, and the term CSR (Cracking Success Rate, the ratio of successful cracking times to total cracking times) is used to evaluate the security. Table 2 shows the security in this scenario. It can be seen that if the user's password is lost, the possibility of our proposed scheme being cracked is small. For Scena.3, assuming the user's original fingerprint information is stolen by the attacker and the user reissues a new vault using another transformation version of his/her fingerprint, i.e., using the original fingerprint minutiae information and a random selected key to crack the new issued vault. In

all 200 trials are conducted on selected fingerprint samples in this branch experiments, and the same term CSR is used to evaluate the result, which is shown in Table 3 with regard to different polynomial degrees. The result shows that if the user's fingerprint is stolen by the attacker and is used to attack the new issued vault, the success possibility is low enough to prevent this case. Combination of Scena.2 and Scena.3's results shows that, if and only if the correct password and the query fingerprints from the same finger as the template fingerprint are present simultaneously, the vault could be decoded successfully.

Table 2. CSR of proposed security-enhanced FV scheme in Scena.2

Degree	**7**	**8**	**9**	**10**	**11**
CSR(%)	0.5	0.5	1.0	0	0.5

Table 3. CSR of proposed security-enhanced FV scheme in Scena.3

Degree	**7**	**8**	**9**	**10**	**11**
CSR(%)	2.0	1.5	1.5	1.0	0.5

4 Security Analysis

In this section, we employ the min-entropy method, adopted in [16], to analyze the security of our proposed scheme. Assuming both minutiae location and orientation are uniformly distributed, the min-entropy of minutiae template M^T given the vault V can be computed as

$$H_\infty(M^T|V) = -log_2\left(\frac{\binom{r}{n+1}}{\binom{r+s}{n+1}}\right), \tag{4}$$

where, r, s and n denote number of minutiae, number of chaff points and polynomial's degree respectively and they are typically 20, 200, 8, respectively, so the typical security of our proposed scheme is approximately 34 bits in normal scenario. In password-stolen scenario and fingerprint-stolen scenario, the CSR is so small that the security under the two scenarios could not be affected. So we can conclude that the security measures of Scena.2 and Scena.3 both are approximately 34 bits. Because of the possibility of cross-matching's existence, previous schemes[7][8][9][10] usually can not achieve their claimed security. We eliminate the possibility of cross-matching to assure the security measure that our proposed scheme can achieve.

5 Conclusion

This paper employs the minutiae's local ridge information to improve the security of fuzzy fingerprint vault by eliminating the possibility of cross-matching between different vaults. Experimental results show that the authentication performance leads a tiny decrease. However, it can still satisfy the typical security need.

References

1. Jain, A.K., Nandakumar, K., Nagar, A.: Biometric Template Security. EURASIP Journal on Advances in Signal Processing 2008, Article ID 579416, 17 pages (2008)
2. Soutar, C., Roberge, D., Stoianov, A., Gilroy, R., Kumar, B.V.K.V.: Biometric Encryption, ICSA Guide to Cryptography. McGraw-Hill, New York (1999), http://www.bioscrypt.com/assets/Biometric_Encryption.pdf
3. Jin, A.T.B., Ling, D.N.C., Goh, A.: Biohashing: Two Factor Authentication Featuring Fingerprint Data and Tokenised Random Number. Pattern Recognition 37, 2245–2255 (2004)
4. Lumini, A., Nanni, L.: An Improved BioHashing for Human Authentication. Pattern Recognition 40, 1057–1065 (2007)
5. Juels, A., Sudan, M.: A Fuzzy Vault Scheme. In: Lapidoth, A., Teletar, E. (eds.) Proceeding of IEEE Int. Symp. Information Theory, p. 408 (2002)
6. Lee, C., Choi, J.Y., Toh, K.A., Lee, S.: Alignment-Free Cancelable Fingerprint Templates Based on Local Minutiae Information. IEEE Transactions on Systems, Man, and Cybernetics, Part B 37(4), 980–992 (2007)
7. Clancy, T.C., Kiyavash, N., Lin, D.J.: Secure Smartcard-based Fingerprint Authentication. In: Proceeding of ACMSIGMM 2003 Multimedia, Biometrics Methods and Applications Workshop, pp. 45–52 (2003)
8. Uludag, U., Pankanti, S., Jain, A.K.: Fuzzy Vault for Fingerprints. In: Proceedings of Fifth International Conference on AVBPA, Rye Town, USA, pp. 310–319 (2005)
9. Uludag, U., Jain, A.: Securing Fingerprint Template: Fuzzy Vault with Helper Data. In: Proceedings of the 2006 Conference on Computer Vision and Pattern Recognition Workshop (2006)
10. Nandakumar, K., Jain, A.K., Pankanti, S.: Fingerprint-Based Fuzzy Vault: Implementation and Performance. IEEE Transactions on Information Forensics and Security 2(4), 744–757 (2007)
11. Nandakumar, K., Nagar, A., Jain, A.K.: Hardening Fingerprint Fuzzy Vault Using Password. In: Lee, S.-W., Li, S.Z. (eds.) ICB 2007. LNCS, vol. 4642, pp. 927–937. Springer, Heidelberg (2007)
12. Chang, E.C., Shen, R., Teo, F.W.: Finding the Original Point Set Hidden among Chaff. In: Proceedings of the 2006 ACM Symposium on Information, computer and communications security, pp. 182–188. ACM Press, New York (2006)
13. Kholmatov, A., Yanikoglu, B.: Realization of Correlation Attack Against the Fuzzy Vault Scheme. In: Proceedings of 2008 SPIE / Biometrics, Security, Forensics, Steganography, and Watermarking of Multimedia Contents, vol. 7, pp. 68190O–68190O-7 (2008)
14. Li, Q., Liu, Z., Niu, X.: Analysis and Problems on Fuzzy Vault Scheme. In: Proceedings of 2006 International Conference on Intelligent Information Hiding and Multimedia Signal Processing, pp. 244–250 (2006)
15. Tico, M., Kuosmanen, P.: Fingerprint Matching Using An Orientation-based Minutia Descriptor. IEEE Transactions on Pattern Analysis and Machine Intelligence 25(8), 1009–1014 (2003)
16. Nagar, A., Nandakumar, K., Jain, A.K.: Securing Fingerprint Template: Fuzzy Vault with Minutiae Descriptor. In: International Conference of Pattern Recognition (to appear) (2008)
17. Li, J., Tian, J., Yang, X., Shi, P., Li, P.: Topological Structrue based Fuzzy Vault Alignment Method. In: International Conference of Pattern Recognition (to appear) (2008)

Systematic Construction of Iris-Based Fuzzy Commitment Schemes*

Christian Rathgeb and Andreas Uhl

University of Salzburg, Department of Computer Sciences, A-5020 Salzburg, Austria
{crathgeb,uhl}@cosy.sbg.ac.at

Abstract. As a result of the growing interest in biometrics a new field of research has emerged entitled *Biometric Cryptosystems*. Only a small amount of work, which additionally tends to be custom-built according to the specific application context, has been published in this area. This work provides a systematic treatment of how to construct biometric cryptosystems based on iris biometrics. A cryptographic primitive called *Fuzzy Commitment Scheme* is adopted to different types of iris recognition algorithms to hide and retrieve a cryptographic key in and out of a biometric template. Experimental results confirm the soundness of the approach.

1 Introduction

Taking into account today's ever-increasing demand on high security standards, in order to secure any kind of crucial information, the science of cryptography has become even more important. While in generic cryptographic systems authentication is possession based [1], key management is performed introducing alternative authentication mechanisms such as password or PIN.

By introducing biometrics to replace password-based authentication the security of cryptographic systems is improved. Several approaches have been made to combine biometric authentication with key management systems to build up so-called "biometric cryptosystems", which are classified by the way biometric authentication is merged with the respective cryptosystem. The trivial way of introducing biometric authentication into a generic key management system, replacing password/PIN-based authentication through biometric authentication, is called "key release scheme". Key release schemes are easy to implement, still these are not frequently used. Within such schemes biometric templates as well as cryptographic keys, which are not secure, are stored in a database separately. This is a very critical issue because biometric templates and cryptographic keys can be stolen or compromised. Thus, a biometric cryptosystem based on a key release scheme is not appropriate for high security applications. The second class of biometric cryptosystems includes "key generation schemes" and "key binding schemes". Key generation schemes directly derive cryptographic keys from biometric data. However, within key generation schemes a cryptographic key cannot

* This work has been supported by the Austrian Science Fund, project no. L554-N15.

M. Tistarelli and M.S. Nixon (Eds.): ICB 2009, LNCS 5558, pp. 940–949, 2009.
© Springer-Verlag Berlin Heidelberg 2009

be changed if it is compromised once. The most promising types of biometric cryptosystems are key binding schemes. By seamlessly binding a cryptographic key with biometric information via a key binding algorithm, secure templates are provided which do not reveal any information about the biometric data, nor about the cryptographic key. With an appropriate key retrieval algorithm keys are released again.

Juels and Wattenberg [2] proposed a theoretical basis for biometric key binding schemes that they refer to as "fuzzy commitment scheme" (FCS). Since the iris is one of the most accurate biometric characteristic [3,4] it is desirable to apply the fuzzy commitment approach to iris biometrics. However, until now only little literature has been published concerning iris-based biometric cryptosystems. This work will provide a systematic approach of how to build up iris-based FCSs. Furthermore, two different iris recognition algorithms [5,6] will be used to demonstrate the construction of different types of iris-based FCSs.

This paper is organized as follows: first a short summary of previous work concerning iris-based biometric cryptosystems will be given, where the fundamentals of a FCS are examined in detail (Sect. 2). Subsequently a generic approach of how to construct iris-based FCS is presented (Sect. 3) which is then demonstrated by applying it to two different types of iris recognition algorithms (Sect. 4). Finally experimental results are presented and discussed (Sect. 5, 6).

2 Iris-Biometric Cryptosystems

In the past several years some key-papers have been published concerning biometric cryptosystems [7,8,9,10,11,12,13,14,15]. Several biometric characteristics, including fingerprints, voice, etc., have been examined for the extraction of cryptographic keys. Still, only a few of these approaches focus on iris biometrics [8,9,10].

Davida *et al.* [8,9] were the first to create a key generation scheme that they refer to as "private template scheme". Within their approach a hashed value of preprocessed iris codes and user specific attributes serves as a cryptographic key. The result of these preprocessed iris codes is concatenated with check digits which are part of a linear ECC. This ECC is capable of correcting a fixed number of errors defined at system setup. At the time of authentication the error correction information, which is stored as part of the template, is used to correct faulty bits in the acquired biometric data. Finally the same hash function as in the registration step is applied to generate a hash which can be used as cryptographic key. Unfortunately, performance measurements and test results are renounced.

According to the idea of the private template scheme Wu *et al.* [16] proposed a system in which 256-dimensional feature vectors are extracted out of preprocessed iris images using a set of 2-D Gabor filters. A hash function is applied to this vector to generate a cryptographic key. Additionally, an ECC is generated. During authentication another feature vector is extracted from a biometric input.

This feature vector is error correction decoded and the same hash function like in the encryption phase is used to generate a cryptographic key. The extracted cryptographic key is suggested to be used in a symmetrical cryptosystem. For a total number of over 100 persons a FRR of approximately 5.55% and a zero FAR are reported.

In order to construct biometric cryptosystems based on the key binding approach Juels and Wattenberg [2] combined well-known techniques from the areas of ECCs and cryptography to achieve a type of cryptographic primitive called FCS. Fuzzy commitment is the analogon to "fuzzy logic" in artificial intelligence. In their definition a FCS consists of a function F, which is used to commit a codeword $c \in C$ and a witness $x \in \{0,1\}^n$. The set C is a set of error correcting codewords c of length n and x represents a bitstream of length n termed witness (in a biometric cryptosystem x represents the biometric data). To enhance security only the difference vector of the codeword and the biometric measurement, $\delta \in \{0,1\}^n$ where $x = c+\delta$, and a hash value $h(c)$ are stored as the commitment. The commitment, which is nothing else then these two values is termed $F(c,x)$. To deal with the fuzziness of x it is proposed that every x', which is sufficiently "close" to x should be able to reconstruct c. If the system is presented with a witness x' that is near x, the difference vector δ is used to translate x' in direction of x. If the correct codeword c is reconstructed with the use of error correction the hash of c', $h(c')$ will match the stored hash value resulting in a successful authentication. The enrollment and authentication process within a FCS operates as follows: during enrollment a user U presents a witness x to the authentication system S. The system selects a codeword $c \in C$ (in a biometric cryptosystem c represents a cryptogaphic key prepared with error correction information), calculates the fuzzy commitment $F(c,x)$ (the difference vector δ and the hash value of the codeword c, $h(c)$) and stores it in a database. At the time of authentication a user purporting to be U presents a witness x' to S. The system looks up the commitment of user U and checks whether x' yields a successful decommitment, which would lead to a successful authentication. In Fig. 1 the basic working flow of a FCS, with respect to the use in a biometric cryptosystem, is illustrated.

Hoa *et al.*[10] applied the FCS to iris biometrics. In their approach a 140-bit cryptographic key is encoded with a concatenation of ECCs and subsequently `XOR`ed with a 2048-bit iris code to generate a secure template. During authentication another iris code is extracted from the person and `XOR`ed with the template. Finally error correction decoding is performed on the resulting bitstream to regenerate the cryptographic key. By applying a concatenation of ECC a remarkable FRR of 0.47% and a zero FAR were reported for a total number of 70 different persons. Still their approach does not clarify the construction of a generic FCS with respect to required precondition as well as the choice of ECCs.

Up to the present there are no other achievements concerning iris-based biometric cryptosystem which are worth mentioning. Thus the motivation of this work is to provide a systematic approach of how to construct iris-based key

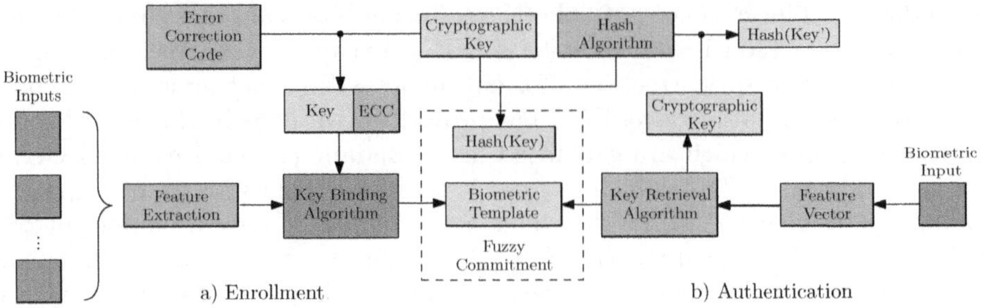

Fig. 1. a) The basic enrollment procedure in a fuzzy commitment scheme. b) The basic authentication procedure in a fuzzy commitment scheme.

binding schemes based on FCSs. In the following section the theoretical basis of constructing iris-based FCSs will be defined.

3 Construction of Iris-Based FCSs

Building up iris-based FCSs first of all preconditions have to be declared: to create a FCS for the use in a biometric cryptosystem the applied iris recognition algorithm should produce an order invariant bitstream (FCSs cannot handle order variant bitstreams). Furthermore, this bitstream should be as long as the cryptographic key concatenated with error correction bits, where the error correction bits provide the information to correct the estimated number of errors between iris data of the same person.

Since ECCs should be able to correct the number of errors occurring between legitimate persons, the next step is to analyze the maximal intra class distance (ICD) of the applied iris recognition algorithm, according to the particular bit block sizes (by analysing ICD of typical and large data sets). Errors between biometric measurements are not distributed uniformly random which implies the usage of block-level ECCs as first layer of error correction is inevitable to handle burst errors. Block level ECCs are codes capable of correcting blocks of bits in which errors occur, while the error correction information is provided by other bit blocks. Experience has shown Reed-Solomon codes are a suitable choice for block-level error correction. Named after I. Reed and G. Solomon, a Reed-Solomon code is defined as $RS(l, n)$ which means that l blocks of of length m are encoded by adding $l - n$ blocks of length m resulting in a total number of $n \cdot m$ bits, where $n \leq 2^{m-1}$ is required. Redundant information is added by oversampling a polynomial constructed from the l information blocks. If less than $(n - l)/2$ blocks are faulty after transmission the polynomial can be reconstructed which means the l information bit blocks can be reconstructed. Further details about Reed-Solomon codes can be found in [18,19].

If the maximal ICD of the applied iris recognition algorithm lies beyond the number of bit blocks, which the applied block-level ECC is capable to correct, another layer of error correction has to be inserted, since simply adding more

redundant bit blocks is not possible (the resulting bitstream must have the same size as the extracted iris code). Bit-level ECCs are capable of correcting single bit errors, while the error correction information lies within each codeword. In practical use one type of bit-level ECC has proved worth, namely Hadamard codes. Hadamard codes, which are generated using Hadamard matrices, are ECCs of the type $[2^n, n+1, 2^{n+1}]$, which means bitstreams of length $n+1$ are mapped to codewords of length 2^n while the whole code consists of a total number of 2^{n+1} codewords. A Hadamard matrix H_n of dimension $n \times n$ generates a Hadamard code consisting of $2n$ codewords, each of length n, capable of correcting up to $n/4 - 1$ errors. Further details about Hadamard codes and Hadamard matrices can be found in [17].

In Fig. 2 the encoding/decoding flow of concatenated error correction is illustrated. It is essential that the block-level ECC and the bit-level ECC operate on the same bit blocks, otherwise one faulty block in the bit level could cause several faulty blocks in the block level during decoding. In the encoding step first a block-level ECC is applied and subsequently a bit-level ECC. Thus, in the decoding step, the bit-level ECC corrects single bit errors and the block-level ECC corrects remaining burst errors. If the the maximal ICD is still too large after bit-level error correction, decoding the block-level ECC will not be able to regenerate a hidden key. If this is the case the applied iris recognition algorithm is not adequate to be used in a FCS. Otherwise the parameters of the applied bit-level ECC and the applied block-level ECC have to be adjusted.

In the following section two different iris recognition algorithms will be analyzed and ECCs will be adapted to build up FCSs which are capable of hiding and retrieving cryptographic keys, sufficiently long to be used in generic cryptosystems.

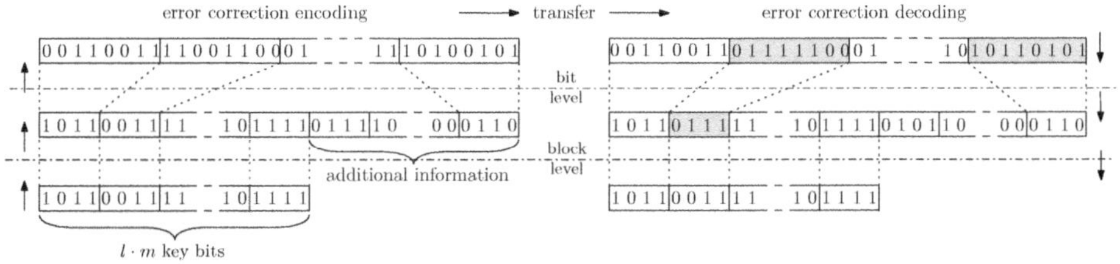

Fig. 2. The concatenation of block-level ECCs and bit-level ECCs (faulty bit blocks are marked gray)

4 Proposed Schemes

For the proposed FCSs two different implementations of iris recognition algorithms are applied: The first implementation is based on a algorithm published by Ma *et al.* which is invariant to translation, scale and rotation. In this approach the iris texture is treated as a kind of transient signal which is processed using wavelet transform. The local sharp variation points, which denote important properties of transient signals, are recorded as features to extract binary iris codes of 1280 bytes. Further details about this algorithm can be found in [5].

The second implementation is based on a algorithm published by Ko *et al.* which uses cumulative sum based change analysis to analyze preprocessed iris textures. Enhanced iris textures are divided into cells out of which mean gray scale values are calculated and furthermore, an iris code $I \in \{0, 1, 2\}^{1000}$ is extracted, using the suggested parameters for the calculation of cumulative sums. Further details about this algorithm can be found in [6].

It will be shown the both types of extracted iris codes are sufficiently long to be used in a FCS. For bit-level error correction Hadamard codes are applied and for block-level error correction Reed-Solomon codes are applied.

To choose an adequate bit block size for block-level error correction the maximal ICDs of both algorithms are estimated, according to the respective block sizes, which are summarized in Table 1. Within the iris code of the algorithm of Ko *et al.* a sequence of 1s indicates an upward slope of cumulative sums of gray scale values and a sequence of 2s indicates a downward slope. This code is simply mapped to a binary code of twice the length. Subsequently the resulting bitstream is rearranged so that the first half of the iris code contains all upward slopes and the second half of the iris code contains all downward slopes. Thus for each part of the resulting code a sequence of 1s suffices to indicate the respective slope. Thereby the number of block errors between different iris codes is minimized. In contrast, Fig. 5 shows the distribution of genuine persons and imposters without the arrangement of the bitstream, according to a blocksize of 8 bit. Analyzing the maximal ICDs of the applied algorithms it is assumable that for the algorithm of Ko *et al.* a single layer of block-level error correction suffices, while for the algorithm of Ma *et al.* a second layer of bit-level error correction is added. This is because within the algorithm of Ko *et al.* mostly burst errors ocurr (the maximal ICD increases only slightly).

Table 1. The maximal intra class distances of the iris codes, using the algorithm of Ma *et al.* and Ko *et al.*, according to the size of faulty bit blocks

Bit Block Size	Max. ICD Ma *et al.* (%)	Max. ICD Ko *et al.* (%)
1	48.3	14.2
2	53.3	21.2
4	67.2	31.0
6	79.5	38.5
8	85.1	43.6

In view of the maximal ICDs of both algorithms, according to the block sizes, ECCs are adapted. As mentioned above, for the algorithm of Ma *et al.* a concatenation of ECCs is applied. It will be shown that a blocksize of $m = 8$ is a suitable choice. To bind and retrieve a sufficiently long key K, the size of the key is set to $|K| = l \cdot m$, where $l = 16$, so that the key consists of 128 bits (note that the key size has to be a multiple of m). Since the entire encoding procedure should produce a bitstream of length $1280 \cdot 8$ bits output size of the Reed-Solomon code, denoted by RS_{res} which is calculated by $|RS_{res}| = 1280 \cdot 8/2^{m-1}$ because the

Hadamard code maps bit blocks of size m to bit blocks of size 2^{m-1}. Thus, for a cryptographic key K, of length $l \cdot m$, Reed-Solomon block-level error correction is defined by $RS(l, 1280 \cdot 8/2^{m-1})$ which means l m-bit information blocks are encoded using a total number of $1280 \cdot 8/2^{m-1}$ bits, which are mapped to $1280 \cdot 8$ bits applying Hadamard encoding. In summary, for the algorithm of Ma *et al.* first a $RS(16, 80)$ block-level ECC is applied and afterwards a Hadamard code which maps the resulting 80 8-bit blocks to 80 128-bit blocks (=1280·8 bits).

For the algorithm of Ko *et al.* the encoding step is trivial. According to Table 1, the maximal ICD is still less than 44% for a block size of $m = 8$. In other words a single layer of block-level error correction does suffice. By mapping the extracted iris code to a binary code and rearranging it, a bitstream of 2000 bits is extracted. By setting the key size to 128 bit, to generate 250 8-bit blocks (=2000 bits), the block-level ECC is defined by $RS(l, 2000/m)$. A single use of bit-level error correction makes no sense for both schemes. While in the algorithm of Ma *et al.* the maximal ICD lies far beyond 25% in the algorithm of Ko *et al.* occurring errors are not distributed uniformly random. This means, applying Hadamard codes to the algorithm of Ko *et al.* the maximal number of occurring bit errors within 128-bit blocks has to be corrected, since a block size of $m = 8$ would be suitable for mapping a 128-bit key to a sufficiently long bitstream. However, the maximal number of bit errors within 128-bit blocks lies beyond $128/4 - 1 = 31$ bits as illustrated in Fig. 4. For the algorithm of Ma *et al.* two layers of error correction are adequate because, according to Table 1, the maximal ICD lies far beyond 50% for suitable block sizes. In contrast, for the algorithm of Ko *et al.* the maximal ICD is clearly below 50% implying the introduction a layer of bit-level error correction is unnecessary.

In the enrollment procedure of the proposed schemes three iris images are preprocessed as described in [4]. The resulting 512×64 pixel iris textures are slitted from the right side [45° to 315°] and from the left side [135° to 225°] to get rid of most of the eyelids and eyelashes, according to the idea in [6]. The three extracted iris codes are majority decoded and gaps of 1s and 0s of the resulting bitstream are filled. The entire enrollment procedure is illustrated in Fig. 3. A cryptographic key is error correction encoded and XORed with the enrollment iris code to create the commitment. At the time of authentication a single iris image is preprocessed, slitted and the according iris recognition algorithm is applied. The resulting iris code is XORed with the stored commitment and error correction decoding is performed. If the error correction decoding succeeds a correct key is returned, which is sufficiently long to be used in a generic cryptosystem.

5 Experimental Results

For the performance evaluation of the proposed schemes a subset of the CASIA-IrisV3-Interval database [20] is used, where for each person at least 8 iris images are available, which makes a total number of about 100 different persons. The first three images of each person were used for the enrollment procedure and the remaining five images were tested against the stored templates.

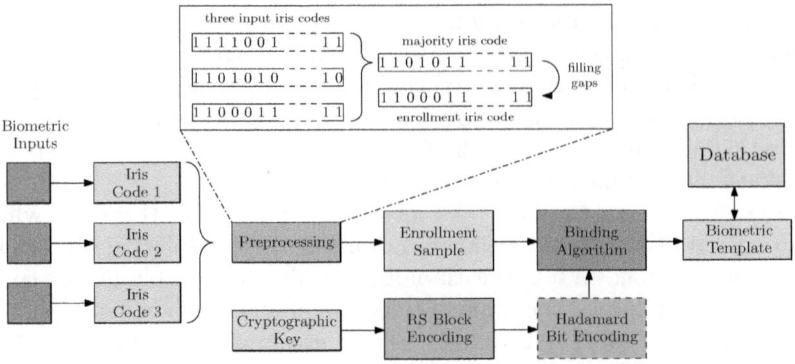

Fig. 3. The entire enrollment procedure: Iris codes are extracted from preprocessed iris images according to the applied algorithm and bound with encoded cryptographic keys

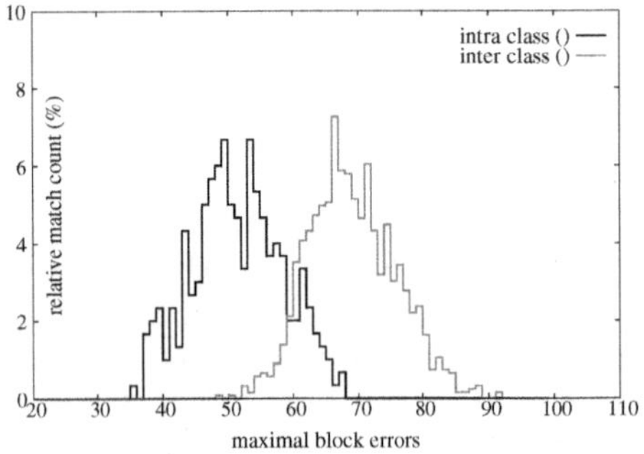

Fig. 4. The distribution of the maximal number of errors within 128-bit block for the algorithm of Ko *et al.*

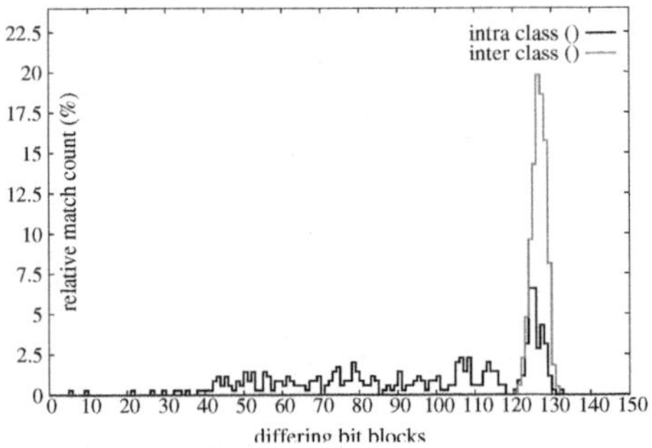

Fig. 5. The distribution of the total number of 8-bit block level errors without rearranging the bits for the algorithm of Ko *et al.*

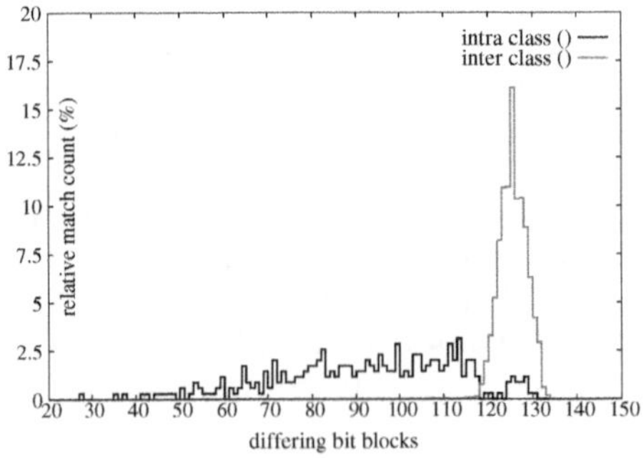

Fig. 6. The distribution of the total number of 8-bit block level errors of the proposed fuzzy commitment scheme, using the algorithm of Ko *et al.*

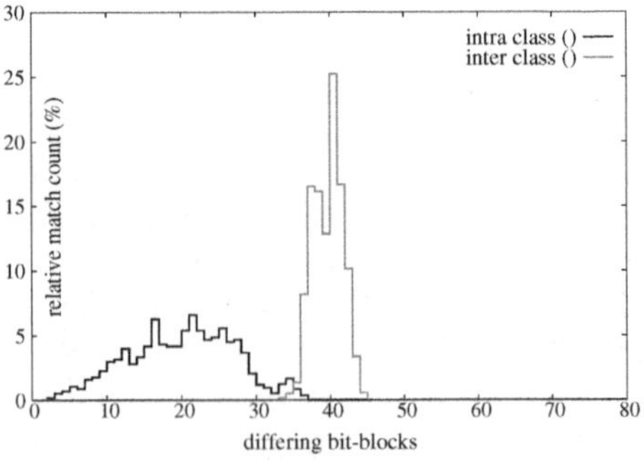

Fig. 7. The distribution of the total number of 8-bit block level errors remaining after Hadamard decoding of the proposed fuzzy commitment scheme, using the algorithm of Ma *et al.*

The performance of each proposed schemes is described by its FRR and FAR. In contrast to common recognition systems, in biometric cryptosystems the FRR is the ratio between truly matching samples for which faulty keys are generated and the total number of tests. By analogy, the FAR describes the percentage of truly non-matching samples for which correct keys are returned. In order to avoid returning faulty keys a hash of the constructed key could be tested against a previously stored hash of the correct key.

In Fig. 7 the intra class and the inter class distance of the FCS, which uses the algorithm of Ma *et al.* is shown, according to the differing bit blocks after Hadamard decoding. While the Hadamard code corrects bit blocks containing less that 25% bit errors, the applied Reed-Solomon code is capable of correcting a total number of (80-16)/2=32 block-level errors resulting in a zero FAR and a FRR of 4.64%. The intra class and the inter class distance according to the differing bit blocks of the FCS, for which the algorithm of Ko *et al.* is applied is shown in Fig. 6 (compare Fig. 5 for the results without bit rearrangement). The Reed-Solomon code is capable of correcting (250-16)/2=117 block-level errors which results in a of FAR 0.08% and a FRR of 6.57%.

6 Summary

Until now only little work has been published concerning biometric cryptosystems which use the iris as biometric characteristic. Additionally, the published literature is mostly custom-built according to the area of application. In contrast, this work shows a generic approach of how to build up iris-based biometric cryptosystems by applying a cryptographic primitive called FCS. Different types of iris recognition algorithms, for which defined preconditions are fulfilled, are used to construct two FCSs in a systematic manner. Experimental results demonstrate the soundness of the proposed approach.

References

1. Uludag, U., Pankanti, S., Prabhakar, S., Jain, A.K.: Biometric cryptosystems: issues and challenges. Proceedings of the IEEE 92(6), 948–960 (2004)
2. Juels, A., Wattenberg, M.: A FCS. In: Sixth ACM Conference on Computer and Communications Security, pp. 28–36 (1999)
3. Bowyer, K., Hollingsworth, K., Flynn, P.: Image understanding for iris biometrics: a survey. Computer Vision and Image Understanding 110, 281–307 (2008)
4. Daugman, J.: How Iris Recognition Works. IEEE Trans. CSVT 14(1), 21–30 (2004)
5. Ma, L., Tan, T., Wang, Y., Zhang, D.: Efficient Iris Recogntion by Characterizing Key Local Variations. IEEE Transactions on Image Processing 13(6), 739–750 (2004)
6. Ko, J.-G., Gil, Y.-H., Yoo, J.-H., Chung, K.-I.: A Novel and Efficient Feature Extraction Method for Iris Recognition. ETRI Journal 29(3), 399–401 (2007)
7. Soutar, C., Roberge, D., Stoianov, A., Gilroy, R., Kumar, B.V.: Biometric Encryption - Enrollment and Verification Procedures. In: Proc. SPIE, Optical Pattern Recognition IX, vol. 3386, pp. 24–35 (1998)

8. Davida, G., Frankel, Y., Matt, B.: On enabling secure applications through off-line biometric identification. In: Proc. of IEEE Symp. on Security and Privacy, pp. 148–157 (1998)
9. Davida, G., Frankel, Y., Matt, B.: On the relation of error correction and cryptography to an off line biometric based identication scheme. In: Proc. of WCC 1999, Workshop on Coding and Cryptography, pp. 129–138 (1999)
10. Hao, F., Anderson, R., Daugman, J.: Combining Cryptography with Biometrics Effectively. IEEE Transactions on Computers 55(9), 1081–1088 (2006)
11. Monrose, F., Reiter, M.K., Wetzel, S.: Password hardening based on keystroke dynamics. In: Proceedings of sixth ACM Conference on Computer and Communications Security, CCCS, pp. 73–82 (1999)
12. Monrose, F., Li, Q., Reiter, M.K., Wetzel, S.: Cryptographic Key Generation from Voice. In: SP 2001: Proceedings of the 2001 IEEE Symposium on Security and Privacy, 12 pages (2001)
13. Dodis, Y., Ostrovsky, R., Reyzin, L., Smith, A.: Fuzzy Extractors: How to Generate Strong Keys from Biometrics and Other Noisy Data. In: Cachin, C., Camenisch, J.L. (eds.) EUROCRYPT 2004. LNCS, vol. 3027, pp. 523–540. Springer, Heidelberg (2004)
14. Ratha, N.K., Connell, J.H., Bolle, R.M.: Enhancing security and privacy in biometrics-based authentication systems. IBM Systems Journal 40, 614–634 (2001)
15. Vielhauer, C., Steinmetz, R., Mayerhöfer, A.: Biometric hash based on statistical features of online signatures. In: ICPR 2002: Proceedings of the 16th International Conference on Pattern Recognition (ICPR 2002), vol. 1, p. 10123 (2002)
16. Wu, X., Qi, N., Wang, K., Zhang, D.: A Novel Cryptosystem based on Iris Key Generation. In: Fourth International Conference on Natural Computation (ICNC 2008), pp. 53–56 (2008)
17. Agaian, S.S.: Hadamard Matrix and Their Applications. Lect. notes in math., vol. 1168. Springer, Heidelberg (1985)
18. Reed, I., Solomon, G.: Polynomial codes over certain finite fields. Journal of the Society for Industrial and Applied Mathematics 8, 300–304 (1960)
19. Berlekamp, E.: Factoring Polynomials Over Finite Fields. Bell Systems Technical Journal 46, 1853–1859 (1967)
20. The Center of Biometrics and Security Research, CASIA Iris Image Database, `http://www.sinobiometrics.com`

Parallel versus Serial Classifier Combination for Multibiometric Hand-Based Identification

Andreas Uhl and Peter Wild

Department of Computer Sciences,
University of Salzburg, A-5020 Salzburg, Austria

Abstract. This paper presents an approach for optimizing both recognition and processing performance of a biometric system in identification mode. Multibiometric techniques facilitate bridging the gap between desired performance and current unimodal recognition rates. However, traditional parallel classifier combination techniques, such as Score sum, Borda count and Highest rank, cause further processing overhead, as they require a matching of the extracted sample with each template of the system for each feature. We examine a framework of serial combination techniques, which exploits ranking capabilities of individual features by reducing the set of possible matching candidates at each iteration, and we compare its performance with parallel schemes. Using this technique, both a reduction of misclassification and processing time in identification mode will be shown to be feasible for a single-sensor hand-based biometric system.

Keywords: Multibiometrics, serial combination, hand biometrics.

1 Introduction

Recently, a number of biometric systems have exploited advantages of multi-instance and multi-algorithm approaches to biometric recognition based on single-sensor input. In particular, such systems benefit of increased matching accuracy, more difficult biometric attacks, reduced enrollment errors in case of failures to extract single biometric traits and, finally, a less complex acquisition procedure [1,2,3]. For single-sensor fusion of iris and face a combined feature achieved 99.75% Genuine Acceptance Rate (GAR) compared to 97.35% GAR for face and 95.42% GAR for iris at 0.1% False Acceptance Rate (FAR) in [1]. A similar result was obtained for fusion of hand-based modalities with reported perfect classification compared to 91.5% GAR at 0.01% FAR for palmprints only and 95-99.9% GAR for individual fingers in [2] and good separation with 0.08% Total Error Rate (TER) compared to 0.24% TER for the best single feature (Minutiae) in [3]. All these studies combined matching scores by employing min-max score normalization and (weighted) sum of scores. While this technique seems to be a good choice outperforming many other alternatives [4,5], processing time requirements are not optimized.

In a system with m biometric matchers M_i, $1 \leq i \leq m$ in identification mode, each sample B is matched against the whole system database $D =$

M. Tistarelli and M.S. Nixon (Eds.): ICB 2009, LNCS 5558, pp. 950–959, 2009.
© Springer-Verlag Berlin Heidelberg 2009

$\{T_1, T_2, \ldots T_t\}$. In case of parallel fusion methods based on the outcome of comparisons $M_i(B, T_j)$, an identification involves $m \cdot t$ matches, i.e. for the total processing time T_p in a single-processor environment we have:

$$T_p = \sum_{j=1}^{t} (C_p + \sum_{i=1}^{m} P_{ij}) \tag{1}$$

where P_{ij} denotes the processing time required for the comparison $M_i(B, T_j)$ and C_p is the time needed for parallel combination. Even if C_p is insignificant, still this term increases linear in the number of users t with a potentially large factor m. You et al. [6] identified this problem and proposed a 4-level hierarchical multifeature coding scheme to facilitate coarse-to-fine matching for palmprints. In this scheme, levels 1-3 reduce the initial set $D_1 := D$ of possible candidates (leading to candidate sets $D_{i+1} \subset D_i \; \forall i < m$) and level 4 performs final matching on D_m. With the preclassification capabilities of fast geometry-based features and final decision by an accurate (but costly) texture-based feature, processing time compared to sequentially executed parallel schemes were reduced from 6.7 seconds to 2.8 seconds in [6]. A similar idea was used in [7] for face detection, where simpler, yet fast classifiers rejected a majority of sub-windows before more complex, but accurate cascaded classifiers guaranteed a low false detection rate.

We extend the approach in [6] and formulate a framework of serial combinations, which, apart from the reduction of candidates, introduces additional information flow between levels by accumulating matching information according to the fusion rule. Serial processing time T_s can be estimated as follows:

$$T_s = \sum_{i=1}^{m} (C_s(i) + \sum_{j \in I(D_i)} P_{ij}) \tag{2}$$

where $I(D_i)$ denotes the index set of templates in D_i and $C_s(i)$ is the time needed for serial combination at level i. With this modification, as will be shown, both a reduction of processing time and increased accuracy can be achieved at the same time. Another contribution of this work is a performance comparison with respect to processing time and accuracy of well-known fusion rules Sum, Borda count and Highest rank in parallel form and (to be defined) serial application using a hand-based multibiometric system. The third contribution of this work is an assessment of individual Minutiae and Eigenfinger-based classifiers in identification mode when operating on single-sensor data, as motivated in [8]. Finally, we present an optimal solution for the employed hand-biometric system minimizing a cost function $C = \omega_1 * (1 - RR) + \omega_2 * T$ with RR being the Rank-1 Recognition Rate, T the processing time, and ω_1, ω_2 weights for misclassification and processing costs.

This paper is structured as follows: An introduction to system design for both parallel and serial setups and employed features is given in Sect. 2. Section 3 presents an overview of implemented parallel and serial combination methods. Employed biometric databases, performance indicators and experimental results are summarized and discussed in Sect. 4. Finally, Sect. 5 forms the conclusion.

2 System Setup

In order to evaluate both serial and parallel classifier combination, our multi-biometric identification system [3] employs exclusive modules for each of the examined combination schemes, see Fig. 1.

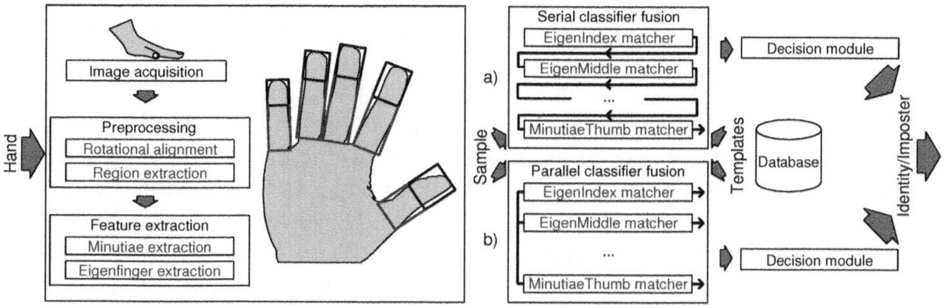

Fig. 1. Experimental setup using a) serial and b) parallel combinations of classifiers

2.1 Proprocessing

At image acquisition stage, each palmar hand is sensed using a HP Scanjet 3500c flatbed scanner sensor. After binary segmentation using Otsu's method and localization of the hand within the image, visible parts of the arm are removed. Translational and rotational invariance are achieved using an iterative algorithm, which aligns each hand based on a hand coordinate system. Its origin is defined as the inter-finger valley between ring and middle finger and the direction of the least-squares approximated outer palm boundary is used for alignment. This method provides better results than traditional moment-based best-ellipse fitting [9], as it is less degraded by different spreadings of fingers, partially visible arms or jewelery in the image. However, if applied to individual fingers, moment-based fitting can improve inter-finger valley and finger tip detection rates.

2.2 Feature Extraction

Results of the preprocessing stage are rectilinear finger images at 100 dpi resolution and contrast-limited adaptive-histogram equalized high-resolution 500 dpi images of the finger tips, illustrated as bounding boxes aligned to each finger in Fig. 1. The task of feature extractors, operating each on a single region, is to extract a feature vector representation. Each extractor contributes its representation to a common augmented template, individual matchers are designed to extract the corresponding features from the common representation and return a scalar value for each match, as defined in Sect 1. Finally, the first rank of the combined ranking vector is used for identity determination within the decision module.

Minutiae Feature Extraction and Matching. Ridge bifurcation and termination points (position, orientation and quality) are extracted from finger-tip images using NIST's *mindtct* [10]. Main processing steps for Minutiae extraction comprise [11]: generation of image maps (e.g. ridge orientation maps), Minutiae detection (by local binarization and pattern matching), and filtering (removing false Minutiae in specific configurations). As Minutiae sets are not pre-aligned by this feature extractor, an optimal pairing between Minutiae sets has to be performed at matching stage (we use NIST's *bozorth3* [10] matcher for this task). Unfortunately this task requires significantly more processing time than just distance-based matching. The processing overhead becomes even more visible in our experiment, as we employ *bozorth3* in 1:1 comparison mode.

Eigenfingers Feature Extraction and Matching. For the Eigenfingers feature, which is based on principal component analysis, we adopt the algorithm proposed by Turk and Pentland [12], which also works well for fingers [13]. Feature extraction requires a pre-calculation of a so-called eigenspace for each finger type, spanned by the most significant eigenvectors obtained from the covariance matrix of mean-normalized training samples (we used a separate dataset of 25 hand-prints and also selected 25 as feature size). The feature vector is composed of the projection coefficients of the normalized finger images projected onto their eigenspace. Therefore, it represents low-resolution 128×384 index, middle, ring fingers and 128×256 thumb and little fingers in an optimal way. In order to obtain matching scores, Manhattan distance is employed in the feature space.

2.3 Matching

Matchers can be distinguised into different groups depending on the type of information provided [14]:

1. **Measurements:** $M_i(B, D) = S_i \in \mathbb{R}^t$ is a measurement vector with its j-th element $S_i[j]$, $1 \leq j \leq t$, indicating the degree that B and T_j correspond to the same identity.
2. **Ranks:** $M_i(B, D) = I_i \in \mathbb{N}^t$ is a rank index vector, where $I_i[j] = x$ indicates, that the template $T_x \in D$, $1 \leq x \leq t$ is ranked at the j-th position. Closely related to the index vector is its corresponding rank vector R_i satisfying:

$$\forall j, x \ with \ 1 \leq j, x \leq t : I_i[j] = x \Leftrightarrow R_i[x] = j. \tag{3}$$

3. **Labels:** in case of fully-automatic systems, only the top first rank or class with the highest score is relevant.

In this paper, we focus on combination schemes at rank and measurement level (see also [15]). Furthermore, we conduct both open-set (with *reject* option [16]) and closed-set identification (system users are identified only) experiments using different ranking dimensions. Since the reject decision is facilitated by thresholding single matching results and returning only classes where the corresponding matching score exceeds a predefined threshold, we interpret a rank index vector with dimensionality zero as a reject.

3 Parallel and Serial Classifier Combinations

In our biometric identification experiments, individual classifiers are *distance classifiers* in the context of [14], i.e. each matcher calculates a distance or similarity between the given template B and each enrolment-template T_j of the system database, see Sect. 1. Following the transformation-based fusion approach [15], distance scores $y = M_i(B, T_j)$ are at first converted into similarity scores by subtraction from an empirically found maximum. Resulting values y' are then mapped to the unit interval $[0, 1]$ using *min-max* normalization [15]. Since normalized scores may be interpreted as some kind of posterior probability $P(B \equiv T_j|B)$ [14], individual matchers may be incorporated in each of the introduced combination strategies *measurement* (using the normalized score vector), *rank* (by simply ranking measurements) and *abstract*-level (using the Bayes-rule estimating the class with the highest posterior probability [14]).

Combining measurements, ranks or labels of individual classifiers has long tradition in pattern recognition. A comprehensive overview of classifier combination techniques including Average Bayes Classifiers, Voting-based schemes and classifiers based on Dempster-Shafer Theory is given in [14]. Combination rules are derived in [5] using Bayesian statistics. A framework for generalized classifier combinations is introduced in [17], with Borda count, Logistic regression and Highest rank combination methods as special cases. According to [18], classifier combination schemes can be divided into *fixed rules* (with all classifiers assumed to exhibit equal performance), *trained rules* (introducing weights to account different strengths) and *adaptive rules* (weights depend on environmental factors). We apply traditional fixed parallel rank and measurement combination rules [15], as they do not need separate training datasets, and compare their serial and parallel implementation with respect to accuracy and performance.

1. **Borda count** (BC): With this technique, rank index vectors I_i of different classifiers M_i are combined to form a consolidated rank index vector I_{BC} by estimating the sum of all ranks of individual classifiers as follows:

$$I_{\mathrm{BC}} \in \mathcal{S}_t : \forall k, l \textit{ with } 1 \leq k < l \leq t \Rightarrow \sum_{i=1}^{m} R_i[I_{\mathrm{BC}}[k]] \leq \sum_{i=1}^{m} R_i[I_{\mathrm{BC}}[l]], \quad (4)$$

where $\mathcal{S}_t$ is the set of all permutations of $\{1, \ldots, t\}$. If multiple such vectors exist, one is chosen randomly [15]. This method assumes equal performance of different matchers. Strengths of BC according to [17] are simplicity and no necessity for training. However, ties constitute a problem [18], especially for open set identification, where the sum of ranks for the rank-1 entry, $\sum_{i=1}^{m} R_i[I_{\mathrm{BC}}[0]]$, is thresholded.

2. **Highest rank** (HR): This method assigns each user its minimum rank value (i.e. highest rank) of all classifiers, again with collisions removed randomly.

$$I_{\mathrm{HR}} \in \mathcal{S}_t : \forall k, l \textit{ with } 1 \leq k < l \leq t \Rightarrow \min_{i=1}^{m} R_i[I_{\mathrm{HR}}[k]] \leq \min_{i=1}^{m} R_i[I_{\mathrm{HR}}[l]]. \quad (5)$$

Since neither rank index vectors reveal information for successful rejection of unseen objects, nor do any other measurements (like the sum of ranks for the rank-1 entry as employed for the BC) indicate genuine identification attempts for HR, we exclusively tested this mode in closed set identification.

3. **Score sum** (SS): This measurement combination method averages scores of matchers M_i, i.e. the common score vector is defined as $S_{\text{SS}} := \frac{1}{m} \sum_{i=1}^{m} S_i$.

   If this measurement vector is to be converted into a rank vector, again we have to break up ties randomly. However, as the set of scores is dense, we further expect strengths of this matcher in open set identification.

For serial classifier combination, these rules remain unchanged, but input and output interfaces for single matchers change. The order of matchers, their dimensionality and the combination rule define a serial classifier combination.

1. **Sequential order of matchers:** The matcher M_i is applied to a subset of templates in D specified by a characteristic function c_i, i.e. $M_i(B, D, c_i)$. In order to leave introduced rules unchanged, for elements j with $c_i(j) = 0$ we technically set $R_i[j] := \infty$ in case of BC and HR and $S_i[j] := 0$ in case of SS. Let R be the rank vector of the chosen combination rule using matchers $M_1, \ldots, M_i$ and $d[i]$ be the desired output dimension, then we define:

$$\forall j \; c_i(j) := \begin{cases} 1 \; \textit{if} \; i = 1 \vee (i > 1 \wedge c_{i-1}(j) = 1 \wedge R[j] \leq d[i-1]), \\ 0 \; \textit{otherwise}. \end{cases} \tag{6}$$

2. **Fixed dimensionality of each matcher's output in the chain:** We define decreasing dimensionalities $d[i], 1 \leq d[i] \leq t$ for each matcher M_i, such that $\forall i : i > 1 \Rightarrow d[i] < d[i-1]$. Finally, the resulting score or index vector is a transformed version of the outcome of the combination technique using all matchers $M_1, \ldots, M_m$ accounting the final dimensionality reduction to $d[m]$ ranks. Note, that the combination method for serial classifiers is applied m times (after each matcher) instead of just once for parallel combination.

4 Experiments

In our identification mode assessments we address absolute and relative performance differences between serial and parallel combination schemes with respect to accuracy and computational cost in both closed set and open set configurations, see Table 1. We try to identify how parameter choice (e.g. rule, order of matchers) for serial combinations influences results. In order to be able to compare performance of parallel or serial schemes relative to individual matchers, we also give a short summary of identification results for single classifiers.

As performance indicators we estimate *Cumulative Match Curves* (CMC) indicating for each rank r the relative frequency of genuine test templates (i.e. templates sharing the same identity with an enrolled template) having their corresponding identity-sharing enrolled template ranked in $\{1, \ldots, r\}$ [16]. For direct comparison, we use the RR measure (evaluating the CMC at rank 1).

Table 1. Comparison of parallel and serial combination results with individual classifiers in identification mode: RR = Rank-1 Recognition (%), RR_{FAR} = Rank-1 Recognition (%) at a specific FAR (%), T = Processing time (ms)

	Parallel classifier fusion			Serial classifier fusion						Minutiae		Eigenf.	
	BC	HR	SS	BC1	BC2	HR1	HR2	SS1	SS2	Index	Mid.	Mid.	Ring
Closed set identification													
RR	99.7	90.0	99.9	99.2	99.4	91.1	95.6	99.9	99.8	99.1	98.8	91.6	89.7
T	10338	10337	10338	2561	8110	2557	8108	2550	8109	5225	5111	< 1	< 1
Open set identification													
$RR_{0.1}$	95.9	–	97.4	89.6	95.0	–	–	97.6	97.4	95.6	92.6	60.1	32.6
RR_{10}	99.1	–	98.9	97.7	99.1	–	–	99.4	99.4	98.0	97.3	85.1	79.0
T	5189	–	5189	1235	4016	–	–	1232	4017	2623	2564	< 1	< 1

As the RR refers to correct rankings of genuine samples (and therefore includes false rejects, in which case the length of the ranking vector is zero), there is a tradeoff between RR and FAR (false acceptance rate of imposter samples) in threshold-based open set identification. We visualize this tradeoff by plotting both rates against each other, varying the threshold over its entire range.

For testing purposes we use a dataset of 443 palmar scans at 4250×5850 resolution (71 males and 15 females, at least 4 different templates per person), captured in a single session at arbitrary rotations in a controlled environment (direct exposure to ambient light was avoided by placing the sensor in a box with a hole for hand insertion). A total of 4 samples were rejected due to enrollment errors and further excluded from experiments. For closed set identification tests we set up 4 different enrollment databases from this dataset using the first, second, third and fourth template of each user as enrollment template. This yields a total of 353 possible comparisons of remaining templates with each system database. For open set identification, again we set up four system databases with randomly selected 43 users for each database using the first, second, third and fourth given template, respectively, as enrolment sample. In this configuration, a total of 177 genuine and 219 imposter identification attempts can be executed per system database. The presented identification scenarios are intended to reflect biometric applications for small-sized companies, e.g. presence detection for employees (closed-set) or access control applications (open set).

4.1 Individual Classifiers

Since not all combination techniques necessarily increase overall performance, we first assess closed set (86 enroled users) and open set (43 enroled users) identification performance of unimodal Minutiae matchers (*M-Thumb, M-Index, M-Middle, M-Ring, M-Little*) and Eigenfingers matchers (*E-Thumb, E-Index, E-Middle, E-Ring, E-Little*) in order to obtain reference rates. Closed-set recognition results in form of an average CMC (averaging results over all 4 system databases) are illustrated in Fig. 2. Presumably, the most accurate feature turned out to be *M-Index* with 99.1% RR, followed by *M-Middle* (98.8%),

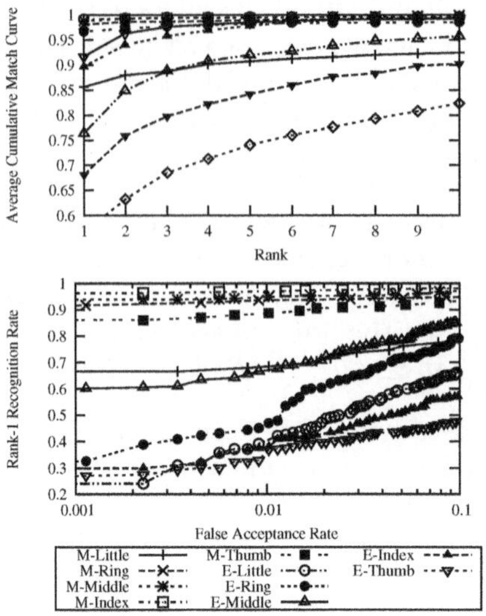

Fig. 2. Closed set Average CMC (top), Open set RR-FAR Tradeoff (bottom)

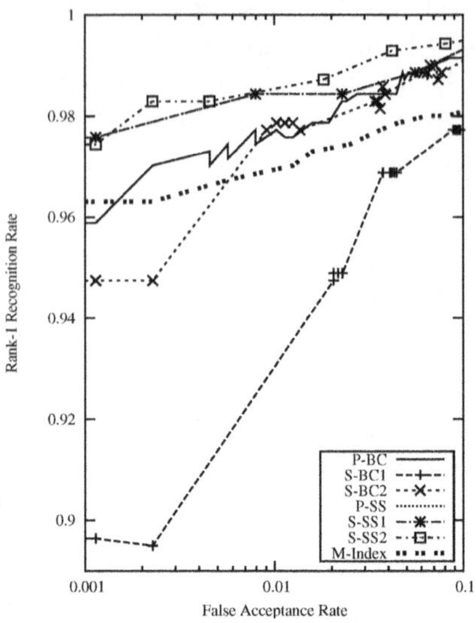

Fig. 3. Open set Recognition-Security Tradeoff for parallel/serial classifiers

M-Ring (98.0%), *M-Thumb* (96.7%), *E-Middle* (91.6%), *E-Ring* (89.7%), *M-Little* (85.6%) *E-Little* (76.4%), *E-Index* (68.1%), and *E-Thumb* (56.7%). However, whereas Minutiae features exhibit processing times of 2478 ms (*M-Little*), 4090 ms (*M-Thumb*), 4205 ms (*M-Ring*), 5111 ms (*M-Middle*) and 5225 ms (*M-Index*), all Eigenfinger-based features could create a ranking vector in almost less than 1 ms. More accurate Minutiae-based matchers tend to exhibit longer average processing times, as they usually operate on larger sets. *M-Little*'s CMC increases slowly with additional ranks, an undesired behavior for serial combination, as a dimensionality reduction may exclude the genuine identity.

Fig. 3 illustrates the tradeoff between RR and FAR for all features in open set identification. The best RR performance at 0.1% FAR is 95.6% (*M-Index*), while the worst performing feature (*E-Little*) only achieves 18.4%. Generally, in this configuration, Eigenfingers perform significantly worse than Minutiae-based features. If FAR is relaxed to 10%, RR reaches 48.2 – 98%. Regarding performance, single Minutiae-based classifiers require up to 2623 ms, whereas Eigenfingers remain at processing times of less than 1 ms.

4.2 Parallel versus Serial Classifier Combinations

The first question arising naturally when serial and parallel classifier combination techniques are compared is the choice of parameters for the serial combination scheme. Recall, that serial combinations are defined by rule, order, count and dimensionality of each matcher. As both the number of possible choices for dimensionalities and order of matchers grow exponentially in the number of matchers, an exhaustive search within a training set to find the best parameters becomes unfeasible if the number of matchers is large. For the comparison of serial and

parallel techniques, we restrict the set of possible individual classifiers to the best two Minutiae features *M-Index*, *M-Middle* and the two best Eigenfinger features *E-Middle* and *E-Ring*. In order to be independent of training datasets, we conservatively define a class reduction factor of two, i.e. $d[0] := \frac{t}{2}, d[i+1] := \frac{d[i]}{2}$, and examine two natural choices for the order of matchers:

- **Configuration 1:** In order to exploit the intrinsic property of class reduction in serial combinations, we select fast classifiers at lower ranks followed by more costly features, i.e. the order is *E-Middle, E-Ring, M-Index, M-Middle.*
- **Configuration 2:** Here we select matchers in the order of accuracy to avoid early loss of potential matches: *M-Index, M-Middle, E-Middle, E-Ring.*

Results of all employed parallel and serial combination experiments can be found in Table 1. In a first test we assessed closed set identification performance of the introduced combination rules Borda count (BC), Highest rank (HR) and Score sum (SS) in its parallel (*P-BC, P-HR, P-SS*) and serial forms (*S-BC1, S-HR1, S-SS1* for Configuration 1, and *S-BC2, S-HR2, S-SS2* using Configuration 2). Whereas for BC serial classifiers exhibit slightly worse accuracy (99.2 – 99.4% instead of 99.7%), serial SS combinations perform almost equally well (99.8 – 99.9%) and serial HR even outperforms parallel techniques (up to 95.6% instead of 90%). However, serial combination techniques could reduce required processing time of parallel schemes significantly (up to a factor of 4 using the fast configuration, and at least by over 20% using the conservative configuration). When comparing different combination rules, SS combination performs best and is at the same time least affected by serial class reduction, followed by BC and the HR method, which is no good choice for the selected classifiers, as overall accuracy even degrades compared to individual classifiers.

When assessing open set identification mode performance of the BC and SS methods in Fig. 3, again SS turns out to be highly accurate. The faster serial combination technique *S-SS1* outperforms the best individual matcher (with 97.6% RR at 0.1% FAR) and at the same time even requires less processing time (1232 ms instead of 2623 ms). Serial combination S-BC1 however seems to be more affected by the worse performance of Eigenfingers. Finally, we have applied exhaustive search using a separate (randomized) training system database, to find the best serial combination with respect to the cost function C introduced in Sect. 1 using weights $\omega_1 = 1, \omega_2 = 10^{-5}$. The best serial combination turned out to be SS considering *E-Middle, E-Ring, M-Index* with dimensions 9, 4, 1 yielding a total accuracy of 98.7% at 616 ms (99.7% at 591 ms on training set).

5 Conclusions and Future Works

We have examined parallel and serial combination techniques on classifiers of a single-sensor hand-based multimodal biometric system. Serial classifiers turned out to decrease processing time significantly at almost the same or even higher accuracy than their parallel combination counterparts. The Score sum method was found to produce the best results for both parallel and serial combination,

followed by Borda count and Highest rank. The ability to improve both accuracy and processing performance simultaneously makes serial combination techniques an excellent choice for classifier fusion. Still, the choice of parameters for serial classification influences recognition accuracy and performance and deserves further attention in future work.

References

1. Zhang, Z., Wang, R., Pan, K., Li, S., Zhang, P.: Fusion of near infrared face and iris biometrics. In: Lee, S.-W., Li, S.Z. (eds.) ICB 2007. LNCS, vol. 4642, pp. 172–180. Springer, Heidelberg (2007)
2. Rowe, R., Uludag, U., Demirkus, M., Parthasaradhi, S., Jain, A.: A multispectral whole-hand biometric authentication system. In: Proc. Biometric Symp. (2007)
3. Uhl, A., Wild, P.: Personal recognition using single-sensor multimodal hand biometrics. In: Elmoataz, A., Lezoray, O., Nouboud, F., Mammass, D. (eds.) ICISP 2008. LNCS, vol. 5099, pp. 396–404. Springer, Heidelberg (2008)
4. Nandakumar, K., Chen, Y., Jain, A., Dass, S.: Quality-based score level fusion in multibiometric systems. In: Proc. 18th Int. Conf. Pattern Recogn (ICPR), pp. 473–476 (2006)
5. Kittler, J., Hatef, M., Duin, R., Matas, J.: On combining classifiers. IEEE Trans. Pattern Anal. Mach. Int. 20(3), 226–239 (1998)
6. You, J., Kong, W.K., Zhang, D., Cheung, K.: On hierarchical palmprint coding with multiple features for personal identification in large databases. IEEE Trans. Circ. Syst. Video Techn. 14(2), 234–243 (2004)
7. Viola, P., Jones, M.: Robust real-time face detection. Int. J. Comput. Vision 57(2), 137–154 (2004)
8. Kumar, A., Zhang, D.: Combining fingerprint, palmprint and hand-shape for user authentication. In: Proc. 18th Int. Conf. Pattern Recogn (ICPR), pp. 549–552 (2006)
9. Kumar, A., Wong, D., Shen, H., Jain, A.: Personal verification using palmprint and hand geometry biometric. In: Kittler, J., Nixon, M.S. (eds.) AVBPA 2003. LNCS, vol. 2688, pp. 668–678. Springer, Heidelberg (2003)
10. NIST: Fingerprint image software 2 (2004), `http://fingerprint.nist.gov/NFIS`
11. Maltoni, D., Maio, D., Jain, A., Prabhakar, S.: Handbook of Fingerprint Recognition. Springer, New York (2003)
12. Turk, M., Pentland, A.: Eigenfaces for recognition. J. Cognitive Neuroscience 3(1), 71–86 (1991)
13. Ribaric, S., Fratric, I.: A biometric identification system based on eigenpalm and eigenfinger features. IEEE Trans. Pat. Anal. Mach. Int. 27(11), 1698–1709 (2005)
14. Xu, L., Krzyzak, A., Suen, C.: Methods of combining multiple classifiers and their applications to handwriting recognition. IEEE Trans. Systems, Man, and Cybernetics 22(3), 418–435 (1992)
15. Ross, A., Nandakumar, K., Jain, A.: Handbook of Multibiometrics. Springer, Secaucus (2006)
16. Bolle, R., Connell, J., Pankanti, S., Ratha, N., Senior, A.: Guide to Biometrics. Springer, New York (2004)
17. Melnik, O., Vardi, Y., Zhang, C.H.: Mixed group ranks: preference and confidence in classifier combination. IEEE Trans. Pat. Anal. Mach. Int. 26(8), 973–981 (2004)
18. Faundez-Zanuy, M.: Data fusion in biometrics. IEEE Aerosp. Electr. Syst. Mag. 20(1), 34–38 (2005)

Robust Multi-modal and Multi-unit Feature Level Fusion of Face and Iris Biometrics

Ajita Rattani and Massimo Tistarelli

Computer Vision Laboratory,
University of Sassari,
Alghero, Italy
ajita.rattani@diee.unica.it, tista@uniss.it

Abstract. Multi-biometrics has recently emerged as a mean of more robust and efficient personal verification and identification. Exploiting information from multiple sources at various levels i.e., feature, score, rank or decision, the false acceptance and rejection rates can be considerably reduced. Among all, feature level fusion is relatively an understudied problem. This paper addresses the feature level fusion of multi-modal and multi-unit sources of information. For multi-modal fusion the face and iris biometric traits are considered, while the multi-unit fusion is applied to merge the data from the left and right iris images. The proposed approach computes the *SIFT* features from both biometric sources, either multi-modal or multi-unit. For each source, feature selection on the extracted *SIFT* features is performed via spatial sampling. Then these selected features are finally concatenated together into a single feature *super-vector* using serial fusion. This concatenated super feature vector is used to perform classification.

Experimental results from face and iris standard biometric databases are presented. The reported results clearly show the performance improvements in classification obtained by applying feature level fusion for both multi-modal and multi-unit biometrics in comparison to uni-modal classification and score level fusion.

1 Introduction

Biometrics refers to the use of physiological, biological or behavioural characteristics to establish the identity of an individual. These characteristics are unique to each individual and remain partially un-altered during the individual's lifetime [1].

In the recent years biometric authentication has gained a considerable improvement in both reliability and accuracy. Nonetheless, the best biometric systems to date present several drawbacks, such as limited applicability, vulnerability to spoofing attacks, less discriminant features, performance degradation due to noisy data, and others. Some of these limitations are inherent in the sensor technology or in the nature of the biometric trait itself. Consequently, a mono-modal biometric system rarely can cope with the variety of requirements in real applications. This is especially true in non-ideal scenarios, like outdoor environments, or highly demanding applications, such as in large-scale systems.

M. Tistarelli and M.S. Nixon (Eds.): ICB 2009, LNCS 5558, pp. 960–969, 2009.
© Springer-Verlag Berlin Heidelberg 2009

By grouping evidence from multiple sources of information, multi-biometric systems [2] allow to overcome some of the drawbacks of the mono-modal biometric systems. The combination of multiple traits provide a better population coverage and provide a platform with wider usability. Multi-biometrics also offer an efficient counter-measure to spoofing, because it would be difficult for an impostor to simultaneously spoof multiple biometric traits of a genuine user [2].

Multi-biometric systems can rely on multi-modalities like face and fingerprint, multiple units like two or more fingerprints, multiple sensors like optical and capacitive sensors, multiple representations like LDA and PCA features for face [2]. The resulting multiple sources of information can be integrated at various levels. Ross and Jain [3] presented a wide overview of multi-modal biometric solutions describing different levels at which fusion can be performed, i.e. sensor level, feature extraction level, matching score level and decision level.

As reported in the literature [3] a biometric system that integrates information at an earlier stage of processing is expected to provide better performances than systems that integrate information at a later stage, because of the availability of more and richer information. Therefore, fusion at sensor and feature level is expected to be more efficient than fusion at matching score, rank and decision levels. Despite of the abundance of research papers related to multi-biometrics [4–6], fusion at feature level is a relatively understudied problem [7-9]. One possible reason is that fusion at feature level is relatively difficult to achieve in practice because different modalities may have incompatible feature sets and the correspondence among different feature spaces may be unknown. Moreover, fusing feature vectors may lead to the problem of the *curse of dimensionality*: due to the large dimension of the fused feature vector, a very complex matcher may be required to correctly classify the fused data. Nevertheless, the fused feature vector may contain noisy or redundant data thus leading to degradation in the performance of the classifier [7].

Among many biometric traits, face and iris posses a number of useful characteristics for recognition. For example, face sensing is non-invasive and friendly. Moreover, it is relatively easy to acquire face images of good quality. On the other hand, iris-based identification systems has proved to be among the best performing biometrics in many application domains. Nonetheless, both face and iris have their own limitations too. The accuracy of face recognition is often affected by illumination, pose, shelter and facial expression, while current iris recognition systems require the active cooperation of the user. Thus in a sense the two modalities are highly complementary. As face data is easy to acquire but delivers low to medium performances, iris data is more difficult to acquire but delivers very high performances. However, as they both belong to the same biological trait (the head) they can be easily sampled at the same time and with the same sensing technology. Therefore, it is expected that fusion of face and iris may overcome the intrinsic limitations of each modality, while delivering the following advantages [10-11]:

1) The total error rate (the combination of the false accept rate and the false reject rate) is decreased with respect to the best of the two modalities [11].
2) The probability of successful spoofing attacks is reduced, because of the increased difficulty in manufacturing (or reproducing in a simulated video) an artificial head with both the face and irises faithfully reproducing the genuine user's appearance. On the contrary, it also makes more difficult for an individual, in a surveillance scenario, to hide his/her identity.

3) The population coverage is larger in comparison to uni-modal biometric system. The adoption of the face modality can allow people with imperfect iris images to enroll, reducing the enrolment failure rate.
4) Conversely to other multi-biometric systems, which require multiple sensors, the combination of face and iris allows for simultaneous acquisition of all the data with a single device. Therefore, multi-modal and multi-unit fusion can be performed without introducing any additional cost for the hardware.

Few work related to face and iris fusion have been reported in the literature. The combination of iris and face at score level have been reported in [10][11]. References [12] [13] are the only one in literature, in our knowledge, describing a methodology for feature level fusion of face and iris. Both of the referred work [12-13] have applied projection to lower dimensional space for feature fusion and reduction, using prior training of the classifier and have used nearest neighbour distance for the classification of fused feature vector.

This paper proposes a new approach for feature level fusion of face and both left and right iris for an individual. In the proposed system, SIFT features are extracted from the face image and both irises. A spatial sampling technique is applied to select a subset of SIFT features individually from features extracted from each of the three biometric traits. The selected features are then combined into a single, higher dimensional fused feature set. Both the multi-modal (face and iris) and multi-unit (left and right irises) sources of information are combined to obtain a single super SIFT feature vector set which is used for matching and classification.

The main novelty of the work stems from the introduction of SIFT [14-16] features to process iris images and the combination of the features in a global SIFT feature vector set. The proposed approach has the advantage over [12][13] of adopting of common feature extractor which delivers more stable features. The applied feature reduction technique is very simple and does not require a preliminary training. Also for classification, a direct matching technique of the fused feature vectors is adopted. Moreover, the work also presented the fusion of both multi-modal and multi-unit sources of information at feature level which further enhanced the results.

Experimental results on the database, composed of a subset of iris images from the CASIA version 3 database [18] and its chimerical combination with the Equinox database [19] are reported. The results obtained demonstrate the performance enhancement of feature level fusion in comparison to uni-modal systems and score level fusion. Section 2, describes the Scale Invariant Features Transform and its application for face and iris. Section 3, describes the feature level fusion of face with iris, both left and right irises. Experimental results are presented in section 4. Conclusions are drawn in Section 5.

2 Scale Invariant Feature Transform (*SIFT*)

David Lowe proposed a method to extract distinctive and invariant features from images applied to general 3D object recognition [14]. These features known as *Scale Invariant Feature Transform* or *SIFT* features are invariant to image scale and rotation. Thus they provide a basis for object representation to perform a robust matching across a substantial range of affine distortion, change in 3D viewpoint, addition of

noise, and change in illumination. SIFT features are well localized in both the spatial and frequency domain, reducing the probability of disruption due to occlusion, clutter, or noise. In addition, SIFT features are highly distinctive. This property allows a single feature to be correctly matched with high probability against a large database of features, providing a basis for robust recognition. The cost of extracting these features is minimized by adopting a cascade filtering approach in which the more time consuming operations are applied only at image locations which have been selected at the initial phase [14].

Due to the stability and robustness of these features, they have been recently been applied for face and fingerprint biometrics [15][16]. Each extracted SIFT features can be defined as $\boldsymbol{S} = (s_1, s_2,...,s_m)$, where each feature $s_i = (x ,y ,\theta, Keydesc)$ includes the spatial location (x, y), the local orientation θ and the key descriptor of size *1x128*. Key descriptor i.e., *Keydesc* part of each SIFT feature is invariant to affine transformations and is adopted for experiments in [15][16].

This work investigates, the applicability of *SIFT* features for iris biometrics and its fusion with the face biometric at feature level fusion. The extracted SIFT features from face and iris images as shown in Fig. 1. In this work also, keypoint descriptors (*Keydesc*) are considered for image representation and experiments.

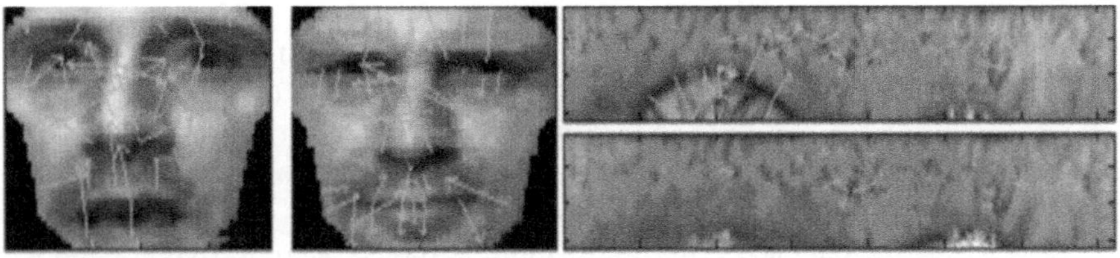

Fig. 1. Shows extracted SIFT Features from the face and iris images. Even with the strong intra-class variations many common SIFT Features can be easily noticed for both the biometrics.

3 Feature Level Fusion

Feature level fusion is performed by the simple concatenation of the feature sets obtained from different sources. Let $\boldsymbol{X} = (x_1, x_2,...,x_m)$ and $\boldsymbol{Y} = (y_1, y_2,...,y_m)$ denote the feature vectors extracted from two different sources. The vector $\boldsymbol{Z} = (z_1, z_2,...,z_m)$ is formed by the concatenation of the two feature sets $\boldsymbol{X}$ and $\boldsymbol{Y}$. Generally, before performing the concatenation, the two vectors $\boldsymbol{X}$ and $\boldsymbol{Y}$ are normalized to ensure the same range and scales of values. In order to reduce the dimensionality of the fused feature space, generally a feature selection technique is also applied to the individual or fused features. The similarity between two individuals is determined by matching the instances of the fused vector $\boldsymbol{Z}$ obtained from database and query sources of information.

By applying SIFT features to represent both the face and iris traits, the entire process of feature level fusion is greatly simplified without compromising recognition ability. In fact, the feature normalization process is not required, because the features from both sets are already commensurable and the matching step is reduced to the computation of the Euclidean distance metric and pairing of the nearest neighboring features between two concatenated database and query feature vectors.

The entire matching process is described by the following steps:

a) **Image Pre-processing and Feature Extraction:** Both face and iris images are pre-processed to extract only the relevant part of the image containing useful information. For face, a template mask is applied to the image to crop the central region of the face. This mask is registered and scaled to each face image on the basis of the position of the eyes and then the face portion is cropped. In the reported experiments, the eyes position are manually extracted but can be equally well extracted using a eye position extractor. It is worth noting that no further preprocessing is applied to the face images to compensate for illumination changes between samples. We deliberately chose not to perform an intensity normalization of the images to fully validate the robustness of the SIFT-based representation and the fused classifier. The iris images are segmented by using the technique proposed by Ross et al. [17] which is based on the implementation of Geodesic Curves. SIFT features [14] are extracted from the segmented and normalized iris images. As a result of which, a set of SIFT features is obtained from the face image and from left and right irises for each subject (Fig. 1).

b) **Feature Selection:** The extracted SIFT features are then selected from both biometric traits by (Fig. 2)
   i. dividing each image (face, left and right iris) into small windows of size *3x2* for the iris and *5x5* for the face. The total number of windows are *256* for face and *3780* for each iris image. The window size has been determined after several trials and taking the size which best captures the best discriminative information content in the data set. After performing several tests on different data, it was established experimentally that the optimal size only depend on the size of the input image.
   ii. For each window the *SIFT* descriptor with average minimum Euclidean distance from all other descriptors within the same window, is selected. In case there is only one descriptor in the window, it is selected for inclusion into the fused feature vector.

c) **Feature Concatenation:** The selected *SIFT* features from the face and the two irises are combined into a single super feature vector set, of dimension *Nx128*, where *N* is equal to sum of all the features selected from different sources ,as a result of spatial sampling, for inclusion in the fused vector.

$$\textbf{\textit{Fused feature set}} = \begin{matrix} \textit{face SIFT} \\ \\ \textit{left iris SIFT} \\ \\ \textit{right iris SIFT} \end{matrix} \begin{pmatrix} 1 \text{ X } 128 \\ \cdots\cdots\cdots \\ 1 \text{ X } 128 \\ \cdots\cdots\cdots \\ 1 \text{ X } 128 \end{pmatrix}$$

d) **Feature Matching:** In order to match two fused feature sets f_1 and f_2 . Given a feature element p_{11} in f_1 , the distances d_1 and d_2 between the first closest element p_{11} and the second closest element p_{21} in f_2 are computed. If the ratio

$\frac{d_1}{d_2}$ is below 60%, then the vector element p_{11} is tagged to match with element p_{21}. The matching score between the two fused feature vectors is equal to the number of matching elements [16]. It is worth noting that this procedure maximizes the separation between matching and non-matching vector elements, thus reducing the probability of false matches between vector pairs.

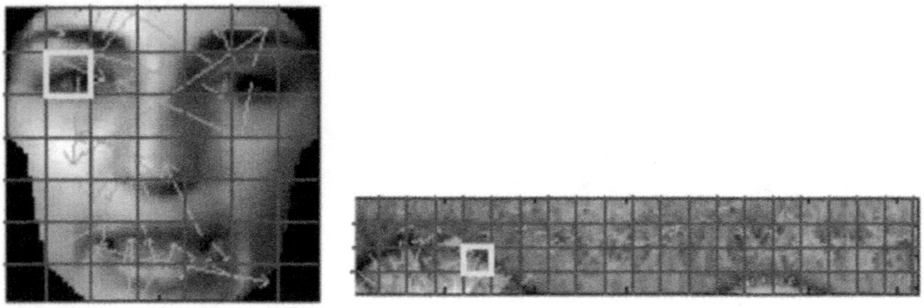

Fig. 2. Example of feature selection by partitioning the face and iris images into a set of small windows and selecting one SIFT feature per window. The yellow boxes show an example of more than one SIFT feature in a single partition. Windows with no features are ignored.

4 Experimental Evaluation

CASIA iris database version 3 [18] and the Equinox face database [19] have been used to evaluate the performance of the proposed system. A total of *57* subjects with ten instances per each client have been taken from the Equinox database. The left and right normalized iris images of *125* clients from the CASIA version 3 database have been used for experiments.

4.1 Protocol for Performance Evaluation

The following evaluation procedure has been applied for mono-modal and multi-biometric matching, with *N* subjects and processing *10* samples for each biometric trait per subject:

Training: one template is built for each subject using one image for each modality, i.e. one face image and one image for each of the two irises. The SIFT features are extracted from each image and both uni-modal and multi-modal feature vectors are built for the face, irises and the combination of the two irises and the face with one or two irises. In the performed experiments, the matching scores were computed to determine the discrimination capability of the single and fused representation.

Testing: Nine samples per person are used for testing and to generate the client matching scores. For the mono-modal systems, the impostor scores are generated by

matching the representation of each client against the first sample of all the other individuals in the dataset. In case of multimodal testing the client is tested against the first face and iris samples of the remaining of the chimerical users. Therefore, in total *9xN* client scores and *9xNx(N-1)* impostor scores for each of the uni-modal and multimodal representations are generated.

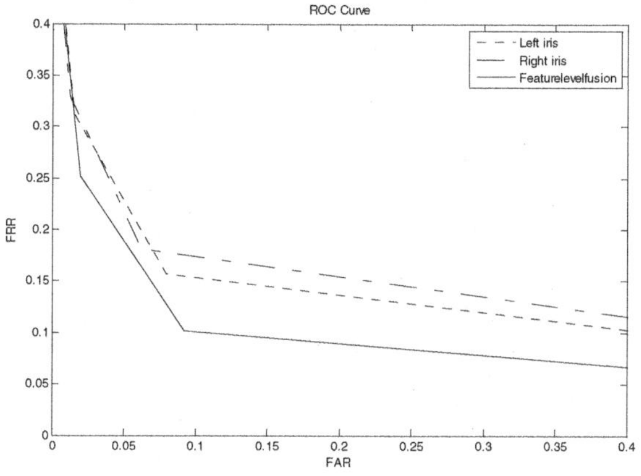

Fig. 3. ROC curves representing the matching performance of individuals based on the SIFT representation extracted from the left and right iris and the feature level fusion of the two irises

4.2 Multi-unit Fusion: Left and Right Irises

For testing the application of SIFT representation to the iris biometric, *125* clients from the CASIA version 3 database [18] with ten left and ten right iris images are used as mentioned in 4.1. The iris images are normalized and segmented as discussed in [17]. Left and Right iris are confirmed to contain discriminatory information for each individual so their combination is a source of complementary information [20]. Thus feature level fusion of left and right iris using SIFT is performed as described in section 3 and it was evaluated against left and right uni-modal traits.

The performance is evaluated as proposed in section 4.1. In total, *125x9=1125* client scores and *125x124x9=139500* impostor scores are generated for each iris. The ROC curves obtained from the matching scores of the left and right iris images using SIFT features and feature level fusion are shown in Fig. 3. As it can be noted from the ROC curves, the feature level fusion of these complementary information sources considerably enhances the identification capabilities.

4.3 Multi-modal with Multi-unit Fusion: Face, Left and Right Irises

To test the performance of matching the fused SIFT representation based on face and iris, a chimerical database is obtained by coupling 57 clients images from CASIA [18] and Equinox face database [19]. The SIFT features are extracted from the normalized face and iris images. The SIFT representation of the face and iris biometric traits are fused at feature level as detailed in section 3. The performance is evaluated as

reported in section 4.1. In total *57x9=513* client scores and *57x56x9=28728* impostor scores are generated for the uni-modal face matching and the face and iris fused vector matching. The ROC curves representing the error distributions of the feature level fused matching are shown in Fig 4. The curves are generated from the scores obtained by matching the mono-modal face representations and the multi-modal feature level fusion with left and right iris images.

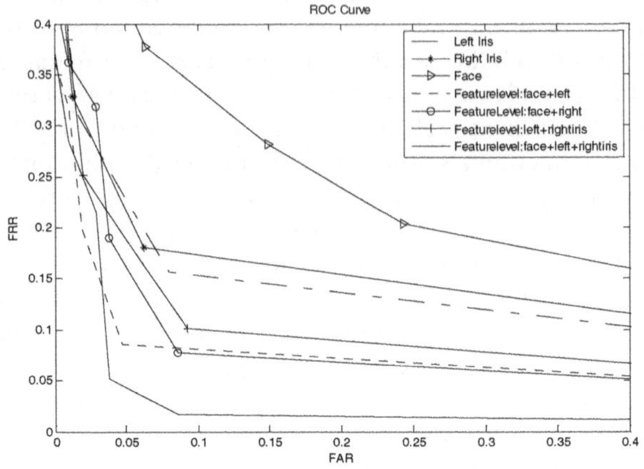

Fig. 4. ROC curves representing the matching performance of individuals based on the SIFT representation extracted from the face, the left and the right iris and the feature level fusion of the three traits

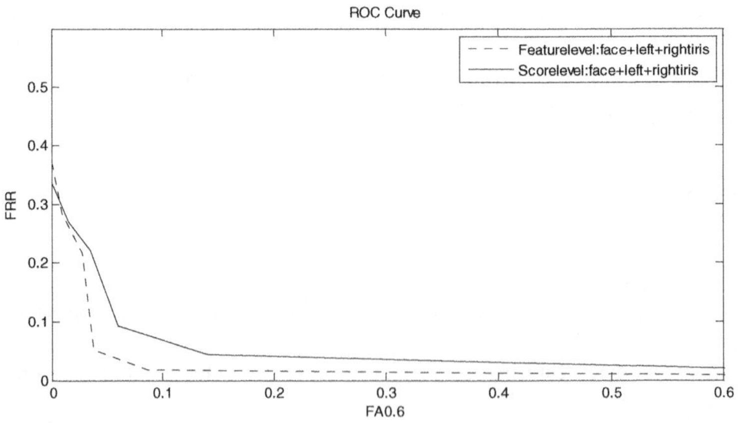

Fig. 5. Comparison of the feature level fusion, of the SIFT representations of the face, the left and the right irises, with the score level fusion of the matching scores of the individual modalities

It is worth noting that the error distributions are lowered by combining more sources of information at the feature level. The performance of the feature level fusion of face and the two irises is compared with the score level fusion in Fig 5. As it can be noted, the data fusion performed at the earlier stage, i.e. feature level, produces better performances in terms of error distributions than the fusion at the score level.

5 Conclusion

The fusion of face and iris biometrics is a natural process in a multi-biometric system which, in principle, can be implemented without introducing extra additional cost for acquisition. At the same time, the two biometrics are strongly complementary. In fact, while face biometric samples are easier to acquire with a conventional camera, iris has demonstrated very superior performances. This suggests an image acquisition system with a camera acquiring face images at a high resolution to provide shape and textural information for both the two irises and the face. Current camera devices allow to sample up to 15 M pixels images at a rate of 30 frames per second and deliver the data stream over a fast Ethernet channel. By properly controlling the image acquisition with a fast face detection algorithm to keep the head within one third of the total image area, the resulting images can provide iris images up to 64 K pixels. Several systems to acquire iris from a distance have been proposed and remarkably demonstrated the possibility to acquire both the face and the iris at the same time [21].

In this paper a novel approach to feature level fusion of face and iris has been proposed. The main advantages of the proposed fusion method are the ease of implementation and the robustness of the resulting representation. Due to the scale-space analysis, SIFT features proved to be very stable and almost insensitive to illumination variations while providing a scale and translation invariant representations. At the same time, the adoption of a common feature representation greatly simplifies the normalization, concatenation and matching processes in the feature level fusion, which is generally an issue related to fusion at this level. From the experiments performed on a mixed database obtained by combining face images from the Equinox database and iris images from the CASIA v.3 database the representation based on feature level fusion demonstrate superior matching performance with respect to unimodal systems and score level fusion. Some issues are still under investigation, such as the optimal feature selection scheme based on quality driven analysis and the evaluation on large multi-modal database.

References

1. Hong, L., Jain, A., Pankanti, S.: Can Multi-biometrics Improve performance. In: Proc. of AutoID 1999, pp. 59–64 (1999)
2. Jain, A.K., Ross, A.: Multi-biometric systems. Communications of the ACM 47(1), 34–40 (2004)
3. Ross, A., Jain, A.K.: Information Fusion in Biometrics. Pattern Recognition Letters 24, 2115–2125 (2003)
4. Chibelushi, C.C., Mason, J.S., Deravi, F.: Integration of acoustic and visual speech for speaker recognition. In: EUROSPEECH 1993, pp. 157–160 (1993)
5. Duc, B., Maître, G., Fischer, S., Bigün, J.: Person authentication by fusing face and speech information. In: Bigün, J., Borgefors, G., Chollet, G. (eds.) AVBPA 1997. LNCS, vol. 1206. Springer, Heidelberg (1997)
6. Hong, L., Jain, A.: Integrating Faces and Fingerprints for Personal Identification. IEEE Transactions on Pattern Analysis and Machine Intelligence 20(12), 1295–1307 (1998)

7. Ross, A., Govindarajan, R.: Feature Level Fusion Using Hand and Face Biometrics. In: Proc. of SPIE Conference on Biometric Technology for Human Identification II, Orlando, USA, pp. 196–204 (2005)
8. Zhou, X., Bhanu, B.: Feature fusion of face and Gait for Human Recognition at a distance in video. In: International Conference on Pattern Recognition, Hong kong, (2006)
9. Singh, S., Gyaourova, G., Pavlidis, I.: Infrared and visible image fusion for face recognition. In: SPIE Defense and Security Symposium, pp. 585–596 (2004)
10. Wang, Y., Tan, T., Jain, A.K.: Combining Face and Iris Biometrics for Identity Verification. In: Kittler, J., Nixon, M.S. (eds.) AVBPA 2003. LNCS, vol. 2688, pp. 805–813. Springer, Heidelberg (2003)
11. Zhang, Z., Wang, R., Pan, K., Li, S.Z., Zhang, P.: Fusion of near infrared face and iris biometrics. In: Lee, S.-W., Li, S.Z. (eds.) ICB 2007. LNCS, vol. 4642, pp. 172–180. Springer, Heidelberg (2007)
12. Son, B., Lee, Y.: Biometric Authentication System Using Reduced Joint Feature Vector of Iris and Face. In: Kanade, T., Jain, A., Ratha, N.K. (eds.) AVBPA 2005. LNCS, vol. 3546, pp. 513–522. Springer, Heidelberg (2005)
13. Gan, J., Liang, Y.: A Method for Face and Iris Feature Fusion in Identity Authentication. IJCSNS, 6 (2B) (2006)
14. Lowe, David, G.: Object recognition from local scale invariant features. In: International Conference on Computer Vision, Corfu, Greece, pp. 1150–1157 (September 1999)
15. Bicego, M., Lagorio, A., Grosso, E., Tistarelli, M.: On the use of SIFT features for face authentication. In: Proc. of Int. Workshop on Biometrics, in association with CVPR (2006)
16. Park, U., Pankanti, S., Jain, A.K.: Fingerprint Verification using SIFT Features. In: Proc. of SPIE Defense and Security Symposium, Orlando, Florida (2008)
17. Ross, A., Shah, S.: Segmenting Non-ideal Irises Using Geodesic Active Contours. In: Proc. of Biometrics Symposium (BSYM), Baltimore, USA (2006)
18. `http://www.cbsr.ia.ac.cn/english/IrisDatabase.asp`
19. `http://www.equinoxsensors.com/products/HID.html`
20. Wu, X., Wang, K., Zhang, D., Qi, N.: Combining left and right irises for personal authentication. In: Yuille, A.L., Zhu, S.-C., Cremers, D., Wang, Y. (eds.) EMMCVPR 2007. LNCS, vol. 4679, pp. 145–152. Springer, Heidelberg (2007)
21. Matey, J.R., Naroditsky, O., Hanna, K., Kolczynski, R., LoIacono, D.J., Mangru, S., Tinker, M., Zappia, T.M., Zhao, W.Y.: Iris on the Move: Acquisition of Images for Iris Recognition in Less Constrained Environments. Proc. of the IEEE 94(11), 1936–1947 (2006)

Robust Human Detection under Occlusion by Integrating Face and Person Detectors

William Robson Schwartz[1], Raghuraman Gopalan[2], Rama Chellappa[2], and Larry S. Davis[1]

[1] University of Maryland, Department of Computer Science
College Park, MD, USA, 20742
[2] University of Maryland, Department of Electrical and Computer Engineering
College Park, MD, USA, 20742

Abstract. Human detection under occlusion is a challenging problem in computer vision. We address this problem through a framework which integrates face detection and person detection. We first investigate how the response of a face detector is correlated with the response of a person detector. From these observations, we formulate hypotheses that capture the intuitive feedback between the responses of face and person detectors and use it to verify if the individual detectors' outputs are true or false. We illustrate the performance of our integration framework on challenging images that have considerable amount of occlusion, and demonstrate its advantages over individual face and person detectors.

1 Introduction

Human detection (face and the whole body) in still images is of high interest in computer vision. However, it is a challenging problem due to the presence of variations in people's poses, lighting conditions, inter- and intra- person occlusion, amongst others. Occlusion, in particular, poses a significant challenge due to the large amount of variations it implies on the appearance of the visible parts of a person.

There are many human detection algorithms in the literature. In general, they fall into two categories: subwindow-based and part-based approaches. In the former category, features extracted from subwindows located within a detection window are used to describe the whole body. Subwindow-based approaches can be based on different types and combinations of features, such as histograms of oriented gradients (HOG) [1], covariance matrices [2], combination of several features [3], and multi-level versions of HOG [4]. On the other hand, part-based approaches split the body into several parts that are detected separately and, finally, the results are combined. For instance, Wu and Nevatia [5] use edgelet features and learn nested cascade detectors for each body part. Mikolajczyk et al. [6] divide the human body into seven parts, and for each part a cascade of detectors is applied. Shet and Davis [7] apply logical reasoning to exploit contextual information augmented by the output of low level detectors.

M. Tistarelli and M.S. Nixon (Eds.): ICB 2009, LNCS 5558, pp. 970–979, 2009.
© Springer-Verlag Berlin Heidelberg 2009

Fig. 1. Image where occlusion is present and fusion of detectors can increase detection accuracy. Face of person *b* is occluded. Once the legs and torso are visible, results from a part-based person detector can be used to support that a human is present at that location. On the other hand, the legs of person *c* are occluded, in such a case, face detector results can be used to reason that there is a person at that particular location since the face of person *c* is perfectly visible.

Subwindow-based person detectors present degraded performance when parts of the body are occluded; part-based approaches, on the other hand, are better suited to handle such situations because they still detect the un-occluded parts. However, since part-based detectors are less specific than whole body detectors, they are less reliable and usually generate large numbers of false positives. Therefore, to obtain more accurate results it is important to aggregate information obtained from different sources with a part-based detector. For this, we incorporate a face detector.

Face detection is an extensively studied problem, and the survey paper [8] provides a comprehensive description of various approaches to this problem. For example, Viola and Jones [9] use large training exemplar databases of faces and non-faces, extract feature representations from them, and then use boosting techniques to classify regions as face or non-face. Other algorithms, for instance Rowley et al. [10], uses a neural network to learn how the appearance of faces differ from non-faces using training exemplars, and then detect faces by seeing how well the test data fits the learned model. Another class of approaches, exemplified by Heisele et al. [11], uses a part-based framework by looking for prominent facial components (eyes, nose etc), and then uses their spatial relationship to detect faces. Although such methods are more robust to image deformations and occlusions when compared with holistic approaches, the choice of feature representations and accurate characterization of the relationships between the facial components is still a challenge.

The question that arises naturally is then, how to fuse these two sources to improve overall detection performance. Specifically, is it possible to use the response profiles of the two separate detectors, to reinforce each other, as well as provide a basis to resolve conflicts? This is the question we address in our work.

Figure 1 motivates the utility of combining face and person detectors. First, while the lower half of person *c* is occluded, the face detector can still detect the face of the person, whereas the person detector might fail. Nevertheless, we can try to *explain* the response of the person detector based on the response of the face detector, and conclude that a person is present. Another case is the reverse situation such as *b* and *d* in Figure 1 whose faces are partially occluded while the body parts are completely visible. Such situations occur often in real-world scenarios, and motivates exploring *feedback* between face and people detectors.

2 Face and Person Detection

In this section we give a synopsis of our algorithms for face detection and person detection. We also provide detection results of applying the individual algorithms on standard datasets, showing that these detectors individually achieve results comparable to state-of-art methods. However, a point to keep in mind is that these standardized datasets do not have considerable amounts of occlusion, which is the main problem that we address in our work.

2.1 Face Detection

We use a feature-based approach to detect faces from still images. Our approach, motivated by [12], is based on using an optimal step edge operator to detect shapes (here, the facial contours are modeled as ellipses). The crux of the algorithm is then to obtain the edge map of the image using a derivative of double exponential (DODE) operator, and fit various sized ellipses to the edge map. Image regions that have high response to ellipse fitting signify locations that likely contain faces.

We then conduct post-processing on these short-listed regions by computing three different cues - color [13], histogram of oriented gradients [1], and eigenfaces [14], and combine the three feature channels using support vector machines [15] to decide whether a face is present or not. The motivation behind the choice of these descriptors is: (i) the human face has a distinct color pattern which can be characterized by fitting Gaussian models for the color pattern of face regions, and non-face regions; (ii) the histogram of oriented gradients capture the high interest areas in faces that are rich in gradient information (eyes, nose and mouth) that are quite robust to pose variations, and (iii) eigenfaces captures the holistic appearance of the human face. These three feature channels capture a mix of global and local information about the face, and are robust to variations in pose.

Our algorithm was tested on the MIT+CMU face dataset [10]. This dataset has two parts. The first part (A) has 130 frontal face images with 507 labeled faces, the second part (B) has 208 images containing 441 faces of both frontal and profile views. The results of our algorithm are presented in Figure 2(a). Most other algorithms that are evaluated on this dataset do not provide the full ROC, but rather provide certain points on the ROC. Since Viola and Jones [9]

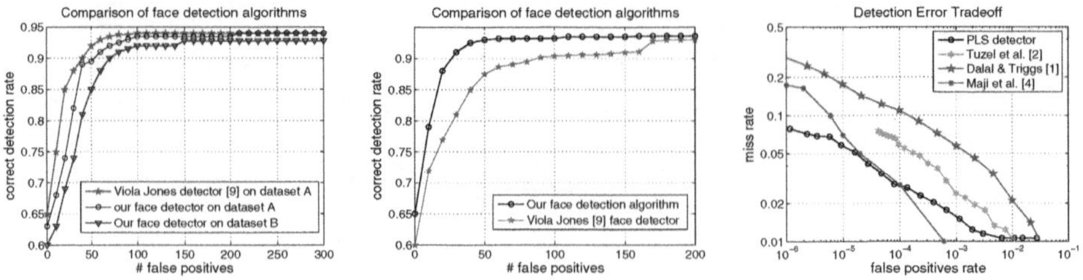

(a) MIT+CMU dataset (b) Maritime dataset (c) INRIA person database

Fig. 2. Experimental results for face and person detection

quote their ROC for part A of this dataset, we have compared our ROC with theirs; even otherwise, it can be observed that our performance is comparable to the ROC points of other algorithms (like Rowley et al. [10]). Since we are interested in detecting partially occluded faces we also compare our approach to the OpenCV implementation of Viola and Jones [9] method on the internally collected maritime dataset in Figure 2(b).

2.2 Person Detection

For person detection we use a method that combines HOG [1] and features extracted from co-occurrence matrices [16]. For each detection window, features extracted from HOG and co-occurrence matrices are concatenated and projected onto a set of latent vectors estimated by the partial least squares (PLS) method [17] in order to reduce the dimensionality of the input vector. The vector obtained after dimensionality reduction is used as the feature vector to describe the current detection window. Finally, the feature vector is classified by a quadratic classifier as either human or non-human sample. As a result, we obtain a probability estimate. Figure 2(c) shows comparisons using the INRIA person dataset [1]. Like face detection, the person detection approach used also achieves results comparable to state-of-art person detectors [1,2,4].

Since part-based approaches are better suited to handle situations of occlusion, we split the person detector into seven different detectors, which consider the following combinations of regions of the body: (1) top, (2) top-torso, (3) top-legs, (4) torso, (5) torso-legs, (6) legs, and (7) full body, as illustrated in Figure 3. Therefore, at each position in the image the person detector estimates a set of seven probabilities. The training for these detectors was performed using the training set of the INRIA person dataset.

As discussed in the literature survey, part-based approaches for person detection have been employed previously. Here, we use a part-based approach in tandem with a face detector creating a small number of intuitive case-based models for overall person detection.

Although the face and person detectors present results comparable to the state of the art on these datasets, these algorithms face difficulties when there is significant occlusion. To this end, we explore how to overcome this problem by combining the responses of the individual detectors.

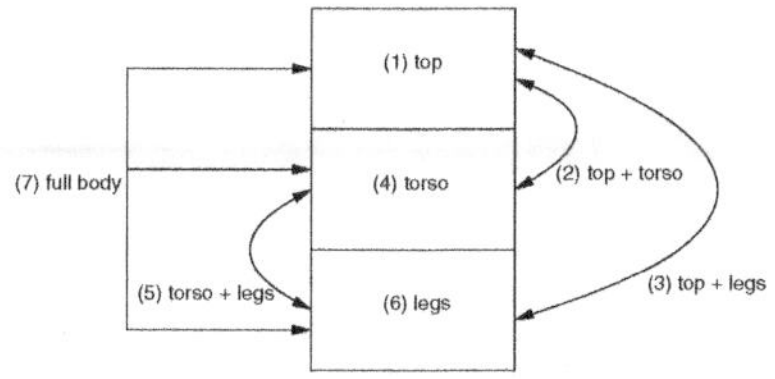

Fig. 3. Parts of a detection window used to train multiple detectors

3 Integrating Face and Person Detection

In this section we present our algorithm for integrating the response profiles of face and person detectors. We model observations of the individual detectors, and generate hypotheses that capture intuitive relationships between the responses of the face detector and the person detector. Specifically, we describe a set of situations where the output of one detector can be logically combined with the other detector's output to eliminate false alarms or confirm true positives.

3.1 Modeling the Response Profiles of the Individual Detectors

To integrate person and face detectors' output we first create models according to the probability profile resulting from individual detectors (the seven probabilities from part-composition person detector and one from the face detector).

For the person detector, we summarize the probability profile obtained by the seven probabilities into a set of four models that inherently capture situations in which various combinations of face and person parts are detected with *high* probability. Specifically,

Model M_1: all body parts are visible
Model M_2: top is visible, torso and legs may or may not be visible. This corresponds to the typical situation in which a person's legs are occluded by some fixed structure like a desk, or the railing of a ship.
Model M_3: top is invisible, whereas torso and legs are visible
Model M_4: all body parts are invisible

Given the set of seven probabilities estimated by the person part-combination detectors, we define probability intervals that characterize each model. The estimation of the intervals for models M_1 and M_4 can be done automatically by evaluating probability of training samples from standard person datasets. However, probability intervals for models M_2 and M_3 only can be estimated if a training set containing partially occluded people were available. Due to the absence of such dataset, we define the probability intervals for M_2 and M_3 manually.

Figure 4 shows the probability intervals for each model. A model M_i fits a detection window if all seven estimated probabilities fall inside the probability intervals defined by M_i. We also estimate a degree of fit of a detection window to each model by simply counting the number of probability intervals satisfied by the response profile:

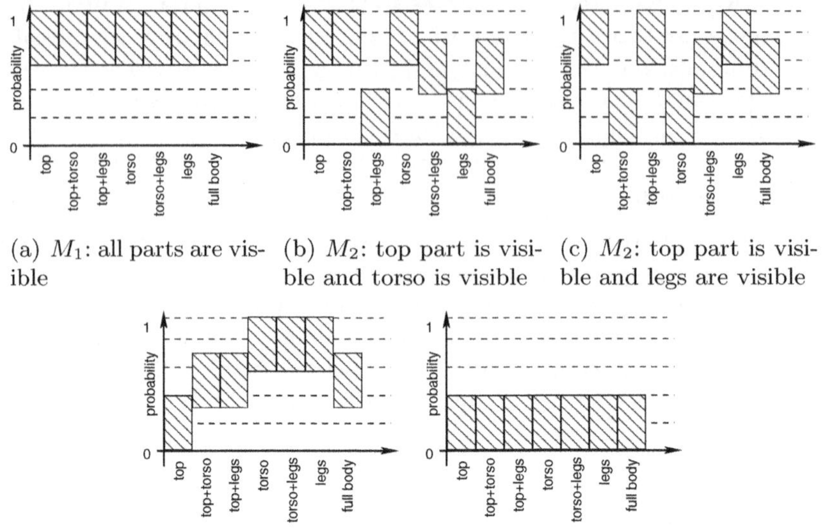

(a) M_1: all parts are visible

(b) M_2: top part is visible and torso is visible

(c) M_2: top part is visible and legs are visible

(d) M_3: top part is invisible

(e) M_4: all parts are invisible

Fig. 4. Models designed considering the output profile of the person detector. The x-axis has the seven detectors and the y-axis the probability interval for each one according to the model. Note that M_2 has two sub-cases, shown in (b) and (c).

$$f(M_i) = \sum_{j=1}^{7} \begin{cases} 1 & \text{if } u_{i,j} \leq P_j \leq l_{i,j} \\ 0 & \text{otherwise} \end{cases} \tag{1}$$

where P_j denotes the probability estimated by the j-th detector, $u_{i,j}$ is the upper bound for the j-th interval defined for M_i and $l_{i,j}$ denotes the lower bound. Therefore, we can rank the models according to how well they fit a given detection window. We say that a model M_i has a rank higher than M_j when $f(M_i) > f(M_j)$.

For the face detector, the observations are characterized by the probability values indicating the presence of face for a given detection window. According to this probability we define three models. We say that a face is present if the probability exceeds a certain threshold (model F_1). We also consider the case when the probability is smaller than the threshold but not negligible (i.e. face might be partially occluded), we refer to this as model F_2. Model F_2 is interesting when the person detector gives a response that supports the low (but not negligible) confidence of the face detector. Finally, we say that a sample fits model F_3 if the probability of face detector is very low.

3.2 Generating Hypotheses to Integrate Detectors

Now that we have designed models according to the response profiles to capture occlusion situations, we create a set of hypotheses (rules) to characterize the

relation between the detector responses so that these different sources of information can be used to *verify* each other's output. We separate the possibilities into five different hypotheses. The first two hypotheses describe the scenario where the person detector (PD) is used to *verify* the output of the face detector (FD), and the remaining three hypotheses deal with the alternate scenario of using face detector to verify the person detector outputs. The hypotheses are described in the form of conditional rules as follows.

H_1 : $[(f(M_1) \wedge f(M_2)) > (f(M_3) \wedge f(M_4))|F_1]$ Given that the face detector provides high response for a detection window, we look at the models that characterize the person detector output. Since the face is visible, the output of PD should better fit models M_1 or M_2 than M_3 and M_4 since we expect the top (head and shoulder) features to be detected by the person detector. If that is the case, then PD output verifies that the FD output is correct. Thus, a person is present at that location.

H_2 : $[(f(M_3) \vee f(M_4)) > (f(M_1) \wedge f(M_2))|F_1]$ The alternate case is, given high response for the face detector, if the output of PD fits either M_3 or M_4, then PD indicates that the face is not visible, and hence the output of the FD is a false alarm.

H_3 : $[(F_1|(f(M_1) \vee f(M_2)) > (f(M_3) \wedge f(M_4))]$ Given that the rank of M_1 or M_2 is greater than M_3, if FD gives a high response, then the face detector is reinforcing the output of the person detector. Thus, we conclude that a person is present at the corresponding location.

H_4 : $[(F_2|f(M_3) > (f(M_1) \wedge f(M_2) \wedge f(M_4))]$ A slightly different case from H_3 is when FD has low response, but still has some probability higher than 0 but not high enough to conclude the presence of face. In this case, if for the person detector the rank of M_3 is higher than M_1, M_2, and M_4, then we still decide that there is a person whose face is partially occluded. This is because M_3 captures the situation where the face is occluded, while the torso and legs are visible.

H_5 : $[F_3|(f(M_1) \vee f(M_2) \vee f(M_3)) > f(M_4)]$: This final hypothesis deals with the case where the output of person detector fits either M_1, M_2, or M_3, and the probability outputted by the face detector is negligible, so that it cannot come under H4. In such a case, since the face is completely invisible, we decide that the PD output is a false alarm.

Essentially, the above hypotheses are built on the fact that the presence of the face implies the presence of a person and vice-versa. We do need some confidence value for the presence of face to make decisions on the output of the person detector. This is based on our observation that the presence of just the torso and legs with no information regarding the face is not a strong cue to detect a person. This condition gives rise to many false alarms.

4 Experimental Results

In this section, we demonstrate with experiments how our integration framework improves detection under occlusion, as well as reduces the false alarms. We tested

(a) (b) (c) (d) (e) (f)

Fig. 5. Results on images from maritime dataset (better visualized in colors)

our algorithm on challenging images taken from an internally collected maritime dataset. It contains images of 3008×2000 pixels, which is suitable for face and person detection, unlike standard datasets used for person detection, which in general contain images with resolution too low to detect faces. This dataset is a good test-bed since it provides challenging conditions wherein the individual face/person detector might fail, thereby emphasizing the need to fuse information obtained by these detectors.

We now present several situations where the integration framework helps to detect humans. In the image shown in Figure 5(b) a person detector would fail to detect people seated since the lower body is occluded. However, our framework combines face information with the presence of the top part of the body (head and shoulders) captured by the person detector. Therefore, it concludes that a person is present. Additionally, Figures 5(c), (e), and (f) contain people who are partially occluded. Such conditions would reduce significantly the probability estimated by an independent person detector, whereas the integration helps resolve this problem.

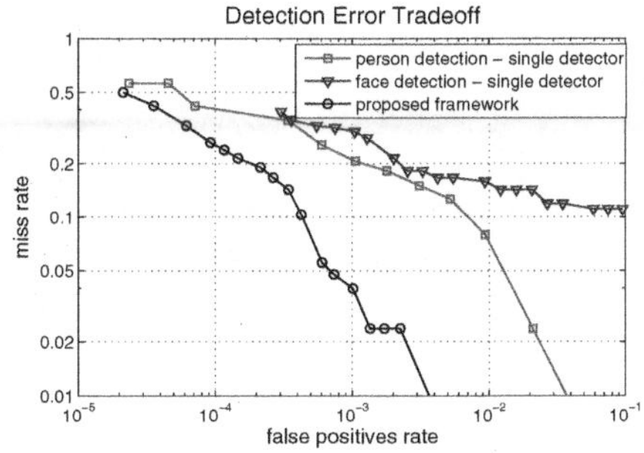

Fig. 6. Detection error tradeoff comparing the integration to individual face and person detectors. The proposed framework outperforms the individual detectors for all points on the curve.

Next, if the face is partially occluded, then the person detector output will belong to model M_3, whereas face detector's output will have some small value that is not very high and not negligible either. In this case, the person detector results can be used to identify the presence of the face. For example, Figures 5(d) and (f) contain people whose faces are occluded. In these cases a face detector would fail to give a high response, but the proposed framework overcomes this problem by aggregating information from body parts.

Essentially, since we are using two separate detectors, if the observations of the person detection and face detection provide conflicting information, then our framework mitigates false positives. A typical example is when hypothesis H_2 is satisfied, which can be used to correct the false alarm of the face detector, and when hypothesis H_5 is satisfied, that helps in reducing the false alarms of the person detector. Additionally, if both individual detectors denote the presence of a person, detection is more reliable than when relying on only one detector.

We tested our algorithm on 20 images containing 126 people. Figure 6 presents the detection error tradeoff of our integration method and compares its results to individual detectors. It can be seen that the use of the proposed method results in a substantial improvement in detection accuracy/false alarm suppression. To generate the curve for the our algorithm, we fix the threshold for the face detector and for the person detector we measure how well each model fits a sample by

$$g(M_i) = \frac{1}{7} \sum_{j=1}^{7} \begin{cases} |P_j - u_{i,j}| & \text{if } P_j > u_{i,j} \\ |P_j - l_{i,j}| & \text{if } P_j < l_{i,j} \\ 0, & \text{otherwise} \end{cases} . \tag{2}$$

With this equation we obtain values of $g(M_i)$ for every sample. Then, varying a threshold value from zero to one we are able to evaluate which hypotheses are satisfied at each step.

5 Conclusions

We have described a framework that combines the observations of face and person detector into different models, and makes decisions based on the hypotheses derived from those models. We then demonstrated our algorithm on several challenging images with considerable occlusion, which illustrates the advantages of exploiting feedback between the response profiles of face and person detectors.

Acknowledgements

This research was partially supported by an ONR MURI Grant N00014-08-10638. W.R. Schwartz acknowledges "Coordenação de Aperfeiçoamento de Pessoal de Nível Superior" (CAPES - Brazil, grant BEX1673/04-1).

References

1. Dalal, N., Triggs, B.: Histograms of oriented gradients for human detection. In: CVPR, vol. 1, pp. 886–893 (2005)
2. Tuzel, O., Porikli, F., Meer, P.: Human detection via classification on riemannian manifolds. In: CVPR, pp. 1–8 (2007)
3. Wu, B., Nevatia, R.: Optimizing discrimination-efficiency tradeoff in integrating heterogeneous local features for object detection. In: CVPR, pp. 1–8 (2008)
4. Maji, S., Berg, A., Malik, J.: Classification using intersection kernel support vector machines is efficient. In: CVPR, pp. 1–8 (2008)
5. Wu, B., Nevatia, R.: Detection of multiple, partially occluded humans in a single image by bayesian combination of edgelet part detectors. In: ICCV, pp. 90–97 (2005)
6. Mikolajczyk, K., Schmid, C., Zisserman, A.: Human detection based on a probabilistic assembly of robust part detectors. In: Pajdla, T., Matas, J(G.) (eds.) ECCV 2004. LNCS, vol. 3021, pp. 69–82. Springer, Heidelberg (2004)
7. Shet, V., Neumann, J., Ramesh, V., Davis, L.: Bilattice-based logical reasoning for human detection. In: CVPR, pp. 1–8 (2007)
8. Yang, M.H., Kriegman, D.J., Ahuja, N.: Detecting Faces in Images: A Survey. IEEE Transactions on Pattern Analysis and Machine Intelligence, 34–58 (2002)
9. Viola, P., Jones, M.: Robust Real-Time Face Detection. International Journal of Computer Vision 57, 137–154 (2004)
10. Rowley, H., Baluja, S., Kanade, T.: Neural network-based face detection. IEEE Transactions on Pattern Analysis and Machine Intelligence 20, 23–38 (1998)
11. Heisele, B., Serre, T., Poggio, T.: A Component-based Framework for Face Detection and Identification. IJCV 74, 167–181 (2007)
12. Moon, H., Chellappa, R., Rosenfeld, A.: Optimal edge-based shape detection. IEEE Transactions on Image Processing 11, 1209–1227 (2002)
13. Hsu, R., Abdel-Mottaleb, M., Jain, A.: Face detection in color images. IEEE Transactions on Pattern Analysis and Machine Intelligence 24, 696–706 (2002)
14. Belhumeur, P., Hespanha, J., Kriegman, D.: Eigenfaces vs. Fisherfaces: Recognition Using Class Specific Linear Projection. PAMI, 711–720 (1997)
15. Osuna, E., Freund, R., Girosit, F.: Training support vector machines: an application to face detection. In: CVPR, pp. 130–136 (1997)
16. Haralick, R., Shanmugam, K., Dinstein, I.: Texture features for image classification. IEEE Transactions on Systems, Man, and Cybernetics 3 (1973)
17. Wold, H.: Partial least squares. In: Kotz, S., Johnson, N.L. (eds.) Encyclopedia of Statistical Sciences, pp. 581–591. Wiley, New York (1985)

Multibiometric People Identification: A Self-tuning Architecture

Maria De Marsico[2], Michele Nappi[1], and Daniel Riccio[1]

[1] Universitá Degli Studi di Salerno,
via Ponte Don Melillo, 84084, Fisciano, Salerno, Italy
{mnappi,driccio}@unisa.it
[2] Universitá Degli Studi di Roma - La Sapienza,
via Salaria 113, 00198, Roma, Italy
demarsico@di.uniroma1.it

Abstract. Multibiometric systems can solve a number of problems of unimodal approaches. One source for such problems can be found in the lack of dynamic update of parameters, which does not allow current systems to adapt to changes in the working settings. They are generally calibrated once and for all, so that they are tuned and optimized with respect to standard conditions. In this work we propose an architecture where, for each single-biometry subsystem, parameters are dynamically optimized according to the behaviour of all the others. This is achieved by an additional component, the *supervisor module*, which analyzes the responses from all subsystems and modifies the degree of reliability required from each of them to accept the respective responses. The paper explores two integration architectures with different interconnection degree, demonstrating that a tight component interaction increases system accuracy and allows identifying unstable subsystems.

1 Introduction

Present systems generally rely on a single biometry. The main drawback is that they are singly vulnerable to possible attacks, as well as little robust with respect to a number of problems. Examples are acquisition errors, possibly due to bad hardware (e.g. a dirty sensor), as well as to an actual distortion of the biometric feature (e.g. a voice altered by a cold or a dimly lit face). In the present work we will consider the combination of iris-, face- and ear-based biometric identification systems. We chose these biometries because they are contact-less, fast and fairly reliable, as we will briefly discuss in the following.

The spatial patterns in the human iris make up a texture information which is highly distinctive of an individual, due to differences in the development of anatomical structures. In particular, biomedical literature [1] suggests that iris is as distinct as patterns of retinal blood vessels, but an iris image can be more easily obtained than a retina image. On the other hand, face is the most immediate characteristic through which we use to recognize a known person, so that many researchers' attention focuses on face recognition [2]. Moreover, people

M. Tistarelli and M.S. Nixon (Eds.): ICB 2009, LNCS 5558, pp. 980–989, 2009.
© Springer-Verlag Berlin Heidelberg 2009

are especially used to be photographed and this makes this automatic recognition modality sufficiently accepted. Face recognition, however, raises a number of non-easy to solve issues, especially when dealing with pose, illumination or expression changes. This reduces reliability of related methods, that are still far from the recognition rates usually provided by DNA or fingerprints. Ear shows some advantages when compared to face. Iannarelli's in-depth medical investigations [3] demonstrated that also the ear meets the basic principle of distinctiveness, so that it is a biometry in every respect. The limited surface allows a faster processing than face, while the lack of expressive variations reduces intra-class variations. However, when the ear in not perfectly frontal with respect to the acquisition device, small variations can modify the obtained image even significantly. Moreover, occlusions due to earrings or hair might be more serious than an occlusion on a face, because of the limited exploitable surface.

A multibiometric system provides an effective alternative, as flaws of an individual system can be compensated by the availability of a higher number of cooperating biometries [4]. In [5] System Response Reliability (SRR) indices have been introduced to evaluate the reliability of each response of single subsystems; responses are considered for fusion only if the corresponding SRR is higher than a given threshold *th*. This improves global system performance, but we argue that we can go even further by considering the "'history"' of the system. We assume the existence of a *supervisor module* exploiting not only single subsystems responses and their reliability, but also the final global response, to evaluate the overall system state and update reliability thresholds of single subsystems. Such module would allow overcoming the invariance of present multibiometric architectures, by implementing an algorithm converging to an optimal parameters configuration independently from the starting one. Each subsystem takes as input an image, pertaining to the corresponding biometry, and extracts salient features; the resulting feature vectors are used to compute similarity scores, to be inputted to the supervisor module, after a normalization step. As all the algorithms exploited for iris, face and ear rely on fractal theory, and in particular on PIFS (Partitioned Iterated Function System) [6], we briefly introduce some related core topics, and then describe the feature extraction procedures.

2 The Integration Scheme

We propose two multibiometric architectures with increasing complexity, a Parallel Protocol (PP) and a Supervised Protocol (SP). The latter significantly differs from the state of art in literature. The algorithm implemented by the recognition systems to classify the single biometry is always the same presented above. Therefore, the reported different performances can be unequivocally ascribed to the peculiar features of the different architectures. This consideration adds value to the experimental results in terms of consistency and readability.

2.1 The Parallel Protocol

In this first solution, PIFS are applied independently to each trait. The three corresponding subsystems work in parallel, each producing a list of subjects from

its specific database of enrolled subjects (gallery); list elements include the ID of a database subject and a numeric score expressing its similarity with the input. Each list is ordered by such similarity, then the three lists are processed by a fusion module to obtain the global response. Two aspects have been especially investigated in designing this system: *reliability margins* and *fusion policies.*

Subsystems might not be equally reliable, e.g. due to the possibly different quality of input. An unreliable response can trigger a further check. A *reliability* measure is then crucial for fusion. Some solutions use margins, measuring the "risk" associated to a single subsystem response after observing its scores. Poh and Bengio [7] introduce a confidence margin based on False Acceptance Rate (FAR) and False Rejection Rate (FRR). Many responses are marked as reliable, as the margin relies on an estimate of the actual distribution of genuine/impostor subjects' scores. This might be inappropriate when very high security is required. Moreover, frequentist approaches assume that the scores of the testing and development sets always originate from similar distributions. We adopt the new System Response Reliability (SRR) index [5], based on a system/gallery dependent metric, and measuring the ability of separating genuine subjects from impostors on a single probe basis. SRR values always fall in the range $[0, 1]$. Each subsystem T_k computes, for each $s_{k,i}$, i=1, ..., in its set of responses, a *reliability measure* $srr_{k,i}$. Moreover, each subsystem T_k is characterized by an estimated threshold th_k, possibly updated over time [2], such that a response $s_{k,i}$ is considered as reliable only if $srr_{k,i} \geq th_k$.

Different policies exist for fusion. Veeramachaneni et al [8] investigate decision level fusion in a binary hypotheses-testing problem, and in particular the selection of the best subset from different fusion rules, to optimize FAR or FRR. Apart from the limitations of the approach, the interesting observations are that, for 2 classifiers fusion, AND and OR rules constitute a very important set (out of 16 possible ones), and that, as the number of classifiers increases, the optimal fusion rules are constructed from these two, giving better results than Averaged Sum Rule. This suggests that, despite the variety of fusion policies, we can assume that AND and OR generally represent a suitable choice. In our case, we have a system S composed by 3 subsystems T_k, $(k = 1, 2, 3)$, processing iris, face and ear respectively, each exploiting a gallery G_k, $(k = 1, 2, 3)$ of images for the same set H of enrolled identities. Notice that each subject can have more than one image in each single gallery. Our fusion rules also exploit the above System Response Reliability (SRR), since each T_k returns a reliability value $srr_{k,i}$ for its i-th response. As for OR, the combined response is valid if at least one subsystem response is reliable; the system returns the identity retrieved by the subsystem with the higher reliability above the corresponding threshold. In AND policy, the combined response is valid only if all responses are reliable; the system returns the identity with the minimum weighted sum of distances from the probe, where weights are the reliability indexes. For sake of consistency, it is important to normalize distances before combining them.

Thresholds th_k can be fixed in advance, and remain fixed in time, or they can be computed and updated according to obtained responses. A compromise

between the number of reliable responses and the system error rate must be obtained. Too high thresholds make the system too restrictive, with a low error rate but also a low number of acceptances, while too low ones might cancel the advantages of the reliability measure. Assume that T_k has executed M times, producing the set $\{s_{k,1}, \ldots, s_{k,M}\}$ of responses. The corresponding reliability measures are combined in a set $RH = \{srr_{k,1}, \ldots, srr_{k,M}\}$ containing the history of system behaviour (reliability history). The value to assign to th_k is strictly correlated to RH features, in particular to the mean and to the variance [5] of its elements. A high $\bar{RH}$ means that system responses are generally reliable so that its threshold can be proportionally high. On the other hand, the variance $\sigma^2[RH]$ measures the stability of T_k. The best situation is then when RH elements have a high mean and a low variance, so that it is possible to fix a high value for th_k. We can summarize the above observations in the formula:

$$th_k = \frac{\bar{RH}^2 - \sigma^2[RH]}{\bar{RH}}, \tag{1}$$

Our Parallel Protocol system, in particular th_k thresholds, has been adjusted according to a set of samples with similar features to those of the set used for identification tests; the two sets are different and disjoint.

2.2 A Self-tuning Architecture: The Supervised Protocol

A revisitation of the classical multibiometric schema takes to system self-tuning, with a much more flexible and robust architecture. The main limit of the architecture in Section 2.1 is that it does not seize the main advantage of exploiting information coming from other subsystems. Each component works independently and final results give no feedback for the overall system. Formula 1 is good for computing and updating the th_k threshold for the reliability index of a single subsystem, according to its behaviour; however, it only accounts for that subsystem. On the contrary, assume the existence of a *supervisor module* (Supervised Protocol - SP) still exploiting single subsystem responses and their reliability to compute the final global response, but also using the latter to evaluate the overall system state and update its parameters (Fig. 1). Such module would implement an algorithm to update single thresholds also according to the behavior of the other subsystems, so converging to an optimal configuration independently from the starting $\{th_1, th_2, th_3\}$ configuration. The algorithm distinguishes two cases:

Case I: More identities I_j, $j \in \{1, 2, \ldots, |H|\}$ share the same maximum number of votes, e.g. when retrieved identities are all different with 1 vote each. If at least one T_k in any such group has $srr_k > th_k$, the system returns the identity retrieved by the subsystem with the higher $srr_k > th_k$, and the response is marked as reliable, otherwise the response is marked as unreliable.

Case II: One identity I_j gets more votes than the others. I_j is returned and the response is marked as reliable.

In both cases, if the response is reliable, each subsystem T_k voting for the returned identity is rewarded by lowering its threshold th_k by an updating step us, unless its current srr_k is already above its th_k. Each other subsystem T_k is penalized by increasing its threshold th_k by us, unless its current srr_k is already below its th_k. In this way the supervisor module lowers thresholds of subsystems voting in agreement, considering such behavior a confirmation of reliability, and increases thresholds of discordant ones, compensating possible distortions (local persistent distortions like lighting variations, dirt on lens).

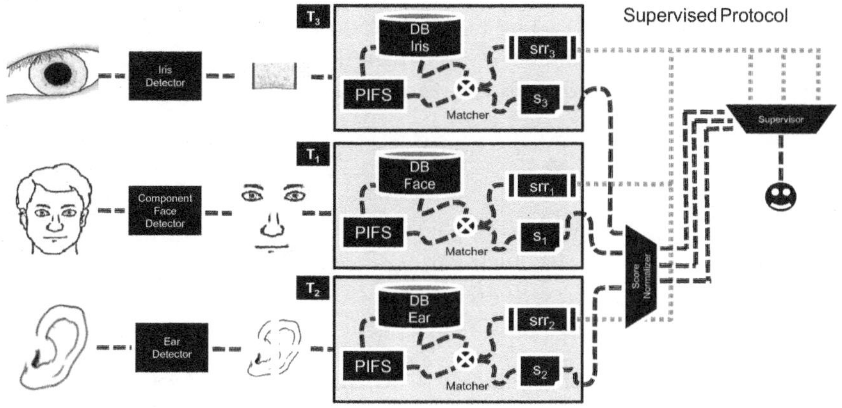

Fig. 1. The architecture of the supervised protocol

Such an architecture does not need an adjustment phase, since the system can start from a default configuration of its parameters and converge in any case towards an optimal one. The speed to reach such latter configuration is a significant system feature, so that it is important to define how to measure it. As we want to simulate the dynamic behavior of an online identification system, we assume that system time is beaten by performed recognition operations; we define a *probe sequence* $P = \{p_1, p_2, \ldots, p_n\}$ as a series of n probes presented to the system, sharing the same acquisition characteristics (normal conditions, right light, earrings, dirty lens). A subsystem equilibrium state (*steady state*) is given by the consecutive instants when threshold fluctuations are lower than a fixed μ, while *convergence speed* λ_k of a subsystem is defined as the ratio between the total variation of its threshold between two consecutive steady states, and the number of instants needed to obtain such transition. Total system convergence speed is defined as the minimum speed among all its subsystems, i.e. $\lambda = min_k \lambda_k$, $k \in \{1, 2, 3\}$.

3 Experimental Results

A suitable standard image database was chosen for each biometry: a) CASIA v 3.0 for iris [9], AR-Faces for face [10] and Notre-Dame for ear [11]. 100 different

chimeric users were enrolled, by extracting the first 100 subjects from each database. In particular, 100 images of iris, face and ear (gallery set) were used for enrolling; further 100 images (tuning) for the same subjects, but distinct from those in the gallery, always in normal conditions, were used to compute Parallel Protocol (PP) parameters, namely th_k thresholds corresponding to reliability index *SRR*. Two different sets of images have been considered for testing (probe set). The first set *probe-I* consists of normal conditions images different from those in the gallery and tuning sets; the second set *probe-II* contains the same images as *probe-I* except that a gaussian filter was applied to them, with kernel 8×8 and variance $\sigma = 3$, to simulate an out of focus situation. Images were manually segmented. As for irises, the center of the pupil, the minor radius and the major radius of the useful region were selected. Once extracted, such iris region was linearized and resized to a rectangle of 256×360 pixels. For each face, eyes, nose and mouth were located, while the useful region was resized to a square of 256×256 pixels. As for the ears, the upper left and the bottom right angle of the rectangle containing the ear were selected, in order to resize it to 200×256 pixels. Adopted measures for accuracy were Recognition Rate (RR), Equal Error Rate (EER) and Number of Reliable Responses (NRR). For each subsystem, the feature extraction process rely on PIFS (Partitioned Iterated Function System). PIFS are largely exploited in image compression and indexing, and they have also been investigated in the field of face and ear recognition [6].

The first experiment consisted in analyzing the performances of the different subsystems, which are summarized in Table 1. We can notice that, in normal conditions, the face shows the better performances, followed by iris and finally ear. This can be ascribed to two reasons: a) quality of iris images is lower than face images b) the exploited fractal techniques were especially optimized for face recognition. This is also confirmed by the fact that with "out of focus", when simulated through gaussian blur, performances of face recognition only drop of 0.07, while those of iris and ear drop of 0.50 and 0.19 respectively. The inversion between ear and iris with blurred images might be due to the fact that the thin characteristic patterns in iris are much more affected by such operation.

3.1 Comparisons among the Two Architectures

In examining results, bear in mind that we are applying a dynamic architecture, while a static multibiometric scheme is usually adopted. We chose different test configurations, with probe sequences varying in the biometry or biometries bearing blurring distortion (e.g. normal iris, normal face and blurred ear). System performances are summarized in Table 2. To clarify the interpretation of the experiments, we notice that each of the selected image sets can be considered as a sequence P of test images, sequentially submitted to the system. The initial thresholds configuration for SP is set at $\{th_1 = 0.0, th_2 = 0.0, th_3 = 0.0$, i.e. all responses are initially considered as reliable. The updating step *us* is fixed at 0.05. Table 2 shows that RR obtained with SP is comparable, or in some cases even better, than with PP, also obtaining a higher Number of Reliable Responses (NRR). More detailed observations pertaining the obtained RR and NRR can

Table 1. Performance of single subsystems on *probe-I* and *probe-II*

SubSets		Single Subsystems Performances	
		RR	EER
Iris	Normal	0.80	0.16
	Blurred	0.30	0.22
Face	Normal	0.97	0.03
	Blurred	0.90	0.07
Ear	Normal	0.72	0.08
	Blurred	0.53	0.14

also result from Table 2. NRR for PP never is never above 52, even guaranteeing a mean RR of 0.9967. On the other hand, SP always succeeds in providing a mean NRR of 99.5, with and RR of 0.97 in the worst case. This can be ascribed to the fact that PP implements a rigid AND policy, strongly limiting NRR to such cases when SRR indices of all voting systems are above the corresponding thresholds. On the contrary, SP shows a comparable mean value for RR of 0.985, yet providing a mean NRR of 99.5. While in PP the drop in performances of one or more subsystems weighs on the global result by compromising the overall AND outcome, in SP the increment in the corresponding thresholds causes to rule out unreliable subsystems, so avoiding to decrease NRR though preserving an high RR. A further relevant observation is related to the way the two protocols set up th_k thresholds for reliability indexes. PP computes such thresholds once and for all using an equivalent tuning set, according to the formula (1), so that the face, which is characterized by an higher average reliability (high mean), presents an high value for its threshold th_2. On the contrary, SP, which also takes into account what happens in the other two subsystems, rewards the face recognizer with a lower threshold. In other words, it implements an even opposite philosophy, taking to better performances. If a subsystem is generally in agreement with the others, we can accept also "less" reliable responses from it. Finally, notice how, in SP, a decrease in input quality in the face subsystem causes a corresponding increase in the reliability threshold. This does not undermine the correctness of the formula (1), but demonstrates its inadequacy in the case of multibiometric systems.

3.2 Further Experiments on the Supervised Protocol

The first experiment aimed at highlighting how the threshold configuration towards which the system converges is independent from the starting one, and how the threshold updating algorithm is stable. We considered 100 probe sequences of 1000 images randomly extracted among the 100 of sets *probe-I* (face and ear) and *probe-II* (iris). For each system run, the initial values for thresholds were randomly chosen (all values being equally probable) in the interval $[0, 1]$. Results of this experiment are reported in Fig. 2, showing that curves generated by the different probe sequences tend to always concentrate in a relatively small final interval. This confirms the convergence of the updating procedure. We can

Table 2. Performance comparison on different input configurations: Parallel Protocol (PP) and Supervised Protocol (SP)

Configurations			Multi-Biometric Architectures							
			Parallel Protocol (PP)				Supervised Protocol (SP)			
			Perf.	th_1	th_2	th_3	Perf.	th_1	th_2	th_3
Iris	Norm	RR	1.00				1.00			
Face	Norm	EER	0.00	0.06	0.15	0.02	–	0.04	0.02	0.10
Ear	Norm	NRR	43				100			
Iris	Norm	RR	1.00				1.00			
Face	Norm	EER	0.01	0.06	0.15	0.16	–	0.10	0.04	0.62
Ear	Blur	NRR	36				100			
Iris	Norm	RR	1.00				0.99			
Face	Blur	EER	0.00	0.06	0.09	0.02	–	0.06	0.13	0.06
Ear	Norm	NRR	45				100			
Iris	Blur	RR	1.00				0.97			
Face	Norm	EER	0.00	0.44	0.15	0.02	–	0.84	0.02	0.12
Ear	Norm	NRR	52				99			
Iris	Norm	RR	1.00				0.98			
Face	Blur	EER	0.04	0.06	0.09	0.16	–	0.06	0.10	0.40
Ear	Blur	NRR	39				100			
Iris	Blur	RR	0.98				0.97			
Face	Norm	EER	0.02	0.44	0.15	0.16	–	0.94	0.02	0.64
Ear	Blur	NRR	41				98			

observe that the threshold th_2 (face) tends to assume lower values than th_1 and th_3 (iris and ear). Notice that the curves pertaining to subsystems iris and ear span a larger interval; this fact underlines that, as they are less reliable on average, they are more easily influenced by the other subsystems. The darker line (in black) in Fig. 2 is the mean value of the 100 computed curves and represents the mean trend for thresholds variation. For each threshold, the mean ({0.92, 0.02, 0.12}) and variance ({0.08, 0.02, 0.09}) of values in the 100 iterations were computed; the same for RR (mean: 0.981, variance: 0.004) and NRR (mean: 980, variance: 1.09). Fig. 3 shows the thresholds variation over a probe sequence composed of 5 subsequences of 200 images each. For each subsequence, we simulate a malfunctioning for one or more subsystems by randomly extracting their input images from *probe-II*. The other subsystems will randomly extract input images from *probe-I*. In Fig. 3, we show this by using "N" for normal images (*probe-I*) and "B" for blurred ones (*probe-II*). Each subsequence is composed of randomly extracted images from sets *probe-I* (Normal Images - N) and *probe-II* (Blurred Images - B). Notice that for the first sequence the thresholds th_1 e th_3 reach higher values than th_2, consistently with the fact that on set *probe-I* face presents the higher reliability. The configuration dramatically changes in the third sequence, in which iris images come from set *probe-II*, where the presence of blur lowers reliability of subsystem T_1. In sequence 4, the blur effect on input ear images lowers performances of T_3 subsystem, while the only one to remain more reliable is T_1. Fig. 3 shows some examples of the computation of the subsystems convergence speed, namely for T_3, and of the identification of equilibrium states with the corresponding value of μ, for T_1; it is also to notice a case when all the subsystems are in equilibrium, so that we can speak about *system equilibrium*.

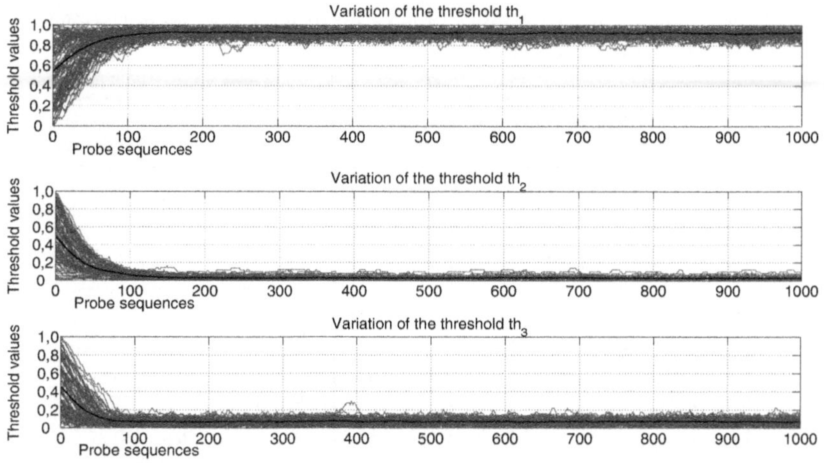

Fig. 2. Variation of thresholds of subsystems T_1, T_2 and T_3 for 100 probe sequences of 1000 images of 100 subjects according to the initial thresholds configuration ($\{th_1, th_2, th_3\}$); the black curve represents the mean trend

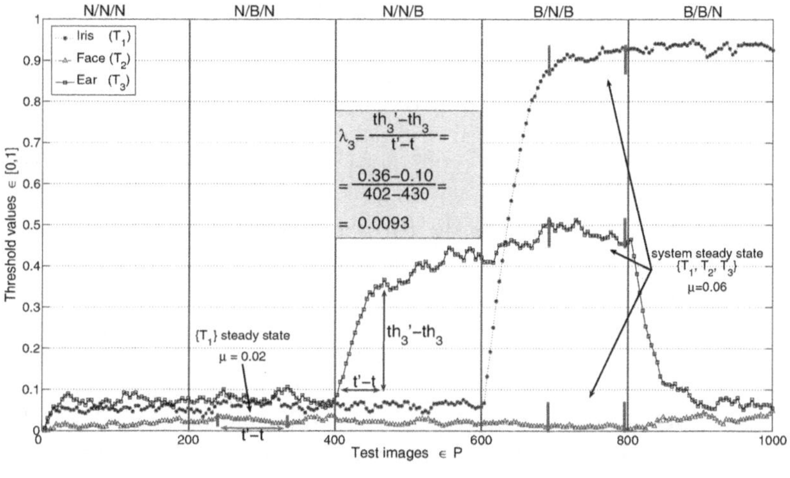

Fig. 3. Example of thresholds variation ($T_1/T_2/T_3$) over a probe sequence of 1000 images from set *probe-I* (Normal Images - N) and *probe-II* (Blurred Images - B)

4 Conclusions

Multibiometric systems, though resolving a number of problems of single-biometry ones, are generally somehow penalized from the invariance of their parameters, making them unable to adapt to changes in the conditions of their working environment. In this paper, we propose an architecture aiming at overcoming such limitation, through the introduction of a supervisor module. Such

additional component collects information from the different subsystems and exploits them to modify the internal conditions (parameters) of each subsystem, aiming at improving the global response. This work opens a line to further investigations, where aspects such as a deeper action of the supervisor module on the internal subsystems state, or gallery updating, represent potential integrations to the architecture presented herein.

Acknowledgements

This work has been partially supported by the Provincia di Salerno Administration, Italy.

References

1. Adler, F.H.: Physiology of the eye: Clinical Application, 4th edn. The C.V. Mosby Company, London (1965)
2. Abate, A.F., Nappi, M., Riccio, D., Sabatino, G.: 2d and 3d face recognition: A survey. Pattern Recogn. Lett. 28(14), 1885–1906 (2007)
3. Iannarelli, A.: Ear identification. In: Calif, Paramont Publishing Fremont (eds.) Forensic Identification Series, pp. 67–74 (1989)
4. Ross, A., Jain, A.K., Qian, J.-Z.: Information fusion in biometrics. In: Bigun, J., Smeraldi, F. (eds.) AVBPA 2001. LNCS, vol. 2091, pp. 354–359. Springer, Heidelberg (2001)
5. Riccio, D., De Marsico, M., Abate, A.F., Nappi, M.: Data normalization and fusion in multibiometric systems. In: International Conference on Distributed Multimedia Systems, DMS 2007, pp. 87–92 (2007)
6. Riccio, D., Tortora, G., Abate, A.F., Nappi, M.: Rbs: A robust bimodal system for face recognition. International Journal of Software Engineering and Knowledge Engineering 17(4), 497–514 (2007)
7. Poh, N., Bengio, S.: Improving fusion with margin-derived confidence in biometric authentication tasks. In: Kanade, T., Jain, A., Ratha, N.K. (eds.) AVBPA 2005. LNCS, vol. 3546, pp. 474–483. Springer, Heidelberg (2005)
8. Veeramachaneni, K., Osadciw, L., Varshney, P.K.: An adaptive multimodal biometric management algorithm. IEEE Transaction on Systems, Man, and Cybernetics-Part C: Applications and Reviews 35, 344–356 (2005)
9. Chinese Academy of Sciences. Note on casia-iris v.3. Website (2008), `http://www.cbsr.ia.ac.cnIrisDatabase.htm`
10. Martinez, A.M.: Recognizing imprecisely localized, partially occluded, and expression variant faces from a single sample per class. IEEE Transaction on Pattern Analisys and Machine Intelligence 24, 748–763 (2002)
11. Flynn, P.: Notre dame ear database, und principal investigator: Professor patrick flynn, computer science and engineering, university of notre dame, notre dame. Website, 2007-01-07 (2007)

Covariate Analysis for View-Point Independent Gait Recognition

I. Bouchrika, M. Goffredo, J.N. Carter, and M.S. Nixon

ISIS, Department of Electronics and Computer Science
University of Southampton, SO17 1BJ, UK
{ib04r,mg2,jnc,msn}@ecs.soton.ac.uk

Abstract. Many studies have shown that gait can be deployed as a biometric. Few of these have addressed the effects of view-point and covariate factors on the recognition process. We describe the first analysis which combines view-point invariance for gait recognition which is based on a model-based pose estimation approach from a single un-calibrated camera. A set of experiments are carried out to explore how such factors including clothing, carrying conditions and view-point can affect the identification process using gait. Based on a covariate-based probe dataset of over 270 samples, a recognition rate of 73.4% is achieved using the KNN classifier. This confirms that people identification using dynamic gait features is still perceivable with better recognition rate even under the different covariate factors. As such, this is an important step in translating research from the laboratory to a surveillance environment.

1 Introduction

In recent years, automated visual surveillance has received considerable interest in the computer vision community. This is largely due to the vital need to provide a safer environment. Because of the rapid growth of security cameras and the need for automated analysis, the deployment of biometric technologies becomes important for the development of automated visual surveillance systems. The suitability of gait recognition for surveillance systems emerges from the fact that gait can be perceived from a distance as well as its non-invasive nature. Although gait recognition is not sufficiently mature to be used in real world applications such as visual surveillance, it overcomes most of the limitations that other biometrics suffer from such as face, fingerprints and iris recognition which can be obscured in most situations where serious crimes are involved.

Gait can be affected by different covariate factors including footwear, clothing, injuries, age, walking speed, and much more akin with other biometrics. In fact, the effects of the different covariates for gait analysis and recognition have not been investigated much by medical and other researchers [1], This is mainly due to the lack of availability for databases, as well as the availability of automated systems which would help for the extraction of gait features. Moreover, the complexity of earlier model-based approaches has precluded their deployment for this analysis. The covariate factors can be related either to the subject as for the

M. Tistarelli and M.S. Nixon (Eds.): ICB 2009, LNCS 5558, pp. 990–999, 2009.
© Springer-Verlag Berlin Heidelberg 2009

case when a subject smiles for face recognition, or related to the environmental conditions such as lighting, nature of the ground or camera setup.

Much research for gait recognition has been done into identifying subjects recorded walking from the side-view. The effects of covariate factors on the performance of gait recognition have been investigated by only a few recent research studies. Sarkar *et al.* [2] described a baseline algorithm for gait recognition based on the temporal correlation of silhouette data. The algorithm is evaluated on a set of twelve experiments in order to examine the effects of the different covariates including viewpoint, footwear, walking surface, time and carrying conditions. However, their work lacks exploratory analysis of the different gait features under covariate data due to the use of the silhouette approach. Tan *at al.* presented an averaging silhouetted-based approach that was tested on the CASIA-B gait dataset with three main variations including clothing, carrying conditions and view angles [3]. Their experimental results showed that the performance of gait recognition is much affected at worst dropping to a recognition rate of just 1% for covariate dataset.

In this paper, a markerless model-based approach is used to investigate of the effects of the covariate factors including, clothing and carrying conditions for view-point independent gait recognition. This extends recent research studies by on covariate analysis [4] and view-point invariant gait recognition [5]. A novel reconstruction method is being employed to rectify and normalize gait features recorded from different view-point into the side-view plane and therefore exploit such data for recognition. As such, we show for the first time that covariate analysis and viewpoint invariance can be combined, thus handling important practical factors in the translation of gait from laboratory to surveillance analysis. This paper is structured as follows: the next section is devoted to the discussion of markerless method used for extracting gait features as well as the normalization approach used for reconstructing gait angular data into the side-view plane. Section 3 describes the gait recognition approach including the derivation of gait signatures and the classification process. Finally, the experimental results on a set of processed videos from CASIA dataset are drawn in the fourth section.

2 Automated Markerless Extraction of Gait Features

2.1 Estimation of the Joint Positions

To extract the gait features of walking subjects from the covariate dataset, we applied the model-based method described in [6] to automate the extraction process of the joint trajectories. To extract candidate joint positions, the Distance Transform is performed on the silhouettes of walking subjects. Spatial motion templates describing the motion of the joints are derived by manual gait analysis and used to aid the markerless extraction of the joint positions. A recursive evidence gathering algorithm is employed for the markerless extraction process whereby spatial model templates for the human motion are presented in a parameterized form invariant to scaling and rotation using the Elliptic Fourier Descriptors described in equation (1):

$$\begin{bmatrix} x(t) \\ y(t) \end{bmatrix} = \begin{bmatrix} a_0 \\ b_0 \end{bmatrix} + \begin{bmatrix} cos(\alpha) & -sin(\alpha) \\ sin(\alpha) & cos(\alpha) \end{bmatrix} \begin{bmatrix} F_x(t) \cdot s_x \\ F_y(t) \cdot s_y \end{bmatrix} \tag{1}$$

where $t \in [0, 2\pi]$, α is the rotation angle, s_x and s_y are the scaling factors across the horizontal and vertical axes respectively. a_0 and b_0 define the position of the shape's centre. $F_x(t)$ and $F_y(t)$ are computed using equation :

$$\begin{aligned} F_x(t) &= \sum_{k=1}^{n} a_{x_k} cos(kt) + b_{x_k} sin(kt) \\ F_y(t) &= \sum_{k=1}^{n} a_{y_k} cos(kt) + b_{y_k} sin(kt) \end{aligned} \tag{2}$$

where a_{x_k},a_{y_k}, b_{x_k} and b_{y_k} are the set of the elliptic phasors which can be computed by Riemann summation [7]. Gait knowledge is exploited via heel strike extraction to reduce the the parameter space dimensionality and therefore reduce the computational load of the evidence gathering algorithm. The Hough Transform is employed to determine the free parameters through the matching process of feature points across the whole sequence of frames to the parametric function, and increase votes in the accumulator space accordingly. The parameters are then determined as the index or key of the accumulator space with the largest value. In the latter phase of the evidence gathering process, an exhaustive local search is performed within every frame to locate the features (i.e., joint positions) whereby, the local search is guided by the motion pattern extracted during the first stage to limit the search area. To more accurately extract the joint positions and reduce the search space, the lower limbs pose estimation algorithm uses as a filtering process the proportions of the human body segments.

2.2 View-Point Rectification

The rectification method is applied to normalise gait features extracted from any viewpoint into the side-view plane. The method is based on four main assumptions: the nature of human gait is cyclic; subjects walk along a straight line during two gait cycles; the distances between the bone joints are constant; and the articulated leg motion is approximately planar.

Considering a subject walking along a straight line, multiple periods of linear gait motion appear analogous to a single period viewed from many cameras related by linear translation. Following this rationale, the positions of the points of interest, i.e. the leg joints, lie in an auto-epipolar configuration consistent with the imaged motion direction. The epipole is thus estimated by computing the intersection of the set of lines formed by linking the correspondent points of interest in each phase of the gait cycle. In order to find these correspondences, the gait periodicity is calculated by applying the stereopsis transformation that maps the epipole $\mathbf{e_0}$ to the ideal point $[1, 0, 0]^T$ and then by computing the cost based on dot product between matching limb segment vectors.

Let $\mathbf{j}_i^\ell$ be the set of joints positions for each leg $\ell = \{1, 2\}$ at the ith frame in the image reference system. After estimating the periodicity of gait, assuming linear velocity between consecutive frames, the set of points of interest $\mathbf{j}_i^\ell$ are

recomputed in order to lie on straight lines starting from the epipole. At first the set of points and the epipole are mapped to the unit square and re-normalized to the unit norm $\|\mathbf{e_0}\| = 1$ respectively. Subsequently, the optimal points are found by estimating the positions $\mathbf{j}_i^\ell$ that lie on the epipolar line and that satisfies the condition

$$\mathbf{j}_i^{\ell T} [\mathbf{e_0}] \times \mathbf{j}_i^\ell = 0 \tag{3}$$

Therefore the back projected rays, formed from a set of optimal points, intersect in a single worldspace point: the epipole. The back projection of all sets of points generates the cluster of 3D points for an assumed single period of reconstructed gait motion. The Direct Linear Transform, DLT, is then used in order to triangulate each worldspace point $\mathbf{J}^\ell$

$$\mathbf{j}_i^\ell \times \mathbf{P}_i \cdot \mathbf{J}^\ell = 0 \tag{4}$$

with the set of camera projection matrices

$$\mathbf{P}_i = \left[\mathbf{R}_{\mathbf{e}}^T, -i\mathbf{e}_0\right] \tag{5}$$

$\mathbf{R}_{\mathbf{e}}^T$ is the rotation matrix for aligning the epipolar vector $\mathbf{e}_0$ with the horizontal axis X. Then,

$$\mathbf{j}_i^\ell = \mathbf{P}_i \begin{pmatrix} 1 & 0 \\ 0 & \mathbf{H_V}^{-1} \end{pmatrix} \begin{pmatrix} 1 & 0 \\ 0 & \mathbf{H_V} \end{pmatrix} = \mathbf{H} \cdot \mathbf{J}^\ell \tag{6}$$

having expressed the limb plane transformation matrix with $\mathbf{H_V}$ so that the two cross section plane lines are centred and normalised respect to Y and Z axes and parallel with Y. By assuming the lengths of the articulated limbs $\mathbf{D}_\ell^2 = \Delta\mathbf{j}_i^{\ell \mathbf{T}} \Delta\mathbf{j}_i^\ell$ are constant over all the frames, the pose difference vectors for the limb segments at two consecutive frames, $\Delta\mathbf{j}_i^\ell$ and $\Delta\mathbf{j}_{i+1}^\ell$, are related by

$$\Delta\mathbf{j}_i^{\ell\mathbf{T}} \cdot \mathbf{H^T} \cdot \mathbf{H} \cdot \Delta\mathbf{j}_i^\ell = \Delta\mathbf{j}_{i+1}^{\ell\mathbf{T}} \cdot \mathbf{H^T} \cdot \mathbf{H} \cdot \Delta\mathbf{j}_{i+1}^\ell \tag{7}$$

After recovering the fronto-parallel structure of subject gait, the representation of the leg joints function $\left[\mathbf{J}_x^\ell(t), \mathbf{J}_y^\ell(t)\right]$ is found by fitting a modified Fourier series to the data with fixed fundamental frequency f_0 and period T:

$$\mathbf{J}_x^\ell(t) = v_x t + \sum_{k=1}^{n} A_k \cos\left(2\pi k f_0 \left(t + \frac{(\ell - 1) T}{2}\right) + \phi_k\right) + \mathbf{J}_{x0}^\ell \tag{8}$$

analogously for $\mathbf{J}_y^\ell(t)$. Thus, the projection of the leg joints on the lateral plane is obtained with an optimized procedure in the following way

$$\breve{\mathbf{J}}^\ell(t) = \left[h_1\ h_2\ h_3\right] g\left(t + \frac{(\ell - 1) T}{2} : f_0, \mathbf{D}_\ell, v_x, v_y, F\right) \tag{9}$$

where $g(t)$ is the bilateral Fourier series function with coefficients F and h are the values of the inverse normalization transform matrix. Therefore, starting from a video sequence from a single camera and without any calibration, the proposed markerless system estimates the gait parameters projected on the lateral plane.

3 Gait Recognition

The processing and derivation of good gait features from this trajectory-based data is a challenging problem due to the compound nature of gait motion inherent in the numerous variables associated with it including kinematics, kinetics and anthropometrics [8]. An important issue in gait recognition is the derivation of appropriate features that can capture the discriminative individuality from a subject's gait. Such features should respond to crucial criteria such as robustness and invariance to weather conditions, clothing and operating conditions.

In order to identify a subject by their gait, we derive the angular measurements, anthropometric measurements as well as the trunk spatial displacement which best describe the gait kinematics. The use of angular motion is very common in model-based gait analysis and recognition. The angles of the joints including the hip and the knee; are considered the most important kinematics of the lower limbs [9]. The anthropometric measurements include the subject height and lengths of the lower limbs. Feature selection is employed to derive as many discriminative cues as possible whilst removing the redundant and irrelevant gait features which may degrade the recognition rate. It is practically infeasible to run an exhaustive search for all the possible combinations of features in order to obtain the optimal subset for recognition due to the high dimensionality of the feature space. For this reason, we employed the Adaptive Sequential Forward Floating Selection (ASFFS) search algorithm [10]. The algorithm uses a validation-based evaluation criterion which is proposed to find the subset of features that minimises the classification errors as well as ensure good separability between the different classes. In contrast to the voting scheme used in the KNN, the evaluation function uses different weights w to signify the importance of the most nearest neighbours. The probability score for a sample s_c to belong to class c is expressed in the following equation (10):

$$f(s_c) = \frac{\sum_{i=1}^{N_c-1} z_i w_i}{\sum_{i=1}^{N_c-1} w_i} \tag{10}$$

where N_c is the number of instances in class c, and the weight w_i for the ith nearest instance is related to proximity as:

$$w_i = (N_c - i)^2 \tag{11}$$

The value of z_i is defined as:

$$z_i = \begin{cases} 1 \text{ if } nearest(s_c, i) \in c \\ 0 \qquad \text{otherwise} \end{cases} \tag{12}$$

such that the $nearest(s_c, i)$ function returns the ith nearest instance to the sample s_c. The Euclidean distance metric is employed to find the nearest neighbours.

The Correct Classification Rate (CCR) is computed using the K-nearest neighbour (KNN) classifier with the Leave-one-out cross-validation rule. The KNN rule is applied at the classification phase due to its low complexity and

hence fast computation besides the ease of comparison to other methods. In the leave-one-out validation, every instance from the original sample is used for testing and is validated against the remaining observations. This is repeated for all the observations in the dataset. The recognition rate is computed as the average of all validations.

4 Experimental Results

The view-invariant gait analysis has been tested on real data from a subset of CASIA-B database [3] with subjects walking along straight lines with 6 different camera orientations (36°, 54°, 72°, 90°, 108°, 126°). The 90° view corresponds to the side view walking direction as shown in Figure 1. The video sequences have a spatial resolution and frame rate of 320x240 pixels and 25fps respectively with an approximate subject height of 90 pixels. Subjects are instructed to walk in three different scenarios; normal walking, wearing a coat and carrying a bag.

4.1 Performance Analyis of Gait Feature Extraction

The markerless algorithm has been applied to the video sequences, the limbs pose has been estimated frame by frame and the hip and knee angles have been extracted for each camera position and for each subject. Figure 1 shows an example of the limbs pose estimation for different camera positions for an example subject carrying a bag. The algorithm allows for estimation of the limbs pose also in such conditions and appears robust with respect to low-resolution, clothing and carrying conditions.

A quantitative validation of the proposed method has been obtained in a virtual environment (Poser 7®) with a humanoid walking for one gait cycle. The results reported in [5] for estimating the leg's angles have a mean value of 2.63 ± 2.61 deg and are particularly encouraging since they present same magnitude to the ones obtained with 3D markerless systems and 2D complex model based methods [11].

Figure 2(a) shows an example of the variations of hip angular motion during two gait cycle for the six different camera positions in the real experimental tests. Predictably, the angles trends, extracted in the image reference system, are influenced by the subject pose respect to the camera and they cannot be used directly for identification. For this reason, the view point correction algorithm is applied and the angle trends after the correction are shown in figure 2(b).

4.2 Gait Recognition and Impact of Covariate Factors

In order to assess the performance of the proposed gait recognition algorithm from different viewpoints using a single uncalibared camera, a set of 1037 video sequences with 20 different subjects recorded at 6 viewpoints are taken from the CASIA-B gait database. To investigate the effects of the viewpoint, an initial experiment is carried out to measure the recognition rate using the non-rectified

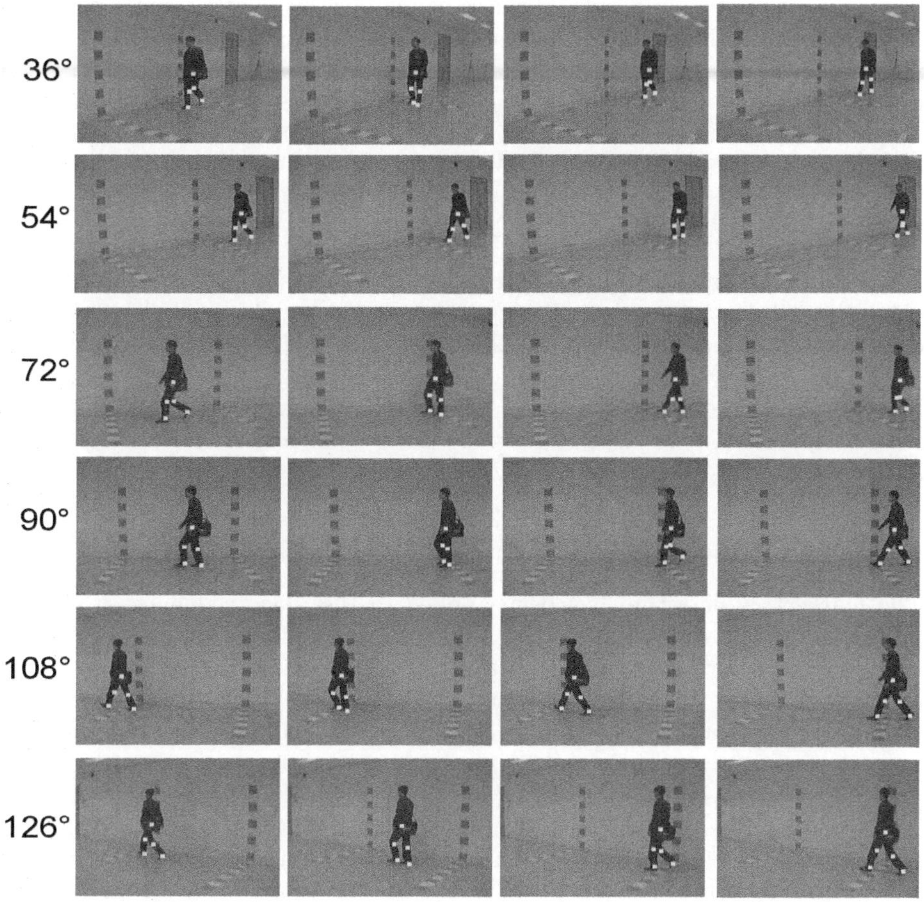

Fig. 1. Joints extraction in different viewpoints for subject carrying a bag

gait data. The CCR is first computed for all the data combined together where a low CCR of 34% is observed based on leave-one-out cross validation.

A set of experiments are carried out to compute the recognition rates for every viewpoint separately after applying the view-rectification approach. This is done based on probing various datasets of different and similar viewpoints. Table 1 shows the variation of the CCRs with respect to the different viewpoints for the achieved results along with comparative results reported by Yu *et al.* [3] in their silhouette-based approach applied on the CASIA-B dataset. The performance of gait recognition largely increases with an average CCR of 73.4% and better classification rates compared to the baseline silhouette-based approach. For both model and silhouette-based methods, the recognition rates along the diagonal for probing dataset against galleries of similar viewpoints, are observed to be higher with an average CCR of 80.8% and 30.15% for our method and the silhouette approach respectively. For the non-diagonal cases, the classification rates

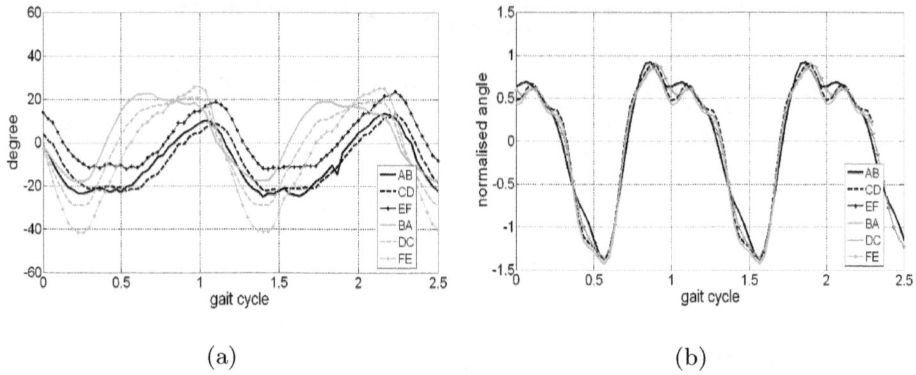

Fig. 2. Hip Angular Motion from different View-points: (a) Unrectified Data. (b) Rectified Angular Data.

drop largely to an average of 9.6% for the silhouette-based due the changes of the silhouette shape when varying the viewpoint which affected the recognition performance. In contrast for the proposed model-based approach, a reported average CCR of 64.48% for probing datasets against probes of different viewpoints. This shows the benefit of using model-based approach combined with the rectification algorithm that can handle the effects of viewpoint. Clearly, the new approach allows for viewpoint invariant analysis and are which handles practical factors in human movement analysis.

Table 1. CCR (%) CASIA-B, Set A: 1) rectified data, 2) results of Yu *et al.* [3]

	Our Method (Rectified Data)							**Yu *et al.* (Silhouette-Based)**						
		Probe Angle							Probe Angle					
		36°	**54°**	**72°**	**90°**	**108°**	**126°**		**36°**	**54°**	**72°**	**90°**	**108°**	**126°**
Gallery Angle	**36°**	67.8	60.8	58.9	50.1	48.0	40.8	**36°**	30.2	16.5	1.2	1.2	1.6	6.9
	54°	57.1	75.3	76.0	65.3	67.9	60.3	**54°**	10.1	30.6	5.6	4.4	7.7	14.1
	72°	52.3	63.3	83.4	81.5	79.0	72.7	**72°**	5.6	7.7	31.0	21.8	14.9	8.9
	90°	5.9	65.1	71.0	88.1	86.5	82.3	**90°**	4.0	6.0	20.6	32.7	16.5	6.0
	108°	44.7	61.0	68.9	79.6	86.6	81.9	**108°**	2.4	4.8	17.7	27.8	30.2	9.3
	126°	38.8	58.2	68.8	65.3	72.6	79.3	**126°**	1.6	4.4	10.1	10.1	18.5	26.2

4.3 Covariate Analysis of Gait with Fixed View-Point

To further describe the covariate effects, an experimental analysis was carried out on the SOTON Covariate database independently from the view-point rectification. A gallery dataset of 160 video sequences is taken from the SOTON gait database consisting of 20 different walking subjects with 8 sequences for every individual recorded without covariate effects. Further, a probe dataset of 440 video sequences for 10 subjects is collected from the Southampton Covariate

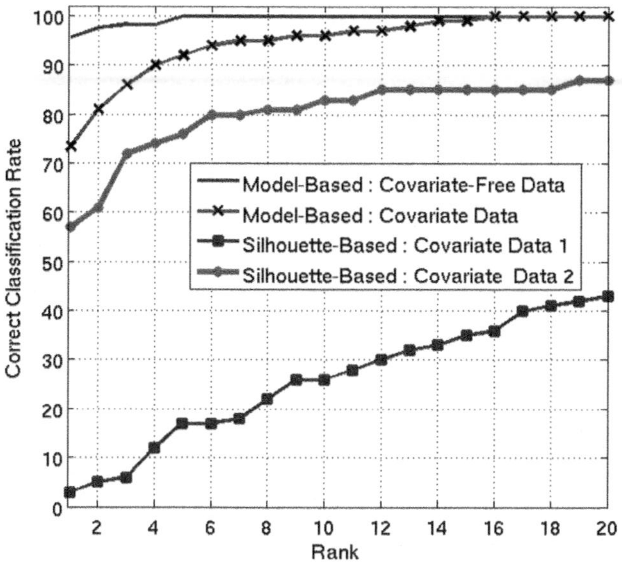

Fig. 3. The Cumulative Match Score Curves for the Classification Results

Database. The covariate factors includes clothing, footwear, carrying conditions as well as walking speed. Based on the subset of features derived using the Feature Selection algorithm, we have achieved a high recognition rate of 95.75% for the value of $k = 5$ using the training covariate-free dataset. This is achieved using solely features describing purely the dynamics of the locomotion process.

Furthermore, we have probed 440 samples from the covariate dataset against the gallery database. A recognition rate of 73.4% is achieved for all the covariate factors which is higher when compared to the low recognition rates reported by Phillips *et al.* [2] using a silhouette-based method. The Cumulative Match Score curves showing the comparative results are shown in Figure (4.3). Phillips reported a CCR of 57% for Data (I) with load carriage and footwear covariates whilst a CCR of 3% is achieved for Data (II) with the following covariates : time, footwear, and clothing. Time has been shown [2] to play a major part in reducing recognition capability by gait. Using a silhouette based approach Veres[12] showed that this could be redressed by fusing those parts of the gait signature which are invariant with time.

5 Conclusions and Future Work

We have taken an important step in deploying gait biometrics for the analysis of surveillance video. A view-invariant markerless model-based approach for gait biometrics is described. Gait features are derived based on pose estimation of the joint positions of walking subjects. A novel reconstruction method is being employed to rectify and normalize gait features recorded from different viewpoint into the side-view plane and therefore exploit such data for recognition.

The method is used is used to investigate of the effects of the covariate factors including clothing and carrying conditions for view-point independent gait recognition. Based on covariate-based probe datasets , a high recognition rate of 73.4% is achieved using the KNN classifier with $k = 5$. This suggests that people identification using dynamic gait features is still perceivable with better recognition rate even under the different covariate factors.

References

1. Nixon, M.S., Tan, T.N., Chellappa, R.: Human Identification Based on Gait. The Kluwer International Series on Biometrics. Springer, New York (2005)
2. Sarkar, S., Phillips, P.J., Liu, Z., Vega, I.R., Grother, P., Bowyer, K.W.: The humanID Gait Challenge Problem: Data Sets, Performance, and Analysis. IEEE Transactions on Pattern Analysis and Machine Intelligence 27(2), 162–177 (2005)
3. Yu, S., Tan, D., Tan, T.: A Framework for Evaluating the Effect of View Angle, Clothing and Carrying Condition on Gait Recognition. In: Proceedings of the 18th International Conference on Pattern Recognition, pp. 441–444 (2006)
4. Bouchrika, I., Nixon, M.: Exploratory factor analysis of gait recognition. In: 8th IEEE Int Conference on Automatic Face and Gesture Recognition (2008)
5. Goffredo, M., Seely, R.D., Carter, J.N., Nixon, M.S.: Markerless view independent gait analysis with self-camera calibration. In: IEEE International Conference on Automatic Face and Gesture Recognition 2008 (2008)
6. Bouchrika, I., Nixon, M.S.: Model-Based Feature Extraction for Gait Analysis and Recognition. In: Proceedings of Mirage: Computer Vision / Computer Graphics Collaboration Techniques and Applications, pp. 150–160 (2007)
7. Aguado, A.S.V., Nixon, M.S., Montiel, M.E.: Parameterizing Arbitrary Shapes via Fourier Descriptors for Evidence-Gathering Extraction. Computer Vision and Image Understanding 69(2), 202–221 (1998)
8. Chau, T.: A review of analytical techniques for gait data. Part 1: fuzzy, statistical and fractal methods. Gait Posture 13(1), 49–66 (2001)
9. Wagg, D.K., Nixon, M.S.: On Automated Model-Based Extraction and Analysis of Gait. In: Proceedings of the Sixth IEEE International Conference on Automatic Face and Gesture Recognition, pp. 11–16 (2004)
10. Somol, P., Pudil, P., Novovičová, J., Paclík, P.: Adaptive Floating Search Methods in Feature Selection. Pattern Recognition Letters 20(11-13), 1157–1163 (1999)
11. Agarwal, A., Triggs, B.: Recovering 3d human pose from monocular images. IEEE TPAMI 28, 44–58 (2006)
12. Veres, G.V., Nixon, M.S., Middleton, L., Carter, J.N.: Fusion of Dynamic and Static Features for Gait Recognition over Time. In: Proceedings of 7th International Conference on Information Fusion, vol. 2 (2005)

Dynamic Texture Based Gait Recognition

Vili Kellokumpu[1], Guoying Zhao[1], Stan Z. Li[2], and Matti Pietikäinen[1]

[1] Machine Vision Group
University of Oulu, P.O. Box 4500, Finland
{kello,gyzhao,mkp}@ee.oulu.fi
[2] Center for Biometrics and Security Research &
National Laboratory of Pattern Recognition,
Institute of Automation, Chinese Academy of Sciences
95 Zhongguancun Donglu, Beijing 100080, China
szli@nlpr.ia.ac.cn

Abstract. We present a novel approach for human gait recognition that inherently combines appearance and motion. Dynamic texture descriptors, Local Binary Patterns from Three Orthogonal Planes (LBP-TOP), are used to describe human gait in a spatiotemporal way. We also propose a new coding of multiresolution uniform Local Binary Patterns and use it in the construction of spatiotemporal LBP histograms. We show the suitability of the representation for gait recognition and test our method on a popular CMU MoBo dataset. We then compare our result to the state of the art methods.

Keywords: Gait recognition, LBP-TOP, multiresolution LBP, uniform patterns.

1 Introduction

The visual analysis of human movements has become an important area of research in computer vision. It has gained a lot of interest because of its many application domains, such as video indexing, surveillance, human computer interaction, sport video analysis, intelligent environments and gait recognition. Especially gait recognition has recently become a topic of great interest. Compared to other biometrics, gait has an advantage as it can be perceived from a distance and it does not require contact or subject cooperation. Furthermore, gait is difficult to hide or replicate.

Many approaches for human gait recognition have been proposed in the literature [1,2]. Typical approaches are to use either motion [3] or shape information [4,5,6]. However, it may not be a good idea to rely on a single modality. For example, shape based methods are naturally sensitive to changes in appearance like carrying a backpack, whereas motion based methods could have trouble in cases where the pace of walking changes.

Interestingly, very good performance in another human motion analysis application area, human action recognition, has recently been reported by using spatiotemporal analysis that combines both motion and shape cues [7,8]. There has been interest to spatiotemporal analysis in gait recognition as well. BenAbdelkaber et al. [9] used Self similarity plots for gait recognition. Another spatiotemporal pattern proposed for

M. Tistarelli and M.S. Nixon (Eds.): ICB 2009, LNCS 5558, pp. 1000–1009, 2009.
© Springer-Verlag Berlin Heidelberg 2009

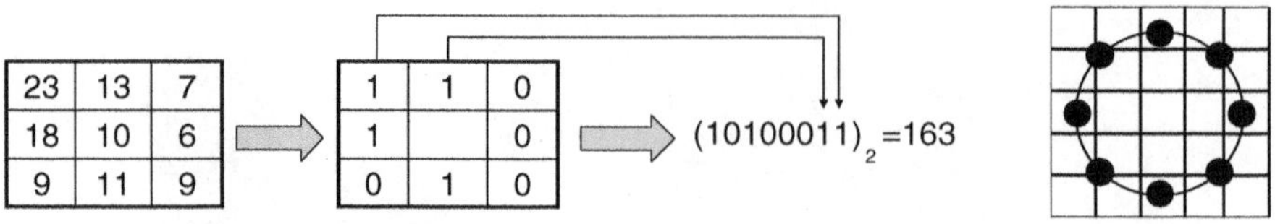

(a) (b)

Fig. 1. (a) Illustration of basic LBP operator (b) Circular (8,2) neighborhood. If the sampling point is not in the center of a pixel, the value at that point is bilinearly interpolated from the nearest pixels.

gait recognition is the frieze pattern [10], which was developed into a Shape Variance-Based frieze pattern [11] with good results. Niyogi and Adelson [12] used *xyt* volume to find out the snakes pattern and then to detect walking people from sequence images. Bounding contours for the walker are recovered and used to define a simple stick figure model, from which several features were extracted and used for individual recognition. Huang and Wang [13] made the recognition based on *xt* plane energy images extracted from *xyt* space. Kobyashi and Otsu [14] used Cubic Higher-order Local Autocorrelation to action and person identification.

Dynamic textures are textures with motion, so they inherently combine both appearance and motion. In this paper, we propose a new method for gait recognition that uses dynamic textures for movement description. The method is based on using a dynamic texture descriptor, Local Binary Patterns from Three Orthogonal Planes (LBP-TOP), to represent human movements in a spatiotemporal way. The LBP-TOP features have successfully been used for facial expression [15], visual speech [16] and action recognition [7]. Furthermore, we propose a novel idea for constructing multiresolution LBP histograms and show that it can improve gait recognition performance compared to traditional multiresolution analysis.

The rest of the paper is organized as follows. Section 2 introduces the LBP-TOP features and describes a new method for creating a multiresolution uniform LBP histogram. We show experiment on the CMU database in Section 3 and conclude in Section 4.

2 Gait Description with Spatiotemporal Local Binary Patterns

Dynamic textures are textures with motion. In this section we introduce the LBP-TOP features, which are originally proposed to describe dynamic textures, and propose a novel method for constructing multiresolution uniform LBP histogram. We also show how these can be applied to gait recognition.

2.1 LBP-TOP Features

LBP operator [17] produces a binary code that describes the local texture pattern, which is built by thresholding a neighborhood of pixels by the grey value of its center pixel. The original LBP operator represents a 3x3 pixel neighborhood as a binary number. Figure 1(a) illustrates the basic LBP operator. When LBP operator is applied to an image, the image texture can be described with a histogram of the binary codes.

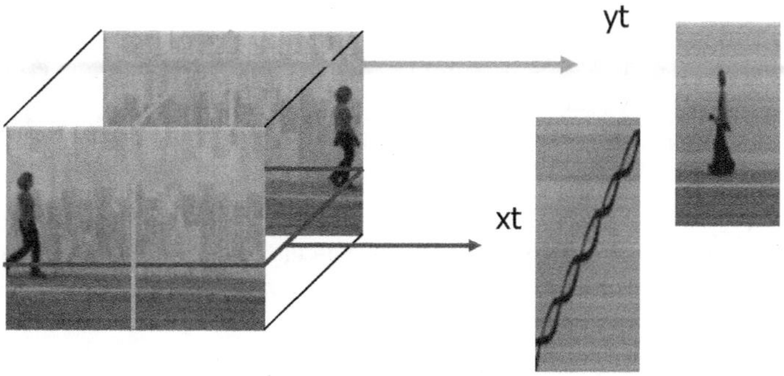

Fig. 2. Illustration of a person walking and the corresponding *xt* and *yt* planes from a single row and column. The different frames correspond to the *xy* planes.

The LBP operator has also been extended to different kinds of neighborhoods. With a circular neighborhood and bilinear interpolation of pixels, any radius and number of sampling points in the neighborhood can be used:

$$LBP(x_c, y_c) = \sum_{p}^{P-1} s(g_p - g_c)2^P, \qquad s(x) = \begin{cases} 1, x \geq 0 \\ 0, x < 0 \end{cases}, \tag{1}$$

where g_c is the gray value of the center pixel (x_c, y_c) and g_p are the gray values at the P sampling points Figure 1(b) shows an example of the circular (8,2) neighborhood that has 8 sampling points and radius of 2.

The LBP operator was extended to a dynamic texture operator by Zhao and Pietikäinen [15], who proposed to form their dynamic LBP description from three orthogonal planes (LBP-TOP) of a space time volume. Figure 2 shows the spatiotemporal volume of a person walking from left to right. It also illustrates the resulting *xt* and *yt* planes from a single row of and column of the volume as well as the first and last *xy* planes that are the frames themselves. The LBP-TOP description is formed by calculating the LBP features from the planes and concatenating the histograms.

The original LBP operator was based on a circular sampling pattern but different neighbourhoods can also be used. Zhao and Pietikäinen proposed to use elliptic sampling for the temporal planes: (x_c-$R_x$sin($2\pi p/P_{xt}$), y_c, t_c-$R_t$cos($2\pi p/P_{xt}$)) for *xt* plane and similarly (x_c, y_c -$R_y$sin($2\pi p/P_{yt}$), t_c-$R_t$cos($2\pi p/P_{yt}$)) for *yt* plane. R_d is the radius of the ellipse to direction of the axis *d* (*x*, *y* or *t*). As the *xy* encodes only the appearance, i.e., both axes have the same meaning, circular sampling is suitable. The values g_p for points that do not fall on pixels are estimated using bilinear interpolation.

2.2 Multiresolution Uniform LBP

Multiple kernels can be used together to improve the LBP description. Traditionally, LBP histograms with different radii are concatenated to form the final histogram, as using a joint histogram is infeasible because of the resulting histogram size. However, a lot of information about the underlying structure is lost when only marginal distributions are used to model the texture statistics.

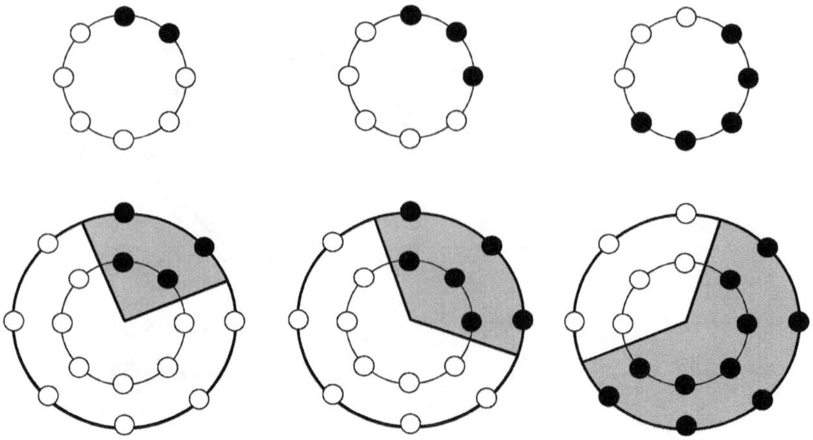

Fig. 3. Illustration of three uniform patterns (top row) and their corresponding multiresolution versions

One way to improve the performance of LBP based methods and shorten the feature histogram is to use the "uniform" subset of LBP codes. A Local Binary Pattern is called uniform if it contains at most two bitwise transitions from 0 to 1 or vice versa when the binary string is considered circular [17]. It is observed that there are a limited number of transitions or discontinuities in the circular presentation of the 3×3 texture patterns; according to Ojala, these uniform patterns are fundamental properties of local image texture and they provide a vast majority amount all patterns.

By the definition, uniform patterns are codes that consist of at most two homogenous regions on a circle. From this definition, they can be thought of as representing sectors on a sampling circle. From this insight a new way of calculating multiresolution LBP representation efficiently can be derived.

Figure 3 illustrates a few uniform patterns and their corresponding multiresolution versions. It can be seen that if the multiresolution sampling points are ordered according to the sampling angle, they will also produce codes that satisfy the bit transition condition. It should also be noted that any number of sampling points can be used on different LBP kernels.

We will not give formulation of the new multiresolution uniform patterns here, but we describe the basic idea and show its applicability to the gait recognition problem. Extensive experiments need be done to verify its generalization capability to different problems. The method does, however, have a lot of potential to increase the performance of the many LBP based methods.

In our preliminary setup, sampling points are ordered by going clockwise around the sampling circle. As the representation of uniform patterns is circular, sampling can be started from any point. Figure 4 illustrates the idea in a case where two eight point sampling kernels are used. It is easy to see that the new method preserves the structural information better than the original method. Also the length of the histogram does not increase too much. For two eight point sampling kernels the histogram length is the same as the regular uniform pattern histogram for a sixteen bit kernel, i.e., 243. If an eight and a sixteen point kernels were used, the histogram length would be 555, the same as for a regular 24bit uniform pattern histogram.

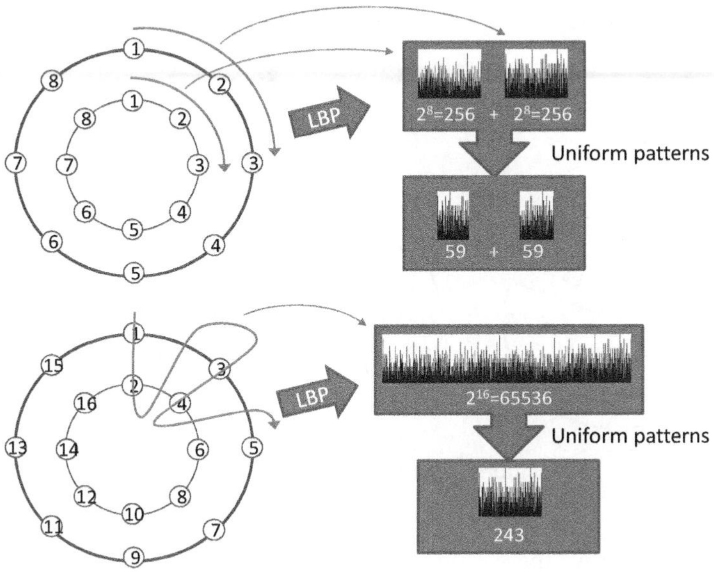

Fig. 4. Illustration of the sampling and formation of the feature histogram. The top image represents the traditional histogram formation where the different kernels are sampled individually. The lower image shows the proposed method. Note the arrows and numbering of the sampling points illustrating the sampling order.

2.3 Gait Description

The dynamic LBP features calculated over the whole *xyt* volume encode the local properties of the movements without any information about their local or temporal locations. For this reason the volume is partitioned into subvolumes and the feature histogram ís formed by concatenating the subvolume histograms. Using the subvolume representation, motion and shape are encoded on three different levels: pixel-level (single bins in the histogram), region-level (subvolume histogram) and global-level (concatenated subvolume histograms).

To obtain a rough spatial definition of human movements, we divide the *xyt* volume into four regions through the centroid of the silhouette. This division roughly separates the hands and legs of the person. Using more blocks would of course allow a more detailed description but would also produce more local histograms and make the whole histogram longer. Using too many blocks could also possibly make the system too sensitive for natural intraperson variance in the gait.

The subvolume division and the formation of our feature histogram are illustrated in Figure 5. The LBP-TOP features are calculated from the whole duration of a sequence and concatenate all the subvolume histograms on each plane.

Furthermore, as we use silhouette data in the experiments, we prune the histogram by removing the bins that represent monotonous areas (all black/white) from the histogram of each plane. We also observed that removing the collection bin, the bin that represents the non-uniform patterns, improves the performance when uniform patterns are used.

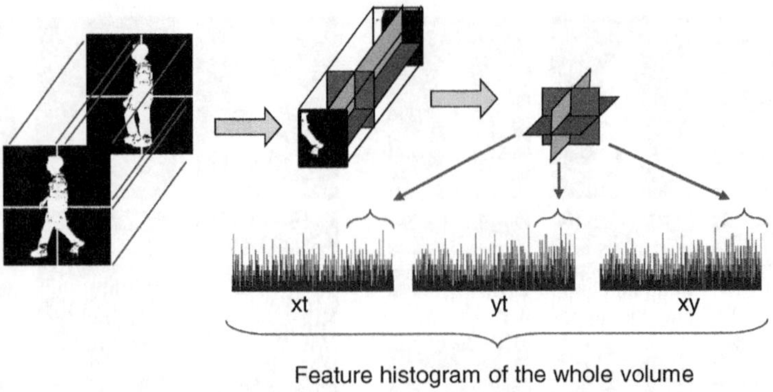

Fig. 5. Formation of the feature histogram

As there is only one appearance plane *xy* and two temporal planes *xt* and *yt*, the motion will have more weight in the concatenated histogram. If histogram of each plane is normalized individually and double weight is given to the appearance plane, no bias is given to either appearance or motion.

Various methods can be used to measure the similarity of LBP-TOP histograms. In this work we have chosen to use the histogram intersection:

$$Similarity = \sum \min(h_i, h_j) \quad , \tag{2}$$

where the summation is taken over the bins. In classification, an unknown gait example is assigned as being from the same person as the most similar example in the training set.

3 Experiments

We demonstrate the performance of our method by experimenting with the popular CMU MoBo database [18]. We use the LBP-TOP methodology described in the previous section and verify the performance of the new multiresolution uniform patterns on gait recognition.

3.1 CMU Database

To show the suitability of the dynamic texture based approach to gait recognition, we experimented with the CMU database. The reason for choosing this database is that we can easily compare our performance to others described in the literature [9, 11, 19, 20, 21]. Furthermore, this database was chosen because it contains examples of people walking with different speeds. As we are using dynamic textures, we want to see how this change in dynamics affects our performance.

The CMU Mobo Database contains 25 different people (23 male, 2 female) walking on a treadmill. Each subject performs walking under four different conditions: walking slow, walking slow holding a ball, walking fast and walking on an inclined surface. Each sequence is recorded at 30 frames per second and is 11 seconds long in duration.

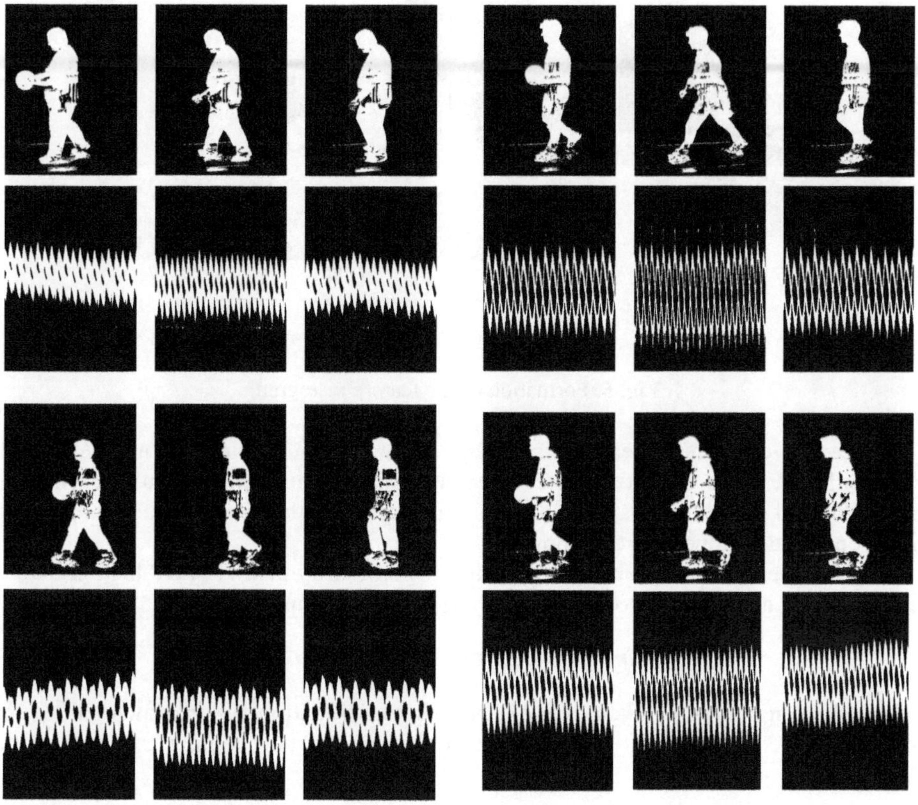

Fig. 6. Illustration of the data in the CMU Mobo database under the three different conditions: walking with a ball, walking fast and walking slow. The first silhouette of each sequence is shown together with a single *xt* slice from whole the duration of the sequence. Notice the change in dynamics and appearance between the conditions.

The database is ideal for testing our method as it contains both appearance (walking and holding a ball) and motion (walking slow vs. fast) changes. The incline walking is excluded from our experiments. Figure 6 gives examples of the database.

3.2 Results

The experiments were run on all the combinations of walking slow (S), walking slow holding a ball (B) and walking fast (F). The test was also performed separately and jointly for appearance and motion features. Table 1 summarizes our results using two LBP kernels (LBP-TOP with R_x=R_y=2, R_t=5, P_x=P_y=P_t=8, and with R_x=R_y=4, R_t=6 P_x=P_y=P_t=8) with the traditional and the proposed multiresolution approaches.

It can be seen that results using just appearance are better than using just motion. Interestingly, the proposed multiresolution method works better on appearance, whereas the original one is better on the temporal planes. When all planes are used together, the new proposed multiresolution uniform LBP performs slightly better.

Table 1. The recognition results on the CMU database. Tests were performed using appearance and motion features separately and jointly. The use of uniform patterns is marked with u2.

Appearance = XY		Hist	B/S	S/B	B/F	F/B	S/F	F/S
Original Multiresolution	u2	228	75 %	79 %	63 %	79 %	76 %	76 %
	-	1020	79 %	79 %	67 %	79 %	76 %	80 %
New Multiresolution	u2	964	71 %	83 %	75 %	79 %	88 %	84 %
	-	262136	79 %	83 %	79 %	83 %	84 %	84 %
Motion = XY+XT								
Original Multiresolution	u2	456	63 %	83 %	54 %	58 %	68 %	56 %
	-	2040	88 %	88 %	33 %	42 %	48 %	40 %
New Multiresolution	u2	1928	67 %	75 %	50 %	54 %	56 %	56 %
	-	524272	96 %	88 %	33 %	42 %	56 %	44 %
Appearance 50% + motion 50%								
Original Multiresolution	u2	684	75 %	83 %	71 %	79 %	80 %	80 %
	-	3060	83 %	88 %	63 %	63 %	76 %	64 %
New Multiresolution	u2	2892	79 %	88 %	75 %	79 %	80 %	76 %
	-	786408	88 %	92 %	67 %	79 %	80 %	72 %
Appearance 65% + motion 35%								
XY = new multiresolution **XT+YT = original multiresolution**	**u2** **u2**	**1420**	**75 %**	**83 %**	**75 %**	**83 %**	**88 %**	**88 %**

Table 2. Results reported in the literature for the CMU database

	S/B	B/S	F/B	B/F	S/F	F/S
CMU [19]	**92 %**	-	-	-	76 %	-
UMD [20]	48 %	68 %	48 %	48 %	80 %	84 %
MIT [21]	50 %	-	-	-	64 %	-
SSP [9]	-	-	-	-	54 %	32 %
SVB frieze [11]	77 %	**89 %**	61 %	73 %	82 %	80 %
LBP-TOP	75 %	83 %	**75 %**	**83 %**	**88 %**	**88 %**

Another interesting thing to note is that when uniform patterns are not used, the new sampling method is always better. This indicates that the more detailed structure information is important and kept by the proposed method. The downside is of course the huge histogram size, but with the proposed multiresolution uniform coding the histogram can be shortened to a reasonable size. The new coding of multiresolution LBP and uniform patterns are definitely worth further investigating.

It can also be seen that using just appearance gives better result when appearance and motion are given equal weight. The best overall result is obtained when the new multiresolution sampling is used without the uniform patterns, though this comes with a cost of a very long feature vector (262136 bins). We then gave appearance more weight (65%) and observed that by taking a combination of the new multiresolution uniform method on *xy* plane and the original multiresolution method on *xt* and *yt* planes, we can match the result of using just appearance with a feature vector that is only 0.5% in length (1420 bins). Also, the performance is better than using the new multiresolution uniform method on *xy* alone. It can be deduced that using both the

appearance and motion is important, though it seems appearance should be given more weight. We are currently working on boosting methods to further improve the results and learn the importance of different features.

The test results are encouraging as in all test combinations exists variation in either shape or motion, or both. The best recognition results are achieved on the F/S and S/F scenarios. In these cases the appearance of the person is roughly the same, though there is variation in the pace of walking. In the other scenarios the recognition result is slightly lower due to the change in appearance. Overall, our result is very good. Table 2 shows that our results are comparable to the state of the art methods reported in the literature. It can be seen that compared to others, we achieve a very good result in the most difficult scenarios F/B and B/F where there is variation in both appearance and dynamics.

4 Conclusions and Future Work

We have proposed a novel dynamic texture based method for human gait recognition. We extract LBP-TOP features in spatiotemporal space and use them to describe human walking movements. The method is computationally simple. We show excellent results on a popular benchmark database.

We have also proposed a new way of constructing multiresolution uniform LBP and we show slightly better results on gait recognition compared to the traditional multiresolution analysis. The method has potential to increase the performance of various LBP based applications. Our current work focuses on more exact formulation of the method and extensive analysis on real image data.

As the used LBP-TOP features can also be calculated from the frame data without segmentation, we are also currently investigating gait recognition from image data rather than silhouettes. This extension makes the method more suitable for many applications.

Acknowledgements

This work was supported by European Regional Development Fund, the Finnish Funding Agency for Technology and Innovation, the Graduate School in Electronics, Telecommunication and Automation (GETA), and the Academy of Finland.

References

[1] Nixon, M.S., Tan, T., Chellappa, R.: Human Identification Based on Gait. Springer, Heidelberg (2005)

[2] Boyd, J., Little, J.: Biometric Gait Recognition. In: Tistarelli, M., Bigun, J., Grosso, E. (eds.) Advanced Studies in Biometrics. LNCS, vol. 3161, pp. 19–42. Springer, Heidelberg (2005)

[3] BenAbdelkader, C., Cutler, R., Davis, L.: Motion based recognition of people in eigengait space. In: Proc. FG, pp. 267 – 272 (2002)

[4] Veres, G.V., Gordon, L., Carter, J.N., Nixon, M.S.: What image information is important in silhouette based gait recognition? In: Proc. CVPR, vol. 2, pp. 776–782 (2004)

[5] Liu, Z., Sarkar, S.: Improved Gait Recognition by Dynamics Normalization. In: PAMI, vol. 2, pp. 863–876 (2006)
[6] Zhao, G., Cui, L., Li, H.: Gait Recognition Using Fractal Scale. PAA 10(3), 235–246 (2007)
[7] Kellokumpu, V., Zhao, G., Pietikäinen, M.: Human activity recognition using a dynamic texture based method. In: Proc. BMVC, 10 p. (2008)
[8] Blank, M., Gorelick, L., Shechtman, E., Irani, M., Basri, R.: Actions as Space-Time Shapes. In: Proc ICCV, vol. 2, pp. 1395–1402 (2005)
[9] BenAbdelkaber, C., Cutler, R.G., Davis, L.S.: Gait Recognition Using Image Self-Similarity. EURASIP Journal on Applied Signal Processing 24(4), 572–585 (2004)
[10] Liu, Y., Collins, R., Tsin, Y.: Gait Sequence Analysis Using Frieze Patterns. In: Heyden, A., Sparr, G., Nielsen, M., Johansen, P. (eds.) ECCV 2002. LNCS, vol. 2351, pp. 657–671. Springer, Heidelberg (2002)
[11] Lee, S., Liu, X., Collins, R.: Shape Variation-Based Frieze Pattern for Robust Gait Recognition. In: Proc CVPR 2007, 8 p. (2007)
[12] Niyogi, S.A., Adelson, E.H.: Analyzing and Recognizing Walking Figures in XYT. In: Proc. CVPR, pp. 469–474 (1994)
[13] Huang, G., Wang, Y.: Human Gait Recognition Based on X-T Plane Energy Images. In: Proc. ICWAPR 2007, pp. 1134–1138 (2007)
[14] Kobyashi, T., Otsu, N.: Action and Simultaneous Multiple-Person Identification Using Cubic Higher-order Auto-Correlation. In: Proc. ICPR, vol. 4, pp. 741–744 (2004)
[15] Zhao, G., Pietikäinen, M.: Dynamic Texture Recognition Using Local Binary Patterns With an Application to Facial Expressions. PAMI 29(6), 915–928 (2007)
[16] Zhao, G., Pietikäinen, M., Hadid, A.: Local Spatiotemporal Descriptors for Visual Recognition of Spoken Phrases. In: Proc. HCM 2007, pp. 57–65 (2007)
[17] Ojala, T., Pietikäinen, M., Mäenpää, T.: Multiresolution Gray-Scale and Rotation Invariant Texture Classification with Local Binary Patterns. PAMI 24(7), 971–987 (2002)
[18] Gross, R., Shi, J.: The CMU Motion of body (MoBo) database. Technical Report CMU-RI-TR-01-18, Robotics Institute, Carnege Mellon University (2001)
[19] Collins, R., Gross, R., Shi, J.: Silhouette Based Human Identification from Body Shape and Gait. In: International Conference on Face and Gesture, pp. 366–371 (2002)
[20] Veerarghavan, A., Chowdhury, A.R., Chellappa, R.: Role of Shape and Kinematics in Human Movement Analysis. In: Proc. CVPR, vol. 1, pp. 730–737 (2004)
[21] Lee, L., Grimson, W.: Gait Analysis for Recognition and Classification. In: International Conference Automatic Face and Gesture, pp. 155–162 (2002)

Gender Recognition Based on Fusion of Face and Multi-view Gait

De Zhang and Yunhong Wang

Intelligent Recognition and Image Processing Laboratory,
School of Computer Science and Engineering, Beihang University,
Beijing 100191, China
zhangde@cse.buaa.edu.cn, yhwang@buaa.edu.cn

Abstract. In this paper, we consider the problem of gender recognition based on face and multi-view gait cues in the same walking sequence. The gait cues are derived from multiple simultaneous camera views. Meanwhile, the face cues are captured by a camera at front view. According to this setup, we build a database including 32 male subjects and 28 female subjects. Then, for face, we normalize the frame images decomposed from videos and introduce PCA to reduce image dimension. For gait, we extract silhouettes from videos and employ an improved spatio-temporal representation on the silhouettes to obtain gait features. SVM is then used to classify gender with face features and gait features from each view respectively. We employ three fusion approaches involving voting rule, weighted voting rule and Bayes combination rule at the decision level. The effectiveness of various approaches is evaluated on our database. The experimental results of integrating face and multi-view gait show an obvious improvement on the accuracy of gender recognition.

1 Introduction

It is known to all that human gender is a typical social category. Therefore, gender identification is an important requirement for different tasks including access control, vital statistics and commercial analysis etc. Gender recognition has received much attention in the computer vision literature. In [4,5,6], we can find some gender identification systems based on the voice of speaker. In [1,2,3], the problem of classifying gender from facial images is described. Furthermore, gender can be recognized from human walking, as shown in [7,8,9]. If a person is far away from the camera, it is hard to get face information at a high enough resolution for recognition tasks. However, when available, it is very powerful for recognition. The modality which can be detected and measured when the subject is far away from the camera is human gait or the style of walking. It will yield optimal performance if the system use as many cues as possible and combine them in meaningful ways.

There have been some attempts on fusion of face and gait for human identification. In [10], Shakhnarovich et al. compute an image based visual hull from a set of monocular views which is then used to render virtual canonical views for tracking and gait recognition. Also, this method is used to render frontal face images. They studied

M. Tistarelli and M.S. Nixon (Eds.): ICB 2009, LNCS 5558, pp. 1010–1018, 2009.
© Springer-Verlag Berlin Heidelberg 2009

the decision fusion of face and gait cues for a multi-camera indoor environment. Later, they further investigate different approaches of classifier combination for face and gait recognition in [11]. Zhou and Bhanu [12] present an approach to the fusion of gait and face which uses side view of non cooperating subjects, making use of PCA and MDA for gait and face features and combining them at the score level. In addition, Shan et al. applied the fusion of face and gait on gender recognition in [14]. They explored Canonical Correlation Analysis (CCA) to fuse the two modalities at feature level on CASIA Gait Database [15]. In their experiments, they used video sequences from two views: front view for face and side view for gait. They extracted facial part from the gait image sequences at front view. Therefore, these low-resolution facial images may do not contain enough information.

In this paper, we are interested in the further study of recognizing human gender by combining face and gait cues. Considering in surveillance environment it is possible to obtain a person's walking videos recorded by far-placed cameras at several views and the person's face images captured by near-placed cameras or the far-placed cameras when he/she comes near, we investigated the fusion of face and multi-view gait information from the same walking sequence. Our first work is to build a database consisting of gait data and face data captured simultaneously for each subject. Second, we carried out the task of gender recognition based on face or gait of each single view separately. Then, for the purpose of fusion, we tried with three schemes: voting rule, weighted voting rule and Bayes combination rule. The Bayes combination rule derives from the Bayes formula and for our case we designed a calculation method to obtain the fusion results.

The remainder of this paper is organized as follows. Section 2 describes our self-building database, and Section 3 briefly introduces the feature extraction for face and gait. The details of fusion strategies are presented in Section 4. Experiments and results are shown in Section 5. Section 6 concludes the paper.

2 Data Collection

We intend to investigate the effectiveness of a two-modality fusion of face and gait information acquired simultaneously for gender recognition. Especially, the gait information is expected to be captured from multiple views. Hence, we need to build such a database which not only contains face and gait data recorded simultaneously but also includes the similar number of males and females for the task of gender recognition.

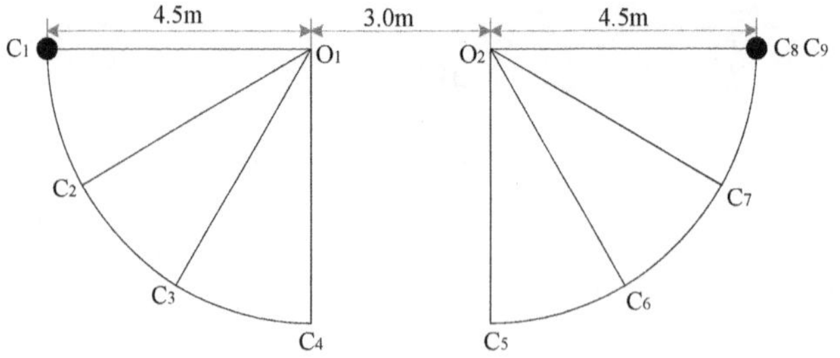

Fig. 1. Cameras setup for data collection. Cameras from C_1 to C_8 are used to record human gait at different views. Camera C_9 records human face.

Our data are captured in an indoor laboratory scenario. Eight cameras are placed at different angles recording the walking of a person. These cameras are divided into two groups each of which comprises four cameras and forms a 1/4 circle. The face of the person is captured by another camera from the front view. The setup of these nine cameras is illustrated in Fig. 1.

In the collecting process, there are 60 volunteers in all including 32 male subjects and 28 female subjects aged between 22~28. During the course of data collection, every participant was asked to walk along the straight line between camera C_1 and C_8, which are denoted by two black points in Fig. 1, from left to right and then return, repeating five times. Thus, every camera recorded five left-to-right and five right-to-left walking video sequences for each person. Meanwhile, we label camera C_8 with the 0^0 view, C_7 with the 30^0 view, till C_1 with the 180^0 view. Camera C_4 and C_5 have the same view angle.

It is not enough in terms of the number of total subjects. We will continue to enlarge our database. Now we name it as IRIP-GF Database 1.0.

3 Gender Recognition

3.1 Face-Based Gender Recognition

For the face data in our database, we crop the facial part from original 24-bit frame images decomposed from face videos. Then, these facial images are normalized to 64-by-64 pixels. Finally, histogram equalization is used to generate a standard 8-bit gray image. Fig. 2 shows an example of a frame image and its preprocessing result.

Fig. 2. The left image is an original frame and the right is the result of preprocessing

Principle Component Analysis (PCA) has long been used for the face-based human recognition. PCA can reduce the dimension of feature space while retaining the most discriminatory information. Let $X \in R^N$ denote a vector representing an image, where N is the dimensionality. The covariance matrix of X is defined as $\Sigma_x = E([x - E(x)][x - E(x)]^T)$, where $E(\hbar)$ is the expectation operator and T

denotes the transpose operation. The covariance matrix Σ_x can be factorized into the following form:

$$\Sigma_x = \Phi\Lambda\Phi \tag{1}$$

where $\Phi = [\Phi_1\Phi_2...\Phi_N] \in R^{N\times N}$ is the orthogonal eigenvector matrix of Σ_x ; $\Lambda = \{\lambda_1\lambda_2...\lambda_N\} \in R^{N\times N}$ is the diagonal eigenvalue matrix of Σ_x with diagonal elements in descending order ($\lambda_1 \geq \lambda_2 \geq ... \geq \lambda_N$). A useful property of PCA is the minimum mean square error (MSE) of signal reconstruction with using only a subset of principal components. The application of this property is the dimensionality reduction:

$$Y = P_{pca}^T X \tag{2}$$

where $P_{pca} = [\Phi_1\Phi_2...\Phi_m]$, $m < N$. The lower dimensional vector $Y \in R^m$ contains the most expressive features of the original data X.

Our 8-bit gray facial images are projected into a low dimensional feature space by using PCA. Then, we select Support Vector Machine (SVM) to classify gender.

3.2 Gait-Based Gender Recognition

For the independence of cloth color or texture information, silhouettes of walking humans are extracted by background subtraction firstly. Then, we need a gait representation that is able to be applied on different views. The spatio-temporal gait representation proposed in [13] for side-view is sensitive to silhouette deformations and robust to spurious pixels. So we take this representation into account and extend it to other views except the side-view. Additionally, we make an improvement to reduce data redundancies.

In [13], for a sequence of silhouette images $b(x, y, t)$ indexed spatially by pixel location (x, y) and temporally by time t, the author forms two new 2D images $F_R(x,t)$ and $F_C(y,t)$. Considering the periodicity of gait, we form these two images only from a gait cycle of silhouettes instead of the whole sequence to reduce redundancies. This is our enhancing point for this method. Hence, the end value of t is the number of silhouettes included in one gait cycle, denoted by N_{gait}.

The silhouettes are normalized to be the same size and centered before feature acquisition. Define a silhouette as $s[i, j],\ i = 0,1,..., M-1,\ j = 0,1,..., N-1$, where M and N denote the number of rows and columns of the silhouette, respectively. Let

$$s[i, j] = \begin{cases} 1 & \text{if } (i, j) \text{ belongs to the foreground} \\ 0 & \text{otherwise.} \end{cases} \tag{3}$$

With the above definitions, the horizontal and vertical projection of silhouettes can be expressed as:

$$p_h[i] = \sum_{j=0}^{N-1} s[i, j], \quad i = 0,....,M-1 \tag{4}$$

$$p_v[j] = \sum_{i=0}^{M-1} s[i, j], \quad j = 0,....,N-1 \tag{5}$$

Consequently, two projection vectors can be defined as follows:

$$H = \{p_h[0], p_h[1],......, p_h[M-1]\} \tag{6}$$

$$V = \{p_v[0], p_v[1],......, p_v[N-1]\} \tag{7}$$

Here, one gait cycle has N_{gait} silhouettes. Therefore, the aforementioned image $F_R(x,t)$ is formed by N_{gait} H vectors arranged by columns. Likewise, N_{gait} V vectors form the image $F_C(y,t)$. Fig. 3 shows these images F_R and F_C at different view angles from a sample.

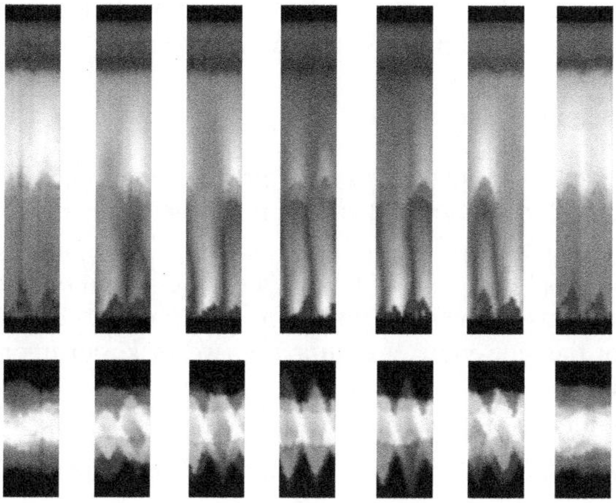

Fig. 3. An example of images F_R and F_C. Columns from left to right represent different view angles: 0^0, 30^0, 60^0, 90^0, 120^0, 150^0, 180^0. F_R images are listed on top row and their corresponding F_C images are shown on bottom row.

N_{gait} is calculated through the autocorrelation of the foreground sum signal. As different people may have different walking period, we resize all F_R and F_C images using bicubic interpolation to the same width (the end value of t i.e.). PCA is then used to retain only the important elements of F_R and F_C images. For each pair of F_R and F_C images, we combine their PCA results into one vector and use it as the feature vector which we intend to extract from a gait sequence. Finally, as the aforementioned face-based gender recognition, we also select Support Vector Machine (SVM) to classify gender.

4 Fusion Strategies

Generally, fusion can be done at two levels: feature level and decision level. Fusion on the feature level consists of combining the observations and mapping them into a single data point to be classified. As shown in [14], CCA is a powerful tool to relate two sets of measurements. Alternatively, one can treat all the input features as separate independent data and perform fusion on the decision level. In our case, each view of gait can be looked as an independent classifier. Adding the face modality, we have eight classifiers in all. Since our classifiers yield decision possibilities as results, the main interest is in fusion on decision level. As discussed in [16], we are interested in general fusion schemes. Particularly, we design a fusion strategy for our problem based on Bayesian inference. In the following paragraphs we will describe these methods.

4.1 Voting Rule

This strategy is motivated by the way humans are making decisions, especially when there is a group of people involved in the decision process. Each classifier is in the position of a human expert with one vote. The resulting class is determined by the majority of votes. Let $v(n)$ denote the number of classifiers with class n on the first rank:

$$v(n) = \sum_{n=n_1} C_i, n_1 \in R_{C_i} \tag{8}$$

where n_1 is the first rank of the i-th classifier C_i.

4.2 Weighted Voting Rule

The weighted voting method is no more than a variant of voting strategy, where each expert has its individual weight.

$$v(n) = \sum_{n=n_1} wC_i, n_1 \in R_{C_i} \tag{9}$$

The weights w need to be defined before the fusion process is applied. The recognition rates of each classifier can be used as weights.

4.3 Bayes Combination Rule

The Bayesian theory is known to all. Given m possible classes $\omega_1, ..., \omega_m$, we assume that both the prior probabilities $P(\omega_j)$ and the conditional densities $p(x \mid \omega_j)$ for an observation x are known. Bayes formula shows how to calculate the posteriori probability:

$$P(\omega_j \mid x) = \frac{p(x \mid \omega_j) P(\omega_j)}{\sum_{k=1}^{m} p(x \mid \omega_k) P(\omega_k)} \tag{10}$$

This theorem can be used for the inference of the joint probability of the input classifiers.

In our case, we intend to combine the results from multiple classifiers on decision level. For a person, there are five walking sequences at one view in our database. We use the voting rule to decide the person's gender, assigning one vote to each sequence. Let $P_m(v_{ij})$ be the possibility of the i-th person being classified as male from the j-th classifier. $P_m(v_{ij})$ can be defined as:

$$P_m(v_{ij}) = \frac{N_m}{N_t} \tag{11}$$

where N_m denotes the number of votes on male and N_t is the number of total votes which is 5 in our database. Likewise, $P_f(v_{ij})$ denotes the possibility being classified as female. Then, the fusion can result in the possibility $p_m(w_i)$ of recognizing a walker as a male with the following expression:

$$p_m(w_i) = \frac{P(m)\sum_j P_m(v_{ij})}{P(m)\sum_j P_m(v_{ij}) + P(f)\sum_j P_f(v_{ij})} \tag{12}$$

where $P(m)$ is the percent that males take up in our database and $P(f)$ is the percent of females. They can be looked as the prior probabilities of the two classes.

5 Experiments

The first part of our experiments is to carry out gender recognition with our face data and gait data. There are seven different views available for gait in our database. Because camera C_4 captured the walking videos in the same view as camera C_5, we leave these videos from C_4 for later experiments. Now, every subject has five walking sequences from one view angle and five corresponding face sequences. We trained and tested SVM on face features and gait features of each view separately. Hence, there are eight gender classifiers comprising front-view face and seven different views' gait. For making the best of our samples, we used the leave-one-out validation method. There are 60 subjects in all, 32 of which are male and the remainder is female. The five sequences of one person were chosen as probe data in turn and all the other sequences were used as gallery data. The correct recognition rates(CRR) are listed in Table 1. We ran SVM with different kernels among which linear kernel performed the best.

Table 1. Results of gender recognition using SVM with linear kernel

modality	gait							face
	0^0	30^0	60^0	90^0	120^0	150^0	180^0	
CRR	78.7%	90.3%	91.7%	90.3%	88.3%	84.0%	82.3%	90.0%

The second part is our fusion experiments. The three combination strategies mentioned in above section lead to an obvious improvement on CRR, as shown in Table 2. The proposed Bayes combination rule achieved the same results compared with the other schemes. Additionally, we wondered whether the face cue gave its contribution in this combination. So, another set of fusion experiments of integrating only the seven different views of gait were done and the results can also be found in Table 2. The results show the fact that there is an obvious increase when face cues are added in the fusion.

Table 2. Results of fusion

Fusion scheme	CRR by fusing face and multi-view gait	CCR by fusing of multi-view gait
Voting rule	98.3%	96.7%
Weighted voting rule	98.3%	96.7%
Bayes combination rule	98.3%	96.7%

In [14] Shan et al. exploit CCA to fuse face and gait at the feature level and their experiments achieve a superior recognition performance of 97.2% on CASIA Gait Database. However, this database doesn't include separate face data and Shan had to crop face images from front-view gait image sequences. Our self-building database can provide an advantage for the study of fusing face and gait and we will do a comprehensive investigation in this field.

6 Conclusions and Future Work

We have presented an effective two-modality fusion system for gender recognition. An important componet of this system is a self-build database including front-view face data and gait data from seven different view. For every person in this database, these data are acquired in the same walking sequences. PCA is used to reduce facial image dimension and an improved spatio-temporal gait representation is introduced to extract the discriminating features from different gait view angles. Then, we employ SVM to implement gender classification based on the different modal features respectively. The final fusion is carried out according to three strategies: voting, weighted voting and Bayes combination. All of them reached the encouraging results.

In the future, we will apply different feature extraction methods to face and gait. In addition, more fusion schemes are waiting for us to make comparison with what we have done.

Acknowledgments

This work was supported by Program of New Century Excellent Talents in University, National Natural Science Foundation of China (No. 60575003, 60332010, 60873158), Joint Project supported by National Science Foundation of China and Royal Society of UK (60710059), and Hi-Tech Research and Development Program of China (2006AA01Z133).

References

1. Moghaddam, B., Yang, M.: Learning Gender with Support Faces. IEEE Transactions on Pattern Analysis and Machine Intelligence 24(5), 707–711 (2002)
2. Shakhnarovich, G., Viola, P.A., Moghaddam, B.: A Unified Learning Framework for Real Time Face Detection and Classification. In: Proc. of IEEE International Conference on Automatic Face and Gesture Recognition, pp. 14–21 (2002)
3. Lu, H., Lin, H.: Gender Recognition using Adaboosted Feature. In: The 3rd International Conference on Natural Computation, vol. 2, pp. 646–650 (2002)
4. Parris, E.S., Carey, M.J.: Language Independent Gender Identification. In: Proc. of IEEE International Conference on Acoustics, Speech and Signal Processing, vol. 2, pp. 685–688 (1996)
5. Harb, H., Chen, L.: Gender Identification Using a General Audio Classifier. In: Proc. of IEEE International Conference on Multimedia and Expo, vol. 2, pp. 733–736 (2003)
6. Huang, T., Yang, Y., Wu, Z.: Combining MFCC and Pitch to Enhance the Performance of the Gender Recognition. In: Proc. of IEEE International Conference on Signal Processing, vol. 1, pp. 16–20 (2006)
7. Davis, J.W., Gao, H.: An Expressive Three-mode Principle Components Model for Gender Recognition. Journal of Vision 4, 362–377 (2004)
8. Lee, L., Grimson, W.: Gait Analysis for Recognition and Classification. In: Proc. of IEEE International Conference on Automatic Face and Gesture Recognition, pp. 148–155 (2002)
9. Yoo, J., Hwang, D., Nixon, M.S.: Gender Classification in Human Gait with SVM. In: Blanc-Talon, J., Philips, W., Popescu, D.C., Scheunders, P. (eds.) ACIVS 2005. LNCS, vol. 3708, pp. 138–145. Springer, Heidelberg (2005)
10. Shakhnarovich, G., Lee, L., Darrell, T.: Integrated face and gait recognition from multiple views. In: Proc. of IEEE International Conference on Computer Vision and Pattern Recognition, vol. 1, pp. 439–446 (2001)
11. Shakhnarovich, G., Darrell, T.: On probabilistic combination of face and gait cues for identification. In: Proc. of IEEE International Conference on Automatic Face and Gesture Recognition, pp. 169–174 (2002)
12. Zhou, X., Bhanu, B.: Integrating face and gait for human recognition. In: Proc. of IEEE workshop on Computer Vision and Pattern Recognition, pp. 55–62 (2006)
13. Liu, Y., Collins, R., Tsin, Y.: Gait Sequence Analysis using Frieze Patterns. Technical Report CMU-RI-TR-01-38, Robotics Institute, Carnegie Mellon University (2001)
14. Shan, C., Gong, S., McOwan, P.W.: Learning Gender from Human Gaits and Faces. In: IEEE Conference on Advanced Video and Signal Based Surveillance, pp. 505–510 (2007)
15. Center for Biometrics and Security Research, CASIA, `http://www.cbsr.ia.ac.cn`
16. Bernard, A., Horst, B.: Combination of Classifiers on the Decision Level for Face Recognition. Tech. Rep. IAM-96-002, IAM, University Bern (1996)

Unsupervised Real-Time Unusual Behavior Detection for Biometric-Assisted Visual Surveillance

Tsz-Ho Yu and Yiu-Sang Moon

The Chinese University of Hong Kong

Abstract. This paper presents a novel unusual behaviors detection algorithm to acquire biometric data for intelligent surveillance in real-time. Our work aims to design a completely unsupervised method for detecting unusual behaviors without using any explicit training dataset. To this end, the proposed approach learns from the behaviors recorded in the history; such that the definition of unusual behavior is modeled according to previous observations, but not a manually labeled dataset. To implement this, pyramidal *Lucas-Kanade algorithm* is employed to estimate the optical flow between consecutive frames, the results are encoded into flow histograms. Leveraging the correlations between the flow histograms, unusual actions can be detected by applying *principal component analysis* (PCA). This approach is evaluated under both indoor and outdoor surveillance scenarios. It shows promising results that our detection algorithm is able to discover unusual behaviors and adapt to changes in behavioral pattern automatically.

1 Introduction

Visual surveillance is ubiquitous in modern societies: *closed-circuit televisions* (CCTV) and other visual surveillance systems are common in public places like airports, train stations and banks. Traditional visual surveillance systems have provided valuable forensic evidence in crime and terror investigations, yet they are not as effective for crime prevention. Because of the rapid growth of surveillance cameras installed and inadequate manpower to supervise them, discovering suspicious activities manually from the surveillance video data becomes inefficient. This has led to an urgent demand for automatic methods to analyze vast quantities of surveillance videos.

Therefore, the fusion of biometric technologies and visual surveillance has attracted much attention recently. Several *biometric-assisted* visual surveillance systems has been proposed, identifying people using gait or face recognition. Besides recording videos for forensic investigations, biometric-assisted surveillance are designed to detect unusual events, identify the people involved and give corresponding warning immediately. For instance, Zhou et al. presented a master-slave camera network to acquire biometric imagery [1]; Nishiyama et al.

M. Tistarelli and M.S. Nixon (Eds.): ICB 2009, LNCS 5558, pp. 1019–1029, 2009.
© Springer-Verlag Berlin Heidelberg 2009

Fig. 1. CCTV images from London bombing on 7 July 2005

designed a surveillance system which recognizes faces in a cluster of moving people [2], and Chan et al.[3] used a Pan-Tilt-Zoom (PTZ) camera to capture face images in a corridor for recognition.

Figure 1 illustrates the motivation for detecting unusual behavior, and the importance of applying biometrics in visual surveillance. This is an excerpt of the CCTV footages recorded during the London bombing terrorist attack on 7 July 2005. Interestingly, the suspects were actually monitored by security cameras, however it still took months for police to examine the 2500 tapes of surveillance videos [4,5]. If the surveillance system could detect the unusual behaviors of the suspects, they would be identified in advance, number of victims could have been minimized.

Unfortunately, due to the inevitable mechanical delays in PTZ cameras and other computational limitations, every biometric-assisted system can only handle a certain amount of people simultaneously. Hence, it is technically infeasible to identify every person using biometircs. To tackle this problem, people are monitored selectively — we are just interested in keeping an eye on the potential troublemakers who behave abnormally. Once an unusual activity is detected, biometric features of the target person is captured and identified [3]. Being the first step for biometric-assisted surveillance, unusual behavior detection directly determines the reliability of the whole system.

In this paper, we present a novel algorithm to detect unusual behaviors for acquiring biometric data in surveillance videos. Leveraging the large correlation between normal behaviors, this technique is specialized for a deployable real-time surveillance system. Utilizing optical flow for motion segmentation and eigenspace transformation for classification, our proposed algorithm is able to detect unusual behavior in real-time without providing any training data. Hence, the surveillance system is able to identify the misbehaving person immediately by means of face recognition or gait analysis.

The rest of the paper is structured as follows: Section 2 reviews the related researches and highlights the contribution of our work. Section 3 describes the feature extraction techniques using pyramidal optical flow from surveillance videos [6,7]. Section 4 explains the unsupervised detection algorithm in order to acquire biometrics data for identity verification. Experiments have been performed to justify our proposed method in various real life scenarios, evaluation results are reported in Section 5. Finally, we draw the conclusion in Section 6.

2 Detecting Unusual Behavior in Video

Techniques for unusual behavior detection has been considered as a challenging topic in computer vision for years. For detecting irregularities in surveillance

video, Xiang et al. developed a sophisticated event-based technique for classifying unusual behavior in video footages incrementally [8,9]. Zhong et al. took a different perspective on detecting unusual activity in videos, motivated by a similar problem in document keyword analysis [10]. Other approaches for detecting unusual behaviors include optical flow [11], silhouette contours [12] or Bayesian inference [13].

Although the importance of behavioral analysis in visual surveillance system has long been recognized, some issues remain largely unsolved under realistic and uncontrolled circumstances. First, it is difficult to give an explicit definition for unusual behaviors, even it is very easy to be recognized manually. A suspicious behavior in one scene can be regarded as normal in other environments. Currently, much work relies on a supervised learning approach, the "unusual behaviors" are required to be well-defined and labeled in the dataset for classifier training. These approaches is not feasible in real surveillance, as the definition of "unusual behaviors" changes over time. In addition, some of detection algorithms rely on detection of local features [12,8,9]. These approaches can acheieve very high accuracy when the background is well modeled. Still, they require accurate segmentation of body silhouettes, which can be affected by occlusions or clutter background. Moreover, many techniques on unusual behavior detection cannot performed in real-time, as they require the complete video for normal/abnormal action modeling [10,12,13].

The main contribution of our work is to present a simple unsupervised algorithm for detecting unusual behavior in real-time, such that the target can be identified immediately using biometric features. Instead of using an explicit training set with well-defined models of "abnormal behaviors", an unsupervised approach is utilized to learn previously-recorded behaviors as prototypes. No training data is required as the proposed algorithm adapts to new behavior pattern dynamically. By learning live samples captured from the actual environment, the definition of unusual behavior can be adjusted incrementally, so the classification criteria gradually changes adaptively. This algorithm is effective yet inexpensive to be performed in real-time, making promptly response to capture biometric features like face images or gait animations. Features of behaviors are extracted using optical flow field, then the recorded behavior samples are transformed into eigenspace by PCA.

3 Feature Extraction and Representation

Framework for the proposed unusual behavior detection is illustrated in Figure 2. For each frame recorded from the visual surveillance system, moving objects are

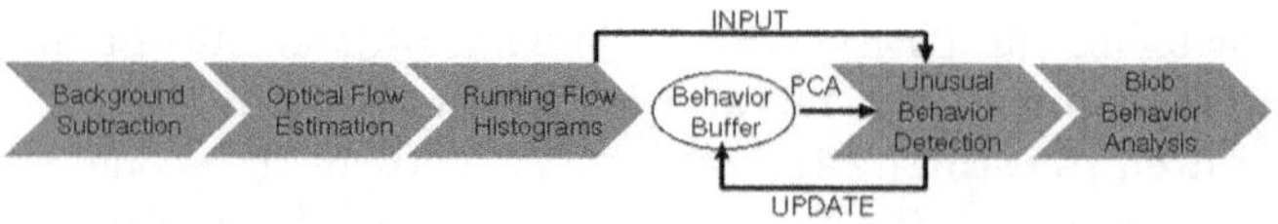

Fig. 2. Framework of the proposed detection algorithm

extracted by using background subtraction. Pyramidal optical flow is applied on the movie objects to analyze their motion direction and velocity. The overall movements in the scene are represented in the running flow histograms which will be discussed later in this section. The histograms samples are stored and transformed using PCA; unusual behaviors in the videos is then detected and located using previous observations.

3.1 Preprocessing

Surveillance videos captured from security cameras are available in gray-scale image sequences. At a specific time, the surveillance system has captured a video segment consisting T frames. This video segment $\mathcal{I}$ can be represented as a series of integer matrices: $\mathcal{I} = \{I_0, I_1, I_2, ..., I_i, ..., I_T\}$; where I_n is the nth frame in the video, and $I_n(x, y)$ represent the *intensity* at frame coordinate (x, y). Before extracting moving objects from the frames, it is necessary to classify stationary frames and frames with actions, this is done by computing the image difference between two consecutive frames:

$$Id_i(x, y) = \begin{cases} 1 & \text{when } I_i(x, y) - I_{i-1}(x, y) \leq thresh_{m1} \\ 0 & \text{otherwise} \end{cases} \tag{1}$$

$$Id_{i,sum} = \sum_x \sum_y Id_i(x, y) \tag{2}$$

$$\text{Frame } I_i \text{ is stationary when } Id_{i,sum} \leq thresh_{m2}. \tag{3}$$

The thresholds $thresh_{m1}$ and $thresh_{m2}$ can be adjusted according to the surveillance environment, they determines the sensitivity of the motion detection. Motionless frames are discarded from detection, assuming that no behavior is captured in the frames.

Foreground pixels are subtracted from the background by employing *mixture of Gaussian* method proposed by Stauffer and Grimson [14]. Figure 3 shows the foreground regions obtained from a frame. It is interesting to note that the foreground region subtracted is not always as precise as expected. In spite of that, our approach can still tolerate with inaccuracies in background subtraction, because the shapes of foreground regions will not be treated as a feature for detection.

3.2 Running Flow Histograms

Instead of measuring the trajectories of moving objects [12] or motion-history image (MHI) [15], optical flow method is chosen because only two consecutive frames are required to estimate the movement fields in the scene, which enables faster responses for a real-time surveillance system. In addition, optical flow methods is capable of finding different motion vectors within one object, offering more information for unusual behavior detection.

Fig. 3. Results of background subtraction using mixture of Gaussian method: (a) original frame; (b) foreground regions of frame (a); (c) original frame; (d) foreground regions of frame (c)

Fig. 4. Number of feature points are reduced using the foreground pixels as mask: (a) original frame; (b) positions of original feature points; (c) the foreground mask; (d) remaining feature points after masking

To ensure that the features points are evening distributed, a dense-feature optical flow field estimation is employed to compute the movements of foreground objects. The amount of feature points are reduced by using a grid mask, which is illustrated in Figure 4, to prune away the inaccurate flows caused by shaking backgrounds. Assuming the movements are small enough, the deformation between two consecutive frames can be estimated using the Lucas-Kanade algorithm.

Addressing the problem of detecting fast movements by Lucas-Kanade algorithm, the flow estimation is augmented by constructing an image pyramid [7]; Figure 5 shows the result of the optical flow algorithm. In an array of flows F_n in frame I_n, each feature point is represented as a four-dimensional feature vector f_n:

$$\begin{aligned} F_i &= \{f_{i1}, f_{i2}, ..., f_{ij}, ..., f_{iJ}\} \\ f_{ij} &= [X_{ij}, Y_{ij}, dx_{ij}, dy_{ij}], \end{aligned} \tag{4}$$

where (dx_{nj}, dy_{nj}) and (X_{nj},Y_{nj}) denote the velocity and location of the feature respectively. For each individual frame with motion detected, a flow histogram $h_i'(d,v)$ is constructed, with $N_d \times N_v$ bins in direction and velocity respectively, as described in 5 and 6.

$$\begin{aligned} \theta_{ij} &= \tan^{-1}\left(\frac{dy_{ij}}{dx_{ij}}\right) \\ S_{ij} &= \sqrt{(dy_{ij})^2 + (dx_{ij})^2} \end{aligned} \tag{5}$$

$$h_i'(d,v) = \text{no. of vector that satisfy} \begin{cases} d\theta * (d-1) \leq \theta_{ij} \leq d\theta * d \\ ds * (v-1) \leq S_{ij} \leq ds * v \end{cases} \tag{6}$$

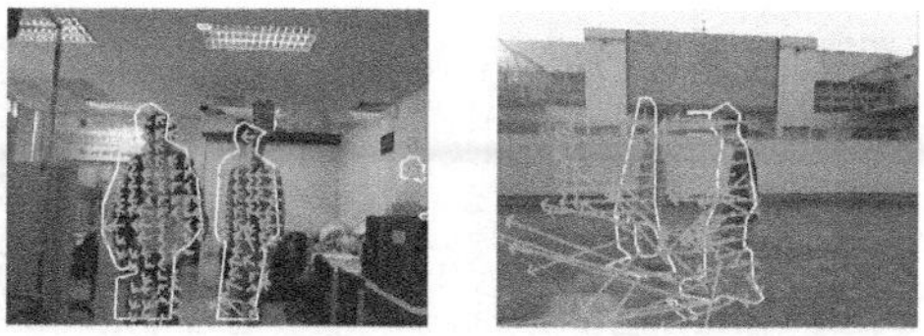

Fig. 5. Optical flow estimations using pyramidal Lucas-Kanade algorithm

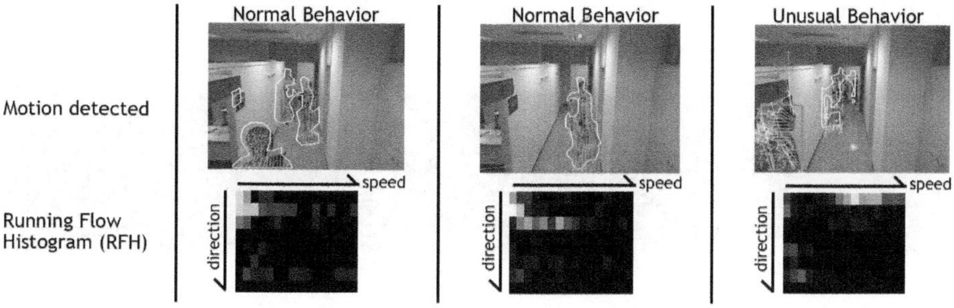

Fig. 6. The *Running Flow Histograms* computed from both normal and unusual behaviors. Note that the histograms of normal behaviors look similar.

For smoothing the transition and balancing the bin values between flow histograms, *running flow histograms (RFH)* are computed on-the-fly in 8.

$$h_i''(d,v) = \alpha_h \times Normalize(h_i'(d,v)) + (1-\alpha_h) \times h_{i-1}(d,v) \quad (7)$$

$$h_i(d * N_v + v) = h_i''(d,v). \quad (8)$$

It is a running average of normalized histograms along the time-line, the parameter α_h determines how much the histograms are smoothed with the previous histogram. Figure 6 visualizes the RFHs computed from different behaviors. In order to perform PCA in the detection stage, every RFH computed is reshaped to a column vector with size $N_d N_v \times 1$.

4 Unsupervised Detection Algorithm

The key concept of the proposed method is that normal behaviors possess high correlation between each others. Therefore, unusual behaviors can be detected by comparing with normal behaviors recorded in the history. If the current behavior presents a large deviation from the observed behaviors in the history, it is classified as unusual. To implement this, a buffer of behavioral data is kept in the system to model the normal behaviors dynamically.

4.1 Eigenspace Transformation

As described in Figure 2, a behavior buffer is kept in the program to record samples of RFHs observed previously. When a new ith frame is acquired from a

surveillance camera, the corresponding RFH h_i is computed. On the other hand, a set of eigenvectors and eigenvalues are obtained by applying PCA on the behavior buffer. Assume that there are S_b RFHs stored in the behavior buffer, the RFHs are concatenated to form a $S_b \times N_d * N_v$ matrix: $H^* = \left[h_1^* \; h_2^* \; h_3^* \ldots h*_{S_b} \right]$. The matrix of eigenvectors V and diagonal matrix of eigenvalues E can be calculated by solving 9:

$$V^{-1}Cov(H^*)V = E. \tag{9}$$

The eigenvectors are sorted by their corresponding eigenvalues in descending order, the first N_e eigenvectors sorted are kept in a matrix V' for unusual behavior detection.

4.2 Detecting Unusual Behavior

Once matrix V' is obtained, the current RFH is projected onto eigenspace and back-projected immediately in 11:

$$\text{Projection: } \mathbf{h}_i = (h_i - \overline{H^*}) * V'^T \tag{10}$$

$$\text{Back-projection: } h_i^r = \mathbf{h}_i * V' + \overline{H^*}, \tag{11}$$

where $\overline{H^*}$ is the element-wise average vector of H^*. As an unusual behavior do not have high correlation with normal behaviors, its RFH cannot be back-projected completely using the eigenvectors computed from normal behaviors. Leveraging this property, the recovery rate R determines if there is unusual behavior in the current frame:

$$R = 1 - \frac{||h_i^r - h_i||}{||h_i||}. \tag{12}$$

A frame is classified as "unusual" when the recovery rate R is smaller than a pre-defined threshold $thresh_r1$. This threshold can be adjusted to control the sensitivity of the detection algorithm:

$$\text{The frame } I_i \text{ contains unusual behavior(s) if its recovery rate } R_i \leq thresh_r1 \tag{13}$$

4.3 Locating Objects of Unusual Behavior

The location of the unusual behavior can be reconstructed using the difference between the original and back-projected RFH:

$$W_i = |h_i^r - h_i|. \tag{14}$$

The vector W represents how "unusual" a flow vector is in the current frame, the difference bewteen two RFHs represents the flow vectors that cannot be recovered by PCA. A "anomaly weight" $W_i(d * N_v + v)$ is assigned to all flow vectors in the scene, according to its corresponding bin in h_i. For each connected component B extracted from the foreground, the average "anomaly" of

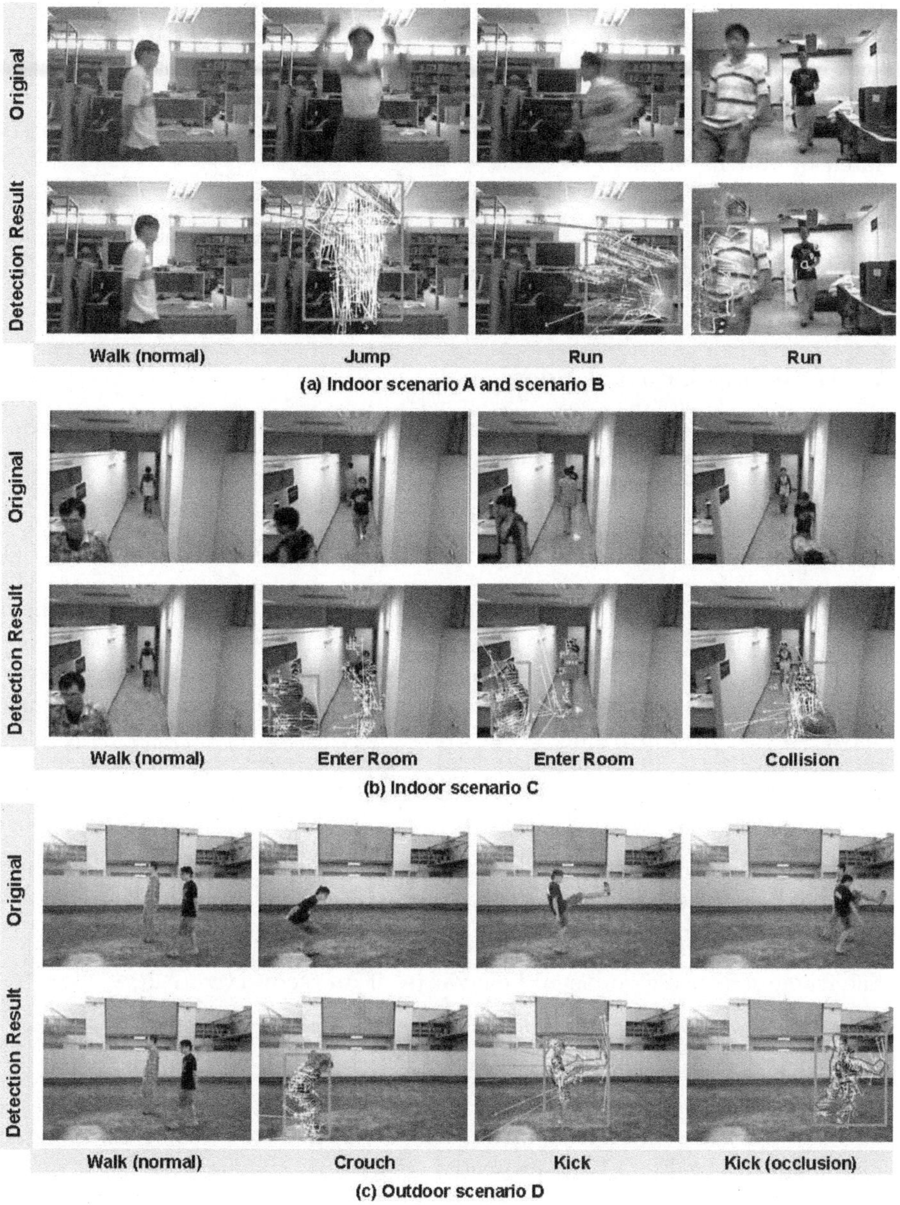

Fig. 7. Sample Detection Result: (a) Detection results in scenario A and B; (b) Detection results in scenario C; (c) Detection results in scenario D

a foreground region can be calculated by summation of all weights inside the connected component divided by its own area:

$$A(B_{ik}) = \frac{\sum\limits_{F_i \in B_{ik}} W_i}{area(B_{ik})}. \tag{15}$$

Consequently, each connected component detected in the scene has been assigned an average "anomaly". This value indicates how the object's behaviors is differ from the normal behaviors in the history. A blob B_{ik} can be regarded as abnormal if $A(B_{ik}) \geq thresh_r 2$.

4.4 Behavior Model Update Scheme

In the unusual behavior detection algorithm, a behavior buffer is used as the prototype of normal behaviors. In real situations, the behavioral pattern changes gradually over time, therefore the behavior buffer needs to be updated continuously with new behaviors.

In our work, the behavior buffer is implemented using a circular linked list with size S'_b where $S'_b \leq S_b$. The behavior buffer employs a "first-in, first-out" design: when a new frame enters the system, it will be evaluated by its recovery rate R; if R is greater than the replacement threshold $thresh_h$, the RFH of the new frame will replace the oldest frame inside the behavior buffer. During the detection process, only the oldest S_b entries are selected to compute the eigenvectors. The most recent frames are not used in detection because they can achieve very high correlations with the current frame, over-weighing the effect of unusual behaviors. As a result, the behavior buffer adapts to new behavior pattern automatically, hence minimizing human intervention.

5 Evaluation

5.1 Experimental Setup

Effectiveness and Robustness of our approach were evaluated under both indoor and outdoor surveillance environments. Three different indoor scenarios were set up for experiments. In scenario A, a webcam was mounted on the wall of a office as the security camera, monitoring the behavior of people passing by the aisle. Scenario B is similar with scenario A, but the camera was mounted at the end of the aisle instead. Scenario C simulated the arrangement of a real indoor CCTV system, the camera was mounted on the ceiling of the entrance/exit corridor — most of the people would just walk straight and enter the office, a few of them entered the control room on one side of the corridor, which is regarded as an unusual behavior.

In the outdoor scenario D, the security camera was assigned to monitor the entrance of a building. The objective of this outdoor experiment is to evaluate the effect of illumination and occlusion on the detection performance. All the experiments described above were conducted on an dual-core 2 GHz platform, with video resolution 640×480pixels, $N_d = 8$, $N_v = 16$, $S_b = 400$, $S'_b = 600$.

5.2 Analysis

The proposed detection algorithm demonstrates satisfactory results: in both indoor and outdoor case, this approach can successfully detect unusual activities in real-time with high robustness, while making only a few false negatives.

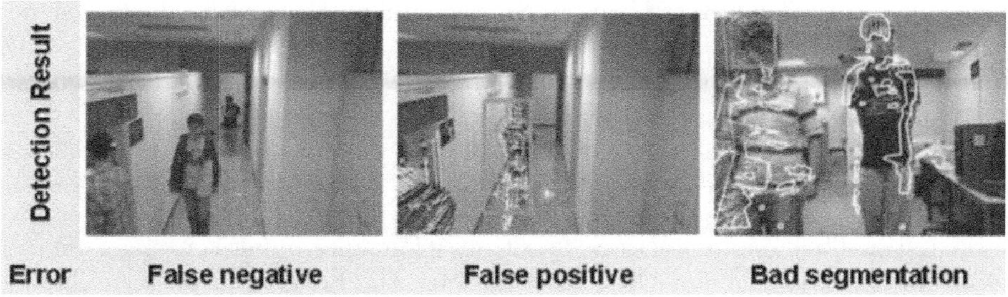

Fig. 8. Examples of incorrect detections

Figure 7(a) illustrates some detection results obtained from the first two indoor environments. In scenario A, without providing any labeled samples, the algorithm is able to discover unusual behaviors automatically. However, some unusual behaviors in the scenario B cannot be detected completely because of bad background/foreground segmentation as depicted in Figure 8. It is expected that the bad segmentation is due to the slow movement of objects in the scene, misinterpreting some foreground regions as background. Some experimental results in scenario C are shown in Figure 7(b). It is shown that the proposed algorithm can also detect unusual behaviors which happen suddenly: on the rightmost column in Figure 7(b), the man on the right hand side was bumped by the man in the middle suddenly, however the algorithm was able to notice the change and made the right response.

Some evaluation results of scenario D are recorded in Figure 7(c). Besides detecting unusual movements, abnormal actions can also be detected by the proposed approach. In addition, this approach also shows high robustness against occlusions, unusual behavior can still be detected when the silhouette is greatly distorted by occlusions.

6 Conclusion and Future Works

We presented a novel unsupervised approach to detect unusual behavior for surveillance. This method is specialized for biometric-assisted surveillance as it is able to perform detection in real-time, in order to identify the target immediately by means of face recognition or gait analysis. This approach needs neither a labeled training set, nor a well-defined behavioral definition to detect unusual behavior automatically and incrementally. In future work, we intend to improve the segmentation accuracy by designing a new background model. We also seek to investigate the possibility of introducing *independent component analysis* (ICA) or *supporting vector machine* (SVM) into the current detection algorithm.

Acknowledgement

The work described in this paper was substantially supported by a grant from the Research Grants Council of the Hong Kong Special Administrative Region, China (Project No. 415207).

References

1. Zhou, X., Collins, R.T., Kanade, T., Metes, P.: A master-slave system to acquire biometric imagery of humans at distance. In: IWVS 2003: First ACM SIGMM international workshop on Video surveillance, pp. 113–120. ACM, New York (2003)
2. Nishiyama, M., Yuasa, M., Shibata, T., Wakasugi, T., Kawahara, T., Yamaguchi, O.: Recognizing faces of moving people by hierarchical image-set matching. In: IEEE Conference on Computer Vision and Pattern Recognition, 2007. CVPR 2007, pp. 1–8 (June 2007)
3. Chan, F., Chen, J., Moon, Y.S.: Fast construction of object correspondence in stereo camera system: an example to human face capturing system. In: IEEE Workshop on Motion and video Computing, 2008. WMVC 2008, pp. 1–6 (January 2008)
4. CNN: Police to give bomb probe details (July 2005), `http://www.cnn.com/2005/WORLD/europe/07/11/london.attacks/index.html`
5. BBC: 7/7 bombers surveillance shown (April 2008), `http://news.bbc.co.uk/1/hi/england/london/7338301.stm`
6. Lucas, B.D., Kanade, T.: An iterative image registration technique with an application to stereo vision. In: IJCAI 1981, pp. 674–679 (1981)
7. Bouguet, J.Y.: Pyramidal implementation of the lucas kanade feature tracker: Description of the algorithm. Jean-Yves Bouguet (2002)
8. Xiang, T., Gong, S.: Activity based surveillance video content modelling. Pattern Recogn. 41(7), 2309–2326 (2008)
9. Xiang, T., Gong, S.: Video behavior profiling for anomaly detection. IEEE Transactions on Pattern Analysis and Machine Intelligence 30(5), 893–908 (2008)
10. Zhong, H., Shi, J., Visontai, M.: Detecting unusual activity in video. In: Proceedings of the 2004 IEEE Computer Society Conference on Computer Vision and Pattern Recognition, 2004. CVPR 2004, vol. 2, June 2-July 2004, pp. II–819–II–826 (2004)
11. Andrade, E., Blunsden, S., Fisher, R.: Hidden markov models for optical flow analysis in crowds. In: 18th International Conference on Pattern Recognition, 2006. ICPR 2006, vol. 1, pp. 460–463 (2006)
12. Wu, X., Ou, Y., Qian, H., Xu, Y.: A detection system for human abnormal behavior. In: 2005 IEEE/RSJ International Conference on Intelligent Robots and Systems, 2005 (IROS 2005), pp. 1204–1208 (August 2005)
13. Li, Y., Xu, C., Liu, J., Tang, X.: Detecting irregularity in videos using kernel estimation and kd trees. In: MULTIMEDIA 2006: Proceedings of the 14th annual ACM international conference on Multimedia, pp. 639–642. ACM, New York (2006)
14. Stauffer, C., Grimson, W.: Adaptive background mixture models for real-time tracking. In: IEEE Computer Society Conference on Computer Vision and Pattern Recognition, vol. 2, p. 2246 (1999)
15. Bobick, A., Davis, J.: The recognition of human movement using temporal templates. IEEE Transactions on Pattern Analysis and Machine Intelligence 23(3), 257–267 (2001)

Multilinear Tensor-Based Non-parametric Dimension Reduction for Gait Recognition

Changyou Chen, Junping Zhang*, and Rudolf Fleischer

Shanghai Key Lab of Intelligent Information Processing
School of Computer Science, Fudan University, Shanghai 200433, China
cchangyou@gmail.com, jpzhang@fudan.edu.cn, rudolf@cse.ust.hk

Abstract. The small sample size problem and the difficulty in determining the optimal reduced dimension limit the application of subspace learning methods in the gait recognition domain. To address the two issues, we propose a novel algorithm named *multi-linear tensor-based learning without tuning parameters (*MTP*)* for gait recognition. In MTP, we first employ a new method for automatic selection of the optimal reduced dimension. Then, to avoid the small sample size problem, we use multi-linear tensor projections in which the dimensions of all the subspaces are automatically tuned. Theoretical analysis of the algorithm shows that MTP converges. Experiments on the USF Human Gait Database show promising results of MTP compared to other gait recognition methods.

Keywords: Subspace learning, multi-linear tensor, small sample size problem, dimension reduction, gait recognition.

1 Introduction

Dimension reduction techniques play a crucial role in the discovery of the intrinsic low-dimensional structure in high-dimensional data such as gait sequences. Principal Component Analysis (PCA) attempts to find a subspace maximizing the total variance of the data and minimizing the mean squared errors. Linear Discriminant Analysis (LDA) constructs an optimal discriminative subspace for classification using label information. Furthermore, by incorporating spectral graph theory, algorithms such as LPP [1] and DNE [2] were developed for discovering the intrinsic low-dimensional manifolds in high-dimensional data. Recently, a unified framework of dimension reduction algorithms was proposed by Yan et al. [3] from the viewpoint of graph embedding.

The above-mentioned algorithms have in common that they are based on a vector representation. The disadvantages of using vector representation are 1) the small sample size problem for high-dimensional data, and 2) unfolding of natural data such as images into vector representation may lose some important information of the original data. To overcome these problems, tensor based learning algorithms [4] were proposed for extending vector based algorithms into tensor based counterparts. For example, He et al. [5] generalized LPP into tensor representation, while the LDE algorithm was extended by Dia et al. [6]. More results about tensor based learning can be found in the work of Yan et al. [7] and Xu et al. [8].

* Corresponding author.

M. Tistarelli and M.S. Nixon (Eds.): ICB 2009, LNCS 5558, pp. 1030–1039, 2009.
© Springer-Verlag Berlin Heidelberg 2009

Most of the subspace learning algorithms mentioned above have been applied for gait recognition. Wang et al. [9] used PCA to reduce the dimension of gait frames for gait recognition. Han et al. [10] enhanced recognition performance by the combination of LDA with some additional gait data. Furthermore, by integrating random transformations and LDA, Boulgouris et al. [11] proposed a new feature extraction process for gait recognition. While manifold learning methods such as LLE [12] and LPP in [13] had been introduced for gait recognition, the results were not very satisfactory. To obtain higher recognition rates, tensor-based methods such as MMFA [13], TLLE [12] and GTDA [14] were proposed.

It is quite obvious that the effectiveness of the algorithms above, which employ either vector representation or tensor representation, heavily depends on empirically tuned parameters. As a result, it is difficult to determine the intrinsic dimension and choose the corresponding subspace for a given problem instance. While DNE [2] tried to solve this problem by introducing a special Laplacian matrix, the nearest k-neighbor factor has to be predefined, and the automatic selection of the optimal projection subspace remains unanswered in [3].

To tackle the above mentioned problems, we now propose an extended tensor based parameter-free version of the DNE algorithm, named MTP (multi-linear tensor-based learning without tuning parameters). The advantage of DNE and MTP is that the intrinsic dimension can be automatically chosen by solving an eigenvalue problem. Unlike in DNE, we suggest that for the automatical selection of the optimal dimension of a subspace, the Laplacian matrix should be non-positive semi-definite, which was not required in DNE [2]. Furthermore, the relationship matrix is defined without the need of a predefined neighborhood factor. With these two improvements, we propose an objective function that preserves the relationship of pairwise data points in the automatically selected optimal subspace. Since the objective is to preserve pairwise relationships but not to achieve good discriminant ability, we enhance the classification ability by using the LDA technique without the singular matrix problem. We then generalize the proposed algorithm to a multi-linear tensor version and prove its convergence.

The rest of this paper is organized as follows. In Section 2 we present our new MTP algorithm together with some theoretical analysis. We report on some experiment in Section 3, and we conclude the paper in Section 4.

2 The MTP Algorithm

In this section, we first state an important property of the Laplacian matrix. Based on this property, we can define a relationship preserving projection on either vector data or tensor data, followed by LDA for enhancing its discriminant ability. Furthermore, we propose an iterative learning algorithm, and show its convergence. In the following, we denote the training data as $X \in \mathbb{R}^{k_1 \times k_2 \times \cdots \times k_n \times N}$, where k_i is the dimension of the training data space in the i-th order of the tensor, m is the dimension of the projection space, and N is the number of training samples. Let $Y \in \mathbb{R}^{m \times N}$ be the projected vector data of each *core tensor* $\mathcal{Y}$, as defined in Section 2.2.

2.1 Non-positive Semi-definite Laplacian Matrix

Yan et al. [3] observed that most dimension reduction techniques can be unified in the graph embedding framework. Its objective is given by

$$F = \min_{Y}\{tr\{YLY^T\}\} \tag{1}$$

subject to some constraints (in the following, the constraints are usually omitted), where L is the Laplacian matrix that we want to find. However, it is not clear from Eq. (1) how to automatically obtain the optimal dimension of Y, i.e., the value of m. The following lemma addresses this problem.

Lemma 1. *We can calculate the optimal dimension of Y in Eq. (1) to find its minimum value if the Laplacian matrix L is not positive semi-definite, i.e., there exist some vectors $y \in \mathbb{R}^N$ such that $y^T Ly < 0$.*

Proof. $\Longrightarrow$: Let $Y = (y_1^T, y_2^T, \cdots, y_m^T)^T$. Then Eq. 1 can be rewritten as

$$F = \min \sum_{i=1}^{m} y_i L y_i^T . \tag{2}$$

Note that L is not positive semi-definite and m is to be determined, we can choose those y_i such that $y_i L y_i^T < 0$ to reconstruct Y. That is, $Y = (y_1^T, y_2^T, \cdots, y_{m^*}^T)^T$ with all $y_i (0 \leq i \leq m^*)$ satisfying $y_i L y_i^T < 0$. In this way, Eq. (1) achieves its minimum value and we can automatically choose the optimal subspace.

In general, the Laplacian matrix L can be decomposed into two symmetric matrices, D and S, such that $L = D - S$ and $D(i,i) = \sum_j S(i,j)$ is a diagonal matrix. We call S the *relationship matrix*, because it reflects the relationship between data points, and the goal of the projection is to preserve these relations. According to Lemma 1, to construct a non-positive semi-definite Laplacian matrix L, we define the relationship matrix S as follows.

$$S(i,j) = \begin{cases} 1 & \text{if } x_i, x_j \text{ are in the same class} \\ -1 & \text{otherwise} \end{cases} \tag{3}$$

Obviously, the matrix L is then non-positive semi-definite, because Eq. (1) can be rewritten as

$$F = \min \sum_{i,j} \|y_i - y_j\|^2 S(i,j) . \tag{4}$$

By analyzing the relationship matrix defined in Eq. (3), we can see that F is not always positive, and thus L is not positive semi-definite. According to Lemma 1, we can get an optimal subspace for the projection automatically. Note that this definition is different from the definition in Zhang et al. [2], where it is necessary to predefine a nearest neighbor factor which is not necessary in our new parameter free version.

2.2 Automatic Subspace Learning Using Multi-linear Tensor Projections

In this section, we employ multi-linear tensor projections in the framework proposed in Section 2.1 to learn an optimal subspace which can preserve the relationships of the original data pairs.

Tensors are an extension of vectors and matrices. We write an order-n tensor as $X \in \mathbb{R}^{N_1 \times N_2 \cdots \times N_n}$, which means an order-$n$ tensor has n-ways varying dimensions. Following the SVD decomposition of a second order tensor (i.e., a matrix), we can decompose an order-n tensor into $X = \mathcal{Y} \times_1 U_1^T \times_2 \cdots \times_n U_n^T$, where $\mathcal{Y}$ is called the *core tensor* of X, the U_k are the order principal components of the k-th mode, and $\times_k$ means the k-mode product. A more detailed survey of tensors and tensor operations can be found in [15]. Intuitively, a vector is viewed as an order-1 tensor, a matrix as an order-2 tensor, and a video stream as an order-3 tensor, etc. Fig. 1 shows some visual descriptions from order-1 to order-4 tensors.

In the following, we assume that the data are given as order-n tensors. Our goal is to learn multi-linear projections U_k, for $k = 1, \cdots, n$, such that the core tensors of the original tensors can achieve the objective defined in Section 2.1, namely to preserve the relationships between data point pairs. To obtain an optimal decomposition of the tensor, we can get the optimal order principal component U_k of the k-th mode of the data tensor X iteratively, similarly to the SVD decomposition of tensors. Suppose one order principal component U_f is unknown, while all other U_i, $i \neq f$, are known. To rewrite the objective function of Eq. (1) in multi-linear tensor projection form with the unknown order principal component U_f, we first project the data using the known order principal components U_i, $i \neq f$, onto a lower subspace, then unfold the projected data into a matrix on the f-mode, using the f-mode unfolding defined in [15].

$$\begin{aligned} F_t(U_f) &= \min \left\{ \sum_{i,j,i \neq j} \| x_i \prod_{k=1}^{n} \times_k U_k - x_j \prod_{k=1}^{n} \times_k U_k \|^2 S_{ij} \right\} \\ &= \min \left\{ tr \left\{ U_f \left(\sum_{i,j} \left[\left(x_i \prod_{k \neq f} \times_k U_k \right)^f - \left(x_j \prod_{k \neq f} \times_k U_k \right)^f \right] \right. \right. \right. \\ &\quad \left. \left. \left. \left[\left(x_i \prod_{k \neq f} \times_k U_k \right)^f - \left(x_j \prod_{k \neq f} \times_k U_k \right)^f \right]^T S_{ij} \right) U_f^T \right\} \right\} \\ &\quad \text{s.t. } U_f^T U_f = I \end{aligned} \tag{5}$$

where y^f means the f-mode unfolding of tensor y, and S_{ij} is the relationship matrix in Eq.3). Let

$$A_f = \sum_{i,j} \left[\left(x_i \prod_{k \neq f} \times_k U_k \right)^f - \left(x_j \prod_{k \neq f} \times_k U_k \right)^f \right]$$

$$\left[\left(x_i \prod_{k \neq f} \times_k U_k \right)^f - \left(x_j \prod_{k \neq f} \times_k U_i \right)^f \right]^T S_{ij} \,. \tag{6}$$

Since the Laplacian matrix corresponding to the relationship matrix S is not positive semi-definite, it is easy to see by Lemma 1 that the optimal projection matrix U_f consists of the eigenvectors of the matrix A_f with negative eigenvalues. In this way, we can iteratively optimize the objective function defined in Eq. (5) to obtain the optimal projection matrix U_f given all other $n-1$ projection matrices U_i, $i \neq f$). Note that when the data are in vector form, there is only a single projection matrix, so we can derive a closed-form solution immediately.

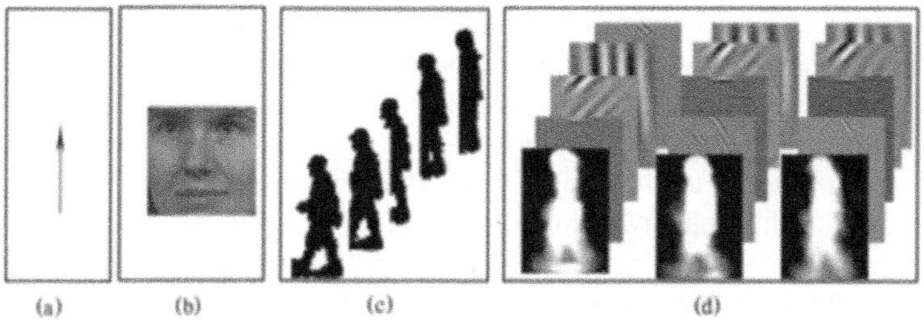

Fig. 1. An order-1 tensor, order-2 tensor, order-3 tensor, and order-4 tensor, from left to right

2.3 Convergence of MTP

To obtain all the projection matrices for tensor based learning, we use the following iterative procedure for searching an optimal solution.

1. We randomly initialize the projection matrices U_i^0, for $i = 1, \cdots, n$. In our experiments, we set $U_i^0 = E_{k_i}$, where E_{k_i} denotes the k_i-dimensional identity matrix.
2. For each U_f, by fixing the other $n-1$ projection matrices U_i, $i \neq f$, we calculate the eigenvalues and eigenvectors of A_f defined in Section 2.2, and choose the eigenvectors corresponding to negative eigenvalues to form the projection matrix U_f.
3. Let ε be a small positive number. We stop if $\|(U_f^{t-1})^T U_f^t - E_{k_f}\|_F^2 < \varepsilon$ for all U_f, otherwise we continue with the next iteration in step (2).

The following theorem shows that the algorithm always converges. The experiments in Section 3 indicate that it usually converges in four to five iterations.

Theorem 2. *The subspace learning procedure of* MTP *converges to a local optimum.*

Proof. Let

$$g(U|_{l=1}^n) = \sum_{i,j} \|(x_i - x_j) \prod \times_k U_k\|^2 S_{ij} \tag{7}$$

Let U_k^t denote the k-th projected matrix learned in the t-th iteration, and g_k^t the minimum objective function value of Eq. 7 in the current state. Then,

$$g_k^t \triangleq \min g(U_l^t|_{l=1}^k, U_l^{(t-1)}|_{l=k+1}^n) \leq g_{k-1}^t \tag{8}$$

which means U_k gets updated with a better projection matrix. Thus,

$$g_0^0 \geq \cdots \geq g_i^{t-1} \geq g_j^{t-1} \geq \cdots \geq g_i^t \geq g_j^t \geq \cdots , \tag{9}$$

where $0 < i < j < n$. Note that Eq. (7) has a lower bound because each x_i and U_i in Eq. (7) has finite norm, suggesting that all g_i^t have a lower bound. As a result, the iterative procedure will stop if $\|(U_l^t)^T U_l^{t-1} - I\| < \varepsilon$.

2.4 Improving MTP

Note that in Eq. (3), a relationship matrix S is defined to give us a non-positive semi-definite Laplacian matrix in the MTP algorithm. From the objective function in Eq. (5) and the relationship matrix S, we can conclude

Remark 3. The objective function in Eq. (5) with the relationship matrix S defined in Eq. (3) will result in data projection onto an optimal low dimensional subspace in which the discriminant ability is not strong.

This can easily be seen from the definition of the objective function in Eq. (5), since it only preserves the pairwise relationships of data points. The advantage is that MTP can automatically find an optimal lower dimensional subspace to represent the data. In this subspace, the traditional LDA technique can be used without the singular matrix problem. Of course, the time complexity is also much lower when we use LDA in the subspace. Thus, we can enhance the discriminant ability of MTP by adding a post-processing step, namely to use LDA in the learned subspace. Experiments show that this step is necessary for a satisfactory discriminant ability.

3 Experiments

3.1 The Gait Database

We evaluated MTP on the USF HumanID gait database [16] version 2.1. In this database, gait sequences of 122 subjects are sampled with varying view points, shoe type, surface type, briefcase and elapse time. Selecting sequences like "Grass, Shoe TypeA, Right Camera, No Briefcase, Time t_1" as the gallery set, Sarkar et al. [16] developed a set of 12 benchmark experiments under different conditions.

Since Gait Energy Image (GEI) [17] is insensitive to noise [10], we employ the same GEI techniques as in [16] and [10] for the representation of gait features. Then the distance between any two gait sequences using GEIs is calculated as

$$D_L(G_i, P) = \frac{1}{R_p} \sum_{k=1}^{R_p} \min_{j=1,\ldots,R_i} d_L(g_{ij}, p_k) , \tag{10}$$

where G_i represents the i-th gait sequence in the gallery set, and P is a specific probe sequence. Here, G_i and P contains R_i and R_p GEIs, respectively, that is, $G_i = \{g_{i_1}, \cdots, g_{iR_i}\}$, and $P = \{p_1, \cdots, p_{R_p}\}$. Furthermore, $d_L(r_1, r_2)$ is the distance between a GEI r_1 in the gallery set and a GEI r_2 in the probe set as computed by our algorithm. Fig. 2 shows some examples of the GEIs of this database. More details can be found in [16,17].

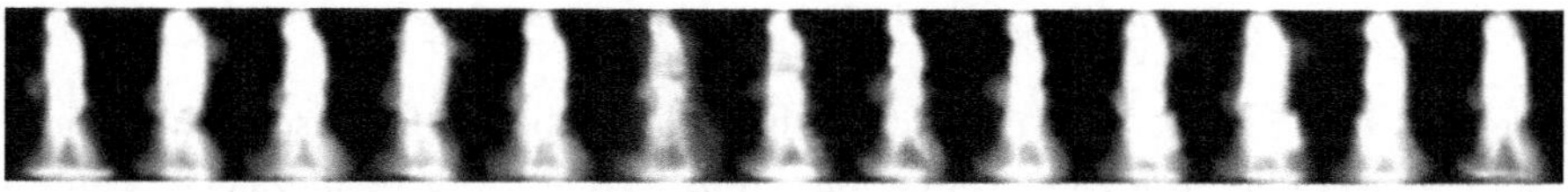

Fig. 2. The GEIs of one individual in the USF gait database. The leftmost one is in the gallery set, while the others are in the probe sets A-L, correspondingly.

3.2 Experimental Results

To evaluate the performance of MTP, we compared it with five different methods: 1) a baseline approach proposed by Sarkar et al. [16] which represents the similarities between frames using the ratio of the number of pixels in their intersection to the number of pixels in their union; 2) two recently developed and popularly used methods, LDA^1 and LDA^2 [10]. LDA^1 refers to performing LDA on the original gait data, while LDA^2 fuses the results of LDA on the original gait data and some generated virtual gait data using LDA; 3) the DNE algorithm by Zhang et al. [2]; 4) 2DLDA [18]. Note that there exist some tensor based dimension reduction algorithms such as [14] whose performance depends on the manual tuning of some parameters. For this reason, we don't compare such kinds of algorithms with our proposed algorithm.

Theoretically, MTP does not need to tune any parameters to obtain an optimal learned projection dimension. However, in the experiments, we observed that some negative eigenvalues of the corresponding matrix were close to zero, which we did not expect to happen. To eliminate this noise, we selected only the prominent negative eigenvalues. More exactly, the number of negative eigenvalues is the minimal m satisfying the following criterion:

$$\gamma = \frac{\sum_{i=1}^{m} |\lambda_i|}{\sum_{i=1}^{n} |\lambda_i|} > 0.98 \,, \tag{11}$$

where n is the total number of negative eigenvalues, and λ_i is the i-th smallest eigenvalue of the corresponding matrix.

Furthermore, we used MTP for gait recognition using both vector data and tensor data (order-2 tensors). We denote the algorithm dealing with vector data as MTP_V while the one working on tensor data as MTP_T. When we compared it to 2DLDA [18], we set the dimension of their projection matrices to be the same as MTP's dimension for a fair comparison. Tables 1 and 2 show the results of Rank-1 and Rank-5 performances of MTP compared with other methods as in [16].

We can see in Tables 1 and 2 that MTP improves on the average the recognition rate compared to other benchmark algorithms. Because MTP does not need any pre-defined parameters, it actually improves greatly the performance of other tensor learning algorithms.

Table 1. Comparison of Rank-1 performances of different algorithms on the USF gait database version 2.1, with the best performance highlighted by boldface. The projection dimension automatically learned by MTP was 20×17.

	Rank-1 Performance						
Experiments	baseline [16]	LDA[1] [10]	LDA[2] [10]	DNE[2]	2DLDA[18]	MTP_V	MTP_T
A	73%	86%	88%	88%	78%	**90%**	**90%**
B	78%	90%	**94%**	87%	89%	89%	91%
C	48%	79%	**83%**	74%	69%	**83%**	**83%**
D	32%	29%	31%	21%	23%	35%	**37%**
E	22%	33%	35%	25%	30%	42%	**43%**
F	17%	17%	22%	14%	9%	22%	**23%**
G	17%	23%	26%	18%	15%	**28%**	25%
H	61%	58%	55%	60%	**74%**	60%	56%
I	57%	51%	61%	58%	**67%**	60%	59%
J	36%	53%	48%	40%	53%	56%	**59%**
K	3%	3%	6%	**9%**	3%	**9%**	**9%**
L	3%	9%	**12%**	0%	6%	6%	6%
Average	40.95%	47.05%	49.15%	44.16%	45.85%	51.30%	**51.57%**

Table 2. Comparison of Rank-5 performances of different algorithms on the USF gait database version 2.1, with the best performance highlighted by boldface. The projected dimension automatically learned by MTP was 20×17.

	Rank-5 Performance						
Experiments	baseline [16]	LDA[1] [10]	LDA[2] [10]	DNE[2]	2DLDA[18]	MTP_V	MTP_T
A	88%	**94%**	**94%**	93%	88%	**94%**	**94%**
B	93%	**94%**	**94%**	93%	**94%**	**94%**	93%
C	78%	88%	90%	87%	81%	**93%**	91%
D	**66%**	57%	53%	51%	54%	64%	64%
E	55%	58%	60%	57%	52%	67%	**68%**
F	42%	37%	42%	36%	30%	47%	**51%**
G	38%	43%	50%	37%	48%	**52%**	**52%**
H	85%	85%	86%	82%	**92%**	89%	88%
I	78%	**85%**	**85%**	80%	**85%**	**85%**	83%
J	62%	77%	75%	75%	73%	78%	**82%**
K	12%	15%	15%	30%	12%	**21%**	18%
L	15%	18%	18%	18%	18%	**21%**	15%
Average	64.54%	66.71%	67.54%	64.7%	64.46%	71.28%	**71.38%**

In Theorem 2 we saw that MTP will converge. To experimentally justify this claim, we evaluated the recognition rates of experiments A, B, H and J with increasing iteration numbers from 0 to 10 in Fig. 3(a). Fig. 3(b) and (c) show the differences of the two projection matrices between consecutive iterations, which are defined as $dif = \|(U_i^t)^T U_i^{t-1} - I\|$, for $i = 1, 2, \ldots$. In Fig. 3 we can see that MTP has a very satisfactory convergent property. Normally, it will converge in as few as 4 iterations.

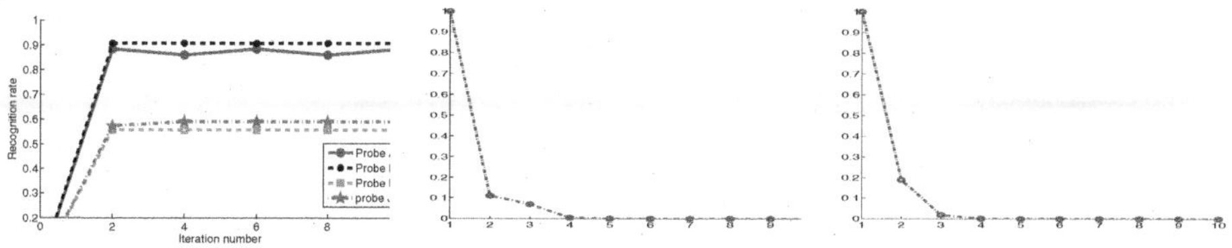

(a) (b) (c)

Fig. 3. Convergence of MTP: (a) the recognition rates of experiments A, B, H and J with increasing iteration numbers from 0 to 10; (b) and (c) the differences of the first and second projection matrices between consecutive iterations

4 Conclusions and Future Work

In this paper, we proposed a parameter-free tuning algorithm, `MTP`, to learn an optimal low-dimensional subspace for high-dimensional gait data. In contrast to `DNE`, `MTP` can automatically determine the optimal dimensions of the projection matrices and can obtain the optimal solution of the objective functions at the same time, which improves on other tensor based learning algorithms. As a result, it has the potential for real applications such as surveillance systems because it needs little human help to obtain an optimal solution. Moreover, we prove convergence of `MTP`. However, its performance for more challenging classifications such as elapse time gait recognition still needs improvement. We should also try to further improve `MTP`'s discriminant ability and extend it with kernel tricks.

Acknowledgements

This paper is sponsored by 863 Project (2007AA01Z176) and NSFC (60635030, 60505002, 60573025).

References

1. He, X., Niyogi, P.: Locality preserving projections. In: Advances in Neural Information Processing Systems (2004)
2. Zhang, W., Xue, X., Sun, Z., Guo, Y., Lu, H.: Optimal Dimensionality of Metric Space for Classification. In: International Conference on Machine Learning (2007)
3. Yan, S., Xu, D., Zhang, B., Zhang, H.J., Yang, Q., Lin, S.: Graph Embedding and Extensions: A General Framework For Dimensionality Reduction. IEEE Transactions on Pattern Analysis and Machine Intelligence 29(1), 40–51 (2007)
4. Tao, D.C., Li, X.L., Hu, W.M., Maybank, S., Wu, X.D.: Supervised Tensor Learning. Knowledge and Information Systems 13(1), 1–42 (2007)
5. He, X.F., Cai, D., Niyogi, P.: Tensor Subspace Analysis. In: Advances in Neural Information Processing Systems (2005)
6. Dai, G., Yeung, D.Y.: Tensor Embedding Methods. In: Proceedings of the National Conference on Artificial Intelligence (2006)

7. Yan, S., Xu, D., Lin, S., Huang, T., Chang, S.: Element Rearrangement for Tensor-Based Subspace Learning. In: IEEE Conference on Computer Vision and Pattern Recognition (2007)
8. Xu, D., Yan, S., Zhang, L., Lin, S., Zhang, H.J., Huang, T.: Reconstruction and Recognition of Tensor-Based Objects With Concurrent Subspaces Analysis. IEEE Transactions on Circuits and Systems for Video Technology 18(1), 36–47 (2008)
9. Wang, L., Tan, T.N., Ning, H.Z., Hu, W.M.: Silhouette analysis-based gait recognition for human identification. IEEE Transactions on Pattern Analysis and Machine Intelligence 25(12), 1505–1518 (2003)
10. Han, J., Bhanu, B.: Individual recognition using gait energy image. IEEE Transation on Pattern Analysis and Machine Intelligence 28(2), 316–322 (2006)
11. Boulgouris, N.V., Chi, Z.X.: Gait recognition using radon transform and linear discriminant analysis. IEEE Transactions on Image Processing 16(3), 731–740 (2007)
12. Li, X., Lin, S., Yan, S., Xu, D.: Discriminant Locally Linear Embedding With High-Order Tensor Data. IEEE Transactions on Systems, Man, and Cybernetics, Part B 38(2), 342–352 (2008)
13. Xu, D., Yan, S., Tao, D., Lin, S., Zhang, H.: Marginal Fisher Analysis and Its Variants for Human Gait Recognition and Content Based Image Retrieval. IEEE Transactions on Image Processing 16(11), 2811–2821 (2007)
14. Tao, D.C., Li, X.L., Wu, X.D., Maybank, S.J.: General Tensor Discriminant Analysis and Gabor Features for Gait Recognition. IEEE Transactions on Pattern Analysis and Machine Intelligence 29(10), 1700–1715 (2007)
15. Lathauwer, L.D.: Signal Processing Based on Multilinear Algebra, PhD thesis, Katholike Universiteit Leuven (1997)
16. Sarkar, S., Phillips, P.J., Liu, Z., Vega, I.R., Grother, P., Bowyer, K.W.: The Humanid Gait Challenge Problem: Data Sets, Performance, and Analysis. IEEE Transactions on Pattern Analysis and Machine Intelligence 27(2), 162–177 (2005)
17. Liu, Z., Sarkar, S.: Simplest Representation yet for Gait Recognition: Averaged Silhouett. In: IEEE International Conference on Pattern Recognition (2004)
18. Ye, J., Janardan, R., Li, Q.: Two-Dimensional Linear Discriminant Analysis. In: Advances in Neural Information Processing Systems (2005)

Quantifying Gait Similarity: User Authentication and Real-World Challenge

Marc Bächlin, Johannes Schumm, Daniel Roggen, and Gerhard Töster

Wearable Computing Laboratory, ETH Zürich, Switzerland
{baechlin,schumm,roggen,troester}@ife.ee.ethz.ch
http://www.ife.ee.ethz.ch

Abstract. Template-based approaches using acceleration signals have been proposed for gait-based biometric authentication. In daily life a number of real-world factors affect the users' gait and we investigate their effects on authentication performance. We analyze the effect of walking speed, different shoes, extra load, and the natural variation over days on the gait. Therefore we introduce a statistical Measure of Similarity (MOS) suited for template-based pattern recognition. The MOS and actual authentication show that these factors may affect the gait of an individual at a level comparable to the variations between individuals. A change in walking speed of 1km/h for example has the same MOS of 20% as the in-between individuals' MOS. This limits the applicability of gait-based authentication approaches. We identify how these real-world factors may be compensated and we discuss the opportunities for gait-based context-awareness in wearable computing systems.

1 Introduction

Biometric authentication is the process of verifying the claimed identity of an individual by his or her physiological or behavioral characteristics. Characteristics include e.g. fingerprint, iris, retina, face, voice, handwriting. This paper presents results on gait-based biometrics. Most of the previous work in gait based user authentication is based on machine vision techniques e.g. [7,11]. An inherent advantage of vision-based gait recognition systems is to capture the gait of a person from a distance. Another, relative new direction, is the user authentication based on miniature acceleration sensors to capture the gait profile. Gait-based authentication using accelerometer sensor was demonstrated in [1,4,9,10] using template-based approaches. Primary advantage of on-body biometric authentication systems is that it excludes the need of ambient technology. Potential applications include smart clothes/bags, or electronic devices that are able to detect who is carrying them to adapt their functionality, or that can detect and report that they were stolen.

Intrinsic gait variability is challenging in authentication systems. For accelerometer based authentication systems these intrinsic gait variability has not been analyzed over long periods of time, nor when parameters typical for real-world scenarios are varied (such as e.g. which pair of shoes is worn or how much

M. Tistarelli and M.S. Nixon (Eds.): ICB 2009, LNCS 5558, pp. 1040–1049, 2009.
© Springer-Verlag Berlin Heidelberg 2009

weight is carried). In order to characterize gait in real-world scenarios we introduce a method to statistically quantify similarity between gait data sets. We use this method to characterize gait variability on a dataset composed of 5 subjects, recorded on 5 consecutive days. This data set includes variations in walking speed, weight carried and shoes worn. We report the results and analyze the implications for gait-based authentication.

2 Related Work

Identifying users from gait patterns using accelerometers is based on the assumption that the gait acceleration profile ("template") is unique to some extent for every person. During training, a *feature template* vector that represents characteristics of the gait of the person to authenticate is computed. The same feature vector is computed during the authentication process and compared to the feature template. The feature vector components commonly used are coming from the temporal domain (e.g. acceleration profile before/after the heel strike), and frequency domain (e.g. power spectrum of a footstep). Usually multiple feature vectors are available (e.g. when the feature vectors are derived from individual steps or out of a sliding window), and the feature template is the average of the feature vectors.

Ailisto and Mäntyjärvi et al. [1,9] were the first to report on acceleration sensor based gait authentication. They recorded the acceleration of the waist of 36 subjects on two different days while the subjects were walking 20 m at their normal, slow and fast walking speed. The performances of three methods were analyzed: (i) a correlation method using a step template in the time domain; (ii) a frequency-domain method using the first 40 FFT coefficients as feature template in the frequency domain; (iii) a data distribution statistics, using the data distribution as feature vector. Using the correlation method they achieved best results with an Equal Error Ratio (EER) of 7%. Further work supported their results. Gafurov et al. [5,6] recorded the acceleration of the right ankle of 21 subjects with a 16 Hz sampling frequency during a 70 m walk. They achieved an EER of 9% and 5% respectively. Gafurov and Buvarp [3] further showed that it is not possible for a person to imitate the gait of another person. A combination of a sensor-based gait and voice biometric has been proposed by Vildjiounaite et al. [12]. The sensor fusion increases the performance of the recognition system in a noisy or loud environment. Liu Rong et al. [10] recorded 5 acceleration data sets at the hip of 35 participants. With a dynamic time warping (DTW) method to normalize the step lengths they achieved an EER of 6,7%.

The literature differs in the number of acceleration signals that are used as well as the sensor positions, however no reported results clearly outperforms the others. The common points of all presented works are short recordings, no variability and the use of template based matching or comparison techniques. However, in real-world scenarios a number of parameters may affect gait, including: walking speed, shoe type, ground, drunkenness and injuries. The effects of these parameters has not been studied, despite their likely influence on "feature templates" and thus on the authentication accuracy.

3 Statistical Template Similarity Measure

The authentication performance depends on how accurately a *feature template* represents the actual feature vectors extracted from the underlying signal. Performance also depends on how dissimilar are the feature vectors of two different datasets. In order to objectively assess this (dis)similarity we devise a statistical measure of similarity. This measure is (i) generic; (ii) classifier independent; (iii) and it reveals specific locations in the feature vector with high entropy.

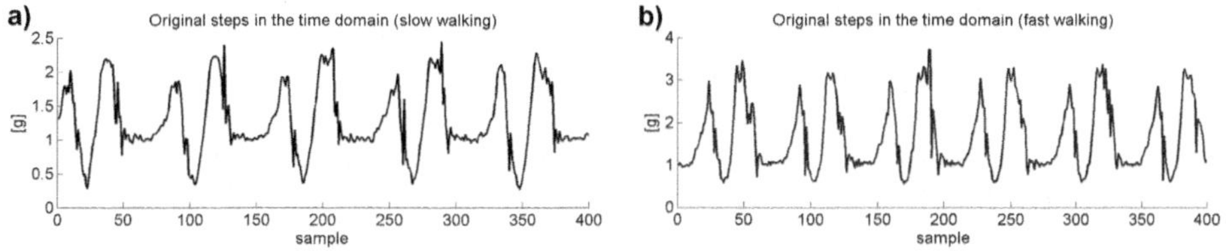

Fig. 1. Two acceleration signal examples: a) 4km/h walking b) 6km/h walking

We measure the gait acceleration at 64 *Hz* with a three-axis MEMS acceleration sensor placed at the ankle (see [2] for sensor details). This three-axis sensor gives an acceleration vector $\boldsymbol{n} = (x, y, z)$ with following axis orientation: $x =$ horizontal forward, $y =$ vertical, $z =$ horizontal lateral. In figure 1 we show the norm of the acceleration signal with walking speeds $4km/h$ and $6km/h$. The signal is then converted into a feature vector that encores characteristics of the gait style. We compare four feature types (two in the time domain and two in the frequency domain) applied to 3 sensor axis combinations (signal types).

Feature type A: A window of 64 samples is aligned on the heel strike. Each sample in the window is one component of the feature vector, yielding a feature vector with 64 components. This feature vector is sensitive to walking speed.

Feature type B: In order to be independent of walking speed, the step length is normalized by dynamic time warping (DTW) to 64 samples. The acceleration magnitudes are normalized to values between 0 and 1 [10]. After the normalization all steps have the same length and amplitude (see figure 2). The normalized acceleration samples form the feature vector.

Feature type C: A jumping window of length 256 samples and jump length of 64 samples is shifted over the signal. A 256 point FFT is computed on the window and the lower 40 FFT coefficients form the feature vector.

Feature type D is identical to feature type C except that the window is centered on the heel strike rather than placed at regular intervals.

Signal type a: The features are computed on the acceleration norm $|n| = \sqrt{x^2 + y^2 + z^2}$.

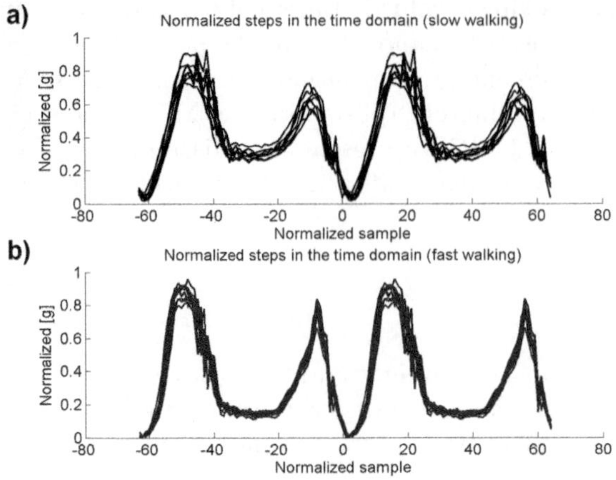

Fig. 2. Normalized steps superimposed on each other a) Normalized steps of 4km/h walking b) Normalized steps of 6km/h walking

Signal type b: The features are computed on the vertical axis of the acceleration.

Signal type c: The features are computed on two signals: the vertical and forward/backward acceleration axes. The feature vector is twice the size as with signal type a or b. The first half are features derived from the vertical acceleration; the second half are features derived from the forward/backward acceleration.

The combination of 4 feature types and 3 signal types results in 12 kinds of feature vectors. We denote the feature vector kind by $Feature_{signal}$ with $Feature \in \{A, B, C, D\}$ and $signal \in \{a, b, c\}$. For each window, the feature vector is computed. Statistical quantification of the similarity between the multiple feature vectors extracted from two datasets is described below. For authentication, multiple feature vectors are averaged into a single *feature template* over a recording period.

Assume two sets $\boldsymbol{F}_X$ and $\boldsymbol{F}_Y$ of N feature vectors of length L computed on datasets S_X and S_Y: $\boldsymbol{F}^i_X = [f^i_{X,0}, ..., f^i_{X,L-1}]$ and $\boldsymbol{F}^i_Y = [f^i_{Y,0}, ..., f^i_{Y,L-1}]$, with $i = 0...N-1$. We introduce a statistical measure of similarity between feature vector set $\boldsymbol{F}_X$ and $\boldsymbol{F}_Y$. For each feature $l = 0, 1, ..., L-1$ we compute the one-way analysis of variance (ANOVA) between datasets S_X and S_Y, i.e. $p_l = Anova(\{f^0_{X,l}, ..., f^{N-1}_{X,l}\}; \{f^0_{Y,l}, ..., f^{N-1}_{Y,l}\})$. p_l is the p-value under the null hypothesis that all data samples at position l in the feature vector are drawn from populations with the same mean. If p_l is near zero, it suggests that one sample mean is significantly different than the other sample mean.

For example, with the feature type B, a feature vector that represents the gait acceleration profile is generated at each heel strike (figure 2). Values near zero in $\boldsymbol{p} = [p_0, ..., p_{L-1}]$ indicate that the acceleration value at position l in the feature vector is statistically different between the gait datasets S_X and S_Y. As feature types C and D are composed in the frequency domain, $\boldsymbol{p}$ indicates statistical similarity of the frequency components.

In order to obtain a single valued statistical similarity measure between the feature vectors of dataset S_X and S_Y, we define the *Measure Of Similarity (MOS)* value as the percentage of positions in the feature vector that are not statistically significantly different ($p > 0.05$). The MOS represents the similarity of two feature sets, and 1-MOS represents the dissimilarity of two feature sets. It allows to statistically compare different gait datasets when real-world factors are varied.

4 Experiment

We analyze the effect of shoes, weight and time (day to day changes) on gait; we compare that to the differences in gait styles from persons to persons. Concretely we compare the set of feature vectors obtained from one recording to the set obtained from another recording using the MOS.

Data is recorded while subjects walk on a treadmill. This allows to control the walking speed and ensures that the behavior of participants is similar among themselves and from one day to the next. The participants performed 8 walking tasks on 5 days over 2 weeks. The 8 walking tasks are walking with a main pair of shoes at different speeds (3,4,5 and 6 km/h), then at 4km/h with additional weight (backpack with 6 and 10 kg) as well as without shoes and with an alternate pair of shoes. Each task lasts at least 1 min every day. This results in 8 min walking data per day and participant. 5 male subjects with age between 25-28 years, weight between 65-75 kg and height between 1.75-1.85 m participated in the experiments, resulting in overall 200 min (about 12000 steps) of walking data.

5 Results and Discussion

In figure 3 we present the results of the statistical similarity analysis between gait datasets using feature type A_a (steps in the time domain, without normalization using the norm of all axes). The bars are the MOS value, the lines indicate the standard deviation (std) over all used data sets.

Figure 3 a) depicts the results of the statistical similarity analysis between the first 30 sec and the second 30 sec of each 1 min data set. The standard deviation reflects the variation between the tasks and days. This is a baseline measure that indicates the highest MOS one is likely to expect (about 80%). It represents the intrinsic gait similarity (or variability) typical for human walking. Figure 3 b) shows that the MOS is about 20% for two different subjects doing the same task. As expected the intrinsic gait similarity of one subject is higher than the similarity between subjects. Thus, at identical tasks different persons have different gait patterns that can be used for authentication.

However, the tasks affect the gait. Figure 3 c) shows the influence of changing weight (+6,+10 kg), kind of shoes (shoes) as well as speed (-1,+1,+2 km). The additional weight has the smallest influence on the gait signal. Changing or removing shoes reduces the MOS by more than 50%. A change of speed as small

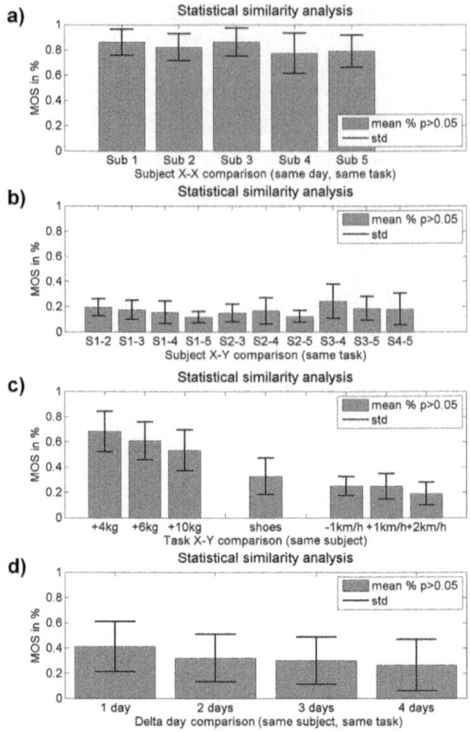

Fig. 3. MOS values with feature type Aa: a) baseline measure b) between different subjects doing the same task c) between different tasks performed by the same subject d) between different days with identical task performed by the same subject

as $\pm 1\ km/h$ reduces the MOS to about 25%. A change of $+2\ km/h$ reduces the MOS to about 20% that is comparable to the MOS between persons. Given that speed modifies the gait of one person (fig 3 c) in a range similar to what the gait of two persons at identical speeds differs (fig 3 b) - seen by the similar MOS values - means that it is challenging to distinguish the two cases.

Figure 3 d) shows the MOS of datasets taken on different days, with 1 to 4 days between the recordings. Surprisingly there is a low level of similarity between different days and the MOS tends to decrease with increasing time distance between the datasets. Future work is needed to assess whether there is a natural tendency for gait patterns to change over time, whether this is cyclic, or whether this may be caused by the user getting used to walk on the treadmill. Analysis of longer recordings will clarify this.

Results obtained using other feature kinds in the time domain show similar results to those presented in figure 3. On the left side of figure 4 we show the result of the similarity analysis in the time domain using feature type B_a (normalized steps using the acceleration norm). The normalization in time domain slightly improves the similarity measure of the intrinsic gait similarity (fig. 4 a_1), and minimizes the effect of weight (fig. 4 c_1), whereas other MOS values are similar to

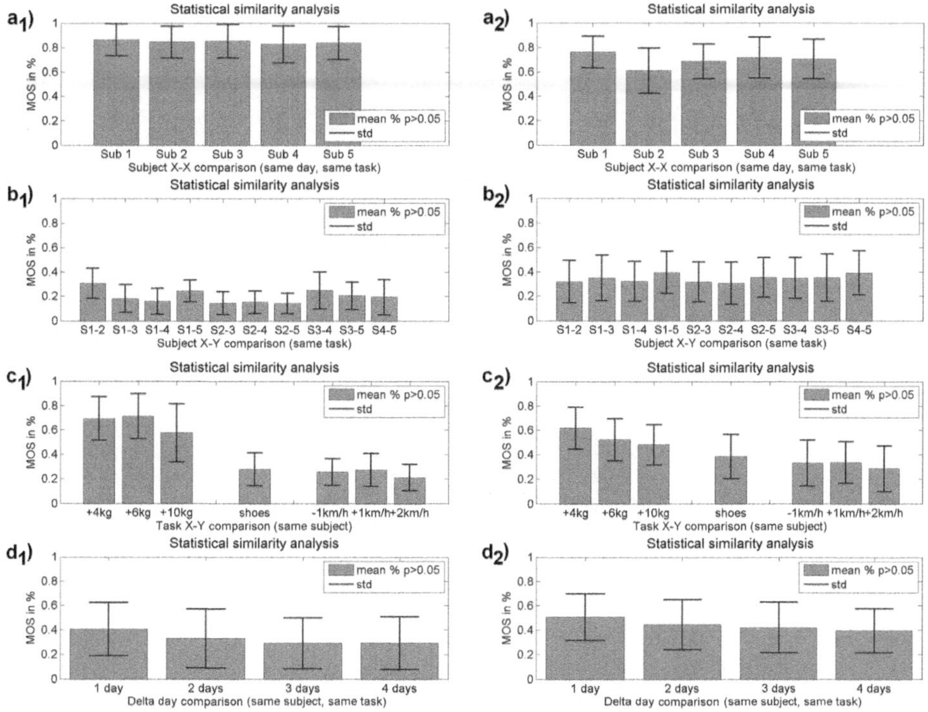

Fig. 4. Results of the statistical similarity analysis: **Left side**: using feature type *Ba* (Normalized steps using the norm); **Right side**: using feature type *Da* (step triggered FFT using the norm)

those obtained with feature type A_a. Despite time and amplitude normalization change in speed leads to MOS values in a range similar to those obtained with feature type A_a. This means that a change in speed does not only affect gait frequency or amplitude but changes the *nature* of the gait acceleration profile itself. This is visible in fig. 2: although acceleration peaks are aligned and of similar amplitude, the acceleration profile nevertheless differs. Similarly changes in shoes also affects the acceleration profile, not only its amplitude.

On the right side of figure 4 we show the result with frequency domain features D_a. Compared to previous results, the intrinsic gait similarity is lower, while the similarity between subjects is greater (fig 4 b_2). Therefore this feature kind is less discriminative than the time domain features. The influence of weight, shoes and speed and variations over time on the MOS is lesser though than with time domain features (higher MOS) (fig 4 c_2,d_2). However overall trends are comparable. Results with other frequency domain features are similar.

To verify the link between the MOS values and authentication performance we apply the authentication scheme described in [3,10] on our dataset. It distinguishes an impostor dataset from a genuine dataset. A *reference* feature template $T^R_{X,D,T}$ is obtained from the person to authenticate X. It is computed on the first half of the 1 min gait recording. The reference template is the average

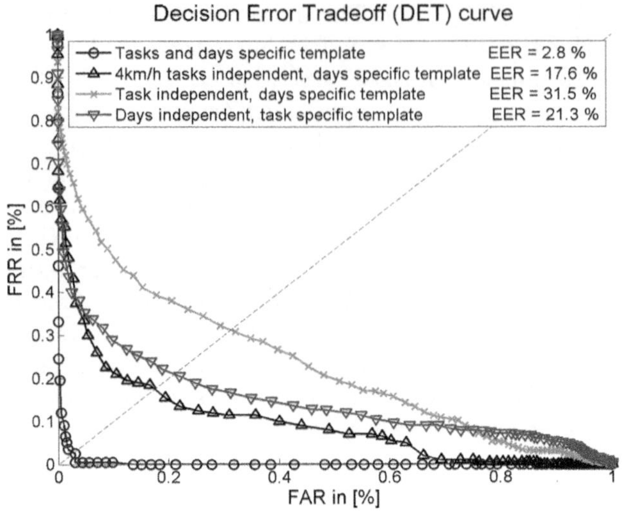

Fig. 5. Person authentication DEF curve for different data set combinations

feature vectors computed on the recording of a single task T on a given day D (task/day-dependent authentication) or the average of feature vectors computed on recordings of multiple tasks $T = t_1..t_n$ or on multiple days $D = d_1..d_n$ (task/day-independent authentication). A *genuine* $T^G_{X,D,T}$ template is obtained from the same person at a later timepoint. It is used to assess the system acceptation performance. It is computed on the second half of the gait recording of each task. An *impostor* template $T^I_{Y,D,T}$ is obtained from the gait of another person Y. It is used to assess the system rejection performance. The authentication scheme compares the normalized Euclidian distance between the genuine/impostor templates and the reference templates. It makes a decision (acceptation/rejection) based on a distance threshold that controls the tradeoff between False Acceptance Ratio (FAR) and False Rejection Ratio (FRR). This is represented by a Decision Error Tradeoff (DET) curve (see figure 5). The Equal Error Rate is the point in the curve where the FAR is equal to the FRR (lower EER is better). We denote the DET curve by $DET(T^R_{X,D,T}, T^G_{X,D',T''}, T^I_{Y,D'',T''})$. Unless specified below, it is computed and averaged on all combinations of 5 persons $X = x_1..x_5$, $Y = y_1..y_5$, $X \neq Y$; 5 days $D = d_1..d_5$, $D' = d_1..d_5$, $D'' = d_1..d_5$; and 8 tasks $T = t_1..t_8$, $T' = t_1..t_8$, $T'' = t_1..t_8$.

The first curve shows a very low EER (2.8%) when the reference template is day and task specific, and the genuine/impostor template come from recordings of the same day and task: $DET(T^R_{X,d_i,t_j}, T^G_{X,d_i,t_j}, T^I_{Y,d_i,t_j})$ with $i \in 1..5$ and $j \in 1..8$. The second DET curve shows the 4km/h walking speed specific performance (weight and shoe independent). The reference template combines the 5 tasks with 4km/h: $DET(T^R_{X,d_i,T_{4km/h}}, T^G_{X,d_i,T_{4km/h}}, T^I_{Y,d_i,T_{4km/h}})$, with $i \in 1..5$ and $T_{4km/h} = t_1, t_2, t_4, t_7, t_8$. The third DET curve shows the task-independent performance. The reference template is day dependent but combines all 8 tasks:

$DET(T^R_{X,d_i,,T}, T^G_{X,d_i,,T'}, T^I_{Y,d_i,,T''})$. The more the template is task-independent, the lesser the authentication performance, due to the template being less specific to a particular task. The fourth DET curve with $DET(T^R_{X,D,t_j}, T^G_{X,D',t_j}, T^I_{Y,D'',t_j})$ shows the day-independent performance. Performance is lower than with day-dependent templates, which implies that day to day changes affect the gate much in the same way as tasks affect the gait, making it difficult to capture in a day-independent reference template. As expected the authentication performance decreases with decreasing MOS difference. Comparing the MOS values of figure 3 a) with the MOS values of figure 3 b) there is a large difference, therefore the EER is very low (2.8%). The MOS values of the first 4 bars in figure 3 c) are closer to the MOS values of figure 3 b) than to the values of figure 3 a). This gives a higher EER of 17.6% (second curve in fig 5). Including the different speed lowers the distance between genuine trials and impostor trials even more, resulting in the worst EER of 31.5% (third curve in fig 5).

6 Conclusion

We analyzed biometric user authentication in real-world scenarios using gait acceleration signals. We introduced a statistical measure which we called Measure Of Similarity (MOS) to characterize the similarity of two gait data sets. Using this MOS value we showed that variations due to weight, shoes, speed and day to day variability reduces the distance between genuine and impostor data sets, making the gait-based authentication in real-life environments challenging. In particular we showed that walking speed changes the nature of the gait acceleration profile and cannot be compensated by frequency and amplitude normalization. Change in shoes also strongly affects the gait acceleration profile. A change in weight has less effect on the acceleration profile, and can be minimized by normalization. We evidenced large day to day gait variations. Longer recordings are required to assess whether this was due to users learning to walk on the treadmill or due to natural variability. In the latter case this may question the real-world applicability of gait-based authentication using the simple features used here. Actual user authentication supports the results of the statistical similarity analysis. Authentication thus must be performed in conditions similar to those of training.

Improvements to gait-based authentication include analyzing the gait only when the actual walking conditions match that of training. This may be easily applied to walking speed and weight (e.g. with pressure sensitive floors). Several training sets may be collected (e.g. various speeds) and used accordingly automatically. The new statistical measure of similarity allows to identify the feature vector components with the highest information content (e.g. regions of the acceleration profile or set of frequency components). Giving more weight to these components when matching a feature vector to a template may better results. Finally, since the gait acceleration profile is affected despite normalization by walking speed, shoes or weight, new features beyond simple acceleration profile or frequency components must be investigated that better capture the unique characteristics of gait.

Finally we note that gait analysis may be used to identify the user's *context*, which is especially important in the field of pervasive computing and wearable computing [8]. Our results show that the gait contains information beyond walking speed, in particular about the weight that a person carries or the shoes that are worn. This information may be automatically interpreted and presents an opportunity for gait-based context-aware systems, as a complement to other sensors used in context-aware systems. Application scenarios include smart devices or clothing that adapt to their user.

Acknowledgment. This project is funded by the EU projects DAPHNet, contract number: 018474-2 (www.daphnet.eu) and SEAT, contract number: 030958 (www.seat-project.org). All views here reflect the authors' opinion and not that of the commission.

References

1. Ailisto, H., Lindholm, M., Mäntyjärvi, J., Vildjiounaite, E., Mäkelä, S.-M.: Identifying people from gait patterns with accelerometers. Proceedings of Biometric Technology for Human Identification II 5779, 7–14 (2005)
2. Bächlin, M., Roggen, D., Tröster, G.: Context-aware platform for long-term life style management and medical signal analysis. In: Proceeding of the 2nd Sensation International Conference, Chania, Greece (2007)
3. Snekkenes, E., Gafurov, D., Buvarp, T.E.: Robustness of biometric gait authentication against impersonation attack. In: 1st Int. Workshop on Information Security, On The Move Federated Conferences, France (2006)
4. Gafurov, D., Snekkenes, E., Bours, P.: Spoof attacks on gait authentication system. IEEE Trans. on Information Forensics and Security 2(3), 491–502 (2007)
5. Gafurov, D., Helkala, K., Søndrol, T.: Biometric gait authentication using accelerometer sensor. Journal of computers 1(7), 51–59 (2006)
6. Gafurov, D., Helkala, K., Sondrol, T.: Gait recognition using acceleration from mems. In: Proc. of 1st Int. Conf. on Availability, Reliability and Security, USA, pp. 432–439 (2006)
7. Han, J., Bhanu, B.: Individual recognition using gait energy image. IEEE Transactions on Pattern Analysis and Machine Intelligence 28(2), 316–322 (2006)
8. Mann, S.: Wearable computing as means for personal empowerment. In: 1st Int. Conf. on Wearable Computing. IEEE Computer Society Press, Los Alamitos (1998)
9. Mäntyjärvi, J., Lindholm, M., Vildjiounaite, E., Makela, S.-M., Ailisto, H.A.: Identifying users of portable devices from gait pattern with accelerometers. In: Int. Conf. on Acoustics, Speech, and Signal Processing, vol. 2, pp. 973–976 (2005)
10. Rong, L., Zhiguo, D., Jianzhong, Z., Ming, L.: Identification of individual walking patterns using gait acceleration. In: 1st Int. Conf. on Bioinformatics and Biomedical Engineering, pp. 543–546 (2007)
11. Sarkar, S., Phillips, P.J., Liu, Z., Robledo, I., Grother, P., Bowyer, K.W.: The human id gait challenge problem: Data sets, performance, and analysis. IEEE Transactions on Pattern Analysis and Machine Intelligence 27(2), 162–177 (2005)
12. Vildjiounaite, E., Mäkelä, S.-M., Lindholm, M., Riihimäki, R., Kyllönen, V., Mäntyjärvi, J., Ailisto, H.: Unobtrusive multimodal biometrics for ensuring privacy and information security with personal devices. In: Fishkin, K.P., Schiele, B., Nixon, P., Quigley, A. (eds.) PERVASIVE 2006. LNCS, vol. 3968, pp. 187–201. Springer, Heidelberg (2006)

40 Years of Progress in Automatic Speaker Recognition

Sadaoki Furui

Department of Computer Science
Tokyo Institute of Technology, Japan
2-12-1 Ookayama, Meguro-ku, Tokyo, 152-8552 Japan
furui@cs.titech.ac.jp

Abstract. Research in automatic speaker recognition has now spanned four decades. This paper surveys the major themes and advances made in the past 40 years of research so as to provide a technological perspective and an appreciation of the fundamental progress that has been accomplished in this important area of speech-based human biometrics. Although many techniques have been developed, many challenges have yet to be overcome before we can achieve the ultimate goal of creating human-like machines. Such a machine needs to be able to deliver satisfactory performance under a broad range of operating conditions. A much greater understanding of the human speech process is still required before automatic speaker recognition systems can approach human performance.

Keywords: Speaker recognition, speaker identification, speaker verification, speaker diarization, text-dependent, text-independent, robust recognition.

1 Introduction

Speech is the primary means of communication between humans. Speaker recognition is the process of automatically recognizing who is speaking by using the speaker-specific information included in speech waves [16, 17, 18]. Many applications have been considered for speaker recognition. These include secure access control by voice, customizing services or information to individuals by voice, indexing or labeling speakers in recorded conversations or dialogues, surveillance, and criminal and forensic investigations involving recorded voice samples. Currently, the most frequently mentioned application is access control. Access control applications include voice dialing, banking transactions over a telephone network, telephone shopping, database access services, information and reservation services, voice mail, and remote access to computers. Speaker recognition technology, as such, is expected to create new services in smart environments and make our daily lives more convenient.

Recently speaker diarization has been actively investigated, in which an input audio channel is automatically annotated with speakers. Diarization can be used for helping speech recognition, facilitating the searching and indexing of audio archives, and increasing the richness of automatic transcriptions, making them more readable.

This paper reviews major highlights during the last four decades in the research and development of automatic speaker recognition so as to provide a technological perspective. Although considerable technological progress has been made, there still remain many research issues that need to be tackled.

M. Tistarelli and M.S. Nixon (Eds.): ICB 2009, LNCS 5558, pp. 1050–1059, 2009.
© Springer-Verlag Berlin Heidelberg 2009

2 40 Years Progress

Topics related to the progress of automatic speaker recognition technology in the past 40 years can be summarized as follows:

2.1 1960s and 1970s

(1) ***Early systems*:** The first attempts at automatic speaker recognition were made in the 1960s, one decade later than automatic speech recognition. Pruzansky at Bell Labs [37] was among the first to initiate research by using filter banks and correlating two digital spectrograms for a similarity measure. Pruzansky and Mathews [38] improved upon this technique; and, Li et al. [26] further developed it by using linear discriminators. Doddington at Texas Instruments (TI) [9] replaced filter banks by formant analysis. Intra-speaker variability of features, one of the most serious problems in speaker recognition, was intensively investigated by Endres et al. [11] and Furui [13].

(2) ***Text-independent methods*:** For the purpose of extracting speaker features independent of the phonetic context, various parameters were extracted by averaging over a long enough duration or by extracting statistical or predictive parameters. They include averaged auto-correlation [5], instantaneous spectra covariance matrix [25], spectrum and fundamental frequency histograms [3], linear prediction coefficients [44], and long-term averaged spectra [14].

(3) ***Text-dependent methods*:** Since the performance of text-independent methods was limited, time-domain and text-dependent methods were also investigated [1, 2, 15, 42]. In time-domain methods, with adequate time alignment, one can make precise and reliable comparisons between two utterances of the same text, in similar phonetic environments. Not surprisingly, as a result text-dependent methods were shown to perform significantly better than text-independent methods.

(4) ***Texas Instruments system*:** TI built the first fully automated large scale speaker verification system providing high operational security. Verification was based on a four-word, randomized utterance built from a set of 16 monosyllabic words. Digital filter banks were used for spectral analysis, and the decision strategy was sequential, requiring up to 4 utterances for each trial. Several million tests were made over a period of 6 years, for several hundred speakers.

(5) ***Bell Labs system*:** Bell Labs built experimental systems aimed to work over dialed-up telephone lines. Furui [15] proposed using the combination of cepstral coefficients and their first and second polynomial coefficients, now called Δ and $\Delta\Delta$cepstral coefficients, as frame-based features to increase robustness against distortions by the telephone system. He implemented an online system and tested it for a half year with numerous calls from 120 users. He also proposed methods for updating templates and thresholds for speaker verification decision. The cepstrum-based features later became standard, not only for speaker recognition, but also for speech recognition.

(6) ***Parameter-domain normalization*:** As one typical normalization technique in the parameter domain, spectral equalization, the so-called "blind equalization" method, was confirmed to be effective in reducing linear channel effects and long-term spectral variation. In this method, cepstral coefficients are averaged over the duration of

an entire utterance, and the averaged values are subtracted from the cepstral coefficients of each frame (CMS; cepstral mean subtraction) [2, 15]. This method can compensate fairly well for additive variation in the log spectral domain. This method is especially effective for text-dependent speaker recognition applications using sufficiently long utterances. It has also been shown that Δcepstral coefficients are resistant to linear channel mismatches between training and testing [15].

2.2 1980s

(1) ***Statistical modeling*:** Speaker recognition research in the 1980s was characterized by a shift in methodology from the more intuitive template-based approach (a straightforward pattern recognition paradigm) towards a more rigorous statistical modeling framework. Today, most practical speaker recognition systems are based on the statistical framework developed in the 1980s and their results, with significant additional improvements having been made in the 1990s.

(2) ***HMM*:** One of the key technologies developed in the 1980s is the hidden Markov model (HMM) approach [12, 39]. This is a doubly stochastic process in that it has an underlying stochastic process that is not observable (hence the term hidden), but can be observed through another stochastic process that produces a sequence of observations. Although the HMM was well known and understood in a few laboratories (primarily IBM, Institute for Defense Analysis (IDA) and Dragon Systems), it was not until widespread publication of the methods and theory of HMMs in the mid-1980s that the technique became widely applied in virtually every speech recognition research laboratory in the world.

(3) ***HMM-based text-dependent methods*:** As an alternative to the template-matching approach for text-dependent speaker recognition, the HMM technique was introduced in the same way as for speech recognition. HMMs have the same advantages for speaker recognition as they do for speech recognition. Remarkably robust models of speech events can be obtained with only small amounts of specification or information accompanying training utterances. Speaker recognition systems based on an HMM architecture used speaker models derived from a multi-word sentence, a single word, or a phoneme. Typically, multi-word phrases (a string of seven to ten digits, for example) were used, and models for each individual word and for "silence" were combined at a sentence level according to a predefined sentence-level grammar [34].

(4) ***VQ/HMM-based text-independent methods*:** Nonparametric and parametric probability models were investigated for text-independent speaker recognition. As a nonparametric model, vector quantization (VQ) was investigated [43]. A set of short-time training feature vectors of a speaker can be efficiently compressed to a small set of representative points, a so-called VQ codebook. A matrix quantizer encoding multi-frame was also investigated [23, 47]. As a parametric model, the HMM was investigated. Pritz [36] proposed using an ergodic HMM (i.e., all possible transitions between states are allowed). An utterance was characterized as a sequence of transitions through a 5-state HMM in the acoustic feature space. Tishby [48] expanded Poritz's idea by using an 8-state ergodic autoregressive HMM represented by continuous probability density functions with 2 to 8 mixture components per state, which had a higher spectral resolution than the Poritz's

model. Rose et al. [40] proposed using a single-state HMM, which is now called a Gaussian mixture model (GMM), as a robust parametric model.

2.3 1990s

(1) ***Robust recognition*:** Research on increasing robustness became a central theme in the 1990s. Matsui et al. [29] compared the VQ-based method with the discrete/continuous ergodic HMM-based method, particularly from the viewpoint of robustness against utterance variations. They found that the continuous ergodic HMM method is far superior to the discrete ergodic HMM method and that the continuous ergodic HMM method is as robust as the VQ-based method when enough training data is available. They investigated speaker identification rates using the continuous HMM as a function of the number of states and mixtures. It was shown that speaker recognition rates were strongly correlated with the total number of mixtures, irrespective of the number of states. This means that using information about transitions between different states is ineffective for text-independent speaker recognition and, therefore, the GMM achieves almost the same performance as the multiple-state ergodic HMM.

(2) ***Combination of spectral envelope and fundamental frequency features*:** Matsui et al. [28] tried a method using a VQ-codebook for long feature vectors consisting of instantaneous and transitional features calculated for both cepstral coefficients and fundamental frequency. Since the fundamental frequency cannot be extracted from unvoiced speech, they used two separate codebooks for voiced and unvoiced speech for each speaker. A new distance measure was introduced to take into account intra- and inter-speaker variability and to deal with the problem of outliers in the distribution of feature vectors. The outlier vectors correspond to intersession spectral variation and to the difference between phonetic content of the training texts and the test utterances. It was confirmed that, although the fundamental frequency achieved only a low recognition rate by itself, the recognition accuracy was greatly improved by combining the fundamental frequency with cepstral features.

(3) ***HMM adaptation for noisy conditions*:** In order to increase the robustness of speaker recognition techniques against noisy speech, Rose et al. [41] applied the HMM composition (PMC) method [20, 27] to speech recorded under noisy conditions. The HMM composition is a technique to combine a clean speech HMM and a background noise HMM to create a noise-added speech HMM. In order to cope with the problem of variation in the signal-to-noise ratio (SNR), Matsui et al. [32] proposed a method in which several noise-added HMMs with various SNRs were created and the HMM that had the highest likelihood value for the input speech was selected. A speaker decision was made using the likelihood value corresponding to the selected model. Experimental application of this method to text-independent speaker identification and verification in various kinds of noisy environments demonstrated considerable improvement in speaker recognition.

(4) ***Text-prompted method*:** Matsui et al. proposed a text-prompted speaker recognition method, in which key sentences are completely changed every time the system is used [30]. The system accepts the input utterance only when it determines that the registered speaker uttered the prompted sentence. Because the vocabulary is unlimited, prospective impostors cannot know in advance the sentence they will

be prompted to say. This method not only accurately recognizes speakers, but can also reject an utterance whose text differs from the prompted text, even if it is uttered by a registered speaker. Thus, a recorded and played back voice can be correctly rejected.

(5) ***Score normalization***: How to normalize intra-speaker variation of likelihood (similarity) values is one of the most difficult problems in speaker verification. Variations arise from the speaker him/herself, from differences in recording and transmission conditions, and from noise. Speakers cannot repeat an utterance precisely the same way from trial to trial. Likelihood ratio- and a posteriori probability-based techniques were investigated [22, 31]. In order to reduce the computational cost for calculating the normalization term, methods using "cohort speakers" or a "world model" were proposed.

(6) ***Relation to other speech research***: Speaker characterization techniques were related to research on improving speech recognition accuracy by speaker adaptation [16], improving synthesized speech quality by adding the natural characteristics of voice individuality, and converting synthesized voice individuality from one speaker to another. Studies on speaker diarization, that is, automatically extracting the speech periods of each person separately ("who spoke when") from a dialogue/conversation/meeting involving more than two people appeared as an extension of speaker recognition technology [21, 45, 49]. Speaker segmentation and clustering techniques have been used to aid in the adaptation of speech recognizers and for supplying metadata for audio indexing and searching. This allows for searching audio by speaker and makes speech recognition results easier to read.

2.4 2000s

(1) ***Score normalization***: A family of new normalization techniques has been proposed, in which the scores are normalized by subtracting the mean and then dividing by standard deviation, both terms having been estimated from the (pseudo) imposter score distribution. Different possibilities are available for computing the imposter score distribution: Znorm, Hnorm, Tnorm, Htnorm, Cnorm and Dnorm [4]. State-of-the-art text-independent speaker verification techniques associate one or more parameterization level normalizations (CMS, feature variance normalization, feature warping, etc.) with a world model normalization and one or more score normalizations.

(2) ***Model adaptation***: Various model adaptation and compensation techniques have been investigated for GMM based speaker recognition methods. McLaren et al. [33] proposed two techniques for GMM mean supervector SVM classifiers: intersession variability modeling and nuisance attribute projection. Petri et al. [35] proposed an unsupervised model adaptation technique including a weighting scheme for the test data, based on the *a posteriori* probability that a test utterance belongs to the target customer model.

(3) ***Combination of audio and visual features***: There has been a lot of interest in audio-visual speaker verification systems, in which a combination of speech and image information is used. As visual information, lip movement is widely used. The audio-visual combination helps improve system reliability. For instance, while background noise has a detrimental effect on the performance of voice, it

does not have any influence on lip information. Conversely, although the performance of lip recognition systems depends on lighting conditions, lighting does not have any effect on voice quality.

The method of combining two information sources (audio-visual fusion) can be treated as either a classifier combination problem or pattern classification problem. For example, for those systems that can only provide decisions, a majority voting method can be used. If the output of classifiers are compatible (e. g., in the form of posterior probabilities), they can be linearly combined (sum rule) or multiplied together (product or log-linear rule). In addition to these combination methods, researchers have also proposed treating the outputs of individual classifiers as feature vectors and using a classifier such as support vector machines, binary decision trees, and radial basis function networks to classify the vectors [7, 8].

(4) ***High-level features***: High-level features such as word idiolect, pronunciation, phone usage, prosody, etc. have been successfully used in text-independent speaker verification. Typically, high-level-feature recognition systems produce a sequence of symbols from the acoustic signal and then perform recognition using the frequency and co-occurrence of symbols. In Doddington's idiolect work [10], word unigrams and bigrams from manually transcribed conversations were used to characterize a particular speaker in a traditional target/background likelihood ratio framework. The use of support vector machines for performing the speaker verification task based on phone and word sequences obtained using phone recognizers has been proposed [6]. The benefit of these features was demonstrated in the "NIST extended data" task for speaker verification; with enough conversational data, a recognition system can become "familiar" with a speaker and achieve excellent accuracy. These methods require utterances of at least several minutes long, much longer than those used in conventional speaker recognition methods.

(5) ***MLLR features***: MLLR (Maximum Likelihood Linear Regression) model adaptation [24] has been widely used in supervised as well as unsupervised HMM adaptation for increasing the robustness of speech recognition. The MLLR has also been widely used in creating text-independent speaker-specific GMMs by adapting speaker-independent GMM (world model). Stolcke [46] proposed using the MLLR adaptation matrix itself as a speaker characterizing feature and reported good experimental results.

3 Discussions

3.1 Summary of the Technology Progress

In the last 40 years, research in speaker recognition has been intensively carried out worldwide, spurred on by advances in signal processing, algorithms, architectures, and hardware. The technological progress in the last 40 years can be summarized by the following changes [19]:

(1) from template matching to statistical modeling, e.g. HMM and GMM,
(2) from filter bank/spectral resonance to cepstral features (cepstrum + Δcepstrum + $\Delta\Delta$cepstrum),

(3) from heuristic time-normalization to DTW/DP matching,
(4) from "distance"-based to likelihood-based methods,
(5) from raw scores to normalized scores,
(6) from acoustic features to high-level features,
(7) from maximum likelihood to discriminative approach, e.g. MCE and SVM,
(8) from clean speech to noisy/telephone speech,
(9) from single-modality (audio signal only) to multimodal (audio/visual) recognition,
(10) emergence of application/combination with speech recognition and speech synthesis, e.g. speaker diarization and voice conversion,

Many of these advances have taken place in both the fields of speech recognition and speaker recognition. The majority of technological changes have been directed toward the purpose of increasing robustness of recognition, including many other additional important techniques not noted above.

Although we have witnessed many new technological promises, we have also encountered a number of practical limitations that hinder a widespread deployment of applications and services.

3.2 How to Achieve Better Speaker Recognition Performance

There are many outstanding issues and problems in the area of speaker recognition. The most pressing issues, providing challenges for implementing practical and uniformly reliable systems for speaker recognition, are rooted in problems associated with variability and insufficient data. Variability is associated with trial-to-trial variations in recording and transmission conditions and speaking behavior. The most serious variations occur between enrollment sessions and subsequent test sessions, resulting in models that are mismatched to test conditions. Most applications require reliable system operation under a variety of environmental and channel conditions and require that variations in speaking behavior will be tolerated. Insufficient data refers to the unavailability of sufficient amounts of data to train representative models and accurate decision thresholds. Insufficient data is a serious and common problem because most applications require systems that operate with the smallest practicable amounts of training data recorded in the fewest number of enrollment sessions, preferably one.

The challenge is to find techniques that compensate for these deficiencies. A number of techniques have been proposed which provide partial solutions, such as cepstral subtraction techniques for channel normalization and spectral subtraction for noise removal. An especially effective technique for combating both variability and insufficient data is updating models with data extracted from test utterances. Studies have shown that model adaptation, properly implemented, can improve verification performance significantly with a small number of updates. It is difficult, however, for model adaptation to respond to large, precipitous changes. Moreover, adaptation provides for the possibility that customer models might be updated and possibly captured by impostors.

A desirable feature for a practical speaker recognition system is reasonably uniform performance across a population of speakers. Unfortunately, it is typical to observe in a speaker recognition experiment a substantial discrepancy between the best performing individuals, the "sheep", and the worst, the "goats". This additional

problem in variability has been widely observed, but there are virtually no studies focusing on its origin. Speakers with no observable speech pathologies, and for whom apparently good reference models have been obtained, are often observed to be "goats". It is possible that such speakers exhibit large amounts of trial-to-trial variability, beyond the ability of the system to provide adequate compensation.

4 Conclusion

Although many important scientific advances have taken place, we have also encountered a number of practical limitations which hinder a widespread deployment of application and services. We still have many research issues as described in the previous section. What we know about human speech processing is very limited. Significant advances in speaker recognition are not likely to come solely from research in statistical pattern recognition and signal processing. Although these areas of investigation are important, the significant advances will come from studies in acoustic-phonetics, speech perception, linguistics, and psychoacoustics.

References

1. Atal, B.S.: Text-independent speaker recognition: J.A.S.A. 52(181) (A), 83th ASA (1972)
2. Atal, B.S.: Effectiveness of linear prediction characteristics of the speech wave for automatic speaker identification and verification: J.A.S.A. 55(6), 1304–1312 (1974)
3. Beek, B., et al.: Automatic speaker recognition system: Rome Air Development Center Report (1971)
4. Bimbot, F.J., et al.: A tutorial on text-independent speaker verification. EURASIP Journ. on Applied Signal Processing, 430–451 (2004)
5. Bricker, P.D., et al.: Statistical techniques for talker identification. B.S.T.J. 50, 1427–1454 (1971)
6. Campbell, W.M., Campbell, J.P., Reynolds, D.A., Jones, D.A., Leek, T.R.: High-level speaker verification with support vector machines. In: Proc. IEEE Int. Conf. Acoust., Speech, Signal Processing, pp. I-73–76 (2004)
7. Campbell, W.M., Campbell, J.P., Reynolds, D.A., Singer, E., Torres-Carrasquillo, P.A.: Support vector machines for speaker and language recognition. Computer Speech and Language 20(2-3), 210–229 (2006)
8. Cheung, M.-C., Mak, M.-W., Kung, S.-Y.: A two-level fusion approach to multimodal biometric verification. In: Proc. IEEE Int. Conf. Acoust., Speech, Signal Processing, pp. V-485-488 (2005)
9. Doddington, G.R.: A method of speaker verification. J.A.S.A. 49(139) (A) (1971)
10. Doddington, G.R.: Speaker recognition based on idiolectal differences between speakers. In: Proc. Eurospeech, pp. 2521–2524 (2001)
11. Endres, W., et al.: Voice spectrograms as a function of age, voice disguise, and voice imitation. J.A.S.A. 49, 6(2), 1842–1848 (1971)
12. Ferguson, J. (ed.): Hidden Markov models for speech, IDA, Princeton, NJ (1980)
13. Furui, S.: An analysis of long-term variation of feature parameters of speech and its application to talker recognition. Electronics and Communications in Japan 57-A, 34–41 (1974)
14. Furui, S., et al.: Talker recognition by long time averaged speech spectrum. Electronics and Communications in Japan 55-A, 54–61 (1972)

15. Furui, S.: Cepstral analysis technique for automatic speaker verification. IEEE Trans. Acoustics, Speech, Signal Processing ASSP-29, 254–272 (1981)
16. Furui, S.: Speaker-independent and speaker-adaptive recognition techniques. In: Furui, S., Sondhi, M.M. (eds.) Advances in Speech Signal Processing, pp. 597–622. Marcel Dekker (1991)
17. Furui, S.: Recent advances in speaker recognition. In: Proc. First Int. Conf. Audio- and Video-based Biometric Person Authentication, Crans-Montana, Switzerland, pp. 237–252 (1997)
18. Furui, S.: Digital Speech Processing, Synthesis, and Recognition, 2nd edn. Marcel Dekker, New York (2000)
19. Furui, S.: Fifty years of progress in speech and speaker recognition. In: Proc. 148th ASA Meeting (2004)
20. Gales, M.J.F., Young, S.J.: HMM recognition in noise using parallel model combination. In: Proc. Eurospeech, Berlin, pp. II-837-840 (1993)
21. Gish, H., Siu, M., Rohlicek, R.: Segregation of speakers for speech recognition and speaker identification. In: Proc. ICASSP, S13.11, pp. 873–876 (1991)
22. Higgins, A., et al.: Speaker verification using randomized phrase prompting. Digital Signal Processing 1, 89–106 (1991)
23. Juang, B.-H., Soong, F.K.: Speaker recognition based on source coding approaches. In: Proc. ICASSP, vol. 1, pp. 613–616 (1990)
24. Leggetter, C.J., Woodland, P.C.: Maximum likelihood linear regression for speaker adaptation of continuous density hidden Markov models. Computer Speech and Language 9, 171–185 (1995)
25. Li, K.P., Hughes, G.W.: Talker differences as they appear in correlation matrices of continuous speech spectra. J.A.S.A. 55, 833–837 (1974)
26. Li, K.P., et al.: Experimental studies in speaker verification using a adaptive system. J.A.S.A. 40, 966–978 (1966)
27. Martin, F., Shikano, K., Minami, Y.: Recognition of noisy speech by composition of hidden Markov models. In: Proc. Eurospeech, Berlin, pp. II-1031–1034 (1993)
28. Matsui, T., Furui, S.: Text-independent speaker recognition using vocal tract and pitch information. In: Proc. Int. Conf. Spoken Language Processing, Kobe, vol. 5.3, pp. 137–140 (1990)
29. Matsui, T., Furui, S.: Comparison of text-independent speaker recognition methods using VQ-distortion and discrete/continuous HMMs. In: Proc. ICSLP, pp. II-157–160 (1992)
30. Matsui, T., Furui, S.: Concatenated phoneme models for text-variable speaker recognition. In: Proc. ICASSP, pp. II-391–394 (1993)
31. Matsui, T., Furui, S.: Similarity normalization method for speaker verification based on a posteriori probability. In: Proc. ESCA Workshop on Automatic Speaker Recognition, Identification and Verification, pp. 59–62 (1994)
32. Matsui, T., Furui, S.: Speaker recognition using HMM composition in noisy environments. Computer Speech and Language 10, 107–116 (1996)
33. McLaren, M., Vogt, R., Baker, B., Sridharan, S.: A comparison of session variability compensation techniques for SVM-based speaker recognition. In: Proc. Interspeech, pp. 790–793 (2007)
34. Naik, J.M., et al.: Speaker verification over long distance telephone lines. In: Proc. ICASSP, pp. 524–527 (1989)
35. Petri, A., Bonastre, J.-F., Matrouf, D., Capman, F., Ravera, B.: Confidence measure based unsupervised target model adaptation for speaker verification. In: Proc. Interspeech, pp. 754–757 (2007)

36. Poritz, A.B.: Linear predictive hidden Markov models and the speech signal. In: Proc. ICASSP, vol. 2, pp. 1291–1294 (1982)
37. Pruzansky, S.: Pattern-matching procedure for automatic talker recognition. J.A.S.A. 35, 354–358 (1963)
38. Pruzansky, S., Mathews, M.V.: Talker recognition procedure based on analysis of variance. J.A.S.A. 36, 2041–2047 (1964)
39. Rabiner, L.R., Juang, B.H.: Fundamentals of Speech Recognition. Prentice-Hall, Englewood Cliffs (1993)
40. Rose, R., Reynolds, R.A.: Text independent speaker identification using automatic acoustic segmentation. In: Proc. ICASSP, pp. 293–296 (1990)
41. Rose, R.C., Hofstetter, E.M., Reynolds, D.A.: Integrated models of signal and background with application to speaker identification in noise. IEEE Trans. Speech and Audio Processing 2(2), 245–257 (1994)
42. Rosenberg, A.E., Sambur, M.R.: New techniques for automatic speaker verification. IEEE Trans. Acoustics, Speech, Signal Proc. ASSP-23(2), 169–176 (1975)
43. Rosenberg, A.E., Soong, F.K.: Evaluation of a vector quantization talker recognition system in text independent and text dependent models. Computer Speech and Language 22, 143–157 (1987)
44. Sambur, M.R.: Speaker recognition and verification using linear prediction analysis. Ph. D. Dissert., M.I.T (1972)
45. Siu, M., et al.: An unsupervised, sequential learning algorithm for the segmentation of speech waveforms with multiple speakers. In: Proc. ICASSP, pp. I-189–192 (1992)
46. Stolcke, A., Ferrer, L., Kajarekar, S., Shriberg, E., Venkataraman, A.: MLLR transforms as features in speaker recognition. In: Proc. Interspeech 2005, pp. 2425–2428 (2005)
47. Sugiyama, M.: Segment based text independent speaker recognition. In: Proc. Acoust., Spring Meeting of Soc. Japan, pp. 75–76 (1988) (in Japanese)
48. Tishby, N.: On the application of mixture AR hidden Markov models to text independent speaker recognition. IEEE Trans. Acoust., Speech, Signal Processing ASSP-30(3), 563–570 (1991)
49. Wilcox, L., et al.: Segmentation of speech using speaker identification. In: Proc. ICASSP, pp. I-161–164 (1994)

Robust Biometric Key Extraction Based on Iris Cryptosystem

Long Zhang[1], Zhenan Sun[2], Tieniu Tan[2], and Shungeng Hu[1,3]

[1] Center for Intelligence Science and Technology Research,
Beijing University of Posts and Telecommunications, Beijing 100876, China
[2] Center for Biometrics and Security Research, National Laboratory of Patten Recognition,
Institute of Automation, Chinese Academy of Sciences, Beijing 100190, China
[3] China Netcom Group Labs, Beijing 100140, China
zl.bupt@gmail.com, {znsun,tnt}@nlpr.ia.ac.cn,
hushungeng@chinaunicom.cn

Abstract. Biometric cryptosystem can not only provide an efficient mechanism for template protection, but also facilitate cryptographic key management, thus becomes a promising direction in information security field. In this paper, we propose a robust key extraction approach which consists of concatenated coding scheme and bit masking scheme based on iris database. The concatenated coding scheme that combines Reed-Solomon code and convolutional code is proposed so that much longer keys can be extracted from the iris data, while the bit masking scheme is proposed to minimize and randomize the errors occur in the iris codes, making the error pattern more suitable for the coding scheme. The experiment results show that the system can achieve a FRR of 0.52% with the key length of 938 bits.

Keywords: Biometrics, cryptography, iris code, key binding, concatenated coding scheme, Reed-Solomon code, convolutional code, bit masking scheme.

1 Introduction

Biometrics provides a convenient and reliable way for user authentication. Compared with the traditional cryptosystem which identifies people based on passwords, biometric authentication has its natural advantages: the user's physiological or behavioral characteristics are something he/she was born with, they are unique, stable and portable, thus can overcome the shortage of the passwords which could be easily guessed or forgotten.

Generally, there are two phases in biometric system for user authentication: during *enrollment* phase, for each registered user, a biometric template is created and stored in the database; during *authentication* phase, a user to be authenticated present his/her biometric sample, the system then compares it to the templates, only a successful match can the user enter the system. In such a system, biometric templates are not stored in any encrypted form, attacks on the templates can pose a fatal threat to the whole system because unlike the password, compromised biometric templates cannot

M. Tistarelli and M.S. Nixon (Eds.): ICB 2009, LNCS 5558, pp. 1060–1069, 2009.
© Springer-Verlag Berlin Heidelberg 2009

be revoked or reissued, once compromised, they are compromised forever [1]. In order to gain higher security level and more public acceptance of biometric system, an efficient biometric template protection scheme should be proposed.

On the other hand, traditional cryptosystem suffers from the key managing problem [2]: cryptographic keys are long and random in order to meet the secure requirement of the system, but they are hard to remember. Instead, the cryptographic keys are stored somewhere in the system and released based on other authentication mechanism such as passwords, which are neither convenient nor secure. Biometrics would be a good alternative, but different from passwords, biometric data are noisy. The existence of intra-user variation among multiple acquisitions of the same biometric identifier makes the cryptographic key extraction a challenging and interesting task.

Biometric cryptosystem, which takes the advantage of both biometrics and cryptography, is developed with the purpose that a repeatable cryptographic key can be extracted from the error-prone biometric data in an error-tolerant way. In such a system, the biometric component plays the role of user authentication while the extracted key can be used in any traditional cryptosystem. Under a careful design, biometric cryptosystem can not only provide an efficient mechanism for template protection, but also facilitate the key management, thus opens new applications in information security field. The existing biometric cryptosystems in the literature can be classified into three categories [1, 2]: *key release* system, *key binding* system and *key generation* system. In this paper, we designed a key binding biometric cryptosystem based on iris database where error-control code is used to handle intra-user variation. The existing key binding methods have two main shortages: firstly, the cryptographic keys extracted from the biometric data are not long enough, which narrows the use of key binding method in more cryptographic applications; secondly, the error-correction capability of the error-control code is not strong enough to handle large intra-user variation of the real biometric data, which may result in poor recognition performance (measured by *FRR* and *FAR*). To solve these problems, for the former, we propose a *concatenated coding scheme* which combines the RS and convolutional codes to increase the key length, and for the latter, we propose a *bit masking scheme* which reduces the intra-user variations of iris codes as well as to randomize them to the acceptable level in accordance with the capability of the error correcting code.

The remaining sections are organized as follows. In section 2, we give a brief review of the existing methods of biometric cryptosystem. Then we describe our proposed method, especially the concatenated coding scheme and the bit masking scheme in section 3. The experiment results and analysis are discussed in section 4. Section 5 is the conclusion.

2 Related Works

There are mainly three approaches to implement a biometric cryptosystem: key release approach, key binding approach and key generation approach. The key release approach is the most straightforward way to combine biometrics and cryptography. Soutar [13] developed such a system where the biometric template and the key are stored together as a user record in an encrypted form. During verification, the biometric subsystem decrypt the user record and performs biometric matching, and only after a successful match will the cryptographic key be released to the security subsystem.

Although the segregation between verification process in the biometric subsystem and the rights and privileges which the user is assigned by the security subsystem will help the revocation of a user's rights and privileges, the exposure of biometric template during verification would make the system vulnerable to attacks. The key generation approach, on the other hand, may be the most difficult and attractive method with the purpose of generating a cryptographic key directly from the biometric data. Dodis *et al.* [10] proposed the fuzzy extractors to solve the problem. The secure sketch is firstly constructed in order to recover the biometric data to the template, and then the fuzzy extractor is built based on the secure sketch where a strong extractor is used to extract a uniformly random key from the biometric template. The drawback of this approach is that the cryptographic key cannot be revoked or reissued because the same biometric template always generates the same key.

The key binding approach provides a feasible way to generate uniformly distributed random keys from the biometric data as well as to protect biometric template securely. Through this approach, a random key is chosen for a special user and then bound with his/her biometric template to form public information, e.g. *helper data*, then the biometric sample which is close enough to the template will be able to extract the key from the helper data.

Davida *et al.* [3] were among the first to apply error-control code to construct an off-line biometric authentication system. Given a canonical iris template, check bits are computed using a selected error-control code to form a codeword, then the check bits along with the hashed codeword are stored as the helper data. If user's biometric sample is close to the template to some extent, the check bits will eliminate the intra-user variations and make the authentication success, meanwhile, the hash function secures the template. Digital signature is also used to bind the biometric data to user attributes. To make the method more resilient and secure, Jules and Wattenberg [4] made some improvements and proposed fuzzy commitment scheme. Instead of regarding the template as the information bits, this scheme treats the template itself as a corrupt codeword, which can commit the cryptographic key in a fuzzy way. Along this line, there are many implementations to realize such integration of biometrics and cryptography. Hao *et.al* [5] designed a key binding system based on iris database using two-layer error-control code that combines Reed-Solomon and Hadamard codes. The length of the key extracted from the iris code is 140 bits with acceptable recognition performance, and can meet the requirement of key length in some cryptographic applications (such as 128-bit AES). Tuyls *et al.* [6] proposed the reliable components scheme based on fingerprint database in order to reduce the intra-user variation. The fuzzy vault scheme proposed by Jules and Sudan [7] may be regarded as an order-invariant version of [4] which is based on set difference. The secret, e.g. the cryptographic key is hidden in a selected polynomial as its coefficient, the vault is then constructed with the help of both genuine points and chaff points. The sample feature set which overlaps the template feature set substantially will be able to reconstruct the polynomial, and thus extract the key. Clancy *et al.* [8] and Uludag *et al.* [9] developed such a system based on fingerprints. Draper *et al.* [11] applied the distributed source coding to ensure the secure storage of fingerprint template. Besides, Itakura and Tsujii [12] proposed a method to apply biometric cryptosystem to PKI environment where the cryptographic key extracted from the biometric data is used as the user private key.

The feature representation of the selected biometric identifier determines which approach should be chosen to design a biometric cryptosystem. If the feature set contains deletions and insertions of its element such as fingerprint, the fuzzy vault scheme is a good choice. And since the iris code is represented as a fix length string, we follow the method proposed in [4] in this paper.

3 Our Proposed Method

In this section, we present the detailed design of our iris cryptosystem, where the verification mode is considered. The iris code is a fixed length feature vector representation of the iris data which can be compared directly using Hamming distance. And the recognition performance of the iris cryptosystem is measured by the False Rejection Rate (FRR) and the False Acceptance Rate (FAR). The FRR is the error rate that the cryptographic key is not correctly extracted with the iris data from the same person, whereas the FAR is the error rate that the cryptographic key is mistakenly extracted with the iris data from different persons. There is always a tradeoff between FRR and FAR where the error-correction capability is the threshold. The system can adjust this threshold to meet different requirements of the applications.

Figure 1 shows the block diagram of the proposed iris cryptosystem. The enrollment phase is comprised of three steps: firstly, the bit masking component is performed to add the masking bits to original iris code I_T to form a masked iris string $I_{TM} \in \{0,1\}^n$, where n denotes the length of the iris code; secondly, a uniformly distributed key $K \in \{0,1\}^k$ is randomly generated for a special user, and then encoded into a codeword $C \in \{0,1\}^n$ through the concatenated encoding component; thirdly, the helper data P that binds the cryptographic key and the iris template together is generated by $P = C \oplus I_{TM}$, where $\oplus$ denotes bitwise XOR. The information about neither the iris template nor the cryptographic key is revealed to the adversary given the helper data P.

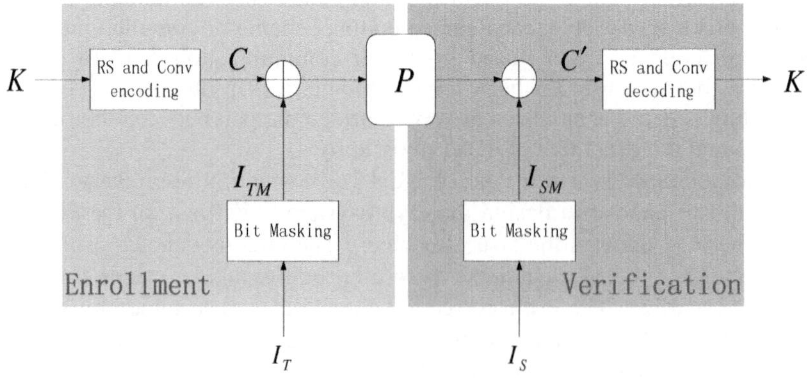

Fig. 1. Overview of the iris cryptosystem

Accordingly, the verification phase comprises three steps: firstly, the system captures and extracts the sample iris code I_S from the user to be verified, and then masked it in the same way as it did during the enrollment phase to form $I_{SM} \in \{0,1\}^n$; secondly, codeword $C' \in \{0,1\}^n$ is calculated based on the helper data, $C' = P \oplus I_{SM} = C \oplus (I_{TM} \oplus I_{SM})$; and thirdly, cryptographic key $K' \in \{0,1\}^k$ is extracted through the concatenated decoding, and if the iris codes I_{TM} and I_{SM} are close enough to each other, e.g. they are from the same person, the exacted key K' will be identical to K, which indicates a successful verification. As shown in figure 1, the concatenated coding scheme and the bit masking scheme are the main factors which directly determine the recognition performance of the iris cryptosystem.

3.1 Concatenated Coding Scheme

Concatenated coding is a particular and efficient method for constructing long codes from short codes, as well as for increasing the error-correction capability of the error-control codes. Applying the concatenated coding scheme to the iris cryptosystem, much longer keys can be embedded into the iris templates, which can then be extracted for legitimate users and used in more cryptographic applications.

There are two types of errors between different iris codes from the same eye: the random errors mainly caused by the precision of the acquisition device, and the burst errors mainly caused by undetected eyelashes and specular reflections. Based on the error characteristic of iris codes, Hao *et.al* [5] designed a two-layer concatenated coding scheme, in which Hadamard code is firstly used to correct the random errors, and then Reed-Solomon code is used to correct the burst errors.

The concatenated coding scheme in this paper handles the errors in iris codes in a different way. The burst errors are firstly transformed into random ones using the bit masking scheme which will be detailed in section 3.2. Then convolutional code and Reed-Solomon code are concatenated to control the errors. Convolutional code is chosen as the inner code because it is suitable to correct random errors, and compared with block coding, convolutional coding with the same code rate can achieve a better error-correction performance. Since burst errors often occur during the decoding phase of the convolutional code, Reed-Solomon code is used as its outer code to correct them at block level. The concatenated coding scheme of convolutional code and Reed-Solomon code is widely used in digital communication field to control the channel noise, but never had it been used in biometrics. In this paper, we will show that this coding scheme is an efficient way to bridge the gap between the fuzziness of the iris codes and the exactness of the cryptography.

Reed-Solomon code is a sub type of BCH code which is non-binary. The Reed-Solomon code we choose to design the cryptosystem is defined on the Galois Field $GF(2^m)$, where m denotes the size of a block. The code is denoted as $RS(n_1, k_1, t)$, where k_1 stands for the block number before encoding, and n_1 stands for the block number after encoding. The error-correction capability is determined by t, which is the number of error blocks that can be corrected, and $n_1 - k_1 = 2t$ must be specified. During verification, Berlekamp-Messey algorithm is used for decoding.

Different from block code, convolutional code operates on serial data where its encoder comprises a group of memory registers. The convolutional code can be denoted as $Conv(n_2, k_2, v)$, where k_2 stands for the bit length of the encoder input stream, and

n_2 stands for the encoder output stream. The error-correction capability is determined by v, which is the number of registers of the encoder. During verification, Viterbi algorithm is used for decoding.

The parameters of our design are set as follows. The length of the iris codes n is fixed at 4096 bits. The code rate of convolutional code $R_2 = k_2 / n_2 = 1/4$, and the register number $v = 6$, which indicates that there should be a tail of 6 zero-bits to flush the registers after convolutional encoding. The block size m of the RS code is 8 bits, then the length of user-specific key can be calculated.

$$k = k_1 \times m - v = (\frac{n_1}{m} - 2t) \times m - v = (\frac{n \times R_2}{m} - 2t) \times m - v = (128 - 2t) \times 8 - 6 \quad (1)$$

The number of blocks that can be corrected by the Reed-Solomon code t is the threshold of the iris cryptosystem. By adjusting the value of t, the key length and the system recognition performance can be changed in light of requirements for different applications.

3.2 Bit Masking Scheme

In iris cryptosystem, the errors of iris codes are always too many to be handled by the error-control code, which is the main factor that degrades the system recognition performance. Moreover, there exist both random errors and burst errors in iris codes from the same eye, which is not a suitable error pattern for our coding scheme. If some positions of the iris codes to be compared can be replaced with the same bits, then the Hamming distance can be reduced. And if these positions are selected uniformly, the burst errors will be randomized. Bit masking scheme is proposed to deal with the errors in this way.

During enrollment, the template iris code $I_T = \{I_{Ti}\}, i = 1, 2, \ldots, n$, where I_{Ti} denotes the i-th bit of the template code, is uniformly masked by the *masking bits* $M_T = \{M_{Ti}\}$, where M_{Ti} denotes the masking bit at the i-th position of the template code. Then during verification, the sample iris code $I_S = \{I_{Si}\}$, $i = 1, 2, \ldots, n$, where I_{Si} denotes the i th bit of the sample code, is masked at the same positions by the masking bits $M_S = \{M_{Si}\}$, where M_{Si} denotes the masking bit at the i-th position of the sample code. The masking bits at the same positions in both the template and the sample codes should keep the same, e.g. $M_{Ti} = M_{Si}$.

Furthermore, we define the *bit masking rate*, denoted as R_M, to describe the extent to which the iris codes are masked. For example, $R_M = 1/4$ means the iris codes are masked at the positions $i = 4j$, $j = 1, 2, \ldots, n/4$. Let $I_{TM} = \{I_{TMi}\}$, $i = 1, 2, \ldots, n$ denotes the masked template code, and $I_{SM} = \{I_{SMi}\}$, $i = 1, 2, \ldots, n$ denotes the masked sample code, they are computed as follows:

$$I_{TMi} = \begin{cases} I_{Ti} & , i \neq 4j \\ M_{Ti} & , i = 4j \end{cases}, \qquad I_{SMi} = \begin{cases} I_{Si} & , i \neq 4j \\ M_{Si} & , i = 4j \end{cases} \quad (2)$$

Since $M_{Ti} = M_{Si}$, we can get $I_{TMi} \oplus I_{SMi} = 0$, $i = 4j$, $j = 1, 2, \ldots, n/4$ .Similarly, $R_M = 1/2$ means the iris codes are masked at the positions $i = 2j$, $j = 1, 2, \ldots, n/2$, and we get $I_{TMi} \oplus I_{SMi} = 0$, $i = 2j$, $j = 1, 2, \ldots, n/2$. $R_M = 0$ stands for the situation that the iris codes are acquired without bit masking scheme.

Obviously, the bit masking scheme can reduce the errors between iris codes, as well as to randomize them. After the iris codes are masked, the proposed concatenated coding scheme can be used to control errors efficiently. But the bit masking rate R_M should be carefully selected. That is because with the increase of the bit masking rate, both the intra-user variations and inter-user variations fall down. If R_M is set too low, the intra-user variations may not be sufficiently reduced, which results in high FRR. On the other hand, if R_M is set too high, the inter-user variations may decrease, which results in high FAR. We suggest that the bit masking rate should be set in accordance with the error-correction capability of the coding scheme.

4 Experimental Results and Analysis

The iris database in the experiment consists of the iris codes from 128 different eyes, and each with 3 measurements. The length of each iris code is 4096 bits. Figure 2(a) shows the Hamming distance distribution of these iris codes. The intra-class Hamming distances are less than 30%, while the inter-class Hamming distances are more than 40%. Then the bit masking schemes with the rates $R_M = 1/4$ and $R_M = 1/2$ are applied to the iris codes respectively, the Hamming distance distributions are changed accordingly as shown in Figure 2(b) and Figure 2(c). With the increase of the masking rate, the curves of intra-class distribution and inter-class distribution become sharper and get closer to each other.

The database contains 3 iris codes per person, and we randomly choose one of them during enrollment. To calculate FRR, we use the remaining two iris codes during verification, thus totally $128 \times C_3^2 = 384$ comparisons should be made. To calculate FAR, we use the iris codes from the other persons during verification, which results in totally $(128 \times C_3^1) \times (127 \times C_3^1)/2 = 73152$ comparisons.

The Reed-Solomon code chosen in our scheme is the shortened code on Galois Field $GF(2^8)$, and the convolutional code is the optimal code with 1/4 code rate given in [14]. The concatenated coding scheme is performed on the iris codes with different bit masking rates $R_M = 0$, $R_M = 1/4$ and $R_M = 1/2$, respectively. With the increase of bit masking rate, the errors occur in the iris codes are reduced and become more random, which can be corrected more easily by the concatenated code. As a result, the FRR decreases while the FAR increases. Figure 3(a) shows the FRR curves for different masking rate, and since the FAR values with the masking rate $R_M = 0$ and $R_M = 1/4$ identically equal to 0 at all the thresholds, we do not show the comparison of the FAR curves. Generally, smaller value of the threshold t results in longer keys with lower FAR but higher FRR, and vice versa. In our design, $R_M = 1/2$ is a suitable value and the FAR and FRR curves are shown in Figure 3(b). The system can extract a user-specific key of 938 bits, with FRR=0.52% and FAR=0. The key extracted from the iris codes is much longer than the existing methods in literatures [5, 6, 8, 9]. And the key can meet many cryptographic applications, such as 128-bit, 192-bit and 256-bit AES. Besides, in the public key cryptosystem, the extracted key

from the iris data can also be used as the user's private key, from which the public-key certificate can be derived. The certificate with a PKI signature can then be used for remote authentication in the network environment.

By our proposed key binding iris cryptosystem, the iris database should meet the following two requirements to ensure the robustness of long biometric key extraction. Firstly, the iris code itself should be long, and if the iris code is not long enough, we can insert the same bits uniformly in both template and sample codes instead of the bit masking scheme, which can not only increase the length of the iris code, but also randomize the errors. Secondly, the captured iris images should be of high quality, which is helpful for our scheme to achieve a better recognition performance.

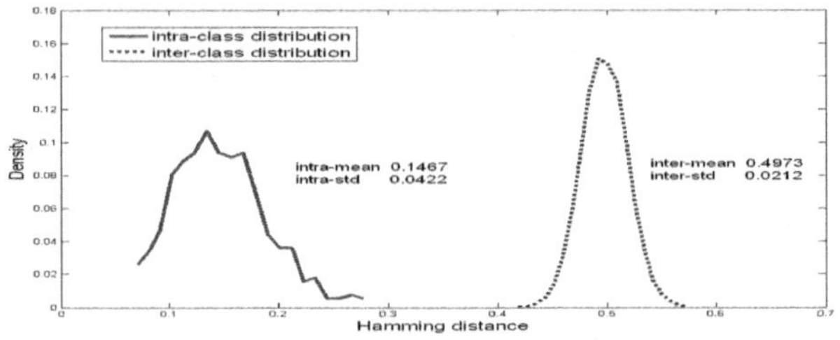

(a)

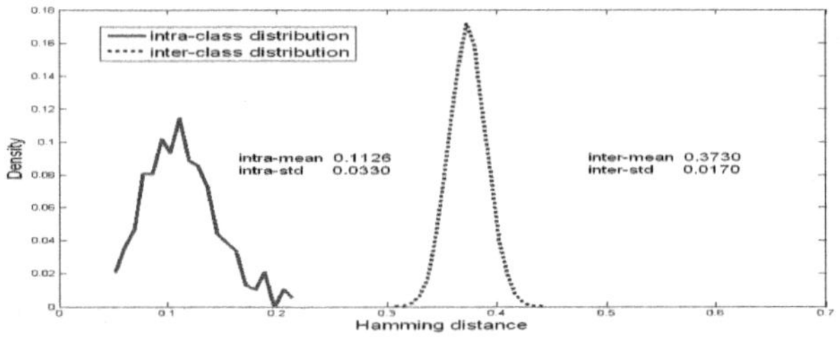

(b)

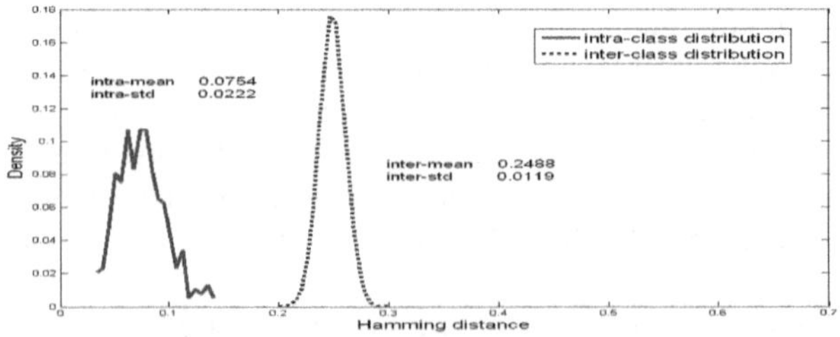

(c)

Fig. 2. Hamming distance distributions: (a) with bit masking rate $R_M = 0$; (b) with bit masking rate $R_M = 1/4$; (c) with bit masking rate $R_M = 1/2$

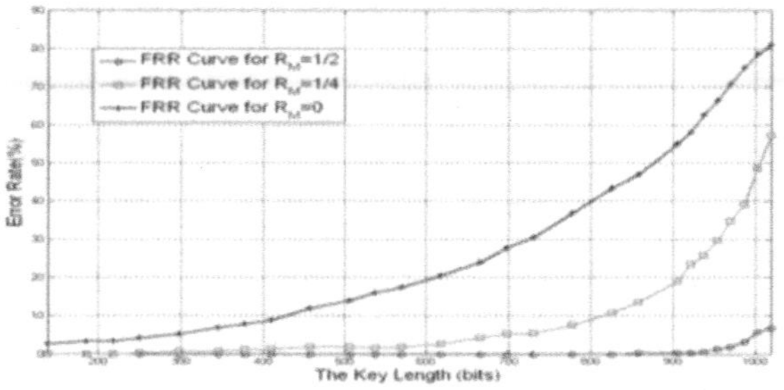

(a)

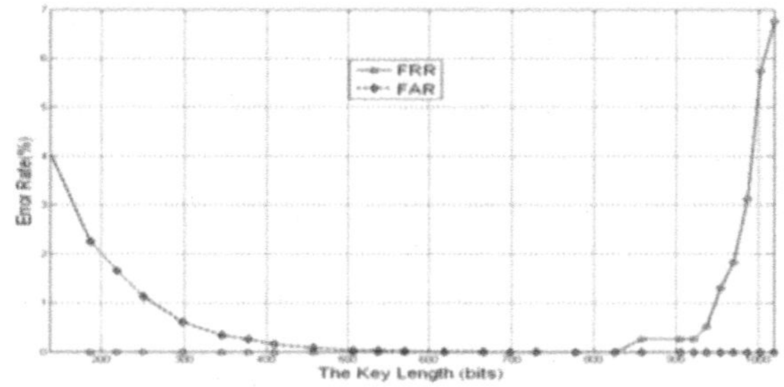

(b)

Fig. 3. Recognition performance of proposed iris cryptosystem: (a) FRR curves for different masking rates; (b) FAR and FRR curves for $R_M = 1/2$

5 Conclusions

In this paper, concatenated coding scheme and bit masking scheme are proposed to construct an iris cryptosystem. The concatenated coding scheme is proposed in order to embed long keys into the iris data. We apply the concatenated code that combines Reed-Solomon code and convolutional code, which is a prevalent channel coding technique in digital communication field, to control the errors in the iris codes. Then the bit masking scheme is proposed to minimize and randomize the errors to make the error pattern more suitable for the coding scheme. The experiment results prove that our iris cryptosystem is robust and efficient since much longer keys can be extracted from the iris data while the system recognition performance remains acceptable. The iris cryptosystem achieves a FRR of 0.52% with the key length of 938 bits.

Acknowledgments. This work is supported by National 973 Program Project (Grant No. 2004CB318100), National Natural Science Foundation Projects (Grant No. 60736018, 60335010, 60702024, 60723005), and National 863 Program Projects (Grant No. 2006AA01Z193, 2007AA01Z162).

References

1. Jain, A.K., Nandakumar, K., Nagar, A.: Biometric Template Security. EURASIP Journal on Advances in Signal Processing 2008, Article ID 579416, 17 (2008), doi:10.1155/2008/579416
2. Uludag, U., Pankanti, S., Prabhakar, S., Jain, A.K.: Biometric cryptosystems: issues and challenges. Proceedings of the IEEE 92(6), 948–960 (2004)
3. Davida, G.I., Frankel, Y., Matt, B.J.: On enabling secure applications through off-line biometric identification. In: Proceedings of the IEEE Computer Society Symposium on Research in Security and Privacy, Oakland, Calif, USA, pp. 148–157 (May 1998)
4. Juels, A., Wattenberg, M.: A fuzzy commitment scheme. In: Proceedings of 6th ACM Conference on Computer and Communications Security (ACM CCS 1999), Singapore, pp. 28–36 (November 1999)
5. Hao, F., Anderson, R., Daugman, J.: Combining crypto with biometrics effectively. IEEE Transactions on Computers 55(9), 1081–1088 (2006)
6. Tuyls, P., Akkermans, A.H.M., Kevenaar, T.A.M., Schrijen, G.-J., Bazen, A.M., Veldhuis, R.N.J.: Practical biometric authentication with template protection. In: Kanade, T., Jain, A., Ratha, N.K. (eds.) AVBPA 2005. LNCS, vol. 3546, pp. 436–446. Springer, Heidelberg (2005)
7. Juels, A., Sudan, M.: A fuzzy vault scheme. In: Proceedings of the IEEE International Symposium on Information Theory, Piscataway, NJ, USA, p. 408 (June-July 2002)
8. Clancy, T., Lin, D., Kiyavash, N.: Secure smartcard-based fingerprint authentication. In: Proceedings of the ACMSIGMM Workshop on Biometric Methods and Applications, Berkley, Mich, USA, pp. 45–52 (November 2003)
9. Uludag, U., Pankanti, S., Prabhakar, S., Jain, A.K.: Fuzzy vault for fingerprints. In: Kanade, T., Jain, A., Ratha, N.K. (eds.) AVBPA 2005. LNCS, vol. 3546, pp. 310–319. Springer, Heidelberg (2005)
10. Dodis, Y., Reyzin, L., Smith, A.: Fuzzy extractors: how to generate strong keys from biometrics and other noisy data. In: Cachin, C., Camenisch, J.L. (eds.) EUROCRYPT 2004. LNCS, vol. 3027, pp. 523–540. Springer, Heidelberg (2004)
11. Draper, S.C., Khisti, A., Martinian, E., Vetro, A., Yedidia, J.S.: Using distributed source coding to secure fingerprint biometrics. In: Proceedings of the IEEE International Conference on Acoustics, Speech and Signal Processing (ICASSP 2007), Honolulu, Hawaii, USA, vol. 2, pp. 129–132 (April 2007)
12. Itakura, Y., Tsujii, S.: Proposal on a multifactor biometric authentication method based on cryptosystem keys containing biometric signatures. International Journal of Information Security, 288–296 (2005)
13. Soutar, C.: Security Considerations for the Implementation of Biometric Systems. In: Ratha, N., Bolle, R. (eds.) Automatic Fingerprint Recognition Systems, pp. 415–431. Springer, Heidelberg (2004)
14. Lin, S., Costello, D.J.: Error Control Coding Second Edition. China Machine Press, Beijing (2007)

Iris Matching by Local Extremum Points of Multiscale Taylor Expansion

Algirdas Bastys, Justas Kranauskas, and Rokas Masiulis

Department of Computer Science II,
Faculty of Mathematics and Informatics,
Vilnius University, Lithuania

Abstract. Random distribution of features in iris image texture allows to perform iris-based personal authentication with high confidence. We propose to use the most significant local extremum points of the first two Taylor expansion coefficients as descriptors of the iris texture. A measure of similarity that is robust to moderate inaccuracies in iris segmentation is presented for the proposed features. We provide experimental results of verification quality for four commonly used iris data-sets. Strong and weak aspects of the proposed approach are also discussed.

1 Introducation

Person identification based on iris recognition gives one of the most reliable results. Iris texture features provides a unique high dimensional information that explains why iris recognition based verification has the lowest false acceptance rate among all types of biometric verification systems [1], [2]. A transparent and simple distribution of calculated distances between iris pairs allows to make a fine prediction of verification quality for a chosen distance threshold [3]. Large-scale application of iris recognition for border-crossing control showed up robustness of the technology even in case of more than half billion different iris images spanning about 150 nationalities [4]. Appearance of new technologies that enable to capture iris images even at distance of 20 meters significantly reduced requirements for subject cooperation with the automated identification system [5].

In this paper, we focus on iris texture description and comparison, thus iris segmentation algorithm is left aside. However, our implementation of fully automatic segmentation algorithm, which approximates iris inner and outer boundaries by Fourier series and unwraps iris texture into polar like representation (similar to Daugman's rubber sheet method), was used in all presented experiments. Iris local features as local extrema of multiscale Taylor expansion are presented in section 2. Only approximations of Taylor coefficients associated with the first and second order derivatives averaged on different scales are used. The most significant local extrema of the expansions shows where the most significant iris texture variations are. In section 3, a metric for comparison of feature sets formed from the local extrema is presented. Quality of the proposed iris recognition algorithm is analyzed in section 4, which is followed by a discussion and conclusion sections.

M. Tistarelli and M.S. Nixon (Eds.): ICB 2009, LNCS 5558, pp. 1070–1079, 2009.
© Springer-Verlag Berlin Heidelberg 2009

2 Local Extrema of Multiscale Taylor Expansion

We developed our scheme of iris representation by local features considering the following constraints: number of local features should be easily controlled and their attributes should be as compact and simple as possible. The main goal of our representation is to control memory consumption and iris matching speed, thus leading to the fast and robust iris-based verification and identification.

2.1 Local Descriptors

Let's consider one dimensional iris signal having a fixed pseudopolar radius $r = const$ as an analytic one-dimensional signal $u = u(x)$, where x is pseudopolar angle. It is well known that any signal can be expressed in classic Taylor series expansion centered on any fixed $x = x_i$. The zero coefficient of the expansion equals to $u(x_i)$ and is strongly influenced by iris lighting conditions thus can not be a robust iris descriptor. The second two coefficients equal to $\frac{u'(x_i)}{1!}$ and $\frac{u''(x_i)}{2!}$. The first derivative $u'(x)$ estimates transition rate of the signal around point x, i.e. the big positive value of $u'(x)$ means a fast increasing transition of the signal around point x and a big negative $u'(x)$ means a fast decay around the same point. The sign of the second derivative $u''(x)$ provides information about the type of concavity of graph of the transition around x, i.e. if $u''(x) > 0$ then graph of transition of the signal is concave upward and $u''(x) < 0$ gives that graph of the signal that is concave downward around point x. If transition of x values corresponds to transition of $u = u(x)$ from left to right through a dark or bright blob one can expect the following transitions of the signs:

Blob	Derivative	Transition of signs
Dark	$u'(x)$	$-+$
	$u''(x)$	$-+-$
Bright	$u'(x)$	$+-$
	$u''(x)$	$+-+$

Left and right boundary of a dark blob can be marked by local negative minimum and positive maximum of the first derivative. The second derivative will have zero crossings around these local extrema. If we expand such features to two-dimensional case, only a fraction of $u_{xx}(x, y)$ zero crossings will correspond to $u_x(x, y)$ local extrema. Therefore local extrema of both two-dimensional functions $u_x(x, y)$ and $u_{xx}(x, y)$ were used as primary source for derivation of iris local features. Notice that $u'(x)$ and $u''(x)$ derivatives can be estimated only numerically in the context of digital iris data and the result of approximations significantly depends on scale at which increments of the signal are estimated. $u''(x)$ approximation will have a positive maximum close to a dark blob center and a negative minimum close to a bright blob center if the chosen scale resonates with blob size.

The well known Taylor series expansion

$$u(x) = \sum_{n=0}^{\infty} \frac{u^{(n)}(a)}{n!}(x-a)^n \tag{1}$$

tells us that sufficiently regular function can be reconstructed from its Taylor coefficients $\frac{u^{(n)}(a)}{n!}$. In theory Taylor coefficients can be derived from $u(x)$ values known at any narrow surrounding of the center point $x = a$. Let us consider some details of estimation of local extrema of the first two derivatives of iris texture and conditions under which a particular extremum point is included in the set of iris local features. Estimation technique of derivatives at different scales is similar to the wavelet multiresolution analysis. Let us fix any symmetric (even) smooth filter $p = p(x) \geq 0$ which defines rule of derivatives averaging. Let us fix some scale $\sigma > 0$ and consider convolutions

$$(u' * D_\sigma p)(x) = 1/\sigma \int_{-\infty}^{\infty} p(\xi/\sigma)u'(\xi - x)\, d\xi \tag{2}$$

and

$$(u'' * D_\sigma p)(x) = 1/\sigma \int_{-\infty}^{\infty} p(\xi/\sigma)u''(\xi - x)\, d\xi. \tag{3}$$

In the bigger scale σ the more blurred version of $u'(x)$ and $u''(x)$ represents $(u' * D_\sigma p)(x)$ and $(u'' * D_\sigma p)(x)$. One the other hand, if σ approaches zero then $(u' * D_\sigma p)(x)$ and $(u'' * D_\sigma p)(x)$ approaches to $u'(x)$ and $u''(x)$ respectively.

We have discrete periodic sequence $u_i = u(i)$ of length NX in case of iris texture. Defining $u(x)$ by linear interpolation $u(x) = u_i + (x - i)(u_{i+1} - u_i)$, $x \in [i, i + 1)$, we will have

$$u''(x) = \sum_{i=-\infty}^{\infty} (u_{i+1} - 2u_i + u_{i-1})\delta(x - i) \tag{4}$$

and

$$(u'' * D_\sigma p)(x) = \sum_{i=-\infty}^{\infty} (u_{i+1} - 2u_i + u_{i-1})p(\frac{x+i}{\sigma})/\sigma \tag{5}$$

$$= \sum_{i=0}^{NX-1} \Delta u_i p_\sigma^{NX}(x + i) \tag{6}$$

$$= \sum_{i=0}^{NX-1} u_i \Delta p_\sigma^{NX}(x + i). \tag{7}$$

Here $\delta(x)$ is Dirac's delta function,

$$\Delta u_i = u(i + 1) - 2u(i) + u(i - 1), \tag{8}$$

and

$$p_\sigma^{NX}(x) = \frac{\sum_{i=-\infty}^{\infty} p(\frac{x+i*NX}{\sigma})}{\sigma}. \tag{9}$$

Similarly, for the first derivative we have

$$(u' * D_\sigma p)(j) = - \sum_{i=0}^{NX-1} \Delta u_i q^{NX} \sigma(j+i), \tag{10}$$

$$q_\sigma^{NX}(x) = \sum_{i=-\infty}^{\infty} 2 \operatorname{arctg}(\frac{x + i * NX}{\sigma}). \tag{11}$$

From symmetry of the filter function follows that (5) defines a filtration of 1D data $\{u_i\}$ by an even-symmetric filter and (10) defines an odd-symmetric filter. In wavelet terminology the even and odd filters has two and one vanishing moments respectively. The proposed filters has more freedom in choosing basic function $p^\sigma = p^\sigma(x)$ in comparison with dyadic wavelet filtration and do not have dyadic restrictions for the scale σ. To regularize the filtration results that at different scales σ approximate second (even filters) or first (odd filters) order derivatives we used some smoothing in orthogonal direction by applying the same filter $p^\sigma = p^\sigma(y)$ with smaller scale value σ.

Table 1 presents scales, type of symmetry and direction which were used to calculate differences. The differences calculated along x and y directions correspond respectively to horizontal and vertical type filters. Derivatives along x direction were estimated at 3 different scales while derivatives along y direction were estimated only at 1 scale.

Table 1. Derivatives and averaging parameters used in derivation of multiscale Taylor expansion and local features

Scale and filter type index s	Blurred derivative	$\sigma_x = \sigma$ of horizontal direction	$\sigma_y = \sigma$ of vertical direction
0	u_x	$\frac{6}{8}$	$\frac{1}{2}$
1	u_x	$\frac{9}{8}$	$\frac{1}{2}$
2	u_x	$\frac{15}{8}$	$\frac{1}{2}$
3	u_{xx}	$\frac{9}{8}$	$\frac{1}{2}$
4	u_{xx}	$\frac{15}{8}$	$\frac{1}{2}$
5	u_{xx}	$\frac{24}{8}$	$\frac{1}{2}$
6	u_y	$\frac{1}{2}$	$\frac{9}{8}$
7	u_{yy}	$\frac{1}{2}$	$\frac{15}{8}$

2.2 Significant Local Descriptors

Blurred derivative may be considered as the coefficients of Taylor series expansion beside $(x - a)/1!$ and $(x - a)^2/2!$ terms estimated at different scales or coefficients of wavelet expansion calculated at different scales with asymmetric or symmetric mother wavelets that have one or two vanishing moments. It is well known that coefficients with highest squared values give the main contribution in approximation of original signal by the expansions. Thus, extrema or

more exactly, positive local maximums and negative local minimums of blurred derivatives, are good candidates for local descriptors of iris texture. Additionally, we required that the local extremum at particular scale σ_0 exceeded the same derivative at the same position but with rougher scales $\sigma > \sigma_0$ (that conditioned more stable extrema). And the very last condition for including the extremum point to the set of local iris features was its magnitude. However, we decided to avoid thresholds in this place and used the following procedure for choosing extrema points (see fig. 1):

1. Divide iris texture annual along polar angle direction into 16 equal sectors.
2. Enumerate sectors clockwise starting at "0" hour.
3. Eliminate local extrema which are found in upper or lower lids. Two different approaches are used:
   (a) Find a convex domain that includes only iris with pupil and excludes lids.
   (b) Ignore sectors 0, 1, 2, 13, 14, and 15 because iris texture in these sectors will likely be hidden by upper lid. Ignore sectors 7 and 8 because iris texture in these sectors will likely be hidden by lower lid.
4. Choose no more than K biggest local extrema from each scale and each active sector.

(a) Original image. (b) A convex domain that includes only iris. (c) Iris division into sectors.

Fig. 1. Elimination of local extrema which are found in upper or lower lids. In (c), active sectors are marked in white and inactive sectors are marked in light gray.

Figure 2 illustrates local extrema points that were found in active sectors for 242116.tiff and 241643.tiff iris images of NIST ICE2005 iris data-set. These two images constitute genuine pair. The two presented derivatives u_{xx} and u_{yy} are calculated with averaging parameters defined by the 6th ($s = 5$) and 8th ($s = 7$) row of the table 1. White "+" and "-" mark local maximum and minimum points of the averaged Taylor expansion. The presented iris texture has no clearly expressed freckles, furrows, stripes or coronas in active sectors. However there are blurred dark and bright blobs where local maximum and minimum points indicate the blobs of the size that resonates with the averaging scale σ.

3 Similarity Metric

Let us fix local features sets A and B formed from Taylor local extrema estimated at different scales. A particular feature $f_i = \{x_i, y_i, s_i, z_i\}$ has the following attributes:

- (x_i, y_i) - local extremum position defined by two integer numbers x_i and y_i.
- $s_i \in \{0, 1, \cdots, 7\}$ - scale and filter type index (see table 1 for details).
- $z_i \in \{0, 1\}$ - extremum type marker. 0 was used for local maximum points and 1 for local minimum points.

The similarity metric between two local features sets is defined by a normalized sum of similarity scores of feature pairs. A particular pair (f_i^A, f_j^B) can gain non-zero similarity scores only if f_i^A and f_j^B correspond to the same filter and extremum type. Such features will be further called *congeneric* features. Two congeneric features can gain positive similarity only if their positions are sufficiently close. Distance between two affine (x_i^A, y_i^A) and (x_j^B, y_j^B) points is estimated considering possible alignment of two irises by means of an angular rotation and radial translation. Angular rotation and radial translation correspond to some shift along x and y directions respectively[1]. With a fixed α and β shifts we use the following scores similarity expression:

$$s(f_i^A, f_j^B; \alpha, \beta) = \begin{cases} 0, \text{if } s_i^A \neq s_j^B \text{ or } z_i^A \neq z_j^B; \\ |DX - |x_i^A - x_j^B - \alpha||_+ \times \\ \times |DY - |y_i^A - y_j^B - \beta||_+, \text{otherwise.} \end{cases} \quad (12)$$

Here DX and DY are two positive integer parameters that define maximal acceptable distance along x and y directions and $|a|_+$ operator assign 0 for negative a values. The similarity scores between A features set and α rotated and β radially shifted B set is defined as sum of the similarity scores of local features pairs, i. e.

$$score(A, B; \alpha, \beta) = \sum_{i=1}^{NA} \sum_{j=1}^{NB} s(f_i^A, f_j^B; \alpha, \beta). \quad (13)$$

Similarity scores between A and B features sets is defined by the following maximum:

$$score^*(A, B) = \max_{-KX \leq \alpha \leq KX, -KY \leq \beta \leq KY} score(A, B; \alpha, \beta). \quad (14)$$

Here KX and KY are integer parameters defining $2KX + 1$ rotations and $2KY + 1$ radial shifts under which similarity scores of shifted unwarped irises are optimized. They can be adjusted according to the database characteristics, for example with irises rotated in plane for up to 35°, $KX = 25$ parameter value was used. Additionally, warped similarity technique from [6] was used. Finally, similarity score should be normalized according to the number of compared local features, for example divided by $NA + NB$.

[1] Shift along the x direction inherits angular periodicity.

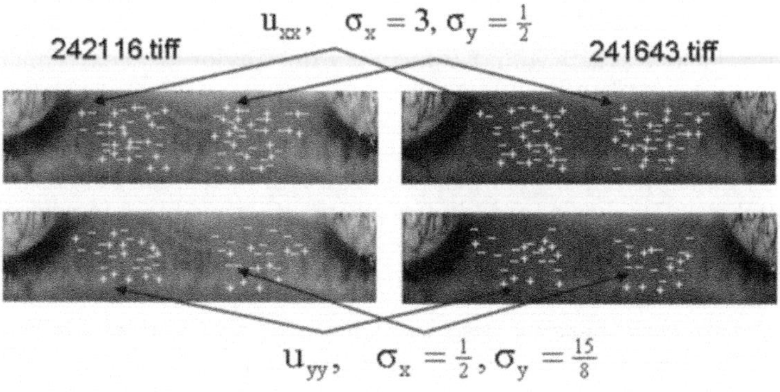

Fig. 2. Unwarped 242116 and 241643 iris images of the NIST ICE2005 data-set. "+" and "-" mark local maximums and minimums of averaged u_{xx} and u_{yy} derivatives (u_{xx} - above, u_{yy} - below). Arrows points to the examples of extrema points that have correspondences in both images.

4 Experimental Results

We used four publicly available iris databases in all experiments: Chinese Academy of Sciences Casia 2.0 (device1) [8] (1200 images of 60 eyes), Casia 3.0 (interval) [8] (2655 images of 396 eyes), and US National Institute of Standards and Technology (NIST) "Iris Challenge Evaluation", experiment 1, (ICE-1) [9] (1425 images of 128 eyes). Our implementation of fully automatic iris segmentation was applied to iris images in all databases. Experiments showed that verification quality increases as value of parameter K (maximum number of congeneric features in one sector) increases. We used $K = 7$ to present the results on all iris databases.

For Casia 2.0 (device 1) iris database ${}_{20}C_2 \times 60 = 11400$ genuine and ${}_{60}C_2 \times 20^2 = 708000$ impostor similarity scores were evaluated. It is recognized that Casia 2.0 iris database contains iris images of lower quality than Casia 3.0 [12]. Best result that can be found in the literature $EER = 0.58\%$ is achieved by [12]. We obtained $EER = 0.13\%$, $ZeroFAR = 1.63\%$, and $ZeroFRR = 5.52\%$ with proposed local extrema of averaged Taylor expansion coefficients as features (filled diamond marked curve in fig. 3).

For Casia 3.0 (interval) iris database 9023 genuine and 3514162 impostor similarity scores were evaluated. Impostor and genuine pairs are completely separable that's why direct estimation of Equal Error Rate (EER) is not possible. We approximated distributions of genuine and impostor empirical similarity scores by Gaussian distributions and concluded that proposed iris verification algorithm has $EER = 4.7 \times 10^{-2}\%$. Complete separation of impostor and genuine pairs was also received on Casia 1.0 [7] database.

For the NIST "Iris Challenge Evaluation", experiment 1, (ICE-1) iris database local extrema based verification algorithm relatively produced the worst results.

The image database contains very difficult and corrupted examples, sometimes with big eyelid occlusion, off-angle or with the iris partly outside the image frame. Poor focus of a part of the iris images especially degrade the quality of the verification that uses only information coming from local extrema. Out of focus iris images lost all subtle details of the iris texture that results in a significant decay of number of Taylor decomposition local extrema at tiny scales. For ICE-1 iris database 12214 genuine and 1002386 impostor similarity scores were evaluated. We obtained $EER = 0.25\%$, $frr@far0.1 = 0.42\%$, $frr@far0.01 = 0.83\%$ (filled triangle marked curve in fig. 3).

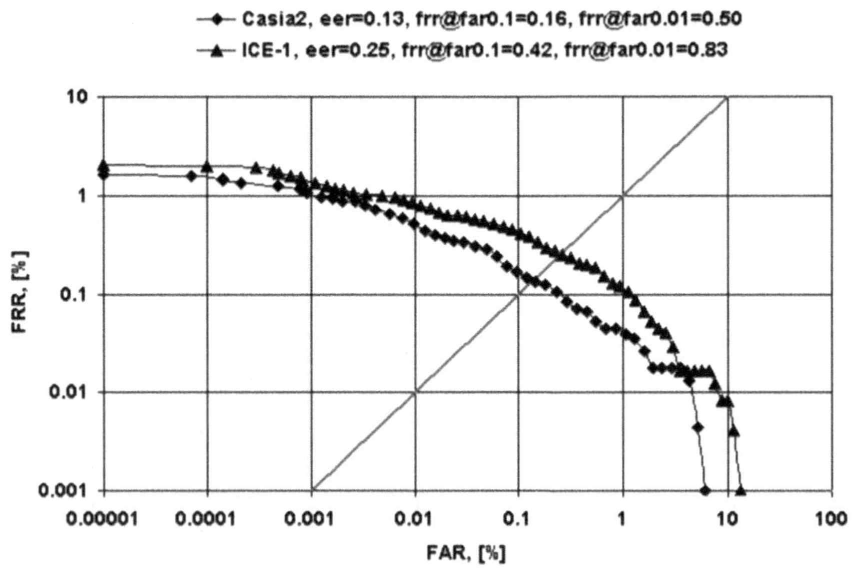

Fig. 3. DET curves of the proposed algorithm for Casia 2.0 (device 1) and ICE-1 iris databases

5 Discussion

The idea to use local extrema of wavelet, Gabor or other expansions for iris recognition is not new [13], [14], [15]. Unlike the 1D local extrema, which where used in [15], we use 2D local extremum points of the multiscale expansion. Such approach reduces the size or iris template and eliminates correlation of the local extremum points along radial direction. We chose the multiscale Taylor expansion since the first two coefficients of the expansion have a transparent interpretation and are associated with very computationally economic filters.

Although the authors of [16] recommends not to use Casia 1.0 data set for iris biometrics research because the images are edited "so that the pupil area is replaced by a circular region of uniform intensity", the images are still usable to

test iris recognition performance (even if the segmentation of edited images is much easier). Our matching algorithm gives complete separation on the Casia 1.0 database. Additionally, we tested the proposed method on larger unedited Casia 3.0 (interval) database and received the same perfect separation of impostor and genuine pairs. About 10 state-of-the-art algorithms were tested on the NIST ICE-1 database with the results available at [9]. Daugman's algorithm 2 [17] is one of the top performers of the ICE [9]. The proposed iris recognition algorithm is very close to Daugman's algorithm 1 [17] and is better than any other published algorithm that uses local extrema.

6 Conclusion

Iris features that can be extracted rapidly were selected on the base of Taylor multiscale expansion. A similarity measure that is robust to moderate iris segmentation inaccuracies for the chosen features was proposed. The proposed features demonstrate good verification performance on publicly available iris databases: Casia 2.0 (device 1), Casia 3.0 (interval) and NIST ICE-1.

Our iris template size is variable and in average contains 1.7 Kb (for $K = 7$). Blurred, out of focus or simple texture containing iris images have reduced number of local extremum points. Therefore, DET curves can have increased ZeroFAR (i.e. FRR at the point where FAR equals to 0) that leads to increased slope near the ZeroFAR point, in the opposite to flat DET curves that are observed for phase based iris matching technique [18]. One of the possible solutions could be additional attributes of the local extremum points that would decrease their matching probability.

Further research will be focused on fusing the phase based iris recognition algorithm (in our implementation) with the proposed method. The proposed method uses different features and different matching approach and is expected to be complimentary to phase based methods. Preliminary experiments showed extraordinary recognition quality results of the fused algorithm but other properties of the algorithm (speed, template size, fusing strategy etc.) must and will be addressed. Additionally, more publicly available databases will be used for testing iris recognition performance.

References

1. Daugman, J., Dowing, C.: Epigenetic randomness, complexity, and singularity of human iris patterns. In: Proceedings of the Royal Society, B, 268, Biological Sciences, pp. 1737–1740 (2001)
2. Daugman, J.: Statistical richness of visual phase information: update on recognizing persons by iris patterns. Int. J. Comput. Vis. 45(1), 25–38 (2001)
3. Daugman, J.: High confidence visual recognition of persons by a test of statistical independence. IEEE Transactions on Pattern Analysis and Machine Intelligence 15(11), 1148–1161 (1993)

4. Daugman, J.: Results from 200 billion iris cross-comparisons. Technical Report UCAM-CL-TR-635 ISSN 1476-2986 (2005), `http://www.cl.cam.ac.uk/techreports/UCAM-CL-TR-635.pdf`
5. Aoptix: Breakthrough in iris recognition (2007), `http://www.aoptix.com/pdf/BreakthroughBio.pdf`
6. Bastys, A., Kranauskas, J., Masiulis, R.: Iris recognition by local extremum points of multiscale Taylor expansion (2008), `http://dx.doi.org/10.1016/j.patcog.2008.12.004`
7. Chinese Academy of Sciences - Institute of Automation Iris Database 1.0 (2003), `http://www.sinobiometrics.com`
8. Chinese Academy of Sciences - Institute of Automation Iris Database 3.0 (2005), `http://www.cbsr.ia.ac.cn/IrisDatabase.htm`
9. National Institute of Science and Technology (NIST): Iris Challenge Evaluation (2005), `http://iris.nist.gov/ICE`
10. Metz, C.E.: Basic principles of ROC analysis. Semin. Nucl. Med. 8, 283–298 (1978)
11. Monro, D.M., Rakshit, S., Zhang, D.: DCT-Based Iris Recognition. IEEE Transactions on Pattern Analysis and Machine Intelligence 29(4), 586–595 (2007)
12. Miyazawa, K., Ito, K., Aoki, T., Kobayashi, K., Nakajima, H.: A Phase-Based Iris Recognition Algorithm. In: Zhang, D., Jain, A.K. (eds.) ICB 2006. LNCS, vol. 3832, pp. 356–365. Springer, Heidelberg (2005)
13. Bae, K., Noh, S., Kim, J.: Iris feature extraction using independent component analysis. In: Proc. 4th Int. Conf. Audio- and Video-Based Biometric Person Authentication, pp. 838–844 (2003)
14. Ma, L.: Person identification based on iris recognition, Ph.D dissertation, Inst. Automation, Chinese Academy of Sciences, Beijing, China (2003)
15. Ma, L., Tan, T., Wang, Y., Zhang, D.: Efficient Iris Recognition by Characterizing Key Local Variations. IEEE Transactions on Image Processing 13(6) (2004)
16. Phillips, P.J., Bowyer, K.W., Flynn, P.J.: Comments on the CASIA version 1.0 Iris Data Set. IEEE Transactions on Pattern Analysis and Machine Intelligence 29(10), 1869–1870 (2007)
17. Daugman, J.: New Methods in Iris Recognition. IEEE Transactions on Systems, Man, and Cybernetics - Part B: Cybernetics 37(5), 1167–1175 (2007)
18. Daugman, J.: Flat ROC Curves, Steep Predictive Quality Metrics: Response to NISTIR-7440 and FRVT/ICE2006 Reports (2007), `http://www.cl.cam.ac.uk/~jgd1000/Response_2_NIST_7440.pdf`

Efficient Iris Spoof Detection via Boosted Local Binary Patterns

Zhaofeng He, Zhenan Sun, Tieniu Tan, and Zhuoshi Wei

Center for Biometrics and Security Research
National Laboratory of Pattern Recognition, Institute of Automation
Chinese Academy of Sciences, P.O. Box 2728, Beijing, P.R. China, 100190
{zfhe,znsun,tnt,zswei}@nlpr.ia.ac.cn

Abstract. Recently, spoof detection has become an important and challenging topic in iris recognition. Based on the textural differences between the counterfeit iris images and the live iris images, we propose an efficient method to tackle this problem. Firstly, the normalized iris image is divided into sub-regions according to the properties of iris textures. Local binary patterns (LBP) are then adopted for texture representation of each sub-region. Finally, Adaboost learning is performed to select the most discriminative LBP features for spoof detection. In particular, a kernel density estimation scheme is proposed to complement the insufficiency of counterfeit iris images during Adaboost training. The comparison experiments indicate that the proposed method outperforms state-of-the-art methods in both accuracy and speed.

1 Introduction

With the increasing demands of security in our daily life, iris recognition has rapidly become a hot research topic for its potential values in personal identification [1,2,3,4]. As shown in Fig. 1(a) and (b), the iris of a human eye is the annular part between the black pupil and white sclera. It displays rich texture that is commonly thought to be highly discriminative between eyes and stable over individuals' lifetime, which makes iris particularly useful for personal identification.

However, it must be aware that, as any other authentication technique, iris recognition is also possibly forged and illegally used [4]. Several potential iris counterfeits have already been considered, e.g., printed iris, re-played video, fake glass/plastic eye, printed contact lens, etc. Among these counterfeits, printed contact lens is commonly thought to be particularly dangerous [5,6]. For the sake of convenience, more and more people wear contact lens. Once they enrolled into an iris system without taking off their contact lens, a big concern rises: anyone who wears the same contact lens can be possibly mistaken as the authorized user even the imposter does not intend to (whereas other counterfeits, e.g. the printed iris, usually take an illegal initiative). Figure 1(c) shows one example contact lens and the resultant iris images by different eyes wearing it. We can see that although the individual live iris textures are quite different from each other, the contact lens wearing iris images look almost the same. It has been reported by several researchers [7,8] that "it is actually possible to spoof some iris recognition systems with

M. Tistarelli and M.S. Nixon (Eds.): ICB 2009, LNCS 5558, pp. 1080–1090, 2009.
© Springer-Verlag Berlin Heidelberg 2009

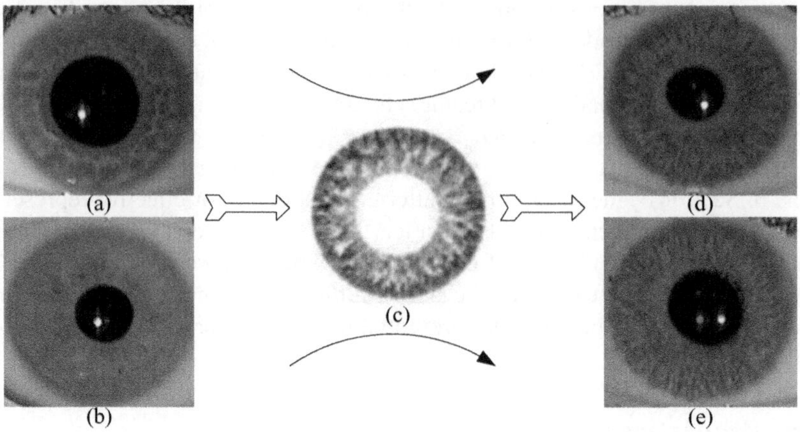

Fig. 1. Example of contact lens and the resultant iris images on different eyes. (a)-(b): The live iris images from two different eyes. (c) The contact lens. (d)-(e): The resultant iris images of (a) and (b) after wearing (c). Although the genuine live iris textures of the two eyes are different, their contact lens wearing iris images seem almost the same.

well-made contact lens" [5]. It is therefore important and desirable to detect the counterfeit iris (especially the contact lens) before recognition to avoid unnecessary losses.

A few researchers have contributed several methods for iris spoof detection. Some of them suggested spoof detection via turning on/off the illuminators followed by checking the responses on the resultant images (e.g., the pupil hippus [4], the specular spots on the cornea [8], and so on). For example, Lee et al. [8] proposed a fake iris detection scheme via investigating the specular spots of collimated IR-LED. Such illuminator based methods can be useful for printed iris or glass/plastic eyes, but tend to fail for contact lens. Moreover, they require additional hardware and have to capture a series of iris images for analysis, which inevitably increase the hardware cost and the recognition time.

Several software-based methods have also been proposed. Daugman [9] and Tan et al. [4] suggested detecting the printed iris via frequency analysis. The basic idea is to utilize the frequency characteristics of the printed iris due to the periodic dot printing. Obviously, this method is limited to printed iris detection. Recently, He et al. [5] proposed a contact lens detection method via statistical texture analysis. Four distinctive features based on gray level co-occurrence matrix (GLCM) are extracted. Support vector machine is used for classification. In [6], Wei et al. also proposed a texture analysis based scheme for contact lens detection. In their work, Iris-Textons are learned and used for texture representation. Both of the texture based methods achieved encouraging performance.

From the above description, we can conclude that the hardware-assisted methods utilize the physical characteristics of the pupil and eye, and are effective for printed iris, replayed video or glass eyes. While the texture based methods focus on textural differences between the live and counterfeit iris images, and are effective for contact lens detection. In this paper, we propose a more efficient texture based method for iris spoof detection (especially the contact lens detection). Firstly, we divide the valid part

of the iris into sub-regions according to the properties of iris textures. Local binary patterns (LBP) are then adopted for representing the statistical texture characteristics of each sub-region. Finally, the well-known Adaboost learning algorithm is performed to select the most discriminative LBP features for spoof detection.

The rest of the paper is organized as follows: in Section 2, we describe how the iris image is preprocessed for spoof detection purpose according to the properties of iris textures. In Section 3, the local binary patterns are adopted for effective representation of the iris texture. In Section 4, Adaboost learning is performed to select the most discriminative LBP features for spoof detection. In particular, we present a novel kernel density estimation scheme to tackle the lack of sufficient training counterfeit iris images. The experiments and discussions are presented in Section 5 prior to the conclusions in Section 6.

2 Iris Properties and Preprocessing

Although iris textures are commonly thought to be highly discriminative between eyes, they (including the contact lens wearing iris textures) still present several desirable common properties [10], such as:

1) *The radial distribution:* Even within an iris, the scale of the iris micro-structures varies a lot along the radius. Usually the larger the radius is, the bigger the iris micro-structures will be (see Fig. 2(a)).
2) *The angular self-similarity:* Although different angular regions remain discriminative, their texture patterns display a certain degree of consistence/correlation as shown in Fig. 2(b).

These properties suggest dividing the iris into multiple regions as shown in Fig. 2(c). We can see that each sub-region contains a particular texture pattern. Via such division, more specific representation of iris can be obtained, and hence makes it easier to discriminate live iris textures from counterfeit iris textures.

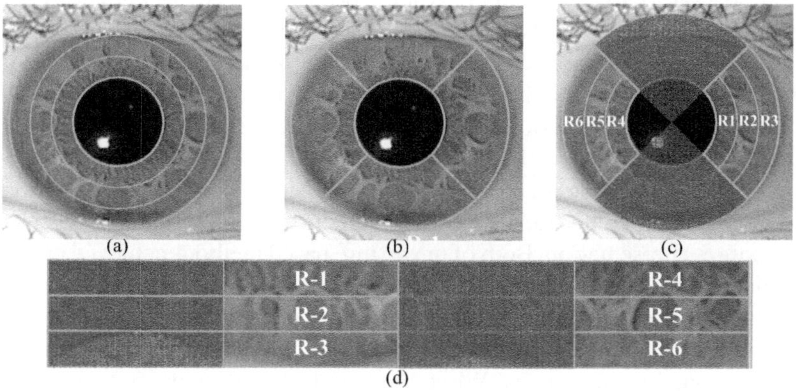

Fig. 2. The properties of iris textures and preprocessing of the iris image. After excluding the upper and lower quarters, the ROI of the iris is divided into six sub-regions.

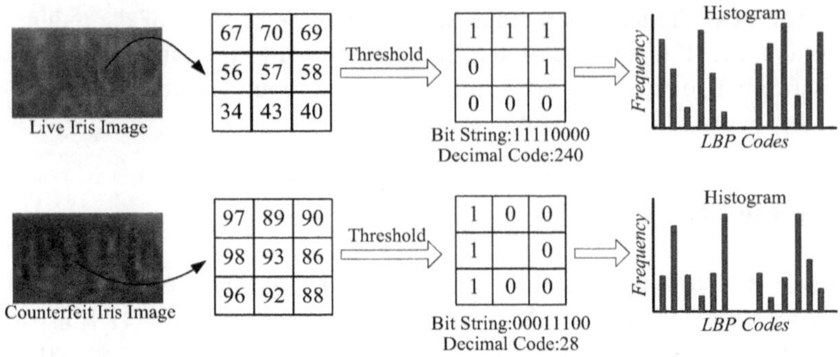

Fig. 3. Local binary patterns encoding. The LBP histograms of different textures are different, and therefore can be used as a texture descriptor.

Moreover, the upper and lower parts of the iris are almost always occluded by eyelids or eyelashes [11]. It is therefore straightforward to exclude the upper and the lower quarters from feature extraction for a more concise representation. And in order to achieve translation and scale invariance, the iris is normalized to a rectangular block of a fixed size of 64*512 [4]. The image preprocessing scheme is illustrated in Fig. 2. As can be seen, the whole iris image is divided into three annular sections along the radial direction, and two sectors along the angular direction. In total, we get six sub-regions.

3 Local Binary Pattern Encoding

As mentioned in Section 1, the textures of counterfeit iris images and live iris images are different in appearance and can be used for spoof detection. But how to encode such differences remains an open problem. Recently, the local binary patterns (LBP) [12] has emerged as an effective texture descriptor. Basically, LBP is defined for each pixel by thresholding its 3*3 neighborhood pixels with the center pixel value, and considering the result as a binary bit string, see Fig. 3. Each LBP code represents a type of micro image structure, and the distribution of them can be used as a texture descriptor [12]. The original LBP is later extended to multi-scale LBP (denoted by $LBP_{P,R}$) and uniform LBP (denoted by LBP^{u2}). $LBP_{P,R}$ is calculated by thresholding P equally spaced points on a circle (whose radius is R) with the center pixel value. A LBP code is called uniform if its bit string contains at most two bit-wise transitions from 0 to 1 or vice versa. LBP based methods have been proved to be successful in biometric texture representation, such as face [13] and iris [14].

In this work, we adopt multi-resolution LBPs (namely, $LBP^{u2}_{8,1}$, $LBP^{u2}_{8,2}$, $LBP^{u2}_{8,5}$, $LBP^{u2}_{8,7}$, $LBP^{u2}_{12,2}$, $LBP^{u2}_{12,3}$, $LBP^{u2}_{12,5}$, $LBP^{u2}_{16,3}$, $LBP^{u2}_{16,5}$, $LBP^{u2}_{16,7}$) for texture representation of each sub-region obtained in Section 2. The number of bins of them is (59, 59, 59, 59, 135, 135, 135, 243, 234, 234) respectively. As described in Section 2, we have 6 sub-regions, hence we will totally get (59+59+59+59+135+135+135+243+243+243) * 6=8220 possible LBP bins. Each bin represents the frequency of one type of micro image structures on one sub-region, and is considered as a candidate texture feature. A large pool of regional LBP features (LBP bins) is therefore generated. Definitely, this

feature pool must contain much redundant information because of the redundancy between different LBP features as well as that between different sub-regions. To learn the most discriminative regional LBP features from the redundant feature pool, we turned to the following Adaboost algorithm.

4 Adaboost Learning

Adaboost is a well-known machine learning algorithm that can select a small set of the most discriminative features from a candidate feature pool [10,15]. It is particularly efficient for binary (two-class) problems, and therefore is suitable for selecting the best LBP features for iris spoof detection.

4.1 Adaboost Learning

Given that $\{x_i, y_i\}_{i=1}^N, (x \in R^d, y \in \{+1, -1\})$ is N labeled training samples with associated weights $\{w(x_i)\}_{i=1}^N$; $\Phi = \{\phi_m(\cdot): R^d \rightarrow R\}_{m=1}^M$ is a candidate feature pool of x; our goal is to automatically learn a small set of the most discriminative features $\{\phi_t\}_{t=1}^T$ from the feature pool, and construct an ensemble classifier:

$$H(x) = \text{sign}\left(\sum_{t=1}^{T} h_t(\phi_t(x))\right) \quad (1)$$

where $h_t(s) : R \rightarrow R$ is a component classifier that can output a 'confidence' of x being a positive when $\phi_t(x)$ equals s [10]. In this work, x is an iris image and $\{\phi_m(x)\}_{m=1}^M$ corresponds to the LBP features (i.e., LBP bins).

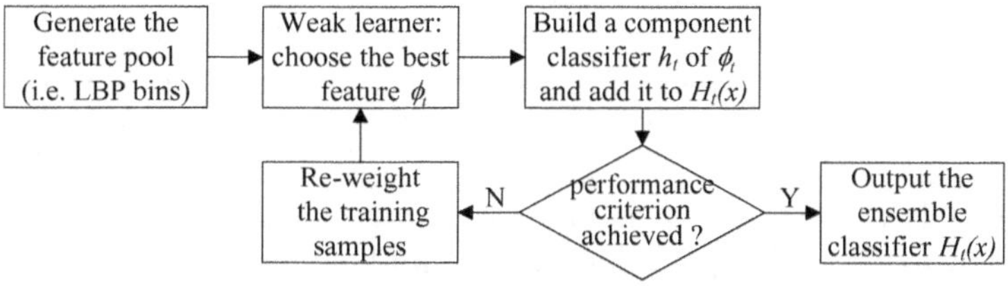

Fig. 4. A generalized framework of Adaboost Learning

The flowchart of Adaboost learning is depicted in Fig. 4 [10]. It begins with generating the feature pool on the training samples. After that, Adaboost repeatedly learns the component classifiers $h_t(\phi_t(\cdot))$ on the weighted versions of the training samples until the performance criterion is satisfied. Clearly, there are three key modules involved in Adaboost: the weak learner, the component classifier and the re-weighting function.

The weak learner is essentially the criterion for choosing the best feature (e.g., $\phi_t(\cdot)$) on the weighted training set. The component classifier h_t outputs a confidence score of x being a positive based on its ϕ_t value. The re-weighting function maintains a distribution over the training samples and updates it in such a way that the subsequent component

classifier can concentrate on the hard samples by giving higher weights to the samples that are wrongly classified by previous classifier.

Among various Adaboost algorithms, we choose the confidence-rated Adaboost learning [15] for its efficiency and simplicity. In confidence-rated Adaboost, the weak learner tries to find the feature that can maximizes the following criterion:

$$\phi_t = \arg\min_{\phi \in \Phi} 2 \sum_{j=1}^{N} \sqrt{P_w^+(\phi^j(x))\, P_w^-(\phi^j(x))} \tag{2}$$

where $P_w^+(\phi_m(x))$, $P_w^-(\phi_m(x))$ are the positive and negative probability distributions of $\phi_m(x)$ on the weighted training set (see Fig. 5). The corresponding component classifier is constructed as follows:

$$h_t(\phi_t) = \frac{1}{2} \ln \frac{P_w^+(\phi_t)}{P_w^-(\phi_t)} \tag{3}$$

The re-weighting function is as follows.

$$w_{t+1}(x_i) \leftarrow w_t(x_i) \exp(-y_i h_t(\phi_t(x_i))) \tag{4}$$

Please refer to [10] and [15] for more details of Adaboost learning.

4.2 Kernel Density Estimation for Counterfeit Iris Images

From Eq. 2 and Eq. 3 we can see that both the weak learner and the component classifier are dependent on the density distributions of positive and negative samples, which are estimated by histograms. More bins in the histogram give a more refined representation of the feature density distribution. However, when the training samples of one class are insufficient (e.g., due to the difficulty of collection), samples dropped into each bin is not enough for a stable estimation of the distribution of this class but just an ad-hoc one of the current training samples. As a result, the classifier learned based on the limited training samples will be ad-hoc, i.e., has low generalization capability and is sensitive to possible noise (see Eq. 3).

A possible solution to this problem is the kernel density estimation (KDE) [16]. The basic idea of kernel density estimation is that: if the feature value of one training sample is x, it is highly possible that there exist several similar samples whose feature values are around x. Suppose $p(x)$ is the probability density of a LBP feature, and we wish to estimate $p(x)$ via a random sample set $x_1, x_2, \ldots, x_N$. In KDE, $p(x)$ is estimated as follows:

$$\hat{p}(x) = \frac{1}{N} \sum_{n=1}^{N} \frac{1}{h} k\left(\frac{x - x_n}{h}\right) \tag{5}$$

where $k\left(\frac{x-x_n}{h}\right)$ is a kernel of width h. From Eq. 5, we can see that the density on x is interpreted as the sum of N local kernels centered on the N data points x_n.

For the sake of smoothness, a popular choice for $k\left(\frac{x-x_n}{h}\right)$ is the Gaussian, which gives rise to the following kernel density model [16]:

$$\hat{p}(x) = \frac{1}{N} \sum_{n=1}^{N} \frac{1}{(2\pi h^2)^{1/2}} exp\left\{-\frac{|x - x_n|^2}{2h^2}\right\} \tag{6}$$

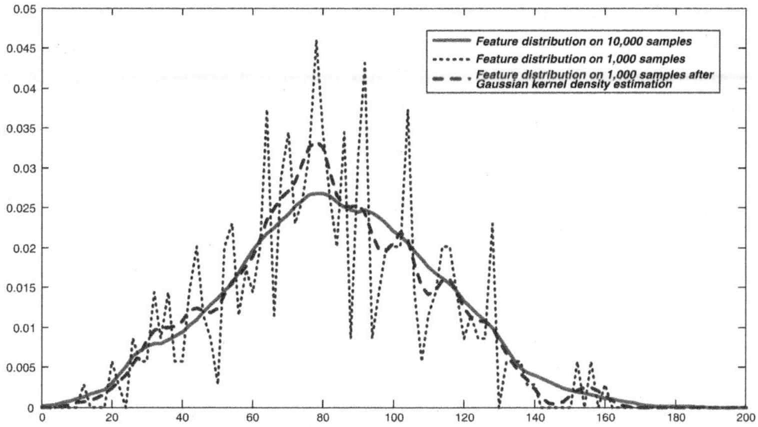

Fig. 5. The distributions of $LBP_{8,2}^{u2}('1001011')$ on 10, 000 iris images, 1, 000 iris images, and the Gaussian kernel density estimation (KDE) version of the 1, 000 one respectively. We can see that although the 1, 000 distribution is quite different to the 10, 000 one, the Gaussian KDE one looks much similar with the 10, 000 one.

where h denotes the standard deviation of the Gaussian kernel. Thus our density model is obtained by placing a Gaussian over each data point and then adding up the contributions over the whole data set. Clearly, the contribution (weight) of x_n decreases while its distance from x increasing.

The usefulness of Gaussian kernel density estimation is illustrated in Fig. 5, where the distribution obtained on 10,000 iris images is surprisingly similar with the distribution obtained on only 1,000 iris images with Gaussian kernel density estimation. This indicates that: although we cannot obtain the genuine distribution of 10, 000 iris images (due to the difficulty of collection), at least we can estimate a much closer one to it via Gaussian kernel density estimation, which, as demonstrated in our experiments, efficiently complements the lack of sufficient training counterfeit iris samples. A notable point in Gaussian KDE is the setting of h which controls the trade-off between having $\hat{p}(x)$ close to the data (at small h) and having $\hat{p}(x)$ smooth (at large h). Experimental results show that $h = 1.8$ is a good choice for this work.

5 Experimental Results

5.1 The Data Set

Experiments are performed to evaluate the usefulness of the proposed method. Due to the absence of public counterfeit iris image database, we manually collected 600 counterfeit iris images. While a minority of these iris images come from printed iris and glass eye, the majority of them are printed color contact lens iris images since this work focuses on contact lens detection. In detail, this counterfeit iris database consists of 20 kinds of different contact lens, with varying textures printed onto them. Some of the counterfeit iris images are shown in Fig. 6. We can see that some of them are difficult even for human to decide.

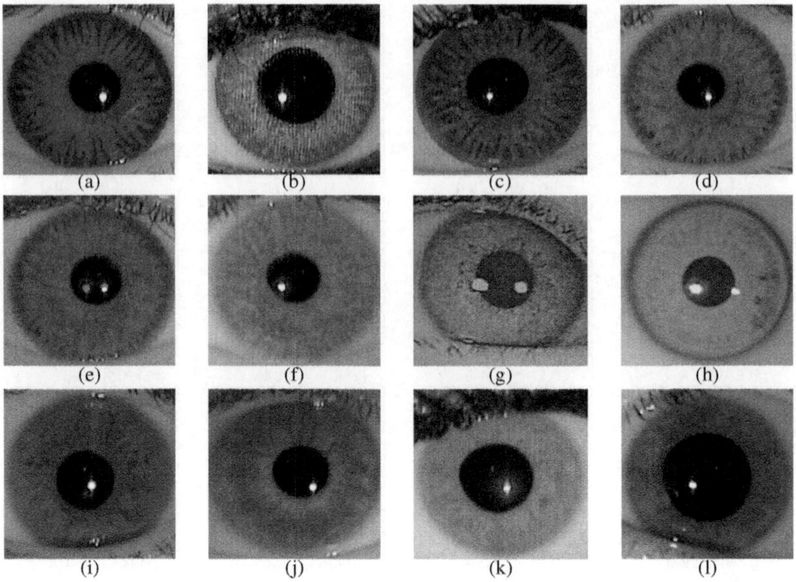

Fig. 6. Examples of training samples. (a)-(f): Contact lens wearing iris images. (g) Printed iris. (h) Glass eye. (i)-(l): Live iris images. We can see that some of the samples are difficult even for human to decide.

Although there are not sufficient counterfeit iris images, fortunately, we can collect sufficient live iris images. This partly complements the potential performance degradation due to the absence of sufficient counterfeit iris images. The live iris images are randomly selected from two well-known iris image databases, namely CASIA-Iris-V3 [17] and the ICE v1.0 [18]. In total, about 10, 000 live iris images are collected, which can cover almost all kinds of textures of the live irises.

5.2 Adaboost Learning

300 counterfeit iris images and 6000 live iris images are randomly selected from the above data set for Adaboost training. As described in Section 3, 8220 LBP features (i.e., LBP bins) are generated for Adaboost learning (during which Gaussian kernel density estimation is applied to complement the insufficiency of counterfeit iris images). The learned Adaboost classifier contains only 85 features, i.e., 85 LBP bins. The first twelve selected LBP features are shown in Fig. 7. We can see that the selected LBP features are in different sub-regions and different scales, which indicates the successfulness of iris division and multi-resolution LBP analysis. Moreover, many different LBP patterns are learned to represent the rich textures of the live and counterfeit iris images. We achieve 100% correction on the training set thanks to the aggressive learning ability of Adaboost [10]. But a big concern is its generalization capability on the test set.

5.3 Test Results and Discussions

The learned Adaboost classifier is tested on the remaining iris images. It is interesting to compare our method with the methods of He [5] and Wei [6]. Table 1 shows the

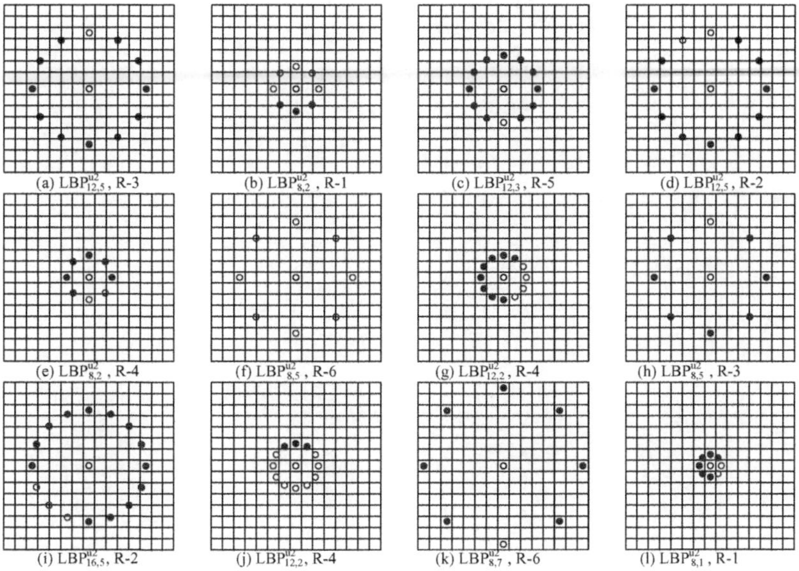

Fig. 7. The first twelve LBP features learned by Adaboost. Note that the learned LBP patterns are in different sub-regions and different scales.

experimental results, where FAR (False Accept Rate), FRR (False Reject Rate) and speed are presented. We can see that the proposed method outperforms the other methods both in accuracy and speed. The encouraging accuracy is perhaps due to four reasons:

1. The local binary patterns are effective in representing the iris textures.
2. The division of the iris into sub-regions enables a more specific representation of the iris textures.
3. The Adaboost learning is efficient in learning the most discriminative features for spoof detection.
4. The proposed Gaussian kernel density estimation scheme partly complements the insufficiency of counterfeit iris images, and increases the generalization capability of the learned Adaboost classifier.

Clearly, the fast execution is due to the computational simplicity of the LBP compared with the calculation of GLCM [5] or iris-textons [6].

Table 1. Overall performance of the learned classifiers via [5], [6] and the proposed method

Algorithm	FAR (%)	FRR (%)	Speed(ms)
He [5]	4.33	6.84	230
Wei [6]	3.67	6.91	340
Proposed	0.67	2.64	160

6 Conclusions

In this paper, we propose a texture analysis based method for efficient iris spoof detection (especially for contact lens detection). The basic idea is the textural differences between counterfeit iris images and the live iris images. Local binary patterns are adopted for representing the textural characteristics of local sub-regions, and Adaboost learning (together with Gaussian kernel density estimation) is performed to select the most discriminative LBP features for spoof detection. Extensive experiments indicate that the proposed method can be well adapted for iris spoof detection.

Acknowledgement

This work is supported by research grants from the National Basic Research Program (Grant No. 2004CB318110), the Natural Science Foundation of China (Grant No. 607-23005, 60736018, 60702024), NLPR 2008NLPRZY-2, the National Hi-Tech Research and Development Program of China (2006AA01Z193, 2007AA01Z162).

References

1. Jain, A.K., Ross, A., Prabhaker, S.: Jain, Arun Ross, and Salil Prabhaker. An introduction to biometric recognition. IEEE Trans. on Circuits and Systems for Video Technology 14(1), 4–20 (2004)
2. Daugman, J.: How iris recognition works. IEEE Trans. On Circuits and Systems for Video Technology 14(1), 21–30 (2004)
3. Wildes, R.: Iris recognition: An emerging biometric technology. Proceedings of the IEEE 85, 1348–1363 (1997)
4. Ma, L., Tan, T., Wang, Y., Zhang, D.: Personal identification based on iris texture analysis. IEEE Trans. on Pattern Analysis and Machine Intelligence 25(12), 1519–1533 (2003)
5. He, X., An, S., Shi, P.: Statistical texture analysis based approach for fake iris detection using support vector machine. In: Proc. of Int'l Conf. on Biometrics 2007, pp. 540–546 (2007)
6. Wei, Z., Qiu, X., Sun, Z., Tan, T.: Counterfeit iris detection based on texture analysis. In: Proc. of IEEE Int'l Conf. on Pattern Recognition (ICPR 2008) (2008)
7. Daugman, J.: Iris recognition and anti-spoof countermeasures. In: Proc. of the 7th Int'l Biometrics Conference (2004)
8. Lee, E.C., Park, K.R., Kim, J.: Fake iris detection by using purkinje image. In: Zhang, D., Jain, A.K. (eds.) ICB 2005. LNCS, vol. 3832, pp. 397–403. Springer, Heidelberg (2005)
9. Daugman, J.: Demodulation by complex-valued wavelets for stochastic pattern recognition. Intl. Journal of Wavelets, Multi-resolution and Information Processing 1, 1–17 (2003)
10. He, Z., Tan, T., Sun, Z., Qiu, X.C.: Boosting ordinal features for accurate and fast iris recognition. In: Proc. of IEEE Conf. on Computer Vision and Pattern Recognition (CVPR 2008) (2008)
11. He, Z., Tan, T., Sun, Z., Qiu, X.C.: Towards accurate and fast iris segmentation for iris biometrics. IEEE Trans. on Pattern Analysis and Machine Intelligence (accepted, 2008)
12. Ojala, T., Pietikäinen, M., Mäenpää, T.: Multiresolution gray-scale and rotation invariant texture classification with local binary patterns. IEEE Trans. on Pattern Analysis and Machine Intelligence 24(7), 971–987 (2002)

13. Ahonen, T., Hadid, A., Pietikäinen, M.: Face description with local binary patterns: Application to face recognition. IEEE Trans. on Pattern Analysis and Machine Intelligence 28(12), 2037–2041 (2006)
14. Sun, Z., Tan, T., Qiu, X.: Graph matching iris image blocks with local binary pattern. In: Proc. of 1st Int'l Conf. on Biometrics, Hong Kong, pp. 366–372 (2006)
15. Schapire, R.E., Singer, Y.: Improved boosting algorithms using confidence-rated predictions. Machine Learning 37, 297–336 (1999)
16. Bishop, C.M.: Pattern Recognition and Machine Learning. In: Information Science and Statistics, ch. 6, pp. 291–324. Springer, New York (2006)
17. Chinese Academy of Sciences Institute of Automation. CASIA Iris Image Database Version 3.0, `http://www.cbsr.ia.ac.cn/IrisDatabase.htm`
18. Ice v1.0 iris image database, iris challenge evaluation (ice), `http://iris.nist.gov/ice/`

Custom Design of JPEG Quantisation Tables for Compressing Iris Polar Images to Improve Recognition Accuracy*

Mario Konrad[1], Herbert Stögner[1], and Andreas Uhl[1,2]

[1] School of Communication Engineering for IT, Carinthia Tech Institute, Austria
[2] Department of Computer Sciences, Salzburg University, Austria
uhl@cosy.sbg.ac.at

Abstract. Custom JPEG quantisation matrices are proposed to be used in the context of compressing iris polar images within iris recognition. These matrices are obtained by employing a Genetic algorithm for the corresponding optimisation. Superior matching results in iris recognition in terms of average Hamming distance and improved ROC are found as compared to the use of the default JPEG quantisation table.

1 Introduction

With the increasing usage of biometric systems the question arises naturally how to store and handle the acquired sensor data. In this context, the compression of these data may become imperative under certain circumstances due to the large amounts of data involved. Among other possibilities (e.g. like compressed template storage on IC cards), compression technology may be applied to sample data in two stages of the processing chain in classical biometric recognition for example: Transmission of sample data after sensor data acquisition and optional storage of (encrypted) reference data in template databases.

The distortions introduced by lossy compression artifacts usually interfere with subsequent feature extraction and may degrade the matching results. In particular, FRR or FNMR will increase (since features of the data of legitimate users are extracted less accurately from compressed data) which in turn affects user convenience and general acceptance of the biometric system. In extreme cases, even FAR or FMR might be affected.

In this work, we will focus on the lossy compression of iris polar images using the JPEG standard. We discuss the use of custom quantisation matrices in order to reflect the specific properties of iris imagery. We will apply a biometric iris recognition systems to the compressed sensor data to evaluate the effects of compression on recognition accuracy.

In Section 2, we will review and discuss the available literature on biometric sample data compression with focus on iris data storage. Section 3 is the main part of this work where we discuss properties of iris imagery and present several

* This work has been supported by the Austrian Science Fund, project no. L554-N15.

M. Tistarelli and M.S. Nixon (Eds.): ICB 2009, LNCS 5558, pp. 1091–1101, 2009.
© Springer-Verlag Berlin Heidelberg 2009

variants of custom JPEG quantisation tables (designed in order to hopefully improve recognition accuracy). In section 4 we first describe the employed iris recognition system and the data this algorithm are applied to. Subsequently we discuss our experimental results with respect to the observed improvements of recognition accuracy. Finally, we describe the Genetic algorithm approach which has been employed for optimising the desired quantisation matrices. Section 5 concludes the paper.

2 Iris Image Compression

ISO/IEC 19794-6 allows iris image data to be stored in lossy manner in the JPEG and JPEG2000 formats. Two types of iris image data are considered: rectilinear images (i.e. images of the entire eye) and polar images (which are basically the result of iris detection and segmentation), the latter much smaller in terms of storage requirement (e.g. 2kB vs. 25-30kB for rectilinear images).

In previous work on compressing iris imagery, rectangular [1,2,3,4] as well as polar [5] iris sample data has been considered. With respect to employed compression technology, we find JPEG [2,3,4], JPEG2000 [1,5,2,3,4], and other general purpose compression techniques [3,4] being investigated. Superior compression performance of JPEG2000 over JPEG is seen especially for low bitrates, however, for high and medium quality JPEG is still an option to consider. While the data formats specified by the ISO/IEC 19794 standard are fixed at present state, their customised use tailored to a specific target modality and the corresponding impact on recognition accuracy as compared to the default settings has not been investigated. In the subsequent study we apply JPEG as covered by ISO/IEC 19794-6 to polar iris images and propose to use custom quantisation tables (Q-tables) adapted to properties of iris imagery.

In some application settings, the requirement for compression technology is caused by low power (mobile) sample acquisition devices which are too weak to conduct feature extraction on board and therefore need to transmit sample data to a remote feature extraction (and matching) module. In this context, it of course makes more sense to apply JPEG instead of JPEG2000 due to its much lower computational demand. In addition to this, applying compression to polar iris images minimises the amount of data to be transmitted (since polar images are smaller by several orders of magnitude as compared to rectangular iris images even without compression applied). Also, this strategy avoids the iris detection process being fooled by compression artifacts as it may be the case when the iris needs to be detected in compressed rectangular iris images. A drawback of the approach relying on polar images is that the acquisition device needs to perform iris detection and the generation of the iris texture patch (i.e. the polar iris image) which involves data interpolation or extrapolation. In any case, the bandwidth required for transmission of sample data is minimised by employing compressed polar iris data.

In [6] compression algorithms tuned for application in the pattern recognition context are proposed, which are based on the modification of the standard compression algorithms: This is done by emphasising middle and high frequencies

and discarding low frequencies (the standard JPEG Q-table is rotated by 180 degrees). JPEG Q-table optimisation has already been considered in biometrics – [7] employ a rate/distortion criterion in the context of face recognition and achieve superior recognition performance as compared to the standard matrix.

3 Custom JPEG Quantisation

The JPEG still image compression standard [8] allows to use custom Q-tables in case image material with special properties is subject to compression. These tables are signalled in the header information. The default Q-tables (see Table 1) have been designed with respect to psychovisual optimality employing large scale experimentation involving a high number of test subjects. There are two reasons which suggest to use different Q-tables as the default configuration: First, iris imagery might have different properties as compared to common arbitrary images, and second, a pleasant viewing experience as being the aim in designing the default tables, might not deliver optimal matching results in the context of biometric recognition (e.g. sharp edges required for exact matching could appear appealing to human observers).

Therefore, as a first stage, we have investigated iris imagery in more detail. 8x8 pixel image blocks have been subjected to DCT transform and the resulting coefficients are averaged for a large number of blocks (i.e. 2000 and 525 blocks for the two types of imagery, respectively). As a first class of blocks, we have used arbitrary images and blocks are extracted randomly. The second class of blocks is taken from polar iris images. Fig. 1 displays the result of both classes where the DC and the largest AC coefficient are set to white, zero is set to black and the remaining values are scaled in between (note that the logarithm is applied to the magnitude of all coefficients before this scaling operation).

The arbitrary blocks (Fig. 1.a) show the typical expected behaviour with decreasing coefficient magnitude for increasing frequency and symmetry with

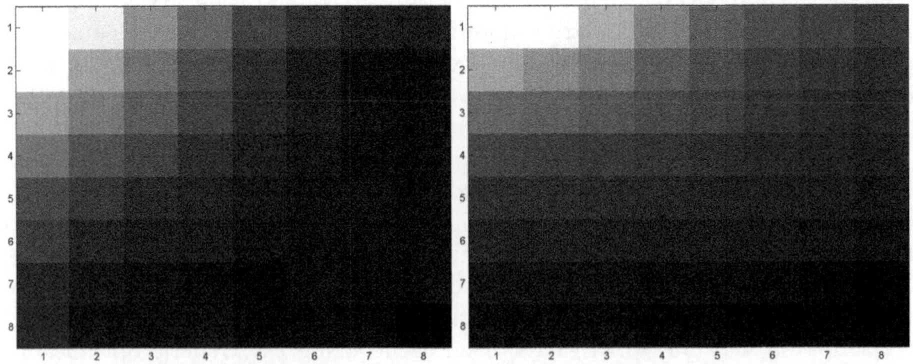

(a) arbitrary blocks (b) polar iris blocks

Fig. 1. Averaged 8x8 DCT blocks

respect to the coordinate axes. Fig. 1.b reveals that in polar iris images there is more energy in the higher frequencies in horizontal direction as compared to vertical direction. This is to be expected since luminance fluctuations in iris texture are more pronounced in radial direction as compared to perpendicular direction.

While we may exploit the direction bias of iris texture in compression directly, we additionally conjecture that the highest and medium frequencies might not be required for the matching stage due to the coarse quantisation used for template generation while at least medium frequencies are required for pleasant viewing. Table 1 displays the Q-tables used in our experiments.

Table 1. JPEG Quantisation tables: STQ, Qtable22, Qtable24 (first line), QTOptk05, and QTOptk10 (second line)

16	11	10	16	24	40	51	61
12	12	14	19	26	58	60	55
14	13	16	24	40	57	69	56
14	17	22	29	51	87	80	62
18	22	37	56	68	109	103	77
24	35	55	64	81	104	113	92
49	64	78	87	103	121	120	101
72	92	95	98	112	100	103	99

10	10	76	255	255	255	255	255
85	112	255	255	255	255	255	255
151	255	255	255	255	255	255	255
255	255	255	255	255	255	255	255
255	255	255	255	255	255	255	255
255	255	255	255	255	255	255	255
255	255	255	255	255	255	255	255
255	255	255	255	255	255	255	255

16	11	10	16	255	255	255	255
12	12	14	255	255	255	255	255
255	255	255	255	255	255	255	255
255	255	255	255	255	255	255	255
255	255	255	255	255	255	255	255
255	255	255	255	255	255	255	255
255	255	255	255	255	255	255	255
255	255	255	255	255	255	255	255

16	11	10	16	24	246	255	255
12	12	14	29	26	255	255	250
14	13	16	24	255	255	255	224
14	17	22	255	255	255	242	255
18	255	255	255	255	255	255	255
24	247	255	255	255	255	255	255
255	255	255	255	255	255	255	255
255	255	255	241	255	241	255	244

15	6	17	19	255	255	255	255
5	15	13	255	255	255	250	255
255	255	247	255	248	255	255	255
255	250	248	255	255	250	255	255
255	222	237	255	251	255	255	250
255	252	251	250	220	249	229	232
254	246	255	251	255	255	255	248
255	255	247	252	255	255	248	255

The first matrix shows the standard case (STQ) where the entries exhibit a steady increase from low frequencies to high frequencies following the well-known zig-zag pattern [8] (which results in more severe quantisation applied to middle and high frequencies).

Qtable22 and Qtable24 have been obtained by large scale trial and error experimentation, setting a large amount of entries to 255 (which causes the corresponding coefficients to be divided by 255 and results in most of them being quantised to zero). Both matrices are asymmetric in the sense that they "protect" more coefficients in horizontal direction (which have been shown to carry more energy as their vertical counterparts in Fig. 1.b), Qtable24 is more pronounced in this respect and retains the values of STQ at the positions not set to 255. The rationale behind the selection of these matrices is to investigate the importance of medium frequency information in the iris recognition process (high frequency information is assumed to be not useful in any case) and to reflect the specific properties of polar iris images.

QTOptk05 and QTOptk10 have been found using the Genetic optimisation approach as described in Section 4.3 using Qtable22 and Qtable24 as individuals of the initial population in addition to randomly generated tables. These tables have been specifically optimised for application with compression rates 5 and 10, respectively.

4 Experimental Study

4.1 Setting and Methods

Iris Recognition System. The employed iris recognition system is Libor Masek's Matlab implementation[1] of a 1-D version of the Daugman iris recognition algorithm. First, this algorithm segments the eye image into the iris and the remainder of the image. Iris image texture is mapped to polar coordinates resulting in a rectangular patch which is denoted "polar image". In case compression is required, it is applied to the polar image in this stage of the procedure. After extracting the features of the iris (which are strongly quantised phase responses of complex 1-D Gabor filters in this case), considering translation, rotations, and disturbed regions in the iris (a noise mask is generated), the algorithm outputs the similarity score by giving the Hamming distance between two extracted templates.

Sample Data. For all our experiments we considered 320x280 pixel images with 8-bit grayscale information per pixel from the CASIA[2] 1.0 iris image database. Note that the fact that the pupil area has been manipulated in these data [9] does not affect our results since we restrict compression to the iris texture area only by compressing polar iris images. We applied the experimental calculations on the images of 50 persons using 3 - 4 images for each eye (i.e. 334 images).

Fig. 2 shows examples of iris templates extracted from uncompressed (first line) and JPEG compressed iris texture patches of one person. Second and third patch (second line) are compressed with rate 10 using STQ and QTOptk10, respectively. By analogy, fourth and fifth patch (third line) are compressed with rate 15 using STQ and Qtable22, respectively.

Note that the images have been scaled in y-direction for proper display, the original dimension is 240×20 pixels.

In the iris texture data (not shown), compression artifacts are clearly visible for both rates displayed, however, the STQ compressed variants are visually more close to the original and seem to have conserved the texture details much better. However, when computing the Hamming distance between both variants having applied compression rate 15 with the uncompressed second image of the same eye in the database, we result in 0.327 for STQ but only 0.317 for Qtable22. Obviously, the Hamming distance between templates does not reflect the visual appearance at all. The smoothing achieved by compression seems to play an important role indeed. More fine grained differences seem to get introduced by the STQ quantisation, while the other two matrices tend to produce rather smooth templates as well.

Compression can be used in various stages of the recognition/matching process. Either the stored reference data may be in compressed form, or the sample data acquired for verification may be compressed (e.g. for efficient transfer), or

[1] `http://www.csse.uwa.edu.au/~pk/studentprojects/libor/sourcecode.html`

[2] `http://www.sinobiometrics.com`

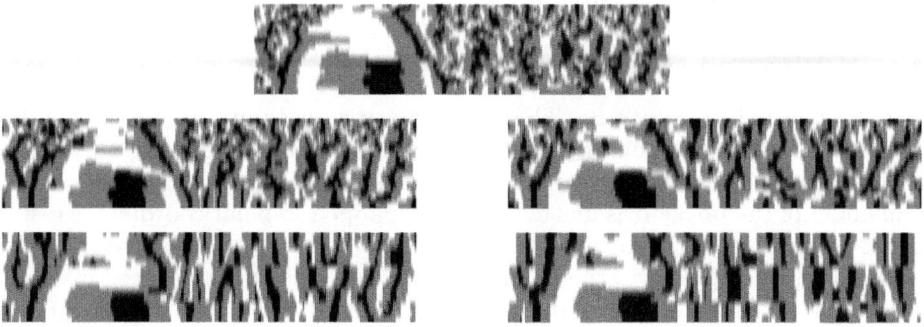

Fig. 2. Comparison of iris templates extracted from uncompressed (top line)/ compressed polar iris images with compression rates 10 (left column) and 15 (right column)

both. Therefore, we use two settings in our experiments: either both images are compressed and matched against each other or only one image is compressed in the matching stage. For investigating correct matches (matches from legitimate users enrolled in the database), we rely on more than 12000 generated images (i.e. 826 images for genuine user matches per compression rate; considering the 15 different compression rates applied (rates 2 - 16), we finally result in 12390 overall images considered). This is only true in the scenario with only 1 compressed image, for 2 compressed images this number is half-ed due to symmetry reasons. For investigating matches between different persons (imposter matches), far more data is available of course (109304 impostor matches are considered for a single rate).

4.2 Experimental Results

First, we investigate the impact of compression on the matching score (i.e. obtained Hamming distance (HD)). Fig. 3 shows the plots of the HD after applying the iris recognition algorithm if iris polar images have been JPEG compressed in the case of genuine user matches. The x-axis shows the compression rates, whereas the y-axis shows the averaged Hamming distance. For reference, we have included the average HD for the case of uncompressed images as horizontal dashed line with circles (as it is the case for all subsequent plots).

The mean value of the HD in the uncompressed case is approximately 0.3. First we consider the standard Q-table (labelled STQ). For increasing compression rate the average HD increases steadily and crosses the suggested matching threshold of 0.34 at compression rates 12 for both cases (one or two images compressed, respectively). Note that the reported numbers refer to averaged HD values which implies the occurrence of a significant number of false negative matches at this compression rate.

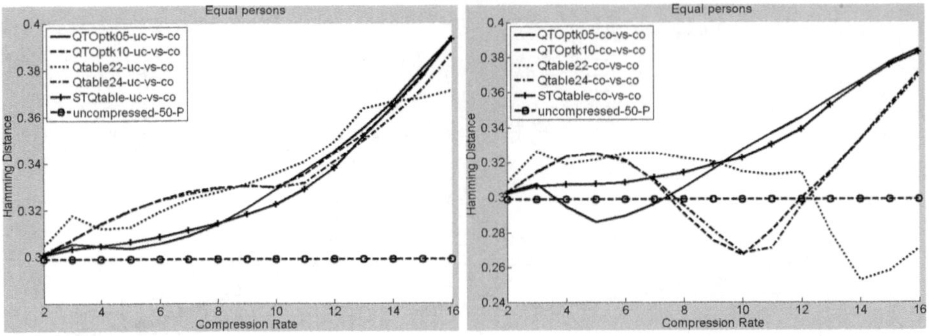

(a) one image compressed (b) both images compressed

Fig. 3. Impact of varying compression rate on HD of genuine users' matches

Concerning the one compressed image scenario (Fig. 3.a), STQ is beaten by QTOptk05 (rates around 6) and Qtable24/Qtable22 (rates 14 and higher), but only by a very small amount.

The situation is different when regarding the two compressed images scenario (Fig. 3.b). QTOptk05 is clearly better than STQ between rate 4 and 8 (and even beats the uncompressed case between rate 4 and 7). Qtable24 and QTOptk10 are better than STQ for rates higher than 7 and also beat the uncompressed case between rate 8 and 11. Finally, Qtable22 beats STQ for rates higher than 10 and is also superior to the uncompressed case for rates 13 and higher.

Next, we focus on rate/distortion performance in terms of PSNR. Figure 4.a shows the averaged rate distortion comparison of JPEG applied to all iris images for the five Q-tables considered. As it is the case for all subsequent plots, the solid graph depicted with crosses shows the results of the standard matrix (STQ). Some interesting results may be observed. First, Qtable24 behaves similarly to QTOptk10 and both exhibit PSNR values clearly above STQ for compression rates larger than 9 – up to 2dB difference may be observed. QTOptk05 is slightly above STQ between rate 5 and rate 8, but the improvement seen is only up to 0.4 dB. Qtable22 shows very fluctuating behaviour for low compression rates, but significantly outperforms STQ for rates larger than 11, up to 2.5dB improvement is found especially for higher rates.

Interestingly, we find that PSNR behaviour is highly dependent on the rate considered and all investigated quantisation matrices are able to outperform STQ in a certain range considerably.

These results indicate that PSNR is indeed a good predictor for matching performance with two compressed iris images in terms of average Hamming distance, but **NOT** in the case of only one image being compressed. The claim that compression up to a rate of 16 even improves the matching scores of not compressed images [5] can be supported at least for the 2 compressed images case for certain "better" Q-tables in distinct ranges of compression rate. This fact is remarkable and may be explained by the fact that compression acts as a kind of low pass filter resulting in denoised and smoothened images which can be matched better as compared to the original noisy counterparts.

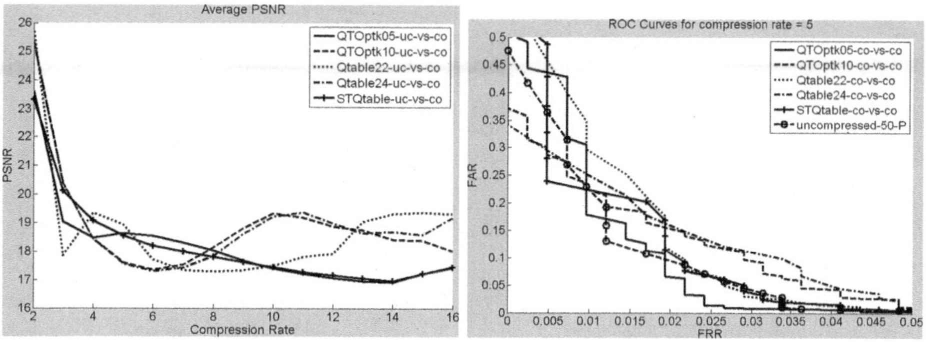

(a) PSNR (b) ROC, Rate 5

Fig. 4. Rate/distortion and ROC performance

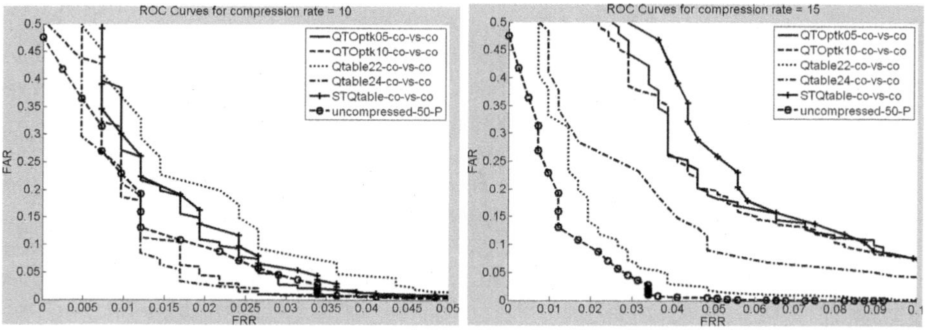

(a) rate 10 (b) rate 15

Fig. 5. ROC at different rates

In order to consider the hidden statistical outliers in the comparisons and to use a quantity often employed in the assessment of biometric system performance, we focus on the receiver operating characteristic (ROC) by computing and plotting the false rejection rate (FRR) against the false acceptance rate (FAR) for different compression rates.

Figs. 4.b to 5 compare the ROC of the different Q-tables for compression rates 5, 10, and 15 (it does not seem to be realistic to operate the iris recognition system at a higher compression rate due to the low visual quality of the images – see Fig. 2). We focus on the two compressed image scenario since the effects observed are identical to the one compressed image case but are seen in a more pronounced manner.

For compression rate 5 (see Fig. 4.b), the proposed QTOptk05 is able to improve the uncompressed and STQ ROC at $FAR < 0.1$ and $FRR > 0.02$. Note also that STQ does hardly outperform the uncompressed case, while this is observed for QTOptk05.

In the case of compression rate 10 (see Fig. 5.a), the situation changes drastically. Now, Qtable24 and QTOptk10 improve over the uncompressed case at $FAR < 0.1$ and $FRR > 0.015$, while QTOptk05 now performs almost equally to STQ and clearly inferior to the uncompressed case.

Finally, when turning to compression rate 15 the situation is again different (Fig. 5.b): now the uncompressed ROC is better as all compressed variants. However, Qtable22 is rather close to the corresponding curve. When comparing ROC to the STQ case, we clearly observe that the customised tables significantly improve over STQ in the entire range displayed in the plots.

There is one more interesting thing to note: STQ performs worst of all investigated matrices. With this rather high compression rate, Qtable22 offers the possibility to actually use the recognition algorithm whereas for STQ the ROC behaviour is actually too poor to be applied in any practical setting.

4.3 Genetic Algorithm Optimisation Approach

We have employed a Genetic algorithm (GA) to generate the two matrices QTOptk05 and QTOptk10 as follows. The Q-table entries (we have restricted the values to be integers from the interval [0, 255]) constitute the genes of each individual, where an individual represents a distinct Q-table. The population size is set to 10 individuals and we limit the number of computed generations to 40 due to reasons of computational demand. Additionally, the optimisation is stopped if no improvement in terms of best individual fitness function is found for 10 generations. In each generation, the two best individuals are selected to be the "elite" and are kept for the subsequent generation. Six individuals are selected for crossover, while two individuals of the subsequent generation are created by mutation.

As the cost function to be evaluated for determining which individuals are to be kept, we compute the sum of the following items for a fixed compression rate: averaged genuine users Hamming distance, average of FAR over a selected

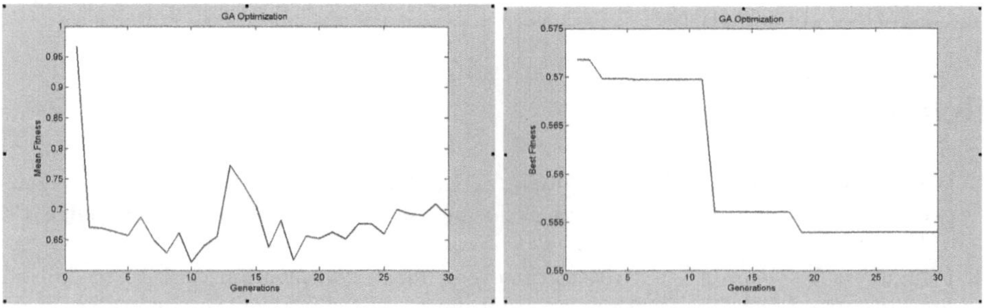

(a) mean per generation (b) best in generation

Fig. 6. Cost function development in the GA (compression rate 10)

number of thresholds, and average of FRR over the same set of thresholds. This costfunction has to be minimised of course.

Figure 6 shows the development of the cost function values for two cases: the mean of the fitness computed over each generation and the best fitness value in each generation. The mean fitness function value is not further improved after an initial "jump", while the best value is improved in several steps until saturation is reached after 30 generations and the GA stops. Note that we have used well performing tables like Qtable22 and Qtable24 as parts of the initial population in addition to randomly generated individuals.

We speculate that a higher mutation rate and a more disruptive crossover strategy might lead to even better results and will conduct further experiments in this direction.

5 Conclusion and Future Work

We have found that custom designed quantisation tables in JPEG can improve matching results in terms of average HD and ROC behaviour significantly as compared to the default tables. This effect is more pronounced for higher compression rates and for the scenario where both images involved in matching are compressed. Moreover it has turned out that these custom matrices need to be optimised with respect to a specific target bitrate – significant improvements are only found within the bitrange the table has been optimized for.

In future work we will consider additional alternative iris recognition algorithms in order to identify possible interference between compression technique and iris recognition system. Furthermore we will further optimise GA parameters in order to determine the ideal configuration.

Acknowledgements

Most of the work described in this paper has been done in the scope of the "Project I Lab" in the master program on "Communication Engineering for IT" at Carinthia Tech Institute.

References

1. Ives, R.W., Broussard, R.P., Kennell, L.R., Soldan, D.L.: Effects of image compression on iris recognition system performance. Journal of Electronic Imaging 17, 011015 (2008)
2. Daugman, J., Downing, C.: Effect of severe image compression on iris recognition performance. IEEE Transactions on Information Forensics and Security 3(1), 52–61 (2008)
3. Matschitsch, S., Tschinder, M., Uhl, A.: Comparison of compression algorithms' impact on iris recognition accuracy. In: Lee, S.-W., Li, S.Z. (eds.) ICB 2007. LNCS, vol. 4642, pp. 232–241. Springer, Heidelberg (2007)

4. Jenisch, S., Lukesch, S., Uhl, A.: Comparison of compression algorithms' impact on iris recognition accuracy II: revisiting JPEG. In: Proceedings of SPIE, Security, Forensics, Steganography, and Watermarking of Multimedia Contents X, San Jose, CA, USA, vol. 6819, p. 68190M (January 2008)
5. Rakshit, S., Monro, D.: Effects of sampling and compression on human iris verification. In: Proceedings of the IEEE International Conference on Acustics, Speech, and Signal Processing (ICASSP 2006), Tolouse, France, pp. II–337–II–340 (2006)
6. Chen, M., Zhang, S., Karim, M.: Modification of standard image compression methods for correlation-based pattern recognition. Optical Engineering 43(8), 1723–1730 (2004)
7. Jeong, G.M., Kim, C., Ahn, H.S., Ahn, B.J.: JPEG quantization table design for face images and its application to face recognition. IEICE Transactions on Fundamentals of Electronics, Communications and Computer Science E69-A(11), 2990–2993 (2006)
8. Pennebaker, W., Mitchell, J.: JPEG – Still image compression standard. Van Nostrand Reinhold, New York (1993)
9. Philips, P., Bowyer, K., Flynn, P.: Comments on the CASIA version 1.0 iris data set. IEEE Transactions on Pattern Analysis and Machine Intelligence 29(10), 1869–1870 (2007)

Improving Compressed Iris Recognition Accuracy Using JPEG2000 RoI Coding

J. Hämmerle-Uhl, C. Prähauser, T. Starzacher, and A. Uhl

Department of Computer Sciences, Salzburg University, Austria
uhl@cosy.sbg.ac.at

Abstract. The impact of using JPEG2000 region of interest coding on the matching accuracy of iris recognition systems is investigated. In particular, we compare the matching scores as obtained by a concrete recognition system when using JPEG2000 compression of rectilinear iris images with and without region of interest coding enabled. The region of interest is restricted to the iris texture area plus the pupil region. It turns out that average matching scores can be improved and that the number of false negative matches is significantly decreased using region of interest coding as compared to plain JPEG2000 compression.

1 Introduction

Iris recognition is claimed to be the most secure biometric modality exhibiting practically 0% FAR and low FRR. An interesting fact is that the iris recognition market is strongly dominated by Iridian Inc. based technology which is based on algorithms of J. Daugman [1]. Certain application scenarios may require compression technology being applied to sample data: In distributed biometric systems, the data acquisition stage is often dislocated from the feature extraction and matching stage (this is true for the enrolment phase as well as for authentication). In such environments the sensor data have to be transferred via a network link to the respective location, often over wireless channels with low bandwidth and high latency. Therefore, a minimisation of the amount of data to be transferred is highly desirable, which is achieved by compressing the data before transmission.

Having found that compression of the raw sensor data can be advantageous in certain applications, we have to identify techniques suited to accomplish this task in an optimal manner. In order to maximise the benefit in terms of data reduction, lossy compression techniques have to be applied. However, the distortions introduced by compression artifacts may interfere with subsequent feature extraction and may degrade the matching results. In particular, FRR or FNMR will increase (since features of the data of legitimate users are extracted less accurately from compressed data) which in turn affects user convenience and general acceptance of the biometric system. In extreme cases, even FAR or FMR might be affected. Therefore, careful optimisation of compression techniques employed is highly desirable.

M. Tistarelli and M.S. Nixon (Eds.): ICB 2009, LNCS 5558, pp. 1102–1111, 2009.
© Springer-Verlag Berlin Heidelberg 2009

The certainly most relevant standard for compressing iris image data is the recent ISO/IEC 19794-6 standard on Biometric Data Interchange Formats which allows iris image data to be stored in lossy manner in the JPEG and JPEG2000 formats. Two types of iris image data are considered: rectilinear images (i.e. images of the entire eye) and polar images (which are basically the result of iris detection and segmentation), the latter much smaller in terms of storage requirement (e.g. 2kB vs. 25-30kB for rectilinear images). It is important to note that with this standardisation it might become more attractive for companies to investigate alternatives to Iridian products due to the available common data format iris recognition systems can rely on.

Only recently, first results and techniques are available on iris image compression and its impact on recognition performance. Ives et al. [4, 3] apply JPEG2000 up to a compression rate of 20 to rectilinear image data (the CASIA database and a proprietary image collection is used) and investigate the effects on FAR and FRR of a 1-D version of the Daugman algorithm (the same system which is used in this study). Rakshit and Monro [9] again use JPEG2000 to compress polar iris images up to a compression rate of 80 and study the impact on verification accuracy of three iris recognition systems (including the Daugman algorithm, the CASIA database is used). Daugman and Downing [2] apply JPEG and JPEG2000 to rectilinear image data (the NIST ICE database is used) and remove image background (i.e. parts of the image not being part of the eye like eye-lids are replaced by constant average gray) before compression is applied. A more compact way of representing the Daugman IrisCode is discussed in [11], however, these results refer to template compression and are only valid for the techniques related to Iridian products.

Five general purpose compression algorithms (including JPEG and JPEG2000) are compared in [6, 5] with respect to their impact on iris recognition accuracy of three different recognition schemes (the CASIA and UBIRIS databases have been used). In accordance to [2] significantly superior compression performance of JPEG2000 over JPEG especially for low bitrates have been found. So far, compression algorithms including JPEG2000 have been applied to iris imagery with their respective standard settings.

In the subsequent experimental study we apply JPEG2000 to rectangular iris data. JPEG2000 region of interest (RoI) coding is applied to the iris texture in order to preserve more textural detail in this area of the images and the corresponding iris recognition results are compared to those obtained by applying plain JPEG2000 compression.

Section 2 introduces the proposed JPEG2000 RoI iris coding framework. Section 3 is the experimental part of this work where we first describe the iris recognition system used to evaluate the JPEG2000 based approach, and the data these algorithms are applied to. Subsequently, we present and discuss our experimental results where we compare iris recognition results using JPEG2000 compression with and without RoI coding enabled. Section 4 concludes the paper.

2 JPEG2000 Region of Interest Coding of Iris Images

The rational for our approach is as follows: Since iris recognition accuracy relies only on the iris texture data, we propose a selective compression approach for iris image data. After detection of the quasi-circular iris region in a rectangular iris image, we pass the information about the identified circular Region of Interest (RoI) to the JPEG2000 encoder which allocates higher quality to this RoI as compared to the rest of the image employing its RoI compression capability. Using this strategy, better iris recognition accuracy is expected as in the case of standard JPEG2000 compression, where the bitrate is distributed evenly among the different regions in the image.

2.1 Iris Detection

Our iris detection approach employs a sequence of standard image processing operations:

1. Histogram equalisation: low contrast images are enhanced to give better results in the subsequent processing stages.
2. Canny edge detection.
3. Canny edges are fed into a Hough transform for circle detection. The circle with the largest radius in a specific range is our detected outer iris boundary.

After the circle detection was successful, we select the detected circle in the image as Region of Interest which is passed to the JPEG2000 encoder.

2.2 JPEG2000 RoI Coding

The JPEG2000 [10] image coding standard uses the wavelet transform as energy compaction method. JPEG2000 may be operated in lossy and lossless mode (using a reversible integer transform in the latter case) and also the wavelet decomposition depth may be defined. The major difference between previously proposed zerotree wavelet-based image compression algorithms such as EZW or SPIHT is that JPEG2000 operates on independent, non-overlapping blocks of transform coefficients ("codeblocks"). After the wavelet transform, the coefficients are quantized and encoded on a codeblock basis using the EBCOT scheme, which renders distortion scalability possible. Thereby the coefficients are grouped into codeblocks and these are encoded bitplane by bitplane, each with three coding passes (except the first bitplane). The codeblock size can be chosen arbitrarily with certain restrictions.

main header	packet header	packet data	... ...	packet header	packet data

Fig. 1. JPEG2000 bitstream structure

The final JPEG2000 bitstream (see Fig. 1) is organised as follows: The main header is followed by packets of data (packet bodies) each of which is preceded by a packet header. A packet body contains CCPs (codeblock contribution to packet) of codeblocks that belong to the same image resolution (wavelet decomposition level) and layer (which roughly stand for successive quality levels). Depending on the arrangement of the packets, different progression orders may be specified. Resolution and layer progression order are the most important progression orders for grayscale images.

In our experiments, we have used the JAVA reference implementation JJ2000[1] which also supports RoI coding which is described in the following.

RoI coding is a feature that enables a non-uniform distribution of the image quality. Parts of the image can be defined as RoI, which are usually areas that contain objects which are of particular interest to the application. An image may contain multiple disjoint RoIs. Although it is possible to define RoIs with arbitrary shapes, the more common cases are rectangles, circles or other simple geometric forms. The image data where a lower (relative) quality is acceptable is called the *Background* (BG). The BG is comprised of all image parts that do not belong to a RoI. RoIs are discussed in Annex H of the JPEG2000 standard [10].

The technical solution to achieve RoI coding is described next. RoIs are often described by simple geometrical shapes. In the JPEG2000 encoding process, the first step is then to derive from this geometric objects in the spatial domain appropriate areas in the wavelet domain. These areas, called *RoI Masks*, contain the wavelet coefficients that belong to RoIs. The method used for priorizing RoI coefficients during the encoding process is called *MaxShift*. MaxShift exploits a property of the EBCOT coding algorithm [10]: The entropy coding for a code block always starts at the most significant bit plane of the coefficients. Thus, information contained in higher bit planes is encoded before the information in lower bit planes. MaxShift simply scales the coefficients belonging to the BG in such a way, that the highest bit plane to which any BG coefficient contributes is lower than the lowest bit plane to which any RoI coefficient contributes.

The exact process at the encoder is described as follows:

1. The encoder determines a scaling value s with the following property:

$$\text{Let } k_{RoImin} = 2^s \tag{1}$$

$$\forall k_{BG} \in BG : k_{RoImin} > k_{BG} \tag{2}$$

2. All coefficients of the Background are scaled down (shifted right) by the scaling value s.
3. The scaling value s is inserted in the resulting JPEG2000 code stream.

At the decoder side, the process is reversed:

1. The decoder extracts the scaling value s from the bit stream.
2. Based on the scaling value the decoder can determine if a coefficient k belongs to a Region of Interest:

[1] `http://jj2000.epfl.ch/`

$$k \geq k_{RoImin} \Longrightarrow k \in RoI \quad (3)$$

$$k < k_{RoImin} \Longrightarrow k \notin RoI \Longrightarrow k \in BG \quad (4)$$

3. All coefficients belonging to the BG are now scaled up to their original value using the scaling value s.

3 Experimental Study

3.1 Iris Recognition System

The employed iris recognition system is Libor Masek's Matlab implementation[2] of a 1-D version of the Daugman iris recognition algorithm. First, this algorithm segments the eye image into the iris and the remainder of the image. After extracting the features of the iris (which are strongly quantized phase responses of complex 1-D Gabor filters in this case), considering translation, rotations and disturbed regions in the iris (a noise mask is generated), the algorithm outputs the similarity score by giving the Hamming distance between two extracted templates. The range of the Hamming distance reaches from zero (ideal matching of two iris images of the same person) to 0.5 (ideal mismatch between two iris images of different persons).

3.2 Sample Data

We use images from two databases for our tests (see table 1). The first images are taken from the Chinese Academy of Sciences V Institute of Automation (CASIA[3]). Each image from the CASIA database is stored as a 8-bit (indexed) gray-scale with a resolution of 320 by 280 pixels. The pictures are taken with a camera using near-infrared light for illumination. This avoids reflections (especially in the dark black pupil area) and may yield a finer structure of the iris. Note that the fact that the pupil area has been manipulated in these data [7] is not relevant in our context since templates for matching are extracted from the polar iris patch only. Moreover, these data come in uncompressed format which is important for the rating of subsequent compression.

The second database we use is UBIRIS [8]. The images are taken with a "normal" (non-infrared) camera and have a size of 200 by 150 pixels in 8-bit gray-scale format (JPEG compressed).

Figure 2 shows an example of a non-RoI JPEG2000 compressed (compression rate 96) iris image of one person (CASIA data), which was used in our calculations together with the extracted iris template data and the noise masks (template and noise mask have been scaled in y-direction by a factor of 4 for proper display).

The differences in the templates are more significant as those in the noise masks which turn out to be very similar. Overall also iris detection is very

[2] `http://www.csse.uwa.edu.au/~pk/studentprojects/libor/sourcecode.html`

[3] `http://www.sinobiometrics.com`

Table 1. Eye-Image Databases used for Tests

	CASIA	UBIRIS
Width	320	200
Height	280	150
Pixels	89600	30000
Format	8-bit gray-scale BMP	8-bit gray-scale JPEG
Number of Image Sets	40	20
Number of Images per Image Set	3	4

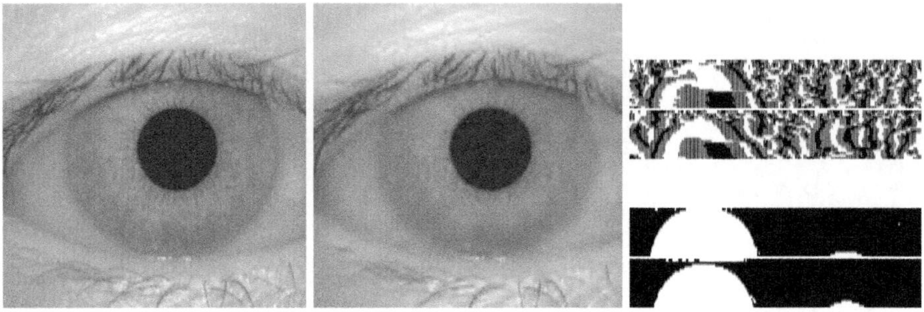

Fig. 2. Comparison of uncompressed/compressed iris image and the corresponding iris templates and noise masks

robust to compression – all matching results shown in the following section have been generated without the software failing to detect the iris.

Compression can be used in various stages of the matching process. Either the stored reference data may be in compressed form, or the sample data acquired for verification may be compressed (e.g. for efficient transfer), or both. In this study, we consider the case of one image out of two involved in matching is employed in compressed manner.

3.3 Experimental Results

For the CASIA images, on average 38 % of the image area belongs to the iris. For UBIRIS, the average size of the iris is 20 %. We performed the tests for JPEG2000 coding with rates of 0.3, 0.2, 0.1 and 0.08 bits per pixel (bpp) on the CASIA image sets and coding rates of 0.8, 0.7,0.6 and 0.5 bpp for the UBIRIS image sets. Note that in the case of UBIRIS images, we apply a re-compression to JPEG-compressed imagery where results are different as compared to compressing raw image data [5]. We measure Hamming distance (HD) and plot on the y-axis $1 - HD$ as the matching score.

Figure 3 shows the average matching results over all image sets considered, we compare the application of JPEG2000 with default settings (J2K) to the employment of the proposed RoI functionality (J2KwROI). For the CASIA image sets (Fig. 3.a), the average matching score for the uncompressed case (original) is

slightly above 0.71. For the four bitrates considered, we find the variant exploiting the RoI functionality being consistently superior to the default JPEG2000 case, the matching score is constantly higher by approximately 0.02 and reaches almost uncompressed performance at 0.3 bpp.

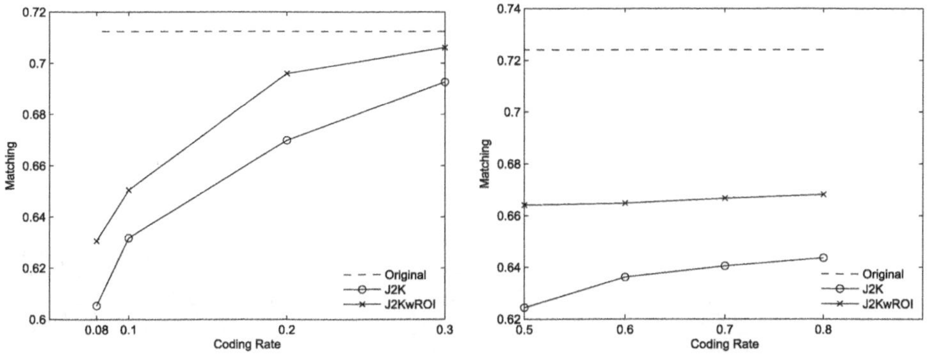

(a) CASIA (b) UBIRIS

Fig. 3. Average matching scores over all image sets (Coding Rate in bpp)

For the UBIRIS image sets, average matching score is slightly above 0.72 in the "uncompressed" case (recall that UBIRIS images already come as JPEG files). Here, superiority of the RoI approach is even more pronounced, the observed matching score gain is in the range 0.02 – 0.04 and decreases for increasing bitrate.

Having documented the gain in terms of average matching score, it is not yet clear in how far this gain will impact on actual recognition performance. In order to investigate this in more depth, we provide averaged matching score results for each image set separately. Fig. 4 shows the corresponding results for the CASIA image sets compressed at 0.2 bpp. For reference, also a hypothetical decision

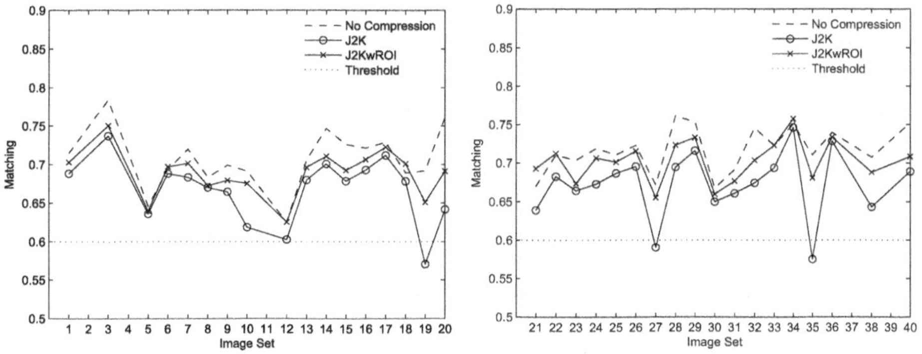

(a) Image Sets 1 to 20 (b) Image Sets 21 to 40

Fig. 4. CASIA: Matching scores at coding rate 0.2 bpp

threshold value is included in the plot at matching score 0.6 (corresponding to $HD = 0.4$ accordingly).

We notice, that for applying JPEG2000 with RoI functionality, matching score values stay well above 0.6 (even above 0.65 for all but sets 5 and 12), while for the standard JPEG2000 case we find values below the threshold for sets 19, 27, and 35.

Fig. 5 shows corresponding results for the CASIA sets 1 - 20 at 0.3 bpp and the UBIRIS sets at 0.5 bpp. For the CASIA sets, at the higher bitrate now no set is blow the decision threshold, while the superior performance of J2KwRoI is still visible. For the UBIRIS sets (Fig. 5.b), again the superiority of J2KwRoI is clearly visible. In addition to the better overall values, only 1 image set results in a matching score below the threshold while 3 set are affected in the default J2K case.

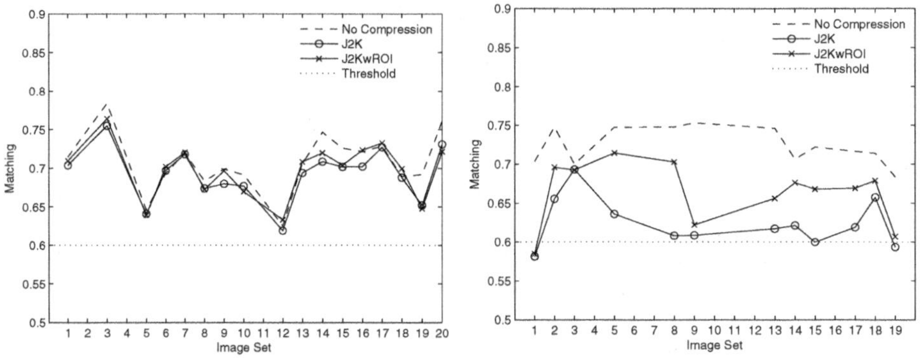

(a) CASIA Sets 1 to 20: Rate 0.3 bpp (b) UBIRIS Sets 1 to 20: Rate 0.5 bpp

Fig. 5. Matching scores at coding rates 0.3 bpp and 0.5 bpp for CASIA and UBIRIS

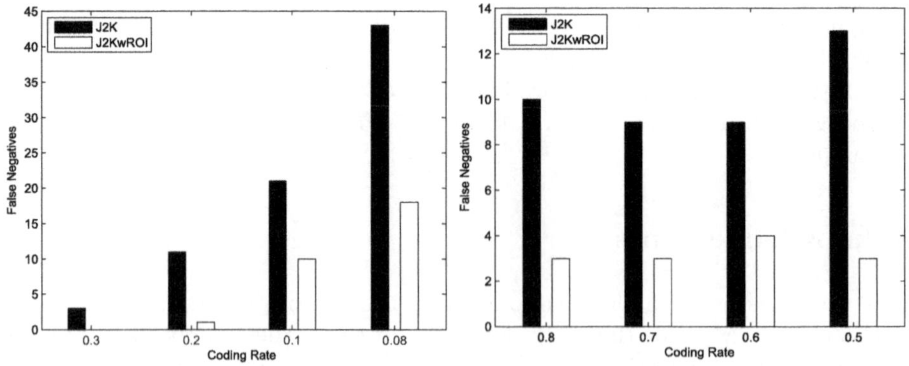

(a) CASIA (b) UBIRIS

Fig. 6. False Negatives ($T_{Match} = 0.6$)

In order to get entirely rid of averaging effects in the comparisons (note that HD-results of each image set are still averaged in the latter results), we provide the number of false negative matches setting the decision threshold to 0.6 (which can be used to compute the false non-match rate (FNMR)). Fig.6 shows corresponding results. For both types of images, the number of false negatives is significantly lower in case of using the RoI approach as compared to the default usage of JPEG2000.

This means, that obviously the ROC is improved by employing the proposed approach, especially in the area of improving user convenience by significantly reducing the number of false negative matches.

4 Conclusion and Future Work

We have investigated the impact of using JPEG2000 region of interest coding on the matching accuracy of iris recognition systems. The region of interest is restricted to the iris texture area plus the pupil region. We find that average matching scores can be improved and that the number of false negative matches is significantly decreased using region of interest coding as compared to plain JPEG2000 compression. This implies an increased user convenience at an at least equal level of security (we have not found evidence for false positives using either technique).

In a certain sense, our approach is related to [2] since both techniques allocate a lower bitrate to the image regions not containing the iris. While the latter technique entirely removes image parts not containing the iris by creating an artificial image containing only pupil, iris texture, and uniform gray pixels, our approach retains the image background with a lower quality and maintains the image integrity.

As a consequence, our approach features the following advantage over [2]: in case the iris detection process fails, our approach retains the entire iris texture (though with reduced quality in the areas which have been falsely detected as being non-iris) while using [2], the corresponding parts of the iris are cut off and replaced by constant gray pixels.

In future work we will compare both approaches with respect to their impact on recognition accuracy and will focus on sensible manners to combine both techniques.

Acknowledgements

This work has been partially supported by the Austrian Science Fund, project no. L554-N15. Most of the work described in this paper has been done in the scope of a semester project (seminar on "Visual Computing and Multimedia") in the master program on "Applied Computer Science" at the University of Salzburg.

References

[1] Daugman, J.: How iris recognition works. IEEE Transactions on Circiuts and Systems for Video Technology 14(1), 21–30 (2004)

[2] Daugman, J., Downing, C.: Effect of severe image compression on iris recognition performance. IEEE Transactions on Information Forensics and Security 3(1), 52–61 (2008)

[3] Ives, R.W., Broussard, R.P., Kennell, L.R., Soldan, D.L.: Effects of image compression on iris recognition system performance. Journal of Electronic Imaging 17, 011015 (2008)

[4] Ives, R.W., Bonney, B.L., Etter, D.M.: Effect of image compression on iris recognition. In: IMTC 2005 – Instrumentation and Measurement Technology Conference (May 2005)

[5] Jenisch, S., Lukesch, S., Uhl, A.: Comparison of compression algorithms' impact on iris recognition accuracy II: revisiting JPEG. In: Proceedings of SPIE, Security, Forensics, Steganography, and Watermarking of Multimedia Contents X, San Jose, CA, USA, vol. 6819, p. 68190M (January 2008)

[6] Matschitsch, S., Tschinder, M., Uhl, A.: Comparison of compression algorithms' impact on iris recognition accuracy. In: Lee, S.-W., Li, S.Z. (eds.) ICB 2007. LNCS, vol. 4642, pp. 232–241. Springer, Heidelberg (2007)

[7] Philips, P.J., Bowyer, K.W., Flynn, P.J.: Comments on the CASIA version 1.0 iris data set. IEEE Transactions on Pattern Analysis and Machine Intelligence 29(10), 1869–1870 (2007)

[8] Proenca, H., Alexandre, L.A.: UBIRIS: a noisy iris image database. In: Roli, F., Vitulano, S. (eds.) ICIAP 2005. LNCS, vol. 3617, pp. 970–977. Springer, Heidelberg (2005)

[9] Rakshit, S., Monro, D.M.: Effects of sampling and compression on human iris verification. In: Proceedings of the IEEE International Conference on Acustics, Speech, and Signal Processing (ICASSP 2006), Tolouse, France, pp. II–337–II–340 (2006)

[10] Taubman, D., Marcellin, M.W.: JPEG2000 — Image Compression Fundamentals, Standards and Practice. Kluwer Academic Publishers, Dordrecht (2002)

[11] von Seelen, U.: IrisCode template compression and its effects on authentication performance. In: Biometrics Consortium Conference 2003 (September 2003)

Image Averaging for Improved Iris Recognition

Karen P. Hollingsworth, Kevin W. Bowyer, and Patrick J. Flynn

University of Notre Dame

Abstract. We take advantage of the temporal continuity in an iris video to improve matching performance using signal-level fusion. From multiple frames of an iris video, we create a single average image. Our signal-level fusion method performs better than methods based on single still images, and better than previously published multi-gallery score-fusion methods. We compare our signal fusion method with another new method: a multi-gallery, multi-probe score fusion method. Between these two new methods, the multi-gallery, multi-probe score fusion has slightly better recognition performance, while the signal fusion has significant advantages in memory and computation requirements.

1 Introduction

The field of iris recognition is an active and rapidly expanding area of research [1]. Many researchers are interested in making iris recognition more flexible, faster, and more reliable. Despite the vast amount of recent research in iris biometrics, all past published work has relied on still iris images. Zhou and Chellapa [2] reported that using video can improve face recognition performance. We postulated that employing similar techniques for iris recognition could also yield improved performance. There is some prior research in iris recognition that uses multiple still images; for example, [3, 4, 5, 6]. However, no researchers have published on how to use additional information available in iris video.

One problem with single still images is that they usually have a moderate amount of noise. Specular highlights and eyelash occlusion reduce the amount of iris texture information present in a single still image. With a video clip of an iris, however, a specular highlight in one frame may not be present in the next. Additionally, the amount of eyelash occlusion is not constant throughout all frames. It is possible to obtain a better image by using multiple frames from a video to create a single, clean iris image.

Zhou and Chellapa suggested averaging to integrate texture information across multiple video frames to improve face recognition performance. By combining multiple images, noise is smoothed away, and relevant texture is maintained. This paper presents experiments demonstrating how signal-level fusion of multiple frames in a video improves iris recognition performance.

2 Related Work

Video has been used effectively to improve face recognition. A recent book chapter by Zhou and Chellapa [2] surveys a number of methods to employ video in

M. Tistarelli and M.S. Nixon (Eds.): ICB 2009, LNCS 5558, pp. 1112–1121, 2009.
© Springer-Verlag Berlin Heidelberg 2009

face biometrics. As an example of a specific algorithm, Thomas et al. [7] use multiple frames from a video clip to improve performance of a face matcher. Thomas et al. use temporal continuity in video to improve face detection accuracy. They use both a rank-based average and score-based average when combining matching scores from different frames.

In contrast, literature in iris biometrics reports results exclusively on still images. Some iris biometric research has used multiple still images, but all such research uses score-level fusion, not signal-level fusion. The information from multiple images has not been combined to produce a better image. Instead, these experiments typically employ multiple enrollment images of a subject, and combine matching results across multiple comparisons.

Du et al. [3] showed that using three enrollment images instead of one increased their rank-one recognition rate from 98.5% to 99.8%. Ma et al. [4] also used three templates of a given iris in their enrollment database, and took the average of three scores as the final matching score. Krichen et al. [5] performed a similar experiment, but used the minimum match score instead of the average. Schmid et al. [6] present two methods for fusing Hamming distance scores. They compute average Hamming distance and also log-likelihood ratio. They find that in many cases, the log-likelihood ratio outperforms the average Hamming distance. In all of these cases, information from multiple images was not combined until after two stills were compared and a score for the comparison obtained. Thus, these researchers use score-level fusion.

Another method of using multiple iris images is to use them to train a classifier. Liu et al. [8] used multiple iris images for a linear discriminant analysis algorithm. Roy and Bhattacharya [9] used six images of each iris class to train a support vector machine. Even in training these classifiers, each still image is treated as an individual entity, rather than being combined with other still images to produce an improved image.

3 Data

We used an Iridian LG EOU 2200 camera to acquire iris data. The LG 2200 takes video data of a subject's eye, and the accompanying software selects specific images from the video to use for iris recognition. During acquisition, the eye is illuminated by one of three infrared LEDs: one above the lens, one to the left, and one to the right. While taking data, the camera cycles through the LEDs, picking one eye image when the top LED is lit, one image when the left LED is lit, and one image while the right LED is lit. Next, the software selects the best of the three selected images to use for recognition. The chosen frame must pass Iridian's quality tests. Our LG 2200 camera uses special software that allows us to save the entire set of three images, rather than just saving the best of the three.

Rather than processing still images we wanted to use multiple frames from video to improve performance. We split the NTSC video signal from the camera so that the analog video feed could be sent both to the computer running the Iridian software, but also to a separate computer so that we could digitize and

save the entire video clip instead of six images. These videos were digitized using a Daystar XLR8 USB digitizer and the resulting videos were stored in a high bit rate (nearly lossless) compressed MP4 format.

We collected a data set of 1061 videos during the spring of 2008. Each video clip was long enough for the Iridian software to capture two sets of three images. We captured videos of both left and right eyes for each subject; however, we treat each individual eye as a separate "subject". We have not yet considered how fusing left and right iris information adds to performance. We captured videos of 296 different eyes. The first video of each subject was used as the gallery, and the remaining videos from each subject were used as the probe. For each subject, we have between one and five iris videos in the probe set. Our total probe set contains 765 videos. For any two videos from the same subject in our data set, they were acquired at least one week apart, and possibly up to two months apart.

LG sells iris cameras that are newer than the LG 2200. However, all of those cameras were designed for saving single images, not video. The videos used in this research were captured with the same camera as the iris videos released in 2008 for the Multiple Biometrics Grand Challenge (MBGC) evaluation [10]. We are not aware of any other data sets containing iris videos instead of iris stills.

4 Creating Average Images and Templates

4.1 Selecting Frames and Preprocessing

The auto-focus on the LG 2200 camera continually adjusts the focus in attempts to find the best view of the iris. Thus, some frames have good focus, while others suffer from severe blurring. To simplify our experiments, we refrained from developing our own quality metrics to select frames. Instead, we decided to use the six frames from the video that corresponded to the six frames that the Iridian software selected to save.

The raw video frames were not pre-processed like the still images that the Iridian software saved. We do not know what preprocessing is done by the Iridian system, although it appears that the system does contrast enhancement and possibly some deblurring. Some differences between the stills and the video frames may be due to differences in the digitizers used to save the signals. We used the Matlab imadjust function to enhance the contrast in each frame. This function scales intensities linearly such that 1% of pixel values saturate at black (0), and 1% of pixel values saturate at white (255).

Our next step was to segment each frame. Our segmentation software uses a Canny edge detector and a Hough transform to find the iris boundaries. The boundaries are modeled as two non-concentric circles. A description of the segmentation algorithm is given in [11].

We did not want segmentation problems to bias our results, so we discarded a few images where the circles clearly did not match the boundaries of the iris. In addition, we discarded a few images because the iris was not visible in the image (e.g. the frame showed a blink, or the bridge of the nose).

4.2 Signal Fusion

For each video, we now had about six frames selected and segmented. We wanted to create an average image consisting only of iris texture. In order to align the irises in the six frames, we transformed the raw pixel coordinates of the iris area in each frame into dimensionless polar coordinates. In the polar coordinates, the radius r ranged from zero (adjacent to the pupillary boundary) to one (adjacent to the limbic boundary). The angle θ ranged from 0 to 2π. This yielded an "unwrapped" iris image for each video frame selected.

In order to combine the six unwrapped iris images, we wanted to make sure they were aligned correctly with each other. Rotation around the optical axis induces a horizontal shift in the unwrapped iris texture. We tried three methods of alignment. First, we identified the shift value that maximized the correlation between the pixel values. Second, we tried computing the iris codes and selecting the alignment that produced the smallest Hamming distance. Third, we tried the naive assumption that people would not actively tilt their head while the iris video was being captured. The first two approaches did not produce any better recognition results than the naive approach. Since the naive approach worked well, we used it in our subsequent experiments.

Parts of the unwrapped images contained occlusion by eyelids and eyelashes. Our iris software implements some eyelid and eyelash detection, and masks those areas. We computed an average unwrapped image from unmasked iris data in the six original images, using the following algorithm. For each (r, θ) position, first find how many of the corresponding pixels in the six unwrapped images are unmasked. If zero or one of the corresponding pixels from the six images are unmasked, mask that position in the average image. Otherwise, compute an average pixel value based on unmasked pixel values of the corresponding frames. (Therefore, the new pixel value could be an average of between two and six pixel intensities, depending on mask values.) To compute an average, we first summed original pixel values, then divided by the number of pixels, then rounded to the nearest unsigned, 8-bit integer.

Using this method, we obtained 296 average images from the gallery videos. We similarly obtained 765 average images from the probe videos. An example average image is shown in Figure 1. On the top of the figure are the six original images, and on the bottom is the average image fused from the six original signals.

4.3 Creating an Iris Code Template

Our software uses one-dimensional log-Gabor filters to create the iris code template. The log-Gabor filter is convolved with rows of the image, and the corresponding complex coefficients are quantized to create a binary code. Each complex coefficient corresponds to two bits of the binary iris code – either "11", "01", "00", or "10" – depending on whether the complex coefficient is in quadrant I, II, III, or IV of the complex plane.

Complex coefficients near the axes of the complex plane do not produce stable bits in the iris code, because a small amount of noise can shift a coefficient from

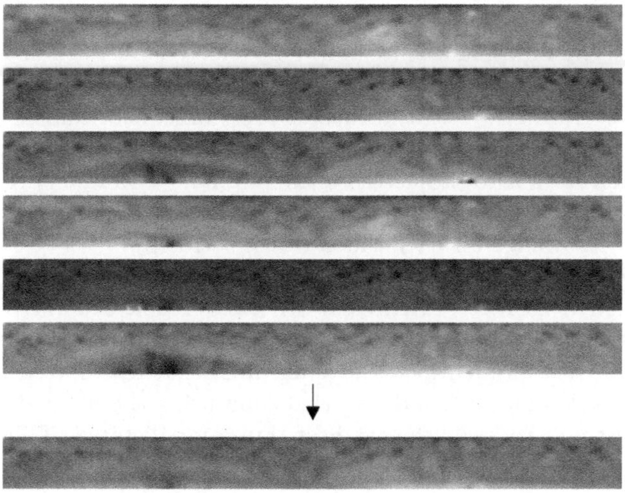

Fig. 1. From the six original images on the top, we created the average image shown on the bottom

one quadrant to the next. We use fragile-bit masking [12,13] to mask out complex coefficients near the axes, and therefore improve recognition performance.

5 Experiments

5.1 Comparison to Previous Work

In biometrics, it has been found that enrolling multiple images improves performance [14, 15, 16]. Iris recognition is no exception. Many researchers [4, 5, 6] enroll multiple images, obtain multiple Hamming distance scores, and then fuse the scores together to make a decision. However, the different researchers have chosen different ways to combine the information from multiple Hamming distance scores.

Let N be the number of gallery images for a particular subject. Comparing a single probe image to the N gallery images gives N different Hamming distance scores. To combine all of the N scores into a single score, Ma et al. [4] took the average Hamming distance. We will call this type of experiment an N-to-1-average comparison. Krichen et al. [5] also enrolled N gallery images of a particular subject. However, they took the minimum of all N different Hamming distance scores. We call this type of experiment an N-to-1-minimum comparison.

Schmid et al. [6] enrolled N gallery images of a particular subject and also took N images of a probe subject. The N gallery images and N probe images were paired in an arbitrary fashion and compared. Thus they obtained N different Hamming distance scores. Finally, they averaged the N scores into one single score. We will call this type of experiment an N-to-N-average comparison. Schmid

Table 1. Signal-fusion Compared to Previous Methods

First Author	Method	d'	EER	FRR@FAR=0.001
Baseline	no fusion	4.37	1.64×10^{-2}	4.51×10^{-2}
Ma	score fusion: N-to-1 average	5.08	6.93×10^{-3}	1.40×10^{-2}
Krichen	score fusion: N-to-1 minimum	5.44	6.63×10^{-3}	1.43×10^{-2}
This work	signal fusion: 1-to-1	6.06	3.88×10^{-3}	7.61×10^{-3}

also combined the N different Hamming scores using the log-likelihood ratio. We will call this type of experiment an N-to-N-loglikelihood comparison.

In our signal-fusion method, we take N frames from a gallery video and do signal-level fusion, averaging the images together to create one single average image. We then take N frames from a probe video and average them together to create a single average image. Thus, we can call our proposed method a signal fusion-1-to-1 comparison.

One automatic advantage of the signal fusion method is that storing a single, average-image iris code takes only a fraction of the space of the score-fusion methods. Instead of storing N gallery templates per subject, the proposed method only requires storing one gallery template per subject.

In order to compare our method to previous methods, we have implemented the N-to-1-average and the N-to-1-minimum methods. Implementing the N-to-N-average and N-to-N-loglikelihood methods remains future work. For our experiments, we let N = 6. For each of these methods, we used the same data sets. Figure 2 shows a decision error threshold curve with results from these experiments. As an additional baseline, we graph the DET curve for a single-gallery, single-probe experiment (No Fusion). The proposed signal fusion method has the lowest false accept and false reject rates for the majority of the decision error threshold curve. There is a small region on the lower right part of the graph where the score fusion average method performs about the same as the proposed method. In this region, the false reject rate is smaller than 2.6×10^{-3}. False reject rates that small are based on 2 or fewer of the 765 possible genuine comparisons in our data set. Thus, we would need a larger amount of data to decide on performance in that region of the graph. (Since we have 225675 impostor comparisons in our data set, we would also need more data if we wanted to determine performance for false accept rates less than 4.4×10^{-6}.)

We conclude that on our data set, the signal-fusion method generally performs better than the previously proposed N-to-1-average or N-to-1-minimum methods. In addition, the signal fusion takes $1/N^{th}$ of the storage and $1/N^{th}$ of the matching time.

5.2 Comparing Signal-Fusion to Large Multi-gallery, Multi-probe Methods

The score-fusion experiments listed in subsection 5.1 compared our signal-fusion method to previously-published methods. Ma's and Krichen's methods both used

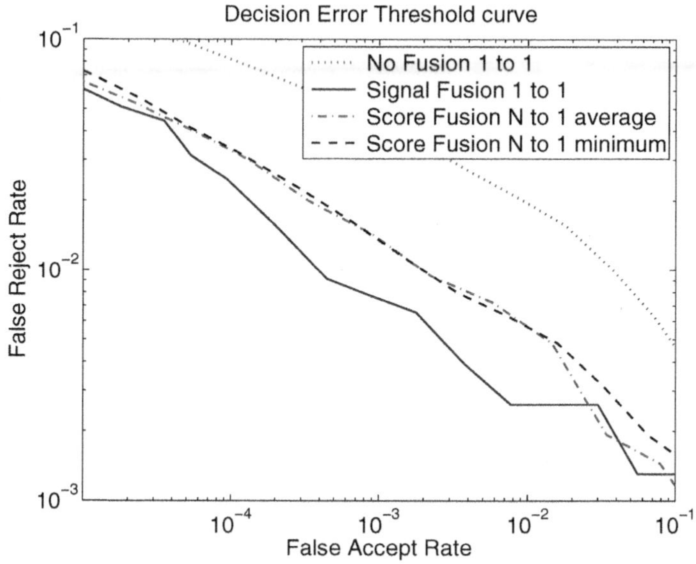

Fig. 2. The proposed signal-fusion method has better performance than using a multi-gallery approach with either an "average" or "minimum" score-fusion rule

multiple gallery images of a subject compared to a single probe image of a subject. Each of those score-fusion methods fused N Hamming distance scores to create the final score. We also wished to consider the situation where for a single comparison, there are N gallery images available, N probe images available, and therefore N^2 Hamming distance scores to fuse. We would expect that the fusion of N^2 scores would perform better than the fusion of N scores. Although this multi-gallery, multi-probe fusion is a simple extension of the methods listed in subsection 5.1, we do not know of any published work that uses this idea for iris recognition.

We tested two ideas: we took the average of all N^2 scores, and also the minimum of all N^2 scores. We call these two methods the (1) multi-gallery, multi-probe, average method (MGMP-average) and the (2) multi-gallery, multi-probe, minimum method (MGMP-minimum). The MGMP-average method produces genuine and impostor Hamming distance distributions with small standard deviations. Since this method has the lowest standard deviation for the impostor distribution, this method is good for reducing false accepts. Therefore, this particular method has the lowest FRR at 0.001 FAR. Using the "minimum" rule for score-fusion produces smaller Hamming distances than the "average" rule. However, both the genuine and impostor distributions are affected. We found that for these two multi-gallery, multi-probe methods, the "average" score-fusion rule works better than the "minimum" rule.

We next compared the MGMP-average method to the signal fusion method. The signal-fusion method presented in this subsection is unchanged from the previous subsection, but we are presenting the results again, for comparison

Table 2. Signal-fusion Compared to a Multi-gallery, Multi-probe Method

Method	d'	EER	FRR @ FAR=0.001
MGMP-average	5.53	2.98×10^{-3}	4.69×10^{-3}
Signal fusion	6.06	3.88×10^{-3}	7.61×10^{-3}

purposes. The comparison between the MGMP-average method and the signal fusion method had mixed results. MGMP-average method had lower EER and a lower FRR at FAR=0.001. However, the signal-fusion method had a better value of d', suggesting that the signal-fusion method does achieve good separation between the genuine and impostor distributions of scores [17]. Statistics for both methods are shown in table 2.

Based on the equal error rate and false reject rate, we must conclude that the multi-gallery, multi-probe average method that we present in this section achieves the best recognition performance of all of the methods considered in this paper. However, the signal-fusion performs well, while taking only $1/N^{th}$ of the storage and $1/N^2$ of the matching time.

5.3 Computation Time

In this subsection, we compare the different methods presented in this paper in terms of processing time. We have three types of methods to compare: (1) the multi-gallery, multi-probe approaches (both MGMP-average and MGMP-minimum) which require N^2 iris code comparisons before fusing values together to create a single score; (2) the multi-gallery approaches (Ma and Krichen) which compare N gallery iris codes to one probe before fusing scores together; and (3) the signal-fusion approach which first fuses images together, and then has a single iris code comparison.

For this analysis, we first define the following variables. Let P be the preprocessing time for each image, I be the iris code creation time, and C be the time required for the XOR comparison of two iris codes. Let N be the number of images of a subject in a single gallery entry for the multi-gallery methods. Let A be the time required to average N images together (to perform signal-fusion). Finally, suppose we have an application such as in the United Arab Emirates where each person entering the country has his or her iris compared to a watch-list of one million people [18]. For this application, let W be the number of people on the watchlist. Expressions for the computation times for all three methods are given in terms of these variables in Table III.

From Daugman's work [17], we can see that typical preprocessing time for an image is 344 ms. He also notes that iris code creation takes 102 ms and an XOR comparison of two iris codes takes 10 μs. Throughout this paper, we have used six images for all multi-gallery experiments. The time to compute an average image from six preprocessed images is 3 ms. Lastly, we know that the United Arab Emirates watchlist contains one million people. By substituting

Table 3. Processing times for different methods

Method	Gallery Preprocessing	Probe Preprocessing	Comparison to Watchlist	Total Time
MGMP	NP+NI = 2.676 s	NP+NI = 2.676 s	WCN^2 = 360 s	365.35 s
Multi-gallery	NP+NI = 2.676 s	P+I = 0.446 s	WCN = 60 s	63.12 s
Signal fusion	NP+A+I = 2.169 s	NP+A+I = 2.169 s	WC = 10 s	14.34 s

these numbers in for our variables, we found the processing time for all of our three types of methods. These numeric values are also presented in table 3.

From this analysis it is clear, that although a multi-gallery, multi-probe method may have some performance improvements over the signal fusion method, it comes at a high computational cost.

6 Conclusions

We perform fusion of multiple biometric samples at the signal level. Our signal fusion approach utilizes information from multiple frames in a video. This is the first published work to use video to improve iris recognition performance. Our experiments show that using average images created from six frames of an iris video performs very well for iris recognition. Average images perform better than (1) experiments with single stills and (2) experiments with six gallery images compared to single stills. Our proposed multi-gallery, multi-probe method achieves slightly better recognition performance than our proposed signal-fusion method. However, the matching time and memory requirements are lowest for the signal-fusion method, and the signal-fusion method still performs better than previously published multi-gallery methods.

Acknowledgement

This research is supported by the National Science Foundation under grant CNS01-30839, by the Central Intelligence Agency, by the Intelligence Advanced Research Projects Activity, and by the Technical Support Working Group under US Army contract W91CRB-08-C-0093. The opinions, findings, and conclusions or recommendations expressed in this publication are those of the authors and do not necessarily reflect the views of our sponsors.

References

1. Bowyer, K.W., Hollingsworth, K.P., Flynn, P.J.: Image understanding for iris biometrics: A survey. Computer Vision and Image Understanding 110(2), 281–307 (2008)

2. Zhou, S.K., Chellapa, R.: Beyond one still image: Face recognition from multiple still images or a video sequence. In: Zhao, W., Chellapa, R. (eds.) Face Processing: Advanced Modeling and Methods, ch. 17, pp. 547–567. Elsevier, Amsterdam (2006)
3. Du, Y.: Using 2D log-Gabor spatial filters for iris recognition. In: SPIE 6202: Biometric Technology for Human Identification III, p. 62020:F1–F8 (2006)
4. Ma, L., Tan, T., Wang, Y., Zhang, D.: Efficient iris recognition by characterizing key local variations. IEEE Transactions on Image Processing 13(6), 739–750 (2004)
5. Krichen, E., Allano, L., Garcia-Salicetti, S., Dorizzi, B.: Specific texture analysis for iris recognition. In: Kanade, T., Jain, A., Ratha, N.K. (eds.) AVBPA 2005. LNCS, vol. 3546, pp. 23–30. Springer, Heidelberg (2005)
6. Schmid, N.A., Ketkar, M.V., Singh, H., Cukic, B.: Performance analysis of iris-based identification system at the matching score level. IEEE Transactions on Information Forensics and Security 1(2), 154–168 (2006)
7. Thomas, D., Bowyer, K.W., Flynn, P.J.: Strategies for improving face recognition from video. In: Workshop on Motion and Video Computing (Feburary 2007)
8. Liu, C., Xie, M.: Iris recognition based on DLDA. In: Int. Conf. on Pattern Recognition, August 2006, pp. 489–492 (2006)
9. Roy, K., Bhattacharya, P.: Iris recognition with support vector machines. In: Proc. Int. Conf. on Biometrics, January 2006, pp. 486–492 (2006)
10. Phillips, P.J., Scruggs, T., Flynn, P.J., Bowyer, K.W., Beveridge, R., Givens, G., Draper, B., O'Toole, A.: Overview of the multiple biometric grand challenge. In: Proc. Int. Conf. on Biometrics (ICB 2009) (2009)
11. Liu, X., Bowyer, K.W., Flynn, P.J.: Experiments with an improved iris segmentation algorithm. In: Proc. Fourth IEEE Workshop on Automatic Identification Technologies, October 2005, pp. 118–123 (2005)
12. Hollingsworth, K.P., Bowyer, K.W., Flynn, P.J.: All iris code bits are not created equal. In: Proc. IEEE Int. Conf. on Biometrics: Theory, Applications, and Systems (September 2007)
13. Hollingsworth, K.P., Bowyer, K.W., Flynn, P.J.: The best bits in an iris code. In: IEEE Transactions on Pattern Analysis and Machine Intelligence (accepted for publication)
14. Bowyer, K.W., Chang, K.I., Yan, P., Flynn, P.J., Hansley, E., Sarkar, S.: Multi-modal biometrics: an overview. In: Second Workshop on Multi-Modal User Authentication (May 2006)
15. Chang, K.I., Bowyer, K.W., Flynn, P.J.: An evaluation of multi-modal 2D+3D face biometrics. IEEE Transactions on Pattern Analysis and Machine Intelligence 27(4), 619–624 (2005)
16. Phillips, P.J., Flynn, P.J., Scruggs, T., Bowyer, K.W., Worek, W.: Preliminary Face Recognition Grand Challenge results. In: Int. Conf. on Automatic Face and Gesture Recognition (FG 2006) (April 2006)
17. Daugman, J.: How iris recognition works. IEEE Transactions on Circuits and Systems for Video Technology 14(1), 21–30 (2004)
18. Daugman, J.: United Arab Emirates deployment of iris recognition, `http://www.cl.cam.ac.uk/~jgd1000/deployments.html` (accessed January 2009)

Iris Recognition Using 3D Co-occurrence Matrix

Wen-Shiung Chen[1], Ren-Hung Huang[1], and Lili Hsieh[2]

[1] VIP-CCLab., Dept. of Electrical Engineering, National Chi Nan University, Taiwan
[2] Dept. of Information Management, Hsiuping Institute of Technology, Taiwan

Abstract. This paper presents a biometric recognition based on the iris of a human eye using gray-level co-occurrence matrix (GLCM). A new approach of GLCM, called 3D-GLCM, which is expanded from the original 2D-GLCM is proposed and used to extract the iris features. The experimental results show that the proposed approach gains an encouraging performance on the UBIRIS iris database. The recognition rate up to 99.65% can be achieved.

Keywords: Personal Authentication, Biometrics, Iris Recognition, Co-occurrence Matrix.

1 Introduction

Biometrics [1] refers to automatic identity authentication of a person on a basis of one's unique physiological or behavioral characteristics and is inherently more suitable for discriminating between an authorized person and an impostor than traditional methods. To date, many biometric features, including fingerprint, hand geometry or palmprint, face, and iris [2]-[10], have been studied. In recent years, the human iris keeps receiving growing interest due to its high uniqueness, high permanence, and high circumvention.

The iris, a kind of physiological biometric feature with genetic independence, contains extremely information-rich physical structure and unique texture patterns, and thus is highly complex enough to be used as a biometric signature [2]. Statistical analysis of faker and authentic scores reveals that irises have an exceptionally high degree-of-freedom up to 266 (much more than fingerprint, showing about 78) [2], and thus are the most mathematically unique feature of the human body. The highly randomized appearance of the iris makes its use as a biometric well recognized.

Compared with other biometric features such as hand and fingerprint, iris patterns are more stable and reliable. Each person's irises are unique, and they are stable with age [2]. In addition, several studies have shown that normal variations in coloring and structure of the tissues of an iris are so multitudinous that no two irises are ever alike - not even our own or identical twins [2]. As a result, the human iris promises to deliver a high level of uniqueness to authentication applications that other biometrics cannot match. Furthermore, iris recognition systems can be non-invasive to their end-users. In this paper, we investigate and design an automatic biometric iris recognition system using the three-dimensional gray level co-occurrence matrix (3D-GLCM) for the discriminating texture features.

M. Tistarelli and M.S. Nixon (Eds.): ICB 2009, LNCS 5558, pp. 1122–1131, 2009.
© Springer-Verlag Berlin Heidelberg 2009

2 Related Work

Nowadays, iris has a high potential in becoming the focus of a very new biometric means of authentication. In 1993, Daugman developed a successful system by using the 2D Gabor wavelet transform [2]. In this system, the visible texture of a person's iris in a real-time video image is encoded into a compact sequence of multi-scale quadrature 2D Gabor wavelet coefficients, which most significant bits consist of a 256-byte "iris code." In 1996, Wildes *et al.* developed a prototype system based on an automatic iris recognition that uses a very computationally demanding image registration technique [3]. This system exploits normalized correlation over small tiles within the Laplacian pyramid bands as a goodness of match measure. Boles and Boashash [4] proposed an iris identification in which zero-crossing of the wavelet transform at various resolution levels is calculated over concentric circles on the iris, and the resulting 1D signals are compared with the model features using different dissimilarity functions. Zhu *et al.* [5] proposed an algorithm for global iris texture feature extraction using multi-channel Gabor filtering and wavelet transform. The mean and standard deviation of each sub-image at different resolution levels are extracted as the features. Subsequently, Ma *et al.* [6] modified the work in [5] by using a bank of circular symmetric filters to extract much more local texture information. The work in [7] also used wavelet multi-resolution analysis based on Gabor filtering for iris feature extraction. Lim *et al.* [8] used wavelet transform to make a feature vector compact and efficient. In summary, all of the previous studies are based upon the multi-resolution analysis technique. Recently, Monro *et al.* [9] proposed a DCT-based iris recognition that the differences of DCT coefficients of overlapped angular patches from normalized iris images are encoded. Thornton *et al.* [10] presented a Bayesian framework for iris pattern matching that experience in-plane nonlinear deformations.

The original gray level co-occurrence matrix (GLCM) [11] proposed by Haralick *et al.* in 1973, one of the most known texture analysis methods, estimates image properties related to second-order statistics. It is based on counting the order of co-occurrence of brightness value of pixels pairs at a certain distance and a direction, and successfully applied to some works in image processing and biometrics. The texture analysis based on the GLCM method was proposed to measure the texture features created from a natural image for classification [12], segmentation [13] and retrieval [14]. The GLCM technique has also applied to biometrics, such as fingerprint [15], face [16], signature [17] and iris [18]-[21]. The GLCM methods used in iris recognition are two-dimensional. In this paper, a novel 3D approach modified from 2D version was designed and then applied to extract the iris features more efficiently.

3 System Overview and Pre-processing

3.1 System Overview

The proposed framework consists of three modules: image pre-processing, feature extraction, and recognition modules, as shown in Fig. 1. First, the iris image pre-processing (IIP) module employs some image processing algorithms to demarcate the region of interest (i.e., iris zone) from the input image containing an eye. It performs three major tasks including iris localization, iris segmentation and coordinate

transformation, and enhancement for the input iris image. Next, the feature extraction (IFE) module produces a newly designed three-dimensional gray level co-occurrence matrix (3D-GLCM), gathers static properties features, and applies appropriate coding methods on these features to generate the iris feature code. Finally, the minimum distance classifier according to Euclidean distance metric is employed in the iris pattern recognition (IPR) module.

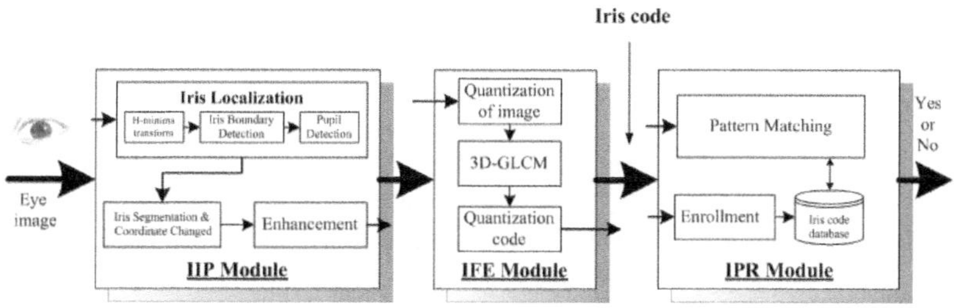

Fig. 1. Structure diagram of the proposed iris recognition system

3.2 Pre-processing Module

Input image does contain not only useful information from iris zone but also useless data derived from the surrounding eye region. Before extracting the features of an iris, the input image must be pre-processed to localize, segment and enhance the region of interest (i.e., iris zone). The system normalizes the iris region to overcome the problem of a change in camera-to-eye distance and pupil's size variation derived from illumination. Furthermore, the brightness is not uniformly distributed due to non-uniform illumination; the system must be capable of removing the effect and further enhancing the iris image for the following feature extraction. Hence, the image pre-processing module is composed of three units: iris localization, iris segmentation and coordinate transform, and enhancement units, as shown in Fig. 2.

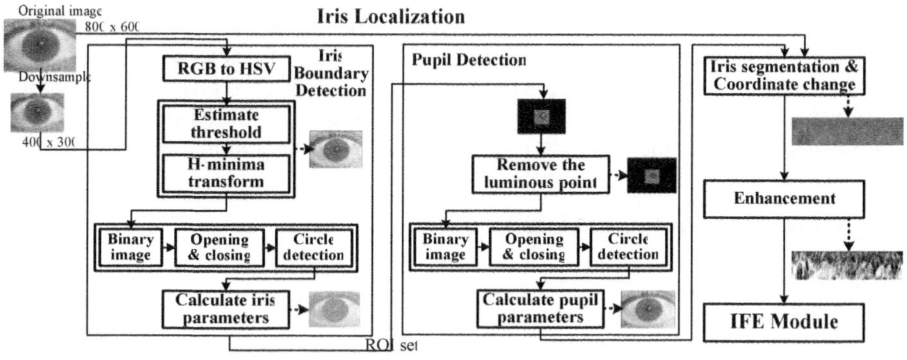

Fig. 2. Pre-processing module

A. Iris Localization Unit

In this unit, we must first determine the useful part (i.e., iris) from an input image. The iris is an annular region between the pupil and the sclera. The inner and outer boundaries of an iris can be treated as non-concentric circles approximately. In order to make the iris localization efficient, the system performs an operation of enhancing principal edges and blurring useless edges on the copied and down-sample image instead of the original one. Following that, the system estimates the center coordinates of the iris first. Since the iris typically is darker than the Sclera and its gray level distribution has a small variance, the system uses Extended-Minima (EM) morphology operator [23]. EM transform is the regional minima of the H-minima transform. H-Minima transform suppresses all minima in the intensity image which depth is less than a scalar. Regional minima are connected with components of pixels with the same intensity value which external boundary pixels all have a greater value than a scalar. We use 8-connected neighborhoods in this process. By choosing an appropriate scalar in EM transform, a perfect edge of outer boundary is gotten. The value of threshold is decided according to the histogram.

Because of inside of iris of pupil, we use a radius 1.2 times as large as outer circle radius to set a ROI area. We regard it as the pupil and search the range (see Fig. 4(a)). Since the pupil typically is darker than the iris and its gray level distribution has a small variance, the system uses a 30×30 mean filter (see Fig. 4(d)) to find the location which has the minimum mean value in the image. This location is considered as an internal point of the iris region, namely the fitting point. Then the system can determine a certain region starting from the fitting point using the iris localization procedure. Since there are obvious intensity differences around each boundary, a proper edge detection method can be applied to find the edge points in this region, and evaluate the exact circle parameters (including the inner and outer boundaries of iris) for all possible point triplets of the edge points.

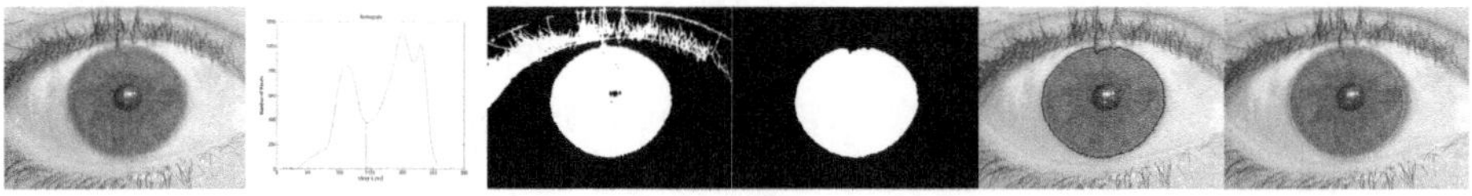

(a) (b) (c) (d) (e) (f)

Fig. 3. (a) Original image, (b) image histogram, (c) EM transform, (d) exact outer edge, (e) iris edge and (f) outer localized image

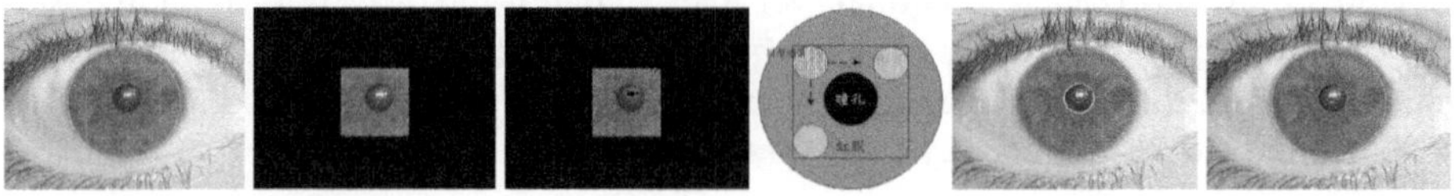

(a) (b) (c) (d) (e) (f)

Fig. 4. (a) Set ROI. (b) ROI image. (c) Remove the luminous point. (d) 30×30 mean filter. (e) Exact inner edge. (f) Iris localized image.

According to Euclidean geometry, by calculating two perpendicular bisectors of any edge point triplet, the system can easily obtain the crossover point of two perpendicular bisectors. We select the center point as the most frequently crossed point. After finding the central points of two circles, we determine the inner radius and the outer radius of the iris by computing the distances between the central points and the inner edge points, and outer edge points, respectively. Eventually, an iris zone can be located by these parameters, as shown in Fig. 4(f).

B. Iris Segmentation and Coordinate Transform Unit
The localized iris zone demarcated from the human eye is transformed from rectangular into polar coordinate systems so as to facilitate the following feature extraction module. When acquiring the human eye image, the iris zone images may be in different sizes (outer and inner boundaries of iris) due to the variations of camera-to-eye distance and/or environmental illumination. The variations will change the patterns of iris texture to some extent. To solve this problem, it is necessary to compensate for the iris deformation. Daugman's system uses radial scaling to compensate for overall size as well as a simple model of pupil variation based on linear stretching [2]. This scaling serves to map Cartesian image coordinates to dimensionless polar image coordinates. In addition, eyelids and eyelashes generally obscure the upper limbus of the iris, so the procedure needs to cut out the obscured area. The system generates a rectangular iris image of a fixed size by linear interpolation. The size of iris image is 512×128.

4 The Proposed Method

4.1 Two-Dimensional Gray Level Co-occurrence Matrix (2D-GLCM)

Co-occurrence matrix, proposed by Haralick *et al.* in 1973 [11], is based on counting the order of co-occurrence of brightness value of pixels pairs at certain distance and direction, which can be described in the following equation:

$$P_{COM}(i,j,d,\theta)=N\left\langle ((x_{n1},y_{n1}),(x_{n2},y_{n2}))\in M_1\times M_2 \left| \begin{array}{l} \max(|x_{n1}-x_{n2}|,|y_{n1}-y_{n2}|)=d \\ \Theta((x_{n1},y_{n1}),(x_{n2},y_{n2}))=\theta \\ I(x_{n1},y_{n1})=i, I(x_{n2},y_{n2})=j \end{array} \right. \right\rangle \tag{1}$$

where θ and d are the direction and distance between two pixels in images, N is the number of elements in the set, and I is an image with the dimension $M_1 \times M_2$ and the range of brightness value N of the processing space. The direction θ is quantized to four directions: horizontal, diagonal, vertical, and anti-diagonal (corresponding to 0°, 45°, 90° and 135°, respectively). Traditionally, Haralick's feature vectors [11] are computed from the co-occurrence matrices. Specifically, the features of *CON*, *HOM*, *ASM*, *H*, and *COR* were usually used for texture analysis.

4.2 Three-Dimensional Gray Level Co-occurrence Matrix (3D-GLCM)

A. 3D-GLCM Generation
The edge structures in iris images are often the most important features for pattern recognition and clear at a variety of scales. To capture the spatial details of an image,

it is advantageous to make use of co-occurrence matrix. Since the 2D-GLCM is unable to fully represent the texture features of the space domain images, we propose a new approach of GLCM, called 3D-GLCM, which is expanded from the original 2D-GLCM and thus can strengthen and demonstrate the texture features of the space domain images. It can be described in the following equation:

$$P_{COM}(i,j,k,d,\theta) = N\left\langle ((x_{n1},y_{n1}),(x_{n2},y_{n2}),(x_{n3},y_{n3})) \in M_1 \times M_2 \times M_3 \;\middle|\; \begin{array}{l} \max(|x_{n1}-x_{n2}|,|y_{n1}-y_{n2}|)=d \\ \max(|x_{n2}-x_{n3}|,|y_{n2}-y_{n3}|)=d \\ \Theta((x_{n1},y_{n1}),(x_{n3},y_{n3}))=\theta \\ I(x_{n1},y_{n1})=i \\ I(x_{n2},y_{n2})=j \\ I(x_{n3},y_{n3})=k \end{array} \right\rangle \quad (2)$$

Fig. 5(a) shows the spatial pixel relation where R is the referenced pixel, N_1 is the 1st neighbour, and N_2 is the 2nd. The structure of 3D-GLCM is shown in Fig. 5(b).

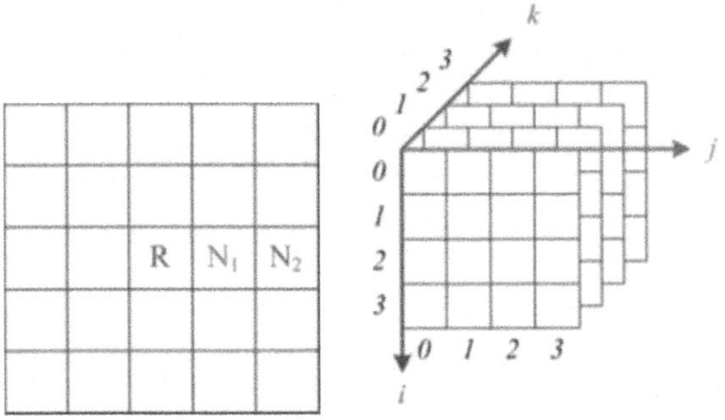

(a) (b)

Fig. 5. (a) Spatial pixel relation. (b) Structure of 3D-GLCM.

B. Feature Extraction using 3D-GLCM Feature Indexes

By modifying the original 2D-GLCM feature indexes, we design the corresponding new 3D-GLCM feature indexes. These new mathematical operators are listed below.

Features	Definitions	
contrast	$CON = \sum_{i,j,k=0}^{Q-1} P_{i,j,k}\left[(i-j)^2+(i-k)^2+(j-k)^2\right]$	(3)
homogeneity	$HOM = \sum_{i,j,k=0}^{Q-1} \frac{P_{i,j,k}}{1+\left[(i-j)^2+(i-k)^2+(j-k)^2\right]}$	(4)
angular second moment	$ASM = \sum_{i,j,k=0}^{Q-1} P_{i,j,k}^2$	(5)
entropy	$H = \sum_{i,j,k=0}^{Q-1} P_{i,j,k}\left(-\ln P_{i,j,k}\right)$	(6)
image pixel correlation	$COR = \sum_{i,j,k=0}^{Q-1} P_{i,j,k}\left[\frac{(i-\mu_i)(j-\mu_j)(k-\mu_k)}{\sigma_i\sigma_j\sigma_k}\right]$	(7)

C. Feature Extraction via Projection

Since we cannot obtain good recognition results by using the 3D-GLCM feature indexes to compute the features of texture, we introduced the projection method in which the 3D co-occurrence matrix is projected to three 2D planes for computing the feature vectors. The proposed 3D-GLCM projection method consists of several steps, as shown in Fig. 6. First, we segment the iris image into n blocks and compute 3D-GLCM for each of the image blocks. Next, we project the 3D-GLCM into three 2D planes and compute the feature vectors from each of these 2D planes using the 2D feature indexes.

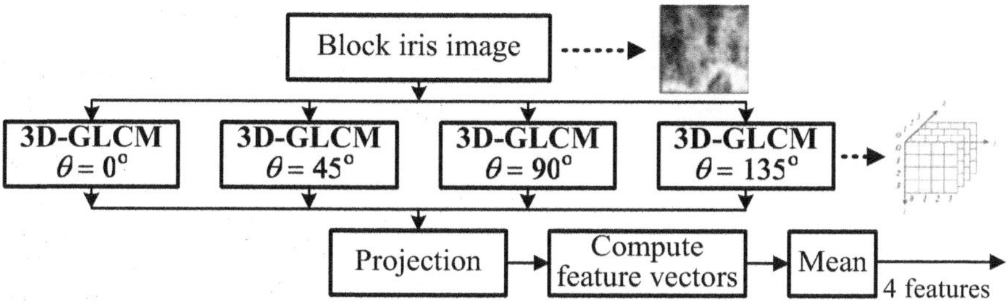

Fig. 6. Feature extraction

4.3 Pattern Recognition Module

In this module, the feature code vector extracted from the claimant iris image is compared against those of the enrolled feature code vectors in our iris database. Here for simplicity, we adopt the mean vector as the prototype of each pattern class in the enrollment phase and utilize the minimum distance classifier to check the approach in the recognition phase. When the feature code vector is compared, we calculate the normalized Euclidean distance D of feature vectors g_1 and g_2, which is defined as

$$D(g_1, g_2) = \frac{1}{255 \cdot N} \min_{rad} \left(\sum_{m=1}^{N} \left(g_1(m+rad) - g_2(m) \right)^2 \right)^{\frac{1}{2}}, \qquad D \in [0,1] \tag{8}$$

where m denotes the position of restructured signal and *rad* correction of the rotation effect of the input image. Both minimum distance classifiers are normalized to [0, 1].

5 Experimental Results

To evaluate the performance of the proposed human iris recognition system, we implemented and tested the proposed schemes on the UBIRIS iris image database [22]. The database comprises 1,205 iris images captured from 241 different eyes (hence, 241 classes).

5.1 Results of Image Pre-processing

We check the accuracy of the boundaries (including pupil, iris, and lower eyelid) subjectively and the proposed system obtains the success rate of 87.22% (1,051

images) from 1,205 images in the experiments for the pre-processing module. It is worth noting that the two main causes of failure come from occlusion by eyelids and non-uniform illumination. Among those 1,051 images obtained successfully from the image pre-processing module, we select 975 images (195 classes) out of them for testing (enrollment and recognition). We train the system by selecting 3 images as the training images set for each person from the authorized users in the enrollment phase. Hence, there are 390 images for testing.

5.2 Results of the 2D-GLCM Method

In this experiment, we test the recognition performance for the 2D-GLCM. To obtain a threshold for separating FRR and FAR, we perform two tests: one is for false rejection test and the other is for false acceptance rate test. For the case of FRR, we can obtain the distribution of non-matching distance between the unknown classes and the registered classes. For the case of FAR, we also obtain the distribution of non-matching between the unknown classes for impostors and the registered classes. Fig. 7(a) shows the distributions of the above two experiments. In this figure, the x-axis and y-axis indicate the degree of distance and the number of data, respectively. Fig. 7(b) shows the plot of the variation of FRR and FAR according to the distribution of non-matching distance by selecting a proper distance threshold. When we set the threshold to be 0.1117, the system obtains the recognition performance of about EER=1.55%. And when the FAR is set to be 0%, the system can obtain FRR=21.79% at a threshold of 0.04. As a result of the bad performance of FAR, we adopt our proposed approaches to improve further.

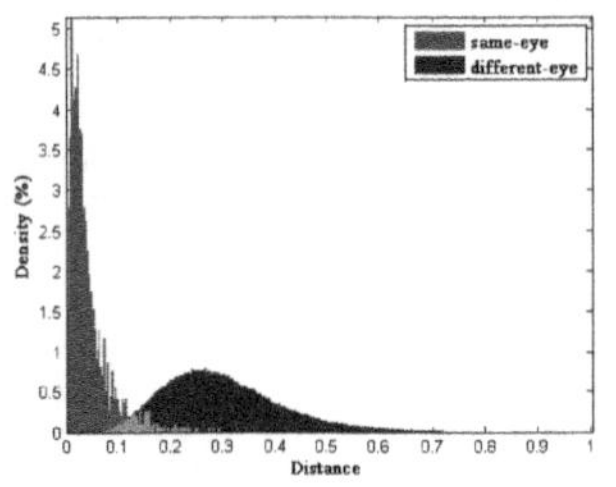

(a)

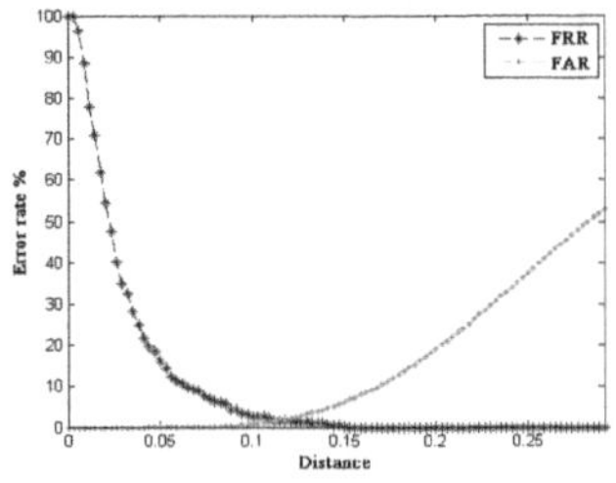

(b)

Fig. 7. The results of the 2D-GLCM method

5.3 Results of the 3D-GLCM Method

In this experiment, we test the performance for the 3D-GLCM method. The distribution of non-matching distance for FRR and FAR experiments is shown in Fig. 8(a). Fig. 8(b) shows the plot of the variation of FRR and FAR in the DM method by selecting a proper distance threshold. By selecting the threshold of 0.1217, the system obtains the system performance of EER=1.13%. Similarly, if the FAR is set to be 0%, the system can obtain FRR=8.72% at a threshold of 0.061. In particular, if we use a code vector of 192 bytes instead of 64 bytes, the recognition performance of the proposed system will be reduced to 0.35% only. The experimental results show that the proposed system performs well.

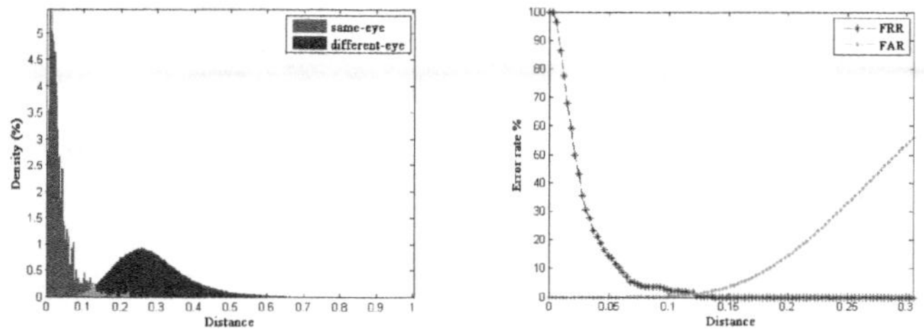

(a) (b)

Fig. 8. The results of the 3D-GLCM method

Finally, we make a summary of the experimental results in Table 1. We may observe that 3D-GLCM methods have more superior performance than the 2D-GLCM method in the results. On the other hand, the 3D-GLCM method can perform superiorly in the case of FAR=0%. Consequently, the 3D-GLCM method provides a securer system than the 2D-GLCM methods.

Table 1. Identification accuracy of the GDC, DM, and CFDM methods

Mode	IFE Module	Size of Features	RA (%)	AA (%)	AF (%)	RF (%)
EER	Projection	256 bytes	0.51	99.49	0.51	99.49
	Projection	192 bytes	0.35	99.65	0.35	99.65
	Projection	128 bytes	0.59	99.41	0.59	99.41
	Projection	64 bytes	1.13	98.87	1.13	98.87
	3D-GLCM	64 bytes	5.17	94.83	5.17	94.83
	2D-GLCM	128 bytes	1.28	98.72	1.28	98.72
	2D-GLCM	64 bytes	1.55	98.45	1.55	98.45
FAR = 0%	Projection	256 bytes	5.38	94.62	0	100
	Projection	192 bytes	5.64	94.36	0	100
	Projection	128 bytes	8.97	91.03	0	100
	Projection	64 bytes	8.72	91.28	0	100
	3D-GLCM	64 bytes	90.77	9.23	0	100
	2D-GLCM	128 bytes	16.15	83.85	0	100
	2D-GLCM	64 bytes	21.79	78.21	0	100

6 Conclusion

In this paper, a personal identification with iris recognition using three-dimensional gray level co-occurrence matrix has been proposed. We use two different methods, including 2D-GLCM and 3D-GLCM, in the feature extraction. The 3D-GLCM method can obtain a good recognition rate. The experimental result shows that the proposed iris recognition is suitable for the environment at high security level.

References

1. Miller, B.: Vital Signs of Identity. IEEE Spectrum 31, 22–30 (1994)
2. Daugman, J.G.: High Confidence Visual Recognition of Persons by a Test of Statistical Independence. IEEE Trans. Pattern Analysis and Machine Intell. 15, 1148–1161 (1993)
3. Wildes, R.P., et al.: A machine-Vision System for Iris Recognition. Machine Vision and Applications 9, 1–8 (1996)
4. Boles, W.W., Boashash, B.: A Human Identification Technique using Images of the Iris and Wavelet Transform. IEEE Trans. on Signal Processing 46, 1185–1188 (1998)
5. Zhu, Y., Tan, T., Wang, Y.: Biometric Personal Identification based on Iris Patterns. In: Proc. of Int. Conf. on Pattern Recognition, vol. II, pp. 801–804 (2000)
6. Ma, L., Wang, Y., Tan, T.: Iris Recognition using Circular Symmetric Filters. In: Proc. of Int. Conf. on Pattern Recognition, vol. II, pp. 414–417 (2002)
7. Ma, L., Tan, T., Wang, Y., Zhang, D.: Personal Identification based on Iris Texture Analysis. IEEE Trans. on Pattern Analysis and Machine Intelligence 25, 1519–1533 (2003)
8. Lim, S., Lee, K., Byeon, O., Kim, T.: Efficient Iris Recognition Through Improvement of Feature Vector and Classifier. ETRI Journal 23, 61–70 (2001)
9. Monro, D., Rakshit, M.S., Zhang, D.: DCT-based Iris Recognition. IEEE Trans. on Pattern Analysis and Machine Intelligence 29, 586–595 (2007)
10. Thornton, J., Savvides, M., Vijaya Kumar, B.V.K.: A Bayesian Approach to Deformed Pattern Matching of Iris Images. IEEE Trans. on Pattern Analysis and Machine Intelligence 29, 596–606 (2007)
11. Haralick, R.M., Shanmugam, K., Dinstein, L.: Textural Features for Image Classification. IEEE Trans. on Systems, Man, and Cybernetics 3, 610–621 (1973)
12. Valkealahti, K., Oja, E.: Reduced Multidimensional Co-occurrence Histograms in Texture Classification. IEEE Trans. on Pattern Analysis and Machine Intelligence 20, 90–94 (1998)
13. Zwiggelaar, R.: Texture based Segmentation: Automatic Selection of Co-occurrence Matrices. In: 17th Int. Conf. on Pattern Recognition, vol. 1, pp. 588–591 (2004)
14. Partio, M., Cramariuc, B., Gabbouj, M., Visa, A.: Rock Texture Retrieval using Gray Level Cco-occurrence Matrix. In: 6th Nordic Signal Processing Symposium, Norway (2002)
15. Yazdi, M., et al.: Novel Ridge Orientation based Approach for Fingerprint Identification using Co-occurrence Matrix. In: Proc. of World Academy of Science, Engineering and Technology, vol. 26, pp. 371–375 (2007)
16. Dabbah, M.A., Woo, W.L., Dlay, S.S.: Secure Authentication for Face Recognition. In: Proc. of IEEE Symp. on Comput. Intell. In: Image and Signal Proc., pp. 121–126 (2007)
17. Busch, A., et al.: Texture for Script Identification. IEEE Trans. on Pattern Analysis and Machine Intelligence 27, 1720–1732 (2005)
18. Szewczyk, R., et al.: Automatic People Identification on the Basis of Iris Pattern - extraction features and classification. In: Proc. Int. Conf. on Microelectronics, pp. 691–694 (2002)
19. Gupta, G., Agarwal, M.: Iris Recognition using Non Filter-based Techniques. In: Biometric Consortium Conference (2005)
20. Bachoo, A.K., Tapamo, J.R.: Texture Analysis and Unsupervised Clustering for Segmenting Iris Images. In: ACM Int. Conf. Proc. Series, vol. 150, pp. 236–243 (2005)
21. Zaim, A., et al.: A New Method for Iris Recognition using Gray-Level Co-occurrence Matrix. In: IEEE International Conference on Electro/information Technology, pp. 350–353 (2006)
22. Proenc, H., Alexandre, L.A.: Ubiris Iris Image Database, `http://iris.di.ubi.pt`
23. Poursaberi, A., Araabi, B.N.: A Novel Iris Recognition System using Morphological Edge Detector and Wavelet Phase Features. ICGST Int. J. on Graphics, Vision and Image Processing 5, 9–15 (2005)

A New Fake Iris Detection Method

Xiaofu He[1], Yue Lu[1], and Pengfei Shi[2]

[1] Department of Computer Science and Technology, East China Normal University, Shanghai 200241, China
{xfhe,ylu}@cs.ecnu.edu.cn
[2] Institute of Image Processing and Pattern Recognition, Shanghai Jiao Tong University, Shanghai 200240, China
pfshi@sjtu.edu.cn

Abstract. Recent research works have revealed that it is not difficult to spoof an automated iris recognition system using fake iris such as contact lens and paper print etc. Therefore, it is very important to detect fake iris as much as possible. In this paper, we propose a new fake iris detection method based on wavelet packet transform. First, wavelet packet decomposition is used to extract the feature values which provide unique information for discriminating fake irises from real ones. Second, to enhance the detecting accuracy of fake iris, Support vector machine (SVM) is used to characterize the distribution boundary based on extracted wavelet packet features, for it has good classification performance in high dimensional space and it is originally developed for two-class problems. The experimental results indicate the proposed method is to be a very promising technique for making iris recognition systems more robust against fake iris spoofing attempts.

1 Introduction

With the increasing requirements for higher security level, biometric systems have been widely used for many applications [1-3]. Biometric recognition or, simply, biometrics refers to the automatic recognition of individuals based on physiological or behavioural characteristics. Biometrics including face, iris, fingerprints, voice, palms, hand geometry, retina, handwriting, gait etc. have been used for the security applications and have many advantages compared to the traditional security systems such as identification tokens, password, personal identification numbers (PINs) etc. Iris recognition is one of the most promising methods because the iris has the great mathematical advantage that its pattern variability among different persons is enormous [4-5]. In addition, as an internal (yet externally visible) organ of the eye, the iris is well protected from the environment and stays unchanged as long as one lives [6-11]. However, biometric recognition systems are vulnerable to be spoofed by fake copies [12], for instance, fake finger tips made of commonly available materials such as clay and gelatine. Iris is no exception. There are potential threats for iris-based systems. The main potential threats are [12-14]: 1) Eye image: Screen image, Photograph, Paper print, Video signal. 2) Artificial eye: Glass/plastic etc. 3) Natural eye (user): Forced

M. Tistarelli and M.S. Nixon (Eds.): ICB 2009, LNCS 5558, pp. 1132–1139, 2009.
© Springer-Verlag Berlin Heidelberg 2009

use. 4) Capture/replay attacks: Eye image, IrisCode template. 5) Natural eye (impostor): Eye removed from body, Printed contact lens. Recently, the feasibility of some attacks have been reported by some researchers [12-16]: they showed that it is actually possible to spoof some iris recognition systems with photo iris, printed iris and well-made colour iris lens. Therefore, it is important to detect the fake iris as much as possible.

In previous research, Daugman introduced the method of using FFT (Fast Fourier Transform) in order to check the printed iris pattern [12-14]. His method detects the high frequency spectral magnitude in the frequency domain, which can be shown distinctly and periodically from the printed iris pattern because of the characteristics of the periodic dot printing. However, if the input counterfeit iris is defocused and blurred purposely, the counterfeit iris may be accepted as live one. Some iris camera manufacturer also proposed counterfeit iris detection method by using the method of turning on/off illuminator and checking the specular reflection on a cornea. Whereas, such method can be easily spoofed by using the printed iris image with cutting off the printed pupil region and seeing through by attacker's eye, which can make corneal specular reflection [15]. Lee et al. [16] proposed a new method of detecting fake iris attack based on the Purkinje image by using collimated IR-LED (Infra-Red Light Emitting Diode). Especially, they calculated the theoretical positions and distances between the Purkinje images based on the human eye model. However, this method requires additional hardware and need the user's full cooperation. To some extent, this interactive mode demands cooperation of the user who needs to be trained in advance and will eventually increase the time of iris recognition.

In this paper, we propose a new fake iris detection method based on wavelet packet transform together with SVM, which can detect the paper printed iris effectively. Wavelet packet transform is firstly used to extract the features. Then SVM is used to classify fake irises from real ones. The remainder of this paper is organized as follows: the proposed method is described in section 2. Section 3 reports experiments and results. Section 4 concludes this paper.

2 Proposed Approach

2.1 Feature Extraction

Wavelet transform is a mathematic tool for hierarchical decomposing functions. Wavelet packets transform (WPT) is a generalization of Wavelet transform that offers a richer signal analysis, which enables us to zoom into any desired frequency channels for further decomposition [17-18]. At each stage in the decomposition part of a WPT, four output subimages are generated, which contain approximation (A), horizontal detail (H), vertical detail (V) and diagonal detail (D) coefficients respectively. For instance, after 2-level WPT, an image has a quadtree with 20 output subimages, each representing different frequency channels, shown in Fig. 1. Therefore, wavelet packet analysis can fully make use of more information of the source image than wavelet analysis. The subimages which exclude approximation are suitable candidates for feature extraction.

In this paper, we present a new fake iris feature extraction method by using WPT. The proposed scheme of feature extraction is to use the n-level coefficients of decomposition parts of iris image via WPT. Since the differences between the fake and

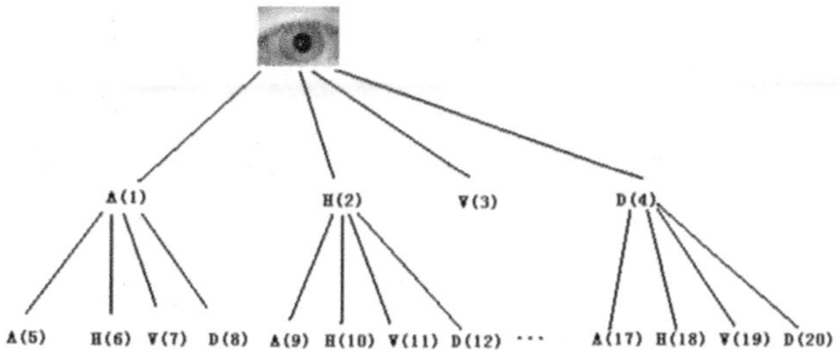

Fig. 1. The structure of 2-level WPT

live irises are located in the high and middle frequency channels, we only select horizontal detail (H), vertical detail (V) and diagonal detail (D) coefficients for discrimination between the fake and live irises. Each iris image was decomposed into n levels using WPT which resulted in 4^n components from wavelet packet tree structure. The iris feature vector consists of high frequency decomposition coefficients except the low frequency. For instance, for n equals to 2, there are totally 18 subimages except A(1) and A(5). Then, the standard deviations of those subimages are arranged to form an m-dimensional iris feature vector.

$$V = [std_1, std_2, std_3, ..., std_m]^T \tag{1}$$

Where $std_i (i = 1, 2..., m)$ denotes the standard deviation of the number i sub iris image after the WPT decomposition.

2.2 Classification

SVM has been recently proposed as a new technique for solving pattern recognition problems [19-20] which is originally developed for two-class problems. It performs pattern recognition between two classes by finding a decision surface determined by certain points of the training set, termed as Support Vectors. At the same time, the decision surface found tends to have the maximum distance between two classes. Therefore, in this paper, we select SVM as fake iris classification.

After feature extraction, an iris image is represented as a feature vector of length m. The features extracted are used for classification by SVM. In this paper, radial basis functions (RBF) kernel function of SVM is used as,

$$K(x, x_i) = \exp\{-\frac{|\mathrm{x} - x_i|^2}{\sigma^2}\} \tag{2}$$

Where, $\mathrm{x_i}$ comprises the input features, and σ is the standard deviation of the RBF kernel, which is three in our experiments.

The input to the SVM texture classifier comes from a feature vector of length m. The sign of the SVM output then represents the class of the iris. For training, +1 was assigned to the live iris class and -1 to the fake iris class. As such, if the SVM output for an input pattern is positive, it is classified as live iris.

3 Experimental Results

In this work, experiment is performed in order to evaluate the performance of the proposed method, which is implemented using Matlab 7.1 on an Intel Pentium IV 3.0G

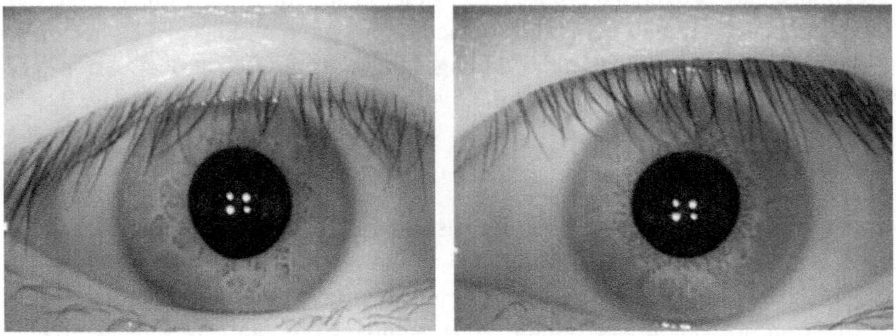

Fig. 2. Samples of live iris

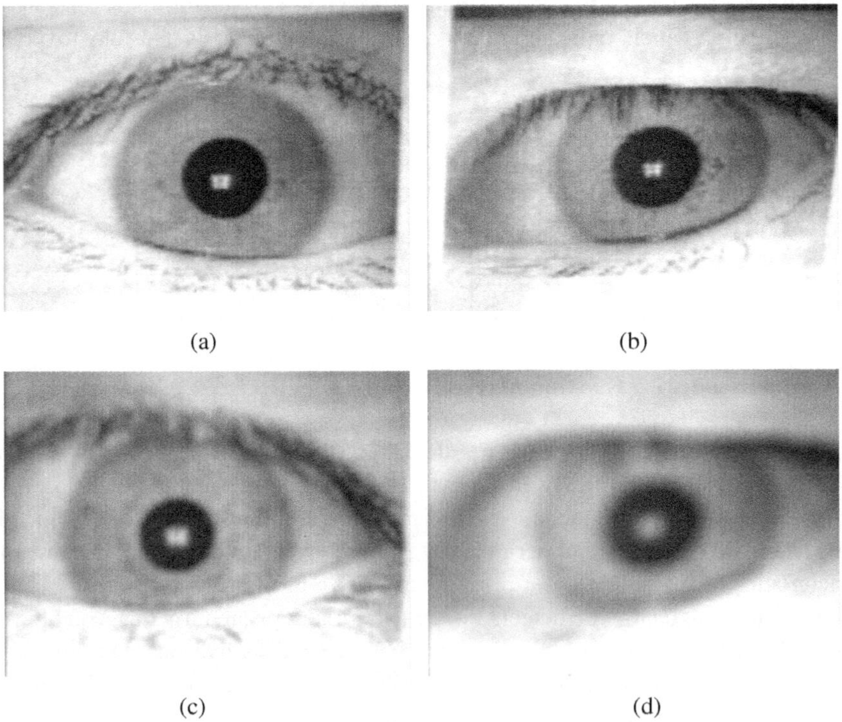

Fig. 3. Samples of printed fake iris. (a) and (b) are clear fake iris. (c) and (d) are defocused fake iris.

processor PC with 512MB memory. We manually collect 1000 live iris images, 220 defocused and motion blurred printed iris images and 140 clear printed iris images. Half of those iris images are used for training and the rest for testing. The positive samples (the live iris images) come from the SJTU iris database version 2.0 (Iris database of Shanghai Jiao Tong University, version 2.0) which is created by using contact iris capture device. Live iris images are printed using Laser Jet printer (The type of the printer is HP LaserJet 1020) and then are captured using the contactless iris capture device. The negative samples (fake iris images) come from those images that are captured at one session. The size of eye images is 320×240. Samples of the live and fake iris are shown in Fig. 2 to Fig. 3.

3.1 Testing Result

By investigating the training results, the iris feature vector consists of a feature vector of length eighteen, which reduces the size of the feature vector and results in an improved generalization performance and classification speed. The parameters of RBF kernel function are set: upper bound is 10, standard deviation is 3 respectively. The correct classification rate (CCR) results of the non-clear (defocused or motion blurred printed iris images) and clear fake irises are showed in Table1. The average execution time for feature extraction and classification (for testing) is 150ms and 14.6 ms respectively, which indicates that the proposed scheme is feasible to practical applications.

Table 1. Comparison of CCR results

Fake iris	Proposed	Traditional
Printed non-clear iris	98.18%	80%
Printed clear iris	98.57%	98.57%

3.2 Comparison with Existing Method

Among previous methods for fake iris detection, the method proposed by Daugman [12-14], is probably the most well-known. He proposed the method of using FFT in order to check the high frequency spectral magnitude in the frequency domain, which can be observed distinctly and periodically from the printed iris pattern because of the characteristics of the periodic dot printing, as shown in Fig. 4. However, the high frequency component cannot be detected in case that input printed iris image is blurred or defocused purposely and the fake iris may be accepted as live one consequently, as shown in Fig. 5. Therefore, there are two problems concerned, i.e. non-clear (e.g. defocused, motion blurred) and clear printed iris. A system that employs fixed-focus optical lens tends to result in defocused iris images. Motion blurred images are often happens if imitator wobbles purposely when spoofing the iris system.

Here, we will present a comparison between the current method and Daugman method described in [12-14] on the same iris database. For the purpose of comparison, we implement his method according to the published paper. Table 1 shows the comparison results of CCR.

Also, we calculated the time consumed of fake iris detection and compared the time consumed of it with traditional detection method, which have been implemented in the same environment, i.e. using Matlab 7.1 on an Intel Pentium IV 3.0G processor PC with 512MB memory. The average time is about 164.6 ms, whereas is about 92 ms with traditional detection method. The reason of it is that Wavelet packets transform is more complex than FFT. Although it is a little slower than traditional method, fake iris still can be detected at real time in practice use.

Based on the comparison results, we can conclude that the proposed method is encouraging comparing to the traditional fake detection method though the speed is a little slower than traditional method. In the case of that iris is defocused or motion blurred on purpose by attacker, our method seems to be more advantageous than the traditional method.

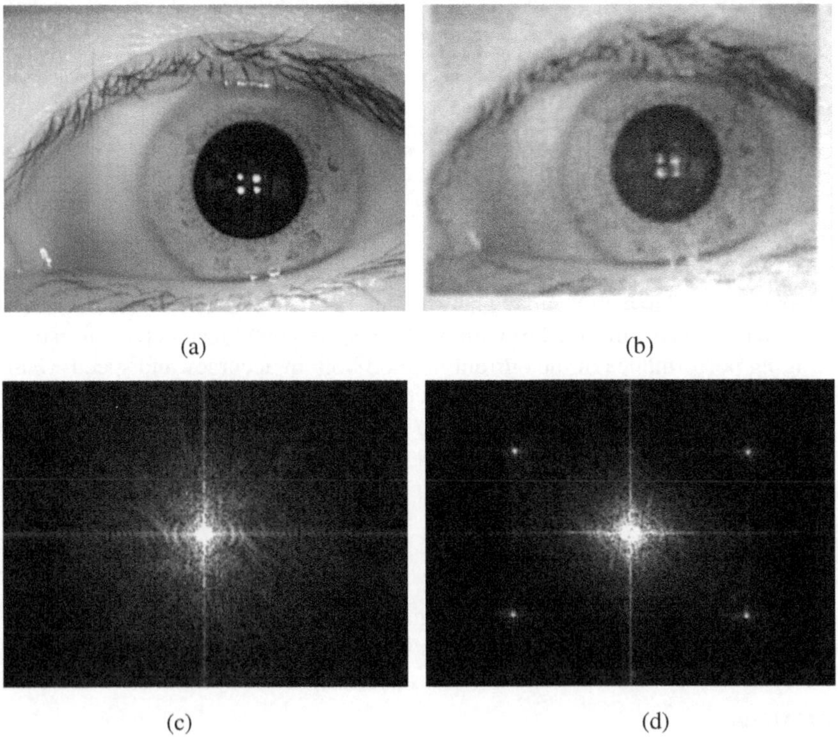

Fig. 4. Comparison of live iris and printed iris. (a) Live iris. (b) Fake iris printed on a paper. (c) 2D Fourier spectrum of live iris. (d) 2D Fourier spectrum of fake iris.

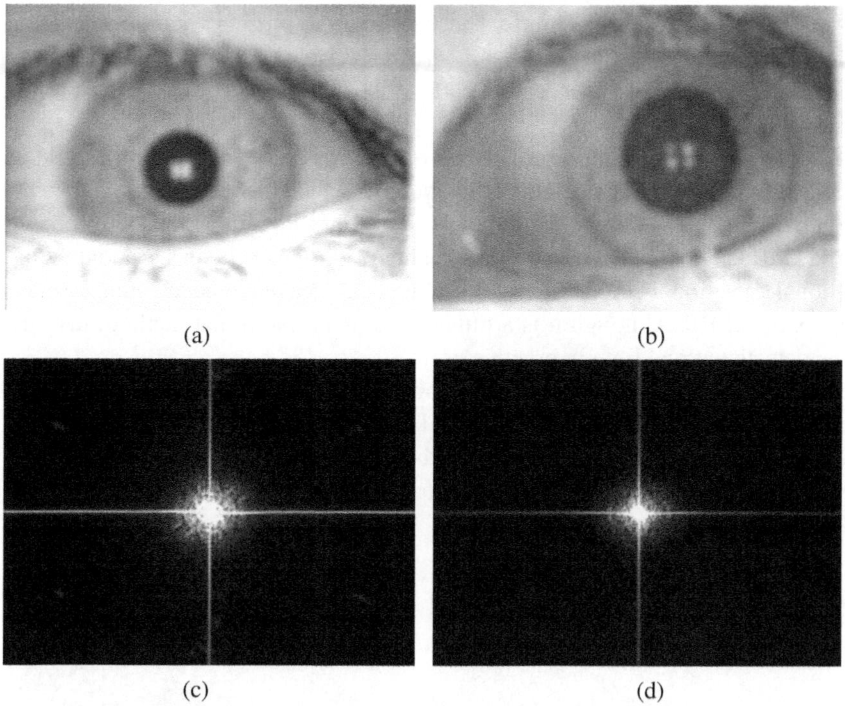

Fig. 5. Defocused printed iris. (a) (b) are defocused printed iris. (c) and (d) are Fourier spectrum of defocused printed iris.

4 Conclusion

In this paper, we have presented an efficient fake iris detection method based on wavelet packet transform together with SVM. Experimental results have illustrated the encouraging performance of the current method both in accuracy and speed. Using this method, paper printed iris can be well detected. It can help to further increase the robust of the iris recognition system. In the future work, we will extend the fake iris database and conduct experiments on a large number of iris databases in various environments to evaluate the stability and reliability of the proposed method.

Acknowledgements

This work is funded by the National 863 Program of China (Grant No. 2006AA01Z119) and Open Fund of National Laboratory of Pattern Recognition (NLPR) (Grant No. 08-2-13).

References

1. Jain, A.K., Bolle, R.M., Pankanti, S. (eds.): Biometrics: Personal Identification in Networked Society. Kluwer, Norwell (1999)
2. Zhang, D.: AutomatedBiometrics: Technologies and Systems. Kluwer, Norwell (2000)

3. Prabhakar, S., Kittler, J., Maltoni, D., O'Gorman, L., Tan, T.: Introduction to the Special Issue on Biometrics: Progress and Directions. IEEE Trans. Pattern Anal. Mach. Intell. 29(4), 513–516 (2007)
4. Daugman, J.: The importance of being random: Statistical principles of iris recognition. Pattern Recognition 36(2), 279–291 (2003)
5. Daugman, J.: How iris recognition works. IEEE Trans. on Circuits and Systems for Video Technology 14(1), 21–30 (2004)
6. Wildes, R.P.: Iris recognition: An emerging biometric technology. Proc. IEEE 85(9), 1348–1363 (1997)
7. Ma, L., Tan, T., Wang, Y., Zhang, D.: Personal identification based on iris texture analysis. IEEE Trans. Pattern Anal. Mach. Intell. 25(12), 1519–1533 (2003)
8. Sun, Z., Wang, Y., Tan, T., Cui, J.: Improving iris recognition accuracy via cascaded classifiers. IEEE Trans. on Systems, Man and Cybernetics, Part C 35(3), 435–441 (2005)
9. Park, K.R., Kim, J.: A real-time focusing algorithm for iris recognition camera. IEEE Trans. on Systems, Man and Cybernetics, Part C 35(3), 441–444 (2005)
10. Wei, Z., Tan, T., Sun, Z.: Nonlinear Iris Deformation Correction Based on Gaussian Model. International Conference on Biometrics, pp. 780–789 (2007)
11. Feng, X., Ding, X., Wu, Y., Wang, P.S.P.: Classifier combination and its application in iris recognition. International Journal of Pattern Recognition and Artificial Intelligence 22(3), 617–638 (2008)
12. Daugman, J.: Iris Recognition and Anti-Spoofing Countermeasures. In: The 7th International Biometrics Conference, London (2004)
13. Daugman, J.: Recognizing Persons by their Iris Patterns: Countermeasures against Subterfuge. In: Jain, et al. (eds.) Biometrics. Personal Identification in a Networked Society, pp. 103–121 (1999)
14. Daugman, J.: Demodulation by complex-valued wavelets for stochastic pattern recognition. International Journal of Wavelets, Multiresolution, and Information Processing 1(1), 1–17 (2003)
15. `http://www.heise.de/ct/english/02/11/114/`
16. Lee, E.C., Park, K.R., Kim, J.: Fake iris detection by using purkinje image. In: Zhang, D., Jain, A.K. (eds.) ICB 2006. LNCS, vol. 3832, pp. 397–403. Springer, Heidelberg (2006)
17. Daubechies, I.: Orthonormal bases of compactly supported wavelets. Commun. Pure Appl. Math. XLI, 909–996 (1988)
18. Laine, A., Fan, J.: Texture classification by wavelet packet signatures. IEEE Trans P. A. M. I 15(11), 1186–1191 (1993)
19. Burges, C.J.C.: A tutorial on support vector machines for pattern recognition. Data Mining Knowledge Discovery 2, 955–974 (1998)
20. Vapnik: Statistical Learning Theory. Wiley-Interscience publication, Hoboken (1998)

Eyelid Localization in Iris Images Captured in Less Constrained Environment

Xiaomin Liu, Peihua Li, and Qi Song

College of Computer Science and Technology, Heilongjiang University
Harbin, Hei Long Jiang Province, 150080, China
peihualj@hotmail.com

Abstract. Eyelid localization plays an important role in an accurate iris recognition system. In less constrained environment where the subjects are less cooperative, the problem becomes very difficult due to interference of eyelashes, eyebrows, glasses, hair and diverse variation of eye size and position. To determine upper eyelid boundary accurately, the paper proposes an integro-differential parabolic arc operator combined with a RANSAC-like algorithm. The integro-differential operator works as a parabolic arc edge detector. During search process of the operator, the potential candidate parabolas should near at least certain percentage of edgels of upper eyelid boundary, detected by 1D edge detector. The RANSAC-like algorithm functions as a constraint that not only makes eyelid localization more accurate, but also enables it more efficient by excluding invalid candidates for further processing. Lower eyelid localization is much simpler due to very less interference involved, and a method is presented that exploits 1D edgels detection and an RANSAC algorithm for parabolic fitting. Experiments are made on UBIRIS.v2 where images were captured at-a-distance and on-the-move. The comparison shows that the proposed algorithm is quite effective in localizing eyelids in heterogeneous images.

Keywords: Eyelid localization, Integro-differential parabolic arc operator, RANSAC, 1D Signal.

1 Introduction

Iris recognition has received increasing interest [12,7,3] among many biometrics technologies, such as face, fingerprint and speech recognitions etc. The characteristics that distinguish iris recognition from others include enormous statistical variability of iris patterns among different persons, stability over time of iris structure, non-invasiveness to human user and high recognition rate [5,6].

Accurate iris segmentation is a prerequisite step for a successful iris recognition system. In iris segmentation, after limbic boundary and pupillary boundary are determined, it is necessary to further localize upper and lower eyelids so that it could be determined whether or not iris is occluded. The straightforward idea is either to simply exclude the top or bottom parts of iris [5] or to use straight lines to fit eyelid boundaries [10,11] and invisible iris regions are

M. Tistarelli and M.S. Nixon (Eds.): ICB 2009, LNCS 5558, pp. 1140–1149, 2009.
© Springer-Verlag Berlin Heidelberg 2009

removed if obstruction occurs. Wildes [12] proposed a two-stage method for both limbic and eyelid boundaries localization, which consists of edge detection followed by Hough transform, tuned for circular and parabolic arcs respectively. In [4], eyelash region is firstly segmented by using high frequency information based on wavelet transform. Next raw eyelid contour, modeled by a parabolic arc, is achieved with least square fitting technique, and at last the eyelid position is fined. Jang et al. [9] introduced a robust eyelid detection approach that could deal with both defocused iris images and rotated iris images. Their method used different masks for eyelid boundary points detection according to focus assessment, and adopted a Hough transform to fit a parabola.

In less constrained environment [1], iris images are captured at-a-distance and on-the-move and eyelid boundary localization becomes more difficult, because of impact of eyelashes, eyebrows, hair, glasses and diverse variation of eye size and position. Fig. 1 shows some challenging examples of iris images.

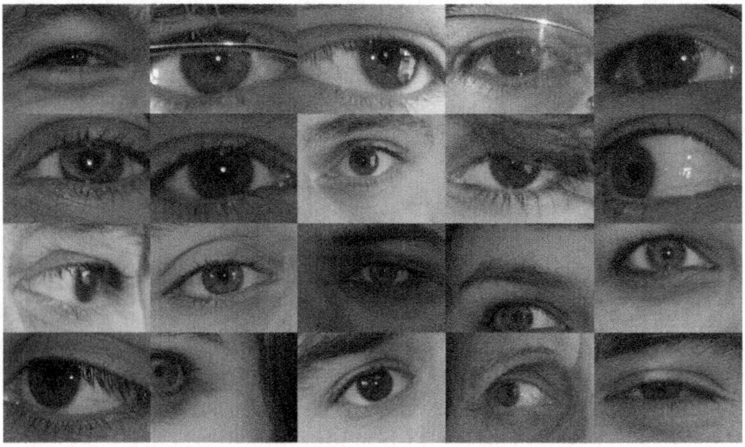

Fig. 1. Some iris images in UBIRIS.v2 database [1]

We focus on eyelid detection in these heterogeneous images. The limbic boundary is firstly localized using an improved Daugman's method [5] that is not discribed here due to page limit. Then we use an integro-differential parabolic arc operator for upper eyelid detection. During searching process of the ideal parabola, we introduce a RANSAC-like [8] algorithm to constrain the candidate parabolas. The constraint is that a valid potential parabola should near at least certain percentage of eyelid edgels detected by 1D edge detector, considering the fact that edgels may be noisy and not accurate. The resulting algorithm becomes more accurate and in the meantime more efficient thanks to exclusion of invalid candidates. In contrast to localization of upper eyelid, that of lower eyelid is much simpler and a straightforward method is briefly described.

2 Our Approach

In this section, we first consider problem of upper eyelid detection in sections 2.1 to 2.3, and then describe an approach for lower eyelid detection in section 2.4.

2.1 Integro-Differential Operator for Upper Eyelid Detection

The upper eyelid is modeled by a parabola of the form

$$\Gamma(a,b,c): \quad y = a(x-b)^2 + c \tag{1}$$

Given the outer circle, characterized by the center (x_c, y_c) and the radius R, of the iris localized by a method that is not described here due to page limit, we can determine the ranges of three parameters a, b and c. The parabola should open to the bottom and its curvature should be less than that of the iris, thus a should be in the interval $[0, 1/(2R)]$. According to geometric relationship of eyelid and iris, we can further constrain the domains of the parabola vertex (b, c): $b \in [x_c - R/2, x_c + R/2]$, $c \in [y_c - 3R/2, y_c + 2R/3]$.

Motivated by Daugman's paper [5], an integro-differential operator for parabolic location is proposed as follows

$$\operatorname*{argmax}_{a,b,c} f(a,b,c) = G_\sigma(c) * \frac{\partial}{\partial c} \int_{\Gamma(a,b,c)} \frac{I(x,y)}{L(a,b,c)} ds \tag{2}$$

where $I(x, y)$ is the iris image, the symbol $*$ stands for convolution operation, $G_\sigma(c)$ is a Gaussian with standard deviation σ, and $L(a, b, c)$ denotes the arc-length of the parabola $\Gamma(a, b, c)$. The integro-differential operator can be interpreted as a parabolic arc edge detector, which searches the parameter spaces of the parabola along which the smoothed integral derivative is maximum. Notice that when occlusion occurs the eyelid boundary above the iris may be weak and difficult to distinguish, thus the segment of the parabolic arc within the region as shown in Fig. 2 is excluded when calculating curve integral with Eq. (2).

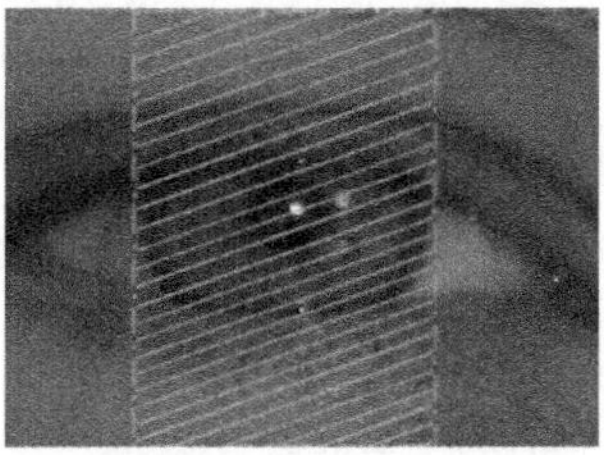

Fig. 2. The segment of a parabolic arc that falls within the shaded region is removed when evaluating curve integral with Eq. (2)

2.2 1D Edgels Detection of Upper Eyelid Boundary

Given the iris position (x_c, y_c) and radius R, we can detect eyelid boundary edgels along line segments

$$\begin{aligned} &x = d \\ &y_c - 2R \leq y \leq y_c + R/2 \end{aligned} \tag{3}$$

where d is constant varying in intervals $[x_c - 3R, x_c - R/2]$ or $[x_c + R/2, x_c + 3R]$. Firstly 1D signal is extracted according to Eq. (3), which is then smoothed with Gaussian, finally the gradient is computed and the point of maximum gradient that is larger than a prescribed threshold ε_e (ε_e=3.0 in our case) is regarded as an edgel of the eyelid boundary. Fig. 3 shows an example, where Fig. 3(a) gives search region of 1D signal. Figs. 3(b), 3(c) and 3(d) provide respectively a line segment along which 1D signal is extracted, filtered 1D signal and the gradient of the filtered 1D signal. The Edgels, $\mathbf{p}_i$, $i = 1, \cdots, N_e$, and the final parabolic arc denoting the upper eyelid boundary are given in Figs. 3(e) and 3(f) respectively. Note that only a portion of detected edgels is accurate, please refer to Fig. 3(e).

2.3 The Proposed Algorithm for Upper Eyelid Boundary Localization

The proposed algorithm is illustrated in Algorithm 1. The search ranges of parameters a, b, c are discretized by *step_a*, *step_b*, *step_c* respectively. For each three-tuple (a, b, c) in the search spaces, a RANSAC-like algorithm is used to determine whether or not the three-tuple is accepted as a potential candidate or just rejected. Given (a, b, c), the number N_p of detected edgels $\mathbf{p}_i$, $i = 1, \cdots, N_e$, is counted for which the distance of the edgel to the parabola is below a threshold ϵ_d. If the percentage of N_p over N_e is above another threshold ϵ_N, the parabola characterized by the three-tuple is a valid candidate for which $f(a, b, c)$ is computed according to Eq. (2). Finally we can get the parabola characterized by the three-tuple (a^*, b^*, c^*) with maximum f. In our experiments, $\varepsilon_d = 15.0$ pixels, $\epsilon_N = 0.5$.

The underlying idea is that at least a portion ϵ_N of edgels, detected as described in section 2.2, is *accurate* that functions as constraint through a RANSAC-like algorithm. The *accurate* here means that the distance between a pixel and the parabolic arc is within a tolerance ε_d pixels. The constraint imposed in this way makes eyelid boundary localization more accurate. An additional advantage is that the proposed algorithm becomes more efficient, because the parabolas that do not satisfy the constraint condition are excluded from further processing.

2.4 Locating the Lower Eyelid

Compared to upper eyelid detection, it is much simpler to detect lower eyelid due to less interference, please refer to Fig. 1. At the beginning, we perform 1D

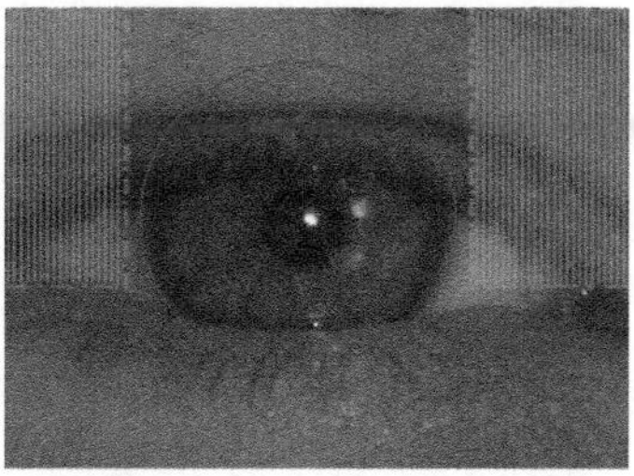

(a) Search region of 1D signal

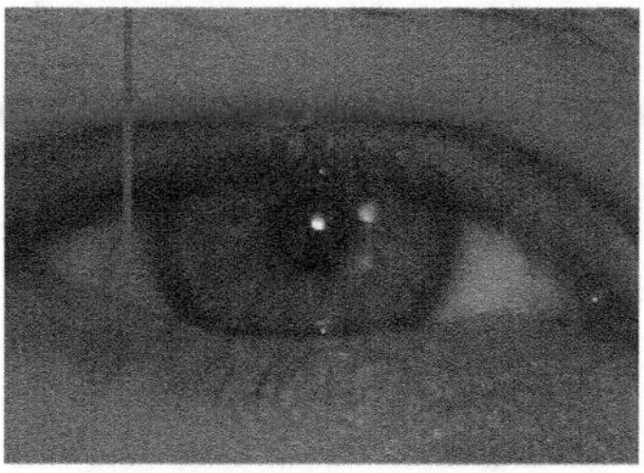

(b) Line segment to extract 1D signal

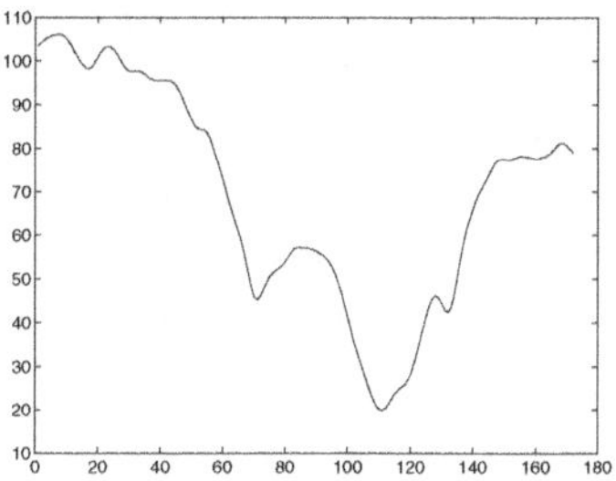

(c) 1D signal filtered with Gaussian

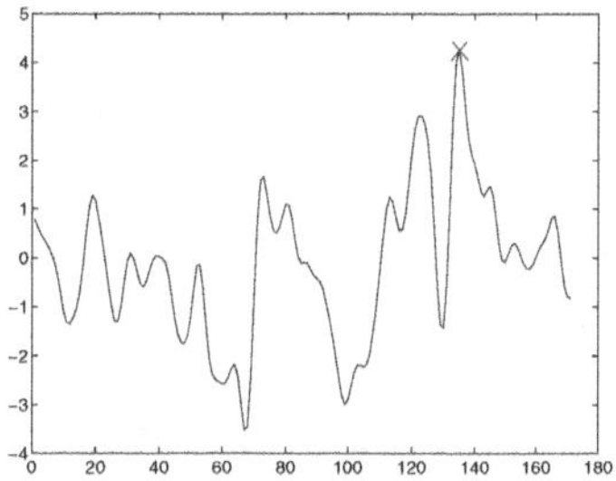

(d) Gradient of 1D signal

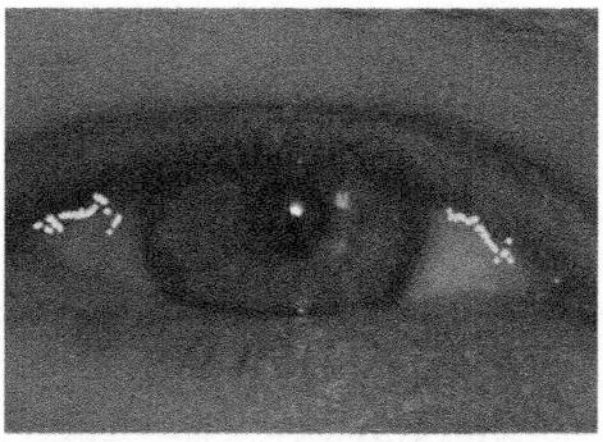

(e) Edgels obtained with 1D search

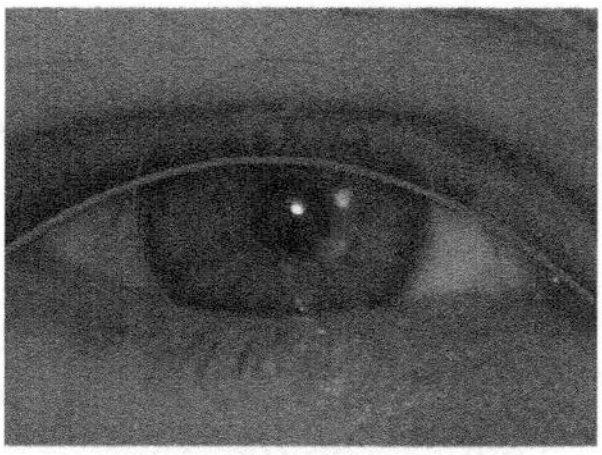

(f) Localization result

Fig. 3. Upper eyelid localization

search to detect edgels of eyelid boundary. The 1D search is along line segments of the form

$$\begin{aligned} &x = d \\ &y_c \leq y \leq y_c + 3R/2 \end{aligned} \tag{4}$$

where d is constant varying in the interval $[x_c - R/2, x_c + R/2]$. Please note the difference of Eq. (4) from Eq. (3). Because the edgels detected in this way may be noisy, RANSAC algorithm [8] is used to fit parabolic arc. Fig. 4 shows two examples of lower eyelid boundary detection, where the dashed rectangle indicates searching area of 1D signal, and the yellow dots represent edgels detected. Please note that when iris is not obstructed by lower eyelid, we actually detect lower boundary of iris, as demonstrated in Fig. 4(c). In both cases, we can determine whether or not occlusion happens and then remove occlusion regions, if any, as

Given searching steps *step_a*,*step_b*,*step_c* of parameters a, b, c, $f_{max} = -10000.0$
Set initial values of a, b, c, $a = step\_a$, $b = x_c - R/2$, $c = y_c - 3R/2$.
while $a < 1/(2R)$ **do**
  while $b < x_c + R/2$ **do**
    while $c < y_c + 2R/3$ **do**
      Count the number N_p of edgels $\mathbf{p}_i = (x_i, y_i)$, $i = 1, \cdots, N_e$, for which $d_i < \epsilon_d$, where $d_i = |y_i - (a(x_i - b)^2 + c))|$.
      if $N_p/N_e > \epsilon_N$ **then**
        Compute $f(a, b, c)$ according to Eq. (2)
        if $f_{max} < f(a, b, c)$ **then**
          $f_{max} = f(a, b, c)$
          $a^* = a, b^* = b, c^* = c$
        end if
      end if
      c += $step\_c$
    end while
    b += $step\_b$
  end while
  a += $step\_a$
end while

Algorithm 1. The pseudocode of the proposed algorithm to locate upper eyelid

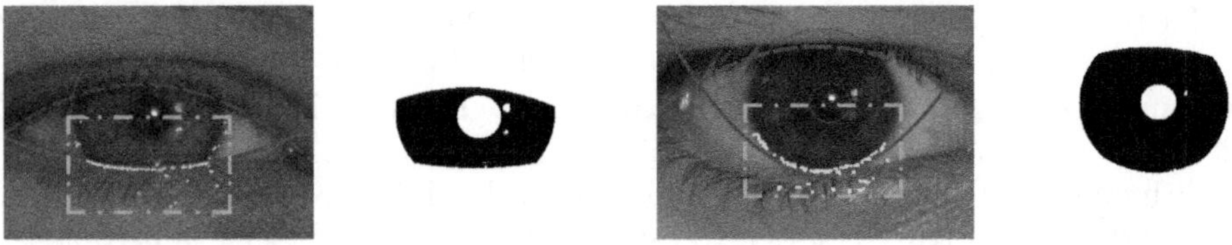

(a) (b) (c) (d)

Fig. 4. Eyelid detection and occlusion removal. Explanation please refer to text.

shown in Figs. 4(b) and 4(d). Note that in the latter case (as shown in Figs. 4(c) and 4(d)) there is no occlusion indeed that can be determined correctly.

3 Experiments

The algorithm is implemented with C++ on a PC with 3.0GHz Intel Pentium(R) 4 CPU and 2G Memory, and we do not make any code optimization. We primarily evaluate the performance of the proposed upper eyelid localization algorithm, comparing it with several other methods as described in Table 1. Because lower eyelid is much simpler to localize, in all algorithms the same method as described in section 2.4 is used for lower eyelid detection.

Table 1. Description of algorithms to be compared with the proposed one

	Algorithm description
Algorithm A	Algorithm A depends only on brute force search with integro-differential operator as described in section 2.1
Algorithm B	Algorithm B is dependent merely on edgels obtained with 1D methods as explained in section 2.2, and uses least square fitting to get the parabola
Algorithm C	Algorithm C counts only on edgels obtained with 1D methods as explained in section 2.2 as well, but, unlike algorithm B, uses RANSAC-like algorithm to fit the parabola
Algorithm D	Algorithm D relies on Canny edge detection followed by Hough transform for parabolic detection

3.1 Database and Ground Truth

We use UBIRIS.v2 iris database [1] in which images were captured at-a-distance and on-the-move. Due to the less constrained environment, the iris images are very heterogeneous and difficult for segmentation due to severe specular highlight, hair and glasses occlusion, considerable variation of eye size and position, eyelashes and eyebrows.

There are a total of 500 iris images of size 400x300 in UBIRIS.v2, in which 29 images are excluded that contains either closed-eyes or rotated-eyes, and the remaining 471 images are used for experiment. The thresholds involved in our algorithm are tuned on 100 images randomly selected in the set of 471 images. For the ith one, $i = 1, 2, \cdots, 471$, we hand-labeled the upper-eyelid boundary points and least square fitting is taken to estimate a parabolic arc that is used as the ground truth $y_{t,i}$, and the interval $[x_1, x_2]$ of the ground truth $y_{t,i}$ is determined simultaneously by the experimenter. The L^2-norm [2] in the following form is adopted to measure the error between the ground truth and an observation

$$E_i = \|y_{t,i} - y_{o,i}\| \triangleq \left(\frac{1}{x_2 - x_1} \int_{x_1}^{x_2} (y_{t,i} - y_{o,i})^2 dx\right)^{\frac{1}{2}} \tag{5}$$

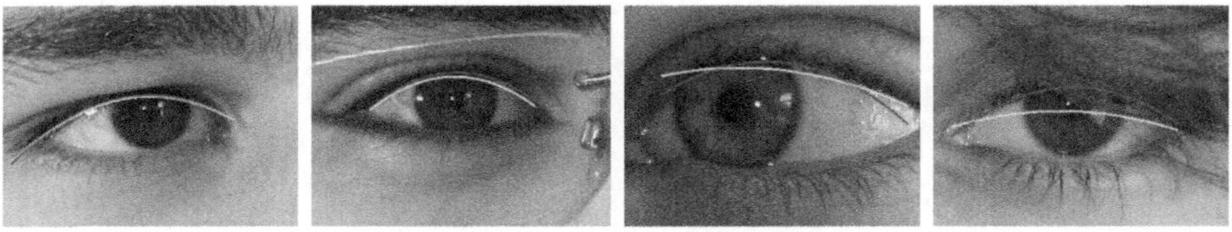

(a) 1.0336 (b) 5.1518 (c) 10.7277 (d) 23.6786

Fig. 5. Four examples of upper eyelid localization error between the ground truth (the red parabolic arc) and the observation (the yellow one), measured by Eq. (5). From left to right, the errors are 1.0336, 5.1518,10.7277 and 23.6786 pixels respectively.

Table 2. Comparison of upper eyelid localization

Algorithm	Average error (pixels)	Average time (s)
Proposed algorithm	12.09	0.424
Algorithm A	15.27	1.848
Algorithm B	34.39	0.037
Algorithm C	99.98	0.037
Algorithm D	24.85	0.746

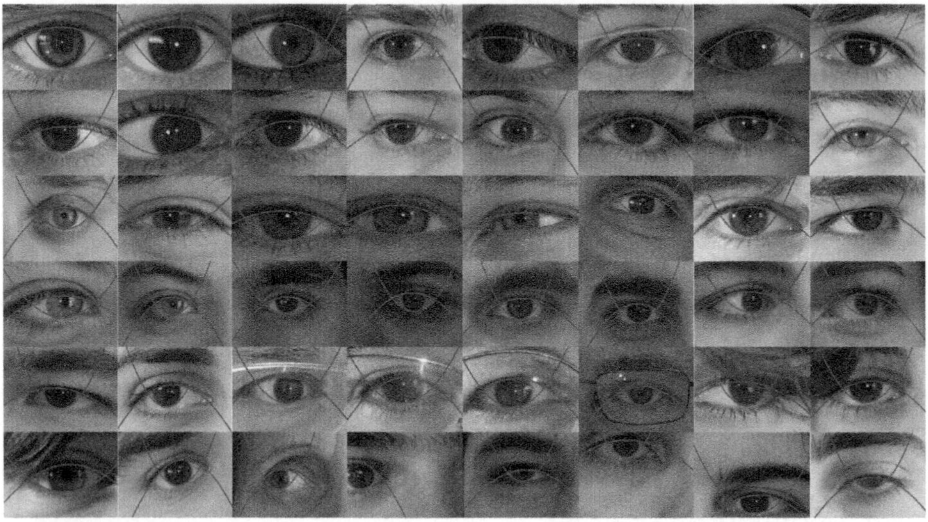

(a) Examples of good localization

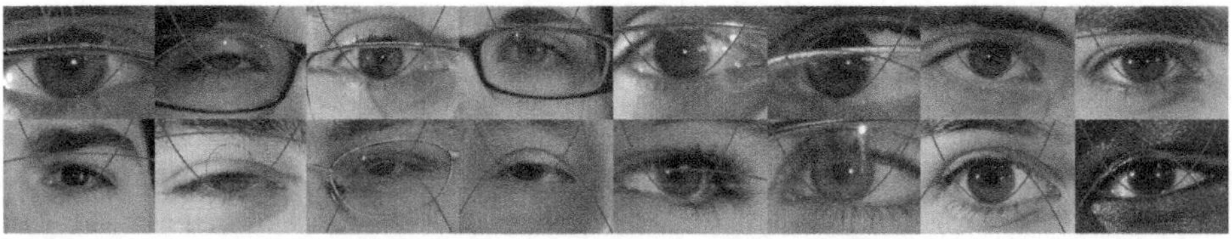

(b) Examples of bad localization

Fig. 6. Examples of eyelid localization with the proposed algorithm

where $y_{o,i}$ denotes the parabola detected in the ith iris image with the proposed algorithm. Then the average error can be computed

$$\bar{E} = \frac{1}{N} \sum_{i=1}^{N} \|y_{t,i} - y_{o,i}\|$$

where N is the total number of images. Fig. 5 presents four examples of upper eyelid localization error, where the red and the yellow parabolic arcs denote the ground truth and the observation respectively. From left to right, the errors measured by Eq. (5) are 1.0336, 5.1518,10.7277 and 23.6786 pixels respectively.

3.2 Comparison Result

Comparison of the proposed algorithm with other ones for upper eyelid localization is presented in Table 2. Above all, we notice that the proposed algorithm is significantly better than Algorithm D–Algorithm of Canny edge detection plus Hough transform, in both accuracy and computing time. Compared to the proposed algorithm, Algorithm A, which depends only on exhaustive search with integro-differential operator, deteriorates in localization accuracy but consumes much more time. Algorithm B that is dependent solely on least square fitting of eyelid edgels demonstrates bad results though it is quite efficient. Finally, Algorithm C, relying only on RANSAC-like fitting of eyelid edgels, exhibits surprisingly large errors. The reason is that, according to our analysis, the number of edges is small due to threshold ε_t we used, and the RANSAC-like algorithm does not work in most cases.

Some typical examples of eyelid localization with the proposed algorithm is shown in Fig. 6, where Fig. 6(a) provides examples of good localization while Fig. 6(b) gives those of bad localization. Boundaries of glasses or eyebrows, inaccurate localization of outer iris, low contrast between sclera region and eyelid can account for most of failure localization.

4 Conclusions

The paper addresses eyelid detection problem in heterogeneous images captured at-a-distance and on-the-move. In this type of less constrained environment, eyelid boundary localization becomes more difficult, due to interference of eyelashes, eyebrows, hair, glasses and more diverse variation of eye size and position. The proposed method for upper eyelid detection consists in an integro-differential parabolic arc operator combined with a RANSAC-like algorithm that imposes constraint of edgels on the operator. Lower eyelid detection is much simpler due to few interference involved, and a straightforward approach is developed. Current work can not deal with cases of closed-eye and rotated-eye that is left for our future research. We would also like to test the performance of our algorithm on other well-known iris databases.

Acknowledgements

The work was supported by the National Natural Science Foundation of China under Grant 60673110 and Natural Science Foundation of Heilongjiang Province (F200512), supported in part by Program for New Century Excellent Talents of

Heilongjiang Province (1153-NCET-002), Sci. & Tech. Research Project of Educational Bureau of Heilongjiang Province (1151G033), the Scientific Research Foundation for the Returned Overseas Chinese Scholars, State Education Ministry and Ministry of Personnel of China, Sci. and Tech. Innovation Research Project (2006RFLXG030) of Harbin Sci. & Tech. Bureau.

References

1. UBIRIS.v2 database (2008), `http://iris.di.ubi.pt/ubiris2.html`
2. Blake, A., Isard, M.: Active Contours. Springer, London (1998)
3. Bowyer, K.W., Hollingsworth, K., Flynn, P.J.: Image understanding for iris biometrics: a survey. Comp. Vis. Image Understanding 110(2), 281–307 (2008)
4. Cui, J., Wang, Y., Tan, T., Sun, Z.: A fast and robust iris localization method based on texture segmentation. In: Proc. of SPIE, pp. 401–408 (2004)
5. Daugman, J.G.: High confidence visual recognition of persons by a test of statistical independence. IEEE Trans. Pattern Anal. Mach. Intell. 15(11), 1148–1160 (1993)
6. Daugman, J.G.: The importance of being random: Statistical principles of iris recognition. Pattern Recognition 36(2), 279–291 (2003)
7. Daugman, J.G.: New methods in iris recognition. IEEE Trans. Systems, Man, Cybernetics B 37(5), 1167–1175 (2007)
8. Fischler, M.A., Bolles, R.C.: Random sample consensus: A paradigm for model fitting with applications to image analysis and automated cartography. Comm. of the ACM 24, 381–395 (1981)
9. Jang, Y.K., Kang, B.J., Park, K.R.: A study on eyelid localization considering image focus for iris recognition. Pattern Recognition Letters 29(1), 1698–1704 (2008)
10. Liu, X., Bowyer, K.W., Flynn, P.J.: Experiments with an improved iris segmentation algorithm. In: IEEE Workshop on Automatic Identification Advanced Technologies (AutoID), pp. 118–123 (2005)
11. Masek, L.: Recognition of human iris patterns for biometric identification. Bachelors Thesis (2003)
12. Wildes, R.P.: Iris recognition:an emerging biometric technology. Proc. IEEE 85(9), 1348–1363 (1997)

Noisy Iris Verification: A Modified Version of Local Intensity Variation Method

Nima Tajbakhsh[1], Babak Nadjar Araabi[1,2], and Hamid Soltanian-zadeh[1,2,3]

[1] Control and Intelligent Processing Center of Excellence, School of Electrical and Computer Engineering, University of Tehran, Iran
[2] School of Cognitive Sciences, IPM, Tehran, Iran
[3] Radiology Image Analysis Lab., Henry Ford Health System, Detroit, MI, USA
n.tajbakhsh@ece.ut.ac.ir, {a.araabi,hszadeh}@ut.ac.ir

Abstract. In this paper, a modified version of local intensity variation method is proposed to enhance the efficiency of identification system while dealing with degradation factors presented in iris texture. Our contributions to improve the robustness and performance of local intensity variation method consist of defining overlapped patches to compensate for deformation of texture, performing a de-noising strategy to remove high frequency components of intensity signals, proposing to add a coding strategy, and combining the dissimilarity values obtained from intensity signals. Experimental results on UBIRIS database demonstrate the effectiveness of proposed method when facing low quality images. To assess the robustness of proposed method to noise, lack of focus, and motion blur, we simulate these degradation factors that may occur during image acquisition in non-ideal conditions. Our results on a private database show that verification performance remains acceptable while the original method [11] suffers from a dramatic degradation.

Keywords: Noisy iris recognition, Score fusion, Robustness evaluation.

1 Introduction

Iris as the most promising biometric has attracted attention in recent years. Stability of iris texture along lifetime and its stochastic structure suggest the iris recognition system as a reliable authentication technology. Although several commercial iris recognition systems have been developed, they are to be used under constraints associated with lighting condition and subjects' behavior. In many cases, subjects have to participate in a verification process several times until the image acquisition process manages to capture a high quality iris image. This mainly comes from the fact that the efficiency of the current commercial iris recognition systems is highly affected by the quality of captured images and consequently the detailed information of iris texture. Therefore, this is typical of such systems to ask the subjects to participate in image acquisition process several times. Regardless of hardships that subjects face, sensitivity of the system performance to the quality of the acquired images restricts application of iris recognition in situations where we do not want to make the subjects aware of acquisition of their biometric data or where the aim is to make use of the system in outdoor conditions.

M. Tistarelli and M.S. Nixon (Eds.): ICB 2009, LNCS 5558, pp. 1150–1159, 2009.
© Springer-Verlag Berlin Heidelberg 2009

Recent research efforts in iris recognition have aimed to relax mentioned constraints. Many recent works with aim of addressing the challenges caused by low quality images have been evaluated on the UBIRIS database [1]. In the following, we briefly point to them and discuss about their results and achievements.

In [2], color Information instead of texture information is proposed to make more reliable decisions dealing with the low quality images. Although the color information on the first session of the UBIRIS outperforms the methods that are merely based on texture information, its performance declines significantly when facing the images of the second session of the UBIRIS. The main reason for lack of performance of color-based information may originate from the fact in uncontrolled lighting conditions, the color content of iris texture influenced and altered by the source of environmental illumination.

In [3], the contour of the shapes of the iris stroma has been exploited to obtain a signature robust to noise factors. Their method especially in small training samples does not provide high efficiency. In [4], an iris coding robust to noisy environment has been proposed in which the entropy of consecutive overlapped angular patches of normalized iris is computed to form a one-dimensional signature for each captured iris image. In spite of effectiveness of this approach, comparison of two signatures in this algorithm is a time-consuming process. In [5], to give a robust representation of iris texture, a division strategy similar to that proposed in [6] has been adopted to reduce the bad effects of eyelids and eyelashes to a minimum in feature extraction stage. Then, an independent feature extraction strategy is performed on each sub-region in which the Local Edge Pattern (LEP) operator is applied to extract various edge or corner patterns. In spite of promising performance, this method has not been evaluated on the second session of the UBIRIS database which is comprised of a large number of blurred or highly occluded images.

In [7], a novel prototype iris recognition system (Eagle-EyesTM) for long-range human identification is developed. Eagle-EyesTM is a multi-biometric system that acquires, a face and two iris images from subjects presenting anywhere within its large capture volume. In our previous works [8,9], texture information of visible and near infrared iris images are combined to enhance verification and recognition results obtained from poor quality images.

In this work, our aim is to extract local intensity variation in a way that its information content less affected by noisy regions. In this way, authentication system will be capable of facing highly occluded images even those captured in uncontrolled lighting conditions. The main ideas are proposing a de-noising strategy prior to feature extraction and a decentralized decision-making process in matching phase.

The rest of this paper is organized as follows: Section 2 describes the local intensity variation followed by a brief analysis on its disadvantages. A detailed description of the proposed feature extraction method is given in Section 3. Section 4 reports the experimental results and, finally, Section 5 presents the conclusions.

2 Local Intensity Variation Method

The idea of analyzing local intensity variations along angular direction is introduced in [10,11]. Since the most discriminative information of iris texture tends to run in the

radial direction, examining variations in the angular direction provides desirable features of iris texture [11]. Ma et al. [10,11] propose two well-known feature extraction techniques based on the local intensity variation. In both of them, the main idea is to extract sharp variations of intensity signals although different approaches are employed. However, the sharp variations of intensity signals are highly affected by noise factors presented in the iris texture. In [10], key local sharp variation points are extracted to represent the appearance or disappearance of important image structures. Determining the positions of key local points is carried out through finding local extreme, minimum or maximum in detail components of decomposed 1-D intensity signals. However, based on the quality of captured images it is possible that local extreme points vanish or some other spurious ones created along intensity signals. In [11], to capture details of the iris texture, Gaussian-Hermite moments of intensity signals are calculated. Due to high frequency components of intensity signals, the Gaussian-Hermite moment kernels are adopted to reflect random shape information of the iris. However, efficiency of the shape information extracted from the intensity signals is greatly influenced by those regions occluded by eyelids and eyelashes. Regardless of this issue, sharp variations of intensity signals playing a crucial role in the feature extraction strategy, perhaps, cannot be obtained from the images captured with poor focus.

3 Proposed Method

In this work, we make four contributions to local intensity variation including defining overlapped patches to compensate for deformation of texture, performing a de-noising strategy for removing high frequency component of intensity signals, proposing to add a coding strategy, and combining the dissimilarity values obtained from intensity signals to make more reliable decisions. This section begins by describing preprocessing stage and then the feature extraction is introduced.

3.1 Preprocessing

In [12], it is shown that isolating the eyelids and eyelashes and replacing the missing iris information through interpolation of the noise-free regions will not result in a significant increment in the separability between the intra- and inter-class comparisons. On the other hand, eyelids and eyelashes can transform the shape of intensity signals in a way that even removing the high frequency components –as a de-noising strategy– will not restore the main information of signals in those regions. Therefore, to eliminate such adverse effects, the upper half of iris that is usually obstructed by noisy regions is discarded [13]. As a result, all of the normalized images are of resolution 256x64 pixels.

3.2 Feature Extraction

The main idea is to extract a number of intensity signals that are capable of facing deformed texture and noisy regions. The intensity signals are constructed by averaging the gray levels of overlapped patches along radial direction. Selecting overlapped patches makes it possible to handle deformed texture. Here, instead of

sharp variations, our goal is to extract the soft variations or in other words, the main behavior of intensity signals. In this way, the captured information is less affected by noisy regions of the texture. Accordingly, we apply the wavelet de-noising to eliminate sharp variations of intensity signals. Moreover, this process enables the information content to be concentrated at a specific level of wavelet decomposition. Using Discrete Wavelet Transform (DWT), one can obtain scale of wavelet function with most informative coefficients. Since the main information of de-noised signals lies in low frequency components; and regarding to the fact that DWT cannot provide enough discriminative coefficients at large scales, the Continues Wavelet Transform (CWT) is adopted to extract features from de-noised signals. Thereafter, a coding strategy based on the signs of the coefficients is performed. That is, the positive coefficients are replaced by one and the negative values by zeros. Therefore, a binary code corresponding to every intensity signal is generated. In view of the fact that texture information is not uniformly distributed over the texture, the generated binary codes which reflect iris information in different regions have to be incorporated into the decision making process in a weighted manner. Furthermore, such decentralized decision-making minimizes the effects of noise factors (iris obstructions and reflections) through splitting the noisy areas into some sub-regions covered by different intensity signals. The way of combining the dissimilarity values is determined based on the power of discrimination of every intensity signal and their associated probability of corruption by noisy regions. Therefore, this strategy can serve as a complementary scheme to cope with noisy and degraded iris images. Combination of dissimilarities can be performed in several ways. Two common approaches that suggested for combining dissimilarity values are (1) performing a classification task in distance space [14, 15, and 16] and (2) estimating density of dissimilarities to form a fusion rule in probability space [17, 18]. In this work, we restrict ourselves to the first approach and adopt Support Vector Machines (SVM) as a candidate of linear and nonlinear classifiers to perform classification in distance space. This is due to the fact that mapping observations from distance space into probability space may result in missing information. For example, it is likely to have a meaningful manifold in distance space that enables us to reduce dimension of distance vectors while improving separability of genuine and imposter classes. However, such valuable property may not exist in probability space. An overview of the feature extraction and decision-making processes is depicted in Figure 1.

4 Experiments

To enable the effective test of proposed iris recognition method, we evaluate the efficiency of proposed method on the both sessions of UBIRIS database, while the training stage is carried out on a private database (UTIRIS) consists of two sessions with 1540 noisy images, 770 captured in Visible Light (VL) and 770 in Near Infrared (NIR) illumination. There are few images in UBIRIS database which are discarded during the evaluation of proposed method. Some of them failed in image segmentation process and others were not of quality to be involved in verification process. At last, we used 1740 out of 1877 images for performing evaluation of proposed method.

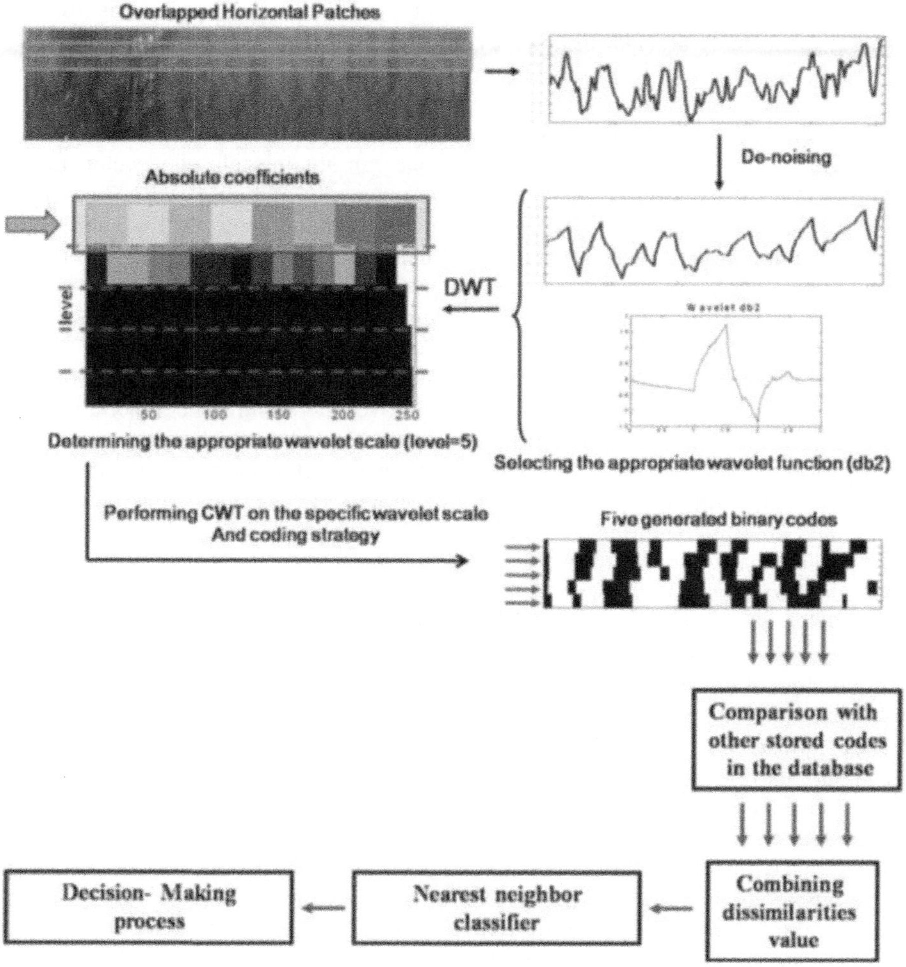

Fig. 1. An overview of proposed method

4.1 Results

The training of proposed method is carried out in two stages. The first one is determining the parameters involved in feature extraction which is performed on VL images of the UTIRIS database, and the other is defining the optimal fusion rule which is accomplished by cross validation test on UBIRIS database.

After extensive experiments, horizontal overlapped patches with 15 pixels width and 10 overlapped pixels were chosen. This enabled extraction of five intensity signals from a given iris pattern. The de-noising strategy was accomplished by performing wavelet decomposition at level 3 and reconstructing the signal just based

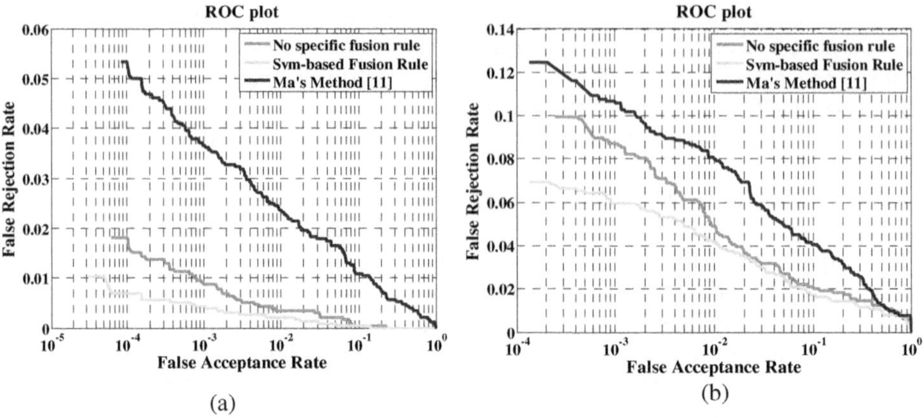

Fig. 2. This figure shows ROC plots of (a) first and (b) second session of UBIRIS. ROC curves are obtained from the proposed method without fusion rule, proposed method with SVM-based fusion, and Ma's method.

Table 1. Obtained results from both sessions of UBIRIS database

Method	Obtained results from			
	First session		Second session	
	EER (%)	FRR, FAR=0.001%	EER (%)	FRR, FAR=0.001%
Poursaberi and Araabi [19]	2.1	7.6	5.0	16.3
Ma et al. [11]	1.9	4.7	5.0	12.4
Rakshit et al. [20]	1.2	5.5	3.8	8.8
Ahmadi et al. [21]	1.9	5.4	8.0	21.6
Proposed method with SVM-based fusion rule	0.4	1.8	3.0	6.8

on approximation coefficients of third level and removing the detailed coefficients of other frequency sub-bands. Based on our experiments, the useful information of de-noised intensity signals were concentrated at the fifth dyadic scale of decomposition. Accordingly, we took the wavelet coefficients obtained through performing CWT on the specified scale and generate the corresponding binary code based on the signs of extracted coefficients. As a result, associated with an intensity signal, an iris code comprised of 256 bits was generated. Thus, a binary matrix consisting of five generated codes was created for a given iris image.

To obtain fusion rules, five-fold cross validation was adopted. That is, database was divided to five subsets, and each time, one of the 5 subsets is used as the test set and the other 4 subsets are put together to form a subset for training of fusion rule. The final ROC is obtained from averaging five resulting ROCs. We followed abovementioned strategy to assess efficiency of proposed method in both sessions of the UBIRIS database.

Figure 2 shows ROCs of proposed method when (1) no specific fusion rule is adopted, (2) using a SVM-based fusion rule, and ROC of method suggested in [11]. This way of comparison enables us to highlight the role of fusion rule in our achievements. It should be noted that utilizing no specific fusion rule is the same as summing dissimilarities or to measure dissimilarity of iris patterns when all five generated codes simultaneously are incorporated in matching process and as a result in this case only one dissimilarity value is obtained when two iris patterns compared. As it is seen, while our contributions including de-noising strategy and defining overlapped patches make a significant prominence in comparison with Ma's method in both sessions, enhancement achieved by utilizing fusion rule is more outstanding in the second session of UBIRIS. This is due to less noise content of images contained in the first session of database and substantial uniform distribution of information over European iris textures captured in visible light illumination. Accordingly, combination of texture information coming from different areas of iris cannot provide a significant enhancement in the first session of UBIRIS.

To make a more comprehensive comparison, we also implemented the methods suggested in [19, 20, and 21] based on their publications and tuned their pertaining parameters to get best possible results. To have a quantitative comparison between our method and other implemented methods, EER and FRR (@ Far=.001%) were computed and tabulated in Table 1. As it is seen, proposed method not only gives a superior performance in comparison with local intensity variation method [11] but also achieves highest efficiency compared with other implemented methods. This prominence is more highlighted when one compares obtained results from second session of UBIRIS. This implies the effectiveness of our method facing highly degraded images especially those suffering from reflections of cornea.

4.2 Robustness Evaluation

In the previous subsection, we demonstrated efficiency of proposed method. In the following, our aim is to assess robustness of proposed approach dealing with degradation factors like Lack of focus owing to a large stand-off distance, blurring as a result of relaxing constraints imposed on the motion of the subject, deformation of texture caused by pupil expansion/ contraction.

Due to the subject of security, being invariant to such degradations is of critical importance. To analyze how our proposed method deals with these unconstraint conditions, we gathered a small but noisy iris database. It is comprised of 200 iris images from 20 subjects that severely suffer from deformation of texture. In order to approach non-cooperative recognition -as an extreme case of noisy recognition, lack of focus and subjects' movement are simulated in our experiments. We also model defects in CCD camera and degradations during transmission of images, by injecting Salt and Pepper noise into captured images.

In each experiment, only one degradation factor is added to the iris images. In the case of blurring, one iris image of quality form per subject is selected and blurred with filters of 10 and 20 pixels length in four different directions (0°, 45°, 90°, and 135°). In the next experiment, Salt and Pepper noise with two noise densities (0.05 and 0.1) are injected to all captured images. We also model lack of focus in all iris images using circular averaging filters with point spread functions in the form of disks with 10 and 15 pixels radii. Verification results based on our experiments are shown in Table 2. As it is seen, except for motion blur of 10-pixel length, the proposed method exhibits the least performance degradation. Although our method has relatively small privilege over [11] in slight noisy condition, there exists a significant difference between achieved results and that of [11] in more noisy conditions. Furthermore, high robustness of feature extraction method versus the lack of focus and motion blur establishes our approach as an effective one to face highly degraded images that captured in non-ideal conditions.

Table 2. Obtained results through robustness evaluation of proposed method

Method	Degradation (%)											
	Motion Blur								Defocus		Noise (Salt & Pepper)	
	Length of the blur 10 pix				Length of the blur 20 pix				Radius of Disk		Density	
	Degree of the blur				Degree of the blur							
	0	45	90	135	0	45	90	135	10	15	5%	10%
Ma et al. [11]	0.5	1.7	0.7	1.7	6.6	12.7	8.7	11.8	3.6	8.2	1.9±0.8	3.8±1.0
Proposed	1.2	1.0	1.3	2.0	3.4	6.4	4.5	5.1	1.0	3.7	1.2±0.4	2.3±1.0

5 Conclusion

One of the main issues in iris recognition is coping with low quality iris images captured in uncontrolled lighting conditions and without active cooperative subject's behavior. In this paper, a modified version of local intensity variation method capable of facing mentioned challenges was proposed. Experimental results on the UBIRIS demonstrated the efficiency of proposed method in comparison with state-of-the-art algorithms. Our algorithm also provided more robustness in comparison with local intensity variation method.

References

[1] Proença, H., Alexandre, L.A.: UBIRIS: A noisy iris image database. In: Roli, F., Vitulano, S. (eds.) ICIAP 2005. LNCS, vol. 3617, pp. 970–977. Springer, Heidelberg (2005), `http://iris.di.ubi.pt`

[2] Krichen, E., Chenafa, M., Garcia-Salicetti, S., Dorizzi, B.: Color-Based Iris Verification. In: Lee, S.-W., Li, S.Z. (eds.) ICB 2007. LNCS, vol. 4642, pp. 997–1005. Springer, Heidelberg (2007)

[3] Hosseini, S.M., Araabi, B.N., Soltanian-Zadeh, H.: Shape Analysis of Stroma for Iris Recognition. In: Lee, S.-W., Li, S.Z. (eds.) ICB 2007. LNCS, vol. 4642, pp. 790–799. Springer, Heidelberg (2007)
[4] Proença, H., Alexandre, L.A.: Iris Recognition: An Entropy-Based Coding Strategy Robust to Noisy Imaging Environments. In: Bebis, G., Boyle, R., Parvin, B., Koracin, D., Paragios, N., Tanveer, S.-M., Ju, T., Liu, Z., Coquillart, S., Cruz-Neira, C., Müller, T., Malzbender, T. (eds.) ISVC 2007, Part I. LNCS, vol. 4841, pp. 621–632. Springer, Heidelberg (2007)
[5] Lee, J., Huang, P.S., Chang, C., Tu, T.: A Novel Approach for Iris Recognition Using Local Edge Patterns. In: Bebis, G., Boyle, R., Parvin, B., Koracin, D., Paragios, N., Tanveer, S.-M., Ju, T., Liu, Z., Coquillart, S., Cruz-Neira, C., Müller, T., Malzbender, T. (eds.) ISVC 2007, Part II. LNCS, vol. 4842, pp. 479–488. Springer, Heidelberg (2007)
[6] Proença, H., Alexandre, L.A.: Toward Noncooperative Iris Recognition: A Classification Approach Using Multiple Signatures. IEEE Transaction on Pattern Analysis and Machine Intelligence 29(4), 607–612 (2007)
[7] Bashir, F., Casaverde, P., Usher, D., Friedman, M.: Eagle-Eyes^{TM}: A System for Iris Recognition at a Distance. In: Proceeding of IEEE Conference on Technologies for Homeland Security, pp. 426–431 (2008)
[8] Tajbakhsh, N., Araabi, B.N., Soltanian-zadeh, H.: An Intelligent Decision Combiner Applied to Noncooperative Iris Recognition. In: IEEE Proceedings of the 11th International Conference on Information Fusion, 6 pages (2008)
[9] Tajbakhsh, N., Araabi, B.N., Soltanian-zadeh, H.: Feature Fusion as a Practical Solution toward Noncooperative Iris Recognition. In: IEEE Proceedings of the 11th International Conference on Information Fusion, 7 pages (2008)
[10] Ma, L., Tan, T., Wang, Y., Zhang, D.: Efficient Iris Recognition by Characterizing Key Local Variations. IEEE Transactions on Image Processing 13(6), 739–750 (2004)
[11] Ma, L., Tan, T., Wang, Y., Zhang, D.: Local intensity variation analysis for iris recognition. Pattern recognition 37(6), 1287–1298 (2004)
[12] Proença, H.: Towards Non-Cooperative Biometric Iris Recognition. University of Beira Interior, Phd Thesis (January 2007) ISBN: 978-972-99648-2-4
[13] Poursaberi, A., Araabi, B.N.: A Half-Eye Wavelet Based Method for Iris Recognition. In: ISDA 2005, Proceeding of 5th International Conference on Intelligent Systems Design and Applications, pp. 262–267 (2005)
[14] Brunelli, R., Alavigna, D.: Person Identification Using Multiple Cues. IEEE Transactions on Pattern Analysis and Machine Intelligence 17(10), 955–966 (1995)
[15] Fierrez-Aguilar, J., Ortega-Garcia, J., Gonzalez-Rodriguez, J., Bigun, J.: Discriminative Multimodal Biometric Authentication based on Quality Measures. Pattern Recognition 38(5), 777–779 (2005)
[16] Nandakumar, K., Chen, Y., Dass, S.C., Jain, A.K.: Likelihood Ratio-Based Biometric Score Fusion. IEEE Transactions on Pattern Analysis and Machine Intelligence 30(2), 342–347 (2008)
[17] Prabhakar, S., Jain, A.K.: Decision-level Fusion in Fingerprint Verification. Pattern Recognition 35(4), 861–874 (2002)
[18] Griffin, P.: Optimal Biometric Fusion for Identity Verification. Identix Research, Technical Report RDNJ-03-0064 (2004)

[19] Poursaberi, A., Araabi, B.N.: Iris Recognition for Partially Occluded Images: Methodology and Sensitivity Analysis. EURASIP Journal on Advances in Signal Processing 2007(1), 12 pages (2007)
[20] Rakshit, S., Monro, D.M., Zhang, D.: DCT-Based Iris Recognition. IEEE Transactions on Pattern Analysis and Machine Intelligence 29(4), 586–595 (2007)
[21] Ahmadi, H., Pousaberi, A., Azizzadeh, A., Kamarei, M.: An Efficient Iris Coding Based on Gauss-Laguerre Wavelets. In: Lee, S.-W., Li, S.Z. (eds.) ICB 2007. LNCS, vol. 4642, pp. 917–926. Springer, Heidelberg (2007)

An Automated Video-Based System for Iris Recognition

Yooyoung Lee[1,2], P. Jonathon Phillips[1], and Ross J. Micheals[1]

[1] NIST, 100 Bureau Drive, Gaithersburg, MD 20899, USA
{yooyoung,jonathon,rossm}@nist.gov
[2] Department of Computer Engineering, Chung-Ang University, Seoul, Korea

Abstract. We have successfully implemented a Video-based Automated System for Iris Recognition (VASIR), evaluating its successful performance on the MBGC dataset. The proposed method facilitates the ultimate goal of automatically detecting an eye area, extracting eye images, and selecting the best quality iris image from video frames. The selection method's performance is evaluated by comparing it to the selection performed by humans. Masek's algorithm was adapted to segment and normalize the iris region. Encoding the iris pattern and then completing the matching followed this stage. The iris templates from video images were compared to pre-existing still iris images for the purpose of the verification. This experiment has shown that even under varying illumination conditions, low quality, and off-angle video imagery, that iris recognition is feasible. Furthermore, our study showed that in practice an automated best image selection is nearly equivalent to human selection.

Keywords: Biometrics, Iris recognition, Eye detection, Image quality measurement, VASIR, Iris segmentation.

1 Introduction

Biometrics is the umbrella term for methods that recognize an individual based on physiological or behavioral characteristics. The human iris is a highly distinctive feature of an individual to establish identity of an individual with very high accuracy.

Still-image to still-image comparison for iris recognition is routinely covered by existing research studies. However, video-image to still-image comparison is a relatively new research subject which needs to overcome a number of challenges before the discipline can receive widespread acceptance. Challenges include recognizing a person in infrared image sequences, coping with high and low resolution, processing video sequences of people walking through a portal, matching to still face images, etc. The U.S. government sponsored the Multiple Biometric Grand Challenge (MBGC) program, which provides a standardized dataset as an aid to find solutions and to advance the current state of iris recognition technology[1].

Video-based Automated System for Iris Recognition (VASIR) is a method that successfully overcomes current limitations caused by external influences such as inconsistent lighting conditions, or low image quality. Its performance and practical feasibility have been evaluated using the above-mentioned MBGC dataset.

M. Tistarelli and M.S. Nixon (Eds.): ICB 2009, LNCS 5558, pp. 1160–1169, 2009.
© Springer-Verlag Berlin Heidelberg 2009

This paper also proposes a method for the selection of the best quality iris image from video frames using an automated algorithm. This method is a confirmation of its viability when compared against manual human selection.

This paper is divided into seven sections. This section served to introduce the notion of automated system for iris recognition in video streams. Section 2 gives an overview on the VASIR procedure. Sections 3-5 describe the eye detection/ extraction, image quality measurement, and the iris recognition algorithm. The experimental results are illustrated in section 6. Section 7 contains results and makes suggestions for future work.

2 Overview

Generally, still-image to still-image iris recognition proceeds in four distinct steps. The first is the image acquisition, which yields an image of the subject's eye region. The second step is the iris localization/normalization, which segments the iris from the rest of the acquired image. As a third step the iris image is encoded in binary. The last step compares an existing iris pattern (gallery) with the generated iris pattern (probe), to provide a decision on whether two irises are the same.

On the other hand, matching a video-based template to a still-based template needs more components for its iris verification procedure. Fig.1 illustrates the basic function of the VASIR process.

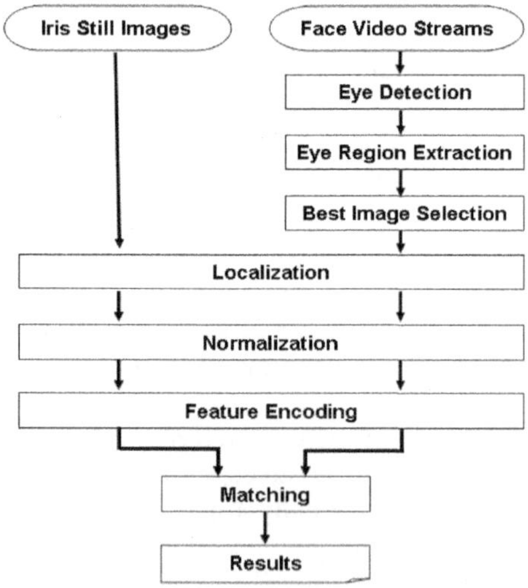

Fig. 1. VASIR Procedure

Before the localization (previously step 2), three additional steps were added in our implementation of the iris verification procedure. The new initial step is the automatic detection of the eye region within video frames; followed by the eye region extraction. In the last additional step, we select the "best" quality image from video.

3 Eye Detection and Extraction

The term "eye detection" will be used to refer to the process of the continuous eye area detection in video streams. We employed a method called "object detector", initially proposed by Viola-Johns[2]and later improved by Lienhart-Maydt[3] through the addition of rotated Haar-like features as described below. The integral image, an intermediate image representation, utilized in the method allows a rapid computation of features. The yielded features are then filtered by classifiers which are based on the boost learning algorithm. Combining multiple classifiers in a cascade allows for the background regions of the image to be quickly rejected, enabling more computation on promising eye pair like regions. A cascade of classifiers has a decision tree; each node is trained to identify objects-of-interest and to reject unrelated objects. In combination with a search window moving across the frame, the classifier can be used to locate the eye region. The classifier is resizable in order to be able to find objects-of-interest independent of varying size, i.e. the scan procedure needs to be repeated several times to find objects of differing size.

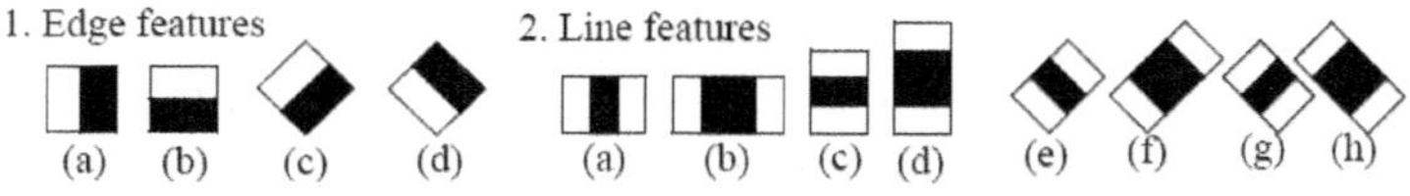

Fig. 2. Used Haar-like Features (captured from [3])

The edge features measure the difference in intensity between the eye regions and the region across the upper cheeks. The features make use of the fact that the eye region is often darker than the cheeks. The line features compare the intensities in the eye regions to the intensity across the bridge of the nose.

The OpenCV community[4] shared a collection of public domain classifiers and the performance of these classifiers were evaluated by On-Santana et al[5]. To detect both the left and right eye region – also called eye-pair or two-eye – in a face video stream, we chose the named eye-pair classifier[6][7] due to its fitting properties, speed and size; The eye-pair classifier is applied to each frame to find the position and size of both eye regions. The classifier in the cascade is 45x11 pixels in size and consists of 19 stages.

We defined a minimum size for the eye area in the frames to lower the number of false positives. The detected region is saved into three separate files: two-eye, left-eye-only and right-eye-only automatically for analysis purposes. Detailed performance results are discussed in section 6.

4 Image Quality Measurement

Normally, present commercial iris recognition systems use images acquired under constrained conditions on account of their image quality requirements. Near-infrared illumination is used to light the face, and the user is prompted with visual and/or auditory feedback to correct the eye position. Due to this the image is often in focus and of sufficient size for iris recognition.

Being able to measure the eye image quality in video streams under general unconstrainted conditions would permit a more accurate determination when to match two iris templates. Several groups[8] have studied how to determine the focus of an image (e.g. analyzing the sharpness of the pupil and iris boundary, calculating the total high frequency power in the 2D Fourier spectrum, a variation of the sum modulus difference (SMD) filter, etc).

This paper proposes two methods of selecting the best image from eye images obtained using video streams. The first method is named the "Human Image Quality Measurement" (HIQM), where the best quality eye images are manually selected via human vision. The other method is an application of the "Automatic Image Quality Measurement Algorithm" (AIQM) via the edge density, which has been proposed for predicting the face image quality by Beveridge et al[9]. Fig.3 illustrates the processes.

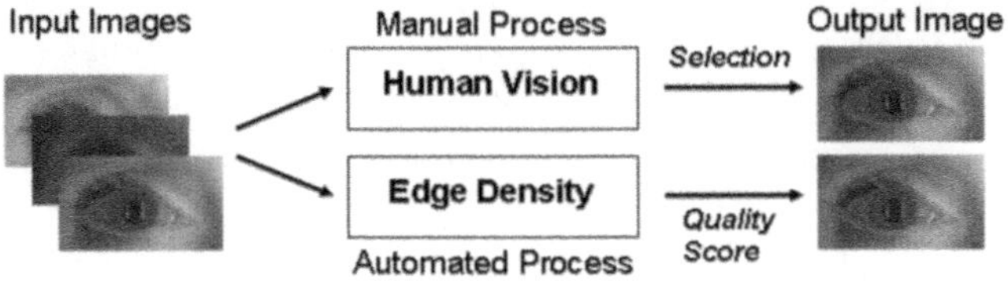

Fig. 3. Image Quality Measurement (via Human Vision and Edge Density)

Out of the set of eye images, and for each eye individually, two images are selected by human vision in HIQM. Afterwards, the two selected eye images are segmented to isolate the iris region, and the best one-eye images (left, right) are chosen corresponding to segmentation results. The detailed segmentation process will be addressed in section 5.

For AIQM, the average edge density in an image is calculated, to determine the level of focus. The sobel edge detector is applied to the image to derive the edges and the average sobel edge is quantified within the size of the eye region.

The performance of VASIR is evaluated as to whether the AIQM is equivalent to HIQM for best image selection. Performance values will be provided in section 6.

5 Iris Recognition Algorithm

The iris verification components in VASIR were implemented using the algorithm initially developed by Masek[10], re-implemented for irisBEE[11].

In the *iris localization* step, the iris is segmented by detecting the inner circle, which is the boundary between the pupil and the iris, and the outer circle, which is the boundary between the iris and the sclera. An automatic segmentation algorithm based on the circular Hough Transform initially employed by Wildes et al[12] is used. The center coordinates (x_c, y_c) and the radius r are able to define any circle according to the equation;

$$x_c^2 + y_c^2 - r^2 = 0 \quad (1)$$

Masek's algorithm uses circular Hough Transform to detect the iris and pupil boundaries by employing Canny edge [10] to generate the edge map. The eyelids were removed using a linear Hough Transform by inserting horizontal lines attached to the eyelids itself. The eyelashes are detected and ignored using threshold manipulation. VASIR applies the segmentation to extract the iris region from the eye images selected by the image quality measurement algorithm.

For *normalization* step, the outcome is a standardized 2D representation of the iris pattern, regardless of the pupil dilation, source image size, non-concentric pupil displacement, or the distance of the eye to the capture device. Daugman[13]'s polar coordinate-based system effectively eliminates scaling issues caused by linear stretching.

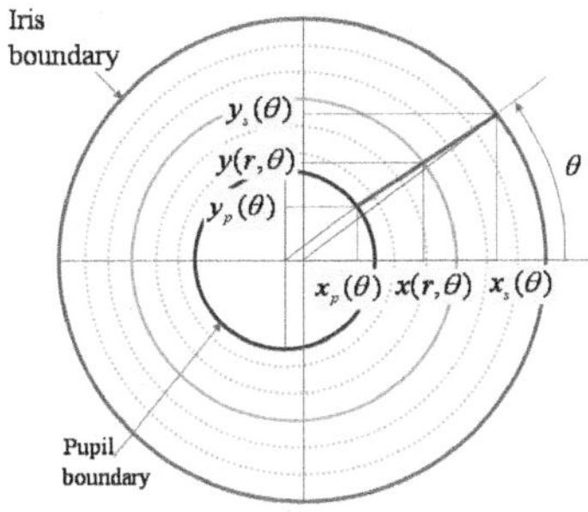

Fig. 4. Doubly Dimensionless Projected Polar Coordinate System[1]

The homogeneous rubber sheet model assigns to each point within the iris region a pair of real coordinates (r,θ) where r lies on the unit interval [0,1] and θ is the angle over $[0,2\pi]$. The remapping of the iris image I(x,y) from Cartesian coordinate (x,y) to polar coordinate (r,θ) is classically represented as;

$$\begin{aligned} &I(x(r,\theta), y(r,\theta)) \rightarrow I(r,\theta) \\ &x(r,\theta) = (1-r)x_p(\theta) + rx_s(\theta) \\ &y(r,\theta) = (1-r)y_p(\theta) + ry_s(\theta) \end{aligned} \qquad (2)$$

$(x_p(\theta), y_p(\theta))$ and $(x_s(\theta), y_s(\theta))$ are the coordinates of the pupil and iris boundaries along the θ direction. In Masek's algorithm, the pupil center is used as reference point, and the iris region is scaled based on the radial vectors. Finally, an iris pattern image and the noise mask using occlusion information produced in the localization stage are generated containing the size of the angle and the radial.

Many different filters have been suggested to extract the features from the iris region (e.g. Laplacian-of-Gaussian filter, Wavelet Transform, Discrete Cosine Transform, etc.) [8]. Masek's algorithm employed a 1D Log-Gabor filter which was introduced by Yao et al[14], in order to process the *feature encoding* from normalized iris images. The normalized iris patterns are then coded by filters that have Gaussian

[1] The figure was modified from the presentation "Analysis of Daugman's High Confidence Visual Recognition of Persons by a Test of Statistical Independence" by Minhwan Kim.

transfer functions when viewed on the logarithmic frequency scale. The frequency response of a Log-Gobor filter is given as;

$$G(w) = e^{(-\log(w/w_0)^2)/(2\log(\sigma/w_0)^2} \tag{4}$$

Where w_0 represents the filter's centre frequency, and σ gives the filter's bandwidth. Each isolated iris pattern is demodulated to extract its phase information using Daugman's method[15].

For *matching*, Masek employed the Hamming Distance (HD) that incorporates noise masking. The HD measure can be used to make a decision whether the iris pattern is of the same person or a different person. The noise mask helps to use only the significant bits in calculating the HD between two iris templates. For rotational inconsistencies between two iris templates, one template is shifted left and right bit-wise and selects values from successive shifts[15]. If an angular resolution of 180 is used, each shift will correspond to a rotation of 2 degrees in the iris region.

6 Experimental Results

6.1 Dataset

VASIR is evaluated using the datasets collected by the MBGC[1]. Two of the goals of the MBGC are to promote and to advance iris recognition technology. The MBGC dataset includes eye images on varying illumination conditions, low quality, and off-angle or occluded images on both still and video imagery.

This paper uses two types of the datasets which we will call 'face visible videos' and 'iris still images'. The face visible videos were captured with a camera by Sarnoff with the Iris On the Move (IOM) system; each frame is 2048x2048 pixels in resolution. An LG2200 EOU camera was used to capture the still images in 640x480 resolutions. There are a total of 149 video sequences and a total of 1,668 iris still images. Of the 149, there is no eye visible in just one of the video sequences, and,in 18 other video sequences only one eye is visible. Since our purpose is to detect both left and right eyes (two-eye) within video frames, we conducted our experiment using only the 130 videos where both eyes are visible.

6.2 Performance for Eye Detection and Extraction

The number of successful two-eye detections in videos was 127 out of the total of 130 videos; in 97.69% both eyes were correctly detected. By defining a minimum size for the eye pair region, the false positives were reduced, resulting in just 3.15%.

Fig.5 depicts examples of successful two-eye detections and false positive detections. For analysis purposes, VASIR automatically extracted 1,083 left-eye and 1,083 right-eye images out of the total of 2,534 frames from the 130 video sequences. The extracted eye images were used to determine the best quality image; the selection is explained in the next section. The sizes of the extracted eye regions are very diverse due to the fact that the subject keeps moving in the video – effectively changing the distance from the camera. The average dimension of the saved images for the left and right is about 500 (width) x ¾ of the width (height) pixels in resolution.

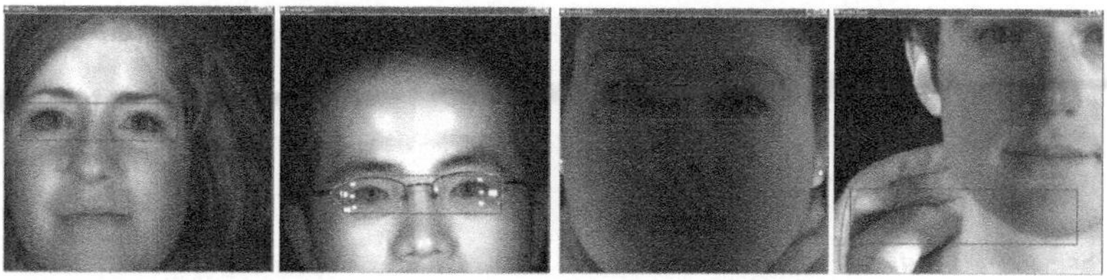

(a) (b)

Fig. 5. Examples of the Eye Detection (a) Successful Two-Eye Detection (b) False Positive Two-Eye Detection

6.3 Best Iris Image Selection and Segmentation

As stated in section 4, the best quality image was selected by two methods: using the HIQM and through the AIQM. Fig. 6 shows an example of the original video frames (6a), the extracted left eyes (6b), and the best-left-eye image selection by AIQM.

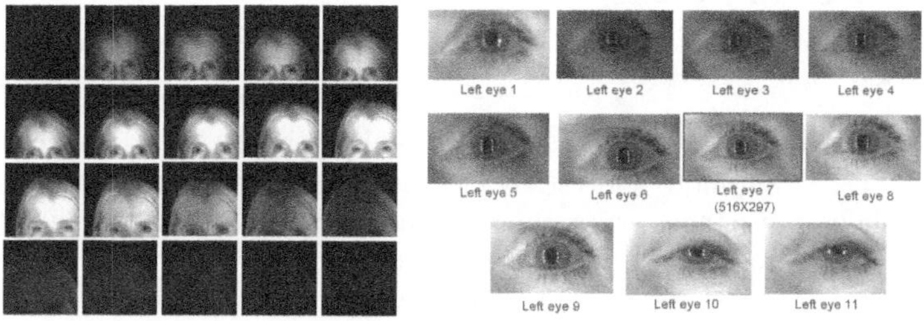

(a) (b)

Fig. 6. An Illustration of the Best Image Selection Process (a) Original Video Frames (b) Best-Left-Eye Selected by AIQM

In the Fig. 6-b, 11 left eyes were extracted in a total 20 frames, and then the seventh left-eye was selected as the best iris image through Edge Density algorithm.

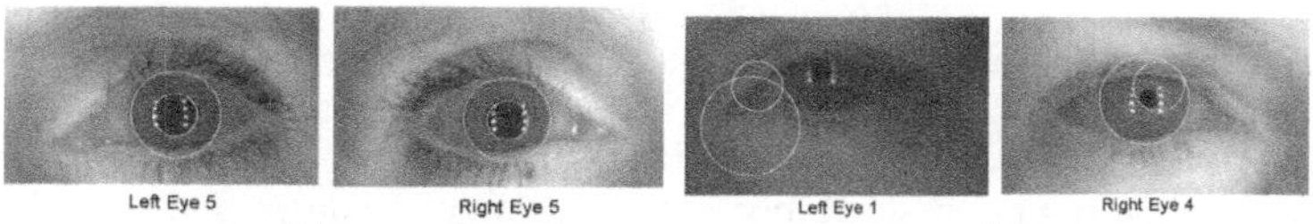

(a) (b)

Fig. 7. Examples of the Iris Region Segmentation (a) Successful Segmentations (b) Incorrect Segmentations

All selected images were segmented in order to extract the iris region. Fig. 7 shows examples of successful and incorrect segmentations.

In the 254 (127 x 2 eyes/image) eye images, the segmentation rate for HIQM was 81.5% and 74.8% for AIQM; it shows the near-equivalency of AIQM and HIQM.

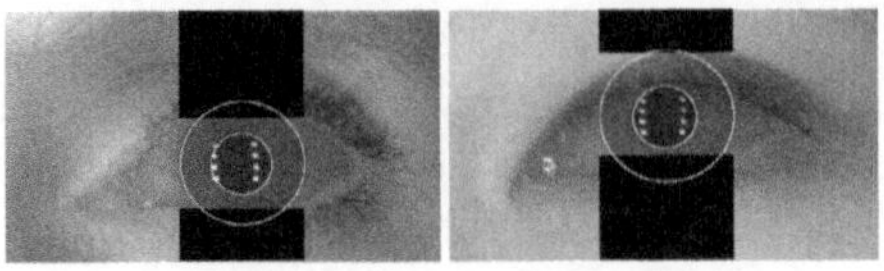

(a) (b)

Fig. 8. Examples of the Occlusion Deletion from Eyelids and Eyelash (a) Succeed to detect occlusions (b) Failed to detect occlusions

Fig. 8 illustrates some examples of the occlusion detection to remove eyelids, eyelashes, and reflection. However, there has been a high rate of false occlusion detection (over 45%) which influenced the failed matching results.

6.4 Normalization

The iris region is rescaled in the normalization process, to work with fixed dimension. Plus, noise masking filters out bits insignificant to the matching process. Fig. 9 shows an illustration of the normalization.

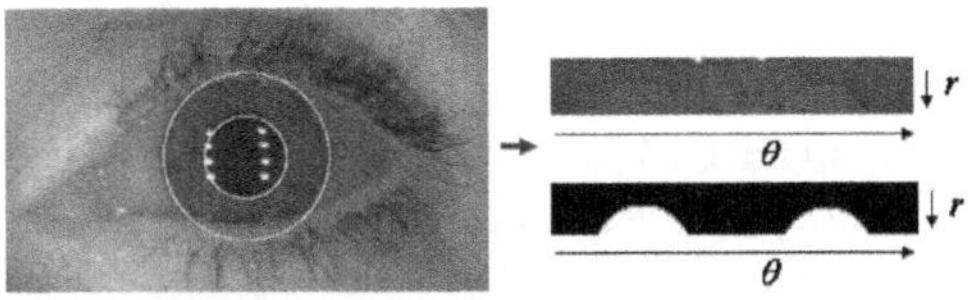

Fig. 9. Normalization Result with the Noise Mask

6.5 Matching Result for Verification (1:1)

For 4 video-images out of the total 127 videos, there were no corresponding still-images. We thus used a total of 123 video-images and the corresponding still-images; 25 persons were featured twice in the 123 datasets. There were either 12 or 6 left and right images per each person in the still-image datasets. In terms of verification, we compared each iris template from video-images with all eye images from the same person. We chose the still-image closest to the video-image according to the computed match score (Hamming Distance, HD). Each and all of the 123 video-images are compared to the 123 still-images – excluding images of the same person, to compute the non-match score. Based on the Match and Non-Match HD distributions, we can choose a separate point which allows a decision to be made when comparing two templates. However, the Match and Non-match distributions may overlap, which would result in a number of incorrect matches or false accepts, and a number of mismatches or false rejects.

The False Accept Rates (FAR) and False Reject Rates (FRR) can be calculated by the overlap area between the two distributions. The Fig. 10 illustrates the Match and Non-match HD distributions with overlap. The generated template is 240x20 pixels in resolution and 10 bits in the template were shifted left and right for rotational inconsistencies. When using a Hamming Distance value of 0.39 for the separation point,

the FAR is 0.80% and the FRR is 43.90% in HIQM; the FAR is 0.83% and the FRR is 45.53% in AIQM. Given this, if two iris templates have a Hamming Distance lower than the separation point then it is concluded that the iris templates belong to the same person; otherwise they are from a different person.

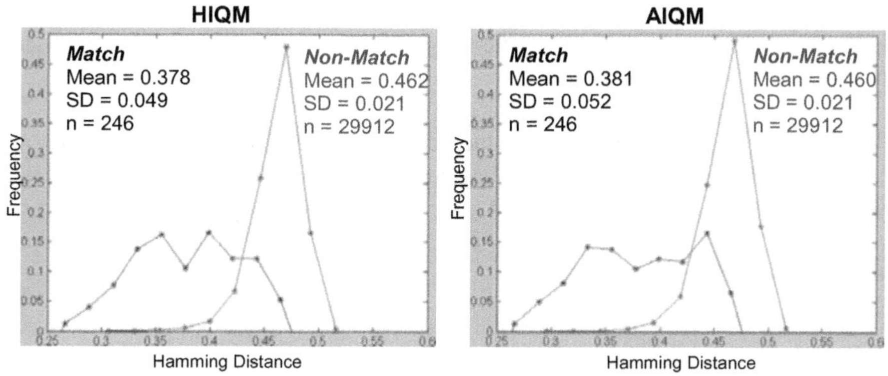

Fig. 10. Match and Non-match Hamming Distance Distributions with Overlap

The experiment resulted in a corresponding matching rate between still-images and video-images of 56.09% in HIQM and 54.47% in AIQM. Comparing HIQM with AIQM, this experiment confirmed their near-equivalency. Regarding system feasibility, the matching rate is relatively low, but system improvements are in development.

7 Conclusion

We implemented the Video-based Automated System for Iris Recognition (VASIR) and evaluated its performance on the MBGC datasets. VASIR used algorithms employing from the OpenCV library for two-eye detection, adapting human and edge density methods for the best image selection, and applying Masek's algorithm for iris recognition. A cross-platform Qt toolkit used to implement the graphical user interface. VASIR successful rate was 97.69% in two-eye detection, 81.50% in the iris region segmentation. The matching rate between a still-image template and video-image template is 56.09% in HIQM and 54.47% in AIQM.

We have shown that even under varying illumination conditions, low quality, and off-angle on both still and video imagery, iris recognition is nevertheless feasible. Furthermore, our study showed that in practice an automated best image selection is nearly equivalent to human selection.

However, VASIR still has opportunities for further improvement. There are a number of VASIR enhancements underway. We are working on new methods to improve each step as shown in Fig.1. In particular, the localization needs to be improved, by substituting the occlusion detection algorithm. Furthermore, enhancing the image quality before the localization should prevent incorrect segmentation of the iris region, caused by false focus, wrong brightness level, and moving objects.

Acknowledgments. The Authors would like to thank James J. Filliben who was extremely helpful in analyzing datasets and plotting the graphs. The identification of any commercial product or trade name does not imply endorsement or recommendation by the National Institute of Standards and Technology.

References

1. Multiple Biometric Grand Challenge, http://face.nist.gov/mbgc/
2. Viola, P., Jones, M.J.: Robust real-time face detection. International Journal of Computer Vision 57(2), 151–173 (2004)
3. Lienhart, R., Maydt, J.: An Extended Set of Haar-like Features for Rapid Object Detection. IEEE International Conference on Image Processing, vol. 1, pp. 900–903 (2002)
4. Intel Open Source Computer Vision Library, v1.0 (2006), http://sourceforge.net/projects/opencvlibrary/
5. Castrillón-Santana, M., Déeniz-Suárez, O., Antón-Canalís, L., Lorenzo-Navarro, J.: Face and Facial Feature Detection Evaluation. In: International Conference on Computer Vision Theory and Applications (2008)
6. Castrillón-Santana, M., Déeniz-Suárez, O., Tejera, M.H., Artal, C.G.: ENCARA2: Real-time detection of multiple faces at different resolutions in video streams. Journal of Visual Communication and Image Representation, 130–140 (2007)
7. Reimondo, A.: Haar cascades repository (2007), http://alereimondo.no-ip.org/OpenCV/34
8. Bowyer, K.W., Hollingsworth, K., Flynn, P.J.: Image Understanding for Iris Biometrics: a Survey. Computer Vision and Image Understanding 110(2), 281–307 (2008)
9. Beveridge, J.R., Givens, G.H., Phillips, P.J., Draper, B.A., Yui Man Lui, C.: Focus on Quality, Predicting FRVT 2006 Performance. In: 8th IEEE International Conference on Automatci Face and Gesture Recogniton, FG (2008)
10. Masek, L.: Recognition of Human Iris Patterns for Biometric Identification, The University of Western Australia, http://www.csse.uwa.edu.au/~pk/studentprojects/libor/
11. Phillips, J., Bowyer, K., Flynn, P., Liu, X., Scruggs, T.: The Iris Challenge Evaluation 2005. In: IEEE 2nd International Conference on Biometrics Theory, Applications and Systems (2008)
12. Richard, P., Wildes, C.: Iris Recogniton: An Emerging Biometric Technology. Proceedings of the IEEE 85(9) (1997)
13. Daugman, J.G.: High Confidence Visual Recognition of Persons by a Test of statistical Independence. IEEE Transactions on Pattern Analysis and Machin Intelligence 15(11) (1993)
14. Yao, P., Li, J., Ye, X., Zhuang, Z., Li, B.: Iris Recognition Algorithm using Modified Log-Gabor Filters. In: Proceedings of International Conference on Pattern Recognition, pp. 461–464 (2006)
15. Daugman, J.G.: How iris recognition works. In: Proceedings of International Conference on Image Processing, vol. 1 (2002)

Empirical Evidence for Correct Iris Match Score Degradation with Increased Time-Lapse between Gallery and Probe Matches

Sarah E. Baker, Kevin W. Bowyer, and Patrick J. Flynn

University of Notre Dame
{sbaker3,kwb,flynn}@cse.nd.edu

Abstract. We explore the effects of time lapse on iris biometrics using a data set of images with four years time lapse between the earliest and most recent images of an iris (13 subjects, 26 irises, 1809 total images). We find that the average fractional Hamming distance for a match between two images of an iris taken four years apart is statistically significantly larger than the match for images with only a few months time lapse between them. A possible implication of our results is that iris biometric enrollment templates may undergo aging and that iris biometric enrollment may not be "once for life." To our knowledge, this is the first and only experimental study of iris match scores under long (multi-year) time lapse.

Keywords: Iris biometrics, enrollment template, template aging, time-lapse, match distribution stability.

1 Introduction

The iris biometrics research community has accepted the premise that the appearance of the iris is highly stable throughout most of a person's life. Daugman stated the assumption this way-"As an internal (yet externally visible) organ of the eye, the iris is well protected and stable over time" [1]. The assumption is repeated in similar form in recent academic references: "[the iris is] stable over an individual's lifetime'" [3], "the iris is highly stable over a person's lifetime" [5], "[the iris is] essentially stable over a lifetime" [4]. While the basic assumption is broadly accepted as valid and commonly re-stated, we know of no experimental work that establishes its validity. This paper describes our experimental evaluation of the extent to which this assumption is true in terms of practical application in biometrics.

We formulate an experimental test of the long-term stability of iris texture in iris biometrics as follows. Assume that a person has an iris image acquired at one point in time for enrollment, and at a later point in time has another image acquired for recognition. The result of matching the two iris images is reported as a fractional Hamming distance, a value between 0 and 1 that indicates the fraction of iris code bits that do not match. A fractional Hamming distance of 0 indicates a perfect match, and a distance of 0.5 indicates random agreement.

M. Tistarelli and M.S. Nixon (Eds.): ICB 2009, LNCS 5558, pp. 1170–1179, 2009.
© Springer-Verlag Berlin Heidelberg 2009

The "stable over a lifetime" concept can be tested by comparing the Hamming distance of image pairs acquired with different time lapses.

To investigate this question experimentally, we use a set of iris images acquired at the University of Notre Dame [6][10][8], and a modified version of the open source "ICE baseline" iris code matcher[7][9][13]. Comparing matching scores between images taken a few months apart with scores between images taken approximately four years apart, we find that there is a statistically significant difference in the average Hamming distance between short-time-lapse matches and long-time-lapse matches. This suggests that the "lifetime enrollment" concept may not be valid. This would also suggest that time lapse between images should factor into a decision about match quality, and that guidelines are needed for time between re-enrollment.

1.1 Related Work

Gonzalez et al. report an effect of time separation on iris recognition [11] that may initially seem similar to this paper. However, their work is based on comparing matches between images acquired at the same acquisition session with those acquired with at most three months time lapse. They report a higher match statistic for images from the same session than those across sessions. They note little change in match statistics when comparing matches with short time lapses, between two weeks and three months. In this paper, we eliminate matches between images acquired at the same session as we expect they would be unfairly similar. Additionally, we focus on the effect of time-lapse between gallery and probe images and same-session images are not used as both the gallery and the probe in a real world scenario. We do not note significant differerences in average Hamming distance for images with a few months time lapse. However, at four years time lapse, we do observe a significant difference.

2 Experimental Methods and Materials

2.1 Experimental Materials

The iris images analyzed in this study were acquired using an LG 2200[2], and the acquisition protocol is the same as that used in the collection of images for the Iris Challenge Evaluation[8]. A small subset of people have participated in data collections from spring of 2004 through spring of 2008. We know of no other iris image data set that has four years of time-lapse data available.

Our data set consists of images acquired approximately weekly during each academic semester. At each acquisition session, six images of each iris are acquired from each subject. Some images were discarded from our data set due to poor quality.

We compare two types of matches: (1) matches between two images both acquired in the same semester (but not on the same day) and (2) matches between one image from spring 2004 and one image from spring 2008. We found 13 subjects in the data set with both spring 2004 and spring 2008 images of each iris.

For these 26 "iris subjects", we used 1236 images from 2004 and 573 images from 2008 for a total of 1809 images. This data set contains eight males and five females between the ages of 24 and 56. Three of these subjects are Asian and ten are Caucasian. Four of these subjects wore contacts and nine did not; no subjects wore glasses for this acquisition.

All images used in our experiments were acquired by the same LG2200 camera. They were also acquired in the same studio using the same acquisition procedure, computer system, digitizer board, driver software, and application software[6][10].

Our iris segmentation technique employs encoding and matching, we used software based on the open source IrisBEE[8]. This software uses one dimensional log-Gabor wavelets to create a 240x20x2-bit iris code and contains improvements to the segmentation as described in [6].

2.2 Experimental Method

Our null hypothesis and alternative hypothesis are stated as follows.

Null Hypothesis: The fractional Hamming distance for iris code matches between images taken a longer time apart is not greater than that for matches between images taken a shorter time apart.

Alternative Hypothesis: The fractional Hamming distance for iris code matches between images taken a longer time apart is greater than that for matches between images taken a shorter time apart.

We consider two experimental scenarios to test the null hypothesis, an "all-irises" test and an "iris-level" test. The experimental results and conclusions are similar for both formulations. The "all-irises" scenario combines the set of images from all 26 "iris-subjects" and is explained as follows.

For each iris we have multiple images. Each such image is considered as a gallery image in succession. For each gallery image, all other images are considered as probe images. Each match between a gallery and a probe image results in a Hamming distance. This HD is placed in either a short-time-lapse set or a long-time-lapse set, depending on the time elapsed between the gallery and the probe image. The process is repeated for every image of that iris subject, yielding a set of short-time-lapse HDs and a set of long-time-lapse HDs. These sets are each averaged, yielding a short-time-lapse mean HD and a long-time-lapse mean HD for that iris subject.

We introduce the following notation:

We have a set of iris images: $\mathcal{I} = \{I_1, I_2, \ldots I_n\}$

Each image in our set has a subject ID including a left-right indicator and a date:

$$\begin{aligned} \forall\, I \in \mathcal{I},\ & I.\text{id} = \text{SubjectID (i.e. 02463L)} \\ & I.\text{date} = \text{Date of Image} \end{aligned}$$

For each unique subject S,

$$\mathcal{I}_S = \{I \in \mathcal{I} \mid I.id = S\}$$

For each $I \in \mathcal{I}_S$, we obtain the set of images within a short-time lapse, $\mathcal{I}_{S*S}$:

$$\mathcal{I}_{S*S} = \{I' \in \mathcal{I}_S \mid |I'.date - I.date| < T_d\}$$

and we obtain the set of images taken after a long-time lapse, $\mathcal{I}_{S*L}$:

$$\mathcal{I}_{S*L} = \mathcal{I}_S - \mathcal{I}_{S*S}$$

where T_d is a time difference threshold. We use $T_d = 6$ months.
We also define sets of Hamming distances as follows:

$$\mathcal{D}_{S*S} = \{HD(I, I') \mid I \in \mathcal{I}_S,\ I' \in \mathcal{I}_{S*S}\}$$
$$\mathcal{D}_{S*L} = \{HD(I, I') \mid I \in \mathcal{I}_S,\ I' \in \mathcal{I}_{S*L}\}$$
$$\mu_{S*S} = \frac{\sum \mathcal{D}_{S*S}}{||\mathcal{D}_{S*S}||} \text{(mean short-time-lapse match score)}$$
$$\mu_{S*L} = \frac{\sum \mathcal{D}_{S*L}}{||\mathcal{D}_{S*L}||} \text{(mean long-time-lapse match score)}$$

The difference between the means $(\mu_{S*L} - \mu_{S*S})$ is computed, and the process is repeated for every iris subject, yielding a set of differences between mean HDs.

We consider two tests of the null hypothesis using these differences. For the sign test, we consider the null hypothesis that a positive difference occurs equally as often as a negative difference. The alternative hypothesis is that the more prevalent, a positive difference, occurs more often. Using a one-tailed Student's t test on the difference of means, we consider the null hypothesis that the mean of the N differences is zero. The alternative hypothesis is that the mean of the differences is greater than zero.

The "iris-level" scenario involves tests performed on each iris separately, yielding 26 different p values. For each iris subject, S, the short-time-lapse set, $\mathcal{D}_{SS}$, and the long-time-lapse set, $\mathcal{D}_{SL}$, used in the "all irises" experiment give two samples of HDs. To test our null hypothesis, we test if these two samples are from a distribution with $\mu_{SS} = \mu_{SL}$.

2.3 Possible Sources of Change in Match Quality

We consider four factors other than time-lapse that itself could conceivably cause poorer quality matches with longer time lapse.

1. The number of bits used in comparisons can affect the match distribution. If two images are masked in such a way that few bits are left to be used in the comparison, the Hamming Distance may be lower than it ought to be[12]. To control for differences in the number of bits used in a match, we implemented score normalization as suggested by Daugman[12]. Across our data, 5400 was the average number of bits used and was used as the scaling parameter in the normalization step.
2. It has been shown that the pupil to iris ratio affects the match distribution [13]. When two images of irises with largely dilated pupils are compared, the Hamming distance is greater than two irises with less dilated pupils. Similarly, as the difference in dilation between the two irises increases, the match distribution shifts in the positive directiont[13]. To account for any effects of pupil dilation, we consider the difference in pupil dilation between irises as a factor in the experiment below.

3. The presence of contact lenses can adversely affect match quality[15]. We performed a manual, retrospective check for contact lenses in all images used in this study. Four subjects wore contact lenses in both years and nine did not wear them in either year. No subjects appear to have begun to wear contacts in 2008 when they did not in 2004, or to have changed the type of contacts they wore.
4. Poor image quality and segmentation affect match quality[16]. We manually inspected every image and its segmentation produced by our segmenter. Approximately 7% of the images acquired for these subjects were discarded due to poor quality and an additional 24% were discarded due to poor segmentation.

3 Results

For every iris-subject, we computed the mean Hamming distance and the standard deviation for the short-time-lapse matches and the long-time-lapse matches. In 23 of the 26 irises, μ_{SL} was greater than μ_{SS} matches. The difference in mean HDs for the two sets of time lapse, $\mu_{diff} = \mu_{S*L} - \mu_{S*S}$, was computed for each iris. We also found the difference in the average number of bits used where $Bits_{diff} = B_{S*L} - B_{S*S}$, where B_{S*L} is the average number of bits used in long-time-lapse matches and B_{S*S} is the average number of bits used in short-time-lapse matches. This data is shown in Table 1.

We found the average pupil to iris ratio in 2004 and the average ratio in 2008 for each of the irises. In 23 of the 26 irises, the average ratio was smaller in 2008 than in 2004. For every match, we computed the difference in the pupil to iris ratio of the two matched images. For each iris we determined the average ratio difference of short-time-lapse matches, $\Delta\mathcal{P}_{S*S}$ and the average ratio difference of long-time-lapse matches, $\Delta\mathcal{P}_{S*L}$. We found $\Delta\mathcal{P}_{S*L}$ was greater than $\Delta\mathcal{P}_{S*S}$ in 22 of the 26 irises. This change in pupil to iris ratio difference may account for an increase in the HD for long-time-lapse matches. However, we observe no correlation between $\Delta\mathcal{P}_{S*L} - \Delta\mathcal{P}_{S*S}$ and $\mu_{S*L} - \mu_{S*S}$ (see Table 1.)

Across all matches, we determined the mean Hamming distance for a long-time-lapse was 0.230, whereas the mean HD for a short-time-lapse was 0.212. However, we found the nonmatch mean HD was 0.447 for a long-time-lapse and 0.446 for a short-time-lapse. These results indicate a time-lapse effect on match scores, but a negligible effect on nonmatch scores. Fig. 2 clearly indicates the shift in the match distribution for long-time-lapse matches and the consistency within the nonmatch distributions.

3.1 All Irises Test

The difference, $\mu_{S*L} - \mu_{S*S}$, was positive for 23 of the 26 irises with an average difference of 0.0165. In a random sample, we would expect the average HD for long-time-lapse matches to be worse for 13 irises and better for 13 irises. We applied a sign test to test the null hypothesis that the number of positive

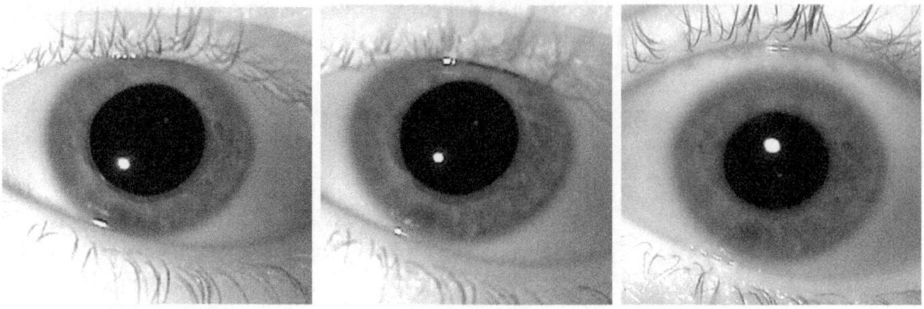

(a) Enrollment Image from Spring 2004 (b) Verification Image from Spring 2004 (c) Verification Image from Spring 2008

Fig. 1. Subject 04233 Left iris- HD for spring 2004 gallery versus spring 2004 probe was 0.156. HD for spring 2004 gallery versus spring 2008 probe was 0.285.

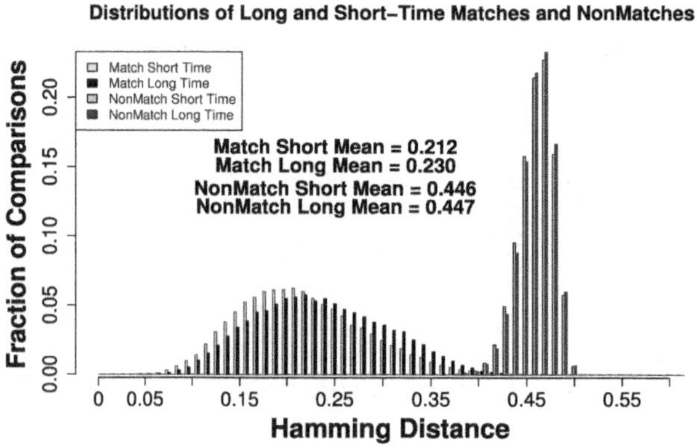

Fig. 2. We observe no change in the non-match distribution, but a significant shift to the right for long-time-lapse matches

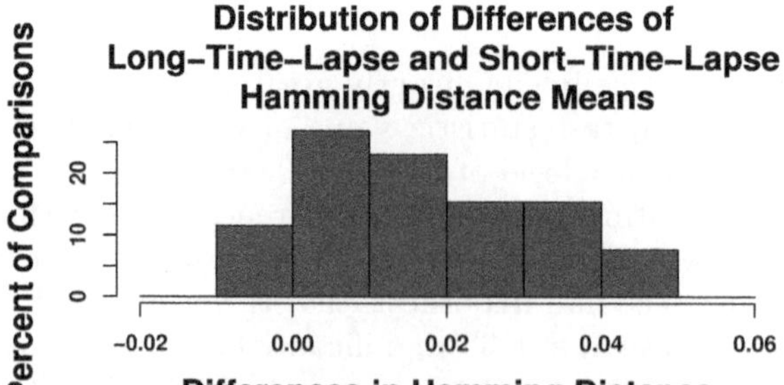

Fig. 3. Distribution of difference of long-time-lapse means and short-time-lapse means

Table 1. Average Hamming distance and standard deviation for short-time-lapse and long-time-lapse matches and the change in mean Hamming distances, bits used, and pupil to iris ratio for all 26 irises

Iris	μ_{S*L}	*std*	$\|\|\mathcal{D}_{S*L}\|\|$	μ_{S*S}	*std*	$\|\|\mathcal{D}_{S*S}\|\|$	μ_{diff}	$Bits_{diff}$	$\Delta\mathcal{P}_{S*L} - \Delta\mathcal{P}_{S*S}$
02463*L*	0.1843	0.0404	1419	0.1847	0.0418	1219	−0.0004	61.5	0.0044
02463*R*	0.2056	0.0377	987	0.1952	0.0375	1008	0.0104	41.0	−0.0014
04233*L*	0.1977	0.0402	2254	0.1795	0.0420	2108	0.0183	46.3	0.0281
04233*R*	0.1752	0.0353	2372	0.1712	0.0398	2080	0.0040	33.8	0.0034
04261*L*	0.1584	0.0378	156	0.1463	0.0403	224	0.0121	−87.7	0.0318
04261*R*	0.1408	0.0255	127	0.1381	0.0302	176	0.0027	−8.1	0.0168
04385*L*	0.2316	0.0409	2676	0.1999	0.0383	1628	0.0317	−91.0	0.0289
04385*R*	0.2288	0.0387	960	0.2140	0.0455	960	0.0148	−74.8	0.0335
04397*L*	0.1398	0.0363	1983	0.1365	0.0355	1323	0.00330	21.8	0.0219
04397*R*	0.1441	0.0266	2311	0.1380	0.0290	1496	0.0061	1.2	0.0087
04470*L*	0.2470	0.0524	479	0.2377	0.0499	572	0.0093	8.9	0.0010
04470*R*	0.2401	0.0468	689	0.2403	0.0521	518	−0.0002	97.2	0.0084
04537*L*	0.2131	0.0463	733	0.1991	0.0433	825	0.0140	−24.2	0.0530
04537*R*	0.1933	0.0387	805	0.1829	0.0442	864	0.0104	−69.0	0.0578
04629*L*	0.2947	0.0440	1246	0.2691	0.0485	1056	0.0255	−97.2	−0.0170
04629*R*	0.2994	0.0438	964	0.2678	0.0476	840	0.0316	−171.8	−0.0072
04815*L*	0.2769	0.0491	2236	0.2336	0.0555	1474	0.0433	−51.6	0.0531
04815*R*	0.2524	0.0493	2922	0.2263	0.0517	2294	0.0260	−26.8	0.0392
04851*L*	0.3101	0.0364	1356	0.2619	0.0478	1122	0.0481	−213.8	0.0203
04851*R*	0.3307	0.0357	2229	0.3092	0.0485	1755	0.0215	−47.5	0.0099
04870*L*	0.2460	0.0594	1477	0.2432	0.0564	880	0.0028	43.1	0.0228
04870*R*	0.2607	0.0539	1556	0.2647	0.0520	855	−0.0041	57.1	0.0267
04888*L*	0.2594	0.0373	147	0.2376	0.0454	162	0.0218	−74.8	0.0268
04888*R*	0.2261	0.0371	216	0.2168	0.0407	252	0.0093	6.0	0.0007
04917*L*	0.2380	0.0397	2022	0.2050	0.0406	1386	0.0330	−250.6	−0.0095
04917*R*	0.2342	0.0395	2419	0.2000	0.0347	1768	0.0343	−246.8	0.0007
All	0.2302	0.0663	28845	0.2118	0.0632	36741	0.0184	−41.3	

differences is not statistically significantly greater than the number of negative differences. With a sign test statistic value of z = 4.1184, we reject the null hypothesis at a significance level of 5% (p = 0.0001).

A histogram representing this sample of differences of mean Hamming distances is shown in Fig. 3. We applied a chi-square goodness-of-fit test to the sample of 26 differences of means. The null hypothesis that this sample is from a normal distribution cannot be rejected at a 5% significance level. Since this data is approximately normal, we can use a t-test to compare the difference of means.

We applied a one-tailed paired Student's t test to test the null hypothesis that this difference-of-means sample comes from a distribution with a mean of zero. The alternative hypothesis is that the difference distribution has a mean greater than zero, which would mean that the long-time-lapse HDs are on average

greater than the short-time-lapse HDs. The null hypothesis was rejected at a 5% significance level (p=0.00001.)

To confirm that there is no significant effect from the number of bits used in matches, we applied a Student's t test to the distribution of $Bits_{diff}$. The null hypothesis was that the mean of this sample was zero. We failed to reject the null hypothesis at a 1% significance level (p = 0.0285). Thus, across all irises, there was no significant change in the number of bits used.

3.2 Iris-Level Test

For each iris subject, we have two samples of Hamming distances, one from long-time-lapse matches, $\mathcal{D}_{S*L}$, and one from short-time-lapse matches, $\mathcal{D}_{S*S}$. These samples were approximately normal, so we applied a one-tailed Student's t test to test the null hypothesis that these two samples of matches come from the same distribution with equal means. The alternative hypothesis is that $\mathcal{D}_{S*L} > \mathcal{D}_{S*S}$. The null hypothesis was rejected for 21 of the 26 irises at a significance level of 0.05.

3.3 Sensor Tests

We have observed that the Hamming distance for long-time-lapse matches is on average larger than that for short-time-lapse matches. One possible cause for this observation would be that there is some subtle change in iris texture over time. However, it is important to note that this is not the only possible cause. For example, if the sensor properties changed over time, this could also produce a change in the imaged texture even if there is no change in the true iris texture.

We performed an experiment with images from the original LG2200 camera used in the acquisition for all images in this paper and a different, rarely-used, LG2200 camera. We tested images from both cameras to determine if the original, well-used, camera and sensor have a degrading effect on match quality. To perform this test, we used two sets of images from Fall 2008 acquired with the original camera as the gallery set and the first probe set, and a third set of images from Fall 2008 acquired with the new, rarely-used camera as the second probe set. We found the matches produced from the two different probe sets were not significantly different. Therefore, we do not see any evidence that the sensor properties have changed enough to explain the time-lapse conclusions we have presented.

4 Discussion and Future Work

We observe an approximate 0.018 increase in Hamming distance for matches with a four years time-lapse. HDs are between 0 and 0.5, so our result represents an approximate $3-4\%$ increase over a four year period. Additionally, at a false accept rate of 0.01%, the false reject rate increases by 75% for long-time-lapse.

The basic results and conclusion presented here run counter to conventional wisdom about iris biometrics. However, we know of no experimental study that has previously tested the "one enrollment for a lifetime" assumption. The previous time variability study referenced in the introduction compared images with

less than three months time lapse. Their results show better performance for images acquired in the same session than images acquired across sessions. They also note no significant differences between two weeks to four weeks to two months time lapse. Our results are based on images of the same iris imaged with time lapse as long as four years. With this long-term time lapse we note statistically significant changes in the iris match quality. Upon visual examination of the irises with the largest difference in Hamming distance, we observed no drastic changes in iris textures, suggesting that if the iris aging affect is real, it is based on subtle differences.

In this study, we use the same iris imaging system, and control for contact lenses, pupil dilation, and number of bits in a match. We noted no apparent trend in the change in the number of bits in a match. In 22 of the 26 irises, the difference in the pupil dilation between the images of a match was greater for matches of long-time-lapse than matches of short-time-lapse. However, this change in pupil dilation difference does not correlate with the change in Hamming distance across the two sets of time-lapse. We have considered the major potential complicating factors for an experimental study of this type. However, it is still important for our result to be replicated by other research groups using different and larger data sets with more subjects.

Future work includes investigation into textural changes and pinpointing the location of such changes. Predicting textural or pupil dilation changes may aid in accounting for degradation in the match statistic. While we have observed an increase in Hamming distance and the false reject rate over a four year period, we do not know if this trend is linear or how the match quality will change with eight years, or longer, time lapse.

Even if the "lifetime enrollment" concept is disproved, it is not necessarily a major barrier to practical deployment of iris biometrics systems. It would mean that consideration should be paid to the time-lapse between image acquisitions in quantifying a match statistic. One possible reconciliation for match quality degradation is to re-enroll a subject with every verification scenario. However, this requires routine verifications as a long time lapse between enrollment and verification will result in an increased false reject rate. Another possibility is to require a re-enrollment session for every subject after a set time frame. The necessary time frame may be difficult to determine. If the time frame is too long the iris match quality may degrade beyond the accept rate before re-enrollment. A third possibility is to report the time lapse between the enrolled and the verification images as well as the match statistic. If further research shows a possible prediction of changes in the match statistic with increased time lapse, we may be able to normalize the statistic based upon this lapse. We suggest these possible considerations but recognize that much further research is needed before making a recommendation.

Acknowledgement

This work is supported by the National Science Foundation under grant CNS01-30839, by the Central Intelligence Agency, by the Intelligence Advanced Research

Projects Activity and by the Technical Support Working Group under US Army contract W91CRB-08-C-0093. The opinions, findings, and conclusions or recommendations expressed in this publication are those of the authors and do not necessarily reflect the views of our sponsors.

References

1. Daugman, J.: How Iris Recognition Works. IEEE Trans. Circuits and Sys. for Video Tech. 14(1), 21–30 (2004)
2. LG, `http://www.lgiris.com/` (accessed, August 2008)
3. Thornton, J., Savvides, M., Kumar, V.: A Bayesian Approach to Deformed Pattern Matching of Iris Images. IEEE Trans. Pattern Anal. and Mach. Intell. 29(4), 596–606 (2007)
4. Miyazawa, K., Ito, K., Aoki, T., Kobayashi, K., Nakajima, H.: An Effective Approach for Iris Recognition Using Phase-Based Image Matching. IEEE Trans. Pattern Anal. and Mach. Intell. 30(10), 1741–1756 (2008)
5. Monro, D., Rakshit, S., Zhang, D.: DCT-Based Iris Recognition. IEEE Trans. Pattern Anal. and Mach. Intell. 29(4), 586–595 (2007)
6. Liu, X., Bowyer, K., Flynn, P.: Experiments with an improved iris segmentation algorithm. In: Fourth IEEE Workshop on Automatic Identification Technologies, Oct. 2005, pp. 118–123 (2005)
7. Liu, X.: Optimizations in Iris Recognition. PhD Dissertation, University of Notre Dame (2006)
8. National Institute of Standards and Technology. Iris Challenge Evaluation (2006) `http://iris.nist.gov/ice`
9. Hollingsworth, K., Bowyer, K., Flynn, P.: The Best Bits in an Iris Code. IEEE Trans. Pattern Anal. and Mach. Intell (in press)
10. Phillips, J., Bowyer, K., Flynn, P., Liu, X., Scruggs, T.: The Iris Challenge Evaluation 2005. In: 2008 IEEE Conf. on Biometrics: Theory, Applications, and Systems,
11. Tome-Gonzalez, P., Alonso-Fernandez, F., Ortega-Garcia, J.: On the Effects of Time Variability in Iris Recognition. In: 2008 IEEE Conf. on Biometrics: Theory, Applications and Systems (2008)
12. Daugman, J.: New Methods in Iris Recognition. IEEE Trans. Sys., Man, and Cyber. 37(5), 1167–1175 (2007)
13. Hollingsworth, K., Bowyer, K., Flynn, P.: Pupil Dilation Degrades Iris Biometric Performance. In: Computer Vision and Image Understanding (in press)
14. Bowyer, K.W., Hollingsworth, K.P., Flynn, P.J.: Image Understanding for Iris Biometrics: A Survey. Computer Vision and Image Understanding 110(2), 281–307 (2008)
15. Ring, S., Bowyer, K.: Detection of Iris Texture Distortions by Analyzing Iris Code Matching Results. In: IEEE Conf. on Biometrics: Theory, Applications, and Systems (2008)
16. Kalka, N., Zui, J., Schmid, N., Cukic, B.: Image Quality Assessment for Iris Biometric. SPIE 6202: Biometric Technology for Human Identification III, D1–D11 (2006)

Practical On-Line Signature Verification*

J.M. Pascual-Gaspar, V. Cardeñoso-Payo, and C.E. Vivaracho-Pascual

ECA-SIMM, Dpto. Informática, Universidad de Valladolid,
Campus Miguel Delibes s/n, 47011 Valladolid, Spain
pascual@gaspar.com, valen@infor.uva.es, cevp@infor.uva.es

Abstract. A new DTW-based on-line signature verification system is presented and evaluated. The system is specially designed to operate under realistic conditions, it needs only a small number of genuine signatures to operate and it can be deployed in almost any signature capable capture device. Optimal features sets have been obtained experimentally, in order to adapt the system to environments with different levels of security. The system has been evaluated using four on-line signature databases (MCYT, SVC2004, BIOMET and MyIDEA) and its performance is among the best systems reported in the state of the art. Average EERs over these databases lay between 0.41% and 2.16% for random and skilled forgeries respectively.

1 Introduction

Handwriting signature has been a mean of personal identification used for centuries and its usage in experimental scenarios as a dynamic biometric modality has proved to be as efficient as many physiological traits [1]. However, this biometric modality faces big challenges when applied to real practical scenarios, far away from ideal laboratory conditions. Working with a small number of genuine signatures for user enrolment is one of those, which motivates the Dynamic Signature Verification (DSV) system which is described and evaluated in this work. The following design guidelines have been applied in order to ensure good practical usability characteristics:

- The system shall use a reduced number of genuine signatures without compromising accuracy and no extra development data sets will be required.
- The signature feature sets should be easily computable and cross-device compatible. Small storage requirements would be desirable.
- The verification algorithm should allow easy and efficient implementation using freely available programming environments.
- The system must be flexible enough as to be adapted to different security restrictions.

* This work has been partially supported by Consejería de Educación de la Junta de Castilla y León under research contract VA077A08.

M. Tistarelli and M.S. Nixon (Eds.): ICB 2009, LNCS 5558, pp. 1180–1189, 2009.
© Springer-Verlag Berlin Heidelberg 2009

2 System Design

This section describes the four building blocks that make up our system, according to Jain's classification of a biometric system in [2]: sensor module, feature extraction module, matcher module and system database.

2.1 Sensor Module

On-line signatures can be obtained through a variety of devices (digital pens, personal digital assistants (PDAs), Tablet-PCs, ...). Digital pen tablets usually provide higher spatial and temporal resolution at affordable prices. That's why many public available signature databases have been acquired using pen tablets, and in particular the four databases used in this work. As will be shown later, our experiments show that the results we report here could be similarly obtained using other devices with almost no modifications.

Pen tablets record temporal and writing gesture information into a sequence of vector samples, at a typical fixed rate of around 100Hz. Raw features f_k can be classified into positional and ergonomical: a) *(p)ositional features* $f_k^p \epsilon \{x_k, y_k\}$, a 2D point in the path followed by the pen; b) *(e)rgonomical features* $f_k^e = \{p_k, a_k, i_k\}$, resulting from the hand-pen interaction. They include the *pressure* p exerted by the pen on the tablet and two pen orientation angles (*azimuth* a and *inclination* i).

2.2 Feature Extraction Module

A raw on-line signature S is represented by a discrete time sequence of N vectors of features, which number is calculated in terms of signature duration and sampling rate:

$$S = \{(x_t, y_t, p_t, a_t, i_t)\}_{t=1..N} = \{f_{k,t}\}_{t=1..N}^{k=1..5} \tag{1}$$

One of three alternative strategies is usually used by DSV systems to select the final feature set from the raw signature parameters:

1. Direct use of the raw features provided by the sensor [3,4]. Although this approach does not exclude some simple preprocessing tasks, emphasis usually focus on the matcher module.
2. Generate larger sets of derived features and empirically select the most effectives using commonly accepted previous results [5,6]. This approach can lead to high-accuracy systems but it is more computationally intensive and the results could be database-dependent [7].
3. Select a combined set of raw and simple derived features which gives optimal performances [8]. This could be the best alternative for real-time, cross-database and cross-device practical applications when statistical classification techniques are applied to feature selection, as we have done in this work.

In our system the basic feature set $F = (x, y, p, a, i)$ was expanded to include first and second time derivatives, to make a final raw feature vector with 15 components: $\widehat{F} = \{F, \Delta F, \Delta\Delta F\} = \{f_k, df_k, ddf_k\}$ with $df_k = (f_{k,t+1} - f_{k,t})/\Delta t$ and $ddf_k = (df_{k,t+1} - df_{k,t})/\Delta t$.

Two normalization process were applied to get the final set of features from this raw set: a *geometrical translation* $N_1 : f_k^{N_1} = f_k^p - \mu_k^p$ to locate the geometric center of the signature at the origin of coordinates, and a *statistical normalization* based on z-norm scaling $N_2 : f_k^{N_2} = (f_k - \mu_k)/\sigma_k$ was applied to all features so that zero mean and unit variance was ensured for each feature in the vector, where $\mu_k = (\sum_{t=1}^{N} f_{k,t})/N$ and $\sigma_k = \sqrt{(\sum_{t=1}^{N} (f_t - \mu_k)^2)/(N-1)}$ are the mean and standard deviation of kth-feature respectively.

2.3 Matcher Module

The two most common alternatives used to determine the similarity between time series associated to on-line signatures are the *reference-based* and *model-based* approaches. Reference-based systems require storing several instances of genuine signatures in order to evaluate the intra-class variability (Fig.1c). Model-based systems do not need to store signature specimens, but just a compact representation of the parameters of the model. Both alternatives have been sucessfully used with similar performance results in the state-of-art systems [1]. In our system, we have chosen a reference-based approach using Dynamic Time Warping (DTW) for time series alignment. This approach combines high accuracy results [9,8] with efficient implementation under a wide spectrum of practical scenarios, which is a goal of our final DSV system.

DTW provides the optimal nonlinear alignment of two sequences of vectors, through a minimization of the overall accumulated distance along the aligning sequence. The distance between a reference signature $S_R = \{r_i\}_{i=1..N}$ and a test signature $S_T = \{t_j\}_{j=1..M}$ is calculated by filling a $DTW_{N+1 \times M+1}$ matrix following equation (2), after initialization of $DTW[0,0] = 0$ and $DTW[i,0] = DTW[0,j] = \infty \forall i, j \epsilon [1, N]$:

$$DTW[i,j] = \overbrace{dist(i,j)}^{\text{current cost}} + \overbrace{\begin{cases} dist(i-1,j) \\ dist(i,j-1) \\ dist(i-1,j-1) \end{cases}}^{\text{accumulated cost}} \tag{2}$$

The distance between reference and test signatures will be stored at the upper right corner of the DTW matrix: $Dist(S_R, S_T) = DTW[N, M]$. The local distance *dist* in equation 2 was the usual Euclidean vector distance $dist(f_i^R, f_j^T) = \sqrt{\sum_{k=1}^{5} (f_{i,k}^R - f_{j,k}^T)^2}$.

To deal with intra-class variability, inherent to the signing process, a number of genuine signature samples should be stored for each user. Previous results show that five signatures is a reasonably low number and could still provide

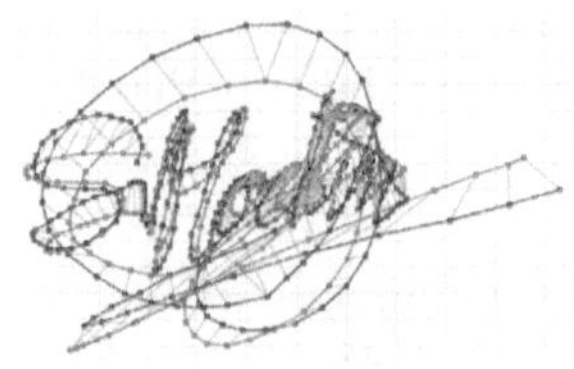

(a) Genuine-genuine

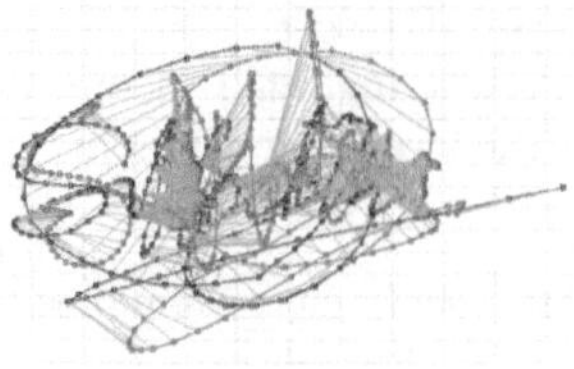

(b) Genuine-forgery

(c) Intra-class variability

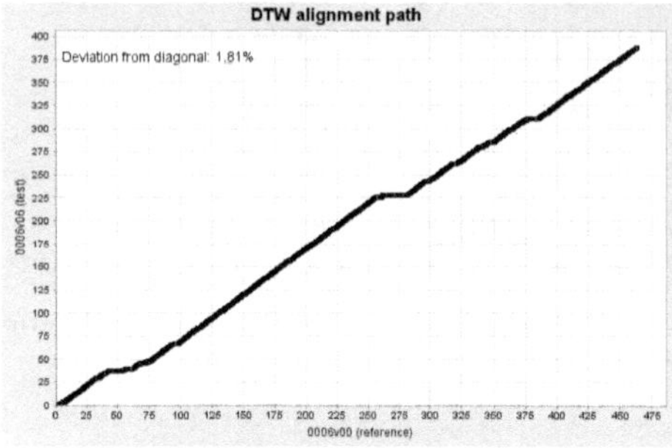

(d) Gen-Gen DTW path

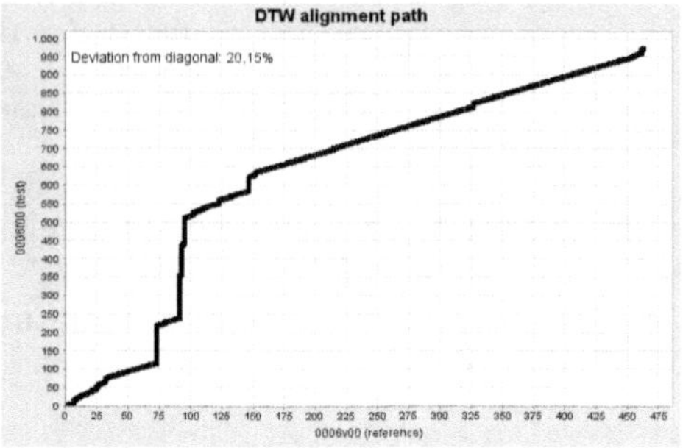

(e) Gen-Forg DTW path

Fig. 1. Figures a) and b) illustrates the DTW alignment between pairs of genuine-genuine and genuine-forgery signatures and shows their respective alignment paths in d) and e). Visual appearance of the alingment path looks closer to a diagonal straight line when genuine samples are compared. Figure c) shows an example of intra-class variability.

good results in practical scenarios [10]. DTW distances between a test signature and the five reference signatures were combined using arithmetic mean and ten-fold cross-validation was carried out to compute a reliable average error value. Five different reference signatures were randomly chosen in each fold and all the remaining genuine signatures of the same user were used for testing.

The performance of our system will be presented using the typical EER measure, both for random and skilled forgery scenarios. This allows easy comparison with results obtained in referenced works for the same databases.

2.4 System Database Module

Four on-line signature databases compiled by different research groups have been used in this work (table 1). These databases were acquired using similar models of pen tablets, providing a common set of signature features. All these databases include skilled forgeries, which is mandatory to test the system in practical scenarios subjected to 'professional forgers' attacks. Another criterion to select these databases had to do with the number of experiments reported on them in the literature.

The signature databases set was split in two disjoint data sets: a) a *Development Data* (*DD*) set, including signatures from the 50 first users of MCYT-100 [11] (hereby MCYT-A), was used to get the optimal features sets, and b) a *Test*

Table 1. Main figures of the two data sets used in this work. Users for whom there were no forgeries available were not included in our experiments. Further details of each database can be found in their reference papers.

Dataset	Database	Users	Signatures		
			Genuines	Forgeries	Total
DD	MCYT-A	50	25	25	**2500**
TD	MCYT-B	50	25	25	**2500**
	SVC2004	40	20	20	**1600**
	BIOMET	84	15	17	**2688**
	MYIDEA	69	18	36	**3726**
Total		**293**	**5802**	**7212**	**13014**

Data (*TD*) set, with the remaining signatures of all other databases, was used to test the system. *TD* set contains the signatures of the last 50 users of MCYT-100 (hereby MCYT-B) and all signatures of SCV2004 [1] [1], BIOMET [12] and MyIDEA [13].

3 Selection of Optimal Feature Sets

The tablets used to acquire the signatures we used in our experiments register both ink (visible) and pen-up strokes. This behaviour is hardware-dependent and many signature capturing devices could not provide hidden strokes. Thus, we removed them from each signature for the experiments. First challenge we faced to increase system performance was to properly and efficiently select an optimal set of features for further experimentation. To this end, both individual and combined feature evaluations have been carried out.

In a first stage, the 15 available raw features were evaluated individually on *DD* set to reduce the dimension of the combined features space. Fig. 2 proves that pen angle features (azimuth a and inclination i) show significantly poorer performance than the rest, for the three signal domains. These results suggest removal of these features, reducing feature space dimension from 15 to 9, which implies a 98% reduction of the number of possible feature combinations (from $2^{15} - 1 = 32767$ to $2^9 - 1 = 511$) and brings a computationally cheaper feature selection process at no relevant performance loss. Additionally, the remaining features (x, y coordinates and pressure p) made up a more common set of features available in the vast majority of ink capturing devices.

After an initial set of isolated features was determined, next step was to select the way to combine them into an almost-optimal final set of features. Three classical feature selection techniques were analyzed [14]: a) *Sequential Forward Selection* (SFS), b) *Sequential Backward Selection* (SBS) and c) *Plus l-take away r Selection* (PTA(l,r)).

SFS progressively incorporates most promising features into larger subsets in such a way that once a feature is added to the set, it cannot be discarded later. In terms of computational cost, this is the most attractive solution, because the

[1] Only the public available development set part was used -40 users-.

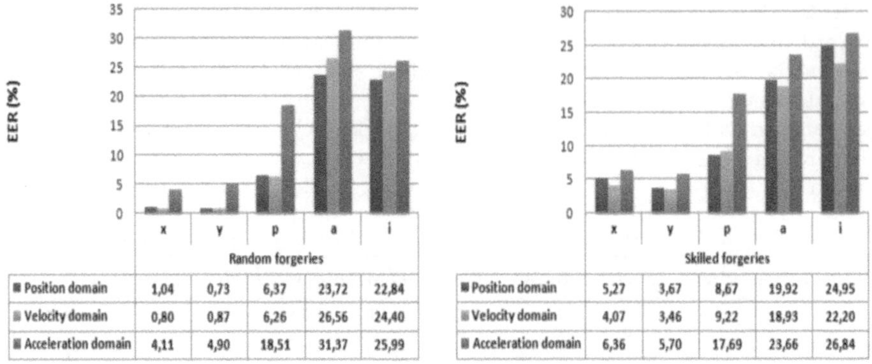

(a) Random forgeries (b) Skilled forgeries

Fig. 2. Evaluation results for individual features. By averaging the random and skilled EER's, the best individual feature is the y-coordinate (3.22%) followed by the x-coordinate (3.61%). Pressure provides intermediate performance (11.12%) but azimuth (24.03%) and inclination (24.54%) perform worse than others and were taken out from the feature selection process.

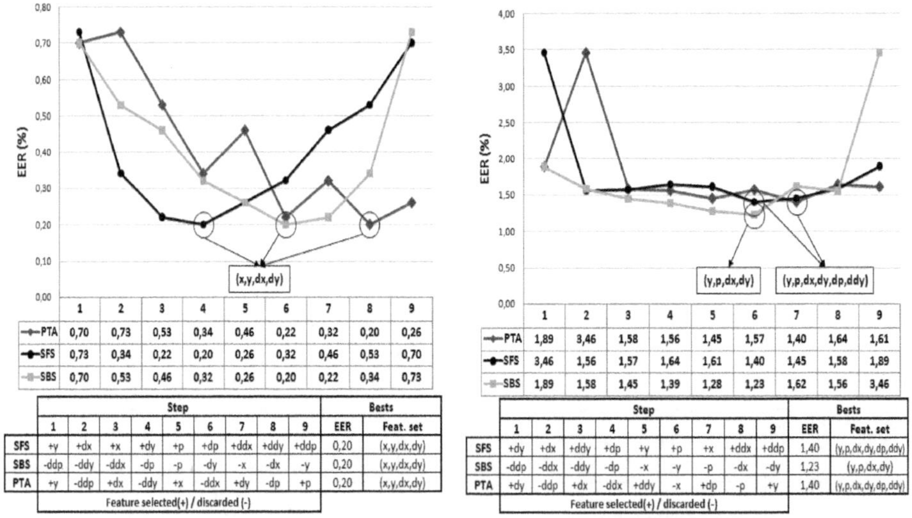

(a) Random forgeries (b) Skilled forgeries

Fig. 3. Evaluation results for combined features. *DD* set for random and skilled scenarios. Curves plot EER evolution at each step of the feature selection process (middle tables below the curves show their numeric values). Tables at the bottom show the order in which features are incorporated or discarded. In both cases, minimum values are significantly lower than extremals. Error is reduced up to 73% (from 0.73% to 0.20%) for random forgeries. Error reduction is 64% (from 3.46% to 1.23%) for skilled forgeries.

Table 2. Standard vs optimal features sets for both scenarios. Minimum values for standard and optimal feature sets in each column are emphazised. Columns in the middle show average over all databases. Lighter shadowed cells show the best feature set for standard and optimal sets. Last row displays relative EER reduction between the best standard feature set and the optimal set; darker cells show average EER reduction of the average best standard set of features ($P + V$) versus the optimal set for each scenario.

		Random forgeries					Skilled forgeries				
	Feat. set	**mcyt-b**	**svc04**	**biomet**	**myidea**	**mean**	**mcyt-b**	**svc04**	**biomet**	**myidea**	**mean**
Stand.	P	5.64	0.78	7.58	*2.45*	**4.11**	6.53	4.70	5.41	*2.89*	**4.88**
	$P + V$	*3.25*	*0.40*	*4.47*	3.52	**2.91**	*4.21*	*4.15*	*3.69*	3.25	**3.83**
	$P + V + A$	4.82	1.01	6.58	6.14	**4.64**	4.23	6.14	4.43	4.10	**4.73**
Opt.	F^o_{rd}	*0.38*	*0.00*	*0.33*	*0.92*	**0.41**	1.16	3.70	*1.25*	2.94	**2.26**
	F^o_{sk}	0.46	0.32	0.96	2.39	**1.03**	*1.06*	*3.38*	1.48	*2.72*	**2.16**
	EER red.	**88.3%**	**100.0%**	**92.6%**	**62.4%**	**86.0%**	**74.8%**	**18.6%**	**59.9%**	**5.9%**	**40.8%**

Table 3. Comparison with selected reference systems. In bold are the bests EER(%) results for each database/scenario. Overall, our system provides better performance in all cases, using similar number of training signatures. System 8 provides better results for the random forgery scenario on MCYT.

			MCYT		SVC2004		Biomet		MyIDea		
	Author	Year	rd	sk	rd	sk	rd	sk	rd	sk	Comments
1	Hennebert [16]	2007							2.7	7.3	- 6 training signatures - GMM-based algorithm - Results of the signature expert with time variability.
2	Humm [17]	2007							2.6	7.3	- 6 training signatures - HMM-based algorithm - Results with time variability schema.
3	Garcia-Salicetti [18]	2007	1.22	3.40							- 5 training signatures - HMM + Distance based algorithm - Best results from individual systems combination - Random test also included skilled forgeries
4	Van-Bao [6]	2007		3.37		4.83		2.33			- 5 training signatures - HMM-based algorithm - Viterbi path and likelihood fusion intra-algorithm
5	Pascual-Gaspar [19]	2007	2.09	6.14							- 3 training signatures - HMM-based algorithm - HMM with user-dependent structure
6	SVC2004 official results [1]	2004			3.02	6.90					- 5 training signatures - Best skilled system: DTW; Best random system: HMM - Results on development set (40 users) for Task2
7	Fierrez-Aguilar [20]	2005			0.15	6.91					- 5 training signatures - Local (DTW) and Regional (HMM) fusion - Bests results on development set (40 users) for Task2
8	Fierrez-Aguilar [5]	2005	**0.24**	2.12							- 5 training signatures - Global (Parzen WC) and local (HMM) experts fusion
9	Fierrez-Aguilar [21]	2007	0.05	0.74							- 10 training signatures* - HMM-based algorithm
	Our system	2008	0.29	**1.23**	**0.00**	**3.38**	**0.33**	**1.48**	**0.92**	**2.72**	- 5 training signatures - DTW-based algorithm - Results with scenario-dependent optimal features - Results on MCYT-100

* Although system 9 uses a more training signatures, it is included for future comparisons due to its excellent results.

size of the evaluated sets of features are kept low up to the final steps of the procedure. SBS deletes one feature at a time, so that it cannot be brought back to the locally optimal subset once it has been discarded. Although this method is computationally expensive, defenders argue that it better takes into account inter-features dependencies [15].

A combined and more sophisticated technique, named *Plus l-take away r Selection* was also evaluated (using $l = r = 1$). This approach aims a balance between computational cost and more adequate treatment of inter-features dependencies. Additionally, this method avoids the nesting problems which arise in SFS and SBS feature selection solutions [14].

Figure 3-a) shows that the three described feature selection methods resulted in the same optimal feature set for the random forgeries scenario (x, y, dx, dy). This indicates the robustness of this combination of features for low FRR (false rejection rate) scenarios. It also suggests that pressure feature could be simply ignored in low to medium security environments, which leads to a widening in the range of capturing devices which could be used for signature acquisiton in those situations. For skilled forgeries, best results are obtained using the SBS method, which provides the optimal combination (y, dx, dy, p), made up of geometric and pressure values. In this case, both SFS and PTA find a slightly worse solution than SBS, as would be initially expected, but with less computational cost. This result indicates that for high-security environments pressure information should be present, even when it has the side effect of a small increase of false rejections.

4 Benchmark Results

Two benchmark tests were carried out on the development data (DD) set to evaluate our system. Table 2 shows error results when evaluating with different features sets. Optimal sets for random (F^o_{rd}) and skilled (F^o_{sk}) scenarios are compared with other standard features sets which combine the three signal domains ([*P*]*osition*, [*V*]*elocity* and [*A*]*cceleration*). Using optimal sets for each scenario drastically improves the accuracy of the system for the verification task, specially in low and medium security scenarios, where no genuine user rejections are desired. Except for the BIOMET database, the optimal set of features for a given scenario (e.g. random or skilled) outperforms the results obtained with the set of features of the other one (e.g. skilled or random) in all cases.

Finally, table 3 compares performance results of our system with the ones in other recently published systems which used the same signature databases we used.

5 Conclusions

We described a new DTW-based on-line signature verification system specially designed to be used in practical scenarios. It does not need special hardware features to get good performance results, just geometric coordinates and, optionally, pen pressure. The system needs only a reduced number of signatures

from the user to bring excellent verification results. Depending on the security requirements, different feature sets could be selectively chosen. Benchmark experiments carried out over four popular on-line signature databases (MCYT, SVC2004, BIOMET and MYIDEA) prove that our system provides excellent results in terms of EER, specially for the skilled forgery scenario, where the system clearly outperforms other up-to-date systems in the literature under similar testing conditions.

References

1. Yeung, D., Chang, H., Xiong, Y., George, S., Kashi, R., Matsumoto, T., Rigoll, G.: SVC2004: First international signature verification competition. In: Zhang, D., Jain, A.K. (eds.) ICBA 2004. LNCS, vol. 3072, pp. 16–22. Springer, Heidelberg (2004)
2. Jain, A.K., Ross, A., Prabhakar, S.: An introduction to biometric recognition. IEEE transactions on circuits and systems for video technology 14(1) (2004)
3. Faundez-Zanuy, M.: On-line signature recognition based on VQ-DTW. Pattern Recognition 40(3), 981–992 (2007)
4. Vivaracho-Pascual, C., Faundez-Zanuy, M., Pascual, J.M.: An efficient low cost approach for on-line signature recognition based on length normalization and fractional distances. Pattern Recognition 42(1), 183–193 (2009)
5. Fierrez-Aguilar, J., Nanni, L., Lopez-Peñalba, J., Ortega-Garcia, J., Maltoni, D.: An on-line signature verification system based on fusion of local and global information. In: Kanade, T., Jain, A., Ratha, N.K. (eds.) AVBPA 2005. LNCS, vol. 3546, pp. 523–532. Springer, Heidelberg (2005)
6. Ly, V.B., Garcia-Salicetti, S., Dorizzi, B.: On using the viterbi path along with hmm likelihood information for online signature verification. IEEE Transactions on Systems, Man, and Cybernetics, Part B 37(5), 1237–1247 (2007)
7. Jain, A.K., Bolle, R.M., Pankanti, S.: Biometrics: Personal Identification in Networked Society. Springer, Heidelberg (2005)
8. Kholmatov, A., Yanikoglu, B.: Identity authentication using improved online signature verification method. Pattern Recognition Letters 26(15), 2400–2408 (2005)
9. Jain, A.K., Griess, F.D., Connell, S.D.: On-line signature verification. Pattern Recognition 35(12), 2963–2972 (2002)
10. Fierrez, J., Ortega-Garcia, J.: 12: On-line signature verification. In: Handbook of Biometrics, pp. 189–209. Springer, Heidelberg (2008)
11. Ortega-Garcia, J., Fierrez-Aguilar, J., Simon-Zorita, D., Gonzalez-Rodriguez, J., Hernaez, I., Igarza, J.J., Vivaracho, C., Escudero, D., Moro, Q.I.: MCYT baseline corpus: a bimodal biometric database. IEE Proc. Visual Image Signal Processing 150(6), 395–401 (2003)
12. Garcia-Salicetti, S., Beumier, C., Chollet, G., Dorizzi, B., les Jardins, J.L., Lunter, J., Ni, Y., Petrovska-Delacretaz, D.: BIOMET: A multimodal person authentication database including face, voice, fingerprint, hand and signature modalities. In: Kittler, J., Nixon, M.S. (eds.) AVBPA 2003. LNCS, vol. 2688, pp. 845–853. Springer, Heidelberg (2003)
13. Dumas, B., Pugin, C., Hennebert, J., Petrovska-Delacretaz, D., Humm, A., Evequoz, F., Ingold, R., von Rotz, D.: MyIDea - Multimodal biometrics database, description of acquisition protocols. In: Proceedings of Third COST 275 Workshop (COST 275), Hatfield (UK), October 27-28, 2005, pp. 59–62 (2005)

14. Jain, A.K., Duin, R.P.W., Mao, J.: Statistical pattern recognition: A review. IEEE Transactions on Pattern Analysis and Machine Intelligence 22(1), 4–37 (2000)
15. Guyon, I., Elisseeff, A.: An introduction to variable and feature selection. Journal of Machine Learning Research 3, 1157–1182 (2003)
16. Hennebert, J., Humm, A., Ingold, R.: Modelling spoken signatures with gaussian mixture model adaptation. In: IEEE International Conference on Acoustics, Speech and Signal Processing (ICASSP 2007) (2007)
17. Humm, A., Hennebert, J., Ingold, R.: Hidden markov models for spoken signature verification. In: First IEEE International Conference on Biometrics: Theory, Applications, and Systems, 2007. BTAS 2007, pp. 1–6 (September 2007)
18. Garcia-Salicetti, S., Fierrez-Aguilar, J., Alonso-Fernandez, F., Vielhauer, C., Guest, R., Allano, L., Trung, T.D., Scheidat, T., Van, B.L., Dittmann, J., Dorizzi, B., Ortega-Garcia, J., Gonzalez-Rodriguez, J., di Castiglione, M.B., Fairhurst, M.: Biosecure reference systems for on-line signature verification: A study of complementarity. Annals of Telecommunications, Special Issue on Multimodal Biometrics 62(1-2), 36–61 (2007)
19. Pascual-Gaspar, J., Cardenoso-Payo, V.: On-line signature verification using hidden markov models with number of states estimation from the signature duration (September 2007)
20. Fierrez-Aguilar, J., Krawczyk, S., Ortega-Garcia, J., Jain, A.K.: Fusion of local and regional approaches for on-line signature verification. In: Li, S.Z., Sun, Z., Tan, T., Pankanti, S., Chollet, G., Zhang, D. (eds.) IWBRS 2005. LNCS, vol. 3781, pp. 188–196. Springer, Heidelberg (2005)
21. Fierrez, J., Ramos-Castro, D., Ortega-Garcia, J., Gonzalez-Rodriguez, J.: Hmm-based on-line signature verification: feature extraction and signature modeling. Pattern Recognition Letters 28(16), 2325–2334 (2007)

On-Line Signature Matching Based on Hilbert Scanning Patterns

Alireza Ahrary[1,2], Hui-ju Chiang[3], and Sei-ichiro Kamata[4]

[1] Fukuoka Industry, Science and Technology Foundation, Japan
[2] Information, Production and Systems Research Center, Waseda University, Japan
[3] Department of Computer Science, National Tsing Hua University, Taiwan
[4] Graduate School of Information, Production and Systems, Waseda University, Japan

Abstract. Signature verification is a challenging task, because only a small set of genuine samples can be acquired and usually no forgeries are available in real application. In this paper, we propose a novel approach based on Hilbert scanning patterns and Gaussian mixture models for automatic on-line signature verification. Our system is composed of a similarity measure based on Hilbert scanning patterns and a simplified Gaussian mixture model for decision-level evaluation. To be practical, we introduce specific simplification strategies for model building and training. The system is compared to other state-of-the-art systems based on the results of the First International Signature Verification Competition (SVC 2004). Experiments are conducted to verify the effectiveness of our system.

Keywords: Hilbert scanning patterns, Gaussian mixture model, Hilbert scanning distance.

1 Introduction

Biometric user recognition techniques are currently undergoing large-scale deployment. Among biometrics, they can be categorized into two classes of methods: physiological and behavioral biometrics. Physiological traits including hand and face geometry, are based on the measurement of biological properties of users by devices like digital cameras and scanners. In contrast to physiological systems, behavioral biometrics, such as recognition of subjects by voice or signatures, require an explicit action to be performed by individuals and thus imply awareness [1]. Signature verification is one of behavioral biometrics which has been an intense research area because of the social and legal acceptance and widespread use of the written signature as a personal authentication method.

Depending on the acquisition process, automatic signature verification systems are divided into two categories: on-line and off-line. In early off-line cases, signatures are captured until the writing process is finished, thus only static images are available. Recently, more researches focus on the online signature verification, where signatures are acquired during the writing process with a special instrument, such as digital tablet. In fact, dynamic information is available

M. Tistarelli and M.S. Nixon (Eds.): ICB 2009, LNCS 5558, pp. 1190–1199, 2009.
© Springer-Verlag Berlin Heidelberg 2009

in the case of online signature verification, such as velocity, acceleration and pressure which is more difficult to imitate than the static shape of signature. Therefore, online signature verification can usually achieve better performance than the offline instance. Many different approaches have been considered in the literature in order to extract discriminative information from on-line signature data. The existing methods can broadly be divided into two classes; i) Feature-based approaches, in which a holistic vector representation consisting of a set of global features is derived from the signature trajectories [2]. ii) Function-based approaches, in which time sequences describing local properties of the signature are used for recognition [3].

Function-based approaches can be classified into local and regional methods. In local approaches, the time functions of different signatures are directly matched by using elastic distance measures such as Dynamic Time Warping (DTW) [4]. In regional methods, the time functions are converted to a sequence of vectors describing regional properties. One of the most popular regional approaches is the method based on Hidden Markov Models (HMM) [5]. Generally, the DTW is regarded as a popular method, but it usually suffers from the following two drawbacks; Heavy computational load and Warping forgeries [6]. The first one makes the DTW time-consuming while the second makes the verification more difficult. On the other hand, the HMM has the capability to perform stochastic matching for a model of signature using a sequence of probability distributions of the features along the signature. Practically, the HMM has been employed in the filed of online signature verification for two decades and has achieved some success. However, the HMM also has its limitations. Its poor discriminative power is fatal which limits its application on the signature verification [7]. These models focus on local properties of signatures such as local moving direction and shape curvature. Global features like writing time or signature length may be incorporated to speed up verification procedure and improve accuracy.

In the practical applications, users may feel uncomfortable if required to provide lots of samples. Only small set of signature samples can be acquired [8]. This demands the ideal signature verification system to be simple because complex system usually needs large training sample set. On the other hand, many existing systems require false samples, i.e., forgery signatures, for system training. But it is actually hard to collect forgery signatures for every user in practice. So the practical system would be more applicable if it uses forgery samples as few as possible.

Due to the above-mentioned two considerations, we proposed a novel method for automatic signature verification system based on Hilbert Scanning (HS) patterns and GMM. The paper is organized as follows. Section 2 and Section 3 detail the theoretical background of the Hilbert Scanning patterns and the statistical modeling based on GMM. In Section 4 gives an overview of the proposed system and its main components. The evaluation of decision-level combination strategies are presented in Section 5. Experimental procedure and results are given in Section 6. Finally in Section 7 we present our conclusions and suggestions for future work.

2 Hilbert Scanning Patterns

Space-filling curve is one-to-one mapping between N-dimensional (N-D) space and one-dimensional (1-D) space. By mapping each point in a N-D space into a 1-D space for a single point, the complex multidimensional access methods can be transformed into simple 1-D processing methods [9]. For this advantage, a lot of research works have been done and there have been many space-filling curves proposed. In 1980, the mathematician G. Peano presented a family of curves [10], such as raster curve, Z curve and Hilbert curve. The Hilbert curve is a locus of points in N-D space. For its special attributes, it becomes an analytical solution of a space-filling curve. In 1891 Hilbert made a curve having the space-filling property in 2-D space (Hilbert curve).

The N-D Hilbert scanning is regarded as a mapping from a point in N-D space to a point in 1-D space using the Hilbert curve. Quinqueton et al. [11] proposed an algorithm for computing the scan by a recursive procedure. However, in general the recursive procedure presents a drawback for high speed computation and hardware implementation. On the other hand, a continuous one-to-one mapping algorithm is considered by Butz [12] but this algorithm is complex. Kamata first discussed a fast sequential algorithm with look-up tables requiring little memory [13]. He also proposed an effective method for constructing look-up tables of N-dimensional Hilbert scanning derived from reflected binary gray codes in order to improve the performance, and also a simple non-recursive algorithm using the look-up tables which expands the algorithm for 2-D space to one for N-dimensional space. Fig. 1 shows how the Hilbert scan transforms the two-dimensional space to one-dimensional sequence.

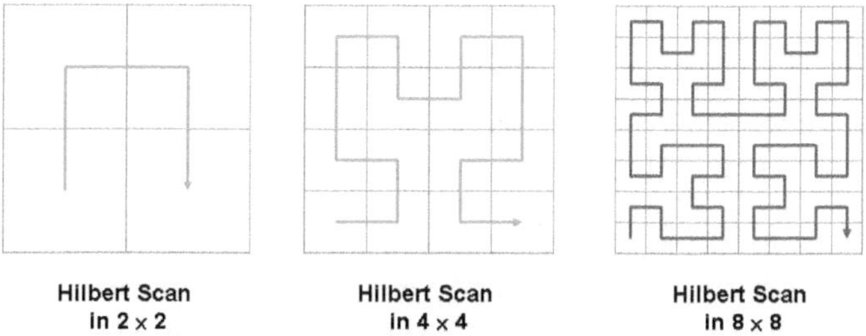

Fig. 1. The Hilbert scanning sequence

The Hilbert scan has two important features; first, extracting a group of neighboring pixels having similar intensities is easier by using Hilbert scan than other scan techniques such as raster scan, row-prime scan, Morton scan, and second is Hilbert scan preserves the coherence in a 2-D space. That is, if the Euclidean distance between two points in the 2-D space is small, then the scanning length between the same pair of points in the 1-D sequence is also small.

The HS patterns using in this paper is the 1-D sequence transformed from the 2-D space, and we use the patterns to compute the similarity measure.

3 Gaussian Mixture Models

GMM is a well known and so much referenced technique for pattern recognition. GMM theory has been known for ages, but it was not till Expectation - Maximization algorithm was developed [16] that it become a useful technique for pattern recognition.

GMM can be thought of as a single state HMM, which means HMM training methods can be applied to GMM with very little or no modification. GMM are widely-used statistical models in many pattern recognition applications. They are a versatile modeling tool as they can be used to approximate any probability density function given a sufficient number of components, and impose only minimal assumptions about the modeled random variables. A GMM is a weighted combination of multiple Gaussians. A distinct property of GMM is its power to approximate any probability density. Let us denote a component GMM by $\Theta = \{(\alpha_l, \mu_l, \Sigma_l) : l = 1, ..., M\}$. It has the general form as follows;

$$p(x \mid \Theta) = \sum_{l=1}^{M} \alpha_l \cdot p(x \mid \mu_l, \Sigma_l) \quad (1)$$

where x is an n-dimensional feature vector, α_l is the weighted coefficient with $\alpha_l = p(l \mid \Theta)$, $\Sigma\alpha_l = 1$, μ_l and Σ_l are the mean vector and the covariance matrix of the l-th Gaussian component, respectively. Each component's distribution is an n-variable Gaussian function. To avoid GMM learning from being stuck into local extreme, K-means clustering is usually applied for "good" initialization. We may choose the best combination with small intra-class distance and large inter-class distance by exhaustive search. Note that the feature vector does not have the same unit in each dimension. We should normalize each dimension before computing the distance matrix between two feature vectors.

For getting this model for each user, Expectation - Maximization (EM) [16] algorithm has been used. EM provides an easy way to estimate the three elements for GMM in an iterative mode. The elements of the models have to be initialized. Random selection from the training data have been used to set up the mean vectors, covariance matrix has been initialized as the unit matrix, and each weighted coefficient is defined as 1/M. After initialization, EM iterative algorithm is run until getting the threshold.

4 System Overview

4.1 Preprocessing

On-line signature always contains associated dynamic information [5], such as: (i) Position in x-axis. (ii) Position in y-axis. (iii) Pressure applied by the pen. (iv) Azimuth angle of the pen with respect to the tablet. (v) Altitude angle of the pen with respect to the tablet.

A signature acquisition process on a restricted size frame is assumed [15]. As a result, users are supposed to be consistent in size and writing dynamics. Moreover, a geometric normalization consisting of position normalization followed by

rotation alignment is applied. In this paper, each signature is normalized on the position and scale firstly.

4.2 Feature Points Extraction

Only coordinate trajectories are considered in the feature extraction process. The feature points are defined as the peaks and valleys of the signal [6]. We first define a rise-distance, denoted by "r", as the amplitude from a valley to the following peak. Similarly we define a drop-distance, denoted by "d", as the amplitude from a peak to the following valley. For any peak (or valley), a rise distance can be computed at one side of the curve, while a drop-distance can be computed at the other side of the curve. The peak or valley is marked as a function point only if $r \geq h_o, d \geq h_0$ condition is satisfied; where h_0 is defined as a threshold. Small ripples are not considered as feature points. This is because small ripples are unreliable most of the time. In the project, the threshold h_0 is chosen as one pixel. Hence any ripples with rise or drop distance less than h_0 will not the counted as feature points. Our simple feature points extracting function can identify the important peaks and valleys along the signal, while excluding the small ripples, as shown in Fig. 2.

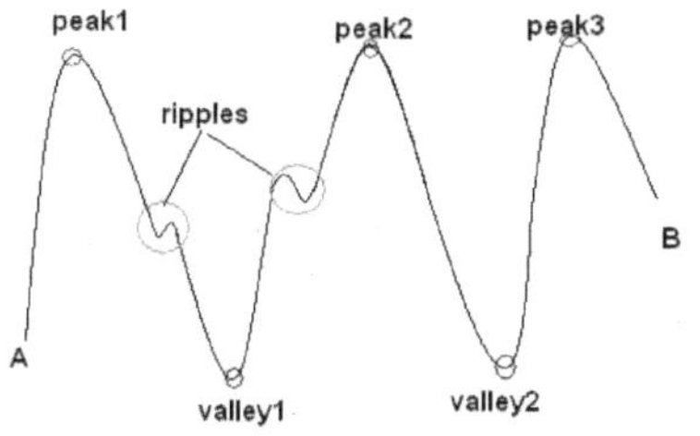

Fig. 2. Examples of peaks, valleys and ripples

4.3 Data Training Procedure

We have two training procedure for Hilbert scanning patterns and Gaussian mixture models respectively. First, we use the Hilbert scanning algorithm to convert the feature points in 2-D space into 1-D sequence. Then, the training procedure of Hilbert scanning patterns consist first 10 genuine signatures for each user and compute the Hilbert scanning distance between each two signatures to represent the Hilbert threshold of each user. In the GMM training, we also apply first 10 signatures of each user for modeling. Every single user will have his/her own model, we will use Θ as the notation to reference each user GMM as defined in eq.1.

5 Evaluation

5.1 Hilbert Scanning Distance

In this paper, we use the Hilbert Scanning Distance which is a distance measure using the Hilbert curve to measure the similarity. Hilbert Scanning Distance

can be viewed as a special Hausdorff distance extended to the 1-D space. We use the Hilbert curve to convert the two-dimensional (2-D) searching space into a one-dimensional (1-D) sequence. An important feature of Hilbert scanning is that it can preserve the coherence in the 2-D image. Thus, the distance between two points in the 2-D space can be migrated into a 1-D sequence. In the 1-D sequence, we only need to consider front and back neighborhood instead of multi-neighborhood in the 2-D space, which means searching many fewer points when computing the distance.

Assume that we are given two finite point sets: $A = \{a_1, ..., a_I\}$ and $B = \{b_1, ..., b_J\}$ such that each point $a \in A$ and $b \in B$ has integer coordinates in the 2-D space. We firstly use Hilbert scanning to convert them to new sets $S = \{s_1, ..., s_I\}$ and $T = \{t_1, ..., t_J\}$ in the 1-D sequence, respectively. Then, the directed Hilbert scanning distance from A to B, $h_{hsd}(A, B)$ is computed by;

$$h_{hsd}(A, B) = \frac{1}{I} \sum_{i=1}^{I} \rho(\min_{j} \|s_i - t_j\|) \tag{2}$$

where $\|\cdot\|$is the Euclidean norm distance in the 1-D space and function ρ is defined as;

$$\rho(x) = \begin{cases} x \ (x \leq \tau) \\ \tau \ (x > \tau) \end{cases} \tag{3}$$

where ρ is called threshold elimination function and τ is a threshold predefined. We also can obtain the directed Hilbert scanning distance from B to A, $h_{hsd}(B, A)$ similarly and Hilbert scanning distance is defined by;

$$H_{hsd}(A, B) = \max(h_{hsd}(A, B), h_{hsd}(B, A)) \tag{4}$$

5.2 GMM Similarity

GMM is a weighted combination of multiple Gaussians. A distinct property of GMM is its power to approximate any probability density. Denote a component GMM by the notation. $\Theta = \{(\alpha_l, \mu_l, \Sigma_l) : l = 1, ..., M\}$. It has the general form as follows;

$$p(x \mid \Theta) = \sum_{l=1}^{M} \alpha_l \cdot p(x \mid \mu_l, \Sigma_l) \tag{5}$$

Every single user will have his/her own model, and we will compute the similarity between the model and training data to get a GMM threshold of each user for the testing procedure.

6 Experiments

6.1 Database Description

There are not many signature databases publicly available at the moment for research purposes. As a result, the common practice in on-line signature recognition research is to evaluate the proposed recognition strategies on small data sets

acquired at the different research laboratories. In this environment, the First International Signature Verification Competition (SVC 2004) has been organized [14] providing a common reference for system comparison on the same signature data and evaluation protocol.

Development corpus of the extended task (including coordinate and timing information, pen orientation and pressure) is used in the experiments that follow. The database 1 used in this experiment consists 5 sets of signatures, the database 2 consists of 40 sets of signatures (Task1 in SVC2004), and the database 3 includes 40 sets of signatures (Task2 in SVC2004). Each set contains 20 genuine signatures from one contributor (acquired in two separate sessions) and 20 skilled forgeries from five other contributors.

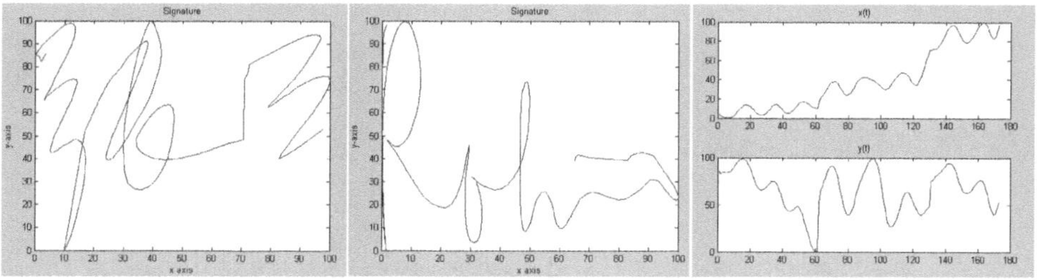

(a) (b) (c)

Fig. 3. (a) shows the Chinese sample of the dataset, (b) shows the English sample of the dataset and (c) shows the trajectories of x(t) and y(t) of the dataset

The signatures are mostly in either English or Chinese. Some examples are shown in Fig. 3 for two different targets of the data set. Plots of the coordinate trajectories are also given.

6.2 Experimental Results

The following tables show the experimental results of the system. The results in terms of the equal error rate (EER) obtained by GMM and proposed method are presented in Table 1. This experiment uses 10 genuine signatures for training, the last 10 genuine signatures and 20 skilled forgery signatures for testing. For skilled signatures, the proposed method acquires the better equal error rate than GMM method.

Random forgeries are also taken into consideration here. We randomly select 20 signatures from other subjects in database for each signer as imposter data. Table 2 shows another experimental result which uses 10 genuine signatures for training, the last 10 genuine signatures and 20 random forgery signatures for testing.

Comparison between our system and the best 5 system competing in SVC2004 task1 and task2, which are following the same experimental protocol, is also given in Table 3 and Table 4. From Table 3 and Table 4, we can see that proposed

Table 1. Error rate of methods for skilled forgeries (Training data=10 samples, testing data=10 genuine + 20 skilled forgeries)

Method\EER	Database1	Database2	Database3
GMM method	16%	15.58%	14.41%
Proposed method	4.6%	6.08%	5.83%

Table 2. Error rate of methods for random forgeries (Training data=10 samples, testing data=10 genuine + 20 random forgeries)

Method\EER	Database1	Database2	Database3
GMM method	12.6%	9.0%	11.1%
Proposed method	4.6%	3.98%	3.25%

Table 3. Error rate of methods using database2 (Training data=10 samples, testing data=10 genuine + 20 forgeries)

Team id	10 genuine signatures + 20 skilled forgeries	10 genuine signatures + 20 random forgeries
106	5.50%	3.65%
Proposed method	6.08%	3.98%
126	6.45%	3.49%
124	7.33%	2.93%
115	9.80%	2.90%

Table 4. Error rate of methods using database 3 (Training data=10 samples, testing data=10 genuine + 20 forgeries)

Team id	10 genuine signatures + 20 skilled forgeries	10 genuine signatures + 20 random forgeries
Proposed method	5.83%	3.25%
219b	6.90%	3.02%
219c	6.91%	3.47%
206	6.96%	2.93%
229	7.64%	4.45%

method gives a better performance in database 2 (Task 1 in SVC 2004) and also performs well both in skilled forgeries and random forgeries in database 3 (Task 2 in SVC 2004).

7 Conclusion

In this paper, we proposed a novel algorithm based on Hilbert scanning patterns and Gaussian mixture models for signature verification system. Proposed method

has two main procedures including training procedure and testing procedure. For training procedure, we compute the Hilbert threshold and GMM threshold representing the similarity. In testing procedure, we proposed a two-stage model to verify the signatures by Hilbert threshold and GMM threshold. The system shows that we can use the fewer features to get the better performance, yielding not only high recognition rate but also less processing time on public database SVC2004.

The possible improvement on this work mainly lies in the combination of the decision. The decision making based on the combined verification is straightforward which need more investigation for improvement. We consider that the threshold variation of every parameter is related to this problem. Examination of a policy to equalize a threshold variation every such parameter is a problem.

Acknowledgment

This work was supported by a grant of Knowledge Cluster Initiative 2nd stage implemented by Ministry of Education, Culture, Sports, Science and Technology (MEXT).

References

1. Vielhauer, C.: A behavioural biometric. Public Service Review: EUROPEAN UNION, 113–115 (2005)
2. Ketabdar, H., Richiardi, J., Drygajlo, A.: Global feature selection for on-line signature verification. In: Proc. of 12th Int'l Graphonomics Society Conf. (2005)
3. Fairhurst, M.C.: Signature verification revisited: Promoting practical exploitation of biometric technology. IEE Electronics and Communication Engineering Journal 9(6), 273–280 (1997)
4. Faundez-Zanuy, M.: On-line signature recognition based on VQ-DTW. Pattern Recognition 40, 981–992 (2007)
5. Fierrez, J., Ortega-Garcia, J., Ramos, D., Gonzalez-Rodriguez, J.: HMM-based on-line signature verification: feature extraction and signature modeling. Pattern Recognition Letters 28(16), 2325–2334 (2007)
6. Feng, H., Wah, C.C.: Online Signature Verification Using a New Extreme Points Warping Technique. Pattern Recognition Letters 24(16), 2943–2951 (2003)
7. Trentin, E., Gori, M.: A Survey of Hybrid HMM/ANN Models for Automatic Speech Recognition. Neurocomputing 37, 91–126 (2001)
8. Zou, M., Tong, J., Liu, C., Lou, Z.: On-line signature verification using local shape analysis. In: Proc. of the 7th Int'l Conf. Document Analysis and Recognition, pp. 314–318 (2003)
9. Kamata, S., Kawaguchi, E.: A method of computing Hilbert curves in two and three dimensional spaces. Int'l Conf. Pattern Recognition, 93–97 (1994)
10. Peano, G.: Sur une courbe qui remplit touteune aire plane. Math. Ann. 36, 157–160 (1890)
11. Quinqueton, J., Berthod, M.: A locally adaptive peano scanning algorithm. IEEE Trans. Pattern Anal. Mach. Intell., PAMI 3(4), 409–412 (1981)

12. Butz, A.: Convergence with Hilbert's space filling curve. Computer and System Science 3(2), 128–146 (1969)
13. Kamata, S., Kawaguchi, E.: A method of computing Hilbert curves in two and three dimensional space. Trans. IEICE J74-D-II, 1217–1226 (1991)
14. Yeung, D.Y., Chang, H., Xiong, Y., George, S., Kashi, R., Matsumoto, T., Rigoll, G.: SVC2004: First international signature verification competition. In: Proc. Intl. Conf. Biometric Authentication, pp. 16–22 (2004)
15. Fierrez-Aguilar, J., Alonso-Hermira, N., Moreno-Marquez, G., Ortega-Garcia, J.: An off-line signature verification system based on fusion of local and global information. In: Maltoni, D., Jain, A.K. (eds.) BioAW 2004. LNCS, vol. 3087, pp. 295–306. Springer, Heidelberg (2004)
16. Bilmes, J.: A Gentle Tutorial of the EM Algorithm and its Application to Parameter Estimation for Gaussian Mixture and Hidden Markov Models. Technical Report ICSI-TR-97-021 (1998)

Static Models of Derivative-Coordinates Phase Spaces for Multivariate Time Series Classification: An Application to Signature Verification

Jonas Richiardi[1], Krzysztof Kryszczuk[2], and Andrzej Drygajlo[1]

[1] Speech Processing and Biometrics Group, Laboratory of IDIAP
Swiss Federal Institute of Technology Lausanne (EPFL)
[2] IBM Zurich Research Laboratory

Abstract. Multivariate time series are sequences, whose order is provided by a time index; thus, most classifiers used on such data treat time as a special quantity, and encode it structurally in a model. A typical example of such models is the hidden Markov model, where time is explicitely used to drive state transitions. The time information is discretised into a finite set of states, the cardinality of which is largely chosen by empirical criteria. Taking as an example task signature verification, we propose an alternative approach using static probabilistic models of phase spaces, where the time information is preserved by embedding of the multivariate time series into a higher-dimensional subspace, and modelled probabilistically by using the theoretical framework of static Bayesian networks. We show empirically that performance is equivalent to state-of-the-art signature verification systems.

1 Introduction

Multivariate time series appear in a vast array of engineering applications, such as computational finance, biosignal processing (e.g. EEG, EMG), bioinformatics, or brain-machine interfacing (which can have tens of data channels). Automatic classification of these time series into pattern classes is often of interest, for example for earthquake detection (which typically is based on two or more data channels) or behavioural biometrics (e.g. speech, signature, gait).

Probabilistic state-space methods such as hidden Markov models (HMMs) offer several advantages over classical time-series models (e.g. ARMAX), for instance handling of hidden variables, continuous and discrete spaces for random variables, and specification of priors [10]. HMMs are top-performing classifiers for signature verification [1, 3]. However, while they are very commonly used, HMMs suffer some important drawbacks, for example feature vectors are deemed conditionally independent given the sequence of states. To relax this unrealistic assumption, techniques have been proposed to model the "trajectories" of the time series across states, for instance in speech recognition [8]. It is also possible to use more general dynamic probabilistic models such as dynamic Bayesian

M. Tistarelli and M.S. Nixon (Eds.): ICB 2009, LNCS 5558, pp. 1200–1208, 2009.
© Springer-Verlag Berlin Heidelberg 2009

networks, and to model the inter-state relationships between the variables of the time series.

Conversely, many discriminative classification methods that perform very well with static[1] multivariate data, such as random forests [2] or multilayer perceptrons, do not yield good results when applied directly to the modelling of time series, since variability in onset or duration is not handled by default.

In this paper, we propose a middle way between static discriminative classifiers and dynamic generative models, by using a static generative model together with specific feature extraction steps. We use a specific topology of static Bayesian network that is equivalent to a Gaussian mixture model (GMM) to perform classification of multivariate time series, as applied to a handwritten signature verification task[2]. We show that, by using appropriate features extracted from the time series, the framework of static Bayesian networks offers an elegant approach to classification of multivariate time series. The underlying principle is to consider that the observable time series data available are in fact a projection from the phase space of higher dimension of an original dynamic system.

The remainder of this paper is organised as follows: In Section 2, we briefly review the theory of phase space reconstruction, drawing attention to the work of Povinelli et al. [14]. In Section 3 we propose a general framework for classifying features that can be extracted from a multivariate time series, linking in more specifically with signature verification. In Section 4, we propose the use of delta features for phase space reconstruction, and show that they posess significant advantages over the more classically used method of delays. Section 5 presents experimental results on two signature databases, which are discussed in Section 6.

2 Phase Space Reconstruction

Phase space reconstruction can be approached through the Takens theorem [18]: Given a (regularly sampled, univariate) time series, it states that, under certain conditions, it is possible to reconstruct a multidimensional phase space which corresponds to the same dynamical system as the one which originally generated the time series, but in a different system of coordinates: there is a diffeomorphism between the two spaces. In other words, the values of the (univariate) time series are seen as a one-dimensional projection of the trajectory in multivariate phase space of the dynamical system under consideration.

One sufficient condition entailed by the Takens theorem is that, for a map to exist between the original and the reconstructed space, we must have

$$D \geq 2\mathcal{D} + 1, \tag{1}$$

where D is the embedding dimension of the reconstructed space, and $\mathcal{D}$ is the box-counting dimension of the trajectory in the original space. The principle

[1] In this paper, *static* data means that there is no time dependence between feature vectors. For models, it means that time is not part of the model structure.

[2] This paper offers a theoretical explanation for the performance of our original proposal of using Gaussian mixture models for signature verification [15].

behind this theorem is that two points should lie close together in D dimenions not because D is too small, but because their proximity is a property of the set of points. The box-counting dimension is difficult to compute in some cases, when not much data is available and/or dimensionality is high [5], and can be approximated by methods such as the false nearest-neighbour algorithm [9] or the correlation integral approach [5].

The coordinates used for embedding must also be chosen. Typically, the coordinates used are the time series $o(t)$ itself, and integer multiples of the lagged time series $(o(t+\tau), o(t+2\tau), \ldots)$. While in principle, with an infinite number of noiseless datapoints, the delay τ could be chosen almost arbitrarily [18], in practice heuristics are applied to select this parameter. Typically, either the first zero of the autocorrelation function (corresponding to linear independence of the coordinates) or the first minimum of the automutual information function (a more general notion of independence) are used to select the optimal τ.

Recently, phase space reconstruction has been applied to statistical classification of time series. Imhoff et al. [6] proposed using statistical models of reconstructed phase spaces in medicine, and Povinelli et al. [14] used the method of delays for ECG classification and speech recognition. As we will show in Section 4, delay coordinates have several shortcomings which can be addressed by the use of other coordinates.

3 Features for Multivariate Time Series Classification

Approaching multivariate time series classification from a statistical pattern recognition viewpoint, an important family of methods is to transform the original signal into another feature space before attempting classification.

For many applications, the goal is to build the most discriminative classifier possible: the semantics associated to states (e.g. for evolution of a disease over time) are of no interest. Indeed, it is even possible that the time series be transformed from a sequence into an unordered set, while maintaining or improving discriminative abilities in comparison to dynamic classifiers.

To recognise patterns in time series, the signal representation and/or the classifier must be made resilient to changes in the aspect of the pattern, for example in amplitude, onset, or duration. To this end, several feature extraction schemes are possible:

Local features are extracted at the same rate as the sampling of the incoming signal: that is, each input sample data vector corresponds to a local feature vector. *Segmental features* are extracted once the time series has been cut into segments or windows. The segmentation paradigms vary, but a segment typically consists of a sequence of points for which some definition of coherence holds. *Global features* summarise some property of the complete observed time series; for instance the total duration of the signal.

In all three cases, features can be computed from single or multiple variables in the multivariate time series. In signature verification, all three types of features have been used [17], and they can be combined in multiple-classifier systems [16].

4 Delta Features for Phase Space Reconstruction

The first and second time derivative of local features, referred to in speech processing as delta (Δ) and delta-delta ($\Delta\Delta$) features, can be used to improve the discriminative ability of the base feature set.

Since the signal is sampled at discrete time intervals, we use a numerical approximation of a first order derivative. By definition:

$$df(\cdot) = lim_{\epsilon \to 0} \frac{f(\cdot + \epsilon) - f(\cdot)}{\epsilon}, \tag{2}$$

which we replace by second order regression using the central difference approximation. Boundary effects are avoided by switching to a forward difference approximation when necessary. Thus, for the tth term in the vector sequence of length T we obtain:

$$df(\cdot_t) \approx \frac{f(\cdot_{t+1}) - f(\cdot_{t-1})}{2} + \frac{f(\cdot_{t+2}) - f(\cdot_{t-2})}{4}. \tag{3}$$

We posit that the key to the effectiveness of *static* probabilistic models for classifying multivariate time series lies in the use of delta features and features that are functions of derivatives of the raw sampled data.

Support for this hypothesis is to be found in a particular method of phase space reconstruction: [18, 12] have shown that, in addition to the method of delays, phase space reconstruction can be performed by using first, second, and successive derivatives of a (univariate) time series $\frac{do}{dt}, \frac{d^2o}{dt^2}, \ldots$. Under condition that the Takens theorem is satisfied, this set of coordinates yields an embedding.

Thus, by using delta features, we are in fact reconstructing a phase space $\mathcal{S}$. This means that the dynamics of the time series (e.g. signature) can asymptotically be fully modelled without the need to resort to dynamic models. Because we also use other features (e.g. pressure), we obtain a feature space $\mathcal{S}'$ that is of a higher dimension than the corresponding reconstructed phase space $\mathcal{S}$, and embeds it. We can then view the training of a generative model on the vectors in this space as the computation of a statistical model of distribution of features in $\mathcal{S}'$. This approach was proposed in [14], using delay coordinates.

Reconstructing the phase space using delta features overcomes some original restrictions of the method of delays to univariate time series. Indeed, in order to compute an embedding into phase space for multivariate time series using the method of delays, it is necessary to estimate the optimal embedding lag (optimal delay coordinate) $\hat{\tau}_d$ for each of the dimension d in the multivariate time series independently. If these optimal embedding lags are not equal, the result is that the phase spaces reconstructed for each of the variables of the original time series will differ in length. Then, the problem of combining the different phase spaces is not trivial to solve [4]. It is also possible to consider (potentially suboptimal) equal time lags for each of the variables in the time series [14]. Using delta features, all the phase spaces reconstructed have the same length.

Also, if the signal available is small (e.g. 60 samples), and the optimal lag computed is large with respect to the signal duration, higher-order lags (e.g. 3τ

or 4τ) may actually not be computable. Using delta coordinates alleviates this problem: phase space coordinates can be computed on small amounts of data.

One potential disadvantage of delta coordinates is noise amplification with successive derivatives, but in practice this can be overcome by low-pass filtering and/or robust derivative extraction (e.g. by fitting Legendre polynomials and taking an analytic derivative).

5 Experiments and Results

5.1 Signature Databases and Verification Systems

We use two different signature databases for our experiments.

The MCYT database [11] contains signature and fingerprint data for 330 users. We use a 100-users subset of this database, called MCYT-100.

The BMEC 2007 development database contains 50 users and is part of the larger BioSecure DS3 dataset. Signatures are acquired on a low-power mobile platform (Ipaq PDA). For both databases, skilled forgeries are used (between levels 6 and 8 [7]).

The classifier used is a static Bayesian network functionally equivalent to the Gaussian mixture model proposed in [15], and is denoted BN/GMM. The number of diagonal covariance Gaussian mixture components is varied, but the initialisation is always performed using the k-means method, followed by the expectation-maximisation algorithm until convergence. The feature set depends on the experiment and database. The first 5 authentic signatures of the first session are used as training samples for each user.

5.2 Classification with Delta Features

In this first set of experiments the goal is to empirically demonstrate the benefits of modelling reconstructed phase spaces with delta coordinates instead of only the base feature vector, and to show the superior performance when compared to delay coordinates.

For the first experiment, the base feature spaces considered is a univariate feature space using only the coordinate x_t (univariate time series). The second feature space consists of the phase-space reconstruction $(x, \dot{x}, \ddot{x})$, which is an embedding dimension of 3. For both feature spaces, the parameters of a BN/GMM classifier, using 40 Gaussian mixture components, are trained using expectation-maximisation. As can be seen in Figure 1, training a statistical classifier on the reconstructed phase space yields better significantly better results (McNemar test on classification decisions, thresholded *a posteriori*, $p < 0.01$) than classification on the initial time series itself, for both phase space reconstruction coordinates used. While not reported here, the results obtained with other univariate coordinates (e.g. y) and on other databases are comparable.

For the second experiment, the classification performance over a derivative-coordinates phase space $(x, \dot{x}, \ddot{x})$ reconstructed from a univariate time series is compared to the classification performance over a delay-coordinates phase space

$(x_t, x_{t+\tau}, x_{t+2\tau})$. The lag is chosen as the mode of the distribution of lags for each user's training, respectively testing, signatures, as per [14]. Figure 1 shows that on this task, a substantial reduction in error rate can be obtained by using phase space reconstruction, and preferring delta coordinates to delay coordinates (statistically significant at $p < 0.01$).

The third experiment reconstructs a phase space from a multivariate time series (x_t, y_t), resulting in a phase space $(x, \dot{x}, \ddot{x}, y, \dot{y}, \ddot{y})$ for derivative coordinates, and $(x_t, x_{t+\tau}, x_{t+2\tau}, y_t, y_{t+\tau}, y_{t+2\tau})$ for delay coordinates. Again, it can be observed from Figure 1 that both coordinate systems yield improvements over the base feature space, and that more improvement results from the use of derivative coordinates (statistically significant at $p < 0.01$).

Our experiments on time series datasets not related to biometrics or signature verification (e.g. ECG, gestures, machine control) show that these trends are generally confirmed: that is, training a BN/GMM classifier in the reconstructed phase space tends to yield significantly better classification accuracy than what is obtained in the base space, and delta coordinates yield better results than lag coordinates.

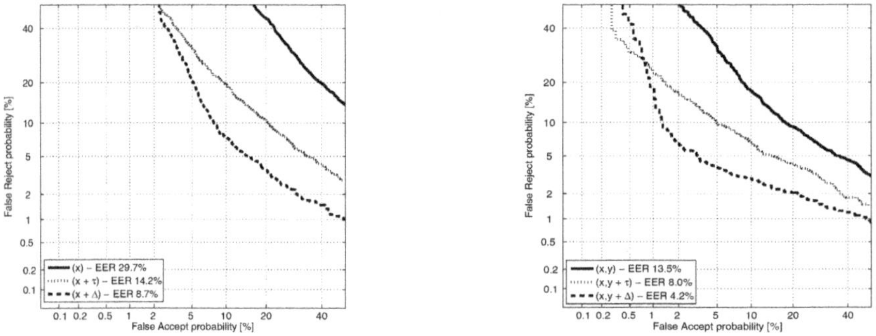

(a) Univariate time series (b) Multivariate time series

Fig. 1. DET curves for classification of univariate (x, left panel) and multivariate ((x, y), right panel) time series in the phase space with delay coordinates (τ) and delta coordinates (Δ)

5.3 Comparison with Dynamic Models

In this series of experiments we compare our static BN/GMM classifier with hidden Markov models of equivalent model complexity: the number of free parameters is kept in the same range. It was chosen to compare 5- and 2- states HMMs to the BN/GMM baseline system, using diagonal covariance matrices. The HMMs have a strict left-to-right topology. Both the BN/GMM and the HMM models are initialised using k-means clustering.

In order to assess statistical significance of the difference between classifiers, we compute the EER threshold a posteriori on the classifier outputs, apply it to the output, yielding decision vectors on which we then perform the McNemar significance test with $p = 0.05$.

On MCYT-100, the feature vector used is

$$\mathbf{o}_t = [x_t, y_t, p_t, \theta_t, v_t + \Delta + \Delta\Delta]' = [x_t, y_t, p_t, \theta_t, v_t, \dot{x}_t, \dot{y}_t, \dot{p}_t, \dot{\theta}_t, \dot{v}_t, \ddot{x}_t, \ddot{y}_t, \ddot{p}_t, \ddot{\theta}_t, \ddot{v}_t]', \quad (4)$$

where x_t and y_t are instantaneous position on the horizontal plane, p_t is instantaneous pressure, θ_t is instantaneous pen trajectory tangent, v_t is instantaneous velocity.

The classifiers compared are a BN/GMM model with 30 Gaussian mixture components (929 free parameters), a two-states, 15-Gaussian components HMM (930 free parameters), and a five-states, 6-Gaussian components HMM (933 free parameters). Figure 2 shows that results are very slightly worse for the BN/GMM than the HMM classifiers at EER, but not statistically significantly so ($p = 0.05$).

On BMEC 2007, the feature vector used is $\mathbf{o}_t = [x_t, y_t, \dot{x}_t, \dot{y}_t, \ddot{x}_t, \ddot{y}_t]'$. The classifiers compared are the BN/GMM model with 20 Gaussian mixture components(259 free parameters), a two-states, 10-Gaussian components HMM (260 free parameters), and a five-states, 4-Gaussian components HMM (263 free parameters). Figure 2 shows that the difference in error rates between these models is not statistically significant ($p = 0.05$). As a comparison point, the BioSecure reference system (based on HMMs) on the same data achieves 15% EER.

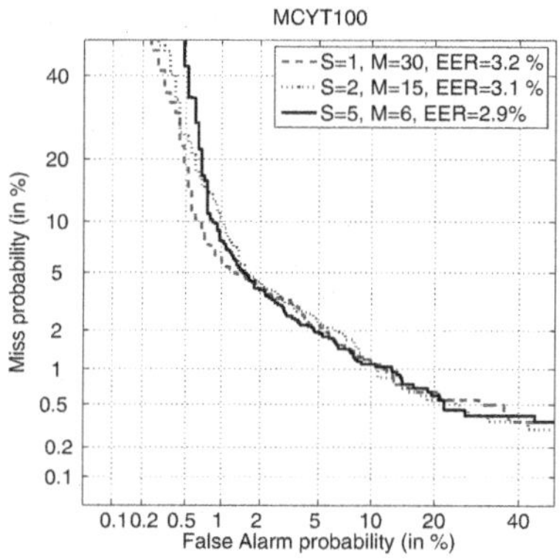

(a) MCYT-100

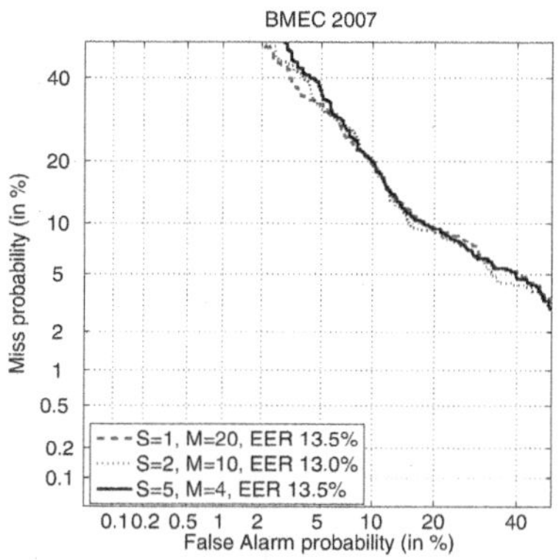

(b) BMEC2007

Fig. 2. Comparison between the BN/GMM model and HMM models with equivalent number of parameters (S, number of states, and M, number of Gaussian components in the mixture) on MCYT-100 and BMEC 2007

The same trends are observed on SVC 2004 database [19], where the comparison of a 50-mixture components BN/GMM to a 2-states, 25-components per state and a 5-states, 10 components per state HMM yield insignificantly different results ($p = 0.05$);

6 Discussion

The use of delta features in speech and signature verification using HMMs was originally motivated by the need to relax the Markov assumption between feature vectors of close states. Delta features have been a common occurence in static models of speech such as used in speaker verification. However, we think that considering the use of delta features in the light of phase space reconstruction theory yields interesting insights into their impressive effectiveness in improving discrimination between classes.

In signature verification, this can be attributed to the fact that the dynamics of the signature are unfolded by embedding into a higher-dimensional subspace, where the specific topology of each user's signature is more apparent in the phase space than when projected down to a single dimension. Since a handwritten signature is the result of a high-order graphomotor process controlled by several parameters [13], it is reasonable to attempt this reconstruction.

The better results obtained for delta coordinates over delay coordinates do not seem to stem from the choice of method used for selecting the optimal lag. We have used both the "first minimum of automutual information" and the "first zero crossing ot the autocorrelation function" heuristics to estimate the optimal lag, and arrived at generally better results with the automutual information. However, both tend to yield small lags, and the resulting reconstructed orbit in phase space is generally stretched along the identity line: the representation in phase space may be needlessly complicated. Secondly, the heuristics mentioned are not necessarily optimal for classification, as observed in [14].

Finally, comparing results with HMMs shows that, for classification tasks on signature data, static models using phase-space signal representations may perform as well as dynamical models. This can be attributed to the fact that time information is embedded in the orbit of the signal in reconstructed phase space. Indeed, for signature time series, static BN/GMM models do not perform on-par with HMMs when no time-dependent information (such as delta features or other features based on derivatives) is included in the feature space.

Acknowledgements

This work was partly funded by the European COST 2101 action and by Swiss National Science Foundation grant 118049. The MCYT-100 database was provided by the Autonomous University of Madrid. The BMEC 2007 development database was provided by Institut Télecom in Paris. The authors wish to thank Laurent Uldry and Jean-Marc Vésin of EPFL-IEL-LTS1 for the stimulating discussions and insights into the use of phase space representations in biosignal processing.

References

[1] Van Bao, L., Garcia-Salicetti, S., Dorizzi, B.: On using the Viterbi path along with HMM likelihood information for online signature verification. IEEE Trans. on Systems, Man, and Cybernetics, Part B 37(5), 1237–1247 (2007)

[2] Breiman, L.: Random forests. Machine Learning 45(1), 5–32 (2001)

[3] Fierrez, J., Ortega-Garcia, J., Ramos, D., Gonzalez-Rodriguez, J.: HMM-based on-line signature verification: Feature extraction and signature modeling. Pattern Recognition Letters 28(16), 2325–2334 (2007)
[4] Garcia, S.P.: Phase space reconstruction of multivariate time series from molecular and field Biology. PhD thesis, Instituto de Tecnologia Quimica e Biológica, Universidade Nova de Lisboa, Oeiras, Portugal (2005)
[5] Grassberger, P., Procaccia, I.: Characterization of strange attractors. Physical Review Letters 50(5), 346–349 (1983)
[6] Imhoff, M., Bauer, M., Gather, U., Löhlein, D.: Statistical pattern detection in univariate time series of intensive care on-line monitoring data. Intensive Care Medicine 24(12), 1305–1314 (1998)
[7] ISO/IEC JTC 1/SC 37 Biometrics. TR 19795-3, biometric performance testing and reporting, part 3: Modality specific testing. Technical report, International Standards Organization (2007)
[8] Iyer, R., Gish, H., Siu, M.-H., Zavaliagkos, G., Matsoukas, S.: Hidden Markov models for trajectory modeling. In: Proc. Int. Conf. on Spoken Language Processing (ICSLP), pp. 891–894 (1998)
[9] Kennel, M.B., Brown, R., Abarbanel, H.D.I.: Determining embedding dimension for phase-space reconstruction using a geometrical construction. Phys. Rev. A 45(6), 3403–3411 (1992)
[10] Murphy, K.: Dynamic Bayesian networks: representation, inference and learning. PhD thesis, University of California at Berkeley (July 2002)
[11] Ortega-Garcia, J., Fierrez-Aguilar, J., Simon, D., Gonzalez, J., Faundez-Zanuy, M., Espinosa, V., Satue, A., Hernaez, I., Igarza, J.-J., Vivaracho, C., Escudero, D., Moro, Q.-I.: MCYT baseline corpus: A bimodal biometric database. In: IEE Proceedings - Vision, Image and Signal Processing, vol. 150, pp. 395–401 (2003)
[12] Packard, N.H., Crutchfield, J.P., Farmer, J.D., Shaw, R.S.: Geometry from a time series. Physical Review Letters 45(9), 712–716 (1980)
[13] Plamondon, R., Maarse, F.J.: An evaluation of motor models of handwriting. IEEE Trans. on Systems, Man and Cybernetics 19(5), 1060–1072 (1989)
[14] Povinelli, R.J., Johnson, M.T., Lindgren, A.C., Roberts, F.M., Ye, J.: Statistical models of reconstructed phase spaces for signal classification. IEEE Trans. on Signal Processing 54(6), 2178–2186 (2006)
[15] Richiardi, J., Drygajlo, A.: Gaussian mixture models for on-line signature verification. In: Proc. ACM SIGMM Multimedia, Workshop on Biometrics methods and applications (WBMA), Berkeley, USA, Nov. 2003, pp. 115–122 (2003)
[16] Richiardi, J., Drygajlo, A.: Promoting diversity in gaussian mixture ensembles: an application to signature verification. In: Schouten, B., Jul, N.C., Drygajlo, A., Tistarelli, M. (eds.) BIOID 2008. LNCS, vol. 5372, pp. 140–149. Springer, Heidelberg (2008)
[17] Richiardi, J., Ketabdar, H., Drygajlo, A.: Local and global feature selection for on-line signature verification. In: Proc. IAPR 8th International Conference on Document Analysis and Recognition (ICDAR 2005), Seoul, Korea, vol. 2, pp. 625–629 (August-September 2005)
[18] Takens, F.: Detecting strange attractors in turbulence. In: Proc. Dynamical Systems and Turbulence. Lecture Notes in Mathematics, vol. 898, pp. 366–381. Springer, Heidelberg (1980)
[19] Yeung, D.-Y., Chang, H., Xiong, Y., George, S., Kashi, R., Matsumoto, T., Rigoll, G.: SVC2004: First international signature verification competition. In: Proceedings 2004 Biometric Authentication: First International Conference (ICBA 2004), Hong Kong, China, pp. 16–22 (July 2004)

Feature Selection in a Low Cost Signature Recognition System Based on Normalized Signatures and Fractional Distances*

C. Vivaracho-Pascual[1], J. Pascual-Gaspar[2], and V. Cardeñoso-Payo[1]

[1] Dep. Informática, U. de Valladolid
{cevp,valen}@infor.uva.es
[2] Grupo ECA-SIMM, Universidad de Valladolid
pascualgaspar@gmail.com

Abstract. In a previous work a new proposal for an efficient on-line signature recognition system with very low computational load and storage requirements was presented. This proposal is based on the use of size normalized signatures, which allows for similarity estimation, usually based on DTW or HMMs, to be performed by an easy distance calcultaion between vectors, which is computed using fractional distance. Here, a method to select representative features from the normalized signatures is presented. Only the most stable features in the training set are used for distance estimation. This supposes a larger reduction in system requirements, while the system performance is increased. The verification task has been carried out. The results achieved are about 30% and 20% better with skilled and random forgeries, respectively, than those achieved with a DTW-based system, with storage requirements between 15 and 142 times lesser and a processing speed between 274 and 926 times greater. The security of the system is also enhanced as only the representative features need to be stored, it being impossible to recover the original signature from these.

1 Introduction

Although the evolution of the technology leads to a continuous increase in the data storage and processing capacity of computers while their price decreases, the reduction in costs will always be an important matter in commercial applications and in large-scale systems, but also in resource-limited systems like smart-cards or PDA terminals.

In on-line signature recognition, this reduction usually lies in the characteristic extraction stage. The different methods that can be found in the literature to reduce the signature representation can be grouped into: (1) The use of global features, (2) Size reduction of the time series representing the signature.

The first option achieves worse results than using local features. Then, it is usually used in a first, quick but weak decision [16], or fusing with local

* This work has been partially supported by Consejería de Educación de la Junta de Castilla y León under research project VA077A08.

M. Tistarelli and M.S. Nixon (Eds.): ICB 2009, LNCS 5558, pp. 1209–1218, 2009.
© Springer-Verlag Berlin Heidelberg 2009

information [4]. So most of the research effort focuses on the second option, where the methods used to reduce the time series size can be classified as follows:

- Selection of representative or critical points and information extraction [10][16].
- Signature resampling [7][9][11], generally performed before feature extraction. Theoretically, sampling rates higher than 40-60 Hz would produce redundant information [14]. However, the results are not always in accordance with that affirmation as, while some authors report better results [7][11] with the application of resampling, some others report worse performance [8].
- Signature resampling + representative points [1][7][12].

In [15] we have applied resampling techniques to normalize the signature size, being possible, then, that the similarity between signatures can be computed as a simple distance measurement between vectors. Normally, this distance calculation is based on the use of the Euclidean one. When data are high dimensional, however, *fractional* p-norms (Minkowski-like norms with an exponent p less than one) are, in general, less sensitive to he concentration phenomenon [6], performing better, as can be seen in our work [15].

Here, we present a first approximation to the idea shown in the third item of the previous list. The goal is to select the signatory's most representative points/features from the normalized signatures. In this first approximation we have focused on the feature selection, considering the signature as a vector of p (points) $\times$ f (features per point) components.

This problem is usually approached by statistical methods, such as, for example, Principal Component Analysis (PCA) or Linear Discriminant Analysis (LDA). Here, we have defined the most representative features as the most stable in the training set, that is, those that show less intra-signatory variability. This selection method is similar to that shown in [2]. Each feature variability/dispersion is measured by its standard deviation (σ). Then, once a threshold, $\sigma_{threshold}$, is fixed, only features with $\sigma < \sigma_{threshold}$ will be used to compute the distance calculation (see section 3).

This approach has the additional advantage of increasing the system's security, since only the representative features need to be stored, it being impossible to recover the original signature from these.

The rest of the paper is organized as follows. In Sec. 2 the main characteristics of the baseline system, as shown in [15], will be described. The feature selection method can be seen in Sec. 3. After describing the experimental environment (Sec. 5), the results achieved will be shown, comparing them with those achieved both with the baseline system and the reference system [3]; the computational load and storage requirements will also be compared. Finally, the conclusions and future work can be seen in Section 6.

2 Baseline System

The approach shown in [15] is based on the use of normalized signatures, that is, the signatures are resampled (using a linear interpolation method) to get a fixed

size (number of points). This idea is based on the property of the "diagonal path" alignment of genuine signatures in DTW. That is, as is shown in the above reference work, the alignment path is more linear (diagonal) for genuine signatures than for those that are not.

Once the signatures have the same length, the problem of measuring similarity between signatures can be computed in a more straightforward way as a simple distance measurement between vectors. Due to the high dimensionality of these, to alleviate the concentration phenomena of the Euclidean distance, the use of *fractional* p-norms is proposed, achieving recognition rates comparable with the state-of-the-art classifiers.

This proposal can be seen as a simplification of the DTW-based systems.

In [15] the proposed system was tested with regard to several different configurations, proposing the final following two for the verification task:

- **Baseline_Ver_Sytem.** p value of the p-norm equals 0.3 and normalized signature size equals 50 points. The n signatures $\{tr_1, \ldots, tr_n\}$ of the training set are used as signatory model. Then, given a test sample (te), the distance with regard to each of the n training signatures ($d(tr_i, te)$) is calculated, these n distances achieve a final measured similarity by means of the min function ($\min_i(d(tr_i, te))$).
- **Baseline_Ver_Sytem_Soft.** p value of p-norms equals 0.3 and normalized signature size equals 25 points. The n training signatures $\{tr_1, \ldots, tr_n\}$ are fused to create a prototype signature, that is called *averaged template (AT)*. The AT is the mean of the training signatures, that is, if a training signature is composed of m feature vectors ($tr^i = \{x_1^i, \ldots, x_m^i\}$), each of them with t components ($x_j^i = (x_{j1}^i, \ldots, x_{jt}^i)$), then the template signature will be composed of m vectors ($AT = \{\overline{x}_1, \ldots, \overline{x}_m\}$) calculated as shown in Eq. 1. The computational load and storage requirements of this option are lower than those in the *Baseline_Ver_Sytem*.

$$\overline{x}_j = \left(\frac{\sum_{i=1}^{n} x_{j1}^i}{n}, \ldots, \frac{\sum_{i=1}^{n} x_{jt}^i}{n}\right) \tag{1}$$

3 Feature Selection

The procedure proposed to select the features with less variability over the training set is shown here.

In the feature extraction stage, a set X^i of m (m being the size of the normalized signature) feature vectors (normalized signature points) of t components, $x_j^i = (x_{j1}^i, ..., x_{jt}^i)$, are extracted from the signature i. Then, this stage output can be seen as a matrix where each row is a feature vector:

$$X^i = \begin{pmatrix} x_{11}^i & \cdots & x_{1t}^i \\ \cdots & & \cdots \\ x_{m1}^i & \cdots & x_{mt}^i \end{pmatrix} \tag{2}$$

The selection method is similar to that shown in [2], being performed over each column as follows.

Given n training signatures, each feature x^i_{jl} distribution is modeled by means of a gaussian, with mean μ_{jl} and standard deviation σ_{jl} calculated as shown in Eq. 3.

$$\mu_{jl} = \frac{\sum_{i=1}^{n} x^i_{jl}}{n}$$
$$\sigma_{jl} = \sqrt{\frac{\sum_{i=1}^{n}(x^i_{jl} - \mu_{jl})^2}{n}} \tag{3}$$

The set of standard deviations achieved as seen in Eq. 3 can also be seen as a matrix:

$$\Sigma = \begin{pmatrix} \sigma_{11} & \cdots & \sigma_{1t} \\ \cdots & & \cdots \\ \sigma_{m1} & \cdots & \sigma_{mt} \end{pmatrix} \tag{4}$$

The standard deviation is a measure of dispersion, then, the smaller σ_{jl} is, the more stable the feature jl will be. Using a threshold, $\sigma_{threshold}$, only features with standard deviation less than this threshold will be used for distance calculation.

This threshold is fixed by column, such that the same number, N_f, of features per column will be selected, not necessarily the same for each vector (row/point). That is, N_f is fixed a priori and, then, the column l threshold, $\sigma^l_{threshold}$, is calculated such that the number of σ_{jl} that comply with Eq. 5 will be N_f.

$$\sigma_{jl} \leq \sigma^l_{threshold} \qquad \forall\, 1 \leq j \leq m \tag{5}$$

Once each $\sigma^l_{threshold}$ is calculated, the distance (using p-norm) between two normalized signatures X^a and X^b is calculated as shown in Eq. 6.

$$d(X^a, X^b) = \left(\sum_{j=1}^{m}\sum_{l=1}^{t} C_{jl} * |x^a_{jl} - x^b_{jl}|^p\right)^{\frac{1}{p}} \quad with\ C_{jl} = 1\ if\ \sigma_{jl} \leq \sigma^l_{threshold} \quad C_{jl} = 0\ if\ \sigma_{jl} > \sigma^l_{threshold} \tag{6}$$

A feature selection per matrix was also tested, that is, using the N_f features with the lesser σ. The threshold, $\sigma_{threshold}$, is calculated such that the number of σ_{jl} that comply with Eq. 7 will be N_f.

$$\sigma_{jl} \leq \sigma_{threshold} \qquad \forall\, 1 \leq j \leq m\ \ and\ \ 1 \leq l \leq t \tag{7}$$

The distance was calculated in a similar way to that shown in Eq. 6, but replacing $\sigma^l_{threshold}$ with $\sigma_{threshold}$.

This selection technique showed a worse performance, so the results have not been included in this work. Therefore, in the following, "feature selection" will only refer to the first technique described, that is, feature selection per column.

4 Experimental Environment

In [3] the performance of the main classifiers used in signature recognition were compared. For this reason, this work was used as the reference in [15]. Then, the same experimental conditions as those in both works will be followed here.

4.1 Experiments

The features selection proposal has been tested for the system configurations shown in *Baseline_Ver_Sytem* and *Baseline_Ver_Sytem_Soft* (see Sec. 2).

Several values of N_f have been tested for each system, being calculated as a percentage of the normalized signature size. The following percentages have been tested: 20%, 40%, 60% and 80%. For these percentages, the corresponding N_f values for each system are:

- **Baseline_Ver_Sytem:** 10, 20, 30 and 40 features per column. For this system the normalized signature has 50 vectors ($m = 50$).
- **Baseline_Ver_Sytem_Soft:** 5, 10, 15 and 20 features per column. For this system the normalized signature has 25 vectors ($m = 50$).

In the following, to identify the systems where the feature selection has been applied, these will be called:

- **FS_Ver_System_N_f:** is the *Baseline_Ver_Sytem* with N_f features selected per column.
- **FS_Ver_System_Soft_N_f:** idem to the previous but for the *Baseline_Ver_Sytem_Soft*.

4.2 Database and Trials

We have used MCYT database [13]. The information acquired for each sampling instant (X, Y, pressure, azimuth and altitude) is used directly as features ($t = 5$).

Samples of 330 different people were acquired. Each target user produced 25 genuine signatures, and 25 skilled forgeries were also captured for each user. These skilled forgeries were produced by the 5 subsequent target users by observing the static images of the signature to imitate, trying to copy them (at least 10 times). Shape-based natural dynamics of highly skilled forgeries are obtained.

The center of mass of each signature was calculated and this point was displaced to the origin of coordinates.

We carried out verification experiments, using 5 signatures per person for training and 5 different genuine signatures per person for testing. In addition, we used the 25 forgeries. This implies a total number of 330×5 genuine tests plus 330×25 skilled impostor tests and $330 \times 329 \times 5$ random impostor tests. This procedure is similar to SVC 2004 [17], although our database is larger.

4.3 Performance Measure

To compare the performance with the reference work [3], the same measure has been used, the minimum value of the Detection Cost Function (DCF): $DCF = C_{miss} * P_{miss} * P_{true} + C_{fa} * P_{fa} * P_{false}$, where C_{miss} is the cost of a miss (rejection), C_{fa} is the cost of a false alarm (acceptance), P_{true} is the a priori probability of the target, and $P_{false} = 1 - P_{true}$. $C_{miss} = C_{fa} = 1$.

Due to the values of C_{miss} and C_{fa}, the minimum value of the DCF is similar, for a great number of tests (which is our case), to the Equal Error Rate (EER).

The minimum DCF or EER can be evaluated with a different threshold for each user or with the same for all, which is the approach used in this work. The latter is the less favorable case, due to the variability of the users scores.

4.4 Score Normalization in Verification Task

To improve the performance of the systems tested with common threshold, score normalization techniques can be applied. A detailed study of this problem is far beyond the scope of this work, so only the technique with the best performance in our baseline work [15] will be used:

- **Target-Centric (TC) Method.** Only target training signatures are used. Given a system score S, the normalized one is calculated as: $S_Z = S - \mu_T$, where μ_T is the mean of the target scores. To generate these scores with the training samples, we have used the rotation sampling method of error estimation [5], implemented by the leave-one-out technique (5 scores).

5 Results: Performance Comparison

Table 1 shows the results achieved with our feature selection proposal, compared with the baseline and reference systems. With regard to the last, only the best results achieved with each classifier tested are shown. In the column "Parameters" of the table, the main characteristics of each system are shown:

- For the VQ system, the size of the model (*cluster* · *vector dimension*).
- For the HMM-based system the number of states (Q) and mixtures (M).
- For the DTW-based system, the function used to integrate the n distances to the n training templates.
- For the baseline systems and with feature selection ones if the score normalization has been applied (*TC2 Rotat*) or not (*No ScorNor*).

From the results shown in Table 1, it can be seen that the best results are achieved with the score normalization application. However, it is interesting to notice that when no score normalization is used, the feature selection improves the system performance, achieving better results than the DTW-based system.

Focusing on the results achieved with score normalization, the use of feature selection allows us to achieve results close to, and even better than, those achieved with the *baseline system*, but with fewer requirements. For example, the **FS_Ver_Sys_30** system (which has the best performance) improves the skilled forgeries result, this being similar to random ones, while the number of vectors is 40% smaller. The "soft" approach is very interesting, since the system has a fairly good performance, but with very low requirements. The feature selection in this case shows that, even here, it is possible to decrease the number of vectors, without decreasing the performance. For example, the **FS_Ver_Sys_Soft_15** system outperforms the random forgeries recognition rates, decreasing by only 3% those with the skilled ones, while the number of vectors is also 40% smaller.

Table 1. Comparison of our feature selection proposal performance and that of the baseline and reference systems in verification. The best performance (best rate random/skilled)is bold face emphasized.

Method	Parameters	Minimum DCF (%)	
		Random	Skilled
VQ	**128x5**	4.3	11.9
HMM	**Q=12,M=1**	5.4	16.3
DTW	**min(·)**	2.4	8.9
Baseline_Ver_Sys	No ScorNor	2.4	8.6
FS_Ver_Sys_10	No ScorNor	2.8	9.4
FS_Ver_Sys_20	No ScorNor	2.1	7.3
FS_Ver_Sys_30	No ScorNor	2.1	7.2
FS_Ver_Sys_40	No ScorNor	2.2	7.9
Baseline_Ver_Sys	TC2 Rotat	1.8	6.6
FS_Ver_Sys_10	TC2 Rotat	2.5	7.8
FS_Ver_Sys_20	TC2 Rotat	2.0	6.2
FS_Ver_Sys_30	TC2 Rotat	**1.9**	**6.2**
FS_Ver_Sys_40	TC2 Rotat	1.8	6.4
Baseline_Ver_Sys_Soft	No ScorNor	2.9	8.4
FS_Ver_Sys_Soft_5	No ScorNor	3.5	9.4
FS_Ver_Sys_Soft_10	No ScorNor	2.9	7.9
FS_Ver_Sys_Soft_15	No ScorNor	2.6	8.1
FS_Ver_Sys_Soft_20	No ScorNor	2.5	8.3
Baseline_Ver_Sys_Soft	TC2 Rotat	2.3	6.4
FS_Ver_Sys_Soft_5	TC2 Rotat	3.2	8.1
FS_Ver_Sys_Soft_10	TC2 Rotat	2.6	6.8
FS_Ver_Sys_Soft_15	TC2 Rotat	2.1	6.6
FS_Ver_Sys_Soft_20	TC2 Rotat	2.3	6.5

In order to get a more complete comparison, it is interesting to compare computational load and storage requirements. With regard to the algorithms tested in the reference work [3], we shall focus this study on the best: DTW.

We will use the following nomenclature:

- J_i is the i signature reference template length.
- I is the candidate's signature length.
- K is the number of reference templates per user.
- m is the normalized signature length.

In our experiments, we have set $K = 5$, and in MCYT database the average length per signature is 454 vectors, so for the calculations we can fix the following $J_i = \overline{J} = 454$ and $I = \overline{I} = 454$.

The DTW algorithm applied in [3] requires the computation of about $\frac{KI\sum_{i=1}^{K} J_i}{3}$ distance measurements to be computed.

In the baseline systems, the number of distances computed is Km per candidate signature, with m the normalized signature length. The time necessary to normalize this signature must be added (an operation that must not be performed with DTW).

When feature selection is applied, we must calculate the standard deviations matrix. This must be performed only once, and can be done off-line, then this time will not be taken into account for the computational load estimation. The number of distances computed is KN_f per candidate signature.

As for the storage requirements, with DTW, it is necessary to save the original signature, that is, $K \sum_{i=1}^{K} J_i$ vectors. In the baseline proposal, the number of vectors to be stored is Km in *Baseline_Ver_Sys* and m in *Baseline_Ver_Sys_Soft*. When feature selection is applied, the number of "vectors" (the same number is selected for each feature, then, we can also use the "vector" unit to compute storage requirements) to be stored is kN_f in *FS_Ver_Sys_*N_f and N_f in *FS_Ver_Sys_Soft_*N_f. In this case we must also store the $m * t$ coefficients C_{jl} matrix. These coefficients have a value 0 or 1, then we need $m*t$ bits. Let us suppose that the data is codified in floating point IEEE 754, single-precision, that is, 32 bits per number (feature). Under this assumption, we need $m*t/32/t = m/32$ "vectors" to store the coefficients.

Table 2 shows a summary of the storage requirements (row *StReq*) and computational load comparison. For the first, the number of vectors needed to store the template/s for each system is shown. As for the computational load, a theoretical and practical estimation is shown. For the theoretical estimation (row *DisNr*) the previous calculations of computed distances have been used replacing I and J_i for their averaged values. The practical comparison has been performed with real processing times, using the same computer and experimental environment. Matlab has been used, performing 1000 verification operations. The following measurements were carried out:

1. Processing time comparison of the algorithms, that is, the input/output operations were excluded. The results of each comparison can be seen in the last four rows of Table 2. The comparison between the systems of the corresponding row and column appears in each cell. The value to the left of the ":" symbol is the processing time of the row system, and the value to the right is the processing time of the column system. Since these processing times are computer and software dependent, the relative values are shown, fixing as 1 the value of the quickest in each case.
2. Productiveness, i.e, the number of operations performed per second, including all the necessary operations: to open and read the test and template signature files, to normalize the test signature in our proposal and get the final score. The results can be seen in row *Op/s* of Table 2.

From the results shown in table 2, and focusing on the comparison with the DTW-based system, it can be seen that the storage requirements, that were 9 times lesser with the *Baseline_Ver_Sys* and 91 with *Baseline_Ver_Sys_Soft*, are 15 and 142, respectively, times lesser now, with the use of feature selection. With regard to the computer load, the processing speed (including I/O operations) that was 181 and 713 times greater with the *baseline systems*, it is 274 and 926, respectively, times greater now. The computer load ratios without I/O are clearly shown in the table.

A comparison with other approaches is far from the scope of this work. In [15] a comparison can be seen between the *baseline systems* and some of the most recent published work with the MCYT database, showing that our proposal is, in general, in the state-of-the-art scores.

Table 2. Computer load and storage requirements comparison between DTW and our systems proposals. In row **StReq (Storage Requirements)**, the number of vectors (5 features each) necessary to store the template/s is shown (the mean in the DTW column). The number of distances calculated per test sample is shown in row **DisNr (Distance Numbers)**; in the DTW column the mean also appears. The number of verification operations per second with each system can be seen in row **Op/s**, and the processing time comparison of the algorithms in the rest of the rows.

	Ba_Ver_Sys	**Ba_Ver_Sys_Soft**	**FS_Ver_Sys_30**	**FS_Ver_Sys_Soft_15**	**DTW**
StReq	250	25	152	16	2270
DisNr	250	25	150	15	343527
Op/s	98	385	148	500	0.54
Ba_Ver_Sys		5:1	1.6:1	8:1	1:328
Ba_Ver_Sys_Soft			1:3	1.6:1	1:1670
FS_Ver_Sys_30				4.8:1	1:557
FS_Ver_Sys_Soft_15					1:2662

6 Conclusions

In this work we have shown a new proposal of representative feature selection. This proposal has been successfully applied to an on-line signature recognition system based on size normalized signatures and fractional distances.

The proposed goal has been achieved, since it has been shown that the system storage requirements and computer load can be reduced with the use of representative points/feature selection, without a noticeable decrease in system performance, and the results are even better with skilled forgeries.

This reinforces the workline shown in our *baseline work* and other related ones, that shows that it is possible to reduce the signature points (undersampling, representative/critical points selection, etc.) without performance loss.

We have achieved a final system with a cost even lower than that of the *baseline system*, which is very interesting for large-scale systems and in resource-limited systems like smart-cards, but with a state-of-the-art performance of systems based on a single classifier, and even in some of those based on classifiers or features fusion. Besides, the security of the system is improved as only the representative features have to be stored, it being impossible to recover the original signature from these.

Although the score normalization is not a goal in this work, we think that it is interesting to remark the good performance of the Target-Centric Methods.

Finally, we should comment that what is shown in this work is a first approximation to the feature selection problem with normalized signatures. The good results achieved shows that it is an interesting workline, with interesting future work, as for example, testing with other normalized sizes and with other proposals of features/points selection.

References

1. Connell, S.D., Jain, A.K.: Template-based online character recognition. Pattern Recognition 34(1), 1–14 (2001)
2. Fabregas, J., Faundez-Zanuy, M.: Biometric dispersion matcher. Pattern Recognition (2009)

3. Faundez-Zanuy, M.: On-line signature recognition based on vq-dtw. Pattern Recognition 40(3), 981–992 (2007)
4. Fiérrez-Aguilar, J., Nanni, L., Lopez-Peñalba, J., Ortega-Garcia, J., Maltoni, D.: An on-line signature verification system based on fusion of local and global information. In: Kanade, T., Jain, A., Ratha, N.K. (eds.) AVBPA 2005. LNCS, vol. 3546, pp. 523–532. Springer, Heidelberg (2005)
5. Fierrez-Aguilar, J., Ortega-Garcia, J., Gonzalez-Rodriguez, J.: Target dependent score normalization techniques and their application to signature verification. IEEE Transactions on Systems, Man and Cybernetics, Part C 35(3), 418–425 (2005)
6. François, D., Wert, V.: The concentration of fractional distances. IEEE Transactions on Knowledge and Data Engineering 19(7), 873–886 (2007)
7. Jain, A.K., Griess, F.D., Connell, S.D.: On-line signature verification. Pattern Recognition 35(12), 2963–2972 (2002)
8. Kholmatov, A., Yanikoglu, B.: Identity authentication using improved online signature verification method. Pattern Recognition Letters 26(15), 2400–2408 (2005)
9. Li, B., Wang, K., Zhang, D.: Minimizing spatial deformation method for online signature matching. In: Lee, S.-W., Li, S.Z. (eds.) ICB 2007. LNCS, vol. 4642, pp. 646–652. Springer, Heidelberg (2007)
10. Li, B., Zhang, D., Wang, K.: Improved critical point correspondence for on-line signature verification. International Journal of Information Technology 12(7), 45–56 (2006)
11. Martinez-Diaz, M., Fierrez, J., Freire, M.R., Ortega-Garcia, J.: On the effects of sampling rate and interpolation in hmm-based dynamic signature verification. In: Proc. Intl. Conf. on Document Analysis and Recognition, ICDAR, vol. 2, pp. 1113–1117 (September 2007)
12. Ohishi, T., Komiya, Y., Matsumoto, T.: Online signature verification using pen-position, pen-pressure and pen-inclination trajectories. In: Proceedings of 15th International Conference on Pattern Recognition, 2000, September 3-7, vol. 4, pp. 547–550 (2000)
13. Ortega-Garcia, J., Fierrez, J., Simon, D., Gonzalez, J., Faundez-Zanuy, M., Hernaez, I., Espinosa, V., Satue, A., Igarza, J.J., Vivaracho, C., Escudero, D., Moro, Q.I.: Mcyt baseline corpus: a bimodal biometric database. IEE Proc. Visual Image Signal Processing 150(6), 395–401 (2003)
14. Plamondon, R., Lorette, G.: Automatic signature verification and writer identification: The state of the art. Pattern Recognition 22(2), 107–131 (1989)
15. Vivaracho-Pascual, C., Faundez-Zanuy, M., Pascual, J.M.: An efficient low cost approach for on-line signature recognition based on length normalization and fractional distances. Pattern Recognition 42(1), 183–193 (2009)
16. Wirotius, M., Ramel, J.-Y., Vincent, N.: Selection of points for on-line signature comparison. In: Ninth International Workshop on Frontiers in Handwriting Recognition, 2004. IWFHR-9 2004, October 26-29, pp. 503–508 (2004)
17. Yeung, D.Y., Chang, H., Xiong, Y., George, S., Kashi, R., Matsumoto, T., Rigoll, G.: Svc2004: First international signature verification competition. In: Zhang, D., Jain, A.K. (eds.) ICBA 2004. LNCS, vol. 3072, pp. 16–22. Springer, Heidelberg (2004)

Feature Selection and Binarization for On-Line Signature Recognition

Emanuele Maiorana, Patrizio Campisi, and Alessandro Neri

Dipartimento di Elettronica Applicata, Università degli Studi "Roma Tre",
Via della Vasca Navale 84, 00146 Roma, Italy
{maiorana,campisi,neri}@uniroma3.it

Abstract. The representation of a biometric trait through a set of parametric features is commonly employed in many biometric authentication systems. In order to avoid any loss of useful information, large sets of features have been defined for biometric characteristics such as signature, gait or face. However, the proposed sets often contain features which are irrelevant, correlated with other features, or even unreliable. In this paper we propose two different approaches for the selection of those features which guarantee the best recognition performances. Moreover, we also face the problem of the binary representation of the selected features. Specifically, an algorithm which selects the minimum number of bits which should be assigned to a given feature, in order to not affect the recognition performances, is here proposed. The effectiveness of the proposed approaches is tested considering a watermarking based on-line signature authentication system, and employing the public MCYT on-line signature corpus as experimental database.

Keywords: Biometrics, On-line Signature, Feature Selection, Feature Binarization, Watermarking.

1 Introduction

Loosely speaking, biometric systems are essentially pattern recognition applications, which perform user authentication by analyzing people physiological or behavioral characteristics. As in the classical framework of a pattern recognition system, a generic biometric system can be represented as a cascade of four main modules performing data acquisition, feature extraction, template storing and template matching [1]. Specifically, the feature extraction module processes the acquired biometric data, namely the pattern, to extract a set of discriminatory features, taken as representative of the acquired data.

Two types of features can be distinguished: parameters and functions. When parameters are used, the considered biometrics is characterized by a vector of elements, each representative of a feature. When functions are used, the given biometrics is characterized in terms of a set of time or space dependent sequences.

However, the sets of parametric features proposed in literature often contain features which are irrelevant or correlated with other features, or even unreliable. This typically happens because many redundant features are included

M. Tistarelli and M.S. Nixon (Eds.): ICB 2009, LNCS 5558, pp. 1219–1229, 2009.
© Springer-Verlag Berlin Heidelberg 2009

in the employed sets, in order to minimize the chances to loose useful information. It is worth pointing out that the representation of biometric data in high-dimensional feature spaces involves several issues: by considering features with high intra-user variability and low inter-user variability, the performance of the employed classification method can be negatively affected, thus resulting in sub-optimal recognition performances [2]. Moreover, the use of a large number of features usually increases the processing and the matching time. High dimensional data also impose severe requirements for storage space and communication bandwidth, which are crucial factors in the actual deployment of a real-time system. Therefore, the selection of the most discriminative parametric features is an issue of paramount importance for many biometric systems. Specifically, the feature selection process holds a significant relevance when considering behavioral biometrics, due to the high intra-class variability which can be experienced by extracting features from different acquisitions.

An algorithm for the selection of those features which offer the best representation of a speech sample has been proposed in [3], where genetic algorithms are employed in conjunction with feed forward neural networks. An approach combining genetic algorithms and particle swarm optimization has been proposed in [4] for the selection of relevant keystroke dynamics features. Moreover, mutual information is exploited in [5] for human gait feature selection. As far as signature recognition is concerned, the problem of feature selection has been discussed in [6], where the selection of a unique set of features for each individual has been suggested, and in [7], where a modified version of the Fisher ratio has been employed as cost function for the selection of local and global signature parametric features. In [8] a statistical distance, based on the difference between the mean values of different users' features, has been defined. In [9], a cost function based on the Equal Error Rate (EER) has been employed to select a subset of signature features, while a consistency measure has been introduced in [10] to rank signature features according to their importance. An algorithm employed to rank a set of on-line signature parametric features has been also presented in [11], while the use of genetic algorithms has been proposed in [12] to select an optimal set of on-line signature features.

In this contribution, a feature selection procedure, based on the distance between distributions of genuine and forgers matching scores, is introduced. Moreover, an approach which allows to binarize a feature set, using only the bits needed to not degrade the system performance, is proposed. The proposed feature selection and binarization procedures are then applied to signature based authentication using watermarking techniques proposed by the authors in [13], as an application example.

2 Feature Selection

In this Section, the proposed procedures for the selection of the most representative features belonging to a given set are detailed. Specifically, the presented approaches perform feature selection on the basis of the distances between genuine and forgers matching scores distributions. In fact, it is well known from

pattern classification that a properly defined feature selection stage should be designed using the matcher employed for the classification. Therefore, the presented approaches can be applied for the selection of the best parametric features extracted from any biometrics. Instead of defining different sets of features for different users as proposed in [6], the proposed approaches define a unique set of features which has to be employed for each enrolled user.

In the proposed approach we assume that a training set of biometric acquisitions, taken from U different users, is available and can be employed to analyze the contribution of each feature to the recognition process. Specifically, we assume that, for each user u, T genuine biometric samples are acquired, and that J forgery samples are also available. These latter samples can be either biometric acquisitions taken from different users, or skilled forgeries which mimick the biometrics of the considered user. Moreover, we assume that the matching module is based on the computation of the Mahalanobis distance between feature vectors extracted from different biometric samples.

Specifically, K features are extracted, for each user, from each of E genuine samples, selected among the T samples available in the training data set. The mean and the standard deviation vectors $\boldsymbol{\mu}^u$ and $\boldsymbol{\sigma}^u$, of the K features are then computed. The feature vectors $\mathbf{g}_i^u$, with $i = 1, \ldots, I$, where $I = T - E$ represents the remaining genuine acquisitions, are evaluated. The feature vector extracted from the j-th forged sample, with $j = 1, \ldots, J$, of the user u is indicated as $\mathbf{f}_j^u$.

An approach to determine a feature ranking, and an iterative approach to determine the best feature to add to an already given set, are described in Sections 2.1 and 2.2, respectively.

2.1 Features Selection: Ranking Approach

The features selection approach here proposed assigns a reliability measure to each feature, which can then be ranked accordingly: the features with the L highest reliability values are then selected to efficiently represent the considered biometric. Specifically, given a feature k, the following distances are computed for each user u belonging to the training set, $u = 1, 2, \ldots, U$:

$$\mathbf{G}^u[i,k] = \frac{|\mathbf{g}_i^u[k] - \boldsymbol{\mu}^u[k]|}{\boldsymbol{\sigma}^u[k]}; \quad \mathbf{F}^u[j,k] = \frac{|\mathbf{f}_j^u[k] - \boldsymbol{\mu}^u[k]|}{\boldsymbol{\sigma}^u[k]} \tag{1}$$

and collected in the matrices $\mathbf{G}^u$ and $\mathbf{F}^u$ of dimensions $I \times K$ and $J \times K$ respectively, with $u = 1, 2, \ldots, U$.

Then, for each user u the distance $\boldsymbol{\vartheta}^u[k]$, for the k-th feature, between $\mathbf{G}^u[i,k]$ and $\mathbf{F}^u[j,k]$ is evaluated, employing one of the distances defined in Section 2.3. The reliability of the k-th feature is then computed by taking the median value $\boldsymbol{\lambda}[k]$ among the U values $\{\boldsymbol{\vartheta}^u[k]\}$. The features with the L highest values of $\boldsymbol{\lambda}[k]$ are considered to be the most representative ones for recognition purposes.

2.2 Features Selection: Incremental Approach

The approach described in Section 2.1 analyzes the reliability of each considered feature, without taking into account the correlation between different features.

However, in most cases the considered features are correlated. Therefore the selection of the best feature which can be added to a given set should be performed on the basis of the already selected ones. In order to take into account possible dependencies between the considered features, in this paper we present an incremental approach.

Specifically, if a subset comprising the L most reliable features, out of the available K, has to be identified, the proposed procedure has to be run by iterating L times the algorithm detailed in the following.

Let us define with $\mathcal{L}_l$ the set of the selected features at step l, with $l = 1, \cdots, L$ and $\mathcal{L}_0$ defines the empty set at the initialization step. Let $\boldsymbol{\Delta} G_0^u[i,k]$ and $\boldsymbol{\Delta} F_0^u[j,k]$ the zero matrices of dimensions $I \times K$ and $J \times K$ respectively.

The generic l-th step of the proposed algorithm is defined as follows:

1. for each user u, with $u = 1, 2, \cdots, U$, computation of $\mathbf{G}_l^u[i,k]$ and $\mathbf{F}_l^u[j,k]$, for $k \in \mathcal{K} \setminus \mathcal{L}_{l-1}$, where "$\setminus$" is the difference operator, as:

$$\mathbf{G}_l^u[i,k] = \sqrt{\left(\frac{\mathbf{g}_i^u[k] - \boldsymbol{\mu}^u[k]}{\boldsymbol{\sigma}^u[k]}\right)^2 + \boldsymbol{\Delta} G_{l-1}^u[i,k]}, \; i = 1, \ldots, I;$$

$$\mathbf{F}_l^u[j,k] = \sqrt{\left(\frac{\mathbf{f}_j^u[k] - \boldsymbol{\mu}^u[k]}{\boldsymbol{\sigma}^u[k]}\right)^2 + \boldsymbol{\Delta} F_{l-1}^u[j,k]}, \; j = 1, \ldots, J; \quad (2)$$

2. for each user u, with $u = 1, 2, \cdots, U$, computation of the distances $\boldsymbol{\vartheta}_l^u[k]$ between $\mathbf{G}_l^u[i,k]$ and $\mathbf{F}_l^u[j,k]$ for $k \in \mathcal{K} \setminus \mathcal{L}_{l-1}$. The measures detailed in Section 2.3 are employed to evaluate the required distances;
3. for each feature $k \in \mathcal{K} \setminus \mathcal{L}_{l-1}$, evaluation of the median values $\boldsymbol{\lambda}_l[k]$ of the U values $\{\boldsymbol{\vartheta}_l^u[k]\}$;
4. selection of the feature $\tilde{k} \in \mathcal{K} \setminus \mathcal{L}_{l-1}$ which possesses the highest reliability value $\boldsymbol{\lambda}_l[\tilde{k}]$;
5. update the set of selected features as $\mathcal{L}_l = \mathcal{L}_{l-1} \bigcup \tilde{k}$;
6. for each user u, with $u = 1, 2, \cdots, U$, update the matrices $\boldsymbol{\Delta} G_l^u$ and $\boldsymbol{\Delta} F_l^u$ as follows:

$$\boldsymbol{\Delta} G_l^u[i,k] = \boldsymbol{\Delta} G_{l-1}^u[i,k] + \left(\frac{\mathbf{g}_i^u[\tilde{k}] - \boldsymbol{\mu}^u[\tilde{k}]}{\boldsymbol{\sigma}^u[\tilde{k}]}\right)^2 \cdot \mathbf{o}[k]$$
$$\boldsymbol{\Delta} F_l^u[j,k] = \boldsymbol{\Delta} F_{l-1}^u[j,k] + \left(\frac{\mathbf{f}_j^u[\tilde{k}] - \boldsymbol{\mu}^u[\tilde{k}]}{\boldsymbol{\sigma}^u[\tilde{k}]}\right)^2 \cdot \mathbf{o}[k] \quad (3)$$

   being $\mathbf{o}[k]$ a row vector with only ones, and "$\cdot$" the column-by-row multiplication operator, for every $k \in \mathcal{K} \setminus \mathcal{L}_l$.
7. if $l < L$ then $l = l + 1$, otherwise the iterations stop.

2.3 Distribution Distances

As already pointed out, both the proposed procedures for feature selection compute a distance between two distributions. In the proposed implementations, two different measures have been employed to evaluate the distance between the probability mass function P_X and P_Y of the score sets X and Y, respectively.

EER Based Distance. The distance $\Delta_{EER}(P_X, P_Y)$ is defined using the EER. Specifically, being P_X the score probability mass function obtained from the matching of authentic biometric samples, and P_Y the score probability mass function obtained from the matching of forgeries, the resulting EER is then computed. We then set $\Delta_{EER}(P_X, P_Y) = 1 - EER$.

Kullback-Leibler Based Distance. The distance $\Delta_{KL}(P_X||P_Y)$ is based on the estimation of the score probability mass function P_X, obtained from the matching of authentic biometric samples, and the score probability mass function P_Y, obtained from the matching of forgeries. The Kullback-Leibler (KL) divergence between the two estimated densities is eventually computed. Specifically, the Parzen window estimator [15] with a Gaussian kernel is employed to estimate P_X and P_Y. The distance $D_{KL}(P_X||P_Y)$ is then obtained as

$$D_{KL}(P_X||P_Y) = \sum_i P_X(i) \ln \frac{P_X(i)}{P_Y(i)}. \tag{4}$$

In Section 5 it is shown that the proposed feature selection approaches lead to recognition rates, expressed in terms of EER, better than those obtained using the approach in [11].

3 Features Binarization

In this Section, the method employed to determine the minimum number of bits necessary to represent a given feature, without affecting the achievable recognition performances, is detailed. As for the feature selection algorithms described in Section 2, it is assumed that a training set of biometric acquisitions can be analyzed. Using the notations introduced in Section 2, we first estimate the minimum and maximum value for a given feature k, with $k = 1, 2, \cdots, K$.

Specifically, the maximum and minimum allowable score values, $\mathbf{M}[k]$ and $\mathbf{m}[k]$, for the k-th feature are computed as follow. The $\alpha\%$ highest score values are discarded thus obtaining as maximum allowable value for the k-th feature $\boldsymbol{\zeta}[k]$. Then, the maximum allowable score value, $\mathbf{M}[k]$ is set as $\mathbf{M}[k] = \boldsymbol{\zeta}[k] + \eta|\boldsymbol{\zeta}[k]|$, where $\eta > 0$. Similarly, the minimum allowable score value $\mathbf{m}[k]$ for the k-th features is computed as $\mathbf{m}[k] = \boldsymbol{\varsigma}[k] - \eta|\boldsymbol{\varsigma}[k]|$, being $\boldsymbol{\varsigma}[k]$ the minimum score value obtained by discarding the $\alpha\%$ lowest scores. The parameters α and η are selected in order to limit the effects of possible outliers in the estimation of the maximum and minimum values of the given distribution. Once the allowed range for the score of the k-th feature has been determined, the following procedure is employed to verify whether a given number of bits b can be used to represent the mean value of the k-th feature, without affecting the achievable verification performances.

For each user u, the matrices $\mathbf{G}^u$ and $\mathbf{F}^u$, whose values are given in (1) are evaluated. The values $\boldsymbol{\mu}^u[k]$ are then binarized employing b bits for each element, taking into account that the k-th feature assumes values into the interval $[\mathbf{m}[k]; \mathbf{M}[k]]$. The decoded version of the binarized k-th feature mean vector is indicated as $\boldsymbol{\nu}^u[k]$, and employed to determine the distances:

$$\mathbf{Gb}^u[i,k] = \frac{|\mathbf{g}_i^u[k] - \boldsymbol{\nu}^u[k]|}{\boldsymbol{\sigma}^u[k]}; \quad \mathbf{Fb}^u[j,k] = \frac{|\mathbf{f}_j^u[k] - \boldsymbol{\nu}^u[k]|}{\boldsymbol{\sigma}^u[k]}. \tag{5}$$

For each user u, a measure representing the goodness of the employed binarization is then computed as:

$$\gamma_b^u[k] = \frac{1}{I}\sum_{i=1}^{I}(\mathbf{G}^u[i,k] - \mathbf{Gb}^u[i,k]) - \frac{1}{J}\sum_{j=1}^{J}(\mathbf{F}^u[j,k] - \mathbf{Fb}^u[j,k]). \tag{6}$$

In fact, it can be observed that when the term $\frac{1}{I}\sum_{i=1}^{I}(\mathbf{G}^u[i,k] - \mathbf{Gb}^u[i,k])$ is positive, the binarization of the k-th feature results in an improvement of the recognition performances. Similarly, an improvement of the recognition accuracy can be obtained when the term $\frac{1}{J}\sum_{j=1}^{J}(\mathbf{F}^u[j,k] - \mathbf{Fb}^u[j,k])$ is negative. Therefore, once the terms $\gamma_b^u[k]$ have been computed for each available user u, a quality measure for the binarization of the k-th feature with b bits can be obtained as $\boldsymbol{\Gamma}_b[k] = \sum_{u=1}^{U}\gamma_b^u[k]/U$. If $\boldsymbol{\Gamma}_b[k] < 0$, the binarization of the mean values of feature k with b bits, for each enrolled user, will result in a performance degradation. When $\boldsymbol{\Gamma}_b[k] \simeq 0$ the recognition performances are not affected by the binarization, and could be even improved if $\boldsymbol{\Gamma}_b[k] > 0$. The minimum number of bits which should be employed to binarize the k-th feature is therefore obtained as the lowest value b for which steadily results $\left|\boldsymbol{\Gamma}_b[k]\right| < \epsilon$, where ε is a predefined threshold, with $\varepsilon \simeq 0$.

4 Feature Selection and Binarization: An Application to Signature Watermarking

The proposed approaches for feature selection and binarization have been applied to the system presented in [13], which is here briefly summarized.

In [13] a multi-level signature based authentication system, where watermarking is employed to hide a set of dynamic signature features into an image of the signature itself, has been described. Specifically, during the enrollment stage E signatures are acquired for each user through an electronic pad, and some relevant dynamic features are extracted from them. The selected features are then embedded into the signature images, by means of watermarking techniques based on the Radon transform. The images are represented using signatures' pressure values, being thus grey-scale images instead of simply binary ones. Among the enrolled signature of a given user, the one whose static features are the closest to the estimated mean is selected to host the chosen dynamic features.

The Malahanobis distance is used to compare both dynamic and static features in the authentication stage. The employed authentication system allows to obtain two levels of security, according to the considered application: when a low security level is needed, the authentication can be performed using only some selected static features, extracted from the signature images. When a higher level of security is required, authentication is accomplished using the embedded dynamic features, by themselves or together with the selected static features.

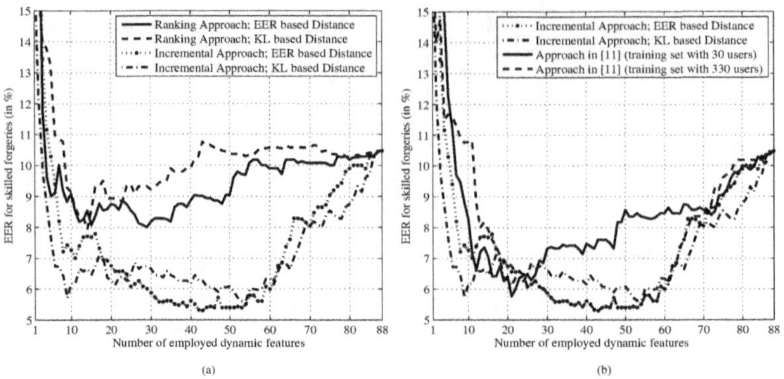

Fig. 1. Application of the proposed feature selection approaches to dynamic signature features. (a): Comparison between the approaches in Section 2.1 and 2.2; (b): Comparison between the approaches in Section 2.2 and in [11].

In this latter case, score fusion techniques [16] can be employed to combine the information extracted from static and dynamic signature features.

As explained in [13], the considered on-line signature based authentication system requires a set of dynamic features to be selected, and also a binarization procedure of the chosen features. Specifically, in order to provide resilience against errors during the mark extraction phase, the binary strings which are embedded into the signature images have to be protected with an error correcting code (ECC). It is therefore important to generate the binary marks employing as more dynamic features as possible, while containing the final dimension of the binary string to be encoded by the employed ECC, and later embedded in the considered signature image. In [13] eleven dynamic features are selected for the embedding, and binarized using a non-optimal procedure, leading to 92 bits. Then, the BCH code $(127, 92)$ is employed to provide an error correction capability of 5 bits. In Section 5 it is shown that, employing the proposed binarization approach, 17 dynamic features can be represented using a binary string of 92 bits, thus leading to an optimal bit allocation, which allows to increase the number of the embedded features.

5 Experimental Results

An extensive set of experimental tests has been performed using the public MCYT on-line signature corpus [14], comprising 100 users, for each of which 25 genuine signatures and 25 skilled forgeries have been captured during five different sessions. The employed database has been divided in two disjoint data sets: a training set, which comprises the genuine and forged signatures of the first 30 users, and a test set, which includes the genuine and forged signatures of the remaining 70 users. The training set has been employed to select the best dynamic and static features, and also to determine the number of bits which has

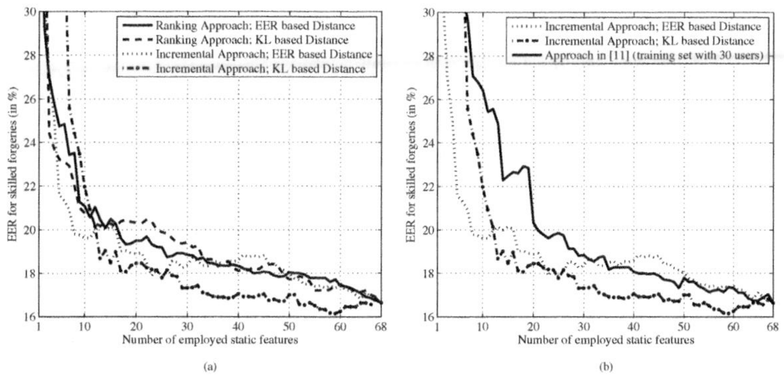

Fig. 2. Application of the proposed feature selection approaches to static signature features. (a): Comparison between the approaches in Section 2.1 and 2.2; (b): Comparison between the approaches in Section 2.2 and in [11].

to be assigned for the binarization of each selected dynamic feature. Specifically, the feature vectors $\mathbf{f}_j^u$, $j = 1, \ldots, 25$ and $u = 1, \ldots, 30$, employed in the proposed methods of Section 2, are extracted from the skilled forgeries available in the training set. The effectiveness of the proposed methods has been tested:

- by evaluating the recognition rates in terms of EER, and by comparing them with those obtained employing the method described in [11];
- by applying the proposed feature selection and binarization approaches to the watermarking based system described in [13].

We have tested the proposed feature selection algorithms employing a set of 88 dynamic features, derived from the set presented in [11] by discarding the parameters related to static information. The results obtained when taking $E = 10$ signatures for the enrollment phase are shown in Figure 1. Specifically, Figure 1(a) shows the comparison between the methods described in Section 2.1 and 2.2, employing both the distances defined in Section 2.3. The results are displayed in terms of the EER, evaluated over the test set with 70 users and considering skilled forgeries, versus the number of employed features. The obtained results illustrate that the incremental procedure of Section 2.2 performs better than the ranking based one. This means that the employed features are highly correlated, being then not possible to perform efficiently the selection of a feature regardless of the already selected ones. Figure 1(b) shows the comparison between the incremental approach presented in Section 2.3 and the feature selection method presented in [11]. Specifically, this latter is applied by performing the feature selection procedure over the training set with $U = 30$ users, and then evaluating the recognition performances over the test set. It can be seen that the proposed incremental methods performs better than the one in [11]. Moreover, the proposed approach is also compared with the feature ranking presented in [11], which has been estimated employing the whole MCYT database with 330 user as training set. The incremental procedure described in Section 2.2 still

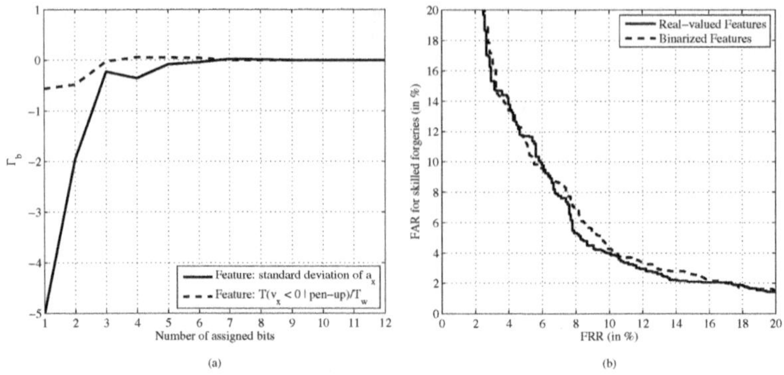

Fig. 3. Analysis of the proposed feature binarization approach. (a): Quality measure $\boldsymbol{\Gamma}_b$ with respect of the number of employed bits, for two considered dynamic features; (b): Effectiveness of the proposed binarization method, verified over the test data set.

performs better when tested for low values of selected features, and allows to reach the same lowest EER which can be achieved employing the feature ranking presented in [11], that is equal to 5.4%. The same comparisons between different feature selection methods have also been carried out considering the set of 68 static features employed in [13]. The obtained results are illustrated in Figure 2, which still validates the effectiveness of the proposed approaches.

As for the second experiment, the feature selection and binarization approaches proposed in this paper are applied to the watermarking based system briefly outlined in Section 4. Specifically, we have selected 17 dynamics features out of the considered 88 ones, employing the proposed incremental feature selection approach with the distribution distance based on the KL divergence. The feature binarization approach described in Section 3 has been employed to represent the 17 selected dynamics features with only 92 bits. The necessary bit-depths have been estimated over the training set with 30 users. The employed parameters are $\alpha = 0.01$ and $\eta = 0.2$. The behavior of the quality measure $\boldsymbol{\Gamma}_b$ introduced in Section 3 is illustrated in Figure 3(a). According to what has been explained in Section 3, the feature "standard deviation of a_x" (defined in [11]) can be binarized employing 7 bits, while feature "$T(v_x < 0|\text{pen-up})/T_w$" (defined in [11]) can be represented employing only 4 bits. The effectiveness of the proposed binarization approach has been verified by analyzing the recognition performances computed over the test data set, while employing 17 dynamics features. Specifically, the performances obtained employing real-valued vectors $\boldsymbol{\mu}^u$ to represent the intra-class mean are compared in Figure 3(b) with those achievable employing vectors $\boldsymbol{\nu}^u$, generated from the binary representation of the considered dynamic features. It is evident that the recognition performances, expressed in terms of False Recognition Rate (FRR) and False Acceptance Rate (FAR) for skilled forgeries, are not affected by the proposed binarization, even if the employed 17 dynamic features are represented with 92 bits, whereas the same amount of bits were employed in [13] to represent only 11 dynamic features.

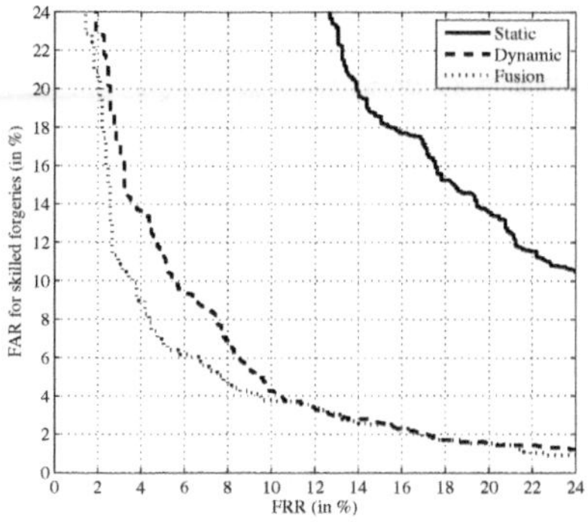

Fig. 4. Performances of the proposed on-line based signature authentication system

The recognition performances achievable for the considered multi-level signature based authentication system of Section 4 are displayed in Figure 4. The proposed incremental feature selection with the KL distribution distance is employed to select the best 50 static features. The achieved EER for static features, computed over the test data set, is equal to 17.07%. The EER achievable employing 17 dynamic features, selected according to the proposed incremental approach, is equal to 7.41%. Combining both static and dynamic features through a *min-max* normalization technique [16], together with a *sum* rule, results in a EER of 6.16%. The parameters needed for the implementation of the *min-max* normalization technique are estimated using the training data set, and later applied to fuse the scores obtained over the test data set.

References

1. Jain, A.K., Ross, A., Prabhakar, S.: An Introduction to Biometric Recognition. IEEE Trans. on Circuits and Systems for Video Technology 14(1) (2004)
2. Hughes, G.: On the mean accuracy of statistical pattern recognizers. IEEE Transactions on Information Theory 14(1), 55–63 (1968)
3. Quixtiano-Xicohtencatl, R., Flores-Pulido, L., Reyes-Galaviz, O.F.: Feature Selection for a Fast Speaker Detection System with Neural Networks and Genetic Algorithms. In: 15th International Conference on Computing (CIC) (2006)
4. Azevedo, G.L.F., Cavalcanti, G.D.C., Filho, E.C.B.: An approach to feature selection for keystroke dynamics systems based on PSO and feature weighting. In: IEEE Congress on Evolutionary Computation (CEC), pp. 3577–3584 (2007)
5. Baofeng, G., Nixon, M.S.: Gait Feature Subset Selection by Mutual Information. In: IEEE Conference on Biometrics: Theory, Applications, and Systems (BTAS) (2007)
6. Brittan, P., Fairhurst, M.C.: Feature selection in automatic signature verification. In: IEE Colloquium on Image Processing for Biometric Measurement (1994)
7. Richiardi, J., Hamed, K., Drygajlo, A.: Local and global feature selection for on-line signature verification. In: Conference on Document Analysis and Recognition (2005)

8. Lee, L., Berger, T., Aviczer, E.: Reliable on-line human signature verification systems. IEEE Trans. on PAMI 18(6), 643–647 (1996)
9. Crane, H., Ostrem, J.: Automatic signature verification using a three-axis force-sensitive pen. IEEE Trans. on Systems, Man, and Cybernetics (1983)
10. Lei, H., Govindaraju, V.: A study on the consistency of features for on-line signature verification. In: IAPR SSPR (2004)
11. Fiérrez-Aguilar, J., Nanni, L., Lopez-Peñalba, J., Ortega-Garcia, J., Maltoni, D.: An on-line signature verification system based on fusion of local and global information. In: Kanade, T., Jain, A., Ratha, N.K. (eds.) AVBPA 2005. LNCS, vol. 3546, pp. 523–532. Springer, Heidelberg (2005)
12. Galbally, J., Fierrez, J., Freire, M.R., Ortega-Garcia, J.: Feature Selection Based on Genetic Algorithms for On-Line Signature Verification. In: IEEE AIAT (2007)
13. Maiorana, E., Campisi, P., Neri, A.: Biometric Signature Authentication using Radon Transform-based Watermarking techniques. In: IEEE BSYM (2007)
14. Ortega-Garcia, J., et al.: MCYT baseline corpus: A bimodal biometric database. IEE Proceedings Vision, Image and Signal Processing 150(6) (2003)
15. Parzen, E.: On Estimation of a Probability Density Function and Mode. Annals of Mathematical Statistics 33, 1065–1076 (1962)
16. Ross, A., Nandakumar, K., Jain, A.K.: Handbook of Multibiometrics. Springer, Heidelberg (2006)

Writer Identification of Chinese Handwriting Using Grid Microstructure Feature

Xin Li and Xiaoqing Ding

State Key Laboratory of Intelligent Technology and Systems
Department of Electronic Engineering, Tsinghua University, Beijing 100084, China
{lixin,dxq}@ocrserv.ee.tsinghua.edu.cn

Abstract. This paper proposes a histogram-based feature to Chinese writer identification. It is called grid microstructure feature. The feature is extracted from the edge image of the real handwriting image. The positions of edge pixel pairs are used to describe the characteristics in a local grid around every edge pixel. After global statistic, the probability density distribution of different pixel pairs is regarded as the feature representing the writing style of the handwriting. Then the similarity of two handwritings is measured with the improved weighted visions of some original metric. On the HIT-MW Chinese handwriting database involving 240 writers, the best Top-1 identification accuracy is 95.0% and the Top-20 accuracy reaches 99.6%.

Keywords: Writer identification, Chinese handwriting, grid microstructure feature, improved weighted metric.

1 Introduction

Automatic writer recognition has been promoted by some valuable researches in recent years. Western handwriting recognition technology has been experimented on large handwriting database and has shown the practical effectiveness. But eastern writer recognition still has some challenges. The basic reason is that western writing (like English) is alphabetic writing while eastern writing (like Chinese) is ideographic writing. English letters are simple and Chinese characters have complex stroke crossing. Western writing has lots of ink connection between individual letters. But eastern writing, no matter Chinese, Japanese or Korean, is composed of characters in separate block structure. The existing methods of Chinese writer recognition mainly divide into two classes: the identical character based text-dependent approaches and the text-independent approach based on texture analysis [1]. Wang [2] proposed a method using directional element features and linear transform, and got rather high accuracy by combining several characters. However, there are tens of thousands of different characters in Chinese and even three to six thousands characters are in common use. If handwritings have only few identical characters even none, the text-dependent methods cannot work. Moreover, the text-independent metheds has no limits to text content. M. Yoshimura [3] counted relative frequencies of some model patterns from Japanese sentences to identify writers. Some texture analysis based

M. Tistarelli and M.S. Nixon (Eds.): ICB 2009, LNCS 5558, pp. 1230–1239, 2009.
© Springer-Verlag Berlin Heidelberg 2009

methods are also introduced in Chinese writer recognition. Zhu [4] used multi-channel Gabor features to Chinese handwritings. Wavelet-based HMT models were utilized by He [5]. These methods use character patching to create texture blocks. But actually the order of character arrangement influences the texture and the complex structures of Chinese characters enhance the influence. So the methods based on texture analysis do not perform as well on Chinese handwritings as on English handwritings. Bulacu and Schomaker [6] proposed a number of very effective histogram-based features for western writer recognition.

In this paper, we propose a histogram-based feature, the grid microstructure feature, for Chinese writer identification. The idea of the feature comes from the contour-hinge feature, which outperforms the other textural and allographic features in [6]. Differently, the proposed feature is generated from a size-adjustable grid and focuses on the positions of edge pixel pairs, while the contour-hinge feature examines the directions of two linked edge fragments. To improve the performance of writer identification, we try different distance metrics for similarity measurement and introduced the improved weighted metrics. The experimental results show the effectiveness of the proposed method. The remaining part of this paper is organized as follows: Section 2 describes the novel histogram-based feature, the grid microstructure feature and the methods of the similarity measurement. Then section 3 introduces the Chinese handwriting database and the experiment process. Section 4 shows and analyzes the results of the experiments. Finally, the conclusions are given in Section 5.

2 Grid Microstructure Feature

The grid microstructure feature is extracted from the edge image of the handwriting image. Edge image is seen as a primary representation of handwritten document. The preprocessing step of our method is to gain the edge image. Real handwritings are scanned to gray images firstly. After binarization using Otsus method, edge detection is carried out. Among many edge detection algorithms, the Sobel operator is used in our implementation. In the edge image, Bulacu et al. saw the two contour fragments extending from an edge pixel as a rotatable hinge and used the direction combination of the fragment pair to describe the small region around the center edge pixel [7]. Maaten fused multi-scale contour-hinge features, whose contour fragment length takes different values, to improve the performance [8]. Our proposed feature is to record the positions of some special edge pixel pairs in every local grid, to calculate the appearance probability of different position pairs and to express the writing style by the probability density distribution.

2.1 Feature Generation

We use a small part of an edge image, which contains a single handwritten character, to demonstrate the process of feature generation as Fig. 1 shows. We assume that a floating grid is moving along the edge traces in the image

and its center square will traverse all edge pixels. Assume that the horizontal and vertical distances between the border squares and the center square of the grid are both L, so the size of the floating grid is $(2L + 1) \times (2L + 1)$. For the diagrammatic sketch in Fig. 1, $L = 4$. Every square position of the grid is marked with a fixed symbol i_m, where m denotes the bigger distance in the horizontal and vertical distances between the square and the center. For each value of m, there are $8m$ square positions and they are assigned from 0_m to $(8m - 1)_m$. In Fig. 1, all square positions except the center are signed with $0_1, \ldots, 7_1, 0_2, \ldots, 15_2, 0_3, \ldots, 23_3, 0_4, \ldots, 31_4$. If there are two edge pixels in some pair of squares, the symbols of the squares are used to record the positions of the edge pixel pair, such as $< i_m, j_l >$. When the center of the grid is moving along the edge traces, we use $h(i_m, j_l)$ to record the occurrence number of edge pixels in each square pair of the local grid. In the beginning, we set $h(i_m, j_l) = 0$ $(1 \leq m, l \leq L, 0 \leq i \leq 8m - 1, 0 \leq j \leq 8l - 1)$. When the grid center is located at a new edge pixel and there are edge pixels in two squares like i_m and j_l, we make

$$h(i_m, j_l) = h(i_m, j_l) + 1. \tag{1}$$

But not all edge pixel pairs have the effectiveness to represent the writing style. Firstly we record the edge pixel pairs which satisfy the set of conditions

$$\begin{cases} 1 \leq m = l \leq L, i < j \\ i_m \text{ and } j_l \text{ are the edge pixels} \\ k_m \text{ is not the edge pixel, } i < k < j \end{cases} \tag{2}$$

Under the conditions, the positions of the edge pixel pairs, which are the same distance away from the grid center, are recorded. We use the continuous distance values $1 \leq m \leq L$ to describe the change of the same edge trace in the local grid, such as $< 1_1, 6_1 >$ and $< 3_2, 11_2 >$. Also the relations between different edge traces are recorded, such as $< 11_2, 14_2 >$ and $< 4_3, 9_3 >$. For $i < j$, we do not record the pairs like $< 31_4, 6_4 >$. Maatten used the implementation to reduce the complexity in his writer identification software [9].

In addition, the extend trend and the turning of a edge trace is more important. A Chinese character always includes much more strokes and has a much finer structure than a western letter. There are many turning points on the edge traces of Chinese characters, such as $< 14_2, 23_3 >$ in Fig.1. Different writers have different writing manners of stroke turnings. We need to record the pairs of pixels which have different distances from the grid center. So we define two other sets of conditions as

$$\begin{cases} 1 \leq m = l - 1 \leq L - 1 \\ i_m \text{ and } j_l \text{ are the connected edge pixels} \\ i_m \text{ is the nearest to } j_l \end{cases} \tag{3}$$

and

$$\begin{cases} 1 \leq m = l - 2 \leq L - 2 \\ i_m \text{ and } j_l \text{ are the connected edge pixels} \\ i_m \text{ is the nearest to } j_l \end{cases} \tag{4}$$

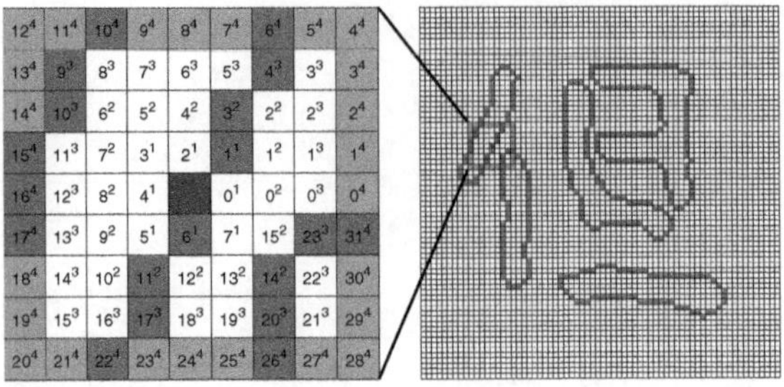

Fig. 1. The example of feature generation when the floating grid lies at a edge pixel. The edge pixel pairs satisfying C_1: $< 1_1, 6_1 >$, $< 3_2, 11_2 >$, $< 11_2, 14_2 >$, $< 4_3, 9_3 >$, $< 9_3, 10_3 >$,..., $< 20_3, 23_3 >$, $< 6_4, 10_4 >$, $< 10_4, 15_4 >$, ..., $< 26^4, 31_4 >$. The pairs satisfying C_2: $< 1_1, 3_2 >$, $< 6_1, 11_2 >$, $< 3_2, 4_3 >$, $< 11_2, 17_3 >$, $< 14_2, 20_3 >$, $< 14_2, 23_3 >$, $< 4_3, 6_4 >$, ..., $< 23_3, 31_4 >$. The pairs satisfying C_3: $< 1_1, 4_3 >$, $< 6_1, 17_3 >$, $< 3_2, 6_4 >$, $< 11_2, 22_4 >$, ..., $< 14_2, 31_4 >$.

Eq. 2, Eq. 3 and Eq. 4 define three sets of conditions. Let C_1, C_2 and C_3 denote them. In Fig. 1, the positions of the edge pixel pairs which satisfy the three sets of conditions are indicated with $< i_m, j_l >$ in the caption. We implement Eq. 1 for $h(i_m, j_l)$ where i_m and j_l in accord with these conditions.

After the center of the floating grid has traversed all edge pixels in the edge image, $h(i^m, j^l)$ gives the occurrence number of that the edge pixels appear at the square pairs $< i_m, j_l >$ in local regions. The sum of all occurrence number is $H = \sum i, j, m, lh(i_m, j_l)$. Then the appearance probability of the pixel pair $< i_m, j_l >$ can be calculated by $p(i_m, j_l) = \frac{h(i_m, j_l)}{H}$. The 2D probability density distribution is called the grid microstructure feature.

Although each handwriting consists of different characters, every character is made up of some elementary sorts of strokes and their deformations. A writer does not write one sort of strokes just the same every time. The edge pairs will appear at different square positions of the moving grid, even around the same sort of strokes. After global statistic, the appearance probabilities of different pairs are variant. However, the writing style of a writer tends to be uniform and the writing tendencies of different writers are discriminating. The variant probability distribution still reflects the writing tendency of the writer because the writer writes in his common manner unconsciously. Different writers write each sort of strokes with different lengths, different breadths, different angles and different curvatures. These characteristics are all considered and computed by the statistic of the grid microstructure feature.

Fig. 2 and Fig 3 show four Chinese handwritings written by two writers (A and B) and their microstructure features under all condition sets, C_1, C_2 and C_3. Each writer has two handwritings. Fig. 4 gives the diagrams of the absolute feature differences between two handwritings which are written by the same writer or different writers. It is observed intuitively that the difference between handwritings of different writers is distinctly bigger than the difference between handwritings of the same writer. The grid microstructure feature embodies the difference of writing style.

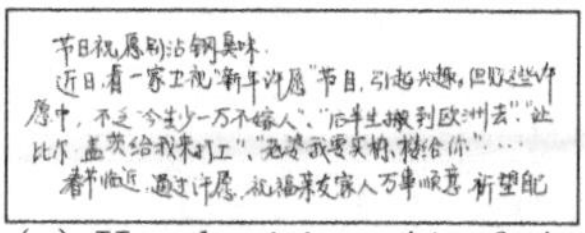

(a) Handwriting $A1$ of A.

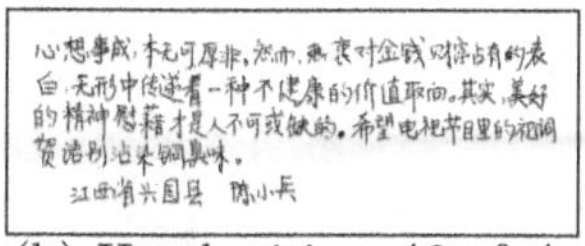

(b) Handwriting $A2$ of A.

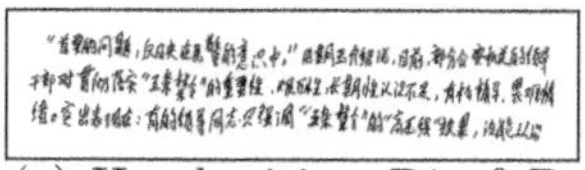

(c) Handwriting $B1$ of B.

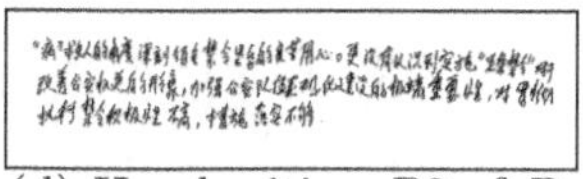

(d) Handwriting $B2$ of B.

Fig. 2. Four handwritings written by two writers

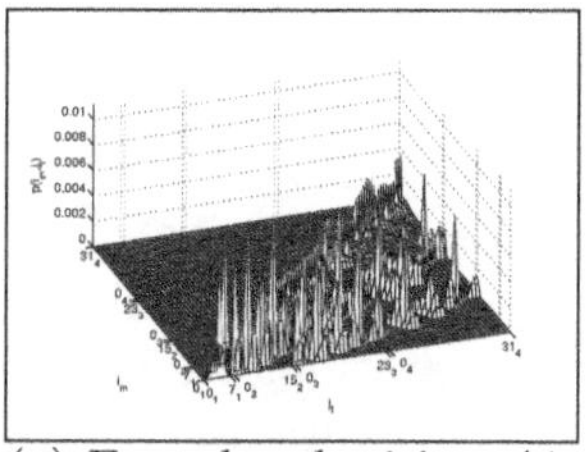

(a) From handwriting $A1$.

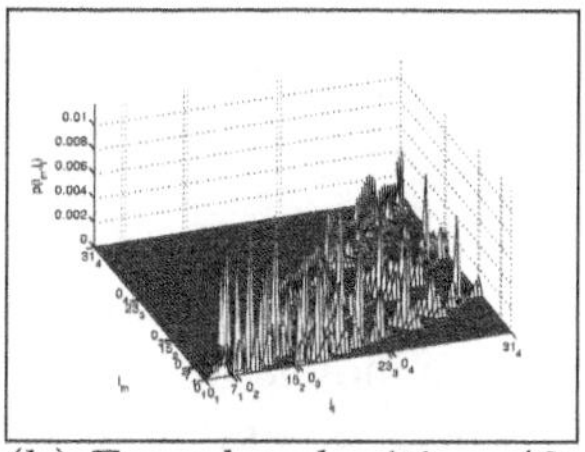

(b) From handwriting $A2$.

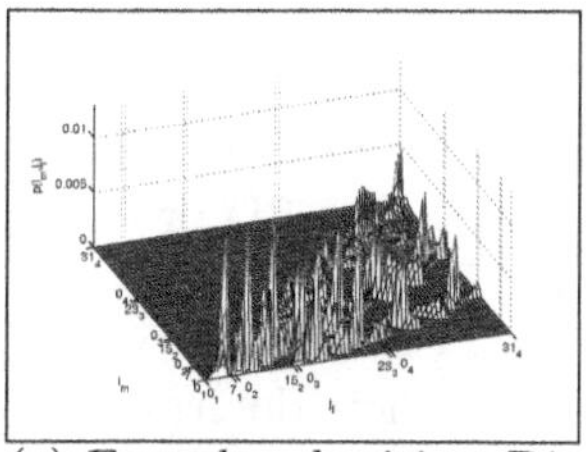

(c) From handwriting $B1$.

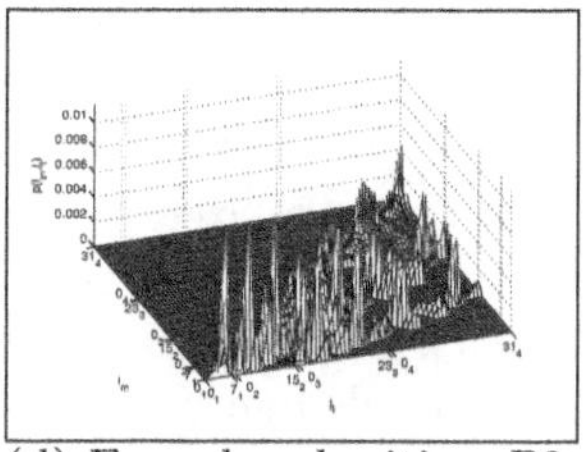

(d) From handwriting $B2$.

Fig. 3. Grid microstructure features extracted from four handwritings

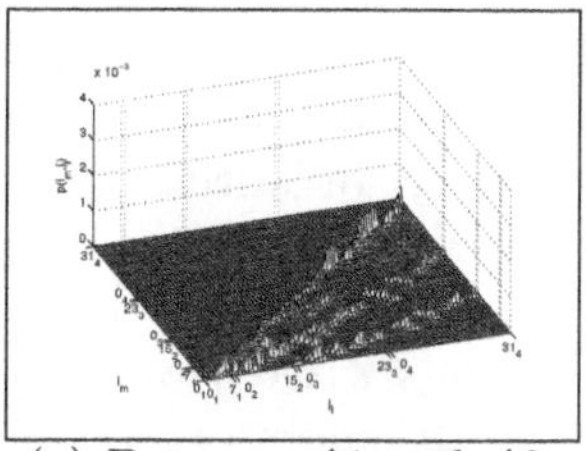

(a) Between $A1$ and $A2$.

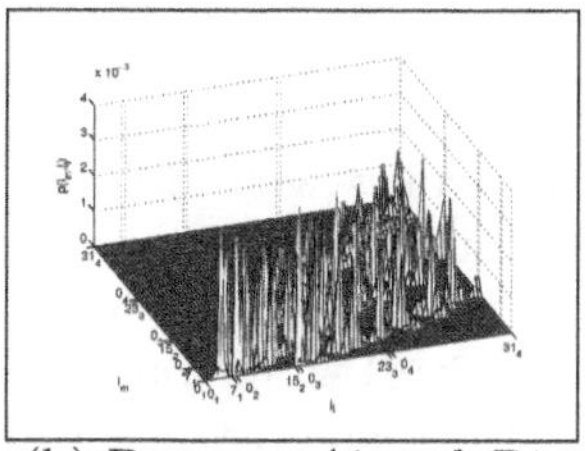

(b) Between $A1$ and $B1$.

Fig. 4. The absolute differences between the features of two handwritings

2.2 Similarity Measurement

To compare a handwriting to another, the difference of writing style should be quantified. We need to measure the similarity of two microstructure features. By arranging all probabilities in a high-dimensional vector, the similarity measurement can be transformed into the distance measurement. The most common distance metric is Euclidean metric though it may not be the best. Belongie used the Chi-square metric (χ^2 distance) after extracting shape contexts for object recognition [10]. Schomaker and Bulacu also considered that Chi-square metric, in which differences of the low probabilities are weighed more importantly, is better than Euclidean metric and Bhattacharya metric [6]. And Zhu [4] used weighted Euclidean distance measure. They used the standard deviation of handwritings written by the corresponding reference writer to weight the difference of each component. If each reference writer has only one handwriting, the weighted method is disable. However, there are also many different reference writers in writer identification. The reference handwritings of all writer give the range of each feature component. The differences of the components with smaller ranges should be given more weight. So we use the standard deviation of all reference handwritings to weight each component. We define two new distance metrics: the improved weighted Euclidean metric (IWED) and the improved weighted Chi-square metric (IWCD). Assume that there are a query handwriting and N reference handwritings, and their microstructure features are $\mathbf{q}$ and $\mathbf{r}_n (1 \leq n \leq N)$. The improved weighted Euclidean distance between $\mathbf{q}$ and $\mathbf{r}_n$ is

$$d_{IWED}(\mathbf{q}, \mathbf{r}_n) = \sqrt{\sum_k \left(\frac{q^{(k)} - r_n^{(k)}}{\sigma^{(k)}} \right)^2}, \tag{5}$$

and the improved weighted Chi-square distance is

$$d_{IWCD}(\mathbf{q}, \mathbf{r}_n) = \sum_k \frac{(q^{(k)} - r_n^{(k)})^2}{(q^{(k)} + r_n^{(k)})\sigma^{(k)}}, \tag{6}$$

where $q^{(k)}$ is kth component of feature vector $\mathbf{q}$, $r_n^{(k)}$ is kth component of $\mathbf{r}_n$ and $\sigma^{(k)} = \sqrt{\frac{1}{N-1} \sum_{n=1}^{N} (r_n^{(k)} - m^{(k)})^2}$, $m^{(k)} = \frac{1}{N} \sum_{n=1}^{N} r_n^{(k)}$. Then in writer identification, the writer of the reference handwriting corresponding to $\underset{\mathbf{r}_n}{\operatorname{argmin}}\, d(\mathbf{q}, \mathbf{r}_n)$ is considered as the most possible writer of the query handwriting.

3 Handwriting Database and Experiment Process

We use a database of handwritten texts, the HIT-MW database, in our experiments. HIT-MW is collected by Artificial Intelligence Lab, Harbin Institute of Technology, China [11]. The database is built for Chinese character recognition. The texts in HIT-MW are different and sampled from China Daily. We got the writer information of a part of the database involving 240 writers. Most of the

writers have only one page of handwritten text. We have to use only one text page for each writer and segment each page into two commensurate parts by manual work. Each part contains about 100 characters and is seen as an independent handwriting. The handwritings $A1$ and $A2$ in Fig. 2 also come from the same text page in HIT-MW and so do $B1$ and $B2$. In this case, the within-class variability is limited and it helps get higher identification accuracy. However, HIT-MW is the best choice in the public Chinese handwriting databases, because it has a large number of writers and provides the writer information.

We carry out writer identification using the segmented handwritings of 240 writers in HIT-MW. One segmented handwriting of each writer is used as the query handwriting and the other one is used as the reference handwriting. So there is only one handwriting sample for each writer in reference handwriting database. For every query handwriting, the distances between it and all reference handwritings are calculated. The writer of the reference handwriting corresponding to the minimal distance is the first candidate. If the candidate is just the writer of the query handwriting, the identification is correct. The correct rate of all 240 query handwritings is the identification accuracy of our experiment. We also make a longer list of candidates. The correct hit rate increases with the length of the candidate list. When there are M candidates in the list, the correct hit rate is called Top-M recognition accuracy.

4 Experimental Result

To begin with, the size of floating grid has influence on the effectiveness of our grid microstructure feature. The information is limited when the grid size is small, and the oversize grid affects the description of local particulars. A best selection of the grid size exists and relates to the scanning resolution of the handwriting samples. HIT-MW samples are all scanned with the resolution 300 dpi. We change the grid size by choosing different values of L in order to get the best grid size. Here the improved weighted Euclidean metric is used. Fig. 5 gives four curves of the correct hit rates when the grid size is changed. The curves are

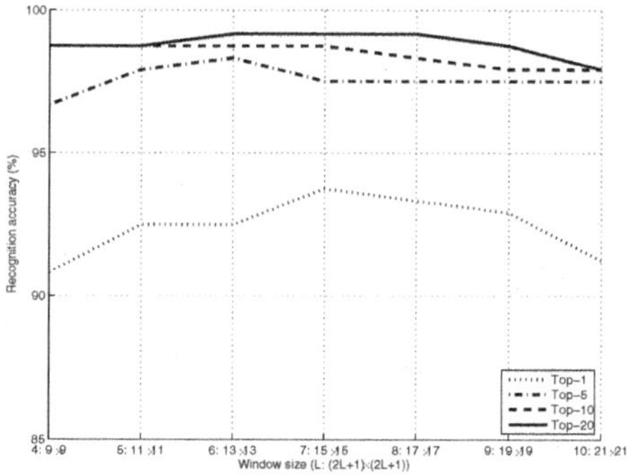

Fig. 5. Accuracy curves of the microstructure feature for different grid sizes

Table 1. Identification performance of microstructure feature using different distance metrics

Distance metric	ED	CD	$IWED$	$IWCD$
Top-1 recognition accuracy (%)	80.4	88.8	93.8	**95.0**

Table 2. Recognition accuracies of writer identification using different features

Features	Recognition accuracy (%) Top-1	Top-5	Top-10	Top-20
CH	84.6	95.4	96.7	98.3
MCH	92.5	97.1	97.5	99.2
GMS^*	94.6	97.5	98.8	99.6
GMS	**95.0**	**98.3**	**98.8**	**99.6**

Table 3. Identification performance of our method and some other methods

Method	Database size (writer number)	Top-1 recognition accuracy (%)
Our	240	95.0
Zhu's [4]	17	95.7
He's [5]	500	36.4
Yoshimura's [3]	20	94.2

obtained when the length of the candidate list is 1, 5, 10 and 20. We find that when L takes value from 6 to 8, the recognition accuracies are the highest. So we make $L = 7$ and choose 15×15 for the best grid size. In fact, the handwritten characters in the 300 dpi resolution are about from 40 to 70 pixels high. It is reasonable to use the 15×15 grid to represent the stroke-level information.

Furthermore, similarity measurement is also an important problem. We test four different distance metrics, Euclidean distance (ED), Chi-square distance (CD), improved weighted Euclidean distance (IWED) and improved weighted Chi-square distance (IWCD), by comparison. The microstructure features with $L = 7$ are extracted. Table 1 shows the identification performance (Top-1 recognition accuracy) changing with the distance metric. The best result is gotten by the improved weighted Chi-square metric.

Finally, we compare the proposed feature with two existing histogram-based features, the contour-hinge feature (CH for short, proposed by Bulacu et al. [7]) and the multi-scale contour-hinge (MCH for short, suggested by Maaten [8]) in the Chinese handwriting database. We also compare two versions of grid microstructure feature, GMS^* and GMS. GMS^* is the feature only satisfying the condition sets $C1$ and GMS is obtained under the case of all three condition sets. The length of edge fragment in CH is 8 and the value set of fragment length in

MCH is 2, 4, 6, 8. The grid size of grid microstructure features is 15×15. In the comparison, the improved weighted Chi-square distance is computed for similarity measurement. Table 2 gives the Top-1, Top-5, Top-10 and Top-20 recognition accuracies using the four features. The grid microstructure features outperform the other two features and GMS is better than GMS^*. It is obvious that our histogram-based feature is more effective for identifying Chinese handwritings.

5 Conclusions

For reference, we list some results of the literatures. Table 3 includes the performance of our method and some other text-independent methods for Chinese writer identification and one method for Japanese handwriting[1]. The recognition accuracy of our method using the grid microstructure feature and the improved weighted Chi-Square metric seems better than the existing text-independent methods for Chinese writer identification. Because of using different databases and different conditions, the results are incomparable. However, Gabor and wavelet features used in the traditional text-independent methods of Chinese writer identification are affected greatly by the normalization and the arrangement of characters in texture blocks. Differently, our histogram-based feature uses original handwriting images and is independent on the order of character arrangement. The grid microstructure feature tries to find out the writing tendency of the writer in local regions. It discriminates the writers using the difference of the tendency. Although the number of Chinese characters is huge and their structures are variant, the sorts of strokes composing characters are limited. If characters in the handwritings are enough to make the appearance percentages of different sorts of strokes balanced, the grid microstructure feature will succeed in denoting the difference of writing style. The effectiveness of our method for Chinese writer identification is proved by the experiments. It is expectable that the feature can be used for multilingual handwritings including western handwritings. As a future work, the effect on the recognition rate caused by the difference of text contents need to be studied. Besides, how to extract writing style information from a smaller number of characters is still a hard problem.

Acknowledgments. This work is supported by National Basic Research Program of China (973 program) (2007CB311004) & National Natural Science Foundation of China (project 60772049, 60872086).

References

1. Plamondon, R., Lorette, G.: Automatic Signature Verification and Writer Identification - the State of the Art. Patt. Rec. 22(2), 107–131 (1989)
2. Wang, X., Ding, X., Liu, H.: Writer Identification Using Directional Element Features and Linear Transform. In: 7th International Conference on Document Analysis and Recognition, pp. 942–945. IEEE Press, New York (2003)

[1] We do not cite the best result of [3] because it was obtained in the text-dependent case that some query text rows and reference rows have the same text.

3. Yoshimura, M., Yoshimura, I., Kim, H.B.: A Text-Independent Off-line Writer Identification Method for Japanese and Korean Sentences. IEICE Trans. Inf. and Syst. E76-D(4), 454–461 (1993)
4. Zhu, Y., Tan, T., Wang, Y.: Biometric Personal Identification Based on Handwriting. In: 15th International Conference on Pattern Recognition, vol. 2, pp. 797–800. IEEE Press, New York (2000)
5. He, Z., You, X., Tang, Y.: Writer Identification of Chinese Handwriting Documents using Hidden Markov Tree Model. Patt. Rec. 41(4), 1295–1307 (2008)
6. Bulacu, M., Schomaker, L.: Text-Independent Writer Identification and Verification using Textural and Allographic Features. IEEE Trans. on Patt. Ana. and Mach. Inte. 29(4), 701–717 (2007)
7. Bulacu, M., Schomaker, L., Vuurpijl, L.: Writer Identification using Edge-Based Directional Features. In: 7th International Conference on Document Analysis and Recognition, pp. 937–941. IEEE Press, New York (2003)
8. Maaten, L., van der Postma, E.: Improving Automatic Writer Identification. In: 17th Belgium-Netherlands Conference on Artificial Intelligence, pp. 260–266 (2005)
9. WRIDE: Maaten's writer identification software, `http://www.cs.unimaas.nl/l.vandermaaten/Laurens_van_der_Maaten/Software.html`
10. Belongie, S., Malik, J., Puzicha, J.: Shape Matching and Object Recognition using Shape Contexts. IEEE Trans. on Patt. Ana. and Mach. Inte. 24(4), 509–522 (2002)
11. Su, T., Zhang, T., Guan, D.: Corpus-Based HIT-MW Database for Offline Recognition of General-Purpose Chinese Handwritten Text. Int. J. of Doc. Ana. and Rec. 10(1), 27–38 (2007)

Enhancement and Registration Schemes for Matching Conjunctival Vasculature*

Simona Crihalmeanu[1], Arun Ross[1], and Reza Derakhshani[2]

[1] West Virginia University, Morgantown WV 26506, USA
[2] University of Missouri, Kansas City, MO 64110, USA

Abstract. Ocular biometrics has made significant strides over the past decade primarily due to the rapid advances in iris recognition. Recent literature has investigated the possibility of using conjunctival vasculature as an added ocular biometric. These patterns, observed on the sclera of the human eye, are especially significant when the iris is off-angle with respect to the acquisition device resulting in the exposure of the scleral surface. In this work, we design enhancement and registration methods to process and match conjunctival vasculature obtained under non-ideal conditions. The goal is to determine if conjunctival vasculature is a viable biometric in an operational environment. Initial results are promising and suggest the need for designing advanced image processing and registration schemes for furthering the utility of this novel biometric. However, we postulate that in an operational environment, conjunctival vasculature has to be used with the iris in a bimodal configuration.

1 Introduction

The eyes have played a significant role in biometrics primarily due to the unique and stable features that can be extracted from the the iris, a circular diaphragm forming the colored portion of the eye. It is well established that the iris *texture* is a reliable biometric, with numerous forms of variability across individuals and is believed to be stable over time when imaged in the near-IR spectrum [3]. However, when the iris becomes "off-angle" with respect to the focal plane of the camera, the sclera (also known as the white of the eye) is revealed. Sclera, the external layer of the eye, is a firm dense membrane comprising of a white and opaque fibrin connective tissue, organized in many bands of parallel and interlacing fibrous tissue bundles. Its outer surface, called episclera, contains the blood vessels nourishing the sclera. The anterior part of the sclera is covered by the conjuctival membrane - a thin membrane containing secretory epithelium that helps lubricate the eyes for eyelid closure. The focus of this work is on the rich vasculature pattern revealed in the episclera and conjunctiva membrane hereafter referred to as the conjunctival vasculature. In a previous study [2], images of 12 eyes pertaining to the left and right eyes of 6 subjects were used to study the possibility of using this pattern as a biometric. In this study, we

* This work was funded by the Center for Identification Technology Research (CITeR).

M. Tistarelli and M.S. Nixon (Eds.): ICB 2009, LNCS 5558, pp. 1240–1249, 2009.
© Springer-Verlag Berlin Heidelberg 2009

conduct a more elaborate experiment involving 50 subjects in order to determine the suitability of using the conjunctival vasculature as a biometric. The primary contribution is the design of image enhancement and registration schemes to match conjunctival vasculature obtained in a non-ideal environment containing ambient lighting.

The paper is divided as follow: section 2 describes data acquisition; section 3 discusses sclera segmentation; section 4 describes specular reflection detection and removal; section 5 describes methods to enhance the image; section 6 discusses the registration method, and section 7 describes the experimental results. The block diagram of the proposed system is shown in Fig.1.

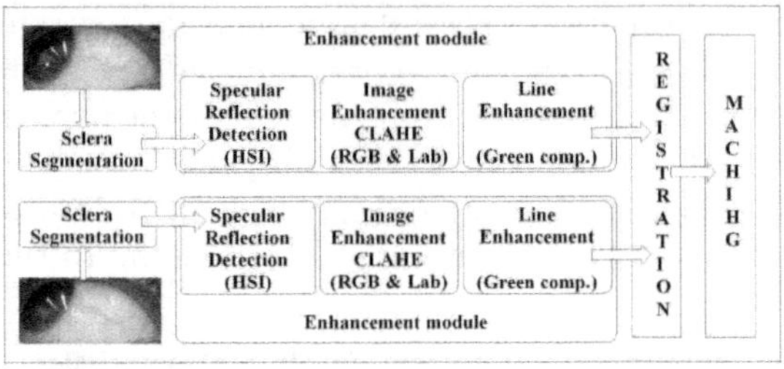

Fig. 1. Block diagram for enhancement and registration of conjunctival vasculature

2 Image Acquisition

The SONY CyberShot DSC F717(5 megapixels) was used to capture color images of the eye. Each subject was asked to move their eyes in the following manner with respect to the optical axis of the camera: frontal, upward, to the left, and to the right. Thus, different regions of the sclera were represented in the ensuing pictures. These color images (RGB) were collected in two sessions. The first session had 2400 images from 50 subjects, and the second session had 816 images from 17 of the original 50 subjects. Images were captured from both eyes at three different distances: 1 feet considered as *near distance* images, 9 feet as *medium distance* images and 12 feet as *far distance* images. For each eye, 2 images were collected per gaze at each distance (Fig. 2).

The paper presents the results on a subset of images belonging to the "left-eye-looking-left near-distance" category that contains 100 images from 50 subjects in the first session, and 34 images from 17 of the 50 subjects in the second session.

3 Sclera Region Segmentation

Image segmentation refers to the procedure that partitions the image into regions of interest according to pixel intensity values, based on either *discontinuities* (abrupt changes in intensity values such as edges) or *similarities* (such as grouping pixels based on predefined criteria). Accurately segmenting the sclera from

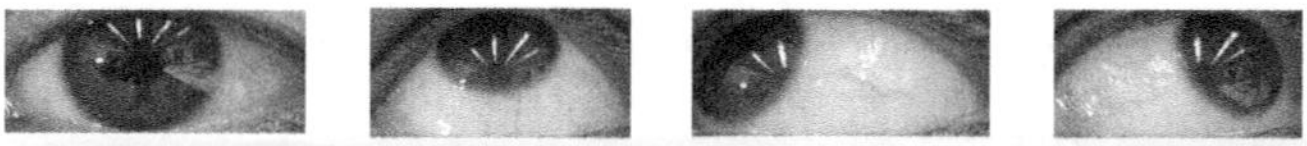

(a) (b) (c) (d)

Fig. 2. *Near* images of the eye where the subject is: a) looking straight ahead, b) looking up, c) looking left, and d) looking right with respect to the camera

the eye image is very important for further processing. A semi-automated technique is used for this purpose. The proposed technique first applies an automated clustering method whose output is subsequently refined by manual intervention. Each pixel is represented as a three-dimensional point in a Cartesian coordinate system based on its primary spectral components of red, green and blue. The k-means clustering algorithm is used to partition the pixels into three categories. Since the sclera is typically whiter than the rest of the eye, such a procedure is expected to work well in separating the scleral pixels from the rest of the image. The output of clustering is a binary mask, which when imposed on the original image, will identify the region of interest corresponding to the sclera.

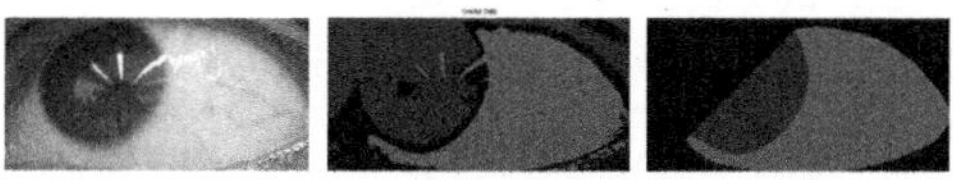

(a) (b) (c)

Fig. 3. Segmenting the sclera from two different eye images, displayed by column: a) Original image, b) Segmented sclera region based on RGB values (red = sclera region, blue = iris region, black = the background) c) Convex hull of the sclera (blue+red) containing a portion of the iris (blue)

The sclera pixel cluster is determined as that cluster which has the largest Euclidean distance from the origin of the coordinate system to its centroid. The iris pixel cluster is determined as that cluster with the smallest Euclidean distance from the origin of the coordinate system to its centroid. Entries marked as '1' in the mask denote the pixels assigned to the sclera cluster. Due to the image characteristics of illumination and specular reflection, it is possible for some pixels from the sclera to not be assigned to the proper cluster, thereby appearing as holes in the mask. In order to eliminate these holes, the convex hull of the pixels in the sclera cluster is considered Fig. 3 (c). This, however, means that pixels pertaining to the iris cluster may be included in the mask. To address this, we first remove those pixels within the convex hull belonging to the iris cluster. Next, we remove the convex hull of the iris pixels included in the mask. Fig. 4 (c) shows the segmented color image, after imposing the mask on the image.

On examining the segmented sclera region, we observed that in some images, a small portion of the lower eyelid was erroneously included. To address this issue, the mask is manually corrected for such images, Fig. 4 (d), thereby eliminating the lower eyelashes, Fig. 4 (e).

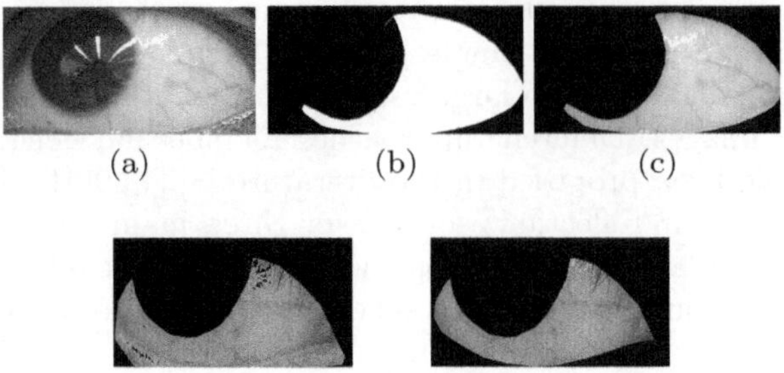

(a) (b) (c)

(d) (e)

Fig. 4. 1) Sclera region segmentation: a) Original image, b) Sclera mask, c) Segmented sclera region. 2) Manual correction of segmented sclera: d) Original image, e) Segmented sclera region after manual correction.

Table 1. Manual correction statistics of segmented sclera

Gaze	Distance	Left eyes		Right eyes	
		automatic segmentation and manual correction	only automatic segmentation	automatic segmentation and manual correction	only automatic segmentation
Looking left	near	61	73	42	92
	medium	49	85	36	98
	far	43	91	39	95
Looking right	near	54	80	56	78
	medium	53	81	51	83
	far	35	99	44	90

Table 1 records the number of images for which manual correction of the segmented sclera was needed.

4 Specular Reflection

4.1 General Considerations

Specular reflections may provide valuable information about the shape of the object and its location with respect to the light source. However, they can cause problems for image processing algorithms that may erroneously consider these specularities as pixels of interest during the process of segmentation resulting in spurious results. Localization of specularities in images is very important and requires a good understanding of the reflection of light, a complicated process that depends on the material of the object under consideration, the roughness of its surface, the angle of illumination, the angle of viewing, and the wavelength of light.

Specular reflections on the sclera have different topologies, sizes and shapes that cannot be described by a single pattern. Their pixel intensity values are distinctively high, and exhibit a large variation both within the same image and across multiple images. Different approaches for specular reflection detection and removal have been proposed in the literature [8],[4],[9],[11],[10],[6],[1]. The algorithm for specular reflection, consists of three main steps: detection and localization of specular reflection, construction of specular reflection mask, and exclusion of the region containing the specular reflection from the sclera region.

4.2 Detection of Specular Reflection

If original images of sclera containing specular reflection were to be further processed, as explained in the following sections, the edges of the specular reflection region may appear as spurious blood vessels in the enhanced image. Our algorithm to detect specular reflections is based on the power law transformation as applied to pixel intensities in the color image. Power law transformations have the basic form:

$$S = c * R^{\gamma}, \tag{1}$$

where c, γ are positive constants, R is the input pixel intensity, and S is the output intensity. As shown in Fig. 5, by simply varying γ we obtain a family of possible transformation curves. For $\gamma > 1$, the power - law curves map a narrow range of light input values into a wider range of output values. For $\gamma < 1$, the power - law curves map a narrow range of dark input values into a wider range of output values. In order to detect specularities, we consider γ an integer in the range $[1, 10]$.

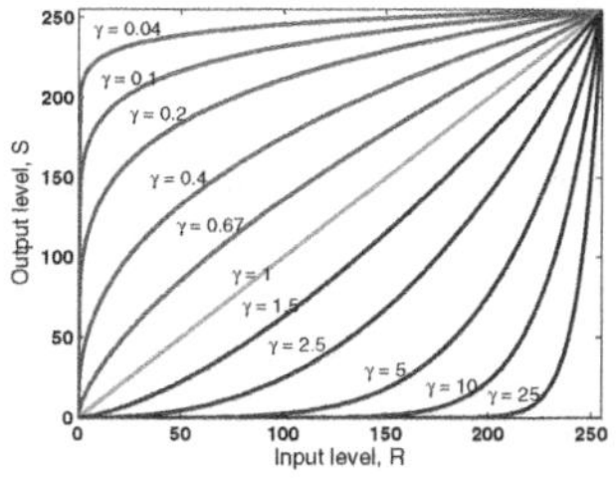

Fig. 5. Plots of the power-law transform for various values of γ; c = 1 in all cases

The algorithm is described by the following steps:

1) Convert the RGB image to the HSI (hue, saturation, illumination) color space.
2) Consider the illumination component of the HSI space as the input image R in equation 1.
3) Compute the output image S for different γ values using equation 1. Fig. 6 (a) displays results for $\gamma = 3$.

4) Compute the histogram for each image S as seen in Fig. 6 (b).
5) Compute the filtered histogram for each image S using the moving average [1/3 1/3 1/3] filter as seen in Fig. 6 (c).
6) Compute the slope θ of the filtered histogram.
7) For the filtered histogram corresponding to each γ, find the first negative θ (θ_γ) and its corresponding intensity value, I_γ, as a potential threshold value for detecting specular reflection.
8) Examine the distribution of θ_γ as a function of γ to select γ_{opt}, $\gamma_{opt} = \arg\max (| \theta_\gamma - \theta_{\gamma-1} |)$. Fig. 7 shows $\gamma_{opt} = 5$; for near distance images, the threshold to detect specular reflection is selected as the mean of all thresholds values found for γ, $5 \leq \gamma \leq 10$.
9) Use the threshold value found to obtain a binary mask for isolating specular reflection.

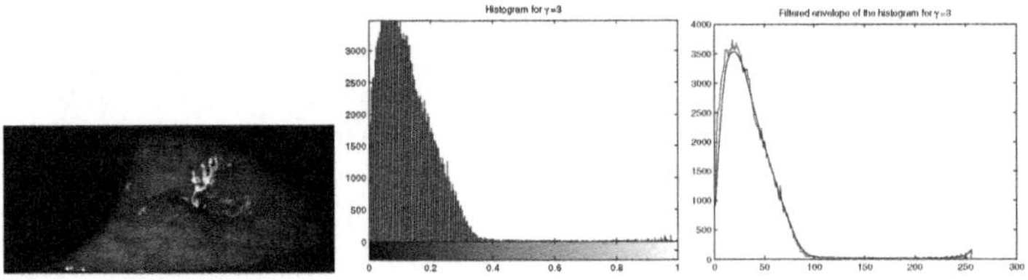

(a) (b) (c)

Fig. 6. Detection of specularities. Examples for $\gamma = 3$: (a) Illumination component of HSI sclera image; (b) Histogram of the illumination component; (c) Filtered envelop of the histogram.

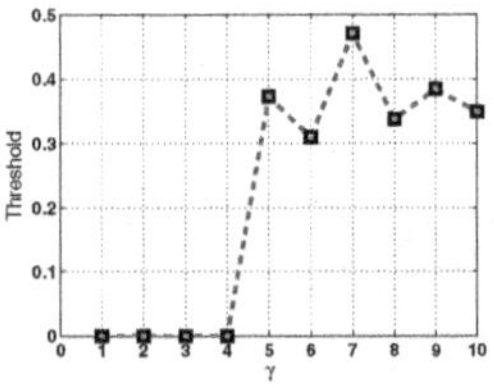

Fig. 7. Threshold values as a function of γ for an image.

Fig.8 shows the results of specular reflection detection.

4.3 Segmented Sclera Image without Specular Reflection

The segmented sclera image without specular reflection is obtained as follows:

1) Use sclera mask and specular reflection mask to obtain the final sclera mask without specular reflection.
2) Superimpose the final mask on the RGB image to obtain segmented sclera without specular reflection (Fig.9).

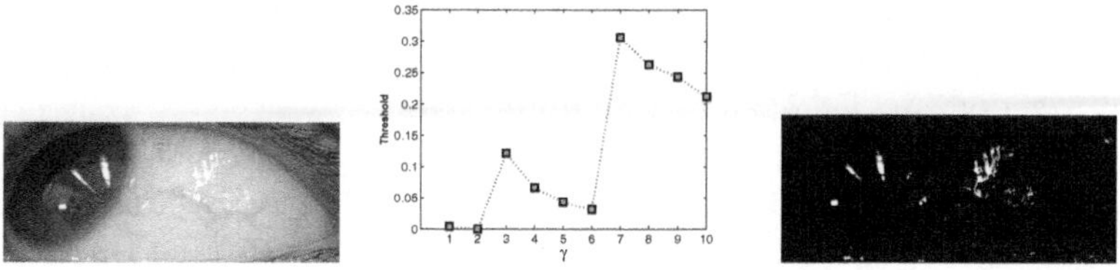

(a) (b) (c)

Fig. 8. Detecting specularities: a) Original image, b) Threshold values for $1 \leq \gamma \leq 10$ c) Specular reflection mask

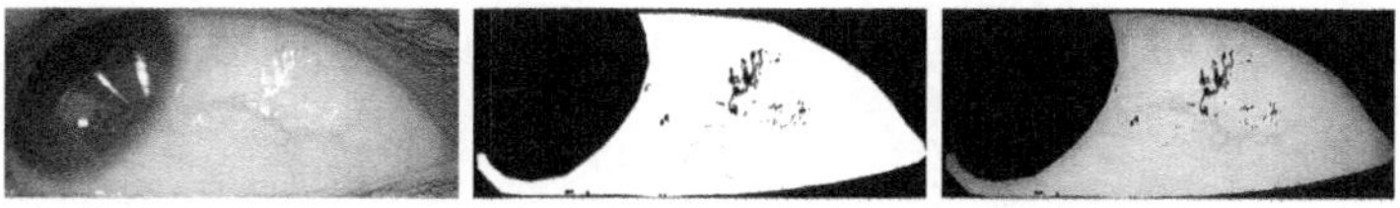

(a) (b) (c)

Fig. 9. Segmenting the sclera after removing specularities: a) Original image, b) Specular reflection mask c) Segmented sclera without specular reflection

5 Image Pre-processing

Images of the conjunctival vasculature present a large variation in intensity values and offer a low contrast between vessels and the background. Therefore sclera veins are pre-processed in two consecutive steps. The first one involves color image enhancement using contrast-limited adaptive histogram equalization (CLAHE) that operates and enhances the contrast on small regions of the color image called tiles and then combines neighboring tiles using bilinear interpolation to eliminate artificially induced boundaries (Fig. 10 (b)).

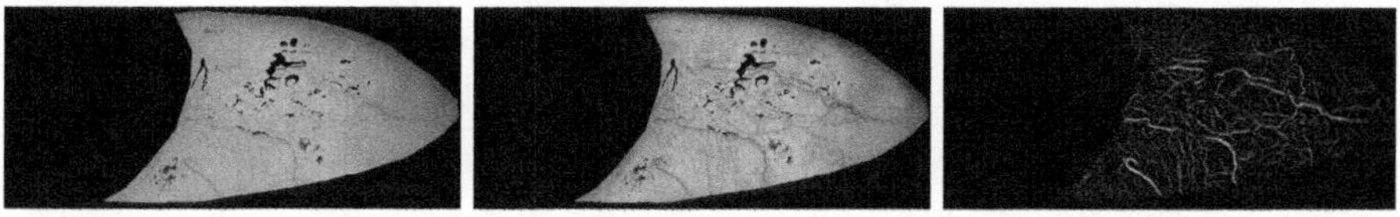

(a) (b) (c)

Fig. 10. Image enhancement: (a) Original sclera vein image, (b) Enhanced sclera vein image; Line enhancement: (c) Example of pre-processed sclera vein image

In order to obtain the best results for vein segmentation, an examination of the three components (red, blue and green) of the enhanced colored sclera images, suggests the use of the green component that has the best contrast between the blood vessels and the background. In order to improve sensitivity to vein detection and segmentation, we use a selective enhancement filter for lines, as described in [5]. The result is a normalized image with enhanced blood vessels as seen in Fig. 10 (c).

6 Sclera Image Registration

Image registration is the process of finding a transformation that aligns one image to another. The method used here, described in [7], models a local affine and a global smooth transformation. It also accounts for contrast and brightness variations between the two images that are to be registered. The registration between two images, the source f(x,y,t) and the target f(x̂,ŷ,t-1), is modeled by the transformation $\boldsymbol{m} = (m_1, m_2, m_3, m_4, m_5, m_6, m_7, m_8)$:

$$m_7 f(x, y, t) + m_8 = f(m_1 x + m_2 y + m_5, m_3 x + m_4 y + m_6, t - 1). \tag{2}$$

where m_1, m_2, m_3, and m_4 are the linear affine parameters, m_5, m_6 are the translation parameters, and m_7, m_8 are the contrast and brightness parameters. A multi-scale approach is employed by using a Gaussian pyramid to downsample the images to be registered. From the coarse-to-fine level, the transformation $\boldsymbol{m}$ is determined globally as well as locally at each level of the pyramid and the estimated parameters are used to warp the source image. Fig. 11 shows results for the registration of two pre-processed sclera images.

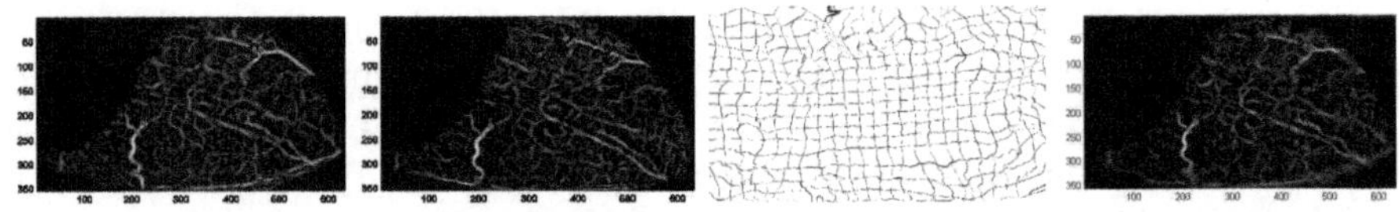

(a) (b) (c) (d)

Fig. 11. Registering images: (a) Target image, (b) Source image, (c) Rectified warp, and (d) Registered source image

Using the linear affine parameters m_1, m_2, m_3, and m_4, and the translation parameters m_5, m_6, the sclera mask of the source image is also registered. The result is a grayscale image that is transformed into a binary image via a threshold, $Th = (M + m)/2$ where M is the maximum value of the registered mask and m is its minimum value.

7 Matching Results

The similarity between two sclera images is assessed using cross-correlation between regions of the sclera that do not include the specular reflections from both images. Results are displayed in Fig.12.

This initial analysis indicates that an EER of $\sim 25\%$ has been obtained on the database of 50 subjects. An assessment of the results suggests that the large intra-class variation is caused by the following factors: (a) the large range of viewing angles; (b) harsh ambient lighting resulting in significant specularities; and (c) the curved ocular surface that cannot be modeled using a linear surface. In our previous paper [2] we manually cropped portions of the sclera and performed matching. In this paper, the entire segmented region of the sclera was considered. Further, the issue of specular reflection and image enhancement were addressed.

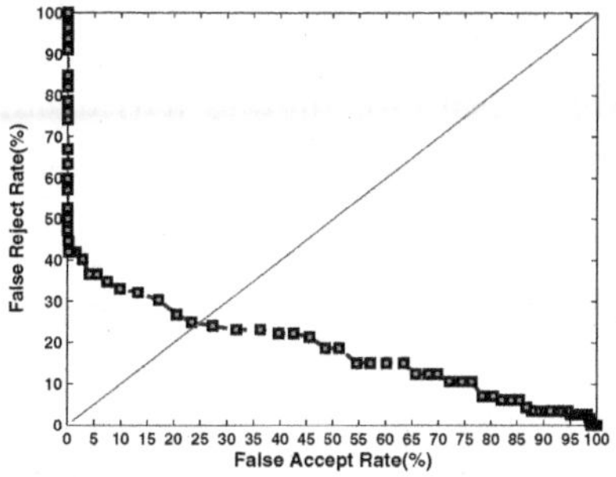

Fig. 12. ROC curve indicating the matching performance for the *left-eye looking-left* images obtained at a *near distance*

8 Summary and Future Work

The use of conjunctival vasculature in the biometric field is a fairly new idea. In this work, we designed image enhancement and registration schemes for processing the conjunctival vasculature information pertaining to 50 subjects. The images used in this work were taken without any constraints on lighting thereby exhibiting a wide range of irregularities including blurring and specularities. An EER of 25% was obtained on the eye images labeled as *left-eye looking-left* obtained at the near distance. We are currently conducting experiments on the remaining eye images. There are several challenges associated with processing these images. These issues are related to: (a) the curved surface of the eye, (b) the glossy surface of the sclera that reflects light, (c) the presence of less prominent veins that can degrade performance, and (d) eyelashes that obstruct the sclera region and can be incorrectly perceived as vasculature patterns. We are currently developing methods to handle these challenging images. We are also looking at ways to combine the sclera with the iris for enhanced biometric recognition. We conclude this paper by noting that more work is essential to make conjunctival vasculature a practical biometric that can operate in non-ideal environments.

References

1. Brelstaff, G., Blake, A.: Detecting Specular Reflections Using Lambertian Constraints. In: Second International Conference on Computer Vision, pp. 297–302 (1988)
2. Derakhshani, R., Ross, A., Crihalmeanu, S.: A New Biometric Modality Based On Conjunctival Vasculature. Artificial Neural Networks in Engineering, St. Louis, USA (2006)
3. Daugman, J.: How Iris Recognition Works. IEEE Transactions on Circuits and Systems for Video Technology 14(1), 21–30 (2004)
4. Klinker, G.J., Shafer, S.A., Kanade, T.: Using a Color Reflection Model to Separate Highlights from Object Color. In: First International Conference on Computer Vision, pp. 145–150 (1987)

5. Li, Q., Sone, S., Doi, K.: Selective enhancement filters for nodules, vessels, and airway walls in two-and three dimentional CT scans. Medical Physics 30(8) (2003)
6. Park, J.B.: Detection of Specular Highlights in Color Images using a New Color Space Transformation. In: International Conference on Robotics and Biomimetics, pp. 737–741 (2004)
7. Periaswamy, S., Farid, H.: Elastic Registration in the Presence of Intensity Variations. IEEE Transactions on Medical Imaging 22(7), 865–874 (2003)
8. Shafer, S.A.: Using Color to Separate Reflection Components. Color Research and Application 10(4), 210–218 (1985)
9. Stoyanov, D., Yang, G.Z.: Removing Specular Reflection Components for Robotic Assisted Laparoscopic Surgery. In: International Conference on Image PRocessing, vol. 3, pp. 632–635 (2005)
10. Tan, R.T., Ikeuchi, K.: Separating Reflection Components of Textured Surfaces using a Single Image. IEEE Transactions on Pattern Analysis and Machine Intelligence 27(2), 178–193 (2005)
11. Tan, R.T., Nishino, K., Ikeuchi, K.: Illumination Chromaticity Estimation using Inverse-Intensity Chromaticity Space. In: Computer Vision and Pattern Recognition Conference, vol. 1, pp. 673–680 (2003)

Entropy of the Retina Template

A. Arakala[1], J.S. Culpepper[1], J. Jeffers[1], A. Turpin[1], S. Boztaş[1], K.J. Horadam[1], and A.M. McKendrick[2]

[1] RMIT University, Melbourne, Australia
[2] University of Melbourne, Parkville, Australia
{arathi.arakala,shane.culpepper,jason.jeffers,andrew.turpin, serdar.boztas,kathy.horadam}@rmit.edu.au, allisonm@unimelb.edu.au

Abstract. We compare two vessel extraction methods for creation of a retina template, using a database of 20 images of normal retinas. Each vessel in a well defined region is represented by a three dimensional feature, from which a retina template is built. Based on the sample distributions, we propose a preliminary theoretical model to predict the entropy of a retina template. We analyse by experimental and theoretical means the entropy present, and infer that entropy from our retina template compares sufficiently favourably with that of a minutia-based fingerprint template to warrant further study.

1 Introduction

Over the past two decades there has been increasing research and community interest in biometric identification, with a growing installed base of biometric devices and a general community acceptance of biometric based authentication. As research in the area has grown, so has the breadth of the biometric characteristics considered for authentication purposes. Interestingly, despite retina-based authentication being one of the early techniques to be explored and successfully commercialized, the volume of published research in the field has been relatively low, though research interest in it has increased noticeably in the last few years. The lack of a large standard database of retinal images for evaluating matching algorithms has meant that it is difficult to compare the performance of different algorithms, or to understand properly the error performance an algorithm in isolation, or even to decide whether further investment in the area is justified. It is expected that matching algorithms will improve significantly over time. In this paper we use an alternative approach to evaluate the information content captured by one of the key algorithms of interest in the field and compare the result to mature fingerprint and iris based techniques.

Section 2 provides background on retina-based authentication and in Section 3 we compare two medical retina image-processing techiques to determine a preferred feature extraction method. In Section 4 we adopt a slight modification of the construction of Farzin, Abrishami-Moghaddam, and Moin [4] to form the retina template. We describe the experiments conducted to study retinal entropy and the theoretical model of the template distribution we develop, giving our results in Section 5.

M. Tistarelli and M.S. Nixon (Eds.): ICB 2009, LNCS 5558, pp. 1250–1259, 2009.
© Springer-Verlag Berlin Heidelberg 2009

2 Background

The retina has many desirable characteristics as the basis of authentication. Retinal blood vessel patterns are known to be very distinctive [12], even between identical twins [14]. The blood vessel structure is very stable over time, well shielded from outside environmental impacts, and believed to be difficult to spoof [6]. From a privacy perspective, a retinal image is unlikely to be captured from an individual without their knowledge, unlike many leading biometrics such as face, fingerprint, voice, and even iris.

Retina-based authentication is perceived as intrusive and user-unfriendly, ever since the first commercially successful retina scanner (Hill [7,6]) required subjects to place their eye very close to the reader and focus on an alignment target as a focussed beam of light was shone in their eye. This process was also the reason retina-based authentication traditionally was not appropriate for individuals with poor eyesight. Now, however, the technological limitations which required an intrusive process no longer exist.

Retinal images appear to contain abundant identifying information and four broad approaches have emerged to capture this information: using general signal and image processing techniques on the raw retina image; matching of the branching blood vessel structure as a whole; matching crossing and bifurcation points of the blood vessel structure; and matching the pattern of vessels traversing a well defined region.

We will concentrate on the last approach but briefly discuss the first three. Tabataee, Fard and Jafariani [13] employ a retinal identification system based on features extracted from the raw image using an analytic Fourier-Mellin transform and image moments. In [9], Marino et al. propose a matching algorithm that focusses on the overall network structure of the blood vessel pattern. The authors store the entire image as the template and perform a cross correlation between the query image and the stored image, but the images are processed in such a way that it is the blood vessel network that is actually being correlated. A template based on the vessel network would be expected to be a natural evolution of this approach. Several research groups have used the points where blood vessels cross one another or bifurcate as the basis of a feature set to match individuals [15,16,10]. This is analogous to successful fingerprint matching techniques.

2.1 Matching Vessels Traversing a Region

The original retina scanner [8] may be classified as using this type of approach. The author used a focussed beam to scan around the fovea, in effect measuring the the blood vessels that traversed a narrow annulus centered on the fovea.

More recently Farzin et al [4] used image processing to locate the optical disk and then extract the blood vessels that traverse an annulus centered on the optical disk, sort them into three different width ranges and match them to an enrolled template by measuring their angular position and angle relative to the border of the annulus. The authors test their algorithm on a database with 60 images, but rotate images randomly to provide more samples for matching. They

do not state whether these rotated images are used to simulate multiple samples from the same subject, as would be obtained in a true enrolment situation.

3 Retinal Image Processing

Our database consists of single images of 20 clinically normal right eyes. The images typically show uneven illumination and poor contrast of minor vessels. These images were obtained from the reflectance images of a Scanning Laser Topographer (HRT, Heidelberg Engineering) which uses a low power laser to scan 384×384 pixel images at 32 different depths of the optic nerve head.

Following the approach of Farzin et al. [4], we extracted vessels that traversed an annulus described by two circles; the inner of radius 80 pixels and an outer of radius 130 pixels from the centre of the image. The center was located as the centroid of the lowest 150 pixels as given by the topographic map that accompanied each image. In order to extract the vessels we compared two methods. The first, due to Can et al. [2,5], uses aligned edge detection filters to locate parallel lines and label them as vessel boundaries. An example appears in Figure 1(a). The second, due to Sofka and Stewart [11], uses multi-scale matched filtering and assigns each pixel a "vesselness" score, and then uses training data to produce a likelihood ratio of the pixel being part of the vessel. An example appears in Figure 1(b). Both methods are implemented in publicly available software at `www.cs.rpi.edu/~sofka/vessels_exec.html`. Once the vessels have been labelled in the images, we extracted the annulus crossings by hand. If a vessel splits while traversing the annulus, we follow only the most anticlockwise vessel. This is an arbitrary choice, but as long as the choice is consistent it should not affect a matching process based on this extraction. After this stage, each vessel has four (x, y) co-ordinates, one for each of the four corners of its crossing of the inner and outer circles. The final step is to "flatten" these vessel segments into two dimensions using polar co-ordinates, as shown in Figure 2. On average, the number of vessel segments extracted from images by Method 1 was 13, while

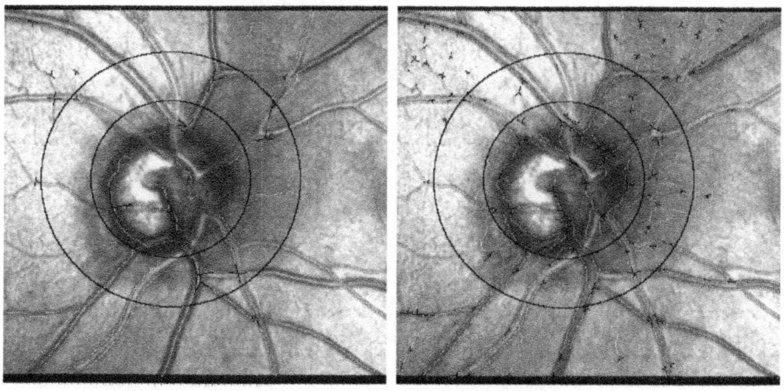

(a) Method 1 (b) Method 2

Fig. 1. Image and labelling typical of those used in this study

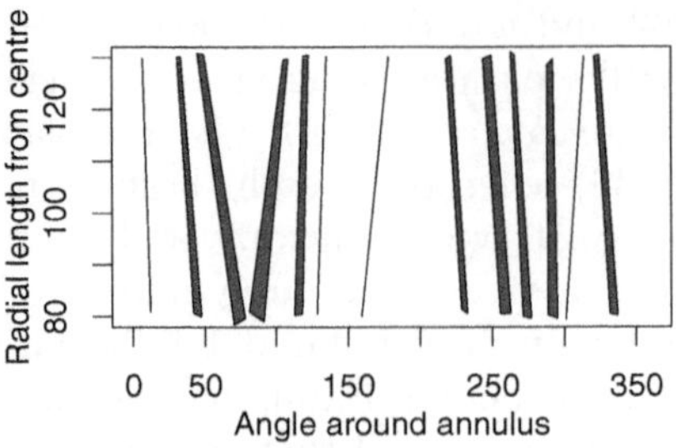

Fig. 2. Flattened extracted features typical of those used in this study

that for Method 2 was 18. We conclude that Method 2 has superior performance in vessel extraction from the annulus centred on the optic disc, confirming the approach in [11]. Consequently we used the vessel segments extracted by Method 2 to build our biometric templates.

4 Entropy of a Retina Template

The *entropy* contained in a biometric template is a measure of the randomness or distinctiveness between the biometric templates of different people. The template itself can have a very different (usually lower) entropy when compared to the original biometric image, because template creation uses only a selection of the significant biometric image features that distinguish one person from another. The more significant features that are used in building a template, the greater will be the template entropy.

Entropy can be determined in two ways. One method is to fit the distribution of the features to existing theoretical models. The other is to determine empirically the probability p that two templates from different people will match, by testing over a very large database of biometric samples. The entropy (in bits) of the biometric template is then $-\log_2 p$.

Owing to the small size of our database, we adopt the first method to estimate the entropy of a retina template.

4.1 Entropy of Fingerprint and Iris Templates

The entropy in a fingerprint has been analysed by Zhu, Dass and Jain [17], who proposed a family of mixture models to simulate the occurrence of minutiae locations and orientations in a fingerprint. They simulated imposter matches and determined an equation for the probability p that fingerprints from different people would match on certain numbers of minutiae [17]. Their model was a close representation of their empirical results (from the FVC2000 DB1) when the match threshold was greater than 7 minutiae. Arakala, Horadam and Boztas [1] applied a simple bounding box technique to quantise the measurements of minutiae location and orientation with respect to the core for fingerprints obtained

from the FVC2002 DB2 and found that the average number of quantised elements (resulting from well separated minutiae) in a fingerprint template was 38. Using Zhu et al's results gives $p = 9.3 \times 10^{-4}$ for templates having a comparable number of minutiae (≈ 36) when the match theshold is 12 minutiae. This corresponds to approximately 10 bits of entropy, and approximates the maximum information that can be contained in a purely minutiae-based template.

On similar lines, Daugman [3] compared 1 billion irises from different people and at the most relaxed relative Hamming Distance threshold of 0.317 between the digital iris templates, found that the probability of obtaining a false match was roughly 0.000001, corresponding to an entropy of 20 bits. These tests show that the specific iris template tested (IrisCode) has extremely high distinctiveness across populations. This allows it to be used to design iris based authentication systems with negligible false match rates.

4.2 The Retina Template and Feature Distribution

After the image processing steps described in Section 3, the vessel segments crossing the annulus are used to create the retina template as an ordered set of three-dimensional vectors. The ith feature (vessel segment) is represented by the triplet (p_i, t_i, c_i), where p_i is the *position* (in degrees around the annulus) of the segment, t_i is its *thickness* in degrees and c_i is its *crossing angle* (the angle in degrees the segment makes with a radial line drawn from the annulus, passing through the centroid of the segment).

These characteristics are the ones selected in [4]. Further research is needed to determine if other characteristics, such as the number of vessel segments, are also important. In our sample, this number had mean 18.45 and standard deviation ≈ 3.245, but the sample is too small to suggest a model distribution.

We analyse the templates obtained from our database to determine the parameters of the *retinal entropy model* developed in Section 4.3. For this model, it is necessary to assume that the retinal images are roughly pre-aligned. The results of Section 5.2 show this assumption may be too strong for current alignment and registration technology.

First we determine the sample distribution of each of the three components of a retinal feature independently. The frequency of occurrence of every possible value of position, thickness and crossing angle across every feature in every template of the database up to a resolution of 1° was noted. The frequency distributions are given in Figure 3.

4.3 The Retinal Entropy Model

We develop a theoretical model of the biometric templates informed by the empirical data. The key simplifying assumption we make is that the position, thickness, and crossing angle are all statistically independent of one another. This assumption appears reasonable given the data available. There are particular circumstances where this assumption is expected to break down. For example, we expect the thickest blood vessels to show a distinctive position distribution.

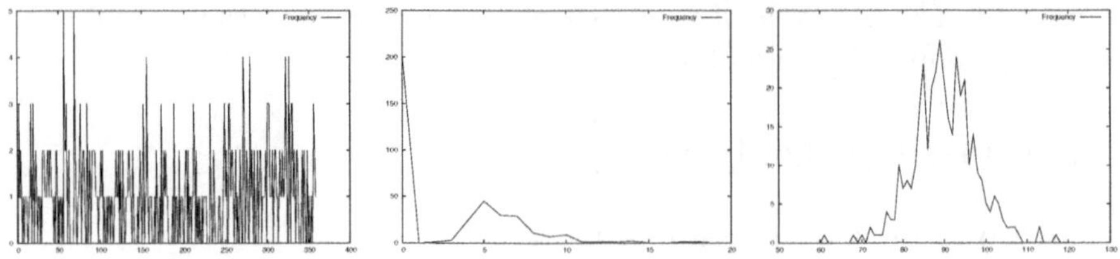

(a) Position (b) Thickness (c) Crossing Angle

Fig. 3. Frequency distributions of features

Figure 3(a), (b) and (c) show our observed distribution of position, thickness and crossing angle respectively. We chose to model the position as a uniform distribution and the crossing angle as a normal distribution. The crossing angles had a mean of 90° and a standard deviation of 7.5°. For the thickness we chose to combine a spiked distribution for the thin vessels containing 55% of the vessels, and a normal distribution with a mean of 6.4 and a standard deviation of 2.4 for the remaining 45% of medium to thick vessels. The overall distribution of features is then modelled as the product of the three distributions.

Templates from the same retina will vary due to movement of the head during retina image capture and variation in the number of features extracted from each capture. To compensate, tolerances have to be specified in each dimension of a feature vector. Each dimension is separately quantised. The resulting number of quanta controls the tradeoff between accuracy and leniency in a matching process. The vector of resolution levels for the position, thickness and crossing angle that generate the quantised space is the *quantisation triplet* $(p_{tol}, t_{tol}, c_{tol})$. It is important for any matching system to determine the smallest permissible tolerances that minimise intra-sample variation and still retain distinctiveness between different retinas.

The model product distribution described above is used to generate a large number of synthetic templates, randomly. This allows us to match a much larger population of templates than can be obtained from current sample databases. Note that the number of features per synthetic template varies according to the model product distribution rather than the sample distribution. The synthetic templates are matched, and their match characteristics provide the entropy predicted by the model. The general procedure is as follows:

- A quantization triplet $(p_{tol}, t_{tol}, c_{tol})$ is chosen.
- A match threshold s is chosen, where two templates are regarded as a match if they share at least s quanta containing a feature.
- A large number of pairs of templates N are generated and the number of matches m is recorded.

Provided N is large enough to result in actual matches, the entropy as predicted by the model is then

$$\log_2 N - \log_2 m.$$

For most of our tests N is of the order of $100,000,000$ and the match number m is recorded for all values of s.

5 Results and Analysis

5.1 Results for Comparison with [4]

We first chose a quantization that matched the algorithm described by the authors in [4] as closely as possible, and was expected to result in an acceptable false non-match rate (FNMR). For position we chose $p_{tol} = 1^\circ$ exactly as in [4]. Fortunately, most of the entropy results from variation in position and consequently this is the most critical choice. For vessel thickness we chose 4 quanta of width 5 pixels, where one of the quanta is too far down the tail of the normal distribution to contribute meaningfully. The remaining three effectively partition the vessels into *thin, medium* and *thick* as in [4]. Crossing angle is less directly comparable because in [4] the authors use a cosine-based correlation that cannot be directly related to our model. We chose 4 quanta of width 25° based around the 90° mean as the closest match to their correlation, given the slow decay of the cosine compared to the normally distributed crossing angle.

The results of running the model on this quantization choice over the range of values of s is shown in Figure 4 as the line labelled by quantisation triplet $(1, 5, 25)$. In [4] the authors find acceptable FNMR up to a threshold of 9 as measured in their matching algorithm. Due to the correlation matching and weighting they employ there is no direct relationship with our model. As the cosine-based correlation decays slowly and thick vessels are weighted by a factor of 2 we should expect that this is equivalent in our model to an s threshold of 4.5. At these threshold values our model predicts an entropy value in the range of at least 17.0 bits for this type of template. Such an entropy value would compare very favorably with fingerprints and quite well with iris (see Section 4.1).

As noted above, the choice of quantization of position is critical to ultimate entropy predicted by the model and the 1° accuracy used in [4] is particularly strict. Having worked with our data, we doubt that acceptable FNMR would result from so strict a choice. The authors of [4] stated that extra images were generated by rotating existing images and this may explain why they could use such a strict choice without harming the FNMR. It is unlikely this would be replicated on a more representative database.

5.2 Intra Sample Variation

An estimate of intra sample variation is required to determine the most suitable quantisation triplet. As our original database contained only one image per individual, we obtained 5 image samples of a different retina, extracted the features using Method 2 and created their templates. For various quantisation triplets the number of quantised features that matched between two samples and the average number of features that matched across the $\binom{5}{2} = 10$ different intra-sample comparisons were noted. Table 1 lists some of these quantisation triplets and the

corresponding average number of matched features between templates. Observe that the quantisation triplet $(10, 10, 10)$ gives the highest average number, about half, of feature matches between samples of the same retina.

5.3 Results and Analysis

The probabilities of finding a feature in each quantum was computed from our database. In this small dataset, only $3 - 5\%$ of quanta had non-zero probabilities and so were unable to use this distribution in the model directly. Consequently, we input each of the quantisation triplets in Table 1 into the theoretical model and used the assumed probability distributions in Section 4.3 to compute the entropy of a retina template at different match thresholds s. The results are given in Figure 4. They show a marked improvement in entropy at finer quantisation triplets. The feature distribution of position in Figure 3(a) confirms that position has a close to uniform distribution and it contributes significantly to retina template entropy. A finer quantisation in position will greatly enhance the maximum retina template entropy possible, but a practical limitation to this is the degree to which alignment and registration accuracy can be achieved. Farzin et al. [4] have performed their matching tests at $1°$ resolution for position at a match threshold of $6 - 7$ features. In Table 1 (quantisation triplet $(1, 5, 25)$) we show that such a fine quantisation on position may only allow an average of 2 features in common between samples of the same retina.

However, if alignment and registration technology improves sufficiently that position can be aligned to $1°$ accuracy, Figure 4 indicates that we should achieve over 20 bits of entropy, which will then be comparable to that of an iris template.

The third column in Table 1 shows the retina template entropies from Figure 4, for different quantisation triplets sampled at a threshold s corresponding to the average number of intra sample feature matches. Choosing the average number of matches as the threshold will imply a FNMR of close to 50%. This might overestimate the FNMR as the average number of matches has been obtained by conducting experiments on multiple samples of one retina only.

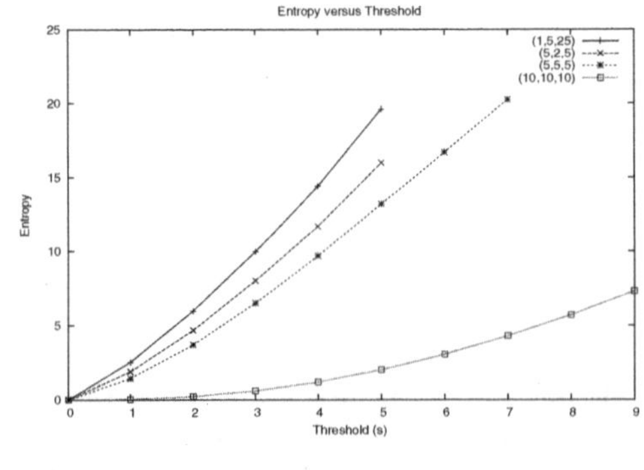

Fig. 4. Entropy (vertical) vs threshold (horizontal) for varying quantisation triplets

Table 1. Quantisation Triplets, Intra Sample Match Average and Predicted Entropies

Quantisation Triplet $(p_{tol}, t_{tol}, c_{tol})$	Average Number of Intra Sample Matches	Predicted Entropy
$(10, 10, 10)$	9	7.3
$(5, 5, 5)$	6	16.7
$(5, 2, 5)$	4	11.7
$(1, 5, 25)$	2	6.0

The quantisation triplet $(5, 5, 5)$ gives the highest entropy of 16.7 bits at the match threshold of $s = 6$. This is significantly higher than that of fingerprint minutiae-based templates and close to that of iris templates.

5.4 Conclusions and Future Work

We have proposed a preliminary theoretical model, with three degrees of freedom, to compute the entropy of retina templates. Choosing quantisation triplets from empirical studies, we determined that retina templates having entropies greater than 16 bits can be created. This compares favourably with minutiae-based fingerprint templates but the model must be tested against much larger retina databases before general conclusions can be drawn. We have identified that position is the principle contributor to retina template entropy. This implies that the maximum possible retina template entropy is limited only by the accuracy of the registration and alignment of intra-retina samples and the limitations introduced by image processing.

Currently we are exploring the use of larger databases to obtain an empirical joint probability distribution to tune our theoretical model and compare with these results. We expect tests on larger databases to exhibit higher entropy, more comparable to that in an iris template. We may incorporate other characteristics, such as number of features, into the model. We also intend to construct retina templates using the three other methods of Section 2 and evaluate their template entropies using a similar modelling technique.

Acknowledgement. This research was supported by an RMIT College of SET Seeding Grant.

References

1. Arakala, A., Horadam, K.J., Boztas, S.: Practical Considerations for Secure Minutiae Based Templates. In: Proc. 2008 Biometrics Symposium, Tampa, Florida, September 23-25, pp. 53–58. IEEE Press, Los Alamitos (2008)
2. Can, A., Shen, H., Turner, J.N., Tanenbaum, H.L., Roysam, B.: Rapid automated tracing and feature extraction from live high-resolution retinal fundus images using direct exploratory algorithms. IEEE Trans. Inform. Technol. Biomed. 3(2), 125–138 (1999)

3. Daugman, J.: Probing the Uniqueness and Randomness of IrisCodes: Results From 200 Billion Iris Pair Comparisons. Proc. IEEE 94(11), 1927–1935 (2006)
4. Farzin, H., Abrishami-Moghaddam, H., Moin, M.: A Novel Retinal Identification System. EURASIP Journal on Advances in Signal Processing, Article ID 280635, 10 (2008), doi:10.1155/2008/280635
5. Fritzsche, K., Can, A., Shen, H., Tsai, C., Turner, J., Tanenbuam, H., Stewart, C., Roysam, B.: Automated model based segmentation, tracing and analysis of retinal vasculature from digital fundus images. In: Suri, J.S., Laxminarayan, S. (eds.) State-of-The-Art Angiography, Applications and Plaque Imaging Using MR, CT, Ultrasound and X-rays, pp. 225–298. Academic Press, London (2003)
6. Hill, R.: Retinal Identification. In: Jain, A.K., Bolle, R.M. (eds.) Biometrics: Personal Identification in Networked Society, pp. 123–142. Kluwer Academic Press, Boston (1999)
7. Hill, R.: Apparatus and method for identifying individuals through their retinal vasculature patterns, US Patent No. 4109237 (1978)
8. Hill, R.: Rotating beam ocular identification apparatus and method, US Patent No. 4393366 (1983)
9. Marino, C., Penedo, M., Carreira, M., Gonzalez, F.: Retinal angiography based authentication. In: Sanfeliu, A., Ruiz-Shulcloper, J. (eds.) CIARP 2003. LNCS, vol. 2905, pp. 306–313. Springer, Heidelberg (2003)
10. Ortega, M., Penedo, M., Marino, C., Carreira, M.: Similarity metrics analysis for feature point based retinal authentication. In: Campilho, A., Kamel, M.S. (eds.) ICIAR 2008. LNCS, vol. 5112, pp. 1023–1032. Springer, Heidelberg (2008)
11. Sofka, M., Stewart, C.V.: Retinal Vessel Centerline Extraction Using Multiscale Matched Filters, Confidence and Edge Measures. IEEE Transactions on Medical Imaging 25(12), 1531–1546 (2006)
12. Simon, C., Goldstein, I.: A new scientific method of identification. New York State Journal of Medicine 35(18), 901–906 (1935)
13. Tabataee, H., Fard, A., Jafariani, H.: A novel human identifier system using retina image and fuzzy clustering approach. In: Proceedings of the 2nd IEEE International Conference on Information and Communications Technologies (ICTTA 2006), Damascus, Syria, pp. 1031–1036 (April 2006)
14. Tower, P.: The fundus oculi in monozygotic twins: report of six pairs of identical twins. Archives of Ophthalmology 54, 225–239 (1955)
15. Xu, Z., Guo, X., Hu, X., Cheng, X.: The blood vessel recognition of ocular fundus. In: Yeung, D.S., Liu, Z.-Q., Wang, X.-Z., Yan, H. (eds.) ICMLC 2005. LNCS, vol. 3930, pp. 4493–4498. Springer, Heidelberg (2006)
16. Xu, Z., Guo, X., Hu, X., Chen, X., Wang, Z.: The identification and recognition based on point for blood vessel of ocular fundus. In: Zhang, D., Jain, A.K. (eds.) ICB 2006. LNCS, vol. 3832, pp. 770–776. Springer, Heidelberg (2006)
17. Zhu, Y., Dass, S.C., Jain, A.K.: Statistical models for assessing the individuality of fingerprints. IEEE Transactions on Information Forensics and Security 2(3), 391–401 (2007)

Lips Recognition for Biometrics

Michał Choraś

Institute of Telecommunications
University of Technology & Life Sciences, Bydgoszcz, Poland
chorasm@utp.edu.pl

Abstract. One of the most interesting emerging method of human identification, which originates from the criminal and forensic practice, is human lips recognition. In this paper we consider lips shape features in order to determine human identity. The major contribution of this paper are novel geometrical parameters developed in order to describe human lips shape for biometric applications.

1 Rationale

Biometric methods of human identification have gained much attention recently, mainly because they easily deal with most problems of traditional identification. In biometric human identification systems users are identified by who they are, not by something they have to remember or carry with them.

Nowadays, there are many well known, already implemented methods of human identification (face, iris, retina, etc.), but still novel and innovative solutions are proposed and needed. Such situation is most apparent due to a large-scale biometrics problem, where even very successful methods fail in real-life security systems, such as access control in airports.

Even strong methods with errors of $FNMR = 2\%$ cause problems when deployed in a large scale: for 100 000 passengers a day, such error effects in 2000 false rejects per day!

The problem is that 99,99% accuracy in laboratory conditions does not result in the same accuracy and efficiency in a large scale and non-controlled, real-life environments. So far the scaling issues in biometrics are not formalized or solved.

Therefore, to cover up for possible errors, other biometric modalities should be used, so that the overall accuracy and reliability of the hybrid systems is higher. Even though, these new emerging methods do not give results comparable with face or fingerprint systems, still such methods may improve accuracy of the hybrid systems, especially for people for whom standard systems fail to be effective.

Current trends in searching for novel modalities to identify humans are:

- 3D imaging (face, finger, ear),
- thermal imaging (face) and
- forensic science.

M. Tistarelli and M.S. Nixon (Eds.): ICB 2009, LNCS 5558, pp. 1260–1269, 2009.
© Springer-Verlag Berlin Heidelberg 2009

Some of the methods recently termed as "emerging" have now matured, to mention ear biometrics (motivated by forensic science).

Other new and emerging biometric modalities are [1][2][3][4]:

- human scent recognition,
- EEG biometrics,
- skin spectroscopy,
- knuckles texture,
- finger-veins and
- finger-nails recognition.

The major contribution of this paper are novel geometrical parameters developed in order to describe human lips shape for biometrics.

In our approach we calculate color features and we merge them with shape features of the binarized lips. We calculate color statistics and moments as well as a set of standard geometrical parameters and the moments of Hu and Zernike. Moreover, we developed original lips shape descriptors that enhance the effectiveness of our lips recognition method for biometric applications.

Lips recognition as biometrics is introduced and motivated in Section 2. In Section 3 related research developments are briefly overviewed. In Section 4.1 our algorithm for lips detection with sample results is presented. Original shape features developed for lips biometrics are proposed in Section 4.3. Experimental results, future work and conclusions are given thereafter.

2 Lips Biometrics

One of the most interesting emerging method of human identification, which originates from the criminal and forensic practice (*cheiloscopy*), is human lips recognition [5][6].

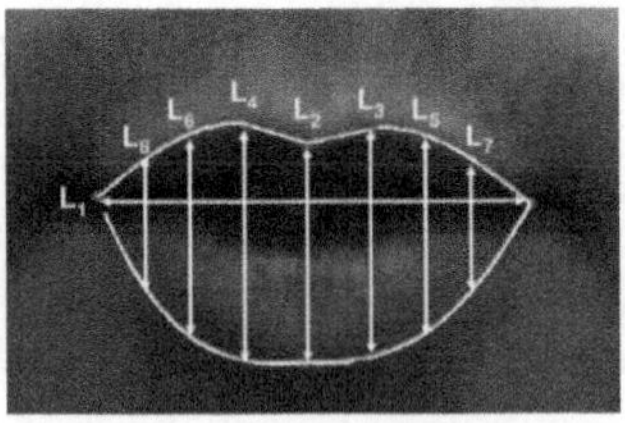

Fig. 1. Lips shape parameters proposed by Cetingul et al. [11]

The fact that lips features may be unique for humans have been confirmed by Yasuo Tsuchihasi and Kazuo Suzuki in their studies at Tokio University (1968-1971). They examined 1364 subjects from 3 to 60 years of age of both genders. Their research proved that lips characteristics are unique and unchangeable (stable) for each examined person [5][7].

In another research lip-prints have been used to support the sex determination of the examined subject [8].

Lip-prints characteristics have been also used in forensics experts and criminal police practice. Such approach to human identity confirmation is used by the police and have been included as a subdiscipline of dactyloscopy.

Human lips features may be implemented in various scenarios, such as:

- speech recognition,
- multimodal audio-video speech recognition,
- speaker identification,
- multimodal audio-video speaker identification,
- lip-reading and finally,
- lips based human identification based on the static mouth/face images.

In general, in accordance to the application scenario, lips features can be divided into three different categories: lips texture features, lips shape features and lips motion features.

Hereby we consider human identification based on the static mouth/face images.

Using lips as modality for human identification has the following advantages:

1. Lips biometrics is passive biometrics - user/subject interaction is not needed. Images may be acquired from the distance without the knowledge of the examined person.
2. Lips biometrics is anatomical - better results are expected than in behavioral biometrics.
3. Lips are usually visible - not hidden/overcast by anything.
4. Lips may be implemented in hybrid lips-face or lips-voice biometric systems.

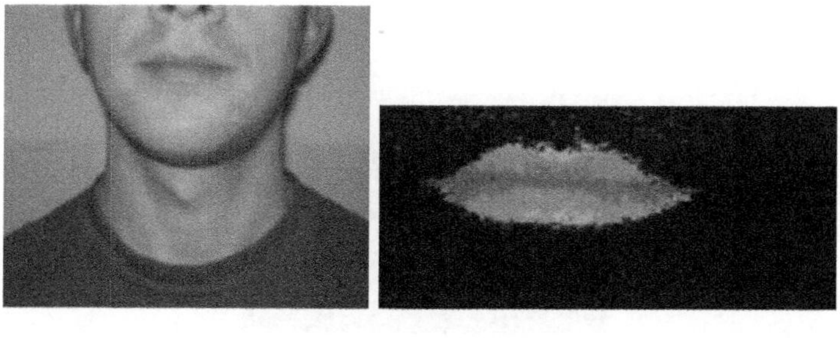

Fig. 2. Example of lower face and extracted corresponding lips area

3 Related Work

While manually examining human lips characteristics, most often, the anatomical patterns on the lips are taken into account. For example the pioneer of *cheiloscopy*, professor J. Kasprzak defined 23 lip patterns [5]. Such patterns

(lines, bifurcations, bridges, pentagons, dots, lakes, crossings, triangles etc.) are similar to fingerprint, iris or palmprint patterns.

However, in biometric systems based on image analysis techniques, such line-topology features cannot be used since it is hard to extract them from acquired images.

Therefore in our approach we do not use lip-prints features, but we focus on characteristics extracted from lips/face images. Such images can be acquired from a certain distance by a standard camera in controlled or uncontrolled environment.

Automated lips biometrics have not been extensively researched so far, but there are some published approaches to human lips feature extraction.

Gomez et al. achieved some very promising results using geometrical parameters, HMM and PCA methods [9]. They reported 96.2% recognition rate on a database of 500 images (50 subjects with 10 images per person).

Cetingul et al. proposed to explore $2d - DCT$ coefficients as lips texture features and used them in multimodal speaker/speech recognition system [10].

They also developed eight lips shape parameters based on the lips geometry. The maximum horizontal distance and the 7 vertical distances (from the Cupid's bow to the lower lip boundary characteristic points) are calculated (Figure 1)[11].

4 A Geometrical Approach to Lips Biometrics

4.1 Lips Detection

In the first step lips are detected from face images. Then we perform segmentation, binarization and size normalization. Most of known lips detection methods proposed so far, had been designed for speech recognition, lips tracking and lip-reading [11][12][13][14][15][16].

We modified a method based on color discriminates proposed by Kubanek [17] and we applied it to the HSV color space.

The condition for lips detection is presented in Eq.1.

$$\begin{cases} H \geq 334 \vee H \leq 10 \\ S \leq 25 \end{cases} \tag{1}$$

The sample results of our lips detection algorithm from lower face images are presented in Figure 2.

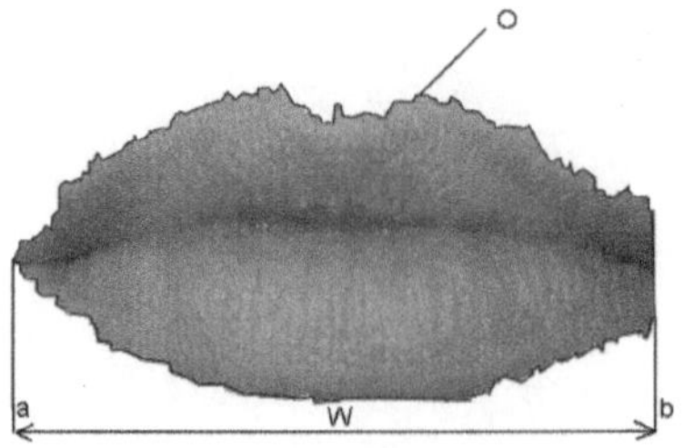

Fig. 3. Lips width to perimeter ratio

4.2 Lips Shape Feature Extraction

After lips detection stage, shape features of the binarized lips images are calculated.

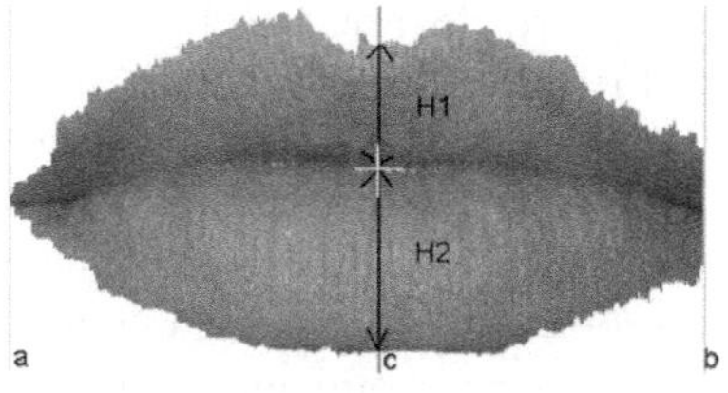

Fig. 4. Upper to lower lip height ratio

We decided to calculate geometrical parameters of detected lips area. In our previous work, we calculated central moments, Zernike moments and Hu moments [18]. Moreover, we used standard geometrical shape parameters to describe human lips [19]:

- Malinowska ratio
- Feret ratio
- Blair-Bliss ratio
- Danielsson ratio
- Haralick ratio
- Lp1 ratio
- Lp2 ratio.

4.3 Original Lip Shape Parameters

Furthermore, in this paper we present our own original parameters developed for the lips biometrics.

9 novel parameters, developed especially for lips biometrics and describing lips' shape are as follows:

1. lips width to perimeter ratio WO given by: $WO = W/O$ where W is the lips width and O is lips perimeter, as presented in Figure 3.

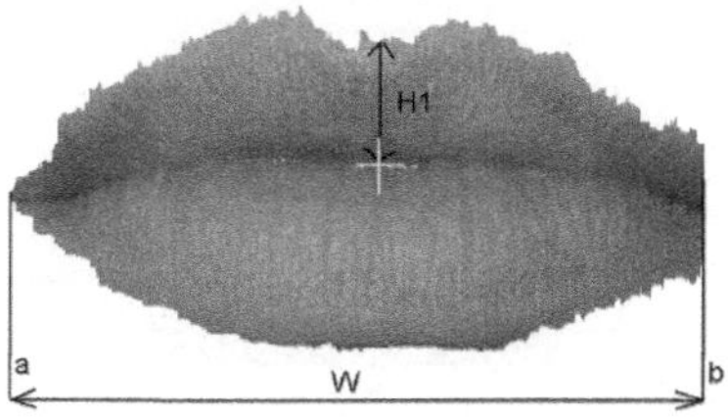

Fig. 5. Upper lip height to width ratio

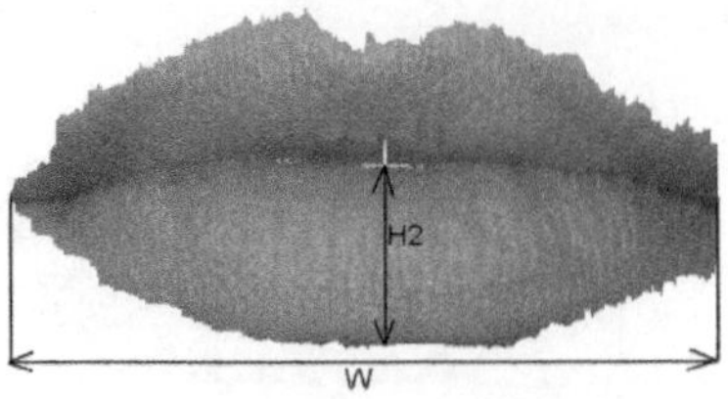

Fig. 6. Lower lip height to width ratio

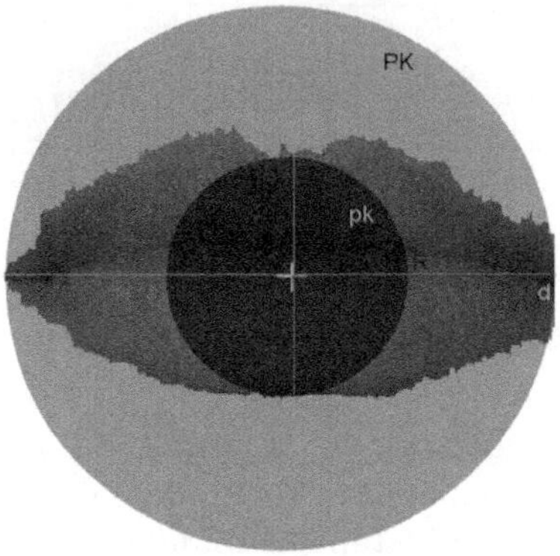

Fig. 7. Inner to outer circle ratio

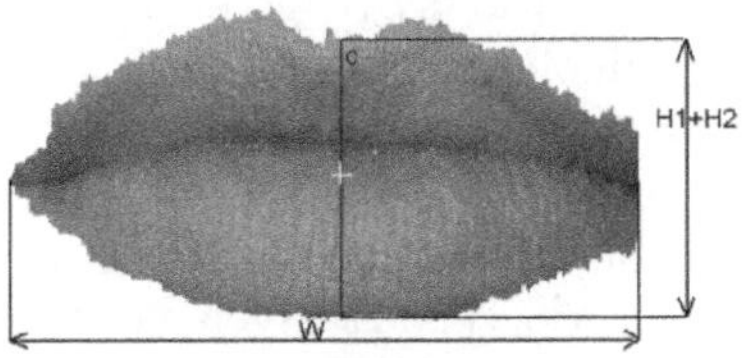

Fig. 8. Width to middle height ratio

Fig. 9. Left side upper to lower lip convexity ratio

Fig. 10. Right side upper to lower lip convexity ratio

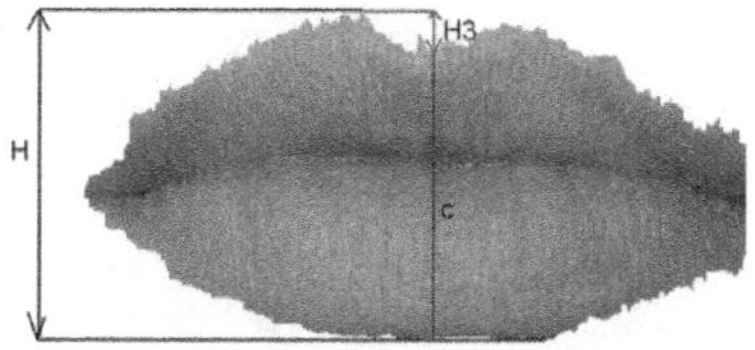

Fig. 11. Indent ratio

2. Upper to lower lip height ratio ULH given by: $ULH = H1/H2$ where $H1$ is the upper lip height calculated in the middle column of the lips, and $H2$ is the lower lip height calculated in the middle column of the lips (Figure 4).
3. Upper lip height to width ratio ULW given by: $ULW = H1/W$ where $H1$ is the upper lip height calculated in the middle column of the lips and W is the lips width (Figure 5).
4. Lower lip height to width ratio LLW given by: $LLW = H2/W$ where $H2$ is the upper lip height calculated in the middle column of the lips and W is the lips width (Figure 6).
5. Inner to outer circle ratio $C2C$ given by: $C2C = pk/PK$ where pk is the inner circle and PK is the outer circle as shown in Figure 7.
6. Width to middle height ratio WMH given by: $WMH = W/(H1 + H2)$ as presented in Figure 8.
7. Left side upper to lower lip convexity ratio LC given by: $LC = m1/m2$ where $m1$ is the upper lip left side convexity and $m2$ is lower lip left side convexity as presented in Figure 9.
8. Right side upper to lower lip convexity ratio RC given by: $RC = n1/n2$ where $n1$ is the upper lip left side convexity and $n2$ is lower lip left side convexity as presented in Figure 10.
9. Indent ratio IR given by: $IR = H3/H$ where H is the lips height and $H3$ is the height of the indent calculated according to the Figure 11.

5 Experimental Results

In the experiments, we used our own database of lower face images. We acquired 3 images per person. The lips database used in our experiments consists of 3 images from 38 people (114 images).

After experiments and evaluation it turned out that adding novel geometrical parameters to the feature vector increased system's efficiency (in comparison to our previous work (without novel geometrical parameters) [18][19]).

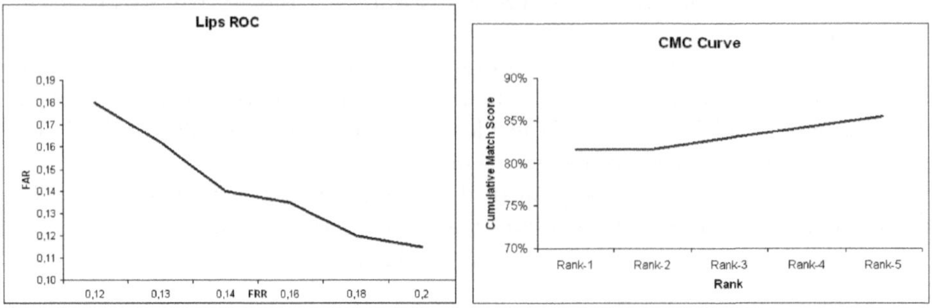

Fig. 12. ROC-Curve (left) and CMC-curve (right) of lips biometric system

Human lips recognition results without and with the novel developed geometrical parameters are shown in the Table 1.

Table 1. Lips Rank-1 recognition rate results

method	number of tests	correct acceptances	false rejections	Rank-1
without novel parameters	76	58	18	76%
with novel parameters	76	62	14	82%

So far we have achieved the recognition rate (Rank-1) of 82%.

The ROC Curve of our lips biometrics system is presented in Figure 12-left.

The CMC curve describing our system is presented in Figure 12-right. The Rank-4 and Rank-5 recognition rate increased up to 86%.

At this moment we work on using symmetry as a feature which may help to better describe human lips.

We use horizontal and vertical symmetry to obtain 2 symmetrical images and 2 'delta' images (difference between the originally acquired subject's image and corresponding symmetry images).

Now we work on selecting efficient features from symmetry and *delta* images. We also plan to use Discrete Symmetry Transform to find lips *fiducial* points to which Gabor Jets may be applied [20].

6 Conclusions

In this paper we presented feature extraction methods and introductory results of automated human lips biometric system.

In our research novel lips shape descriptors had been used to determine human identity.

We have achieved promising recognition results for well detected lips images, which motivate our further research in this area.

We plan to extend our research with the development of new lips features. We are investigating the texture and symmetry features of the human lips. Moreover, we try to improve the lips detection algorithm with the Active Shape Models approach.

Lips biometrics could be used to enhance effectiveness of other well-known biometrics, by its implementation in multimodal systems. Since most of the methods have some drawbacks, the idea of building multimodal (hybrid) biometric systems is gaining lot of attention [21].

Lips biometrics seems to be a natural choice to support well known methods like voice and face recognition in emerging applications such as access control, border security and recognition for human-vehicle interaction [22].

References

1. Prabhakar, S., Kittler, J., Maltoni, D., O'Gorman, L., Tan, T.: Introduction to the Special Issue on Biometrics: Progress and Directions. IEEE Trans. on PAMI 29(4), 513–516 (2007)
2. Goudelis, G., Tefas, A., Pitas, I.: On Emerging Biometric Technologies. In: Proc. of COST 275 Biometrics on the Internet, Hatfield, UK, pp. 71–74 (2005)
3. Morales, A., Ferrer, M.A., Travieso, C.M., Alonso, J.B.: A knuckles texture verification method in a transformed domain. In: Proc. of 1st Spanish Workshop on Biometrics, Girona, Spain (2007) (on CD)
4. Choraś, M.: Emerging Methods of Biometrics Human Identification. In: Proc. of ICICIC 2007 - Kummamoto, Japan. IEEE CS Press, Los Alamitos (2007)
5. Kasprzak, J., Leczynska, B.: Cheiloscopy. Human Identification on the Basis of Lip Prints (in Polish). CLK KGP Press, Warsaw (2001)
6. Kasprzak, J.: Forensic Otoscopy (in Polish). University of Warmia and Mazury Press (2003)
7. Tsuchihasi, Y.: Studies on Personal Identification by Means of Lip Prints. Forensic Science 3(3) (1974)
8. Sonal, V., Nayak, C.D., Pagare, S.S.: Study of Lip-Prints as Aid for Sex Determination. Medico-Legal Update 5(3) (2005)
9. Gomez, E., Travieso, C.M., Briceno, J.C., Ferrer, M.A.: Biometric Identification System by Lip Shape. In: Proc. of Carnahan Conference on Security Technology, pp. 39–42 (2002)
10. Cetingul, H.E., Yemez, Y., Erzin, E., Tekalp, A.M.: Multimodal speaker/speech recognition using lip motion, lip texture and audio. Signal Processing 86, 3549–3558 (2006)
11. Cetingul, H.E., Yemez, Y., Erzin, E., Tekalp, A.M.: Discriminative Analysis of Lip Motion Features for Speaker Identification and Speech-Reading. IEEE Trans. Image Processing 15(10), 2879–2891 (2006)
12. Ouyang, H., Lee, T.: A New Lip Feature Representation Method for Video-based Bimodal Authentication. In: Proc. of NICTA-HCSNet Multimodal User Interaction Workshop, Sydney, Australia, vol. 57, pp. 33–37 (2005)
13. Leung, S.H., Wang, S.L., Lau, W.H.: Lip Image Segmentation Using Fuzzy Clustering Incorporating an Elliptic Shape Function. IEEE Trans. Image Processing 13(1), 51–62 (2004)

14. Nowak, H.: Lip-reading with Discriminative Deformable Models. Machine Graphics and Vision 15(3-4), 567–576 (2006)
15. Yau, W.H., Kumar, D.K., Chinnadurai, T.: Lip-Reading Technique Using Spatio-Temporal Templates and Support Vector Machines. In: Ruiz-Shulcloper, J., Kropatsch, W.G. (eds.) CIARP 2008. LNCS, vol. 5197, pp. 610–617. Springer, Heidelberg (2008)
16. Joumun, F.E., Gnanayutham, P., George, J.: Multimedia Interfaces for BSL Using Lip Readers. In: Miesenberger, K., Klaus, J., Zagler, W.L., Karshmer, A.I. (eds.) ICCHP 2008. LNCS, vol. 5105, pp. 663–669. Springer, Heidelberg (2008)
17. Kubanek, M.: Method of Speech recognition and Speaker Identification with Use Audio-Visual of Polish Speech and Hidden Markov Models. In: Saeed, K., et al. (eds.) Biometrics, Computer Security Systems and Artificial Intelligence Applications, pp. 45–55. Springer, Heidelberg (2006)
18. Choraś, M.: Human Lips Recognition. In: Kurzyñski, M., et al. (eds.) Computer Recognition Systems 2. Advances in Soft Computing, pp. 838–843. Springer, Heidelberg (2007)
19. Choraś, M.: Human Lips as Emerging Biometrics Modality. In: Campilho, A., Kamel, M.S. (eds.) ICIAR 2008. LNCS, vol. 5112, pp. 994–1003. Springer, Heidelberg (2008)
20. Choraś, M., Andrysiak, T.: Symmetry-based Salient Points Detection in Face Images. In: Rutkowski, L., Tadeusiewicz, R., Zadeh, L.A., Żurada, J.M. (eds.) ICAISC 2006. LNCS, vol. 4029, pp. 758–767. Springer, Heidelberg (2006)
21. Ross, A., Nandakumar, K., Jain, A.K.: Handbook of Multibiometrics. International Series on Biometrics. Springer, Heidelberg (2006)
22. Erzin, E., Yemez, Y., Tekalp, A.M., Ercil, A., Erdogan, H., Abut, H.: Multimodal Person Recognition for Human-Vehicle Interaction. IEEE Multimedia 13(2), 18–31 (2006)

Biometrics Method for Human Identification Using Electrocardiogram

Yogendra Narain Singh[1] and P. Gupta[2]

[1] Institute of Engineering & Technology, Lucknow, India
singhyn@gmail.com
[2] Indian Institute of Technology Kanpur, Kanpur, India
pg@iitk.ac.in

Abstract. This work exploits the feasibility of physiological signal electrocardiogram (ECG) to aid in human identification. Signal processing methods for analysis of ECG are discussed. Using ECG signal as biometrics, a total of 19 features based on time interval, amplitudes and angles between clinically dominant fiducials are extracted from each heartbeat. A test set of 250 ECG recordings prepared from 50 subjects ECG from Physionet are evaluated on proposed identification system, designed on template matching and adaptive thresholding. The matching decisions are evaluated on the basis of correlation between features. As a result, encouraging performance is obtained, for instance, the achieved equal error rate is smaller than 1.01 and the accuracy of the system is 99%.

1 Introduction

Poets for long have been using it as a parameter to measure human emotions, where as for doctors it has been a health indicator. But now research also identifies heartbeat as a biometrics measure which can be used for human verification and identification.

Biometrics technology is an automated use of unique patterns of physiological (e.g., face, fingerprint, iris and hand-geometry) and behavioral (e.g., signature, gait) characteristics present in human beings to determine or verify the identity of individuals. Security is a prime concern of the modern society. From local in-house setting to more global scope the concerns of identity theft problems are growing in today's interconnected world. To ensure safe and secure environment, biometrics technology is now being used in many commercial, government and forensic applications. The success of using biometrics technology is that the biometrics characteristics exhibit unique patterns which are difficult to copy, share or distribute. The technology is more reliable and user friendly. However, every biometrics has its strengths and weaknesses. None of them meets the requirements of all the applications efficiently. This research shows the biometrics use of ECG can aid to more accurate identification results among all biometrics characteristics used at present.

The Electrocardiogram (ECG) is a physiological signal generated by electrical activity of heart. It is a non-invasive tool used by cardiologists to diagnose

M. Tistarelli and M.S. Nixon (Eds.): ICB 2009, LNCS 5558, pp. 1270–1279, 2009.
© Springer-Verlag Berlin Heidelberg 2009

cardiac diseases such as to recognize electrolyte abnormalities and electrical or structural cardiac dysfunctions [1]. In the recent past, only few studies treating ECG as biometrics can be found in the literature. Biel *et al.* [2] have conducted ECG comprising biometrics experiment on 20 subjects using twelve features of their heartbeats. Shen *et al.* [3] have investigated the feasibility of ECG as a new biometrics for identity verification. They have evaluated 20 individuals on seven heartbeat features, mostly extracted from QRS complex. Template matching and decision-based neural network techniques have been used for identity verification. Issues on these studies are mainly the extraction of ECG features and accuracy, selection of consistent features and investigation of ECG to change in physiology of heart. In a recent study, Israel *et al.* [4] have focused on more physiologically found features describing the characteristic timings of ECG signal. On 29 subjects, they have investigated the influence of different anxiety states on the identification of humans by their heartbeat characteristics.

The objectives of this work are two fold: firstly, it discusses the signal preprocessing methods for the automatic delineation of ECG and secondly, it quantifies the potential of ECG as a biometrics for individual identification. ECG delineation process usually concerns signal analysis and diagnostic classification [5] that carries in a series of steps. These steps are: (1) signal acquisition, includes digitization, sampling and filtering, (2) data transformation, includes finding of complexes, (3) data representation, a process of waveform delineation and (4) feature extraction, the measurement of amplitude and interval features between different diagnostic fiducials.

Once the ECG is delineated, features of classes time interval, amplitude and angle between different diagnostic fiducials are extracted from each heartbeat. The extracted features are normalized and to make them independent from changes in heart rate. The biometrics experiment is conducted on the consistent features using proposed identification model, designed on pattern recognition and adaptive thresholding techniques.

The reminder of this paper is organized as follows. The schematic of biometrics system for individual identification on ECG is presented in Section 2. A high level summary of the techniques utilized for automatic ECG delineation to prepare ECG feature set including normalization of features are given in Section 3. A detailed description is next given for proposed identification model, in Section 4. In order to evaluate the performance of identification system, biometrics results are summarized in Section 5. Finally, concluding remarks are presented in Section 6.

2 Biometrics Method

The schematic description of individual identification system on ECG is shown in Figure 1. The method is implemented in a series of steps: starting from ECG signal acquisition, preprocessing that include correction of signal from noise and noise-artifacts, ECG delineation that include detection of waveforms and their dominant fiducials from each heartbeat, feature extraction that include

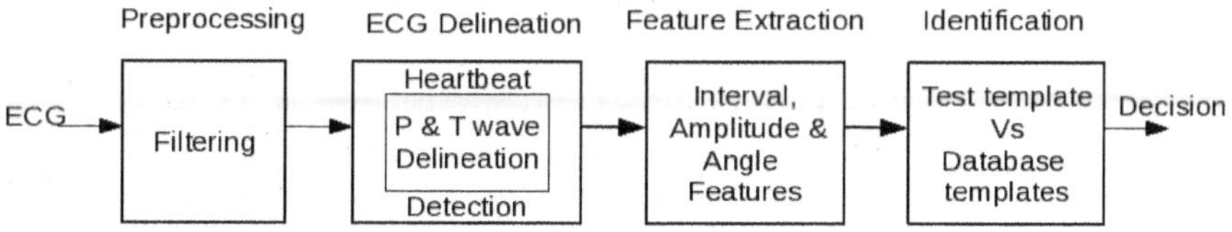

Fig. 1. Schematic of individual identification system on ECG

extraction of time interval, amplitude and angle features from dominant fiducials, followed by identification and decision making. The identification model utilizes template matching and adaptive thresholding technique which is evaluated on the basis of correlation between features.

3 Automated Feature Set Detection

The ECG data is acquired from individuals and subsequently it is digitized. The preprocessing of ECG signal involves correction from low and high frequency noises. Low frequency noise is resulted from baseline oscillations, body movements and respiration while high frequency noise is caused from power line interferences and digitization of analog potential [6]. Digital filters of linear phase characteristics are employed in this experiment. Regarding ECG delineation, it involves the detection of dominant complexes in a heartbeat such as QRS complex, P and T waves from the signal. The heartbeats are detected using a QRS complex delineator. Once a heartbeat is detected a temporal search windows are set before and after a QRS complex to search for other waveforms. The delineation of P and T waveforms are performed by their respective delineators. Found dominant fiducials are later used in preparing the feature set.

3.1 Automated ECG Delineation

QRS Complex Delineation. The QRS complex delineator is implemented using the technique proposed by Pan and Tompkins [7] with some improvements. It employs digital analysis of slope, amplitude and width information of ECG waveforms. The fiducials of QRS complex such as QRS_{onset} and QRS_{offset} are delineated according to the location and convexity of R peak.

P Wave Delineation. P wave is a low amplitude wave and has low signal to noise ratio. It shows atrium function in heart. The delineation technique utilized in this study works directly over the digitized ECG without compromising the accuracy of detected fiducials. It uses first derivative approach for P wave delineation. The first derivative, y_{nT} at time instant T is calculated using the following time difference equation,

$$y_{nT} = -2 * x_{(n-2)T} - x_{(n-1)T} + x_{(n+1)T} + 2 * x_{(n+2)T} \quad (1)$$

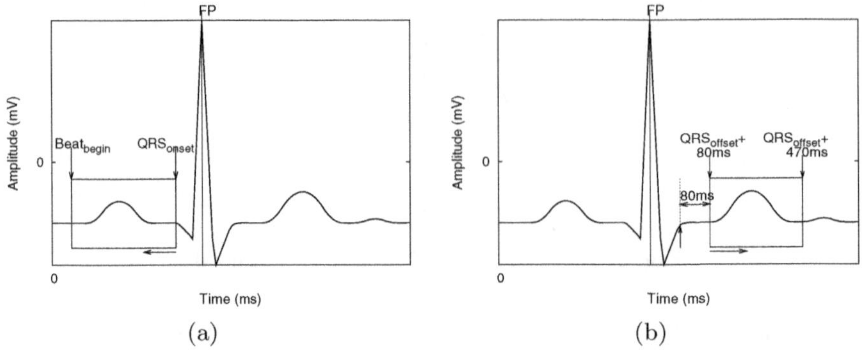

Fig. 2. Setting of search windows for: (a) P wave delineation and (b) T wave delineation

where x_{nT} represent the data sample of size n at discrete instance of time T which is set to $10ms$ at sampling rate of $100Hz$.

In order to determine P wave and its end fiducials, a search window is set prior to the beginning of QRS complex (QRS_{onset}). The search window that approximately contained P wave is set heuristically and extended from QRS_{onset} to the beginning of heartbeat ($Beat_{begin}$) as shown in Figure 2 (a). The $Beat_{begin}$ fiducial can be determined by searching of first isoelectric sample prior to the start of atrium deflection.

The detection of P wave is performed using adaptive thresholding technique. It dynamically adjusts the slope threshold with the consideration of high frequency noise present in the beat. The dominant fiducials of P wave: P_{peak} is determined by finding the local maximum in the surrounding region while P_{onset} and P_{offset} are determined by finding the local minimum at the ends within the search region. Some adjustments are needed if there are the existence of inflections near to the ends of the waveform.

T Wave Delineation. T wave is concerned to ventricles repolarization. The problem with T wave delineation is its repolarization cycle which terminates faster and comprising lower stimulation in comparison to the noise artifacts present in the beat. This makes detection of T wave end fiducial (T_{offset}) more cumbersome. In order to achieve the reliable performance of T wave delineation the signal is first corrected from oscillatory patterns of reference potential. A recursive lowpass filter is of following time difference equation,

$$y_{nT} = 2 * y_{(n-1)T} - y_{(n-2)T} + x_n - 2 * x_{(n-4)T} + x_{(n-8)T} \tag{2}$$

is utilized, where x_n represents the data sample of size n at discrete instant of time T. At sampling frequency of 100 Hz, T is found to be 10 ms and processing delay caused by the filter is nearly 30 ms.

Prior to start the delineation process, a search window is defined that approximately contained T wave. The boundaries of search window are set heuristically relative to QRS_{offset} fiducial, extended from $QRS_{offset} + 80ms$ to $QRS_{offset} +$

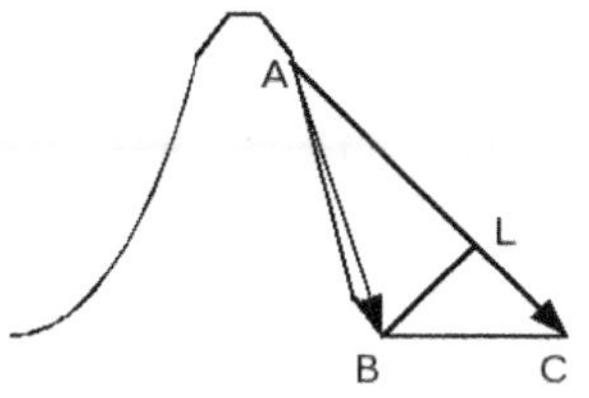

Fig. 3. Determination of T wave end fiducials

470ms as shown in Figure 2 (b). It is observed from the morphology of T wave that a time segment of 80ms just after QRS_{offset} fiducial usually concerns the time prior to repolarization of ventricles that shows the negligible stimulation. Similarly, the right boundary of the search window is set according to the duration of depolarization to repolarization of the ventricles. The dominant fiducial corresponding to peak of T wave is determined using the technique of time derivative and adaptive thresholding, as similar to P wave with some adjustments.

The end fiducials of T wave are determined efficiently using the analysis of its waveform curvature. This analysis is based on assumption that the portion of curvature near to T wave ends is convex. The end fiducials of T wave are determined by tracking the signal downhill and finding the location of minimum radius of curvature as shown in Figure 3. By fixing of time difference between A, B and A, C the minimum radius of curvature is found by maximizing BL using vector cross product between two directed line segments i.e., $BL = \frac{|\vec{AC} \times \vec{AB}|}{|\vec{AC}|}$. Using this technique, fiducials are found more robust to local noise present in the beat.

3.2 Feature Set Preparation

In order to carry out the biometrics experiment a feature set is prepared from extracted dominants fiducials of P, QRS complex and T waves. The attributes of feature set are grouped into interval features, amplitude features and angle features. A total of 19 attributes listed in Table 1 are extracted from each heartbeat. The positions of different attributes are shown in Figure 4.

Prior to process the feature set for biometrics experiment, it is required to account for changes in these individual attributes with changes in heart rate. The heart rate varies due to changes in pressure inside heart and ventricular volume. The changes in heart rate consequently changes the duration of P wave, PR interval and QT interval. Thus, attributes related to P and T waves are normalized by dividing the beat length ($PR_I + QT_{CI}$) while the RQ and RS are used as raw attributes. Finally, normalized attributes are represented relative positions of fiducials within a heartbeat. The attributes related to peak fiducials of different waveforms are fairly invariant with changes in heart rate. As a result, raw attributes of amplitude and angle features are used in the feature set preparation.

Table 1. Considered groups of attribute features are selected from ECG dominant fiducials ($\star$: RR is used for QT interval correction)

Group Label	Features	Representation
Interval Features	PR interval	PR_I
	PR segment	PR_S
	corrected-QT interval	QT_{CI}
	ST segment	ST_S
	ST interval	ST_I
	R_{peak} to T_{onset} segment	RT_L
	R_{peak} to P_{peak} segment	RP
	R_{peak} to P_{offset} segment	RP_R
	R_{peak} to Q_{peak} segment	RQ
	R_{peak} to S_{peak} segment	RS
	R_{peak} to P_{onset} segment	RP_L
	R_{peak} to T_{offset} segment	RT_R
	RR interval*	RR
Amplitude Features	RQ amplitude	RQ_A
	RS amplitude	RS_A
	RP amplitude	RP_A
	RT amplitude	RT_A
Angle Features	Angle Q	$\angle Q$
	Angle R	$\angle R$
	Angle S	$\angle S$

4 Identification Model

In order to carry out the biometrics experiment a new identification model is proposed. The model is designed on pattern matching and adaptive thresholding technique which is evaluated on the basis of the correlation between corresponding attributes of the feature vectors. Two ECG records are declared to be matched if their feature sets are matched. Alternatively, if the correlation between corresponding attributes of the feature sets lies above than a threshold then matching decision of two records could be taken. The detection threshold for each attribute of feature set is estimated separately on the basis of interbeat correlation between corresponding attributes of different feature sets extracted from ECG recordings.

The computation procedure of threshold is as follows: Firstly, an ECG dataset (recording of T sec) of kth individual is chipped into n subdata sets (recordings of S sec each, where $S < T$). From each subdata set a pattern of m features, called attributes is extracted. Let P_k be a pattern matrix of size $n \times m$.

$$P^k_{n,m} = \begin{pmatrix} a^k_{1,1} & a^k_{1,2} & \cdots & a^k_{1,m} \\ a^k_{2,1} & a^k_{2,2} & \cdots & a^k_{2,m} \\ \vdots & \vdots & \ddots & \vdots \\ a^k_{n,1} & a^k_{n,2} & \cdots & a^k_{n,m} \end{pmatrix} \tag{3}$$

where $a^k_{i,j}$ represents the jth attribute of ith subdata set of an individual k.

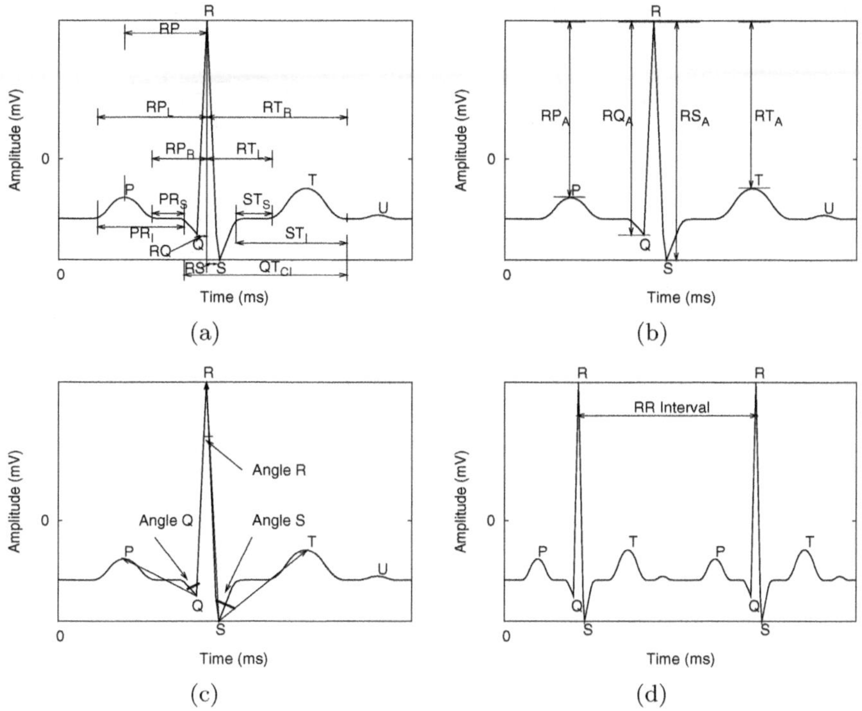

Fig. 4. Considered feature positions of different classes: (a) Time interval, (b) Amplitude and (c) Angles. The RR interval shown in (d) is used for computing QT_I using Bazett's formula [8].

The purpose of ECG data set partitioning is to measure the correlation between corresponding collection of attributes of different ECGs. The degree of relationship between attributes of feature vectors are evaluated on the basis of Pearson product-moment correlation coefficient which is computed as follows,

$$R = \left(\left(r_{X_1^i X_1^j} \right) \left(r_{X_2^i X_2^j} \right) \cdots \left(r_{X_m^i X_m^j} \right) \right) \tag{4}$$

where, $X_1^i = [a_{1,1}^i, a_{2,1}^i, \cdots a_{n,1}^i]^T$; $X_2^i = [a_{1,2}^i, a_{2,2}^i, \cdots a_{n,2}^i]^T \cdots X_m^i = [a_{1,m}^i, a_{2,m}^i, \cdots a_{n,m}^i]^T$. These are the column vectors of the pattern matrix for an individual i for $1 \leq i \leq k$. The correlation coefficient $r_{X_1^i X_1^j}$ is calculated using the formula,

$$r_{X_1^i X_1^j} = \frac{n \sum X_1^i X_1^j - \sum X_1^i \sum X_1^j}{\sqrt{n \sum {X_1^i}^2 - (\sum X_1^i)^2} \sqrt{n \sum {X_1^j}^2 - (\sum X_1^j)^2}}. \tag{5}$$

The correlation matrix of size $k \times k$ is generated for each column of the matrix R. The maximum is selected from each correlation matrix which shows the optimum degree of relationship between the attributes of feature vectors of different ECGs. Let $\delta_l = Max(r_{X_l^i X_l^j})$ for $l = 1, 2, \ldots, m$, be the maximum

correlation coefficient computed for each column l of matrix R. Then, maximum of correlation for all attributes are obtained as follows,

$$[\delta] = (\delta_1\ \delta_2 \cdots \delta_m) \tag{6}$$

The detection threshold for each attribute of the feature set is estimated on the computed value of corresponding δ. This model estimates detection threshold according to the change in ECG. The power of the proposed model is that all attributes of feature vector participate equally for deciding the threshold level. In decision making, one can set the decision limit on the number of matched attributes according to the detected threshold.

5 Biometrics Results

The performance of identification system on ECG is evaluated on Physionet QT database [9]. The database provides an evaluation tool for validation of experiments based on ECG delineation. The experiment is conducted on 50 first lead individual ECG recordings of the database. Among them 32 recordings are taken from MIT-BIH Arrhythmia database that contains the ECG of mostly inpatient men and women of age between 47 to 84 years. Remaining, 18 recordings are taken from MIT-BIH Normal Sinus database of subjects age between 20 to 50 years those have no significant arrhythmias. During experiment first 8 seconds of each recording is used for training and setting of system parameters.

Table 2. Performance of identification system over test database of size 50 at different threshold levels [GMR is the genuine matching rate, computed as $GMR(\%) = 100 - FNMR(\%)$]

Threshold Level	# FNM	FNMR (%)	GMR (%)	# FM	FMR (%)	Accuracy (%)
14	0	0	100	2	4	98
15	0	0	100	1	2	99
16	0	0	100	1	2	99
17	1	2	98	0	0	99
18	2	4	96	0	0	98
19	2	4	96	0	0	98

Biometrics performance of the identification system is measured on the parameters of false matching rate (FMR) and false non-matching rate ($FNMR$). From the correlation FMR and $FNMR$ are calculated for different threshold values to generate the detection error tradeoff (DET) curve of the identification process. The accuracy (Acc) of the system is also determined using FMR and $FNMR$ as, $Acc(\%) = 100 - \left(\frac{FMR+FNMR}{2}\right)$.

During experiment, five subdata sets of nearly 100 beats are selected randomly from each individual ECG and the feature sets are generated. Thus, a

total of 250 (50×5) samples are prepared in the database for training the identification system. The test of positive identification is conducted on different data samples of an individual ECG recording while the test of negative identification is conducted on data samples of different ECG recordings. The ECG recordings of European ST-T database are used for testing of negative identification.

In order to test the identity of an individual, a test template which is a collection of feature sets consist of five subdata sets of nearly 100 beats, is compared with templates stored in database, e.g. *one-to-many* matching. The decision of best match is taken on the basis of threshold whose level can be set on the number of attributes matched. For example, a test template can be declared matched with its counterpart stored in database if 17 out of 19 attributes are matched. The performance of identification system is examined on different settings of threshold level of attributes between 14 to 19 and results are given in Table 2.

It shows that at threshold level 15, only one individual (out of 50) is matched wrongly while none of them is non-matched wrongly. Alternatively, the system achieved $GMR = 100\%$ at $FMR = 2\%$. At threshold level 17, no one is matched wrongly while only one individual is non-matched wrongly. Thus, 99% of genuine subjects are matched correctly while none of them is forged to system. Therefore, system achieved the accuracy of 99% at this setting of threshold.

The detection error tradeoff (DET) curve of the identification process is shown in Figure 5. It is worthwhile to mention that an equal error rate (EER) smaller than 1.01 is obtained. The obtained results are encouraging and demonstrate the potential of heartbeat signals for human identification.

The concern of identification system on ECG is the changes of ECG dominant fiducials during aging. Study shows that these changes are only seen upto the age of adolescence (∼14 years) [10]. After adolescence the ECG dominant features are relatively consistent. This may be a minor concern because biometrics applications are mainly employed to identify the individuals those have passed the adolescence age.

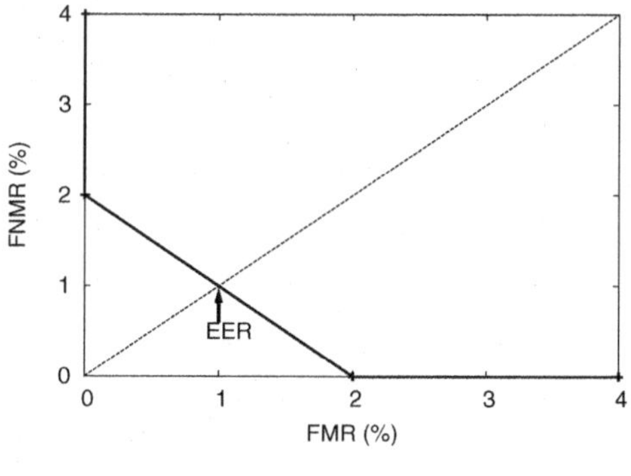

Fig. 5. Detection error trade-off curve (DET) for identification of individuals using ECG. The location of equal error rate (ERR) is indicated.

6 Conclusions

In this paper, it has shown that dominant fiducials delineated from ECG recording exhibits features that are unique to an individual. A series of experiments have been conducted for individual identification on ECG on bench mark database. The database consists of normal and inpatient men and women of age 50 ± 23 years. Biometrics results have shown that the ECG features are useful for distinguishing different individuals. From these observations it has been concluded that each individual ECG has a unique set of heartbeat features that hold sufficient information which can be used as potential biometrics for individual identification.

References

1. Kligfield, P.: The Centennial of the Einthoven Electrocardiogram. Journal of Electrocardiology 35, 123–129 (2002)
2. Biel, L., Pettersson, O., Lennart, P., Peter, W.: ECG Analysis: A New Approach in Human Identification. IEEE Transaction on Instrumentation and Measurement 50(3), 808–812 (2001)
3. Shen, T.W., Tompkins, W.J.: One-Lead ECG for Identity Verification. In: Proceedings of the Second Joint EMBS/BMES Conference, pp. 62–63 (2002)
4. Israel, S.A., Irvine, J.M., Andrew, C., Mark, D.W., Brenda, K.W.: ECG to Identify Individuals. Pattern Recognition 38(1), 133–142 (2005)
5. Kors, J.A., Bemmel, J.H., Zywietz, C.: Signal analysis for ECG interpretation. Methods Inf. Med. 29(4), 317–329 (1990)
6. Friesen, G.M., Thomas, C.J., Manal, A.J., Stanford, L.Y., Stephen, R.Q., Troy, N.: A Comparison of the Noise Sensitivity of Nine QRS Detection Algorithms. IEEE Transaction on Biomedical Engineering 37(1), 85–98 (1990)
7. Pan, J., Tompkins, W.J.: A Real Time QRS Detection Algorithm. IEEE Transactions on Biomedical Engineering 33(3), 230–236 (1985)
8. Bazett, H.C.: An Analysis of the Time-Relations of Electrocardiograms. Heart 7, 353–370 (1920)
9. Laguna, P., Mark, R.G., Goldberger, A.L., Moody, G.B.: A Database for Evaluation of Algorithms for Measurement of QT and Other Waveform Intervals in the ECG. In: Computers in Cardiology, pp. 673–676 (1997)
10. Rijnbeek, P.R., Witsenburg, M., Schrama, E., Hess, J., Kors, J.A.: New Normal Limits for the Pediatric Electrocardiogram. European Heart Journal 22, 702–711 (2001)

Real-Time Model-Based Hand Localization for Unsupervised Palmar Image Acquisition

Ivan Fratric and Slobodan Ribaric

University of Zagreb, Faculty of Electrical Engineering and Computing,
Unska 3, 10000 Zagreb, Croatia
{Ivan.Fratric,Slobodan.Ribaric}@fer.hr

Abstract. Unsupervised and touchless image acquisition are two problems that have recently emerged in biometric systems based on hand features. We have developed a real-time model-based hand localization system for palmar image acquisition and ROI extraction. The system operates on video sequences and produces a set of palmprint regions of interest (ROIs) for each sequence. Hand candidates are first located using Viola-Jones approach and then the best candidate is selected using model-fitting approach. Experimental results demonstrate the feasibility of the system for unsupervised palmar image acquisition in terms of speed and localization accuracy.

Keywords: Hand localization, real-time, model-based, unsupervised biometrics.

1 Introduction

In biometric systems based on hand features, two problems have recently emerged: unsupervised and touchless image acquisition. In previously developed systems [1-8], the image is taken under strictly controlled (light and position) conditions. In most of the systems the user is required to place a hand on the sensor. However, this is a problem for large-scale scenarios because many people refuse to touch the biometric sensor for sanitary and other reasons. Also, touchless hand-based biometric systems in combination with unsupervised image acquisition are more convenient to use.

The motivation of our work is the development of a touchless palmprint recognition system. The system should aslo be unsupervised in the sense that there is no need to guide a user during the acquisition procedure; it is sufficient for the user to wave the hand in front of a sensor or a group of space-distributed sensors.

The system should be robust enough to be able to work in uncontrolled environment conditions with varying lighting and a cluttered background, as required by most real-world applications. The robust real-time detection and localization of the hand are crucial in order to be able to develop such a system.

In this paper we describe the development of a real-time model-based hand localization system for palmar image acquisition. The system operates on video sequences taken in a real environment and produces a sequence of images containing the palmprint regions of interest (ROIs). This sequence of ROIs is intended to be the input for a biometric verification system, which would select the best ROIs in terms of feature extraction and use them for transparent user verification.

M. Tistarelli and M.S. Nixon (Eds.): ICB 2009, LNCS 5558, pp. 1280–1289, 2009.
© Springer-Verlag Berlin Heidelberg 2009

Scanner-based hand biometric systems have been proposed in [1, 2, 3]. However, scanners are slow and require the hand to be placed on the scanning surface. Zhang et al. [4] developed an online palmprint identification system in which a hand image is captured using a CCD camera, but the user is still required to place a hand on the device and pegs are used to constrain the position of the hand. Papers [5, 6] describe biometric systems based on the hand's geometry. Both systems use cameras as the input devices and require the user to place the hand on the surface, where it is constrained by pegs. The multimodal biometric systems described in [7, 8] use a camera-based input device and do not restrain the position of the hand, but the hand is still required to be placed on a flat, uniformly colored surface.

In a recently published paper, Ong et al. [9] describe a touchless palmprint verification system. Skin-color-based segmentation is used in order to segment the hand in the image, which may not work in environments with skin-colored backgrounds or poor lighting. Their system is set to extract a single ROI every 2 seconds, which is then used for the verification.

Most of the work related to hand detection in real environments is in the area of natural human-computer interaction. A review of techniques used in this area is given in [10]. Many of the techniques described use motion information and/or heuristical approaches for hand detection. However, in our scenario there could be multiple moving objects on the image, so this information alone would not be sufficient for robust hand detection.

Little work has been done on finding hands in images based on their appearance and shape. Kölsch and Turk [11] used an object detector proposed by Viola and Jones to locate hands. Their approach is fast, but finds only hands in a pre-defined pose and is unable to determine the contour on the hand in the image. Stenger et al. [12] use a hierarchical model-based approach for finding and tracking hands. Their approach finds the hand shape and orientation, but takes approximately 2 seconds per frame to execute on a 1-GHz Pentium IV.

2 System Organization

Fig. 1 shows the overview of the developed hand localization system for palmar image acquisition and ROI extraction.

A camera is used to capture a video sequence of the hand at a resolution of 640x480 pixels, which is a trade-off between image quality and processing time. Each of the video frames is then processed as follows.

Firstly, the frame is scaled to a resolution of 320x240 pixels in order to decrease the time required for the image processing. This reduced image (converted into gray-scale with 256 levels) is used for the detection of the hand candidates. The Viola-Jones approach is used for hand-candidate detection because of its speed and good detection rate [13].

Once the positions of the hand candidates have been located on the image, each of the candidates is verified by fitting it to the large, predefined set of hand models. Two measures are used for the correspondence of the hand candidate and the model [12]: the edge distance and the skin-color likelihood. These two measures are combined in a unique measure that determines which of the candidates, if any, should be used for the feature extraction.

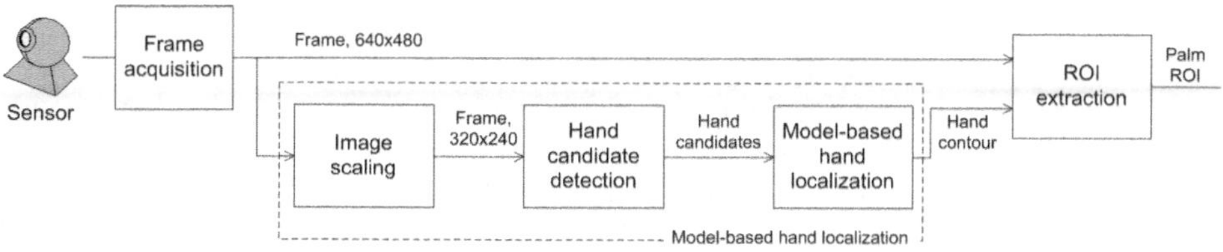

Fig. 1. Overview of the hand localization system for palmar image acquisition and ROI extraction

The output of the model-based hand localization stage is a hand contour with marked stable points, which are used in the ROI extraction stage to determine the position of the palm ROI on the original (non-scaled) image.

3 Hand-Candidate Detection

The goal of the hand-candidate detection process is to obtain a relatively small set of image locations where the presence of the hand is possible. This process should be as fast as possible so as to be able to process the entire frame in real time.

The Viola-Jones approach was selected for the candidate detection [13]. This approach has become very popular in computer vision because of its high speed and good accuracy. It has been primarily used for face detection [13]; however, it has also been used for hand detection [11].

The Viola-Jones object detector operates on grayscale images. A sliding-window approach is used to examine each image location for the presence of a hand. The classifier itself is built as a cascade of weak classifiers. These weak classifiers are constructed using AdaBoost on a large set of 2D Haar-like features, which can be computed quickly by using integral images (each image element on the integral image contains the sum of all the pixels to its upper left on the original image). Weak classifiers are trained to have a very high detection rate (99.5% in our case) and a high false positive rate (50% in our case). This means that at each stage of the classification most of the real hands will be passed to the subsequent classifiers in the cascade, while a part of the non-hands will be rejected. Thus, most of the non-hands will be rejected in the first several stages of the cascade, which, together with fast feature computation, makes real-time object detection possible.

Usually, after the detection, the results are filtered by grouping overlapping detection windows and eliminating groups with a small number of members. However, we skipped this step because we wanted to pass all of the detection to the model-fitting stage in order to achieve the best possible fitting.

We used the OpenCV [14] implementation of the Viola-Jones object detector.

4 Model-Based Hand Localization

In the model-based hand localization phase, each of the candidates for the hand position, found in the previous stage, were examined by fitting a large number of hand models over that position on the image. The goal of this process is to further verify the hand candidates, remove the non-hands, and only select the candidate with the highest matching score. The output of this stage is an approximate hand contour – that of the best-fitting model.

The model-matching process is taken from [12]. Each model is represented as (a) a contour, (b) a binary mask, where white points indicate the inside of the hand and black indicate the outside of the hand.

Unlike [12], where multiple hand models were obtained by rotating a small set of models covering basic hand gestures, in our system models are generated from the database of 1872 hand images of 243 people, obtained using a table scanner (180 dpi). The hands in the database images are cropped and rescaled to the resolution of 64x64 pixels to form the models. This set of models contains significant variations in hand positions and rotations. Fig. 2 shows some of the models used in our system.

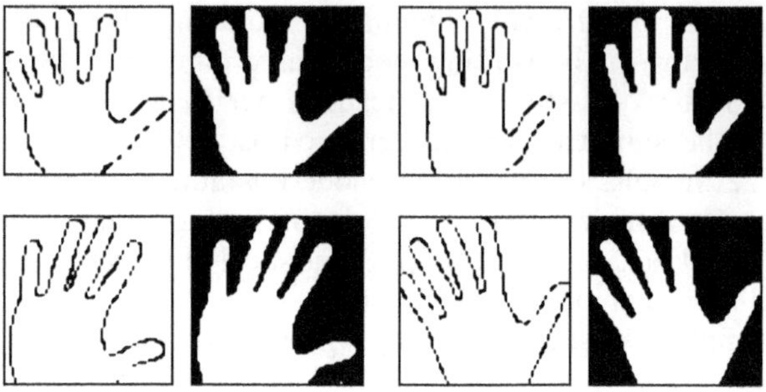

(a) (b) (a) (b)

Fig. 2. Some of the hand models used in the model-fitting stage: (a) the hand contours, (b) the corresponding binary masks

Two measures are used for the correspondence between the model and the hand candidate: the edge distance and the skin-color likelihood. The edge distance is computed as the quadratic chamfer distance between the model contour and the edges of the image. For two sets of points, *A* and *B*, it is given as

$$d(A,B) = \frac{1}{N_a} \sum_{a \in A} \min_{b \in B} \left\| a - b \right\|^2 \tag{1}$$

where N_a denotes the number of points in the set *A*. In our system *A* is the set of points on the model contour, while *B* is the set of the images' edge elements. The edges on an image are obtained using the Canny edge detector [15]. The computation speed of this distance can be increased if a distance transform of the edge image is computed prior to the model-fitting stage.

The hand's skin-color likelihood, given the model x, is computed as

$$p(hand \mid Model\ x) = \prod_{k \in S(Model\ x)} p^s(I(k)) \prod_{k \in \bar{S}(Model\ x)} p^{bg}(I(k)) \tag{2}$$

where $S(Model\ x)$ denotes the set of points inside the hand in the model *x*, where $\bar{S}(Model\ x)$ denotes the set of points outside the hand in the model *x*, *I(k)* is the color of the image at location *k* and p^s and p^{bg} are the skin and the background color distributions.

The skin and the background color distributions are taken from [16]. A non-parametric color model is used, where the skin and background color distributions are

computed based on RGB histograms with R, G and B resolutions of 32x32x32 obtained from a large database of images with manually marked skin regions.

This likelihood in (2) is easier to compute as a log likelihood

$$\log(p(hand \mid Model\,x)) = \sum_{k\in S(Model\,x)} \log(p^{s}(I(k))) + \sum_{k\in \bar{S}(Model\,x)} (\log p^{bg}(I(k))) \quad (3)$$

The hand matching score between the model and a hand candidate is obtained by first normalizing the edge distance and skin-color likelihood using the min-max normalization [17] and then summing the two normalized scores.

If speed was not one of the requirements of the system, the process of model fitting for a hand candidate would involve iteratively matching all the models to the hand candidate and selecting the best-fitting model based on the matching score. However, this process would be very slow, so instead, in our system, each hand candidate is matched only to the subset of models arranged as a 3-level tree. This 3-level tree contains all 1872 models, in which each node contains a group of models (and its prototype). Each child-node in the tree is obtained by grouping models contained in its parent node based on a modified simple heuristical clustering algorithm. The nodes at the third level represent individual models.

A modified simple heuristical clustering algorithm is given below:

```
Set the first model as the prototype for the first
cluster
For each model m
|     If the distance between model m and any prototype
|     is more than characteristic distance
|     |     Create a new cluster and set m as its proto-
|     |     type;
For each model m
|     For each cluster C
|     |     If the distance between the model m and the
|     |     prototype of C is less than characteristic
|     |     distance
|     |     |     Put m in C;
```

Note that in the above algorithm one model can end up in different clusters. The edge distance between the model contours computed according to (1) is used as the measure of the distance between models. The characteristic distance is selected experimentally as 8 at the level 1 of the tree and 4 at the level 2 of the tree.

Once the tree is constructed, the fitting process for each candidate can be performed as follows

```
1. Set current node to tree root
2. Compare hand candidate to the prototype of every
   child of the current node
3. Set current node to the child with the highest hand
   matching score
4. If current node has children go to step 2.
5. Best matching model is the one contained in the
   current node
```

This process is repeated for all the hand candidates and the best candidate in terms of the hand matching score is selected. This hand matching score is then compared to the threshold T, set to 1 (50% of the maximum expected hand matching-score value), which was selected based on preliminary experiments on the training set. The purpose of this thresholding is to eliminate the non-hands and also hands with strong shape deformations (e.g. closed or half-closed hand). If the score is lower than T, it is assumed that the matching is not good enough and there is no hand in the current frame. Otherwise, the contour of the best-matching model is assumed to be the contour of the hand on the image.

We also experimented with using snakes [18], initially positioned as the contour of the best-matching model, to further adapt the contour to the hand in the image. However, it was observed that this step was not robust enough to be used in environments with a cluttered background, and that better fitting is achieved if it is skipped.

5 ROI Set Extraction

The inputs to the ROI extraction module are: (a) the hand contour of the best model with marked stable points, and (b) the original frame (640x480 pixels).

A simplified procedure described in [3] was used for determining stable points on the contour of each model (Fig. 3). Based on these stable points, the location of the palm ROI in the current frame is determined.

In our system the palm ROI is defined as a square region with two of its corners placed on the middle-points of the line segments P1-B2 and B4-P2 (see Fig. 3).

This region is extracted from the current frame and geometry normalization is applied in order to obtain a ROI image of fixed size and orientation.

Any additional normalization, such as lighting normalization, can be applied at this stage as well.

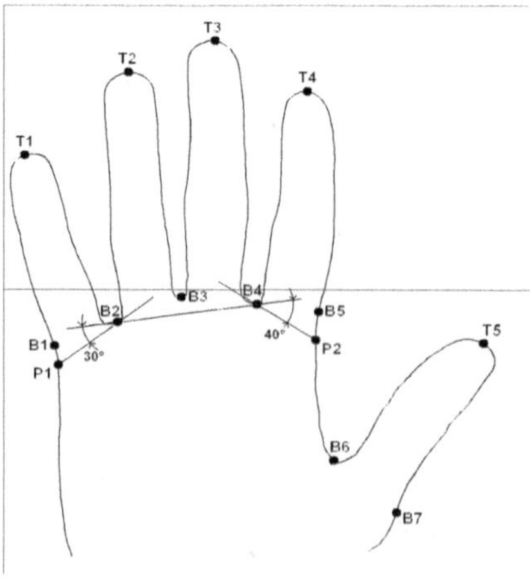

Fig. 3. Hand contour with the stable points marked

6 Experimental Results

The experiments were conducted on a database of video sequences. The database was divided into a training set and a test set. The training set consists of 32 video sequences of 10 people. A total of 2138 frames containing hands were extracted and used for the training. The test set consists of 10 video sequences of 5 people. The people in the test set are not the same as the people in the training set. A total of 362 frames were randomly selected for testing from the test sequences.

The database was collected using two cameras: a high-quality CCD camera and a low-quality CMOS web camera. The sequences were captured under varying lighting conditions and all of them feature a cluttered background.

We first tested the detection using the Viola-Jones hand detector only, and then using the full model-based detector. The results of the detection using only the Viola-Jones object detector can be seen in the ROC curve (detail) in Fig. 4.

The results are shown for two sizes of detection window. As can be seen from the ROC curve, high detection rates can be achieved using the Viola-Jones detector. Because the results for different sizes of detection windows were similar in performance, a smaller window of 20x20 pixels was selected for further experiments. A detection rate of 98.3% can be achieved with 19 false positives, and a detection rate of 99.7% can be achieved with 713 false positives. This point in the ROC curve was selected as the working point of our system. This maximizes the detection rate, and the false positives can be eliminated in the model-fitting stage.

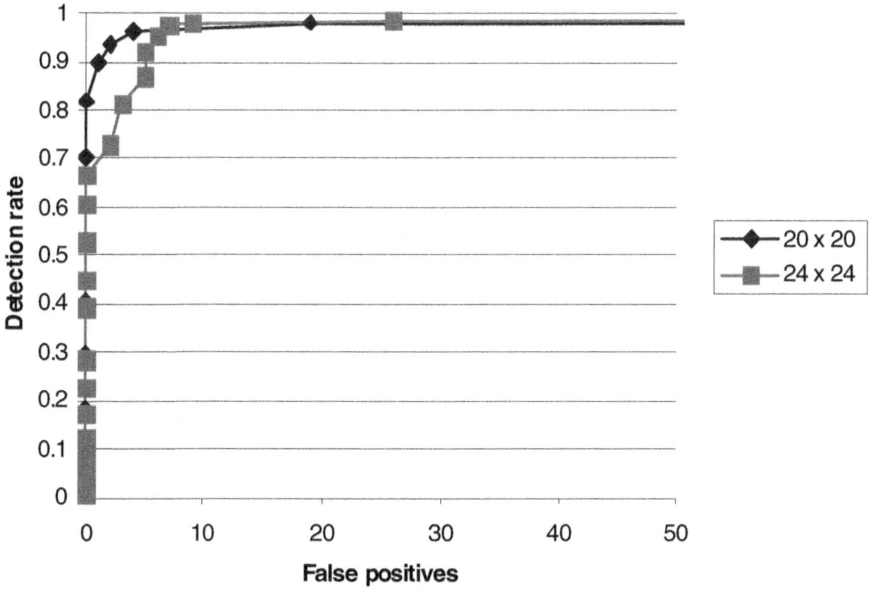

Fig. 4. ROC curve for hand detection using the Viola-Jones object detector

In the second experiment the model-based localization and ROI extraction procedure, as given in the previous sections, were applied to all the frames in the test set. Fig. 5 shows some of the successful and non-successful ROI localization results.

As can be seen from Fig. 5, the system can operate successfully in scenes with a cluttered background, even if the hand is placed over other skin-colored regions, such as the face. The errors are mostly due to skin-colored regions with edge elements resembling, in terms of shape, edges typically found on a hand, or a hand being incorrectly positioned on the image.

To measure the accuracy of the system, we defined the successful ROI localization as localization where the ROI falls inside the palm on the actual frame. On our test set, with the accuracy measure as described above, the system accurately locates the palm ROI in 96.9% of cases.

Fig. 5. Some of the results of the hand and ROI localization (successful and unsuccessful)

The average processing time for a single frame on a single processor with 1177 SPECint_base2000 was 0.27 seconds. However, hand detection in the verification system is intended to be run on a newer, quad-core processor, where each core would be assigned one frame. In this setup, taking into account the ability of the detection system to independently process frames, the average processing time could be reduced by up to four times.

7 Conclusion

We have developed a hand-localization system for palmar image acquisition and ROI extraction. The system first locates the hand candidates using the Viola-Jones approach and then selects the best candidate using a model-fitting approach.

The Viola-Jones approach is very fast and highly accurate for hand detection (a 98.3% detection rate with 19 false positives). The model-fitting process selects the best candidate and finds a palm ROI that falls inside the palm on the image in almost all frames (96.9% in our test set). This demonstrates the feasibility of our system, even in environments with cluttered or skin-colored backgrounds. The system is designed to operate in real-time on a multi-core processor.

In the future we plan to use this system as part of a touchless palmprint verification system. To avoid any contact, the identity could be presented to the system by RFID or a similar unsupervised technology. The unsupervised characteristic of the system could be achieved by using a group of space-distributed (intelligent) sensors.

The system would be operating on better equipment than that used in the experiments described in this paper, such as a high-speed multi-core processor together with a high-quality camera, which would enable us to obtain high-quality features usable for the identity verification. Techniques for selecting the best ROIs for verification will be developed and tested in terms of the FAR and the FRR.

References

1. Han, C., Cheng, H.L., Fan, K.C., Lin, C.L.: Personal Authentication Using Palm-print Features. Pattern Recognition 36, 371–381 (2003)
2. Lin, L., Chuang, T.C., Fan, K.C.: Palmprint verification using hierarchical decomposition. Pattern Recognition 38(12), 2639–2652 (2005)
3. Ribaric, S., Fratric, I.: A Biometric Identification System Based on Eigenpalm and Eigenfinger Features. IEEE Trans. PAMI 27(11), 1698–1709 (2005)
4. Zhang, D., Kong, W.K., You, J., Wong, M.: Online Palm Print Identification. IEEE Trans. PAMI 25(2), 1041–1050 (2003)
5. Jain, A.K., Ross, A., Pankanti, S.: A prototype hand geometry-based verification system. In: Proc. 2nd Intl. Conf. on Audio- and Video-Based Biometric Person Authentication, Washington DC, USA, pp. 166–171 (1999)
6. Sanchez-Reillo, R., Sanchez-Avila, Gonzalez-Marcos, A.: Biometric identification through hand geometry measurements. IEEE Trans. PAMI 22(10), 1168–1171 (2000)
7. Kumar, A., Wong, D.C.M., Shen, H., Jain, A.K.: Personal verification using palmprint and hand geometry biometric. In: Proc. Intl. Conf. Audio- and Video-based Person Authentication, pp. 668–675 (2003)

8. Kumar, A., Zhang, D.: Personal Authentication using Multiple Palmprint Representation. Pattern Recognition 38(10), 1125–1129 (2005)
9. Ong, M.K.G., Connie, T., Teoh, A.B.J.: Touch-less palm print biometrics: Novel design and implementation. Image and Vision Computing 26(12), 1551–1560 (2008)
10. Erol, A., Bebis, G., Nicolescu, M., Boyle, R., Twombly, X.: A Review on Vision-Based Full DOF Hand Motion Estimation. In: Proc. IEEE Workshop on Vision for Human-Computer Interaction, USA, pp. 75–82 (2005)
11. Kölsch, M., Turk, M.: Robust Hand Detection. In: Proc. IEEE Intl. Conference on Automatic Face and Gesture Recognition, pp. 614–619 (2004)
12. Stenger, B., Thayananthan, A., Torr, P.H.S., Cipolla, R.: Model-based hand tracking using a hierarchical Bayesian filter. IEEE Trans. PAMI 28(9), 1372–1384 (2006)
13. Viola, P., Jones, M.: Robust Real-Time Object Detection. Intl. Journal of Computer Vision 57(2) (2004)
14. Open Computer Vision Library, `http://sourceforge.net/projects/opencvlibrary/`
15. Canny, J.: A Computational Approach To Edge Detection. IEEE Trans. PAMI 8, 679–714 (1986)
16. Jones, M.J., Rehg, J.M.: Statistical color models with application to skin detection. Intl. Journal of Computer Vision 46(1), 81–96 (2002)
17. Jain, A.K., Nandakumar, K., Ross, A.: Score Normalization in Multimodal Biometric Systems. Pattern Recognition 38(12), 2270–2285 (2005)
18. Kass, M., Witkin, A., Terzopoulos, D.: Snakes: active contour models. Intl Journal of Computer Vision 1(4), 259–268 (1988)

Palm Vein Verification System Based on SIFT Matching

Pierre-Olivier Ladoux[1], Christophe Rosenberger[2], and Bernadette Dorizzi[1]

[1] Institue TELECOM
Télécom & Management SudParis
9 Rue Charles Fourier, 91011 Évry Cedex, France
pierre-olivier.ladoux@it-sudparis.eu,
bernadette.dorizzi@it-sudparis.eu
[2] Laboratoire GREYC
ENSICAEN – CNRS – Université de Caen Basse-Normandie
6 boulevard Maréchal Juin, 14000 Caen, France
christophe.rosenberger@ensicaen.fr

Abstract. We present in this communication a new biometric system based on the use of hand veins acquired by an infrared imager. After the preprocessing stage and binarization, the vein image is characterized by specific patterns. One originality of the proposed system is to use SIFT descriptors for the verification process. The developed method only necessitates a single image for the enrollment step allowing a very fast verification. The experimental results on a database containing images of 24 individuals acquired after two sessions show the efficiency of the proposed method.

Keywords: Hand vein modality, IR acquisition, normalization preprocessing, SIFT descriptors, key points matching.

1 Introduction

Hand vein is a biometric modality that seems promising as it is acquired in Near Infrared light (NIR), which implies that skin variations and dirtiness are less sensible than in visible light [1]. Moreover, the hemoglobin which flows in the veins is sensible to NIR light, this way allowing a good quality of acquisition of the hand veins. It is possible to use either the back of the hand or the hand palm. A recent study [2] using back hand vein data and tested with 5 sessions per person and 50 persons showed promising results. The main problem of this database is the low resolution of the images (images at resolution 132x124 pixels).

The first commercialized products have been produced by Hitachi [3] on the back and Fujitsu [4] on the palm. They have been patented but only little information is available on them. These companies claim a very low FRR at very low FAR on a huge database – close to 0% on 140000 hands.

Unfortunately at this moment, there is no public database allowing verifying these figures. In general, in the various papers present in the literature, after the acquisition phase, some preprocessing algorithms are used such as histogram equalization, low pass filtering. Then, some binarization is performed and for verification using this

M. Tistarelli and M.S. Nixon (Eds.): ICB 2009, LNCS 5558, pp. 1290–1298, 2009.
© Springer-Verlag Berlin Heidelberg 2009

image, two main streams can be found: global matching of the reference and test images through a pixel to pixel superposition [5] or after some frequency analysis as in Laplacian palm processing [8]. Another direction consists in performing local matching of specific feature points extracted in the two references and test images [6]. The matching algorithms in this last approach are similar to those used for fingerprint verification [7].

Our approach developed in this paper falls in the last category. However, we use for the matching of the feature points, the well known SIFT [9] algorithm which, to our knowledge has never been used so far for hand vein verification. This algorithm developed for graph matching was proven to be very efficient for face verification [10]. We tested the proposed system on a small home database of videos of the hand palm of 24 persons containing two sessions.

We present in Section 2 the details of our approach including a synopsis of the developed system, the acquisition device we used for acquiring the palm database as well as the preprocessing we made on these images. We also explain our binarization procedure. In the following subsection, the detection of the feature points and the SIFT procedure is briefly described. Due to the fact that some translations and rotations are present between the two sessions, we also propose a post processing allowing a better separation of genuine and impostor's scores. Finally, section 3 presents the experimental work, namely the protocols defined on each database and the results in terms of EER and ROC curves.

2 Description of the Proposed System

The proposed scheme of the proposed method is shown in Figure 1.

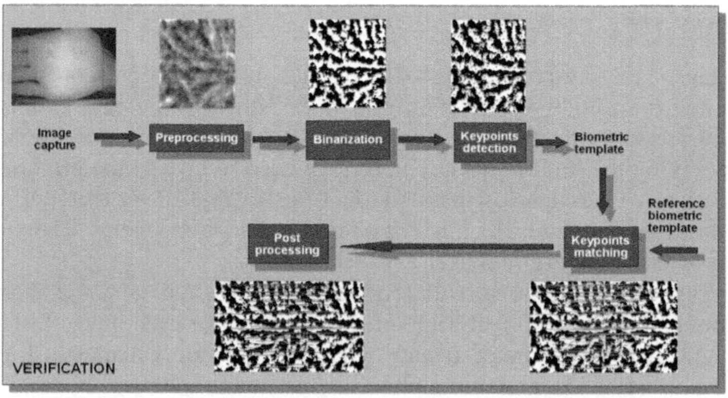

Fig. 1. Synopsis of the developed method (verification step)

It is standard in biometrics with an enrollment and a verification step. One important characteristic of our method is that only one image is needed for the enrollment step. Moreover, a post-processing is added after the matching phase in order to tackle translation and rotation problems.

2.1 Acquisition

Image acquisition is done by 48 infrared LEDs and a CCD camera, which sensitive range is between 400nm and 900nm but the interesting wavelength is around 850nm. We added to this system one support in order to help the persons positioning their hand and to therefore limit translations and rotations. The hand is approximately at 20cm of the camera's lens (see Figure 2).

Fig. 2. NIR acquisition device

At each acquisition, we acquired a short video which provided us a set of 30 grey-scale pictures of size 640x320 pixels. We this way acquired the data of 24 persons in two sessions.

2.2 Pre-processing

The first step of the pre-processing is the extraction of the region of interest (ROI). Due to limited translation and rotation, this is eased. Sizes are cut down to 232x280 pixels. Then, we apply the 5x5 box filter on the ROI in order to reduce the noise.

After removing the high frequency noise, we need to correct the brightness, which is not uniform. A Gaussian low-pass 51x51 filter is applied on the ROI in order to obtain the brightness image which is considered as low frequencies. Then, the brightness is subtracted of the original ROI.

At this step, the contrast is still too bad. We therefore apply a normalization method commonly used in fingerprint verification [7]. For each image I of size NxM, the mean and variance (denoted μ and σ respectively) are calculated. Equation describes the normalization process applied on the image with μ_d and σ_d, the desired value of the mean and variance. For each pixel, we modify its gray-level with the following formula:

$$I'(x,y) = \begin{cases} \mu_d + \sqrt{\dfrac{\sigma_d^2 g (I(x,y)-\mu)^2}{\sigma^2}} & I(x,y) > \mu \\ \mu_d - \sqrt{\dfrac{\sigma_d^2 g (I(x,y)-\mu)^2}{\sigma^2}} & Otherwise \end{cases} \quad (1)$$

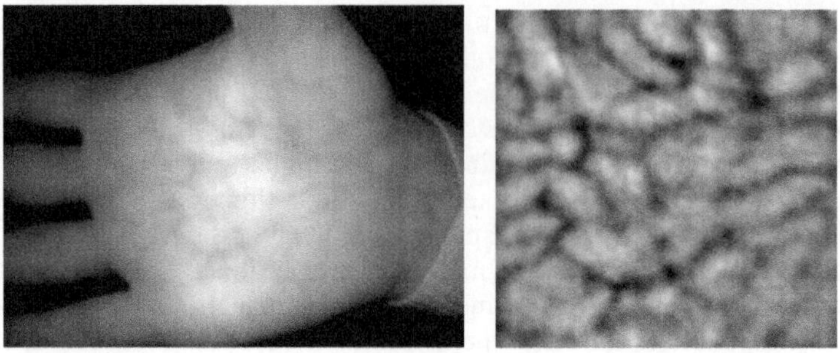

Fig. 3. (left) NIR image of palm, (right) Image after pre-processing

Where *I(x,y)* corresponds to the gray-level for the pixel located at *(x,y)* for the original image and *I'(x,y)* for the resulting one after pre-processing. Figure 3 shows the original image of the palm acquired with our sensor and the corresponding image after pre-processing. For our experiments, we set empirically the values of μ_d to 128 and σ_d to 40^2.

2.3 Vein Pattern Extraction

After noise reduction and contrast normalization, the quality of the image is improved as can be seen in Figure 3. To obtain the vein pattern, it is necessary to extract the veins from the background. In fact, the grey level is low where the hemoglobin absorbs the NIR. Therefore, the chosen extraction algorithm is a local thresholding one depending on the mean value of the neighborhood of each pixel.

Figure 4 shows two processing results for two images for the same individual. If we would ask somebody to decide if these images correspond to the same individual, he would probably try to find similar areas between these two images. This is the idea of the proposed methodology defined in the next section.

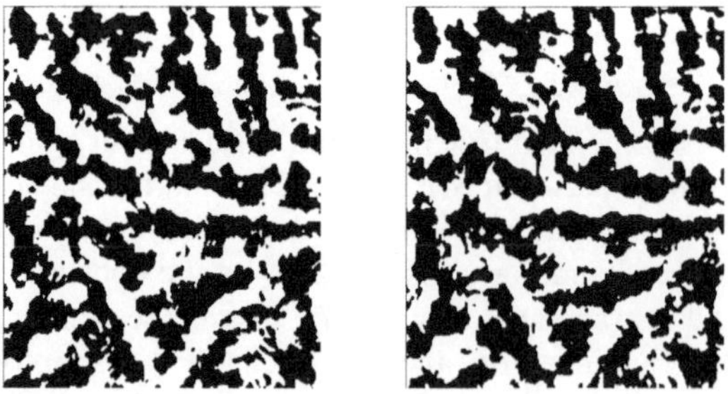

Fig. 4. Examples of two binarized images corresponding to the same individual

2.4 Pattern Definition

We chose to use a local description of the vein image in order to facilitate the verification step. We used the SIFT descriptor, following of a comparative study [11] showing that it is one of the most efficient local ones.

The invariant descriptor developed in the SIFT algorithm described in [9] is applied locally on key-points and is based on the image gradients in a local neighborhood. The

SIFT detector and descriptor are constructed from the Gaussian scale space of the source image. The algorithm makes use of another scale space too, called difference of Gaussian (DOG), which can be considered as the scale derivative of the Gaussian scale space. Extracted key-points are defined as points of local extremum of the DOG scale space.

The descriptor is created by sampling the magnitudes and orientations of the image gradients in a neighborhood of each key-point and building smoothed orientation histograms that contain the important aspect of the neighborhood. Each local descriptor is composed on a 4x4 array (histogram). To each coordinate of this array, an 8 orientation vector is associated. A 128-elements vector is then built for each key-point.

We used in this article the implementation provided by Lowe [9]. As illustration, we obtained an average value of 800 detected key-points for the vein images at hand.

2.5 Matching Similarity

Each individual is described by a set of invariant features $Y(I) = \{k_i=(s_i, x_i, y_i)\}$, $i=1:N(I)$ where s_i is the 128-elements SIFT invariant descriptor computed near key-point k_i, (x_i,y_i) its position in the original image I and $N(I)$ the number of detected keypoints for image I.

The verification problem for an individual given the set $Y(I)$ corresponding to the biometric model of an individual in our case, is to measure the similarity with another set of keypoints computed on the supposed vein image of the individual. We thus have to compute a similarity between two sets of points $Y(I_1)$ and $Y(I_2)$. We thus use the following matching method which is a modified version of a decision criterion first proposed by Lowe [9]:

Given two points $x \in Y(I_1)$ and $y \in Y(I_2)$, we say that x is associated to y iff :

$$d(x,y) = min_{\{z \in Y(I2)\}} d(x,z) \text{ and } d(x,y) \leq C\, d(x,y') \quad (2)$$

Where C is an arbitrary threshold, $d(.,.)$ denotes the Euclidean distance between the SIFT descriptors and y' denotes any point of $Y(I_2)$ whose distance to x is minimal but greater than $d(x,y)$:

$$d(x,y')=min_{\{z \in Y(I2),\ d(x,z)>d(x,y)\}}\, d(x,z) \quad (3)$$

In other words, x is associated to y if y is the closest point from x in $Y(I_2)$ according to the Euclidean distance between SIFT descriptors and if the second smallest value of this distance $d(x,y')$ is significantly greater than $d(x,y)$. The significance of the necessary gap between $d(x,y)$ and $d(x,y')$ is encoded by the constant C. In the same way, we say that $y \in Y(I_2)$ is associated to $x \in Y(I_1)$ iff x is the closest point from y among $Y(I_1)$ according to the Euclidean distance between SIFT descriptors and if the second smallest value of this distance $d(y,x')$ satisfies $d(y,x)<C\, d(y,x')$. Then, we will say that x is matched to y iff x is associated to y and y is associated to x. If $Y(I_1)$ and $Y(I_2)$ do not define any matching point, we consider that both sets have an infinite distance.

Figure 5 shows an example of matching results for an impostor and a genuine. We obtained 23 detected associations for an impostor (left result) and 63 for the genuine one. In the whole database, the average value for impostors equals 6 (we show in Figure 5 the comparison result having the larger number of associations for an impostor) and 30 for genuine ones.

The number of associations is used here as a similarity measure.

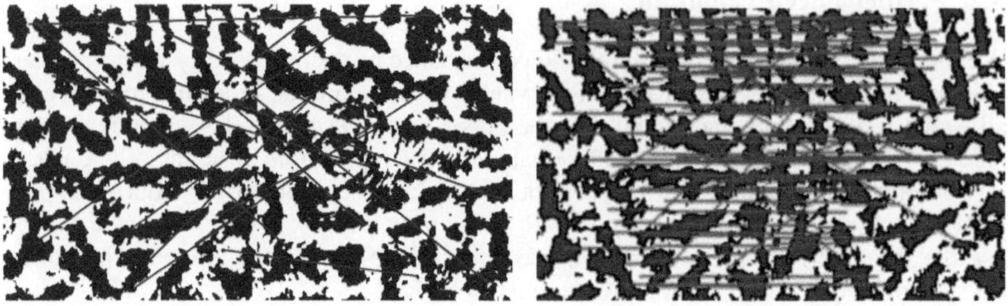

Fig. 5. Example of matching result for an impostor (on the left) and with a genuine (on the right)

2.6 Post-processing

The post-processing has for objective to select among the detected associations the relevant ones. The acquisition can provide mainly small translation and rotation artifacts. That means that the associations for genuine images corresponding to displacement vectors in Figure 5 shall be relatively collinear. In order to determine relevant ones, we compute the direction of each displacement vector by considering the keypoints location in each image to match. We compute the histogram of orientation of these vectors (with a step of 0.1 radian). The updated similarity value is the number of associations for the highest peak of the histogram.

Figure 6 shows the post-processing result of the association result given in Figure 5 (right). Note that we obtained a similarity value equals to 6 instead 23 and 51 instead 63 for the results presented in Figure 3. This shows that we significantly increase the difference between genuine and impostor samples.

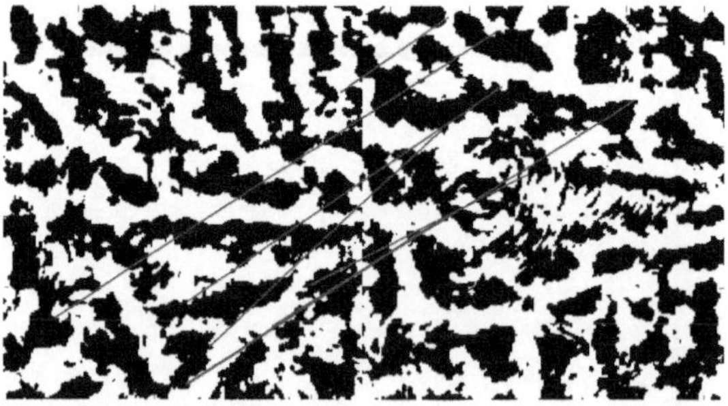

Fig. 6. Post processing of the result of Figure 5 (right)

3 Experimental Results

3.1 Protocol

We realized a home database with 24 individuals. Two sessions have been made and videos of hand have been acquired. We extracted from videos, 30 images for each session. The total amount of images used for this experiment is 24x30x2 = 1440 images.

We use one image for the enrollment and 59 images for the test for each individual.

3.2 Performance Evaluation

We used several measures to quantify the performance of the proposed system. In order to compute the False Rejection Rate (FRR) value, we compare the biometric reference with all samples of the same individual. For the False Acceptance Rate (FAR) value, we compare the biometric reference of an individual with all samples from different individuals.

For out test database, the FRR value is therefore computed considering 24x59= 1416 matching procedures and the FAR value uses 24x59x23= 32568 ones. We will use the Equal Error Rate (EER) measure that corresponds to the value of the biometric system threshold for which FAR equals to FRR.

We will plot the disparities distribution between intraclass and interclass matching results and the ROC (Receiver Operating Characteristic) curve.

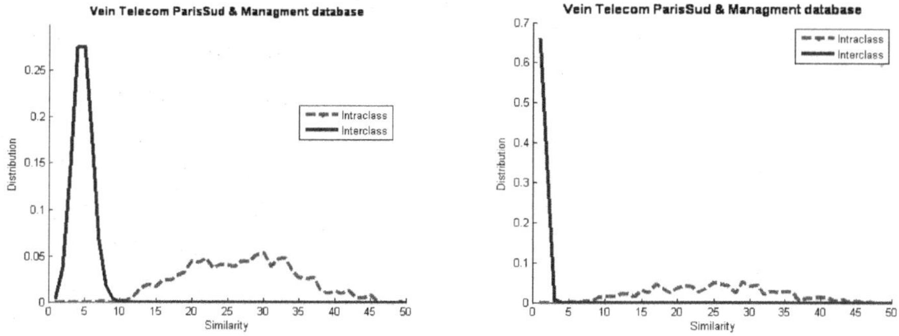

Fig. 7. Disparities distribution for the proposed method on the test database (left) without any post-processing (right) with post-processing

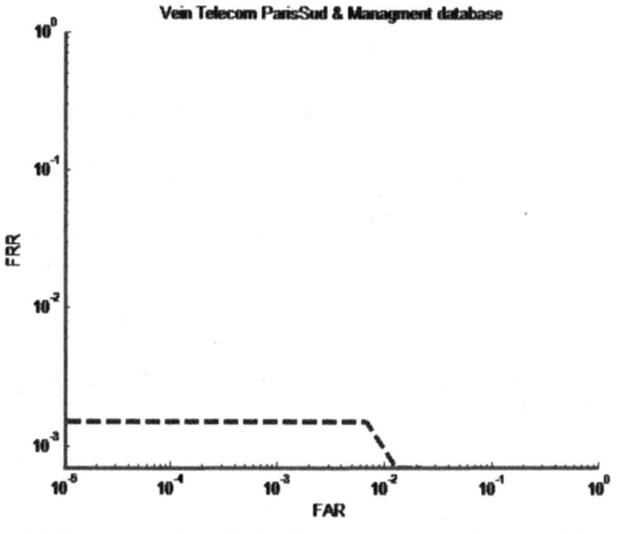

Fig. 8. DET curve of the proposed method without any post-processing (blue dotted line) and with processing (red line)

3.3 Results

First, we plot the distribution of the intraclass and interclass disparities on the test database using the proposed method without and with post-processing. The benefit of the post-processing is clear. It allows a good separation between genuine and impostors.

Computing the DET curve of the proposed method without any post-processing (see Figure 8) we obtain an EER value equals to 0.14%. After using the proposed post-processing, we obtained an ideal EER value that is to say 0%.

The average computation time for verification with images of size 232x280 pixels and using a Matlab implementation is about 2.5 seconds.

4 Conclusions and Perspectives

Hand vein verification is a promising technique for the authentication of an individual as the expected performance is very good. The performance we can expect is very good. This biometric modality is also difficult to copy that makes it a good candidate for many applications such as authentication method for e-banking operations.

In this paper, a complete biometric system based palm veins has been developed. We proposed an original method based on the use of SIFT descriptors for the enrollment and the verification steps. One of the main benefit of the proposed approach is that only one image is needed for the enrollment.

The efficiency of the proposed method is promising even if the test database used in the experiments is quite small. In the future, we intend to define a more significant database confirming our first preliminary results.

References

1. Im, S.-K., Park, H.-M., Kim, Y.-W., Han, S.-C., Kim, S.-o., Kang, C.-H.: An Biometric Identification System by Extracting Hand Vein Patterns. Journal of the Korean Physical Society 38(3) (March 2001)
2. Shahin, M., Badawi, A., Kamel, M.: Biometric Authentication Using Fast Correlation of Near Infrared Hand Vein Patterns. International journal of Biomedical sciences 2 (2007)
3. Zhang, Z., Ma, S., Han, X.: Multiscale Feature Extraction of Finger-Vein Patterns Based on Curvelets and Local Interconnection Structure Neural Network, Pattern Recognition. In: 18th International Conference on Pattern Recognition, vol. 4, pp. 145–148 (September 2006)
4. Watanabe, M., Endoh, T., Shiohara, M., Sasaki, S.: Palm vein authentication technology and its applications. In: Proceedings of the Biometric Consortium Conference, September 19-21 (2005)
5. Im, S.K., Choi, H.S., Kim, S.W.: A Direction-Based Vascular Pattern Extraction Algorithm for Hand Vascular Pattern Verification. ETRI J. 25(2), 101–108 (2008)
6. Wang, L., Leedham, G.: A Thermal Hand Vein Pattern Verification System. In: Singh, S., Singh, M., Apte, C., Perner, P. (eds.) ICAPR 2005. LNCS, vol. 3687, pp. 58–65. Springer, Heidelberg (2005)
7. Hong, L., Wan, Y., Jain, A.: Fingerprint Image Enhancement: Algorithm And Performance Evaluation. IEEE Transactions on Pattern Analysis and Machine Intelligence 20(8), 777–789 (1998)

8. Wang, J.-G., Yau, W.-Y., Suwandy, A., Sung, E.: Person recognition by fusing palmprint and palm vein images based on 'Laplacianpalm' representation. Pattern Recognition 41, 1531–1544 (2008)
9. Lowe, D.: Distinctive image features from scale-invariant key-points. International Journal of Computer Vision 60(2), 91–110 (2004)
10. Rosenberger, C., Brun, L.: Similarity-Based Matching for Face Authentication. In: International Conference on Pattern Recognition (ICPR) (2008)
11. Mikolajczyk, K., Schmid, C.: A performance evaluation of local descriptors. IEEE Transactions on Pattern Analysis and Machine Intelligence 27(10), 1615–1630 (2005)

Author Index

GPSR Compliance

The European Union's (EU) General Product Safety Regulation (GPSR) is a set of rules that requires consumer products to be safe and our obligations to ensure this.

If you have any concerns about our products, you can contact us on ProductSafety@springernature.com

In case Publisher is established outside the EU, the EU authorized representative is:

Springer Nature Customer Service Center GmbH
Europaplatz 3
69115 Heidelberg, Germany

Batch number: 10761599

Printed by Printforce, the Netherlands